Manual of
ENVIRONMENTAL MICROBIOLOGY

Manual of
ENVIRONMENTAL MICROBIOLOGY

Editor in Chief

Christon J. Hurst
U. S. Environmental Protection Agency
Cincinnati, Ohio

Editors

Guy R. Knudsen
Department of Plant, Soil, and Entomological Sciences
University of Idaho
Moscow, Idaho

Michael J. McInerney
Department of Botany and Microbiology
University of Oklahoma
Norman, Oklahoma

Linda D. Stetzenbach
Harry Reid Center for Environmental Studies
University of Nevada, Las Vegas
Las Vegas, Nevada

Michael V. Walter
Texaco E & P Technology Department
Bellaire, Texas

ASM PRESS
Washington, D.C.

Manual of environmental
microbiology

Copyright © 1997
American Society for Microbiology
1325 Massachusetts Ave., N.W.
Washington, DC 20005

Library of Congress Cataloging-in-Publication Data

Manual of environmental microbiology / editor-in-chief Christon J.
    Hurst . . . [et al.].
        p.   cm.
    Includes bibliographical references and index.
    ISBN 1-55581-087-X
    1. Microbial ecology—Laboratory manuals.   2. Sanitary
microbiology—Laboratory manuals.   I. Hurst, Christon J.
QR100.M36   1996
576′.15—dc20   96-18805
                                                               CIP

10 9 8 7 6 5 4 3 2 1

Contents

Editorial Board

Contributors

MARTIN ALEXANDER
Department of Soil, Crop and Atmospheric Sciences,
Cornell University, Ithaca, NY 14853

CHARLES W. BACON
Toxicology and Mycotoxin Research Unit, U.S. Department
of Agriculture, Agricultural Research Service, Russell
Research Center, Athens, GA 30613

MORTON A. BARLAZ
Department of Civil Engineering, North Carolina State
University, Raleigh, NC 27695-7908

J. W. BENNETT
Department of Cell and Molecular Biology, Tulane
University, New Orleans, LA 70118

ROBERT F. BONSALL
USDA-ARS, Root Disease and Biological Control Research
Unit, Washington State University, Pullman,
WA 99164-6430

INGEBORG D. BOSSERT
Department of Chemical and Biochemical Engineering,
Rutgers University, New Brunswick, NJ 08855

PAUL M. BRADLEY
U. S. Geological Survey, 720 Gracern Road, Columbia,
SC 29210

STEPHEN N. BRADLEY
University of Idaho Center for Hazardous Waste
Remediation Research, University of Idaho, Moscow, ID
83844-1052, and Department of Microbiology, Molecular
Biology, and Biochemistry, University of Idaho,
Moscow, ID 83844-3052

CORALE L. BRIERLEY
VistaTech Partnership, Ltd., Salt Lake City, UT 84171

JAMES A. BRIERLEY
Newmont Metallurgical Services, Salt Lake City, UT 84108

A. E. BROWN
Department of Botany and Microbiology, Auburn University,
Auburn, AL 36849

SUSAN BROWN
Culture Collection of Algae and Protozoa, Institute of
Freshwater Ecology, The Windermere Laboratory, Far
Sowrey, Ambleside, Cumbria, LA22 OLP, United Kingdom

ROBERT S. BURLAGE
Environmental Sciences Division, Oak Ridge National
Laboratory, Oak Ridge, TN 37831-6036

MARK P. BUTTNER
Harry Reid Center for Environmental Studies, University of
Nevada, Las Vegas, Las Vegas, NV 89154-4009

DOUGLAS E. CALDWELL
Department of Applied Microbiology and Food Science,
University of Saskatchewan, Saskatoon, Saskatchewan
S7N 5A8, Canada

DOUGLAS G. CAPONE
Chesapeake Biological Laboratory, Center for Environmental
and Estuarine Studies, University of Maryland, Solomons,
MD 20688-0038

DAVID A. CARON
Biology Department, Woods Hole Oceanographic
Institution, Woods Hole, MA 02543

CARL E. CERNIGLIA
Microbiology Division, National Center for Toxicological
Research, Food and Drug Administration, Jefferson,
Arkansas 72079

FRANCIS H. CHAPELLE
U. S. Geological Survey, 720 Gracern Road, Columbia,
SC 29210

GERARDO CHIN-LEO
Environmental Studies, The Evergreen State College,
Olympia, WA 98505

ROBERT R. CHRISTIAN
Biology Department, East Carolina University, Greenville,
NC 27858

ROBERT C. COOPER
BioVir Laboratories, Inc., Benicia, CA 94510

DON L. CRAWFORD
Department of Microbiology, Molecular Biology, and
Biochemistry, University of Idaho, Moscow, ID 83844-3052

RONALD L. CRAWFORD
University of Idaho Center for Hazardous Waste
Remediation Research, University of Idaho, Moscow, ID
83844-1052, and Department of Microbiology, Molecular
Biology, and Biochemistry, University of Idaho,
Moscow, ID 83844-3052

A. J. CROSS-SMIECINSKI
Harry Reid Center for Environmental Studies, University of
Nevada, Las Vegas, Las Vegas, NV 89154

STEPHEN J. CULLEN
Vadose Zone Research Laboratory, Institute for Crustal
Studies, University of California Santa Barbara, Santa
Barbara, CA 93106, and Geraghty and Miller, Inc., 3700
State Street, Santa Barbara, CA 93105

LOUISE-MARIE DANDURAND
Department of Plant, Soil and Entomological Sciences,
University of Idaho, Moscow, ID 83844-2339

F. DANE
Department of Horticulture, Auburn University, Auburn,
AL 36849

RICHARD E. DANIELSON
BioVir Laboratories, Inc., Benicia, CA 94510

JOHN G. DAY
Culture Collection of Algae and Protozoa, Institute of
Freshwater Ecology, The Windermere Laboratory, Far
Sowrey, Ambleside, Cumbria, LA22 OLP, United Kingdom

RICARDO DE LEON
Metropolitan Water District of Southern California, Water
Quality Division, La Verne, CA 91750

JODY W. DEMING
School of Oceanography, University of Washington, Seattle,
WA 98195

LEE A. DEOBALD
Department of Microbiology, Molecular Biology, and
Biochemistry, University of Idaho, Moscow, ID 83844-3052

RICHARD DEVEREUX
Gulf Ecology Division, U. S. Environmental Protection
Agency, Gulf Breeze, FL 32561

NICHOLAS J. E. DOWLING
IRSID, BP 320, Voie Romaine, 57214 Maizières-les-Metz,
France

B. D. FAISON
Chemical Technical Division, Martin Marietta Energy
Systems, Oak Ridge National Laboratories, Oak Ridge,
TN 37831-6194

XIUHONG FENG
Department of Crop and Soil Sciences, Washington State
University, Pullman, WA 99164-6420

BARRY S. FIELDS
Division of Bacterial and Mycotic Diseases, National Center
for Infectious Diseases, Centers for Disease Control and
Prevention, Atlanta, GA 30333

D. D. FOCHT
Department of Soil and Environmental Sciences, University
of California, Riverside, CA 92921

JAMES K. FREDRICKSON
Pacific Northwest National Laboratory, P. O. Box 999,
Richland, WA 99352

JED A. FUHRMAN
Department of Biological Sciences, University of Southern
California, Los Angeles, CA 90089-0371

ROGER S. FUJIOKA
Water Resources Research Center, University of Hawaii,
Honolulu, HI 96822

JAMES R. FUXA
Department of Entomology, Louisiana Agricultural
Experiment Station, Louisiana State University Agricultural
Center, Baton Rouge, LA 70803

MARK O. GESSNER
Center for Ecosystem Research, University of Kiel,
Schauenburgerstraße 112, 24118 Kiel, Germany

SERGEY A. GRINSHPUN
Aerosol Research Laboratory, Department of Environmental
Health, University of Cincinnati, Cincinnati,
OH 45267-0056

ETHAN L. GROSSMAN
Department of Geology and Geophysics, Texas A&M
University, College Station, TX 77843

JEAN GUEZENNEC
IFREMER Centre de Brest, BP 70, 29280 Plouzané, France

TERRY B. HAMMILL
University of Idaho Center for Hazardous Waste
Remediation Research, University of Idaho, Moscow, ID
83844-1052, and Department of Microbiology, Molecular
Biology, and Biochemistry, University of Idaho,
Moscow, ID 83844-3052

GWEN LEE HARDING
Texaco Inc., P.O. Box 509, Beacon, NY 12508

RONALD W. HARVEY
Water Resources Division, U. S. Geological Survey, 3215
Marine Street, Boulder, CO 80303-1066

MARK E. HINES
Department of Biological Sciences, University of Alaska
Anchorage, Anchorage, AK 99508

DOROTHY M. HINTON
Toxicology and Mycotoxin Research Unit, U.S. Department
of Agriculture, Agricultural Research Service, Russell
Research Center, Athens, GA 30613

WILLIAM E. HOLBEN
Division of Biological Sciences, University of Montana,
Missoula, MT 59812

CHRISTON J. HURST
National Risk Management Research Laboratory, U. S.
Environmental Protection Agency, Cincinnati, OH 45268

M. KHALID IJAZ
H. H. Shaikh Khalifa Scientific Centre for the Racing
Camel, P.O. Box 17292, Al-Ain, United Arab Emirates

AMIEL G. JARSTFER
Division of Natural Sciences, LeTourneau University,
Longview, TX 75607

LEE-ANN JAYKUS
Department of Food Science, North Carolina State
University, Raleigh, NC 27695-7624

PAUL A. JENSEN
U.S. Department of Health and Human Services, Public Health Service, Centers for Disease Control and Prevention, National Institute for Occupational Safety and Health, Cincinnati, OH 45226

ECKARDT JOHANNING
Department of Community Medicine, Mt. Sinai School of Medicine, New York, NY 10029, and Eastern New York Occupational Health Program, Albany, NY 12210

RONALD D. JONES
Southeast Environmental Research Program and Department of Biological Sciences, Florida International University, University Park, Miami, FL 33199

GARY M. KING
Darling Marine Center, University of Maine, Walpole, ME 04573

GUY R. KNUDSEN
Department of Plant, Soil and Entomological Sciences, University of Idaho, Moscow, ID 83844-2339

DARREN R. KORBER
Department of Applied Microbiology and Food Science, University of Saskatchewan, Saskatoon, Saskatchewan S7N 5A8, Canada

ROGER A. KORUS
Department of Chemical Engineering, University of Idaho, Moscow, ID 83844

DAVID S. KOSSON
Department of Chemical and Biochemical Engineering, Rutgers University, New Brunswick, NJ 08855

JOHN H. KRAMER
Condor Earth Technologies, Inc., 21663 Brian Lane, Sonora, CA 95370-3905, and Vadose Zone Research Laboratory, Institute for Crustal Studies, University of California Santa Barbara, Santa Barbara, CA 93106

YASUHISA KUNIMI
Department of Applied Biological Sciences, Faculty of Agriculture, Tokyo University of Agriculture and Technology, Saiwai, Fuchu-shi, Tokyo 183, Japan

JOHN R. LAWRENCE
National Hydrology Research Institute, Saskatoon, Saskatchewan S7N 3H5, Canada

EDWARD R. LEADBETTER
Graduate Program in Microbiology, Department of Molecular and Cell Biology, University of Connecticut, Storrs, CT 06269-2131

ESTELLE LEVETIN
Faculty of Biological Science, University of Tulsa, Tulsa, OK 74105

STEVEN E. LINDOW
Department of Environmental Science, Policy and Management, University of California at Berkeley, Berkeley, CA 94720-3110

KATHLEEN L. LONDRY
Department of Botany and Microbiology, University of Oklahoma, Norman, OK 73019-0245

JOYCE E. LOPER
Horticultural Crops Research Laboratory, Agricultural Research Service, U. S. Department of Agriculture, Corvallis, OR 97330

RICHARD G. LUTHY
Department of Civil and Environmental Engineering, Carnegie Mellon University, Pittsburgh, PA 15213

EUGENE L. MADSEN
Section of Microbiology, Division of Biological Sciences, Cornell University, Ithaca, NY 14853-8101

W. MAHAFFEE
Department of Plant Pathology, Auburn University, Auburn, AL 36849

AARON B. MARGOLIN
Department of Microbiology, Biological Sciences Center, University of New Hampshire, Durham, NH 03824

KEVIN C. MARSHALL
School of Microbiology and Immunology, The University of New South Wales, Sydney NSW 2052, Australia

TIMOTHY R. McDERMOTT
Department of Plant, Soil, and Environmental Sciences, Montana State University, Bozeman, MT 59717-0312

GORDON A. McFETERS
Department of Microbiology, Montana State University, Bozeman, MT 59717

MICHAEL J. McINERNEY
Department of Botany and Microbiology, University of Oklahoma, Norman, OK 73019-0245

AARON L. MILLS
Laboratory of Microbial Ecology, Department of Environmental Sciences, University of Virginia, Charlottesville, VA 22903

CHRISTINE L. MOE
Department of Epidemiology, School of Public Health, University of North Carolina, Chapel Hill, NC 27599-7400

ALAN JEFF MOHR
Aerosol & Environmental Technology Branch, U.S. Army, Dugway Proving Ground, Dugway, UT 84022-5000

MATTHEW J. MORRA
Soil Science Division, University of Idaho, Moscow, ID 83844-2339

DAVID D. MYROLD
Department of Crop and Soil Science, Oregon State University, Corvallis, OR 97331-7306

STEVEN Y. NEWELL
Marine Institute, University of Georgia, Sapelo Island, GA 31327

SHIRLEY F. NISHINO
U. S. Air Force, Armstrong Laboratory, Tyndall AFB, FL 32403

ANDREW OGRAM
Soil and Water Science Department, University of Florida, Gainesville, FL 32611

STEPHEN A. OLENCHOCK
National Institute for Occupational Safety and Health, Morgantown, WV 26505

HANS W. PAERL
University of North Carolina at Chapel Hill, Institute of Marine Sciences, Morehead City, NC 28557

PIERRE PAYMENT
Centre de Recherche en Virologie, Institut Armand-Frappier, Université du Québec, Laval, Québec H7N 4Z3, Canada

IAN L. PEPPER
Department of Soil, Water, and Environmental Science, University of Arizona, Tucson, AZ 85721

TOMMY J. PHELPS
Oak Ridge National Laboratory, P. O. Box 2009, Oak Ridge, TN 37831-6036

HOLLY C. PINKART
Center for Environmental Biotechnology, University of Tennessee/Oak Ridge National Laboratory, Knoxville, TN 37932, and Department of Microbiology, University of Tennessee, Knoxville, TN 37996

WESLEY O. PIPES
Department of Civil Engineering and the Environmental Studies Institute, Drexel University, Philadelphia, PA 19104

ANURADHA RAMASWAMI
Environmental Science and Engineering Division, Colorado School of Mines, Golden, CO 80401

DAVID B. RINGELBERG
Center for Environmental Biotechnology, University of Tennessee/Oak Ridge National Laboratory, Knoxville, TN 37932

SYED A. SATTAR
Department of Microbiology and Immunology, Faculty of Medicine, University of Ottawa, Ottawa, Ontario K1H 8M5, Canada

FRANK W. SCHAEFER III
Human Exposure Research Division, National Exposure Research Laboratory, U.S. Environmental Protection Agency, Cincinnati, OH 45268

STEVEN K. SCHMIDT
Department of Environmental, Population, and Organismic Biology, University of Colorado, Boulder, CO 80309-0334

KATE M. SCOW
Department of Land, Air, and Water Resources, University of California, Davis, CA 95616

JOE J. SHAW
Department of Botany and Microbiology, Auburn University, Auburn, AL 36849

BARRY F. SHERR
College of Oceanic and Atmospheric Sciences, Corvallis, OR 97331-5503

EVELYN B. SHERR
College of Oceanic and Atmospheric Sciences, Corvallis, OR 97331-5503

KAY L. SHUTTLEWORTH
Microbiology Division, National Center for Toxicological Research, Food and Drug Administration, Jefferson, Arkansas 72079

N. P. SINGH
Molecular Biology Institute, University of Scranton, Scranton, PA 18510

K. SMALLA
Biologische Bundesanstalt, Institute for Biochemistry, Messeweg 11/12, W-3300 Braunschweig, Germany

RICHARD L. SMITH
U. S. Geological Survey, 3215 Marine Street, Boulder, CO 80303

JIM C. SPAIN
U. S. Air Force, Armstrong Laboratory, Tyndall AFB, FL 32403

DAVID A. STAHL
Environmental Health Engineering, Department of Civil Engineering, Northwestern University, Evanston, IL 60208-3109

LINDA D. STETZENBACH
Harry Reid Center for Environmental Studies, University of Nevada, Las Vegas, Las Vegas, NV 89154-4009

G. STOTZKY
Laboratory of Microbial Ecology, Department of Biology, New York University, New York, NY 10003

KERRY L. SUBLETTE
Department of Chemical Engineering, University of Tulsa, Tulsa, OK 74101

JOSEPH M. SUFLITA
SEC Institute for Energy and the Environment and Department of Botany and Microbiology, University of Oklahoma, Norman, OK 73019-0245

CURTIS A. SUTTLE
Departments of Oceanography, Botany, and Microbiology, University of British Columbia, Vancouver, British Columbia V6T 1Z4, Canada

DAVID M. SYLVIA
Soil and Water Science Department, University of Florida, Gainesville, FL 32611-0290

RALPH S. TANNER
Department of Botany and Microbiology, University of Oklahoma, Norman, OK 73019-0245

LINDA S. THOMASHOW
USDA-ARS, Root Disease and Biological Control Research Unit, Washington State University, Pullman, WA 99164-6430

GARY A. TORANZOS
Environmental Microbiology Laboratory, Department of Biology, P.O. Box 23360, University of Puerto Rico, Rio Piedras, PR 00931-3360

J. T. TREVORS
Department of Environmental Biology, University of Guelph, Guelph, Ontario N1G 2W1, Canada

GLENN A. ULRICH
Department of Botany and Microbiology, University of Oklahoma, Norman, OK 73019-0245

J. D. VAN ELSAS
Institute for Plant Protection IPO-DLO, Binnenhaven 5, P. O. Box 9060, 6700 GW Wageningen, The Netherlands

PIETER T. VISSCHER
Department of Marine Sciences, University of Connecticut, Groton, CT 06340

LAWRENCE P. WACKETT
Department of Biochemistry and Institute for Advanced Studies in Biological Process Technology, University of Minnesota, St. Paul, MN 55108

MICHAEL V. WALTER
E & P Technology Department, Texaco Inc., Bellaire, TX 77402

ALAN WARREN
Department of Zoology, The Natural History Museum, Cromwell Road, London SW7 5BD, United Kingdom

DAVID M. WELLER
USDA-ARS, Root Disease and Biological Control Research Unit, Washington State University, Pullman, WA 99164-6430

DAVID C. WHITE
Center for Environmental Biotechnology, University of Tennessee/Oak Ridge National Laboratory, Knoxville, TN 37932; Environmental Science Division, Oak Ridge National Laboratory, Oak Ridge, TN 37831; and Department of Microbiology, University of Tennessee, Knoxville, TN 37996

KLAUS WILLEKE
Aerosol Research Laboratory, Department of Environmental Health, University of Cincinnati, Cincinnati, OH 45267-0056

GIDEON M. WOLFAARDT
Department of Applied Microbiology and Food Science, University of Saskatchewan, Saskatoon, Saskatchewan S7N 5A8, Canada

S.-J. WU
Department of Botany and Microbiology, Auburn University, Auburn, AL 36849

CHIN S. YANG
P & K Microbiology Services, Inc., 1950 Old Cuthbert Road, Cherry Hill, NJ 08034

MARYLYNN V. YATES
Department of Soil and Environmental Sciences, University of California, Riverside, CA 92521

L. Y. YOUNG
Center for Agricultural Molecular Biology and Department of Environmental Sciences, Cook College, Rutgers, the State University of New Jersey, New Brunswick, NJ 08903

JAMES T. ZMUDA
Research and Development Laboratory, The Metropolitan Water Reclamation District of Greater Chicago, Schaumburg, IL 60193

INTRODUCTION TO ENVIRONMENTAL MICROBIOLOGY

I

VOLUME EDITOR
MICHAEL J. McINERNEY

Introduction to Environmental Microbiology

CHRISTON J. HURST

1

Environmental microbiology is the study of those microorganisms which exist in natural or artificial environments. The origin of scientific research in this field rests in the observations of Antony van Leewenhoeck that were published in 1677 (1). van Leewenhoeck used a microscope of his own creation to discover what he termed "animalcula," or the "little animals" which lived and replicated in rainwater, well water, seawater, and water from snowmelt. During the intervening centuries, the expansion of our knowledge regarding environmental microorganisms has been based on increasingly detailed observations and experimentation, in which we have been aided by advancements in microscopy and the development of biochemical and mathematical tools. Yet many of the phenomena which we have since come to better recognize and characterize in environmental microbiology were initially described in van Leewenhoeck's publication.

Many of van Leewenhoeck's observations were based on the examination of environmental samples which he variously maintained in a wineglass, porcelain dish, and glass bottle in his house. In current terminology, we would say that he had created small artificial environments, or microcosms. While we often tend to think of artificial environments as taking the form of small laboratory systems, the term "artificial environment" also encompasses both production-scale fermentations and anthropogenic structures emplaced in the natural world. Field-emplaced anthropogenic structures are often open to the environment and thus conceptually merge with the spectrum which represents natural environments.

We have continued to study both the presence and the metabolic activities of microorganisms found in surface waters, and since the studies of van Leewenhoeck, we have discovered that microorganisms in fact cover the planet, living even in the fumaroles of surface volcanoes and in sedimentary rocks within the dry valleys of Antarctica and providing the biological basis for the life that surrounds seafloor thermal vents. We have learned that the microbial colonization of environments is a process which involves a succession of organisms which differ in their ecological requirements. The microorganisms chemically interact with their physical environment, and their most notable effect has been the creation of an oxidizing atmosphere on this planet. By way of these chemical interactions, the microbes remain crucial to the biogeochemical cycling which

supports the continuance of life on our planet, turning over those elements that represent the basic ingredients of life such as carbon, hydrogen, nitrogen, oxygen, phosphorus, and sulfur.

Over the course of millenia, we have learned to use for our benefit naturally occurring microbial products, including petroleum and some of the fungal metabolites which have antimicrobial properties. While expanding the extent to which we understand microbial processes, during the last few decades we have begun learning how to harness microbial biosynthetic and degradative activities. This harnessing, and in some instances the intentional manipulation of microbial activities, constitutes the basis of microbial biotechnology, whereby we direct the activity of microorganisms within both natural and artificial environments for a variety of purposes. As one example, we utilize microorganisms as tools to help us achieve goals such as the production of materials which are beneficial to our existence, among which are antibiotics, vitamins, and fuels such as biogas and ethanol. Microorganisms also are used as tools to help us intentionally degrade both natural and anthropogenic materials in wastewater digestors, composters, landfills, natural terrestrial environments, and natural or artificial aquatic environments. Sometimes we use microorganisms as tools in achieving agricultural goals such as protecting plants from insect damage. Furthermore, microbial processes have been employed as a means for reducing the use of hazardous chemicals in geochemical recovery operations, by using microorganisms to leach metals from ores and to enhance the recovery of petroleum from wells. Just as we sometimes use our knowledge of microbial processes to our benefit, at other times we try to prevent natural microbial activities such as those which contribute to the biofouling, corrosion, and decay of objects exposed to the environment.

Microbes represent the very origin of life on Earth, and they comprise the basis of our biological legacy. Presently, we divide microorganisms into five major taxonomic groups. Four of these, the algae, bacteria, fungi, and protozoans, are considered to be cellular, meaning that they possess cell membranes. The fifth group, viruses, are acellular. This classification scheme is very traditional, and biochemically based phylogeny studies constantly provide us with suggestions for revising the groupings. What matters more, from the viewpoint of environmental microbiology, is our increasing level of knowledge regarding the fact that in ecosystems, these groups of microorganisms interact both with one another and with the macroorganisms on this planet. These interactions occur and can be studied on several levels: spatially, biochemically, and even genetically. We have tended to divide our descriptions of these interactions into categories based on whether the relationships are neutral, positive, or negative with regard to the organisms involved. Neutral relationships are ones in which there is neither perceived harm nor advantage for the involved organisms, as when two populations coexist, ignoring the presence of one another either because their numbers are sparse or because they occupy different ecological niches. Commensal relationships occur when one organism is perceived to benefit while the other is unaffected, as when microbes which colonize the external surfaces of animals or plants derive energy by metabolizing exudates produced by their hosts. There are also mutually advantageous relationships in which both populations benefit, examples of which are synergism and mutualism. Synergism is a voluntary interaction wherein both populations of organisms can exist on their own but do better when living together. The existence of certain intestinal bacteria within human hosts is an example of synergism. In return for being sheltered, these bacteria produce vitamins that the host cannot manufacture on its own and would otherwise have to acquire from its diet. Mutualism or symbiosis occurs when neither organism seems capable of living on its own in a particular environment. An example of symbiosis is the interrelationship between certain chemoauttotrophic bacteria and the marine tube worms (*Riftia* sp.) which inhabit seafloor thermal vents. Lastly, there are relationships in which one population may benefit while the other is harmed, such as the parasitism represented by viral or protozoan diseases. Competition describes relationships in which both populations of organisms suffer by attempting to coexist within a common ecological niche when their combined numbers are too great to be supported by that niche. These interactions between organisms are not fixed in nature, and a relationship which normally is benign may become detrimental, as is the case with some opportunistic pathogens which normally exist as benign skin surface microflora but can cause disease if they gain access to the interior of an immunologically weakened host.

Not only have we tried to elucidate the natural fate of microorganisms in the environment, but often we have attempted to outright eliminate from the environment those microorganisms which are pathogenic either to humans or to the plants and animals upon which we depend for our sustenance. Some of these pathogens are indigenous environmental organisms, while various others are of human origin or from animals and plants and are released into the environment through natural processes. The study of their persistence in environmental media, including survival and transport in soil, water, and air, provides clues that help us in anticipating and controlling their populations.

Readers of this manual will find that the core sections are structured with regard to the type of environmental medium being discussed. The subject of water, the hydrosphere, has been divided into two sections; section III contains chapters which address the fact that water often serves as a vehicle in the transmission of pathogenic microbes, and the chapters in section IV address general aquatic ecology. The terrestrial environments of the lithosphere have been divided into soil and plant zone interactions (section V); section VI covers the microbiology of deeper subsurface environments and landfills. While microbes are not known to colonize the atmosphere, air serves as one of the vehicles by which both they and their often toxic by-products are transported (section VII). The subject of microbially mediated chemical transformations bridges the hydrosphere and lithosphere by virtue of relatedness of the involved microbial metabolic processes, which often are performed by either the same or related genera of organisms. For this reason, the topics addressing biotransformation and biodegradation have largely been grouped into one section (section VIII). Likewise, the basic principles of environmental microbiology (section I) and general analytical methodologies (section II) tend to be common across the range of environments that we study, and these latter two subject areas have been placed in the front of the manual because of their primary importance.

I am not sure whether Antony van Leewenhoeck could have foreseen where his discoveries would lead: to the diversity of environmental microbiology subjects that we now study and the wealth of knowledge that we have accumulated. But just as I have always enjoyed reading his account of environmental microorganisms, I feel that he would enjoy our efforts at once again summarizing all of environmental microbiology in a single document as represented by this manual. I thank the numerous microbiologists who have collaborated in creating this manual, many as contributors and some as editors. I also thank those giants in microbiology upon whose shoulders we have stood, for we could not have accomplished the creation of this manual without the advantage that those giants have afforded us.

REFERENCE

1. **van Leewenhoeck, A.** 1677. Observations communicated to the publisher by Mr. Antony van Leewenhoeck, in a Dutch letter of the 9th of October, 1676. Here Englished: Concerning little animals by him observed in rain- well- sea- and snow-water; as also in water wherein pepper had lain infused. *Philos. Trans. R. Soc. Lond.* **11:**821–831.

Microbial Communities and Interactions: a Prelude

MARTIN ALEXANDER

2

The author of an introduction to microbial communities and the interactions among microbial populations is immediately faced with three significant problems. First, an enormous number of different microorganisms exist in a wide array of highly dissimilar environments. These microorganisms include bacteria, fungi, algae, protozoa, and viruses as well as invertebrates that frequently consume microorganisms, plants that provide readily available organic nutrients, and higher animals that have intriguing but frequently highly different types of associations with microbial populations and communities. Second, the interactions among microbial populations are not only numerous and dissimilar but often poorly characterized or totally undefined. Furthermore, a difficulty in considering such interactions is the frequent lack of understanding of the role of animal and plant populations that have a major impact on microorganisms but are not within the usual scope of interest of microbiologists. Third, a woeful lack of agreement exists about the various definitions that are appropriate for characterizing microbial populations, communities, environments, and interactions.

This chapter is not designed to be encompassing. The topic of microbial communities and interactions would easily fill an entire monograph. Rather, the text will serve as a prelude, that is, an introduction that precedes a discussion of the principle issues of environmental microbiology. A prelude is of value only if it serves as a suitable introduction, and it is the chapters that follow that will serve as the meat of the monograph. An analogy to a musical composition is not inappropriate, for many such compositions have preludes or overtures; they are designed to start the performance and give the audience time to become settled but rarely have intrinsic value or intrinsic interest.

THE PLEASURES OF MISUSAGE

"When I use a word," Humpty Dumpty said, "it means just what I choose it to mean—neither more nor less—adjectives you can do anything with." "The question is," said Alice, "whether you can make words mean so many different things." (Lewis Carroll, *Through the Looking Glass*, 1871).

Environmental microbiology is blessed, or possibly cursed, with a multitude of terms that are often contradictory, sometimes misused, and frequently misunderstood even by the specialists. Semantic sensitivity is frequently not the

hallmark of the experimentalist, and this lack of sensitivity often is the basis for confusion and disagreement. Inasmuch as no language academy exists to serve as the final arbiter of technical definitions, it is incumbent upon an author to define clearly the terms to be used so that the reader will understand the frame of reference and the approach employed by the author, even if the reader does not agree, totally or partially, with the definitions used. Terms like environment, niche, ecosystem, community, population, consortium, parameter, and secretion fill the literature of environmental microbiology, and these words are used in so many ways and often in such conflicting manners that the reader frequently is unsure of the specific issue under discussion.

Population is one example of a frequently misused word. A population is an assemblage of individual organisms having common characteristics. They all have common traits, indicating a similar or identical ancestry. In microbiology, a population may represent all individuals of a single species, subspecies, variety, race, or other subspecies designation. Within a population may be a large number of individual cells, or a population may be composed of a multicellular filament, such as is found among the fungi and algae. An appropriate term for the individual is *propagule*.

All of the organisms that occupy a particular site represent a *community*. However, many specialists use the term community to encompass a particular category of organisms, so that an environment may be considered to have a community of bacteria, fungi, algae, plants, or animals. The community of a soil is somewhat different from that of the subsoil or underlying aquifer, and the community in the water column is not the same as that inhabiting the underlying sediment. A term synonymous with community but not widely used in English-speaking countries is *biocoenosis*. Although clear distinctions thus exist between communities and populations, the words are frequently and mistakenly used as synonyms.

By *habitat* is meant a site having some uniformity in properties that appear to be of ecological importance. Some species are restricted to a single and frequently unique habitat, whereas others are cosmopolitan and reside in a wide variety of dissimilar habitats. An individual habitat may be soil, the intestinal tract, the outer surfaces of a plant root, or lake bottom sediments. Frequently, however, owing to the small size of microorganisms, the concern of the microbial ecologist is the microhabitat, inasmuch as individual propagules, filaments, or cells often are restricted to a site that is no larger than several millimeters or sometimes even micrometers. *Niche* is another word that is frequently misused. Among ecologists, the niche of an organism is what it does; that is, its function in its natural habitat. Thus, an organism that exists in the water column of a lake has that aquatic environment as its habitat, but its niche may, for example, be the fixation of atmospheric N_2 or the feeding upon bacteria. Such semantic sensitivity may appear to be pedantic, but when these insensitivities create confusion or prevent understanding of technical matters, they are not mere pedantries.

Consortium is a term used more commonly for bacteria in associations than for assemblages of other organisms (11). It is a useful word when it refers to a collection of organisms that have some functional association with one another. For example, one species may provide growth factors for a second or serve to eliminate inhibitory compounds that affect a second species. To some degree, a consortium may represent a commensal or protocooperative interaction that has not yet had the basis for the interaction established. Unfortunately, however, mixtures of bacteria that have yet to be separated into pure cultures are frequently designated consortia; in these instances, the mixture does not represent a functional association but rather reflects the lack of isolation of individual components of the mixture.

The ability to use language properly is an art. The literature of microbiology is ample testimony to the fact that the use of language is not a science. The misuse of *niche* and *habitat* is an old example of semantic insensitivity. More recently, use of the word *secrete* has become another addition to the confusion of nomenclature. The etymology of the word shows its relationship with *secret*: to secrete is to hide or to conceal. This is well illustrated in its usage for secretion of precious heirlooms to prevent theft or, in animal physiology, for the secretions of glands. However, the word has now come to mean, in microbiology at least, exactly the opposite, and a microorganism that "secretes" a molecule is making it available for all to see. It is no longer hidden or concealed but rather available outside the organism. Even Samuel Johnson, centuries ago in his *A Dictionary of the English Language*, defined "secrete" as "hide" and defined excretion as "ejecting somewhat quite out of the body." Presumably, some reluctance exists among 20th century microbiologists to use the more appropriate term *excretion*, possibly because of its etymological relationship with excrement. Other examples of words that have completely opposite meanings are not difficult to find: to cleave means either to separate or to adhere, and the verb *doctor* may be used to mean either treat medically or alter deceptively. At some time, the words become accepted as standard usage, but until that time, care should be taken to avoid use of terms that might be taken as having different meanings by different readers.

THE VIRTUES OF COMMUNITY

Although microbiologists are prone to work with pure cultures, communities having only one species are rare in nature, except in those monospecific communities associated with disease processes or in environments so extreme that few species can survive or proliferate. Clearly, therefore, there is, in nature at least, a major benefit gained by one species by coexisting with other species. Although the benefits are unquestionable, the basis for those benefits and the

mechanisms associated with the interactions in natural communities are rarely understood and rarely studied (12, 13).

Most communities typically have several, many, or an uncounted number of species. The degree of such species diversity varies enormously. To be a member of a community, the propagules of a species not only must reach that environment but must successfully cope with the detrimental interactions occurring there, and the fact that some environments are characterized by high species diversity indicates that their habitat is accessible to many different types of microorganisms and that these arrivals are able to survive and occasionally proliferate. Soils, surface waters, sediments, sites rich in decaying organic materials, and other environments support a multitude of species of bacteria, fungi, protozoa, and sometimes algae. On the other hand, sites with high salinity, low pH, or high light intensity are characterized by low species diversity.

An old truism states that microorganisms are potentially everywhere. That truism is little more than a platitude if it is taken to mean that some organism will alight on any exposed and previously uncolonized site. However, the statement takes on a degree of ecological inaccuracy when it is used as the basis for assuming that species or genera that are widely disseminated are also widely established. Potentially cosmopolitan species frequently are able to grow, in culture at least, very rapidly, so that a few propagules or a small biomass can multiply to reach, in the absence of other organisms, a high cell density or large biomass in a matter of hours or days. Such largely unchecked proliferation rarely occurs in nature, however, and effective checks exist to prevent the unbridled multiplication of widely disseminated propagules. The operation of these various checks is the basis for the appropriate second half of the truism, namely, that the environment selects. Hence, the full and ecologically appropriate statement is, Microorganisms are potentially everywhere, but the environment selects.

The impact of that selection is evidenced by the characteristic communities of individual habitats. Many of the bacteria, fungi, and protozoa present in surface waters are thus not characteristic of soils, and the types of heterotrophs found in soils are often quite different from those that would be found on the surfaces of leaves. The forces of selection are often nonbiological, and an area that has a low pH, is exposed to high light intensity, has no available oxygen, or contains low concentrations of readily available carbon sources supports a community that is able to cope with these restraints or limitations. Such abiotic factors are often reasonably easy to demonstrate, but more difficult to establish and scientifically more interesting are those biotic stresses that are operating in environments in which major abiotic stresses do not determine community composition.

The indigenous populations that make up the community are responsible for the biotic balance that is maintained. They regulate the population densities or biomass of the individual component species of the community, and they act to prevent the establishment of invading species. These various biotic interactions determine the stability of established microbial communities, and they reflect what is designated *homeostasis*. From an ecological standpoint, homeostasis is the capacity of a community to maintain its stability and integrity in an environment subject to abiotic and potentially biotic modifications. Should there be some modification associated with an external nonbiological perturbation or should a nonindigenous organism arrive in the environment, mechanisms associated with homeostasis operate to maintain the integrity of the community and often to eliminate the nonindigenous species. These homeostatic mechanisms are constantly operating, and although they may be modified somewhat, they are not altered appreciably unless an enormous perturbation occurs. Perturbations that may modestly or drastically upset the community are characteristically abiotic, but the mere introduction of propagules of a nonindigenous microorganism is rarely sufficient to alter the community to an appreciable extent or, in many instances, to allow the arrival to become established.

Ignoring the potency, as well as the limitations, of the homeostatic mechanisms has led to a number of viewpoints that run counter to observations in nature. For example, individuals desiring to introduce bacteria or fungi into soils, subsoils, or aquifers for the purposes of bioremediation assume that the introduced species will become established and bring about the destruction of a contaminant of concern (4). It is assumed that because the microorganism has a beneficial trait, in this case the capacity to grow on and thus destroy an organic pollutant, it will become established and perform what is desired of it. However, the capacity to use a substrate for growth, although of paramount importance in culture, is necessary but not sufficient for ecological success: the organism also must be able to cope with the various factors associated with homeostasis. An attribute that is a necessary requirement for growth is not an attribute sufficient for establishment, and both necessary and sufficient traits must be present. Thus, an organism must be able to compete effectively for limited resources other than the organic compound that can serve as its carbon source, and it must be able to cope with the stresses associated with predation and parasitism, which are of great importance in many environments. Farmers know that sowing a new variety of plant, one that has major beneficial attributes, is not sufficient to obtain high yields of that variety. The introduced crop species invariably is unable to cope with competition by weeds, parasitism by insects and plant pathogens, poor soil structure, and other stresses. Few introduced species become established, whether the introduction has the capacity to do good or is potentially injurious. Yet in some instances, an alien species does become successfully established, as evidenced by the major upsets that have occurred because of invasion by plants and animals. Attempts have been made to predict the capacity of an introduction to become established (21), and studies have been directed to establishing the traits that might be used to predict the outcome of an introduction. To date, the outcome of that research is modest (22).

In contrast with vendors of microorganisms who expect success from their introductions are those individuals concerned with the fate of genetically engineered microorganisms. Many of these individuals assume that all introductions will fail and that modest changes in the genotype of an existing organism will not result in its establishment. Here, too, it is necessary to consider the strengths and the limitations of the homeostatic mechanisms, and the likely consequence of their operation would eliminate most introductions. Nevertheless, homeostatic mechanisms are not omnipotent, and some introduced organisms succeed and become established. The failure to predict the success or failure of introductions is a reflection of the incomplete knowledge of the various components of homeostasis.

THE STRUGGLE FOR EXISTENCE

The very fact that species can be distinguished is a priori evidence that each has unique biochemical, physiological, or morphological properties. Some of these traits serve as the basis for the presence of a species in one environment but not another and for the relative abundance and activity of the various inhabitants of those environments. Because of the large number of microorganisms arriving at exposed sites, one or more of those traits are the basis for natural selection, the geography of the organism, and its role in the environment.

The abundant literature on the biochemistry, physiology, and morphology of microorganisms is derived from studies of those organisms under artificial conditions in liquid media in the absence of any other species. Pure, or axenic, cultures have provided a wealth of information, but it is far from certain which of those biochemical, physiological, or morphological properties are of ecological significance. Many facile extrapolations have been made from in vitro to in vivo conditions, but rarely have these extrapolations been verified as being ecologically relevant.

To exist in an environment, a bacterium, fungus, alga, or protozoan must be able to endure all of the abiotic stresses characteristic of that environment. These stresses include pH, high or low temperature, occasional drying or freezing in some environments, intense solar radiation, high pressures deep in the ocean, or salinity in certain terrestrial or aquatic ecosystems. If an organism in culture is unable to survive and occasionally grow when exposed to these stresses, even under artificial conditions, it is unlikely to be an inhabitant of an environment in which those stresses occur. These are the stresses that are easy to establish, and it is simple to show which may be important in determining the absence of an organism. Yet many species are transported to environments in which they are able to tolerate all the abiotic stresses, but they still do not become established. The homeostatic mechanisms operating in the community in which the new arrival alights are effective in eliminating many of the arrivals, yet the reasons for the elimination or, conversely, the reasons for the presence of the inhabitants are rarely understood. Although it is little more than a platitude to state that the biochemical, physiological, or morphological properties of the organism determine its ecological success, the information necessary to predict or explain the ecological success or failure is largely lacking.

An obvious need that must be satisfied for an organism to become established is the presence of all nutrients that it requires. Many natural ecosystems contain most and sometimes all inorganic nutrients in concentrations sufficient to maintain a reasonably large community, but the supply of the energy source, and particularly the carbon sources for heterotrophs, is frequently inadequate. Hence, to some degree, the availability of a suitable carbon source is essential in determining the distribution of an organism. In most environments containing readily or slowly available organic molecules, a variety of dissimilar propagules able to use those carbon sources arrive, yet only a few become established. In this instance, as in so many cases in environmental microbiology when causation is being sought, the presence of an energy source is a *necessary* but not *sufficient* requirement for establishment. It is necessary to look further.

Natural selection, or selection in nature, is not easy to explain on the basis of existing knowledge. It may be simple to show what is necessary for a positive outcome in selection but not what is sufficient. The issue is to identify those attributes of organisms that are necessary and sufficient for them to survive and occasionally multiply in particular environments. After showing the importance of tolerance to the abiotic factors that are detrimental to one or another group of organisms and the need for a supply of inorganic nutrients and a carbon or energy source, then what? The methodologies associated with the enrichment or elective culture technique are ideal for natural selection under the artificial conditions of the enrichment solution, but the use of such methods typically results in an organism that grows fastest under those conditions. In contrast, many inhabitants of natural communities grow slowly and do not appear under the usual artificial conditions imposed by the enrichment medium. From an environmental viewpoint, it is the fitness trait or set of fitness traits that underlies ecological success and sometimes dominance. These traits are the specific biochemical, physiological, or morphological characteristics that determine an organism's habitat and its niche.

The fitness traits of the initial colonists in environments that are largely free of microorganisms or that have been drastically disturbed are often reasonably simple to predict. Such environments are not unknown, and they exist, for example, on the previously uncolonized surfaces of roots moving through soil, plant materials that become bruised, waters that receive sudden influxes of organic materials, etc. Their fitness traits are commonly associated with successful dispersal to the site and those characteristics that are the basis for their capacity to use the organic nutrients present. Those nutrients may be either compounds excreted by the emerging root or constituents of the tissues that become bruised and thus accessible for microbial utilization. Frequently, the initial colonist preempts the site so that a propagule that arrives later but has the same enzymatic capacity is unable to multiply. Preemptive colonization appears to be important in many habitats that suddenly become accessible for microbial growth, and in these instances, the key fitness trait is frequently associated with dispersal.

With time, the initial colonizing species are displaced and new organisms become abundant. The initially dominant organisms may become of less significance as successions proceed, and they may be eliminated totally. The trend with time of succession is increasing species diversity. As succession thus proceeds and species diversity increases, the identities of the fitness traits become increasingly less certain. Among the factors that contribute or determine selection during succession are the availability of nutrients that are synthesized by the preceding species, the alteration in concentration of inorganic nutrients, the formation of toxic products by the initial colonists, competition for limiting resources (especially the supply of organic carbon), predation by protozoa and invertebrates, and the appearance of organisms that parasitize the pioneering species or the subsequent colonists.

The displacement of organisms during colonization and succession ultimately leads to the *climax community*. This is the assemblage of organisms that are characteristic of many habitats and the communities of chief concern to environmental microbiologists. The climax community tends to reproduce itself and remain similar in composition with the passage of time. The organisms making up the climax community interact in a variety of ways, but the component populations are not eliminated. At this stage, the nature of the interactions is difficult to unravel, and the specific fitness traits that underlie an organism's position in the community are frequently uncertain.

Undoubtedly, competition is one of the major interactions in climax communities. As microorganisms usually grow readily, one or more factors in the environment become limiting. That limiting factor serves as the basis for competition. For communities dominated by heterotrophic bacteria and fungi, the limiting factor is frequently the supply of available carbon. Many environments, soils and sediments for example, contain large amounts of organic matter, but much of that organic matter is not readily available. Thus, the organisms that are successful are often those that are able to make use of the less readily available organic materials. In communities containing chemoautotrophs, the limiting factor is frequently the supply of the inorganic compound or ion that serves as their energy source. In surface waters, the limiting factor for the algae or cyanobacteria is often the concentration of phosphorus or nitrogen. In some instances, the community is dominated by a species that has preemptively colonized the site, and its role is associated with its presence at the site before other organisms arrived. In other instances, however, a variety of organisms endowed with appropriate physiological capacities have reached the site, and they are thus competing for the factor that they have in common but which is in limited supply.

What determines the outcome of competition? It is tempting to suggest that the successful competitor is the organism that grows most rapidly. However, the fact that many of the dominant organisms in natural environments do not grow quickly suggests that it is imprudent to extrapolate readily from pure culture studies on growth rates under noncompetitive circumstances to natural environments. More is involved than simply growth rate per se because of other stresses, the possible need to be transported to sites where there is an additional supply of the limiting nutrient, the need to avoid predation, etc.

THE BLESSING OF TRAVEL: DISPERSAL

An interesting question to put to an audience of intelligent lay individuals is, What are the major global pollutants? Most of the answers can be predicted. Typically, similar answers would be obtained if one addressed the same question to many ecologists, microbiologists, and other scientists. The answers probably would include air pollutants, ozone, pesticides, polychlorinated biphenyls, heavy metals, etc. However, the available morbidity and mortality data give answers that are quite unexpected. Indeed and, to many scientists and lay individuals, surprisingly, the major pollutants on the basis of morbidity and mortality data are microorganisms. Pneumonic infections, diarrhea, tuberculosis, and malaria are among the major causes of human death and suffering. This may not seem to be true for individuals in developed countries, in which the psychological issues associated with chemical pollutants and some of the physiological changes associated with air pollutants are of prime concern, but the actual data show that microorganisms are the chief environmental stressors, at least on a global basis.

What does the importance of microorganisms as major pollutants have to do with dispersal? Simply put, for most of these major causes of human misery, prophylaxis has some impact and chemotherapy has a significant role in reducing the incidence or effects, although not in remote or poor areas of developing countries. The major means of controlling or preventing these diseases is by interfering with microbial dispersal. Dispersal is thus not only important for the human population, as well as animals and plants, but also critical for microorganisms to maintain their own existence. Dissemination is a key factor, moreover, not only for pathogenic microorganisms but for all bacteria, fungi, protozoa, and algae.

Some of the scientific interest in microbial dispersal comes from basic research. Thus, there have been major breakthroughs in studies of chemotaxis and in the use of genetic and biochemical techniques to understand such movement in response to a chemical stimulus (19). In addition, a number of very useful mathematical models have been developed in aerobiology and for use in predicting microbial dissemination through aquifers and soils (3). Interest in microbial dispersal also comes from concern with bacteria and viruses causing diseases of importance to humans (9) and with fungi that, as a consequence of their aerial movement, are contributors to decline in food and feed production. The disposal on land of agricultural and urban wastes containing pathogens has also resulted in considerable research and monitoring of microbial movement.

Potentially new habitats regularly become available for microbial colonization. Thus, seedlings that emerge from seeds, growing roots, damaged fruits, the newborn infant, and bruised tissues contain sterile sites potentially inhabited by a variety of microorganisms that reach those sites. Healthy humans, animals, and plants exposed to pathogens that are being transported also represent potentially hospitable habitats. Nutrients appear at new locations in water bodies because of vertical and horizontal mixing of water, and new environments are also created by soil erosion and by the building of dams and ponds. The earlier statement that microorganisms are everywhere, or at least potentially everywhere, is clearly a platitude when one considers that the issue is not whether a microorganism will appear but whether a particular species will reach a site in which it can potentially grow.

Many microbial communities completely consume the supply of the limiting nutrient element or other limiting resource, and if there is no further input or regeneration of that limiting nutrient or resource, the species that make up those communities will disappear. Should this happen in all ecosystems in which a particular species resides, it will be eliminated unless it somehow can find new circumstances. Thus, unless a species forms a resistant structure that allows it to endure until conditions become favorable, its survival dictates the finding of a new environment. It must escape either in space (via a dispersal mechanism) or in time (via a persistence mechanism).

Microorganisms have developed many means for dispersal, and each species must have one or more methods for accomplishing its migration. Among these mechanisms are specialized structures to launch cells into air, vectors to carry propagules unerringly to new locations, mechanisms that cause changes in animal or plant hosts that result in dispersal of the parasite, and photo- or chemotaxis. Some of these means of dispersal have a high risk of failure, as is common with organisms dispersed through the atmosphere. To be successful, such an organism must produce enormous numbers of propagules to have one or several alight in an environment in which the organism can grow. Species with more efficient dispersal mechanisms require fewer propagules because of the greater likelihood of successfully encountering a hospitable site. Bacteria, viruses, or fungi that are transmitted by living vectors typically have efficient dispersal mechanisms and require fewer propagules for the species to be maintained.

Active dispersal, in which the physiology of the microorganism controls its transport, has been the subject of considerable research, probably because it is more comforting to the microbiologist to have his or her pet control its own fate. As a consequence, considerable research has been done on chemotaxis, although because of the energy requirement for movement toward or away from a chemical stimulus, such dispersal is restricted to short distances. Movement, as by motility or by growth of filaments, may be somewhat random rather than result from a taxis or tropism, but the motility of protozoa and the growth of the filaments of fungi in soil and algae in water often results in their colonization of new environments.

Passive dispersal, in contrast with the limited extent of spread associated with active dispersal, can result in an organism being disseminated to locations meters, kilometers, or hundred of kilometers away from its original reservoir. Aerial dispersal is frequently of considerable importance to fungi (17). A propagule that is transported through the air must have mechanisms to overcome three major hazards: radiation, desiccation, and extremes of temperature. In not all instances is it clear what physiological adaptations are responsible for the resistance to such environmental hazards. However, the presence of thick walls and dark pigments in many fungi transported through the air and of carotenoid pigments in many aerially dispersed bacteria suggests that these features are important adaptations for this mode of transport. The extent of migration of some of these organisms is truly impressive. For example, *Helminthosporium maydis,* a pathogen of corn, spread in a single year over areas of thousands of kilometers, doing enormous financial damage to the corn crop of North America. Similarly, the spores of the fungal genus *Hemileia* apparently spread from Angola to Brazil with tradewinds in the Atlantic Ocean, a distance of greater than 1,000 km.

Passive dispersal in water and soils has also been the subject of considerable inquiry, in part because of public health problems. For example, algae associated with red tides move for some distances and then suddenly create a bloom, a population explosion that may have major consequences for aquatic fauna and people consuming shellfish present in the area of the algal bloom. The interest in passive dispersal through soil is often a result of concern with the vertical migration of bacteria or viruses that cause human disease, but appreciable research has also been conducted on the vertical movement that is of importance in colonization of roots or the entry of microorganisms in aquifers underlying soils into which the microorganisms were inadvertently or deliberately introduced.

THE PLAGUE OF GEOGRAPHY

Geography? Yes, there is a microbial geography. Restricted distributions on a macroscale as well as on a microscale characterize all groups of organisms. The truism that microorganisms are everywhere or potentially everywhere has almost no meaning in the context of biogeography. The literature dealing with the geography of microbial groups is often unknown to most laboratory scientists, but an investigation of microbial communities in natural environments quickly shows marked and sometimes extreme localization of microbial groups. Biogeography, which deals with the distribution of organisms and the basis for the distribution, is evident among aquatic and terrestrial algae, free-living and pathogenic fungi, protozoa in marine waters and freshwaters as well as in soil, and bacteria in countless habitats. Many genera, species, and often subspecies are cosmopolitan, but the fact of their widespread distribution does not mask the restricted nature of their occurrence within particular regions, sites, or microenvironments (10, 23, 25).

Well known to phycologists is the limited distribution

of diatoms, some of which are present only in subtropical or tropical waters, whereas others characteristically are found in the Arctic or Antarctic. Other algae have snowfields as their habitats. In a more restricted but still large area are the algae that exist in patches in lakes and oceans. Some of these patches are no more than a few centimeters across, whereas others extend over areas of greater than 300 km. Bacteria such as *Beijerinckia* species similarly have a restricted distribution, being commonly but not solely present in soils of the tropics or subtropics (2). Often, the biogeography of a microorganism that is transmitted by a living vector or that is an obligate parasite is determined largely or exclusively by the biogeography of its vector or host. This is true of both *Plasmodium* and *Trypanosoma* species, for example, and the explanations for their distributions thus are quite simple. Not quite as simple to explain is the distribution of microorganisms that have no vector or are not obligately associated with a particular host organism, at least in their distribution. A notable example is *Coccidioides immitis*, which has a unique distribution in the Western Hemisphere (25). It is characteristically found in the soils of certain semiarid regions that typically are exposed to high temperatures, receive little rainfall, and frequently have high salinities. The organism causes disease in humans, but though the host is largely global, the fungus is restricted to certain localized sites. Although *C. immitis* is transmitted by wind, wind movement alone cannot account for its geography. Even if the fungus is cold intolerant, that possible sensitivity would not explain why it is not present in warm, humid areas to which it may be carried by the wind.

Zonation is also evident at a microenvironmental level, and marked horizontal and vertical differences in distribution or occurrence of organisms are evident in waters, sediments, and soils. Many studies have shown that algae, bacteria, and fungi have restricted distributions even at a microscale, and these clearly reflect differences in the physical and chemical characteristics of the environment that are biologically important. However, in only a few cases have the causes of this highly localized microenvironmental distribution been established. Among the factors known or postulated to be important in microscopic biogeography are nutrient concentration and type, temperature, pH, oxygen, grazing by zooplankton, mechanical barriers, and inhibitory substances.

THE BENEFITS OF ALTRUISM

To the general public, microorganisms are frequently considered to be solely harmful. This is evident in the use of the word *germ*. Yet altruism is a widespread attribute among all major categories of microorganisms. Admittedly, competition, parasitism, and predation are of great importance, and disease-producing microorganisms are widespread. Nevertheless, the good that many heterotrophic and autotrophic populations do to their neighbors should not be overlooked. Such beneficence is commonly categorized, from the ecological viewpoint at least, by the terms commensalism, protocooperation, and symbiosis. Although semantically sensitive readers might expect a clear distinction among these terms, in fact a continuum exists so that the range of each of these types of beneficial interaction merges into the next. Indeed, it is likely that there is a continuous evolution so that organisms which were, at one time, commensals evolve to exist in protocooperative relationships and those that are involved in protocooperation will, with

time, evolve into highly dependent and mutually beneficial symbioses.

In a commensal association, one species benefits a second but the first gets no good in return. The second is deemed the commensal. Such interactions are evident between heterotrophic bacteria and algae excreting photosynthetically fixed carbon (7) and apparently between populations in a biodegradative process (27). The mechanisms underlying these commensal association are numerous, but only a few have been characterized. In one such type of commensalism, one population converts a compound unavailable to a second population into a product that can be used as a nutrient source by that second species. The first gets no benefit from the association, but the second is provided with something essential for its replication. Similarly, if the second population is auxotrophic, it will not grow in the absence of the growth factors that are required, yet an organism that excretes a growth factor may allow for the development of its commensal. Studies of water and soils indicate that frequently as many as three-fourths of the indigenous species are able to excrete one or more vitamins or amino acids. Another type of commensalism is evident in environments containing organic or inorganic inhibitors, and a species that destroys an organic inhibitor or somehow detoxifies an inorganic ion will serve an altruistic role for its sensitive neighbor. Because terrestrial and aquatic communities contain a high percentage of microorganisms that are auxotrophic, often considerably more than half requiring one to many growth factors, and because algae and cyanobacteria excrete carbon into organic carbon-poor surface waters, commensalism must be of considerable importance.

An organism that relies on its neighbors for the synthesis of a carbon source, growth factors, or detoxifying enzymes does not need to expend energy for those purposes, so that in an environment in which the neighbor is present, that commensal has a competitive advantage, all other factors being equal, over an organism that must expend energy rather than being somewhat dependent. If evolution thus benefits a commensal, it would also likely favor a second species that itself is not fully self reliant. The second species would be more competitive with similar populations that must develop the full armament of enzymes and physiological mechanisms. In this way, two populations that rely on one another will have, with time, a selective advantage over partially or fully independent populations as long as the two species coexist. This is the basis for protocooperation, an association in which each of the two interactants needs and benefits the other. This two-member association, when functioning as a unit, has greater fitness than the two species functioning independently. The association, however, is not obligatory or is not specific, and the identities of the associates are not fixed. Synergism is often the result of protocooperation, synergism referring to a process in which two species cause a change that neither could perform alone or would carry out slowly. The two interactants may function because one provides a carbon or energy source, a growth factor, O_2, or detoxifying enzymes that serve as the basis for the benefit. Well-studied protocooperative relationships include those in which there is interspecies H_2 transfer (8) and the fermentation of polysaccharides (20).

In a simple protocooperation, only a single benefit needs to be transferred from one population to the second. However, if a simple protocooperation gives the interactants additional fitness, so too would greater degrees of integration. Each population may thus contribute more than one requisite to the second population. When this occurs, the association becomes somewhat less flexible, and the identities of the two interactants become somewhat more restricted because each must provide the full complement of benefits required by the second. This seems to be the basis for the tight interactions that characterize symbiotic or mutualistic associations.

Symbioses are evident among many microbial groups and in many different environments. Some involve two species of microorganisms, and some involve one microbial species and a plant or animal. In the orchid symbiosis, by contrast, there are three species, the orchid, a fungus, and a tree; the first two are symbionts, but the fungus is a parasite of the tree. Among the intermicrobial symbioses, a few have been the subject of considerable scrutiny, and sometimes the contributions of the symbionts to their partners have been well defined. The lichen, an association of an alga with a fungus, has long been of interest to botanists, although to few microbiologists, and the nature of the symbiosis is becoming ever more clear. In contrast, the symbiotic associations between bacteria residing within protozoa and protozoa relying on their internal inhabitants have received scant attention. Several of the symbioses involving plants, by contrast, have been intensively investigated and remain the subject of active research in many countries. That research has been in part driven by the benefits to human society of such relationships as the fungus-root symbiosis termed the mycorrhiza and the association between legumes and members of the genera *Rhizobium* and *Bradyrhizobium*. In both instances, the microorganisms are heterotrophs, although photosynthetic rhizobia are now known to occur, and the microorganism must thus get carbon from the plant. In turn, the root-nodule bacterium provides nitrogen in a usable form for the higher plant, but the precise contribution of the mycorrhizal fungus to the plant is still the subject of considerable controversy. Nevertheless, the simple provision of carbon by the plant to the microorganism and of nitrogen by the rhizobia to the legume does not adequately explain the symbiosis, because nonlegumes also provide carbon through their root excretions and many bacteria are able to fix nitrogen yet do not enter into symbiotic relationships.

THE HAZARDS FROM THE HUNTERS: PREDATION AND PARASITISM

Microorganisms may feed upon other organisms. In turn, microorganisms serve as prey and hosts for other microorganisms. A truly vast literature exists on some of these relationships, particularly on the actions of bacteriophages on bacteria in culture media. Far less sizable is the literature on the ecological role of the species that parasitize or prey upon microorganisms in nature. Too often, the abundance of literature and the ease of working with some of the organisms leads the nonecologist to conclude that a particular group of parasites or predators is indeed important in nature. The chief reason for believing in importance, implicit but not explicit in the research, is the interest of the researcher but not necessarily the data that are available. Unfortunately, it is not quite as easy to establish the significant interactions in highly heterogeneous microbial communities as it is to carry out studies with only two organisms in culture media.

Among the microbial predators are protozoa, myxobacteria, acrasiomycetes, and a number of chlorophyll-containing flagellates that are often but not always classified as algae. Of these various groups, incontrovertible evidence exists for the importance only of protozoa in determining the structure or function of microbial communities (1, 5, 6). Wastewaters, many surface waters, and the rhizosphere of actively growing plants often have large numbers of protozoa, and a high percentage of the individual protozoa are present in the trophic stage that is associated with active feeding on bacteria. In many aquatic environments containing an inflow of readily available organic compounds, bacterial numbers increase rapidly as the organic substrates they utilize are metabolized, but then there is a sharp decline in bacterial abundance as the protozoa begin to proliferate. Since a single protozoan cell, upon division, consumes 10^3 to 10^4 bacteria, a community with 10^3 protozoa per ml or per g is characterized by active predation. By use of eukaryotic inhibitors, moreover, it has been shown that inhibition of protozoan growth is associated with the maintenance of large numbers of potential prey individuals and often the existence of populations that otherwise would have been suppressed.

Nevertheless, despite their abundance and activity, protozoa do not eliminate their prey. They, and most other predators, are prudent, and the elimination of prey by an obligate predator would result in the subsequent elimination of the predator species itself. Prudence appears to be related to the density dependence of feeding, and a predator that is able to markedly reduce the size of a large population is unable to effectively destroy a small population. This has often been interpreted in terms of the balance between energy needed for hunting and energy gained by the hunt; thus, when a predator population is getting more energy by feeding on the prey, as would occur at high prey densities, than it needs to hunt for those prey, it will continue to graze. On the other hand, at low prey densities, the predator must use considerable energy to move to the few surviving prey cells or to bring those survivors to itself for consumption, and it thus will not continue to feed and may either die or encyst (3). This is not to say that the predator cannot eliminate a single prey species, which it will do if there are sufficient alternative prey to maintain a bacterial density sufficient for feeding; under such circumstances, the alternative prey would provide sufficient energy to the protozoan that it can eliminate a small population of a nongrowing or slowly growing bacterial species (18). Thus, the effects of protozoa may be viewed in terms of reducing the size of the bacterial community and also, under some circumstances, possibly eliminating individual species when the bacterial community is large. Still, soils contain higher bacterial densities, at least by total counts, than are predicted on the basis of the capacity of the indigenous protozoa to reduce the bacterial densities to lower numbers. For example, the total count of bacteria in soil may be in the vicinity of $10^9/cm^3$, but in solution, the protozoa may reduce the bacterial density to approximately $10^6/cm^3$. Therefore, it is likely that a refuge exists in soil, a refuge that possibly results from the inability of protozoa to penetrate pores large enough for bacteria but too small for the predator or from the inability of protozoa to feed on bacteria that are sorbed to particulate surfaces.

Less certain and still subject to controversy is the role of bacteriophages and *Bdellovibrio* species in nature. Because these organisms rely on bacteria for their development and even existence in nature, the ecology of their host or prey must be important in the ecology of the bacteriophages and *Bdellovibrio* species. However, does the virus or the vibrio have an appreciable impact on the population that sustains it? Both the virus and the vibrio exhibit density dependence (15, 24, 26), and most evidence suggests that they are unable to have an impact on small populations and thus do not eliminate the species on which they feed. The view of density dependence of viral replication has been challenged, however (16). Moreover, inasmuch as viruses are species or strain specific and individual species or strains rarely are present at very high cell densities in nature, the likelihood is small that the bacteriophages will do more than occasionally eliminate a sufficient number of host bacteria for the virus itself to be maintained. A possible exception is in monospecific algal or cyanobacterial blooms. The vibrios, by contrast, act on a broader range of hosts, so that even if the density of an individual prey species is low, the vibrio may be maintained and have an impact on a community with a high bacterial cell density. Nevertheless, although there are no data suggesting that either the bacteriophages or vibrios have a significant impact and data to the contrary exist (1), further study is required to establish their role and ecological significance.

THE ORDEAL OF RELEVANCY

An ecologist has been euphemistically defined as an individual whose feet are firmly planted in midair. Admittedly, much of ecology, including environmental microbiology, is concerned with basic science. This is as it should be. However, it is also clear that environmental microbiology is partly an applied science and that it has much to offer to our knowledge and approaches to public health, prevention of the transmission of diseases of animals and plants, the maintenence or restoration of environmental quality, and a variety of technologies. The epidemiology of communicable diseases is, to a significant degree, an extension of evaluations of microbial dispersal. The spread of viruses, bacteria, fungi, and protozoa thus has a tremendous impact on human and animal health and plant protection. Many aspects of chemotherapy rely on knowledge of upsets and restorations of microbial communities, and preventive medicine likewise requires information from environmental microbiology.

Plant pathologists have long recognized the significance of information on the ecology not only of the disease-producing fungi, bacteria, and viruses but also of other microorganisms that reside in the same habitats. Aerobiologists and soil microbiologists frequently interact with plant pathologists, and much of the research is interdisciplinary. Many plant diseases are not effectively controlled by chemical agents or by sanitation procedures, and it is the activity of other, nonpathogenic members of microbial communities that is the basis for effective control of particular diseases.

The maintenance of environmental quality relies on the role of microorganisms in preventing pollution, destroying organic materials before the concentration becomes objectionable, or destroying toxic chemicals before they have an impact on humans, animals, or plants. Similarly, environmental microbiologists are actively engaged in bioremediation technologies and in designing ways to enhance microbial activities to bring about the destruction of chemicals that already exist as pollutants. Therefore, knowledge of microbial community structure as well as function has a key role in both basic and applied science, and this knowledge will aid in furthering our understanding of both natural ecosystems and ways of maintaining environments in which humans wish to live.

REFERENCES

1. **Acea, M. J., C. R. Moore, and M. Alexander.** 1988. Survival and growth of bacteria introduced into soil. *Soil Biol. Biochem.* **20:**509–515.

2. **Alexander, M.** 1977. *Introduction to Soil Microbiology.* John Wiley & Sons, Inc., New York.

3. **Alexander, M.** 1981. Why microbial predators and parasites do not eliminate their prey and hosts. *Annu. Rev. Microbiol.* **35:**113–133.

4. **Alexander, M.** 1994. *Biodegradation and Bioremediation.* Academic Press, Inc., San Diego, Calif.

5. **Berninger, U.-G., B. J. Finlay, and P. Kuuppo-Leinikki.** 1991. Protozoan control of bacterial abundance in freshwater. *Limnol. Oceanogr.* **36:**139–147.

6. **Bloem, J., F. M. Ellenbroek, M. J. B. Bar-Gilissen, and T. E. Cappenberg.** 1989. Protozoan grazing and bacterial production in stratified Lake Vechten estimated with fluorescently labeled bacteria and by thymidine incorporation. *Appl. Environ. Microbiol.* **55:**1787–1795.

7. **Cole, J. J.** 1982. Interactions between bacteria and algae in aquatic ecosystems. *Annu. Rev. Ecol. Syst.* **13:**291–294.

8. **Conrad, R., H.-P. Mayer, and M. Wüst.** 1989. Temporal change of gas metabolism by hydrogen-syntrophic methanogenic bacterial associations in anoxic paddy soil. *FEMS Microbiol. Ecol.* **62:**265–274.

9. **Cox, C. S.** 1989. Airborne bacteria and viruses. *Sci. Prog.* (Oxford) **73:**469–499.

10. **Ducklow, H. W.** 1984. Geographical ecology of marine bacteria: physical and chemical variability at the mesoscale, p. 22–31. *In* M. J. Klug and C. A. Reddy (ed.), *Current Perspectives in Microbial Ecology.* American Society for Microbiology, Washington, D.C.

11. **Ferry, J. G., and R. S. Wolfe.** 1976. Anaerobic degradation of benzoate to methane by a microbial consortium. *Arch. Microbiol.* **107:**33–40.

12. **Fletcher, M., T. R. G. Gray, and J. G. Jones (ed.).** 1987. *Ecology of Microbial Communities.* Cambridge University Press, Cambridge.

13. **Hairston, N. G.** 1959. Species abundance and community organization. *Ecology* **40:**404–416.

14. **Hurst, C. J. (ed.).** 1991. *Modeling the Environmental Fate of Microorganisms.* American Society for Microbiology, Washington, D.C.

15. **Keya, S. O., and M. Alexander.** 1975. Regulation of parasitism by host density: the *Bdellovibrio-Rhizobium* interrelationship. *Soil Biol. Biochem.* **7:**231–237.

16. **Kokjohn, T. A., G. S. Sayler, and R. V. Miller.** 1991. Attachment and replication of *Pseudomonas aeruginosa* bacteriophages under conditions simulating aquatic environments. *J. Gen. Microbiol.* **137:**661–666.

17. **Leonard, K. J., and W. E. Fry (ed.).** *Plant Disease Epidemiology,* vol. 1. McGraw-Hill Book Co., New York.

18. **Mallory, L. M., C. S. Yuk, L.-N. Liang, and M. Alexander.** 1983. Alternative prey: a mechanism for elimination of bacterial species by protozoa. *Appl. Environ. Microbiol.* **46:**1073–1079.

19. **Manson, M. D.** 1992. Bacterial motility and chemotaxis. *Adv. Microb. Physiol.* **33:**277–346.

20. **Murray, W. D.** 1986. Symbiotic relationship of *Bacteroides cellulosolvens* and *Clostridium saccharolyticum* in cellulose fermentation. *Appl. Environ. Microbiol.* **51:**710–714.

21. **Simberloff, D.** 1989. Which insect introductions succeed and which fail?, p. 61–75. *In* J. A. Drake, H. A. Mooney, F. di Castri, R. H. Groves, F. J. Kruger, M. Rejmanek, and M. Williamson (ed.), *Biological Invasions: a Global Perspective.* John Wiley & Sons, Inc., Chichester, England.

22. **Simberloff, D., and M. Alexander.** 1994. Biological stressors, p. 6-1–6-60. *In Ecological Risk Assessment Issue Papers.* EPA/630/R-94/009. U.S. Environmental Protection Agency, Washington, D.C.

23. **Tett, P.** 1987. Modelling the growth and distribution of marine microplankton, p. 387–425. *In* M. Fletcher, T. R.

G. Gray, and J. G. Jones (ed.), *Ecology of Microbial Communities*. Cambridge University Press, Cambridge.

24. **Varon, M., and B. P. Ziegler.** 1978. Bacterial predator-prey interaction at low prey density. *Appl. Environ. Microbiol.* **36:**11–17.

25. **Weitzman, I.** 1991. Epidemiology of blastomycosis and coccidiomycosis, p. 51–74. *In* D. K. Arora (ed.), *Handbook of Applied Mycology*, vol. 2. Marcel Dekker, New York.

26. **Wiggins, B. A., and M. Alexander.** 1985. Minimum bacterial density for bacteriophage replication: implication for significance of bacteriophages in natural ecosystems. *Appl. Environ. Microbiol.* **49:**19–23.

27. **Zeikus, J. G.** 1983. Metabolic communication between biodegradative populations in nature, p. 423–462. *In* J. H. Slater, R. Whittenbury, and J. W. T. Wimpenny (ed.), *Microbes in Their Natural Environments*. Cambridge University Press, Cambridge.

Prokaryotic Diversity: Form, Ecophysiology, and Habitat

EDWARD R. LEADBETTER

3

Although the meaning of the word "diversity" is ordinarily quite clear, this is not always so when it is used in connection with living organisms. The connotations of the word with regard to large organisms ("macrobes") are usually different than when applied to smaller ones ("microbes"). One obvious difference is that with the former group it is ordinarily organisms with distinctive anatomical and morphological features visible to the eye that are the objects of the habitat census and it is these traits that lead to enumeration or identification of different genera and species present. The sustained *presence* of an organism is taken as evidence of its significance—the ability to thrive—in the habitat. Usually there is little regard to, or indeed knowledge of, the organism's precise functions in that habitat (6). By contrast, establishing the diversity of microorganisms present is not really possible by relying on morphological features alone; instead, specific physiological capabilities need also to be deciphered in order to establish the different types of microbes present. Invariably this has required the isolation of pure cultures and determination of cell traits.

The specific biochemical and physiological activities of microbial cells are of profound significance for the habitat. The high surface-to-volume ratio of these very small cells endows them with high biochemical and physiological activities, and this in turn marks them as having major impacts on chemical aspects of their habitats (16). This impact of nutrient consumption and product formation as a consequence of microbial growth is such that the organisms are properly regarded as determinants of their habitat environments and not merely, as is the case for so many macrobes, organisms that either survive, or do not, in response to environmental changes.

Thus, diversity in the context of the macrobe is more an index of the suitability or degree of habitat health for existing, or desired, macrobiota, while in the microbial context the diversity revealed by a census leads to establishment of the bases for and maintenance of the chemical as well as many physical features of the habitat.

As used in this Manual, the adjective "environmental" refers to microbiology that has other than a clinical emphasis and thus sets this Manual in contrast to the *Manual for Clinical Microbiology*, which is oriented in large part to microbes associated with diseases of animals, especially humans. This chapter not only discusses the diversity of microbes and their functions in the more usual environmental sense (association with soils and waters, for example), but also includes a focus on microbes associated, either extracellularly or intracellularly, with macrobes, including animals, insects, and plants. This seems appropriate for several reasons. One is that the alimentary tract of many macrobes has long been considered a tube within the organism, with openings at each end to the exterior environment in which the macrobe exists; invariably the lumen, wall, and other anatomical aspects of such a tube are colonized by microbes. The factors that affect initial microbial colonization and subsequent survival are, at least in part, those of the environment exterior to the macrobe. A second reason is that, for example, the bacterial invasion of plant root cells that results in formation of the oft-described "nodule" associated with symbiotic assimilation of dinitrogen, or the formation of the light-emitting organs of some marine biota, is, in either instance, a reflection of the existence and persistence of microbes both in the macrobes and the environment in which the latter live. For similar reasons, other significant and often not yet well understood interactions of microbes and macrobes are included in this discussion of the diversity of prokaryotes and of the habitats in which they are present.

RECOGNITION OF PROKARYOTE DIVERSITY: BACKGROUND

Although it has been more than three centuries since what we now call prokaryotes were first seen, it has been only slightly more than a century that the organisms termed bacteria, in particular, began to be studied extensively. As many have pointed out in general terms, especially Stanier (53) in an environmental microbiological context, our perceptions that microbes as a group possess extremely diverse characteristics and live and function in quite diverse habitats (many of which seem "extreme" to those acquainted only with macrobiota) had their origins in the extraordinary, novel, and penetrating observations and studies of two particular individuals—the Russian S. Winogradsky and the Hollander M. Beijerinck. As van Niel noted (56), those who studied with Beijerinck and his successors in Holland and elsewhere (the "Delft School") were among the major contributors to both the breadth and depth of our comprehension of prokaryote diversity, the crucial functional roles these microbes play in the biosphere, and, accordingly, the recognition that in the absence of these diverse microbial

14

activities, other forms of life on the planet would promptly cease to exist!

Even though our present understanding of prokaryotic diversity, in terms of form, function, and habitats occupied, is enormous when compared with that enjoyed some 30 or 40 years ago, we are nonetheless regularly made aware of the real limits of that understanding. The enormous success of enrichment culture and related isolation approaches and considerations (28, 53, 56), along with application of new analytical techniques (2, 13, 38, 59), has enabled us to see (literally, in some instances) that we understand very little about so many of the presumptive prokaryotes present in nearly every habitat we explore. The "humbling experience" (60) that resulted from the study of natural populations with then state-of-the-art light microscopes has become ever more common and daunting as newer approaches to light, electron, and other forms of microscopy have evolved and as tools of a molecular biological nature (e.g., analysis of RNA sequences, use of RNA-targeted oligonucleotide probes) have been brought to bear in the examination of mixed microbial populations of many habitats (4, 47, 58).

LIMITATIONS IN COMPREHENSION OF PROKARYOTE DIVERSITY

The past several decades have indeed been heady times in what might be termed the "Golden Age of Environmental Microbiology" (by analogy to the label applied to the period of about a century past in which there were stellar accomplishments establishing specific microbes as causes of specific diseases). It is a rare issue of a major journal that does not contain a report of the cultivation and characterization of an organism that had been refractory to study in the laboratory, or the presence of one with apparently novel features detected in a habitat that had seemingly already been well explored, or that the microbial content of some novel habitat is shown to be worthy of inspection. Even so, it is not possible at this time to describe completely either the enormous diversity in features of the many prokaryotes presently recognized, the wide range of habitats in which they are known to occur, or the microbial composition of the populations present in those habitats.

There is no little irony in the fact that as we learn more and more about ever more members of the *Bacteria* and *Archaea*, we come to realize that we comprehend less and less, both qualitatively and quantitatively, the diverse features of different prokaryotes and of environmental factors affecting, and reflecting, their growth and persistence. Twenty years ago, some might have surmised that perhaps as much as 40% of the prokaryotic world was recognized and understood; today it is probably overly optimistic to suggest that the figure is more than 5%. Such has been the progress in the more than 300 years since van Leeuwenhoek not only discovered bacteria, but described some diversity in morphology (e.g., cocci, rods), in function (e.g., motile versus nonmotile cells), and in habitat (e.g., a pepper-grain infusion, tooth and gum surfaces, sputum). Pasteur, when he observed the "vibrion butyrique" some 200 years later, laid the foundations of the biochemical, physiological, and genetic diversity that continues to be studied and enlarged today.

Readers are reminded that in this assessment of prokaryotic diversity and the range of habitats in which these organisms are known to multiply (or at least survive), the phrase "as known at the present," even though it is not regularly stated, needs to be kept uppermost in mind. This constitutes, then, a challenge to those interested in enlarging our understanding of the microbes' contribution to life on the planet. In the space available for it, this chapter was not intended to be so much a vade mecum as it was to be a descriptive outline of the *Bacteria*, the *Archaea*, and their habitats.

SOME FACTORS AFFECTING LIMITATIONS IN OUR UNDERSTANDING

Many different factors are responsible for the fact that we are yet unable to describe the prokaryotic world in its entirety. As in any of the sciences, it is quite often major advances in the development and application of new technologies that permit significant increases in our comprehension. In addition, a regular impediment is the limited ability of humans to apply new findings in an integrative way to the microhabitat dimensions and features that are so difficult to describe adequately. Among the factors which continue to be hurdles that we must overcome are those noted below.

Only in the most rare instances does the size or shape alone, invariably deduced by microscopic examination of a prokaryote, permit one to identify it and equate its properties with those of an organism already described. This limitation is no less serious if one encounters a novel or "new" organism. The quite considerable range in sizes and shapes sometimes makes difficult the determination of whether an object is in fact an organism or is instead some inanimate material. In addition, many cells are sufficiently small so as to be at or below the limits of resolution of the light microscope and hence not visible or visible only with special optics (e.g., dark-field microscopy, as for some spirochetes). Still another important consideration is that for a single cell or even a few cells to be detected in a typical field of view using light microscopy, on the order of 10^6 cells per ml needs to be present. Clearly, it would be easy to overlook the presence of an organism less abundantly present.

It is a fact that no single laboratory growth environment (e.g., nutrients, other chemical and physical factors) permits the multiplication and subsequent isolation and culture of even a small subset of the *Bacteria* or *Archaea* that we know at present. In addition, in many instances cells are attached to or localized on surfaces or form biofilms from which it is difficult to remove the cells without inflicting damage to them. Accordingly, for many habitats it is not readily possible either to enumerate the prokaryote population present or to isolate and then describe thoroughly the types present.

Pure cultures, as such, are rarely found in habitats other than the laboratory. Even though we continue to explore the metabolic and physiologic properties of pure (as well as some simple, mixed) cultures in controlled laboratory environments and to use these insights to make reasoned suggestions about functional attributes that *might* occur in natural habitats, we obviously lack such information about the capabilities of those organisms not amenable to cultivation. Of no little importance are the many demonstrations (e.g., references 1 and 33) of the interactions of different prokaryotes with each other, or with other biota, with the result that the behaviors of members of mixed populations (mixed cultures) are seen to differ from those predicted on the basis of study of component pure cultures.

These are but a few reasons why a particular organism's presence, in and of itself, in a habitat does not necessarily provide evidence concerning the precise in situ activities of that organism and its contributions to habitat functions. Another complication is that many of the *Bacteria* exhibit an enormous degree of metabolic flexibility (e.g., nutrients consumed, metabolic end products, mode of energy conservation); these traits are often regulated by numerous and complex environmental factors that are difficult to ascertain with accuracy for the microscale environments occupied. In contrast, not all prokaryotes possess such broad spectra; instead, many possess a specialized metabolism or physiology. Detection of significant numbers of these types of organisms in a habitat may allow a more confident prediction of their roles in the habitat than is possible for the more metabolically flexible ones.

Highly significant contributions to our present understanding of the relationships between particular microbes and cause-effect relationships in many habitats have resulted from use of the enrichment culture approach. Many who have carefully employed this approach to ask (i) if organisms with already known attributes were present in samples from a habitat being examined or (ii) whether a novel transformation of compound "C" to compound "D," or some other alteration of a habitat feature, could in fact be mediated by a microbe have recognized the need to quantify the numbers of the particular organisms present as an important aspect of establishing causal relationships (57). Such quantification often involves use of most-probable-number (MPN) studies employing particular enrichment culture media and environmental conditions as well as determination of biotransformation rates under, for example, conditions closely approximating those of the habitat in question.

Even though application of relatively newly described "molecular probe" technologies is argued (e.g., references 4 and 59) to obviate the need for bringing organisms into culture in order to describe the microbial composition of the habitat studied, we need to recognize some caveats associated with this methodology (2). One is that, in general, the probes are designed on the basis of our knowledge of molecular signatures of well-described and understood organisms, that is, organisms whose properties have been discerned from study of pure cultures. As long, then, as the habitats being examined using these probe methodologies contain no heretofore novel, unrecognized organisms with, accordingly, previously undescribed molecular signatures, the "there is no longer a need to culture" phrase is to a degree correct. On the other hand, because of the inherent metabolic/physiologic flexibility noted above of many prokaryotes, even if probe studies reveal the presence of, for example, sulfate-reducing bacteria in a habitat, this information alone does not establish whether it is sulfate, sulfite, thiosulfate, some other sulfur-containing electron acceptor, or even nitrate that is being anaerobically respired by the population or whether it is lactate or some other electron donor that is serving as the energy source. Still, the potential and utility of the molecular probe approaches, when used in combination with those of classic enrichment culture for habitat analysis, are very likely to lead to greater comprehension, as demonstrated by a recent elegant study of the ecology of methanotrophs (23), a physiological group with a notably limited nutritional spectrum for growth.

But perhaps the most significant factors responsible for the limitations in our understanding of bacterial diversity relate, first and foremost, to the relatively small numbers of analysts who have devoted themselves to these tasks, and second, to the fact that often their fascinating revelations led many of them to a type of "reductionism," namely, the exploration and elucidation of the properties of the organism(s) newly isolated. Because the elucidation of novel properties of such isolates was and continues to be in itself both stimulating and rewarding, individuals with interests in surveys and elucidation of microbial diversity have had their attentions refocused. There is probably no better explanation for the enormous numbers of newly described prokaryotes possessing newly recognized biochemical, morphological, and physiological properties other than that there has been a real increase over the past two to three decades in the numbers of investigators who have pursued this type of scholarship. Even so, the total number of practitioners of this art and science is not large, and their goals, significant as they are, often have not been widely appreciated or understood, nor has the research been well funded; "a fishing expedition" or "there is no hypothesis" is a term too often seen in reviews of requests for funding of promising projects.

Many thoughtful students of prokaryotic diversity have long recognized that only a fraction of the presumed prokaryotes seen in habitat samples or appearing in enrichment cultures have ever been brought into pure culture (or into stable mixed culture) so that traits could be elucidated and the organisms could be identified (44). Seemingly insignificant modifications of the culture medium or incubation conditions often resulted in pronounced changes in the populations that developed. Even when similar modifications were made in isolation media in attempts to further cultivate such organisms, the efforts were often unsuccessful. Clearly, the organisms' nutritional needs (in the broad sense of these words) remained unmet. Possible explanations are legion. An oft-noted situation is that the populations developing in an enrichment medium with or without agar present as a solidifying agent (an example of the accumulative versus the separative enrichment culture approach) are recognizably different. Many investigators note that some organisms that do appear in accumulative (liquid) enrichments are unable to grow in an otherwise presumably identical agar-solidified medium (53, 57).

This discordance between what can be seen and what can be cultivated has been brought into renewed view, literally speaking, as a result of current analytical approaches based on extraction of nucleic acids from populations collected from native habitats to permit amplification and cloning of genes for rRNA. In many instances, rRNA with sequences unlike those already characterized from cultivated organisms is found, thereby indicating the presence of a novel, undescribed organism(s) in the habitat sampled. Thus, many factors that had long been recognized by those seeking to enlarge the understanding of different prokaryotes present in different habitats have now become more widely appreciated and will lead to new emphases and approaches in pursuit of long-sought goals.

MORPHOLOGIES AND SIZES OF PROKARYOTIC CELLS

We no longer consider prokaryotic cells to have shapes once described simply as spheres, rods, vibrioids, or spirals, for in addition we recognize cells that are essentially square and others that are nearly rectangular; some that may approximate a triangle, or others a star. Many of these different morphological types are characterized as well by the presence of constrictions, protuberances, lobes, or other geometrically irregular aspects of their surfaces. Cells may exist as single entities or as units forming chains, clumps, or fila-

TABLE 1 Examples of diversity in morphology of selected prokaryotes

Morphology or other cell anatomical feature	Representative genus (genera)
Coccus	
Single	*Acidianus, Megasphaera*
In chains	*Lactococcus, Streptococcus*
In other groupings	*Pediococcus, Sarcina*
Coccus-rod, lobed	*Sulfolobus*
Coccus ↔ rod[a]	*Arthrobacter*
Circular (or nearly so)	*Cyclobacterium*
Cyst (or microcyst) formed[a]	*Azotobacter, Sporocytophaga*
Endospore formed[a]	*Bacillus, Clostridium, Acetonema, Sporosarcina, Thermoactinomyces*
Exospore formed[a]	*Methylocystis*
Irregular	*Nocardia, Mycobacterium, Streptomyces*[a]
Myxospore formed[a]	*Myxococcus, Stigmatella*
Rectangle	*Methanopyrus*
Rod	
Long, single	*Bacillus*
Short, single	*Pseudomonas*
Often in chains	*Bacillus, Lactobacillus*
Spiral, small	*Bdellovibrio, Desulfovibrio, Rhodospirillum, Methanospirillum*
Sheathed	*Thermotoga, Sphaerotilus*
Square	Not yet named
Stalked[a]	*Asticcaulis, Caulobacter*
Triangle	*Pyrodictium*
Vibrioid	*Bdellovibrio, Marinomonas, Vibrio*
Wall-less	*Mycoplasma, Thermoplasma*

[a] Cell undergoes morphogenesis.

TABLE 2 Diversity in prokaryote cell size

Size (μm)[a]	Representative genus
<0.2	Some "picoplankton"
0.3–0.5	*Veillonella*
2–3	*Megasphaera*
5–25	*Thiovulum*
0.6–1.2 × 2.5–5.8	*Bacillus*
1.1–1.5 × 2–6	*Escherichia*
5–6 × 8–12	*Chromatium*
2.5–4 × 40–100	*Thiospirillum*
1–100 × 50–200	*Beggiatoa*
80 × 600	*Epulopiscium*

[a] Approximate diameter; × length, where appropriate.

ments. Table 1 provides a selected list of the diverse morphological traits of representative organisms.

This morphological diversity is a contributing factor to the difficulty sometimes encountered in distinguishing cells from inanimate material. Of additional significance is the fact that the morphology of many prokaryotes has an inherent degree of inconstancy, that is, that the morphology of a given organism can undergo change depending upon the extracellular environment or the stage of growth, for example. In addition, cells subjected to drying for purposes of staining, for instance, sometimes appear discernibly different from cells in the living, unstained state.

The range of sizes for different prokaryotic cells is quite large. There is an overlap in the size range of pro- and eukaryotic cells. One unicellular organism may be barely sufficiently large to be resolved by and thus seen in the light microscope while another, conversely, can be just large enough to be seen with the unaided human eye (5). Table 2 provides a selected listing of the cell sizes currently recorded for prokaryotes.

PHYSIOLOGICAL DIVERSITY AMONG THE PROKARYOTES

In the biological world, there are no known parallels to the abilities of the *Bacteria* to utilize an enormous array of energy sources to support growth. Presumably this represents the consequences of mutation and adaptation that occurred during the lengthy evolutionary history of this life on the planet. The recognition that beneath the widely disparate nutritional and environmental needs for the growth of different bacteria there was an underlying unity in their physiological attributes was a major conceptual contribution (28) that had a marked practical influence on the development, nature, and extent of our understanding of the significance of prokaryotic diversity (29).

This encompassing view was that energy was conserved and made available for life processes as a result of cellularly mediated, coupled oxidation-reduction reactions and that as a group the *Bacteria* exploited in this way nearly every conceivable source of energy available. Then novel, this concept of an extraordinary versatility, along with the wide range of environments in which different prokaryotes were shown to grow, led to the notion that prokaryotes are unique in their ability to inhabit and thrive in environmental "extremes" (i.e., those not conducive for life for macrobes) (31).

One outgrowth of these considerations was the recognition that every naturally occurring organic compound was subject to attack (usually meaning utilization as a growth-supporting nutrient) by one or another bacterium and min-

TABLE 3 Some nutritional aspects of physiological diversity

Electron donors utilized (examples)	Electron acceptor, *reduced end product(s)*

In aerobic respiration
  Organic molecules
    Carbohydrates, amino acids, purines, pyrimidines, lipids, fatty acids, alcohols, hydrocarbons (both aliphatic and aromatic), sulfonic and aromatic acids
  Inorganic molecules or ions
    Carbon monoxide, molecular hydrogen, metallic sulfides, ammonium, nitrite, ferrous and manganous salts, elemental sulfur
	Molecular oxygen, *water*

In anaerobic respiration
  Organic molecules: much as above; possible exceptions are some sulfonic and aromatic acids, gaseous hydrocarbons
 	
  Inorganic molecules or ions: much as above
	Nitrate, nitrite; *nitrite, nitrogen gas, "ammonium"*
	
	Sulfate, sulfite, elemental sulfur; *sulfite, sulfide*
	
	Fumarate, *succinate*; dimethyl sulfoxide, *dimethylsulfide*; ferric salts, *ferrous salts*; trimethylamine oxide, *trimethylamine*

But methanogens typically utilize primarily hydrogen gas, formate, or acetate	Carbonate, *methane*
In fermentation Organic molecules: carbohydrates, purines, pyrimidines	Organic molecules, protons; *alcohols, fatty acids, ketones, hydrogen gas*

In phototrophy
  Organic molecules: alcohols, fatty acids, organic acids (e.g., malate, succinate, benzoates)
  Inorganic compounds or ions: hydrogen gas, ferric, sulfide, elemental sulfur, thiosulfate
  But, for cyanobacteria, water
	Carbonate, *cellular components*

eralized as a result of participation in the biogeochemical cycles. We have come to realize that not all compounds synthesized by animals and plants are subject to biodegradation at identical rates (the slow attack on lignocellulosics is one example) and that the initial degradative steps of some biosynthesized materials are carried out not by *Bacteria* or *Archaea*, but instead by other microbes. The initial contributions of Beijerinck and Winogradsky (53) that established the utilization of inorganic ions and molecules for energy conservation coupled to growth have since been extended in terms both of the scope of the *Bacteria* and *Archaea* involved and of the range of inorganic entities able to serve in this capacity (3, 26, 35).

It is customary (8, 24, 42, 48) to categorize and contrast the several ways prokaryotes employ the coupled oxidation-reduction reactions listed below. Table 3 provides examples of some additional aspects of these processes.

- Aerobic respiration: molecular oxygen serves as the oxidant in a redox reaction and appears in reduced form as water, one end product of this metabolism.
- Anaerobic respiration: in environmental conditions where molecular oxygen is absent or in limited supply, an inorganic ion such as nitrate, sulfate, or carbonate serves as the terminal oxidant and becomes reduced to dinitrogen (or ammonia), sulfide, or methane, respectively. It is now recognized that a variety of other ions

such as oxidized (ferric) iron, or organic molecules such as fumarate, trimethylamine oxide, or dimethyl sulfoxide, are also able to serve as terminal oxidants for a variety of anaerobically respiring prokaryotes.

- Fermentation: an organic compound, usually a metabolic intermediate resulting from oxidation of the organic energy source, serves as the terminal oxidant and, accordingly, a more reduced organic molecule is a metabolic end product(s).
- Phototrophy: radiant energy is absorbed by chlorophyll-, bacteriochlorophyll-, or "accessory pigment"-containing pigment complexes, resulting in an excitation of electrons present in the complex and leading to an oxidation and charge separation. When water is employed as the ultimate reductant for the sequential reactions, the processes are characterized by evolution of molecular oxygen ("oxygenic photosynthesis"); when either inorganic or organic compounds replace water's role, molecular oxygen is not formed and the processes are termed, instead, anoxygenic photosynthesis.

We may summarize, then, one aspect of the prokaryotes' physiological diversity by noting that some are able to inhabit and multiply only in habitats that are in regular contact with the Earth's atmosphere and the molecular oxygen it contains, employing aerobic respiration (i.e., are strictly or obligatorily aerobic). Others function only in the absence of air, employing fermentation or anaerobic respiration, and thus pursue other modes of energy acquisition and conservation; we term such organisms strictly or obligatorily anaerobic (but might also describe them as strictly fermentative or as living strictly by anaerobic respiration). Another group may possess, for example, the ability either to live by aerobic *or* anaerobic respiration; the adjective "facultative" is added to describe such an organism's respiration. The same adjective is used in describing the ability of an organism to live either by aerobic respiration or by fermentation (e.g., facultatively fermentative) and in the description of a bacterium able to live either by anoxygenic phototrophy or by aerobic respiration. We recognize that evolution has resulted in a continuum of traits, rather than a set of neatly packaged ones. The inadequacy of words such as "facultative" to describe the physiologies of a bacterium able to live either by fermentation, aerobic respiration, or anoxygenic phototrophy, for example, becomes readily apparent.

It is also customary to categorize physiological traits in yet other ways. One such way refers to the source(s) of carbon assimilated for biosynthesis: "autotrophy" describes the ability of an organism to utilize carbon dioxide as the principal source of carbon (save perhaps the need for vitamins or an amino acid[s]), while "heterotrophy" or "organotrophy" describes the use of carbon atoms of organic molecules as the principal carbon source.

When categorization is focused on the source of energy to be conserved, an organism doing so at the expense of reduced inorganic ions, or molecules such as hydrogen gas, would be regarded as a "lithotroph." The term "organotroph" would describe an organism utilizing organic molecules; note the use of the identical word to describe both the carbon and the energy source. Table 4 lists examples of descriptive terms often used in categorizing what are indeed different lifestyles of prokaryotes.

Prokaryote physiological diversity is not, however, restricted only to relationships to molecular oxygen or to the ability to utilize radiant energy to capture energy. Optimal growth under conditions of low pH ("acidophiles") or high pH ("alkaliphiles") is characteristic of many different prokaryotes. Similar diversity exists in temperature optima for growth: cold-loving organisms ("psychrophiles") contrast with those that are unable to grow at temperatures less than ca. 80°C, some of which ("hyperthermophiles") have been shown to grow at temperatures in excess of 110°C (54). Among other traits in which prokaryotes show remarkable variation from organism to organism are the ability to tolerate (17) or the need for inorganic salts (as for the strictly "halophilic" subgroup of the *Archaea*) and the ability of some organisms ("oligotrophs") to grow only in environments with low nutrient concentrations while others ("copiotrophs") either tolerate or require much higher nutrient levels. Yet another trait noted for some but not all organisms is a requirement for pressure in excess of that on the Earth's surface ("barophiles"). Table 5 summarizes ranges of selected traits.

As noted earlier in the chapter, the scope of nutritional diversity among the prokaryotes is not only impressive but of important consequence in affecting the ability of these microbes to colonize and thrive in nearly every imaginable habitat. Some of the bacteria are able to oxidize one-carbon compounds (those containing no carbon-to-carbon bonds) and to assimilate the oxidized moiety for synthesis of the small and macromolecules characteristic of nearly any prokaryotic (or, for that matter, eukaryotic) cell. While many of these organisms are also able to oxidize and assimilate

TABLE 4 Some terms used in relation to bacterial growth

Acidophiles: organisms with growth optima at ca. pH 1–5

Alkaliphiles: organisms with growth optima at ca. pH > 8

Anaerobes: organisms that are unable to use (i.e., consume) molecular oxygen
  Obligate: those that cannot grow in the presence of molecular oxygen
  Oxyduric: those not killed by (i.e., tolerant of) molecular oxygen
  Oxylabile: those killed by the presence of molecular oxygen
  Aerotolerant: those able to grow in the presence of molecular oxygen even though they do not use it

Aerobes: organisms that use molecular oxygen in redox reactions coupled to energy conservation
  Obligate: those that are unable to grow in the absence of molecular oxygen
  Facultative (or euryoxic): those that are able to grow in the absence of molecular oxygen

Microaerophiles: organisms that require molecular oxygen for growth, but can tolerate its presence only when present at low levels (often ca. 10% of atmospheric levels)

Mixotrophs: organisms that utilize both autotrophic and heterotrophic means, usually simultaneously, of conserving energy and assimilating nutrients for growth

Phototrophs: organisms that use radiant energy (light) as source of energy for growth
  Obligate: those that are unable to grow in the absence of light
  Facultative: those able to grow by gaining (conserving) energy in the absence of light

Symbiosis: two or more dissimilar organisms that interact and live together

Syntrophy: the relationship between proton-reducing organisms and other organisms that consume hydrogen gas
  Obligate: the relationship when the proton-reducing organism is unable to grow in the absence of the hydrogen-consuming one

methanol carbon, they are generally incapable of assimilating the carbon of other simple or complex organic molecules as sole carbon and energy source for growth. By contrast with these "methanotrophs," another group, the "methylotrophs," differ in two ways: they are unable to utilize methane for growth but will grow at the expense of methanol, formate, and methylated amines, but more strikingly, they can grow at the expense of acetate, other organic acids, amino acids, etc. Most of the known methanotrophs, then, are quite specialized in terms of nutrients utilizable for growth, while the methylotrophs are much less so.

Another example of nutritional specialization is shown by at least one *Bacillus* species which is able to grow readily with urate, a purine, or compounds such as allantoin or allantoate, both of which are intermediates in the degradative pathway of urate oxidation and assimilation. However, in contrast to scores of other *Bacillus* species, *B. fastidiosus* is unable to grow at the expense of a range of sugars, polysac-

TABLE 5 Environmental extremes in which prokaryotes are thought to multiply

Characteristic	Value
Pressure	1 to ca. 1,000 atm[a]
Temp	−1.2 to 110–113°C
Depth	2,000 m (terrestrial subsurface)
	ca. 10,000 m (ocean floor)
Salt concn	Up to ca. 4–5 M
Acidity/alkalinity	pH of 1–2 to 11–12
Available water	As low as 0.6 a_w (water activity)

[a] 1 atm = 101.29 kPa.

charides, amino acids, simple proteins, etc. In similar fashion, some *Cytophaga* species, unable to grow with amino acids serving as carbon and energy sources when first isolated from a natural habitat, are strictly dependent upon polymers such as cellulose or chitin as a source of carbon and energy for growth. Other cytophagas, by contrast, are unable to utilize cellulose or chitin for growth, but are able to grow on complex media containing peptides, amino acids, yeast extract, or simple sugars. These specificities contrast to those of some pseudomonads and relatives which are able to use any of scores of organic compounds including simple organic acids, the range of amino acids, and benzenoid and polycyclic compounds to support cell multiplication.

Such nutritional specificity occurs not only in regard to utilization of organic compounds by these organotrophs. Autotrophs—organisms able to synthesize the majority, if not all, of the carbon skeletons of their cellular compounds by reducing carbon dioxide to the "level" of carbohydrate (representative of the overall ratio of C:H:O of cells)—vary in the reductant they employ for this purpose. Some are able to oxidize hydrogen gas or may use a compound such as hydrogen sulfide from which to capture energy for growth, or they may be able to grow as organotrophs by use of organic compounds. However, not all chemolithotrophs are so versatile; some "sulfur-oxidizing" chemolithotrophs will use any of several reduced sulfur compounds to acquire energy, while others display much more selectivity and specificity.

For another group of autotrophs (e.g., *Nitrosomonas*), ammonia serves as an energy source, while nitrite does not, yet *Nitrobacter* spp. will oxidize nitrite to form nitrate but are unable to gain energy from ammonia. Reduced forms of iron and manganese are suitable electron donors in energy-conserving processes for yet other groups of autotrophs.

At least two other aspects of physiological diversity are significant for the discussions that follow. One of these deals with the fact that although a particular organic molecule may not function as a sole source of carbon and energy for growth of a particular organism, the compound may be metabolizable (i.e., it may be either oxidized or reduced in whole or in part, and all or part of the molecule may be assimilated into cellular components) so long as the organism is utilizing a different molecule as a carbon and energy source. This phenomenon, first termed "co-oxidation" when it was described in reference to oxidation of ethane

and other gaseous hydrocarbons by a methanotroph, has been shown to be of more general significance and distribution in prokaryotes and is now considered an important aspect of many bioremediation processes. A second trait, which has become of more widespread interest and significance since it was initially recognized, is syntrophy (33). In anoxic environments the degradation of reduced organic compounds (e.g., simple alcohols, fatty acids, certain aromatic acids), resulting in the accumulation of acetate, carbon dioxide, hydrogen gas, and formate (among other end products), is energetically unfavorable. On the other hand, if the concentrations of these products are kept sufficiently low, the energetics for growth become somewhat more favorable. Methanogens and sulfate-reducing bacteria (sulfidigens), commonly present in anoxic habitats, function to consume hydrogen gas and thus lower its partial pressure in the microhabitat. Thus, such cocultures function to permit the biodegradation of compounds that would otherwise be refractory to attack by pure cultures. The term "interspecies hydrogen transfer" has been used to describe such interactions as well.

DIVERSITY: INSIGHTS FROM MACROMOLECULAR ANALYSES

Because it appears that prokaryotes have been present on this planet for perhaps four-fifths of its age, there has been ample time for repeated mutations, their accumulation and natural selection, and an introduction of widely varied traits in these organisms. For many years, the lack of a substantial traditional fossil record impeded critical considerations of evolutionary change in the microbial world. It was not until the Zuckerkandl and Pauling (62) proposal of macromolecular traits as indicators of evolutionary changes that we could ponder events of the past. Data from both protein (15) and rRNA (22, 39) sequences are now used widely for this purpose and for reaching inferences about phylogenetic relationships. The evolution of prokaryote diversity has become a subject of intense interest and significance, regularly revisited and reviewed; a recent international symposium (46) is notable for its scope.

The implications of sequence information suggest strongly that life on the planet should be considered in three major categories, *Archaea*, *Bacteria*, and *Eucarya*. A widely accepted phylogenetic tree (Fig. 1) is based on the inferences that the groups termed *Archaea* and *Eucarya* diverged from ancestors of *Bacteria*, first as a single lineage and only later diverging and becoming separately recognizable entities. The timing of this divergence, and particularly the estimates for the divergence point of the eukaryotes and prokaryotes, have been considered often and in different ways, but most recently and carefully by Doolittle and colleagues (15). The potential for lateral gene transfer in natural populations may remain a matter of concern in using sequence information for phylogenetic considerations (21, 30).

In the line leading to *Bacteria*, an initial bifurcation leads to thermophilic cells now represented by genera such as *Aquifix*, *Fervidobacterium*, and *Thermotoga*. It is a matter of curiosity and concern that *Aquifix*, regularly placed at the bottom of the tree, is nonetheless known as an aerobic, respiring chemolithotroph; this represents an anomaly, as the planet presumably was devoid of molecular oxygen at the time of this evolutionary event (21).

Many but not all practitioners consider that the green non-sulfur phototrophs were among the next group(s) to diverge. Perhaps at about the same time a group emerged

Bacteria   Archaea   Eucarya

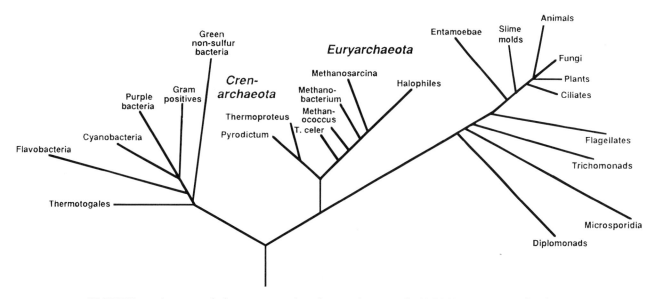

FIGURE 1   A recent phylogenetic tree based on evaluation of 16S RNA sequences. The three major lineages of life (Archaea, Bacteria, Eucarya) are shown. From reference 39, with permission.

that possessed cell walls devoid of peptidoglycan: the *Planctomycetales*, including genera such as *Gemmata*, *Isosphaera*, *Pirellula*, and *Planctomyces* (21, 55).

The remaining well-studied and recognized bacterial groupings designated the flavobacteria, cyanobacteria, gram-positive bacteria (of which there appear to be at least two major subgroups), and purple or proteobacteria (a group itself considered to warrant five subdivisions: alpha, beta, gamma, delta, and epsilon) appear to have emerged later as a single radiation which then underwent divergence.

Even before this refined evidence about likely events in bacterial evolution, it was deemed likely that thermophilic anaerobes were likely candidates as the most ancient of living beings. Arguments over whether autotrophs preceded heterotrophs or vice versa abounded but were based more on conjecture than they were on anything approaching persuasive evidence. Among the candidates for the earliest of organisms were the methanogenic bacteria, prokaryotes which are now considered not to be *Bacteria* per se, but instead members of the *Archaea*, appearing later on the evolutionary scene.

A current view of the *Archaea* is that they are as unlike the *Bacteria* as they are unlike the *Eucarya*. Their possession of unusual ether-linked lipids and the lack of peptidoglycan in their cell walls are among their distinguishing characteristics. The group is now considered in subcategories as methanogens, thermo- and hyperthermophiles, and strict halophiles. Some limits of our understanding of the *Archaea* are demonstrated by the diversity recognized after the surprising discovery of new types of these organisms in cold, marine environments (13, 37).

Although the *Eucarya* are usually cited as characteristically containing a membrane-bound nucleus, they are also usually considered to possess subcellular organelles termed chloroplasts or mitochondria. It may be significant for evo-

lutionary considerations that in the *Eucarya*, three groups at the base of the tree—the diplomonads, the microsporidians, and the trichomonads—are devoid of mitochondria.

One of the many surprises arising from such approaches to the study of phylogeny, and of special interest for considerations of prokaryotic diversity, has been the demonstration that some biochemical and physiological traits appear in quite distantly related organisms. For example, sulfate reduction occurs both in *Bacteria* and in *Archaea*; the same is true for autotrophic assimilation of carbon dioxide as well as the trait of hyperthermophily.

SOME ASPECTS OF BEHAVIORAL DIVERSITY AMONG PROKARYOTES

When samples of mixed microbial populations are obtained from either natural habitats or laboratory cultures, or when pure cultures are examined using any one of several types of light microscopy, we quickly note that not all organisms have identical aspects of behavior. Some organisms swim through liquids with ease and rapidity, others regularly do so more slowly; some dart about seemingly randomly as if in a frenzy, while others move for longer distances and appear to be swimming smoothly; in some, no swimming motility is ever discerned. We now recognize that, in nearly every instance, such motility reflects the presence and function of either a single flagellum or multiple flagella and the arrangement of these on the cell surface(s). Many of the *Bacteria* termed spirochetes are seen to swim much better through viscous liquids than they do in typical water samples or laboratory media; this may be a reflection of the position of their flagella (axial filaments; endoflagella) between the cell wall and an external sheath of the cell. This location clearly does not permit the flagellum to rotate in the extracellular milieu and propel the cell in a manner similar to that of, for example, the well-studied enteric bacteria.

Usually the behavior of swimming flagellated bacteria has been described as alternating "runs" and "tumbles." The observations made with the enteric bacteria may be of limited application, however. *Azospirillum* spp. (61), pseudomonads, and rhizobia are among organisms observed not only to behave differently, but to have different intracellular regulatory mechanisms involved.

A group of bacteria with rather diverse physiological properties move in a different manner, not by swimming through liquids, but rather by migrating over insoluble substrates which provide solid surfaces (e.g., agar, cellulose, chitin, other cells) on which to move. This motility, for which locomotor organelles have as yet not been visualized and for which, accordingly, the mechanism for motility remains unknown, has often been termed "gliding." Cells can be seen to move (glide) on a solid surface, and the colony can spread a considerable distance (often more than a centimeter) beyond the point of cell deposition (inoculation). Such colonies usually are characterized as much thinner (very little vertical height) than those typical of enteric bacteria or pseudomonads, for example, growing on an identical nutrient-poor medium. Nutrient concentration usually exerts a significant effect on the ability of cells to glide and, accordingly, to form spreading colonies; the ability of a cell to glide is ordinarily not expressed on a nutrient-rich medium such as "nutrient broth," and the gliding ability of an organism is thus not noted when such a medium is used for growth. Among the best-studied *Bacteria* known to possess gliding ability are those of genera such as *Beggiatoa*, *Cytophaga*, *Flexibacter*, and *Myxococcus*; this is far from an inclusive list.

The ability of a prokaryote to swim through liquids in the absence of flagella is known in at least one cyanobacterium (a *Synechococcus* sp.); no locomotor organelles have yet been discerned. The organism is not known to glide.

It has already been suggested that cells are able to alter their physiological phenotypic properties, within limits imposed by their genotype, as indicated by the fact that some bacteria are able to live by either phototrophy or aerobic respiration, others by fermentation or aerobic respiration, and still others by aerobic or anaerobic respiration. Such environmental effects are myriad, but occur also with respect to anatomical features and motility phenomena. One such example is the motility of some flagellated *Bacteria* on solid surfaces, a behavior termed "swarming." Certain *Serratia*, *Proteus*, and *Vibrio* species, among others, synthesize additional, specialized flagella as a result of contact with a solid surface, and it is these additional flagella that enable the organism to spread. Surface features, among them the concentration of agar used in solidifying the medium, have significant effects on events in the cell's interior that in turn affect synthesis of these appendages for motility.

The ability of cells to respond (chemotaxis) to concentration gradients of attractants (often nutrients) or repellants by employing temporal sensing mechanisms has been extensively studied as one category of behavioral responses. Although phototrophic eukaryotic cells have long been known to exhibit responses to light gradients (phototaxis), no true example of this behavioral phenomenon among the anoxygenic, phototrophic prokaryotes was noted until a recent persuasive study (45) in the non-sulfur phototroph *Rhodospirillum centenum*. It is important to note that studies of the in situ behavior of prokaryotic populations have established phenomena such as "diel migration" (20) which undoubtedly reflect both chemo- and phototactic (and possibly other types of) behavior of members of these communities. Such behavior is, then, not an artifact of pure culture study.

Another sort of response is reflected in the ability of some cells to undergo either cellular or colonial morphogenesis in connection with entrance into a dormant or nongrowing ("resting") state. The formation of the remarkably resistant endospore in the cell interior of the organisms grouped in genera such as *Acetonema*, *Bacillus*, *Clostridium*, *Desulfotomaculum*, *Sporomusa*, *Sporosarcina*, and *Thermoactinomyces*, for example, generally is regarded as a response to an encounter with nutrient-poor conditions. That seems to be true as well for myxospores and microcysts formed by myxobacters and cytophagas, where the entire cell shortens, thickens, and becomes spherical (or nearly so), and is probably the explanation accounting for exospore formation by some methanotrophs. Other examples of changes in cell morphology resulting in specialized functions are the formation of heterocysts in some cyanobacteria and akinetes in others and stalk formation in prosthecate bacteria as typified by *Caulobacter*. These particular morphogenetic (behavioral?) responses clearly make it more difficult to distinguish a potential cell from debris in samples taken from the environment, as noted earlier. Application of specific molecular probes is likely to be of enormous aid in identifying organisms of these types in situ and in assessing their growth (versus dormancy) state in natural habitats.

DIVERSITY OF HABITATS EXPLOITED FOR GROWTH OF PROKARYOTES

Everything is everywhere; but the milieu selects . . . in nature and in the laboratory. (7)

Why is it that the prokaryotes have been known to occur in so many different habitats and that our comprehension of "new places" in which they are found is increasing? Certainly, as may be obvious from the degree of physiological diversity outlined above, prokaryotes have the potential to exploit (to multiply in) habitats judged "extreme" when compared with those that support the existence of animals and plants. Yet another factor must be reflected in the ability of these metabolically diverse and very active cells to enter a state of inactivity, or dormancy, and be able to survive for periods of time far longer than the generation times deduced for them in studies of laboratory cultures. Although dormancy is a trait often equated with cysts or endospores, for example, it is well recognized (even though it is more difficult to study inactive, dormant cells than it is to examine traits of exponentially growing ones!) that dormancy is a significant phenomenon in the persistence of an enormous variety of cells (27).

If we accept, as we should, the dictum that prokaryotes are found virtually everywhere, it follows that the entirety of habitats and the microbes therein cannot be enumerated, let alone described, in a limited space. Nonetheless, given the dynamic and exciting state of the study of microbial diversity, selected recent descriptions, reflecting either novel revelations or reconsiderations of earlier findings, are noted very briefly below.

• The biodiversity in hot spring microbiota, bacterial as well as archaeal, was judged to be far greater, as deduced from sequence analysis of macromolecules, than had been concluded from isolation and cultivation of microbiota from these environments (9, 58).

- The prokaryote *Buchnera*, thus far uncultivated intracellular symbionts of aphids, has recognizable phylogenetic affinities and provides important nutrients to the insect host (10).
- Purple sulfur bacteria exist in the form of readily visible aggregates ("berries") in salt marsh ponds (50).
- Sulfate-reducing bacteria, some not yet cultivable, play significant roles as members in deep ocean sediments (41).
- Luminescent bacteria, long thought to occupy primarily marine habitats, occur in significant numbers and roles in terrestrial macrobes (36).
- Archaea, many seemingly novel, are present in abundance in marine picoplankton, including those from cold waters (14, 19, 49). Thermophilic bacteria occur in cold marine sediments (25).
- Ammonium ion can be oxidized under denitrifying conditions (34).
- Barophilic responses occur in archaeal hyperthermophiles isolated from hydrothermal vent areas (43).
- Salt-excreting plants produce halophilic minihabitats in soils (51).
- So-called uncultivable microbiota from marine environments are in large part bacteria that have lost their nucleoids (63).
- The phenomenon of magnetotaxis is a trait more widely distributed than had been recognized previously (52).
- Modern microbial mats found in hypersaline environments are characterized by unusual features (11, 12).
- Mats of *Thioploca*, a sulfur-oxidizing bacterium, play unanticipated roles in concentrating and transporting nitrate in a marine environment (18).
- Dense accumulations ("plates") of nearly pure culture of a phototrophic bacterium may develop in a meromictic saline lake (40).
- Control of growth of a single organism in the mixed population of activated sludge is possible (32).

REFERENCES

1. **Achtnich, C., A. Schuhmann, T. Wind, and R. Conrad.** 1995. Role of interspecies H_2 transfer to sulfate and ferric iron-reducing bacteria in acetate consumption in anoxic paddy soil. *FEMS Microbiol. Ecol.* **16:**61–70.
2. **Akkermans, A. D. L., M. S. Mirza, H. J. M. Harmsen, J. H. Biok, P. R. Herron, A. Sessitsch, and W. M. Akkermans.** 1994. Molecular ecology of microbes: a review of promises, pitfalls and true progress. *FEMS Microbiol. Rev.* **15:**185–194.
3. **Albrechtsen, H.-J., G. Heron, and T. H. Christensen.** 1995. Limiting factors for microbial Fe(III)-reduction in landfill leachate polluted aquifer (Vejen, Denmark). *FEMS Microbiol. Ecol.* **16:**233–248.
4. **Amann, R. I., W. Ludwig, and K.-H. Schleifer.** 1995. Phylogenetic identification and in situ detection of individual microbial cells without cultivation. *Microbiol. Rev.* **59:**143–169.
5. **Angert, E. R., K. D. Clements, and N. R. Pace.** 1993. The largest bacterium. *Nature* (London) **362:**239–241.
6. **Atlas, R. M.** 1986. Applicability of general ecological principles to microbial ecology, p. 339–370. *In* J. S. Poindexter and E. R. Leadbetter (ed.), *Bacteria in Nature*, vol. 2. *Methods and Special Applications in Bacterial Ecology*. Plenum Press, New York.
7. **Baas Becking, L. G. M.** 1934. Geobiologie ov inleiding tot de milieukunde. Stockum und Zoon N. V., Den Haag, Netherlands. (Translation from the Dutch by C. B. van Niel.)
8. **Balows, A., H. G. Truper, M. Dworkin, W. Harder, and K. H. Schleifer (ed.).** 1992. *The Prokaryotes*, 2nd ed. A *Handbook on the Biology of Bacteria: Ecophysiology, Isolation, Identification, Applications*, vol. 1–4. Springer-Verlag, New York.
9. **Barns, S. M., F. E. Fundyga, M. W. Jeffries, and N. R. Pace.** 1994. Remarkable archaeal diversity detected in a Yellowstone National Park hot spring environment. *Proc. Natl. Acad. Sci. USA* **91:**1609–1613.
10. **Baumann, P., L. Baumann, C.-Y. Lai, and D. Rouhbakhsh.** 1995. Genetics, physiology, and evolutionary relationships of the genus *Buchnera*: intracellular symbionts of aphids. *Annu. Rev. Microbiol.* **49:**55–94.
11. **Caumette, P., R. Matheron, N. Raymond, and J.-C. Relexans.** 1994. Microbial mats in the hypersaline ponds of Mediterranean salterns (Salins-de-Giraud, France). *FEMS Microbiol. Ecol.* **13:**273–286.
12. **Conrad, R., P. Frenzel, and Y. Cohen.** 1995. Methane emission from hypersaline microbial mats: lack of aerobic methane oxidation activity. *FEMS Microbiol. Ecol.* **16:**297–306.
13. **DeLong, E. F.** 1992. Archaea in coastal marine environments. *Proc. Natl. Acad. Sci. USA* **89:**5685–5689.
14. **DeLong, E. F., K. Y. Wu, B. B. Prezelin, and R. V. M. Jovine.** 1994. High abundance of archaea in antarctic marine picoplankton. *Nature* (London) **371:**695–697.
15. **Doolittle, R. F., D.-F. Feng, S. Tsang, G. Cho, and E. Little.** 1996. Determining divergence times of the major kingdoms of living organisms with a protein clock. *Science* **271:**470–477.
16. **Ehrlich, H. L.** 1985. The position of bacteria and their products in food webs, p. 199–220. *In* E. R. Leadbetter and J. S. Poindexter (ed.), *Bacteria in Nature*, vol. 1. *Bacterial Activities in Perspective*. Plenum Press, New York.
17. **Fortin, D., G. Southam, and T. J. Beveridge.** 1994. Nickel sulfide, iron-nickel sulfide and iron sulfide precipitation by a newly isolated *Desulfotomaculum* species and its relation to nickel resistance. *FEMS Microbiol. Ecol.* **14:**121–132.
18. **Fossing, H., V. A. Gallardo, B. B. Jorgensen, M. Huttel, L. P. Nielsen, H. Schulz, D. E. Canfield, S. Forster, R. N. Glud, J. K. Gundersen, J. Kuver, N. B. Ramsing, A. Teske, B. Thamdrup, and O. Ulloa.** 1995. Concentration and transport of nitrate by the mat-forming sulphur bacterium *Thioploca*. *Nature* (London) **374:**713–717.
19. **Fuhrman, J. A., K. McCallum, and A. A. Davis.** 1992. Novel major archaebacterial group from marine plankton. *Nature* (London) **356:**148–149.
20. **Garcia-Pichel, F., M. Mechling, and R. W. Castenholz.** 1994. Diel migrations of microorganisms within a benthic, hypersaline mat community. *Appl. Environ. Microbiol.* **60:**1500–1511.
21. **Giovannoni, S. J., M. S. Rappe, D. Gordon, E. Urbach, M. Suzuki, and K. G. Field.** Ribosomal RNA and the evolution of bacterial diversity. *Symp. Soc. Gen. Microbiol.*, in press.
22. **Gutell, R. R., N. Larsen, and C. R. Woese.** 1994. Lessons from an evolving rRNA: 16S and 23S rRNA structures from a comparative perspective. *Microbiol. Rev.* **58:**10–26.
23. **Holmes, A. J., N. J. P. Owens, and J. C. Murrell.** 1995. Detection of novel marine methanotrophs using phylogenetic and functional gene probes after methane enrichment. *Microbiology* **141:**1947–1955.
24. **Hungate, R. E.** 1985. Anaerobic biotransformations of organic matter, p. 39–96. *In* E. R. Leadbetter and J. S. Poindexter (ed.), *Bacteria in Nature*, vol. 1. *Bacterial Activities in Perspective*. Plenum Press, New York.

25. **Isaksen, M. F., F. Bak, and B. B. Jorgensen.** 1994. Thermophilic sulfate-reducing bacteria in cold marine sediment. *FEMS Microbiol. Ecol.* **14**:1–8.

26. **Jannasch, H. W.** 1995. Microbial interactions with hydrothermal fluids, p. 273–296. *In* S. E. Humphris (ed.), *Sea Floor Hydrothermal Systems: Physical, Chemical, Biological, and Geological Interactions, Geophysical Monograph 91.* American Geophysical Union, Washington, D.C.

27. **Kaprelyants, A. S., J. C. Gottschal, and D. B. Kell.** 1993. Dormancy in non-sporulating bacteria. *FEMS Microbiol. Rev.* **104**:271–286.

28. **Kluyver, A. J.** 1931. *The Chemical Activities of Micro-Organisms.* University of London Press, Ltd.

29. **Kluyver, A. J., and C. B. van Niel.** 1956. *The Microbe's Contribution to Biology.* Harvard University Press, Cambridge, Mass.

30. **Lorenz, M. G., and W. Wackernagel.** 1994. Bacterial gene transfer by natural genetic transformation in the environment. *Microbiol. Rev.* **58**:563–602.

31. **Lowe, S. E., M. K. Jain, and J. G. Zeikus.** 1993. Biology, ecology, and biotechnological applications of anaerobic bacteria adapted to environmental stresses in temperature, pH, salinity, or substrates. *Microbiol. Rev.* **57**:451–509.

32. **Madoni, P., and D. Davoli.** 1993. Control of *Microthrix parvicella* growth in activated sludge. *FEMS Microbiol. Ecol.* **12**:277–284.

33. **McInerney, M. J.** 1986. Transient and persistent associations among prokaryotes, p. 293–338. *In* J. S. Poindexter and E. R. Leadbetter (ed.), *Bacteria in Nature*, vol. 2. *Methods and Special Applications in Bacterial Ecology.* Plenum Press, New York.

34. **Mulder, A., A. A. van de Graaf, L. A. Robertson, and J. G. Kuenen.** 1995. Anaerobic ammonium oxidation discovered in a denitrifying fluidized bed reactor. *FEMS Microbiol. Ecol.* **16**:177–184.

35. **Nealson, K. H., and D. Saffarini.** 1994. Iron and manganese in anaerobic respiration: environmental significance, physiology, and regulation. *Annu. Rev. Microbiol.* **48**:311–343.

36. **Nealson, K. H., T. M. Schmidt, and B. Bleakley.** 1990. Biochemistry and physiology of Xenorhabdus, p. 271–285. *In* R. R. Gaugler and H. K. Kaya (ed.), *Entomopathogenic Nematodes in Biological Control.* CRC Press, Boca Raton, Fla.

37. **Olsen, G. J.** 1994. Archaea, archaea, everywhere. *Nature* (London) **371**:657–658.

38. **Olsen, G. J., D. L. Lane, S. J. Giovannoni, N. R. Pace, and D. A. Stahl.** 1986. Microbial ecology and evolution: a ribosomal RNA approach. *Annu. Rev. Microbiol.* **40**:337–366.

39. **Olsen, G. J., and C. R. Woese.** 1993. Ribosomal RNA: a key to phylogeny. *FASEB J.* **7**:113–123.

40. **Overmann, J., J. T. Beatty, and K. J. Hall.** 1994. Photosynthetic activity and population dynamics of *Amoebobacter purpureus* in a meromictic saline lake. *FEMS Microbiol. Ecol.* **15**:309–320.

41. **Parkes, R. J., B. A. Cragg, S. J. Bale, J. M. Getliff, K. Goodman, P. A. Rochelle, J. C. Fry, A. J. Weightman, and S. M. Harvey.** 1994. Deep bacterial biosphere in Pacific Ocean sediments. *Nature* (London) **371**:410–413.

42. **Pfennig, N.** 1985. Stages in the recognition of bacteria using light as a source of energy, p. 113–129. *In* E. R. Leadbetter and J. S. Poindexter (ed.), *Bacteria in Nature*, vol. 1. *Bacterial Activities in Perspective.* Plenum Press, New York.

43. **Pledger, R. J., B. C. Crump, and J. A. Baross.** 1994. A barophilic response by two hyperthermophilic, hydrothermal vent archaea: an upward shift in the optimal temperature and acceleration of growth rate at supra-optimal temperatures by elevated pressure. *FEMS Microbiol. Ecol.* **14**:233–242.

44. **Poindexter, J. S., and E. R. Leadbetter.** 1986. Enrichment cultures in bacterial ecology, p. 229–260. *In* J. S. Poindexter and E. R. Leadbetter (ed.), *Bacteria in Nature*, vol. 2. *Methods and Special Applications in Bacterial Ecology.* Plenum Press, New York.

45. **Ragatz, L., Z.-Y. Jiang, C. E. Bauer, and H. Gest.** 1995. Macroscopic phototactic behavior of the purple photosynthetic bacterium *Rhodospirillum centenum.* *Arch. Microbiol.* **163**:1–6.

46. **Roberts, D.** Evolution of microbial life. *Symp. Soc. Gen. Microbiol.*, in press.

47. **Salyers, A. A., and D. D. Whitt.** 1994. *Bacterial Pathogenesis: a Molecular Approach.* ASM Press, Washington, D.C.

48. **Schlegel, H. G.** 1993. *General Microbiology*, 7th ed. Cambridge University Press, London.

49. **Schmidt, T. M., E. F. DeLong, and N. R. Pace.** 1991. Analysis of a marine picoplankton community by 16S rRNA gene cloning and sequencing. *J. Bacteriol.* **173**:4371–4378.

50. **Seitz, A. P., T. H. Nielsen, and J. Overmann.** 1993. Physiology of purple sulfur bacteria forming macroscopic aggregates in Great Sippewissett Salt Marsh, Massachusetts. *FEMS Microbiol. Ecol.* **12**:225–236.

51. **Simon, R. D., A. Abeliovich, and S. Belkin.** 1994. A novel terrestrial halophilic environment: the phylloplane of atriplex halimus, a salt-excreting plant. *FEMS Microbiol. Ecol.* **14**:99–110.

52. **Spring, S., and K.-H. Schleifer.** 1995. Diversity of magnetotactic bacteria. *Syst. Appl. Microbiol.* **18**:147–153.

53. **Stanier, R. Y.** 1951. The life-work of a founder of bacteriology. *Q. Rev. Microbiol.* **26**:35–37.

54. **Stetter, K. O.** 1995. Microbial life in hyperthermal environments. *ASM News* **61**:285–290.

55. **van de Peer, Y., J.-M. Neefs, P. de Rijk, P. de vos, and R. de Wachter.** 1994. About the order of divergence of the major bacterial taxa during evolution. *Syst. Appl. Microbiol.* **17**:32–38.

56. **van Niel, C. B.** 1949. The "Delft school" and the rise of general microbiology. *Bacteriol. Rev.* **13**:161–174.

57. **van Niel, C. B.** 1955. Natural selection in the microbial world. *J. Gen. Microbiol.* **13**:201–217.

58. **Ward, D. M., M. J. Ferris, S. C. Nold, M. M. Bateson, E. D. Kopczynski, and A. L. Ruff-Roberts.** 1994. Species diversity in hot spring microbial mats as revealed by both molecular and enrichment culture approaches—relationship between biodiversity and community structure, p. 33–44. *In* L. J. Stal and P. Caumette (ed.), *Microbial Mats.* NATO ASI Series, vol. G35. Springer-Verlag, Berlin.

59. **Woese, C. R.** 1994. Microbiology in transition. *Proc. Natl. Acad. Sci. USA* **91**:1601–1603.

60. **Wolfe, R. S.** 1992. Foreword, p. v–vi. *In* A. Balows, H. G. Truper, M. Dworkin, W. Harder, and K.-H. Schleifer (ed.), *The Prokaryotes*, 2nd ed. Springer-Verlag, New York.

61. **Zhulin, I. G., and J. P. Armitage.** 1993. Motility, chemokinesis, and methylation-independent chemotaxis in *Azospirillum brasilense.* *J. Bacteriol.* **175**:952–958.

62. **Zuckerkandl, E., and L. Pauling.** 1965. Molecules as documents of evolutionary history. *J. Theor. Biol.* **8**:357–366.

63. **Zweifel, U. L., and A. Hagstrom.** 1995. Total counts of marine bacteria include a large fraction of non-nucleoid-containing bacteria (ghosts). *Appl. Environ. Microbiol.* **61**:2180–2185.

GENERAL METHODOLOGY

VOLUME EDITOR
MICHAEL J. McINERNEY

SECTION EDITOR
DAVID A. STAHL

Overview: General Methodology

MICHAEL J. McINERNEY AND DAVID A. STAHL

4

Microbial ecology is a dynamic field with new technologies being developed that will allow the investigator to detect, identify, and quantify microorganisms in their natural ecological settings, to assess the microenvironments in which these microorganisms reside, and to determine the activity and physiological state of microorganisms in their natural habitats. Additionally, methods for the culture of microbial communities are being developed which will allow the study of important community-level questions such as the rules that govern the development of communities and how population interactions affect community functions such as nutrient cycling and homeostasis in a controlled and reproducible manner. Coupled with the development of these new technologies is an ever-increasing ability to culture microorganisms from physiologically and phylogenetically diverse populations.

Most of the chapters in this section deal with ways to quantify the numbers and kinds of microorganisms within a community. The ability to obtain this type of quantitative information is fundamental to the understanding of the structure and function of any ecosystem. Because many of the chapters in this section can be topics for entire books themselves, it was not possible to provide detailed protocols of all approaches or in-depth discussions of modifications that may be needed for investigation of a specific habitat. Rather, the emphasis is on the advantages and disadvantages of a particular method so that the investigator will appreciate when and why a procedure is used. Specific modifications needed for studying a given habitat will be discussed in other sections of the manual. By surveying the information provided in the different chapters in this section, the investigator can choose the method that is most applicable for the system under study.

Microscopy has been one of the primary approaches of bacteriology for hundreds of years. However, the development of fluorescent probes, highly sensitive charge-coupled device cameras, and relatively inexpensive computer software and hardware for digital image acquisition and analysis has made it possible to obtain quantitative information on the taxonomic affiliation and the physiological state of individual cells in complex associations of organisms that might occur in their natural environments. Chapter 5 provides a detailed discussion of the utility of scanning confocal laser microscopy. That chapter will allow even the most novice investigator to enter the digital world of microscopy. An important advantage of scanning confocal laser microscopy is that it allows nondestructive, in situ, optical sectioning of biological materials without fixation. Thus, many of the artifacts encountered with light or electron microscopic approaches can be avoided. The last part of the chapter discusses how this equipment is used for the enumeration and identification of cells, determination of their metabolic state, and analysis of the architecture of microbial assemblages.

Pure culture approaches have traditionally been an important part of any microbiological investigation, providing a relatively inexpensive assessment of the numbers, kinds, and metabolic activities of microorganisms. Also, pure culture methodology is still the definitive approach to demonstrate that an activity is microbially mediated. Even a cursory review of the literature quickly reveals that a great number and variety of media have been used to cultivate prokaryotic and eukaryotic microorganisms, which makes it appear that culturing is an empirical art. However, an understanding of the basic concepts of medium design and the physical conditions that limit microbial growth provides a rational approach to devising media and cultural conditions for growth of many different kinds of microorganisms. Chapter 6 discusses how a basal medium constructed from a few mineral solutions can be easily modified to grow many different kinds of microorganisms.

Cultivation of microeukaryotes (chapter 7) uses many of the approaches developed for the cultivation of prokaryotes. The cultivation of algae most closely mirrors that of bacteria, while for the cultivation of protozoa, one must first consider whether the organism uses a dissolved or particulate form of organic carbon. For the latter, the choice of the appropriate particulate food or prey organism is important, since protozoa are often selective feeders. Continued refinements of procedures have led to an increasing number of microeukaryotes available in laboratory culture as either monoprotist, monoaxenic, or axenic cultures. This has allowed a more thorough understanding of the evolution and taxonomy of microeukaryotes than could have been obtained solely from morphological characteristics.

The developments of cell culture and concentration methodologies have led to the ability to detect very low levels of environmentally important animal and human viruses (chapter 8). The viruses most often encountered in environmental samples include bacteriophages and animal and human enteric viruses. The presence of the latter in potable water sources is a major public health concern. Thus, it is important to be able to detect small numbers of infectious units. Since viruses cannot be propagated in artificial medium, the enumeration of an infectious animal virus requires its propagation in a living host cell. Cell cultures have replaced the need for the use of laboratory animals, and the optimization of cell culture methodologies combined with the development of concentration or extraction procedures allows the detection of very low numbers of infectious units from environmental samples.

With the exception of certain diseases, pure cultures of microorganisms rarely exist in natural environments. Indeed, the activities of free-living microorganisms are most effective only when organisms are present in association with other groups of organisms. Laboratory culture of these organisms in association with one another is essential in order to understand the community-level mechanisms involved in specific processes and the effects of perturbations on these associations. Such microbial assemblages can be maintained in laboratory systems by controlling the substrate(s), its concentration, and the rate of flux, often using systems that provide for both temporal and spatial variations of these parameters (chapter 9). These assemblages have many of the characteristics of natural communities. They arise spontaneously, remain relatively constant over time, and exhibit homeostatic behavior. Also, distinct assemblages of microorganisms with clearly differentiated transition zones (ecotones) form in response to nutrient gradients. These systems should become important tools for many areas of environmental microbiology, including bioremediation, wastewater treatment, composting, landfills, and development of biological control agents.

Viable plate count or other cultural methods for the determination of biomass have a strong bias in that these methods detect only those organisms that grow under the selected set of conditions. For many environments, the physiological status of an organism or the lack of understanding of its nutritional requirements may prevent its successful detection. An effective and quantitative way to measure microbial biomass in situ without the need to culture or to distinguish individual cells in an environmental sample is to measure the amount of certain cellular components of the microorganism. If the cellular component is universally distributed, has a short residence time in detrital pools after the death of the cell, and is expressed at relatively constant levels and throughout the growth cycle, then this component can be used as a measure of biomass. A number of components could potentially be used for biomass estimation (chapter 10). However, because all intact cells contain polar lipids, primarily as phospholipids, and because organisms without membranes are not viable, phospholipid analysis has become an important method for the determination of microbial biomass. In addition, phospholipid analysis can provide information on the abundance of specific groups of microorganisms and their physiological status. One problem with the interpretation of biochemical biomass data is that the units of measure are different from those traditionally used by microbiologists, i.e., micromoles of a component rather than the number of cells per unit amount of sample. However, conversion factors provide a mechanism to relate these different measures of biomass.

There is no doubt that the development of methods based on nucleic acid sequence analysis will revolutionize how environmental microbiology studies are conducted in the future. For example, molecular approaches based on 16S RNA sequence analysis allow the direct measure of the abundance, diversity, and phylogeny of microorganisms in almost any environment. Also, the use of nucleic acid probes allows the determination of the taxonomic status of individual cells in complex microbial assemblages. Although the molecular approaches described in chapter 11 provide the framework for describing the phylogeny of microorganisms present in an ecosystem, it is not yet clear how this information relates to current taxonomic ranks. Also, the proportional recovery of specific sequences cannot be directly equated with traditional measures of species abundance.

Most of the methods described above for use in determining the numbers and kinds of microorganisms in a given environmental sample result in the destruction or alteration of the sample, and there is also a significant time delay before the results are obtained. The development of bioreporter genes and biosensors can provide, in certain instances, information on the workings of individual cells or populations within a community in real time (chapter 12). These methods also allow the analysis of the microenvironment in which the cell resides. Bioreporter gene technology allows the analysis of the expression of almost any gene by providing a gene product that can be easily detected. To date, these technologies are limited to the study of aerobic habitats because of the need for molecular oxygen for bioluminescence or for aerobic conditions for the formation of the chromophore present in the fluorescent protein. This chapter also discusses microprobes and biosensors which allow the detection of a specific physical or chemical property in a microenvironment. These technologies define the physical and/or chemical gradients that both reflect and control the development of biofilms and other multispecies assemblages.

The final chapter in this section (chapter 13) summarizes the management and administrative procedures needed to ensure the quality of the data generated by research activities. The reliability of scientific information is undergoing increasing scrutiny in the public sector. This is particularly true in environmental microbiology, where this information is often used for risk assessment and liability decisions. Thus, we must all be aware of the procedures needed to ensure the continuous reliability of the data and the quality of the measurement process.

Analytical Imaging and Microscopy Techniques

J. R. LAWRENCE, D. R. KORBER, G. M. WOLFAARDT, AND D. E. CALDWELL

5

Examination of microorganisms in their natural habitat may be achieved most effectively through the application of a variety of microscopic techniques. As indicated by Murray and Robinow (77), microscopy provides the primary approach to the study of bacteria. In ecological studies, a variety of methods ranging from buried slides (27), soil peels (114), and capillary methods (81) have been used in conjunction with microscopy to examine the diversity of microorganisms and their associations in natural systems. These approaches have provided a clear sense of the diversity of microorganisms, the presence of unculturable organisms, and the existence of microenvironments and thus a natural basis for many fundamental questions in microbial ecology. However, without the capacity to couple microscopy and image analysis to other analytical techniques such as labeling with fluorescent reporter molecules, it has not been possible to assess, for example, the microenvironment, the state of the organism, and its taxonomic affiliation. There have been several important developments relevant to analytical imaging, including fluorescent probes whose fluorescence is influenced by pH, E_h, etc., fluorescent oligonucleotide probes, highly sensitive charge-coupled device (CCD) cameras, and computer software and hardware required for digital image acquisition and analysis, particularly by scanning confocal laser microscopy (SCLM). These developments provide the basis for the application of scientific visualization in environmental microbiology. In the increasingly digital world in which we live, it is not surprising that traditional microscopic observation is developing into the discipline of analytical imaging. Any image which may be obtained with a microscope can now be digitized, after which a wide range of potential analyses are possible. Thus, the emphasis in this school of microscopy is on the quantitative extraction of the information contained in the image. To a large degree, this approach remains full of potential and is far from fully exploited.

SPECIMEN PREPARATION AND HANDLING SYSTEMS FOR MICROSCOPIC ANALYSIS

A major goal of the use of microscopic techniques in microbial ecology is to achieve minimum disturbance of the system under observation. Although the option always exists to take the microscope and place it in the habitat, only Staley (105) has taken the plunge. Thus, for microscopy studies, a plethora of devices and methods that are amenable to the culture and study of bacteria and bacterial communities have been developed. These systems all represent attempts to recreate nature in the laboratory, and all succeed and fail to various degrees for observation, enumeration, or biodiversity measurements.

Details of the preparation of living cell suspensions as wet mounts and hanging block and drop mounts and the use of agar-gelatin slide cultures are discussed by Murray et al. (76). These authors as well as Marshall (72), Brock (16), and Wimpenny (118) provide selective guides to the original literature on these approaches, which are fundamental to many light microscopic studies of microorganisms.

Extraction and embedding of soil materials for microscopic examination have been described by Bae et al. (8). Further developments in embedding have been described by Yu et al. (125). Waid (114) described a method for obtaining soil films which can be examined by light and electron microscopy. This technique may result in introduction of various artifacts such as collapse of structures and distortions. However, this problem can be minimized with SCLM, for which drying or embedding of the peels is not required. Culture methods include the use of glass slide incubation (13, 27, 42, 105), capillary techniques (81), and agar block microculture (108).

A variety of flow cells and microperfusion chambers have been used to study attached growth by microorganisms (21, 33, 34, 121, 122). These systems offer the advantages of controlled but easily changed conditions, ease of use with in situ observation, microscopy, and microelectrodes (useful for monitoring community development), application of fluorescent probes, and in situ antibody studies. The approach also allows easy examination of spatial and temporal changes in the material under examination, particularly when used in conjunction with computer-controlled stages (see Computer Control of the Microscope Stage below). Various designs of gel-stabilized systems have been described. Early development of this approach, as well as its advantages and limitations, have been reviewed by Wimpenny et al. (119). Chapter 9 in this volume provides information regarding culture techniques suitable for use with light microscopy. Some of these devices, gel-stabilized gradient plates (34, 121), and flow cells (122) are commercially available from Koh Industries, Ann Arbor, Mich.

LIGHT MICROSCOPY

The Limits to Light Microscopy

In addition to proper object illumination and precise alignment of the microscope's optical elements, perhaps the most important factors limiting primary image quality are the selection of the microscope objective lens, matching condenser elements, and appropriate interference filters best suited to a particular application. A number of factors influence the selection of an ideal lens, including the resolution and magnification required, the type of material to be examined and how the sample is prepared, the nature of the supporting material (glass slide; wet or dry mount; flow cell; oil, water, or glycerol immersion; thickness of coverslips; etc.), and what type of information is desired (e.g., enumeration, cellular structure, specific fluor binding patterns, or gross morphology).

When determination of image detail is a concern, remember that resolution is dependent not on lens magnification but rather on the numerical aperture (NA) of the lens, according to the relationship $d = 0.61 \times \lambda/NA$, where λ is the light source wavelength and d represents the minimum distance between two dots which can be identified. Usually a combination of high magnification ($\times 60$ or $\times 100$) and high NA (1.3 or 1.4) is desirable for microscopic analysis of cellular and subcellular detail. For larger objects or when gross morphological structure and arrangement are of interest, excellent lower-magnification lenses ($50\times$, 1.0-NA, and $<40\times$ lenses) may also be obtained from the major microscope manufacturers.

Depth of field, the distance between the closest and farthest objects in focus within a field of view, is also a consideration for light microscopy applications. For high-NA lenses, the depth of focus is approximately calculated by using the following equation: depth of focus (in micrometers $= \lambda(\mu m)/2(NA)^2$.

The working distance or depth of focus (the range of distances within which the image formed by the lens is clearly focused) is also a critical factor in considering objective lenses. The working distances for various lenses showing the impact of NA on this parameter are shown in Table 1.

Brock (17) concluded that direct microscopy is rarely able to reveal the presence of small microorganisms at the densities found in natural systems and that overestimation of the significance of aggregates is likely to occur from uncritical use of the light microscope. However, direct microscopic examination of various habitats by using the techniques outlined below reveals considerably more detail than was previously thought possible. This fact does not, however, excuse uncritical application of the technology.

Analog Methods for Image Acquisition

The modern microscopist has an array of digital tools which extend the range and extent of possible analyses; however, the quality of the extracted data is still dependent on the fidelity of the primary microscopic image. It thus follows that the principles governing basic light microscopy play a critical role in most advanced microscopic applications. The various microscope techniques, dark-field, phase-contrast, interference, fluorescence, and the others described below, provide various means to enhance the contrast between microorganisms and their environment, thus enabling the investigator to visualize them. The reader is referred to references 77 and 102 for reviews of the fundamentals of microscopy.

TABLE 1 Objective lenses for fluorescence and SCLM applications

Lens	Magnification/ NA/immersion	Working distance (mm)
Leica		
NPL Fluotar	10/0.45/oil	0.33
NPL Fluotar	25/0.75/oil	0.14
NPL Fluotar	40/1.3/oil	0.21
NPL Fluotar	50/1.00/water	0.68
NPL Fluotar	50/1.00/oil	0.18
NPL Fluotar	63/1.3/oil	
EF L (PHACO)	20/0.32/air	6.83
NPL FL L (PHACO)	32/0.40/air	6.55
NPL Fl L (PHACO)	40/0.60/air	1.65
	25/0.60/water	
	100/1.2/water	
Zeiss		
Plan-Neofluor	16/0.50/multi	0.22
Plan Apochromat	25/0.8/multi	0.32
Plan-Neofluor	40/0.9/multi	0.13
Plan-Neofluor	40/1.3/oil	0.12
Plan Apochromat	40/1.00/oil	0.38
	63/1.0/water	
Plan Apochromat	63/1.4/oil	0.09
Plan-Neofluor	100/1.3/oil	0.08
Achro	40/0.85/oil	0.35
Achro	40/0.75/water	1.6
Olympus		
D-Plan Apo	20/0.70/dry	
D-Plan Apo	40/0.85/dry	
D-Apo	40/1.3/oil	
D-Apo	100/1.3/oil	
S Plan Apo	10/0.40	0.55
S Plan Apo	20/0.70	0.55
S Plan Apo	40/0.95	0.13
S Plan Apo	60/1.40/oil	0.12
ULWD-CD Plan	20/0.40/air	10.5
LWD-CD Plan	40/0.60/1.92	
Nikon		
CF Fluor	20/0.75/dry	0.66
CF UVF	40/1.3/glycerol	0.10
CF Plan Apo	40/1.0/oil	0.12
CF N ELWD CFM Plan Achro	20/0.40/dry	6.0
CF N ELWD/CFM Plan Achro	40/0.55/dry	5.08–6.84
CF N LWD CFM Plan Achro	100/0.40/dry	0.66
CFM Plan Achro	60/0.70/dry	4.9
CFDB Plan Achro	100/0.90/dry	1.0
CFDB Plan Achro	100/0.80/dry	2.0
CF N Plan Apochromat	60/1.4/oil	0.17
CF N Plan Apochromat	100/1.4/oil	0.10

Dark-Field Microscopy

Contrast between organisms and their surrounding liquid can be achieved economically and easily by the use of dark-field microscopy. The nature of dark-field microscopy and its applications are covered by Murray and Robinow (77).

Dark-field microscopy has been applied in conjunction with image analysis in a number of studies. Dark-field microscopy is based on the scattering of light from a point source. Accurate size measurements cannot be performed on dark-field images unless some conversion factor is included in the areal calculation. However, this fact is generally not a problem for particle enumeration (e.g., for bacteria, it actually increases the sensitivity of cell detection) and also does not preclude monitoring changes in dark-field light scattering over time as a function of growth rate (56, 65). In recent years, dark-field image analysis has been used to quantify rates of bacterial deposition, monitor the growth and development of bacteria and bacterial microcolonies, and document pathways of bacterial microcolony formation (56, 57, 66). Dark-field image analysis has also been used to quantify the motile behavior of both bacteria and surface-associated protozoans (56, 65). Modified dark-field apparatuses have been developed to provide illumination of objects positioned on opaque or solid materials, such as enamel, steel, or plastics. Sjollema et al. (99) used an externally mounted incandescent system to illuminate (using a very low illumination angle) bacteria attached to the lower surfaces of a flow cell, creating a reflected dark field that provided high-quality images for image analysis.

Phase-Contrast Microscopy

Because of the high resolving power and magnification afforded by moderately expensive lenses and accessory equipment, phase-contrast microscopy remains the most routinely used microscopic technique. Procedures and theory for phase-contrast microscopy have been well detailed in a number of excellent publications and manuals (77). It is important to note that Kohler illumination (a method of providing centered, uniform illumination focused to the plane of analysis that achieves maximum resolution) must be achieved for optimum results. In addition, chromatic aberration, which can blur portions of the object being viewed, may be minimized by using a green interference filter. Spherical aberration may also be reduced by using curvature-corrected lenses, such as planapochromat lenses. With high NA, planapochromat lenses will result in the highest-quality high-resolution images, albeit at the highest cost.

A number of workers have combined digital imaging and analysis with phase-contrast microscopy. This approach has proven useful for estimating the growth rates and quantifying the morphologies of attached bacteria growing as biofilms within glass flow cells, as cells are maintained in a single focal plane at the glass-liquid interface, allowing acquisition of excellent primary images for both photography and digitization (21, 47, 63, 64, 68). Enumeration studies have also been conducted by using low-magnification (40×) extra-long-working-distance (ELWD) phase-contrast objectives in conjunction with image analysis (99, 100). Image analysis of low-magnification (×16) phase-contrast video recordings of diatoms responding to gradients of mannose has also been used to document diatom chemotactic responses (53). Although use of low magnification and low-NA lenses may reduce image quality, it may provide an image with the desired information content at lower cost.

Interference Microscopy

In interference contrast microscopy, two beams, one carrying the image and a reference beam causing interference patterns, are produced optically. Details explaining the setup and operation of interference optics are found in references 15, 48, 76, and 103. This method is very useful for resolving fine subcellular detail without halo effects observed in phase-contrast microscopy and is also less subject to interference from the optical effects of other cell materials than phase-contrast microscopy. If white light is used, the image will contain interference colors which are useful for microstructure studies as well as the analysis of crystals or mineral thin sections. The image may also attain a pseudo-three-dimensional (3-D) appearance.

Despite the apparent benefits, relatively few microbiological studies have used interference microscopy in conjunction with analytical imaging methods. Keevil and Walker (49) described the use of episcopic Nomarski differential interference contrast (DIC) microscopy in conjunction with epifluorescence microscopy, which allowed the simultaneous analysis of hydrated biofilm structure and the binding patterns of acridine orange (AO) and a vital stain in microbial films which exceeded 100 μm in total thickness.

Fluorescence Microscopy

In general, high-quality microscopes may be equipped for fluorescence microscopy. This involves the fitting of an appropriate light source, including quartz-halogen-tungsten filament lamps, mercury vapor arc lamps, and xenon arc lamps. The latter two are the most widely used for fluorescence microscopy. Mercury lamps have major lines at 365, 405, 436, and 546 nm, in contrast to the xenon lamps, which lack prominent spectral lines and provide more potential for excitation in the UV range than mercury lamps. For successful application, any of the light sources must be used in combination with the correct filters (heat-adsorbing filters, narrow-band-pass filters, barrier filters) which permit excitation of the specimen at optimal wavelengths, detection of reemitted fluorescence, and blockage of harmful UV wavelengths. Epifluorescence microscopy (whereby incident light passes through an objective lens which functions as the condenser and objective) is the most common type of fluorescence microscopy system. Epifluorescence provides a major advantage in that it permits visualization of bacteria on opaque surfaces. However, transmitted and dark-field illumination fluorescence also may be used (77). See references 40 and 41 for reviews on the development of fluorescent probes. Additional technical information is also available in reference 115. Rost (88, 89) has prepared two excellent volumes on fluorescence microscopy that cover all major aspects of the area.

Fluorescence microscopy dictates special care in specimen handling and observation. Many of these factors apply equally well to SCLM. Light sources for fluorescence microscopy are generally more energetic and cause photooxidation and bleaching of the sample. However, a number of fade-retardant formulations are commercially available and can substantially increase working time with various fluors. Most fade-retardant solutions are toxic, and we consider that their purchase from suppliers such as Molecular Probes Inc. (Eugene, Ore.) or Citifluor (London, England) is a safe and simple option. Fluorescence yield may be very weak, requiring long exposure times to obtain either photographs or video images. Consequently, high-speed (e.g., 1600 pushed to 3200 ASA) films that produce high-contrast photographs are desirable, as are highly light sensitive video systems (10^{-3}-lx or greater sensitivity). Because silicate glasses transmit UV wavelengths poorly, use of UV-transparent coverslips (either quartz or high-quality zinc titanium; G.M. Associates, Inc., Oakland, Calif., or Corning Glass Works, Corning, N.Y.) will increase the strength of excitation wavelengths and substantially improve fluorescence emission. To reduce bleaching, samples should be protected from unnecessary exposure to any light source. The room lighting should be subdued to aid in viewing the specimen. Similarly, prolonged viewing under epi-illumination should also be avoided for the same reasons.

Fluorescence microscopy is the technique used most frequently in combination with image analysis, primarily for the enumeration of samples stained with a fluorescent probe (58, 98, 111). Provided that the strength of fluorescence emission is strong enough, this technique can be used to visualize bacteria in a wide range of environmental samples. The use of fluorescence microscopy for accurate determination of cell sizes has limitations as a result of the undefined cell boundary resulting from the fluorescent halo effect; however, this problem can be minimized by using specialized object recognition imaging routines (19, 58). Excellent examples of the application of fluorescence microscopy and image analysis in microbiology are provided in references 11, 75, 96, and 97.

Other Microscopic Approaches

Typically, images have been obtained by using transmitted, incident, or polarizing light microscopy techniques, as outlined by Marshall (72). Casida (25) described the use of color infrared photography for microscopic visualization of nonstained microorganisms in soil and other habitats. Casida (23) also detailed the use of a modified metallurgical reflected light microscope with a floating stage for observation of microorganisms in natural habitats. The continuously variable phase and amplitude-contrast system microscope was also shown to work for observations in soil habitats (26), as was interval scanning (24). Polarizing microscopy, which has been used for soil fabric analyses in soil science, has been proposed for use in combination with phase-contrast microscopy for examination of relationships between soil particles and microorganisms (72). The technique may also be combined with epifluorescence techniques to achieve the same end. Interference reflectance microscopy may also provide information regarding attached microorganisms on transparent substrata (36).

These microscopic techniques have been little used by microbial ecologists or environmental microbiologists. However, when coupled to current camera and image analysis systems, these approaches may have considerable potential for studies of microorganisms in their natural habitats.

ANALYTICAL MICROSCOPY

The essential tools for analytical imaging include a high-quality photomicroscope with the capacity to mount both a standard photographic camera and a light-sensitive video camera. The system should also be capable of phase-contrast microscopy, dark-field microscopy, and epifluorescence microscopy, and the images should be transferable to a computer with as much random-access memory as can be provided and a video card or frame grabber to allow digital acquisition of images. A wide variety of software packages are available from suppliers in both DOS and Macintosh formats; each must be assessed to ensure that it is capable of image analysis and processing and is flexible (e.g., allowing users to program their own processing routines). The ultimate choices in microscope, computer, and software packages should be dictated by the applications routinely performed, microscope resolution required, and ease and speed of image handling.

The development of SCLM has provided a tool with increased resolution and optical sectioning capability. The system allows imaging of fully hydrated living materials particularly in combination with fluorescence techniques (19). Scanning confocal laser microscopes may be configured to combine all of the required elements for analytical imaging. A variety of systems are available from several manufacturers; they come in various configurations, inverted and upright, and cost from $80,000 to $450,000, depending on the options chosen or required. It is essential when setting up a digital imaging and laser microscopy facility to assess the equipment offered by each manufacturer with your own samples. The technology continues to evolve and improve; therefore, although one manufacturer may have an advantage at the time of writing, substantial technological and software innovation could alter this position. Thus, we make no recommendations regarding specific instrumentation or software but do provide examples of applications. The descriptions of light microscopy, analytical imaging steps, suggested probes, and laser microscopy are general in nature and not limited to equipment available from a specific manufacturer.

Digital Image Acquisition (SCLM)

Description

SCLM is a fusion of traditional light microscope hardware with a laser light source and computerized digital imaging. In conventional microscopy, all light passing through a specimen is viewed simultaneously by eye or by camera. With SCLM, a laser beam, positioned by means of a galvanometrically driven oscillating mirror (or the less common scanning stage approach), continuously scans point by point and line by line in a user-defined region of the specimen. With beam-scanning SCLM systems, points in the specimen excited by the incident laser light fluoresce and emit light which is digitally accumulated by a photomultiplier detector (i.e., in a 768 by 512 matrix) over the scan interval. Scan rates are variable (1/4 to 4 s); however, frame scan speed is increased at the expense of horizontal image resolution for the BioRad MRC 600, for example, scanning at 1/4 s per frame is performed at 128 lines horizontal resolution, whereas scanning at 4 s per frame is performed at 512 lines horizontal resolution. This means that the 1/4-s-per-frame image is of much lower (25%) resolution. Magnification is enhanced (by zooming) when the scanning is performed over user-defined image areas of decreasing size. Because the scanned image is accumulated in a pointwise fashion, image degradation from objects positioned laterally from the beam point is precluded. More importantly, pinholes confocally positioned in front of the incident laser light and in front of the photomultiplier detector prevent fluorescent signals originating from above, below, or beside the point of focus from reaching the photodetector (as the angles are not correct for the light to pass through the detector aperture). This effectively enhances image quality by eliminating out-of-focus light from thick specimens, resulting in what essentially is an optical-digital thin section with a thickness approaching the theoretical resolution of the light microscope (approximately 0.2 μm), depending on the lenses used and the size of the confocal pinhole. *Handbook of Biological Confocal Laser Microscopy* is an excellent source of information on nearly all aspects of confocal microscopy (80).

Choice of Lenses

The principal factors defining the quality of SCLM images are the NA of the objective lens in combination with the small size of the pinhole. For practical purposes, images cease to be confocal in nature when the NA of the lens is less than 0.5; greater confocality is obtained with high-NA (1.4) objective lenses. When lenses having an NA of 0.2 or less are used, the thickness of the optical thin section is approximately 10 μm; however, this value drops rapidly to less than 1 μm when objectives having an NA of >0.6 are used. However, the recommended objective lenses for best performance (high resolution) are those which provide a flat field and a high NA, such as 1.4-N.A. 60 or 100× planapochromat lenses. Electronic enlargement of objects up to ~100× can be achieved with the 1.4-NA 60× objective without a decrease in resolution. Table 1 provides a selected listing of lenses with high NA and some with long-working-distance characteristics. The reader is referred to reference 80 for an in-depth discussion of the choice of objective lenses for SCLM applications. Shotton (95) also provides an excellent review of laser microscopy fundamentals and applications.

However, resolution is not the complete story; images with high information content can be obtained with a wide range of lenses. Special-purpose lenses used in traditional microscopy include ELWD lenses and water-immersible and water immersion lenses. Excellent 150× water immersion lenses with high NA (1.3) values (Olympus) can be obtained. Zeiss provides a tapered 63× 1.0-NA water-immersible lens designed specifically for use with microelectrodes. The Zeiss lens provides an excellent combination of working distance and resolution for $2,000. The Nikon 40 and 20× water-immersible lenses have proven very useful in our hands for examination of stream materials and rock surfaces. Nikon is also marketing a water immersion 60× 1.2-NA lens, although the cost is about $9,000 and the benefits of increased working distance are lost. A similar 60× water immersion lens is also available from Olympus for a similar price. Although ELWD lenses sacrifice NA to achieve the longer working distances, they can be usefully applied to studies of the rhizosphere and of soil and sediment samples, for which increased working distances can be advantageous.

SCLM Setup and Operation

Alignment

The setup and operating procedures of SCLM systems vary with the model of the microscope; however, a number of fundamental elements are common to all systems. Following SCLM installation, the internal mirrors and optical elements of the unit should be aligned by a company representative; however, working alignment must be confirmed prior to each usage. The ability of the user to check or modify alignment varies greatly with instrumentation provided by the manufacturers. Desirable alignment features include centering the incident laser light with respect to the objective lens by using an aligning prism and optimizing the emitted fluorescent or reflected light with respect to the photodetector. This is achieved by using specific tools or procedures outlined by the manufacturer to adjust and align the various galvanometric mirrors.

It is essential to ensure that the correct filters are installed. For example, Bio-Rad utilizes a 488-nm excitation, 515-nm long-pass emission filter for green excitation and emission and a 514-nm excitation, 550-nm long-pass emission filter for red fluorescence excitation and emission. In most SCLM systems, the user may collect the images obtained in green and red wavelengths separately or simultaneously. Simultaneous dual-wavelength imaging involves the use of dual photomultiplier detectors and merging of the images by using software provided by the manufacturer. Reflectance imaging may be achieved through installation or positioning of a special filter set available from most manufacturers. Special blocks and filter sets for use with specific fluorescent probes may also be obtained. Always ensure that the laser shutter is not open when filter blocks are changed. Avoid touching any of the optical elements during this procedure. The filter blocks or wheels are precision aligned at the factory and are generally not user serviceable. Always store filter blocks in a dust-free container.

After the filters are aligned, the sample is placed on the microscope and imaged by using a low zoom factor, ensuring that the PMT (photomultiplier tube) gain is set to manual and the detector aperture is about half open. Focus on a portion of the sample, which should provide an even signal response once the SCLM is aligned. Scanning in real time at 1/4 frame per s, the gain of the PMT should be adjusted such that the image is about at half brightness (no part of the image should be saturated) but of reasonable quality. The mirrors should then be adjusted so that the image is evenly illuminated. If the system was misaligned, this operation will result in a significant increase in brightness and image clarity. The gain may need to be reduced to prevent saturation as image quality is enhanced. This process should be repeated with the PMT detector aperture (pinhole) reduced incrementally until it is at its smallest size. If dual-PMT detection is required, ensure that the correct filter blocks are in place in PMT1 and PMT2 filter holders and repeat alignment procedures as outlined above.

It is good practice to periodically check the alignment of the system during extended use and always at the beginning of each session. The system should always be realigned when filter blocks are changed. The advent of fiber optic connections in the new generation of laser microscopes has greatly simplified these operations.

SCLM Fluorescence Yield

Assuming that the SCLM system is properly aligned, numerous other factors can influence fluorescence yield from the specimen, transfer efficiency within the microscope, and subsequent detection by the photomultiplier. Sample-related factors include the strength of fluorescence emitted from the sample, which in turn is dependent on the fluorochrome used or the inherent reflectance of the material. There are a number of staining options whereby the fluorescent signal may be optimized (see Applications of Digital Microscopy, below). The wide array of applications and different fluorescent compounds dictates that all or most of the SCLM settings and associated hardware be optimized for each set of experiments. Of significance are the following. (i) Care must be taken to ensure that the objective lens used for a particular analysis is of sufficient magnification and resolving potential. (ii) The filter sets (and laser source) must be matched to the excitation and emission spectra of the fluorochrome. (iii) The intensity of the incident laser light will not always be sufficient to provide a strong fluorescent signal, especially when the excitation wavelengths are not ideal for a particular fluorochrome. Higher-intensity laser illumination (adjustable between 1 and 100% transmission by using neutral density filters) may be required to produce a detectable fluorescence emission). (iv) The size of the confocal aperture at the PMT detector must be correct. However, this parameter also regulates the thickness of the optical thin section, with larger apertures producing thicker optical sections.

Optimizing the fluorescence yield from a given sample often involves compromising on one or more of these four factors. However, under ideal conditions, the lowest-intensity laser settings (usually 1% transmission) and the smallest detector pinhole aperture size should be used to minimize photobleaching or photodamage of the specimen while optimizing image quality. Transfer efficiency is the fraction of light from the specimen detected by the photomultiplier detector. Within the various laser microscopes, this factor has changed considerably with time; for example, the Bio-Rad MRC 500 had a transfer efficiency of 0.32, the MRC 600 has an improved efficiency of 0.70, and the current Leica model lists an optical transfer efficiency of 0.92. Transfer efficiency is an important factor to be considered during purchase of laser microscopy equipment, but one must consider how it was measured; for example, the Leica value is an optical bench measurement, whereas others are working measurements. Optical transfer efficiency will vary with sample, oil versus air, fluor, etc. Thus, users should always provide their own materials in any evaluation of equipment.

Detection of Fluorescence by SCLM

In addition to being affected by mirror alignment and pinhole optimization, the quality of the digital image observed on the computer screen is influenced by the adjustment of the scan head or photodetector(s) controls as well as adjustment of the computer monitor. Improper adjustment of the photodetector(s) in the SCLM scan head is a very common cause of low-quality SCLM images. Before the SCLM is used, the black level should be adjusted to facilitate image detection and display. The point where the boundary between scans just becomes invisible against the dark background is the correct setting for the black level. This setting is critical for detecting darker parts of the image, the details of which may be irretrievably lost if the black level is set too high.

The photodetector gain level must also be optimized after the black level has been set; in general, this setting controls the brightness of the image. To optimize the range of brightness available for any image, it is usually desirable to increase the brightness of the image until the brightest regions (ideally only a few pixels) of the image approach saturation (a grey-level setting of 255 on a scale of 0 to 255, where 0 = black and 255 = white). When one is attempting to locate a specimen, high gain settings may be used, but as the image is brought into crisp focus, the gain level should be reduced to utilize the entire grey-level range. For details on the function and operation of additional controls influencing image collection, users are directed to their SCLM user manuals.

Adjustment of the computer monitor can influence how the user perceives the quality of the image. An improperly adjusted monitor can result in an over- or underestimation of what combination of gain and black levels results in image saturation, decreasing the amount of information contained in the final digital images. Under working lighting conditions, view the monitor test image and adjust the monitor brightness and contrast controls until all intensities are visible and discrete. The image should then be centered and sized with the aid of a test image. Before using any SCLM system, one should become completely familiar with the user manuals for associated SCLM hardware and software.

Laser Sources

SCLM systems may function confocally in either the fluorescence (using a dichroic beam splitter) or reflectance (using a 50% reflecting/transmitting beam splitter) mode. In terms of fluorescence operation, a consideration of importance is matching the fluorochrome (see Application of Digital Microscopy, below) with the major lines of emission from the laser source. SCLM systems may be outfitted with a number of different types of lasers, each having characteristic lines of light emission. The most common, least expensive, and most reliable (5,000- to 10,000-h life span) is the argon ion laser, which when operated at maximum power (50 mW) emits two main lines at 488 (the blue line) and 514 (the green line) nm (as well as minor lines ranging from 274 to 528 nm). It is important to ensure that the laser is operated in the full-power mode as opposed to the standby mode when a green line is required, as the green line generated in standby mode is very weak. The major drawbacks to the argon laser are the limited range of fluors excited by the laser and the lack of separation of green and red fluorescence for dual labeling of samples.

Helium-neon and krypton-argon mixed-gas lasers may also be obtained and have the advantage of providing a greater number of lines of excitation for use with a wider range of fluorochromes and the potential for simultaneous excitation of three fluors. For example, the krypton-argon mixed-gas laser provides lines of excitation at 488 (blue), 568 (yellow), and 647 (red) nm from a single laser beam. The principal advantage of the mixed-gas laser is that there is little spectral emission overlap between fluorescein-based fluors and red fluors such as Texas red or rhodamine, thus minimizing bleed-through of fluorescein into the red detection channel. Dual lasers include the argon and helium-neon pairing, which mainly provides a strong green line (from the helium-neon laser) while maintaining the blue and light-green line of the argon laser. Once again, low bleed-through is prevented; however, the two beams cannot be used simultaneously, as the 543-nm line of the helium-neon laser overlaps with that of fluorescein emission. Bleed-through in dual-channel imaging is also controllable through application of software to electronically remove overlap. Specialized applications may also necessitate the use of alternate laser sources, including the krypton ion and helium-cadmium lasers (providing a strong red line and 325/442-nm lines, respectively). Lastly, UV–visible-UV excimer lasers (157 to 351 nm) may now be obtained for use with SCLM systems; however, these continue to be quite expensive (the result of quartz optical elements and special lenses). However, UV-based systems change the range of fluorescent probes which may be used.

PMTs

The range of PMTs offered in commercial SCLM systems is somewhat limited, but potential exists for application of PMTs with various spectral ranges, from 185 to 930 nm. Combinations of these PMTs with various lasers, filters, and probes could allow creation of multichannel, multiparameter imaging. For example, in a simple case, cell types may be indicated by using polyclonal or monoclonal antibodies or nucleotide probes labeled with green and red fluors, while polymers may be visualized by using lectins conjugated to colloidal gold and visualized by reflectance.

Image Acquisition

To acquire an image by SCLM entails optimizing the instrument as described above. After these steps have been carried out and a sample has been prepared with the appropriate fluors or probes, the area of interest must be defined. This may be carried out either through viewing the sample by conventional epifluorescence or on-line scanning with the laser; the latter method becomes preferred with experience, since bleaching of the sample and handling time are reduced. The user then has the choice of acquiring a single *xy* image from a selected optical plane or a series of optical planes (i.e., a *z* series). The nature of confocal microscopy allows the system to be operated as though it had infinite depth of field, and thus a series of in-focus, blur-free images in perfect register may be sequentially collected. Figure 1 shows a series of xy images taken at the 2-, 8-, and 14-μm depths in a biofilm community, using negative staining so that cells appear as dark objects. One of the most novel and interesting imaging options is to use the z-axis control function of the stage to scan in the x and z axes to collect a sagittal section of the material in the vertical (xz) axis. Some typical examples of this type of imaging are shown in Fig. 2. These images show both the xy (bottom) and xz (top) images separated by a black line (the coverslip) of two biofilms and illustrate the information content, i.e., surface roughness and depth variations, that can be seen in images obtained through both xy and xz optical sectioning techniques.

These images may be acquired either as direct images or through a mathematical filter. For example, Kalman filtration (a running average filtration that mathematically averages sequentially collected images, reducing noise level in the image but maintaining edge features) is extremely effective in reducing random noise and producing a clean image. The general equation for Kalman filtration is as follows: $(P_n = I/j + P_{j1} (1 - 1/j)$, where Pn is the new pixel value, I is the input value, and j is the current frame value.

However, collection through mathematical filtration reduces the information content of the image for purposes such as ratiometric analysis. Thus, the user should define the end use of the images prior to collection. Various manufacturers offer other mathematical collection filters that allow for image enhancement through summation of signals, collection to a preset maximum signal, local contrast enhancement, median filters, linear smoothing filters, and gradient filters (80). Each of these filters attempts to reduce noise levels in images, smooth the data set, and enhance the edge of objects within the image. For example, the median filter establishes a box of pixels and calculates the median intensity within the box; this value is then substituted for the central pixel value, the box is moved 1 pixel, and the process is repeated. This process tends to remove small objects, smoothing the data but retaining the edges of objects of interest. Deconvolution filters may also be applied to the confocal image to further enhance their clarity. The Bio-Rad systems also offer a low-signal-detection or photon-counting mode that may be applied to weak signals.

A number of factors limit the optical sectioning of biological materials, most significantly the ability of the laser beam to penetrate the sample and the quenching of the fluorescence emission. The staining method chosen will also influence the maximum depth from which optical sections may be obtained. For example, negative staining (fluorescence exclusion [the sample is flooded with fluor]) results in much greater quenching of the laser beam and auto-quenching of emitted signal than the use of a positive stain such as fluorescein isothiocyanate (FITC) or tetramethyl rhodamine isothiocyanate (TRITC). However, the former method is not subject to bleaching and loss of contrast (20).

The objective lens selected for imaging positively or negatively stained samples is also a major factor in successful optical sectioning. In general, the high-NA lenses are not suitable for collection of images in thick specimens. We have found that we can effectively section through 50 to 100 μm of biofilm materials by using 60$\times$ 1.4-NA lenses, several hundred micrometers with 63$\times$ 1.0-NA water-immersible lenses, and up to 1 mm and more with 20 or 40$\times$ ELWD lenses or 40$\times$ water-immersible lenses. The 40$\times$ 1.0-NA objectives have been reported to give working dis-

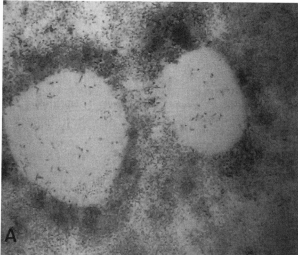
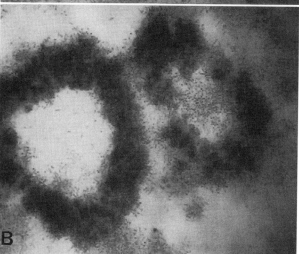
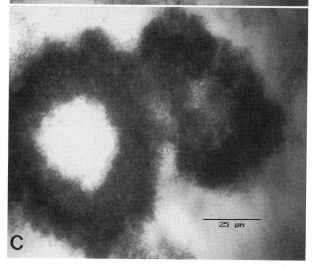

FIGURE 1 A series of three xy optical thin sections taken at 2-, 8-, and 14-μm depths within a 40-μm-thick degradative biofilm community. These sections were created with a Bio-Rad MRC 600 scanning confocal laser microscope mounted on a Nikon SA photomicroscope and interfaced with a Northgate 486 computer. The staining procedure used was negative staining (20).

FIGURE 2 SCLM photomicrographs of *xz* or sagittal sections (top images marked with vertical scale bar) through a biofilm community. The black horizontal lines separating the *xz* and *xy* images are the coverslip that the biofilm is attached to. Both images were taken with a Bio-Rad MRC 600 laser microscope and subjected to negative staining. The location of the *xz* image is shown as a dotted line in the bottom *xy* image. The *xz* view is one of the unique features of SCLM imaging and provides a valuable view of the fully hydrated biofilm community.

tances of 300 to 400 μm. Shotton (95) and Rost (88, 89) provide additional information on effective working distances for lenses.

In fluorescence confocal imaging, simultaneous detection of two or more fluorochromes is of benefit. A number of parameters may be collected for the same field (see Applications of Digital Microscopy, below). One of the great benefits of dual or multichannel imaging is that the images obtained are in exact spatial register, which facilitates dual channel display, fluorescence ratio imaging, stereoscopic imaging, and 3-D image reconstruction. The application of pseudocoloring can also be useful for display of information from ratiometric images of pH, E_h, or ion concentrations or to represent results from the application of multiple fluors.

Reflection imaging may also be performed with SCLM systems in the confocal mode. This has great significance for quality control in the microprocessor industry but also has microbiological applications in that differences in refractive index within the specimen result in the reflection of laser light. Confocal reflection contrast images may be

obtained by SCLM and provide superior images to conventional interference microscopy. The combination of reflectance imaging with colloidal gold labeling also has been used in cell biology (95).

It is also possible to equip the SCLM system with a nonconfocal detector system for scanned transmitted light imaging. Provided that the host microscope is properly equipped, dark-field, phase-contrast, and Nomarski DIC images may be digitized of the same microscope field. This provides the microscopist with the opportunity to combine confocal fluorescence (including multichannel imaging), confocal reflectance, transmitted phase-contrast, dark-field, and Nomarski DIC, and conventional epifluorescence imaging of the same sample materials.

IMAGE PROCESSING AND ANALYSIS

Image Input Devices

To perform image analysis, a device to create an analog video signal is needed. This signal is subsequently converted (digitized) to a primary digital image via a dedicated image-processing computer or an add-on video-digitizing device (video board or frame grabber). Used in conjunction with a photomicroscope, the video camera represents the most common route for obtaining an analog signal of a microscopic image. Image fidelity is crucial, as the success of all image-processing and analysis steps is basically contingent on the quality of the primary video signal. It follows that the camera system must therefore be matched with the intended end application (primarily in terms of the sensitivity, resolution, and linearity of response) and also with the equipment available.

The microscope must be outfitted with a video tube which, as a standard, accepts a C-mount collar threaded into the video camera. The video camera is connected to a computer capable of digitizing an image by means of coaxial video cable and BNC (baby N connector)- or UHF (ultra-high frequency)-type connectors. For grey-scale images, a single cable (RS 170; most commonly 8 bit and 256 levels of grey) may be used to transfer images. Four cables are required for a color RGB (red, green, blue) signal transfer, which involves a 24-bit signal at 8 bits for each of the three color channels and a separate synchronization channel.

Scientific video cameras vary with respect to horizontal resolution, light sensitivity, intrascene dynamic range, blooming characteristics, geometric distortion, and spectral response (46). Manual adjustment of the camera gain-and-voltage control is preferred, as self-adjusting cameras alter their light sensitivity in response to changes in inter- or intrascene brightness, limiting the quantitative use of the data. Low- to mid-sensitivity CCD (10^{-3} to 10^{-1} lx) cameras are not as susceptible to electronic noise and geometric distortion as are the tube-based cameras; in recent years they have become less expensive and now dominate the laboratory market. Monochrome and color models vary somewhat in their cost, grey-level sensitivity, and horizontal resolution; however, most models provide resolutions of 480 by 640 pixels (National Television System Committee [North American] standard) and 512 by 768 pixels (phase alternating line [European] standard). The generally flexible characteristics of CCD cameras make them applicable for phase, bright-field, DIC, and dark-field imaging needs and are also useful for fluorescence applications if specimen fluorescence emission is strong and tends to resist photobleaching. The relatively large intrascene dynamic range and lack of blooming characteristics of CCD cameras are also desirable for high-contrast applications such as dark-field or fluorescence image analysis. Cooled CCD cameras, based on thermoelectrically (Peltier) cooled circuitry, now offer RGB color imaging (Optronics Engineering, Goleta, Calif.). These systems provide a very "quiet" (noise-free) image as a result of extremely low dark current and perform on-chip integration (summation of image information) over extended exposure times (up to 4 min). These systems also have a broad exposure range (from 0.0002 to 13,000 lx) and on-board digital contrast control; thus, weakly fluorescent and bright-field images can be digitized with the same basic system.

Low-light scientific requirements may also be satisfied by using either silicone intensification target tube (10^{-3} lx) or intensified silicone intensification target tube (10^{-5} lx) devices; however, these are generally more expensive than plumbicon or newvicon tube-based units or CCD cameras of similar sensitivity. Drawbacks of using silicone intensification cameras are that they usually have a lower intrascene dynamic range and response linearity and a lower overall exposure range (upper and lower limits of illumination), are more prone to blooming artifacts, and generally have a higher signal-to-noise ratio.

Extremely light sensitive, liquid-cooled ($-40°C$) CCD cameras (Photometrics Ltd., Tucson, Ariz.; Astromed Ltd., Cambridge, England) having a wider range of grey-level intensities (12 bits 4,096 grey levels) may also be obtained. Cameras with photon-counting capabilities (limiting low-level illuminance, 10^{-11} lx) are suitable only for highly specialized applications (e.g., bio- or chemiluminescence or *lux* gene expression), require specialized support equipment and darkrooms, and are not generally applicable for routine laboratory use. A list of scientific camera manufacturers is provided in Table 2. Additional useful information on camera systems may be found in reference 46.

Image Analysis Systems and Digitizing Boards

A number of image analysis programs (e.g., Ultimage, Image Analyst, NIH Image, VoxView, Micro-Tome, and IBAS) are currently available. These systems vary in terms of system architecture (parallel or pipeline [pipeline architecture allows the execution of several parallel or repetitive processing tasks at high speed without input from the central computer]) and whether the system is a dedicated image processor, is mainframe based, or is an add-on unit configured for personal computers (PCs). Other variables include grey-level resolution, programming languages, user interface (menu or command driven), capacity for software modification, system memory configurations, and capacity. Reductions in the cost of computer random-access memory and image-archiving platforms and increases in microprocessor operating speeds have resulted in more powerful add-on image-processing capabilities. Consequently, add-on digitizer boards used in conjunction with high-speed, PC platform imaging software now dominate the entry-level image-processing market primarily because imaging hardware may be interfaced with existing computer equipment. Most current-generation digitizing boards also permit the digital capture of color video signals from RGB cameras and thus provide the option of performing true color video image analysis.

The Ultimage image analysis system (GTFS Inc., Santa Rosa, Calif.) permits the analysis of various object parameters based on size, shape, or densitometry and also allows separate analyses to be performed in the different color bands (red, green, or blue) on a Macintosh platform. NIH Image is Macintosh-based freeware image analysis software programmed by Wayne Rasband of the National Institutes of Health (wayne@helix.nih.gov) available over the Internet via anonymous FTP (file transfer protocol) at the address zippy.nimh.nih.gov (to acquire this program over the Internet via FTP, see Digital Transmission of Information, below). Integration of NIH Image with laboratory computers requires the acquisition of a frame grabber or digitizing board; however, digital images may be acquired over computer networks or on disk. To maintain the range of features offered by NIH Image, Scion AG5 or LG3 boards are recommended. Different versions of NIH Image are available for various Macintosh computers, including a PowerPC version (Image 1.57) which offers excellent performance as well as a wide range of analysis, enhancement, and user-programmable functions. Gel-scanning systems (Alpha Innotech Corp., San Leandro, Calif.; PDI, Huntington Station, N.Y.; Scanalytics, Billerica, Mass.) for the analysis of 1- and 2-D electrophoretic gels further extend the range of computer-assisted video analyses that may be routinely

TABLE 2 Manufacturers of equipment for digital imaging

Manufacturer	Address	Phone and/or fax no.
Chinon America, Imaging Division	1065 Bristol Rd., Mountainside, NJ 07092	Phone: (908) 654-0404 Fax: (908) 654-6656
Cohu Inc.	Box 85623, San Diego, CA 92186	Phone: (619) 277-6700
Control Vision	Box 596, Pittsburg, KS 66762	Phone: (316) 231-6647 Fax: (316) 231-5816
Creative Micro Inc.	P.O. Box 4477, Englewood, CO 80155-4477	Phone: (800) 771-1295 Fax: (303) 771-1136
Dage-MTI, Inc.	701 N. Roeske Ave., Michigan City, IN 46360	Phone: (219) 872-5514
Detection Dynamics	4700 Loyola Lane, Suite 119, Austin, TX 78723	Phone: (512) 345-8401 Fax: (512) 926-0940
Fotodyne Inc.	Hartland, Wisc.	Phone: (800) 362-3642 Fax: (414) 369-7013
Gateway Electronics	8123 Page Blvd. St., Louis, MO 63130	Phone: (314) 427-6116 Fax: (314) 427-3147
EPIX, Inc.	381 Lexington Dr., Buffalo Grove, IL 60089	Phone: (708) 465-1818 Fax: (708) 465-1919
Hammamatsu Corp.	360 FootHill Rd., Bridgewater, NJ 08807	Phone: (908) 231-0960 Fax: (908) 231-1539
ImageNation Corp.	P.O. Box 276, Beaverton, OR 97075	Phone: (800) 366-9131 Fax: (503) 643-2458
Imagraph	11 Elizabeth Dr., Chelmsford, MA 01824	Phone: (508) 256-4624 Fax: (508) 250-9155
Industrial Video Source	1220 Champion Circle #100, Carrollton, TX 75006	Phone: (800) 627-6734 Fax: (214) 280-9668
ISIS Inc.	323 Love Place, Suite D, Goleta, CA 93117	Phone: (805) 692-2390 Fax: (805) 692-2391
Marshall Electronics	P.O. Box 2027, Culver City, CA 90230	Phone: (310) 390-6608 Fax: (310) 391-8926
Micro Video Products	16201 Osborne St., Westminster, CA 92683	Phone: (800) 473-0538 Fax: (714) 847-4486
MIKROMAK GmbH	Am Weichselgarten 4, D-91058, Erlangen, Germany	Phone: 49-9131-77939-0 Fax: 49-9131-77939-8
Perceptics Corp.	Knoxville, Tenn.	Phone: (615) 966-9200 Fax: (615) 966-9330
Photometrics Ltd.	Tucson, Ariz.	Phone: (602) 889-9933 Fax: (602) 573-1944
Polaris Industries	141 West Wieuca Rd., 300-B, Atlanta, GA 30342-3219	Phone: (800) 752-3571 Fax: (404) 252-8929
Sierra Scientific	605 West California Ave., Sunnyvale, CA 94086	Phone: (800) 397-7332 Fax: (408) 773-5672
Sharp Microelectronics	Camas, WA 98607	Phone: (206) 834-2500
Super Circuits	13552 Research Blvd. #B, Austin, TX 78750	Phone: (800) 335-9777 Fax: (512) 335-1925
Toshiba America	1010 Johnson Dr., Buffalo Grove, IL 60089-6900	Phone: (800) 253-5429 Fax: (708) 541-1927
Wintriss Engineering Corp.	6342 Ferris Square, San Diego, CA 92121	Phone: (800) 733-8089
Xillix Technologies Corp.	Suite 200, 2339 Colombia St., Vancouver, B.C. V5Y 3Y3, Canada	Phone: (604) 875-6161 Fax: (604) 872-3356

conducted. Software packages range from excellent freeware such as NIH Image to basic packages (approximately $5,000) to dedicated image analysis systems ($60,000+).

SCLM systems provide a series of perfectly registered optical thin sections that can be used to produce rendered 3-D reconstructions, which can be rotated around various axes to aid in the interpretation of the image or for volumetric analyses. PC-based 3-D imaging and image analysis systems include Micro Voxel (Indee Systems, Sunnyvale, Calif.), VoxelView (Vital Images, Fairfield, Iowa), and VoxBlast (VayTek, Inc., Fairfield, Iowa). It should be noted that most 3-D imaging programs are add-ons for use with SCLM systems and cost between $5,000 and $20,000.

In 3-D imaging, voxels (volume elements) rather than pixels (picture elements) are the basic imaging units. Using a series of image sections which consist of pixels, the 3-D imaging program calculates (from the location and intensity of 2-D image pixels) the corresponding space-filled model of the image slices (a 3-D array of voxels). A good example of this technique is the imaging of individual biofilm bacteria, whereby most information is obtained near a solid-liquid interface. For instance, sampling a 3-D matrix of 20 by 20 by 20 μm should provide excellent results (note that the 3-D image could be extracted from 20 optical thin sections of normal xy size [768 by 512 pixels] collected at 1-μm increments). However, there is a limit as to how many thin

sections (some suggest 12 or fewer) may be included before the 3-D image becomes too complex; however, these problems are also specimen dependent. In addition, if there is a large space between optical sections, the computer software may simply interpolate the missing information. Since this is undesirable, sectioning must be performed at increments scaled in accordance with the required 3-D resolution.

Examples of red-green anaglyph projections (a pair of superimposed projections, colored red and green, that when viewed through complementary-colored glasses appears stereoscopic) of SCLM optical thin sections may be seen in the reviews by Caldwell et al. (19) and Amann et al. (2). In addition, stereo pairs have been published in several articles, including that by Lawrence et al. (67). In Fig. 3 we provide two stereo pairs to be viewed with or without the aid of stereo viewing glasses. These images are projections of a series of optical thin sections of a biofilm community (40 μm thick) subjected to negative staining (top) and positive staining (bottom). The user may also vary or exaggerate the apparent thickness of the projection by changing the offset of the two images; thus, some sense of actual scale should be provided. It is important to spend considerable time viewing the images to ensure that a 3-D image is seen and that all the detail is observed. Note that 10 to 20% of the population has difficulty viewing stereo pairs. Serial displays of images or sequential

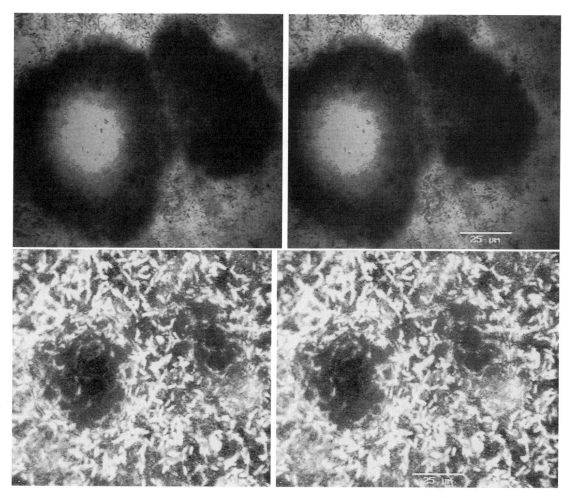

FIGURE 3 Two stereo pairs to be viewed with or without the aid of stereo viewing glasses. These images are projections of a series of optical thin sections (selected sections are shown in Fig. 1) of a 40-μm-thick biofilm community subjected to negative staining (20) (top) and positive staining (bottom). It is important to spend considerable time viewing the images to ensure that a 3-D image is seen and that all detail is observed. In each case, the reader should see structures projecting out of the page.

rotations of a projection will also provide the reader with a 3-D display of the confocal data set (80).

Basic Image Processing Steps

Image analyses involve a series of functions linked to form a sequence of computer operations common to most if not all image analysis procedures. These steps include image acquisition, image processing, image segmentation, object measurement, and data output. Readers should consult references 37, 90, and 91 for further detail on image processing and analyses.

Primary Image

Initially, the analog signal, which has continuous amplitudes, must be converted into a digital image which has discrete steps, either 1 or 0, before storage in computer random-access memory or disk memory. The manner by which the primary image is digitized depends on the intended analysis and the software and hardware capabilities. If the specimen is stationary, image averaging improves the quality of the primary image by reducing nonimage electrical interfer-

ence (electronic noise or snow, pixel dropout) inherent to all television cameras and by eliminating image degradation by objects which move during the image-averaging process. Advances in image integration such as on-chip camera integration and Kalman filtration function like simple image averaging but result in images with improved signal-to-noise ratios. Images that are of sufficient quality for processing without any enhancement may not require image averaging. In addition, for many applications, information can be lost as a consequence of the mathematical filtration process.

Mathematical Filtration and Image Enhancement

Background

Microscope misalignment and optical imperfections frequently result in the uneven illumination of the primary image, a factor which will make reliable object recognition over the whole field of analysis difficult and inconsistent. As a general step for high-quality image analysis, some form of shading correction should be used. Shading corrections involve the digitization and storage of a reference image; this image may be subtracted from all subsequent images, removing factors such as uneven illumination. When specimens are subsequently analyzed, the computer will automatically correct the image by subtracting the grey level of the imperfections from the grey level of the image, thereby eliminating localized bright spots or surface imperfections without affecting the fidelity of objects of interest. Buyers of image-processing and analysis systems should recognize the fundamental role that either shading correction or background subtraction functions play in digital imaging.

Image analysis systems can detect only object pixels having grey levels different from those of neighboring pixels. Ideally, the distribution of grey levels contained in a high-contrast image will span the entire available range (0 to 255 for 8-bit system); however, this is seldom the case. A number of image-processing steps can facilitate the analysis of images which have poor contrast or are "noisy." It should be noted that these steps cannot add information in excess of that contained within the primary image but can improve the ease of object recognition and specificity and accuracy of object measurement. A number of mathematical enhancement operations have been used to facilitate the analysis of microbiological systems, including histogram analysis in combination with grey-level transformation and/or normalization (53); contrast enhancement (68); filtration (median, low pass, Gaussian, Laplacian, etc.); image subtraction (65), addition (68), and multiplication (68); and object erosion and/or dilation (53, 65, 68). Examples of multiple-step image analysis applied to studies in microbiology are provided in references 11, 53, 58, 68, 75, 97, and 100. Digital images may also be manually edited, permitting the interactive modification of the image. For details on the use and applicability of functions available for image processing and enhancement, readers should refer to the user manual or to reference 38, 90, or 91.

Deconvolution

Deconvolution refers to the sharpening, or deblurring, of an image obtained with an optical microscope by removal of out-of-focus information, thereby mathematically producing an optical thin section. A series, or stack, of high-quality primary images of the specimen from different sectioning depths at the same xy location is required. A computer then calculates which information results from objects located in other focal planes and removes these objects from the image. The most common algorithm used for deconvolution is the nearest-neighbor deconvolution algorithm. While mathematical approaches for deblurring microscopic images are less expensive than hardware-based optical-sectioning methods, there are limitations to the number of depths that may be used during the calculation as well as to the complexity of the image. For example, a structurally simple, three-slice image may be deconvolved rapidly and with excellent results. However, 10- or 20-slice images are much more complex, and the time for iterative calculations becomes quite long. In addition, if only neighboring image slices are used during image deconvolution, the blurring caused by all other images at different depths (as well as those images not collected but between image slices) will not be removed.

Deconvolution approaches are also usable with SCLM and can be used to enhance the quality of individual optical thin sections. Shaw and Rawlins (94) used high-NA, high-magnification lenses in conjunction with a BioRad MRC 500 to develop deconvolution algorithms based on the 3-D point-spread function of the microscope lenses. The authors demonstrated that losses in image confocality, caused by use of large detector aperture sizes, could be reversed if the 3-D data were processed by using their algorithm. Deconvolution software, such as Micro-Tome, is available either through commercial sources (VayTek) or in the open literature (94).

Object Recognition and Thresholding

Images must be thresholded (segmented) to inform the computer as to which objects are to be measured during the analysis. Thresholding may be performed subjectively (with the user manually defining the grey-level cutoff value) or by using automatic thresholding functions (e.g., with a probabilistic model). During manual thresholding, objects with a grey value below a user-set level (0 to 255) are considered background (grey value reset to 0), whereas objects with grey values above this level are redefined as objects (grey level reset to 255), creating a binary image. While detection and placement of the cell boundary may accurately be estimated by using phase-contrast or bright-field illumination, a number of factors influence the computer-measured size of fluorescent objects: the length of exposure to light (long times result in bleaching and an apparent reduction in object size), the halo effect (which makes the fluorescent object seem larger), and user subjectivity (58).

For critical applications, it may be necessary to use automated edge-seeking algorithms. Sieracki et al. (97) evaluated nine different automatic thresholding methods, using fluorescent microspheres with known diameters as controls, to position the boundaries of fluorescent objects. It was found that the grey-level histogram analysis of individual cells present within images containing a number of cells could be used to construct a circular cell profile. This profile was then analyzed to determine the maximum in the second derivative and used to calculate the best thresholding level for the entire image. Viles and Sieracki (112) subsequently found that this approach, suitable for nanoplankton analyses, did not function well with smaller picoplankton and thus used a combination of edge detection and adaptive edge strengthening. Møller et al. (75) recently used Cellstat (product on-line information: http://www.lm.dtu.dk/cellstat/index.html) to obtain isointensity-thresholded images of fluorescently stained cells with the cell boundary positioned at a point equivalent to 20% of the smoothed cell intensity maximum grey-level value. It should be noted that this approach should be required only when accurate details on the area of the cell or average grey value are required and is not necessary for simple object enumeration.

Difference Imagery

Difference imagery is a technique which permits the selective identification of changes which occur or exist between two digital images and consequently facilitates the visualization and measurement of either microbial growth or movement over a defined interval. Both approaches (measurement of microbial growth and measurement of movement) have been used to facilitate the study of microbiological growth in the presence of an inorganic matrix (18) and bacterial motility (56, 65). These applications have been extensively reviewed by Caldwell et al. (19). Difference imagery may be used for temporal and spatial analyses conducted by using SCLM. For example, the development of chemical microzones, visualized by using either pH- or E_h-sensitive molecular probes, could be quantified by observing changes in the fluorescent signal over a time course.

Pseudocolor

The human eye is significantly less sensitive to changes in grey level than it is to changes in color. Color lookup tables (a user-defined function in most image analysis programs) may be applied to grey-level images, thereby improving the capacity to interpret the intensity differences contained within an image. It is also important to note that in SCLM imaging, the output of the PMTs is scaled and converted to a grey-scale image. Thus, all color images created with SCLM equipment (e.g., merged images or 3-D red-green anaglyph projections) are falsely colored so that, for example, information from the green and red channels is appropriately colored. Applications of pseudocolor to show pH gradients (20) and for production of 3-D projections (2) have been described.

Analysis and Data Output

Once the objects in an image have been defined by thresholding, a number of user-defined parameters may be measured. These include the area, morphology, densitometry, number, and/or distribution and position of objects. Any of these parameters may also be used to specifically exclude or include objects in subsequent processing steps, permitting a high degree of selectivity in the analysis. Most image analysis programs also offer flexible programming languages which may be integrated, as subroutines, with the basic program to provide for specialized analysis and measurements. Excellent programs have been developed for automatic determination of biomass (10), cell number, cell volumes, and the number of dividing cells (11); quantification of growth parameters (19, 21, 47, 65, 75); and determination of bacterial size and morphology (52, 58, 65, 96, 97).

APPLICATIONS OF DIGITAL MICROSCOPY
Enumeration

Only a small fraction of the microorganisms in natural environments can be cultured (14). Thus, significant underestimations may occur when culturing techniques are used to enumerate microorganisms from terrestrial or aquatic samples. Direct microscopic analyses, especially epifluorescence microscopy, are therefore often used to assess the number of cells present in these environments. A recent review by Kepner and Pratt (50) provides a comprehensive overview of the use of epifluorescence microscopy for enumeration.

The method described by Hobbie et al. (44), and modifications of it, have been widely used to enumerate microbial populations. In principle, the method involves placing a known amount of an environmental sample on a known area of a slide or filter and staining it with a fluorescent dye; the cells are visualized by epi-illumination with UV or blue light. Fluorochromes often used in this procedure include AO, 4′,6-diamidino-2-phenylindole (DAPI), and FITC. In addition, several DNA stains, including the SYTO (Molecular Probes) series, are available. Fluorochromes are particularly useful for fieldwork because the samples can be preserved until returned to the laboratory for analysis. Preservatives include formaldehyde at a final concentration of 2% (44, 84) and Lugol's iodine solution (82). Storage for up to 4 years in the dark at room temperature with Lugol's iodine solution without any significant change in bacterial numbers has been reported (82). Another advantage of this procedure is its simplicity.

Direct count procedures may be used in combination with other methods to assess the ratio of specific cells or populations in relation to the whole community, for instance in conjunction with autoradiography to determine the uptake of specific substrates by individuals from a community (74, 106) or in conjunction with oligonucleotide probes to monitor the presence of defined populations in heterogeneous communities (3). The direct count method is also frequently used in combination with other fluorescent probes, such as fluorescein diacetate (FDA) (28) and 5-cyano-2,3-ditotyl tetrazolium chloride (CTC) (86), for estimation of the percentage of actively metabolizing cells in microbial communities (see below).

The actual counting may be performed by eye, although this is considered more subjective than counting with automated systems and does not facilitate measurements of cell length and width for calculation of bacterial biovolume and biomass C. Attempts to automate counting have achieved various degrees of success (10, 35, 96, 97). Although the major application of image analysis has been for water samples, the approach has also been applied to soil smears. In this application, the system has involved operator decisions regarding removal of detritus and separation of cell aggregates. The application of SCLM to these enumeration problems has resulted in the development of a fully automatic system for determination of not only cell number but also cell volumes and frequency of dividing cells (11). Reference 11 provides an excellent example of coupling image analysis with SCLM techniques and an overview of the general approach.

Determination of Growth, Viability, and Metabolic Condition

Knowledge of the metabolic state of microbial cells is often important because it largely determines the effects that the microbes have on their environment. As a result, much effort has been made to develop methods for the identification of viable cells within heterogeneous communities. Many of these methods are based on fluorescent probes specifically designed (or derivatized) to link one or more physiological facets of cellular condition to the overall cell viability of individual bacteria. These fluorescent viability indicators have been based on cytoplasmic redox potential, electron transport chain activity, enzymatic activity, cell membrane potential, membrane integrity, and various combinations thereof (5, 7, 9, 14, 22, 73, 79, 86, 106).

It is questionable whether absolute determination of cell viability is possible with these probes. For example, Back and Kroll (7) pointed out that the relationship between differential fluorescence of AO when bound to RNA and to DNA, used to differentiate between active and inactive cells, and cell viability is not well understood. Nevertheless, the studies conducted to date have indicated the potential of viability probes. For instance, oxidized CTC is soluble and nonfluorescent; however, when reduced by the electron transport chain of an active bacterial population, it becomes an insoluble CTC-formazan crystal which accumulates intracellularly and fluoresces in the red wavelengths. Rodriguez et al. (86) used CTC to visualize actively respiring cells in mixed bacterial populations. These workers also compared the total number of bacteria obtained from environmental samples by using DAPI with those which were metabolically active and found that CTC counts were less than the total counts but equal to or greater than plate count results. A practical application of CTC was also reported by Yu and McFeters (126), who used CTC for estimating cell injury resulting from chlorination of biofilms in a model pipeline system and found that the results obtained with CTC were in general agreement with those of Rodriguez et al. (86). Others (e.g., Sieracki et al. [97]) have used 2-(p-iodophenyl)-3-(p-nitrophenyl)-5-phenyltetrazolium chloride (INT). Reduction of INT leads to the deposition of INT-formazan, which is recognized as opaque deposits in cells stained with AO. Caldwell et al. (19) used resorufin as an indicator of cytoplasmic reducing potential to differentiate between living and killed yeast cells. When observed with SCLM, actively metabolizing cells appeared negatively stained, whereas dead cells stained white. Tetrazolium reduction, in conjunction with image analysis, has also been used to assess the metabolic condition of nonculturable cells of *Helicobacter* and *Vibrio* species (38). One should be aware, however, that the variable permeability of cell membranes in natural (heterogeneous) assemblages to these redox-dependent fluors may have a significant impact on the results, and therefore suitable preliminary studies are required.

The integrity of the cell membrane is generally considered a dependable indicator of the viability of bacterial cells. Probes which have been used for determining the integrity of bacterial membranes include nucleic acid stains such as propidium iodide. Because these probes are relatively membrane impermeable, they do not cross functional cell membranes (40). However, they easily penetrate cells with a ruptured membrane, and once inside the cell cytoplasm, they bind to nucleic acids and undergo a large Stoke's shift (to the red wavelengths), resulting in an increase in fluorescence emission intensity and easier detection with fluorescence microscopy. A weakness of this method is that it cannot indicate dead cells which have already lost their DNA, for instance by the action of degradative enzymes.

The multiple-fluor probe Live/Dead Baclite (Molecular Probes) is also a viability indicator based on membrane integrity. This mixture (the composition is proprietary) contains a green-fluorescing probe which penetrates both the membrane and cytoplasm and a red-fluorescing probe which does not penetrate the cell unless there is a loss of membrane integrity. Thus, living cells appear green and dead cells appear red. Figure 4 shows a soil biofilm community stained with the commercial preparation; the live cells are shown on the left and the nonviable cells are shown on the right of this dual-channel image. A potential weakness of this mixture is that the red probe sometimes accumulates on the surface of living cells, producing a faint red "ghost." This could be a problem when epifluorescence microscopy is used. With laser microscopy, subtraction of green fluorescence bleed-through from the red channel (and vice versa) can be done to improve image quality. Additional pitfalls of the method include the inability of the green probe (which is presumably hydrophobic) to penetrate microcolonies of living cells that were embedded in exopolymer while the red probe (which is presumably hydrophilic) accumulated at the surfaces of these cells. Thus, it is possible to underestimate the number of viable cells. In contrast, cells which have lost their cytoplasmic contents following death may fail to fluoresce red, although they are thoroughly penetrated by the viability probe. Alternative methods such as difference imaging (18) or physical viability assays may be

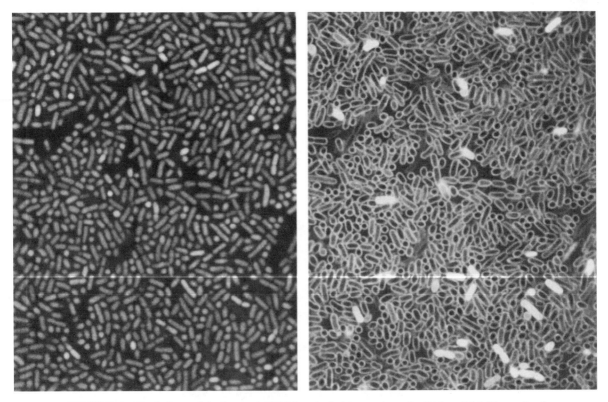

FIGURE 4 Dual-channel (red-green) SCLM optical thin section of a 24-h soil biofilm stained with a Molecular Probes Live/Dead Baclite kit. The left channel shows living cells which emit green fluorescence, whereas the right channel (corrected for green fluorescence bleed-through by using the program D-Bleed, which mathematically reduces the overlap in red-green signal) shows a number of dead cells emitting intense red fluorescence. Note that the fluorescent indicator can be seen to accumulate outside the membranes of living cells in the red channel, resulting in a ring of low-intensity fluorescence.

required to verify the results obtained with these chemical probes.

Plasmolysis (contraction of the cell contents in a living cell) of cells when pulsed with a hypertonic solution such as NaCl is an inexpensive physical viability assay that is relatively simple and amenable to a wide range of microscopic techniques (22). Cells with a semipermeable membrane plasmolyze, while those with a damaged membrane do not. Another simple physical method for determining the functionality of the cell membrane and protoplast is use of fluorescein. This compound fluoresces maximally at pH 9 and is quenched at low pH (<5) (40). Cells with a functional membrane maintain their intracellular pH at approximately 7.0. Thus, by reducing the pH in the ambient environment in the presence of 0.01% fluorescein, viable cells become fluorescent against a dark background because a small amount of fluorescein diffuses through the membrane of viable cells and fluoresces.

FDA, a nonfluorescent derivative of fluorescein, is also used to detect active cells. FDA is transported across the cell membrane and hydrolyzed by a variety of esterases. The resulting intracellular release of fluorescein allows direct visualization of actively metabolizing cells by epifluorescence microscopy or SCLM. This approach has been used to detect metabolic activity in bacteria (28) and fungi (101). Other compounds with potential for microbiological studies include fluorogenic enzymatic substrates (carboxy-FDA and calcein AM), probes sensitive to DNA damage (Hoechst 33342), fluorescent Gram stains (Baclite), and stains sensitive to membrane potential (rhodamine 123, RH-795, and carbocyanine derivatives).

It is sometimes useful to apply fluorescent probes which reflect the type of inactivation that is likely to occur. Korber et al. (52) applied fleroxacin (a fluoroquinolone DNA gyrase inhibitor which alters nucleic acid conformation and inhibits cell replication) in combination with AO to *Pseudomonas fluorescens* biofilms. The sensitivity of AO to DNA conformation (30) allowed the detection of fleroxacin-affected cells by SCLM and dual-wavelength fluorescence microscopy. Use of AO was much more appropriate for this study than use of an indicator of redox potential, as cells not affected by fleroxacin initially sustained high metabolic activity.

Møller et al. (75) also used AO staining to obtain simultaneous estimates of cellular RNA and DNA contents of single cells. Direct cell growth rates may be determined from rRNA determinations based on quantitative hybridization with ribosomal probes (32, 75, 85). These determinations may be made very effectively by using the combination of SCLM and image analysis of the fluorescence signals.

Microstructural Analyses

The form and arrangement of naturally occurring bacteria and bacterial communities have most commonly been assessed by using either electron microscopy (EM) or fluorescence microscopy. Electron microscopy involves extensive sample preparation and potential for development of artifacts; the major limitation for fluorescence is out-of-focus haze, which has made it difficult to accurately determine the spatial arrangements of cells and other components in living, fully hydrated biofilms or other microbial systems. The development of SCLM allows nondestructive, in situ optical sectioning of biological materials (67), enabling detailed analysis of microbes in their habitat without fixation.

A variety of fluorescent probes have been used in conjunction with SCLM to study the form and arrangement of

bacteria, exopolysaccharides, and voids in attached communities. For instance, fluorescence exclusion (negative staining) has been used to yield high-resolution optical sections of biofilms (20). Fluorescein solutions (fluorescein salt) are probably most frequently used as negative stains, although conjugated forms of fluorescein (FITC, TRITC, and rhodamine) may be used. This procedure, in which bacteria are imaged as dark objects against a bright background, has been used to visualize biofilms formed by both pure cultures and communities (54, 67, 122). Other negative stains include resazurin and fluor-dextran conjugates. A number of workers applied positive-staining techniques to study biofilms (55, 98, 122, 125). Common probes which have been used for positive staining include AO, FITC, TRITC, and Nile red. Both positive and negative staining of the same biofilm location is shown in Fig. 5. These photomicrographs clearly show the contrasting views provided by these techniques; negative staining shows the cell-cell boundary, and positive staining shows cell-polymer boundaries.

For a better understanding of the role of exopolysaccharide in microbial processes, it is necessary to characterize the exopolysaccharide in terms of spatial distribution within biofilms, density, chemical composition, and charge. Commercially available fluorescent dextrans bearing a defined charge (e.g., polyanionic, cationic, or neutral) have been used by Korber et al. (52) and Lawrence et al. (69) to study the charge distribution and polymer permeability within pure culture biofilms as well as in biofilm communities.

Fluor-conjugated lectins, characterized by their highly specific binding with saccharide moieties, can be used to map the chemical compositions of the exopolymer materials associated with bacterial cells (19, 78). Wolfaardt et al. (124) applied a panel of fluorescent lectins [specific for either α-D-Man, β-D-Gal(1–4)-D-GlcNAc, β-D-Gal(1–3)-D-GalNAc, or α-D-Man; α-D-Glc, (D-GlcNAc)$_3$, α-L-fucose, sialic acid, D-GalNAc, or D-Man; or D-Glc] to a herbicide-degrading community to study the vertical and horizontal distribution of these moieties within the biofilms formed by this community.

Microenvironmental Analyses

Protocols for defining the microbial microenvironment have focused on biofilms because (i) the large number of cells in biofilms tends to concentrate any chemical or physical microenvironmental effect, (ii) biofilm bacteria generally remain stably juxtaposed, (iii) the parameters may be viewed temporally, and (iv) the biofilm growth habit is ubiquitous (29, 117). Perhaps the most significant feature of biofilm bacteria and their exuded exopolymers is that the resultant cell-polymer matrix tends to trap metabolic end products and hinder the migration of substrates (e.g., O_2) from the bulk phase to bacteria located at the base of the biofilm.

The extent of hindered diffusion (effective diffusion coefficient [D_e]) can be measured by using nonreactive (nonbinding) fluorescent probes and imaging procedures based on either SCLM (51, 69) or epifluorescence microscopy and image analysis. D_e may be measured by methods based on fluorescence recovery after photobleaching (FRAP) or diffusion monitoring. In FRAP, a high-intensity light source (usually a laser) is used to bleach out an area of fluorescence at a specific biofilm location; the critical assumption is made that molecules in the bleached area are rendered permanently nonfluorescent. The gradual return to initial fluorescence values, occurring only as a function of lateral fluor diffusion (commonly fluorescently labeled

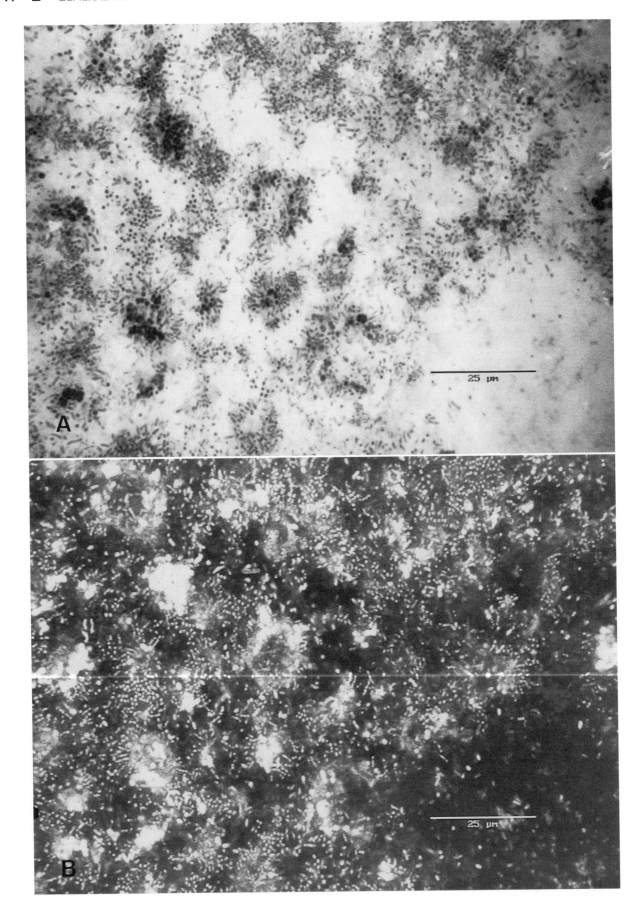

dextrans), is then measured temporally and used in conjunction with the appropriate equation to calculate D_e (6, 12). For FRAP studies, 1-D, 2-D, and 3-D equations have been used to calculate D_e values.

Because of questions surrounding the application of FRAP techniques to biofilms, Lawrence et al. (69) developed an in situ fluorescence-monitoring method for the estimation of 1-D effective diffusion coefficients at defined locations within biofilm matrices. In diffusion monitoring, a pulse of fluor is introduced into a biofilm system grown in rectangular flow cells, after which the migration of the fluor through the biofilm matrix to the biofilm-substratum interface is monitored (using SCLM) at a defined xy location under static flow conditions. In this case, diffusion occurs primarily along the vertical axis (from the bulk liquid phase through the biofilm); subsequently, 1-D formulae are deemed sufficient. Either the FRAP or the diffusion-monitoring method may be used with dextran-conjugated fluors of various molecular weights or any other defined, fluorescently conjugated molecule (DNA, antibodies, etc.). Difficulties may arise as a consequence of specific and/or nonspecific binding of the tracer fluor to the biofilm matrix.

Many direct measurements (e.g., using microelectrodes) have indicated that microbial activity can radically alter the chemistry of a aqueous system (29, 31, 70). A number of fluorescent probes (fluorescein, 5- and 6-carboxyfluorescein, etc.) exhibit pH- and E_h-sensitive fluorescence; consequently, these probes have potential for in situ measurements of microbe-associated chemical microenvironments. To date, little quantitative data (other than intriguing images showing relative changes in fluorescence intensity) have been amassed for microbiological (biofilm) systems, despite the potential benefits of this approach (19). Pseudocolor images of these gradients may be seen in the review by Caldwell et al. (19). The primary limiting factors for this approach are the fact that biofilms lack discrete boundaries, the presence of multiple variables which influence fluorescence, and the difficulty of achieving adequate controls. With eukaryotic systems, in contrast, much work has been performed on pH, calcium ion concentration, and magnesium ion concentration (43, 61, 62, 92). Ratiometric methods (40, 71, 109, 110) generally involve the application of a fluorescent compound that has an excitation/emission line that is sensitive to its environment and a corresponding excitation/emission line that is insensitive to the environment. Thus, the ratio of the images at these two wavelengths removes variations due to concentration, and the resulting image contains variations in fluorescence due to the parameter of interest. This is an approach that may benefit future studies of microenvironmental chemistry, as may the use of microelectrodes for probe calibration. The reader is directed to the extensive cell biology literature for additional information and examples that may find application in microbial systems (43, 61, 62, 92, 109, 110).

Quantification of Biodiversity

One approach for defining and quantifying the antigenic diversity contained in microbiological systems is the use of immunological (fluor-conjugated poly- or monoclonal antibodies) fluorescent probes (60). These methods were originally adapted for ecological studies (93) but also have broad potential for distinguishing between morphologically similar but functionally different bacteria present within complex biofilms (87). Details on preparation of antibodies are provided by Lam and Mutharia (60); however, optimization for use in natural systems (biofilms, soil samples, etc.) will inevitably be required.

The biodiversity of natural microbial populations may also be determined phylogenetically, allowing the identification of specific bacteria from complex systems (104, 120). Given the relatedness of slowly evolving (conserved) RNA sequences (120) and the synthesis of complementary oligonucleotides to conserved RNA regions, this method does not require the isolation and cultivation of community members. Thus, organisms which are hard to culture or of unknown ecological significance may now be ascertained often to the species (and in some cases, subspecies) level (1–3, 104, 116). Fluorescent oligonucleotide conjugates have previously been used in conjunction with SCLM and epifluorescence microscopy to document microbial diversity in a range of environments, including sewage sludge and the rhizosphere (2, 4, 39, 113). The application of in situ PCR (45) provides a powerful tool for visualization of genes and gene products. An excellent review dealing with the synthesis and application of oligonucleotide probes for microbiological systems has been provided by Amann et al. (2). This review also includes a series of very good SCLM optical thin sections and red-green anaglyph projections of rRNA-probed materials.

Many biotechnology companies will also custom-prepare fluor-conjugated oligonucleotides if provided the appropriate genetic sequence and wavelengths of fluorescent emission as well as the amount required. Ribosomal oligonucleotide sequences are also freely available over the Internet. A good starting point is the site http://specter.dcrt.nih.gov:8004/Desc/loci__toc__by__desc.html.

Commercial fluors (Molecular Probes) for the identification of fungi, as well as Bac-Lite Gram stain, a fluorescent stain for bacteria, may also be used to provide broad indications of system biodiversity. These probes have been developed and tested with pure batch cultures; therefore, the user must apply them with caution to biofilms and natural communities.

Computer Control of the Microscope Stage

Computer-programmable microscope stages may be obtained from any major microscope manufacturer (e.g., Zeiss, Bio-Rad, Nikon, or Leica). Using software executed by a PC, the user may return to sites of previous analysis with excellent accuracy (within ± 1 to 3 μm). This approach permits the microscopist to conduct temporal studies at a large number of sampling sites (e.g., down multiple transects), thereby precluding the need to resample randomly. This approach also facilitates microbiological analyses using molecular probes with different emission spectra, as the user may observe a number of sites stained with one compound and then observe additional sites after the second probe is added.

FIGURE 5 Optical thin sections (xy) at the same biofilm location showing the effect of negative (top) and positive (bottom) staining on the resulting image. These are clearly contrasting views of the biofilm, with negative staining (top) showing the cells (black objects) and polymer boundaries (dark grey). The positive stain defines some cells very well, but cells in the large microcolonies are obscured by the fluor binding to extensive exopolymer.

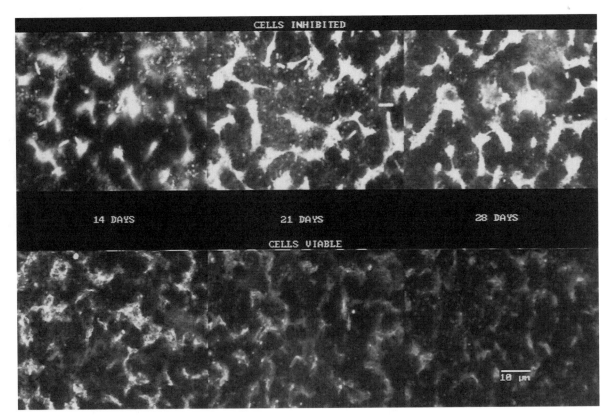

FIGURE 6 Time series of SCLM images illustrating the visualization of metabolic activity within a biofilm community. In these images, the metabolism of the herbicide diclofop methyl, which is fluorescent, is shown as a decrease in fluorescence intensity in uninhibited biofilm bacteria. Images were collected at various times and locations, taking xy optical thin sections in inhibited (top) and noninhibited (bottom) biofilms. The change in fluorescence intensity was measured by using image analyses to assess change in grey value. (Reprinted from reference 123.)

Wolfaardt et al. (123) used a monitoring approach to examine differences in the apparent utilization of the fluorescent herbicide diclofop methyl in viable and potassium cyanide-treated biofilms at various intervals over a 28-day time course. An example of such a time series of SCLM images is shown in Fig. 6. In these images, the metabolism of the herbicide diclofop methyl, which is fluorescent, is shown by the decrease in fluorescence intensity in uninhibited biofilm bacteria. This utilization was also examined through pulse-labeling of the biofilm with [^{14}C]diclofop and monitoring the appearance of $^{14}CO_2$ in parallel with the decrease in fluorescence (123).

SCLM systems typically come equipped with xz computer-controlled stages. Thus, when used in conjunction with computer-controlled xy-positioning systems, routine 3-D positioning may be performed accurately and precisely. In addition to saving positions of previous analysis, most stage-controlling programs also allow construction of analysis grids. To construct an analysis grid, the xy area to be analyzed as well as the area of the microscopic field (this is usually performed on-line), must be defined. This allows the analysis of large, contiguous areas of interest as well as the reconstruction of large montages consisting of multiple images. This capability allows for the application of geostatistical techniques to the analysis of microbial communities or systems (55) (see below).

Statistical Considerations

Image processing and analysis allow detailed measurements to be performed on a range of parameters; however, most microscopy studies have relied on relatively small areas of analysis. Data thus obtained may be prone to errors based on rare events resulting from insufficient or selective sampling (e.g., reference 17). The question of how large an area must be sampled for representative data has previously been addressed (108). Computer-controlled stages have subsequently aided in the development of statistical methods for biofilm analysis. Korber et al. (54, 55) used a computer-controlled microscope stage and SCLM imaging to construct large "maps" of contiguous regions of *P. fluorescens* biofilms. In addition to providing a unique 3-D view of the biofilm architecture (55), elements, or subsets, of the large biofilm montage were analyzed to define the extent of analysis required to provide representative biofilm biomass data. It was found that the spatial variability contained within pure-culture *P. fluorescens* biofilms was large, requiring analysis areas exceeding 10^5 μm^2 for statistically representative results.

HANDLING DIGITAL IMAGES
Image Formats and Conversion

A wide and seemingly confusing variety of image formats (e.g., TIFF, GIF, JPEG, RAW, PICT, EPS, BioRadTIFF) exists. However, all of these formats have advantages and disadvantages in terms of their use by specific manufacturers, their degrees of image fidelity, and their ability to compress images for archiving or exchange over the Internet. In addition, many are interconvertible with the correct software. A program such as Adobe Photoshop (Adobe Systems Inc., Mountain View, Calif.) will handle conversion of many image types. However, the user should be cautious in the application of conversions and compression, since for some formats the original image and information content may not be retrievable (always maintain backup copies).

Image Reproduction

The standard approach for decades has been the use of photographic reproduction techniques. These methods are discussed in detail by Popkin (83) and will not be addressed in this chapter. Digital imaging opens many options in terms of image reproduction. In addition, the digital record may be extensively altered in a variety of software packages such as Adobe Photoshop, Quark XPress, or Digital Darkroom.

Adobe Photoshop is generally considered the "gold standard" for the import, manipulation, and enhancement of digital images and photographs. Importantly, a method for obtaining a digital version of the desired image (e.g., slide or page scanners, frame grabber or digitizer boards coupled with analog imaging devices, on-line or network connections with digitizing facilities, or high-capacity, transferable storage media) is still required for a fully functional system. Adobe Photoshop and similar applications also have significant flexibility in being able to import and handle a wide variety of image formats, facilitating the exchange of digital information between different acquisition systems or platforms. Once the digital photograph is within the application, the user may merge, enhance, adjust the contrast, create montages, add text, or delete extraneous material en route to the development of presentation or publication graphics. The utility of this type of program as a scientific presentation tool, as opposed to one of analysis, has only recently been exploited.

Images may then be printed for publication by a variety of means, including video printers, dye sublimation printers, and slide printers. Quality output can be expensive; however, these devices put complete production control in the hands of the researcher, which can be efficient and cost-effective.

Archiving and Storage

The handling of image data has evolved considerably over the last decade. This evolution has essentially involved the transition from the physical storage and manipulation of photographs to digital or electronic means of achieving the same ends. Associated with this transition is an expanded ability to organize, archive, manipulate, enhance, transfer, and analyze visual data with speed and efficiency.

Video recorders may be used to store images and have the advantage of allowing retrieval and playback to an image-processing computer at a later time to test new hypotheses, measure new parameters, or increase the frequency of data collection to the highest resolution of the particular video system used (1/30 s or greater). Time-lapse recorders are also desirable for specific applications, such as the analysis of bacterial growth and motion, and often allow the insertion of time-date reference information.

Digital storage devices are basically the same as those used for computing applications, with the exception that high-capacity devices are generally preferred because of the large size of most digital images and the relative ease of acquiring numerous files. High-capacity (1 gigabyte [GB] or more) hard drives have significant application, with the advantage of being erasable and having fast read/write speeds. Optical drives have recently become quite dependable, provide high performance, and have the advantage of being relatively stable platforms for long-term image archiving. Included in this category are WORM (write once/read many) drives, read/write units, and recordable compact disks (CDs). At present, WORMS and read/write units can store approximately 470 MB of information per disk side (providing approximately 1 GB of storage space). A standard CD may store 640 MB; newer formats allow substantial increases in storage space. WORM and read/write optical disks typically cost around $200, whereas CDs cost around $20, making them a highly cost-effective means for data storage. All of the optical storage systems suffer from the problem of not playing back on equipment made by competing manufacturers, thus limiting their use as shuttle disks. Typical imaging systems usually utilize a high-capacity hard drive for rapid data collection and are interfaced with optical devices for storage and archiving of select information.

Digital Transmission of Information

File Moving and Transfer

The rapid expansion of the Internet clearly demonstrates the future trend of information transfer. By using the Internet, information may be transferred easily and rapidly from one location to another, usually without loss or distortion of information. This is achieved by means of protocols, or rules, which govern the operation of the networks. Internet protocols and transmission control protocols work in conjunction with specialized computer applications to provide a standard which allows different computers to communicate easily and without the usual hurdles which accompany interfacing computers of different makes or with different operating systems. Typically, these systems function to provide computer electronic mail, remote log-ins, or the transfer of computer files.

The digital transfer of data files is one element of the Internet which is particularly appealing to those individuals conducting scientific research. FTP is a format for the moving and/or copying of files located at a remote site. Both interfacing systems in question must have access to the Internet and have the appropriate FTP capabilities. To transfer documents via FTP, one user must also have access to the remote computing facility. Often, access may be gained anonymously (without the need for a formal account). Many computing department mainframes have public files which may be freely accessed by using FTP programs such as Fetch (Dartmouth College, Hanover, N.H.), which provides a hierarchical folder-based interface (in UNIX, files may be viewed by using "dir" [directory] or "ls" [list] commands). Refer to reference 59 or 107 for more details on log-in and copy procedures.

Using the Internet To Obtain Information

The Internet represents a vast source of information of scientific use, including (in much abbreviation) details on current research, granting agencies and deadlines, news and discussion groups, manufacturer listings, and various industrial and academic institutes. Once a computer is connected to the Internet, a multitude of navigation software and utilities may be used to assist in successful searching. For example, Archie, a utility which allows the search of public files located on the Internet, may be used to locate a needed piece of software. Finger, an on-line UNIX utility, identifies

TABLE 3 Internet sites containing information on digital imaging and microscopy

Internet address	Topic(s)
http://www.microscopy-online.com/	Microscopy and related information
http://rsb.info.nih.gov/nih-image/	NIH Image home page and links to image analysis information
http://outcast.gene.com/ae/RC/ASM/index.html	American Society for Microbiology home page
http://www.edt.com/dvc/dvc.html	Video camera information
http://zebu.uoregon.edu/images/universe.au	CCD video camera information
http://www.robotics.com/video.html	Various video camera manufacturers

electronic mail addresses of system users throughout the world. Whois is both an electronic e-mail directory and an application for accessing it. Gopher may also be used to locate and retrieve various indexed resources by using a menu-driven interface. Two routes to the World Wide Web (high-level, hypertext-driven Internet search sites) include NCSA Mosaic (University of Illinois) and Netscape (Netscape Communications Corporation, Mountain View, Calif.). However, additional programs exist or are in development. These programs incorporate powerful, efficient search "engines" as well as cross-referenced resource material accessed by "clicking" on highlighted text, which facilitates the location of specific information, software, or digital images. Highly graphics-oriented and menu- or hypertext-driven software represents perhaps the most user-friendly entry into the Internet. Key resource material will enable the exploitation of the incredible array of resource material currently available (59, 107). Table 3 provides a sampling of specific sites for image analysis and microbiology on the Internet.

SOURCES OF EQUIPMENT, FLUORESCENT MOLECULAR PROBES, AND OTHER REAGENTS

A good starting point for the acquisition of equipment and fluorescent probes is the site http://corn.eng.buffalo.edu/www/CommercialInst/readme.html. The user will find in this directory the names of SCLM manufacturers, including Bio-Rad, Leica, Meridian, Molecular Dynamics, and Zeiss, image-processing equipment manufacturers, and a comprehensive list of fluorochromes. The annual *American Laboratory Buyers Guide* is an invaluable source of information and suppliers. Molecular Probes (see reference 40) provides a comprehensive list of available fluors applicable to a range of molecular and environmental applications.

REFERENCES

1. **Amann, R. I., L. Krumholz, and D. A. Stahl.** 1990. Fluorescent-oligonucleotide probing of whole cells for determinative, phylogenetic, and environmental studies in microbiology. *J. Bacteriol.* **172:**762–770.
2. **Amann, R. I., W. Ludwig, and K. H. Schleifer.** 1995. Phylogenetic identification and in situ detection of individual microbial cells without cultivation. *Microbiol. Rev.* **59:**143–169.
3. **Amann, R. I., J. Stromley, R. Devereux, R. Key, and D. A. Stahl.** 1992. Molecular and microscopic identification of sulfate-reducing bacteria in multispecies biofilms. *Appl. Environ. Microbiol.* **58:**614–623.
4. **Assmus, B., P. Hutzler, G. Kirchhof, R. I. Amann, J. R. Lawrence, and A. Hartmann.** 1995. In situ localization of *Azospirillum brasilense* in the rhizosphere of wheat using fluorescently labeled rRNA-targeted oligonucleotide probes and scanning confocal laser microscopy. *Appl. Environ. Microbiol.* **61:**1013–1019.
5. **Autio, K., and T. Mattila-Sandholm.** 1992. Detection of active yeast cells (*Saccharomyces cerevisiae*) in frozen dough sections. *Appl. Environ. Microbiol.* **58:**2153–2157.
6. **Axelrod, A., D. E. Koppel, J. Schlessinger, E. Elsen, and W. W. Webb.** 1976. Mobility measurement by analysis of fluorescence photobleaching recovery kinetics. *Biophys. J.* **16:**1055–1069.
7. **Back, J. P., and R. G. Kroll.** 1991. The differential fluorescence of bacteria stained with acridine orange and the effects of heat. *J. Appl. Bacteriol.* **71:**51–58.
8. **Bae, H. C., E. H. Cota-Robles, and L. E. Casida, Jr.** 1972. Microflora of soil as viewed by transmission electron microscopy. *Appl. Microbiol.* **23:**637–648.
9. **Betts, R. P., P. Bankes, and J. G. Banks.** 1989. Rapid enumerations of viable micro-organisms by staining and direct microscopy. *Lett. Appl. Microbiol.* **9:**199–202.
10. **Bjornsen, P. K.** 1986. Automatic determination of bacterioplankton biomass by image analysis. *Appl. Environ. Microbiol.* **51:**1199–1204.
11. **Bloem, J., M. Veninga, and J. Shepherd.** 1995. Fully automatic determination of soil bacterium numbers, cell volumes and frequency of dividing cells by confocal laser scanning microscopy and image analysis. *Appl. Environ. Microbiol.* **61:**926–936.
12. **Blonk, J. C. G., A. Don, H. van Aalst, and J. J. Birmingham** 1993. Fluorescence photobleaching recovery in the confocal scanning laser microscope. *J. Microsc.* **169:**363–374.
13. **Bott, T. L., and T. D. Brock.** 1970. Growth and metabolism of periphytic bacteria: methodology. *Limnol. Oceanogr.* **20:**191–197.
14. **Bottomley, P. J., and S. P. Maggard.** 1990. Determination of viability within serotypes of a soil population of *Rhizobium leguminosarum* bv. *trifolii. Appl. Environ. Microbiol.* **56:**533–540.
15. **Bradbury, S.** 1970. *Additional Notes on the Use of the Microscope,* p. 17. Royal Microscopical Society, Oxford.
16. **Brock, T. D.** 1971. Microbial growth rates in nature. *Bacteriol. Rev.* **35:**39–58.
17. **Brock, T. D.** 1984. How sensitive is the light microscope for observations on microorganisms in natural habitats? *Microb. Ecol.* **10:**297–300.
18. **Caldwell, D. E., and J. J. Germida.** 1985. Evaluation of difference imagery for visualizing and quantitating microbial growth. *Can. J. Microbiol.* **31:**35–44.
19. **Caldwell, D. E., D. R. Korber, and J. R. Lawrence.** 1992. Confocal laser microscopy and digital image analysis in microbial ecology. *Adv. Microb. Ecol.* **12:**1–67.
20. **Caldwell, D. E., D. R. Korber, and J. R. Lawrence.** 1992. Imaging of bacterial cells by fluorescence exclusion using scanning confocal laser microscopy. *J. Microbiol. Methods* **15:**249–261.
21. **Caldwell, D. E., and J. R. Lawrence.** 1986. Growth kinetics of *Pseudomonas fluorescens* microcolonies within

the hydrodynamic boundary layers of surface microenvironments. *Microb. Ecol.* **12**:299–312.

22. **Caldwell, D. E., and J. R. Lawrence.** 1989. Microbial growth and behavior within surface microenvironments, p. 140–145. *In* T. Hattori, Y. Ishida, Y. Maruyama, R. Y. Morita, and A. Uchida (ed.), *Proceedings of ISME-5.* JSS Press, Tokyo.

23. **Casida, L. E.** 1969. Observation of microorganisms in soil and other natural habitats. *Appl. Microbiol.* **18**:1065–1071.

24. **Casida, L. E.** 1972. Interval scanning photomicrography of microbial cell populations. *Appl. Microbiol.* **23**:190–192.

25. **Casida, L. E.** 1975. Infrared color photomicrography of soil microorganisms. *Can. J. Microbiol.* **21**:1892–1893.

26. **Casida, L. E.** 1976. Continuously variable amplitude contrast microscopy for the detection and study of microorganisms in soil. *Appl. Environ. Microbiol.* **31**:605–608.

27. **Cholodny, N.** 1930. Uber eine neue Method zur Untersuchung der Bodenflora. *Arch. Mikrobiol.* **1**:620–652.

28. **Chrzanowski, T. H., R. D. Crotty, J. G. Hubbard, and R. P. Welch.** 1984. Applicability of the fluorescein diacetate method of detecting active bacteria in freshwater. *Microb. Ecol.* **10**:179–185.

29. **Costerton, J. W., Z. Lewandowski, D. DeBeer, D. E. Caldwell, D. R. Korber, and G. A. James.** 1994. Biofilms: the customized microniche. *J. Bacteriol.* **176**:2137–2142.

30. **Daley, R. J.** 1979. Direct epifluorescence enumeration of native aquatic bacteria: uses, limitations, and comparative accuracy, p. 29–45. *In* J. W. Costerton and R. R. Colwell (ed.), *Native Aquatic Bacteria: Enumeration, Activity, and Ecology.* STP 695. American Society for Testing and Materials, Philadelphia.

31. **DeBeer, D., P. Stoodley, F. L. Roe, and Z. Lewandowski.** 1994. Effects of biofilm structures on oxygen distribution and mass transport. *Biotechnol. Bioeng.* **43**:1131–1138.

32. **DeLong, E. F., G. S. Wickham, and N. R. Pace.** 1989. Phylogenetic stains: ribosomal RNA-based probes for identification of single cells. *Science* **243**:1360–1362.

33. **Duxbury, T.** 1977. A microperfusion chamber for studying the growth of bacterial cells. *J. Appl. Bacteriol.* **42**:247–251.

34. **Emerson, D., R. M. Worden, and J. A. Breznak.** 1994. A diffusion gradient chamber for studying microbial behavior and separating microorganisms. *Appl. Environ. Microbiol.* **60**:1269–1278.

35. **Estep, K. W., F. MacIntyre, E. Hjorleifson, and J. M. Sieburth.** 1986. MacImage: a user-friendly image analysis system for the accurate mensuration of marine organisms. *Mar. Ecol. Prog. Ser.* **33**:243–253.

36. **Fletcher, M.** 1988. Attachment of *Pseudomonas fluorescens* to glass and influence of electrolytes on bacterium substratum separation distance. *J. Bacteriol.* **170**:2027–2030.

37. **Gonzalez, R. C., and P. Wintz.** 1977. *Digital Image Processing.* Addison-Wesley, Reading, Mass.

38. **Gribbon, L. T., and M. R. Barer.** 1995. Oxidative metabolism in nonculturable *Helicobacter pylori* and *Vibrio vulnificus* cells studied by substrate-enhanced tetrazolium reduction and digital image processing. *Appl. Environ. Microbiol.* **61**:3379–3384.

39. **Hahn, D., R. I. Amann, W. Ludwig, A. D. L. Akkermans, and K.-H. Schleifer.** 1992. Detection of microorganisms in soil after *in situ* hybridization with rRNA-targeted, fluorescently labelled oligonucleotides. *J. Gen. Microbiol.* **138**:879–887.

40. **Haugland, R. P.** 1992. *Molecular Probes: Handbook of Fluorescent Probes and Research Chemicals.* Molecular Probes Inc., Eugene, Ore.

41. **Haugland, R. P., and A. Minta.** 1990. Design and application of indicator dyes, p. 1–20. *In* J. K. Foskett and S. Grinstein (ed.), *Noninvasive Techniques in Cell Biology.* Wiley-Liss, Inc., New York.

42. **Henrici, A. T.** 1933. Studies of freshwater bacteria. I. A direct microscopic technique. *J. Bacteriol.* **25**:277–286.

43. **Hernandez-Cruz, A., F. Sala, and P. R. Adams.** 1990. Subcellular calcium transients visualized by confocal microscopy in a voltage-clamped vertebrate neuron. *Science* **247**:858–862.

44. **Hobbie, J. E., R. J. Daley, and S. Jasper.** 1977. Use of Nucleopore filters for counting bacteria by fluorescence microscopy. *Appl. Environ. Microbiol.* **33**:1225–1228.

45. **Hodson, R. E., W. A. Dustman, R. P. Garg, and M. A. Moran.** 1995. In situ PCR for visualization of microscale distribution of specific genes and gene products in prokaryotic communities. *Appl. Environ. Microbiol.* **61**:4074–4082.

46. **Inoué, S.** 1986. *Video Microscopy,* p. 263–307. Plenum Press, New York.

47. **James, G. A., D. R. Korber, D. E. Caldwell, and J. W. Costerton.** 1995. Digital image analysis of growth and starvation responses of a surface-colonizing *Acinetobacter* sp. *J. Bacteriol.* **177**:907–915.

48. **James, J.** 1976. *Light Microscopic Techniques in Biology and Medicine.* Martinus Nijhoff Medical Division, Amsterdam.

49. **Keevil, C. W., and J. T. Walker.** 1992. Nomarski DIC microscopy and image analysis of biofilms. *Binary* **4**:93–95.

50. **Kepner, R. L., and J. R. Pratt.** 1994. Use of fluorochromes for direct enumeration of total bacteria in environmental samples: past and present. *Microbiol. Rev.* **58**:603–615.

51. **Korber, D. R., D. E. Caldwell, and J. W. Costerton.** 1994. Structural analysis of native and pure-culture biofilms using scanning confocal laser microscopy, p. 347–353. *In Proceedings of the National Association of Corrosion Engineers (NACE) Canadian Region Western Conference.*

52. **Korber, D. R., G. A. James, and J. W. Costerton.** 1994. Evaluation of fleroxacin activity against established *Pseudomonas fluorescens* biofilms. *Appl. Environ. Microbiol.* **60**:1663–1669.

53. **Korber, D. R., J. R. Lawrence, K. E. Cooksey, B. Cooksey, and D. E. Caldwell.** 1989. Computer image analysis of diatom chemotaxis. *Binary* **1**:155–168.

54. **Korber, D. R., J. R. Lawrence, M. J. Hendry, and D. E. Caldwell.** 1992. Programs for determining representative areas of microbial biofilms. *Binary* **4**:204–210.

55. **Korber, D. R., J. R. Lawrence, M. J. Hendry, and D. E. Caldwell.** 1993. Analysis of spatial variability within mot$^+$ and mot$^-$ *Pseudomonas fluorescens* biofilms using representative elements. *Biofouling* **7**:339–358.

56. **Korber, D. R., J. R. Lawrence, B. Sutton, and D. E. Caldwell.** 1989. Effect of laminar flow velocity on the kinetics of surface recolonization by mot$^+$ and mot$^-$ *Pseudomonas fluorescens. Microb. Ecol.* **18**:1–19.

57. **Korber, D. R., J. R. Lawrence, L. Zhang, and D. E. Caldwell.** 1990. Effect of gravity on bacterial deposition and orientation in laminar flow environments. *Biofouling* **2**:335–350.

58. **Krambeck, C., H.-J. Krambeck, D. Schröder, and S. Y. Newell.** 1990. Sizing bacterioplankton: a juxtaposition of bias due to shrinkage, halos, subjectivity in image interpretation and symmetric distributions. *Binary* **2**:5–14.

59. **Krol, E.** 1992. *The Whole Internet: User's Guide and Catalog,* p. 396. O'Reilly & Associates, Inc., Sebastopol, Calif.

60. **Lam, J. S., and L. M. Mutharia.** 1994. Antigen-antibody reactions, p. 104–132. In P. Gerhardt, R. G. E. Murray, W. A. Wood, and N. R. Krieg (ed.), *Methods for General and Molecular Bacteriology.* American Society for Microbiology, Washington, D.C.

61. **Lattanzio, F. A., Jr.** 1990. The effects of pH and temperature on fluorescent calcium indicators as determined with chelex-100 and EDTA buffer systems. *Biochem. Biophys. Res. Commun.* **171:**102–108.

62. **Lattanzio, F. A., Jr., and D. K. Bartschat.** 1991. The effect of pH on rate constants, ion selectivity and thermodynamic properties of fluorescent calcium and magnesium indicators. *Biochem. Biophys. Res. Commun.* **177:** 184–191.

63. **Lawrence, J. R., and D. E. Caldwell.** 1987. Behavior of bacterial stream populations within the hydrodynamic boundary layers of surface microenvironments. *Microb. Ecol.* **14:**15–27.

64. **Lawrence, J. R., P. J. Delaquis, D. R. Korber, and D. E. Caldwell.** 1987. Behavior of *Pseudomonas fluorescens* within the hydrodynamic boundary layers of surface microenvironments. *Microb. Ecol.* **14:**1–14.

65. **Lawrence, J. R., D. R. Korber, and D. E. Caldwell.** 1989. Computer-enhanced darkfield microscopy for the quantitative analysis of bacterial growth and behavior on surfaces. *J. Microbiol. Methods* **10:**123–138.

66. **Lawrence, J. R., D. R. Korber, and D. E. Caldwell.** 1992. Behavioral analysis of *Vibrio parahaemolyticus* variants in high- and low-viscosity microenvironments using computer-enhanced microscopy. *J. Bacteriol.* **174:** 5732–5739.

67. **Lawrence, J. R., D. R. Korber, B. D. Hoyle, J. W. Costerton, and D. E. Caldwell.** 1991. Optical sectioning of microbial biofilms. *J. Bacteriol.* **173:**6558–6567.

68. **Lawrence, J. R., J. A. Malone, D. R. Korber, and D. E. Caldwell.** 1989. Computer image enhancement to increase depth of field in phase contrast microscopy. *Binary* **1:**181–185.

69. **Lawrence, J. R., G. M. Wolfaardt, and D. R. Korber,** 1994. Monitoring diffusion in biofilm matrices using confocal laser microscopy. *Appl. Environ. Microbiol.* **60:** 1166–1173.

70. **Lens, P. N. L., D. De Beer, C. C. H. Cronenberg, F. P. Houwen, S. P. P. Ottengraf, and W. H. Verstraete.** 1993. Heterogeneous distribution of microbial activity in methanogenic aggregates: pH and glucose microprofiles. *Appl. Environ. Microbiol.* **59:**3803–3815.

71. **Luby-Phelps, K., F. Lanni, and D. L. Taylor.** 1988. The submicroscopic properties of cytoplasm as a determinant of cellular function. *Annu. Rev. Biophys. Chem.* **17:** 369–396.

72. **Marshall, K. C.** 1986. Microscopic methods for the study of bacterial behaviour at inert surfaces. *J. Microbiol. Methods* **4:**217–227.

73. **Marxsen, J.** 1988. Investigations into the number of respiring bacteria in groundwater from sandy and gravelly deposits. *Microb. Ecol.* **16:**65–72.

74. **Meyer-Reil, L.-A.** 1978. Autoradiography and epifluorescence microscopy combined for the determination of number and spectrum of actively metabolizing bacteria in natural waters. *Appl. Environ. Microbiol.* **36:**506–512.

75. **Møller, S., C. S. Kristensen, L. K. Poulsen, J. M. Carstensen, and S. Molin.** 1995. Bacterial growth on surfaces: automated image analysis for quantification of growth rate-related parameters. *Appl. Environ. Microbiol.* **61:** 741–748.

76. **Murray, R. G. E., R. N. Doetsch, and C. F. Robinow.** 1994. Determinative and cytological light microscopy, p. 21–41. In P. Gerhardt, R. G. E. Murray, W. H. Wood, and N. R. Krieg (ed.), *Methods for General and Molecular Bacteriology.* American Society for Microbiology, Washington, D.C.

77. **Murray, R. G. E., and C. F. Robinow.** 1994. Light microscopy, p. 8–20. In P. Gerhardt, R. G. E. Murray, W. A. Wood, and N. R. Krieg (ed.), *Methods for General and Molecular Bacteriology.* American Society for Microbiology, Washington, D.C.

78. **Neu, T., and K. C. Marshall.** 1991. Microbial "footprints"—a new approach to adhesive polymers. *Biofouling* **3:**101–112.

79. **Nix, P. G., and M. M. Daykin.** 1992. Resazurin reduction tests as an estimate of coliform and heterotrophic bacterial numbers in environmental samples. *Bull. Environ. Contam. Toxicol.* **49:**354–360.

80. **Pawley, J. B. (ed.).** 1990. *Handbook of Biological Confocal Microscopy.* Plenum Press, New York.

81. **Perfil'ev, B. V., and D. R. Gabe.** 1969. *Capillary Methods of Investigating Micro-Organisms* (J. M. Shewan, transl.). University of Toronto Press, Toronto.

82. **Pomroy, A. J.** 1984. Direct counting of bacteria preserved with Lugol iodine solution. *Appl. Environ. Microbiol.* **47:** 1191–1192.

83. **Popkin, T. J.** 1994. Photography, p. 735–761. In P. Gerhardt, R. G. E. Murray, W. A. Wood, and N. R. Krieg (ed.), *Methods for General and Molecular Bacteriology.* American Society for Microbiology, Washington, D.C.

84. **Porter, K. G., and Y. S. Feig.** 1980. The use of DAPI for identifying and counting aquatic microflora. *Limnol. Oceanogr.* **25:**943–948.

85. **Poulsen, L. K., G. Ballard, and D. A. Stahl.** 1993. Use of rRNA fluorescence in situ hybridization for measuring the activity of single cells in young and established biofilms. *Appl. Environ. Microbiol.* **59:**1354–1360.

86. **Rodriguez, G. G., D. Phipps, K. Ishiguro, and H. F. Ridgway.** 1992. Use of a fluorescent redox probe for direct visualization of actively respiring bacteria. *Appl. Environ. Microbiol.* **58:**1801–1808.

87. **Rogers, J., and C. W. Keevil.** 1992. Immunogold and fluorescein immunolabelling of *Legionella pneumophila* within an aquatic biofilm visualized by episcopic differential interference contrast microscopy. *Appl. Environ. Microbiol.* **58:**2326–2330.

88. **Rost, F. W. D.** 1992. *Fluorescence Microscopy,* vol. I. Cambridge University Press, Cambridge.

89. **Rost, F. W. D.** 1992. *Fluorescence Microscopy,* vol. II. Cambridge University Press, Cambridge.

90. **Russ, J. C.** 1990. *Computer-Assisted Microscopy: the Measurement and Analysis of Images.* Plenum Press, New York.

91. **Russ, J. C.** 1995. *The Image Processing Handbook,* 2nd ed. CRC Press, Inc., Boca Raton, Fla.

92. **Saavedra-Molina, A., S. Uribe, and T. M. Devlin.** 1990. Control of mitochondrial matrix calcium: studies using fluo-3 as a fluorescent calcium indicator. *Biochem. Biophys. Res. Commun.* **167:**148–153.

93. **Schmidt, E. L.** 1972. Fluorescent antibody techniques for the study of microbial ecology, p. 67–76. In T. Rosswall (ed.), *Modern Methods in the Study of Microbial Ecology,* vol. 17. Swedish Natural Science Research Council, Stockholm.

94. **Shaw, P. J., and D. J. Rawlins.** 1991. The point-spread function of a confocal microscope: its measurement and use in deconvolution of 3-D data. *J. Microsc.* **163:** 151–165.

95. **Shotton, D. M.** 1989. Confocal scanning optical microscopy and its applications for biological specimens. *J. Cell Sci.* **94:**175–206.

96. **Sieracki, M. E., P. W. Johnson, and J. M. Sieburth.** 1985. Detection, enumeration and sizing of planktonic bacteria by image-analyzed epifluorescence microscopy. *Appl. Environ. Microbiol.* **49:**799–810.

97. **Sieracki, M. E., S. E. Reichenbach, and K. L. Webb.** 1989. Evaluation of automated threshold selection methods for accurately sizing microscopic fluorescent cells by image analysis. *Appl. Environ. Microbiol.* **55:**2762–2772.

98. **Singh, A., F.-P. Yu, and G. A. McFeters.** 1990. Rapid detection of chlorine-induced bacterial injury by the direct viable count method using image analysis. *Appl. Environ. Microbiol.* **56:**389–394.

99. **Sjollema, J., H. J. Busscher, and A. H. Weerkamp.** 1988. Deposition of oral streptococci and polystyrene lattices onto glass in a parallel plate flow cell. *Biofouling* **1:**101–112.

100. **Sjollema, J., H. J. Busscher, and A. H. Weerkamp.** 1989. Real-time enumeration of adhering microorganisms in a parallel plate flow cell using automated image analysis. *J Microbiol. Methods* **9:**73–78.

101. **Söderström, B. E.** 1977, Vital staining of fungi in pure cultures and in soil with fluorescein diacetate. *Soil Biol. Biochem.* **9:**59–63.

102. **Spencer, M.** 1982. *Fundamentals of Light Microscopy.* Cambridge University Press, Cambridge.

103. **Stahl, D. A., B. Flesher, H. R. Mansfield, and L. Montgomery.** 1988. Use of phylogenetically based hybridization probes for studies of ruminal microbial ecology. *Appl. Environ. Microbiol.* **54:**1079–1084.

104. **Stahl, D. A., D. J. Lane, G. J. Olson, and N. R. Pace.** 1984. Analysis of hydrothermal vent-associated symbionts by ribosomal RNA sequences. *Science* **224:**409–411.

105. **Staley, J. T.** 1971. Growth rates of algae determined *in situ* using an immersed microscope. *J. Phycol.* **7:**13–17.

106. **Tabor, P. S., and R. A. Neihof.** 1982. Improved microautoradiographic method to determine individual microorganisms active in substrate uptake in natural waters. *Appl. Environ. Microbiol.* **36:**945–953.

107. **Tolhurst, W. A., M. A. Pike, K. A. Blanton, and J. R. Harris.** 1994. *Using the Internet*, p. 1188. Que Corp., Indianapolis.

108. **Torrella, F., and R. Y. Morita.** 1981. Microcultural study of bacterial size changes and microcolony and ultramicrocolony formation by heterotrophic bacteria in seawater. *Appl. Environ. Microbiol.* **41:**518–527.

109. **Tsien, R. Y.** 1989. Fluorescent indicators of ion concentrations. *Methods Cell Biol.* **30:**127–156.

110. **Tsien, R. Y., and A. Waggoner.** 1990. Fluorophores for confocal microscopy: photophysics and photochemistry, p. 169–178. *In* J. B. Pawley (ed.), *Handbook of Confocal Microscopy.* Plenum Press, New York.

111. **Tumber, V. P., R. D. Robarts, M. T. Arts, M. S. Evans, and D. E. Caldwell.** 1993. The influence of environmental factors on seasonal changes in bacterial cell volume in two prairie saline lakes. *Microb. Ecol.* **26:**9–20.

112. **Viles, C., and M. E. Sieracki.** 1992. Measurement of marine picoplankton cell size by using a cooled, charge-coupled device camera with image-analyzed fluorescent microscopy. *Appl. Environ. Microbiol.* **58:**584–592.

113. **Wagner, M., R. Amann, H. Lemmer, and K.-H. Schleifer.** 1993. Probing activated sludge with oligonucleotides specific for proteobacteria: inadequacy of culture-dependent methods for describing microbial community structure. *Appl. Environ. Microbiol.* **59:**1520–1525.

114. **Waid, J. S.** 1973. A method to study microorganisms on surface films from soil particles with the aid of the transmission electron microscope. *Bull. Ecol. Res. Commun.* (Stockholm) **17:**103–108.

115. **Wang, Y.-L., and D. L. Taylor (ed.).** 1989. *Fluorescence Microscopy of Living Cells in Culture*, part A. *Fluorescent Analogs, Labeling of Cells and Basic Microscopy.* Academic Press, Inc., New York.

116. **Ward, D. M., M. M. Bateson, R. Weller, and A. L. Ruff-Roberts.** 1992. Ribosomal RNA analysis of microorganisms as they occur in nature. *Adv. Microb. Ecol.* **12:**219–286.

117. **Wilderer, P. A., and W. G. Characklis.** 1989. Structure and function of biofilms, p. 4–17. *In* W. G. Characklis and P. A. Wilderer (ed.), *Structure and Function of Biofilms.* John Wiley & Sons, New York.

118. **Wimpenny, J. W. T. (ed.).** 1988. *Handbook of Laboratory Model Systems for Microbial Ecosystems*, vol. 1 and 2. CRC Press, Inc., Boca Raton, Fla.

119. **Wimpenny, J. W. T., P. Waters, and A. Peters.** 1988. Gel-plate methods in microbiology, p. 229–251. *In* J. W. T. Wimpenny (ed.), *Handbook of Laboratory Model Systems for Microbial Ecosystems*, vol. 1. CRC Press, Inc., Boca Raton, Fla.

120. **Woese, C. R.** 1987. Bacterial evolution. *Microbiol. Rev.* **51:**221–271.

121. **Wolfaardt, G. M., J. R. Lawrence, M. J. Hendry, R. D. Robarts, and D. E. Caldwell.** 1993. Development of steady-state diffusion gradients for the cultivation of degradative microbial consortia. *Appl. Environ. Microbiol.* **59:**2388–2396.

122. **Wolfaardt, G. M., J. R. Lawrence, R. D. Robarts, and D. E. Caldwell.** 1994. Multicellular organization in a degradative biofilm community. *Appl. Environ. Microbiol.* **60:**434–446.

123. **Wolfaardt, G. M., J. R. Lawrence, R. D. Robarts, and D. E. Caldwell.** 1995. Bioaccumulation of the herbicide diclofop in extracellular polymers and its utilization by a biofilm community during starvation. *Appl. Environ. Microbiol.* **61:**152–158.

124. **Wolfaardt, G. M., J. R. Lawrence, R. D. Robarts, and D. E. Caldwell.** *In situ* characterization of biofilm exopolymers involved in the accumulation of chlorinated organics. Submitted for publication.

125. **Yu, F. P., G. M. Callis, P. S. Stewart, T. Griebe, and G. A. McFeters.** 1994. Cryosectioning of biofilms for microscopic examination. *Biofouling* **8:**85–91.

126. **Yu, F. P., and G. A. McFeters.** 1994. Physiological response of bacteria in biofilms to disinfection. *Appl. Environ. Microbiol.* **60:**2462–2466.

Cultivation of Bacteria and Fungi

RALPH S. TANNER

6

A brief review of the literature in applied and environmental microbiology quickly reveals a great number and variety of media used to culture bacteria and fungi in the laboratory. These range from very simple formulations of commercially available media to media which require a very meticulous and skillful assembly to prepare (43, 54). Thousands of microbiological medium formulations are available from just a handful of sources (2, 5, 6, 15, 22). It may appear that cultivation of bacteria and fungi is a complicated and empirical art, requiring some skill, observation, and attention to detail. However, with knowledge of the basic concepts of medium composition and of the physical conditions which may limit microbial growth, one can enhance one's ability to culture bacteria and fungi available in pure culture and to enrich for, isolate, and culture many microorganisms of interest from the environment (31). Many of the basic aspects of the culture of microorganisms have been outlined previously (13, 21, 26, 29). The major nutritional requirements for microbial growth, most of which were outlined by Gottschal et al. (25), that must be considered are sources of carbon; sources of energy (also called electron or hydrogen sources); electron acceptors; nitrogen sources; sources of other major mineral nutrients such as sulfur, phosphate, potassium, magnesium, and calcium; vitamin requirements; and trace metal requirements. Some physicochemical factors affecting growth include temperature, pH, requirement for oxygen, and salinity.

Several points should be kept in mind for use of the material in this chapter. The first is the importance and utility of considering culture conditions when working with bacteria and fungi. The research on some problems, such as the novel growth factor requirement of *Methanomicrobium mobile* (47) or the biochemistry of dehalogenation in *Desulfomonile tiedjei* (14), was not really feasible until the finer points of microbial culture were determined. Consideration of this issue has also been useful for environmental studies, such as the recovery of sulfate-reducing bacteria from environmental samples (46) or bacteria from potable water (44). While important, however, the rationale for design of a medium or use of a specific culture condition is discussed in the literature infrequently. Second, many of the problems with the culture of microorganisms are due to incorrect or failure to control incubation temperature or culture pH, lack of a required nutrient, less than ideal provision of a carbon or energy source, and even excess of some medium component.

Finally, it is not possible to thoroughly examine all of the specifics for the culture of all of the different bacteria in a single chapter. The relevant literature should be consulted for specific groups of organisms, such as thermophiles (17, 31), phototrophs (5, 19, 38), and fungi (6, 20, 31, 50). However, the basics of microbial culture are reviewed in this chapter, and this information can be used in conjunction with that available from the literature.

BASIC MEDIUM PREPARATION

A recipe for a basic basal medium for cultivating microorganisms is given in Table 1. It includes a mineral solution which provides the major inorganic chemicals required by many microorganisms, a vitamin solution, a trace metal solution, and a buffer for control of the culture pH. This basal medium will support the growth of many types of microorganisms with the appropriate selection of carbon sources, energy sources, and electron acceptors (1). The selection of the various components included in this basal medium and their roles are discussed below.

Mineral Solution

The composition of a mineral solution useful for preparing medium for many different microorganisms is given in Table 2. This mineral solution provides a source of sodium, chloride, ammonium as a nitrogen source, potassium, phosphate, magnesium, sulfate as a sulfur source, and calcium. While a mineral solution with this composition is useful for many different cultures, it may require modification for certain groups or species of microorganisms.

A requirement for sodium by most microorganisms other than marine or halophilic species has not been demonstrated. However, some phototrophs and anaerobic bacteria do require sodium for metabolism and growth (25). Many commercially prepared media contain some sodium chloride. The presence of a low concentration of sodium chloride in a medium does not appear to inhibit the culture of microorganisms. In contrast, potassium is required for the culture of essentially all microorganisms.

The use of ammonium as a source of available nitrogen will be examined below under the heading of nitrogen sources. It needs to be noted here that nitrate may be the preferred nitrogen source for the culture of cyanobacteria and aerobic soil microorganisms, for the enrichment and

TABLE 1 Basal microbiological medium[a]

Component	Amt/liter
Mineral solution[b]	10 ml
Vitamin solution[c]	10 ml
Trace metal solution[d]	0.5–5 ml
Buffer[e]	1–20 g
Adjust pH with NaOH[f]	

[a] A general-purpose medium adaptable to pure culture or ecological studies with the appropriate selection of carbon sources, energy sources, and electron acceptors. A general source of growth factors, such as yeast extract (0.1 to 2.0 g/liter), must be included for many studies.
[b] Described in Table 2.
[c] Described in Table 3.
[d] Described in Table 4.
[e] The buffer used is selected for each medium: TES (pK$_a$ 7.4) for neutral pHs, NaHCO$_3$ for CO$_2$-containing gas phases, MES (pK$_a$ 6.1) for acidic pHs, phosphate for lactic acid bacteria or thiobacilli, TAPS (pK$_a$ 8.4) or glycine (pK$_a$ 9.8) for alkaline pHs, etc.
[f] The final pH of many media is 7.0 to 7.5. This should be adjusted as required for individual cultures or experiments.

isolation of fungi, etc. (6, 19). Both ammonium and nitrate would be omitted from a medium designed to enrich for and isolate potential nitrogen-fixing bacteria.

The phosphate concentration in the basal medium described here is lower than that found in many other media but is more than enough to support the growth of most microorganisms. It was observed while labeling cells with radioactive phosphate for molecular systematics studies that 20 to 50 μM phosphate would support the culture of most bacteria. Notable exceptions to this observation were the lactic acid bacteria and thiobacilli, which required millimolar levels of phosphate for culture. Phosphate at higher concentrations can function as a buffer, but there are other compounds which can serve this function without the potential problem of precipitate formation or sequestering of trace metals (see below).

Magnesium is also required for the culture of bacteria and fungi. Sulfate can be used by many species as a sulfur source, but some groups or species may require sulfide or sulfur-containing amino acids as a sulfur source. Calcium may be considered more a trace metal requirement. However, it is readily incorporated in the mineral solution and may be required at a higher concentration for the growth of some species. For example, increasing the final concentration of calcium in medium 5- to 10-fold over what is in the basal medium presented in Table 1 can greatly increase the cell yield of *Methanobacterium thermoautotrophicum* from a large-scale culture. Some marine bacteria require levels of magnesium, calcium, and iron higher than those provided in the basal medium described in Table 1.

TABLE 2 Mineral solution[a]

Component	Amt (g)/liter
NaCl	80
NH$_4$Cl	100
KCl	10
KH$_2$PO$_4$	10
MgSO$_4$·7H$_2$O	20
CaCl$_2$·2H$_2$O	4

[a] A solution containing the major inorganic components required for microbial growth. Add and dissolve each component in order. The mineral solution can be stored at room temperature.

Issues that should be considered for the formulation of a mineral solution or a set of stock solutions for medium preparation include precipitation, stability upon storage, and convenience. The mineral solution in Table 2 does not form a precipitate, is stable at room temperature for at least 1 year, and is useful for many microorganisms. The fewer the number of stock solutions needed to constitute a medium, the easier its preparation. The mineral solution in Table 2 was adapted from the mineral solutions 1 and 2 for the culture of methanogens (3), with the idea of reducing the number of solutions and eliminating the need to weigh ammonium chloride for medium preparation.

Mineral stock solutions can be developed for more specific groups of bacteria and fungi. For example, a 20× mineral solution for preparation of medium to recover sulfate-reducing bacteria from aquatic or terrestrial samples (46) would contain (grams per liter) NaCl, 100; (NH$_4$)$_2$SO$_4$, 10; MgSO$_4$·7H$_2$O, 4; KH$_2$PO$_4$, 6; and CaCl$_2$·2H$_2$O, 0.8. Ammonium sulfate replaced ammonium chloride so that the medium would contain more sulfate as an electron acceptor, and the amount of phosphate was increased to the point that the medium would contain a precipitate after the addition of ferrous ammonium sulfate (46). A 100× mineral solution for culture of marine microorganisms could contain (grams per liter) KCl, 200; NH$_4$Cl, 100; MgSO$_4$·7H$_2$O, 40; and KH$_2$PO$_4$, 20. Sodium chloride was omitted from the stock solution since it could not be added at a concentration high enough for this stock to meet the requirements of a marine medium (20 to 28 g of sodium chloride per liter). A calcium salt was omitted from this marine mineral stock solution because it would produce a precipitate.

Omission of a calcium salt, and sometimes a phosphate salt, from a mineral stock solution may be required to avoid a precipitate. Use monobasic potassium phosphate (slightly acidic) rather than dibasic potassium phosphate (alkaline) to reduce problems with precipitate formation or microbial contamination upon room temperature storage in a stock solution. The use of hydrated salts rather than anhydrous salts, such as for magnesium sulfate or calcium chloride, eases preparation of mineral stock solutions.

Vitamin Solution

A solution of water-soluble vitamins which will support or stimulate the growth of many microorganisms was adapted from reference 55 and is given in Table 3. The changes from the previous water-soluble vitamin solution are the

TABLE 3 Vitamin solution[a]

Component	Amt (mg)/liter
Pyridoxine-HCl	10
Thiamine-HCl	5
Riboflavin	5
Calcium pantothenate	5
Thioctic acid	5
p-Aminobenzoic acid	5
Nicotinic acid	5
Vitamin B$_{12}$	5
MESA[b]	5
Biotin	2
Folic acid	2

[a] A solution designed to meet the water-soluble vitamin requirements of many microorganisms. Store at 4°C in the dark.
[b] Mercaptoethanesulfonic acid.

inclusion of mercaptoethanesulfonic acid, which is required by *Methanobrevibacter ruminantium* (4), and elevation of the concentration of vitamin B_{12}, which is required for the growth of *Clostridium aceticum* and *Methanomicrobium mobile* (47).

Other vitamins which may be required to culture microorganisms are hemin (8, 26), 1,4-naphthoquinone (14, 23), and vitamin K_1 (2-methyl-3-phytyl-1,4-naphthoquinone) (26). A $100\times$ stock solution of hemin can be prepared by dissolving 50 mg of hemin in 1 ml of 1 N sodium hydroxide and taking the dissolved hemin to a final volume of 100 ml with water for the stock solution (26). A $5,000\times$ stock solution of vitamin K_1 can be prepared by dissolving 0.15 ml in 30 ml of 95% ethanol (26). It is possible that a microorganism will be stimulated by other chemical forms of a vitamin, such as niacinamide instead of nicotinic acid or pyridoxamine rather than pyridoxine. Some strains, probably host associated, may require a preformed coenzyme such as pyridoxal phosphate or thiamine pyrophosphate for growth, but these organisms would be encountered rarely by most investigators. A recent novel vitamin, 7-mercaptoheptanoylthreonine phosphate, was reported in 1991 (30). It is likely that other vitamins will be discovered in the future.

Trace Metal Solution

A stock solution which can supply the trace metal requirements of many microorganisms was adapted from reference 55 and is given in Table 4. Changes from the previous formulation include deletion of the sodium and magnesium salts (supplied by the mineral solution); deletion of the aluminum and borate salts (probably included since these are found in soil extracts and bacteria were once classified with plants); addition of sources of selenium, nickel, and tungsten (required for hydrogenases, formate dehydrogenases, etc.); and doubling the concentration of metals. In general, aerobic microorganisms grow well with only a small addition of the trace metal solution (0.2 to 0.5 ml/liter of culture), while anaerobes seem to grow better with larger additions (more than 10 ml/liter for large-scale cultures). Borate (1 to 5 mg of boron per liter) is required for the culture of cyanobacteria (19).

Carbon and Energy Sources

The carbon and energy requirements of a microorganism can range from carbon dioxide and hydrogen for the chemolithotrophic methanogens and acetogens to very complex

media containing amino acids, carbohydrates, nitrogen bases, and undefined components (e.g., yeast extract, meat extract, or ruminal fluid) for some bacteria. Carbon and energy requirements for microbial culture are considered together since many microorganisms that many microbiologists work with are chemoorganotrophs which often use a single organic substrate to satisfy both needs. The great number of potential carbon and energy sources used for microbial culture prohibits a detailed examination of them. Indeed, many of these compounds have probably not been examined in detail, and surprises regarding their use still await. For example, enrichments containing valerate as the carbon and energy source will probably yield just bacteria, but enrichments containing *n*-pentanol, the corresponding alcohol, may yield fungi or grazing protozoa. Some particular items with regard to the culture of chemolithotrophs and phototrophs and the use of xenobiotic or toxic substrates will be considered in this section. However, there are useful generalities in regard to carbon and energy sources which may be examined briefly.

Essentially any organic compound may be considered as a carbon and energy source for the culture of microorganisms. Some compounds, such as glucose, may be used by a great number of species, while others, such as allantoin under anoxic conditions, are used by only a few species. Lower levels of substrates are often used for the culture of aerobes (0.1 to 2 g/liter) compared with substrate levels used for the culture of anaerobes (20 g/liter for some substrates in large-scale cultures). The medium recipes given for the culture of known microorganisms are very useful as a guide for the substrates that a species may use. Some ideas as to other substrates that a strain might utilize, or which may be used to improve the culture of a strain, are given below.

A good way to classify substrates is by biochemical and chemical categories: amino acids and related compounds; carboxylic acids, volatile fatty acids, and long-chain fatty acids and lipids; alcohols; nitrogen bases; carbohydrates; biological polymers; aromatic compounds, methoxylated aromatic compounds; etc. The ability of a single species to utilize any single substrate can be very specific, thus the continued role of carbohydrate and other substrate utilization patterns for phenotypic characterization of microorganisms.

Pyruvate and intermediates in the tricarboxylic acid cycle are often overlooked as substrates to support or improve the culture of microorganisms, but these have proved very useful in my own and others' hands for this purpose (14, 44). For example, the use of α-ketoglutarate as a substrate has greatly facilitated the culture of *Legionella* species (42).

Another issue with regard to carbon and energy sources is substrate metabolism without apparent growth. The monitoring of the pH of substrate utilization tests in cultures that do not exhibit visible growth may indicate substrate utilization. A decrease in pH can mean the formation of acidic end products of metabolism. An increase in pH can mean the release of ammonia from a substrate or the effect of the cations remaining after the utilization of an organic acid. The pH change alone may inhibit the growth of a strain, and conditions may need to be varied for the compound in question to support growth.

Some compounds may be toxic or insoluble at the levels at which certain substrates, such as formaldehyde, hydrocarbons, xenobiotics, etc. (31), are often added to media. Insoluble and volatile substrates are frequently introduced into cultures via a vapor phase (11, 31, 52). It may be necessary

TABLE 4 Trace metal solution[a]

Component	Amt (g)/liter
Nitrilotriacetic acid	2.0
Adjust pH to 6 with KOH	
$MnSO_4 \cdot H_2O$	1.0
$Fe(NH_4)_2(SO_4)_2 \cdot 6H_2O$	0.8
$CoCl_2 \cdot 6H_2O$	0.2
$ZnSO_4 \cdot 7H_2O$	0.2
$CuCl_2 \cdot 2H_2O$	0.02
$NiCl_2 \cdot 6H_2O$	0.02
$Na_2MoO_4 \cdot 2H_2O$	0.02
Na_2SeO_4	0.02
Na_2WO_4	0.02

[a] A solution designed to meet the trace metal requirements of many microorganisms. Store at 4°C.

with some toxic substrates to add them to a culture at low, nontoxic levels, monitor utilization and reamend the culture as the toxic substrate is consumed.

Substrates which may be inhibitory if used as pure compounds, such as oleic acid, may be added in a less available form, such as polyoxyethylenesorbitan monooleate (Tween 80). Tween 80 can be highly stimulatory for some microorganisms, both aerobes and anaerobes, for reasons not fully delineated, at a concentration of 0.2 g/liter (3, 10). Some media, included those commercially prepared, may have Tween 80 added at as high a level as 1 g/liter.

Important energy sources for chemolithotrophs include hydrogen, sulfide, and other reduced sulfur compounds. Other inorganic energy sources are carbon monoxide and elemental metals (33), which may be regarded essentially as hydrogen sources, and ammonium or nitrite (18). Many chemolithotrophs can use carbon dioxide as a sole carbon source, but others may require organic carbon as a carbon source or for specific biosynthetic reactions, and organic compounds may be taken up and utilized by autotrophic microorganisms. One of the most common of these is acetate, which is required by a number of methanogens and other anaerobic microorganisms (47). If it does not interfere with a study, it may be a good idea to include some acetate (2 to 5 g/liter) in media for chemolithotrophs (1). Some methanogens and ruminal bacteria require the branched-chain fatty acids isobutyrate, isovalerate, and 2-methylbutyrate for the synthesis of branched-chain amino acids (8, 47). The actual branched-chain amino acids may satisfy this requirement for some microorganisms (53).

Light may be considered the energy source for the growth of phototrophic bacteria (19, 31, 38). These organisms constitute physiologically and phylogenetically a very diverse group of bacteria which includes photoautotrophs (which can use carbon dioxide as the sole carbon source) and photoheterotrophs (which use simple organic compounds such as acetate, pyruvate, malate, and ethanol as sources of reducing equivalents and cell carbon). The anoxygenic phototrophs, which includes those able to use reduced sulfur compounds as sources of reducing potential, need to be cultured anaerobically (see below). I prefer to culture cyanobacteria by using the techniques required for anaerobes, since these may be gas exchanged for culture under an enriched carbon dioxide and reduced oxygen gas phase. Specific wavelengths of light, such as near infrared or infrared for the purple sulfur and nonsulfur purple bacteria, may be used to enrich for certain phototrophs. Some of the phototrophs can be grown under low light intensities, as little as 10 lx, and it is often recommended that light intensities be moderated (100 to 400 lx, for example) to enrich for, isolate, and culture many phototrophic bacteria (19, 38). However, I have successfully enriched for and grown cyanobacteria and nonsulfur purple bacteria under plant growth lamps at 6,000 lx.

Nitrogen Sources

Ammonium is the inorganic nitrogen source for the basal medium described in this chapter. A lower concentration of ammonium may be used for the routine culture of aerobes, but the growth of many anaerobes is stimulated by the recommended level of ammonium (8). Nitrate may be a better source of nitrogen for the culture of many soil microorganisms and the cyanobacteria and for the enrichment and isolation of fungi. If nitrate is provided as a source of nitrogen, the possibility that it can also serve as an electron acceptor should be considered. Nitrogen gas can serve as a

sole nitrogen source for strains with the ability to fix nitrogen.

Urea is an organic source of nitrogen which may have use in field projects outside the laboratory. Urea needs to be filter sterilized for laboratory evaluation because it can hydrolyze to ammonia and carbon dioxide during steam sterilization. Some microorganisms require organic nitrogen compounds such as nitrogen bases or, more frequently, amino acids for their culture. Vitamin assay Casamino Acids (Difco Laboratories, Detroit, Mich.) provides a semidefined, if incomplete, source of amino acids if this requirement needs to be examined (40). Two grams of Casamino Acids per liter of medium is a useful concentration for examining the nitrogen requirements of an isolate, but 10 to 20 g/liter would be used if the goal was to generate cell tissue. Peptone, enzymatic digests of casein or soybean meal, yeast extract, etc., provide more complex nitrogen sources for microorganisms which require peptides or other compounds. Microorganisms, such as some of the lactic acid bacteria, may require nitrogen bases (adenine, cytosine, guanine, thymine, and/or uracil) as trace nutrients in medium. Addition of each nitrogen base at a concentration of 1 to 10 mg/liter may fulfill this requirement.

Electron Acceptors

There are a number of electron acceptors which will support microbial respiration, the more common being oxygen (reduced to water), nitrate (reduced to nitrite, ammonia, or dinitrogen), sulfate (reduced to sulfide), and carbon dioxide (reduced to methane or acetate). Recently there has been an increasing awareness of the importance of iron reduction (ferric to ferrous iron) in the environment (34). Electron acceptors are usually added to media at a concentration of 5 to 20 mM.

There are a number of less frequently considered or more unusual electron acceptors that can be reduced by microorganisms, and many of these reductions can support metabolism and growth. These include sulfite; thiosulfate; selenate; Mn(IV); U(VI); oxyanions of tellurium, europium, and rhodium; dimethyl sulfoxide; trimethylamine n-oxide; and arsenate (32, 34, 35, 39, 41). Organic compounds may also be used as electron acceptors, a common one being fumarate (reduced to succinate). For instance, *Escherichia coli* is capable of anaerobic respiration using fumarate, malate, or aspartate as an electron acceptor (36). Unsaturated bonds in other compounds, such as ferulic acid, may be used as electron acceptors.

Undefined Media and Medium Components

Microorganisms will frequently be cultured on undefined media or on media which contain undefined components, either because the exact nutritional requirements of strains are unknown or simply because of convenience (many undefined media are commercially available). Probably the most commonly used undefined medium component is yeast extract (Difco). Yeast extract is a complex mixture of acetate, amino acids, peptides, nitrogen bases, vitamins, trace metals, phosphate, etc., that would be required to support the growth of most known microbial strains. It even contains some fermentable carbohydrate, ribose. Yeast extract in a concentration as low as 50 mg/liter can enhance the recovery of microorganisms from environmental samples in an otherwise defined medium. It is commonly used at 0.5 to 5 g/liter in a number of culture media.

All microbiologists are familiar with nutrient broth, which contains peptone and beef extract. However, nutrient broth is not a good medium for the culture of many microorganisms, primarily because although it does contain amino acids and peptides, it lacks many important nutrients, especially carbohydrate. Two undefined media which are more complete and useful than nutrient broth are plate count broth and tryptic soy broth (15). Both contain glucose (2 to 2.5 g/liter). Plate count broth contains yeast extract and tryptone (Trypticase) as additional components. Tryptic soy broth contains tryptone and soytone (a digest of soy meal) and is probably more commonly used than plate count broth. One should note when preparing solid medium that commercial tryptic soy agar is different from a solid medium prepared from tryptic soy broth in that glucose is omitted from tryptic soy agar (15).

An undefined medium commonly used for the culture of fungi is Sabouraud dextrose agar, which contains a peptone and glucose. Two other undefined media which are useful for the culture of fungi are potato dextrose broth, which contains glucose and potato infusion (starch and trace nutrients), and malt extract broth, which contains glucose, maltose, malt extract, and yeast extract (15).

It has been noted that a higher number of cells may be recovered from environmental samples if media containing low concentrations of substrate and nutrients are used instead of rich media (37, 56). The undefined media may be used at $\frac{1}{2}$ to $\frac{1}{10}$ of their label-indicated concentrations for environmental studies without loss of cell recovery. A medium was formulated for potable water samples on the basis of the principle that use of low concentrations of a wide variety of components could result in higher recovery of different heterotrophic bacteria (44).

Medium Preparation

Basic medium preparation consists of constituting most or all of the medium components at the proper concentration in solution, adjusting the pH (see below), and then adding components which may interfere with pH adjustment (calcium carbonate for detection of acid production or sodium bicarbonate for medium under a carbon dioxide-containing gas phase) or a gelling agent for solid medium. The most common gelling agent is agar, which is available in a variety of purities and forms. The extra expense of purified agar (BBL, Becton Dickinson, Cockeysville, Md.) can be justified on the basis of a greater and more consistent agar strength, higher clarity, and relative absence of interfering organic substances. Agar is commonly used in medium at a concentration of 7 to 20 g/liter. If a very pure agar is desired, Oxoid Ionagar no. 2 (Oxford, Columbia, Md.) or a high-gel-strength, electrophoretic-grade agarose may be used. Another gelling agent which is particularly useful for the culture of thermophiles and extreme thermophiles is Gelrite (Scott Laboratories, Carson, Calif.). This gellan gum (5 to 12 g/liter) requires divalent cations, such as 1 g of $MgCl_2 \cdot 6H_2O$ per liter, for solidification. A mixture of $CaCl_2 \cdot 2H_2O$ and $MgCl_2 \cdot 6H_2O$ (0.8 g of each per liter) may yield a harder medium. Gelling agents should be melted in the medium prior to steam sterilization.

The major means of medium sterilization are steam sterilization (autoclaving) and filtration. For steam sterilization, it is generally considered necessary to heat the medium to 121°C at 100-kPa (15-lb/in^2) gauge pressure for 15 to 20 min. In general, the shorter the time required to sterilize a medium the better, although as the volume of medium to be sterilized increases, so does the sterilization time. Clean laboratory conditions, the use of chemically clean glassware, and the use of analytical reagent-grade or better medium

components facilitate the preparation of sterile medium and materials. For some media tubed at 10 ml or less, I have been able to use routinely a 5-min steam sterilization (46).

There are media or medium components which break down under the conditions of steam sterilization. These should be filter sterilized. For small volumes, the use of a prefilter-sterile filter (pore size of 0.8 to 0.2 μm) combination, such as an Acrodisc PF syringe filter (Gelman Sciences, Ann Arbor, Mich.), is convenient. Disposable, easy-to-use filter units, such as Nalgene disposable filterware (Nalge, Rochester, N.Y.), are useful for volumes up to 100 ml. Filter capsules which have good flow rates and filtration capacity, such as Gelman minicapsules, are available for filter sterilization of many liters of medium.

PHYSICOCHEMICAL FACTORS

Physicochemical factors which should be considered for the cultivation of bacteria and fungi include temperature, pH, relationship to oxygen, and, for some microorganisms, salinity and ionic strength. Common problems in culture of bacteria and fungi are incorrect or uncontrolled temperature or pH. If one is having a problem with the culture of a microorganism, these factors should be checked first. Microorganisms can be particular about growth conditions. For example, strains of the same (phylogenetic) species can have 5 to 7°C differences in optimal growth temperatures.

The concept and control of incubation temperature are straightforward and will not be elaborated upon. One point that should be considered is that the recovery of microorganisms from many common environmental samples will be inhibited by incubation at 37°C, the temperature of many laboratory incubators. Room temperature (20 to 23°C) incubation is better for many environmental samples. Few viable colonies are lost if these samples are incubated at 28 to 30°C (44, 46). Of course, different temperatures would be required for samples from a hot spring or alpine stream. All of the physicochemical factors need to be considered and adjusted for the recovery of microorganisms from different environmental samples.

pH

The importance of control of pH for the culture of microorganisms is often overlooked. Many bacteria, especially those associated with a host or recovered from an otherwise constant or protected environment, have fairly narrow pH optimum ranges. The pH range of *Legionella* species can be only ±0.05 pH units from optimum (42). pH can change during culture of microorganisms. As an organic acid (added as a salt) is used as a substrate, as nitrate is reduced to ammonia or dinitrogen gas, or as carbon dioxide is consumed, culture pH may increase. The production of organic acid from fermentation or incomplete oxidation of substrate is a common reason for a decrease in pH in culture.

Some common buffers used for microbial culture are phosphate and Tris, which have often proven unsatisfactory (24). Good's buffers are zwitterionic organic buffers with pK_as in the range useful to most microbiologists. They were developed specifically for biological research (24), and one of these, HEPES (N-2-hydroxyethylpiperazine-N'-2-ethanesulfonic acid), proved useful for tissue culture. My laboratory has found TES [N-tris(hydroxymethyl)methyl-2-aminoethanesulfonic acid (pK_a 7.4)] a better buffer for culture at neutral pHs than HEPES or PIPES [piperazine-N-N'-bis(2-ethanesulfonic acid)] (1, 46). MES [2-(N-morpholino)ethanesulfonic acid (pKa 6.1)] is useful for the culture of bacteria requiring slightly acidic pHs. TAPS [N-tris(hydroxymethyl)methyl-3-aminopropanesulfonic acid (pKa 8.4)] can be used for those microorganisms with slightly alkaline pH optima. In general, changes in pH with growth are less for aerobic cultures than for anaerobic cultures. A Good's buffer at 1 to 2 g/liter may control the pH of an aerobic culture, while 5 to 20 g/liter may be required for anaerobic cultures.

Acetate (pK$_a$ 4.75) or citrate (pK$_{a1}$ 3.2, pK$_{a2}$ 5.05, pK$_{a3}$ 6.4) is a good buffer for those organisms requiring more acidic conditions. Remember that acetate and citrate may also serve as carbon and energy sources. Sodium bicarbonate is a good buffer at pH 6 to 8 for cultures incubated with a carbon dioxide gas phase. The concentration of bicarbonate used will depend on the partial pressure of carbon dioxide in the gas phase and the incubation temperature. For example, 4 to 5 g of sodium bicarbonate would be added per liter of medium for methanogens under a gas phase containing 20% carbon dioxide at 250-kPa gauge pressure (1, 3, 4, 47). This concentration of bicarbonate can be added to medium after pH adjustment and before steam sterilization, obviating the need for sterile additions of carbonate to media formulated for 100% carbon dioxide gas phases (26).

Relationship to Oxygen

Bacteria and fungi have varying requirements and tolerances for oxygen for culture and may be classified as aerobes (e.g., pseudomonads), microaerophiles (e.g., some strains of *Legionella* and *Campylobacter*), facultative anaerobes (e.g., enteric organisms such as *E. coli*), or anaerobes (e.g., clostridial species). Small-scale culture of aerobes is relatively straightforward—media are simply (aseptically) exposed to air. The culture of microaerophiles and aerotolerant anaerobes can be relatively easy with the use of a candle jar, which also provides an atmosphere enriched with carbon dioxide, or the use of semisolid media (29). There are several systems developed for the culture of microaerophiles and many anaerobes, particularly those of clinical interest, such as the GasPak (BBL). These use water for the production of the desired gas phase, which sometimes leads to problems with plate culture of environmental samples, with actively motile cells swarming over a plate's surface. Waterless gas-generating systems for the culture of microaerophiles and clinically important anaerobes have been developed (7).

The challenges of the general culture of strictly anaerobic microorganisms were alleviated in the late 1970s with the introduction of syringe transfer technique, specialized glassware, improved anaerobic glove bags and chambers, improved (handmade) plate incubators, and other methods under the umbrella classification of "Balch technique" (3, 4). The flexible chamber manufactured by Coy (Grass Lake, Mich.) can easily maintain an oxygen level of 0 to 2 ppm (compared with around 200,000 ppm in the atmosphere) and is probably the easiest system to use and maintain. One can handle strict anaerobes in such an environment just like aerobes on an open bench. The author incorporates a bed of activated charcoal just below the catalyst tray in the chamber to absorb compounds which could inactivate the oxygen-scrubbing catalyst and shallow jars of calcium chloride (anhydrous) to remove water from the chamber gas phase, which increases the interval between catalyst "activation" (drying). It seems to be easier to maintain an anaerobic chamber if plates and cultures are not incubated inside but are incubated in plate holders (3) outside of the chamber.

The best way to prepare anaerobic medium is to boil the medium under an oxygen-free gas stream, which can reduce any dissolved oxygen below a detection limit (20 ppb), seal the medium, transfer the medium into an anaerobic chamber, dispense the medium and finally seal the tubed medium in the chamber and remove it for further manipulation or sterilization. Many anaerobes, such as sulfate-reducing bacteria and most clostridia, can be cultured in "Hungate-type" anaerobic culture tubes (no. 2047; Bellco Glass, Vineland, N.J.) (1, 46). These culture tubes are relatively inexpensive. The butyl rubber stopper used with the Hungate-type tubes is thin enough that skin testing syringes can be used for transfers. These inexpensive syringes are available sterile in multipack trays (no. 5539; Becton Dickinson, Rutherford, N.J.).

Strict anaerobes and those using gases as substrates, such as methanogens and acetogens, should be cultured in aluminum seal tubes (no. 2048; Bellco Glass) (1, 3, 4). It may be necessary to use glass rather than plastic syringes for the transfer of the strict anaerobes, but these can be cleaned and reused. A gassing station (3) can replace the anaerobic chamber atmosphere which was sealed in the tubes with a desired gas phase. This system (3, 4) is very useful for any experiment for which control of the culture gas phase is required.

The culture of many strict anaerobes requires poising the redox potential of the medium lower than what mere elimination of oxygen will achieve. Resazurin (50 μg/liter) can be incorporated into medium as a redox indicator. Reducing agents for the culture of anaerobes include cysteine, sulfide, thioglycolate, dithionite, glutathione, yeast extract, and dithiothreitol (28). Ascorbic acid, elemental iron, or titanium(III) citrate may be used if sulfur compounds need to be avoided (28, 46). The use of titanium(III) may result in a lower recovery of anaerobes from environmental samples (51). A good general cysteine-sulfide reducing agent which can be stored in an anaerobic chamber for at least a year can be prepared as follows. Take 2 g of L-cysteine and 2 g of washed and dried crystals of Na$_2$S·9H$_2$O into an anaerobic chamber. Boil 100 ml of water under nitrogen, take it into an anaerobic chamber, dissolve the cysteine and sulfide, and dispense and seal the reducing agent. This mixture is added at 1 to 10 ml per liter of anaerobic medium, after oxygen has been removed (1, 3, 4, 14, 47).

Salinity and Ionic Strength

Many microorganisms from marine or saline environments require salt or salts for their culture. For many marine bacteria, this requirement can be met simply by adding 10 to 20 g of sodium chloride per liter of medium. Seawater contains other ions, particularly magnesium. A complete seawater medium can be constituted from the recipe for Turks island salt (9) or based on commercially available formulations such as Instant Ocean (Aquarium Systems, Mentor, Ohio). Some marine media require filter sterilization to avoid the formation of precipitates during steam sterilization. Many extreme halophiles can be cultured in medium containing elevated concentrations of sodium chloride, though some have an additional requirement for magnesium ions (49).

CULTIVATION OF FUNGI

Most microbiologists are familiar with the methods to culture bacteria, and most of the material presented above was derived from the literature on bacteria. The fundamentals of the culture of fungi are generally the same as for the bacteria, and there are good basic references in this area (6, 12, 20).

General trends in the media used to culture fungi include use of nitrate as a nitrogen source and use of slightly acidic pH in the culture medium (6, 20). The recovery of fungi from environmental samples can be facilitated in general, in addition to the two factors above, by the inclusion of antibacterial agents, such as streptomycin, in the medium and by incubation at low (4 to 10°C) temperatures, which permits a number of fungi to outcompete bacteria (6, 12, 20). Many fungi can also be enriched for and isolated on common bacteriological media. Isolated strains of fungi can be cultured on the basal medium described above, which contains ammonium as the nitrogen source. Once yeast cells are in pure culture, growth in liquid medium is fairly easy. Molds, on the other hand, may prefer to grow in shallow culture in tissue culture flasks, a technique that is useful for liquid culture of actinomycetes as well.

One problem with the culture of fungi is the possibility of contamination of the laboratory with fungal spores. Working in a hood will alleviate some of these problems. Remember to keep the hood clean and disinfected. A useful trick for working with a mold on a plate on an open bench is to first place a lint-free towel (Kimwipes; Kimberly-Clark Corp., Roswell, Ga.) moistened with a disinfectant on the working area. Change the towel each time a different culture is manipulated.

As in the rest of biology, there are a number of exceptions to every generality given above about the cultivation of fungi. The fungi may continue to hold surprises for microbiologists as different techniques are used for their culture. Examples include the recent discovery of anaerobic fungi (16, 50) and the culture of a fungus with toluene as the sole carbon and energy source (52).

LARGER-SCALE CULTIVATION

Larger-scale (0.1- to 100-liter) culture of bacteria and fungi is, again, in principle similar to small-scale culture, with some attention given to the problems of sterilization of larger volumes of medium, use of specialized equipment such as fermentors, provision of gaseous substrate such as oxygen, safe handling of large volumes of very hot liquids, etc. Oxygen is provided to larger cultures through shaking, stirring, and/or injection of sterile air in as small bubbles as possible, all to force as much oxygen as possible into solution. The supply of oxygen to aerobic cultures is a prime consideration for the scale-up of these cultures. Foaming and the use of antifoam should be considered when fermentation equipment is used. It is relatively easier to scale up the culture of many anaerobes by using Pyrex bottles (Corning Glass, Corning, N.Y.) (3) in place of fermentors, but it will probably be necessary to have a mechanism to control the culture pH during growth. If sealed glassware is used, provision must be made to prevent a buildup of pressure due to microbial production of gases as end products of metabolism.

Large-scale culture of a microorganism may point out nutritional requirements or other factors limiting the growth of the culture that were not apparent in smaller cultures. The experience with M. thermoautotrophicum is a good illustration of this. The medium used for tube culture of M. thermoautotrophicum is very simple and defined (3). When the culture of this methanogen was scaled up to obtain enough cell material for biochemical studies, it was noted that production of cell mass was limited by trace metals, such as iron, cobalt, and molybdenum (45, 48). A surprising

discovery was made that M. thermoautotrophicum required nickel as a result of scaling up its culture (45), and it was noted in our laboratory that cell yields were improved if the concentration of calcium in the medium was increased 5- to 10-fold. M. thermoautotrophicum can take up acetate from its medium, and the cell yield was enhanced by the inclusion of 2 g of sodium acetate per liter of medium. One liter of gas (H₂/CO₂, 80:20) per min was sparged through a 100-liter fermentor of M. thermoautotrophicum. This gassing rate stripped some nutrients such as ammonia and sulfide, and periodic additions to the fermentor of sodium sulfide and ammonium hydroxide improved cell yield.

Like most other aspects of microbiological research, the cultivation of bacteria and fungi rests on a general foundation of basic principles but does require some mastery of a number of details. The literature is a good guide to the specifics of culture of known microorganisms. One should also rely on colleagues for advice when starting to work with unfamiliar microorganisms, such as cyanobacteria or anaerobes, or to work with new techniques, such as fermentation equipment. Two important things to consider for the culture of microorganisms are to not shy away from trying something new and to closely observe (for such phenomena as macroscopic, microscopic, and chemical characteristics) the culture. Simple, yet novel and perhaps important, discoveries can still be made on the basis of these tenets for the culture of bacteria and fungi.

REFERENCES

1. **Adkins, J. P., L. A. Cornell, R. S. Tanner.** 1992. Microbial composition of carbonate petroleum reservoir fluids. Geomicrobiol. J. **10:**87–97.
2. **Atlas, R. M.** 1993. Handbook of Microbiological Media. CRC Press, Boca Raton, Fla.
3. **Balch, W. E., L. J. Magrum, G. E. Fox, R. S. Wolfe, and C. R. Woese.** 1979. Methanogens: reevaluation of a unique biological group. Microbiol. Rev. **43:**260–296.
4. **Balch, W. E., and R. S. Wolfe.** 1976. New approach to the cultivation of methanogenic bacteria: 2-mercaptoethanesulfonic acid (HS-CoM)-dependent growth of Methanobacterium ruminantium in a pressurized atmosphere. Appl. Environ. Microbiol. **32:**781–791.
5. **Balows, A., H. G. Trüper, M. Dworkin, W. Harder, and K.-H. Schleifer (ed.).** 1992. The Prokaryotes, 2nd ed., vol. I–IV. Springer-Verlag, New York.
6. **Booth, C.** 1971. Fungal culture media. Methods Microbiol. **4:**49–94.
7. **Brazier, J. S., and V. Hall.** 1994. A simple evaluation of the AnaeroGen™ system for the growth of clinically significant anaerobic bacteria. Lett. Appl. Microbiol. **18:**56–58.
8. **Bryant, M. P., and I. M. Robinson.** 1962. Some nutritional characteristics of predominant culturable ruminal bacteria. J. Bacteriol. **84:**605–614.
9. **Budavari, S. (ed.).** 1989. Merck Index, 11th ed. Merck & Co., Rahway, N.J.
10. **Cato, E. P.** 1983. Transfer of Peptostreptococcus parvulus (Weinberg, Nativelle, and Prevot 1937) Smith 1957 to the genus Streptococcus: Streptococcus parvulus (Weinberg, Nativelle, and Prevot 1937) comb. nov., rev., emend. Int. J. Syst. Bacteriol. **33:**82–84.
11. **Claus, D., and N. Walker.** 1964. The decomposition of toluene by soil bacteria. J. Gen. Microbiol. **36:**107–122.
12. **Collins, C. H., P. M. Lyne, and J. M. Grange.** 1989. Microbiological Methods, 6th ed. Butterworths, Boston.
13. **Cote, R. J., and R. L. Gherna.** 1994. Nutrition and media, p. 155–178. In P. Gerhardt, R. G. E. Murray, W. A. Wood,

and N. R. Krieg (ed.), *Methods for General and Molecular Bacteriology*. American Society for Microbiology, Washington, D.C.

14. DeWeerd, K. A., L. Mandelco, R. S. Tanner, C. R. Woese, and J. M. Suflita. 1990. *Desulfomonile tiedjei* gen. nov. and sp. nov., a novel anaerobic, dehalogenating, sulfate-reducing bacterium. *Arch. Microbiol.* **154**:23–30.

15. Difco Laboratories. 1984. *Difco Manual*, 10th ed. Difco Laboratories, Detroit.

16. Dore, J., and D. A. Stahl. 1991. Phylogeny of anaerobic rumen *Chytridiomycetes* inferred from small subunit ribosomal RNA sequence comparisons. *Can. J. Bot.* **69**:1964–1971.

17. Edwards, C. (ed.). 1990. *Microbiology of Extreme Environments*. McGraw-Hill Publishing Co., New York.

18. Focht, D. D., and W. Verstraete. 1977. Biochemical ecology of nitrification and denitrification. *Adv. Microb. Ecol.* **1**:135–214.

19. Fogg, G. E., W. D. P. Stewart, P. Fay, and A. E. Walsby. 1973. *The Blue-Green Algae*. Academic Press, New York.

20. Garraway, M. O., and R. C. Evans. 1984. *Fungal Nutrition and Physiology*. John Wiley & Sons, New York.

21. Gerhardt, P., and S. W. Drew. 1994. Liquid culture, p. 224–247. *In* P. Gerhardt, R. G. E. Murray, W. A. Wood, and N. R. Krieg (ed.), *Methods for General and Molecular Bacteriology*. American Society for Microbiology, Washington, D.C.

22. Gherna, R., P. Pienta, and R. Cote (ed.). 1992. *American Type Culture Collection Catalogue of Bacteria and Phages*, 18th ed. American Type Culture Collection, Rockville, Md.

23. Gomez-Alarcon, R. A., C. O'Dowd, J. A. Z. Leedle, and M. P. Bryant. 1982. 1,4-Naphthoquinone and other nutrient requirements of *Succinovibrio dextrinosolvens*. *Appl. Environ. Microbiol.* **44**:346–350.

24. Good, N. E., G. D. Winget, W. Winter, T. N. Connolloy, S. Izawa, and R. M. M. Singh. 1966. Hydrogen ion buffers for biological research. *Biochemistry* **5**:467–477.

25. Gottschal, J. C., W. Harder, and R. A. Prins. 1992. Principles of enrichment, isolation, cultivation, and preservation of bacteria, p. 149–196. *In* A. Balows, H. G. Trüper, M. Dworkin, W. Harder, and K.-H. Schleifer (ed.), *The Prokaryotes*, 2nd ed., vol. I. Springer-Verlag, New York.

26. Holdeman, L. V., E. P. Cato, and W. E. C. Moore (ed.). 1977. *Anaerobe Laboratory Manual*, 4th ed. Virginia Polytechnic Institute and State University, Blacksburg.

27. Holt, J. G., and N. R. Krieg. 1994. Enrichment and isolation, p. 179–215. *In* P. Gerhardt, R. G. E. Murray, W. A. Wood, and N. R. Krieg (ed.), *Methods for General and Molecular Bacteriology*. American Society for Microbiology, Washington, D.C.

28. Jones, G. A., and M. D. Pickard. 1980. Effect of titanium(III) citrate as reducing agent on the growth of rumen bacteria. *Appl. Environ. Microbiol.* **39**:1144–1147.

29. Krieg, N. R., and P. Gerhardt. 1994. Solid, liquid/solid, and semisolid culture, p. 216–223. *In* P. Gerhardt, R. G. E. Murray, W. A. Wood, and N. R. Krieg (ed.), *Methods for General and Molecular Bacteriology*. American Society for Microbiology, Washington, D.C.

30. Kuhner, C. A., S. S. Smith, K. M. Noll, R. S. Tanner, and R. S. Wolfe. 1991. 7-Mercaptoheptanoylthreonine phosphate substitutes for heat-stable factor (mobile factor) for growth of *Methanomicrobium mobile*. *Appl. Environ. Microbiol.* **57**:2891–2895.

31. Labeda, D. P. (ed.). 1990. *Isolation of Biotechnological Organisms from Nature*. McGraw-Hill Publishing Co., New York.

32. Laverman, A. M., J. S. Blum, J. K. Schaefer, E. J. P. Phillips, D. R. Lovley, and R. S. Oremland. 1995. Growth of strain SES-3 with arsenate and other diverse electron acceptors. *Appl. Environ. Microbiol.* **61**:3556–3561.

33. Lorowitz, W. H., D. P. Nagle, Jr., and R. S. Tanner. 1992. Anaerobic oxidation of elemental metals coupled to methanogenesis by *Methanobacterium thermoautotrophicum*. *Environ. Sci. Technol.* **26**:1606–1610.

34. Lovley, D. R. 1991. Dissimilatory Fe(III) and Mn(IV) reduction. *Microbiol. Rev.* **55**:259–287.

35. Lovley, D. R., and E. J. P. Phillips. 1992. Bioremediation of uranium contamination with enzymic uranium reduction. *Environ. Sci. Technol.* **26**:2228–2234.

36. Macy, J., H. Kulla, and G. Gottschalk. 1976. H_2-dependent growth of *Escherichia coli* on L-malate. *J. Bacteriol.* **125**:423–428.

37. Mallory, L. M., B. Austin, and R. R. Colwell. 1977. Numerical taxonomy and ecology of oligotrophic bacteria isolated from the estuarine environment. *Can. J. Microbiol.* **23**:733–750.

38. Mann, N. H., and N. G. Carr (ed.). 1992. *Photosynthetic Prokaryotes*. Plenum Press, New York.

39. Moore, M. D., and S. Kaplan. 1994. Members of the family *Rhodospirillaceae* reduce heavy-metal oxyanions to maintain redox poise during photosynthetic growth. *ASM News* **60**:17–23.

40. Nolan, R. A. 1971. Amino acids and growth factors in vitamin-free casamino acids. *Mycologia* **63**:1231–1234.

41. Oremland, R., J. S. Blum, C. W. Culbertson, P. T. Visscher, L. G. Miller, P. Dowdle, and F. E. Strohmaier. 1994. Isolation, growth, and metabolism of an obligately anaerobic, selenate-respiring bacterium, strain SES-3. *Appl. Environ. Microbiol.* **60**:3011–3019.

42. Pasculle, A. W. 1992. The genus *Legionella*, p. 3281–3303. *In* A. Balows, H. G. Trüper, M. Dworkin, W. Harder, and K.-H. Schleifer (ed.), *The Prokaryotes*, 2nd ed., vol. IV. Springer-Verlag, New York.

43. Rabus, R., R. Nordhaus, W. Ludwig, and F. Widdel. 1993. Complete oxidation of toluene under strictly anoxic conditions by a new sulfate-reducing bacterium. *Appl. Environ. Microbiol.* **59**:1444–1451.

44. Reasoner, D. J., and E. E. Geldreich. 1985. A new medium for the enumeration and subculture of bacteria from potable water. *Appl. Environ. Microbiol.* **49**:1–7.

45. Schonheit, P., J. Moll, and R. K. Thauer. 1979. Nickel, cobalt and molybdenum requirement for growth of *Methanobacterium thermoautotrophicum*. *Arch. Microbiol.* **123**:105–107.

46. Tanner, R. S. 1989. Monitoring sulfate-reducing bacteria: comparison of enumeration media. *J. Microbiol. Methods* **10**:19–27.

47. Tanner, R. S., and R. S. Wolfe. 1988. Nutritional requirements of *Methanomicrobium mobile*. *Appl. Environ. Microbiol.* **54**:625–628.

48. Taylor, G. T., and S. J. Pirt. 1977. Nutrition and factors limiting the growth of a methanogenic bacterium (*Methanobacterium thermoautotrophicum*). *Arch. Microbiol.* **113**:17–22.

49. Tindall, B. J. 1992. The family *Halobacteriaceae*, p. 768–808. *In* A. Balows, H. G. Trüper, M. Dworkin, W. Harder, and K.-H. Schleifer (ed.), *The Prokaryotes*, 2nd ed. Springer-Verlag, New York.

50. Trinci, A. P. J., D. R. Davies, K. Gull, M. I. Lawrence, B. B. Nielsen, A. Rickers, and M. K. Theodorou. 1994. Anaerobic fungi in herbivorous animals. *Mycol. Res.* **98**:129–152.

51. Wachenheim, D. E., and R. B. Hespell. 1984. Inhibitory effects of titanium(III) citrate on enumeration of bacteria from rumen contents. *Appl. Environ. Microbiol.* **48**:444–445.

52. **Weber, F. J., K. C. Hage, and J. A. M. de Bont.** 1995. Growth of the fungus *Cladosporium sphaerospermum* with toluene as the sole carbon and energy source. *Appl. Environ. Microbiol.* **61:**3562–3566.

53. **Whitman, W. B., E. Ankwanda, and R. S. Wolfe.** 1982. Nutrition and carbon metabolism of *Methanococcus voltae.* *J. Bacteriol.* **149:**852–863.

54. **Widdel, F., and F. Bak.** 1992. Gram-negative mesophilic sulfate-reducing bacteria, p. 3352–3378. *In* A. Balows, H. G. Trüper, M. Dworkin, W. Harder, and K.-H. Schleifer (ed.), *The Prokaryotes,* 2nd ed., vol. IV. Springer-Verlag, New York.

55. **Wolin, E. A., M. J. Wolin, and R. S. Wolfe.** 1963. Formation of methane by bacterial extracts. *J. Biol. Chem.* **238:**2882–2886.

56. **Yanagita, T., T. Ichikawa, T. Tsuji, Y. Kamata, K. Ito, and M. Sasaki.** 1978. Two trophic groups of bacteria, oligotrophs and eutrophs: their distributions in fresh and sea water areas in the central northern Japan. *J. Gen. Appl. Microbiol.* **24:**59–88.

Cultivation of Algae and Protozoa

ALAN WARREN, JOHN G. DAY, AND SUSAN BROWN

Traditionally, algae and protozoa have been regarded as primitive plants and animals, respectively. Differentiation between the two groups is based primarily on the possession of chloroplasts for photoautotrophic nutrition in the algae and a reliance on phagotrophic or osmotrophic nutritional strategies in the protozoa. However, distinctions between the algae and the protozoa have never been entirely clear, and certain groups, such as the euglenids and dinoflagellates, have long been claimed by both phycologists and protozoologists. During the last 20 years, studies on the systematics and evolution of unicellular eukaryotes (algae, protozoa, and lower fungi) have been in a state of great of activity. Over this period, many taxonomic boundaries, including those between the algae and the protozoa, have been broken down and new relationships have been established. As a result, the constituent organisms are now grouped together as protists, reviving the term originally coined by Haekel (22), or protoctists (37). Nevertheless, although the concept of algae and protozoa no longer has evolutionary or systematic validity, these are still useful terms in a functional or ecological sense, defining photoautotrophic and heterotrophic protists, respectively.

Algae and protozoa are essentially aquatic organisms, although many examples of both groups are found in terrestrial environments; for example, algae are present as symbionts with fungi in lichens, while both algae and protozoa are common soil inhabitants. However, even in soil they are active only when there is sufficient moisture present. The ecological importance of both groups in the aquatic environment is becoming increasingly recognized. Algae, for example, are the principal primary producers in many aquatic ecosystems, contributing up to 80% of the biomass and primary productivity in open ocean waters (reference 2 and references therein). Algae are also important bioindicators of water quality and are used to assess and monitor the health of aquatic systems. On the other hand, algae cause much concern when they reach densities sufficient to form surface scums or impart off-flavors to potable water supplies. Algal blooms, particularly those of *Microcystis*, *Oscillatoria*, and other genera of potential toxin-producing cyanobacteria, are a serious problem to those charged with supplying drinking water and may significantly impair the amenity value of a water body.

Protozoa have been identified as major consumers of primary producers, both algal and bacterial, in pelagic food chains and are a vital component of the microbial loop (3). They, in turn, constitute an important food resource for metazoan plankters. Likewise, protozoa often dominate as consumers of microalgae and bacteria in sands and sediments, in organically polluted waters, and in biological aerobic wastewater treatment processes (12, 13, 18). They may also be used as indicators of organic pollution and effluent quality (13). In addition, marine protozoa, such as the ciliate *Mesodinium rubrum* and the dinoflagellate *Ptychodiscus brevis*, cause red tides in many parts of the world, with their attendant toxicological problems.

In all of these cases, accurate identification of the organisms involved is essential, and this usually relies on an ability to culture the organisms in the laboratory. Stable, pure cultures are the bedrock of many aspects of fundamental and applied research, as well as the commercial exploitation of algae and protozoa. In addition, the ex situ maintenance of organisms is essential for the preservation of microbial diversity. Conservation of organisms, including algae and protozoa, has been adopted as the basis of international agreements, including article 9 of the Convention on Biological Diversity (Rio Convention, 1992) and Microbiological Diversity 21, as well as a range of other international initiatives and programs (23). Therefore, successful cultivation is an important first step in many types of study of algae and protozoa.

IDENTIFICATION

Algae

Algae are generally assumed to be photoautotrophs, and in most cases cultures will grow in the absence of organic carbon. However, many species are capable of utilizing organic substrates (17, 24, 43), and some require trace quantities of vitamins or other organic molecules for normal growth (60). Achlorophyllous mutants or colorless organisms closely related to pigmented algae require an organic carbon source. This is generally provided in culture media by yeast extract, acetate, glucose, soil extract, or other undefined medium components (see Protozoa, below). Media and culture conditions may be designed to suit individual strains, although in general, standard media which are satisfactory for a range of organisms are used; these are discussed in greater detail below.

Protozoa

Protozoa are generally assumed to be heterotrophic, although some may be autotrophic or both (mixotrophic). They utilize a wide range of sources of carbon and other nutrients. Many are obligately parasitic, endocommensal, or epibiotic. These types fall outside the scope of this review, and some are dealt with elsewhere (chapter 16). Information on their cultivation may also be found elsewhere in the literature (references 34 and 37 and references therein). Free-living protozoa may obtain carbon and other nutrients in either dissolved or particulate form. In most cases, the nutrient source is external to the cell and often takes the form of other (prey) organisms. These may be eukaryotes, prokaryotes, or viruses (57). Alternatively, protozoa may feed on detritus and other particulate or dissolved organic matter in the environment. Since most protozoa are selective feeders, cultures must seek to provide appropriate food. Some protozoa have plastids (e.g., dinoflagellates) or harbor endosymbionts (e.g., *Zoochlorella* species in *Paramecium bursaria*) which may provide internal sources of nutrients. In these cases, the culture methods used for algae (see below) might be more appropriate for their maintenance in the laboratory.

Traditionally, free-living protozoa have been divided into three main groups according to their morphology: flagellates, amoebae, and ciliates. Of these, only the ciliates are a truly natural (monophyletic) group; the flagellates and amoebae are both polyphyletic and include groups that may be only distantly related. Nevertheless, from a practical viewpoint it is still sometimes useful to refer to these groupings since the isolation and culture methods used are often the same within each group.

To maintain cultures of protozoa long term, it is necessary to provide a medium that suits each species and a supply of appropriate food. Various publications provide comprehensive information or refer to medium preparations for protozoa (34, 37, 44, 59). However, certain isolation techniques, growth media, and culture conditions suit a wide range of organisms. Some of these are discussed below.

ISOLATION METHODS

Isolation and purification of algae and protozoa are broadly similar. The standard objective is to produce a representative uniprotistan culture, preferably clonal, derived from a single protozoan or algal unit: cell, filament, trichome, or colony. For some applications, use of axenic strains is preferable, in which case all contaminating organisms must be removed and the culture must be maintained in an axenic state. However, for many protozoa, a more realistic goal is to produce a monoxenic culture in which only one other species, usually the food organism, is present in addition to the one that has been purposefully isolated.

The methods employed are comparable to those used for other microorganisms and can be broadly classified as enrichment methods, dilution methods, or physical and chemical methods. In all cases, good microbiological practices and aseptic techniques should be employed. Furthermore, where possible, all initial manipulations and transfers should be performed in media similar in pH and osmotic potential to the site of isolation. It is also optimal to ensure that suitable temperature, light, and oxygen tension regimes are maintained during transportation and during initial manipulations of field samples. It should be noted, however, that some protozoa fail to grow in fresh culture media even when precautions such as these are taken. In these cases, it may be necessary to use a conditioned culture medium, i.e., one that has been previously used for culturing other protozoa and that has been conditioned by the exudation of substances by these protozoa. Likewise, for subculturing of certain protozoa, greater success may be attained if a significant amount of spent medium, which may contain key nutrients and/or growth factors, is transferred to the new culture.

Enrichment

Enrichment is the inoculation of a field sample into an equal or greater volume of suitable medium and incubation under favorable conditions. By inoculating parallel cultures in a range of media, different organisms will be selected. Transfer to mineral medium with no available nitrogen (e.g., BG 11 without NaNO₃ [see Table 2]) will select for nitrogen-fixing cyanobacteria. For bacterivorous protozoa, the simplest way to enrich a sample is to add boiled grains of barley, wheat, or rice, which will promote the growth of bacteria and thereby produce a food source for the protozoa. Enrichment does not guarantee the production of a uniprotistan culture but may be useful in increasing the numbers of cells of the desired organism for further isolation.

Dilution

Dilution methods are most effective for use on preponderantly uniprotistan samples. Material is sequentially diluted in appropriate medium and incubated under favorable conditions. The greatest dilution in which growth occurs is likely to be uniprotistan. The chances of success with this technique are obviously increased if the species in question is one of the most abundant, since unwanted organisms will be diluted out more quickly. For very small or abundant protists, a dipping-wire method in which the number of cells transferred at each dilution step is significantly reduced has been developed (10). Other variations of serial dilution techniques have been described elsewhere (56). Dilution methods cannot guarantee that the cultures obtained are clonal, and it is most unlikely that an axenic strain will be produced.

Physical Methods

Physical methods involve the selection of individual protistan units and their transfer to appropriate sterilized medium and environmental conditions. This may result in an axenic and clonal culture. Pipetting using thin capillary pipettes, working under a dissecting microscope, can be used for a wide range of organisms, particularly those which are relatively large and/or slowly moving. This method, however, requires dexterity and patience and is not applicable to small protists. Other methods, such as those described below, should be tried first. The additional sequential transfers of an isolated cell through a series of washes may result in the generation of an axenic culture. An alternative that may be successfully employed is the use of a micromanipulator (58), while for small flagellates, Cowling (9) describes an agar cavity-channel isolation technique which may be used in combination with such micromanipulatory methods.

Other methods by which protists may be physically isolated include selective filtration, silicone oil plating, density gradient centrifugation, and flow cytometry. The silicone oil plating technique may be used for a wide range of protists and relies on the isolation of clone-founding cells within microdroplets formed from vortex-mixed oil-culture emulsions (53). Density gradient centrifugation (5, 21, 62) and flow cytometry (6, 11) have good potential as automated means of discriminatory cell sorting and strain isolation on the basis of cell size, density, or other cell attributes.

Many anaerobic protists are sensitive to oxygen and must be isolated and maintained in an oxygen-free environment. Cowling (9) devised a microscope slide ring chamber which permits observation of anaerobic protists and their withdrawal by micropipetting without compromising the oxygen-free conditions within the chamber. A somewhat more complicated apparatus is described elsewhere (64).

Agar Plating

Agar plating methods rely on discerning colony growth of isolated clones on agar surfaces or within agar tubes. These techniques have been used mostly for algae, amoebae, and some flagellates. Certain algae may be isolated by streaking a drop of algal culture onto the surface of an agar plate, using a sterile loop or capillary pipette; discrete colonies may arise from single cells or groups of cells. After an appropriate incubation interval (generally in the range of 7 to 21 days), cells should be aseptically transferred from a colony and resuspended in liquid medium, after which the process is repeated. A variation of this technique that may be employed for motile strains involves the use of pour plates (i.e., the addition of molten agar to aliquots of a dilution series of the algal suspension). The resulting colonies are excised and transferred to fresh medium, and the process is repeated. This technique can give rise to clonal, axenic isolates of many algal species. Spray-plating techniques, which achieve separation of individual cells by directing a fine spray of cell suspension onto an agar surface, have also been used successfully (63).

Amoebae are commonly isolated by placing one or two drops of sample onto a nonnutrient agar plate that has been streaked with a suitable food organism and incubating the plate (see Amoebae, below). Sufficient room (>1 cm) should be allowed to ensure that the amoebae migrate away fom their contaminants. As the amoebae exhaust the local food supply, they migrate over the agar surface, thereby isolating themselves from other microorganisms in the sample.

Isolated amoebae may then be picked off with a sterile scalpel blade and subcultured.

In many cases, the procedures described above are used in conjunction with other physical and chemical techniques, some of which are listed below (Table 1).

These methods are discussed in detail elsewhere; for additional information, see references 4, 20, 34, 47, 51, and 56.

MAINTENANCE METHODS
Algae

The choices of medium and maintenance regime are dependent on the growth requirements for the algal isolate(s) as well as the resources which are available locally. The trend in larger collections is to minimize the number of media and growth regimes utilized and, in addition, to maximize the subculture interval, without adversely effecting culture quality. To this end, many planktonic strains commonly isolated from aquatic environments are maintained on solidified medium.

Media

Although standardization of methods and medium has been achieved for some purposes, particularly ecotoxicity testing (45), in general, medium composition depends both on the requirements of the algae (e.g., diatoms require the inclusion of a silica source) and the preferences of the researcher. Therefore, a wide variety of media have been developed and are in common use; details of medium composition and their suitability for particular algae are available in specialist literature (4, 41, 44, 55, 59, 61). Many isolates may be successfully maintained on or in relatively broad spectrum media. Some of the most commonly used are detailed below. In general, isolates will be most successfully cultured and manipulated in liquid medium. However, for some applications (and in order to increase the time between subcultures), solidified medium may be more appropriate. Solidified medium is produced by the addition of 15 g of bacteriological

TABLE 1 Additional techniques used to isolate and purify algae and protozoa

Process	Rationale and outline of method
Use of antibiotics	To kill contaminating bacteria and cyanobacteria
Use of nonspecific bactericides	To kill contaminants. Methods include 2% (vol/vol) formaldehyde for 2 min or 1% (vol/vol) sodium lauryl sulfate for 12 h (for the purification of cyst- or aplanospore-forming algae). Sequential sodium hypochlorite washes (0.5, 1, 2, 5, and 10% [vol/vol]) for 5 min each.
Use of germanium dioxide	Incubation in medium containing 5 to 10 mg of germanium dioxide per liter kills diatoms.
UV irradiation	Low doses will kill bacterial contaminants. Irradiation may also damage or kill the algal and protozoan cells and induce mutations.
Fragmentation and ultrasonication	Disruption of clumps of cells and dislodging of adhering contaminants to produce a uniform dispersion
Low-speed centrifugation	Centrifugation at 100 to 200 × *g* for 10 min should not affect gas vacuolate cyanobacteria; most contaminants are removed by centrifugation.
Migration response	Separation from contaminants by unidirectional migration may be greatly enhanced by strong phototactic, chemotactic, or geotropic responses.
High incubation temperature	Selection of thermophiles

TABLE 2 BG 11 (blue-green algal [cyanobacterial] medium)[a]

Stock solution	Amt (g)/liter
1. NaNO$_3$	15.0[b]
2. K$_2$HPO$_4$·3H$_2$O	4.0
3. MgSO$_4$·7H$_2$O	7.5
4. CaCl$_2$·2H$_2$O	3.6
5. Citric acid	0.6
6. Ferric ammonium citrate	0.6
7. EDTA (disodium salt)	0.1
8. Na$_2$CO$_3$	2.0
9. Trace metal mixture	
H$_3$BO$_3$	2.86
MnCl$_2$·4H$_2$O	1.81
ZnSO$_4$·7H$_2$O	0.222
Na$_2$MoO$_4$·2H$_2$O	0.39
CuSO$_4$·5H$_2$O	0.079
Co(NO$_3$)$_2$·6H$_2$O	0.0494

[a] Add 100 ml of solution 1, 10 ml each of solution 2 to 8, and 1 ml of solution 9 to distilled or deionized water to obtain a total volume of 1 liter. pH is adjusted to 7.8 prior to sterilization. Based on reference 54.
[b] May be omitted for nitrogen-fixing cyanobacteria.

agar per liter to standard medium formulations before the medium is dispensed into tubes, universal bottles, or other appropriate vessels. After sterilization, vessels are placed at an acute angle to maximize the surface area of the agar upon gelation.

BG 11 (Table 2), a medium developed for culturing cyanobacteria (which generally prefer alkaline conditions), is also suitable for a wide range of eukaryotic microalgae.

Jaworski's medium (Table 3) is used for a wide range of freshwater and terrestrial eukaryotic algae (59). It may also be used to maintain diatoms upon addition of 57 mg of NaSiO$_3$·9H$_2$O per liter. In addition, Jaworski's medium supplemented with organic material is the medium most commonly used at the Culture Collection of Algae and Protozoa to maintain axenic algal strains (59). The additional organic material in the form of sodium acetate and complex nitrogen sources increases the potential yield of cultures which can grow mixotrophically or photoheterotrophically. Also, any

TABLE 3 Jaworski's medium[a]

Stock solution	Amt (g)/200 ml
1. Ca(NO$_3$)$_2$·4H$_2$)	4.0
2. KH$_2$PO$_4$	2.48
3. MgSO$_4$·7H$_2$O	10.0
4. NaHCO$_3$	3.18
5. FeNaEDTA	0.45
Na$_2$EDTA	0.45
6. H$_3$BO$_3$	0.496
MnCl$_2$·4H$_2$O	0.278
(NH$_4$)$_6$Mo$_7$O$_{24}$·4H$_2$O	0.2
7. Cyanocobalamin (vitamin B$_{12}$)	0.008
Thiamine HCl (vitamin B$_1$)	0.008
Biotin	0.008
8. NaNO$_3$	16.0
9. NaHPO$_4$·12H$_2$O	7.2

[a] Add 1 ml each of solutions 1 to 9 to distilled or deionized water to obtain a total volume of 1 liter of medium. Reprinted from reference 15a with permission. © 1995 Springer-Verlag GmbH.

TABLE 4 Guillard's f/2 medium[a]

Stock solution	Amt/liter
1. NaNO$_3$	7.5 g
2. NaH$_2$PO$_4$·2H$_2$O	5.65 g
3. Na$_2$EDTA	4.36 g
FeCl$_3$·6H$_2$O	3.15 g
CuSO$_4$·5H$_2$O	0.01 g
ZnSO$_4$·7H$_2$O	0.022 g
CoCl$_2$·6H$_2$O	0.01 g
MnCl$_2$·4H$_2$O	0.18 g
Na$_2$MoO$_4$·2H$_2$O	0.006 g
4. Cyanocobalamin (vitamin B$_{12}$)	0.5 mg
Thiamine HCl (vitamin B$_1$)	100.0 mg
Biotin	0.5 mg
5. Na$_2$SiO$_3$·9H$_2$O	40.18

[a] Add 10 ml of solution 1 and 1 ml each of solutions 2 to 4 to filtered natural seawater to obtain a total volume of 1 liter. Adjust pH to 8.0 prior to sterilization. If the medium is to be used for diatoms, 1 ml of solution 5 is added to the medium prior to pH adjustment and sterilization. Reprinted from reference 15a with permission. © 1995 Springer-Verlag GmbH.

bacterial contaminants in previously axenic cultures rapidly become obvious, as the medium will appear cloudy within 2 days.

Guillard's f/2 medium (Table 4) is used for a wide spectrum of marine microalgae (59). It may be used for marine diatoms upon addition of 40 mg of Na$_2$SiO$_3$·9H$_2$O per liter. Like Jaworski's medium, it is suitable for maintaning axenic cultures. The addition of 1.0 g of yeast extract and 1.0 g of glucose or sodium acetate per liter increases the potential yield of cultures which can grow mixotrophically or photoheterotrophically.

Once an algal culture is established, it requires periodic subculturing by the aseptic transfer of an inoculum (generally 0.5 to 10% of the original culture) to sterile medium. The duration of intervals between subculturing depends on the growth characteristics of the isolate as well as the medium and culture conditions used. The duration of intervals generally ranges between 2 weeks and 6 months. Use of senescent material as an inoculum is avoided when possible.

Maintenance Conditions

Optimal temperatures depend on the environment from which the cultures were isolated. Most commonly cultured strains can be maintained at 15 to 20°C, although some organisms from extreme environments, both hot and cold, cannot survive within this temperature range. These cultures should be maintained at as close to the temperature of their original habitat as possible. Care should be taken to avoid exceeding the temperature maximum, and it is advisable to use illuminated incubators with good temperature control (e.g., ± 2°C). Light regimes, in many cases, depend on the facilities available. Usually cool white fluorescent tubes are appropriate with a light-dark regime of 16 h-8 h. A wide range of light-dark periods may be used with reasonable success. Alternatively, natural light from a north-facing window is suitable. In general, light levels of approximately 50 μmol of photon m^{-2} s^{-1} are satisfactory, although some organisms, particularly cyanobacteria, will grow better at light levels of <25 μmol of photon m^{-2} s^{-1}.

Protozoa

For flagellates and ciliates, the isolation techniques, growth media, and culture conditions are often identical or similar.

Therefore, these two groups will be dealt with together, while the amoebae will be discussed separately.

Flagellates and Ciliates

The choice of culture medium will depend largely on what the protozoan feeds upon. Many flagellates and ciliates feed upon bacteria, and in these cases nonselective media, designed to encourage the growth of bacterial populations, may be used. For many species, isolates may be cultured in the presence of a mixed bacterial flora which coexisted with the organism in its original habitat. Alternatively, selective cultures may be obtained by incubating the protozoa in an inorganic salt solution along with an appropriate food organism. For bacterivorous forms, common, nonpathogenic laboratory cultures of bacteria may be used. In some cases, however, it may be necessary to isolate, characterize, and identify bacteria from the original sample and then use one or more of these bacteria as selected food organisms in an axenified culture.

For omnivores and carnivores, an examination of the contents of the food vacuoles of the isolate may give an indication of its preferred food. Attempts should then be made to feed the isolate on something similar in shape, size, and chemical composition to its normal prey. If all else fails, it may be necessary to carry out a replicated feeding experiment using a range of food organisms in order to determine which, if any, will support the growth of the isolate.

For some species, axenic cultures may be established. In these cases, the culture medium, which may be defined or semidefined, is usually rich in organics.

Culture media may be categorized into four main types: plant infusions, soil extract-based media, inorganic salt solutions, and specific (organic-rich) media.

Plant Infusions (Table 5)

Plant infusions have long been used for the culture of bacterivorous flagellates (e.g., *Bodo* species) and ciliates (e.g., *Colpidium* and *Paramecium* species). The principle is that organic compounds leach out of, or are extracted from, plant material, and these compounds support bacterial growth. The simplest to use are polished rice, boiled wheat, or barley grains. Likewise hay or lettuce infusions have been used successfully for many years. Dehydrated, powdered cereal leaf preparations (Sigma product C7141) are commercially available. These may be prepared as 0.1 to 0.25% (wt/vol) infusions either in an inorganic salt solution or in distilled water. Care should be taken not to add amounts of infusion larger than the recommended 0.1 to 0.25%, since this could lead to excessive growth of bacteria, which will inhibit protozoa. Plant infusions may also be used for culturing marine flagellates and ciliates. Here, alternative sources of enrichment may be used, such as the seaweeds *Ulva* and *Enteromorpha* species, which should be autoclaved before use. For marine species, the infusions are made in either natural or artificial seawater (59).

Soil Extract-Based Media

As with plant infusions, the underlying principle is to extract from the soil organic compounds that will support the growth of bacterial food organisms. There are several ways of doing this, one of which is as follows. Sterilize (autoclave for 1 h at 15 lb/in^2) 1 part air-dried, sieved soil—preferably a sandy loam at pH 7.0—with 2 parts distilled water to produce a supernatant extract stock. Allow the contents to settle for at least 1 week before use. The medium can be used with the soil remaining as a so-called biphasic (soil-water) culture medium. A biphasic tube culture method (51) has been used as an effective and convenient means of maintaining many flagellate and ciliate strains in long-term batch culture. Alternatively, the supernatant may be decanted and filtered for use as a liquid culture medium, either directly or diluted and with inorganic salts added. The addition of calcium carbonate or cereal grains to culture dishes or tubes of soil extract media may promote faster

---

TABLE 5 Culture media used for maintenance of nonmarine flagellates and ciliates[a]

Type of medium	Preparation
Plant infusions	
Hay	1–10 g of timothy hay or similar. Boil and filter.
Lettuce	1–2 g of dried lettuce leaves. Boil and filter.
Cereal grains	Boil barley or wheat grains (5 min). Add 1 grain per 10 ml of medium. Further sterilization is not required.
Cereal leaves	Boil 0.1–0.2 g of dehydrated cereal leaves (Sigma C7141) (10 min). Filter. Adjust pH to 7.0–7.2.
Soil extract medium with added salts	Soil extract stock[b], 100 ml; K_2HPO_4, 20 mg; $MgSO_4 \cdot 7H_2O$, 20 mg; KNO_3, 200 mg
Inorganic salt solutions (Prescott's and James's)	$CaCl_2 \cdot 2H_2O$, 43.3 μg; KCl, 16.2 μg; K_2HPO_4, 51.2 μg; $MgSO_4 \cdot 7H_2O$, 28.0 μg
Specific media	
Proteose peptone yeast extract	Proteose peptone (Oxoid L85), 10 g; yeast extract (Oxoid L21), 2.5 g. Boil until dissolved.
Euglena medium	Sodium acetate (trihydrate), 1 g; Lab Lemco (beef extract) powder (Oxoid L29), 1 g; tryptone (Oxoid L42), 2 g; yeast extract (Oxoid L21), 2 g; $CaCl_2$, 10 mg

[a] Components are given for 1 liter (final volume) of medium made with glass-distilled water unless otherwise indicated. All media are sterilized by autoclaving for 1 h at 15 lb/in^2 unless otherwise indicated. Based mainly on references 9 and 20.
[b] See text for method of preparing soil extract stock.

bacterial growth and improve culture longevity of sapro-phytic flagellates and ciliates. As with plant infusions, care must be taken not to allow excessive growth of bacteria, including actinomycetes, which might inhibit protozoa. Soil extracts are therefore normally diluted to 3 to 10% (vol/vol).

Inorganic Salt Solutions

Inorganic salt solutions provide a balanced medium, the ionic composition of which is suitable for the growth of many protozoa. Such solutions, however, contain negligible quantities of dissolved organic matter; therefore, the addition of food organisms, and/or a carbon source for the growth of food organisms, is essential. Carnivorous ciliates (e.g., *Didinium* species) generally grow well in inorganic salt solutions if they have sufficient prey (e.g., *Paramecium* species). Inorganic salt solutions may either be natural or artificial. Water can be regarded as a weak inorganic salt solution. For some species, filtered water from the sample site is the best medium. Most freshwater and soil species will also grow in commercially available, noncarbonated bottled mineral waters (e.g., Volvic [Perrier UK Ltd.]) Artificial inorganic salt solutions have the advantage that their chemical composition is known precisely. Perhaps the most commonly used is Prescott's and James's solution (Table 5).

Marine protozoa generally grow best in dilute media, and for many species the most convenient medium is seawater itself, filtered and autoclaved. However, defined marine media such as artificial seawater medium, sold commercially as Ultramarine Synthetic Salt Solution (Waterlife Research Industries), and supplemented seawater medium may also be used for several flagellate and ciliates species. Formulations for both are given elsewhere (59).

Specific Media

For certain applications, the ultimate objective is to produce an axenic culture. Many protozoa, including the flagellates *Astasia*, *Euglena*, and *Chilomonas* species and the ciliates *Tetrahymena* and *Paramecium* species, have been grown axenically. To achieve axenic growth, a defined or semidefined medium must be used. Such media invariably contain relatively high concentrations of dissolved organic carbon so that the protozoa can feed osmotrophically. The ingredients of such media are usually derived from animal sources (powdered whole or defatted liver, proteose peptone, beef extract, etc.), although yeast extracts are often added as nutritional supplements. Commonly used formulations include proteose peptone yeast extract medium (for *Tetrahymena* species) and *Euglena* medium (for acetate flagellates, e.g., euglenids and cryptomonads). Formulations for these media are given in Table 5. Gradual release of nutrients for prolonged axenic cultivation of tetrahymenids and phagotrophic chrysomonads has been achieved by using large screw-cap tubes containing powdered liver (0.4 to 0.8% [wt/vol]) overlaid with water (29). An alternative inexpensive method to achieve long-term maintenance of *Tetrahymena* species is to use as the culture medium a garbanzo bean autoclaved in a test tube of water.

Maintenance Conditions

The main culture variables to be considered are temperature, pH, oxygen tension, food type and concentration, culture medium type and concentration, culture vessel type and capacity, and inoculum age, size, and density (9). As a general rule, conditions should be maintained as close as possible to those from which the organism was isolated, al-though most species will grow more quickly at slightly elevated temperatures (18 to 25°C is optimal for many) and in the presence of a plentiful food supply. One of the greatest perils is overgrowth by bacteria, which can normally be avoided by diluting the culture medium or by adding a suspension of nongrowing bacteria, generally heat treated, to nonnutrient medium. On the other hand, maintenance of strains at low or suboptimal temperatures can save labor and time. Sterile plastic tubes, flasks with screw caps, or petri dishes are all suitable for maintaining cultures of flagellates and ciliates.

Some protozoa are algivorous, while others are omnivores and feed on algae facultatively. In order to maintain an adequate food supply, cultures of these types should be kept in conditions of illumination as described in the section on maintenance of algae. By contrast, other protozoan cultures may best be incubated in the dark in order to control algal contaminants.

Many species of protozoa are known to produce cysts. These may be resting cysts, which are typified by their highly resistant cyst walls, or reproductive cysts, the walls of which are usually thinner and more permeable. The factors which cause encystment vary in different species, and the literature contains many descriptive accounts of this process (for reviews, see references 8 and 52). Usually these factors are adverse environmental features such as shortage or excess of food, extreme pH, increase in salt concentration, high temperatures, overcrowding, lack of oxygen, accumulation of metabolic by-products following rapid growth, or desiccation. Interestingly, there are far fewer accounts of excystation. Like encystation, excystation has been ascribed to a variety of factors, including osmotic phenomena, high food and oxygen concentrations, enzymatic dissolution of cystic membranes, and presence of protein degradation products (8). Often, simply transferring cysts to a fresh culture medium will stimulate excystation.

The techniques described above were designed principally for aerobic protozoa. Anaerobic protozoa require special techniques, since one of the most important considerations is the exclusion of oxygen from the culture. Cultures of anaerobic flagellates and ciliates have been routinely maintained by employing a modification of the Hungate technique. This modification uses pH-adjusted, degassed (N_2/CO_2, 95%:5%) cereal infusion or soil extract medium in serum bottles. The headspace is replaced with nitrogen, and the bottles are sealed with butyl rubber bungs held in place by crimped aluminum caps (40). With this technique, a species of the ciliate *Trimyema* has been maintained in culture for 4 years (19).

Amoebae

Amoebae may be cultured in liquid media, on agar, or in biphasic media. Information concerning the maintenance of individual species and genera is available in the literature (28, 44, 48–50, 59). In general, the larger amoebae (e.g., those belonging to the family Amoebidae) are grown in liquid culture, and smaller amoebae (e.g., those belonging to the family Vahlkampfiidae) are grown on agar. Freshwater and soil amoebae may persist for some time in dilute salt solutions, mineral water, or even distilled water. Marine amoebae may persist in seawater diluted to 75% strength or less. These inorganic liquids form the basis of all culture media suitable for amoebae. Growth and cell division require additional nutrients either included in the media or from prey organisms, selected according to the needs of the strain. Tables 6 and 7 provide a guide to combinations of

TABLE 6 Culture media for freshwater and soil amoebae

Representative family or genus	Culture medium	Food organisms
Vahlkampfiidae, Thecamoebidae, Paramoebidae, Hartmannellidae, Vexilliferidae, Vannellidae, Leptomyxidae, Acanthamoebidae, Echinamoebidae, Flabellulidae, Cochliopodiidae, *Stachyamoeba, Deuteramoeba, Dactylamoeba, Rosculus, Cryptodifflugia, Nuclearia, Dictyostelium*	Agar (e.g., nonnutrient amoeba saline agar[a] or Sigma cereal leaf-Prescott agar[a])	Bacteria (e.g., *E. coli* or *K. aerogenes*)
Acrasidae, Cavosteliidae, Protosteliidae	Agar (e.g., cornmeal glucose agar[a] or hay infusion agar[b])	Bacteria (e.g., *E. coli*) or yeasts (e.g., *R. mucilaginosa*)
Axenic–Hartmannelliidae, Vahlkampfiidae, Acanthamoebidae	Organic liquid medium (e.g., proteose peptone glucose[a] or PYNFH medium[c])	None
Euglyphidae, *Mayorella*	Biphasic (e.g., Sigma cereal leaf-Prescott liquid and Sigma cereal leaf-Prescott agar[a])	Bacteria in coculture

[a] See reference 59.
[b] See reference 46.
[c] See reference 33.

culture media and food organisms appropriate for the culture of a wide diversity of freshwater and marine amoebae, the majority of which are classified in the phylum Rhizopoda (for classification, see references 48 to 50).

Organic nutrients can be included in the medium itself either by the inclusion of nutrients such as glucose, serum, proteose peptone, and yeast extract or by using plant infusions (see above). However, apart from axenically cultured amoebae, for which dissolved nutrients are the sole source of food, the principal reason for increasing the organic content of a medium is to support the growth in the culture of bacteria on which amoebae can feed. Bacterial growth can also be stimulated by the addition of one or two unpolished rice grains which have been surface sterilized by rapid passage through a flame. Care must be taken not to make a medium too rich, or the bacteria may overgrow the amoebae. Amoebae may be feeding on bacteria which were isolated with them from their original habitat, but if suitable bacteria are not present in coculture with the amoebae, species such as *Escherichia coli* or *Klebsiella aerogenes* can be cultured separately and added directly to the culture vessel. Yeasts (e.g., *Rhodotorula mucilaginosa*) may be suitable for slime molds.

Although bacteria are an appropriate food source for many amoebae (particularly smaller amoebae), larger amoebae can also feed on other protists. By using an inorganic liquid medium with rice grains, it is possible to maintain cultures in which an adequate growth of bacteria supports a population of small protists (e.g., *Colpidium* sp.) which are in turn preyed upon by amoebae. Alternatively, the rice grains (and bacteria) can be omitted and the protists can be added directly at regular intervals (2 to 7 days). In this case, the food organism is cultured separately in a rich organic medium but is transferred, after centrifugation, into an inorganic medium before being added to the amoeba culture. It is important not to overfeed large amoebae. Some amoebae will feed on algae, including cyanobacteria. Again, the food organism is cultured separately and added to the amoeba culture containing an inorganic liquid medium.

Maintenance Conditions

The main culture variables are temperature, light, and moisture. Cultures of most free-living, nonpathogenic amoebae are best maintained at room temperature, about 18 to 22°C. Although some species will grow well at temperatures up to and above 30°C, temperatures above approximately 24°C may be lethal. Potential pathogens belonging to the genera *Naegleria* and *Acanthamoeba* are cultured at temperatures in the range of 37 to 45°C. Many small freshwater and soil amoebae form cysts; these amoebae can be stored at 7°C. Cultures should never be placed in direct sunlight, and most amoebae are best maintained in the dark. Amoebae which require a light-dark cycle include those

TABLE 7 Culture media for marine amoebae

Representative family or genus	Culture medium	Food organisms
Vahlkampfiidae, Hartmannellidae, Flabellulidae, Paramoebidae, Thecamoebidae, Leptomyxidae, *Stygamoeba, Nolandella*	Agar (e.g., malt and yeast extract–75% seawater agar[a])	Bacteria in coculture or (e.g.) *E. coli*
Gruberella, Korotnevella, Parvamoeba, Mayorella, Trichosphaerium	Liquid medium (e.g., 75% seawater, Sigma cereal leaf–75% seawater,[a] modified Føyns Erdschreiber medium,[a] or biphasic)	Bacteria in coculture

[a] See reference 59.

which feed on algae, those with algal endosymbionts, and the acrasids. To prevent a detrimental increase in the concentration of solutes during incubation of a culture and/or the drying out of agar, the junction between the lid and base of the culture vessel can be sealed with a narrow strip of polyethylene film. This also reduces the risk of contamination.

Subculturing Methodology

The frequency with which a strain requires transfer to fresh medium depends on the species and the culture method. Subculturing is preceded by microscopic examination to ensure that cultures contain many viable cells and have not become contaminated. The risk of contamination can be minimized by subculturing under aseptic conditions, using standard microbiological techniques. Information concerning the safe handling of potential pathogens can be found in reference 1.

Agar cultures. Amoebae can be transferred to a new plate on an agar block cut from a dense parent culture. A suitable area is located microscopically, and its location is marked on the inverted base of the parent plate. A small block (about 9 mm^2) is excised with a scalpel and placed amoeba-side down near the edge of the fresh agar. If bacteria or yeasts are required as food organisms, they are streaked across the agar first and the block is placed at one end of the streak. Amoebae from less dense cultures can be transferred by washing the cells from the parent plate with an inorganic liquid by using a Pasteur pipette and distributing the suspension over a fresh agar surface, previously spread with bacteria or yeasts, if necessary. Each time a strain is subcultured, two to three new cultures should be created. Amoeba strains vary widely in how rapidly they multiply on an agar surface; subculturing intervals range from 7 days to 6 weeks. Cyst-forming amoebae are subcultured onto a slope of agar (with a bacterial streak) in a universal bottle and are incubated at room temperature for 2 weeks before being transferred to 7°C. At this temperature, they may remain viable for 6 to 12 months. Excystment is induced by transferring cysts to an agar plate streaked with bacteria, either by washing (see above) or on a small block of agar. The new culture is incubated at room temperature. Some cysts may remain viable for considerably longer; however, the proportion of the cysts capable of excystment and the probability of generating an active culture will decrease over time.

Liquid cultures. Suitable culture vessels include dishes with lids and tissue culture flasks made of glass or plastic. To ensure adequate oxygenation, the depth of liquid should not exceed 10 to 12 mm. Axenic cultures can be maintained in tubes. Subculturing intervals range from 2 to 4 weeks, and three to four new cultures should be generated each time. Strains in liquid culture can be subcultured and transferred either by pouring or by pipette. Amoebae readily attach to plastic and glass surfaces but can be resuspended by agitating the media immediately prior to transfer, either with a pipette or by swirling the culture vessel. In most cases, between half and the total volume of parent culture is transferred to the new culture vessel containing fresh medium, and then mixed culture is transferred back to the parent culture to equalize the volumes. To subculture axenic strains, only 5 to 10% (vol/vol) of culture is transferred aseptically to fresh medium.

QUALITY CONTROL

In most cases, assessment of quality control is based on cell morphology and culture appearance. Obvious growth is the first factor noted. The gross morphology of the culture and the individual cells are observed microscopically. If no major differences between the parent culture and the subculture are observed, then quality is generally assumed to be satisfactory. The presence of any contaminating organism in previously axenic cultures is checked by phase-contrast microscopy. In addition, a small aliquot of culture (0.1 to 0.5 ml) is spread onto rich organic medium such as nutrient or tryptone soya agar. These plates are incubated at 20 to 25°C for 48 to 72 h and then examined, by eye, for the presence of any bacterial or fungal growth.

Biochemical and molecular methods could easily be adopted for quality control. A standard "fingerprint" of information on chemical composition, biochemical characteristics (e.g., isoenzyme analysis), or genetic sequences could be maintained on a database. These data would be compared with those for the maintained organism or new isolates. Many of the tests that could be used lend themselves to automation, and such a system is certainly technically feasible, at a cost. A number of problems currently prevent the adoption of this type of system. First, for many of the above-mentioned tests, strains should be axenic, which is often difficult to achieve and requires a great deal of time. In reality, the vast majority of algal and protozoan cultures either have one protist present or are dominated by one organism. Second, the stability of the measured characteristic should be known, and it should not be affected by the maintenance regime; this parameter would have to be established for each strain. In addition, the cost of developing and running such a quality control system makes it unlikely to be adopted other than in specialist units or major collections.

Documentation should be maintained on each isolate. This should include name to species level and authority if appropriate; any strain number or culture collection code or number; whether it is a type culture; origin; isolator's name and date of isolation; medium and maintenance regime; preservation methods; any unique or representative biochemical or other characteristics; and any applicable regulatory conditions (e.g., quarantine, containment levels, and patent status). In addition, a maintenance schedule giving details on culture conditions, dates of subculture, and status of culture at each transfer should be recorded. This documentation can most easily be maintained as a computerized database, although hard copy in the form of card indices, catalogs, etc., can be used as an alternative.

LIMITATIONS TO PROCEDURES

Although serial subculture is the standard method for maintaining laboratory algal and protozoan strains, it is widely recognized that this technique is suboptimal. By its nature it is selective, as only those strains which can be cultured ex situ can be maintained. Some organisms obviously change when cultured on artificial medium or under axenic conditions. This is most commonly observed in cyanobacterial cultures, where loss of ability to produce colonies, heterocysts, gas vacuoles, or akinetes can occur. In addition, changes in pigmentation and loss of toxicity have been observed. Eukaryotic microalgae may also change morphology in culture. These changes include loss of ability to produce spines in *Micractinium* species and apparently irreversible shrinkage in many diatom cultures. Cyst-forming protozoa have been observed to lose the ability to form viable cysts, and vahlkampfiid amoebae have been found to lose the ability to form flagellates. In some of these examples, the changes are apparently irreversible; in others, the lost attribute can be regained by changing the medium or other environmental factors. There is as yet little evidence that there are any major shifts in the genetic complement of cultures which have been maintained under laboratory conditions for prolonged periods. Also, most commonly observed or measured characteristics, in the vast majority of organisms retained in microbial collections, appear to be stable upon serial subculture. The alternative to serially subculturing laboratory strains, which is the reisolation and purification of specific algal or protozoan strains each time they are required for a laboratory-based study, is obviously not possible. However, the use of long-term preservation methods has the potential to prevent any major changes in phenotype or genotype of cultured specimens.

In addition to the foregoing limitations, there are the dangers of mislabeling, contamination of specimens, and other handling problems (e.g., the transfer of an inoculum to an inappropriate medium). The likelihood of these latter problems may be minimized by ensuring rigid adherence to maintenance protocols and good documentation and quality control procedures.

OTHER USEFUL OR ALTERNATIVE PROCEDURES

To conserve genetic integrity and to minimize the maintenance requirements, long-term preservation methods have been developed for a wide range of biological materials (15, 31, 32). In general, these methods depend on the reduction, removal, or rendering unavailable of extra- and intracellular water. Protocols which have been developed involve drying, freeze-drying, and freezing. These processes remove water or make it biologically unavailable, resulting in the reduction, or in some cases complete cessation, of metabolic activity.

Drying, generally air drying, may be used successfully for a wide range of cyst-forming protozoa, and many strains are commonly transported as dried material on filter paper (1a, 44). Some algal strains, particularly those which produce resistant aplanospores, have the potential for storage in dried form. Dried *Haematococcus pluvialis* may remain viable for at least 27 years (35). However, drying has not been widely applied as a method of long-term conservation of algae, primarily because of the low levels of recovery for some organisms and the short shelf life of stored material (16). More recent research using a controlled drying protocol demonstrated that the method has some potential for a number of green algae (36).

Freeze-drying/lyophilization is the method of choice by which many scientists preserve bacterial and fungal strains. For algae, viability levels may be extremely low in some cases, and these levels may decrease further upon prolonged storage (16, 25); this may lead to difficulties in regenerating an active culture. In addition, it is theoretically possible that extremely low viability levels obtained may result in the selection of a nonrepresentative, freeze-drying-tolerant subpopulation. Although freeze-drying has not been widely adopted for the conservation of algae or protozoa, it has been used to preserve cyanobacterial cultures for up to 5 years with no reduction in viability (26). In addition, it is used at the American Type Culture Collection to successfully preserve a wide range of organisms (1a, 44).

Freezing or cryopreservation is the optimal method of long-term storage, since high postthaw viability can be guaranteed. At low subzero temperatures ($< -135°C$), no further deterioration of stored material can occur and viability is effectively independent of storage duration (42). A range of freezing protocols has been developed. Most utilize a two-step system with controlled or semicontrolled cooling from room temperature to an intermediate holding temperature ($-30°C$ being commonly used), allowing cryodehydration of the cells to occur before the cells are plunged into liquid nitrogen ($-196°C$) (1a, 14, 27, 30, 39). The frozen material is stored in either liquid or vapor-phase nitrogen in an appropriate liquid nitrogen storage system. Although some organisms can be successfully cryopreserved and stored at higher subzero temperatures, viability levels rapidly fall during storage (7). It is therefore necessary to maintain frozen cultures at extremely low temperatures, optimally in liquid nitrogen at $-196°C$.

The methods described above all have the advantage of requiring little routine maintenance of the preserved material and, in the case of cryopreservation, guarantee genetic stability. The obvious disadvantage is the cost of specialist equipment and personnel. In addition, the preserved cultures require a period of recovery and growth before they are usable; this period may be as little as a couple of days for fast-growing amoebae (7) but may be in excess of 4 weeks for some strains (59). The main limitation to the implementation of these techniques in large culture collections, where the equipment for preservation is available, is the recalcitrance of many organisms to conventional preservation protocols. Research to improve techniques and to expand the range of organisms which can be maintained by long-term preservation methods is ongoing.

REFERENCES

1. **Advisory Committee on Dangerous Pathogens.** 1995. *Categorisation of Biological Agents. According to Hazard and Categories of Containment,* in press. Her Majesty's Stationery Office, London.

1a. **Alexander, M., P.-M. Daggett, R. Gherna, S. Jong, and F. Simione.** 1980. *American Type Culture Collection Methods. 1. Laboratory Manual on Preservation Freezing and Freeze-Drying as Applied to Algae, Bacteria, Fungi and Protozoa.* American Type Culture Collection, Rockville, Md.

2. **Anderson, R. A.** 1992. Diversity of eukaryotic algae. *Biodiversity Conserv.* **1:**267–292.

3. **Azam, F., T. Fenchel, J. G. Field, J. S. Gray, L. A. Meyer-Reil, and F. Thingstad.** 1983. The ecological role of water-column microbes in the sea. *Mar. Ecol. Prog. Ser.* **10:** 257–263.

4. **Belcher, H., and E. Swale.** 1988. *Culturing Algae: a Guide for Schools and Colleges.* Culture Collection of Algae and Protozoa, Ambleside, England.

5. **Berk, S. G., P. Guerry, and R. R. Colwell.** 1976. Separation of small ciliate protozoa from bacteria by sucrose gradient centrifugation. *Appl. Environ. Microbiol.* **31:** 450–452.

6. **Bertz, J. W., W. Aretz, and W. Hartel.** 1984. Use of flow cytometry in industrial microbiology for strain improvement programs. *Cytometry* **5:**145–150.

7. **Brown, S., and J. G. Day.** 1993. An improved method for the long-term preservation of *Naegleria gruberi. Cryo-Letters* **14:**347–352.

8. **Corliss, J. O., and S. C. Esser.** 1974. Comments on the role of the cyst in the life cycle and survival of free-living protozoa. *Trans. Am. Microsc. Soc.* **93:**578–593.

9. **Cowling, A. J.** 1991. Free-living heterotrophic flagellates:

methods of isolation and maintenance, including sources of strains in culture, p. 477–491. *In* D. J. Patterson and J. Larson (ed.), *The Biology of Free-Living Heterotrophic Flagellates.* Systematic Association special volume 45. Clarendon Press, Oxford.

10. **Cowling, A. J., and H. G. Smith.** 1987. Protozoa in the microbial communities of maritime Antarctic fellfields. Deuxieme Colloque sur les Ecosistemes Terrestres Sunantarctiques, 1986, Paimpoint. *Com. Natl. Francais Rech. Antarct.* **58:**205–213.

11. **Cunningham, A., and J. W. Leftley.** 1986. Application of flow cytometry to algal physiology and phytoplankton ecology. *FEMS Microbiol. Rev.* **32:**159–164.

12. **Curds, C. R.** 1973. The role of protozoa in the activated-sludge process. *Am. Zool.* **13:**161–169.

13. **Curds, C. R.** 1992. *Protozoa in the Water Industry.* Cambridge University Press, Cambridge.

14. **Day, J. G., and M. M. DeVille.** 1995. Cryopreservation of algae, p. 81–89. *In* J. G. Day and M. R. McLellan (ed.), *Cryopreservation and Freeze-Drying Protocols.* Humana Press Inc., Totowa, N.J.

15. **Day, J. G., and M. R. McLellan (ed.).** 1995. *Cryopreservation and Freeze-Drying Protocols.* Humana Press Inc., Totowa, N.J.

15a. **Day, J. G., and M. R. McLellan.** 1995. Conservation of algae, p. 75–98. *In* B. Grout (ed.), *Genetic Preservation of Plant Cells in Vitro.* Springer-Verlag, Heidelberg, Germany.

16. **Day, J. G., I. M. Priestley, and G. A. Codd.** 1987. Storage reconstitution and photosynthetic activities of immobilized algae, p. 257–261. *In* C. Webb and F. Mavituna (ed.), *Plant and Animal Cells. Process Possibilities.* Ellis Harwood Ltd., Chichester, England.

17. **Droop, M. R.** 1974. Heterotrophy of carbon, p. 530–559. *In* W. D. P. Stewart (ed.), *Algal Physiology and Biochemistry.* Blackwell, London.

18. **Fenchel, T.** 1969. The ecology of marine microbenthos. IV. Structure and function of the benthic ecosystem. *Ophelia* **6:**1–182.

19. **Finlay, B. J., T. M. Embley, and T. Fenchel.** 1993. A new polymorphic methanogen, closely related to *Methanocorpusculum parvum,* living in stable symbiosis within the anaerobic ciliate *Trimyema* sp. *J. Gen. Microbiol.* **139:**371–378.

20. **Finlay, B. J., A. Rogerson, and A. J. Cowling.** 1988. *A Beginners Guide to the Collection, Isolation, Cultivation and Identification of Freshwater Protozoa.* Culture Collection of Algae and Protozoa, Freshwater Biological Association, Ambleside, England.

21. **Griffiths, B. S., and K. Ritz.** 1988. A technique to extract, enumerate and measure protozoa from mineral soils. *Soil Biol. Biochem.* **20:**163–173.

22. **Haekel, E.** 1866. *Generelle Morphologie der Organismen.* G. Reimer, Berlin.

23. **Hawksworth, D. L., and B. Aguirre-Hudson.** 1994. International initiatives in microbial diversity, p. 65–72. *In* B. Kirsop and D. L. Hawksworth (ed.), *The Biodiversity of Microorganisms and the Role of Microbial Resource Centres.* World Federation for Culture Collections, United Nations Environment Programme, England.

24. **Hellebust, J. A., and J. Lewin.** 1977. Heterotrophic nutrition, p. 169–197. *In* D. Werner (ed.), *The Biology of Diatoms.* Botany Monographs 13. Blackwell, London.

25. **Holm-Hansen, O.** 1967. Factors affecting viability of lyophilized algae. *Cryobiology* **4:**17–23.

26. **Holm-Hansen, O.** 1973. Preservation by freezing and freeze-drying, p. 195–206. *In* J. Stein (ed.), *Handbook of Phycological Methods: Culture Methods and Growth Measurements.* Cambridge University Press, Cambridge.

27. **James, E. R.** 1991. Maintenance of parasitic protozoa by cryopreservation, p. 209–226. *In* B. Kirsop and A. Doyle (ed.), *Maintenance of Microorganisms and Cultured Cells.* Academic Press Ltd., London.

28. **Kalinina, L. V., and F. C. Page.** 1992. Culture and preservation of naked amoebae. *Acta Protozool.* **31:**115–126.

29. **Keenan, K., E. Erlich, K. H. Donnelly, M. B. Basel, S. H. Hutner, R. Kassoff, and S. A. Crawford.** 1978. Particle-based axenic media for tetrahymenids. *J. Protozool.* **25:**385–387.

30. **Kilvington, S.** 1995. Cryopreservation of pathogenic and nonpathogenic free-living amoebae, p. 63–70. *In* J. G. Day and M. R. McLellan (ed.), *Cryopreservation and Freeze-Drying Protocols.* Humana Press Inc., Totowa, N.J.

31. **Kirsop, B., and A. Doyle (ed.).** 1991. *Maintenance of Microorganisms and Cultured Cells.* Academic Press Ltd., London.

32. **Kirsop, B. E., and J. J. S. Snell (ed.).** 1984. *Maintenance of Microorganisms.* Academic Press, London.

33. **Laverde, A. V., and M. M. Brent.** 1980. Simplified soluble media for the axenic cultivation of *Naegleria. Protistologica* **16:**11–15.

34. **Lee, J. J., and A. T. Soldo (ed).** 1992. *Protocols in Protozoology.* Society of Protozoologists, Lawrence, Kans.

35. **Leeson, E. A., J. P. Cann, and G. J. Morris.** 1984. Maintenance of algae and protozoa, p. 131–160. *In* B. E. Kirsop and J. J. S. Snell (ed.), *Maintenance of Microorganisms.* Academic Press, London.

36. **Malik, K. A.** 1993. Preservation of unicellular green-algae by liquid-drying. *J. Microbiol. Methods* **18:**41–46.

37. **Margulis, L., J. O. Corliss, M. Melkonian, and D. J. Chapman (ed.).** 1989. *Handbook of Protoctista.* Jones and Bartlett, Boston.

38. **McLachlan, J.** 1973. Growth media–marine, p. 25–52. *In* J. Stein (ed.), *Handbook of Phycological Methods: Culture Methods and Growth Measurements.* Cambridge University Press, Cambridge.

39. **McLellan M. R., A. J. Cowling, M. F. Turner, and J. G. Day.** 1991. Maintenance of algae and protozoa, p. 183–208. *In* B. Kirsop and A. Doyle (ed.), *Maintenance of Microorganisms and Cultured Cells.* Academic Press Ltd., London.

40. **Miller, T. L., and M. J. Wolin.** 1974. A serum bottle modification of the Hungate technique for cultivating obligate anaerobes. *Appl. Microbiol.* **27:**985–987.

41. **Miyachi, S., O. Nakayama, Y. Yokohama, Y. Hara, M. Ohmori, K. Komogata, H. Sugawara, and Y. Ugawa (ed.).** 1989. *World Catalogue of Algae.* Japan Scientific Societies Press, Tokyo.

42. **Morris, G. J.** 1981. *Cryobiology.* Institute of Terrestrial Ecology, Cambridge.

43. **Neilson, A. H., and R. A. Lewin.** 1974. The uptake and utilization of organic carbon by algae: an essay in comparative biochemistry. *Phycologia* **13:**227–264.

44. **Nerad, T. A.** 1991. *ATCC Catalogue of Protists.* American Type Culture Collection, Rockville, Md.

45. **OECD.** 1984. *Guidelines for Testing Chemicals.* Section 2. *Effects on Biotic Systems.* no. 201. *Algal Growth Inhibition Test.* Organization for Economic Cooperation and Development, Geneva.

46. **Olive, L. S.** 1967. The Prostelida—a new order of the Mycetozoa. *Mycologia* **59:**1–29.

47. **Packer, L., and A. N. Glazer (ed.).** 1988. Cyanobacteria. *Methods in Enzymology,* vol. 167. Academic Press Inc., London.

48. **Page, F. C.** 1983. *Marine Gymnamoebae.* Institute of Terrestrial Ecology, Cambridge.

49. **Page, F. C.** 1988. *A New Key to Freshwater and Soil Gymnamoebae.* Freshwater Biological Association, Ambleside, England.

50. **Page, F. C.** 1990. Nackte Rhizopoda, p. 1–160. *In* D. Matthes (ed.), *Protozoenfauna 2.* Gustav Fischer Verlag, Stuttgart, Germany.
51. **Pringsheim, E. G.** 1946. *Pure Cultures of Algae.* Cambridge University Press, Cambridge.
52. **Sleigh, M. A.** 1989. *Protozoa and Other Protists.* Edward Arnold, London.
53. **Soldo, A., and S. A. Brickson.** 1980. A simple method for plating and cloning ciliates and other protozoa. *J. Protozool.* **27:**328–331.
54. **Stanier, R. Y., R. Kunisawa, M. Mandel, and G. Cohen-Bazire.** 1971. Purification and properties of unicellular blue-green algae (order *Chroococales*). *Bacteriol. Rev.* **35:** 171–205.
55. **Starr, R. C., and J. A. Zeikus.** 1993. UTEX—the culture collection of algae at the University of Texas at Austin. *J. Phycol.* **29:**1–106.
56. **Stein, J. (ed.).** 1973. *Handbook of Phycological Methods: Culture Methods and Growth Measurements.* Cambridge University Press, Cambridge.
57. **Suttle, C. A., and F. Chen.** 1992. Mechanisms and rates of decay of marine viruses in seawater. *Appl. Environ. Microbiol.* **58:**3721–3729.
58. **Throndsen, J.** 1973. Special methods—micromanipulators, p. 139–144. *In* J. Stein (ed.), *Handbook of Phycological Methods: Culture Methods and Growth Measurements.* Cambridge University Press, Cambridge.
59. **Tompkins, J., M. M. DeVille, J. G. Day, and M. F. Turner (ed.).** 1995. *Culture Collection of Algae and Protozoa Catalogue of Strains.* Culture Collection of Algae and Protozoa, Ambleside, England.
60. **Turner, M. F.** 1979. Nutrition of some marine microalgae with special reference to vitamin requirements and utilization of nitrogen and carbon sources. *J. Mar. Biol. Assoc. UK* **59:**535–552.
61. **Watanabe, M. M., and H. Nozaki.** 1994. *NIES—Collection List of Strains.* National Institute for Environmental Studies, Tsukuba, Japan.
62. **Whitelam, G. C., T. Lanaras, and G. A. Codd.** 1983. Rapid separation of microalgae by density gradient centrifugation in percoll. *Br. Phycol. J.* **18:**23–28.
63. **Wiedeman, V. E., P. L. Walne, and F. R. Trainor.** 1964. A new technique for obtaining axenic cultures of algae. *Can. J. Bot.* **42:**958–959.
64. **Zhukov, B. F., and A. P. Mylnikov.** 1983. Cultivation of free-living colourless flagellates from waste water treatment plants. *Protozoologiya* **8:**142–152. (In Russian.)

Cultivation and Assay of Viruses

PIERRE PAYMENT

8

Viruses from infected individuals can be disseminated by fecal material, saliva, nasal secretions, skin lesions, blood, etc. They can thus be found in most environments; water, soil, and air as well as surfaces can become contaminated. Because viruses are obligate parasites, they cannot be propagated in artificial medium as is often done with bacteria and fungi. Methods for their detection are thus more limited. Their presence can be detected by immunological or nucleic acid methods, but demonstration of their infectivity requires the use of live animals or cell cultures. Laboratory animals are of little use in the routine environmental laboratory, and cell culture has become the method of choice for cultivating and detecting viruses. Environmental virologists have been able to select cell cultures that are susceptible to many commonly encountered viruses and have developed methods for their detection in the environment. This chapter will review these methods and their practical use and describe some of the quality control procedures required to maximize their level of sensitivity.

Table 1 lists the viruses that through one route or another may be found in the environment. Almost every known virus is potentially present in air, soil, water, or fomites. In contrast to samples used in clinical virology (i.e., fecal material, sputum, secretions, etc.), in which large numbers of viruses are often present, in environmental samples, viruses have been diluted or disseminated and are relatively scarce. These viruses must be concentrated or extracted from samples as large as several thousand liters of water and air or several hundred grams of soil or sludge (13, 30, 43). These methods are described in other chapters of this manual describing the specific environments under study. The discussion in this chapter assumes that a concentrate has been obtained. The main objective of the cultivation and assay of viruses is to optimize detection methods to a level at which even a single infectious unit can be detected with confidence (2, 13).

PRECAUTIONARY NOTE

All viruses must be considered pathogenic for humans or animals. Any attempt to manipulate contaminated samples, to concentrate viruses, or to assay them should be performed under the proper biosafety conditions and by trained personnel only. The reader is referred to institutional, national, or international biosafety guidelines and other appropriate literature (45).

DETECTION METHODS

Methods for detection of viruses in the environment can be divided in three types: detection of viral antigen, detection of viral nucleic acid, and detection of viral infectivity. In most cases, the detection of viral infectivity is the main objective: it will enable the researcher to estimate a health risk, a level of contamination, the efficacy of disinfection or sterilization procedures, etc. If the objective is solely to determine whether a given environment has been contaminated by viruses, the detection of viral antigens or nucleic acids could be sufficient. The latter two methods have the advantage of being available for a larger number of viruses than the detection of infectivity. However, they are not precise and quantitation is difficult.

Electron Microscopy

The visualization of viral particles by electron microscopy is the basic method. It is not very sensitive, as a concentration of at least 1 million viral particles per ml is needed for viral particles to be observed. Electron microscopy is most often used on samples obtained after the virus has been grown in cell culture; viral morphology can then provide valuable information of the type of virus present, and use of immunoelectron microscopy may even provide specific identification (17, 36).

Immunoassays

Antigen detection methods based on immunoassays can also be used. Enzyme-linked immunosorbent assays are available for many viruses and can be used when large amounts of viral antigens are present in a sample. They have been used successfully in clinical microbiology for the detection of viruses in clinical material (17) and for the detection of enteric viruses in water samples (8, 12, 41). These methods are the least sensitive and very susceptible to interference from extraneous material in environmental concentrates, and they do not provide information on the infectivity of the viruses detected.

Nucleic Acid Probes

Nucleic acid probes are the basis for the most specific methods for the detection and identification of viruses (37). They

TABLE 1 Viruses that could potentially be found in the environment (air, water, soil)

Family	Genus	Common species
Picornaviridae	*Enterovirus*	Polioviruses; coxsackieviruses, group A; coxsackieviruses, group B; echoviruses; enteroviruses 68 through 71
	Heparnavirus	Hepatitis A virus
	Rhinovirus	Virus types infecting humans
	Aphthovirus	Foot-and-mouth disease viruses
Caliciviridae	*Calicivirus*	Norwalk gastroenteritis virus
Reoviridae	*Reovirus*	Reovirus
	Orbivirus	17 subgroups
	Rotavirus	Human rotaviruses
Flaviviridae	*Flavivirus*	Hepatitis C virus
Orthomyxoviridae	*Influenzavirus*	Influenza virus
Paramyxoviridae	*Paramyxovirus*	Parainfluenza virus, mumps virus
	Morbillivirus	Measles virus
	Pneumovirus	Human respiratory syncytial virus
Coronaviridae	*Coronavirus*	Human coronavirus
Bunyaviridae	*Hantavirus*	Hantaan virus
Lentivirinae	*Lentivirus*	Human immunodeficiency virus
Parvoviridae	*Parvovirus*	Human parvovirus B19
Papovaviridae	*Papillomavirus*	Human papillomaviruses (warts)
Adenoviridae	*Mastadenovirus*	Human adenoviruses
Hepadnaviridae	*Hepadnavirus*	Human hepatitis B virus
Herpesviridae	*Simplexvirus*	Human herpes simplex virus
	Varicellovirus	Varicella-zoster virus (herpesvirus 3)
	Cytomegalovirus	Human cytomegalovirus (herpesvirus 5)
	Lymphocryptovirus	Epstein-Barr virus (herpesvirus 4)
Poxviridae	*Orthopoxvirus*	Vaccinia virus, smallpox virus (variola)

are currently being developed for a large number of viruses and are being applied with some success to the detection of viruses in environmental samples (11, 14, 22, 24, 38, 44). These methods are the most sensitive but are also susceptible to various inhibitors found in environmental concentrates; false-negative results and sample contamination during laboratory manipulations are the major problems encountered with these methods. Finally, use of nucleic acid probes provides no information on the infectivity of the viruses detected.

Cell Culture

The detection of infective viruses in environmental samples relies mainly on cell culture as the method of choice. The method does not allow detection of all viruses, but neither do other methods; any method used for this purpose will thus provide only an incomplete answer. Detection of viruses on cell culture is not an easy process: concentrates from environmental samples often contain organic and inorganic compounds which can be toxic to cell culture. Microbial contaminants such as bacteria and fungi, many types of planktonic microorganisms, as well as some invertebrates are always present in water and can still be present in the concentrate. When concentrated from samples as large as several thousands liters, they can account for a significant problem that must be resolved before one attempts to even deposit the sample on a cell culture, the bacteriological sterility of which is of utmost importance.

Isotonicity and pH

To be applied to cell culture without any deleterious effect, samples must be isotonic, physiological, and free of toxic compounds and microbial contaminants. Viruses adsorbed to surfaces, flocs, or collection filters are often eluted by using alkaline proteinaceous solutions such as beef extract or buffered solutions (13). These eluates must be neutralized (pH 7.0 to 7.5) immediately to prevent viral inactivation; this can be achieved by dropwise addition of 1 N hydrochloric acid. Balanced salt solutions or tissue culture medium can be used to maintain isotonicity and the correct pH.

Decontamination and Detoxification

Samples must be decontaminated and detoxified before being applied to cell cultures (13). A preliminary test on cell culture with a small amount of the sample will rapidly indicate the presence of toxic substances: toxic samples will usually destroy cells in less than 24 h (cytotoxic effect), while bacterial contaminants will rapidly take over the cell culture and destroy all cells in a few days (cloudy appearance). Microbial contamination is easily recognized by microscopy: bacilli, cocci, yeasts, or fungi can be seen in the culture medium at a magnification of ×400.

Samples can be decontaminated by simple mechanical means such as centrifugation at 4,000 to 10,000 × g for 15 to 30 min or filtration on sterile 0.45- or 0.2-μm-pore-size membrane filters. Viruses are smaller than the smallest bacteria and will not be retained by membranes with the pore sizes used for bacteria. However, precautions should be taken to ensure that viruses are not retained by the membrane via adsorption to the matrix. To prevent this from occurring, membranes can be pretreated with a small amount of fetal calf serum (2 to 3 ml of a 10% solution) or a 1% bovine serum albumin solution. When filtration is not feasible, decontamination can be achieved through the addition to the sample of a cocktail of antibiotics at high concentrations (penicillin, 1,000 U/ml; streptomycin, 1,000 μg/ml; gentamicin, 50 μg/ml; amphotericin B [Fungizone],

2.5 μg/ml). A combination of these three methods is often used to achieve optimal results for problematic samples.

Both filtration and centrifugation can result in significant loss of viruses and should be avoided if possible. If such processing is unavoidable, laboratory experiments with seeded viruses should help establish the consequent losses.

Detoxification can be achieved by extraction of the sample with trichlorofluoroethane. Half a volume of the chemical is added, and the sample is mixed vigorously for several minutes and centrifuged at 1,000 $\times$ g to obtain two phases. The aqueous phase is collected for viral assay. Toxicity can be reduced by removing the sample after the adsorption period on the cells and washing the cells twice with phosphate-buffered saline (PBS) before adding fresh culture medium. While this practice can slightly reduce the efficiency of the method, it is preferable to total loss of the sample.

Choice of Cell Cultures

There is no single cell system for the detection of viruses. Primary human cell cultures are very sensitive to virus infection and have been a substrate of choice for virologists (23, 25). The difficulty in obtaining human cell cultures has been partly solved by the use of monkeys as the source of cells. Even this source has become more and more erratic and expensive and is now infrequently used for routine work, especially as these cultures can be contaminated by adventitious agents.

Comparisons of the sensitivities of various cell lines to viruses have been reported on many occasions (6–9, 15, 16, 20, 28, 35, 40, 44, 46). The appropriate cell line for detection of the virus under study should be determined by using these references as guidelines. For viruses of animal species other than humans, homologous cell cultures are often the best choice (29). Detailed procedures for cell culture and the required solutions are described elsewhere (2, 23, 25, 33, 34). Environmental virologists have debated, and are still debating, which cell line is the most suitable for detecting with the greatest sensitivity the viruses of concern to environmental virology. Cultures from established kidney cell lines from rhesus or green monkeys such as BSC-1, MA-104, Vero, and BGM (5–7, 9) are widely used, but human cell cultures such as HeLa and RD (rhabdomyosarcoma) are also used (5, 15, 28, 36, 40, 46). It is up to the researcher to select the most suitable cell culture. There are several sources for cell cultures that can be used in virology. The American Type Culture Collection (Bethesda, Md.) maintains a large selection of cell culture sources and provides a quality-controlled source of cells. Each cell type has its own specific nutritional requirements, which are determined not only by the experimental conditions but also by the adaptation of cell populations to the medium used. There are almost as many variations in the composition and the use of culture media as there are cell types in culture. Any researcher using cell cultures must maintain culture conditions at an optimal level and apply the ultimate care in their maintenance; failure to do so will lead to genetic modifications of the cells and often a decrease in their sensitivity to virus growth. This decrease would lead to a significant decrease in the efficiency of virus recovery, which is usually detected through proper quality assurance procedures.

Cell cultures can occasionally be obtained through researchers and colleagues. However, cell cultures bearing the same name can in fact be entirely different cells. Cell cultures can easily be contaminated by another cell type when both cells are being manipulated at the same time; a good practice is to handle only one cell type at a time in the working area. Cell cultures can also be "selected" by the researcher's practices: different multiplication patterns, growing temperature, medium, serum concentration, etc. Such practices account for the fact the same cell line at the same passage level in two different laboratories will differ in sensitivity to viruses (i.e., have different growth abilities). Interlaboratory comparisons should therefore be performed with the same cells and according to strict protocols of quality control and quality assurance.

Cell cultures must also be tested for the presence of adventitious agents (i.e., microorganisms naturally infecting these cells). Mycoplasmas are a major problem, as they are easily introduced in cell cultures by insufficiently tested biologicals (serum or trypsin) as well as poor laboratory practices such as mouth pipetting. Detailed methods on using cell culture for the detection of viruses can be found elsewhere (2, 10, 23, 25, 34).

Material for Cell Culture and Virus Assays

All procedures are preferably performed by a trained virologist using a class II type I biological safety cabinet for the protection of both cell cultures and analysts (45). All material used for cell culture and virus assays must be sterile and clean. All glassware for solutions or media that will come in contact with cells must be washed, rinsed with high-quality distilled water, and sterilized in an autoclave or sterilizing oven. Any trace of detergent or residues will be toxic to the cells and should be removed. Cells can be grown on almost any surface; glass and disposable plasticware have been used in a wide variety of formats. Plasticware has become popular, and specially treated plastics are available for cell culture work. Most commonly used are the 25-, 75-, and 150-cm^2 flasks sterilized by irradiation by the manufacturer. For virus enumeration, it is often convenient to use multiwell plastic plates commercially available in 4-, 12-, 24-, and 96-well configurations. Only material labeled as "Sterile, for cell culture" should be used.

Most cell culture media used today are synthetic and are commercially available. The chemically defined (synthetic) media are often supplemented with fetal bovine or calf serum which contains nutrients not available otherwise. Serum is added to the medium to a final concentration of 0.1 to 10%, the highest concentrations being used for cell growth and the lowest concentrations being used for cell culture maintenance or virus detection. When the viruses to be detected are susceptible to inhibitors or antibodies present in serum, the media can be supplemented with 1 to 10% lactalbumin hydrolysate in place of serum.

Cell cultures can be easily maintained if a few simple rules are followed: store all solutions at the optimal temperature (follow the manufacturer's recommendation) and give preference to freshly prepared solutions, store all solutions in the dark, check the chemical quality of the distilled water on a regular basis, and always prewarm to 37°C any medium or solution that will be in contact with cell cultures. However, do not leave medium or solutions at 37°C for several hours.

Use of Antibiotics

Antibiotic and antimycotic solutions may be incorporated into media to reduce contamination introduced by environmental samples. Their use should serve not to mask bad laboratory practices but to maintain the cell cultures in good condition. The recommended concentrations should not be increased for cell culture maintenance. In fact, it is preferable not to use antibiotics to maintain cell culture stocks; any contamination will be detected earlier and will not jeopardize experiments or sample analysis. Their use should be restricted to decontaminating samples and preventing contamination of experimental cell cultures. Penicillin (100 to 1,000 U/ml), streptomycin (100 to 1,000 μg/ml), gentamicin (50 μg/ml), and amphotericin B (2.5 μg/ml) are generally used. Stock solutions in water (100$\times$) should be prepared in 5-ml aliquots and can be stored at $-20°C$ for a year. Combinations of antibiotics at concentrations of up to 10 times the normal concentration are used to suppress or control bacterial contamination found in environmental samples. These high antibiotic concentrations can be toxic for some cell lines; prior testing is imperative.

Quality Control

A regular quality control and quality assurance control program should be in place (42). Both cell cultures and viruses can occasionally behave erratically. It is important to test on a regular basis (according to the number of tests performed) the cell cultures that are used for the virus isolation and enumeration. Suspensions of selected viruses are prepared and frozen in small aliquots at $-70°C$ with a protecting agent (glycerol, albumin, or serum). The quality of the reagent grade water (type I) used in the laboratory must be checked frequently. Simple tests such as pH and conductivity determination will rapidly identify deteriorating water quality and failing purifying apparatus. A routine program could include monthly verification of the susceptibility of cells to viruses, using standardized viral suspensions frozen at $-70°C$ in aliquots (poliovirus type 1, reovirus, echovirus 11, or any other virus with the same characteristics as the viruses being sought).

VIRUS QUANTITATION

The endpoint that is used for the quantitation of viruses will affect the number of viruses detected. Cytopathic effect, plaque formation, and detection of viral products are the usual means used by the environmental virologist.

When a virus multiplies, it can either destroy the cell it has infected or use the cell metabolism to replicate itself without affecting these cells. Destruction of the cells is readily assessed by microscopic observation: the dead cells are dislodged from the flask surface often after rounding up or showing characteristic morphological changes (cytopathic effect). When the cells are covered by a light agar medium to prevent them from being dislodged, the virus will migrate from one cell to another, producing characteristic round holes in the cell monolayer. These holes, called plaques, are produced by a single initial virus particle. If the virus does not destroy the cells, indirect methods of detection must be used.

Plaque Assay

Some viruses, such as the enteroviruses, produce plaques or zones of lysis in cell monolayer overlaid with solidified nutrient medium. These plaques originate from a single infectious virus particle; thus, the titer of virus may be estimated very precisely. This method and the required solutions are described elsewhere (2, 33, 34). Plaque assay is used for a very limited number of viruses and has the lowest sensitivity of the available methods (32). Cultivable enteroviruses will generally produce plaques, but other enteric viruses such as adenoviruses and reoviruses will not. The main advantage of plaque assay is that each individual virus (or

aggregate) will form a single plaque and each plaque is rarely a mixture of several virus types. To detect plaque-forming viruses, serial 10-fold dilutions of the concentrate in maintenance medium (changing the pipette at each dilution) are inoculated to 2 to 10 flasks (25 or 75 cm^2) per dilution, using not more than 1 ml of inoculum per flask. Flasks are incubated at 37°C for 1 h, gently rocking the flasks every 15 min to promote virus adsorption. An agar overlay nutrient medium is then added to each flask. After solidification, the flasks are incubated at 37°C in an inverted position (cell monolayer up) in the dark. The number of plaques is counted daily for 3 to 5 days, or until the monolayers starts degrading. The number of viruses is the average number of plaques per flask, corrected for a volume of 1 ml and multiplied by the dilution factor, if any. The result is expressed as PFU per unit of volume tested. The viral origin of a plaque must often be confirmed. Pick up the plaque with a curved Pasteur pipette and transfer it to a confluent cell monolayer; if a cytopathic effect develops, the viral origin of the plaque is confirmed. The virus can also be identified at this step.

Some viruses will not form plaques except under a semisolid medium. It is suggested that one use 1% (wt/vol) methylcellulose, which is semisolid at 37°C and semiliquid at 4°C. This substance is sterilized by autoclaving (it then has a thick pasty appearance) and can be solubilized by placing it at 4°C or on ice for several hours. Other viruses produce a cytopathic effect but unfortunately one that is slow or subtle. The presence of neutral red in the culture medium makes it toxic for cells during the long incubations. Therefore, the preferred method is to proceed in two steps: a first incubation period without neutral red, followed by a second incubation for 24 to 48 h with the stain in the medium. Proceed as described earlier, omitting neutral red from the agar. Incubate the samples for 7 to 10 days, add fresh nutrient agar containing neutral red, or remove the agar and stain the monolayer with crystal violet (34).

Quantal Assays

Enteroviruses (polioviruses; coxsackievirus types B1 to B6, A7, and A9; echovirus types 1 to 34) induce a characteristic cytopathic effect: rounding of the cells is followed by their detachment from the cell monolayer. Adenoviruses produce rounding of the cells, an increase in refractivity, and a strong tendency to aggregate. Titers of these viruses can be determined either by placing the sample on an established cell monolayer or by mixing the virus with cells in suspension before seeding into flasks. Sensitivity can be increased by the addition of trypsin to the medium to promote infection (13). The microtechnique is performed in multiwell plates and is useful when large number of viruses are expected (in wastewaters, heavily polluted surface waters, etc.). The macrotechnique is used for samples containing low levels of viruses (treated waters, minimally polluted surface waters, groundwaters, etc.) when the volume of sample to be inoculated to the cell cultures is larger. The maximum volume of sample that can be applied to a cell culture is 1 ml/25 cm^2; larger volumes will result in loss of sensitivity.

Some laboratories may have access to an electron microscope to detect directly the presence of viruses in the supernatants of infected flasks or wells. This method is expensive but has the advantage of detecting virus types on the basis of morphology and detecting growth of multiple viruses. To enhance the detection of slowly growing viruses, part of the infected cells or the supernatant can be placed onto fresh cell cultures. This is usually done after freezing the original culture at $-20°C$ to break apart the cells and then transferring 0.1 to 0.5 ml to the fresh cell monolayers. This can be performed several times to increase the number of viruses present and demonstrate a cytopathic effect, and the virus is identified by serum neutralization using specific antisera (25). Most detection and identification methods are based on immunological reactions with specific or polyvalent antisera that are fluorescein or peroxidase tagged.

Macrotechnique (in Flasks)

A few days in advance, prepare the appropriate cell culture in growth medium (i.e., Eagle's minimal essential medium with 10% serum) in 25-cm² flasks so that the cells are confluent when they are used. Prepare serial dilutions of the virus (or sample) to be titrated in maintenance medium containing antibiotics (change the pipette at each dilution to prevent virus carryover). Remove the culture medium from each flask and inoculate not more than 1 ml of each dilution per 25-cm² flask; the larger the number of flasks, the more precise the titer (31). Use 10 to 20 flasks if there are few viruses in the sample. If a higher number of viruses is expected, the sample can be diluted and inoculated to five flasks per dilution. Preferably place the flasks on a rocking platform to slowly distribute the inoculum on the full monolayer for 60 min, add maintenance medium (i.e., Eagle's minimal essential medium with 1% serum), and incubate the flasks at 37°C. Observe the flasks daily and record the number showing a cytopathic effect. Once the cytopathic effect has stopped progressing or when the controls begin to show aging, estimate the viral titer by using an estimation of the most probable number (MPN). The MPN approach has been in use for many years in bacteriology, using predefined tables for a number of experimental settings (3, 5, or 10 tubes). Its use in virology has evolved from the use of tables to simplified formulas for the calculation of the virus density in a sample (3, 4), and with the advent of microcomputers, it is now possible to use a more precise statistical approach of the number of viruses in a sample. The computer programs for the calculations of the MPNIU (MPN of infectious units) and their confidence intervals are available from the literature (18, 19, 21, 26, 39, 47).

The simplest form of the MPNIU estimation equation for a single dilution is

$$MPNIU = -\ln(q/n)$$

where q is the number of negative cultures and n is the total number of cultures inoculated.

For more than one dilution, apply the following formula, using results from dilutions with both positive and negative results (i.e., exclude dilutions which are 100% positive):

$$MPNIU/ml = P/(NQ)^{1/2}$$

where P is the total number of cultures from all dilutions, N is the total volume (milliliters) of sample inoculated for all dilutions, and Q is the total volume (milliliters) of samples in negative cultures.

Microtechnique (in Multiwell Plates)

A few days in advance, seed the cells in multiwell plates and incubate them until a confluent monolayer is obtained (note that multiwell plates are not tight fitting and must be incubated in an incubator with a 5% CO₂ atmosphere. Prepare serial dilutions of the sample to be titrated in maintenance medium (change the pipette at each dilution to prevent virus carryover). Remove the medium and inoculate 25 μl of each dilution into appropriate wells (with cells) of a flat-bottom 96-well plate. Inoculate each dilution to at

least four wells; the larger the number of wells, the more precise the titer estimation by the MPN approach. If the sample contains very few viruses, all wells from a plate can be inoculated with 25 μl of undiluted sample or after dilution 1:2 or 1:5 in maintenance medium. The use of 24-well plates is an alternative that has also been shown useful in many laboratories; their larger surface culture area increases the volume of sample that can be inoculated. When the cytopathic effect has stopped progressing or the controls begin to show aging, estimate the virus concentration by the MPN approach.

Immunoassays

The detection of viral products in infected cells can also be achieved by using specific or polyvalent antisera that are available commercially or can be prepared in house (1). Both immunofluorescence and immunoperoxidase assays can be used (1, 33, 34). In the immunofluorescence assay, the antibodies fixed on viral products in the cells are detected by using a commercial preparation of antibodies conjugated to a fluorochrome (usually fluorescein) that can be observed under a microscope equipped with UV light. In the immunoperoxidase assay, the antibodies fixed on viral products in the cells are detected by using a commercial preparation of antibodies conjugated to an enzyme, peroxidase, which can be detected with an appropriate substrate and observed with a regular microscope.

The first step in these assays is to inoculate the sample or dilutions of the sample to an appropriate cell culture (BGM or MA-104, for example) in 25-cm² flasks (not more than 1 ml/25 cm²) or directly in 24- or 96-well plates if high virus concentrations are expected. Incubate the samples for 7 to 10 days at 37°C. Observe the cells after 24 h to detect possible toxic effects; note wells in which cytotoxicity is observed. After the incubation period, freeze-thaw the cell cultures, transfer 0.25 ml of the supernatant of each flask in the corresponding well of a new tissue culture plate containing the same cells, and incubate the samples again for 7 to 10 days to increase the titer of virus, if present. Drain the supernatant, wash it twice with PBS, fix the cells by covering them for 15 min with absolute methanol containing 1% hydrogen peroxide, and finally rinse the cell twice with PBS. Add to each well 0.3 ml of a 1:200 dilution of the human immune serum globulin preparation (from Cutter Biologicals or equivalent from another source). Use only immune serum globulins and not purified gamma globulins. Incubate the samples for 1 to 2 h at 37°C and then wash them twice with PBS.

For detection by the immunoperoxidase assay, add to each well 0.3 ml of an optimally diluted protein A-peroxidase conjugate, incubate the samples for 1 to 2 h at 37°C, and wash them four times with PBS. Prepare a fresh 0.025% solution of substrate (3,3'-diaminobenzidine tetrahydrochloride) in 100 ml of 0.05 M Tris-HCl (pH 7.6) filtered on a Whatman filter paper to remove debris. Just before use, add 30 μl of hydrogen peroxide (30% solution) per 100 ml of substrate. Substrate reactivity can be checked by adding a small drop of the diluted conjugate to 1 ml of substrate in a separate tube; the dark brown reaction should appear rapidly, indicating that the reagents are properly prepared. Add 0.5 ml of the freshly prepared substrate to each well, and allow the reaction to proceed for 5 to 15 min. The reaction can be observed under the microscope and stopped by washing with tap water when a maximum difference is observed between the infected and noninfected controls and the background level is still low. Wash the monolayers with tap water and observe them wet with an inverted microscope. Infected cells will appear as dark brown, often with clearly visible nuclear or cytoplasmic inclusions. All wells in which a specific reaction or a cytopathic effect is observed are considered positive; record both the presence of a positive immunoperoxidase reaction and a cytopathic effect.

For the immunofluorescence test, replace the protein A-peroxidase conjugate with a protein A or an anti-specific immunoglobulin fluorescein conjugate. There is then no need for a substrate; simply wash the monolayers carefully and examine them with a UV light microscope (1, 33, 34).

The number of viruses in the original sample can be evaluated by the MPN approach, using the volume of sample inoculated at the first passage and the number of wells at each dilution that have finally been shown to contain viruses.

Identification of Viruses

Viruses isolated on cell culture can be propagated further in cell culture and identified by a variety of methods: electron microscopy (33, 34), serum neutralization (1, 25, 33, 34), molecular methods (27), and immunoassays (1, 33, 34). In a specific outbreak situation, it is useful to identify the causative agent. It is often not worth the cost of identifying such viruses in monitoring environmental samples, since these viruses are pathogens and should never be present in the environment unless a major fecal contamination has occurred. However, this remains a decision to be taken by each researcher according to the objective of the ongoing research.

The viruses that are most commonly found in the environment are the polioviruses, coxsackieviruses (group B), adenoviruses, and reoviruses. By using molecular biology methods and gene probe technology, all viruses will eventually be detected.

REFERENCES

1. Balows, A., W. J. Hausler, K. L. Hermann, H. D. Isenberg, and H. J. Shadomy (ed.). 1991. *Manual of Clinical Microbiology*, 5th ed. American Society for Microbiology, Washington, D.C.
2. Berg, G., R. S. Safferman, D. R. Dahling, D. Berman, and C. J. Hurst. 1983. *USEPA Manual of Methods for Virology*. EPA-600/4-84-013. Environmental Protection Agency, Cincinnati, Ohio.
3. Chang, S. L. 1965. Statistics of the infective units of animal viruses, p. 219–234. *In* G. Berg (ed.), *Transmission of Viruses by the Water Route*. Interscience Publishers, New York.
4. Chang, S. L., G. Berg, K. A. Busch, R. E. Stevenson, N. A. Clarke, and P. W. Kabler. 1958. Application of the most probable number method for estimating concentration of animal viruses by the tissue culture technique. *Virology* 6:27–42.
5. Dahling, D. R., G. Berg, and D. Berman. 1974. BGM, a continuous cell line more sensitive than primary rhesus and African green kidney cells for the recovery of viruses from water. *Health Lab. Sci.* 11:275–282.
6. Dahling, D. R., R. S. Safferman, and B. A. Wright. 1984. Results of a survey of BGM cell culture practices. *Environ. Int.* 10:309–311.
7. Dahling, D. R., and B. A. Wright. 1986. Optimization of the BGM cell line culture and viral assay for monitoring viruses in the environment. *Appl. Environ. Microbiol.* 51:790–812.
8. Dahling, D. R, B. A. Wright, and F. P. Williams. 1993. Detection of viruses in environmental samples—suitability of commercial rotavirus and adenovirus test kits. *J. Virol. Methods* 45:135–147.
9. Davis, P. M., and R. J. Phillpots. 1974. Susceptibility of the Vero line of African green monkey kidney cells to human enteroviruses. *J. Hyg.* (London) 72:23–30.
10. Freshney, R. 1987. *Culture of Animal Cells. A Manual of Basic Techniques*, 2nd ed. Alan R. Liss, Inc., New York.
11. Genthe, B., M. Gericke, B. Bateman, N. Mjoli, and R. Kfir. 1995. Detection of enteric adenoviruses in South African waters using gene probes. *Water Sci. Technol.* 31:345–350.
12. Genthe, B., G. K. Idema, R. Kfir, and W. O. K. Grabow. 1991. Detection of rotavirus in South African waters: a comparison of a cytoimmunolabeling technique with commercially available immunoassays. *Water Sci. Technol.* 24:241–244
13. Gerba, C. P., and S. M. Goyal. 1982. *Methods in Environmental Virology*. Marcel Dekker, Inc., New York.
14. Girones, R. M. Puig, A. Allard, F. Lucena, G. Wadell, and J. Jofre. 1995. Detection of adenovirus and enterovirus by PCR amplification in polluted waters. *Water Sci. Technol.* 31:351–357.
15. Guttman-Bass, N. 1987. Cell cultures and other host systems for detecting and quantifying viruses in the environment, p. 195–228. *In* G. Berg (ed.), *Methods for Recovering Viruses from the Environment*. CRC Press, Inc., Boca Raton, Fla.
16. Hasler, P., and R. Wigand. 1978. The susceptibility of Vero cell cultures for human adenoviruses. *Med. Microbiol. Immunol.* 164:267–276.
17. Herrmann, J. E. 1995. Immunoassay for the diagnosis of infectious diseases, p. 110–122. *In* P. R. Murray, E. J. Baron, M. A. Pfaller, F. C. Tenover, and R. H. Yolken (ed.), *Manual of Clinical Microbiology*, 6th ed. ASM Press, Washington, D.C.
18. Hugues, C., and C. Pietri. 1985. Influence du volume d'inoculum dans la quantification des virus selon deux techniques comparées: plagues et N.P.P. utilisant un grand nombre d'inoculum par dilution. *Chemosphere* 14:149–153.
19. Hugues, B., C. Pietri, and M. André. 1985. Estimation of virus density in sewage effluents by 2 counting techniques. Comparison of precision as a function of inoculum volume. *Zentralbl. Bakteriol. B* 181:409–417.
20. Hurst, C. J. 1986. Evaluation of mixed cell types and 5-iodo-2′-deoxyuridine treatment upon plaque assay titers of human enteric viruses. *Appl. Environ. Microbiol.* 51:1036–1040.
21. Husson van Vliet, J., and P. Roussel. 1986. Estimating viral concentrations: a reliable computation method programmed on a pocket calculator. *Comput. Methods Programs Biomed.* 21:167–172.
22. Jehl-Pietri, C. B. Hugues, M. Andre, J. M. Diez, and A. Bosch. 1993. Comparison of immunological and molecular hybridization detection methods for the detection of hepatitis-A virus in sewage. *Lett. Appl. Microbiol.* 17:162–166.
23. Jones Brando, L. V. 1995. Cell culture systems, p. 158–165. *In* P. R. Murray, E. J. Baron, M. A. Pfaller, F. C. Tenover, and R. H. Yolken (ed.), *Manual of Clinical Microbiology*, 6th ed. ASM Press, Washington, D.C.
24. Leguyader, F., V. Apaire-Marchais, J. Brillet, and S. Billaudel. 1993. Use of genomic probes to detect hepatitis A virus and enterovirus RNAs in wild shellfish and relationship of viral contamination to bacterial contamination. *Appl. Environ. Microbiol.* 59:3963–3968.
25. Lennette, E. H., and R. W. Emmons. 1989. *Diagnostic Procedures for Viral and Rickettsial and Chlamydial Infections*, 6th ed. American Public Health Association, Washington, D.C.

26. **Macdonell, M. T., E. Russek, and R. R. Colwell.** 1984. An interactive microcomputer program for the computation of most probable number. *J. Microbiol. Methods* **2:**1–7.

27. **Margolin, A. B., C. P. Gerba, K. J. Richardson, and J. E. Naranjo.** 1993. Comparison of cell culture and a poliovirus gene probe assay for the detection of enteroviruses in environmental water samples. *Water Sci. Technol.* **27:**311–314.

28. **Morris, R.** 1985. Detection of enteroviruses—an assessment of 10 cell lines. *Water Sci. Technol.* **17:**81–88.

29. **Payment, P., F. Affoyon, and M. Trudel.** 1988. Detection of animal and human enteric viruses in water from the Assomption river and its tributaries. *Can. J. Microbiol.* **34:**967–972.

30. **Payment, P., and E. Franco.** 1993. *Clostridium perfringens* and somatic coliphages as indicators of the efficiency of drinking water treatment for viruses and protozoan cysts. *Appl. Environ. Microbiol.* **59:**2418–2424.

31. **Payment, P., and M. Trudel.** 1985. Influence of inoculum size, incubation temperature and cell density on virus detection in environmental samples. *Can. J. Microbiol.* **31:**977–980.

32. **Payment, P., and M. Trudel.** 1987. Detection and quantitation of human enteric viruses in waste waters: increased sensitivity using a human immune serum globulin-immunoperoxidase assay on MA-104 cells. *Can. J. Microbiol.* **33:**568–570.

33. **Payment, P., and M. Trudel.** 1989. *Manuel de Techniques Virologiques,* 2nd ed. Presses de l'Université du Québec/AUPELF, Québec, Canada.

34. **Payment, P., and M. Trudel.** 1993. *Methods and Techniques in Virology.* Marcel Dekker Inc., New York.

35. **Pietri, C., and B. Hugues.** 1985. Influence du système cellulaire sur la quantification des virus dans les eaux usées. *Microbios Lett.* **30:**67–72.

36. **Pietri, C., B. Hugues, and D. Puel.** 1988. Immune electron microscopy in the detection of viruses other than enteroviruses on cell culture in untreated sewage. *Zentralbl. Bakteriol. Mikrobiol. Hyg.* **186:**67–72.

37. **Podzorsky, R. P., and D. H. Persing.** 1995. Molecular detection and identification of microorganisms, p. 130–157. *In* P. R. Murray, E. J. Baron, M. A. Pfaller, F. C. Tenover, and R. H. Yolken (ed.), *Manual of Clinical Microbiology,* 6th ed. ASM Press, Washington, D.C.

38. **Puig, M., J. Jofre, F. Lucena, A. Allard, G. Wadell, and R. Girones.** 1994. Detection of adenoviruses and enteroviruses in polluted waters by nested PCR amplification. *Appl. Environ. Microbiol.* **60:**2963–2970.

39. **Russek, E., and R. R. Colwell.** 1983. Computation of most probable numbers. *Appl. Environ. Microbiol.* **45:**1646–1650.

40. **Schmidt, N. J., H. H. Ho, and E. H. Lennette.** 1975. Propagation and isolation of group a coxsackieviruses in RD cells. *J. Clin. Microbiol.* **2:**183–185.

41. **Sellwood, J., and P. Wynjones.** 1995. A novel method for the detection of infectious rotavirus from water. *Water Sci. Technol.* **31:**367–370.

42. **Sewel, D. L., and R. B. Schifman.** 1995. Quality assurance: quality improvement, quality control and test validation, p. 55–66. *In* P. R. Murray, E. J. Baron, M. A. Pfaller, F. C. Tenover, and R. H. Yolken (ed.), *Manual of Clinical Microbiology,* 6th ed. ASM Press, Washington, D.C.

43. **Simard, C., M. Trudel, G. Paquette, and P. Payment.** 1983. Microbial investigation of the air in an apartment building. *J. Hyg.* (Cambridge) **91:**277–286.

44. **Sobsey, M. D., L. R. Sangermano, and C. J. Palmer.** 1993. Simple method of concentrating enteroviruses and hepatitis A virus from sewage and ocean water for rapid detection by reverse transcriptase-polymerase chain reaction. *Appl. Environ. Microbiol.* **59:**3488–3491.

45. **Strain, B. A., and D. H. M Gröschel.** 1995. Laboratory safety and infectious waste management, p. 75–85. *In* P. R. Murray, E. J. Baron, M. A. Pfaller, F. C. Tenover, and R. H. Yolken (ed.), *Manual of Clinical Microbiology,* 6th ed. ASM Press, Washington, D.C.

46. **Tougianidou, D., M. Jacob, K. Herbold, T. Hahn, B. Flehmig, and K. Botzenhart.** 1989. Assessment of various cell lines (including mixed-cell cultures) for the detection of enteric viruses in different water sources. *Water Sci. Technol.* **21:**311–314.

47. **Wyshak, G., and K. Detre.** 1972. Estimating the number of organisms in quantal assays. *Appl. Environ. Microbiol.* **23:**784–790.

Cultivation of Microbial Consortia and Communities

DOUGLAS E. CALDWELL, GIDEON M. WOLFAARDT, DARREN R. KORBER, AND JOHN R. LAWRENCE

9

Most medical microbiologists work exclusively with pure cultures. However, environmental microbiologists often cultivate consortia and communities in addition to pure cultures. Laboratory cultivation of these associations is necessary to gain an understanding of the community-level mechanisms for specific environmental processes. It is also needed to determine the potential effects of environmental contaminants on microbial systems and to make effective use of them as inoculants in industry, agriculture, and the environment (9, 10, 16, 71). Applications include bioremediation, industrial and municipal wastewater treatment, composting, solid waste landfills, biological control agents, and biofertilizers.

Koch's postulates require that each organism be maintained in isolation to demonstrate that it is a causative agent, and this works well when a single organism slips past a host's defenses and causes a disease. However, most aquatic and terrestrial environments contain communities and not isolated cell lines. Some hot springs were once thought to be "virtually pure cultures," but it is now known that there are no natural environments in which the product of evolution is either a pure culture or a single race of organisms. Evolution has resulted in interactive systems with multiple levels of organization, adapted to optimize the conversion of abiotic resources to biotic resources. Germ theory and pure culture are inappropriate in this context. They constrain cause-and-effect relationships to the organismal level when the causative mechanism of environmental processes is commonly at the community level or at multiple levels of organization.

Many free-living microorganisms are most effective only when they are present in association with other groups of organisms. This is the case in the degradation of pesticides, polychlorinated biphenyls, and other halogenated organic compounds. The same is true for the organisms involved in interspecies hydrogen transfer within methanogenic systems. It is also true in the case of nitrification, which requires a set of ammonia-oxidizing organisms and another set of nitrite-oxidizing organisms for complete oxidation of ammonia to nitrate.

CRITERIA FOR THE ISOLATION OF PURE CULTURES AND COMMUNITIES

A community is not an isolated cell line, nor is it a mixture of unrelated organisms. Thus, the criteria for obtaining and characterizing communities are different from those used to obtain isolated cell lines or mixed cultures and enrichments. Each community is an individual network of organisms, the members of which are normally adapted to propagate both as individuals and as part of one or more associations. Individuals associate through specific behavioral adaptations which allow them to position themselves within an association and to interact physically, chemically, and biologically (34, 46, 50, 51). These adaptive positioning mechanisms often involve motility, chemoreceptors, chemotaxis, chemoadherence, production of exopolymers, and phagocytosis. Cultivation of these communities sometimes requires that community inputs and outputs be maintained. This is necessary to prevent the depletion of substrates and the accumulation of products as well as to sustain the quasi-steady-state dynamics of many interactive associations. Methods used for community culture are described later in this chapter. Some of the criteria used for isolating communities and for other types of cultures are discussed briefly below.

Pure Cultures

Pure cultures are normally obtained by isolating individual cells in dilution tubes or spread plates. Aseptic technique is then used to maintain them in isolation while studying their properties. Until the development of fluorescent molecular probes and confocal laser microscopy, this was the only way to obtain enough homogeneous cell material to perform the chemical analyses (14) necessary for identification and study.

Isolated cell lines are used to produce the fluorescent molecular probes used to detect and enumerate specific microorganisms in environmental samples (fluorescein isothiocyanate-conjugated mono- and polyclonal antibodies, 16S rRNA probes, etc.). They are also useful as test organisms to predict the effects of environmental contaminants on key organisms involved in nutrient cycling. Although generally less effective than degradative microbial consortia, pure cultures have also been used as inoculants in bioremediation of aquatic and terrestrial environments contaminated with halogenated organics.

There are two criteria normally used to confirm that a pure culture has been obtained. One is that streak plates on several different media yield a single colony type. The other is that phase microscopy and Gram staining reveal a single cell type. However, there are situations in which microbial associations have been mistaken for pure cultures despite these precautions. This occurs when the organisms involved are so tightly associated that they cannot readily be cloned unless both organisms are present within each colony, as was once the case with *Methanobacterium omelianskii* (8).

Enrichment Cultures

Enrichments are frequently used as the first step in obtaining pure cultures of organisms involved in biogeochemical cycles or in the degradation of environmental contaminants. Enrichment culture methods, which have been reviewed by Holt and Krieg (38), make it easier to isolate organisms in dilution tubes and streak plates. For example, some enrichment media provide no biologically available form of nitrogen and thus enrich for nitrogen-fixing organisms. Others provide only a single halogenated organic compound as the energy source, and these enrich for specific degradative organisms.

An effective enrichment results in the predominance of a single bacterial species (1). However, enrichments have also been used to obtain degradative consortia (16), although they were not specifically designed for this purpose. The variability in environmental conditions during the time course of enrichment (depletion of substrate and accumulation of end products) generally produces corresponding variability in the species composition and spatial arrangement of organisms within consortia and communities (9). Cultivation of defined communities thus requires that community culture methods be used instead of enrichment culture to better define the environment in terms of the concentration and flux of both substrates and products.

The difference between community culture (9, 10, 16) and enrichment culture (38) is that an enrichment culture uses environmental stress to enrich for a specific species and reduce biological diversity (1), while community culture often results in an increase in diversity. To be effective, an enrichment culture must impose an environmental stress that selects only those organisms with the desired adaptations. In the words of Beijerinck, "Enrichment culture experiments can be called perfect or imperfect. In a perfect experiment a single species is isolated . . ." (1). In contrast, a community culture supports a diverse community of organisms and never leads to the isolation of a single species. Consequently, in community culture the conditions are not necessarily selective, whereas in an enrichment culture they must always be selective.

Community Cultures

Community culture refers to the methods used to isolate and cultivate consortia, biofilm communities, bioaggregates, microecosystems, and other networks of interacting microorganisms growing in association with one another. The primary difference between community culture and most other methods of cultivation is that the culture is defined by defining the environment rather than by isolating cells through the use of aseptic technique. The species composition of the resulting culture is defined through the formation of self-organizing microbial associations. To confirm that a community culture has been obtained, it is thus important to demonstrate that the association of organisms meets criteria of autopoiesis, synergy, communality, and homeostasis.

Meeting the criterion of autopoiesis requires that the presumptive community be self-organizing. This implies that the structure (species composition) and architecture (spatial arrangement) of the association arise spontaneously and remain relatively constant over time (or vary in a regular and reproducible way) if the environment is constant and adequately controlled. The primary test of autopoiesis is to confirm that the cultured association forms spontaneously under septic conditions. It then remains stable even when challenged by pure cultures which are foreign to the association. The only exception is when one organism periodically displaces another from its functional role (niche) in the network as a result of subtle environmental variations or the lack of an optimally adapted organism when the community initially formed. The mechanisms of autopoiesis include interconnections between species populations based upon chemoreceptors and exopolymers involved in intrageneric and intergeneric coaggregation. These mechanisms also include physiological and behavioral adaptations.

The primary test of synergy is that the cultured association proliferate and convert abiotic resources to biotic resources more effectively than its component members when cultivated individually. This can be demonstrated if the association has a broader habitat range than its individual members, if it is more resilient to environmental stress, or if it has more favorable collective growth constants (maximum specific growth rate, half-saturation constant, or cell yield). The criterion of synergy is thus met by comparing the performance of individuals isolated from the community with that of the cultured community itself and with that of mixed cultures of the isolates. If the performance of the native association is superior to the performance of the individual associates, and if this performance is partially regained by recombining isolated members in mixed culture, the criterion of synergy has been adequately met. It is never possible to fully test the importance of all community members in synergy because of the difficulty of testing for synergy under every conceivable environmental challenge. A member which seems insignificant or detrimental in mixed culture may be essential for successful in situ proliferation of the community. The role of predators is a good example. In the laboratory, a mixed culture containing predators might not perform as well as one without. However, when the predator-free culture is used as an environmental inoculant, its members may not have retained the adaptations necessary to avoid predation and thus be unable to function effectively in situ.

The primary test of communality is to determine the response of the presumptive community to various environmental gradients. If ecotones occur, then they define boundaries between communities and confirm the existence of a communal association. If ecotones do not occur, then the organisms are not necessarily organized into recognizable associations whose interactions are significant at the community level. Ecotones are present if the discontinuities in cell distribution along environmental gradients are sharp and distinct for organisms in association compared with individuals cultured along the same gradients but in isolation. Ecotones arise because of the need for biological systems to reorganize in response to environmental variation. As environmental constraints change, one set of relationships must undergo a transformation to another set of relationships. The transition from one network of relationships to another is normally sharper than the transition from one set of unrelated organisms to another. This is due to the difficulties associated with generating a gradual community response to changing environmental conditions as opposed to generating gradual individual responses. A gradual change from one community to another requires a coordinated transition in the interactions linking community members (including chemoreceptors, exopolymers, and behavior). This type of coordination is difficult. However, a gradual transition from one set of unrelated organisms to another requires only that the individuals respond independently. No coordinated reorganization is necessary, which makes a more uniform and gradual transition possible.

Homeostasis implies that the presumptive community creates a favorable and stable microenvironment within an unfavorable macroenvironment. It is not enough to show that an association is protected by producing an exopolymer which shields it from periodic environmental stresses. Meeting the criterion of homeostasis requires that, as in the case of synergy, the homeostasis of the association is greater than the homeostasis of its members and is partially regained by recombining isolated associates in mixed culture. Not all microbial associations fulfill the criterion of homeostasis, and homeostasis should thus be considered only one of many potential mechanisms of synergy.

Most microbial associations can be considered as either consortia, communities, or microecosystems. Consortia consist of two or more organisms associated through physical attachment in a specific spatial arrangement (37). The term is most commonly used at present to describe the mixtures of organisms which occur in degradative enrichments. However, this usage is presumptive until the presence of consortia has been confirmed. Communities consist of two or more cell lines associated through any of several mechanisms, including attachment, exchange of primary or secondary metabolites, and behavioral adaptations. Communities may also include one or more consortia as well as subcommunities. A microecosystem consists of a community (or set of communities and consortia) with multiple levels of biological organization and well-defined boundaries, exhibiting internal cycling of C, N, S, and other elements between their oxidized and reduced forms, possessing multiple trophic levels, spatial organization, and a favorable internal microenvironment(s) within a relatively unfavorable macroenvironment, and in a quasi-steady-state condition (changes in the system are relatively small compared with the rate of internal and/or external nutrient cycling). Microbial associations thus often involve layered networks, with the possibility of functional proliferation strategies arising at any layer or level.

Community culture should be used preferentially over other culture methods when the objective is to produce a microbial association to be used as an inoculant in an in situ application or when the objective is to understand the mechanisms of community-level environmental processes and the effects of toxicants on those processes. Community cultures are inherently stable and durable under septic conditions. They contain predators and parasites to which pure and mixed cultures often become prey. Specific applications include degradative biofilm communities for use in bioremediation, bioaggregates and biofilms for use in restoring the activity of municipal and industrial wastewater systems following periodic bulking or exposure to toxic chemicals, inoculants for use as biological control agents, inoculants for plant growth promotion, as well as starter cultures for silage and composting. Community inoculants are important because environments which support the growth of the communal associations may not necessarily be adequate to support the formation of these communities.

Community cultures are also useful in understanding the mechanisms of environmental processes and the effects of toxicants on these processes. It has been postulated that nitrification is inhibited in climax ecosystems as a result of the effects of allelopathic plant compounds (66), and this hypothesis has been questioned (6, 7). The nitrification process is always mediated by associations of ammonia-oxidizing (nitrosofying) and nitrite-oxidizing (nitrifying) organisms. Consequently, the effects of allelopathic compounds should be tested not only for pure cultures of the organisms involved but also for community cultures, and this has not yet been done. This analysis would permit the detection of community-level inhibition as well as mechanisms of inhibition at the organismal level. Potential mechanisms of community-level inhibition include increasing the susceptibility of nitrifying organisms to predators, blocking chemoreceptors necessary for coaggregation of the organisms involved in nitrifying associations, and inhibiting the production of exopolymers that serve as the matrix within which many microbial associations form and which protect them from periodic desiccation, salt stress, and other environmental stresses.

It is also important to test the effects of new pesticides and herbicides on the microbial associations involved in nutrient cycling. Testing for inhibition only at the cellular level may result in false-positive results due to the relative susceptibility of isolated cell lines compared with the durability of microbial associations. It may also result in false-negative tests if the mechanism of inhibition is targeted at community-level biological adaptations which inhibit the mechanisms of interaction between organisms without necessarily inhibiting the individual member of the association when cultured in isolation.

Mixed Cultures

A mixed culture is produced by combining two or more pure cultures and excluding other organisms by using aseptic technique. Mixed cultures are not necessarily microbial communities. If the organisms within a mixed culture did not all evolve within the same adaptive association, then the culture may be inherently unstable, with one or more organisms gradually displacing others.

The primary value of mixed culture is to demonstrate synergy among the isolated members of consortia, communities, and microecosystems as well as to establish the mechanisms of synergy. If the members of a community are synergistic, then various mixtures of isolated individuals should proliferate more effectively than the isolates themselves. If a specific trait of a single community member is postulated to be involved in the formation of an association (for example, a chemoreceptor or exopolymer), then the performance of a synthetic community (mixed culture of isolates from a native community) containing an adaptation-negative mutant can be compared with that of a synthetic community (mixed culture) containing the parental strain. The difference in performance between the individual cell lines, the adaptation-negative cell line, the native community, the synthetic community (mixed culture of community members), and the synthetic community with the adaptation-negative mutant substituted for its parental strain can be used to elucidate the specific mechanisms of community-level interactions. In many cases, a community-level adaptation may have a detrimental effect on the proliferation of the isolated individual and a positive effect on the proliferation of its community (i.e., the adaptation-negative mutant outperforms its parent when the two are compared as isolated cell lines but not when they are compared as members of an association). Experiments of this type are needed to establish a design theory for the construction of degradative and other types of microbial communities for use in agriculture, industry, and the environment.

COMMUNITY THEORY IN ENVIRONMENTAL MICROBIOLOGY

Both reductive and nonreductive experimental information is required to adequately understand microbial communities. As suggested by Marshall (56), the analysis of communities includes defining population dynamics in living communities, defining the physicochemical characteristics of the microenvironments within communities, and defining the metabolic processes carried out by individual bacteria in communities. It is also important to characterize them as competing networks and as mechanisms of larger environmental systems (16). This process involves defining the stability of specific community networks under defined laboratory conditions, the habitat range of organisms alone and in association with their community network, the role of each organismal trait in proliferation of the organism versus proliferation of the community (habitat range of parental strains and adaptation-negative mutants alone and in association with other organisms), whether the final endpoint of biological adaptation in a specific environment is a single species or a community, whether there is only one community network that can become established in a specific environment or several, whether the presence of one network precludes the development or encroachment of another, whether there are sharp ecotones between communities, whether the community occupying a specific environment has been optimized through evolution in terms of its ability to propagate and to most efficiently convert abiotic resources to biotic resources, whether spatial or temporal pathways are necessary for the development of specific microbial associations, whether favorable community microenvironments are homeostatic (stable despite periodically unfavorable changes in the macroenvironment), and whether the species composition and spatial architecture of the community depend on the timing of immigration or emigration by specific species.

These questions are of importance not only to environmental microbiology but also to ecology and evolution in general (16). Microbial ecologists thus have a unique opportunity to reexamine the assumptions of ecology and evolution from an experimental perspective as opposed to descriptions of the fossil record, genetic sequences, and natural communities. Prokaryotic communities are more likely to form and are easier to study than plant and animal communities because of the large populations which can be contained within an extremely small space (10^{12} cells per ml), the time period over which bacterial communities have evolved (3.5 billion years for unicellular organisms, compared with 0.6 billion years for marine animals and 0.4 billion years for vertebrates), and the limited physical capacity of prokaryotes for storing large quantities of genetic information, which thus requires a greater degree of interaction in optimizing proliferation. The small size of prokaryotes also prevents the formation of diffusion gradients around individual cells (42a), and this necessitates the formation of multicellular associations if they are to effectively control their microenvironment and optimize the probability of reproductive success.

SPECIFIC EXAMPLES OF COMMUNAL ASSOCIATIONS

Examples of cultured consortia and communities include heterotrophic-phototrophic associations, interspecies hy-

drogen transfer, and degradative consortia. In each case, the association of two or more lineages produces a group of organisms that becomes a unit of proliferation (and hence a unit of evolution) in itself.

The *Chlorochromatium aggregatum* Consortium

The formation of the C. *aggregatum* association occurs through attachment via small cups located on the surface of the sulfate reducer which connect it with a sulfide oxidizer (19, 20, 25, 64). However, connections between individual organisms can also arise through endosymbiosis, behavioral adaptations, antibiotic production, and numerous other chemical, physical, and behavioral mechanisms in addition to attachment. Once the individuals have connected, they acquire new characteristics as an association which they lacked as individuals. For example, the association may have collective growth constants which are different from the growth constants of its individuals. In this case, the organismal half-saturation constants for sulfate and sulfide would be lower than the same constants for the association because of internal recycling of sulfide and sulfate. This would allow the organisms to proliferate most effectively under low-sulfide and low-sulfate conditions only when in association.

In addition to the metabolic synergy which results from this association, there is also a physical synergy. The photosynthetic symbiont is the only photosynthetic bacterium found in anaerobic hypolimnia which does not have either gas vacuoles or flagella. These adaptations are necessary for hypolimnetic organisms to avoid being sedimented out. They allow the organisms to position themselves along shifting vertical gradients of sulfide, oxygen, and light (19, 20). Consequently, the photosynthetic symbiont is completely dependent on the sulfate reducer to properly position the association to optimize the proliferation of both organisms.

The *Anabaena* sp.-*Zoogloea* sp. Consortium, a Heterotrophic-Phototrophic Interaction

The association between *Anabaena* and *Zoogloea* species was originally observed in the epilimnia of eutrophic lakes and subsequently cultivated in continuous culture (11, 70). The metabolism of the phototroph and that of the heterotroph were complementary not only in terms of recycling of carbon but also in terms of minimizing photorespiration and preventing pH inhibition. Numerous other similar associations have also been demonstrated (39, 48, 53, 58–60), including the stimulation of nitrogen fixation in cyanobacteria through the preferential attachment of bacteria to heterocysts as opposed to vegetative cells. These and related consortia have been reviewed by Paerl and Pinckney (61).

The *M. omelianskii* Consortium, an Interaction Based on Interspecies Hydrogen Transfer

Methanobacillus omelianskii is a symbiotic association of two bacterial species originally thought to be a single pure culture (8). Many associations of this type break down during isolation as a result of loss of the steady-state inputs and outputs that connect them with their macroenvironment. In this case, however, the metabolic connection is strong enough to sustain the association even in batch culture.

The association is based on the transfer of molecular hydrogen from one organism to the other (interspecies hydrogen transfer). One associate produces acetate and H_2 from ethanol, while the other produces methane from H_2 and CO_2. This relationship is in contrast to the S- and C-cycling association mentioned in the examples cited above, in which electron donors and acceptors are continuously provided by repeatedly cycling an element through various oxidation states. *M. omelianskii* is one example of numerous methanogenic associations based on interspecies hydrogen transfer or formate transfer (2, 4, 23, 24, 42, 54, 67, 74, 77).

Diclofop-Degrading Communities: Interactions Based on the Degradation of Halogenated Hydrocarbons

Studies of diclofop degradation have shown that there is a temporal and spatial order to the development of degradative communities (85–89). These associations contain several structured consortia which degenerate when the biofilm is irrigated with more labile substrates but not when it is irrigated with refractory compounds. Some members of the consortia concentrate diclofop in their exopolymers prior to degradation, while others assimilate it directly. Diclofop, like most other compounds containing aromatic structures, is fluorescent, and its distribution within degradative biofilms (as well as the distribution of many of its degradation products) can thus be monitored directly by using confocal laser microscopy. Use of fluorescein isothiocyanate-conjugated lectins shows that diclofop-binding sites consist of polymers with a relatively high fucose density. The fluorescence of diclofop can also be used to monitor its movement up the food chain into protozoa feeding on diclofop-containing cells and feeding on bacterial exopolymers with sorbed diclofop.

Diclofop degradation is one of many situations in which a community network is required for complete degradation of refractory molecules (27–29, 40, 47, 49, 68, 71, 72, 76). Degradative networks may contribute to chemical communication and cometabolism between different microbial species, a larger and more flexible genetic resource pool from which metabolic pathways may evolve, formation of favorable microenvironments as in the case of reductive dechlorination, and protection within the biofilm exopolymer matrix against perturbations (5, 31, 51).

COMMUNITY CULTURE METHODS

Describing a community in situ does not necessarily ensure that it is well enough understood to be controlled or effectively utilized in environmental applications. If a community network has been adequately elucidated at the community level, then it should be possible to cultivate it in the laboratory under defined conditions, complete with predators and parasites, and without using aseptic technique. This is a criterion of success in environmental microbiology, just as the isolation of a pure culture is a criterion of success in understanding and controlling disease using the germ theory and Koch's postulates.

When one is culturing communities, the environment, rather than aseptic technique, defines the culture (10, 16). Accurately controlling the environment, in terms of substrate concentration and flux as well as spatial and temporal variations, is thus crucial. As stated by Senior et al. (71), "It is reasonable to postulate that interacting microbial communities are common in nature; however, very few microbial associations of this type have been defined, principally because inappropriate enrichment and selection techniques have been used to isolate them from natural environments." Defining the environment requires the provision of a steady-state system which is capable of supplying the necessary inputs and outputs to sustain a dynamic microbial community (71). Hence, in addition to defining substrate concentrations, it may also be necessary to define substrate avail-

ability in terms of flux as well as spatial and temporal variability. Similarly, it is important that the flux of end products moving from the microenvironment to the macroenvironment be continually removed as they would be in situ. Community-level studies thus involve microbiological systems which provide multidimensional environmental gradients for studies of habitat range (11, 12, 17, 26, 52, 81–84) as well as providing the spatial (9, 10, 12, 14, 86) and temporal (69) pathways necessary for communities to develop.

Batch culture systems are normally inadequate for defined studies of community cultures because of the lack of environmental control as the culture develops. Consequently, the discussion below includes only those culture systems which provide adequate environmental control of substrate flux and concentration throughout the time course of cultivation. Although batch systems may be used to cultivate microbial associations, they should be used with caution and subcultured frequently.

The chemostat, nutristat, microstat, and continuous-flow slide culture (CFSC) are among the methods commonly used to cultivate consortia and communities. The chemostat provides control of substrate flux and growth rate (assuming that there is no wall growth). However, the concentration of substrate cannot be directly controlled and must be determined empirically once a quasi-steady-state condition has been obtained. The nutristat provides control of the concentration and temporal variation of substrate, although the growth rate and flux must be allowed to vary so that a quasi steady state can be obtained. CFSC provides control of the flux, concentration, and temporal variation of substrate (but not direct control of growth rate). The microstat provides control of substrate flux, concentration, temporal variation, and spatial variation of substrate (but not direct control of growth rate).

The Chemostat

Chemostats and other continuous cultures are among the most widely used systems for cultivating microbial communities and consortia (33, 35, 49, 62, 65, 69, 71, 73, 79, 80, 87). They provide a quasi-steady-state system with inputs and outputs necessary to sustain dynamic networks of interacting microorganisms (16, 71). The volume of the culture is constant, and medium continually flows into the culture at the same rate that effluent flows out. The growth of organisms within the culture gradually reduces the concentration of the limiting nutrient until a quasi steady state is obtained in which the dilution rate (flow rate/volume) equals the specific growth rate (57). If a community is cultivated, the mean growth rate of each member within the community is equal to the dilution rate once the culture is substrate limited and has had sufficient time to reach a quasi-steady-state condition.

Chemostat systems are particularly effective in providing a continuous supply of refractory compounds over the long time periods necessary for degradative consortia and communities to form (49, 65, 71, 87). They are also used to cultivate algal-bacterial bioaggregates and anaerobic digestor granules. These applications generally require a modified outlet which ensures that flocs, granules, and bioaggregates are diluted at the same rate as suspended cells (30).

The Nutristat and Gradostat

Rutgers et al. (69) devised a new continuous-culture device, the nutristat, providing defined concentrations of pentachlorophenol (PCP) and gradually increasing the PCP concentration. This approach, based on a temporal substrate gradient, resulted in higher growth rates and shorter acclimation times in the selection of PCP degraders than batch or traditional chemostat systems. In most continuous-culture systems, the substrate concentration is allowed to vary while the growth rate (dilution rate and flux of substrate) is held constant. The concentration of substrate is continuously monitored and controlled via an automated system. Gradually increasing the concentration of substrate in this system provides sufficient time for organisms and associations to gradually adapt.

The gradostat (36, 52, 84) is a multistage chemostat that is used to provide spatial gradients of concentration as opposed to the temporal gradients of the nutristat. It consists of a series of bidirectionally linked chemostats which provide a physical pathway for community development. The gradostat normally consists of five separate continuous-culture vessels connected in series. A medium containing one substrate is pumped into the first stage, and a second substrate is pumped into the last stage. Effluent is removed at the same rate from both ends. This produces opposing gradients in one dimension, and all possible proportions of the two substrates are represented. Representing all concentrations, as well as all proportions of the two substrates (as in the case of the microstat described below), would require a five-by-five array of linked chemostats, with substrates supplied from two adjacent edges of the array and with effluent being removed from the two opposing edges.

Dual-Dilution Continuous Culture

Although mathematical models of continuous culture often assume that no wall growth occurs, some chemostat studies have focused on biofilms of attached organisms (21, 32, 41, 55, 78, 89). A dual-dilution continuous culture was used by Caldwell and Lawrence (18) as a behavioral enrichment to select an aggressive surface colonizer from soil that was not only effective at colonizing surfaces but also highly efficient at emigrating from surface microenvironments and colonizing new surface environments. In this culture system both the aqueous and the solid phase of the culture were continually diluted (replaced) at a constant rate. This system was subsequently used by Korber et al. (45) to evaluate the role of motility in colonization by comparing the performance of nonmotile mutants with that of parental strains. Thus, the primary value of this approach compared with other methods is that one obtains communities which are capable of rapidly emigrating and colonizing new surfaces as they become available. This adaptation would be of value in the colonization of growing root hairs and tips in the rhizosphere or in the recovery of biofilm communities from periodic grazing by protozoa and invertebrates.

The Rototorque Annular Bioreactor

One of the best characterized continuous-culture systems for studying surface-associated organisms is the Rototorque annular bioreactor (22). The Rototorque consists of two (one fixed and one rotatable) concentrically mounted cylinders. The resistance to rotation caused by biofilm growth provides data on friction and shear. If biofilm accumulation is of interest, then rates of dilution are set to exceed the growth rate of the organisms. In this case, only sloughed cells are detected in the planktonic phase. Alternatively, if both biofilm and planktonic growth are of interest, the dilution rate is set below the maximum specific growth rate and planktonic cells are able to grow more rapidly in the aqueous phase than they are diluted out. Biofilms are normally stud-ied by removing and observing coupons (75, 78). This system is useful in cultivating biofilm communities from rotating biological contactors and other systems in which the biofilm community is several millimeters or more thick and subject to sloughing. It also provides sufficient cell material for direct chemical analysis of biofilm material by standard chemical analysis.

Continuous-Flow Slide Culture (CFSC)

The primary advantage of CFSC over other methods of community culture is that it provides the opportunity to view the development of communities by using various forms of microscopy. This ability, coupled with the use of fluorescent molecular probes, provides a very effective means of studying community-level processes nondestructively. Another advantage is that the concentration and flux of substrate are precisely and directly controlled. In the chemostat, the concentration of substrate varies until it reaches a quasi-steady-state concentration which must be empirically measured. In CFSC, substrate concentrations can also be changed almost instantaneously, permitting the use of pulses of fluorescent probes and test compounds.

However, this culture system is appropriate only for study of biofilm communities. Nonadherent communities are more readily studied in chemostat systems. A drawback of the approach is that the quantity of biomass available for conventional chemical and biochemical analyses is very small and fluorescent molecular probes are not sufficient to quantify many important processes, including the consumption and production of most substrates. This drawback can sometimes be overcome by using a column of glass beads to provide sufficient cell material for traditional analysis while using CFSC, connected in series or parallel, for nondestructive analysis using fluorescent molecular probes.

CFSC has been particularly helpful in expanding the scope of laboratory studies to include microbial communities. It allows the study of microorganisms at multiple levels of organization, including the plasmid, cell, consortium, community, and microecosystem (10, 14, 16). Studies in CFSC have shown that spatial organization, including intrageneric and intergeneric coaggregation, is necessary to optimize the activity of degradative and other biofilm communities (43, 85, 87, 89). A glass flow cell mounted on a microscope stage is irrigated with stream water, groundwater, wastewater, the effluent from a chemostat, a defined medium, etc. The community is analyzed during the time course of its development by using fluorescent molecular probes, epifluorescence microscopy, confocal laser microscopy, and digital image analysis (14). These methods allow quantification of biochemistry, genetic sequences, genetic expression, growth, metabolism, etc., as well as the corresponding three-dimensional relationships among them. Information concerning confocal laser microscopy, digital imaging, and fluorescent molecular probes is provided in chapter 5 of this volume. Prior to development of these technologies, cells within communities were cloned as isolated cell lines to produce sufficient material for conventional chemical analyses. Although confocal laser microscopy is optimal, epifluorescence microscopy can also be used quantitatively if brilliant black (32) is used as a quenching agent in the irrigation solution. This provides a physical method of obtaining an optical thin section by removing light originating from beneath the x-y (horizontal) plane.

Square glass capillaries similar to those used by Perfil'iev and Gabe (63) are available commercially (Friedrich and Dimmock Inc., P.O. Box 230, Millville, NJ 08332; phone:

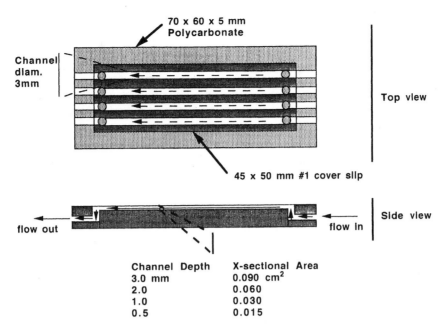

FIGURE 1 Schematic diagram of the multichannel flow cell. The use of flow cells containing multiple channels simplifies multiple comparisons, replication, and controls.

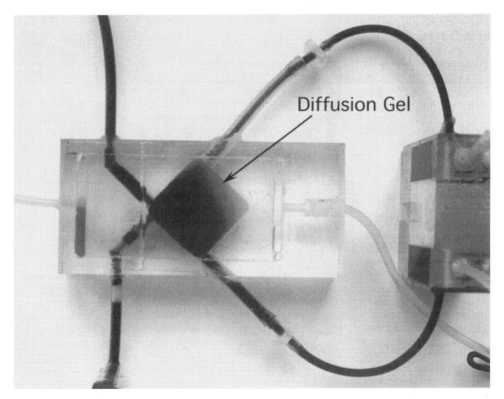

FIGURE 2 Photograph of the Wolfaardt microstat. Two-dimensional steady-state diffusion gradients are created by diffusion through an agar or acrylamide gel. Biofilm communities are cultivated on the surface of the gradient which forms within the gel. The response of communities to the two-dimensional gradients is referred to as an environmental response surface (ecogram) (shown in Fig. 3).

[609] 825-0305) in sizes as small as an internal width of 50 μm with a 25-μm wall thickness. However, conventional coverslips assembled with silicone adhesive provide a larger and more uniform viewing area (18). In most cases, multi-channel flow cells are required for experimental treatments, duplicates, and controls. These are constructed using a polycarbonate base containing several irrigation channels as shown in Fig. 1. Coverslips are bonded to cover the base such that the gaps between the coverslips are located between channels (44, 85). UV transparent coverslips (zinc titanium glass; Corning Glass code 0211) and oversize coverslips are available by special order from Corning (Corning Glass Works, P.O. Box 5000, Corning, NY 14830; phone: [607] 737-1632).

Microstats

Microstats are used for the cultivation of biofilm communities as opposed to planktonic communities. They do not lend themselves to microscopy, as in the case of CFSC, but they do provide spatial pathways which allow the formation of communities that might not otherwise be able to develop. Spatial gradients also allow ecotones to form. Ecotones confirm that communities have been obtained, define the habitat range of a specific community, and can be used to determine whether the same community is responsible for the degradation of two different compounds. Although pulses of test compounds or fluorescent molecular probes can be used in the irrigation solution, the gradients cannot be turned on and off. Thus, the environment cannot be changed quickly as in the case of CFSC.

The microstat provides a two-dimensional gradient of steady-state environmental conditions within a surface microenvironment (9, 10, 13, 15, 16, 86). This makes it possible to quantify the effect of environmental variation on the formation of communities and to quantify the distribution of various community networks. It also provides spatial pathways of community development. For example, a community degrading toxicants under unfavorable conditions may be unable to form unless a gradient of toxicant allows the community to form initially at low concentrations and gradually expand toward higher concentrations (once it is optimally configured in terms of community structure and biofilm architecture). In some cases a primer gradient, consisting of a labile toxicant analog, may be necessary perpendicular to the primary toxicant gradient. Primer gradients provide a physical pathway along which the community may reorganize and in which labile compounds permit the cometabolism of more refractory compounds. Temporal gradients, as described by Rutgers et al. (69), are also effective in producing degradative consortia. However, use of two-dimensional gradients allows the community to determine its own time frame and pathway for development. This avoids the possibility that the development of the community will not be able to keep up with the pace of change or that the sequence of changes selected will be inadequate.

The Wolfaardt microstat, shown in Fig. 2, uses a gel as the diffusion medium to generate two-dimensional steady-state diffusion gradients. The response of the community is measured by using a fluorescent probe that is normally quantified by scanning confocal laser microscopy (see chapter 5 of this volume) as shown in Fig. 3.

The Laminar Wedge Microstat

The laminar wedge microstat does not rely on diffusion through a gel to generate a two-dimensional gradient. Instead, the gradient is produced by diffusion through "wedge

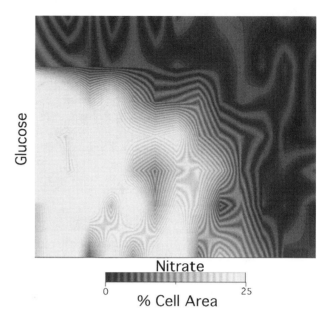

FIGURE 3 The response surface for a 5-day-old biofilm community. The surface of the gel was continuously irrigated with a minimal salts medium containing all growth requirements except carbon and nitrogen. Carbon and nitrogen gradients originated at two edges of the gel matrix as shown. The biofilm was stained with Nile red (5 μg of Nile red ml^{-1} in 50% aqueous glycerol; Eastman Kodak Co., Rochester, N.Y.) A total of 91 scanning confocal laser microscope images were taken at 2-mm intervals along the width and length of the gel surface. These were analyzed to determine the extent of biofilm coverage at each location. The ecogram shown here was prepared by using SpyGlass Transform (banded gray-scale legend) and shows an increase in surface coverage at locations near the carbon and nitrogen sources.

lamina" a few hundred micrometers in thickness. This greatly decreases the time necessary to obtain steady-state gradients and allows the gradients to be turned on and off during the incubation of a biofilm community. However, it does not produce perpendicular gradients containing all combinations and proportions of substrate concentrations as in the case of the original microstats.

The laminar wedge microstat shown in Fig. 4 produces a two-dimensional steady-state gradient within a 100-μm flow lamina which moves across the surface of a coverslip. The laminar wedges are created by using wedge-shaped inlet tubes. The upper wedge contains one test compound, and the lower wedge contains the other. The wedges provide one-dimensional opposing gradients that represent all proportions of the two test compounds. As the wedge lamina moves downstream beneath the coverslip, both test compounds gradually diffuse downward into the bulk phase of the irrigation solution. This gives the gradients a second dimension in which the various proportions of the test compounds are supplied at gradually decreasing concentrations.

Biofilm communities are cultivated on the irrigated surface of the coverslip and form in response to the environmental gradients. Thicker biofilm communities can thus develop without distorting the gradients. It also allows quantification of fluorescent molecular probes by confocal microscopy (see chapter 5 of this volume for more information) without disassembly, as well as allowing the gradients

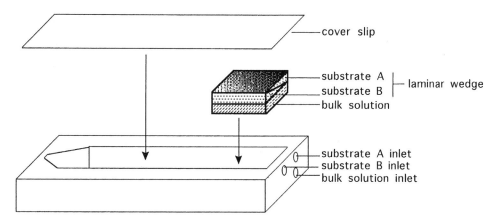

FIGURE 4 Diagram of the laminar wedge microstat. Two-dimensional steady-state gradients are produced by a $100\text{-}\mu$m flow lamina which moves across the surface of a coverslip. The wedges are created by using wedge-shaped inlets for each of the two test compounds. A third inlet supplies the bulk irrigation solution. Biofilm communities are cultivated on the irrigated surface of the coverslip and form in response to the environmental gradients. An epifluorescence microscope or scanning confocal laser microscope is used to quantify the environmental response surface for the community, using the appropriate fluorescent molecular probes. A quenching agent may be added to a carrier solution (within the bulk phase beneath the gradient lamina) so that the response surface can be quantified by using a UV light and video camera. The advantage of the laminar wedge microstat over the Wolfaardt microstat is that the community can be observed continually and nondestructively on both macro- and microscales, using the appropriate fluorescent molecular probes. It also allows the development of thicker biofilms without significantly perturbing the gradients. However, it does not provide as broad a range of concentrations at all proportions of the test compounds (see text for explanation).

to be turned on and off during a time course. This permits response surfaces to be mapped more quickly than would be possible with confocal laser microscopy or epifluorescence microscopy. For example, quantifying 10 transects with 10 points per transect involves a test grid with 100 measurements, each of which would require at least 1 min because of the time required for the laser to scan each field several times and to average the resulting images. This process takes more than an hour, and the value for the last measurement may be considerably different from that for the first measurement. This discrepancy can result in the distortion of response surfaces. In addition, some measurements made by using microscopy are highly dependent on positioning of the objective lens at the same distance beneath the coverslip during each of the measurements. This results in random variability between measurements. These problems are avoided if a television camera is used to produce a response surface which is then calibrated by using microscopy (if necessary) at selected points along the isopleths of the surface. The time for digitization with a television camera normally ranges from 1/60 of a second to 10 s, depending on the detector and digital averaging method used.

We acknowledge the U.S. Department of Energy, Environment Canada (National Hydrology Research Institute and GASReP), the Saskatchewan Agricultural Development Fund, and the National Science and Engineering Research Council of Canada for financial support.

REFERENCES

1. **Beijerinck, M. W.** 1901. Enrichment culture studies with urea bacteria. *Zentralbl. Bakteriol. Part II* **7:**33–61.

2. **Bochem, H. P., S. M. Schoberth, B. Sprey, and P. Wengler.** 1982. Thermophilic biomethanation of acetic acid: morphology and ultrastructure of a granular consortium. *Can. J. Microbiol.* **28:**500–510.

3. **Bochner, B.** 1989. "Breathprints" at the microbial level. *ASM News* **55:**536–539.

4. **Boone, D. R., R. L. Johnson, and Y. Liu.** 1989. Diffusion of the interspecies electron carriers H_2 and formate in methanogenic ecosystems and its implications in the measurement of K_m for H_2 and formate uptake. *Appl. Environ. Microbiol.* **55:**1735–1741.

5. **Bouwer, H.** 1989. Transformations of xenobiotics in biofilms, p. 251–267. *In* W. G. Characklis and P. H. Wilderer (ed.), *Structure and Function of Biofilms*. John Wiley & Sons, Toronto.

6. **Bremner, J. M., and G. C. McCarthy.** 1988. Effects of terpenoids on nitrification in soil. *Soil Sci. Soc. Am. J.* **52:**1630–1633.

7. **Bremner, J. M., and G. C. McCarthy.** 1993. Inhibition of soil nitrification by allelochemicals derived by plants and plant residues. *Soil Biochem.* **8:**181–218.

8. **Bryant, M. P., E. A. Wolin, M. J. Wolin, and R. S. Wolfe.** 1967. *Methanobacillus omelianskii*, a symbiotic association of two species of bacteria. *Arch. Mikrobiol.* **59:**20–31.

9. **Caldwell, D. E.** 1993. The microstat: steady-state microenvironments for subculture of steady-state consortia, communities, and microecosystems, p. 123–128. *In* R. Guerrero and C. Pedros-Alio (ed.), *Trends in Microbial Ecology.* Spanish Society for Microbiology, Barcelona.

10. **Caldwell, D. E.** 1995. Cultivation and study of biofilm communities, p. 1–15. *In* H. Lappin-Scott and J. W. Costerton (ed.), *An Introduction to Bacterial Biofilms.* Cambridge University Press, Cambridge.

11. **Caldwell, D. E., and S. J. Caldwell.** 1978. A *Zoogloea* sp. associated with blooms of *Anabaena flos-aquae. Can. J. Microbiol.* **24:**922–931.

12. **Caldwell, D. E., S. J. Caldwell, and J. M. Tiedje.** 1975.

An ecological study of sulfur-oxidizing bacteria from the littoral zone of a Michigan lake and a sulfur spring in Florida. *Plant Soil* **43**:101–114.

13. **Caldwell, D. E., and P. Hirsch.** 1973. Growth of microorganisms in two-dimensional steady-state diffusion gradients. *Can. J. Microbiol.* **19**:53–58.

14. **Caldwell, D. E., D. R. Korber, and J. R. Lawrence.** 1992. Confocal laser microscopy and computer image analysis. *Adv. Microb. Ecol.*, **12**:1–67.

15. **Caldwell, D. E., D. R. Korber, and J. R. Lawrence.** 1993. Analysis of biofilm formation using 2-D versus 3-D digital imaging, p. 525–665. *In* L. B. Quesnel, P. Gilbert, and P. S. Handley (ed.), *Microbial Cell Envelopes: Interactions and Biofilms.* Blackwell Scientific Publications, Oxford.

16. **Caldwell, D. E., D. R. Korber, G. M. Wolfaardt, and J. R. Lawrence.** 1996. Do bacterial communities transcend Darwinism? *Adv. Microb. Ecol.* **15**:1–72.

17. **Caldwell, D. E., S. H. Lai, and J. M. Tiedje.** 1973. A two-dimensional steady-state diffusion gradient for ecological studies. *Bull. Ecol. Res. Commun.* (Stockholm) **17**:151–158.

18. **Caldwell, D. E., and J. R. Lawrence.** 1986. Growth kinetics of *Pseudomonas fluorescens* microcolonies within the hydrodynamic boundary layers of surface microenvironments. *Microb. Ecol.* **12**:299–312.

19. **Caldwell, D. E., and J. M. Tiedje.** 1975. The structure of anaerobic bacterial communities in the hypolimnia of several michigan lakes *Can. J. Microbiol.* **21**:377–385.

20. **Caldwell, D. E., and J. M. Tiedje.** 1975. A morphological study of anaerobic bacteria from the hypolimnia of two Michigan lakes. *Can. J. Microbiol.* **21**:362–376.

21. **Characklis, W. G.** 1988. Model biofilm reactors, p. 155–174. *In* J. W. T. Wimpenny (ed.), *CRC Handbook of Laboratory Model Systems for Microbial Ecology Research*, vol. 1. CRC Press, Boca Raton, Fla.

22. **Characklis, W. G., G, A. McFeters, and K. C. Marshall.** 1990. Physiological ecology in biofilm systems, p. 341–393. *In* W. G. Characklis and K. C. Marshall (ed.), *Biofilms.* John Wiley & Sons, New York.

23. **Chartrain, M., and J. G. Zeikus.** 1986. Microbial ecophysiology of whey biomethanation: intermediary metabolism of lactose degradation in continuous culture. *Appl. Environ. Microbiol.* **51**:180–187.

24. **Chartrain, M., and J. G. Zeikus.** 1986. Microbial ecophysiology of whey biomethanation: characterization of bacterial trophic populations and prevalent species in continuous culture. *Appl. Environ. Microbiol.* **51**:188–196.

25. **Croome, R. L., and P. A. Tyler.** 1984. The microanatomy and ecology of *Chlorochromatium aggregatum* in two meromictic lakes in Tasmania Australia. *J. Gen. Microbiol.* **130**:2717–2724.

26. **Emerson, D., R. M. Worden, and J. A. Breznak.** 1994. A diffusion gradient chamber for studying microbial behavior and separating organisms. *Appl. Environ. Microbiol.* **60**:1269–1278.

27. **Federle, T. W., and G. M. Pastwa.** 1988. Biodegradation of surfactants in saturated subsurface sediments: a field study. *Groundwater* **26**:761–770.

28. **Federle, T. W., and B. S. Schwab.** 1989. Mineralization of surfactants in microbiota of aquatic plants. *Appl. Environ. Microbiol.* **55**:2092–2113.

29. **Ferry, J. G.** 1976 Anaerobic degradation of benzoate to methane by a microbial consortium. *Arch. Microbiol.* **107**:33–40.

30. **Feuillade, J., and M. Feuillade.** 1979. A chemostat device adapted to planktonic *Oscillatoria* cultivation. *Limnol. Oceanogr.* **24**:562–574.

31. **Fletcher, M.** 1984. Comparative physiology of attached and free-living bacteria, p. 223–232. *In* K. C. Marshall

(ed.), *Microbial Adhesion and Aggregation.* Springer-Verlag, New York.

32. **Food and Agricultural Organization of the United Nations.** 1963. *Food Colors*, vol. II, p. 136. Food and Agricultural Organization of the United Nations, Rome.

33. **Gottschal, J. C., and L. Dijkhuizen.** 1988. The place of continuous culture in ecological research, p. 19–49. *In* J. W. T. Wimpenny (ed.), *CRC Handbook of Laboratory Model Systems for Microbial Ecology Research*, vol. 1. CRC Press, Boca Raton, Fla.

34. **Haefele, D. M., and S. E. Lindow.** 1987. Flagellar motility confers epiphytic fitness advantages upon *Pseudomonas syringae. Appl. Environ. Microbiol.* **53**:2528–2533.

35. **Harder, W., J. G. Kuenen, and A. Matin.** 1977. Microbial selection in continuous culture. *J. Appl. Bacteriol.* **43**:1–24.

36. **Herbert, R. A.** 1988. Bidirectional compound chemostats: applications of compound diffusion-linked chemostats in microbial ecology, p. 99–115. *In* J. W. T. Wimpenny (ed.), *CRC Handbook of Laboratory Model Systems for Microbial Ecology Research*, vol. 1. CRC Press, Boca Raton, Fla.

37. **Hirsch, P.** 1984. Microcolony formation and consortia, p. 373–393 *In* K. C. Marshall (ed.), *Microbial Adhesion and Aggregation.* Springer-Verlag, New York.

38. **Holt, J. G., and N. R. Krieg.** 1994. Enrichment and isolation, p. 179–204. *In* P. Gerhardt, R. G. E. Murray, W. A. Wood, and N. R. Krieg (ed.), *Methods for General and Molecular Bacteriology.* American Society for Microbiology, Washington, D.C.

39. **Humenik, F. J.** 1970. Respiratory relationships of a symbiotic algal-bacterial culture for wastewater nutrient removal. *Biotechnol. Bioeng.* **12**:541–560.

40. **Jiménez, L., A. Breen, N. Thomas, T. W. Federle, and G. S. Sayler.** 1991. Mineralization of linear alkylbenzene sulfonate by a four-member aerobic bacterial consortium. *Appl. Environ. Microbiol.* **57**:1566–1569.

41. **Kieft, T. L., and D. E. Caldwell.** 1984. Chemostat and in-situ colonization kinetics of *Thermothrix thiopara* on calcite and pyrite surfaces. *Geomicrobiol. J.* **3**:217–229.

42. **Kinner, N. E., D. L. Balkwill, and P. L. Bishop.** 1983. Light and electron microscope studies of microorganisms growing in rotating biological contactor biofilms. *Appl. Environ. Microbiol.* **45**:1659–1669.

42a. **Koch, A. I.** 1991. Diffusion: the crucial process in many aspects of the biology of bacteria. *Adv. Microb. Ecol.* **11**:37–70.

43. **Kolenbrander, P. E., and J. London.** 1992. Ecological significance of coaggregation among oral bacteria. *Adv. Microb. Ecol.* **12**:183–217.

44. **Korber, D. R., G. A. James, and J. W. Costerton.** 1994. Evaluation of fleroxacin activity against established *Pseudomonas fluorescens* biofilms. *Appl. Environ. Microbiol.* **60**:1663–1669.

45. **Korber, D. R., J. R. Lawrence, and D. E. Caldwell.** 1994. Effect of motility on surface colonization and reproductive success of *Pseudomonas fluorescens* in dual-dilution continuous culture and batch culture systems. *Appl. Environ. Microbiol.* **60**:1421–1429.

46. **Korber, D. R., J. R. Lawrence, B. Sutton, and D. E. Caldwell.** 1989. Effects of laminar flow velocity on the kinetics of surface recolonization by mot$^+$ and mot$^-$ *Pseudomonas fluorescens. Microb. Ecol.* **18**:1–19.

47. **Lang, E., H. Viedt, J. Egestorff, and H. H. Hanert.** 1992. Reaction of the soil microflora after contamination with chlorinated aromatic compounds and HCH. *FEMS Microbiol. Ecol.* **86**:275–282.

48. **Lange, W.** 1971. Enhancement of algal growth in cyanophyta-bacteria systems by carbonaceous compounds. *Can. J. Microbiol.* **17**:303–314.

49. **Lappin, H. M., M. P. Greaves, and J. H. Slater.** 1985.

Degradation of the herbicide mecoprop 2-(2-methyl-4-chlorophenoxy)propionic-acid by a synergistic microbial community. *Appl. Environ. Microbiol.* **49:**429–433.

50. **Lawrence, J. R., D. R. Korber, and D. E. Caldwell.** 1992. Behavioral analysis of *Vibrio parahaemolyticus* variants in high and low viscosity microenvironments using digital image processing. *J. Bacteriol.* **174:**5732–5739.

51. **Lawrence, J. R., D. R. Korber, G. M. Wolfaardt, and D. E. Caldwell.** 1995. Surface colonization strategies of biofilm-forming bacteria. *Adv. Microb. Ecol.* **14:**1–75.

52. **Lovitt, R. W., and J. W. T. Wimpenny.** 1981. Physiological behavior of *Escherichia coli* grown in opposing gradients of oxidant and reductant in the gradostat. *J. Gen. Microbiol.* **127:**269–284.

53. **Mackenthun, K. M.** 1973. *Environmental Phosphorus Handbook*, p. 613–631. John Wiley & Sons, Toronto.

54. **MacLeod, F. A., S. R. Guiot, and J. W. Costerton.** 1990. Layered structure of bacterial aggregates produced in an upflow anaerobic sludge bed reactor. *Appl. Environ. Microbiol.* **56:**1598–1607.

55. **Maigetter, R. Z., and R. M. Pfister.** 1974. A mixed bacterial population in a continuous culture with and without kaolinite. *Can. J. Microbiol.* **21:**173–180.

56. **Marshall, K. C.** 1994. Microbial ecology: whither goest thou?, p. 5–8. *In* R. Guerrero and C. Pedros-Alio (ed.), *Trends in Microbial Ecology.* Spanish Society for Microbiology, Barcelona.

57. **Monod, J.** 1949. The growth of bacterial cultures. *Annu. Rev. Microbiol.* **3:**371–394.

58. **Montgomery L., and T. M. Vogel.** 1992. Dechlorination of 2, 3, 5, 6 tetrachlorobiphenyl by a phototrophic enrichment culture. *FEMS Microbiol. Lett.* **94:**247–250.

59. **Nalewajko, C.** 1976. Kinetics of extracellular release in axenic algae and in mixed algal-bacterial cultures: significance in estimation of total (gross) phytoplankton excretion rates. *J. Phycol.* **12:**1–5.

60. **Paerl, H. W.** 1976. Specific associations of the bluegreen algae *Anabaena* and *Aphanizomenon* with bacteria in freshwater blooms. *J. Phycol.* **12:**431–447.

61. **Paerl, H. W., and J. L. Pinckney.** 1995. Microbial consortia: their roles in aquatic production and biogeochemical cycling. *Microb. Ecol.* **12:**1–20.

62. **Parkes, R. J., and E. Senior.** 1988. Multistage chemostats and other models for studying anoxic ecosystems, p. 51–71. *In* J. W. T. Wimpenny (ed.), *Handbook of Laboratory Model Systems for Microbial Ecosystems*, vol. 1. CRC Press, Boca Raton, Fla.

63. **Perfil'iev, B. V., and D. R. Gabe.** 1969. *Capillary Methods of Investigating Microorganisms.* J. M. Shewan, trans. University of Toronto Press, Toronto.

64. **Pfennig, N.** 1980. Syntrophic mixed cultures and symbiotic consortia with phototrophic bacteria: a review, p. 127–131. *In* G. Gottschalk, N. Pfennig, and H. Werner (ed.), *Anaerobes and Anaerobic Infections.* Fischer-Verlag, Stuttgart.

65. **Prosser, J. I.** 1989. Modeling nutrient flux through biofilm communities, p. 239–250. *In* W. G. Characklis and P. A. Wilderer (ed.), *Structure and Function of Biofilms.* John Wiley & Sons, Toronto.

66. **Rice, E. L., and S. K. Pancholy** 1972. Inhibition of nitrification by climax ecosystems. *Am. J. Bot.* **59:**1033–1040.

67. **Robinson, R. W., D. E. Akin, R. A. Nordstedt, M. V. Thomas, and H. C. Aldrich.** 1984. Light and electron microscopic examinations of methane-producing biofilms from anaerobic fixed-bed reactors. *Appl. Environ. Microbiol.* **48:**127–136.

68. **Rozgaj, R., and M. Glancer-Soljan.** 1992. Total degradation of 6-aminonaphthalene-2-sulphonic acid by a mixed culture consisting of different bacterial genera. *FEMS Microbiol. Ecol.* **86:**229–235.

69. **Rutgers, M., J. J. Bogte, A. M. Breure, and J. G. van Andel.** 1993. Growth and enrichment of pentachlorophenol-degrading microorganisms in the nutristat, a substrate concentration-controlled continuous culture. *Appl. Environ. Microbiol.* **59:**3373–3377.

70. **Schiefer, G. E., and D. E. Caldwell.** 1982. Synergistic interaction between *Anabaena* and *Zoogloea* spp. in carbon dioxide limited continuous cultures. *Appl. Environ. Microbiol.* **44:**84–87.

71. **Senior, E., A. T. Bull, and J. H. Slater.** 1976. Enzyme evolution in a microbial community growing on the herbicide Dalapon. *Nature* (London) **263:**476–479.

72. **Slater, J. H.** 1988. Microbial population and community dynamics, p. 51–74. *In* J. M. Lynch and J. E. Hobbie (ed.), *Micro-organisms in Action: Concepts.* Blackwell Scientific Publication, Inc., Palo Alto, Calif.

73. **Slater, J. H., and D. J. Hartman.** 1982. Microbial ecology in the laboratory: experimental systems, p. 255–274. *In* R. G. Burns and J. H. Slater (ed.), *Experimental Microbial Ecology*, Blackwell Scientific Publications, Oxford.

74. **Stams, A. J. M., J. T. C. Grotenhuis, and A. J. B. Zehnder,** 1989. Structure-function relationship in granular sludge, p. 440–445. *In* T. Hattori, Y. Ishida, Y. Maruyama, R. Y. Morita, and A. Uchida (ed.), *Recent Advances in Microbial Ecology.* Japan Scientific Societies Press, Tokyo.

75. **Stewart, P. S., B. M. Peyton, W. J. Drury, and R. Murga.** 1993. Quantitative observations of heterogeneities in *Pseudomonas aeruginosa* biofilms. *Appl. Environ. Microbiol.* **59:**327–329.

76. **Tagger, S., N. Truffaut, and J. Le Petit.** 1990. Preliminary study on relationships among strains forming a bacterial community selected on naphthalene from a marine sediment. *Can. J. Microbiol.* **36:**676–681.

77. **Thiele, J. H., C. M. Thartrain, and J. G. Zeikus.** 1988. Control of interspecies electron flow during anaerobic digestion: role of floc formation in syntrophic methanogenesis. *Appl. Environ. Microbiol.* **54:**10–19.

78. **Trulear, M. G., and W. G. Characklis.** 1982. Dynamics of biofilm processes. *J. Water Pollut. Control Fed.* **54:**1288–1301.

79. **Veldkamp, H.** 1977. Ecological studies with the chemostat. *Adv. Microb. Ecol.* **1:**59–94.

80. **Veldkamp, H., and H. W. Jannasch.** 1972. Mixed culture studies with the chemostat. *J. Appl. Chem. Biotechnol.* **22:**105–123.

81. **Wimpenny, J. W. T.** 1992. Microbial systems: patterns in space and time. *Adv. Microb. Ecol.* **12:**469–522.

82. **Wimpenny, J. W. T., H. Gest, and J. L. Favinger.** 1986. The use of two-dimensional gradient plates in determining the responses in non-sulphur purple bacteria to pH and NaCl concentration. *FEMS Microbiol. Lett.* **37:**367–371.

83. **Wimpenny, J. W. T., and P. Waters.** 1984. Growth of microorganisms in gel-stabilized two-dimensional diffusion gradient systems. *J. Gen. Microbiol.* **130:**2921–2936.

84. **Wimpenny, J. W. T., P. Waters, and A. Peters.** 1988. Gel-plate methods in microbiology, p. 229–251. *In* J. W. T. Wimpenny (ed.), CRC *Handbook of Laboratory Model Systems for Microbial Ecology Research*, vol. 1. CRC Press, Boca Raton, Fla.

85. **Wolfaardt, G. M., J. R. Lawrence, J. V. Headley, R. D. Robarts, and D. E. Caldwell.** 1994. Microbial exopolymers provide a mechanism for bioaccumulation of contaminants. *Microb. Ecol.* **27:**279–291.

86. **Wolfaardt, G. M., J. R. Lawrence, M. J. Hendry, R. D. Robarts, and D. E. Caldwell.** 1993. Development of steady-state diffusion gradients for the cultivation of degra-

dative microbial consortia. *Appl. Environ. Microbiol.* **59:** 2388–2396.

87. **Wolfaardt, G. M., J. R. Lawrence, R. D. Robarts, and D. E. Caldwell.** 1994. The role of interactions, sessile growth and nutrient amendment on the degradative efficiency of a bacterial consortium. *Can. J. Microbiol.* **40:** 331–340.

88. **Wolfaardt, G. M., J. R. Lawrence, R. D. Robarts, and D. E. Caldwell.** 1995. Bioaccumulation of the herbicide diclofop in extracellular polymers and its utilization by a biofilm community during starvation. *Appl. Environ. Microbiol.* **61:**152–158.

89. **Wolfaardt, G. M., J. R. Lawrence, R. D. Robarts, S. J. Caldwell, and D. E. Caldwell.** (1994). Multicellular organization in a degradative biofilm community. *Appl. Environ. Microbiol.* **60:**434–446.

Biomass Measurements: Biochemical Approaches

DAVID C. WHITE, HOLLY C. PINKART, AND DAVID B. RINGELBERG

10

RATIONALE

Determination of biomass is of great importance in microbial ecology, as microbes form the base of the food web. As used in biochemical approaches to biomass, the microbes include the prokaryotes and microeukaryotes that pass through a 0.5-mm sieve. It is particularly useful to differentiate between the viable biomass, which manifests the potential for the metabolic activities in the environment, and the nonviable biomass, which can still be part of the food web but has little or no potential metabolic activity. Classical biochemical and microbiological techniques successfully used in public health for the isolation and culture of clinical specimens have proven less than adequate for the determination of biomass and community structure in environmental samples. It has been repeatedly documented in reviews of the literature that viable counts of bacteria from various environmental samples may represent only a very small proportion of the extant microbial community (85, 86). Furthermore, microbes may be metabolically active and potentially infectious even though they are not culturable (98). Finally, classical microbial tests are time-consuming and provide little indication of the nutritional status or evidence of toxicity which can affect metabolic activities and can be crucial in studies of microbial ecology.

TRADITIONAL BIOMASS MEASURES

Microbiologists have traditionally quantitated the biomass of microbiota in a sample to the number of viable cells detected by viable count. While this is sufficient for monocultures which are readily cultured in the laboratory, it is not satisfactory for most environmental samples, such as soils and sediments, in which the viable counts often represent 0.1 to 10% of the cells detected by using acridine orange direct counts (AODC) or biochemical methods. In soils and sediments, the problems are intensified, as the often sparse and heterogeneously distributed microbial community can significantly interfere with the accuracy of the determination of microbial abundance biomass by AODC. Direct microscopic counts measure abundance and not biomass. The relationship between biomass and abundance requires insight into the biovolume, and conversion factors are not simple and show variations of 500 between different size classes of cells (65). Direct microscopic counts present problems when autofluorescence of sediment clay granules or

opacities in the sediments obscure detection of bacteria in situ (22). This problem is often ameliorated by inducing detachment of the microbes from soil granules by using solutions containing multicharged ions like polyphosphate followed by recovery on membrane filters for microscopic counting. We have evidence from signature lipid biomarker analysis that in subsurface sediments, the detachment of the microbes is selective and often not quantitative. With in situ direct counting of bacteria at densities of less than 10^4 cells per g (dry weight) of sediment, the reproducibility of the counts is very low and the error is large, even with counting of 20 fields with adequate numbers of cells and replicate subsampling (53).

BIOCHEMICAL BIOMASS MEASURES

An effective and quantitative way to measure the biomass of the microbiota in situ without the requirements of culture or finding each organism in microscopic fields is to measure cellular components of the microbes. If cellular components are universally distributed, have a short (in terms of the process being studied) residence time in detrital pools after excretion by death, and are expressed at relatively constant levels among the microbial community and throughout the growth cycle, then they can be used as a measure of the biomass. A number of cellular components, such as muramic acid and other cell wall components (23), have been used as measures of bacterial biomass. Lipopolysaccharide (LPS) components have been used as measures of the gram-negative bacteria (71). ATP is a universal measure of metabolizing cells, provided that it does not persist in soil following cell death (42). If the assay is combined with a treatment of added extracellular ATPase, then assay of all adenosine-containing components can be a measure of the energy charge (ATP/AMP + ADP + ATP). The ratio of adenosine to the energy charge is an exquisitely sensitive indicator of stress in bacteria, as it measures a key homeostatic mechanism to maintain the energy charge, which is essential for metabolic functioning (14).

LIPID ANALYSIS

All intact cells contain polar lipids. Polar lipids in microbes are primarily phospholipids. Determination of lipid phosphate (LP) or phospholipid ester-linked fatty acids (PLFA) provides a quantitative measure of the microbial biomass containing intact cellular membranes. Either determination is a measure of the viable microbial biomass because organisms without intact cellular membranes are not viable. With cell death, exogenous and endogenous phospholipases rapidly transform the polar lipids in the cell membranes to nonpolar neutral lipid diglycerides by removing polar phosphate-containing head groups (97). The diglyceride-to-PLFA ratio increases in many subsurface sediments from <0.2 at the surface to over 2.0 at >200 m (95).

BIOMASS-TO-CELL NUMBER CONVERSIONS

One of the major problems with biochemical biomass measures is that the results are in micromoles of component per gram of soil or sediment. Since microbiologists traditionally think of biomass as the number of cells in a gram of soil or sediment, a simple conversion could be made by determining the biomarker content of monocultured cells and then counting the cells to determine a value per cell. Problems in equating AODC measurements, with their accuracy problems (22), to estimates from the PLFA or LP derive from the lack of a universally applicable conversion factor for estimating the PLFA or LP per bacterial cell and the number of cells per gram (dry weight) of bacteria (11, 21, 22, 85, 92). This problem results from observations that most environments harbor microbes of widely differing volumes and shapes. Bacterial biovolumes can vary over 3 orders of magnitude (31, 65). The volume of a viable cell can also vary with nutritional status. In bacterial enrichments and in isolates or mixed cultures from the sea, the LP content can vary between 34 and 380 μmol of LP per g of carbon for aerobic organisms, compared with contents of between 118 and 250 μmol of LP per g of carbon for anaerobic cultures (11). *Arthrobacter crystallopoietes* showed a 30% decrease in PLFA per cell after 2 weeks of starvation (48), and *Vibrio cholerae* showed up to a 99.8% decrease in PLFA per cell after 7 days of starvation, with loss of culturability but not membrane integrity (39). Brinch-Iverson and King (11) stated that the conversion factor for bacteria of 100 μmol of LP per g of carbon (100 μmol of PLFA per g [dry weight]) based on earlier work (90) was reasonably applicable to sediments with a significant proportion of anaerobes. Bacterial cell volume varies between 0.01 and 7 μm^3, with smaller bacteria having a higher dry weight-to-volume ratio than larger bacteria (65, 66). These authors developed an allometric relationship between dry weight and volume whereby biomass equals a conversion factor times the volume raised to an exponential scaling factor. Scaling factors and conversion factors show a dependence on size classes (65).

EQUIVALENCE OF BIOMASS MEASURES

In one specific environment, a comparison of methods for determination of microbial biomass showed equivalence. Subsurface sediment samples containing sparse prokaryote communities of minicells (demonstrated microscopically) were used in these experiments. In these sediments, the viable biomass determined by PLFA was equivalent (but with a much smaller standard deviation) to that estimated by intracellular ATP, cell wall muramic acid, LP, and very carefully done AODC measurements (6). It was assumed that there were 2.5×10^{12} cells per g (dry weight) and 100 μmol of PLFA per g (dry weight) (6). Generally, environmental bacteria growing in dilute media in the laboratory

or as mixed bacterial populations have average volumes of 0.48 ± 0.2 μm^3 (10). It has been our experience from a wide variety of environmental samples that the relationship between the PLFA and estimated number of cells varies by a factor of at least 4.

VIABLE BIOMASS

The viable microbial biomass can be determined by quantifying organic phosphate from the polar lipid fraction of the lipid extract, using a relatively simple colorimetric analysis (11, 22, 29, 92). The sensitivity of the classical colorimetric analysis for LP as initially proposed for environmental samples (92) has been improved considerably with a dye-coupled reaction to sensitivities of 1 nmol of LP, corresponding to about 10^7 bacteria (22). Higher sensitivities (about 10 fmol with gas chromatography [GC]-mass spectrometry [MS] and single-ion monitoring of negative ions) and specificity with a concomitant determination of the nonviable cell biomass, the community composition, and the nutritional-physiological status can be obtained through the application of the various GC-MS methods (66).

COMMUNITY COMPOSITION

Information obtained from the lipid analysis provides insight into the community composition as well. The PLFA patterns derived from environmental microbial communities are much like the infrared spectra of complex molecules in that the PLFA patterns provide quantitative analysis, but the interpretation in terms of specific components may be obscured because of overlapping compositions among constituents. Quantitative comparisons of total community PLFA patterns accurately mirror shifts in community composition but may not provide definitive analysis of shifts in specific microbial groups. Some specific groups of microbes contain characteristic fatty acid profiles. Several examples are listed in Table 1. However, it must be kept in mind that signature lipid biomarker analysis cannot detect every species of microorganism in an environmental sample, as many species have overlapping PLFA patterns. Further analysis of other lipids such as the sterols (for the microeukaryotes such as nematodes, algae, and protozoa) (57, 78, 91), glycolipids (phototrophs and gram-positive bacteria), or hydroxy fatty acids (OHFA) in the LPS of the lipid A of gram-negative bacteria (LPS-OHFA) (18, 21, 46, 71) can provide a more detailed community structure analysis.

EXTENSION TO NUCLEIC ACID ANALYSIS

The solvent extraction utilized in signature lipid biomarker analysis has recently been shown to liberate cellular nucleic acids which can be used for gene probing (43). Over 50% of the gene *nahA* present in intact *Pseudomonas fluorescens* cells added to soil was recovered by using the lipid extraction protocol compared with recovery by the standard techniques (68). The DNA recovered from the lipid extraction was of high quality and suitable for enzymatic amplification. The combined lipid extraction and recovery of nucleic acids can be very useful in biomass and community composition determinations. The DNA probe analysis offers powerful insights because of the exquisite specificity in the detection of genes. Concomitant DNA-lipid analysis readily provides quantitative recoveries independent of the ability to isolate or culture the microbes. The lipid analysis gives evidence of the phenotypic properties of the community that indicates ex-

TABLE 1 Examples of signature lipids and their cellular locations[a]

Genus, organism, or group	Lipid biomarker	Cellular localization	Reference(s)
Desulfovibrio	i17:1ω7c, i15:1ω7c, i19:1ω7c	PLFA	18
Desulfobacter	10Me16, cy18:0(ω7,8)	PLFA	17
Desulfobulbus	17:1ω6c, 15:1	PLFA	71
Francisella tularensis	20:1ω11, 22:1ω13, 24:1ω15, 26:1ω17	PLFA	61
Nostoc commune	i15:1ω11, br18:1, 18:3ω3	PLFA	74
Flexibacter	i5:1ω5, i15:1ω6, β-OH-i15:0, β-OH17:0	PLFA and LPS	32, 63
Vibrio cholerae	11Me19:1, 18:2ω6.9	PLFA	32
Archaea	Ether-linked lipids, diphytanyl glycerol diethers, bidiphytinyl glycerol ethers	Membrane	60
Methanotrophs, type I	16:1ω8c, 16:1ω5c	PLFA	62
Methanotrophs, type II	18:1ω8c, 18:1ω8t	PLFA and LPS	62
Thiobacillus	i17:1ω5, 10Me18:1ω6, 11Me18:1ω6, hydroxy cyclopropane, methoxy, mid-chain branched OHFA	PLFA and LPS	45, 46
Actinomycetes	Mid-chain branched fatty acids	PLFA	47, 48
Frankia	cy18:0(ω8,9), i16:1ω6	PLFA	82
Planctomyces	19:1ω10, 3-OH20:0	PLFA and LPS	44
Desulfomonile	br3-OH19:0, br3-OH21:0, br3-OH22:0	LPS	76
Legionella	3-OHi14:0, 2.3diOHi14:0, 27-oxo28:0, 27:2,3 OH22:0	LPS	83
Geobacter	14:1ω7, i17:1ω8,3-OH15:0, 9-OH16:0, 10-OH16:0, 11-OH16:0, 3-OH17:0	PLFA and LPS	50
Mycobacteria	Micocerosic acids, 2M3,3-OHFA, 2-OH alcohols	Neutral lipid	2
Bacillus or *Arthrobacter*	i15:0/a15:0, i17:0/a17:0 <0.2	PLFA	48
Fungi	18:2ω6, 18:3ω6, 18:3ω3, sterols	PLFA	57, 91
Clostridia	Plasmalogen-derived dimethyl acetals	Polar lipids	51
Diatoms	16:1ω13t, 16:2ω4, 16:3ω4, 20:5ω3	PLFA	9, 70, 96
Higher plants	18:1ω11, 18:3ω3, 20:5ω3, 26:0	PLFA	81, 99
Protozoa	20:2ω6, 20:3ω6, 20:4ω6	PLFA	85, 86

[a] Fatty acids are designated by the total number of carbon atoms followed by the number of double bonds, with the position of the double bond indicated from the methyl end (ω) of the molecule. Configuration of the double bonds is indicated as either *cis* (c) or *trans* (t). For example, 16:1ω7c is a PLFA with 16 total carbons with one double bond located seven carbons from the ω end in the *cis* configuration. Branched fatty acids are designated *iso* (i) or *anteiso* (a) if the methyl branch is one or two carbons, respectively, from the ω end (e.g., i15:0) or by the position of the methyl group from the carboxylic end of the molecule (e.g., 10Me16:0). Methyl branching at undetermined positions in the molecule is indicated by the prefix "br". Cyclopropyl fatty acids are designated by the prefix "cy" followed by the total number of carbons (e.g., cy17:0). The position of a hydroxyl group is numbered from the carboxyl end of the fatty acid, with OH as a prefix (e.g., 3-OH16:0).

tant microbial activity by providing in situ indications of starvation, growth rate, exposure to toxicity, unbalanced growth, deficiencies of specific nutrients, and the aerobic/ anaerobic metabolic balance, while DNA probes define the physiological potential of the microbial community. The combined DNA-lipid analysis overcomes some deficiencies in microbial ecology studies involving only nucleic acid analysis (87).

COMMUNITY PHYSIOLOGICAL STATUS

As already mentioned, it is possible to assess the physiological status of the microbial community by using lipid analysis. Many subsets of the microbial community respond to specific conditions in their microenvironment with shifts in lipid composition. The proportion of poly-β-hydroxyalkanoic acid (PHA) in bacteria (64) or triglyceride (in the microeukaryotes) (30) relative to the PLFA provides a measure of nutritional-physiological status. Some bacteria undergo unbalanced growth and cannot divide when exposed to adequate carbon and terminal electron acceptors but lack some essential nutrient such as phosphate, nitrate, and trace metals. These bacteria form PHA. When the essential component becomes available, these bacteria catabo-

lize PHA and form PLFA as they grow and divide. For example, the PHA/PLFA ratio in rhizosphere microbes from *Brassica napus* planted in sand and recovered from roots was <0.0001, compared with 6.6 for bacteria not associated with the rootlets (81).

Specific patterns of PLFA can also indicate physiological stress in certain bacterial species (32). Starvation and stationary-phase growth lead to conversion of monoenoic PLFA to the cyclopropane PLFA. Exposure to solvents, alcohols, and acids induces changes in PLFA (77). Starvation can lead to minicell formation and a relative increase in specific *trans*-monoenoic PLFA compared with the *cis* isomers (32). It has been shown that for increasing concentrations of phenol, *Pseudomonas putida* P8 forms increasing proportions of *trans*-unsaturated fatty acids (38). Increasing the proportions of *trans*-monoenoic PLFA is not the critical feature of solvent resistance in *P. putida*. Comparison of a solvent-sensitive strain with the Idaho strain, which is resistant to saturating concentrations of solvents and surfactants, showed that although both exhibited increases in *trans*-monoenoic PLFA, the resistant strain also shifts its lipid composition, decreases the proportion of monoenoic PLFA to saturated PLFA, increases the level of LPS-OHFA, and exhibits decreased permeability to the hydrophobic antibiotic difloxacin not detected in the solvent-sensitive strain (73).

Phospholipid patterns also change in response to environmental stress. Some *Pseudomonas* species form acylornithine lipids in lieu of phospholipids when growing with limited bioavailable phosphate (52). Respiratory quinone structure indicates the degree of aerobic activity in gram-negative heterotrophic facultative bacteria (37). Anaerobes are usually associated with high ratios of *iso*-branched to *anteiso*-branched PLFA which are typical of *Desulfovibrio*-type sulfate-reducing bacteria. Gram-positive aerobes like *Micrococcus* or *Arthrobacter* species have low ratios of *iso*- to *anteiso*-branched saturated PLFA.

METABOLIC ACTIVITY IN ESTIMATING BIOMASS

There are also biomass measures which are based on detecting the activity of environmental microbiota. Assays dependent on enzymatic activities (80), growth, or respiration after chloroform fumigation (40) have been used. The major problem with activity-based assessments of microbial biomass is that microbial community metabolic activity does not necessarily correlate with microbial biomass (high activity does not mean an actively growing, dividing microbial community). Measurements of the viable microbial biomass by PLFA determination in marine sediments recovered from the Antarctic, the deep sea, and neotropical marine mud flats are remarkably constant at about 10 nmol of PLFA per g (dry weight) of sediment ($\sim 10^9$ equivalent cells per g [dry weight]) (87). Metabolic activities (measured with injected substrates in situ), however, showed neotropical sediments to be at least 300-fold more active than those in the Antarctic in terms of DNA synthesis determined from [^{3}H]thymidine incorporation rates (96).

DISTURBANCE ARTIFACTS IN ACTIVITY MEASUREMENTS

Great care must be taken with metabolic activity measurements to avoid the generation of disturbance artifacts. Lipid analysis has been used to overcome the problem of disturbance artifacts generated in determining microbial activity (16, 24, 25, 54). The facile determination of the ratio of [^{14}C] acetate incorporation into PHA and PLFA has proven especially valuable in ecological studies (27).

COMMUNITY COMPOSITION FROM ISOLATED BACTERIA

Community composition based on distinctive patterns of ester-linked fatty acids released from bacteria (largely from the phospholipids and LPS of clinical isolates) is currently used in identifying cultured microbes. Patterns of the prominent fatty acids of isolated microbes after growth on standardized media are used to differentiate over 2,000 species of organisms by using the MIDI microbial identification system (MIDI, Newark, Del.). (84). Utilization of MIDI requires isolation and culture of the microbes prior to analysis. As a result, the unculturable microbes which may represent the vast majority of the environmental microbes are not detected. Utilization of fatty acids from cultured microbes from a soil or sediment does not indicate in situ biomass, nor is it as accurate as other methods in reflection of the viable community composition, as often the viable counts represent 0.1 to 10% of the cells detected by using AODC or biochemical measures (1, 4, 69).

BIOLOG

A new automated microbial identification system based on aerobic metabolic activities, Biolog, has been used for community microbiological composition. The system is based on differential activities among 92 various substrates and has been shown to show differences in community metabolism that paralleled those provided by the lipid analysis in differentiating microbial communities in drilling fluids, makeup waters, and deep subsurface cores (49). Unfortunately, this form of analysis requires a transparent carbon-free inoculum. Although many groundwater samples can be assayed directly, soils and subsurface sediments need to be blended, extracted with sodium pyrophosphate, and incubated without added carbon and nutrients for 24 h with agitation, and the supernatant must be flocculated with a mixture of calcium and magnesium salts before assay (49). The Biolog assay cannot be used in a quantitative determination of microbial biomass but can provide community activity comparisons between biomes. With the Biolog system, it was possible to detect considerable variation in the substrate utilization of microbial communities of soils taken from six different plant communities and compare differences in functional diversity (100). Patterns of substrate utilization were reproducible for model communities, but the extents of substrate utilization were not reflected in comparisons of responses of isolates, model communities of known composition, and soils (35). Replicate soil communities from the same pots varied considerably when the community activity analysis was used.

WHEN TO UTILIZE BIOCHEMICAL BIOMASS MEASURES

Biochemical biomass determinations have been successfully applied to a multitude of environments. Since the measurement involves extraction, concentration, purification, fractionation, derivatization, and analysis by GC-MS, with structural identification of each signature component, there are few environments to which it cannot be applied. The assay has even been used to determine the biomass of microbes in sludges of petroleum storage tanks, although extra purification steps were included to remove the neutral lipid hydrocarbon components. Samples from soils (4, 29, 40, 47, 80), rhizospheres (81, 82, 99), ocean abyss (5), stream periphyton (33), clinical specimens (2, 67), pus (61), mummies (unpublished data), ice cores (70), mongoose anal sacs (15), sediments (1, 22, 25, 31, 54, 72, 78, 88), subsurface materials (6, 24, 28, 94, 95), membrane filter retentates from groundwater (50), bioprocessing (36, 51), biofouling films (89), concrete (46), detritus (55), sponge spicule mats (96), drinking water biofilms (83), rocks (3), fungal biomass (91), grazed detritus (56), substratum biodegradability (8), microbially influenced corrosion (41), predation (93), pollution (79), anaerobic digestors (36), and microcosm microbial community comparison with the field (20) have all been characterized by using lipid analysis.

COLLECTION OF SAMPLES FOR ANALYSIS

Proper sample collection techniques are essential for obtaining a representative sample from the environmental matrix. The analysis is most satisfactory when at least 10^8 bacteria are analyzed, since some of the more interesting signatures can be found as trace components in the lipid extract. It has proved possible though, to generate lipid profiles of deep subsurface sediments which contain only 1.0 pmol of PLFA per g of sediment (equivalent to 2.5×10^4 bacteria) (94). Signature lipid biomarker analysis also can provide a nested analysis for the determination of appropriate sample sizes for experimental plots to determine the appropriate quadrat size (19). The analysis provides a quantitative estimation of sample-to-sample or within-sample heterogeneity in the determination of the microbial biomass and community composition in estuarine mud flats (19).

SAMPLE HANDLING

It is important that once the environmental samples are recovered, they be frozen (at least $-20°C$ or below) or lyophilized as quickly as possible. If this is not possible, wet weights are recorded and the sample is cooled as rapidly as possible. To minimize community compositional changes, it is necessary that microbial activity be stalled as soon as possible following sample collection. If the samples have been grown in culture, the medium should be centrifuged and the resulting cell pellet should be rinsed twice with 0.05 μM phosphate buffer (pH 7.5) before lyophilization. Samples should not be held on ice ($\sim 4°C$) any longer than absolutely necessary. Dry ice is satisfactory for holding frozen samples. Rock samples held at 4°C showed rapid and significant changes in biomass and community composition (3). Preserving samples with buffered formaldehyde or gluteraldehyde is not as satisfactory, as these preservatives can damage some of the less stable lipids.

PERFORMING THE SIGNATURE LIPID BIOMARKER ANALYSIS

Meticulous technique must be practiced to ensure contaminant-free analyses. Scrupulous cleaning of all glassware is absolutely necessary. Once glassware is used, it is immediately fully immersed in a washtub full of hot water and detergent. The cleaning process is so effective that phosphate-containing detergent can be used. The glassware is scrubbed with a brush and rinsed five times each with cold tap water and then deionized water. Glassware is allowed to dry completely before being wrapped in aluminum foil and heated in a clean muffle furnace for a minimum of 4 h at 450°C. Disposable glassware such as pipettes and silicic acid columns need not be washed but is also baked in the muffle furnace. No materials other than fired glass and acetone-rinsed Teflon may come into contact with lipid solvents. Lipids from fingers, hair, stopcock grease, oils, and hydrocarbons are all potential contaminants. Plasticware cannot be used in lipid analysis. Samples can be extracted at room temperatures but should be protected from light, especially fluorescent light, if photosensitive lipids (such as quinones) are to be analyzed. The extractant consists of a single-phase chloroform-methanol mixture (1:2, vol/vol), generally called a Bligh and Dyer (7), which can be modified to accept a phosphate buffer (88). Investigators have found that modification of the buffer can increase the recovery of PLFA from soils with high clay content (29). Samples can be extracted in glass centrifuge bottles and then centrifuged at $6,000 \times g$ for 30 min, with the liquid phase decanted into a separatory funnel or analyzed directly in a separatory funnel or in a test tube or other suitable container. With sandy sediments and sufficient one-phase extractant volume, it is usually not necessary to wash the sediment for a quantitative recovery. For bacterial samples, approximately 10^8 bacterial

cells are sufficient to achieve a good signal-to-noise ratio during GC or GC-MS analysis. To obtain the lipid fraction of the extracted sample, equal volumes of chloroform and distilled water (or buffer) are added and the emulsion is shaken. With time a split phase develops, which is then centrifuged or allowed to separate passively overnight. The lower organic phase (containing the bacterial lipids) is collected and filtered through a fluted Whatman 2V filter that has been preextracted with $CHCl_3$. The organic phase is removed by rotary evaporation at 37°C. The dried total lipid extract is dissolved in chloroform and then transferred to a silicic acid column and separated into neutral lipid, glycolipid, and polar lipid fractions (31) by elution with solvents of increasing polarity. The neutral lipid fraction is analyzed for lipids such as free fatty acids, sterols, respiratory quinones, triglycerides, and diglycerides (47, 78). The glycolipid fraction can be analyzed for PHA (64). The polar lipid fraction, containing the phospholipids, is subjected to a transesterification by a mild alkaline methanolysis protocol (32), resulting in fatty acid methyl esters which are then separated, quantified, and tentatively identified by capillary GC. Individual components can then be definitively identified by their mass spectra. Monoenoic PLFA double-bond positions are determined by GC-MS analysis of the dimethyl disulfide adducts (59).

The LPS-OHFA from the lipid A of gram-negative bacteria can be recovered from the lipid-extracted residue. This residue is hydrolyzed in acid, and the lipid components released by the hydrolysis are reextracted (71). After centrifugation at $6,000 \times g$ for 30 min, the chloroform phase is recovered, evaporated to dryness, and methylated by using "magic" methanol (methanol-chloroform-concentrated HCl [10:1:1, vol/vol/vol]) (61). The methylated OHFA are recovered, and the solvent is removed under a stream of nitrogen. The OHFA are purified by thin-layer chromatography (developed in hexane-diethyl ether [1:1, vol/vol]), recovered in chloroform-methanol (1:1, vol/vol), and then derivatized by using bis(trimethylsilyl)trifluoroacetamide prior to GC-MS analysis.

Results are reported with fatty acids designated by the total number of carbon atoms followed by the number of double bonds, with the position of the double bond indicated from the methyl end (ω) of the molecule (see footnote to Table 1).

PLFA and other lipid profiles can be entered into spreadsheet formats and subjected to statistical analysis. In addition to analysis of variance, two multivariate statistical applications have grown in popularity in addressing similarities between PLFA profiles. Dendrograms from a hierarchical cluster analysis are generally constructed from arcsine-transformed PLFA mole percent values, with similarities based on modified Euclidean distances. The two-dimensional plots generated from a principal-components analysis not only illustrate profile similarities (or differences) but also identify which PLFA contribute to the formation of the plots and to what extent (i.e., coefficients of loadings) (2, 76, 94, 95).

INTERPRETATION

Viable biomass is estimated from the total amount of PLFA detected in a sample. Phospholipids are an essential part of the intact cell membranes; thus, this biomass is a measure of the viable or potentially viable cells. Viable biomass can also be determined as organic LP by colorimetric methods (22, 92). The great majority of microbial phospholipids are diacyl, with a molar ratio of LP to PLFA of 1:2. Environmental samples showed LP-to-PLFA ratios of 1:2 to 1:1.5 with a mean of 1:1.7 in 200 samples (unpublished data). Our laboratory uses a cell equivalent value calculated from experiments performed with subsurface bacteria. It is based on the assumptions that there are 2.5×10^{12} cells per g (dry weight) and 100 μmol of phospholipid per g (dry weight) of cells (6). This equivalent yields 2.5×10^4 cells per pmol of PLFA. It is important to note that the number of cells per gram (dry weight) can vary by up to an order of magnitude, as summarized by Findlay and Dobbs (21). With cell death, or as the cell ruptures, phospholipids are attacked by enzymes, resulting in a lipid molecule called a diglyceride which is not present in the membranes of viable cells. A rapidly growing microbial community will show a diglyceride fatty acid/PLFA ratio of μ~0, while in less ideal environments the ratio can exceed 3 (95).

COMMUNITY COMPOSITION

Microbial community composition can be characterized from the pattern and types of PLFA identified in a sample. Some examples of signature lipids are shown in Table 1. When one is determining community structure by using a signature lipid approach, it is crucial to consider the environment from which the sample was retrieved when interpreting results. Terminally branched saturated PLFA are common to gram-positive bacteria but also to some gram-negative anaerobic bacteria, such as the sulfate-reducing bacteria. Monoenoic PLFA are found in most all gram-negative microorganisms and many types of microeukaryotes. Specific groups of bacteria form monoenoic PLFA with the unsaturation in an atypical position, such as 18:1ω8c in the type II methane-oxidizing bacteria (62). Polyenoic PLFA generally indicate the presence of microeukaryotes but have also been sparingly reported in some bacteria (34). The PLFA 18:2ω6 is prominent in fungi but is also found in algae and protozoa. Polyenoic PLFA with the first unsaturation in the ω6 position are classically considered to be of animal origin, whereas organisms with the first unsaturation in the ω3 position are generally considered to be of either plant or algal origin. Normal saturated PLFA longer than 20 carbons are typical of the microeukaryotes. There are exceptions to these generalizations. Sterol types and patterns are very helpful in identifying microeukaryotes, especially when combined with PLFA results. For example, cholesterol has been found to be prominent in protozoans such as *Cryptosporidium* species, ergosterol is found in many fungi (58), and algae (12) contain a diversity of sterols in patterns which have proven to be useful in forming taxonomic relationships. Branched-chain monoenoic PLFA are common in the anaerobic *Desulfovibrio*-type sulfate-reducing bacteria both in culture and in manipulated sediments (18, 72). They are also found in certain actinomycetes, which as a group contain mid-chain branched saturated PLFA, in particular 10Me18:0, with lesser amounts of other 10 methyl-branched homologs. Environments with 10Me16:0 > > 10Me18:0 often feature anaerobic gram-negative *Desulfobacter*-type sulfate-reducing bacteria (17, 72). Although normal (straight-chain) saturated PLFA are found in both prokaryotes and eukaryotes, bacteria generally contain greater amounts of the 16-carbon moiety (16:0), whereas the microeukaryotes contain greater amounts of the 18-carbon moiety (18:0). Methylotrophs are an exception to this rule, generally making more 18:0 than 16:0.

PHYSIOLOGICAL STATUS

Insight into the nutritional and physiological status of the microbial community can be determined through application of signature lipid biomarker analysis. The monoenoic PLFA 16:1ω7c and 18:1ω7c are increasingly converted to the cyclopropyl fatty acids cy17:0 and cy19:0, respectively, in gram-negative bacteria as the microbes move from a logarithmic to a stationary phase of growth. This ratio varies from organism to organism or environment to environment but usually falls within the range of 0.05 (log phase) to 2.5 or greater (stationary phase) (51, 85). An increase in cyclopropyl PLFA formation has also been associated with increased anaerobic metabolism in facultative heterotrophic bacteria in monoculture studies. Bacteria make *trans*-monounsaturated fatty acids as a result of changes in the environment, usually as a result of stress (i.e., toxicity or starvation). For example, gram-negative bacteria make 16:1ω7t or 18:1ω7t fatty acids in the presence of toxic pollutants such as phenol (38). In addition *trans*/*cis* ratios of greater than 0.1 have been shown to indicate starvation in bacterial isolates (32). This value is usually 0.05 or less in healthy, nonstressed populations. A ratio of storage lipid (PHA) to membrane lipid (PLFA) can be interpreted as a measure of unbalanced growth. When bacteria are in the presence of a carbon source and a terminal electron acceptor but lack an essential nutrient, they do not undergo cell division but instead form storage compounds such as PHA. When growing vigorously, they do not form PHA but instead show an increase in total PLFA. PHA/PLFA ratios can range anywhere from 0 (dividing cells) to over 40 (carbon storage). Ratios greater than 0.2 usually indicate the beginnings of unbalanced growth in at least part of the microbial community.

It is sometimes useful to determine in situ proportions of aerobic and anaerobic metabolism within a microbial community. Benzoquinones (ubiquinones, coenzyme Q) are produced by aerobic and facultative gram-negative bacteria. Terminal electron acceptors in the membrane-bound electron transport chain are either oxygen or nitrate, both of which carry high potentials (37). Naphthoquinones (menaquinones, dimethylmenaquinones) are produced by aerobic gram-positive bacteria, extreme halophiles, and gram-negative facultative or obligately anaerobic bacteria. These organisms use succinate, CO_2, or other low-potential electron acceptors in the electron transport chain. Fermentative anaerobic growth by facultative or obligate anaerobes generally produces no respiratory quinones. A ratio of total benzoquinones to total naphthoquinones provides an indication of the extent of aerobic versus anaerobic microbial respiration. In gram-negative bacteria, respiratory quinones are usually 10 to 100 times less in content than the PLFA. Sometimes proportions of isoprenologs of the respiratory quinones can be helpful in identifying species. Benzoquinone, with 13 isoprenolog units in the side chain, is found uniquely in *Legionella pneumophila*. When plasmologens (lipids typical of clostridia) are subjected to a mild acid methanolysis, fatty aldehydes are formed, which can then be converted into dimethyl acetals. With increasing proportions of obligate anaerobes and anaerobic metabolism, the dimethy acetal/PLFA ratio will increase. In certain situations, anaerobic metabolism can be estimated from the ratio of *iso*-branched to *anteiso*-branched saturated PLFA. The gram-positive aerobes (*Arthrobacter* and *Micrococcus* species) have i17:0/a17:0 ratios of approximately 0.2, whereas the gram-negative anaerobes (*Desulfovibrio*) have i17:0/a17:0 ratios of greater than 5 (17).

VALIDATION

The use of signature lipid biomarker analysis in determining the in situ viable microbial biomass, community composition, and nutritional-physiological status has been validated in a series of experiments (86). The induction of microbial community compositional shifts by altering the microenvironment resulted in changes that were often predictable, given past experience with microbial communities. For example, biofouling communities incubated in seawater at altered pH in the presence of antibiotics and specific nutrients resulted in a community dominated by fungi, while other conditions resulted in a community dominated almost exclusively by bacteria (91). Similar experiments showed that light-induced shifts which occurred within microbial communities were matched by expected shifts in signature lipid biomarkers and in terminal electron acceptors (9). These community compositional shifts resulting from specific perturbations have been reviewed (86). A second validation was the isolation of a specific organism or groups of organisms, with subsequent detection of the same organisms by signature lipid analysis in consortia under conditions in which their growth was induced. It was possible to induce a "crash" in methanogenesis in a bioreactor by inducing the growth of sulfate-reducing bacteria (51) or by adding traces of chloroform or oxygen (36). These crashes were accompanied by shifts in the signature lipid biomarkers that were correlated with the changes in the microbial populations. Specific sulfate-reducing bacterial groups can also be "induced" in estuarine muds (52), as can methane-oxidizing populations (60) or propane-oxidizing actinomycetes (75), through the addition of appropriate substrates. Again, all of these community shifts were evidenced by measurable changes in lipid signatures and in lipid patterns. A third validation was the induction of shifts in microbial community nutritional status by generating conditions of unbalanced growth in which cell growth but not cell division was possible. This was accomplished by chelating trace metals in the presence of tannins on epiphytic microbiota (58) and by disturbing anaerobic sediments with oxygenated seawater (25). Under these conditions, the ratio of PHA to PLFA biosynthesis increased dramatically just as it does in monocultures of appropriate bacteria under laboratory conditions. A fourth validation was the detection of specific shifts in microbial communities as a result of specific grazing by predators. The sand dollar *Mellita quinquiesperforata* was shown to selectively remove nonphotosynthetic microeukaryotes from sandy sediments. Examination of the morphology of the organisms in its feeding apparatus and of the signature lipid biomarker patterns before and after grazing by the echinoderm demonstrated the specific loss of nonphotosynthetic microeukaryotes (26). Another example involved the amphipod *Gammarus mucronatus*, which exhibited a relatively nonspecific grazing of the estuarine detrital microbiota. This organism removed the microeukaryotes, which were then replaced, to a large extent, by bacteria (55, 56). Results of the signature lipid biomarker analysis agreed with the cellular morphologies present as shown by scanning electron microscopy.

LIMITATIONS

Determination of microbial biomass with a colorimetric analysis of the organic phosphate of the phospholipids is straightforward and requires little specialized equipment other than a spectrophotometer (92). However, this analysis is relatively insensitive, with limits of detection in the micromolar range ($\sim10^{10}$ bacteria with the stable colorimetric analysis [92] or $\sim10^7$ bacteria the size of *Escherichia coli* with the dye-coupled assay [22]).

UNITS

A major problem with signature lipid biomarker analysis in determining environmental microbial biomass is that the results are not presented in the traditional units. Biomass is measured as picomoles of PLFA or micromoles of LP per sample instead of cells per sample. Although this value can be related to the number of specific organisms present, the conversion is problematic because of the variety of shapes and sizes organisms maintained in nature. The analysis of fungi based on sterol content also presents a problem since mycelia often exist as large multinucleated cells with a huge biomass, much of which is not active.

In the determination of signature lipid biomarkers in environmental samples, the lipid profiles will not result in the definition of each individual species present. Some species are readily defined since they contain either unique lipid components or unique lipid patterns. However, in environmental analyses, overlapping patterns may necessitate less specific interpretations, i.e., at the functional group level. Since DNA suitable for gene probing can be recovered with signature lipid biomarker analysis, the combination of signature lipid biomarker analysis with DNA gene probe technology greatly expands the specificity and scope of community compositional determinations.

Analysis of lipid components requires special analytical skills and entails expenses for extractions, processing, and GC-MS equipment for analysis. Scrupulous attention must be paid to the purity of solvents, reagents, and glassware since signatures at 1 part in 10^{14} are commonly detected by these analyses. Once any difficulties in performing the analyses have been overcome, the interpretation of community composition and nutritional-physiological status requires an extensive familiarity with widely scattered literature. Research toward automating and accelerating the speed of the analysis has been initiated in a number of laboratories. In the not too distant future, it is likely that signature lipid biomarker analysis will be fully automated and accomplished in a matter of hours instead of the current time frame of days. In the meantime, lipid analysis provides significant insight into the microbial biomass, community structure, and physiological status of environmental samples and provides a quantitative means for obtaining this type of information.

REFERENCES

1. **Albrechtsen, H.-J., and A. Winding.** 1992. Microbial biomass and activity in subsurface sediments from Vejen, Denmark. *Microb. Ecol.* **23:**303–317.
2. **Almeida, J. S., A. Sonesson, D. B. Ringelberg, and D. C. White.** 1995. Application of artificial neural networks (ANN) to the detection of *Mycobacterium tuberculosis*, its antibiotic resistance and prediction of pathogenicity amongst *Mycobacterium* spp. based on signature lipid biomarkers. *Binary Comput. Microbiol.* **7:**53–59.
3. **Amy, P. A., D. L. Halderman, D. Ringelberg, and D. C. White.** 1994. Changes in bacteria recoverable from subsurface volcanic rock samples during storage at 4°C. *Appl. Environ. Microbiol.* **60:**2679–2703.

4. **Bååth E., Å. Frostegård, and H. Fritze.** 1992 Soil bacterial biomass, activity, phospholipid fatty acid pattern, and pH tolerance in an area polluted with alkaline dust deposition. *Appl. Environ. Microbiol.* **58:**4026–4031.

5. **Baird, B. H., and D. C. White.** 1985. Biomass and community structure of the abyssal microbiota determined from the ester-linked phospholipids recovered from Venezula Basin and Puerto Rico Trench sediments. *Mar. Geol.* **68:**217–231.

6. **Balkwill, D. L., F. R. Leach, J. T. Wilson, J. F. McNabb, and D. C. White.** 1988. Equivalence of microbial biomass measures based on membrane lipid and cell wall components, adenosine triphosphate, and direct counts in subsurface sediments. *Microb. Ecol.* **16:**73–84.

7. **Bligh, E. G., and W. J. Dyer.** 1959. A rapid method of total lipid extraction and purification. *Can. J. Biochem. Physiol.* **31:**911–917.

8. **Bobbie, R. J., S. J. Morrison, and D. C. White.** 1978. Effects of substrate biodegradability on the mass and activity of the associated estuarine microbiota. *Appl. Environ. Microbiol.* **35:**179–184.

9. **Bobbie, R. J., J. S. Nickels, G. A. Smith, S. D. Fazio, R. H. Findlay, W. M. Davis, and D. C. White.** 1981. Effect of light on biomass and community structure of estuarine detrital microbiota. *Appl. Environ. Microbiol.* **42:**150–158.

10. **Bratbak, G., and I. Dundas.** 1984. Bacterial dry matter content and biomass estimations. *Appl. Environ. Microbiol.* **48:**755–757.

11. **Brinch-Iverson, J., and G. M. King.** 1990. Effects of substrate concentration, growth state, and oxygen availability on relationships among bacterial carbon, nitrogen and phospholipid phosphorous content. *FEMS Microbiol. Ecol.* **74:**345–356.

12. **Canuel, E. A., J. E. Cloen, D. B. Ringelberg, J. B. Guckert, and G. H. Rau.** 1995. Molecular and isotopic tracers used to examine sources of organic matter and its incorporation into the food webs of San Francisco Bay. *Limnol. Oceanogr.* **40:**67–81.

13. **Collins, M. D., R. M. Keddie, and R. M. Kroppenstedt.** 1983. Lipid composition of *Arthrobacter simplex, Arthrobacter tumescens,* and possibly related taxa. *Syst. Appl. Microbiol.* **4:**18–26.

14. **Davis, W. M., and D. C. White.** 1980. Fluorometric determination of adenosine nucleotide derivatives as measures of the microfouling, detrital and sedimentary microbial biomass and physiological status. *Appl. Environ. Microbiol.* **40:**539–548.

15. **Decker, D. M., D. B. Ringelberg, and D. C. White.** 1992. Lipid components in anal scent sacs of three mongoose species (*Helogale parvula, Crossarchus obscurus, Suricatta suricatta*). *J. Chem. Ecol.* **18:**1511–1524.

16. **Dobbs, F. C., and R. H. Findlay.** 1993. Analysis of microbial lipids to determine biomass and detect the response of sedimentary microorganisms to disturbance, p. 347–358. *In* P. F. Kemp, B. F. Sherr, E. B. Sherr, and J. J. Cole (ed.), *Handbook of Methods in Aquatic Microbial Ecology.* Lewis Publishers, Boca Raton, Fla.

17. **Dowling, N. J. E., F. Widdel, and D. C. White.** 1986. Phospholipid ester-linked fatty acid biomarkers of acetate-oxidizing sulfate reducers and other sulfide forming bacteria. *J. Gen. Microbiol.* **132:**1815–1825.

18. **Edlung, A., P. D. Nichols, R. Roffey, and D. C. White.** 1985. Extractable and lipopolysaccharide fatty acid and hydroxy acid profiles from *Desulfovibrio* species. *J. Lipid Res.* **26:**982–988.

19. **Federle, T. W., M. A. Hullar, R. J. Livingston, D. A. Meeter, and D. C. White.** 1983. Spatial distribution of biochemical parameters indicating biomass and community composition of microbial assemblies in estuarine mud flat sediments. *Appl. Environ. Microbiol.* **45:**58–63.

20. **Federle, T. W., R. J. Livingston, L. E. Wolfe, and D. C. White.** 1986. A quantitative comparison of microbial community structure of estuarine sediments from microcosms and the field. *Can. J. Microbiol.* **32:**319–325.

21. **Findlay, R. H., and F. C. Dobbs.** 1993. Quantitative description of microbial communities using lipid analysis, p. 271–284. *In* P. F. Kemp, B. F. Sherr, E. B. Sherr, and J. J. Cole (ed.), *Handbook of Methods in Aquatic Microbial Ecology.* Lewis Publishers, Boca Raton, Fla.

22. **Findlay, R. H., G. M. King, and L. Watling.** 1989. Efficiency of phospholipid analysis in determining microbial biomass in sediments. *Appl. Environ. Microbiol.* **55:**2888–2895.

23. **Findlay, R. H., D. J. W. Moriarty, and D. C. White.** 1983. Improved method of determining muramic acid from environmental samples. *Geomicrobiol. J.* **3:**135–150.

24. **Findlay, R. H., P. C. Pollard, D. J. W. Moriarty, and D. C. White.** 1985. Quantitative determination of microbial activity and community nutritional status in estuarine sediments: evidence for a disturbance artifact. *Can. J. Microbiol.* **31:**493–498.

25. **Findlay, R. H., M. B. Trexler, J. B. Guckert, and D. C. White.** 1990. Laboratory study of disturbance in marine sediments: response of a microbial community. *Mar. Ecol. Prog. Ser.* **61:**121–133.

26. **Findlay, R. H., M. B. Trexler, and D. C. White.** 1990. Response of a benthic microbial community to biotic disturbance. *Mar. Ecol. Prog. Ser.* **61:**135–148.

27. **Findlay, R. H., and D. White.** 1987. A simplified method for bacterial nutritional status based on the simultaneous determination of phospholipid and endogenous storage lipid poly beta-hydroxy alkanoate. *J. Microbiol. Methods* **6:**113–120.

28. **Fredrickson, J. K, J. P. McKinley, S. A. Nierzwicki-Bauer, D. C. White, D. B. Ringelberg, S. A. Rawson, S.-M. Li, F. J. Brockman, and B. N. Bjornstad.** 1995. Microbial community structure and biogeochemistry of miocene subsurface sediments: implications for long-term microbial survival. *Mol. Ecol.* **4:**619–626.

29. **Frostegaard, A., A. Tunlid, and E. Baath.** 1991. Microbial biomass measured as total lipid phosphate in soils of different organic content. *J. Microbiol. Methods* **14:**151–163.

30. **Gehron, M. J., and D. C. White.** 1982 Quantitative determination of the nutritional status of detrital microbiota and the grazing fauna by triglyceride glycerol analysis. *J. Exp. Mar. Biol.* **64:**145–158.

31. **Guckert, J. B., C. P. Antworth, P. D. Nichols, and D. C. White.** 1985. Phospholipid, ester-linked fatty acid profiles as reproducible assays for changes in prokaryotic community structure of estuarine sediments. *FEMS Microbiol. Ecol.* **31:**147–158.

32. **Guckert, J. B., M. A. Hood, and D. C. White.** 1986. Phospholipid, ester-linked fatty acid profile changes during nutrient deprivation of *Vibrio cholerae*: increases in the *trans/cis* ratio and proportions of cyclopropyl fatty acids. *Appl. Environ. Microbiol* **52:**794–801.

33. **Guckert, J. B., S. C. Noid, H. L. Boston, and D. C. White.** 1991. Periphyton response along an industrial effluent gradient: lipid-based physiological stress analysis and pattern recognition of microbial community structure. *Can. J. Fish. Aquat. Sci.* **49:**2579–2587.

34. **Guckert, J. B., D. B. Ringelberg, D. C. White, R. S. Henson, and B. J. Bratina.** 1991. Membrane fatty acids as phenotypic markers in the polyphasic taxonomy of methylotrophs within the proteobacteria. *J. Gen. Microbiol.* **137:**2631–2641.

35. **Haack, S. K., H. Garchow, M. J. Klug, and L. J. Forney.** 1995. Analysis of factors affecting the accuracy, reproducibility, and interpretation of microbial community carbon source utilization patterns. *Appl. Environ. Microbiol.* **60:** 1458–1468.

36. **Hedrick, D. B., J. B. Guckert, and D. C. White.** 1991. The effect of oxygen and chloroform on microbial activities in a high-solids, high-productivity biomass reactor. *Biomass Bioenergy* **1:**207–212.

37. **Hedrick, D. B., and D. C. White.** 1986. Microbial respiratory quinones in the environment: a sensitive liquid chromatographic method. *J. Microbiol. Methods* **5:** 243–254.

38. **Heipieper, H. J., R. Diffenbach, and H. Keweloh.** 1992. Conversion of *cis* unsaturated fatty acids to *trans*, a possible mechanism for the protection of phenol-degrading *Pseudomonas putida* P8 from substrate toxicity. *Appl. Environ. Microbiol.* **58:**1847–1852.

39. **Hood, M. A., J. B. Guckert, D. C. White, and F. Deck.** 1986. Effect of nutrient deprivation on the levels of lipid, carbohydrate, DNA, RNA, and protein levels in *Vibrio cholerae. Appl. Environ. Microbiol.* **52:**788–793.

40. **Horwath, W. R., and E. A. Paul.** 1994. Microbial biomass, p. 753–774. *In* R. W. Weaver, S. Angle, P. Bottomley, D. Bezdicek, S. Smith, A. Tabatabai, A. Wollum, S. H. Mickelson, and J. M. Bigham (ed.), *Methods of Soil Analysis, Microbiological and Biochemical Properties*, part 2. Soil Science Society of America, Madison, Wis.

41. **Jack, R. F., D. B. Ringelberg, and D. C. White.** 1992. Differential corrosion of carbon steel by combinations of *Bacillus* sp., *Hafnia alvei*, and *Desulfovibrio gigas* established by phospholipid analysis of electrode biofilm. *Corros. Sci.* **32:**1843–1853.

42. **Karl, D. M.** 1993. Total microbial biomass estimation derived from the measurement of particulate adenosine-5'-triphosphate, p. 359–368. *In* P. F. Kemp, B. F. Sherr, E. B. Sherr, and J. J. Cole (ed.), *Handbook of Methods in Aquatic Microbial Ecology.* Lewis Publishers, Boca Raton, Fla.

43. **Kehrmeyer, S. R., B. M. Appelgate, H. C. Pinkart, D. B. Hedrick, D. C. White, and G. S. Sayler.** Combined lipid/DNA extraction method for environmental samples. *J. Microbiol. Methods*, in press.

44. **Kerger, B. D., C. A. Mancuso, P. D. Nichols, D. C. White, T. Langworthy, M. Sittig, H. Schlessner, and P. Hirsch.** 1988. The budding bacteria, *Pirellula* and *Planctomyces*, with a typical 16S-rRNA and absence of peptidoglycan, show eubacterial phospholipids and unusually high proportions of long-chain beta-hydroxy fatty acids in the lipopolysaccharide lipid A. *Arch. Microbiol.* **149:** 255–260.

45. **Kerger, B. D., P. D. Nichols, C. P. Antworth, W. Sand, E. Bock, J. C. Cox, T. A. Langworthy, and D. C. White.** 1986. Signature fatty acids in the polar lipids of acid-producing *Thiobacilli*: methoxy, cyclopropyl, alpha-hydroxy-cyclopropyl and branched and normal monoenoic fatty acids. *FEMS Microbiol. Ecol.* **38:**67–77.

46. **Kerger, B. D., P. D. Nichols, W. Sand, E. Bock, and D. C. White.** 1987. Association acid-producing *Thiobacilli* with degradation of concrete: analysis by "signature" fatty acids from the polar lipids and lipopolysaccharide. *J. Ind. Microbiol.* **2:**63–69.

47. **Kieft, T. L., D. B. Ringelberg, and D. C. White.** 1994. Changes in ester-linked phospholipid fatty acid profiles of subsurface bacteria during starvation and desiccation in a porous medium. *Appl. Environ. Microbiol.* **60:** 3292–3299.

48. **Kostiw, L. L., C. W. Boylen, and B. J. Tyson.** 1972. Lipid composition of growing and starving cells of *Arthrobacter crystallopoietes. J. Bacteriol.* **94:**1868–1874.

49. **Lehman, R. M., F. S. Colwell, D. B. Ringelberg, and D. C. White.** 1995. Microbial community-level analyses based on patterns of carbon source utilization and phospholipid fatty acid profiles for quality assurance of terrestrial subsurface cores. *J. Microbiol. Methods* **22:**263–281.

50. **Lovely, D. R., S. J. Giovannoni, D. C. White, J. E. Champine, E. J. P. Phillips, Y. A. Gorby, and S. Goodwin.** 1992. *Geobacter metallireducens* gen. nov. sp. nov., a microorganism capable of coupling the complete oxidation of organic compounds to the reduction of iron and other metals. *Arch. Microbiol.* **159:**363–344.

51. **Mikell, A. T., Jr., T. J. Phelps, and D. C. White.** 1987. Phospholipids to monitor microbial ecology in anaerobic digesters, p. 413–444. *In* W. H. Smith and J. R. Frank (ed.), *Methane from Biomass, a Systems Approach.* Elsevier Publishing Co., New York.

52. **Minnikin, D. E., and H. Abdolrahimzadeh.** 1974. The replacement of phosphatidylethanolamine and acidic phospholipids by ornithine-amide lipid and a minor phosphorus-free lipid in *Pseudomonas fluorescens* NCMB129. *FEBS Lett.* **43:**257–260.

53. **Montagna, P. A.** 1982. Sampling design and enumeration statistics for bacteria extracted from marine sediments. *Appl. Environ. Microbiol.* **43:**1366–1372.

54. **Moriarty, D. J. W., D. C. White, and T. J. Wassenberg.** 1985. A convenient method for measuring rates of phospholipid synthesis in seawater and sediments: its relevance to the determination of bacterial productivity and the disturbance artifacts introduced by measurements. *J. Microbiol. Methods* **3:**321–330.

55. **Morrison, S. J., J. D. King, R. J. Bobbie, R. E. Bechtold, and D. C. White.** 1977. Evidence for microfloral succession on allochthonous plant litter in Apalachicola Bay, Florida, U.S.A. *Mar. Biol.* **41:**229–240.

56. **Morrison, S. J., and D. C. White.** 1980. Effects of grazing by estuarine gammaridean amphipods on the microbiota of allochthonous detritus. *Appl. Environ. Microbiol.* **40:** 659–671.

57. **Nes, W. R.** 1977. The biochemistry of plant sterols. *Adv. Lipid Res.* **15:**233–324.

58. **Newell, S. Y.** 1993. Membrane-containing fungal mass and fungal specific growth rate in natural samples, p. 579–586. *In* P. F. Kemp, B. F. Sherr, E. B. Sherr, and J. J. Cole (ed.), *Handbook of Methods in Aquatic Microbial Ecology.* Lewis Publishers, Boca Raton, Fla.

59. **Nichols, P. D., J. B. Guckert, and D. C. White.** 1986. Determination of monounsaturated fatty acid double-bond position and geometry for microbial monocultures and complex consortia by capillary GC-MS of their dimethyl disulphide adducts. *J. Microbiol. Methods* **5:**49–55.

60. **Nichols, P. D., C. A. Mancuso, and D. C. White.** 1987. Measurement of methanotroph and methanogen signature phospholipids for use in assessment of biomass and community structure in model systems. *Org. Geochem.* **11:**451–461.

61. **Nichols, P. D., W. R. Mayberry, C. P. Antworth, and D. C. White.** 1985. Determination of monounsaturated double bond position and geometry in the cellular fatty acids of the pathogenic bacterium *Francisella tularensis. J. Clin. Microbiol.* **21:**738–740.

62. **Nichols, P. D., G. A. Smith, C. P. Antworth, R. S. Hanson, and D. C. White.** 1985. Phospholipid and lipopolysaccharide normal and hydroxy fatty acids as potential signatures for the methane-oxidizing bacteria. *FEMS Microbiol. Ecol.* **31:**327–335.

63. **Nichols, P. D., B. K. Stulp, J. G. Jones, and D. C. White.** 1986. Comparison of fatty acid content and DNA homol-

ogy of the filamentous gliding bacteria *Vitreoscilla, Flexibacter, Filibacter. Arch. Microbiol.* **146:**1–6.

64. **Nickels, J. S., J. D. King, and D. C. White.** 1979. Poly-beta-hydroxybutyrate accumulation as a measure of unbalanced growth of the estuarine detrital microbiota. *Appl. Environ. Microbiol.* **37:**459–465.

65. **Norland, S.** 1993. The relationship between biomass and volume of bacteria, p. 303–307. *In* P. F. Kemp, B. F. Sherr, E. B. Sherr, and J. J. Cole (ed.), *Handbook of Methods in Aquatic Microbial Ecology.* Lewis Publishers, Boca Raton, Fla.

66. **Norland, S., M. Heldal, and O. Tumyr.** 1987. On the relationship between dry matter and volume of bacteria. *Microb. Ecol.* **13:**95–101.

67. **Odham, G., A. Tunlid, G. Westerdahl, L. Larsson, J. B. Guckert, and D. C. White.** 1985. Determination of microbial fatty acid profiles at femtomolar levels in human urine and the initial marine microfouling community by capillary gas chromatography-chemical ionization mass spectrometry with negative ion detection. *J. Microbiol. Methods* **3:**331–344.

68. **Ogram, A., G. S. Sayler, and T. Barkay.** 1987. The extraction and purification of microbial DNA from sediments. *J. Microbiol. Methods* **7:**57–66.

69. **Olsen R. A., and L. R. Bakken.** 1987. Viability of soil bacteria: optimization of plate-counting technique and comparison between total counts and plate counts within different size groups. *Microb. Ecol.* **13:**59–74.

70. **Palmisano, A. C., M. P. Lizotte, G. A. Smith, P. D. Nichols, D. C. White, and C. W. Sullivan.** 1988. Changes in photosynthetic carbon assimilation in Antarctic sea-ice diatoms during a spring bloom: variations in synthesis of lipid classes. *J. Exp. Mar. Biol. Ecol.* **116:**1–13.

71. **Parker, J. H., G. A. Smith, H. L. Fredrickson, J. R. Vestal, and D. C. White.** 1982. Sensitive assay, based on hydroxy-fatty acids from lipopolysaccharide lipid A for gram-negative bacteria in sediments. *Appl. Environ. Microbiol.* **44:**1170–1177.

72. **Parkes, R. J., N. J. E. Dowling, D. C. White, R. A. Herbert, and G. R. Gibson.** 1992. Characterization of sulfate-reducing bacterial populations within marine and estuarine sediments with different rates of sulfate reduction. *FEMS Microbiol. Ecol.* **102:**235–250.

73. **Pinkart, H. C., J. W. Wolfram, R. Rogers, and D. C. White.** 1996. Cell envelope changes in solvent-tolerant and solvent-sensitive *Pseudomonas putida* strains following exposure to O-xylene. *Appl. Environ. Microbiol.* **62:**1129–1132.

74. **Potts, M., J. J. Olie, J. S. Nickels, J. Parsons, and D. C. White.** 1987. Variations in phospholipid ester-linked fatty acids and carotenoids of desiccated *Nostoc commune* (cyanobacteria) from different geographic locations. *Appl. Environ. Microbiol.* **53:**4–9.

75. **Ringelberg, D. B., J. D. Davis, G. A. Smith, S. M. Pfiffner, P. D. Nichols, J. B. Nickels, J. M. Hensen, J. T. Wilson, M. Yates, D. H. Kampbell, H. W. Reed, T. T. Stocksdale, and D. C. White.** 1988. Validation of signature polarlipid fatty acid biomarkers for alkane-utilizing bacteria in soils and subsurface aquifer materials. *FEMS Microbiol. Ecol.* **62:**39–50.

76. **Ringelberg, D. B., T. Townsend, K. A. DeWeerd, J. M. Suflita, and D. C. White.** 1994. Detection of the anaerobic declorinator *Desulfomonile tiedjei* in soil by its signature lipopolysaccharide branched-long-chain hydroxy fatty acids. *FEMS Microbiol. Ecol.* **14:**9–18.

77. **Sikkema J., J. A. M. deBont, and B. Poolman.** 1995. Mechanisms of membrane toxicity of hydrocarbons. *Microbiol. Rev.* **59:**201–222.

78. **Smith, G. A., P. D. Nichols, and D. C. White.** 1989. Triglyceride and sterol and composition of sediment mi-croorganisms from McMurdo Sound, Antarctica. *Polar Biol.* **9:**273–279.

79. **Smith, G. A., J. S. Nickels, B. D. Kerger, J. D. Davis, S. P. Collins, J. T. Wilson, J. F. McNabb, and D. C. White.** 1986. Quantitative characterization of microbial biomass and community structure in subsurface material: a prokaryotic consortium responsive to organic contamination. *Can. J. Microbiol.* **32:**104–111.

80. **Tabatabai, M. A.** 1994. Soil enzymes, p. 775–834. *In* R. W. Weaver, S. Angle, P. Bottomley, D. Bezdicek, S. Smith, A. Tabatabai, A. Wollum, S. H. Mickelson, and J. M. Bigham (ed.), *Methods of Soil Analysis: Microbiological and Biochemical Properties*, part 2. Soil Science Society of America, Madison, Wis.

81. **Tunlid, A., B. H. Baird, M. B. Trexler, S. Olsson, R. H. Findlay, G. Odham, and D. C. White.** 1985. Determination of phospholipid ester-linked fatty acids and poly beta hydroxybutyrate for the estimation of bacterial biomass and activity in the rhizosphere of the rape plant *Brassica napus* (L). *Can. J. Microbiol.* **31:**1113–1119.

82. **Tunlid, A., N. A. Schultz, D. R. Benson, D. B. Steele, and D. C. White.** 1989. Differences in the composition between vegetative cells and nitrogen-fixing vesicles of *Frankia* spp. strain Cp11. *Proc. Natl. Acad. Sci. USA* **86:**3399–3403.

83. **Walker, J. T., A. Sonesson, C. W. Keevil, and D. C. White.** 1993. Detection of *Legionella pneumophila* in biofilms containing a complex microbial consortium by gas chromatography-mass spectrometric analysis of genus-specific hydroxy fatty acids. *FEMS Microbiol. Lett.* **113:**139–144.

84. **Welch, D. F.** 1991 Applications of cellular fatty acid analysis. *Clin. Microbiol. Rev.* **4:**422–438.

85. **White, D. C.** 1983. Analysis of microorganisms in terms of quantity and activity in natural environments. *Symp. Soc. Gen. Microbiol.* **34:**37–66.

86. **White, D. C.** 1988, Validation of quantitative analysis for microbial biomass, community structure, and metabolic activity. *Adv. Limnol.* **31:**1–18.

87. **White, D. C.** 1994. Is there anything else you need to understand about the microbiota that cannot be derived from analysis of nucleic acids? *Microb. Ecol.* **28:**163–166.

88. **White, D. C.** 1995. Chemical ecology: possible linkage between macro- and microbial ecology. *Oikos* **74:**177–184.

89. **White, D. C., and P. H. Benson.** 1984. Determination of the biomass, physiological status, community structure and extracellular plaque of the microfouling film, p. 68–74. *In* J. D. Costlow and R. C. Tipper (ed.), *Marine Biodeterioration: an Interdisciplinary Study.* U. S. Naval Institute Press, Annapolis, Md.

90. **White, D.C., R. J. Bobbie, J. S. Herron, J. D. King, and S. J. Morrison.** 1979. Biochemical measurements of microbial mass and activity from environmental samples, p. 69–81. *In* J. W. Costerton and R. R. Colwell (ed.), *Native Aquatic Bacteria: Enumeration, Activity and Ecology.* ASTM STP 695. American Society for Testing and Materials, Philadelphia.

91. **White, D. C., R. J. Bobbie, J. S. Nickels, S. D. Fazio, and W. M. Davis.** 1980. Nonselective biochemical methods for the determination of fungal mass and community structure in estuarine detrital microflora. *Bot. Mar.* **23:**239–250.

92. **White, D. C., W. M. Davis, J. S. Nickels, J. D. King, and R. J. Bobbie.** 1979. Determination of the sedimentary microbial biomass by extractible lipid phosphate. *Oecologia* **40:**51–62.

93. **White, D. C., and R. H. Findlay.** 1988. Biochemical markers for measurement of predation effects on the bio-

mass, community structure, nutritional status, and metabolic activity of microbial biofilms. *Hydrobiologia* **159:** 119–132.

94. **White, D. C., and D. B. Ringelberg.** Utility of signature lipid biomarker analysis in determining in situ viable biomass, community structure, and nutritional/physiological status of the deep subsurface microbiota. *In* P. S. Amy and D. L. Halderman (ed.), *The Microbiology of the Terrestrial Subsurface,* in press. CRC Press, Boca Raton, Fla.

95. **White, D. C., and D. B. Ringelberg.** Monitoring deep subsurface microbiota for assessment of safe long term nuclear waste disposal. *Can. J. Microbiol.,* in press.

96. **White, D. C., G. A. Smith, and G. R. Stanton.** 1984. Biomass, community structure, and metabolic activity of the microbiota in benthic marine sediments and sponge spicule mats. *Antarct. J. U.S.* **29:**125–126.

97. **White, D. C., and A. T. Tucker.** 1969. Phospholipid metabolism during bacterial growth. *J. Lipid Res.* **10:** 220–233.

98. **Xu, H.-S., N. Roberts, F. L. Singelton, R. W. Atwell, D. J. Grimes, and R. R. Colwell.** 1982. Survival and viability of nonculturable *Escherichia coli* and *Vibrio cholerae* in the estuarine and marine environment. *Microb. Ecol.* **8:**313–323.

99. **Zac, D. R., D. B. Ringelberg, K. S. Pregitzer, D. L. Randlett, D. C. White, and P. S. Curtis.** Soil microbial communities beneath *Populus grandidentata* Michx grown under elevated atmospheric CO_2. *Ecol. Appl.,* in press.

100. **Zak, J. C., M. R. Willig, D. L. Moorhead, and H. G. Wildman.** 1994. Functional diversity of microbial communities: a quantitative approach. *Soil Biol. Biochem.* **26:** 1101–1108.

Molecular Approaches for the Measurement of Density, Diversity, and Phylogeny

DAVID A. STAHL

11

INTRODUCTION: THE MEANING OF MICROBIAL DIVERSITY

This chapter considers the use of molecular methods for direct measures of abundance, diversity, and phylogeny of environmental populations of microorganisms. In part, the methods used to make these measurements cannot be separated from those of microbial systematics. Although issues in microbial systematics are fundamental to the characterization of environmental diversity and phylogeny, full coverage of them is beyond the scope of this chapter, and they are discussed only as necessary to provide background to specific methods. This emphasis also results in the exclusion of certain molecular techniques from detailed discussion, since they have not been productively applied to the description of natural systems.

Measures of biological diversity generally consider both the total number of species and the partition of abundance (29). Species richness (species abundance) is used to describe the number of species present, and species evenness (species equitability) is used to describe how evenly individuals are distributed among these species (29, 36). There are two major problems associated with the use of this measure. First, a variety of diversity indices have been proposed and debated, and the utility of a single index has been questioned (36). However, a critical discussion of diversity index formulations is beyond the scope of this chapter. The second problem is the fundamental unit measured. By default, this unit is species, the basic unit of evolution and ecology. This is an essential consideration for interpretation of the molecular techniques. Although the taxonomic level of species is common to the vocabulary of all biologists, the boundaries separating individual species are not always well defined. Microbiologists in particular remain undecided about the utility of the species concept in the description of microorganisms (61). As yet there is no satisfactory species concept for bacteria or fungi. This has resulted in much debate among microbial systematists and some sidestepping among microbial ecologists. Although the development of molecular systematics provided an essential phylogenetic framework to microbial ecology and a basis for describing natural microbial diversity, it has not resolved the species question. Thus, a necessary prelude to the following discussion of techniques is a brief discussion of species concepts, from the perspectives of both macroecologists and microecologists. The discussion is brief, and the reader is referred to a recent overview presented by King (40) for more complete coverage.

Biological Species Concepts

The biological species concept as defined and revised by Mayr (54) is summarized below (40).

1. Species are defined by distinctness rather than by differences.
2. Species consist of populations rather than unconnected individuals.
3. Species are not defined by the fertility of individuals but by the reproductive isolation of populations.
4. Species are reproductive communities of populations that occupy a specific niche in nature.
5. The species is an ecological unit that, regardless of the individuals composing it, interacts as a unit with other species with which it shares the environment.
6. The species is a genetic unit, a gene pool, whereas the individual is a temporary vessel holding a portion of the gene pool for a short time.

Reproductive isolation is a central feature of the biological species concept and has always created great difficulty in the treatment of organisms that reproduce asexually, microorganisms in particular. An attempt at synthesis was the introduction of the niche concept in the revised definition presented above (54). Although this attempt provided the foundation for including microorganisms, it was severely criticized for lack of a clear conceptual framework for the definition of niche (40).

Evolutionary Species Concepts

The evolutionary species concept as introduced by Simpson also provides the foundation for including uniparental organisms, such as microorganisms: "An evolutionary species is a lineage (an ancestor-descendant sequence of populations), evolving separately from others and with its own unitary evolutionary role and tendencies" (78). This definition was subsequently modified by Wiley to incorporate the dichotomous branchings clearly evident from molecular systematics, defining a species as "a single lineage of ancestral descendent populations of organisms which maintains its identity from other such lineages and which has its own evolutionary tendencies and historical fate" (112). A key feature of the evolutionary species concept, relevant to microorganisms, is that it accommodates the capacity to share genetic information between populations without the loss of distinct evolutionary roles. This definition is consistent with the clear genealogy of microorganisms now revealed by comparative sequence analyses (64, 113), in contrast to a promiscuity evidenced by the movement of broad-host-range plasmids between widely divergent populations (72). However, a weakness of this definition is the same as for the modified biological species concept: the use of the niche (unitary evolutionary role) as a defining characteristic. Although the evolutionary species concept is appealing, considering both niche and evolution, it does not serve practicing biologists.

Microbiologists also have also sought an acceptable operational definition of species. As discussed in the following section, the recognized definition of species is based on chromosomal DNA similarity (37, 108): "The phylogenetic definition of species generally would include strains with approximately 70% or greater DNA-DNA relatedness and with 5°C or less ΔT_m." (108). Although the standard, this definition does not provide a basis for relating microorganisms to other described species of metazoa and metaphyta, for which the species definition is anchored in reproductive isolation. This represents a fundamental discontinuity between macroecology and microecology that must be considered in the context of this chapter's coverage and also relates to current efforts to survey global biological diversity.

There are a couple of other fundamental features that distinguish studies of the microbial world from those of the visible world. These differences arise from the character of past studies in microbial ecology and microbial systematics. The current microbial taxonomy is based primarily on organisms isolated in pure culture and described on the basis of observable phenotype. This has had two major consequences. First, the pure culture collection is now recognized to represent only a small fraction of environmental diversity (6, 45). Thus, a sense of the scope of existing environmental diversity, a necessary backdrop for interpretation of described diversity, is virtually lacking. Second, the selection of phenotypic attributes upon which the description of a microorganism is based often is arbitrary (e.g., use of an established panel of general biochemical tests) and may not fully address the ecology of the organism. Consequently, there is little understanding of the niche occupancy of microorganisms. Most studies have been performed in the laboratory (pure culture) without observing other, possibly more relevant, activities in nature. Thus, microbiologists appear to be even less able than other biologists to evaluate microbial diversity within the framework of niche. Although molecular methods described in this chapter have provided the framework for describing the phylogeny of microorganisms, this has yet to be clearly integrated into microbial systematics or ecology. For example, there is no rule for depth of phylogenetic relationship defining species, genus, or higher taxonomic ranks. As a consequence, operational taxonomic units based on molecular characterization of environmental populations are sometimes used as a means to sidestep taxonomic definition. These aspects of the discipline have a direct bearing on the application and interpretation of the molecular techniques described in this chapter. Although further elaboration is not possible here, the reader should bear these considerations in mind while reading the following descriptions of methods.

MOLECULAR SYSTEMATICS: THE FOUNDATION FOR COMPREHENSIVE DIVERSITY SURVEYS

The conceptual foundation for a molecular systematics was formally introduced in 1965 by Zuckerkandl and Pauling (114). Biopolymers (DNA, RNA, or polypeptides) were classified according to the degree to which they reflected the genealogy of an organism. The central model of molecular evolution is one of random evolutionary changes occurring at a stochastically constant rate. This model was introduced in the form of the "molecular clock," which requires a constant rate of change in the sequence of a common biopolymer, the "molecular chronometer." Useful chronometers must display certain properties, including (i) molecular clock-like behavior, (ii) range (rates of change must be commensurate with the spectrum of evolutionary distances measured), and (iii) size (the molecule must be large enough to provide an adequate amount of information) (113). A good molecular chronometer should also fulfill a number of additional criteria, including universal distribution, functional constancy, and the absence of lateral transfer.

The first generally applied measure of sequence divergence was the use of DNA-DNA hybridization to evaluate total genomic similarity. As discussed below, this technique remains an important tool in microbial systematics and also has been applied to evaluating environmental diversity. However, the ease with which nucleic acid sequence can be determined has resulted in its nearly routine use in microbial systematics and increasing application to direct environmental studies. Proteins so far used in phylogenetic studies include ATPases, protein elongation factors, and cytochromes (25, 38). However, the comparative sequencing of the rRNAs has had the most far-reaching application. The establishment of a robust phylogeny based on comparative sequencing of the rRNAs provided the first consistent taxonomic description of microorganisms. This event was an essential prelude to the unambiguous description of their communities. By far, this has become the most commonly used measure of environmental diversity (65, 83, 85, 106).

The rRNAs—the Ultimate Molecular Chronometers?

Three types of rRNA are common to the ribosomes of prokaryotes and eukaryotes: the 16S (and 16S-like), the 23S (and 23S-like), and the 5S rRNAs. Since the larger rRNAs of many eukaryotes and some prokaryotes differ significantly in size, the terms "16S-like," and "23S-like" have been used to refer to the two larger rRNAs. This will be the convention used in this chapter. Certain attributes of the rRNAs favor their use as molecular chronometers. The sequences that code for rRNA are among the most highly conserved (113). The rRNAs can be viewed as composed of structural domains within which sequence variation differs with respect to increasing phylogenetic distance. Regions that vary sufficiently slowly allow inference of relationships between members of the three domains (Bacteria, Eucarya, and Archaea) (113), whereas the most variable regions provide for discrimination between organisms of approximate genus and species rank differences. Regional differences in sequence conservation have provided the basis for designing nucleic acid probes varying in specificity; group- and species-specific oligonucleotide probes have been used for direct assessment of environmental diversity (see below).

The accuracy of phylogenetic inference is dependent on not only the number of bases compared but also the particular region(s) of the molecule compared. The 16S-like rRNA (ca. 1,500 nucleotides) provides a large amount of information useful for phylogenetic inference. Although the 23S-like rRNA (ca. 3,000 nucleotides) contains about twice as much information as the 16S-like rRNA, and therefore should provide greater accuracy of phylogenetic inference, the smaller molecule, because it is much easier to sequence, has become the established reference. Thus, the 23S-like rRNA has been used primarily as a supplement to 16S-like rRNA data for resolving closely spaced evolutionary branching (50). Well over 5,000 16S-like rRNA sequences are now available. These have provided the most encompassing of available molecular frameworks with which to explore natural microbial diversity and phylogeny (65, 83, 106). Although the 5S rRNA molecule provides relatively little sequence (ca. 120 nucleotides) and has been less useful in the study of distant phylogenetic relationships (82), it has been used to provide "fingerprints" of microbial population structure in complex environments (see below).

Relationship between DNA and RNA Similarity

The most widely accepted molecular criterion for defining a microbial species is based on DNA-DNA hybridization (37). Organisms demonstrating genomic DNA sequence similarities of approximately 70% or greater are generally accepted as representatives of the same genomic species (37, 82, 108). It is important to distinguish the experimental basis for determining genomic sequence similarity in contrast to similarity values calculated on the basis of sequence comparisons (percentage identity). Genomic DNA similarity is estimated on the basis of two heterologous DNA strands reassociating to form a stable duplex structure. Under standard hybridization conditions, stable structure requires approximately 80% matching. Thus, there is not a direct relationship between similarity values determined by hybridization and percentage identity determined by sequence comparison. The reader is referred to a review by Stackebrandt and Liesack for more detailed discussion of the use of nucleic acids for classification (82). Although the correlation is somewhat arbitrary, the genomic species corresponds reasonably well with phenotypic attributes defining the group (species) (25, 82). The classification of organisms demonstrating DNA-DNA hybridization values between about 30 and 65% is less certain, but such organisms often are assigned to the same genus (37). In addition to DNA-DNA similarity, hybridization of the more highly conserved rRNAs was used to establish relationships at the intergeneric level and to define several genera (14, 39). Now, with increasing use of rRNA sequence comparisons in ecology, phylogeny, and taxonomy, the relationship between genomic DNA hybridization and 16S rRNA sequence similarity is important to establish. Unfortunately, there have been relatively few systematic comparisons of these two molecular measures of relationship.

Although identical 16S rRNA sequences are generally sufficient to confirm genomic species relationship (2, 82), the relative divergence of the two measures appears to vary for different microbial groups. For example, an analysis of the genus *Fibrobacter* demonstrated that strains related at approximately 50% DNA-DNA similarity (40 to 52%) shared 16S rRNA sequence similarity values of between 98.7 and 99.8% (2). Strains of *Bacillus globisporus* and *B. psychrophilus* related at 99.5% 16S similarity demonstrated less than 50% DNA reassociation (22). An estimation of this relationship based on previously published data suggested a value of approximately 98% 16S similarity (15).

Regardless, there appears to be little overlap between genomic hybridization- and rRNA sequence-based measures of relationship. Assemblages closely related by 16S rRNA sequence (ca. 95 to 98% similarity) would be usually undetectable with techniques now used to measure genomic DNA similarity (81a).

Yet another complicating consideration is sequence variation between the different rRNA operons of an organism. This variation is also found between organisms and has yet to be systematically evaluated. Inspection of 16S rRNA sequences published in GenBank suggested that interoperon difference may vary from 0 to 5% (12). It is uncertain to what extent sequencing errors contribute to this variation. Although interoperon variation is less a concern for defining taxonomic rank at levels higher than species, there is no consensus on evolutionary distance values (phylogenetic depth), as inferred from 16S sequence divergence (90), for assignment of organisms to genera or higher ranks. Thus, although 16S rRNA sequence divergence increasingly serves as a primary criterion in microbial classification and environmental surveys, the definition of higher taxons remains entirely subjective, and different described genera vary markedly in phylogenetic depth. This is a major criticism of phylogenetically based classifications, since they may not distinguish between the theoretical definitions and objects in nature, and is of great significance to the study of environmental microbial diversity and ecology, to which diversity measures based on rRNA sequence relationship are increasingly applied. The ecological underpinnings remain unknown.

Qualitative versus Quantitative Assessments of Community Structure

An important consideration in the discussion and evaluation of the various molecular techniques is the suitability of a method for identifying and quantifying environmental populations. Methods suitable for identification may not be well suited to quantification. For example, DNA restriction fragment length polymorphism is not easily interpreted when applied to environmental systems of unknown complexity and population composition. Techniques that rely on the recovery of relatively intact DNA (e.g., for PCR amplification, cloning, or restriction digestion) generally must use less disruptive extraction techniques than methods that analyze RNA. This is because DNA is much more sensitive to mechanical shearing than is RNA. Methods of RNA analysis can be combined with the most disruptive of extraction methods, for example, mechanical breakage by reciprocal shaking with zirconium beads (85, 87). Even so, the efficiency and representativeness of nucleic acid recovery from environmental samples remains a fundamental concern in all studies. Selected methods for evaluating representative recovery of nucleic acids from environmental samples are discussed below in the section on rRNA-targeted DNA probes. In addition to unresolved issues of representative DNA recovery, biases of PCR amplification and DNA cloning are well recognized (69). Although PCR amplification of conserved biopolymers (primarily rRNAs) is increasingly used to describe the environmental diversity of populations, the proportional recovery of specific sequences cannot be equated with abundance. And even though hybridization to extracted nucleic acids should provide more direct information of abundance, the meaning of abundance as defined in molecular terms is very different from established microbiological criteria based on CFU or direct cell counts. Thus, comparison of different measures of total biomass should be very informative. For example, phospholipid analyses might provide a relatively independent quantification of biomass that could be related to nucleic acid-based measures (27).

RECOVERY OF NUCLEIC ACID SEQUENCES FROM ENVIRONMENTAL SAMPLES

There are three basic formats now used to recover DNA sequence information, isolated from either pure culure or environmental samples: chain termination sequencing of cloned (or PCR-amplified) DNA templates, restriction enzyme digestion, and DNA probe hybridization. The latter two methods are used to identify relatively short sequence elements. For example, restriction enzymes commonly recognize 6- to 8-nucleotide sequence elements, and DNA probes, with the capacity for single-nucleotide mismatch discrimination, are usually around 20 nucleotides in length. Although longer DNA probes are commonly used to identify homologous targets, they do not provide defined sequence information.

Nucleic Acid Probes
General

Either DNA or RNA can serve as a nucleic acid probe. However, for a variety of technical reasons (e.g., ease of synthesis and stability), most studies have used DNA probes. There are two basic categories of DNA probes: group specific (phylogenetic or taxonomic) and functional. Group-specific probes generally target conserved biopolymers that can be used to infer phylogenetic relationships among the host organisms (3, 85). Today the most widely used target molecule is the small subunit rRNA (16S-like rRNA), and probes can be designed to target phylogenetic groups varying in evolutionary diversity ("phylotypes"). Phylogenetic probes therefore have the potential to provide explicit measures of community structure at different levels of resolution. Another category of group-specific probes consists of total genomic DNA probes used for species-level identification (e.g., reverse sample genome probing [RSGP]).

An important distinction is made between probes designed to identify phylogenetic or taxonomic groups and probes designed to monitor specific metabolic functions. There has been considerable development of probes targeting genes encoding specific enzymes to evaluate specific chemical transformations or potential activity of environmental populations. These are functional gene probes (e.g., for catabolism of aromatic compounds) (21, 74). Although they will not be specifically addressed in this chapter, functional probes provide an essential connection between the different measures of environmental diversity (phylogeny) and ecology. Also in this regard, the use of phylogenetic groups (phylotypes) as a measure of environmental diversity ultimately must include better understanding of unifying phenotypic characteristics of circumscribed groups. If certain traits are found to reflect membership within a group, these traits should serve to better relate community structure and function. Some examples of traits, and corresponding genes, conserved within phylogenetically defined groups include genes for nitrogen fixation (28), [NiFe] hydrogenase (107), and cellulases within some cellulolytic assemblages (48). Although the emphasis of this chapter is molecular, it is the phenotypic reflection of molecular diversity that must remain the central consideration in environmental microbiology.

Phylogenetic Probes

The scope of this review limits discussion of many of the technical aspects of rRNA probe-based analyses of natural systems. The reader is directed to recent reviews (3, 85, 106) and references therein for a more complete description of methods. A full presentation or review of rRNA-targeted probes would require extensive referencing of recent applications to environmental, diagnostic, and determinative research. I apologize for any exceptional omissions in this regard. The following discussion is intended to provide an overview of key considerations.

Phylogenetic Probe Design

The essential attribute of the rRNAs with regard to probe design is the regional conservation of nucleotide sequence. Although overall conserved in sequence, these biopolymers exhibit great variation in regional sequence conservation. Some nucleotide positions and locales have remained virtually unchanged since the divergence of all existing life (universal sequences), whereas other regions vary so quickly that they can be used to differentiate among species of bacteria. In addition, the generally high copy number of rRNA per cell lends greater sensitivity to direct detection using hybridization techniques.

Probes designed to complement the rRNAs are generally 15 to 25 nucleotides in length (3, 85). The assemblage of organisms encompassed by a probe varies according to the region of the molecule selected as the hybridization target. Species-specific probes complement the most variable regions. More general probes, identifying phylogenetic groups of rank greater than species, target more conserved regions of the molecule (85, 87, 106). The publications describing the development of 16S rRNA probes for clinical or environmental application are much too numerous to address here. Thus, I will not discuss the more specialized applications but rather focus on the more general class of probes. Since these probes are designed to encompass larger phylogenetic groupings (Fig. 1), they can be used to provide a phylogenetic overview to community structure.

Probe Characterization: Phylogenetic Nesting of Probes

An essential aspect of probe development is the demonstration of target group specificity. To some extent this can be demonstrated empirically, using a collection of target and nontarget group nucleic acids or fixed reference cells for studies using whole-cell hybridization (3). My laboratory and others have routinely used a panel of diverse rRNAs ("phylogrid") to characterize new probes (16, 47, 68). Prior to an evaluation using the reference panel, it is essential that the temperature of dissociation (T_d) of the probe-target complex be experimentally determined (3, 85). It is generally insufficient to use one of the available formulas to predict T_d (85). Also, the temperature interval over which probe dissociates from the target RNA varies considerably for probes having comparable T_ds. Knowledge of the temperature range over which dissociation occurs is essential for adjusting hybridization conditions as needed to discriminate between closely related nontarget species. Thus, the initial T_d characterization should include closely related nontarget species rRNA. An independent T_d evaluation must be used to characterize probes used for whole-cell hybridization since there may be a significant difference between the T_d values and transition temperature ranges determined for the same probe by using different formats, membrane and whole-cell hybridization.

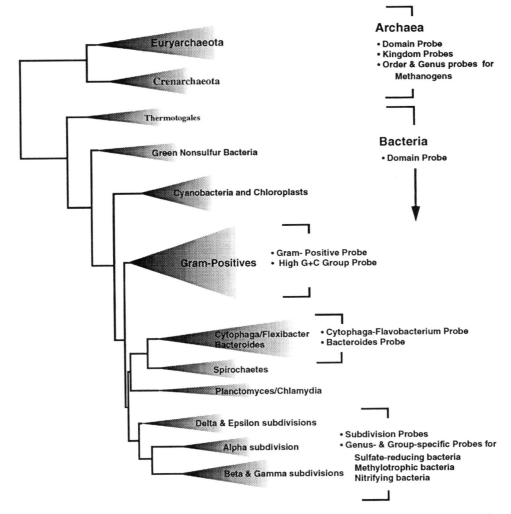

FIGURE 1 Partial listing of group-specific probes in relationship to prokaryote phylogeny. The phylogeny was adapted by Stahl (84), using the maximum-likelihood analysis of Olsen et al. (64) to provide the relative branching order of the major prokaryotic lineages. The probes for the indicated target groups have been described in the indicated references: archaeal domain (24, 68, 86), bacterial domain (24, 85), archaeal kingdoms (11), gram-positive organisms (97), high- G + C gram-positive organisms (71), cytophaga-flavobacterium and bacteroides (53), proteobacteria subdivisions (52), sulfate-reducing bacteria (16), methylotrophic bacteria (10, 98), and nitrifying bacteria (55, 104, 105). The reader can consult these references for more complete descriptions of probe design and characterization. Reprinted from reference 84 with permission from Blackwell Science Ltd.

There is a limitation to an empirical characterization of probe specificity. If we have only a limited appreciation of microbial diversity, it is impossible to construct a reference panel to unequivocally demonstrate specificity. However, there are a couple of additional methods to further evaluate specificity. The first is to use multiple probes, each having the same target group specificity, to quantify a single target population. For example, two probes for the archaeal domain (68) were recently used by DeLong and coworkers to independently confirm estimates of high archaeal abundance in Antarctic waters (13). Both probes hybridized to the same fraction of 16S rRNA extracted from these waters.

The second approach to probe validation takes advantage of the phylogeny. As already noted, it is generally possible to construct probes for phylogenetic groups (phylotypes) differing in evolutionary depth. These probes are of a hierarchical specificity and have been described as being nested. The use of a nested set of probes to characterize environmental diversity provides yet another consistency check. If the more specific probes fully represent the larger phylogenetic group, then the sum of the specific probe hybridization values should equal that obtained by using the more general probe. For example, the sum quantification obtained by using a complete set of species-specific probes should equal that of the corresponding genus-level probe. This approach was used to identify a novel lineage of cellulolytic bacteria in the equine cecum (47) and frequently used to evaluate consistency between domain-probe summation and total population abundance determined by using a universal probe (67).

General caveats relating to the use of phylogenetic probes include the following. The probes are tools subject to refinement through experimentation. Only through general application and combined use with other methods of community analysis will they be fully evaluated or, as necessary, refined. Also, the resolution of the 16S rRNA probes is approximately at the level of species. Questions relating to the abundance and distribution of subspecies and strains will require the combination of different approaches and methods (e.g., fluorescent antibody techniques). Another concern relates to the extraction of RNA from environmental samples. Although resistant to mechanical breakage, RNA is subject to degradation during and following extraction, generally as a consequence of endogenous nucleases or nuclease contamination. One consequence of partial degradation of sample is variable destruction of different probe target sites. For example, one of the regions used as a target site for hybridization to a universal probe is very sensitive to degradation (70). Thus, for methods of quantification using extracted rRNA, it is essential that sample integrity be evaluated. This is most conveniently accomplished by using acrylamide gel electrophoresis to demonstrate recovery of high-molecular-weight species (1, 73).

Quantification of Populations or Single Cells

There are two basic formats for using phylogenetic probes to study the environmental distribution of microorganisms: hybridization to total rRNA extracted from the environment and hybridization to whole cells for subsequent microscopic visualization and enumeration. The application of phylogenetic probes to single-cell identification and enumeration is discussed in recent reviews (3, 85). Both approaches have associated technical difficulties, but I will restrict my examples and discussion to analyses of extracted nucleic acid (66, 67, 87).

The central concern is recovery of rRNA from a variety of different environmental matrices and organism types. There are two aspects of nucleic acid recovery that are distinguished by the analytical approach reviewed here. The first is the efficiency of extraction. What fraction of total nucleic acid is recovered from the environmental matrix? For example, although the breakage technique may disrupt all microorganisms present, recovery might be reduced by degradation or adsorption of nucleic acids to matrix material (e.g., clays). The second consideration is representative rRNA recovery. Does the fractional recovery correspond to the environmental abundance of the corresponding nucleic acids present in the environment? For example, a population resistant to breakage would be fractionally underrepresented; conversely, an exceptionally easy-to-break microorganism would be overrepresented. The use of a universal hybridization probe to evaluate relative breakage efficiency of different groups is discussed below.

These are aspects of the method that must be more fully resolved before any nucleic acid technology can be applied routinely to environmental studies. Although the identification of novel species is relatively straightforward, quantifying their abundance is far from straightforward. The technical aspects of nucleic acid (both RNA and DNA) recovery from the environment have been discussed by a number of investigators (1, 35, 62, 66, 96) and will not be addressed further here.

Use of a Universal Probe to Normalize Hybridization Results

The near-universal conservation of certain regions of the rRNAs can be used to quantify total rRNA recovered from an environmental sample. For example, hybridization to a probe complementary to a region of sequence near position 1400 in the 16S rRNA has been used to quantify total 16S rRNA present in nucleic acid extracted from a variety of environments (67, 87). This value can then be used to express the abundance of more specific 16S rRNA target groups as a fraction, or percentage, of the total 16S rRNA recovered. This method was first used to evaluate population changes in the bovine rumen following the addition of antibiotic to the diet (87). Normalization offers several advantages. Experimental variation in hybridization, as might result from the presence of substances interfering with hybridization, is accounted for. Also, a measure of total rRNA recovery provides an internal reference for evaluating the recovery of specific rRNA target groups, for example, by comparing different breakage conditions. We have used normalization with a universal probe to compare the recovery of 16S rRNA from easy-to-break and hard-to-break organisms by using different mechanical breakage conditions (66).

Restriction Enzymes

The most common format for using restriction enzymes to define environmental diversity consists of combining digestion and fractionation of DNA extracted from an environmental sample with hybridization using nucleic acid probes complementary to conserved gene sequences common to all or many of the organisms present in the sample. The general format was first described by Southern and is often referred to as Southern blotting (80). The rRNAs are optimal targets for hybridization (ribotypes) (26), but other conserved elements (e.g., nitrate reductase and formyltetrahydrofolate synthetase) also have been used for environmental studies (49, 79). In application, restricted DNA is separated by size on an agarose gel and transferred to a membrane support for hybridization with a radiolabeled probe or a label appropriate for use with one of a variety of nonradioactive detection agents (e.g., digoxigenin) (85). The resulting population of different-size DNA fragments hybridizing to the probe is then used to infer a relationship between individual isolates or resolve different environmental populations. The separation of genes derived from different populations requires that they differ in sequence at the sites of DNA restriction or differ in length of DNA flanked by common restriction sites. For this reason, more than one restriction enzyme is generally used for restriction enzyme analysis, and the resulting size distribution patterns (banding patterns) are compared. This technique has been used primarily as a method to relate different microbial isolates on the basis of restriction fragment length polymorphism, with lesser direct application to environmental samples.

Direct Sequence Analyses

The now well-established techniques of PCR amplification, DNA cloning, and nucleic acid sequencing have become general tools for the study of environmental microbiology. These basic methods are well described in a variety of methods manuals (73) and will not be specifically addressed here. By far, the sequence information most commonly extracted from natural systems is that for the RNAs. As discussed above, for a variety of technical and practical considerations (size, information content, ease of sequencing), the 16S-like rRNA has become the standard measure for defining phylogenetic affiliation. Many thousands of sequences are now available in nucleic acid sequence databases, and associated tools for analysis are available (51, 99). Although the

23S-like rRNAs would provide the basis for more precise phylogenetic assignments and improved design of phylogenetic probes, it is not currently feasible to supplement the 16S-like rRNA sequence collection with corresponding sequences for the larger rRNA. The general approach as developed for community-level analyses based on rRNA sequence content could be applied to any biopolymer of appropriate conservation and community representation.

There are three basic methods to recover rRNA sequence information from nucleic acid extracted from environmental samples: (i) shotgun cloning, screening, and sequencing (63, 65, 75), (ii) cDNA cloning and sequencing of rRNA (109), and (iii) PCR amplification, cloning, and sequencing (4, 23, 106). The PCR-based methods can use either DNA or RNA as the template, the latter requiring the use of reverse transcriptase to generate cDNA from rRNA. These sequencing approaches all require the screening and analysis of large clone collections. Screening of a shotgun library derived from total environmental DNA is the more difficult approach since only about 0.125 to 0.3% of the clones contain part or all of the rRNA gene (75). These clones can be identified by hybridization, for example by using total rRNA derived from the environmental sample as a probe. Alternatively, DNA probes targeting highly conserved regions of the molecule (phylogenetic probes) may be used. This first screening step is generally not necessary for PCR-based recovery methods since the majority of clones will contain rRNA sequence. The second phase of the analysis is the elimination of redundant clones, which is essential to avoid expense and time associated with unnecessary sequence determinations. A variety of approaches have been used to identify redundant clones, using or combining the following strategies: complete or single-nucleotide sequencing of a small variable region, species- or group-specific phylogenetic probe screening, and restriction analysis (63, 106). The resolution of the different screening techniques must be balanced against the time and expense associated with each strategy. This determination can be made only with consideration of habitat and research objectives.

The caveats of sequencing approaches relate primarily to previously discussed biases of nucleic acid extraction, PCR amplification, and cloning. Experience in my laboratory has shown that even when nucleic acids extracted from pure cultures are used, PCR results are sometimes inconsistent. Amplification of rRNA sequences by using general primers has been shown in some cases to exclude important environmental populations (4, 81). The issues of PCR-generated sequence hybrids (chimeras) and the extent of sequence variation between rRNA operons of individual organisms remain to be fully evaluated.

GENOMIC DNA HYBRIDIZATION MEASURES OF COMMUNITY STRUCTURE

The general category of genetic complexity is used to classify methods of analysis that provide limited information of specific sequence content. The extent to which these methods can be used to estimate species diversity and individual population identity varies with the method, community complexity, and the aforementioned questions of microbial species definition.

DNA Reassociation

DNA reassociation kinetics was initially used to evaluate genomic sequence complexity, revealing repetitive DNA elements in the genomes of higher eukaryotes (9). More recently it has been used to assess the diversity of natural microbial communities (93, 95). Community-level DNA sequence complexity, as inferred from the rate of DNA reassociation, is related to population complexity. This measurement of complexity is a function of the concentration of complementary strands. Under defined conditions, strand reassociation follows second-order kinetics. Thus, the rate of reassociation is proportional to the square of the nucleotide concentration of homologous DNA strands. At a given concentration of total DNA (molar concentration of nucleotides in single-stranded DNA), increasing genomic complexity (larger genomes, larger numbers of genomes, fewer repeat elements per genome) results in a reduced concentration of complementary strands and a correspondingly reduced rate of reassociation. For example, as microbial community diversity (heterogeneity) increases (e.g., there are a greater number of unique genomes), the rate of reassociation of DNA extracted from the community decreases.

Experimentally, DNA reassociation is measured over time and the fraction of reassociated DNA (C/C_0) is expressed as a function of C_0t, where C_0 is the initial molar concentration of nucleotides in single-stranded DNA and t is the time in seconds. The plot of this relationship is referred to as a C_0t curve. The reaction rate constant can be expressed as $1/C_0t_{1/2}$, where $t_{1/2}$ is the time required for 50% reassociation. Under defined conditions, most importantly temperature and monovalent ion concentration, $C_0t_{1/2}$ is proportional to the complexity (e.g., number of unique genomes) of the DNA. The practical and theoretical considerations of DNA reassociation are well developed (8, 9, 111), and the reader can consult the cited references for more complete theoretical and practical treatments.

The interpretation of DNA reassociation kinetics is made in the context of information theory, as has been developed for other diversity indices and briefly discussed above (94). It is a measure of the total amount of information in a system (richness, number of unique genomes) and the distribution of that information (evenness, abundance of individual genomes). Torsvik and coworkers have expressed diversity as the number of "standard" genomes with no homology (93, 94). Although this method has been applied to relatively few communities, the results are notable in that they suggest far greater diversity than anticipated. In an initial study of a soil sample taken from a beech forest (in Seim, Norway), reassociation kinetics suggested the presence of approximately 4,000 genomes (93). Here, the system is defined as the soil specimen from which the DNA was extracted. A more recent study examined the change in diversity of marine sediment populations associated with pollution from an associated fish farm (94). Reassociation rates were determined for DNA isolated from the bacterial fraction isolated from the 10-cm top layer of the sediments. The bacterial fraction of the unpolluted sediment had a $C_0t_{1/2}$ equivalent to 10,000 *Escherichia coli*-size genomes with no homology. In contrast, the diversity of the polluted sediment corresponded to about 50 *E. coli*-size genomes. Thus, organic pollution apparently resulted in a dramatic reduction (ca. 250-fold) in bacterial diversity.

Although DNA reassociation should provide a generally useful measure of community structure, a variety of parameters must be considered, and all have yet to be systematically evaluated. One concern in the interpretation of DNA reassociation estimates is a reduction in the rate of reassociation resulting from impurities in the DNA sample. For example, Torsvik et al. showed that the rate of reassociation increased with repeated purification of the sample DNA.

This can be evaluated in part by the addition of exogenous DNA to serve as an internal control (93). However, it is also important to more fully evaluate changes in reassociation kinetics that might result from the use of different extraction and purification techniques (92). Another consideration for any DNA-based analysis is the source of the DNA. The persistence of "inactive" DNA, either in the environment or entrained within dead or moribund cells, is essentially unanswered and is a concern in interpreting any data obtained solely from DNA.

RSGP

Reverse sample genome probing (RSGP) makes use of the observation that the entire genome of a microorganism can be used as a specific probe for its detection in the environment. Whole-genome probes have been used to detect *Mycobacterium*, *Mycoplasma*, *Chlamydia*, *Bacteroides*, and *Campylobacter* species (5, 17, 18, 56, 110). RSGP reverses the usual relationship of sample DNA and probe. The genomic DNA from different reference organisms is denatured and immobilized on a membrane support, the reference panel. DNA extracted from the environment, containing an unknown diversity of organisms, is randomly labeled and hybridized to the reference panel. Under conditions of high stringency, whole-genome probes hybridize only to identical or closely related genotypes. For example, when a reference panel of DNAs from different sulfate-reducing bacteria was hybridized with randomly labeled genomic DNA from any species represented on the panel, only self-hybridization or hybridization with nearly identical isolates was observed (102, 103).

The following overview addresses only key technical considerations (100–103). The method requires the antecedent isolation of reference organisms from the environment. Chromosomal DNAs are then isolated from the different strains, measured amounts are applied to membranes, and cross-hybridization among strains is evaluated by using stringent hybridization conditions. Strongly cross-hybridizing DNA preparations are combined and treated as the same standard (either a single species or a set of closely related species). Different reference standards have genomes that generally cross-hybridize less than 1%. The standards are then used to prepare a master filter, using bacteriophage lambda as an internal control. The amount of genomic DNA applied to the membrane varies with analytical need. For example, 20 ng is sufficient for analysis of DNAs obtained from pure or enrichment cultures, and 200 ng is appropriate for analysis of total community DNAs. A reference concentration series of bacteriophage DNA (e.g., 10, 20, 50, and 100 ng) is applied on the same membrane. Sample DNA (ca. 100 ng) and lambda DNA (ca. 200 pg) are combined, boiled, and placed on ice. A probe is prepared by random hexamer labeling, using $[\alpha\text{-}^{32}P]dCTP$ and Klenow polymerase. Following denaturation, the probe is hybridized to a master filter under stringent conditions. Following washing of the filter under defined conditions, bound probe is quantified (e.g., by autoradiography), and the fraction of community DNA composed of individual component genomes (represented by the individual reference DNAs on the master membrane) is calculated from the hybridization to individual DNA standards relative to the lambda reference series (101). This analysis assumes that the lambda and environmentally derived DNAs present in the probe mixture are labeled to the same specific activity and hybridize with comparable efficiency. One possible concern is that impurities associated with the environmental DNA

may influence both relative labeling and the extent of hybridization.

To date, the technique has been used primarily to evaluate microbial populations associated with oil fields (101–103). RGSP revealed a significant difference between planktonic and biofilm-associated populations in oil recovery systems. Planktonic populations were more diverse and dominated by organochemotrophs. In contrast, biofilm populations were typically dominated by one to three populations of sulfate-reducing bacteria from the family *Desulfovibrionaceae*, with much lower representation by organochemotrophs. More recently this technique was used to demonstrate the effects of pollutants (e.g., benzene and toluene) on microbial communities (100).

Advantages of the RSGP technique are that once an appropriate microbial survey of the target environment has been completed, master filters can be prepared rapidly and economically in large numbers. These filters can be stored indefinitely for immediate use when new sample DNA become available for analysis. DNA from newly isolated standards can be spotted on side strips that are hybridized with the sample DNA probe together with the master filter. Since as many as 20 sample DNA preparations can be labeled simultaneously and subsequently hybridized, routine screening of sample DNAs in terms of a large number of different standards is feasible. An advantage is also that a single set of highly stringent hybridization conditions can be used.

A disadvantage of the technique is that although the actual assay does not involve culturing, the microbial community is described only in terms of its culturable component. Although RSGP has good precision, the calculated fractions can be subject to systematic errors. For example, the calculated fractions are sensitive to label allocation to sample and internal standard DNA (i.e., differences in specific activity). Also, detection sensitivity is defined by the extent of cross-hybridization of the standard DNAs. Fractional representation of individual populations cannot be reproducibly determined below the threshold defined by cross-hybridization.

Reciprocal Hybridization of Community DNA

An approach to assess change in community structure in time and space by using reciprocal hybridization of total community DNA was described by Lee and Fuhrman (42, 43). The method provides a relatively rapid format to reveal significant population differences or shifts. However, interpretation is complicated by uncertainties about relative population abundances and hybridization kinetics and by possible differences in labeling efficiencies of DNAs comprising different DNA pools. Nonetheless, the method provides a relatively clear assessment of community relationship when the communities are nearly identical (high reciprocal hybridization similarity) or distinct (low reciprocal hybridization). Intermediate cross-hybridization values are more difficult to interpret. Thus, the primary utility of the technique is to reveal significant differences or shifts in community structure.

A consideration in the interpretation of any of the DNA-DNA hybridization methods described above is the resolution of the technique. Significant hybridization is observed only between populations that have significant DNA similarity. Recall that 50% DNA similarity, as measured by hybridization techniques, is anticipated for organisms that have approximately 98 to 99% rRNA sequence similarity (2). Thus, although little apparent similarity in structure may be evidenced by a DNA hybridization-based measure, significant relationship at the approximate level of genus could nonetheless be present.

NUCLEIC ACID FINGERPRINTS OF COMMUNITY STRUCTURE

The term "molecular fingerprint" is reserved for methods of analysis that generate a pattern-based characterization of community structure, most commonly represented by a banding pattern of nucleic acid fragments resolved by gel electrophoresis, but provide little or no direct information of specific microbial population identity. The techniques of low-molecular-weight (LMW) RNA pattern analysis and denaturing gradient gel electrophoresis (DGGE) are specifically addressed. The restriction fragment length polymorphism method previously described has also been referred to as DNA fingerprinting and could be included in this category. The following discussion of methods does not address sample collection or extraction of nucleic acid.

RNA Pattern Analyses

The method is based on the analytical separation of LMW RNAs (5S rRNA and RNAs) to generate a profile (fingerprint) characteristic for individual organisms (31, 33). High-resolution gel electrophoresis has been most commonly used to separate the different RNA species, primarily on the basis of size, but other possible approaches include high-performance and fast-performance liquid chromatography. The resulting fractionation pattern of resolved 5S rRNAs and class 1 and class 2 tRNAs is a stable feature of the organism. It does not vary with different growth phases or medium composition and has been used as a chemotaxonomic character set for microbial classification (32). The resolution is sufficient to differentiate organisms classified at the approximate taxonomic ranks of species and above. For example, profiling of five *Alcaligenes* species revealed 5S rRNA bands of identical length (116 nucleotides) but significantly different tRNA patterns. A survey of typical genera from five of the major bacterial lineages demonstrated a unique tRNA pattern for each genus and also revealed a wide range of 5S rRNA sizes (109 to 114 nucleotides) (33). A comparison of organisms previously characterized by DNA-DNA homology showed that those of identical RNA profiles had homology values of 80 to 100% (32). These values are well within the accepted DNA homology range for inclusion in a single species.

The method also provides for direct analyses of natural microbial populations (30, 34). LMW RNA species extracted from naturally available biomass are fractionated to generate a community profile that can be related at some level to pure culture isolates previously characterized by this criterion. The pattern alone serves as a direct measure of community structure and change in structure, as might result from change in environmental context or perturbation of the system (30, 34). However, the resolution of evaluated separation techniques imposes an upper limit on the number of organisms that can be clearly resolved. Thus, the method as now developed is most applicable to environments dominated by a few species (32). If individual bands are sufficiently well resolved, sequencing of individual 5S rRNA species provides a more direct identification of population member identity. Although the information content of the 5S rRNA is insufficient for precise phylogenetic placement, its sequence does serve for identification when one or more closely related reference organism sequences are available (41, 88, 89).

Quantification of individual natural populations depends on the above-noted issues of resolution and the method of detection. The most sensitive technique currently available is end labeling with radioactive ^{32}P (41, 88, 89); however, quantitative inference is compromised by variable incorporation of radiolabel. Structural differences at either the 5' or 3' termini of different LMW RNAs influence the kinetics of label incorporation using either polynucleotide kinase or RNA ligase. This is a well-recognized problem for end labeling of 5S rRNAs (89) but is suggested to be much less severe for end labeling of tRNAs. The tRNAs label with high and more uniform efficiency when RNA ligase is used to append a radiolabeled nucleotide (e.g., cytidine 3',5'-[5'-^{32}P] bisphosphate) to the CCA$_{OH}$ tail comprising the 3' terminus of all mature species (32). Even so, a remaining concern is the partial or differential loss of 3' termini as a result of degradation during sample collection or extraction, resulting in nonuniform labeling. An alternative detection method uses silver staining for visualizing 5S rRNA and tRNA bands. The advantage of silver staining is that the intensity of staining is proportional to the amount of RNA present. However, this detection method is 1 to 2 orders of magnitude less sensitive than radioactive end labeling and so requires large amounts (ca. 10 μg) of environmental RNA.

Although there are a number of caveats associated with the use of LMW RNA fingerprinting, it is the one nucleic acid-based method that provides an all-encompassing and direct measure of population diversity and abundance. All other nucleic acid-based approaches use techniques that are well recognized to introduce bias (e.g., PCR and cloning) (69) or rely on indirect detection (e.g., nucleic acid hybridization) of a population subset of the community.

DGGE

The DGGE method of analysis is based on the analytical separation of DNA fragments identical or nearly identical in length but different in sequence composition. The method was first developed to detect single-base changes in genes for diagnosis of human genetic diseases and in genetic linkage studies (60). More recently DGGE has been extended to resolve environmental populations of microorganisms by separating PCR amplification products generated by using primers targeting conserved genes. Common flanking sequences within highly conserved genes (e.g., rRNA and hydrogenase genes) serve as general priming sites for amplification. PCR primers designed for the amplification and cloning of 16S-like rRNA genes were first used to demonstrate the technique as applied to environmental microbiology (58). A more recent study used the same general approach to resolve natural populations of sulfate-reducing bacteria that share a highly conserved hydrogenase (107).

Separation is based on changes in electrophoretic mobility of DNA fragments migrating in a gel containing a linearly increasing gradient of DNA denaturants (urea and formamide). Changes in fragment mobility are associated with partial melting of the double-stranded DNA in discrete regions, the so-called melting domains. Since the temperature of the gel is held constant, the melting temperature for each domain varies according to concentration of denaturant and therefore by position in the gel. When the DNA enters a region of the gel containing sufficient denaturant, a transition of helical to partially melted molecules occurs, and migration is severely retarded. Sequence variation within such domains alters their melting behavior, and sequence variants of the different amplification products stop migrating at different positions in the denaturing gradient (44).

By using the method as first developed, approximately 50% of sequence variants studied could be detected in DNA fragments up to 1,000 bp in length. However, virtually all variants can be resolved by the attachment of a GC-rich sequence to the DNA fragment. This terminal appendage, the GC clamp, acts as a high-temperature-melting domain. The GC clamp is added by cloning (60) or by PCR performed with a primer that contains a 40-bp GC-rich sequence at the 5' end (76, 77).

The use of DGGE to characterize a population of DNA amplified from a particular environment generally requires a preliminary study to establish appropriate running conditions. Amplification products are first characterized by using perpendicular gels with an increasing gradient of denaturants from left to right (perpendicular to the direction of electrophoresis). The sample is applied across the entire width of the gel, electrophoresed for about 3 h, and visualized by ethidium bromide staining. DNA molecules at the side of the gel containing a low concentration of denaturants migrate as double-stranded DNA, whereas those to the other side melt upon entry and stop. At intermediate concentrations of denaturants, the molecules have different degrees of melting, as reflected in different mobilities. This preliminary analysis provides some information of population complexity and defines a narrower range of denaturants for use in subsequent higher-resolution studies. The later studies generally use a parallel gradient gel with an increasing concentration of denaturants from top to bottom. Analyses of multiple samples are generally conducted with parallel gradient gels. These studies are generally preceded by a time travel experiment to establish the appropriate electrophoresis run time(s). This determination involves simply loading the same sample repeatedly to different lanes of a parallel gradient gel at a defined time interval between sample loadings.

DGGE analysis of PCR-amplified 16S rDNA fragments provides a rapid method to characterize community population structure, with consideration of some of the caveats discussed below. The initial study by Muyzer et al. demonstrated the presence of several distinguishable bands (between 5 and 10) in the gel separation pattern, which were most likely derived from the predominant species within those communities characterized (58). Recent studies have examined microbial mats (7, 19), deep-sea hydrothermal vent samples, and a stratified marine water column (91). More specific information of population composition can be obtained by secondary analysis of the DGGE banding pattern via sequencing or hybridization. Sequences of individual bands (fragments) are determined following their extraction from the gel, a second round of PCR amplification, and sequencing (direct or after cloning). Group- and species-specific DNA hybridization probes have been used to identify specific populations within the pattern of resolved bands following transfer of the DNA to nylon membranes (58, 59).

In addition to community structure information, the DGGE method is amenable to semiquantitative assessments of activity. Although detailed discussion of this topic is beyond the scope of this chapter, brief mention is appropriate. One of the most general measures of cellular activity is ribosome content. The ratio of rRNA to rDNA increases with increasing growth rate (activity). DGGE has been used to evaluate this ratio among different natural populations by comparing patterns and intensities of bands derived from

using either rDNA or rRNA (using reverse transcriptase to generate cDNA) as the template (91). A similar approach was used to evaluate the expression of a NiFe hydrogenase conserved among natural populations of desulfovibrios (57).

The caveats of the DGGE analysis method include the following. Beyond the usual concerns of representative DNA extraction, the questions of representative PCR amplification of individual populations within the target collection and formation of chimerical sequences between populations remain mostly unanswered (46, 69). Separation of the many fragments amplified from a highly diverse bacterial community is not possible with available technology, although resolution may be improved by using a narrower range of denaturants or two-dimensional electrophoresis (20). The phylogenetic information obtained from sequencing of individual bands is limited, because only fragments of up to approximately 500 bp can be well separated. However, this amount of information is substantially more than that provided by 5S rRNA sequencing (see LMW RNA Pattern Analyses, above). Another concern associated with the technique is the a priori assignment of individual bands to individual populations. As discussed in the introductory considerations of diversity assessments, sequences between rRNA operons of an individual organism can vary significantly, so individual organisms could potentially contribute to multiple bands on a DGGE gel.

I thank Gerrit Voordouw for providing a detailed description of the RSGP technique. Thanks is also given to Gerard Muyzer, Manfred Höfle, and Vigdis Torsvik for providing supporting papers and information.

Research originating in my laboratory and discussed in this chapter was supported by grants from NSF, ONR, and USDA.

REFERENCES

1. **Alm, E., and D. A. Stahl.** Extraction of DNA from sediments. *In* A. K. L. Akkermans, J. D. van Elsas, and F. J. de Bruijn (ed.), *Molecular Microbial Ecology Manual*, in press. Kluwer Academic Publishers, Dordrecht, The Netherlands.

2. **Amann, R. I., C. Lin, R. Key, L. Montgomery, and D. A. Stahl.** 1992. Diversity among *Fibrobacter* isolates: towards a phylogenetic classification. *Syst. Appl. Microbiol.* **15:**23–31.

3. **Amann, R. I., W. Ludwig, and K.-H. Schleifer.** 1995. Phylogenetic identification and in situ detection of individual microbial cells without cultivation. *Microbiol. Rev.* **59:**143–169.

4. **Amann, R. I., J. Stromley, R. Devereux, R. Key, and D. A. Stahl.** 1992. Molecular and microscopic identification of sulfate-reducing bacteria in multispecies biofilms. *Appl. Environ. Microbiol.* **58:**614–623.

5. **Andrew, P. W., and G. J. Boulnois.** 1990. Early days in the use of DNA probes for *Mycobacterium tuberculosis* and *Mycobacterium avium* complexes, p. 179–198. *In* A. J. L. Macario and E. Conway de Macario (ed.), *Gene Probes for Bacteria.* Academic Press, Inc., San Diego, Calif.

6. **Barns, S. M., R. E. Fundyga, M. W. Jeffries, and N. R. Pace.** 1994. Remarkable archaeal diversity detected in a Yellowstone National Park hot spring environment. *Proc. Natl. Acad. Sci. USA* **91:**1609–1613.

7. **Bateson, M. M., and D. M. Ward.** 1995. Analysis of *Chloroflexus* diversity in hot spring microbial mats by molecular methods and extincting dilution enrichment, abstr. N-53, p. 341. *In Abstracts of the 95th General Meeting of the American Society for Microbiology 1995.* American Society for Microbiology, Washington, D.C.

8. **Britten, R. J., D. E. Graham, and B. R. Neufeld.** 1974. Analysis of repeating DNA sequences by reassociation. *Methods Enzymol.* **29:**363–418.

9. **Britten, R. J., and D. E. Kohne.** 1968. Repeated sequences in DNA. *Science* **161:**529–540.

10. **Brusseau, G. A., E. S. Bulygina, and R. S. Hanson.** 1994. Phylogenetic analysis and development of probes for differentiating methylotrophic bacteria. *Appl. Environ. Microbiol.* **60:**626–636.

11. **Burggraf, S., T. Mayer, R. Amann, S. Schadhauser, C. R. Woese, and K. O. Stetter.** 1994. Identifying members of the domain *Archaea* with rRNA-targeted oligonucleotide probes. *Appl. Environ. Microbiol.* **60:**3112–3119.

12. **Clayton, R. A., G. Sutton, P. S. Hinkle, Jr., C. Bult, and C. Fields.** 1995. Intraspecific variation in small-subunit rRNA sequences in GenBank: why single sequences may not adequately represent prokaryotic taxa. *Int. J. Syst. Bacteriol.* **45:**595–599.

13. **DeLong, E. F., K. Y. Wu, B. B. Prézelin, and R. V. M. Jovine.** 1994. High abundance of archaea in Antarctic marine picoplankton. *Nature* (London) **371:**695–697.

14. **De Smedt, J., and J. De Ley.** 1977. Intra- and intergeneric similarities of *Agrobacterium* ribosomal ribonucleic acid cistrons. *Int. J. Syst. Bacteriol.* **27:**222–240.

15. **Devereux, R., S.-H. He, C. L. Doyle, S. Orkland, D. A. Stahl, J. LeGall, and W. B. Whitman.** 1990. Diversity and orgin of *Desulfovibrio* species: phylogenetic definition of a family. *J. Bacteriol.* **172:**3609–3619.

16. **Devereux, R., M. D. Kane, J. Winfrey, and D. A. Stahl.** 1992. Genus- and group-specific hybridization probes for determinative and environmental studies of sulfate-reducing bacteria. *Syst. Appl. Microbiol.* **15:**601–609.

17. **Dular, R.** 1990. Gene probe detection of human and cell culture mycoplasmas, p. 417–449. *In* A. J. L. Macario and E. Conway de Macario (ed.), *Gene Probes for Bacteria*. Academic Press, Inc., San Diego, Calif.

18. **Dutilh, B., C. Bebaer, and P. A. D. Grimont.** 1990. Detection of *Chlamydia trachomatis* with DNA probes, p. 46–64. *In* A. J. L. Macario and E. Conway de Macario (ed.), *Gene Probes for Bacteria*. Academic Press, Inc., San Diego, Calif.

19. **Ferris, M. J., G. Muyzer, and D. Ward.** 1995. Analysis of native bacterial populations in a hot spring microbial mat using denaturing gradient gel electrophoresis (DGGE) of 16S rDNA fragments, abstr. N-26, p. 337. *In Abstracts of the 95th General Meeting of the American Society for Microbiology 1995.* American Society for Microbiology, Washington, D.C.

20. **Fischer, S. G., and L. S. Lerman.** 1979. Length-independent separation of DNA restriction fragments in two dimensional gel electrophoresis. *Cell* **16:**191–200.

21. **Fleming, J., J. Sanseverino, and G. S. Sayler.** 1993. Quantitative relationship between naphthalene catabolic frequency and expression in predicting PAH degradation in soils at town gas manufacturing sites. *Environ. Sci. Technol.* **27:**1068–1074.

22. **Fox, G. E., J. D. Wisotzkey, and J. R. Jurtshuk.** 1992. How close is close: 16S ribosomal RNA sequence identity may not be sufficient to guarantee species identity. *Int. J. Syst. Bacteriol.* **42:**166–170.

23. **Giovannoni, S. J., T. B. Britschgi, C. L. Moyer, and K. G. Field.** 1990. Genetic diversity in Sargasso Sea bacterioplankton. *Nature* (London) **345:**60–63.

24. **Giovannoni, S. J., E. F. DeLong, G. J. Olsen, and N. R. Pace.** 1988. Phylogenetic, group-specific oligonucleotide probes for identification of single microbial cells. *J. Bacteriol.* **170:**720–726.

25. **Goodfellow, M., and A. G. O'Donnell.** 1993. Roots of bacterial systematics, p. 3–54. *In* M. Goodfellow and A.

G. O'Donnell (ed.), *Handbook of New Bacterial Systematics*. Academic Press, London.

26. **Grimont, F., and P. A. D. Grimont.** 1991. DNA fingerprinting, p. 249–279. *In* E. Stackebrandt and M. Goodfellow (ed.), *Sequencing and Hybridization Techniques in Bacterial Systematics.* John Wiley & Sons, Chichester, England.

27. **Hedrick, D. B., T. White, J. B. Guckert, W. J. Jewell, and D. C. White.** 1992. Microbial biomass and community structure of a phase separated methanogenic reactor determined by lipid analysis. *J. Ind. Microbiol.* **9:**193–199.

28. **Hennecke, H., K. Kaluza, B. Thony, M. Fuhrmann, W. Ludwig, and E. Stackebrandt.** 1985. Concurrent evolution of nitrogenase genes and 16S rRNA in *Rhizobium* species and other nitrogen fixing bacteria. *Arch. Microbiol.* **142:**342–348.

29. **Hill, M. O.** 1973. Diversity and evenness: a unifying notation and its consequences. *Ecology* **54:**427.

30. **Höfle, M., and I. Brettar.** 1995. Taxonomic diversity and metabolic activity of microbial communities in the water column of the Central Baltic. *Limnol. Oceanogr.* **40:**868–874.

31. **Höfle, M. G.** 1988. Identification of bacteria by low molecular weight RNA profiles: a new chemotaxonomic approach. *J. Microbiol. Methods* 8:235–248.

32. **Höfle, M. G.** 1990. RNA chemotaxonomy of bacterial isolates and natural microbial communities, p. 129–159. *In* J. Overbeck and R. J. Chróst (ed.), *Aquatic Microbial Ecology: Biochemical and Molecular Approaches.* Springer-Verlag, New York.

33. **Höfle, M. G.** 1990. Transfer RNAs as genotypic fingerprints of eubacteria. *Arch. Microbiol.* **153:**299–304.

34. **Höfle, M. G.** 1992. Bacterioplankton community structure and dynamics after large-scale release of nonindigenous bacteria as revealed by low-molecular-weight-RNA analysis. *Appl. Environ. Microbiol.* **58:**3387–3394.

35. **Holben, W. E.** 1994. Isolation and purification of bacterial DNA from soils. *In Methods of Soil Analysis*, part 2. *Microbiological and Biochemical Properties.* Soil Science Society of America, Madison, Wis.

36. **Hurlbert, S. H.** 1971. The nonconcept of species diversity: a critique and alternative parameters. *Ecology* **52:**577–586.

37. **Johnson, J. L.** 1984. Nucleic acids in bacterial classification, p. 8–11. *In* N. R. Krieg and J. G. Holt (ed.), *Bergey's Manual of Systematic Bacteriology.* Williams & Wilkins, Baltimore.

38. **Jones, D., and N. R. Krieg.** 1984. Serology and chemotaxonomy, p. 15–18. *In* N. R. Krieg and J. G. Holt (ed.), *Bergey's Manual of Systematic Bacteriology.* Williams & Wilkins, Baltimore.

39. **Kilpper-Balz, R.** 1991. DNA-rRNA hybridization, p. 45–68. *In* E. Stackebrandt and M. Goodfellow (ed.), *Sequencing and Hybridization Techniques in Bacterial Systematics.* John Wiley & Sons, Chichester, England.

40. **King, M.** 1993. *Species Evolution: the Role of Chromosome Change.* Press Syndicate of the University of Cambridge, Cambridge.

41. **Lane, D. J., D. A. Stahl, G. J. Olsen, D. Heller, and N. R. Pace.** 1985. Phylogenetic analysis of the genera *Thiobacillus* and *Thiomicrospira* by 5S rRNA sequences. *J. Bacteriol* **163:**75–81.

42. **Lee, S., and J. A. Fuhrman.** 1990. DNA hybridization to compare species compositions of natural bacterioplankton assemblages. *Appl. Environ. Microbiol.* **56:**739–746.

43. **Lee, S., and J. A. Fuhrman.** 1991. Spatial and temporal variation of natural bacterioplankton assemblages studied by total genomic DNA cross-hybridization. *Limnol. Oceanogr.* **36:**1277–1287.

44. **Lerman, L. S., S. G. Fischer, I. Hurley, K. Silverstein, and N. Lumelsky.** 1984. Sequence determined DNA separations. *Annu. Rev. Biophys. Bioeng.* **13:**399–423.

45. **Liesack, W., and E. Stackebrandt.** 1992. Occurrence of novel groups of the domain *Bacteria* as revealed by analysis of genetic material isolated from an Australian terrestrial environment. *J. Bacteriol.* **174:**5072–5078.

46. **Liesack, W., H. Weyland, and E. Stackebrandt.** 1991. Potential risks of gene amplification by PCR as determined by 16S rDNA analysis of a mixed-culture of strict barophilic bacteria. *Microb. Ecol.* **21:**199–209.

47. **Lin, C., B. Flesher, W. C. Capman, R. I. Amann, and D. A. Stahl.** 1994. Taxon specific hybridization probes for fiber-digesting bacteria suggest novel gut-associated *Fibrobacter. Syst. Appl. Microbiol.* **17:**418–424.

48. **Lin, C., and D. A. Stahl.** 1995. Comparative analyses reveal a highly conserved endoglucanase in the cellulolytic genus *Fibrobacter. J. Bacteriol.* **177:**2543–2549.

49. **Lovell, C. R., and Y. Hui.** 1991. Design and testing of a functional group-specific probe for the study of natural populations of acetogenic bacteria. *Appl. Environ. Microbiol.* **57:**2602–2609.

50. **Ludwig, W., G. Kirchhof, N. Klugbauer, M. Weizenegger, D. Betzl, M. Ehrmann, C. Hertel, S. Jilg, R. Tatzel, H. Zitzelsberger, S. Liebl, M. Hochberger, J. Shah, D. Lane, P. Wallnöfer, and K. H. Scheifer.** 1992. Complete 23S ribosomal RNA sequences of gram-positive bacteria with a low DNA G + C content. *Syst. Appl. Microbiol.* **15:**487–501.

51. **Maidak, B. L., N. Larsen, M. J. McCaughey, R. Overbeek, G. J. Olsen, K. Fogel, J. Blandy, and C. R. Woese.** 1994. The ribosomal database project. *Nucleic Acids Res.* **22:**3485–3487.

52. **Manz, W., R. Amann, W. Ludwig, M. Wagner, and K.-H. Schleifer.** 1992. Phylogenetic oligodeoxynucleotide probes for the major subclasses of proteobacteria: problems and solutions. *Syst. Appl. Microbiol.* **15:**593–600.

53. **Manz, W., R. Amann, M. Vancanneyt, and K. H. Schleifer.** Whole cell hybridization probes for members of the cytophaga-flexibacterium-bacteroides (CFB) phylum. *Syst. Appl. Microbiol.*, in press.

54. **Mayr, E.** 1982. *The Growth of Biological Thought: Diversity, Evolution, and Inheritance.* The Belknap Press of Harvard University Press, Cambridge, Massachusetts.

55. **Mobarry, B. K., M. Wagner, V. Urbain, B. E. Rittmann, and D. A. Stahl.** 1996. Phylogenetic probes for analyzing abundance and spatial organization of nitrifying bacteria. *Appl. Environ. Microbiol.* **62:**2156–2162.

56. **Morotomi, M., T. Ohno, and M. Mutai.** 1988. Rapid and correct identification of intestinal *Bacteroides* spp. with chromosomal DNA probes by whole cell dot blot hybridization. *Appl. Environ. Microbiol.* **54:**1158–1162.

57. **Muyzer, G.** Personal communication.

58. **Muyzer, G., E. C. DeWall, and A. G. Uitterlinden.** 1993. Profiling of complex microbial populations by denaturing gradient gel electrophoresis analysis of polymerase chain reaction-amplified genes coding for 16S rRNA. *Appl. Environ. Microbiol.* **59:**695–700.

59. **Muyzer, G., S. Hottenträger, A. Teske, and C. Wawer.** Denaturing gradient gel electrophoresis of PCR-amplified 16S rDNA—a new molecular approach to analyse the genetic diversity of mixed microbial communities. *In* A. D. L. Akkermans, J. D. van Elsas, and F. J. de Bruijn (ed.), *Molecular Microbial Ecology Manual,* in press. Kluwer, Dordrecht, The Netherlands.

60. **Myers, R. M., T. Maniatis, and L. S. Lerman.** 1987. Detection and localization of single base changes by denaturing gradient gel electrophoresis. *Methods Enzymol.* **155:**501–527.

61. **O'Donnell, A. G., M. Goodfellow, and D. L. Hawksworth.** 1994. Theoretical and practical aspects of the quantification of biodiversity among microorganisms. *Phil. Trans. R. Soc. Lond. Ser. B Biol. Sci.* **345:**65–73.

62. **Ogram, A., G. S. Sayler, and T. Barkay.** 1987. The extraction and purification of microbial DNA from sediments. *J. Microbiol. Methods* **7:**57–66.

63. **Olsen, G. J., D. J. Lane, S. J. Giovannoni, N. R. Pace, and D. A. Stahl.** 1986. Microbial ecology and evolution: a ribosomal RNA approach. *Annu. Rev. Microbiol.* **40:**337–365.

64. **Olsen, G. J., C. R. Woese, and R. Overbeek.** 1994. The winds of (evolutionary) change: breathing new life into microbiology. *J. Bacteriol.* **176:**1–6.

65. **Pace, N. R., D. A. Stahl, D. J. Lane, and G. J. Olsen.** 1986. The analysis of natural microbial populations by ribosomal RNA sequences. *Adv. Microb. Ecol.* **9:**1–55.

66. **Raskin, L., W. C. Capman, R. Sharp, and D. A. Stahl.** Molecular ecology of gastrointestinal ecosystems. *In* R. I. Mackie, B. A. White, and R. E. Isaacson (ed.), *Ecology and Physiology of Gastrointestinal Microbes: Gastrointestinal Microbiology and Host Interactions,* in press. Chapman and Hall, London.

67. **Raskin, L., L. K. Poulsen, D. R. Noguera, B. E. Rittmann, and D. A. Stahl.** 1994. Quantification of methanogenic groups in anaerobic biological reactors using oligonucleotide probe hybridizations. *Appl. Environ. Microbiol.* **60:**1241–1248.

68. **Raskin, L., J. M. Stromley, B. E. Rittmann, and D. A. Stahl.** 1994. Group-specific 16S rRNA hybridization probes to describe natural communities of methanogens. *Appl. Environ. Microbiol.* **60:**1232–1240.

69. **Reysenback, A.-L., L. J. Giver, G. S. Wickham, and N. R. Pace.** 1992. Differential amplification of rRNA genes by polymerase chain reaction. *Appl. Environ. Microbiol.* **58:**3417–3418.

70. **Risatti, J. B., W. C. Capman, and D. A. Stahl.** 1994. Community structure of a microbial mat: the phylogenetic dimension. *Proc. Natl. Acad. Sci. USA* **91:**10173–10177.

71. **Roller, C., R. Wagner, R. Amann, W. Ludwig, and K.-H. Schleifer.** 1994. In situ probing of Gram-positive bacteria with a high DNA G + C content by using 23S rRNA-targeted oligonucleotides. *Microbiology* **140:**2849–2858.

72. **Salyers, A. A., and N. B. Shoemaker.** 1994. Broad host range gene transfer: plasmids and conjugative transposons. *FEMS Microbiol. Ecol.* **15:**15–22.

73. **Sambrook, J., E. F. Fritsch, and T. Maniatis.** 1989. *Molecular Cloning: a Laboratory Manual,* 2nd ed. Cold Spring Harbor Laboratory Press, Cold Spring Harbor, N.Y.

74. **Sanseverino J. C. Werner, J. Fleming, B. Applegate, J. M. H. King, and G. S. Sayler.** 1993. Molecular diagnostics of polycyclic aromatic hydrocarbon biodegradation in manufactured gas plant soils. *Biodegradation* **4:**303–321.

75. **Schmidt, T. M., E. F. DeLong, and N. R. Pace.** 1991. Analysis of a marine picoplankton community by 16S rRNA gene cloning and sequencing. *J. Bacteriol.* **173:**4371–4378.

76. **Sheffield, V. C., J. S. Beck, E. M. Stone, and R. M. Myers.** 1992. A simple and efficient method for attachment of a 40-base pair, GC-rich sequence to PCR-amplified DNA. *BioTechniques* **12:**386–387.

77. **Sheffield, V. C., D. R. Cox, L. S. Lerman, and R. M. Myers.** 1989. Attachment of a 40-base pair G + C-rich sequence (GC-clamp) to genomic DNA fragments by the polymerase chain reaction results in improved detection of single-base changes. *Proc. Natl. Acad. Sci. USA* **86:**232–236.

78. **Simpson, G. G.** 1961. *Principles of Animal Taxonomy.* Columbia University Press, New York.

79. **Smith, G. B., and J. M. Tiedje.** 1992. Isolation and characterization of a nitrate reductase gene and its use as a

probe for denitrifying bacteria. *Appl. Environ. Microbiol.* **58:**376–384.

80. **Southern, E. M.** 1975. Detection of specific sequences among DNA fragments separated by gel electrophoresis. *J. Mol. Biol.* **98:**503–517.

81. **Spring, S., R. Amann, W. Ludwig, K.-H. Schleifer, and N. Peterson.** 1992. Phylogenetic diversity and identification of non-culturable magnetotactic bacteria. *Syst. Appl. Microbiol.* **15:**116–122.

81a.**Stackebrandt, E., and B. M. Goebel.** 1994. Taxonomic note: a place for DNA-DNA reassociation and 16S rRNA sequence analysis in the present species definition in bacteriology. *Int. J. Syst. Bacteriol.* **44:**846–849.

82. **Stackebrandt, E., and W. Liesack.** 1993. Nucleic acids and classification, p. 151–194. *In* M. Goodfellow and A. G. O'Donnell (ed.), *Handbook of New Bacterial Systematics.* Academic Press, London.

83. **Stahl, D. A.** 1986. Evolution, ecology and diagnosis: unity in variety. *Bio/Technology* **4:**623–628.

84. **Stahl, D. A.** 1995. Application of phylogenetically based hybridization probes to microbial ecology. *Mol. Ecol.* **4:** 535–542.

85. **Stahl, D. A., and R. Amann.** 1991. Development and application of nucleic acid probes in bacterial systematics, p. 205–248. *In* E. Stackebrandt and M. Goodfellow (ed.), *Sequencing and Hybridization Techniques in Bacterial Systematics.* John Wiley & Sons, Chichester, England.

86. **Stahl, D. A., R. Devereux, R. I. Amann, B. Flesher, C. Lin, and J. Stromley.** 1989. Ribosomal RNA based studies of natural microbial diversity and ecology, p. 669–673. *In* T. Hattori et al. (ed.), *Recent Advances in Microbial Ecology.* Japan Scientific Societies Press, Tokyo.

87. **Stahl, D. A., B. Flesher, H. Mansfield, and L. Montgomery.** 1988. Use of phylogenetically based hybridization probes for studies of ruminal microbial ecology. *Appl. Environ. Microbiol.* **54:**1079–1084.

88. **Stahl, D. A., D. J. Lane, G. J. Olsen, and N. R. Pace.** 1984. Analysis of hydrothermal vent-associated symbionts by ribosomal RNA sequences. *Science* **224:**409–411.

89. **Stahl, D. A, D. J. Lane, G. J. Olsen, and N. R. Pace.** 1985. Characterization of a Yellowstone hot spring microbial community by 5S rRNA sequences. *Appl. Environ. Microbiol.* **49:**1379–1384.

90. **Swofford, D. L., and G. J. Olsen.** 1990. Phylogeny reconstruction, p. 411–501. *In* D. M. Hillis and C. Moritz (ed.), *Molecular Systematics.* Sinauer Associates, Inc., Sunderland, Mass.

91. **Teske, A., C. Wawer, G. Muyzer, and N. B. Ramsing.** 1996. Distribution of sulfate-reducing bacteria in a stratified fjord (Mariager Fjord, Denmark) as evaluated by most probable-number counts and denaturing gradient gel electrophoresis of PCR-amplified ribosomal DNA fragments. *Appl. Environ. Microbiol.* **62:**1405–1415.

92. **Torsvik, V., F. L. Daae, and J. Goksoyr.** 1995. Extraction, purification, and analysis of DNA from soil bacteria. p. 29–48. *In* J. T. Trevors and J. D. van Elsas (ed.), *Nucleic Acids in the Environment: Methods and Applications.* Springer-Verlag, New York.

93. **Torsvik, V., J. Goksoøyr, and F. L. Daae.** 1990. High diversity in DNA of soil bacteria. *Appl. Environ. Microbiol.* **56:**782–787.

94. **Torsvik, V., J. Goksøyr, F. L. Daae, R. Sørheim, J. Michalsen, and S. Salte.** 1993. Diversity of microbial communities determined by DNA reassociation technique p. 375–378. *In* R. Guerrero and C. Pedrórs-Alió (ed.), *Trends in Microbial Ecology.* Spanish Society for Microbiology, Madrid.

95. **Torsvik, V., K. Salte, R. Sørheim, and J. Goksøyr.** 1990.

Comparison of phenotypic diversity and DNA heterogeneity in a population of soil bacteria. *Appl. Environ. Microbiol.* **56:**776–781.

96. **Torsvik, V. L.** 1980. Isolation of bacterial DNA from soil. *Soil Biol. Biochem.* **12:**12–21.

97. **Toze, S., and D. A. Stahl.** Unpublished data.

98. **Tsien, H. C., B. J. Bratina, K. Tsuji, and R. S. Hanson.** 1990. Use of oligodeoxynucleotide signature probes for identification of physiological groups of methylotrophic bacteria. *Appl. Environ. Microbiol.* **56:**2858–2865.

99. **Van de Peer, Y., I. Van de Broeck, P. De Rijk, and R. De Wachter.** 1994. Database on the structure of small ribosomal subunit RNA. *Nucleic Acids Res.* **22:** 3488–3494.

100. **Voordouw, G.** Personal communication.

101. **Voordouw, G., Y. Shen, C. S. Harrington, A. J. Telang, T. R. Jack, and D. W. S. Westlake.** 1993. Quantitative reverse sample genome probing of microbial communities and its application to oil field production waters. *Appl. Environ. Microbiol.* **59:**4101–4114.

102. **Voordouw, G., J. K. Voordouw, T. R. Jack, J. Foght, P. M. Fedorak, and D. W. S. Westlake.** 1992. Identification of distinct communities of sulfate-reducing bacteria in oil fields by reverse sample genome probing. *Appl. Environ. Microbiol.* **57:**3542–3552.

103. **Voordouw, G., J. K. Voordouw R. R. Karkhoff-Schweizer, P. M. Fedorak, and D. W. S. Westlake.** 1991. Reverse sample genome probing, a new technique for identification of bacteria in environmental samples by DNA hybridization, and its application to the identification of sulfate-reducing bacteria in oil field samples. *Appl. Environ. Microbiol.* **57:**3070–3078.

104. **Wagner, M., G. Rath, R. Amann, H.-P. Koops, and K.-H. Schleifer.** 1995. In situ identification of ammonia-oxidizing bacteria. *Syst. Appl. Microbiol.* **18:**251–264.

105. **Wagner, M., G. Rath, H.-P. Koops, J. Flood, and R. Amann.** In situ analysis of nitrifying bacteria in sewage treatment plants. *Water Sci. Technol.*, in press.

106. **Ward, D. M., M. M. Bateson, R. Weller, and A. L. Ruff-Roberts.** 1992. Ribosomal RNA analysis of microorganisms as they occur in nature. *Adv. Microb. Ecol.* **12:** 219–286.

107. **Wawer, C., and G. Muyzer.** 1995. Genetic diversity of *Desulfovibrio* spp. in environmental samples analyzed by denaturing gradient gel electrophoresis of [NiFe] hydrogenase gene fragments. *Appl. Environ. Microbiol.* **61:** 2203–2210.

108. **Wayne, L. G., D. J. Brenner, R. R. Colwell, P. A. D. Grimont, O. Kandler, M. I. Krichevsky, L. H. Moore, W. E. C. Moore, R. G. E. Murray, E. Stackebrandt, M. P. Starr, and H. G. Trüper.** 1987. Report of the ad hoc committee on reconciliation of approaches to bacterial systematics. *Int. J. Syst. Bacteriol.* **37:**463–464.

109. **Weller, R., and D. M. Ward.** 1989. Selective recovery of 16S rRNA sequences from natural microbial communities in the form of cDNA. *Appl. Environ. Microbiol.* **55:** 1818–1822.

110. **Wetherall, B. L., and A. M. Johnson.** 1990. Nucleic acid probes for *Campylobacter* species, p. 255–285. *In* A. J. L. Macario and E. Conway de Macario (ed.), *Gene Probes for Bacteria.* Academic Press, Inc., San Diego, Calif.

111. **Wetmur, J. G., and N. Davidson.** 1968. Kinetics of renaturation of DNA. *J. Mol. Biol.* **31:**349–370.

112. **Wiley, E. O.** 1978. *Phylogenetics: the Theory and Practice of Phylogenetic Systematics.* John Wiley & Sons, New York.

113. **Woese, C. R.** 1987. Bacterial evolution. *Microbiol. Rev.* **51:**221–271.

114. **Zuckerkandl, E., and L. Pauling.** 1965. Molecules as documents of evolutionary history. *J. Theor. Biol.* **8:**357–366.

Emerging Technologies: Bioreporters, Biosensors, and Microprobes

ROBERT S. BURLAGE

12

The study of microbial communities is dependent on the ability to discern individual bacterial species and microbial activities in a complex matrix. Classical and molecular methods are very useful tools that have provided scientists with great insight into the microbial world. For example, microbial communities can be examined by using the molecular techniques of DNA hybridization (40), PCR (1), and in situ hybridization. Lipid analysis has been used to describe microbial communities (47). However, these techniques have important limitations: they require destruction of a sample (e.g., DNA extraction) as well as significant delays in obtaining results. The development of new techniques for examining microbial communities nondestructively, in an on-line, real-time manner, has provided opportunities to observe the workings of these communities in great detail. Ideally, these techniques would supply data on a continuous basis and be able to resolve individual bacterial cells or environmental conditions at the submillimeter scale.

Biosensors of environmental conditions and bioreporter genes of genetic activity have been developed to augment our ability to detect, identify, and quantify microorganisms in complex ecological settings and to analyze their microniches. This chapter describes several techniques that have been developed to provide this level of resolution and that have proven useful for studies in microbial ecology. These techniques are likely to become more generally applied as the technology matures.

BIOREPORTER TECHNOLOGY

The analysis of genetic expression is made more difficult by the lack of suitable assays for most gene products. Bioreporter genes fill a surrogate role, supplying an assayable gene product when an assay for the gene product of interest is not available or is very difficult to perform. Bioreporter genes have been used extensively with pure cultures to demonstrate expression of specific genes. However, many of the most useful bioreporter genes, such as *lacZ* (encoding the β-galactosidase enzyme) and *xylE* (encoding catechol 2,3-oxygenase), are usually unsuitable for ecological studies. While any single bioreporter gene might not be present in a particular species, the presence of the gene in other mem-

bers of a community is not unlikely, potentially creating a significant background problem. Another drawback of conventional bioreporters is that they are often dependent on the destruction of a sample of the community for performance of a biochemical assay. Nondestructive assays allow repeated experiments to be performed on the same sample, and so changes (development) of a microbial community can be observed. Biochemical assays for conventional bioreporters require time for sampling, processing, and signal development, and these steps introduce a significant time delay between the microbial event and the analysis of the data. There is a distinct advantage in having an assay that gives results in real time, i.e., as they happen.

There are now two methods which use a nondestructive, noninvasive means of detecting gene expression, both of which depend on the efficient collection of light. Light can be measured with great sensitivity and precision, allowing the detection of a single cell under some conditions and with appropriate light-gathering equipment (44). Bioreporters for bioluminescent and fluorescent gene products are described below.

Bioluminescence

Bioluminescence is the production of visible light by a biochemical process. Unlike most chemical reactions, which produce heat as the main by-product, these reactions also generate enough light to be detected by conventional photodetectors. This phenomenon is easily observed in fireflies, although many other species are capable of producing light. The genes for these light-producing reactions have been isolated by researchers, and they are now available on cloning vectors. When genetically fused (by using either a transcriptional or a protein fusion) to appropriate genes from a host bacterium, these strains will produce light under defined conditions. The cloned firefly luciferase gene (*luc*) has been used to observe gene expression in animals (36), plants (21, 29), and bacteria (30). Firefly luciferase is a powerful tool for genetic analysis, although it is difficult to use for microbial ecology experiments since the reaction requires the substrate luciferin, which must be added exogenously. Often these assays use extracts from samples, rather than whole cells, and therefore destructive sampling is required.

Several genera and species of bioluminescent bacteria are known, although *Vibrio harveyi* and *V. fischeri* have received the most attention. These bacteria contain *lux* genes,

Publication no. 4557, Environmental Sciences Division, Oak Ridge National Laboratory.

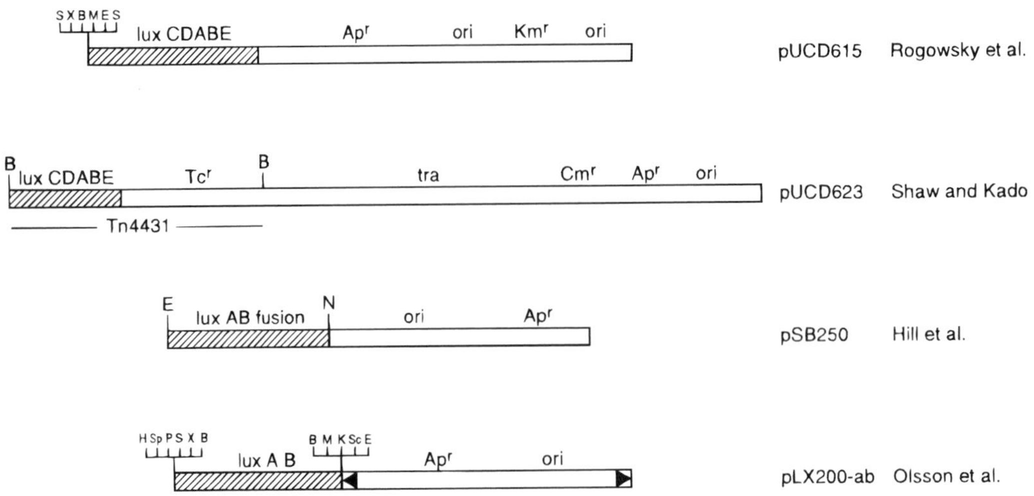

FIGURE 1 Plasmids useful for constructing *lux* fusions. Plasmids pUCD615 and pUCD623 (containing the *lux* transposon Tn*4431*) use the *lux* genes from *V. fischeri*. Plasmids pSB250 and pLX200-ab use the *lux* genes from *V. harveyi*. All plasmids have been linearized for comparative purposes. Abbreviations: Ap, ampicillin; Cm, chloramphenicol; Km, kanamycin; Tc, tetracycline; ori, origin of replication; tra, conjugative transfer genes; B, *Bam*HI; E, *Eco*RI; H, *Hind*III; K, *Kpn*I; N, *Nae*I; P, *Pst*I; Sc, *Sac*I; S, *Sal*I; M, *Sma*I; Sp, *Sph*I; X, *Xba*I. Other useful constructions are reviewed by Stewart and Williams (45).

which are responsible for bioluminescence. The *lux* operon is a complex pathway of five genes, *luxCDABE*, and efficient expression of all of these genes in the host is required for appropriate functioning of the bioreporter. Only two genes, *luxA* and *luxB*, encoding the heterodimeric luciferase enzyme, are needed for the actual bioluminescent reaction. The *luxCDE* genes have been implicated in the recycling of the required aldehyde substrate, so that a pool of substrate is continuously available. The *lux* genes, comprising about 7 kbp of DNA, have been cloned and sequenced and are available for analysis of microbial consortia. Several recent reviews describe the genetics and physiology of bacterial bioluminescence (24, 25) and the use of these fusions (4). Examples of genetic fusion vectors are shown in Fig. 1.

Advantages and Disadvantages

The advantages of bioluminescent bioreporters lie primarily in the relative ease of light measurement. Light can be measured accurately and with great sensitivity. Since light radiates in all directions from a point source, light detection can be performed in three dimensions, giving a more sophisticated analysis of an object's position in space. It can be measured quickly (in real time) and without perturbing or destroying the sample. For instance, the light detector can be introduced into the sample and left there for an extended period, or it can detect light that passes through the glass wall of a bioreactor vessel. Bacterial interactions can thus be examined in real time, for example, to study predator-prey or symbiotic relationships. There is usually no need to add any substrates or reagents for the bioluminescence assay (although see below regarding the requirement for an aldehyde substrate). In most environments, bioluminescence is a rare trait, and therefore a background problem is unlikely.

The *lux* genes are especially useful if a qualitative analysis is sufficient, i.e., for determining whether light is being produced at a given time. For these experiments, light output can be expressed as relative light units. Expression as relative light units is acceptable when all work is performed with the same light-measuring apparatus and under identical conditions (distance from detector, temperature, composition of the vessel holding the sample) and if no comparisons with published results are made. If quantitation is desired, photodetectors can be calibrated (see below).

The advantages described above must be weighed against several disadvantages that are inherent in the bioluminescent bioreporters. The bioluminescent reaction requires molecular oxygen, and its absence will render the bioreporter inoperative. Oxygen limitation results in a lower intensity of light generated, perhaps to a level that would escape detection. If the host strain does not have a suitable aldehyde substrate for the bioluminescent reaction, an aldehyde must be added exogenously. Few strains make sufficient aldehyde for prolonged light production, and strains utilizing *luxAB* vectors must always have aldehyde supplemented in the medium. The substrate, usually *n*-decanal at a final concentration of 0.1 to 1.0% (vol/vol), penetrates the cells readily, although it can be toxic at relatively low doses. Finally, the luciferase enzyme of *V. fischeri* is heat labile and is not recommended for use above 30°C. However, the luciferase from *V. harveyi* is stable at 37°C.

Applications

Many genetic constructions involving the *lux* genes have been prepared, and most use the genes from a *Vibrio* species. Examples of genetic constructions which are useful for *lux* fusions are shown in Fig. 1. The intact *luxCDABE* cassette is available on a plasmid cloning vector, pUCD615, so that expression of bioluminescence can be placed under the control of a host promoter (38). The full cassette ensures that the aldehyde substrate will be regenerated for continuous availability, although the host cell must have a suitable long-chain aldehyde (e.g., decanal) present as substrate for the luciferase reaction. The presence of such an aldehyde can be determined only empirically, and fluctuations of the aldehyde concentration are always possible. If such fluctuations are suspected, n-decanal can be added at the concentrations listed above to ensure an adequate supply.

The inclusion of the full cassette (i.e., the luxA and luxB luciferase genes and the luxCDE genes) obviates the need for exogenously added aldehyde, at least in those strains that have a suitable aldehyde substrate to start with. The luciferase enzyme uses molecular oxygen to convert the aldehyde substrate to a carboxylic acid, with the resulting light a by-product of the reaction. The action of the luxCDE genes is to recycle the carboxylic acid product of the light-product reaction, giving a continuous supply of aldehyde. It has been observed that at peak light emissions, the light intensity of such a strain can be boosted, although not greatly, by the addition of aldehyde. This finding suggests that there is always an aldehyde limitation in the cell.

Alternatively, a construction containing only the luxA and luxB genes can be used if the aldehyde substrate is added exogenously. The luxA and luxB genes have been fused into a single open reading frame by Escher et al. (14), although the resulting luxAB fusion luciferase is more temperature sensitive than the native enzyme. Transposons carrying the lux genes, such as Tn4431 (43), are also available. This construct carries the intact luxCDABE cassette and a gene for antibiotic selection of transposon insertion. This construct is a valuable means of generating transposon mutants quickly. The luxAB genes are available on a Tn5 derivative (3) and on a mini-Tn5 derivative (12).

Light Measurement

Unlike many assays, in which a standard procedure is used to describe the results, the measurement of light can be accomplished by a variety of means. Visualization of bacterial colonies may be sufficient for screening of clones during genetic construction, although the observer must be in a darkened room and the technique is not at all quantitative. Photographic film can be exposed to the light emitted from colonies, although this technique is usually cumbersome. Several types of electronic equipment are suitable for the measurement of light. ATP photometers or luminometers, which are used for measurement of ATP concentrations by the luciferase assay, are common in laboratories. Liquid scintillation counters are also common. Liquid scintillation counters must be very sensitive to detect photons resulting from radioactive decay, and so they are good photodetectors for bioluminescence, although the coincidence channel should be disconnected prior to use. The coincidence channel eliminates background during its measurement of radiation but is a hindrance for measurement of bioluminescence, since light emanating from a single cell might not be detected by both photodetectors simultaneously. In bioluminescence work, a background sample can be tested, and therefore all light from samples should be measured. These methods are sensitive but are not designed specifically for measurement of bioluminescence. Accordingly, there are problems in introducing representative samples to the photodetectors as well as problems with incubation conditions for the samples. That is, the samples would have to fit inside ordinary scintillation vials, which might not provide adequate aeration or mixing.

Commercial photomultipliers (e.g., Oriel, Stratford, Conn.) are recommended for remote sampling of light, including that emitted by bioreactors and soil microcosms. These devices usually include flexible fiber-optic cables which have a high efficiency of light transmittance, an important feature in measuring low amounts of light. For extremely low amounts of light, such as would be expected from single bacterial cells, charge-coupled devices (CCDs) (e.g., those made by Hamamatsu) can be used. The added sensitivity is reflected in the increased cost of this equipment, and few laboratories have access to one. A CCD can be used, however, to visualize signals that are seen through a microscope and thus has the potential to describe the physiological response of single cells, although integration of weak signals can delay output for several minutes. Accordingly, samples that move or drift during the integration time will give a blurred image, if the image is detected at all.

The lack of standardization is a major shortcoming of bioluminescent reporter work and has its greatest impact on the quantification of results. The output of a bioluminescent strain must be expressed in terms of specific activity to allow comparisons between laboratories. Units of light production would ideally be expressed as photons (quanta) of light per minute per milligram of total protein. However, each photodetector system has a different efficiency of light detection (for instance, different sensitivities for different wavelengths) as well as a different geometry of its detector window. A method to standardize photodetectors by using a light-producing biochemical reaction has been described (27), and this method should be applied more generally. Calibration of a photodetector by using a standard light source is possible, although the equipment is expensive and not generally available in laboratories.

Fluorescence

The Green Fluorescent Protein (GFP) is a relatively new bioreporter that meets many of these requirements and is developing into a versatile and valuable tool. GFP addresses many of the disadvantages of the bioluminescent bioreporters: (i) the bioluminescence reaction requires oxygen and is unreliable under conditions of reduced oxygen tension; (ii) bioluminescence requires functioning of the luciferase enzyme, which requires correct synthesis and folding of the protein; (iii) the luciferase enzyme requires a substrate, which might not be available in the cell; and (iv) the luciferase from V. fischeri is heat labile and is useless at 37°C. Increased stability at high temperatures would be an asset for all luciferases. GFP has the advantage that it is measured on the basis of its intrinsic properties and not on the basis of its biological activity in a certain milieu.

The gene for GFP is found in the jellyfish Aequorea victoria. GFP converts the blue bioluminescent light of the jellyfish to a green color; the advantage to the jellyfish of shifting the color from blue to green is not known. The GFP gene has been cloned and sequenced, and the protein has been extensively characterized (31, 32). The protein that is synthesized from the GFP gene autocyclizes (8), producing a chromophore that is brightly fluorescent (Fig. 2). When the GFP gene is expressed in a cell (either prokaryotic or eukaryotic), it fluoresces a bright green after cyclization of the chromophore (6). The fluorescence makes the cell easy to detect by using UV light (excitation, 395 nm) and conventional light-gathering equipment.

Advantages and Disadvantages

Like bioluminescence, fluorescence can be measured accurately and with great sensitivity. Detection is dependent on the ability of the researcher to expose the GFP molecule to the excitation wavelength; this can be performed with flexible fiber-optic cables that are introduced into a microbial ecosystem. Measurement is rapid, and there is no need

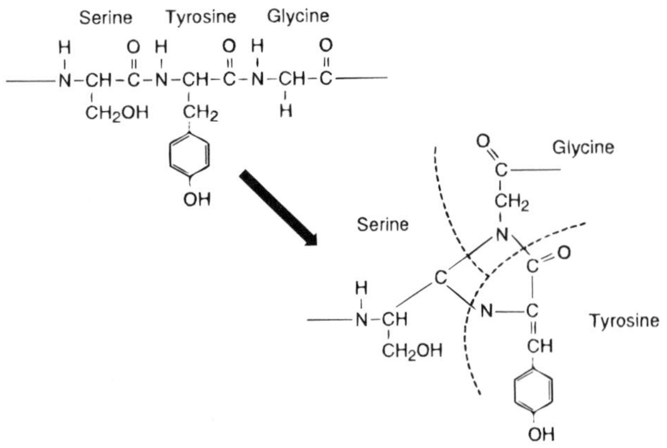

FIGURE 2 The chromophore of *Aequorea* GFP. Amino acids 65, 66, and 67 of GFP form a cyclical structure by an autocatalytic reaction. This chromophore is the source of the bright fluorescence seen with this protein. The dotted lines delineate the separate amino acids in the chromophore.

to add any substrates or reagents. The problems of sample perturbation and destruction are therefore avoided.

Fluorescence of GFP is very bright, and individual bacterial cells can easily be seen by epifluorescence microscopy. GFP appears to be very slow in forming the chromophore (typically taking several hours), and the speed at which it forms seems to vary with different organisms and different growth conditions, although a comprehensive analysis of this phenomenon is lacking. The protein is extremely stable, being largely unaffected by treatment with detergents, proteases, glutaraldehyde, or organic solvents. It is also very stable over a pH range of 6 to 12 and in high (65°C) temperatures. GFP may be useful in genetic analysis of thermophiles and other extremophiles. Once the protein is made, it does not degrade quickly in the cell, and therefore assays on the dynamics of gene expression, such as have been performed with the *lux* genes, are probably not possible with GFP. On the other hand, its stability makes it ideal for some applications, such as for tagging bacteria for a transport experiment (5). Bacteria expressing GFP were ideal for this experiment because the measure of interest was the concentration of bacteria in a transported sample versus the input concentration (i.e., C/C_0). Samples were collected from the sand column through which the bacteria migrated and were immediately analyzed with a fluorescence spectrometer. The fluorescent signal was proportional to the number of bacteria present in the sample. Throughout the experiment, the fluorescence of representative samples remained constant, as demonstrated by comparison of the fluorescence signal with conventional CFU counts on selective agar plates.

It is hypothesized that A. *victoria* utilizes an enzyme that is responsible for degrading GFP and that the gene can be cloned onto the same vector that contains the GFP gene. This could result in a more transient production of GFP. Production of secondary metabolites in some species, such as *Pseudomonas*, may result in the creation of other fluorescent compounds that may be mistaken for GFP. The formation of the GFP chromophore requires molecular oxygen, although not in great amounts, and therefore is unsuitable for completely anaerobic conditions. No fluorescence is seen when cells are grown in an anaerobic environment, although once the chromophore is formed, it will continue to fluoresce in an anaerobic environment. It is unknown

how much oxygen must be present for efficient formation of the chromophore. The presence of GFP in bacteria does not appear to have deleterious effects on the host, although a comprehensive analysis has not been performed.

Applications

Use of GFP is still a relatively new technique, and the construction of convenient cloning vectors is continuing. However, the number of applications of GFP is impressive, guaranteeing that more vectors will soon become available. Plasmid vectors containing GFP are available from Clontech Laboratories (Palo Alto, Calif.) and Life Technologies (Gaithersburg, Md.). These plasmids contain the GFP gene within a polylinker region, allowing convenient manipulation of the gene for transcriptional fusions. The intact GFP gene has been inserted into a derivative of Tn5, and therefore random mutations with GFP are possible (5). This transposon, Tn5GFP1, can be introduced into a variety of gram-negative species by using electroporation.

Mutations have been introduced into the GFP gene in order to produce fluorescent signals with altered properties. The red-shifted GFP was isolated in this manner (11). The name refers to the shift of the excitation wavelength toward the red end of the spectrum. The protein fluoresces at approximately the same wavelength (the maximum is at 505 nm instead of 510 nm) but excites at 490 nm instead of 395 nm. This shift is expected to be helpful, since the 490-nm excitation wavelength is beyond the wavelengths of excitation at which cellular proteins fluoresce (because of their aromatic amino acids). The red-shifted GFP gene is available on a plasmid, pTU58K (Clontech). As this construct becomes more widely available, it will certainly be incorporated into broad-host-range plasmids for use as cloning vectors and in transposons for mutagenesis methods. A mutant GFP developed by Heim et al. (17) results in the production of a blue color instead of green. Relevant characteristics of GFP and its derivatives are presented in Table 1. It is anticipated that many more fluorescent protein genes, such as the GFP gene from *Renilla reniformis* (22, 39), will be isolated or created in the near future. A spectrum of excitation and emission wavelengths might soon be available that would allow the use of several bioreporter genes in one species or the use of bioreporters to distinguish individual species in a community.

Fluorescence Measurement

Bacterial colonies expressing GFP can be easily detected upon exposure to UV light. This is easily accomplished by using the UV supply that is used in most molecular biology laboratories to visualize DNA in agarose gels, although an inexpensive handheld UV light will work just as well. Fluorescent bacteria can also be easily visualized by epifluorescence microscopy. An appropriate filter set should be used; the filter for fluorescein detection has proved to be very useful for this purpose. A xenon or mercury lamp can be used as a source of UV excitation.

Fluorescence spectrometry facilitates detection of GFP fluorescence. Fluorescence spectrometers vary in sensitivity and versatility, although in general they should be able to detect GFP expression in bacteria. Quantification of bacteria in the sample is possible when a standard is examined contemporaneously. Digital imaging spectroscopy (15, 50) is an excellent means of detecting and characterizing fluorescent signals, although the expense of the system makes it unavailable to all but a few researchers.

TABLE 1 Comparison of GFPs from several sources[a]

Source	Amino acid at position:						Excitation (nm)	Emission (nm)	Reference
	64	65	66	67	68	69			
Aequorea GFP	Phe	Ser	Tyr	Gly	Val	Gln	395	510	6
Red-shifted GFP	Met	Gly	Tyr	Gly	Val	Leu	490	520	11
P4	Phe	Ser	His	Gly	Val	Gln	382	448	17
Renilla GFP	Phe	Ser	Tyr	Gly	Asp	Arg	498	508	39

[a] Amino acids critical to formation of the chromophore are presented, as are excitation and emission wavelengths for detection. Although not yet available for bioreporter purposes, the *Renilla* GFP is presented for comparison.

MICROPROBES AND BIOSENSORS

Microprobes

A microprobe is a device that measures a specific physical or chemical property in a microenvironment. For instance, microprobes can be devised to test for pH, temperature, or the concentration of chloride ion. The quality that makes these probes different from other probes is their small size, which makes them more suitable for analysis at submillimeter resolution. The development of microprobes for the examination of microbial environments has proceeded at a rapid rate thanks to innovative construction techniques. Microprobes have been described for ammonium (9), nitrate (19, 46), oxygen (34), denitrification (by nitrous oxide production) (7), and sulfate reduction (33). Conventional technology is used for measurement of the analyte; the critical development is the miniaturization of the electrode. The probe tip can be in the range of 1 to 10 μm in diameter, although 10 to 50 μm is more common. The signal is usually reported as a current on an ammeter. Delicate construction of the probe is required, as is manipulation of the probe in three dimensions by using a micromanipulator and dissecting microscope. Microprobes are especially useful for the study of microecology of biofilms, including formation and activity at various depths inside the biofilm. An excellent review is available (35). The main obstacle to the general use of microprobes is that they are not commercially available (with a few notable exceptions, such as Diamond Electro-Tech [Ann Arbor, Mich.] and Microelectrodes [Londonderry, N.H.]), and the microprobe must be handmade by the researcher. Techniques for construction are available, although they require skill and patience. As the usefulness of microprobes becomes better appreciated, the number of manufacturers will certainly increase.

The oxygen microprobe described by Revsbech (34) provides a good example of the current microprobe technology. This microprobe has a tip that is approximately 10 μm in diameter and is sensitive to oxygen concentrations in the micromolar range. It incorporates a guard cathode that removes oxygen diffusing toward the sensor tip from the electrolyte solution, permitting stable signal acquisition from the sample. This microprobe is suitable for examination of biofilm ecology or aquatic microbiology, with the ability to discern microbial processes at the water interface. Although the microprobe is extraordinarily small for an analytical instrument, it should not be forgotten that its presence is likely to disturb or influence the surrounding environment, however slightly. The prudent researcher will be attentive to possible effects from the use of these tools.

Biosensors

A biosensor is a type of probe in which a biological component, such as an enzyme, antibody, or nucleic acid, interacts with an analyte, which is then detected by an electronic component and translated into a measurable (electronic) signal (Fig. 3). Biosensor probes are possible because of a fusion of two technologies, microelectronics and biotechnology. Biosensors are useful for measuring many analytes (e.g., gases, ions, and organic compounds) or even bacteria and are suitable for studies of complex microbial environments. The technology of the biological component is usually based on a conventional system. For example, nucleic acid biosensors rely on hybridization of DNA, much the way that Southern blot hybridization works. Biosensors have been used extensively for clinical applications (49), and a thorough review of them is not possible here. A recent review on the application of biosensors for environmental study is available (37).

Each biosensor has a biological and an electronic component. A variety of substances, including nucleic acids, proteins (particularly antibodies and enzymes), lectins (plant

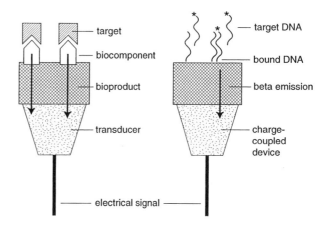

FIGURE 3 Biosensors. (Left) A generalized scheme for a biosensor. Interaction of the target analyte with the biological component results in a signal, which is transmitted to the transducer. The transducer senses the signal and converts it to an electrical signal. (Right) A DNA biosensor. The hybridization event brings the labeled DNA in contact with the transducer, a CCD camera. The CCD camera detects beta emission from ^{32}P decay and converts it to an electrical signal.

proteins that bind sugar moieties), and complex materials (organelles, tissue slices, microorganisms), can be used as the biological component. In each case, it is the specificity of the biological component for an analyte (or group of related analytes) that makes the biomolecule attractive for sensing technology. For example, a single strand of DNA will hybridize only to its complementary strand under the appropriate conditions. The conditions of the assay are very important, especially if reversibility of binding is a key factor. This is apparent in antibody-based biosensors, in which the affinity of the antibody for the antigen affects the sensitivity of the biosensor.

Isolation of the biological component is necessary to ensure that only the molecule of interest is bound or immobilized on the electronic component (also called the transducer). In some cases, such as the isolation of DNA, this is easy. It is possible to have a commercial vendor make a specific oligonucleotide that is pure and already labeled. Antibody or enzyme extraction and purification are much more complex procedures, although crude extracts can sometimes be used. The stability of the biological components is also critical, since they are being used outside their usual biological environment. A labile protein usually makes a poor candidate for a biosensor.

A variety of materials can also be used to construct the transducer. Generally, the transducers fall into distinct categories: electrochemical, optical, piezoelectric, and calorimetric (42). Electrochemical transducers report on changes in voltage when current is held constant (potentiometric) or report changes in current when voltage is held constant (amperometric). These are by far the most common electrochemical transducers, although transducers based on conductance and capacitance have also been described (42). In each case, the interaction of the analyte with the biological component will cause a change in potential that is detected by the sensor. Optical biosensors utilize a fiber-optic probe to receive specific wavelengths of light. The versatility of fiber-optic probes is due to their capacity to transmit signals that report on changes in wavelength, wave propagation time, intensity, distribution of the spectrum, or polarity of the light. In general, acquisition of the signal from these devices is accomplished through flexible cables, which can transmit light to the biological component (such as an excitation light for fluorescence) and receive light back from the sample (such as for light generation by the sample or for light absorption or reflection). Light conductance can be accomplished with great efficiency (less than 1% over short distances), and loss of signal is usually not a problem. Acquisition of the light signal by the detector can be a considerable problem if the light source is very weak, and the difference in light signals may be small. Light propagation over longer distances (greater than 1 m) is usually accompanied by a loss of conductance efficiency unless laser light is used.

Piezoelectric biosensors measure changes in mass. A piezoelectric material, such as a quartz crystal, will oscillate at a certain frequency when a potential is applied across its surface. If the mass at the surface changes because, for example, a bound antibody complexes with a specific antigen, the frequency of oscillation will change, and this change is detectable. Calorimetric transducers are comparable in function to optical transducers except that heat generation is measured instead of light.

Attachment of the biological component to the electronic component is vital for the success of these devices. If the biological component is destroyed in the process of binding or if it binds with the active site unavailable to the analyte, the biosensor will not function. Attachment techniques that are inefficient will diminish the sensitivity of the instrument. Attachment can be accomplished in a variety of ways, such as covalent binding of the molecule to the detector (usually through a molecular cross-bridge), adsorption onto the surface, entrapment in porous material, or microencapsulation. Ultrathin applications of biological material are usually deposited on transducers by using the Langmuir-Blodgett (2) or molecular self-assembly (26) technique. The need to deposit organic surfaces on transducers in a predictable manner has been addressed by the work of Decher et al. (10). This group demonstrated the sequential construction of layers on typical transducer surfaces, such as glass, silicon wafers, and quartz. This technique should improve biosensors by increasing uniformity of results.

The essence of the biosensor is matching the appropriate biological and electronic components to produce a relevant signal during analysis. For example, antibodies can be attached to a piezoelectric transducer, so that the binding of the antigen will be recorded as a change in the attached mass. Alternatively, the antibodies can be attached to an optical fiber, and the antigen binding is recorded by evanescent wave detection (2, 26). This technique is particularly suitable for immunoassays because the evanescent waveform is operational only a short distance from the surface of the optical electrode, and this distance is approximately equal to the size of the immune complex (42). Immunosensors that utilize other detection systems have been described (48).

Both electrochemical and optical electrodes are very useful for the detection of signals from attached enzymes. The enzymatic reaction can cause a potential change that is detected by the electrochemical electrode or a change in one of the components of the enzyme system that can be detected by the optical electrode. Scheper et al. (41) utilized the latter concept in their biosensor. They attached the glucose-fructose oxidase from *Zymomonas mobilis* to a fiber-optic cable attached to a fluorimeter. The enzyme complex contained bound $NADP^+$. When the enzyme oxidizes glucose, it reduces $NADP^+$ to NADPH, which is a fluorescent molecule. The change in fluorescence is therefore proportional to the concentration of glucose. There is also a critical weakness associated with this assay, in that it is dependent on the availability of $NADP^+$. When the supply is exhausted, the biosensor will no longer function. Resupplying the biosensor with an essential cofactor is often technically challenging but is required for long-term monitoring of an environment.

Nucleic acid biosensors depend on the ability of a single-stranded nucleic acid to hybridize with another fragment of DNA by complementary base pairing. Technological innovation is introduced in the manner in which the nucleic acid oligomer is attached to the surface of the detector and the manner in which the hybridized nucleic acid is detected and transduced into a measurable signal. Amino-derivatized oligonucleotides can be attached to glass (SiO_2) surfaces, such as fiber-optic cables, glass beads, or microscope slides, through covalent bonding with a chemical linker (Fig. 4). Some techniques result in the nonspecific attachment of oligonucleotide to the surface, which is an impediment to the hybridization that will be detected by the transducer. Losses of efficiency such as this can be avoided through careful attachment of the oligonucleotide via a modified 3' or 5' end of the oligomer. Graham et al. (16) used a simple procedure to attach oligomers to an evanescent wave bio-

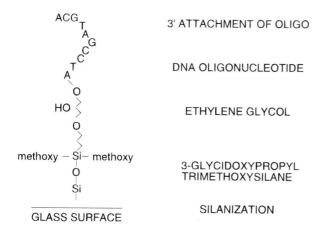

3' ATTACHMENT OF OLIGO

DNA OLIGONUCLEOTIDE

ETHYLENE GLYCOL

3-GLYCIDOXYPROPYL TRIMETHOXYSILANE

SILANIZATION

FIGURE 4 The attachment of a single-stranded oligonucleotide to a glass surface through a series of chemical modifications of the glass surface. The oligonucleotide must be amino derivatized. Based on the procedure of Maskos and Southern (23).

sensor array. Maskos and Southern (23) described the synthesis of oligonucleotides on derivatized glass bead supports. Glycol spacers of various dimensions can be added between the oligonucleotide and the glass support, which facilitates hybridization. Although not described specifically for biosensor development, this technique should be easily adaptable to biosensors.

A nucleic acid biosensor that utilizes evanescent wave technology has been described by Graham et al. (16). They used short fragments of nucleic acids that are small enough to reside within the field of the evanescent wave. They were able to detect fluorescein-labeled DNA hybridizing to their complementary immobilized probes in a flow cell. Fluorescence was monitored and reported as a change in the output voltage. Nucleic acid biosensors are potentially useful in the field of rapid DNA sequencing as well as in clinical applications. The biosensor described by Eggers et al. (13) integrates microelectronics, molecular biology, and computational science in an optical electrode format. Their device can detect hybridization and report on the spatial address of the hybridization signal on a glass surface (13) or a silicon wafer (20), to which the DNA probes are attached. Several different DNA oligomers can be attached to the optical electrode at different locations. The DNA on the biosensor is then hybridized to DNA that is free in solution. The free DNA must be labeled, usually with a fluorescent, luminescent, or radioisotope decay ^{32}P signal. The signal is detected by a CCD, which is extremely sensitive. The computer identifies the location of the affected pixels and forms the signals into a recognizable array. Not only is this technology suitable for rapid DNA sequencing, but it is also applicable to the rapid detection of many different gene sequences from DNA extracted from a consortium. This technology will require further development before it is suitable for routine analysis of samples.

SUMMARY

The need to examine microbial communities in detail has led to the development of a variety of tools that, if used properly, can yield data on the response of microorganisms to their environment. Bioreporter genes that allow detection of genetic expression without disturbing or destroying the sample are available. The development of additional innovative bioreporter genes is likely. Microprobes have demonstrated their efficacy under complex environmental conditions. Biosensors that were originally developed for clinical applications are rapidly being adapted for environmental purposes. Development of these tools is expected to continue, resulting in more accurate and versatile measurement systems.

Creation of additional bioreporter genes that are based on light or fluorescence production would be beneficial. The mutants of GFP that have been described are a testament to the power of genetic engineering to create new tools for research. A combination technology (e.g., *lux* bioreporter gene and optical electrode biosensor) has excellent prospects for success. Such a device would combine the power of the bioreporter to respond to a biological event and the power of the biosensor to detect the resulting increased light output. In essence, the bioreporter would gain portability, while the biosensor would gain the uniqueness of an individual bacterial strain.

Among the developments that can be expected for biosensors is the attachment of complex multimeric enzymes to transducers. The spatial arrangement of the several separate proteins is critical to the efficacy of the enzymatic reaction. Genetic engineering may produce enzymes and antibodies that are more stable, giving the biosensors a greater shelf life. DNA-based biosensors will probably be used in rapid DNA sequencing, which has broad implications for the analysis of microbial communities. For instance, it may soon be possible to achieve a census of microorganisms in a community that is based on DNA extraction and not on plating efficiency. All of the described technologies will benefit from a further reduction in size, making them applicable to field work and at a spatial scale that reflects cell-cell interactions.

This chapter is based on work supported by the Laboratory Directed Research and Development Program and by Defense Programs/U.S. Department of Energy under Cooperative Research and Development Agreement DOE 92-0077 between Lockheed Martin Energy Systems, Inc., and the National Center for Manufacturing Sciences. Oak Ridge National Laboratory is managed for the U.S. Department of Energy under contract DE-AC05-84OR21400 with Lockheed Martin Energy Systems, Inc.

REFERENCES

1. **Atlas, R. M., and R. J. Steffan.** 1991. Polymerase chain reaction: applications in environmental microbiology. *Annu. Rev. Microbiol.* **45:**137–161.
2. **Blodgett, K. B., and I. Langmuir.** 1937. Build-up films of barium stearate and their optical properties. *Phys. Rev.* **51:** 964–982.
3. **Boivin, R., F. P. Chalifour, and P. Dion.** 1988. Construction of a Tn5 derivative encoding bioluminescence and its introduction in *Pseudomonas, Agrobacterium,* and *Rhizobium. Mol. Gen. Genet.* **213:**50–55.
4. **Burlage, R. S., and C. Kuo.** 1994. Living biosensors for the management and manipulation of microbial consortia. *Annu. Rev. Microbiol.* **48:**291–309.
5. **Burlage, R. S., Z. Yang, and T. Mehlhorn.** A Tn5 derivative labels bacteria with green fluorescent protein for transport experiments. *Gene,* in press.
6. **Chalfie, M., Y. Tu, G. Euskirchen, W. W. Ward, and D. C. Prasher.** 1994. Green fluorescent protein as a marker for gene expression. *Science* **263:**802–805.
7. **Christensen, P. D., L. P. Nielsen, N. P. Revsbech, and**

J. Sorensen. 1989. Microzonation of denitrification activity in stream sediments as studied with a combined oxygen and nitrous oxide microsensor. *Appl. Environ. Microbiol.* **55:**1234–1241.

8. **Cody, C. W., D. C. Prasher, W. M. Westler, F. G. Prendergast, and W. W. Ward.** 1993. Chemical structure of the hexapeptide chromophore of the *Aequorea* green-fluorescent protein. *Biochemistry* **32:**1212–1218.

9. **DeBeer, D., and J. C. Van den Heuvel.** 1988. Response of ammonium-selective microelectrodes based on the neutral carrier nonactin. *Talanta* **35:**728–730.

10. **Decher, G., B. Lehr, K. Lowack, Y. Lvov, and J. Schmitt.** 1994. New nanocomposite films for biosensors: layer-by-layer adsorbed films of polyelectrolytes, proteins or DNA. *Biosens. Bioelectronics* **9:**677–684.

11. **Delagrave, S., R. E. Hawtin, C. M. Silva, M. M. Yang, and D. C. Youvan.** 1995. Red-shifted excitation mutants of the green fluorescent protein. *Bio/Technology* **13:**151–154.

12. **DeLorenzo, V., M. Herrero, U. Jakubzik, and K. N. Timmis.** 1990. Mini-Tn5 transposon derivatives for insertion mutagenesis, promoter probing, and chromosomal insertion of cloned DNA in gram-negative bacteria. *J. Bacteriol.* **172:**6568–6572.

13. **Eggers, M., M. Hogan, R. K. Reich, J. Lamture, D. Ehrlich, M. Hollis, B. Kosicki, T. Powdrill, K. Beattie, S. Smith, R. Varma, R. Gangadharan, A. Mallik, B. Burke, and D. Wallace.** 1994. A microchip for quantitative detection of molecules utilizing luminescent and radioisotope reporter groups. *BioTechniques* **17:**516–524.

14. **Escher, A., D. J. O'Kane, J. Lee, and A. A. Szalay.** 1989. Bacterial luciferase $\alpha\beta$ fusion protein is fully active as a monomer and highly sensitive *in vivo* to elevated temperature. *Proc. Natl. Acad. Sci. USA* **86:**6528–6532.

15. **Goldman, E. R., and D. C. Youvan.** 1992. An algorithmically optimized combinatorial library screened by digital imaging spectroscopy. *Bio/Technology* **10:**1557–1561.

16. **Graham, C. R., D. Leslie, and D. J. Squirrell.** 1992. Gene probe assays on a fibre-optic evanescent wave biosensor. *Biosens. Bioelectronics* **7:**487–493.

17. **Heim, R., D. C. Prasher, and R. Y. Tsien.** 1994. Wavelength mutations and posttranslational autoxidation of green fluorescent protein. *Proc. Natl. Acad. Sci. USA* **91:** 12501–12504.

18. **Hill, P. J., S. Swift, and G. S. A. B. Stewart.** 1991. PCR based gene engineering of the *Vibrio harveyi lux* operon and the *Escherichia coli trp* operon provides for biochemically functional native and fused gene products. *Mol. Gen. Genet.* **226:**41–48.

19. **Jensen, K., N. P. Revsbech, and L. P. Nielsen.** 1993. Microscale distribution of nitrification activity in sediment determined with a shielded microsensor for nitrate. *Appl. Environ. Microbiol.* **59:**3287–3296.

20. **Lamture, J. B., K. L. Beattie, B. E. Burke, M. D. Eggers, D. J. Ehrlich, R. Fowler, M. A. Hollis, B. B. Kosicki, R. K. Reich, S. R. Smith, R. S. Varma, and M. E. Hogan.** 1994. Direct detection of nucleic acid hybridization on the surface of a charge coupled device. *Nucleic Acids Res.* **22:** 2121–2125.

21. **Langridge, W. H. R., K. J. Fitzgerald, C. Koncz, J. Schell, and A. A. Szalay.** 1989. Dual promoter of *Agrobacterium tumefaciens* mannopine synthase genes is regulated by plant growth hormones. *Proc. Natl. Acad. Sci. USA* **86:** 3219–3223.

22. **Lorenz, W. W., R. O. McCann, M. Longiaru, and M. J. Cormier.** 1991. Isolation and expression of a complementary DNA encoding *Renilla reniformis* luciferase. *Proc. Natl. Acad. Sci. USA* **88:**4438–4442.

23. **Maskos, U., and E. M. Southern.** 1992. Oligonucleotide hybridizations on glass supports: a novel linker for oligonucleotide synthesis and hybridization properties of oligonucleotides synthesized in situ. *Nucleic Acids Res.* **20:** 1679–1684.

24. **Meighen, E. A.** 1991. Molecular biology of bacterial bioluminescence. *Microbiol. Rev.* **55:**123–142.

25. **Meighen, E. A.** 1994. Genetics of bacterial bioluminescence. *Annu. Rev. Genet.* **28:**117–139.

26. **Moaz, R., L. Netzer, J. Gun, and J. Sagiv.** 1988. Self-assembling monolayers in the construction of planned supramolecular structures and as modifiers of surface properties. *J. Chim. Phys.* **85:**1059–1065.

27. **O'Kane, D. J., M. Ahmad, I. B. C. Matheson, and J. Lee.** 1986. Purification of bacterial luciferase by high-performance liquid chromatography. *Methods Enzymol.* **133:** 109–128.

28. **Olsson, O., C. Koncz, and A. A. Szalay.** 1988. The use of a *luxA* gene of the bacterial luciferase operon as a reporter gene. *Mol. Gen. Genet.* **215:**1–9.

29. **Ow, D. W., K. V. Wood, M. DeLuca, J. R. deWet, D. R. Helinski, and S. H. Howell.** 1986. Transient and stable expression of the firefly luciferase gene in plant cells and transgenic plants. *Science* **234:**856–859.

30. **Palomares, A. J., M. A. DeLuca, and D. R. Helinski.** 1989. Firefly luciferase as a reporter enzyme for measuring gene expression in vegetative and symbiotic *Rhizobium meliloti* and other Gram-negative bacteria. *Gene* **81:**55–64.

31. **Perozzo, M. A., K. B. Ward, R. B. Thompson, and W. W. Ward.** 1988. X-ray diffraction and time-resolved fluorescence analyses of *Aequorea* green fluorescent protein crystals. *J. Biol. Chem.* **263:**7713–7716.

32. **Prasher, D. C., V. K. Eckenrode, W. W. Ward, F. G. Prendergast, and M. J. Cormier.** 1992. Primary structure of the *Aequorea victoria* green-fluorescent protein. *Gene* **111:** 229–233.

33. **Ramsing, N. B., M. Kuhl, and B. B. Jorgensen.** 1993. Distribution of sulfate-reducing bacteria, O_2, and H_2S in photosynthetic biofilms determined by oligonucleotide probes and microelectrodes. *Appl. Environ. Microbiol.* **59:** 3840–3849.

34. **Revsbech, N. P.** 1989. An oxygen microsensor with a guard cathode. *Limnol. Oceanogr.* **34:**474–478.

35. **Revsbech, N. P., and B. B. Jorgensen.** 1986. Microelectrodes: their use in microbial ecology. *Adv. Microb. Ecol.* **9:**293–352.

36. **Rodriguez, J. F., D. Rodriguez, J. Rodriguez, E. B. McGowan, and M. Esteban.** 1988. Expression of the firefly luciferase gene in vaccinia virus: a highly sensitive gene marker to follow virus dissemination in tissues of infected animals. *Proc. Natl. Acad. Sci. USA* **85:**1667–1671.

37. **Rogers, K. R.** 1995. Biosensors for environmental applications. *Biosens. Bioelectronics* **10:**533–541.

38. **Rogowsky, P. M., T. J. Close, J. A. Chimera, J. J. Shaw, and C. I. Kado.** 1987. Regulation of the *vir* genes of *Agrobacterium tumefaciens* plasmid pTiC58. *J. Bacteriol.* **169:** 5101–5112.

39. **San Pietro, R. M., F. G. Prendergast, and W. W. Ward.** 1993. Sequence of chromogenic hexapeptide of *Renilla* green-fluorescent protein. *Photochem. Photobiol.* **57:**63S.

40. **Sayler, G. S., and A. C. Layton.** 1990. Environmental application of nucleic acid hybridization. *Annu. Rev. Microbiol.* **44:**625–648.

41. **Scheper, T., C. Muller, K. D. Anders, F. Eberhardt, F. Plotz, C. Schelp, O. Thordsen, and K. Schugerl.** 1994. Optical sensors for biotechnological applications. *Biosens. Bioelectronics* **9:**73–83.

42. **Sethi, R. S.** 1994. Transducer aspects of biosensors. *Biosens. Bioelectronics* **9:**243–264.

43. **Shaw, J. J., and C. I. Kado.** 1987. Direct analysis of the invasiveness of *Xanthomonas campestris* mutants generated

by Tn4431, a transposon containing a promoterless luciferase cassette for monitoring gene expression, p. 57–60. *In*
D. P. S. Verma and N. Brisson (ed.), *Molecular Genetics
of Plant-Microbe Interactions*. Martinus Nijhoff, Dordrecht,
The Netherlands.

44. **Silcock, D. J., R. N. Waterhouse, L. A. Glover, J. I.
Prosser, and K. Killham.** 1992. Detection of a single genetically modified bacterial cell in soil by using charge coupled
device-enhanced microbiology. *Appl. Environ. Microbiol.*
58:2444–2448.

45. **Stewart, G. S. A. B., and P. Williams.** 1992. *lux* genes
and the applications of bacterial bioluminescence. *J. Gen.
Microbiol.* **138:**1289–1300.

46. **Sweerts, J. R., and D. DeBeer.** 1989. Microelectrode measurements of nitrate gradients in the littoral and profundal
sediments of a meso-eutrophic lake (Lake Vechten, The
Netherlands). *Appl. Environ. Microbiol.* **55:**754–757.

47. **Vestal, J. R., and D. C. White.** 1989. Lipid analysis in
microbial ecology. *Bioscience* **39:**535–541.

48. **Vo-Dinh, T., M. J. Sepaniak, G. D. Griffin, and J. P.
Alarie.** 1993. Immunosensors: principles and applications.
Immunomethods **3:**85–92.

49. **Wise, D. L., and L. B. Wingard (ed.).** 1991. *Biosensors
with Fiberoptics.* Humana Press, Clifton, N.J.

50. **Youvan, D. C.** 1994. Imaging sequence space. *Nature*
(London) **369:**79–80.

Quality Assurance

A. J. CROSS-SMIECINSKI

13

The purpose of this chapter is to discuss in terms of environmental microbiology, when possible, the principles of quality assurance (QA). Starting with the definition of QA and its counterparts, quality control (QC) and quality improvement, the chapter then moves into a description of general requirements, with an emphasis on those QA requirements that will most likely be encountered in a sponsored project. A primary goal of this chapter is to describe requirements in familiar terms and in the process illustrate how an effective QA program can benefit an environmental microbiology project.

QA

Taylor defines QA as "a system of activities whose purpose is to provide to the producer or user of a product or a service the assurance that it meets defined standards of quality with a stated level of confidence" (18). In general terms, QA is a concept used to ensure the continuous reliability of data. QA articulates a program of planning, control, and improvement efforts coordinated to meet data quality requirements for diverse studies.

Today most organizations sponsoring scientific research and even laboratory production require the implementation of a QA program. As a taxpayer or other type of customer, each person understands the reason for this. Consider the definition of QA—to ensure that a product will satisfy specific needs. The QA program is designed to "ensure that the *money invested* in the project will yield scientifically valid conclusions related to a principle hypothesis . . ." (emphasis added) (6).

QC

The quality of the measurement process (performance) is evaluated by comparison against standards (7). QC is the set of materials and practices used throughout the measurement process to ensure that each discrete part contributes a minimum amount of error to the results. QC requirements may be applied to any field, sampling, and measurement aspect of the project that affects the data quality. The choice of QC requirements is dependent on the study or project and usually is tied closely to the field and measurement methods. QC is a required part of all laboratory QA programs.

Note: The terms QA and QC are often used interchangeably. However, they have different meanings. QC, like QA, is used to ensure the reliability of data, but on a different level. Where QA is concerned with the quality of the information provided by the data, QC is concerned with the quality of the measurement process.

QUALITY IMPROVEMENT

Quality improvement is the evolutionary process used to improve or prevent a stagnant quality program. Often in the form of quality circles, or small groups of employees, quality improvement utilizes the ingenuity of the personnel to solve specific problems in the workplace, thereby promoting a process of continuous improvement. Quality improvement works on the premise that employees have a positive interest in efficiency and want to improve their work practices, products, and environment (11).

QUALITY COSTS

A QA program typically requires 10 to 20% of an investigator's time and generally 15% of total measurement time (6). The cost depends on the extent of the specific QA requirements. Some programs, for example those supporting enforcement, litigation, or nuclear licensing activities, require extensive documentation, procurement regulation, and record keeping.

Other programs supporting basic or applied research emphasize careful record keeping practices, controls on the measurement process, and a review to ensure validity of the work. Small organizations can encounter the situation of receiving funding from an agency with a QA program that has extensive requirements. The sponsoring agency that has a large QA staff may find it difficult to understand why the two-person research group is overwhelmed by its QA requirements. For this reason, the ideal solution is to work under a graded approach to QA, whereby the QA requirements are tailored to the needs or importance of the project. A graded approach to QA is one way of achieving the maximum benefits of a QA program at the least cost.

Whether an organization is establishing its own QA program or implementing that of a sponsor, the added cost of the time required to implement the program must be covered. In the case of a sponsored study, it is necessary to be fully aware of the extent of the QA required for the study.

QA AND ENVIRONMENTAL MICROBIOLOGY

QA is not entirely new to environmental microbiology (1, 4). However, with the emphasis in environmental sciences being the detection of chemical contaminants and resulting engineered solutions (e.g., Superfund), environmental QA program requirements have focused on chemistry and engineering fields. The environmental microbiologist therefore often finds it necessary to establish a QA program for which the requirements are written in terms of chemical measurements.

The subsections of this chapter are headed with the most common terms that an environmental microbiologist will be confronted with when implementing a sponsor's QA requirements. The discussion following each heading defines the requirement and provides some examples of fulfillment.

MANAGEMENT AND ORGANIZATION

A goal of the QA program is to give management the opportunity to provide input and take responsibility at the planning, implementation, and assessment of environmental microbiology projects. The organization's policies and commitment regarding QA should be clarified as to the importance placed on QA and the reasons. The requirement to incorporate QA in the initial planning stages of a study, or that management and subordinates be involved in certain QA decisions, is an example of an organization's QA policies.

All components of an organization, especially the QA staff and their relationships to the management and technical staff, are important to ensure coordinated efforts and logical flow of information. Specifying the differences between informal and formal communications is important. Divisions of responsibility among the QA staff, management, supervisory staff, and contractors should be clear.

The responsibility for quality lies with all personnel, and it is supportive to illustrate in management and organization policies just how personnel are to interact to comply with the QA program in a manner that maximizes the flow and quality of work. Most importantly, management should ensure that the program's QA requirements are understood and complied with throughout all studies.

QA PROGRAM

It is convenient to base the QA program on a national standard such as ANSI E-4, U.S. Environmental Protection Agency Standard R-2 (23), ASME NQA-1 (3), and U.S. Department of Energy orders 5700.6c (20). In this way, the organization can get an efficient start toward establishment of its own QA program. Standard requirements can be embellished to create a program tailored to the organization's philosophy. The QA program is established and maintained through a number of processes and documents, among them (i) management reviews to evaluate the effectiveness of the QA program in achieving adequate data quality, (ii) QA plans or procedures that describe the QA requirements which each study must meet, (iii) technical plans and procedures (stepwise descriptions of work, including how QA requirements are met), and (iv) audits or assessments to evaluate performance and/or compliance with plans and procedures.

PERSONNEL QUALIFICATIONS AND TRAINING

Qualifications and training for various personnel should be identified to ensure that the work is performed safely and efficiently. Management and technical training requirements for project personnel may be specified. In addition, documentation of personnel training may need to be established and maintained. It should be noted here that providing a training program need not be an overwhelming effort for a small environmental microbiology group. While large organizations may be able to provide employees with a training facility and staff, etc., small organizations may be able to accomplish training with documentation of presentations and/or assigned reading.

PROCUREMENT OF ITEMS AND SERVICES

The organization's procurement system should ensure that suppliers provide the items and services required. To document this, the organization's procurement procedure may require suppliers to have a QA program consistent with the organization's own or another accepted QA program. For example, organizations within one agency will sometimes accept a supplier on each others' approved-supplier lists. A thorough and effective procurement process should define, in the following order, procurement planning, procurement document preparation, selection of procurement sources, proposal and bid evaluation and award, purchaser evaluation of supplier performance, acceptance of the item or service, and control of nonconformances.

For most QA programs, items ordered by catalog number, sometimes called commercial grade items (3), require only a procedure for recording inspection and acceptance of the items received.

DOCUMENT CONTROL

Document control includes a system for handling documents and records to ensure that only the most recent version of a procedure is being used and to ensure that the documents accurately reflect completed work (23).

- Documents that specify quality requirements or prescribe work activities should be reviewed, approved, and controlled.
- Reviews should address technical adequacy, compliance with requirements, correctness, and completeness prior to approval and issuance.
- Effective dates are placed on documents.
- The disposition of obsolete documents is ensured to avoid their inadvertent use.
- Procedures are established and maintained to provide the current revision of a controlled document.
- Procedures defining the distribution of controlled documents are established.

Who designates the documents that are to be controlled for each study, and which types of documents qualify to be controlled, should be established. As an illustration, in most studies the principal investigator would probably designate which documents should be controlled and would designate those documents by title and revision number. For example, QA plans and technical procedures are often controlled.

QA RECORDS

A QA record is a completed document that furnishes evidence of the quality of items or activities (3) or study results. Records are prepared, reviewed, and maintained to reflect the processes resulting in the completed work. Examples of QA records include plans and procedures used during the study, raw data, procurement documents, computer files, and chain-of-custody forms. Which specific documents and data become QA records and when should be clarified. Normally, documents and data become records when personnel complete a segment of work and no longer need to refer to the documents. Controls should be sufficient to ensure that records are legible, accurate, complete, and secure. Work performed should be sufficiently documented to verify that all data collected are scientifically sound and defensible as well as repeatable. Provisions should be made for storage (including proper storage of electronic data, considering the longevity of the various media), length of storage, preservation, and disposition of records.

Some agencies' QA programs require records to be stored in fireproof cabinets or in duplicate. One way to avoid the cost that either measure would incur is to submit the records to that agency's document control center as soon as possible after they become QA records. Requirements for transfers of records to other users should be defined.

COMPUTER SOFTWARE

A QA program should require establishment of procedures that are designed to prevent defects in environmental microbiology data due to computer software. Software QA requirements may be extensive depending on the software used and whether the software was written or purchased. Software used for graphics, for presentations, or in other ways that do not affect data quality do not normally require QA. Such software is used and documented with practices consistent with its intended use.

Certification from the manufacturer indicates that the software has been thoroughly tested and that the manufacturer stands behind the quality of its product. However, software used to reduce (process into a usable form) environmental data needs to be verified to ensure that it accomplishes its intended purpose accurately. A hand-held calculator can be used to check calculations. Another way to check the software is to use it to process a known data set and determine whether it produces the correct response. If the software is used with equipment or instrumentation, this can be accomplished during standardization steps. Database and spreadsheet software usually does not need QA documentation. However, to prevent inadvertent modification of data, software QA requirements may specify use of a limited access system, such as a password. Computer programs may need to meet extensive QA requirements if the computer code was written by an individual in the organization or procured or if the program or software is being used on a sensitive study, such as one affecting national health or related to nuclear licensing. QA requirements for these types of programs may include software specifications, such as program capabilities, data source, definition of valid input, and error checking on the data input (14); individual(s) responsible for software QA; a description of the approach to and documentation of software, such as development, acquisition, testing and use; and software life cycle processes and controls, such as classification, requirements development, design, code development, and other details (10, 19).

QUALITY PLANNING

Certain activities should be included in the planning process for any project. The sponsor's technical and quality requirements need to be translated into project specifications to achieve the desired results. Costs and time constraints need to be considered during planning. The requirement to document planning is often accomplished via a study plan. Planning elements should include, as appropriate, general overview and background of the project; scope of work, study purpose, objective or hypotheses to be tested, and a listing of tasks involved, including estimated resource requirements, deliverables, and their schedules; scientific approach or technical method used to collect, evaluate, analyze, or study results; standards and acceptance criteria; applicable procedures; equipment, instruments, and software; provisions for documenting the work performed and providing required records; prerequisites, special controls, specific environmental conditions, processes, skills, or staff training and certification requirements; data to be collected; flowcharts; summary tables; specification of how results will be presented; time line of project tasks and subtasks; and uses and limitations of the results (23).

WORK PROCESSES

Work processes must be documented to ensure that work is performed as planned and/or can be repeated if necessary. Work processes for most environmental studies are controlled by or recorded in technical procedures manuals, the scientific notebook, or a combination of the two. If scientific notebooks are used, establish some controls concerning proper record keeping practices and perhaps the types and assignments of notebooks to be used.

For example, most guidelines for controlling scientific notebooks have specific requirements (12) for the manner in which errors are corrected, for prohibiting the use of pencils and correction fluid, for ensuring traceability to reagent lots, QC materials, and raw and electronic data, and other record keeping practices that affect report preparation, data evaluation, and repeatability.

Generally, a process which is performed the same way each time is a work process that is appropriate for conversion to a written standard operating or technical procedure. Two formats for written procedures that illustrate the differences in QA program requirements are as follows. (i) Documentation for a work process written to a national QA standard (3) includes technical, QA, or other requirements, including reference to the applicable method, standards, and criteria; a description of the laboratory equipment, including manufacturer, operating manuals, model, and serial number; identification of data acquisition computer hardware and software, including software title and version; special controls, precautions, environmental conditions, and hold points; QC determinations of accuracy, precision, and bias; a sequential description of the work to be performed, including controls for altering the sequence of steps; quantitative or qualitative acceptance criteria; information to be recorded and where; provision for recording deviations from the procedure; and records to be generated by use of the procedure. (ii) Documentation for a work process written for a production laboratory, such as one that tests water samples for pathogens, includes parameter, reference, method, reagents and equipment, procedure, QC requirements, and comments.

AUDITS AND ASSESSMENTS

Audits and assessments are defined according to the individual needs and philosophy of the QA program. Audit and assessment activities can be designed to be a review of routine data or may involve visits to the laboratory or field for an on-site evaluation. Auditors may use a general check sheet that applies to similar laboratory or field operations or possibly the study's QA plan and written procedures.

Audits or assessments emphasize to project personnel the importance of complying with the QA program and documents. The assessment process provides the opportunity to review good laboratory practices and make changes such as updating procedures and adding equipment or personnel. Such changes serve to benefit the work team and data quality.

Audits and assessments fall generally into two major categories, internal and external. Internal audits or self-assessments are performed within the organization. External audits are performed by personnel external to the defined organization such as the sponsor's QA personnel or the project leader's corporate office auditors. An environmental microbiology group may be either the subject of an audit or an auditor, as in the case of having to evaluate the QA program of a service or supplier.

Terms for some specific types of audits and assessments are defined as follows. An *audit* is inspection of objective evidence (evidence that is written, not oral) for compliance to written instructions, plans, and procedures. Auditors request and inspect records and documents for evidence of compliance. A report detailing specific instances of noncompliance found by the auditors is written. During *surveillance*, auditors observe the work as it is being performed (22) for the purpose of process control or product acceptance (3). A report detailing specific instances of noncompliance found by the surveyors is written. In *performance evaluation*, a system, process, or work group is evaluated on the basis of actual performance, usually against a performance sample. A sample is submitted to the laboratory for measurement. The laboratory's sample results are evaluated statistically in comparison with the results of other participants in the study or the known value of the sample. A written report detailing this comparison is sent to the participant. If the laboratory has produced an unacceptable result, corrective action is needed and often requested by the organization that submitted the performance sample to the laboratory. In *data quality assessment* (22), a set of data is evaluated against specific criteria, contained in a check sheet. A data package is evaluated against technical, QA, and QC requirements to establish whether the data are valid and whether the laboratory that produced the data is in compliance with requirements.

Since many scientists have not experienced an audit or surveillance, it is useful for the project leader and/or in-house QA staff to prepare the study personnel. A description, with as much detail as possible, of what the audit will involve should be provided. Although good things often come out of an audit, the process usually presents a tense situation for those being audited. It is a valuable exercise to prepare personnel for an external audit or assessment by having a preaudit or by performing a self-assessment.

NONCONFORMANCE AND CORRECTIVE ACTION

Nonconformance is defined as a deficiency in characteristic, documentation, or procedure that renders the quality of an item or activity unacceptable or indeterminate (3). A non-conformance, such as measuring a sample with an instrument that does not meet calibration requirements, is therefore a serious deficiency.

Discovery of a nonconformance is followed by tagging, segregating, or otherwise isolating the nonconforming item or condition so that it is not used again until it is corrected. An example is to label the data produced on the nonconforming instrument so that they are not used and to indicate with a label that the instrument does not meet calibration requirements.

Corrective action is defined as the measure(s) taken to rectify conditions adverse to quality and, where necessary, to preclude repetition (3). A corrective action would be to remeasure the samples described after the instrument has met calibration requirements. The nonconformance and resulting corrective action should be recorded.

To get from the nonconforming item or condition to the corrective action is a process that can vary greatly depending on the QA program. In one extreme, upon discovery of a nonconformance, the corrective action is chosen by the appropriate person, executed, and recorded. The cycle could be repeated depending on the success of the initial corrective action.

In the other extreme, a QA program may require that a nonconformance be reviewed for recommended solutions. Following this, recommendations are approved and documented by a responsible party. The corrective action is then carried out, and its effectiveness is evaluated and documented. Any items potentially affected by the nonconformance must be reviewed for any adverse effects from the nonconformance.

An example of an extensive nonconformance/corrective procedure can be described for a laboratory whose balance has passed its required calibration due date. It is tagged to be unacceptable for the studies requiring scheduled calibration. Personnel prepare a nonconformance report that includes recommended corrective actions: that the calibration service be contacted for an emergency calibration visit; that the balance continue to be used, because the daily calibration checks are meeting the posted acceptance limits; or that the balance in the next building be used.

The principal investigator reviews the suggested corrective actions and assigns an individual to execute the following corrective action: get the calibration service to the laboratory. The balance can be used as long as it meets daily acceptance limits to accomplish the current workload. The assignee is also to see that future balance calibration recalls are scheduled appropriately.

When the calibration service calibrates the balance, it is found that the balance is still within tolerances. The service documents the calibration and method used, provides certificates, tags the balance calibrated until the next recall, and schedules the next recall for a more convenient date. The principal investigator documents that the data produced by the uncalibrated balance were unaffected, since the balance maintained measurement tolerances, and that the corrective action was effective.

SAMPLING

Sampling is an integral part of the study design. The selection of sampling sites and sampling strategies is based on the study objectives and hypotheses. Therefore, the sampling strategy should be one that ensures the achievement of the quality of data needed to reach conclusions about the experiment. Entire documents have been prepared and published (21) regarding the subject of sampling (2,4,5,9,13,17).

Statisticians are valuable in planning the sampling activity. Study questions are most effectively supported by a predetermined number and variety of samples to be measured and/or collected. In turn, the proper statistical evaluation of resulting data lends confidence to the study conclusions. Statistical support, if available, or use of a good practical text (6,18) is strongly suggested in statistical planning of sampling activities.

With respect to the sampling site, specify the requirements that will ensure that the study objectives are met. If sampling does not involve a geographic location, it may have other requirements, such as age, sex, or species-sample population. Consider specifying the following when planning the sampling activity: (i) sample matrices (e.g., air, water, and soil or body fluids) and target organisms and (ii) the useful measurement levels expected or required from the site.

If the site is mandated rather than selected for the study, determine the effect that the site will have on data and the project objectives: the reason (statistical or scientific) for choosing the sampling site and sampling frequencies; physical, behavioral, and other factors at the site that may affect sampling (e.g., elevation, permits, or physiological obstacles) and sample quality (e.g., sources of contamination or extreme weather); measurements that may need to be taken in the field (e.g., pH and temperature of fluid samples); and site modifications that may be necessary before sampling can occur (e.g., use of special transportation or construction).

With respect to sampling procedures, make specific plans for collection, preservation, storage, record keeping, and transportation: measurement parameters and sample volumes to be collected for each; sampling methods to be used (e.g., composite or grab) (use validated or [preferably] standard methods); preparation and cleaning of sampling equipment, containers, reagents, and supplies; calibration of field equipment (e.g., pH meter and vacuum pump); sample processing, which may include preservation techniques (e.g., use of formalin or azide), transportation (e.g., overnight delivery), and storage (e.g., refrigeration) of samples before laboratory use; provision for the control of field failures (e.g., extra batteries, tubing, and sampling containers); holding times, which are important in planning collection of microbiological samples (consider the maximum length of time samples can be kept before the measurement integrity is affected); and forms, notebooks, and procedures to be used to record sample collection, sampling conditions, and measurements.

Chain of custody is the use of signatures to record individuals responsible for the sample at all times during its life. Sample control in some form will be helpful, especially when there are a large number of samples or more than a few sampling sites and/or sampling teams.

QC can be used to aid in determining variability due to sampling. Field sampling QC includes the use of a combination of laboratory and field or travel blanks (travel blanks carried or shipped with samples to detect contamination during the trip), splits, duplicates, and other QC sample types as well as the requirement that all sample personnel use the same procedure and be proficient.

MEASUREMENT METHODS

Measurement methods are chosen by the environmental microbiologist for the analytical parameters or target organisms in the specific matrix at the anticipated levels. Methods need to agree with the qualitative and quantitative objectives established for the study. Investigators may be limited in choices by the sponsor requirements. In planning measurement methods, consider the following topics: sample preparation methods, decontamination procedures, reagents and other materials needed (their cost and availability), performance requirements, predictable corrective actions, and measurement confirmation.

CALIBRATION

An investigator must demonstrate that the measurement systems are operating properly. The equipment and instrumentation are calibrated and checked at a prescribed frequency commensurate with the type and history of the measurement system as well as the accuracy and precision needs of the project.

Instruments such as colony counters and analytical balances have internally automated or computerized calibrations. It is necessary, however, to verify that the internal calibration is accurate by using a calibration check. For measurements that need a multipoint calibration to establish a concentration curve (initial calibration), a periodic calibration check can be used as an alternative to full recalibration to ensure that the curve remains valid.

Calibration methods and frequencies should be planned for instruments and for equipment used to obtain measurement data (e.g., pH meters, analytical balances, pipettes, and counters) that may affect the quality of the project data (22). Consider including the following items and actions when planning calibrations:

- Specific calibration procedures.
- When possible and practical, certify equipment and/or standards with respect to source, traceability, and purity. If a nationally recognized standard (e.g., one issued by the National Institute of Standards and Technology) does not exist, designate the standards to be used. Records will be maintained and traceable to the item being calibrated (22).
- Record keeping system for achieving traceability from the calibrated item (instrument or equipment) to the standard used.
- Frequency of initial calibrations (multipoint concentration curves), calibration checks (single-point checks), or recalibrations (criteria for establishing new curve).
- Control limits and specific acceptance criteria for all calibration measurements (17). Example: correlation coefficient requirement of 0.995 for initial calibration curve, or calibration check must agree within 10% of its original concentration.

PREVENTIVE MAINTENANCE

It is critical to a project that field and laboratory instruments and equipment be operational and reliable. It can be useful to clarify exactly where each scientist's responsibility for instruments and equipment maintenance begins and ends. Stating the division of responsibility is important in a setting where there are multiple users and/or service contracts on instruments.

Preventive maintenance procedures are best addressed by reference to instrument or equipment procedures. These are often the manufacturer-supplied manuals and instructions. In addition to the following list, any maintenance issues (e.g., specific instrument problems or parts or service needs) that the investigator has noted previously should be considered in an effective preventive maintenance program.

- A schedule of preventive maintenance performed in-house and by a service representative.
- An inventory of critical spare parts and supplies that must be kept in-house and who is responsible for this supply.
- Maintenance contract information for critical measurement systems (e.g., current phone numbers and contract numbers) needed to facilitate service. These may be placed in documentation of procedures, in manuals, or in the instrument log or even attached to the instrument.
- Archiving requirements and locations of one-of-a-kind manuals and instructions, including all versions of computer programs used to reduce data.
- Inspections and acceptance testing performance and record keeping.
- Training of users.
- Continuity in case of a breakdown.
- Record keeping requirements, such as what details of maintenance are recorded, by whom, when, and where.

QUALITY CONTROLS

The investigator must decide which parameters, measurement systems, and sets of data are critical and to what extent detection of problems can be predicted by using various indicators. The investigator then determines what type of QC can be instituted in the study to monitor each parameter or data set. Plan QC frequency, acceptance limits, corrective actions, and record keeping.

Results of the QC determinations must fall within certain limits of acceptance to support the quality of specific measurements. If QC measurements do not meet acceptable limits, they become real-time indicators flagging unacceptable data. Immediate corrective action can then be performed, preventing the collection of more unacceptable data. For some studies, measurement methods may not be standardized initially. Consequently, routine QC requirements may not be practical and a certain amount of trial and error may be necessary to pinpoint the most effective QC regime.

Often a standard or published method will provide QC requirements; however, existing methods often do not provide adequate QC. Therefore, the investigator needs to thoroughly review the method to ensure that all critical parameters are controlled and supplement QC measurements where necessary.

Standards versus Subjective Determinations

Unfortunately, for many aspects of environmental microbiology there are no standards or reference materials to be used for QC and often measurements are subjective, such as the assignment of color to microbial colonies and visual identification of individual organisms. In these cases, the scientist must be creative in developing QC standards. Ornithologists use a standard color chart for checking visual accuracy (6).

Industry uses reliability indices in several ways to treat subjective measurements. However, two techniques can be used to deal with almost any subjective situation. For classifications of nonquantitative (attribute) data, kappa tech-

niques (8) are appropriate. For ratings based on a scale, such as 1 to 10, methods based on the intraclass correlation (8) are a good choice.

Other QC Methods

Reference materials include a substance with one or more of its properties established well enough to be used for checking the accuracy of a measurement method. An example is the use of a standard culture (e.g., American Type Culture Collection strain). These work well as reference materials (6).

A second measurement method should be used to confirm a measurement that is not completely selective. An example is the use of phase-contrast microscopy to verify the identification of a *Giardia* cyst detected with epifluorescence assay (6).

Other QC methods used include the use of control charts to indicate trends in instrument or system stability; proficiency testing of analysts in order to hone and document individual skills; and multilaboratory performance programs, a form of proficiency testing usually associated with certification programs.

For each QC check that is planned, consider frequency of use, the location (field or laboratory) of use, the organisms or analytes to be used, the levels, the control limits, any forms or bench sheets used, and the corrective actions.

DATA REDUCTION

Considerable time and effort go into planning a study and collecting data in a manner that meets a predetermined set of objectives. It is important to ensure that errors are not introduced into the data during the last phases of the study: data reduction, verification, validation, and reporting. Although it may seem labor-intensive to plan these activities, an investigator does not want to jeopardize data quality this late in the game. Consider the following.

(i) Data reduction is changing raw data into a more useful form (15). It includes calculations, tabulations, and transfers that result in interpretation and presentation of data. A listing of personnel, procedures, statistical approaches, and products that result from data reduction should be established to include data reduction procedures, including software or computer programs; statistical approach for reducing data; sample data sheets or formats; treatment of QC measurements; and products of data reduction.

(ii) Data verification determines if measurements have been made in accordance with prescribed, approved procedures and whether data have been quantified, recorded, and transcribed accurately. Completeness and consistency are assessed. Data verification is a routine activity performed on small sets of data as they are produced (16). Among the items to be included are criteria used to accept, reject, or qualify data (22); critical control points (points at which results are so important that they require compliance prior to further progress) (6); and verification forms and checklists to be used.

(iii) The data validation process determines if the data are adequate for their intended use by comparison against a predetermined set of criteria (15). The process includes procedures for determining outliers and criteria for flagging data and acceptance criteria used to validate data (16).

CONCLUSION

QA, QC, and quality improvement are integral parts of a valid scientific study. Although QA is not necessarily new to microbiologists, the emphasis of QA requirements in environmental science has been on chemical measurements and engineering. The chapter has defined and provided examples, when possible, of typical QA requirements that the environmental microbiologist will probably encounter when implementing a QA program.

REFERENCES

1. **American Public Health Association, American Water Works Association, and Water Pollution Control Federation.** 1981. *Standard Methods for the Examination of Water and Wastewater*, 15th ed. American Public Health Association, Washington, D.C.

2. **American Public Health Association, American Water Works Association, and Water Pollution Control Federation.** 1992. *Standard Methods for the Examination of Water and Wastewater*, 17th ed. American Public Health Association, Washington, D.C.

3. **ASME Nuclear Quality Assurance Committee.** 1994. *Quality Assurance Requirements for Nuclear Facility Applications*. ASME NQA-1. The American Society of Mechanical Engineers, New York.

4. **Bordner, R., and J. Winter.** 1978. *Microbiological Methods for Monitoring the Environment, Water, and Wastes*. EPA 600/8-78-017. U.S. Environmental Protection Agency, Washington, D.C.

5. **Britton, L. J., and P. E. Greeson (ed.).** 1989. Methods for collection and analysis of aquatic biological and microbiological samples, p. 16–158. *U.S. Geological Survey Techniques of Water Resources Investigations*, book 5. U.S. Government Printing Office, Washington, D.C.

6. **Cross-Smiecinski, A. J., and L. D. Stetzenbach.** 1994. *Quality Planning for the Life Science Researcher*. CRC Press, Inc., Boca Raton, Fla.

7. **Deming, S., and S. L. Morgan.** 1993. *Experimental Design: a Chemometric Approach*, 2nd ed., vol. 11. Elsevier Science Publishing Co., New York.

8. **Futrell, D.** 1995. When quality is a matter of taste, use reliability indexes, p. 81–86. *In* B. Stratton (ed.), *Quality Progress*. American Society of Quality Control, Milwaukee, Wis.

9. **Ghiorse, W. C., and D. L. Balkwill.** 1983. Enumeration and morphological characterization of bacteria indigenous to subsurface environments. *Dev. Ind. Microbiol.* **24:** 213–224.

10. **Harris, S. D.** 1992. Software quality assurance requirements of a nuclear program written to meet current software quality assurance standards. p. 6.A.3.1–6.A.3.10. *In Proceedings of the Nineteenth Annual National Energy & Environmental Division Conference*. American Society for Quality Control, Milwaukee, Wis.

11. **Juran, J. M., F. M. Gryna, Jr., and R. S. Bingham (ed.).** 1979. *Quality Control Handbook*, 3rd ed. McGraw-Hill, Inc., New York.

12. **Kanare, H. M.** 1985. *Writing the Laboratory Notebook*, p. 29–52. American Chemical Society, Washington, D.C.

13. **Labeda, D. P. (ed.).** 1990. *Isolation of Biotechnological Organisms from Nature*. McGraw-Hill, New York.

14. **Morris, C. R.** 1993. Computer validation. *Network News* **7:**1.

15. **Parker, S. P.** 1984. *McGraw-Hill Dictionary of Scientific and Technical Terms*, 4th ed. McGraw-Hill Book Co., New York.

16. **Ramos, S. J.** 1992. A project manager's primer on data validation, p. 4.C4.1–4.C4.12. *In Proceedings of the Nineteenth Annual National Energy & Environmental Division Conference*. American Society for Quality Control, Milwaukee, Wis.

17. **Simes, G.** 1991. *Preparation Aids for the Development of Category I-IV Quality Assurance Project Plans*. EPA 600/8-91/003-006. U.S. Environmental Protection Agency, Risk Reduction Engineering Laboratory, Cincinnati.

18. **Taylor, J. K.** 1987. *Quality Assurance of Chemical Measurements*. Lewis Publishers, Inc., Chelsea, Mich.

19. **U.S. Department of Energy.** 1990. *Quality Assurance Requirements Document*. DOE/RW-021n4, revision 4. U.S. Office of Civilian Radioactive Waste Management, Washington, D.C.

20. **U.S. Department of Energy.** 1991. Order 57006.c. Offices of Nuclear Energy and Environment, Safety, and Health, Washington, D.C.

21. **U.S. Environmental Protection Agency.** 1989. *Soil Sampling Quality Assurance User's Guide*. EPA/600/8-89/046. U.S. Environmental Monitoring Systems Laboratory, Las Vegas.

22. **U.S. Environmental Protection Agency, Quality Assurance Management Staff.** 1992. *Revised Draft EPA Requirements for Quality Assurance Project Plans for Environmental Data Operations*. EPA QA/R-5. U.S. Environmental Protection Agency, Washington, D.C.

23. **U.S. Environmental Protection Agency, Quality Assurance Management Staff.** 1992. *Interim Draft EPA Requirements for Quality Management Plans*. EPA QA/R-2. U.S. Environmental Protection Agency, Washington, D.C.

WATER MICROBIOLOGY IN PUBLIC HEALTH

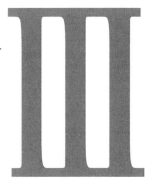

VOLUME EDITOR
CHRISTON J. HURST

SECTION EDITOR
GARY A. TORANZOS

Overview of Water Microbiology as It Relates to Public Health

CHRISTON J. HURST

14

One of the most important aspects of water microbiology, from a human perspective, is the fact that we acquire numerous diseases from microorganisms found in water. The magnitude of human morbidity and mortality associated with waterborne infectious diseases has led to the development of epidemiological surveillance studies (3, 8, 9). Some of these diseases represent intoxications, which we acquire from ingesting microbial toxins in foods. One type of toxin comes from the metabolic activities of bacteria such as *Proteus* and *Klebsiella* species, which grow on the surfaces of some fish and generate the histidine associated with scombroid poisoning. Dinoflagellates such as *Gambierdiscus, Gonyaulax,* and *Ptychodiscus* species are a major source of neurotoxins which can become biologically concentrated in the tissues of reef fish and shellfish.

The majority of human diseases associated with microbially contaminated water are, however, infectious in nature (7), and the associated pathogens include numerous bacteria, viruses, and protozoa. These water-related infectious hazards can be characterized according to various schemes. One such approach divides the hazards into categories based on the source of the involved pathogen and the route by which human recipients contact that pathogen (see chapter 15). These categories could be defined as follows.

(i) Water-washed infections are caused by pathogens which are not acquired from water but rather are acquired either by contact with or ingestion of microbially contaminated material such as feces. These disease hazards could be ameliorated by using clean water for sanitation purposes, including bodily cleansing and the washing of plates and drinking glasses.

(ii) In infections which have water-related insect vectors, the associated pathogens are not acquired from water but rather are acquired as a consequence of humans being bitten by invertebrate vectors (usually mosquitos) whose life cycles depend on access to water either on the land surface or in uncovered containers. Thus, the incidence of these diseases may increase during periods of land surface flooding and during monsoon seasons.

(iii) Water-based infections are caused by pathogenic worms which must spend part of their life cycle within intermediate vertebrate or invertebrate hosts that reside in aquatic environments. These diseases are acquired either by physical contact with water that is contaminated by the pathogenic organisms or by inadvertent ingestion of infested intermediate host animals.

(iv) Waterborne infections are recognized as resulting either from physical contact with microbially contaminated water, as may occur during bathing and recreational activities, or from ingestion of contaminated water, ice that has been made from contaminated water, or food items which have come into contact with contaminated water.

The chapters in this section address various issues related to the routes by which humans acquire waterborne infections. These potential routes of infection are diagrammed in Fig. 1.

The reservoirs for the pathogenic microorganisms found in environmental waters can be either humans, animals, or the environment itself (5), as summarized in Table 1. However, it commonly is presumed that many of the human pathogens found in our aquatic resources originate from human reservoirs. This human-related contamination can occur during either defecation in water or recreational activities conducted in water (4). Additionally, domestic wastewater seems to have particular importance as a contrib-

TABLE 1 Examples of infectious disease hazards associated with microorganisms in water

Reservoir for microorganisms	Disease	Causative genus or genera
Human	Cholera	*Vibrio*
	Encephalitis	*Enterovirus*
	Entamoebiasis	*Entamoeba*
	Gastroenteritis	*Astrovirus, Calicivirus, Coronavirus, Rotavirus*
	Hepatitis	*Calicivirus, Hepatovirus*
	Meningitis	*Enterovirus*
Animal	Campylobacteriosis	*Campylobacter*
	Cryptosporidiosis	*Cryptosporidium*
	Giardiasis	*Giardia*
	Leptospirosis	*Leptospira*
Environmental	Encephalitis	*Naegleria*
	Cholera	*Vibrio*
	Legionellosis	*Legionella*

133

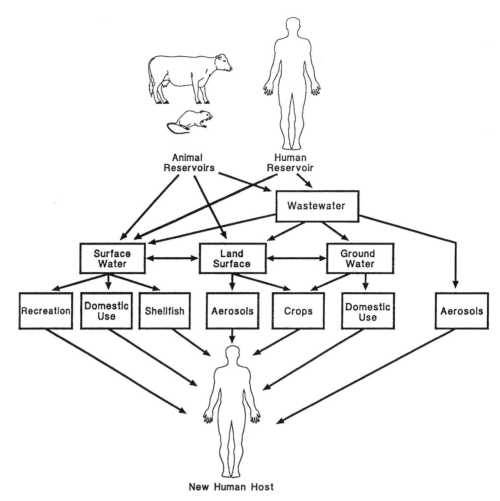

FIGURE 1 Water-related environmental routes by which infectious agents are transmitted to susceptible individuals. The animals symbolically represented are cows as a source of cryptosporidiosis and rodents as a source of giardiasis and campylobacteriosis (from reference 7 with permission).

utor of the pathogenic contaminants found in aquatic environments (7, 10), and the attendant public health concerns have resulted in the development of methods for studying and reducing the levels of pathogens in wastewater (1). Wastewater treatment efforts may help reduce the incidence of problems which can follow the discharge of wastewaters into environmental surface water, including illness among swimmers (2), contamination of drinking water, and disease that can occur when bivalve molluscan shellfish harvested from contaminated waters are subsequently consumed by humans. Wastewater treatment will also reduce the contamination of aquifers, which can either result indirectly following the percolation of surface-applied wastewaters into the subsurface or occur directly during the subsurface injection of wastewater. The treatment of wastewater also is intended to reduce the contamination of crops that may occur when wastewater is eventually discharged onto land surfaces (1, 6, 10).

REFERENCES

1. **Acher, A. J., E. Fischer, and Y. Manor.** 1994. Sunlight disinfection of domestic effluents for agricultural use. *Water Res.* **28:**1153–1160.

2. **Corbett, S. J., G. L. Rubin, G. K. Curry, D. G. Kleinbaum, and Sydney Beach Users Study Advisory Group.** 1993. The health effects of swimming at Sydney beaches. *Am. J. Public Health* **83:**1701–1706.

3. **Glass, R. I., and R. E. Black.** 1992. The epidemiology of cholera, p. 129–154. *In* D. Barua and W. B. Greenough III (ed.), *Cholera.* Plenum Medical Book Company, New York.

4. **Grabow, W. O. K.** 1991. New trends in infections associated with swimming-pools. *Water S A* (Pretoria) **17**(2): 173–177.

5. **Grimes, D. J.** 1991. Ecology of estuarine bacteria capable of causing human disease: a review. *Estuaries* **14:**345–360.

6. **Hopkins, R. J., P. A. Vial, C. Ferreccio, J. Ovalle, P. Prado, V. Sotomayor, R. G. Russell, S. S. Wasserman, and J. G. Morris, Jr.** 1993. Seroprevalence of *Helicobacter pylori* in Chile: vegetables may serve as one route of transmission. *J. Infect. Dis.* **168:**222–226.

7. **Hurst, C. J., and P. A. Murphy.** The transmission and prevention of infectious disease, p. 3–54. *In* C. J. Hurst (ed.), *Modeling Disease Transmission and Its Prevention by Disinfection,* in press. Cambridge University Press, Cambridge.

8. **Moore, A. C., B. L. Herwaldt, G. F. Craun, R. L. Calde-**

ron, A. K. Highsmith, and D. D. Juranek. 1993. Surveillance for waterborne disease outbreaks—United States, 1991–1992. *Morbid. Mortal. Weekly Rep.* **42**(SS-5):1–22.

9. Payment, P., L. Richardson, J. Siemiatycki, R. Dewar, M. Edwardes, and E. Franco. 1991. A randomized trial to evaluate the risk of gastrointestinal disease due to consumption of drinking water meeting current microbiological standards. *Am. J. Public Health* **81:**703–708.

10. Straub, T. M., I. L. Pepper, and C. P. Gerba. 1993. Hazards from pathogenic microorganisms in land-disposed sewage sludge. *Rev. Environ. Contam. Toxicol.* **132:** 55–91.

Waterborne Transmission of Infectious Agents

CHRISTINE L. MOE

15

Waterborne transmission is a highly effective means for spreading infectious agents to a large portion of the population. Large quantities of enteric organisms can be introduced into the aquatic environment through fecal contamination from infected persons (or animals) that is discharged into sewers or unprotected waterways. In contrast to person-to-person transmission, infectious agents even from bedridden infected persons can play a role in waterborne disease transmission, since pathogens in soiled bedding and clothing may be released into water during washing. When such fecal contamination mixes with unprotected and/or inadequately treated drinking water, large numbers of susceptible hosts can be exposed and become infected. Historically, this process has been documented repeatedly since the time of John Snow in the mid-19th century. Waterborne disease outbreaks infecting tens to hundreds of thousands of people have been reported, and Ewald has observed that those pathogens that are more frequently waterborne tend to be the most virulent enteric pathogens (50).

Numerous infectious agents have been transmitted by ingestion, contact, or inhalation of water (Tables 1 and 2). Disease outcomes associated with waterborne infections include mild to life-threatening gastroenteritis, hepatitis, skin infections, wound infections, conjunctivitis, respiratory infections, and generalized infections. Some waterborne microorganisms are frank pathogens, some are opportunistic pathogens, and some are toxigenic. The total number of potentially pathogenic microorganisms is unknown but may number in the thousands. New infectious agents of disease continue to be recognized.

There are five critical elements in the transmission of infectious agents through water: (i) source of the infectious agents, (ii) specific water-related modes of transmission, (iii) specific attributes of the organism that allow it to survive and possibly multiply and to move into and within the aquatic environment, (iv) infectious dose and virulence factors of the organism, and (v) host susceptibility factors.

Some agents of waterborne diseases, such as *Legionella* species, *Vibrio* species, *Aeromonas hydrophila,* and *Pseudomonas aeruginosa,* are indigenous aquatic organisms. Control of these infections may depend on controlling exposure to water containing such organisms or, when possible, treating the water to remove or inactivate the infectious agents. Most microbial waterborne pathogens of concern originate in the enteric tracts of humans or animals and enter the aquatic environment via fecal contamination. The concentration of these pathogens in a community water supply will depend in part on the number of infected persons and/or animals in the community and the opportunities for feces from these individuals to enter the water supply. Control of these diseases rests on sanitation measures and wastewater treatment to prevent the introduction of feces containing these organisms into drinking water supplies or recreational waters and on adequate water treatment to remove or inactivate these organisms in drinking water.

There are several water-related modes of transmission of infectious agents, which are discussed in the following section. Many enteric pathogens transmitted via ingestion of fecally contaminated water can also be transmitted person to person by contact with fecally contaminated hands or fomites or by consumption of fecally contaminated food. In situations of endemic disease due to poor sanitation and hygiene, the attributable risk due to water may be difficult to determine because of the risks from many other transmission routes of infection.

Different classes of organisms have specific attributes, such as size and charge, which determine their movement and survival in the aquatic environment and their susceptibility to various water and wastewater treatment processes. Knowledge of these attributes can aid in the design of effective barriers or control strategies.

The infectious dose is the minimum number of organisms required to cause infection and varies considerably by type of organism. In general, enteric viruses and protozoa have low infectious doses, typically between 1 and 50 tissue culture infectious units, PFU, cysts, or oocysts. Bacterial pathogens tend to require a larger dose to cause infection. Observed median infectious doses (numbers of organisms resulting in a 50% infection rate) for enteric bacteria range from 10^2 to 10^8. Because such data are derived from studies of healthy, adult human volunteers with controlled conditions (vehicle, gastric acidity), care must be taken when these findings are extrapolated to other populations, such as malnourished children in developing countries or human immunodeficiency virus-infected persons. For example, studies of *Vibrio cholerae* indicate that 10^6 organisms when ingested with water by fasting volunteers caused no illness. However, the same inoculum ingested with food or sodium bicarbonate caused illness in 90 to 100% of the exposed

TABLE 1 Illnesses acquired by ingestion of water

Agent	Source	Incubation period	Clinical syndrome	Duration
Viruses				
Astrovirus	Human feces[a]	1–4 days	Acute gastroenteritis	2–3 days, occasionally 1–14 days
Calicivirus	Human feces[a]	1–3 days	Acute gastroenteritis	1–3 days
Enteroviruses (polioviruses, coxsackieviruses, echoviruses)	Human feces	3–14 days (usually 5–10 days)	Febrile illness, respiratory illness, meningitis, herpangina, pleurodynia, conjunctivitis, myocardiopathy, diarrhea, paralytic disease, encephalitis, ataxia	Variable
Hepatitis A virus	Human feces	15–50 days (usually 25–30 days)	Fever, malaise, jaundice, abdominal pain, anorexia, nausea	1–2 wk to several months
Hepatitis E virus	Human feces	15–65 days (usually 35–40 days)	Fever, malaise, jaundice, abdominal pain, anorexia, nausea	1–2 wk to several months
Norwalk and Norwalk-related viruses (Snow Mountain, Hawaii, Southampton, Taunton, and Toronto agents)	Human feces	1–2 days	Acute gastroenteritis with predominant nausea and vomiting	12–48 h
Group A rotavirus	Human feces[a]	1–3 days	Acute gastroenteritis with predominant nausea and vomiting	5–7 days
Group B rotavirus	Human feces[a]	2–3 days	Acute gastroenteritis	3–7 days
Bacteria				
Aeromonas hydrophila	Freshwater		Watery diarrhea	Avg, 42 days
Campylobacter jejuni	Human and animal feces	3–5 days (1–7 days)	Acute gastroenteritis, possible bloody and mucoid feces	1–4 days, occasionally >10 days
Enterohemorrhagic *Escherichia coli* O157:H7	Human and animal feces	3–8 days	Watery then grossly bloody diarrhea, vomiting, possible hemolytic uremic syndrome	1–12 days (usually 7–10 days)
Enteroinvasive *E. coli*	Human feces	1–3 days	Possible dysentery with fever	1–2 wk
Enteropathogenic *E. coli*	Human feces	1–6 days	Watery to profuse watery diarrhea	1–3 wk
Enterotoxigenic *E. coli*	Human feces	12–72 h	Watery to profuse watery diarrhea	3–5 days
Plesiomonas shigelloides	Fresh surface water, fish, crustaceans, wild and domestic animals?	1–2 days	Bloody and mucoid diarrhea, abdominal pain, nausea, vomiting	Avg, 11 days
Salmonellae	Human and animal feces	8–48 h	Loose, watery, occasionally bloody diarrhea	3–5 days
Salmonella typhi	Human feces and urine	7–28 days (avg, 14 days)	Fever, malaise, headache, cough, nausea, vomiting, abdominal pain	Weeks to months
Shigellae	Human feces	1–7 days	Possible dysentery with fever	4–7 days
Vibrio cholerae O1	Human feces	9–72 h	Profuse, watery diarrhea, vomiting, rapid dehydration	3–4 days
Non-O1 *V. cholerae*	Human feces	1–5 days	Watery diarrhea	3–4 days
Yersinia enterocolitica	Animal feces and urine	2–7 days	Abdominal pain, mucoid occasionally bloody diarrhea, fever	1–21 days (avg, 9 days)

(*Table continued on next page*)

TABLE 1 Illnesses acquired by ingestion of water (*Continued*)

Agent	Source	Incubation period	Clinical syndrome	Duration
Parasites				
Balantidium coli	Human and animal feces	Unknown	Abdominal pain, occasional mucoid or bloody diarrhea	Unknown
Cryptosporidium species	Human and animal feces	1–2 wk	Profuse, watery diarrhea	4–21 days
Entamoeba histolytica	Human feces	2–4 wk	Abdominal pain, occasional mucoid or bloody diarrhea	Weeks to months
Giardia lamblia	Human and animal feces	5–25 days	Abdominal pain, bloating, flatulence, loose, pale, greasy stools	1–2 wk to months and years
Algae				
Cyanobacteria (*Anabaena*, *Aphanizomenon*, and *Microcystis* species)	Cyanobacterial blooms in marine water or freshwater	Few hours	Toxin poisoning (blistering of mouth, gastroenteritis, pneumonia)	Variable
Helminth				
Dracunculus medinensis (guinea worm)	Larvae discharged from worms protruding from skin of infected person	8–14 mo (usually 12 mo)	Blister, localized arthritis of joints adjacent to site of infection	Months

[a] Animal strains are believed to be not pathogenic for humans.

volunteers because these vehicles decreased the protective gastric acidity (116). Virulence of microorganisms varies by type and strain and route of infection.

Infection and development of clinical symptoms depend on a number of specific and nonspecific host factors, such as age, immune status, gastric acidity, nutritional status, and vitamin A deficiency. With hepatitis A virus, symptomatic infection is rare in children and greatly increases with age. For enteric viruses like group A rotavirus and astrovirus, symptomatic infections are common among children under 2 years of age.

CLASSIFICATION OF WATER-RELATED DISEASES

The classification of water-related diseases by Bradley (10) provided a valuable framework for understanding the relationship between infectious disease transmission and water. This classification system facilitates planning effective prevention and control measures for a variety of water-related diseases, depending on the type of agent and type of transmission route involved. Bradley described four main categories of water-related infections: waterborne infections, water-washed infections, water-based infections, and infections with water-related insect vectors.

The waterborne infections are those classically recognized as waterborne disease, such as typhoid and cholera, whereby an enteric microorganism enters the water source through fecal contamination and transmission occurs by ingestion of contaminated water. Transmission by this route

depends on (i) the concentration of pathogens in water, which is determined by the number of infected persons in the community, the amount of fecal contamination in the water, and the survival of the organism in water, (ii) the infectious dose of the organism, and (iii) individual ingestion (exposure) of the contaminated water. Control of these infections is generally through improvement of microbiological water quality, by water treatment and/or source protection.

Water-washed infections are diseases due to poor personal and/or domestic hygiene. These diseases are not due to the presence of infectious agents in water but rather are due to the lack of readily accessible water, which limits washing of hands and utensils and thus permits transmission of infectious agents, such as *Shigella* species, by fecally contaminated hands and utensils. Transmission is again related to the presence of feces from an infected individual, the infectious dose, the amount of fecal contamination on the hand or surface, and the survival of the organism on surfaces. Diseases that affect the eye and skin, such as trachoma, conjunctivitis, and scabies, that are related to lack of water for bathing are also included in this category. Control of these diseases is through provision of greater quantities of water, closer, easier access to water, and education to improve personal and domestic hygiene.

Water-based infections are worm infections in which the pathogen must spend a part of its life cycle in the aquatic environment. This category is further subdivided into those diseases acquired by ingestion of water and those diseases acquired by contact with water. The prototype illnesses of

TABLE 2 Illnesses acquired by recreational contact with water[a]

Agent	Source	Incubation period	Clinical syndrome	Duration
Viruses				
Adenovirus (serotypes 1, 3, 4, 7, 14)	Humans	4–12 days	Conjunctivitis, pharyngitis, fever	7–15 days
Bacteria				
Aeromonas hydrophila	Fresh and brackish water	8–48 h	Wound infections	Weeks to months
Legionellae	Freshwater, soil	Legionnaires' disease: 2–14 days (usually 5–6 days) Pontiac fever: 5–66 h (usually 24–48 h)	Legionnaires' disease: pneumonia with anorexia, malaise, myalgia and headache, rapid fever and chills, cough, chest pain, abdominal pain and diarrhea Pontiac fever: fever, chills, myalgia, headache	Legionnaires' disease: variable (usually weeks to months) Pontiac fever: 2–7 days
Mycobacterium spp. (M. marinum, M. balnei, M. platy, M. kansasii, M. szulgai)	Marine or brackish waters, freshwater	2–4 wk	Lesions of skin or subcutaneous tissues	Months
Pseudomonas spp.	Water		Dermatitis, ear infections, conjunctivitis	
Vibrio spp. (V. alginolyticus, V. parahaemolyticus, V. vulnificus, V. mimicus)	Marine water	V. vulnificus: 24 h V. parahemolyticus: 4–48 h	V. vulnificus: acute gastroenteritis, wound infections, septicemia V. parahaemolyticus: acute gastroenteritis, wound infections Ear infections	V. vulnificus: septicemia fatal in 2–4 days V. parahemolyticus: usually 3 days
Other				
Cyanobacteria (Anabaena, Aphanizomenon, and Microcystis species)	Cyanobacterial blooms in marine or freshwater	Few hours	Dermatitis	
Naegleria fowleri	Freshwater in warm climates, soil, decaying vegetation	3–7 days	Meningoencephalitis, headache, anorexia, fever, nausea, and vomiting; usually fatal	10 days
Acanthamoeba species	Water		Subcutaneous abscesses, conjunctivitis	8 days to several months
Schistosoma species	Feces and urine of infected animals and birds	Few minutes to hours	Dermatitis, prickly sensation, itching	Years

[a] Agents acquired through ingestion of water are not included.

this category are dracontiasis, which is due to ingestion of water contaminated with guinea worm, and schistosomiasis, which is transmitted by contact with water contaminated with species of the trematode genus *Schistosoma*. The original source of the guinea worm is larvae that are discharged from the female worm, which lies in a vesicle usually on the lower leg or foot of an infected human. The larvae are discharged when the vesicle is immersed in water and are then ingested by a copepod (genus *Cyclops*), in which they develop into the infective stage. Humans become infected when they ingest water containing the copepods (10). The eggs of schistosome worms enter the aquatic environment from the urine or feces of an infected human. The eggs hatch in the water to produce miracidia, which infect snails, develop into the infective stage, and are shed by snails into the water over a period of months. Humans become infected when the free-swimming infective larvae penetrate the skin

during water contact (10). Although schistosomiasis is typically considered a tropical disease, there are reports of schistosomal dermatitis (swimmer's itch) in the United States that appear to be associated with *Schistosoma* species (26). Control of dracontiasis and schistosomiasis is through protection of the water source and the user by limiting skin contact with water and by eradication of intermediate hosts.

The types of water contact diseases most frequently encountered in developed countries are those associated with recreational water exposure to contaminated marine water, freshwater lakes, ponds, creeks, or rivers or to possibly treated water in swimming pools, wave pools, hot tubs, and whirlpools. While many recreational water outbreaks are associated with ingestion of water, there are some diseases of the ear, eye, and skin that are associated with actual water contact, as well as systemic illness associated with penetration of a pathogen through an open wound or abrasion. Illness from recreational water contact can be due to enteric organisms in fecally contaminated water. Reports of recreational water outbreaks have involved *Giardia* and *Cryptosporidium* species, *Shigella sonnei*, and *Escherichia coli* O157:H7 that presumably entered the gastrointestinal tract via ingestion (26). Other recreational water outbreaks have involved ingestion, contact, or inhalation of indigenous aquatic organisms such as *Naegleria*, *Pseudomonas*, and *Legionella* species (26), several *Vibrio* species, and several *Mycobacterium* species (41). Epidemiological and microbiological studies indicate that *Staphylococcus aureus* skin and ear infections are often associated with recreational use of water, and the source of these organisms may be other bathers or the water (27, 28). *Vibrio vulnificus* can cause serious wound infections when an injury to skin occurs in marine water or from contact of preexisting wounds with marine water (92). Cyanobacterial toxins have been associated with contact irritation after bathing in marine waters or freshwaters (32). An additional cause of recreational water infections are the *Leptospira* species, which are neither enteric organisms nor aquatic organisms but enter water via the urine of infected domestic and wild animals (41).

Water-vectored infections are those transmitted by insects which breed in water, such as mosquito vectors of malaria, or insects which bite near water, like the tsetse flies that transmit sleeping sickness. Control of these infections is through the application of pesticides, destruction of breeding grounds, and construction of piped water supplies.

Two additional modes of transmission which are water related are the transmission of infectious agents by inhalation of water aerosols and by the consumption of raw or undercooked shellfish or contaminated fish. The major pathogens associated with aerosol transmission are *Legionella* species, especially *Legionella pneumophila*, the etiologic agent of Legionnaires' disease and Pontiac fever. *Legionella* species are ubiquitous in water and soil and are capable of prolonged survival and reproduction in the aquatic environment. Growth within free-living amoebae appears to enhance survival and provide protection from routine disinfection (160). Outbreaks of legionellosis have been associated with aerosols from cooling towers and evaporative condensers of large buildings or with hot and cold water systems in hospitals, hotels, and other institutions. *Legionella* species can proliferate in hot water tanks maintained at 30 to 54°C (152), and exposure occurs from aerosols created from showerheads. Control of these infections is through minimizing exposure to contaminated aerosols and routine cleaning and disinfection of water systems with adequate doses of chlorine or ozone (114, 160).

The potential for aerosol transmission of *Mycobacterium avium* and other nontuberculous mycobacteria and the risk to the immunocompromised population continues to be a concern. Like legionellae, these organisms are frequently isolated in environmental and treated water systems and are able to colonize and propagate within water distribution systems (79). Wendt et al. (158) reported the isolation of nontuberculous mycobacteria from aerosol samples near the James River in Virginia. The isolates (mostly *Mycobacterium intracellulare*) were biochemically similar to those recovered from human clinical specimens, suggesting that airborne mycobacteria derived from freshwater may be a significant source of infection.

Bivalve molluscan shellfish have served as vehicles of enteric disease transmission because of their ability to concentrate enteric organisms from fecally contaminated water in their tissue. Numerous outbreaks have been attributed to the consumption of raw or undercooked oysters, clams, and mussels (112). Many pathogens, including hepatitis A and E viruses, Norwalk-related viruses, pathogenic *E. coli*, *Salmonella typhi*, and species of *Shigella*, *Vibrio*, *Plesiomonas*, and *Aeromonas*, have been implicated in shellfish-borne disease (62, 63, 78). Shellfish and some species of fish may also serve as vehicles for algal toxins. Toxic species of *Gonyaulax* and *Gymnodinium* are concentrated by filter-feeding mollusks and can cause paralytic shellfish poisoning among shellfish consumers (17). Reef-feeding fish can concentrate toxic dinoflagellates of the genus *Gambierdiscus* that cause ciguatera seafood poisoning among consumers (17).

STUDIES OF WATERBORNE DISEASE

Epidemic Waterborne Disease

Most of the information on the risk factors and etiologic agents of waterborne disease comes from investigations of waterborne disease outbreaks by state and local health departments and the surveillance program maintained by the Centers for Disease Control and Prevention and the Environmental Protection Agency. Because most infections due to waterborne agents are not reportable diseases, it is difficult to recognize disease clusters. This surveillance system is based on voluntary reporting by state health departments and clearly represents only a fraction of the true incidence of waterborne disease outbreaks. Data on waterborne outbreaks in the United States have been summarized by Craun (34, 35) and others (21, 23, 24, 26). These data indicate that since the early 1980s, parasitic agents have become the major etiologic agents associated with waterborne disease outbreaks in the United States and that recently recognized etiologic agents, such as *E. coli* O157:H7 and *Cryptosporidium* species, are being reported more frequently and from new settings, such as outbreaks associated with recreational water use (26).

Still, the vast majority of waterborne outbreaks are classified by this surveillance system as acute gastrointestinal illness of unknown etiology. Stool examinations by hospital laboratories typically include culture for *Salmonella*, *Shigella*, and *Campylobacter* species. In addition, many laboratories will test for group A rotavirus in specimens from young children, *Giardia*, and, more recently, *Cryptosporidium* species at the request of a physician. Clinical symptoms suggest that many of the outbreaks of acute gastroenteritis illness of unknown etiology may be due to viral agents, such as Norwalk virus (NV) and related small round structured viruses. However, inadequate diagnostic technology has limited the detection of these agents in clinical and environmental samples. In addition, for a number of newly recognized etiologic agents, there is some evidence of an association with waterborne disease, such as *Cyclospora cayetanensis* (163), cyanobacteria (17), *Helicobacter pylori* (91), mycobacteria (79), and aeromonads (134). These agents are discussed in the section on emerging waterborne pathogens.

Endemic Waterborne Disease

The primary concern with infectious agents in drinking water is acute gastrointestinal illness. In industrialized countries, there have been few studies of endemic gastrointestinal disease associated with the consumption of drinking water. A randomized intervention trial was conducted in Montreal, Canada, to examine the risk of gastrointestinal illness associated with the consumption of conventionally treated municipal drinking water that met current microbiological standards (126). A total of 606 households were recruited into the study, and 299 of these households were supplied with reverse-osmosis filters that provided additional in-home water treatment. Gastrointestinal symptoms were recorded in family health diaries. Water samples from the surface water source, treatment plant, distribution system, and study households were analyzed for several indicator bacteria and culturable viruses. Over a 15-month period, a 35% higher rate of gastrointestinal symptoms was observed in the 307 study households drinking municipal tapwater without in-home treatment compared with the 299 study households supplied with reverse-osmosis filters. Symptoms and serologic evidence suggested that much of this increased illness may have been due to low levels of enteric viruses in the municipal water supply, which originated from a river contaminated by human sewage.

A longitudinal study of French alpine villages that used untreated groundwater for their drinking water supplies observed a weak relationship between rates of acute gastrointestinal disease and the presence of fecal streptococcal indicator bacteria in the public water system over a 15-month study period (166). Illness data were collected through active surveillance by physicians, pharmacists, and schoolteachers. Weekly water samples were collected from frequently used taps in the distribution system of each village and were analyzed for several bacterial indicator organisms.

Ecological studies have attempted to find a relationship between rates of endemic hepatitis A infection and municipal source water quality and/or level of water treatment (5). Several case-control studies have found the consumption of unfiltered municipal water or shallow well water to be a risk factor for endemic giardiasis (8, 30, 40).

MICROBIAL AGENTS COMMONLY ASSOCIATED WITH WATERBORNE DISEASE

The commonly recognized waterborne pathogens consist of several groups of enteric and aquatic bacteria, enteric viruses, and three enteric protozoa (Table 1).

Enteric and Aquatic Bacteria

V. cholera and S. typhi, the first waterborne pathogens to be recognized, were identified in the 19th century and have been responsible for tremendous morbidity and mortality worldwide. Vibrio species are aquatic bacteria that are well adapted to both the estuarine environment and the intestinal tract. Waterborne enteric bacteria include both human-associated and zoonotic species. Campylobacter and Salmonella species are found in the intestinal tracts of numerous domestic and wild animals. Therefore, contamination of water from animal feces also poses a human health risk. For other enteric bacteria, such as S. typhi and Shigella species, infections are generally limited to humans. The infectious dose for enteric bacteria depends on several host factors, including gastric acidity and the vehicle of transmission. Volunteer studies have shown that when the bacteria are ingested with milk, the median infectious doses are approximately 10^2 CFU for Shigella species, approximately 10^7 CFU for S. typhi, and approximately 10^8 CFU for enterotoxigenic E. coli (116). Diagnosis of these infections is typically by culturing the microorganisms from clinical specimens, detecting bacterial antigens or antibodies by enzyme immunoassays, or detecting specific genes by using molecular biology methods such as nucleic acid probe hybridization.

In the developing world, classic waterborne bacterial in-fections due to V. cholerae and S. typhi continue to be a problem. In the United States, recent (1985 to 1992) drinking water outbreaks due to bacterial agents have been due predominantly to Shigella species (nine outbreaks), followed by Campylobacter species (four outbreaks). The majority of recreational water outbreaks have been dermatitis infections caused by Pseudomonas species (41 outbreaks), followed by shigellosis (11 outbreaks) and Legionella infection (5 outbreaks) (21, 23, 24, 26).

Data on the occurrence of pathogenic bacteria in water have been summarized by Emde et al. (46). However, much of the data on microbial occurrence in raw water and treated water are believed to be obsolete because they were gathered by insensitive analytical methods with selective recovery media that underestimate the bacterial levels by as much as 100- to 1,000-fold (141). The isolation of bacterial pathogens from water is difficult because of low concentrations, their stressed condition, interference by competing aquatic microorganisms, and the fastidious nutrient requirements of some pathogens. Enrichment procedures are often used to promote growth of the organisms before they are transferred to selective media for identification. However, this limits the ability to quantify the pathogens in the sample (48).

The persistence of enteric bacteria in the aquatic environment depends on the species and on a variety of environmental factors (temperature, pH, sunlight, predation or competition by indigenous aquatic microorganisms, dissolved organics, attachment to particulates, association with vectors such as amoebae, protozoa or copepods, presence of salts and other solutes) (reviewed in references 56 and 120). The classic study by McFeters et al. (104) demonstrated that the time for a 50% reduction in the population of several enteric bacteria in well water at 20°C ranged from 2.4 h for some Salmonella species to 26.8 h for some Shigella species. More recently, it was recognized that some enteric bacteria have the ability to enter a dormant state, referred to as viable but nonculturable, in which they can survive for long periods of time and can still be infectious at high doses (153). Difficulty recovering stressed bacteria from water samples may lead to false-negative results when microbial water quality is evaluated. The public health significance of enteric bacteria in this state is unknown. Some enteric bacteria have the ability to multiply in the aquatic environment under favorable nutrient and temperature conditions (68).

The majority of bacterial pathogens are removed or inactivated by standard water treatment practices. Removal capabilities of specific processes have been reviewed (99). Most of the recent drinking water outbreaks (1985 to 1992) associated with bacterial pathogens have been due to consumption of untreated groundwater. Enteric bacteria are relatively susceptible to disinfection. The CT (product of disinfectant concentration in milligrams per liter and contact time in minutes) values for 99% inactivation by free chlorine (pH 6 to 7) vary from 0.034 to 0.05 for enteric bacteria like E. coli at 5°C (31) to 15 and 28 for aquatic bacteria such as L. pneumophila at 20°C and Mycobacterium fortuitum, respectively (140).

Enteric Viruses

The enteric viruses are more recently recognized waterborne pathogens. Human rotaviruses, NV (and related viruses like Snow Mountain agent and Hawaii agent), human astroviruses, and human caliciviruses were first described in the early to mid-1970s. Depending on the virus, diagnosis of these infections may be by commercially available enzyme immunoassays for antigen or antibodies (group A rotavirus, enteric adenovirus types 40 and 41, hepatitis A virus), electron microscopy, tissue culture (*Enterovirus, Adenovirus, Astrovirus, Rotavirus*), or molecular methods such as PCR and probe hybridization (Norwalk-related viruses, *Astrovirus, Rotavirus, Adenovirus, Enterovirus*, hepatitis E virus [HEV]). Unlike the case for the enteric bacteria, concern about waterborne transmission of enteric viruses is generally limited to the strains that have humans as their natural reservoir. Although there are animal strains of many of these viruses, animal-to-human transmission is believed to be uncommon. The infectious dose of these agents is low, typically in the range of 1 to 10 infectious units. Human volunteer studies with group A human rotavirus estimated the median infectious dose to be between 5 and 6 focus-forming units (156).

Enteric viruses tend to be more persistent in the aquatic environment than most enteric bacteria; however, their survival depends on numerous physical, chemical, and microbial characteristics of the water as well as the virus type. The estimated time for about a 50% reduction of hepatitis A virus in groundwater at 25°C was approximately 14 days (142). Their prolonged survival times and small sizes enable viruses to move greater distances in soil and water. Unlike aquatic or enteric bacteria, enteric viruses cannot multiply in the environment. Cultivable enteric viruses have been detected in surface waters, groundwaters, and treated drinking waters in concentrations ranging from 647 PFU/liter (135) to 1 PFU/1,000 liters (124). Virus recovery from water samples is relatively poor, and currently many enteric viruses cannot be cultured in vitro. However, the development of new molecular amplification techniques and nucleic acid hybridization has improved the sensitivity of virus detection in water (76).

Inactivation and/or removal of enteric viruses by water treatment processes (reviewed in references 75, 124, and 140) varies by virus type and treatment conditions. Reported CT values for 99% inactivation of poliovirus 1 and *Rotavirus* at 5°C with free chlorine (pH 6 to 7) are 1.1 to 2.5 and 0.01 to 0.05, respectively (31). Experiments by Sobsey et al. (143) demonstrated that the CT values for 99.99% inactivation of hepatitis A virus (pH 6, 5°C) were 2.3 for dispersed virus and 29 for cell-associated virus. A human volunteer study that examined the infectivity of NV in water treated with different doses of chlorine indicated that NV remained infectious after a 3.75-mg/liter dose of chlorine and 30-min contact time (88).

Inadequate diagnostic technology has limited the detection of many enteric viruses in both clinical and environmental samples. Consequently, enteric viruses have not been frequently identified as the etiologic agents of waterborne disease outbreaks. From 1983 to 1992, there were 10 reported drinking water outbreaks associated with viral agents (hepatitis A or Norwalk-related viruses) in the United States (21, 23, 24, 26). These outbreaks involved over 6,000 total cases, and all were attributed to the ingestion of untreated or inadequately treated groundwater. In addition, group A rotavirus was implicated in a large waterborne outbreak in Colorado due to the mechanical breakdown of a chlorinator (72). Many outbreaks currently reported as acute gastroenteritis of unknown etiology are likely due to viral agents. Recent advances in the diagnosis and detection of enteric viruses through the application of molecular techniques should increase the recognition of waterborne outbreaks of viral infections.

Enteric Protozoa

Since 1981, enteric protozoa have become the leading cause of waterborne disease outbreaks for which an etiologic agent was determined. The enteric protozoa are relatively recently recognized waterborne pathogens. Human cryptosporidiosis was first described in 1976 (105, 118), and the first reported waterborne outbreak occurred in 1984 (38). Recent evidence indicates that *Cryptosporidium* species is the third most common enteric pathogen worldwide (37). Diagnosis is typically by immunofluorescence microscopy or enzyme immunoassay; however, cryptosporidiosis is probably greatly underdiagnosed. A prospective surveillance study in one state indicated that only 0.5 to 1% of all fecal samples submitted were examined for *Cryptosporidium* species and that 3.7 to 8.5% of stool samples contained oocysts (7). Although both trophozoites and cysts or oocysts are shed in feces, the cysts are the infective form, and as for the enteric viruses, the infectious dose for these agents is low. A recent study of *Cryptosporidium parvum* in human volunteers determined that the median infectious dose was 132 oocysts (42). Water contamination by both human and animal feces is an important mode of transmission for *Giardia* and *Cryptosporidium* species. Humans are the only host for *Entamoeba histolytica*.

The thick-walled protozoan cysts and oocysts are environmentally resistant, but there is little information on their survival in water (132). Their sizes (7 to 14 μm for *Giardia* cysts and 4 to 6 μm for *Cryptosporidium* oocysts) tend to limit extensive migration through soil; however, groundwater has become contaminated by protozoan cysts during flooding. Methods for the recovery and detection of *Giardia* and *Cryptosporidium* cysts or oocysts in water are cumbersome and relatively inefficient and do not distinguish between viable and nonviable cysts or oocysts. Surveys of raw and treated water supplies indicate that the occurrence of *Cryptosporidium* oocysts is widespread. Two surveys of almost 300 surface water supplies in the United States revealed that 55 to 77% of surface water samples had *Cryptosporidium* oocysts (130, 133). Water analysis at 66 U.S. and Canadian surface water treatment plants demonstrated that up to 27% of treated drinking water samples had low levels of *Cryptosporidium* oocysts (98). However, current methods may underestimate the occurrence of these organisms in water because of poor recovery efficiency or overestimate the risk to human health because of inability to assess the viability of the cysts or oocysts in a way that relates to their potential for human infectivity (132). Studies by LeChevallier and Norton suggest that approximately 90% of cysts and 66% of oocysts detected in raw water may not be viable (97). Like enteric viruses, enteric protozoa cannot multiply in the environment.

The cysts and oocysts of enteric protozoa are relatively resistant to chlorine disinfection. CT values for *Giardia lamblia* cysts for 99% inactivation at pH 6 to 7 by free chlorine range from 47 to 150 (31). The CT value for 99% inactivation of *Cryptosporidium parvum* by free chlorine (pH 7) at 25°C is estimated to be 7,200 (31). Therefore, removal of these organisms by water treatment processes depends on effective coagulation-flocculation, sedimentation, and filtration. A recent case study of three surface water treatment facilities that used coagulation-flocculation, sedimentation, and filtration observed log_{10} reductions of 2.24 to 2.78 for *G. lamblia* and 2.3 to 2.45 for *Cryptosporidium parvum* (97).

As diagnostic methods for *Giardia* and *Cryptosridium* species improved, the recognition of their role in waterborne disease outbreaks increased dramatically. From 1991 to 1992, 21% of outbreaks associated with drinking water were attributed to parasitic agents (26). In 1993, a *Cryptosporidium* sp. was implicated in the largest recorded waterborne disease outbreak in the United States, in which it was estimated that over 400,000 people were infected (100). The majority of drinking water outbreaks attributed to enteric protozoa have been associated with surface water supplies that were either unfiltered or subject to inadequate flocculation and filtration processes. *Cryptosporidium* outbreaks tend to have high attack rates, and fecal contamination from cattle is usually suspected to be the source of the oocysts (132). There have been no reported waterborne outbreaks of *Entamoeba histolytica* in the United States since 1971 (34, 35).

EMERGING WATERBORNE PATHOGENS

Emerging infectious diseases have been defined as those whose incidence in humans has increased within the past two decades or threatens to increase in the near future (77). There are a number of newly recognized infectious agents that have recently been associated with outbreaks of waterborne disease or appear to have the potential for waterborne transmission.

Recently Recognized Pathogens Associated with Waterborne Disease Outbreaks
Norwalk-Like Viruses
NV and related small round structured viruses are the leading cause of epidemic viral gastroenteritis in older children and adults in the United States. Numerous NV and small round structured virus outbreaks linked to drinking water, recreational water, ice, shellfish, various food items, and environmental contamination have been documented (70, 85, 86, 89, 110). NV has been proposed as a member of the *Caliciviridae* family (83), and recent sequence analyses indicate that this family includes Snow Mountain agent, Hawaii agent, and Taunton agent (2). Insufficient diagnostic technology has limited the study of the role of these viruses in both epidemic and endemic gastroenteritis. The recent cloning and sequencing of NV (80) and Snow Mountain agent (155) has allowed the development of sensitive new molecular diagnostic methods (39, 81–83).

Seroprevalence data suggest that epidemic and endemic NV-associated infection may be common. In the United States, 50 to 70% of adults have NV antibodies by the fifth decade of life (9, 59). A recent study in England reported antibody prevalence rates of 48% among children 12 to 23 months old, 70% among those 5 to 9 years old, 81% among adults 20 to 29 years old, and over 90% by the fifth decade of life (57). These high rates of antibody prevalence indicate that much of the population is frequently exposed to NV and related viruses and may experience recurrent NV-associated gastroenteritis, since antibodies do not appear to confer protection from illness.

Over 13,500 documented cases of waterborne disease associated with these viruses were reported between 1971 and 1990 (1). Most small round structured virus outbreaks, however, are probably not recognized. In 1982, Kaplan et al. (86) estimated, on the basis of serologic evidence from 74 outbreaks investigated by the (then) Centers for Disease Control from 1976 to 1980, that 42% of acute nonbacterial gastroenteritis may be due to NV and related viruses. Examining outbreak characteristics, Kaplan et al. (85) suggested that NV and related viruses may be responsible for 23% of waterborne outbreaks of acute gastroenteritis in the United States.

E. coli O157:H7
Enterohemorrhagic *E. coli* O157:H7 is a pathogenic strain of *E. coli* that produces two potent toxins. This organism causes bloody diarrhea, and 2 to 7% of infections result in hemolytic uremic syndrome, in which the erythrocytes are destroyed and the kidneys fail (25). In many parts of the United States and Canada, *E. coli* O157:H7 is the second or third most commonly isolated enteric bacterial pathogen (58). Children and the elderly are most susceptible to hemolytic uremic syndrome complications. This severe disease was first recognized in 1982 and has the highest mortality rate of all waterborne diseases in the United States. The infectious dose is believed to be low, as for shigellae, and the incubation period is 12 to 60 h. The reservoir appears to be healthy cattle, and transmission often occurs by ingestion of undercooked beef or raw milk. Person-to-person transmission is important among families and in child care centers. Diagnosis is by culture followed by a commercially available latex agglutination test for *E. coli* O157 antigen (58).

Two waterborne outbreaks of *E. coli* O157:H7 have been reported in the United States. One outbreak was associated with drinking water in a Missouri community in 1989. Of the 243 people affected, one-third had bloody diarrhea, 32 were hospitalized, 2 had hemolytic uremic syndrome and 4 died (24). The *E. coli* strain was isolated from clinical specimens but not from water samples. Unchlorinated well water and breaks in the water distribution system were considered to be contributing factors. The other waterborne outbreak, with 80 cases, was associated with recreational water exposure to a lake in Oregon in 1991 (26). Fecal contamination from swimmers and poor water exchange in the lake were believed to be contributing factors. Prolonged survival of *E. coli* O157:H7 in water has been reported by Geldreich et al. (55), who observed only a 2-log-unit reduction after 5 weeks at 5°C.

Cyanobacteria
Cyanobacteria (blue-green algae) occur naturally in fresh and brackish waters worldwide. Although these are not infectious agents, some species produce toxins during algal blooms, which are triggered by nutrient enrichment from natural waters, agricultural fertilizer runoff, or domestic or industrial effluents (32). Approximately 25 species of cyanobacteria have been associated with adverse health effects (6). Marine toxic forms are in the genera *Lyngbya*, *Schizothrix*, and *Oscillatoria*. Toxic freshwater cyanobacteria are members of the genera *Microcystis*, *Anabaena*, *Aphanizomenon*, *Nodularia*, and *Oscillatoria* (17). Toxic blooms have been reported in many parts of the United States, Canada, Europe, southern Africa, Asia, Australia, and New Zealand (17, 32). The conditions that influence the toxicity of a bloom are not known (43). However, temperature is believed to be one key factor (17). One survey in the United Kingdom found that 75% of cyanobacterial blooms contained toxins (6).

Cyanobacterial toxins are of three main types: lipopolysaccharide endotoxins, hepatotoxins, and neurotoxins. Acute health effects in humans include gastroenteritis, liver damage, nervous system damage, pneumonia, sore throat, earache, and contact irritation of skin and eyes (32, 146). The potential chronic health effects of long-term exposure to cyanobacterial toxins in drinking water is unknown. It has been suggested that high rates of liver cancer in parts of China may be linked to cyanobacterial hepatotoxins in drinking water (16). A 1976 outbreak of intestinal illness in Pennsylvania was associated with a cyanobacterial bloom in municipal water supply and affected 62% of the population (17). In addition, there are several reports of adverse health effects related to contact with recreational water in the United States (17).

Although cyanobacterial poisoning has been long recognized, in recent years there have been numerous reports of cyanobacterial poisoning associated with surface water ingestion and contact in Australia (44, 67, 144). One outbreak of hepatoenteritis in the Palm Island Aboriginal settlement on the Australian northeast coast in 1979 affected 138 children and 10 adults, the majority of whom required hospitalization (32, 66). The outbreak was linked to a dense algal bloom in the drinking water reservoir for the island that was treated with copper sulfate, which caused lysis of the cyanobacteria and consequent release of toxins. A recent case-control study in southern Australia reported that ingestion of chlorinated river water was a significant risk factor for gastrointestinal symptoms compared with ingestion of rain water and that the weekly number of gastroenteritis cases in the study area was correlated with mean log cyanobacterial cell counts in the river water (45). An earlier epidemiological study in New South Wales, Australia, examined levels of hepatic enzymes in routine blood specimens submitted to a local hospital and found elevated levels of gamma-glutamyltransferase (indicating damage to liver cell membranes) among city residents during a heavy algal bloom (*Microcystis aeruginosa*) in the municipal water reservoir (51). These levels were significantly higher than those measured in specimens collected from city residents 1 month before and after the bloom and compared with levels in country residents who used other water supplies.

Control of cyanobacteria is problematic since several studies indicate that the toxins can remain potent for days after the organisms have been destroyed by copper sulfate or chlorination (45, 66). On the basis of toxicity data from mouse bioassays, the Engineering and Water Supply Department of South Australia developed interim guidelines for acceptable numbers of cyanobacteria in water supplies (45). However, further research is needed on the acute and chronic toxicity of cyanobacterial toxins, and suitable methods need to be developed for monitoring the types and concentrations of cyanobacterial toxins in natural and treated water supplies (43).

Cyclospora cayetanensis

C. cayetanensis, formerly called cyanobacterium-like bodies or big cryptosporidium, was identified as a new protozoan pathogen of humans in 1993 (121). A member of the family *Eimeriidae*, it was first identified by Soave and colleagues in 1986 (139) in stools from patients who had lived or traveled in developing countries and is frequently found in AIDS patients with prolonged diarrhea (121). Under light microscopy, it appears as a refractile sphere 8 to 10 μm in diameter and has been associated with prolonged, self-limited watery diarrhea with an average duration of 40 days. This organism occurs in tropical climates worldwide and has been identified as the cause of diarrheal disease outbreaks in North, Central, and South America, Caribbean countries, Southeast Asia, and eastern Europe (93).

Evidence for waterborne transmission comes from a case-control study of a diarrhea outbreak among foreign residents in Nepal, where consumption of untreated water was identi-fied as a risk factor (137). The Nepal study and two prospective cohort studies of Peruvian children (121) indicate a seasonal pattern of infection, with peak incidence occurring in warm summer months. In the United States, an outbreak of *Cyclospora* diarrhea occurred among the house staff of a Chicago hospital in the summer of 1990 (24). Contaminated open-air, rooftop water storage tanks that supplied the house staff dormitory were believed to the source of the outbreak. However, *Cyclospora* species could not be detected in water samples from the tanks. A recent case report described a *Cyclospora* infection in an immunocompetent host who developed prolonged diarrhea after exposure to sewage-contaminated water in his home (64). Sewage from a neighboring dairy backed up into the basement of the patient's home, and *Cyclospora* organisms were detected in sewage effluent samples taken from the pipe that served the dairy.

Recently Recognized Waterborne Pathogens in Developing Countries

HEV

HEV is the only known agent of enterically transmitted non-A, non-B infectious hepatitis, although some evidence suggests that there may be others (145). A single serotype of virus that has single-stranded, positive-sense RNA with three open reading frames has been described (145). HEV is believed to be either a new RNA virus or a member of the *Caliciviridae* family. The recent development of serologic tests for HEV antibody based on recombinant DNA technology has enabled further study of the epidemiology of this virus, which is clinically and epidemiologically similar to hepatitis A virus (145). However, in contrast to hepatitis A virus, the majority of cases in epidemics occur among young adults (15 to 40 years of age), and there is a high case fatality rate (up to 30%) among pregnant women (36). Large outbreaks, involving thousands of cases, have been reported in developing areas in Africa, Asia, and Mexico and have been linked to fecally contaminated water and inadequate chlorination. In 1991, the largest documented waterborne HEV outbreak affected an estimated 79,000 persons in Kanpur, India, and was associated with contaminated surface water (115). Person-to-person transmission has also been observed. HEV RNA has been detected in samples from sewage treatment plants in India by using reverse transcriptase PCR and nucleic acid probe hybridization (84). Currently, there are no data on the inactivation of this virus in water by disinfection.

Group B Rotavirus

Group B rotavirus was first reported in connection with a waterborne outbreak in China in 1984 (74) and differs from the group A rotavirus strains that commonly cause pediatric diarrhea. Group B rotavirus infections occur more frequently in adults than children, are associated with severe, cholera-like illness, and have been reported mainly in China. By electron microscopy, the virus is morphologically similar to group A rotavirus but is antigenically distinct. Diagnosis is by electron microscopy and genome electrophoresis or by enzyme immunoassay. Seroprevalence surveys for group B antibody have reported rates of up to 41% positive in China, 5 to 18% positive in Thailand, Burma, and Hong Kong, 1 to 12.5% positive in the United Kingdom, United States, and Canada, and 10% positive in Kenya (11). More than 1 million cases were reported in China in 1982 to 1983 (73). Many outbreaks involving tens of thousands of cases were attributed to fecally contaminated water (74). However, the detection of group B rotaviruses in water has not yet been reported.

V. cholerae O139

V. cholerae O139 is a new toxigenic strain of epidemic V. cholerae which has been called the Bengal strain. This is the first non-O1 V. cholerae ever reported to be associated with epidemic cholera, and it is causing a new pandemic in Asia which began in India in October 1992 (29) and is currently spreading into the Middle East. The clinical illness is indistinguishable from cholera caused by V. cholerae O1. However, this new epidemic is affecting persons of all ages in areas where most of the population, except for young children, have acquired immunity to V. cholerae O1. As with other strains of V. cholerae, transmission occurs through fecally contaminated water and food.

Recently Recognized Frank or Opportunistic Pathogens Suspected To Cause Waterborne Disease

Enteric Viruses

Several recently recognized enteric viruses are known to be transmitted by drinking water and contaminated shellfish, as well as via cold foods and person to person. Human astroviruses, first described in 1975 by Appleton and Higgins (3), and human caliciviruses, first described in 1976 by Madeley and Cosgrove (101), are both RNA viruses characterized as small round structured viruses on the basis of their appearance by electron microscopy. Both have been associated with acute gastroenteritis, mainly in children and the elderly, and have a worldwide distribution. The predominant clinical features are vomiting and diarrhea. For both of these viruses, the incubation period is between 1 and 3 days, and symptoms last 1 to 4 days. Outbreaks typically occur in institutional settings with children or the elderly (22). Diagnosis for both of these viruses is generally by electron microscopy. However, immunofluorescence (96), enzyme immunoassay (69, 109), nucleic acid probe hybridization (109), and PCR (119) techniques have recently been developed to detect astrovirus in fecal specimens. There are no reports of the detection of these viruses in water. Evidence of waterborne transmission is from epidemiologic analyses of outbreaks.

Several other viruses are recognized enteric pathogens or putative pathogens; however, to date there is no evidence of waterborne transmission of these organisms. Enteric adenoviruses (serotypes 40 and 41, also known as subgenus F) are DNA viruses associated with about 5 to 12% of pediatric diarrhea and median durations of 8.6 and 12.2 days, respectively (151). Diagnosis is by commercially available enzyme immunoassay or by PCR (151). There is no evidence of waterborne transmission of enteric adenoviruses. However, nonenteric adenoviruses (serotypes 3, 7, 1, 4, and 14) have been associated with recreational water outbreaks of pharyngoconjunctival fever and have been isolated from a sewage outlet at a lake and from a swimming pool (53).

Coronaviruses are pleomorphic, enveloped RNA viruses that are well established causes of diarrhea in animals. They were first observed in feces of persons with gastroenteritis by electron microscopy in 1975 (19, 20, 103) but since then have also been frequently detected in the feces of healthy persons. Hence, there continues to be doubt about their etiologic role in human diarrhea. Except for one strain, these viruses have not been propagated in vitro, and diagnosis depends on observation of characteristic morphological features by electron microscopy (18). Epidemiologic evidence suggests that fecal-oral transmission and personal hygiene may be key factors in transmission, since several studies

noted that the highest prevalence rates were among populations with low socioeconomic status and poor personal hygiene (18). One study in Lesotho reported prevalence in children of 30 to 68% but found no difference between the prevalence in a village with an "improved water supply" and a village with a "traditional contaminated water" supply (90).

Toroviruses are enveloped, single-stranded, positive-sense RNA viruses in the Coronaviridae family that are recognized enteric pathogens of cattle and horses (161). New diagnostic methods have made it possible to detect toroviruses in stool samples from children and adults with diarrhea (94, 95), but they are not established diarrhea pathogens in humans. Picobirnaviruses are RNA viruses that are also known to be etiologic agents of diarrhea in animals. Picobirnaviruses have been detected in fecal specimens from human cases of diarrhea in Brazil (127) and from a cohort of AIDS patients in the United States who had diarrhea (60). Pestiviruses, single-stranded, positive-sense RNA viruses in the Flaviviridae family, were identified in 23% of stool specimens from Arizona Indian children under 2 years of age with diarrhea, compared with 3% of healthy controls (164). Seroprevalence studies of serum from Arizona, Maryland, and Peru indicate that 30 to 50% of children and adults had antibodies to pestivirus and that peak exposure occurred before 2 years of age (165).

There are no reports of waterborne transmission of enteric adenovirus, coronavirus, torovirus, picobirnavirus, or pestivirus. Methods to detect them in water have not been developed, and there is no information about the survival of these viruses in water.

Enteric and Aquatic Bacteria

Aeromonas species are commonly found in water and soil. Reported densities in water range from 10^2 to 10^3/ml in river water to 1 to 100/liter in groundwater. High population densities appear to be related to fecal pollution and temperature, and aeromonads proliferate in domestic and industrial wastewaters (134). There is some evidence to suggest that a high proportion of environmental isolates may produce enterotoxins (102), and several reports have suggested an association between gastroenteritis and aeromonads in drinking water (14, 15, 134). A 2-year clinical study in Iowa concluded that three strains of Aeromonas were capable of causing diarrhea and that consumption of untreated water was a risk factor for Aeromonas infection (113). Studies in London showed a correlation between water and fecal isolates of Aeromonas sobria (117). High summer prevalence of nosocomial A. hydrophila infection was linked to high counts in hospital water storage tanks in France (128). However, other studies in The Netherlands and London found little similarity between aeromonads isolated from diarrheal feces and those found in drinking water (65, 107). Concern about the possible health effects from these organisms in The Netherlands has led to the establishment of drinking water guidelines of <20 CFU/100 ml for drinking water leaving the treatment plant and <200 CFU/100 ml for drinking water in the distribution system (149).

A wide range of atypical (nontuberculosis) mycobacteria occur in the environment. Generally, these organisms cause illness mainly in immunosuppressed populations. Although opportunistic mycobacteria can infect almost any site in the body, they are most commonly associated with pulmonary disease, cervical lymphadenopathy and localized skin and soft tissue infections (79). Cutaneous lesions have been associated with recreational water exposure to Mycobacterium *marinum* (also called *Mycobacterium balnei*) (41, 79). The incidence of *Mycobacterium avium* complex infection, which typically causes pulmonary and disseminated disease, in human immunodeficiency virus-infected patients has dramatically increased, causing further speculation about the possible role of water in the transmission of this agent (138). A recent study that compared clinical mycobacteria isolates with environmental isolates from hospital hot water systems suggested that exposure to these water sources was the transmission route for several groups of AIDS patients (150). Other species (*Mycobacterium kansasii* and *Myobacterium xenopi*) have been isolated from domestic and hospital water supplies; however, the health significance of this finding is uncertain (79).

H. pylori, formerly referred to as *Campylobacter pylori*, is associated with indigestion and abdominal pain. Chronic infection may result in peptic ulcers and gastric cancer. *H. pylori* infections occur throughout the world, and the prevalence of infection increases with age. Individuals in developing countries are more likely to become infected earlier in life, and by adulthood, infection rates range from approximately 40% in developed countries to 80% in developing countries. If untreated, infection becomes chronic and probably persists for life (33).

Fecal-oral transmission of *H. pylori* infection has been suggested by several studies that implicated crowding, socioeconomic status, and consumption of raw, sewage-contaminated vegetables as risk factors for infection (71, 106, 108). Studies in Peru have identified the type of water supply (municipal or community wells) as a risk factor for infection and found that water source appeared to be a more important risk factor than socioeconomic status (91). A seroprevalence survey of 245 healthy children in Arkansas did not find a relation between *H. pylori* seropositivity and type of water supply (municipal or well) (52). However, it is likely that the levels of fecal contamination in the Peruvian water sources were substantially higher than those in Arkansas.

H. pylori has not yet been isolated from an environmental source. Laboratory studies (136, 159) demonstrated that *H. pylori* can survive in freshwater and sterile, distilled water (7°C) for 10 and 14 days, respectively. These studies also found evidence that *H. pylori* can survive for prolonged periods as viable, nonculturable coccoid bodies. Klein et al. (91) reported difficulty detecting *H. pylori* from water samples because the culture plates were overgrown before the 3 to 4 days required to isolate *H. pylori*. Enroth and Engstrand (47) have recently reported the development of PCR methods to detect *H. pylori* in water.

Protozoa, Fungi, and Algae

Microsporidia is a general term that describes a large group of primitive, obligate, intracellular protozoa commonly found in animals. Species of five genera, *Nosema*, *Encephalitozoon*, *Pleistophora*, *Enterocytozoon*, and *Septata*, are known to cause human infections (54). *Enterocytozoon bieneusi* and *Septata intestinalis* primarily parasitize the intestine and can cause persistent diarrhea. The vast majority of reported cases of microsporidial infections have occurred among persons infected with human immunodeficiency virus (13), and recent epidemiological studies suggest that *Enterocytozoon bieneusi* is an important cause of chronic diarrhea in patients with AIDS (157).

Infection is acquired through ingestion of small (1- to 2-μm) spores with thick walls that make them environmentally resistant and allow them to remain infective for up to 4 months in the environment (154). Transmission of microsporidia is believed to be primarily fecal-oral (13). Although there is currently no evidence of waterborne transmission, there is one report of repeated isolation of microsporidia spores from ditch water samples collected in Florida during a 1-year period (4). Spore concentrations of up to 3,000/ml of water were observed, and two of the five genera identified in the samples (*Nosema* and *Pleistophora*) were groups that include human pathogens.

Waterborne Disease of Unknown Etiology

Waterborne disease surveillance in the United States indicates that no etiologic agent can be determined for more than 50% of reported waterborne disease outbreaks (21, 23, 24, 26), possibly in part because of lack of appropriate clinical and/or environmental samples or limited diagnostic techniques for many of the newly recognized pathogens. However, there are some outbreaks that, despite thorough investigation, could not be attributed to any known etiologic agent and suggest the existence of unrecognized agents of waterborne disease.

"Brainerd diarrhea" was first described in an outbreak with 122 cases that occurred in Brainerd, Minnesota, in 1983 (122). The illness was characterized by chronic diarrhea with acute onset, marked urgency, lack of systemic symptoms, and failure to respond to antimicrobial therapy. The distinctive feature of the illness is the average duration of 12 to 18 months. In the Brainerd outbreak, transmission was linked to raw milk that had passed through hoses rinsed with poor-quality water. A subsequent outbreak in Illinois in 1987 was clearly associated with the consumption of untreated well water which had sporadic problems with fecal coliform counts (123). Several other possible outbreaks have been suspected in Texas in 1985 and five other states, as well as among over 200 tourists who participated in group tours to the Galapagos Islands in 1991 and 1992. Despite intense microbiological analyses of several outbreaks, no etiologic agent for this syndrome has been identified.

PREVENTION AND CONTROL OF WATERBORNE DISEASE

Historical surveillance data (35) and epidemiological and microbiological studies of water and health have documented the relationship between improvements in water quality and water treatment practices and a reduction in morbidity and mortality associated with waterborne diseases. Reviewing studies of water quality, water quantity, and hygiene, Esrey et al. (49) concluded that improvements in microbiological water quality resulted in a 16% median reduction of diarrhea morbidity (range, 0 to 90%) and that improvements in both water quality and availability resulted in a 37% median reduction of diarrhea morbidity (range, 0 to 82%). Prevention and control of waterborne diseases require accurate and rapid methods to measure microbiological water quality and to identify and evaluate risk factors for waterborne disease.

Measurements of Microbiological Water Quality

Detection of Infectious Agents in Water

Many waterborne pathogens are difficult to detect and/or quantify in water, and for most of the newly recognized agents, methods to detect them in environmental samples have still to be developed. For enteric organisms, the concentrations are much lower in water than in clinical specimen; thus, their detection in water starts with some type of concentration process such as filtration. This is followed by a process to recover the pathogen from the filter and then by an enrichment or amplification process either by culture or molecular biology methods. Typically the recovery efficiencies of these procedures are low, making it difficult to estimate the original concentration of the infectious agent in the water. Also, some methods, such as those used to detect *Giardia* cysts and *Cryptosporidium* oocysts, do not give an indication of the viability or infectivity of the organisms. Furthermore, many of these laboratory techniques are limited to specialized research or reference laboratories and are not done on a routine basis.

In most investigations of waterborne disease outbreaks, water is identified as the vehicle of transmission by epidemiological evidence rather than by the detection of the infectious agent in water samples. Of the 32 waterborne infectious disease outbreaks reported in the United States between 1991 and 1992, the etiologic agent was detected in the water in only four outbreaks: *Shigella* species (one outbreak), *Giardia* species (two outbreaks), and *Cryptosporidium* species (one outbreak) (26). This may be because the contamination of the water supply was temporary and the infectious agent died off or was flushed out of the water system before the outbreak was recognized and appropriate water samples were collected. The longer the incubation period of the infectious agent, the longer it may be before the outbreak is recognized and water is suspected as the vehicle. The size and timing of the contamination event, the lag time until recognition of the outbreak, the survival characteristics and transport patterns of the agent, and the sensitivity and efficiency of the laboratory methods will affect the likelihood that an infectious agent will be detected in water (131).

Microbial Indicator Organisms

Environmental Protection Agency and World Health Organization standards for microbiological water quality are expressed in terms of total coliforms and fecal coliforms. These are groups of bacteria excreted by healthy humans and animals that serve as indicators of fecal contamination (148, 162). Laboratory tests for total and fecal coliforms in water are much easier to perform than tests to detect pathogenic microorganisms in water. Because these indicator organisms are excreted in high numbers by all individuals, the concentrations of these indicators are likely to be higher and a more constant fraction of the community fecal waste pool. By contrast, specific pathogens are excreted only by infected individuals, and their numbers in the community fecal waste pool depend on the excretion level of the particular pathogen and on the number of infected individuals in the community.

Ideally, microbial indicators should provide a measure of health risk associated with the ingestion or contact with water. Total and fecal coliforms have many limitations as predictors of risk of waterborne disease. Because of their shorter survival times in water and their greater susceptibility to water treatment processes, these indicator organisms tend to be poor models for enteric protozoa and viruses. Outbreaks of waterborne disease, especially protozoal outbreaks such as the Milwaukee epidemic (100), have been associated with water which met total and fecal coliform standards. Moreover, there are nonfecal sources for these indicator organisms, and in contrast to most enteric pathogens, total and fecal coliforms may multiply in aquatic environments with sufficient nutrients and optimal temperatures. Such characteristics may result in false-positive

reports of water contamination. The performance of total and fecal coliforms as reliable indicators of tropical water quality is especially problematic and has been investigated and reviewed by Hazen (68). This is of particular concern in tropical developing countries where there are a greater number of untreated, contaminated water sources and high morbidity and mortality associated with waterborne disease (68).

Many alternative indicator organisms have been investigated, and their advantages and limitations have been reviewed (87). *E. coli* and enterococci were included in the recent revision of the Environmental Protection Agency recreational water standards (147). *Clostridium perfringens* and male-specific coliphage have been proposed as potential indicators of drinking water quality that may better model the survival and disinfection resistance of enteric protozoa and viruses (125). Epidemiological and microbiologic studies have demonstrated that *E. coli* and enterococci are better indicators of tropical drinking water quality than fecal coliforms (111). There is increasing recognition that no single organism can serve as an adequate indicator for all types of water and all routes of exposure.

Multiple-Barrier Approach

In developed countries, waterborne disease prevention and control is based on a multiple-barrier approach that involves source water protection, water treatment, and distribution system management and protection (1). Waterborne disease outbreaks in the United States usually involve (i) source contamination and the breakdown of one or more of the treatment barriers (disinfection or filtration), (ii) contamination of the distribution system, or (iii) the use of untreated water. The multiple-barrier approach emphasizes identification of all available barriers, assessing the degree of vulnerability of each barrier to the passage of pathogens, recognizing and anticipating conditions under which pathogen risk increases, and maintaining barriers at high levels of effectiveness (1). For waterborne diseases associated with aquatic microorganisms, approaches to prevention and control may depend on treatment technology or controlling exposure to untreated waters.

Risk Assessment Approaches

Risk assessment approaches have been useful to systematically identify, analyze, quantify, and characterize the risk of specific waterborne illnesses (141). These models are based on field data on the occurrence of specific microorganisms in raw and treated water supplies, experimental data on removal or inactivation by various water treatment processes, and experimental dose-response data (61, 129). However, the shape of the dose-response curve, especially in the low-dose region representative of waterborne exposure, is ill defined. Furthermore, it is difficult to model variation in microbial virulence factors and host-specific characteristics, such as age and immune status, that may affect individual exposure and susceptibility to infection and disease. Finally, for infectious agents with multiple transmission routes, it may be difficult to determine the attributable risk associated with waterborne transmission compared with other routes of transmission, especially in areas where waterborne diseases are endemic (12).

Recommendations and Research Needs

Although the incidence of waterborne diseases has been greatly reduced in areas with effective water treatment, control of waterborne transmission of infectious agents continues to be an important challenge for public health research. Methods to detect many of these agents in clinical and environmental samples need to be developed or improved. Information on the occurrence and persistence of enteric and aquatic pathogens in various types of water supplies needs to be updated, and the factors that contribute to virulence and waterborne transmission need to be better characterized. For many waterborne pathogens, we need to improve our understanding of the effectiveness of various water treatment processes and disinfectants to remove and/or inactivate these microorganisms. Finally, better surveillance of waterborne diseases would lead to earlier recognition and investigation of epidemic and endemic waterborne disease. Infection with G. *lamblia* and *E. coli* O157:H7 are now reportable diseases in many states. It has been recommended that testing for *Cryptosporidium* species become part of routine stool examination and that public health officials make *Cryptosporidium* infection a reportable condition (100). Increased awareness among health professionals of the symptoms associated with more recently described waterborne illnesses, such as *Cyclospora* or cyanobacterial poisoning, may lead to greater reporting of these conditions and provide opportunities to study their waterborne transmission. Because the health risks posed by many waterborne infectious agents are still unknown, water utilities and public health professionals need to work together to maintain vigilance for any indication of waterborne transmission of disease in their communities.

REFERENCES

1. **American Water Works Association.** 1994. Preventing waterborne disease: is your system at risk? Satellite Teleconference, April 8.
2. **Ando, T., S. S. Monroe, J. R. Gentsch, Q. Jin, D. C. Lewis, and R. I. Glass.** 1995. Detection and differentiation of antigenically distinct small round-structured viruses (Norwalk-like viruses) by reverse transcription-PCR and Southern hybridization. *J. Clin. Microbiol.* **33:**64–71.
3. **Appleton, H., and P. G. Higgins.** 1975. Viruses and gastroenteritis in infants. *Lancet* **i:**1297.
4. **Avery, S. W., and A. H. Undeen.** 1987. The isolation of microsporidia and other pathogens from concentrated ditch water. *J. Am. Mosq. Control Assoc.* **3:**54–58.
5. **Batik, O., G. F. Craun, R. W. Tuthill, and D. F. Kraemer.** 1980. An epidemiologic study of the relationship between hepatitis A and water supply characteristics and treatment. *Am. J. Public Health* **70:**167–168.
6. **Baxter, P. J.** 1991. Toxic marine and freshwater algae: an occupational hazard? *Br. J. Ind. Med.* **49:**505–506.
7. **Binford, L. R. F., M. A. Pentella, and B. H. Kwa.** 1991. A survey to determine the presence of *Cryptosporidium* infection associated with diarrhea in west central Florida, 1985–1992. *Fla. J. Public Health* **3:**18–21.
8. **Birkhead, G., and R. L. Vogt.** 1989. Epidemiologic surveillance for endemic *Giardia lamblia* infection in Vermont. *Am. J. Epidemiol.* **129:**762–768.
9. **Blacklow, N. R., G. Cukor, M. K. Bedigian, P. Echeverria, H. B. Breenber, D. S. Schrieber, and J. S. Trier.** 1979. Immune response and prevalence of antibody to Norwalk enteritis virus as determined by radioimmunoassay. *J. Clin. Microbiol.* **10:**903–909.
10. **Bradley, D. J.** 1977. Health aspects of water supplies in tropical countries, p. 3–17. *In* R. G. Feachem, M. G. McGarry, and D. D. Mara (ed.), *Water, Wastes and Health in Hot Climates.* John Wiley & Sons, London.
11. **Bridger, J. C.** 1994. Non-group A rotaviruses, p.

369–407. *In* A. Z. Kapikian (ed.), *Viral Infections of the Gastrointestinal Tract.* Marcel Dekker, New York.
12. **Briscoe, J.** 1984. Intervention studies and the definition of dominant transmission routes. *Am. J. Epidemiol.* **120:**449–455.
13. **Bryan, R. T., A. Cali, R. L. Owen, and H. C. Spencer.** 1991. *Microsporidia*: opportunistic pathogens in patients with AIDS, p. 1–26. *In* T. Sun (ed.), *Progress in Clinical Parasitology.* Field and Wood, Philadelphia.
14. **Burke, V., J. Robinson, M. Gracey, D. Peterson, N. Meyer, and V. Haley.** 1984. Isolation of *Aeromonas* spp. from an unchlorinated domestic water supply. *Appl. Environ. Microbiol.* **48:**367–370.
15. **Burke, V., J. Robinson, M. Gracey, D. Peterson, and K. Partridge.** 1984. Isolation of *Aeromonas hydrophila* from a metropolitan water supply: seasonal correlation with clinical isolates. *Appl. Environ. Microbiol.* **48:**361–366.
16. **Carmichael, W. W.** 1994. The toxins of cyanobacteria. *Sci. Am.* **270:**78–86.
17. **Carmichael, W. W., C. L. A. Jones, N. A. Mahmood, and W. C. Theiss.** 1985. Algal toxins and water-based diseases. *Crit. Rev. Environ. Control* **15:**275–313.
18. **Caul, E. O.** 1994. Human coronaviruses, p. 603–625. *In* A. Z. Kapikian (ed.), *Viral Infections of the Gastrointestinal Tract.* Marcel Dekker, New York.
19. **Caul, E. O., and S. K. R. Clarke.** 1975. Coronavirus propagated from patient with non-bacterial gastroenteritis. *Lancet* **ii:**953–954.
20. **Caul, E. O., W. K. Paver, and S. R. K. Clarke.** 1975. Coronavirus particles in faeces in patients with gastroenteritis. *Lancet* **i:**1192.
21. **Centers for Disease Control.** 1988. Water-related disease outbreaks, 1985. *Morbid. Mortal. Weekly Rep.* **37:**15–24.
22. **Centers for Disease Control.** 1990. Viral agents of gastroenteritis: public health importance and outbreak management. *Morbid. Mortal. Weekly Rep.* **39:**1–24.
23. **Centers for Disease Control.** 1990. Waterborne disease outbreaks, 1986–1988. *Morbid. Mortal. Weekly Rep.* **39:**1–13.
24. **Centers for Disease Control.** 1991. Waterborne-disease outbreaks, 1989–1990. *Morbid. Mortal. Weekly Rep.* **40:**1–21.
25. **Centers for Disease Control and Prevention.** 1993. *Preventing Foodborne Illness: Escherichia coli O157:H7.* Centers for Disease Control and Prevention, Atlanta.
26. **Centers for Disease Control and Prevention.** 1993. Surveillance for waterborne disease outbreaks—United States, 1991–1992. *Morbid. Mortal. Weekly Rep.* **42:**1–22.
27. **Charoenca, N., and R. S. Fujioka.** 1993. Assessment of *Staphylococcus* bacteria in Hawaii's marine recreational waters. *Water Sci. Technol.* **27:**283–289.
28. **Charoenca, N., and R. S. Fujioka.** 1995. Association of staphylococcal skin infections and swimming. *Water Sci. Technol.* **31:**11–17.
29. **Cholera Working Group, International Centre for Diarrhoeal Disease Research, Bangladesh.** 1993. Large epidemic of cholera-like disease in Bangladesh caused by *Vibrio cholerae* O139 synonym Bengal. *Lancet* **342:**387–390.
30. **Chute, C. G., R. P. Smith, and J. A. Baron.** 1987. Risk factors for endemic giardiasis. *Am. J. Public Health* **77:**585–587.
31. **Clark, R. M., C. J. Hurst, and S. Regli.** 1993. Costs and benefits of pathogen control in drinking water, p. 181–198. *In* G. F. Craun (ed.), *Safety of Water Disinfection: Balancing Chemical and Microbial Risks.* ILSI Press, Washington, D.C.
32. **Codd, G. A., S. G. Bell, and W. P. Brooks.** 1989. Cyanobacterial toxins in water. *Water Sci. Technol.* **21:**1–13.

33. **Cover, T. L., and M. J. Blaser.** 1995. *Helicobacter pylori*: a bacterial cause of gastritis, peptic ulcer disease and gastric cancer. *ASM News* **61:**21–26.

34. **Craun, G. F.** 1986. Recent statistics of waterborne disease outbreaks (1981–1983), p. 161–168. *In* G. F. Craun (ed.), *Waterborne Diseases in the United States.* CRC Press, Boca Raton, Fla.

35. **Craun, G. F.** 1986. Statistics of waterborne outbreaks in the US (1920–1980), p. 73–159. *In* G. F. Craun (ed.), *Waterborne Diseases in the United States.* CRC Press, Boca Raton, Fla.

36. **Cubitt, W. D.** 1994. Caliciviruses, p. 549–568. *In* A. Z. Kapikian (ed.), *Viral Infections of the Gastrointestinal Tract.* Marcel Dekker, New York.

37. **Current, W. L., and L. S. Garcia.** 1991. Cryptosporidiosis. *Clin. Microbiol. Rev.* **4:**325–358.

38. **D'Antonio, R. G., R. E. Winn, J. P. Taylor, T. L. Gustafson, W. L. Current, M. M. Rhodes, G. W. Gary, and R. A. Zajac.** 1985. A waterborne outbreak of cryptosporidiosis in normal hosts. *Ann. Intern. Med.* **103:**886–888.

39. **DeLeon, R., S. M. Matsui, R. S. Baric, J. E. Herrmann, N. R. Blacklow, H. B. Greenberg, and M. D. Sobsey.** 1992. Detection of Norwalk virus in stool specimens by reverse transcriptase-polymerase chain reaction and nonradioactive oligoprobes. *J. Clin. Microbiol.* **30:**3151–3157.

40. **Dennis, D. T., R. P. Smith, J. J. Welch, C. G. Chute, B. Anderson, J. L. Herndon, and C. F. von Reyn.** 1993. Endemic giardiasis in New Hampshire: a case-control study of environmental risks. *J. Infect. Dis.* **167:**1391–1395.

41. **Dufour, A. P.** 1986. Diseases caused by water contact, p. 23–41. *In* G. F. Craun (ed.), *Waterborne Diseases in the United States.* CRC Press, Boca Raton, Fla.

42. **DuPont, H. L., C. L. Chappell, C. R. Sterling, P. C. Okhuysen, J. B. Rose, and W. Jakubowski.** 1995. The infectivity of *Cryptosporidium parvum* in healthy volunteers. *N. Engl. J. Med.* **332:**855–859.

43. **Elder, G. H., P. R. Hunter, and G. A. Codd.** 1993. Hazardous freshwater cyanobacteria (blue-green algae). *Lancet* **341:**1519–1520.

44. **El Saadi, O., and A. S. Cameron.** 1993. Illness associated with blue-green algae. *Med. J. Aust.* **158:**792–793.

45. **El Saadi, O., A. J. Esterman, S. Cameron, and D. M. Roder.** 1995. Murray River water, raised cyanobacterial cell counts, and gastrointestinal and dermatological symptoms. *Med. J. Aust.* **162:**122–125.

46. **Emde, K. M. E., H. Mao, and G. R. Finch.** 1992. Detection and occurrence of waterborne bacterial and viral pathogens. *Water Environ. Res.* **64:**641–647.

47. **Enroth, H., and L. Engstrand.** 1995. Immunomagnetic separation and PCR for detection of *Helicobacter pylori* in water and stool samples. *J. Clin. Microbiol.* **33:**2162–2165.

48. **Ericksen, T. H., and A. P. Dufour.** 1986. Methods to identify waterborne pathogens and indicator organisms, p. 195–214. *In* G. F. Craun (ed.), *Waterborne Diseases in the United States.* CRC Press, Boca Raton, Fla.

49. **Esrey, S. A., R. G. Feachem, and J. M. Hughes.** 1985. Interventions for the control of diarrhoeal diseases among young children: improving water supplies and excreta disposal facilities. *Bull. W.H.O.* **63:**757–772.

50. **Ewald, P. W.** 1994. When water moves like a mosquito, p. 67–86. *In* P. W. Ewald (ed.), *Evolution of Infectious Disease.* Oxford University Press, New York.

51. **Falconer, I. R., A. M. Beresford, and M. T. C. Runnegar.** 1983. Evidence of liver damage by toxin from a bloom of the blue-green alga, *Microcystis aeruginosa. Med. J. Aust.* **1:**511–514.

52. **Fiedorek, S. C., H. M. Malaty, D. L. Evans, C. L. Pumphrey, H. B. Casteel, D. J. Evans, and D. Y. Graham.** 1991. Factors influencing the epidemiology of *Helicobacter pylori* infection in children. *Pediatrics* **88:**578–582.

53. **Foy, H. M.** 1991. Adenoviruses, p. 77–94. *In* A. S. Evans (ed.), *Viral Infections of Humans: Epidemiology and Control.* Plenum Medical Book Company, New York.

54. **Garcia, L. S., R. Y. Shimizu, and D. A. Bruckner.** 1994. Detection of microsporidial spores in fecal specimens from patients diagnosed with cryptosporidiosis. *J. Clin. Microbiol.* **32:**1739–1741.

55. **Geldreich, E. E., K. R. Fox, J. A. Goodrich, E. W. Rice, R. M. Clark, and D. L. Swerdlow.** 1992. Searching for a water supply connection in the Cabool, Missouri, disease outbreak of *Escherichia coli* O157:H7. *Water Res.* **26:**1127–1137.

56. **Gerba, C. P., and G. Bitton.** 1984. Microbial pollutants: their survival and transport pattern to groundwater, p. 65–88. *In* G. Bitton and C. P. Gerba (ed.), *Groundwater Pollution Microbiology.* John Wiley & Sons, New York.

57. **Gray, J. J., X. Jiang, P. Morgan-Capner, U. Desselberger, and M. K. Estes.** 1993. Prevalence of antibodies to Norwalk virus in England: detection by enzyme-linked immunosorbent assay using baculovirus-expressed Norwalk virus capsid antigen. *J. Clin. Microbiol.* **31:**1022–1025.

58. **Gray, L. D.** 1995. *Escherichia, Salmonella, Shiegella,* and *Yersinia,* p. 450–452. *In* P. R. Murray, E. J. Baron, M. A. Pfaller, F. C. Tenover, and R. H. Yolken (ed.), *Manual of Clinical Microbiology,* 6th ed. ASM Press, Washington, D.C.

59. **Greenberg, H. B., J. Valdesuso, A. Z. Kapikian, R. M. Chanock, R. G. Wyatt, W. Szmuness, J. Larrick, J. Kaplan, R. H. Gilman, and D. A. Sack.** 1979. Prevalence of antibody to the Norwalk virus in various countries. *Infect. Immun.* **26:**270–273.

60. **Grohmann, G. S., R. I. Glass, H. G. Pereira, S. S. Monroe, A. W. Hightower, R. Weber, R. T. Bryan, and Enteric Opportunistic Infections Working Group.** 1993. Enteric viruses and diarrhea in HIV-infected patients. *N. Engl. J. Med.* **329:**14–20.

61. **Haas, C. N.** 1993. Quantifying microbiological risks, p. 389–398. *In* G. F. Craun (ed.), *Safety of Water Disinfection: Balancing Chemical and Microbial Risks.* ILSI Press, Washington, D.C.

62. **Hackney, C. R., and M. E. Potter.** 1994. Animal-associated and terrestrial bacterial pathogens, p. 172–209. *In* C. R. Hackney and M. D. Pierson (ed.), *Environmental Indicators and Shellfish Safety.* Chapman and Hall, New York.

63. **Hackney, C. R., and M. E. Potter.** 1994. Human-associated bacterial pathogens, p. 154–171. *In* C. R. Hackney and M. D. Pierson (ed.), *Environmental Indicators and Shellfish Safety.* Chapman and Hall, New York.

64. **Hale, D., W. Aldeen, and K. Carroll.** 1994. Diarrhea associated with cyanobacterialike bodies in an immunocompetent host. *JAMA* **271:**144–145.

65. **Havelaar, A. H., F. M. Schets, A. van Silfhout, W. H. Jansen, G. Wieten, and D. van der Kooij.** 1992. Typing of *Aeromonas* strains from patients with diarrhoea and from drinking water. *J. Appl. Bacteriol.* **72:**435–444.

66. **Hawkins, P. R., M. T. C. Runnegar, A. R. B. Jackson, and I. Falconer.** 1985. Severe hepatotoxicity caused by the tropical cyanobacterium (blue-green alga) *Cylindrospermopsis raciborskii* (Woloszynska) Seenaya and Subba Raju isolated from a domestic water supply reservoir. *Appl. Environ. Microbiol.* **50:**1292–1295.

67. **Hayman, J.** 1992. Beyond the Barcoo—probable human

tropical cyanobacterial poisoning in outback Australia. *Med. J. Aust.* **157**:794–796.

68. **Hazen, T. C.** 1988. Fecal coliforms as indicators in tropical waters: a review. *Toxic. Assess.* **3**:461–477.

69. **Herrmann, J. E., N. A. Nowak, D. M. Perron-Henry, R. W. Hudson, W. D. Cubitt, and N. R. Blacklow.** 1990. Diagnosis of astrovirus gastroenteritis by antigen detection with monoclonal antibodies. *J. Infect. Dis.* **161**:226–229.

70. **Herwaldt, B. L., J. F. Lew, C. L. Moe, D. C. Lewis, C. D. Humphrey, S. S. Monroe, E. W. Pon, and R. I. Glass.** 1994. Characterization of a variant strain of Norwalk virus from a food-borne outbreak of gastroenteritis on a cruise ship in Hawaii. *J. Clin. Microbiol.* **32**:861–866.

71. **Hopkins, R. J., P. A. Vial, C. Ferreccio, J. Ovalle, P. Prado, V. Sotomayor, R. G. Russell, S. S. Wasserman, and J. G. Morris.** 1993. Seroprevalence of *Helicobacter pylori* in Chile: vegetables may serve as one route of transmission. *J. Infect. Dis.* **168**:222–226.

72. **Hopkins, R. W., G. B. Gaspard, T. B. Williams, R. J. Karlin, C. Cukor, and N. R. Blacklow.** 1984. A community waterborne gastroenteritis outbreak: evidence for rotavirus as the agent. *Am. J. Public Health* **74**:263–265.

73. **Hung, T., G. Chen, C. Wang, R. Fan, R. Yong, J. Chang, R. Dan, and M. H. Ng.** 1987. Seroepidemiology and molecular epidemiology of the Chinese rotavirus, p. 49–62. *In* G. Bock and J. Wheelan (ed.), *Novel Diarrhoea Viruses. CIBA Foundation Symposium 128.* John Wiley & Sons, Chichester, England.

74. **Hung, T., G. M. Chen, C. A. Wang, and X. Chang.** 1984. Waterborne outbreak of rotavirus diarrhoea in adults caused by novel rotavirus. *Lancet* **i**:1139–1142.

75. **Hurst, C. J.** 1991. Presence of enteric viruses in freshwater and their removal by the conventional drinking water treatment processes. *Bull. W.H.O.* **69**:113–119.

76. **Hurst, C. J., W. H. Benton, and R. E. Stetler.** 1989. Detecting viruses in water. *J. Am. Water Works Assoc.* **81**:71–80.

77. **Institute of Medicine.** 1992. *Emerging Infections: Microbial Threats to Health in the United States.* National Academy Press, Washington, D.C.

78. **Jaykus, L.-A., M. T. Hernard, and M. D. Sobsey.** 1994. Human enteric pathogenic viruses, p. 92–153. *In* C. R. Hackney and M. D. Pierson (ed.), *Environmental Indicators and Shellfish Safety.* Chapman and Hall, New York.

79. **Jenkins, P. A.** 1991. Mycobacteria in the environment. *J. Appl. Bacteriol.* **70**:137S–141S.

80. **Jiang, X., D. Y. Graham, K. Wang, and M. K. Estes.** 1990. Norwalk virus genome cloning and characterization. *Science* **250**:1580–1583.

81. **Jiang, X., D. O. Matson, G. M. Ruiz-Palacios, J. Hu, J. Treanor, and L. K. Pickering.** 1995. Expression, self-assembly, and antigenicity of a Snow Mountain agent-like calicivirus capsid protein. *J. Clin. Microbiol.* **33**:1452–1455.

82. **Jiang, X., J. Wang, D. Y. Graham, and M. K. Estes.** 1992. Detection of Norwalk virus in stool by polymerase chain reaction. *J. Clin. Microbiol.* **30**:2529–2534.

83. **Jiang, X., M. Wang, D. Y. Graham, and M. K. Estes.** 1992. Expression, self-assembly and antigenicity of the Norwalk virus capsid protein. *J. Virol.* **66**:6527–6532.

84. **Jothikumar, N., K. Aparna, S. Kamatchiammal, R. Paulmurugan, S. Saravanadevi, and P. Khanna.** 1993. Detection of hepatitis E virus in raw and treated wastewater with the polymerase chain reaction. *Appl. Environ. Microbiol.* **59**:2558–2562.

85. **Kaplan, J. E., R. Feldman, D. S. Campbell, C. Lookabaugh, and G. W. Gary.** 1982. The frequency of a Norwalk-like pattern of illness in outbreaks of acute gastroenteritis. *Am. J. Public Health* **72**:1329–1332.

86. **Kaplan, J. E., G. W. Gary, R. C. Baron, N. Singh, L. B. Schonberger, R. Feldman, and H. B. Greenberg.** 1982. Epidemiology of Norwalk gastroenteritis and the role of Norwalk virus in outbreaks of acute nonbacterial gastroenteritis. *Ann. Intern. Med.* **96**:756–761.

87. **Kator, H., and M. Rhodes.** 1994. Microbial and chemical indicators, p. 30–91. *In* C. R. Hackney and M. D. Pierson (ed.), *Environmental Indicators and Shellfish Safety.* Chapman and Hall, New York.

88. **Keswick, B. H., T. K. Satterwhite, P. C. Johnson, H. L. DuPont, S. L. Secor, J. A. Bitsura, G. W. Gary, and J. C. Hoff.** 1985. Inactivation of Norwalk virus in drinking water by chlorine. *Appl. Environ. Microbiol.* **50**:261–264.

89. **Khan, A. S., C. L. Moe, R. I. Glass, S. S. Monroe, M. K. Estes, L. E. Chapman, X. Jiang, C. Humphrey, E. Pon, J. K. Iskander, and L. B. Schonberger.** 1994. Norwalk virus-associated gastroenteritis traced to ice consumption aboard a cruise ship in Hawaii: comparison and application of molecular method-based assays. *J. Clin. Microbiol.* **32**:318–322.

90. **Kidd, A. H., S. A. Esrey, and M. J. Ujfalusi.** 1989. Shedding of coronavirus-like particles by children in Lesotho. *J. Med. Virol.* **27**:1164–1169.

91. **Klein, P. D., Gastrointestinal Physiology Working Group, D. Y. Graham, A. Gaillour, A. R. Opekun, and E. O. Smith.** 1991. Water source as risk factor for *Helicobacter pylori* infection in Peruvian children. *Lancet* **337**:1503–1506.

92. **Klontz, K. C., S. Lieb, M. Schreiber, H. T. Janowski, L. M. Baldy, and R. A. Gunn.** 1988. Syndromes of *Vibrio vulnificus* infections: clinical and epidemiologic features in Florida cases, 1981–1987. *Ann. Internal. Med.* **109**:318–323.

93. **Knight, P.** 1995. One misidentified human parasite is a cyclosporan. *ASM News* **61**:520–522.

94. **Koopmans, M., A. Herrewegh, and M. C. Horzinek.** 1991. Diagnosis of torovirus infection. *Lancet* **337**:859.

95. **Koopmans, M., M. Petric, R. I. Glass, and S. S. Monroe.** 1993. Enzyme-linked immunosorbent assay reactivity of torovirus-like particles in fecal specimens from humans with diarrhea. *J. Clin. Microbiol.* **31**:2738–2744.

96. **Kurtz, J. B., and T. W. Lee.** 1984. Human astrovirus serotypes. *Lancet* **ii**:1405.

97. **LeChevallier, M. W., and W. D. Norton.** 1993. Treatments to address source water concerns: protozoa, p. 145–164. *In* G. F. Craun (ed.), *Safety of Water Disinfection: Balancing Chemical and Microbial Risks.* ILSI Press, Washington, D.C.

98. **LeChevallier, M. W., W. D. Norton, and R. G. Lee.** 1991. *Giardia* and *Cryptosporidium* spp. in filtered drinking water supplies. *Appl. Environ. Microbiol.* **57**:2617–2621.

99. **Logsdon, G. S., and J. C. Hoff.** 1986. Barriers to the transmission of waterborne disease, p. 255–274. *In* G. F. Craun (ed.), *Waterborne Diseases in the United States.* CRC Press, Boca Raton, Fla.

100. **MacKenzie, W. R., N. J. Hoxie, M. E. Proctor, M. S. Gradus, K. A. Blair, D. E. Peterson, J. J. Kazmierczak, D. G. Addiss, K. R. Fox, J. B. Rose, and J. P. Davis.** 1994. A massive outbreak in Milwaukee of *Cryptosporidium* infection transmitted through the public water supply. *N. Engl. J. Med.* **331**:161–167.

101. **Madeley, C. R., and B. P. Cosgrove.** 1976. Caliciviruses in man. *Lancet* **i**:199.

102. **Mascher, F., F. F. Reinthaler, D. Stunzner, and B. Lamberger.** 1988. *Aeromonas* species in a municipal water supply of a central European city: biotyping of strains and

detection of toxins. *Zentralbl. Bakteriol. Mikrobiol. Hyg.* **186**:333–337.

103. **Mathan, M., V. I. Mathan, S. O. Swaminathan, S. Yesudoss, and S. J. Baker.** 1975. Pleomorphic virus-like particles in human faeces. *Lancet* **i**:1068–1069.

104. **McFeters, G. A., G. K. Bissonnette, J. J. Jezeski, C. A. Thomson, and D. G. Stuart.** 1974. Comparative survival of indicator bacteria and enteric pathogens in well water. *Appl. Microbiol.* **27**:823–829.

105. **Meisel, J. L., D. R. Perera, C. Meligro, and C. E. Rubin.** 1976. Overwhelming watery diarrhea associated with a *Cryptosporidium* in an immunosuppressed patient. *Gastroenterology* **70**:1156–1160.

106. **Mendall, M. A., P. M. Goggin, N. Molineaux, J. Levy, T. Toosy, D. Strachan, and T. C. Northfield.** 1992. Childhood living conditions and *Helicobacter pylori* seropositivity in adult life. *Lancet* **339**:896–897.

107. **Millership, S. E., J. R. Stephenson, and S. Tabaqchali.** 1988. Epidemiology of *Aeromonas* species in a hospital. *J. Hosp. Infect.* **11**:169–175.

108. **Mitchell, H. M., Y. Y. Li, P. J. Hu, Q. Liu, M. Chen, G. G. Du, Z. J. Wang, A. Lee, and S. L. Hazell.** 1992. Epidemiology of *Helicobacter pylori* in Southern China: identification of early childhood as the critical period for acquisition. *J. Infect. Dis.* **166**:149–153.

109. **Moe, C. L., J. R. Allen, S. S. Monroe, H. Gary, C. D. Humphrey, J. E. Herrmann, N. R. Blacklow, C. Carcamo, M. Koch, K. H. Kim, and R. I. Glass.** 1991. Detection of astrovirus in pediatric stool samples by immunoassay and RNA probe. *J. Clin. Microbiol.* **29**:2390–2395.

110. **Moe, C. L., J. Gentsch, T. Ando, G. Grohmann, S. S. Monroe, X. Jiang, J. Wang, M. K. Estes, Y. Seto, C. Humphrey, S. Stine, and R. I. Glass.** 1994. Application of PCR to detect Norwalk virus in fecal specimens from outbreaks of gastroenteritis. *J. Clin. Microbiol.* **32**:642–648.

111. **Moe, C. L., M. D. Sobsey, G. P. Samsa, and V. Mesolo.** 1991. Bacterial indicators of risk of diarrhoeal disease from drinking-water in the Philippines. *Bull. W.H.O.* **69**:305–317.

112. **Morse, D. L., J. J. Guzewich, J. P. Hanrahan, R. Stricof, M. Shayegani, R. Deibel, J. C. Grabau, N. A. Nowak, J. E. Herrmann, G. Cukor, and N. R. Blacklow.** 1986. Widespread outbreaks of clam- and oyster-associated gastroenteritis: role of Norwalk virus. *N. Engl. J. Med.* **314**:678–681.

113. **Moyer, N. P.** 1987. Clinical significance of *Aeromonas* species isolated from patients with diarrhea. *J. Clin. Microbiol.* **25**:2044–2048.

114. **Muraca, P. W., V. L. Yu, and A. Goetz.** 1990. Disinfection of water distribution systems for Legionella: a review of application procedures and methodologies. *Infect. Control Hosp. Epidemiol.* **11**:79–88.

115. **Naik, S. R., R. Aggarwal, P. N. Salunke, and N. N. Mehrotra.** 1992. A large waterborne viral hepatitis E epidemic in Kanpur, India. *Bull. W.H.O.* **70**:597–604.

116. **Nataro, J. P., and M. M. Levin.** 1994. Bacterial diarrheas, p. 697–752. *In* A. Z. Kapikian (ed.), *Viral Infections of the Gastrointestinal Tract.* Marcel Dekker, New York.

117. **Nazer, H., E. Price, G. Hunt, U. Patel, and J. Walker-Smith.** 1990. Isolation of *Aeromonas* spp. from canal water. *Indian J. Pediatr.* **57**:115–118.

118. **Nime, F. A., D. L. Page, M. A. Holscher, and J. H. Yardley.** 1976. Acute enterocolitis in a human being infected with the protozoan *Cryptosporidium. Gastroenterology* **70**:592–598.

119. **Noel, J. S., T. W. Lee, J. B. Kurtz, R. I. Glass, and S. S. Monroe.** 1995. Typing of human astroviruses from clinical isolates by enzyme immunoassay and nucleotide sequencing. *J. Clin. Microbiol.* **33**:797–801.

120. **Olson, B. H.** 1993. Pathogen occurrence in source waters: factors affecting survival and growth, p. 83–97. *In* G. F. Craun (ed.), *Safety of Water Disinfection: Balancing Chemical and Microbial Risks.* ILSI Press, Washington, D.C.

121. **Ortega, Y. R., C. R. Sterling, R. H. Gilman, V. A. Cama, and F. Diaz.** 1993. Cyclospora species—a new protozoan pathogen of humans. *N. Engl. J. Med.* **328**:1308–1312.

122. **Osterholm, M. T., K. L. MacDonald, K. E. White, J. G. Wells, J. S. Spika, M. E. Potter, J. C. Forfang, R. M. Sorenson, P. T. Milloy, and P. A. Blake.** 1986. An outbreak of a newly recognized chronic diarrhea syndrome associated with raw milk consumption. *JAMA* **256**:484–490.

123. **Parsonnet, J., S. C. Trock, C. A. Bopp, C. J. Wood, D. G. Addiss, F. Alai, L. Gorelkin, N. Hargrett-Bean, R. A. Gunn, and R. V. Tauxe.** 1989. Chronic diarrhea associated with drinking untreated water. *Ann. Intern. Med.* **110**:985–991.

124. **Payment, P., and R. Armon.** 1989. Virus removal by drinking water treatment processes. *Crit. Rev. Environ. Control* **19**:15–31.

125. **Payment, P., and E. Franco.** 1993. *Clostridium perfringens* and somatic coliphages as indicators of the efficiency of drinking water treatment for viruses and protozoan cysts. *Appl. Environ. Microbiol.* **59**:2418–2424.

126. **Payment, P., L. Richardson, J. Siemiatycki, R. Dewar, M. Edwardes, and E. Franco.** 1991. A randomized trial to evaluate the risk of gastrointestinal disease due to consumption of drinking water meeting current microbiological standards. *Am. J. Public Health* **81**:703–708.

127. **Pereira, H. G., A. M. Fialho, T. H. Flewett, J. M. S. Teixeira, and Z. P. Andrade.** 1988. Novel viruses in human faeces. *Lancet* **ii**:103–104.

128. **Picard, B., and P. Goullet.** 1987. Seasonal prevalence of nosocomial *Aeromonas hydrophila* infection related to aeromonas in hospital water. *J. Hosp. Infect.* **10**:152–155.

129. **Regli, S., J. B. Rose, C. N. Haas, and C. P. Gerba.** 1991. Modeling the risk from *Giardia* and viruses in drinking water. *J. Am. Water Works Assoc.* **83**:76–84.

130. **Rose, J. B.** 1988. Occurrence and significance of *Cryptosporidium* in water. *J. Am. Water Works Assoc.* **80**:53–58.

131. **Rose, J. B.** 1990. Environmental sampling for waterborne pathogens: overview of methods, application limitations and data interpretation, p. 223–234. *In* G. F. Craun (ed.), *Methods for the Investigation and Prevention of Waterborne Disease Outbreaks.* Office of Research and Development, U. S. Environmental Protection Agency, Washington, D.C.

132. **Rose, J. B.** 1993. Enteric waterborne protozoa: hazard and exposure assessment, p. 115–126. *In* G. F. Craun (ed.), *Safety of Water Disinfection: Balancing Chemical and Microbial Risks.* ILSI Press, Washington, D.C.

133. **Rose, J. B., C. P. Gerba, and W. Jakubowski.** 1991. Survey of potable water supplies for *Cryptosporidium* and *Giardia. Environ. Sci. Technol.* **25**:1393–1400.

134. **Schubert, R. H. W.** 1991. Aeromonads and their significance as potential pathogens in water. *J. Appl. Bacteriol.* **70**:131S–135S.

135. **Sellwood, J., and J. V. Dadswell.** 1991. Human viruses and water, p. 29–45. *In* P. Morgan-Capner (ed.), *Current Topics in Clinical Virology.* The Laverham Press, Salisbury, England.

136. **Shahamat, M., C. Paszko-Kolva, H. Yamamoto, U. Mia, A. D. Pearson, and R. R. Colwell.** 1989. Ecological studies of *Campylobacter pylori. Klin. Wochenschr.* **67**:62–63.

137. **Shlim, D. R., M. T. Cohen, M. Eaton, R. Rajah, E. G.**

Long, and B. L. P. Ungar. 1991. An alga-like organism associated with an outbreak of prolonged diarrhea among foreigners in Nepal. *Am. J. Trop. Med. Hyg.* **45:**383–389.

138. **Singh, N., and V. L. Yu.** 1994. Potable water and *Mycobacterium avium* complex in HIV patients: is prevention possible? *Lancet* **343:**1110–1111.

139. **Soave, R., J. P. Dubey, L. J. Ramos, and M. Tummings.** 1986. A new intestinal pathogen? *Clin. Res.* **34:**533A.

140. **Sobsey, M. D.** 1989. Inactivation of health-related microorganisms in water by disinfection processes. *Water Sci. Technol.* **21:**179–195.

141. **Sobsey, M. D., A. P. Dufour, C. P. Gerba, M. W. LeChevallier, and P. Payment.** 1993. Using a conceptual framework for assessing risks to health from microbes in drinking water. *J. Am. Water Works Assoc.* **85:**44–48.

142. **Sobsey, M. D., P. A. Shields, F. S. Hauchman, A. L. Davis, V. A. Rullman, and A. Bosch.** 1988. Survival and persistence of hepatitis A virus in environmental samples, p. 121–124. *In* A. J. Zuckerman (ed.), *Viral Hepatitis and Liver Disease.* Alan R. Liss, New York.

143. **Sobsey, M. D., T. Fuji, and R. M. Hall.** 1991. Inactivation of cell-associated and dispersed hepatitis A virus in water. *J. Am. Water Works Assoc.* **83:**64–67.

144. **Soong, F. S., E. Maynard, K. Kirke, and C. Luke.** 1992. Illness associated with blue-green algae. *Med. J. Aust.* **156:**67.

145. **Ticehurst, J.** 1995. Hepatitis E virus, p. 1056–1067. *In* P. R. Murray, E. J. Baron, M. A. Pfaller, F. C. Tenover, and R. H. Yolken (ed.), *Manual of Clinical Microbiology,* 6th ed. ASM Press, Washington, D.C.

146. **Turner, P. C., A. J. Gammie, K. Hollinrake, and G. A. Codd.** 1990. Pneumonia associated with contact with cyanobacteria. *Br. Med. J.* **300:**1440–1441.

147. **U.S. Environmental Protection Agency.** 1986. *Ambient Water Quality Criteria for Bacteria—1986.* EPA 44015-84-002. Office of Regulations and Standards, U. S. Environmental Protection Agency, Washington, D.C.

148. **U.S. Environmental Protection Agency.** 1994. *National Primary Drinking Water Standards.* EPA 810-F-94-001A. Office of Water, U.S. Environmental Protection Agency, Washington, D.C.

149. **van der Kooij, D.** 1993. Importance and assessment of the biological stability of drinking water in the Netherlands, p. 165–179. *In* G. F. Craun (ed.), *Safety of Water Disinfection: Balancing the Chemical and Microbial Risks.* ILSI Press, Washington, D.C.

150. **von Reyn, C. F., J. N. Maslow, T. W. Barber, J. O. Falkinham, and R. D. Arbeit.** 1994. Persistent colonisation of potable water as a source of *Mycobacterium avium* infection in AIDS. *Lancet* **343:**1137–1141.

151. **Waddell, G., A. Allard, M. Johansson, L. Svensson, and I. Uhnoo.** 1994. Enteric adenoviruses, p. 519–547. *In* A. Z. Kapikian (ed.), *Viral Infections of the Gastrointestinal Tract.* Marcel Dekker, New York.

152. **Wadowsky, R. M., R. B. Yee, L. Mezmar, E. J. Wing, and J. N. Dowling.** 1982. Hot water systems as sources of *Legionella pneumophila* in hospital and nonhospital plumbing fixtures. *Appl. Environ. Microbiol.* **43:**1104–1110.

153. **Walch, M., and R. R. Colwell.** 1994. Detection of nonculturable indicators and pathogens, p. 258–273. *In* C. R. Hackney and M. D. Pierson (ed.), *Environmental Indicators and Shellfish Safety.* Chapman and Hall, New York.

154. **Waller, T.** 1979. Sensitivity of *Encephalitozoon cuniculi* to various temperatures, disinfectants and drugs. *Lab. Anim.* **13:**227–230.

155. **Wang, J., X. Jiang, H. P. Madore, J. Gray, U. Desselberger, T. Ando, Y. Seto, I. Oishi, J. F. Lew, K. Y. Green, and M. K. Estes.** 1994. Sequence diversity of small, round-structured viruses in the Norwalk virus group. *J. Virol.* **68:**5982–5990.

156. **Ward, R. L., D. I. Bernstein, and E. C. Young.** 1986. Human rotavirus studies in volunteers: determination of infectious dose and serological response to infection. *J. Infect. Dis.* **154:**871–880.

157. **Weber, R., R. T. Bryan, R. L. Owen, C. M. Wilcox, L. Gorelkin, G. S. Visvesvara, and Enteric Opportunistic Infections Working Group.** 1992. Improved light-microscopical detection of microsporidia spores in stool and duodenal aspirates. *N. Engl. J. Med.* **326:**161–166.

158. **Wendt, S. L., K. L. George, B. C. Parker, H. Graft, and J. O. Falkinham.** 1980. Epidemiology of infection by nontuberculous mycobacteria. *Am. Rev. Respir. Dis.* **122:**259–263.

159. **West, A. P., M. R. Millar, and D. S. Tompkins.** 1990. Survival of *Helicobacter pylori* in water and saline. *J. Clin. Pathol.* **43:**609.

160. **Winn, W. C.** 1995. Legionella, p. 533–544. *In* P. R. Murray, E. J. Baron, M. A. Pfaller, F. C. Tenover, and R. H. Yolken (ed.), *Manual of Clinical Microbiology,* 6th ed. ASM Press, Washington, D.C.

161. **Woode, G. N.** 1994. The toroviruses: bovine (Breda virus) and equine (Berne virus) and the torovirus-like agents of humans and animals, p. 581–602. *In* A. Z. Kapikian (ed.), *Viral Infections of the Gastrointestinal Tract.* Marcel Dekker, New York.

162. **World Health Organization.** 1993. *Guidelines for drinking-water quality,* 2nd ed., vol. 1. *Recommendations.* World Health Organization, Geneva.

163. **Wurtz, R.** 1994. *Cyclospora:* a newly identified intestinal pathogen of humans. *Clin. Infect. Dis.* **18:**620–623.

164. **Yolken, R., F. Leister, J. Almeido-Hill, E. Dubovi, R. Reid, and M. Santosham.** 1989. Infantile gastroenteritis associated with excretion of pestivirus antigens. *Lancet* **i:**517–519.

165. **Yolken, R., M. Santosham, R. Reid, and E. Dubovi.** 1988. Pestiviruses: major etiological agents of gastroenteritis in human infants and children? *Clin. Res.* **36:**80A. (Abstract.)

166. **Zmirou, D., J. P. Ferley, J. F. Collin, M. Charrel, and J. Berlin.** 1987. A follow-up study of gastro-intestinal diseases related to bacteriologically substandard drinking water. *Am. J. Public Health* **77:**582–587.

Detection of Protozoan Parasites in Source and Finished Drinking Waters

FRANK W. SCHAEFER III

16

Protozoa are unicellular microorganisms which, unlike bacteria and viruses, possess membrane-bound genetic material or nuclei and other assorted cellular organelles. They exhibit various forms of locomotion and reproduction, differences that have been used to categorize them into broad groups; however, the taxonomy of the protozoa is a dynamic, evolving process about which there is little agreement.

Free-living protozoa are ubiquitous in natural waters and moist soils and fall into three broad groups: amoebae, flagellates, and ciliates. They are often abundant in both surface and groundwater supplies as part of the normal aquatic community. Free-living protozoa are of little concern to water treatment plant operators, since they cause no treatment problems and have no effect on health. Identification of free-living protozoa to genus and species requires a great deal of skill and training. Keys, line drawings, and pictures in *An Illustrated Guide to the Protozoa* (22) aid in identifying unknown organisms at least to ordinal and familial levels. Included as representatives of free-living protozoa found in source waters are *Hartmanella* spp., *Acanthamoeba* spp., *Echinamoeba* spp., *Naegleria fowleri*, *Euglena* spp., *Tetrahymena* spp., and *Paramecium* spp.

Parasitic protozoa, which live in or on another organism such as a plant or animal, may or may not be of concern to the water treatment plant operator, depending on the organism on which they live. This chapter deals primarily with human enteric protozoa. Within the human small and large intestines, a number of protozoa, including amoebae, flagellates, coccidians, and ciliates, can be found. Many of these organisms are not recognized as pathogenic and consequently are dismissed as commensals. *Entamoeba coli* and *Trichomonas hominis* are recognized as examples of commensal organisms which live in another organism without harming it. Parasitic protozoa, on the other hand, are known to be harmful when living in or on another organism. Included as representatives of enteric protozoa are *Entamoeba histolytica*, *Giardia lamblia*, *Cryptosporidium parvum*, *Cyclospora* spp., *Enterocytozoon bieneusi*, *Isospora belli*, *I. hominis*, and *Balantidium coli*.

Amoebae belong to the phylum Sarcomastigophora and generally possess a vesicular nucleus and pseudopodia, which are retractable cytoplasmic protrusions used both for locomotion and feeding. Thecate amoebae are distinguished from athecate or naked forms, the other main subgroup, by a closely fitting envelope or shell secreted by the trophozoite. Many of the athecate amoebae are able to withstand adverse environmental conditions by transforming from a vegetative trophozoite form to a dormant, resistant cyst form. Cysts of certain amoebae have been isolated from the air in association with dust particles (30, 41). *Hartmanella* spp., *Acanthamoeba castellanii*, *Echinamoeba* spp., and *N. fowleri* are athecate examples which can produce cysts. *Hartmanella* spp., *A. castellanii*, and *Echinamoeba* spp. are known to harbor and amplify *Legionella* spp. (12). *Acanthamoeba* spp. and *N. fowleri* are normally free-living forms but under the right conditions become opportunistic pathogens which are responsible for primary amoebomeningoencephalitis in humans. Furthermore, *Acanthamoeba* spp. also are well-documented causes of corneal keratitis in contact lens wearers.

Flagellates, also members of the phylum Sarcomastigophora, have vesicular nuclei and move by hair- or whip-like cylindrical organelles which are approximately 0.25 μm wide, called flagella. Flagellar length and number in this group are variable. Members of this group may also possess chloroplasts, thecal plates, basal bodies, and collars around the flagellum. Some genera exhibit multicellular colonial forms. *Euglena* spp. are examples of free-living flagellates which contain chloroplasts. *G. lamblia* is an example of a parasitic flagellate.

Ciliates, which belong to the phylum Ciliophora, are distinguished from the other protozoan groups by their unique nuclei. They have both a large macronucleus which regulates cellular metabolism and a small micronucleus which is involved in genetics and sexual recombination. Many ciliates are phagotrophic; i.e., they ingest nutrients through a mouth or cytostome. These organisms are covered with cilia, which are used in both feeding and movement. Cilia are organelles similar in structure to flagella. They differ from flagella in being generally shorter and in being interconnected through their intracytoplasmic basal structure. As a result of the interconnection, ciliary movement can be coordinated. *Tetrahymena* spp. and *Paramecium* spp. are examples of free-living ciliates, while *Balantidium coli* is an example of a parasitic form.

The phylum Apicomplexa is sometimes referred to as the phylum Sporozoa. This group is exclusively parasitic and is responsible for both human and veterinary disease. Malaria and coccidian diseases are well known examples from humans and animals, respectively. The apicomplexans are characterized by possession of an anterior apical complex. In addition, they have complicated life cycles. Unlike members of the phyla mentioned above, they possess no obvious locomotor organelles. Most members of this group have a vesicular nucleus, a Golgi apparatus, and mitochondria. Moreover, there is usually a sexual phase to the life cycle in this group.

BIOLOGY AND ECOLOGY

Most enteric protozoa have two stages in their life cycles. The trophozoite is an actively feeding, growing, and reproducing stage. Stimuli in the host's intestinal tract induce most of the enteric protozoa to produce a resistant, dormant transmission form, which is referred to as either a cyst, oocyst, or spore. Whether commensal or parasitic, these protozoa have simple, direct life cycles and are transmitted as fecal contaminants of food and/or water. Furthermore, person-to-person transmission is known to occur. Trophozoites generally do not survive outside their hosts unless they are propagated in a specialized culture medium (29, 36). Cysts, oocysts, and spores are different in that they are known to survive for long periods outside the host, especially under cool, moist conditions. Ambient temperatures around 25°C or higher and desiccation are known to rapidly reduce the time a cyst, oocyst, or spore can survive outside the host. Unlike clinical specimens, only cysts, oocysts, and spores of parasitic forms are of concern in environmental samples. Most troubling for the water treatment industry is the fact that many of these forms are more resistant to chemical disinfection than enteric bacteria and viruses.

Enteric parasitic protozoa are known to produce gastrointestinal distress including diarrhea, flatulence, cramps, anorexia, and weight loss. However, they can produce a continuum of pathologies ranging from no symptoms to extremes of illness requiring hospitalization.

Presently, *Giardia* and *Cryptosporidium* species are of great concern to the water treatment industry because they are known to be the etiologic agents responsible for a number of episodes of waterborne gastroenteritis. Their significance is also increased because they can use a number of mammalian reservoir hosts besides humans. As a result, there is an amplification in the numbers of cysts and/or oocysts which are potential challenges to water impoundments and treatment plants.

Free-living forms are found in air, water, and soil. Large numbers of bacterivorous ciliates and amoebae can be found in sewage and other organic wastes. Furthermore, they may be found on the surface, bottom, banks, and slopes of most bodies of freshwater and marine water. Free-living forms may withstand temperatures ranging from freezing up to 28°C. Certain ecological stimuli may cause amoebae to encyst or to differentiate into a flagellated form.

DETECTION AND QUANTITATION

Whether one is looking for free-living or parasitic forms, no one collection method suffices for all protozoa. Free-living forms are usually in sufficient density, if present at all, and can be collected by a variety of grab sample techniques from either sediment or water above the sediment. Parasitic forms, on the other hand, are usually very sparse and most likely are suspended in the water. Consequently, they must be concentrated from large volumes of water.

Free-Living Protozoa

After collection, samples of free-living forms may sometimes be observed directly under the microscope, or they may be cultured. Each free-living form has optimal cultural requirements, which cannot be covered in detail here. Many of the media used by the American Type Culture Collection to culture free-living forms are listed in the back of the catalog (9). In addition, older protozoology books by Jahn and Jahn (15) and Kudo (18) contain valuable information on methods for collecting and culturing free-living protozoa. Parasitic forms, like *Giardia* and *Cryptosporidium* species, can be cultured by using very exacting aseptic procedures (29, 36) which do not lend themselves to environmental samples. Whatever the organism of interest, the investigator must pay close attention to pH, nutrients present, toxic substances present (like hydrogen sulfide), temperature, osmotic effects, and other organisms in the sample.

Initial identification of free-living protozoa is done with taxonomic keys; however, a basic knowledge of protozoan morphology is required to use them. Fundamental keys have been developed (15) but now are somewhat out of date taxonomically. A more recent, taxonomically accepted key has been published (22) but requires considerable knowledge of protozoa on the part of the user. In many cases, the only way to identify some free-living protozoa to genus and species is to collaborate with experts in the field.

Detection and quantitation of protozoa usually require microscopic techniques including epifluorescence, brightfield, phase-contrast, and differential interference contrast (DIC) microscopy. Moreover, some protozoa are identified on the basis of exacting staining techniques. Very large forms may require observation with a dissecting microscope. Many free-living forms are very motile and must be slowed down to be seen at all. Substances like Protoslo (Carolina Biological Supply, Burlington, N.C.), 10% methylcellulose, Detain (1% aqueous solution of a polyether; Ward's Natural Science Establishment, Inc.), and 3% polyacrylamide hydrazide (Sigma Chemical Co.) have been used to increase the viscosity of the medium and slow protozoan movement. Whenever a microscope slide is to be observed for a protracted period, the coverslip must be sealed to the slide with either nail polish, Vaspar, or petroleum jelly to prevent evaporation. Air bubbles under a sealed coverslip can provide a limited source of oxygen for a living preparation. For photomicrography of motile living material, a flash attachment for the microscope is essential.

Free-living amoebae, like *Acanthamoeba* spp., *Hartmanella* spp., *Naegleria* spp., and others, are usually grab sampled. Volumes of the samples are generally small, ranging from 1 to 4 liters of fluid or 10 to 100 ml of sediment. Fluid samples are concentrated by centrifugation at $250 \times g$ for 5 to 8 min when necessary, while sediment samples are used directly. The fluid concentrates and sediments are then cultured on nonnutrient agar lawns of *Escherichia coli*, which the amoebae use as a food source. Detection, purification, and cloning of the free-living amoebae are done by isolating organisms from plaques, places where the amoebae have consumed all of the bacteria (17). Identification is based on cyst and trophozoite morphology. Collaboration with an expert in the field is required in many instances. Antibodies and genetic primers and probes can be obtained to confirm preliminary identifications.

Parasitic Protozoa

Detection of *G. lamblia* and *C. parvum* in source and finished waters has been and continues to be of great interest.

G. *lamblia* is one of the most frequently reported parasitic waterborne pathogens (8). Furthermore, over the past 10 years *C. parvum* also has been responsible for numerous waterborne outbreaks of gastroenteritis, with the most notable being in Carrollton, Georgia, in 1987 (14), Talant, Oregon, in 1992 (38), and Milwaukee, Wisconsin, in 1993 (23), in which 13,000, 80,000, and 403,000 individuals were affected, respectively. The method used to detect these protozoan pathogens in water is an immunofluorescence detection procedure (3) after concentration of large volumes of water. Currently, *Giardia* cysts and *Cryptosporidium* oocysts are concentrated from water by retention on a yarn-wound filter. Retained particulates are eluted from the filter and reconcentrated by centrifugation. The pelleted *Giardia* cysts and *Cryptosporidium* oocysts are separated to some extent from other particulate debris by flotation on a Percoll-sucrose solution with a specific gravity of 1.1. A portion of the water layer/Percoll-sucrose interface is placed on a membrane filter to form a monolayer, indirectly stained with fluorescent antibody, and examined under a UV microscope. Cysts and oocysts are classified according to specific criteria (immunofluorescence, size, shape, and internal morphological characteristics), and the results are reported in terms of presumptive and confirmed cysts and oocysts per 100 liters. Presumptive cysts and oocysts are defined as having the right size and shape as well as fluorescence characteristics. Confirmed cysts and oocysts are those in which internal morphological characteristics like nuclei, axonenes, and median bodies can be demonstrated in *Giardia* cysts and sporozoites can be seen in oocysts by DIC microscopy (Fig. 1).

This method is continually evolving as the result of the efforts of many researchers who are evaluating and attempting to improve various steps of the procedure (21, 27).

Sampling

Sample collection usually requires an apparatus consisting of an inlet hose fitted with a pressure regulator and pressure gauge. The inlet hose connects to a plastic filter holder containing a polypropylene fiber depth filter of 1-μm nominal porosity. The effluent hose has a water meter in the middle and a limiting orifice flow control valve with a flow rate of 4 liters/min on the end (Fig. 2). A fluid proportioner or proportioning injector is needed for chlorinated disinfectant-treated waters (Fig. 3). To neutralize disinfectant, sodium thiosulfate solution is added in excess via the proportioning system. The sampling apparatus does not have to be sterile, but it must be clean and uncontaminated by cysts or oocysts. This can be accomplished by thoroughly flushing the system with the water to be sampled. Quasi-aseptic technique is used during sample collection to prevent cross-contamination between samples. The volume of water to be sampled depends on the water quality. Minimum sample sizes of 100 liters of source water and 1,000 liters of finished water are required. Records of the time, start and stop water meter readings, location, operator, and turbidity must be noted at the beginning and end of collection. After the sample collection is complete, the apparatus is carefully drained. During this step, the inlet hose is kept above the level of the outlet hose to prevent filter backwashing. The plastic filter holder is opened, and the filter is aseptically transferred to a labeled Zip-Loc bag. To guard against bag leaks, the first bag is placed inside a second Zip-Loc bag. Samples are stored and transported under refrigeration. Freezing of filters is never permitted, because cysts and oocysts are disrupted by ice crystals. Consequently, they do not respond to the flotation step that follows and are lost in the pellet. This method has the advantage of being reasonably portable. It is known that some organisms of interest may pass through this filter. An additional disadvantage is that organisms may enter the filter matrix and not be recovered. This approach to sampling, even with its limitations, is currently the only method available to sample large volumes of turbid water. Since this sampling apparatus, along with a small gas-powered pump, is reasonably portable, it can be carried into a remote protected watershed sampling site. The steps below are followed when the fiber depth filter is used for raw water sample collection.

1. Put on a pair of latex gloves.
2. Before connecting the sampling apparatus (Fig. 2) to the tap or source to be sampled, turn on the tap and allow the water to purge residual debris from the line for 2 to 3 min, or until the turbidity of the water becomes uniform.
3. Connect the apparatus minus the filter to the tap, and allow 20 gal (76 liters) to flush the system. If a pressurized source is not available, use a pump, following the manufacturer's instructions, to get water through the sampling apparatus. While the flushing of the apparatus is being done, adjust the pressure regulator so that the adjacent pressure gauge reads no more than 30 lb/in^2.
4. Turn off the water flow when the flushing of the apparatus is complete. Open the filter housing, and pour all of the water out. Put the filter in, and close and tighten the filter housing.
5. Use a water-resistant marking pen to record the start time, meter reading, name of the person collecting the sample, turbidity, date, and sampling location on the filter holder label.
6. Start the water flow through the filter. Check the pressure gauge adjacent to the pressure regulator to make sure that the reading is no more than 30 lb/in^2. Readjust the regulator, if necessary.
7. After the 100 liters of raw water has passed through the filter, shut off the water flow and record the stop time, final meter reading, and turbidity of the water at the end of filtration on the filter holder label.
8. Disconnect the sampling apparatus while maintaining the inlet hose level above the level of the opening on the outlet hose in order to prevent backwashing and the loss of particulate matter from the filter.
9. After allowing the apparatus to drain, open the filter housing and pour the residual water remaining in the filter holder into a plastic sample bag.
10. Aseptically remove the filter from the holder and transfer the filter to the plastic sample bag containing the residual water.
11. Seal the bag and place it inside a second plastic sample bag. Transfer the label or label information from the filter holder to the outside of this second (outer) bag.
12. Transport the sample to the laboratory on wet ice or with, but not on, cold packs. When the filters arrives at the laboratory, it should be immediately stored at 2 to 5°C. *Do not freeze the filter during transport or storage.*

The following steps are used for finished water sample collection. If the water must be neutralized, add sodium thiosulfate solution via the proportioner system. A proportioning injector is assumed in the steps that follow. For each 100 liters of finished water sampled, 250 ml of 2.0% (wt/vol) sodium thiosulfate solution will be needed.

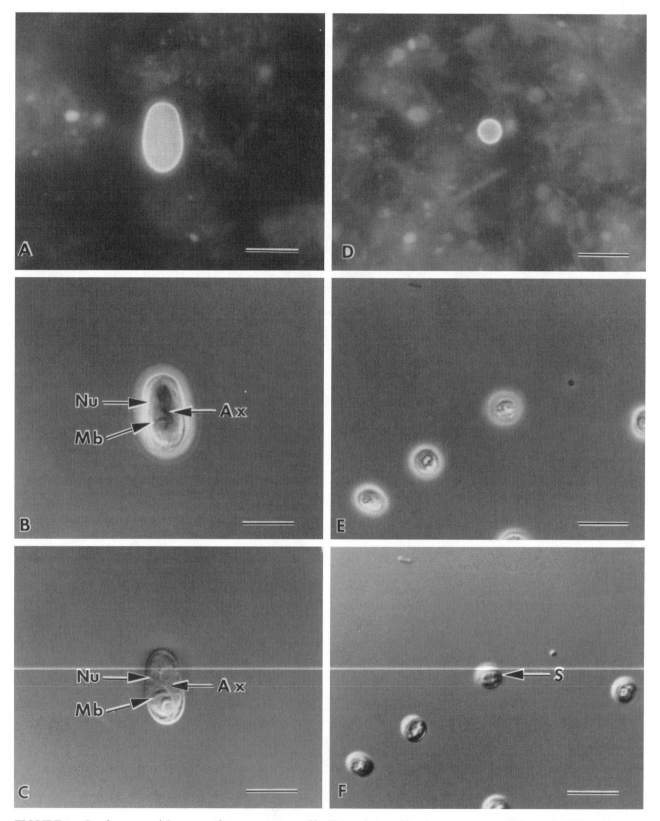

FIGURE 1 *Giardia* cysts and *Cryptosporidium* oocysts stained by IFA and viewed by phase-contrast and Nomarski DIC microscopy. (A) IFA-stained *G. lamblia* cyst. (B) Phase-contrast photomicrograph of *G. lamblia* cyst. Nu, nucleus; Mb, median body; Ax, axonemes. (C) Nomarski DIC photomicrograph of *G. lamblia* cyst. Nu, nucleus; Mb, median body; Ax, axonemes. (D) IFA-stained *C. parvum* oocyst. (E) Phase-contrast photomicrograph of *C. parvum* oocyst. (F) Nomarski DIC photomicrograph of *C. parvum* oocyst. S, sporozoite, one of two showing. Bars = 10 μm.

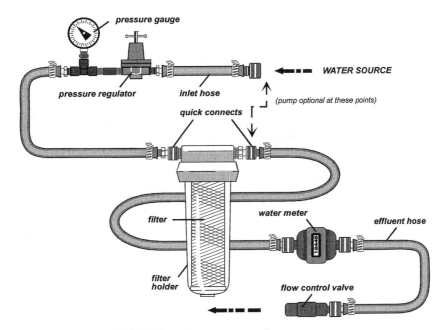

FIGURE 2 Raw water sampling apparatus.

1. Put on a pair of latex gloves.

2. Before connecting the sampling apparatus (Fig. 3) to the tap or source to be sampled, turn on the tap and allow the water to purge residual debris from the line for 2 to 3 min, or until the turbidity of the water becomes uniform.

3. Connect the apparatus minus the filter to the tap, and allow 20 gal (76 liters) to flush the system. If a pressurized source is not available, use a pump, following the manufacturer's instructions, to get water through the sampling apparatus. While the flushing is being done, adjust the pressure regulator so that the adjacent pressure gauge reads no more than 30 lb/in^2. Pour the 2% sodium thiosulfate solu-

tion into a graduated cylinder. Place the injector tube into the solution, and adjust the larger top (vacuum) screw on the injector so that the pressure on the pressure gauge following the injector reads no more than 19 lb/in^2. Now adjust the smaller bottom (flow) screw on the injector so that the flow rate of the thiosulfate solution is 10 ml/min. A hose cock clamp on the injector tube may be required to achieve the correct thiosulfate flow rate. After this adjustment is complete, transfer the injector tube to a graduated carboy of thiosulfate solution.

4. Turn off the water flow when the flushing of the apparatus is complete. Open the filter housing, and pour all of

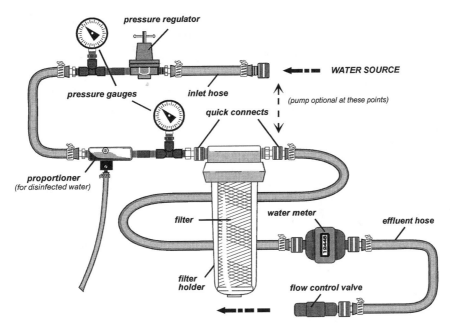

FIGURE 3 Finished water sampling apparatus.

the water out. Put the filter in, and close and tighten the filter housing.

5. Use a water-resistant marking pen to record the start time, meter reading, name of the person collecting the sample, turbidity, date, and sampling location on the filter holder label.

6. Start water flow through the filter. Check the pressure gauge adjacent to the pressure regulator to make sure the reading is no more than 30 lb/in^2. Also check to make sure that the thiosulfate solution is being drawn into the sampling apparatus. Readjust the regulator and injector if necessary.

7. After the 1,000 liters of finished water has passed through the filter, shut off the water flow and record the stop time, final meter reading, and turbidity of the water at the end of filtration on the filter holder label.

8. Disconnect the sampling apparatus while maintaining the inlet hose level above the level of the opening on the outlet hose in order to prevent backwashing and the loss of particulate matter from the filter.

9. After allowing the apparatus to drain, open the filter housing and pour the residual water remaining in the filter holder into a plastic sample bag.

10. Aseptically remove the filter from the holder and transfer the filter to the plastic sample bag containing the residual water.

11. Seal the bag and place it inside a second plastic sample bag. Transfer the label or label information from the filter holder to the outside of this second (outer) bag.

12. Transport the sample to the laboratory on wet ice or with, but not on, cold packs, and refrigerate it at 2 to 5°C. *Do not freeze the filter during transport or storage.*

A polycarbonate membrane filter sampling has been used in place of the cartridge depth fiber filter (13). When this approach to sampling is used, 10 to 20 liters of water is filtered through a 1- or 2-μm-absolute-porosity polycarbonate filter, using negative pressure ranging from 250 to 270 mm Hg (ca. 33 to 36 kPa). Following filtration, the polycarbonate filter is removed from the filter holder and placed on a glass plate larger than the filter. The glass plate, in turn, is positioned on an edge in a rectangular plastic trough (a kitchen drawer organizer). The filter is rinsed with distilled water, and the particulates, which collect in the plastic trough, are scraped off with a squeegee. This sampling method does not lend itself to collecting large volumes of turbid water. Moreover, the sampling apparatus is not portable to remote sampling sites.

Another approach to sampling raw water employs an electronegative fiber filter with a porosity of either 1 or 3 μm (28). Using an electronegatively charged filter allowed for collection of culturable human enteric viruses in addition to *Giardia* and *Cryptosporidium* species, coliphages, and *Clostridium perfringens*. The 100 liters of water to be sampled is conditioned with 0.001 M aluminum chloride, and the pH is adjusted to 3.5 by using an in-line injector system monitored by a pH meter to allow for virus adsorption. Microorganisms are eluted by using a slow backwash with 1.8 liters of 1% beef extract solution containing 1% Tween 80 at pH 10. A 400-ml aliquot of backwash is used for the *Clostridium perfringens* analysis. After centrifugation of the remaining 1,400 ml of backwash for 20 min at 3,000 × g, 1,000 ml of the supernatant is used for human enteric virus analysis and the remaining 400 ml of the supernatant is used for coliphage analysis. The pellet is resuspended in a solution containing 2% formalin and 0.1% Tween 80 and then used

for *Giardia* cyst and *Cryptosporidium* oocyst analysis. From this published account, one cannot tell whether the protozoa were identified on the basis of fluorescence alone or whether DIC optics were used to confirm internal morphological characteristics. A major question is, What effect on the protozoa and their morphological characteristics do the pH extremes have? Since this sampling method has been used by only one researcher, it is hard to evaluate how efficient and utilitarian it is. Perhaps a clue to the utility of this sampling procedure is given in the paper by the following statement: "We do not advocate the use of large volume concentration methods for routine monitoring of drinking water treatment; we have used these methods only to obtain a sufficient number of positive samples to establish correlations among the selected pathogens and indicators."

Recoveries of the organisms of interest by the depth fiber filter and polycarbonate membrane filter have been reported to range between 5 and 59%. Another sampling method is based on carbonate precipitation of particulates in a 10-liter sample (40). In this method, 100 ml of 1 M calcium chloride and 100 ml of 1 M sodium bicarbonate are mixed into a 10-liter sample. The pH of the mixture is adjusted to 10 with 3 M sodium hydroxide. The resulting floc of calcium carbonate crystals and particulates is allowed to settle overnight at room temperature. In the morning, the supernatant is discarded and the calcium carbonate precipitate is dissolved with 10% (wt/vol) sulfamic acid. Recovery of *C. parvum* oocysts seeded into deionized, tap, and Thames River water is reported to be in excess of 68%. This sampling method has now been applied to *Giardia* cysts (39). While this sampling method enables greater recovery of the organisms of interest, it suffers from some of the same problems of the polycarbonate membrane sampling method: it is not amenable to sampling large volumes of water, and a volume of 10 liters of water is not easy to transport from remote sites to laboratories.

Elution and Concentration

Filter elution is the next stage of the immunofluorescence detection procedure for *Giardia* and *Cryptosporidium* species. The initiation of sample collection and elution from the collection filter should be performed within 96 h. Elution solution is made by mixing 100 ml of 1% sodium dodecyl sulfate, 100 ml of 1% Tween 80, 100 ml of 10× phosphate-buffered saline, and 0.1 ml of Sigma antifoam A with 500 ml of reagent water. Adjust the final volume to 1 liter with additional reagent water after adjusting the pH to 7.4. Two approaches to eluting the particulates from the filter may be used: either washing by hand or use of a stomacher. For hand washing, the following steps are used.

1. Pour the residual solution in the bag into a beaker, rinse the bag with eluting solution, add the rinse solution to the beaker, and discard the bag.

2. Using a razor knife or other appropriate disposable cutting instrument, cut the filter fibers lengthwise down to the core. Discard the blade after the fibers have been cut. Divide the filter fibers into a minimum of six equal portions, with one-sixth consisting of the cleanest fibers (those nearest the core), the second one-sixth being the second layer of fibers, and so on until the final one-sixth consists of the outermost filter fibers (the dirtiest fibers).

3. Beginning with the cleanest fibers (the one-sixth nearest the core), hand wash the fibers in three consecutive 1.0-liter volumes of eluting solution. Wash the fibers by kneading them in the eluting solution contained in either

a beaker or a plastic bag. Wring the fibers to express as much of the liquid as possible before discarding. Maintain the three 1.0-liter volumes of eluate separate throughout the washing procedure. An additional beaker or two of eluting solution may be required for extremely dirty filters.

4. Using the three 1.0-liter volumes of eluate used in step 3, repeat the washing procedure on the second one-sixth layer of fibers, and then continue sequentially with the remaining one-sixth layers of fibers.

5. The minimum total wash time of fibers should be 30 min. After all fibers have been washed, combine the three 1.0-liter volumes of eluate with the residual filter water from step 1 and discard the fibers.

For stomacher washing, the following steps are used.

1. Use a stomacher with a bag capacity of 3,500 ml. Using a razor knife or other appropriate disposable cutting instrument, cut the filter fibers lengthwise down to the core. Discard the blade after the fibers have been cut.

2. After loosening the fibers, place all of the filter fibers in a stomacher bag. To ensure against bag breakage and sample loss, place the filter fibers in the first stomacher bag into a second stomacher bag.

3. Add 1.75 liters of eluting solution to the fibers. Homogenize the samples for two 5-min intervals. Between each homogenization period, hand knead the filter material to redistribute the fibers in the bag.

4. Pour the eluted particulate suspension into a 4-liter pooling beaker. Wring the fibers out to express as much of the liquid as possible into the beaker as well.

5. Put the fibers back into the stomacher bag, add 1.0 liter more of eluting solution, and homogenize the samples for two 5-min intervals. Between each homogenization period, hand knead the filter material to redistribute the fibers in the bag.

6. Add the eluted particulate suspension to the 4-liter pooling beaker. Wring the fibers out to express as much of the liquid as possible into the beaker as well. Discard the fibers. Rinse the stomacher bag with eluting solution into the pooling beaker.

Particulate concentration, the next stage in the method, is done as follows.

1. Concentrate the combined eluate and residual water into a single pellet by centrifugation at 1,050 × g for 10 min, using a swinging-bucket rotor and plastic conical centrifuge bottles.

2. Carefully aspirate and discard the supernatant fluid, and resuspend the pellet in sufficient elution solution by vortexing.

3. After pooling the particulates in one conical bottle, centrifuge them once more at 1,050 × g for 10 min and record the packed pellet volume.

4. Carefully aspirate and discard the supernatant fluid, and resuspend the pellet by vortexing it in an equal volume of 10% neutral buffered formalin solution. If the packed pellet volume is less than 0.5 ml, bring the pellet and solution volume to 0.5 ml with eluting solution before adding enough 10% buffered formalin solution to bring the resuspended pellet volume to 1.0 ml.

At this point, a break may be inserted if the procedure is not going to progress immediately to the flotation purification procedure described below. If a break is inserted at this point, be sure to store the formalin-treated sample at 4°C for not more than 72 h.

Flotation

The flotation step in the protocol for detecting *Giardia* and *Cryptosporidium* species has included zinc sulfate at a specific gravity of 1.18 to 1.2, Percoll-sucrose at a specific gravity of 1.1, and a Percoll-Percoll step gradient with steps at specific gravities 1.05 and 1.09. Zinc sulfate was abandoned because of its high osmotic stress on the protozoa. If cysts and oocysts are not removed from it fast enough, their buoyant density changes and they sink from the interface. Moreover, osmotic stress crenates the morphological characteristics to the point of making them indistinguishable. Percoll-sucrose has been found to be less osmotically stressful than zinc sulfate and has been widely used by a number of investigators. However, Percoll-Percoll has been shown to recover more cysts and oocysts than Percoll-sucrose (27). The limitations to Percoll-Percoll flotation are that it is expensive to make and it recovers more algal cells and other debris with the protozoa. The steps that follow assume use of Percoll-sucrose as the flotation medium. To make Percoll-sucrose, mix together 45 ml of Percoll (specific gravity, 1.13), 45 ml of reagent water, and 10 ml of 2.5 M sucrose. Check the specific gravity, which should be between 1.09 and 1.1, with a hydrometer.

1. In a clear plastic 50-ml conical centrifuge tube(s), vortex a volume of resuspended pellet equivalent to not more than 0.5 ml of packed pellet volume with a sufficient volume of eluting solution to make a final volume of 20 ml.

2. Using a 50-ml syringe and 14-gauge cannula, underlay the 20-ml vortexed suspension of particulates with 30 ml of Percoll-sucrose flotation solution (specific gravity, 1.10).

3. Without disturbing the pellet suspension/Percoll-sucrose interface, centrifuge the preparation at 1,050 × g for 10 min, using a swinging-bucket rotor. Slowly accelerate the centrifuge over a 30-s interval up to the speed at which the tubes are horizontal in order to avoid disrupting the interface. Similarly, at the end of centrifugation, decelerate slowly. *Do not use the brake.*

4. Using a polystyrene 25-ml pipette rinsed with eluting solution, draw off the top 20 ml of particulate suspension layer, the interface, and 5 ml of the Percoll-sucrose below the interface. Place all of these volumes in a plastic 50-ml conical centrifuge tube.

5. Add additional eluting solution to the plastic conical centrifuge tube (step 4) to a final volume of 50 ml. Centrifuge the sample at 1,050 × g for 10 min.

6. Aspirate and discard the supernatant fluid down to 5 ml (plus pellet). Resuspend the pellet by vortexing, and save this suspension for further processing with fluorescent antibody reagents.

Immunofluorescent Staining

Before the immunofluorescent staining can be done, the amount of particulate material to be stained must be determined. In this procedure, immunofluorescent staining is done with membrane filters and a filter manifold. The volume of sample concentrate from the flotation procedure described above that may be applied to each 25-mm-diameter membrane filter can be determined as follows.

1. Vortex the sample concentrate, and apply 40 μl to one 5-mm-diameter well of a 12-well red heavy Teflon-coated slide.

2. Allow the sample to sit for approximately 2 min at room temperature.

3. Examine the flooded well at a total magnification of ×200. If the particulates are distributed evenly over the

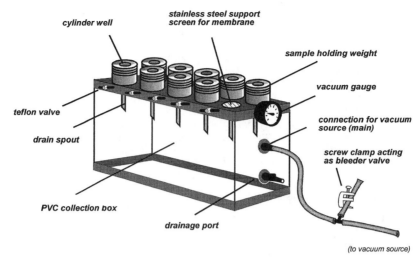

FIGURE 4 Filtration manifold assembly.

well surface area and are not crowded or touching, then apply 1 ml of the undiluted sample to each 25-mm-diameter membrane filter to be prepared.

4. Adjust the volume of the sample accordingly if the particulates are too dense or are widely spread. Retest on another well. Always adjust the sample concentrate volume so that the density of the particulates is just a little sparse. If the layer of sample particulates on the membrane filters is too dense, any cysts or oocysts present in the sample may be obscured during microscopic examination. Make sure that the dilution factor, if any, from this step is recorded.

Prepare the filtration manifold (Fig. 4 shows a diagram of the filtration manifold assembly) as follows (this phase of the procedure assumes use of a Hoefer model FH 255V manifold).

1. Connect the filtration manifold to the vacuum supply, using a vacuum tube containing a T-shaped tubing connector.

2. Attach a Hoffman screw clamp to 4 to 6 cm of latex tubing, and then attach the latex tubing to the stem of the T connector. The screw clamp is used as a bleeder valve to regulate the vacuum to 2 to 4 in. (5 to 10 cm) of Hg.

3. Close all manifold valves and open the vacuum all the way. Using the bleeder valve on the vacuum tubing, adjust the applied vacuum to 2 to 4 in. (5 to 10 cm) of Hg. Once the bleeder valve is adjusted, do not readjust it during filtration. If necessary, turn the vacuum on and off during filtration at the vacuum source.

One Sartorius 25-mm-diameter cellulose acetate filter (0.2-μm pore size) and one 25-mm-diameter ethanol-compatible membrane support filter (any porosity) are required for each 1 ml of adjusted suspension obtained in the procedure described above for determining sample volume per filter. The membrane filters are prepared in the following fashion.

1. Soak the required number of each type of filter separately in petri dishes filled with 1× phosphate-buffered saline. Drop the filters, handling them with blunt-end filter forceps, one by one flat on the surface of the buffer.

2. Once the filters are wetted, push the filters under the fluid surface with the forceps.

3. Allow filters to soak for at least 1 min before use.

4. Turn the filtration manifold vacuum source on. Leaving all of the manifold well support valves closed, place one support filter on each manifold support screen. This filter ensures even distribution of sample.

5. Place one Sartorius 25-mm-diameter cellulose acetate filter on top of each support filter. Use a rubber policeman to adjust the cellulose acetate filter if necessary.

6. Open the manifold well support valves to flatten the filter membranes. Make sure that no bubbles are trapped and that there are no creases or wrinkles on any of the filter membranes.

7. Use as many filter positions as there are sample volumes to be assayed. Record the number of sample 25-mm-diameter membrane filters prepared and the volume of floated pellet represented by these membranes. In addition, include at least one positive control for *Giardia* cysts and *Cryptosporidium* oocysts and one negative control each time the manifold is used.

8. Position the 1-lb (454-g) stainless steel wells firmly over each filter.

9. Label each sample and control well appropriately with little pieces of tape on the top of the stainless steel wells, and/or use a manifold membrane labeling diagram (Fig. 5) to keep track of each sample and control.

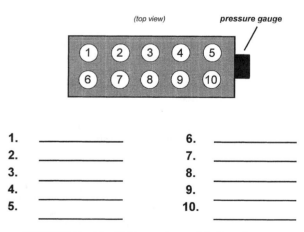

FIGURE 5 Manifold membrane labeling diagram.

TABLE 1 Composition of ethanol-glycerol dehydration series

95% ethanol (ml)	Glycerol (ml)	Reagent water (ml)	Final vol (ml)	Final % ethanol
10	5	80	95	10
20	5	70	95	20
40	5	50	95	40
80	5	10	95	80
95	5	0	100	90.2

Sample application and immunofluorescent staining are done by using the following steps. This section of the method assumes the use of the Ensys Hydrofluor Combo kit for immunofluorescent staining. The reagents in the kit are diluted according to the manufacturer's instructions, using phosphate-buffered saline. In addition, an ethanol-glycerol dehydration series, DABCO (1,4-diazabicyclo-[2.2.2]-octane)-glycerol mounting medium, and 1% bovine serum albumin are required. The ethanol-glycerol dehydration series is prepared as shown in Table 1.

To prepare DABCO-glycerol mounting medium, prewarm 95 ml of glycerol, using a magnetic stir bar on a heating stir plate. Add 2 g of DABCO (Sigma catalog no. D-2522) to the warm glycerol with continuous stirring until it dissolves. (*Caution:* DABCO is hygroscopic and causes burns; avoid inhalation, as well as skin and eye contact.) Adjust the final volume to 100 ml with additional glycerol. Bovine serum albumin (1%) is made by sprinkling 1.0 g of bovine serum albumin crystals over 85 ml of $1 \times$ phosphate-buffered saline (pH 7.4). Allow crystals to fall before stirring them into solution with a magnetic stir bar. After the bovine serum albumin is dissolved, adjust the volume to 100 ml with phosphate-buffered saline.

1. Open the manifold support valve for each well containing filters.
2. Rinse the inside of each stainless steel well and membrane filter with 2 ml of 1% bovine serum albumin applied with a Pasteur pipette. Drain the bovine serum albumin completely from the membrane.
3. Close the manifold valves under each membrane filter.
4. For the positive controls, add 500 to 1,000 G. *lamblia* cysts and 500 to 1,000 C. *parvum* oocysts, or use the Ensys positive control antigen as specified by the manufacturer.
5. For a negative control, add 1.0 ml of $1 \times$ phosphate-buffered saline to one well.
6. Add 1.0 ml of the vortexed, adjusted water sample to a well.
7. Open the manifold valve under each membrane filter to drain the wells. Rinse each stainless steel well with 2 ml of 1% bovine serum albumin. *Do not touch the pipette to the membrane filter or to the well.* Close the manifold valve under each membrane filter.
8. Pipette 0.5 ml of the diluted primary antibody onto each membrane, and allow it to remain in contact with the filter for 25 min at room temperature.
9. At the end of the contact period, open the manifold valve to drain the antisera.
10. Rinse each well and filter five times with 2 ml of $1 \times$ phosphate-buffered saline. *Do not touch the tip of the pipette to the membrane filter or to the stainless steel wells.* Close all manifold valves after the last wash is completed.

11. Pipette 0.5 ml of labeling reagent onto each membrane, and allow it to remain in contact with the filter for 25 min at room temperature. Cover all wells with aluminum foil to shield the reagents from light and to prevent photobleaching of the fluorescein isothiocyanate dye during the contact period.
12. At the end of the contact period, open the manifold valves to drain the labeling reagent.
13. Rinse each well and filter five times with 2 ml of $1 \times$ phosphate-buffered saline. *Do not touch the tip of the pipette to the membrane filter or to the stainless steel wells.* Close all manifold valves after the last wash is completed.
14. Dehydrate the membrane filters in each well by sequentially applying 1.0 ml of 10, 20, 40, 80, and 90.2% ethanol solutions containing 5% glycerol. Allow each solution to drain thoroughly before applying the next in the series.

Filter Mounting

1. Label glass slides for each filter, and place them on a slide warmer or in an incubator calibrated to 37°C.
2. Add 75 μl of 2% DABCO-glycerol mounting medium to each slide on the slide warmer or in the incubator, and allow the slides to warm for 20 to 30 min.
3. Remove the top cellulose acetate filter with fine-tip forceps, and layer it over the correspondingly labeled DABCO-glycerol mounting medium-prepared slide. *Make sure that the sample application side is up.* If the entire filter is not wetted by the DABCO-glycerol mounting medium, pick up the membrane filter with the same forceps and add a little more DABCO-glycerol mounting medium to the slide under the filter.
4. Use a clean pair of forceps to handle each membrane filter. Soak used forceps in a beaker of diluted detergent cleaning solution.
5. After a 20-min clearing period on the slide warmer, the filter should become transparent and appear drier. After clearing, if the membrane starts to turn white, apply a small amount of DABCO-glycerol mounting medium *under* the filter.
6. After the 20-min clearing period, apply 20 μl of DABCO-glycerol mounting medium to the center of each membrane filter, and cover it with a cover glass (25 mm by 25 mm). Tap out air bubbles with the handle end of a pair of forceps. Wipe off excess DABCO-glycerol mounting medium from the edge of each cover glass with a slightly moistened Kimwipe.
7. Seal the edge of each cover glass to the slide with clear fingernail polish.
8. Store the slides in a "dry box." A dry box can be constructed from a covered Tupperware-type container to which a thick layer of anhydrous calcium sulfate has been added. Cover the desiccant with paper towels, and lay the slides flat on the top of the paper towels. Place the lid on the dry box, and store the box at 4°C.
9. Examine the slides microscopically as soon as possible but within 5 days of preparation, because they may become opaque if stored longer, and DIC or Hoffman modulation optical examination would then no longer be possible.

Microscopic Examination

The procedure described below assumes the use of a microscope capable of epifluorescence and DIC or Hoffman modulation optics, with stage and ocular micrometers and 20× (numerical aperture = 0.6) to 100× (numerical aperture

= 1.3) objectives. The epifluorescence portion of the microscope should be equipped with appropriate excitation and band pass filters for examining fluorescein isothiocyanate-labeled specimens (exciter filter, 450 to 490 nm; dichroic beam-splitting mirror, 510 nm; barrier or suppression filter, 515 to 520 nm).

1. Remove the dry box from 4°C storage and allow it to warm to room temperature before opening.

2. Adjust the microscope to assure that the epifluorescence and Hoffman modulation or DIC optics are in optimal working order. Make sure that the fluorescein isothiocyanate cube is in place in the epifluorescence portion of the microscope.

The immunofluorescent staining controls are examined first. The purpose of these controls is to ensure that the assay reagents are functioning, that the assay procedures have been properly performed, and that the microscope has been adjusted and aligned properly.

3. Using epifluorescence, scan the positive control slide at a total magnification of no less than ×200 for apple-green fluorescence of *Giardia* cyst and *Cryptosporidium* oocyst shapes. Background fluorescence of the membrane should be either very dim or nonexistent. *Cryptosporidium* oocysts may or may not show evidence of oocyst wall folding, which is characterized under epifluorescence by greater concentrations of fluorescein isothiocyanate along surface fold lines, depending on the manner in which the oocysts have been treated and the amount of turgidity they have been able to maintain (31). If no apple-green fluorescing *Giardia* cyst or *Cryptosporidium* oocyst shapes are observed, then the fluorescent staining did not work or the positive control cyst preparation was faulty. Do not examine the water sample slides for *Giardia* cysts and *Cryptosporidium* oocysts. Recheck reagents and procedures to determine the problem.

4. If apple-green fluorescing cyst and oocyst shapes are observed, change the microscope from epifluorescence to the 100× oil immersion Hoffman modulation or DIC objective. At a total oil immersion magnification of no less than ×1,000, examine *Giardia* cyst shapes and *Cryptosporidium* oocyst shapes for internal morphology. The *Giardia* cyst internal morphological characteristics include one to four nuclei, axonemes, and median bodies. *Giardia* cysts should be measured to the nearest 0.5 μm with a calibrated ocular micrometer. Record the length and width of cysts. Also record the morphological characteristics observed. Continue until at least three *Giardia* cysts have been detected and measured in this manner. The *Cryptosporidium* oocyst internal morphological characteristics include one to four sporozoites. Examine the *Cryptosporidium* oocyst shapes for sporozoites, and measure the oocyst diameter to the nearest

0.5 μm with a calibrated ocular micrometer. Record the size of the oocysts. Also record the number, if any, of the sporozoites observed. Sometimes a single nucleus is observed per sporozoite. Continue until at least three oocysts have been detected and measured in this manner.

5. Using epifluorescence, scan the negative control membrane at a total magnification of no less than ×200 for apple-green fluorescence of *Giardia* cyst and *Cryptosporidium* oocyst shapes. If no apple-green fluorescing cyst or oocyst shapes are found and if background fluorescence of the membrane is very dim or nonexistent, continue with examination of the water sample slides. If apple-green fluorescing cyst or oocyst shapes are found, discontinue examination, since possible contamination of the other slides is indicated. Clean the equipment, recheck the reagents and procedure, and repeat the assay with additional aliquots of the sample.

Scan each slide in a systematic fashion, beginning with one edge of the mount and covering the entire coverslip. An up-and-down or side-to-side scanning pattern may be used. Figure 6 illustrates two alternatives for systematic slide scanning.

Presumptive counts and confirmed counts are determined as follows. (i) When appropriate responses have been obtained for the positive and negative controls, use epifluorescence to scan the entire coverslip from each sample at a total magnification of not less than ×200 for apple-green fluorescence of cyst and oocyst shapes.

(ii) When brilliant apple-green fluorescing round to oval objects (8 to 18 μm long by 5 to 15 μm wide) with brightly highlighted edges are observed, switch the microscope to either Hoffman modulation or DIC optics. Look for external or internal morphological characteristics atypical of *Giardia* cysts (spikes, stalks, appendages, pores, one or two large nuclei filling the cell, red fluorescing chloroplasts, crystals, spores, etc.). If these atypical structures are not observed, then categorize such apple-green fluorescing objects of the aforementioned size and shape as presumptive *Giardia* cysts. If *Giardia* cysts with two or more internal structures (nuclei, axonemes, and median bodies) are found, categorize them as confirmed *Giardia* cysts. Confirmed *Giardia* cysts are a subset of the presumptive *Giardia* cyst set. Record the shape and measurements (to the nearest 0.5 μm at a total magnification of ×1,000) for each such object. Record the internal structures observed.

(iii) When brilliant apple-green fluorescing ovoid or spherical objects (3 to 7 μm in diameter) with brightly highlighted edges are observed, switch the microscope to either Hoffman modulation or DIC optics. Look for external or internal morphological characteristics atypical of *Cryptosporidium* oocysts (spikes, stalks, appendages, pores, one or

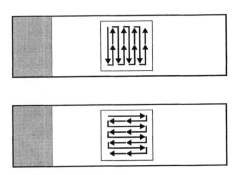

FIGURE 6 Illustration of two alternatives for systematic slide scanning.

two large nuclei filling the cell, red fluorescing chloroplasts, crystals, spores, etc.). If these atypical structures are not observed, then categorize such apple-green fluorescing objects of the aforementioned size and shape as presumptive *Cryptosporidium* oocysts. If *Cryptosporidium* oocysts with internal structure (one to four sporozoites per oocyst) are observed, record them as confirmed *Cryptosporidium* oocysts. Record the shape and measurements (to the nearest 0.5 μm at a total magnification of $\times 1,000$) for each such object. Although not a defining characteristic, surface mold folds may be observed in some specimens. Confirmed *Cryptosporidium* oocysts are a subset of the presumptive *Cryptosporidium* oocyst set. Record the number of sporozoites observed.

Calculation

Record the percentage of floated sediment examined microscopically. Calculate this value from the total volume of floated pellet obtained, the number of 25-mm-diameter membrane filters prepared together with the volume of floated pellet represented by these membrane filters, and the number of membrane filters examined. The following values are used in calculations: V = volume (liters) of original water sample; P = eluate packed pellet volume (milliliters); F = fraction of eluate packed pellet volume (P) subjected to flotation, determined as milliliters of P subjected to flotation/total milliters of P; R = percentage (expressed as a decimal) of floated sediment examined; PG = presumptive *Giardia* immunofluorescent antibody (IFA) stained cyst count; CG = count of confirmed *Giardia* cysts with two or more internal structures; PC = presumptive *Cryptosporidium* IFA-stained oocyst count; and CC = count of confirmed *Cryptosporidium* oocysts with one to four sporozoites.

For positive samples, calculate the number of cysts or oocysts per 100 liters of sample as follows:

$$\frac{X}{100 \text{ liters}} = \frac{(\text{PG, CG, PC, or CC}) (100)}{FVR}$$

For samples in which no cysts or oocysts are detected, (for example, PG or CC or CG or PC), the value is <1. Calculate the detection limit as follows:

$$\frac{<X}{100 \text{ liters}} = \frac{(<1) (100)}{FVR}$$

Limitations

Overall limitations associated with the concentration and immunofluorescence detection methods include the following. (i) Organism recovery is low. The recovery efficiency is influenced by water quality and in particular by the amount of particulates in the water being sampled. Generally, the greater the turbidity, the lower the recovery efficiency (26, 27, 39, 40). (ii) The procedure is time-consuming and requires specialized, expensive equipment. (iii) Nonspecificity of the monoclonal antibodies allows for a number of cross-reactions and potential false-positive results for both *Giardia* and *Cryptosporidium* samples. Algal cells of the same size and shape as *Giardia* cysts and *Cryptosporidium* oocysts react with the monoclonal antibodies used in this procedure (33). Furthermore, the *Cryptosporidium* and *Giardia* monoclonal antibodies react with other *Cryptosporidium* and *Giardia* species. (iv) The method does not determine viability, much less infectivity of the detected organisms. (v) The procedure does not identify the host of origin. Consequently, it is not known whether the particular isolate is a significant human threat. (vi) Large amounts of algae and

debris, which can physically obscure the observation of the cysts and oocysts on the microscope slides, are also recovered from water samples during the procedure. (vii) Many algae and other animate and inanimate objects have pigments that autofluoresce at the excitation wavelength for fluorescein isothiocyanate and other common fluorochromes. This autofluorescence interferes with cyst and oocyst detection and makes reading of slides more time-consuming and tedious. (viii) The method is dependent on the skill, experience, and training of the microscopist. After detecting objects of the right size, shape, and fluorescence characteristic, the microscopist must demonstrate internal morphological characteristics by DIC or Hoffman modulation microscopy. Unfortunately the microscopic portion of the analysis can be quite subjective even among experts. Moreover, this type of microscopic work is intense, demanding, and tiring. When the microscopist becomes too tired, the reliability of results becomes questionable.

While these methods have been used successfully on some occasions to confirm epidemiological evidence that a waterborne outbreak has occurred and in surveys of raw (20) and finished (19) waters, they do not lend themselves to easy, routine monitoring (16). In an effort to address the limitations of the monitoring methods, the U.S. Environmental Protection Agency convened a protozoa, virus, and coliphage monitoring workshop of experts in Cincinnati, Ohio, in August 1993. Additional workshops on the protozoan method occurred in Cocoa Beach, Florida, in January 1994 and in Cincinnati in May 1995. Out of these workshops came the recognition that a negative count and low detection limit do not ensure pathogen-free water. In the report from the 1993 workshop, it was recommended that results obtained by using this method be interpreted with extreme caution. The report also concluded that a training course of 3 to 4 days is required before a laboratory can begin the laboratory approval process, that microscopic work must be supervised by a senior analyst, and that microscopic work by a single analyst should not exceed 4 h/day or more than 5 consecutive days/week. Intermittent rest periods during the 4 h/day are encouraged.

Alternate Approaches

Recognition of the limitations of the *Giardia-Cryptosporidium* immunofluorescence assay has stimulated research using other approaches for detection and identification, including fluorescence-activated cell sorting (FACS), enzyme-linked immunosorbent assay (ELISA), confocal microscopy, PCR, electrorotation assay (ERA), use of a cooled charge-coupled device (CCD), and fluorescence in situ hybridization (FISH). Presently, none of these approaches except ELISA has been validated with environmental and treated water samples. In addition, a great deal of emphasis is being placed on determining viability by using various fluorochromes like 4',6-diamino-2-phenylindole (DAPI) (6), propidium iodide (34), and fluorescein diacetate (35). At best, fluorochrome viability testing appears to be unreliable and variable between laboratories.

A number of clinical ELISA kits are commercially available and have been evaluated for the ability to detect spiked *Giardia* cysts and *Cryptosporidium* oocysts in environmental water samples (10). Fewer than 10 cysts or oocysts can be detected by this technique. Like the immunofluorescence assay for these organisms, detection by ELISA suffers from cross-reactions with some algae and is adversely impacted by turbidity. However, since this technique can be done more rapidly than the immunofluorescence procedure and does not require a skilled microscopist to read the result, it has potential for screening water samples which might require more rigorous testing.

Confocal microscopy in conjunction with FISH has been used to detect *Giardia* cysts and identify them to the species level (11). Oligomeric probes to 16S rRNA which were conjugated to a carboxymethylindocyanine dye (Cy3 or Cy5) or fluorescein were used to distinguish among G. *lamblia*, G. *muris*, and G. *ardeae* cysts. When the fluorochrome fluorescein isothiocyanate was conjugated to the detection monoclonal antibody and the rRNA gene probes were used as a counterstain, it was possible to detect cysts of G. *lamblia* in fecal as well as environmental samples taken from a sewage lagoon. The preparations were observed with a krypton-argon laser-equipped confocal microscope which allowed the visualization of all of the fluorochromes. While nothing has been published on using FISH for *Cryptosporidium* detection, there is likely to be something on this approach soon. The major advance in using this approach is that false-positive results are reduced and the need for DIC optics is eliminated. The drawback, however, is the great expense of and skill needed to operate a confocal microscope. If the investigator can forgo the simultaneous observation of two or more fluorochrome labels, then all that is needed is a conventional epifluorescence microscope equipped with the requisite number of filters for the various fluorochrome labels. If a procedure exploiting FISH in conjunction with fluorescent antibodies could be developed for use in a FACS apparatus, then the microscope time required for water samples would be greatly reduced if not eliminated.

A FACS apparatus is a very elaborate laser-based particle counter. Besides being able to sort particles, it is able to sense the fluorescence of labeled particles or cells as well as determine their size. Presently, this instrument is being used to detect *Giardia* and *Cryptosporidium* in water samples in the United Kingdom and Australia (40). In combination with calcium carbonate flocculation, recoveries of *Giardia* cysts and *Cryptosporidium* oocysts range from 92 to 104% in spiked samples of wastewater, reservoir, and river water. A FACS instrument is capable of looking at much more of the sample than an immunofluorescence microscope can. Because fluorescent antibodies are used in the FACS procedure and cross-reactions with other fluorescent objects can occur, it is recommended that sorted fluorescent objects of the right size and shape be confirmed on an epifluorescence microscope equipped with DIC optics. Microscopic observation of the sample is much less fatiguing after FACS processing, as most of the contaminating debris and background fluorescence are eliminated. This approach allows most of the sample to be sorted in 5 to 15 min. Moreover, since more of the sample is observed than in the immunofluorescence procedure, the probability of finding the organisms of interest is greatly increased. The major disadvantages of this technique are the skill required of the FACS operator and/or microscopist and the high cost of the instrument.

Fluorescence imaging of *Cryptosporidium* oocysts and the sporozoites inside by using a CCD attached to a fluorescence microscope has been reported (5, 7). Originally a CCD was attached to a telescope and was designed to capture and intensify low light levels from distant celestial bodies, after cooling to 140 K with liquid nitrogen. Since a CCD also is able to capture a broad spectral range, it has the ability to detect fluorochromes of low intensity when attached to an epifluorescence microscope. Oocysts labeled with surface detection monoclonal antibodies and sporozoites labeled with DAPI, rhodamine 6-G, and/or propidium iodide were

detected and measured. Perhaps the most remarkable feature reported for the slow-scan CCD is its ability to detect C. *parvum* oocysts at a magnification of only ×20. With the immunofluorescence procedure, a 100× objective is required to visualize internal oocyst structures like sporozoites and their nuclei. The potential exists for easily determining sporozoite/oocyst ratios by using this highly experimental device. At present, the technique is limited because the sophisticated software needed to run it is not commercially available. Furthermore, since the CCD is in development, one does not know what effect autofluorescing algae and other naturally occurring organisms in a water sample will have.

The detection of *Giardia* cysts and *Cryptosporidium* oocysts by amplifying specific regions of DNA via PCR has been reported (24, 25, 42). A single *Giardia* cyst was detected by using giardin DNA as the target for PCR. Moreover, total mRNA increased significantly after induction of excystation in living *Giardia* cysts, making this the first successful molecular viability determination reported for this organism. Discrimination between *Giardia* species which are pathogenic to humans from those which are not was also possible by using PCR. Primers and probes have been reported for detecting C. *parvum* oocysts as well. Theory suggests that this procedure is just what is needed for detecting organisms in environmental samples that are usually low in density. However, analyses of environmental samples are compromised by the presence of inhibitors such as humic acids. Consequently, PCR works well only in highly purified samples. To overcome this problem, it has been suggested that water sample concentrates should be purified by chromatography over Sephadex G-100 and Chelex-100 spin columns (1). To date, successful application of this purification procedure has been reported only for enteric viruses. Attempts to use column purification and sample dilution in conjunction with PCR detection of *Giardia* cysts have not been very successful (32), and it has been concluded that other inhibitors of PCR besides organic compounds like humic acids are present. Other problems with PCR are that it is not quantitative and is susceptible to false-positive reactions from dead, deteriorated organisms (24).

Because of licensing agreements and proprietary concerns, little has been published other than in the popular press on the ERA (4). Like the FACS procedure, the ERA does not use the conventional immunofluorescence assay sampling and processing steps. Water samples of 10 to 20 liters are taken by using a foam depth filter. After being eluted from the filter, the particulates are exposed to anti-*Giardia* cyst and anti-*Cryptosporidium* oocyst antibodies which have been conjugated with paramagnetic beads. The resultant mixture is passed through an affinity immunocapture device which captures and washes the organisms of interest. The final step in this assay employs a microscope that has a special stage attachment composed of a well around which electrodes are strategically place. The electrodes are in turned connected to a power source. When an electric field is induced by the electrodes, the organism-bead complexes on the stage rotate at characteristic rates and directions. Observations are reported to be done at a magnification of ×400. No information is available regarding the percent recovery, specificity, sensitivity, etc., of this procedure.

Microscopic particulate analysis (MPA) (37) is a method that has evolved from the immunofluorescence microscopy methods used to detect *Giardia* and *Cryptosporidium* species.

Initially this method was developed to determine if ground-water was being impacted by surface water. In this protocol, 500 gal (1,893 liters) of water are sampled by using a depth fiber filter over an 8- to 24-h period. The particulates caught by the filter fibers are extracted, using particle-free water containing Tween, either by hand washing or with a stomacher blender (Tekmar catlog no. 10-0610-00). The particulates are concentrated by centrifugation. Samples with particulate concentrations of less than 20 μl/100 gal (379 liters) are observed directly, while samples with particulate concentrations of greater than 20 μl/100 gal are floated on Percoll-sucrose (specific gravity, 1.15). Unlike the parent immunofluorescence procedure, the particulates are not stained with monoclonal antibodies. Staining with Lugol's iodine may or may not be done, depending on whether bright-field microscopy is being used. The particulates are observed by either bright-field, phase-contrast, or DIC microscopy. All microorganisms in the size range of 1 to 400 μm are counted and measured with a calibrated ocular micrometer. These microorganisms, referred to as bioindicators in this method, include diatoms, algae, Giardia cysts, coccidian oocysts, plant debris, pollen, rotifers, crustaceans, amoebae, nematodes, insect parts, and insect larvae. The bioindicators are identified at least to taxonomic phylum and hopefully to class. In addition, inanimate particulates are tallied. Identification of Giardia cysts, coccidian oocysts, and/or enteric helminths is indicative of groundwater under the influence of surface water. The occurrence of pigment-bearing diatoms and algae may also be indicative of groundwater under the influence of surface water. While this list of bioindicators is long, there is little agreement among individuals using this method as to what level of risk they represent. To overcome this problem, this consensus MPA protocol categorizes each bioindicator into relative risk factor groups. On the basis of each bioindicator's density and risk factor category, the water is mathematically rated as to the level of risk for being under the influence of surface water. Recently the MPA method has been adapted to evaluate filtration plant efficiency. This is done by looking at pre- and post-filter water samples. In experienced hands, this method has great value. However, it requires much greater expertise than the immunofluorescence method from which it evolved and is susceptible to extreme subjectivity, especially in the hands of the inexperienced. Analysts need, besides extensive experience, a strong background in limnology, parasitology, phycology, and invertebrate zoology. Even though this method is very laborious, it still provides a more detailed analysis of the water than is possible from more automated approaches like particle counting.

EMERGING PROTOZOA OF POTENTIAL HEALTH SIGNIFICANCE IN WATER

Of all cases of waterborne gastroenteritis, only about 50% are attributable to a specific microorganism, toxin, or chemical. Twenty-five years ago G. lamblia was thought to be an innocuous commensal and C. parvum was basically unknown. With vigilant epidemiology and the ability to monitor these protozoa, we have come to appreciate their significance. As for Cryptosporidium species 15 years ago, there are now clinical reports of gastroenteritis caused by microsporidia, Cyclospora and Isospora species, and Blastocystis hominis in both immunocompetent and immunocompromised individuals. Since these organisms are enteric, are these emerging protozoans of potential health significance in water? At present, there are no methods other than clinical approaches

for detecting them in environmental samples. This does not preclude adaptation of existent techniques for Giardia and Cryptosporidium detection to look for them in water.

Microsporidia (a nontaxonomic term) are obligate intracellular parasites belonging to the class Microsporididea of the protozoan phylum Microspora. They are ubiquitous parasites infecting a variety of vertebrate and invertebrate hosts. From an anthropocentric point of view, insects of commercial significance like honeybees and silkworms are affected by this group. Moreover, moving up the phylogenetic tree, snails and commercial fish such as salmon, flounder, and monkfish are prone to microsporidian diseases. Only with the advent of the AIDS epidemic was this group of pathogens recognized as a cause of human disease. Six genera of microsporidia have been recovered in humans: Encephalitozoon, Nosema, Pleistophora, Enterocytozoon, Septata, and "Microsporidium," a category for all forms as yet unclassified. Although rare, microsporidian infections are now being found in immunocompetent people. Presently, classification is done on the basis of spore size (1.5 to 5 μm), nuclear configuration, the number of polar tube coils within the spore and developing forms, and the host cell-parasite relationship. Infection by these parasites is via either ingestion, inhalation, or inoculation. Two enteric forms, Enterocytozoon bieneusi and Septata intestinalis, may be transmitted by the water route. Furthermore, Encephalitozoon cuniculi and E. hellem have been recovered from the central nervous system and eye, respectively, but can disseminate to the urinary tract, from which spores can be released to the environment. Consequently, Encephalitozoon spp. seem to be likely candidates for waterborne transmission. Since microsporidia have been detected in clinical specimens with indirect fluorescent poly- and monoclonal antibodies (2), this approach could be adapted for detecting them in water samples in a fashion similar to the immunofluorescence assay for Giardia and Cryptosporidium detection.

Cyclospora sp., which is an apicomplexan, is being incriminated as a cause of prolonged diarrhea in humans in the developing world and more recently in developed countries. The oocysts are 8 to 10 μm in diameter and are similar to Cryptosporidium and Isospora species, since all have an acid-fast staining characteristic. However, Cyclospora oocysts have the unique characteristic of autofluorescing when viewed by UV microscopy.

Three species of Isospora are reported to be parasitic in humans: I. belli, I. hominis, and I. natalensis. Confusion exists regarding the status of I. belli and I. hominis, as some investigators believe them to be the same organism. There is even a report suggesting that I. hominis is really a species of Sarcocystis. I. natalensis, which is rare and occurs only in South Africa, will not be discussed. Unlike C. parvum oocysts, I. belli oocysts are not fully sporulated at the time of evacuation. When sporulation, which requires about 48 h at room temperature, is complete, the oocyst has two sporocysts, each containing four crescent-shaped sporozoites. Overall, I. belli oocysts are elongated oval structures, measuring 20 to 30 μm in length by 10 to 19 μm in width, with both ends being somewhat narrow. Oocysts of I. belli can be detected on the basis of their shape and ability to be acid fast stained. Like Cryptosporidium oocysts and Giardia cysts, I. belli oocysts can be viewed with phase-contrast and Nomarski DIC microscopy. Dogs have been suspected of being a reservoir host for I. belli.

I. hominis oocysts, which are ovoid and 25 to 33 μm in length, are fully sporulated and contain two sporocysts before being evacuated from the host. At the time of passage, the oocyst has broken open, and ripe sporocysts containing four sporozoites are passed in the feces. The sporocysts measure around 14 μm in length. Visualization can be by acid-fast staining and bright-field microscopy or by either phase-contrast or Nomarski DIC microscopy with unstained material.

Blastocystis hominis is a yeast-like organism of questionable taxonomic status. Depending upon the source or expert consulted, *B. hominis* may or may not be classified as a pathogen. Generally, those classifying it as a pathogen do so because drug treatment eliminates this organism, as well as the mild gastroenteritis that is present, when no other etiologic agent can be detected. Often confused with amoebic cysts, *B. hominis* varies in size and has a large vacuole surrounded peripherally by granules in addition to a single nucleus. Visualization can be by trichrome stain.

SUMMARY

Protozoa are eukaryotic organisms with either a free-living or parasitic existence. Some free-living forms, under the right conditions, can become opportunistic pathogens. Enteric pathogenic protozoa such as *Giardia* and *Cryptosporidium* species, which are now known to be transmitted by water, have been responsible for numerous waterborne outbreaks of gastroenteritis. The primary means for detection, since density levels in water are low, involves concentrating a large volume of water by filtration, extracting the particulates from the filter, separating the organisms from the particulates, and assaying for the pathogens. The most widely used method for detecting these protozoa has been the indirect immunofluorescence assay. Besides being extremely labor-intensive and highly dependent on the skill of the microscopist, this technique is known to have a number of deficiencies, including false-positive results, inability to determine the viability of the detected organisms, and poor recovery efficiency.

Various attempts have been made to improve the immunofluorescence detection method. Rather than sampling by filtration and buoyant density centrifugation to purify the organisms, carbonate flocculation is reported to improve recoveries. PCR, CCD, ERA, and FACS are currently being evaluated as alternate test procedures. As each of these approaches is relatively new and much more research is needed, it remains to be seen whether they will be equal to or better than the current IFA procedure.

REFERENCES

1. **Abbaszadegan, M., M. S. Huber, C. P. Gerba, and I. L. Pepper.** 1993. Detection of enteroviruses in groundwater with the polymerase chain reaction. *Appl. Environ. Microbiol.* **59:**1318–1324.

2. **Aldras, A. M., J. M. Orenstein, D. P. Kotler, J. A. Shadduck, and E. S. Didier.** 1994. Detection of microsporidia by indirect immunofluorescence antibody test using polyclonal and monoclonal antibodies. *J. Clin. Microbiol.* **32:**608–612.

3. **American Society for Testing and Materials.** 1992. *Annual Book of ASTM Standards 11.01,* p. 925–935. American Society for Testing and Materials, Philadelphia.

4. **Butt, J. P. H., R. Pethig, A. Parton, and D. Dawson.** Electrorotation assay for *Cryptosporidium* and *Giardia* from raw water supplies. *In Proceedings of the 9th International Conference on Electrostatics,* in press.

5. **Campbell, A. T., R. Haggart, L. J. Robertson, and H. V. Smith.** 1992. Fluorescent imaging of *Cryptosporidium* using

a cooled charge couple device (CCD). *J. Microbiol. Methods* **16:**169–174.

6. **Campbell, A. T., L. J. Robertson, and H. V. Smith.** 1992. Viability of *Cryptosporidium parvum* oocysts: correlation of in vitro excystation with inclusion/exclusion of fluorogenic vital dyes. *Appl. Environ. Microbiol.* **58:**3488–3493.

7. **Campbell, A. T., L. J. Robertson, and H. V. Smith.** 1993. Novel methodology in the detection of *Cryptosporidium parvum*: a comparison of cooled charge couple device (CCD) and flow cytometry. *Water Sci. Technol.* **27:**89–92.

8. **Craun, G. F.** 1990. Waterborne giardiasis, p. 267–293. *In* E. A. Meyer (ed.), *Giardiasis.* Elsevier Science Publishers, Amsterdam.

9. **Dagget, P.-M.** 1982. Algae and protozoa, p. 31–56, 601–656. *In* H. D. Hall (ed.), *Catalog of Strains,* 15th ed., vol. I. American Type Culture Collection, Rockville, Md.

10. **de la Cruz, A. A., and M. Sivaganesan.** 1995. Detection of *Giardia* and *Cryptosporidium* spp. in source water samples by commercial enzyme-immunosorbent assay, p. 543–554. *In Proceedings of the 1994 Water Technology Conference.* American Water Works Association, Denver.

11. **Erlandsen, S. L., H. van Keulen, A. Gurien, W. Jakubowski, F. W. Schaefer III, P. Wallis, D. Feely, and E. Jarroll.** 1994. Molecular approach to speciation and detection of *Giardia*: fluorochrome-rDNA probes for identification of *Giardia lamblia, Giardia muris,* and *Giardia ardeae* in laboratory and environmental samples by in situ hybridization, p. 64–66. *In* R. C. A. Thompson, J. A. Reynoldson, and A. J. Lymbery (ed.), *Giardia: from Molecules to Disease.* CAB International, Oxon, England.

12. **Fields, B. S., G. N. Sanden, J. M. Barbaree, W. E. Morrill, R. M. Wadowsky, E. H. White, and J. C. Feeley.** 1989. Intracellular multiplication of *Legionella pneumophila* in amoebae isolated from hospital hot water tanks. *Curr. Microbiol.* **18:**131–137.

13. **Hansen, J. S., and J. E. Ongerth.** 1991. Effect of time and watershed characteristics on concentration of *Cryptosporidium* oocysts in river water. *Appl. Environ. Microbiol.* **57:**2790–2795.

14. **Hayes E. B., T. D. Matte, T. R. O'Brien, T. W. McKinley, G. S. Logsdon, J. B. Rose, B. L. P. Ungar, D. M. Word, P. F. Pinsky, M. L. Cummings, M. A. Wilson, E. G. Long, E. S. Hurwitz, and D. D. Juranek.** 1989. Large community outbreak of cryptosporidiosis due to contamination of a filtered public water supply. *N. Engl. J. Med.* **320:**1372–1376.

15. **Jahn, T. L., and F. F. Jahn.** 1979. *How To Know the Protozoa.* Wm. C. Brown Publishers, Dubuque, Iowa.

16. **Jakubowski, W.** 1984. Detection of *Giardia* cysts in drinking water, p. 263–286. *In* S. L. Erlandsen and E. A. Meyer (ed.), *Giardia and Giardiasis: Biology, Pathogenesis, and Epidemiology.* Plenum Press, New York.

17. **Krogstad, D. J., G. S. Visvesvara, K. W. Walls, and J. W. Smith.** 1985. Blood and tissue protozoa, p. 612–630. *In* E. H. Lennette, A. Balows, W. J. Hausler, Jr., and H. J. Shadomy (ed.), *Manual of Clinical Microbiology,* 4th ed. American Society for Microbiology, Washington, D.C.

18. **Kudo, R. R.** 1966. *Protozoology,* 5th ed. Charles C Thomas, Springfield, Ill.

19. **LeChevallier, M. W., W. D. Norton, and R. G. Lee.** 1991. *Giardia* and *Cryptosporidium* spp. in filtered drinking water supplies. *Appl. Environ. Microbiol.* **57:**2617–2621.

20. **LeChevallier, M. W., W. D. Norton, and R. G. Lee.** 1991. Occurrence of *Giardia* and *Cryptosporidium* spp. in surface water supplies. *Appl. Environ. Microbiol.* **57:**2610–2616.

21. **LeChevallier, M. W., W. D. Norton, J. E. Siegel, and M. Abbaszadegan.** 1995. Evaluation of an immunofluorescent procedure for detection of *Giardia* cysts and *Cryptosporidium* oocysts in water. *Appl. Environ. Microbiol.* **61:**690–697.

22. Lee, J. J., S. H. Hutner, and E. C. Lee. 1985. *An Illustrated Guide to the Protozoa.* Society of Protozoologists, Lawrence, Kans.
23. MacKenzie, W. R., N. J. Hoxie, M. E. Proctor, M. S. Gradus, K. A. Blair, D. E. Peterson, J. J. Kazmierczak, D. G. Addiss, K. R. Fox, J. B. Rose, and J. P. Davis. 1994. A massive outbreak in Milwaukee of *Cryptosporidium* infection transmitted through the public water supply. *N. Engl. J. Med.* **331:**161–167.
24. Mahbubani, M. H., A. K. Bej, M. Perlin, F. W. Schaefer III, W. Jakubowski, and R. M. Atlas. 1991. Detection of *Giardia* cysts by using polymerase chain reaction and distinguishing live from dead cysts. *Appl. Environ. Microbiol.* **57:**3456–3461.
25. Mahbubani, M. H., A. K. Bej, M. H. Perlin, F. W. Schaefer III, W. Jakubowski, and R. M. Atlas. 1992. Differentiation of *Giardia duodenalis*, and other *Giardia* spp. by using polymerase chain reaction and gene probes. *J. Clin. Microbiol.* **30:**74–78.
26. Musial, C. E., M. J. Arrowood, C. R. Sterling, and C. P. Gerba. 1987. Detection of *Cryptosporidium* in water by using polypropylene cartridge filters. *Appl. Environ. Microbiol.* **53:**687–692.
27. Nieminski, E. C., F. W. Schaefer III, and J. E. Ongerth. 1995. Comparison of two methods for the detection of *Giardia* cysts and *Cryptosporidium* oocysts in water. *Appl. Environ. Microbiol.* **61:**1714–1719.
28. Payment, P., and E. Franco. 1993. *Clostridium perfringens* and somatic coliphages as indicators of the efficiency of drinking water treatment for viruses and protozoan cysts. *Appl. Environ. Microbiol.* **59:**2418–2424.
29. Radulescu, S., and E. A. Meyer. 1990. In vitro cultivation of *Giardia* trophozoites, p. 99–110. *In* E. A. Meyer (ed.), *Giardiasis.* Elsevier Science Publishers, New York.
30. Rivera, F., G. Roy-Ocotla, I. Rosas, E. Ramirez, P. Bonilla, and F. Lares. 1987. Amoeba isolated from the atmosphere of Mexico City and environs. *Environ. Res.* **42:**149–154.
31. Robertson, L. J., A. T. Campbell, and H. V. Smith. 1993. Induction of folds or sutures on the walls of *Cryptosporidium parvum* oocysts and their importance as a diagnostic feature. *Appl. Environ. Microbiol.* **59:**2638–2641.
32. Rodgers, M. R., C. M. Bernardino, and W. Jakubowski.

1993. A comparison of methods for extracting amplifiable *Giardia* DNA from various environmental samples. *Water Sci. Technol.* **27:**85–88.
33. Rodgers, M. R., D. J. Flanigan, and W. Jakubowski. 1995. Identification of algae which interfere with the detection of *Giardia* cysts and *Cryptosporidium* oocysts and a method for alleviating this interference. *Appl. Environ. Microbiol.* **61:**3759–3763.
34. Sauch, J. F., D. Flanigan, M. L. Galvin, D. Berman, and W. Jakubowski. 1991. Propidium iodide as an indicator of *Giardia* cyst viability. *Appl. Environ. Microbiol.* **57:**3243–3247.
35. Schupp, D. G., and S. L. Erlandsen. 1987. A new method to determine *Giardia* cyst viability: correlation of fluorescein diacetate and propidium iodide staining with animal infectivity. *Appl. Environ. Microbiol.* **53:**704–707.
36. Upton, S. J., M. Tilley, and D. B. Brillhart. 1994. Comparative development of *Cryptosporidium parvum* (Apicomplexa) in 11 continuous host cell lines. *FEMS Microbiol. Lett.* **118:**233–236.
37. U.S. Environmental Protection Agency. 1992. *Consensus Method for Determining Groundwaters under the Direct Influence of Surface Water Using Microscopic Particulate Analysis (MPA).* EPA 910/9-92-029. U.S. Environmental Protection Agency, Washington, D.C.
38. Vasconcelos, J. (U.S. Environmental Protection Agency, Region X.) 1992. Personal communication.
39. Vesey, G., P. Hutton, A. Champion, N. Ashbolt, K. L. Williams, A. Warton, and D. Veal. 1994. Application of flow cytometric methods for the routine detection of *Crystosporidium* and *Giardia* in water. *Cytometry* **16:**1–6.
40. Vesey, G., J. S. Slade, M. Byrne, K. Shepard, and C. R. Fricker. 1993. A new method for the concentration of *Cryptosporidium* oocysts from water. *J. Appl. Bacteriol.* **75:**82–86.
41. Walker, P. L., P. Prociv, W. G. Gardiner, and D. E. Moorhouse. 1986. Isolation of free-living amoebae from air samples and an air conditioner filter in Brisbane. *Med. J. Aust.* **145:**3–4.
42. Webster, K. A., J. D. E. Pow, M. Giles, J. Catchpole, and M. J. Woodward. 1993. Detection of *Cryptosporidium parvum* using a specific polymerase chain reaction. *Vet. Parasitol.* **50:**35–44.

Detection of Viruses in Environmental Waters, Sewage, and Sewage Sludges

CHRISTON J. HURST

17

Many different groups of viruses can be found in environmental waters. These include the many types of viruses whose hosts are natural aquatic organisms (17). There also are groups of viruses present in environmental waters which represent exogenous contaminants, whose hosts are nonaquatic plants and animals. The water-associated viruses of greatest concern from the viewpoint of human public health are those that replicate in cells of the human gastrointestinal tract. These are referred to as human enteric viruses, and they are shed in fecal material. The principal transmission route for enteric viruses is fecal-oral, meaning that they can cause illness when a susceptible host subsequently ingests fecally contaminated water or food. Septic effluents and wastewater from human populations doubtlessly contribute many of the infectious human enteric viruses that are found in surface waters and groundwaters (13, 15). The presence of human enteric viruses in surface water can also result from human recreational activities performed either in or around those bodies of water.

At least some of the human enteric viruses can cross-infect, meaning that they are capable of also replicating within animals other than humans. As an example, acquisition of the disease poliomyelitis, caused by a member of the genus *Enterovirus*, has been documented in a wild population of chimpanzees living near a human community (16). Other genera of enteric viruses which should be considered to have the potential for cross-infection of related mammalian host species are *Orthoreovirus* (5), *Rotavirus* (5), and *Mastadenovirus* (14). Likewise, some enteric viruses of animals likely can cross-infect into humans. Sources of such viruses are fecal wastes from animals, including farm wastewater which drains into surface waters, overland runoff containing animal fecal wastes, and fecal wastes directly deposited in the water by animals. It is difficult, and in some instances may be impossible, to determine the host species from which many of these potentially cross-infective enteric viruses originate. Thus, for the purpose of protecting public health, the presence in water of any viruses capable of replicating in mammalian host animals or cultured mammalian cells is presumed to represent a pathogenic hazard to humans.

Enteric viruses are capable of surviving improperly operated conventional drinking water treatment processes and may therefore be found in potable water (6). This knowledge has resulted in the development of methods for detecting both human enteric viruses and fecally associated bacteriophages in water (3, 7). These bacteriophages are not considered to be enteric in nature because they infect the intestinal bacteria rather than the animal host, and thus they themselves do not pose a public health threat. Instead, these bacteriophages have been studied as potential indicators of the presence of fecal material and on the premise that they may prove useful as indicators by which to gauge the fate of human enteric viruses within the natural environment. They also may prove useful as indicators of the removal or destruction of human enteric viruses during water treatment processes (6).

The levels of bacteriophages in environmental waters, particularly those that receive sewage effluent, are often sufficiently high that the water can be directly assayed for the presence of these bacteriophages without need for resorting to concentration techniques. This represents a desirable advantage, since viral concentration techniques have a measure of inherent inefficiency which results in some loss of virus. Also, some virus inactivation may occur during concentration processes as a result of such factors as changes in pH and the addition of chelating chemicals like EDTA. The levels of human enteric viruses present in environmental waters often are so low that the viruses must be concentrated from the water before a successful, cost-effective assay can be performed. This chapter describes techniques which can be used for concentrating human enteric viruses from environmental water, drinking water, raw wastewater, wastewater sludges, and wastewater effluents.

MECHANISMS INVOLVED IN VIRAL CONCENTRATION METHODS
Descriptions

The numerous approaches that have been used for concentrating viruses contained in water samples (7, 12) can be separated into five major categories based on the mechanisms involved (7).

(i) *Passive adsorption* relies upon unaided entrapment of viruses via adsorption or absorption into pads of gauze and cotton, used in stationary or flowthrough configurations. This procedure is followed by expression of the entrapped fluids, either with or without the supplemental use of an eluant solution, to recover the viruses.

(ii) *Directed adsorption* entails adsorption of viruses to the

168

surface of filter materials or granular solids, facilitated in many instances by pretreatment of the water. Processes which have been evaluated as pretreatments include removing soluble organics from the water by using charged resins, adjustment of the water sample pH, and addition of salts to the water. Efforts have also been made to pretreat the adsorbent material itself in order to enhance the efficiency of viral adsorption. Examples of the latter approach include the binding of metal precipitates or charged polymers to the matrix of microporous filters. The bound viruses are then recovered from the filter materials or granular solids used as adsorbents by exposing the adsorbents to an eluant solution.

(iii) *Ultrafiltration* involves retention of viruses in the original water sample during a reduction of its volume by pore size exclusion. This generally is accomplished by recirculating the water, under pressure, through hollow fiber filters or, in either recirculating or nonrecirculating systems, over the surface of flat-sheet filters. In some cases, an eluant is subsequently passed through the same concentration unit to facilitate virus recovery.

(iv) *Direct physicochemical flocculation* or *phase separation* is a process by which viruses are concentrated either by association with a generated precipitate, by a selective partitioning of the base water sample during processes which yield a polymeric phase separation, or by retention in the original water sample during a reduction in its volume through hydroextraction.

(v) *Affinity chromatography* is the process whereby viruses are retained during passage of the water sample through a column of polysaccharide gel particles which bear covalently linked antibodies that specifically recognize and bind a particular virus type. Viruses are then released from the column by alterations in the composition or ionic strength of the buffer fluid used for maintaining hydration of the polysaccharide gel.

Theoretical Explanation of Directed Adsorption, Elution, and Reconcentration

Directed adsorption has generally replaced the other types of concentration techniques for use in recovering viruses from large sample volumes of water. The different types of solid adsorbents that have successfully been used can be divided into two groups, filters and granular solids. The types of filters that have been tested are composed of yarn fiber wound around a hollow core to form a depth filtration cartridge, sheet filter materials configured either as flat layers or as cartridges composed of pleated sheets of filter material, and cartridges prepared as hollow tubes of filter material. Viruses adsorb onto the filter matrix during the passage of virus-containing water samples through these filters. The adsorbed viruses are normally recovered by subsequently passing an eluant through the filters or by dissolution of the filter if it is made of alginate. Granular solids generally are used as viral adsorbents in one of three modes: batch utilization, whereby the granules are mixed into the water sample and then recovered by using either filtration or magnetic attraction, the latter requiring that the granules be magnetic; as layers of granules either supported on or sandwiched between sheets of non-virus-adsorbing flat filter material through which the water sample is passed; or as columns or fluidized beds of granules exposed to the water sample by using a flowthrough configuration. Subsequent desorption of viruses from the granular solids occurs during exposure of the granules to an eluant, using one of these same three modes (7).

Of the many different types and configurations of virus adsorbents, the ones that currently seem to be preferred for use in recovering viruses from large volumes of water are wound yarn cartridge filters, pleated sheet cartridge filters composed of either glass microfiber or nylon, including sheet filter material types that are modified to have a more electropositive charge, and columns of glass powder. Filters seem to offer some advantage over glass powder columns in that the filters are easily transported to the field, even by postal shipment, and then can be readily used in the field and shipped back to the processing laboratory with very little concern about detrimental effects on either the filter or the adsorbed viruses. This chapter describes the use of cartridge filter-based methods for concentrating viruses from environmental water, drinking water, and sewage effluent. The methods presented for isolating viruses from raw sewage (raw wastewater) and wastewater sludges effectively utilize a process of directed adsorption and elution but differ in that they use the wastewater solids as an in situ adsorbent.

The adsorption of viruses onto filters, and presumably other solid adsorbents as well, is governed both by electrostatic interactions, which predominate at lower pH levels, and hydrophobic interactions, which predominate at higher pH levels. The chemical composition of the solid adsorbent and the fluid in which the viruses are suspended will influence the extent to which either attraction or repulsion occurs between the viruses and adsorbent. Two other important factors that relate to virus adsorption from water include the relative proximity and amount of contact time allowed between the adsorbent and the viruses in the water sample. These latter factors are influenced by (i) the rate at which the water being sampled flows either past or through the solid adsorbent and (ii) the ratio of the diameter of the virus particle to the diameter of any pores in the adsorbent material through which the water may pass. Virus adsorption can be inhibited by the presence of added proteins in the input virus suspension, possibly the result of competition between the proteins and viruses for adsorption sites extant on the surface of the solid matrix. Other types of soluble organics such as humic and fulvic acids may also interfere with virus adsorption. Means of facilitating the adsorption of viruses onto solid matrices include removing dissolved organic materials from the water samples by passing the water through a resin column and adding salts, including chlorides of either sodium, magnesium, or, more effectively, aluminum. Virus adsorption to filters can also be facilitated by adjusting (usually lowering) the pH of a water sample, a process that can be performed by adding HCl either through a batch operation or by using an in-line injector configured ahead of the filter (2). A pH of approximately 3.5 to 4.0 seems preferred for use with negatively charged filters (1, 7), whereas a pH of near neutrality appears preferable for the more positively charged filters (7). There may also be optimal pH levels for adsorption of viruses to the different types of granular solids.

The adsorption process is reversible, and the reverse process is termed elution. Elution is accomplished by exposing the adsorbent to an aqueous solution termed an eluant, following which the eluant is termed an eluate and contains at least some of the viruses which had been adsorbed to the solid matrix. Eluants can be divided into at least two categories based on their modes of action (7). The first category consists of proteinaceous materials which simply compete with the proteins of the virus particles for binding sites on the adsorbent; the second category consists of compounds that alter the favorability of adsorption. Beef extract is now the most widely used eluant material of the first category and is recommended in this chapter, although other protein products may prove to be suitable substitutes. The second category of eluants includes solutions that contain various active substances, among which are chaotropic agents like glycine or trichloroacetic acid, detergents like Tween 80, and EDTA as a chelating agent for metal cations. Both categories of eluant may be used at an elevated pH, which would help to decrease electrostatic attraction between the viruses and adsorbents. Eluants containing viruses which have been desorbed from filters generally are reduced in volume by a secondary, or second-step, concentration technique before being assayed for the presence of viruses. The secondary concentration method customarily used for beef extract-based eluants is organic flocculation as introduced by Katzenelson and coworkers (10), performed by lowering the pH of the eluate, during which the proteinaceous material supplied by the beef extract spontaneously precipitates. Viruses contained in the beef extract become associated with this precipitate, which can subsequently be collected by centrifugation and dissolved in a small amount of higher-pH salt-based buffer such as sodium phosphate or glycine. A pH of approximately 3.5 appears optimal for recovering viruses from beef extract eluates via organic flocculation (10).

SAMPLING APPARATUS AND PROCESSING EQUIPMENT

A variety of types of sampling equipment may be needed for detecting viruses in environmental waters. Such equipment includes not only the usual sampling jars or bottles but often also buckets, carboys, pumps, cartridge filters and their holders, and hoses. If it is necessary to store a volume of sample water which is too large to fit in a standard jug or carboy, then a new sterilized plastic garbage can with a tightly fitting cover can be used as a water storage container. It is preferable that containers used for water samples be made of polypropylene, which can reduce unintentional adsorption of viruses to container walls (11). Alternatively, it is suggested that the apparatus be made of either stainless steel, brass, or a chemically resistant polymer, which will aid in terms of ruggedness for use in field operations, cleanability, and resistance to the chemical corrosion that can be caused by salt or extremes of pH.

If chlorinated tap waters or wastewater effluents will be sampled, then a particularly helpful item to have available in the field is a compact field-type N, N-diethyl-p-phenylene diamine (DPD) chlorine test kit. This kit can be used to confirm that by adding the prescribed quantity of sodium thiosulfate solution, you have successfully neutralized all free chlorine or other chlorine compounds present in the water sample. Disinfectant neutralization must be accomplished to prevent disinfectant-related die-off of the viruses before the water sample is processed on site or transported for off-site (i.e., laboratory) processing. If viruses will be concentrated from water samples in the field by using cartridge filters, a portable pH meter may also be needed in the event that the pH of the waters to be sampled must be adjusted prior to filtration. When virus-adsorbing filters that are considered electropositive are used, pH adjustment would likely be required only for recovery of viruses from a water whose pH was above approximately 7.5. Alternatively, negatively charged filters can be used; they require that the sample water first be adjusted to a pH of approximately 3.5 and supplemented with aluminum chloride prior to filtration.

Cartridge Filtration Apparatus

The types of apparatus used for concentrating viruses from large volumes of environmental waters, drinking water, or sewage effluent by means of cartridge filtration are not standard equipment for most environmental microbiology laboratories. Thus, the equipment used with the cartridge filters will be described in some detail. Figure 1 presents the type of equipment configuration used for concentrating viruses by means of cartridge filters. A lightweight portable electric or gasoline-powered water pump fitted for use with either brass or stainless steel quick-disconnect plumbing adaptors or garden hose couplings can be used to supply pressure for passing water or sewage effluent through cartridge filters. Such a pump will not be needed if the water or wastewater is being supplied directly via a tap from a pressurized source. Figure 1 also shows two filter holders designed for use with standard 10-inch (25.4-cm)-long cartridge water filters, fitted for use with either quick-disconnect plumbing adaptors or garden hose couplings. The first holder contains a non-virus-adsorbing prefilter that may be needed to prevent clogging of the virus-adsorbing filter, which is contained in the second filter holder. If it is necessary to modify the water

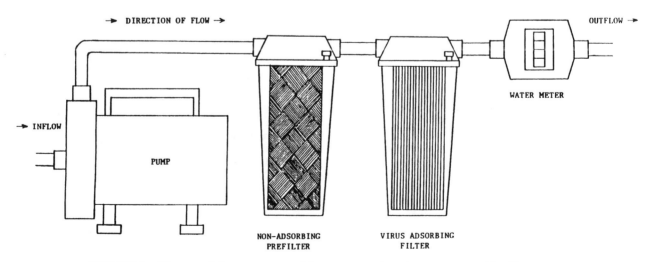

FIGURE 1   Diagram of the apparatus used for concentrating viruses onto cartridge filters.

pH or to add aluminum chloride to aid viral adsorption, then an in-line injector can be configured between the pump (or pressurized tap) and the first filter holder. An alternative to using an injector is to first pump the water sample into a large temporary storage tank, make any necessary chemical adjustments (supplementations) to the water while it is in that tank, and then pump the adjusted water from the tank through the filters. The filter holders should have clear base sections so that the filtration process can be visually monitored, particularly for air pockets, which occasionally develop within the filter holders and tend to impede the water flow. Suitable holders should be available from the manufacturers and distributors from which the cartridge filters are obtained. Water being processed through these filters flows from the outside of the filter, through the filter material, and into the hollow core. It is helpful if the top of each filter holder is fitted with a stainless steel finger-operated air pressure release valve connecting to the inflow side of the filter holder. The installation of air pressure release valves in this manner allows for easy elimination of air pockets from the filter holders. Figure 1 also shows a portable water meter fitted for use with either quick-disconnect plumbing adaptors or garden hose couplings attached to the outlet end of the second filter holder. Suitable meters are available from plumbing suppliers and used for metering the amount of water which is passed through the filters. Four lengths of fiber-reinforced garden hose, likewise fitted for use with either quick-disconnect plumbing adaptors or garden hose couplings, are used for connecting the pump, filter holders containing their cartridge filters, and water meter to form the virus concentration apparatus shown in Fig. 1. The fourth piece of hose is used to direct outflow from the water meter. Also, one length of strong-walled water supply hose, fitted for use with either quick-disconnect plumbing adaptors or garden hose couplings, is required for leading water into the pump. Standard garden hose may not be suitable for this latter usage since many types of garden hose tend to collapse inward during operation of the pump. It may be helpful to have a stainless steel or brass wire strainer fitted to the intake end of the water supply hose when one is concentrating environmental waters by pumping directly from their source through the filters. Use of a strainer in this instance will help to prevent the water supply hose from clogging. The suggested maximum force for driving a water sample through these filters is 30 lb/in^2. If the pump being used is particularly powerful, it may be necessary to install a flow-restricting device ahead of the first filter holder. An ice chest containing either wet ice or commercial ice packs is needed for transport (or shipment via overnight delivery service) of the used virus-adsorbing filter, inside a polypropylene container of eluant, to the laboratory where the elution and secondary concentration steps are to be performed.

Sterilization Requirements

All reagents that will be used during any of these methods must be sterilized prior to use. Solutions of HCl or NaOH which are at least 1 M in strength are considered to be self-sterilizing, as are solutions of at least 0.5% (wt/vol) sodium hypochlorite or calcium hypochlorite. Containers used to store water or wastewater samples should be presterilized. All parts of equipment that will come into contact with the samples should likewise be visibly clean and sterile. The only parts of a pump or fluid injector which need to be sterile are those that come in contact with the water sample. These parts can be treated by a hypochlorite solution sterilization technique. The choice of methods to use for steriliz-

ing different items of equipment will depend on the size of the equipment and the item's resistance to heat. Many types of apparatus can be sterilized in the laboratory before their use in the field, either by autoclaving, treatment with ethylene oxide gas, or, if accessible, gamma irradiation. All of the cartridge filters described in this study can be presterilized within their holders by ethylene oxide gas treatment before they are transported to the field. Alternatively, some filters are sufficiently resistance to heat and steam that they can be sterilized by autoclaving.

Techniques for Field Sterilization of Sampling Equipment

Field sterilization of metal or borosilicate glass objects can be performed by using flame from a portable gas torch (such as propane or butane) to heat the sample collection objects to red incandescence. Objects less resistant to high heat, such as nonborosilicate glass and some polymer materials, can be surface sterilized by dousing or immersing them in commercial 95% ethanol and then igniting the alcohol with a flame. Nearly all objects described in this chapter except filters and liquids can be subjected to field sterilization techniques which consist of chlorine-based chemical treatment with solutions such as calcium hypochlorite or sodium hypochlorite followed by the use of sodium thiosulfate to neutralize the residual disinfectant activity.

If the water sample is to be placed into a storage container for any reason, then the container must first be treated by a sterilization technique. A method for field sterilizing water storage containers is described below. The container used for storage of a water sample must be visibly clean, should be sterile, and preferably should not previously have been used for any purpose other than the storage of water samples. Large (≥50-gal [1 gal = 3.785 liters] capacity) polyethylene containers with lids are available from scientific supply companies and can be used for storage of water samples. In an emergency situation, as mentioned earlier, a new plastic garbage container with a lid can be used for this purpose. One can field sterilize containers by first completely filling them with clear water, and then adding calcium hypochlorite solution (0.5% [wt/vol]) to the water at a rate of 3.8 ml/gal. A 10% solution of standard household liquid bleach (sold as 5.25% [wt/wt] sodium hypochloride) can be used as a substitute for the calcium hypochlorite solution. The hypochlorite solution is throughly mixed into the water, which is then allowed to remain in the storage container for at least 15 min. The chlorinated water is then emptied from the storage container, and the container is rinsed thoroughly with clear water and again completely filled with clear water. Sterile sodium thiosulfate solution (50% [wt/vol]) should be added to this second filling of water at a rate of 10 ml/gal. The sodium thiosulfate solution is thoroughly mixed into the water, and the water is then allowed to remain in the storage container for at least 15 min. The container is then emptied and filled with the intended water sample.

If necessary, all of the equipment used in the virus concentration apparatus (Fig. 1) except the filters can be subjected to this chlorine-based field sterilization technique. The apparatus shown in Fig. 1 should be disassembled prior to treatment by the field sterilization technique to ensure that all surfaces that may later come into contact with the sample water will have received thorough exposure to the chlorine disinfectant and that all residual hypochlorite is then neutralized before the water sample is processed through the system. Hoses and assembled filter holders (without filters), with their air pressure release valves open, should be carefully submerged in a solution of clear water containing hypochlorite prepared as described above for treatment of storage containers. Be certain that there are no pockets of trapped air in either the hoses or filter holders, as this may preclude thorough treatment. Keep the hoses and filter holders completely submerged in this chlorinated water for at least 15 min, and then remove and drain the hoses and filter holders. Next, similarly treat the hoses and filter holders by submersion for at least 15 min in clear water containing sodium thiosulfate solution prepared as described above for treatment of sample storage containers. The hoses and filter holders should be drained and can then be used immediately; if they are to be used at some later time, their openings should be wrapped with sterile aluminum foil or some other type of sterile covering to prevent contamination. This chlorine-based field sterilization technique can be carried out on pumps and injectors by first continuously recirculating clear water containing hypochlorite solution, prepared as described above, through them for at least 15 min. Residual chlorinated water should then be drained from the pump or injector and its necessarily attached hoses. Clear water containing sodium thiosulfate solution, likewise prepared as described above, should then be recirculated through the pump or injector and attached hoses for at least 15 min. Residual water containing thiosulfate solution should then be drained from the pump or injector and attached hoses. The pump, injector, and hoses can then immediately be used to process a water sample; if they are not used immediately, their openings should receive a sterile covering as described above to prevent contamination. When properly performed, this chlorine-based field sterilization technique should be sufficient to destroy the infectiousness of viral contaminants on the treated surfaces of the equipment. This technique may not be sufficient to destroy some encysted protozoans, which can have exceptional resistance to chlorine.

DETECTION OF VIRUSES

Detecting viruses in water involves a three-stage process. The first stage consists of either collecting a grab sample for processing in the laboratory or performing a field operation in which the water being sampled is passed through cartridge filters, which are then usually returned to the laboratory for processing. The grab sampling approach is generally used when either the level of viruses or the content of solids is relatively high, as is the case for raw wastewater or wastewater sludges. If the level of viruses and the concentration of solids is expected to be relatively low, as is the case with most environmental surface water, groundwater, treated drinking water, and wastewater effluents, then the cartridge filtration approach is likely to be used. The second stage consists of laboratory elution processing, either concentrating viruses onto the solids in the grab sample and then eluting those solids or eluting the cartridge filters. Usually these eluates are further reduced in volume through a subsequent concentration process. The third stage consists of assaying viruses contained in the eluates.

Viral Concentration Techniques

The viral concentration techniques presented in this chapter are categorized according to the type of environmental sample being examined: environmental waters (fresh or saline), treated drinking water, wastewater effluent, raw wastewater, and wastewater sludges.

Environmental Water, Drinking Water, and Sewage Effluent

The procedure described below can be used to concentrate suspected human enteric viruses from environmental surface waters, including brackish and marine waters, groundwater, tap water, and sewage effluents. The produced concentrated sample materials can then be examined for the presence of viruses. The viruses are first adsorbed onto cartridge membrane filters. A beef extract-based eluant fluid is then used to desorb viruses from the filters. Viruses contained in the eluate are then secondarily concentrated by low-pH organic flocculation. The volume of water sample that is processed for detection of human enteric viruses is not fixed by convention and depends on the level of viruses likely to be present in the water. For sampling sewage effluents, the minimum useful sample volume may range from 15 to 30 liters. For general surface waters, a minimum useful volume would probably be 100 liters, with 150 to 200 liters preferred. For water that has very low turbidity, the apparatus described here could be used for processing volumes as great as 1,000 liters, as may be necessary for examination of either groundwater or tap water that has been processed by sand filtration.

If the intended water sample consists of wastewater effluent, tap water, or some other water that may have received a chemical disinfectant, then the collected water sample is dosed with sodium thiosulfate solution at a rate of 10 ml/gal, and a sample of the dosed water is checked to ensure that all residual disinfectant activity has been neutralized. This neutralization of residual disinfectant activity must be done before the water is either stored as a bulk sample for processing off site or processed on site for concentrating any viruses which the water may contain. If possible, the water sample should be processed by the virus concentration technique immediately after collection. If it is necessary to store a water sample before it is processed, then the storage container must first be adequately treated by a sterilization technique; after the sample water is added, the container should be kept covered to prevent contamination. The sample should also be kept in a cool location to reduce viral inactivation until such time as the water sample can be processed. Prolonged storage of water samples should be done in a refrigerated room (1 to 4°C) and preferably for no longer than 2 days.

For each water sample to be processed by cartridge filtration, a virus-adsorbing filter and possibly also a prefilter are needed. Neither of these filter types should be reused. The prefilter should be a 3-μm-pore-size (nominal porosity) wound polypropylene yarn filter of the type available as stock no. M27R10S from Commercial Filters Division, Parker Hannifin Company (Lebanon, Ind.). When viruses from environmental waters are concentrated by using cartridge filters that are considered to be positively charged, it may not be necessary to chemically modify the pH of the waters prior to passage through the filters except for lowering the pH of naturally alkaline waters to below the range of 7.5 to 8.5. This is the maximum pH range which should be used for adsorbing viruses onto the positively charged filters. The recommended positively charged virus-adsorbing filters are a pleated 0.45-μm-pore-size charged glass fiber membrane type, presently available as a 1-MDS filter cartridge from Cuno Inc., Division of Commercial Shearing Inc. (Meriden, Conn.), and a pleated 0.20-μm-pore-size charged nylon membrane type, presently available as a Posidyne N-66 filter cartridge from Pall Trinity Micro Corporation (Cortland, N.Y.) (7). Both of these virus-adsorbing filters are considered to be positively charged relative to untreated cellulose ester or glass microfiber filters and are therefore suitable for concentrating viruses from water at most ambient pH levels (approximately 5 to 7.5). If the 1-MDS type of charged glass fiber filter is used, then it is preferable to use dilute hydrochloric acid to adjust the pH of the any alkaline water samples to less than 7.5 before filtration (7). The charged nylon filter type may be preferable for use with water samples of moderately alkaline pH level, as these filters can be used effectively for concentrating viruses from water at pH levels up to approximately 8.5 (7). A more electronegatively charged glass microfiber filter, of the type available as a 0.25-μm-pore-size Duo-Fine series filter from Memtec America Corporation, Filterite Division (Timonium, Md.), can also be used. In fact, this type of filter may be preferred for concentrating viruses from either brackish or marine water. When the latter type of virus-adsorbing filter is used, it will be necessary to adjust the pH of the sampled water to 3.5 by addition of hydrochloric acid and to add aluminum chloride before passing the sample water through the filter. The recommended final concentration of aluminum chloride is 0.0005 to 0.0015 M (12).

The various reagents (aluminum chloride, hydrochloric acid, and sodium thiosulfate) which may need to be added to the water sample before the sample passes through the filters (1, 7, 12) can be prepared as stock reagent solutions and injected into the flowing sample between the pump (or pressurized source) and the prefilter (2). Alternatively, a batch adjustment technique can be used. For this technique, the water sample is collected into a sterilized temporary storage container, chemical supplements are added to the water while the sample is in the storage container, and the water is then pumped through the filters.

Assemble the cartridge filter virus concentration apparatus as shown in Fig. 1, making certain to properly align the pump, filter holders containing their respective filters, and water meter with respect to their indicated directions of water flow. If injectors will be used for chemical supplementation, they should be installed ahead of the prefilter. Be certain that the tops of the filter holders are fully tightened to their bottom sections, that the filter holders contain the correct filters, and that the air pressure release valves on the filter holders are closed. Connect one end of the water supply hose to the pump inlet, and place the other end of this hose into the water being tested, whether it is being taken directly from an environmental water source, a tap water source, or a storage container prepared as described above. Note the reading on the water meter (flowmeter), as these meters usually cannot be reset; to determine how much water sample has been filtered, it will be necessary to read the meter both before and after the filtration. Turn on the pump and let the sample water pass through the filters and meter. It may be necessary to occasionally release trapped air pockets from the filter holders. The presence of air pockets within the filter holders can easily be detected by looking at the clear bottom parts of the holders. Such air pockets should be released, as they may reduce the rate at which water is able to flow through the filters. To release air, first make sure that the holder in question is being held upright so that the trapped air can pass through the release valve. While keeping the holder upright, slightly open the valve until the air has passed through, and then close the valve to keep the water sample from leaking out. If the holder does not contain a release valve, it sometimes is possible to eliminate the trapped air by rapidly inverting the filter holder. This maneuver is done by first placing the

filter holder into an upright vertical position, then quickly inverting the filter holder, and rapidly restoring the holder to its normal upright position. It may be necessary to repeat this "flipping" maneuver. Alternatively, with the filter holder held upright, the top of the filter holder can be loosened slightly where the top connects with the bottom by means of screw threads, to release trapped air. The top and bottom sections should then be tightened together when water rather than air begins to seep out between the screw threads.

There are several procedures available for eluting viruses from the virus-adsorbing cartridge filter. This chapter describes one which allows elution to begin in the field, with eluant presterilized ahead of time in the laboratory. For each water sample to be processed, you will need 1,600 ml of sterile, pH 7, 3% beef extract solution contained in a separate watertight, 1-gal-capacity, wide-mouth, screw-cap autoclavable container. A polypropylene container is preferred for this purpose because viruses do not readily adsorb to polypropylene. The eluant solution can be sterilized by autoclaving it inside the polypropylene containers. Containers of this type are available as stock no. 2121-0010 from Nalge Company, Division of Sybron Corporation (Rochester, N.Y.). The beef extract solution should consist of 48 g of microbiological-grade beef extract powder, of a type which produces an easily visible amount of precipitate when adjusted to pH 3.5, dissolved in distilled water. It may be helpful if at least one-half of this powdered beef extract is of the type sold as Beef Extract V by BBL Microbiology Systems (Cockeysville, Md.). This will help to ensure that an easily visible precipitate is produced during subsequent secondary concentration of the filter eluant.

After the water sample has been filtered, disassemble the apparatus and allow any remaining water to drain from the filter holders. The filter holders should then be opened carefully. The virus-adsorbing membrane filter should aseptically be placed into the container of beef extract solution (eluant). If a prefilter was used, both it and one-half of the eluate should be placed into an additional sterile 1-gal-capacity polypropylene container. These filters should be kept in their tightly capped containers of eluant and, along with wet ice or ice packs, immediately placed into the insulated container. The insulated container of samples should then immediately be transported to the laboratory or shipped there by overnight delivery service.

The virus-adsorbing filter (and prefilter, if one was used) and filter eluant should be processed immediately upon receipt by the laboratory, using the elution and secondary concentration processes described below. If immediate elution is not possible, then the filter (and prefilter, if one was used) should be stored refrigerated at 1 to 4°C until such time as the elution and secondary concentration processes can be performed. This period of storage should not be longer than 2 days so as to avoid excessive bacterial growth in the filter and container of eluant. These filters should be kept in their containers of eluant until such time as the filters are processed by the virus elution technique. If the filters must be stored for a longer period of time, then storage should be at −70°C, although it must be understood that some viruses will lose viability when frozen. In particular, members of family *Inoviridae* seem extremely susceptible to loss of infectivity when frozen (personal observation).

To begin the virus elution procedure, open the jar of eluant, aseptically remove the virus-adsorbing filter from its container of eluant, and place the filter into a filter holder that has been presterilized. The top of the filter holder should be fully tightened to its bottom section and the air pressure release valve should be closed. The filter holder is then connected by hoses to either the pump or the pressure can which will hold the eluant. If a prefilter was used, then the prefilter likewise is placed into a presterilized holder, and both holders (respectively containing the virus-adsorbing filter and prefilter) are connected together in series so that the filters which they contain will be eluted simultaneously. Next, while continuously stirring the full 1,600-ml volume of filter eluant, adjust it to a pH of 9.5 by dropwise addition of 1 M sodium hydroxide and hydrochloric acid as necessary. The eluant is then passed through the filter (and prefilter, if one was used) three times sequentially, with the flow of eluant being opposite that used during the virus adsorption process, and the eluate is collected into a sterile container. This represents a reversal of the virus adsorption technique described above. The filters should then be discarded. Any eluant remaining in the filter holder (or both holders, if a prefilter was used), pump chamber, pressure can, or hoses should then be combined with the eluate in the sterile container.

At this point, the pH of the eluate should immediately by neutralized by the dropwise addition of 1 M hydrochloric acid. A portion of the pH-neutralized filter eluate can be removed to a separate container, and this portion of the eluate is then saved and subsequently assayed for the presence of bacteriophages. The enteric viruses contained in the filter eluate tend to have a greater resistance to brief low-pH exposure than do bacteriophages. Enteric viruses contained in the filter eluate therefore can be subjected to a secondary, or second-step, concentration procedure. This second concentration step, as published by Katzenelson et al. (10), represents a 20-fold concentration of the filter eluate. To begin the Katzenelson procedure, lower the pH of the filter eluate to 3.5 by dropwise addition of 1 M hydrochloric acid, using continuous stirring. During this pH adjustment, the eluate should become cloudy as an organic precipitate forms. The pH 3.5 eluant is stirred for 30 min and then centrifuged for 10 min at $3,000 \times g$ to collect the precipitate. The supernatant resulting from this centrifugation is discarded, and the precipitate from all 1,600 ml of eluant is dissolved in place within the centrifuge bottles, using a total volume of 80 ml of sterile 0.15 M Na_2HPO_4 prepared in distilled water. The pH of the dissolved precipitate, now called the final virus concentrate, is then checked and, if necessary, adjusted to between 7.0 and 7.2 by adding 1 M sodium hydroxide or 1 M hydrochloric acid. Antibiotics can be added to the final virus concentrate if desired. Suggested final antibiotic concentrations are 200 U of penicillin G per ml, 200 μg of streptomycin sulfate per ml, and 2.5 μg of amphotericin B per ml. These concentrates can then be stored at $-70°C$ prior to assay if the viruses of interest are stable when frozen in this solution. Otherwise, the concentrates should be stored under refrigeration (1 to 4°C). These same storage conditions should be used for any pH-neutralized filter eluate that was kept for bacteriophage analysis.

The types of filter material used in fabricating the cartridge filters are usually available commercially as circular flat-sheet filters which can be used with other types of commercial filter holders. These flat-sheet filters and filter holders are often suitable for processing smaller volumes of sample and can be eluted by passing a correspondingly smaller volume of the same type beef extract-based eluant through the filters (1, 7).

Raw Wastewater

The procedure described below was developed for the detection of viruses in 8-liter volume samples of raw sewage (9). After the sample is collected, adjust it to a final concentration of 0.05 M $MgCl_2$ by adding a 4 M $MgCl_2$ stock solution. After storage of the sample overnight at 1°C to allow the solids to settle, approximately the top one-half of the supernatant volume is decanted and discarded. The remainder of the sample is then centrifuged in aliquots for 20 min at $9,500 \times g$ to pellet the solids. All of the pelleted solids from the sample are then combined and resuspended to approximately 120 ml with phosphate-buffered saline (pH 7.2, 0.01 M phosphate, prepared with sodium phosphate) containing 2% (vol/vol) fetal bovine serum. The resuspended solids are then extracted with an equal volume of Freon (1,1,2-trichloro-1,2,2-trifluoroethane) and centrifuged at $1,200 \times g$ for 30 min to separate the two phases. The upper (aqueous) phase from this extraction is collected and centrifuged for 20 min at $9,500 \times g$ to remove residual suspended solids. Then, the supernatant from this last $9,500 \times g$ centrifugation step is adjusted to a final concentration of 3.0% (wt/vol) beef extract by addition of a 30% (wt/vol) beef extract solution and subjected to concentration by organic flocculation at pH 3.5, using the Katzenelson procedure (10) as described earlier in this chapter, and the low-pH precipitate is dissolved in 10 ml of 0.15 M Na_2HPO_4. The pH of the dissolved precipitate should be adjusted to 7.0 to 7.2 if it is not already within that range. Also, the small amount of residual suspended solids which had been present in the upper phase following Freon extraction should be eluted by mixing the solids in a 5-ml volume of buffered 10% beef extract elution solution (containing, per liter, 100 g of commercial powdered beef extract, 13.4 g of $Na_2HPO_4 \cdot 7H_2O$, and 1.2 g of citric acid); the sample is then subjected to centrifugation for 10 min at $2,800 \times g$ to repellet the solids, and the resulting supernatant is filtered through a sterile 0.25-μm-pore-size (absolute filtration) microbiological filter, producing a filtered eluate. To obtain a final concentrated sample, the dissolved precipitate resulting from organic flocculation of the Freon-extracted supernatant should be combined with the filtered eluate from the residual solids. The volume of the final concentrated sample ranges from approximately 11 to 14 ml, representing an overall concentration factor of approximately 6,000-fold. These concentrated samples can then be stored at $-70°C$ prior to assay if the viruses of interest are stable when frozen in this solution. Otherwise, the concentrated samples should be stored under refrigeration (1 to 4°C). The beef extract used should be of a type that produces a visible amount of precipitate when adjusted to pH 3.5.

Wastewater Sludge

Viruses can be isolated from wastewater sludges by using the following procedure (18). An appropriate sample size would be approximately 500 ml of primary wastewater sludge or 2,000 ml of secondary (mixed liquor-activated) wastewater sludge. Each sludge sample is adjusted to pH 3.5 with 1 M HCl, and then 0.05 M $AlCl_3$ is added to a final concentration of 0.0005 M. This can be done while stirring the sludge in a large beaker. The adjusted sample is then centrifuged at $1,400 \times g$ for 15 min, and the supernatant is discarded. Resuspend the solids pellet in an amount of buffered 10% beef extract elution solution equal to five times the pellet volume. The beef extract used should be of a type that produces a visible amount of precipitate when adjusted to pH 3.5. The solids pellet should be stirred in the eluant solution for 30 min. The resuspension is then centrifuged at 7,000 × g for 30 min, and the retained supernatant is passed in series through sterilized filters of 3.0-, 0.45-, and 0.25-μm nominal porosity (a suggestion being Duo-Fine material, obtained in sheets and cut into 142-mm-diameter discs) to remove remaining sludge solids and contaminating bacteria. The resulting filtrate can be assayed for bacteriophages. If the goal is to detect enteric viruses, then the filtrate should be diluted with a volume of distilled water equal to 2.3 times the filtrate volume (to achieve a final effective beef extract concentration of 3% by weight) and concentrated by the Katzenelson technique (10) as described earlier in this chapter. The supernatant resulting from the Katzenelson procedure is then discarded, and the precipitate is resuspended in a volume of 0.15 M Na_2HPO_4 equal to 1/20 the volume of the diluted filtrate. These concentrates can be stored at −70°C prior to assay if the viruses of interest are stable when frozen in this solution. It has been noted that some types of viruses will die off even when stored at −70°C in these processed eluates (8). If there is suspicion regarding the stability of viruses when they are frozen in any type of sample, then storage should be done under refrigeration (1 to 4°C) and for the least possible amount of time. The colder the storage temperature is, the greater is the stability of the viruses. Alternate sludge processing methods that were collaboratively tested have been published by Goyal et al. (4).

Virus Assay Techniques

In general, the infectivity of enteric viruses contained in environmental samples is examined by inoculating the sample into cultures of either human or animal cells that are prepared in the laboratory, as opposed to inoculating live animals. This is often done by using either a cytopathogenicity assay, looking for virus-induced cytopathogenic effects in the inoculated cells, or a plaque formation assay technique, looking for cell death. Methods for performing these two types of assays for detecting enteric viruses contained in environmental samples are described in chapter 43. It may prove necessary to use toxicity reduction methods when assaying concentrated environmental samples in cultured human or animal cells. The cell monolayer washing procedure described in chapter 43 can be used for reducing cytotoxic effects. Methods for using plaque formation assays to detect bacteriophages contained in environmental samples are likewise described in chapter 43. General approaches used for alternative types of viral assay techniques are described in chapter 8.

REFERENCES

1. **Biziagos, E., J. Passagot, J.-M. Crance, and R. Deloince.** 1989. Hepatitis A virus concentration from experimentally contaminated distilled, tap, waste and seawater. *Water Sci. Technol.* **21:**255–258.
2. **Dahling, D. R., and B. A. Wright.** 1987. Comparison of the in-line injector and fluid proportioner used to condition water samples for virus monitoring. *J. Virol. Methods* **18:**67–71.
3. **El-Abagy, M. M., B. J. Dutka, and M. Kamel.** 1988. Incidence of coliphage in potable water supplies. *Appl. Environ. Microbiol.* **54:**1632–1633.
4. **Goyal, S. M., S. A. Schaub, F. M. Wellings, D. Berman, J. S. Glass, C. J. Hurst, D. A. Brashear, C. A. Sorber, B. E. Moore, G. Bitton, P. H. Gibbs, and S. R. Farrah.** 1984. Round robin investigation of methods for recovering human enteric viruses from sludge. *Appl. Environ. Microbiol.* **48:**531–538.
5. **Holmes, I. H., G. Boccardo, M. K. Estes, M. K. Furuichi, Y. Hoshino, W. K. Joklik, M. McCrae, P. P. C. Mertens, R. G. Milne, K. S. K. Samal, E. Shikata, J. R. Winton, I. Uyeda, and D. L. Nuss.** 1995. Family *Reoviridae*, p. 208–239. *In* F. A. Murphy, C. M. Fauquet, D. H. L. Bishop, S. A. Ghabrial, A. W. Jarvis, G. P. Martelli, M. A. Mayo, and M. D. Summers (ed.), *Virus Taxonomy (Sixth Report of the International Committee on Taxonomy of Viruses)*. Springer-Verlag, Vienna.
6. **Hurst, C. J.** 1991. Presence of enteric viruses in freshwater and their removal by the conventional drinking water treatment process. *Bull. W.H.O.* **69:**113–119.
7. **Hurst, C. J., W. H. Benton, and R. E. Stetler.** 1989. Detecting viruses in water. *J. Am. Water Works Assoc.* **81(9):**71–80.
8. **Hurst, C. J., and T. Goyke.** 1986. Stability of viruses in waste water sludge eluates. *Can. J. Microbiol.* **32:**649–653.
9. **Hurst, C. J., K. A. McClellan, and W. H. Benton.** 1988. Comparison of cytopathogenicity, immunofluorescence, and in situ DNA hybridization as methods for the detection of adenoviruses. *Water Res.* **22:**1547–1552.
10. **Katzenelson, E., B. Fattal, and T. Hostovesky.** 1976. Organic flocculation: an efficient second-step concentration method for the detection of viruses in tap water. *Appl. Environ. Microbiol.* **32:**638–639.
11. **Moore, R. S., D. H. Taylor, L. S. Sturman, M. M. Reddy, and G. W. Fuhs.** 1981. Poliovirus adsorption by 34 minerals and soils. *Appl. Environ. Microbiol.* **42:**963–975.
12. **Nestor, I.** 1983. Methodologie de detection des virus enteriques dans l'eau de mer. *Rev. Epidemiol. Santé Publique* **31:**21–38.
13. **Paul, J. R., J. D. Trask, and S. Gard.** 1940. II. Poliomyelitic virus in urban sewage. *J. Exp. Med.* **71:**765–777.
14. **Russell, W. C., T. Adrian, A. Bartha, K. Fujinaga, H. S. Ginsberg, J. C. Hierholzer, J. C. de Jong, Q. G. Li, V. Mautner, I. Nasz, and G. Wadell.** 1995. Family *Adenoviridae*, p. 128–133. *In* F. A. Murphy, C. M. Fauquet, D. H. L. Bishop, S. A. Ghabrial, A. W. Jarvis, G. P. Martelli, M. A. Mayo, and M. D. Summers (ed.), *Virus Taxonomy (Sixth Report of the International Committee on Taxonomy of Viruses)*. Springer-Verlag, Vienna.
15. **Tambini, G., J. K. Andrus, E. Marques, J. Boshell, M. Pallansch, C. A. de Quadros, and O. Kew.** 1993. Direct detection of wild poliovirus circulation by stool surveys of healthy children and analysis of community wastewater. *J. Infect. Dis.* **168:**1510–1514.
16. **van Lawick-Goodall, J.** 1971. *In the Shadow of Man*, p. 214–224. Houghton Mifflin, Boston.
17. **Watanabe, R. A., J. L. Fryer, and J. S. Rohovec.** 1988. Molecular filtration for recovery of waterborne viruses of fish. *Appl. Environ. Microbiol.* **54:**1606–1609.
18. **Williams, F. P., Jr., and C. J. Hurst.** 1988. Detection of environmental viruses in sludge: enhancement of enterovirus plaque assay titers with 5-iodo-2'-deoxyuridine and comparison to adenovirus and coliphage titers. *Water Res.* **22:**847–851.

Indicators of Marine Recreational Water Quality

ROGER S. FUJIOKA

18

INTRODUCTION

Value of Clean Coastal Recreational Waters

In most countries, populations are concentrated near coastal waters because coastal waters provide food, jobs, recreation, and aesthetic appeal. For islands such as Hawaii, the value of coastal water is economic as much as social and cultural (17). To ensure their value, coastal waters must not be polluted since clean water and attractive beaches guarantee a healthy ecosystem as well as a strong tourist industry. People will travel long distances to spend time at beaches suitable for swimming, sunbathing, surfing, and fishing. As a result, government agencies and private industries support tourism by building homes, hotels, restaurants, shops, parks, harbors, and other facilities. Unfortunately, these activities plus the commensurate increase in population in the area inevitably result in increased production of point source pollution such as the discharge of sewage into nearby streams or estuaries or directly into coastal waters. These activities also increase nonpoint sources of pollution such as streams and storm drains which have not been regulated and cause excessive contamination of coastal waters (35). Today, one can hardly find a popular beach that has not been affected by pollution, with consequent degradation of the suitability of water for swimming. Diseases acquired from recreational use of waters are of the greatest concern to public health agencies and the tourism industry. Thus, management of coastal water quality becomes the key to maintaining this resource. Water quality is usually maintained by establishing water quality standards and implementing a water monitoring program to ensure that the quality of water is safe for primary contact uses, especially swimming.

Establishment of Microorganisms as Indicators of Water Quality

Coastal waters are susceptible to contamination with numerous types of microbial pathogens and can thus serve as a vehicle for the transmission of diseases to people by contact with the water (water contact disease) or by ingestion of the water (waterborne disease). The microbial pathogens which can potentially be transmitted to people by these two means and the most common disease symptoms that they cause are summarized in Table 1. It should be noted that diseases associated with recreational use of water have been difficult to document because infections with pathogens are events which cannot be easily determined and the percentages of infected people who develop the various degrees of observable clinical symptoms vary with the age and health status of each individual, the concentrations of the pathogens in the water, and the physiological status and virulence of the pathogen. These observations are complicated by the fact that many symptoms (coughing, vomiting, headache, skin rash) associated with recreational use of water may be caused by factors unrelated to microbial infections. Moreover, some pathogens giving rise to infections may originate from the bodies of swimmers and not from the contaminated water.

Most health officials recognize the complicating factors discussed above, accept the fact that most of the infections acquired as a result of recreational use of water cannot be documented, and accept the premise that by preventing the pollution of recreational waters, they can reduce the incidence of water-related infections and diseases. To implement this goal, health officials must operate under conservative and practical guidelines to develop a simple, routine monitoring test to determine the hygienic quality of water. The greatest risk to human health occurs during recreational use of water involving primary contact (wading, swimming, surfing, snorkeling, scuba diving) with water, and thus the greatest public health threat is present when these waters are contaminated with sewage. Since the pathogens in sewage are transmitted by ingestion, the first responsibility of health officials is to protect the public against waterborne diseases. It is this reasoning that has led to the establishment of recreational water quality standards which are targeted specifically to waterborne diseases.

The need to determine when a body of water is contaminated with sewage was recognized long ago, when classical waterborne diseases such as typhoid, shigellosis, amebiasis, and cholera were everyday threats and it was well known that sewage was the source of these pathogens. At that time, methods for the detection of many bacteria were crude and isolation of any pathogen from water was not possible. Four important discoveries led to our present-day system of assessing the hygienic quality of water based on concentrations of fecal bacteria. First, Escherich in 1885 determined that coliform bacteria were numerous and always detected in feces and sewage. Second, Klein and Houston (33) reported that they could dilute sewage more than 10,000-fold and still detect the presence of

TABLE 1 Pathogen transmitted through recreational water

Organism	Disease
Bacteria	
Water contact	
Aeromonas hydrophila	Wound infection
Citrobacter spp.	Wound infection
Leptospira icterohemorrhagia	Leptospirosis
Mycobacterium marinum	Otitis externa
M. balnei	Otitis externa
Pseudomonas spp.	Otitis externa
Staphylococcus aureus	Wound infection
Vibrio spp.	Wound infection, otitis externa
Waterborne	
Campylobacter jejuni	Gastroenteritis
Escherichia coli serotypes	Gastroenteritis
Salmonella typhi	Typhoid fever
Salmonella spp.	Gastroenteritis
Shigella dysenteriae	Dysentery
Shigella spp.	Gastroenteritis
Vibrio cholerae	Gastroenteritis
Vibrio spp.	Gastroenteritis
Viruses	
Water contact	
Adenovirus	Pharyngitis, eye infections
Waterborne	
Adenovirus	Enteritis
Coxsackieviruses and echoviruses	Meningitis, myocarditis
Hepatitis A and E viruses	Hepatitis
Norwalk virus	Enteritis
Polioviruses	Poliomyelitis
Rotaviruses	Enteritis
Caliciviruses and astroviruses	Enteritis
Protozoa	
Water contact	
Naegleria spp.	Meningoencephalitis
Waterborne	
Giardia lamblia	Giardiasis
Entamoeba histolytica	Dysentery
Cryptosporidium spp.	Dysentery

coliform bacteria. Under the same conditions, all known physical or chemical tests could no longer detect the presence of sewage. Thus, Klein and Houston coined the term "indicator of sewage" for coliform bacteria. Third, Scott (43) in 1932 took a pragmatic approach and proposed that concentrations of coliforms (<1,000/100 ml) be used as Connecticut's marine bathing water quality standards. Fourth, Stevenson (46) in 1953 conducted the first epidemiological study in the United States and reported a correlation between elevated concentrations (>2,000/100 ml) of coliform bacteria at freshwater beaches and increased incidences of illnesses in swimmers. It should be noted that this correlation was not observed at saltwater beaches. The results of these four key studies led public health officials to accept the level of coliform bacteria as the best criterion for assessing the hygienic quality of water. As a result, the word "indicator" has been used interchangeably with coliforms to indicate water pollution.

Initial U.S. Water Quality Standards Based on Detectable Risk

In the early history of the United States, the responsibility for protecting the public's health was vested in each state; therefore, water quality guidelines and standards varied among states and even among counties within a state. However, primarily on the basis of the findings of Scott (43) and Stevenson (46), the predominant recreational water quality guideline used was 1,000 total coliforms per 100 ml. In 1972, Congress passed the Clean Water Act (Public Law 92-500), which established the formation of the U.S. Environmental Protection Agency (USEPA) and the formulation of national standards for all states to follow. The newly established USEPA recognized that many bacteria which tested positive by the total coliform test were environmental (soil, plants) and not specifically associated with feces of warm-blooded animals. Fecal coliforms or thermotolerant coliforms constitute a subgroup of coliform bacteria which can grow at 44.5°C and were shown to be more specifically related to coliform bacteria growing in the intestines of humans and warm-blooded animals. Since it was previously established that approximately 20% of total coliforms are fecal coliforms, USEPA in 1976 recommended that the national recreational water quality standards be established at a geometric mean of 200 fecal coliforms per 100 ml. A serious limitation in the use of total coliform and fecal coliform levels to establish recreational water quality standards was the absence of data correlating the increasing concentrations of coliforms in a body of water with increasing numbers of pathogens in the water or with increasing incidences of diarrheal diseases among swimmers using the water. However, since the concentration of coliforms in recreational water is taken as evidence that the water is contaminated with feces or sewage, and since feces or sewage is the source of most pathogens transmitted by ingestion of water, an increased concentration of coliform bacteria in the water is concluded to signal an increase in detectable risk to swimmers. Although coliform standards are limited to determining detectable but nonpredictable risk levels, these standards appear to be conservative and over a number of years have served to maintain water quality and prevent waterborne diseases among swimmers

Current U.S. Recreational Water Quality Standards Based on Acceptable Risk

Recognizing the limitations in the coliform standards, USEPA as early as 1972 initiated a remarkable 10-year project (4) initially to develop more specific tests for various indicators, including *Escherichia coli*, enterococci, and *Clostridium perfringens*, which were required to complete a comprehensive microbiological and epidemiological study to determine the concentrations of nine different microbial water quality indicators in marine waters from beaches in New York City, Boston, and New Orleans; the researchers also interviewed people who swam and those who did not swim at those beaches. With that experimental design, the USEPA study for the first time was able to determine the incidence of diseases among swimmers and nonswimmers in waters with various measured concentrations of microorganisms. The results showed that of all the microorganisms measured, only the concentrations of enterococci in marine waters correlated with incidences of diarrheal diseases among swimmers. On the basis of the short incubation period and the symptoms, the USEPA scientists conducting the study concluded that most of the diarrheal diseases were attributable to viral infections (5). Initially USEPA recommended that the marine recreational water quality standard be set at a geometric mean of three enterococci per 100 ml, which related to an acceptable risk level of six diarrheal diseases per 1,000 swimmers. However, after the public comment period, USEPA (47) recommended that the marine recreational standard be set at a geometric mean of 35 enterococci per 100 ml, a recommendation based on analysis of five water samples from a swimming site over a 30-day period. This recommended standard correlates to an expected rate of 19 diarrheal diseases per 1,000 swimmers.

It should be noted that a major contribution of this USEPA study was the confirmation that concentrations of indicators such as total coliform and fecal coliform which were previously used in establishing water quality standards in the United States, as well as other proposed indicators such as fecal streptococci (23), *C. perfringens*, and *Staphylococcus aureus*, could not be used to predict incidences of diarrheal diseases among swimmers in marine waters. The value of the epidemiological-water microbiological study conducted by USEPA can be summarized as follows. First, it was the first well-designed epidemiological and water quality study. Second, the findings answered many questions related to the usefulness of indicators and the limitations of the traditional recreational water quality standards based on concentrations of coliform bacteria. Third, it resulted in recommending new recreational water quality standards based on measurable and acceptable risk levels. Fourth, it recognized the need for standards which are specific for freshwaters and for marine waters. Fifth, it developed new membrane filtration (MF) methods which reliably measured the concentrations of several useful indicator bacteria. Sixth, it provided evidence that human enteric viruses were the pathogens most likely responsible for marine recreational waterborne diseases.

Implementation of USEPA recreational water quality standards is the responsibility of individual states (47). Some states such as Hawaii readily acted on the USEPA recommendations and the application of acceptable risk levels associated with the concentrations of enterococci in the recreational water. The State of Hawaii in 1990 (11) determined that the USEPA's recommended marine recreational water quality standard of 35 enterococci per 100 ml correlated to an unacceptable risk level of 19 diseases per 1,000 swimmers and therefore established a much more stringent standard of 7 enterococci per 100 ml, which corresponded to a lower acceptable risk level of 10 diseases per 1,000 swimmers. Many states have yet to change their recreational water quality standards. Others, such as California, rely on separate agencies to establish recreational standards; as a result, in California there are coastal water segments with three different marine recreational water quality standards (1,000 total coliforms per 100 ml, 200 fecal coliforms per 100 ml, and 35 enterococci per 100 ml).

Marine Recreational Water Quality Standards in Other Countries

Canada is the only country outside the United States which has adopted the USEPA marine recreational standard of a geometric mean of 35 enterococci per 100 ml. However, Canada also retains its older standard of 200 fecal coliforms per 100 ml. It should be noted that some scientists (16, 32) have reevaluated the USEPA study and have reported flaws in the design, assumptions, or calculations used. Thus, the reliability and applicability of the USEPA study to all beach sites are still controversial. This controversy in the newly

recommended USEPA recreational water quality standard was fueled by a more recent study by USEPA (6) which reported that when the source of fecal indicators in the recreational waters is from a nonpoint source rather than sewage, the concentrations of fecal indicators in water are no longer a reliable predictor of diseases among swimmers. This more recent finding affects each state's monitoring program because the source of fecal indicator in coastal waters is generally never known, and recent evidence indicates that nonpoint source pollution (21, 24, 27) rather than sewage is the source of most coastal water pollution (35).

Pike (40) and Salas (42) recently reviewed the marine recreational water quality standards and guidelines used in various countries in the world. They reported that European countries use guidelines which state that 80% of the samples should not exceed 500 total coliforms per 100 ml or 100 *E. coli* bacteria per 100 ml. Other countries use guidelines for recreational waters based on the standard of either 1,000 total coliforms or 200 fecal coliforms per 100 ml.

THE MOST PROBABLE NUMBER (MPN) METHOD FOR INDICATOR BACTERIA

Proper Collection and Preparation of Sample

Considerable technician skill, time, and use of expensive supplies and equipment are required for the microbiological analyis of water. Moreover, the results of these assays are used in important management decisions such as the posting of warning signs for swimmers and actual closing of beaches. It is well known that the value of a water analysis is only as good as the sample to be tested. Other sections of this manual as well as the latest version of *Standard Methods for the Examination of Water and Wastewater (Standard Methods)* (1) should be consulted for proper collection and preparation of samples as well as the approved methods for determining whether marine water meets marine recreational water quality standards. The following are some relevant factors to consider in sampling marine recreational waters: (i) samples should be collected at the time and site where maximum bathing occurs, (ii) they should be collected every 6 days over a 30-day period so that a valid geometric mean can be obtained to fulfill marine recreational water quality standards, (iii) they should be immediately protected from sunlight exposure, since microorganisms in marine water samples are especially vulnerable to inactivation by sunlight, (iv) they should be analyzed as soon as possible, preferably within 6 h of collection, because marine water conditions injure most enteric bacteria, and (v) they must be well mixed just prior to microbiological assay to disperse aggregated microorganisms and thus enable better quantitation of the concentrations of microorganisms they contain.

Theory and Multistep Nature of the MPN Method

The method uses probability statistics to determine the mean concentration of bacteria as the MPN per 100 ml. The MPN test is generally conducted in three sequential phases (presumptive, confirmatory, and completed), each phase requiring 1 to 2 days of incubation. In the initial or presumptive phase, three volumes of samples (10, 1, and 0.1 ml) are inoculated into 3, 5, or 10 tubes containing bacteriological medium constituted to allow the target bacteria to grow. In this test, it is assumed that any single viable target bacterium in the inoculum will result in growth or a positive reaction in the tubed medium. For example, to recover coliform bacteria, tubes containing fermentation tubes and lactose-containing lauryl tryptose broth are inoculated; after 1 to 2 days of incubation at 35 ± 0.5°C, tubes showing turbidity, acidity, and gas (bubbles in fermentation tubes) are considered presumptively positive for coliform bacteria. These readings are considered presumptive because false-positive and false-negative reactions often occur as a result of growth and interference by nontarget bacteria in the growth medium (36). As a result, subsamples from all presumptively positive tubes must be inoculated into a more selective medium to confirm the presence of the target bacteria. For total coliforms, the confirmation test is the growth of target bacteria in brilliant green lactose bile broth at 35°C within 1 to 2 days; the confirmed test for fecal coliforms is the growth of target bacteria in *E. coli* (EC) medium at 44.5°C within 1 to 2 days (1). The confirmed test is reliable evidence but not proof that the target bacteria have been detected. Therefore, subsamples of the confirmed positive reactions should be inoculated onto a selective agar medium, and the target bacterium should be physically recovered and Gram stained. This completed test is generally conducted on 10% of the positive confirmed samples as a quality control measure. For practical purposes, the number of positive and negative tubes in the confirmed phase of the test is usually used to determine the MPN of the target bacteria by using the following formula, credited to H. A. Thomas (1):

$$\text{MPN per 100 ml} = \frac{\text{no. of positive tubes} \times 100}{\sqrt{\substack{(\text{ml of sample in negative tubes}) \\ \times (\text{ml of sample in all tubes})}}}$$

However, in practical terms, most people use the tables of positive and negative tube reactions as shown in *Standard Methods* (1) to determine the MPN.

In summary, the disadvantages of the MPN method are (i) the total time, labor, material, and costs required to analyze one sample, (ii) the substantial increase in reagents, tubes, incubation space, and cleanup requirements when multiple samples need to be analyzed or when the sample volume must be increased to 100 ml, (iii) the multiphase nature of the method, each phase requiring an overnight incubation period and therefore requiring 2 to 4 days before a final reading can be made, and (v) the fact that MPN is a single estimated number, while the true number (95% confidence limit) may show extreme variation from the MPN. The major advantages of the MPN method are as follows: (i) it will accept both clear and turbid samples, (ii) it inherently allows for the resuscitation and growth of injured bacteria, (iii) the endpoint can be easily read by personnel with minimal skill, and (v) minimal preparation time and effort are required to start the test, and therefore processing of samples can be initiated at any time of the day.

Applicability of the MPN Method to Monitoring of Marine Waters

Historically, the five-tube MPN test and the single MPN figure have most often been used to determine whether marine waters meet the recreational water quality standards. The inherent imprecision of the MPN method has generally been ignored and can be illustrated by the following results of a confirmed five-tube MPN test for fecal coliform when the combination of positives resulted in 5/5 in the 10-ml sample, 3/5 in the 1-ml sample, and 1/5 in the 0.1-ml sample. The final reading of 5-3-1 is then matched to Table 9221.IV in *Standard Methods* (1), which results in a single MPN index

of 110 target bacteria per 100 ml but with a 95% confidence limit that the true value may be as low as 40 or as high as 300 target bacteria per 100 ml. On the other hand, if the combination of reading were 5-4-1 (only one more positive reaction), the MPN index would be 170/100 ml with a 95% confidence limit from a low of 70/100 ml to a high of 480/100 ml. These results show that a water sample assayed for fecal coliform by the five-tube MPN method and determined to have an MPN index of 110 or 170/100 ml may in reality be in violation of the standard of 200 fecal coliforms per 100 ml. It should be noted that the possible range of numbers reflecting the true concentration is much wider when a 3-tube test is performed and much narrower when a 10-tube test is performed.

When the MPN method is used to monitor marine waters, the precision of the method must be balanced with the resources required to use it. The first decision of the analyst is to select the 3-tube (9 tubes per test), the 5-tube (15 tubes per test), or the 10-tube (30 tubes per test) test format. The 3-tube test requires less material, labor, and time but is the least accurate; the 10-tube test is most accurate, but its requirement for material, labor, and time per test often makes its use prohibitive when a large number of samples must be processed. When the marine recreational water quality standards are established with high concentrations of indicator bacteria such as 1,000 total coliforms or 200 fecal coliforms per 100 ml, the imprecision of the MPN method and the fact that the five-tube test analyzes only a total of 55.5 ml of sample notwithstanding, the five-tube method can usually be successfully used as a monitoring assay. However, the imprecision of the five-tube MPN format would make this method unreliable in determining the geometric mean for a recreational water quality standard set at a low level of seven enterococci per 100 ml as is the situation in Hawaii. This is illustrated in Table 9221.IV in *Standard Methods* (1), which indicates that when all tubes are negative, the MPN is <2 but the upper and lower limits of this assay cannot be determined. Reliability and precision of the determined concentrations of indicator bacteria in marine recreational waters are required to support a legal decision such as closing a beach. Under these conditions, the 10-tube MPN test should be used.

A valuable feature of the MPN method is its ability in determine the concentrations of indicator bacteria and pathogens in sediments from environmental waters. This method should be used to analyze sediments at marine beaches, since the results will provide information on the history of contamination at a given site. Unfortunately, there are no standards for indicators and pathogens in sediments at recreational beaches, and therefore this useful application of the MPN test has largely been ignored.

Improvements in the MPN method to shorten the time needed to complete the test have been achieved primarily by incorporating into the bacterial growth medium a substrate which will detect the presence of a specific enzyme in the target bacteria, resulting in a specific reaction which can be detected by fluorescence. Assays using substrates such as 4-methylumbelliferyl β-D-glucuronide for *E. coli* and 4-methylumbelliferyl-β-D-glucoside for enterococci are being prepared commercially (37) and can be devised in miniaturized format to save on materials, time, and space (28). However, these improvements do not address the inherent imprecision of the MPN method and the observation that marine waters contain many types of bacteria which have enzyme systems similar to those of the target indicator bacteria and thus yield false-positive reactions (9, 36, 37). These limitations must be seriously considered if the numbers determined by the MPN method are to be used in a regulatory fashion to open and close beaches.

MF METHOD FOR INDICATOR BACTERIA
Theory and Single-Step Nature of MF

The procedure for the MF test is to pass the water sample containing the target bacteria through a membrane with a pore size (e.g., 0.45 μm) small enough to physically retain the bacterium on the filter surface while allowing the water to filter through the membrane. Generally, a vacuum is used to pull the water sample through the filter; this technique generally works very well because most marine water designated for swimming is not turbid and 100-ml samples can be easily filtered through a standard 47-mm-diameter membrane. This filter is then placed onto a solid growth medium (agar) which is selective or differential for growth of the target bacteria. The membrane filter used in this assay must be made of material, such as cellulose nitrate, which will allow nutrients from the agar medium to pass from the underside to the surface of the medium and thus allow the viable and trapped bacteria on the surface of the membrane to grow and to form a visible colony. After incubation, usually for 1 day at a favorable temperature, the target bacteria retained on the membrane will grow into a characteristic, visible colony which can be counted. The number of target colonies counted is directly related to the concentration of the bacteria in the water. For example, if 100 ml of water is assayed and 100 target colonies are counted, the concentration of the target bacteria will be determined as 100 CFU/100 ml of water. CFU is used because although this test assumes that each colony resulted from the growth of one viable bacterium, the discrete unit or particle from which the colony formed may be aggregates of several bacteria.

The reliability of the number of target colonies counted is dependent on the selectivity of the bacteriological medium and the conditions of the test. At the minimum, the target colony count after 1 day of incubation is nearly equivalent to the count in the confirmed phase of the MPN method and in some tests is nearly equivalent to the count in the completed test phase. It is recommended that 10% of the target colonies counted be streaked for isolation on another selective medium and that the isolate be Gram stained to verify that the target bacteria were recovered. Since reliable counts are obtained after one major step and after 1 day of incubation, the MF method is considered a one-step method, as opposed to the multistep MPN method.

In summary, the advantages of the MF method include (i) savings in terms of time, labor, and cost compared with the MPN method, (ii) direct determination of the concentrations of bacteria with high precision and accuracy, (iii) the formation of the target bacteria as colonies which can be purified for further identification and characterization, and (iv) the ability to process large volumes of water samples to greatly increase the sensitivity of this method. The disadvantages include (i) inapplicability of the method to turbid samples, which can clog the membrane or prevent the target bacteria from growing on the membrane, (ii) false-negative results due to the inability of viable but injured or stressed bacteria in environmental waters to grow with standard MF methods, and (iii) false-positive results when nontarget bacteria form colonies similar to the target colonies (15, 22, 29). These results caution anyone analyzing marine waters to be aware that such waters contain a diversity of bacteria which can interfere with tests that had been found to yield excellent results when fresh water samples were assayed.

Applicability of the MF Method to Monitoring of Marine Water

The MF method is preferred over the MPN method to determine whether marine water meets the recreational water quality standards primarily because the results are obtained within a day and the precision of the resulting counts is very high. Moreover, since each colony represents one viable unit of bacteria, the counts can easily be related to the concentrations of viable bacteria to which bathers are exposed when swallowing a unit volume of marine water.

To test whether marine waters meet recreational standards based on total coliform and fecal coliform, the traditional one-step method is used. For total coliforms, the processed filter is placed onto M-Endo agar medium and incubated 1 day at 35°C (1). The target colonies, characterized as red with a metallic green sheen, are counted. For fecal coliforms, the processed filter is placed onto m-FC agar in a tight-fitting petri plate, which is incubated submerged in a water bath at 44.5°C for 1 day (1). The target colonies, characterized as various shades of blue, are counted. Since coliform bacteria in marine waters are often in a stressed or injured state, they often will not grow when incubated directly onto selective medium. As a result, a resuscitation step is often recommended. This may involve initially placing the processed membrane onto nonselective medium (phenol red lactose agar) and incubating it for 4 h at 35°C to allow the injured or stressed bacteria to recover. The filter is then placed onto m-FC agar for overnight incubation at 44.5°C (1).

To test marine waters for the new marine recreational standard based on enterococci, the two-step MF method is used. In this method, the processed filter is placed onto mE agar and incubated for 2 days at 41 ± 0.5°C. The first step involves counting all pink to red colonies on the filter, which represent fecal streptococci bacteria. The entire filter is then transferred to an esculin iron agar medium and incubated for 20 min at 41°C. The formation of pink to red colonies which develop a black or reddish brown precipitate on the underside of the filter demonstrates that the colonies are able to hydrolyze esculin and therefore confirms that the fecal streptococci are truly enterococci. Since this method was used in the USEPA microbiological-epidemiological study to establish the new marine recreational standard based on enterococci, it should be used to determine whether the marine water meets the new USEPA marine recreational water quality standard.

As mentioned earlier, it is desirable to analyze the sediments at marine recreational beaches. Although the MF method has not been recommended for analysis of sediments, a modification of the method has been developed primarily to measure bacteria in foods (38), using the principle of the MPN method. This method uses special membrane filters which contain numerous hydrophobic grids; each grid on the membrane represents a separate culture chamber which is equivalent to a tube in the MPN test. This method allows particles to distribute and bacterial colonies to grow within that grid. The usefulness of this method is that it allows one to use a number of different selective media to isolate the various types of bacteria which are associated with particles. This method can be used to characterize the microbial content of sediments and can provide valuable information on the history of chronic contamination at different recreational water sites.

PROPOSED ALTERNATIVE INDICATORS OF MARINE WATER QUALITY

Current Assessment of Indicators of Recreational Waters

There is a worldwide consensus among scientists that the bacterial indicators used in establishing marine water quality standards have been helpful but are interim measures until better methods and more complete studies are conducted to identify better indicators or to monitor for pathogens. This consensus is based on the agreement that human enteric viruses are the pathogens most likely to be transmitted by recreational use of marine waters, and it is well known that these viruses are much more stable in marine waters than any of the bacterial indicators. Dufour (12) summarized the five criteria for selecting an ideal indicator of water quality as follows: (i) the indicator should be consistently present in feces and at higher concentrations than the pathogens, (ii) it should not multiply outside the human intestinal tract, (iii) it should be as resistant as or more resistant than the pathogens to environmental conditions and to disinfection, (iv) it must be assayed by means of a simple and reliable test, and (v) its concentrations in water should correlate with concentrations of feces-borne pathogens or with a measurable health hazard.

None of the current bacterial indicators used in establishing marine recreational standards fulfill all of the desirable criteria of an ideal indicator, and most are more susceptible to environmental conditions (19) than viruses. As a result, many other microorganisms are being evaluated as alternative indicators of water quality.

Promising Alternative Bacterial Indicators

C. perfringens

C. perfringens is useful as a water quality indicator because (i) it is consistently present in moderate concentrations in human feces as well as in sewage, (ii) it is more stable in environmental waters than most pathogens, (iii) the method for its recovery can be used by most laboratories because it is similar to most bacterial MF assays, and (iv) it has been successfully used to monitor sewage-contaminated streams (13, 20, 45), ocean water (18), and sewage sludges in ocean environments (30). One reported limitation in the use of C. perfringens is the extreme stability of this bacterial spore to environmental conditions and the possibility that its detection indicates a pollution event that occurred a long time ago. In summary, C. perfringens is a conservative indicator of fecal pollution. If monitoring results show no or low levels of C. perfringens, one can be assured that the quality of that recreational water is good.

An excellent two-step MF method developed by Bisson and Cabelli (3) is available to assay for concentrations of C. perfringens. In this assay, the water samples are filtered through membranes by using the procedure used to assay for enterococci or fecal coliform. The processed membranes are then placed onto m-CP agar (3) and incubated at 44°C under anaerobic conditions. After 1 day of incubation, the membranes are immediately exposed to fumes of ammonium hydroxide. All yellow colonies which turn pink to red are acid phosphatase positive, confirming that the colonies are C. perfringens.

S. aureus

The results of most epidemiological studies show higher incidences of nondiarrheal diseases such as eye, ear, nose, and skin infections. In this regard, many swimmers acquire staphylococcus skin infections after swimming in marine waters (7), and there is a need for an indicator of water quality which addresses skin infections rather than diarrheal infections. Evidence to support the use of S. aureus as indicator of water quality includes the following: (i) the bacterium is stable in marine waters, (ii) its concentrations in water was shown to represent the load of microorganisms being shed by swimmers (8, 14), and (iii) an effective MF method is now available to recover it from marine waters (8). The limitations in the use of this indicator system are the multistep nature of the method to recover and to enumerate the concentrations of S. aureus and the still unresolved question as to whether recreational water serve as a vector for the transmission of this pathogen to uninfected humans (a public health problem) or whether the bacterium is present on the human skin and recreational use of marine waters causes opportunities for self-infection (a personal problem).

An MF method to assay for S. aureus from marine waters was recently reported by Charoenca and Fujioka (8). This is a simple one-step method in which the processed membrane is placed onto Vogel-Johnson agar supplemented with 0.005% sodium azide. After 1 to 2 days of incubation at 35°C, the target black colonies representing total staphylococci were counted; 10% of the black colonies from a given site were tested for coagulase to determine the concentrations of S. aureus in the water sample.

Promising Alternative Virus Indicators

Coliphage (FRNA Phage)

Coliphages are viruses which infect coliform bacteria. A review of the literature indicates that the most likely group of coliphage for use as a water quality indicator is the F (male)-specific RNA (FRNA) coliphage because (i) it represents viruses similar in size, shape, and genetic makeup to human enteric viruses, which are responsible for most of waterborne diseases, (ii) it represents viruses which are more stable than human enteroviruses such as poliovirus in environmental waters and more resistant to disinfection, (iii) its concentrations in environmental waters have been reported to correlate with sewage contamination (25, 26), and (iv) two bacterial hosts are available to selectively grow this coliphage (10, 26). The method to recover coliphages from environmental waters is relatively simple and within the capability and resources of most water quality laboratories. The limitations in the use of this method include (i) the need to maintain the host cells in the mutated state to maintain the method's efficiency to recover FRNA virus, (ii) the observation that this virus is a good marker of sewage but not human feces and the assumption that it must be multiplying under environmental conditions, and (iii) the need to develop simple and reliable methods to concentrate the virus from large volumes of water.

The virus assay is similar to the MF method in that the concentration of virus is based on the formation of PFU. Each plaque is a visible area of infection and is similar to a bacterial CFU because it is assumed that each plaque results from the infection of one virus. However, since viruses are specific, intracellular parasites, a live cell culture is required. For the assay of FRNA bacterial virus, two reliable strains of piliated bacteria which can support the growth of FRNA virus can be used. The first strain is a mutant of Salmonella typhimurium (WG-49) developed by Havelaar and Pot-Hogeboom (26); the second, an E. coli mutant called HS(pFamp)R, has been developed by Debartolomeis and Cabelli (10).

Phages of *Bacteroides fragilis*

B. fragilis is a species of bacteria found in higher concentrations than *E. coli* in the human intestinal tract. However, because this bacterium is anaerobic, it dies rapidly when discharged into environmental waters. A virus or phage which infects *B. fragilis* has been proposed as a good indicator of water quality because (i) the virus multiplies only in the intestinal tracts of humans and therefore shares a major characteristic with human enteric virus, (ii) it is not found in feces of most animals, and therefore its presence is evidence of human and not animal fecal contamination, and (iii) it survives well in the environment and can be detected by using *B. fragilis* bacteria as a host.

The method used to recover this phage is similar in principle to the method used to recover FRNA phage except that a specific strain (HSP40) of *B. fragilis* is required for recovery and enumeration. Use of *B. fragilis* as a host has resulted in promising data that this bacteriophage can be related to fecal contamination in marine waters (31, 34). It is curious that this phage has not been readily isolated from sewage and environmental waters in the United States (44). Thus, the method or the host for this group of virus must be further developed so it can be used under all conditions.

Human Enteric Virus

Since human viruses are the primary waterborne pathogens, it is logical that they would serve as good water quality indicators. This potential is supported by data which show that all bacterial indicators are unreliable predictors for the presence of human viruses in environmental waters. The limitations in the use of human viruses as indicators of water quality are the complexity of the methods to assay for human enteric viruses in terms of cost, time, equipment, and trained personnel. For example, animal cell cultures must be maintained, and methods to concentrate viruses from large volumes of water are difficult and inefficient (41). As a result, most routine monitoring laboratories will have to scale up considerably to be able to assay for human viruses. Although the methods for the recovery of human viruses can be made more efficient, the procedure will always be difficult. The best approach may be to establish virus standards but to designate and fund only select laboratories nationwide to monitor waters for viruses. By this approach, information on virus loads in environmental waters in different regions of the United States will be obtained; such information is currently lacking and is required to determine risks of acquiring waterborne diseases.

USE OF GENE PROBES TO MONITOR WATERS FOR PATHOGENS

The primary goal of monitoring marine recreational waters is to determine the presence and absence of pathogens rather than indicators of pathogens. Since human viruses are the most likely waterborne pathogens, this is the primary group of pathogens which should be monitored for. However, monitoring recreational waters for the many different human viruses by using the standard tissue culture method is not practical because different groups of virus require that different cells be cultured, and some viruses are not culturable. As a result, scientists are actively developing gene probe methods to detect the presence of pathogens such as viruses. The recently developed PCR method is capable of selecting specific gene sequences and amplifying them until enough of the products are available to measure the presence of a pathogen by more standard laboratory means such as gel electrophoresis. This method has already been used to detect the presence of most pathogens in clinical samples (39). Moreover, microorganisms can be concentrated from water on membranes, and their presence can be detected by applying PCR to the membrane (2). This method is fast and sensitive, and many viruses can be measured simultaneously by combining several gene probes for the different groups of virus. The current limitation of this approach is that it cannot differentiate between live and dead cells (only live microorganisms are of public health significance). Thus, pathogens in effluents which have been properly disinfected before being discharged into environmental waters may still be detected as positive by the PCR method. In a current application of PCR, tissue culture is used to start the infection of pathogenic viruses and PCR is used to rapidly detect and identify the virus infection in that tissue culture.

In summary, PCR is a powerful tool, and many scientists are convinced that science is the study of how to overcome limitations. In this regard, many scientist are working to improve the PCR method so that it is quantitative and will be able to distinguish between the detection of gene sequences from dead and from live cells. When this occurs, the quality of recreational waters will be determined within a few hours of sampling, and thus decisions affecting the health of swimmers will be able to be made in a timely manner.

REFERENCES

1. **American Public Health Association, American Water Works Association, Water Pollution Control Federation.** 1995. *Standard Methods for the Examination of Water and Wastewater*, 19th ed. American Public Health Association, Washington D.C.
2. **Bej, A. K., M. R. Mahbubani, J. L. Dicesare, and R. M. Atlas.** 1991. Polymerase chain reaction-gene probe detection of microorganisms by using filter-concentrated samples. *Appl. Environ. Microbiol.* **57:**3529–3534.
3. **Bisson, J. W., and V. J. Cabelli.** 1979. Membrane filtration enumeration method for *Clostridium perfringens. Appl. Environ. Microbiol.* **37:**55–66.
4. **Cabelli, V. J.** 1983. *Health Effects Criteria for Marine Recreational Waters.* EPA-600/1-80-031. U.S. Environmental Protection Agency, Washington, D.C.
5. **Cabelli, V. J., A. P. Dufour, L. J. McCabe, and M. A. Levin.** 1982. Swimming-associated gastroenteritis and water quality. *Am. J. Epidemiol.* **115:**606–616.
6. **Calderon, R. L., E. W. Mood, and A. P. Dufour.** 1991. Health effects of swimmers and nonpoint sources of contaminated water. *Int. J. Environ. Health Res.* **1:**21–31.
7. **Chang, W. J., and F. D. Pien.** 1986. Marine-acquired infections: hazards of the ocean environment. *Postgrad. Med.* **80:**30–33.
8. **Charoenca, N., and R. Fujioka.** 1991. Assessment of staphylococcus bacteria in Hawaii's marine recreational waters. *Water Sci. Technol.* **27:**283–289.
9. **Davies, C. M., S. C. Apte, and S. M. Peterson.** 1995. Possible interference of lactose-fermenting marine vibrios in coliform β-D-galactosidase assays. *J. Appl. Bacteriol.* **78:**387–393.
10. **Debartolomeis, J., and V. J. Cabelli.** 1991. Evaluation of an *Escherichia coli* host strain for enumeration of F male-specific bacteriophages. *J. Appl. Environ. Microbiol.* **57:**1301–1305.
11. **Department of Health, State of Hawaii.** 1990. *Hawaii Ad-*

ministrative Rules on Water Quality Rules, chapters 11–54. Department of Health, State of Hawaii, Honolulu.

12. **Dufour, A. P.** 1984. Bacterial indicators of recreational water quality. *Can. J. Public Health* **75:**49–56.

13. **Easterbrook, T. J., and P. A. West.** 1987. Comparison of most probable number and pour plate procedures for isolation and enumeration of sulphite-reducing *Clostridium* spores and group D fecal streptococci from oysters. *J. Appl. Bacteriol.* **62:**413–419.

14. **Fattal, B., E. Peleg-Oleevsky, Y. Yoshpe-Purer, and H. I. Shuval.** 1986. The association between morbidity among bathers and microbial quality of seawater. *Water Sci. Technol.* **18:**59–69.

15. **Figueras, M. J., F. Polo, I. Inza, and J. Guarro.** 1994. Poor specificity of m-Endo and m-FC culture media for the enumeration of coliform bacteria in sea water. *Lett. Appl. Microbiol.* **19:**446–450.

16. **Fleischer, J. M.** 1991. A re-analysis of data supporting US Federal bacteriological waters quality criteria governing marine recreational waters. *Res. J. Water Pollut. Control Fed.* **63:**259–265.

17. **Fujioka, R.** 1992. Value of coastal water quality for island communities, p. 53–68. *In 92 Annual Joint Symposium on Water Resources and Quality Management, and Development and Conservation of Groundwater Resources in Che-Ju Do.* Chung Buk National University, Cheong Ju, Korea.

18. **Fujioka, R., C. Fujioka, and R. Oshiro.** 1992. Application of *Clostridium perfringens* to assess the quality of environmental and recreational waters. WRRC project completion report to Department of Public Works, City and County of Honolulu.

19. **Fujioka, R. S., H. H. Hashimoto, E. B. Siwak, and R. H. F. Young.** 1981. Effect of sunlight on survival of indicator bacteria in seawater. *Appl. Environ. Microbiol.* **41:**690–696.

20. **Fujioka, R. S., and L. K. Shizumura.** 1985. *Clostridium perfringens*, a reliable indicator of stream water quality. *J. Water Pollut. Control Fed.* **57:**986–992.

21. **Fujioka, R. S., K. Tenno, and S. Kansako.** 1988. Naturally occurring fecal coliforms and fecal streptococci in Hawaii's freshwater streams. *Toxic. Assess.* **3:**613–630.

22. **Fujioka, R. S., A. A. Ueno, and O. T. Narikawa.** 1990. Unreliability of KF agar to recover fecal streptococcus from tropical marine waters. *Res. J. Water Pollut. Control Fed.* **62:**27–33.

23. **Geldreich, E. E., and B. A. Kenner.** 1969. Concepts of fecal streptococci in stream pollution. *J. Water Pollut. Control Fed.* **41:**R336–R352.

24. **Hardina, C. M., and R. S. Fujioka.** 1991. Soil: the environmental source of E. coli and enterococci in Hawaii's streams. *Environ Toxicol. Water Qual.* **6:**185–195.

25. **Havelaar, A. H., K. Furuse, and W. H. Hogeboom.** 1986. Bacteriophages and indicator bacteria in human and animal faeces. *J. Appl. Bacteriol.* **60:**255–262.

26. **Havelaar, A. H., and W. M. Pot-Hogeboom.** 1988. F-specific RNA-bacteriophages as model viruses in water hygiene: ecological aspects. *Water Sci. Technol.* **20:**399–407.

27. **Hazen, T. C.** 1988. Fecal coliforms as indicators in tropical waters: a review. *Toxic. Assess.* **3:**461–477.

28. **Hernandez, J. F., J. M. Guibert, J. M. Delattre, C. Oger, C. Charriere, B. Hughes, R. Serceau, and F. Sinegre.** 1991. Miniaturized fluorogenic assays for enumeration of E. coli and enterococci in marine water. *Water. Sci. Technol.* **24:**137–141.

29. **Hernandez-Lopez, J., and F. Vargas-Albores.** 1994. False-positive coliform readings using membrane filter techniques for seawater. *Lett. Appl. Microbiol.* **19:**483–485.

30. **Hill, R. T., L. T. Knight, M. S. Anikis, and R. R. Colwell.** 1993. Benthic distribution of sewage sludge indicated by *Clostridium perfringens* at a deep-ocean dump site. *J. Appl. Environ. Microbiol.* **59:**47–51.

31. **Jofre, J., A. Bosch, F. Lucena, R. Girones, and C. Tartera.** 1986. Evaluation of *Bacteroides fragilis* bacteriophages as indicators of the virological quality of water. *Water Sci. Technol.* **18:**167–177.

32. **Jones, F., D. Kay, R. Stanwell-Smith, and M. D. Wyer.** 1990. An appraisal of the potential public health impacts of sewage disposal to UK coastal waters. *J. Inst. Water Environ. Man* **4:**295–303.

33. **Klein, E., and A. C. Houston.** 1899. Further reports on bacteriological evidence of recent and therefore dangerous sewage pollution, p. 498–504. *In Metropolitan Water Supply.* 28th Annual Report of Local Government Board. Supplementary Report of Medical Officer for 1898–1899. Appendix B. London Government Office, London.

34. **Lucena, F., J. Lasobras, D. McIntosh, M. Forcadell, and J. Jofre.** 1994. Effect of distance from the polluting focus on relative concentrations of *Bacteroides fragilis* phages and coliphages in mussels. *Appl. Environ. Microbiol.* **60:**2272–2277.

35. **Novotny, V.** 1988. Diffuse (nonpoint) pollution—a political, institutional, and fiscal problem. *J. Water Pollut. Control Fed.* **60:**1404–1413.

36. **Olson, B. H.** 1978. Enhanced accuracy of coliform testing in seawater by a modification of the most-probable-number method. *Appl. Environ. Microbiol.* **36:**438–444.

37. **Palmer, C. J., Y. Tsai, A. L. Lang, and L. Sangermano.** 1993. Evaluation of Colilert-Marine Water detection of total coliforms and *Escherichia coli* in the marine environment. *Appl. Environ. Microbiol.* **59:**786–790.

38. **Parrington, L. J., A. N. Sharpe, and P. I. Peterkin.** 1993. Improved aerobic colony count technique for hydrophobic grid membrane filters. *Appl. Environ. Microbiol.* **59:**2784–2789.

39. **Persing, D. H., T. F. Smith, F. C. Tenover, and T. J. White (ed.).** 1993. *Diagnostic Molecular Microbiology: Principles and Applications.* American Society for Microbiology, Washington, D.C.

40. **Pike, E. B.** 1994. Recreational use of coastal waters: development and application of health-related standards, p. 189–199. *In G. Eden and M. Haigh (ed.), Water and Environmental Management in Europe and North America: a Comparison of Methods and Practices.* Ellis Horwood Press, Chichester, Great Britain.

41. **Reynolds, K. A., C. P. Gerba, and I. L. Pepper.** 1995. Detection of enteroviruses in marine waters by direct RT-PCR and cell culture. *Water Sci. Technol.* **31:**323–328.

42. **Salas, H. J.** 1986. History and application of microbiological water quality standards in the marine environment. *Water Sci. Technol.* **18:**47–57.

43. **Scott, W. J.** 1932. Survey of Connecticut's shore bathing waters. *Public Health Eng.* **19:**316.

44. **Sobsey, M.** Personal communication.

45. **Sorensen, D. L., S. G. Eberl, and R. A. Dicksa.** 1989. *Clostridium perfringens* as a point source indicator in nonpoint polluted streams. *Water Res.* **23(2):**191–197.

46. **Stevenson, A. H.** 1953. Studies of bathing water quality and health. *Am. J. Public Health.* **43:**529–538.

47. **U.S. Environmental Protection Agency.** 1986. Ambient water quality criteria for bacteria—1986. EPA 440/5-84-002. U.S. Environmental Protection Agency, Washington, D.C.

Detection of Indicator Microorganisms in Environmental Freshwaters and Drinking Waters

GARY A. TORANZOS AND GORDON A. McFETERS

19

Indicator organisms are used globally as a warning of possible contamination and as an index of water quality deterioration. Heavy reliance has been placed on the coliform group of bacteria to determine the safety of potable water, while coliforms and other organisms are also used to signal conditions in recreational and shellfish-harvesting water that might lead to adverse health consequences. Although this practice is not perfect and there is considerable variety in the ways that different indicator microorganisms are applied in various geographical areas and situations, public health concerns have generally been well served. The presence of indicator organisms will likely continue to be used as a criterion of water quality that will be of value if attention is given to the development and use of optimal methods for the recovery of these microorganisms.

Gastroenteritis is the most common affliction associated with waterborne pathogens. Although for most of the population in developed countries minor gastroenteritis may simply mean several hours of discomfort, in developing countries up to 10 million people die every year as a direct result of the consumption of contaminated water (26). However, the continuing increase in the proportion of immunologically compromised and elderly in developed countries has focused more attention on possible waterborne outbreaks. The seriousness of this reality is compounded by the fact that determining the microbiological quality of waters is more complicated than previously thought.

The presence of enteric pathogens in drinking and recreational waters is of great concern. As a result of the danger to public health due to the presence of pathogens, it is extremely important to determine the microbiological safety of these waters. The ideal manner for doing this would be to analyze the waters for the presence of specific pathogens of concern. However, hundreds of different microorganisms have been shown to be involved in waterborne disease outbreaks; thus, it would be impractical to look for every pathogen potentially present in water samples. Culture methods are usually used for bacteria and cell culture techniques for the detection of viruses, while microscopic methods are used for protozoa. However, the target bacteria may not grow in culture media since they are frequently injured as a result of exposure to environmental stressors such as disinfectants used during water treatment. Additionally, several enteric viruses cannot be cultured in the laboratory, and methods for the detection of protozoan pathogens are notoriously inaccurate.

As a result, analyses for the presence of waterborne pathogens becomes extremely complicated and do not ensure complete safety to the consumer. Thus, the following groups of microorganisms are used to determine the biological safety of the waters.

COMMONLY USED INDICATORS
Total Coliforms

The traditional definition of the coliform group of bacteria specifies that they are aerobic and facultatively anaerobic, gram-negative, nonsporeforming, rod-shaped bacteria that ferment lactose with gas and acid production in 24 to 48 h at 35°C (14). Hence, these criteria are not strictly taxonomic, although coliform bacteria belong to the family *Enterobacteriaceae* and usually include *Escherichia coli* as well as various members of the genera *Enterobacter, Klebsiella,* and *Citrobacter* when these criteria are applied in the microbiological analysis of water quality. That definition of the coliform bacteria has classically been translated into specific biochemical reactions or the appearance of characteristic colonies on commonly used media. However, the more recent advent of enzyme-specific media and tests has allowed the application of cytochrome oxidase (negative) and β-galactosidase (positive) as additional criteria for the coliform group (14).

These bacteria are classically used as indicators of fecal contamination or water pollution from sewage and thus are of sanitary significance, although it might also be pointed out that while these bacteria can originate from the intestinal tracts of homeothermic animals, other bacteria numerically dominate that type of microbial community (20). In addition, experience has demonstrated that some members of the coliform group can originate from nonenteric environments such as wastes from the wood industry and surfaces of redwood water tanks (75), biofilms within drinking water distribution systems (21, 38), and epilithic algal mat communities in pristine streams (5, 47). The persistence of these bacteria in aquatic systems is comparable to that of some of the waterborne bacterial pathogens (51), although they are much less persistent than enteric viruses and protozoa. Likewise, bacteria belonging to the coliform group are somewhat like many of the waterborne bacterial pathogens with respect to disinfection susceptibility but unlike the more persistent viruses and protozoa. These factors illustrate why experience and discretion are often vital in the correct interpretation of coliform data obtained from some aquatic environments.

The coliform bacteria that comprise what has traditionally been termed the "total coliform" group form the primary standards for potable water in North America and indeed most of the world. This designation is usually interpreted as the totality of bacteria that conform to the classical, nontaxonomic coliform definition, although the inclusion of the term "total" has often led to the false assumption that all viable coliform bacteria are detected by a given method specified for the analysis of coliform bacteria.

Fecal Coliforms (Thermotolerant Coliforms)

The subset of the more comprehensive coliform, or total coliform, group that is more definitive as an indicator of homeothermic fecal contamination consists of what are termed the fecal coliforms (18). However, the term "thermotolerant coliform" may be more accurate for this group (73) and is beginning to be accepted (11). These bacteria conform to all of the criteria used to define total coliforms plus the requirement that they grow and ferment lactose with the production of gas and acid at $44.5 \pm 0.2°C$. Bacteria in this coliform subgroup have been found to have an excellent positive correlation with fecal contamination from warm-blooded animals (52). The physiological basis of the elevated temperature phenotype in the fecal coliforms has been described as a thermotolerant adaptation of proteins to, and their stability at, the temperatures found in the enteric tracts of animals (10) that are both constant and higher than temperatures in most aquatic and terrestrial environments. However, fecal coliform bacteria which conform to this definition can belong to the genus *Klebsiella* (13, 35) and have been isolated from environmental samples in the apparent absence of fecal pollution. Such observations have been made on water receiving high levels of carbohydrate-rich industrial effluent and in contact with plant material. Similarly, in the last 10 years there have been reports that other members of the fecal coliform group, including *E. coli*, were detected in some pristine areas of the world (58) and associated with regrowth events in potable water distribution systems (38). Thus, caution needs to be exercised when one is trying to decide whether the presence of indicator microorganisms does indeed represent fecal contamination and thus a threat to public health.

E. coli

Among the fecal coliforms, *E. coli* deserves further discussion. This bacterium not only satisfies all of the criteria of the total and fecal coliforms in most cases but has additional characteristics that make it a useful microbiological indicator of water quality. In particular, *E. coli* has been demonstrated to be a more specific indicator for the presence of fecal contamination than the fecal coliform group of bacteria (13). In addition, *E. coli* conforms to taxonomic as well as functional identification criteria and is enzymatically distinguished by the lack of urease and presence of β-glucuronidase, enzymes that form the basis for recently developed differential methods that will be discussed later in this chapter. One possible disadvantage of this organism as an indicator in water is that it has been consistently found in pristine tropical rain forest aquatic and plant systems and so may not be a reliable signal of fecal contamination in those environments (26). However, *E. coli* has been effectively used

for some time in Europe and has recently been incorporated into U.S. drinking water regulations as a specific indicator of fecal contamination.

Fecal Streptococci and Enterococci

Fecal streptococci and enterococci, which are gram-positive bacteria, are useful as indicators of microbiological water quality since they are common inhabitants of the intestinal tracts of humans and lower animals (20). Like the coliforms, some of these organisms have persistence patterns that are similar to those of a range of potential waterborne pathogenic bacteria (57). That is particularly true of the enterococci, including *Streptococcus faecalis* and *S. faecium*, which are thought to be more human specific than the other fecal streptococci. In addition, other members of that group such as *S. bovis*, *S. equinus*, and *S. avium* are somewhat characteristic of fecal contaminants from specific animals and birds (18), although the latter species do not survive in water as well as do many of the other classical indicator bacteria and some pathogens (19, 47).

OTHER COMMONLY USED INDICATORS
Sulfite-Reducing Clostridia

The use of the sulfite-reducing members of the genus *Clostridium* (*C. perfringens* and *C. welchii*) as indicators of fecal pollution was originally proposed in the late 1800s by Houston. The presence of these microorganisms in the feces of all warm-blooded animals is the basis for this practice, although they are considered ubiquitous in aquatic sediments and the spore form explains their persistence. As a result of the longevity of the spores, these bacteria can be considered indicators of remote fecal pollution (73).

Pseudomonas spp.

Members of the genus *Pseudomonas* are possibly the microorganisms isolated most often from bodies of water. However, contrary to the previously discussed indicators, their presence does not necessarily indicate a possible risk to public health. Some species have been linked to infections associated with exposure to recreational waters and thus have been proposed as indicators of recreational water quality (5). *Pseudomonas aeruginosa* was found to be more resistant than acid-fast bacteria during ozonation processes (23), which demonstrates its resistance to chemical disinfection and thus its usefulness in analysis of recreational waters such as swimming pools, which receive chemical disinfection.

HPC

The heterotrophic plate count (HPC) method takes into consideration the enumeration of all aerobic bacteria capable of growing in some commonly used media such as R_2A, HPC agar, or plate count agar. The level of bacteria determined by HPC indicates the overall microbiological status (i.e., distribution) of the system and not necessarily the possibility of risk to public health, although such bacteria have been recently implicated as potential pathogens in drinking water (42, 57).

Fecal Sterols

Alternative analytical approaches have been sought in lieu of classically used indicator microorganism as indicators of fecal pollution. This interest continues because of inadequacies associated with the failure of cultivation-based methods to detect these organisms as well as other concerns. The use of specific saturated sterols such as 5β-cholestan-3β-ol (coprostanol) has been investigated as a molecular signature of fecal contamination in water (14, 75). Although not frequently employed, this approach has been useful as a marker of fecal pollution in sediments because the compound is specific to the feces of higher animals, including humans, and is biodegradable, and there is a relationship between its concentration and the degree of fecal pollution (70, 74, 75). However, since this is a chemical method, it is beyond the scope of this chapter and will not be discussed further.

FC/FS Ratio

It is often desirable to differentiate between humans and lower animals as the probable source of fecal pollution in studies of contaminated aquatic systems. As noted above with regard to the fecal streptococci, different animals can have somewhat distinctive signatures of fecal microflora. Although both fecal coliforms and fecal streptococci are numerous in the feces of humans and animals, the ratio of fecal coliforms to fecal streptococci (FC/FS ratio) has been proposed as an indicator of the origin of the contamination (21). The research has indicated that a ratio of greater than 4 is characteristic of human fecal contamination whereas a ratio of less than 0.7 suggests animal waste. Although that index has been used with success in many applications, it is complicated by the differential die-off kinetics of the two bacterial groups as well as differences among individual species within those categories. Specifically, some reports have indicated that the fecal streptococci have a greater survival than fecal coliforms in water and that certain enterococci such as S. bovis and S. equinus die off much more rapidly than the other fecal streptococci (47). For that reason, Geldreich and Kenner (21) suggested that the ratio would be valid only for less than 24 h following the discharge of feces into the water being tested. Interpretation of the FC/FS ratio is further complicated by the observation that the ratio is also influenced by the efficiency of methods and media used to detect fecal streptococci. Hence, the FC/FS ratio is not generally recommended as a totally unambiguous means of differentiating human from animal fecal pollution (5, 52), although it has value in some instances if care is used in interpreting the results.

SIGNIFICANCE OF INDICATORS TO PUBLIC HEALTH

Analysis for the presence of indicators is a shortcut attempt to determine the microbiological quality and public health safety of waters. Very few natural water bodies are completely free of indicator microorganisms. As mentioned above, high concentrations of total coliforms can be found associated with plant material. Thus, the presence of high numbers of coliforms in surface waters is not necessarily a cause for concern. Under some circumstances, even the presence of fecal coliforms (Klebsiella spp., E. coli) can be expected in surface waters. In these instances, other indicators should be used to determine the possible threat to public health. However, the difference between untreated and treated waters should be emphasized. The detection of fecal (thermotolerant) coliforms in treated drinking waters should be a cause for concern, since current drinking water treatment processes successfully eliminate indicator microorganisms.

Current Regulations Regarding Drinking and Surface Waters

The U.S. Environmental Protection Agency and the World Health Organization as well as the European Economic Community standards allow for a maximum number of microorganisms that can be present in a given volume of drinking water (22, 56). Table 1 gives some of the U.S. and international standards for drinking and bathing waters. These standards are believed to provide optimal protection of public health. In the United States, the maximum contaminant levels (MCL) are the maximum permissible concentrations of these organisms and are legally enforceable. In Canada, the maximum acceptable concentrations (MAC) are nonenforceable guidelines.

ALTERNATE INDICATORS

Bacteriophages

The use of bacteriophages as indicators is not a new concept (12, 36). More specifically, bacterial viruses that use E. coli as the host bacterium (coliphages) have been proposed as

TABLE 1 Bacteriological drinking water and recreational freshwater standards or guidelines

Standards established by:	No./100 ml					Turbidity (NTU[b])
	Total coliforms[a]		Fecal (thermotolerant) coliforms		Enterococci (recreational)	
	Drinking	Recreational[c]	Drinking	Recreational[c]		
World Health Organization	1–10		0			<1–5
Canada	<10		0	200[d]	35[e]	<1–5
European Economic Community	0	<10,000[f]		<2,000[f]		0–4
United States	0	200[g]				1 (monthly)

[a] In systems analyzing <40 samples per month, the maximum contaminant level specifies that no more than one sample per month may be total coliform positive. In systems analyzing >40 samples per month, the maximum contaminant level specifies that no more than 5% of the monthly samples may be total coliform positive (22).

[b] NTU, nephelometric turbidity units.

[c] Recreational refers to primary contact (swimming) waters.

[d] Geometric mean of at least five samples (27) when experience has shown that greater than 90% of the fecal coliforms are E. coli.

[e] Geometric mean of at least five samples taken during a period not to exceed 30 days (28).

[f] Compulsory limit. If exceeded in more than 20% of samples with at least 14 days of sampling, then bathing is prohibited (56).

[g] This is a U.S. Environmental Protection Agency criterion. Since no uniform national standard exists, it may vary from state to state.

alternate indicators of fecal contamination and as possible model organisms for the presence of human enteric viruses in water. The term "coliphage" can be rather misleading, since it is used to refer to bacterial viruses such as phage lambda, which can also infect *Salmonella* spp., which are not, by definition, coliforms. Nonetheless, there are several hundred (if not thousands of) different coliphages present in sewage which can serve as alternate indicators of fecal contamination for some applications (66, 68). The icosahedral coliphages (such as the MS2, or male-specific, phage) are very similar in morphology and diameter to the human enteroviruses. Thus, there has been much discussion concerning the use of these microorganisms as alternate indicators. However, as Kott very succinctly expressed it "When introducing a new indicator, it should be borne in mind that superiority [of the new indicator] is obligatory" (36). Reports have also shown that coliphages, although similar to enteric viruses in morphology and size, behave differently under environmental conditions. Thus, bacteriophages could possibly be used as indicators of fecal pollution rather than as specific indices of the presence of certain pathogens (25, 30). Extreme care should be taken, however, during enumeration of coliphages in the environment, since the numbers obtained depend on the type of host bacterium being used. Some of the hosts are permissive only to somatic phage or male-specific phage replication, whereas others are permissible for both.

Other bacteriophages (such as *Bacteroides fragilis* phages) have been proposed as alternate indicators of human pollution (64, 65). These phages are quite promising; however, the host is anaerobic, which presents problems for the easy implementation of this technique in water utility laboratories. More research needs to be done on this group of viruses.

H$_2$S Producers

Although most of the members of the total and fecal coliform groups are not H$_2$S producers, Manja et al. (43) developed a simple method which has been tested in developing countries (37) and rural areas of North America (61). Although no correlation was observed between the presence of coliforms and H$_2$S producers (63), other studies have shown that the presence of the H$_2$S producers does correlate with other criteria of microbiological water quality. Thus, this test may be a good alternative for rural areas, as well as remote areas of the world, since it does not necessitate incubators, sophisticated equipment, or trained personnel (67).

GENERAL METHODS
Sample Collection, Transport, and Storage
It should be noted that it is just as important to use appropriate care in obtaining the proper sample as it is to analyze the sample correctly. The sample should be collected in sterile glass or polypropylene bottles or bags. Whenever it is suspected that a disinfectant is present, the sample should be amended with a solution of sodium thiosulfate (Na$_2$S$_2$O$_3$), which inactivates any residual halogen compounds present in the sample. An Na$_2$S$_2$O$_3$ concentration of 18 mg/liter will neutralize up to 5 mg of free (residual) chlorine per liter (4).

Some environmental samples may also contain high concentrations of zinc and copper. In such cases, it is recommended that the sampling containers be amended with a solution of the disodium salt of EDTA (Na$_2$EDTA) at a concentration of 372 mg/liter. The presence of this chelating agent will reduce metal toxicity and is especially important if more than 4 h elapses between collection and analysis of the sample (4) to avoid any possible metal toxicity to the microorganisms.

Both compounds (Na$_2$S$_2$O$_3$ and Na$_2$EDTA) can be added together or separately to the containers before sterilization by autoclaving. It is recommended that the reader consult *Standard Methods for the Examination of Water and Wastewater* (4) for more detailed information.

Additionally, the amount of elapsed time allowed between sample collection and analysis should not exceed 24 h. The samples should be kept in a refrigerated container (or icebox) at a temperature below 10°C during transport and storage. It should be noted that the sample will not necessarily reach this temperature. If sample analysis within the 24-h time limit is not possible, in situ analysis using portable equipment should be considered.

Water samples can be analyzed for total coliform bacteria by a range of techniques, including most probable number (MPN; multiple-tube fermentation test), presence-absence (P-A), and membrane filtration (MF), as discussed below. Because of this variety of analytical approaches, there are a number of media that are commonly used.

MPN (Multiple-Tube Technique)
MPN analysis is a statistical method based on the random (Poisson) dispersion of microorganisms in a given sample. The results are expressed in terms of the MPN of microorganisms detected per volume of sample. Classically, this assay has been performed as a multiple-tube fermentation test. Although the technique is rather time-consuming (to perform the presumptive, confirmed, and sometimes completed phases, a process which takes several days), most laboratory technicians around the world are fully trained in this technique and prefer it to other methods of water analysis. The MPN technique is also recommended for high-turbidity waters.

Classically, lauryl tryptose broth is used with the MPN method; with this technique, a small inverted tube is included to facilitate the detection of gas formation, and 0.01 g of bromcresol purple per liter (final concentration) is used to determine acid production. The concentration of this medium is prepared so that the addition of water sample volumes of 10, 20, or 100 ml to the liquid medium will not reduce the ingredient concentration below that of the specified medium. Typically, five identical (20-ml) aliquots of drinking water are analyzed as a set. Each tube is vigorously mixed with the medium about 25 times, taking care not to introduce air bubbles into the inverted tubes. These samples are then incubated at 35°C and examined for growth and for gas and acid production after 24 and 48 h. The production of gas or acidic growth after 48 h constitutes a positive presumptive reaction. Confirmation of presumptive reactions is done by inoculating the positive aliquots into brilliant green lactose bile broth tubes containing an inverted tube, incubating the samples for 24 and 48 h at 35°C, and scoring tubes with gas as representing confirmed samples that contain coliform bacteria. Bacterial density and the 95% confidence limits can be estimated with the use of MPN tables for the volumes and number of aliquots used (4).

The MPN method can also be used for the direct detection of coliforms and *E. coli* by using enzyme-specific tests as mentioned below. The characteristic color and/or fluorescent endpoints specified by the manufacturer should be used

TABLE 2 Analysis of drinking water or surface water (including treated wastewater) by the multiple-tube (MPN) technique

Type of analysis	Medium	Vol (ml) inoculated into 5- or 3-tube series	Positive reaction
Total coliforms (incubate at 35 ± 0.5°C)	Lactose broth or	10, 1, 0.1	Gas and/or acid
	lauryl tryptose broth	10, 1, 0.1	Gas and/or acid
Fecal coliforms			
Incubate at 44.5 ± 0.2°C	EC medium	10, 1, 0.1	Gas and/or acid
Incubate at 35 ± 0.5°C for 3 h and transfer to	A-1[a]	10, 1, 0.1	Gas
44.5 ± 0.2°C for 21 h			
Enterococci (incubate at 35 ± 0.5°C)	Azide dextrose broth	10, 1, 0.1	Growth

[a] Although A-1 medium is recommended by the American Public Health Association only for analysis of marine waters and treated wastewater, several laboratories have tested it for analysis of surface and drinking waters with excellent results (13a). Additionally, A-1 medium does not require a confirmation test.

to determine the presence or level of the target bacteria. The U.S. Environmental Protection Agency has indicated that confirmation is not needed with the commercially available 4-methylumbelliferyl-β-D-glucuronide-based media. For the examination of surface waters, inoculate a series of medium tubes with a 10-fold dilution series of the sample. Using five- or three-tube series, inoculate each series with one dilution. If low concentrations of indicator bacteria are present, inoculate 100 ml of water into flasks containing 100 ml of the appropriate medium at double strength or, alternatively, 10-ml volumes directly into tubes containing 10 ml of the appropriate medium at double strength. If high concentrations of bacteria are suspected, a 10-fold dilution series should be performed. A sterile solution such as phosphate buffer (pH 7.2) or peptone water (0.1% final solution, pH 6.8) should be used as the diluent.

Inoculate media as indicated in Table 2 for the presumptive test. Subsequently, all positive tubes should be subjected to the confirmation test by inoculation into a second medium as indicated in Table 3.

MF

MF is possibly the most widely used method in North America and Europe. It is a simple test and lends itself well to the in situ analysis of samples as a result of the portability of the necessary equipment. The technique is based on the entrapment of the bacterial cells by a membrane filter (pore size of 0.45 μm). After the water is filtered, the membrane is placed on an appropriate medium and incubated. Discrete colonies with typical appearance are counted after 24 to 48 h. This technique is more precise than the multiple-tube (MPN) technique. The greatest limitation of the MF test is that it is useful only for low-turbidity waters and for waters that have low concentrations of nontarget (i.e., background) microorganisms. Samples of water that possess high turbidity clog the filter, and high concentrations of nontar-

get microorganisms mask the presence of the target colonies. Tables 4 and 5 outline the basic MF method, including confirmation procedures.

Total Coliforms

The MF analysis of water for total coliform bacteria can involve the use of LES Endo agar (Difco), m-Endo broth (Difco), or m-Coliform broth (BBL) as well as other media. If liquid medium is used, ca. 2 ml of the broth can be added to certified absorbent pads, available from commercial sources, in small sterile petri dishes, or 1.5% agar-agar is added to the broth to make it semisolid. Details for carrying out the MF procedure are given elsewhere (5). Characteristic colonies appear pink to dark red with a unique metallic green sheen (viewed at a magnification of ×10 to ×15 if necessary). Tergitol agar has been used to detect coliform bacteria as a means to follow the movement of the plume resulting from a marine sewage outfall (46). Only plates containing between 20 and 80 typical colonies should be counted. Confirmation can be done as described earlier by the selection of both representative typical and atypical colonies. The population density of the target bacteria, usually described as CFU/100 ml, in the original sample can be calculated from the volume filtered and dilutions used, if any. The percentage of the typical colonies that are confirmed as positive can be used to determine the verified coliform density.

Fecal Coliforms (Thermotolerant Coliforms)

M-FC medium is used to quantify fecal coliforms in water samples when the MF technique is used. The addition of 1% rosolic acid (dissolved in 0.02 N NaOH) to the medium is highly recommended. However, care should be taken since rosolic acid decomposes if sterilized by autoclaving. It should be stored refrigerated (4 to 6°C) and discarded after 2 weeks or sooner if its color changes from dark red to muddy

TABLE 3 Confirmation procedures for the MPN technique

Type of analysis	Medium	Positive reaction
Total coliforms (incubate at 35 ± 0.5°C)	Brilliant green lactose bile broth and	Gas and/or acid
	lauryl tryptose broth	Gas and/or acid
Fecal coliforms (incubate at 44.5 ± 0.2°C)	EC medium	Gas and/or acid
Enterococci (incubate at 35 ± 0.5°C)	Pfizer selective *Enterococcus* medium[a]	Growth
		Brownish colonies (with halos)[b]

[a] No longer commercially available and therefore must be prepared separately.
[b] Brownish colonies confirm the presence of fecal streptococci. Colonies are transferred to brain heart infusion broth containing 6.5% NaCl. Growth in the latter medium confirms the presence of enterococci.

TABLE 4 MF procedures for analysis of drinking water or surface water (including treated wastewater)

Type of analysis	Medium	Vol (ml) filtered (depending on type of sample)[a]	Positive reaction
Total coliforms (incubate at 35 ± 0.5°C)	LES Endo agar or	100, 10, 1.0	Colonies with metallic green sheen
	M-Endo medium	100, 10, 1.0	Colonies with metallic green sheen
Fecal coliforms (incubate at 44.5 ± 0.2°C)	M-FC	100, 10, 1.0	Pale to deep-blue colonies
Fecal streptococci (incubated at 35 ± 0.5°C)	M-Enterococcus	100, 10, 1.0	Pink to deep-red colonies
Enterococci (incubate at 41°C for 24–48 h, then transfer membrane to EIA medium and incubate for 20 min at 41°C)	mE	100, 10, 1.0	Pink to red colonies with a black or reddish brown precipitate on the underside after transfer of the membrane to EIA medium

[a] When treated wastewater or waters affected by sewage are being sampled, several dilutions may be analyzed, and the 100- and 10-ml volumes may be skipped altogether, with higher dilutions used instead.

brown. M-FC can be used as an agar-based medium (1.5% agar added to the broth), or broth can be added to certified absorbent pads. Petri dishes containing filters are incubated at 44.5 ± 0.2°C. Incubation with this degree of accuracy can best be achieved by using incubators or water baths for this specific purpose. Dishes can be submerged in a water bath with weights after being placed inside two tightly sealed plastic bags. Typical fecal coliform colonies will appear various shades of blue, although atypical *E. coli* may be pale yellow whereas non fecal coliform colonies are grey to cream colored. Again, magnification of ×10 to ×15 can be used to observe colonies, and the desired range is 20 to 60 colonies per plate.

Typical colonies sometimes lose the characteristic appearance (green sheen and bluish color for total and fecal coliforms, respectively) after 24 h; thus, reading all plates at 24 h postinoculation is highly recommended.

Enterococci

The medium designated mE is typically used for the detection of enterococci in freshwater and marine waters (the latter topic is covered in chapter 18). Although KF medium was used in the past, it has been found to be susceptible to false-positive results (3, 17, 66) and is no longer recommended. The mE agar is prepared by heating the basal ingredients (10 g of peptone, 15 g of NaCl, 30 g of yeast extract, 1 g of esculin, 0.05 g of cycloheximide, 0.15 g of NaN$_3$, and 15 g of agar) and cooling the mixture to <46°C. Then 0.25 g of nalidixic acid is mixed in 5 ml of reagent-grade water, a few drops of 0.1 N NaOH are added to dissolve the antibiotic, and the whole volume is added to the basal medium. Finally, 0.15 g of 2,3,5-triphenyl tetrazolium chloride is added and dissolved, and the entire mixture is dispensed into petri dishes. It should be noted that the quantities specified

above are for the preparation of a 1-liter volume of medium. After processing of the sample, the membrane filter is placed onto the solidified medium, incubated at 41°C, and observed after 24 and 48 h. After incubation, the membrane is transferred very carefully to a petri dish containing EIA medium (1 g of esculin, 0.5 g of ferric citrate, 15 g of agar, 1 liter of water; dissolve and sterilize by autoclaving) and incubated at 41°C for 20 min. All pink to red colonies developing a black or reddish brown precipitate on the underside of the filter should be counted.

For the detection of fecal streptococci in freshwaters and marine waters, m-Enterococcus medium (20 g of tryptone, 5 g of yeast extract, 2 g of glucose, 4 g of K$_2$HPO$_4$, 0.4 g of NaN$_3$, 0.1 g of 2,3,5-triphenyl tetrazolium chloride, 10 g of agar, 1 liter of water) is used. The sample is processed, and the petri dishes are incubated at 35°C for 24 to 48 h. Colonies should be counted after 24 and 48 h, as some of the pinpoint colonies are not readily visible after 24 h.

P-A

The microbiological monitoring of drinking water has historically relied on MPN and MF approaches to estimate population densities of indicator organisms. Although those methods have been useful, disadvantages such as space and time requirements prompted the idea of simply testing for the presence or absence of indicator bacteria in a standardized volume (i.e., 100 ml) of water, using a liquid medium resembling many employed in the MPN approach, as described by Clark (10). One of the major advantages of this approach is that it is easier to perform and hence more useful in small potable water systems where microbiological problems are more frequent, and it permits larger systems to analyze greater numbers of samples to gain a more comprehensive microbiological evaluation of their distribution

TABLE 5 Confirmation procedures for the MF technique

Type of analysis	Medium	Positive reaction
Total coliforms (incubate at 35 ± 0.5°C)	Brilliant green lactose bile broth and	Gas and/or acid
	lauryl tryptose broth	Gas and/or acid
Fecal coliforms (incubate at 44.5 ± 0.2°C)	EC medium	Gas and/or acid
Fecal streptococci[a]	Bile esculin (incubate at 35 ± 0.5°C)	Growth
	BHI–6% NaCl (incubate at 35 ± 0.5°C)	Growth
	BHI (incubate at 45 ± 0.5°C)	Growth

[a] Growth in brain heart infusion broth (BHI) at 45 ± 0.5°C and bile esculin confirms the presence of fecal streptococci. Growth in bile esculin and brain heart infusion broth containing 6.5% NaCl confirms the presence of enterococci.

network. Therefore, the presence-versus-absence concept of indicator organism occurrence has replaced the specified monthly average for compliance with U.S. regulations in the current total coliform rule (16). Although it can yield valuable information on the prevalence of microbiological problems within a system, a weakness is that this approach fails to provide data on the magnitude of such occurrences. In addition, the common connotation of "absence" can be misleading in the case of injured bacteria that are frequently present in treated drinking water systems and fail to produce a positive test on established media (45), as discussed in more detail below.

The P-A analysis of drinking water for total coliforms can entail 100 ml of sample added to 50 ml of triple-strength P-A broth (single strength is as follows: 13 g of lactose broth, 17.5 g of lauryl tryptose broth, 0.0085 g of bromcresol purple, 1 liter of water) in 250-ml bottles. Bottles containing aliquots of the water sample to be tested are incubated, and resulting endpoints are determined as described above. A distinct yellow color results from the fermentation of lactose and gas formation can be detected as bubbles with gentle shaking. Confirmation of such presumptive results can be done as described above and in Table 3.

Enzyme-Specific Tests

New criteria have been added to the traditional definition of coliform bacteria, presented earlier, which had been based on classical microbiological characteristics and phenotypes. This new approach uses the presence of characteristic enzymatic activities to permit the differentiation of the coliform group of bacteria and *E. coli* in the determination of microbiological water quality. It should be realized that the definition of a coliform or a fecal coliform basically relies on the activity of a single enzyme (β-galactosidase). The new enzymatic definition of total coliform bacteria is based on the presence of β-galactosidase, and that of *E. coli* is based on the enzymatic action of β-glucuronidase. Although the common use of media that are based on these specific enzymatic activities to identify indicator bacteria in water has gained wide acceptance only in the past few years and is listed as an alternative only in the current (18th) edition of *Standard Methods for the Examination of Water and Wastewater* (5), enzymatic criteria have been accepted since the 1985 edition of that book (3) and microbiological media based on the same principle have been available for over two decades. In addition, a verification method using cytochrome oxidase and β-galactosidase has proved to be superior to the more traditional technique for confirming the presence of total coliforms in water (39). The endpoint characteristically incorporated into media designed for the enzymatic detection of coliforms and *E. coli* in water is the development of a specific color or fluorescence. The activity of the enzyme β-galactosidase results in the hydrolysis of substrates such as *ortho*-nitrophenyl-β-D-galactoside, which is colorless, to a colored product indicating the presence of coliform bacteria. *E. coli* is characterized by the production of a fluorescent end product following the hydrolysis of 4-methylumbelliferyl-β-D-glucuronide by the action of the enzyme β-glucuronidase. A wide range of other chromogenic and fluorogenic substrates acted upon by β-glucuronidase that have been incorporated into various media are also available. Exhaustive comparative testing of at least one commercially available enzyme-based medium indicated that it performed as well as accepted media in the detection of both total coliforms and *E. coli* from drinking

water (15) and *E. coli* following chlorine injury, using mixed, natural bacterial suspensions (44). However, chlorine-mediated injury results in the somewhat delayed development of both enzymatic endpoints in two of the commercially available media (unpublished results) because of the extended lag phase that is characteristic of stressed bacteria (45).

Currently, a number of different media based on this principle have been developed for use in the MF, MPN, and P-A techniques. Commercially available media include Colisure (Millipore Corp., Bedford, Mass.) (48), Colilert (Idexx, Westbrook, Maine) (15, 34), m-ColiBlue (Hach Co., Loveland, Colo.), ColiComplete (BioControl, Bothell, Wash.), and MicroSure (Gelman Sciences, Ann Arbor, Mich.). Similar media to detect coliforms and/or *E. coli* in water have also been described in the literature; representative examples of the new enzyme-specific media are MI agar (7), mX (9), m-LGA (59), and mLGA (71). All of these media can be used in a P-A or MPN manner, depending on the needs of the analyst.

Bacteriophages

The enumeration of bacteriophages is one of the simplest methods. This technique utilizes a host strain of *E. coli*, since viruses (phages) are obligate intracellular parasites and thus need a metabolically active host in which to replicate. Several host strains can be used; each has its limitations. One of the most widely used hosts is *E. coli* C (ATCC 13706). This host strain allows for the replication of somatic phages (those phages having their receptor sites located on the bacterial cell wall). However, *E. coli* C3000 (ATCC 15597) is a strain which has sex pili and thus allows for the replication of male-specific coliphages (those phages having the receptor sites on the sex pilus) as well as somatic phages. A third host is a genetically manipulated *Salmonella* strain that has the ability to produce sex pili but does not contain the receptor sites for coliphages on its cell wall. Thus, only male-specific (F^+, or F-specific) coliphages should replicate with this *Salmonella* strain used as the host (32).

For the analysis of surface and drinking waters, it is simpler to use a single-layer, direct plaque assay as outlined by Grabow and Coubrough (24). In addition, the method allows for the analysis of 100-ml volumes, which increases the sensitivity of the technique. The medium consists of 14 g of meat extract, 4 g of yeast extract, 4 g of NaCl, 12 g of peptone, 1 g of sodium carbonate, 1 g of magnesium chloride, 12 g of agar, and 1,000 ml of water; a 13% (wt/vol) $CaCl_2$ solution is prepared separately and autoclaved. The medium is autoclaved in 100-ml volumes and kept liquid at ca. 48°C, and 1 ml of the $CaCl_2$ solution is added. Five milliliters of an overnight culture of the appropriate host (ATCC 13706 or ATCC 15597) is added to the medium, and finally a 100-ml volume of the sample is mixed carefully and thoroughly (taking care not to create air bubbles) and poured into large petri dishes. The plates are incubated at 35°C for 6 to 24 h and read for the presence of small clear areas which represent PFU. Care should be taken to include negative controls (i.e., plates which contain only the medium and the bacterial host). The total number of viral plaques is counted and expressed as PFU/100 ml.

Rapid Tests

The need for tests to quickly determine the possible existence of water contamination of public health significance is longstanding. This need exists because the time required to perform the current tests is greater than the mean residence time for water within many potable distribution networks. Significant factors associated with that dilemma in drinking water and other relatively high quality aquatic environments and systems include the low densities of ambient bacteria plus the need for at least 18 h of bacterial growth before an observable endpoint is obtained with the classical techniques. The constraints associated with other systems containing higher concentrations of bacteria are clearly less demanding. However, for most applications, the rapid methods available should be viewed as research tools since at present they either lack sufficient sensitivity or are unsuitable for use in routine monitoring laboratories. However, a variety of rapid methods are in the developmental stage, and some appear very promising. Those seeking such a method need to critically evaluate which of the available methods are compatible with their specific needs. A variety of analytical approaches have been proposed for the rapid detection of bacteria in water and wastewater (5), although most are limited by sensitivity with respect to analysis of water of good microbiological quality. The following is a list of some of the more attractive rapid methodological options that have been described in the literature and a brief statement about potential applications and noteworthy limitations of each.

(i) The direct total microbial count or acridine orange direct count is often used to determine the total bacterial population in an aquatic system (5). This analytical approach requires a fluorescence microscope with appropriate optical filters and is of limited value in determining bacterial viability unless known organisms are used under defined conditions (50). The use of 5-cyano-2,3-ditolyl tetrazolium chloride (a fluorescent compound) and a contrasting counterstain allows the determination of the total bacterial population as well as the fraction respiring in water (60) and the assessment of biofilm disinfection (31). This method provides little information of value in bacterial identification and is tedious, although image analysis technology has proven useful in making the technique less problematic and allows statistical concerns to be addressed more easily (62).

(ii) A rapid test that detects fecal coliform bacteria after 7 h of incubation is available (54). This method has proven of value in the examination of surface waters as well as unchlorinated sewage and might serve as an emergency test for detection of sewage or fecal contamination in potable water. However, the presence of injured bacteria in such systems might represent an explanation for false-negative results because of the prolonged lag that is characteristic of stressed bacteria in environmental samples (45).

(iii) ATP detection assays have been useful in the determination of bacterial population density. This approach is rather insensitive, requiring the presence of at least 10^3 cells, and does not identify organisms, but it has been used successfully in ecological and wastewater studies (33, 34).

(iv) Radiometry using labeled substrates can be both rapid and sensitive in the detection of organisms that metabolize the labeled nutrient to CO_2. This method does not discriminate bacteria phylogenetically, although it provides information of functional significance. Despite the potential of this method, very few reports document its application in environmental analysis (55).

(v) Fluorescently labeled antibodies have been used for the detection and identification of bacteria in environmental studies for some time. This method also requires a fluorescence microscope and is tedious unless image analysis is used, and it is constrained by the specificity of the antibodies used and does not indicate viability or physiological activity.

However, this approach can be used in a variety of applications, including tracking a specific organism in the environment or detecting enteric bacteria (41, 72). The lack of sensitivity of this method prohibits its use for the direct detection of pathogens in waters. Large volumes of water need to be concentrated before this technique becomes useful for pathogen detection.

(vi) A hybrid method that incorporates the use of fluorescent antibodies to identify bacteria along with 5-cyano-2,3-ditolyl tetrazolium chloride to determine cellular respiratory activity has recently been introduced (53). Although this method requires the use of a fluorescence microscope and can be tedious, it is unique in that it allows the rapid, simultaneous detection of specific bacteria and the discrimination of respiratory activity at the cellular level.

Recovery of Injured Bacteria

Indicator bacteria become injured in water and wastewater following sublethal exposure to a wide variety of chemical and physical environmental stressors, including disinfectants, metals, and UV irradiation (45). Such bacteria are unable to form colonies on most selective media, and between 10 and 90% of the coliform bacteria in treated drinking water may be injured (45, 49). As a consequence, injured cells are undetected in water. This can lead to an underestimation of a contamination event, and injured bacteria may also pass undetected into finished drinking water distribution networks (8), where they may eventually recuperate and colonize the system. A medium, m-T7, was developed to detect injured bacteria in drinking water and is commercially available (8). Other guidelines associated with the detection of injured bacteria are given in *Standard Methods for the Examination of Water and Wastewater* (5). Although the use of media and methods designed to detect injured bacteria does not always result in the enumeration of greater numbers of indicator bacteria in all systems, it provides a more complete view of the water quality plus guidance in the diagnosis of problems within water distribution systems experiencing unexplained occurrences of excessive indicator bacteria.

Molecular Methods

(i) The PCR technique provides a powerful and sensitive option for detection of microorganisms in aquatic environments (57). This approach has been used for the detection of coliforms and *E. coli* in water (6) as well as sewage and sludge (69). Caution is needed in the interpretation of PCR results since this test allows for the amplification and detection of DNA sequences without regard for bacterial viability (2). In addition, materials in natural samples can interfere with the required reactions.

(ii) Phylogenetic identification of specific bacteria or groups of organisms can be achieved without cultivation through the use of fluorescently labeled oligonucleotide probes for rRNA sequences. This approach has gained wide application in ecological studies (2) and has been used to detect an opportunistic pathogen (*Vibrio vulnificus*) in waters after concentration on membrane filters (29).

SURFACE FRESHWATERS

It should be kept in mind that the microbiology of untreated waters will be drastically different from that of treated drinking waters. Thus, strong emphasis should be placed on the use of the appropriate indicators. In the case of recreational waters (excluding swimming pools, which receive fecal contamination from bathers), it should be determined if the recreational areas are being subjected to fecal pollution. If so, then use the appropriate indicators to determine if the bathers are at risk of enteric disease. Risk can be determined most accurately if the area is known thoroughly. Indicators such as total coliforms may not be of much value in analyses of recreational waters. *E. coli* and coliphages may be much more reliable indicators of the presence of fecal pollution (and thus possible risk), unless the recreational waters are part of a tropical rain forest (26). In the latter case, the use of coliphages may be more advisable. Analyses for the presence of total coliforms will not shed much information, in spite of possible statistical correlations.

DRINKING WATERS

Traditional treatment is based on both physical and chemical barriers. Thus, indicator microorganisms should not be present in finished waters. The presence of any indicator (total or fecal coliforms, coliphages, or *E. coli*) in the finished waters suggests that one of the barriers is not functioning properly. The distribution network may be susceptible to contamination, and therefore the presence of fecal coliforms in treated drinking water should be a cause for concern. However, coliforms have been shown to be able to grow in potable water distribution systems as a result of the bacterial biofilms on pipe surfaces. The analyst should be aware of the different types of indicators, and proper use of these tools should be emphasized.

SUMMARY

We have attempted to give an overall view of the currently used methods as well as those that are now being developed for the microbiological analysis of drinking waters and surface freshwaters. The practice of using groups of bacteria or specific organisms such as *E. coli* as indicators of the possible presence of fecal contamination or as indicators of the possible presence of pathogenic microorganisms is and has been extremely useful to protect public health. However, it should be kept in mind that indicators, regardless of which ones are used, are only tools and have limitations. There exist numerous misunderstandings regarding the use of indicators. For example, analysis of levels of coliforms and fecal coliforms is considered by many to be a simplistic method for determining and guaranteeing water quality. Such a dogmatic approach to water quality is counterproductive.

There are exciting new developments in the area of rapid methods as well as the direct detection of pathogenic microorganisms. The reader is exhorted to become familiar with the advantages and disadvantages of these new advances. However, there is no better way of approaching water quality management and analysis than being familiar with the watershed or distribution system under investigation. This knowledge and the correct and timely application of suitable analytical methods allow for early detection of many anomalous conditions within the system. The resulting data can then be used to better manage the system and thus protect the public health.

REFERENCES

1. **Alvarez, A. J., E. A. Hernandez-Delgado, and G. A. Toranzos.** Advantages and disadvantages of traditional and molecular techniques applied to the detection of pathogens in waters. *Water Sci. Technol.* **27:**253–256.

2. **Amann, R. I., W. Ludwig, and K.-H. Schleifer.** 1995. Phylogenetic identification and in situ detection of individual microbial cells without cultivation. *Microbiol. Rev.* **59:**143–169.

3. **American Public Health Association.** 1985. *Standard Methods for the Examination of Water and Wastewater*, 17th ed. American Public Health Association, Washington, D.C.

4. **American Public Health Association.** 1989. *Standard Methods for the Examination of Water and Wastewater*, 18th ed., p. 9–31, 9–32. American Public Health Association, Washington, D.C.

5. **American Public Health Association.** 1992. *Standard Methods for the Examination of Water and Wastewater*, 19th ed. American Public Health Association, Washington D.C.

6. **Bej, A. K., S. W. McCarty, and R. M. Atlas.** 1991. Detection of coliform bacteria and *E. coli* by multiple PCR: comparison with defined substrate and plating methods for water quality monitoring. *Appl. Environ. Microbiol.* **57:** 2429–2432.

7. **Brenner, K. P., C. C. Rankin, Y. R. Roybal, G. N. Stelma, P. V. Scarpino, and A. P. Dufour.** 1993. New medium for the simultaneous detection of total coliforms and *E. coli* in water. *Appl. Environ. Microbiol.* **59:** 3534–3544.

8. **Bucklin, K. E., G. A. McFeters, and A. Amirtharaja.** 1991. Penetration of coliforms through municipal drinking water filters. *Water Res.* **25:**1013–1017.

9. **Chang, G. W, and R. I. Lum.** 1994. mX, a simple, economical membrane filter medium for *E. coli* and total coliforms, p. 1397–1400. *In Proceedings of the AWWA/WQTC*. American Water Works Association.

10. **Clark, J. A.** 1990. The presence-absence test for monitoring water quality, p. 399–411. *In* G. A. McFeters (ed.), *Drinking Water Microbiology*. Springer-Verlag, New York.

11. **Comité Coordinador Regional de Instituciones de Agua Potable y Saneamiento de Centroamérica, Panama y Republica Dominicana (CAPRE).** 1993. Normas de Calidad del Agua para Consumo Humano, 1st ed. CAPRE, San José, Costa Rica.

12. **Dienert, F.** 1944. Hygiene of bathing places. *Bull. Acad. Med.* **128:**660–665.

13. **Dufour, A. P.** 1977. Escherichia coli: the fecal coliform, p. 48–58. *In* A. W. Hoadley and B. J. Dutka (ed.), *Bacterial Indicators/Health Hazards Associated with Water*. ASTM STP 635. American Society for Testing and Materials, Philadelphia, Pa.

13a. **Dutka, B. J.** Personal communication.

14. **Dutka, B. J., A. S. Y. Chau, and J. Coburn.** 1974. Relationship between bacterial indicators of water pollution and fecal sterols. *Water Res.* **8:**1047–1055.

15. **Edberg, S. C., M. J. Allen, and D. B. Smith.** 1988. National field evaluation of a defined substrate method for the simultaneous detection of total coliforms and *Escherichia coli* from drinking water: comparison with the standard multiple-tube fermentation method. *Appl. Environ. Microbiol.* **54:**1559–1601.

16. **Federal Register.** 1989. Drinking water; national primary drinking water regulations; total coliforms; final rule. *Fed. Regist.* **54:**27544.

17. **Fujioka, R. A., A. A. Ueno, and O. T. Narikawa.** 1984. Recovery of false positive fecal streptococcus on KF agar from marine recreational waters. Technical Report no. 168. Water Resources Research Center, University of Hawaii at Manoa, Honolulu.

18. **Geldreich, E. E.** 1967. Fecal coliform concepts in stream pollution. *Water Sewage Works* **114:**98–110.

19. **Geldreich, E. E.** 1970. Applying bacteriological parameters

to recreational water quality. *J. Am. Water Works Assoc.* **62:**113–120.

20. **Geldreich, E. E.** 1978. Bacterial pollution and indicator concepts in feces, sewage, stormwater and solid wastes, p. 51–97. *In* G. Berg (ed.), *Indicators of Viruses in Water and Food.* Ann Arbor Science, Ann Arbor, Mich.

21. **Geldreich, E. E., and B. A. Kenner.** 1969. Concepts of fecal streptococci in stream pollution. *J. Water Pollut. Control Fed.* **41:**R355–R352.

22. **Gleick, P. H.** 1993. Water quality and contamination, p. 225. *In* P. H. Gleick (ed.), *Water in Crisis, a Guide to the World's Fresh Water Resources.* Oxford University Press, Oxford.

23. **Grabow, W. O. K., J. S. Burger, and E. M. Nupen.** 1980. Evaluation of acid-fast bacteria, *Candida albicans,* enteric viruses and conventional indicators for monitoring wastewater reclamation systems. *Prog. Water Technol.* **12:** 803–817.

24. **Grabow, W. O. K., and P. Coubrough.** 1986. Practical direct plaque assay for coliphages in 100-milliliter samples of drinking water. *Appl. Environ. Microbiol.* **52:**430–433.

25. **Havelaar, A. H.** 1987. Bacteriophages as model organisms in waste treatment. *Microbiol. Sci.* **4:**362–364.

26. **Hazen, T. C., and G. A. Toranzos.** 1990. Tropical source water, p. 32–54. *In* G. A. McFeters (ed.), *Drinking Water Microbiology.* Springer-Verlag, New York.

27. **Health and Welfare Canada.** 1992. *Guidelines for Canadian Recreational Water Quality.* Canadian Governement Publishing Centre, Ottawa, Canada.

28. **Health and Welfare Canada.** 1993. *Guidelines for Canadian Drinking Water Quality.* Canadian Governement Publishing Centre, Ottawa, Canada.

29. **Heidelberg, J. F., K. R. O'Neil, D. Jacobs, and R. R. Colwell.** 1993. Enumeration of *Vibrio vulnificus* on membrane filters with a fluorescently labeled oligonucleotide probe specific for kingdom-level 16S rRNA sequences. *Appl. Environ. Microbiol.* **59:**3474–3476.

30. **Hernandez-Delgado, E. A., M. L. Sierra, and G. A. Toranzos.** 1991. Coliphages as alternate indicators of fecal contamination in tropical waters. *Environ. Toxicol. Water Qual.* **6:**131–143.

31. **Huang, C.-T., F. P. Yu, G. A. McFeters, and P. S. Stewart.** 1994. Non-uniform spatial patterns of respiratory activity within biofilm during disinfection. *Appl. Environ. Microbiol.* **61:**2252–2256.

32. **IAWPRC Study Group on Health Related Water Microbiology.** 1991. Bacteriophages as model viruses in water quality control. *Water Res.* **25:**529–545.

33. **Jorgensen, P. E., T. Eriksen, and B. K. Jensen.** 1992. Estimation of viable biomass in wastewater and activated sludge by determination of ATP, oxygen utilization rate and FDA hydrolysis. *Water Res.* **11:**1495–1501.

34. **Karl, D. M.** 1980. Cellular nucleotide measurements and applications in microbial ecology. *Microbiol. Rev.* **44:** 739–796.

35. **Knittel, M. D., R. J. Seidler, C. Eby, and L. M. Cabe.** 1977. Colonization of the botanical environment by *Klebsiella* isolates of pathogenic origin. *Appl. Environ. Microbiol.* **34:**557–563.

36. **Kott, Y.** 1977. Current concepts of indicator bacteris, p. 3–14. *In* A. W. Hoadley and B. J. Dutka (ed.,), *Bacterial Indicators/Health Hazards Associated with Water.* American Society for Testing and Materials, Philadelphia.

37. **Kromoredjo, P., and R. S. Fujioka.** 1991. Evaluating three simple methods to assess the microbial quality of drinking water in Indonesia. *Environ. Toxicol. Water Qual.* **6:** 259–270.

38. **LeChevallier, M. W.** 1990. Coliform regrowth in drinking water: a review. *J. Am. Water Works Assoc.* **82:**74–86.

39. **LeChevallier, M. W., S. C. Cameron, and G. A. McFeters.** 1983. Comparison of verification procedures for the membrane filtration total coliform technique. *Appl. Environ. Microbiol.* **45:**1126–1128.

40. **LeChevallier, M. W., S. C. Cameron, and G. A. McFeters.** 1983. New medium for the improved recovery of coliform bacteria from drinking water. *Appl. Environ. Microbiol.* **45:**484–492.

41. **Levasseur, S., M.-O. Husson, R. Leitz, F. Merlin, F. Laurent, F. Peladan, J.-L. Drocourt, H. Leclerc, and M. Van Hoegaerden.** 1992. Rapid detection of members of the family *Enterobacteriaceae* by a monoclonal antibody. *Appl. Environ. Microbiol.* **58:**1524–1529.

42. **Lye, D. E., and A. P. Dufour.** 1991. A membrane filter procedure for assaying cytotoxic activity in heterotrophic bacteria isolated from drinking water. *J. Appl. Bacteriol.* **70:**89–94.

43. **Manja, E. S., M. S. Maurya, and D. M. Rao.** 1982. A simple field test for the detection of fecal pollution in drinking water. *Bull. W.H.O.* **60:**797–801.

44. **McCarty, S. C., J. H. Standridge, and M. C. Stasiak.** 1992. Evaluating commercially available define-substrate test for recovery of *E coli. J. Am. Water Works Assoc.* **84**(May):91–97.

45. **McFeters, G. A.** 1990. Enumeration, occurrence and significance of injured indicator bacteria in drinking water, p. 478–492. *In* G. A. McFeters (ed.), *Drinking Water Microbiology.* Springer-Verlag, New York.

46. **McFeters, G. A., J. P. Barry, and J. Howington.** 1993. Distribution of enteric bacteria in Antarctic seawater surrounding a sewage outfall. *Water Res.* **27:**645–650.

47. **McFeters, G. A., G. K. Bissonnette, J. J. Jezeski, C. A. Thomson, and D. G. Stewart.** 1974. Comparative survival of indicator bacteria and enteric pathogens in well water. *Appl. Microbiol.* **27:**823–829.

48. **McFeters, G. A., S. C. Broadaway, B. H. Pyle, M. Pickett, and Y. Egozy.** 1995. Comparative performance of Colisure and accepted methods in the detection of chlorine-injured total coliforms and *E. coli. Water Sci. Technol.* **31:** 259–261.

49. **McFeters, G. A., J. S. Kippin, and M. W. LeChevallier.** 1986. Injured coliforms in drinking water. *Appl. Environ. Microbiol.* **51:**1–5.

50. **McFeters, G. A., A. Singh, S. Byun, S. Williams, and P. R. Callis.** 1991. Acridine orange staining reaction as an index of physiological activity in *E. coli. J. Microbiol. Methods* **13:**87–97.

51. **Payment, P., L. Richardson, J. Siemiatycki, R. Dewar, M. Edwardes, and E. Franco.** 1991. A randomized trial to evaluate the risk of gastrointestinal disease due to consumption of drinking water meeting current microbiological standards. *Am. J. Public Health* **81:**703–708.

52. **Pourcher, A.-M., L. A. Devriese, J. F. Hernandez, and J. M. Delattre.** 1991. Enumeration by a miniaturized method of *E. coli, S. bovis* and enterococci as indicators of the origin of faecal pollution of waters. *J. Appl. Bacteriol.* **70:**525–530.

53. **Pyle, B. H., S. C. Broadaway, and G. A. McFeters.** 1995. A rapid, direct method for enumerating respiring enterohemorrhagic *E. coli* O157:H7 in water. *Appl. Environ. Microbiol.* **61:**2614–2619.

54. **Reasoner, D. J., J. C. Blannon, and E. E. Geldreich.** 1979. Rapid seven-hour fecal coliform test. *Appl. Environ. Microbiol.* **38:**229–236.

55. **Reasoner, D. J., and E. E. Geldreich.** 1989. Detection of fecal coliforms in water using [^{14}C] mannitol. *Appl. Environ. Microbiol.* **55:**907–911.

56. **Rheinheimer, G.** 1992. *Aquatic Microbiology,* 4th ed., p. 279–282. John Wiley & Sons, New York.

57. **Richardson, K. J., M. H. Stewart, and R. L. Wolfe.** 1991. Application of gene probe technology to the water industry. *J. Am. Water Works Assoc.* **83**(Sept.):71–81.

58. **Rivera, S. C., T. C. Hazen, and G. A. Toranzos.** 1988. Isolation of fecal coliforms from pristine sites in a tropical rain forest. *Appl. Environ. Microbiol.* **54:**513–517.

59. **Sartory, D. P., and L. Howard.** 1992. A medium detecting β-glucuronidase for the simultaneous membrane filtration enumeration of *E. coli* and coliforms from drinking water. *Lett. Appl. Microbiol.* **15:**273–276.

60. **Schaule, G., H.-C. Flemming, and H. F. Ridgway.** 1993. Use of CTC for quantifying planktonic and sessile respiring bacteria in drinking water. *Appl. Environ. Microbiol.* **59:** 3850–3857.

61. **Seidl, P.** 1990. Microbiological investigations of drinking and recreational waters from an Indian reserve in Canada, p. 59–65. *In* G. Castillo, V. Campos, and L. Herrera (ed.), *Proceedings of the Second Biennial Water Quality Symposium.* Editorial Universitaria, Santiago, Chile.

62. **Singh, A., B. H. Pyle, and G. A. McFeters.** 1989. Rapid enumeration of bacteria by image analysis epifluorescence microscopy. *J. Microbiol. Methods* **10:**91–102.

63. **Sivaborborn, K.** 1988. *Abstr. First Biennial Water Qual. Symp.* Banff, Canada.

64. **Tartera, C., and J. Jofre.** 1987. Bacteriophages active against *Bacteroides fragilis* in sewage-polluted waters. *Appl. Environ. Microbiol.* **53:**1632–1637.

65. **Tartera, C., F. Lucena, and J. Jofre.** Human origin of *Bacteroides fragilis* bacteriophages present in the environment. *Appl. Environ. Microbiol.* **55:**2696–2701.

66. **Toranzos, G. A.** Unpublished observation.

67. **Toranzos, G. A.** 1991. Current and possible alternate indicators of fecal contamination in tropical waters: a short review. *Environ. Toxicol. Water Qual.* **6:**121–130.

68. **Toranzos, G. A., C. P. Gerba, and H. Hanssen.** 1988. Enteric viruses and coliphages in Latin America. *Tox. Assess.* **5:**491–510.

69. **Tsai, Y.-L., C. J. Palmer, and L. R. Sangermano.** 1993. Detection of *E. coli* in sewage and sludge by PCR. *Appl. Environ. Microbiol.* **59:**353–357.

70. **Venkattsen, M. I., and I. R. Kaplan.** 1990. Sedimentary coprostanol as an index of sewage addition in the Santa Monica Basin, California. *Environ. Sci. Technol.* **24:** 208–213.

71. **Walter, K. S., E. J. Fricker, and C. R. Fricker.** 1994. Observations on the use of a medium detecting β-glucuronidase activity and lactose fermentation for the simultaneous detection of *E. coli* and coliforms. *Lett. Appl. Microbiol.* **19:**47–49.

72. **Winkler, J., K. N. Timmis, and R. A. Snyder.** 1995. Tracking the response of *B. cepacia* G4 5223-Prl. *Appl. Environ. Microbiol.* **61:**448–455.

73. **World Health Organization.** 1989. *Guidelines for Drinking-Water Quality,* 2nd ed., p. 8–29. World Health Organization, Geneva.

74. **Writer, J. H., J. A. Leenheer, L. B. Barber, G. A. Amy, and S. C. Chapra.** 1995. Sewage contamination in the upper Mississippi River as measured by the fecal sterol, coprostanol. *Water Res.* **29:**1427–1436.

75. **Wun, C. K., R. W. Walker, and W. Litsky.** 1976. The use of XAD-2 resin for the analysis of coprostanol in water. *Water Res.* **10:**955–959.

Control of Microorganisms in Source Water and Drinking Water

AARON B. MARGOLIN

20

INTRODUCTION

"There is no problem in science more definitely settled, than the one pertaining to the drinking of contaminated water as a cause of enteric disease" (5). This quote appeared in *JAMA* more than 100 years ago. The importance of potable water has been known for over 4,000 years. A Sanskrit document dictates that "foul water be boiled, kept in the sun, have a piece of hot copper dipped into it, and then cooled in a container in the earth" (8). Today we are still dealing with some of the same issues concerning the ingestion of contaminated water.

Over the last several decades, increases and shifting demographics, as well as a decline in economic prosperity, have stressed and tested our ability to maintain and deliver potable water. Increases in technology and the ability to detect pathogens have created new challenges for our existing water treatment practices. Classical disinfectants such as chlorine have been shown to be partially or completely ineffective against viral and protozoal pathogens (44). In a time when we are asked to do more for less, creating water treatment practices which protect individuals from disease becomes one of our greatest challenges. Understanding water treatment processes, knowing how to protect our source waters, and understanding the natural environmental conditions which affect source waters must be coupled to our knowledge of water disinfection. As our society changes and its members grow older, the control of traditional and new pathogens remains a top priority.

Microbial Contamination

Microbial populations in source waters can be divided into distinct categories: nonpathogens, opportunistic pathogens, and frank pathogens. While nonpathogens do not cause disease, they are of major concern with respect to water quality. Water which is contaminated with high counts of nonpathogenic bacteria or algae can change the water quality, making it less desirable as potable water because of problems such as taste and odor associated with the organisms. In addition to decreasing the potability of the water, having a high burden of nonpathogens can greatly affect the disinfection efficiency of a treatment process, permitting the passage of pathogens into the distribution system (44).

Nonpathogens

Water quality, with respect to nutrients and microbial populations, can be divided into three categories or trophic levels: (i) oligotrophic (low nutrients, minimal microbiological activity), (ii) mesotrophic (moderate nutrients, moderate microbiological activity), and (iii) eutrophic (high nutrients, high microbiological activity). Water quality effects which can be associated with eutrophication include anoxic conditions, high microbial populations, high turbidity and color, and even the formation of trihalomethanes (THMs) (36). In addition to detracting from the overall quality of the water, these conditions can decrease the efficiency of treatment process by clogging filters and placing increased organic demands on the system, resulting in the survival of pathogens. The most common indicator of eutrophication is a high level of algae (36).

Pathogens

Microbial pathogens which are of major concern are those bacteria, viruses, or protozoans which originate from fecal contamination, known collectively as the enterics. Some bacterial and protozoan pathogens have reservoirs in human as well as certain animal populations. For example, the protozoan *Cryptosporidium parvum* is found in cows and several other animals. Surface waters which are affected by runoff or shallow wells which are easily infiltrated by tainted water may become contaminated with *Cryptosporidium* oocysts. Controlling the entrance of these pathogens into water sources is complex and requires good watershed management. Viral pathogens of interest reside predominantly in humans; hence, control can be somewhat easier.

Sources of contamination can be classified as point or nonpoint. Point sources are known and can be documented, facilitating their control under good management practices. The primary point source of contamination for fecal pathogens comes from wastewater discharges or effluent reuse. In either type of point source, controlling the introduction of fecal pathogens into source waters relies on efficient maintenance of treatment facilities and strict adherence to disinfection policies. Failure to eliminate pathogens prior to discharge into receiving waters can cause subsequent problems in drinking water treatment processes, resulting in the distribution of contaminated water (36).

Nonpoint sources are sources of contamination which are not known. They include, but are not limited to, agricultural runoff, livestock, urban runoff, landfills, land development, recreational activities, inadequate septic systems, and illegal dumpings or discharges. Nonpoint sources of contamination are much more difficult to control than point sources and hence in some aspects present a greater public health threat (36).

CHOOSING WATER SOURCES AND TREATMENT PROCESSES

Selection of a water treatment process is a complicated task that ultimately must result in delivery of potable water which is acceptable to the consumer at a reasonable cost. Factors which must be considered include (22) water supply source quality, desired finished water quality, reliability of process equipment, operational requirements and personnel capabilities, flexibility in dealing with changing water quality and equipment malfunctions, available space for construction of treatment facilities, waste disposal constraints, and capital and operating costs (including chemical availability).

Water Source Selection

Water source selection should be based on several factors which will enable the continued delivery of potable water. Factors which must be considered include the yield quantity, source water quality, ability to protect the watershed from either human or animal impact or both, collection and treatment requirements, and the distribution process that will be required. Additional consideration should be given to whether surface water or groundwater should be chosen as the source water. Surface waters are usually easier to locate and assess than groundwater supplies. However, surface water is more readily affected by alterations in the surrounding environment and may require modification of treatment processes to meet the changing characteristics of the water (22).

Surface Water Sources

Surface water is made up from rivers, streams, lakes, ponds, and runoff from rain events. One problem associated with surface water is its ability to rapidly change as stresses are placed on the surrounding environment. Treatment processes associated with surface water sources must be able to respond to changes in the water quality. Runoff from rain can rapidly increase turbidity, thus decreasing efficiency in existing disinfection practices. Spills and runoff of surface contaminants such as fertilizers or insecticides can inhibit treatment plant operations. Additional problems can arise from biological contamination, such as algal blooms, insects, and animal intrusion. Once a surface water source is chosen, consideration must be given to natural process in relation to placement of intake pipes. Spring and fall turnover can turn pristine waters with low nutrient loads into turbid waters with high concentrations of organics. During summer months, surface waters may stratify into distinct layers such that warmer water stays near the surface and cooler water is trapped below with little intermixing, resulting in anoxic conditions and solubilization of iron and manganese as well as the production of sulfur from anaerobic bacteria (29). Treatment plants must be flexible and able to respond with additional processes, such as increased sedimentation or flocculation times and perhaps the use of ion-exchange resins to remove manganese ions, as well as additional disinfection in order to maintain potable water quality which is acceptable to the consumer and free of risk.

There are several advantages to using surface water as a source water, if they can be exploited. The location and elevation of the water sources may offer the advantage of gravity flow to the treatment facility. Visualization of the water supply can yield a daily evaluation. Water levels and sources of contamination can be less costly to identify compared with those of groundwater (29). Finally, surface waters which are properly managed, and this can be an enormous task, can also be used for recreational purposes.

Groundwater

Groundwater supplies are derived primarily from wells. Water from shallow wells or "hand-dug" wells is not considered groundwater because it is usually under the influence of surface water via runoff and infiltration and hence can have many of the characteristics of surface water. "True" groundwater, in contrast, is relatively constant from season to season and is usually not affected by changes in the surface environment. This does not mean that variation in water quality from well to well does not exist. Depending on hydrological influences, groundwater quality can vary greatly between wells. In general, groundwater is more expensive than surface water, in large part because of the pumping requirements, but also groundwater usually requires less treatment than most surface water with respect to microbial pathogens, organic matter, and turbidity (36). The microbial pathogens of concern in groundwater are the enteric viruses. Viruses, unlike bacteria and protozoan cysts, are not always retained by the soil matrix and, if environmental conditions (i.e., soil type and pH, rain events, and cation concentration) are correct, can migrate to underground aquifers. Additionally, viruses which are complexed with particulate matter and kept in a cold environment have been documented to migrate great distances from the source of contamination and survive long periods outside the human host (42).

Watershed Protection

All too often, the burden of controlling pathogens in drinking water has been placed on treatment plant operations. A more comprehensive approach is the use of multiple barriers. One of the first barriers in place should be control of the entrance of microbes into source waters via watershed protection. Regulatory protection of public water supply sources is multifactorial, involving federal, state, and local programs. At the national level, the Safe Drinking Water Act and the Clean Water Act both directly regulate the introduction of contaminants into the nation's surface and groundwater. At the state and local levels, regulatory programs can vary but are usually designed to protect source waters by restricting and/or regulating certain activities which could degrade the quality of the source water. Such activities may include the curtailment of building permits until treatment facilities exist to handle the human waste. The ultimate goal of any watershed protection should be to limit or exclude the introduction of contaminants into a source water (36).

In addition to physically protecting source waters, other activities such as monitoring and sanitary surveys can help elucidate pathogen contamination. Monitoring programs should be designed to include all of the parameters that need to be evaluated. The monitoring program design should include frequencies for sampling analysis, protocols, and the types of analysis needed (36). Additionally, monitoring programs must be interpreted with caution, especially when things such as indicator organisms are used as predictors of pathogens. To help determine the true sanitary quality of source water and to determine which type of pathogen may be a contaminant, a multipronged approach to monitoring is much more valuable than use of a single type of organism. While bacteria have long been used as the sole indicator for microbiological water quality, enough evidence has been established over the years to indicate that bacteria, while able to predict the presence of other bacterial contaminants or the introduction of fecal material, are not adequate predictors for the presence of viruses or protozoans. To accurately assess the presence of coliform or fecal coliform bacteria, bacteriophages, human enteric viruses, or enteric protozoa, each should be evaluated separately. Also, since any source water is a dynamic body and because pathogens usually occur in very low numbers, monitoring cannot be a one-time exercise. For successful monitoring, a schedule covering at least 1 year, and preferably more, must be adhered to, and then results must be evaluated at the end of that period. The monitoring program must be designed to take into account the changing properties of the source water. For instance, if the water is used for recreational activities during the summer months, the frequency of monitoring should reflect the increased usage. The same may hold true for dry versus wet weather events. Discharges or runoff may occur only during heavy rains, and the pathogens that they bring may not be detected if monitoring is done only during the dry period.

TREATMENT AND DISINFECTION

Physical Methods

RO

Reverse osmosis (RO) is the physical process of separating particles contained in water from the aqueous component. RO systems have been used primarily for the desalination of seawater for the production of potable water. This has limited the use of RO systems to the production of potable water for coastal communities and for desert communities which have no source of water other than seawater. RO has several advantages over other physical methods such as distillation. RO is a relatively low energy consumption process. The principal energy consumers are the pumps necessary to run the RO system. The major disadvantages to an RO system are the presence of particulate and colloidal matter in feed water and the precipitation of soluble salts. All RO membranes can become clogged. This is particularly true for spiral-wound and hollow-fiber modules, especially when submicron and colloidal particles are present in the feed water (13).

Other problems associated with RO membranes include the chemical composition and the pH of the feed water. Hydrolysis of cellulose acetate membranes can occur if the feed water is too acidic or too alkaline. Other compounds such as phenols or free chlorine can be soluble in the membrane and result in membrane failure (13). Because of the low-molecular-weight cutoff of some RO membranes, water which is passed through the membrane may be considered free of microorganisms. One problem with using RO as a means of disinfecting water is that failure of membrane integrity can potentially allow for the passage of pathogens.

Distillation

Distillation is a process that uses heat to separate the aqueous phase of water from the solid phase or particulates. Distillation, by its nature, provides the purest water and water which is free from all pathogens. There are several disadvantages to using distillation as a means of producing potable water, but by far the greatest disadvantage is the energy required to run the process. Distillation processes are usually confined to those areas where ultrapure water is required and is not a practical alternative as a means for the production of potable water (13). Distillation processes have been tried in areas that can rely on solar energy as a means of providing electricity. However, even under these conditions, the capital investment is large and the process of producing potable water is slow, making this system much less attractive than RO.

Chemical Methods

Chlorine

Chlorination has been the major line of defense against waterborne disease outbreaks in the United States since 1908 (9). Chlorination is probably the oldest and most widely used form of drinking water disinfection. It has several advantages which make it appealing and have bolstered its popularity. Chlorine is relatively inexpensive and easily obtained, and it provides residual protection in distribution systems. Chlorine is a strong oxidizing agent and can be used to modify the chemical character of water. In water, chlorine gas hydrolyzes to form hypochlorous acid (HOCl) according to the following reaction:

$$Cl_2 + H_2O \leftrightarrows HOCl + H^+Cl^-$$

The hypochlorous acid undergoes further ionization to form hypochlorite ions (OCl^-) according to the following reaction:

$$HOCl \leftrightarrows H^+ + OCl^-$$

Equilibrium concentrations of HOCl and OCl^- depend on the pH of the water. Alkaline pH shifts the equilibrium to the right, causing the formation of higher concentrations of HOCl. Both HOCl and OCl^- are commonly referred to as free available chlorine. Upon chlorination of water, a portion of the chlorine is reacted with compounds in the water, such as organics. Chlorine which remains available for further reaction is known as free chlorine. The difference between the chlorine concentration applied to water and the free chlorine is known as the chlorine demand (i.e., the concentration of chlorine added minus the concentration of free chlorine resulting) and is one of the physical attributes used to define the needed level of water treatment.

One of chlorine's greatest attributes is ease of use. While chlorine gas requires special handling, there are two forms of chlorine which are readily available and easy to use. One form is calcium hypochlorite [$Ca(OCl)_2$], which is the predominant dry form and when dissolved in water contains approximately 70% available chlorine. Sodium hypochlorite (NaOCl) is available in liquid form at concentrations of between 5 and 15%.

When chlorine reacts with ammonia in water, the resulting reactions can form chloramines according to the following reactions:

$HOCl + NH_3 \rightarrow H_2O + NH_2Cl$ monochloramine

$HOCl + NH_2Cl \rightarrow H_2O + NHCl_2$ dichloramine

$HOCl + NHCl_2 \rightarrow H_2O + NCl_3$ trichloramine

The specific reaction products that are formed depend on the pH of the water, temperature, reaction time, and the initial chlorine-to-ammonia ratio. In general, below pH 4.4, trichloramine is produced; above pH 8.5, monochloramine usually exists alone; and between pH 4.5 and 8.5, both mono- and dichloramine exist. Chloramines are thought to be less effective as disinfectants than hypochlorite (11).

The ability of hypochlorite to destroy microbial life is predominantly due to the ability of HOCl to oxidize proteins and other structures found on bacteria, viruses, and protozoa. Since HOCl is neutral and has a relatively low molecular weight, it is able to penetrate the organism with relative ease and hence has a high germicidal activity. The cidal activity of chlorine is greatly reduced at high pH, probably because at an alkaline pH, the predominant species of chlorine is OCl^-. The negative charge of this molecule may prevent it from penetrating or coming in close contact with surface proteins as a result of electrostatic repulsive forces that exist between the chlorine and the carboxyl ends of proteins. Organisms that are relatively large and contain a "waxy" outer coat, such as some protozoan cysts, may be more resistant to the oxidizing potential of chlorine or require much longer contact times and hence not be adequately disinfected by chlorine CT values which have been established for the inactivation of bacteria (CT values are defined by the concentration of the disinfectant and the contact time). Leahy et al. (28) showed that 2.80 mg of chlorine per liter achieves a 99% inactivation of *Giardia muris* cysts in 16 min of contact time at pH 7.0 and 25°C. Studies by Korich et al. (27) demonstrated that 80.0 mg of chlorine per liter requires more than 90 min of contact time for a 90% inactivation of *C. parvum* oocysts. All of these studies were done at the bench level, in static systems. Work by Tilton (45) has demonstrated that there is very little correlation between the CT developed with a static bench-scale system and that developed with a dynamic pilot plant system, the latter usually requiring a much greater CT than is indicated by static bench-scale systems.

Advances in the microbiological examination of water have shown that hypochlorite is also less effective at inactivating viruses than at killing total and fecal coliforms, the historical indicator organisms. This may be in part because viruses are not living organisms and do not rely on any metabolic (enzymatic) functions for survival in the environment. Harakeh (23) reported that of six viruses tested, coxsackievirus B5 was the most resistant, with 99.99% inactivation achieved at a dose of 18.0 mg of free chlorine per liter after 5 min of contact, whereas simian rotavirus was the most sensitive, with a dose of 5.0 mg of chlorine per liter required to obtain 99.99% inactivation.

While chlorine has been widely used over the decades, it is not without its drawbacks. Since the early 1970s it has been known that undesirable disinfection by-products (DBPs) can be produced when free chlorine interacts with selected precursors such as humic substances and other organics. This interaction produces a group of halogen-substituted single-carbon compounds referred to as total THMs (16). Because of the potential carcinogenicity of the THMs, chloroform in particular, the U.S. Environmental Protection Agency has established a maximum contaminant level of 80 μg/liter for total THMs. Approaches to control THM formation are the removal of organic precursors that react with the chlorine, removal of THMs from finished water, and the use of alternative disinfectants (43).

Chlorine Dioxide

Chlorine dioxide (ClO_2) was first used in 1921, by the German chemist Erick Schmidt, as a bleaching agent for cellulose fibers. However, the dangers and high costs of generating chlorine dioxide delayed the widespread use of it until the late 1930s, when Mathieson (now Olin) Chemical Co. developed an economical ClO_2 manufacturing process.

The application of chlorine dioxide to water treatment arose from the work of Aston (6), who demonstrated that it could be used to remove taste- and odor-causing compounds such as chlorophenols. However, since the cost of using chlorine dioxide for large-scale water treatment was significantly higher than the cost of using chlorine, its usage was restricted to special situations until the early 1970s. In 1975, the U.S. Environmental Protection Agency, faced with increasing public concern over THMs in drinking water, sponsored research into alternative disinfection techniques to control THMs. The research of Granstrom and Lee (21) indicated that chlorine dioxide did not form THMs, particularly chloroform, when applied at dosages similar to those used for chlorine. The implication of those studies was that ClO_2 was a viable alternative to chlorination. Preliminary studies have suggested that ClO_2 may be more effective than chlorine at disinfecting cysts and viruses (24, 41, 46), and it is now well documented that ClO_2 forms far lower amounts of THMs than Cl_2 does (30).

The generation and reaction of chlorine dioxide in aqueous systems are poorly understood and are a topic for significant debate. Presently, ClO_2 for water treatment is generated primarily by the reaction of chlorine with sodium chlorite in acid solution, as described by Granstrom and Lee (21):

$$Cl_2 + 2NaClO_2 - 2ClO_2 + 2NaCl$$

The optimum pH for this reaction is considered to be less than or equal to 3.5. At this pH a theoretical yield of 2.6 mol of ClO_2 per mol of Cl_2 applied will be produced. Under typical pH conditions (6.5 to 8.5) encountered in water treatment, chlorine dioxide has been shown to readily disproportionate to chlorate and chlorite ion (21):

$$2ClO_2 + 2OH_2 \leftrightarrows ClO_2^- + ClO_3^- + H_2O$$

However, the detection of chlorate in chlorine dioxide-treated waters has not been achieved, possibly because of analytical problems (47).

Chlorite has been detected in treated water and is of concern, since it has been shown to react with hemoglobin, causing hemolytic anemia at levels as low as 50 mg/liter (1). Additionally, studies with rats indicated a decrease in sperm motility at levels above 100 mg/liter (12). The production of chlorite ions led the U.S. Environmental Protection Agency to set a maximum contaminant level of residual chlorine dioxide together with its inorganic by-products, chlorite and chlorate ions, at 1.0 mg/liter in drinking water (48).

The oxidizing power, and therefore germicidal effectiveness, of chlorine dioxide was once thought to be 2.5 times that of chlorine, on the basis of its full oxidation potential, $E_0 = -1.95$ V, in the following reaction:

$$ClO_2 + 4H^+ + 5e^- \leftrightarrows Cl^- + 2H_2O$$

However, more recent work has shown that the full oxidation potential of chlorine dioxide can be reached only under acidic conditions. At pH 6 to 8, the predominant reaction is the reduction of chlorine dioxide to chlorite ion, which has an oxidation potential of -1.16 V, which is comparable to that of chlorine (34):

$$ClO_2 + e^- \leftrightarrows ClO_2^-$$

Several studies comparing the bactericidal (10, 31, 37) and virucidal (17) qualities of chlorine dioxide with those of chlorine have concluded that chlorine dioxide is as effective as hypochlorous acid and more effective than hypochlorite ion or chloramines in disinfection. Other significant findings show that the effectiveness of chlorine dioxide as a disinfectant is much higher than that of chlorine in waters above pH 8.

The physiological mode of inactivation of bacteria by chlorine dioxide has been shown to be the disruption of protein synthesis. The mechanism by which viruses are inactivated was partially elucidated by Noss (35), who showed that phage f2 was inactivated by chlorine dioxide reacting with tyrosine residues on the capsid protein and/or the A protein, resulting in the inhibition of viral adsorption to host bacteria.

UV Light

Downes and Blount (18) were the first to recognize the effects that UV light in sunlight have on bacteria. Basic UV technology was established by 1910, when Hewitt developed the mercury vapor lamp enclosed in a quartz sheath. The quartz sheath was used to dampen the effects of temperature changes on the light. Over the next three decades, the optimum UV wavelength for disinfection was narrowed to a range of 250 to 266 nm. Sensitivity to UV light was found to be species as well as strain dependent (18).

UV radiation (A and B) is a type of low-energy electromagnetic radiation with poor penetrating power. UV causes excitation rather than ionization of atoms by raising electrons to a state of higher energy without removing them. The UV region of the electromagnetic spectrum that is most optimal for the destruction of bacteria is between 220 and 300 nm. This is often known as the abiotic region (25).

In the 1960s, it was found that the mechanism for UV disinfection was the creation of thymine-thymine dimers or of breaks in the nucleic acid, such as in DNA. More specifically, it seems that UV energy has its greatest effect on pyrimidines and is much less effective on purines. While thymine-thymine dimers are the predominant dimers, dimerization is not limited to these bases; cytosine-cytosine and cytosine-thymine dimers have also been identified in UV-irradiated bacterial cells. It has been speculated that inactivation of RNA viruses, i.e., the enteroviruses and rotavirus, occurs because of uracil dimerization (38).

One problem associated with UV irradiation is its absorbance by proteins, and hence those viruses with a tight outer capsid, such as bacteriophage MS-2, or those viruses containing a double-shell capsid, such as rotavirus, are usually more resistant to UV inactivation than other single-stranded RNA viruses, such as poliovirus or hepatitis A virus. A similar observation can be made for bacteria: the gram-negative rods are much more susceptible to inactivation than the gram-positive cocci, the latter requiring upwards of 5 to 10 times the dose of the former.

There are many factors which influence the efficiency of UV disinfection. In theory, any water can be disinfected to any degree required, using the following equation (14):

$$P/P_0 = e^{-Et/Q}$$

where P is the average number surviving, P_0 is the original number before irradiation, E is the intensity of germicidal energy, t is the time of exposure in minutes, and Q is the exposure, termed a unit lethal exposure, approximately 40 μW-min/cm^2.

Factors which affect the cidal activity of UV light include physical blockage of the light or absorbance of the UV energy. These factors can range from water turbidity, water color, organics, iron, nitrates, and dissolved cations such as manganese, calcium, sodium, and aluminum.

One particular advantage of UV disinfection is that it is able to destroy microbial life without adding anything to the water. This same characteristic is also a disadvantage, since UV disinfection leaves no residual, and water disinfected by this means remains open to subsequent contamination during distribution. Hence, most UV water treatment facilities add a small amount of chlorine after UV treatment to aid in controlling regrowth of bacteria in the distribution lines and to eliminate any pathogens present as a result of posttreatment contamination.

Ozone

Ozone was first discovered by the Dutch philosopher Van Marum in 1783 and used in France to sterilize polluted water. To this day, ozone disinfection for drinking water is much more popular in Europe than in other areas, evidenced by the more than 1,000 European plants which utilize ozone for disinfection of drinking water. There are several factors which have limited the use of ozone compared with chlorination in the United States, primarily cost and the requirement to generate O_3 on site (19).

Chemically, ozone is the strongest oxidant of all the classical disinfectants. As a strong oxidant, ozone reacts with a wide variety of organics and, unlike chlorine, does not form DBPs. Additionally, ozone does not impart taste and can help control organoleptic properties (odor, taste, color). When ozone is added to water, it reacts with hydroxide ions to form hydroxyl radicals and organic radicals. Ozone attacks organic compounds by adding oxygen atoms to the unsaturated carbon-carbon bonds, yielding carboxylic acids, ketones, and aldehydes (7).

Ozonation is not pH dependent, but its cidal activity decreases as the water temperature increases. This is in part because the solubility of ozone in water is a function of temperature, and as the temperature increases, the solubility decreases. Ozone does not leave any residual, and hence water disinfected by ozone remains accessible to contamination by posttreatment introduction of microbes. It is for this reason that in the United States, most ozonation processes do not stand alone but are usually only the first step, which is then followed by the addition of chlorine.

Commercially, ozone is generated by producing a high-voltage corona discharge in a purified oxygen-containing feed gas. The ozone is then contacted with the water, and the feed gas is recycled or discharged. A typical ozone treatment plant consists of three basic subsystems: feed gas preparation, ozone generation, and ozone-water contact (15).

Ozonation has been shown to be an effective process for the destruction of bacteria and viruses. France has adopted a standard for the use of ozone to inactivate viruses. When an ozone residual of 0.4 mg/liter can be measured 4 min after the initial ozone demand has been met, viral inactivation is satisfied (15). Kim et al. (26) demonstrated that ozone breaks the protein capsid of phage f2 into several subunits, which liberates the viral RNA and disrupts the adsorption of the virion to the host pili.

Ozonation is not without its own problems. Bromide ions (Br^-) in drinking water which is disinfected by ozone can lead to the formation of DBPs such as bromate ion (BrO_3^-). Bromide ions can enter water sources from geologic sources, saltwater intrusions, crop and soil fumigation, and salt that is applied to pavement during winter months. Unfortunately, if bromide ion is present in source waters, there are no known treatment techniques available for economically removing it (40). Additionally, complex organics in the source water which are poorly assimilated by bacteria may be broken down by ozone into more readily assimilated organics, resulting in posttreatment bacterial regrowth.

Biological Treatment

Biological treatment of water has been classically used for the conversion of organic material to biomass in wastewater. Drinking water is usually rendered potable by the addition of one or more disinfectants. As cited previously, DBPs have been shown to cause cancers and other abnormalities in laboratory animals; thus, treatment of drinking water by biological filters is one way to avoid the DBPs produced by more classical disinfection methods. In pilot- and full-scale biological filters, contact time is typically measured in minutes, and the water is treated in a single-pass, flowthrough mode (39). Studies by Shukairy et al. (39) have also shown that biological treatment of water can be effectively used to produce a good quality biostabilized water low in DBPs. Aldehydes and keto acids have also been shown to be removed by biological filters at the pilot-scale level (32).

MEASUREMENTS OF DISINFECTANTS

Chlorine

Whenever chlorine is added for disinfection, effluent chlorine residual should be continually monitored. This can be done with automated, continuous, total chlorine analyzers which measure the total chlorine residual and do not differentiate between species (33). In treatment plants without continuous monitors or those wishing to manually measure different chlorine species, residual chlorine can be measured by several different techniques which are detailed in the 19th edition of *Standard Methods for the Examination of Water and Wastewater* (4). In brief, for natural and treated waters, the iodometric method is good for measuring total chlorine at concentrations greater than 1 mg/liter. Amperometric titration is good for free or combined chlorine, and this method is not affected by common oxidizing agents. For measurements of free chlorine, which is unaffected by significant concentrations of monochloramine, dichloramine, nitrate, and nitrite, use of syringaldazine is recommended. For waters which are heavily laden with organic material, either the use of *N,N*-diethyl-*p*-phenylenediamine or some other method which specifically is not affected by interfering compounds should be chosen (4).

Chlorine Dioxide

Accurate determination of the applied ClO_2 dosage as well as the residual concentration of ClO_2 and its by-products is vital to the control of microorganisms. Several reviews of ClO_2 measurement techniques (2, 20) have concluded that the amperometric method for ClO_2 (3) and the sequential iodometric method for ClO_2^- and ClO_3^- (3) are the best choices at present. These methods are detailed in reference 4.

UV Light

UV measurement is more difficult than measurement of chlorine or chlorine dioxide since no chemical is added to the water. Additionally, since UV adsorption and correlation with UV adsorption are site specific, calculations made at one site may not be comparable to those made at other water sources with different water characteristics (4). Presently, there is a *Standard Methods* Joint Task Group which has been charged with developing a UV absorbance standard method. Inactivation of microorganisms by UV is a function of both time and intensity. Light intensity can be determined by a UV sensor, and retention time can be determined by tracer studies. Factors which can alter UV efficiency include water matrix changes, UV lamp coating, UV intensity sensor coating, or lamp age.

Ozone

The accepted method for measuring ozone in water is the indigo colorimetric method. The method is relatively simple and quantitative and can be used on many types of natural waters, such as surface water, groundwater, and biologically treated domestic wastewaters (4).

NOVEL METHODS FOR THE CONTROL OF ORGANISMS IN SOURCE WATER AND DRINKING WATER

Electromagnetic Waves

Electromagnetic radiation is the propagation of energy through space. The name assigned to this treatment technology, electromagnetic waves, refers to the fact that travel of the photons emitted by the source of electromagnetic radiation can be treated as a wave-like function. The order of the electromagnetic energy spectrum from low to high is as follows: visible, low-end UV, X rays, and gamma rays. Microwave ovens are an example of the use of photons to heat water by exciting the water molecules. For water treatment, electromagnetic waves at the low end of the UV range will result in heating of the water. The temperature generated by electromagnetic waves at the low end is usually not sufficient for the destruction of most bacterial spores, viruses, and cysts or oocysts. In the visible range, some photochemical reactions such as dissociation and increased ionization may take place, but here too, not enough energy is available for the inactivation of most waterborne pathogens. At the other end of the spectrum are the X rays and the gamma rays. They are of high energy and will tend to ionize almost anything with which they collide (14). These types of energies have been successful in destroying *Cryptosporidium* oocysts (unpublished data); however, their efficiency is highly dependent on good quality water, and all require great amounts of energy and so are usually not cost effective.

Electromagnetism

Electromagnetism itself has no germicidal effects, but since microbes are charged and do associate with particles in the water, electromagnetism has been used to remove the particles containing the attached organisms. This technique by itself is not sufficient for removal of all pathogens but does reduce turbidity and hence can enhance the disinfection capabilities of more traditional disinfectants (14).

Electron Beams

Electron beams are another means of using ionizing radiation for the destruction of microbes. The energetic electrons dissociate water into free radicals H^+ and OH^-. These may combine to form hydrogen, peroxides, and ozone. These highly active molecules attack living structures to promote their oxidation, reduction, dissociation, and degradation. Bacteria, viruses, and protozoan cysts are usually affected by these secondary ionization products (14).

SUMMARY

In summary, one of the greatest challenges facing societies today is providing drinking water that is free of any pathogens and that is both palatable and affordable. Approaches for the production of potable water have traditionally centered on treatment practices. Because of economic restraints or the production of harmful by-products, this is no longer a plausible solitary option. Development of a single alternative treatment which is economical, reliable, and safe is not likely to occur in the next decade. In fact, as we are faced with the emergence of a population whose individuals have an older average age and subpopulations that have special needs, such as the immunocompromised, compounded by declining economics, we will face mandates that require our skill in the production of potable water to become more adept than ever before.

One approach to a solution would be the use of multiple barriers. While this concept is not new, its practice has been very lax. In order to maximize the resources at hand, enhance traditional treatment practices which now dictate lower concentrations of disinfectants, and be able to remove emerging pathogens, it will be important to address all of the different facets that are involved in preventing and removing microbial pathogens. This effort will have to start with the selection of good quality source waters, which, in turn, cannot happen unless we start protecting these sources from further pollution. Once a water source has been identified, it should be protected from human and industrial impact. When the water enters the treatment process, there should be an allowance of multiple sites for the addition of disinfectants, the addition of coagulant or flocculent, and different areas for sedimentation and filtration. With careful planning for the protection of our source waters and good management practices at our treatment facilities, we can shift the burden that has long been placed solely on disinfection.

REFERENCES

1. **Abdul-Rahman, M. S., D. Couri, and R. J. Bull.** 1980. Kinetics of ClO_2 and effect of ClO_2 in drinking water on blood glutathione and hemolysis in rat and chicken. *J. Environ. Pathol. Toxicol.* **3:**431–449.
2. **Aieta, E. M., and J. D. Berg.** 1986. A review of chlorine dioxide in drinking water treatment. *J. Am. Water Works Assoc.* **78:**6–62.
3. **Aieta, E. M., and P. V. Roberts.** 1981. Chlorine dioxide chemistry: generation and residual analysis, p. 429–452. *In* W. J. Cooper (ed.), *Chemistry in Water Reuse*, vol. 1. Ann Arbor Science Publishers, Ann Arbor, Mich.
4. **American Public Health Association.** 1995. Chlorine, p. 36–57. *In* M. A. H. Franson (ed.), *Standard Methods for the Examination of Water and Wastewater*, 19th ed. American Public Health Association, Washington, D.C.
5. **Anonymous.** 1891. The water we drink. JAMA **17:**819–820.
6. **Aston, R. N.** 1947. Chlorine dioxide use in plants on the Niagara border. *J. Am. Water Works Assoc.* **39:**687–690.
7. **Baily, P. S.** 1975. Reactivity of ozone with various organic functional groups important to water purification, p. 263–277. *In First International Symposium on Ozone for Water and Wastewater Treatment Proceedings.* Waterbury, Conn.
8. **Baker, M. N.** 1930. *The Quest for Pure Water.* American Water Works Association, New York.
9. **Belohlav, L. R., and E. T. McBee.** 1962. Discovery and early work, p. 137–156. *In* J. S. Sconce (ed.), *Chlorine: Its*
10. **Bernarde, M. A., W. B. Snow, V. P. Olivieri, and B. Davidson.** 1967. Kinetics and mechanism of bacterial disinfection by chlorine dioxide. *Appl. Environ. Microbiol.* **15:**257–262.
11. **Bitton, G.** 1980. Fate of viruses in land disposal of wastewater effluents, p. 201–241. *In Introduction to Environmental Virology.* John Wiley & Sons, New York.
12. **Carlton, B. D., and M. K. Smith.** 1986. Reproductive effects of alternative disinfectants and their by-products, p. 295–300. *In* R. L. Jolley, R. J. Bull, W. P. Davis, S. Katz, M. H. Roberts, Jr., and V. A. Jacobs (ed.), *Water Chlorination: Chemistry, Environmental Impact and Health Effects.* Lewis Publishers, Inc., Mich.
13. **Cheremisinoff, N. P., and P. N. Cheremisinoff,** 1993. Physical separation processes, p. 1–30. *In* WordCrafters Editorial Services Inc. (ed.), *Water Treatment and Waste Recovery Advanced Technology and Applications.* Prentice-Hall, Inc., Englewood Cliffs, N.J.
14. **Cheremisinoff, N. P., and P. N. Cheremisinoff.** 1993. Chemical technologies for water disinfection, p. 31–94. *In* WordCrafters Editorial Services Inc. (ed.), *Water Treatment and Waste Recovery Advanced Technology and Applications.* Prentice-Hall, Inc., Englewood Cliffs, N.J.
15. **Cheremisinoff, N. P., and P. N. Cheremisinoff.** 1993. Sterilization by radiation, p. 140–158. *In* WordCrafters Editorial Services Inc. (ed.), *Water Treatment and Waste Recovery Advanced Technology and Applications.* Prentice-Hall, Inc., Englewood Cliffs, N. J.
16. **Craun, G. F.** 1988. Surface water supplies and health. *J. Am. Water Works Assoc.* **80:**40–52.
17. **Cronier, S., P. V. Scarpino, and M. L. Zink.** 1978. Chlorine dioxide destruction of viruses and bacteria in water, p. 651–658. *In* R. L. Jolley, H. Gorchev, and D. H. Hamilton, Jr. *Water Chlorination: Environmental Impacts and Health Effects*, vol. 2. Ann Arbor Science, Ann Arbor, Mich.
18. **Downes, A., and T. Blount.** 1877. Research on the effect of light upon bacteria and other organisms. *Proc. R. Soc. Lond.* **26:**488.
19. **Glaze, W. H.** 1987. Drinking water treatment with ozone. *Environ. Sci. Technol.* **21:**224.
20. **Gordon, G.** 1988. Methods of measuring disinfectant residuals. *J. Am. Water Works Assoc.* **80:**9–94.
21. **Granstrom, M. L., and G. F. Lee.** 1957. Rates and mechanisms of reaction involving oxychloro compounds. *Public Works* **22:**209–217.
22. **Hamann, C. L., Jr., J. B. McEwen, and A. G. Myers.** 1990. Guide to selection of water treatment processes, p. 157–188. *In* F. W. Pontius (ed.), *Water Quality and Treatment: a Handbook of Community Water Supplies.* McGraw-Hill, Inc., New York.
23. **Harakeh, M. S.** 1986. Factors influencing chlorine disinfection of wastewater effluent contaminated by rotaviruses, enteroviruses, and bacteriophages, p. 681–690. *In* R. L. Jolley, R. J. Bull, V. A. Jacobs, W. P. Davis, S. Katz, and M. H. Roberts, Jr. (ed.), *Water Chlorination: Chemistry, Environmental Impact and Health Effects*, vol. 5. Lewis Publishers, Mich.
24. **Hoff, J. C., and E. E. Geldreich.** 1981. Comparison of the biocidal efficiency of alternative disinfectants. *J. Am. Water Works Assoc.* **73:**40–44.
25. **Huff, C. B.** 1965. Study of ultraviolet disinfection of water and factors in treatment efficiency. *Public Health Rep.* **80:**695.
26. **Kim, C. K., D. M. Gentils, and O. J. Sproul.** 1980. Mechanism of ozone inactivation of bacteriophage f-2. *Appl. Environ. Microbiol.* **39:**210–218.

27. **Korich, D. G., J. R. Mead, M. S. Mador, N. A. Sinclair, and C. R. Sterling.** 1990. Effects of ozone, chlorine dioxide, chlorine, and monochloramine on *Cryptosporidium parvum* oocyst viability. *Appl. Environ. Microbiol.* **56:**1423–1428.

28. **Leahy, J. G., A. J. Rubin, and O. J. Sproul.** 1987. Inactivation of *Giardia muris* cysts by free chlorine. *Appl. Environ. Microbiol.* **53:**1448–1453.

29. **Levenspiel, O.** 1972. *Chemical Reaction Engineering.* John Wiley & Sons, New York.

30. **Malley, J. P., Jr., J. K. Edzwald, and N. M. Ram.** 1988. Pre-oxidant effects on organic halide formation and removal of organic halide precursors. *Environ. Technol. Lett.* **9:**1089.

31. **Malpas, J. F.** 1973. Disinfection of water using chlorine dioxide. *Water Treat. Exam.* **22:**209–217.

32. **Miltner, R. J., H. M. Shukairy, and R. S. Summers.** 1992. Disinfection by-product formation and control by ozonation and biological treatment. *J. Am. Water Works Assoc.* **84:**11–53.

33. **Montgomery, J. M.** 1985. Disinfection, p. 560–561. *In* J. M. Montgomery (ed.), *Water Treatment Principles and Design.* John Wiley & Sons, New York.

34. **Myhrstad, J. A., and J. E. Samdal.** 1969. Behavior and determination of chlorine dioxide. *J. Am. Water Works Assoc.* **4:**205–208.

35. **Noss, C. I.** 1986. Chlorine dioxide reactivity with proteins. *Water Res.* **20:**351–356.

36. **Reinert, P. E., and J. A. Hroncich.** 1990. Source water quality management, p. 189–268. *In* F. W. Pontius (ed.), *Water Quality and Treatment: a Handbook of Community Water Supplies.* McGraw-Hill, Inc., New York.

37. **Ridenour, G. M., and R. S. Ingols.** 1947. Bactericidal properties of chlorine dioxide. *J. Am. Water Works Assoc.* **39:**561–567.

38. **Roessler, W. P., E. V. Dellen, Amway Corporation, ADA Michigan, M. Abbaszadegan, and C. P. Gerba.** 1992. Coliphage MS-2 as a UV water disinfection efficacy test surrogate for bacterial and viral pathogens. American Water Works Association Water Quality Technology Conference poster presentation.

39. **Shukairy, H. M., R. J. Miltner, and R. S. Summers.** 1995. Bromide's effect on DBP formation, speciation, and control, part 2. Biotreatment. *J. Am. Water Works Assoc.* **87:**71–82.

40. **Siddiqui, M. S., G. L. Amy, and R. G. Rice.** 1995. Bromate ion formation: a critical review. *J. Am. Water Works Assoc.* **87:**58–70.

41. **Sobsey, M.** 1988. *Detection and Chlorine Disinfection of Hepatitis A in Water.* EPA CR-813-024. U.S. Environmental Protection Agency, Washington, D. C.

42. **Sobsey, M. D., and J. S. Glass.** 1981. Improved electropositive filter for concentrating viruses from large volumes of water, p. 239–245. *In* M. Goddard and M. Butler (ed.), *Viruses and Wastewater Treatment.* Pergamon Press, New York.

43. **Stetler, R. E., R. L. Ward, and S. C. Waltrip.** 1984. Enteric virus and indicator bacteria levels in a water treatment system modified to reduce trihalomethane production. *Appl. Environ. Microbiol.* **47:**319–324.

44. **Tate, C. D., and K. F. Arnold.** 1990. Health and aesthetic aspects of water quality, p. 83–156. *In* F. W. Pontius (ed.), *Water Quality and Treatment: a Handbook of Community Water Supplies.* McGraw-Hill, Inc., New York.

45. **Tilton, K. S.** 1995. Comparison of chlorine and chlorine dioxide as disinfectants for surface water. Ph.D. dissertation. University of New Hampshire, Durham.

46. **U.S. Environmental Protection Agency.** 1989. Drinking water; national primary drinking water regulations; filtration, disinfection; turbidity, *Giardia lamblia*, viruses, *Legionella*, and heterotrophic bacteria. *Fed. Regist.* **54:**27486–27491.

47. **Weber, W. J.** 1972. *Physicochemical Processes for Water Quality.* Wiley-Interscience, New York.

48. **Werdehoff, K. S., and P. C. Singer.** 1987. Chlorine dioxide's effect on THMFP, TOXFP, and the formation of inorganic by-products. *J. Am. Water Works Assoc.* **79:**107–113.

Detection of the Presence of Bacteria and Viruses in Shellfish

RICARDO DE LEON AND LEE-ANN JAYKUS

21

Bacterial and viral disease outbreaks have been associated with all major types of edible bivalve mollusks, and human enteric pathogens have been isolated from shellfish obtained from both opened and closed harvesting areas (35, 89). This chapter contains information which pertains to those shellfish which are edible bivalve mollusks of the class Pelecypoda and include the species commonly referred to as oysters, mussels, clams, and cockles. Since these organisms are filter feeders, they use siphoning organelles and mucous membranes to sieve suspended food particles from the aquatic environment as a source of food. If their surrounding water is contaminated by bacteria or viruses, these mucous membranes may entrap the pathogens and transfer them to the digestive tract. Since shellfish are usually consumed whole and raw, they may act as passive carriers of human pathogens.

STANDARDS FOR PREVENTION OF SHELLFISH-ASSOCIATED DISEASE

The National Shellfish Sanitation Program was developed in 1925 to target the prevention of shellfish-associated disease caused primarily by enteric bacteria (33). In the United States, this program has established bacteriological standards for shellfish and their harvesting waters based on the fecal coliform index. For approved harvesting waters, the most probable number of coliforms must not exceed 70/100 ml, with no more than 10% of the samples exceeding 230/100 ml. Shellfish meat may contain no more than 230 fecal coliforms per 100 g. While these standards have proved to be very effective in preventing outbreaks of disease due to enteric bacterial pathogens (90), there is little or no relationship between the levels of coliform indicator bacteria in water and shellfish and the presence of human enteric viruses or naturally occurring marine *Vibrio* species. Consequently, shellfish-associated outbreaks and cases of disease due to both human enteric viruses and *Vibrio* species continue to occur in the United States (43, 48, 61). Human enteric viruses appear to be the major cause of shellfish-associated disease, while the *Vibrio* species cause rarer but more severe disease syndromes. Microbiological standards for enteric virus or *Vibrio* contamination in shellfish have not been established, primarily because of the lack of standard detection methodology.

ETIOLOGY OF SHELLFISH-ASSOCIATED OUTBREAKS OF ILLNESS
Pathogenic Bacteria Transmitted by Shellfish

Two general groups of pathogenic bacteria may be transmitted by shellfish. The first group represents those organisms which are indigenous to the marine environment, predominantly members of the family *Vibrionaceae*, including the genera *Vibrio*, *Aeromonas*, and *Plesiomonas*. The presence of these organisms is unrelated to fecal pollution and therefore cannot be monitored by using coliform indices. The second group, called nonindigenous bacterial pathogens, are not natural marine inhabitants, and their presence in shellfish arises either from direct fecal contamination by human or animal reservoirs or from poor general sanitation during harvesting, processing, or preparation of the food animals. While bacteria account for only about 2% of shellfish-associated outbreaks of illness (2), both the incidence and public awareness of shellfish-associated human *Vibrio* infections have increased during the last decade (92).

Enteric Bacteria

The presence of *Salmonella* and *Shigella* species in shellfish is predominantly due to fecal contamination of harvesting sites or cross-contamination by human carriers, particularly food handlers. The primary source of *Staphylococcus aureus* in shellfish is also food handlers, and shellfish can readily support the growth of this organism when contaminated and stored at an improper temperature. In a study of staphyloccocal food poisoning in the United Kingdom between 1969 and 1990, Weineke et al. (93) found that up to 7% of staphylococcal food poisoning incidents were attributable to fish and shellfish.

The Vibrios

The vibrios are part of the normal estuarine microflora, have excellent survival capabilities in both marine and estuarine environments, and may be accumulated by shellfish during feeding. While the transmission of *Vibrio cholerae* by seafood is well documented throughout the world, the United States has not had a major outbreak since 1911, although sporadic cases have occurred. Most documented sporadic cases have been attributed to the consumption of crustaceans and shrimp; only a few have been associated with bivalve mollusks. However, sporadic cases in both Texas and Florida in the 1970s were linked to raw shellfish. More recently, seafoods imported from South American locations where cholera is endemic have been associated with disease in industrialized nations.

V. *parahaemolyticus* outbreaks are rare in the United States and generally due to gross mishandling of contaminated seafood, particularly improper refrigeration, insufficient cooking, or cross-contamination. V. *vulnificus* has been the cause of greater concern in the United States in recent years, since epidemiological evidence indicates that enteric infection with V. *vulnificus* can result in a syndrome characterized by gastrointestinal disease followed by primary septicemia, with mortality rates approaching 50%. Individuals with underlying liver dysfunction or those who are immunocompromised are particularly at risk (11, 86). At least 16 states have reported V. *vulnificus* infections, with seawater and raw shellfish as the primary source of infection. Hlady et al. (43) have reported V. *vulnificus* as a leading cause of reported food-borne illness-related deaths in Florida. Levine and Griffin (61) have also reported that raw shellfish may contribute significantly to *Vibrio*-related gastroenteritis in the Gulf of Mexico region. Furthermore, this organism is capable of establishing a nonculturable but viable state, which means that routine examination by conventional cultural methods may result in negative results although viable and potentially virulent cells may be present in high numbers (92). Rippey (74) provides an excellent review of shellfish-associated disease due to marine *Vibrio* species.

Enteric Viruses Transmitted by Shellfish

Enteric viruses are parasites of animals and humans believed to be transmitted primarily by the fecal-oral route. They are excreted in large numbers (10^6 to 10^{10} virus particles per g) in the feces of infected individuals and can almost always be detected in domestic sewage effluents (75, 80). Since mammalian viruses are extremely specific for both species and tissue, the sources of human enteric virus contamination of shellfish harvesting beds is always human fecal pollution rather than general animal waste. Enteric viruses may enter the ocean or estuaries directly through the discharge of domestic sewage, sewage-contaminated rivers and streams, ocean disposal of domestic sewage sludge, and boat wastes. Indirect sources of contamination occur through land application and sludge burial, with subsequent runoff via rivers and perhaps groundwater to estuaries and the coastline (39).

Enteric viruses are more resistant than bacterial pathogens to common sewage treatment processes, including chlorination (8). In laboratory studies, enteric viruses have been reported to survive for 2 to 130 days in seawater. These survival periods far exceed those reported for coliform bacteria in similar environments (65). For these reasons and others, the fecal coliform index is inadequate for monitoring the presence of viral contamination in shellfish. Numerous studies have been conducted on the survival of viruses in marine water, and reviews on the subject have been published by Gerba and Goyal (35), Goyal (39), and Jaykus et al. (48).

There are over 110 known enteric viruses which are excreted in human feces and ultimately find their way into domestic sewage (65, 94). These viruses cause a wide variety of illnesses such as hepatitis, fever, diarrhea, gastroenteritis, paralysis, meningitis, and myocarditis. The list of known enteric viruses has grown rapidly over the last two decades as better methods have become available for their detection (65). Regardless of the large number of enteric viruses, only hepatitis A virus (HAV), non-A, non-B hepatitis (hepatitis

E) virus, small round structured viruses including Norwalk virus and the Snow Mountain agent, astroviruses, typical caliciviruses (non-small round structured viruses), and small round (featureless) viruses (SRVs) have been epidemiologically linked in various degree to shellfish-associated viral disease. Nonetheless, sporadic cases of illness that are not detected as outbreaks may potentially be attributable to shellfish consumption (34). Reviews of shellfish-borne viral disease epidemiology have been published by Mosley (67), Gerba and Goyal (35), Richards (73), Appleton (2), De Leon and Gerba (22), and Jaykus et al. (48).

HAV

Infectious hepatitis or hepatitis A is perhaps the most serious viral illness transmitted by the ingestion of shellfish (73). HAV is currently classified as a member of the genus *Hepatovirus* of the family *Picornaviridae* and has several basic characteristics in common with the enterovirus group, i.e., single-stranded RNA genome, 28-nm diameter, and fecal-oral transmission. HAV causes a higher incidence of symptomatic infections than do enteroviruses, up to 95% during outbreaks (59). Common symptoms of HAV infection are dark urine, nausea, vomiting, malaise, fever, chills, and jaundice. Fulminant HAV infection, although rare, has a very high mortality rate (85). The high rate of HAV asymptomatic infections in children may contribute greatly to the nonepidemic prevalence of the disease in persons of other ages (85).

Outbreaks of shellfish-associated infectious hepatitis in the United States were not documented until the early 1960s. Since then, nearly 1,500 cases of shellfish-associated HAV have been reported (73). Shellfish-associated hepatitis A is of greater concern in countries where raw shellfish is widely consumed. For example, Xu et al. (96) reported 300,000 cases of hepatitis A attributable to shellfish over a 2-month period in 1988 in Shangai, China. While the number of sporadic cases of HAV infection is difficult to ascertain, they are nonetheless considered to be more significant than outbreak-related cases, and some may be associated with shellfish (69). The significance of shellfish in the transmission of hepatitis A in the United States and Europe has been summarized by Richards (73) and Appleton (2), respectively.

Non-A, Non-B Hepatitis (Hepatitis E)

Enteral non-A, non-B hepatitis is transmitted mainly by sewage-contaminated water and person-to-person contact. The etiological agent of this disease is a calicivirus, now designated hepatitis E virus (13, 72). The disease caused by hepatitis E virus can be more severe than that caused by HAV, with a high incidence of cholestasis and significant mortality (>10%) in pregnant women. Enteral non-A, non-B hepatitis is endemic in the Middle East, Africa, and India but not in the United States (16). The role of shellfish in the transmission of this disease is suspected but currently unconfirmed epidemiologically.

Caliciviruses

Human Caliciviruses

Human caliciviruses have been isolated from the feces of infected individuals. These viruses have a characteristic morphology of 32 "capped" depressions in icosahedral symmetry and are 28 to 34 nm in diameter. Some human calicivirus strains more commonly affect children, causing vomiting and diarrhea, whereas other strains affect all age groups, with more flu-like symptoms (i.e., fever, malaise, and nausea) (18). Attack rates are generally high. Cubbitt (19) provides a recent review of the caliciviruses. Shellfish are an important vehicle in the transmission of human caliciviruses, and molecular techniques have recently provided much-needed information and opportunities for the detection and study or these noncultivable viruses (1, 23). Several important agents of viral disease transmitted by shellfish, generally referred to as Norwalk-like viruses or SRVs, have been recently shown to share genomic characteristics with the caliciviruses. Members of this group include Norwalk virus; Snow Mountain, Southhampton, Hawaii, Taunton, and Parramatta agents; and many other viruses which are yet unnamed (52). Shared morphological and genomic characteristics of these viruses, such as a single-stranded RNA genome, single type of capsid protein, and similar genomic organizations and sequences have led scientists to refer to these agents as the Norwalk-like group, with Norwalk virus itself as the prototype strain. The genome of Norwalk virus was first sequenced in 1990 (51), and partial sequences from various other Norwalk-like viruses have been reported recently (1, 91).

Norwalk and Norwalk-Like Viruses
The Norwalk agent, the first recognized human gastroenteritis virus of clinical importance, was discovered in 1972 in Norwalk, Ohio, by immunoelectron microscopy of an infectious stool filtrate (54). The infectivity and pathogenicity of the virus have since been studied in adult human volunteers (20). The attack rate in adult human volunteers is 50%, with typical gastroenteritis symptoms and reinfection occurring even in the presence of detectable serum antibody levels (10, 95). A short-term resistance to reinfection of 4 to 14 weeks has been reported (9, 53). Forty-two percent of the outbreaks of nonbacterial gastroenteritis investigated by the Centers for Disease Control in 1976 to 80 were attributed to the Norwalk agent (55).

The first documented outbreak of shellfish-associated Norwalk virus gastroenteritis involved over 2,000 persons throughout Australia (68). In the United States, shellfish-associated gastroenteritis attributed to Norwalk virus was first reported in 1980 after individuals consumed oysters from Florida (40). Since that time, other U.S. outbreaks have been reported (41, 66). The importation of depurated English clams has led to outbreaks of enteric illness due to Norwalk virus (73), and Norwalk viral illness associated with shellfish is a continuing problem in the United States (73).

Other SRVs
A group of SRVs have been reported to be the cause of numerous outbreaks of shellfish-associated gastroenteritis (5, 36). These viruses do not appear to be serologically related to Norwalk virus or HAV. SRV infection is diagnosed by electron microscopy, and information on the characteristics of the viruses is virtually nonexistent. The SRV group most likely represents more than one virus type. While some, such as the Snow Mountain agent (1, 30, 88, 92) and Hawaii agent (1, 87), may be genetically related to the Norwalk-like viruses, others, such as the W-Ditchling agent (3) and the cockle agent (5), are less well characterized. Gill et al. (36) investigated an outbreak involving small round structured virus illness due to consumption of Pacific oysters, with a particularly high attack rate of 79%.

Some of the SRVs may belong to the genus *Parvovirus*. Currently, it is not clear whether parvoviruses are involved in gastroenteritis associated with shellfish consumption. It is suspected that involvement of some parvovirus-like particles in dual infections with Norwalk virus may originate from enhanced replication of persistent parvoviruses during the Norwalk virus infection (15). Additional research is needed to differentiate the parvoviruses as a separate group of shellfish-associated gastroenteritis viruses. Nonetheless, small round parvovirus-like particles have been found in shellfish incriminated in the outbreaks (2).

Astroviruses
An astrovirus was initially detected in the feces of children with gastroenteritis (4) and given its name because of its characteristic five- to six-pointed star-like form seen under electron microscopy (63). Symptomatic infection has been found in 80% of babies infected with astroviruses (62). Infection of volunteers who have detectable serum antibody does not result in diarrhea (58).

At least one outbreak of astrovirus gastroenteritis has been associated with the consumption of oysters (14). The outbreak, which occurred at an officers' dinner on a naval base, had two phases. The first phase was caused by an SRV, and the second phase occurred 4 days later after recovery from the SRV illness. The second phase consisted of another bout of diarrhea accompanied by the shedding of large numbers of astroviruses.

METHODS FOR THE DETECTION OF PATHOGENS IN SHELLFISH
General Steps in Isolation and Detection
While the methods for the recovery and detection of microorganisms in shellfish differ for bacteria and viruses, they both follow five to six general steps from sample collection to final identification (Table 1). In this chapter, only the steps which significantly affect the assay method will be discussed. Sample collection and shucking as well as meat homogenization are common to both contaminant types and are straightforward techniques provided that they are done aseptically. For viruses, the approach used for pathogen extraction and concentration will affect not only the efficiency of recovery of targeted viruses but also the sample toxicity or inhibition of the assay. While bacterial cultural enrichment steps are fairly standard, they can be complicated by fastidious organisms or those which enter the viable but nonculturable state. For molecular detection methods, the nucleic acid extraction-concentration approach will affect detection accuracy. In all cases, assay inhibitors must be removed before application of molecular methods.

Various methods have been developed for the recovery

TABLE 1 General steps in the isolation of pathogens from shellfish

1. Sample collection and shucking

2. Meat homogenization

3. Virus extraction, cultural enrichment of bacteria, or nucleic acid extraction and concentration

4. Removal of assay inhibitors

5. Assay by conventional or molecular methods

of enteric viruses from shellfish and other seafoods. These methods have been summarized by Sobsey (81–83) and reviewed by Bouchriti and Goyal (12) and Jaykus et al. (48). Two general schemes, designated extraction-concentration and adsorption-elution-concentration, have proven successful. The goal in both cases is to separate viruses from shellfish meats, provide a low-volume aqueous solution that is free of cytotoxic material, and recover most of the viruses present in the shellfish sample. Both schemes employ conditions favoring the separation of viruses from shellfish tissues, primarily through the use of filtration, precipitation, polyelectrolyte flocculation, and solvent extraction. However, the adsorption-elution-precipitation methods have been used more commonly in recent years. A modified adsorption-elution-precipitation procedure (84) is recommended for virus concentration. Briefly, 50 g of oysters is diluted sevenfold with cold distilled water and vigorously blended, and the viruses are adsorbed to the meat by reducing conductivity to <2,000 ppm and pH to 5.0. After centrifugation at 1,700 × g and 4°C for 15 min, the solids-adsorbed viruses are eluted by resuspension in 350 ml of 0.05 M glycine–0.14 M NaCl, with subsequent pH adjustment to 7.5. Shellfish solids are removed by centrifugation at 1,700 × g and 4°C for 15 min, and viruses in the supernatant are concentrated by acid precipitation at pH 4.5. The resulting floc is sedimented by centrifugation at 1,700 × g and 4°C for 15 min and resuspended in 10 ml of 0.1 M disodium phosphate buffer (pH 9.3 to 9.5), and the pH is then adjusted 7.0 to 7.5. Cytotoxic components of the sample are precipitated by addition of Cat-Floc T to a final concentration of 0.1%, and the precipitate is removed by centrifuging at 3,000 × g and 4°C for 20 min. Recovered supernatants are supplemented with gentamicin and kanamycin to final concentrations of 50 and 250 μg/ml, respectively, and incubated at room temperature for 2 h. The final concentrates can be stored at −80°C until assay or further treatment. Various modifications have included further clarification of concentrates by polyelectrolyte flocculation, ultrafiltration, or a freeze-thaw cycle followed by centrifugation. The adsorption-elution method is easily adapted to the different species of shellfish and has even been used for freshwater clams (24).

Conventional Methods for the Detection of Bacteria in Shellfish

Examination of food products for bacterial pathogens differs from the method used for clinical samples because of the potentially low numbers of contaminants and the frequently poor physiological state of the organisms. Conventional methods for the detection of bacteria in foods rely on the ability of the organism to multiply in specific selective enrichment media, followed by defined growth and evidence of morphology or other distinguishing features on selective plating media. Therefore, detection of pathogens is dependent on both the availability of media specific for their isolation and the ability of the organisms to grow. Consequently, conventional cultural methods will fail to detect bacteria when they are in the viable but nonculturable state. Nonetheless, successful methods have been developed for the detection of culturable forms of V. cholerae, V. parahaemolyticus, and V. vulnificus in shellfish (56). The general approach involves preparation of a shellfish homogenate by 1:10 dilution in phosphate-buffered saline (PBS) (V. vulnificus), PBS supplemented with 2 to 3% NaCl (V. parahaemolyticus), or alkaline peptone water (APW) (V. cholerae). APW dilutions of the resulting homogenate are made for enrichment

purposes. For V. cholerae, enrichment is performed by incubation of the homogenate and APW dilutions at 42°C for 6 to 24 h; for V. parahaemolyticus and V. vulnificus, enrichment is done in a three-tube, multiple-dilution most-probable-number format at 35 to 37°C for 16 to 18 and 12 to 16 h, respectively. Selective plating is done with thiosulfate-citrate-bile salts-sucrose agar and/or modified cellobiose-polymyxin B-colistin agar. Typical colony color and morphology are described elsewhere (32). Preliminary biochemical tests for the identification of V. cholerae include tests using oxidase, triple sugar iron agar, Kligler's iron agar, 1% tryptone broth supplemented with 1 to 3% NaCl, gelatin and gelatin salt agars, and Hugh-Leifson glucose broth. For V. vulnificus and V. parahaemolyticus, biochemical confirmation is done by using the oxidase test, motility test, arginine glucose slant, triple sugar iron agar, O/129 disc assay, and the O-nitrophenyl-β-D-galactopyranoside test. Other biochemical tests may be appropriate and are commercially available. Standard serological tests are available for V. cholerae and V. parahaemolyticus, and various forms of immunological and/or DNA probe technologies are in developmental phases (32).

Conventional Methods for the Detection of Enteric Viruses in Shellfish

The shellfish extracts obtained after extraction-concentration or adsorption-elution-concentration may be assayed for viruses by using conventional mammalian cell culture techniques (chapter 8). There are two basic procedures for detection and quantification of viruses in cell culture: quantal methods (50% tissue culture infective dose or most probable number) and enumerative methods (plaque assay). Quantal methods rely on the development of widespread cell monolayer deterioration (cytopathic effects), while enumerative methods detect localized cell monolayer damage under an agar overlay (a plaque). Selection of the cell culture type depends almost entirely on the viruses that one wishes to detect. Unfortunately, there is no single universal cell culture system that will detect all or even a majority of the more than 100 different human enteric viruses that can potentially contaminate shellfish. Among the several convenient and sensitive cell lines commonly used for the detection of human enteroviruses are the BGMK (buffalo green monkey kidney-derived), MA-104 (rhesus monkey kidney-derived), RD (human rhabdomyosarcoma-derived), and AGMK (primary African monkey kidney-derived) cell lines. Because of their ease of detection in mammalian cell culture, most studies on the incidence of enteric viral contamination in shellfish have been limited to the enterovirus family. Likewise, because of its ease of detection, Richards (73) has proposed the use of poliovirus, a model human enterovirus, as an alternative indicator of virus contamination of shellfish and their harvesting waters.

However, the epidemiologically important shellfish-borne enteric viruses do not replicate (Norwalk and Norwalk-like viruses and SRVs) or replicate poorly (HAV and astroviruses) in all currently used mammalian cell cultures. While the FRhK-4 (fetal rhesus monkey kidney-derived) cell line has proven useful in the propagation of a laboratory-adapted strain of HAV (17, 71), its use in the detection of wild-type HAV is of limited value. There are presently no mammalian cell culture lines available for the detection of Norwalk virus, Norwalk-like viruses, SRVs, or astroviruses. Furthermore, lengthy assay time and high cost seriously limit the value of this detection methodology.

TABLE 2 Advantages and disadvantages of molecular techniques

Advantages	Disadvantages
1. More rapid detection of fastidious organisms	1. No distinction between viable and nonviable organisms can be made.
2. Detection of nonculturable pathogens	2. Environmental samples often need extensive preparation and cleaning.
3. Simultaneous detection of multiple pathogens	3. Results are mostly qualitative.
4. Greater selectivity for the detection of groups of organisms or specific pathogens within a group	4. Quantitative techniques, when available, are usually very target specific.

Molecular Methods for the Detection of Pathogens in Shellfish

The advent of recombinant DNA technology has made it possible to use a segment of cloned DNA or RNA as the complementary sequence for the detection of viral genomes in clinical samples, in infected cells, and recently in environmental samples, as reviewed by De Leon and Sobsey (26). Of particular value have been nucleic acid hybridization and amplification techniques. Gene probe hybridization to genomic viral RNA has been reported to detect 10^4 to 2.5×10^5 physical virus particles (29, 50) or 500 to 1,000 PFU of virus (21, 79). Margolin et al. (64) have reported the detection of 1 to 10 PFU of poliovirus with a probe of high specific activity (2×10^9 cpm/μg of DNA). In side-by-side comparisons of DNA and RNA probes, RNA probes were five- to eightfold more sensitive than DNA probes (31, 49, 79). Although most reported detection sensitivities are 100-fold less than a desired goal of 1 to 10 infectious units, several investigators have reported detection of enteric viruses in environmental samples (31, 79).

The sensitivity of gene probes in environmental samples can be increased by the use of other molecular techniques in conjunction with gene probe hybridization. Enzymatic amplification of target nucleic acids is the most feasible alternative for achieving the sensitivity limits needed for routine assay of viral contamination in environmental and food samples. The most promising nucleic acid amplification method is PCR, an in vitro enzymatic enrichment method

which results in amplification of a specific DNA sequence up to 10^6-fold within a few hours (70, 77). The method is readily adapted to the detection of RNA viruses by reverse transcribing the RNA before PCR amplification (reverse transcriptase PCR [RT-PCR]) (25, 38, 76). Research on the PCR amplification of enteric pathogens in shellfish is now in progress (Tables 2 to 4).

PCR Methods for the Detection of *Vibrio* Species in Shellfish

V. vulnificus is an organism which can enter a viable but nonculturable state and is thus a prime candidate for detection methods which do not rely on conventional culture. Hill et al. (42) used guanidinium isothiocyanate, chloroform extraction, and ethanol precipitation to isolate DNA sequences of *V. vulnificus* from oyster homogenates seeded at the 100-CFU/g level (Table 3). While preincubation for 24 h was required prior to detection, this may not necessarily be a drawback since it can save painstaking pathogen extraction procedures and provides some indication that the organisms may be viable. Koch et al. (57) used a short-term (6 to 8 h) enrichment period to detect 10 CFU of seeded *V. cholerae* per gram of oyster meat via direct PCR from crude homogenates. No methods specifically targeting pathogens in the viable but nonculturable state, or foregoing cultural enrichment procedures, have been reported.

PCR Methods for the Detection of Enteric Viruses in Shellfish

Shellfish meat represents a chemically complex matrix, containing many compounds which can interfere with the performance of PCR. Two general approaches have been undertaken to reduce the level of interfering compounds and thereby facilitate the application of PCR to the detection of viral pathogens in shellfish. In the first approach, viral nucleic acids are isolated and purified from the shellfish meat before application of RT-PCR. The second approach involves the concentration and purification of intact virus particles from the complex matrix with simultaneous reduction in sample volume and removal of PCR inhibitors, followed by subsequent heat release of the viral nucleic acid from the viruses and RT-PCR (Table 3). Although direct extraction of nucleic acids is usually a simpler procedure, destruction of the integrity of the virions during this process exposes naked viral RNA to potential degradation and destroys viral infectivity, thus precluding direct comparison with cell culture infectivity.

TABLE 3 Reported sensitivities of detection of pathogens in shellfish by PCR

Pathogen(s)	Sample(s)	Sensitivity	Reference
HAV	Clams	2,000 particles/g (10 PFU/g)	37
Poliovirus	Oysters	38 PFU/20 g (2 PFU/g)	6
V. vulnificus	Oysters	100 CFU/g	42
Poliovirus, HAV	Oysters	10 PFU/50 g (0.02 PFU/g)	46
Norwalk virus	Oysters	4,500 PCRU^a/50 g (90 PCRU/g)	46
HAV	Clams and oysters	100 PFU/1.5 g (67 PFU/g)	7
Norwalk virus	Clams and oysters	5–10 PCRU/1.5 g (3–7 PCRU/g)	7
Poliovirus	Oysters and mussels	10 PFU/5 g (2 PFU/g)	60
HAV	Clams and oysters	Not specified	27
V. cholerae	Oysters	10 CFU/g	57

^{a} PCRU, PCR-amplifiable units.

TABLE 4 Removal of enzymatic inhibitors from shellfish

Pathogen(s)	Sample(s)	Target	Method	Reference
HAV	Clams	RNA	Sodium acetate washes	37
Poliovirus	Oysters	RNA	CTAB[a] precipitation	6
V. vulnificus	Oysters	DNA	Guanidinium isothionate-chloroform extraction	42
Poliovirus, HAV	Oysters	Virions	ProCipitate and PEG	46
	Oysters	Virions	CTAB precipitation	47
HAV	Clams and oysters	Virions	Antibody capture	27
Poliovirus	Oysters and mussels	RNA	Guanidinium isothiocyanate-glass beads	60
V. cholera	Oysters	Cells	Heat lysis	57

[a] CTAB, cetyl trimethyl ammonium bromide.

Removal of PCR Inhibitors from Shellfish Homogenates

Adaptation of PCR amplification techniques to shellfish has proven to be a significant challenge, predominantly because of large concentrations of uncharacterized enzymatic inhibitors associated with shellfish tissue. Nevertheless, several investigators have devised methods either for direct RNA extraction of viral RNA or for isolation of virions from shellfish followed by detection of the RNA by RT-PCR (Table 4). The most popular approach for detection of viruses in environmental and shellfish samples has involved direct RNA extraction. A frequently used method for the preparation of nucleic acid for hybridization and RT-PCR involves extracting RNA directly from shellfish extracts or environmental concentrates by using sodium dodecyl sulfate-proteinase K procedures, phenol-chloroform extraction, and ethanol precipitation (6, 37, 64). Further removal of inhibitors has been accomplished by treatment of crude nucleic acid extracts with the cationic detergent cetyl trimethyl ammonim bromide (6, 7, 97). Briefly, nucleic acids from oyster extracts are obtained by initial digestion with proteinase K (50 μg/ml) in 10 mM Tris HCl (pH 7.5)–5 mM EDTA–0.5% sodium dodecyl sulfate at 56°C for 30 min. The digest is extracted twice with an equal volume of phenol-water-chloroform (68:18:14), and the aqueous phase is precipitated in ethanol. The resulting pellet is suspended in water, to which cetyl trimethyl ammonium bromide and NaCl are added to final concentrations of 1.4% and 0.11 M, respectively. The mixture is incubated at room temperature for 30 min, and the nucleic acids are pelleted by centrifugation at 12,100 × g for 30 min. The pellet is resuspended in 1 M NaCl and precipitated in ethanol. The precipitated nucleic acids are suspended in 100 μl of water and used in RT-PCR in 1- to 20-μl aliquots.

Other investigators found that the guanidinium isothiocyanate-chloroform extraction system produced nucleic acid extracts low in enzymatic inhibitors (42, 60). Although the resulting RNA is usually of relatively high purity after such extractions, disadvantages to this method include incomplete recovery and/or potential degradation of RNA during the extraction procedure and difficulty in making direct correlations with infectivity because of loss of virion integrity. Shieh et al. (79) combined traditional cell culture and molecular techniques by amplifying environmental water concentrates in mammalian cell culture prior to RNA extraction of cell monolayers and probing with single-stranded RNA probes. This approach accomplishes the dual purpose of increasing the number of copies of target nucleic acid and incorporating an infectivity assay. Potential disadvantages to such an approach include the lack of susceptible cell lines for epidemiologically important human enteric viruses and cytotoxicity of shellfish extracts.

An alternative to RNA extraction would involve direct isolation of viruses from environmental and shellfish extracts. Jansen et al. (44) detected HAV in stool specimens by using an antibody capture–RT-PCR approach. Desenclos et al. (28) implicated HAV in oyster outbreak specimens by immunocapture of virus, heat release of RNA, and subsequent RT-PCR, a method that has been further investigated by Deng et al. (27). De Leon et al. (23) demonstrated the direct PCR amplification of Norwalk virus in 10 to 20% of stool samples initially clarified by fluorocarbon extraction and centrifugation and further purified by spin column chromatography with Sephadex G-200. Spin columns were prepared as described by Sambrook et al. (78) except that Silane-treated glass wool was used as the column support. The columns were prepared in 3-ml syringes and spun at 400 × g and room temperature for 4 min. Approximately 75 to 85 μl of sample was passed through a 1-ml column bed (1:12 sample-to-resin ratio) that had been equilibrated with 0.1× Taq buffer II. Viruses were recovered in the excluded volume, which averaged 70 to 80 μl per column; 10 μl was used in each subsequent RT-PCR.

Another direct virion isolation method seeks to concentrate the virions from oyster extracts in a small volume, with removal of enzymatic inhibitors (45–47). In this method, RT-PCR inhibitors were initially removed by two consecutive extractions with equal volumes of Freon. After centrifugation at 6,000 × g and 5°C for 20 min, the resulting supernatant was adjusted to pH 7.3 to 7.4 and 0.3 M NaCl, supplemented with 4 to 12% polyethylene glycol (PEG) 6000, and incubated overnight at 4°C to precipitate viruses. The precipitated viruses were recovered by centrifugation at 6,000 × g and 4°C for 20 min and resuspended in one-seventh the original sample volume, using 50 mM Tris–0.2% Tween 20 (pH 9.0). Resuspended precipitates were held at room temperature for 30 min to aid virus elution and then centrifuged for 15 min at 8,200 × g and room temperature to remove extraneous particulates. The resulting supernatant of 1.5 to 2.0-ml volume was a 10-fold concentrate of the initial extract volume.

Viruses in 1-ml volumes of the PEG eluants were precipitated for 15 min by addition of an equal volume of Pro-Cipitate (Affinity Technology, Inc., New Brunswick, N.J.). The solid phase was recovered by centrifugation for 15 min at room temperature and 13,800 × g. To elute adsorbed viruses, 4 ml of 50 mM Tris–0.2% Tween 20 (pH 9.0) was added, and the sample was gently rotated for 1 h at room temperature and then centrifuged at 6,000 × g and 15°C for 20 min to remove excess Pro-Cipitate. Viruses in the eluant were further concentrated by a secondary PEG precipitation using 5 to 10% PEG and 0.3 M NaCl. After 2 h at 4°C and gentle rotation, the solutions were centrifuged at 6,000 × g and 4°C for 20 min. The virus-containing precipitate was resuspended in one-seventh the original volume, using 50 mM Tris–0.2% Tween 20 (pH 8.0). Ten-microliter aliquots of final concentrates, representing 5 to 10% of the total sample volume, were used in RT-PCR. Using this approach, processed samples could be readily assayed by both RT-PCR and cell culture infectivity.

BARRIERS TO THE APPLICATION OF MOLECULAR METHODS FOR DETECTION OF PATHOGENS IN SHELLFISH

Although prototype molecular methods for the detection of human pathogens in shellfish have been reported, there are still several barriers to overcome before these methods are applicable to routine monitoring (Table 5). To obtain sample representation and detection sensitivity adequate for the low infectious doses of most enteric pathogens, in particular the viruses, large samples of shellfish will need to be processed. While some of the viral detection methods reported in the literature begin with large (50-g) sample sizes, most effectively involve extraction from very small volumes of sample aliquots, which limits the sensitivity of the assay procedure from the very beginning. The approaches then applied to concentrate and purify the pathogens from the homogenate not only need to be reasonably efficient but also need to produce a concentrate that is low in both volume and enzymatic inhibitors. To complicate matters further, the prototype extraction and detection procedures differ for bacteria and viruses, as well as for different shellfish species. It appears that shellfish have extremely high levels of enzymatic inhibitors, with complex carbohydrates (particularly glycogen) being the major culprit. These compounds have remained recalcitrant to almost all removal processes applied. In addition, the type of enzymatic inhibitors may differ with the shellfish species, potentially restricting the use of some methods to a single species of shellfish. Type and concentration of enzymatic inhibitors may also differ with harvesting season and geographic location.

RESEARCH NEEDS FOR THE DETECTION OF PATHOGENS IN SHELLFISH

Research is needed in order to develop and refine the prototype protocols into collaboratively tested methods which could be used routinely and expeditiously to evaluate the

TABLE 5 Barriers to the application of molecular techniques to shellfish

1. Large sample volumes of shellfish are required to obtain the necessary level of detection sensitivity.

2. Extraction procedures differ for each pathogen group.

3. Diverse enzymatic inhibitors are present at high levels.

4. Enzymatic inhibitors may differ with shellfish species and harvesting location.

5. Concentrations of inhibitors may vary with seasons.

TABLE 6 Research needs for application of molecular methods to shellfish extracts

1. Simple and rapid pathogen (or nucleic acid) extraction methods

2. Simple and reliable methods for the removal of enzymatic inhibitors

3. Standardized methods which are not specific to shellfish species or harvesting location

4. Quantitative methods for assessing relative levels of contamination

microbiological safety of shellfish products (Table 6). In general, future research needs for the routine application of PCR methods to the detection of microbiological contamination in shellfish include development of the following: (i) simple, rapid, and cost-effective pathogen extraction procedures; (ii) simple and reliable methods for the removal of enzymatic inhibitors; (iii) methods which are not restricted by shellfish species, harvesting location, or season; and (iv) quantitative approaches for assessing the relative levels of contamination. Furthermore, additional experimentation is needed to establish the relationship between detection by molecular amplification techniques and the presence of infective virus particles in food and environmental samples. With sufficient developmental effort, effective methods for the rapid detection of human pathogenic microorganisms in shellfish at naturally occurring levels of contamination should become available.

REFERENCES

1. **Ando T., S. S. Monroe, J. R. Gentsch, Q. Jin, D. C. Lewis, and R. I. Glass.** 1995. Detection and differentiation of antigenically distinct small round-structured viruses (Norwalk-like viruses) by reverse transcription-PCR and southern hybridization. *J. Clin. Microbiol.* **33:**64–71.

2. **Appleton, H.** 1987. Small round viruses: classification and role in food-borne infections. *Ciba Found. Symp.* **128:**108–125.

3. **Appleton, H., M. Buckley, B. T. Thorn, J. L. Cotton, and S. Henderson.** 1977. Virus-like particles in winter vomiting disease. *Lancet* **i:**409–411.

4. **Appleton, H., and P. G. Higgins.** 1975. Viruses and gastroenteritis in infants. *Lancet* **i:**1297.

5. **Appleton, H., and M. S. Pereira.** 1977. A possible virus aetiology in outbreaks of food-poisoning from cockles. *Lancet* **ii:**780–781.

6. **Atmar, R. L., T. G. Metcalf, F. H. Neill, and M. K. Estes.** 1993. Detection of enteric viruses in oysters by using the polymerase chain reaction. *Appl. Environ. Microbiol.* **59:**631–635.

7. **Atmar, R. L., H. F. Neill, J. L. Romalde, F. Le Guyader, C. M. Woodley, T. G. Metcalf, and M. K. Estes.** 1995. Detection of Norwalk virus and hepatitis A virus in shellfish tissues with the PCR. *Appl. Environ. Microbiol.* **61:**3014–3018.

8. **Bitton, G.** 1980. *Introduction to Environmental Virology.* John Wiley & Sons, New York.

9. **Blacklow, N. R., G. Cukor, M. K. Bedigian, P. Echeveria, H. B. Greenberg, D. S. Schreiber, and J. S. Trier.** 1979. Immune response and prevalence of antibody to Norwalk enteritis virus as determined by radioimmunoassay. *J. Clin. Microbiol.* **10:**903–909.

10. **Blacklow, N. R., R. Dolin, D. S. Fedson, H. DuPont, R. S. Northrop, R. B. Hornick, and R. M. Chanock.** 1972. Acute infectious nonbacterial gastroenteritis: aetiology and pathogenesis. *Ann. Intern. Med.* **76:**993–1008.

11. **Blake, P. A., M. H. Merson, R. E. Weaver, D. G. Hollis, and P. C. Heublin.** 1979. Disease caused by a marine Vibrio. *N. Engl. J. Med.* **300:**1–5.

12. **Bouchriti, N., and S. M. Goyal.** 1993. Methods for the concentration and detection of human enteric viruses in shellfish: a review. *Microbiologica* **16:**105–113.

13. **Bradley, D. W., A. Andjaparidze, and E. H. Cook.** 1988. Etiologic agent of enterically transmitted non-A, non-B hepatitis. *J. Gen. Virol.* **69:**731–738.

14. **Caul, E. O.** 1987. Discussion. Astroviruses: human and animal. *Ciba Found. Symp.* **128:**102–107.

15. **Caul, E. O.** 1987. Discussion. Small round viruses: classification and role in food-borne infections. *Ciba Found. Symp.* **128:**120–125.

16. **Cliver, D. O.** 1988. Virus transmission via foods. *Food Technol.* **42:**241–247.

17. **Cromeans, T., M. D. Sobsey, and H. A. Fields.** 1987. Development of a plaque assay for a cytopathic, rapidly replicating isolate of hepatitis A. *J. Med. Virol.* **22:**45–56.

18. **Cubitt, W. D.** 1987. The candidate caliciviruses. *Ciba Found. Symp.* **128:**126–143.

19. **Cubitt, W. D.** 1994. Caliciviruses, p. 549–568. *In* A. Z. Kapikian (ed.), *Viral Infections of the Gastrointestinal Tract.* Marcel Dekker, Inc. New York.

20. **Cukor, G., and N. R. Blacklow.** 1984. Human viral gastroenteritis. *Microbiol. Rev.* **48:**157–179.

21. **De Leon, R.** 1989. Use of gene probes and an amplification method for the detection of rotaviruses in water. Ph.D. dissertation. University of Arizona, Tucson.

22. **De Leon, R., and C. P. Gerba.** 1990. Viral disease transmission by seafood, p. 639–662. *In* J. O. Nraigu and M. S. Simmons (ed.), *Food Contamination from Environmental Sources.* John Wiley & Sons, Inc., New York.

23. **De Leon, R., S. M. Matsui, R. S. Baric, J. E. Herrmann, N. R. Blacklow, H. B. Greenberg, and M. D. Sobsey.** 1992. Detection of Norwalk virus in stool specimens by reverse transcriptase-polymerase chain reaction and nonradioactive oligoprobes (RT-PCR-OP). *J. Clin. Microbiol.* **30:**3151–3157.

24. **De Leon, R., H. A. Payne, and C. P. Gerba.** 1986. Development of a method for poliovirus detection in freshwater clams. *Food Microbiol.* **3:**345–349.

25. **De Leon, R., C. Shieh, R. S. Baric, and M. D. Sobsey.** 1990. Detection of enteroviruses and hepatitis A virus in environmental samples by gene probes and polymerase chain reaction. Presented at the Water Quality Technology Conference, American Water Works Association, San Diego, Calif.

26. **De Leon, R., and M. D. Sobsey.** 1991. Virus detection in water: a review of current and proposed methods, p. 400–421. *In* J. R. Hall and D. Glysson (ed.), *Monitoring Water in the 1990's: Meeting New Challenges.* ASTM STP 1102. American Society for Testing and Materials, Philadelphia.

27. **Deng, M. Y., S. P. Day, and D. O. Cliver.** 1994. Detection of hepatitis A virus in environmental samples by antigen-capture PCR. *Appl. Environ. Microbiol.* **60:**1927–1933.

28. **Desenclos, J. C. A., K. C. Klontz, M. H. Wilder, O. V. Nainan, H. S. Margolis, and R. A. Gunn.** 1991. A multistate outbreak of hepatitis A caused by the consumption of raw oysters. *Am J. Public Health Assoc.* **81:**1268–1272.

29. **Dimitrov, D. J., D. Y. Graham, and M. K. Estes.** 1985. Detection of rotavirus by nucleic acid hybridization with cloned DNA of simian rotavirus SA-11 genes. *J. Infect. Dis.* **152:**293–300.

30. **Dolin, R., C. Reichman, K. D. Roessner, T. S. Tralka, R. T. Scooley, W. Gary, and D. Morens.** 1982. Detection by immune electron microscopy of the Snow Mountain Agent of acute viral gastroenteritis. *J. Infect. Dis.* **146:**184–189.

31. **Dubrou, S., H. Kopecka, J. M. Lopez-Pila, J. Marechal, and J. Prevot.** 1990. Detection of hepatitis A virus and other enteroviruses in wastewater samples by gene probe assays. Presented at the International Symposium on Health Related Microbiology, Tübingen, Germany.

32. **Elliot, E. L., C. A. Kaysner, and M. L. Tamplin.** 1992. *V. cholerae, V. parahaemolyticus, V. vulnificus,* and other Vibrio spp., p. 111–140. *In FDA Bacteriological Analytical Manual,* 7th ed. AOAC International, Arlington, Va.

33. **Frost, H. W.** 1925. Report of committee on the sanitary control of the shellfish industry in the United States. *Public Health Rep. Suppl.* **53:**1–17.

34. **Gerba, C. P.** 1988. Viral disease transmission by seafoods. *Food Technol.* **42:**99–103.

35. **Gerba, C. P., and S. M. Goyal.** 1978. Detection and occurrence of enteric viruses in shellfish: a review. *J. Food Prot.* **41:**743–754.

36. **Gill, O. N., W. D. Cubitt, O. A. McSwiffan, B. M. Watney, and C. L. R. Bartlett.** 1983. Epidemic of gastroenteritis caused by oysters contaminated with small round structured viruses. *Br. Med. J.* **287:**1532–1534.

37. **Goswami, B. B., W. H. Koch, and T. A. Cebula.** 1993. Detection of hepatitis A in *Mercenaria mercenaria* by coupled reverse transcription and polymerase chain reaction. *Appl. Environ. Microbiol.* **59:**2765–2770.

38. **Gouvea, V., R. I. Glass, P. Ward, K. Taniguchi, H. F. Clark, B. Forrester, and Z. Y. Fang.** 1990. Polymerase chain reaction amplification and typing of rotavirus nucleic acid from stool specimens. *J. Clin. Microbiol.* **28:**276–282.

39. **Goyal, S. M.** 1984. Viral pollution of the marine environment. *Crit. Rev. Environ. Control* **14:**1–32.

40. **Gunn, R. A., H. T. Janowski, S. Lieb, E. C. Prather, and H. Greenberg.** 1982. Norwalk virus gastroenteritis following raw oyster consumption. *Am. J. Epidemiol.* **115:**348–351.

41. **Guzewich, J. J., and D. L. Morse.** 1986. Sources of shellfish in outbreaks of probable viral gastroenteritis: implications for control. *J. Food Prot.* **49:**389–394.

42. **Hill, W. E., S. P. Keasler, M. W. Truckess, P. Feng, C. A. Kayser, and K. A. Lampel.** 1991. Polymerase chain reaction identification of *Vibrio vulnificus* in artificially contaminated oysters. *Appl. Environ. Microbiol.* **57:**707–711.

43. **Hlady, W. G., R. C. Mullen, and R. S. Hopkins.** 1993. *Vibrio vulnificus* from raw oysters: leading cause of reported deaths from illness in Florida. *J. Fla. Med. Assoc.* **80:**536–538.

44. **Jansen, R. W., G. Seigl, and S. M. Lemon.** 1990. Molecular epidemiology of human hepatitis A virus defined by antigen-capture polymerase chain reaction. *Proc. Natl. Acad. Sci. USA* **87:**2867–2871.

45. **Jaykus, L., R. DeLeon, and M. D. Sobsey.** 1993. Application of RT-PCR for the detection of enteric viruses in oysters. *Water Sci. Technol.* **27:**49–53.

46. **Jaykus, L., R. DeLeon, and M. D. Sobsey.** A virion concentration method for detection of human enteric viruses in oysters by PCR and oligoprobe hybridization. Submitted for publication.

47. **Jaykus, L., R. DeLeon, and M. D. Sobsey.** 1995. Development of a molecular method for the detection of enteric viruses in oysters. *J. Food Prot.* **58:**1357–1362.

48. **Jaykus, L., M. T. Hemard, and M. D. Sobsey.** 1994. Human enteric pathogenic viruses, p. 92–153. *In* C. R.

Hackney and M. D. Pierson (ed.), *Environmental Indicators and Shellfish Safety*. Chapman and Hall, New York.

49. **Jiang, X., M. K. Estes, and T. G. Metcalf.** 1987. Detection of hepatitis A virus by hybridization with single-stranded RNA probes. *Appl. Environ. Microbiol.* **53:**2487–2495.

50. **Jiang, X., M. K. Estes, T. G. Metcalf, and J. L. Melnick.** 1986. Detection of hepatitis A virus in seeded estuarine samples by hybridization with cDNA probes. *Appl. Environ. Microbiol.* **52:**711–717.

51. **Jiang, X. N., D. Y. Graham, K. N. Wang, and M. K. Estes.** 1990. Norwalk virus genome cloning and characterization. *Science* **250:**1580–1583.

52. **Kapikian, A. Z., and R. M. Chanock.** 1990. Norwalk group of viruses, p. 671–693. *In* B. N. Fields (ed.), *Virology*. Raven Press, Ltd., London.

53. **Kapikian, A. Z., H. B. Greenberg, W. L. Cline, A. R. Kalica, R. G. Wyatt, H. J. James, Jr., N. L. Lloyd, R. M. Chanock, R. W. Ryder, and H. W. Kim.** 1978. Prevalence of antibody to the Norwalk agent by a newly developed immune adherence haemagglutination assay. *J. Med. Virol.* **2:**281–294.

54. **Kapikian, A. Z., R. G. Wyatt, R. Dolan, T. Thornhill, A. R. Kalica, and R. M. Chanock.** 1972. Visualization by immune electron microscopy of a 27-nm particle associated with acute infectious nonbacterial gastroenteritis. *J. Virol.* **10:**1075–1081.

55. **Kaplan, J. E., G. W. Gary, R. C. Baron, N. Singh, L. B. Schonberger, R. Feldman, and H. B. Greenberg.** 1982. Epidemiology of Norwalk gastroenteritis and the role of Norwalk virus in outbreaks of acute nonbacterial gastroenteritis. *Ann. Intern. Med.* **96:**756–761.

56. **Kaysner, C. A., M. L. Tamplin, and R. M. Twedt.** 1992. Vibrio, p. 451–473. *In* C. Vanderzant and D. F. Splittstoesser (ed.), *Compendium of Methods for the Microbiological Examination of Foods*. American Public Health Association, Washington, D.C.

57. **Koch, W. E., W. L. Payne, B. A. Wentz, and T. A. Cebula.** 1993. Rapid polymerase chain reaction method for detection of *Vibrio cholerae* in foods. *Appl. Environ. Microbiol.* **59:**556–560.

58. **Kurtz, J. B., T. W. Lee, J. W. Craig, and S. E. Reed.** 1979. Astrovirus infection in volunteers. *J. Med. Virol.* **3:**321–330.

59. **Lednar, W. M., S. M. Lemon, J. W. Kirkpatrick, R. R. Redfield, M. L. Fields, and P. W. Kelley.** 1985. Frequency of illness associated with epidemic hepatitis A virus infections in adults. *Am. J. Epidemiol.* **122:**226–233.

60. **Lees, D. N., K. Henshilwood, and W. J. Dore.** 1994. Development of a method for detection of enteroviruses in shellfish by PCR with poliovirus as a model. *Appl. Environ. Microbiol.* **60:**2999–3005.

61. **Levine, W. C., and P. M. Griffin.** 1993. Vibrio infections on the Golf Coast: results of first year of regional surveillance. Gulf Coast Vibrio Working Group. *J. Infect. Dis.* **167:**479–483.

62. **Madeley, C. R.** 1979. Viruses in the stools. *J. Clin. Pathol.* **32:**1–10.

63. **Madeley, C. R., and B. P. Cosgrove.** 1975. 28 nm particles in faeces in infantile gastroenteritis. *Lancet* **ii:**451–452.

64. **Margolin, A. B., M. J. Hewlett, and C. P. Gerba.** 1986. Use of a cDNA dot-blot hybridization technique for detection of enteroviruses in water, p. 87–95. *In Water Quality Technology Conference Proceedings*. American Water Works Association, Denver. Colo.

65. **Melnick, J. L., and C. P. Gerba.** 1980. The ecology of enteroviruses in natural waters. *Crit. Rev. Environ. Control* **10:**65.

66. **Morse, D. L., J. J. Guzewich, J. P. Hanrahan, R. Stricof, M. Shayegani, R. Deibel, J. C. Grabau, N. A. Nowak,** J. E. Herrmann, G. Cukor, and N. R. Blacklow. 1986. Widespread outbreaks of clam- and oyster-associated gastroenteritis. *N. Engl. J. Med.* **314:**678–681.

67. **Mosley, J. W.** 1967. Transmission of viral disease by drinking water, p. 5–23. *In* G. Berg (ed.), *Transmission of Viruses by the Water Route*. Interscience, New York.

68. **Murphy, A. M., G. S. Grohmann, P. J. Christopher, W. A. Lopez, G. R. Davey, and R. H. Millsom.** 1979. An Australian-wide outbreak of gastroenteritis from oysters caused by Norwalk virus. *Med. J. Aust.* **2:**329–333.

69. **O'Mahony, M. C., C. D. Gooch, D. A. Smyth, A. J. Thrussel, C. L. R. Bartlett, and N. D. Noah.** 1983. Epidemic hepatitis A from cockles. *Lancet* **i:**518.

70. **Ou, C. Y., S. Kwok, S. W. Mitchell, D. H. Mack, J. J. Sninsky, J. W. Krebs, P. Feorino, D. Warfield, and G. Schochetman.** 1988. DNA amplification for direct detection of HIV-1 in DNA of peripheral blood mononuclear cells. *Science* **239:**295–297.

71. **Provost, P. J., and M. R. Hilleman.** 1979. Propagation of human hepatitis A virus in cell culture in vitro. *Proc. Soc. Exp. Biol. Med.* **160:**213–221.

72. **Reyes, G. R., M. A. Purdy, and J. P. Kim.** 1990. Isolation of DNA from the virus responsible for enterically transmitted non-A, non-B hepatitis. *Science* **247:**1335–1339.

73. **Richards, G. P.** 1985. Outbreaks of shellfish-associated enteric virus illness in the United States: requisite for development of viral guidelines. *J. Food Prot.* **48:**815–823.

74. **Rippey, S. R.** 1994. Infectious diseases associated with molluscan shellfish consumption. *Clin. Microbiol. Rev.* **7:**419–425.

75. **Rodgers, F. G.** 1981. Concentration of viruses in faecal samples from patients with gastroenteritis, p. 15–18. *In* M. Goddard and M. Buttler (ed.), *Viruses and Wastewater Treatment*. Pergamon Press, New York.

76. **Rotbart, H. A.** 1990. Enzymatic amplification of the enteroviruses. *J. Clin. Microbiol.* **28:**438–442.

77. **Saiki, R. K., D. H. Gelfand, S. Stoffel, S. J. Scharf, R. Higuchi, G. T. Horn, K. B. Mullis, and H. A. Erlich.** 1988. Primer-directed enzymatic amplification of DNA with a thermostable DNA polymerase. *Science* **239:**487–491.

78. **Sambrook, J., E. F. Fritsch, and T. Maniatis.** 1989. *Molecular Cloning: a Laboratory Manual*, 2nd ed. Cold Spring Harbor Laboratory Press, Plainview, N.Y.

79. **Shieh, Y. S. C., R. S. Baric, M. D. Sobsey, J. Ticehurst, T. A. Miele, R. DeLeon, and R. Walter.** 1991. Detection of hepatitis A virus and other enteroviruses in water by RNA probes. *J. Virol. Methods* **31:**119–136.

80. **Slade, J. S., and B. J. Ford.** 1983. Discharge to the environment of viruses in wastewater, sludges and aerosols, p. 3–18. *In* G. Berg (ed.), *Viral Pollution of the Environment*. CRC Press, Inc., Boca Raton, Fla.

81. **Sobsey, M. D.** 1982. Detection of viruses in shellfish, p. 243–259. *In* C. P. Gerba and S. M. Goyal (ed.), *Methods in Environmental Virology*. Marcel Dekker, New York.

82. **Sobsey, M. D.** 1985. Procedures for the virological examination of seawater, shellfish and sediment, p. 81–118. *In* A. E. Greenberg and D. A. Hunt (ed.), *Laboratory Procedures for the Examination of Seawater and Shellfish*. American Public Health Association, Washington, D.C.

83. **Sobsey, M. D.** 1987. Methods for recovering viruses from shellfish, seawater and sediments, p. 77–108. *In* G. Berg (ed.), *Methods for Recovering Viruses from the Environment*. CRC Press, Inc., Boca Raton, Fla.

84. **Sobsey, M. D., R. J. Carrick, and H. R. Jensen.** 1978. Improved methods for detecting enteric viruses in oysters. *Appl. Environ. Microbiol.* **36:**121–130.

85. **Tabor, E.** 1984. Clinical presentation of hepatitis A, p.

47–53. *In* A. R. J. Gerety (ed.), *Hepatitis A*. Academic Press, New York.

86. **Tacket, C. O., F. Brenner, and P. A. Blake.** 1984. Clinical features and an epidemiological study of *Vibrio vulnificus* infections. *J. Infect. Dis.* **149:**558–561.

87. **Thornhill, T. S., R. G. Wyatt, A. R. Kalica, R. Dolin, R. M. Chanock, and A. Z. Kapikian.** 1977. Detection by immune electron microscopy of 26 to 27 nm virus-like particles associated with two family outbreaks of gastroenteritis. *J. Infect. Dis.* **135:**20–27.

88. **Truman, B. L., H. P. Madore, M. A. Menengus, J. L. Nitzkin, and R. Dolin.** 1987. Snow Mountain agent gastroenteritis from clams. *Am. J. Epidemiol.* **126:**516–525.

89. **Vaughn, J. M., E. F. Landry, M. Z. Thomas, T. J. Vicale, and W. F. Panello.** 1980. Isolation of naturally occurring enteroviruses from a variety of shellfish species residing in Long Island and New Jersey marine embayments. *J. Food Prot.* **43:**95–98.

90. **Verber, J. L.** 1984. *Shellfish Borne Disease Outbreaks*, p. 6–14. U.S. Public Health Service, Food and Drug Administration, Davisville, R.I.

91. **Wang, J., X. Jiang, H. P. Madore, J. Gray, U. Desselberger, T. Ando, Y. Set, I. Oishi, J. F. Lew, K. Y. Green, and M. K. Estes.** 1994. Sequence diversity of small, round-structured viruses in the Norwalk virus group. *J. Virol.* **68:**5982–5990.

92. **Ward, D. R., and C. R. Hackney (ed.).** 1991. *Microbiology of Marine Food Products*. Van Nostrand Reinhold, New York.

93. **Wieneke, A. A., D. Roberts, and R. J. Gilbert.** 1993. Staphylococcal food poisoning in the United Kingdom, 1969–90. *Epidemiol. Infect.* **110:**519–31.

94. **World Health Organization.** 1974. *Fish and Shellfish Hygiene*, p. 8. Technical Report Series No. 550. World Health Organization, Geneva.

95. **Wyatt, R. G., R. Dolin, N. R. Blacklow, H. DuPont, R. Buscho, T. S. Thornhill, A. Z. Kapikian, and R. M. Chanock.** 1974. Comparison of three agents of acute infectious nonbacterial gastroenteritis by cross-challenge in volunteers. *J. Infect. Dis.* **129:**709–714.

96. **Xu, Z. Y., Z. H. Li, J. X. Wang, Z. P. Xiao, and D. X. Dong.** 1992. Ecology and prevention of a shellfish-associated hepatitis A epidemic in Shangai, China. *Vaccine* Suppl. **1:**S67–S68.

97. **Zhou, Y., M. K. Estes, X. Jiang, and T. G. Metcalf.** 1991. Concentration and detection of hepatitis A virus and rotavirus from shellfish by hybridization tests. *Appl. Environ. Microbiol.* **57:**2963–2968.

Modeling the Fate of Microorganisms in Water, Wastewater, and Soil

CHRISTON J. HURST

22

The natural environment is filled with microorganisms, most of which are natural residents and colonize various ecological niches. These microorganisms either live independently within the environment or live in association with various host organisms. There also are places where and times when microorganisms are released into the environment as the result of human activities. Sometimes these microorganisms are released with the goal that they will degrade noxious or toxic wastes. At other times, microorganisms are intentionally released with the goal that they will compete against natural microflora which cause disease in plants or animals. Humans also intentionally release microorganisms such as *Bacillus thuringiensis* and polyhedrosis viruses that are intended to reduce and control populations of insects. Also released are microorganisms that are pathogenic to humans, as often manifested through normal bodily functions such as sneezing, coughing, and fecal excretion. There also is a fear that human pathogens may intentionally be released for purposes of biological warfare.

Those released microorganisms which are associated with human fecal wastes often are directly deposited on the land surface. They also may enter the terrestrial environment as infiltrates or permeates from outhouses and septic tanks, and they are present in raw municipal wastewaters, wastewater effluents, and wastewater sludges which may be discharged onto the land surface for fertilizing and irrigating crops. Wastewater effluents sometimes are applied to the land surface with the intention that they will eventually percolate into and recharge aquifers, while at other times treated wastewater effluents are directly injected into subsurface aquifers. Just as these surface-applied wastewaters will percolate into groundwaters, so too can contaminated groundwater find its way back to the surface via springs and seeps as a result of subsurface flow patterns and pressure gradients. We also discharge wastewater directly into surface waters. The end result is that regardless of where we apply the contaminants, water flow patterns can distribute them to other locations which may be far distant. Included among the microorganisms in fecal wastes are many that are pathogenic for humans, and if these pathogenic organisms survive their environmental exposure long enough, they may be encountered by new human hosts and thereby complete an environmental cycle of disease transmission: from humans, through the environment, and back to humans.

There is an interest in understanding the aquatic and terrestrial fate of naturally resident microorganisms. Likewise, we as humans also have an interest in monitoring the fate of those microorganisms released into the environment as a consequence of our own activities. Understanding the fate of microorganisms following either accidental or deliberate release into aquatic and terrestrial environments is of key importance when we try to understand and address the potential impact of those microorganisms on the environment. The goal of this chapter is to explain some of the quantitative approaches which are applicable to understanding the fate of microorganisms after their release into water and soil.

DILUTION AND ENVIRONMENTAL TRANSPORT OF CONTAMINANTS

Once microorganisms are released into the environment, their fate becomes subject to the whims of that environment. The most immediately noticeable consequence is that organisms become dispersed as they are added into water or soil. To some extent, microorganisms can be transported through the environment along with the soil, either as wind-carried particles or as material that is carried downslope in landslides. The movement of released organisms is greatly facilitated by the flow of water, which can occur either on the surface (2, 4) or in the subsurface (20, 25).

Because microbes have an intrinsic electrostatic charge, they will adsorb to the surface of charged environmental particulates (6, 17). This is an important factor, because adsorption to particulates can enhance the ability of microorganisms to survive in the environment (12). In the subsurface, adsorption to the matrix acts to retard the transport of microorganisms (19). Microbes will also become adsorbed to the matrix as they are transported by flowing subsurface waters. Additionally, microorganisms can adsorb to gas-water interfaces by hydrophobic mechanisms, a process which happens in unsaturated subsurface zones (23). In surface flows, the tendency of microorganisms to adsorb onto suspended particulates facilitates the sedimentation of those organisms during periods of low water flow. These sediments can then serve as a reservoir of microbes, with subsequent souring of stream bottoms during times of turbulent water flow resulting in resuspension of the particulate-associated microbes (24). Microbes contained in surface water flows may also be removed by impaction onto the sides or bottom of the channel.

Microorganisms released into the environment become susceptible to inactivation by a variety of physical and chemical factors such as desiccation (13) and temperature- or pH-related effects on their biomolecules (12, 18), which may include denaturation, radiation from sunlight (1, 21), and effects of inorganic ions (12). Released populations of microbes may also be subjected to biochemical antagonism by microbial products (8) such as enzymes and subjected to predation (7) by natural environmental microorganisms. Some released microorganisms can replicate within the natural environment, while others cannot. Under appropriate conditions, microbes may persist for several months and perhaps even years following their accidental or deliberate release into aquatic and terrestrial environments.

DEVELOPING AND PERFORMING MICROBIAL SURVIVAL STUDIES

From a biological and ecological standpoint, it is always best to study an indigenous population which is openly exposed within the natural environment and free to move around within that environment. This approach often can be used for macroorganisms. Individual plants can be identified, and reidentified as necessary, because they are held stationary by their roots; animals are identifiable either by natural or artificially applied bodily markings or by the use of tagging devices which emit electronic signals. However, microbes are generally too small to be monitored as individuals, especially when allowed unrestricted movement within the natural environment. Thus, we often must resort to compromises when it comes to experimental design.

Designing the Experiments

A series of decisions must be made when one is planning a microbial survival study. The major decisions are whether to study the fate of seeded organisms or an indigenous population and whether the organisms will be studied in place within the natural environment, in environmental material brought into the laboratory, or in an entirely artificial system.

Physical Considerations

The most important aspect of studies designed to evaluate the fate of microorganisms is to define the study population, which implies controlling or understanding any possible movement of individuals into or out of that population. Studies of microbial persistence are most manageably performed in the laboratory, although even when this is done with natural environmental materials that have been brought into the laboratory, the results are not guaranteed to be identical to findings that would be obtained from observation of natural organisms freely moving in nature and interacting with other organisms within their ecosystem. The other popular alternative is to study natural or seeded organisms under a constrained natural exposure. This can be done by placing organisms into the environment under conditions which restrain their movement, such as by using membrane filter chambers (15, 16, 22) or membrane filter sandwiches (18), artificial basins (13), or even intravenous bags (3). However, it is important to understand the possible impact of any imposed barriers which would limit the metabolic and physiologic activities of the test microbial population. Barriers may affect exposure of the test population to environmental factors such as the movement or loss of moisture, gas exchange, movement of chemicals, and radiation from sunlight. Conversely, one of the most important aspects of studies which are intended to mimic environmental conditions within laboratory confines is to first study the natural environment in an attempt to assess and duplicate within the laboratory the effects of natural factors.

Statistical Considerations

When the ultimate goal is to mathematically model the observed fate of microorganisms, it is best to design the experiments in such a way that the data will be amenable to statistical analyses. Also, it must be recognized that variability within the data set being modeled is a crucial factor which can cause even the best-intended analyses to produce spurious or possibly even erroneous results. Some sources of variability can be eliminated by the use of appropriate techniques in designing and performing the experiments. Techniques for eliminating unwanted error include, whenever possible, monitoring concurrently and in a single place the fate of the subject microorganisms in each of the candidate sample materials being examined and under all possible combinations of the test conditions. This approach provides a full factorial design, which ensures the acquisition of a complete set of data upon which to base statistical models, and also helps to eliminate some of the variation which would occur if the combinations of conditions were examined at different times or in different places. If it is not possible to test all combinations of conditions simultaneously, then the order in which the conditions are tested should be chosen at random. Also, whenever possible, it is preferable to keep the experimental conditions unchanged during the course of the observation (experimentation) period. For example, if at the start of a survival study the conditions are established as being a series of time point observations for monitoring the survival of organisms in soil, performed at 23°C and with 30% relative humidity, then both the relative amount of moisture in the system and the temperature should be held constant for the entire period of observation. Otherwise, even if the total amount of water is held constant, changes in temperature will vary the relative humidity. Maintaining a constancy of conditions during the course of the observation period also greatly simplifies the subsequent task of modeling the data. Another important means of eliminating error at the design stage of experiments consists of reducing the amount of contaminating materials, such as deleterious chemicals or even potential microbial nutrients, that inadvertently may be added to the experimental system as a part of the suspending fluid in which the microorganisms are introduced.

Many sources of variability cannot be eliminated, although their impact on the body of data may be controlled to some extent through randomization. One such source of variability is sampling error. The material being studied should be thoroughly mixed before each sample is removed. Alternatively, if the material has been aliquoted before the survival study is begun or if the quantity of material being studied is very large, then selection of aliquots or sampling locations should be done in a randomized manner. All samples must be treated identically, including the performance of any elutions or special handling processes. Daily variability in the sensitivity of sample processing and microbial assay techniques can contribute a substantial amount of variation to the body of data. For this reason, the order in which related samples are processed and assayed may become critical. It is preferable that all samples be processed and assayed either as a single batch or else on a completely randomized schedule. Alternatively, all samples representing the same sampling date should be processed and assayed simultaneously. For example, if three sets of experimental conditions are being studied, with each sampled at time points of 0, 1, 2, 3, 4, and 5, then ideally the samples from all 18 time points should be processed and assayed simultaneously to help reduce day-to-day variation associated with the processing and assay procedures. Alternatively, the 18 samples could be processed and assayed independently of one another to help ensure that variation due to reproducibility (or sensitivity) of the processing and assaying techniques is randomly inserted within the data set. Another possibility is to process and assay all time zero samples as a batch and likewise to batch all 1's together, all 2's together, etc. This latter approach can result in artificially high and at times apparently erratic estimates of the changes in microbial titer that occur between sequential sampling times. However, this approach gives a truer estimation of the statistical differences between the sets of conditions. The worst alternative would be to process and assay all samples from the first combination of conditions as one batch, all samples from the second set of conditions as another batch, and all samples from the third set of conditions as yet another batch. While this last alternative yields aesthetically pleasing graphs of the observed titer changes with respect to observation time, it is faulted because it hopelessly intermixes the variability caused by differences between experimental conditions and the batch variability associated with the processing and assay techniques.

Designing the Modeling Analysis

Science is, of course, based on careful experimentation and observation of produced results. Model equations can be of great value by helping us to understand the interactions which led to the observed results and then providing a basis for anticipating the future outcome of similar situations. The most important thing to remember in developing models is that they are only estimates and have validity within specific ranges. One must also keep in mind the fact that models can only approximate the relationship between their dependent and independent variables, and the possibility that variability which exists within the data set can interfere with attempts at successful modeling efforts. While all of the models presented in this chapter are empirical in nature, it still must be recognized that there are questionable assumptions involved in making statements of cause-and-effect relationships based on these models. With these caveats in mind, this section will focus on the use of regression analysis to develop statistical models that describe temporal changes in microbial titer as a function of experimental parameters and water or soil characteristics.

A microbial population can either increase, remain constant, or decrease in numbers during the course of an observation period, depending on environmental factors such as temperature, desiccation, nutrients, and environmental transport. Because of the small physical size of most microorganisms, the results of microbial fate studies are commonly presented and discussed in terms of the microbial population taken as a whole. Most often, those populations of microorganisms that are released into the natural environment are fated to die there. Free-living microorganisms, including algae plus most bacteria, fungi, and protozoa, that are released into the environment usually cannot successfully compete for nutrients with the populations of organisms that already are established in that environment. Other microorganisms, such as viruses and some protozoa, whose rep-

lication is dependent on finding a suitable host organism may succeed in establishing an ongoing infection within the host population, but even then, most of the microorganisms released from an infected host are fated to die before they establish suitable contact with a new host. The remainder of this chapter is devoted to mathematically examining the rate at which populations of microorganisms die away following their release into the natural environment. The die-off of a microbial population can be described as analogous to the decay rates associated with radioisotopes, in that the frequency of individual death or decay events is assumed to occur at a statistically calculable rate. Because of this analogy, the rapidity with which the members of a microbial population die off is often expressed as a rate function, and the results can be termed microbial population decay rates (14).

The Basis of Survival Calculations

The following is a general exponential decay equation in which N represents the number, or concentration, of microorganisms, t is time, and K is a rate constant:

$$\frac{dN}{dt} = -KN \qquad (1)$$

The sign associated with the rate constant is negative, indicating a net loss in titer with respect to time. According to this equation, the amount of microbial titer decrease observed within a given time period is dependent on the number of organisms present at the start of that time period. Integration of equation 1 yields

$$\frac{N_t}{N_0} = e^{-Kt} \qquad (2)$$

where N_0 is the titer at time zero (the outset of the experiment or series of observations) and N_t represents the titer at some subsequent observation time. Performing a logarithmic transformation of equation 2 yields equation 3:

$$\log_n\left(\frac{N_t}{N_0}\right) = -Kt \qquad (3)$$

Equation 3 suggests that the rate of microbial die-off during the interval between times zero and t is log linear with a constant value equal to K.

Linear Regression

The relationship shown in equation 3 can be examined by using simple linear regression as defined by equation 4:

$$Y = B_0 + B_1X_1 \qquad (4)$$

Whenever simple linear regression is used, Y represents the dependent, or response, variable; X represents the independent variable; B_1 is the coefficient associated with X_1, which results from regressing Y with respect to X_1; and B_0 represents the y-axis intercept value, which can also be viewed as an error term. Linear regression is a concept which is simple to understand and produces results that can be readily visualized and easily compared. It must be understood that linear regression imposes constraints on the developed model equations and will not always provide the best fit in terms of describing the relationship between the dependent and independent variables. Nevertheless, linear regression can be a very valuable tool in the study of microbial activity. Interrelations between the independent variables should either be eliminated from the models or else recognized and

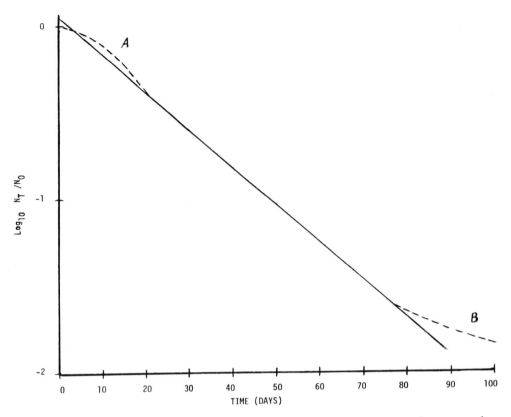

FIGURE 1 Visual presentation of equation 5. $\log_{10}$-transformed titer ratio values are used as the dependent variable; these values are regressed linearly with respect to time, t, as the independent variable. The solid line is the slope, B_t, which represents a rate expressed as $[\log_{10}(N_t/N_0)]/t$. B_0 is the point where the solid line intercepts the y axis. The dashed lines demonstrate deviations from log linearity as a result of shouldering (A) and tailing (B).

their influence understood within the general framework of the developed models.

Linear regression analysis of microbial survival studies is usually performed with the dependent variable being log base$_{10}$ reduction in titer since the initial time of observation versus elapsed time as the independent variable, and the regression is performed by a least-squares technique. This application of simple linear regression to data from microbial survival studies is shown in equation 5:

$$\log_{10} \frac{N_t}{N_0} = B_0 + B_t t \qquad (5)$$

This equation expresses Y in terms of titer change following a $\log_{10}$ transformation [presented as $\log_{10}(N_t/N_0)$] as a function of time, t. Figure 1 shows how this equation appears when graphed. The slope of the solid line, B_1, is expressed in terms of $[\log_{10}(N_t/N_0)]/t$ and is analogous to K. The dashed lines shown in Fig. 1 illustrate the effects described as shouldering, where a lag time exists before the die-off rate becomes exponential, and tailing, where die-off becomes less than the exponential rate at the end of the observation period. Individual plots of data from microbial die-off experiments may express shouldering, tailing, both effects, or neither effect.

Figure 2 presents the use of linear regression as defined by equation 5 to describe the loss of titer for indigenous viruses contained in aerobically digested wastewater sludge following land surface disposal. The individual points shown

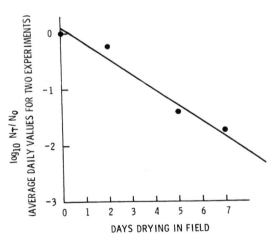

FIGURE 2 Graph of equation 5 applied to data points representing averages from two experiments that followed the die-off of natural virus populations in aerobically digested wastewater sludge solids after land surface disposal (11).

on the graph represent average values from two independent trials each for days 1, 2, and 7 and a single value for day 5. This figure shows that when replicate survival experiments are performed and analyzed, the experimental results may be approximated reasonably well by using the log-linear approach represented by equation 5.

Figure 2 shows that when microbial survival is expressed as the $\log_{10}$ of the titer ratio N_t/N_0 (called the survival ratio), the titer at the beginning of the experiment (time zero) is plotted as an N_t/N_0 value of zero. A $\log_{10} N_t/N_0$ value of -1 then equals 10% of the initial titer, a value of -2 equals 1% of the initial titer, and a value of -3 equals 0.1% of the initial titer. On this particular graph, the units of time are presented as days. However, depending on the duration of any given experiment, the appropriate time units might range from seconds to even years.

The graphed function represented by the linear regression line shown in Fig. 2 does not pass through any of the graphed points. In fact, these graphed points possibly suggest a sigmoidal equation, the result of both shouldering and tailing. However, using the simple linear regression approach for modeling microbial die-off often suffices to describe the data and can yield a great deal of information regarding those environmental factors and processes which led to the observed experimental findings. A measure of the appropriateness of using this type of linear regression approach can be derived by assessing the values of r and P calculated for the regressions performed on the $\log_{10}$-transformed data from individual experiments. Very often, the calculated values for r are greater than 0.9 and can range up to greater than 0.98; the corresponding P values are often less than 0.0001 (10). In most cases, the calculated B_0 value is not zero. This probably often reflects the result of random variation in the data points and also the outcome of at-tempting to use a linear equation format to accommodate data that demonstrate nonlinear traits such as shouldering and tailing. Also, when the change in titer of the population is very small or the titer does not change at all with respect to time, then the sign of the value B_t can sometimes be positive. This presumably results from an impact of random variation obscuring any clear trend in titer decrease.

Figure 3 presents an example of what microbial survival data can look like when the results of individual survival experiments are plotted in this fashion. Notice that the actual graphed values are connected sequentially in a dot-to-dot manner. The only time when a line that does not connect the graphed points in a dot-to-dot manner should be drawn is when that line represents the plot of a calculated mathematical function. In Fig. 2, the rate of viral inactivation reflects the influence of numerous uncontrolled environmental factors such as drying, thermal inactivation, and microbial antagonism. These factors can be sorted out if the incubation conditions are experimentally controlled. Sets of slope values derived from linearly regressing the results of individual survival experiments, conducted under controlled survival conditions, can be compared by using t tests and analysis of variance to determine whether differences in experimental conditions had a statistically significant effect on the rate of microbial die-off (8, 10).

The type of linear regression shown in equation 5 can also be used to determine whether data sets derived from two different studies are comparable and thus can potentially be combined into a single data set. Table 1 shows such an example, comparing the data set represented in Fig. 2 (11) with a data set generated by Farrah and coworkers (5). Both of these studies examined the fate of indigenous viruses in aerobically digested sludge solids following land surface disposal. These two studies were performed during the same

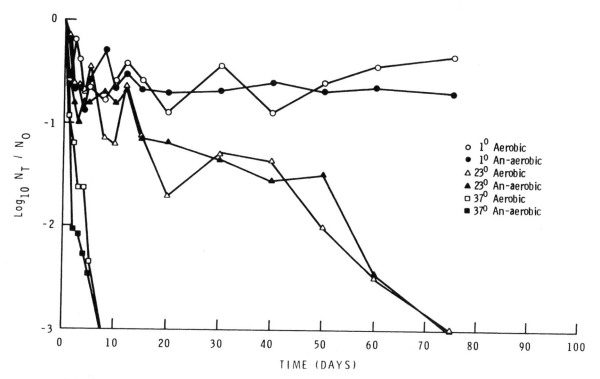

FIGURE 3 Plot of data points for the comparative survival of seeded viruses incubated in sterile soil (12).

TABLE 1 Regression equations describing data for inactivation of indigenous viruses following land surface spreading of sludge[a]

Description	Regression equation		r value for the regression	P value for the regression
	Slope	y intercept		
Data from study by Farrah et al. (5)	0.24325	0.20603	0.984	0.002
Data from study by Hurst et al. (11)	0.28769	− 0.10325	0.950	0.001
Both sets of data analyzed as a single group	0.26601	0.03114	0.959	<0.00001

[a] Linear regression analysis was performed with $\log_{10}$ reduction in titer since day of spreading as the dependent variable versus number of days since sludge was placed on land as the independent variable (9).

season of the year in geographical areas that have roughly similar environmental conditions. The first two regressions in Table 1 represent changes in titer for the indigenous virus populations as the dependent variable (expressed as $\log_{10}$-transformed titer ratio values, $\log_{10} N_t/N_0$) against days of drying in the field as the independent variable. The individual data set from the study by Farrah et al. (5) is shown in the first regression. The data set from Hurst et al. (11) is shown in the second regression. The third regression listed in Table 1 shows the result obtained by pooling the sets of viral titer data from these two different studies and then regressing the $\log_{10}$-transformed titer ratio values in that pool as the dependent variable against time of drying in the field. This table also presents the respective Pearson correlation coefficients (r values) and statements of probability (P values) that the regression values expressed are different from zero. It can be seen that the regression values for both slope and y intercept generated from the pooled data set were intermediate between those corresponding values yielded when the two sets of data were regressed individually. In addition, it can be seen that regression of the combined set of data still generated a strong r value and produced a lower (better) P value than did regression of the two sets of data independently. Thus, these two sets of data appear to be compatible (9).

Thus far, the examples of data regressions presented in this chapter have utilized the titer ratio as the dependent variable versus time as the independent variable. Linear regressions as shown in equation 4 can also be performed by using the slope, B_t, from equation 5 as the dependent variable. This requires a two-step regression technique (12). The first step consists of deriving inactivation rate values in the form of $[\log_{10}(N_t/N_0)]/t$, accomplished by regressing $\log_{10}$-transformed titer ratios $[\log_{10}(N_t/N_0)]$ from various sampling dates as the dependent variable versus the length of time that the organisms were incubated in the test material as the independent variable (equation 5). The slope values (B_t) derived from this first step of the regression technique are the inactivation rate values, and these are then used as the dependent variable in a second step of analysis, where they are linearly regressed (using equation 4) against a single independent variable, typically either an environmental factor such as soil moisture level or soil pH, insolation, or a soil or water characteristic.

When an individual environmental factor, or a single characteristic of the soil or water, is regressed as the independent variable against the inactivation rate values as the dependent variable, the equation used to model the relationship takes the form of equation 4. That individual factor or characteristic would be represented as X_1 in equation 4, with the relationship between the rate values ($[\log_{10}(N_t/N_0)]/t$, derived as B_t in equation 5 and then inserted as Y in equation

4) and that individual factor or characteristic being represented as B_1 in equation 4 (12). The resulting value for B_1 is analogous to the rate constant, K, as shown in equation 1.

The inactivation rate values ($[\log_{10}(N_t/N_0)]/t$) can also be regressed against a multiple number of independent variables that represent various environmental factors or characteristics of the soil or water that is being studied. One approach for doing this type of comparison is to use a multiple linear regression as presented in equation 6 (9, 12):

$$Y = B_0 + B_1X_1 + B_2X_2 + \ldots + B_nX_n \quad (6)$$

In such equations, Y is the dependent, or response, variable and has the form $[\log_{10}(N_t/N_0)]/t$; B_0 represents the y-axis intercept; X_1 through X_n represent the different independent variables; and B_1 through B_n represent the coefficients assigned to those different independent variables. This approach thus presents the inactivation rate constant, K, as being the sum of several rate constants which reflect the impact of different independent variables, as described in equation 7:

$$K = k_1 + k_2 + k_3 + \ldots + k_n \quad (7)$$

The multiple regression equation is easiest to interpret if the different independent variables are unrelated. However, in practice, this rarely happens. For this reason, when one is selecting candidate independent variables for inclusion into a multiple linear regression model, it is important to first examine the set of all possible independent variables for the presence of any strong cross-correlations between them. This examination can be done quite easily by using Pearson's test (10). If two or more of the candidate variables correlate with one another in a highly significant manner, then it would probably be better to allow only one of those variables to be included in a multiple regression model. Having first examined the candidate independent variables for correlation with one another and then eliminated the potential problem of including highly cross-correlated variables in a model, one must select which of the remaining candidate variables will be incorporated into the actual multiple linear regression model equation. Two approaches can be used to make this determination. The first is to build the regression model by using a stepwise technique beginning with that candidate independent variable which demonstrates the strongest correlation with the dependent variable (inactivation rate). Additional individual independent variables can then be incorporated in a successive manner to improve the model fit. When one is developing such models, it is helpful to establish criteria for deciding when to stop adding additional independent variables to the multiple regression equation. One possible criterion is to include at least one independent variable and to stop adding further variables when the next to be incorporated in sequence would increase the P value for the overall model to above 0.05, would produce no improvement in the overall P value for the model, or would contribute an amount of less than 0.001 to the overall r value for the model. Another approach would be to predetermine which variables one wished to include in the equation on the basis of other analyses, such as the use of Pearson's correlation coefficients to assess the relationship between candidate independent variables and the inactivation rate values.

An example of using the stepwise approach for selecting multiple independent variables is shown in equation 8. This is the best stepwise regression fit for modeling virus survival based on the physical and chemical characteristics of nine soils (12).

$$Y = 0.1005 + (0.0025) \text{ (viral adsorption to soil)}$$
$$- (0.0008) \text{ (extractable phosphorus)}$$
$$- (0.0007) \text{ (exchangeable aluminum)}$$
$$- (0.0510) \text{ (saturation pH)} \qquad (8)$$

In this model equation, Y is represented as $[\log_{10} (N_t/N_0)/\text{day}]$, and four independent variables are included: the extent of viral adsorption to the particles of a test soil and the levels of resin-extractable phosphorus, exchangeable aluminum, and saturation pH for each of nine soils. This equation summarizes results obtained by modeling the survival of seeded viruses incubated aerobically, under nonsterile conditions, at a single temperature (23°C) in nine different characterized soils (12). Both the choice of which soil characteristics were included and the order in which those soil characteristics were inserted as variables into equation 8 were determined by stepwise multiple linear regression.

Another example of this latter approach, using a two-step regression technique to represent inactivation rate values as a function of multiple independent environmental factors or sample characteristics, can be shown for a study in which the survival of seeded viruses was studied in samples of water from five sites. Seeded viruses were incubated independently at different temperatures (−20, 1, and 22°C) in portions of water from those five different sites under aerobic, nonsterile conditions (10). Samples of the water collected from each of the different sites were chemically analyzed. The survival data from that study were first regressed by using equation 5. The viral inactivation rate values, expressed in terms of $\log_{10}(N_t/N_0)/\text{day}$, derived from that first step of regression analysis, were then regressed as the dependent variable (equation 4) versus each of the individual water characteristics. From a list of water characteristics that were evaluated in this way, five characteristics were found to correlate with viral inactivation rates (10): hardness, conductivity, turbidity, suspended solids, and the capacity of indigenous nutrients present in the water samples to support the growth of seed bacteria.

Stepwise multiple linear regression was used to determine the best fit for describing virus survival in these waters at the individual incubation temperatures of −20, 1, and 22°C. Of the five characteristics which, when tested individually, were found to correlate with the inactivation rate values, hardness and conductivity so strongly correlated with one another that only one of the two was subsequently allowed to occur in any given model. Likewise, turbidity and suspended solids so strongly correlated with one another that only one of these two was subsequently allowed to occur in any given model. When a stepwise multiple linear regression

was performed by using the inactivation rate values from incubation at 22°C (equation 9) as the dependent variable versus the five characteristics as potential independent variables, two of those characteristics were included in the model: hardness and the mean number of generations of growth which the water could support for three different bacterial species after the bacteria were independently seeded into sterilized or pasteurized samples of water from those five different sites. This second variable has been abbreviated as "mean gen PEK," since the bacterial species used were *Pseudomonas fluorescens*, *Enterobacter cloacae*, and *Klebsiella oxytoca*. All three of these bacterial species are heterotrophic organisms common to water samples.

$$Y = 2.04 \times 10^{-2} + (1.37 \times 10^{-3}) \text{ (hardness)}$$
$$- (2.93 \times 10^{-3}) \text{ (mean gen PEK)} \qquad (9)$$

In this model equation, Y is represented as $\log_{10} (N_t/N_0)/\text{day}$. This model equation suggests that the factors which significantly affected survival of the test viruses in these waters at 22°C were the level of water hardness and the level of nutrients available to support microbial growth. Stepwise multiple linear regression equations that were similarly derived for the other two incubation temperatures revealed that the factors which affected viral survival at 1°C were turbidity and the level of nutrients available for supporting growth of one of the three organisms, *K. oxytoca*. The only factor which appeared to have affected survival at −20°C was turbidity.

Comparing the Accuracies of Equation Formats

The linear modeling approach requires two major assumptions. The first assumption is that any differences, or distances, between the plotted data points and the graphed equation (equation 5; Fig. 2) are due to error. The second assumption is that the use of linear modeling is appropriate for the data being examined. Sources of possible error include any random variability associated with the development of ecological conditions during the course of the experiment and variability associated with sample collection, sample processing, dilutions, and the assay system. One way of testing this first assumption, and in the process comparing the basic accuracy of the equation format that is used for modeling the microbial population decay rates, is to first use the observed experimental values to generate a model and then use that model to generate a corresponding set of predicted values and to compare the sets of predicted versus observed values by means of simple linear regressions for which Y and X represent the predicted values and observed values, respectively (14).

During the first part of this analysis, when the model equations are generated, Y represents the actual experimentally determined titer ratio values (N_t/N_0). During the second part of the analysis, when those model equations are used to predict the outcome of the study, the predicted titer ratio values represent estimates of the expected value of Y, $E(Y)$. If an ideal equation format existed, then the set of predicted values would be a duplicate of the set of experimentally observed values. This duplication does not occur, in part because it is impossible to identify and include all pertinent environmental variables into a mathematical model and also because it is impossible to predict the extent of those errors in each titer ratio value that would be caused by such factors as random variation, experimental variation, and variation associated with sample processing and assaying.

If an ideal equation format existed, and if it were possible to incorporate and model the effects of all sources of variability, then one could derive an ideal model. In such a case, if the set of values predicted by the model were linearly regressed against the set of experimentally observed values, then the equation yielded by that regression would have a slope of 1.0 and a y-axis intercept of zero. For this particular analysis, we could thus consider a line whose slope is 1.0 with a y-axis intercept of zero to be an ideal line. The absolute numerical difference between the slope of that ideal line and the slope of the line derived by linearly regressing the actual sets of predicted and observed values has been used as an assessment of the basic accuracy of the equation format (14). Hurst et al. (14) published eight equation formats which were compared for their accuracy in modeling a common set of experimental data. According to that assessment, the most accurate equation format is the one which yields the smallest value for the absolute numerical difference between the calculated slope values from regression of predicted versus observed values, and the ideal slope value of 1.0. An equation format termed multiplicative error II proved best in this regard. This equation format is shown in Equation 10:

$$Y = B_0 X_1^{B_1} X_2^{B_2} \ldots X_n^{B_n} t^{B_t} \qquad (10)$$

In that test, Y was the titer ratio expressed as N_t/N_0, X_1 through X_n represented experimental variables, and t represented time. B_1 through B_n represented the corresponding regression coefficients calculated for the variables X_1 through X_n, and B_t represented the coefficient calculated for time, t. Essentially, in this equation, the rate constant K is represented as being a product of several different contributing rates. This can be represented as equation 11:

$$K = k_1 k_2 k_3 \ldots k_n \qquad (11)$$

To develop the models using the multiplicative error II equation format, equation 10 was subjected to a base 10 logarithmic transformation to yield equation 12, and the models were created by using a linear regression procedure.

$$\log_{10} \frac{N_t}{N_0} = \log_{10} B_0 + B_1 \log_{10} X_1 + B_2 \log_{10} X_2 \\ + \ldots + B_n \log_{10} X_n + B_t \log_{10} t \qquad (12)$$

The regression shown in equation 12 can be performed in a stepwise manner provided that the obligatory term $B_t \log_{10} t$ is included before $B_1 \log_{10} X_1$.

The accuracy of the equation format represented by equation 12 indicated that when survival data are plotted by using $\log_{10} (N_t/N_0)$ as the dependent variable and time as the independent variable (equation 5; Fig. 2), the deviation (distance) between the individual graphed points and the linear repression line are not due entirely to random variability. This finding would also suggest that rather than using the linear approach shown in equation 5 as the basis for modeling survival data, it may be better to use the equation

$$\frac{N_t}{N_0} = B_0 t^{B_t} \qquad (13)$$

after it has been $\log_{10}$ transformed to generate

$$\log_{10} \frac{N_t}{N_0} = \log_{10} B_0 + B_t \log_{10} t \qquad (14)$$

Values of B_t derived by regressing data from individual experiments by using equation 14 could be grouped and then compared and contrasted as sets, using analytical techniques such as t tests and analysis of variance in the same manner that was described earlier with respect to B_t values derived from equation 5. That same study (14) also demonstrated that the effects of both temperature and time were best modeled by using temperature and time as independent variables, thus allowing temperature and time to have coefficients which take the form of exponents. Perhaps the most important implication from that study is that neither the influence of thermal effects, as measured by incubation temperature, nor the relationship of titer ratio to time is a linear function; rather, both are exponential functions.

Another, simpler test which can be used to determine whether data should be modeled by using linear regression is a ranking test. If a correlation between titer ratio as the dependent variable and some independent variable such as temperature is statistically significant by rank correlation but not statistically significant by linear regression, then linear regression may be an inappropriate analytical approach for modeling that relationship; instead, the effect of the independent variable upon titer ratio should perhaps be analyzed by a non-linear regression technique. Also, it should be noted that in the equations presented in this chapter, it has been assumed that all of the terms incorporated into a survival model have the same general form (i.e., as $X_1 B_1$ or $X_1^{B_1}$). This is not a requirement. An example of a model in which the incorporated terms take different formats is that of Auer and Niehaus (1).

The technique of using a set of experimentally observed values to generate a model and then comparing values predicted by the model against that same set of experimentally observed values can be used to give an assessment of the accuracy of the equation format which was used to construct the model. This approach is what has been presented both here and in the report by Hurst et al. (14). However, this technique cannot be used to estimate the accuracy of the developed model itself, as defined by the choice of included variables and their corresponding parameter values. Performing a test of the accuracy of the developed model requires instead that the predicted values be compared against a set of experimentally observed values other than those used to generate the model. One possible approach for performing this latter type of assessment would be to divide the original data set into two subsets, or half-sets, before the model is developed. One of these half-sets of experimentally observed values could be used to generate a model, and then the predicted values from that model could be compared against the other half-set of observed values. An even better approach would be to perform two independent sets of experiments. The observed values from the first set of experiments could be used to generate a model, whose accuracy is then assessed by using the observed values from the second set of experiments. An example of the latter approach might be to develop a model by using data from the study by Farrah et al. (5), assessing the model's accuracy by using it to predict what the outcome of the Hurst et al. study (11) might have been, and finally comparing those predicted values against the values actually observed by Hurst et al. (11). Unfortunately, neither of these two studies (5, 11) was considered to have yielded a large enough data set for accurately performing such a test of model accuracy.

REFERENCES

1. **Auer, M. T., and S. L. Niehaus.** 1993. Modeling fecal coliform bacteria. I. Field and laboratory determination of loss kinetics. *Water Res.* **27:**693–701.

2. **Canale, R. P., M. T. Auer, E. M. Owens, T. M. Heidtke, and S. W. Effler.** 1993. Modeling fecal coliform bacteria. II. Model development and application. *Water Res.* **27:** 703–714.
3. **Connolly, J. P., R. B. Coffin, and R. E. Landeck.** 1992. Modeling carbon utilization by bacteria in natural water systems, p. 249–276. *In* C. J. Hurst (ed.), *Modeling the Metabolic and Physiologic Activities of Microorganisms.* John Wiley & Sons, New York.
4. **Dahling, D. R., and R. S. Safferman.** 1979. Survival of enteric viruses under natural conditions in a subarctic river. *Appl. Environ. Microbiol.* **38:**1103–1110.
5. **Farrah, S. R., G. Bitton, E. M. Hoffmann, O. Lanni, O. C. Pancorbo, M. C. Lutrick, and J. E. Bertrand.** 1981. Survival of enteroviruses and coliform bacteria in a sludge lagoon. *Appl. Environ. Microbiol.* **41:**459–465.
6. **Grant, S. B., E. J. List, and M. E. Lidstrom.** 1993. Kinetic analysis of virus adsorption and inactivation in batch experiments. *Water Resour. Res.* **29:**2067–2085.
7. **Harvey, R. W.** 1991. Parameters involved in modeling movement of bacteria in groundwater, p. 89–114. *In* C. J. Hurst (ed.), *Modeling the Environmental Fate of Microorganisms.* American Society for Microbiology, Washington, D.C.
8. **Hurst, C. J.** 1988. Influence of aerobic microorganisms upon virus survival in soil. *Can. J. Microbiol.* **34:**696–699.
9. **Hurst, C. J.** 1991. Using linear and polynomial models to examine the environmental stabiliy of viruses, p. 137–159. *In* C. J. Hurst (ed.), *Modeling the Environmental Fate of Microorganisms.* American Society for Microbiology, Washington, D.C.
10. **Hurst, C. J., W. H. Benton, and K. A. McClellan.** 1989. Thermal and water source effects upon the stability of enteroviruses in surface freshwaters. *Can. J. Microbiol.* **35:** 474–480.
11. **Hurst, C. J., S. R. Farrah, C. P. Gerba, and J. L. Melnick.** 1978. Development of quantitative methods for the detection of enteroviruses in sewage sludges during activation and following land disposal. *Appl. Environ. Microbiol.* **36:** 81–89.
12. **Hurst, C. J., C. P. Gerba, and I. Cech.** 1980. Effects of environmental variables and soil characteristics on virus survival in soil. *Appl. Environ. Microbiol.* **40:**1067–1079.
13. **Hurst, C. J., C. P. Gerba, J. C. Lance, and R. C. Rice.** 1980. Survival of enteroviruses in rapid-infiltration basins during the land application of wastewater. *Appl. Environ. Microbiol.* **40:**192–200.
14. **Hurst, C. J., D. K. Wild, and R. M. Clark.** 1992. Comparing the accuracy of equation formats for modeling microbial population decay rates, p. 149–175. *In* C. J. Hurst (ed.), *Modeling the Metabolic and Physiologic Activities of Microorganisms.* John Wiley & Sons, New York.
15. **McFeters, G. A., and D. G. Stuart.** 1972. Survival of coliform bacteria in natural waters: field and laboratory studies with membrane-filter chambers. *Appl. Microbiol.* **24:** 805–811.
16. **Mezrioui, N., B. Baleux, and M. Troussellier.** 1995. A microcosm study of the survival of *Escherichia coli* and *Salmonella typhimurium* in brackish water. *Water Res.* **29:** 459–465.
17. **Moore, R. S., D. H. Taylor, M. M. M. Reddy, and L. S. Sturman.** 1982. Adsorption of reovirus by minerals and soils. *Appl. Environ. Microbiol.* **44:**852–859.
18. **Pesaro, F., I. Sorg, and A. Metzler.** 1995. In situ inactivation of animal viruses and a coliphage in nonaerated liquid and semiliquid animal wastes. *Appl. Environ. Microbiol.* **61:** 92–97.
19. **Powelson, D. K., and C. P. Gerba.** 1994. Virus removal from sewage effluents during saturated and unsaturated flow through soil columns. *Water Res.* **28:**2175–2181.
20. **Powelson, D. K., C. P. Gerba, and M. T. Yahya.** 1993. Virus transport and removal in wastewater during aquifer recharge. *Water Res.* **27:**583–590.
21. **Sinton, L. W., R. J. Davies-Colley, and R. G. Bell.** 1994. Inactivation of enterococci and fecal coliforms from sewage and meatworks effluents in seawater chambers. *Appl. Environ. Microbiol.* **60:**2040–2048.
22. **Smith, J. J., J. P. Howington, and G. A. McFeters.** 1994. Survival, physiological response, and recovery of enteric bacteria exposed to a polar marine environment. *Appl. Environ. Microbiol.* **60:**2977–2984.
23. **Wan, J., J. L. Wilson, and T. L. Kieft.** 1994. Influence of the gas-water interface on transport of microorganisms through unsaturated porous media. *Appl. Environ. Microbiol.* **60:**509–516.
24. **Wilkinson, J., A. Jenkins, M. Wyer, and D. Kay.** 1995. Modelling faecal coliform dynamics in streams and rivers. *Water Res.* **29:**847–855.
25. **Yates, M. V., and S. R. Yates.** 1991. Modeling microbial transport in the subsurface: a mathematical discussion, p. 48–76. *In* C. J. Hurst (ed.), *Modeling the Environmental Fate of Microorganisms.* American Society for Microbiology, Washington, D.C.

Detection of Bacterial Pathogens in Wastewater and Sludge

ROBERT C. COOPER AND RICHARD E. DANIELSON

23

The number and variety of bacteria present in wastewater and associated solids are legion. Their sources are the excrement of humans and animals, other waste materials that find their way into domestic sewage, and the microbial flora in the source water. This great diversity and associated variety of required growth conditions hamper attempts to isolate, identify, and enumerate most bacterial members of this microcosm. The number of pathogenic bacteria that might be present in wastewater and biosolids is a function of the disease morbidity in the community from which the waste materials are derived and the degree of sewage treatment received. Relative to the total number of bacteria present, the pathogens normally represent but a minor part. In most instances, these pathogens play a passive role in the dynamics of the microbial ecosystem; the waste environment is hostile, and consequently the number of pathogens present tends to decrease over time. The isolation, identification, and enumeration of pathogens from this milieu are replete with all of the difficulties listed above.

Representative genera of bacterial pathogens that can be found in domestic wastewater and sludge include *Salmonella*, *Shigella*, *Vibrio*, *Escherichia*, *Campylobacter*, and *Yersinia*. The genus *Salmonella* includes more than 1,800 species and serovars; the genus *Shigella* is represented by four species encompassing 34 serovars; *Vibrio cholerae* O1 and non-O1 can be present; the pathogenic varieties of *Escherichia coli* include at least 5 serovars; *Campylobacter jejuni* is the most common *Campylobacter species* found in wastewater, followed by *C. coli*; and there are reported to be more than 50 serovars of *Yersinia enterocolitica* (15, 17, 30, 33).

Because of the difficulties in the isolation and detection of bacterial pathogens in wastewater and sludges, the use of surrogate (indicator) bacteria has been standard practice in water quality monitoring. Historically the coliform group, the enterococci, and *Clostridium perfringens* have, in descending priority, been the bacterial indicators of choice. In all of these cases, the indicator bacteria are assumed to be indigenous to feces, and thus their presence in environmental samples is indicative of fecal contamination. The presence, and in some instances the absence, of these indicators is not an absolute indication of the presence of bacterial pathogens. Rather, they indicate the potential for the presence of pathogens because of the likelihood that infectious feces are present in wastewater or sludge. The bacterial indicators' greatest weakness as a public health monitoring tool

for water and wastewater is their greater sensitivity to disinfection relative to viruses and, in particular, the cysts of protozoan parasites. In these instances, the absence of indicator bacteria is not a guarantee that other, more resistant microbial forms are not present. Because of these problems, there have been ongoing efforts to find better indicators for the presence of microbial pathogens in environmental samples. The ideal would be to monitor for the presence of all microbial pathogens that might be present in a liquid or solid sample, an ideal not likely to be realized any time in the foreseeable future.

There are a number of instances in which there is a need to make direct measurements for the presence and number of bacterial pathogens in water and solids. This approach can be most useful in epidemiologic studies of waterborne disease, in the development of the relationship between indicator numbers and specific pathogen concentrations which could be of aid in standard settings, in determining the efficacy of water and solids treatment processes in the reduction of pathogens, and in situations in which the sanitary significance of high indicator numbers is in question. This latter situation has been observed in composting sewage sludge in which coliform after-growth can occur, the significance of which is determined by the direct measurement of *Salmonella* species (16). If a specific pathogen type were to be used as an indicator of pollution, one would still have the problem of selecting which would be the ideal surrogate. The majority of work done thus far on measuring bacterial pathogens in water and solids has been directed toward salmonellae, and these have become standard pathogens by consensus.

TRADITIONAL METHODS FOR THE DETECTION AND ENUMERATION OF BACTERIAL PATHOGENS

To date, methods for the detection and enumeration of bacterial pathogens from wastewater and sludge have used a cumbersome approach which includes enrichment, isolation, and identification. The number of bacterial pathogens found in wastewater and sludges is usually lower than the number of nonpathogens present. For example, the number of salmonellae detected in activated sludge and digested sludge ranges from none detected to a most probable number (MPN) of 400/g (dry weight) of solids (62). The numbers reported in raw sewage range from 7 to 8,000/100 ml (15). Because of these small numbers, some form of enrichment, which may be preceded by a sample concentration step, is required.

Determining the concentrations of bacterial pathogens in wastewater and biosolids is a cumbrous task. The traditional method of choice is to estimate the numbers of bacteria by the MPN method. The MPN is determined by placing a series of dilutions of a wastewater sample, sample concentrate, or sludge-solids extract into tubes of enrichment broth and determining if the target bacterium has grown in any of the inoculated tubes. This requires that each tube (dilution) be screened for the presence of the pathogen by completing the isolation and identification steps.

Sampling Requirements, Transportation, and Preservation

Raw (untreated) sewage does not require the concentration of large-volume samples (typically 250 ml to 1 liter). Sterile glass or polypropylene bottles can be used to collect the samples. All samples should be refrigerated or stored on ice and processed in the laboratory as soon as possible but not more than 24 h after collection.

Treated sewage effluent (primary, secondary, or better) may require the collection of larger sample sizes (1 to 10 liters), depending on the final quality of the wastewater effluent. If the collected wastewater represents an effluent which has been treated by a tertiary or better process, large samples (10 to 20 liters or more) must be collected. For molecular investigations, depending on the final water quality, bacteria may be concentrated directly onto filters in the field for recovery of their nucleic acids (29, 52). Furthermore, if a chlorinated effluent is discharged, the sample must be dechlorinated by adding to the sample bottle an appropriate concentration of sodium thiosulfate, usually enough to give a final concentration of 100 mg/liter, prior to sterilization (18).

Sewage sludge samples should be collected in accordance with reference 16. The tested sample consists of subsamples taken throughout the same pile or source of sludge. All samples should be refrigerated or held on ice and processed within 24 h of collection. Results are reported as MPN per gram (dry weight).

Wastewater samples generally are concentrated by filtration or centrifugation of known volumes of sample whereby the bacteria present are captured on a filter or in a centrifuged pellet. Filtration using 0.45-μm-pore-size membranes is most applicable to "clean" (filterable) water. In the case of sewage, the utility of filtration is restricted because clogging limits the sample volume. If necessary, coarser filters such as spun glass can be used, although they will be less efficient as bacterium collectors (13). In the case of wastewater and biosolids extracts, centrifugation can be used but there will be some obvious limitations in the size of the sample that can be examined. A nonquantitative pseudo-filtration method has been used (34) in which gauze pads (sanitary napkins work well) are suspended in sewage flow for a period of time, after which the entire pad is placed in a suitable enrichment broth. Bacteria can be isolated from these pads up to four times more frequently than from grab samples (27). This method can also be used to detect enteric viruses in wastewater. Other concentration methods using diatomaceous earth, fiberglass, and membrane filtration are described in *Standard Methods for the Examination of Water and Wastewater* (*Standard Methods*; 18). Methods that use concentration prior to enrichment are problematic in that they concentrate all manner of materials and microbes as well as the target pathogen; thus, while concentrating the sample, one also significantly increases the amount of interference that will be encountered in subsequent steps of the isolation and identification process.

Enrichment and Isolation

The enrichment process employs a broth medium that allows the target bacteria to multiply to numbers large enough to facilitate their isolation and identification. In many instances, the medium to be used has been chosen directly from those used in clinical laboratories where isolation would be from the stools of infected individuals. In these samples, one would expect large numbers of pathogens. The enrichments are devised to limit the growth of interfering bacteria, such as coliforms and *Proteus* species, and encourage the growth of the pathogen. Frequently the formulations used were less inhibitory to the target than to the unwanted bacteria and at the same time were not optimal for growth of the pathogen. In the case of stools from infected individuals, and in many instances infected food, the target bacteria are robust enough and in sufficient numbers that they multiply despite the suboptimal growth conditions. In wastewater, sewage solids, and treated effluents, the pathogenic bacteria are in a hostile environment, not robust, and in small numbers. In these instances, the selective enrichment used in the clinical setting may be too restrictive for successful enhancement. Less restrictive or nonrestrictive media can be used to allow the growth of the target pathogen, and other bacteria, followed by secondary enrichment in a restrictive medium. After enrichment for the appropriate time and temperature, material from the broth is streaked or plated onto a solid medium for bacterial isolation. There is a large variety of such media which are selective, differential, or both. Typical colonies are selected and further identified by using standard biochemical (triple sugar iron agar, lysine iron agar, urea agar, etc.) and immunological methods.

Salmonella Detection

Because of the great diversity of pathogens that can be found in wastewater and solids, the diversity of media and incubation regimens required, and in most instances the paucity of information available, the following discussion is limited to salmonellae. Most of the available literature and reported experience in the sphere of wastewater and sludge have been directed toward the detection and enumeration of salmonellae. Much of the methodology used has been adapted from food microbiology, an area of intense activity in the detection of salmonellae.

A representative list of selective enrichment media is presented in Table 1. These media fall into three ingredient

TABLE 1 Selective enrichment media used for the detection of salmonellae in wastewater and sludge

Enrichment medium	Reference
Selenite-F broth	32
Selenite brilliant green	42
Selenite brilliant green sulfa broth	42
Selenite cystine broth[a]	40
Selenite dulcitol broth[a]	28
Tetrathionate broth[a]	38
Tetrathionate brilliant green broth	22
RVB	59

[a] Selected in *Standard Methods* (18).

TABLE 2 Major components of selected *Salmonella* isolation agars

Class	Ingredient	Agar[a]						
		SS	XLD	XLBG	BG	HE	BS	MSRV
Carbohydrate	Dextrose	−	−	−	−	−	+	−
	Lactose	+	+	+	+	+	−	−
	Saccharose	−	+	+	+	+	−	−
	Xylose	−	+	+	−	−	−	−
Amino acid	Lysine	−	+	+	−	−	−	−
Inhibitors	Bile salts	+	−	−	−	+	−	−
	Brilliant green	+	−	+	+	−	+	−
	Deoxycholate	−	+	−	−	−	−	−
	Bismuth sulfate	−	−	−	−	−	+	−
	Malachite green	−	−	−	−	−	−	+
	MgCl$_2$	−	−	−	−	−	−	+
	Novobiocin	−	−	−	−	−	−	+

[a] SS, salmonella-shigella; XLBG, xylene lysine brilliant green; BG, brilliant green; HE, Hektoen enteric; BS, bismuth sulfite.

categories: (i) selenium based, (ii) tetrathionate based, and (iii) malachite green-magnesium salt based (Rappaport-Vassiliadis broth [RVB]). The conditions presented in these enrichment media do not present optimum growth conditions for salmonellae and are selective in that they inhibit the growth of interfering bacteria, such as *Proteus* species and coliforms, and provide opportunity for the ascendancy of any salmonellae that might be present. A number of additional ingredients have been incorporated into *Salmonella* cultivation medium in an attempt to enhance selectivity. These include brilliant green dye, cystine, sulfapyridine, and novobiocin. This latter antibiotic, in concentrations of up to 80 μg/ml, has been found very useful in the suppression of interfering bacteria (48) and is incorporated in the RVB. Because these enrichment media can exert a significant inhibitory effect on stressed salmonellae, some form of pre-enrichment, such as buffered peptone broth, can be used prior to inoculation into a selective medium. This is the procedure followed when the tetrathionate brilliant green broth of Hussong et al. (22) is used.

The incubation temperature can have a significant impact on the recovery efficiency of a selective medium. Harvey and Price (19) in their review of *Salmonella* isolation methods indicate that the enhancement of *Salmonella* recovery by using incubation temperatures of between 40 and 43°C has been recognized since the turn of the century. The use of elevated temperature must be matched with the enrichment system used. These authors point out that results at elevated temperatures can vary among laboratories, probably because of lack of standardization of methods. In general, elevated temperature does not enhance the usefulness of selenite brilliant green broth or tetrathionate-based medium samples. These authors also point out that elevated temperature methods may not detect *Salmonella typhi*. The elevated temperature regimen works well with RVB and is used in the secondary enrichment scheme of Hussong et al. (22) and with the use of selenite dulcitol broth.

Salmonella Isolation Media

There are a number of formulations of isolation agar for the detection of salmonellae. The major components for six of the most common agars are shown in Table 2. We have included a seventh, less common medium, modified semisoft Rappaport-Vassiliadis (MSRV) agar, because of our success in the isolation of salmonellae from wastewater and biosolids

with this formulation. These agars are selective and differential. Selectivity is brought about through the use, singly or in combination, of brilliant green dye, bile salts, or deoxycholate. Variations of this theme include the use of bismuth sulfide agar and the use of malachite green (a homolog of brilliant green), a relatively high concentration of magnesium chloride, and novobiocin in MSRV medium. Differentiation among colony types is, for the most part, based on carbohydrate fermentation and associated pH change, as shown by indicator dyes. For example, the fermentation of lactose, present in five of the seven agars, is an attribute of coliforms but not salmonellae; colonies of the former will assume the characteristic color of the pH indicator used. A good illustration of the use of these activities to differentiate pathogens from nonpathogens is the make-up of xylose lysine deoxycholate (XLD) agar. As shown in Table 2, the medium contains three sugars, lactose, saccharose, and xylose, with the last at a lesser concentration than the first two. The pH indicator used is phenol red, which is yellow in acid. Coliform colonies will be yellow because of fermentation of all of the carbohydrates present; nonpathogenic non-lactose fermenters, such as *Proteus* species, will be detected by the fermentation of the other carbohydrates present, while salmonellae, which can ferment xylose, will neutralize the acid produced through the decarboxylation of lysine, resulting in a colony of red color. The presence of excess lactose and saccharose prevents lysine decarboxylase-active coliforms from neutralizing the acid produced. The production of hydrogen sulfide is indicated by reaction with ferric salts in the medium to produce red-black or yellow-black colonies, the former being common to *Salmonella* species. In contrast, bismuth sulfite agar, which contains only dextrose, relies on the production of hydrogen sulfide, with the concomitant precipitation of iron sulfide producing a black metallic sheen on suspect *Salmonella* colonies. The medium relies on the inhibitory effect of brilliant green to control the growth of interfering coliforms.

MSRV agar does not contain any carbohydrate but relies on the motility of salmonellae as a differential characteristic. The use of motility as a means of separating salmonellae from background bacteria has been reported by numerous authors throughout the years (19). More recently, De Smedt et al. (10) and De Smedt and Bolderdijk (9) have developed a modification of the Rappaport-Vassiliadis enrichment medium in semisoft (0.8%) agar onto which drops of enrichment culture are placed; the sample is incubated at 42°C, and motility (cloudy growth) away from the spot is considered to be an indication of the presence of salmonellae. This limits the test to *Salmonella* isolates that are motile; however, there are very few nonmotile *Salmonella* species, notably *Salmonella pullorum* (30).

The considerable variety of enrichment and isolation media available poses a problem for the investigator who wishes to select the most efficacious method for the isolation of salmonellae from wastewater and biosolids. Part of this decision will be based on the investigator's past experience and sample matrix. Morinigo et al. (35–37) recently evaluated the use of various enrichment media and isolation agars for their effectiveness in the isolation of salmonellae from environmental samples. In one study on river water (35), they examined 10 different enrichment media: RVB with four concentrations of novobiocin, selenite cysteine at two different incubation temperatures, and selenite-F with and without novobiocin at two incubation temperatures. All of the samples were preenriched in buffered peptone broth and isolated onto XLD and brilliant green agars. They concluded that the best recoveries of salmonellae were made with RVB containing novobiocin at 10 μg/ml. In another study, these authors (36) again evaluated this enrichment medium with the inclusion of tetrathionate broth, using laboratory cultures of bacteria including salmonellae and species known to interfere with the isolation of salmonellae. They found that selenite-containing media were less effective at restraining the growth of gram-positive interference bacteria. Salmonellae grew poorly in tetrathionate broth and media containing brilliant green. Stressed (by exposure to seawater) salmonellae did well in selenite-F broth, selenite cysteine broth, and RVB. When natural water samples were used, RVB was deemed the best for enrichment. These authors also evaluated a great variety of isolation agars, excluding MSRV, for the ability to grow salmonellae as well as to inhibit interfering bacteria. Laboratory cultures of bacteria, both stressed and unstressed, were used. The results indicated that XLD agar was equal in selectivity to the best of the others and was best for the recovery of stressed salmonellae. From their work it appears that RVB with novobiocin is the enrichment medium of choice and XLD agar is an acceptable isolation agar for use with polluted water samples.

There has been an increased interest in the ability to isolate salmonellae from sewage sludge, particularly with the adoption of a *Salmonella* standard for treated sludge that can be applied, unrestrictedly, to land (16). Yanko et al. (62) evaluated five different procedures for the isolation and enumeration of salmonellae in biosolids: two standard methods (18) which use dulcitol selenite and tetrathionate enrichments, respectively, followed by the use of XLD and brilliant green isolation agars; Kenner and Clark's method using dulcitol selenite enrichment incubated at 40°C and isolation on XLD agar; the method of Hussong et al. (22), which uses a preenrichment in buffered peptone at 36°C followed by a secondary enrichment in tetrathionate brilliant green broth incubated at 43°C and isolation on xylose lysine brilliant green agar; and a selenite brilliant green enrichment incubated at 37°C, with confirmation on XLD agar. This evaluation was conducted on digested sludge, activated sludge, and composted sludge. The Hussong et al. (22) and selenite brilliant green methods recovered more salmonellae than did the other three, and the first two were determined to be inadequate for use in compliance testing.

In none of the these evaluations was MSRV agar used.

In our laboratory, we have compared three methods for the enumeration (MPN) of salmonellae in biosolids and raw sewage. These methods include the selenite brilliant green enrichment and selenite dulcitol enrichment, both followed by isolation on XLD agar, and a nonselective enrichment in tryptic soy broth (12) followed by isolation on MSRV agar. All of the enrichments were incubated at 35°C for 24 to 48 h. The MSRV agar was then spotted with up to 0.2 ml (one drop per spot) of enrichment and incubated for 24 h at 42°C. At the end of this period, the MSRV agar plates were examined and motility away from the spots was recorded. Material taken from the edge of a motile spot was either restreaked onto XLD agar or used directly for biochemical screening. Our experience has been that this material contains, with few exceptions, a pure culture of salmonellae. In our hands, this procedure is two to three times more effective than the other two methods tested (unpublished data). In no instance have we recovered salmonellae by either of the first two methods and not made a recovery in MSRV agar. One reason for the success of this method may be the use of a large inoculum (0.2 ml) of enrichment culture relative to that used on a streak plate. Harvey and Price (19) point out that salmonellae growing in an enrichment culture tend not to be uniformly distributed and that a streak plate may well come up negative even when salmonellae are present in the enrichment. Our observations seem to confirm this phenomenon, since frequently not all of the six spots on a positive MSRV agar plate are motile. The use of a nonselective enrichment medium may also be of value in resuscitating stressed salmonellae.

Other Bacterial Pathogens

The recently published literature on the detection of bacterial pathogens other than *Salmonella* species in wastewater and sludge is relatively sparse. The cultural method for the detection of *V. cholerae* is well established (18). The use of alkaline peptone water as an enrichment followed by isolation on thiosulfate citrate bile salts sucrose agar and confirmation of typical colonies by biochemical and serological tests is an effective method. In our laboratory, we have had good success in isolating non-O1 *V. cholerae* from estuarine sediments with this procedure.

Reports on the detection of *Shigella* species have not been found. *Standard Methods* (18) suggests that methods similar to those used for the detection of salmonellae should be employed for the detection of shigellae and points out that the methodology is qualitative and low in sensitivity. Reports on the isolation and enumeration of *Yersinia* species in wastewater and sludge were not found. A membrane filter method for the detection of *Yersinia* species in water is described in *Standard Methods* (18).

Methods for the isolation and enumeration of thermophilic *Campylobacter* species from the water environment are reported by Arimi et al. (1) and Jones et al. (24, 25). These authors report successful isolation and MPN enumeration with Preston *Campylobacter* selective enrichment broth (Oxoid USA, Inc., Columbia, Md.) incubated at 43°C and isolation on Preston *Campylobacter* selective agar (Oxoid USA) incubated under microaerophilic conditions.

It is known that *Aeromonas hydrophila* is associated with human enteric disease (21). Because of this understanding, the distribution of this autochthonous aquatic bacterium in surface water, wastewater, and sludge has been reported. The most recent literature contains descriptions of methods that may or may not use an enrichment step. Nishikawa and

Kishi (39) describe a qualitative method using an alkaline peptone broth enrichment followed by isolation on bile salts brilliant green starch agar. Amylase-positive colonies on the latter medium were considered positive presumptive *Aeromonas* isolates. Further identification was based on biochemical differentiation. Poffe and Op de Beek (47) reported good results in the isolation and enumeration of *A. hydrophila* from sewage and sludge by using direct detection methods. In the case of wastewater, they used the membrane filter method with mA agar (50), which contains ampicillin, trehalose, and deoxycholate. In the case of sludge samples, suspensions of various dilutions were spread onto the surface of mA agar. In both instances the suspect colonies were yellow, a result of trehalose fermentation. These were further identified by biochemical differentiation. Ribas et al. (49) compared four similar media for their usefulness in the detection of *Aeromonas* spp. in polluted water. The media were of various compositions that contained ampicillin, a complex carbohydrate, and, in one case, glutamate; mA agar was included. The authors concluded that all of the media gave similar results but that the starch glutamate ampicillin agar was the simplest to use.

MOLECULAR BIOLOGICAL METHODS

The use of molecular techniques is growing rapidly in the environmental microbiology field. In the last 5 years, molecular techniques have moved from the research laboratory to wastewater agencies (31, 45, 56, 57). There can be insurmountable difficulties on the use of culture methods for the isolation of certain pathogenic bacteria from wastewater and sludges.

The primary advantage to the application of molecular methods to the detection of pathogenic microorganisms from environmental samples is the ability to specifically and rapidly detect the organism of interest without having to actually isolate it on growth media. It may even be possible to apply molecular methods to detect those target organisms that exhibited characteristics commonly associated with viability (the presence of RNA species). Given the large background of competing bacteria, these methods hold promise for the future of monitoring for either specific or selected groups of pathogenic bacteria in wastewater and sludges.

PCR

In the section that follows, it is assumed that the reader is familiar with the basic theory behind PCR. The details of the PCR will not be discussed here but can be found in various textbooks (2, 14, 23). The PCR involves a few critical steps; the following are highlights of those procedures.

Primer Selection

PCR primers can be chosen as specific sequences adjacent to predetermined target sequences, or a group of random hexamers can be used as primers to create PCR-amplified molecules that contain the target sequence. If specific primer sequences are to be chosen, one can determine the preferred primer sequences either by searching the literature or nucleic acid libraries or by sequencing areas of DNA or RNA of the bacteria of interest. Primer sets can be highly specific for only one particular gene that, one hopes, is unique to a single organism, or they can be universal, amplifying related sequences that cross taxonomic family or kingdom boundaries. In addition, depending on the amplification conditions, multiple primer sets may be simultaneously run in a nested form or as a suite of primers (5, 6, 57). Target

sequences of high tertiary complexity (e.g., superhelices) may be more difficult to amplify because of the structural geometry of the finished sequence (2, 51). It is important to limit the size of the final target (<3 kbp) because as the PCR is used to replicate longer and longer pieces of DNA, there is greater chance of error (e.g., in base pair matching, the formation of primer dimers, etc.) (2, 51).

PCR Conditions

PCR is performed with a thermal cycler, an instrument that is capable of rapidly cycling temperature changes over a wide range, generally between 4 and 100°C. Thermal cycler settings must be optimized for each primer set and desired product and often are determined empirically through trials. In addition, the various commercial thermal cyclers may perform slightly differently. Therefore, in order to keep thorough quality control, it is imperative that PCR conditions be optimized on one instrument and subsequently reproven if the investigator switches to another brand or type of cycler. The optimal annealing temperature is critical and can be determined either empirically through a series of experiments or by consulting computer-based software which can predict melting and annealing temperatures on the basis of base pair sequences (2).

Molecular Probes and Detection Techniques

There are many methods available for the detection of the PCR product. One of the easiest methods is gel electrophoresis, whereby the products of the PCR amplification are separated into bands based on size. Some drawbacks to this method include the following: (i) if the target was present in an extremely low concentration, even an amplified product may be difficult to visualize; (ii) other amplification products may interfere with the resolution of the electrophoretic bands; and (iii) if only a single band is generated for a target organism, then a single electrophoretic image may not be definitive proof of a successful amplification of specific target. The first two of these problems can be resolved by using specific complementary probes, which can be labeled either isotopically or nonisotopically. One may then visualize extremely low concentrations of product and avoid interfering bands which do not bind to the probe. As an aid in solving the third problem, it is advisable to amplify more than a single target site for any organism of interest, since the presence of multiple products from the same organism provides a more specific test for that organism.

Isotopic versus Nonisotopic Techniques

For many years, isotopically labeled (e.g., ^{32}P-labeled) probes were the method of choice given their sensitivity for detecting nucleic acid sequences. Methods using isotopically labeled probes can measure as little as femtogram quantities of product (4–6). However, because of the necessary precautions associated with handling and cleanup of radioactive material, many investigators now prefer to use nonisotopic detection procedures whenever possible. Biotin-streptavidin, digoxigenin, or fluorescent labels can be used for obtaining colorimetric results (31, 45, 54, 56). Methods for nonisotopically labeled probes may not be as sensitive as those which use isotopically labeled probes, but they are improving. The final product from the reaction can be detected by immobilizing the target sequences onto nylon filters and hybridizing them in place with a known, labeled probe, or the known probe can be immobilized and used to capture the labeled target material through hybridization as the sample is placed into contact with the filter.

The specificity of binding between the nucleic acid probe and the complementary target nucleic acid sequences versus other sequences which are similar, but not identical, is known as stringency and is dictated by the chemical conditions and temperature under which the hybridization is performed (2). The selection of hybridization conditions and the format of test presentation (Southern hybridization, dot blots, etc.) will vary depending on the makeup of the probe and the typical yield of the PCR. Consequently, the best hybridization conditions must be determined empirically for each complementary set of nucleic acid primer sequences.

Direct Probing

If the microorganisms of interest are in sufficient quantity, PCR amplification of the target nucleic acid sequences may not be necessary. Recovery of total nucleic acid from an environmental source such as sewage, sludge, or filter concentrates may yield sufficient material for direct probing. Sommerville et al. (52) and Knight et al. (29) demonstrated that some pathogens may be present in sufficient quantity but may not be culturable. In these instances, total nucleic acids were harvested and known nucleic acid probes specific for the bacterial pathogens were used as a detection system.

Quantification

It is possible to use PCR to quantify the number of bacteria present in the sample. Internal nucleic acid amplification controls can be run simultaneously, and the intensity of the hybridization product between the probe and the PCR amplification can be compared with that of the product of hybridization with known quantities of target nucleic acid (3, 5, 44, 45). Picard et al. (46) prepared multiple PCR tubes containing undiluted and diluted DNA prepared from soil extracts and used the resulting test reactions as a basis for constructing an MPN result. Yamashiro et al. (61) used a technique based on the method of Holland et al. (20), whereby the PCR-amplified product is fluorescently labeled as the PCR proceeds. To arrive at a quantitative result, the increasing concentration of fluorescently labeled target DNA is compared with the increasing concentration of fluorescently labeled control DNA within the same reaction tube.

Quality Control

There are several quality control measures that can be used to test for a successful PCR and hybridization. If the same source of wastewater or sludge is to be repeatedly tested, spiking samples with either the organism in question or target DNA will provide information on the efficiency of sampling techniques and the effects of inhibitory compounds. In addition to a spiked study, every reaction tube must contain a known piece and quantity of control DNA that will indicate whether the PCR had been inhibited and to what extent. Specificity of the hybridization can be measured by using known labeled DNA that differs from the target by one or two base pairs. Therefore, first the level of stringency must be empirically determined for each primer set, and then the established amplification regimen can be rigorously followed. Sensitivity of the test results can be determined by adding either target DNA or the test organism in the desired wastewater or sludge matrix in such a manner as to conduct dilution-to-extinction experiments.

Sample Preparation

Currently, the primary disadvantage of the application of molecular methods to environmental microbiology is the presence of naturally occurring compounds such as humic acids and minerals which may be inhibitory to the molecular procedure (23, 53, 55). Therefore, removing these inhibitory agents is of primary importance when one is designing sample preparation and concentration techniques. The following discussion will highlight a variety of methods that have been used to isolate the DNA or RNA of the target bacteria in question.

DNA is a double-stranded nucleic acid which can form very complex superhelical structures. These structures can be quite recalcitrant to environmental effects (26, 58, 60). Treated wastewater may contain disinfectants which are indiscriminate oxidizing compounds and can damage DNA molecules. Some treated sludges can have extremely high pHs from liming processes which can denature DNA. Therefore, if molecular techniques are to be applied to limed sludges, it may be necessary to adjust (lower) the pH of suspensions before testing.

Use of RNA as a molecular target is important because it may be helpful as a qualitative measure of viability (i.e., by measuring the presence of mRNA or rRNA). In addition, RNA is far more abundant within bacterial cells and will be easier to detect with fewer cycles of amplification in a PCR. Reverse transcriptase PCR has been used for the detection of enteric viruses in environmental waters and wastewaters (89, 57), and techniques used to preserve viral RNA should be applicable to bacterial RNA. Although RNA is more plentiful than DNA, RNA is a single-stranded nucleic acid which has been shown to be quite labile in environmental matrices (58). Therefore, it is important that steps be taken to improve the survival of RNA molecules during sample manipulations (2).

As stated above, one of the major obstacles in amplifying bacterial nucleic acids from environmental sources is the presence of inhibitory agents such as humic acid (55). Since environmental matrices can be complex and undefined, it is difficult to apply a "one cleanup step fits all" procedure. For example, approaches used for the recovery of bacterial nucleic acids from wastewater differs those used for recovery from sludge. Palmer et al. (44, 45) used the following techniques to recover bacterial (*Legionella* spp.) nucleic acids from wastewater and reclaimed wastewater. First, the sample was concentrated onto Teflon filters (Durapore; Millipore Corp., Bedford, Mass.). The filter was then vortexed in a lysis buffer and heated to release the bacterial DNA. The nucleic acids were further purified by centrifugation through a Spin-X filter (Costar, Cambridge, Mass.). A carrier reagent (homopolymer A RNA) was added to aid subsequent precipitation of the DNA in isopropanol. Following centrifugation, the DNA pellet was washed twice in isopropanol. Finally, the pellet was resuspended in PCR water (Perkin-Elmer, Norwalk, Conn.), and an aliquot was submitted to amplification by PCR. Although one manufacturer of an environmental PCR kit (Perkin-Elmer) uses Durapore filters, other investigators (4, 43) reported that greater yields were achieved with Fluoropore filters (both filter types; Millipore).

Environmental water samples may also be centrifuged to concentrate bacteria into a pellet; the pelleted cells are then lysed in a silica-guanidinium thiocyanate lysis buffer and subjected to several washing steps (3, 54). These crude extracts have been shown to be sufficient to add directly to a PCR amplification.

For sludges, Tsai et al. (56) reported that rapid freeze-thawing of samples followed by a phenol-chloroform extraction and cleanup through a Sephadex-200 spin column provided adequate recovery of target bacterial DNA. Typically, small volumes (about 25 to 30 μl) of the extracted nucleic acids are used in 100-μl reaction mixes; therefore, rare targets may be missed. Analysis of larger-volume samples by PCR has been described by Stewart and Abbaszadegan (53), who applied it to the detection of viruses in the environment. The basic volumes were adjusted to accommodate 100 μl of extract for a total volume of 300 μl. Because the larger volumes may introduce more inhibitory compounds, Stewart and Abbaszadegan employed cleanup through a Sephadex G-100 column (for humic acids) and through a column of Chelex-100 (for ionic inhibitors).

Recently, magnetic capture beads have been used in a wide variety of clinical, food, and environmental applications for the recovery of target nucleic acids that were subsequently amplified by PCR (41). This technique holds promise as a mechanism for capturing the target nucleic acid while incorporating wash steps which help to remove potential inhibiting agents.

FA Techniques

Fluorescently labeled antibodies specific for a target organism have been used to screen wastewater and reclaimed wastewater for target microorganisms (11, 44, 45). Since there are many microorganisms in wastewater and sludge, there may be some that share similar antigenic reaction sites (epitopes) with which the antibody stain could cross-react. Therefore, when one is searching through a diverse number of bacteria, it is best to limit the occurrence of cross-reactivity by using monoclonal fluorescent antibodies (FA). The biggest advantage to using monoclonal fluorescent antibody stains is that they can be used to produce a rapid and relatively inexpensive test (once the monoclonal antibody has been made). The disadvantages of the use of monoclonal FA stains for environmental samples are that (i) there remains a probability that cross-reactivity will occur; (ii) the FA stain can be used to detect the presence of the target organism but does not measure viability; (iii) the array of antigens exhibited by a bacterium may differ depending on the conditions in its environment, and for this reason, depending on the physiological state of the bacteria used to prepare the monoclonal cell line (which in turn produces the antibodies used in the detection reaction), the monoclonal antibody(ies) chosen may not not recognize the key target bacterial epitopes that are expressed on bacteria in environmental waters; and (iv) the number of bacteria required for microscopic detection must be substantial, and especially in environmental samples, the level of test organisms may be below this minimum number. A negative test result, therefore, should be reported as "none detected" rather than "none present."

ELISA

Enzyme-linked immunosorbent assays (ELISAs) are also antibody based. Brigmon et al. (7) applied an ELISA to detect the presence of *Salmonella enteritidis* in raw sewage, sludges, and wastewater. The advantage associated with applying the ELISA technique directly to environmental samples is that it can provide a mechanistic measure (via a microtiter dish reader) with only minor manipulation of the sample. However, one of the major disadvantages of this technique is the lack of sensitivity. Brigmon et al. (7) reported that a minimum of 10^5 *S. enteritidis* cells per ml were required to generate a clear response above background. Tamanai-Shacoori et al. (54) demonstrated that PCR was superior to ELISA for screening environmental samples for enterotoxi-

genic *E. coli*. Given the current standard of <3 *Salmonella* MPN per 4 g of sludge (16), this technology has not yet reached the sensitivity necessary for performing the testing necessary for regulatory compliance. Other disadvantages of this technique include those outlined for FA stains.

CONCLUSION

There is a growing interest in the direct measurement of specific bacterial pathogens in environmental samples. This interest is stimulated by concern for the reliability of the use of standard indicator bacteria in setting public health standards, the need for better epidemologic information relative to the use or reclaimed wastewater and the disposal of biosolids, and the evaluation of pathogen reduction in waste treatment processes. Methods for the isolation and enumeration of pathogenic bacteria in wastewater and sludge are not yet well developed. There are myriad difficulties, ranging from the huge diversity of microorganisms involved to the considerable variety of methodological approaches that one can pursue. At this time, development of methods for the detection and enumeration of salmonellae is the most advanced, followed by an increasing interest in the development of cultural methods for *Campylobacter* and *Aeromonas* species, among others. The more recent developments in molecular biological methods offer great possibilities for the future.

REFERENCES

1. **Arimi, S. M., C. R. Fricker, and R. W. A. Park.** 1988. Occurrence of thermophilic campylobacters in sewage and their removal by treatment processes. *Epidemiol. Infect.* **101:**279–286.
2. **Ausubel, F. M., R. Brent, R. E. Kingston, D. D. Moore, J. G. Seidman, J. A. Smith, and K. Struhl (ed.).** 1989. John Wiley & Sons, New York.
3. **Bej, A. K., J. L. DiCesare, L. Haff, and R. M. Atlas.** 1991. Detection of *Escherichia coli* and *Shigella* spp. in water by using the polymerase chain reaction and gene probes for *uid*. *Appl. Environ. Microbiol.* **57:**1013–1017.
4. **Bej, A. K., M. H. Mahbubani, J. L. DiCesare, and R. M. Atlas.** 1991. Polymerase chain reaction-gene probe detection of microorganisms by using filter-concentrated samples. *Appl. Environ. Microbiol.* **57:**3529–3534.
5. **Bej, A. K., M. H. Mahbubani, R. Miller, J. L. DiCesare, L. Haff, and R. M. Atlas.** 1990. Multiplex PCR amplification and immobilized capture probes for detection of bacterial pathogens and indicators in water. *Mol. Cell. Probes* **4:**353–365.
6. **Bej, A. K., S. C. McCarty, and R. M. Atlas.** 1991. Detection of coliform bacteria and *Escherichia coli* by multiplex polymerase chain reaction: comparison with defined substrate and plating methods for water quality monitoring. *Appl. Environ. Microbiol.* **57:**2429–2432.
7. **Brigmon, R. L., S. G. Zam, G. Bitton, and S. R. Farrah.** 1992. Detection of *Salmonella enteritidis* in environmental samples by monoclonal antibody-based ELISA. *J. Immunol. Method.* **152:**135–142.
8. **De Leon, R., C. Shieh, R. S. Baric, and M. D. Sobsey.** 1990. Detection of enteroviruses and hepatitis A virus in environmental samples by gene probes and polymerase chain reaction, p. 833–853. *In Advances in Water Analysis and Treatment.* Proceedings of the Water Quality Technology Conference, San Diego, Calif. American Water Works Association, Denver.
9. **De Smedt, J. M., and R. F. Bolderdijk.** 1987. Dynamics of Salmonella isolation with modified semi-solid Rappaport-Vassiliadis medium. *J. Food Prot.* **50:**658–661.

10. **De Smedt, J. M., R. F. Bolderdijk, H. Rappoid, and D. Lautenschlaeger.** 1986. Rapid Salmonella detection in foods by motility enrichment on a modified semi-solid Rappaport-Vassiliadis medium. *J. Food Prot.* **49:**510–514.

11. **Desmonts, C., J. Minet, R. Colwell, and M. Cormier.** 1990. Fluorescent-antibody method useful for detecting viable but nonculturable *Salmonella* spp. in chlorinated water. *Appl. Environ. Microbiol.* **56:**1448–1452.

12. **Difco Laboratories.** *Difco Manual,* 10th ed. Difco Laboratories, Detroit.

13. **Dutka, J. B., and J. B. Bell.** 1973. Isolation of salmonellae from moderately polluted waters. *J. Water Pollut. Control Fed.* **45:**316–323.

14. **Erlich, H. A. (ed.).** 1989. *PCR Technology, Principles and Applications for DNA Amplification.* Stockton Press, New York.

15. **Feachem, R. G., D. J. Bradley, H. Garelick, and D. D. Mara (ed.).** 1983. *Sanitation and Disease: Health Aspects of Excreta and Wastewater Management.* John Wiley & Sons, New York.

16. **Federal Register.** 1993. Standards for the use of disposal of sewage sludge: final rules. *Fed. Regist.* **58:**32,9387–9404.

17. **Gary, L. D.** 1995. *Escherichia, Salmonella, Shigella,* and *Yersinia,* p. 450–464. *In* P. R. Murray, E. J. Baron, M. A. Pfaller, F. C. Tenover, and R. H. Yolken (ed.), *Manual of Clinical Microbiology.* 6th ed. American Society for Microbiology, Washington, D.C.

18. **Greenberg, A. E., et al. (ed.).** 1992. *Standard Methods for the Examination of Water and Wastewater,* 18th ed. APHA, AWWA and WEF Publishers, Washington, D.C.

19. **Harvey, R. W. S., and T. H. Price.** 1979. A review: principals of Salmonella isolation. *J. Appl. Bacteriol.* **46:**27–56.

20. **Holland, P. M., R. D. Abramson, R. Watson, and D. H. Gelfand.** 1991. Detection of specific polymerase chain reaction product by utilizing the 5′ to 3′ exonuclease activity of *Thermus aquaticus* DNA polymerase. *Proc. Natl. Acad. Sci. USA* **88:**7276–7280.

21. **Holmberg, S. E., and J. J. Farmer.** 1984. *Aeromonas hydrophila* and *Plesiomonas shigelloides* as causes of intestinal infections. *Rev. Infect. Dis.* **6:**633–639.

22. **Hussong, D., W. D. Burge, and N. K. Enkiri.** 1985. Occurrence, growth, and suppression of salmonellae in composted sewage sludge. *Appl. Environ. Microbiol.* **50:**887–893.

23. **Innis, M. A., D. H. Gelfand, J. J. Sninsky, and T. J. White (ed.).** 1990. *PCR Protocols. A Guide to Methods and Applications.* Academic Press, Inc., New York.

24. **Jones, K., M. Betaieb, and D. R. Telford.** 1990. Seasonal variation of thermophilic campylobacters in sewage sludge. *J. Appl. Bacteriol.* **69:**185–189.

25. **Jones, K., M. Betaieb, and D. R. Telford.** 1990. Comparison between environmental monitoring of thermophilic campylobacters in sewage effluent and the incidence of *Campylobacter* infection in the community. *J. Appl. Bacteriol.* **69:**235–240.

26. **Karl, D. M., and M. D. Bailiff.** 1989. The measurement and distribution of dissolved nucleic acids in aquatic environments. *Limnol. Oceanogr.* **34:**543–558.

27. **Kelley, S., and W. W. Sanderson.** 1960. Density of enteroviruses in sewage. *J. Water Pollut. Control Fed.* **32:**1269–1273.

28. **Kenner, B. A., and H. A. Clark.** 1974. Determination and enumeration of *Salmonella* species and *Pseudomonas aeruginosa. J. Water Pollut. Control Fed.* **46:**2163–2171.

29. **Knight, I. T., S. Shults, C. W. Kaspar, and R. R. Colwell.** 1990. Direct detection of *Salmonella* spp. in estuaries by using a DNA probe. *Appl. Environ. Microbiol.* **56:**1059–1066.

30. **Krieg, N. R., and J. G. Holt (ed.).** 1984. *Bergey's Manual*

of Systematic Bacteriology, vol. 1. Williams & Wilkins, Baltimore.

31. **Lang, A. L., Y.-L. Tsai, C. L. Mayer, K. C. Patton, and C. J. Palmer.** 1994. Multiplex PCR for detection of the heat-labile toxin gene and shiga-like toxin I and II genes in *Escherichia coli* isolated from natural waters. *Appl. Environ. Microbiol.* **60:**3145–3149.

32. **Liefson, E.** 1936. New selinite enrichment media for the isolation of typhoid and paratyphoid (salmonella) bacilli. *Am. J. Hyg.* **24:**423–432.

33. **McLaughlin, J. C.** 1995. *Vibrio,* p. 465–476. *In* P. R. Murray, E. J. Baron, M. A. Pfaller, F. C. Tenover, and R. H. Yolken (ed.), *Manual of Clinical Microbiology,* 6th ed. American Society for Microbiology, Washington, D.C.

34. **Moore, B.** 1948. The detection of paratyphoid carriers in towns by means of sewage examination. *Monthly Bull. Minist. Public Health Lab. Serv.* **7:**1241.

35. **Morinigo, M. A., E. Martinez-Mananares, M. A. Munoz, and R. Cornax.** 1989. Evaluation of different plating media used in the isolation of salmonellas from environmental samples. *J. Appl. Bacteriol.* **66:**353–360.

36. **Morinigo, M. A., M. A. Munoz, R. Cornax, D. Castro, and J. J. Borrego.** 1990. Evaluation of different enrichment media for the isolation of Salmonella from polluted seawater samples. *J. Microbiol. Methods* **11:**43–49.

37. **Morinigo, M. A., M. A. Munoz, E. Martinez-Mananares, and J. M. Sanchez.** 1993. Laboratory study of several enrichment broths for the detection of *Salmonella* spp. particularly in relation to water samples. *J. Appl. Bacteriol.* **74:**330–335.

38. **Muller, L.** 1923. Un nouveau milieu d'enrichissement pour la recherche du bacille typhique et des paratyphiques. *C. R. Soc. Biol.* **89:**434–437. (As cited in reference 19.)

39. **Nishikawa, Y., and T. Kishi.** 1987. A modification of bile salts brilliant green agar for isolation of motile *Aeromonas* from food and environmental specimens. *Epidemiol. Infect.* **98:**331–336.

40. **North, W. R., and M. T. Bartram.** 1953. The efficiency of selenite broth of different compositions in the isolation of *Salmonella. Appl. Microbiol.* **1:**130–134.

41. **Olsvik, O., T. Popovic, E. Skjerve, K. S. Cudjoe, E. Hornes, J. Ugelstad, and M. Uhlen.** 1994. Magnetic separation techniques in diagnostic microbiology. *Clin. Microbiol. Rev.* **7:**43–54.

42. **Osborne, W. W., and J. L. Stokes.** 1955. A modified selenite brilliant green medium for the isolation of *Salmonella* from egg products. *Appl. Microbiol.* **3:**295–299.

43. **Oyofo, B. A., and D. M. Rollins.** 1993. Efficacy of filter types for detecting *Campylobacter jejuni* and *Campylobacter coli* in environmental water samples by polymerase chain reaction. *Appl. Environ. Microbiol.* **59:**4090–4095.

44. **Palmer, C. J., G. F. Bonilla, B. Roll, C. Paszko-Kolva, L. R. Sangermano, and R. S. Fujioka.** 1995. Detection of *Legionella* species in reclaimed water and air with the EnviroAmp Legionella PCR kit and direct fluorescent antibody staining. *Appl. Environ. Microbiol.* **61:**407–412.

45. **Palmer, C. J., Y.-L. Tsai, C. Paszko-Kolva, C. Mayer, and L. R. Sangermano.** 1993. Detection of *Legionella* species in sewage and ocean water by polymerase chain reaction, direct fluorescent-antibody, and plate culture methods. *Appl. Environ. Microbiol.* **59:**3618–3624.

46. **Picard, C., C. Ponsonnet, E. Paget, X. Nesme, and P. Simonet.** 1992. Detection and enumeration of bacteria in soil by direct DNA extraction and polymerase chain reaction. *Appl. Environ. Microbiol.* **58:**2717–2722.

47. **Poffe, R., and E. Op de Beek.** 1991. Enumeration of *Aeromonas hydrophila* from domestic wastewater plants and surface waters. *J. Appl. Bacteriol.* **71:**366–370.

48. **Restino, L., G. S. Grauman, W. A. McCall, and W. M.**

Hill. 1977. Effects of varying concentrations of novobiocin incorporated into two *Salmonella* plating media on the recovery of four *Enterobacteriaceae*. *Appl. Environ. Microbiol.* **33:**585–589.

49. **Ribas, F., J. Frias, J. M. Huguet, F. R. Ribas, and F. Lucena.** 1991. Comparison of different media for the identification and quantification of *Aeromonas* spp. in water. *Antonie van Leeuwenhoek* **59:**225–228.

50. **Rippey, S. R., and V. J. Cabelli.** 1979. Membrane filter procedure for enumeration of *Aeromonas hydrophila* in fresh waters. *Appl. Environ. Microbiol.* **38:**108–111.

51. **Saiki, R. K.** 1990. Amplification of genomic DNA, p. 13–20. *In* M. A. Innis, D. H. Gelfand, J. J. Sninsky, and T. J. White (ed.), *PCR Protocols. A Guide to Methods and Applications.* Academic Press, Inc., New York.

52. **Sommerville, C. C., I. T. Knight, W. L. Straube, and R. R. Colwell.** 1989. Simple rapid method for the direct isolation of nucleic acids from aquatic environments. *Appl. Environ. Microbiol.* **55:**548–554.

53. **Stewart, P. W., and M. Abbaszadegan.** 1995. Large volume RT-PCR for the detection of enteroviruses in ground water samples, abstr. Q-394, p. 469. *In Abstracts of the 95th General Meeting of the American Society for Microbiology 1995.* American Society for Microbiology, Washington, D.C.

54. **Tamanai-Shacoori, Z., A. Jolivet-Gougeon, M. Pommepuy, M. Cormier, and R. R. Colwell.** 1994. Detection of enterotoxigenic *Escherichia coli* in water by polymerase chain reaction amplification and hybridization. *Can. J. Microbiol.* **40:**243–249.

55. **Tsai, Y.-L., and B. H. Olson.** 1992. Rapid method for separation of bacterial DNA from humic substances in sediments for polymerase chain reaction. *Appl. Environ. Microbiol.* **58:**2292–2295.

56. **Tsai, Y.-L., C. J. Palmer, and L. R. Sangermano.** 1993. Detection of *Escherichia coli* in sewage and sludge by polymerase chain reaction. *Appl. Environ. Microbiol.* **59:**353–357.

57. **Tsai, Y.-L., M. D. Sobsey, L. R. Sangermano, and C. J. Palmer.** 1993. Simple method of concentrating enteroviruses and hepatitis A virus from sewage and ocean water for rapid detection by reverse transcriptase-polymerase chain reaction. *Appl. Environ. Microbiol.* **59:**3488–3491.

58. **Tsai, Y.-L., B. Tran, and C. J. Palmer.** 1995. Analysis of viral RNA persistence in seawater by reverse transcriptase-PCR. *Appl. Environ. Microbiol.* **61:**363–366.

59. **Vassiliadis, P., D. Trichopoulos, J. Papadakas, V. Kalapothaki, X. Zavitsanos, and C. Serie.** 1981. *Salmonella* isolation with Rappaports enrichment medium of different composition. *Zentralbl. Bakteriol. Mikrobiol. Hyg. Abt. 1 Orig. Reihe B* **173:**382–389.

60. **Weinbauer, M. G., D. Fuks, and P. Peduzzi.** 1993. Distribution of viruses and dissolved DNA along a coastal trophic gradient in the northern Adriatic Sea. *Appl. Environ. Microbiol.* **59:**4074–4082.

61. **Yamashiro, C. T., M. Rogers, W. Jakubowski, and C. Paszko-Kolva.** 1995. Development and evaluation of species-specific PCR detection assays for *Giardia lamblia* and *Cryptosporidium parvum*, abstr. Q-209, p. 436. *In Abstracts of the 95th General Meeting of the American Society for Microbiology 1995.* American Society for Microbiology, Washington, D.C.

62. **Yanko, W. A., A. S. Walker, J. L. Jackson, L. L. Liabo, and A. L. Garcia.** 1995. Enumerating *Salmonella* in biosolids for compliance with pathogen regulations. *Water Environ. Res.* **67:**364–370.

Assessing the Efficiency of Wastewater Treatment

WESLEY O. PIPES AND JAMES T. ZMUDA

24

INTRODUCTION

Wastewater treatment plants usually include a series of processes, and in most cases one of the processes is biological. Each process can be evaluated separately to determine its contribution to the overall effectiveness of the treatment. Various characteristics and parameters of the microbial consortium in individual biological treatment processes can be measured for use as guides for operational control. Thus, there are a large number of physical, chemical, and biological tests which can be applied to wastewater treatment processes, many more tests than are covered herein.

This chapter is concerned with measuring the overall efficiency and effectiveness of the entire treatment plant and not with evaluating the performance of individual processes or determining the functioning of the microorganisms in a biological process. The methods covered are those which can be used to ascertain the strength of the wastewater and the potential of the effluent to cause water pollution. Also, the methods covered are limited to microbiological methods, all of which are types of bioassays. To establish a context for discussing how the methods are used, it is expedient to review the objectives of wastewater treatment, the concepts of efficiency and effectiveness of wastewater treatment, and the differences between wastewater treatment processes and sludge treatment processes.

Objectives of Treatment

The overall objectives of wastewater treatment are to separate the wastes from the water for disposal elsewhere and to produce an effluent which can be discharged to a receiving water body without causing pollution. Various kinds of waste materials can cause different types of water pollution; therefore, evaluation of wastewater treatment requires measurement of several parameters which provide information about a variety of possible pollution effects as well as the measurement of concentrations of specific waste components in the wastewater and effluent. Thus, these overall objectives need to be subdivided into a number of specific objectives related to the different types of water pollution in order to be a useful framework for evaluation of treatment.

One way of subdividing the overall objectives of wastewater treatment into specific objectives would be to list all of the biological and chemical constituents of the wastewater (e.g., typhoid bacteria, ammonia, cyanide, and starch) and then specify limits on the concentration of each constit-

uent in the effluent. However, application of this approach would require massive analytical effort. For reasons of economy, the cost of data collection to evaluate treatment should be only a small fraction of the cost of the treatment itself. A different approach which depends on parameters which are directly related to specific pollution effects is used. Table 1 lists the parameters used for evaluation of wastewater and effluents along with the associated pollution effects. These are aggregate parameters; that is, each measures something about the combined effect of a number of related constituents which may cause a particular pollution effect.

Pathogenic microorganisms which cause waterborne diseases are wastewater constituents, but there are too many different types for detection and quantitation of individual pathogens to be a practical means of assessing wastewater treatment. Indicator organisms are more easily detected than the pathogens and signify that the water has been contaminated with fecal material. Evaluation of the density of one or more indicator groups in an effluent is a well-established method of assessing one type of water pollution.

Depletion of the dissolved oxygen (DO) in receiving waters is very serious because desirable aquatic organisms are aerobic and the solubility of oxygen in water is extremely low, less than 12 mg/liter in the temperature range of significance to aquatic biology. Microbial utilization of only a few milligrams of DO per liter as a result of wastewater discharges can result in the elimination of the desired populations and communities of aquatic organisms in the receiving waters. It is not feasible to measure the concentrations of all biodegradable organic compounds or to determine the densities of all microorganisms which could be involved in depletion of DO. The biochemical oxygen demand (BOD) test was developed as a measure of the potential effects of the activities of aerobic microorganisms in consuming oxygen. There are several versions of the BOD test; the one used for evaluation of wastewater treatment is the 5-day, 20°C BOD (BOD_5) bottle test.

The discharge of effluents with even moderate amounts of nitrogen and phosphorus in forms which are available as nutrients for algae and aquatic plants can result in excessive growths of nuisance organisms in lakes, streams, and estuaries. Chemical analysis for the concentrations of the total amounts of nitrogen and phosphorus in wastewaters and effluents is feasible. However, the amount of growth stimulation in the receiving waters depends on the specific N and

TABLE 1 Parameters of water pollution

Parameter	Pollution effect(s)	Measurement technique(s)
Pathogenic microorganisms	Water unsuitable for human consumption or recreation	Microbiological (see chapters 15, 16, 17, 20, 21, and 23)
Indicator organisms	Indication of fecal contamination and possible presence of pathogens	Microbiological (see chapters 18 and 19)
Biodegradable organic compounds	Depletion of DO in the receiving waters	Microbiological—BOD (this chapter); chemical—COD (not covered); chemical—TOC (not covered)
Plant nutrients (N and P)	Stimulation of the growth of algae and aquatic plants	Microbiological—algal growth assay (this chapter)
Toxic compounds	Killing of aquatic life	Microbiological—toxicity tests (this chapter); biological—fish toxicity assay (not covered); chemical—analyses for specific compounds (not covered)
Settleable solids	Formation of sludge banks, smothering of benthos, and anaerobic benthic conditions	Physical (not covered)
Fats, oils, and grease	Surface films—interference with surface phenomena	Chemical extraction (not covered)

P compounds present in the effluent and on chemical and biological factors of the receiving waters. The algal growth assay provides an integrated assessment of this aspect of the possible pollution effects of the effluent.

Potentially toxic compounds are present in many effluents, but the expression of toxicity depends on the concentration of the particular compound, the sensitivity of the exposed organisms, chemical and biological factors of the receiving waters, and synergistic and antagonistic interactions of the effluent constituents. It is not feasible to measure the concentrations of all possible toxic compounds which might be present. Toxicity bioassays are used to demonstrate the relative degree of toxicity of wastewaters and effluents. Toxicity bioassay techniques using fish, algae, invertebrates, and microorganisms have been used to evaluate wastewaters and effluents. Only the toxicity bioassays using microorganisms are covered in this chapter.

Suspended solids and fats, oils, and grease are also measured by aggregate parameters because it would not be feasible to measure individual components. The analytical procedures for those two parameters are physical and chemical techniques and therefore are not covered in this manual of microbiological methods.

The objectives of wastewater treatment include both separation of the wastes from wastewater and prevention of pollution of the receiving waters. Treatment efficiency depends upon the extent to which the wastes are separated from the wastewater. Treatment effectiveness is evaluated by measuring the values of the parameters of pollution in the effluent without reference to their values in the wastewater. Each of the parameters of pollution is a measure of the aggregate effect of several waste constituents in relation to a specific form of water pollution such as transmission of waterborne disease or deoxygenation of the receiving waters.

Pollutant Removal and Parameter Reduction

When applied to a conservative substance such as phosphorus, the term "efficiency" is used to express information about the fraction of the waste material present in the wastewater which is separated during treatment. Most often, removal efficiency is expressed as a percentage:

$$\% \text{ efficiency} = (100)\left(\frac{C_i - C_e}{C_i}\right) \quad (1)$$

where C_i is the concentration of the waste material in the influent and C_e is the concentration in the effluent. Percent efficiency values are used to compare different treatment processes and to determine if a particular treatment plant is accomplishing the purpose for which it was designed.

The amount of waste material separated from the water is given by

$$\text{waste load reduction} = (C_i - C_e)(Q)(f) \quad (2)$$

where Q is the wastewater flow rate and f is a unit conversion factor; e.g., when concentration is expressed in milligrams per liter and flow is expressed in cubic meters per day, an f of 0.001 kg/g is used to give the waste load reduction in kilograms per day. Waste load reduction expresses how much is accomplished by wastewater treatment and also provides information about the amount of the residual waste material (usually called sludge) for disposal elsewhere.

In wastewater treatment, the terms "removal" and "reduction" are used more or less interchangeably. The context for "removal" is usually the accomplishment of treatment or the production of sludge, while "reduction" is used most often in reference to the effect of treatment on possible pollution of the receiving waters. This distinction causes no conceptual problems in relation to the separation of a conservative material, such as phosphorus, from the wastewater, but when applied to a parameter which is a property rather than a material, use of the term "removal" may be misleading or make no sense at all.

Toxicity is a property, not a material. Although it is possible to measure the relative toxicity of the wastewater and the treated effluent, the term "toxicity removal" does not make sense and "toxicity reduction" is used. If the toxicity of the effluent is 1/100 of the toxicity of the wastewater, the toxicity reduction can be expressed as 99%; however, the 99% in this case does not give any information about the amount of sludge produced.

In the same manner, the exertion of oxygen demand is a property of organic compounds which can be assimilated by heterotrophic microorganisms, but the compounds themselves are not oxygen demand. Also, exertion of oxygen demand is an activity of the microorganisms after the organic compounds have been assimilated. Nevertheless, "BOD_5 removal" is used much more frequently than "BOD_5 reduction" because a quantitative relationship (direct proportionality) between the waste components which cause the exertion of oxygen demand and BOD_5 itself is usually assumed. Reduction of BOD_5 is considered to be a direct measure of the removal of the waste components (organic compounds and microorganisms) which cause BOD. This equating of the property with the waste components causes conceptual problems and confusion about the proper way to evaluate the effectiveness of wastewater treatment.

Water Treatment versus Sludge Treatment

Wastewater treatment plants consist of two series of treatment processes, one used for treatment of the wastewater and one used for treatment of the sludge which results from the separation of the wastes from the water. Wastewater treatment is intended to produce purified water which can be discharged to the receiving water body without causing pollution. Sludge treatment is intended to change the characteristics of the waste materials so that the ultimate disposal (usually to land) is accomplished with less expense and without creating a nuisance.

The series of sludge treatment processes usually includes a biological process, either anaerobic digestion or aerobic digestion. The objectives of sludge digestion are (i) biodegradation of some of the organic matter in the sludge so that the rate of decomposition of the remaining organic matter is very slow and (ii) alteration of the properties of the sludge so that it is more easily thickened and dewatered. The focus for the design and operation of the sludge digestion processes is on the changes which occur in the waste material rather than on changes in water quality. The water which is separated from the sludge (usually called supernatant) is not suitable for discharge but is mixed back in with the incoming wastewater for further treatment.

By and large, the parameters of water pollution (Table 1) are not appropriate for evaluation of the efficiency or effectiveness of sludge digestion processes. The supernatant removed from an anaerobic sludge digester will often have a higher BOD_5 (due to the presence of organic compounds which are assimilated very rapidly) and a greater toxicity (due to the release of nitrogen from proteins) than the sludge which is put into the digester. Other aggregate parameters such as volatile suspended solids tests and determinations of volatile acids are used to assess the functioning of sludge digestion processes. Those methods are not covered in this chapter.

PATHOGENIC AND INDICATOR MICROORGANISMS

The presence of pathogenic microorganisms in wastewater makes it unsuitable for discharge to receiving water. The absence of pathogens which can cause waterborne disease is an important measure of the effectiveness of wastewater treatment. In this case, the efficiency of treatment is never calculated; the total number or percentage of the pathogens removed or inactivated is not pertinent to evaluation of treatment. Instead, wastewater treatment is evaluated in terms of its effectiveness in reaching the ideal of no pathogens in the effluent.

There are several protozoa, bacteria, viruses, and parasitic worms which can cause waterborne diseases. The techniques for detection of these pathogens are quite varied, and some are very complex. Methods for the detection of pathogens in wastewaters, effluents, and receiving waters are covered in chapters 16, 17, and 23 of this Manual.

The methods for detection of most of the waterborne disease organisms are qualitative rather than quantitative. Indicator organisms which are also present in wastewater and which are easier to evaluate quantitatively have been proposed and/or used as surrogates for the pathogens. One aspect of the effectiveness of wastewater treatment is often measured by determining the densities of certain indicator bacteria in the effluent. Methods for evaluating indicator bacteria densities are discussed in chapters 18, 19, 20, and 22 of this Manual.

OXYGEN DEMAND MEASUREMENTS

Municipal wastewater and most industrial wastewaters contain various types of bacteria, fungi, and protozoa which will assimilate organic matter and consume oxygen (if any is present). The identities of many of these heterotrophic organisms are not known and are not considered important. What are important are the possible decrease in the concentration of DO in the receiving waters and the fact that wastewater treatment can be accomplished by making use of the metabolic activities of the heterotrophic microorganisms.

A method for predicting the rate of consumption of DO by microorganisms in a body of water receiving an effluent is an essential part of assessing the potential for deoxygenation of the receiving water. The original approach to making this assessment (developed approximately 100 years ago) was to measure the DO concentration in the receiving water, incubate a sample of the water sealed off from contact with sources of oxygen for a period of time, and then measure the DO remaining to find out how much had been consumed. This approach evolved into the BOD_5 test (bottle determination), which has been in use in more or less its present form for more than 60 years.

The BOD_5 of the wastewater provides an assessment of what the oxygen demand on the receiving water would be if wastewater treatment were not provided, and the difference between that value and the BOD_5 of the effluent is a measure of the reduction in the potential for deoxygenation of the receiving waters. In addition, the BOD_5 of wastewater is also interpreted as a measure of the concentration of organic material which can serve as a substrate to support the growth of microorganisms in biological wastewater treatment. However, the BOD_5 test does not measure microbial substrate directly; it measures the amount of oxygen consumed by microorganisms in a 5-day period. Mathematical models are used to extrapolate BOD_5 results to "deoxygenation of the receiving waters" or to "substrate concentration." The mathematical models are applicable only when certain rather stringent conditions are met and thus may not represent very well either BOD reduction in biological wastewater treatment or the various phenomena of deoxygenation of a stream or lake polluted with biodegradable organic matter.

There is no alternative for estimating water pollution potential, but two other methods have been developed for measuring the strength of wastewater in relation to biological treatment processes. The chemical oxygen demand (COD) test was developed because a BOD_5 test requires 5 days for completion and therefore is not suitable either for real-time evaluation of the efficiency of wastewater treatment or for operational control of the treatment processes.

The total organic carbon (TOC) test was developed as a third alternative for measuring the strength of wastewater on the basis of an assumption that the primary purpose of biological treatment is to reduce the concentration of organic material in the wastewater.

BOD

The BOD_5 test is the primary method for determining the strength of wastewater to be treated by a biological process and the strength of effluents in terms of the load on the oxygen resources of the receiving waters. It is one of the most important tests for regulation of discharges via limits on the BOD_5 of an effluent. It is also used for evaluation of the efficiency of wastewater treatment by comparison of the BOD_5 of the effluent with the BOD_5 of the influent (equation 1), and a minimum percent BOD_5 reduction is often one of the regulatory requirements.

The BOD_5 test, as used for assessing the efficiency of wastewater treatment, is intended to be a measure of some fraction of the carbonaceous oxygen demand, that is, the oxygen consumed by heterotrophic microorganisms which utilize the organic matter of the waste in their metabolism, and not the oxygen demand exerted by autotrophic nitrifying bacteria. Since ammonia is usually present in wastewaters, nitrification inhibitors must be used to suppress the exertion of nitrogenous oxygen demand. Carbonaceous oxygen demand is called first-stage BOD, and nitrogenous oxygen demand is called second-stage BOD.

Measurement of BOD_5

The details of the determination of BOD_5 by the bottle method (illustrated in Fig. 1) are described in section 5010 of *Standard Methods for the Examination of Water and Wastewater* (hereafter referred to as *Standard Methods*) (1). The

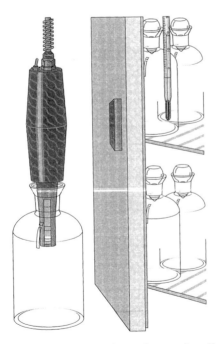

FIGURE 1 The BOD_5 test for evaluating the effectiveness of wastewater treatment is carried out by incubating diluted samples of wastewater and effluent in sealed bottles for 5 days at 20°C and measuring the oxygen depletion. Illustration by Stephen Brayfield.

important characteristics of a BOD bottle are (i) it serves as a culture vessel for the wastewater microorganisms and (ii) it can be filled completely with a diluted sample and then sealed from contact with the atmosphere. The decrease in the DO in a BOD bottle over a period of 5 days provides a measure of the respiration of the microorganisms present because there is no interchange with other sources of oxygen. The conditions for the BOD bottle test were selected to give oxygen consumption measurements which can be used to compare the strength (pollution potential) of wastewaters and effluents in relation to oxygen consumption in the receiving waters.

The BOD_5 test is a bioassay and requires the same careful attention to detail that is essential for the performance of other bioassays. The details of providing the proper conditions for exertion of oxygen demand by microbiological respiration should be readily understood by microbiologists familiar with microbiological cultures and with the respirometric techniques used for many other biochemical measurements. Extensive use of both control tests and quality assurance procedures is essential to obtaining the proper results.

The rate of oxygen consumption varies with temperature, and at a constant temperature, the total amount of oxygen consumed varies with time. The selection of an incubation period of 5 days at 20°C was a compromise from among many possibilities. The temperature of 20°C is, roughly, a median summer temperature for surface waters in temperate climates. Experience has shown that the microbiological oxidation of organic matter in municipal wastewaters can be completed, for all practical purposes, in a period of 20 days at 20°C in a BOD bottle. However, 20 days is too long to wait for results, and the empirically selected period of 5 days allows for exertion of about two-thirds of the 20-day oxygen consumption of municipal wastewaters.

In the United States, municipal wastewaters usually can be expected to have BOD_5 values in the range of 100 to 300 mg/liter, and industrial wastewaters are usually much stronger, with BOD_5 values of thousands and sometimes tens of thousands of milligrams per liter. Since the solubility of oxygen in water at 20°C is about 9 mg/liter, wastewaters have to be diluted 50- to 10,000-fold in order to have some DO remaining after 5 days of incubation. Effluents with BOD_5 values of >7 mg/liter also need to be diluted, usually two- to fivefold.

Since BOD_5 is a parameter (the rate of oxygen consumption averaged over 5 days), not a material which is conserved during dilution, the effect that dilution will have on its value is not obvious. Evidence accumulated over many years demonstrates that if the conditions are proper for the growth and metabolism of heterotrophic microorganisms, the rate of oxygen consumption will be proportional to the remaining concentration of waste components, which will result in the exertion of a BOD. Thus, the decrease in the rate of exertion of oxygen demand caused by dilution is proportional to the dilution factor. When this empirical relationship holds, the BOD_5 of the undiluted wastewater can be calculated from the following equation: BOD_5 of wastewater = (BOD_5 of diluted sample)(dilution factor). In using this relationship, it is important to remember that both the concentrations of the organic compounds, which serve as a microbial substrate, and the cell density of the microorganisms, which exert the oxygen demand, are decreased by dilution.

The DO of the dilution water should be close to saturation before the dilutions are made because the respiration of some aerobic microorganisms is inhibited if the DO drops below 1.0 mg/liter during the BOD test. In order to provide a factor of safety, the residual DO in the BOD bottle after 5 days of incubation should be at least 2.0 mg/liter. Because the dilution water is seeded with organisms capable of assimilating the organic mater of the wastewater, there may be a small depletion of DO in 5 days even if no wastewater is added. To account for this, control bottles containing only dilution water are incubated with the BOD_5 test bottles and a correction is made for the DO depletion in the controls. The depletion of DO in the BOD test bottles should be at least 2 mg/liter during the incubation period in order to be significantly greater than the DO depletion in the control. To be sure of attaining this range of a DO depletion of at least 2 mg/liter with at least 2 mg/liter remaining after the 5-day incubation period, it is necessary to prepare and incubate several dilutions of the original sample.

The composition of the solution used for dilution of samples at the start of the BOD_5 test is an important part of providing the proper conditions for exertion of oxygen demand. The dilution water specified in *Standard Methods* (1) utilizes a phosphate buffer to provide a pH near neutrality during the incubation period. The dilution water also contains the required mineral nutrients in case the wastewater is deficient in any of them. Finally, to avoid confusing nitrogenous oxygen demand with carbonaceous oxygen demand, a compound which inhibits the nitrifying bacteria is added to the dilution water.

It is important that the dilution water not contain any constituents which are inhibitory to the microorganisms which exert the carbonaceous oxygen demand. This needs to be demonstrated by running BOD_5 quality control standards prepared from pure organic compounds, usually glucose and glutamic acid. The controls also protect against the possibility of inhibitory material derived from the system for laboratory purification of water used for making up solutions.

For measurement of the BOD_5 of industrial wastewaters,

the dilution water must be seeded (inoculated) with microorganisms which are capable of degrading the waste organic compounds. The seed may be obtained from a biological treatment process which has been acclimated to the wastewater or from sediments downstream from the point at which the effluent is discharged into the receiving waters. In some cases, development of a laboratory culture of the seed organisms or purchase of a commercial seed may be necessary. Effluents from biological treatment processes and municipal wastewaters are expected to contain the microorganisms required for exertion of the carbonaceous oxygen demand, but dilution water for testing these samples is often seeded anyway.

The Standard BOD Curve and Equation

If a diluted sample of wastewater is incubated under the proper conditions and the oxygen consumption is recorded every day, a curve similar to that shown in Fig. 2 can be plotted to show that the rate of oxygen consumption is high initially but decreases continuously over time, approaching zero asymptotically. The fractional reduction in the amount of oxygen consumed each day is approximately the same, day by day, throughout the entire period of BOD exertion. When the rate of oxygen consumption is so small that it is no longer measurable, exertion of the first-stage BOD has been completed. In Fig. 2, which represents an idealized curve rather than actual data, the rate of oxygen consumption is no longer measurable at 20 days, and the ultimate, first-stage BOD is 5.8 mg/liter. The BOD_5 of the diluted sample is 4.4 mg/liter.

A mathematical equation is used as a generalization of the BOD curve. It is assumed that the rate of oxygen consumption at any time is proportional to the BOD remaining. This is expressed mathematically as $dy/dt = kL = k(L_0 - y)$, which integrates to

$$y = L_0(1 - e^{-kt}) \qquad (3)$$

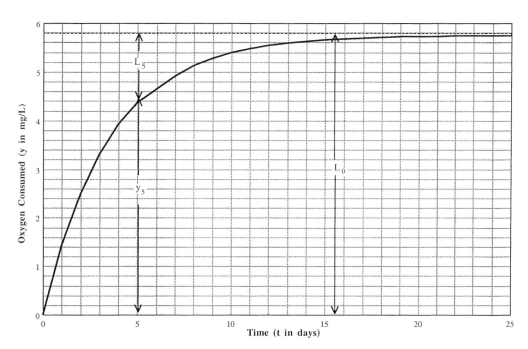

FIGURE 2 Curve representing oxygen consumption versus time during exertion of first-stage BOD.

in which y is oxygen consumed (milligrams per liter), t is time (days), and L is the BOD remaining (milligrams per liter). The parameters are L_0, the ultimate carbonaceous BOD, and k, the BOD rate parameter. The differential equation is analogous to first-order chemical kinetics if the ultimate carbonaceous BOD remaining is regarded as a reactant.

Although there have been several attempts, a convincing theoretical rationale for the assumption that the rate of BOD exertion is proportional to the concentration of ultimate carbonaceous BOD remaining has never been developed. Thus, equation 3 is merely a mathematical description of the curve of oxygen consumption versus time. The description is reasonably accurate if the proper conditions are maintained for the BOD test, but if the proper organisms are absent or toxic substances are present in inhibitory concentrations or an inorganic nutrient is not available in an adequate concentration, the course of BOD exertion can be quite different from that indicated by equation 3. Even though the parameters are empirical, L_0 is considered to be the theoretical amount of oxygen required for complete biodegradation of the organic matter present in the wastewater and k is interpreted as being analogous to a rate parameter for a first-order chemical reaction. Ascribing this type of theoretical significance to L_0 and k focuses attention on the waste organic matter rather than on the microorganisms which consume the oxygen, opening the way for misinterpretations.

Microbiological Phenomena Which Result in Oxygen Consumption

A major difficulty with raising the assumption that the rate of microbial oxygen consumption is proportional (under the proper conditions) to the concentration of ultimate carbonaceous BOD remaining to the level of theory is that there are several different microbiological phenomena which occur in a BOD bottle: (i) assimilation of organic matter, (ii) endogenous respiration, (ii) cryptic growth, (v) oxygen consumption by predators, (v) oxygen consumption by nitrifying bacteria, and (vi) oxygen consumption by algae. One assumption does not describe all of these phenomena, and none of them necessarily result in a rate of oxygen consumption which is proportional to the concentration of remaining ultimate BOD; however, the overall effect is a continuous decline in the rate of oxygen consumption from day to day.

The assimilation of the waste organic compounds by a microbial consortium which is growing in a wastewater results in a rate of oxygen consumption proportional to the rate at which the organic compounds are utilized. The course of oxygen consumption will be first order only if microbial assimilation is limited by the concentration of organic substrates and not by other factors (nutrients, inhibitory conditions, oxygen concentration, etc.). The BOD_5 test methodology is intended to produce the conditions under which concentration of organic substrates is limiting.

The microorganisms which directly assimilate the waste organic matter may be called the primary heterotrophs. While growth is occurring, some of the organic compounds are used to synthesize compounds which are stored in the microbial cells as a food reserve. Assimilation is considered to be the first step of BOD exertion, and when most of the organic compounds have either been oxidized or converted into new microorganisms or stored food reserve, it has been completed. The assimilation phase is followed by a period of endogenous metabolism which is manifest by oxygen consumed as the primary heterotrophs metabolize their stored food reserve. Endogenous metabolism is a mechanism by which the microbial cells meet their maintenance requirements, i.e., replace protein molecules and other cellular components which break down spontaneously. It does not result in an increase in the number of organisms, and after the stored food reserve has been depleted, the number of viable organisms decreases because some cellular components are breaking down and are not being replaced.

As some of the primary heterotrophs die, the cell membranes rupture and the cell contents spill out. This organic material supports the growth of secondary heterotrophs. This growth when there is an overall decrease in the total number of viable cells present is called cryptic growth. Oxygen consumption due to cryptic growth cannot be differentiated from the endogenous respiration of the primary heterotrophs. Secondary heterotrophs also include slowly growing bacteria and/or fungi which are able to assimilate the wastewater organic compounds which are not utilized by the primary heterotrophs.

In addition to the bacteria and fungi which are the primary and secondary heterotrophs, wastewater contains protozoa and microscopic invertebrates (rotifers, microcrustacea, etc.) which feed on the bacteria and fungi. Oxygen consumption by the predators occurs while the heterotrophs are growing as well as when they are in the endogenous phase, but this type of oxygen consumption is quantitatively much more significant during the endogenous phase. The three simultaneous phenomena (endogenous metabolism of the primary heterotrophs, cryptic growth of the secondary heterotrophs, and predation of protozoa and invertebrates on the bacteria and fungi) do not produce a rate of oxygen consumption which is proportional to the BOD remaining, but the rate of oxygen consumption does continue to decline from day to day.

Wastewater contains ammonia and proteinaceous material which will release ammonia when assimilated. Ammonia can serve as an energy source for nitrifying bacteria. Nitrification is an oxygen demand which is independent of the organic content of the wastewater or effluent. The oxygen demand per unit mass of ammonia is very high, theoretically 4.57 mg of oxygen consumed per mg of ammonia oxidized, and the nitrogenous oxygen demand of a treated effluent can be a major factor in the pollution of the receiving waters. However, nitrification in a BOD_5 test used for evaluating wastewater treatment is considered to be an interference with the determination of the carbonaceous oxygen demand, and nitrification inhibitors are added to the dilution water. Nitrogenous oxygen demand is not included in the assumption that the rate of BOD exertion is proportional to the BOD remaining, and thus is not included in equation 3, but its reduction is an important part of the effectiveness of wastewater treatment in preventing water pollution.

If biological wastewater treatment processes are exposed to sunlight, they often will grow significant amounts of algae. This is more common for attached growth processes (e.g., trickling filters), but the effluent weirs of the settling tanks of activated sludge processes sometimes develop long strings of filamentous algae. Any algal cells incorporated into a BOD sample will, when incubated in the dark, consume rather than produce oxygen, and this oxygen consumption is measured as part of the BOD. The algal contribution to the BOD_5 of an effluent is not related to waste organic matter originally present in the wastewater and is not proportional to the ultimate carbonaceous oxygen demand of the effluent.

BOD5 and the Kinetics of Biological Treatment Processes

For kinetic models of biological wastewater treatment processes, BOD$_5$ test results are interpreted as measures of the concentration of organic substrate which will support the growth of heterotrophs, despite of the fact that several phenomena other than assimilation occur in a BOD bottle. BOD$_5$ is considered to be a (more or less) constant fraction of the first-stage BOD (L_0), which is considered to be the oxygen equivalent of the assimilable organic matter. BOD$_5$ "removal" efficiency is calculated by equation 1 on the basis of the rationale that organic matter is a material which can be removed, although strictly speaking, BOD$_5$ is a parameter which can be reduced but not removed.

For many untreated industrial wastewaters, the interpretation of BOD$_5$ as a measure of the concentration of organic substrate which will support microbial growth is a reasonable approximation. Much of the waste material is raw organic matter, and the time of flow in the sewers is so short that little microbial growth occurs before the treatment plant. However, by the time the wastewater has passed through a biological treatment process, the assimilation of organic substrate is complete and the oxygen consumption measured during the effluent BOD$_5$ test is the result of endogenous respiration, the activities of secondary heterotrophs, predation, and possibly algal respiration. Clearly, considering effluent BOD$_5$ to be influent BOD$_5$ which was not "removed" during biological treatment is a misinterpretation.

If the BOD$_5$ is 70% of the ultimate carbonaceous BOD, the rate of BOD reduction is proportional to the ultimate carbonaceous BOD remaining, and the BOD$_5$ reduction is specified as 90%, what percentage of the BOD$_5$ of the raw wastewater has to be removed? The answer to this question is $(100)(0.9/0.7) = 128\%$, if it is assumed that the rate parameter, k, is the same for both the wastewater and effluent. This exercise is trivial in terms of evaluating wastewater treatment; it is presented here only to emphasize the facts that substrate assimilation must be completed during biological treatment and that the BOD$_5$ of the effluent is due to other phenomena.

Kinetic models of biological wastewater treatment circumvent the conceptual problem of the changing BOD phenomena by specifying that the effluent samples should be filtered to remove microorganisms before the BOD$_5$ determination is made. Thus, effluent BOD$_5$ is considered to be a result of the same phenomenon as the wastewater BOD$_5$, i.e., assimilation of organic matter not removed during treatment. However, the whole (not filtered) effluent is discharged to the receiving waters, and evaluating wastewater treatment on the basis of the BOD$_5$ of the filtered effluent uncouples the evaluations of efficiency and effectiveness of the treatment. The kinetic models used for the design of biological treatment processes cannot be used to predict if the effluent will be suitable for discharge without adding back the BOD$_5$ from respiration of microorganisms which were not removed by settling, the nitrogenous oxygen demand, and any algal oxygen demand.

In the case of untreated municipal wastewater, the interpretation of BOD$_5$ as solely a measure of organic substrate to be assimilated during biological treatment is considerably less accurate. The average time of flow in sewers ranges from hours to days, and a great deal of assimilation can occur during that time. The wastewater BOD$_5$ as well as the effluent BOD$_5$ may be due mostly to endogenous respiration, the activities of secondary heterotrophs, and predation. The primary job of the biological treatment process in this case will be the aggregation of the microorganisms which grew in the sewer rather than the assimilation of organic matter.

In spite of the conceptual problems of the "theoretical" interpretation of BOD$_5$ when, in fact, no coherent theory exists, there is no apparent candidate technique to replace it as the primary measure of both the efficiency and the effectiveness of wastewater treatment in relation to the possible deoxygenation of the receiving waters. The use of BOD$_5$ is well established by accumulation of decades of experience and recent regulatory usage. However, there need is for a more widespread understanding of the BOD$_5$ test as a bioassay and a more sophisticated interpretation of BOD$_5$ results as used for evaluation of wastewater treatment.

COD

The COD test was developed as a method for determining, in a short period of time, the oxygen equivalent of the organic content of wastewater. The procedure for determining COD is described in section 5220 of *Standard Methods* (1). The organic compounds are oxidized by dichromate ion in the presence of a catalyst which promotes the breaking of aromatic ring structures. The dichromate solution is standardized in terms of equivalent molecular oxygen. Most organic compounds are oxidized completely, or nearly so, but volatile compounds may be driven out of the sample before they are oxidized, and some aromatic, nitrogen-containing compounds are resistant to the oxidant. There are also interferences with the test caused by some inorganic compounds.

The COD test measures not only the oxygen equivalent of the waste organic matter but also that of the microbial cells. The oxygen demand associated with the microbial cells is only partially exerted during a BOD$_5$ test. Also, some of the organic compounds measured by the COD determination may not be metabolized by the microorganisms in either the BOD bottle or the biological treatment process. The COD is usually higher than the BOD$_5$ both in the wastewater and in the effluent, but in many cases, there is a consistent ratio between COD and BOD$_5$ although the COD/BOD$_5$ ratio for the effluent is different from the COD/BOD$_5$ ratio for the untreated wastewater. COD may be used as an index of BOD$_5$ for adjusting the operation of a biological wastewater treatment process as the strength of the waste varies from day to day. However, after 40 years in general use, the COD test has not replaced the BOD$_5$ test for evaluation of either the efficiency or the effectiveness of wastewater treatment, and there is no indication that it is likely to.

TOC

The TOC test is a direct measure of organic content rather than the oxygen equivalent of the organic content. The procedure for determining TOC is described in section 5310 of *Standard Methods* (1). The inorganic carbon is first driven out by acidifying the sample and purging with a purified gas. Then the carbon in organic compounds is oxidized to carbon dioxide by using heat and oxygen, UV light, chemical oxidants, or combinations of the oxidants, and the amount of CO_2 produced is measured by one of several possible methods.

TOC has not replaced either BOD or COD as a wastewater parameter, but it is used for different purposes. Determination of the TOC of a wastewater or effluent gives a separate measure of the level of contamination. TOC is also used for measuring concentrations of hazardous wastes in water when toxicity or resistance to biodegradation would interfere with the BOD$_5$ test. Neither the BOD$_5$ test nor the COD determination is useful for measuring contaminant levels in the ranges of milligrams or micrograms per liter. Concentrations in those ranges are seldom of consequence in relation to the deoxygenation of receiving waters unless the contaminants are toxic. In recent years, various versions of the TOC test have been used more extensively in the water treatment field than in the wastewater treatment field.

PLANT NUTRIENTS: ALGAL GROWTH TEST

As explained previously, the plant nutrients which may be present in wastewater are compounds of nitrogen and phosphorus. There are suitable chemical methods for determining the total amounts of N and P and also for measuring the concentrations of specific compounds containing one or the other. However, not all of the N and P in an effluent is necessarily in a form which will stimulate algal growth, and there may be effluent–receiving water interactions which affect the availability of the nutrients to the algae. Thus, an algal growth test has been developed to measure the relative potential of an effluent to support the growth of algae and aquatic macrophytes (seed plants). Untreated wastewaters are seldom tested by the algal growth test; it is used as a measure of effectiveness rather than efficiency of wastewater treatment.

The algal growth test can be used to determine what reductions in nitrogen and phosphorus are necessary to protect the water quality downstream of the treatment plant. It can also be used to determine the feasibility of nutrient criteria for the treatment plant effluent, e.g., to establish a 1.0-mg/liter total P limitation. Algal growth test studies can be used to measure the bioavailable nitrogen and phosphorus in the plant effluent as well as to assess the trophic status of the receiving water body upstream and downstream of the point of discharge.

The procedure for performing the algal growth test for nutrient studies, which differs in some details from the test used to measure chronic toxicity, was published by the U.S. Environmental Protection Agency (EPA) (10). Section 8111 of *Standard Methods* (1) also describes algal assays for evaluating the nutritional status of water. These tests are based on Liebig's law of the minimum, which states that the maximum yield is proportional to the concentration of the nutrient which is present and biologically available in the lowest concentration in relation to the growth requirements of the organisms.

The EPA method (10) employs a single test alga, *Selenastrum capricornutum*, while *Standard Methods* (1) allows for the use of any of six different freshwater algae, including *S. capricornutum*, and three marine algae. The use of a single test algal species makes comparison of data obtained with different waters easier, and the understanding of the algal growth restrictions increases as more data become available for a single test species. The benefits of using a standard test organism outweigh the benefits of using an indigenous species, and *S. capricornutum* and indigenous algal species will produce parallel growth yield responses. For these reasons, it is prudent to follow the EPA recommendation and use *S. capricornutum* in all algal assays.

Detailed procedures for conducting the algal assays are found in reference 10, but it is worthwhile to point out several considerations. (i) Sample preparation is dictated by the purpose of the study being conducted, but in all cases, the sample must be filtered to remove indigenous algae. For nutrient studies, the sample can be autoclaved and then filtered (0.45-μm-pore-size filter) to determine the amount of algal biomass that can be grown from all nutrients in the water. For a study of the effects of complex wastes, the sample is simply filtered and not autoclaved. (ii) It is necessary to add disodium EDTA as a chelating agent to algal culture media and to test water at a concentration of 1 mg/liter. It was determined empirically that this concentration makes trace metals, particularly iron, biologically available while not complexing other essential nutrients, such as Ca and Mg, or heavy metals which may be toxic to algae. (iii) The basic test design to determine which nutrient is limiting involves additions of N, P, wastewater, etc., depending on the specific purpose of the test. Adding materials other than the growth-limiting nutrient does not increase the growth yield. (iv) The minimum chemical data needed to evaluate the assay response include initial pH, total P, P$_i$, nitrite-N, nitrate-N, ammonia-N, and total Kjeldahl-N.

Bioavailable N and P are determined from the dry weight of the maximum standing crop of *S. capricornutum* after 14 days of incubation. Dry weight of the algal biomass can be directly determined gravimetrically, or an electronic particle counter can be used to determine biomass indirectly. Each microgram of P will support 0.430 mg (± 20%) (dry weight) of *S. capricornutum* if other constituents are not growth limiting. Each microgram of N will support 0.038 mg (± 20%) (dry weight) of *S. capricornutum* if other constituents are not growth limiting.

The algal growth test is not normally performed on a routine basis on effluents. Results of chemical analyses provide day-to-day information on nutrient loadings on the receiving waters. The algal growth test makes it possible to differentiate the amounts of nutrients available for algal growth from the total concentrations of nutrients determined by chemical analyses. Algal growth tests can be used in studies designed to determine what reductions in the N and P content of effluents would be necessary to improve water quality downstream of a treatment plant. Such studies involve measuring the bioavailable N and P in the effluent as well as assessing the trophic status of the receiving water before and after addition of the effluent.

Those waters containing greater than 0.015 mg of bioavailable P per liter and 0.165 of bioavailable N per liter are considered eutrophic. In general, these values correlate well with the P$_i$ and total soluble N concentrations in the water. The procedure for performing the algal growth test in nutrient studies differs in some detail from the algal bioassay used to detect chronic toxicity (see below).

TESTING FOR EFFLUENT TOXICITY

Toxicity is a property of many wastewater components which may be manifest in the receiving waters when the component is present in the appropriate concentration range. The Federal Water Pollution Control Act Amendments of 1972 (Public Law 92-500), the Clean Water Act of 1977 (PL 95-217), and the Water Quality Act of 1987 (PL 100-4) all state that it is the national policy that the discharge of toxic substances in toxic amounts to the nation's waters is prohibited. Toxic discharges are regulated by National Pollutant Discharge Elimination System (NPDES) permits, which are issued by state authorities and subject to EPA regulations and approval. All dischargers are required to report data as specified in their NPDES permits in order to comply with limitations on the concentrations of specific chemicals in their effluents. However, it is not feasible to specify limits for all the potentially toxic compounds which might be present in an effluent.

The EPA's surface toxics control regulation (11a) established specific requirements that an "integrated approach" be used in water quality-based toxics control. The integrated approach refers to the performance of bioassays or whole effluent toxicity (WET) tests, in addition to the collection of chemical data. The WET approach is a better way to protect aquatic life and human health because it is simply not feasible to analyze effluents for all possible industrial chemicals. Also, chemical data do not provide knowledge about the biological effects of various chemicals acting synergistically. Bioassays provide information about the effects of effluents on living organisms (12).

The EPA's WET monitoring policy (15) states that ". . . where appropriate, the permitting authority should impose WET monitoring condition upon dischargers . . ." Technical Support Document for Water Quality-Based Toxics Control (13) further states that as techniques become available for implementing biologically based criteria, they too will be integrated into the water quality-based toxics control. This document refers to the performance of field surveys to assess the biological integrity of water, an approach that is very important for control of non-point source pollution but has limited applicability to the control of point source pollution. Thus, a triad of approaches (WET testing, biologically based water quality criteria, and field surveys) to prevent the discharge of toxic chemicals to the nation's waterways will be implemented.

The procedures for bioassays to assess toxicity of effluent are described in section 8100 of *Standard Methods* (1) Bioassays using algae, invertebrates, and fish are included in that section, but the bioassay using luminescent bacteria is not a standard method. The bioassays considered in this chapter are those using algae and those using bacteria.

Toxicity bioassays using fish (Fig. 3) or invertebrates cannot be conducted with the same ease and frequency as chemical analyses such as ammonia determination. The performance of bioassays requires the culture and maintenance of test organisms, special laboratory facilities including environmental chambers to house test organisms, and trained technicians working under the supervision of a biologist or aquatic toxicologist. Each test takes from several days (acute tests) to more than a week (chronic tests) to perform. Furthermore, the number of tests that can be conducted concurrently is limited by space and available personnel. The number of bioassay tests that can be performed at one time in a given laboratory is very small compared with the number of chemical test that can be performed in the same amount of space.

Very few dischargers are able to perform conventional bioassays with fish or invertebrates in house. Facilities and personnel costs eliminate that as an option. Yet dischargers should not have to rely completely on outside testing laboratories to investigate the possible toxicity of their effluents. Cost-effective bioassays using microorganisms that can be performed quickly and easily are needed for this purpose (4).

The conditions of NPDES permits for many municipal agencies require that if toxicity is found to be associated with the effluent, the agency may be required to conduct a toxicity identification evaluation (TIE) and a toxicity reduction evaluation (14). The TIE and TRE may involve a large number of bioassays to identify the cause of toxicity in an effluent and how to reduce it. The important information to be obtained from the bioassays for a TIE or TRE is the relative toxicities of various sources of wastewater discharged to the sewers. Bioassays using microorganisms are an efficient and economical way to obtain this information.

Bioassays Using Bacteria

The most widely used WET bioassays using microorganisms are the Microtox tests (Microbics Corporation, 2232 Rutherford Road, Carlsbad, CA 92008; phone: [619] 438-8282). There are three versions of the bioassay: an acute test, a chronic test, and a mutagenicity test (Mutatox). The basic Microtox test protocol is an official standard of the American Society for Testing and Materials (standard D-5660 under ASTM Committee D-34 on Waste Management).

The Microtox acute toxicity test system was developed to provide a standardized means for monitoring aquatic toxicity (6). The Microtox reagent (luminescent bacteria) is stored frozen until the test is to be performed. The luminescent bacteria are exposed to a wastewater sample for 30 min to perform a surrogate bioassay. The test can be conducted by an operator after several days of training, using about 25 ft^2 (ca. 2.3 m^2) of bench space. The cost of each test is a fraction of that for bioassays which utilize either fish or invertebrates. Conventional bioassays and the Microtox test are compared in Table 2.

The concept behind the use of surrogate bioassays is not to replace conventional bioassays but to allow more testing to be done. Chemical analysis results and data from short-term surrogate bioassays such as the Microtox analysis can be gathered and analyzed to provide background comparative information before numerous WET tests using higher organisms are performed. This will make it possible to know when and where to conduct whatever limited number of WET tests can be performed so as to maximize the useful information derived from the WET tests. The results of bacterial bioassays for toxicity provide a way to direct the utilization of limited resources which may be available for performing conventional bioassays.

Kaiser and Palabrica (8) compared EC$_{50}$ values from Microtox acute tests with LC$_{50}$ values from fish toxicity bioassays for individual chemicals (for definitions, see Table 2, footnote b). They found high linear correlations of the Microtox test data with the acute toxicities of a large variety of organic chemicals to several fishes, demonstrating that the Microtox test is, at least to some extent, predictive of acute fish bioassay results, making it a good test for screening. Intralaboratory coefficients of variation of 10 to 20% indicate that the Microtox basic test is reasonably reproducible (6). Intralaboratory coefficients of variation for conventional bioassays with higher organisms reported by EPA (14) were generally much higher.

Comparison of the results of conventional bioassays and Microtox testing on complex effluents have validated the Microtox test as a useful means of identifying toxic effluents, even though in some instances it may be less sensitive (7) or more sensitive (9) than bioassays with higher organisms. It is not necessary that the Microtox test have the same sensitivity as fish and invertebrate bioassays to be a useful tool in monitoring for aquatic toxicity.

The greatest amount of information can and should be gained from running a battery of bioassays with different organisms. Such a battery might include test organisms at three trophic levels: bacteria, *Ceriodaphnia* species, and fish, for example. Bacterial bioassay data should not replace data from fish and invertebrate bioassays, but they are appropriate for screening and may be used in conjunction with other bioassays.

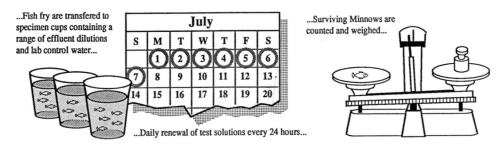

1 Maintenance of Cultures

Minnows are transferred to breeding tanks where eggs are laid on the under-side of tiles...

When eggs develop eyes, tiles are transfered from breeding tanks to beakers where the eggs hatch...

STOCK TANK **BREEDING TANK** **HATCHERY**

2 Conducting Bioassay 7 day Test

...Fish fry are transfered to specimen cups containing a range of effluent dilutions and lab control water...

...Surviving Minnows are counted and weighed...

July						
S	M	T	W	T	F	S
	1	2	3	4	5	6
7	8	9	10	11	12	13
14	15	16	17	18	19	20

...Daily renewal of test solutions every 24 hours...

3 Calculate NOEC's for survival and growth using Dunnet statistic

...80% of controls must survive. Average weight of controls must be 0.25 mg.

FIGURE 3 *Pimephales promelas* larval survival and growth test. Bioassays for toxicity of wastewater and effluents using fish require large amounts of laboratory space and technician time. NOEC, no-observed-effect concentration. Illustration and graphic design by Stephen Brayfield.

The Microtox acute toxicity test employs a specific strain of a marine bacterium which emits light as an end product of its respiration. This strain, *Photobacterium phosphoreum*, also referred to as *Vibrio fischeri*, is deposited with the Northern Regional Research Laboratory in Peoria, Illinois, and designated NRRL B-11177. Microbics cultures these bioluminesent bacteria, harvests and lyophilizes them, and markets them as Microtox reagent. The lyophilized bacteria are rehydrated with Microtox Reconstitution Solution to provide a ready-to-use cell suspension.

In the Microtox test, these organisms are exposed to a test sample to measure the toxic effect of the sample on the organisms. The light output of the luminescent bacteria is measured before and after the bacteria have been challenged by a sample of unknown toxicity. A difference in light output is attributed to the effect of the sample on the organisms. The light of a control (reagent blank) that contains no sample is used to correct for light lost naturally (not as a result of toxicity). The endpoint of the bioassay is expressed in terms of an effective concentration that causes a preselected percent of light loss at a particular time. For example, the $EC_{50}(5)$ is the effective concentration of a sample causing a 50% decrease in the light output in 5 min.

Toxicity Bioassays Using Algae

The algal toxicity bioassay involves exposing a growing culture of *S. capricornutum* to a range of dilutions of an effluent for a 96-h incubation period and measuring the response, in terms of cell density, biomass, chlorophyll content, or absorbance, relative to that of an inoculated control of stock culture medium (Fig. 4). In the absence of materials toxic to the alga, the maximum standing crop will be proportional to the initial concentration of limiting nutrient available. Macro- and micronutrients are added to the effluent being assayed so that, at a minimum, it contains the same concentrations of nutrients as the stock medium. If the maximum

TABLE 2 Comparison of various types of WET tests

Organism (reference)	Type	Duration	Person-h[a]	Endpoint	Expression[b]
Ceriodaphnia dubia (13)	Acute	48 h	7	Survival	LC_{50}
C. dubia (11)	Short term, chronic	96 h	28	Survival	NOEC
				Reproduction	NOEC; IC_{25}
Pimephales promelas (12)	Acute	7 days	24	Survival	LC_{50}
P. promelas (11)	Short term, chronic	7 days	24	Survival	NOEC
				Growth	NOEC; IC_{25}
Selenastrum capricornutum (10, 11)	Short term	96 h	12	Growth	NOEC
Salmonella typhimurium	Genotoxicity	48 h	4	Induced reversions	MR
Photobacterium phosphoreum	Acute	5–30 min	1–2	Light loss	EC_{50}
	Chronic	20–24 h	1–2	Light loss	EC_{50}
	Genotoxicity	24 h	1–2	Light production	MR

[a] Estimates are based on the time needed to conduct the tests in the Metropolitan Water Reclamation District of Greater Chicago Biomonitoring Laboratory. Time spent maintaining cultures is not included.

[b] LC_{50}, point estimate of the concentration required to kill 50% of the test organisms; NOEC, no-observed-effect concentration; IC_{25}, point estimate of the concentration which will cause a 25% reduction in the observed response of the test organism; EC_{50}, point estimate of the concentration which will reduce the light output by 50%; MR, mutagenicity ratio.

[c] Mutatox test.

standing crop is significantly smaller in the cultures containing the effluent than in the stock culture medium, the presence of a toxic material is indicated.

The procedure for performing this assay in 125- or 250-ml flasks was developed and described by the EPA (11). A more economical version of the assay using 96-well microtiter plates has been developed (5). Problems with volatility and cross-contamination of wells have been reported with the microtiter plate version of the assay, but these should be resolved in the near future (3). The algal toxicity bioassay is particularly useful for evaluating effluents of plants with a large percentage of industrial wastes in their influent.

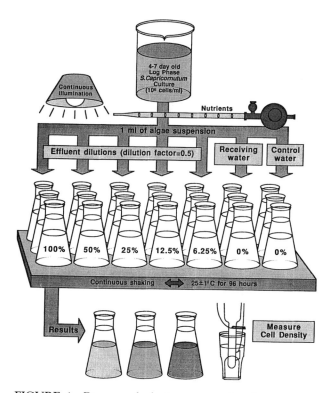

FIGURE 4 Diagram of a basic test procedure for a toxicity bioassay using an alga. Graphic by Stephen Brayfield.

SUMMARY

The efficiency of wastewater treatment is measured by the amount of a waste material separated from the wastewater and often is expressed as percent removal (equation 1) or as waste load reduction (equation 2). Since many different waste materials are carried by water, it is possible to calculate a number of different efficiencies for any wastewater treatment plant. In contrast to efficiency, the effectiveness of wastewater treatment is related to the potential of the effluent to cause pollution without considering how much waste is separated during treatment. Both the efficiency and the effectiveness of wastewater treatment plants need to be evaluated.

Because of their large number and variety, it is not feasible to measure the concentrations of all waste materials, and evaluation of the effectiveness of wastewater treatment depends to a large extent on the use of a few parameters of pollution which are aggregate measures of specific pollution effects, e.g., transmission of waterborne disease (indicator organism density), deoxygenation of the receiving waters (BOD_5), and toxicity to aquatic organisms (bioassays). The individual pollution effects occur independently (although two or more often occur simultaneously), and a thorough evaluation of the accomplishments of wastewater treatment requires the use of several parameters of pollution.

The magnitudes of the parameters of pollution are reduced during wastewater treatment as waste materials are removed. Parameters are properties of the wastewater components, not "things" which can be removed from the water. Confusion between wastewater components and their properties can result in the term "removal" being used in place of "reduction." This, in turn, can lead to misunderstanding of the objectives of, and misinterpretation of the data representing the accomplishments of, wastewater treatment. The major example of this is the widespread use of the term "BOD removal," which indicates confusion between certain waste organic compounds and their property of causing exertion of an oxygen demand. Even though BOD exertion is known to be the result of microbial metabolism and at least five different metabolic phenomena are involved in BOD exertion, BOD_5 is often treated as if it were a material which can be removed from wastewater.

In sum, microbiological techniques are used extensively

for evaluation of the effectiveness of wastewater treatment. Techniques for detecting pathogenic microorganisms and indicator organisms are described in other chapters of this section.

Illustrations were prepared by Stephen Brayfield, Brayfield Graphics.

REFERENCES

1. **American Public Health Association.** 1992. *Standard Methods for the Examination of Water and Wastewater*, 18th ed. American Public Health Association, American Waterworks Association, and Water Environment Federation, Washington, D.C.
2. **Beckman Instruments, Inc.** 1992. *Microtox™ System Operating Manual*. Beckman Instruments, Inc., Carlsbad, Calif.
3. **Blaise, C. R.** 1993. Practical laboratory applications with micro-algae for hazard assessment of aquatic contaminants, p. 83-107. *In* M. Richardson (ed.), *Ecotoxicology Monitoring*. VCH Publishers, Weinheim, Germany.
4. **Environment Canada.** 1992. *Biological Test Method: Toxicity Test Using Luminescent Bacteria (Photobacterium phosphoreum)*. Report EPS 1/RM/24. Environment Canada, Ottawa, Ontario, Canada.
5. **Environment Canada.** 1992. *Biological Test Method: Growth Inhibition Test Using the Freshwater Alga, Selenastrum capricornutum*. Report EPS. Environment Canada, Ottawa, Ontario, Canada
6. **Isenberg, D. L.** 1993. The Microtox™ toxicity test: a developer's commentary, p. 3–15. *In* M. Richardson (ed.), *Ecotoxicology Monitoring*. VCH Publishers, Weinheim, Germany.
7. **Johnson, I., R. Buffer, R. Milne, and C. J. Redshaw.** 1993. The role of Microtox™ in the monitoring and control of effluents, p. 309–317. *In* M. Richardson (ed.), *Ecotoxicology Monitoring*. VCH Publishers, Weinheim, Germany.
8. **Kaiser, K. L. E., and V. S. Palabrica.** 1991. *Photobacterium phosphoreum toxicity data index*. *Water Pollut. Res. J. Can.* **26:**361–431.
9. **Munkittrick, K. R., E. A. Power, and G. A. Sergy.** 1991. The relative sensitivity of Microtox™, daphnid, rainbow trout and fathead minnow acute lethality tests. *Environ. Toxicol. Water Qual.* **6:**3–62.
10. **U.S. Environmental Protection Agency.** 1978. *The Selenastrum capricornutum Printz Algal Assay Bottle Test*. EPA-600/9-78-018. U.S. Environmental Protection Agency, Corvallis, Oreg.
11. **U.S. Environmental Protection Agency.** 1989. *Short-term Methods for Estimating the Chronic Toxicity of Effluents and Receiving Waters to Freshwater Organisms*, 2nd ed. EPA-600/4-89-001. U.S. Environmental Protection Agency, Cincinnati, Ohio.
11a. **U.S. Environmental Protection Agency.** 1989. National pollutant discharge elimination system; surface water toxics control program; final rule. *Fed. Regist.* **54:**23868–23899.
12. **U.S. Environmental Protection Agency.** 1991. *Methods for Measuring the Acute Toxicity of Effluents and Receiving Waters to Freshwater and Marine Organisms*, 4th ed. EPA-600/4-90/027. U.S. Environmental Protection Agency, Washington, D.C.
13. **U.S. Environmental Protection Agency.** 1991. *Technical Support Document for Water Quality-Based Toxics Control*. EPA-505/2-90/001. U.S. Environmental Protection Agency, Washington, D.C.
14. **U.S. Environmental Protection Agency.** 1991. *Methods for Aquatic Toxicity Identification Evaluations: Phase I. Toxicity Characterization Procedures*, 2nd ed. EPA-600/6-91/003. U.S. Environmental Protection Agency, Washington, D.C.
15. **U.S. Environmental Protection Agency.** 1994. *Whole Effluent Toxicity Monitoring Policy*. EPA-833/B-94/002. U.S. Environmental Protection Agency, Washington, D.C.

AQUATIC ENVIRONMENTS

VOLUME EDITOR
CHRISTON J. HURST

SECTION EDITORS
STEVEN Y. NEWELL AND ROBERT R. CHRISTIAN

Overview of Issues in Aquatic Microbial Ecology

ROBERT R. CHRISTIAN AND DOUGLAS G. CAPONE

25

Aquatic microbiology has advanced greatly during the past 25 years concomitantly with the development of new methods and technologies. It has often been stated that the field of microbial ecology in general is methods limited (16, 38, 82). Certainly, we are able to address specific questions now that were not tractable before. We now have estimates of process rates; standing stocks of elements, compounds, and biomass; and densities of organisms that could not have been obtained without the advances in methodology that have occurred. This is a recurring theme throughout this Manual.

To a large extent, the approach used by aquatic microbiologists to determine a particular aspect of microbial community structure or activity is dictated by the research objectives and perspective. Depending on the intent and questions, very different methods may be used. A primary objective of traditional microbiological studies has been, and remains, the identification of the physiological diversity of microorganisms in nature. The field of microbial ecology more specifically seeks to identify the roles and interactions of microorganisms in the environment (9). Environmental microbiology considers the interactions of microbes in nature with respect to interactions, both positive and negative, with human society. Microbial biogeochemistry is interested in the quantitative significance of the microbiota in elemental transformations in the environment. While each perspective has validity within its own realm, the approaches that each uses represent a continuum in terms of our understanding of the role of microbes in the environment, moving from identification and recognition to potential roles to quantification of environmental processes and their regulation. The perspectives of this section are primarily those of microbial ecology and biogeochemistry. However, there are several broader questions that underpin the science and its various subdisciplines which are similar to those that were articulated by Hungate (40) in 1962: "(1) What kinds of organisms occur? (2) What activities do they perform and how are these activities interrelated? (3) In what numbers does each kind of organism occur? (4) What is the magnitude of its activity, and what factors influence this magnitude?"

Today's versions of these questions are similar but not necessarily the same; they have gained maturity. The first question has been given new meaning. Progressing from the 1960s through the 1980s in aquatic microbial ecology, less emphasis was given to identifying species. Physiological groupings, guilds, and communities were often the hierarchical level of interest. As a result of the advances in molecular biology and nucleic acid technologies, we now recognize new groups of organisms to characterize and identify and new ways to do the work. Also, we can add the question, "What is the basis for the evolutionary heritage of an organism?" Direct access to the genotype has truly altered the field.

The second of Hungate's questions was founded in the tradition of laboratory culturing, pure and mixed. Today this question is addressed with both laboratory and in situ studies. But regardless of how the question is addressed, we continue to discover novel and interesting microbial activities with ecological relevance.

When Hungate formulated the third question, quantification of organisms in microbiology was largely through enumeration of "viable" species populations. In the years that have followed, a wide variety of techniques have been developed that either have improved our ability to enumerate or have allowed estimation of cell constituents. Again, emphasis was not necessarily at the species level and has often shifted to estimation of standing stocks of elements or biomass in the microbial community or its components. The "kind of organism" has been at a higher hierarchical level. With the new molecular techniques, there is a resurgence of research designed to address this question at the level originally asked.

If one question has dominated modern aquatic microbial ecology, it is the fourth question, addressing the magnitudes of activities and their control. The other questions have often been subsumed within this larger issue. Thus, the abundances of different kinds of microbes have been determined in the context of their processing of matter or energy. In fact, energy flow and elemental cycling have been two major themes of much of aquatic microbial ecology, as reflected in this section. Growth and the acquisition and metabolic processing of materials have been the activities whose magnitudes need to be measured. Since Hungate's statements, development of methods has continually focused on estimation of in situ rates. With this focus has been continued emphasis on factors that influence these in situ rates through both field observations and experiments done in the field and laboratory.

Coincidently, Hungate's questions were written at the beginning of what Tiedje considered a new "era" in microbial ecology (80a). Tiedje designated the first era the period of "discovery of biochemical diversity (1880–1930)." At this time, the new techniques of enrichment, selection, and pure culture helped direct the field. He considered the period from 1930 to 1960 the era of "enumermania." Using procedures to measure CFU and viable organisms, microbial ecologists enumerated populations in habitats throughout the world. As Hungate wrote his questions, we entered the era of the "process level" study (1960 to 1990). As mentioned, the magnitude of activities and their controls were the topics of many studies using new chemical and microscopical techniques. Now, Tiedje indicates that we have entered a new era—one of "populations and communities" (1990 on). This era has been spurred by the advances of molecular biology. Therefore, the first two qualitative questions posed by Hungate were largely the ones to be addressed during the first era. During the second era, the third question dominated research, although we now know that results were less than complete. The latter two quantitative questions were the focus of the third era, but at a higher level in the hierarchy (e.g., guilds and communities) than originally implied. It is the promise of the new era that the hierarchical level of study can be brought back to the species. In this section, the reader will find a blend of methodologies which bridge these last two eras.

We see, therefore, that the ability to address all of Hungate's questions did in fact await the advent of new methods. Further, the desire to ask new questions or forms of the questions prompted new methods. Kuhn (52) describes this interactive process as part of "normal science." He suggests that most scientific study occurs within the context of a "paradigm" or accepted perceptions and rules concerning the field. "Scientific revolutions" occur when old paradigms are no longer tenable and a new one is introduced. In aquatic microbial ecology, major paradigms have their roots near the turn of the century or before: the dominant roles of microorganisms in decomposition and elemental cycling (58, 83, 84), the ubiquity of microorganisms, the diversity of their physiologies, and the unity of their biochemistry.

One might, however, consider that in the 1970s two topics were novel enough to change the way aquatic microbial ecologists thought about their field. The finding of deep-sea hydrothermal vents opened new venues for research (19, 44). The result has been an expansion of the boundaries of previous knowledge that forced consideration of options not obvious before the discovery. One now may more readily recognize that energy sources other than light can support a community, that life may occur above 100°C, and that mutualisms may take forms not previously acknowledged. Such recognition was possible before but was overshadowed by more traditional views of life both at the Earth's surface and in the deep sea. A second paradigm shift occurred in aquatic ecology that had its roots in microbial ecology. It involved the recognition of the importance of the microbial food web and "loop" (5, 71, 88). This paradigm revised our view of both the trophic structure and nutrient cycling in aquatic ecosystems. Much of the research in aquatic microbial ecology since this recognition has either focused on the topic or alluded to it. If we consider the "revolution" caused by molecular biology, we find ourselves in a time of increased opportunities for answering both old and new questions.

In this section, we describe methods applicable to studying the community structure and activities of microorganisms in aquatic environments. The world's aquatic environment is indeed large and diverse. Microbes live within a variety of habitats and interact with a variety of other microbes and macroorganisms. These varieties influence our concerns about methodology in at least two ways. First, the diversity of microbes has evolved to conduct an assortment of activities requiring diversity in methodology for discernment. Second, the diversity in habitat conditions often precludes the universal use of methods. Modifications for these habitat differences are routine within the field. In this chapter, we discuss some of the issues associated with these diversities and the applications of methods to answering the present-day versions of Hungate's questions.

MECHANISMS ESTABLISHING COMMUNITY STRUCTURE AND ACTIVITY

In this section, we consider three arenas of the aquatic environment: the water column, sediments, and biofilms. Many of the described methods are specifically for one of these environments. Unique properties of the ecosystems in these environments often warrant separate consideration. However, the reader should recognize that overlaps exist among methods for the different environments and indeed among the sections in this manual. Therefore, one may wish to consult multiple chapters in this section and other sections when developing methods.

Within each of the environments mentioned above exist many variations in both abiotic and biotic factors that influence microbial activity and abundance. Herein, we briefly describe these factors and provide references for more information. We consider that any activity has a maximum rate relative to the organism (e.g., per unit organism or per unit cell) or to the sampled system (e.g., per unit volume or unit area). The maximum rate can be the optimum rate for those organisms. For example, the maximum rate of growth for a bacterial species may also be the optimum rate for the continued survival and success of the species. For many microbial activities that are studied this is true, but the two may be separate. The optimum rate may be less than the maximum (74). Environmental factors may be considered to regulate activity relative to the maximum rate. Prevention of the maximum can be through limitation by insufficient supply or through inhibition by overabundance. Therefore, each factor has a distribution over which it may be limiting or inhibiting to an activity. In fact, these distributions are likely to be dependent on other conditions influencing the activities. Hungate's fourth question can be restated as, "What is the magnitude of its activity relative to the distributions of factors which influence the organism?"

A list of abiotic factors that influence microbial activity can be found in nearly any general microbiology text. Also, these abiotic factors and their influences have been reviewed in a variety of publications. (Representative references are cited below.) They are often discussed in the context of extreme environments and include such items as radiation (26, 87); hydrogen ion concentration (31, 42); temperature (7, 22, 35); pressure (60); water availability (10) and ion requirements (27); energy sources, organic matter, and growth factors (24, 30, 59); and terminal electron acceptors, oxygen, and redox potential (61). All entries in this list are known to affect activities and community structure within aquatic environments. They may also have an effect on our ability to measure the activities and structure. We recommend Prosser's text (72) for a general account of the effects of these factors on organisms and organismal adaptation. Books and monographs which address the effects of environmental factors on microbes include those by or edited by Codd (17), Herbert and Codd (36), Edwards (21), Jennings (45), and Lowe et al. (55).

Less attention in aquatic microbial ecology has been given to the physics of movement of water and materials in water relative to microbial activity and migration. Movement of and in water occurs under the control of a variety of mechanisms (e.g., molecular diffusion, tidal exchange, and the Coriolis effect). The dominance of one mechanism over another and the subsequent interaction with chemical and biological activity depend on the spatial and temporal scales considered (46, 51, 63, 77). The time and space scales of various physical, chemical, and biological processes can be ordered or compared to gain insight into the relative importance of each in establishing patterns of growth (89). For example, Sverdrup (80) compared depths of integrated planktonic primary productivity and respiration with the depth of the mixed layer to predict vernal blooms in the ocean. Patchiness and the mode of sampling and analysis of different chemical species in an estuary were related to the relative turnover times from planktonic and benthic exchanges and water movement (41). There have even been attempts to relate the success of phytoplankton to interaction with nutrient patchiness at the microscale (29, 43). Unfortunately, most microbial ecologists lack the skills to evaluate many of the issues related to hydrodynamics. Collaboration with physical scientists or engineers is warranted (63).

In contrast to the relatively few studies on hydrodynamic interactions with microbes, those on biotic interactions are numerous. General microbial ecology texts such as those by Atlas and Bartha (4) and Stolp (79) have substantial sections on biotic interactions. Further, Bull and Slater (11) have edited a book on the topic. Their second chapter classifies interactions "based on effects," as you would find in most books (e.g., mutualism, competition, predation, etc.). As indicated earlier, the importance of mutualisms has been given new meaning with discovery of deep-sea communities based on chemoautotrophy. Also, microbial trophic interactions have been studied considerably in the past decade in association with the microbial loop (88). Bull and Slater (12) discuss limitations to the scheme of classification "based on effects." They provide an alternate classification after Slater (78). Microbial communities are classified according to seven different biological mechanisms: (i) provision of specific nutrients by different members of the community; (ii) the alleviation of growth inhibition; (iii) interactions which result in the alteration of individual population growth parameters, therefore producing more competitive and/or efficient community performance; (iv) a combined metabolic activity not expressed by individual populations alone; (v) cometabolism; (vi) hydrogen transfer reactions; and (vii) the interaction of several primary (i.e., dominant) species. These categories serve as an alternate perspective to evaluating biotic interactions.

APPROACHES TO COMMUNITY STRUCTURE ASSESSMENT

The microbial community represents the composite of the various populations. The actual definition of what is a community is not universally agreed upon by ecologists. Some consider that a community incorporates species that interact (e.g., competition or predation) within a location, whereas others consider it to be the sum of all species of a particular taxon within a location (3, 75). We refer the reader to Ludwig and Reynolds (56) for mathematical methods of indexing and analyzing communities. In this section, we focus on what is necessary for assessing the individual species or physiologically based taxa.

Our knowledge of microbial processes in nature has traditionally begun with the isolation and identification of the organism of interest. The development of procedures for the routine enrichment and enumeration of specific populations or types of microbes, springing directly from laboratory studies of bacterial growth physiology, fostered much of the initial insights gained in microbial ecology. However, there are clear limitations to multiple tube tests with most probable number analysis or CFU types of enumeration procedures (e.g., reference 25) if the question at hand is to gain an accurate and unbiased assessment of the population sizes of relevant physiological groups (2, 76). Conventional population assessment methods suffer, as they are based on selective enrichment of organisms. Alternative immunological and molecular approaches are rapidly emerging, are valuable adjuncts to the classical methods, and avoid the inherent biases of enrichment-based methods. The biomarker approach advanced by White and his colleagues uses the distribution of relevant biochemical markers as a distinct and unbiased means of determining the presence, distribution, and biomass of particular functional classes of organisms in the environment (86; see also reference 62).

APPROACHES TO ACTIVITY ASSESSMENT

Knowledge of the distribution of organisms can provide information only on their potential activities and roles in material cycling. Various approaches may be taken to assess directly the microbial processes in particular environments and estimating their rates. These range from those dependent on observations to more experimental approaches which require some degree of sample perturbation.

Observations of the distributions of chemical and biochemical species in the environment over appropriate scales of space and time have provided important direct evidence of biological or microbiological processes in the environment. (At the grand scale, Lovelock [54] predicted no life on Mars and developed his theory of Gaia from his evaluation of the disequilibrium of atmospheric gases on Earth.) Success of an observational approach often depends on the nature of the physical environment. For example, chemical gradients, say through the water column of a lake or ocean, may develop only where there is a degree of hydrological stability. Transient maxima of indicator chemical species, or anomalies relative to distributions predicted by physical diffusion, often provide important evidence of zones of microbial activity. Also, researchers have begun to study natural isotope abundance distributions intensively. This mode of study continues to develop and show great promise in providing a new level of insight into biological transformations for a range of biogeochemically active elements.

Direct assays of isolated samples are the most familiar means of demonstrating, at least qualitatively, the presence of a particular pathway or process in an environmental sample. Although not always the case, we often find ourselves moving from relatively insensitive or more qualitative methods, and those that may substantially perturb the original system, to those providing more sensitive, quantitative information and less disruption of the sample.

At a relatively simple level, incubations which monitor short-term changes in a relevant substrate or product pool can give us direct information on a particular microbiological process or transformation which impinges upon that pool. However, in the environment, such pools may exist near steady state, in which production and consumption are in balance. Unless equilibrium is disturbed, no change in ambient pool size may be observed. Manipulation of physical factors (e.g., light), inclusion of chemical inhibitors to disrupt either the consumption or production of a relevant pool (65), or addition of limiting substrates to amplify flux through a pathway is often used to disrupt the equilibrium and allow direct detection of a pathway or process.

Alternatively, one may use substrates enriched in either radioactive (e.g., ^3H and ^{14}C) or stable (e.g., ^{13}C and ^{15}N) isotopes (18, 33, 50). Direct tracer procedures generally follow the conversion of an added, labeled substrate to product. While radiotracers may generally be added without significant perturbation of natural pool size, this is generally more difficult in stable isotope-based procedures, and the effects of substrate increases need be considered. When natural substrate concentration is increased by addition of the labeled substrate, the kinetics of uptake may be studied. Both in situ and projected rates may be inferred from kinetic parameters (37, 66).

Isotope dilution methods represent an approach distinct from direct tracer methods (6, 34). One typically labels a product pool and monitors the decrease in isotope ratio of that pool from the initial enrichment. This provides a measure of the rate of production of the substance in the pool, on the assumption that new product will initially be unlabeled. A distinct advantage of isotope dilution approaches is that, apart from a small addition to the product pool, it generally does not significantly affect the natural concentration; rates of increase in a product pool may be estimated even in systems near steady state. Furthermore, when combined with measures over time of the target pool, one can accurately estimate both production and consumption of that pool (28).

Biochemical approaches which involve the isolation of relevant enzymes and direct determination of their activities have also been adapted for the environmental context. Generally, the subject enzyme is extracted from the biological matrix, and its activity is determined. The typical assays of enzymology often provide substrates in excess to maximize activity. The advent of a wide variety of fluorometric derivatives of relevant substrates has prompted a renewed interest in environmental enzymes (57). This approach does not assess an in situ rate. However, attempts have been made to relate rates of activity in extracts to in situ rates (e.g., reference 68).

CONSIDERATIONS IN ASSAYING MICROBIAL ACTIVITIES

A variety of factors need be considered in setting up and conducting assays of microbial activity. Besides the obvious considerations of assay temperature, controls, and pathway-specific considerations such as light, O_2 tension, etc., there are also important issues of containment method, sample size, and time scale of assay, to name a few. Often the goal, from the ecological and biogeochemical perspectives, is determination of in situ rates of activity. However, various factors of logistics, instrumentation, assay sensitivity, and practicality often result in compromises.

The format of the assay is a factor in need of primary consideration, particularly with respect to sediments (15). The simplest way to leave a natural sample undisturbed during assay is to let it remain in its natural setting and rely on observations of relevant constituents. But the sampling of constituents may itself impose some level of perturbation.

Generally (ignoring system-scale experiments), some type of sample enclosure or semienclosure is necessary to observe changes in relevant parameters over reasonable time scales or to contain introduced tracers, inhibitors, or stimulants (e.g., reference 69). For sediments, domes are often used for in situ incubation (e.g., reference 23). As these domes include overlying water, corrections with incubated bottles of water may be necessary to account for activities in the water column. However, even such in situ approaches perturb the sample through isolation, restriction, or prevention of the natural exchange of substrates.

More commonly, the sample of interest is fully enclosed in a container of appropriate size. Depending on the application, various plastics, glass, or metal may be used. Applications requiring containment of gases must consider materials of low or insignificant permeability to the gases of interest. In other cases, free exchange of gases or even low-molecular-weight metabolites (e.g., with dialysis membranes or ultrafilters) may be allowed or desired.

The concerns of bottle effects are inherent in sample containment. Surfaces are well known to provide substrate for many microorganisms (90) and may allow differential proliferation of subpopulations, sorption of substrates or metabolites, or release of unwanted contaminants (e.g., metal ions). As bottle effects are of most concern with the surface, scaling up assay size will diminish the surface area relative to volume. However, this may result in logistic problems (cumbersome samples). Furthermore, with larger container size, consideration must be given to the development of gradients within the container and to the need for sample mixing and homogenization.

With sediments, slurry-type assays are often used; in these assays, samples of sediments are combined with water under appropriate conditions (e.g., under anoxic conditions) in some proportion to produce a homogenate (e.g., reference 85). Slurries offer the advantages of providing homogeneous experimental aliquots while allowing rapid equilibration of substrates and products within a sample during assay. When experimental treatments are to be used to elicit a response from a particular sediment, slurries may allow effects which would otherwise be obfuscated by the natural heterogeneity. However, slurrying of sediments will drastically alter geochemical gradients and, presumably, the relationships and activities of indigenous microorganisms (e.g., references 13 and 15). The alternative approach to slurries is the "intact core" method (47). In this approach, sediment samples are taken by core or by subcoring larger samples (e.g., reference 14). Substrates, stimulants, or inhibitors can then be injected into the sediment or overlain if required. Analysis can then be of the sediment, destructively, or from overlying waters or gas space (e.g., reference 64).

The requirement for a threshold level of change in the relevant analyte over some reasonable time scale, be it a detectable decrease in substrate or increase in product, is often a constraint imposed by the instrumental methods used. Assay procedures which include saturating levels of an otherwise limiting substrate or cofactors are used to achieve a detectable signal. However, such procedures are often termed assays of potential activity because of the likelihood that they may have artifactually stimulated or (ideally) maximized activity. Similarly, sample concentration is often used to increase detectability of activity. Care must be taken to establish the effects of such procedures and whether there is a linear relationship between concentration factor and observed activity.

The use of inhibitors which uncloak a particular process may allow increased sensitivity in detection of that process without the need for additional substrate additions and may more closely approximate in situ activities. However, inhibitors used in this manner may often have undesired or unanticipated effects which compromise this usefulness in regard to detecting in situ rates (65).

Appropriate controls are essential in establishing the biological basis of a particular process or transformation. They are crucial when abiological chemical reactions (e.g., sulfide oxidation) may be important or when physical processes (e.g., sorption of added substrates) may produce artifactual evidence of uptake or release. Brock (8) has discussed in detail approaches to poisoned controls in biogeochemical studies, pointing out the importance of ascertaining the effectiveness of a chosen poison and the absence of its interference in abiological pathways. One may draw firm conclusions when an inhibitor is effective. Without controls, absence of inhibition does not necessarily denote an abiological pathway but may only indicate choice of ineffective inhibitor (8). Perhaps worse, partial but unquantified degree of inhibition may lead to interpretative errors regarding the importance of biological processing. Factors such as the length of time for poisons to act, and the possibility of their being rendered ineffective with time, need to be considered. For instance, mercuric chloride is generally a very effective biological poison, but it may be rendered ineffective in sulfidogenic environments by mercury's precipitation as its sulfidic salt. Tuominen et al. (81) have recently reported on a number of routinely used methods of inhibition of bacterial thymidine or leucine incorporation, including autoclaving and irradiation, which are ultimately compromised in extended assays of bacterial activity in sediments.

The length of assay is also a matter to carefully consider. Extended assay may result in large changes in the microbial assemblages relative to those initially present. The activity of extant enzyme systems may be modified with time as substrates are depleted. Induction of enzyme systems not present in the original sample may account for apparent lags in particular activities. Selection of subdominant microbial populations may occur with depletion of substrates (including gaseous components) and/or buildup of metabolites not originally present. If estimation of natural rates is sought, and the sensitivity of the procedure allows it, brevity of assay is strongly advised. If one wishes to extrapolate back to the in situ rates, it is wise to establish the linearity of the assay through time course sampling.

SUMMARY OF SECTION

Over recent decades there have been several books of methodology which may be of interest and use to the reader. As might be expected, earlier ones focus on methods of culture and light microscopy. Aaronson (1) and Rodina (73) take this approach. These references are still valuable to the microbiologist interested in enriching or isolating particular organisms. In 1979, the American Society for Testing and Materials sponsored two more methods-oriented books: one for sediments (53) and one for aquatic bacteria (20). Poindexter and Leadbetter (70) edited a volume with an expanded discussion of the measurements of biomass, biochemical markers, and metabolic rates. Other contributions have included a book edited by Overbeck and Chrost (67) focused on biochemical and molecular approaches and a volume of *Methods in Microbiology* (32). More recently, Kemp et al.

(48) edited an extensive methods manual for aquatic microbial ecology which complements this section.

Contrary to the experience in public health/environmental or clinical microbiology, there are no "standard methods" in microbial ecology. Although there are methods which may be commonly used, such as the acridine orange direct count (39, 49), no governmental or industrial organization mandates the use of specific protocols. Also, the diversities of purpose and application require modifications of protocols. Given the limitations of space in this section, the reader cannot infer that a method given here can be applied without alteration or that it is the only solution to a problem. Each author has been directed to describe limitations and potential modifications, but the reader is encouraged to examine other appropriate manuals.

This section reflects the recent interest in biogeochemistry and microbial food webs as well as the reinvigoration of interest in community structure and autecology through molecular techniques. The first half of the section contains chapters on productivity, trophic interactions, and community structure. Primary and secondary productivity are addressed in chapters 26 and 27, with fungal productivity discussed in 31. In chapter 26 and chapters 28 to 31 are discussions of methods concerning community structure of primary producers, viruses, bacteria, protozoans, and eukaryotic organoosmotrophs, respectively. Sherr and Sherr describe methods for the study of trophic interactions in chapter 32.

Biogeochemistry and habitat oriented methods are described next. First, King discusses benthic biogeochemistry of carbon in chapter 33. Sulfur, nitrogen, and phosphorus cycling are covered in chapters 34, 35, and 36, respectively. Metal-microbe interactions are then discussed by Mills in chapter 37. In the last two chapters, Marshall describes methods associated with attachment, adhesion, and biofilms (chapter 38) and Deming explains methods applied to unusual and extreme environments (chapter 39).

We thank Joe Montoya, Becky Tarnowski, and Steve Newell for reviewing and commenting on an earlier draft of this chapter.

Support for R. R. Christian through NSF DEB 94-11974 and continued support for D. G. Capone from NSF are acknowledged.

REFERENCES

1. **Aaronson, S.** 1970. *Experimental Microbial Ecology.* Academic Press, New York.
2. **Amman, R. I., W. Ludwig, and K.-H. Schleifer.** 1995. Phylogenetic identication and in situ detection of individual microbial cells without cultivation. *Microbiol. Rev.* **59:** 143–169.
3. **Anderson, D. J., and J. Kikkawa.** 1986. Development of concepts, p. 3–16. *In* J. Kikkawa and D. J. Anderson (ed.), *Community Ecology: Pattern and Process.* Blackwell Scientific Publications, Melbourne, Australia.
4. **Atlas, R. M., and R. Bartha.** 1993. *Microbial Ecology: Fundamentals and Applications.* Benjamin/Cummings Publishing Co., Redwood City, Calif.
5. **Azam, F., T. Fenchel, J. G. Field, J. S. Gray, L.-A. Meyer-Reil, and F. Thingstead.** 1983. The ecological role of water-column microbes in the sea. *Mar. Ecol. Prog. Ser.* **10:**257–263.
6. **Blackburn, T. H.** 1979. A method for measuring rates of NH_4^+ turnover in anoxic marine sediments using a $^{15}NH_4^+$ dilution technique. *Appl. Environ. Microbiol.* **37:** 760–765.
7. **Brock, T. D.** 1978. *Thermophilic Microorganisms and Life at High Temperatures.* Springer-Verlag, New York.

8. **Brock, T. D.** 1978. The poisoned control in biogeochemical studies, p. 717–725. In W. E. Krumbein (ed.), *Environmental Biogeochemistry and Geomicrobiology*, vol. 3. Ann Arbor Science, Ann Arbor, Mich.

9. **Brock, T. D.** 1989. The study of microorganisms in situ: progress and problems, p. 1–17. In M. Fletcher, T. R. G. Gray, and J. G. Jones (ed.), *Ecology of Microbial Communities*. Symposium 41. Cambridge University Press, Cambridge.

10. **Brown, A. D.** 1990. *Microbial Water Stress Physiology: Principles and Perspectives*. John Wiley & Sons, Chichester, England.

11. **Bull, A. T., and J. H. Slater (ed.).** 1982. *Microbial Interactions and Communities*. Academic Press, London.

12. **Bull, A. T., and J. H. Slater.** 1982. Microbial interactions and community structure, p. 13–44. In A. T. Bull and J. H. Slater (ed.), *Microbial Interactions and Community Structure*. Academic Press, London.

13. **Burdige, D.** 1989. The effect of sediment slurrying on microbial processes, and the role of amino acids as substrates for sulfate reduction in anoxic marine sediments. *Biogeochemistry* **8:**1–23.

14. **Capone, D. G., and E. J. Carpenter.** 1982. A perfusion method for assaying microbial activities in estuarine sediments: applicability to studies of N$_2$ fixation by C$_2$H$_2$ reduction. *Appl. Environ. Microbiol.* **43:**1400–1405.

15. **Christian, R. R., and W. J. Wiebe.** 1979. Three experimental regimes in the study of sediment microbial ecology, p. 148–155. In C. D. Litchfield and P. L. Seyfried (ed.), *Methodology for Biomass Determinations and Microbial Activities in Sediments*. ASTM Special Technical Publication 673. American Society for Testing and Materials, Philadelphia.

16. **Christian, R. R., and R. L. Wetzel.** 1991. Synergism between research and simulation models of estuarine microbial food webs. *Microb. Ecol.* **22:**111–125.

17. **Codd, G. A. (ed.).** 1984. *Aspects of Microbial Metabolism and Ecology*. Academic Press, London.

18. **Coleman, D. C., and B. Fry (ed.).** 1991. *Carbon Isotope Techniques*. Academic Press, New York.

19. **Corliss, J. B., J. Dymond, L. I. Gordon, J. M. Edmond, R. P. von Herzen, R. D. Ballard, K. Green, D. Williams, A. Bainbridge, K. Crane, and T. H. van Andel.** 1979. Submarine thermal springs on the Galapagos rift. *Science* **203:**1073–1082.

20. **Costerton, J. W., and R. R. Colwell (ed.).** 1979. *Native Aquatic Bacteria: Enumeration, Activity and Ecology*. ASTM Special Technical Publication 695. American Society for Testing and Materials, Philadelphia.

21. **Edwards, C. (ed.).** 1990. *Microbiology of Extreme Environments*. McGraw-Hill Publishing Co., New York.

22. **Edwards, C.** 1990. Thermophiles, p. 1–32. In C. Edwards (ed.), *Microbiology of Extreme Environments*. McGraw-Hill Publishing Co., New York.

23. **Fisher, T. R., P. R. Carlson, and R. T. Barber.** 1982. Sediment nutrient regeneration in three North Carolina estuaries. *Estuarine Coastal Mar. Sci.* **14:**101–116.

24. **Fry, J. C.** 1990. Oligotrophs, p. 93–116. In C. Edwards (ed.), *Microbiology of Extreme Environments*. McGraw-Hill Publishing Co., New York.

25. **Gerhardt, P. (ed.).** 1981. *Manual of Methods for General Bacteriology*. ASM Press, Washington, D.C.

26. **Gibson, C. E., and D. H. Jewson.** 1984. The utilisation of light by microorganisms, p. 97–128. In G. A. Codd (ed.), *Aspects of Microbial Metabolism and Ecology*. Academic Press, London.

27. **Gilmore, D.** 1990. Halotolerant and halophilic microorganisms, p. 147–177. In C. Edwards (ed.), *Microbiology of Extreme Environments*. McGraw-Hill Publishing Co., New York.

28. **Glibert, P. M., and D. G. Capone.** 1993. Mineralization and assimilation in aquatic, sediment and wetland systems, p. 243–272. In R. Knowles and T. H. Blackburn (ed.), *Nitrogen Isotope Techniques*. Academic Press, New York.

29. **Goldman, J. C.** 1984. Conceptual role for microaggregates in pelagic waters. *Bull. Mar. Sci.* **35:**462–476.

30. **Gottschal, J. C.** 1992. Substrate capturing and growth in various ecosystems. *J. Appl. Bacteriol.* **73:**39S–48S.

31. **Grant, W. D., and B. J. Tindall.** 1986. The alkaline saline environment, p. 25–54. In R. A. Herbert and G. A. Codd (ed.), *Microbes in Extreme Environments*. Academic Press, London.

32. **Grigorova, R., and J. R. Norris (ed.).** 1990. *Methods in Microbiology*, vol. 22. *Techniques in Microbial Ecology*. Academic Press, London.

33. **Harrison, W. G.** 1983. Nitrogen in the marine environment: use of isotopes, p. 763–807. In E. J. Carpenter and D. G. Capone (ed.), *Nitrogen in the Marine Environment*. Academic Press, New York.

34. **Harrison, W. G., and L. R. Harris.** 1986. Isotope-dilution and its effects on measurements of nitrogen and phosphorus uptake by oceanic microplankton. *Mar. Ecol. Prog. Ser.* **27:**253–261.

35. **Herbert, R. A.** 1986. The ecology and physiology of psychrophilic microorganisms, p. 1–24. In R. A. Herbert and G. A. Codd (ed.), *Microbes in Extreme Environments*. Academic Press, London.

36. **Herbert, R. A., and G. A. Codd (ed.).** 1986. *Microbes in Extreme Environments*. Academic Press, London.

37. **Hobbie, J. E.** 1990. Measuring heterotrophic activity in plankton. *Methods Microbiol.* **22:**235–250.

38. **Hobbie, J. E.** 1993. Introduction, p. 1–8. In P. F. Kemp, B. F. Sherr, E. B. Sherr, and J. J. Cole (ed.), *Handbook of Methods in Aquatic Microbial Ecology*. Lewis Publishers, Boca Raton, Fla.

39. **Hobbie, J. E., R. J. Daley, and S. Jasper.** 1977. Use of Nuclepore filters for counting bacteria by fluorescence microscopy. *Appl. Environ. Microbiol.* **33:**1225–1228.

40. **Hungate, R. E.** 1962. Ecology of bacteria, p. 95–120. In I. C. Gunsalus and R. Y. Stanier (ed.), *The Bacteria*, vol. IV. *The Physiology of Growth*. Academic Press, New York.

41. **Imberger, J., T. Berman, R. R. Christian, E. B. Sherr, D. E. Whitney, L. R. Pomeroy, R. G. Wiegert, and W. J. Wiebe.** 1983. The influence of water motion on the distribution and transport of materials in a salt marsh estuary. *Limnol. Oceanogr.* **28:**201–214.

42. **Ingledew, W. J.** 1990. Acidophiles, p. 33–54. In C. Edwards (ed.), *Microbiology of Extreme Environments*. McGraw-Hill Publishing Co., New York.

43. **Jackson, G. A.** 1989. Physical and chemical properties of aquatic environments, p. 213–233. In M. Fletcher, T. R. G. Gray, and J. G. Jones (ed.), *Ecology of Microbial Communities*. Cambridge University Press, Cambridge.

44. **Jannasch, H. W., and C. O. Wirsen.** 1979. Chemosynthetic primary production at east Pacific sea floor spreading centers. *BioScience* **29:**592–598.

45. **Jennings, D. H. (ed.).** 1993. *Stress Tolerance of Fungi*. Marcel Dekker, Inc., New York.

46. **Johnson, B. D., K. Kranck, and D. K. Muschenheim.** 1994. Physiochemical factors in particle aggregation, p. 75–96. In R. S. Wotton (ed.), *The Biology of Particles in Aquatic Systems*. Lewis Publishers, Boca Raton, Fla.

47. **Jorgensen, B. B.** 1978. A comparison of methods for the quantification of bacterial sulfate reduction in coastal marine sediments. I. Measurements with radiotracer techniques. *Geomicrobiol. J.* **1:**11–28.

48. **Kemp, P. F., B. F. Sherr, E. B. Sherr, and J. J. Cole (ed.).**

1993. *Handbook of Methods in Aquatic Microbial Ecology.* Lewis Publishers, Boca Raton, Fla.

49. **Kepner, R. L., and J. R. Pratt.** 1994. Use of fluorochromes for direct enumeration of total bacteria in environmental samples: past and present. *Microbiol. Rev.* **58:**603–615.

50. **Knowles, R., and T. H. Blackburn (ed.).** 1991. *Nitrogen Isotope Techniques.* Academic Press, New York.

51. **Koch, A. L.** 1990. Diffusion: the crucial process in many aspects of the biology of bacteria. *Adv. Microb. Ecol.* **11:** 37–70.

52. **Kuhn, T.** 1970. *The Structure of Scientific Revolutions,* 2nd ed. University of Chicago Press, Chicago.

53. **Litchfield C. D., and P. L. Seyfried (ed.).** 1979. *Methodology for Biomass Determinations and Microbial Activities in Sediments.* ASTM Special Technical Publication 673. American Society for Testing and Materials, Philadelphia.

54. **Lovelock, J. E.** 1987. *Gaia: a New Look at Life on Earth.* Oxford University Press, Oxford.

55. **Lowe, S. E., M. K. Jain, and J. G. Zeikus.** 1993. Biology, ecology and biotechnological applications of anaerobic bacteria adapted to environmental stresses in temperature, pH, salinity, or substrates. *Microbiol. Rev.* **57:**451–509.

56. **Ludwig, J. A., and J. F. Reynolds.** 1988. *Statistical Ecology: a Primer on Methods and Computing.* John Wiley & Sons, New York.

57. **Meyer-Reil, L.-A.** 1991. Ecological aspects of enzymatic sediments, p. 84–95. *In* R. J. Chrost (ed.), *Microbial Enzymes in Aquatic Environments.* Springer-Verlag, New York.

58. **Mills, E. L.** 1989. *Biological Oceanography, an Early History, 1870–1960.* Cornell University Press, Ithaca, N.Y.

59. **Moriarty, D. J. W., and R. T. Bell.** 1993. Bacterial growth and starvation in aquatic environments, p. 25–53. *In* S. Kjelleberg (ed.), *Starvation in Bacteria.* Plenum Press, New York.

60. **Morita, R. Y.** 1986. Pressure as an extreme environment, p. 171–186. *In* R. A. Herbert and G. A. Codd (ed.), *Microbes in Extreme Environments.* Academic Press, London.

61. **Morris, J. G.** 1984. Changes in oxygen tension and the microbial metabolism of organic carbon, p. 59–96. *In* G. A. Codd (ed.), *Aspects of Microbial Metabolism and Ecology.* Academic Press, London.

62. **Newell, S. Y.** 1992. Estimating fungal biomass and productivity in decomposing litter, p. 521–561. *In* G. C. Carroll and D. T. Wicklow (ed.), *The Fungal Community,* 2nd ed. Marcel Dekker, Inc., New York.

63. **Nihoul, J. C. J.** 1980. Marine hydrodynamics at ecological scales, p. 1–12. *In* J. C. J. Nihoul (ed.), *Ecohydrodynamics.* Elsevier Scientific Publishing Co., Amsterdam.

64. **Nishio, T., A. Hattori, and I. Koike.** 1983. Estimation of denitrification and nitrification in coastal and estuarine sediments. *Appl. Environ. Microbiol.* **43:**648–653.

65. **Oremland, R. S., and D. G. Capone.** 1988. Use of specific inhibitors in microbial ecological and biogeochemical studies. *Adv. Microb. Ecol.* **10:**285–383.

66. **Oren, A., and T. H. Blackburn.** 1979. Estimation of sediment denitrification rates at in situ nitrate concentrations. *Appl. Environ. Microbiol.* **37:**174–176.

67. **Overbeck, J., and R. J. Chrost.** 1990. *Aquatic Microbial Ecology: Biochemical and Molecular Approaches.* Springer-Verlag, New York.

68. **Packard, T. T.** 1985. Measurements of electron transport activity of microplankton. *Adv. Aquat. Microbiol.* **3:** 207–261.

69. **Paerl, H. W., J. Rudek, and M. A. Malin.** 1990. Stimulation of phytoplankton production in coastal waters by natural rainfall inputs: nutritional and trophic implications. *Mar. Biol.* **107:**247–254.

70. **Poindexter, J. S., and E. R. Leadbetter (ed.).** 1986. *Bacte-ria in Nature,* vol. 2. *Methods and Special Applications in Bacterial Ecology.* Plenum Press, New York.

71. **Pomeroy, L. R.** 1974. The ocean's food web, a changing paradigm. *BioScience* **24:**499–504.

72. **Prosser, C. L.** 1986. *Adaptational Biology: Molecules to Organisms.* John Wiley & Sons, New York.

73. **Rodina, A. G.** 1972. *Methods in Aquatic Microbiology.* Translated, edited, and revised by R. R. Colwell and M. S. Zambruski. University Park Press, Baltimore.

74. **Roszak, D. B., and R. R. Colwell.** 1987. Survival strategies of bacteria in the natural environment. *Microbiol. Rev.* **51:** 365–379.

75. **Salt, G. W. (ed.).** 1983. A roundtable on research in ecology and evolutionary biology. *Am. Nat.* **122:**583–705.

76. **Schut, F., E. J. DeVries, J. C. Gottschal, B. R. Robertson, W. Harder, R. A. Prins, and D. K. Button.** 1993. Isolation of typical marine bacteria by dilution culture: growth, maintenance, and characteristics of isolates under laboratory conditions. *Appl. Environ. Microbiol.* **59:**2150–2160.

77. **Shimeta, J., P. A. Jumars, and E. J. Lessard.** 1995. Influences of turbulence on suspension feeding by planktonic protozoa: experiments in laminar shear fields. *Limnol. Oceanogr.* **40:**845–859.

78. **Slater, J. H.** 1981. Mixed cultures and microbial communities, p. 1–24. *In* M. E. Bushell and J. H. Slater (ed.), *Mixed Culture Fermentations.* Academic Press, London.

79. **Stolp, H.** 1988. *Microbial Ecology: Organisms, Habitats, Activities.* Cambridge University Press, Cambridge.

80. **Sverdrup, H. U.** 1953. On conditions for vernal blooming of phytoplankton. *J. Cons. Exp. Mar.* **18:**287–295.

80a. **Tiedje, J. M.** 1992. Advances in microbial ecology. *In Abstracts of the 92nd General Meeting of the American Society for Microbiology 1992.* American Society for Microbiology, Washington, D.C.

81. **Tuominen, L., T. Kairesalo, and H. Hartikainen.** 1994. Comparison of methods for inhibiting bacterial activity in sediments. *Appl. Environ. Microbiol.* **60:**3454–3457.

82. **Van Es, F. B., H. J. Laanbroek, and H. Veldkamp.** 1984. Microbial ecology: an overview, p. 1–34. *In* G. A. Codd (ed.), *Aspects of Microbial Metabolism and Ecology.* Academic Press, London.

83. **Waksman, S. A., C. L. Carey, and H. W. Reuszler.** 1933. Marine bacteria and their role in the cycle of life of the sea. I. Decomposition of marine plant and animal residues by bacteria. *Biol. Bull.* **65:**57–79.

84. **Waksman, S. A., C. L. Carey, and M. Hotchkiss.** 1933. Marine bacteria and their role in the cycle of life of the sea. II. *Biol. Bull.* **65:**137–167.

85. **Wellsbury P., R. A. Herbert, and R. J. Parkes.** 1994. Bacterial [methyl-^{3}H] thymidine incorporation in substrate-amended sediment slurries. *FEMS Microbiol. Ecol.* **15:** 237–248.

86. **White, D. C.** 1983. Analysis of microorganisms in terms of quantity and activity in natural environments, p. 37–66. *In* J. H. Slater, R, Whittenbury, and J. W. T. Wimpenny (ed.), *Microbes in Their Natural Environment.* Cambridge University Press, Cambridge.

87. **Whitelam, G. C., and G. A. Codd.** 1986. Damaging effects of light on microorganisms, p. 129–170. *In* R. A. Herbert and G. A. Codd (ed.), *Microbes in Extreme Environments.* Academic Press, London.

88. **Wiebe, W. J., M. A. Moran, and R. E. Hodson (ed.).** 1994. Special issue: the microbial loop. *Microb. Ecol.* **28:** 111–334.

89. **Wimpenny, J. W. T.** 1992. Microbial systems: patterns in time and space. *Adv. Microb. Ecol.* **12:**469–522.

90. **ZoBell, C. E., and D. Q. Anderson.** 1936. Observations on the multiplication of bacteria in different volumes of stored sea water and the influence of oxygen tension and solid surfaces. *Biol. Bull.* **71:**324–342.

Primary Productivity and Producers

HANS W. PAERL

26

THE PROCESS OF PRIMARY PRODUCTION AND RELEVANT MICROORGANISMS

Primary production, the biochemical conversion of inorganic matter to organic matter, is a fundamental measure of the fertility of aquatic systems (5, 11). Except for systems dominated by macroalgae and macrophytes, production of organic matter at the base of planktonic and benthic food webs is generally mediated by the microbial autotrophic processes photoautotrophy or phototrophy (photosynthetically mediated CO_2 fixation) and chemolithotrophy or chemoautotrophy (chemically mediated CO_2 fixation). In these processes, light- and chemical-derived energy, respectively, are used to reduce CO_2 to organic carbon. The basic stoichiometry of CO_2 fixation is $nCO_2 + 2nH_2O \rightarrow n(CH_2O) + O_2 + nH_2O$.

The manner in which autotrophs derive and donate reducing power to drive the carbon stoichiometry differs among physiological groups. Phototrophs utilize light-driven photolysis of simple, readily oxidizable compounds as a source of electrons and protons to generate reducing power (NADPH) and energy (ATP) for CO_2 fixation. The two modes of phototrophy are oxygenic and anoxygenic photosynthesis. In oxygenic photosynthesis, the dominant form of primary production in oxygenated surface waters, water fills this role. The photolysis of H_2O results in the formation of protons, electrons, and molecular oxygen (O_2), as outlined below:

$$2H_2O \xrightarrow{h\nu} 4H^+ + 4e^- + O_2$$

This is the light reaction of photosynthesis. Combining the reducing power of the light reaction with the CO_2 fixation, or dark reaction, of photosynthesis yields the following overall biochemical reaction:

$$6CO_2 + 12H_2O \xrightarrow{h\nu} C_6H_{12}O_6 + 6O_2 + 6H_2O$$

Oxygenic photosynthesis is conducted by all chlorophyll a-containing microorganisms. These include the prokaryotic blue-green algae (cyanobacteria) and eukaryotic microalgal groups listed in Table 1. Anoxygenic photosynthesis is confined to the photosynthetic bacterial groups listed in Table 1. H_2S is the source of protons and electrons, with S being the product:

$$2H_2S \xrightarrow{h\nu} 4H^+ + 4e^- + 2S$$

Anoxygenic photosynthesis is restricted to O_2-devoid water columns or surface sediments. Anoxygenic photosynthesis was, in all likelihood, the prevalent form of aquatic primary production during the O_2-free Precambrian period ($\approx$2.5 billion years ago). Oxygenic photosynthesis is identified with the appearance of cyanobacteria during the late Precambrian (4). Present-day dominance by oxygenic phototrophs has led to confinement of anoxygenic phototrophs to stratified, illuminated bottom waters in lakes, near-surface anoxic sediments, laminated microbial mats and biofilms containing O_2-free microenvironments, and endosymbiotic (O_2-free) habitats.

Chemolithotrophy takes place in oxic and anoxic waters. In either case, the oxidation of a range of inorganic and organic compounds provides reductant and energy for CO_2 fixation. Examples of aerobic chemolithotrophs commonly encountered in the water column and oxic surface sediments include nitrifying bacteria (*Nitrosomonas* and *Nitrobacter* species), which oxidize ammonium (NH_4^+) or nitrite (NO_2^-) and reduce CO_2 (Table 1). Representative facultative anaerobic and microaerophilic chemolithotrophs include the nonpigmented sulfide (S^{2-})-oxidizing bacteria (*Beggiatoa* species) and the sulfur (S^0) oxidizers (*Thiobacillus* species) (Table 1).

METHODS FOR MEASURING PRIMARY PRODUCTION IN AQUATIC HABITATS

Methods for assessing aquatic primary production are based on measuring rates of consumption of reactants or formation of products of autotrophy. The most commonly examined reactants are CO_2 (total or ΣCO_2) and H^+ (i.e., pH), while the products include O_2 and changes in pH. Primary production measurements are frequently supplemented with measurements of nutrient (e.g., N and P) uptake, in order to obtain parallel estimates of biomass production.

Change in biomass over time has also been used to approximate primary production (11, 49). This is accomplished either by microscopic determination of cell numbers or by spectrophotometric or fluorometric determinations of chlorophyll a content. Cell counts or measurements of chlorophyll a are converted to amount of cellular carbon by multiplying these parameters by cellular conversion factors (e.g., number of cells or amount of chlorophyll a × amount of carbon per cell per unit of chlorophyll a or biovolume).

TABLE 1 Major functional groups of aquatic microbial autotrophs, categorized according to environmental requirements and oxic/anoxic characteristics

Oxygenic phototrophs
  Prokaryotic
    Cyanobacteria (planktonic and benthic)
    Photosynthetic bacteria (planktonic and benthic), *Erythrobacter* (O_2 tolerant, contains bacteriochlorophyll *a*)
  Eukaryotic
    Chlorophytes (green algae, planktonic and benthic)
    Chrysophytes (including diatoms, planktonic and benthic)
    Cryptophytes (planktonic)
    Dinoflagellates (colorless and pigmented, planktonic)
    Euglenophytes (colorless and pigmented, planktonic and benthic)
    Prasinophytes (mainly planktonic)

Anoxygenic phototrophs (prokaryotic; also grow photoheterotrophically)
  Sulfide and sulfur as sole electron donor for photosynthesis
    Chromatium (planktonic only in meromictic waters, benthic)
    Chlorobium (planktonic only in meromictic waters, benthic)
    Thiocapsa (planktonic only in meromictic waters, benthic)
    Thiopedia (planktonic only in meromictic waters, benthic)
    Thiospirillum (planktonic only in meromictic waters, benthic)
    Ectothiorhodospira (high salinity and alkaline requirements)
    Prosthecochloris (strictly marine, benthic)
    Rhodobacter (planktonic and benthic)
    Rhodopseudomonas (planktonic and benthic)
    Heliobacter (mainly benthic)
    Chloroflexis (thermophilic, benthic)

Aerobic chemolithotrophs (prokaryotic)
  Colorless sulfur-oxidizing bacteria
    Thiobacillus (mainly benthic)
    Thiovulum (microaerophilic, benthic)
    Beggiatoa (microaerophilic, benthic)
    Thiosphaera (microaerophilic, mainly benthic)
  Iron- and manganese-oxidizing bacteria
    Gallionella (acidophilic, benthic)
    Planctomyces (planktonic and benthic)
    Leptothrix (mainly benthic)
  Nitrifying bacteria
    Nitrosomonas (ammonia oxidizers, planktonic and benthic)
    Nitrosococcus (ammonia oxidizers, planktonic and benthic)
    Nitrobacter (nitrite oxidizers, planktonic and benthic)
    Nitrococcus (nitrite oxidizers, planktonic and benthic)

Biomass has also been estimated directly as C or N by combusting microbial biomass and measuring C and N particulate content with a C,H,N analyzer (36, 44).

Other techniques that have been used to indirectly assess primary production include monitoring changes in the isotopic composition of C, N, and S contained in microbial biomass. This approach is based on the knowledge that microorganisms discriminate against heavy isotopic forms of each element during uptake, assimilatory, and growth reactions (8). Details of indirect techniques for estimating primary production are provided in the literature cited in Table 2.

CO_2 fixation can be assessed by measurements of CO_2 uptake by gas analysis or by using isotopic tracer techniques, specifically measuring $^{14}CO_2$ uptake (i.e., the ^{14}C technique).

CO_2 Uptake by Gas Analysis

CO_2 uptake can be assessed directly by gas chromatographic or infrared (IR) absorption analyses of CO_2 consumption, either in the headspace or in the soluble phase of gastight vessels. This approach is most often used in laboratory or manipulated field experiments (e.g., sealed chambers or mesocosms), for which access to analytical instrumentation is readily available.

Procedure

Populations and communities of interest are dispensed into a gastight, optically transparent vessel (normally borosilicate or Pyrex glass, polycarbonate, or plexiglas [i.e., Lucite]). The vessel aqueous and gas phases may be altered as desired. In the case of the gas phase, the vessel may be flushed with gas mixtures, either initially or continuously. Vessels are incubated under varying-illumination (including a dark treatment) regimes. At prescribed intervals, gas and/or aqueous subsamples (0.2 to several milliliters) are withdrawn through a serum stopper placed in the vessel. Vessels can also be sampled on a continuous-gas-flow basis.

Analysis of total $[CO_2]$ or $[\Sigma CO_2]$ is either by IR absorption (nondispersive IR detector) or gas chromatography (thermal conductivity detector). The sensitivities of these instruments are comparable. If CO_2 or ΣCO_2 is the only species to be analyzed, IR analysis is most practical (especially if flowthrough procedures are used). If additional gases (e.g., O_2 and N_2) are to be analyzed, gas chromatography is preferred. CO_2 is analyzed by sampling the gas phase directly or by sparging the aqueous phase with an inert gas (He, Ar, or N) and passing the evolved CO_2 through the appropriate instruments. To determine ΣCO_2, a 0.2-ml sample of aqueous phase is acidified in 5 to 10 ml of 30% H_3PO_4 to volatilize all inorganic C species (CO_2, HCO_3^{-2}, and CO_3^{-2}), sparged with inert gas, and analyzed by IR absorption (27). Instrument sensitivity and linearity are determined by making up a set of CO_2 or dissolved inorganic carbon (using Na_2HCO_3 or Na_2CO_3) standards covering the range of concentrations encountered in samples.

Discussion of Technique

The gas analysis technique is most applicable to laboratory studies, for which constant access to instrumentation can be ensured. If dense and active cultures (i.e., single populations) or microcosms (i.e., mixed communities) are available, this is a very effective, easily executed, and unequivocal techique. On the other hand, the ^{14}C technique is often

TABLE 2 Commonly used methods for determining primary production in planktonic and benthic aquatic environments

Method	References
^{14}C technique	
Planktonic systems	5, 6, 11, 32, 43, 49, 50
Benthic systems	16, 17, 28, 38
O_2 Winkler method	
Planktonic systems	5, 11, 31, 49, 50
Benthic systems	15, 16, 18, 34, 38
O_2 microelectrode method, benthic and planktonic systems	33, 37, 38

preferred in ecosystem-level (i.e., water column or whole-lake/marine environments) studies.

While both techniques measure CO_2 uptake as an estimate of primary production, they do not necessarily account for identical processes and as a result may yield contrasting results. This is because the measure of CO_2 "uptake" by gas analysis includes simultaneous production (e.g., respiration and decomposition) and autotrophic consumption of CO_2. Thus, like the oxygen method of assessing primary production, this is a net measure of CO_2 flux. The ^{14}C method, on the other hand, measures only consumption of CO_2 (as $^{14}CO_2$). Small amounts of refixation of respired or excreted ^{14}C may be possible, especially during lengthy (>3-h) incubations.

The CO_2 uptake and production technique is suitable for both planktonic and benthic studies. Limitations are largely due to physical (structure and size) constraints. In particular, analyses of benthic samples can be problematic. Maintaining structural integrity of sediments, microbial biofilms, and mats is critically important. Disruption of these samples can lead to altered diffusive regimes and changes in vertical biogeochemical gradients. Such alterations can significantly affect measurements of primary production and respiration, which in turn affect CO_2 flux (30, 38). In addition, laminar flow and other forms of turbulence can profoundly affect CO_2 flux at the sediment-water interface. Therefore, samples must be stirred during measurements.

Uptake of ^{14}C-Labeled CO_2

The radioactive (β-emitting) form of carbon, ^{14}C, is a long-half-life (5,760 years) isotope commonly used to quantify CO_2 uptake and its incorporation into organic matter. The ^{14}C method is based on several assumptions. (i) $^{14}CO_2$ biochemically behaves similarly to the stable, dominant form of carbon, ^{12}C; however, there is a small (6%) isotopic discrimination against the heavier ^{14}C. (ii) The chemical form of ^{14}C administered (usually as ^{14}C-labeled Na_2HCO_3) rapidly (within a few seconds after mixing) equilibrates with other forms of inorganic C (CO_2 and CO_3^{2-}) in accordance with pH and $[\Sigma CO_2]$. ^{14}C should be administered in trace quantities ($[\Sigma CO_2]$ remains unaltered by ^{14}C additions). Because of its high sensitivity, ease of use and deployment in the field, and relatively low cost, the ^{14}C method is widely used for measuring primary production in natural waters (5, 11, 49).

Procedure

Planktonic Samples

Water samples are collected (preferably with a nonmetallic sampler, such as a Van Dorn or Niskin bottle) and dispensed in 100- to 250-ml incubation bottles. Transparent Pyrex, borosilicate, or polycarbonate bottles with gastight seals are preferred. If bottles are incubated outdoors, they should be kept shaded during filling and processing to prevent photoinhibition. For field assessments of primary productivity, at least triplicate light and single dark (opaque bottles) samples should be incubated at each sampling location. Bottles should be stored in a lighttight box until collection is complete. Parallel water samples for alkalinity (ΣCO_2; $CO_2 + HCO_3^{2-} + CO_3^{2-}$) and pH measurements are collected in well-sealed glass bottles or vials (completely filled). All samples should be stored in a cool, dark place until analysis.

After sample collection, $NaH^{14}CO_3$ is added to each bottle, using a preset repeating pipetter or syringe equipped with disposable tips or needles. Then 0.5 to several milliliters of stock ^{14}C solution (3 to 10 μCi ml^{-1}) is added per bottle.

Commercial sources of $NaH^{14}CO_3$ having specific activities of 20 to 100 μCi μmol^{-2} are available. At the above-mentioned dilutions, trace quantities of ΣCO_2 (i.e., <0.1 μM) are added. Stock solutions can be made up and stored in several ways. Commercially available $NaH^{14}CO_3$ solutions (usually shipped in sterile deionized water) can be diluted with sterile water (at pH 7.5 to 8.5) and stored refrigerated in a well-sealed bottle. Alternatively, the diluted stock can be dispensed in glass, break-neck ampoules (i.e., "gold seal" or "blue seal" types), which are then sealed and autoclaved for long-term storage at room temperature.

Standard radioactive protection measures, including wearing disposable gloves and laboratory coats and protecting the work area with disposable protective paper or plastic covers (Benchkote or equivalent material), should be enforced. A well-labeled set of glassware, pipettes, filtration funnels, forceps, etc., should be dedicated for repeated use with ^{14}C. Appropriate disposal and handling of solid and liquid waste (in accordance with state and federal regulations) are required.

After ^{14}C additions, seal bottles tightly and agitate them by repeatedly inverting (5 to 10 times) them to ensure thorough mixing of the isotope. Make sure that dark bottles are truly dark (cover them with layers of foil if necessary) and (for in situ studies) place bottles at locations or depths from which they were collected. Incubate the bottles for 2 to 4 h.

To terminate the incubation, collect bottles and place them in a well-insulated, lighttight box for rapid transportation to the laboratory. If the time between sample collection and processing is longer than 1 h, samples should be placed on ice to restrict biological activity (i.e., respiration and decomposition of ^{14}C-labeled cell constituents). Bottle contents (either partial or entire volume) are filtered, under gentle vacuum (2 to 5 lb/in^2 or ≈ 200 torr), on 0.45-μm-pore-size nitrocellulose (HA Millipore or equivalent) or glass fiber (Whatman GFF or equivalent) filters. Filter funnels are rinsed with small quantities of prefiltered (0.45-μm-pore-size filter) water from which samples were obtained. Filters are then removed and placed face up in a well-ventilated area.

After drying, filters are placed in a well-sealed plastic container having a removable lid (≈ 20-by-30-by-10-cm polyethylene food storage containers are suitable). A small (≈ 50-ml) widemouth beaker half-filled with concentrated (fuming) HCl is placed in the box, and the filters are exposed to HCl fumes for at least 30 min. This treatment eliminates adsorbed and abiotically precipitated inorganic ^{14}C from filtered material. Filters are then allowed to thoroughly vent and dry (usually overnight). They are placed in liquid scintillation (LS) vials to which a chemically compatible, high-efficiency LS cocktail is added. A variety of suitable cocktails are available, the most popular of which are those composed of biodegradable, nontoxic solvents (ICN Ecolume, Cytoscint, Betacount, or equivalent). Typically, ^{14}C counting efficiencies, determined from a set of quenched standards, range from 90 to 95% for ^{14}C in filtered materials.

Water samples saved for alkalinity measurements are analyzed (preferably on the same day) either by IR absorbance or by titration and pH for ΣCO_2.

The primary productivity (CO_2 fixed per unit time) is calculated as

primary productivity (mg of C m^{-3} h^{-1})

$$= \frac{(A_l - A_d) \times {}^{12}C \times 1.06 \times 1,000}{{}^{14}C \text{ added} \times T}$$

where A_l is the mean counts (dpm) in light bottles, corrected for quenching and instrument background; A_d is the quench- and background-corrected counts (dpm) in dark bottles; ^{12}C is the total inorganic carbon (ΣCO_2) available, in milligrams of carbon per liter; 1.06 is a correction for the isotope effect; 1,000 is the conversion factor from liters to cubic meters; ^{14}C added is the total activity of ^{14}C added, in dpm; and T is the incubation time, in hours (5, 49).

Primary productivity can also be calculated per square meter by using integration techniques. It can additionally be normalized per milligram of chlorophyll a and per microeinstein per square meter (or micromoles of photons per square meter) of photosynthetically active radiation (PAR; 400 to 700 nm for oxygenic phototrophs; 400 to 850 nm for anoxygenic phototrophs) flux. More detailed discussions of the ^{14}C technique for freshwater and marine studies are presented by Goldman (11), Wetzel and Likens (49), and Falkowski (5).

Benthic Samples

Primary productivity of benthic microalgal mats and submersed biofilms can also be estimated by the ^{14}C technique. A modification of methods described for planktonic samples is used. Mat or biofilm subsamples are first obtained by cutting or coring small (≈ 1- to 2-cm^2) pieces from natural assemblages. Whole cores, incubated intact, can also be used (28). Because of a high degree of heterogeneity in mats and biofilms, a large number of replicate (minimally triplicates) light samples is recommended. Replicated light and single dark subsamples are placed in cleaned 22-ml borosilicate LS vials having polypropylene-lined caps. Dark vials can be wrapped with foil. Then 20 ml of ambient water and 0.2 ml of [^{14}C]NaHCO$_3$ (2 to 4 μCi of total activity; specific activity of 40 to 100 μCi μmol^{-1}) are added. Vials are sealed, gently mixed, and placed on their sides, with mat surface facing up, under natural illumination and temperatures. Vials are agitated to ensure uniform distribution of isotope, nutrients, and gases. Dissolved inorganic carbon (ΣCO_2) is determined by IR analyses of acidified, sparged ambient water.

In the case of whole cores, [^{14}C]NaHCO$_3$ can be directly injected into cores, which are then left to incubate under conditions described above (38). Cores are normally incubated under natural illumination and temperature conditions.

Incubations are terminated by removing mat and core samples from vials and placing them, face up, in a fuming HCl atmosphere for at least 30 min. Samples are then air dried and processed for LS counting, using a tissue-solubilizing, biodegradable cocktail (e.g., Cytoscint cocktail; ICN Inc.). Quenching (due to pigments and humic and other colored substances in sediments) may reduce LS counting efficiencies. This is corrected for by developing a quench curve whereby various amounts of unlabeled sediments are amended with known quantities of calibrated ^{14}C-labeled hexadecane or toluene (New England Nuclear or Amersham).

The formula described for planktonic samples can be applied to benthic microalgal communities. Results are usually expressed as milligrams of C per unit surface area (square centimeter or square meter) by correcting for the surface area of mat or biofilm samples used) per unit time. Data can be normalized per amount of chlorophyll a and/or PAR flux.

Discussion of Technique

The ^{14}C technique is the technique of choice when large numbers of samples need to be rapidly analyzed and when extensive spatial-temporal (synoptic) sampling of lakes, rivers, estuaries, or oceanic regions is required. Since its introduction to marine and freshwater productivity studies in the 1950s (43, 47), this technique has been the dominant method for assessing primary production in natural waters. Despite its popularity, the ^{14}C method is not without technical and interpretational problems (9, 32, 50), which are summarized below.

Sample Collection and Preparation

Water samples should preferably be collected with nonmetallic samplers (e.g., Van Dorn or Niskin bottles; plastic hoses and buckets) as far away as possible from sampling platforms or vessels by using trace chemical "clean" techniques. Heavy metals, hydrocarbons (i.e., fuel and oil) and other anthropogenic contaminants (lubricants, paints, solvents, etc.) associated with sampling can alter photosynthetic rates of resident microflora (6).

The [^{14}C]NaHCO$_3$ stock solution should be made up of either distilled, deionized water or seawater free of the above-mentioned contaminants. When trace metal contamination is of concern, stock solution water is initially purified by passing it through a Chelex column. The stock pH must be sufficiently alkaline (>pH 8.0) to guard against ^{14}C losses to the atmosphere (as $^{14}CO_2$) during preparation and dispensing.

Once samples are collected and transferred to incubation vessels, they are placed in a light-shielded environment as soon as possible. Photoinhibition, photorespiration, and (under extremely high light conditions) photooxidation of microflora may otherwise result. One should be particularly concerned about this problem in high-irradiance environments, including alpine, low-latitude tropical and subtropical regions, and intertidal and shallow-water benthic regions, and when high reflectivity is encountered (i.e., on ice, snow, sand, etc.). Photoinhibition can lead to underestimates (10 to 40%) of primary productivity (12, 14).

If in situ incubations are undertaken, the investigator is faced with several choices. If samples are incubated at the depth from which they were collected, they are essentially held under static light and turbulence conditions during the incubation period. This is the simplest and most commonly used means of incubating ^{14}C samples. In nature, phytoplankton seldom reside under static conditions. For example, in well-mixed, illuminated near-surface waters (i.e., epilimnia in freshwater and mixed layers in estuarine and marine waters), phototrophs experience a highly dynamic, transient light regime during daytime.

Incubations conducted under transient light regimes yield primary productivity estimates differing from those derived under static conditions (13, 19). This disparity holds true even when multiple-depth static incubations are integrated to yield an areal estimate (square meters) of primary productivity. The differences between static and dynamic incubations are particularly large in highly turbid waters, where light extinction gradients are very steep in the mixed layer. To compensate for this, various transient-light incubation schemes have been devised. These include use of on-deck incubators which rotate samples through variable-light regimes and use of in situ variable-light incubators, such as rotating "wheels" containing a range of light transmittance screens (13, 19). Sedimentation effects in the incubation vessels may also affect productivity measurements. To compensate for this, bottles should be agitated to most closely mimic natural turbulence.

Choosing appropriate incubation periods is also important. A priori, it is stressed that no single incubation length can be prescribed for all aquatic ecosystems. As a rule, 2 to 4 h is recommended, because this time frame should be long enough to allow for sufficient isotopic labeling of cell pools and constituents critical to the primary production process while being short enough to avoid significant ^{14}C losses due to excretion, cell death and lysis, grazing, etc., all of which ultimately affect quantitation of primary productivity. There may also be significant seasonal differences in the above-mentioned sources of error. It is suggested that the investigator start with this time frame and adjust it according to sensitivity needs.

The time(s) of the day during which productivity measurements are made can be critical to the determination of daily and longer-term production. It has been shown that primary productivity correlates with irradiance. Productivity and PAR curves are not necessarily superimposable, however. Studies on a variety of marine and freshwater ecosystems have shown that highest rates of primary productivity occur during mid to late morning, prior to the period of maximum irradiance (7, 9, 14, 32). Fee (7) and others have modeled primary productivity on the basis of knowledge of such deviations from the diel PAR flux curve. In addition to diel variability in the relationship between PAR flux and productivity, there are day-to-day, seasonal, and spatial differences in environmental variables regulating primary productivity.

After incubation, samples should be processed expeditiously. Unless it is necessary for specific applications (i.e., autoradiography, sample vouchers, and radioimmunological and molecular samples), one should avoid the use of preservatives such as formaldehyde, glutaraldehyde, and Lugol's iodide. Preservation steps can lead to loss of radioactivity from samples; this is particularly true if samples are washed or treated with other solutions prior to filtration, drying, and further handling (25). If required, samples can be quick-frozen and freeze-dried, but only if this is the last step prior to determining radioactivity (25).

Planktonic samples are usually concentrated by filtration prior to ^{14}C determination. One should determine appropriate filtration vacuum conditions under which successful concentrations can be achieved without loss of cellular labeled constituents. Appropriate vacuum settings depend on phytoplankton community composition and physiological state. A gentle vacuum, on the order of ≈200 torr, is recommended if the above-mentioned variables are unknown. The choice of filters is also critical. One should choose a filter that quantitatively retains all relevant phototrophs at any time of the year and at any location. From a practical standpoint, it is probably best to choose the smallest-pore-size filter capable of passing the volume of water incubated.

As an alternative to filtration, one can collect 5 to 20 ml of incubation water and then quantitate ^{14}C by direct LS counting. Prior to counting, samples are acidified (0.1 N HCl), sparged with air or N_2 to expel nonassimilated $^{14}CO_2$, and neutralized (with dilute NAOH) prior to counting. This approach is most applicable to productive waters, for which large incubation volumes are not required. Relatively low LS counting efficiencies (<85%) may result, because large amounts of water (which is a strong quenching agent) must be mixed with the LS cocktail. An LS fluor chemically compatible with the sample and yielding maximum ^{14}C counting efficiency is preferred. For filters, there is a wide range of high-efficiency (>95%) organic solvent (toluene or xylene)-based and anionic-surfactant-based (Ecolume or Cytoscint; ICN) fluors available. Fluors of the latter group are gaining popularity because they are nontoxic and biodegradable.

Data Interpretation

Interpretation of data is one of the more critical aspects of the ^{14}C technique. Compared with other methods of assessing productivity, specifically the O_2 method, the ^{14}C technique yields results that range from general agreement (50) to serious underestimates (41, 42). Some of the disagreement can be attributed to various methodological problems, including those for determining cellular ^{14}C losses (40). Other discrepancies may be due to the biochemistry of respective measurements. Because it accounts for the balance between CO_2 uptake, assimilation, and losses via respiration, excretion, and lysis, the ^{14}C method represents net primary production. In contrast, the O_2 method accounts for total photosynthetic production of O_2 minus (only) O_2 consumption via dark-mediated respiration. It does not account for other biomass losses incurred during excretion, lysis, death, grazing, and additional light-mediated respiratory processes such as photorespiration. As such, the O_2 method is a closer estimate of gross primary production.

In practice, when these methods are compared, both agreement and varying results emerge (9, 50). The greatest disparity between these methods occurs when primary productivity is closely linked, in time and space, to higher trophic levels. At the level of microbial interactions, autotrophy is closely linked to heterotrophy. Carbon and other nutrient (N and P) transfers can occur within minutes between host autotrophs to epiphytic heterotrophs (26). Such close metabolic coupling of primary and secondary producer components of the microbial food web is referred to as the microbial loop (2). If significant C transfer occurs within the ^{14}C incubation period, it could lead to underestimates of primary production. Labeled dissolved organic carbon will not be retained on filters. If microheterotrophs consume labeled dissolved organic carbon, this loss may be partially recovered. However, if these microheterotrophs then excrete or respire ^{14}C, an underestimate occurs. Similarly, if micro- or mesozooplankton grazing of phytoplankton with respiration and excretion takes place during the incubation period, transfer of fixed ^{14}C can occur, leading to underestimates of net primary production. Grazers can be eliminated prior to incubation to assess the quantitative significance of trophic transfer during incubations. On the other hand, respiration can lead to the loss of fixed C. If recently synthesized ^{14}C-labeled organic compounds are respired, a portion of the ^{14}C fixed may be lost and/or recycled during incubation. This may lead to underestimates of primary production.

Oxygen-Based Methods

Oxygen evolution has long been used as a benchmark measurement of photosynthesis and, in the case of natural microbial assemblages, primary productivity (34). The O_2 method is often preferred over other, less direct primary production techniques. One of its main advantages is that it measures photosynthetic performance on the basis of a key product of the light reaction of photosynthesis. Hence, measurements can be directly related to the capture of radiant energy, electron flow, and the generation of reductant. The O_2 technique is suitable for determining productivity of complex microbial assemblages, where parallel measurements of O_2 evolution and its utilization serve as useful indicators of community metabolism (17). Oxygen can be easily and accurately measured without need for radioactive materials and sophisticated detection equipment. Diffusion limitations with the ^{14}C technique seriously compromise the use of this method in laminated benthic systems. Such limitations are minimized with the O_2 measurements (especially those obtained by using microelectrodes) outlined below. In moderate- to high-productivity systems, where sensitivity is not limiting, the O_2 method is often preferred.

Oxygen evolution is measured in two basic ways: (i) determination of changes in dissolved oxygen in bottles or other vessels titrimetrically (i.e., the Winkler method) and (ii) electrochemical, macroelectrode, or microelectrode determinations. The two techniques should yield identical results, although in the latter, small amounts of O_2 are consumed during measurement. Supply needs and measuring equipment are relatively simple and portable, making this an excellent shipboard or field technique.

Winkler Method

A divalent manganese solution is added to water samples, and then alkaline iodide is added. Under these conditions, dissolved O_2 is rapidly reduced and precipitated as manganese oxide. This step "fixes" O_2 prior to analysis. The sample is then acidified in the presence of iodide, and the oxidized manganese is converted to its divalent state, while iodine (equivalent to the original amount of dissolved O_2 present) is liberated. The iodine is quantified either titrimetrically or spectrophotometrically.

Procedure

Duplicate 125- or 250-ml Pyrex, ground glass-stoppered light and dark BOD (biological oxygen demand) bottles are normally used to hold samples. Quartz bottles may be substituted if effects of UV radiation are assessed. Bottles are flushed several times with sample water and filled, making sure not to generate bubbles during filling (bubbling will affect O_2 solubility and hence concentration). This is accomplished by inserting a filling tube (Tygon or equivalent material) to the bottom of the bottle and rapidly filling the bottle, allowing for overflow. Quickly insert the ground glass stopper. Store each sample in a cool, lighttight box until all samples are collected.

After incubation, reagents may be added by pipette in the following order: 3 M manganese chloride (dissolve 600 g of $MnCl_2 \cdot H_2O$ in distilled water to make 1 liter) (solution 1); alkaline iodide solution (8 N NaOH–4 N NaI), which is made by dissolving 320 g of NaOH and then 600 g of NaI in distilled water to make a volume of 1 liter (solution 2); and 10 N sulfuric acid (add 280 ml of concentrated H_2SO_4 to 500 ml of distilled water) (solution 3). Allow each solution to cool to room temperature before use.

The following procedure is for 125-ml bottles. Add 1 ml of solution 1 and 1 ml of solution 2. Stopper bottles and mix well (by inverting bottles at least 10 times). Following this fixation step, bottles can be kept cool and dark for at least 1 to 2 h prior to analysis. When analyzing, allow the precipitate to settle at least halfway down the bottle.

For tritrimetric determination of $[O_2]$, the iodine liberated is titrated with a 0.01 N standardized thiosulfate solution, yielding $[O_2]$ in milligrams per liter. Solutions are as follows. For solution A, dissolve 2.9 g of analytical-grade $Na_2S_2O_3 \cdot 5H_2O$ and 0.1 g of Na_2CO_3 in 1 liter of distilled water; add 0.2 ml of CS_2 as a preservative. For solution B, make up a 0.01 N potassium iodide standard by adding 0.3567 g of air-dried (105°C) analytical-grade KIO_3 to 250

ml of distilled deionized water (warm if necessary to dissolve KIO_3). For solution C, make up a 1% starch solution (solubilize in dilute NaOH if necessary and neutralize with dilute HCl).

Add 1 ml of solution 3 (10 N sulfuric acid). Within 1 h of sample acidification, use a pipette to transfer 50 ml into a 125-ml Erlenmeyer flask containing a magnetic stir bar. Titrate immediately with 0.01 N thiosulfate which has previously been dispensed in a burette, while stirring, until the solution is slightly yellow. Then add 0.5 ml of starch solution, which should turn the sample blue. Continue titrating slowly and carefully until the blue color disappears. Titrate N_2-flushed blanks and subtract each blank to obtain a corrected titration volume (V) in milliliters. $[O_2]$ is calculated as follows: mg of $O_2 \cdot$ liter^{-1} = 0.1016 × f × V × 16.

To determine the calibration factor f, add 1 ml of concentrated sulfuric acid (solution 3) to a 125-ml BOD bottle filled with distilled water. Mix thoroughly. Add 1 ml of the $MnCl_2$ solution (solution 1) and mix again. Transfer 50 ml into the titration flask; add 50 ml of the 0.01 N KIO_3 standard (solution B). Mix gently for 2 min and titrate with thiosulfate. Using V as the titration volume, $f = 5.00/V$. Determine the calibration factor in triplicate.

For spectrophotometric determination of $[O_2]$, allow the precipitate formed (after adding solutions 1 and 2 to the sample) to settle halfway down the sample bottle. Add 1 ml of solution 3, stopper, and mix again. Wait at least 5 min for the precipitate to thoroughly dissolve. A small aliquot (5 to 10 ml) is withdrawn from the bottle and diluted to 100 ml with distilled water. Avoid violent mixing to minimize oxidation of iodine by exposure to air. Simply inverting the mixing flask will provide adequate mixing. The extinction of iodine (E_s for samples and E_b for the blank) is measured at 287.5 nm in a 1-cm quartz cuvette, using a UV spectrophotometer. An N_2-flushed (to remove O_2) water sample is run in parallel as a blank. To quantify $[O_2]$ by the spectrophotometric technique, a calibration factor must first be determined by the standard tritration technique discussed above. One can also refer to a table to determine standard O_2 saturations (percentage of milligrams of O_2 per liter) at specific temperatures (at sea level) and salinity conditions (31). This yields an O_2 solubility factor (F_s) based on the following formula: $F_s = [O_2]/(E_s - E_b)$.

Discussion of Technique

The Winkler technique is best suited for moderate to highly productive productive waters, characterized by relatively high O_2 production and consumption rates. Measurements are most easily conducted on planktonic samples, for which sampling volume is not restricted. The technique can also be applied to sediments, biofilms, mats, and epibiotic communities. In these systems, the method is most effectively deployed by placing domes and other enclosures directly over the surface to be examined. Specially designed domes, equipped with electric stirrers and ports for withdrawing subsamples, are routinely deployed for assays of benthic primary production, respiration, and (by subtracting respiration from production) net community metabolism (17). During measurements of benthic metabolism, corrections must be made for metabolism of the overlying water column.

Surficial O_2 production and consumption measurements suffer from methodological shortcomings, some of them similar to those incurred with ^{14}C-based methods. Perhaps most serious are diffusional problems. Oxygen, like CO_2, is a gas, and it is transported along diffusional gradients. In the case of very high rates of O_2 production or consumption, the rate of diffusional transport can control the rate at which O_2 is emitted from or absorbed by surficial and subsurface boundary layers. If the surface is submersed, diffusional barriers are most pronounced. Unless turbulence (as stirring or shaking) can overcome this limitation, underestimates of O_2 production and/or consumption may result. Potential underestimation due to transport of photosynthetically produced oxygen to lower depths in sediments may constitute an additional problem. Diffusional problems can be minimized by stirring or agitating samples. Care must be taken, however, to avoid disturbance and resuspension of sediments.

One way to minimize diffusional problems is to use microelectrode measurements of $[O_2]$ (see below). Microelectrodes are small enough to minimize the impact of diffusional gradients. They are usually mounted on a micromanipulator, which facilitates rapid movement of the microelectrode along diffusional gradients.

An interpretational limitation of the O_2 method is that, unlike the ^{14}C method, it does not account for primary productivity of anoxygenic phototrophs (e.g., photosynthetic bacteria) and chemolithotrophs (e.g., nitrifying bacteria and S-oxidizing bacteria). On the other hand, it does account for respiration by these and all other microorganisms (which the ^{14}C method is incapable of).

Electrochemical Techniques

Procedures

Oxygen can be measured electrochemically, using several types of electrodes. The most widely used are cathode and Clark-style combination electrodes. In the cathode electrode, a membrane-coated platinum cathode, coated with a gold surface (to sense O_2), is embedded in a thin glass or steel cannula housing. A voltage of -0.75 V (relative to a calomel reference electrode) is applied to the cathode. The current resulting from the reduction of O_2 at the gold surface is directly proportional to $[O_2]$ in the sample. Cathode-style electrodes can be constructed with very small ($-5\text{-}\mu m$) sensing tips. Microelectrodes with small sensing tips have a high degree of spatial resolution. Furthermore, the O_2 diffusion path to the tip is fast, leading to rapid response times. The 90% response time (i.e., the time required to approach equilibration of external and internal $[O_2]$ conditions) is typically <0.5 s. The construction and general application of cathode-style microelectrodes in microbial O_2 production and consumption studies are discussed in detail by Revsbech and Jørgensen (37).

In Clark-style or combination O_2 electrodes, the cathode and reference electrode are combined but physically separated in the same electrode. The cathode and electrodes are both immersed in an electrolyte (1 M KCl), which additionally serves to shield the cathode. The tip of the electrode is sealed with an O_2-permeable membrane. Clark-style electrodes vary in size and shape, depending on the scale, sensitivity, stability, response time, and durability desired. The larger BOD electrodes are typically used for environmental measurements (i.e., on a sampling line) or measurements to be made on large containerized water samples (i.e., those in BOD bottles, domes, and other enclosures). Response time is on the order of a few to 30 s. If high resolution, high sensitivity, and rapid response time are desired, smaller-diameter electrodes are called for. Minielectrodes have sensing tips of 0.5 to several millimeters in diameter, while microelectrodes have tips as small as 5 μm. The response time of Clark-style microelectrodes is similar to that for cathode microelectrodes.

$[O_2]$ is determined by measuring changes in the current flowing through the electrode. In large Clark-style electrodes, a digital or analog ammeter serves as the detector. With all microelectrodes, the current changes are very small, and hence a very sensitive ammeter, capable of measuring down to 1 pA (10^{-12}A), is used. Portable picoammeters are commercially available (e.g., Kiethley Inc.) for field studies.

Mini- and microelectrodes are typically used for small-scale pO_2 measurements in biofilms, benthic boundary layers, sediments, and microbial mats, where strong, microscale (micrometer) O_2 gradients are an integral component of primary production and nutrient cycling dynamics. Details of the construction and applications of Clark-style mini- and micro-electrodes are provided by Revsbech and Jørgensen (37). The procedure briefly outlined below is applicable to the use of fabricated mini- and microelectrodes for measuring pO_2 in surficial environments.

Oxygen regimes and changes therein are analyzed by mounting the micro- or minielectrode on a micromanipulator. The micromanipulator is positioned above the sample in such a way that the electrode can be rapidly inserted into the sample and $[O_2]$ measurements can be made at discrete intervals. O_2 profiles are made by advancing the electrode downward into the sample at 0.25- to 0.5-mm intervals. Measurements are usually made under alternating illuminated and dark conditions. Once stable O_2 conditions occur under these conditions, the next interval is measured. The most common way to determine photosynthetic O_2 production is to use the light-dark shift method (37–39). In this method, samples are exposed to the irradiance of interest. The microelectrode is inserted to the sampling location. $[O_2]$ is determined. Photosynthesis is then stopped by darkening the sample. This allows O_2 consumption and diffusional exchange to continue at the presumed original rate (prior to darkening). Since there is no photosynthetic O_2 production to balance consumption, $[O_2]$ decreases in proportion to the photosynthetic rate just prior to darkening. The initial slope (measured on either a recorder or data logger) of the decrease in $[O_2]$ after darkening is equal to the $[O_2]$ produced by photosynthesis.

Electrodes are standardized and calibrated by measuring current changes in O_2-depleted (N_2- or Ar-flushed) and O_2-saturated (by bubbling with air) solutions of ambient water.

Discussion of Techniques

Electrochemical measurements of $[O_2]$ afford obvious advantages over colorimetric Winkler techniques. They are rapid and easily executed and, in the case of mini- and microelectrodes, circumvent diffusional artifacts and limitations. All electrodes require calibration against the Winkler technique. Some microelectrodes are sensitive to illumination, and this needs to be compensated for by measuring current changes in response to light versus dark conditions in media free of microorganisms (e.g., distilled water). Oxygen microelectrodes are additionally sensitive to temperature changes, which also need to be checked using sterile or microorganism-free solutions.

Because they are very thin and constructed in large part of glass, mini- and microelectrodes are prone to breakage. Breakage is a particularly serious problem in sediments and microbial mats, where impenetrable heterogeneous substrates (sand, gravel, mucilage, slimes, etc.) can hinder their use. In addition, microelectrodes are sensitive to fouling, chemical (e.g., sulfide) poisoning, and changes in sensitivity and response time, all of which necessitate extensive and repeated standardization. Unless investigators have ready access to the technology for constructing microelectrodes, replacement of broken electrodes, calibration, and standardization are expensive, time-consuming aspects of these techniques.

Possible interpretive problems include extrapolating or integrating from the short time scales required to make measurements (minutes) to the longer time scales (hours to days) required to assess productivity on community and ecosystem levels. In addition, microelectrodes can potentially disrupt biofilms and mats during measurements, leading to invasion of O_2 into previously hypoxic and anoxic microzones.

Electrochemical techniques require electronic amplification (of current output), detection, and measurement accessories, the degrees of sophistication and costs varying considerably. Cost of ancillary equipment ranges from approximately $1,000 to $1,500 for an O_2 meter (e.g., YSI 54 ARC), in the case of large electrodes, to in excess of $5,000 for microelectrode amplification and measurement equipment (picoammeters and electrochemical sensing equipment). In addition, a micromanipulator (~$1,500) is required to position electrodes.

Complementary Techniques and Their Applications

Standard measurements of primary productivity can be complemented by modifications, ancillary techniques, and procedures which enhance physiological, ecological, and taxonomic interpretations of productivity measurements (1, 15–18, 33). Measurements of photopigments are highly informative in that they can serve as surrogates for rate measurements of productivity and as complementary biomass measurements. In addition, the ^{14}C method lends itself to specific techniques capable of assessing species-, group-, and community-specific rates of primary production. Two such techniques, microautoradiography and radiolabeling of diagnostic microalgal pigments, can enhance the specificity and dimensionality of primary productivity measurements.

Pigment-Based Estimates of Primary Productivity

Changes in net phototrophic biomass over time can be used as an indirect estimate of primary productivity. This approach has been used when direct physiological rate measurements are not available. The simplest and often most effective estimate of phototrophic biomass is one based on determining cellular photopigment content. The most common pigment assessed is chlorophyll a, because it is present in all oxygenic phototrophs. Other pigments, specific for taxonomic groups (carotenoids, phycobilins), can be used to partition production among these groups (see Photopigment Radiolabeling, below). All pigment-based approaches are reliant on the assumption that the cellular ratio of photopigments to dry weight or carbon content is reasonably constant. This assumption may not be valid in communities experiencing environmental change and physiological stresses. Pigment-based measurements are net estimates, since they simultaneously account for synthesis and loss of photopigments. Synthesis is largely due to growth and resultant production of new biomass. Losses and changes in cell pigment synthesis may be attributable to cellular death, consumption, advective processes (sinking, sedimentation, and horizontal transport due to flushing), and changes in

physiological state (i.e., photoadaptation, photobleaching, and photooxidation).

The most common measure of chlorophyll a is by spectrophotometric absorbance of intact cells or extracted pigment (usually in 90% acetone or 90% ethanol or methanol) at either 440 or 663 nm and 750 nm (turbidity blank). Detailed methodologies for microalgal pigment extraction and analysis are given by Strickland and Parsons (44), and applications to aquatic production studies are addressed by Millie et al. (22). Fluorometric analyses of chlorophyll a (excitation maximum at ≈400 nm, emission maximum at ≈660 nm) are also commonly used in limnological or oceanographic studies. Fluorometry is more sensitive than and not as prone to optical interference as spectrophotometry (22). Fluorometry can additionally be applied in situ by placing fluorometers and recording and telemetry devices in the water column.

The above-mentioned methods are applicable to both planktonic and benthic systems. In the case of benthic measurements, accumulations of chlorophyll degradation products (e.g., pheophytin) can be significant. This can lead to interpretational problems, even if acidification corrections for pheophytin are made (44). High-performance liquid chromatography (HPLC) has, however, proven effective in rapidly separating photosynthetic pigments prior to spectrophotometric and fluorometric detection (3, 20, 46). When coupled to real-time multiwavelength rapid-scanning capabilities, such as those provided by a photodiode array spectrophotometer (PDAS), HPLC offers the ability to rapidly and accurately distinguish and quantify pigments and a number of their degradation products within monotypic and mixed algal assemblages (22, 29). In this manner, certain taxa known to contain unique pigments and combinations of pigments (such as diagnostic carotenoids, phycobilins, specific combinations of chlorophylls a, b, and c, and chlorophyll derivatives) may be identified and potentially quantified. When coupled with autoinjectors and driven by state-of-the-art computer software, HPLC-PDAS is very amenable to large-scale production characterization and monitoring programs (National Science Foundation's Global Ocean Flux Studies; U.S. Environmental Protection Agency's Toxics and Lakes Survey).

Photopigment Radiolabeling

Photopigment radiolabeling is a relatively new, potentially useful technique for measuring C-specific growth rates among diverse phototrophic taxa. The conventional ^{14}C method gives a single production value for the entire community. In contrast, the photopigment radiolabeling technique allows for simultaneous measurement of specific growth rates among individual taxa. The method is an extension of the ^{14}C technique. When exposed to ^{14}C, the labeled C is incorporated into 3-phosphoglyceric acid, passes through the Calvin-Benson cycle, and eventually passes into a pool of smaller-molecular-weight intermediates. These labeled compounds enter the various photopigment biosynthetic pathways. After a sufficiently long incubation (2 to 24 h), the specific activity of photopigment C will equal that of the total phytoplankton carbon pool. The C-specific growth rate (μ, where $C_P = C_{p,o}e^{\mu t}$, C_P is C-specific biomass at time t, $C_{p,o}$ is C-specific biomass at time zero, and t is time) is a more useful parameter for describing the physiological responses of functional groups of phototrophs, in terms of C assimilation and allocation into biomass. A comparison of μ values obtained under different experimental conditions allows for predictions of the ultimate fate of community components over time.

Using HPLC coupled to PDAS detection of pigments, photopigments are separated on a Vydac reverse-phase C_{18} column by using a binary gradient system (46). On-line PDAS produces real-time three-dimensional chromatograms of absorbance spectra for each photopigment. Computer software allows data from individual chromatograms to be stored and processed for qualitative analyses and sample comparisons of absorption spectra and peak retention times. C-specific growth rates are determined by published methods (35). The ^{14}C specific activity (dpm) of individual pigments is quantitatively determined by using an in-line radioactivity (β-emission) detector. Radioactivity (dpm) peaks are overlayed onto pigment chromatograms to identify labeled pigments. Values of community-specific growth rate (μ) and C biomass are determined by using equations given by Goericke and Welschmeyer (10) and Redalje (35). The growth rates and C biomass of phylogenetic groups (cyanobacteria, diatoms, dinoflagellates, chlorophytes, etc.) are estimated from the activity of labeled diagnostic pigments.

Microautoradiography

Samples that have been incubated with $^{14}CO_2$ can be analyzed for single cell-, taxon-, and community-specific photosynthetic capabilities and rates by using microautoradiography (45, 48). This method is based on the fact that ^{14}C assimilated by microorganisms can be detected and visualized by radiation-sensitive silver halide emulsions which are placed over radioactive organisms and subsequently processed by standard photographic techniques.

^{14}C-labeled cells are either filtered onto membrane filters or settled onto glass slides, after which radiosensitive liquid or film emulsions are applied (23, 24). Following exposure and processing, both the host microorganism and its radioactivity (which appears as exposed silver grains in the developed emulsion) can be examined microscopically and photographed. Radioactivity can be quantified (either as silver grains or as tracks of grains) if a uniformly thick film or liquid emulsion is applied (30). For details of the technique, the reader is referred to references 9, 24, 25, and 30. Below is a short description of the technique applied to planktonic samples.

Samples that have been incubated with ^{14}C are either fixed in 2% borate-buffered formaldehyde (pH 7 to 8) (in the case of delicate cells) or left unfixed and gently filtered ($\approx$200 torr) on 0.45-μm-pore-size Millipore HA filters. Filters are then rinsed with prefiltered, unlabeled sample water to remove excess radiolabel. In the case of marine samples, salts must be removed from filters, or they will crystallize upon drying and interfere with microscopic viewing. This is accomplished by rinsing filters twice with small quantities (10 to 30 ml) of 0.01 N phosphate-buffered saline, which removes salts while avoiding lysis of fragile cells. Filters are then quick-frozen and/or air dried (for at least 4 h) and optically cleared by placing them, face up, on clean microscope slides and carefully passing them over the mouth of a 250-ml beaker containing boiling (fuming) acetone. This step will simultaneously clear filters and firmly attach them to slides.

The following steps are conducted in a darkroom, either under a dark red 20-W or lesser-wattage safelight (at least 1 m away from samples) or under complete darkness, which ensures the lowest background. Slides are dipped in a nuclear track emulsion which has been melted previously. Kodak NTB-2, which is a moderately sensitive, low-background, reusable emulsion, is widely recommended. The stock emulsion is kept in a lighttight container in a refrigerator at $\approx$4°C. In darkness, it is melted at 40°C and diluted 1:1 with water for use with grain density microautoradiography. If quantitative autoradiography is desired, the application of thick emulsions combined with track autoradiography or stripping film emulsions (Kodak AR-10) is recommended (30). All emulsions should be thoroughly melted at 40°C (use a water bath in a darkroom or heat light-sealed containers with emulsion in an oven) before use. In both thick-layer (undiluted emulsion) track and diluted thin-layer grain density techniques, dipped and jelled slides are air dried for 30 min and exposed in lighttight slide boxes containing packets of silica gel desiccant. Exposure times typically vary from 2 to 10 days.

Exposed track and thin-layer microautoradiographs are developed in Kodak D-19 for 5 and 2 min, respectively, and transferred to a stop bath for 2 min. They are then fixed for 5 min in Kodak rapid fixer or 15% NaS_2O_3 and rinsed in a gently flowing water bath for 15 min. All slides are then air dried overnight. Microautoradiographs are viewed microscopically, usually with phase-contrast oil immersion optics. For thin-layer autoradiographs, immersion oil can be directly applied to the dried emulsion. Track autoradiograph emulsions are allowed to swell by the addition of 30% glycerine followed by the application of a coverslip. Immersion oil can then be applied to the coverslip.

While microautoradiography can overcome and circumvent technical limitations of other techniques, it is time-consuming, requires experience and patience, and is limited by microscopic resolution. It can also be prone to differences in interpretation among investigators.

REGIONAL AND GLOBAL STUDIES OF PRIMARY PRODUCTION: THE UTILITY OF REMOTE SENSING

Increasingly, humans are affecting aquatic primary production. Perhaps the most perceptible and problematic symptoms of this troubling trend are anthropogenic nutrient-enhanced eutrophication and associated increases in harmful algal blooms in lakes, estuaries, and coastal waters. Such ecosystem-level and larger-scale (regional and global) events range from days to several weeks, depending on the timing of and relative enrichments from direct and indirect (runoff) nutrient inputs. In addition, ever-expanding industrial, agricultural, and domestic discharges of toxic xenobiotic substances are strongly suspected of altering rates of primary production and community composition. The functional overlay of growth- and community-altering anthropogenic inputs in the world's aquatic ecosystems over a wide range of spatiotemporal scales necessitates short- as well as long-term detection and characterization of productivity responses, often integrated over large areas.

These informational needs are clearly beyond the scope of limited field surveys using the techniques discussed above. However, they constitute tasks ideally suited for multispectral (to detect diagnostic phytoplankton pigments such as chlorophylls, carotenoids, and phycobilins) aircraft and satellite-based remote sensing (21). The National Aeronautics and Space Administration's SeaWiFS satellite-based remote-sensing platform, to be launched in 1996, will be capable of discrete spectral analyses with sufficient sensitivity and resolution. SeaWiFS includes six narrow visible bands (412, 443, 490, 510, 555, and 670 nm) and two broader near-IR bands (765 and 865 nm). All eight bands are recorded in

10-bit resolution to take advantage of their high signal-to-noise ratios (up to 1,000). This makes this a remote-sensing platform ideal for high-resolution, high-sensitivity analyses of primary production among phygenetically diverse algal groups whose diagnostic pigments can be individually characterized.

The spatial and temporal coupling of nutrient and toxic inputs with patches of pigmented phototrophs is of critical importance in determining primary production, changes in algal biomass, bloom dynamics, and CO_2 flux in these waters. In addition, the contemporaneous impacts of autochthonous (internal; upwelling, regenerated N supplied to surface waters via mixing) and allochthonous (external; runoff, riverine, and atmospheric) nutrient- and toxin-mediated primary production should be assessed. Evolving remote-sensing techniques will help clarify and evaluate these nutrient-production interactions on scales appropriate for each system in question.

Research and logistic support were provided by the National Science Foundation, projects OCE 9012496, 9115706, DEB 921049, and DEB 9220886; the National Sea Grant Office (NOAA) and North Carolina Sea Grant College Program, projects NC R/EHP-1 and R/MER-23; the U.S. Geological Survey/UNC Water Resources Research Institute, project UNC-79-2; the U.S. Department of Agriculture, project 93-37102-9103; and the U.S. Environmental Protection Agency.

Technical assistance was provided by H. Barnes and J. Pinckney. I appreciate the editorial assistance of R. Christian and an anonymous reviewer.

REFERENCES

1. **Archer D., and A. Devol.** 1992. Benthic oxygen fluxes on the Washington shelf and slope: a comparison of in situ microelctrode and chamber flux measurements. *Limnol. Oceanogr.* **37:**614–629.

2. **Azam, F., T. Fenchel, J. G. Filed, J. S. Gray, L. A. Meyer-Reil, and F. Thingstad.** 1983. The ecological role of water-column microbes in the sea. *Mar. Ecol. Progr. Ser.* **10:**257–263.

3. **Bowles, N. D., H. W. Paerl, and J. Tucker.** 1985. Effective solvents and extraction periods employed in phytoplankton carotenoid and chlorophyll determinations. *Can. J. Fish. Aquat. Sci.* **42:**1127–1131.

4. **Cloud, P.** 1976. Beginnings of biospheric evolution and their biogeochemical consequences. *Paleobiology* **2:**351–387.

5. **Falkowski, P. G.** 1980. *Primary Productivity in the Sea.* Plenum Press, New York.

6. **Fitzwater, S. E., G. A. Knauer, and J. H. Martin.** 1982. Metal contamination and its effect on primary production measurement. *Limnol. Oceanogr.* **27:**544–551.

7. **Fee, E. J.** 1975. The importance of diurnal variation of photosynthesis vs. light curves to estimate of integral primary production. *Verh. Int. Ver. Limnol.* **19:**39–46.

8. **Fogel, M. L., and L. A. Cifuentes.** 1993. Isotope fractionation during primary production, p. 73–98. *In* M. H. Engel and S. A. Macko (ed.), *Organic Geochemistry.* Plenum Press, New York.

9. **Gieskes, W. W., and G. W. Kraay.** 1984. State-of-the-art in the measurement of primary production, p. 171–190. *In* M. J. R. Fasham (ed.), *Flow of Energy and Materials in Marine Ecosystems.* Plenum Publishing Co., New York.

10. **Goericke, R., and N. Welschmeyer.** 1993. The chlorophyll-labeling method: measuring specific rates of chlorophyll a synthesis in cultures and in the open ocean. *Limnol. Oceanogr.* **38:**80–95.

11. **Goldman, C. R.** 1986. *Primary Productivity in Aquatic Environments.* University of California Press, Berkeley, Calif.

12. **Goldman, C. R., D. T. Mason, and J. E. Hobbie.** 1967. Two Antarctic desert lakes. *Limnol. Oceanogr.* **12:**295–310.

13. **Harding, L. W., T. R. Fisher, and M. A. Tyler.** 1987. Adaptive responses of photosynthesis in phytoplankton: specificity to time-scale of change in light. *Biol. Oceanogr.* **4:**403–437.

14. **Harris, G. P., and B. B. Piccinin.** 1977. Photosynthesis by natural phytoplankton populations. *Arch. Hydrobiol.* **80:**405–457.

15. **Hickman, M.** 1969. Methods for determining the primary productivity of epipelic and epipsamnic algal associations. *Limnol. Oceanogr.* **14:**936–941.

16. **Hunding, C., and B. Hargrave.** 1973. A comparison of benthic microalgal production measured by ^{14}C and oxygen methods. *J. Fish. Res. Board Can.* **30:**309–312.

17. **Kemp, M., P. A. Sampou, J. Garber, J. Tuttle, and W. R. Boynton.** 1992. Seasonal depletion of oxygen from bottom waters of Chesapeake Bay: roles of benthic and planktonic respiration and physical exchange processes. *Mar. Ecol. Prog. Ser.* **85:**137–152.

18. **Lindeboom, H., A. Sandee, and H. Driessche.** 1985. A new bell jar/microelectrode method to measure changing oxygen fluxes in illuminated sediments with microalgal cover. *Limnol. Oceanogr.* **30:**693–698.

19. **Mallin, M. A., and H. W. Paerl.** 1992. Effects of variable irradiance on phytoplankton productivity in shallow estuaries. *Limnol. Oceanogr.* **37:**54–62.

20. **Mantoura, R. C. F., and C. A. Llewellyn.** 1983. The rapid determination of algal chlorophyll and carotenoid pigments and their breakdown products in natural waters by reverse-phase high performance liquid chromatography. *Anal. Chim. Acta* **151:**297–314.

21. **Millie, D. F., M. C. Baker, C. S. Tucker, B. T. Vinyard, and C. P. Dionigi.** 1992. High-resolution, airborne remote-sensing of bloom-forming phytoplankton. *J. Phycol.* **28:**281–290.

22. **Millie, D. F., H. W. Paerl, and J. P. Hurley.** 1993. Microalgal pigment assessments using high-performance liquid chromatography: a synopsis of organismal and ecological applications. *Can. J. Fish. Aquat. Sci.* **50:**2513–2527.

23. **Paerl, H. W.** 1974. Bacterial uptake of dissolved organic matter in relation to detrital aggregation in marine and freshwater systems. *Limnol. Oceanogr.* **19:**966–972.

24. **Paerl, H. W.** 1978. Microbial organic carbon recovery in aquatic ecosystems. *Limnol. Oceanogr.* **23:**927–935.

25. **Paerl, H. W.** 1984. An evaluation of freeze-fixation as a phytoplankton preservation method for microautoradiography. *Limnol. Oceanogr.* **29:**417–426.

26. **Paerl, H. W.** 1984. Transfer of N_2 and CO_2 fixation products from *Anabaena oscillarioides* to associated bacteria during inorganic carbon sufficiency and deficiency. *J. Phycol.* **20:**600–608.

27. **Paerl, H. W.** 1987. Dynamics of blue-green algal (*Microcystis aeruginosa*) blooms in the lower Neuse River, North Carolina: causative factors and potential controls. Report no. 229. University of North Carolina Water Resources Research Institute, Raleigh, N.C.

28. **Paerl, H. W., B. M. Bebout, S. B. Joye, and D. J. Des Marais.** 1993. Microscale characterization of dissolved organic matter production and uptake in marine microbial mat communities. *Limnol. Oceanogr.* **38:**1159–1161.

29. **Paerl, H. W., and D. F. Millie.** 1991. Evaluations of spectrophotometric, fluorometric and high performance liquid chromatographic methods for algal pigment determinations in aquatic ecosystems. Workshop report. U.S. Environmental Protection Agency region II, Cincinnati.

30. **Paerl, H. W., and E. A. Stull.** 1979. In defense of grain density autoradiography. *Limnol. Oceanogr.* **24:** 1166–1169.

31. **Parsons, T. R., Y. Maita, and C. M. Lalli.** 1992. *A Manual of Chemical and Biological Methods for Seawater Analysis.* Pergamon Press, Oxford.

32. **Peterson, B. J.** 1980. Aquatic primary productivity and the ^{14}C-CO_2 method: a history of the productivity problem. *Annu. Rev. Syst.* **11:**359–385.

33. **Pinckney, J., and R. G. Zingmark.** 1993. Photophysiological responses of intertidal benthic microalgal communities to in situ light environments: methodological considerations. *Limnol. Oceanogr.* **38:**1373–1383.

34. **Pomeroy, L.** 1959. Algal productivity in salt marshes of Georgia. *Limnol. Oceanogr.* **4:**386–397.

35. **Redalje, D.** 1993. The labeled chlorophyll a technique for determining photoautotrophic carbon specific growth rates and carbon biomass, p. 563–572. *In* P. Kemp, B. Sherr, E. Sherr, and J. Cole (ed.), *Handbook of Methods in Aquatic Microbial Ecology.* Lewis Publishing Co., Boca Raton, Fla.

36. **Redfield, A. C.** 1958. The biological control of chemical factors in the environment. *Am. Sci.* **46:**205–221.

37. **Revsbech, N. P., and B. B. Jørgensen.** 1986. Microelectrodes: their use in microbial ecology. *Adv. Microb. Ecol.* **9:**273–352.

38. **Revsbech, N. P., B. B. Jørgensen, and O. Brix.** 1981. Primary production of microalgae in sediments measured by oxygen microprofile, $H^{14}CO_3$ fixation, and oxygen evolution methods. *Limnol. Oceanogr.* **26:**717–730.

39. **Revsbech, N. P., J. Nielsen, and P. K. Hansen.** 1988. Benthic primary production and oxygen profiles, p. 69–83. *In* T. H. Blackburn and J. Sørensen (ed.), *Nitrogen Cycling in Coastal Marine Environments.* John Wiley & Sons, New York.

40. **Sharp, J. H.** 1977. Excretion of organic matter by marine phytoplankton. Do healthy cells do it? *Limnol. Oceanogr.* **22:**381–399.

41. **Sheldon, R. W., and W. H. Sutcliffe.** 1978. Generation times of 3 h for Sargasso Sea microplankton determined by ATP analysis. *Limnol. Oceanogr.* **23:**1051–1054.

42. **Shulenberger, E., and J. L. Reid.** 1981. The Pacific shallow oxygen maximum, deep chlorophyll maximum, and primary productivity, reconsidered. *Deep-Sea Res.* **28A:** 29–35.

43. **Steeman Nielsen, E.** 1952. The use of radioactive carbon (^{14}C) for measuring organic production in the sea. *J. Cons. Cons. Int. Explor. Mer* **18:**117–140.

44. **Strickland, J. D. H., and T. R. Parsons.** 1972. A practical handbook of seawater analysis. *Bull. Fish. Res. Board Can.* **167:**310p.

45. **Stull, E. A., E. De Amezaga, and C. R. Goldman.** 1973. The contribution of individual species of algae to primary productivity of Castle Lake, California. *Int. Ver. Theor. Angew. Limnol. Verh.* **18:**1776–1783.

46. **Van Heukelem, L., A. Lewitus, T. Kana, and N. Craft.** 1994. Improved separations of phytoplankton pigments using temperature-controlled high performance liquid chromatography. *Mar. Ecol. Prog. Ser.* **114:**303–313.

47. **Vollenweider, R. A., and A. Nauwerck.** 1961. Some observations on the ^{14}C method for measuring primary production. *Verh. Int. Ver. Limnol.* **14:**134–139.

48. **Watt, W.** 1971. Measuring the primary production rates of individual phytoplankton species in natural mixed populations. *Deep-Sea Res.* **18:**329–339.

49. **Wetzel, R. G., and G. E. Likens.** 1991. *Limnological Analyses.* Springer-Verlag, New York.

50. **Williams, P. J. L., K. R. Heinemann, J. Marra, and D. A. Purdie.** 1983. Comparison of ^{14}C and O_2 measurements of phytoplankton production in oligotrophic waters. *Nature* (London) **305:**49–50.

Bacterial Secondary Productivity

GERARDO CHIN-LEO

27

Planktonic heterotrophic bacteria (bacterioplankton) are now recognized to be a large and metabolically active group in aquatic systems that contribute significantly to the biomass and to the flow of carbon. Bacterial cell densities often exceed 10^9 cells liter^{-1}. Bacterial biomass in marine systems is generally greater than that of zooplankton and can be 20% of phytoplankton biomass (15, 20). Estimates of bacterial biomass and growth rate show that bacterial biomass turns over rapidly (hours). Comparison of bacterial biomass production rates with corresponding rates of phytoplankton production and biomass indicate that bacteria consume a substantial fraction (20 to 40%) of the carbon fixed by phytoplankton (15, 80). Bacteria are now considered major secondary producers, as they convert dissolved organic matter (DOM) derived from primary producers into an abundant biomass. The consumption of bacterial biomass by specialized predators (ciliates and flagellates) may be an important pathway for the transfer of DOM to metazoan food webs (3, 61, 70).

This view of bacteria has emerged as a result of recent improvements in the methods for measuring the abundance and growth rates of bacteria in environmental samples. Bacterial abundance is currently determined from the direct enumeration of bacterial cells by using nucleic acid-specific fluorochromes and epifluorescence microscopy (35, 59, 62). The contributions of bacteria to overall metabolism in ecosystems have been determined primarily from estimates of bacterial biomass production (BBP). If the assimilation efficiency of the bacterial population is known, BBP rates can be used to calculate the total utilization of organic matter by bacteria and consequently to determine the geochemical significance of bacterial metabolism. The amount of bacterial biomass potentially available to grazers and thus to higher trophic levels can also be determined from rates of BBP. In addition, changes in the rates of BBP can be used as an indicator of the response of bacteria to spatial and temporal fluctuations in environmental conditions. Currently, the most widely used methods to estimate BBP involve the measurement of various aspects of population growth such as the increase in cell numbers when predators are removed, changes in the frequency of dividing cells (FDC), and increases in the rate of synthesis of cellular constituents such as nucleic acids and proteins (64). Systemwide estimates of BBP determined by these independent methods agree in that bacterial production is high and consumes a substantial fraction of the algal primary production (15, 80).

Even though there is a general agreement regarding the importance of bacteria in aquatic ecosystems, the methods to assess BBP have remained controversial. Presently, there is little agreement about the best method to estimate BBP. This disagreement is based largely on uncertainties regarding the aspects of the taxonomy and metabolism of aquatic bacteria necessary to calibrate the methods and thus to determine the aspect of growth which is the best indicator of BBP. We know very little about the species composition and metabolic diversity of natural assemblages of bacteria. Traditional methods of laboratory microbiology which depend on the isolation of single bacterial strains are inadequate to study bacteria in aquatic systems. Only a small fraction of the bacteria in natural aquatic systems can be cultured (34). Therefore, current approaches for estimating BBP have focused on determining the growth of all heterotrophic bacteria rather than attempting to assess the growth rates of individual species. To use this synecological approach, assumptions must be made regarding the overall requirements and response of the entire bacterial assemblage. Furthermore, little is understood about the growth state and the growth environment of aquatic bacteria. For example, methods which measure the rate of synthesis of cellular constituents assume that bacteria have balanced, steady-state growth. Under these conditions, the rates of synthesis of all cellular constituents are equal. However, there is evidence that unbalanced growth is common in natural populations of bacteria (12, 42). During unbalanced growth, rates of synthesis of different macromolecules will yield different estimates of growth. Determinations of BBP based on the utilization of DOM can be inaccurate because bacteria in aquatic systems consume DOM which occurs in very low concentrations (2), and the chemical composition of this DOM is largely unknown (7).

Additional challenges to developing accurate and precise methods of measuring BBP relate to the possible artifacts caused by the manipulation of natural samples. Bacteria occur in association with other organisms and nonliving particles of similar size, making it difficult to isolate them by size fractionation. Bacteria can also respond rapidly to changes in their environment; thus, containment (bottle effects) and long incubation periods may significantly modify species composition and growth (26).

To properly assess the contribution of bacterial metabolism to the flow of organic matter in ecosystems, BBP needs to be converted into units of carbon or nitrogen. Because of the difficulties in isolating bacteria from surrounding debris, these values can seldom be measured directly. Carbon production is generally obtained by applying a conversion factor which translates cell production into carbon production. These conversion factors are typically derived by comparing the volume of cells with carbon content in mixed cultures of bacteria under laboratory conditions (53). Published values of volume to biomass vary by a factor of 14 (20), and thus conclusions derived from BBP estimates are very sensitive to the choice of conversion factor used. This problem is compounded by the need for other conversion factors which translate a given measure of growth (e.g., DNA or protein synthesis) into cells produced.

Given these uncertainties and methodological limitations, it is remarkable that current BBP estimates based on a variety of independent methods show strong correlations to algal production and biomass (15, 20, 80). This suggests that when compared over large time and space scales, these methods reflect actual rates of BBP and thus are adequate to study systemwide contributions by bacteria. However, as our knowledge of the ecology of aquatic bacteria increases, questions concerning the dynamics and controls of BBP have become more sophisticated and demand greater accuracy and precision of the methods. Intercalibration studies have shown that differences between methods can be significant (14), making the study of the dynamics and controls of bacteria over short time and space scales problematic. Because of the various uncertainties of current methods of measuring BBP, their adequacy to accurately estimate bacterial processes and rates has been questioned (39, 45). However, for many ecological questions, the current methods are still valuable tools. For example, understanding the response of bacteria to changing conditions requires high precision, but knowledge of the absolute value of BBP is not needed (16). Additionally, there has been substantial research on the causes for differences among methods, and uncertainties can be reduced by experimental validation of the assumptions of the chosen method in the system being studied (64, 67).

Because current methods of measuring BBP suffer from many uncertainties and limitations, the choice of method must follow careful thought regarding the question being addressed and the degree of accuracy that is satisfactory. For example, estimates of protein synthesis may be a better indicator of the amount of biomass available for consumption by grazers than the rate of cellular division because cells may increase in mass but not divide. The investigator must also be prepared to invest time and effort to test, in a given system, the validity of the assumptions of the method chosen.

A common approach given the difficulties in determining the accuracy of each method has been to simultaneously measure BBP by using independent methods. Agreement between these methods provides confidence that observed variations reflect real changes in rates of bacterial production. Because each method has its own assumptions and targets a different aspect of bacterial growth, differences in BBP estimates may reveal variations in environmental factors or changes in the growth state of bacteria. In this context, current methods of BBP can complement each other and provide useful information regarding the growth state and the environment of bacteria.

Of the various methods available to determine BBP, those that employ radiolabeled precursors to estimate the rate of synthesis of nucleic acids and proteins have become the most widely used and will be the focus of this chapter. The rate of synthesis of these macromolecules closely reflects cell division and cellular growth. Advantages of these procedures include high specificity for bacteria, high sensitivity, small sample volumes (<100 ml), and short incubation periods (<1 h). Disadvantages include the need to test for the specificity of the label for the chosen macromolecule and determine possible sources of external and internal isotope dilution and the need for conversion factors to obtain cell or carbon production estimates from incorporation rates. Currently, the three methods most commonly used are thymidine (TdR) incorporation into DNA, leucine (Leu) incorporation into protein, and adenine incorporation into RNA.

When bacterial growth is balanced, rates of incorporation into RNA, DNA, and protein will give equal estimates of growth. However, uncoupling of macromolecular synthesis (unbalanced growth) occurs when bacteria shift growth rates (38). Periods of unbalanced growth may consequently reflect the response of bacteria to changes in environmental conditions (12). Simultaneous measurement of the rate of synthesis of DNA, RNA, and protein thus can provide information regarding the growth state of bacteria in nature (64, 65).

Methods to determine BBP have been reviewed extensively (56, 64), and a manual with step-by-step instructions of several methods is available (47). In this chapter, the rationale, advantages, and disadvantages of TdR, Leu, and adenine methods will be discussed. Because adenine is also incorporated by algae and size fractionation is required to obtain estimates of BBP, only the procedures for TdR and Leu are described in detail. In addition, methods to determine empirically a conversion factor from TdR or Leu incorporation to cells produced, as well as several procedures designed to test various assumptions of these methods, are presented. Alternative methods to determine BBP which do not rely on the uptake of radiolabeled compounds are also briefly discussed.

TdR INCORPORATION INTO DNA

Rationale, Advantages, and Disadvantages

Of the various methods available to measure bacterial production, TdR incorporation into DNA (28) has become the most widely used in both freshwater and marine systems (15, 20, 80). This method estimates the rate of bacterial DNA synthesis and consequently the rate of cell division. Nucleotides are normally synthesized de novo, but in bacteria they can also be produced via a salvage pathway (56). dTMP is synthesized de novo from dUMP by the enzyme thymidylate synthetase. In the salvage pathway, dTMP is formed through the phosphorylation of exogenous TdR by the enzyme thymidine kinase. The TdR method assumes that TdR is added in sufficient concentration to inhibit de novo biosynthesis of nucleotides and promote the salvage mechanism. The added TdR is also assumed to be taken up rapidly by bacteria, to remain stable during uptake, and to be rapidly converted into nucleotides.

The advantages of the TdR method include (i) high specificity for heterotrophic bacteria, (ii) high sensitivity and precision, and (iii) ease of use in the field. Only bacteria have the transport mechanisms for assimilating TdR and the enzyme thymidine kinase needed for nucleotide synthesis via the salvage pathway. Other organisms such as eukaryotes or cyanobacteria do not appear to incorporate TdR into DNA (8). The TdR method requires minimal manipulation of water samples. Short (typically <1-h) incubation periods are sufficient, and only nanomolar concentrations of TdR are needed. The procedures are relatively simple and do not require complex equipment, and the method can be completed quickly and is easily adapted to field use. Furthermore, the filters containing the incorporated TdR can be stored for subsequent extraction and determination of radioactivity. The basic TdR procedure has been modified for use in other environments such as sediments (27) and marine snow (1).

Extensive use and testing in many aquatic environments has raised concerns related to the accuracy of the method. Major concerns are as follows. (i) TdR is not always incorporated into DNA, and other macromolecules are also labeled. (ii) The conversion factors required to obtain cell or carbon production from TdR incorporation vary widely and are difficult to derive empirically. (iii) De novo synthesis may not be entirely suppressed by added TdR, leading to underestimates of DNA synthesis. (iv) Microautoradiographic studies show that not all of the bacteria present incorporate TdR (19). It is not clear if these unlabeled bacteria are dormant or moribund cells or are actively growing bacteria with unique metabolism, or whether these bacteria lack TdR transport systems (60, 82). (v) Bacteria in anaerobic environments such as chemolithotrophic and sulfate-reducing bacteria do not appear to incorporate TdR (29, 40), and so the TdR method will underestimate BBP in anaerobic waters. The problems associated with the molecular specificity of TdR and with the conversion factor have been extensively studied. Isotope dilution caused by de novo synthesis and the fraction of the bacterial population incorporating TdR in various environments have not been investigated in as much detail.

The ^{3}H radioactive label of TdR often appears in macromolecules other than DNA, such as RNA and proteins, indicating catabolism of the added TdR (36, 66). The fraction of the total incorporated radioactivity which appears in non-DNA macromolecules can be very variable and may lead to serious overestimates of DNA synthesis. To circumvent this problem, several purification procedures have been developed to isolate the DNA fraction (36, 43, 81). Although differences in the degree of TdR catabolism reported can be traced to differences in the extraction method used (78), examination of the degree of TdR catabolism obtained by individual methods suggests that the extent of this metabolism varies predictably with experimental and environmental conditions. For example, nitrogen starvation in seawater cultures of bacteria causes the utilization of TdR for protein synthesis (13). Differences in TdR catabolism have been observed over a salinity gradient (10) and with depth (32). In San Francisco Bay, the patterns of TdR metabolism could be related to differences in the source of carbon to the system and to differences in bacterial species composition (37). Consistent patterns of TdR metabolism within a system may reflect an environmental or physiological basis. Thus, measurements of TdR catabolism can give insights into environmental conditions. Determination of the extent of TdR catabolism has become an important component of the TdR method. The procedures to isolate DNA are relatively simple and should be performed routinely. However, if the percentage of the total TdR that is incorporated into DNA is constant in a given system, it may not be necessary to perform the DNA isolation for every sample.

Another major concern of the TdR method is the need for conversion factors to translate TdR incorporation rates (picomoles per liter per hour) to cells or carbon produced (cells per liter per hour or micrograms of carbon per liter per hour). Rates of TdR incorporation are first converted into cells produced, using a theoretically or empirically derived factor. The theoretical factor is calculated from (i) the percentage of TdR in bacterial DNA, (ii) the DNA content per cell, (iii) assumptions regarding the extent of extracellular isotope dilution, and (iv) the effective inhibition of de novo synthesis (intracellular isotope dilution). The value of this factor is generally obtained from the results of laboratory studies, and it is often referred to as the literature conversion factor. This factor is 0.4×10^{18} to 0.5×10^{18} cells per mol of TdR incorporated. A major drawback of the theoretical conversion factor is that the information used in its computation is derived from studies using laboratory cultures of bacteria. These values may not be representative of natural bacteria. In studies using natural assemblages of bacteria, the amount of DNA per cell was quite variable, with smaller cells generally containing more DNA per unit volume (13, 71). Furthermore, other assumptions related to the theoretical factor, such as the extent of isotope dilution by de novo synthesis, are very difficult to determine directly (57).

The difficulties of using the theoretical conversion factor can be circumvented by deriving an empirical conversion factor. The empirical conversion factor uses natural assemblages of bacteria from the study area to calibrate the TdR method and is derived by comparing, under controlled conditions, TdR incorporation with increases in bacterial numbers (22). An added advantage of an empirical conversion factor is that all possible relationships between TdR incorporation and cell production (such as TdR catabolism) are, in theory, included in the conversion factor. Published empirical conversion factors vary widely and are generally higher than the theoretical factors (20, 64). The most commonly used empirical factor is 2×10^{18} cells per mol of TdR, which is the median of values from 97 studies (20). Variations in the empirical conversion factor may be caused by differences in the method used to analyze the data or to differences in the experimental procedures. For example, high conversion factors relative to the commonly used factor of 2×10^{18} cells per mol of TdR may be due to underestimates of TdR incorporation caused by the lack of uptake saturation (4, 24). If uptake is not saturated, de novo synthesis leads to isotope dilution. Bell (4) reported that variations in the conversion factors could be reduced by using higher concentrations of added TdR. Thus, when the TdR method is used, preliminary experiments should be performed to determine the TdR concentration at which maximal uptake is achieved (saturation concentration). Conversion factors determined by the same method can vary seasonally and spatially. These variations may reflect changes in environmental parameters such as nutrient availability (17).

For many questions regarding the ecological role of bacteria, BBP in units of carbon is required. Once TdR incorporation values are converted to cell production, another factor is applied to translate this information into units of carbon produced. This factor is determined by comparing cell volumes with carbon content in cultures of bacteria. A wide range of biomass conversion factors are available (53, 58). The choice of conversion factors and their combination is likely to significantly affect the calculated magnitude of bacterial carbon production based on any method using incorporation of radiolabeled compounds (20). Because of the various factors available to convert TdR incorporation rates into cell or carbon production, data are generally reported in units of TdR incorporation (typically picomoles per liter per hour).

Procedures for Estimating BBP from Incorporation Rates of TdR

The basic TdR procedure examines [methyl-^{3}H]TdR incorporation into a macromolecular fraction containing DNA, RNA, and protein which is isolated by a cold trichloroacetic acid (TCA) extraction (5). For each water sample, replicate samples (usually 10 ml) are incubated with [methyl-^{3}H]TdR (specific activity of 40 to 80 Ci mmol^{-1}; usually a final concentration of 20 nM). Abiotic absorption of radioactivity is measured in replicate samples in which bacteria have been killed with TCA or formalin. After an incubation period (30 to 60 min), the samples are placed in an ice-cold water bath for 1 min, and then 1 ml of ice-cold 50% TCA is added to a final concentration of 5%. Samples are cooled for an additional 15 min. This step kills the bacteria and lyses cell membranes, allowing the removal of the TdR taken up but not incorporated into macromolecules. Following this extraction, samples are filtered through 0.45-μm-pore-size cellulose acetate membrane filters. Filters with the bacteria are then rinsed twice with 1 ml of ice-cold 5% TCA and once with distilled water. To remove TdR incorporated into cells but not into DNA (e.g., into lipids), the filter is rinsed further with five 1-ml aliquots of 80% ethanol. Radioactivity in the filter is measured with a scintillation counter. Quenching during radioassay caused by the filter is reduced by dissolving the filter with 1 ml of ethyl acetate for 30 min prior to the addition of scintillation solution.

Several procedures are available for the isolation of DNA if incorporation into other macromolecules is an issue. These include an acid-base hydrolysis (43) and a phenol-chloroform extraction (81). To purify DNA by the phenol-chloroform procedure, lysed samples are filtered through cellulose nitrate membrane filters instead of cellulose acetate filters, since the latter are dissolved by chloroform-phenol. Following the incubation period, incorporation is stopped by addition of formaldehyde to a final concentration of 2%. RNA is extracted first by adding 5 N NaOH to a final concentration of 0.25 N and incubating the sample at 20 to 25°C for 60 min. The sample is then acidified with 100% TCA (1 to 2 ml/10-ml sample) to a pH of approximately 1 and stored on ice for 15 min. The sample is then filtered, and any protein remaining on the filter is extracted by rinsing with 5 ml of a 50% (wt/vol) chloroform-phenol solution (50 g of phenol in 100 ml of chloroform), leaving radiolabeled DNA on the filter. A final rinse is done with five 1-ml portions of ice-cold 80% ethanol, and the filter is prepared for radioassay as described above.

Exact sample volumes, incubation times, and the final concentration of TdR may vary depending on the environment (freshwater, salt water, oligotrophic, eutrophic, etc.) and are derived from preliminary experiments. The appropriate concentration of TdR is determined from saturation experiments in which replicate water samples are incubated with various concentrations of TdR (e.g., 5, 10, 15, 20, and 25 nM). The rate of TdR incorporation is then plotted versus added TdR, and the lowest concentration at which incorporation is saturated (i.e., no increase in incorporation observed) is chosen. The length of incubation is determined by incubating replicate samples over various time intervals (e.g., 10, 20, and 30 min) and determining if the relationship between rate of incorporation and incubation time is linear. To avoid bottle effects, the shortest incubation time which yields a linear, measurable rate is chosen.

The extent of extracellular isotope dilution is determined indirectly by a standard additions method. Replicate samples are incubated with a given amount of radioactive TdR (e.g., 5 nM) but with various concentrations of unlabeled TdR (e.g. 0, 5, 15, 25, and 35 nM). Increasing amounts of added unlabeled TdR dilute the radioactive TdR and yield decreasing incorporation rates. In the absence of external concentrations of the substrate, the plot of 1/incorporation rate versus total added substrate (labeled plus unlabeled) should pass through 0. However, if additional TdR is present in excess of what is added experimentally, the plot will not pass through 0 and the ambient concentration of TdR is determined by extrapolating the linear plot to 1/incorporation rate equals 0 ($y = 0$).

The proportion of bacterial cells incorporating TdR can be determined by microautoradiography (9, 60, 74). In this technique, filters containing bacterial cells previously incubated with TdR are placed face down on a radioactivity-sensitive emulsion coating one surface of a microscope slide. The filter is then removed, and the bacteria are embedded in the emulsion. Cells which incorporated TdR "expose" the underlying emulsion, producing silver grains. The entire slide is developed by normal photographic means and additionally stained with DNA-specific fluorochromes so that silver grains and bacteria can be identified simultaneously. The preparation is then observed with bright-field illumination to count the cells with associated silver grains. By using epifluorescence, the total bacterial population is determined by counting the fluorescent cells in the same sample.

LEU INCORPORATION INTO PROTEIN

Rationale, Advantages, and Disadvantages

The Leu approach has not been used as widely as the TdR approach. However, recently it has received much attention as an independent check of or as a promising alternative to the TdR method. The Leu method estimates the rate of bacterial protein synthesis by measuring the incorporation rate of radioactive Leu into proteins which are isolated by a hot TCA extraction (51, 52, 71). Protein synthesis is a good indicator of BBP because proteins account for a large percentage (~60%) of bacterial biomass in natural assemblages (30), and their synthesis consumes a large percentage of cellular energy.

Protein synthesis can be estimated by using $^{35}SO_4$ or [^{3}H]Leu as a radiolabeled precursor. [^{3}H]Leu is the preferred substrate because algae also utilize $^{35}SO_4$ and the high concentration of sulfate in marine environments can result in significant isotope dilution. Kirchman et al. (51) demonstrated that increases in Leu incorporation agreed with increases in cell numbers and protein content. Furthermore, most of the Leu was incorporated into proteins, and little was degraded. Given the small concentrations used (usually 10 to 20 nM) and the short incubation periods (<1 h), bacteria exclusively take up the added Leu. The uptake of other amino acids such as valine has also been used to estimate BBP (41, 68).

Advantages of the Leu method include the same ease of use and application to field work as for the TdR method.

Furthermore, the Leu method has several advantages over the TdR method. It is more sensitive than the TdR method because more Leu (proteins) is needed than TdR (DNA) in the production of new cells. The environmental and intracellular concentration of Leu, and hence the extent of isotope dilution, can be determined by high-performance liquid chromatography (71). Furthermore, Leu incorporation can be used to determine the bacterial carbon content following determination of a protein/carbon conversion factor (71).

When used simultaneously with the TdR method, the Leu method can provide an independent check of the TdR method. In addition, because the TdR and Leu methods estimate the rates of syntheses of different macromolecules (DNA and protein), simultaneous application of these methods also provides complementary information regarding the growth state of bacteria. The Leu method has been used alone and in conjunction with the TdR method for analysis of samples from freshwater (54, 72), marine environments (65, 79), and sediments (77) and to measure epiphytic bacterial production (76). The procedures of the Leu method are very similar to those of the TdR method, and the two methods can be performed with the same equipment and materials. Modifications have been made to reduce the cost of the procedures by using centrifugation rather than filters to concentrate the protein fraction (73). Furthermore, a dual-label approach which allows the simultaneous measurement of Leu and TdR in the same sample has been developed (11).

Leu incorporation can be used to determine bacterial carbon production directly because Leu composes a constant percentage of bacterial protein, and the ratio between protein and carbon over a wide range of bacterial cells is constant (71). With this information, carbon production from Leu incorporation is obtained by solving the following formula (48): BBP (g of C liter^{-1} h^{-1}) = Leu incorporation (mol^{-1} h^{-1}) · 131.2 · (%Leu)$^{-1}$ · (C/protein) · ID, where 131.2 is the formula weight of Leu, %Leu is the fraction of Leu in protein, C/protein is the ratio of cellular carbon to protein, and ID is isotope dilution. With information derived from using natural assemblages of bacteria, %Leu = 0.073, C/protein = 0.86, and ID = 2 results in a conversion factor of 3.1 kg of C mol^{-1} (71).

Concerns associated with the Leu method include isotope dilution by external and internal sources. Simon and Azam (71) found a twofold internal isotope dilution, suggesting that Leu biosynthesis was not effectively inhibited by the added Leu. An advantage, nevertheless, of the Leu method is that this dilution can be measured directly. However, the procedures are not easy and require special equipment (71). Another concern related to the Leu method is that in eutrophic environments, very high concentrations of Leu (up to 200 nM) may be needed to saturate incorporation (79). Addition of very high (millimolar) concentrations of Leu may modify bacterial protein synthesis and lead to the appearance of the radiolabel in phytoplankton or fungi (25), possibly as a result of the presence of low-affinity but specific uptake systems. Finally, if bacterial internal protein turnover is rapid, the Leu method will overestimate BBP. Kirchman et al. (52) reported that bacterial protein turnover in natural assemblages of bacteria did not appear to affect the Leu method, but there have been no further published studies examining this issue.

Procedures for Estimating BBP from Incorporation Rates of Leu

The incubation and filtration procedures of the Leu method are similar to those of the TdR method (11, 51). The protein fraction is isolated following a hot TCA extraction which solubilizes nucleic acids. Replicate samples (5 to 25 ml) are incubated with [4,5-^{3}H]Leu (usually a final concentration of 10 nM [48] but in some cases as high as 35 nM [55]; specific activity of 40 to 60 Ci mmol^{-1}). In eutrophic environments where incorporation rates are expected to be high, a combination of radioactive and nonradioactive leucine may be added (e.g., 1 nM [4,5-^{3}H]Leu and 10 nM nonradioactive Leu) to reduce the expense of the procedure. Abiotic absorption of radioactivity is measured with formalin- or TCA-killed controls. Samples are incubated for an appropriate period of time (30 to 60 min) and then placed in ice-cold water for 1 min to end the incubation. The total macromolecular fraction is isolated by adding 1 ml of 50% TCA to a final concentration of 5% and then heating samples to 80°C for 15 min, which extracts nucleic acids. Upon cooling to room temperature, samples are filtered through 0.45-μm-pore-size cellulose acetate filters to isolate the proteins. As with the TdR procedure, filters are rinsed with 5% cold TCA and with cold 80% ethanol to remove unincorporated Leu and Leu in lipids. Prior to radioassay, the filters are also dissolved with ethyl acetate. The exact sample volumes, incubation times, and final concentration of Leu are derived from preliminary experiments. These experiments are the same as those described above for the TdR method. Leu specificity for proteins can be tested by comparing the radioactivity on the filter with that in the hot-TCA filtrate which contains nucleic acids. The Leu method is very specific for proteins, and nearly all of the Leu taken up is assimilated into the protein fraction (51). If experimental evidence shows that this is the case in the study area, it is possible to simplify this procedure by substituting a cold TCA extraction for the hot TCA extraction. In this case, the procedures are identical to those of the TdR method. Leu incorporation can be translated directly to carbon production by using a theoretical conversion factor of 3.1 kg of C mol^{-1} (71). Alternatively, an empirical conversion factor can be determined by using mixed cultures of bacteria from the study area.

Simultaneous measurements of independent indices of BBP can provide complementary information regarding the growth of bacteria (65, 69). To achieve a simultaneous measurement of TdR and Leu incorporation in the same sample, [^{3}H]TdR and [^{14}C]Leu are added to a single sample and the cold TCA-insoluble material is collected (11). [^{14}C]Leu is chosen because [2-^{14}C]TdR can undergo catabolism and still label DNA because of the retention of the label in uracil. A final concentration of 20 nM Leu (10 nM [^{14}C]Leu; specific activity of about 300 mCi mmol^{-1} and 10 nM nonradioactive Leu) is used. Termination of incubation, cold TCA extraction of macromolecules, subsequent filtration, and preparation of filters for radioassay are the same as those described above. This dual-label approach has been modified and applied successfully to sediments (77).

EXPERIMENTS TO DETERMINE EMPIRICAL CONVERSION FACTORS FOR THE TdR AND LEU METHODS

Cultures of mixed assemblages of bacteria are prepared by inoculating filter-sterilized sample water with small volumes of prefiltered sample water (1 part prefiltered to 9 parts sterilized) (23, 49). Prefiltering is accomplished with a 0.6-μm-pore-size track-etch polycarbonate filter and reduces grazers, phytoplankton, and detritus in relation to bacteria. Sterile filtration is accomplished with 0.22-μm pore-size Nuclepore filters. Adding the prefiltered sample to sterilized water from the same site promotes bacterial growth (2), reduces grazing by bacterivores passing the 1-μm-pore-size filter, and reduces viral encounters (63). Gentle vacuum (<100 mm Hg [13 kPa]) is used throughout to avoid release of DOM due to cell lysis. Inoculated samples (1 to 4 liters) are kept in the dark and incubated at the in situ water temperature. Subsamples (10 to 50 ml) are taken every 2 to 4 h for a total of 24 h and assayed both for bacterial abundance (35, 62) and TdR (and/or Leu) uptake by the methods described above.

There are several methods to compute a conversion factor from these experiments. The simplest and most commonly used method (integrative method) involves dividing the increase in cell numbers by the substrate incorporation integrated over time (49). The change in cellular abundance is simply the final abundance minus the initial abundance. The total substrate incorporated is computed by integrating the area under the curve described by the plot of substrate incorporation versus time. A drawback of this approach is that only two values are used to determined the increase in cell numbers. Alternatively, the data can be analyzed by using an exponential growth model which considers all points (21, 23, 49, 50). The equation used to compute the conversion factor (CF) is $CF = \mu\, e^B/e^b$, where μ is the growth rate determined from changes in cells abundance over time, B is the y intercept of ln(cells) versus time, and b is the y intercept of ln(substrate incorporation) versus time (49). The conversion factor yields values in units of cells. To obtain carbon production, an additional factor must be applied (see above).

ADENINE INCORPORATION INTO DNA AND RNA

Rationale, Advantages, and Disadvantages

The adenine method has been applied to many water column environments (6, 46) and sediments (18) but has not been used as extensively as the TdR method. The adenine method estimates the rate of both DNA and RNA syntheses. Adenine is incorporated into bacteria and into eukaryotic algae and is designed to measure the growth of the entire microbial community. Estimates of BBP obtained by using this method, therefore, require a size fractionation step to isolate the bacteria.

An advantage of this method is the ability to simultaneously measure RNA and DNA synthesis in the same incubation. Relative changes in these rates can yield information regarding the extent of unbalanced growth in mixed assemblages of bacteria. In this method, it is possible to correct for isotope dilution by measuring the specific activity of the nucleic acid precursors. Adenine is incorporated into nucleic acids through various precursor molecules (ATP, dATP, ADP, and AMP). The existing pools of these compounds, in addition to inputs via de novo synthesis, dilute the specific activity of the added radioactive adenine ([2, ³H] adenine). In the adenine method, the immediate nucleotide triphosphate precursors are isolated and assayed for their specific activities.

Briefly, the procedure for the adenine method involves incubating replicate water samples with [³H] adenine (44). Following incubation, subsamples are extracted for isolation of nucleic acids. RNA, DNA, and the precursor ATP are separated, purified, and radioassayed. From this information, rates of RNA and DNA synthesis corrected for isotope dilu-

tion are computed. If a time course experiment in which replicate samples are incubated over different times is performed, the individual rates of DNA and RNA synthesis are determined. DNA synthesis can be converted to BBP if the C/DNA ratio is known.

OTHER METHODS

Several other methods for the estimation of BBP which do not require radiolabeled compounds are available. The FDC method, for example, estimates the growth rate of bacteria from an empirically determined relationship between growth rate, temperature, and the FDC (31, 59). By using epifluorescence microscopy, dividing cells are identified as cells visibly showing invagination. Advantages of this method are the lack of an incubation and the ability to determine the bacterial biomass from the same sample. Disadvantages are the need to determine the relationship between temperature, growth rate, and FDC for each environment investigated, the time and tedium involved in microscopic examination of all samples, and the subjectivity involved in the identification of dividing cells.

Other methods focus on determining increases in bacterial numbers after grazers have been eliminated by size fractionation or by using eukaryotic inhibitors. The overlap of sizes of bacteria and their predators and the possible control of bacteria by virus action complicates this method. Furthermore, metabolic inhibitors and antibiotics which are used to selectively inhibit groups of organisms are not always effective (75). Finally, the long-term incubations that are required to observe changes in bacterial numbers can introduce significant artifacts such as changes in DOM. Some of these bottle effects can be minimized by using dialysis bags, which allow exchange of dissolved materials but maintain the bacterial population isolated from predators (33).

SUMMARY AND CONCLUSIONS

The importance of heterotrophic bacteria as important decomposers of organic matter and as secondary producers in aquatic systems is widely recognized. While many methods have been developed to successfully study the contributions of bacteria at the ecosystem level, these methods have been difficult to apply to the understanding of the dynamics and controls of BBP over short time scales. A major difficulty in refining current methods and in developing new ones has been the dearth of knowledge regarding the species composition of natural assemblages of bacteria. Current methods of BBP must make assumptions regarding the overall characteristics of the community under consideration. Consequently, important information is lost and it is difficult to accurately calibrate the methods. Knowledge of the species composition and metabolic diversity of aquatic bacteria is sorely needed.

Presently, the most commonly used methods for the estimation of BBP are TdR incorporation into DNA, Leu incorporation into protein, and adenine incorporation into DNA and RNA. These methods are easily used to determine BBP in situ, and information derived with these methods has contributed significantly to our current view of bacteria in aquatic systems. Nevertheless, there are several concerns associated with these methods, such as the specificity of the label for bacteria and for the target molecule, dilution of the specific activity of the added substrate by extracellular and intracellular pools, and the need for conversion factors to determine carbon production. These uncertainties introduce into the results ambiguities which make the study of issues related to the controls and fate of bacterial production problematic. For example, the various conversion factors needed to obtain carbon production are not easily derived, and the ranges associated with the choice of factors and their combination may exceed the experimental errors of the methods. It is therefore difficult to compare BBP over short time and space scales and to accurately assess the role of bacteria in carbon flow.

Given the current state of knowledge, the simultaneous measurement of several independent indices of BBP is recommended. Simultaneous measurements help reduce the methodological limitations of individual methods. For example, because the Leu method provides carbon production estimates without the need for a carbon-per-cell conversion factor, simultaneous measurements of TdR and Leu provide estimates of both carbon and DNA synthesis. Furthermore, because balanced growth is probably not common in nature, simultaneous measurements of DNA, RNA, and protein can provide important complementary information regarding the growth state and growth environment of bacteria. Evaluation of the most appropriate combination of methods must follow careful consideration of the question being addressed and thorough testing of methodological assumptions in the study area.

REFERENCES

1. **Alldredge, A. L.** 1993. Production of heterotrophic bacteria inhabiting marine snow, p. 531–536. *In* P. F. Kemp, B. F. Sherr, E. B. Sherr, and J. J. Cole (ed.), *Handbook of Methods in Aquatic Microbial Ecology.* Lewis Press, Boca Raton, Fla.

2. **Ammerman, J. W., J. A. Fuhrman, Å. Hagström, and F. Azam.** 1984. Bacterioplankton growth in seawater. I. Growth kinetics and cellular characteristics in seawater cultures. *Mar. Ecol. Prog. Ser.* **18:**31–39.

3. **Azam, F., T. Fenchel, J. G. Field, J. S. Gray, L. A. Meyer-Reil, and T. F. Thingstad.** 1983. The ecological role of water-column microbes in the sea. *Mar. Ecol. Prog. Ser.* **10:**257–263.

4. **Bell, R. T.** 1990. An explanation for the variability in the conversion factor deriving bacterial cell production from incorporation of [³H]-thymidine. *Limnol. Oceanogr.* **35:**910–915.

5. **Bell, R. T.** 1993. Estimating production of heterotrophic bacterioplankton via incorporation of tritiated thymidine, p. 495–504. *In* P. F. Kemp, B. F. Sherr, E. B. Sherr, and J. J. Cole (ed), *Handbook of Methods in Aquatic Microbial Ecology.* Lewis Press, Boca Raton, Fla.

6. **Bell, R. T., and B. Riemann.** 1989. Adenine incorporation into DNA as a measure of bacterial production in freshwater. *Limnol. Oceanogr.* **34:**435–444.

7. **Benner, R., J. D. Pakulski, M. McCarthy, J. I. Hedges, and P. G. Hatcher.** 1992. Bulk chemical characterization of dissolved organic matter in the ocean. *Science* **255:**1561–1564.

8. **Bern, L.** 1985. Autoradiographic studies of [methyl-³H]thymidine incorporation in a cyanobacterium (*Microcystis wesenbergii*)-bacterium association and in selected algae and bacteria. *Appl. Environ. Microbiol.* **49:**232–233.

9. **Carman, K. R.** 1993. Microautoradiographic detection of microbial activity, p. 397–404. *In* P. F. Kemp, B. F. Sherr, E. B. Sherr, and J. J. Cole (ed.), *Handbook of Methods in Aquatic Microbial Ecology.* Lewis Press, Boca Raton, Fla.

10. **Chin-Leo, G., and R. Benner.** 1992. Enhanced bacterioplankton production and respiration at intermediate sa-

11. **Chin-Leo, G., and D. L. Kirchman.** 1988. Estimating bacterial production in marine waters from the simultaneous incorporation of thymidine and leucine. *Appl. Environ. Microbiol.* **54:**1934–1939.

12. **Chin-Leo, G., and D. L. Kirchman.** 1990. Unbalanced growth in natural assemblages of marine bacterioplankton. *Mar. Ecol. Prog. Ser.* **63:**1–8.

13. **Cho, B. C., and F. Azam.** 1988. Heterotrophic bacterioplankton production measurement by the tritiated thymidine incorporation method. *Arch. Hydrobiol. Beih. Ergebn. Limnol.* **31:**153–162.

14. **Christian, R. R., R. B. Hanson, and S. Y. Newell.** 1982. Comparison of methods for measurement of bacterial growth rates in mixed batch cultures. *Appl. Environ. Microbiol.* **43:**1160–1165.

15. **Cole, J. J., S. Findlay, and M. L. Pace.** 1988. Bacterial production in fresh and saltwater ecosystems: a cross-system overview. *Mar. Ecol. Prog. Ser.* **43:**1–10.

16. **Cole, J. J., and M. L. Pace.** 1995. Why measure bacterial production? A reply to the comment by Jahnke and Craven. *Limnol. Oceanogr.* **40:**441–444.

17. **Coveney, M. F., and R. G. Wetzel.** 1988. Experimental evaluation of conversion factors for the [³H]thymidine incorporation assay for bacterial secondary productivity. *Appl. Environ. Microbiol.* **54:**160–168.

18. **Craven, D. B., and D. M. Karl.** 1984. Microbial RNA and DNA synthesis in marine sediments. *Mar. Biol.* **83:**129.

19. **Douglas, J. D., J. A. Novitsky, and R. O. Fournier.** 1987. Microautoradiography-based enumeration of bacteria with estimates of thymidine-specific growth and production rates. *Mar. Ecol. Prog. Ser.* **36:**91–99.

20. **Ducklow, H. W., and C. A. Carlson.** 1992. Oceanic bacterial production. *Adv. Microb. Ecol.* **12:**113–181.

21. **Ducklow, H. W., and S. M. Hill.** 1985. Tritiated thymidine incorporation and the growth of heterotrophic bacteria in warm core rings. *Limnol. Oceanogr.* **30:**260–272.

22. **Ducklow, H. W., and D. L. Kirchman.** 1983. Bacterial dynamics and distribution during a spring diatom bloom in the Hudson River Plume. *J. Plankon Res.* **5:**333–355.

23. **Ducklow, H. W., D. L. Kirchman, and H. L. Quinby.** 1992. Determination of bacterioplankton growth rates during the North Atlantic spring phytoplankton bloom. *Microb. Ecol.* **24:**125–144.

24. **Ellenbroek, F. M., and T. E. Cappenberg.** 1991. DNA synthesis and tritiated thymidine incorporation by heterotrophic freshwater bacteria in continuous culture. *Appl. Environ. Microbiol.* **57:**1675–1682.

25. **Fallon, R. D., and S. Y. Newell.** 1986. Thymidine incorporation by the microbial community of standing-dead *Spartina alterniflora*. *Appl. Environ. Microbiol.* **52:**1206–1208.

26. **Ferguson, R. L., E. N. Buckley, and A. V. Palumbo.** 1984. Response of marine bacterioplankton to differential filtration and confinement. *Appl. Environ. Microbiol.* **47:**49–55.

27. **Findlay, S.** 1993. Thymidine incorporation into DNA as an estimate of sediment bacterial production, p. 505–508. *In* P. K. Kemp, B. F. Sherr, E. B. Sherr, and J. J. Cole (ed.), *Handbook of Methods in Aquatic Microbial Ecology.* Lewis Press, Boca Raton, Fla.

28. **Fuhrman, J. A., and F. Azam.** 1982. Thymidine incorporation as a measure of heterotrophic bacterioplankton production in marine surface waters: evaluation and field results. *Mar. Biol.* **66:**109–120.

29. **Gilmour, C. C., M. E. Leavitt, and M. P. Shiaris.** 1990. Evidence against incorporating of exogenous thymidine by

sulfate-reducing bacteria. *Limnol. Oceanogr.* **35:** 1401–1409.

30. **Hagström, Å.** 1984. Aquatic bacteria: measurements and significance of growth, p. 495–501. *In* M. J. Klug and C. A. Reddy (ed.), *Current Perspectives in Microbial Ecology.* American Society for Microbiology, Washington, D.C.

31. **Hagström, Å, U. Larsson, P. Hörstedt, and S. Normark.** 1979. Frequency of dividing cells, a new approach to the determination of bacterial growth rates in aquatic environments. *Appl. Environ. Microbiol.* **37:**805–812.

32. **Hanson, R. B., and H. K. Lowery.** 1983. Nucleic acid synthesis in oceanic microplankton from the Drake Passage, Antarctica; evaluation of steady of steady state growth. *Mar. Biol.* **73:**79–89.

33. **Herndl, G. J., E. Kaltenböck, and G. Müller-Niklas.** 1993. Dialysis bag incubation as a nonradiolabeling technique to estimate bacterioplankton production, p. 553–556. *In* P. F. Kemp, B. F. Sherr, E. B. Sherr, and J. J. Cole (ed.), *Handbook of Methods in Aquatic Microbial Ecology.* Lewis Press, Boca Raton, Fla.

34. **Hobbie, J. E.** 1988. A comparison of the ecology of planktonic bacteria in fresh and salt water. *Limnol. Oceanogr.* **33:**750–764.

35. **Hobbie, J. E., R. J. Daley, and S. Jasper.** 1977. Use of Nuclepore filters for counting bacteria by fluorescence microscopy. *Appl. Environ. Microbiol.* **33:**1225–1228.

36. **Hollibaugh, J. T.** 1988. Limitations of the [³H]thymidine method for estimating bacterial productivity due to thymidine metabolism. *Mar. Ecol. Prog. Ser.* **43:**19–30.

37. **Hollibaugh, J. T.** 1994. Relationships between thymidine metabolism, bacterioplankton community metabolic capabilities, and sources of organic matter. *Microb. Ecol.* **28:** 117–131.

38. **Ingraham, J. L., O. Maaloe, and F. C. Neidhardt.** 1983. *Growth of the Bacterial Cell.* Sinauer Associates, Sunderland, Mass.

39. **Jahnke, R. A., and D. B. Craven.** 1995. Quantifying the role of heterotrophic bacteria in the carbon cycle: a need for respiration rate measurements. *Limnol. Oceanogr.* **40:** 436–441.

40. **Johnstone, B., and R. P. Jones.** 1989. A study of the lack of (methyl-³H)thymidine uptake and incorporation by chemolithotrophic bacteria. *Microb. Ecol.* **18:**73–77.

41. **Jørgensen, N. O. G.** 1992. Incorporation of [³H]leucine and [³H]valine into protein of freshwater bacteria: uptake kinetics and intracellular isotope dilution. *Appl. Environ. Microbiol.* **58:**3638–3646.

42. **Karl, D. M.** 1981. Simultaneous rates of ribonucleic acid and deoxyribonucleic acid syntheses for estimating growth and cell division aquatic microbial communities. *Appl. Environ. Microbiol.* **44:**891–902.

43. **Karl, D. M.** 1982. Selected nucleic acid precursors in studies of aquatic microbial ecology. *Appl. Environ. Microbiol.* **42:**802–810.

44. **Karl, D. M.** 1993. Microbial RNA and DNA synthesis derived from the assimilation of [2,³H]-adenine, p. 471–482. *In* P. F. Kemp, B. F. Sherr, E. B. Sherr, and J. J. Cole (ed.), *Handbook of Methods in Aquatic Microbial Ecology.* Lewis Press, Boca Raton, Fla.

45. **Karl, D. M.** 1994. Accurate estimation of the microbial loop processes and rates. *Microb. Ecol.* **28:**147–150.

46. **Karl, D. M., and C. D. Winn.** 1984. Adenine metabolism and nucleic acid synthesis: applications to microbiological oceanography, p. 197–216. *In* J. Hobbie and P. Williams (ed.), *Heterotrophic Activity in the Sea.* Plenum Press, New York.

47. **Kemp, P. F., B. F. Sherr, E. B. Sherr, and J. J. Cole (ed.).** 1993. *Handbook of Methods in Aquatic Microbial Ecology.* Lewis Press, Boca Raton, Fla.

48. **Kirchman, D. L.** 1993. Leucine incorporation as a measure of biomass production by heterotrophic bacteria, p. 509–512. *In* P. F. Kemp, B. F. Sherr, E. B. Sherr, and J. J. Cole (ed.), *Handbook of Methods in Aquatic Microbial Ecology.* Lewis Press, Boca Raton, Fla.

49. **Kirchman, D. L., and H. W. Ducklow.** 1993. Estimating conversion factors for thymidine and leucine methods for measuring bacterial production, p. 513–517. *In* P. F. Kemp, B. F. Sherr, E. B. Sherr, and J. J. Cole (ed.), *Handbook of Methods in Aquatic Microbial Ecology.* Lewis Press, Boca Raton, Fla.

50. **Kirchman, D. L., and M. P. Hoch.** 1988. Bacterial production in the Delaware Bay estuary estimated from thymidine and leucine incorporation rates. *Mar. Ecol. Prog. Ser.* **32:** 169–178.

51. **Kirchman, D. L., E. K'nees, and R. E. Hodson.** 1985. Leucine incorporation and its potential as a measure of protein synthesis by bacteria in natural aquatic systems. *Appl. Environ, Microbiol.* **49:**599–607.

52. **Kirchman, D. L., S. Y. Newell, and R. E. Hodson.** 1989. Incorporation versus biosynthesis of leucine: implications for measuring rates of protein synthesis and biomass production by bacteria in marine systems. *Mar. Ecol. Prog. Ser.* **32:**47–49.

53. **Lee, S., and J. A. Fuhrman.** 1987. Relationships between biovolume and biomass of naturally derived marine bacterioplankton. *Appl. Environ. Microbiol.* **53:**1298–1303.

54. **McDonough, R. J., R. W. Sanders, K. G. Porter, and D. L. Kirchman.** 1986. Depth distribution of bacterial production in a stratified lake with an anoxic hypolimnion. *Appl. Environ. Microbiol.* **52:**992–1000.

55. **Moran, M. A., and R. E. Hodson.** 1992. Contributions of three subsystems of a freshwater marsh to total bacterial secondary productivity. *Microb. Ecol.* **24:**161–170.

56. **Moriarty, D. J. W.** 1986. Measurement of bacterial growth rates in aquatic systems from rates of nucleic acid synthesis. *Adv. Microb. Ecol.* **9:**243–292.

57. **Moriarty, D. J. W.** 1988. Accurate conversion factors for calculating bacterial growth rates from thymidine incorporation into DNA: elusive or illusive? *Arch. Hydrobiol. Beih. Ergebn. Limnol.* **31:**211–217.

58. **Nagata, T., and Y. Watanabe.** 1990. Carbon and nitrogen-to-volume ratios of bacterioplankton grown under different nutritional conditions. *Appl. Environ. Microbiol.* **56:** 99–109.

59. **Newell, S. Y., R. D. Fallon, and P. S. Tabor.** 1986. Direct microscopy of natural assemblages, p. 1–48. *In* J. S. Poindexter and E. R. Leadbetter (ed.), *Bacteria in Nature,* vol. 2. Plenum Press, New York.

60. **Pedrós-Alió, C., and S. Y. Newell.** 1989. Microautoradiographic study of thymidine uptake in brackish waters around Sapelo Island, Georgia, USA. *Mar. Ecol. Prog. Ser.* **55:**83–94.

61. **Pomeroy, L. R.** 1974. The ocean's food web: a changing paradigm. *BioScience* **9:**499–504.

62. **Porter, K. G., and Y. S. Feig.** 1980. The use of DAPI for identifying and counting aquatic microflora. *Limnol. Oceanogr.* **25:**943–948.

63. **Proctor, L. M., and J. A. Fuhrman.** 1990. Viral mortality of marine bacteria and cyanobacteria. *Nature* (London) **343:**60–62.

64. **Riemann, B., and R. Bell.** 1990. Advances in estimating bacterial biomass and growth in aquatic systems. *Arch. Hydrobiol.* **118:**385–402.

65. **Riemann, B., R. T. Bell, and N. O. G. Jørgensen.** 1990. Incorporation of thymidine, adenine and leucine into natural bacterial assemblages. *Mar. Ecol. Prog. Ser.* **65:**87–94.

66. **Robarts, R. D., R. J. Wicks, and L. M. Septhon.** 1986. Spatial and temporal variations in bacterial macromolecule

labeling with [*methyl-*^{3}H]thymidine. *Appl. Environ. Microbiol.* **52:**1368–1373.

67. **Robarts, R. D., and T. Zohary.** 1993. Fact or fiction—bacterial growth rates and production rates as determined by [methyl-^{3}H]thymidine? *Adv. Microb. Ecol.* **13:**371–425.

68. **Servais, P.** 1995. Measurement of the incorporation rates of four amino acids into proteins for estimating bacterial production. *Microb. Ecol.* **29:**115–128.

69. **Servais, P., and G. Billen.** 1991. Bacterial production measured by ^{3}H-thymidine and ^{3}H-leucine incorporation in various aquatic ecosystems. *Arch. Hydrobiol. Beih. Ergebn. Limnol.* **37:**73–81.

70. **Sherr, E. B., and B. F. Sherr.** 1987. High rates of consumption of bacteria by pelagic ciliates. *Nature* (London) **325:**710–711.

71. **Simon, M., and F. Azam.** 1989. Protein content and protein synthesis rates of planktonic marine bacteria. *Mar. Ecol. Prog. Ser.* **51:**201–213.

72. **Simon, M., and B. Rosenstock.** 1992. Carbon and nitrogen sources of planktonic bacteria in Lake Constance studied by the composition and isotope dilution of intracellular amino acids. *Limnol. Oceanogr.* **37:**1496–1511.

73. **Smith, D. C., and F. Azam.** 1992. A simple, economical method for measuring bacterial protein synthesis rates in seawater using [^{3}H]leucine. *Mar. Microb. Food Webs* **6:**107–114.

74. **Tabor, P. S., and R. A. Neihof.** 1982. Improved microautoradiographic methods to determine individual microorganisms active in substrate uptake in natural waters. *Appl. Environ. Microbiol.* **44:**945.

75. **Taylor, G. T., and M. L. Pace.** 1987. Validity of eucaryotic inhibitors for assessing production and grazing mortality of marine bacterioplankton. *Appl. Environ. Microbiol.* **53:**119–128.

76. **Thomaz, S. M., and R. G. Wetzel.** 1995. [^{3}H]leucine incorporation methodology to estimate epiphytic bacterial biomass production. *Microb. Ecol.* **29:**63–70.

77. **Tibbles, B. J., C. L. Davis, J. M. Harris, and M. I. Lucas.** 1992. Estimates of bacterial productivity in marine sediments and water from a temperate saltmarsh lagoon. *Microb. Ecol.* **23:**195–209.

78. **Torréton, J. P., and M. Bouvy.** 1991. Estimating bacterial DNA synthesis from [^{3}H]thymidine incorporation: discrepancies among macromolecular extraction procedures. *Limnol. Oceanogr.* **36:**299–306.

79. **van Looij, A., and B. Riemann.** 1993. Measurement of bacterial production in coastal marine environments using leucine: applications of a kinetic approach to correct for isotope dilution. *Mar. Ecol. Prog. Ser.* **102:**97–104.

80. **White, P. A., J. Kalff, J. B. Rasmussen, and J. M. Gasol.** 1991. The effect of temperature and algal biomass on bacterial production and specific growth rate in freshwater and marine habitats. *Microb. Ecol.* **21:**99–108.

81. **Wicks, R. J., and R. D. Robarts.** 1987. The extraction and purification of DNA labeled with [methyl-^{3}H]thymidine in aquatic bacterial population studies. *J. Plankton Res.* **9:**1159–1166.

82. **Zweifel, U. L., and Å. Hagström.** 1995. Total counts of marine bacteria include a large fraction of non-nucleoid-containing bacteria (ghosts). *Appl. Environ. Microbiol.* **61:**2180–2185.

Community Structure: Viruses

CURTIS A. SUTTLE

28

The realization that viruses are an abundant and ubiquitous component of aquatic ecosystems (2, 33) has stimulated a great deal of interest in the role of viruses in microbial systems (7, 15, 38). Many studies are beginning to include estimates of viral abundance, while others are attempting to directly address the structure and diversity of viral communities. As interest in the roles of viral communities in nature has increased, so has the effort to develop new methods. As a result, a number of approaches have been introduced in the past few years. When these are combined with well-established methods, this provides a suite of tools with which to address questions of community structure in natural virus communities.

The approach used to examine viruses in nature depends on the question being addressed. Nonetheless, it is possible to identify two major approaches. The first is used when one is interested in quantifying the entire viral community in terms of abundance or composition; the other is used when the primary objective is to examine populations of viruses that infect a single host. This chapter discusses methods that can be used to examine the structure of aquatic viral communities, although many of the methods can be adapted for other environments.

COMMUNITY APPROACHES
Determination of Total Viral Abundance in Aquatic Samples by Yo-Pro Staining

There are numerous instances when it is desirable to determine the abundance of viruses in aqueous samples, such as in studies of factors regulating viral abundance in nature and in routine environmental and public health monitoring programs. The latter include waste treatment, public water supplies, groundwater, and bathing areas (see also chapters 8 and 17).

There has been considerable interest in natural viral communities since observations that typical concentrations of viral particles in aquatic systems are in excess of 10^7 ml^{-1}. Overall, the abundance of viruses in marine and freshwaters is positively correlated with the concentration of chlorophyll and with bacterial abundance and production (11, 24, 27). Yet the abundance of viruses in aquatic communities can be extremely dynamic on time scales ranging from hours to weeks. For example, viral abundances in mesocosms filled with coastal seawater have been observed to change rapidly over periods of hours to days (3, 18). Few studies have examined seasonal changes in the total concentration of viruses. In Tampa Bay, a productive estuary on the west coast of Florida (24), viral abundance ranged about 10-fold (from 0.2×10^7 to 2.3×10^7 ml^{-1}). The highest concentrations occurred in the summer and fall, when bacteria and chlorophyll were also most abundant. Similarly, viral abundance in the Adriatic Sea was associated with higher chlorophyll and bacterial concentrations (45). The most extensive freshwater study was done for a large lake. Viral abundance was examined on a weekly basis during 4 months (19) and, although highly variable among sampling dates, was found to covary with the abundance of bacteria.

Most studies on viral abundance in aquatic systems have been done by using transmission electron microscopy (TEM) to count virus-like particles. For aquatic samples, this requires concentrating the viruses about 100- to 1,000-fold. Typically, the viruses are either pelleted directly onto electron microscopy grids by ultracentrifugation (3, 11, 46) or concentrated by ultrafiltration and then spotted and dried onto the surfaces of the grids (31, 33). TEM-based methods have the advantage that viral particles are directly observed; however, no information is obtained on the infectivity of the particles, and the hosts which the viruses infect remain unknown. Moreover, problems arise in samples with high concentrations of particulate material, as nonviral particles can obscure much of the grid surface, making counting difficult. Consequently, estimates of viral abundance made by using TEM often have relatively large coefficients of variation. Despite these difficulties, TEM can be used to estimate the abundance of viral particles in aquatic samples.

The greatest shortcomings of TEM-based methods are the cost of equipment and the time required to prepare samples. Transmission electron microscopes are expensive and tend to be associated with large research facilities. In addition, investigators need access to an ultracentrifuge or ultrafiltration equipment in order to concentrate the viruses. These constraints have motivated the development of other protocols for counting viruses in natural samples.

Epifluorescence microscopy of samples stained with acridine orange (22) or DAPI (4',6-diamidino-2-phenylindole) (32) is routinely used to estimate the abundance of bacteria in aquatic samples. As DAPI is quite specific for double-stranded DNA (dsDNA) and is brighter than acridine orange, it has also been used to estimate viral abundance in aquatic samples (16, 34, 37, 42). Estimates of viral abundance obtained by using DAPI staining can be comparable to those obtained by TEM (34). However, DAPI-stained viruses are near the limit of detection by epifluorescence microscopy, and the accuracy of the counts depends greatly on the optical equipment. Moreover, only dsDNA viruses are visible by DAPI staining. Although marine viral communities appear primarily to be made up of dsDNA viruses, this will not be true for all samples. Further details on counting viruses in aquatic samples by TEM and DAPI staining can be found elsewhere (37).

Recently, a protocol that uses epifluorescence microscopy in combination with a very bright nucleic acid-specific cyanine-based dye (Yo-Pro-1) for counting viruses was developed (20). The precision and accuracy of the counts appear to be higher than those obtained by TEM, and the results from the study suggest that viral abundance can be significantly underestimated using TEM.

There are several constraints of the method. Background fluorescence prohibits its use on samples with a high humic content, and bacteria as well as detritus are stained. In most cases, viruses can be readily distinguished from bacteria and detritus. Detritus is generally larger and fluoresces yellow, whereas the fluorescence of viruses is green. Bacteria can usually be distinguished from viruses by shape, but some very small bacteria may be counted as viruses. Counting bacteria as viruses is more a theoretical problem than a practical one for natural samples. Even if all bacteria in a water sample were counted as viruses, the error that would be introduced into estimates of viral abundance would typically be less than the standard deviation of the estimate of viral abundance (20). The actual error is much less, because only a very small proportion of the bacteria in a water sample are likely to be confused with viruses.

As for TEM-based methods, no information is provided on the number of infective viruses, and the hosts which the viruses infect remain unknown. Furthermore, although sample preparation is rapid and a large number of samples can be processed in parallel, the Yo-Pro procedure requires incubating samples for 2 days before counting. In some applications, it may pose a problem that the samples cannot be preserved prior to staining. Perhaps future modifications will circumvent these problems. Nonetheless, for most situations the Yo-Pro protocol should be suitable for total counts of viruses in aqueous samples. An outline of the method follows; a more complete description can be found elsewhere (20).

Outline of Method

(i) Dilute a stock solution of Yo-Pro-1 (1 mM Yo-Pro-1 in 1:4 dimethyl sulfoxide:water; Molecular Probes, Inc., Eugene, Ore.) to 50 μM in an aqueous solution of 2 mM sodium cyanide.

(ii) Place a series of 80-μl drops of the solution from the preceding step on the bottom of a 10-cm-diameter plastic petri dish. Place a filter paper soaked with 3 ml of an aqueous NaCl solution (0.3% [wt/vol]) in the lid to prevent evaporation of the stain.

(iii) Dilute unpreserved samples (100 μl) with 700 μl of sterile deionized-distilled water and place them on the surface of a 0.02-μm pore-size Al$_2$O$_3$ Anodisc 25 membrane filter (Whatman). The samples should remain within the plastic support ring of the filter by surface tension. The samples cannot be preserved with aldehyde fixatives, which prevent binding of Yo-Pro. Dilution of the samples is necessary so that a reasonable volume (800 μl) can be filtered and

because the high concentration of divalent cations in many natural water samples interferes with binding of the stain.

(iv) Gently filter (15 kPa) each sample, using a premoistened 0.45-μm-pore-size cellulose-nitrate membrane as a backing filter.

(v) Take the moist Anodisc membranes on which the samples have been filtered and lay them (sample side up) on drops of the staining solution. Incubate in the covered petri dish for 2 days in the dark at room temperature.

(vi) Filter two 800-μl aliquots of deionized distilled water through the membrane to rinse it.

(vii) Transfer the damp membranes to glass slides, immediately cover each with a drop of spectrophotometry-grade glycerol and a coverslip, and store them at $-20°$C until processing.

(viii) Count >200 viruses in 20 randomly selected fields at a magnification of $\times 1,000$, using an epifluorescence microscope equipped with an acridine orange filter set (excitation, <490 nm; dichroic filter, 500 nm; barrier filter, >515 nm).

(ix) The number of viruses per milliliter is calculated from $P \times (A/F) \times (1,000/V)$, where P is the number of stained viruses in the microscope field, A is the filtration area of the filter (square micrometers), F is the area of the microscope field (square micrometers), and V is the volume of the original sample prior to dilution or addition of stain (microliters).

Morphological Characterization of Viral Communities by TEM

For many samples, the abundance of viral particles is all that is required, while in other cases, data on the composition of viral communities will be needed. TEM is one approach that has been routinely used to examine the composition of natural virus communities.

Coincident with studies reporting high viral concentrations in natural waters were descriptions of the morphological characteristics of the viruses making up these communities. In general, viruses with capsid diameters of <60 nm dominate marine (2, 6, 45, 48) and freshwater (19) systems. In contrast, most viruses that have been isolated from marine systems are larger than the most abundant size classes in nature (4). Currently, data on changes in the relative abundance of different morphotypes in viral communities can be obtained only by TEM.

Morphological studies of viruses in natural waters are most appropriate for inferring the dynamics of specific components of viral communities. For example, Bratbak et al. (5) observed a substantial increase in the abundance of a large icosahedral virus coincident with the collapse of an induced phytoplankton bloom, suggesting that the virus was responsible for its demise. Other studies have shown that the abundance of small viruses is more dynamic than that of larger viruses (6, 18, 19) and that increases in specific viral morphotypes appear to be associated with the collapse of bacterial communities in natural systems (6, 19).

Typically, viruses are concentrated by ultracentrifugation or ultrafiltration, spotted on an electron microscopy grid, and viewed by TEM. The method can be used to gather data on the size structures of natural viral communities (6, 11, 19, 45, 48) and can even be used to determine the abundance of visibly infected cells and burst sizes for natural bacterial populations (46). Methods for staining, visualizing, and enumerating viral particles by TEM are reviewed in detail elsewhere (3, 17, 37). A protocol that can be used for natural viral communities is briefly outlined below.

Outline of Method

(i) Viruses are pelleted from water samples as soon as possible after collection. It is preferable that samples not be treated with fixatives or preservatives which can cause clumping; however, samples can be stored for 1 to 2 weeks in 1 to 2% electron microscopy-grade glutaraldehyde.

(ii) Using double-sided tape, attach carbon-Formvar-coated electron microscopy grids onto platforms that are flat on the surface and will fit into the bottom of an ultracentrifuge tube that can be placed in a swing-out rotor. Epoxy platforms can be made from resin, which is placed in the bottom of an ultracentrifuge tube and allowed to harden. For most natural water samples, 100-mm centrifuge tubes concentrate the viruses enough so that they can be viewed by TEM.

(iii) Place the platforms with the attached grids into the bottom of the centrifuge tubes, fill the tubes with sample, and load the tubes into the rotor. Centrifuge the samples so that particles of 80S are sedimented with 100% efficiency. For most samples, 3 h at 180,000 $\times$ g is suitable, but for samples of high salinity or which are cold, it may be necessary to increase the time.

(iv) The liquid is aspirated from the tube, the grids are removed, and the excess water is wicked from the surface of the grids with a piece of filter paper. For saltwater samples, it may also be necessary to wash the grids by placing them sample side down on a drop of distilled water. It is important not to let the surface of the grids become dry during this procedure.

(v) The sample is stained by floating the grid face down on a drop of 1% uranyl acetate for several seconds and immediately wicking the stain from the surface of the grid. This procedure results in areas where stain has been deposited around the perimeter of the viruses, yielding a negatively stained image. This technique provides the best preparation for observing the morphology of viruses. If the samples are stained too long, the viruses will be positively stained, which obscures detail and causes inaccurate estimates of viral size. Negatively stained preparations are generally not adequate for estimating the abundance of viruses in aquatic samples; however, by placing several grids at the bottom of the ultracentrifuge tubes, positively stained preparations can be made from the same water sample.

(vi) The grids are observed by TEM at ca. 80 kV and a magnification of ca. $\times$30,000. Photographic images of the preparations will provide much greater resolution than the phosphorescent screen of the microscope.

This approach can be used to obtain estimates on the abundance of viral morphotypes in a sample. However, there are many potential pitfalls, including the fact that viruses do not sediment in parallel paths, staining is uneven, and there is interference from particulates. Before attempting estimates of abundance by TEM, consult more comprehensive descriptions of the protocol (e.g., references 3, 16, 17, and 37).

POPULATION APPROACHES

Community approaches are useful in studies of the total abundance of viral particles or for documentation of the dominant morphotypes in viral communities. However, these methods do not yield information about the infectivity of viruses in natural communities or about the organisms that the viruses infect. Infectivity data are important, as much of the natural viral community may be noninfective

(43), and the organisms which are infected cannot be determined from morphology. For example, cyanophages belong to the same families of dsDNA viruses as those which infect heterotrophic bacteria, yet their effects on natural microbial systems are very different because of the different trophic roles of cyanobacteria and heterotrophic bacteria.

For more than 40 years, studies have documented the presence of viruses in the sea that infect specific isolates of marine bacteria (36), yet only a few groups (30, 35, 49) have made concerted efforts to investigate the occurrence of viruses that infect strains of bacteria and cyanobacteria from natural environments. Concentrations of viruses which infect specific hosts vary from less than a few viruses to $>10^5$ ml^{-1}. Typically, investigations have found relatively low titers of viruses which infect specific isolates of heterotrophic bacteria, although concentrations of $>10^4$ ml^{-1} have been reported in some instances (29). Concentrations of viruses infecting phytoplankton can be higher. For example, viruses which infect strains of the cyanobacteria *Synechococcus* spp. can occur in excess of 250,000 infective units ml^{-1} in coastal seawater and routinely occur at $>10^3$ ml^{-1} (39, 40, 44). Eukaryotic phytoplankton are also subject to infection by lytic viruses, and abundances can also be high; concentrations of $>10^5$ ml^{-1} have been measured in coastal seawater for viruses which infect the unicellular photosynthetic flagellate *Micromonas pusilla* (13).

In some cases, the abundance of viruses is highest when host density is highest. For example, cyanophages are most abundant in productive coastal waters when the concentration of *Synechococcus* spp. and water temperatures are highest (39, 40, 44). In contrast, the levels of viruses which infect the photosynthetic flagellates *M. pusilla* and *Chrysochromulina brevifilum* are highest in winter (13, 41); however, these species are also most likely to be abundant when water temperatures are cooler. In other cases, the highest concentrations of infective viruses have been associated with the collapse of host cell populations (21).

Abundance of Viral Populations

To infer the effects of viruses on specific taxa within the microbial community, it is necessary to measure the abundance of viruses which infect specific hosts. As this requires a bioassay, the organism which is infected by the virus(es) must be in culture. As many microorganisms are difficult to isolate and culture, this is a major limitation of the method. For bacteria and phytoplankton which can be grown on solid media, plaque assays are relatively straightforward and give the highest precision for estimates of the abundances of lytic viruses. Unfortunately, many isolates do not grow well on solid substrates, and a most probable number (MPN) assay in liquid culture must be used. The precision of MPN assays is less than that of plaque assays, but they are extremely sensitive and, if carefully optimized can be used to estimate the abundance of a few infectious viruses per liter. Below are generalized descriptions of plaque and MPN assays. Further details can be found elsewhere (37; chapter 17 of this volume).

Plaque Assays

Plaque assays are based on the fact that viruses form visible plaques on a lawn of host cells. Typically, host cells and the sample to be assayed are mixed in molten agar and poured quickly and evenly over a solidified bottom layer of agar. As the bacteria or algae grow, the lawn will become darker, and areas where lysis of cells has occurred will appear as clearings on the lawn. The number of viruses is quantified as the number of PFU in the volume of water sampled. The method has the advantages that it is relatively sensitive, with a detection limit of about 5 viruses ml^{-1}, only viruses that are infectious for a specific host are enumerated, results are accurate, and the infective agent can be purified relatively easily from individual plaques. A brief protocol for a plaque assay is provided below; additional details can be found in reference 1.

Outline of Method

(i) The water sample is collected as soon as possible before the plaque assay. It may be necessary to prefilter the sample through a 0.2- or 0.45-μm-pore-size filter to prevent other bacteria from overgrowing the lawn and interfering with the assay.

(ii) Combine 500 μl of exponentially growing host cells with 500 μl of the sample to be assayed in a sterile microcentrifuge tube, and mix quickly by gentle vortexing.

(iii) Adsorb the viruses to the host cells. The time to allow for adsorption depends on host cell density as well as the adsorption kinetics of the virus. Adsorption occurs most rapidly at high host densities when contact rates are highest. However, at very high host densities, a significant proportion of the cells may be nonviable, and if adsorption onto nonviable hosts occurs, the titer of infectious viruses will be underestimated. In general, a period of several minutes is adequate to allow for adsorption, but adsorption times should be optimized for each system.

(iv) Add the sample to a tube containing 2 ml of molten top agar, vortex, and quickly pour the contents onto the bottom agar. The host cells and viruses should be tested to ensure that the elevated temperatures required for top agar do not decrease viability.

(v) Count the plaques after an even lawn appears (usually 1 to 7 days).

MPN Assays

When the host organism cannot be grown on solid medium, the number of infective viruses must be estimated by MPN assay. The principle of the method is that the sample to be titered is taken through a series of dilutions and added to liquid cultures of host cells. In those cultures to which one or more infective viruses have been added, lysis of the culture will occur. The estimate of viral titer is obtained from the number of cultures in which lysis occurs at each step in the dilution series. Typically, a series of 10-fold dilutions, with 3 to 10 replicates at each dilution, is added to culture tubes, and the concentration of infective viruses is estimated by comparing the number of replicates at each dilution in which lysis occurs with an MPN table. In many instances, greater accuracy and precision are obtained by conducting the assays in 96-well microtiter plates rather than culture tubes. The protocol works especially well for phytoplankton (13, 39, 40, 44). However, when sensitivity greater than ca. 50 viruses ml^{-1} is required, it is necessary to use larger cultures so that a larger volume of sample can be screened. The microtiter procedure is outlined below; a more detailed discussion of MPN assays is provided elsewhere (26, 37).

Outline of Method

(i) Prepare the medium which will be used to construct the dilution series of the natural water sample that will be titered. The medium should allow vigorous growth of the host organism, be free of any viruses that might cause lysis, and be free of particulate material to which viruses might adsorb. Depending on the host-virus system, suitable media include natural water that has been ultrafiltered, filter sterilized, and autoclaved or artificial media.

(ii) Transfer the assay organism into fresh medium so that it will be near the beginning of exponential growth when it is dispensed into the microtiter plates.

(iii) Collect the sample to be titered in a clean container and prepare a series of 10-fold dilutions. It may be necessary to prefilter the sample through 1.0- or 0.2-μm-pore-size filters to remove other organisms which might overgrow the assay organism.

(iv) Transfer 200 to 250 μl of the exponentially growing culture into each well of a microtiter plate, and for each dilution add ca. 50 μl into each of 8 to 16 wells of the microtiter plate. Include 8 to 16 wells to which only medium and host cells are added to ensure that the medium supports vigorous growth.

(v) After sufficient time has passed to complete ca. 10 lytic cycles, score the plates for the number of wells that have cleared at each dilution series. The plates should be monitored on a regular basis during this interval, as resistant cells can grow in wells in which lysis has occurred, thereby causing incorrect estimates of MPN. Ideally, an aliquot from each well should be transferred into fresh, exponentially growing cultures to ensure that the lytic agent can be propagated or, in case a virus has not been amplified, enough to cause lysis. Calculate the MPN by using tables or preferably a computer program that also provides statistical parameters for the estimates (e.g., reference 23).

Genetic Diversity

Although it is clearly established that viruses are an abundant, dynamic, and ecologically important component of aquatic ecosystems, we are at the earliest stages in defining the genetic diversity within these communities and populations. Furthermore, the mechanisms regulating this diversity are even less well investigated. A major constraint in examining diversity is that most microbes that serve as hosts have not been cultured. Therefore, approaches that require isolation of the viruses likely represent a small portion of the diversity that is present. From results for the few systems that have been examined, it appears that aquatic viral populations are genetically diverse within and among geographic locations. This can be seen in studies which have shown that host range is often very restricted (30, 39, 44). Potentially there is an enormous variety of approaches that can be used to examine genetic diversity in natural virus communities, but few attempts have been made. Consequently, no single approach can be recommended or has a proven track record. Moreover, many approaches require familiarity with molecular techniques, the description of which is beyond the scope of this chapter. Hence, I will restrict discussion to the methods that have been used in the few experiments which have examined genetic diversity in aquatic viruses.

The best approach to quantifying diversity in aquatic viral populations and communities is to directly examine the nucleic acids. One of the simplest approaches for making comparisons among DNA viruses is analysis by restriction fragment length polymorphism (RFLP). RFLP analysis is done by extracting DNA from the viruses, cutting it by using restriction enzymes, and running the cut DNA on an agarose gel. DNAs with different restriction banding patterns have different genotypes. This approach was used to examine the genetic diversity among viruses (MpV) which infect the marine photosynthetic flagellate M. pusilla. These large dsDNA viruses have a genome of about 200 kb, are widely distributed in coastal waters (11), and are not known to infect any other species (28). Although viruses isolated from within and among locations were morphologically indistinguishable by electron microscopy, they were all genotypically different (11). A similar approach was used to examine marine cyanophages, which were also found to be genetically variable (47). Because of the large genetic variation within viral populations, quantitative comparisons of the variation within environmental isolates may require a larger degree of resolution than can be obtained by RFLP. Recently, however, the genetic diversity among vibrio phages isolated from waters around Florida and Hawaii was compared by probing DNA from the viral isolates with a DNA fragment which had been cloned from a marine vibrio phage and which hybridized to each of the viral isolates (25). Using this approach, it was shown that viruses isolated from Hawaiian waters were less diverse than those from Florida.

An ideal way to make direct comparisons of genetic variation among viruses is to select a gene which reflects their evolutionary history. There are many suitable gene targets that can be chosen, depending on the viruses being compared. These include capsid and polymerase genes. For many dsDNA viruses, the DNA polymerase (*pol*) gene is a good target. This gene is found in all organisms as well as in many large dsDNA viruses. In addition, the gene has regions which are highly conserved when the inferred amino acid sequences are compared and other regions which are variable. The conserved regions make it relatively easy to find the gene and align it with other *pol* genes once sequenced, while the variable regions facilitate comparison within groups of viruses which are closely related. Once a *pol* gene is found, it may be possible to assign a virus or a group of viruses to an established family on the basis of the sequence that is obtained. It is also possible to identify regions of the gene which can be used with PCR to specifically amplify segments of the gene from the group of viruses of interest. In this way, it is possible to rapidly obtain sequences from a relatively large number of viral isolates or even from natural viral communities, thereby eliminating the selective and time-consuming steps of isolating and purifying the virus before extracting the DNA. Furthermore, because the sequences are more conserved at the inferred amino acid level than at the DNA level, it is possible to design primers with low degeneracy that amplify DNA from very closely related viruses or to make primers which are quite degenerate and can be used to amplify DNA from groups of more distantly related viruses (e.g., reference 8).

An example of this approach is the examination of the genetic diversity of a group of viruses which infect microalgae. On the basis of sequence analysis of DNA *pol* genes from MpV and from two viruses which infect a microalgal symbiont found in *Paramecium bursaria*, a set of primers specific for this group of viruses was designed (9). These primers were used to amplify DNA from a variety of viral isolates as well as from natural marine viral communities. From phyletic analysis of these sequences, it was possible to compare the genetic diversity of these viruses within and among water samples, to determine the evolutionary relationships among the different viral isolates and those in natural viral communities, and to demonstrate that the closest relatives to these viruses are the herpesviruses (10). Moreover, by comparing the results of the genetic relationships obtained by sequence analysis of the *pol* fragments with those obtained from total genomic DNA hybridization (10, 14), it was possible to show that the relationships inferred from

sequence analysis were representative of the entire viral genomes.

This example shows one approach that can be used, assuming that the virus or group of viruses in which one is interested has an α-like DNA polymerase gene. Other viruses will require identifying other genes which have suitable resolution. The question of the overall genetic diversity of natural communities of viruses has yet to be addressed, and there are no obvious ways in which this problem can be easily tackled. Ultimately, we may have to wait for some new insights and approaches before it is possible to solve this problem.

REFERENCES

1. **Adams M. H.** 1959. *Bacteriophages*. Interscience, New York.
2. **Bergh, O., K. Y. Børsheim, G. Bratbak, and M. Heldal.** 1989. High abundance of viruses found in aquatic environments. *Nature* (London) **340:**476–468.
3. **Børsheim, K., G. Bratbak, and M. Heldal.** 1990. Enumeration and biomass estimation of planktonic bacteria and viruses by transmission electron microscopy. *Appl. Environ. Microbiol.* **56:**352–356.
4. **Børsheim, K. Y.** 1993. Native marine bacteriophages. *FEMS Microbiol. Lett.* **102:**141–159.
5. **Bratbak, G., J. K. Egge, and M. Heldal.** 1993. Viral mortality of the marine alga *Emiliania huxleyi* (Haptophyceae) and termination of algal blooms. *Mar. Ecol. Prog. Ser.* **93:**39–48.
6. **Bratbak, G., M. Heldal, S. Norland, and T. Thingstad.** 1990. Viruses as partners in spring bloom microbial trophodynamics. *Appl. Environ. Microbiol.* **56:**1400–1405.
7. **Bratbak, G., T. F. Thingstad, and M. Heldal.** 1994. Viruses and the microbial loop. *Microb. Ecol.* **28:**209–221.
8. **Chen, F., and C. A. Suttle.** 1995. Nested PCR with three highly degenerate primers for amplification and identification of DNA from related organisms. *BioTechniques* **18:**609–611.
9. **Chen, F., and C. A. Suttle.** 1995. Amplification of DNA polymerase gene fragments from viruses infecting microalgae. *Appl. Environ. Microbiol.* **61:**1274–1278.
10. **Chen, F., and C. A. Suttle.** 1996. Evolutionary relationships among large double-stranded DNA viruses that infect microalgae and other organisms as inferred from DNA polymerase genes. *Virology* **219:**170–178.
11. **Cochlan, W. P., J. Wikner, G. F. Stewart, D. C. Smith, and F. Azam.** 1993. Spatial distribution of viruses, bacteria and chlorophyll a in neritic, oceanic and estuarine environments. *Mar. Ecol. Prog. Ser.* **92:**77–87.
12. **Cottrell, M. T., and C. A. Suttle.** 1991. Wide spread occurrence and clonal variation in viruses which cause lysis of a cosmopolitan, eukaryotic marine phytoplankter, *Micromonas pusilla*. *Mar. Ecol. Prog. Ser.* **78:**1–9.
13. **Cottrell, M. T., and C. A. Suttle.** 1995. Dynamics of a lytic virus infecting the photosynthetic marine picoflagellate *Micromonas pusilla*. *Limnol. Oceanogr.* **40:**730–739.
14. **Cottrell, M. T., and C. A. Suttle.** 1995. Genetic diversity of algal viruses which lyse the photosynthetic picoflagellate *Micromonas pusilla* (Prasinophyceae). *Appl. Environ. Microbiol.* **61:**3088–3091.
15. **Fuhrman, J. A., and C. A. Suttle.** 1993. Viruses in marine planktonic systems. *Oceanography* **6:**50–62.
16. **Hara, S., K. Terauchi, and I. Koike.** 1991. Abundance of viruses in marine waters: assessment by epifluorescence and transmission electron microscopy. *Appl. Environ. Microbiol.* **57:**2731–2734.
17. **Hayat, M. A., and S. E. Miller.** 1990. *Negative Staining*. McGraw-Hill, New York.

18. **Heldal, M., and G. Bratbak.** 1991. Production and decay of viruses in aquatic environments. *Mar. Ecol. Prog. Ser.* **72:**205–212.

19. **Hennes, K. P., and M. Simon.** 1995. Significance of bacteriophages for controlling bacterioplankton growth in a mesotrophic lake. *Appl. Environ. Microbiol.* **61:**333–340.

20. **Hennes, K. P., and C. A. Suttle.** 1995. Direct counts of viruses in natural seawater and laboratory cultures by epifluorescence microscopy. *Limnol. Oceanogr.* **40:**1050–1055.

21. **Hennes, K. P., C. A. Suttle, and A. M. Chan.** 1995. Fluorescently labeled virus probes show that natural virus populations can control the structure of marine microbial communities. *Appl. Environ. Microbiol.* **61:**3623–3627.

22. **Hobbie, J. E., R. J. Daley, and S. Jasper.** 1977. The use of Nuclepore filters for counting bacteria by epifluorescence microscopy. *Appl. Environ. Microbiol.* **33:**1225–1228.

23. **Hurley, M. A., and M. E. Roscoe.** 1983. Automated statistical analysis of microbial enumeration by dilution series. *J. Appl. Bacteriol.* **55:**159–164.

24. **Jiang, S. C., and J. H. Paul.** 1994. Seasonal and diel abundance of viruses and occurrence of lysogeny/bacteriocinogeny in the marine environment. *Mar. Ecol. Prog. Ser.* **104:**163–172.

25. **Kellogg, C. A., J. B. Rose, S. C. Jiang, J. M. Thurmond, and J. H. Paul.** 1995. Genetic diversity of vibrio phages isolated from marine environments around Florida and Hawaii. *Mar. Ecol. Prog. Ser.* **120:**89–98.

26. **Koch, A. L.** 1981. Growth measurement, p. 179–207 *In* P. Gerhardt, R. G. E. Murray, R. N. Costelow, E. W. Nester, W. A. Wood, N. R. Krieg, and G. B. Phillips (ed.), *Manual of Methods for General Bacteriology.* American Society for Microbiology, Washington, D.C.

27. **Maranger, R., and D. F. Bird.** 1995. Viral abundances in aquatic systems: a comparison between marine and fresh waters. *Mar. Ecol. Prog. Ser.* **121:**217–226.

28. **Mayer, J. A., and F. J. R. Taylor.** 1979. A virus which lyses the marine nanoflagellate *Micromonas pusilla. Nature* (London) **281:**299–301.

29. **Moebus, K.** 1992. Laboratory investigations on the survival of marine bacteriophages in raw and treated seawater. *Helgol. Wiss. Meeresunters.* **46:**251–273.

30. **Moebus, K., and H. Nattkemper.** 1983. Taxonomic investigations of bacteriophage sensitive bacteria isolated from marine waters. *Helgol. Wiss. Meeresunters.* **34:**375–373.

31. **Paul, J. H., S. C. Jiang, and J. B. Rose.** 1991. Concentration of viruses and dissolved DNA from aquatic environments by vortex flow filtration. *Appl. Environ. Microbiol.* **57:**2197–2204.

32. **Porter, K., and Y. Feig.** 1980. The use of DAPI for identifying and counting aquatic microflora. *Limnol. Oceanogr.* **25:**943–948.

33. **Proctor, L. M., and J. A. Fuhrman.** 1990. Viral mortality of marine bacteria and cyanobacteria. *Nature* (London) **343:**60–62.

34. **Proctor, L. M., and J. A. Fuhrman.** 1992. Mortality of marine bacteria in response to enrichments of the virus size fraction from seawater. *Mar. Ecol. Prog. Ser.* **87:**283–293.

35. **Safferman, R. S., and M. E. Morris.** 1967. Observations on the occurrence, distribution and seasonal incidence of blue-green algal viruses. *Appl. Microbiol.* **15:**1219–1222.

36. **Spencer, R.** 1955. A marine bacteriophage. *Nature* (London) **175:**690.

37. **Suttle, C. A.** 1993. Enumeration and isolation of viruses, p. 121–134. *In* P. F. Kemp, B. F. Sherr, E. F. Sherr, and J. J. Cole (ed.), *Handbook of Methods in Aquatic Microbial Ecology.* Lewis Publishers, Boca Raton, Fla.

38. **Suttle, C. A.** 1994. The significance of viruses to mortality in aquatic microbial communities. *Microb. Ecol.* **28:**237–243.

39. **Suttle, C. A., and A. M. Chan.** 1993. Marine cyanophages infecting oceanic and coastal strains of *Synechococcus:* abundance, morphology, cross-infectivity and growth characteristic. *Mar. Ecol. Prog. Ser.* **92:**99–109.

40. **Suttle, C. A., and A. M. Chan.** 1994. Dynamics and distribution of cyanophages and their effects on marine *Synechococcus* spp. *Appl. Environ. Microbiol.* **60:**3167–3174.

41. **Suttle, C. A., and A. M. Chan.** 1995. Viruses infecting the marine prymnesiophyte *Chrysochromulina* spp.: isolation, preliminary characterization and natural abundance. *Mar. Ecol. Prog. Ser.* **118:**275–282.

42. **Suttle, C. A., A. M. Chan, and M. T. Cottrell.** 1990. Infection of phytoplankton by viruses and reduction of primary productivity. *Nature* (London) **347:**467–469.

43. **Suttle, C. A., and F. Chen.** 1992. Mechanisms and rates of decay of marine viruses in seawater. *Appl. Environ. Microbiol.* **58:**3721–3729.

44. **Waterbury, J. B., and F. W. Valois.** 1993. Resistance to co-occurring phages enables marine *Synechococcus* communities to coexist with cyanophages abundant in seawater. *Appl. Environ. Microbiol.* **59:**3393–3399.

45. **Weinbauer, M. G., D. Fuks, and P. Peduzzi.** 1992. Distribution of viruses and dissolved DNA along a coastal trophic gradient in the northern Adriatic Sea. *Appl. Environ. Microbiol.* **59:**4047–4082.

46. **Weinbauer, M. G., and P. Peduzzi.** 1994. Frequency, size and distribution of bacteriophages in different marine bacterial morphotypes. *Mar. Ecol. Prog. Ser.* **108:**11–20.

47. **Wilson, W. H., L. R. Joint, N. G. Carr, and N. H. Mann.** 1993. Isolation and characterization of five marine cyanophages propagated on *Synechococcus* sp. strain WH7803. *Appl. Environ. Microbiol.* **59:**3736–3743.

48. **Wommack, K. E., R. T. Hill, M. Kessel, E. Cohen, and R. R. Colwell.** 1992. Distribution of viruses in the Chesapeake Bay. *Appl. Environ. Microbiol.* **58:**2965–2970.

49. **Zachary, A.** 1976. Physiology and ecology of bacteriophages of the marine bacterium *Beneckea natriegens:* salinity. *Appl. Environ. Microbiol.* **31:**415–422.

Community Structure: Bacteria and Archaea

JED A. FUHRMAN

29

Community structure is generally considered to be information related to the types of organisms present in an environment and the relative proportions of those types. The various ways to determine such structure could easily fill a whole book; therefore, this brief chapter will basically provide an outline of the various approaches that are used to determine community structure and some guidelines regarding which methods may be most appropriate for specific aquatic environments and specific scientific questions. Also, it must be realized that this topic is undergoing massive revision as new molecular biological and other highly technical approaches are being brought to the fore. Therefore, I will describe some of these new methods in detail, even though they are still under development, because they are likely to become particularly important in the future.

Community structure analysis might be performed for a number of purposes, and the most appropriate techniques to use will vary with the goals of the study. Certain techniques may allow study of subsets of the microbial community in great detail but at the same time entirely miss other community members. For example, one might be interested in the portion of the community that is performing a certain function, such as photosynthesis, and this might be studied by analyzing extracted photosynthetic pigments (34) or possibly by flow cytometric analysis of pigment contents of individual cells (6). Several of the techniques described below rely on the ability to identify particular preselected components of the microbial community. However, in addition to learning about known types of microorganisms, a study of the general community structure can also potentially find previously unknown microbial types, and some of these may be the most interesting. As might be expected, when one finds novel organisms, it takes considerable time to characterize such organisms.

MICROSCOPY

The oldest method for obtaining information on microbial community structure is to examine the sample with a microscope and characterize the microbes by their morphology. With some aquatic environments, one can in fact identify many types of organisms this way. An example is microbial mats: numerous types have readily recognizable morphologies, particularly by electron microscopy (34). Many colonial microorganisms have distinctive colony morphologies

that can be recognized by standard light microscopy or even with the naked eye. However, this approach is risky in that some types may appear similar by convergence and may thus be misidentified. More to the point, in most aquatic environments, the vast majority of bacteria have nondistinct and/or variable morphologies, so this method is inappropriate for them (33). Related to identification by morphology is the identification of certain kinds of organisms by their natural fluorescence properties as examined by epifluorescence microscopy. For example, methanogens contain the F_{420} cofactor that fluoresces green when excited with light at 420-nm wavelength. More commonly, such identification may be based on photosynthetic pigments (often in conjunction with general size and morphology information) that can yield distinctive colors of fluorescence. For example, marine *Synechococcus* cells containing phycoerythrin have a distinctive golden color fluorescence when excited with blue light and a bright red fluorescence when excited by green light. These pigment "signatures" can be more precisely determined by flow cytometry, which provides quantitative fluorescence (and other) information on many thousands of individual cells in a short time. This approach in fact was used to discover the presence in the ocean of large quantities of tiny marine prochlorophytes that had evaded detection by standard visual epifluorescence microscopy techniques (6).

CULTURING

The most common traditional method of determining community structure involves culturing the organisms from the habitat in question and identifying the cultures by standard techniques (see chapter 6). If the organisms of interest are culturable, then this approach may be suitable. It goes without saying that the culturable organisms must be viable in order to be detected, and such viable counts are often thought to avoid the problem of counting inactive organisms that may be of less interest. However, there are several important caveats. First, even for culturable organisms, it has been suggested that many individuals may be viable but still not culturable (31). A second and bigger concern is that culture conditions that have typically been used to perform such counts usually recover on the order of 1% or less of the total number of organisms in aquatic habitats (11, 19, 21), even though the majority of organisms in such habitats can be shown to be metabolically and synthetically active (12). The reasons for this inability to culture the organisms are somewhat speculative and are generally thought to be the result of culture conditions that do not adequately mimic the natural growth conditions. Possibilities include wrong substrates, trace metal sensitivity, excessive storage product formation (leading to cell damage), and viral infection, among others. Recent attempts to use dilution cultures (i.e., dilute a sample with filtered seawater such that only a few bacteria are in each tube) have greatly increased the percentage of bacteria that can be cultured, but for unknown reasons many of such cultures stop growth when abundances reach about 10^5 cells per ml (4). At such abundances, the classical identification tests cannot be readily performed, but it may be possible to coax such organisms into growth under richer conditions. While this may assist identification, the organisms may have substantially different physiological properties after such a process. In any case, many aquatic microbiologists are concerned about accepting the results of culture-based approaches for determining community structure in a comprehensive manner, and one rarely sees such studies in the literature.

IMMUNOLOGICAL APPROACHES

Immunological approaches have been used primarily to characterize and count nitrifying bacteria (39) and cyanobacteria (5). To prepare the antibodies for such a study, one must first culture the organisms and then use the culture to vaccinate an animal. The antibodies are purified from serum and then labeled with a fluorescent tag, such as fluorescein. The antibodies are mixed with a sample, unattached antibodies are rinsed away, and the cells are then observed by epifluorescence microscopy. Of course, this method presupposes that one must have cultures of the organisms in question, and so this approach would miss organisms that are resistant to cultivation. Also, one must know something about the cross-reactivity of the antibodies to know the range of organisms that they bind to. Often, antibodies can be specific for a particular species or strain. This approach has the potential to identify nonculturable members (viable or not) of a particular serotype, which is functionally defined as the type(s) to which the antibody binds. It has also been found to be useful for organisms that might be cultured, but grow slowly, such as nitrifiers.

LIPID ANALYSIS

Different groups of microorganisms have different types of lipids, and this fact has been used extensively for microbial identification. For environmental work, lipid analysis has been used as an indicator of community structure, with certain classes of lipids (or particular lipids) being used as markers for certain groups. In general, the lipids are extracted in organic solvents and analyzed by gas chromatography. In aquatic systems, this work has been pioneered by White and colleagues (briefly reviewed in reference 41; see chapter 10). The approach provides information not only on cell type but also about nutritional status or stress. Because most such analytical methods require a substantial number of microbes of each type in each sample, this approach has been used primarily for sediments. However, with the appropriate instrumentation, it is now possible to extend such work to planktonic environments.

LOW-MOLECULAR-WEIGHT RNA PROFILES

The molecular size distribution pattern of low-molecular-weight RNA (including tRNAs and 5S rRNA) is thought to be unique within narrow phylogenetic groups of microorganisms. This concept has been applied to natural aquatic planktonic communities by Hofle (17): the whole community RNA profile is examined on an electrophoretic gel and compared with profiles of other communities and of standard cultures. As long as the bands on the gel are well resolved, one can see patterns of similarity and dissimilarity between natural communities. It is also possible to detect patterns that suggest the presence of specific organisms or groups, and bands may be excised and sequenced for partial identification.

DNA-DNA HYBRIDIZATION

There are situations in which one wishes to know the community structure to learn if two microbial communities contain the same or different organisms, yet the quantitative information on species compositions is not of particular interest. That is, one is asking simply if two communities are the same or different. In such situations, it is possible to perform a DNA-DNA hybridization assay with total DNAs extracted from the two communities (20). In this approach, which to date has been performed only with plankton, organisms from each sample are collected on a 0.22-μm-pore-size filter (after prefiltration through a glass fiber filter to remove eukaryotes), and then the DNA is extracted with hot 1% sodium dodecyl sulfate and purified by phenol extraction. A portion of the DNA from each is labeled with ^{35}S by nick translation. Then filter hybridization is performed under stringent conditions (68°C) by spotting a known amount of each DNA onto a filter and hybridizing these two spots with the labeled DNA. Two filters are used, each with both kinds of DNA in separate spots; one filter is hybridized to one labeled DNA, and the other filter is hybridized to the other labeled DNA. The results are scored by expressing the cross-hybrids as a percentage of the self hybrids on the same filter. This percentage is expected to be the sum total of the shared common fractions between the two filters. For example, if sample 1 has 10% species A, 40% species B, and 50% species C, and sample 2 has 25% species A, 30% species C, and 45% species D, the shared common fraction is 10% + 30% = 40%.

Tests with mixtures of pure culture DNAs have shown that the results are usually as expected, with 100% hybridization of identical or nearly identical samples, ranging down to about 5 to 10% hybridization between samples sharing few or no species (the 5 to 10% represents well-conserved DNA sequences that cross-hybridize between distantly related organisms). Each assay yields two numbers, from the two reciprocal hybridizations, and these are usually similar to each other. However, when one sample contains a subset of the species of the other, one hybridization yields a higher number than the other (it is even possible for one number to exceed 100%). This is because a complex probe (many species) can more easily hybridize to a simple (fewer species) subset of itself than it can hybridize to itself. Empirically it was found that when the two reciprocal hybridizations are different, then the lower number of the two is generally correct, and the sample used to make the probe that gives the higher number is more complex than the other. Therefore, asymmetrical results say something about relative species richness of the samples, yet still also yield the original desired information about shared common species. This approach has been used to see broad differences between ocean basins, between depths at stratified locations, and over seasonal scales at the same location, and it has also shown similarities between some communities over time and space (20).

OTHER DNA-BASED APPROACHES

Another type of DNA-DNA hybridization is analysis of reassociation rates of single sample DNA that is melted (strands separated). The rate is related to the complexity (or one may say diversity) of the DNA and thus can be an index of the diversity of the sample. In other words, it indicates the number of species present but not what types they are. Torsvik et al. (35) examined DNA extracted from soil microorganisms in this way, and the very slow reassociation of most of the DNA was interpreted as indicating the presence of about 4,000 completely different genomes in a 30-g sample from a deciduous forest. This was 200 times higher than the diversity from standard plate counts and graphically demonstrates the remarkably high potential overall diversity of microbial communities and its undersampling by conventional means.

A method to "fingerprint" a microbial community based upon the quantitative distribution of genomes with different percent G + C contents was recently presented (18). In this approach, community DNA is centrifuged in a density gradient that separates the DNA based upon percent G + C content, yielding a profile. Although the percent G + C is not an unambiguous identifier for particular groups, the profile of DNA along the percent G + C gradient can strongly indicate the presence of certain groups (prompting probe analysis for verification; see below). Furthermore, profiles from different samples can indicate differences between the microbial communities over time and space.

16S rRNA-BASED APPROACHES

The difficulty of culturing bacteria and the ordinary requirement that a culture be available to identify new species has presented a major dilemma to microbial ecologists. About a decade ago, Pace et al. (28) and Olsen et al. (27) presented the elegant idea that cultures are not necessary to identify the organisms present in a natural habitat. The idea came against the backdrop of the increasing use of molecular phylogeny to help define microbial systematics (42). It was becoming clear that the nucleotide sequences in molecules like 16S rRNA are very powerful tools in determining phylogenies and consequently in microbial systematics. Large databases of such sequences have been made available. The new idea was to use molecular biological techniques to obtain 16S rRNA sequences directly from the organisms freshly collected from the natural habitat without culturing them. These sequences could then be compared with those in the databases and with each other to learn how they fit into the microbial phylogenetic framework. Even if the sequence is unknown from previous work, it can be placed in relation to known organisms and other sequences from nature. A further major benefit is that the sequences can be used to make probes for quantitative composition analyses of microbial communities. This is proving to be a very powerful approach and is being augmented by inclusion of 23S rRNA data as well (see chapter 11). A major aspect of the power of this approach is that the data are in the form of sequences that are universally understood and readily analyzed.

The 16S rRNA sequence analyses can be done in a few

ways. If one is interested in learning what types of organisms are present in a sample, it is best to use the approaches that do not restrict the results to certain groups. The initial method proposed by Pace et al. (28) called for extraction of DNA from all the microorganisms, fragmentation of that DNA with a restriction enzyme, and ligation to DNA from a bacteriophage (e.g., lambda) to make a library containing bits of DNA from all of the original organisms in the sample. That library is then screened for fragments coding for 16S rRNA by low-stringency hybridization to rRNA from a culture (or from cultures representing very broad groups like the domains *Bacteria*, *Archaea*, and *Eucarya*). The low stringency, combined with the moderately high level of conservation of this molecule, is expected to allow virtually any 16S rRNA to be detected. Screening with another rRNA, rather than with universally conserved oligonucleotide probes, is preferable because of the large number of false positives expected as a result of random matches between the oligonucleotide and the myriad genes from the natural community (in our laboratory, we found such false positives were very common). The positive clones are then sequenced, and the sequences are aligned to sequences from a database and analyzed for phylogenetic relationships by a computer. This library approach was used successfully by Schmidt et al. (32) with marine plankton.

Ward et al. (40) took a different approach to cloning 16S rRNA sequences of microorganisms from a well-studied hot spring at Yellowstone National Park. They extracted RNA from the natural sample and performed reverse transcription with universal primers to make cDNA that was then cloned and sequenced. The length of the cloned products was usually a few hundred bases, which is suitable for identification and some general phylogenetic analyses. The results yielded numerous clone types that were not the same as the organisms that had previously been cultured from that hot spring and had been thought to be the dominant types (by microscopical observation as well as culture work). This has been seen as good evidence that the culture-based approach finds only a subset of the natural diversity and that morphological identification of even distinctive organisms can be deceptive.

One major alternative related approach to cloning 16S rRNA genes has also been used with success: PCR. With PCR methods, DNA is extracted from freshly collected organisms, PCR is performed with the primers of choice (see below), the PCR products are ligated into a phage or plasmid vector, these are cloned by standard techniques, and the clones are sequenced and analyzed phylogenetically. This approach is particularly suited to planktonic communities because there is usually very little DNA to work with. For example, with typical bacterial abundances of 10^6 bacteria per ml and typical bacterial DNA content of a few femtograms, there is on the order of a few micrograms of bacterial DNA per liter. The amplification inherent in PCR means that one can begin with 1 ng of total genomic DNA (representing roughly 1 million bacteria) and end up with micrograms of amplified 16S rRNA genes.

The existence of regions of the 16S rRNA molecule that are essentially invariant among all known organisms means that universal primers can be used in PCR; the longest distance between such primers is about 860 bases, which is adequate (but not ideal) for most phylogenetic analyses. The universal nature of the primers allows the broadest coverage, but it can also be a problem. For example, the high copy number of nucleus-encoded 16S rRNA genes in eukaryotes may cause eukaryotic PCR products to swamp those of prokaryotes if they are together in the sample. One solution is to try to remove eukaryotes by filtration, as can be done in marine plankton; Fuhrman et al. (14, 15) used glass fiber filtration that removed essentially all of the eukaryotes but only about 10% of the prokaryotes. Alternatively, one can choose more specific primers to target specific groups. An example of this might be to use *Bacteria*-specific primers to avoid amplifying eukaryotic genes when eukaryotes and prokaryotes may not be easily separated; however, one must remember that chloroplasts and mitochondria contain "bacterial" 16S rRNA sequences, so these primers do not completely avoid interference from eukaryotes. On the positive side, there are nearly universal bacterial primers that allow cloning of almost the whole 16S rRNA gene (nearly 1,500 bases), maximizing the available phylogenetic information (1). However, the so-called universal primers should be used with care, in part because they are not exactly universal. Even with some ambiguous bases, such primers have mismatches with certain known groups, and one should check updated databases before embarking on this work. For example, the nearly universal bacterial primer set does not match the *Planctomyces* sequence well, and therefore one may suspect that other unknown groups may be missed. The commonly used primer at *Escherichia coli* position 1492, often treated as if it is universal, is not really so. While the PCR annealing conditions can be adjusted to allow priming in the presence of some mismatches, there still may be some rRNA genes that do not amplify well, and this possibility must be considered in interpreting the results.

Among results from PCR-based cloning and sequencing of 16S rRNA genes from marine plankton, Giovannoni et al. (16) used moderately specific PCR primers designed primarily for cyanobacteria, yet still found a novel proteobacterial group (SAR11) in addition to a cyanobacterial group. Fuhrman et al. (14, 15) used universal primers with microbial DNA collected from the deep sea and the euphotic zone and found novel groups of both *Archaea* and *Bacteria*. DeLong et al. (8) examined "marine snow" with bacterial primers and found that the clones were distinctly different from those collected from surrounding free-living bacteria. As was found by Schmidt et al. (32), who used a phage library, very few of these marine clones (with the exception of cyanobacteria) were closely related to previously known cultures. The SAR11 group was found at several marine locations and depths. Some of the clones were so distant as likely to be considered separate phyla from those previously studied. DeLong (7) and DeLong et al. (9) used archaeal primers and found two distinct archaeal groups in coastal temperate and polar waters; these were in the same groups as found by Fuhrman et al. (14, 15) at different locations and with universal primers. When specific primers are used, even a relatively minor component can be amplified, detected, and studied. As an example, ammonium-oxidizing bacterial 16S rRNA genes were specifically amplified from plankton by Voytek and Ward (37).

The PCR methods have been used with samples from sediments and microbial mats (2, 10, 25, 29), as well as deep-sea holothurian guts (23), often showing unexpected bacterial and archaeal diversity. Although such material is far more concentrated than plankton (more organisms per unit volume), making it easier to obtain enough material for analysis, there are many substances that can interfere with molecular analyses, so the DNA may need extensive purification. Also, such organisms may be difficult to extract. A further consideration with at least some of such samples is that they should be frozen or extracted immediately upon sampling; this is because it has been found that storage of sediments (especially if the sediments were initially anaerobic but stored in aerobic conditions) can lead to significant and rapid shifts in species compositions (30). Such rapid potential changes seem most likely in rich material with rapid potential growth.

Some studies have used information about the 16S rRNA clones short of partial or full sequences. These include restriction fragment length polymorphism analysis (3, 8, 24), which can be useful in grouping clones together. However, such analysis is based on only a few base positions in the sequence, and so it can lack resolution between closely related groups, particularly when universal primers are used for the initial amplification. On the other hand, it is possible to do the PCR with more specific primers, followed by restriction analysis, to indicate rather rapidly the presence of particular groups or types of interest and to compare different samples with respect to these groups. One could envision some fairly specific analyses by this approach with judicious selection of primers and restriction enzymes. Even with universal primers, the results can have resolution adequate for many types of studies. Another type of analysis is denaturing gradient gel electrophoresis, which is a way to separate similar-length nucleic acid molecules on the basis of small differences in the sequences. Muyzer et al. (26) used denaturing gradient gel electrophoresis analysis of PCR products to separate different components of the PCR mixtures. One can get an idea of the broad diversity (number of different bacterial types) of a sample by examining the number of different bands in such an analysis. It is also possible to use probes to characterize individual bands, or the bands may be cloned and sequenced for a detailed phylogenetic analysis.

A non-16S rRNA method that deserves mention here is PCR from repetitive sequences in DNA to yield electrophoretic banding patterns (genomic fingerprints) that can be distinct for particular groups or strains (36). Such methods may be applicable in the future for characterizing community structure.

There is still a question about possible biases in the molecular methods: Have we replaced culturing biases with unknown biases? Potential biases could arise at a few stages. In the extraction stage, one can check microscopically to see that substantially all the cells have lysed, and this has been done with some of the extraction techniques with aquatic samples (15). The PCR may introduce biases due to variations in primer binding or extension efficiency. There are also possible biases in the cloning step, as it is known that some sequences clone more readily than others. These possible problems indicate that caution is still in order when one is interpreting data on the relative amounts of different clones in libraries, and one cannot assume that clones are found in proportion to their natural abundance. There is also some concern about possible chimeras being formed during PCR amplification, and one needs to check clone libraries for these. However, even if there are biases, they are probably quite unrelated to culturing biases, and so these approaches are yielding much new information on what organisms are present.

PROBES

To avoid the possibility of biases in the clone libraries, probes have been used to quantify certain sequence types. One way this can be done is with oligonucleotide probes hybridized to bulk nucleic acid extracted from the aquatic habitat in question. Probe sequences are determined from sequence databases and can be universal or specific to certain domains, groups, or even some species (1). It is generally preferred to use RNA as the target instead of DNA because DNA is likely to yield far more false positives (unintentional hits) because mixed genomic DNA from innumerable species will have an immense variety of genes. RNA also has the benefit of being present in ribosomes and thus much more abundant as a target and also related to the cellular growth rate (although the exact relationship in natural communities is still uncertain). Quantitative bulk hybridizations to extracted RNA from aquatic habitats have been reported by Giovannoni et al. (16) and DeLong et al. (9), showing the relative abundances of the SAR11 cluster and *Archaea*, respectively. It should be noted that in order to best standardize such probe binding, it is ideal to have a culture of the organism or group in question to determine the relative binding of different probes empirically.

Often, however, one is not interested in knowing the fraction of RNA coming from particular groups but instead wants to know the proportions of individuals in those groups. For such work, it is ideal to tag each cell type with specific probes that allow visual identification. The preferred mode of observation has been epifluorescence microscopy with fluorescent oligonucleotide probes to rRNA (reviewed in reference 1). Flow cytometry has also been used to automate the analysis with success (38). This area has blossomed very recently. However, it has been found to work best with relatively rich environments, probably because slowly growing cells in relatively oligotrophic environments have few ribosomes and thus fluoresce dimly. Nevertheless, some approaches are yielding data with difficult samples like marine plankton. Lee et al. (22) found that multiple probes yield enough fluorescence for standard visual observation of about 75% of the bacteria in marine plankton. Fuhrman et al. (13) found that about 75% of the DAPI (4′,6-diamidino-2-phenylindole)-countable cells from coastal marine plankton may be seen even with single fluorescent probes when video image intensification is used to boost the brightness of the images, and about 8% fluoresced with an *Archaea*-specific probe. Laser confocal microscopy (1) is another promising approach with excellent sensitivity (see chapter 5). This is clearly an area that will progress rapidly. There is little doubt that in the future, such single-cell probe approaches will be common tools in studies of microbial community structure.

I thank N. Pace, S. Giovannoni, E. DeLong, and G. Olsen for helping me get involved in this work; S. Lee and R. Amann for helpful discussions; and R. Hicks and R. Christian for reviewing of the manuscript.
This work was supported by NSF grant DEB-9401110.

REFERENCES

1. **Amann, R. L., W. Ludwig, and K.-H. Schleifer.** 1995. Phylogenetic identification and in situ detection of individual microbial cells without cultivation. *Microbiol. Rev.* **59:**143–169.
2. **Barns, S. M., R. E. Fundyga, M. W. Jeffries, and N. R. Pace.** 1994. Remarkable archaeal diversity detected in a Yellowstone National Park hot spring environment. *Proc. Natl. Acad. Sci. USA* **91:**1609–1613.
3. **Britschgi, T., and S. J. Giovannoni.** 1991. Phylogenetic analysis of a natural marine bacterioplankton population by rRNA gene cloning and sequencing. *Appl. Environ. Microbiol.* **57:**1707–1713.
4. **Button, D. K., F. Schuts, P. Quang, R. Martin, and B.** R. Robertson. 1993. Viability and isolation of marine bacteria by dilution culture: theory, procedures, and initial results. *Appl. Environ. Microbiol.* **59:**881–891.
5. **Campbell, L., E. J. Carpenter, and V. J. Iacono.** 1983. Identification and enumeration of marine chroococcoid cyanobacteria by immunofluorescence. *Appl. Environ Microbiol.* **46:**553–559.
6. **Chisholm, S. W., R. J. Olson, E. R. Zettler, J. Waterbury, R. Goericke, and N. Welschmeyer.** 1988. A novel free-living prochlorophyte abundant in the oceanic euphotic zone. *Nature* (London) **334:**340–343.
7. **DeLong, E. F.** 1992. Archaea in coastal marine environments. *Proc. Natl. Acad. Sci. USA* **89:**5685–5689.
8. **DeLong, E. F., D. G. Franks, and A. A. Alldredge.** 1993. Phylogenetic diversity of aggregate-attached vs. free-living marine bacterial assemblages. *Limnol. Oceanogr.* **38:**924–934.
9. **DeLong, E. F., K. Y. Wu, B. B. Prezelin, and R. V. M. Jovine.** 1994. High abundance of archaea in Antarctic marine picoplankton. *Nature* (London) **371:**695–697.
10. **Devereux, R., and G. W. Mundfrom.** 1994. A phylogenetic tree of 16S rRNA sequences from sulfate-reducing bacteria in a sandy sediment. *Appl. Environ. Microbiol.* **60:**3437–3439.
11. **Ferguson, R. L., E. N. Buckley, and A. V. Palumbo.** 1984. Response of marine bacterioplankton to differential filtration and confinement. *Appl. Environ. Microbiol.* **47:**49–55.
12. **Fuhrman, J. A., and F. Azam.** 1982. Thymidine incorporation as a measure of heterotrophic bacterioplankton production in marine surface waters: evaluation and field results. *Mar. Biol.* **66:**109–120.
13. **Fuhrman, J. A., S. H. Lee, Y. Masuchi, A. A. Davis, and R. M. Wilcox.** 1994. Characterization of marine prokaryotic communities via DNA and RNA. *Microb. Ecol.* **28:**133–145.
14. **Fuhrman, J. A., K. McCallum, and A. A. Davis.** 1992. Novel major archaebacterial group from marine plankton. *Nature* (London) **356:**148–149.
15. **Fuhrman, J. A., K. McCallum, and A. A. Davis.** 1993. Phylogenetic diversity of marine subsurface microbial communities from the Atlantic and Pacific Oceans. *Appl. Environ. Microbiol.* **59:**1294–1302.
16. **Giovannoni, S. J., T. B. Britschgi, C. L. Moyer, and K. G. Field.** 1990. Genetic diversity in Sargasso Sea bacterioplankton. *Nature* (London) **345:**60–63.
17. **Hofle, M. G.** 1992. Bacterioplankton community structure and dynamics after large-scale release of nonindigenous bacteria as revealed by low-molecular-weight RNA analysis. *Appl. Environ. Microbiol.* **58:**3387–3394.
18. **Holben, W. E., and D. Harris.** 1991. Monitoring changes in microbial community composition by fractionation of total community DNA based on G + C content, abstr. Q-222. In *Abstracts of the 91st General Meeting of the American Society for Microbiology 1991.* American Society for Microbiology, Washington, D.C.
19. **Jannasch, H. W., and G. E. Jones.** 1959. Bacterial populations in sea water as determined by different methods of enumeration. *Limnol. Oceanogr.* **4:**128–139.
20. **Lee, S., and J. A. Fuhrman.** 1990. DNA hybridization to compare species compositions of natural bacterioplankton assemblages. *Appl. Environ. Microbiol.* **56:**739–746.
21. **Lee, S. H., and J. A. Fuhrman.** 1991. Species composition shift of confined bacterioplankton studied at the level of community DNA. *Mar. Ecol. Prog. Ser.* **79:**195–201.
22. **Lee, S. H., C. Malone, and P. F. Kemp.** 1993. Use of multiple 16S rRNA-targeted fluorescent probes to increase signal strength and measure cellular RNA from natural planktonic bacteria. *Mar. Ecol. Prog. Ser.* **101:**193–201.

23. **McInerney, J. O., M. Wilkinson, J. W. Patching, T. M. Embley, and R. Powell.** 1995. Recovery and phylogenetic analysis of novel archaeal rRNA sequences from a deep-sea deposit feeder. *Appl. Environ. Microbiol.* **61:**1646–1648.

24. **Moyer, C. L., F. C. Dobbs, and D. M. Karl.** 1994. Estimation of diversity and community structure through restriction fragment length polymorphism distribution analysis of bacterial 16S rRNA genes from a microbial mat at an active, hydrothermal vent system, Loihi Seamount, Hawaii. *Appl. Environ. Microbiol.* **60:**871–879.

25. **Moyer, C. L., F. C. Dobbs, and D. M. Karl.** 1995. Phylogenetic diversity of the bacterial community from a microbial mat at an active, hydrothermal vent system, Loihi Seamount, Hawaii. *Appl. Environ. Microbiol.* **61:**1555–1562.

26. **Muyzer, G., E. C. De Waal, and A. G. Uitterlinden.** 1993. Profiling of complex microbial populations by denaturing gradient gel electrophoresis analysis of polymerase chain reaction-amplified genes coding for 16S rRNA. *Appl. Environ. Microbiol.* **59:**695–700.

27. **Olsen, G. J., D. L. Lane, S. J. Giovannoni, and N. R. Pace.** 1986. Microbial ecology and evolution: a ribosomal RNA approach. *Annu. Rev. Microbiol.* **40:**337–365.

28. **Pace, N. R., D. A. Stahl, D. L. Lane, and G. J. Olsen.** 1986. The analysis of natural microbial populations by rRNA sequences. *Adv. Microb. Ecol.* **9:**1–55.

29. **Reysenbach, A.-L., G. W. Wickham, and N. R. Pace.** 1994. Phylogenetic analysis of the hyperthermophilic pink filament community in Octopus Spring, Yellowstone National Park. *Appl. Environ. Microbiol.* **60:**2113–2119.

30. **Rochelle, P. A., B. A. Cragg, J. C. Fry, R. J. Parkes, and A. J. Weightman.** 1994. Effect of sample handling on estimation of bacterial diversity in marine sediments by 16S rRNA gene sequence analysis. *FEMS Microbiol. Ecol.* **15:**215–226.

31. **Roszak, D. B., and R. R. Colwell.** 1987. Survival strategies of bacteria in the natural environment. *Microbiol. Rev.* **51:**365–379.

32. **Schmidt, T. M., E. F. DeLong, and N. R. Pace.** 1991. Analysis of a marine picoplankton community by 16S rRNA gene cloning and sequencing. *J. Bacteriol.* **173:**4371–4378.

33. **Sieburth, J. M.** 1979. *Sea Microbes.* Oxford University Press, New York.

34. **Stolz, J. F.** 1990. Distribution of phototrophic microbes in the flat laminated microbial mat at Laguna Figueroa, Baja California, Mexico. *BioSystems* **23:**345–357.

35. **Torsvik, V., J. Goksøyr, and F. L. Daae.** 1990. High diversity of DNA of soil bacteria. *Appl. Environ. Microbiol.* **56:**782–787.

36. **Versalovic, J., M. Schnieder, F. J. den Bruijn, and J. R. Lupski.** 1994. Genomic fingerprinting of bacteria using repetitive sequence-based polymerase chain reaction. *Methods Mol. Cell Biol.* **5:**25–40.

37. **Voytek, M. A., and B. B. Ward.** 1995. Detection of ammonium-oxidizing bacteria of the beta subclass of the class *Proteobacteria* in aquatic samples with PCR. *Appl. Environ. Microbiol.* **61:**1444–1450.

38. **Wallner, G., R. Erhart, and R. Amann.** 1995. Flow cytometric analysis of activated sludge with rRNA-targeted probes. *Appl. Environ. Microbiol.* **61:**1859–1866.

39. **Ward, B. B.** 1982. Oceanic distribution of ammonium-oxidizing bacteria determined by immunofluorescent assay. *J. Mar. Res.* **40:**1155–1172.

40. **Ward, D. M., R. Weller, and M. M. Bateson.** 1990. 16S rRNA sequences reveal numerous uncultured microorganisms in a natural community. *Nature* (London) **345:**63–65.

41. **White, D. C.** 1994. Is there anything else you need to understand about the microbiota that cannot be derived from analysis of nucleic acids? *Microb. Ecol.* **28:**163–166.

42. **Woese, C. R.** 1987. Bacterial evolution. *Microbiol. Rev.* **51:**221–271.

Protistan Community Structure

DAVID A. CARON

30

Protistan assemblages of aquatic ecosystems have become the focus of a concerted research effort in aquatic ecology. One stimulus for this work has been the recognition that phototrophic protists (the microalgae) constitute a major fraction of the primary productivity within aquatic ecosystems. Another incentive has been the growing realization that protozoa (heterotrophic protists) play a pivotal role in the flow of energy and elements in these communities (59). Studies of the abundance, biomass, and feeding activity of protozoa have been conducted in a wide range of ecosystems in recent years. In addition, many laboratory studies have examined the general biology and physiology of various protozoa grown under carefully controlled conditions. The synthesis of this information into useful models of how protozoan assemblages are structured and how they function in nature has begun.

Protozoa typically are defined as heterotrophic, eukaryotic, single-celled organisms. It is important to recognize that the term "protozoa" now has more historical significance than phylogenetic or ecological meaning. Protozoa are distinguished from other heterotrophic eukaryotes by their ability to exist as unicells and from the microscopic algae by their inability to photosynthesize. These distinctions are rather ambiguous in many cases. For example, many "protozoan" taxa occur among the "algal" taxa. Numerous species of heterotrophic chrysomonads and dinoflagellates exist. These species are closely related to chloroplast-bearing species on the basis of ultrastructural features and DNA sequence data (3, 44, 65). They are separated from the algae solely on the basis of the absence of a chloroplast. In addition to the existence of apochlorotic "algal" species, chloroplast-bearing genera that are capable of phagotrophy (i.e., heterotrophic nutrition) in addition to photosynthesis exist within the chrysophyte, dinoflagellate, prymnesiophyte, cryptophyte, and euglenophyte algae (62). This "mixotrophic" behavior obscures the distinction between traditional definitions of algae and protozoa.

There are also difficulties in the classification of ciliated protozoa as phototrophs or heterotrophs. Some ciliates ingest and digest algal prey but are able to retain the chloroplasts of these prey in a functional state, thereby providing the ciliates with a photosynthetic capability (62, 73). Photosynthesis in these "green ciliates" contributes significantly to the overall nutrition of the protozoan and also forms a notable fraction of the primary productivity of some planktonic communities (24).

The close phylogenetic affinities of some flagellated protozoa with algal taxa, as well as the mixed nutrition of many protists, indicate the artificiality of the historical distinction between the algal and protozoan taxa (44). For these reasons, consideration of the "purely protozoan" species in an assemblage is not appropriate from a phylogenetic or an ecological perspective. This discussion will, therefore, include references to chloroplast-bearing protists (i.e., algae).

ASSESSING PROTISTAN COMMUNITY STRUCTURE

Defining "community structure" in protistan communities is a difficult task. Some of this difficulty is related to problems associated with identifying species in mixed, natural assemblages, while some of it is related to the fact that the appropriate analysis of protistan community structure depends to some extent on the goals of the investigator.

There are at least three major considerations for assessing protistan species diversity in natural samples: the large number of species present in most aquatic environments, the disparate methodologies that are necessary for sampling these species, and the variety of morphological criteria that are used for identifying them. Free-living protists range in size from approximately 2-μm flagellated protists up to some species of radiolaria that can form cylindrical gelatinous colonies measuring 0.01 by 1 m. This 6-order-of-magnitude size range makes it necessary to apply a variety of sampling techniques in order to adequately sample all of the protistan species in an environment. The unique physical and chemical characteristics of different aquatic environments (e.g., planktonic and benthic environments) also contribute to the varied protocols that are necessary to sample protistan assemblages. In addition, identification keys for the major protistan taxa are based on morphological criteria that differ among the various protistan groups. Taxonomic expertise among protozoologists is often limited to one of the major groupings of protozoa (e.g., flagellates, amoebae, and ciliates) or some portion of one of these major categories. Recognizing the difficulties posed by species number, sampling, and taxonomy is central to understanding the present state of our knowledge concerning the structure of natural assemblages of aquatic protists.

Methods for sampling protistan communities typically are tailored for specific groups of protists and often for specific environments. Very distinct sampling methods typically are used for planktonic and benthic environments. Some of these sampling protocols are necessary because of the wide range of protistan abundances in different ecosystems. Abundances of protists in oxygenated surface sediments typically are several orders of magnitude greater than abundances in an equivalent volume of water above the sediment environment.

Even within an environment, sampling protocols must be adapted for particular groups of protists. For example, enumerating species of benthic protozoa among the sedimentary particles in which they exist has been a long-standing problem in assessing protistan species diversity in these environments. Various methods for extracting and concentrating protozoa from sediments have relied on the mobility of the community in response to changing salinity, extraction by centrifugation, or enrichment culture (1, 32, 61, 72). Such approaches have resulted in reasonable estimates of the protozoan diversity of some sediment environments, but the success of these methods is usually group specific. The extraction of protozoa by the sea ice method may work well for highly mobile species such as benthic ciliates, but this method may be less useful for more slowly moving forms such as small amoebae. For the latter forms, enrichment cultivation appears to be the most appropriate method.

Adjustments to sampling protocols also exist for sampling different protistan groups in plankton communities. Sample volumes of 200 to 500 ml are usually sufficient for flagellated protists (typical abundances are hundreds to thousands per milliliter), and volumes of 0.5 to 2 liters are usually sufficient for ciliated protists (typical abundances are tens to thousands per liter), but sarcodine protozoa (amoebae, actinopods, and foraminifera) usually must be concentrated by using plankton nets or filters. These latter techniques, however, are damaging to delicate species of planktonic ciliates (38). Several common methods for sampling protistan assemblages are reviewed by Sieburth (68).

Preservation, fixation, and other manipulations are prerequisites for the identification of most protistan species once appropriate samples have been collected. Notable exceptions to this generalization are the "naked" amoebae (primarily the Gymnamoebae), in which some of the characteristics that are essential for proper identification are present only in living specimens. For the remaining protist groups, correct preservation is dependent on the protistan group under consideration. For flagellated (and often ciliated) protists, aldehyde fixatives (formaldehyde and glutaraldehyde) are commonly used, often followed by osmium tetroxide when electron microscopy is planned (43, 67). A variety of fixatives have been developed for ciliated protozoa, most of which are usually employed in combination with postfixation staining methods that are used to visualize cytological features of the cells (44, 75).

The preservation of some protistan taxa requires special consideration. A preservative that does not promote dissolution of skeletal structures must be used for those species which possess such structures (e.g., actinopods and foraminifera). Careful adjustment of the pH of the preservative is necessary to prevent dissolution of foraminiferan tests (9), while addition of strontium is necessary to prevent dissolution of acantharian skeletons (50). When these requirements conflict, subsamples must be preserved separately for the different groups. For example, samples for planktonic foraminifera (which require alkaline pH) typically would be preserved differently than samples for planktonic ciliates, which are often preserved in acid Lugol's solution (75).

The identification of protistan species in mixed natural assemblages depends on criteria that are often as different as the methodologies used to sample and preserve these assemblages. Ciliates typically possess morphological features that provide sufficient taxonomic criteria for identifying species by light microscopy. Cell shape, size, location, and characteristics of the oral area, presence of a lorica, and particularly the arrangement of the somatic ciliature are useful features for species identification (44). Ciliates are often easier to identify than many of the flagellated and amoeboid protists because of the presence of these features, and extensive species lists exist for various environments (see reference 57 for a review).

Flagellated protists typically possess fewer morphological features that can serve as useful taxonomic criteria when they are observed by light microscopy. Cell size and shape, chloroplast arrangement, and flagellation are important criteria for identification by light microscopy. Some diagnostic features (e.g., flagellar mastigonemes and body scales), however, are visible only by electron microscopy. Electron microscopy is often necessary for distinguishing the numerous genera and species of small heterotrophic flagellates (<10 μm). The need to establish these features by using electron microscopy makes it difficult to process large numbers of samples. Moreover, many of the latter taxa have not been adequately described. There is considerable uncertainty about the validity of numerous genera (56) and, thus, the true species diversity of small heterotrophic flagellates in many environments.

The amoeboid protists are a polyphyletic collection of species, and the methods of identification applied to these species are heterogeneous. The naked amoebae are identified on the basis of features of the living organisms: cell size and shape during locomotion, arrangement and type of pseudopodia, morphology of the floating form, etc. The requirement for live material for species identification has made the determination of species diversity of natural assemblages of amoebae a difficult topic, but the taxonomy, distribution, and general ecology of these species are slowly emerging (5, 20, 53). Identification of the many types of testate amoeboid protists (testacea, foraminifera, radiolaria, heliozoa, and others) is based on the skeletal structures that are present in these species and on features of cellular organization such as the pseudopodia that characterize this diverse assemblage. The presence of a rigid skeletal structure in many of these species makes it possible to use plankton nets or screens for collecting and concentrating these specimens from the plankton or sediment.

Difficulties associated with sampling and identifying the entire spectrum of protists (as described above) in natural communities hampers the documentation of true protistan species diversity of any natural ecosystem. Very few, if any, studies have attempted to characterize all of the protistan species present in an aquatic ecosystem or have documented even all of the heterotrophic protistan species. Exceptions to this generalization might be found in environments in which protozoan diversity is greatly reduced as a result of severe environmental factors such as anaerobic conditions (36), but it is safe to say that the vast majority of studies of natural communities have underestimated total protistan species diversity.

Analyses of species diversity for particular taxa of protists (i.e., the ciliates, flagellates, or amoebae), however, have been more accurately determined. The most complete information exists for plankton communities, for which extensive lists of ciliated protozoa, chloroplast-bearing flagellates, and skeleton-bearing sarcodines (foraminifera and actinopods) have been obtained (8, 29, 48, 54, 55, 57).

Protozoan Abundance and Biomass

Identification of the protistan species present in an aquatic environment provides useful but limited information on their potential contribution to the structure and function of the total biological community because of the tremendous size range and varied trophic activities of protistan species. A much greater understanding of their importance can be obtained by combining species lists with estimates of abundance and biomass. Most modern methods for collection and identification of protists have been designed with this goal in mind. Loss of protistan cells during collection, enrichment, preservation, and sample processing continues to be a major concern in studies attempting to measure protistan species diversity and abundance, but generally accepted methods which minimize this problem for specific groups of protists and allow accurate estimates of protistan abundance to be obtained are now emerging.

The estimation of population abundances of amoebae presents a particularly difficult problem. Not only must these species be observed alive in order to be accurately identified, but their amorphous cell shape makes them difficult to enumerate in preserved samples. The few abundance estimates that are available for these species have been obtained by using a most probable number culture technique that relies on the growth of the amoebae in serial dilutions of the water or sediment samples (25, 61).

Protistan abundance measurements can be used to calculate total protistan biomass (typically expressed in units of carbon), using measurements of abundance, cell volume, and empirically derived carbon/volume conversion factors. Cell volume measurements obtained from microscopical studies are combined with abundance estimates to calculate the volume of particular protistan taxa, and carbon/volume conversion factors are then applied to calculate the carbon content. Carbon/volume conversion factors must take into account shrinkage due to fixation and the variable vacuolar space of protists. Shrinkage due to fixation can be both taxon and size specific. Recently published values for converting carbon to volume are 160, 240, and 360 fg of C μm^{-3} for flagellated protists of 10^3, 10^2, and 10^1 μm^3, respectively (80), and 190 fg of C μm^{-3} for ciliated protists (58) preserved appropriately. Carbon/volume conversion factors for larger sarcodines (acantharia, radiolaria, and foraminifera) are based on aspects of the cells that are resistant to net collection (51). Methods for estimating the cell volume of naked amoebae that directly relate the diameter of the nucleus to total cell volume have been proposed (60).

Describing Protistan Community Structure

The term "community structure" implies that organized relationships exist between protists and other microorganisms within natural ecosystems. Indeed, the "niche" concept has been applied to protistan assemblages with the implication that the number of protistan species in an environment is indicative of the number of unique ecological roles for protists in these assemblages. Unfortunately, it is unrealistic to consider all protistan species in a community as separate entities at this time because of the great species diversity of these assemblages, the limited ecological information available on the realized niches of many protistan species, and the extreme difficulty in obtaining species identifications

and abundance/biomass information for all protistan species in an assemblage. For these reasons, various simplifying groupings of protists have been used as a way of reducing the complexity of protistan assemblages into manageable (and measurable) quantities.

Various manners of grouping species of protists have been used. The most popular types of these methods have attempted to group protists by trophic mode (phototrophic versus heterotrophic), size, and prey type (for heterotrophs), in keeping with the trophic-level concept of Lindeman (47). In recent years, however, there has been a recognition that these trophic categories must be somewhat more flexible than originally proposed because of the common behavior of mixed nutrition among protists (18, 79). Nevertheless, aggregation of species into "trophospecies" (79) is a useful and necessary procedure for compartmentalizing the assemblage in order to allow investigations of energy and elemental flow through aquatic communities in models of manageable size (22).

For heterotrophic protists, it is common to group species according to the type of prey that they consume. Bacterivorous flagellates and ciliates in plankton communities or bacterivorous flagellates and amoebae on suspended particles may be grouped together to represent a major sink for bacterial biomass in the plankton. Similarly, ciliate species may be grouped into bacterivorous, herbivorous, or predacious species (26). Such classifications based on "feeding guilds" ignore some of the details of protistan feeding behavior (such as omnivory), but they are useful for reducing the complexity of the assemblage. Feeding guilds are often treated as single species in biological or biogeochemical models of ecosystem function.

The organization of protists by size is a logical one for two reasons. Allometric dependence of growth and metabolism can be used to constrain the potential contribution of a particular size range of protists to biogeochemical cycles (7, 19, 34). In addition, predator-prey relationships are typically size dependent, with larger predators consuming smaller prey. This generalization is realistic. Many heterotrophic flagellates 2 to 20 μm in size consume bacteria and cyanobacteria that are <2 μm in size, and many ciliate species 20 to 100 μm in size consume algae and protozoa <20 μm in size.

There are some notable exceptions, however, to size-dependent grazing. Many species of heterotrophic dinoflagellates consume diatom prey that are considerably larger than themselves by employing a pseudopodial "feeding veil" (41). Similarly, some planktonic sarcodines (acantharia, radiolaria, and foraminifera), because they produce a sticky pseudopodial network that entangles and immobilizes prey items, are capable of consuming metazoan prey considerably larger than themselves (76).

Notwithstanding these exceptions, size-dependent grazing models are the most common means of organizing protistan populations into manageable units for inclusion into models of elemental flow in aquatic ecosystems (6, 28, 52). The aggregation of species into groups within models probably reduces the predictive capabilities of these models, but their outcomes thus far appear to be in reasonable agreement with field data. It remains to be seen how the reduction of species diversity in these models will affect predictions of the response of the community to internal and external perturbations, but the gradual disaggregation of these models into more ecologically relevant compartments should provide insight into this issue.

TABLE 1 Species list showing the range of protistan diversity of an oligotrophic oceanic environment[a]

Protistan taxon	Avg size (μm)	Chloroplasts present?	Phagotrophy?	Probable prey	Representative abundance (liter^{-1})
Flagellated and nonmotile protists					
Dinoflagellates					
Protoperidinium sp.	55	No	No	Dia	10^2
Gymnodinium sp.	50	Yes	Yes	C, Din, Dia	2×10^2
Prorocentrum micans	25 by 40	Yes	No		10^3
Ornithocercus magnificus	40	No[b]	Yes	Dia, Sf	10^2
Chrysophytes and chrysomonads					
Paraphysomonas imperforata	7.0	No	No	B, Cc, Sf	10^5
Ochromonas sp.	6.0	Yes	Yes	B, Cc	10^5
Prymnesiophytes					
Chrysochromulina ericina	6.0	Yes	Yes	B, Cc	10^4
Chlorophytes					
Nannochloris atomus	4.0	Yes	No		10^5
Bacillariophytes (diatoms)					
Minutocellus polymorphus	3.0	Yes	No		10^5
Coscinodiscus concinnus	75 by 200	Yes	No		10^{-2}
Ditylum brightwellii	20 by 100	Yes	No		10^2
Rhizosolenia clevei	200 by 500	Yes	No		10^{-3}
Ethmodiscus rex	1,000	Yes	No		10^{-4}
Choanoflagellates					
Diaphanoeca grandis	2.5	No	Yes	B	10^5
Amoeboid protists					
Gymnamoebae (naked amoebae)					
Platyamoeba weinsteini	3 by 12	No	Yes	B	10^0
Flabellula citata	4 by 30	No	Yes	B	5×10^{-1}
Foraminifera[c]					
Globigerina bulloides	700[d]	No	Yes	Omi, Mz	10^{-3}
Globigerinoides sacculifer	700[d]	No[e]	Yes	Omi, Mz	10^{-3}
Acantharea[c]					
Amphilonche elongata	50 by 400	No[e]	Yes	Omi	10^{-1}
Spumellarian radiolaria[c]					
Thalassicolla nucleata	1,000[d]	No[e]	Yes	Omi, Mz	10^{-4}
Collozoum caudatum	200[d,f]	No[e]	Yes	Omi, Mz	10^{-5}
Ciliated protists					
Tintinnids					
Tintinnopsis parva	20 by 40	No	Yes	B, Cc, Sf	10^2
Oligotrichs					
Strombidium sulcatum	25 by 50	Yes[g]	Yes	B, Cc, Sf	10^2
Loboea strobila	50 by 150	Yes[g]	Yes	B, Cc, Sf	10^2
Hypotrichs					
Euplotes woodruffi	65 by 120	No	Yes	C	5×10^{-1}
Hymenostomatids					
Uronema marinum	10 by 20	No	Yes	B, Cc	10^3

[a] Only a partial list of representative species to exemplify the breadth of trophic modes in a real assemblage. Species are organized according to major taxa. Pertinent ecological information and realistic abundances are based on literature values. Dia, diatoms; C, ciliates; Din, dinoflagellates; Sf, small flagellated protists; B, bacteria; Cc, chroococcoid cyanobacteria; Mz, metazoan zooplankton; Omi, omnivorous on prokaryotic and eukaryotic unicells.

[b] Species harboring extracellular, symbiotic cyanobacteria that contribute to the photosynthetic nutrition of the host.

[c] Adult specimens only.

[d] Size does not take into account extensive pseudopodial network.

[e] Species harboring intracellular, symbiotic dinoflagellates that contribute to the photosynthetic nutrition of the host.

[f] Colonial species with colonies up to 1 cm in width and 1 m in length.

[g] Species that retain functional chloroplasts from ingested prey.

An Example from the Plankton

A hypothetical example easily indicates the analytical approaches for examining protistan community structure and the limitations of these approaches. A species list of protists that are typical of an oceanic plankton community is given in Table 1. This assemblage is not meant to be complete but rather indicative of the breadth of protistan sizes and nutritional modes in this type of ecosystem. Pertinent information on cell size, photosynthetic and phagotrophic ability, prey type(s), and typical abundances are also provided.

The species in this assemblage have been arranged according to major taxonomic categories (44).

As shown in Table 1, taxonomic groupings of the protists correspond poorly to the nutritional modes of the species. Reorganization of the same species into groups based on the nutritional modes of the species provides a very different view of this assemblage (Table 2). This reorganization indicates the classical dichotomy between phototrophs and heterotrophs, but it also indicates the more recent realization that many protistan species possess the ability for mixed nutrition. This ability results in some of the species occupying more than one trophic category.

The collection of species in this assemblage also demonstrates the enormous breadth of cell sizes that can be displayed by protistan assemblages (Fig. 1). The size range is not necessarily restricted for any particular trophic mode. In this assemblage, heterotrophs range from 2.5 μm to >1 mm in size, phototrophs range from 3.0 μm to >1 mm, and mixotrophs range from 6.0 μm to >1 mm if one includes symbiont-bearing sarcodines in this last category. Commonly used plankton size class designations are also shown in this figure. These designations correspond to organisms 2 to 20 μm (nanoplankton), 20 to 200 μm (microplankton), 0.2 to 20 mm (mesoplankton), and 20 to 200 mm (macroplankton) in longest dimension (69). Protists occur in all of these size classes, as indicated in Fig. 1, although the majority of these species typically occur in the nano- and microplankton classes.

One generality that is clear from Table 1 is that small planktonic protists typically occur in greater abundances than large species. This relationship is shown clearly by differences of 1 to 2 orders of magnitude in abundances when the species in each plankton size class are summarized (Fig. 2A). The magnitudes of these differences are typical for both phototrophic and heterotrophic protists (and often for mixotrophs, as shown in Fig. 2). However, the abundances of phototrophic, mixotrophic, and heterotrophic protists within a size class are often similar (hatched and stippled columns within each size class).

The large disparity that is apparent when one compares the abundances of protists in different size classes (Fig. 2A) is greatly reduced when the total volumes of living protists are compared (Fig. 2B). Small cell size among nanoplankton is generally balanced by high abundances of these species, while low abundances of the larger protists are balanced by their larger volumes. These general relationships of protistan abundance and biovolume are consistent with data from natural assemblages of nanoplanktonic and microplanktonic protists (17).

The information summarized in Tables 1 and 2 and Fig. 1 and 2 can be used to construct a typical box model depicting the flow of materials from producers through consumers in this hypothetical protistan community (Fig. 3). The species have been grouped according to their nutritional modes and approximate predator-prey relationships. Arrows in the model indicate the presumed direction of energy and material flow (i.e., from producers to consumers and from small organisms to large consumers). Examination of this model exemplifies the advantages and disadvantages of this approach for describing nutritional modes and trophic relationships among protists in a mixed natural assemblage.

The goals for most of the work on modeling "microbial loop" processes have been the development of working models that accurately describe energy and elemental flow within these communities and the incorporation of microbial processes into classical models of aquatic community

TABLE 2 Planktonic protist species described in Table 1 but arranged according to trophic category[a]

Phototrophy
  Gymnodinium sp.
  Prorocentrum micans
  Ochromonas sp.
  Chrysochromulina ericina
  Nannochloris atomus
  Minutocellus polymorphus
  Coscinodiscus concinnus
  Ditylum brightwellii
  Rhizosolenia clevei
  Ethmodiscus rex

Mixotrophy
  Phagotrophic algal species
    Gymnodinium sp.
    Ochromonas sp.
    Chrysochromulina ericina
  Chloroplast-retaining species
    Strombidium sulcatum
    Loboea strobila
  Symbiont-bearing species
    Ornithocercus magnificus
    Hastigerina pelagica
    Amphilonche elongata
    Thalassicolla nucleata
    Collozoum caudatum

Heterotrophy
  Bacterivory
    Paraphysomonas imperforata
    Ochromonas sp.
    Chrysochromulina ericina
    Diaphanoeca grandis
    Platyamoeba weinsteini
    Flabellula citata
    Amphilonche elongata
    Thalassicolla nucleata
    Collozoum caudatum
    Tintinnopsis parva
    Strombidium sulcatum
    Loboea strobila
    Uronema marinum
  Herbivory
    Protoperidinium sp.
    Gymnodinium sp.
    Ornithocercus magnificus
    Paraphysomonas imperforata
    Amphilonche elongata
    Thalassicolla nucleata
    Collozoum caudatum
    Tintinnopsis parva
    Strombidium sulcatum
    Loboea strobila
    Globigerina bulloides
  Carnivory
    Paraphysomonas imperforata
    Globigerina bulloides
    Amphilonche elongata
    Thalassicolla nucleata
    Collozoum caudatum
    Euplotes woodruffi

[a] Note that some species occur in more than one category.

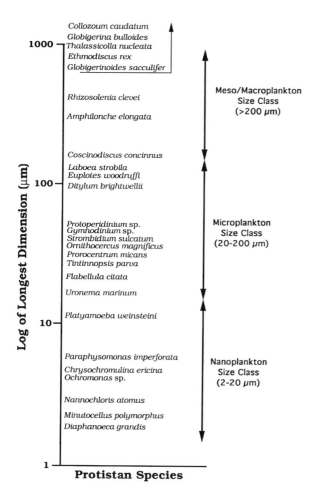

FIGURE 1 Approximate sizes (longest dimension) of the planktonic protistan species listed in Table 1. Commonly used size class designations are shown on the right. Note that the sizes of these protists span more than 3 orders of magnitude. The upward-pointing arrow under *Globigerinoides sacculifer* indicates that the group of five species enclosed by the arrow can be larger than 1,000 μm.

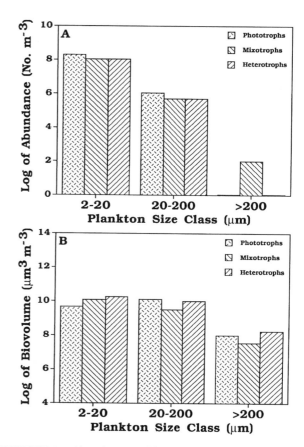

FIGURE 2 Abundance and biovolume relationships of the protistan assemblage listed in Table 1. The species have been grouped according to size class and according to trophic mode (phototrophic, mixotrophic, heterotrophic).

structure and function. Models such as the one shown in Fig. 3 are appropriate for these purposes because they attempt to reduce a complex assemblage of microorganisms to a manageable number of trophic "compartments" and trophic interactions. These models, therefore, are strongly influenced by methodologies available for identifying protistan species (or trophospecies) and for investigating their trophic interactions.

The model in Fig. 3 might adequately describe energy or elemental flow in this hypothetical protistan assemblage if the biomass and flow parameters of the model could be determined. However, this type of depiction of community structure still has some inherent disadvantages. As referred to earlier, predator-prey relationships that are not size dependent are difficult to represent and measure. Energy is depicted as moving from smaller to larger size classes in this model, but this representation is incorrect for species such as *Protoperidinium* sp., which can graze on diatoms larger than itself, and for the sarcodines *Globigerinoides sacculifer*, *Thalassicolla nucleata*, *Collozoum caudatum*, and *Globigerina bulloides*, which can consume metazoan prey. The double-

headed arrows connecting these latter compartments indicate the potential for the flow of energy in either direction. In practice, these measurements are difficult to make.

Similarly, selective grazing and omnivory are difficult to incorporate into this type of model. For example, *Euplotes woodruffi* is a predacious ciliate feeding primarily on other ciliates (in this assemblage, it might feed on *Uronema marinum*). On the other hand, *Tintinnopsis parva* may accept a variety of small protists and other microorganisms as prey. The distinction between these two rather different nutritional modes has been forfeited by placing them into the same trophic compartment. Clearly, if the goal of this modeling exercise was to understand the factors affecting the success or failure of either of these two species in plankton communities, then this model would be unsatisfactory. It is for reasons such as this last example that the appropriate conceptualization and representation of protistan community structure must take into account the goal of the investigator.

TEMPORAL AND SPATIAL CHANGES IN COMMUNITY STRUCTURE

The most significant differences in the species composition and trophic relationships of protistan communities exist between different aquatic environments. However, there is also a rapidly increasing database on changes in community

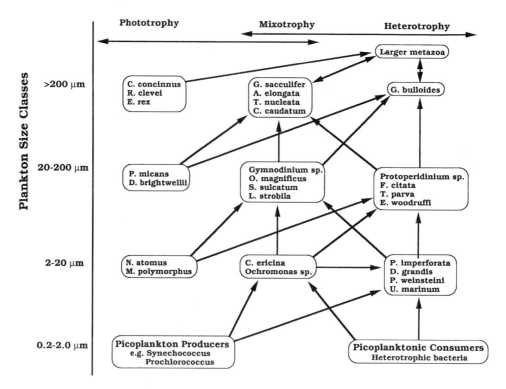

FIGURE 3 A box model approximating the major trophic interactions within the protistan assemblage listed in Table 1. The species have been grouped according to known or presumed size-dependent and trophy-dependent relationships. Arrows indicate the direction of energy flow in predator-prey interactions.

structure over seasonal and shorter time scales. These latter changes appear to be most significant in temperate and polar climates.

Freshwater versus Marine Ecosystems

Probably the most distinct difference between freshwater and marine protistan communities is the restriction of the larger sarcodines (acantharia, radiolaria, and foraminifera) to brackish and marine ecosystems. In tropical and subtropical oceanic plankton communities, adult sarcodines are often the most conspicuous macroscopic organisms in surface water, while swarmer cells and juvenile specimens of these species contribute to the entire size spectrum of protozooplankton (20). In marine sediments, benthic foraminifera constitute an important component of faunal assemblages (39, 45).

In contrast to large differences in the assemblages of larger sarcodine species in freshwater and salt water, there appears to be similarity with respect to types of ciliates and flagellates in these environments. Recent summaries from both environments have indicated that planktonic bacterial biomass and production are related to phytoplankton biomass and primary productivity (23). Likewise, the ecological roles of small protozoa in freshwater and marine plankton communities appear to be similar and related to bacterial production over very broad scales of examination (63). Most ciliates in the freshwater and marine plankton also appear to play grossly analogous ecological roles as consumers of small prokaryotes and eukaryotes (10, 57). Mixotrophic (phagotrophic) algae exist in both freshwater and marine water (4, 12, 13), as do species of chloroplast-retaining ciliates (35, 74). These generalizations do not mean that the same species of flagellates or ciliates occur in both ecosystems, but rather that similar ecological niches have been filled by protistan species in these different environments.

Benthic versus Pelagic Ecosystems

Although there are species of protists that are commonly found in both benthic and pelagic environments, there are clearly numerous species within these assemblages that are uniquely suited for one environment or the other. Morphological adaptations of ciliates to life between sediment particles in the benthos have resulted in the evolution of cell forms that allow movement through this medium. Common adaptations include cylindrical or flattened shapes, flexible cell walls, and patterns of ciliature that allow "crawling" along surfaces and grazing on prey loosely associated with particulate material. Some species permanently attach to surfaces. In contrast, ciliates in pelagic environments (e.g., choreotrichs) tend to have more rounded shapes and patterns of ciliature that afford rapid swimming behavior and feeding on suspended particles.

There is great diversity among and within benthic environments as a consequence of sediment grain size, organic loading, oxygen gradients, etc. The number of microenvironments at one locale may be considerable. For this reason, the remoteness of many benthic environments, and the difficulties of sampling and concentrating protists from these environments, the ciliate (and other protist) fauna of many benthic ecosystems is still poorly characterized. There are extremely few observations of the protistan fauna of the deep-ocean benthos (39, 71, 78).

The amoebae are particularly well suited for existence in benthic environments. Locomotion and feeding of these species take place on particles. Therefore, benthic environments tend to support significant assemblages of amoebae (64). Amoebae occurring in the plankton are generally assumed to be associated with suspended particulate material or with the air-water interface (25, 61).

Among the larger sarcodines, there are clear differences between pelagic and benthic assemblages. Foraminifera occur in both environments, but the species occurring in these two environments are different. The planktonic species are restricted to pelagic, oceanic ecosystems, while benthic species are common from salt marshes to abyssal depths. Most radiolaria (polycystines and phaeodaria) and acantharia are restricted to pelagic, oceanic ecosystems, but most heliozoa are coastal and/or benthic. There are relatively few exceptions to these generalities, making the larger sarcodine fauna of benthic and pelagic ecosystems quite distinct.

Beyond the obvious contribution of phototrophic protists to the flagellated protistan assemblages in surface waters of pelagic ecosystems, the heterotrophic flagellate assemblages of benthic and pelagic environments can also differ in composition. Many flagellated protozoa occur in both environments, but species that are capable of particle attachment or movement along surfaces (e.g., bodonid flagellates) tend to predominate in benthic environments. Forms that feed on suspended bacteria (e.g., chrysomonad flagellates and choanoflagellates) tend to predominate in pelagic ecosystems.

Pelagic environments generally are considered to be more homogeneous than benthic ecosystems, but there are clearly sources of heterogeneity in the plankton. Epibiotic (and possibly enteric) protistan assemblages have not been adequately studied, but they contribute to protistan species diversity in the water column (77). Suspended particles also create unique microhabitats in pelagic ecosystems for some protozoan species that are more characteristic of the benthos. Macroscopic detrital aggregates in marine planktonic ecosystems (so-called marine snow) may create a false benthos for benthic species by creating microenvironments with elevated abundances of bacteria and other prey (16, 70). Similar oases for unique protozoan assemblages in the plankton may be established by using artificial foam substrates (14). It has been demonstrated that the colonization and species succession of protozoa on these natural and artificial substrates may follow a pattern similar to that found for the colonization of oceanic islands by higher organisms (14, 84).

Depth and Seasonal Distributions

The seasonality of algal species composition and abundance in pelagic environments is well known. Distribution of the protistan algae with depth has also received considerable attention. In contrast, changes in total protozoan abundance or biomass with season or depth have been documented, but there is relatively little information on changes in species composition or community structure and function with depth. This paucity of information is not surprising given the difficulties mentioned previously with identification and high species diversity and the logistical problems associated with the collection of long-term data sets or multiple samples from discrete depths.

Most studies to date have been restricted to a particular group of heterotrophic protists because of either methodological approach or taxonomic expertise. Often these investigations have reported only changes in broad taxonomic or ecologically relevant categories (heterotrophic flagellates, mixotrophic flagellates, ciliates, etc.). For example, depth

and seasonal changes in abundance have been reported in a variety of marine and freshwater environments for flagellated and/or ciliated protozoa (11, 21, 27, 42). More detailed data on spatiotemporal distributions are available only for specific taxa for which identification is more straightforward (8, 40).

It is difficult to generalize concerning changes in the community structure of heterotrophic protistan assemblages as a function of season from these scattered reports. For temperate communities, seasonal changes in species composition and winter reductions in the intensity of grazing activity are likely, but the extent of these changes remains largely undetermined for most environments. Temperature is a strong controlling influence on processes within the microbial loop of temperate ecosystems (82), but diverse heterotrophic and phototrophic protistan assemblages abound even in extremely low temperature environments such as sea ice habitats near Antarctica (37).

The vertical distributions of heterotrophic protists typically demonstrate greater overall abundances in surface waters relative to abundances at depth. These distributions of abundance are clearly related to the production of organic material in surface waters. Fine-scale vertical distributions, however, can be complex. Elevated abundances of protozoa have been observed at the air-water interface (25), at oxic-anoxic boundaries within water columns (85), and at subsurface biological features such as deep chlorophyll maxima (30). Vertical distributions of protists in the sediments typically are related to physical and chemical gradients within the benthos (33). The exploitation of these chemical and physical features within the benthos and water column can increase the diversity of protistan assemblages of an environment by providing unique microhabitats for the growth of species able to exist there.

NEW APPROACHES TO STUDYING PROTISTAN COMMUNITY STRUCTURE

The significant problems associated with determining species diversity, abundance/biomass, and trophic activity of protistan assemblages in aquatic ecosystems continue to hamper in-depth analyses of the structures and functions of these communities. As described above, classical methods of identification are time-consuming and often do not provide quantitative information on the occurrence of species. New conceptual and methodological approaches will be necessary to deal with these recalcitrant problems.

Conceptual approaches and methodologies from molecular biology and immunology offer hope for addressing some of these problems. For example, RNA-targeted oligonucleotide probes and PCR-based methods for determining the presence of specific microorganisms are becoming commonplace in medical and environmental bacteriology (2). At present, the application of these approaches to protistan ecology is largely confined to investigations designed to determine the presence or absence of species of interest to human health (49) or species that might have adverse environmental impact (66). Applications to species of purely ecological significance, however, are beginning (31). Development of immunological methods for determining the presence and abundance of specific microbiological taxa also has begun. This approach has been successfully applied to phototrophic protists (15).

The application of molecular and immunological approaches to protistan groups with problematic taxonomic features may be particularly helpful in the future. For example, species of heterotrophic flagellates or amoebae which presently involve nonquantitative or impractical approaches such as electron microscopy or the observation of living specimens might be particularly suited for this approach (46).

Biochemical markers to indicate the presence, abundance, and activity of heterotrophic protists also may provide new methods for examining natural assemblages of protists. Detailed pigment analyses have provided useful insights into the contribution of microalgal taxa to total algal biomass (83). Lipid biomarkers have been applied to obtain information on the biomass and nutritional status of bacterial assemblages (81). Analogous methodologies for assessing community-level features of heterotrophic protistan assemblages would be useful for investigating natural, mixed assemblages of protists.

Support for the preparation of this chapter was provided by National Science Foundation grants OCE-9216270, 9314533, and 9310693.

REFERENCES

1. **Alongi, D. M.** 1986. Quantitative estimates of benthic protozoa in tropical marine systems using silica gel: a comparison of methods. *Estuarine Coastal Shelf Sci.* **23**:443–450.
2. **Amann, R. I., W. Ludwig, and K.-H. Schleifer.** 1995. Phylogenetic identification and in situ detection of individual microbial cells without cultivation. *Microbiol. Rev.* **59**:143–169.
3. **Andersen, R. A., G. W. Saunders, M. P. Paskind, and J. P. Sexton.** 1993. Ultrastructure and 18S rRNA gene sequence for *Pelagomonas calceolata* gen. et sp. nov. and the description of a new algal class, the Pelagophyceae classis nov. *J. Phycol.* **29**:701–715.
4. **Arenovski, A. L., E. L. Lim, and D. A. Caron.** 1995. Mixotrophic nanoplankton in oligotrophic surface waters of the Sargasso Sea may employ phagotrophy to obtain major nutrients. *J. Plankton Res.* **17**:801–820.
5. **Arndt, H.** 1993. A critical review of the importance of rhizopods (naked and testate amoebae) and actinopods (heliozoa) in lake plankton. *Mar. Microb. Food Webs* **7**:3–29.
6. **Azam, F., T. Fenchel, J. G. Field, J. S. Gray, L. A. Meyer-Reil, and F. Thingstad.** 1983. The ecological role of water-column microbes in the sea. *Mar. Ecol. Prog. Ser.* **10**:257–263.
7. **Banse, K.** 1982. Cell volumes, maximal growth rates of unicellular algae and ciliates, and the role of ciliates in the marine pelagial. *Limnol. Oceanogr.* **27**:1059–1071.
8. **Bé, A. W. H.** 1977. An ecological, zoogeographic and taxonomic review of recent planktonic foraminifera, p. 1–100. *In* A. T. S. Ramsay (ed.), *Oceanic Micropaleontology.* Academic Press, London.
9. **Bé, A. W. H., and O. R. Anderson.** 1976. Preservation of planktonic foraminifera and other calcareous plankton, p. 250–258. *In* H. F. Steedman (ed.), *Zooplankton Fixation and Preservation.* UNESCO Press, Paris.
10. **Beaver, J. R., and T. L. Crisman.** 1989. The role of ciliated protozoa in pelagic freshwater ecosystems. *Microb. Ecol.* **17**:111–136.
11. **Bernard, C., and F. Rassoulzadegan.** 1994. Seasonal variations of mixotrophic ciliates in the northwest Mediterranean Sea. *Mar. Ecol. Prog. Ser.* **108**:295–301.
12. **Berninger, U.-G., D. A. Caron, and R. W. Sanders.** 1992. Mixotrophic algae in three ice-covered lakes of the Pocono Mountains, USA. *Freshwater Biol.* **28**:263–272.
13. **Bird, D. F., and J. Kalff.** 1986. Bacterial grazing by planktonic lake algae. *Science* **231**:493–495.

14. **Cairns, J., Jr., D. L. Kuhn and J. L. Plafkin.** 1979. Protozoan colonization of artificial substrates, p. 34–57. *In* R. L. Wetzel (ed.), *Methods and Measurements of Periphyton Communities: a Review.* Special Technical Publication 690. American Society for Testing and Materials, Philadelphia.
15. **Campbell, L., L. P. Shapiro, and E. Haugen.** 1994. Immunochemical characterization of eukaryotic ultraplankton from the Atlantic and Pacific oceans. *J. Plankton Res.* **16**:35–51.
16. **Caron, D. A.** 1991. Heterotrophic flagellates associated with sedimenting detritus, p. 77–92. *In* D. J. Patterson and J. Larsen (ed.), *The Biology of Free-Living Heterotrophic Flagellates,* special vol. 45. Clarendon Press, Oxford.
17. **Caron, D. A., H. G. Dam, P. Kremer, E. J. Lessard, L. P. Madin, T. C. Malone, J. M. Napp, E. R. Peele, M. R. Roman, and M. J. Youngbluth.** 1995. The contribution of microorganisms to particulate carbon and nitrogen in surface waters of the Sargasso Sea near Bermuda. *Deep-Sea Res.* **42**:943–972.
18. **Caron, D. A., and B. J. Finlay.** 1994. Protozoan links in food webs, p. 125–130. *In* K. Hausmann and N. Hüllsmann (ed.), *Progress in Protozoology. Proceedings of the IX International Congress of Protozoology, Berlin 1993.* Gustav Fischer Verlag, Stuttgart.
19. **Caron, D. A., and J. C. Goldman.** 1990. Protozoan nutrient regeneration, p. 283–306. *In* G. M. Capriulo (ed.), *Ecology of Marine Protozoa.* Oxford University Press, New York.
20. **Caron, D. A., and N. R. Swanberg.** 1990. The ecology of planktonic sarcodines. *Rev. Aquat. Sci.* **3**:147–180.
21. **Carrick, H. J., and G. L. Fahnenstiel.** 1990. Planktonic protozoa in Lakes Huron and Michigan: seasonal abundance and composition of ciliates and dinoflagellates. *J. Great Lakes Res.* **16**:319–329.
22. **Christian, R. R.** 1994. Aggregation and disaggregation of microbial food webs. *Microb. Ecol.* **28**:327–329.
23. **Cole, J. J., S. Findlay, and M. L. Pace.** 1988. Bacterial production in fresh and saltwater ecosystems: a cross-system overview. *Mar. Ecol. Prog. Ser.* **43**:1–10.
24. **Crawford, D. W.** 1989. *Mesodinium rubrum:* the phytoplankter that wasn't. *Mar. Ecol. Prog. Ser.* **58**:161–174.
25. **Davis, P. G., D. A. Caron, and J. M. Sieburth.** 1978. Oceanic amoebae from the North Atlantic culture, distribution, and taxonomy. *Trans. Am. Microsc. Soc.* **96**:73–88.
26. **Dolan, J. R.** 1991. Guilds of ciliate microzooplankton in the Chesapeake Bay. *Estuarine Coastal Shelf Sci.* **33**:137–152.
27. **Dolan, J. R., and D. W. Coats.** 1990. Seasonal abundances of planktonic ciliates and microflagellates in mesohaline Chesapeake Bay waters. *Estuarine Coastal Shelf Sci.* **31**:157–175.
28. **Ducklow, H. W.** 1994. Modeling the microbial food web. *Microb. Ecol.* **28**:303–319.
29. **Dworetzky, B. A., and J. J. Morley.** 1987. Vertical distribution of radiolaria in the Eastern Equatorial Atlantic: analysis of a multiple series of closely-spaced plankton tows. *Mar. Micropaleontol.* **12**:1–19.
30. **Fairbanks, R. G., and P. H. Wiebe.** 1980. Foraminifera and chlorophyll maximum: vertical distribution, seasonal succession, and paleoceanographic significance. *Science* **209**:1524–1526.
31. **Fell, J. W., A. Statzell-Tallman, M. J. Lutz, and C. P. Kurtzman.** 1992. Partial rRNA sequences in marine yeasts: a model for identification of marine eukaryotes. *Mol. Mar. Biol. Biotechnol.* **1**:175–186.
32. **Fenchel, T.** 1967. The ecology of marine microbenthos. I. The quantitative importance of ciliates as compared with

metazoans in various types of sediments. *Ophelia* **4**: 121–137.

33. **Fenchel, T.** 1969. The ecology of marine microbenthos. IV. Structure and function of the benthic ecosystem, its chemical and physical factors and the microfauna communities with special reference to the ciliated protozoa. *Ophelia* **6**:1–182.

34. **Fenchel, T., and B. J. Finlay.** 1983. Respiration rates in heterotrophic, free-living protozoa. *Microb. Ecol.* **9**: 99–122.

35. **Finlay, B. J., U.-G. Berninger, L. J. Stewart, R. M. Hindle, and W. Davison.** 1987. Some factors controlling the distribution of two pond-dwelling ciliates with algal symbionts (*Frontonia vernalis* and *Euplotes daidaleos*). *J. Protozool.* **34**:349–356.

36. **Finlay, B. J., K. E. Clarke, E. Vicente, and M. R. Miracle.** 1991. Anaerobic ciliates from a sulfide-rich solution lake in Spain. *Eur. J. Protistol.* **27**:148–159.

37. **Garrison, D. L., and M. M. Gowing.** 1993. Microzooplankton, p. 123–166. *In* E. I. Friedmann (ed.), *Antarctic Microbiology*. Wiley-Liss, New York.

38. **Gifford, D. J.** 1985. Laboratory culture of marine planktonic oligotrichs (Ciliophora, Oligotrichida). *Mar. Ecol. Prog. Ser.* **23**:257–267.

39. **Gooday, A. J., L. A. Levin, P. Linke, and T. Heeger.** 1992. The role of benthic foraminifera in deep-sea food webs and carbon cycling, p. 63–91. *In* G. T. Rowe (ed.), *Deep-Sea Food Chains—and the Global Carbon Cycle.* Kluwer Academic, Dordrecht, The Netherlands.

40. **Gowing, M. M.** 1993. Seasonal radiolarian flux at the VERTEX North Pacific time-series site. *Deep-Sea Res.* **40**: 517–545.

41. **Jacobson, D. M., and D. M. Anderson.** 1986. Thecate heterotrophic dinoflagellates: feeding behavior and mechanisms. *J. Phycol.* **22**:249–258.

42. **Jürgens, K., and G. Stolpe.** 1995. Seasonal dynamics of crustacean zooplankton, heterotrophic nanoflagellates and bacteria in a shallow, eutrophic lake. *Freshwater Biol.* **33**: 27–38.

43. **Leadbeater, B. S. C.** 1993. Preparation of pelagic protists for electron microscopy, p. 241–251. *In* P. F. Kemp, B. F. Sherr, E. B. Sherr, and J. J. Cole (ed.), *Handbook of Methods in Aquatic Microbial Ecology.* Lewis Publishers, Boca Raton, Fla.

44. **Lee, J. J., S. H. Hutner, and E. C. Bovee.** 1985. *An Illustrated Guide to the Protozoa.* Society of Protozoologists, Lawrence, Kans.

45. **Lee, J. J., K. Sang, B. ter Kuile, E. Strauss, P. J. Lee, and W. W. Faber, Jr.** 1991. Nutritional and related experiments on laboratory maintenance of three species of symbiont-bearing, large foraminifera. *Mar. Biol.* **109**:417–425.

46. **Lim, E. E., L. A. Amaral, D. A. Caron, and E. F. DeLong.** 1993. Application of rRNA-based probes for observing marine nanoplanktonic protists. *Appl. Environ. Microbiol.* **59**: 1647–1655.

47. **Lindeman, R. L.** 1942. The trophic-dynamic aspect of ecology. *Ecology* **23**:399–418.

48. **Maeda, M.** 1986. An illustrated guide to the species of the families Halteriidae and Strobilidiidae (Oligotrochida, Ciliophora), free swimming protozoa common in the marine environment. *Bull. Ocean Res. Inst. Univ. Tokyo* **21**: 1–67.

49. **McLaughlin, G. L., S. Montenegrojames, M. H. Vodkin, D. Howe, M. Toro, E. Leon, R. Armijos, I. Kakoma, B. M. Greenwood, M. Hassanking, J. Marich, J. Ruth, and M. A. James.** 1992. Molecular approaches to malaria and babesiosis diagnosis. *Mem. Inst. Oswaldo Cruz* **87**:57–68.

50. **Michaels, A. F.** 1991. Acantharian abundance and symbi-

51. **Michaels, A. F., D. A. Caron, N. R. Swanberg, F. A. Howse, and C. M. Michaels.** 1995. Planktonic sarcodines (Acantharia, Radiolaria, Foraminifera) in surface waters near Bermuda: abundance, biomass and vertical flux. *J. Plankton Res.* **17**:131–163.

52. **Moloney, C. L., and J. G. Field.** 1991. The size-based dynamics of plankton food webs. 1. A simulation model of carbon and nitrogen flux. *J. Plankton Res.* **13**:1003–1038.

53. **Page, F. C.** 1983. *Marine Gymnamoebae.* Institute of Terrestrial Ecology, Cambridge.

54. **Parke, M., and P. S. Dixon.** 1976. Checklist of British marine algae—third revision. *J. Mar. Biol. Assoc. U.K.* **56**:527–594.

55. **Patterson, D. J., and J. Larsen.** 1991. *The Biology of Free-Living Heterotrophic Flagellates.* Clarendon Press, Oxford.

56. **Patterson, D. J., and M. Zölffel.** 1991. Heterotrophic flagellates of uncertain taxonomic position, p. 427–475. *In* D. J. Patterson and J. Larsen (ed.), *The Biology of Free-Living Heterotrophic Flagellates.* Clarendon Press, Oxford.

57. **Pierce, R. W., and J. T. Turner.** 1992. Ecology of planktonic ciliates in marine food webs. *Rev. Aquat. Sci.* **6**: 139–181.

58. **Putt, M., and D. K. Stoecker.** 1989. An experimentally determined carbon:volume ratio for marine "oligotrichous" ciliates from estuarine and coastal waters. *Limnol. Oceanogr.* **34**:1097–1103.

59. **Reid, P. C., C. M. Turley, and P. H. Burkill.** 1991. *Protozoa and Their Role in Marine Processes.* Springer-Verlag, Berlin.

60. **Rogerson, A., H. G. Butler, and J. C. Thomason.** 1994. Estimation of amoeba cell volume from nuclear diameter and its application to studies in protozoan ecology. *Hydrobiologia* **284**:229–234.

61. **Rogerson, A., and J. Laybourn-Parry.** 1992. The abundance of marine naked amoebae in the water column of the Clyde estuary. *Estuarine Coastal Shelf Sci.* **34**:187–196.

62. **Sanders, R. W.** 1991. Mixotrophic protists in marine and freshwater ecosystems. *J. Protozool.* **38**:76–81.

63. **Sanders, R. W., D. A. Caron, and U.-G. Berninger.** 1992. Relationships between bacteria and heterotrophic nanoplankton in marine and fresh water: an inter-ecosystem comparison. *Mar. Ecol. Prog. Ser.* **86**:1–14.

64. **Sawyer, T. K.** 1980. Marine amebae from clean and stressed bottom sediments of the Atlantic Ocean and Gulf of Mexico. *J. Protozool.* **27**:13–32.

65. **Schlegel, M.** 1991. Protist evolution and phylogeny as discerned from small subunit ribosomal RNA sequence comparisons. *Eur. J. Protistol.* **27**:207–219.

66. **Scholin, C. A., and D. M. Anderson.** 1994. Identification of group- and strain-specific genetic markers for globally distributed *Alexandrium* (Dinophyceae). I. RFLP analysis of SSU rRNA genes. *J. Phycol.* **30**:744–754.

67. **Sherr, E. B., and B. F. Sherr.** 1993. Preservation and storage of samples for enumeration of heterotrophic protists, p. 207–212. *In* P. F. Kemp, B. F. Sherr, E. B. Sherr, and J. J. Cole (ed.), *Handbook of Methods in Aquatic Microbial Ecology.* Lewis Publishers, Boca Raton, Fla.

68. **Sieburth, J. M.** 1979. *Sea Microbes.* Oxford University Press, New York.

69. **Sieburth, J. M., V. Smetacek, and J. Lenz.** 1978. Pelagic ecosystem structure: heterotrophic compartments of the plankton and their relationship to plankton size fractions. *Limnol. Oceanogr.* **23**:1256–1263.

70. **Silver, M. W., M. M. Gowing, D. C. Brownlee, and J. O. Corliss.** 1984. Ciliated protozoa associated with oceanic sinking detritus. *Nature* (London) **309**:246–248.

71. **Small, E. B., and M. E. Gross.** 1985. Preliminary observa-

tions of protistan organisms, especially ciliates, from the 21°N hydrothermal vent site. *Biol. Soc. Wash. Bull.* **6:** 401–410.

72. **Starink, M., M.-J. Bär-Gilissen, R. P. M. Bak, and T. E. Cappenberg.** 1994. Quantitative centrifugation to extract benthic protozoa from freshwater sediments. *Appl. Environ. Microbiol.* **60:**167–173.

73. **Stoecker, D., A. E. Michaels, and L. H. Davis.** 1987. Large proportion of marine planktonic ciliates found to contain functional chloroplasts. *Nature* (London) **326:** 790–792.

74. **Stoecker, D., A. Taniguchi, and A. E. Michaels.** 1989. Abundance of autotrophic, mixotrophic and heterotrophic planktonic ciliates in shelf and slope waters. *Mar. Ecol. Prog. Ser.* **50:**241–254.

75. **Stoecker, D. K., D. J. Gifford, and M. Putt.** 1994. Preservation of marine planktonic ciliates: losses and cell shrinkage during fixation. *Mar. Ecol. Prog. Ser.* **110:**293–299.

76. **Swanberg, N. R., and D. A. Caron.** 1991. Patterns of sarcodine feeding in epipelagic oceanic plankton. *J. Plankton Res.* **13:**287–312.

77. **Taylor, F. J. R.** 1982. Symbioses in marine microplankton. *Ann. Inst. Oceanogr.* (Paris) **58**(Suppl.):61–90.

78. **Turley, C. M., and K. Lochte.** 1990. Microbial response to the input of fresh detritus to the deep-sea bed. *Palaeogeogr. Palaeoclimatol. Palaeoecol.* **89:**3–23.

79. **Turner, J. T., and J. C. Roff.** 1993. Trophic levels and trophospecies in marine plankton: lessons from the microbial food web. *Mar. Microb. Food Webs* **7:**225–248.

80. **Verity, P. G., C. Y. Robertson, C. R. Tronzo, M. G. Andrews, J. R. Nelson, and M. E. Sieracki.** 1992. Relationships between cell volume and the carbon and nitrogen content of marine photosynthetic nanoplankton. *Limnol. Oceanogr.* **37:**1434–1446.

81. **White, D. C.** 1994. Is there anything else you need to understand about the microbiota that cannot be derived from analysis of nucleic acids? *Microb. Ecol.* **28:**163–166.

82. **Wiebe, W. J., W. M. Sheldon, Jr., and L. R. Pomeroy.** 1992. Bacterial growth in the cold: evidence for an enhanced substrate requirement. *Appl. Environ. Microbiol.* **58:**359–364.

83. **Wright, S. W., S. W. Jeffrey, F. C. Mantoura, C. A. Llewellyn, T. Bjørnland, D. Repeta, and N. Welschmeyer.** 1991. Improved HPLC method for the analysis of chlorophylls and carotenoids from marine phytoplankton. *Mar. Ecol. Prog. Ser.* **77:**183–196.

84. **Yongue, W. H., Jr., and J. Cairns, Jr.** 1978. The role of flagellates in pioneer protozoan colonization of artificial substrates. *Pol. Arch. Hydrobiol.* **25:**787–801.

85. **Zubkov, M. V., A. F. Sazhin, and M. V. Flint.** 1992. The microplankton organisms at the oxic-anoxic interface in the pelagial of the Black Sea. *FEMS Microbiol. Ecol.* **101:** 245–250.

Bulk Quantitative Methods for the Examination of Eukaryotic Organoosmotrophs in Plant Litter

MARK O. GESSNER AND STEVEN Y. NEWELL

31

Heterotrophic organisms acquire their carbon and nutrients by one of two processes. Animals and protists are capable of ingesting particles, a mode of nutrition known as phagotrophy. Organoosmotrophic organisms, in contrast, assimilate organic carbon and nutrients in dissolved form. The heterotrophic prokaryotes fall into this category. They share this feature with a number of eukaryotes, including the eumycotic fungi (Basidiomycota, Ascomycota, Zygomycota, Chytridiomycota, and Deuteromycota; e.g., references 11 and 25) and the taxonomically unrelated oomycetous protists (Oomycota) (83), thraustochytrids, and labyrinthulids (Labyrinthulomycota), the last of which have also been shown to be partially phagotrophic (96).

Eukaryotic organoosmotrophs, or fungi sensu lato, are ubiquitous in aquatic environments and exhibit a remarkable diversity (9, 24, 40, 47, 53, 54, 60, 72, 103, 108, 109, 114). They occur as pathogens, commensals, or mutualistic symbionts of algae, vascular plants, and animals in both pelagic and benthic systems (45, 59, 119), and they may mediate a variety of biogeochemical transformations (113). Their predominant habitat, however, is plant litter accumulating within land-water transition zones such as coastal marine areas, streams and rivers receiving litter from the streamside vegetation, the littoral zones of lakes, and other freshwater wetlands (40). The implication of this occurrence and the penetrative mycelial growth form (74) is that aquatic eukaryotic organoosmotrophs are primarily found within opaque solid substrates. In consequence, their hyphae are neither immediately visible nor easily separable from the plant tissue without significant destruction taking place (71). Here lies the major challenge for the investigator who wants to assess the community structure, biomass, productivity, or activity of these organisms in their natural environment.

The methods traditionally used in soil mycology have also been applied to the study of fungi in aquatic environments. Essentially two approaches have been adopted. One involves the direct observation of reproductive structures either present in the natural habitat or induced prior to observation; the other is based on the isolation of fungi on nutrient media. Frankland (27), Frankland et al. (28), Parkinson (92), Parkinson and Coleman (93), and others provide overviews of these techniques; Dick (24), Maltby (60), and Webster and Descals (114) address methods more specifically applicable to the study of aquatic assemblages. There is general consensus that dilution plating as commonly used to isolate bacteria has severe drawbacks in ecological studies of fungi. The main argument against the use of this method is its strong bias toward organisms producing large numbers of propagules. Particle plating partly avoids this bias, especially if sample pieces are surface sterilized (82) or serially washed with sterile water (51) before plating. These pretreatments remove or kill many (though not all) of the propagules and hyphal bits loosely attached to the particle surface but not necessarily representative of mycelial development within the substrate. Probably the most realistic results with particle plating techniques are achieved when small standardized pieces are plated (6, 52). However, although more adequate than dilution plating, particle plating also introduces considerable bias in the interpretation of data due to the inevitable selectivity of culture media.

A remedy to the problem of culture medium selectivity is direct observation of fungal fruiting structures in or on the material collected in the field. In streams, for example, many species of aquatic hyphomycetes can be readily identified on the basis of the characteristic shapes of their conidia forming on leaf litter and wood, transported in stream water, and accumulating in foam (9, 109, 114). When conidia are counted, a quantitative picture of fungal presence, community structure, and sporulation activity can be obtained. Likewise, fruiting structures of aeroaquatic hyphomycetes (97, 114) and marine litter fungi (54) can often be induced in damp chambers. To obtain quantitative information on the reproductive activity of such fungi, ascomata have been counted in standing-dead macrophyte leaves (86) and expulsed spores trapped on sticky surfaces in the laboratory (86) and in the field (56). However, all of these methods rely on the presence or formation of fruiting structures. Because fungi differ greatly in their potential to produce such structures and also change their patterns of resource allocation to vegetative and reproductive structures during their life history, direct observational methods only partially reflect the presence and activity of the mycelium located within the solid substrate.

Another, indirect approach for examining fungal assemblages consists in the determination of selected biomolecules characteristic of specific groups of organisms. Lipids, in particular, have been used to this end. The analyses can be done in a qualitative (examination of community structure) or a quantitative (determination of biomass) fashion (69, 71, 112). White et al. (chapter 10, this volume) describe these techniques in some detail, with particular attention to the analysis of microbial community structure. This chapter is primarily concerned with indirect quantitative approaches. Special attention is given to the ergosterol assay, which is a bulk method measuring the biomass of eumycotic fungi in solid substrates (71, 73, 88). The so-called acetate-to-ergosterol method is an extension of the ergosterol assay allowing direct measurements of eumycotic fungal growth rates and productivity in situ (73, 79, 110); it will also be presented in some detail.

FUNGAL BIOMASS IN PLANT LITTER

The Ergosterol Assay

Background Rationale

The rationale underlying the ergosterol assay and similar techniques is that a quantitative relationship exists between the concentration of a selected cell constituent and the amount of organismic mass present. The principle of the assay thus consists in quantifying ergosterol (Fig. 1) in a solid substrate colonized by fungi and relating it to the amount of eumycotic fungal dry mass, ash free dry mass, or a particular cell constituent via an appropriate conversion factor.

Ergosterol is particularly useful as an index molecule of eumycotic fungal mass since it combines a number of advantageous properties. Ergosterol is the major sterol in the majority of eumycotic fungi, including members of the Zygomycota, Ascomycota, Basidiomycota, and Deuteromycota, though some species of ascomycetes may also contain high proportions of brassicasterol (ergosta-5,22-dien-3β-ol) (116). The phytopathogenous rusts (Uredinales) are notable exceptions in that their mycelium contains no ergosterol. Likewise, chytrids (Chytridiomycota) and oomycetous protists (Oomycota) are not included in the assay because their cells do not contain ergosterol either (71, 116). Ergosterol is absent from vascular plants, metazoan animals, and most other organisms, although it may occur, usually in minor amounts, in some bacteria, cyanobacteria, green microalgae, and protozoa (71, 116). Thus, ergosterol is essentially restricted to a broad but taxonomically and ecologically quite well defined group of organisms. For many practical considerations, the critical condition of specificity is therefore met. Nevertheless, when exploring natural systems, investigators should ensure that nontarget organisms potentially present either are lacking in ergosterol, contribute insignificantly to the total pool owing to a low biomass or a low cellular ergosterol content, or can be removed from the sample before analysis. At least one of these criteria is generally met in aquatic litter systems (7, 72, 84, 111), although attached algae might sometimes be abundant enough on decomposing plant material to cause potential interference (33a, 72, 84).

Ergosterol is mainly a membrane component (57, 94, 115, 116). Since membranes are critical in maintaining cell function, ergosterol is likely to be indicative of cytoplasm-containing, metabolically active cells. In addition, ergosterol is prone to oxidation at its conjugated double bond and to other reactions of the labile ring B of the sterol skeleton (Fig. 1). As a result, the molecule is likely to undergo quite rapid degradation upon cell death, suggesting that it is basically an indicator of living fungal biomass (43, 71, 84, 117). However, rigorous tests of this assumption have

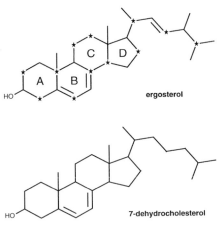

FIGURE 1 Ergosterol (ergosta-5,7,22-trien-3β-ol; provitamin D₂) and the commercially available 7-dehydrocholesterol (cholesta-5,7-dien-3β-ol; provitamin D₃), which because of its structural similarity to ergosterol and strong UV absorbance at 282 nm can be useful as an internal standard in routine applications. All but one of the 28 carbon atoms in ergosterol are derived from acetate during fungal sterol biosynthesis; the 12 carbon atoms originating from C-1 of the acetate molecule (carboxyl carbon) are indicated by stars. The four rings of the steroid skeleton are indicated by A to D.

not been conducted. The conjugated double bond of ergosterol results in a shift in the UV absorbance spectrum toward longer wavelengths, with a maximum occurring at 282 nm (Fig. 2). This allows the unambiguous spectrophotometric distinction from plant sterols such as sitosterol and campesterol, which show no significant absorbance above 240 nm (17, 104, 105). As ergosterol is a quantitatively important constituent of fungal cells (commonly near 0.5% of organic

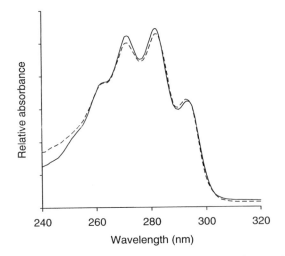

FIGURE 2 UV absorbance spectrum of ergosterol in methanol, showing the characteristic absorbance maxima of the provitamins D at about 262, 271, 282, and 294 nm. The solid line indicates a commercially available ergosterol standard (Fluka; >98% purity); the broken line corresponds to the eluted HPLC fraction of an extract obtained from a plant litter sample colonized by fungi in a stream (33a).

mass [35, 75]), analytical detection does not pose major problems.

Sample Collection and Storage

The importance of adequate collection of field material, avoidance of artificial sample treatments and experimental procedures (e.g., drying and grinding of litter before field exposure), and a realistic setup of experiments cannot be overemphasized (15, 72; chapter 4, this volume). For example, Newell (72) and others made the point that much of the leaf litter in (subtropical) salt marshes decomposes in the standing-dead position. Despite this fact, investigators examining microbial assemblages associated with such litter have commonly employed litter bags placed on the marsh surface and found little evidence for fungal involvement in the decomposition process (58). In contrast, by using a tagging technique, it was demonstrated that fungi were eminently important in the natural, standing-dead decomposition system (72). Similar cases have been made for freshwater litter systems (10, 39, 55, 98), although at least in streams the observed effects on fungal assemblages were found to be less dramatic than in salt marshes (8, 18, 41).

Once samples are collected, precautions must be taken to avoid degradation of ergosterol. Throughout storage and sample processing, samples must be protected from bright sunlight and other sources of strong UV radiation, which rapidly destroys the molecule (1, 77, 104). Samples must not be oven dried or, for some types of samples, even air dried because such procedures likewise can lead to severe losses (76, 77) despite reports to the contrary (14, 17). However, under normal laboratory conditions of light and temperature, pure dissolved ergosterol is stable for several weeks to months (1, 20, 33a), and thus excessive precautions to protect samples from light and to prevent contact with molecular oxygen are unnecessary (77).

A safe way of preserving samples intended for ergosterol analysis involves subsampling of a freshly collected sample, immediate immersion in methanol, and subsequent storage in the dark at 4°C (77) or below. The recommended procedure with plant litter samples is to cut a set of leaf discs or a number of small, otherwise standardized plant pieces, immerse them immediately in methanol, and carry out extractions within a week after collection. A second, identical set of plant pieces is set aside, dried, and, if necessary, ashed in order to relate sample wet weight or surface area to dry mass, ash free dry mass, or a similar parameter of interest. In our experience, methanol-preserved samples can be stored for at least 1 month (see also reference 20), but longer storage is probably possible. Because sample matrix components might chemically or physically interact with the ergosterol molecule, we suggest, however, that investigators planning not to process samples immediately after collection check whether significant degradation occurs with their specific sample type and size over the projected storage period. This can be conveniently done by spiking samples with an ergosterol standard or, perhaps better (20), a known amount of mycelium before storage. 7-Dehydrocholesterol, displaying close structural similarities with ergosterol (Fig. 1) and strongly absorbing light at 282 nm, could also be useful as an internal standard (14, 31), permitting estimation of ergosterol losses during both storage and later extraction. Note also that ergosterol powder used for standard curves should be checked for degradative loss and possibly discarded after several months of cold storage (75).

An alternative way of storing samples is to freeze bulk samples, lyophilize them just before analysis, and take subsamples from the homogenized (e.g., shredded and well-mixed) bulk sample (20, 34, 91). The appeal of this procedure resides in the possibilities of (i) making precise dry mass determinations of the same sample analyzed for ergosterol and (ii) minimizing within-sample heterogeneity. Lyophilization might also be indicated when samples contain much moisture (e.g., >5% of the extraction solvent volume), potentially compromising extraction efficiencies (1, 20). With colonized leaf litter from a stream, Gessner et al. (34) found that ergosterol recoveries from samples placed in a freezer at −18°C and subsequently lyophilized were comparable to those obtained with methanol-preserved samples. Similar results were reported by Davis and Lamar (20) and Anderson et al. (1) for soil samples. However, Newell et al. (77) and Antibus and Sinsabaugh (2) obtained lower recoveries from lyophilized samples (up to 82%) than from methanol-preserved samples, indicating that losses may occur during freezing, storage in the freezer, and/or lyophilization. Losses from aged mycelium appear to be especially pronounced (2). The way in which samples are frozen (instantly in liquid nitrogen, more slowly in a freezer, etc.) is possibly critical (20), but insufficient data are currently available to make clear recommendations. One factor possibly contributing to lower ergosterol recoveries from freeze-dried material is dry storage of samples after lyophilization (33a), which is therefore best avoided.

In conclusion, immersion in methanol is the preferred preservation method when representative subsampling and reasonably accurate weight determinations of subsamples are possible. Deep-freezing followed by lyophilization may give more satisfactory results when dry weight determinations are difficult and within-sample heterogeneity is high, provided that potential losses of ergosterol are either acceptably low or accurately predictable. As different fungal species and/or sample matrices may give inconsistent results with regard to the most appropriate preservation technique (76), investigators should run preliminary tests when examining unfamiliar environmental samples.

Extraction and Saponification

Methods for the analysis of sterols and chemically related compounds have a long tradition in biochemistry and physiology. Methods employing classic lipid extraction together with a wide range of chromatographic methods have been used (46, 120), but these will not be discussed here. Note that because of the relative lability of ergosterol toward oxidizing agents, light, and temperature, some of these methods may not be appropriate for quantitative ergosterol analysis. Current evidence suggests that adequate extraction of ergosterol from fungal mycelium is achieved either by wet homogenization with a high-speed blender followed by repeated extraction in cold methanol (34, 67, 105) or ethanol (87) or by refluxing in a mixture (32) or either of these alcohols (44, 77). With plant litter from marine and freshwater environments, both methods produce satisfactory results, but refluxing is more convenient (34, 77).

A standard method for ergosterol analysis has been established by the French Association for Normalization (AFNOR standard VI8-I12 [17]), and another detailed protocol for the extraction, purification, and quantitation of ergosterol from plant litter has been presented by Newell (73). Samples are refluxed for 2 h in absolute methanol. Subsequently an alcoholic KOH solution is added, and refluxing is continued for another 30 min in order to convert ergosterol esters into free ergosterol. Contrary to previous beliefs (73, 77), it is perhaps advantageous not to remove solids prior to saponification, given the finding that extraction efficiencies are higher when mycelial pellets of marine fungi are retained (75, 91). The same trend has been observed by Nout et al. (87), although differences between treatments were small. In more complex samples such as soil (20) or colonized plant litter, matrix compounds might interact with ergosterol; therefore, investigators should first check whether keeping or removing solids results in higher yields from their particular sample type. However, recoveries even from complex matrices have consistently been found to be high (>85%), implying that physical or chemical interactions leading to ergosterol losses are generally small (43, 117).

In many cases, the 2-h refluxing in methanol can probably be omitted and ergosterol can be concurrently extracted and saponified during a single reflux cycle (generally 0.5 h) in alcoholic base (21, 34, 104, 123). KOH in a methanol-ethanol mixture is as suitable for that purpose as KOH in pure methanol (38; see also reference 14). Gessner et al. (34) found that wet homogenization in cold methanol and subsequent saponification (105), 2-h refluxing in methanol followed by saponification (77), and concurrent extraction and saponification in alcoholic base (104) are equally efficient at extracting ergosterol from leaf litter colonized by fungi in a stream. Consequently, results obtained with these different extraction methods should be directly comparable.

Fatty acid esters make up a substantial but variable portion of total ergosterol in fungal cells (32, 38, 87, 91, 100). Therefore, saponification is an important analytical step if the concentration of total ergosterol in a sample is to be determined. In contrast to other workers, Martin et al. (62) claimed that saponification is not necessary. This conclusion is supported by their finding that ergosterol yields were increased when saponification was omitted. Presumably, however, this result reflects proportionally higher losses during the analysis including saponification or an inefficient extraction of ergosterol esters (see reference 91) rather than the absence of ergosterol esters in the strain of *Laccaria laccata* examined. Salmanowicz and Nylund (100), for example, found that ergosterol yields from their strain of *L. laccata* increased by 24% when alcoholic extracts were saponified. Apart from the analytical considerations, there is a fundamental debate as to whether a saponification step should be included in ergosterol analysis of environmental samples (75). The debate relates to the question of whether free or total ergosterol is the better indicator of fungal biomass. Although the issue is far from being settled, total ergosterol currently appears to be the better indicator (75).

Glycosidic bonds are not cleaved by refluxing in alcoholic base, but this is not a problem because ergosteryl glycosides make up a minute fraction of the total ergosterol pool in three species of fungi examined (14).

Partitioning

After saponification, the extracted lipids including ergosterol are partitioned into a nonpolar phase. Neutralization (34, 44) is not required at this stage, but it is essential to add water to the alcoholic extract in order to get a clearly defined phase boundary between the alcoholic and nonpolar phases. Abundant flocculation may occur with some types of samples at this stage. Addition of an aqueous NaCl solution (14, 20, 34) instead of water may reduce the problem, although it will not be a perfect remedy. However, if in the first partitioning step one collects the clean nonpolar phase while entirely avoiding the intermediate emulsive layer,

flocculation will automatically diminish in subsequent partitioning steps, allowing quantitative recovery of ergosterol.

Newell et al. (77) and others (1, 34) made the case that the type, quality, and relative volume of the nonpolar solvent as well as the intensity of mixing need to be carefully controlled to ensure high and reproducible recoveries (see also reference 20). Pentane (2, 67, 77, 91), hexane (29, 100, 104), and petroleum ether of various boiling ranges (5, 14, 34) are suitable as nonpolar solvents. Each has advantages and disadvantages. Pentane quickly evaporates at a low temperature (36°C) but is more highly flammable and shows less affinity to ergosterol than the other solvents (34). In terms of its partition coefficient, hexane is superior to the other solvents, but it is also the most toxic and takes longer to evaporate (boiling point, 69°C). Petroleum ether is significantly cheaper than both pentane and hexane, but its composition is not chemically defined. In practice, all alternatives will give satisfactory results when the partitioning conditions are carefully chosen. Diethyl ether as used by some workers (44) is better avoided both for safety reasons (explosion hazards) and because of potential oxidation of ergosterol by peroxides.

Modifications of Extraction Procedures

Other published schemes for the extraction of ergosterol from plant tissue and soils differ considerably in regard to solvent type, sample volume, sample-to-solvent ratio, whether extraction is carried out at ambient or elevated temperature, whether or not a saponification step is included, and many procedural details. While many of the reported methods are probably adequate, we feel that simple extraction in cold methanol (88, 100, 104, 123, 124) without simultaneous homogenization by a high-speed blender may lead to inefficient extraction.

To reduce analysis time, radical modifications of the procedures described by Seitz et al. (105) and Newell et al. (77) have been proposed. Johnson and McGill (48, 49) achieve extraction and saponification by brief heating (100°C, 5 min) of samples in methanolic KOH placed in screw-cap test tubes. The extracted lipids including ergosterol are subsequently partitioned into petroleum ether (four times) without prior sample transfer to a clean vessel and without adding water. Martin et al. (62) scaled the original procedure by Seitz et al. (105) down to an extraction volume of 0.5 ml, omitting both the saponification step and the partitioning step, and injecting the primary alcoholic extract directly into the high-pressure liquid chromatography (HPLC) system. Ruzicka et al. (99) used ultrasonication in a methanol-ethanol mixture, also omitting saponification procedures. All three protocols thus reduce analysis time substantially, and although used only for ectomycorrhizae and clay soils so far, these procedures hold potential for general application. Reported recoveries of added ergosterol carried through the procedures are excellent in Johnson and McGill's (48, 49) study (97% ± 2%, n = 4), but Martin et al. (62) tended to obtain recoveries (85% ± 9%, n = 4) that were both lower and more variable than those reported by other workers (e.g., references 20, 34, 87, 100, and 105). An important shortcoming limiting the usefulness of the protocols proposed by Martin et al. (62) and Ruzicka et al. (99) is that they measure only free ergosterol—unless Ruzicka et al. (99) are correct in their hypothesis that ultrasonication can break up ergosteryl compounds. Using microwave radiation, Young (121) has developed another rapid extraction method that includes saponification; an ordinary microwave oven can be used, samples can be as small as

2.5 ml of liquid volume, and extraction, including sample-cooling time, takes only about 15 min.

SPE

Two alternative approaches have been proposed to extract ergosterol from solid samples. Gessner and Schmitt (38) used solid-phase extraction (SPE) to clean up lipid extracts before HPLC analysis. The saponified extracts are acidified and then passed over a reverse-phase extraction column (C_{18}, Sep-Pak Vac). Ergosterol is subsequently eluted with isopropanol and directly injected into an HPLC system for final purification and quantitation. SPE thus replaces the time-consuming and, because large volumes of highly flammable and toxic solvent are used, potentially hazardous partitioning step into nonpolar solvent. With manual operation, sample throughput is significantly greater than for conventional liquid-liquid extraction, and as SPE lends itself to automation, throughput could be increased further. However, the method is not as straightforward as it appears, given that ergosterol tends to be degraded in acidified extracts (38; see also reference 73). With different types of litter from freshwater systems, losses could be prevented by using small sample sizes (<50 mg) and ensuring a strongly alkaline pH in final extracts, but more work is needed before a conclusive assessment of SPE as a generally useful procedure to determine ergosterol in environmental samples is possible.

SFE

Another method, which takes advantage of supercritical fluid extraction (SFE), has even greater potential to simplify and speed up ergosterol analysis (121, 122), especially if extraction and chromatographic isolation are carried out on-line. In the method developed by Young and Games (122), a solid sample is placed in the extraction vessel of an SFE system and extracted with supercritical carbon dioxide under controlled conditions of temperature, fluid density, and flow rate. The extracted analytes including ergosterol are trapped on a short reverse-phase column, eluted, and subjected to final chromatographic separation. The main advantages of the method are the all-in-one extraction, prepurification, and, potentially, final separation and quantitation of the analyte. In addition, the method largely avoids organic solvents and is rapid and minimally labor-intensive (122). Extraction efficiencies and reproducibility are nearly as good as with other procedures. Disadvantages include the fact that only free, not esterified, ergosterol is analyzed and the need for special equipment not readily available in most laboratories at present.

Sample Preparation for Chromatography

The volume of the prepurified extract needs to be reduced before injection into the HPLC system. This is commonly achieved by evaporation in a fume hood under a stream of nitrogen and gentle heating (34, 67, 105). Although ergosterol has a general tendency to be oxidized, evaporation can be carried out under a stream of air, rather than nitrogen, without taking the risk of ergosterol losses (33a, 77). Safety is nevertheless increased when highly flammable solvents such as pentane and hexane (flash points at −49 and −23°C) are evaporated under nitrogen atmosphere.

The dried-down lipids are reconstituted in a known volume (usually 0.5 to 2 ml) of suitable solvent. Dichloromethane (methylene chloride) has generally been used for this purpose (105, 106) but has not always been found to be appropriate (123). Dichloromethane is also quite toxic. It

may be replaced by methanol (77), although the solubility of ergosterol is much poorer in this solvent. Isopropanol (29, 38) or isopropanol-acetone (3:1 [vol/vol] [5]) is probably a good compromise, and ethyl acetate may give satisfactory results as well. When a highly fluid solvent such as dichloromethane or ethyl acetate is used, pipetting with a microsyringe will avoid gross errors in the final sample volume.

Various procedures, both before and after solvent evaporation, have been used to purify extracts prior to HPLC analysis. In general, it has been concluded that these purification steps such as washing, neutralization, and drying (5, 34, 63), passage over columns packed with various sorbent types (17, 34, 106), or thin-layer chromatography (63, 89) are not necessary (14, 34, 63, 104). Extensive cleanup might even result in severe losses of ergosterol (73), as suggested, for example, by the exceedingly low concentrations in fungal mycelium analyzed by Osswald et al. (89). Membrane filtration (5, 77) or centrifugation (34) of the final concentrated extract is nevertheless indicated when particles are present.

Chromatography

Among the techniques that have been used for final purification and quantitation of ergosterol (120) in environmental samples, HPLC is the most widely used and recommended technique. Spectrophotometric quantitation without prior chromatographic separation may be sufficient when pure mycelium of some (but not all) fungi is analyzed, but it suffers from strong interference by nontarget compounds when ergosterol is analyzed in complex matrices (63) such as plant litter. Thin-layer chromatography may be useful for screening purposes (102) and when coupled with UV spectroscopy might even allow ergosterol quantitation (70). Gas-liquid chromatography (GLC) is the traditional method used to analyze sterols. It probably produces satisfactory results also with environmental ergosterol samples (29, 30, 48, 49), although separation from plant sterols may not always be simple (63). Advantages of GLC over HPLC are the lower detection limit and the fact that the gas chromatograph can be readily coupled to a mass spectrometer (29, 121), generally allowing unambiguous peak identification. However, identification by mass spectrometry is not routinely necessary because of the characteristic UV absorbance spectrum of ergosterol between 240 and 300 nm (Fig. 2) and the lack of strong absorbance of other major plant and animal sterols above 240 nm. Therefore, unlike other sterols, ergosterol is readily detected with standard UV absorbance detectors of HPLC systems at a wavelength of 282 nm. When diode array technique is used (104), additional information on peak identity and purity can be obtained, which may even surpass information gained with gas chromatography-mass spectrometry (120). The particular advantages of HPLC over GLC are the greater robustness of both the instrument and separation methods and the fact that derivatization of ergosterol is not needed with HPLC, allowing nondestructive analysis.

The intermediate polarity of ergosterol allows its separation from matrix lipids by both normal-phase and reverse-phase HPLC. The isocratic separation procedures are straightforward. A selection of published protocols is listed in Table 1. With plant litter from aquatic environments, good resolution of ergosterol is obtained under the conditions of the first two entries in Table 1, but other methods may be equally adequate and sometimes required to achieve sufficient separation (1, 104, 121). However, if a reverse-phase method with methanol as the mobile phase gives satis-

TABLE 1 Selection of published methods for the separation of ergosterol from matrix lipids with HPLC

Method	Material	Column type[a]	Column dimensions (length [cm] by diam [mm])	Mobile phase	Flow rate (ml/min)	Temp (°C)	Injection vol (μl)	Retention time (min)	Reference(s)
Reverse-phase chromatography	Salt marsh litter	Pierce RP18	22 by 4.6	Methanol	2	Ambient	50	3.6	77
	Leaf litter from stream	Lichrospher RP18	25 by 4.6	Methanol	1.5	28	10	9.0	34
	Soybeans	10 ODS Spher-column	25 by 4.6	Methanol-water (92:8)	1.5	40	20	12	87
	Wood	Nova-pak C_{18}	15 by 3.9	Methanol-sulfuric acid (2.5 mM; 98:2)	2.2	Ambient		8.3	32
	Ectomycorrhizae	Nova-pak C_{18}	15 by 3.9	KH_2PO_4 (10 mM; pH 2.75) in methanol	1.4	Ambient		10	100
	Activated sludge	Ultrasphere C_{18}	15 by 4.6	Acetonitrile-isopropanol (97:3)	2.0	Ambient	200	8.4	33
	Air filters	Ultramex 3 Phenyl	15 by 4.6	Acetonitrile-water (55:45)	0.7	Ambient		15	121, 121a
Normal-phase chromatography	Cereals, feeds	LiChrosorb Si 60	25 by 4.6	n-Hexane–isoamyl alcohol (95:5)	1.5	Ambient	10–100	3.9	104
	Cereals	μ-Porasil	30 by 4.6	Dichloromethane-isopropanol (99:1)	1.67	Ambient	10–25	5.8	105
	Fungal culture	LiChrosorb Si 60	12.5 by 4.6	n-Hexane–isopropanol (97:3)	1	Ambient		4	124

[a] For details of packing characteristics, see pertinent catalogs of manufacturers or dealers.

factory results, there is no need to resort to other solvents such as acetonitrile, or various mixtures thereof (31, 33, 44), or normal-phase chromatography involving potentially hazardous and/or highly toxic mobile phases such as dichloromethane and hexane (104, 105, 124). Therefore, normal-phase HPLC is recommended only in exceptional cases. An interesting alternative to conventional HPLC is supercritical fluid chromatography (122).

Conversion Factors

Newell et al. (71, 77) made the case that conversion factors relating ergosterol concentrations to biomass need to be established under conditions identical to or at least closely simulating those prevailing in nature. The fundamental dilemma in this regard is that fungi grow within solid substrates, and thus a straightforward reference method to calibrate the ergosterol assay is not readily available. The closest one can get is to analyze extramatrical mycelia of fungi growing for the most part inside their natural substrate. When this approach has been taken, ergosterol concentrations were found, at least in eight strains of three fungal species examined, not to differ significantly from those in mycelium grown in more artificial situations (75), suggesting that data derived from less natural growth conditions are valuable. If so, one can plumb the range of possible variability in mycelial ergosterol contents by subjecting fungi to various (artificial) growth conditions (3, 4, 35) and thus assess the magnitude of potential error when making projections to field samples (35).

Ergosterol concentrations in mycelium of fungi isolated

from aquatic environments and grown on agar plates or in liquid culture have been found to range generally from 2 to 16 mg/g of mycelial dry mass (35, 36, 91, 110, 111, 115) or ash free dry mass (75, 84). A similar range has been found in ecological studies with terrestrial ergosterol-containing fungi (2, 49, 62, 71, 115). Some published values are considerably lower (<1 mg/g of mycelial mass) (13, 14, 29, 89, 115, 124), but in many instances application of these values to field samples is clearly inappropriate since it would result in unrealistically or even impossibly high estimates of fungal biomass (2). For example, in both decomposing leaf litter and ectomycorrhizae, ergosterol contents can exceed 1 mg/g of the total system mass (2, 36, 48, 78, 110). In two cases, mycelial ergosterol concentrations were found to be much higher than 16 mg/g (84, 87). Both estimates are probably inflated because they were made by reference to data obtained by means of hyphal length measurements, a method which is known to result in severe underestimates of mycelial mass in solid substrates (71).

Ergosterol concentrations vary among different fungal species, among isolates of the same species, and even within a strain depending on its physiological state. A number of factors, including age (14, 22, 32, 87; but see references 35 and 71), growth rate, carbon and nutrient availability (3, 4, 22, 49; but see reference 35), temperature (23), and oxygen (17, 21, 87), have been found to affect mycelial ergosterol contents. Newell (71) has discussed this issue in some detail. Generally speaking, extents of variability in mycelial ergosterol contents are smaller than those reported for alternative index molecules and other methods to determine fungal biomass (71). Temperature was generally found not to have a pronounced effect (e.g., references 70a and 87). In contrast, oxygen exerts a strong influence on ergosterol contents in fungi. This is readily explicable by the fact that the synthesis of ergosterol in fungi requires molecular oxygen (57, 64, 65, 115, 116). The oxygen conditions of the habitat in which fungi grow may therefore be expected to influence mycelial ergosterol contents (21, 87); therefore, caution must be used when work is carried out in microaerobic environments (58, 61, 90). Also, there is recent evidence that the cultural maintenance history of a fungal strain can sometimes be a factor leading to unnatural levels of ergosterol content of mycelia (75). If possible, strains used in finding conversion factors should therefore be maintained on the natural substrate and be as newly isolated as practicable.

In studies in which a range of fungal isolates have been analyzed, the average (mean and/or median) mycelial ergosterol concentration was found to be close to 5 mg/g (2, 35, 62, 66, 75, 91), suggesting that a conversion factor derived from this concentration allows rough estimates of fungal biomass in diverse natural systems. Only Bermingham et al. (13) found lower concentrations, but one suspects that incomplete determinations of ergosterol were involved, for the ergosterol contents reported in that study were up to 20-fold lower than literature values (35) for particular species (e.g., *Alatospora acuminata* Ingold). If interspecies differences are a major source of total variability in mycelial ergosterol concentrations, the combination of ergosterol measurements, species-specific conversion factors, and determinations of fungal community structure may enhance the accuracy of fungal mass estimates (35). This is possible in systems in which fungal assemblages are strongly dominated by a single species (48, 73, 81, 90, 110) or the more complex species composition can be determined (35, 36). However, given the possibility of significant intraspecies variability (35, 62, 84, 91; but see reference 75) and fungal responses to environmental conditions in terms of mycelial ergosterol contents (3, 4, 35, 49, 84, 87), there are limits to this approach.

Alternative Approaches to Fungal Biomass Determinations

Hyphal Length Measurement

A number of other methods have been proposed to determine the amount of fungal biomass in aquatic environments. Frankland (27), Frankland et al. (28), Newell (71), Parkinson (92), and Parkinson and Coleman (93) provide detailed reviews of these methods as applied to both soils and plant litter systems. Hyphal length measurements, conversion to hyphal biovolumes, and further conversion into biomass constitute the traditional method for determining fungal biomass in soils. That the method is tedious cannot be a primary argument against its use (93). However, it also suffers from uncertainties associated with the use of appropriate conversion factors at several levels, thus multiplying potential errors, and is likely to result in severe underestimates when applied to fungi growing in solid substrates such as plant litter (71, 81). Unless perhaps mycelium is stained, the latter effect would appear to be especially pronounced with the hyaline aquatic hyphomycetes dominating leaf-associated fungal assemblages in streams.

Hexosamine Assay

The rationale underlying the hexosamine assay is similar to that of the ergosterol assay in that it aims at quantifying a particular cell constituent, chitin in this case, which is characteristic of the eumycotic fungi (71). As in the ergosterol assay, most oomycetes are not included in this assay but chytrids are. Chitin, unlike ergosterol, is a recalcitrant cell wall component and therefore not representative only of living fungal hyphae. In addition, the assay is less sensitive, the analytical procedures are more complicated and time-consuming, and conversion factors are more variable than for the ergosterol method (71). Nevertheless, in combination with the ergosterol assay, it may be useful both as a cross-check and to get an idea of the ratio between living and dead fungal mass (71, 85).

Immunoassay

Quantitative immunological techniques (enzyme-linked immunosorbent assay and radioimmunoassay) are based on the specific recognition of cell constituents (antigens) by antibodies raised against them. If the target molecules are cell wall components, immunological techniques suffer from the same basic handicap as the hexosamine assay when living fungal mass is to be determined. The specific advantage of the approach lies in the potential to distinguish between major taxonomic groups, species, or even strains, especially when monoclonal antibodies can be generated (28). On the other hand, a number of drawbacks have been identified (71). Frankland et al. (28) and Newell (71) discuss the advantages and disadvantages of immunoassays for fungi in an ecological context and provide access to the pertinent literature. To date, only Fallon and Newell (26) and Bermingham et al. (12) have developed quantitative immunoassays for aquatic plant litter systems.

ATP

Another index molecule that has been widely used as a general measure of microbial mass is ATP (50). In terms of selectivity, the assay ranges, relative to the immunological techniques, at the other end of the spectrum of methods to determine fungal biomass, since ATP occurs in all living cells. Despite this major limitation, it can be useful as an indicator of fungal mass in systems in which only the target organisms are known to contribute significantly to the ATP pool in both laboratory and field studies (42, 111). The problem of the appropriate conversion factors is at least as severe as with the ergosterol and hexosamine assays (50).

Prospects

A range of other approaches to measure fungal mass in solid substrates can be envisioned. Some of these have been briefly presented by Newell (71, 74) and White et al. (chapter 10, this volume), but so far none has been applied to aquatic eukaryotic osmotrophs. Both biochemical approaches and DNA and RNA technological approaches (chapter 11, this volume) hold promise in this respect. With the development of new techniques, it may eventually be possible to quantify not only the biomass of ergosterol-containing eumycotic fungi but also that of chytrids and oomycetous protists (71), both of which might be important components in some aquatic ecosystems (41).

FUNGAL GROWTH AND PRODUCTIVITY IN PLANT LITTER

The Acetate-to-Ergosterol Method

Background Rationale

Measurements of fungal biomass largely ignore the dynamics of fungal growth and are hence insufficient to provide answers to ecological questions at various levels of investigation. This recognition led Newell and Fallon (79) to devise the acetate-to-ergosterol method, which allows the direct determination of virtually instantaneous fungal growth rate and productivity in situ. The acetate-to-ergosterol method is an extension of the ergosterol technique described above in that it measures the rate of ergosterol synthesis, which is then related to fungal growth and productivity by means of appropriate conversion factors. The basic rationale underlying the technique is thus similar to that of tracer techniques developed to determine bacterial (68) and microalgal (118) productivity (see chapters 26 and 27, this volume). However, in contrast to those methods, the acetate-to-ergosterol method achieves specificity by isolating the radiolabeled target molecule (i.e., ergosterol) rather than by offering a precursor which is specifically metabolized only by the target organism or by offering the precursor at a concentration not efficiently exploitable by nontarget organisms.

Principle

A detailed protocol of how to determine fungal growth and productivity with the acetate-to-ergosterol method has been described by Newell (73). Briefly, a sample such as pieces of decomposing litter colonized by fungi is placed in a solution containing radiolabeled acetate of a known concentration and specific activity and then incubated for a defined period. The incubation is stopped by rapid rinsing and preservation (111a) or addition of a killing agent such as formaldehyde (37, 79), and the sample is then carried through the ergosterol extraction procedure described above. The ergosterol peak eluting from the HPLC column is collected, and radioactivity is counted. Thus, all points relating to sampling, sample storage, and processing discussed above in connection with the ergosterol assay apply equally to the acetate-to-ergosterol method. Additional, specific points are addressed below.

Choice of Radiolabel

Both tritiated and ^{14}C-labeled acetate can be used as the radiolabeled precursor for ergosterol synthesis, but ^{14}C is probably superior because a number of hydrogen atoms are lost and introduced during sterol synthesis (79). To date only [^{14}C]acetate labeled at C-1 has been used. However, because of the decarboxylation of mevalonic acid during sterol biosynthesis (57, 64, 65, 115, 116), labeling with [1-^{14}C]acetate is theoretically 20% less efficient than labeling with [2-^{14}C]acetate (Fig. 1). Note also that in the equation given by Newell (73) for calculation of specific rate of ergosterol-carbon synthesis, the specific activity of the precursor acetate solution should be adjusted to *effective* specific activity (i.e., ×0.89 if [1-^{14}C] acetate is used) so as to account for the facts that (i) one of the 28 carbon atoms in ergosterol does not directly emanate from acetate and (ii) only 12 atoms are derived from the acetate carboxyl group (C-1). The usefulness of precursors more specific to ergosterol synthesis than acetate (e.g., mevalonate, which is commercially available) has not been tested in ecological applications.

Like animal and plant sterols, ergosterol is synthesized almost entirely from acetate units via the mevalonate pathway (57, 64, 65, 115, 116). Important intermediates are hydroxymethylglutarate, mevalonate, squalene, and lanosterol. Subsequent reactions leading to ergosterol are numerous and variable, but all pathways lead to the incorporation of 12 carbon atoms from C-1 of the acetate molecule and 15 carbon atoms originating from C-2 (Fig. 1; reference 107). Only one carbon is introduced by methylation of the

C-24 position in the sterol side chain and hence is not immediately derived from acetate. Thus, the efficiency of labeling ergosterol with [^{14}C]acetate is particularly high.

Experimental Conditions during Incubation with Radiolabel

Radiotracer techniques to measure in situ growth and productivity of microorganisms are based on the premise that the metabolism of the examined organism remains unchanged after sampling and during incubation with the added radiolabeled precursor. Otherwise, measured synthesis rates would obviously not reflect the rates actually occurring in nature. Although this prerequisite can never be perfectly met, attempts must be made to (i) handle samples as little as possible in order to minimize stress on the examined organisms, (ii) start incubations quickly after sampling, (iii) during incubations simulate natural conditions as closely as possible (note, for example, that some aquatic fungi may be stimulated by aeration and/or agitation [111a], whereas others may be hampered by agitation [73]), and (iv) keep incubation times short in order to prevent fungal adaptation to the experimental situation. Ideally, incubations are carried out in sealed containers in the field immediately after sampling (e.g. reference 101).

Length of Incubation Period

Newell (73) suggested incubating litter samples for 1 h with radiolabel, but that time may be too short to achieve measurable labeling of fungi which grow, for example, in submerged leaf litter in streams at temperatures ≤10°C (33a). Under such conditions, incubation times of 3 to 5 h are probably adequate given that rates of incorporation of acetate into ergosterol have been found to be linear for at least 4 (111a) to 6 (37) h. Note also that the incubation time for measuring productivity of autotrophic eukaryotes (phytoplankton) is commonly 4 h (118). However, because the acetate-to-ergosterol technique has not yet been widely tested, investigators should carry out preliminary time course experiments to determine if incorporation rates are constant over the projected incubation period.

Precursor Concentration

Incubations need to be carried out at a concentration of added acetate high enough to preclude notable fungal synthesis of ergosterol at the expense of other carbon sources. For both marine (73, 79) and freshwater (37, 111a) litter fungi, an added acetate concentration of 5 mM has been found sufficient to maximize rates of incorporation into ergosterol, but again it is recommended that investigators check in preliminary trials whether this concentration is also appropriate in their specific study system.

Isotope Dilution

Even at saturation levels of added precursor, some ergosterol synthesis can occur at the expense of other (unlabeled) extra- or intracellular carbon pools, a phenomenon referred to as isotope dilution (37, 68, 79), resulting in an underestimation of true synthesis rate from apparent incorporation rates. The extent of isotope dilution can be determined by incubating replicate samples with constant amounts of radioactivity but varying precursor concentrations and subsequently plotting the added precursor concentration versus the reciprocal of the incorporated radioactivity (37, 68, 79). Isotope dilution corresponds to the negative value of the concentration that is equivalent to the intercept between the regression line through the plotted datum points and the abscissa. If isotope dilution is high and variable, a complete isotope dilution experiment needs to be conducted for each measurement of growth rate and productivity.

Extraction and Chromatography

Extraction of labeled ergosterol and chromatography are carried out as described for the ergosterol assay. However, in contrast to that assay, complete separation of ergosterol from matrix lipids not absorbing at 282 nm is critical, because any coeluting compounds may interfere with radioactivity counting. Although this may not be a great problem when the coeluting compounds are of fungal origin and chromatographic separation conditions are always the same, the safe and proper solution consists of complete purification of ergosterol. In this respect, a diode array detector can be useful to check the identity and purity of the collected ergosterol fraction. If satisfactory chromatographic resolution cannot be achieved, it may be beneficial to collect only the central portion of the ergosterol fraction and correct scintillation counts mathematically for total peak area (79).

Scintillation Counting

Scintillation counting of the radiolabeled ergosterol fraction is straightforward (19). Background radioactivity and counts of formaldehyde-killed controls are commonly low (<1 Bq or 60 dpm) (79), and counting efficiencies are generally near 90% (33a, 70a). As a rule of thumb, the final specific activity of the precursor and the final volume of the sample extract should be chosen such that expected scintillation counts of samples are at least 10-fold greater than control values. When HPLC separation is achieved with methanol as the mobile phase, solvent evaporation after collection of the ergosterol fraction is not necessary, but care must be taken to add enough fluor for scintillation counting. The tedious and potentially hazardous collection of the ergosterol fraction after chromatographic separation can be circumvented by counting radioactivity with an on-line detector (95); however, on-line radioactivity counting in HPLC systems is less sensitive than standard liquid scintillation counting.

Conversion Factors

There are two possible approaches to obtain conversion factors relating acetate incorporation rates to fungal growth and production. The first, theoretical approach takes advantage of existing biochemical knowledge about ergosterol biosynthesis in fungi, i.e., the fact that the origin of carbon atoms in ergosterol is precisely known (Fig. 1; reference 107). As a consequence, the amount of radiolabel incorporated by a fungus can be directly converted to the amount of ergosterol produced, and if the proportion of ergosterol in fungal mass is known (see references 35, 71, 75, and 91 and others cited above), the latter can be readily calculated. Corrections may be made for isotope dilution (see above) and for isotope discrimination, i.e., the potential disproportionate incorporation of heavy carbon isotopes such as ^{14}C.

The alternative, empirical approach consists of measuring acetate incorporation rates at one or several time points during early exponential growth when losses of fungal mass can be assumed to be insignificant and then relating the incorporation rate to the change in mycelial biomass over a time interval encompassing the time points where acetate incorporation is measured (see chapter 27, this volume). Fungal mass can be expressed in terms of organic mass (or a similar parameter such as dry mass or carbon) (79) or indirectly in terms of ergosterol (110). Expression in terms of organic mass permits the establishment of a direct relationship between acetate incorporation rates and biomass production but requires harvest of pure mycelium, which usually involves culture in unnatural conditions. The specific advantage of relating incorporation rates to ergosterol (rather than to organic mass) lies in the possibility of designing experimental conditions that mimic natural growth within solid substrates. Thus, if the main objective of the study is to determine fungal growth rates, this latter approach is probably preferable, because information on the amount of the fungal mass present and produced is not necessary. If, on the other hand, fungal production is the main parameter to be measured, a second, independently determined conversion factor is required to calculate fungal mass from ergosterol, and so the direct conversion of acetate incorporation rates to fungal biomass production may be more advantageous. A comparison of empirical and theoretical conversion factors reveals that they differ by less than 20% to about sixfold (Table 2).

Problems and Outlook

At present, the acetate-to-ergosterol method has been used in only a rather limited number of studies (37, 79, 85, 101,

TABLE 2 Conversion factors relating rates of [1-^{14}C]acetate incorporation to fungal biomass production

Culture system	Fungal species	Determination method	Conversion factor (μg of fungal biomass/nmol of acetate)	Reference
Extract of smooth cordgrass leaves	Phaeosphaeria spartinicola	Empirical, organic mass	9.3 ± 1.4[a]	79
		Theoretical[b]	7.0	
	Passeriniella obiones	Empirical, organic mass	7.0 ± 2.4[a]	79
		Theoretical[c]	5.4	
Yellow poplar leaves in stream microcosm	Anguillospora filiformis	Empirical, ergosterol	19.2	111a
		Theoeretical[d]	3.0	
Ash leaves in stream microcosm	Articulospora tetracladia	Empirical, ergosterol[e]	5.5	37
		Theoretical[e]	6.6	37

[a] Mean ± standard deviation of three strains.
[b] Assuming a mycelial ergosterol concentration of 4.7 mg/g of organic mass; average of two recently isolated strains (75).
[c] Assuming a mycelial ergosterol concentration of 6.1 mg/g of organic mass (75).
[d] Assuming a mycelial ergosterol concentration of 10.9 mg/g of dry mass (111).
[e] Assuming a mycelial ergosterol concentration of 5.0 mg/g of dry mass (35, 37).

110). When tested, basic requirements necessary for a meaningful interpretation of ergosterol synthesis rates in terms of fungal growth and productivity have been consistently found to be met in these studies: (i) significant labeling of the target organisms, (ii) no incorporation by nontarget organisms (bacteria), (iii) negligible impact of added acetate on fungal growth, (iv) constant incorporation rates over typical periods of incubation with radiolabel, and (v) saturation of incorporation in the lower millimolar range (37, 79, 111a). Furthermore, the growth rates calculated from acetate incorporation rates were generally found to be plausible, suggesting that the method has great potential to fill the current gap in knowledge on in situ fungal growth and productivity in aquatic environments. Nevertheless, considerably more methodological work is necessary before the general usefulness of the method can be evaluated. Possible problem areas that need to be further examined include impaired diffusion of precursor inside the solid matrix of the fungal substrate, inadequate equilibration of radiolabel with internal and external acetate pools (37), rapid response of the target organisms to the (potentially artificial) experimental situation (80), dependence of mycelial ergosterol contents on fungal growth rate (2–4), and reliability of conversion factors.

Alternative Methods for Measuring Fungal Productivity

Differences in Fungal Biomass

There is currently no true alternative available to the acetate-to-ergosterol method for measuring virtually instantaneous in situ fungal growth and productivity. Nevertheless, in studies in which losses in fungal mass can be assumed to be negligible over the time period of interest, growth rate and productivity can be inferred from measured differences in biomass (71). With fungi, this approach is probably less prone to error than with many bacteria, because eukaryotes generally exhibit a slower turnover and because filamentous growth within solid substrates reduces physical losses and provides protection from predation. Also, the output of fungal mycelium, to mycophagous arthropods, for example, can be measured by use of fluorochrome-tagged hyphae (28; for a protozoan analog, see chapter 32, this volume).

Examples of Other Potential Approaches

Other potentially useful but currently untested tracer techniques to measure fungal growth and productivity would involve the measurement of glucosamine incorporation into chitin or of mannose incorporation into the fungal cell wall component mannoprotein (see reference 85). Likewise, determination of the synthesis rate of nucleic acid or protein sequences specific to fungi and directly related to growth may prove valuable in this regard.

CONCLUSION

In recent years, significant strides have been made in the development of bulk methods to measure the biomass and productivity of eukaryotic organoosmotrophic microbes. The ergosterol assay discussed in this chapter appears to be the best method currently available to estimate the biomass of eumycotic fungal assemblages associated with plant litter and other solid substrates. We make this statement well knowing that a number of limitations continue to exist, the most important being the uncertainty associated with the choice of appropriate conversion factors from ergosterol to

biomass or specific cell-constituents such as protein, nitrogen, and phosphorus. Future work will possibly reveal more accurate conversion factors established under (semi)natural conditions, provide insights into the physiological reasons causing variability in mycelial ergosterol contents, and hence give us a better feel for the error likelihood to be expected when conversion factors are applied to ergosterol data from field samples. Notwithstanding these probable advances in the future, we must acknowledge that eukaryotic osmotrophs exhibit considerable plasticity, limiting the chance of success for making exact measurements of mycelial mass with current approaches.

The acetate-to-ergosterol technique described in this chapter is currently the only available method to determine virtually instantaneous growth rates and productivity of eumycotic fungi in their natural environment. Several methodological studies with litter-associated fungi from both marine and freshwater environments have shown that the technique produces plausible results, but because of its novelty, conclusions as to the validity of the method remain preliminary at present.

As regards the purely analytical aspect of ergosterol measurements, current schemes for extraction, purification, and quantitation are basically sound. However, a number of pitfalls need to be avoided. Problems potentially associated with the acetate-to-ergosterol method, in particular, may not even have been identified to date. Moreover, there is considerable potential for further refinement, simplification, and acceleration of analytical procedures. New technologies such as microwave-assisted extraction, SPE, SFE, and supercritical fluid chromatography may prove valuable in this respect.

The future should bring a critical assessment of the validity and an improvement and fine-tuning of existing bulk methods such as those discussed in this chapter. The concurrent development of alternative methods indubitably will prove advantageous as well. Perhaps more importantly, however, methods for the analysis of eukaryotic organoosmotrophs other than the eumycotic, ergosterol-containing fungi need to be developed (74), and eventually attempts must be made to devise group-selective and even species- and strain-specific quantitative assays (12, 26). Techniques borrowed from other fields of biology and biochemistry hold great potential in this respect (see chapter 11, this volume), but the elaboration or adaptation of such techniques for purposes of environmental mycology will not be a quick and easy endeavour. The required effort appears nevertheless to be justified in view of the importance of eukaryotic organoosmotrophs in aquatic ecosystems (40, 72), which hitherto has been largely ignored.

We thank J. Christopher Young for sharing unpublished data and ideas.

REFERENCES

1. **Anderson, P., C. M. Davidson, D. Littlejohn, A. M. Ure, C. A. Shand, and M. V. Cheshire.** 1994. Extraction of ergosterol from peaty soils and determination by high performance liquid chromatography. *Talanta* **41:** 711–720.
2. **Antibus, R. K., and R. L. Sinsabaugh.** 1993. The extraction and quantification of ergosterol from ectomycorrhizal fungi and roots. *Mycorrhiza* **3:**137–144.
3. **Arnezeder, C., and W. A. Hampel.** 1990. Influence of growth rate on the accumulation of ergosterol in yeast-cells. *Biotechnol. Lett.* **12:**277–282.

4. **Arnezeder, C., and W. A. Hampel.** 1991. Influence of growth rate on the accumulation of ergosterol in yeast-cells in a phosphate limited continuous culture. *Biotechnol. Lett.* **13:**97–100.

5. **Arnezeder, C., W. Koliander, and W. A. Hampel.** 1989. Rapid determination of ergosterol in yeast cells. *Anal. Chim. Acta* **225:**129–136.

6. **Bååth, E.** 1988. A critical examination of the soil washing technique with special reference to the effect of the size of the soil particles. *Can. J. Bot.* **66:**1566–1569.

7. **Baldy, V., M. O. Gessner, and E. Chauvet.** 1995. Bacteria, fungi and the breakdown of leaf litter in a large river. *Oikos* **74:**93–102.

8. **Bärlocher, F.** 1992. Effects of drying and freezing autumn leaves on leaching and colonization by aquatic hyphomycetes. *Freshwater Biol.* **28:**1–7.

9. **Bärlocher, F. (ed.).** 1992. *The Ecology of Aquatic Hyphomycetes, Ecological Studies,* vol. 94. Springer-Verlag, Berlin.

10. **Bärlocher, F., and N. R. Biddiscombe.** Geratology and decomposition of *Typha latifolia* and *Lythrum salicaria. Arch. Hydrobiol.,* in press.

11. **Barr, D. J. S.** 1992. Evolution and kingdoms of organisms from the perspective of a mycologist. *Mycologia* **84:**1–11.

12. **Bermingham, S., F. M. Dewey, and L. Maltby.** 1995. Development of a monoclonal antibody-based immunoassay for the detection and quantification of *Anguillospora longissima* colonizing leaf material. *Appl. Environ. Microbiol.* **61:**2606–2613.

13. **Bermingham, S., L. Maltby, and R. C. Cooke.** 1995. A critical assessment of the validity of ergosterol as an indicator of fungal biomass. *Mycol. Res.* **99:**479–484.

14. **Bindler, G. N., J. J. Piadé, and D. Schulthess.** 1988. Evaluation of selected steroids as chemical markers of past and presently occurring fungal infections on tobacco. *Beitr. Tabakforsch. Int.* **14:**127–134.

15. **Boulton, A. J., and P. I. Boon.** 1991. A review of methodology used to measure leaf litter decomposition in lotic environments: time to turn over an old leaf? *Aust. J. Mar. Freshwater Res.* **42:**1–43.

16. **Cahagnier, B., D. Melcion, and D. Richard-Molard.** 1995. Growth of *Fusarium moniliforme* and its biosynthesis of fumonisin B1 on maize grain as a function of different water activities. *Lett. Appl. Microbiol.* **20:**247–251.

17. **Cahagnier, B., D. Richard-Molard, J. Poisson, and C. Desserme.** 1983. Evolution de la teneur en ergostérol des grains au cours de la conservation. Une possibilité d'évaluation quantitative et rapide de leur mycoflore. *Sci. Aliment.* **3:**219–244.

18. **Chergui, H., and E. Pattee.** 1993. Fungal and invertebrate colonization of *Salix* fresh and dry leaves in a Moroccan river system. *Arch. Hydrobiol.* **127:**57–72.

19. **Coleman, D. C., and B. Fry (ed.).** 1991. *Carbon Isotope Techniques.* Academic Press, New York.

20. **Davis, M. W., and R. T. Lamar.** 1992. Evaluation of methods to extract ergosterol for quantitation of soil fungal biomass. *Soil Biol. Biochem.* **24:**189–198.

21. **Delaveau, J., G. Lipus, and M. Moll.** 1983. Importance des stérols au cours du cycle de la levure de brasserie (*Saccharomyces uvarum*). *Bios* **14:**37–40.

22. **Desgranges, C., C. Vergoignan, M. Georges, and A. Durand.** 1991. Biomass estimation in solid state fermentation. I. Manual biochemical methods. *Appl. Microbiol. Biotechnol.* **35:**200–205.

23. **Dexter, Y., and R. C. Cooke.** 1984. Fatty acids, sterols and carotenoids of the psychrophile *Mucor strictus* and some mesophilic *Mucor* species. *Trans. Br. Mycol. Soc.* **83:**455–461.

24. **Dick, M. W.** 1992. Patterns of phenology in populations of zoosporic fungi, p. 355–382. *In* G. C. Carroll and D. T. Wicklow (ed.), *The Fungal Community. Its Organization and Role in the Ecosystem,* 2nd ed. Marcel Dekker, New York.

25. **Durrieu, G.** 1995. Quel statut pour les champignons? *Cryptogam. Mycol.* **16:**27–36.

26. **Fallon, R. D., and S. Y. Newell.** 1989. Use of ELISA for fungal biomass measurement in standing-dead *Spartina alterniflora* Loisel. *J. Microbiol. Methods* **9:**239–252.

27. **Frankland, J. C.** 1990. Ecological methods of observing and quantifying soil fungi. *Trans. Mycol. Soc. Jpn.* **31:**89–101.

28. **Frankland, J. C., J. Dighton, and L. Boddy.** 1990. Methods for studying fungi in soil and forest litter. *Methods Microbiol.* **22:**343–403.

29. **Frey, B., H. R. Buser, and H. Schüepp.** 1992. Identification of ergosterol in vesicular-arbuscular mycorrhizae. *Biol. Fert. Soils* **13:**229–234.

30. **Frey, B., A. Vilariño, H. Schüepp, and J. Arines.** 1994. Chitin and ergosterol content of extraradical and intraradical mycelium of the vesicular-arbuscular mycorrhizal fungus *Glomus intraradices. Soil Biol. Biochem.* **26:**711–717.

31. **Fritze, H., and E. Bååth.** 1993. Microfungal species composition and fungal biomass in a coniferous forest soil polluted by alkaline deposition. *Microb. Ecol.* **25:**83–92.

32. **Gao, Y., T. Chen, and C. Breuil.** 1993. Ergosterol—a measure of fungal growth in wood for staining and pitch control fungi. *Biotechnol. Tech.* **7:**621–626.

33. **Gardner, R. M., G. W. Tindall, S. M. Cline, and K. L. Brown.** 1993. Ergosterol determination in activated sludge and its application as a biochemical marker for monitoring fungal biomass. *J. Microbiol. Methods* **17:**49–60.

33a. **Gessner, M. O.** Unpublished data.

34. **Gessner, M. O., M. A. Bauchrowitz, and M. Escautier.** 1991. Extraction and quantification of ergosterol as a measure of fungal biomass in leaf litter. *Microb. Ecol.* **22:**285–291.

35. **Gessner, M. O., and E. Chauvet.** 1993. Ergosterol-to-biomass conversion factors for aquatic hyphomycetes. *Appl. Environ. Microbiol.* **59:**502–507.

36. **Gessner, M. O., and E. Chauvet.** 1994. Importance of stream microfungi in controlling breakdown rates of leaf litter. *Ecology* **75:**1807–1817.

37. **Gessner, M. O., and E. Chauvet.** Growth and productivity of aquatic hyphomycetes in decomposing leaf litter. Submitted for publication.

38. **Gessner, M. O., and A. L. Schmitt.** 1996. Use of solid-phase extraction to determine ergosterol concentrations in plant tissue colonized by fungi. *Appl. Environ. Microbiol.* **62:**415–419.

39. **Gessner, M. O., and J. Schwoerbel.** 1989. Leaching kinetics of fresh leaf-litter with implications for the current concept of leaf-processing in streams. *Arch. Hydrobiol.* **115:**81–90.

40. **Gessner, M. O., K. Suberkropp, and E. Chauvet.** Decomposition and biomass: freshwater and marine habitats. *In* D. T. Wicklow and B. Söderström (ed.), *The Mycota: a Comprehensive Treatise on Fungi as Experimental Systems for Basic and Applied Research,* vol. IV. *Environmental and Microbial Relationships,* in press. Springer-Verlag, Berlin.

41. **Gessner, M. O., M. Thomas, A.-M. Jean-Louis, and E. Chauvet.** 1993. Stable successional patterns of aquatic hyphomycetes on leaves decaying in a summer cool stream. *Mycol. Res.* **97:**163–172.

42. **Golladay, S. W., and R. L. Sinsabaugh.** 1991. Biofilm development on leaf and wood surfaces in a boreal river. *Freshwater Biol.* **25:**437–450.

43. **Grant, W. D., and A. W. West.** 1986. Measurement of ergosterol, diaminopimelic acid and glucosamine in soil: evaluation as indicators of microbial biomass. *J. Microbiol. Methods* **6:**47–53.

44. **Griffiths, H. M., D. G. Jones, and A. Akers.** 1985. A bioassay for predicting the resistance of wheat leaves to *Septoria nodorum. Ann. Appl. Biol.* **107:**293–300.

45. **Heaney, S. I., J. W. G. Lund, H. M. Canter, and K. Gray.** 1988. Population dynamics of *Ceratium* spp. in three English lakes, 1945–1985. *Hydrobiologia* **161:** 133–148.

46. **Heupel, R. C.** 1989. Isolation and primary characterization of sterols, p. 1–31. *In* W. D. Nes and E. J. Parish (ed.), *Analysis of Sterols and Other Biologically Significant Steroids.* Academic Press, San Diego, Calif.

47. **Hyde, K. D., and S. Y. Lee.** 1995. Ecology of mangrove fungi and their role in nutrient cycling: what gaps occur in our knowledge? *Hydrobiologia* **295:**107–118.

48. **Johnson, B. N., and W. B. McGill.** 1990. Comparison of ergosterol and chitin as quantitative estimates of mycorrhizal infection and *Pinus contorta* seedling response to inoculation. *Can. J. For. Res.* **20:**1125–1131.

49. **Johnson, B. N., and W. B. McGill.** 1990. Ontological and environmental influences on ergosterol content and activities of polyamine biosynthesis enzymes in *Hebeloma crustuliforme* mycelia. *Can. J. Microbiol.* **36:**682–689.

50. **Karl, D. M.** 1986. Determination of in situ microbial biomass, viability, metabolism, and growth, p. 85–176. *In* J. S. Poindexter and E. R. Leadbetter (ed.), *Bacteria in Nature,* vol. 2. Plenum Press, New York.

51. **Kirby, J. J. H.** 1987. A comparison of serial washing and surface sterilization. *Trans. Br. Mycol. Soc.* **88:**559–562.

52. **Kirby, J. J. H., J. Webster, and J. H. Baker.** 1990. A particle plating method for analysis of fungal community composition and structure. *Mycol. Res.* **94:**621–626.

53. **Kohlmeyer, J., and E. Kohlmeyer.** 1979. *Marine Mycology.* Academic Press, New York.

54. **Kohlmeyer, J., and B. Volkmann-Kohlmeyer.** 1991. Illustrated key to filamentous higher marine fungi. *Bot. Mar.* **34:**1–61.

55. **Kok, C. J., G. Van der Velde, and K. M. Landsbergen.** 1990. Production, nutrient dynamics and initial decomposition of floating leaves of *Nymphaea alba* L. and *Nuphar lutea* (L.) Sm. (Nymphaeaceae) in alkaline and acid waters. *Biogeochemistry* **11:**235–250.

56. **Kuehn, K., and K. Suberkropp.** 1994. The fluctuations in fungal activity associated with standing dead litter of the aquatic macrophyte *Juncus effusus. In Abstracts of the 5th International Mycological Congress.*

57. **Kuhn, P. J., A. P. J. Trinci, M. J. Jung, M. W. Goosey, and L. G. Copping (ed.).** 1990. *Biochemistry of Cell Walls and Membranes in Fungi.* Springer-Verlag, Berlin.

58. **Lee, C., R. W. Howarth, and B. L. Howes.** 1980. Sterols in decomposing *Spartina alterniflora* and the use of ergosterol in estimating the contribution of fungi to detrital nitrogen. *Limnol. Oceanogr.* **25:**290–303.

59. **Lichtwardt, R. W., M. J. Huss, and M. C. Williams.** 1994. Biogeographic studies on Trichomycete gut fungi in winter stonefly nymphs of the genus *Allocapnia. Mycologia* **85:**535–546.

60. **Maltby, L.** 1992. Heterotrophic microbes, p. 165–194. *In* P. Calow and G. E. Petts (ed.), *The Rivers Handbook. Hydrological and Ecological Principles,* vol. 1. Blackwell, London.

61. **Mansfield, S. D., and F. Bärlocher.** 1993. Seasonal variation of fungal biomass in the sediment of a salt marsh in New Brunswick. *Microb. Ecol.* **26:**37–45.

62. **Martin, F., C. Delaruelle, and J. L. Hilbert.** 1990. An improved ergosterol assay to estimate fungal biomass in ectomycorrhizas. *Mycol. Res.* **94:**1059–1064.

63. **Matcham, S. E., B. R. Jordan, and D. A. Wood.** 1985. Estimation of fungal biomass in a solid substrate by three independent methods. *Appl. Microbiol. Biotechnol.* **21:** 108–112.

64. **Mercer, E. I.** 1984. The biosynthesis of ergosterol. *Pestic. Sci.* **15:**133–155.

65. **Mercer, E. I.** 1991. Sterol biosynthesis inhibitors: their current status and modes of action. *Lipids* **26:**584–597.

66. **Miller, J. D., Y. E. Moharir, J. A. Findlay, and N. J. Whitney.** 1984. Marine fungi of the Bay of Fundy. VI. Growth and metabolites of *Leptosphaeria oraemaris, Sphaerulina oraemaris, Monodictys pelagica,* and *Dendryphiella salina. Proc. Nova Scotia Acad. Sci.* **34:**1–8.

67. **Miller, J. D., J. C. Young, and H. L. Trenholm.** 1983. *Fusarium* toxins in field corn. I. Time course of fungal growth and production of deoxynivalenol and other mycotoxins. *Can. J. Bot.* **61:**3080–3087.

68. **Moriarty, D. J. W.** 1990. Techniques for estimating bacterial growth rates and production of biomass in aquatic environments. *Methods Microbiol.* **22:**211–234.

69. **Müller, M. M., R. Kantola, and V. Kitunen.** 1994. Combining sterol and fatty acid profiles for the characterization of fungi. *Mycol. Res.* **98:**593–603.

70. **Naewabanij, M., P. A. Seib, R. Burroughs, L. M. Seitz, and D. S. Chung.** 1984. Determination of ergosterol using thin-layer chromatography and ultraviolet spectroscopy. *Cereal Chem.* **61:**385–388.

70a.**Newell, S. Y.** Unpublished data.

71. **Newell, S. Y.** 1992. Estimating fungal biomass and productivity in decomposing litter, p. 521–561. *In* G. C. Carroll and D. T. Wicklow (ed.), *The Fungal Community: Its Organization and Role in the Ecosystem,* 2nd ed. Marcel Dekker, New York.

72. **Newell, S. Y.** 1993. Decomposition of shoots of a saltmarsh grass. *Adv. Microb. Ecol.* **13:**301–326.

73. **Newell, S. Y.** 1993. Membrane-containing fungal mass and fungal specific growth rate in natural samples, p. 579–584. *In* P. F. Kemp, B. F. Sherr, E. B. Sherr, and J. J. Cole (ed.), *Handbook of Methods in Aquatic Microbial Ecology.* Lewis Publishers, Boca Raton, Fla.

74. **Newell, S. Y.** 1994. Ecomethodology for organoosmotrophs: prokaryotic unicellular versus eukaryotic mycelial. *Microb. Ecol.* **28:**151–157.

75. **Newell, S. Y.** 1994. Total and free ergosterol in mycelia of saltmarsh ascomycetes with access to whole leaves or aqueous extracts of leaves. *Appl. Environ. Microbiol.* **60:** 3479–3482.

76. **Newell, S. Y.** 1995. Minimizing ergosterol loss during preanalytical handling and shipping of samples of plant litter. *Appl. Environ. Microbiol.* **61:**2794–2797.

77. **Newell, S. Y., T. L. Arsuffi, and R. D. Fallon.** 1988. Fundamental procedures for determining ergosterol content of decaying plant material by liquid chromatography. *Appl. Environ. Microbiol.* **54:**1876–1879.

78. **Newell, S. Y., and F. Bärlocher.** 1993. Removal of fungal and total organic matter from decaying cordgrass leaves by shredder snails. *J. Exp. Mar. Biol. Ecol.* **171:**39–49.

79. **Newell, S. Y., and R. D. Fallon.** 1991. Toward a method for measuring fungal instantaneous growth rates in field samples. *Ecology* **72:**1547–1559.

80. **Newell, S. Y., R. D. Fallon, R. M. Cal Rodriguez, and L. C. Groene.** 1985. Influence of rain, tidal wetting and relative humidity on release of carbon dioxide by standing dead salt-marsh plants. *Oecologia* **68:**73–79.

81. **Newell, S. Y., R. D. Fallon, and J. D. Miller.** 1989. Decomposition and microbial dynamics for standing, nat-

urally positioned leaves of the salt-marsh grass *Spartina alterniflora. Mar. Biol.* **101:**471–481.

82. **Newell, S. Y., and J. W. Fell.** 1982. Surface sterilization and the active mycoflora of leaves of a seagrass. *Bot. Mar.* **25:**339–346.

83. **Newell, S. Y., and J. W. Fell.** 1995. Do halophytophthoras (marine Pythiaceae) rapidly occupy fallen leaves by intraleaf mycelial growth. *Can. J. Bot.* **73:**761–765.

84. **Newell, S. Y., J. D. Miller, and R. D. Fallon.** 1987. Ergosterol content of salt-marsh fungi: effect of growth conditions and mycelial age. *Mycologia* **79:**688–695.

85. **Newell, S. Y., M. A. Moran, R. Wicks, and R. E. Hodson.** 1995. Productivities of microbial decomposers during early stages of decomposition of leaves of a freshwater sedge. *Freshwater Biol.* **34:**135–148.

86. **Newell, S. Y., and J. Wasowski.** 1995. Sexual productivity and spring intramarsh distribution of a key saltmarsh microbial secondary producer. *Estuaries* **18:**241–249.

87. **Nout, M. J. R., T. M. G. Bonants-Van Laarhoven, P. de Jongh, and P. G. de Koster.** 1987. Ergosterol content of *Rhizopus oligosporus* NRRL 5905 grown in liquid and solid substrates. *Appl. Microbiol. Biotechnol.* **26:**456–461.

88. **Nylund, J. E., and H. Wallander.** 1992. Ergosterol analysis as a means of quantifying mycorrhizal biomass. *Methods Microbiol.* **24:**77–88.

89. **Osswald, W. F., W. Höll, and E. F. Elstner.** 1986. Ergosterol as a biochemical indicator of fungal infection in spruce and fir needles from different sources. *Z. Naturforsch.* **41c:**542–546.

90. **Padgett, D. E., and D. A. Celio.** 1990. A newly discovered role for aerobic fungi in anaerobic salt marsh soils. *Mycologia* **82:**791–794.

91. **Padgett, D. E., and M. H. Posey.** 1993. An evaluation of the efficiencies of several ergosterol extraction techniques. *Mycol. Res.* **97:**1476–1480.

92. **Parkinson, D.** 1994. Filamentous fungi, p. 330–350. *In* R. W. Weaver, S. Angle, P. Bottomley, D. Bezdicek, S. Smith, A. Tabatabai, and A. Wollum (ed.), *Methods of Soil Analysis: Microbiological and Biochemical Properties.* SSSA Book Series 5. Soil Science Society of America, Madison, Wis.

93. **Parkinson, D., and D. C. Coleman.** 1991. Methods for assessing microbial populations, activity and biomass. Microbial communities, activity and biomass. *Agric. Ecosyst. Environ.* **34:**3–33.

94. **Passi, S., C. De Luca, S. Fabbri, S. Brasini, and C. Fanelli.** 1994. Possible role of ergosterol oxidation in aflatoxin production by *Aspergillus parasiticus. Mycol. Res.* **98:**363–368.

95. **Peacock, G. A., and M. W. Goosey.** 1989. Separation of fungal sterols by normal-phase high-performance liquid chromatography: application to the evaluation of ergosterol biosynthesis inhibitors. *J. Chromatogr.* **469:**293–304.

96. **Porter, D.** 1990. Phylum Labyrinthulomycota, p. 388–398. *In* L. Margulis, J. O. Corliss, M. Melkonian, and D. J. Chapman (ed.), *Handbook of Protoctista.* Jones and Bartlett, Boston.

97. **Premdas, P. D., and B. Kendrick.** 1991. Colonization of autumn-shed leaves by four aero-aquatic fungi. *Mycologia* **83:**317–321.

98. **Rogers, K. H., and J. De Bruyn.** 1988. Decomposition of *Paspalum distichum* L.: methodology in seasonally inundated systems. *Verh. Int. Ver. Limnol.* **23:**1945–1948.

99. **Ruzicka, S., M. D. P. Norman, and J. A. Harris.** 1995. Rapid ultrasonication method to determine ergosterol concentration in soil. *Soil Biol. Biochem.* **27:**1215–1217.

100. **Salmanowicz, B., and J. E. Nylund.** 1988. High performance liquid chromatography determination of ergosterol as a measure of ectomycorrhiza infection in Scots pine. *Eur. J. For. Pathol.* **18:**291–298.

101. **Sarvis Weyers, H.** 1994. A comparison of biomass and productivity of fungi and bacteria on decomposing leaves in two streams. M.Sc. thesis. Department of Biology, University of Alabama, Tuscaloosa.

102. **Sashidhar, R. B., V. S. Rao, Y. Ramakrishna, and R. V. Bhat.** 1989. Rapid and specific method for screening ergosterol as an index of fungal contamination in cereal grains. *Food Chem.* **31:**51–56.

103. **Schaumann, K.** 1993. Marine Pilze, p. 144–195. *In* L.-A. Meyer-Reil and M. Köster (ed.), *Mikrobiologie des Meeresbodens.* Gustav Fischer Verlag, Jena, Germany.

104. **Schwadorf, K., and H. M. Müller.** 1989. Determination of ergosterol in cereals, mixed feed components, and mixed feeds by liquid chromatography. *J. Assoc. Offic. Anal. Chem.* **72:**457–462.

105. **Seitz, L. M., H. E. Mohr, R. Burroughs, and D. B. Sauer.** 1977. Ergosterol as an indicator of fungal invasion in grains. *Cereal Chem.* **54:**1207–1217.

106. **Seitz, L. M., D. B. Sauer, R. Burroughs, H. E. Mohr, and J. D. Hubbard.** 1979. Ergosterol as a measure of fungal growth. *Physiol. Biochem.* **69:**1202–1203.

107. **Seo, S., U. Sankawa, H. Seto, A. Uomori, Y. Yoshimura, Y. Ebizuka, H. Noguchi, and K. Takeda.** 1986. Biosynthesis of sitosterol in tissue cultures of *Rabdosia japonica* Hara and ergosterol in yeast from $[2\text{-}^{13}C, 2\text{-}^2H_3]$acetate. *J. Chem. Soc. Chem. Commun.* **1986:**1139–1141.

108. **Shearer, C. A.** 1993. The freshwater Ascomycetes. *Nova Hedwigia* **56:**1–33.

109. **Suberkropp, K.** 1992. Aquatic hyphomycete communities, p. 729–747. *In* G. C. Carroll and D. T. Wicklow (ed.), *The Fungal Community. Its Organization and Role in the Ecosystem,* 2nd ed. Marcel Dekker, New York.

110. **Suberkropp, K.** 1995. The influence of nutrients on fungal growth, productivity, and sporulation during leaf breakdown in streams. *Can. J. Bot.* **73**(Suppl. 1):S1361–S1369.

111. **Suberkropp, K., M. O. Gessner, and E. Chauvet.** 1993. Comparison of ATP and ergosterol as indicators of fungal biomass associated with decomposing leaves in streams. *Appl. Environ. Microbiol.* **59:**3367–3372.

111a.**Suberkropp, K., and H. Weyers.** 1996. Application of fungal and bacterial production methodologies to decomposing leaves in streams. *Appl. Environ. Microbiol.* **62:**1610–1615.

112. **Vestal, J. R., and D. C. White.** 1989. Lipid analysis in microbial ecology. Quantitative approaches to the study of microbial communities. *BioScience* **39:**535–541.

113. **Wainwright, M.** 1992. The impact of fungi on environmental biogeochemistry, p. 601–618. *In* G. C. Carroll and D. T. Wicklow (ed.), *The Fungal Community. Its Organization and Role in the Ecosystem,* 2nd ed. Marcel Dekker, New York.

114. **Webster, J., and E. Descals.** 1981. Morphology, distribution, and ecology of conidial fungi in freshwater habitats, p. 295–355. *In* G. T. Cole and B. Kendrick (ed.), *Biology of Conidial Fungi,* vol. 1. Academic Press, New York.

115. **Weete, J. D.** 1980. *Lipid Biochemistry of Fungi and Other Organisms.* Plenum Press, New York.

116. **Weete, J. D.** 1989. Structure and function of sterols in fungi. *Adv. Lipid Res.* **23:**115–167.

117. **West, A. W., and W. D. Grant.** 1987. Use of ergosterol, diaminopimelic acid and glucosamine contents of soils to monitor changes in microbial populations. *Soil Biol. Biochem.* **19:**607–612.

118. **Wetzel, R. G., and G. E. Likens.** 1991. Primary productivity of phytoplankton, p. 207–226. *In Limnological Analyses,* 2nd ed. Springer-Verlag, New York.

119. **Wigand, C., and J. C. Stevenson.** 1994. The presence and possible ecological significance of mycorrhizae of the

submersed macrophyte, *Vallisneria americana. Estuaries* **17**:206–215.

120. **Xu, S., R. A. Norton, F. G. Crumley, and W. D. Nes.** 1988. Comparison of the chromatographic properties of sterols, select additional steroids and triterpenoids: gravity-flow column liquid chromatography, thin-layer chromatography, gas-liquid chromatography and high-performance liquid chromatography. *J. Chromatogr.* **452:** 377–398.

121. **Young, J. C.** Microwave-assisted extraction of the fungal metabolite ergosterol and total fatty acids. *J. Agric. Food Chem.* **43**:2904–2910.

121a.**Young, J. C.** Personal communication.

122. **Young, J. C., and D. E. Games.** 1993. Supercritical fluid extraction and supercritical fluid chromatography of the fungal metabolite ergosterol. *J. Agric. Food Chem.* **41:** 577–581.

123. **Zelles, L., K. Hund, and K. Stepper.** 1987. Methoden zur relativen Quantifizierung der pilzlichen Biomasse im Boden. *Z. Pflanzenernae hr. Bodenkd.* **150**:249–252.

124. **Zill, G., G. Engelhardt, and P. R. Wallnöfer.** 1988. Determination of ergosterol as a measure of fungal growth using Si 60 HPLC. *Z. Lebensm. Unters. Forsch.* **187:** 246–249.

Phagotrophy in Aquatic Microbial Food Webs

EVELYN B. SHERR AND BARRY F. SHERR

32

TROPHIC INTERACTIONS AMONG AQUATIC MICROBES

The focus on grazing interactions—who eats whom—in aquatic ecosystems has expanded during the past two decades to include, in addition to the usual macroscopic consumers such as fish and copepods, microbial predators: the phagotrophic protists. At the microscale, the macroscale notions about linear food chains, segregated trophic levels, and dynamics of monospecific populations are not entirely applicable for several reasons. (i) Mixotrophy, a combination of trophic modes, is common among the protists. Examples include photosynthetic flagellates that ingest prey (34, 55) and ciliates that harbor endosymbiotic algae (72) or temporarily "enslave" chloroplasts from algal prey (71). (ii) Taxonomy of microbes is more difficult than that of metazoans, which complicates analysis of individual populations of microorganisms. Only recently has the "black box" of taxonomy of aquatic bacteria been opened via molecular genetic techniques (15, 26). Protistan taxonomy has lagged because of lack of funding and lack of experts in this field (14, 50). (iii) Among microbial ecologists, dominant research themes are generally in the area of systems ecology rather than population and community ecology. Heterotrophic microbes are responsible for a large share of overall carbon utilization, respiration, and regeneration of mineral nutrients in ecosystems. Thus, most research has considered functional groups of microbes rather than individual populations. The systems ecology perspective has influenced the types of questions asked as well as the methodologies devised to address them.

Microbial trophodynamics is a rapidly developing field of investigation. Newly discovered classes of aquatic microbes and unexpected trophic modes among long-recognized protists have forced changes in conceptualization of microbial food webs. For instance, in oligotrophic regions of the sea, chlorophyll a- and chlorophyll b-containing bacteria, the prochlorophytes, are an abundant and often significant component of both the phytoplankton (as phototrophic cells) and the bacterioplankton (as <1-μm prokaryotes) (9). In such systems, bacterivorous flagellates must play a role as grazers of primary as well as of secondary microbial production. A second example has to do with dinoflagellates. Long assigned to the phytoplankton, dinoflagellates are now recognized as having an equally impor-

tant role as consumers of a wide range of microbial prey (41, 63). Not only are half of all dinoflagellate species nonpigmented, and thus heterotrophic (21), but many species of autotrophic dinoflagellates are also phagotrophic (5, 42).

The microbial loop concept of Azam et al. (1), in which bacteria, bacterivorous flagellates, and ciliate consumers of the flagellates function as a regenerative adjunct to the classic metazoan food web, has evolved to encompass the emerging new information. We have proposed that an overall microbial food web should include all microorganisms with the functional roles of primary production, primary and secondary consumption, and nutrient regeneration (62). The microbial food web per se, not simply phytoplankton, should thus be considered as the resource base for metazoan food webs.

Developing methods to elucidate specific trophic links within highly complex microbial food webs has been a challenge. We will outline the techniques that have been used to investigate the processes of bacterivory and of herbivory within microbial food webs and then briefly discuss approaches used to study other trophic pathways. We must emphasize, however, that current methods are not entirely satisfactory and that progress in this area remains limited by methodology.

BACTERIVORY

The question of the fate of bacterial production within aquatic food webs has been of interest since first Pomeroy (51) and then Williams (83) and Azam et al. (1) proposed that heterotrophic bacteria process a large fraction (30 to 50%) of total primary production. Phagotrophic protists, predominantly <5-μm heterotrophic flagellates, but also ciliates in some systems, have been identified as major consumers of bacterioplankton in both marine waters and freshwaters (18, 60, 68). Viral lysis may rival protistan bacterivory as a source of bacterioplankton mortality (7), but more work is needed to confirm the significance of this process in aquatic food webs.

As an additional caveat, we stress that the rationale and results of the methods discussed below depend on the traditional methods of enumerating bacteria via epifluorescence counts of cells after staining with the fluorochrome acridine orange or 4′,6′-diamidino-2-phenylindole (DAPI). Zweifel and Hagstrom (85) have presented evidence that the stan-

TABLE 1 Summary of methods used to study protistan bacterivory in natural microbial communities in marine water and freshwater systems

Method (references[s])	Advantages	Disadvantages

Prey growth with and without predation
  Selective filtration (84)
  Selective inhibition (43, 57, 73, 75)
  Progressive dilution (37)
	Not technically difficult, can estimate both growth and grazing rates	Manipulation artifacts, disruption of predator-prey feedbacks, time-consuming microscopy

Uptake of labeled prey
  Fluorescent beads (3, 6, 20)
  FLB, heat inactivated or live stained (17, 40, 47, 58, 60)
  Radiolabeled bacteria (11, 32, 41, 48)
	Short-term incubations, can identify which protists are bacterivores	Preparation of labeled prey, manipulation artifacts, labeled prey not "natural," microscopy, incomplete separation of radiolabeled bacteria from protist grazers

Disappearance of labeled prey
  FLB (12)
  Radiolabeled bacteria (56) or *E. coli* minicells (81, 82)
	Simplified microscopy for FLB, no microscopy for radiolabeled prey	Preparation of labeled prey, manipulation artifacts, labeled prey not "natural"

Acid lysozyme activity Cleavage of peptidoglycan analog MUF-chitotriose by enzymes at pH 4.5 in sonicated seawater samples (28)	In vitro assay, no incubation of live organisms, rapid analysis of samples	New method, uncertainty about specificity of enzyme activity, must be calibrated with other method

dard DAPI staining method produces nonspecifically stained bacterial particles, a large fraction of which do not appear to contain DNA. If their results are confirmed, then perhaps protists graze only specific components of the bacterioplankton assemblage (e.g., intact cells with DNA) and produce other components (e.g., cell ghosts without DNA).

A number of methods have been used to quantify protistan bacterivory (Table 1). Measurement of this parameter is useful since it independently provides a lower bound to rates of bacterial cell production and an upper bound to transfer of bacterial carbon to larger organisms in the food web via bacterivorous protists. Bacterivory techniques segregate generally into manipulation/long-term incubation methods and prey surrogate addition/short- or long-term incubation methods (Table 1). A third approach, recently proposed by Gonzalez et al. (28), is an in vitro method that does not require incubation of live organisms, although it must be calibrated with an incubation method.

Manipulation Methods

The basic rationale for these methods is that grazing impact of bacterivores is eliminated (or greatly minimized) and changes in abundance of bacterial prey are compared in treatments with and without grazing. The earliest of such methods was physical separation of bacterial prey from their predators via size-selective screening (84). Bacterial abundance was monitored over 24 h in unfiltered samples (with grazers) and in samples screened through either 1- or 3-μm-pore-size (hereafter referred to as 3-μm) filters. Wright and Coffin (84) reported that the concentration of bacteria increased linearly in the 1-μm-filtered sample and showed a slower decrease, or decline, in 3-μm-filtered water and in unfiltered samples. Abundant heterotrophic flagellates passed through the 3-μm filter, and some eukaryotic cells also were able to pass through the 1-μm filter (84).

Concerns about physical manipulation of samples via screening and incomplete separation of predator and prey led to attempts to quantify in situ bacterivory via selective metabolic inhibition of either the prokaryotic prey or the eukaryotic predator (19, 46). As in the size fractionation method, changes in bacterial cell abundance in treatments with and without addition of inhibitory chemicals are monitored over time periods of 12 to >24 h. The main assumption inherent in this method is that the chemicals added specifically inhibit only the group of interest and do not affect the growth rate or feeding rate of other components of the planktonic assemblage. Fuhrman and McManus (19) used ampicillin at 5 mg liter^{-1} to inhibit bacterial growth, assuming that eukaryotic predators were not affected. Newell et al. (46) used a combination of cycloheximide and thiram to inhibit bacterivorous protists and assumed that bacteria were unaffected. However, such assumptions were not entirely justified. Working with microbial communities in freshwater sediments, Tremaine and Mills (75) reported apparent incomplete inhibition of protists by cycloheximide and decreased bacterial growth rate. Taylor and Pace (73) examined effects of a number of eukaryotic inhibitors on both phototrophic and heterotrophic protists and concluded that use of metabolic inhibitors to evaluate bacterivory did not produce reliable results. Sherr et al. (61) evaluated a number of prokaryotic and eukaryotic chemical inhibitors and concluded that most were not sufficiently selective. Only combinations of cycloheximide (200 mg liter^{-1}) plus colchicine (100 mg liter^{-1}) for protists and of vancomycin (200 mg liter^{-1}) plus penicillin (1 mg liter^{-1}) for bacteria resulted in acceptable selective inhibition of the two target assemblages (57). Liu et al. (43) recently reported that kanamycin at 1 mg liter^{-1} specifically inhibited the growth of the marine prokaryotic phototrophs *Prochlorococcus* and *Synechococcus* spp. but not of heterotrophic bacteria or of phagotrophic protists. Use of this selective inhibitor allowed estimation of growth and grazing mortality rates of autotrophic prokaryotes in surface waters off Hawaii (43).

A third manipulation approach to bacterivory was adapted from the Landry and Hassett (38) dilution method initially used to quantify microzooplankton grazing on phytoplankton. The dilution method involves establishing a series of mixtures of whole water and 0.2-μm-filtered water in order to progressively decrease grazing impact on prey populations. In theory, prey cells in diluted samples will show proportionally decreased grazing mortality but the same intrinsic growth rate (36). Monitoring growth of prey over 24 h in the dilution series allows calculation of both grazing rate and growth rate (36). The dilution method has been used to estimate bacterivory in Hawaiian coastal waters (37) and protist grazing on coccoid cyanobacteria at several sites in the northwest Atlantic Ocean (8). Tremaine and Mills (76) did a limited test of the assumptions of the dilution method applied to analysis of bacterivory and concluded that the approach was valid for the eutrophic freshwater habitat examined.

Several problems are associated with the manipulation/incubation methods outlined above. (i) Separation of predator and prey varies in degree depending on the microbial community involved but is rarely complete. (ii) Manipulation can cause experimental artifacts in predator and prey assemblages; e.g., screening can cause mechanical damage to protists. (iii) Long-term incubations of small water samples may disrupt the natural species composition of predator and prey assemblages; i.e., growth of some species may be inhibited and growth of others may be favored by the incubation conditions. (iv) Separation of predator and prey or artificial decrease of grazing rates can disrupt positive feedback interactions between microbial assemblages. For example, excretions of bacterivorous protists provide their bacterial prey with dissolved organic substrates (67) and inorganic nitrogen and phosphorus nutrients (10). Sherr et al. (57) found lower bacterial growth rates in experimental manipulations in which protist grazing was reduced via addition of eukaryotic cell inhibitors. Addition of ammonium to such treatments resulted in enhanced bacterial growth, suggesting that bacteria were limited by availability of inorganic nutrients excreted by protists. There is now a substantial literature on requirements of heterotrophic bacteria in both marine water and freshwater systems for inorganic nitrogen and phosphorus (35). Disruption of positive feedback processes is likely to be a more significant problem in oligotrophic than in eutrophic systems.

Prey Addition Methods

Quantifying uptake of added, labeled prey minimizes the experimental artifacts that may result from disruption of predator-prey feedback processes caused by manipulation methods. Prey addition experiments may involve short-term incubations in which rate of appearance of labeled prey in protists is monitored or longer-term incubations in which rate of disappearance of labeled prey is monitored. To prevent alteration of protist clearance rates, labeled prey should be added at tracer concentrations, generally 5 to 50% of natural bacterial abundance (58). Surrogate prey particles used to determine bacterivory have included (i) fluorescently labeled prey, either bacterium-size fluorescent plastic beads (3, 6, 20) or fluorescently labeled bacteria (FLB) (13, 47, 58), and (ii) radioisotopically labeled prey, including natural bacterioplankton assemblages (32, 41, 62, 82).

Hollibaugh et al. (32) pioneered the use of bacteria prelabeled with tritiated thymidine in grazing experiments. Lessard and Swift (41) modified this technique to determine relative grazing by specific components of the >20-μm protistan assemblage on [^{3}H]thymidine-labeled heterotrophic bacteria and on ^{14}C-labeled phototrophic cells. A problem with the use of tritium-labeled bacteria as prey is that retention of the label by bacterivorous protists is lower than expected in incubations lasting longer than 1 to 2 h (11). Servais et al. (56) combined size-selective filtration with long-term disappearance of [^{3}H]thymidine-labeled bacteria in order to discriminate between bacterial mortality due to protist grazing and that due to viral lysis or other factors. The potential problem of growth of labeled natural bacteria in disappearance experiments was addressed by the protocol of Wikner et al. (82), in which nongrowing *Escherichia coli* minicells the size of natural bacteria are labeled with [^{35}S]methionine for use in grazing assays. Nygaard and Hessen (48) subsequently developed a method for ^{14}C labeling of bacterial cells via growth on ^{14}C-labeled protein hydrolysate and reported higher ingestion rates of radiolabeled cells compared with uptake of heat-inactivated FLB. A concern with respect to use of radiolabeled bacterial cells in grazing experiments is effective separation of label ingested by bacterivores from label present in unconsumed bacterial cells.

Advantages of addition methods include less disruption of natural microbial communities, identification of the specific components of the protistan community responsible for bacterivory, and ability to run short-term prey uptake experiments, which minimizes potential artifacts inherent in long-term incubations. In the case of radioisotopically labeled prey, extensive microscopy is not required as it is in the other approaches. Use of fluorescently labeled tracer particles has revealed that ciliates as well as flagellates can be significant bacterivores in some aquatic systems (64, 68) and that some phototrophic flagellates are bacterivorous (3, 35, 55). Specific protocols for estimating bacterivory via addition of fluorescently labeled particles (64) and of radiolabeled minicells (81) have been published.

A major assumption inherent in grazing methods based on addition of labeled prey is that predators do not show significant selection for or against the added prey compared with their rates of feeding on natural, unlabeled prey. This assumption has been tested and found to be invalid in certain respects. Some species of bacterivorous protists have lower feeding rates on plastic microspheres than on FLB (58, 47). Bacterivorous flagellates selectively graze larger bacterial cells (13, 29), dividing cells (59), and motile cells (30, 45). Bacterivorous flagellates may also exhibit differential feeding on individual bacterial strains (44). A criticism of the FLB grazing method of Sherr et al. (58) is that the labeled bacteria are heat inactivated and thus may not be an adequate surrogate for live bacteria (37). Gonzalez et al. (30) concluded that if more than a few percent of natural bacterioplankton were motile, use of heat-inactivated FLB would result in underestimation of rates of bacterivory.

Use of live-stained FLB has been proposed to achieve more accurate estimates of bacterial mortality due to protist grazing. Landry et al. (40) developed a method to live-stain bacteria with fluorescein isothiocyanate by treatment with dithioerythritol, a compound that breaks sulfur bonds in the bacterial cell wall, enhancing binding of the fluorochrome. More recently, Epstein and Rossel (17) have suggested cyanoditolyl tetrazolium chloride (CTC; Polysciences, Inc., Warrington, Pa.) for preparation of live FLB. CTC is a redox compound that can be used in place of oxygen in electron transport systems. In its oxidized state, CTC is nonfluorescent and easily diffuses into bacterial cells; after reduction in the electron transport system, CTC precipitates in the bacterial cytoplasm as a red-fluorescent fluorochrome (53). A drawback in using CTC is that although >90% of cultured bacteria will brightly stain with this compound after 1 to 2 h, in natural samples often <10% of bacteria, presumably only metabolically active cells, will stain even after 10 to 12 h of incubation in 5 mM CTC (24). Thus, live-stained bacteria may also not be representative of the entire bacterioplankton assemblage.

Notwithstanding the problems inherent in using surrogate prey, the FLB uptake/disappearance approach has been widely used to evaluate grazing within natural microbial assemblages on both heterotrophic bacteria and cyanobacteria (4, 12). FLB have also been used to evaluate rates of digestion of bacterial prey by protists (27, 61). Monger and Landry (45) and Keller et al. (34) applied flow cytometry to analysis of FLB uptake by marine bacterivorous flagellates in order to increase the precision of the method.

The Digestive Enzyme Activity Approach

Gonzalez et al. (28) have proposed that in situ bacterivory might be assessed by determining the activity of one or more hydrolytic enzymes present in the food vacuoles of phagotrophic protists that would be specific for bacterial prey. The basis of the method is that protistan digestive enzymes can be distinguished from extracellular hydrolytic enzymes of bacteria or from extracellular or intracellular hydrolytic enzymes of other eukaryotic organisms by pH optima of enzymatic activity. In theory, the digestive enzymes of protists should have peak activity at a pH of 4 to 5, the pH of their early-stage food vacuoles, while other types of hydrolytic enzymes would have optimal activity at higher pH. Gonzalez et al. (28) developed an assay for bacterivory based on measurement of activity of lysozyme, an enzyme that hydrolyzes the cell walls of eubacteria, in sonicated seawater samples at acid pH. Lysozyme activity was assessed by rate of enzymatic cleavage of the fluorochrome methylumbelliferone (MUF) from the substrate MUF-triacetylchitotriose (Sigma Co., St. Louis Mo.), which serves as an analog of peptidoglycan, a major structural component of bacterial cell walls. MUF exhibits low fluorescence when bound to substrates and high fluorescence after cleavage (33).

In cultures of bacterivorous flagellates, the rate of production of MUF fluorescence showed two peaks, one at pH 4.5, presumably due to flagellate digestive enzymes, and one at pH 7 to 8, presumably due to bacterial ectoenzymes, with little overlap (28). Cultured phytoplankton exhibited only minor lysozyme activity at pH 4.5 (27). Acid lysozyme activity was calibrated against rates of bacterial ingestion via the FLB uptake method (58). Results of the two methods were significantly correlated ($r^2 = 0.98$) over a wide range of rates of bacterivory (10^3 to 10^8 bacteria ingested per ml per h). Within a single order of magnitude of bacterivory, positive correlations between the two indices of bacterivory

were also found both for Oregon shelf water (56 km from shore) and for offshore gyre water (380 km from shore) (28). We have since obtained a calibration correlation between acid lysozyme activity and bacterivory assessed via FLB uptake in the Arctic Ocean ($r^2 = 0.96$). These results indicate that the acid lyzozyme assay for bacterivory can be used in a variety of oceanic systems spanning a range of trophic states. However, from our experience, there does not appear to be a universal calibration relationship; thus, it appears necessary to perform calibration experiments for each study area. In addition, the bacterivory assay, e.g., FLB uptake, used to calibrate the acid lysozyme assay is likely to have its own inherent problems, as discussed above. An advantage of the digestive enzyme activity approach is that it allows higher resolution sampling, in time and space, of grazing rates within a particular system compared with the previously described methods used to assess bacterivory.

Simek et al. (69) have proposed an alternate substrate, MUF-β-N-acetylglucosaminide, as an indicator of enzyme activity associated with protistan bacterivory. We have found problems with use of this substrate to assess bacterivory in natural microbial communities: (i) enzyme activity versus pH plots show a peak at pH 6.5 to 7 for protistan enzymes that cleave the substrates, and (ii) most of the species of marine phytoplankton that we tested also had high enzymatic activity for the substrate in this pH range. Thus, outside of controlled laboratory experiments, enzyme activity based on cleavage of MUF-acetylglucosaminide could be attributed to multiple sources, not strictly to bacterivorous protists.

PROTISTAN HERBIVORY

Phagotrophic protists are increasingly recognized as significant grazers of phytoplankton in aquatic ecosystems (2, 63, 65). In marine systems, <200-μm herbivores (ciliates, dinoflagellates, and nanoflagellates) often graze 50 to 100% of daily phytoplankton production (65). Thus, development of methods to assess protistan herbivory is critical in determining the structure and functioning of pelagic food webs. However, less work has been done in this regard (Table 2) compared with the number of approaches developed to investigate protistan bacterivory.

The most commonly used method for quantification of protist grazing on phytoplankton is the original approach proposed: the dilution method of Landry and Hassett (38), discussed previously. The advantages of this approach are that the method is straightforward and that both grazing

TABLE 2 Summary of methods used to study protistan herbivory in natural microbial communities in marine systems

Method (references[s])	Advantages	Disadvantages
Progressive dilution: Change in chlorophyll measured over 24 h in a series of dilutions to decrease grazing impact (22, 36, 38, 39, 79, 80)	Commonly used, can obtain both prey growth and grazing rate, technically simple	Does not always work, manipulation artifacts, disruption of predator-prey feedbacks
Selective filtration: Size fractionation separates phytoplankton from protist grazers (77, 79)	Technically simple, can obtain growth and grazing rates	Manipulation artifacts, incomplete separation of prey and predator, disrupts feedbacks
Uptake and disappearance of labeled prey: Fluorescently labeled (12, 42, 52, 54, 66) or radiolabeled (41) cells	Can obtain cell-specific uptake rates for particular types of predator and prey	Preparation of labeled prey, labeled prey may not be "natural," difficult to get community grazing rates

mortality and intrinsic phytoplankton growth rate are obtained. In addition, by analyzing plant pigments via high-performance liquid chromatography at the beginning and end of the incubations, grazing on specific taxonomic components of the phytoplankton assemblage can be assessed (78, 80). The disadvantages are that the dilution method requires both manipulation, which can disrupt predator-prey feedback processes, and long-term incubation (24 h), which may create experimental artifacts. When the method is used in low-nutrient environments, a separate dilution series with addition of major nutrients is recommended as part of the experimental protocol (36). Setting up each dilution assay is sufficiently time-consuming that generally only one experiment can be performed in a day.

The assumption that the cell-specific clearance rates of phagotrophic protists are not affected by dilution can be violated in eutrophic systems in which protists experience saturated feeding at in situ prey abundance (22). To directly address the problem of variation in cell-specific grazing rate with dilution, Landry et al. (39) have modified the dilution assay protocol by adding FLB or fluorescently labeled algae to the dilution series. Proportional decrease of added labeled prey cells over the duration of the dilution experiment is then used to determine relative grazer activity, and the net phytoplankton growth is regressed against relative grazing rather than against the dilution factor as in the original protocol. This modification of the method allows direct measurement of the effect of dilution on grazing, thus making application of the assay less ambiguous.

Verity (77) used a size fractionation/incubation method to quantify grazing rates of microzooplankton (10 to 202 μm) on <10-μm phytoplankton in coastal marine waters. This approach is limited to situations in which there is a clear size difference between phytoplankton and the dominant protistan herbivores. In a later study in the open North Atlantic, Verity et al. (79) compared estimates of microzooplankton herbivory made via the dilution method and via the size fractionation approach. The dilution method yielded consistently higher estimates of grazing mortality, suggesting an incomplete separation of predator and prey in the size fractionation experiments.

Other approaches to determination of protist grazing within natural planktonic communities appear to be useful only for specific predator-prey interactions. Campbell and Carpenter (8) showed that prokaryotic inhibitors could be used to measure protist grazing on coccoid cyanobacteria. Rublee and Gallegos (54) adapted the heat-inactivated dichlorotriazinyl amino fluorescein staining method of Sherr et al. (58) to prepare fluorescently labeled phytoplankton for in situ grazing experiments in coastal marine waters. Sherr et al. (66) further tested this labeled prey addition method and concluded that for short-term uptake assays, it was necessary to add fluorescently labeled phytoplankton at abundances higher than generally present in mesotrophic systems. Labeled prey could be added at tracer concentrations only in applications of the method in which long-term (>12 to 24 h) disappearance of the prey is analyzed.

The method of Rublee and Gallegos (54) produces fluorescently labeled algal cells which have been shrunk and hardened by heating and which are no longer motile in the case of phytoflagellates. Putt (52) developed a live-staining method for algal cells that uses the fluorochrome hydroethydine. She reported higher ingestion rates of live-stained phytoplankton than of heat-inactivated and stained phytoplankton. However, in practice, hydroethydine-stained cells

lost fluorescence upon fixation and could be visualized only by the pink to brown color of stained algae in transmitted light microscopy (52). Hydroethydine staining also decreased the motility of phytoflagellates (52).

A more promising compound for producing live-stained fluorescently labeled eukaryotic cells, including both phototrophic and heterotrophic protists, is the reagent 5-chloromethylfluorescein diacetate (CellTracker Green CMFDA; Molecular Probes, Eugene, Oreg.) (42). CMFDA is membrane permeable. Once inside the cell, the compound is cleaved by esterases and fluoresces bright green when stimulated with blue light, and it is precipitated by thiol groups so that it is membrane impermeable. Fluorescence is not diminished by preservation. Li et al. (42) demonstrated ingestion of CMFDA-stained ciliates and phytoflagellates by autotrophic dinoflagellates in Chesapeake Bay. We have prepared CMFDA-stained bacterivorous flagellates by incubating cells for 2 to 3 h with 0.5 μM (final concentration) CMFDA and have observed ingestion of the labeled flagellates by a heterotrophic dinoflagellate, *Gymnodinium* sp., by a tintinnid, *Tintinnopsis* sp., as well as by a variety of protists in Oregon coastal waters.

OTHER TROPHIC PATHWAYS IN MICROBIAL FOOD WEBS

Phagotrophic protists may ingest other types of food in addition to bacteria and phytoplankton. Alternate food resources include other heterotrophic protists (70, 78), sperm cells (23), particulate detritus (25), viral particles (31), and high-molecular-weight compounds (74). Fluorescently labeled viral particles (31) and colloidal polysaccharides and proteins (74) have been used to demonstrate ingestion of submicrometer nonliving material. However, there is still virtually no information on the importance of alternate foods for phagotrophic protists in marine water and freshwater environments. Of particular interest is assessing the significance of the protist-protist trophic link in marine water and freshwater food webs. From the early conceptualizations of the role of heterotrophic microbes in marine ecosystems (1, 16, 49) to more recent ideas about the structure of pelagic microbial food webs (62), a major component of the microbial loop concept has been that bacterivorous flagellates are consumed in turn by larger protists. There is supporting evidence for this trophic link. Both Ohman and Snyder (49) and Verity (78) reported that cultured ciliates were able to grow on a diet of heterotrophic flagellates.

Only recently has the bacterivorous flagellate → larger protist trophic link been formally evaluated in a natural system. Working in a shallow bay in the Adriatic Sea, Solic and Krstulovic (70) monitored growth of heterotrophic flagellates during 24-h incubations in seawater samples prefiltered through either 8-μm polycarbonate filters or a 100-μm mesh screen. These authors reported that on an annual basis, phagotrophic protists in the 8- to 100-μm size fraction consumed nearly 100% of the biomass production of <8-μm flagellates in the bay. Use of CMFDA-stained heterotrophic protists in combination with size-selective screening would allow determination of which components of the protistan assemblage were in fact ingesting the flagellates. This area of research is ripe for further investigation.

SUMMARY

Determination of predator-prey interactions and of pathways of energy and elemental flows within microbial food webs is important for understanding the structure and functioning of ecosystems. Methods to assess specific trophic pathways, e.g., bacterivory and herbivory, have been applied to in situ microbial communities in both marine waters and freshwaters. Most approaches used so far entail assumptions that have been demonstrated to be invalid in certain cases. In order for grazing assays to yield interpretable results, it is important for alternate mortality processes to be assessed via judicious control treatments. For example, spontaneous lysis or viral lysis of bacterial cells may confound results in long-term prey disappearance methods used to estimate protistan bacterivory; thus, a nongrazer control treatment should be included in the experimental design. Methods for evaluation of food resources for phagotrophic protists other than bacteria or phytoplankton, e.g., other heterotrophic protists or nonliving organic matter, have only recently been proposed. This field is a rapidly expanding area within aquatic microbial ecology; additional methodological approaches and further application of existing approaches should yield major new insights into microbial trophodynamics in the future.

REFERENCES

1. **Azam F., T. Fenchel, J. G. Field, J. S. Gray, L. A. Meyer-Reil, and F. Thingstad.** 1983. The ecological role of water-column microbes in the sea. *Mar. Ecol. Prog. Ser.* **10:** 257–263.

2. **Beaver, J. R., and T. L. Crisman.** 1989. The role of ciliated protozoa in pelagic freshwater ecosystems. *Microb. Ecol.* **17:**111–136.

3. **Bird, D. F., and J. Kalff.** 1986. Bacterial grazing by planktonic lake algae. *Science* **231:**493–494.

4. **Bloem, J., M. Starink, B. M.-J. Bar-Gilissen, and T. E. Cappenberg.** 1988. Protozoan grazing, bacterial activity, and mineralization in two-stage continuous cultures. *Appl. Environ. Microbiol.* **54:**3113–3121.

5. **Bockstahler, K. R., and W. D. Coats.** 1993. Spatial and temporal aspects of mixotrophy in Chesapeake Bay dinoflagellates. *J. Protozool.* **40:**49–60.

6. **Borsheim, K. Y.** 1984. Clearance rates of bacteria-sized particles by freshwater ciliates, measured with mono-disperse fluorescent latex beads. *Oecologia* **63:**286–288.

7. **Bratbak, G., F. Thingstad, and M. Heldal.** 1994. Viruses and the microbial loop. *Microb. Ecol.* **28:**209–222.

8. **Campbell, L., and E. J. Carpenter.** 1986. Estimating the grazing pressure of heterotrophic nanoplankton on *Synechococcus* spp. using the sea water dilution and selective inhibitor techniques. *Mar. Ecol. Prog. Ser.* **33:**121–129.

9. **Campbell, L., and D. Vaulot.** 1993. Photosynthetic picoplankton community structure in the subtropical North Pacific near Hawaii (Station ALOHA). *Deep-Sea Res.* **40:** 2043–2060.

10. **Caron, D. A.** 1994. Inorganic nutrients, bacteria, and the microbial loop. *Microb. Ecol.* **28:**295–298.

11. **Caron, D. A., E. J. Lessard, M. Voytek, and M. R. Dennett.** 1993. Use of tritiated thymidine (TdR) to estimate rates of bacterivory: implications of label retention and release by bacterivores. *Mar. Microb. Food Webs* **7:** 177–196.

12. **Caron, D. A., E. L. Lim, G. Miceli, J. B. Waterbury, and F. W. Valois.** 1991. Grazing and utilization of chroococcoid cyanobacteria and heterotrophic bacteria by protozoa in laboratory cultures and a coastal plankton community. *Mar. Ecol. Prog. Ser.* **76:**205–217.

13. **Chrzanowski, T. H., and K. Simek.** 1990. Prey-size selection by freshwater flagellated protozoa. *Limnol. Oceanogr.* **35:**1429–1436.

14. **Corliss, J. O.** 1995. The ambiregnal protists and the codes of nomenclature: a brief review of the problems and of proposed solutions. *Bull. Zool. Nomencl.* **52:**11–17.

15. **DeLong, E. F., D. G. Franks, and A. L. Alldredge.** 1993. Phylogenetic diversity of aggregate-attached vs free-living marine bacterial assemblages. *Limnol. Oceanogr.* **38:** 924–934.

16. **Ducklow, H.** 1983. Production and fate of bacteria in the oceans. *BioScience* **33:**494–501.

17. **Epstein, S. S., and J. Rossel.** 1995. Methodology of *in situ* grazing experiments: evaluation of a new vital dye for preparation of fluorescently labeled bacteria. *Mar. Ecol. Prog. Ser.* **128:**143–150.

18. **Fenchel, T.** 1982. Ecology of heterotrophic microflagellates. IV. Quantitative occurrence and importance as bacterial consumers. *Mar. Ecol. Prog. Ser.* **9:**35–42.

19. **Fuhrman, J. A., and G. B. McManus.** 1984. Do bacteria-sized marine eukaryotes consume significant bacterial production? *Science* **224:**1257–1260.

20. **Fuhrman, J. A., and G. B. McManus.** 1986. Bacterivory in seawater studied with the use of inert fluorescent particles. *Limnol. Oceanogr.* **31:**420–426.

21. **Gaines, G., and M. Elbrachter.** 1987. Heterotrophic nutrition, p. 224–268. *In* F. J. R. Taylor (ed.), *The Biology of Dinoflagellates*. Blackwell Scientific, Oxford.

22. **Gallegos, C. L.** 1989. Microzooplankton grazing on phytoplankton in the Rhode River, Maryland: nonlinear feeding kinetics. *Mar. Ecol. Prog. Ser.* **57:**23–33.

23. **Galvao, H. M., A. T. Fritz, and R. Schmaljohann.** 1989. Ingestion of gametes by protists: fate of surplus reproductive energy in the sea. *Mar. Ecol. Prog. Ser.* **51:**215–220.

24. **Gasol, J. M., P. A. del Giorgia, R. Massana, and C. M. Duarte.** 1995. Active versus inactive bacteria: size-dependence in a coastal marine plankton community. *Mar. Ecol. Prog. Ser.* **128:**91–97.

25. **Gast, V.** 1985. Bacteria as a food resource for microzooplankton in the Schlei Fjord and Baltic Sea with special reference to ciliates. *Mar. Ecol. Prog. Ser.* **22:**107–120.

26. **Giovannoni, S., T. B. Britschgi, C. L. Moyer, and K. G. Field.** 1990. Genetic diversity in Sargasso Sea bacterioplankton. *Nature* (London) **345:**60–63.

27. **Gonzalez, J. M., J. Iriberri, L. Egea, and I. Barcina.** 1990. Differential rates of digestion of bacteria by freshwater and marine phagotrophic protozoa. *Appl. Environ. Microbiol.* **56:**1851–1857.

28. **Gonzalez, J. M., B. F. Sherr, and E. B. Sherr.** 1993. Digestive enzyme activity as a quantitative measure of protistan grazing: the acid lysozyme assay for bacterivory. *Mar. Ecol. Prog. Ser.* **100:**197–206.

29. **Gonzalez, J. M., E. B. Sherr, and B. F. Sherr.** 1990. Size-selective grazing on bacteria by natural assemblages of estuarine flagellates and ciliates. *Appl. Environ. Microbiol.* **56:**583–589.

30. **Gonzalez, J. M., E. B. Sherr, and B. F. Sherr.** 1993. Differential feeding by marine flagellates on growing versus starving, and on motile versus nonmotile, bacterial prey. *Mar. Ecol. Prog. Ser.* **102:**257–267.

31. **Gonzalez, J. M., and C. A. Suttle.** 1993. Grazing by marine nanoflagellates on viruses and viral-sized particles: ingestion and digestion. *Mar. Ecol. Prog. Ser.* **94:**1–10.

32. **Hollibaugh, J. T., J. A. Fuhrman, and F. Azam.** 1980. Radioactively labeling of natural assemblages of bacterioplankton for use in trophic studies. *Limnol. Oceanogr.* **25:** 172–181.

33. **Hoppe, H.-G.** 1993. Use of fluorogenic model substrates for extracellular enzyme activity (EEA) measurement of bacteria, p. 423–432. *In* P. F. Kemp, B. F. Sherr, E. B. Sherr, and J. J. Cole (ed.), *Handbook of Methods in Aquatic Microbial Ecology*. Lewis Publishers, Boca Raton, Fla.

34. **Keller, M. D., L. P. Shapiro, E. M. Haugen, T. L. Cucci, E. B. Sherr, and B. F. Sherr.** 1994. Phagotrophy of fluorescently labeled bacteria by an oceanic phytoplankter. *Microb. Ecol.* **28:**39–52.

35. **Kirchman, D.** 1994. The uptake of inorganic nutrients by heterotrophic bacteria. *Microb. Ecol.* **28:**255–272.

36. **Landry, M. R.** 1993. Estimating rates of growth and grazing mortality of phytoplankton by the dilution method, p. 715–722. *In* P. F. Kemp, B. F. Sherr, E. B. Sherr, and J. J. Cole (ed.), *Handbook of Methods in Aquatic Microbial Ecology.* Lewis Publishers, Boca Raton, Fla.

37. **Landry, M. R., L. W. Haas, and V. L. Fagerness.** 1984. Dynamics of microbial plankton communities: experiments in Kaneohe Bay, Hawaii. *Mar. Ecol. Prog. Ser.* **16:**127–133.

38. **Landry, M. R., and R. P. Hassett.** 1982. Estimating the grazing impact of marine micro-zooplankton. *Mar. Biol.* **67:**283–288.

39. **Landry, M. R., J. Kirshtein, and J. Constantinou.** 1995. A refined dilution technique for measuring the community grazing impact of microzooplankton with experimental tests in the central equatorial Pacific. *Mar. Ecol. Prog. Ser.* **120:**53–63

40. **Landry, M. R., J. M. Lehner-Fournier, J. A. Sundstrom, V. L. Fagerness, and K. E. Selph.** 1991. Discrimination between living and heat-killed prey by a marine zooflagellate, *Paraphysomonas vestita* (Stokes). *J. Exp. Mar. Biol. Ecol.* **146:**139–152.

41. **Lessard, E. J., and E. Swift.** 1985. Species-specific grazing rates of heterotrophic dinoflagellates in oceanic waters. *Mar. Biol.* **87:**289–296.

42. **Li, A., D. K. Stoecker, D. W. Coats, and J. Adam.** 1995. Phagotrophy of fluorescently labeled protists by photosynthetic dinoflagellates from Chesapeake Bay, p. a-31. *In Program/Book of Abstracts, American Society of Limnology and Oceanography Meeting.* University of Nevada, Reno, Nev.

43. **Liu, H., L. Campbell, and M. R. Landry.** 1995. Growth and mortality rates of *Prochlorococcus* and *Synechococcus* measured with a selective inhibitor technique. *Mar. Ecol. Prog. Ser.* **116:**227–287.

44. **Mitchell, G. C., J. H. Baker, and H. A. Sleigh.** 1988. Feeding of a freshwater flagellate *Bodo saltans* on diverse bacteria. *J. Protozool.* **35:**219–222.

45. **Monger, B. C., and M. R. Landry.** 1992. Size-selective grazing by heterotrophic nanoflagellates: an analysis using live-stained bacteria and dual-beam flow cytometry. *Arch. Hydrobiol. Beih. Ergebn. Limnol.* **37:**173–185.

46. **Newell, S. Y., B. F. Sherr, E. B. Sherr, and R. D. Fallon.** 1983. Bacterial response to presence of eukaryotic inhibitors in water from a coastal marine environment. *Mar. Environ. Res.* **10:**147–157.

47. **Nygaard, K., K. Y. Borsheim, and T. F. Thingstad.** 1988. Grazing rates on bacteria by marine heterotrophic microflagellates compared to uptake rates of bacterial-sized monodisperse fluorescent latex beads. *Mar. Ecol. Prog. Ser.* **44:**159–165.

48. **Nygaard, K., and D. O. Hessen.** 1990. Use of ^{14}C-protein-labelled bacteria for estimating clearance rates by heterotrophic and mixotrophic flagellates. *Mar. Ecol. Prog. Ser.* **68:**7–14.

49. **Ohman, M. D., and R. A. Snyder.** 1991. Growth kinetics of the omnivorous oligotrich ciliate *Strombidium* sp. *Limnol. Oceanogr.* **36:**922–935.

50. **Patterson, D. J., and J. Larsen.** 1992. A perspective on protistan nomenclature. *J. Protozool.* **39:**125–131.

51. **Pomeroy, L. R.** 1974. The ocean's food web, a changing paradigm. *BioScience* **24:**499–504.

52. **Putt, M.** 1991. Development and evaluation of tracer particles for use in microzooplankton herbivory studies. *Mar. Ecol. Prog. Ser.* **77:**27–37.

53. **Rodriguez, G. G., D. Phipps, K. Ishiguro, and H. F. Ridgeway.** 1992. Use of a fluorescent redox probe for direct visualization of actively respiring bacteria. *Appl. Environ. Microbiol.* **58:**1801–1808.

54. **Rublee, P. A., and C. L. Gallegos.** 1989. Use of fluorescently labelled algae (FLA) to estimate microzooplankton grazing. *Mar. Ecol. Prog. Ser.* **51:**221–227.

55. **Sanders, R. W., and K. G. Porter.** 1988. Phagotrophic phytoflagellates. *Adv. Microb. Ecol.* **10:**167–192.

56. **Servais, P., G. Billen, and J. V. Rego.** 1985. Rate of bacterial mortality in aquatic environments. *Appl. Environ. Microbiol.* **49:**1448–1454.

57. **Sherr, B. F., E. B. Sherr, T. L. Andrew, R. D. Fallon, and S. Y. Newell.** 1986. Trophic interactions between heterotrophic protozoa and bacterioplankton in estuarine water analyzed with selective metabolic inhibitors. *Mar. Ecol. Prog. Ser.* **32:**169–179.

58. **Sherr, B. F., E. B. Sherr, and R. D. Fallon.** 1987. Use of monodispersed, fluorescently labeled bacteria to estimate in situ protozoan bacterivory. *Appl. Environ. Microbiol.* **53:**958–965.

59. **Sherr, B. F., E. B. Sherr, and J. McDaniel.** 1992. Effect of protozoan grazing on the frequency of dividing cells in bacterioplankton assemblages. *Appl. Environ. Microbiol.* **58:**2381–2385.

60. **Sherr, B. F., E. B. Sherr, and C. Pedros-Alio.** 1989. Simultaneous measurement of bacterioplankton production and protozoan bacterivory in estuarine water. *Mar. Ecol. Prog. Ser.* **54:**209–219.

61. **Sherr, B. F., E. B. Sherr, and F. Rassoulzadegan.** 1988. Rates of digestion of bacteria by marine phagotrophic protozoa: temperature dependence. *Appl. Environ. Microbiol.* **54:**1091–1095.

62. **Sherr, E. B., and B. F. Sherr.** 1988. Role of microbes in pelagic food webs: a revised concept. *Limnol. Oceanogr.* **33:**1225–1226.

63. **Sherr, E. B., and B. F. Sherr.** 1992. Trophic roles of pelagic protists: phagotrophic flagellates as herbivores. *Arch. Hydrobiol. Beih. Ergebn. Limnol.* **37:**165–172.

64. **Sherr, E. B., and B. F. Sherr.** 1993. Protistan grazing rates via uptake of fluorescently labeled prey, p. 695–701. *In* P. F. Kemp, B. F. Sherr, E. B. Sherr, and J. J. Cole (ed.), *Handbook of Methods in Aquatic Microbial Ecology.* Lewis Publishers, Boca Raton, Fla.

65. **Sherr, E. B., and B. F. Sherr.** 1994. Bacterivory and herbivory: key roles of phagotrophic protists in pelagic food webs. *Microb. Ecol.* **28:**223–235.

66. **Sherr, E. B., B. F. Sherr, and J. McDaniel.** 1991. Clearance rates of <6 μm fluorescently labeled algae (FLA) by estuarine protozoa: potential grazing impact of flagellates and ciliates. *Mar. Ecol. Prog. Ser.* **69:**81–92.

67. **Sieburth, J. M., and P. G. Davis.** 1982. The role of heterotrophic nanoplankton in the grazing and nurturing of planktonic bacteria in the Sargasso and Caribbean Seas. *Ann. Inst. Oceanogr.* **58:**285–296.

68. **Simek, K., and V. Straskrabova.** 1992. Bacterioplankton production and protozoan bacterivory in a mesotrophic reservoir. *J. Plankon Res.* **14:**773–787.

69. **Simek, K., J. Vrba, and P. Lavrentyev.** 1994. Estimates of protozoan bacterivory: from microscopy to ectoenzyme assay? *Mar. Microb. Food Webs* **8:**71–85.

70. **Solic, M., and N. Krstulovic.** 1994. Role of predation in controlling bacterial and heterotrophic nanoflagellate standing stocks in the coastal Adriatic Sea: seasonal patterns. *Mar. Ecol. Prog. Ser.* **114:**219–235.

71. **Stoecker, D. K., A. E. Michaels, and L. H. Davis.** 1987. Large proportion of marine planktonic ciliates found to

contain functional chloroplasts. *Nature* (London) **326:** 415–423.

72. **Taylor, F. J. R.** 1990. Symbiosis in marine protozoa, p. 323–340. *In* G. M. Capriulo (ed.), *The Ecology of Marine Protozoa.* Oxford University Press, New York.

73. **Taylor, G. T., and M. L. Pace.** 1987. Validity of eucaryotic inhibitors for assessing production and grazing mortality of marine bacterioplankton. *Appl. Environ. Microbiol.* **53:** 119–128.

74. **Tranvik, L. J., E. B. Sherr, and B. F. Sherr.** 1993. Uptake and utilization of 'colloidal DOM' by heterotrophic flagellates in seawater. *Mar. Ecol. Prog. Ser.* **92:**301–309.

75. **Tremaine, S. C., and A. L. Mills.** 1987. Inadequacy of the eukaryotic inhibitor cycloheximide in studies of protozoan grazing on bacteria at the freshwater-sediment interface. *Appl. Environ. Microbiol.* **53:**1969–1972.

76. **Tremaine, S. C., and A. L. Mills.** 1987. Tests of the critical assumptions of the dilution method for estimating bacterivory by microeukaryotes. *Appl. Environ. Microbiol.* **53:** 2914–2921.

77. **Verity, P. G.** 1986. Grazing of phototrophic nanoplankton by microzoo-plankton in Narragansett Bay. *Mar. Ecol. Prog. Ser.* **29:**105–115.

78. **Verity, P. G.** 1991. Measurement and simulation of prey uptake by marine planktonic ciliates fed plastidic and aplastidic nanoplankton. *Limnol. Oceanogr.* **36:**729–750.

79. **Verity, P. G., D. K. Stoecker, M. E. Sieracki, and J. R. Nelson.** 1993. Grazing, growth and mortality of microzooplankton during the 1989 North Atlantic spring bloom at 47°N, 18°W. *Deep-Sea Res.* **40:**1793–1814.

80. **Welschmeyer, N., R. Goericke, S. Strom, and W. Peterson.** 1991. Phytoplankton growth and herbivory in the subarctic Pacific: a chemotaxonomic analysis. *Limnol. Oceanogr.* **36:**1631–1649.

81. **Wikner, J.** 1993. Grazing rate of bacterioplankton via turnover of genetically marked minicells, p. 703–714. *In* P. F. Kemp, B. F. Sherr, E. B. Sherr, and J. J. Cole (ed.), *Handbook of Methods in Aquatic Microbial Ecology.* Lewis Publishers, Boca Raton, Fla.

82. **Wikner J., A. Andersson, S. Normark, and A. Hagstrom.** 1986. Use of genetically marked minicells as a probe in measurement of predation on bacteria in aquatic environments. *Appl. Environ. Microbiol.* **52:**4–8.

83. **Williams, P. J. L. B.** 1981. Incorporation of microheterotrophic processes into the classical paradigm of the planktonic food web. *Kieler Meeresforsch. Sonderh.* **5:**1–28.

84. **Wright, R. T., and R. B. Coffin.** 1984. Measuring microzooplankton grazing on planktonic marine bacteria by its impact on bacterial production. *Microb. Ecol.* **10:**137–149.

85. **Zweifel, U. L., and A. Hagstrom.** 1995. Total counts of marine bacteria include a large fraction of non-nucleoid containing bacteria (ghosts). *Appl. Environ. Microbiol.* **61:** 2180–2185.

Applications of ^{14}C and ^{3}H Radiotracers for Analysis of Benthic Organic Matter Transformations

GARY M. KING

33

Organic matter transformations in benthic ecosystems affect the cycles of a number of elements, including nitrogen, phosphorus, and sulfur, in addition to carbon and oxygen (10). Rates of organic matter transformation contribute to the establishment of oxygen and sulfide gradients in sediments, patterns of animal distribution and abundance, and the extent of coupling between benthic and pelagic ecosystems (10). However, analyses of benthic organic matter transformations have been hampered by the diversity and complexity of the organics present in most systems. Direct measurements of CO_2 production have proven useful for estimates of system metabolism (1, 18) but do not provide insights about transformations of specific organics or the partitioning of metabolism among various heterotrophic processes. Rates of bulk organic matter metabolism have been inferred from rates of oxygen, nitrate, or sulfate utilization (3, 16, 21), but these assays also lack specificity for individual organics.

Although assays of substrate appearance or disappearance have been applied successfully in some cases to measure metabolic rates for specific organics (e.g., reference 6), the most common and powerful methods have been based on the use of radiotracers. In particular, the availability of ^{14}C- and ^{3}H-labeled organics has facilitated rate measurements that would not otherwise have been possible. However, in most studies, radioassays have been restricted to one or a few dissolved, low-molecular-weight organics, the overwhelming predominance of particulate, high-molecular-weight organic forms in situ notwithstanding. This lack of emphasis on particulate organics arises from two facts: (i) relatively few polymeric radiolabeled organics are routinely available from commercial suppliers, and those that are available hardly reflect the diversity of polymeric organics in situ; and (ii) even when available through commercial or noncommercial sources, polymeric organics are not easily introduced to natural systems, especially sediments or soils, in forms comparable to those that occur naturally. Judicious choices of radiolabeled monomeric organics can at least partially offset these limitations, since rates of polymeric organic metabolism can be inferred from rates of monomer metabolism if certain assumptions are valid.

Since the toolkit for benthic biogeochemists and micro-

bial ecologists is largely confined to low-molecular-weight radiolabeled organics (but see reference 3 for examples of exceptions), this summary will focus on glucose and acetate, both of which are central elements of polysaccharide metabolism and organic matter fermentation. Rather than providing detailed method descriptions or recipes, the intent is to briefly highlight approaches and problems common to the use of radiotracers in general. Wood and Paterek (36) provide a brief but useful description of some of the principles of isotope use and processing.

SUMMARY OF KINETIC AND MASS FLOW APPROACHES

Estimation of Kinetic Parameters for Substrate Uptake by the Wright-Hobbie Method (37)

Prior to the advent of sensitive methodologies for determining the absolute concentrations of specific solutes (e.g., amino acids, glucose, and acetate), radiolabeled substrates were used to estimate turnover times (T_t), maximum uptake or oxidation capacity (V_{max}), and the sum of a transport constant and the ambient substrate concentration ($K_t + S_n$). Appropriate methodologies have been summarized in detail by Gocke (12) and will not be reviewed here. Although Gocke (12) emphasizes the water column, the principles described can be readily applied to sediments (34). Variations in T_t have been considered a useful index of substrate dynamics (7), though estimates of mass flow are preferred. Kinetic data have also been useful for comparing temporal and spatial patterns of V_{max}, often considered a surrogate for the biomass that actively consumes a specific substrate (7). $K_t + S_n$ has provided insights about affinities of natural populations for various substrates, since S_n is often small relative to K_t. Problems include inaccuracies in K_t estimates for complex assemblages and the well-known inaccuracies of the double-reciprocal data transformation (29) used for parameter estimation in the Wright-Hobbie mathematical model. In addition, the data from the partitioning of ^{14}C-labeled substrates between fractions nominally considered biomass and $^{14}CO_2$ have sometimes been used to calculate the extent of substrate respiration (13). These estimates are somewhat doubtful though, since the effects of isotopic dilution have been rarely considered (see below and references 20, 22, and 31). In spite of these limitations, the Wright-Hobbie method and variants thereof remain useful

Contribution 286 from the Darling Marine Center.

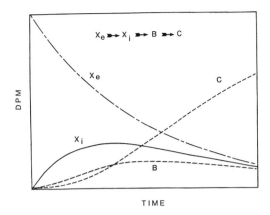

FIGURE 1 Uptake of an exogenous radiolabeled substrate, X_e, added to a pool with a steady-state concentration and time courses of radioactivity in an internal pool of the substrate (X_i), an intermediate metabolite (B), and an end product (C), also with steady-state concentrations. Note that as a result of isotopic dilution, the rate of accumulation of radioactivity in C is not equal to the uptake rate of X_e, nor is C/X_e constant.

in the absence of reliable data on natural substrate concentrations.

Estimation of Turnover Times and Mass Flow

An alternate approach for estimating the turnover times of a given substrate is based on the loss of radiolabeled substrate added as a pulse at tracer levels to a pool with a steady-state concentration during the course of an assay. Under these conditions, the radiotracer disappears exponentially (Fig. 1 and 2) and a time course analysis of tracer concentration can be analyzed to estimate the first-order rate constant in the expression

$$X^*_t = X^*_0 \cdot e^{-kt} \tag{1}$$

where X^*_0 is the radiotracer concentration at time zero, X^*_t is the radiotracer concentration at time t, and k is the uptake rate constant. For a steady-state substrate concentration, the turnover time, T_t, is calculated simply as

$$T_t = 1/k \tag{2}$$

If T_t is a desired parameter, an analysis of radiotracer disap-

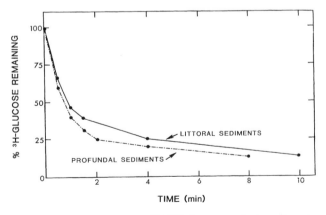

FIGURE 2 Time course of [6-^{3}H]glucose uptake in sediments of Wintergreen Lake, Michigan. Reprinted from reference 22 with permission.

pearance from a given substrate pool is usually preferred over the Wright-Hobbie procedure. In addition, if the uptake of a given substrate at a given concentration conforms to a simple first-order model, e.g., $dQ/dt = -kQ$, then mass flow is simply the product of the estimated uptake rate constant and the substrate concentration. Since many ecologically important substrates can now be measured with sufficient sensitivity for natural concentrations, estimates of mass flow for carbon budgets can be obtained for water column or sediment samples maintained under near in situ conditions.

Some high-performance liquid chromatography (HPLC) and gas chromatographic techniques facilitate analysis of radiotracer specific activity (i.e., kilobecquerels per micromole) as well as substrate concentration. This is especially advantageous since a record of specific activity eliminates the need to assume steady-state conditions during an assay. Variations in substrate concentration and radiotracer specific activity over time during an incubation allow both accurate estimates of mass flow and insights about substrate dynamics (11).

Successful measurement of substrate dynamics by using substrate pool sizes and turnover rates depends to a great extent on the investigators' ability to add radiotracers to samples at true tracer levels, that is, levels very small relative to S_n. In addition, it is obvious that at least ranges of natural substrate concentrations must be known initially, and exact substrate concentrations must be known ultimately for rate estimates. In addition, the comparatively low specific activity of ^{14}C-labeled organics can become problematic, since ^{14}C tracer additions may result in unacceptably low activity during subsequent time courses and sample processing. Tritiated organics have much higher specific activity than ^{14}C-labeled compounds and can provide a suitable alternative when substrate pool sizes are low (8). However, the use of ^{3}H-organics restricts the amount of information that one can obtain, since the fate of carbon cannot be directly measured.

General Caveats and Limitations for Isotope Use

Though powerful, radioisotopic methods have a number of drawbacks. Because of its long half-life and current limitations on low-level radioactive waste, the disposal of ^{14}C wastes has become increasingly problematic and costly. Investigators must consider approaches to minimize waste generation, including the use of biodegradable scintillation fluids, as a routine aspect of experimental design. Regulations and common sense dictate that isotopes must be used in specifically designated and approved laboratory areas, ideally with glassware, pipettors, incubators, and other supplies restricted for isotope use. Contamination must be routinely monitored, typically with a scintillation counter. Access to a scintillation counter is also required for sample assay. Thus, the costs associated with isotope use can become substantial.

Careful attention must be given to experimental design and radioassay. Questions that must be addressed include the following. Is a radiotracer for a given compound available with a specific activity high enough to facilitate tracer studies? What levels of radioactivity are expected in target pools or samples? At what position in a given compound is the radioisotope incorporated, and can transformations at this position be followed unambiguously? Are sample processing and analytical procedures compatible with radioisotopes? Investigators must also consider the effects of isotopic dilution when assaying the movement of radiotracer among multiple pools. Time course analyses of specific activity in precursor pools may be required to establish rates of mass flow into product pools (Fig. 1; see references 9, 15, 17, 32, 33, and 38 for detailed discussions). Proper attention must also be given to errors that arise during scintillation assays when one is counting blanks and samples for a fixed time rather than for a fixed level of counting error. The most reliable results from radioassays are obtained by accumulating 10,000 counts or more per sample, irrespective of counting time. In some instances, counting requirements may necessitate modifications in experimental design. Additional constraints arise from the form and purity of commercially available radiosubstrates. Many ^{14}C- and ^3H-labeled compounds are supplied in solutions of ethanol or buffers that may not be suitable for introduction into samples. Dilution alone is usually insufficient for eliminating potential problems with the carrier solution; even a 1:1,000 dilution of ethanol still results in a solution of about 15 mM. Instead, volatile carrier solutions need to be removed by careful evaporation using a stream of nitrogen; ionizable compounds can be purified by ion exchange. However, the exchange of stock carrier solutions (e.g., ethanol for sterile, deionized water) can result in storage problems, since some radiolabeled compounds, especially those containing ^3H, are susceptible to significant self-decomposition.

Considering all of the limitations of radioisotope use, investigators are well advised to use nonradiotracer methods if possible. The advantages of doing so are illustrated well by comparing chemiluminescence and radiotracer techniques for the visualization of nucleic acids in molecular biology (2). Obviously, there are certain questions or problems for which radiotracer approaches are not just the best choice but the only choice. In other cases, radiotracer approaches should be considered as a last resort.

SUBSTRATE UPTAKE AND TRANSFORMATION
Glucose Uptake Based on [6-^3H]Glucose

Glucose pool sizes in sediments and the water column of lakes and marine systems are typically nanomolar to low micromolar (5, 22). For true tracer studies, these levels require the use of [^{14}C]glucose in amounts that can become limiting for radioassay. Tritiated glucose, because of its high specific activity, provides a suitable alternative, especially if the primary objective of an investigation is to establish turnover times or to determine mass flow by using concentration data as well. The approach is suitable for sediment slurries or undisturbed sediments. Since [^3H]glucose is available at a very high specific activity, direct injections into sediment without mixing are possible. This is often not the case for [^{14}C]glucose. Turnover times are estimated from either direct fits of an exponential curve to time course data of unreacted radioglucose (e.g., Fig. 1 and 2; equations 1 and 2) or a linear regression of log-transformed data. The product of the uptake rate constants thus calculated (k in units of t^{-1}) and measured glucose concentrations provide an estimate of glucose utilization rates: $kh^{-1} \cdot \mu mol\ cm^{-3} = \mu mol\ cm^{-3}\ h^{-1}$.

Availability and Processing of [^3H]Glucose

Tritiated glucose is available in several forms, including double-labeled and single-labeled isomers, as aqueous or ethanolic solutions, and in a chromatographically purified grade. The position of labeling is important, since glucose transformations, especially fermentations, differentially affect hy-

drogen atoms within the glucose molecule. We have typically used [6-^3H]glucose (22, 31) and have chosen aqueous stocks. Although we have not examined chromatographically purified grades, we recommend purifying all stocks prior to use with a strong anion-exchange column. A suitable column can be prepared by loading sterile Dowex-AG1 resin (100/120 mesh), chloride form, or equivalent into sterile 3-cm^3 syringe barrels to create 1-cm^3 bed volumes. Radioglucose stock solutions are passed through the resin and collected in the original isotope container. This procedure removes anionic glucose decomposition products in the stock solution and is essential for accuracy in subsequent radioassays. We have found comparable results with use of either radioglucose solutions preincubated for 24 h in filter-sterilized seawater or solutions diluted in deionized water with no preincubation (31).

Use of Slurries versus Intact or "Undisturbed" Sediments

Tritiated glucose stock solutions are available with concentrations of approximately 20 to 40 μM for activities of 37 TBq ml^{-1} (specific activity = 0.9 to 1.9 TBq mmol^{-1}). High specific activities permit direct injection of tritiated glucose into undisturbed sediments by using protocols analogous to those for measuring sulfate reduction rates (see chapter 34 of this volume). For example, dilution of a stock without added glucose to 0.37 kBq μl^{-1} results in final concentration of 0.2 to 0.4 μM; injection of 10 μl of such a dilute stock into 5 to 10 cm^3 of sediment adds only trace amounts of glucose, with minimal disturbance of glucose concentrations at or near the injection site. A total injection of 0.37 kBq cm^{-3} of sediment provides sufficient radioactivity for accurate analyses of glucose uptake.

Use of sediment slurries as opposed to intact sediments allows additions of even greater radioactivity per cubic centimeter of sediment, since the [^3H]glucose stock is diluted very quickly to suitably low concentrations by mixing. For example, mixing 100 μl of undiluted [^3H]glucose quickly into 10 cm^3 of slurry results in a final added concentration of 0.2 to 0.4 μM with an activity of 370 kBq cm^{-3} of sediment. This level of radioactivity is well above typical minimal requirements for uptake assays.

Design of Uptake Assays and Termination of Activity

Time course analyses using undisturbed sediments can be facilitated by collecting sediment in modified 5-cm^3 syringes (luer tips cut off, and filled syringes stoppered with gray butyl serum bottle closures [Supelco, Inc., Bellafonte, Pa.]). Resolution of 1 to 2 cm for depth profiles is obtained by using ≥6-cm-inner-diameter core tubes with ports bored at intervals into the side of the tube. Two replicate minicores can be obtained (with care) at a given depth from a single core, but one minicore per depth is preferable. Using a minimum of five time points in an assay and triplicates for each time point requires at least 15 minicores per depth interval.

Since glucose turnover times in both freshwater and marine sediments are rapid ($T_t < 10$ min), incubation times must be relatively short. A geometric progression in sampling times (e.g., 1, 2, 4, 8, and 16 min) should be used if possible. Minicores used for the shorter incubation periods (e.g., 1 to 4 min) should be processed individually to ensure precision in the timing of injections and termination. Minicores used for the longer incubation times can be processed in batches by allowing 30 to 60 s between the injections of the replicates for a given time point: inject a series of replicates at 0.0, 0.5, and 1.0 min; terminate at 8.0, 8.5, and 9.0 min. Rapid termination of activity is an obvious necessity. This can be achieved by expressing the contents of a mini-corer into a 15 to 50-ml conical, disposable, plastic centrifuge tube containing a volume of glutaraldehyde (10%) equal to or greater than the volume of sample. Immediate and thorough vortexing results in rapid termination of uptake. The use of minicores terminated immediately after isotope addition provides blank data that can be used to correct the time series results for uptake that occurs during the termination step. Samples treated with glutaraldehyde can be stored frozen, but storage times should be minimized.

Sample Processing and Radioassay

The decrease in activity of unreacted [3H]glucose is measured after centrifugation of the glutaraldehyde-sediment slurry (2,000 to 5,000 × g for 10 min). Triplicate subsamples (0.25 to 1 ml) from each centrifugate are processed as follows.

(i) To determine total dissolved radioactivity, a known volume of centrifugate is transferred to a scintillation vial and assayed by liquid scintillation counting for total dissolved radioactivity.

(ii) To determine total nonvolatile radioactivity, a second volume is transferred to a scintillation vial and amended (400 μl per ml of sample) with a solution of deionized water containing sodium acetate (0.3 M) and a sodium carbonate buffer (0.25 M sodium bicarbonate, 0.1 M sodium carbonate [pH 11]). The resulting solution is dried under an infrared heat lamp in a fume hood to evaporate ^3H_2O; the alkaline acetate solution prevents losses of radiolabeled fatty acid end products from glucose metabolism. After drying, the residue is resuspended and assayed as described above for radioactivity.

(iii) To determine total ^3H_2O and unreacted [3H]glucose, a third subsample is passed through an anion-exchange column similar to that described above for processing of [3H]glucose but with a 5-cm^3 resin in the fluoride form. Nonionic radiolabel (i.e., unreacted [3H]glucose plus ^3H_2O) is eluted from the column with 7 ml of deionized water after loading of the sample. The eluate is assayed as described above for radioactivity. *Cautionary note:* The samples contain glutaraldehyde and should be handled with care. Tritiated water must be evaporated in a fume hood; total radioactivity volatilized must conform to statutory constraints on gaseous tritium release. The very low levels of [3H]glucose required for most analyses should pose few, if any, disposal problems for scintillation fluids, other liquids, or evaporated water.

Processed samples can be assayed readily by liquid scintillation counting using biodegradable scintillation fluids. Since some of these fluids are not compatible with significant volumes of water, comparisons of efficacy are necessary. Quench curves should be prepared for different sample matrices, using procedures recommended for the available counter; standardization is simplified by the availability of certified solutions of tritiated water and toluene. Because of the potentially low activity in some of the above-mentioned fractions and the difference calculations specified below, it is imperative that samples be counted to yield a fixed counting error, not a fixed time. Calculations include the following (after blank correction):

1. ^3H_2O = subsample 1 − subsample 2
2. Unreacted [3H]glucose = subsample 3 − ^3H_2O

Estimating Rate Constants, Pool Sizes, and Uptake Rates

Rate constants can be calculated simply by using nonlinear parameter estimation or linear regression methods for exponentially decreasing unreacted [3H]glucose activity. Departures from an exponential model may indicate non-steady-state behavior in the glucose pool size or variation in uptake rates during incubation. Typically, exponential models fit [3H]glucose data with a high r^2 value (>0.9). Total glucose pool sizes can be estimated by using several approaches, including HPLC (30) and enzyme-based assays (22). The former has the advantage of providing information on multiple monosaccharides, while the latter are relatively inexpensive. In addition, enzyme-based assays presumably avoid problems arising from the presence of chemically detectable but biologically unavailable pools. Regardless of the approach chosen, a separate but parallel set of samples will likely be required unless the method used to terminate activity during the radioassays is compatible with the pool size assays.

Fermentation End Products Assayed with [U-14C]glucose

[3H]glucose is a very suitable probe for measuring glucose uptake, but it provides little useful information on the subsequent specific fate of glucose carbon or on patterns of glucose transformation. [U-14C]glucose and the various specifically labeled glucose isomers serve this purpose well, but not without some caveats. As indicated previously, the specific activities of radioglucose are approximately 100 to 600 times lower than those of tritiated glucose. Although counting efficiencies for ^{14}C are about threefold greater than those for ^3H, it is evident from considerations presented above that true tracer studies may not be possible with [14C]glucose directly injected into undisturbed sediments. True tracer levels of [14C]glucose are readily achieved with slurries, but data from slurries may have little direct relevance for glucose turnover in situ. Sawyer and King (31) examined potential inaccuracies for estimates of the extent and kinetics of glucose oxidation based on the use of slurries versus undisturbed sediments. $^{14}CO_2$ production from slurries with tracer levels of [14C]glucose was compared with results from undisturbed sediments to which various levels of elevated carrier glucose had been added. No significant differences were observed either in time courses or amounts of $^{14}CO_2$ produced. These results suggest that patterns of metabolism might show some short-term independence with respect to elevated glucose concentrations. However, this observation should be verified empirically for each new sediment type examined.

Analyses of carbon transformation patterns must also consider the effects of isotopic dilution. For example, King and Klug (22) showed that conclusions about the fate of glucose carbon differed significantly depending on the incubation time chosen for examining radiocarbon distribution (Fig. 3). Neither $^{14}CO_2$ nor $^4CH_4$ was present at time points comparable to the glucose turnover time (1 to 2 min), even though radioglucose was largely consumed. The true patterns of glucose transformation, e.g., the relative extent of glucose mineralization and methane formation, were evident only after extended incubations of >24 h. The potential problems associated with isotopic dilution have been documented further in studies using pure cultures (20) and marine sediments (31). Smith and Horner (33) have also provided a lucid theoretical and practical treatment of tracer kinetic problems involving multicompartment systems sub-

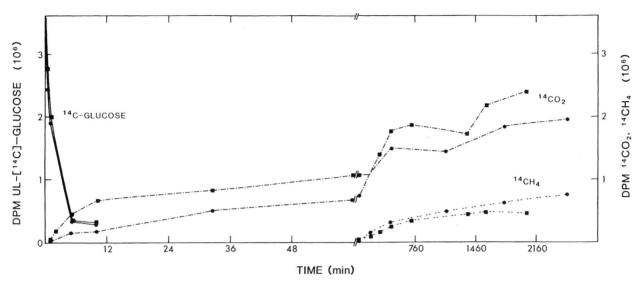

FIGURE 3 Time course of [U-^{14}C]glucose uptake and production of $^{14}CO_2$ and $^{14}CH_4$ in sediments of Wintergreen Lake, Michigan. Reprinted from reference 22 with permission.

jected to pulse-labeling. The tracer kinetic modeling software recommended by Smith and Horner (33) is currently available for desktop and microcomputers using UNIX, DOS, and Macintosh operating systems (SAAM Institute, University of Washington, Seattle). SAAM II has several powerful features, among which are the capability of fitting models to user data sets and providing estimates (with statistics) for rate constants and steady-state compartment sizes. STELLA (High Performance Systems, Inc., Hanover, N.H.) is a rather different software package, also available for Macintosh and DOS operating systems. It provides a better environment for simulating tracer and mass exchanges among compartments but does not allow for statistical fitting of user models to datasets.

[^{14}C]glucose transformations can be analyzed by processing samples as described for [^{3}H]glucose, with the following modifications.

(i) An active or passive distillation apparatus can be used to determine $^{14}CO_2$ production. Briefly, an active distillation system requires flasks or jars with removable closures (e.g., rubber stoppers) that are connected with Teflon or Tygon tubing to a source of compressed gas (e.g., nitrogen) and to vials or tubes containing a liquid CO_2 absorbent (e.g., 0.1 N KOH). After addition of sediment, the flasks are sealed rapidly and acidified by injection of a suitable volume of 1 N HCl. A slow flow of nitrogen (10 ml min^{-1}) is initiated, flushing $^{14}CO_2$ from the flask and into the absorbent. The flask contents can be mixed with a magnetic stirrer during flushing or more simply by passing the nitrogen input through the sediment-acid solution. Flushing periods of 30 to 60 min generally suffice for complete $^{14}CO_2$ absorption; however, trapping efficiencies and optimal conditions can be assessed readily by adding known amounts of sodium [^{14}C]bicarbonate to control sediments. Passive $^{14}CO_2$ trapping systems make use of flasks or tubes that are sealed with stoppers fitted with filter paper traps soaked with a suitable CO_2 absorbent (see reference 20). Such systems are inherently more simple than active distillation and generate less radioactive waste; they are likewise calibrated with sodium [^{14}C]bicarbonate. Regardless of the approach used, $^{14}CO_2$ production provides a measure of apparent, not absolute,

glucose respiration unless the effects of isotopic dilution can be specifically estimated (20). Problems that may arise with both active and passive distillation include chemiluminescence during liquid scintillation assay of the CO_2 traps and excessive degassing for sediments rich in carbonates. The former problem is typically transient and becomes negligible if samples are allowed to stand a few hours after mixing of scintillation fluor and absorbent. The latter problem is solved only through judicious choices of sample size, volume, and concentration of acid solutions; volume of distillation flasks; and volume and concentration of absorbents, all of which must be optimized empirically.

(ii) To measure unreacted [^{14}C]glucose, $^{14}CO_2$-free sediment slurries from the preceding treatment are centrifuged at moderate speeds (1,000 to 3,000 × g) to pellet sediment and other particulates. Known volumes of the centrifugates are neutralized (pH paper provides a suitable indicator), loaded onto anion-exchange columns, and eluted with deionized water (e.g., as described above for determining total ^3H_2O and unreacted [^{3}H]glucose). The radioactivity in the eluate is a measure of the unreacted [^{14}C]glucose. The radioactivity in centrifugates prior to treatment with anion-exchange resins provides a measure of the sum of unreacted [^{14}C]glucose and ^{14}C-labeled fermentation end products (i.e., organic acids).

(iii) A variety of organic acids (e.g., acetate, propionate, and lactate) are produced during glucose utilization in anaerobic sediments. Total ^{14}C-organic acid production can be measured as described above, using anion-exchange resins. Alternatively, pore water from slurries or undisturbed sediments can be analyzed by HPLC, using columns designed for organic acid analysis, and conductimetric or spectrophotometric detection. Column eluates corresponding to the retention times of specific organic acids can be collected manually or by using a fraction collector for radioassay (22, 31). However, the data should be interpreted with caution. Sawyer and King (31) have shown that the radioactivity in specific fractions may not be unambiguously assigned to a given organic acid on the basis of retention times alone. An enzymatic approach has been designed to verify several key organic acids: acetate, lactate, and pyruvate. Briefly, the approach involves parallel assays of samples treated with and without enzymes, i.e., lactate dehydrogenase. The extent of radioactivity lost from a target fraction, i.e., that containing lactate, provides a means of verifying qualitative identities and activity. See reference 31 for methodological details.

(iv) Several approaches are possible for estimating the incorporation of glucose carbon into biomass or particulate carbon. We have used a wet oxidation with active $^{14}CO_2$ distillation for sediments washed with an acetate buffer to remove unreacted [^{14}C]glucose, $^{14}CO_2$, and ^{14}C-organic acids (31). Briefly, sediments are centrifuged (1,000 to 3,000 × g) after terminations of incubations with [^{14}C]glucose. The supernatants are decanted, and the pellets are resuspended in 2 volumes of 0.1 M sodium acetate (pH 7.0). The suspensions are centrifuged, and the acetate wash is repeated twice. After the final supernatants are decanted, the pellets are transferred to two-neck boiling flasks fitted with a condenser and gas inlet tube; the outlet from the condenser is connected to a series of CO_2 traps containing KOH. At least 10 volumes of an acidic potassium persulfate solution is added to each flask; the resulting persulfate-sediment suspensions are boiled for 2 h after the flasks are sealed and a flow of nitrogen is initiated. The method can be standardized conveniently by using cells from a suitable bacterial culture or enrichment grown in the presence of [^{14}C]glucose. A known amount of the radiolabeled cells is collected by filtration; the radioactivity on one set of filters is assayed by liquid scintillation counting. Parallel filters are oxidized as described above in the presence of control sediment. We have obtained conversion efficiencies of about 80% in the latter case (31).

Uptake and Metabolism of Acetate

Acetate is a key intermediate in the anaerobic metabolism of organic matter in both marine and freshwater sediments (10, 25, 26). Acetate turnover and the fate of acetate carbon can be readily analyzed by using the methods described for glucose. Important considerations include the following.

(i) Radioacetate solutions, containing ^3H or ^{14}C, are available at significantly lower specific activities than radioglucose; [^3H]acetate is therefore the best, if not the only, choice for turnover assays.

(ii) Acetate appears to occur in bioavailable and unavailable pools (19, 28, 35); added radiotracers may not equilibrate with the unavailable pools. This can result in overestimates of mass flow if total rather than bioavailable acetate concentrations are measured.

(iii) In freshwater and, to a lesser extent, marine sediments, acetate transformation results in production of $^{14}CH_4$ as well as $^{14}CO_2$. The latter can be measured readily by using active or passive distillation methods as described above. However, assay of $^{14}CH_4$ requires one of several other approaches: a gas proportional counter (22), oxidation in a tube furnace followed by $^{14}CO_2$ trapping (24), or oxidation in the flame of a gas chromatograph equipped with a flame ionization detector and a $^{14}CO_2$ trap (23).

(iv) Acetate concentrations can be measured by two approaches. First, gas chromatography and HPLC with ion or spectrophotometric detection offer relatively sensitive methods that also provide concentrations of other organic acids (14, 27). However, these methods measure both the bioavailable and the unavailable acetate pools. A second, sensitive alternative approach designed to measure bioavailable acetate is based on the use of acetyl coenzyme A synthase with detection of a reaction by product, AMP, by

HPLC (19). A biosensor approach based on the use of *Desulfobacter* species provides yet another alternative, though it is perhaps inconvenient for routine use (35).

CONCLUSIONS

A relatively large number of organic compounds are available with ^3H or ^{14}C radiolabel. Various monosaccharides, amino acids, sugar acids, fatty acids, and nucleic acids, among others, can be purchased in forms suitable for use in assays of benthic biogeochemical transformations. Procedures for using them vary little from those outlined for glucose and acetate. However, in almost all cases, prospective users are encouraged to carefully consider nonisotopic approaches followed by implementation of experimental designs that minimize isotope use. Careful consideration must also be given to the use of time courses, time zero blanks, adsorption or chemical controls, the implications of isotopic dilution for estimates of end product formation, and proper mathematical analysis.

This contribution was supported in part by NSF grants OCE-8700358 and DEB-9107315.

REFERENCES

1. **Andersen, F. O., and B. T. Hargrave.** 1984. Effects of Spartina detritus enrichment on aerobic/anaerobic benthic metabolism in an intertidal sediment. *Mar. Ecol. Prog. Ser.* **16:**161–171.
2. **Ausubel F. M., R. Brent, R. E. Kingston, D. D. Moore, J. G. Seidman, J. A. Smith, and K. Struhl (ed.).** 1992. *Short Protocols in Molecular Biology,* 2nd ed. John Wiley & Sons, Inc., New York.
3. **Benner, R., A. E. Maccubbin, and R. E. Hodson.** 1984. Preparation, characterization, and microbial degradation of specifically radiolabeled [^{14}C]lignocelluloses from marine and freshwater macrophytes. *Appl. Environ. Microbiol.* **47:**381–389.
4. **Berner, R. A.** 1980. *Early Diagenesis.* Princeton University Press, Princeton, N.J.
5. **Burney, C. M.** 1986. Diel dissolved carbohydrate accumulation in coastal waters of South Florida, Bermuda, and Oahu. *Estuarine Coastal Shelf Sci.* **23:**197–203.
6. **Christensen, D.** 1984. Determination of substrates oxidized by sulfate reduction in intact cores of marine sediments. *Limnol. Oceanogr.* **29:**189–192.
7. **Christian, R. R., K. Bancroft, and W. J. Wiebe.** 1978. Resistance of the microbial community within salt marsh soils to selected perturbations. *Ecology* **59:**1200–1210.
8. **Dietz, A. S., L. J. Albright, and T. Tuominen.** 1977. Alternative model and approach for determining microbial heterotrophic activities in aquatic systems. *Appl. Environ. Microbiol.* **33:**817–823.
9. **Dring, M. J., and D. H. Jewson.** 1982. What does ^{14}C uptake by phytoplankton really measure? A theoretical modelling approach. *Proc. R. Soc. Lond. Ser. B* **214:**351–368.
10. **Fenchel, T., and T. H. Blackburn.** 1979. *Bacteria and Mineral Cycling.* Academic Press, London.
11. **Fuhrman, J. A., and R. I. Ferguson.** 1986. Nanomolar concentrations and rapid turnover of dissolved free amino acids in seawater: agreement between chemical and microbiological measurements. *Mar. Ecol. Prog. Ser.* **33:**237–242.
12. **Gocke, K.** 1977. Heterotrophic activity, p. 198–222. *In* G. Rheinheimer (ed.), *Microbial Ecology of a Brackish Water Environment.* Springer-Verlag, New York.

13. **Griffiths, R. P., B. A. Caldwell, and R. Y. Morita.** 1984. Observations on microbial percent respiration values in Arctic and subarctic marine waters and sediments. *Microb. Ecol.* **10:**151–164.
14. **Henderson, M. H., and T. A. Steedman.** 1982. Analysis of C_2-C_6 monocarboxylic acids in aqueous solution using gas chromatography. *J. Chromatogr.* **244:**337–346.
15. **Hobson, L. A., W. J. Morris, and K. T. Pirquet.** 1976. Theoretical and experimental analysis of the ^{14}C technique and its use in studies of primary production. *J. Fish. Res. Board Can.* **33:**1715–1721.
16. **Jørgensen, B. B., and J. Sørensen.** 1985. Seasonal cycles of O_2, NO_3^- and SO_4^{2-} reduction in estuarine sediments: the significance of a NO_3^- reduction maximum in the spring. *Mar. Ecol. Prog. Ser.* **24:**65–74.
17. **Jung, A., and P. Bartholdi.** 1980. Turnover of multicompartment systems. *J. Theor. Biol.* **85:**657–663.
18. **Kepkay, P. E., and F. Ø. Andersen.** 1985. Aerobic and anaerobic metabolism of a sediment enriched with Spartina detritus. *Mar. Ecol. Prog. Ser.* **21:**153–161.
19. **King, G. M.** 1991. Measurement of acetate in marine porewaters by using an enzymatic approach. *Appl. Environ. Microbiol.* **57:**3476–3481.
20. **King, G. M., and T. Berman.** 1984. Potential effects of isotopic dilution on apparent respiration in ^{14}C-heterotrophy experiments. *Mar. Ecol. Prog. Ser.* **19:**175–180.
21. **King, G. M., B. L. Howes, and J. W. H. Dacey.** 1985. Short term endproducts of sulfate reduction: formation of acid volatile sulfides, elemental sulfur, and pyrite. *Geochim. Cosmochim. Acta* **49:**1561–1566.
22. **King, G. M., and M. J. Klug.** 1982. The kinetics of glucose uptake and endproduct formation in Wintergreen Lake sediments. *Appl. Environ. Microbiol.* **44:**1308–1317.
23. **King, G. M., M. J. Klug, and D. R. Lovley.** 1983. Metabolism of acetate, methanol, and methylated amines in intertidal sediments of Lowes Cove, Maine. *Appl. Environ. Microbiol.* **46:**1848–1853.
24. **King, G. M., and W. J. Wiebe.** 1980. Tracer analysis of methanogenesis in salt marsh soils. *Appl. Environ. Microbiol.* **39:**877–881.
25. **Kristensen, E., G. M. King, M. Holmer, G. Banta, M. J. Jensen, K. Hansen, and N. Bussawarvit.** 1994. Sulfate reduction, acetate turnover and carbon metabolism in sediments of the Ao Nam Bor mangrove forest, Thailand. *Mar. Ecol. Prog. Ser.* **109:**245–255.
26. **Lovley, D. R., and M. J. Klug.** 1982. Intermediary metabolism of organic matter in the sediments of a eutrophic lake. *Appl. Environ. Microbiol.* **43:**552–560.
27. **Parkes, R. J., and J. Taylor.** 1983. Analysis of volatile fatty acids by ion-exclusion chromatography, with special reference to marine pore water. *Mar. Biol.* **77:**113–118.
28. **Parkes, R. J., J. Taylor, and D. Jorck-Ramberg.** 1984. Demonstration, using *Desulfobacter* spp., of two pools of acetate with different biological availabilities in marine pore water. *Mar. Biol.* **83:**271–276.
29. **Robinson, J. A.** 1985. Determining microbial kinetic parameters using nonlinear regression analysis: advantages and limitations in microbial ecology. *Adv. Microb. Ecol.* **8:**61–114.
30. **Rocklin, R. D., and C. A. Pohl.** 1983. Determination of carbohydrates by anion exchange chromatography with pulsed amperometric detection. *J. Liq. Chromatogr.* **6:**1577–1590.
31. **Sawyer, T. E., and G. M. King.** 1993. Glucose uptake in an intertidal marine sediment: metabolism and endproduct formation. *Appl. Environ. Microbiol.* **59:**120–128.
32. **Sheppard, C. W., and A. S. Householder.** 1951. The mathematical basis of the interpretation of tracer experiments in closed steady-state systems. *J. Appl. Phys.* **4:**510–520.
33. **Smith, D. F., and S. M. J. Horner.** 1981. Tracer kinetic analysis applied to problems in marine biology. *Bull. Fish. Aquat. Sci. Can.* **210:**113–127.
34. **Toerien, D. F., and B. Cavari.** 1982. Effect of temperature on heterotrophic glucose uptake, mineralization, and turnover rates in lake sediments. *Appl. Environ. Microbiol.* **43:**1–5.
35. **Wellsbury, P., and R. J. Parkes.** 1995. Acetate bioavailability and turnover in an estuarine sediment. *FEMS Microbiol. Ecol.* **17:**85–94.
36. **Wood, W. A., and J. R. Paterek.** 1994. Physical analysis, p. 465–511. *In* P. Gerhardt, R. G. E. Murray, W. A. Wood, and N. R. Krieg (ed.), *Methods for General and Molecular Bacteriology.* American Society for Microbiology, Washington, D.C.
37. **Wright, R. T., and J. E. Hobbie.** 1965. The uptake of organic solutes in lake water. *Limnol. Oceanogr.* **10:**22–28.
38. **Zilversmit, D. B., C. Entenman, and M. C. Fishler.** 1943. On the calculation of "turnover time" and "turnover rate" from experiments involving the use of labeling agents. *J. Gen. Physiol.* **26:**325–331.

Sulfur Cycling

MARK E. HINES, PIETER T. VISSCHER, AND RICHARD DEVEREUX

34

Sulfur (S) is an important element biochemically and geochemically. It constitutes ~1% of the dry mass of organisms, in which it serves many structural and enzymatic functions. S also acts as a significant electron donor and acceptor during many bacterial metabolisms (28). S can be found in a range of valence states from the highly reduced sulfide (-2) to the most oxidized form in sulfate (SO_4^{2-}) ($+6$). There are several intermediate valence forms of S that can act as both electron donors and electron acceptors, depending on environmental conditions, the most notable being elemental sulfur (S^0) and thiosulfate ($S_2O_3^{2-}$) (44). Microbial S transformations are closely linked with the carbon cycle in which S reduction coupled with organic matter utilization is a major mineralization pathway in anaerobic habitats, while S oxidations, some of which are autotrophic and/or phototrophic (28, 65), can occur aerobically and anaerobically. Many S compounds are highly reactive, and microorganisms often must compete with abiotic reactions. This competition makes study of certain physiological types of S-oxidizing bacteria difficult. In addition, the S cycle is complicated further by the reactivity of sulfide with metals and the oxidation of metal sulfides by bacteria. In general, the high chemical and biological reactivity of S compounds results in a tight coupling of the oxidative and reductive portions of the S cycle in aquatic habitats, particularly at the redoxcline, where S cycling can be extremely rapid (28).

Microorganisms of the S cycle are extremely diverse. The anaerobic SO_4^{2-}-reducing bacteria (SRB), which are unique physiologically and genetically, are represented by several genera, most of which were discovered in the last 15 years (15, 67). Once thought to be restricted primarily to SO_4^{2-} respiration or fermentation, SRB have been demonstrated in recent studies to be capable of utilizing iron, manganese, and even O_2 as electron acceptors during the oxidation of reduced S compounds (40). Chemolithotrophic S oxidation is mediated aerobically by colorless S bacteria, some purple S bacteria, and SRB. Anaerobically, nitrate-respiring chemolithotrophs oxidize sulfide, and both oxygenic and anoxygenic phototrophic bacteria use sulfide as an electron donor for photosynthesis (61, 63). Intermediates produced during sulfide oxidation can be either oxidized, reduced, or fermented (disproportionated) (16, 30). Although S reduction and oxidation are often studied separately for convenience, it is becoming clear that in many instances both processes must be considered simultaneously.

SAMPLE COLLECTION AND HANDLING

For enumeration, isolation, and rate measurements, it is important to process samples quickly after they are obtained and to maintain samples near ambient temperatures (26). For anaerobes, collect and handle samples anoxically. Typical water samplers include Niskin bottles and bags. To prevent contamination by O_2 which diffuses through or from plastic, water samples should be transferred to glass bottles by allowing them to overflow before sealing. If rubber stoppers are used, they should be the type that retard O_2 diffusion. Butyl rubber is best, and thick butyl stoppers (e.g., Bellco 2048-11800) can be penetrated many times with needles without O_2 penetration. Since many S compounds are reactive with rubber stoppers, and most stoppers release some S compounds, Teflon-lined butyl septa (e.g., Wheaton 224168) or Teflon Mininert valves (Supelco 3-3305) are a useful alternative. However, they usually cannot withstand repeated needle penetration.

Sediments are obtained by using grabs or corers. Ideally, for rate measurements, use small-diameter (3 cm) whole-core (~15 cm long) incubations in which amendments of dissolved constituents are introduced axially via microliter syringes through predrilled holes filled with silicone sealant. Whole cores should have the overlying water removed prior to incubation to prevent depletion of O_2 in surficial layers due to stagnant water. In aquatic sediments in which O_2 penetration is less than 10 to 15 mm, most of the sulfide-oxidizing activity takes place in the upper 20 to 30 mm; therefore, cores for S oxidation measurements should sample the upper 3 to 5 cm. Sediments removed from cores should be handled in N_2-filled glove bags or boxes. However, most relatively active sediments can withstand short exposures to laboratory air. Cores should be used within 24 h after collection and should not be exposed to temperatures above the ambient temperature, since even short exposures to elevated temperatures will increase rates.

Sedimentary pore water samples for chemical measurements can be obtained in several ways: (i) centrifugation of sediments under N_2 or centrifugation (2 min at 8,000 × g) of small volumes in a microcentrifuge filter unit (0.45-μm pore size; Gelman Sciences, Ann Arbor, Mich.); (ii) squeezing of samples under inert gas flow (N_2), using prefabricated squeezers (48); (iii) use of in situ dialysis samplers referred to as peepers (20), which must equilibrate with pore water for ~2 weeks (these cannot be used when sediments are subjected to tides); and (iv) use of in situ "sippers," which are lysimeters with porous Teflon collars that allow for the removal of pore water in a few minutes under a slight vacuum (23). In all cases, samples must not come in contact with O_2, and except for peeper samples, which do not have to be filtered, pore waters must be filtered anoxically. When sampling sites such as marshes or seagrass beds that contain macrophytes, avoid sampling techniques that disturb the sediments, since damaged roots leak dissolved material (25). Sippers are well suited for vegetated sediments (21).

SULFATE REDUCTION

SRB are a specialized group of anaerobic bacteria that are responsible for the dissimilatory reduction of SO_4^{2-} to sulfide, which is linked to organic matter oxidation. They are important in the anaerobic degradation of organic matter in most aquatic habitats, where they are situated at the terminal end of the anaerobic food chain. SRB are also a major source of sulfide, which is highly reactive, geochemically important, and an important source of sulfide for use by S-oxidizing bacteria described below. Because of the abundance of SO_4^{2-} in seawater, dissimilatory SO_4^{2-} reduction is considered a major process in marine sediments, yet it is also important in the decomposition of organic material in anoxic freshwater habitats. Although the reduction of SO_4^{2-} is considered to be the classic role of SRB in the environment, recent studies have demonstrated that these bacteria are capable of diverse metabolisms including metal and oxygen reduction, metal methylation and demethylation, organic fermentations, S disproportionation, and the utilization of various intermediate redox states of S. It is generally accepted that SRB oxidize products of fermentative bacteria such as fatty acids, alcohols, some aromatic acids, a few amino acids, and hydrogen. The suite of substrates used varies among genera. For detailed information on SRB, various reviews are available (19, 44, 51, 67, 68).

Rate Measurements

Measurements of rates of SO_4^{2-} reduction have become routine in studies of the biogeochemistry of anoxic aquatic environments. Rates can be determined by monitoring the loss of SO_4^{2-} in samples incubated anaerobically over time (jar experiments) or by determining the rate of production of reduced ^{35}S end products generated during the reduction of $^{35}SO_4^{2-}$ in incubated samples. The ^{35}S technique is preferable since it uses a short incubation period (1 day or less) and undisturbed sediment. However, if one is unable to use the radiotracer technique, a jar approach provides an estimate of SO_4^{2-} reduction. Minimize manipulation in all cases, especially the mixing of oxic and anoxic layers in sediments. See reference 26 for a discussion of effects of mixing and dilution on measuring rates of sedimentary SO_4^{2-} reduction.

Jar Technique

Homogenize a sediment sample and either distribute subsamples among several vessels which are sacrificed over time or maintain one vessel in which subsamples are removed over time. For the former, use centrifuge tubes (15 to 50 ml) so that pore water can be extracted easily. When using plastic, preincubate tubes in an inert atmosphere for ~2

weeks and incubate sample-filled tubes similarly, since many plastics are permeable to O_2 and are capable of bleeding O_2 dissolved in plastic. Diffusion-retardant plastic bags can be used without additional precautions (8). At selected time intervals, extract and filter pore waters and measure SO_4^{2-} content (analytical techniques are described below). Determine the wet and dry weights of known volumes of sediment so that rates can be expressed on a sediment volume basis. Incubation times should be kept to a minimum, and a linear decrease in SO_4^{2-} content is desirable since nonlinear losses are difficult to quantify. For nearshore marine sediments, incubation times are from a few days to weeks. Since freshwaters contain much less SO_4^{2-}, incubation times are short (a few days to <2.0 h).

Consumption of SO_4^{2-} is calculated from the slope of the concentration over time ($\partial[SO_4^{2-}]/\partial t$). The consumption rate is expressed in either amount (micromoles or millimoles) per volume (e.g., cubic centimeter of sediment or milliliter of pore water) or per weight (gram of sediment) per unit of time (minute, hour, or day). Rates are usually expressed as nanomoles per cubic centimeter of sediment per day or integrated over depth as millimoles per square meter per day.

^{35}S Technique

The radiotracer technique is the most widely used method for measuring rates of SO_4^{2-} reduction. $^{35}SO_4^{2-}$ is introduced in tracer amounts in small volumes, and ^{35}S end products are isolated. Unlike the jar technique, which measures SO_4^{2-} loss, the accumulation of labeled end products is very sensitive; therefore, incubation times are short and undisturbed samples can be used. Early studies measured the accumulation of acid-volatile sulfides only. However, since the discovery that significant quantities of reduced ^{35}S are recovered as non-acid-soluble phases (i.e., S^0 and FeS_2) (24), these are included by using a chromium reduction procedure (71). The rapid isotopic exchange of ^{35}S among reduced pools (18) precludes the use of this technique as a measure of the rate of production of specific reduced S species. Therefore, rates are determined from the sum of all inorganic reduced S compounds. The following is a one-step distillation procedure for measuring rates of SO_4^{2-} reduction. See reference 17 for additional details.

Radiotracer (2.0 μl of $^{35}SO_4^{2-}$ in weak acid or sterile water, ~1.0 μCi) is introduced, via microliter syringe through longitudinal holes filled with silicone sealant, into small (~3-cm) diameter short (15-cm) cores while the syringe is withdrawn. It is also possible to take horizontal subcores in syringes which are sealed and injected (22). For samples that are difficult to subcore, such as salt marsh peats which contain thick rhizomes, large-diameter cores are sliced into small sections which are introduced into 5-ml syringes without homogenization. To determine the SO_4^{2-} content of the sediments, additional cores are sectioned horizontally and pore water is extracted by centrifugation. After incubating in the dark for 2 to 24 h at ambient temperature, cores are cut into segments which are mixed with 20% zinc acetate and then frozen to fix sulfides and terminate activity. Incubation time depends on the SO_4^{2-} content of the sediment, with shorter incubation periods used when SO_4^{2-} levels are low. Try to avoid a significant loss (>10%) of ambient SO_4^{2-} during incubation. Thawed samples are centrifuged, and the activity of the remaining $^{35}SO_4^{2-}$ is determined by scintillation counting. When individual subcores are used, the ^{35}S activity of this supernatant is not required as long as the activity of the injected ^{35}S is known.

The pellet is washed twice with seawater to remove excess $^{35}SO_4^{2-}$, transferred to a reaction vessel, and mixed with 10 ml of ethanol. The vessel is connected to a condenser and contains ports for introducing reagents and a constant stream of N_2 gas (7). The top of the condenser is connected to two traps in series, each containing 10 ml of a solution of 6% zinc acetate and 10% ammonium hydroxide. Two drops of Antifoam B (Baker Chemicals) are added to the traps to prevent foaming, which can be excessive for samples with relatively high reduced S levels. The reaction vessel is flushed with N_2 for 20 min, and then 40 ml of $CrCl_2$ and 20 ml of concentrated HCl are added. The contents are gently boiled for 45 to 60 min with a constant stream of N_2. The total reduced S is converted to H_2S, which is swept into the traps and collected as ZnS. Over 98% of the ^{35}S is collected in the first trap. Half of the trap material (~5.0 ml) is mixed with a gel-forming scintillation cocktail and counted. If desired, the sulfide content of the remaining ZnS can be determined colorimetrically (9) for a measure of the pool size of total reduced inorganic S.

The reduced chromium solution is prepared by percolating 1.0 M $CrCl_3\cdot6H_2O$ (pH 1.0) through a column filled with amalgamated zinc granules (grain size of 0.5 to 1.5 mm). The zinc is amalgamated by soaking briefly in a solution of $HgCl_2$ (~0.25 M), transferred to a glass column (~2-cm inside diameter), and washed with 3 volumes of 0.1 M HCl. The $CrCl_3$ solution, which is dark green, changes to a bright blue color when reduced ($CrCl_2$) and is collected within syringes or in serum bottles previously flushed with N_2. It is important that only bright blue reduced chromium be used in reaction vessels. If the solution elutes from the column as even slightly greenish blue, it should be discarded. Only freshly prepared $CrCl_3\cdot6H_2O$ solutions should be run through zinc columns, since solutions a few days old will not reduce adequately. Also use only $CrCl_3\cdot6H_2O$ which is a fine green powder. $CrCl_3\cdot6H_2O$ in large chunks that exhibit a slight bluish hue will never completely reduce, even if recently manufactured.

The rate of SO_4^{2-} reduction is calculated as

$$rate = \frac{[SO_4^{2-}]\cdot(^{35}S\text{-TRIS})\cdot1.06}{(^{35}SO_4^{2-}+{}^{35}S\text{-TRIS})\cdot t}\text{ nmol of }SO_4^{2-}\text{ cm}^{-3}$$

(whole sediment) day^{-1}

where $[SO_4^{2-}]$ is the SO_4^{2-} concentration in nanomoles per cubic centimeter (calculated from pore water values and water content of sediment determined from wet and dry weights), ^{35}S-TRIS is the total radioactivity of ZnS, $^{35}SO_4^{2-}$ is the total radioactivity of the supernatant (the $^{35}SO_4^{2-}$ left after incubation), t is the incubation time in days, and 1.06 corrects for isotope fractionation (26).

Cultivation, Enrichment, Enumeration, and Isolation

SRB are readily isolated from many environments, including freshwater and salt water, soils, oil-bearing shales and strata, intestinal contents, sewage, and paper mill effluents (47). In addition to the common mesophilic forms, thermophilic and hyperthermophilic forms have also been isolated. Although SO_4^{2-} reduction has been shown to occur in the oxic layer of microbial mats and SRB can survive brief exposure to air, SRB have so far been cultivated only under anoxic, reducing conditions. The physical removal of O_2 from a medium by boiling or sparging with nitrogen is often sufficient to permit growth of robust strains, since sulfide may be carried over in the inoculum. However, the use of

a prereduced, "poised" medium is usually necessary, particularly for fastidious strains or when SRB are present in low numbers as would be obtained from a dilution series. The greatest variety of SRB has been isolated by using a bicarbonate-buffered medium that contains SO_4^{2-}, an electron donor, and sulfide as the reductant. Background on culturing anaerobic bacteria is provided in chapter 6. Many practical details of cultivating SRB have been described elsewhere (47, 67).

Composition and Preparation of Media

The following general-purpose media can be used to cultivate most types of mesophilic or moderately thermophilic SRB, although no one particular formulation is suited to all types. As with all anaerobes, cultures are incubated in the absence of air by filling the containers as completely as possible or using a headspace filled with anoxic gas. Use septa or stoppers made from butyl rubber. Incubation temperatures are typically 30°C for mesophilic strains and 55°C for thermophilic strains.

Postgate's Media

Postgate's media (47) are useful for routine cultivation of *Desulfovibrio* and *Desulfotomaculum* spp. and are relatively simple to prepare. They may be solidified with 10 g of agar per liter. N_2 may be used as the headspace gas.

Postgate's medium B

KH_2PO_4	0.5 g
NH_4Cl	1.0 g
$CaSO_4$	1.0 g
$MgSO_4\cdot7H_2O$	2.0 g
Sodium lactate	3.5 g
Yeast extract	1.0 g
Ascorbic acid	0.1 g
Thioglycolic acid	0.1 g
$FeSO_4$	0.5 g
Tap water	1.0 liter

Adjust pH to between 7.0 and 7.5 and autoclave under N_2. Add NaCl to 2.5% for marine strains or use seawater. This medium forms a precipitate which is useful for long-term maintenance of cultures, since it is thought to provide a microhabitat. The ingredients may be prepared as stock solutions. However, the reductants thioglycolate and ascorbate must be freshly added, as they readily deteriorate in air. The completed medium should be used as soon as possible; deterioration of the reductants is indicated by the transient appearance of a purple color. The medium diagnostically blackens upon growth of SRB.

Postgate's medium C

KH_2PO_4	0.5	g
NH_4Cl	1.0	g
$NaSO_4$	4.5	g
$CaCl_2\cdot6H_2O$	0.06	g
$MgSO_4\cdot7H2O$	0.06	g
Sodium lactate	6.0	g
Yeast extract	1.0	g
$FeSO_4$	0.004	g
Sodium citrate·$2H_2O$	0.3	g
Distilled water	1.0	liter

Adjust pH to ~7.2 and autoclave under N_2. For marine strains, prepare with NaCl at 2.5%. The medium may be cloudy after autoclaving. However, this is generally a clear medium, the iron being chelated by the citrate, and is useful for large-scale cultures of *Desulfovibrio* or *Desulfotomaculum* spp.

Widdel and Pfennig Medium

Widdel and Pfennig medium, a defined medium described in detail by Widdel and Bak (67), was used to isolate most of the newly identified genera of SRB. All SRB cultivated to date can be grown on a variation of this medium. Adjust the basal salts solution with respect to NaCl, $MgCl_2 \cdot 6H_2O$, and $CaCl_2$ for preparation of freshwater, brackish, and marine media, respectively. Select electron donors according to the physiology of the strain to be grown or the type to be enumerated. Assemble the medium under 80% N_2–20% CO_2 from separately prepared solutions added to the basal salts solution in the order given below. To make a solid medium, it is recommended that the agar be washed three times in distilled water to remove inhibitors and growth substrates (67), added to the basal salts (10 g/liter, final concentration), and then autoclaved. The sterile solution must be kept at ~50°C to keep the agar molten while the other additions are made.

Solution A, basal salts

	Freshwater		Brackish		Marine	
Na_2SO_4	4.0	g	4.0	g	4.0	g
KH_2PO_4	0.2	g	0.2	g	0.2	g
NH_4Cl	0.25	g	0.25	g	0.25	g
NaCl	1.0	g	7.0	g	20.0	g
$MgCl_2 \cdot 6H_2O$	0.4	g	1.2	g	3.0	g
KCl	0.5	g	0.5	g	0.5	g
$CaCl_2 \cdot 2H_2O$	0.1	g	0.1	g	0.15	g
Resazurin (0.1%)	1.0	ml	1.0	ml	1.0	ml
Distilled water to	950	ml	950	ml	950	ml

Adjust the volume of water used for subsequent additions; the amounts of salts given are for 1 liter of the completed medium. Add salts to stirred water to prevent the formation of precipitates. Prepare the solution in a stoppered vessel for making subsequent additions and dispense the completed medium. After autoclaving, cool the solution under N_2-CO_2 and make additions under the anaerobic gas while mixing on a magnetic stir plate. The final pH of the medium should be between 7.0 and 7.3.

Solution B, trace elements (69)

HCl (66%, 8 M).	12.5 ml
$FeSO_4 \cdot 7H_2O$.	2,100 mg
H_3BO_3.	30 mg
$MnCl_2 \cdot 4H_2O$	100 mg
$CoCl_2 \cdot 6H_2O$.	190 mg
$NiCl_2 \cdot 6H_2O$.	24 mg
$CuCl_2 \cdot 2H_2O$.	2 mg
$ZnSo_4 \cdot 7H_2O$.	144 mg
$Na_2MoO_4 \cdot 2H_2O$	36 mg
Distilled water	987 ml

Autoclave under N_2 and add 1.0 ml/liter of medium.

In addition, some strains require additional trace elements (prepared in 100 ml of distilled water): NaOH, 40 mg; $Na_2SeO_3 \cdot 5H_2O$, 0.6 mg; and $Na_2WO_4 \cdot 2H_2O$, 0.8 mg. Autoclave under nitrogen, and add 1.0 ml/liter of medium.

Solution C, bicarbonate (1.0 M)

$NaHCO_3$.	8.4 g
Distilled water.	100 ml

Saturate the solution with CO_2 by sparging or by shaking in a tightly closed bottle with one-third of the capacity left as headspace that is replenished several times with CO_2. The solution may be filter sterilized or autoclaved in bottles with secured seals under CO_2. Add 30 ml/liter of medium.

Solution D, organic substrates

Prepare separately as neutralized solutions (usually 1.0 to 2.0 M) and autoclave under N_2. Select one for use, depending on the physiology of the strain to be grown or the type to be enriched. The ranges of substrate concentrations used (millimolar) are as follows: acetate, 10 to 20; benzoate, 2 to 5; butyrate, 5 to 12; caproate, 2 to 5; ethanol, 10 to 20; formate, 10 to 20; lactate, 10 to 40; propionate, 10 to 20; succinate, 10 to 20; and pyruvate (filter sterilized), 10 to 40. Add 5 to 15 ml of the concentrated stock solution per liter medium.

Solution E, vitamins

Prepare separately to ensure stability, filter sterilize, and keep in the dark.

Mixed vitamins

4-Aminobenzoic acid.	4 mg
D-(+)-biotin	1 mg
Nicotinic acid.	10 mg
Calcium D-(+)-pantothenate.	5 mg
Pyridoxine dihydrochloride	15 mg
Sodium phosphate buffer (10 mM, pH 7.1).	100 ml

Add 1.0 ml/liter of medium.

Thiamine

Thiamine chloride.	10 mg
Sodium phosphate buffer (25 mM, pH 3.4).	100 ml

Add 1.0 ml/liter of medium.

Vitamin B_{12}

Cyanocobalamin.	5 mg
Distilled H_2O.	100 ml

Add 1.0 ml/liter of medium.

Solution F, sulfide

$Na_2S \cdot H_2O$.	4.8 g
Distilled water.	100 ml

Sodium sulfide crystals deteriorate in air. Therefore, select clean crystals or rinse crystals with distilled water if needed. Sparge the water with N_2 and maintain crystals under N_2 while they are dissolving. Autoclave under N_2 and add 7.5 ml/liter of medium.

Dithionite solution

O_2-free distilled water (autoclaved under nitrogen).	10 ml
$Na_2S_2O_4$ (weighted aseptically).	0.2 g

Add 1.0 ml/liter of medium.

The more fastidious species of SRB are stimulated by the addition of sodium dithionite, which further reduces the medium. Add this reductant after the medium has already been reduced with sulfide. Aqueous sodium dithionite rapidly breaks down with exposure to O_2 and must be prepared with O_2-free water under N_2. A small tube with a side arm to introduce a sterile stream of N_2 is recommended for preparation and storage of dithionite, which will remain stable for about 5 days if not exposed to air. Since dithionite crystals are usually sterile because of their toxicity, an estimated solid amount is added directly from the reagent bottle to the medium, using a sterile spatula fashioned from a platinum wire.

Inoculation, Transfer, and Storage

Use a 1 to 10% (vol/vol) inoculum, with the higher amount for more slowly growing strains. Inoculations are made with a syringe needle through septum seals or with a pipette for stoppered cultures. When using a pipette, place the tip beneath the surface of the medium to limit exposure of the inoculum to air and flush the headspace with the anaerobic gas as the stopper is replaced. Perform transfers on a 6- to 12-week basis. A stringiness becomes apparent as a result of lysis of cells in aged cultures. For storage, remove cultures from the incubator before they attain stationary phase and store them at 4 to 6°C in the dark. Care should be taken, as some strains may expire if refrigerated. Postgate's medium B is suited to the storage of *Desulfovibrio* and *Desulfotomaculum* spp. For long-term maintenance, cultures containing 5 to 10% dimethylsulfoxide may be kept under liquid nitrogen.

Enumeration and Isolation, MPN, and Solid Media

Dilution of samples into liquid media can be used to enumerate SRB by the most probable number (MPN) technique (39). Containers should be refrigerated until the samples are processed, and samples are diluted in poised media and shaken to dislodge and separate cells, since many SRB grow in clumps. Growth of SRB is confirmed by blackening of media that contain an iron indicator, such as Postgate medium B, or by growth in the media of Widdel and Pfennig. With the latter, growth is indicated by the appearance of turbidity, which is measured by increase in A_{600} against an uninoculated blank, and should be confirmed by chemically testing for SO_4^{2-} reduction or sulfide production by one of the assays given below. Transfer an inoculum from a culture obtained with the highest dilution of the series to a medium solidified with agar to obtain an isolate in pure culture.

Enumerations are also performed by counting the formation of colonies in solid media. Serially dilute samples in tubes, held at 40°C, that contain the liquefied medium. Use a pipette to make a transfer, and briefly distribute the inoculum while keeping the tip beneath the surface of the medium. Solidify the agar as a layer over the inside of the tube (roll tubes) or as a plug. Incubate tubes in an inverted position so that condensation collects away from the forming colonies and where it can be easily removed. As noted above, colonies of SRB blacken in the presence of the iron indicator. In media without an indicator, the colonies may contain a slight yellowish, reddish, or grayish pigment.

Obtain pure cultures by picking well-separated colonies from the agar with a drawn Pasteur pipette. Suspend the colony mass in a few milliliters of anoxic saline or medium to inoculate a small volume (~10 ml) of medium. Reserve some of the inoculum for microscopic inspection.

Tests for Purity

Microscopically observe the cell suspension for uniform morphology. It may be necessary to repeat selection of a colony. Postgate (47) offers these additional tests. Contamination of a culture by an aerobe can readily be determined by plating onto nutrient agar containing glucose and peptone; no colonies should appear after the plate is incubated in air at 30°C. To test for contaminant anaerobes, prepare 25 ml of molten peptone-glucose agar (pH 7.0 to 7.6) and add sterile $Fe(NH_4)_2(SO_4)_2$ to 0.05%. Cool the medium to 40°C, dispense into tubes, and make a dilution series from

the test culture. Allow the agar to solidify as a plug, and incubate the sample under N_2. Black colonies should appear; colonies that are not black are contaminants.

Molecular Techniques for Density and Diversity Measurement

It is now well understood that only a very small percentage of the bacteria observed in an environmental sample can be cultivated. In recent years, molecular methods based on nucleic acid hybridization probes or gene cloning have been developed to examine natural populations of bacteria directly without cultivation. SRB have historically been difficult to isolate and identify with culture techniques. Most genera have been described fairly recently. Yet it has been determined that culturable numbers of SRB, based on MPN determined by using defined media of Widdel and Pfennig, can account for measured rates of SO_4^{2-} reduction (31, 59). Nonetheless, a morphologically unique and undescribed vacuolated spirillum was obtained by enrichment on acetate. Similarly, 16S rRNA genes cloned from a marine sediment indicated that there are phylogenetic lines of SRB not yet represented in the culture collection (14).

rRNA-Based Methods

The methods are based largely on oligonucleotide probes designed to hybridize to 16S rRNA sequences and take advantage of the agreement between the phylogeny and physiology of the gram-negative mesophilic SRB (11, 12, 15). Details on the utilization of rRNA probes are given elsewhere in this manual (chapter 10). Oligonucleotide probes may be used to measure the rRNA abundance of a targeted group by hybridization to nucleic acids extracted from a natural microbial community and immobilized on a membrane support. The probes may also be used to directly observe and enumerate targeted cells in a sample by whole-cell hybridization (2). The oligonucleotides may also serve as PCR primers for the amplification of 16S rRNA genes of a targeted group from an environmental sample (3, 14, 42).

Seven oligonucleotides have been developed as hybridization probes or PCR primers for SRB (1, 3, 13). Four are genus specific [targeted group, probe number, sequence (*Escherichia coli* sequence numbers)]: *Desulfobacterium*, 221, TGCGCGGACTCATCTTCAAA (221 to 240); *Desulfobacter*, 129, CAGGCTTGAAGGCAGATT (129 to 146); *Desulfobulbus*, 660, GAATTCCACTTTCCCCTCTG (660 to 679); and *Desulfovibrio*, 687, TACGGATTTCACTCCT (687 to 702). Three encompass more diverse assemblages: the phylogenetic lineage composed of *Desulfococcus multivorans*, *Desulfosarcina variabilis*, and *Desulfobotulus sapovorans*, 814, ACCTAGTGATCAACGTTT (814–831); and a probe specific for a broader assemblage composed of *Desulfobacterium* spp., *Desulfobacter* spp., *D. multivorans*, *D. variabilis*, and *D. sapovorans*, 804, CAACGTTTACTGCGTGGA (804 to 821). The remaining probe, SRB385F, CGGCGTCGCTGCGTCAGG (385 to 402), is selective for most, but not exclusive for, gram-negative SRB and has been used primarily as a PCR primer (1, 14).

DNA-DNA Hybridizations

A technique termed reverse sample genome probing (RSGP) has been developed and used to study SRB in oil field production waters, soil, and mining wastes (53, 66). In RSGP, genomic DNAs from pure cultures are applied to membranes and serve as standards against which radiolabeled DNA obtained from a microbial community by either direct extraction or selective enrichment is hybridized. The pure cultures may be obtained from collections or isolated from the environment under investigation. The genomes used as standards are chosen on the basis of having no or limited cross-hybridization. RSGP was shown to generate reproducible fingerprints of microbial communities of SRB. Quantitation of hybridization signals indicated enrichment of SRB in biofilms on metal surfaces that had contact with production water.

SULFUR OXIDATION

Sulfide is produced from degradation of sulfur-containing organic matter and by dissimilatory SO_4^{2-} reduction. Microbial metabolism of sulfide competes with chemical oxidation, by either O_2 or Fe^{2+}, while emissions of H_2S from sediments and the water column present an alternative but small loss factor. During biological and chemical conversion of sulfide (oxidation state = -2) to SO_4^{2-} ($+6$), a variety of intermediates (-1 to $+5$) are formed. Some of these intermediates, such as polysulfides, react chemically with organic matter; during these reactions, organic S compounds can be formed. Metabolic pathways discussed here include those which are of particular environmental importance. Under oxic conditions, chemolithotrophic sulfide oxidation takes place by colorless S bacteria (thiobacilli and filamentous sulfide oxidizers) and some purple S bacteria and SRB. In the absence of O_2, nitrate-respiring chemolithotrophs engage in sulfide oxidation, and both oxygenic and anoxygenic phototrophic bacteria use sulfide as an electron donor for photosynthesis. Intermediates which are produced during sulfide oxidation can be oxidized or sometimes reduced by SRB or, alternatively, fermented (disproportionated). Various reviews provide detailed information on oxidative processes in the sulfur cycle (28, 32, 55).

During chemolithotrophic metabolism, sulfide is respired with O_2 or NO_3^-. Since chemical oxidation competes with microbial oxidation, thiosulfate ($S_2O_3^{2-}$) is often used as a sulfide analog, since the former does not chemically oxidize spontaneously (28). Chemolithotrophic sulfide oxidation plays an important role in hydrothermal vent environments, in certain symbiotic relationships, and, most importantly, in (marine) sediments and water columns of stratified lakes.

Rate Measurements

Rate measurements can be determined by O_2, nitrate, and/or sulfide (or $S_2O_3^{2-}$) consumption. Problems introduced by chemical reactions of sulfide and O_2 or metals such as iron require one to exercise caution. Use of killed controls or inhibitors is necessary to account for abiotic loss factors. When sulfide or $S_2O_3^{2-}$ is added, it is important to use concentrations as close to those in situ as possible. Sulfide can be especially toxic to bacteria at elevated concentrations, and so its use in excess may cause consumption rates which deviate from those in situ.

For water samples, place 5 to 50 ml in a chamber or reaction vessel. Determine the endogenous O_2 consumption while stirring. Add a small volume (10 to 100 μl) of a concentrated sulfide or $S_2O_3^{2-}$ stock solution to the reaction vessel and determine the disappearance of O_2 over time. Alternatively, determine the disappearance of sulfide or $S_2O_3^{2-}$. Treat autoclaved or filter-sterilized controls identically to determine abiotic loss factors. Preparation of stock solutions of sulfide and $S_2O_3^{2-}$ (50 to 500 mM) is described

below. The analytical techniques for measuring these compounds and nitrate and O_2 are also described below.

Intact core measurements are the preferred basis for measuring S transformations (26). A major problem with sediments is abiotic reactions, such as iron and manganese sulfide mineral formation and oxidation of reduced S by O_2. Introduce the substrate (preferably $S_2O_3^{2-}$ to minimize abiotic consumption) with a glass syringe by axial microinjection. Add concentrated stock solutions in 5- to 25-μl amounts to avoid changing pore water content. Incubate multiple cores simultaneously and sacrifice individual ones at each time point. Upon sampling, section the cores and measure the pore water concentration of the substrate as described below. To prevent use of $S_2O_3^{2-}$ by SRB, molybdate (final concentration of Na_2MoO_4, 5 to 10 mM) can be injected simultaneously as a specific inhibitor (see below). Since this method is elaborate and makes the use of replicates impractical, it may be necessary to repeat the experiment two or three times.

Measurements can also be made by using sediments that are slurried (e.g., 1 part sediment plus 1 to 3 parts sterilized water or artificial medium of similar ionic strength) and placed in a vessel (100 to 500 ml) or serum bottle (20 to 100 ml) in which the consumption of sulfide, $S_2O_3^{2-}$, O_2, or nitrate is monitored in aliquots collected over time. The slurry should contain a gas phase (N_2 for anoxic conditions) and should be stirred or shaken, and the temperature should be controlled. In slurry experiments, measure sulfide and O_2 directly, using electrodes placed in the reaction vessel.

Calculate the consumption of sulfide or $S_2O_3^{2-}$ from the initial slope of the concentration over time ($\delta[\text{sulfide}]/\delta t$ or $\delta[S_2O_3^{2-}]/\delta t$). For water samples, this can be normalized easily to volume; for sediments, the consumption rate is expressed either in amount (micromoles or millimoles) per volume (e.g., cubic centimeter of sediment or milliliter of pore water) or per mass (gram of sediment) per unit of time (minute, hour, or day). Another approach is to calculate rates of sulfide or $S_2O_3^{2-}$ consumption by assuming steady state in which the SO_4^{2-} reduction (i.e., sulfide production) rate equals the sulfide consumption rate, or the $S_2O_3^{2-}$ oxidation rate equals $1/2 \times SO_4^{2-}$ reduction rate $\times 0.6$ (which assumes that 60% of the sulfide proceeds through $S_2O_3^{2-}$ [54]).

Radiotracer Studies and Isotope Exchange

The use of [^{35}S]sulfide for measuring S oxidation creates difficulties because, in addition to chemical reaction with O_2 and Fe^{2+}, rapid isotope exchange occurs among reduced S species (i.e., $H^{35}S^-$, $^{35}S^0$, and $Fe^{35}S$), presumably via polysulfide ($^{35}S_n^{2-}$) (18). However, [^{35}S]sulfide synthesized from $^{35}S^0$ has been applied successfully for short-term sediment incubations (31). In general, the use of $^{35}S_2O_3^{2-}$ is preferable (29). Keep incubation times short (<1 h) to prevent metabolism of the S compounds formed. Apply radionuclides as described above for whole-core incubations and terminate reactions by slicing and then homogenizing sections of cores in ice-cold 1% zinc acetate. Retrieve pore water by microcentrifugation (2 min at 8,000 $\times$ g at 4°C) and filtration through a 0.2-μm-pore-size nylon filter (Alltech, Deerfield, Ill.). After dilution in an $S_2O_3^{2-}$ carrier solution (0.1 mM), separate $S_2O_3^{2-}$, polythionates, and SO_4^{2-} by ion chromatography (60). Collect the eluted fractions containing these S compounds in scintillation vials and determine activity by using scintillation cocktails designed for aqueous samples (e.g., Ecolume or Ecolite; ICN Biomedicals, Inc.).

Cultivation, Enrichment, and Isolation

Cultivation and enumeration of S oxidizers are most successful in CO_3^{2-}-buffered media (58). *Thiomicrospira* spp. from marine environments are especially sensitive to elevated phosphate concentrations. Success of isolation procedures can be enhanced by mixing mineral media with filtered (0.45-μm-pore-size filters) and autoclaved water from which the inoculum is taken (1:1 [vol/vol]). Gradient cultures (two-layered agar systems, in which the bottom compartment contains sulfide and the upper compartment consists of mineral agar in which O_2 can freely diffuse) are useful for the isolation and cultivation of filamentous S bacteria (*Beggiatoa* sp. [43]). Cultivation in chemostats is also a powerful tool with which to determine sulfide oxidation rates (56, 58).

Medium for S Oxidizers

Mineral medium base
NH_4Cl	0.2 g
$CaCl_2 \cdot 2H_2O$	0.225 g
KCl	0.2 g
$MgCl_2 \cdot 6H_2O$	0.2 g
KH_2PO_4	0.02 g

Add 1.0 ml of Widdel and Pfennig's trace element solution (listed above for SRB). For *Thiobacillus* and *Thiomicrospira* spp., add $FeSO_4 \cdot 7H_2O$ (0.1 to 1 mg); for marine isolates, add NaCl (25 g). Dissolve salts for 1.0 liter in 800 ml and autoclave. Prepare and autoclave separately (grams per liter) Na_2CO_3 (20), $Na_2S_2O_3$ (248), $Na_2S \cdot 7–9H_2O$ (dissolve 24 g/liter in boiling 2% Na_2CO_3 solution, close cap, and autoclave), CH_3COONa (8.1), 1 M HCl, 1 M NaOH, and, if desired, yeast extract (10). Add Na_2CO_3 stock to medium base (1:9 [vol/vol]) and either $Na_2S_2O_3$ (1:9 [vol/vol]) or Na_2S (1:99 [vol/vol]); if required, add CH^3COONa (1:9 [vol/vol]) and yeast extract (1:99 [vol/vol]). Adjust final volume to 1.0 liter. Adjust pH to 7.5 to 8.0 with sterile HCl or NaOH. For aerobes, transfer complete medium into an Erlenmeyer flask with a cotton plug (do not shake, because this will cause the pH to change since the medium is CO_3^{2-}-buffered). For anaerobes, use bottles or tubes sealed with butyl rubber stoppers or septa. Culturing of phototrophs requires incandescent light (15- or 40-W light bulb placed 10 to 20 cm from the culture).

Inocula and Maintenance

Inocula that have a high success rate can be obtained from a variety of environments, including marine sediments, estuaries, stratified lakes, and sulfur-containing (thermal) springs. Storage time of the inoculum should be kept to a minimum. When sediments are used, allow air exchange for aerobes, restrict exposure to O_2 for microaerophiles (denitrifying *Thiomicrospira* spp.), and keep O_2-free for anaerobes (purple and green sulfur bacteria). Isolation can be direct (sample is transferred to a bottle or Erlenmeyer flask) or after serial 10-fold dilutions. Alternatively, a reaction vessel through which a continuous low flow (dilution rate of 0.005 to 0.02 h^{-1}) of medium is fed (simple chemostat) yields good results for slowly growing species such as certain (denitrifying) thiobacilli and *Thiomicrospira* spp. Isolate phototrophs after dilution in liquefied agar (see above), and pick single colonies for transfer to fresh (liquid) medium (46).

Enumeration

Enumerate S oxidizers by using MPN techniques (39). Dilute a water sample of known size by 10-fold several times to obtain a series ranging from 10^{-3} to 10^{-8} the original concentration. Dilute sediment samples 100 times and sonicate for approximately 30 to 60 s (40 W) prior to the same treatment. Use each dilution step as an inoculum (1:9 [vol/vol]) in fresh medium to which a pH indicator (e.g., bromocresol purple; 0.4 g/liter) is added. Add acetate, or any desired organic substrate, in addition to CO_3^{2-}, to include heterotrophic $S_2O_3^{2-}$ utilizers. Since a wide variety of heterotrophs use $S_2O_3^{2-}$ but produce polythionates, it is important to include a pH indicator. Three or five tube series can be used. Score positive tubes and transfer to fresh medium to confirm growth. Use positives from greatest dilutions to isolate organisms which are likely to be important in the environment. Use epifluorescence microscopy (57) to estimate cell counts in the inoculum.

Phototrophic Processes

Sulfide and $S_2O_3^{2-}$ are electron donors for purple S and green S bacteria (27, 63). Certain cyanobacteria can also use these electron donors, but this is environmentally significant in only a few lakes (29). In addition to sulfide oxidation experiments discussed above, chlorophyll, bacteriochlorophylls, and other pigments can be used as biomass indicators, and light of specific wavelengths can be used to stimulate or inhibit certain species selectively (49, 52). Cyanobacteria have additional pigments, phycobiliproteins, which absorb at 550 to 650 nm. Accessory pigments (carotenoids) of anoxygenic phototrophs typically absorb at between 400 and 550 nm. Purple non-S bacteria contain spirilloxanthin, okenone, or lycopene, while green non-S bacteria contain carotene, chlorobactene, or isonieratene. By using certain light sources, either group of bacteria (oxygenic or anoxygenic) can be selected in enrichment cultures. Photopigments can be measured spectrophotometrically after extraction in methanol with or without hexane (52). In methanol, absorption maxima (nanometers) and coefficients (per gram per liter per centimeter) of chlorophyll *a* and bacteriochlorophylls *a*, *b*, *c*, *d*, *e*, and *g* are 665 and 74.5, 771 and 84.1, 794 (unstable in air), 668 and 86.0, 654 and 82.3, 646 and 82.3, and 765 (coefficient in methanol unknown), respectively. Different bacteriochlorophylls can be further differentiated upon acidification, yielding bacteriopheophytin with different absorption maxima (52). High-performance liquid chromatography with fluorescence detection enables photopigment quantification as well (70). Photopigment concentrations can be used to calculate protein, which may be useful for specific rate measurements in situ. Although the specific photopigment concentration typically increases with decreasing light intensity, a value of 22 to 30 mg of protein per μg of bacteriochlorophyll is typically found (59).

Organic Sulfur Oxidation

Dimethyl sulfide (DMS), dimethyl sulfoxide, and methane thiol (CH_3SH) are consumed by a variety of thiobacilli, anoxygenic phototrophs, and SRB (38, 61, 62, 64). Metabolism can be assessed as described above for other rate measurements. Prepare fresh stock solutions of oxidizable substrates. Prepare DMS gravimetrically. Prepare CH_3SH from dimethyl disulfide by cleaving with tributylphosphine (34). Determine the concentrations of these organosulfur compounds by gas chromatography (GC) with flame ionization detection (FID) or flame photometric detection (FPD) (35). FPD is more sensitive than FID, and measurements of volatile S compounds via GC-FPD are facilitated by doping the FPD fuel line with low levels of an S gas (i.e., carbonyl sulfide, sulfur dioxide, or carbon disulfide [41]). Care should be taken with CH_3SH, which rapidly binds to sediments, glass, and rubber (37), and appropriate controls are necessary to account for abiotic losses. DMS can be oxidized biologically to dimethyl sulfoxide, which is measured as DMS after chemical reduction with a stabilized titanium solution (36). When performing experiments with organosulfur compounds, use Teflon-lined stoppers and heat glassware overnight (300 to 450°C) prior to use.

INHIBITORS

Molybdate (MoO_4^{2-}) has been used as a specific competitive inhibitor of SO_4^{2-} reduction, which is useful for determining the direct role of SRB in biogeochemical transformations (45). In marine samples, MoO_4^{2-} concentrations of 20 to 28 mM are applied, although much lower (5 mM) amounts have almost the same effect (45). For freshwater samples, less MoO_4^{2-} (<2 mM) is required. In addition to SRB, MoO_4^{2-} inhibits S metabolism by thiobacilli and phototrophs 15 and 30%, respectively (59). All of these organisms contain ADP and/or ATP sulfurylases, which may be affected by MoO_4^{2-}. Other group VI oxyanions, such as WO_4^{2-} and SeO_4^{2-}, have also been used to inhibit SO_4^{2-} reduction but are not quite as thorough. However, these latter compounds may prove useful for some applications, since they interact less with free thiols than does MoO_4^{2-} (34).

ANALYSES

Measure dissolved sulfide colorimetrically (9) or with ion-specific electrodes (50). Both techniques have a detection limit of 1 to 10 μM. Alternatively, samples can be acidified in closed serum bottles and the headspace assayed for H_2S by GC-FID or GC-FPD. GC techniques have very low detection limits (nanomolar) and are not necessary for most applications. Determine $S_2O_3^{2-}$ colorimetrically after cyanolysis (33), which can also be used to measure polysulfides (58). The detection limit with this technique for both compounds is 10 μM. It is important to use pore water as a reagent blank to account for possible background color. Alternatively, anion-exchange chromatography and UV detection can be used to measure $S_2O_3^{2-}$, using samples that are filtered through 0.2-μm-pore-size nylon filters (Alltech) (limit of detection, 1 to 5 μM [4]). Measure nitrate colorimetrically after reaction with sulfanilamide (limit of detection, 1 to 5μM [6]) or in freshwater samples with an ion-specific electrode (limit of detection, 30 μM [10]). Nitrate can also be measured by ion chromatography. Measure O_2 with an O_2 monitor consisting of a polarographic electrode (Clark type; Yellow Springs International, Yellow Springs, Ohio) attached to a strip chart recorder (5).

REFERENCES

1. **Amann, R. I., B. Binder, S. W. Chisholm, R. Olsen, R. Devereux, and D. A. Stahl.** 1990. Combination of 16S rRNA-targeted oligonucleotide probes with flow cytometry for analyzing mixed microbial populations. *Appl. Environ. Microbiol.* **56:**1919–1925.
2. **Amann, R. I., W. Ludwig, and K.-H. Schleifer.** 1995. Phylogenetic identification and in situ detection of individual cells without cultivation. *Microbiol. Rev.* **59:** 143–169.
3. **Amann, R. I., J. Stromley, R. Devereux, R. Key, and D. A. Stahl.** 1992. Molecular and microscopic identification of sulfate-reducing bacteria in multispecies biofilms. *Appl. Environ. Microbiol.* **58:**614–623.
4. **Bak, F., A. Schuhmann, and K. H. Jansen.** 1993. Determination of tetrathionate and thiosulfate in natural samples and microbial cultures by a new, fast and sensitive ion chromatographic technique. *FEMS Microbiol. Ecol.* **12:** 257–264.
5. **Beechey, R. B., and D. W. Ribbons.** 1972. Oxygen electrode measurements. *Methods Microbiol.* **6B:**25–53.
6. **Bendschneider, K., and R. J. Robinson.** 1957. A new spectrophotometric method for the determination of nitrite in sea water. *J. Mar. Res.* **11:**87–96.
7. **Canfield, D. E., R. Raiswell, J. T. Westrich, C. M. Reaves, and R. A. Berner.** 1986. The use of chromium reduction in the analysis of reduced inorganic sulfur in sediments and shales. *Chem. Geol.* **54:**149–155.
8. **Canfield, D. E., B. Thamdrup, and J. W. Hansen.** 1993. The anaerobic degradation of organic matter in Danish coastal sediments: iron reduction, manganese reduction, and sulfate reduction. *Geochim. Cosmochim. Acta* **57:** 3867–3883.
9. **Cline, J. D.** 1969. Spectrophotometric determination of hydrogen sulfide in natural waters. *Limnol. Oceanogr.* **14:** 454–458.
10. **de Beer, D., and J. P. R. A. Sweerts.** 1989. Measurement of nitrate gradients with an ion selective microelectrode. *Anal. Chim. Acta* **219:**351–356.
11. **Devereux, R., M. Delaney, F. Widdel, and D. A. Stahl.** 1989. Natural relationships among sulfate-reducing eubacteria. *J. Bacteriol.* **171:**6689–6695.
12. **Devereux, R., S.-H. He, C. L. Doyle, S. Orklnad, D. A. Stahl, J. LeGall, and W. B. Whitman.** 1990. Diversity and origin of *Desulfovibrio* species: phylogenetic definition of a family. *J. Bacteriol.* **172:**3609–3619.
13. **Devereux, R., M. D. Kane, J. Winfrey, and D. A. Stahl.** 1992. Genus- and group-specific hybridization probes for determinative and environmental studies of sulfate-reducing bacteria. *Syst. Appl. Microbiol.* **15:**601–609.
14. **Devereux, R., and G. W. Mundfrom.** 1994. A phylogenetic tree of 16S rRNA sequences from sulfate-reducing bacteria in a sandy marine sediment. *Appl. Environ. Microbiol.* **60:**3437–3439.
15. **Devereux, R., and D. A. Stahl.** 1993. Phylogeny of sulfate-reducing bacteria and a perspective for analyzing their natural communities, p. 131–160. *In* J. M. Odom and J. R. Singleton, (ed.), *Sulfate-Reducing Bacteria: Contemporary Perspectives.* Springer-Verlag, New York.
16. **Elsgaard, L., and B. B. Jørgensen.** 1992. Anoxic transformations of radiolabeled hydrogen sulfide in marine and freshwater sediments. *Geochim. Cosmochim. Acta.* **56:** 2425–2435.
17. **Fossing, H., and B. B. Jørgensen.** 1989. Measurement of bacterial sulfate reduction in sediments: evaluation of a single step chromium reduction. *Biogeochemistry* **8:** 205–222.
18. **Fossing, H., S. Thode-Andersen, and B. B. Jørgensen.** 1992. Sulfur isotope exchange between S-35 labeled inorganic sulfur compounds in anoxic marine sediments. *Mar. Chem.* **38:**117–132.
19. **Gibson, G. R.** 1990. Physiology and ecology of the sulphate-reducing bacteria. *J. Appl. Bacteriol.* **69:**769–797.
20. **Hesslein, R. H.** 1976. An in situ sampler for close interval pore water studies. *Limnol. Oceangr.* **21:**912–914.
21. **Hines, M. E., G. T. Banta, A. E. Giblin, J. E. Hobbie, and J. T. Tugel.** 1994. Acetate concentrations and oxidation in salt marsh sediments. *Limnol. Oceanogr.* **39:**140–148.
22. **Hines, M. E., and G. E. Jones.** 1985. Microbial bio-

geochemistry in the sediments of Great Bay, New Hampshire. *Estuarine Coastal Shelf Sci.* **20:**729–742.

23. **Hines, M. E., S. L. Knollmeyer, and J. B. Tugel.** 1989. Sulfate reduction and other sedimentary biogeochemistry in a northern New England salt marsh. *Limnol. Oceanogr.* **34:**578–590.

24. **Howarth, R. W.** 1979. Pyrite: its rapid formation in a salt marsh and its importance in ecosystem metabolism. *Science* **203:**49–51.

25. **Howes, B. L., J. W. H. Dacey, and S. G. Wakeham.** 1985. Effects of sampling technique on measurements of porewater constituents in salt marsh sediments. *Limnol. Oceanogr.* **30:**221–227.

26. **Jørgensen, B. B.** 1978. A comparison of methods for the quantification of bacterial sulfate reduction in coastal marine sediments. I. Measurements with radiotracer techniques. *Geomicrobiol. J.* **1:**11–27.

27. **Jørgensen, B. B.** 1982. Ecology of the bacteria of the sulphur cycle with special reference to anoxic-oxic interface environments. *Philos. Trans. R. Soc. Lond.* **298:**543–561.

28. **Jørgensen, B. B.** 1988. Ecology of the sulphur cycle: oxidative pathways in sediments, p. 31–63. *In* J. A. Cole and S. J. Ferguson (ed.), *The Nitrogen and Sulphur Cycles.* Cambridge University Press, Cambridge.

29. **Jørgensen, B. B.** 1994. Sulfate reduction and thiosulfate transformations in a cyanobacterial mat during a diel oxygen cycle. *FEMS Microbiol. Ecol.* **13:**303–312.

30. **Jørgensen, B. B.** 1990. A thiosulfate shunt in the sulfur cycle of marine sediments. *Science* **249:**152–154.

31. **Jørgensen, B. B., and F. Bak.** 1991. Pathways and microbiology of thiosulfate transformations and sulfate reduction in a marine sediment (Kattegat, Denmark). *Appl. Environ. Microbiol.* **57:**847–856.

32. **Kelly, D. P.** 1989. Physiology and biochemistry of unicellular bacteria, p. 193–218. *In* H. G. Schlegel and B. Bowien (ed.), *Autotrophic Bacteria.* Springer-Verlag, Berlin.

33. **Kelly, D. P., L. A. Chambers, and P. A. Trudinger.** 1969. Cyanolysis and spectrometric estimation of trithionate in a mixture with thiosulfate and tetrathionate. *Anal. Chem.* **41:**898–901.

34. **Kiene, R. P.** 1991. Evidence for the biological turnover of thiols in anoxic marine sediments. *Biogeochemistry* **13:**117–135.

35. **Kiene, R. P.** 1993. Measurement of diemthylsulfide (DMS) and dimethylsulfoniopropionate (DMSP) in seawater and estimation of DMS turnover rates, p. 601–610. *In* P. F. Kemp, B. F. Sherr, E. B. Sherr, and J. J. Cole (ed.), *Handbook of Methods in Aquatic Microbial Ecology.* Lewis Press, Boca Raton, Fla.

36. **Kiene, R. P., and G. Gerard.** 1994. Determination of trace levels of dimethylsulfoxide (DMSO) in seawater and rain water. *Mar. Chem.* **47:**1–12.

37. **Kiene, R. P., and M. E. Hines.** 1995. Microbial formation of dimethyl sulfide in anoxic *Sphagnum* peat. *Appl. Environ. Microbiol.* **61:**2720–2726.

38. **Kiene, R. P., and P. T. Visscher.** 1987. Production and fate of methylated sulfur compounds from methionine and dimethylsulfoniopropionate in anoxic salt marsh sediments. *Appl. Environ. Microbiol.* **53:**2426–2434.

39. **Koch, A. L.** 1994. Growth measurement, p. 248–277. *In* P. Gerhardt, R. G. E. Murray, W. A. Wood, and N. R. Krieg (ed.), *Methods for General and Molecular Bacteriology.* American Society for Microbiology, Washington, D.C.

40. **Lovley, D. R., and E. J. P. Phillips.** 1994. Novel processes for anaerobic sulfate production from elemental sulfur by sulfate-reducing bacteria. *Appl. Environ. Microbiol.* **60:**2394–2399.

41. **Morrison, M. C., and M. E. Hines.** 1990. The variability of biogenic sulfur flux from a temperate salt marsh on short time and space scales. *Atmos. Environ.* **24:**1771–1779.

42. **Muyzer, G., E. C. de Waal, and A. G. Uitterlinden.** 1993. Profiling of complex microbial populations by denaturing gradient gel electrophoresis analysis of polymerase chain reaction-amplified genes coding for 16S rRNA. *Appl. Environ. Microbiol.* **59:**695–700.

43. **Nelson, D. C., B. B. Jørgensen, and N. P. Revsbech.** 1986. Growth pattern and yield of a chemoautotrophic *Beggiatoa* sp. in oxygen-sulfide microgradients. *Appl. Environ. Microbiol.* **52:**225–233.

44. **Odom, J. M., and R. Singleton, Jr.** 1993. *The Sulfate-Reducing Bacteria: Contemporary Perspectives.* Springer-Verlag, New York.

45. **Oremland, R. S., and D. G. Capone.** 1988. Use of "specific" inhibitors in biogeochemistry and microbial ecology. *Adv. Microb. Ecol.* **10:**285–383.

46. **Pfennig, N.** 1978. *Rhodocyclus purpureus* gen. nov. and sp. nov., a ring-shaped, vitamin B12-requiring member of the *Rhodspirillaceae. Int. J. Syst. Bacteriol.* **28:**283–288.

47. **Postgate, J. R.** 1984. *The Sulphate-Reducing Bacteria*, 2nd ed. Cambridge University Press, Cambridge.

48. **Reeburgh, W. S.** 1967. An improved interstitial water sampler. *Limnol. Oceanogr.* **12:**163–165.

49. **Repeta, D. J., D. J. Simpson, B. B. Jørgensen, and H. W. Jannasch.** 1989. Evidence for anoxygenic photosynthesis from the distribution of bacteriochlorophylls in the Black Sea. *Nature* (London) **342:**69–72.

50. **Revsbech, N. P., and B. B. Jørgensen.** 1986. Microelctrodes: their use in microbial ecology. *Adv. Microb. Ecol.* **9:**293–352.

51. **Skyring, G. W.** 1987. Sulfate reduction in coastal ecosystems. *Geomicrobiol. J.* **5:**295–374.

52. **Stal, L. J., H. van Gemerden, and W. E. Krumbein.** 1984. The simultaneous assay of chlorophyll and bacteriochlorophyll in natural microbial communities. *J. Microbiol. Methods* **2:**295–306.

53. **Telang, A. J., G. Voordouw, S. Ebert, N. Sifeldeen, J. M. Foght, P. M. Fedorak, and D. W. S. Westlake.** 1994. Characterization of the diversity of sulfate-reducing bacteria in soil and mining waste water environments by nucleic acid hybridization techniques. *Can. J. Microbiol.* **40:**955–964.

54. **Thamdrup, B., K. Finster, H. Fossing, J. Würgel Hansen, and B. B. Jørgensen.** 1994. Thiosulfate and sulfite distributions in porewater of marine sediments related to manganese, iron, and sulfur geochemistry. *Geochim. Cosmochim. Acta* **58:**67–73.

55. **Trüper, H. G.** 1989. Chemosynthetically sustained ecosystems in the deep sea, p. 147–166. *In* H. G. Schlegel and B. Bowien (ed.), *Autotrophic Bacteria.* Springer-Verlag, Berlin.

56. **Van Gemerden, H., and H. H. Beeftink.** 1978. Specific rates of substrate oxidation and product formation in autotrophically growing *Chromatium vinosum* cultures. *Arch. Microbiol.* **119:**135–143.

57. **Velji, M. I., and L. J. Albright.** 1993. Improved sample preparation for enumeration of aggregated aquatic substrate bacteria, p. 139–142. *In* P. F. Kemp, B. F. Sherr, E. B. Sherr, and J. J. Cole (ed.), *Handbook of Methods in Aquatic Microbiol Ecology.* Lewis Press, Boca Raton, Fla.

58. **Visscher, P. T., J. W. Nijburg, and H. van Gemerden.** 1990. Polysulfide utilization by *Thiocapsa roseopersicina. Arch. Microbiol.* **155:**75–81.

59. **Visscher, P. T., R. A. Prins, and H. van Gemerden.** 1992. Rates of sulfate reduction and thiosulfate consuption in a marine microbial mat. *FEMS Microbiol. Ecol.* **86:**283–294.

60. **Visscher, P. T., P. Quist, and H. Vangemerden.** 1991. Methylated sulfur compounds in microbial mats: in situ

concentrations and metabolism by a colorless sulfur bacterium. *Appl. Environ. Microbiol.* **57**:1758–1763.

61. **Visscher, P. T., and B. F. Taylor.** 1993. Aerobic and anaerobic degradation of a range of alkyl sulfides by a denitrifying marine bacterium. *Appl. Environ. Microbiol.* **59**: 4083–4089.

62. **Visscher, P. T., and B. F. Taylor.** 1993. Organic thiols as organolithotrophic substrates for growth of phototrophic bacteria. *Appl. Environ. Microbiol.* **59**:93–96.

63. **Visscher, P. T., F. P. Vandenende, B. E. M. Schaub, and H. Vangemerden.** 1992. Competition between anoxygenic phototrophic bacteria and colorless sulfur bacteria in a microbial mat. *FEMS Microbiol Ecol.* **101**:51–58.

64. **Visscher, P. T., and H. Van Gemerden.** 1991. Photoautotrophic growth of *Thiocapsa roseopersicina* on dimethyl sulfide. *FEMS Microbiol. Lett.* **81**:247–250.

65. **Voordouw, G.** 1995. Minireview—the genus *Desulfovibrio:* the centennial. *Appl. Environ. Microbiol.* **61**: 2813–2819.

66. **Voordouw, G., Y. Shen, C. S. Harrington, A. J. Telang, T. R. Jack, and D. W. S. Westlake.** 1993. Quantitative reverse sample genome probing of microbial communities and its application to oil field production waters. *Appl. Environ. Microbiol.* **59**:4101–4114.

67. **Widdel, F., and F. Bak.** 1991. Gram-negative mesophilic sulfate-reducing bacteria, p. 3352–3378. *In* A. Balows, H. G. Trüper, M. Dworkin, W. Harder, and K.-H. Schleifer (ed.), *The Prokaryotes,* 2nd ed. Springer-Verlag, New York.

68. **Widdel, F., and T. A. Hansen.** 1991. The dissimilatory sulfate- and sulfur-reducing bacteria, p. 583–634. *In* A. Balows, H. G. Trüper, M. Dworkin, W. Harder, and K.-H. Schleifer (ed.), *The Prokaryotes,* 2nd ed. Springer-Verlag, New York.

69. **Widdel, F., G.-W. Kohring, and F. Mayer.** 1983. Studies on dissimilatory sulfate-reducing bacteria that decompose fatty acids. III. Characterization of the filamentous gliding *Desulfonema limicola* gen. nov., sp. nov., and *Desulfonema magnum* sp. nov. *Arch. Microbiol.* **134**:286–294.

70. **Yacobi, Y. Z., W. Eckert, H. G. Trüper, and T. Berman.** 1990. High-performance liquid-chromatography detection of phototrophic bacterial pigments in aquatic environments. *Microb. Ecol.* **19**:127–136.

71. **Zhabina, N. N., and I. I. Volkow.** 1978. A method of determination of various sulfur compounds in sea, sediments and rocks, p. 735–745. *In* W. E. Krumbein (ed.), *Environmental Biogeochemistry and Geomicrobiology,* vol. 3. Ann Arbor Science Publishers, Ann Arbor, Mich.

Microbial Nitrogen Cycling

DOUGLAS G. CAPONE

35

Nitrogen (N) availability is a key factor regulating the biological productivity of many aquatic ecosystems (19, 47), and bacteria have long been recognized as important agents affecting N pools through various transformations (63, 100). The assimilation into organic form and subsequent release of inorganic N, as performed by a broad array of prokaryotic and eukaryotic organisms, comprise the inner core of the N cycle in nature. However, it is the uniquely bacterial processes of N_2 fixation, nitrification, and denitrification which define the broader cycle and which can affect directly the availability and form of N within particular ecosystems. The relative importance of each of these processes can vary greatly among systems.

Hence, there has been sustained interest in determining the importance of these processes in the environment, and a variety of procedures are currently in use. This chapter provides an update and synopsis of current procedures for determining N_2 fixation, nitrification, and denitrification in aquatic environments (38, 96).

N_2 FIXATION

Biological N_2 fixation is the reduction of N_2 gas to ammonium as catalyzed by nitrogenase, an enzyme system found in a physiologically diverse array of eubacteria and archaea (79). As such, N_2 fixation provides a primary input of usable N to the biosphere and may be a key process in N-limited environments. Nitrogen fixation occurs throughout a variety of freshwater and marine habitats. Several recent reviews on aquatic N_2 fixation are available (15, 48, 76).

Assessing Populations

Growth on N-free media has long been used as prima facie evidence of N_2 fixation, and early researchers in the field relied on such procedures in attempting to enrich and enumerate N_2-fixing populations in nature. Because of the physiological diversity of N_2 fixers, a variety of media and conditions appropriate for aerobic and anaerobic photoautotrophs, chemoautotrophs, and aerobic and anaerobic heterotrophs need to be used (8), depending on the circumstance. However, because of pervading contamination of many ingredients for bacteriological media with low levels of combined N and the possibility of uptake of atmospheric

NH_3, growth on N-free media is generally not considered definitive.

The nitrogenase enzyme has been highly conserved (79). Antibodies to components of the protein complex which show broad reactivity with nitrogenases from diverse sources have been developed (60a, 119). Such antibodies, linked to fluorescent reporters, have been used to detect and enumerate diazotrophs in marine seagrass and mat systems (23). Nucleotide probes to genes for the structural proteins of nitrogenase (*nifH* and *nifK*) have been developed (120) and used to detect diazotrophs in a variety of marine substrates (121). Both of these approaches hold great promise for detecting and enumerating diazotrophs in aquatic environments.

Assessing Activity

Assessing the rate of N_2 fixation in the environment has been the subject of intensive study and methods development over the past several decades (8). As for the determination of any activity or process in the environment, there are a variety of generic issues and caveats which need to be considered. The reader is referred to chapter 25 for a further discussion of the topic.

The earliest method for detecting active N_2 fixation in natural samples was to monitor increases in organic N, typically by the Kjeldahl method (8). However, this is a very insensitive approach as one seeks to observe small increases above a large background of organic N preexisting in the sample. Furthermore, it cannot clearly discriminate increases in N due strictly to N_2 fixation from those which may result from assimilation of preexisting inorganic forms of N.

$^{15}N_2$ Direct Tracer Methods

When first introduced, the $^{15}N_2$ isotope tracer procedure provided the first sensitive, direct method for quantitatively assessing rates of N_2 fixation in natural samples (14, 28, 95) and remains the procedure of choice for definitively demonstrating and quantitating N_2 fixation.

The assay in outline is very straightforward and follows standard stable isotope tracer protocols (see reference 24 for a detailed treatment). Briefly, a sample is amended with enriched $^{15}N_2$ (up to 99 atom% $^{15}N_2$ available commercially or generated in the laboratory or field from ^{15}N-enriched ammonium salts) and incubated for a period, after which

334

the biological material is collected and analyzed for increases over natural abundance due to the uptake of the enriched $^{15}N_2$. For aquatic samples, assay protocols often call for removal of atmospheric N_2 by sparging or vacuum before introduction of the $^{15}N_2$ tracer. This is done in order to maximize the enrichment of the actual substrate pool and thereby provide a discernible enrichment in product over a relatively brief period of sample exposure and containment (e.g., reference 34).

Enrichment in the sample is determined by either isotope ratio mass spectrometry (IRMS) or emission spectrometry. Biological samples are typically converted to N_2 gas by Dumas combustion for either instrumental method, either off-line, or on-line in carbon-nitrogen analyzer/flowthrough mass spectrometers. Dedicated ^{15}N emission spectrometers are considerably less expensive than IRMS and have the advantage of requiring substantially smaller sample mass (<1 to 10 μg of N) than single-collector mass spectrometers. (The latter are mass spectrometers with a single Faraday cup collector which analyze samples by scanning several masses. The analysis usually requires >500 μg of sample). However, emission spectrometers are typically far less precise than IRMS and therefore require higher enrichment in product pools to discern activity. Multiple-collector IRMS are now available at reasonable cost and can obtain accurate isotope ratios on much smaller samples (down to 10 μg) than single-collector instruments. Furthermore, the high precision of the multiple-collector systems allows reproducible detection of small increases in ^{15}N in natural plankton assemblages {1 to 10 $\delta^{15}N$ units, where $\delta^{15}N$ per mil = $[(^{15}N{:}^{14}N_{sample}/^{15}N{:}^{14}N_{atmosphere} - 1) \times 1,000]$} after 4 h of exposure to 10 atom% $^{15}N_2$, without the need to remove ambient N_2 (64).

Acetylene Reduction Method

The introduction of the acetylene (C_2H_2) reduction method in 1967 (94) prompted a revolution in studies of N_2 fixation by providing a highly sensitive, convenient, and readily available procedure for determining nitrogenase activity. The procedure was designed for field studies, and the original description included assay of lake cyanobacteria (94).

Briefly, a sample of interest is incubated in the presence of C_2H_2, typically at a partial pressure of about 0.1 to 0.2 atm of C_2H_2 (1 atm = 101 kPa). Nitrogenase preferentially reduces C_2H_2 at these levels to C_2H_4. C_2H_2 and C_2H_4 can be easily separated and quantified by flame ionization gas chromatography. In active samples, C_2H_4 production can be detected within minutes of C_2H_2 addition. Relative to $^{15}N_2$ fixation, sample preparation (injection of a gas-phase sample) is trivial and simple. Sample throughput is very rapid (about 1 min per analysis). Portable gas chromatographs suitable for this analysis are readily available and may be operated in remote locations as long as power and the requisite compressed gases (typically N_2, air, and H_2) are available. Acetylene may be obtained in small, portable cylinders or generated in the field from CaC_2. A detailed protocol is provided in reference 17.

For broad-based field studies or experimental studies requiring multiple treatment levels, the C_2H_2 reduction procedure is the method of choice. Moreover, in certain systems, it may be run concurrently with C_2H_2 block assays of denitrification (e.g., reference 116; see Denitrification, below). However, the C_2H_2 reduction procedure is indirect in that it measures the reduction of a substrate analog of N_2 and ideally assumes that the activity observed would have been directed to N_2 reduction in the absence of C_2H_2.

While good correspondence has been found between rates of C_2H_2 reduction and direct determination of $^{15}N_2$ fixation for a broad variety of systems, large divergences have been noted in others. Departures from the theoretical value of 3:1 have often been related to the relative extent of nitrogenase-catalyzed H_2 production, a natural function of nitrogenase (79). Hence, when quantitative estimates are required, intercomparisons with $^{15}N_2$ reduction are strongly advised (17).

Emergent and Innovative Procedures

With the development of molecular tools to detect nitrogenase (*nif*) genes in the environment comes the possibility of detecting mRNA from *nif* genes as well (117, 118). This ability would be useful in providing direct evidence of active synthesis of the genes, rather than of the potential for N_2 fixation. Furthermore, with in situ hybridization, one could in theory determine the location and distribution of cells induced for N_2 fixation in natural samples.

Another approach to assessing N_2 fixation, which has had some use in agricultural systems but has not been considered extensively in aquatic systems, is the use of the natural isotope abundance in different N pools. The nitrogenase reaction, unlike many of the other N transformations, results in very little isotope fractionation (99). Thus, biological systems directly dependent on N_2 gas for N nutrition should have N isotopic signatures very close to that in N_2 (0.3662 atom% ^{15}N, or a $\delta^{15}N$ close to 0). Such anomalies have been observed in studies of marine ecosystems (e.g., reference 1; see reference 18) but have not generally been interpreted with respect to a possible diazotrophic source.

NITRIFICATION

Nitrification is the oxidation of more-reduced species of inorganic N, namely NH_4^+ and NO_2^-, to more-oxidized forms and thus is a crucial component of the N cycle (6, 55, 102). While nitrification has been found in some heterotrophs, in the environment it is generally attributed to two specialized types of aerobic autotrophic eubacteria, the ammonium oxidizers and the nitrite oxidizers (55). Nitrifiers appear to be ubiquitous in soils, as well as in marine waters, freshwaters, and sediments, although their activity is not always simple to demonstrate directly.

Assessing Populations

While nitrifiers have been grown in culture for over a century, they are generally regarded as slow growing and difficult to culture (55, 101, 102). Nitrifier populations in aquatic environments may be enumerated by most-probable-number procedures, using media designed to support dark autotrophic growth in air with NH_4^+ or NO_2^- as the electron donor. However, growth yields are poor, and this approach is generally not held to be a quantitative means of accurately assessing nitrifier populations in nature. Ward (101) pioneered the application of immunofluorescence detection in aquatic systems (earlier developed for soil systems by Schmidt [84]), having obtained antisera to several strains of marine nitrifiers. Her studies detected populations of nitrifiers far more numerous than previously noted by conventional procedures. However, the polyclonal antibodies used, generated against specific cell surface antigens, often have a high degree of strain specificity (106) and may still underestimate in situ populations.

Because of a relatively high phylogenetic coherence of nitrifiers, probes to relevant rRNA sequences may be useful for identification of ammonium oxidizers and nitrite oxidizers (98).

Assessing Activity

At the simplest level, nitrification may be detected by the production of NO_3^- and NO_2^- from NH_4^+, and simple spectrophotometric procedures may suffice (25, 77). In highly oligotrophic environments, such as open ocean waters, high-sensitivity methods (range of 1 to 100 nM) for NO_3^- (32) and NH_4^+ (51) are also available. In freshwater environments, ion chromatographic methods with UV detection are good alternatives for NO_3^- and NO_2^- detection (e.g., reference 43). However, the success of this approach will depend on a relatively high rate of nitrification and will detect only a net rate over any concurrent consumption of the N oxides produced (either by dissimilatory or assimilatory NO_3^- reduction). In a system with balanced nitrification and NO_3^- uptake, changes in the NO_3^- pool may be undetectable and other procedures will be required. Henriksen et al. (41) have described an assay of nitrification potential for sediments in which a high (1 mM) amendment of NH_4^+ is added and NO_3^- production is monitored. N_2O, which may be sensitively detected by electron capture gas chromatography, is also formed during nitrification and has been used as a diagnostic tool for nitrification in certain environments (55, 105).

^{15}N Direct Tracer Methods

Both direct ^{15}N tracer and ^{15}N isotope dilution methods have been applied to obtain quantitative estimates of nitrification in various aquatic ecosystems (65). As mentioned above, while access to appropriate instrumental facilities is often a constraining factor, highly sensitive IRMS have become more affordable and available.

Studies of open ocean (72) and estuarine (46) waters have used the direct approach. Briefly, a small enrichment of the ambient NH_4^+ pool (typically 10% increase) is made with highly enriched $^{15}NH_4^+$. After a period of incubation, samples are taken for analysis of the ^{15}N enrichment in the NO_2^- and NO_3^- pool, along with changes in concentration by appropriate methodology (see above). The $^{15}NO_2^-$ pool may also be enriched, and flux into NO_3^- can be traced.

For isotopic analysis, NO_2^- is typically trapped as a diazo complex and extracted (65). NO_3^- can be reduced to NO_2^- and then trapped as described above, and each sample is dried and combusted on- or off-line to N_2 before introduction into the mass or emission spectrometer (65). Alternatively, NO_3^- and NO_2^- may be reduced to NH_4^+ by Devarda's alloy or $TiCl_3$, with the NH_4^+ captured by microdiffusion (10) or other methods (see below). Methods have recently been developed for direct microbiological conversion of very low levels of NO_3^- or NO_2^- to N_2O prior to mass spectrometry (42, 81).

The isotope dilution approach, which provides simultaneously estimates of both nitrification and nitrate consumption (combined assimilatory and dissimilatory), has also been used in water column (109) and sediment (44, 59, 83) systems. For this method, one labels the product pool, NO_3^-, and monitors the decrease in isotope ratio of that pool relative to the initial enrichment. This provides a measure of the rate of NO_3^- production, on the assumption that new product will initially be unlabeled. A distinct advantage of isotope dilution approaches is that, apart from a small addition (typically 10%) to the product pool that generally does not significantly affect absolute concentra-

tion, rates of increase in the product pool may be estimated even in systems near steady state, where consumption is in close balance with production. Furthermore, when this approach is combined with measures over time of the NO_3^- pool, one can accurately estimate both its production and its consumption (35).

Inhibitor-Based Nitrification Assays

Nitrification has been particularly amenable to the inhibitor approach because of the sensitivity of the steps of nitrification to a variety of metabolic inhibitors (6, 73). Inhibitors, if appropriately applied, can disrupt steady-state conditions and reveal production or consumption of nitrification substrates, intermediates, or end products. Besides relevant N species, these may also include functions specific to autotrophic metabolism (e.g., dark $^{14}CO_2$ uptake). In contrast to ^{15}N methods, which require careful pool separation and sample preparation, inhibitor-based assays are often more rapid and sensitive.

In general, the ammonium oxidation step has received the most attention in the application of inhibitors. One of the earliest compounds to be used as a nitrification inhibitor in an ecological context was N-Serve (nitrapyrin or 2-chloro-6-trichloromethyl pyridine) (e.g., reference 110), a compound developed to minimize ammonium fertilizer loss from agricultural systems. Inhibition of ammonium oxidation should result in a cessation of NO_2^- and NO_3^- production, with a concomitant decrease in NH_4^+ consumption, as was reported in several early studies (40, 110). Typical concentrations used in sediments were about 10 to 20 µg/ml (37, 40). N-Serve requires an organic solvent, typically acetone, and therefore solvent controls should be used in addition to inhibitor-free, solvent-free controls. Furthermore, N-Serve may sorb to particles and is susceptible to degradation, which may result in decreasing effectiveness with time. Hall (37) proposed use of allylthiourea (ATU) as a water-soluble alternative to N-Serve. He found somewhat higher rates of nitrification in a direct comparison, with some evidence of N-Serve and solvent inhibition of nontarget processes. Hall (37) recognized the problems of slow diffusion of these inhibitors to sites of nitrification.

Most recently, gaseous inhibitors of ammonium oxidation have been introduced. Gaseous inhibitors have the advantage for sediment systems of diffusing more rapidly through assay systems than organic compounds such as N-Serve or ATU. Sloth et al. (91) found 1 kPa (0.01 atm) of C_2H_2 to be an effective inhibitor of nitrification in sediments. Greater ammonium efflux from sediments occurring in the presence of C_2H_2 is taken as the measure of nitrification. C_2H_2 is of course also an inhibitor of N_2 fixation and of the final step in denitrification (see below), and this may need to be considered. Similarly, Miller et al. (61) have proposed CH_3F (at 0.1 atm) as an effective and specific inhibitor of nitrification which apparently does not interfere with dissimilatory NO_3^- reduction and denitrification. Moreover, CH_3F has the added benefit of inhibiting nitrifier-specific N_2O production. A comparison of the two methods found insignificant differences with respect to stimulating ammonium flux from sediments. It should be noted that whereas C_2H_2 is readily available or easily generated from CaC_2, CH_3F is relatively difficult to obtain and can be quite costly.

Coupling inhibition of ammonium oxidation with determination of dark $^{14}CO_2$ uptake has been used to detect nitrifier activity (9, 92). This approach provides a very sensitive radioisotopic method but requires knowledge of a conversion factor if N transformation estimates are the objective (reported values of N oxidized to C assimilated range from 4 to over 100) (55). Jones et al. (52) described an alternative radioisotope approach based on the ammonium oxidizer-specific (via ammonium monooxygenase) conversion of ^{14}CO to $^{14}CO_2$ in the presence and absence of N-Serve.

Nitrite oxidation can also be specifically inhibited, and this offers the advantages of monitoring the NO_2^- pool, for which the spectrophotometric procedure is quite sensitive and for which there is generally no appreciable background. Belser and Mays (7) found chlorate at about 10 mM effective and were able to determine NO_2^- oxidation rates in soils and sediments. However, a subsequent report by Hynes and Knowles (49) pointed out several potential problems with this approach.

Emergent and Innovative Approaches

With respect to population assessment, antibodies developed to functional enzymes (e.g., ammonium monooxygenase) may provide a means for more accurately determining the abundance of these organisms. Similarly, nucleotide probes developed to the genes of functional enzymes or to 16S sequences unique to nitrifiers could be similarly used to enumerate and localize nitrifiers in environmental matrices in a less biased manner than traditional procedures (24, 98, 104, 105).

Microprobes which allow very fine (<1 mm) scale resolution of NO_3^- profiles in sediments have recently been developed (50). The dynamic aspects of these profiles in sediments were experimentally demonstrated by the introduction of an inhibitor (C_2H_2).

^{15}N natural abundance studies also show great promise in discerning nitrification rates over integrative time and space scales in aquatic ecosystems (45). Under NH_4^+-replete conditions, nitrification can impose a very large (up to 20 per mil) fractionation between the NH_4^+ and NO_3^- pools. For instance, clear evidence for nitrification has been obtained in Chesapeake Bay waters where increases in subpycnocline $\delta^{15}N$ values of NH_4^+ directly correlated with increases in NO_2^- concentrations. Natural abundance data have also been used to infer the source of N_2O in open ocean waters (114).

DISSIMILATORY NITRATE REDUCTION AND DENITRIFICATION

Dissimilatory nitrate reduction is the reduction of NO_3^- to NO_2^- in anaerobic respiration and is performed by a wide variety of facultative anaerobes when O_2 is unavailable; biological denitrification is the reduction of nitrogen oxides, such as NO_3^- and NO_2^-, as respiratory electron acceptors to gaseous end products, either N_2O or N_2 (57). In general, denitrification is performed by a subset of the facultative anaerobes which reduce nitrate. Denitrification represents a sink for combined nitrogen within ecosystems (85). In aquatic ecosystems, denitrification is generally associated with anaerobic zones or with chemical interfaces where NO_3^- may be supplied from adjacent NO_3^--rich zones or areas sustaining aerobic nitrification (60). In some aquatic ecosystems, it has been implicated in promoting N limitation (70). Dissimilatory NO_3^- reduction to NH_4^+ (NO_3^- fermentation), which conserves combined N within an ecosystem, can at times equal or exceed denitrification (60).

Assessing Populations

As for N_2 fixation and nitrification, enrichment and enumeration of denitrifiers have been long used to evaluate the presence of such populations in situ. Media appropriate for organotrophs, NO_3^- as an electron acceptor, and anaerobic conditions are generally used (31; also see chapter 6). Production of gas in anaerobic enrichments with NO_3^- is often taken as positive for denitrifiers, and such a test may be coupled with most-probable-number analysis (31). However, as for a broad range of bacteria (3), heterotrophic denitrifiers have diverse organic substrate requirements, and it is unlikely that any particular medium would satisfy the specific requirements of all such that a quantitative procedure based on enrichment could be developed.

Immunofluorescence methods based on cell surface antigens of cultured denitrifiers have been developed and applied (107). However, while the antibodies provide information on the strain of interest and closely related serotypes in situ, they appear to be highly strain specific and unsuitable as a means of assessing the broader denitrifying population. Similarly, antibodies generated to nitrite reductase (NiR) are similarly strain specific (108), despite the apparent similarity among denitrifiers (13).

Assessing Activity

Because of the perceived importance of denitrification in many aquatic ecosystems, considerable effort has been made to develop and improve procedures to provide quantitative estimates of this process. As for nitrification, many of our initial insights into denitrification came from observations of spatial and temporal patterns of NO_3^- or NO_2^- with use of standard colorometric procedures (77) or more recently by ion chromatographic approaches (43). Observations of NO_3^- anomalies in waters and sediments, i.e., deficits in NO_3^- concentrations over that predicted by simple diffusion, mixing, and nutrient regeneration (21, 97) or excesses of N_2 relative to Ar (5, 67), provide geochemical evidence for denitrification in various water bodies and sediment systems. Occasionally, transients of NO_2^- or excesses (or deficits) of N_2O have also been taken as evidence of denitrification (39, 105).

Direct N_2 Flux

Field studies have sought to demonstrate short-term rates of denitrification in many aquatic systems, typically by using an incubation format. As for nitrification, and with similar provisos, one may monitor decreases in NO_3^- pools as a measure of nitrate reduction (e.g., reference 26). However, this determination will provide only an estimate of net NO_3^- flux and will not discriminate between assimilatory and dissimilatory consumption.

Determination of N_2 flux has long been used as a quantitative measure of denitrification in respirometric studies of bacterial physiology. Aquatic studies have also used N_2 flux from natural samples as an estimate of in situ denitrification rates (56). N_2 is generally determined on subsamples, after headspace equilibration or sparging, by thermal conductivity after gas chromatographic separation. However, the large natural background of N_2 provides a formidable barrier to detecting small increases. Furthermore, leakage of ambient N_2 into assay vessels or sampling devices (e.g., syringes) may be erroneously construed as denitrification (78). Seitzinger et al. (86, 88) developed procedures for stringent containment of sediment samples, removal of ambient N_2, and incubation of samples in N_2 sparged seawater. Incubation periods of over a week are required for establishing rates of denitrification, and the procedures are not amenable for broad-scale sampling, replication, or experimentation. Nowicki (71) has recently proposed some improvements in the method which provide means of obtaining estimates over shorter (<1-week) time periods. She also introduced the use of a control using deoxygenated water over the sediments, which eliminates NO_3^- arising from in situ nitrification.

^{15}N Tracer Methods

Direct tracer procedures have been widely applied in denitrification studies. The bulk of these studies have used ^{15}N isotopes, although some research on denitrification has been accomplished with the short-lived radioisotope ^{13}N (30, 33). Tracer studies offer the advantage of unambiguous evidence of the presence of a pathway, although quantitation may be confounded by a number of factors, such as artificially elevating substrate pools or disturbing natural gradients (see chapter 25 for a detailed discussion).

Koike and Hattori (58, 59) introduced procedures to determine NO_3^- reduction and nitrification rates in sediments by isotope dilution of the NO_3^- pool (see above) and denitrification and NO_3^- reduction to NH_4^+ by direct tracer studies of $^{15}NO_3^-$ conversion to $^{15}N_2$ and $^{15}NH_4^+$, respectively. The latter direct tracer procedure was estimated to be about 100-fold more sensitive than the isotope dilution approach. NH_4^+ may be captured for subsequent conversion to N_2 and mass spectrometry by steam distillation or direct microdiffusion (35). A variety of alternative procedures, including precipitation, organic complexation and extraction or trapping on nonpolar C-18 adsorbent, or trapping on zeolite resin, are available (35, 65).

The problem of elevated NO_3^- concentrations was addressed by Oren and Blackburn (75), who applied a kinetic approach to the problem. Nishio et al. (68) subsequently described a flowthrough method for determining denitrification near the sediment-water interface. $^{15}NO_3^-$ is fed into the waters overlying sediment contained in a core, and the enrichment of the $^{15}N_2$ pool in the effluent is monitored. However, this can provide only an estimate of denitrification dependent on an overlying water source, which may only be a fraction of total denitrification (87). Parallel assays with $^{15}NH_4^+$ can provide an estimate of nitrifier-dependent denitrification (69). A recent innovation is the use of isotope pairing analysis to estimate from tracer $^{15}NO_3^-$ assays alone the extent of denitrification dependent on external NO_3^- and that derived from in situ NO_3^- production via nitrification (66).

Acetylene Blockage Method

The observation that C_2H_2 blocked the final step of the denitrification pathway (4, 116) led rapidly to development of convenient and sensitive procedures for aquatic denitrification (93, 115). C_2H_2, in blocking N_2O reductase, focuses the products of denitrification into N_2O, easily detected by electron capture gas chromatography and for which there is no appreciable background. C_2H_2 is typically added at about 10 to 20% of the gas phase of the assay vessel or injected into cores in C_2H_2-saturated water. Over time, samples are taken and analyzed for the appearance of N_2O. N_2 fixation and denitrification can be assayed simultaneously (115).

This C_2H_2 blockage procedure has found wide use in diverse habitats. However, under conditions of low ambient NO_3^- or when NO_3^- derives primarily from nitrification (which is inhibited by C_2H_2) or S^{2-} is present, the procedure likely underestimates in situ denitrification (10, 74, 87,

90). In conjunction with C_2H_2 blockage, Joye and Paerl (53) have used additions of millimolar NO_3^- and glucose, along with chloramphenicol to prevent de novo synthesis of denitrifying enzymes, to provide an estimate of potential denitrification. Recent evidence indicates that chloramphenicol may interfere with the activities of some of the denitrifying enzymes (112). In systems in which NO_3^- levels are relatively high (>10 μM) and denitrification over the length of the incubation is unlikely to deplete the NO_3^- pool, the C_2H_2 blockage procedure remains an appropriate and useful approach.

An alternative to the C_2H_2 blockage assay is the N_2O reductase assay first proposed by Sherr and Payne (89) and refined by Miller et al. (62). In this procedure, solutions containing known quantities of N_2O (5 to 20 kPa) are injected into samples, and the consumption of N_2O is determined over time (relative to values for killed, or C_2H_2 inhibited, controls).

Emergent and Innovative Approaches

As mentioned above, current efforts are being directed toward development of molecular methods to detect and enumerate denitrifying populations in environmental samples (104). In contrast to nitrifiers, the diversity of denitrifiers probably precludes use of rRNA-based probes for assessing denitrifying populations at large (103). However, functional genes may provide a useful target for such studies. For instance, two major forms of NiR are known: those containing a c,d_1 heme-type cytochrome and non-heme Cu-containing forms. An oligonucleotide probe developed to a portion of the c,d_1-type NiR of *Pseudomonas stutzeri* was more broadly reactive with other c,d_1-containing strains and may prove more useful in evaluating denitrifier population sizes in environmental samples (103, 108). Recently, Ye et al. (113) have reported on the sequence and development of a preliminary probe for the Cu-type NiR. The potential for discriminating active from inactive denitrifiers by using mRNA also exists (105).

With respect to determining activity, the advent of affordable benchtop quadrapole mass spectrometers or mass selective detectors suitable for gas analysis through membrane inlet systems should provide a stimulus for direct gas flux studies. Kana et al. (54) have reported on an improved N_2/Ar method applicable for real-time analysis of denitrifying sediments. Isotope ratio instruments have also become more available and affordable and will promote both tracer and natural abundance studies (2) of denitrification. Microprobe technology continues to advance, and as mentioned above, microprobes for N_2O (10, 20) and NO_3^- which allow fine-scale determination of profiles and temporal dynamics of each pool and thereby provide an ability to observe the results of nitrification and denitrification with minimal sample perturbation, have been described (50).

OTHER ASPECTS

Bacteria are involved in a variety of other relevant N transformations, including (i) the key processes of uptake of inorganic N and its incorporation into organic form and (ii) the degradation of organic forms returning N to inorganic pools. Considerable research has proceeded on both fronts. Research using tracer methodology with size fractionation (38) and specific inhibitors (73) has demonstrated important roles of heterotrophic bacteria in the uptake of limiting nutrients in lacustrine and marine environments (22, 111).

The $^{15}NH_4{}^+$ isotope dilution procedure has been used widely, and the results indicate that the regeneration of inorganic N is often dominated by heterotrophic bacteria in many systems (38). Shallow sediments are particularly important sites of N regeneration, and procedures for sediment studies are well developed (11, 12).

With the interest in radiatively important trace gases ("greenhouse" gases), there has been renewed interest in sources and sinks of N_2O in aquatic ecosystems (16). N_2O can arise from both nitrification and denitrification (including via nitrifier-catalyzed denitrification [80]). There remains considerable controversy over the relative roles of denitrifiers and nitrifiers in N_2O cycling. Efforts to resolve the source of N_2O in short-term assays with inhibitors have often been confounded by the nonspecificity of commonly used inhibitors (e.g., C_2H_2). In this regard the CH_3F nitrification inhibitor described by Miller et al. (61) reportedly is specific in inhibiting nitrifier, but not denitrifier, N_2O production.

New pathways of N transformations continue to be found, and novel observations are often abetted by innovations in methods or instrumentation. In the late 1970s, with the advent of the $^{15}NO_3{}^-$ isotope dilution procedure, the existence and importance of the dissimilatory pathway to $NH_4{}^+$ were first recognized (58, 59). While N_2O production from nitrification has been known since the 1960s, its importance in particular ecosystems was not appreciated until a decade later, with the widespread use of electron capture gas chromatography. Most recently, direct ^{15}N tracer studies have provided evidence for a novel pathway of anaerobic nitrification (36). Aerobic denitrifiers have been isolated (82), and their importance in the environment remains to be resolved. Similarly, heterotrophic nitrifiers are widely distributed in soils (see chapter 49), but little is known of their contribution to nitrification in aquatic environments.

REFERENCES

1. **Altabet, M. A., W. G. Deuser, S. Honjo, and C. Stienen.** 1991. Seasonal and depth-related changes in the source of sinking particles in the North Atlantic. *Nature* (London) **354:**136–139.
2. **Altabet, M. A., R. Francois, D. W. Murray, and W. L. Prell.** 1995. Climate-related variations in denitrification in the Arabian Sea from sediment $^{15}N/^{14}N$ ratios. *Nature* (London) **373:**506–509.
3. **Amman, R. I., W. Ludwig, and K.-H. Schleifer.** 1995. Phylogenetic identification and in situ detection of individual microbial cells without cultivation. *Microbiol. Rev.* **59:**143–169.
4. **Balderston, W. L., B. Sherr, and W. J. Payne.** 1976. Blockage by acetylene of nitrous oxide reduction in *Pseudomonas perfectomarinus*. *Appl. Environ. Microbiol.* **31:**504–508.
5. **Barnes, R. O., K. K. Bertine, and E. D. Goldberg.** 1975. N_2:Ar, nitrification and denitrification in southern California borderland basin sediments. *Limnol. Oceanogr.* **20:**962–970.
6. **Bedard, C., and R. Knowles.** 1989. Physiology, biochemistry, and specific inhibitors of CH_4, NH_4, and CO oxidation by methanotrophs and nitrifiers. *Microbiol. Rev.* **53:**68–84.
7. **Belser, L. W., and E. L. Mays.** 1980. Specific inhibition of nitrite oxidation by chlorate and its use in assessing nitrification in soils and sediments. *Appl. Environ. Microbiol.* **39:**505–510.
8. **Bergerson, F. J. (ed.).** 1980. *Methods for Evaluating Biological Nitrogen Fixation.* John Wiley & Sons, New York.
9. **Billen, G.** 1976. A method for evaluating nitrifying activity in sediments by dark ^{14}C-bicarbonate incorporation. *Water Res.* **10:**51–57.
10. **Binnerup, S. J., K. Jensen, N. P. Revsbech, M. H. Jensen, and J. Sorensen.** 1992. Denitrification, dissimilatory reduction of nitrate to ammonium, and nitrification in a bioturbated estuarine sediment as measured with ^{15}N and microsensor techniques. *Appl. Environ. Microbiol.* **58:**303–313.
11. **Blackburn, T. H.** 1979. A method for measuring rates of $NH_4{}^+$ turnover in anoxic marine sediments using a ^{15}N-$NH_4{}^+$ dilution technique. *Appl. Environ. Microbiol.* **37:**760–765.
12. **Blackburn, T. H.** 1993. Turnover of $^{15}NH_4{}^+$ tracer in sediments, p. 643–648. *In* P. F. Kemp, B. F. Sherr, E. B. Sherr, and J. J. Cole (ed.), *Handbook of Methods in Aquatic Microbial Ecology.* Lewis Publishers, Boca Raton, Fla.
13. **Bryan, B. A.** 1981. Physiology and biochemistry of denitrification, p. 67–84. *In* C. C. Delwiche (ed.), *Denitrification, Nitrification, and Atmospheric Nitrous Oxide.* John Wiley & Sons, New York.
14. **Burris, R. H., and P. W. Wilson.** 1957. Methods for measurement of nitrogen fixation. *Methods Enzymol.* **4:**355–365.
15. **Capone, D. G.** 1988. Benthic nitrogen fixation: microbiology, physiology and ecology, p. 85–123. *In* H. Blackburn and J. Sorensen (ed.), *Nitrogen Cycling in Coastal Waters.* John Wiley & Sons, New York.
16. **Capone, D. G.** 1991. Aspects of the marine nitrogen cycle with relevance to the dynamics of nitrous and nitric oxide, p. 255–275. *In* J. E. Rogers and W. B. Whitman (ed.), *Microbial Production and Consumption of Greenhouse Gases: Methane, Nitrogen Oxides, and Halomethanes.* American Society for Microbiology, Washington, D.C.
17. **Capone, D. G.** 1993. Determination of nitrogenase activity in aquatic samples using the acetylene reduction procedure, p. 621–631. *In* P. F. Kemp, B. F. Sherr, E. B. Sherr, and J. J. Cole (ed.), *Handbook of Methods in Aquatic Microbial Ecology.* Lewis Publishers, Boca Raton, Fla.
18. **Carpenter, E. J., D. G. Capone, B. Fry, and H. R. Harvey.** Biogeochemical tracers of the marine cyanobacterium *Trichodesmium.* Submitted for publication.
19. **Carpenter, E. J., and D. G. Capone (ed.).** 1983. *Nitrogen in the Marine Environment.* Academic Press, New York.
20. **Christensen, P. B., L. P. Nielsen, N. P. Revsbech, and J. Sorensen.** 1989. Microzonation of denitrification activity in stream sediments as studied with a combined oxygen and nitrous oxide microsensor. *Appl. Environ. Microbiol.* **55:**1234–1241.
21. **Cline, J. D., and F. A. Richards.** 1972. Oxygen-deficient conditions and nitrate reduction in the Eastern Tropical North Pacific Ocean. *Limnol. Oceanogr.* **17:**885–900.
22. **Currie, D. J., and J. Kalff.** 1984. The relative importance of phytoplankton and bacterioplankton in phosphorus uptake in freshwater. *Limnol. Oceanogr.* **29:**311–321.
23. **Currin, C. A., H. W. Paerl, G. Suba, and R. S. Alberte.** 1990. Immunofluorescent detection and characterization of N_2-fixing microorganisms from aquatic environments. *Limnol. Oceanogr.* **35:**59–71.
24. **Degrange, V., and R. Bardin.** 1995. Detection and counting of *Nitrobacter* populations in soil by PCR. *Appl. Environ. Microbiol.* **61:**2093–2098.
25. **D'Elia, C.** 1983. Nitrogen determination in seawater, p. 731–762. *In* E. J. Carpenter and D. G. Capone (ed.), *Nitrogen in the Marine Environment.* Academic Press, New York.
26. **Devol, A. H.** 1991. Direct measurement of nitrogen gas fluxes from continental shelf sediments. *Nature* (London) **349:**319–322.

27. **Dugdale, R. C., and J. J. Goering.** 1967. Uptake of new and regenerated forms of nitrogen in primary productivity. *Limnol. Oceanogr.* **12:**685–695.

28. **Dugdale, R. C., J. J. Goering, and J. H. Ryther.** 1964. High nitrogen fixation rates in the Sargasso Sea and the Arabian Sea. *Limnol. Oceanogr.* **9:**507–510.

29. **Fiedler, R., and G. Proksch.** 1975. The determination of nitrogen-15 by emission and mass spectrometry. *Anal. Chim. Acta* **78:**1–62.

30. **Firestone, M. K., R. B. Firestone, and J. M. Tiedje.** 1980. Nitrous oxide from soil denitrification: factors controlling its biological production. *Science* **208:**749–751.

31. **Gamble, T. N., M. Betlach, and J. M. Tiedje.** 1977. Numerically dominant denitrifying bacteria from world soils. *Appl. Environ. Microbiol.* **33:**926–939.

32. **Garside, C.** 1982. A chemiluminescent technique for the determination of nanomolar concentrations of nitrate and nitrite in seawater. *Mar. Chem.* **11:**159–167.

33. **Gersberg, R., K. Krohn, N. Peck, and C. R. Goldman.** 1976. Denitrification studies with ^{13}N-labeled nitrate. *Science* **192:**1229–1231.

34. **Glibert, P. M., and D. A. Bronk.** 1994. Release of dissolved organic nitrogen by marine diazotrophic cyanobacteria, *Trichodesmium* spp. *Appl. Environ. Microbiol.* **60:**3996–4000.

35. **Glibert, P. M., and D. G. Capone.** 1993. Mineralization and assimilation in aquatic, sediment and wetland systems, p. 243–272. *In* R. Knowles and T. H. Blackburn (ed.), *Nitrogen Isotope Techniques.* Academic Press, New York.

36. **Graaf, V. D. A. A., A. Mulder, P. De Bruijn, M. S. M. Jetten, L. A. Robertson, and J. G. Kuenen.** 1995. Anaerobic oxidation of ammonium is a biologically mediated process. *Appl. Environ. Microbiol.* **61:**1246–1251.

37. **Hall, G.** 1984. Measurements of nitrification rates in lake sediments: comparison of the nitrification inhibitors Nitrapyrin and allylthiourea. *Microb. Ecol.* **10:**25–36.

38. **Harrison, W. G.** 1983. Nitrogen in the marine environment: use of isotopes, p. 763–807. *In* E. J. Carpenter and D. G. Capone (ed.), *Nitrogen in the Marine Environment.* Academic Press, New York.

39. **Hattori, A.** 1983. Denitrification and dissimilatory nitrate reduction, p. 191–232. *In* E. J. Carpenter and D. G. Capone (ed.), *Nitrogen in the Marine Environment.* Academic Press, New York.

40. **Henriksen, K.** 1980. Measurement of in situ rates of nitrification in sediment. *Microb. Ecol.* **6:**329–337.

41. **Henriksen, K., J. I. Hansen, and T. H. Blackburn.** 1981. Rates of nitrification, distribution of nitrifying bacteria and nitrate fluxes in different types of sediment from Danish waters. *Mar. Biol.* **61:**299–304.

42. **Hojberg, O., H. S. Johansen, and J. Sorensen.** 1994. Determination of ^{15}N abundance in nanogram pools of NO_3^- and NO_2^- by denitrification bioassay and mass spectrometry. *Appl. Environ. Microbiol.* **60:**2467–2472.

43. **Hordijak, C. A., M. Snieder, J. J. M. van Engelen, and T. E. Cappenberg.** 1987. Estimation of bacterial nitrate reduction rates at in situ concentrations in freshwater sediments. *Appl. Environ. Microbiol.* **53:**217–223.

44. **Horrigan, S. G., and D. G. Capone.** 1985. Rates of nitrification and nitrate reduction in nearshore marine sediments under varying environmental conditions. *Mar. Chem.* **16:**317–327.

45. **Horrigan, S. G., J. P. Montoya, J. L. Nevins, and J. J. McCarthy.** 1990. Natural isotopic composition of dissolved inorganic nitrogen in the Chesapeake Bay. *Estuarine Coastal Shelf Sci.* **30:**393–410.

46. **Horrigan, S. G., J. P. Montoya, J. L. Nevins, J. J. McCarthy, H. W. Ducklow, R. Goericke, and T. Malone.** 1990. Nitrogenous nutrient transformations in the spring and fall in the Chesapeake Bay. *Estuarine Coastal Shelf Sci.* **30:**369–391.

47. **Howarth, R. W.** 1988. Nutrient limitation of net primary production in marine ecosystems. *Annu. Rev. Ecol.* **19:**89–110.

48. **Howarth, R. W., R. Marino, J. Lane, and J. J. Cole.** 1988. Nitrogen fixation in freshwater, estuarine, and marine ecosystems. 1. Rates and importance. *Limnol. Oceanogr.* **33:**669–687.

49. **Hynes, R. K., and R. Knowles.** 1983. Inhibition of chemoautotrophic nitrification by sodium chlorate and sodium chlorite: a reexamination. *Appl. Environ. Microbiol.* **54:**1178–1182.

50. **Jensen, K., N. P. Revsbech, and L. P. Nielsen.** 1993. Microscale distribution of nitrification activity in sediment determined with a shielded microsensor for nitrate. *Appl. Environ. Microbiol.* **59:**3287–3296.

51. **Jones, R. D.** 1991. An improved fluorescence method for the determination of nanomolar concentrations of ammonium in natural waters. *Limnol. Oceanogr.* **36:**814–818.

52. **Jones, R. D., R. Y. Morita, and R. P. Griffiths.** 1994. Method for estimating in situ chemolithotrophic ammonium oxidation using carbon monoxide oxidation. *Mar. Ecol. Prog. Ser.* **17:**259–269.

53. **Joye, S. B., and H. W. Paerl.** 1993. Contemporaneous nitrogen fixation and denitrification in intertidal microbial mats: rapid response to runoff events. *Mar. Ecol. Prog. Ser.* **94:**267–274.

54. **Kana, T., C. Darkangelo, M. D. Hunt, J. B. Oldham, G. E. Bennett, and J. C. Cornwell.** 1994. Membrane inlet mass spectrometer for rapid high-precision determination of N_2, O_2, and Ar in environmental water samples. *Anal. Chem.* **66:**4166–4170.

55. **Kaplan, W. A.** 1983. Nitrification, p. 139–190. *In* E. J. Carpenter and D. G. Capone (ed.), *Nitrogen in the Marine Environment.* Academic Press, New York.

56. **Kaplan, W. A., I. Valiela, and J. M. Teal.** 1979. Denitrification in a saltmarsh ecosystem. *Limnol. Oceanogr.* **24:**726–734.

57. **Knowles, R.** 1982. Denitrification. *Microbiol. Rev.* **46:**43–70.

58. **Koike, I., and A. Hattori.** 1978. Denitrification and ammonia formation in anaerobic coastal sediments. *Appl. Environ. Microbiol.* **35:**278–282.

59. **Koike, I., and A. Hattori.** 1978. Simultaneous determinations of nitrification and nitrate reduction in coastal sediments by a 15-N dilution technique. *Appl. Environ. Microbiol.* **35:**853–857.

60. **Koike, I., and J. Sorensen.** 1988. Nitrate reduction and denitrification in marine sediments, p. 251–273. *In* T. H. Blackburn and J. Sorensen (ed.), *Nitrogen Cycling in Coastal Marine Environments.* John Wiley & Sons, New York.

60a. **Ludden, P. (University of Wisconsin).** Personal communication.

61. **Miller, L. G., M. D. Coutlakis, R. S. Oremland, and B. B. Ward.** 1993. Selective inhibition of ammonium oxidation and nitrification-linked N_2O formation by methyl fluoride and dimethyl ether. *Appl. Environ. Microbiol.* **59:**2457–2464.

62. **Miller, L. G., R. S. Oremland, and S. Paulsen.** 1986. Measurement of nitrous oxide reductase activity in aquatic sediments. *Appl. Environ. Microbiol.* **51:**18–24.

63. **Mills, E. L.** 1989. Biological oceanography, an early history, 1870–1960. Cornell University Press, Ithaca, N.Y.

64. **Montoya, J. P., M. Voss, P. Kaehler, and D. G. Capone.** 1996. A simple, high-precision, high-sensitivity tracer

assay for N₂ fixation. *Appl. Environ. Microbiol.* **62:** 986–993.

65. **Mosier, A. R., and D. S. Schimel.** 1993. Nitrification and denitrification. p. 181–208. *In* R. Knowles and T. H. Blackburn (ed.), *Nitrogen Isotope Techniques.* Academic Press, New York.

66. **Nielsen, L. P.** 1992. Denitrification in sediments determined from nitrogen isotope pairing. *FEMS Microbiol. Ecol.* **86:**357–362.

67. **Nishio, T., I. Koike, and A. Hattori.** 1981. N₂/Ar and denitrification in Tama Estuary sediments. *Geomicrobiol. J.* **2:**193–209.

68. **Nishio, T., I. Koike, and A. Hattori.** 1982. Denitrification, nitrate reduction, and oxygen consumption in coastal and estuarine sediments. *Appl. Environ. Microbiol.* **43:**648–653.

69. **Nishio, T., I. Koike, and A. Hattori.** 1983. Estimates of denitrification and nitrification in coastal and estuarine sediments. *Appl. Environ. Microbiol.* **43:**648–653.

70. **Nixon, S. W.** 1981. Remineralization and nutrient cycling in coastal marine ecosystems, p. 111–138. *In* B. J. Neilson and L. E. Cronin (ed.), *Estuaries and Nutrients.* Humana Press, Clifton, N.J.

71. **Nowicki, B. L.** 1994. The effect of temperature, oxygen, salinity, and nutrient enrichment on estuarine denitrification rates measured with a modified nitrogen gas flux technique. *Estuarine Coastal Shelf Sci.* **35:**137–156.

72. **Olson, R. J.** 1981. ¹⁵N tracer studies of the primary nitrite maximum. *J. Mar. Res.* **39:**203–226.

73. **Oremland, R. S., and D. G. Capone.** 1988. Use of specific inhibitors in microbial ecological and biogeochemical studies. *Adv. Microb. Ecol.* **10:**285–383.

74. **Oremland, R. S., C. Umberger, C. W. Culbertson, and R. L. Smith.** 1984. Denitrification in San Francisco Bay intertidal sediments. *Appl. Environ. Microbiol.* **47:** 1106–1112.

75. **Oren, A., and T. H. Blackburn.** 1979. Estimation of sediment denitrification rates at in situ nitrate concentrations. *Appl. Environ. Microbiol.* **37:**174–176.

76. **Paerl, H. W.** 1990. Physiological ecology and regulation of N₂ fixation in natural waters. *Adv. Microb. Ecol.* **11:** 305–344.

77. **Parsons, T. R., Y. Maita, and C. M. Lalli.** 1984. *A Manual of Chemical and Biological Methods for Seawater Analysis.* Pergamon Press, New York.

78. **Pearsall, K. A., and F. T. Bonner.** 1980. Analysis of dinitrogen-nitrogen oxide mixtures employing direct vacuum line-gas chromatograph injection. *J. Chromatogr.* **200:**224–227.

79. **Postgate, J. R.** 1978. Nitrogen fixation. *Studies in Biology* no. 92. E. Arnold, London.

80. **Poth, M., and D. D. Focht.** 1985. ¹⁵N kinetic analysis of N₂O production by *Nitrosomonas europaea*: an examination of nitrifier denitrification. *Appl. Environ. Microbiol.* **49:**1134–1141.

81. **Risgaard-Petersen, N., S. Rysgaard, and N. P. Revsbech.** 1993. A sensitive assay for determination of ¹⁴N/¹⁵N isotope distribution in NO₃⁻. *J. Microbiol. Methods* **17:**155–164.

82. **Robertson, L. A., and J. G. Kuenen.** 1984. Aerobic denitrification: a controversy revived. *Arch. Microbiol.* **139:** 351–354.

83. **Rysgaard, S., N. Risgaard-Petersen, L. P. Nielsen, and N. P. Revsbech.** 1993. Nitrification and denitrification in lake and estuarine sediments measured by the ¹⁵N dilution technique and isotope pairing. *Appl. Environ. Microbiol.* **59:**2093–2098.

84. **Schmidt, E. L.** 1974. Quantitative autoecological study

of microorganisms in soil by immuno-fluorescence. *Soil Sci.* **118:**141–149.

85. **Seitzinger, S.** 1988. Denitrification in freshwater and coastal ecosystems: ecological and geochemical significance. *Limnol. Oceanogr.* **33:**702–724.

86. **Seitzinger, S. P.** 1993. Denitrification and nitrification rates in aquatic sediments, p. 633–641. *In* P. F. Kemp, B. F. Sherr, E. B. Sherr, and J. J. Cole (ed.), *Handbook of Methods in Aquatic Microbial Ecology.* Lewis Publishers, Boca Raton, Fla.

87. **Seitzinger, S. P., L. P. Nielson, J. Caffrey, and P. B. Christensen.** 1993. Denitrification measurements in aquatic sediments: a comparison of three methods. *Biogeochem. J.* **23:**147–167.

88. **Seitzinger, S., S. W. Nixon, M. E. Pilson, and S. Burke.** 1980. Denitrification and N₂O production in nearshore marine sediments. *Geochim. Cosmochim. Acta* **44:** 1853–1860.

89. **Sherr, B. F., and W. J. Payne.** 1978. Effect of the Spartina alterniflora root-rhizome system on salt marsh denitrifying bacteria. *Appl. Environ. Microbiol.* **35:**724–729.

90. **Slater, J. M., and D. G. Capone.** 1989. Nitrate requirement for acetylene inhibition of nitrous oxide reduction in marine sediments. *Microb. Ecol.* **17:**143–157.

91. **Sloth, N. P., L. P. Nielsen, and T. H. Blackburn.** 1992. Nitrification in sediment cores measured with acetylene inhibition. *Limnol. Oceanogr.* **37:**1108–1112.

92. **Somville, M.** 1978. A method for the measurement of nitrification rates in water. *Water Res.* **12:**843–848.

93. **Sorensen, J.** 1978. Denitrification rates in a marine sediment as measured by the acetylene inhibition technique. *Appl. Environ. Microbiol.* **36:**139–143.

94. **Stewart, W. D., G. P. Fitzgerald, and R. H. Burris.** 1967. In situ studies on N₂ fixation using the acetylene reduction technique. *Proc. Natl. Acad. Sci. USA* **58:** 2071–2078.

95. **Stewart, W. P. D.** 1965. Nitrogen turnover in marine and brackish habitats. *Ann. Bot.* **29:**229–239.

96. **Taylor, B. F.** 1983. Assays of microbial nitrogen transformations, p. 809–837. *In* E. J. Carpenter and D. G. Capone (ed.), *Nitrogen in the Marine Environment.* Academic Press, New York.

97. **Vanderbroght, J. P., and G. Billen.** 1975. Vertical distribution of nitrate concentration in interstitial water of marine sediments with nitrification and denitrification. *Limnol. Oceanogr.* **20:**953–961.

98. **Voytek, M. A., and B. B. Ward.** 1995. Detection of ammonium-oxidizing bacteria of the beta-subclass of the class *Proteobacteria* in aquatic samples with the PCR. *Appl. Environ. Microbiol.* **61:**1444–1450.

99. **Wada, E., and A. Hattori.** 1991. Nitrogen in the sea: forms, abundances, and rate processes. CRC Press, Boca Raton, Fla.

100. **Waksman, S. A., C. L. Carey, and M. Hotchkiss.** 1933. Marine bacteria and their role in the cycle of life of the sea. *Biol. Bull.* **65:**137–167.

101. **Ward, B. B.** 1982. Oceanic distribution of ammonium-oxidizing bacteria determined by immunofluorescent assay. *J. Mar. Res.* **40:**1155–1172.

102. **Ward, B. B.** 1986. Nitrification in marine environments. *Symp. Soc. Gen. Microbiol.* **20:**157–184.

103. **Ward, B. B.** 1995. Diversity in denitrifying bacteria: limits of rDNA RFLP analysis and probes for the functional gene, nitrite reductase. *Arch. Microbiol.* **163:**167–175.

104. **Ward, B. B.** Functional and taxonomic probes for bacteria in the nitrogen cycle. *In* I. Joint (ed.), *Molecular Ecology of Aquatic Microbes,* in press. NATO.

105. **Ward, B. B.** Nitrification and denitrification: probing the

nitrogen cycle in aquatic environments. *Microb. Ecol.*, in press.

106. **Ward, B. B., and A. F. Carlucci.** 1985. Marine ammonia- and nitrite-oxidizing bacteria: serological diversity determined by immunofluorescence in culture and in the environment. *Appl. Environ. Microbiol.* **50:**194–201.

107. **Ward, B. B., and A. R. Cockcroft.** 1993. Immunofluorescence detection of denitrifying bacteria in seawater and intertidal sediment environments. *Microb. Ecol.* **25:** 233–246.

108. **Ward, B. B., A. R. Cockcroft, and K. A. Kilpatrick.** 1993. Antibody and DNA probes for detection of nitrite reductase in seawater. *J. Gen. Microbiol.* **139:**2285–2293.

109. **Ward, B. B., K. A. Kilpatrick, E. H. Renger, and R. W. Eppley.** 1989. Biological nitrogen cycling in the nitracline. *Limnol. Oceanogr.* **34:**493–513.

110. **Webb, K. L., and W. J. Wiebe.** 1975. Nitrification on a coral reef. *Can. J. Microbiol.* **21:**1427–1431.

111. **Wheeler, P. A., and D. L. Kirchman.** 1986. Utilization of inorganic and organic nitrogen by bacteria in marine systems. *Limnol. Oceanogr.* **31:**998–1009.

112. **Wu, Q., and R. Knowles.** 1995. Effect of chloramphenicol on denitrification in *Flexibacter canadensis* and *Pseudomonas denitrificans*. *Appl. Environ. Microbiol.* **61:** 434–437.

113. **Ye, R. W., M. R. Fries, and S. G. Bezborodnikov.** 1993. Characterization of the structural gene encoding a copper-containing nitrite reductase and homology of this gene to DNA of other denitrifiers. *Appl. Environ. Microbiol.* **59:**250–254.

114. **Yoshida, N., H. Morimoto, M. Hirano, I. Koike, S. Matsuo, E. Wada, T. Saino, and A. Hattori.** 1989. Nitrification rates and ^{15}N abundances of N_2O and NO_3^- in the western North Pacific. *Nature* (London) **342:**895–898.

115. **Yoshinari, T., R. Hynes, and R. Knowles.** 1977. Acetylene inhibition of nitrous oxide reduction and measurement of denitrification and nitrogen fixation in soil. *Soil Biol. Biochem.* **9:**177–183.

116. **Yoshinari, T., and R. Knowles.** 1976. Acetylene inhibition of nitrous oxide reduction by denitrifying bacteria. *Biochem. Biophys. Res. Commun.* **69:**705–710.

117. **Zehr, J., and D. G. Capone.** Problems and promises of assaying the genetic potential for nitrogen fixation in the marine environment. *Microb. Ecol.*, in press.

118. **Zehr, J., S. Braun, Y. Chen, and M. Mellon.** Nitrogen fixation in the marine environment: relating genetic potential to nitrogenase. *In NATO AWI Symposium*, in press.

119. **Zehr, J. P., R. J. Limberger, K. Ohki, and Y. Fujita.** 1990. Antiserum to nitrogenase generated from an amplified DNA fragment from natural populations of *Trichodesmium* spp. *Appl. Environ. Microbiol.* **56:**3527–3531.

120. **Zehr, J. P., and L. A. McReynolds.** 1989. Use of degenerate oligonucleotides for amplification of the *nifH* gene from the marine cyanobacterium *Trichodesmium thiebautii*. *Appl. Environ. Microbiol.* **55:**2522–2526.

121. **Zehr, J. P., M. Mellon, S. Braun, W. Litaker, T. Steppe, and H. W. Paerl.** 1995. Diversity of heterotrophic nitrogen fixation genes in a marine cyanobacterial mat. *Appl. Environ. Microbiol.* **61:**2527–2532.

Phosphorus Cycling

RONALD D. JONES

36

Phosphorus is considered to be a major growth-limiting nutrient in aquatic systems (17, 32, 35), and unlike the case for nitrogen, there is no large atmospheric source that can be made biologically available. All living organisms require phosphorus for growth and metabolism. It is essential in both cellular energetics (ATP) and cellular structure (DNA, RNA, phospholipids). As such, the phosphorus cycle has been extensively studied and is of great interest in the aquatic environment (5, 6, 10, 34, 35).

In contrast to the microbial cycles of nitrogen and sulfur, phosphorus, in general, does not commonly undergo any oxidations or reductions and remains combined with oxygen in its pentavalent state as phosphate molecules (PO_4^{3-}). It is thought that some sediment microorganisms may have the capability to utilize phosphate as a terminal electron acceptor in the absence of oxygen, nitrate, or sulfate and produce phosphine (PH_3) (3). This process is, however, of little ecological significance, and I will limit my discussion to the cycling of the phosphate molecule.

The aquatic cycling of phosphorus is essentially the conversion of phosphorus from the organic to the inorganic state and vice versa, with microorganisms playing a key role in both transformations. There are two pools of phosphorus that must be considered: the large slowly cycled pool of geologic phosphate contained in the earth's crust and sediments and the much smaller but rapidly cycled pool of biologically active phosphate. In the water column, this smaller pool exists predominantly as dissolved organic and inorganic phosphates, living and dead biological materials, and some suspended particulate and sediment inorganic phosphates. Although microorganisms play a key role in the release of phosphorus from the large geologic pool by mineralization and dissolution through the production of organic acids, the primary role of microorganisms is in mediating the cycling of the small biologically active pool (11, 27). It is this pool of phosphate and the methods used to examine it that will be discussed here. In addition, this chapter will be limited to a discussion of phosphorus cycling in the water column. For more information on sediment and intracellular phosphorus cycling, the reader is referred to references 5 and 37, respectively.

Since phosphorus is a limiting nutrient in most of the systems studied, its examination is often difficult because of the small quantities actively being cycled. Many of the traditional analytical methods are not sensitive enough to be used or require such large sample volumes that they are inappropriate. For this reason and because there are readily available radioactive phosphorus compounds, ^{32}P and ^{33}P, many researchers have relied on the use of radioisotopes for the examination of phosphorus (1, 25, 37). In addition, there are several nonradioactive methods for the examination of phosphate availability, based on the production of the enzyme alkaline phosphatase (8) or 5'-nucleotidase (1, 2). The magnesium-induced coprecipitation (MAGIC) technique for the measurement of nanomolar quantities of phosphate recently developed by Karl and Tien (21) may help to solve some of the analytical problems, but it is unlikely to replace the use of radioisotopes.

MEASUREMENT OF PHOSPHORUS UPTAKE AND INCORPORATION BY USING ^{32}P AND DIFFERENTIAL FILTRATION

The method described below relies on the use of radioactive $^{32}PO_4^{3-}$ and requires that certain precautions be taken. Briefly, the procedure consists of the addition of trace amounts of $^{32}PO_4^{3-}$ to a water sample; after various periods of incubation, the sample is filtered through a series of filters with different pore sizes to obtain both turnover time and the incorporation rates of various size fractions. The use of radioactive phosphate allows the determination of phosphate uptake at concentrations at or near ambient concentrations. In general, phosphate concentrations below 0.03 μM are not easily quantifiable by colorimetric chemical analysis, however (21); concentrations of phosphate in aquatic systems are often below this concentration. If this is the case or if one wishes to determine which size fraction is incorporating phosphorus from the water column, then the use of ^{32}P is appropriate.

Because of the speed at which trace quantities of phosphate are taken up by microbial communities, this procedure is generally divided into two phases. In the first phase, the objective is to collect as many samples as possible within the first 10 min for the calculation of phosphate turnover time (as defined by Lean et al. [24]). After this intense sampling, a second series of samples is differentially filtered for size fractionation, if these results are desired. Since turnover times of less than 10 min are common (24), it is important to have all of the materials necessary set up before the procedure is begun. Initially the assay is conducted with a 0.2-μm-pore-size filter only, and results are calculated on the basis of the percentage of the ^{32}P remaining in the filtrate. Since it is easier to count the ^{32}P retained on the filters and calculate the activity remaining in the filtrate, the filtrate is discarded and the filters are retained for counting.

For this procedure, duplicate 250-ml water samples are placed in 500-ml polyethylene, glass, or polycarbonate bottles. Plastic bottles are recommended because of the fragility of glass, even though they can represent a radioactive decontamination problem. Samples are placed in a water bath at field temperature. The initial uptake of phosphate is usually restricted to bacterial activity (9, 10); however, if it is suspected that photosynthetic organisms play a significant role, then incubations should be done in both light and dark conditions.

After a sufficient interval has been allowed for the samples to reach equilibrium with incubation conditions, a calculated amount of carrier-free ^{32}PO$_4$$^{3-}$ (Amersham or DuPont, NEN Research Products) is added to give between 20,000 and 50,000 cpm/ml. This amount should be calculated immediately before the experiment and is based on the decay rate and isotope preparation data provided by the manufacturer. Since ^{32}P decays to form ^{32}S, the volume added is unimportant but should generally be less than 0.1 ml because of the acidity of the phosphate solution (generally as orthophosphoric acid in dilute HCl). After the isotope is added, the bottle is mixed vigorously without shaking and placed in the water bath. Subsamples of 0.5, 1, 2, 4, 8, 10, 20, 40, 60, and 120 min and filtered through a 25-mm-diameter, 0.2-μm-pore-size filter. Before isotope is added, the water to be examined should be checked for its filtration time. As a general rule, the filtration process for a 10-ml sample should take less than 30 s. If a longer time is required, then smaller subsamples should be used. Use of subsamples smaller than 1.0 ml is not recommended, as reproducibility seems to suffer. In general, nuclear track emission (Nuclepore or equivalent) filters work better than depth retention filters. It is critical that accurate times be recorded and that the precise time of the end of filtration be noted. For this purpose, each bottle should have a separate stopwatch assigned to it and the filtration process should be observed so that the time can be recorded as soon as the surface of the filter is dry. The accuracy of the results depends highly on

the precision of the time determination and the reproducibility of subsample collection.

After 1 or 2 h, additional 10-ml subsamples are filtered onto 0.2-, 1.0-, 5.0-, and 12-μm-pore-size (Nuclepore or equivalent) filters to determine the size dependence of phosphate uptake. The use of Nuclepore or equivalent filters ensures that little water will remain in the filter matrix. This is important since any remaining water retained by the filter will give results higher than expected. As the filters are collected, they are placed into 20-ml scintillation vials and covered with a sufficient amount of scintillation cocktail or water. If water is used (recommended, since problems associated with the disposal of waste are minimized), it is important that the filters be placed flat on the bottom of the vial, filtration surface up, and covered with 2 ml of water. The use of water makes the counting efficiency depend on the geometry of the sample (39), since water has no scintillation properties. This reproduction of the geometry is not necessary if a scintillation cocktail is used.

For the determination of total activity, samples along with 2.0 ml of the unfiltered water are placed in a liquid scintillation counter and assayed for radioactivity. For calculation of the turnover time, the counts retained on the filter are subtracted from the total counts, and the natural logarithm of the percent of ^{32}P in the filtrate is plotted against time (Table 1 and Fig. 1) (24). The initial linear portion of the slope is used to approximate the uptake rate constant, and the turnover time of phosphate is the reciprocal of the rate constant (Fig. 1). Usually, rapid turnover times indicate phosphorus limiting conditions and long turnover times indicate that some other factor limits microbial activity or system productivity. For a full discussion of interpretation of data, the reader is referred to reference 24.

Just as turnover time gives information pertaining to the rate of phosphate uptake, the result of the size fractionation gives us information about the type of microorganism responsible for phosphate uptake. To calculate these values, the counts from the sequential filtration steps are subtracted from each other to give the percentage of phosphate incorporated by the 0.2- to 1.0-, 1.0- to 5.0-, 5.0- to 12.0-, and >120-μm fractions. The value for the material collected on the 1.0-μm-pore-size filter is subtracted from the value for the material collected on the 0.2-μm-pore-size filter to give the 0.2- to 1.0-μm fraction, the value for the material collected on the 5.0-μm-pore-size filter is subtracted from

TABLE 1 Rate of ^{32}PO$_4$$^{3-}$ incorporation by an Everglades National Park, Fla., water sample

Sample collection time (min)	cpm[a] (filter)	% Filter [(filter cpm/total cpm) × 100]	% Filtrate (100 − % filter cpm)	ln % filtrate
0.80	14,736	25.23	74.77	4.31
4.20	35,369	60.56	39.44	3.67
7.80	43,992	75.33	24.67	3.21
11.20	47,464	81.27	18.73	2.93
17.80	49,511	84.78	15.22	2.72
30.90	52,675	90.20	9.80	2.28
40.60	54,462	93.26	6.74	1.91
49.30	53,888	92.27	7.73	2.04
68.50	54,369	93.10	6.90	1.93
77.50	55,218	94.55	5.45	1.70
109.00	55,426	94.91	5.09	1.63

[a] Total cpm added was 58,400.

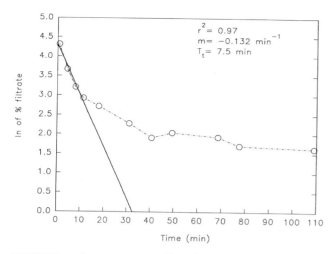

FIGURE 1 Incorporation of $^{32}PO_4{}^{3-}$ into particulate material. Solid line, initial slope for turnover time calculation; m, slope; T, turnover time. Water was collected from Everglades National Park, Fla.

the value for the material collected on the 1.0-μm-pore-size filter to give the 1.0- to 5.0-μm fraction, etc. The counts retained on the 12.0-μm-pore-size filter represent the fraction incorporated into particles of >12 μm. Results are generally presented as the percentage of the total amount of phosphate incorporated (counts on the 0.2-μm-pore-size filter). However, if the initial concentration of phosphorus is known, the result can be expressed in absolute terms. An example of the type of distribution expected is shown in Fig. 2. For interpretation, it is generally assumed that the 0.2- to 1.0-μm fraction represents the bacteria, that the 1.0- to 5.0-μm fraction represents a mix of bacteria and

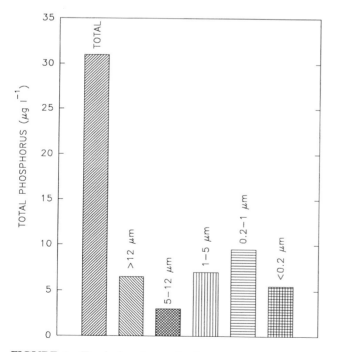

FIGURE 2 Total phosphorus size fractionation in a water sample from Everglades National Park, Fla.

phytoplankton, and that both the 5.0- to 12.0-μm and >12.0-μm fractions are phytoplankton (9, 13). These size classifications are highly dependent on the system being examined, and microscopic examination of the retained material should be made for a more precise determination. In addition, the determination of chlorophyll or other specific pigments can help interpret size classes.

The use of $^{32}PO_4{}^{3-}$ has many other applications, several of which require only minor modification of the foregoing procedures. Rate constants such as K_m (the Michaelis-Menten constant) and V_{max} (the maximum velocity of the reaction at substrate saturation) can be calculated by using several different concentrations of cold (nonradioactive) phosphate to dilute the activity of the $^{32}PO_4{}^{3-}$ (23, 25). For a more complete discussion of enzyme kinetics, the reader is referred to references 7 and 26. Generally, total inorganic phosphate concentrations must be kept below 2.0 μM for this procedure. ^{32}P can also be used as a tracer of phosphate movement through the food web. This can be accomplished by conducting the differential filtration described above at different time intervals and observing the change in the percentage of P contained in each fraction (13, 22). This approach could also be used to examine phosphorus incorporation by larger organisms and macrophytes.

The procedure described above is generally limited in use only by the precautions necessary for the safe handling of ^{32}P (39). Care must be taken so that the samples collected for ^{32}P uptake and incorporation are homogeneous and do not contain large particles or rapidly settling sediments. If the samples used contain these materials, subsamples will not be uniform and the results could be affected. Samples should not be refrigerated and should be assayed as soon as possible after collection. In general, samples more than 6 to 12 h old should not be used. Effects of storage are, however, easily determined by experimentation.

Interpretation of the data can be more difficult if there are high concentrations of particulate inorganic phosphates. Without chemical analysis of the incorporated phosphate, it is not possible to be certain that all incorporation was a consequence of the conversion of inorganic phosphates into organic phosphates. Results from many systems have indicated that much of the material is incorporated into microbial cells, but caution is advised when making this assumption (9, 14, 38).

ALKALINE PHOSPHATASE AS AN INDICATOR OF PHOSPHORUS LIMITATION

The use of $^{32}PO_4{}^{3-}$ as described above yields potential rates for the microbial conversion of inorganic phosphorus to organic phosphorus. It cannot establish if phosphorus is limiting. Alkaline phosphatase, on the other hand, can serve both as an indicator for the potential for conversion of organic phosphorus into inorganic phosphorus and as a test for phosphate limitation (4, 12, 15, 31). The importance of alkaline phosphatase in the cycling of phosphorus in natural waters is paramount. Phosphorus occurs in nature in the form of organic and inorganic phosphates, but almost all osmotrophic organisms utilize only inorganic phosphate for growth (37). All organisms possess alkaline phosphatase, but only bacteria, fungi, and perhaps some algae excrete the enzyme outside of their cells, thus participating in the remineralization and dissolution of organic phosphates (8, 36, 37, 41). Since inorganic phosphate is almost always found in growth-limiting concentrations (17), one of the most important microbial activities is the conversion of organic phosphates to inorganic phosphates through the action of phosphatases.

There are two types of phosphatases, acid and alkaline, identified by their pH optima (28, 29, 37). These enzymes function by hydrolyzing organic phosphate esters as well as inorganic pyrophosphate and certain other inorganic phosphates. Phosphatases are present in all organisms, but only organisms that possess an extracellular (usually periplasmic) phosphatase are able to mineralize external sources of organic phosphate. Without alkaline phosphatase, organic phosphorus would act as a terminal sink for inorganic phosphorus and productivity in most systems would be limited by external phosphorus inputs from continental weathering. Phosphatase production and the various genetic and kinetic control mechanisms are in themselves an area of active study; indeed, some of the mechanisms for cell regulation make phosphatases excellent bioindicators (37). However, for purposes of this chapter, the use of extracellular phosphatases will be discussed as a method here.

What makes alkaline phosphatase so appropriate is that microorganisms do not produce extracellular alkaline phosphatase in the presence of excess dissolved phosphate. This allows the use of alkaline phosphatase as a sensitive indicator for phosphate limitation. Alkaline phosphatase has been used successfully as a bioindicator in several systems (4, 18, 19), but because of the nature of the regulation of alkaline phosphatase, the environments for which it is best suited are oligotrophic (15, 30, 31). Extracellular production of alkaline phosphatase responds to increased levels of dissolved inorganic phosphate on the order of minutes to hours rather than days, making it suitable to the temporal variations of phosphorus concentration often found in natural systems.

In environments in which the concentration of dissolved inorganic phosphate is high and phosphate is not growth limiting, the assay of extracellular alkaline phosphatase may not be appropriate. If, however, there is an active cycling between the organic and inorganic forms of phosphate, alkaline phosphatase activity assays may prove useful (31).

The most commonly used procedures for assaying alkaline phosphatase activity are based on the use of an artificial substrate such as p-nitrophenyl phosphate (PNPP) (33) or 3-o-methylfluorescein phosphate (MFP) (30). PNPP is a colorless compound that is cleaved by alkaline phosphatase to yield PO_4^{3-} and p-nitrophenol, which is yellow in color and can be detected spectrophotometrically. MFP is hydrolyzed by alkaline phosphatase to yield PO_4^{3-} and 3-o-methylfluorescein (MF), which is highly fluorescent and is detected fluorometrically. The use of MFP gives sensitivity nearly 3 orders of magnitude greater than that with use of PNPP, and detection can be in the nanograms-per-liter range. Because of this and several other difficulties pertaining to the use of PNPP, the procedure for the use of MFP will be described here. However, in addition to MFP, another fluorogenic substrate, 4-methylumbelliferyl phosphate, can be used, requiring only a modification of the excitation and emission wavelengths (8). For the MFP procedure, it is very important that caution be taken to ensure that the conditions of the assay are standardized. Temperature and pH both affect the fluorescence of MF as well as the activity of alkaline phosphatase. A saturating concentration of MFP dissolved in a buffer solution is added to a water sample. The sample is incubated, and alkaline phosphatase present in the samples cleaves the phosphate from the MFP, leaving MF. The concentration of MF, assessed fluorometrically, is proportional to the alkaline phosphatase activity of the

sample. In this assay, duplicate 3-ml subsamples are pipetted into disposable cuvettes (four clear sides, polystyrene; Fisher Scientific), and 30 μl of MFP reagent is added to each and mixed with disposable plastic transfer pipettes (Fisher Scientific). The MFP reagent is prepared by dissolving 53 mg of anhydrous MF (Sigma Chemical Company) in 100 mM Tris buffer, pH 8.7 (Sigma). The final concentrations of MFP are 1 mM in the stock reagent and 100 μM in the assay mixture. The final concentration of Tris buffer in the assay mixture is 1 mM. This stock MFP reagent can be stored in a freezer at $-20°C$ in small plastic vials (2 to 3 ml) and thawed prior to each use. The reagent stored in this manner has a useful life of more than 1 year.

The pH of the MFP reagent generally depends on the pH of the sample and the type of phosphatase, acid or alkaline, being assayed. Once again, it is important to remember that the objective of this procedure is to assay enzyme activity under standard conditions and that the greatest sensitivity will be obtained under conditions favoring the activity of alkaline phosphatase, not microbial growth. It is also important to note that the concentration of the Tris buffer may have to be increased or the pH of the samples may have to be adjusted if samples with high acidity or alkalinity are used. For most natural waters, the MFP reagent described here will be adequate.

After the sample is mixed, the fluorescence is immediately measured by using a fluorometer with an excitation wavelength of 430 nm and an emission wavelength of 507 nm and recorded. The subsamples are then incubated for 2 h at 25°C in the dark. The temperature of incubation is unimportant as long as the conditions are standardized. However, again it is important to remember the objective of optimum conditions for enzyme activity. After this period, the cuvettes are again assayed fluorometrically at the same wavelengths, and the results are recorded. The amount of MF produced in 2 h is quantified by subtracting the initial value from the final value, and the result is compared against a standard curve. Different incubation periods can be used, but the amount of MFP converted to MF should never exceed 10% of the initial concentration of MFP (i.e., 10 μM). If it does, then the assayed value for alkaline phosphatase activity will be lower than the actual value since the MFP may no longer be at a saturating concentration.

To prepare the standard MF curve, it is important to ensure that the standards are made up in the same buffer at the same pH, concentration, and temperature as the assay mixture. Working standards are prepared by using a 1 mM solution of MF in methanol (stock stored in freezer at $-20°C$) diluted into 1 mM Tris buffer, pH 8.7. Concentrations of 0, 1, 2.5, 5, and 10 μM are used to prepare the final curve. If care is not taken to ensure that the pH and temperature are the same as in the assay mixture, the results will be incorrect, as MF fluorescence is highly dependent on these factors.

When samples for alkaline phosphatase activity are collected, it is important that sediments and large particles be excluded, as they will cause a positive interference and lead to difficulty in data interpretation. In general, samples should not be refrigerated but rather stored in the dark at ambient or assay temperature. It is important that the samples be assayed as soon as possible, but if necessary, samples can be stored for up to 24 h with little change in activity. An example of the results obtained by this procedure is

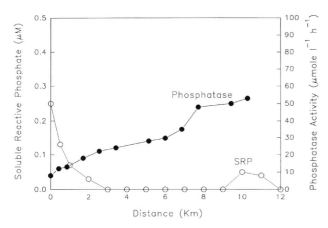

FIGURE 3 Alkaline phosphatase activity and soluble reactive phosphate (SRP) concentrations in water samples from a transect into Everglades National Park, Fla.

shown in Fig. 3. Although the results of alkaline phosphatase assays are often expressed in absolute units (micromoles per liter per hour), these results are generally not strictly comparable from one environment to another. Results are best viewed in the context of relative values when one is working with a series of similar matrix samples. In the example in Fig. 3, the low alkaline phosphatase activity at 0 km indicates an excess of dissolved inorganic phosphate and nonlimiting conditions, while the value of 53 at 11 km indicates a situation in which phosphate is in limited supply. Thus, high alkaline phosphatase activity indicates phosphate limitation, while low activity represents the opposite. In some situations, alkaline phosphatase activity can be affected by the number of bacterial cells present in the sample. This can be of concern when low alkaline phosphatase activities are associated with low cell numbers rather than low concentrations of alkaline phosphatase per cell. When examining waters with low bacterial numbers one should exercise caution when interpreting results. The opposite result can occur with high bacterial numbers, but this is much less common.

In addition to the procedure described above, MFP can be used to estimate the actual rate of organic phosphate mineralization. This is done by assaying the rate of MF formation at several different concentrations of MFP below saturation, applying first-order rate kinetics. The most appropriate method for interpreting these results when the concentration of organic phosphate cannot be measured accurately is that proposed by Wright and Hobbie (40) and discussed by Button (7). MFP can also be applied to sediments and soils with only minor modifications, usually involving centrifugation or filtration to remove interferences for the fluorescence measurements (8). Other enzymes, such as 5'-nucleotidase, have also been shown to be important in the regeneration of phosphate (1, 2). Intracellular production of alkaline phosphatase has also been shown to be related to nutrient limitation and can be used to examine phosphate cycling by larger organisms, particularly the algae associated with surfaces (16, 20). In summary, the use of MFP is a simple, sensitive, and inexpensive method to examine phosphate mineralization and limitation.

REFERENCES

1. **Ammerman, J. W., and F. Azam.** 1985. Bacterial 5'-nucleotidase in aquatic ecosystems: a novel mechanism of phosphorus regeneration. *Science* **227:**1338–1340.

2. **Ammerman, J. W., and F. Azam.** 1991. Bacterial 5'-nucleotidase activity in estuarine and coastal marine waters: characterization of enzyme activity. *Limnol. Oceanogr.* **36:**1427–1436.

3. **Atlas, R. M., and R. Bartha.** 1992. *Microbial Ecology: Fundamentals and Applications,* 3rd ed. Benjamin/Cummings, Redwood City, Calif.

4. **Berman, T.** 1970. Alkaline phosphatases and phosphorus availability in Lake Kinneret. *Limnol. Oceanogr.* **15:**663–674.

5. **Boers, P. C. M., T. E. Cappenberg, and W. van Rarphorst (ed.).** 1993. *Developments in Hydrobiology 84. Proceedings of the Third International Workshop on Phosphorus in Sediments.* Kluwer Academic Publishers, Dordrecht, The Netherlands.

6. **Bostrom, B., G. Persson, and B. Broberg.** 1988. Bioavailability of different phosphorus forms in freshwater systems. *Hydrobiologia* **170:**133–155.

7. **Button, D. K.** 1991. Biochemical basis for whole-cell uptake kinetics: specific affinity, oligotrophic capacity, and the meaning of the Michaelis constant. *Appl. Environ. Microbiol.* **57:**2033–2038.

8. **Chrost, R. J. (ed.).** 1991. *Microbial Enzymes in Aquatic Environments.* Springer-Verlag, New York.

9. **Currie, D. J., and J. Kalff.** 1984. A comparison of the abilities of freshwater algae and bacteria to acquire and retain phosphorus. *Limnol. Oceanogr.* **29:**298–310.

10. **Currie, D. J., and J. Kalff.** 1984. The relative importance of bacterioplankton and phytoplankton in phosphorus uptake in freshwater. *Limnol. Oceanogr.* **29:**311–321.

11. **Doremus, C., and L. S. Clesceri.** 1982. Microbial metabolism in surface sediments and its role in the immobilization of phosphorus in oligotrophic lake sediments. *Hydrobiologia* **91:**261–268.

12. **Fitzgerald, G. P., and T. C. Nelson.** 1966. Extractive and enzymatic analyses for limiting or surplus phosphorus in algae. *J. Phycol.* **2:**32–37.

13. **Francko, D. A.** 1983. Size-fractionation of alkaline phosphatase activity in lake water by membrane filtration. *J. Freshwater Ecol.* **2:**305–309.

14. **Gachter, R., J. S. Meyer, and A. Mares.** 1988. Contribution of bacteria to release and fixation of phosphorus in lake sediments. *Limnol. Oceanogr.* **33:**1542–1558.

15. **Hashimoto, S., K. Fujiwara, and K. Fuwa.** 1985. Relationship between alkaline phosphatase activity and orthophosphate in the present Tokyo Bay. *J. Environ. Sci.* **A20:**781–809.

16. **Healy, F. P., and L. L. Hendzel.** 1979. Fluorometric measurement of alkaline phosphatase activity in algae. *Freshwater Biol.* **9:**429–439.

17. **Howarth, R. W.** 1988. Nutrient limitation of net primary production in marine ecosystems. *Annu. Rev. Ecol. Syst.* **19:**89–110.

18. **Huber, A. L., and D. K. Kidby.** 1984. An examination of the factors involved in determining phosphatase activities in estuarine waters. 1. Analytical procedures. *Hydrobiologia* **111:**3–11.

19. **Huber, A. L., and D. K. Kidby.** 1984. An examination of the factors involved in determining phosphatase activities in estuarine waters. 2. Sampling procedures. *Hydrobiologia* **111:**13–19.

20. **Jones, J. G.** 1972. Studies on freshwater bacteria: association with algae and phosphatase activity. *J. Ecol.* **60:**59–75.

21. **Karl, D. M., and G. Tien.** 1992. MAGIC: a sensitive and precise method for measuring dissolved phosphorus in aquatic environments. *Limnol. Oceanogr.* **37:**105–116.

22. **Lean, D. R. S.** 1973. Phosphorus movement between biologically important compartments. *J. Fish. Res. Board Can.* **30:**1525–1536.

23. **Lean, D. R. S.** 1984. Metabolic indicators for phosphorus limitation. *Verh. Int. Ver. Limnol.* **22:**211–218.

24. **Lean, D. R. S., A. A. Abbott, and F. R. Pick.** 1987. Phosphorus deficiency of Lake Ontario plankton. *Can. J. Fish Aquat. Sci.* **44:**2069–2076.

25. **Lean, D. R. S., and E. White.** 1983. Chemical and radiotracer measurements of phosphorus uptake by lake plankton. *Can. J. Fish. Aquat. Sci.* **40:**147.

26. **Lehninger, A. L.** 1982. *Principles of Biochemistry.* Worth Publishers, New York.

27. **Lijklema, L.** 1977. The role of iron in the exchange of phosphate between water and sediments, p. 313–317. *In* H. L. Golterman (ed.), *Interactions between Sediments and Fresh Water.* Dr. W. Junk Publishers, The Hague, The Netherlands.

28. **Olsson, H.** 1983. Origin and production of phosphatases in the acid lake Gardsjon. *Hydrobiologia* **101:**49–58.

29. **Olsson, H.** 1991. Phosphatase activity in an acid, limed Swedish lake, p. 206–219. *In* R. J. Chrost (ed.), *Microbial Enzymes in Aquatic Environments.* Springer-Verlag, New York.

30. **Perry, M. J.** 1972. Alkaline phosphatase activity in subtropical Central North Pacific waters using a sensitive fluorometric method. *Mar. Biol.* **15:**113–119.

31. **Pick, F. R.** 1987. Interpretation of alkaline phosphatase activity in Lake Ontario. *Can. J. Fish Aquat. Sci.* **44:**2087–2094.

32. **Redfield, A. C.** 1958. The biological control of chemical factors in the environment. *Am. Sci.* **46:**205–222.

33. **Reichardt, W., J. Overbeck, and L. Steubing.** 1967. Free dissolved enzymes in lake waters. *Nature* (London) **216:**1345–1347.

34. **Schindler, D. W.** 1971. Carbon, nitrogen, and phosphorus and the eutrophication of freshwater lakes. *J. Phycol.* **7:**321–329.

35. **Smith, S. V.** 1984. Phosphorus versus nitrogen limitation in the marine environment. *Limnol. Oceanogr.* **29:**1149–1160.

36. **Stewart, A. J., and R. G. Wetzel.** 1982. Phytoplankton contribution to alkaline phosphatase activity. *Arch. Hydrobiol.* **93:**265–271.

37. **Torriani-Gorini, A., F. G. Rothman, S. Silver, A. Wright, and E. Yagil (ed.).** 1987. *Phosphate Metabolism and Cellular Regulation in Microorganisms.* American Society for Microbiology, Washington, D.C.

38. **Vadstein, O., Y. Olsen, H. Reinertsen, and A. Jensen.** 1993. The role of planktonic bacteria in phosphorus cycling in lakes—sink and link. *Limnol. Oceanogr.* **38:**1539–1544.

39. **Wang, C. H., D. L. Willis, and W. D. Loveland.** 1975. *Radiotracer Methodology in the Biological, Environmental and Physical Sciences.* Prentice-Hall, Englewood Cliffs, N.J.

40. **Wright, R. T., and J. E. Hobbie.** 1966. The use of glucose and acetate by bacteria and algae in aquatic ecosystems. *Ecology* **47:**447–464.

41. **Wynne, D., B. Kaplan, and T. Berman.** 1991. Phosphatase activities in Lake Kinneret phytoplankton, p. 220–226. *In* R. J. Chrost (ed.), *Microbial Enzymes in Aquatic Environments.* Springer-Verlag, New York.

Metal Requirements and Tolerance

AARON L. MILLS

37

INTRODUCTION

Definition of Metals

Metals are defined chemically as all elements on the left side of the periodic table, including the alkali and alkaline earth metals (groups 1A and 2A) and the transition metals (groups 3B through 2B). Also included in the transition metals are the heavier elements of groups 3A through 6A, specifically those elements including and below Al, Ge, Sb, and Po. Elements to the right of and above the aforementioned are defined as nonmetals. This definition includes trace elements often included with metals in the environment; for example, Se and As are often considered along with heavy metals, even though they are not metals. Discussion of these elements will be undertaken here only as it enhances discussion of the metals. In the context of metals of biological and environmental importance, the alkali and alkaline earth metals are usually not considered metals. The term "heavy metals" has become popular for group 3B and heavier metals. These are sometimes required by biological systems but are often toxic if present at concentrations only slightly higher than that at which they are required. It is the heavy metals that will be the focus of this chapter, primarily because of the toxicity of many of the elements included in this grouping.

It is impossible to offer a comprehensive view of all facets of metal interactions with microorganisms in this chapter, given the diversity of the metals, the breadth of their functions, and the number of problems associated with quantitative measures of individual species in microbiological systems, combined with the resultant fact that standard methods for examining metals in those microbiological systems are not well developed. Instead, this presentation will endeavor to provide general information to aid the decision-making process about how to examine metal-microbe interactions. Approaches will be suggested, but detailed methods cannot be specified. Most importantly, several pitfalls associated with metal-related questions will be identified, with appropriate cautions for the investigator. Perhaps the best advice to one whose curiosity motivates research into metal-microbe interactions in an environmental context is to collaborate with a geochemist who shares curiosity about the behavior of metals in a biogeochemical setting. Such a collaboration can yield results that have meaning beyond the confines of the experimental laboratory system and will produce information that can be extrapolated to the environment in a quantitative manner.

Biological Importance

Metals are critical components of many biochemical systems (11, 14, 17). The lighter metal ions (Na^+, K^+, Ca^{2+}, and Mg^{2+}) are used as enzyme cofactors and are important components of ion pumps that drive oxidative phosphorylation and help to maintain osmotic balance within cells. Mg^{2+} is also involved in the stabilization of ribosomes, nucleic acids, and cell membranes. Several anionic polymers contain K^+, Ca^{2+}, and Mg^{2+}. These ions are abundant in nature and at most environmentally encountered concentrations represent little threat to cellular function. Very high concentrations, such as may be present in brines, disrupt the osmotic conditions of cells, causing them to lose water rapidly. Death usually follows quickly from osmotic shock. Additionally, high concentrations of these ions can disrupt the electrical stability of cell membranes, causing them to cease function and rupture.

Many of the heavy metals are also important components of biological systems; they are required at low concentrations for growth and metabolic function. For example, microorganisms have evolved mechanisms that vary in specificity to accumulate Cu, Zn, Fe, Ni, Mn, and Co from the surrounding environment (14). The essential metals serve several functions: (i) they represent prosthetic groups in many proteins and dictate the configuration of the active sites of enzymes; (ii) they serve as cofactors for some enzymatic reactions; (iii) they serve as multidentate centers for porphyrin molecules; and (iv) they serve as redox centers, transferring electrons in important redox reactions in cells. Note that these functions are not mutually exclusive—the cytochrome system has a series of porphyrin-like components that have iron as their center (Fig. 1). The iron in the cytochromes serves as an electron transfer (redox) site in the electron transport system, and it changes from Fe(III) to Fe(II) and back as electrons are moved along the cytochrome chain. Examples of the main metal functions in biochemical processes are given in Table 1. Most organisms require iron, manganese, zinc, copper, and cobalt, and several organisms also require molybdenum. As a result, many microbiological media are formulated to contain small amounts of these elements (Table 2). For example, the trace element solution HO-LE, which is used for the enrichment

349

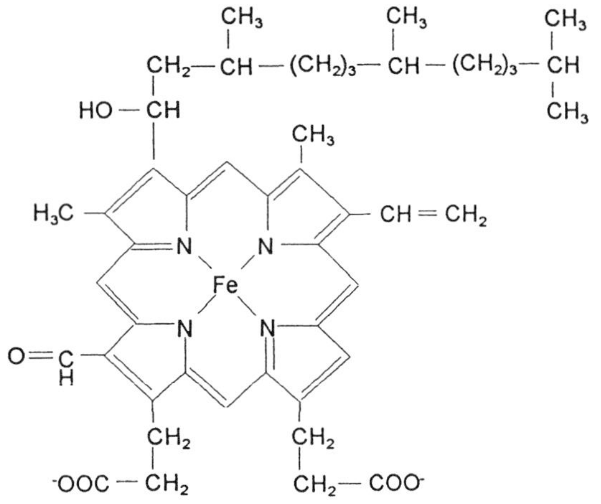

FIGURE 1 Heme A, the prosthetic group of the class A cytochromes. Note the porphyrin structure that contains iron as the redox center.

of other media requiring added trace metals, contains (per liter of solution; dilute salts to 1 liter with distilled water) H_3BO_3, 2.85 g; $MnCl_2 \cdot 4H_2O$, 1.8 g; sodium tartrate, 1.77 g; $FeSO_4 \cdot 7H_2O$, 1.36 g; $CoCl_2 \cdot 6H_2O$, 0.04 g; $CuCl_2 \cdot 2H_2O$, 0.027 g; $NaMoO_4 \cdot 2H_2O$, 0.025 g; and $ZnCl_2$, 0.020 g (2). This solution can then be added to a medium to ensure delivery of required trace elements to the microbes. A typical dilution of HO-LE solution would place 1 ml in 1,000 ml of medium solution (to give final metal concentrations in the range of micrograms per liter) (8).

In addition to their roles as described above, metals can serve as energy sources and electron acceptors. Some bacteria have developed efficient mechanisms for linking the energy given off by the exergonic oxidation reactions involving metal elements. On the other hand, other bacteria have

TABLE 1 Some enzymes that require metal ions as cofactors[a]

Ion	Enzyme(s)
Zn^{2+}	Alcohol dehydrogenase
	Carbonic anhydrase
	Carboxypeptidase
Mg^{2+}	Phosphohydrolases
	Phosphotransferases
	Hexokinase
Mn^{2+}	Arginase
	Phosphotransferases
Fe^{2+} or Fe^{3+}	Cytochromes
	Peroxidase
	Catalase
	Ferridoxin
Cu^{2+} (Cu^{+})	Tyrosinase
	Cytochrome oxidase
K^{+}	Pyruvate phosphokinase (also requires Mg^{2+})
Na^{+}	Membrane-bound ATPase (also requires K^{+} and Mg^{2+})
Co^{2+} (Co^{+})	Cobalamin (vitamin B_{12})
Mo^{4+} (Mo^{5+})	Nitrogen fixation
	Nitrate reduction

[a] Based on a table in reference 17.

developed mechanisms for transferring the electrons produced during the oxidation of reduced organics (or possibly even some other metals) to oxidized metal ions. Despite a great deal of work on iron- and sulfur-oxidizing bacteria and the recent findings of abundant enzymatic iron reduction in anaerobic environments, this area of inquiry is still in its infancy. Only a few standard techniques exist, and many of those are being reconsidered in light of the newer discoveries.

Environmental Importance

Metals are a natural part of most environments; however, in some cases certain ions may be present in high concentrations because of an enhanced abundance in a particular mineral body. More often, high concentrations of metals arise from anthropogenic activities, especially those which involve the processing of metallic substances, i.e., mining, smelting, printing, electroplating, and battery manufacture and disposal (Table 3). Metal ions, for example, Cu^{2+}, have seen extensive use as herbicides and fungicides, and many antifouling coatings act by leaching metal ions or metal-containing substances from the coating surface. Many of the same elements required by microbes at low concentrations are toxic to the same organisms when present at concentrations found in many contaminated environments (14). Additionally, some elements (e.g., Al, Cd, Hg, Pb, and Sn) seem to have no essential biological function but are also taken up and accumulated by microorganisms. These metals are also often toxic at concentrations found in environmental contamination (14, 27).

The presence of elevated concentrations of heavy metals has been associated with increased numbers of metal-resistant microbes; however, resistant bacteria are also often found in environments with metals present at background concentrations (e.g., 9, 12, 21). The presence of metal-resistant bacteria, even in large numbers, cannot be used as direct evidence for metal contamination. In areas where the metals themselves may present no great risk to the environment or to plant, animal, or human use of the resources in that environment, metals may play a role in maintaining retention of antibiotic resistance once resistance is attained. Many of the antibiotic resistance genes are plasmid borne, and in clinical isolates, they tend to be present on the same plasmids as are the resistance factors for metals such as mercury (14). Nonclinical environmental isolates tend not to have linked antibiotic and mercury resistance. If the plasmids from the clinical isolates are transferred in the environment, however, elevated concentrations of metals could, by providing selective pressure for metal resistance, also provide an indirect selective pressure for antibiotic resistance.

DIFFICULTIES IN WORKING WITH METALS
Speciation

Metals tend to exist as polyvalent ions (frequently as more or less oxidized species such as Hg^{2+} [mercuric] and Hg^{+} [mercurous] or Cu^{2+} and Cu^{+} [cupric and cuprous, respectively]), and they also tend to form a variety of aqueous-phase complexes with dissociated water molecules. In the presence of species such as HS^{-} or S^{2-}, metals often form soluble complexes with ions other than OH^{-}. The identity of the dominant species is, therefore, a function of both pH and the redox potential (E_h; sometimes expressed as the negative logarithm $-p\epsilon$, which is the hypothetical electron

TABLE 2 Trace elements commonly added to microbiological media[a]

Element	Medium formulation								
	Trace element mixture	HO-LE	SL-6	SL-7	SL-8	SL-10	A-5	Metals "44"	Wolfe's
Al									+
B	+	+	+	+	+	+	+	+	+
Ca[b]	+								+
Co	+	+	+	+	+	+	+	+	+
Cu	+	+	+	+	+	+	+	+	+
Fe	+	+		+	+	+		+	+
Mn	+	+	+	+	+	+	+	+	+
Mo	+	+	+	+	+		+		+
Ni			+	+	+	+		+	
Zn	+	+	+	+	+	+	+	+	+

[a] Data from reference 2. For specific formulary, consult that reference.
[b] Most media are formulated to contain Na, K, Ca, and Mg, although those elements are often specified as cations for addition of major anionic nutrients (NaNO₃ or MgSO₄, for example).

activity at equilibrium). The relationship between acid-base equilibrium and redox equilibrium for solutions of given bulk composition is often expressed by means of E_h-pH diagrams; a diagram for Hg is shown in Fig. 2. By examining the diagram, one can determine the dominant species of a given metal in a solution of the strength and composition specified for the calculations used to generate the diagram. For a detailed discussion of this subject, refer to appropriate

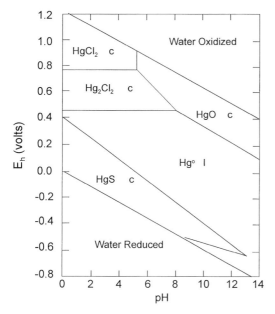

FIGURE 2 An example of an E_h-pH diagram for equilibrium speciation calculations for Hg. The zones represent regions of stability for the phases shown. Boundaries are points at which the dominant species in adjacent regions will be present in equal concentrations. Centers (c) of regions represent conditions under which the species indicated in that region will be at the highest concentration and those surrounding it will be at their lowest respective concentrations. Specific conditions for these calculations are 25°C and 1 atm (101.29 kPa) of pressure in a system containing 1 mM Cl^- and 1 mM total sulfur as SO_4^{2-}.

geochemistry references such as that by Stumm and Morgan (33). The multivalency and the complex formation exhibited by metals make identifying the species to which microbes are sensitive a more complex task than simply determining the total concentration of the element in the bulk solution.

Microbial processes are sensitive to metals and metal speciation (see reference 11 for an early review). As a single example, Tatara et al. (41) observed that ferric iron (added at 1 to 20 μM) and copper (0.25 mg of CuSO₄·5H₂O liter^{-1}) inhibited the mineralization of CCl₄ by a *Pseudomonas* strain at neutral pH, but the inhibition was relieved when the pH was increased to 8.2, thus causing the precipitation of the iron (and presumably the copper). The effect of the metal on the mineralization process changed with the metal species, even though the total amount of metal added to the system (as CuSO₄) did not change. The combined metal concentration-environmental condition spectrum must be accounted for in determining the exact effect of metals on any microorganism or microbial process.

The species of metals present (and therefore the bioavailability of each metal) change with pH, E_h, and other constituents in a complex manner. The equilibria are not simple, two-component reactions. To assist with determining speciation of inorganic ions, a number of computer programs which carry out the large number of calculations quickly and efficiently have been developed. Routine use by microbiologists is not easily accomplished, however, and the models suffer from some severe disadvantages. The best of the equilibrium speciation models is the WATEQ family (5). Models such as WATEQ are not designed to be used by inexperienced individuals; the entry of data into the model for calculation is not always straightforward. Moses and Herman (24) have written a small program to serve as a data entry module for WATEQ, but changes in the WATEQ code itself are often required to make the program work on a specific computer platform.

Included in the models is a large table of thermodynamic data associated with the solubility and appropriate equilibria for a given element. In general the models work well for speciation calculations in dilute inorganic solutions, and they usually contain options for calculating speciation in more concentrated solutions (seawater and brines). However, in order to use the options properly, the user must

TABLE 3 Comparison of anthropogenic inputs of some heavy metals with inputs from natural sources[a]

Metal	Metric tons (10^3)/yr derived from:			Ratio of anthropogenic to weathering sources
	Weathering	Mining	Industrial emissions	
Antimony	15	55	41	6.4
Cadmium	4.5	19	24	9.6
Chromium	810	6,800	1,010	9.6
Copper	375	8,114	1,048	24
Lead	180	3,100	565	20
Mercury	0.9	6.8	11	20
Molybdenum	15	98	98	13
Nickel	255	778	356	4.4
Zinc	540	6,040	1,427	14

[a] Data are from reference 27.

understand them, and few microbiologists will be familiar with the specifics of these calculations. The thermodynamic data are not complete, and for all but a few compounds they do not exist at all. The most recent version of WATEQ will allow entry of humate and fulvate, and reasonable complexation estimates can be obtained from the model if concentrations of these specific substances are known. Thus, speciation models are not helpful in determining the most likely form of a metal in a complex, organic-rich microbiological medium. Furthermore, the calculations are based on equilibrium assumptions. Microorganisms are well known for their proclivity to move a thermodynamic system away from equilibrium. Additionally, the models assume no kinetic constraints; that is, they assume that the system being modeled is at equilibrium and that any change (requiring a recalculation) to a new equilibrium state occurs instantly. While there are many drawbacks to the use of equilibrium speciation models, microbiologists collaborating with geochemists will find the results of their use helpful in interpreting results of experiments in terms of realistic solution metal concentrations. The models can be especially useful in studies involving organic-poor solutions or natural waters. In such cases, a complete chemical analysis is possible (in fact, the entire solution composition may often be defined by the investigator) so that speciation of the metal can be determined, and changes in concentrations of the various species with changing conditions of pH, E_h, temperature, ionic strength, solid mineral phases, etc., can be related to changes in behavior of the microbe or assemblage being examined.

Analytical Considerations

Not all metal species can be readily analyzed with some techniques. Atomic absorption spectroscopy (AA) measures the total amount of a metal in a digest or in solution, but it cannot differentiate among the dissolved species present. The digests or extracts often include material that was in the solid phase in the original sample. For aqueous samples in which interest is placed in total metal concentration, AA is an excellent analytical tool. The simplest AA techniques involve aspiration of a small amount of sample into the flame of the instrument. Detection limits for flame AA vary with the metal but are generally in the range of 0.01 to 0.1 mg liter^{-1} (1). Modifications such as the use of a graphite furnace (electrothermal method) instead of the flame burner can lower the detection limit 20- to 1,000-fold. Many elements can be determined at levels as low as 1.0 μg liter^{-1} (1).

Plasma emission spectroscopy is also an excellent method

for the quantitative analysis of metals at low concentrations (1). The detection limits for the heavy metals compare favorably with those obtained from graphite furnace AA, although the latter is slightly more sensitive in most cases. The choice of methods for microbiological applications will depend more on the availability of instrumentation than on other factors.

Specific ion electrodes can detect ionic forms of the metals and are quite specific as to what forms are sensed. A Cu^{2+} electrode, for example, can be used to determine the amount of cupric ion in a given solution. In many applications, however, there are interfering materials that preclude accurate determination of the metal ions, which may be present in extremely low concentrations. Ion chromatography can sometimes be adapted to differentiate among metal species (23), but this approach is not as sensitive as other techniques, such as graphite furnace AA, in which larger samples may be concentrated prior to the actual analysis.

METAL REQUIREMENTS

Determination of the requirement of any given metal ion by a microbial process is most appropriately determined in synthetic media, in which case the potential for inclusion of minute amounts of material found in environmental systems can be eliminated or controlled. No standard procedures for this type of determination exist, but some general guidelines can be suggested. In most cases, the organisms of interest will be heterotrophs. Thus, organics in the medium are required. The concentrations of organics should be kept as low as possible to avoid rendering the added metals unavailable because of complexation with the medium components. Complexation is likely to be a more serious problem in undefined, complex, organic-rich media (nutrient broth, peptone, yeast extract, etc.). However, effects of complexation on metal requirements in culture are less likely to be a problem than complexation when one is examining metal toxicity because the investigator often wishes to know how much of a supplement must be added to the medium to produce optimum growth. To understand the form of the metal being used, it would be best to avoid a highly rich, complex medium, and it should be noted that the amounts of metals necessarily added to a complex medium may not represent the amounts required to support microbial growth and metabolism in water or soils.

Examine all medium components, organic and inorganic, for trace metal contamination. Microbes can often obtain minute but adequate amounts of trace elements that occur as contaminants in medium ingredients. For this type of assay, use only the purest media and salts obtainable. The use of double-distilled water (with the second distillation in scrupulously clean glass) is recommended. Deionized water may be extremely low in total ions but may contain just enough of the target ion to support growth of the test organism. Whenever possible, conduct the tests in a liquid medium with as low an organic content as practical. If the medium is a simple one (a single carbon source consisting of a carbohydrate, for example), the speciation model approach may be used to estimate the amount of a given metal species present in the medium. Diffusion of metals in agar can be a kinetic limitation on availability when metal concentrations are extremely low, and agar is known to complex with a variety of metal ions (30), lowering the actual concentration of metal delivered to the microbial cell.

RESISTANCE TO METALS

Most metal toxicity arises from the reaction of metal ions with proteins. For example, toxicity of mercuric ions arises from the aggressive binding of the ions to sulfhydryl groups in organic molecules. This action can inhibit macromolecular synthesis and enzyme activity. Proteins often contain R-SH groups that control the tertiary and quaternary structure of the molecule. Binding by Hg^{2+} can destroy that structure and inactivate the enzyme. Disulfide bonds are also sensitive to Hg^{2+} binding, and transcription and translation are particularly sensitive (13, 34). An increase in mercury resistance is thought to have been mediated by the use of compounds such as phenyl mercury and thimerosal as disinfectants (15). This concept is supported by the observation of a decreasing resistance to mercury coincident with abandonment of mercurials as disinfectants (29).

Metal resistance is frequently determined by plasmid-carried genes in bacteria. Mechanisms of resistance to metals generally involve either a transformation mechanism or an efflux mechanism. A transformation-type mechanism is exemplified by the mercury reductase system, the enzymatic reduction of the Hg^{2+} ion to elemental mercury (Hg^0). The elemental form is much less toxic than the mercurous ion (35, 36), and the physical properties of Hg^0 greatly decrease its availability to the microbes. Hg^0 is virtually insoluble in water, and it has a high vapor pressure, which leads to rapid evaporation. Cadmium resistance, on the other hand, is considered to occur by an efflux system in which it is pumped out of the cell by a Cd^{2+}-H^+ antiport system (13). Other plasmid-associated Cd and Cu resistance mechanisms involve metallothionein-type molecules (18, 28) or siderophore production (7).

General Considerations of Resistance Testing

For many compounds, standard protocols exist for the determination of resistance to the compound. Antibiotics have disk sensitivity assays that represent a comparative standard among investigators. Such standards do not exist for metals. Obviously, there are some commonalities in practice that allow statements to be made about the presence or absence of metal resistance, and to some extent a quantitative response to metal concentrations may be made. Problems associated with unknown speciation of the metals as inorganic complexes, and especially as organometallic complexes with components of growth media, make difficult an exact determination of how much of what metals the bacteria can actually resist.

Resistance to specific medium components including metals is often used as a method for the selection or isolation of organisms known to be resistant to a certain metal or determination of some degree of resistance. The degree of resistance is usually expressed on the basis of the total amount of metal added to the medium (9, 16, 19, 21, 31). Enumeration of metal-tolerant organisms was often accomplished by plating water samples or dilutions of sediment or soil samples on agar-solidified media to which some concentration of metal salt had been added. Mills and Colwell (21) reported very high concentrations of metals (e.g., 250 mg liter of medium^{-1}) needed to be added to agar media to achieve reasonable reductions of organisms in estuarine water. Ramamoorthy and Kushner (30) noted that the actual concentration of free metal ion in microbiological media was substantially lower than the concentration added as a result of complexation of the metal ions with organic components of the media.

The main problem with inclusion of metals in organic-rich media is that the concentration of metal added to the medium, and therefore the concentration to which growing organisms are considered resistant, is generally much greater than the active concentration of metal in, for example, lake water or seawater. The true MIC, i.e., the level of true resistance, is therefore generally much lower than that demonstrated through a medium-supplement-based assay. For example, Sunda et al. (38–40) used specific ion electrodes to demonstrate that the actual toxic copper species was free Cu^{2+}, which was present in only small amounts in comparison with the total copper present; the strength of the relationship for some bacteria was so great that the suggestion was made that the response of the organism might be used as a measure of the cupric ion activity (38). Similarly, free-metal (ionic) cadmium was also shown to be the toxic species in a study of shrimp (37).

Inhibition of Growth

Several methods have been suggested to determine the resistance of bacteria to metals in solution. No method has been adopted as a standard procedure, and each of the methods presented here has some significant disadvantages. Nevertheless, each could be used to give a relative measure of the resistance of an organism to various levels of metal, of the resistance of an organism to different metals at the same or different relative concentrations, or of the resistance of different organisms to one or more concentrations of one or more metals. Obviously, the more complex the relationships being examined, the more the complexity of the procedure becomes important. As always, it is important to restate that in complex media containing organics and many other ions, the exact concentration of the toxic species will rarely, if ever, be known accurately.

Agar Dilution Method

An agar dilution method similar to that used for antibiotic resistance assays (44) has been used by Nieto et al. (25, 26) to evaluate metal tolerance in halophilic bacteria. This technique simply involves the use of agar-solidified media with a range of metal concentrations included. Nieto et al. (25) placed 21 spots of liquid inocula (with 10^4 to 10^5 cells) on each plate with an automatic pipettor, incubated the plates for 2 days, and looked for visible growth. Resistance was reported as the highest concentration of metal (actually metal added to the medium) that allowed growth of the organisms.

Continuous Culture

Mayfield et al. (20) used a continuous culture system to examine the differences in growth kinetics of a target organism in a dilute growth medium with or without metals present. Such a system has a distinct advantage in that by careful tuning of the system, organisms can be grown in media very similar to natural conditions (assuming an aquatic lifestyle). Use of dilute organic substrates in chemostats also allows for better control of metal speciation. Effects of different metal concentrations can be determined by adjusting the concentration and examining the response in the steady-state growth parameters. Both lethal and sublethal effects on growth can be observed. There is a distinct disadvantage in terms of the number of strains that can be examined in this fashion and the amount of time necessary to obtain results. Continuous culture systems can also be used to select slower-growing metal-resistant strains from a general assemblage.

Agar Diffusion Method

In determining the resistance of chemoorganotrophic bacterial isolates to antibiotic chemicals, a rapid assay that utilizes the diffusion of the antibiotics from paper disks into the agar-solidified medium to achieve a radial concentration gradient from the disk outward into the medium has been developed (6). This method was adapted for metal elements by Smith et al. (32) and further modified by Thompson and Watling (42, 43). The ensuing description follows closely the method of Thompson and Watling (42), but because there is no standard, modifications could be made to improve the technique or to make it more suitable for answering a specific question.

Preparation

1. Reconstitute and sterilize nutrient agar in 100-ml volumes. (Small volumes make pouring uniform plates easier; if an automatic filler is used, larger batches of medium may be prepared.) Nutrient agar was suggested for use by Thompson and Watling (42), who indicated that metals diffused uniformly through the emulsion. A less rich medium (e.g., one-half-strength nutrient agar, R2A agar [1], or dilute peptone-tryptone-yeast extract-glucose agar) may cause less metal complex formation with medium components, thus delivering more metal to the organisms.

2. Prepare tubes containing 5 ml of a suitable liquid medium for growing log-phase cultures of the organisms being tested. Thompson and Watling (42) used a tryptone medium in their examination of metal resistance in *Escherichia coli*.

3. Sterilize Whatman AA disks (13-mm diameter) or equivalents in a 160°C oven for at least 1 h in glass petri dishes.

4. Prepare stock solutions (1 mg of metal ion ml^{-1}) of metal salts. (Minor contamination with other metals can produce artificial results. AA standards are commercially available at this concentration, and they provide uniformly prepared solutions with no, or minimal but known, contamination.)

Procedure

1. Grow cultures of the isolates to be tested on suitable agar plates (presumably the same medium as that to be used for the test).

2. Inoculate a few representative colonies into a tube of the liquid medium, and incubate them for 18 h (or a suitable

length of time for appearance of turbidity; the object is to obtain cells in the logarithmic phase of growth).

3. Make a standard inoculum by transferring 0.1 ml of the 18-h culture to a fresh tube containing 5 ml of liquid medium (which serves as a transfer diluent) immediately prior to the test.

4. Melt the agar, cool it to 50°C, and pour it into plates containing 10 ml of the medium. Uniform plates are important, as thicker plates will tend to reduce the amount of diffusion of the metal away from the disk. When the agar has solidified, spread 0.1 ml of the inoculum from step 3 evenly over the surface of the plate.

5. Make the metal-impregnated disks by dropping 0.1 ml of the stock metal solution onto each disk (100 μg of metal per disk). Use of an automatic pipettor will facilitate the requisite uniformity in the disks. Place the disk on the inoculated plate, and incubate the plate at a relevant temperature for a period suitable for the organisms being examined.

6. After incubation, measure the distance (in millimeters) from the edge of the disk to the edge of the clearing zone.

7. To determine an apparent MIC, compare the distance between the edge of the clearing zone and the edge of the disk with a concentration determined by the diffusion gradient for that metal in that medium. Figures 3 and 4 show examples from the Thompson and Watling publications (42, 43). The concentrations needed to determine a diffusion gradient may be measured as follows. Carefully cut strips of agar 15 by 2 mm thick (to the bottom of the agar layer) from the edge of the disk to the edge of the plate. Weigh each strip, digest it with 2 ml of concentrated HNO$_3$, and evaporate the material to dryness. Add 2 ml of 10% (vol/vol) HNO$_3$ to dissolve the sample residue. The metal concentration in these samples can then be determined by AA, and the concentration of metal (in micrograms per gram) is plotted against distance from the disk to obtain the diffusion gradient plots. From these plots, the total metal concentration at the point of inhibition can be determined and reported as the MIC. Keep in mind that this is an apparent value, in that the concentration of the true toxic species of the metal is not determined by this assay.

This procedure provides a fundamental approach to metal resistance testing similar to that commonly used for antibiotic resistance testing. Modifications can easily be made: the medium can be changed; the metal salts used in the filters can be altered to determine the effects of different forms, e.g., anion components, of metal contaminants; and metals can also be mixed in the solutions in different proportions to examine synergistic interactions in metal resistance. Despite the several advantages that are offered by an approach such as the diffusion assay, all of the problems associated with cultivation of microbes are manifested; not the least is the limitation on the type of organism (chemoorganotrophs) to which the technique can be applied easily.

Inhibition of Growth Determined by Uptake of [³H]Thymidine

A method for determining the effects of dissolved metals on growth of bacterial communities in soil has been proposed by Bååth and coworkers (3, 4, 10). The method is based on the assumption that bacterial growth is proportional to the amount of incorporation of thymidine into macromolecules in the bacterial cell. For use in soils, the bacterial cells are first extracted from the soil, and the metal of interest is

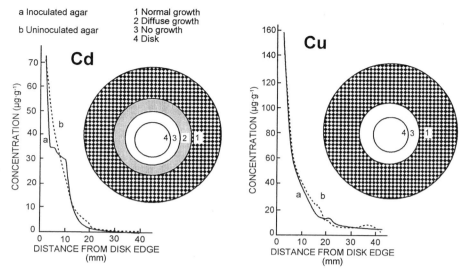

FIGURE 3 Diffusion gradients associated with Cd- and Cu-impregnated disks in the agar diffusion assay using nutrient agar as the medium. The figure also illustrates the appearance of the clearing zones surrounding the disks. Note the alteration of the gradient in the inoculated Cd-containing plates that arises as a result of Cd uptake by the bacterial cells at the edge of the clearing zone. Reprinted from Thompson and Watling (42) with permission.

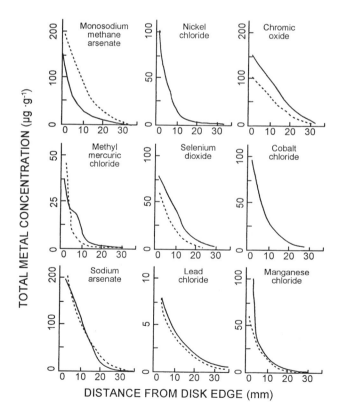

FIGURE 4 Diffusion gradients associated with several trace elements used in the agar diffusion assay reported by Thompson and Watling (43). Reprinted from Thompson and Watling (43) with permission.

added to the cell suspension. Tritiated thymidine is then added, and the suspension is incubated. At the end of the incubation period, the macromolecules (fraction soluble in cold trichloroacetic acid) are extracted from the suspension, and the amount of ^{3}H incorporated is measured. For studies of the effect of the metal on the entire community, this method does not select for those organisms that can grow on the isolation or enumeration medium. Because the suspension does not need to be supplemented with large amounts of complex organic mixtures (as found in many culture media), determination of a reasonable metal concentration and speciation are much more straightforward, yielding results that are more environmentally relevant. Although the published procedure is for microbes extracted from soil, the same approach could easily be adapted to water samples, which would not require extraction prior to testing.

The method, as presented here, outlines the report of Bååth (4). For details that would be useful in adapting the method to water or to a specific soil situation, the reader is referred to the original articles. The extraction efficiency for bacterial cells reported by Bååth (4) was from 10 to 30% and depended on the soil type.

Procedure

1. Homogenize 10 g of soil (the amount necessary may vary depending on the numbers of organisms and their activity) with 200 ml of distilled water in a high-speed blender or homogenizer for 10 min.

2. Sediment the solids in a centrifuge at 750 × g for 10 min.

3. Decant the supernatant through glass wool, and place 1- to 4-ml portions into plastic vials.

4. Add solutions containing the necessary concentration of the metal to be tested. For this assay, a convenient final volume is 5 ml. Be sure to add distilled water or an appropriate salt solution to controls that contain no added metal.

5. Hold the vials at room temperature for 10 min before adding [*methyl-³H*]thymidine.

6. Add [³H]thymidine, and incubate the mixture for 1 to 2 h. (The amount of thymidine added and the incubation time used will vary from study to study. All considerations appropriate for use of the [³H]thymidine method for assaying bacterial productivity apply here. For a specific and detailed discussion, see chapter 27.) It is also recognized that various individuals conduct the thymidine uptake procedure with different protocols. The Bååth (4) procedure is presented as an approach that can be modified as necessary to suit the needs and practices of individual laboratories.

7. At the end of the incubation period, terminate all reactions by the addition of 1 ml of 5% formaldehyde solution.

8. Filter the suspension through a 0.45-μm-pore-size membrane filter prewashed three times with 1% sodium hexametaphosphate.

9. Wash the filter with cells in place with three 5-ml portions of 80% ethanol and three 5-ml portions of ice-cold 5% trichloroacetic acid.

10. Trim away the edges of the filters, and place each into a liquid scintillation vial.

11. Add 1 ml of 0.1 M NaOH, and heat the vials to 90°C for 1 to 2 h to solubilize macromolecules.

12. Allow the vials to cool, add liquid scintillation cocktail, and count the radioactivity with a liquid scintillation counter.

13. Compare the levels of thymidine incorporation in samples with the metals and concentrations added (including the controls).

For water samples, the cells may be collected directly on the filter without any extraction.

Inhibition of Specific Metabolic Processes

A number of studies have examined the effects of added metals on specific enzymatic activities or on microbial processes. In many cases, the assay is often quite simple: addition of a concentration of metal ion to a solution used to test enzymatic activity. The assays can suffer from the same speciation-complexation problems that plague other approaches as well, although use of dilute solutions can minimize the problem. Many times, speciation modeling can be accomplished for the assay solutions. If whole cells are used for the test, complexation of the metal with the purified or semipurified enzyme (protein) may be avoided if it is not the mechanism of toxicity in the environment.

Examples of enzyme activity assays for testing metal toxicity include that of Montuelle et al. (22), who examined dehydrogenase activity and β-glucosidase activity in bacteria isolated from the sediments of polluted and unpolluted rivers in solutions in the presence or absence of added metals. As mentioned earlier, Tatara et al. (41) examined the mineralization of CCl_4 as affected by the presence of iron and copper and the effect of environmental conditions on the degree of toxicity. Other methods of measuring microbial growth and activity may be readily adapted to examine the effects of metals on the organisms simply by including different concentrations of metal ions in the test solutions. In all cases, care must be taken to use only the purest of chemicals and, whenever reasonable, to determine the dominant metal species in the test solution.

Activity assays also have the ability to provide direct evidence of synergistic relationships (also possible at a more crude level with classical culture methods) by allowing the mixing of various combinations of metals at different concentrations. Additionally, sublethal effects may be readily observed as inhibition of an activity. Such an observation may lead to the documentation of chronic inhibition in the environment.

The determination of metal requirement and resistance is not always a straightforward procedure. Metal loadings for culture media can easily be determined, but these results are operational definitions for the culture conditions and may have no relationship to the actual toxic levels in an environmental situation. Thus, caution should always be used in the quantitative extrapolation of laboratory results involving metals to the environment. Laboratory studies will continue to be of great value in the interpretation of field-derived observations, and experiments that examine the molecular basis of metal requirement and resistance should continue apace. Qualitative or mechanistic conclusions may follow directly from laboratory studies, but quantitative relationships, and especially regulatory guidelines, based on such procedures should be viewed with some skepticism.

REFERENCES

1. **American Public Health Association.** 1989. *Standard Methods for the Examination of Water and Wastewater*, 17th ed. American Public Health Association, Washington, D.C.

2. **Atlas, R. M.** 1993. *Handbook of Microbiological Media*. CRC Press, Inc., Boca Raton, Fla.

3. **Bååth, E.** 1989. Effects of heavy metals in soil on microbial processes and populations (a review). *Water Air Soil Pollut.* **47:**335–379.

4. **Bååth, E.** 1992. Measurement of heavy metal tolerance of soil bacteria using thymidine incorporation into bacteria extracted, after homogenization-centrifugation. *Soil Biol. Biochem.* **11:**1167–1172.

5. **Ball, J. W., D. K. Nordstrom, and D. W. Zachman.** 1987. WATEQ4F—a personal computer FORTRAN translation of the geochemical model WATEQ2 with revised data base. USGS Open File Report 87-50. U.S. Geological Survey, Menlo Park, Calif.

6. **Bauer, A. W., W. M. M. Kirby, J. C. Sherris, and M. Turck.** 1965. Antibiotic susceptibility testing by a standardized single disc method. *Am. J. Clin. Pathol.* **45:**493.

7. **Clarke, S. E., J. Stuart, and J. Sanders-Loehr.** 1987. Induction of siderophore activity in *Anabaena* spp. and its moderation of copper toxicity. *Appl. Environ. Microbiol.* **53:**917–922.

8. **Cote, R. J., and R. L. Gherna.** 1994. Nutrition and media, p. 155–178. *In* P. Gerhardt, R. G. E. Murray, W. A. Wood, and N. R. Kreig (ed.), *Methods for General and Molecular Bacteriology.* American Society for Microbiology, Washington, D.C.

9. **Dean-Ross, D., and A. L. Mills.** 1989. Bacterial community structure and function along a heavy metal gradient. *Appl. Environ. Microbiol.* **55:**2002–2009.

10. **Diaz-Ravina, M., E. Bååth, and A. Frostegard.** 1994. Multiple heavy metal tolerance of soil bacterial communities and its measurement by a thymidine incorporation technique. *Appl. Environ. Microbiol.* **60:**2238–2247.

11. **Duxbury, T.** 1985. Ecology of heavy metal responses in microorganisms. *Adv. Microb. Ecol.* **8:**185–235.

12. **Duxbury, T., and B. Bicknell.** 1983. Metal-tolerant bacterial populations from natural and metal-polluted soils. *Soil Biol. Biochem.* **15:**243–250.

13. **Foster, T. J.** 1983. Plasmid-determined resistance to anti-

microbial drugs and toxic metal ions in bacteria. *Microbiol. Rev.* **47**:361–409.

14. **Gadd, G. M.** 1990. Metal tolerance, p. 178–210. *In* C. Edwards (ed.), *Microbiology of Extreme Environments.* McGraw-Hill Publishing Co., New York.

15. **Hall, B. M.** 1970. Mercury resistance of *Staphylococcus aureus. J. Hyg.* **68**:121–129.

16. **Harnett, N. M., and C. L. Gyles.** 1984. Resistance to drugs and heavy metals, colicin production, and biochemical characteristics of selected bovine and porcine *Escherichia coli* strains. *Appl. Environ. Microbiol.* **48**:930–935.

17. **Lehninger, A. L.** 1993. *Principles of Biochemistry,* 2nd ed. Worth Publishers, New York.

18. **Lerch, K.** 1980. Copper metallothonein, a copper-binding protein from *Neurospora crassa. Nature* (London) **284**: 368–370.

19. **Loutit, M. W., J. Aislabie, P. Bremner, and C. Pillidge.** 1988. Bacteria and chromium in marine sediments. *Adv. Microb. Ecol.* **10**:415–437.

20. **Mayfield, C. I., W. E. Inniss, and P. Sain.** 1980. Continuous culture of mixed sediment bacteria in the presence of mercury. *Water Air Soil Pollut.* **13**:335–349.

21. **Mills, A. L., and R. R. Colwell.** 1977. Microbiological effects of metal ions in Chesapeake Bay water and sediment. *Bull. Environ. Contam. Toxicol.* **18**:99–103.

22. **Montuelle, B., X. Latour, B. Volat, and A.-M. Gounot.** 1994. Toxicity of heavy metals to bacteria in sediments. *Bull. Environ. Contam. Toxicol.* **53**:753–758.

23. **Moses, C. O., A. T. Herlihy, J. S. Herman, and A. L. Mills.** 1988. Ion chromatographic analysis of mixtures containing ferrous and ferric iron. *Talanta* **35**:15–22.

24. **Moses, C. O., and J. S. Herman.** 1986. Watin—a computer program for generating input files for WATEQF. *Ground Water* **24**:83–89.

25. **Nieto, J. J., R. Fernandez-Castillo, and M. C. Marquez.** 1989. Survey of metal tolerance in moderately halophilic eubacteria. *Appl. Environ. Microbiol.* **55**:2385–2390.

26. **Nieto, J. J., A. Ventosa, and F. Ruiz-Berraquero.** 1987. Susceptibility of halobacteria to heavy metals. *Appl. Environ. Microbiol.* **53**:1199–1202.

27. **Nriagu, J. O.** 1990. Global metal pollution: poisoning the biosphere? *Environment* **32**:6–11, 218–233.

28. **Olafson, R. W., K. Abel, and R. S. Sim.** 1979. Prokaryotic metallothionein: preliminary characterization of a blue-green alga heavy metal-binding protein. *Biochem. Biophys. Res. Commun.* **89**:36–43.

29. **Porter, F. D., S. Silver, C. Ong, and H. Nakahara.** 1982. Selection for mercurial resistance in hospital settings. *Antimicrob. Agents Chemother.* **22**:852–858.

30. **Ramamoorthy, S., and D. Kushner.** 1975. Binding of mer-

cury and other heavy metals by microbial growth media. *Microb. Ecol.* **2**:162–176.

31. **Schmidt, T., R.-D. Stoppel, and H. G. Schlegel.** 1991. High-level nickel resistance in *Alcaligenes xylosoxydans* 31A and *Alcaligenes eutrophus* KT02. *Appl. Environ. Microbiol.* **57**:3301–3309.

32. **Smith, G. W., A. M. Kozuchi, and S. S. Hayasaka.** 1982. Heavy metal sensitivity of seagrass rhizoplane and sediment bacteria. *Bot. Mar.* **25**:19–24.

33. **Stumm, W., and J. J. Morgan.** 1981. *Aquatic Chemistry, an Introduction Emphasizing Chemical Equilibria in Natural Waters,* 2nd ed. John Wiley & Sons, Inc., New York.

34. **Summers, A. O.** 1992. Untwist and shout: a heavy metal-responsive transcriptional regulator. *J. Bacteriol.* **174**: 3097–3101.

35. **Summers, A. O., and S. Silver.** 1972. Mercury resistance in a plasmid-bearing strain of *Escherichia coli. J. Bacteriol.* **112**:1228–1236.

36. **Summers, A. O., and S. Silver.** 1978. Microbial transformations of metals. *Annu. Rev. Microbiol.* **32**:637–672.

37. **Sunda, W. G., D. W. Engel, and R. M. Thuotte.** 1978. Effect of chemical speciation on toxicity of cadmium to grass shrimp, *Palaemonetes purgio:* importance of free cadmium ion. *Environ. Sci. Technol.* **12**:409–413.

38. **Sunda, W. G., and P. Gillespie.** 1979. The response of a marine bacterium to cupric ion and its use to estimate cupric ion activity in seawater. *J. Mar. Res.* **37**:761–777.

39. **Sunda, W. G., and R. R. Guillard.** 1976. The relationship between cupric ion activity and the toxicity of copper to phytoplankton. *J. Mar. Res.* **34**:511–529.

40. **Sunda, W. G., and J. M. Lewis.** 1978. Effect of complexation by natural organic ligands on the toxicity of copper to a unicellular alga, *Monochrysis lutheri. Limnol. Oceanogr.* **23**:870–878.

41. **Tatara, G. M., M. J. Dybas, and C. S. Criddle.** 1993. Effects of medium and trace metals on kinetics of carbon tetrachloride transformation by *Pseudomonas* sp. strain KC. *Appl. Environ. Microbiol.* **59**:2126–2131.

42. **Thompson, G. A., and R. J. Watling.** 1983. A simple method for the determination of bacterial resistance to metals. *Bull. Environ. Contam. Toxicol.* **31**:705–711.

43. **Thompson, G. A., and R. J. Watling.** 1984. Comparative study of the toxicity of metal compounds to heterotrophic bacteria. *Bull. Environ. Contam. Toxicol.* **33**:114–120.

44. **Washington, J. A., II, and V. L. Sutter.** 1980. Dilution susceptibility test: agar and macrobroth dilution procedures, p. 453–458. *In* E. H. Lennette, A. Ballows, W. J. Hausler, Jr., and J. P. Truant (ed.), *Manual of Clinical Microbiology,* 3rd ed. American Society for Microbiology, Washington, D.C.

Colonization, Adhesion, and Biofilms

KEVIN C. MARSHALL

38

RELATIONSHIPS BETWEEN COLONIZATION, ADHESION, AND BIOFILMS

Rationale for Surface Colonization by Microorganisms

Many natural habitats are characterized by insufficient energy substrates and/or nutrients (i.e., oligotrophic conditions) to support microbial growth. As a result, most microbes in these habitats exist in a state termed starvation-survival (59). Macromolecules and smaller hydrophobic molecules, originating as excretions from living organisms or as lytic products from dead organisms, tend to partition at interfaces and, particularly, to adsorb at solid surfaces to form conditioning films (CF). These CF alter the charge (60) and free energy (2) of the surface and act as concentrated energy substrates for microbes associated with the surface (53). In flowing systems, there is a continuous input of nutrients which encourage rapid growth and reproduction of colonizing bacteria and the eventual buildup of biofilms.

Adhesion to Surfaces

Microorganisms in nature encounter a wide range of solid surfaces that may markedly alter their physiological and ecological behavior (49). Movement of microorganisms toward surfaces may result from water flow, from the organism's motility, or from a combination of these processes. Bacteria, in particular, may be considered as living colloidal particles (49); when in the vicinity of a surface, they are subjected to various attraction and repulsion forces (50) that can lead to rapid but passive adhesion, due solely to physicochemical interactions between the bacterium and the surface, or to a time-dependent but active adhesion involving an intermediate reversible phase of adhesion followed by physiological responses induced in the bacterium by the presence of the surface (34). Firm adhesion to surfaces is generally mediated by bacterial exopolymers, a process termed polymer bridging (56). The nature and extent of adhesion depend on the physicochemical properties of the substratum, the source of the CF molecules, and the species and physiological status of the organisms involved (74).

General Aspects of Colonization

Bacteria are the primary colonizing organisms on surfaces, with small, starved cells predominating (19). Successful colonization of the surface depends on continued microbial adhesion accompanied by growth of the adherent organisms, with a regular succession of bacterial types (57) leading ultimately to the formation of a mature biofilm containing a wide variety of microbial types depending on the surfaces and environmental conditions encountered (16, 58).

Growth at Surfaces

Use of radiolabeled substrates and time-lapse video techniques have revealed that primary colonizing starved bacteria are able to metabolize surface-bound substrates, grow to normal size, and commence cellular reproduction (42, 66). Observations have indicated a range of strategies of growth at and cell detachment from surfaces in different bacteria (53), thereby ensuring effective colonization of the surface along with a return of a proportion of the cells to the aqueous phase (52). Growth at surfaces is accompanied by the production of copious quantities of secondary exopolymers (13), which ultimately provide the polymer matrix of a biofilm.

Biofilm Formation

A biofilm forms when multilayers of bacteria embedded in a polymer matrix develop at a surface. The modern concept of a biofilm, as revealed by scanning confocal laser microscopy (SCLM), is of discrete columns of mixed bacterial species embedded in a polymer matrix surrounded by water-filled voids (8). Microelectrode studies have revealed that oxygen penetrates to the base of the voids but not very far into the columnar bacterial matrices (20). These findings have enormous implications regarding the mass transport of gases, nutrients, and metabolic by-products to and from the bacterial biomass and in terms of the spatial distribution of microbial populations within the biofilm community.

A wide range of techniques have been used in studies of adhesion, colonization, and biofilm formation related to many different ecosystem, industrial, and medical applications. It is not possible to give full details of all of these techniques in this chapter, but appropriate references are provided along with advantages and disadvantages of the techniques in most instances. The emphasis in this chapter is on basic laboratory techniques; aspects of biofilm development related to practical problems of biofouling and biocorrosion are treated in chapter 66.

SUBSTRATUM PROPERTIES AND CF

Solid substrata possess characteristic surface properties that are important in microbial adhesion and colonization, particularly surface free energy, surface charge, and surface roughness. These properties can be monitored by a range of physical and chemical techniques, as can modifications resulting from the adsorption of organic molecules (CF) at the solid-liquid interface.

Contact Angle Measurements

The angle of contact made by a droplet of liquid at a solid surface is a reflection of the surface free energy of the substratum. It may spread on the surface (completely wettable) or remain as a discrete drop making a finite angle of contact (Θ) with the surface. As Θ increases, the degree of wettability decreases. Θ may be measured using a contact angle meter (e.g., Erma Contact Angle Meter, Tokyo, Japan), which is essentially a low-power microscope set to view the droplet at an oblique angle. The reflection of the droplet is seen on the illuminated, smooth substratum to provide a sharp delineation of the actual contact point of the droplet; the angle is accurately measured with the aid of a goniometer (rotatable protractor) attached to the microscope. Several precautions must be observed: (i) the droplet size should be constant and small (no more than 2 mm in diameter) to avoid gravitational effects, (ii) measurements should be made in an enclosed chamber (containing a beaker of water in which water contact angles are being measured) to ensure minimal evaporation of the droplet, and (iii) the substratum surface needs to be scrupulously cleaned (by vigorous buffing or, better still, by radiofrequency glow discharge in a high-vacuum chamber) to ensure the removal of contaminating organics derived from adsorption of volatile chemicals abundant in any laboratory.

The surface tension (γ) at the solid-air interface is given by the equation $\gamma_{sa} = \gamma_{sl} + \gamma_{al} \cos \Theta$, where s, a, and l refer to the solid, air, and liquid phases, respectively. The terms γ_{sa} and γ_{sl} cannot be measured accurately. On the other hand, contact angles obtained on a surface by using an homologous series of liquids may be used to determine the critical surface tension (γ_c), which is a measure of the highest surface tension that a liquid can have and still spread over a given solid surface by molecular attraction and is defined as the intercept of the horizontal axis at cos $\Theta = 1$ with the extrapolated straight line plot of cos Θ against the surface tension (γ_o) of the various liquids (2). Similarly, a homologous series of liquids has been used to determine the dispersion (γ^d_s) and polar (γ^p_s) components for surfaces (7).

Substratum wettability has also been expressed in terms of water contact angles (30), the work of adhesion (W_A) between water and the substratum (67), bubble contact angles (31), and the surface free energy (γ_s) determined from a single contact angle measurement by using the equation-of-state approach (62). For comparative purposes, particularly when clean and CF surfaces are being considered, most of these expressions are probably satisfactory. Busscher and van Pelt (6) have provided evidence to suggest that the use of a single liquid is not suitable for comparisons of surface free energies of a range of solid surfaces. In addition, Fletcher and Marshall (31) have criticized the use of contact angles on dried substrata for the measurement of wettability of surfaces with CF because of the denaturation of, and hence structural changes to, macromolecules at the surface. They proposed the use of bubble contact angles on wetted surfaces yet recognized the problems arising from changes to the liquid surface tension upon the addition of the macromolecules.

Surface Physicochemistry Methods

Baier (see reference 2, Fig. 3.15, p. 90) gives an excellent schematic representation of a range of nondestructive physicochemical techniques applied to the characterization of clean and CF surfaces. Using infrared (IR)-transmitting, reflective prisms, such as germanium, the nature of CF may be characterized by contact angle measurements to give γ_c, by attenuated total reflectance (ATR)-IR spectrometry to give macromolecular properties, by ellipsometry to give optical thickness, by X-ray diffraction to give crystal structure, by scanning electron microscopy (SEM) for surface texture, and by energy dispersive X-ray analysis for elemental composition. These techniques are limited by the need to dry the prisms prior to examination, with the result that macromolecules are denatured and the true nature of groups that may interact with adhesive bacteria may be masked (31). This problem has been overcome, at least for IR analysis, by the application of Fourier transform-IR (FTIR) spectrometry (63) combined with ATR in a circle cell assembly (4), which allows a time course study of macromolecular adsorption from a flowing system by means of subtraction of the water signal to reveal the IR absorption peaks of the macromolecules.

Other techniques applied to the study of CF include Auger electron spectroscopy and X-ray photoelectron spectroscopy (XPS) (44) and pyrolysis-chemical ionization combined with mass spectrometry (83). Again, these techniques require dried specimens and are destructive, resulting in fragment analysis, and may not identify the precise groups involved in interactions with bacteria adhering to the CF surface.

Surface Roughness

Much of the research on bacterial adhesion and colonization ignores the important parameter of surface roughness, yet apparently smooth surfaces possess surface undulations equal to or greater than the size of an average bacterial cell (see reference 10, Fig. 3, p. 147). In flowing systems, consequently, bacteria deposited in these microcrevices are more protected from shear forces than those deposited on high points of the undulations or on smooth surfaces and thus have a greater chance of colonizing the surface. Surface tracing equipment is available to provide a measure of the surface roughness of various substrata (74).

SURFACE CHARACTERIZATION OF ADHESIVE BACTERIA

Electron Microscopy

Techniques of electron microscopy directly related to the study of adhesive properties of bacteria are presented elsewhere (36, 37). Transmission electron microscopy (TEM) usually provides the most useful information on surface structures likely to be involved in adhesion and colonization processes. Negative staining and shadowing techniques have proved valuable in revealing external structures such as flagella, fimbriae, pili, patterned surfaces, and exopolymers, especially adhesive "footprints" following removal of the bacterial cells from the surface (61). Ultrathin sectioning provides the most convincing evidence for the actual mode of adhesion and the surface structures involved (14, 15).

SEM is very useful for indicating the numbers and distribution of microorganisms adhering to surfaces (15) but lacks sufficient resolution to give any convincing idea of structures involved in the process of adhesion. Extravagant claims on the form of adhesive polymers, as viewed by SEM, pay little regard to the fact that the specimens are viewed from above and that hydrated polymers become highly condensed during dehydration procedures. The use of ruthenium red and antibody treatments (15) help to preserve exopolymers in SEM preparations but do not overcome the problem of satisfactorily viewing the specimens.

Cell Surface Hydrophobicity

A variety of methods have been used to determine the relative hydrophobicities of microorganisms, and comparisons of the methods have revealed certain advantages of some methods under some circumstances (see reference 71 for full details of methods and their applicability). Certainly hydrophobicity plays an important role in the adhesion of some bacteria to particular substrata, but caution is urged in interpreting the results of single methods of testing for hydrophobicity, because (i) different mechanisms can exist for the adhesion of the same organism to different substrata (64); (ii) some authors have reported using "very" hydrophobic bacteria, and yet the organisms give uniform aqueous suspensions (e.g., reference 71, Fig. 4, p. 375)—thus, they must be relatively hydrophilic; and (iii) some bacteria possess a localized hydrophobic area on their surface, but most of the cell is hydrophilic—these cells show as hydrophobic by many of the tests (54). It is important to note that the standard methods for determining cell surface hydrophobicity give only a net result for the whole surface and do not measure localization of hydrophobic sites (54). A technique that may prove valuable in determining such localization is the judicious use of colloidal gold labeling at high pH (36).

Surface Charge Properties

The surface charge properties of small particles, including bacteria, may be determined by the technique of microelectrophoresis. Basically, the apparatus consists of a chamber of uniform thickness, with reversible electrodes at each end of the chamber and a facility for flushing electrolyte (buffer) and bacterial suspensions in electrolyte through the chamber, mounted on a suitable microscope (39). The velocity of the particles is determined following the application of an electrical potential, with negatively charged particles moving toward the cathode. Reversing the polarity of the electrodes results in the particles moving in the opposite direction so that a mean velocity is obtained. Modern instruments based on laser detection of particles in the chamber and computer evaluation of electrophoretic mobilities are available.

The electrophoretic mobility is given by $m = v\chi_s q \cdot i^{-1}$, where m is the electrophoretic mobility (micrometers per second per volt per centimeter), v is the velocity of the particles (micrometers per second), χ_s is the specific conductivity of the electrolyte (mhos [1 mho = 1 Siemens] per centimeter), q is the cross-sectional area of the chamber (square centimeters), and i is the current (amperes). It is not recommended that the surface charge density be calculated from the electrophoretic mobility because of underestimation of the surface charge owing to problems of surface conductivity in the electrophoresis chamber (33). The adsorption of CF or colloidal material to particles, including bacteria, results in significant changes in the electrophoretic mobilities of the particles (48, 60).

Although different bacteria and mutants with altered surface structures possess different surface charges, caution is needed in interpreting such measurements, particularly in relation to variable natural environments. The apparent charge on a cell surface is very dependent on electrolyte concentration and valency (49). Most measurements are made in a single electrolyte at a fixed concentration, whereas in nature, the electrolyte will consist of ions of mixed valencies and the concentrations will vary with circumstances. Electrophoretic mobility measurements provide a net value for the whole bacterial surface and do not measure localization of positively or negatively charged sites on the cell surface. Marshall and Cruickshank (54) determined the distribution of positively charged sites on *Flexibacter* strain CW7 by reacting the cells with negatively charged silver iodide sol (colloidal solution) and viewing them under TEM to determine the distribution of electron-dense AgI particles on the bacterial cell surfaces.

Chemical Characterization

The complex chemistry of bacterial exopolymers is discussed by Christensen and Characklis (13), and the detection of extracellular polysaccharides by electron microscopy is considered by Handley (37). Conventional methods for exopolymer isolation mainly involve precipitation of polymers from culture supernatants (13). Such polymers are not necessarily related to extracellular components involved in adhesion, as in the following instances: (i) when fimbriae are implicated in the adhesion process and subsequent production of excess polymer results in masking these fimbriae and the desorption of bacteria from the surface (70); (ii) when cell surface hydrophobicity is involved in adhesion and, again, excess polymer production results in desorption of the cells (24); (iii) when different mechanisms exist for adhesion of the same organism to different surfaces (64); and (iv) when an organism passes through a time-dependent reversible phase of adhesion (56) prior to the irreversible adhesion phase, suggesting changes in gene expression specifically induced by contact with a surface (34). As a consequence of these problems, attempts have been made to remove bacteria from the surface, following short-term adhesion, and to analyze the resultant adhesive polymeric footprints remaining on the surface (61). Analyses of such footprint material may be achieved in the future by the application of newer surface chemistry techniques, such as FTIR (4, 63), XPS (44), and pyrolysis-chemical ionization combined with mass spectrometry (83).

In addition to polymer analysis, some of the physicochemical analytical techniques are being applied directly to structural analyses of the outer surfaces of microbial cells. FTIR has been used for such direct analysis of bacteria attached to germanium ATR crystals (63), and XPS has been employed extensively in surface analyses of bacteria and yeasts (72). The method of cell preparation for XPS, whereby bacteria are suspended in distilled water and freeze-dried, has been criticized (55) on the basis that artifacts may result from adsorption to cell surfaces of internal macromolecules leaking under these conditions.

COLONIZATION STUDY SYSTEMS

Static Systems

Static adhesion and colonization methods are applicable to environments such as soils, sediments, and nonflowing lake systems. A convenient method for studying adhesion and colonization of surfaces under static conditions involves adding bacterial suspensions to petri and/or tissue culture dishes, allowing them to stand for variable times, decanting the suspension, rinsing thoroughly with a suitable diluent (depending on the environment being simulated), staining (generally with crystal violet), and reading the absorbance at a number of sites on each dish in a spectrophotometer to give a comparative measure of bacterial numbers adhering to the surface (25). This method allows rapid setting up and reading of many replicates and, generally, very good reproducibility between replicates. The method is limited by the availability of suitable transparent substrata capable of containing bacterial suspensions, by the variability of different batches of petri and tissue culture dishes, and by the influence of gravitational effects on the movement of cells to the surfaces.

Other static systems include the classical slide immersion technique used in soils (12) and in lake waters (38), whereby the slides are recovered, washed, stained, and examined by direct light microscopy. The technique has been modified by the use of nontransparent metal coupons and observation by epifluorescence or scanning electron microscopy after suitable staining or coating (58).

Flowing Systems

Flowing systems are relevant to studies on flowing streams, on saliva flow in the mouth, and on flow past ship hulls, pipelines, and rocks in streams. Transport of microorganisms is more efficient under dynamic conditions, where it is dominated by advection and diffusion, than under static conditions, where it is subject only to diffusion (23). In many dynamic systems, the flow is uncontrolled and often leads to turbulent flow past the test surface. This is well illustrated by the use of water flow through a chamber containing glass coverslips mounted at right angles to the flow, to study manganese deposition by bacteria colonizing the coverslip surfaces (79). An indication of the uneven flow occurring in the system was the accumulation of bacteria and manganese deposits mainly at the edges of the coverslips. A simpler system involved the use of pure cultures suspended in seawater in a large evaporating dish in which glass slides were suspended and the liquid was rotated by means of a magnetic stirrer (56). These methods give useful results but do not provide even flow over the deposition surface, and it is not possible to calculate shear forces operative at any point on the surface.

In terms of colonization data useful in real flowing systems, more realistic information can be obtained with the use of laminar flow chambers, in which flow rates can be controlled and shear forces at the surface can be calculated. Laminar flow refers to the movement of a fluid along a tube in laminae, or sheets, as opposed to turbulent flow, in which case eddies exist within the flow. The measure of flow is the dimensionless Reynolds number (Re), which is the ratio of inertial forces to viscous forces in the flow. If Re is <2,100, then the flow is laminar; if Re is >2,100, then turbulent flow is occurring (see reference 11 for a full discussion of this concept). Laminar flow chambers can be constructed to allow continuous observation under a phase-contrast microscope and have been used extensively in recent times for controlled studies of adhesion and colonization (45).

Culture Conditions

The source of microorganisms in adhesion studies may be natural waters (45, 57), suspensions of pure cultures (56), or mixtures of cultures (75). Although batch cultures are often used, these are variable in their adhesion behavior because of the range of physiological ages of cells within the culture at any particular stage of growth (26, 74). The use of continuous cultures, particularly when grown under conditions of controlled nutrient limitation, gives more reliable results because of the physiological uniformity of cells within the culture (74).

COLONIZATION AND BIOFILM FORMATION: LIGHT MICROSCOPY
General Considerations

Comparisons of the use of transmitted, incident, and polarizing light microscopic techniques in studies on the behavior of bacteria at surfaces have been presented by Marshall (51), and the topic of microscopy (enumeration) is dealt with in chapter 5. Observation by phase-contrast microscopy is generally the most convenient method because it avoids the need for drying and staining, it is nondestructive and slides can be returned to a test system if necessary, and it can be used readily in continuous-observation systems. Most observers in the past have relied on the extraction of substratum samples from nonflowing or flowing systems at regular intervals to obtain some measure of the types of primary colonizers, the rate of colonization of the surface, and the succession of organisms at the surface. This approach has provided useful preliminary data on these various parameters but suffers from the lack of continuous observation of the sequence of events occurring at the surface. Appropriate precautions must be observed in all aspects of enumeration of total numbers or of particular types of microorganisms adhering to and colonizing solid substrata, as personal bias is very readily introduced in such enumeration. Fields for enumeration must be chosen at random, and sufficient replicates must be taken to ensure valid statistical analyses of the results obtained. Manual counting of bacteria in many fields is very laborious and time-consuming. More satisfactory methods include direct counting by on-line image analysis (8) or by taking photographs of the random fields to be counted and determining the counts by subjecting the photographs to image analysis.

Growth Rates at Surfaces

Staley (76) used a partly submerged microscope to observe the succession of bacteria and algae and to determine algal growth rates at the surface of a glass slide, and Brock (5) described a number of methods for determining growth rates of bacteria in nature. Bott and Brock (3) devised methods to distinguish between passive attachment and in situ growth of bacteria at surfaces, and they used the exponential growth equation to determine specific growth rates. Subsequently, this equation was shown to overestimate growth rates, and a more satisfactory equation has been derived to describe colonization in terms of simultaneous growth and attachment of bacteria at surfaces (9).

Reversible and Irreversible Adhesion

Under a phase-contrast microscope, reversible adhesion may be distinguished by the presence of Brownian motion in cells associated with the surface and by their removal by a moderate shear force, such as water flow or flagellar motion (56). Irreversible adhesion, on the other hand, is characterized by a lack of Brownian motion, and cells are not removed from the surface by a moderate shear force. Some bacteria appear to adhere almost immediately to a surface without an intermediate reversible phase, a process which probably involves only physicochemical interactions with the surface and is termed passive adhesion (28). Those bacteria exhibiting a two-phase (reversible and a time-dependent irreversible) adhesion almost certainly show some physiological response to the surface (i.e., altered gene expression [34]) in addition to physicochemical interactions (50).

The numbers of bacteria adhering at surfaces depends on complex interactions between the surface free energy of the substrata, the surface free energy of the bacteria, and the surface tension of the liquid phase (1). Removal of calcium ions from the system assists in the detachment of bacteria and biofilms from surfaces, probably by preventing divalent bridging between adhesive polymer fibrils (78). Interference reflection microscopy has proved useful in defining the distance of separation from the surface, in terms of nanometers, of adherent bacteria at different electrolyte concentrations (29). Zvyagintsev et al. (84) have attempted to estimate adhesive strength of attached bacteria by determining the force necessary to remove the cells by centrifugation; similarly, Duddridge et al. (22) have quantitated removal rates of adhering bacteria from surfaces under defined fluid shear forces. The true adhesive strength of attached bacteria would not be obtained by removal procedures, however, because these generally break the adhesive polymers between the cell and the surface and merely give a measure of the cohesive strength of the polymers and leave a polymer footprint at the surface (61).

Colonization Behavior at Surfaces

Continuous observations of bacterial surface colonization have been made in flow cells, and the behavior of the bacteria at surfaces has been revealed either by computer-enhanced microscopy, including image analysis techniques (45), or by time-lapse video techniques (18, 66). The latter technique has the advantages of being relatively inexpensive in terms of the equipment required and providing a permanent record of behavioral events that, when played at normal speed, gives a spectacular view of the often unusual events. Both techniques have shown that various bacteria possess very different colonization behavioral patterns. Published observations include cellular growth and subsequent reproduction of starving adhesive and nonadhesive bacteria, a "mother-daughter" relationship between attached cells and their progeny, a "packing" maneuver, a "spreading" or "slow migration" maneuver following cell division, a "rolling" maneuver, the response of gliding bacteria to inhibitors and, in some instances, their recovery following removal of the inhibitor, and the outgrowth of distinctly different morphological types of one bacterium on different substrata (details summarized in references 43a and 53).

Standard forms of light microscopy are not very suitable for observing biofilms more than one or two cells thick because of the translucent nature of the films.

SCLM

The use of SCLM has revolutionized our understanding of the structure of biofilms, because its use enables the optical sectioning of a biofilm and, by capturing images at various levels within the biofilm, allows a computerized three-dimensional image to be developed (8). The technique is nondestructive and can be used to monitor biofilms. It has been employed to demonstrate the excretion of acids by bacteria into the interconnected, water-filled voids (8), the diffusion of oxygen into the voids and columnar bacterial masses of biofilms (20), the heterogeneity and flow characteristics within biofilms (77), the changes in morphology of marine bacterium SW5 biofilms induced by different substrata (18), and the multicellular organization of biofilms as it relates to biodegradative activity (82).

Specific Probes

A major problem in studying natural biofilms is to identify particular microorganisms in situ. Although distinctive morphological differences between two species have been exploited in monitoring colonization in binary population biofilms (75), this approach cannot be implemented in more complex biofilms. Rogers and Keevil (69) used episcopic differential interference microscopy to locate *Legionella pneumophila*, labeled by immunogold and fluorescein-linked antibodies, in aquatic biofilms.

Recently developed 16S and 23S rRNA-directed fluorescent oligonucleotide probes (see details in chapter 11 of this manual) have been used with various degrees of specificity to detect particular bacteria in biofilms (47), to measure the activity of single cells in biofilms (65), and to monitor the isolation and enrichment of specific bacteria from samples, such as biofilms, using sequences derived from environmental sources (41). Another approach is the use of *algD*-bioluminescent reporter plasmids to monitor alginate production in biofilms (80). The combination of specific probes and SCLM should provide opportunities to obtain three-dimensional relationships between interacting species within biofilms.

COLONIZATION AND BIOFILM FORMATION: ELECTRON MICROSCOPY
TEM

Suspension of nickel or gold electron microscope grids directly in aquatic environments and examination by TEM following negative staining or shadowing have provided some indication of the nature of primary colonizers and, in the short term, of successional sequences of colonizing microorganisms (57). TEM examination of ultrathin sections of embedded, surface-colonized materials has given a clear indication of the heterogeneity of bacterial types and their spatial distribution, potential for interactions, and physical separation by the polymer matrix within natural biofilms (16). Observations of such sections, however, did not reveal the void spaces between bacterial masses in biofilms that were detected by optical sectioning with SCLM (8).

SEM

SEM has proved very useful in enumeration of microorganisms colonizing surfaces, the variety of types involved, and the succession of types with time of colonization (58). On the other hand, SEM is of doubtful value in studies of the structure of biofilms, even when attempts are made to stabilize the polymer matrix of the biofilm (15), because of the collapse of the overall matrix on drying. This problem is exemplified by a comparison between biofilm material prepared in the normal way for examination by conventional SEM and similar material maintained in moist conditions for examination by environmental SEM (46).

COLONIZATION AND BIOFILM FORMATION: OTHER TECHNIQUES
Radiolabeling Experiments

The numbers of bacteria colonizing various substrata over time have been determined by growing cells in an appropriately labeled substrate, centrifuging the cells to remove excess label, resuspending the cells in an appropriate diluent, exposing substrata to the labeled cell suspension, removing substratum samples at regular intervals, rinsing the samples to remove loosely associated cells, placing the samples in a liquid scintillant, and determining the level of radioactivity, as disintegrations per minute, present at the surface (43). Total cell numbers and radioactivity of dilutions of the original supernatant can be determined to give a standard curve of total cell numbers against disintegrations per minute. Radiolabeled substrates have also been used in autoradiographic techniques and to measure assimilation rates as indications of the activity of bacteria at surfaces (27) and to demonstrate the scavenging of surface-bound fatty acids (42) and proteins (73).

Fatty Acid Profiles

An indirect measure of changes in biofilm community structure with time has been obtained by the application of chemical methods to determine fatty acid profiles to reveal the presence of significant numbers of specific groups of bacteria (35).

Spectroscopic Techniques

FTIR techniques have been used to monitor the appearance of colonizing bacteria on germanium surfaces and the production of exopolymers leading to the formation of biofilms (4, 63) and to examine the role of exopolymers in complexing copper ions, resulting in corrosion of thin copper films deposited on germanium crystals (32). Other spectroscopic techniques used in studies on the involvement of exopolymers in copper corrosion include Auger electron spectroscopy and XPS (40).

Microelectrodes

The introduction of microelectrodes into biofilm studies (68) has greatly increased our knowledge of the importance of pH and diffusion gradients in controlling the distribution of different microbial populations in these complex communities. For instance, microelectrodes have been used to study the distribution of *Beggiatoa* species in microbial mats (68), denitrification activity in trickling filter biofilms (17), nitrification activity in biofilms (21), and the role of mass transport and diffusion in voids in biofilms (20).

Molecular Methods of Biofilm Community Analysis

Molecular methods are dealt with in detail in chapters 10 and 11 of this manual. Because natural biofilms, including microbial mats, consist of a complex mixture of different microbial populations, they have been the subject of some early attempts to describe the community structure by extraction and analysis of the naturally occurring nucleic acid material (81). It is expected that the application of such methods eventually will provide details on the diversity and phylogeny of microorganisms present in a wide range of biofilms.

REFERENCES

1. **Absolom, D. R., F. V. Lamberti, Z. Policova, W. Zingg, C. J. van Oss, and A. W. Neumann.** 1983. Surface thermodynamics of bacterial adhesion. *Appl. Environ. Microbiol.* **46:**90–97.
2. **Baier, R. E.** 1980. Substrate influence on adhesion of microorganisms and their resultant new surface properties, p. 59–104. *In* G. Bitton and K. C. Marshall (ed.), *Adsorption of Microorganisms to Surfaces.* Wiley-Interscience, New York.
3. **Bott, T. L., and T. D. Brock.** 1970. Growth and metabolism of periphytic bacteria: methodology. *Limnol. Oceanogr.* **15:**333–342.
4. **Bremer, P. J., and G. G. Geesey.** 1991. An evaluation of biofilm development utilizing non-destructive attenuated total reflectance Fourier transform infrared spectroscopy. *Biofouling* **3:**89–100.
5. **Brock, T. D.** 1971. Microbial growth rates in nature. *Bacteriol. Rev.* **35:**39–58.
6. **Busscher, H. J., and A. W. J. van Pelt.** 1987. The possibility of deriving solid-surface free energies from contact-angle measurements with one liquid. *J. Materials Sci. Lett.* **6:**815–816.
7. **Busscher, H. J., A. W. J. van Pelt, H. P. de Jong, and J. Arends.** 1983. Effect of spreading pressure on surface free energy determinations by means of contact angle measurements. *J. Colloid Interface Sci.* **95:**23–27.
8. **Caldwell, D. E., D. R. Korber, and J. R. Lawrence.** 1992. Confocal laser microscopy and computer image analysis in microbial ecology. *Adv. Microb. Ecol.* **12:**1–67.
9. **Caldwell, D. E., J. A. Malone, and T. L. Kieft.** 1983. Derivation of a growth rate equation describing microbial surface colonization. *Microb. Ecol.* **9:**1–6.
10. **Characklis, W. G.** 1984. Biofilm development: a process analysis, p. 137–157. *In* K. C. Marshall (ed.), *Microbial Adhesion and Aggregation.* Springer-Verlag, Berlin.
11. **Characklis, W. G., N. Zelver, and M. H. Turakhia.** 1990. Transport and interfacial transfer phenomena, p. 265–340. *In* W. G. Characklis and K. C. Marshall (ed.), *Biofilms.* Wiley-Interscience, New York.
12. **Cholodny, N.** 1930. Über eine neue Methode zur Untersuchung der Bodenflora. *Arch. Mikrobiol.* **1:**409–420.
13. **Christensen, B. E., and W. G. Characklis.** 1990. Physical and chemical properties of biofilms, p. 93–130. *In* W. G. Characklis and K. C. Marshall (ed.), *Biofilms.* Wiley-Interscience, New York.
14. **Corpe, W. A.** 1980. Microbial surface components involved in adsorption of microorganisms onto surfaces, p. 105–144. *In* G. Bitton and K. C. Marshall (ed.), *Adsorption of Microorganisms to Surfaces.* Wiley-Interscience, New York.
15. **Costerton, J. W.** 1980. Some techniques involved in study of adsorption of microorganisms to surfaces, p. 403–432. *In* G. Bitton and K. C. Marshall (ed.), *Adsorption of Microorganisms to Surfaces.* Wiley-Interscience, New York.
16. **Costerton J. W., R. J. Irvin, and K.-J. Cheng.** 1981. The bacterial glycocalyx in nature and disease. *Annu. Rev. Microbiol.* **35:**299–324.
17. **Dalsgaard, T., and N. P. Revsbech.** 1992. Regulating factors of denitrification in trickling filter biofilms as measured with the oxygen/nitrous oxide microsensor. *FEMS Microbiol. Ecol.* **101:**151–164.
18. **Dalton, H. M., L. K. Poulsen, P. Halasz, M. L. Angles, A. E. Goodman, and K. C. Marshall.** 1994. Substrata-induced morphological changes in a marine bacterium and their relevance to biofilm structure. *J. Bacteriol.* **176:**6900–6906.
19. **Dawson, M. P., B. A. Humphrey, and K. C. Marshall.** 1981. Adhesion: a tactic in the survival strategy of a marine vibrio during starvation. *Curr. Microbiol.* **6:**195–199.
20. **de Beer, D., P. Stoodley, F. Roe, and Z. Lewandowski.** 1994. Effects of biofilm structures on oxygen distribution and mass transport. *Biotechnol. Bioeng.* **43:**1131–1138.
21. **de Beer, D., J. C. van den Heuvel, and S. P. P. Ottengraph.** 1993. Microelectrode measurements of the activity distribution in nitrifying bacterial aggregates. *Appl. Environ. Microbiol.* **59:**573–579.

22. **Duddridge, J. E., C. A. Kent, and J. F. Laws.** 1982. Effect of surface shear on the attachment of *Pseudomonas fluorescens* to stainless steel under defined flow conditions. *Biotechnol. Bioeng.* **24:**153–164.

23. **Escher, A., and W. G Characklis.** 1990. Modeling the initial events in biofilm accumulation, p. 445–486. *In* W. G. Characklis and K. C. Marshall (ed.), *Biofilms.* Wiley-Interscience, New York.

24. **Fattom, A., and M. Shilo.** 1984. Hydrophobicity as an adhesion mechanism of benthic cyanobacteria. *Appl. Environ. Microbiol.* **47:**135–143.

25. **Fletcher, M.** 1976. The effects of proteins on bacterial attachment to polystyrene. *J. Gen. Microbiol.* **94:**400–404.

26. **Fletcher, M.** 1977. The effects of culture concentration and age, time, and temperature on bacterial attachment to polystyrene. *Can. J. Microbiol.* **23:**1–6.

27. **Fletcher, M.** 1979. A microautoradiographic study of the activity of attached and free-living bacteria. *Arch. Microbiol.* **122:**271–274.

28. **Fletcher, M.** 1980. The question of passive versus active attachment in nonspecific bacterial adhesion, p. 197–210. *In* R. C. W. Berkeley, J. M. Lynch, J. Melling, P. R. Rutter, and B. Vincent (ed.), *Microbial Adhesion to Surfaces.* Ellis Horwood Publishers, Chichester, England.

29. **Fletcher, M., J. M. Lessmann, and G. I. Loeb.** 1991. Bacterial surface adhesives and biofilm matrix polymers of marine and freshwater bacteria. *Biofouling* **4:**129–140.

30. **Fletcher, M., and G. I. Loeb.** 1979. Influence of substratum characteristics on the attachment of a marine pseudomonad to solid surfaces. *Appl. Environ. Microbiol.* **37:**67–72.

31. **Fletcher, M., and K. C. Marshall.** 1982. Bubble contact angle method for evaluating substratum interfacial characteristics and its relevance to bacterial attachment. *Appl. Environ. Microbiol.* **44:**184–192.

32. **Geesey, G. G., L. Lang, M. R. Hankins, T. Iwaoka, and P. R. Griffiths.** 1988. Binding of metal ions by extracellular polymers of biofilm bacteria. *Water Sci. Technol.* **20:**161–165.

33. **Gittens, G. J., and A. M. James.** 1963. Some physical investigations of the behaviour of bacterial surfaces. VII. The relationship between zeta potential and surface charge as indicated by microelectrophoresis and surface conductance measurements. *Biochim. Biophys. Acta* **66:**250–263.

34. **Goodman, A. E., and K. C. Marshall.** 1995. Genetic responses of bacteria at surfaces, p. 80–98. *In* H. M. Lappin-Scott and J. W. Costerton (ed.), *Microbial Biofilms.* Cambridge University Press, Cambridge.

35. **Guckert, J. B., C. B. Antworth, P. D. Nichols, and D. C. White.** 1985. Phospholipid, ester-linked fatty acid profiles as reproducible assays for changes in prokaryotic community structure of estuarine sediments. *FEMS Microbiol. Ecol.* **31:**147–158.

36. **Handley, P. S.** 1991. Negative staining, p. 63–86. *In* N. Mozes, P. S. Handley, H. J. Busscher, and P. G. Rouxhet (ed.), *Microbial Cell Surface Analysis. Structural and Physicochemical Methods.* VCH Publishers, Inc., New York.

37. **Handley, P. S.** 1991. Detection of cell surface carbohydrate components, p. 87–107. *In* N. Mozes, P. S. Handley, H. J. Busscher, and P. G. Rouxhet (ed.), *Microbial Cell Surface Analysis. Structural and Physicochemical Methods.* VCH Publishers, Inc., New York.

38. **Henrici, A. T.** 1933. Studies of freshwater bacteria. I. A direct microscopic technique. *J. Bacteriol.* **25:**277–286.

39. **James, A. M.** 1991. Charge properties of microbial cell surfaces, p. 221–262. *In* N. Mozes, P. S. Handley. H. J. Busscher, and P. G. Rouxhet (ed.), *Microbial Cell Surface Analysis. Structural and Physicochemical Methods.* VCH Publishers, Inc., New York.

40. **Jolley, J. G., G. G. Geesey, M. R. Hankins, R. B. Wright, and P. L. Wichlacz.** 1988. Auger electron spectroscopy and X-ray photoelectron spectroscopy of the biocorrosion of copper by gum arabic, bacterial culture supernatant and *Pseudomonas atlantica* exopolymer. *J. Surf. Interface Anal.* **11:**371–376.

41. **Kane, M. D., L. K. Poulsen, and D. A. Stahl.** 1993. Monitoring the enrichment and isolation of sulfate-reducing bacteria by using oligonucleotide hybridization probes designed from environmentally derived 16S rRNA sequences. *Appl. Environ. Microbiol.* **59:**682–686.

42. **Kefford, B., S. Kjelleberg, and K. C. Marshall.** 1982. Bacterial scavenging: utilization of fatty acids localized at a solid-liquid interface. *Arch. Microbiol.* **133:**257–260.

43. **Kefford, B., and K. C. Marshall.** 1984. Adhesion of *Leptospira* at a solid-liquid interface: a model. *Arch. Microbiol.* **138:**84–88.

43a. **Korber, D. R., J. R. Lawrence, H. M. Lappin-Scott, and J. W. Costerton.** 1995. Growth of microorganisms on surfaces, p. 15–45. *In* H. M. Lappin-Scott and J. W. Costerton (ed.), *Microbial Biofilms.* Cambridge University Press, Cambridge.

44. **Kristoffersen, A., G. Rolla, T. Sonju, and E. Jantzen.** 1982. The organic film developed on metal surfaces exposed to seawater: chemical studies. *J. Colloid Interface Sci.* **90:**191–196.

45. **Lawrence, J. R., and D. E. Caldwell.** 1987. Behavior of bacterial stream populations within the hydrodynamic boundary layers of surface microenvironments. *Microb. Ecol.* **14:**15–27.

46. **Little, B., P. Wagner, R. Ray, R. Pope, and R. Scheetz.** 1991. Biofilms: an ESEM evaluation of artifacts introduced during SEM preparation. *J. Ind. Microbiol.* **8:**213–222.

47. **Manz, W., U. Szwezyk, P. Ericsson, R. Amann, K.-H. Schleifer, and T.-A. Stenstrom.** 1993. In situ identification of bacteria in drinking water and adjoining biofilms by hybridization with 16S and 23S rRNA-directed fluorescent oligonucleotide probes. *Appl. Environ. Microbiol.* **59:**2293–2298.

48. **Marshall, K. C.** 1968. Interaction between colloidal montmorillonite and cells of Rhizobium species with different ionogenic surfaces. *Biochim. Biophys. Acta* **156:**179–186.

49. **Marshall, K. C.** 1976. *Interfaces in Microbial Ecology.* Harvard University Press, Cambridge, Mass.

50. **Marshall, K. C.** 1985. Mechanisms of bacterial adhesion at solid-liquid interfaces, p. 131–161. *In* D. C. Savage and M. Fletcher (ed.), *Bacterial Adhesion: Mechanisms and Physiological Significance.* Plenum Press, New York.

51. **Marshall, K. C.** 1986. Microscopic methods for the study of bacterial behaviour at inert surfaces. *J. Microbiol. Methods* **4:**217–227.

52. **Marshall, K. C.** 1992. Planktonic versus sessile life of prokaryotes, p. 262–275. *In* A. Balows, H. G. Trüper, M. Dworkin, W. Harder, and K.-H. Schleifer (ed.), *The Prokaryotes. A Handbook on the Biology of Bacteria: Ecophysiology, Isolation, Identification, Applications,* 2nd ed. Springer-Verlag, New York.

53. **Marshall, K. C.** Adhesion as a strategy for access to nutrients. *In* M. Fletcher and D. Savage (ed.), *Molecular and Ecological Diversity of Bacterial Adhesion,* in press. John Wiley & Sons, New York.

54. **Marshall, K. C., and R. H. Cruickshank.** 1973. Cell surface hydrophobicity and the orientation of certain bacteria at interfaces. *Arch. Mikrobiol.* **91:**29–40.

55. **Marshall, K. C., R. Pembrey, and R. P. Schneider.** 1994. The relevance of X-ray photoelectron spectroscopy for the analysis of microbial cell surfaces: a critical view. *Colloids Surf. B Biointerfaces* **2:**371–376.

56. **Marshall, K. C., R. Stout, and R. Mitchell.** 1971. Mecha-

nism of the initial events in the sorption of marine bacteria to surfaces. *J. Gen. Microbiol.* **68:**337–348.

57. **Marshall, K. C., R. Stout, and R. Mitchell.** 1971. Selective sorption of bacteria from seawater. *Can. J. Microbiol.* **17:**1413–1416.

58. **Marzalek, D. S., S. M. Gerchakov, and L. R. Udey.** 1979. Influence of substrate composition on marine microfouling. *Appl. Environ. Microbiol.* **38:**987–995.

59. **Morita, R. Y.** 1982. Starvation-survival of heterotrophs in the marine environment. *Adv. Microb. Ecol.* **6:**171–198.

60. **Neihof, R. A., and G. I. Loeb.** 1972. The surface charge of particulate matter in seawater. *Limnol. Oceanogr.* **17:** 7–16.

61. **Neu, T. R., and K. C. Marshall.** 1991. Microbial "footprints"—a new approach to adhesive polymers. *Biofouling* **3:**101–112.

62. **Neumann, A. W., R. J. Good, C. J. Hope, and M. Sejpal.** 1974. An equation of state approach to determine surface tensions of low energy solids from contact angles. *J. Colloid Interface Sci.* **49:**291–304.

63. **Nichols, P. D., J. M. Henson, J. B. Guckert, D. E. Nivens, and D. C. White.** 1985. Fourier-transform infrared spectroscopic methods for microbial ecology: analysis of bacteria, bacteria-polymer mixtures and biofilms. *J. Microbiol. Methods* **4:**79–94.

64. **Paul, J. H., and W. H. Jeffery.** 1985. Evidence for separate adhesion mechanisms for hydrophilic and hydrophobic surfaces in *Vibrio proteolytica*. *Appl. Environ. Microbiol.* **50:** 431–437.

65. **Poulson, L. K., G. Ballard, and D. A. Stahl.** 1993. Use of rRNA fluorescence in situ hybridization for measuring the activity of single cells in young established biofilms. *Appl. Environ. Microbiol.* **59:**1354–1360.

66. **Power, K., and K. C. Marshall.** 1988. Cellular growth and reproduction of marine bacteria on surface-bound substrate. *Biofouling* **1:**163–174.

67. **Pringle, J. H., and M. Fletcher.** 1983. Influence of substratum wettability on attachment of freshwater bacteria to solid surfaces. *Appl. Environ. Microbiol.* **45:**811–817.

68. **Revsbech, N. P., and B. B. Jorgensen.** 1986. Microelectrodes: their use in microbial ecology. *Adv. Microb. Ecol.* **9:**293–352.

69. **Rogers, J., and C. W. Keevil.** 1992. Immunogold and fluorescein immunolabeling of *Legionella pneumophila* within an aquatic biofilm visualized by using episcopic differential interference microscopy. *Appl. Environ. Microbiol.* **58:** 2326–2330.

70. **Rosenberg, E., A. Gottlieb, and M. Rosenberg.** 1983. Inhibition of bacterial adherence to epithelial cells and hydrocarbons by emulsan. *Infect. Immun.* **39:**1024–1028.

71. **Rosenberg, M., and S. Kjelleberg.** 1986. Hydrophobic interactions: role in bacterial adhesion. *Adv. Microb. Ecol.* **9:**353–393.

72. **Rouxhet, P. G., and M. J. Genet.** 1991. Chemical composition of the microbial cell surface by X-ray photoelectron spectroscopy, p. 173–220. *In* N. Mozes, P. S. Handley, H. J. Busscher, and P. G. Rouxhet (ed.), *Microbial Cell Surface Analysis. Structural and Physicochemical Methods.* VCH Publishers, Inc., New York.

73. **Samuelsson, M.-O., and D. L. Kirchman.** 1991. Degradation of adsorbed protein by attached bacteria in relationship to surface hydrophobicity. *Appl. Environ. Microbiol.* **56:**3643–3648.

74. **Schneider, R. P., and K. C. Marshall.** 1994. Retention of the Gram-negative bacterium SW8 on surfaces—effects of microbial physiology, substratum nature and conditioning films. *Colloids Surf. B Biointerfaces* **2:**387–396.

75. **Siebel, M. A., and W. G. Characklis.** 1991. Observations of binary population biofilms. *Biotechnol. Bioeng.* **37:** 778–789.

76. **Staley, J. T.** 1971. Growth rates of algae determined in situ using an immersed microscope. *J. Phycol.* **7:**13–17.

77. **Stoodley, P., D. de Beer, and Z. Lewandowski.** 1994. Liquid flow in biofilm systems. *Appl. Environ. Microbiol.* **60:** 2711–2716.

78. **Turakhia, M. H., K. E. Cooksey, and W. G. Characklis.** 1983. Influence of a calcium-specific chelant on biofilm removal. *Appl. Environ. Microbiol.* **46:**1236–1238.

79. **Tyler, P. A., and K. C. Marshall.** 1967. Microbial oxidation of manganese in hydro-electric pipelines. *Antonie van Leeuwenhoek J. Microbiol. Serol.* **33:**171–183.

80. **Wallace, W. H., J. T. Fleming, D. C. White, and G. S. Sayler.** 1994. An *algD*-bioluminescent reporter plasmid to monitor alginate production in biofilms. *Microb. Ecol.* **27:** 225–239.

81. **Ward, D. M., M. M. Bateson, R. Weller, and A. L. Ruff-Roberts.** 1992. Ribosomal RNA analysis of microorganisms as they occur in nature. *Adv. Microb. Ecol.* **12:** 219–286.

82. **Wolfaardt, G. M., J. R. Lawrence, R. D. Robarts, S. J. Caldwell, and D. E. Caldwell.** 1994. Multicellular organization in a degradative biofilm community. *Appl. Environ. Microbiol.* **60:**434–446.

83. **Zsolnay, A., and B. Little.** 1983. Characterization of fouling films by pyrolysis and chemical ionization mass spectrometry. *J. Anal. Appl. Pyrol.* **4:**335–341.

84. **Zvyagintsev, D. G., A. F. Pertsovskaya, E. D. Yakhnin, and E. I. Averbach.** 1971. Adhesion value of microorganism cells to solid surfaces. *Microbiology* (USSR) **40:** 889–893.

Unusual or Extreme High-Pressure Marine Environments

JODY W. DEMING

39

ENVIRONMENTS CONSIDERED

In general, the best-known microorganisms are opportunists that thrive in the "usual or mild" environments provided to them in laboratory settings that meet our own sense of favorable conditions—body temperature, atmospheric pressure, plentiful organic material for food, neutral pH, and abundant oxygen and water. By contrast, "unusual or extreme" environments have come to mean any that deviate from conditions intolerable to *Homo sapiens*, regardless of spatial scale, from acidic and anoxic pockets in a digestive tract to volcanic hot springs to near-freezing waters in a pressurized ocean. Given the volume of all such "intolerable" habitats, the vast majority of microorganisms derive from unusual or extreme environments. Some combination of extreme temperature, pressure, food supply, acidity, redox condition, or water activity describes the norm, not the rarity, in microbial environments.

The database for any single type of extreme environment has expanded greatly since Kushner's 1978 compilation of chapters by experts on individual environments (52). This chapter provides some history, current trends, and specific examples of strategies and experimental protocols for studying microorganisms that inhabit the largest of accessible extreme environments on the planet—the pressurized deep ocean. The pressure class of microbe addressed most frequently is the barophilic group, composed of organisms that require elevated hydrostatic pressure to achieve optimal growth rate. Methods for studying both extremely psychrophilic and thermophilic prokaryotes are considered, excepting molecular protocols. Few if any specific molecular methods have been developed for deep-sea members of the domain *Bacteria* (eubacteria), since general protocols have proven adequate (9). Complete molecular details for *Archaea*, the microbial domain best known for its extremophiles, including inhabitants of the deep sea, can be found elsewhere (69). Sampling and experimental information is provided according to two broadly defined temperature and pressure regimes in the deep sea: (i) the cold abyssal and hadal zones in the ocean, where cold means <4°C, abyssal means >3.7 km (the average depth of the ocean), and hadal means >6 km (as in island-arc trenches); and (ii) the hot habitats associated with deep-sea hydrothermal vent systems, where hot means >90°C and deep-sea means >2 km in depth. The cold and hot deep seas have in common ele-

vated levels of hydrostatic pressure, which increases approximately 100 bar (1.01 bar = 1.00 atm = 14.7 lb/in^2 = 10.1 MPa) for every kilometer increase in depth. It increases at a greater rate as a result of lithostatic pressure below the seafloor in the marine subsurface biosphere (27), a heretofore inaccessible extreme environment that may prove quite significant volumetrically and scientifically in the future.

The overall goal of this chapter is to enlighten researchers new to the study of the pressurized ocean about the availability of simple, affordable laboratory equipment and methods for culturing and studying its prokaryotic inhabitants. In spite of increasing interest in "extremophiles" (69) and related biodiversity issues, the vast pool of microbial inhabitants of the deep ocean has barely been tapped. Here I hope to dispel some myths regarding the remoteness or limited feasibility of deep-sea microbiology that may be limiting further exploration of high-pressure marine environments.

BACKGROUND RATIONALE FOR SAMPLING PROCEDURES

Although few scientists have ready access to submersible operations in the deep sea or even to standard surface-ship sampling expeditions, the new investigator can find established researchers forthcoming with expertise, field samples, or strains from existing culture collections. Many research goals can be met through collaboration without ever going to sea (9, 35, 45). However, the conditions under which the original sample or culture was obtained frequently figure in the conclusions that can be drawn from subsequent research. Even the views of a molecular biologist working with a well-established deep-sea microorganism will be sharpened by an understanding of the rationale and protocols applied in obtaining the test strain.

Temperature Considerations

For studies of the cold deep sea, temperature appears to be the essential parameter to consider and control. Barophiles were not readily obtained and maintained in pure culture until insulated sampling gear was used and a silica gel culturing medium that could be inoculated and solidified exclusively at cold temperatures was developed (34). All subsequent barophiles from the cold deep sea have proved sensitive to warming at atmospheric pressure in laboratory culture (reference 28 and citations therein), though notably less so under elevated pressures (32). The fact that obligately barophilic bacteria (those undergoing stasis or mortality at atmospheric pressure) have been isolated from samples warmed (and decompressed) during gear recovery (33) suggests that strict attention to in situ temperature during shipboard enrichments may be more critical than maintenance of in situ temperature (or pressure) during sample retrieval procedures.

If the goal is not to isolate in culture but rather to study metabolic activities of the barophilic component of microbial assemblages, then strict adherence to in situ temperature during all operations appears essential (31). Common sense and existing data (83) dictate that sample warming will eliminate some of the psychrophilic barophiles, allowing shallow-water "intruders" present but typically dormant in the deep sea to assume dominance in incubation experiments. A systematic use of different recovery and incubation temperatures to assess indigenous versus shallow-water microbes has not been attempted, although a trend in the literature toward that goal is apparent (62, 63, 78). Niche-specific mixes of thermal and pressure classes of microorganisms in the cold deep sea have been predicted (29).

Isolating thermophilic microorganisms from the hot deep sea presents an opposing situation to psychrophilic barophiles. Samples from the cold deep sea are brought from <4°C to warmer surface waters (except at high latitudes) over the course of several hours, while samples from hot vents cool in situ, from superheated temperatures (>100°C and up to 420°C) to ambient seawater temperature (<4°C), within minutes as samples are pulled from hot sources into the massive cold-water bath of the deep sea. In the first case, some psychrophilic barophiles may succumb to thermal stress during transit through even a mild thermal gradient. In the second case, the potential loss of hot vent inhabitants due to thermal shifts is not known, but rapid cooling has not prevented the establishment of extensive collections of hyperthermophilic microorganisms (maximal growth temperature of >90°C; typically anaerobic heterotrophic or methanogenic members of the domain *Archaea*) native to deep-sea smoker vents (references 1 and 7 and citations therein).

In laboratory tests, hyperthermophiles have proved tolerant of cold temperatures, even under oxygenated conditions, for long periods of time (months to years). These observations have compelling implications for novel stress responses and long-term survival in the cold deep sea after dispersal from a vent (29). The stability of hyperthermophiles at low temperatures also means that sample chilling (immersing a hot container into an ice bath) can be a simple way to end an experimental growth or reaction period in the laboratory.

Pressure Considerations

With rare exception (e.g., gas-vacuolate bacteria), marine prokaryotes exist in pressure equilibrium with their fluid surroundings: they do not implode upon decompression. Decompression-sensitive points in enzymatic pathways or membrane processes, however, may lead to time-dependent (17) impaired metabolic activities, morphological aberrations, or cell lysis (38). For microscopic assessments of deep-sea microbial biomass, samples need only be fixed as soon as possible after retrieval shipboard (and protected against temperature increases, as discussed above), not protected against decompression during gear ascent. When temperature is minded, brief decompression periods during exponential growth phase in laboratory culture do not appear to

cause obvious or irreversible damage (33, 85). Bacteria isolated in the complete absence of decompression (43) have proved no different in their (lack of) sensitivity to brief periods of decompression from those obtained via traditional sampling and culturing means.

For metabolic assessments of microbial community activity in the cold deep sea that rely on samples recompressed shipboard, the effects of decompression during retrieval remain poorly known. Discriminating decompression effects from other factors has proved elusive. The problem lies with the near-universal distribution of microorganisms in the ocean: microorganisms of widely diverse origins and requirements for activity exist throughout the deep sea, even though inactive under in situ conditions (29). Thus, higher activity rates in decompressed samples can signal the release of shallow-water bacteria from the dormancy-inducing effects of deep-sea conditions, while lower rates can mean a substrate or temperature limitation on the same organisms (11, 42). If the goal is to record true in situ rates of activity, the surest route would be to protect against both temperature and pressure changes, either by using decompression-free samplers or by conducting incubation experiments on the seafloor (15, 39, 80). Much has and can be learned, however, from studies of various treatment effects on cold, recompressed deep-sea samples (28, 54).

In the hot deep sea, methods for studying hyperthermophilic activity in situ are still largely in the developmental phase. Whereas sophisticated samplers are available for studies of undecompressed samples from the cold deep sea (see below), means to maintain both in situ temperature and pressure on hot vent samples during retrieval procedures do not exist. Instead, the emphasis has been on developing means to measure hyperthermophilic microbial activity directly on the seafloor (see below).

Whether one is studying pressure effects on microorganisms from the cold or the hot deep sea, it is critical to realize that hydrostatic (liquid) pressures affect organisms differently than hyperbaric (gas) pressures. To generalize, microorganisms are more sensitive to gas pressure, regardless of the thermal class of organisms under study (60, 76). Pressurized sources of gas are sometimes used to generate hydrostatic pressures in the laboratory, but the organism is exposed directly only to liquid under pressure. Pressure information provided in this chapter pertains to hydrostatic pressure. Some microorganisms in subsurface regions of hydrothermal vents may be exposed to gases under pressure during phase separation. Hyperbaric pressure thus becomes a consideration in the design of high-temperature, high-pressure equipment for in situ smoker vent studies (7), but these are highly specialized cases (25).

Substrate and Redox Considerations

In early studies of temperature or pressure effects on deep-sea bacteria, neither substrate concentration nor oxygen availability was a limiting experimental factor (42, 89). With the development of decompression-free sampling devices and the use of different levels of substrate by different groups of investigators, it became apparent that substrate concentration is a critical parameter (22). Renewed interest in oligotrophy and the nonculturable state of many marine bacteria (29, 61) further alerts a new researcher to the importance of substrate levels in the design of microbial experiments.

Few researchers studying psychrophilic barophiles have concerned themselves with oxygen limitation: the cold deep sea from which all barophiles have been cultured is saturated with oxygen. A relatively unexplored area of deep-sea microbiology is the realm of anaerobic psychrophilic barophily, which may pertain to buried sediments and animal guts (47). Means to preserve redox conditions during sampling and experimentation are available (see below).

In contrast, attention to reducing conditions has come to the forefront in studies of the hot deep sea. Superheated geothermal fluids contain no free oxygen and can be heavily laden with reduced gases and forms of sulfur and other mineral species (7). In keeping with such highly reduced environments, virtually all known vent hyperthermophiles have proved anaerobic (1, 7). Their isolation, however, has not required anaerobic conditions during sampling; many have derived from submersible-recovered samples exposed to oxygenated seawater during retrieval. Although known hyperthermophiles can survive oxygenated conditions if the temperature is suboptimal for growth, taking pains to subsample from vent samplers (sealed in situ) by using anaerobic techniques (5, 26) may be contributing to the continued isolation of new strains of anaerobic hyperthermophiles from the deep sea.

Low Water Activity

Low water activity expressed as desiccation is rarely a consideration in microbial studies of the deep ocean, although terrestrial spore-forming bacteria are known to persist in deep-sea sediments (28). Low water activity as a result of extreme saltiness, however, is relevant to the microbiology of brine formations, such as occur in the Gulf of Mexico (77) or in subsurface regions of hydrothermal vents, where liquid-gas phase separation leads to high salt concentrations in the fluids (27). Theoretical arguments have been put forward to link high salinity with microbial stability under high temperatures and hydrostatic pressures, while some experimental data from the Juan de Fuca vent field are explained by invoking microbial stability during brine formation by phase separation (7, 27). On the other hand, no hyperthermophilic halophiles (salt-loving microorganisms) have been obtained in culture, either from the hot deep sea or from other hot or salty environments. Whether a concerted effort to detect such organisms has been made is not clear from the literature: hyperthermophilic halophiles may be a new class of microorganisms awaiting discovery (7). Sampling approaches and culturing methods for known (mesophilic) halophiles from shallow and terrestrial habitats are well established and were reviewed quite recently elsewhere (10, 71).

Physical Disassociation of Organisms

The physical orientations of microorganisms to each other and to surfaces in the environment can be critical determinants of in situ activity and success in detecting activity or culturing targeted organisms. This generality has specific application to the cold deep sea, where attachment to particles has different effects on microbial activity, depending on the pressure requirements of the test organism, its starvation state, and the incubation temperature (reference 68 and citations therein). Barophilic bacteria are predicted to favor association with surfaces, while barotolerant or barosensitive bacteria may tend to exist in suspension (29). Selection of deep-sea inocula replete with particles (e.g., sediments and animal guts) may thus promote success in isolating new barophiles (22). The same appears true at hydrothermal vents, where hot sediments, flanges, particulate sulfides, and animal guts have all yielded hyperthermophilic members of the domain Archaea well adapted to deep-sea pressures (27).

The thermal and chemical gradients established in porous matrices of sulfidic deposits at hydrothermal vents can help guide the cultivation of members of the domain Bacteria versus hyperthermophilic Archaea (7, 35).

In assessing the metabolic activities of microbial communities, the effects of physical disassociations during sample treatment should be considered carefully. The most common experimental treatment of deep-sea sediments is to slurry the sediment with ambient, typically prefiltered (0.2-μm-pore-size filter) seawater (24). The benefit of slurrying is homogeneity in labeling and distribution of subsamples for a time course experiment. The cost is disruption of physical associations that may determine in situ activity. An activity rate measured in slurried sediment can be expected to differ from that which occurs in situ; e.g., for aerobic substrate utilization, rates are higher in oxygenated slurries than in undisturbed cores (66). For hydrolytic enzyme activity, some evidence suggests that slurried rates will be lower the greater the slurry dilution factor (64). To circumvent some of these problems, methods for injecting the substrate of interest directly into undisturbed sediment cores have been developed (56) and adapted for pressurized incubations (66). The disadvantages of the whole-core injection approach include diffusional constraints on the distribution of substrate within the injected horizon of sediment and reliance on separate subcores rather than homogeneously mixed subsamples for replication and killed controls. Nevertheless, the whole-core injection approach seems promising for estimating in situ sediment activity if seafloor experiments are not feasible (66).

Other examples of the importance of considering physical associations of bacteria with surfaces in the deep sea include the discovery of high densities of barophilic bacteria on the hindgut lining of abyssal invertebrates (30), of barophilic manganese-oxidizing (19) and methane-oxidizing (20) bacteria likely associated with particles raining from hydrothermal plumes, and of surface-colonizing prokaryotes of unusual morphologies in both the cold (28) and the hot (41) deep sea.

Contamination Problems

Because of an inability to sterilize most sampling gear and the near-universal distribution of a wide variety of microorganisms in seawater, contamination takes on special meaning in deep-sea research. Contaminants can include not only the typical airborne and skin-transferred microbes but also marine bacteria naturally present but not the target of study. Novel types of microorganisms can be enriched selectively from a deep-sea sample regardless of the presence of contaminants, but aseptic technique should be used as much as possible throughout all procedures as a matter of good practice. For example, to sample uncompromised gut contents, deep-sea animals should be kept chilled, handled minimally, and dissected with sterile tools as soon as possible after retrieval. Sediment cores should be subsampled from the interior of the recovered mass by using a sterile device. Fluid sampling devices, from Niskin bottles to sophisticated pressure-retaining seawater samplers to thermally tolerant titanium vent samplers, have been designed to wash liberally with the desired sample prior to capturing it. Flushing with hot geothermal fluids should be particularly effective against contaminants in situ, since only very unusual indigenous microorganisms are expected to survive exposure to superheated temperatures. To guard against the unlikely, sample analyses should be as selective as possible for the molecular or metabolic types under study.

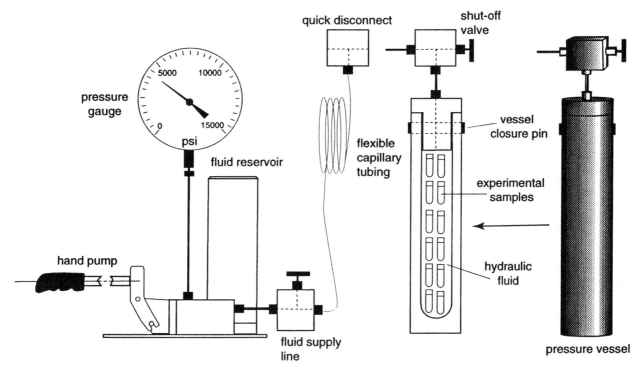

FIGURE 1 Typical equipment for conducting microbial studies under elevated hydrostatic pressure: stainless steel hand pump, gauge, and fluid reservoir (Enerpac; Division of Applied Power, Inc., Butler, Wis.); high-pressure valves and flexible capillary tubing (HIP, Erie, Pa.); and quick-disconnect, threadless pressure vessel and threadless cap (custom built by Tem-Pres Division of Leco Corporation, Blanchard, Pa., after Yayanos [82] and Yayanos and Van Boxtel [86]). See text for additional details.

SAMPLING EQUIPMENT
For Cold Deep Environments

Pressure-retaining samplers for retrieving undecompressed water samples have been developed independently by three laboratories. The multichambered stainless steel device developed in the 1970s by Jannasch and colleagues (44) collected seawater, recovered it shipboard, and allowed for injection of (radiolabeled) substrate and later removal of subsamples, all without a pressure change. Protection against pressure changes depended on gas pressure in a chamber separated from the sample by a piston floating in a buffer chamber of sterile water. This design was later refined to allow sample concentration during collection (40) and subculturing on an agar plate in a hyperbaric chamber (43).

A second pressure-retaining water sampler was developed by Colwell and colleagues in the same time frame (75) and is still in use. It operates entirely on the basis of hydrostatic pressure, generated by a hand pump (such as shown in Fig. 1), thus obviating the extra precautions required to work with gas (potentially explosive) pressure. Use of hydrostatic pressure also demands attention to safety issues, including the potential mechanical failure of a pressure fitting or user release of pressure too quickly, which can cause a spray of fluid from the pressurized chamber. The former has not occurred in research laboratories to my knowledge; pressure equipment is built to withstand pressures twice those anticipated by the user. The latter causes little damage to the operator (you get wet), though protective eye gear is advised.

A third pressure-retaining water sampler is newly devel-

oped by A. Bianchi and colleagues and described by Bianchi and Garcin (11). It borrows from the designs of its predecessors, being simpler in operation than the first ones but still dependent on gas pressure. It has not yet been deployed in the cold deep sea, but at warm (13°C) Mediterranean depths it has enabled detection of pressure-adapted microbial activity (12) (Table 1).

For microbial studies of cold deep-sea sediments without a pressure change, researchers are dependent on free-vehicle benthic landers (39, 80) or submersible operations (15). As a result, few data are available and some results remain ambiguous. The situation for studying microbial aspects of undecompressed deep-sea animals is better. A pressure-retaining trap for abyssal crustacea was developed and used successfully in the 1970s to recover and study live specimens under pressure (81) and, subsequently, to obtain the first barophilic bacterium (84). Others developed remote means to study the responses of invertebrate intestinal flora in situ (79). Methodological difficulties in discriminating microbial from eukaryotic activity, however, have apparently slowed research in this direction.

For research on decompressed deep-sea samples, a wide variety of sampling devices described in the oceanographic literature (reference 72 and citations therein) have been used by microbiologists. These include Niskin samplers for seawater (12), moored sediment traps for sinking particulates (31), large and small box corers (14) and multicorers (4) for sediments (21, 24, 54), and a variety of trawls, dredges, and tethered and free-vehicle traps for animals (30, 59).

TABLE 1 Examples of recent[a] studies of microbial community activity in the cold deep sea under in situ conditions or simulated in situ temperature and pressure

Process	Approach	Sample container[b]	Investigator(s)
Biomass production	Increase in cell number by epifluorescence microscopy	A	Deming (21); Deming and Baross (28)
		E	Lochte and Turley (54)
	Incorporation of [³H]thymidine	A	Alongi (2)
		E	Turley (78)
		E	Poremba (63)
Protein production	Incorporation of [³H]leucine	E	Turley (78)
		E	Poremba (63)
Dissolved organic carbon consumption	Incorporation and/or respiration of ¹⁴C-amino acids, glucose, acetate, etc.	A	Deming and Colwell (31)
		Corer[c]	Wirsen and Jannasch (80)
		Corer[d]	Cahet and Sibuet (15)
		Pressure-retaining sampler	Bianchi and Garcin (11, 12)
Organic carbon hydrolysis	Hydrolysis of fluorogenic substrate analogs	A	Meyer-Reil and Koster (57)
		E	Boetius and Lochte (13)
		E	Poremba (64)
		D	Deming and Baross (29)
Particulate organic carbon consumption	Incorporation and/or respiration of ¹⁴C-phytodetritus	Corer[d]	Cahet and Sibuet (15)
	Disappearance of unlabeled phytodetritus	E	Lochte and Turley (54)
		E	Poremba (62)

[a] Since 1985; see text for discussion and citations of earlier work.
[b] Letters refer to those in Fig. 2.
[c] Experiment conducted in situ by remote free-vehicle lander.
[d] Experiment conducted in situ by manned submersible.

For Hot Deep Environments

To date, the small-scale sampling of hot deep environments at submarine hydrothermal vents has required manned submersible operations. The number of microbial sampling expeditions since the discovery of hot vents in 1979 is limited (see Table 3 in reference 27). The capability for small-scale sampling via remotely operated vehicles exists but has not been used by microbiologists studying hyperthermophiles. Gross-scale sampling from a surface ship may prove rewarding, especially for tapping into the deep hot subsurface via deep-sea drilling cores (70).

Thermally tolerant sampling gear for collection of superheated geothermal fluids under deep-sea pressure was first developed by geochemists. The basic sampling instrument, a titanium syringe, draws in hot geothermal fluid in situ and returns it shipboard cooled and decompressed. Diagrams of the original titanium syringe and an expanded sampling system incorporating multiple syringes, a flowthrough system, and rigorous temperature checks on the collected samples are provided and reviewed elsewhere (7). The latter system has proved useful for the recovery of hot fluids minimally diluted with seawater (27, 73). To collect particles that may be present in superheated geothermal fluids, microbiologists have developed instruments for emplacement directly in the flow of a smoker: (i) the smoker poker, a titanium tube with a spring-loaded closure mechanism and glass mesh filling that can be inserted deep (decimeters) into the throat of a smoker (50); and (ii) the vent cap, a conical flowthrough device that is positioned atop a smoker and offers colonizing surfaces to microorganisms in the smoker plume (50). Solid sulfidic materials from smoker walls and flanges (horizontal expansions of a smoker wall) are collected directly by the submersible manipulator. More precise coring devices for smoker walls are recently available (18) but await use by microbiologists. Nevertheless, numerous hyperthermophilic

isolates and evidence for whole communities of hyperthermophilic members of the domain *Archaea* have been obtained from grab samples of smoker structures (reference 7 and citations therein). Standard submersible push-cores into the hot sediments of Guaymas Basin vents have yielded evidence for viable communities of the first known hyperthermophilic sulfate-reducing microorganisms (46).

A few sophisticated titanium samplers that tolerate smoker extremes in temperature, pressure, and chemistry and double as in situ incubators have been developed by microbiologists for smoker experiments on the seafloor. The first was designed to capture superheated fluids, release a tracer compound into the sampling chamber, and incubate the sample in the continuous flow of the smoker. Results of deployments in 1988 at Juan de Fuca smokers (7, 73) yielded indications of microbial stability (particulate DNA) under the conditions of the experiment. However, the sampler was not reliable against leaks, returning only 4 of 39 deployments with uncompromised samples. A new in situ titanium sampler-incubator with an improved closure mechanism has since been designed and built (25); it was scheduled for test dives in November 1995.

SIMULATING EXTREME ENVIRONMENTS IN THE LABORATORY

At Cold Temperatures and Elevated Pressures

Pressure-retaining water samplers also double as laboratory chambers for simulating in situ temperature and pressure conditions in the cold deep sea. With constant pressure maintained mechanically, the challenge is frequently to keep sampler temperature constant during instrument retrieval shipboard and transfer from one setting (ship deck) to the next (temperature-controlled laboratory). Temperature control during transport is typically accomplished in unsophisticated fashion, using ice-filled jackets around the sampler. The possible negative effects of temperature shifts are avoided by limiting the time the sampler spends outside a temperature-controlled chamber.

For the majority of researchers who do not have access to pressure-retaining deep-sea samplers, previously decompressed samples and cultures can be studied under in situ conditions simulated in simple, stainless steel pressure vessels (Fig. 1) immersed in temperature-controlled water baths. The earliest pressure vessels used by microbiologists involved thick threaded caps sealed to the main vessel via multiple large bolts (16, 89). Pressure was applied through tubing that also required threaded fittings to the cap. This configuration greatly limited the rate at which a vessel could be sealed or opened; many early experiments under simulated deep-sea conditions were thus based on endpoint incubations.

Important advances in pressure vessel design that streamlined culture work and enabled time course experiments were published by Yayanos (82) and Yayanos and Van Boxtel (86). The threaded cap and vessel design was replaced by a smooth-sided cap (and vessel) held in place by a smooth-sided crossbar penetrating the walls of the vessel (Fig. 1). After pressurization of the closed vessel, via introduction of a small volume of hydraulic fluid through flexible capillary tubing (Fig. 1), the pin and cap become immobilized by the pressure differential. With no pressure in the vessel, the pin slides out easily, followed by the cap. Pressurization or decompression is further facilitated by use of a quick-disconnect interface between pump and pressure vessel (Fig. 1), as described by Yayanos and Van Boxtel (86). To facilitate my pressure research in the 1980s, I worked with engineers at the Tem-Pres Division of the Leco Corporation (Blanchard, Pa.) to develop such pressure vessels and quick-disconnect devices; these can be ordered directly from the company.

The simplest means of pressurizing a vessel involves a hand pump, equipped with a high-pressure gauge for monitoring the rise or fall of pressure during operations, such as that produced by Enerpac (Division of Applied Power, Inc., Butler, Wis.) and featured in Fig. 1. However, more sophisticated electric pumps have also been used to create deep-sea hydrostatic pressures in the laboratory (87). Although early pressure studies involved the use of various hydraulic oils and risked sample contamination, distilled water (frequently sterilized, depending on the chances of sample contact) has since replaced oil as a simple, clean, effective hydraulic fluid for the type of operation depicted in Fig. 1.

Advancing beyond the approach of incubating one or more pressure vessels in a temperature-controlled water bath is the pressure-temperature gradient apparatus described by Yayanos et al. (88). This instrument accommodates eight pressure vessels housed in parallel in an insulated aluminum block. A temperature gradient is established in the block via a thermostatted heater at one end and a thermostatted cooler at the other. Large multifactored data sets can be generated relatively quickly with this approach. To date, it has been used to score bacterial growth in small tubes of liquid media placed along the temperature gradient and colony formation in long glass tubes filled with solid media (32, 83, 88). The potential for additional types of assays seems promising.

When a microbial sample is pressurized, it must be isolated from the surrounding hydraulic fluid in a pressure-responsive container to avoid container collapse. Either the entire container or some part of the container, such as a mobile piston or pliant cap, must be flexible in response to pressure. In all cases, air bubbles must be expelled from the container prior to sealing. Their presence can cause hyperbaric gas toxicity to the test organisms and leaks in containers as the gas expands and contracts during sample pressurization and decompression (unlike pressurizing gas, pressurizing a liquid reduces its volume by only a small degree). Examples of containers used by a wide cross section of researchers working on microbial issues in the cold deep sea include (i) a modified plastic syringe with a mobile, rubber-capped syringe plunger (Fig. 2A); (ii) a plastic test tube, sealed with stretched Parafilm, or a glass test tube, sealed with a rubber stopper that is free to move into the tube (Fig. 2B); (iii) an open-ended glass tube sealed at both ends with a rubber stopper free to move into the tube (Fig. 2C); (iv) an open-ended plastic tube, sealed as in the model in Fig. 2C, with predrilled injection ports sealed individually (with a silicon-based glue) or collectively (by heat shrinking a plastic sleeve around the injection core) (Fig. 2D); (v) a polyethylene bag, heat sealed after sample loading (Fig. 2E); and (vi) an agar-filled plastic pouch, heat sealed after inoculation (Fig. 2F).

Modified syringes were the early choice and have been used extensively by many researchers (Table 1), especially for work at sea: they are calibrated and can double as the subsampling device. Note that if a Becton Dickinson syringe is used, the air space in the rubber cap of the syringe plunger should be displaced with sterile water (by injecting from the back end of the syringe) to guard against cap collapse and leakage under pressure. Glass or plastic test tubes, which do not double as subsampling devices at sea, have proved a simpler alternative in laboratory work with pure cultures in liquid media (32, 68, 88). A sterile glass bead added to a modified syringe or test tube enables sample mixing without decompression: the pressurized vessel can be rocked automatically (82) or inverted manually to effect the mixing. Test tubes are the container of choice when the solid silica gel medium of Dietz and Yayanos (34) is used. Open-ended glass tubes have also been used effectively with solid media and pure cultures in the pressure-temperature gradient apparatus (88) mentioned above. Open-ended plastic tubes with predrilled, sealed injection ports and a beveled edge at one end, much like those developed by Meyer-Reil (56) for coastal work, have been used at sea as both subsampling devices for abyssal sediments and subsequent experimental containers (29, 66). Radiolabeled or fluorogenic substrates can be injected into sediment horizons, and the whole subcore can be incubated under simulated deep-sea conditions. Prior to pressurization, the injected subcore is wrapped its full length with Parafilm to protect against accidental leakage through an injection port under pressure.

Polyethylene bags have been used to good advantage by European researchers for both field work (Table 1) and pure culture (36). In this approach, the entire container flexes under pressure. An adaptation of the bag approach was developed recently by Yoshida's group in Japan (59). In this elaborate method, used to isolate barophiles from deep-sea fish, a plano-convex agar lens is inoculated on the planar surface, overlaid with a second agar lens, and heat sealed in a plastic pouch for subsequent pressurization. Colonies form at the interface of the two lenses.

Other clever sample containers not featured in Fig. 2 have also been developed (53, 74). For example, in some morphological or rapid reaction studies in which the sample may need to be fixed or poisoned under pressure, the Lan-

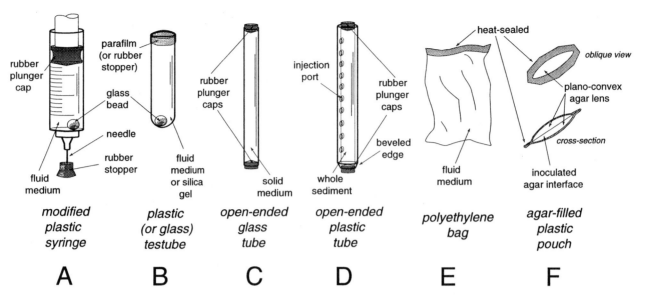

FIGURE 2 Examples of sample containers used under elevated hydrostatic pressure, i.e., with pressure-responsive (movable) parts (A to D) or a completely flexible design (E and F). A and D double as sample collectors in the field for fluid and sediment, respectively. B and C (when filled with solid media [34, 88]) and F (59) allow for colony formation under pressure. C, which can be of any length, is used in the pressure-temperature gradient instrument of Yayanos et al. (88). See Table 1 for examples of container use in selected studies and text for additional details.

dau-Thibodeau method originally devised for the study of pressure-sensitive nonmarine microorganisms (53) is available. A two-chambered metal device allows the process of interest to occur in one chamber separated from fixative or poison in the other chamber by a glass coverslip. The pressurized vessel is inverted sharply, causing a small steel ball in the second chamber to break the glass partition and mix fixative or poison with the culture or reaction solution. Samples are then decompressed for processing. The use of this approach to examine ultrastructural effects of decompression on an obligately barophilic hadal bacterium revealed that morphological aberrations due to decompression are time dependent (17); i.e., there are no immediate effects.

At High Temperatures and Pressures

Fewer laboratories worldwide are equipped to study microbial processes under simulated temperatures and pressures of the hot deep sea. The material complications at high temperatures are greatly magnified compared with work at cold temperatures. Pressure vessels and sampling containers must be heat resistant and frequently resistant to acidic or reducing conditions. Pressure relief valves become essential safety features, since pressure increases rapidly in a closed system as the temperature is increased. Liquids also vaporize under pressure, depending on the temperature and chemistry of the solution; possible phase transitions must be anticipated and either avoided or monitored. All existing high-temperature, high-pressure systems are custom built; researchers must rely on knowledgeable engineers to build effective systems and establish appropriate safety protocols. The field of high-temperature, high-pressure research appears to be expanding in parallel with keen communitywide interests in the phylogeny, biochemistry, molecular genetics, and biotechnological applications of hyperthermophiles (1, 7, 27, 46, 60).

Two types of laboratory systems have been constructed and used in studies of microorganisms from the hot deep sea: (i) a main pressure vessel incubated at one combination of temperature and pressure conditions and subsampled over time without altering incubation conditions of the sample and (ii) multiple pressure vessels incubated at different temperatures (in one or more ovens) for endpoint experiments. The first of the time course systems was developed by Yayanos and colleagues (87) and used by others (6, 8) to culture hyperthermophiles at superheated temperatures. A second time course system was developed by myself and colleagues (51). It has been used primarily to test the suitability (thermal stability) of selected organic molecules (51, 65) for experiments at superheated temperatures on the seafloor (25). A third, developed by J. A. Baross and colleagues, is in use with cultures of deep-sea hyperthermophiles (37). These three systems are designed to accommodate experimental temperatures to 350°C and pressures to about 600 bar. A fourth system, developed by D. S. Clark and colleagues, has been used very effectively in the study of methanogenesis under milder (<120°C) high-temperature conditions (reference 60 and citations therein). Of these systems, the second and third were built commercially, by the Tem-Pres Division of Leco Corporation; design specifications and models are available directly from Tem-Pres.

The most rapid progress in the study of pressure effects on cultures of deep-sea heterotrophic hyperthermophiles has come from the use of individual pressure vessels and ovens for endpoint experiments (see Table 2 in reference 27 and citations therein). The same vessels used in cold deep-sea research (as manufactured by Tem-Pres) can be used to temperatures of about 200°C. Any standard oven large enough to accommodate the vessels can provide the desired temperature. A setup similar to that shown in Fig. 1, but with pump and gauge mounted outside of the oven, was used to document colony formation at 120°C under deep-sea pressure (see reference 26 for further operational details). An important improvement on this system is the "wine rack," designed by Baross and colleagues and described by Baross and Deming (7). The device accommodates eight pressure vessels in separately insulated and temperature-regulated chambers, such that a matrix of temperatures and pressures can be tested in a single experiment. Pressure is created and monitored as in Fig. 1, but temperature is computer controlled by independent thermoregulators inserted into each vessel. This system has been used to establish barophily in a deep-sea hyperthermophile, quantify pressure-induced upward shifts in optimal temperature and growth rate, and document novel stress responses enabling hyperthermophilic survival under deep-sea pressures (references 7, 27, and 37 and citations therein). The key to success in these studies was use of a wider range of test pressures than usual, including pressures expected in the subsurface biosphere, well below vent features on the seafloor (27).

Pressure-responsive sample containers for use at high temperatures resemble designs shown in Fig. 2A and E but are constructed of heat- and acid-tolerant materials. If only modest (relative to smoker) temperatures (<120°C) are tested, then modified (autoclavable) plastic syringes (Fig. 2A) are still usable, though not repeatedly. Glass syringes, with silicon grease to facilitate movement of the frosted glass plunger under pressure, can be used to about 200°C. Higher temperatures require custom-built titanium syringes with thermally tolerant O-rings on the plunger. The flexible bag approach (Fig. 2E) can be used at the highest temperatures achievable in the existing laboratory systems if gold is substituted for plastic (51).

SAMPLE ANALYSIS
For Isolation and Culturing

A variety of culturing media and techniques have been developed specifically for high-pressure deep-sea microbial research. For the cold deep sea, the repertoire is somewhat limited to enrichment media selective for aerobic (or facultatively anaerobic) heterotrophs. Other nutritional types of psychrophilic barophiles have not yet been isolated, although their existence from community analyses is clear (19, 20). Development of the solid silica gel medium for

heterotrophs (34) marked a turning point for cold deep-sea research, since it enabled colony formation under pressure and isolation of the first barophile (84). The advantage of the medium is that it can be inoculated as a liquid (salt-free silica sol) and solidified later by adding sea salts. The whole process can be conducted at a cold temperature so that the inoculum is never exposed to warming, as would be the case with an agar-based medium. This medium was used to advantage in characterizing cardinal growth temperatures of new barophiles in the years following its development (32, 33, 83, 88), but researchers today appear to be either developing new solid-medium approaches for isolations by colony formation under pressure (59) or studying existing cultures in standard marine broth media (reference 29 and citations therein). A typical marine broth is based on peptone and yeast extract (Difco); the silica gel medium can be used with a variety of organic substrates but has been optimized for tryptone, glucose, and yeast extract (34).

A wider menu of culturing media and protocols has been developed for hyperthermophiles from the hot deep sea, in keeping with a more intense interest in the diversity of known hot reducing habitats. Several recent publications provide complete details on medium preparation, inoculation, and general anaerobic technique at high temperatures, as well as culture maintenance protocols for the range of known deep-sea hyperthermophiles (5, 7, 69). Each nutritional group and many individual species have very specific trace nutrient requirements, including metals, sulfur compounds, and organic compounds. For example, the new researcher should be aware of unusual requirements for tungsten at concentrations much higher than would be provided in a standard trace metal mixture (5, 7). The range of hyperthermophiles isolated from the hot deep sea include sulfur-dependent heterotrophs, non-sulfur-dependent heterotrophs, and methanogens. All are marine in origin, as indicated by a strict salt requirement; none to date are obligate barophiles, but each shows some form of positive response to in situ pressure (27). Because of thermostability problems, few studies of deep-sea hyperthermophiles rely on solid media. However, a solid medium based on the use of thermally stable GELRITE (Kelco Div. of Merck & Co., Inc., San Diego, Caif.) has been described and used to temperatures of 120°C under elevated pressure (26).

TABLE 2 Examples of microbial community measurements in the hot (>90°C) deep sea under in situ conditions or simulated in situ temperature and pressure[a]

Process	Approach	Sampling device or incubator	Investigator(s)
Cell survival[b]	ATP assay[c]	Smoker poker	Karl et al. (49)
	Surface colonization[c]	Vent cap	Karl et al. (50)
	DNA extraction and microscopy	Ti syringe and in situ Ti sampler-incubator	Straube et al. (73); Baross and Deming (7)
	Lipid analysis and subculturing	Grab sample of flange	Hedrick et al. (35); Baross and Deming (7)
	[35S]sulfate reduction[d]	Push-cores	Jorgensen et al. (46)
Biomass production	Increase in hydrolyzable protein	Ti syringe in pressure vessel	Baross and Deming (6); Baross et al. (8)
	Incorporation of [3H]thymidine	Glass syringes in pressure vessels	Deming (23)
	Increase in extractable DNA	In situ Ti sampler-incubator	Baross and Deming (7); Deming (25)

[a] See reference 7 for a comparative review of experimental details and results.
[b] Incubation not required.
[c] Negative results at superheated temperatures.
[d] Measured at high temperature but atmospheric pressure after sample recovery.

For Community Measurements

The direct analysis of microbial communities in samples from the cold deep sea (without sample incubation) has involved most of the obvious techniques available to microbial ecologists. For example, epifluorescence microscopy (reference 29 and citations therein), the firefly luciferase ATP assay (48), and lipid extractions (3) have been used to assess microbial biomass in deep-sea sediments. Measurements of microbial activity requiring sample manipulation and an experimental incubation period have also involved most of the standard techniques regularly applied to milder environments. Recent examples of processes measured, analytical approaches taken, and types of sample containers used during pressurized incubations are summarized in Table 1, along with citations providing further methodological details. Inspection of authorship (Table 1) reveals a European revival of this type of experimental research on microbial communities from the cold deep sea.

Attempts to analyze microbial biomass or measure microbial activity in hot deep-sea samples and, by implication, in a deep hot subsurface biosphere have been plagued by technical difficulties and controversy over interpretation of results. The difficulties and debate stem from problems of contamination, as discussed earlier and in detail by Karl et al. (50), Straube et al. (74), and Deming and Baross (27), but also from the prevailing wisdom that microorganisms should not be capable of withstanding, much less responding favorably to, the superheated temperatures of submarine geothermal fluids. The issues surrounding the debate have been addressed recently (7). Here is provided instead, for practical interest, a list of the several experiments that have been attempted with hot deep-sea samples, either in situ or under simulated in situ conditions, according to the analytical approaches taken, types of sample containers used during pressurized incubations, and relevant citations (Table 2).

LIMITATIONS YET TO BE OVERCOME

The outstanding challenge to all microbial ecologists is to find ways to measure in situ microbial activity noninvasively. This goal has not been achieved for any environment, much less for extreme ones. The problem lies with available approaches to activity measurements and not with sampling restrictions. Sophisticated pressure-retaining devices have been developed for deep-sea microbiology, yet the samples obtained have been analyzed only by traditional means, e.g., viable plate counts or radiolabeled tracer experiments, both of which are invasive and selective. The clever investigator can still learn much about the microbiology of an extreme environment by using existing equipment and (invasive) laboratory techniques. This is especially true for the understudied deep sea, as evidenced by the recent work of new researchers in the field, especially on the topic of extracellular enzyme activity (Table 1).

The future predictably lies with the noninvasive application of molecular techniques to deep-sea samples. The ideal might involve fixing samples in situ (no chance for negative or selective effects of shifts in temperature or pressure during recovery) and then analyzing them creatively, using molecular approaches that reveal not only the phylogenetic diversity of the microbial community but also the dominant functional genes being expressed in situ at the time of sample fixation. New discoveries on the phylogenetic diversity of both the cold and hot deep sea are happening now (55, 58, 67). Measures of functional gene expression, all-important to understanding ecological aspects of the deep sea, await

further developments in molecular environmental biotechnology.

Preparation of this chapter was supported in part by an NSF Presidential Young Investigator award.

I am grateful to Shelly Carpenter for preparation of the figures and to John Baross for input on an earlier draft.

REFERENCES

1. **Adams, M. W. W.** 1994. Biochemical diversity among sulfur-dependent hyperthermophilic microorganisms. *FEMS Microbiol. Rev.* **15:**261–277.
2. **Alongi, D. M.** 1990. Bacterial growth rates, production and estimates of detrital carbon utilization in deep-sea sediments of the Solomon and Coral Seas. *Deep-Sea Res.* **37:**731–746.
3. **Baird, B. H., D. E. Nivens, J. H. Parker, and D. C. White.** 1985. The biomass, community structure, and spatial distribution of the sedimentary microbiota from a high-energy area of the deep sea. *Deep-Sea Res.* **32:**1089–1099.
4. **Barnett, P. R. O., J. Watson, and D. Connelly.** 1984. A multiple corer for taking virtually undisturbed samples from shelf, bathyal and abyssal sites. *Oceanol. Acta* **7:**399–408.
5. **Baross, J. A.** 1993. Isolation and cultivation of hyperthermophilic bacteria from marine and freshwater habitats, p. 21–30. *In* P. F. Kemp, B. F. Sherr, E. B. Sherr, and J. J. Cole (ed.), *Handbook of Methods in Aquatic Microbial Ecology.* Lewis Publishers, Boca Raton, Fla.
6. **Baross, J. A., and J. W. Deming.** 1983. Growth of "black smoker" bacteria at temperatures of at least 250°C. *Nature* (London) **303:**423–426.
7. **Baross, J. A., and J. W. Deming.** 1995. Growth at high temperatures: isolation and taxonomy, physiology, and ecology, p. 169–217. *In* D. M. Karl (ed.), *The Microbiology of Deep-Sea Hydrothermal Vents.* CRC Press, New York.
8. **Baross, J. A., J. W. Deming, and R. R. Becker.** 1984. Evidence for microbial growth in high pressure, high temperature environments, p. 186–195. *In* M. J. Klug and C. A. Reddy (ed.), *Current Perspectives in Microbial Ecology: Third International Symposium on Microbial Ecology.* American Society for Microbiology, Washington, D.C.
9. **Bartlett, D., M. Wright, A. A. Yayanos, and M. Silverman.** 1989. Isolation of a gene regulated by hydrostatic pressure in a deep-sea bacterium. *Nature* (London) **342:**572–574.
10. **Betlach, M. C., and R. F. Shand.** 1995. Growth of halophilic archaebacteria under conditions of low oxygen tension and high light intensity, p. 17–20. *In* F. T. Robb, K. R. Sowers, S. DasSarma, A. R. Place, H. J. Schreier, and E. M. Fleischmann (ed.), *Archaea, a Laboratory Manual.* Cold Spring Harbor Laboratory Press, Plainview, N.Y.
11. **Bianchi, A., and J. Garcin.** 1993. In stratified waters the metabolic rate of deep-sea bacteria decreases with decompression. *Deep-Sea Res.* **40:**1703–1710.
12. **Bianchi, A., and J. Garcin.** 1994. Bacterial response to hydrostatic pressure in seawater samples collected in mixed-water and stratified-water conditions. *Mar. Ecol. Prog. Ser.* **111:**137–141.
13. **Boetius, A., and K. Lochte.** 1994. Regulation of microbial enzymatic degradation of organic matter in deep-sea sediments. *Mar. Ecol. Prog. Ser.* **104:**299–307.
14. **Boland, G. S., and G. T. Rowe.** 1991. Deep-sea benthic sampling with the GOMEX box corer. *Limnol. Oceanogr.* **36:**1015–1020.
15. **Cahet, G., and M. Sibuet.** 1986. Activité biologique en domaine profond: transformations biochimiques in situ de composes organiques marques au carbone-14 a l'interface eau-sediment par 2000 m de profondeur dans le golfe de Gascogne. *Mar. Biol.* **90:**307–315.

16. **Certes, A.** 1884. Sur la culture, a l'abri des germes atmospheriques, des eaux et des sediments rapportes par les expeditions du Traveilleur et du Talisman, 1882–1883. *C. R. Acad. Sci.* **98:**690–693.

17. **Chastain, R. A., and A. A. Yayanos.** 1991. Ultrastructural changes in an obligately barophilic marine bacterium after decompression. *Appl. Environ. Microbiol.* **57:**1489–1497.

18. **Cook, T. L., and D. S. Stakes.** 1995. Biogeological mineralization in deep-sea hydrothermal deposits. *Science* **267:** 1975–1979.

19. **Cowen, J. P.** 1989. Positive pressure effect on manganese binding by bacteria in deep-sea hydrothermal plumes. *Appl. Environ. Microbiol.* **55:**764–766.

20. **de Angelis, M. A., J. A. Baross, and M. D. Lilley.** 1991. Enhanced microbial methane oxidation in water from a deep-sea hydrothermal vent field at simulated in situ hydrostatic pressures. *Limnol. Oceanogr.* **36:**565–570.

21. **Deming, J. W.** 1985. Bacterial growth in deep-sea sediment trap and boxcore samples. *Mar. Ecol. Prog. Ser.* **25:** 305–312.

22. **Deming, J. W.** 1986. Ecological strategies of barophilic bacteria in the deep ocean. *Microbiol. Sci.* **3:**205–211.

23. **Deming, J. W.** 1987. Thermophilic bacteria associated with black smokers along the East Pacific Rise, p. 325–332. *In Deuxieme Colloque International de Bacteriologie Marine, IFREMER, Actes de Colloques 3.* Centre National de la Recherche Scientifique, Brest, France.

24. **Deming, J. W.** 1993. ^{14}C tracer method for measuring microbial activity in deep-sea sediments, p. 405–414. *In P. F. Kemp, B. F. Sherr, E. B. Sherr, and J. J. Cole (ed.), Handbook of Methods in Aquatic Microbial Ecology.* Lewis Publishers, Boca Raton, Fla.

25. **Deming, J. W.** 1994. The LAREDO sampler: last attempt at remote experimentation in the deep ocean, abstr. R-07. *In Abstracts of the Oceanography Society Meeting.* The Oceanography Society, Virginia Beach, Va.

26. **Deming, J. W., and J. A. Baross.** 1986. Solid medium for culturing black smoker bacteria at temperatures to 120°C. *Appl. Environ. Microbiol.* **51:**238–243.

27. **Deming, J. W., and J. A. Baross.** 1993. Deep-sea smokers: windows to a subsurface biosphere? *Geochim. Cosmochim. Acta* **57:**3219–3230.

28. **Deming, J. W., and J. A. Baross.** 1993. The early diagenesis of organic matter: bacterial activity, p. 119–144. *In M. H. Engel and S. A. Macko (ed.), Organic Geochemistry.* Plenum Press, New York.

29. **Deming, J. W., and J. A. Baross.** Survival, dormancy and non-culturable cells in extreme deep-sea environments. *In R. R. Colwell and D. J. Grimes (ed.), Non-Culturable Microorganisms in the Environment,* in press. Chapman & Hall, New York.

30. **Deming, J. W., and R. R. Colwell.** 1981. Barophilic bacteria associated with deep-sea animals. *BioScience* **31:** 507–511.

31. **Deming, J. W., and R. R. Colwell.** 1985. Observations of barophilic microbial activity in samples of sediment and intercepted particulates from the Demerara Abyssal Plain. *Appl. Environ. Microbiol.* **50:**1002–1006.

32. **Deming, J. W., H. Hada, R. R. Colwell, K. R. Luehrsen, and G. E. Fox.** 1984. The ribonucleotide sequence of 5S rRNA from two strains of deep-sea barophilic bacteria. *J. Gen. Microbiol.* **130:**1911–1920.

33. **Deming, J. W., L. K. Somers, W. L. Straube, D. G. Swartz, and M. T. MacDonell.** 1988. Isolation of an obligately barophilic bacterium and description of a new genus, *Colwellia* gen. nov. *Syst. Appl. Microbiol.* **10:**152–160.

34. **Dietz, A. S., and A. A. Yayanos.** 1978. Silica gel media for isolating and studying bacteria under hydrostatic pressure. *Appl. Environ. Microbiol.* **36:**966–968.

35. **Hedrick, D. B., R. J. Pledger, D. C. White, and J. A. Baross.** 1992. In situ microbial ecology of hydrothermal vent sediments. *FEMS Microb. Ecol.* **101:**1–10.

36. **Helmke, E., and H. Weyland.** 1986. Effect of hydrostatic pressure and temperature on the activity and synthesis of chitinases of Antarctic Ocean bacteria. *Mar. Biol.* **91:**1–7.

37. **Holden, J. F., and J. A. Baross.** 1995. Enhanced thermotolerance by hydrostatic pressure in the deep-sea hyperthermophile Pyrococcus strain ES4. *FEMS Microb. Ecol.* **18:** 27–34.

38. **Jaenicke, R.** 1987. Cellular components under extremes of pressure and temperature: structure-function relationship of enzymes under pressure, p. 257–272. *In H. W. Jannasch, R. E. Marquis, and A. M. Zimmerman (ed.), Current Perspectives in High Pressure Biology.* Academic Press, London.

39. **Jahnke, R. A., and M. B. Christiansen.** 1989. A free-vehicle benthic chamber instrument for sea floor studies. *Deep-Sea Res.* **36:**625–637.

40. **Jannasch, H. W., and C. O. Wirsen.** 1977. Retrieval of concentrated and undecompressed microbial populations from the deep sea. *Appl. Environ. Microbiol.* **33:**642–646.

41. **Jannasch, H. W., and C. O. Wirsen.** 1981. Morphological survey of microbial mats near deep-sea thermal vents. *Appl. Environ. Microbiol.* **41:**528–538.

42. **Jannasch, H. W., and C. O. Wirsen.** 1982. Microbial activities in undecompressed and decompressed deep-seawater samples. *Appl. Environ. Microbiol.* **43:**1116–1124.

43. **Jannasch, H. W., C. O. Wirsen, and C. D. Taylor.** 1982. Deep-sea bacteria: isolation in the absence of decompression. *Science* **216:**1315–1317.

44. **Jannasch, H. W., C. O. Wirsen, and C. L. Winget.** 1973. A bacteriological, pressure-retaining, deep-sea sampler and culture vessel. *Deep-Sea Res.* **20:**661–664.

45. **Jones, W. J., J. A. Leigh, F. Mayer, C. R. Woese, and R. S. Wolfe.** 1983. *Methanococcus jannaschii* sp. nov., an extremely thermophilic methanogen from a submarine hydrothermal vent. *Arch. Microbiol.* **136:**254–261.

46. **Jorgensen, B. B., M. F. Isaksen, and H. W. Jannasch.** 1992. Bacterial sulfate reduction above 100°C in deep-sea hydrothermal vent sediments. *Science* **258:**1756–1757.

47. **Jumars, P. A., L. M. Mayer, J. W. Deming, J. A. Baross, and R. A. Wheatcroft.** 1990. Deep-sea deposit-feeding strategies suggested by environmental and feeding constraints. *Philos. Trans. R. Soc. Lond. Ser. A* **331:**85–101.

48. **Karl, D. M.** 1978. Distribution, abundance, and metabolic states of microorganisms in the water column and sediments of the Black Sea. *Limnol. Oceanogr.* **23:**936–949.

49. **Karl, D. M., D. J. Burns, K. Orrett, and H. W. Jannasch.** 1984. Thermophilic microbial activity in samples from deep-sea hydrothermal vents. *Mar. Biol. Lett.* **5:**227–231.

50. **Karl, D. M., G. T. Taylor, J. A. Novitsky, H. W. Jannasch, C. O. Wirsen, N. R. Pace, D. J. Lane, G. J. Olsen, and S. J. Giovannoni.** 1988. A microbiological study of Guaymas Basin high temperature hydrothermal vents. *Deep-Sea Res.* **35:**777–791.

51. **Kelly, R. M., and J. W. Deming.** 1988. Extremely thermophilic archaebacteria: biological and engineering considerations. *Biotechnol. Prog.* **4:**47–62.

52. **Kushner, D. J. (ed.).** 1978. *Microbial Life in Extreme Environments.* Academic Press, New York.

53. **Landau, J. V., and L. Thibodeau.** 1962. The micromorphology of *Amoeba proteus* during pressure-induced changes in the sol-gel cycle. *Exp. Cell Res.* **27:**591–594.

54. **Lochte, K., and C. M. Turley.** 1988. Bacteria and cyanobacteria associated with phytodetritus in the deep sea. *Nature* (London) **333:**67–69.

55. **McInerney, J. O., M. Wilkinson, J. W. Patching, T. M. Embley, and R. Powell.** 1995. Recovery and phylogenetic

analysis of novel archaeal rRNA sequences from a deep-sea deposit feeder. *Appl. Environ. Microbiol.* **61**:1646–1648.

56. **Meyer-Reil, L.-A.** 1986. Measurement of hydrolytic activity and incorporation of dissolved organic substrates by microorganisms in marine sediments. *Mar. Ecol. Prog. Ser.* **31**:143–149.

57. **Meyer-Reil, L.-A., and M. Koster.** 1992. Microbial life in pelagic sediments: the impact of environmental parameters on enzymatic degradation of organic material. *Mar. Ecol. Prog. Ser.* **81**:65–72.

58. **Moyer, C. L., F. C. Dobbs, and D. M. Karl.** 1995. Phylogenetic diversity of the bacterial community from a microbial mat at an active, hydrothermal vent system, Loihi Seamount, Hawaii. *Appl. Environ. Mcriobiol.* **61**:1555–1562.

59. **Nakayama, A., Y. Yano, and K. Yoshida.** 1994. New method for isolating barophiles from intestinal contents of deep-sea fishes retrieved from the abyssal zone. *Appl. Environ. Microbiol.* **60**:4210–4212.

60. **Nelson, C. M., M. R. Schuppenhauer, and D. S. Clark.** 1992. High-pressure, high-temperature bioreactor for comparing effects of hyperbaric and hydrostatic pressure on bacterial growth. *Appl. Environ. Microbiol.* **58**:1789–1793.

61. **Oliver, J. D.** 1993. Formation of viable but nonculturable cells, p. 239–272. *In* S. Kjelleberg (ed.), *Starvation in Bacteria.* Plenum Press, New York.

62. **Poremba, K.** 1994. Simulated degradation of phytodetritus in deep-sea sediments of the NE Atlantic (47°N, 19°W). *Mar. Ecol. Prog. Ser.* **105**:291–299.

63. **Poremba, K.** 1994. Impact of pressure on bacterial activity in water columns situated at the European continental margin. *Netherlands J. Sea Res.* **33**(10):29–35.

64. **Poremba, K.** 1995. Hydrolytic enzymatic activity in deep-sea sediments. *FEMS Microb. Ecol.* **16**:213–222.

65. **Qian, Y., M. H. Engel, S. A. Macko, S. Carpenter, and J. W. Deming.** 1993. Kinetics of peptide hydrolysis and amino acid decomposition at high temperature. *Geochim. Cosmochim. Acta* **57**:1271–1274.

66. **Relexans, J.-C., J. W. Deming, A. Dinet, J.-F. Gaillard, and M. Sibuet.** Sedimentary organic matter and micro-meiobenthos with relation to trophic conditions in the northeast tropical Atlantic. *Deep-Sea Res.*, in press.

67. **Reysenbach, A.-L., S. Rehm, N. R. Pace, P. S. Kessler, S. L. Carpenter, and J. W. Deming.** 1994. Molecular phylogeny of hyperthermophilic microorganisms from seafloor and inferred subsurface smoker environments, abstr. The Oceanography Society, Virginia Beach, Va. R-09. *In Abstracts of the Oceanography Society Meeting.* The Oceanography Society, Virginia Beach, Va.

68. **Rice, S. A., and J. D. Oliver.** 1992. Starvation response of the marine barophile CNPT-3. *Appl. Environ. Microbiol.* **58**:2432–2437.

69. **Robb, F. T., K. R. Sowers, S. DasSarma, A. R. Place, H. J. Schreier, and E. M. Fleischmann (ed.).** 1995. *Archaea, a Laboratory Manual.* Cold Spring Harbor Laboratory Press, Plainview, N.Y.

70. **Rochelle, P. A., J. C. Fry, R. J. Parkes, and A. J. Weightman.** 1992. DNA extraction for 16S rRNA gene analysis to determine genetic diversity in deep sediment communities. *FEMS Microbiol. Lett.* **100**:59–66.

71. **Rodriguez-Valera, F.** 1995. Cultivation of halophilic Archaea, p. 13–16. *In* F. T. Robb, K. R. Sowers, S. Das-Sarma, A. R. Place, H. J. Schreier, and E. M. Fleischmann (ed.), *Archaea, a Laboratory Manual.* Cold Spring Harbor Laboratory Press, Plainview, N.Y.

72. **Rowe, G. T., and M. Sibuet.** 1983. Recent advances in instrumentation in deep-sea biological research, p. 81–95. *In* G. T. Rowe (ed.), *The Sea*, vol. 8. *Deep-Sea Biology.* John Wiley & Sons, New York.

73. **Straube, W. L., J. W. Deming, C. C. Somerville, R. R. Colwell, and J. A. Baross.** 1990. Particulate DNA in smoker fluids: evidence for existence of microbial populations in hot hydrothermal systems. *Appl. Environ. Microbiol.* **56**:1440–1447.

74. **Straube, W. L., M. O'Brien, K. Davis, and R. R. Colwell.** 1990. Enzymatic profiles of 11 barophilic bacteria under in situ conditions: evidence for pressure modulation of phenotype. *Appl. Environ. Microbiol.* **56**:812–814.

75. **Tabor, P. S., J. W. Deming, K. Ohwada, H. Davis, M. Waxman, and R. R. Colwell.** 1981. A pressure-retaining deep ocean sampler for measurement of microbial activity in the deep sea. *Microb. Ecol.* **7**:51–65.

76. **Taylor, C. D.** 1979. Growth of a bacterium under a high-pressure oxyhelium atmosphere. *Appl. Environ. Microbiol.* **37**:42–49.

77. **Tuovila, B. J., F. C. Dobbs, P. S. LaRock, and B. Z. Siegel.** 1987. Preservation of ATP in hypersaline environments. *Appl. Environ. Microbiol.* **53**:2749–2753.

78. **Turley, C. M.** 1993. The effect of pressure on leucine and thymidine incorporation by free-living bacteria and by bacteria attached to sinking oceanic particles. *Deep-Sea Res.* **40**:2193–2206.

79. **Wirsen, C. O., and H. W. Jannasch.** 1983. In situ studies on deep-sea amphipods and their intestinal microflora. *Mar. Biol.* **78**:69–73.

80. **Wirsen, C. O., and H. W. Jannasch.** 1986. Microbial transformations in deep-sea sediments: free-vehicle studies. *Mar. Biol.* **91**:277–284.

81. **Yayanos, A. A.** 1978. Recovery and maintenance of live amphipods at a pressure of 580 bars from an ocean depth of 5700 meters. *Science* **200**:1056–1059.

82. **Yayanos, A. A.** 1982. Deep-sea biophysics, 409–416. *In Subseabed Disposal Program Annual Report January to September 1981*, vol. II. *Appendices (Principal Investigator Progress Reports)*, part 2 of 2. Sandia National Laboratories, Albuquerque, N. Mex.

83. **Yayanos, A. A., and E. F. DeLong.** 1987. Deep-sea bacterial fitness to environmental temperatures and pressures, p. 17–32. *In* H. W. Jannasch, R. E. Marquis, and A. M. Zimmerman (ed.), *Current Perspectives in High Pressure Biology.* Academic Press, London.

84. **Yayanos, A. A., A. S. Dietz, and R. Van Boxtel.** 1979. Isolation of a deep-sea barophilic bacterium and some of its growth characteristics. *Science* **205**:808–810.

85. **Yayanos, A. A., A. S. Dietz, and R. Van Boxtel.** 1981. Obligately barophilic bacterium from the Mariana Trench. *Proc. Natl. Acad. Sci. USA* **78**:5212–5215.

86. **Yayanos, A. A., and R. Van Boxtel.** 1982. Coupling device for quick high-pressure connections to 100 MPa. *Rev. Sci. Instrum.* **53**(5):704–705.

87. **Yayanos, A. A., R. Van Boxtel, and A. S. Dietz.** 1983. Reproduction of *Bacillus stearothermophilus* as a function of temperature and pressure. *Appl. Environ. Microbiol.* **46**:1357–1363.

88. **Yayanos, A. A., R. Van Boxtel, and A. S. Dietz.** 1984. High-pressure-temperature gradient instrument: use for determining the temperature and pressure limits of bacterial growth. *Appl. Environ. Microbiol.* **48**:771–776.

89. **ZoBell, C. E., and R. Y. Morita.** 1957. Barophilic bacteria in some deep sea sediments. *J. Bacteriol.* **73**:563–568.

SOIL, RHIZOSPHERE, AND PHYLLOSPHERE

VOLUME EDITOR
GUY R. KNUDSEN

SECTION EDITOR
LINDA THOMASHOW

Introduction: Soil, Rhizosphere, and Phyllosphere

GUY R. KNUDSEN

40

We are only beginning to comprehend the abundance and diversity of the soil microbiota. Yet although soil microorganisms remain unseen and underappreciated by most, all life on earth depends on their activities. As the human species continues to have a profound and sometimes devastating impact on this planet, it is becoming clear that good stewardship of agricultural, forest, and natural resources will require a deeper understanding of soil microbial communities. The 18 chapters in this section present a diversity of general methodologies and specific techniques suitable for the study of those microorganisms associated with soil and with plants. It is hoped that this volume will prove useful to researchers and students of soil science, plant pathology, agronomy, soil biochemistry, biological control, and numerous related disciplines. Because microbial ecology is advancing rapidly not only in its technology but also conceptually (2), we have endeavored to provide some linkage between the well-established techniques and those that have only recently become available.

Where possible, attempts have been made to avoid parochialism and to integrate concepts and methods from soil microbiology and plant microbiology, disciplines which have been too much isolated from one another. Historically, methodologies for the study of soil microbes have evolved somewhat separately from the study of plant-associated microbes, often deriving from a different scientific perspective and with different research goals. As Paul and Clark (3) noted, the primary emphasis of soil microbiology has been on the metabolic activities of soil-inhabiting organisms, especially their roles in the energy flow and nutrient cycling associated with primary productivity. Much of the research approach has focused on biochemical processes, including microbial roles in the processes of biogeochemical cycles, i.e., carbon mineralization, nitrogen fixation, nitrification and denitrification, and sulfur oxidation. In contrast, approaches to the study of rhizosphere and phyllosphere microbes have, quite naturally, tended to focus on plant-microbe interactions, especially mutualistic relationships (e.g., mycorrhizae and *Rhizobium*-legume symbioses) and the vast array of plant diseases that are of microbial origin. By definition, the rhizosphere and phyllosphere do not lend themselves to pure culture studies, and there has been less attention to measurement of metabolic activities of plant-associated microorganisms than to measurement of activities of microorganisms in soil. A wide array of traditional and novel methods is now available to investigate the activities of microbes in soil and plant habitats. The chapters that follow are intended to provide critical analyses of the advantages and disadvantages of most of the important techniques in soil and plant microbiology, to introduce new technological applications to researchers in these fields, and to provide a foundation for interpretation of experimental results. Because space limitations frequently preclude detailed protocols, extensive citation of the most recent and relevant literature has been encouraged.

Chapters 41 and 42 address the current state of methodologies for isolating and culturing the microbial components of soil (chapter 41) and rhizosphere and phyllosphere (chapter 42) communities. In chapter 43, techniques for sampling viruses from soil are considered. Chapter 44 addresses methods for detecting, isolating, and culturing mycorrhizal fungi, and chapter 45 considers the isolation and characterization of endophytic fungi and bacteria. It has long been known that the majority of soil microbes are not amenable to recovery by standard culture techniques. Recent advances in molecular biology have provided new tools to detect and characterize soil microbes and to assess their abundance, genetic makeup, and environmental functions. The use of molecular biology techniques for soil microbial community structure analysis is covered in chapter 46. Techniques for the recovery of soil bacterial community DNA are described in chapter 47, and a discussion of applications of PCR techniques to amplify gene sequences from plant and soil microbes is presented in chapter 48.

Several chapters address aspects of microbial activity and chemical transformations in soil. In chapter 49, quantitative approaches to the study of nitrogen transformations are presented. Chapter 50 covers methods to measure the metabolic activity of soil- and plant-associated microbes, with emphasis on quantification of microbial respiration. Bacteria and fungi modify soil and rhizosphere environments in part by producing extracellular enzymes that digest soluble materials, providing products that are absorbed back into the cell; chapter 51 presents methods for assessment of this extracellular enzymatic activity.

Recombinant DNA technology offers the possibility of developing genetically engineered bacteria for purposes such as degradation of toxic compounds or biological control of plant pathogens. Particularly exciting advances have been made in our ability to observe, in situ, many aspects of microbial physiology that previously could be demonstrated only with pure cultures under highly controlled conditions. Various new strategies are available for detection of bacteria, their genes, and their gene products in natural environments. These strategies are equally useful for tracking the fate of introduced organisms and their genetic information in soil and the rhizosphere. In chapter 52, monitoring techniques involving use of antibiotic resistance, chromogenic, and luminescent markers are discussed; the use of fluorescent antibodies for studying the ecology of soil- and plant-associated microbes is covered in chapter 53. The ability to elucidate microbial gene expression in natural environments is becoming an exciting possibility, and chapter 54 describes the use of reporter genes to assess gene expression of soil- and plant-associated bacteria. Although much is known about the genetics, regulation, and biosynthesis of antibiotics, the ecological significance of antibiotic production is speculative unless antibiotic production can be demonstrated and quantified in situ; methods to accomplish this are the subject of chapter 55.

Potential environmental release of engineered bacteria has raised concerns about the possible spread of recombinant DNA sequences to other organisms present in the same environment, and knowledge about the potential of introduced organisms to transfer genetic information to other bacteria present in a given habitat is essential for assessment of possible environmental risk. Bacterial plasmids and transposons occur in most prokaryotes, and they carry the genetic determinants of many important phenotypic properties, in-

cluding antibiotic and heavy metal resistance, production of antibiotics, bacteriocins, and toxins, catabolism of xenobiotic compounds, nodulation and nitrogen fixation, plant tumorogenesis, and insecticidal toxin production (1). Chapter 56 summarizes available methods to quantify rates of transfer of mobile genetic material in soil and rhizosphere environments. Finally, one cannot ignore the importance in soil and plant habitats of the most abundant members of the animal kingdom, the arthropods. Chapter 57 focuses on methods to assess microbial interactions with terrestrial invertebrates, with emphasis on the entomopathogens.

The choice of topics in this section, although broad, is necessarily selective. We have attempted to emphasize those subjects and methods that have wide application to quantitative ecological investigations of soil and plant microbial habitats. It is hoped that these chapters will also stimulate greater cooperation and understanding among members of the various scientific disciplines studying terrestrial microorganisms.

REFERENCES
1. **Chater, K. F., and D. A. Hopwood.** 1989. Diversity of bacterial genetics, p. 23–52. *In* K. F. Chater and D. A. Hopwood (ed.), *Genetics of Bacterial Diversity.* Academic Press, New York.
2. **Lynch, J. M., and J. B. Hobbie (ed.).** 1988. *Micro-Organisms in Action: Concepts and Applications in Microbial Ecology.* Blackwell Scientific Publications, Oxford.
3. **Paul, E. A., and F. E. Clark.** 1989. *Soil Microbiology and Biochemistry.* Academic Press, New York.

Methods for Sampling Soil Microbes

J. D. VAN ELSAS AND K. SMALLA

41

Representative sampling of soils is crucial for assessing microbiological soil parameters such as the functioning of the soil ecosystem in relation to the presence of specific microbial groups, the survival and distribution of unmodified or genetically modified microorganisms (GMOs) following a release, and the presence of plant-pathogenic microbes after a plant disease outbreak. The determination of microbial diversity in soil also depends heavily on how samples are collected and evaluated, since spatial aspects play a role in determining the interactions between microorganisms and shifts in microbial populations in soil (10, 40). Hence, without a sampling plan that takes into account the spatial localizations of, and relationships between, microorganisms in soil, any data on soil microbial diversity may have little meaning.

Soil is a highly complex environment dominated by the soil solid phase. In contrast to water systems, in which mixing to a certain extent is possible, soil is relatively recalcitrant to mixing. Nevertheless, many soils are slowly and unevenly mixed over a long period of time as a result of such processes as soil surface erosion and/or earthworm soil turnover, which can amount to 10% of topsoil per year (53). Furthermore, soluble constituents of the soil solid matrix may dissolve in soil water and reprecipitate at other sites.

The soil microbiota, including bacteria, fungi, and protozoa, is localized in close association with soil particles, mainly clay-organic matter complexes (10). Often, microbes can be found as single cells or as microcolonies, embedded in a matrix of polysaccharides (10, 40, 53). Their activities and interactions with other organisms and with soil particles depend to a large extent on conditions at the microhabitat level, which may differ between microhabitats even over very small distances (48). The microhabitats for soil microorganisms include the interior as well as exterior surfaces of soil aggregates of various sizes (<250 μm, 250 μm to 2 mm, and >2 mm) and compositions. Soil can therefore be regarded as highly heterogeneous with respect to the distribution of soil matter and organisms, at both the microscale (micrometer to millimeter) and macroscale (meter) levels. The level of heterogeneity or variability varies for each soil, and variability may fall into two classes: (i) random variation, caused by the inherent characteristics of a relatively unmixed system, and (ii) variation due to natural or human-induced environmental gradients. The presence of plant roots, resulting in the establishment of a rhizosphere soil compartment, is an example of gradient-type variation due to a natural process. Agricultural soils usually exhibit vertical gradients in texture, structure, composition, and biota, due in part to maximal input of nutrients (e.g., carbon, nitrogen, and phosphorus) and oxygen at the surface. Variation along horizontal gradients may be less extreme (at least in the top layer) as a result of tillage, which increases homogeneity. On the other hand, a forest soil may have a degree of vertical variability similar to that of an agricultural soil but also may show substantial horizontal variability, even in the top layer, since little horizontal mixing takes place.

Heterogeneity and consequent variability in soil pose particular problems for any soil sampling strategy. Nevertheless, the samples collected often are expected to provide information that is meaningful and representative (averaged) for the entire field or area that is being sampled. Depending on the purpose of the experiment or analysis, the sampling strategy used should take into account the level of precision (defined as accuracy with which the real mean value of the parameter to be assayed is determined) of the data needed in relation to the commonly observed variability. Considerations on how to establish a sampling scheme based on a statistical evaluation of variability are discussed by Gilbert (11). Also, principles of geostatistics, i.e., how to go about sampling given that populations often show a spatial distribution, are well treated by Isaacs and Srivastava (25). Chapter 42 of this manual also includes some considerations on the use of geostatistical considerations in sampling of the rhizosphere. In any case, representativeness and/or randomness of sampling is important in order to facilitate meaningful comparisons and characterizations of population fluctuations, as outlined in the following sections of this chapter.

The International Organization for Standardization (ISO) (Geneva, Switzerland) has developed standards for adequate soil sampling, which have been described in a series of ISO norms (20–24). These norms can be obtained from the ISO. This chapter describes soil sampling strategies in conformance with these and other norms, since we feel it crucial that standardized procedures be used for adequate and comparable sampling. Figure 1 shows the different steps needed in soil sampling as outlined below. Table 1 provides a protocol for soil sampling and processing.

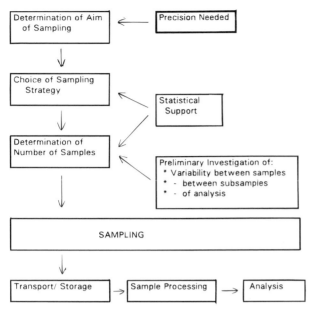

FIGURE 1 Schematics of soil sampling.

DEVELOPING SOIL SAMPLING STRATEGIES

Microscale versus Macroscale Variability

Variability at the microhabitat scale, even though a fact of life for microorganisms in soil, is not commonly regarded as relevant in soil sampling schemes which aim to characterize microbial populations or activities across whole fields or areas. In such sampling strategies, microscale variability tends to be overlooked or averaged out because of the sizes of the samples collected and analyzed. Rather, a description of variability (distribution) over larger distances relevant to the field (macroscale) is regarded as the crucial issue. We

TABLE 1 Protocol for soil sampling and processing

Question	Step in the procedure
What kind of field variation is expected?	Inspect and characterize field
How should an experiment be designed to meet the study objective?	Define precise objectives of study and establish block design
What type of samples is to be used?	Select sample type
Where should samples be taken?	Define strategy related to study objective
How many samples should be taken?	Calculate number of samples needed
How is sampling to be done?	Define utensils, prevention of cross-contamination
How are samples to be transported, processed, and analyzed?	Define processing (storage) and analysis methods

discuss below strategies for sampling soil across fields in which variability at or exceeding the scale of common sample size (e.g., 100 g to several kg) is addressed. It is important to note that whenever relevant to the question of interest, microscale variability can be taken into account in locally obtained intact field cores by modifying the level of resolution of sampling and analysis, i.e., by applying a soil sampling plan at a millimeter-to-centimeter scale instead of, for instance, a meter scale. Classical soil fractionation or separation techniques, which separate soil aggregates or layers into fractions, are useful tools for this purpose.

Experimental (Block) Designs

Sampling strategies must be designed in accordance with experimental objectives in order to produce data sets useful for statistical analyses (13, 48). Often, experiments are designed to test the effects of experimental factors or treatments on predetermined variables (13). To evaluate these effects, a statistically sound experimental design is needed to minimize variation due to random factors in the field (random variability) while revealing differences caused by the treatments (experimental factors). Various experimental designs, all of which are based on dividing the field site into experimental units, have been developed for this purpose.

Designs for field experiments are based on the way the experimental units are arranged in the field. Fields can be blocked by using completely randomized, randomized complete block, randomized incomplete block, Latin square, or split plot designs (7). Each design serves a specific experimental goal; the statistical considerations underlying the different experimental designs will not be further addressed here but can be found in recent literature (7, 48). A field soil with just one treatment does not need to be blocked, but sampling is often performed by the completely randomized approach (see Sampling Strategy, below). Sampling in fields with more than one treatment is guided by the blocking scheme applied.

Measurements of potentially confounding variables should be taken at different sites in the field before the actual experiment is performed, in order to determine how to divide the field into blocks. These measurements provide information about the level of parameter variation over the field, indicating an optimal experimental and sampling setup. To enable comparisons, one should choose an area of study that is as homogeneous as possible, avoiding areas which might cause aberrations in measurements due to unusually extreme characteristics (e.g., areas with unusual rockiness, water input, or weeds or those with missing plants). Alternatively, blocking should be done to take such variability into account.

As a rule of thumb, the best experimental setup is the simplest design available that enables the assessment to be made with the desired precision. Totsche (48) stated there is no clear "decision recipe" which will easily provide the best design for a given problem. However, three criteria by which to choose from the above experimental designs were provided. Formulated as questions, these are as follows: (i) will the experimental design selected provide the information required to achieve the experimental objective, (ii) will the proposed sample population be sufficient to answer the problem with accuracy and reliability, and (iii) is the proposed experimental design sufficient and economical. Totsche also provided a series of subcriteria which illustrate the questions (48).

Sampling Strategy

The objective of soil sampling is normally to obtain a collection of sampling units from a population in order to assay a characteristic of that population without the need to collect and analyze the entire population. To provide meaningful data, the samples should be representative either of the whole (global estimation) or of a selected subarea (local estimation) (25). It is also important to decide whether a mean parameter value is sufficient or if the complete distribution of data values is desired. If necessary, variability at the microscale in addition to that at macroscale also can be considered in the sampling strategy, but this often will be limited to situations involving a local extreme parameter value. Also, in conventional soil sampling, the spatial arrangement of soil aggregates, soil organisms, and enzymes is commonly disturbed, which, depending on the purpose of the experiment, may blur the data at the microscale level. Designing a sampling strategy thus requires taking into consideration the objective of the study and the level of precision of the data needed, in addition to the practicality of meeting the sampling requirements (sampling sites, numbers and types of samples, size of sampling units, sample replicates versus subsampling within replicates, composite samples if necessary, etc.) to reach the objective of the study. Hence, the sampling strategy chosen usually is a compromise between theoretical/statistical and purely practical considerations. The practical considerations include how many soil samples of what mass can be handled in the laboratory in a reasonable time and with reasonable investment of labor hours and instruments. The theoretical/statistical considerations relate to the degree of variability expected and the degree of precision required. A statistical basis for deciding on the allocation of resources between replication of sampling and subsampling within replicates is provided by Snedecor and Cochran (41).

A decision on what constitutes the basic sampling unit also is required. For sampling schemes used to study variables much less numerous per unit mass of soil than microorganisms (e.g., plants, insects, or nematodes), this issue is of paramount importance, since selecting too small a sampling unit may result in unacceptably high variation or error. However, for most microbiological assays, commonly used sampling units of 100 g to several kilograms probably are adequate, since each sample will contain a sufficient amount of the population assayed.

Before sampling, one must also decide on the types of samples and the sampling strategy to be used. Several sample types are distinguished, and each type serves a specific purpose (7, 52), as outlined below. The underlying sampling strategies are shown in Fig. 2. The choice of the sample type and strategy thus depends entirely on the purpose of the study.

Judgment samples are inherently biased and are selected by the investigator to suit the experimental needs. Judgment samples are nonrandom, obviating the application of any statistics, since they are meant to be nonindicative for the whole field. They can therefore be recommended only to serve as a source for, e.g., the isolation of microorganisms.

Simple random samples are collected randomly over a field or site to be studied. Randomness is a prerequisite in this approach; i.e., each possible sample must have an equal opportunity of being selected. Simple random samples allow for a statistical treatment of the data produced and are appropriate for such purposes as the characterization of fields by mean parameter values, variation, and spatial distribution. Simple random samples are, for instance, obtained by dividing the field into a grid-like pattern and taking samples at randomly selected intersections of the grid.

Stratified random samples are in principle comparable to simple random samples. However, subareas are established in the plot on the basis of some selected (physical) property of the soil. Stratified random sampling allows for more precise statements about each of the subareas sampled and increases the precision of estimates over the entire population. However, it requires that a greater number of samples be collected.

Systematic sampling ensures that the entire area to be studied is well represented by the samples. In practice, predetermined points are used, e.g., according to a regular grid pattern. Choice of the grid size (e.g., 0.1 by 0.1 m, 1 by 1 m, or 10 by 10 m) depends on the level of refinement in the data required and on the practicality of processing large numbers of samples. Systematic samples are comprehensive yet nonrandom. They are useful for initial systematic characterization of the spatial variability of a parameter across a whole field. Furthermore, systematic samples can be very precise in measuring an effect. For instance, if effects of plants on adjacent soil are studied, soil may systematically be sampled only at one side (downwind or down-groundwater-movement) of the plants.

The simple random sampling approach (2) can also be combined with a block design, as shown in Fig. 2. In this approach, random samples are taken from replicate blocks in a homogeneous field site. Analysis of replicate blocks as independent experimental tests results in estimates of reproducibility of observations.

Often, multiple individual samples are obtained, which are then bulked and mixed (composited). This approach reduces observed variability in the composite units and is often applied entirely for this purpose. For instance, in a field study aimed at assessing variations in the sizes of kanamycin-resistant bacterial populations, we found that the variance between samples was as high as 50% of the mean, whereas that between composites (made up of eight subsamples) was only about 15% of the mean (unpublished data). Also, to characterize fluctuations in total viable microbial communities in grassland soil over a growing season, randomly taken replicate samples from different depths in the field were composited for each depth (17). Composite samples can be compared only if they are similarly constructed and if there are no interactions between the sampling units. When comparing composites, one will know only the variance between them, not that between the samples making up the composites.

As noted above, the sampling scheme selected will depend on the study objective, which can change in the course of experimentation. An initial question might be: What are the mean value and variation, as well as spatial distribution (posting), of a parameter (e.g., the number of fluorescent

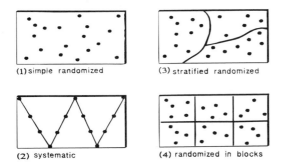

FIGURE 2 Schematics of soil sampling strategies.

pseudomonads) across a 100- by 100-m field? Global denominators over the field are required for the parameter selected. Depending on the outcome of the initial screening, it may then be of interest to study how certain high values (local denominators) are spatially structured in the field in relation to soil or plant aspects. A sensible sampling plan may be to first collect systematic samples, using a series of composite samples collected in a grid (e.g., 10 by 10 m), which will provide data about spatial variation of the parameter studied. In a second phase, areas around sampling points with high values may be sampled by using a smaller grid size, in order to construct contour maps and possibly relate the high values to some local soil or plant variable. This approach, i.e., initial systematic sampling followed by directed sampling of "interesting spots," is a commonly used geostatistical principle (25).

Distribution of Populations and Determination of the Number of Samples Needed

Population Distributions

Since there are limits to the amount of soil that can be handled and processed for analysis, a well-chosen sampling strategy should allow for robust inferences to be made about the (statistical) population from assays of the subpopulations represented in the samples. It is therefore important to have an idea of the expected distribution of the populations under study. Populations can be randomly distributed in space, and such distributions are mathematically described by parameters describing distributions with central tendency, e.g., the normal (Gaussian) distribution (25, 36). Parameters characterizing the normal distribution are the mean, the median, and the variance. The median and mean are usually not very far apart. However, populations are often nonrandomly distributed over fields, which results in a distribution that cannot be described with parameters of the normal distribution (25, 30, 36). Such distributions may result from, for instance, microorganisms occurring in high numbers in some clusters ("hot spots") and in substantially lower numbers in surrounding areas, and they are typically characterized by data sets which contain a large number of "average" values and a few extreme ones (36). The resulting distributions can be highly (positively) skewed, and the mean exceeds the median. Often, the median provides the best estimate of the central tendency of the data. However, since a normal distribution offers great advantages with respect to statistical tests to be performed (25), it is often advantageous to make the data approach normality by using logarithmic, arcsine, or other mathematical transformations (31). Many positively skewed distributions can, after logarithmic transformation, be well described by the lognormal distribution, and a variable is said to be lognormally distributed if the distribution of the logarithm of the variable is normal. Thus, parametric tests can be used in the case of populations with distributions approaching normality, whereas the less powerful nonparametric tests are used for other distributions. For further information about population distributions and the use of parametric versus nonparametric tests, the reader is referred to texts by McSpadden and Lilley (31) and Isaacs and Srivastava (25).

In practice, the distribution of the parameter under study will often be unknown. From a limited number of samples, one can obtain an idea of the most likely spatial distribution of the whole population, but this approach represents just a first approximation at characterizing this distribution. As it is often assumed that the variations in populations of microorganisms over a field result in lognormal distributions (33), a lognormal distribution is commonly taken as most likely; however, this assumption can be challenged.

It is further important to relate the distribution of a population across a field to expectations with respect to localization, i.e., whether extreme values are expected at certain local sites. Collecting too many samples in such areas may lead to a sample mean exceeding the actual population mean. Hence, the sampling strategy selected should take such expected local variability into account, e.g., by undersampling such localities or by processing the data separately.

Determination of the Number of Samples Needed

A general statistical rule for sampling is, the larger the number of samples, the better the estimate of the population mean. However, sample numbers are commonly restricted by limitations in the laboratory. An important consideration when determining the minimum number of samples needed is how close the sample mean must be to the population mean. The number of samples needed can be determined by statistical methods, as outlined by McIntosh (30), Wollum (52), Lamé and Defize (28), and McSpadden and Lilley (31). The expected variation in the parameter to be assayed first has to be determined via analysis of the parameter in a small number of samples collected in the field. For instance, the mean of the sample pool can be required to be within 10% of the population mean. A statistical formula, $n = Z^2 S^2/d^2$ (n = number of samples, $Z = Z$ with probability [$Z_{0.05} = 1.96$], S^2 = error variance of samples, and d^2 = margin of error for the plot), allows for calculation of the number of samples required (30, 35). For instance, to estimate the mean population size of a test organism to within 5×10^3 CFU (d) at the 0.05 level of confidence ($Z = 1.96$), the sampling error variance is estimated as 25×10^6 based on the samples collected. The equation presented above then provides the minimum sample number as 3.8, rounded off to 4. This formula is one of the most commonly used, but it may underestimate n (27). Hence, the number of samples may be increased by 4 if alpha = 0.01 or by 2 if alpha = 0.05 or 0.10. Alpha is the probability of unfair rejection of the null hypothesis, which is known as making a type I error (30). For instance, if the null hypothesis is that two population means are equal, the erroneous conclusion that the two means are different would represent a type I error. Alternatively, the number of samples needed can be inferred from tables provided by Beal (5).

ISO norm 10381-4 (23) provides general rules for sampling soil for microbiological analyses. According to this norm, fields with homogeneous utilization of up to 2 ha (most agricultural practice) are well sampled with one composite sample composed of 15 subsamples per replicate to yield an average whole field estimate. Homogeneous fields of 2 to 5 ha require two such composite samples, fields of 5 to 10 ha require three, fields of 10 to 20 ha require five, etc. Less homogeneous fields such as forest soils, large-area fruit culturing soils, and vineyards need special instructions for sampling, as discussed above and outlined in ISO norm 10381-4 (23).

SOIL SAMPLING IN PRACTICE: SAMPLE COLLECTION, TRANSPORT, STORAGE, AND PROCESSING

Soil (Site and Condition) Characterization

By using a preestablished sampling plan that is based on the objectives of the study, the level of precision required, and the microorganism(s) or process(es) of interest, bulk soil samples are obtained from the field site. The field under study should be well characterized as to its history, topography, type of soil, degree of homogeneity, type of and variability in vegetation and slope, and presence of water streams. Field history is important, since management or disturbance will certainly affect microbial activity and diversity. For instance, previous use of a fungicide may leave fungicide residues that can affect microbial populations in soil. Cropping history also is important for soil microflora and processes such as nitrogen fixation. Knowledge of the site's topography and surroundings is key to understanding possible influences from the surroundings, such as via water movement along slopes.

The scale of heterogeneity also must be considered in determining the experimental design and the sampling strategy (7). For instance, extensive or less extensive sampling in depth may be needed, depending on soil heterogeneity, when a depth profile is required. Also, it should be decided whether entire soil cores (useful to take into account localized soil heterogeneity) or smaller homogenized samples are needed. The purpose of the study is key to this decision.

In characterizing the experimental site, valuable information may be obtained from (in the United States) the Soil Science Society of America, U.S. Geological Survey topographic maps, and/or U.S. Department of Agriculture soil survey maps. Comparable organizations and national bodies may be consulted in other countries.

Since climate as well as seasonal fluctuations (wet/dry, high/low temperatures) can exert a great influence on soil microorganisms and processes, it is important to decide when and under which conditions to sample the soil (9). If, for example, the value of a soil parameter over a growing season or a year is required, the sample values over this time span should be averaged and the range should be determined. Variation across time can also be presented. Unless one is particularly interested in post-drying/rewetting events, sampling immediately following a dry/wet cycle is not recommended, since major microbiological shifts are known to take place in this process (9).

Practical Soil Sampling: Bulk and Rhizosphere Soil

Depending on statistical considerations (see Sampling Strategy, above), the purpose of the experiment, and the practical requirements of the assays to be performed, small samples (up to 100 g), medium-size samples (100 g to several kilograms), or large samples (over several kilograms) may be required. Most microbiological, biochemical, and soil chemical assays require small to medium sample sizes, whereas large samples may be needed when undisturbed soil cores are to be used or for certain soil industrial applications. Small to medium-size samples can be obtained for each soil horizon by using presterilized tools (hand auger, sample corer, spade, shovel, or trowel). A cutting frame is especially useful for forest soils. Sampling of rather small soil volumes from different depths can be performed with a hand auger, which is thoroughly cleaned or disinfected between handlings so as to prevent sample cross-contamination. Sampling of larger soil volumes or of intact soil cores requires the use of a specialized technique (e.g., drilling, boring, or preparation of trial pits) and mechanical equipment (boring tools) (for details, see ISO norm 10381-4 [23]).

Rhizosphere soil and the rhizoplane (surface of plant roots) are sampled by carefully excavating plants from soil with a sterile shovel or trowel. To minimize the introduction of artifacts due to sampling, plant roots as well as other plant parts should be left as intact as possible. A substantial amount of rhizosphere soil should be left around the roots, since the rhizosphere, including root hairs, may extend quite far into soil. Alternatively, an entire soil core con-

taining the plant can be carefully dug out. Sampling of the rhizosphere is further treated in chapter 42 of this manual.

Sampling depth is defined by the type of soil and the experimental requirements. Commonly, the plough layer (0 to 25 cm deep) is sampled in agricultural soils, whereas in grassland, the most densely rooted layer (0 to 10 cm) is taken (9). Forest soils as well as soil profiles are often sampled with consideration of the soil horizons of interest. For instance, the experimental design may dictate that each discernible soil horizon be sampled separately if effects of the different horizons are to be studied.

Bulk and rhizosphere soil samples are commonly placed in thin-walled plastic bags for transport to the laboratory. Bulk soil samples can be thoroughly mixed in these bags, and root/soil samples are dissected in the laboratory. Thin plastic bags are quite useful because they allow for slow gas (O_2, CO_2) exchange to take place while preventing excessive soil drying. Soil cores with or without plants may be placed in boxes for transport, either directly or packed in plastic film. Drying of bulk or rhizosphere soil or roots or exposure to strong sunlight (resulting in drying) should be avoided.

With a laboratory nearby (within 1 to 2 h of travel time), there is often no need to perform extensive sample processing on site. However, if soil samples must be transported for several hours or more, it may be preferable to prepare composite samples or rhizosphere subsamples on site, packing them in plastic bags for transport to the laboratory. Also, samples may be processed on site in order to validly assess a measured variable, such as respiratory capacity.

Sample Transport

Field soil samples often have to be obtained at locations far from the laboratories where analyses are performed. Hence, soil and plant root samples in plastic bags must be transported by a means that will ensure that they do not change substantially before analysis. Soil cores which are sampled to obtain information on small-scale soil heterogeneity should be transported with minimal disturbance of soil structure. Precautions during transport may include keeping the sample temperature low (e.g., 4°C or on ice) and shielding samples from direct sunlight. This practice may be appropriate for soils in temperate climate regions, where soil temperatures of top layers generally vary between slightly above 0°C (winter) to 15 to 20°C (summer). However, in tropical climates, where topsoil temperatures may reach extremes of up to 50 to 60°C, transport at ambient temperature may be more appropriate if analysis takes place shortly after transport.

Following sampling and, if needed, on-site compositing of subsamples, transport to the laboratory should be as quick as possible to prevent unnecessary lags between soil sampling and sample processing and analysis.

Sample Storage

Storage of soil samples should in principle be avoided if microbiological parameters such as CFU counts or enzymatic activities are to be assessed, since storage invariably alters the microbiology of the soil (1). Storage, however, may sometimes be unavoidable, especially if the same soil sample is used in sequential experiments, e.g., with microorganisms and/or plants. During storage, air drying of soil should be avoided, since it results in a drop in bacterial numbers and a concomitant shift in microbial population structure (43). In dried soils, for instance, yeasts instead of bacteria and fungi were shown to become dominant (43). Soil drying also results in increased soil surface acidity, reduced Mn, and increased solubility and oxidizability of soil organic matter (4). Remoistening of dried soil might bring it back to its original physical, chemical, and/or microbiological composition; however, this proposition has been challenged. For instance, denitrifying bacteria may be more active in dried and remoistened soil than in soil kept constantly at the same moisture content (34); rewetted soil suffers a microbiological population explosion, and the microbiological "behavior" of the soil 1 month after rewetting may be anomalous or unpredictable between replicates or when reproduced in time (4). Therefore, storage of soil for long periods of time is preferably done in a moist condition at 4°C. Room temperature may be most appropriate for short-term storage. However, at 4°C, several problems may arise. Conditions in plastic bags may become anaerobic because of decreased permeability of the plastic to O_2 at this low temperature. According to Gordon (12), plastic bags of 0.025-mm thickness are optimal. Sample drying may also take place upon long-term storage at 4°C (45). To prevent this, double bags are sometimes used, but this practice may promote anaerobism and thus change prevailing soil conditions.

It is questionable whether soil stored even under optimal conditions maintains its properties. For instance, Stotzky et al. (45) found that after storage for 3 months, the numbers of microorganisms except actinomycetes decreased. Hence, if the microbiota is to be assessed, long-term storage is not recommended. Obviously, storage of soil might be acceptable if assays are based on cell material that is supposedly preserved in soil under proper conditions, e.g., at 4 or even −20°C (for DNA). Wollum (52) recently found that short-term storage (7 to 21 days) of soils did not result in appreciable changes in soil properties, including total microbial biomass, counts of specific microbial groups, available N, and enzymatic activities; some properties (e.g., arylsulfatase activities), however, changed. Hence, storage of soil at 4°C for a short period (up to 3 weeks) in the dark may be acceptable, especially if storage is unavoidable in a practical sense. If stored samples have to be used, it is important that some quality control and quality assurance measures be taken, e.g., via the assessment of selected critical soil microbiological parameters (52). In a recent treatise of soil microbiology (47), Tate provides very similar recommendations.

Sample Processing

For most microbiological or biochemical analyses, soil samples usually are processed mechanically in the laboratory before the assays are performed. Soil processing in this case means that the soil sample is made accessible for analyses. In most cases, the spatial relationships between microorganisms in soil (microscale organization) are thus lost (see below). During processing, soil samples should never be allowed to dry out excessively or be exposed to high temperatures. Soil is sometimes sieved through a 2- to 4-mesh sieve to remove roots, stones, etc., but this should only be done if a greater homogeneity is required. If soil is too wet to allow for proper sieving, it should be dried to not below 30% of water-holding capacity to facilitate handling without greatly affecting the microbial biomass (42).

To obtain rhizosphere soil, root segments with attached soil may be cut into pieces with sterile scissors. However, any destruction of plant material should be minimized and postponed until just prior to analysis. Adhering soil should be kept in close contact with the roots as much as possible. Processed samples should be used immediately for analyses. Sampling of the rhizosphere and sample processing are further described in chapter 42 of this manual. Further details on experimental procedures are provided by Postma et al.

(37), Van Overbeek and van Elsas (51), and Van Overbeek et al. (50).

Recovery of the Bacterial Fraction

For some so-called in situ techniques (for instance, in situ hybridization [14]), it is important to carefully remove a small part of the sample to assess the spatial relationships between microorganisms and soil or plant material. Most other analyses are performed by extractive techniques, in which the physical structure of soil and its spatial arrangements are not preserved. For such studies of microorganisms from soil via culturing, immunofluorescence (IF), or DNA-based techniques, microbial cells must be efficiently dislodged from soil particles. Ideally, a large and representative part of the soil microflora is thus recovered in a form suitable for analysis. However, efficient recovery may be problematic because of (i) the heterogeneous distribution of microbial cells throughout soil and (ii) the adsorptive interactions between cells and soil particles.

Strategies to obtain primarily bacterial cells from soil in suspension are discussed below. It is possible to modify these procedures in order to also obtain fungi and protozoa. The bacterial fraction can be obtained in suspension by using the following steps: dispersion of soil aggregates in liquid, separation of soil particles from the liquid fraction with microorganisms, and purification of the microbial fraction.

Dispersion can be brought about by shaking, stomacher blending, or ultrasonication of soil in a liquid containing soil dispersion agents. Stomacher blending is based on mechanical action imposed on soil slurries in plastic bags, resulting in efficient dislodgment of cells from soil. The efficiency of ultrasonication depends on the energy introduced; excessive energy may impede the culturability of cells (38). Blending in a stomacher or in a Waring blender or shaking in shake flasks is a suitable alternative, with cell recovery values on the same order of magnitude (2, 8, 29, 38). An alternative approach is based on the use of ion-exchange resins, which, by taking up polyvalent cations from soil, break bonds between soil particles as well as between cells and soil particles. Resins often are used in combination with detergents such as sodium deoxycholate in order to break bonds which hold cells and soil particles together (19, 26, 29). The use of a density gradient with Nycodenz resin (Nycomed Pharma AS, Torshov, Oslo, Norway) has been reported to increase the recovery of bacterial cells by dislodging an enhanced number of cells from soil (3). Sodium pyrophosphate is a good soil-dispersing agent (46). In our laboratories, either shaking (for 10 to 30 min) of 10 g of soil in 95 ml of 0.1% sodium pyrophosphate in water or stomacher blending in phosphate buffer has given consistently high recoveries (same order of magnitude as estimated in soil) of total and specific bacteria measured as CFU on plates and/or via total and specific microscopic cell counts.

Following dispersion, the bacterial fraction may be separated from soil particles by low-speed centrifugation (2, 8, 44). Except for soil of high organic matter content, a short (5- to 15-min) centrifugation step at 500 to 1000 × g often permits the precipitation of soil particles and most fungal hyphae, yielding a cleared supernatant containing bacterial cells. The recovery of bacterial cells is dependent on their previous dislodgment from the soil as well as the potential for interactions with soil particles. Separation of bacterial cells from a sandy soil is easier than separation from a clayey one because clay particles tend to strongly adsorb bacterial cells in the presence of bivalent cations. However, repeated centrifugation and homogenization steps were found not to release all bacterial cells from three organic Norwegian soils into the supernatant (8). It has been estimated that by established methods, often only up to 30% of microscopically detectable bacteria are recovered from soil (44). This recovery rate may be acceptable when coupled with results of sensitive analytical methods such as selective plating for CFU counts of specific culturable organisms. However, it may not be adequate for less sensitive techniques such as IF or DNA-based techniques used for monitoring populations, since it enhances the limit of detection. An exception is PCR (6), which by its nature can overcome this drop in sensitivity. A more important problem is caused by the fact that current recovery methods may have inherent bias, the precise nature of which is often difficult to ascertain. For example, Hattori (15) found that simply shaking and dislodging bacteria from soil led to a bias toward better recovery of bacteria from the outer shells of soil aggregates than of those from internal soil aggregate sites.

After separation of the bacterial fraction from soil particles, further purification can be performed via density centrifugation over a Percoll gradient (2, 29). However, such methods are very laborious and quite difficult to perform, and they may give rise to additional losses of cells.

SOIL SAMPLING AND ANALYSES OF BACTERIA AND THEIR CHARACTERISTICS

Once samples have been obtained from field soil, immediate analysis is often required; otherwise, sample material should be preserved in such a way that crucial characteristics are not lost or changed. This is the case when a release of unmodified bacteria or GMOs in soil is monitored by assessing CFU numbers, when cellular activity such as the detection of in situ expression of a *lacZ* reporter gene is measured, and when the physiological status of cells is assessed via cell stressor experiments (50). When the GMO DNA is targeted, samples should be either processed for analysis as quickly as possible or immediately stored at low temperature (for instance, −20°C) until analysis. IF analysis of specific cells in soil also is best performed on fresh samples, whereas analyses performed on frozen samples should be carefully controlled.

An important point to consider in the analyses is whether and how sample-to-sample variation, e.g., across a field site, affects the assays performed with the samples. Often, such interference is nil, but it is important to check for this potential problem. For instance, microscopic observation of specific bacterial cells via IF is done by using conventional soil flocculation techniques (37). Samples from homogeneous field sites are expected to flocculate with similar efficiencies, and hence data between samples (sites, treatments) can be compared. However, less homogeneous sites may require special attention with respect to a possible effect of soil on the efficiency of the IF assay. DNA-based techniques (e.g., PCR) either use soil suspended in a buffer directly for cell lysis and DNA extraction (32, 39) or first separate the bacterial fraction via differential centrifugation steps (18). Interference with DNA extraction and purification efficiency as a result of sample-to-sample variation may be small unless soil samples differ greatly across the field site sampled. However, PCR amplification is extremely sensitive to inhibitory compounds still present in soil DNA extracts, and undetected differences between samples can result in major differences in final PCR data. Hence, a careful check on this possible bias, e.g., via amplification of pure target DNA added to the different soil DNA extracts, is recommended.

In spite of these possible caveats, these and other methods can adequately describe bacterial fate in field soils as affected by experimental treatments. Albeit performed on different samples, a statistical treatment of the data can pro-

vide a consistent picture of inoculant fate in the field. CFU enumerations over time thus provide information on the dynamics of a GMO culturable population, whereas concomitant IF counts describe the fate of the introduced GMO cell population. Activity measurements, either of specific genes as evidenced by, for example, the *lacZ* reporter (51) or overall via reaction of bacterial cells with a redox dye such as CTC (16), can elucidate the degree of in situ activity of cells subjected to field conditions. Finally, assessment of specific target DNA provides information about their possible persistence in soil. The use of both direct cell lysis and cell extraction-lysis procedures provides evidence about the dynamics of total (cell-associated plus extracellular) versus cell-associated target DNA molecules. Both data sets are important for assessing the persistence of GMOs and their specific genes in field soils.

This work was supported by a grant from the EU-BIOTECH program (BIO2-CT92-0491).

We thank Brian McSpadden and Leo S. van Overbeek for critically reviewing the manuscript.

REFERENCES

1. **Anderson, J. P. E.** 1987. Handling and storage of soils for pesticide experiments, p. 45–60. *In* L. Somerville and M. P. Greaves (ed.), *Pesticide Effects on Soil Microflora.* Taylor and Francis, London.
2. **Bakken, L. R.** 1985. Separation and purification of bacteria from soil. *Appl. Environ. Microbiol.* **49:**1482–1487.
3. **Bakken, L. R., and V. Lindahl.** 1995. Recovery of bacterial cells from soil, p. 9–27. *In* J. T. Trevors and J. D. van Elsas (ed.), *Nucleic Acids in the Environment; Methods and Applications.* Springer-Verlag, Heidelberg, Germany.
4. **Bartlett, R., and B. James.** 1980. Studying dried, stored soil samples—some pitfalls. *Soil Sci. Soc. Am. J.* **44:**721–724.
5. **Beal, S. L.** 1989. Sample size determination for confidence intervals on the population mean and on the difference between two population means. *Biometrics* **45:**101–105.
6. **Briglia, M., R. Eggen, W. M. De Vos, and J. D. van Elsas.** 1996. Rapid and sensitive method for the detection of Mycobacterium chlorophenolicum PCP-1 in soil based on 16S rDNA gene-targeted PCR. *Appl. Environ. Microbiol.* **62:** 1478–1480.
7. **Environmental Protection Agency.** 1992. *Monitoring Small-Scale Field Tests of Microorganisms.* Environmental Protection Agency, Washington, D.C.
8. **Faegri, A., V. L. Torsvik, and J. Goksoyr.** 1977. Bacterial and fungal activities in soil: separation of bacteria and fungi by a rapid fractionated centrifugation technique. *Soil Biol. Biochem.* **9:**105–112.
9. **Forster, J. C.** 1995. Soil sampling, handling, storage and analysis, p. 49–51. *In* K. Alef and P. Nannipieri (ed.), *Methods in Applied Soil Microbiology and Biochemistry.* Academic Press, London.
10. **Foster, R. C.** 1988. Microenvironments of soil microorganisms. *Biol. Fertil. Soils* **6:**189–203.
11. **Gilbert, R. O.** 1987. *Statistical Methods for Environmental Pollution Monitoring.* Van Nostrand Reinhold Co., New York.
12. **Gordon, A. M.** 1988. Use of polyethylene bags and films in soil incubation studies. *Soil Sci. Soc. Am. J.* **52:**1519–1520.
13. **Green, R. H.** 1979. *Sampling Design and Statistical Methods for Environmental Biologists.* John Wiley & Sons, New York.
14. **Hahn, D., R. I. Amann, W. Ludwig, A. D. L. Akkermans, and K.-H. Schleifer.** 1992. Detection of microorganisms in soil after in situ hybridization with rRNA-targeted, fluorescently labelled oligonucleotides. *J. Gen. Microbiol.* **138:**879–887.
15. **Hattori, T.** 1988. Soil aggregates as microhabitats of microorganisms. *Rep. Inst. Agric. Res. Tohoku Univ.* **37:**23–36.

16. **Heijnen, C. E., S. Page, and J. D. van Elsas.** 1995. Metabolic activity of *Flavobacterium* strain P25 during starvation and after introduction into bulk soil and the rhizosphere of wheat. *FEMS Microbiol. Ecol.* **18:**129–138.

17. **Higashida, S., and K. Takao.** 1985. Seasonal fluctuation patterns of microbial numbers in the surface soil of a grassland. *Soil Sci. Plant Nutr.* **31:**113–121.

18. **Holben, W. E., J. K. Jansson, B. K. Chelm, and J. M. Tiedje.** 1988. DNA probe method for the detection of specific microorganisms in the soil bacterial community. *Appl. Environ. Microbiol.* **54:**703–711.

19. **Hopkins, D. W., and A. G. O'Donnell.** 1992. Methods for extracting bacterial cells from soil, p. 104–112. *In* E. M. H. Wellington and J. D. van Elsas (ed.), *Genetic Interactions among Microorganisms in the Natural Environment.* Pergamon Press, London.

20. **International Organization for Standardization.** 1992. *Soil Quality—Sampling,* part 1. *Guidance on the Design of Sampling Programmes.* ISO/CD 10381-1-1992. International Organization for Standardization, Geneva.

21. **International Organization for Standardization.** 1992. *Soil Quality—Sampling,* part 2. *Guidance on Sampling Techniques.* ISO/CD 10381-2. International Organization for Standardization, Geneva.

22. **International Organization for Standardization.** 1992. *Soil Quality—Sampling,* part 3. *Guidance of Safety.* ISO/CD 10381-3. International Organization for Standardization, Geneva.

23. **International Organization for Standardization.** 1992. *Soil Quality—Sampling,* part 4. *Guidance on the Procedure for the Investigation of Natural and Cultivated Sites.* ISO/CD 10381-4. International Organization for Standardization, Geneva.

24. **International Organization for Standardization.** 1992. *Soil Quality—Sampling,* part 6. *Guidance on the Collection, Handling and Storage of Soil for the Assessment of Aerobic Microbial Processes in the Laboratory.* ISO/DIS 10381-6. International Organization for Standardization, Geneva.

25. **Isaacs, E. H., and R. M. Srivastava.** 1989. *Applied Geostatistics.* Oxford University Press, Oxford.

26. **Jacobsen, C. S., and O. F. Rasmussen.** 1992. Development and application of a new method to extract bacterial DNA from soil based on separation of bacteria from soil with cation-exchange resin. *Appl. Environ. Microbiol.* **58:** 2458–2462.

27. **Kupper, L. L., and K. B. Hafner.** 1989. How appropriate are sample size formulas? *Am. Stat.* **43:**101–105.

28. **Lamé, F. P. J., and P. R. Defize.** 1993. Sampling of contaminated soil: sampling error in relation to sample size and segregation. *Environ. Sci. Technol.* **27:**2035–2044.

29. **MacDonald, R. M.** 1986. Sampling soil microfloras: dispersion of soil by ion exchange and extraction of specific microorganisms from suspension by elutriation. *Soil Biol. Biochem.* **18:**399–406.

30. **McIntosh, M. S.** 1990. Statistical techniques for field testing of genetically engineered microorganisms, p. 219–239. *In* M. Levin and H. Strauss (ed.), *Risk Assessment in Genetic Engineering—Environmental Release of Organisms.* McGraw-Hill Inc., New York.

31. **McSpadden, B. M., and A. Lilley.** Application of common statistical tools. *In* J. D. van Elsas, E. M. H. Wellington, and J. T. Trevors (ed.), *Modern Soil Microbiology,* in press. Marcel Dekker, Inc., New York.

32. **Ogram, A., G. S. Sayler, and T. Barkay.** 1987. The extraction and purification of microbial DNA from sediments. *J. Microbiol. Methods* **7:**57–66.

33. **Parkin, T. B., and J. A. Robinson.** 1994. Statistical treatment of microbiological data, p. 15–39. *In* R. W. Weaver, S. Angle, P. Bottomley, D. Bezdicek, S. Smith, A. Tabata-bai, and A. Wollum (ed.), *Methods of Soil Analysis,* part 2. *Microbiological and Biochemical Properties.* Soil Science Society of America, Inc., Madison, Wis.

34. **Patten, D. K., J. M. Bremner, and A. M. Blackmer.** 1980. Effects of drying and air-dry storage of soils on their capacity for denitrification of nitrate. *Soil Sci. Soc. Am. J.* **44:** 67–70.

35. **Peterson, R. G., and L. D. Calvin.** 1986. Sampling, p. 35–51. *In Methods of Soil Analysis,* part 1, 2nd ed. Agronomy Monograph 9. American Society of Agronomy and Soil Science Society of America, Madison, Wis.

36. **Pielou, E. C.** 1983. *Population and Community Ecology, Principles and Methods,* 4th ed. Gordon and Breach Science Publishers, New York.

37. **Postma, J., J. D. Van Elsas, J. M. Govaert, and J. A. van Veen.** 1988. The dynamics of *Rhizobium leguminosarum* biovar *trifolii* introduced into soil as determined by immunofluorescence and selective plating techniques. *FEMS Microbiol. Ecol.* **53:**251–260.

38. **Ramsay, A. J.** 1984. Extraction of bacteria from soil: efficiency of shaking or ultrasonication as indicated by direct counts and autoradiography. *Soil Biol. Biochem.* **16:** 475–481.

39. **Smalla, K., N. Cresswell, L. C. Mendonca-Hagler, and J. D. van Elsas.** 1993. Rapid DNA extraction protocol from soil for polymerase chain reaction mediated amplification. *J. Appl. Bacteriol.* **74:**78–85.

40. **Smiles, D. E.** 1988. Aspects of the physical environment of soil organisms. *Biol. Fertil. Soils* **6:**204–215.

41. **Snedecor, G. W., and W. G. Cochran.** 1980. *Statistical Methods,* 7th ed. Iowa State College Press, Ames.

42. **Somerville, L., and M. P. Greaves.** 1987. *Pesticide Effects on Soil Microflora.* Taylor and Francis, London.

43. **Sparling, G. P., and M. V. Cheshire.** 1979. Effects of soil drying and storage on subsequent microbial growth. *Soil Biol. Biochem.* **11:**317–319.

44. **Steffan, R. J., J. Goksoyr, A. K. Bej, and R. M. Atlas.** 1988. Recovery of DNA from soils and sediments. *Appl. Environ. Microbiol.* **54:**2908–2915.

45. **Stotzky, G., R. D. Goos, and M. I. Timonin.** 1962. Microbial changes in soil as a result of storage. *Plant Soil* **16:** 1–18.

46. **Strickland, T. C., P. Sollins, D. S. Schimel, and E. A. Kerle.** 1988. Aggregation and aggregate stability in forest and range soils. *Soil Sci. Soc. Am. J.* **52:**829–833.

47. **Tate, R. L.** 1995. *Soil Microbiology.* John Wiley & Sons, New York.

48. **Totsche, K.** 1995. Quality control and quality assurance in applied soil microbiology and biochemistry, p. 5–24. *In* K. Alef and P. Nannipieri (ed.), *Methods in Applied Soil Microbiology and Biochemistry.* Academic Press, London.

49. **Van Elsas, J. D., and L. S. van Overbeek.** 1993. Bacterial responses to soil stimuli, p. 55–79. *In* S. Kjelleberg (ed.), *Starvation in Bacteria.* Plenum Press, New York.

50. **Van Overbeek, L. S., L. Eberl, M. Givskov, S. Molin, and J. D. van Elsas.** 1995. Survival of, and induced stress resistance in, carbon-starved *Pseudomonas fluorescens* cells residing in soil. *Appl. Environ. Microbiol.* **61:**4202–4208.

51. **Van Overbeek, L. S., and J. D. van Elsas.** 1995. Root exudate induced promoter activity in *Pseudomonas fluorescens* mutants in the wheat rhizosphere. *Appl. Environ. Microbiol.* **61:**890–898.

52. **Wollum, A. G., II.** 1994. Soil sampling for microbiological analysis, p. 1–14. *In* R. W. Weaver, S. Angle, P. Bottomley, D. Bezdicek, S. Smith, A. Tabatabai, and A. Wollum (ed.), *Methods of Soil Analysis,* part 2. *Microbiological and Biochemical Properties.* Soil Science Society of America, Inc., Madison, Wis.

53. **Wood, M.** 1989. *Soil Biology.* Chapman and Hall, London.

Sampling Microbes from the Rhizosphere and Phyllosphere

LOUISE-MARIE C. DANDURAND AND GUY R. KNUDSEN

42

RHIZOSPHERE AND PHYLLOSPHERE HABITATS

Populations

Plant surfaces and interiors are important habitats for microorganisms; indeed, some microorganisms are able to grow only in association with plants. We will use the terms rhizosphere and phyllosphere as nouns to describe plant-associated habitats and as adjectives to describe the microbial components of those habitats. Although Hiltner (35) first coined the term rhizosphere specifically to designate the zone of interaction between bacteria and the roots of legumes, we follow the more recent convention of classifying plant-associated microbes as occupying rhizosphere habitats (i.e., the plant root surface and the surrounding soil in which microbes are influenced by root processes) or phyllosphere habitats (the surfaces of aerial plant parts). As Lynch (65) noted, the rhizosphere environment actually is composed of the interacting trinity of the plant, the soil immediately adjacent to the root, and the organisms associated with the roots. Microbial habitats within the rhizosphere are sometimes further classified as endorhizosphere (within the root tissue itself) or rhizoplane (on the root surface). Analogous terms, with similar etymology, have been applied to aerial plant parts, or the phyllosphere; i.e., habitats may include plant interiors and plant surfaces (phylloplane). An extensive discussion of endophytic microbes is presented in chapter 45. Our discussion here will be limited to plant surface microbes and, in the case of the rhizosphere, those in the immediately surrounding soil.

Populations of microbes in the rhizosphere differ quantitatively and qualitatively from those in the bulk soil; their numbers are generally higher, and different populations commonly are represented. Plant roots release compounds, including simple sugars and amino acids, that encourage growth of specific microbial communities. Because rhizosphere inhabitants rely heavily on organic exudates for their energy requirements, their metabolism may be closely tied to that of the plant (65). Similarly, numerous microbes colonize the aerial parts of plants. These are sometimes called epiphytic microbes (from the Greek *epi* [on top of] and *phytos* [plant]), but unlike the epiphytic plants which derive no nutrition from the plant surfaces that they inhabit, epiphytic

microbes actively utilize compounds exuded from leaves and other aerial plant parts. Noncolonizing microbes frequently can be found in the phyllosphere, but there also is substantial evidence of defined microbial succession and development of specific phyllosphere communities.

Although many rhizosphere and phyllosphere microbes form apparently commensal relationships with plant roots, there are numerous well-known relationships involving mutualism and/or parasitism. Roots of many plants form an intimate relationship with certain fungi; the resulting mycorrhizal association results in enhanced uptake of certain mineral nutrients, including phosphorus. Legumes and some other plants form symbioses with nitrogen-fixing bacteria. The mutualistic relationship between legumes and *Rhizobium* bacteria allows the plants to grow in otherwise nitrogen-deficient soil and enhances the soil's nitrogen status for other plants. Other species of nitrogen-fixing bacteria inhabit foliar surfaces. Some microbes in the rhizosphere or the phyllosphere are plant pathogens. With the exception of vectored or seed-borne organisms, most plant pathogens first contact the plant at the phylloplane or the rhizoplane. While some fungal plant pathogens are well equipped to mechanically and enzymatically breach these surfaces, many others spend much of their life cycles as nonparasitic rhizosphere or phyllosphere inhabitants and enter a parasitic phase only when conditions are conducive to passive entry and subsequent colonization of plant interiors. In contrast, certain microbes in the rhizosphere or the phyllosphere are antagonistic to plant pathogens or insect pests and thus serve as potential biological control agents for agriculture and forestry.

Environmental Attributes

Plant roots modify the surrounding soil environment in various ways. Almost all chemical components of the plant have been found to be lost from roots. Root deposition includes exudates that contain water-soluble compounds such as sugars, amino acids, organic acids, hormones, and vitamins; actively secreted polymeric carbohydrates and enzymes; lysates which include cell walls, sloughed whole cells, and whole roots; and volatile compounds. It is generally believed that the zone just behind the root tip is the site of maximum root exudation (18). As roots elongate through soil, cells are sloughed from the root cap, and root hairs and cortical cells are sloughed from the surface. Deposition from

Paper no. 96714 of the Idaho Agricultural Experiment Station.

roots provides a readily used energy source for microbes, and spatial variation in the components of rhizodeposition over the root surface greatly influences variability in the composition of the rhizosphere community (97). Other plant influences on the rhizosphere community include the removal of soil water by plant uptake, release of extracellular enzymes, and reduced oxygen tension and the release of CO_2 as a result of aerobic respiration, with an accompanying localized increase in soil acidity. In turn, microbes associated with plant roots modify the rhizosphere environment by themselves producing extracellular enzymes and plant growth factors and by forming parasitic or mutualistic relationships with the plant. As a consequence, the quantities and types of available substrates in the rhizosphere differ from those in the bulk soil, resulting in different compositions of the respective microbial communities. This so-called rhizosphere effect reflects the influence of plant roots on the size and composition of the surrounding soil microbial community. One numerical measure of rhizosphere effect is the rhizosphere/soil (R/S) ratio, which compares the total number of microbes in a rhizosphere habitat with the number in the root-free soil; rhizosphere/soil ratios range from as low as 5 to 100 or more but most frequently are from 10 to 20 (18, 32). Absolute size estimates of the rhizosphere vary greatly, depending in part on the plant species and spatial location on the root, which is reflective of physiological maturity (20), as well as soil temperature and moisture content.

The environment similarly determines aspects of epiphytic microbial populations and communities, as do plant host type and the age and position of leaves (42, 48). On new leaves, bacteria predominate in the microbial community, while yeasts and filamentous fungi succeed them and become dominant later in the growing season (6). The important environmental factors affecting phyllosphere populations include ambient and leaf surface temperature, relative humidity and leaf wetness, and solar radiation. Generally, environmental factors fluctuate more rapidly in the phyllosphere than in soil, and thus microbial numbers can differ greatly even between samples taken over a short time period.

SAMPLING AND ISOLATION TECHNIQUES

There are two general approaches to the quantitative estimation of microbial populations in the rhizosphere or the phyllosphere: direct methods such as visualization techniques and indirect methods such as dilution plating of root or leaf washings or most-probable-number (MPN) estimates (15).

Indirect methods have variously been reported to recover only about 0.01 to 10% of the numbers of bacteria enumerated by direct counts (9, 16, 24, 77). However, direct counts may overestimate the number of bacteria in soil because it is difficult to differentiate between living and dead cells unless a viability stain is used. In addition, a high level of concentration is needed by the observer; direct counts are labor-intensive if large numbers of samples need to be processed. Another disadvantage with direct counts is that different groups of organisms are difficult to identify unless their morphologies differ.

Direct Visualization via Microscopy

Most of the wide range of potential microscopic techniques have been attempted for visualization of microbes on root and foliar surfaces; such techniques include bright-field, phase-contrast and differential interference, epifluores-

cence, scanning and transmission electron, and atomic force microscopy. In general, direct visualization for quantitative purposes can be difficult because of the opacity of the substrate and the small size of microorganisms; spatial associations are difficult to describe because of the short working distance and the high magnification needed to see them. Phase-contrast or interference microscopy of fresh material is possible, but associated soil and root material can give a confusing background for observation of microorganisms. Fluorochromes and other stains are commonly used to observe microbes on plant surfaces. However, autofluorescence of plant tissue can be a problem when fluorochromes are used to detect microbes on surfaces of plants. Although a discussion of general microscopic techniques is beyond the scope of this chapter, Table 1 lists methods and references for microscopic visualization of plant surface microbes, chapter 5 provides a general discussion of microscopic methods for environmental microbiology, and chapter 53 discusses immunofluorescence techniques.

Indirect Methods: General Requirements

General microbiological laboratory equipment for the isolation and culture of bacteria or fungi from plant surfaces includes an autoclave for sterilization of glassware and preparation of media, glassware, balance, glass or disposable plastic petri dishes, test tubes (or microcentrifuge tubes) and racks, and Bunsen burner or alcohol lamp. When possible, it is desirable to work in a closed room with minimum airflow. A laboratory benchtop that has been swabbed with ethanol is usually suitable as a work surface, although use of a laminar flow hood will further guard against contamination of samples. If it is likely that animal pathogens are present in samples, a biological safety hood should be used. Necessary tools and general materials include dissecting needles, scalpel or single-edged razor blades, 70% ethanol solution, and bleach (sodium hypochlorite) solution (e.g., 10 parts household bleach, 10 parts 95% ethanol, 80 parts water) for surface sterilization of plant material when microbes are to be isolated from within plant tissues.

Collection and Transport of Samples

Field samples should be transported to the laboratory as soon as possible after they are collected. Although some authors advocate maintaining samples at or near their temperature at the time of collection (13), more commonly samples are transported on ice or in an insulated container. If ice is used, care must be taken not to allow water from the melting ice to contact the samples. It is important to avoid exposing samples to excessive heat or desiccation, since these conditions will adversely affect most microbial populations (99). Ambient environmental conditions at the time of collection should be noted for later reference.

Removing Microbes from the Rhizosphere or Phyllosphere

Since precise identification of what constitutes rhizosphere soil is difficult or impossible, researchers often are forced to rely on an operational definition of the rhizosphere. In many studies, the rhizosphere has been defined as the thin layer of soil adhering to the root system after loose soil has been removed by shaking. However, depending on soil texture and moisture content, the amount of adhering soil may vary considerably. Parke et al. (72) reported that the amount of adhering soil (from 0 to 1,200 mg/cm of root) on roots and the amount of adhering soil mass (25 to 1,000 mg) did not significantly affect estimates of root populations of a rhizo-

TABLE 1 Microscopic visualization of rhizosphere and phyllosphere microbes

Method	Purpose	Stain(s) or technique	Selected reference(s)
Visible light microscopy	General techniques	Various stains	22
Fluorescence microscopy	Separation of bacteria from soil by dispersion, dilution, and filtration	Acridine orange	8
	Observation of soil microbes	Various fluorochromes 4', 6-Diamidino-2-phenylindole	22 8
	Determination of metabolically active bacteria and fungi	Tetrazolium dye	8, 70, 83
		Fluorescein diacetate	8, 64, 89
	Enumeration of bacteria; disadvantage is that soil bacteria have less RNA because of low metabolic activity, and thus intensity of fluorescence may be low	Fluorescently labeled oligonucleotides	1, 8, 30, 94
	Direct observation of microorganisms on roots; autofluorescence from root tissue may prevent observation of microbes on root surfaces	Various fluorochromes (acridine orange, auramine O, berberinsulfate, coriphosphine, coriphosphine O, Congo red)	96
	Dual-stain fluorescence	Fluorescein isothiocyanate, tetraethylrhodamine isothiocyanate; europium chelate and fluorescent brightener	2, 43, 74
	Applications in microbial ecology	Immunofluorescence	7
Electron microscopy	Observation of microbes on the rhizoplane or rhizosphere; advantage is that the spatial patterns of microbes can be retained		10, 11, 17, 25–28, 60, 81
Atomic force microscopy	Ecological applications		56

bacterial strain of *Pseudomonas fluorescens*. If microbial numbers are to be expressed as CFU per gram of rhizosphere soil, a subsample of the rhizosphere soil should be weighed, then oven-dried, and weighed again to determine soil moisture content. For phyllosphere samples there are fewer difficulties, as generally the phyllosphere will be more or less synonymous with the phylloplane (in all cases, we are excluding the plant interior from our definitions).

A variety of methods have been used to separate rhizosphere microbes from root surfaces and to separate phyllosphere microbes from leaf surfaces. These methods include washing (in flasks or tubes on a platform or wrist action shaker), vortexing (with or without additional agitation, for example, with added glass beads), sonication, and blending or maceration. The method used for removal of microbes from either the rhizosphere or the phyllosphere affects the recovery of populations (23, 51). Table 2 presents a list of common techniques, some advantages and disadvantages of each method, and references to their experimental use. Washing or homogenizing diluents commonly include sterile tap or distilled water, saline solution, or buffer solutions. Regardless of the method, one must consider whether rhizo-

sphere or phyllosphere microbes may be killed by the process, resulting in artificially low counts, and/or whether growth in the diluent may occur, possibly resulting in inflated counts. Freezing leaf samples for more than 3 days prior to processing is not recommended, as recovery of bacteria has been shown to be adversely affected (23).

Dilution Plating

The dilution plate count is one of the most widely used methods for enumeration of microbes. However, it is well established that indirect counts take into account only culturable microbes, overlooking viable but nonculturable organisms, and thus underestimate actual microbial populations (16). Skinner et al. (86) listed several reasons for inadequacies of the plate count procedure for bacterial isolation: (i) clumps of bacterial cells may give rise to single colonies on a plate, (ii) antibiosis and competition on plates cause some colonies to fail to develop, and (iii) even so-called general growth media are usually selective in some respect. Many cells that are metabolically active do not produce colonies on solid media. Plate counts estimate only those cells that are tolerant of the medium and incubation

TABLE 2 Methods for removal of rhizosphere and phyllosphere microorganisms from plant surfaces

Method	Comments	Selected reference(s)
Root or leaf washings (platform or wrist action shaker)	Numerous samples can be processed at once; time-consuming (30 min–several hours); may not remove tenacious microbes from surfaces; aqueous medium may promote microbial growth	23, 51, 53, 61, 76
Vortexing	Vigorous agitation helps remove tenacious microbes from leaf or root surfaces; each sample is processed rapidly (30 s–2 min), but samples usually must be processed individually	
Stomacher homogenizer	Effective at removing tenacious microbes from surfaces, but only one sample may be processed at a time; processed samples may contain endophytic microbes and leaf tissues; rapid processing of individual samples is relatively specialized and expensive	3, 23, 51, 61
Blender or trituration with a mortar and pestle	Effective at removing tenacious microbes from surfaces, but samples may also contain endophytic microbes and leaf tissues; equipment is relatively inexpensive; processing of individual samples is relatively rapid, but only one sample may be processed at a time	23, 38, 51
Ultrasonication		23, 33, 76
Root or leaf imprints on agar	Allows visualization of the in situ spatial patterns of microbes	90
Adhesive (e.g., cellophane tape) removal	Removes fungal spores and hyphae and some bacterial cells; retains spatial information and can be viewed microscopically; inexpensive; serial dilution is not possible, and it may be difficult to culture removed microbes	53

conditions to which they are subjected (99). In contrast, the plate count procedure may provide an apparent overestimation of fungal or actinomycete biomass if clusters of spores are present, since each spore may give rise to a colony on the plate.

To accommodate the enormous size of most microbial populations in environmental samples, as well as the microscopic size of individuals, the main principle of enumeration is dilution of the population to countable numbers. There are three common versions of the method. In the spread plate method, aliquots of dilutions are spread over the surface of agar medium. In the drop plate method, aliquots (10 μl) are spotted onto the surface of agar medium. Each drop plate can contain four drops of four concentrations of the samples, which greatly reduces the number of plates needed. Although the level of detection is less than in the spread plate method (the drop plate uses a smaller amount of sample), no difference in enumeration of bacteria from leaf surfaces was found between the spread plate and drop plate methods (23). In the pour plate method, aliquots are placed in sterile petri dishes and then molten (40°C) agar is added to the petri dish, which is swirled to mix the agar with the sample.

For all three methods, the initial sample, whether leaf washing or homogenate or rhizosphere soil suspension, is serially diluted in sterile distilled or tap water, saline, or buffer solution (e.g., 1.2 mM KH_2PO_4 [pH 7.2]). Serial dilutions are done by pipetting the sample through a series of 1:10 diluent blanks (typically, a 1-ml sample is added to a 9-ml test tube blank or 100 μl is added to 900 μl of diluent in a microcentrifuge tube) until an appropriate dilution has been reached. Typically, for fungi, this dilution is approximately 1:10³, whereas for rhizosphere bacteria, a dilution of 1:10⁶ to 1:10⁷ is generally suitable (18). When possible, it is advisable to perform a preliminary isolation to estimate

the best dilution level and also to plate out one dilution level on either side of the expected optimum level. From dilution tubes, samples (0.1 ml) are pipetted and spread onto appropriate agar medium. Two or three replicate spread plates per dilution tube are recommended. For further information on preparation of serial dilutions and on plating techniques, see reference 99.

Colony counts usually are made after 2 to 3 days (for bacteria) or 4 to 7 days (for fungi), and by taking into account the dilution factor, an estimated number of organisms in the original sample is obtained. Several different measures have been used to express rhizosphere population numbers from dilution plate counts: (i) number of bacteria per gram of rhizosphere soil, (ii) number of bacteria per unit weight of root (5, 39, 95), (iii) number of bacteria per unit length or area of root (4, 29, 52, 62, 91), and (iv) number of bacteria per total root system (62). Because the amount of root surface area available for bacterial colonization is not directly related to either root weight or root length, use of these measures tends to make comparison of samples, even from within the same root system, difficult (21). Davies and Whitbread (21) advocated estimating the entire surface area of the root system as a method that could adequately distinguish differences in root colonization and still be practical for sampling a large number of replicate root samples. Their method involved the assumption that roots are made up of a series of cylinders, so that the surface area of each cylinder could be calculated from measurements of its diameter and length. Cylinder radii were calculated by observing root thin sections microscopically and calculating means of 10 to 50 diameter measurements. Lengths of lateral roots were obtained by using the grid intersect technique (67, 92). The authors determined that estimation of root surface area as a linear function of dry root weight was relatively accurate

TABLE 3 Media for culture of rhizosphere and phyllosphere microbes

Medium	Description
Potato dextrose agar	Rich, general fungal growth medium
Cornmeal agar	Rich, general fungal growth medium
Czapek's agar	Rich, general fungal growth medium
Unamended agar	Useful for selection of some rapidly growing fungi, e.g., *Pythium* spp.
Tryptic soy agar	General bacterial growth medium
Nutrient agar	General bacterial growth medium
King's medium B agar	Diagnostic medium for fluorescent pseudomonads (nonselective)
Malt extract agar	Fungal growth medium, especially for basidiomycetes

TABLE 4 Selected microbial inhibitors used in the culture of rhizosphere and phyllosphere microbes

Inhibitor(s) (concn)	Description
Cycloheximide, nystatin (50 μg/ml)	For general fungal inhibition
Rose bengal (30–700 μg/ml)	Fungal growth retardant; aids detection of slower-growing fungi; may be fungitoxic if incubated in light
Benomyl fungicide (10–50 μg/ml)	For selective fungal inhibition
Lactic acid (0.1%), malic acid (0.5%)	For general bacterial inhibition; add to medium after sterilizing, or agar may not solidify
Rifampin	For general bacterial inhibition
Kanamycin, streptomycin, neomycin (100–200 μg/ml)	For general inhibition of gram-negative bacteria; not heat stable; filter sterilize and add to agar just before pouring
Crystal violet (10 μg/ml)	For inhibition of gram-positive bacteria; may be autoclaved with medium
Chloramphenicol, tetracycline (25–200 μg/ml)	For general bacterial inhibition; not heat stable; filter sterilize and add just before pouring

and much quicker than measurement of diameters and lengths. Root surface area can also be estimated by a gravimetric method (12).

From leaf or root surfaces, it is possible to directly isolate fungal and bacterial propagules by pressing the surface of the leaf or root segment onto an agar surface and then removing it. Best results are obtained with a slightly softer agar concentration (e.g., 1% [by weight]), and it is important to ensure that there is no free water on the agar surface; otherwise, discrete colonies cannot be obtained. This method has the advantage of providing some information about the microbial spatial distribution on the leaf surface. It is also possible to pour a small volume of molten agar onto the leaf surface, allow it to solidify, and then transfer it to fresh medium (57). Cellophane tape may similarly be used, first pressing it to the leaf surface and then plating it out on agar medium. If double-sided tape is used, the tape may be placed on a microscope slide, after which the sampling surface is stained for microscopic visualization (53).

Table 3 presents several general and special-purpose media that are appropriate for culturing rhizosphere and phyllosphere microbes. Usually, an investigator will add components which inhibit either fungi or bacteria; a list of commonly used inhibitors is given in Table 4. More detailed methods suitable for culture and/or selective inhibition of specific rhizosphere and phyllosphere microbes may be found for bacteria (including actinomycetes) in reference 55, for yeasts in reference 75, and for filamentous fungi in reference 85. General information on cultivation of bacteria and fungi is given in chapter 6 of this volume.

MAINTENANCE OF MICROBIAL CULTURES

In general, standard microbiological methods may be used for storage and maintenance of bacterial and fungal isolates from rhizosphere and phyllosphere habitats. Most bacterial isolates can be stored at 4°C on agar slants or even plastic-wrapped plates for relatively short periods of time, e.g., several weeks to a few months. For longer-term storage, lyophilization or storage at ultralow temperature (-70 to -80°C) is desirable. Late-logarithmic- to early-stationary-phase

broth cultures are diluted 1:1 with 30% glycerol in microcentrifuge tubes or special cryostorage tubes and then placed in racks in the ultralow-temperature freezer. To recover cells, a small sample from the upper surface of the frozen culture can be scraped with a sterile wooden toothpick onto (or into) fresh growth medium, taking care not to let the remainder of the culture thaw out.

Most fungi can be stored at 4°C on agar slants, but they should be subcultured at approximately 6- to 12-month intervals. Because repeated subculturing of fungal strains may result in genotypic and/or phenotypic changes over time, methods such as freeze-drying or liquid nitrogen storage are preferable for long-term maintenance (87).

Other methods to store cultures include storage in sterile soil, in sterile water, or under mineral oil. Some bacteria and many fungi can be stored in sterile water for many years. Fungi can be stored in sterile soil or in a sterile sand–1% cornmeal mixture. Spore suspensions in sterile water are added to sterile soil or sand, allowed to grow for several days at room temperature, and then left to dry while stored at 4°C. The fungi can be retrieved by sprinkling small amounts of soil or sand onto a suitable medium. This method is especially useful for *Fusarium*, *Aspergillus*, *Rhizopus*, and *Alternaria* species.

STATISTICAL CONSIDERATIONS
Special Considerations

General aspects of microbial sampling and data analysis are covered in chapter 4 of this volume. However, some special considerations arise when one is sampling microbes from the rhizosphere and the phyllosphere. Hirano and Upper (37), Kinkel et al. (49), Kloepper and Beauchamp (50), and Parberry et al. (71) have reviewed some of the difficulties inherent in sampling the phyllosphere and the rhizosphere. Problems include the nonnormal distribution of microorganisms and the selection of sampling strategy, sample scale, and sample unit.

The choices of sampling strategy, sample scale, and sample unit can all have a significant effect on the results of an experiment. The type of sampling strategy used in sampling the rhizosphere and the phyllosphere often depends on the question that is being asked. It is generally recommended that samples be taken at random so that the assumption of independence of samples for standard statistical tests (e.g., analysis of variance) is met. However, sampling of microbes from the rhizosphere in many cases cannot be done randomly. For example, colonization is often determined by taking root segments at a prespecified location (e.g., 1-cm segments at 2 cm below the soil line). Another example is sequential sampling of roots for assessment of the root-colonizing ability of a microbe, which can result in autocorrelated data. In these situations, one must verify that the data meet the assumptions of the statistical analysis before conducting the tests (31, 88).

Microbial populations in the rhizosphere or the phyllosphere frequently exhibit a high degree of autocorrelation. Problems arise when the assumption of sample independence is not met because the data may be autocorrelated. When data are autocorrelated, the knowledge of a variable's value at some location gives the observer some prior knowledge of the value that the variable will have at another location. Thus, each observation does not bring one full degree of freedom to the analysis. Autocorrelation may cause results of standard statistical tests to be declared significant when they are not (58, 78, 88). When spatial autocorrelation is present, spatial dependence among observations can be removed by using trend surface analysis or spatial variate differencing so that the usual statistical tests can be done (58). Legendre (58) provides for further information on correcting standard statistical methods when autocorrelation is present.

The tendency for both phyllosphere and rhizosphere microbial populations to conform to a lognormal distribution (log-transformed sample colony counts are normally distributed) has been noted (4, 36, 37, 63, 68), and transformation is routinely used. However, Kinkel et al. (49) and Ishimaru et al. (41) reported that distributions of epiphytic bacteria were not consistently described by the lognormal distribution. The appropriateness of the lognormal distribution should be verified before its application to a data set.

Selection of sample scale and unit is based on biological features of the plants, reduction of sample variance, ease of processing, and other methodological constraints (49). When epiphytic bacterial populations were sampled, a high level of variability was reported among sample units at every scale investigated (leaf segments, leaflets, or whole leaves) (49). In contrast, variability of rhizobacteria was significantly less when whole roots were sampled than when root segments were sampled (51). Thus, for phyllosphere sampling, Kinkel et al. (49) suggested that selection of sample scale and unit should not be based only on the assumption that variance is scale dependent but may be more appropriately based on methodological and biological considerations. Although sample variance may be reduced by sampling whole root systems rather than root segments, this sample scale may not always be feasible (e.g., for sampling of mature plants).

Application of Spatial Statistics, Including Geostatistics

Most natural environments are spatially structured by various energy inputs that result in patchy structures or gradients (59). Thus, biological organisms are rarely distributed in a random or uniform manner. The rhizosphere is a good example of this, since energy input is largely due to root deposition. Since the quantity and composition of root deposition vary along the root, spatial variability of deposition from seeds and roots may influence sites of colonization by rhizosphere microbes. Sites may be preferentially colonized by some microbes, such as bacteria in cell junctions (25, 26, 80) or zoospore encystment in the zone of root cell elongation (34, 66, 98). Although the tendency for both phyllosphere and rhizosphere microbial populations to conform to lognormal or similar frequency distributions has been noted (4, 63, 68), the actual spatial variability of these populations is less well documented. Characterization of spatial variability allows us to improve quantification of the process under study as well as to identify the driving variables (73).

Spatial statistical analysis provides a mechanism to explore processes that generate different patterns of organisms over time and determine the sensitivity of patterns to variations in these processes. Spatial analysis is defined here as quantitative evaluation of variations or changes in spatial orientation of an entity or population within a defined area or volume. Such an analysis requires that the spatial integrity (spatial coordinate framework) of the observations be maintained. Frequency distribution methods (e.g., indices of dispersion) are commonly based on mean/variance ratios, which do not provide reliable interpretations of spatial structure, since information on the location of each sample is ignored. Although such indices are useful for estimation of the population mean, they do not maintain spatial integrity of samples, making spatial analysis impossible (45, 69, 82). One method for spatial analysis, geostatistics, determines the degree of autocorrelation among samples on the basis of the direction and distance between them (40, 93) and can be used to interpolate values between measured points on the basis of the degree of autocorrelation encountered (78). Geostatistics detects spatial dependence among neighboring samples and defines the degree of dependence by giving quantifiable parameters. For readers interested in an introductory text on geostatistical theory and practice, Isaaks and Srivastava's text (40) is highly recommended. Other recommended reading includes publications by Rossi et al. (79), Legendre (58), and Robertson (78). Geostatistics has proven highly applicable to biological systems; for example, geostatistics has been effectively used to evaluate insect spatial distributions (47, 84) and a spatial simulation model (54) as well as plant disease patterns (14, 44, 69) and patterns of zoospore encystment on roots (19).

REFERENCES

1. **Amman, R. I., L. Krumholz, and D. A. Stahl.** 1990. Fluorescent-oligonucleotide probing of whole cells for determinative, phylogenetic and environmental studies in microbiology. *J. Bacteriol.* **172:**762–770.
2. **Anderson, J. R., and J. M. Slinger.** 1975. Europium chelate and fluorescent brightener staining of soil propagules and their photomicrographic counting. I. Methods. II. Efficiency. *Soil Biol. Biochem.* **7:**205–215.

3. **Armstrong, J. L., G. R. Knudsen, and R. J. Seidler.** 1987. Survival of recombinant bacteria associated with plants and herbivorous insects in a microcosm. *Curr. Microbiol.* **15:**229–232.

4. **Bahme, J. B., and M. N. Schroth.** 1987. Spatial-temporal colonization patterns of a rhizobacterium on underground organs of potato. *Phytopathology* **77:**1093–1100.

5. **Bennett, R. A., and J. M. Lynch.** 1981. Colonization potential of bacteria in the rhizosphere. *Curr. Microbiol.* **6:** 137–138.

6. **Blakeman, J. P.** 1981. *Microbial Ecology of the Phylloplane.* Academic Press, London.

7. **Bohlool, B. B., and E. L. Schmidt.** 1980. The immunofluorescence approach in microbial ecology. *Adv. Microb. Ecol.* **4:**203–241.

8. **Bottomley, P. J.** 1994. Light microscopic methods for studying soil microorganisms, p. 81–105. *In* R. W. Weaver, J. S. Angle, and P. S. Bottomley (ed.), *Methods of Soil Analysis*, part 2. *Microbiological and Biochemical Properties.* Soil Science Society of America, Madison, Wis.

9. **Campbell, R., and M. P. Greaves.** 1990. Anatomy and community structure of the rhizosphere, p. 11–34. *In* J. M. Lynch (ed.), *The Rhizosphere.* John Wiley & Sons, Chichester, England.

10. **Campbell, R., and R. Porter.** 1982. Low-temperature scanning electron microscopy of microorganisms in soil. *Soil Biol. Biochem.* **14:**241–245.

11. **Campbell, R., and A. D. Rovira.** 1973. The study of the rhizosphere by scanning electron microscopy. *Soil Biol. Biochem.* **5:**747–752.

12. **Carley, H. E., and R. D. Watson.** 1966. A new gravimetric method for estimating root-surface area. *Soil Sci.* **102:** 289–291.

13. **Casida, L., Jr.** 1968. Methods for the isolation and estimation of activity of soil bacteria, p. 97–122. *In* T. R. G. Gray and D. Parkinson (ed.), *The Ecology of Soil Bacteria.* University of Toronto Press, Toronto.

14. **Chellemi, D. O., K. G. Rohrbach, R. S. Yost, and R. M. Sonoda.** 1988. Analysis of the spatial pattern of plant pathogens and diseased plants using geostatistics. *Phytopathology* **78:**221–226.

15. **Cochran, W. G.** 1950. Estimation of bacterial densities by means of the most probable number. *Biometrics* **6:**105–116.

16. **Colwell, R. R., P. R. Brayton, D. J. Grimes, D. B. Roszak, S. A. Huq, and L. M. Palmer.** 1985. Viable but nonculturable *Vibrio cholerae* and related pathogens in the environment: implications for release of genetically engineered microorganisms. *Biol. Tech.* **3:**817–820.

17. **Costerton, J. W., J. C. Nickel, and T. I. Ladd.** 1986. Suitable methods for the comparative study of free-living and surface-associated bacterial populations, p. 49–84. *In* J. S. Poindexter and E. R. Leadbetter (ed.), *Bacteria in Nature*, vol. 2. *Methods and Special Applications in Bacterial Ecology.* Plenum Press, New York.

18. **Curl, E. A., and B. Truelove.** 1986. *The Rhizosphere.* Springer-Verlag, Berlin.

19. **Dandurand, L. M., G. R. Knudsen, and D. J. Schotzko.** 1995. Quantification of *Pythium ultimum var. sporangiiferum* zoospore encystment patterns using geostatistics. *Phytopathology* **85:**186–190.

20. **Darbyshire, J. F., and M. R. Greaves.** 1970. An improved method for the study of the interrelationships of soil microorganisms and plant roots. *Soil Biol. Biochem.* **2:**63–71.

21. **Davies, K. G., and R. Whitbread.** 1989. A comparison of methods for measuring the colonisation of a root system by fluorescent Pseudomonads. *Plant Soil* **116:**239–246.

22. **Dhingra, O. D., and J. B. Sinclair.** 1994. *Basic Plant Pathology Methods*, 2nd ed. CRC Press, Inc., Boca Raton, Fla.

23. **Donegan, K., C. Maytac, R. Sekler, and A. Porteous.** 1991. Evaluation of methods for sampling, recovery and enumeration of bacteria applied to the phylloplane. *Appl. Environ. Microbiol.* **57:**51–56.

24. **Faegri, A., V. L. Torsvik, and J. Goksoyr.** 1977. Bacterial and fungal activities in soil: separation of bacteria and fungi by rapid fractionated centrifugation techniques. *Soil Biol. Biochem.* **9:**105–112.

25. **Foster, R. C.** 1986. The ultrastructure of the rhizoplane and rhizosphere. *Annu. Rev. Phytopathol.* **24:**211–234.

26. **Foster, R. C., and A. D. Rovira.** 1973. The rhizosphere of wheat roots studied by electron microscopy of ultra-thin sections. *Bull. Ecol. Res. Comm.* (Stockholm) **17:**93–102.

27. **Foster, R. C., and A. D. Rovira.** 1976. The ultrastructure of the wheat rhizosphere. *New Phytol.* **76:**343–352.

28. **Fyson, A., P. Kerr, J. N. A. Lott, and A. Oaks.** 1988. The structure of the rhizosphere of maize seedling roots, a cryogenic scanning electron microscopy study. *Can. J. Bot.* **66:**2431–2435.

29. **Geels, F. P., and B. Schippers.** 1983. Reduction of yield depressions in high frequency potato cropping soil after seed tuber treatments with antagonistic fluorescent pseudomonad spp. *Phytopathol. Z.* **108:**207–214.

30. **Giovannoni, S. J., E. F. DeLong, G. J. Olsen, and N. R. Pace.** 1988. Phylogenetic group-specific oligonucleotide probes for identification of single microbial cells. *J. Bacteriol.* **170:**720–726.

31. **Gomez, K. A., and A. A. Gomez.** 1984. *Statistical Procedures for Agricultural Research.* John Wiley & Sons, New York.

32. **Gray, T. R. G., and D. Parkinson.** 1968. *The Ecology of Soil Bacteria.* University of Toronto Press, Toronto.

33. **Haefele, D., and R. Webb.** 1982. The use of ultrasound to facilitate the harvesting and quantification of epiphytic and pathogenic microorganisms. *Phytopathology* **72:**947.

34. **Hickman, C. J., and H. H. Ho.** 1966. Behavior of zoospores in plant-pathogenic phycomycetes. *Annu. Rev. Phytopathol.* **4:**195–220.

35. **Hiltner, L.** 1904. Uber neuere Erfahrungen und Problem auf dem Gebiet der Bodenbakteriologie und unter besonderer Berucksichtigung der Grundungung und Brache. *Arb. Dtsch. Landwirtsch. Ges.* **98:**59–78.

36. **Hirano, S. S., E. V. Nordheim, D. C. Arny, and C. D. Upper.** 1982. Lognormal distribution of epiphytic bacterial populations on leaf surfaces. *Appl. Environ. Microbiol.* **44:** 695–700.

37. **Hirano, S. S., and C. D. Upper.** 1983. Ecology and epidemiology of foliar bacterial plant pathogens. *Annu. Rev. Phytopathol.* **21:**243–269.

38. **Hirano, S. S., and C. D. Upper.** 1993. Dynamics, spread, and persistence of a single genotype of *Pseudomonas syringae* relative to those of its conspecifics on populations of snap bean leaflets. *Appl. Environ. Microbiol.* **59:**1082–1091.

39. **Howie, W. J., and E. Echandi.** 1983. Rhizobacteria: influence of cultivar and soil type on plant growth and yield of potato. *Soil Biol. Biochem.* **15:**127–132.

40. **Isaaks, E. H., and R. H. Srivastava.** 1989. *Applied Geostatistics.* Oxford University Press, Oxford.

41. **Ishimaru, C., K. M. Eskridge, and A. K. Vidaver.** 1991. Distribution analysis of naturally occurring epiphytic populations of *Xanthomonas campestris p.v. phaseoli* on dry beans. *Phytopathology* **81:**262–268.

42. **Jacques, M.-A., L. L. Kinkel, and C. E. Morris.** 1995. Population sizes, immigration, and growth of epiphytic bacteria on leaves of different ages and positions of field-grown endive (*Cichorium endiva* var. *latifolia*). *Appl. Environ. Microbiol.* **61:**899–906.

43. **Johnen, B. G.** 1978. Rhizosphere microorganisms and roots stained with europium chelate and fluorescent brightener. *Soil Biol. Biochem.* **10:**495–502.

44. **Johnson, D. A., J. R. Alldredge, J. R. Allen, and R. Allwine.** 1991. Spatial pattern of downy mildew in hop yards during severe and mild disease epidemics. *Phytopathology* **81:**1369–1374.

45. **Jumars, P. A., D. Thistle, and M. L. Jones.** 1977. Detecting two-dimensional spatial structure in biological data. *Oecologia* **8:**109–123.

46. **Katznelson, H.** 1965. Nature and importance of the rhizosphere, p. 187–209. *In* K. J. Baker and W. C. Snyder (ed.), *Ecology of Soil Borne Plant Pathogens—Prelude to Biological Control.* University of California Press, Berkeley.

47. **Kemp, W. P., T. M. Kalaris, and W. F. Quimby.** 1989. Rangeland grasshopper (Orthoptera: Acrididae) spatial variability: macroscale population assessment. *J. Econ. Entomol.* **82:**1270–1276.

48. **Kinkel, L. L.** 1991. Fungal community dynamics, p. 253–270. *In* J. H. Andrews and S. S. Hirano (ed.), *Microbial Ecology of Leaves.* Springer-Verlag, New York.

49. **Kinkel, L. L., M. Wilson, and S. E. Lindow.** 1995. Effect of sampling scale on the assessment of epiphytic bacterial populations. *Microb. Ecol.* **29:**283–297.

50. **Kloepper, J. W., and C. J. Beauchamp.** 1992. A review of issues related to measuring colonization of plant roots by bacteria. *Can. J. Microbiol.* **29:**1219–1232.

51. **Kloepper, J. W., W. F. Mahafee, J. A. McInroy, and P. A. Backman.** 1991. Comparative analysis of five methods for recovering rhizobacteria form cotton roots. *Can. J. Microbiol.* **37:**953–957.

52. **Kloepper, J. W., M. N. Schroth, and T. D. Miller.** 1980. Effects of rhizosphere colonization by plant growth promoting rhizobacteria on potato plant development and yield. *Phytopathology* **70:**1078–1082.

53. **Knudsen, G. R., and G. W. Hudler.** 1987. Use of a computer simulation model to evaluate a plant disease biocontrol agent. *Ecol. Model.* **35:**45–62.

54. **Knudsen, G. R., and D. J. Schotzko.** 1991. Simulation of Russian wheat aphid movement and population dynamics on preferred and non-preferred host plants. *Ecol. Model.* **57:**117–131.

55. **Labeda, D. P., and M. C. Shearer.** 1990. Isolation of actinomycetes for biotechnological applications, p. 1–19. *In* D. P. Labeda (ed.), *Isolation of Biotechnological Organisms from Nature.* McGraw-Hill, New York.

56. **Lal, R., and S. A. John.** 1994. Ecological applications of atomic force microscopy. *Am. J. Physiol.* **266:**C1–C21.

57. **Langvad, F.** 1980. A simple and rapid method for qualitative and quantitative study of the fungal flora of leaves. *Can. J. Microbiol.* **26:**666–670.

58. **Legendre, P.** 1993. Spatial autocorrelation: trouble or new paradigm? *Ecology* **74:**1659–1673.

59. **Legendre, P., and M.-J. Fortin.** 1989. Spatial pattern and ecological analysis. *Vegetatio* **80:**107–138.

60. **Levanony, H., Y. Bashan, B. Romano, and E. Klein.** 1989. Ultrastructural localization and identification of *Azospirillum brasilense* Cd on and within wheat root by immunogold labeling. *Plant Soil* **117:**207–218.

61. **Lindow, S. E., G. R. Knudsen, R. J. Sekler, M. V. Walter, V. W. Lambou, P. S. Amy, D. Schmedding, B. Prince, and S. Hern.** 1988. Aerial dispersal and epiphytic survival of *Pseudomonas syringae* during a pretest for the release of genetically engineered strains into the environment. *Appl. Environ. Microbiol.* **54:**1557–1563.

62. **Loper, J. E., C. Haack, and M. N. Schroth.** 1985. Population dynamics of soil pseudomonads in the rhizosphere of potato (*Solanum tuberosum* L.). *Appl. Environ. Microbiol.* **49:**416–417.

63. **Loper, J. E., T. V. Suslow, and M. N. Schroth.** 1984. Lognormal distribution of bacterial populations in the rhizosphere. *Phytopathology* **74:**1454–1460.

64. **Lundgren, B.** 1981. Fluorescein diacetate as a stain for metabolically active bacteria. *Oikos* **36:**17–22.

65. **Lynch, J. M.** 1987. Soil biology—accomplishments and potential. *Soil Sci. Soc. Am. J.* **51:**1409–1412.

66. **Mitchell, R. T., and J. W. Deacon.** 1986. Differential (host specific) accumulation of zoospores of Pythium on roots of graminaceous and non-graminaceous plants. *New Phytol.* **102:**113–122.

67. **Newman, E. I.** 1966. A method of estimating the total length of a root in a sample. *J. Appl. Ecol.* **3:**139–145.

68. **Newman, E. I., and H. J. Bowen.** 1974. Patterns of distribution of bacterial on root surfaces. *Soil Biol. Biochem.* **6:**205–209.

69. **Nicot, P. C., D. I. Rouse, and B. S. Yandell.** 1984. Comparison of statistical methods for studying spatial patterns of soilborne plant pathogens in the field. *Phytopathology* **74:**1399–1402.

70. **Norton, J. M., and M. K. Firestone.** 1991. Metabolic status of bacteria and fungi in the rhizosphere of Ponderosa pine seedlings. *Appl. Environ. Microbiol.* **57:**1161–1167.

71. **Parberry, I. H., J. F. Brown, and V. J. Bofinger.** 1981. Statistical methods in the analysis of phylloplane populations, p. 47–65. *In* J. P. Blakeman (ed.), *Microbial Ecology of the Phylloplane.* Academic Press, London.

72. **Parke, J. L., C. M. Liddell, and M. K. Clayton.** 1990. Relationship between soil mass adhering to pea taproots and recovery of *Pseudomonas fluorescens* from the rhizosphere. *Soil Biol. Biochem.* **22:**495–499.

73. **Parkin, T. B.** 1993. Spatial variability of microbial processes in soil—a review. *J. Environ. Qual.* **22:**409–417.

74. **Pfender, W. F., L. G. King, and J. R. Rabe.** 1991. Use of dual-stain fluorescence microscopy to observe antagonism of *Pyrenophora tritici-repentis* by *Limonomyces roseipellis* in wheat straw. *Phytopathology* **81:**109–112.

75. **Phaff, H. J.** 1990. Isolation of yeasts from natural sources, p. 53–79. *In* D. P. Labeda (ed.), *Isolation of Biotechnological Organisms from Nature.* McGraw-Hill, New York.

76. **Ramsay, A. J.** 1984. Extraction of bacteria from soil: efficiency of shaking or ultasonication as indicated by direct counts and autoradiography. *Soil Biol. Biochem.* **16:**475–481.

77. **Richaume, A., D. Steinberg, L. Jocteur-Monrozier, and G. Faurie.** 1993. Differences between direct and indirect enumeration of soil bacteria: the influence of soil structure and cell location. *Soil Biol. Biochem.* **25:**641–643.

78. **Robertson, G. P.** 1987. Geostatistics in ecology: interpolating with known variance. *Ecology* **68:**744–748.

79. **Rossi, R. E., D. J. Mulla, A. G. Journel, and E. H. Franz.** 1992. Geostatistical tools for modeling and interpreting ecological spatial dependence. *Ecol. Monogr.* **62:**277–314.

80. **Rovira, A. D.** 1956. A study of the development of the root surface microflora during the initial stages of plant growth. *J. Appl. Bacteriol.* **19:**72–79.

81. **Rovira, A. D., and R. Campbell.** 1974. Scanning electron microscopy of microorganisms on the roots of wheat. *Microb. Ecol.* **1:**15–23.

82. **Sawyer, A. J.** 1989. Inconstancy of Taylor's B: simulated sampling with different quadrant sizes and spatial distributions. *Res. Popul. Ecol.* **31:**11–24.

83. **Schmidt, E. L., and E. A. Paul.** 1982. Microscopic methods for soil microorganisms, p. 1027–1042. *In* A. L. Page, R. H. Miller, and D. R. Keeney (ed.), *Methods of Soil Analysis: Chemical and Microbiological Methods.* Agronomy Society of America and Soil Science Society of America, Madison, Wis.

84. **Schotzko, D. J., and G. R. Knudsen.** 1992. Use of geostatistics to evaluate a spatial simulation of Russian wheat aphid (Homoptera: Aphididae) movement behavior on preferred and nonpreferred hosts. *Environ. Entomol.* **21:**1271–1282.

85. **Seifert, K. A.** 1990. Isolation of filamentous fungi, p. 21–57. *In* D. P. Labeda (ed.), *Isolation of Biotechnological Organisms from Nature.* McGraw-Hill, New York.

86. **Skinner, F. A., P. C. T. Jones, and J. E. Mollison.** 1952. A comparison of a direct- and a plate-counting technique for the quantitative estimation of soil micro-organisms. *J. Gen. Microbiol.* **6:**261–271.

87. **Smith, D., and A. H. S. Onions.** 1983. A comparison of some preservation techniques for fungi. *Trans. Br. Mycol. Soc.* **81:**535–540.

88. **Snedecor, G. W., and W. G. Cochran.** 1967. *Statistical Methods.* The Iowa State University Press, Ames.

89. **Stamatiadis, S., J. S. Doran, and E. R. Ingham.** 1990. Use of staining and inhibitors to separate fungal and bacterial activity in soil. *Soil Biol. Biochem.* **22:**81–88.

90. **Stanghellini, M. E., and S. L. Rasmussen.** 1989. Root prints: a technique for the determination of the in situ spatial distribution of bacteria on the rhizoplane of field-grown plants. *Phytopathology* **79:**1131–1134.

91. **Suslow, T. V., and M. N. Schroth.** 1982. Rhizobacteria of sugar beets: effects of seed application and root colonization on yield. *Phytopathology* **72:**199–206.

92. **Tennant, D.** 1975. A test of a modified line intersect method of estimating root length. *J. Ecol.* **63:**995–1001.

93. **Trangmar, B. B., R. S. Yost, and G. Uehara.** 1985. Application of geostatistics to spatial studies of soil properties. *Adv. Agron.* **38:**45–94.

94. **Tsien, H. S., B. J. Bratina, K. Tsuji, and R. S. Hanson.** 1990. Use of oligonucleotide signature probes for identification of physiological groups of methylotrophic bacteria. *Appl. Environ. Microbiol.* **56:**2858–2865.

95. **Turner, S. M., and E. I. Newman.** 1984. Growth of bacteria on roots of grasses: influence of mineral nutrient supply and interactions between species. *J. Gen. Microbiol.* **130:**505–512.

96. **van Vuurde, J. W. L., and P. F. M. Elenbaas.** 1978. Use of fluorochromes for direct observation of microorganisms associated with wheat roots. *Can. J. Microbiol.* **24:**1272–1275.

97. **Whipps, J. M.** 1990. Carbon economy, p. 59–97. *In* J. M. Lynch (ed.), *The Rhizosphere.* John Wiley & Sons, Chichester, England.

98. **Zentmyer, G. A.** 1961. Chemotaxis of zoospores for root exudates. *Science* **133:**1595–1596.

99. **Zuberer, D. A.** 1994. Recovery and enumeration of viable bacteria, p. 119–158. *In* R. W. Weaver, J. S. Angle, and P. S. Bottomley (ed.), *Methods of Soil Analysis*, part 2. *Microbiological and Biochemical Properties.* Soil Science Society of America, Madison, Wis.

Sampling Viruses from Soil

CHRISTON J. HURST

43

Human enteric viruses are common microbial contaminants of domestic drainage and sewage systems. As such, they are present in permeates from outhouses, as well as effluents from septic tanks and related waste disposal facilities that generally are discharged directly into soil. They are also present in municipal sewage, which is often discharged onto land following various levels of formal treatment. One of the major public health concerns associated with human enteric viruses contained in domestic drainage and sewage is that they may survive following introduction into the soil environment (11) and subsequently cause illness in susceptible persons who consume virally contaminated groundwaters or crops (15). Research has been done on several topics that are pertinent to this concern, including the extent of viral adsorption to soil, the ability of viruses to persist in soil, migration of viruses through soil in association with the movement of water or wastewater, viral persistence in groundwater, and the survival of viruses on vegetables that might be contaminated by "night soil" or wastewater irrigation. In addition to concerns regarding human enteric viruses in soil, other types of viruses also will be present in the terrestrial environment. These viruses include those which naturally infect other categories of microorganisms: bacteria, algae, fungi, and protozoans. Viruses which infect other categories of microorganisms, often termed phages, can interact with their microbial hosts even within the pores of consolidated rock (4). The soil will also contain viruses of vascular plants and viruses whose natural hosts are vertebrate or invertebrate animals which exist in terrestrial environments.

This chapter focuses on the methodology used for detecting animal viruses in samples of soil. This methodology generally relies on elution and subsequent concentration of viruses from the soil, with subsequent use of either cytopathogenicity or plaque formation assays to detect the viruses. These assays are based on the use of cultured animal cells as hosts for supporting viral replication. This chapter also discusses the use of plaque formation methodology for detecting bacteriophages, or viruses which infect bacteria. Other types of assay procedures, such as those based on PCR (24), have also been developed for detecting viruses in soil samples. Two good general reviews on viruses in soil are those by Duboise et al. (5) and Williams et al. (27).

VIRUSES IN SOIL SAMPLES
Background

Viruses have a strong tendency to bind to soil and mineral particles (11, 17, 19, 25), and therefore it can be expected that most of the viruses in environmental soil samples will be adsorbed to the surface of soil particles. The adsorption of viruses to soil particles occurs by charged colloidal particle-charged surface interactions (18). This adsorption is reversible, and the favorability of adsorption depends on the energy of activation for the system that is being studied (21). The electrostatic interactions which are important to viral adsorption are pH dependent, being stronger at lower pH levels (26). For this reason, the pHs of eluant solutions used to elute (desorb) viruses from soil normally range from neutral to alkaline. Proteinaceous materials such as beef extract are often used as the basis for such eluant solutions in the hope that protein compounds in the eluant will help to displace the adsorbed viruses (3, 16, 24). The success of eluting viruses from soil depends not only on the physical and chemical characteristics of the soil (10, 20) but also on the physiological and morphological characteristics of the viruses (20).

Soil Sampling and Processing Equipment

Many different types of apparatus, ranging from spoons and spatulas to shovels and powered augers, can be used to collect soil samples. Occasionally, even powered earth-moving equipment may be used for clearing away adjacent or overlying material before the soil that is to be sampled can be reached. It is important that those surfaces on the sampling apparatus which will come into direct contact with the soil to be analyzed for viruses not be contaminated with the types of viruses for which the soil is being examined, and it is certainly preferable that these surfaces be sterile. Collection of the final soil sample should be done in an aseptic manner, and this can be facilitated by using such devices as presterilized liners for core samplers and wearing presterilized, disposable gloves. Soil samples should be kept in sterile containers until either assayed for viruses by use of direct inoculation or subjected to an elution process to separate the viruses from the soil particles. The available options for suitable sample containers include wide-mouthed screw-cap plastic jars and zipper closure plastic bags. Soil samples should be kept chilled to reduce thermal inactivation of the viruses. This chilled storage can be done by packing the

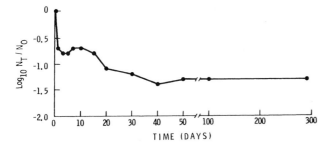

FIGURE 1 Stability of *Enterovirus human poliovirus 1* when frozen in soil at $-70°C$. There was a detrimental freezing effect evidenced as a large initial loss in titer, followed by relative stability. N_0 is the titer at time zero, the outset of freezing (in this case, 1.4×10^4 PFU/g [dry weight] of soil); N_T is the titer at any subsequent elapsed time T. From reference 8a.

sample containers in wet ice (H_2O) or by refrigerating them. A dramatic loss in viral titer may occur upon freezing soil samples (Fig. 1). Thus, soil samples should neither be stored in a freezer nor packed in dry ice (CO_2). Following the use of eluant solutions to elute viruses from the soil, the solutions are termed eluates. The eluates are frequently subjected to concentration techniques for reducing the volume of material which subsequently must be assayed. These eluates can be stored under refrigeration (1 to 4°C) prior to being assayed or even stored frozen at or below $-70°C$ if the viruses of interest are stable when frozen in the eluate. It should be noted that viral infectivity is not necessarily stable in frozen beef extract-based eluates (13).

Many types of equipment used for soil sampling can be sterilized in the laboratory before use in the field. Depending on the composition of the equipment, this sterilization may be done by autoclaving, by using ethylene oxide gas, or by gamma irradiation. Field sterilization of metal or borosilicate glass objects can be performed by using the flame from a portable propane or butane gas torch to heat the objects to red incandescence. Objects less resistant to high heat, such as nonborosilicate glass and some polymer materials, can be surface sterilized by dousing or immersing them in 95% ethanol and then igniting the alcohol with a flame. Objects made from many types of polymers can be subjected to field sterilization techniques which use chlorine solutions, followed by the use of a sodium thiosulfate solution to neutralize the residual disinfectant activity, as described in chapter 17 of this manual.

Reporting Results

The results from the different types of viral assay techniques presented in this chapter are reported as viral infectious units, expressed in terms of either PFU (plaque-forming units), MPN (most probable number; method 9510-G in reference 6), or tissue culture infectious dose as a 50% endpoint (23), depending on the nature of the viral titration assay being performed. These assay results should be reported per unit dry weight of the analyzed soil sample, expressed in either grams or kilograms of soil. Expressing the results in terms of dry weight of the soil will avoid the problem of artifactual differences which may result from comparisons of viral titer reports that are based on the wet weight of soil samples.

DETECTION OF ANIMAL VIRUSES

It is possible to directly assay soil-associated enteric viruses by inoculating dilute soil suspensions onto cultures of animal host cells. Likewise, it is possible to assay soil-associated bacteriophages by directly inoculating dilute soil suspensions into cultures of bacterial host cells. The advantage of this direct inoculation approach is that it eliminates the additional time and materials which would be needed for processing the soil by an elution technique. However, three potential problems must be considered when one is directly inoculating host cell cultures with suspensions of soil particles. First, the soil particles can obscure visualization of the assay results. Second, the soil particles may contain associated chemicals which prove toxic to the host cells. Third, other microorganisms associated with the soil particles may result in contamination that kills or overgrows the host cells. To minimize the adverse effects associated with these problems, the concentration of viruses contained in a soil sample must be high enough to allow for dilution of a suspension of the soil prior to direct inoculation. Otherwise, it will be necessary to elute the viruses from the soil. It also may be necessary to concentrate the resulting eluates before they are assayed.

Titers resulting from the assay of either soil eluates or concentrated eluates cannot be considered equivalent to viral titers obtained from assaying soil suspensions by direct inoculation, since processes used for viral elution and for subsequent concentration of the eluates are not completely efficient. Since elution methods always involve some loss of virus, direct inoculation may, in some ways, be considered to yield a more accurate estimation of the level of viruses present in a soil sample.

Techniques for Elution and Subsequent Concentration of Viruses from Soil Samples

The following methods were developed for detecting human enteric viruses; with appropriate modification, they might also be usable for detecting bacteriophages and other types of viruses.

Elution of Viruses from Soil

For processing by this technique, 50 g (wet weight) of soil should be suspended in 50 ml of a buffered, 10% beef extract solution (containing, per liter, 100 g of commercial powdered beef extract, 13.4 g of $Na_2HPO_4 \cdot 7H_2O$, and 1.2 g of citric acid). The pH of this formulation is approximately 7.0. The beef extract used should be of a type which produces a visible amount of flocculant precipitate when solutions of the beef extract are lowered to pH 3.5. One suitable type is Beef Extract V, available from BBL Microbiology Systems, Cockeysville, Md. The suspensions of soil and buffer should be agitated with a magnetic stirrer for 30 min, during which time the pH of the suspensions should be maintained at 7.0 ± 0.1 by addition of 1 M HCl or 1 M NaOH dropwise as needed. The suspensions should then be centrifuged for 30 min at 2,500 × g, after which the supernatant material is collected and the soil pellet is discarded. The supernatant, now termed an eluate, should then be passed through a stacked sandwich of 3.0-, 0.45-, and 0.25-μm-pore-size (nominal porosity) glass microfiber filters (Duo-Fine series; Filterite Division, Memtec America Corporation, Timonium, Md., or equivalent) to remove remaining soil particles and reduce the level of contaminating soil bacteria. The filtered supernatant (filtered eluate) may then be either directly assayed for the presence of viruses or further concentrated prior to assay, using the concentration method described later in this chapter. If the viruses of interest are stable when frozen, then the filtered eluates prepared from soil samples should be stored either at or below $-70°C$.

Otherwise, the filtered eluates should be stored at 1 to 4°C. This soil elution procedure comes from a collaborative testing study (16).

Concentration of Viruses from Soil Eluates

If the filtered eluate needs to be concentrated, then it should first be diluted to 3% [wt/vol] beef extract by addition of 2.3 equivalent volumes of sterile distilled water. While the diluted eluate is agitated with a magnetic stirrer, the pH of the diluted eluate should be adjusted to 3.5 ± 0.1 by dropwise addition of 1 M HCl. Stirring should be continued until either a visible flocculant precipitate has formed or 30 min has passed. The pH of the solution should be periodically checked during the course of stirring and readjusted as necessary to 3.5 ± 0.1 by addition of either 1 M HCl or 1 M NaOH. The solution should then be centrifuged at 1,000 × g for 5 min at 4°C, following which the resulting supernatant fluid should carefully be removed by pipetting and discarded. A magnetic stirring bar should then be added to the centrifuge bottle containing the precipitate, and the precipitate should be resuspended in 8 to 10 ml (total volume) of 0.05 M glycine, pH 11.0. The pH of the precipitate and glycine buffer mixture should be checked periodically during this stirring and, if necessary, increased to 9.5 by addition of 1 M NaOH. After dissolution of the precipitate, centrifuge the solution at 1,000 × g for 10 min at 4°C and collect the resulting supernatant. This supernatant is now considered a concentrated sample and should be adjusted to pH 7.5 by addition of 1 M HCl (with a caution not to lower the pH below 7.4 to avoid formation of a troublesome precipitate), supplemented with antibiotics if desired (suggested final concentrations: 200 U of penicillin G per ml, 200 μg of streptomycin sulfate per ml, and 2.5 μg of amphotericin B per ml), and made isotonic by addition of 0.5 ml of 3.0 M NaCl per 10 ml of concentrated sample. If the viruses to be studied are stable when frozen in this concentrated sample material, then the concentrated sample should be stored frozen either at or below −70°C until assayed for the presence of viruses. If the target viruses are not stable when frozen in this concentrated sample material, then the concentrated sample should be stored under refrigeration at 1 to 4°C. This procedure comes from a collaborative testing study (16). Not all viruses are stable when exposed to the pH 3.5 and pH 9.5 conditions that are incorporated in this concentration procedure. Thus, caution is necessary when this method is used for concentrating viruses whose pH stability is unknown.

Techniques for Assaying Animal Viruses

The assay techniques described in this chapter were developed for human enteric viruses and use explanted cells that have been propagated in the laboratory to serve as hosts for supporting replication of the viruses. With appropriate modification, these methods might be usable for assaying other types of animal or plant viruses. Methods are presented for performing both a cytopathogenicity assay and a plaque formation assay. Cytopathogenicity assays are based on the appearance of characteristic visible changes in the morphology of living host cells that result from the process of viral infection. Such changes, termed cytopathogenic effects (CPE), can be assessed by microscopic examination of the infected cells. Plaque assays detect the production of focal areas of cell death (plaques) that may appear within a virally infected culture of host cells. Detection of these plaques is often facilitated by exposing the cultures of cells to a vital stain such as neutral red. Neutral red causes live cells to appear pink, while dead cells appear colorless. At modest concentrations, this dye does not seem to interfere with cellular viability. The plaques normally can be detected by visual examination of the infected cell cultures without aid of a microscope. Further background information on cytopathogenicity and plaque assay procedures, along with specific information on their use with environmental samples and comparisons of their relative effectiveness for this purpose, can be found in chapter 8 of this manual and reference 14. These viral assays should always be performed with monolayer cultures of cells that have been grown to confluency. Stock virus suspensions for use as positive controls in animal virus assays can be prepared as described in chapter 8 of this manual.

Method for Performance of Viral CPE Assays

A cell line which can be used for detecting a large number of different human enteric virus species by CPE assays is BGM, a continuous line originating from African green monkey kidneys. A cell line useful for detecting human adenoviruses is HEp-2, a continuous human epidermoid carcinoma line. The BGM cell line can be obtained from Daniel R. Dahling, U.S. Environmental Protection Agency (Cincinnati, Ohio), or else purchased as BGMK (Buffalo green monkey kidney) from Bio-Whittaker (Walkersville, Md.). The HEp-2 cell line can be purchased from the American Type Culture Collection (Rockville, Md.). Production of individual tube cultures of either BGM or HEp-2 cells for use in CPE assays can be done in screw-cap 16- by 125-mm borosilicate glass or polystyrene cell culture tubes. Individual cultures of cells for use in CPE assays can also be prepared in multiple-well polystyrene cell culture plates. Plates which contain 24 wells work well for this type of assay. Both of these cell lines should be incubated at 37°C. Culture tubes should be tightly capped during incubations. Multiple-well plates should be sealed with transparent sealing tape during incubations.

The growth medium used for routine cultivation of the BGM cell line consists of equal parts Eagle's minimum essential medium (MEM) and Leibovitz medium L-15, supplemented with 10% (vol/vol) fetal bovine serum (FBS) and 5 ml of a 7.5% (wt/vol) stock sodium bicarbonate solution per liter. This same medium can also be used for preparing individual tube or plate cultures of BGM cells for use in virus CPE assays. The growth medium used both for routine cultivation of the HEp-2 cell line and for preparation of tube or plate cultures of the HEp-2 line should consist of MEM supplemented with 10% (vol/vol) FBS and 5 ml of sodium bicarbonate solution per liter. Either Earle's balanced salt solution or MEM, supplemented with FBS at a final level of 5% (vol/vol) and 15 ml of a 7.5% (wt/vol) stock sodium bicarbonate solution per liter, can be used for both cell lines as a maintenance medium for maintaining the viability of confluent cell culture monolayers prior to performance of the CPE assay. Both the growth and maintenance media can be supplemented with N-(2-hydroxyethyl)-piperazine-N'-2-ethanesulfonic acid (HEPES) buffer at a level of 20 mM, although addition of HEPES may require a corresponding change in the amount of sodium bicarbonate that is added to the medium. Antibiotics can be added to growth and maintenance media to suppress microbial contamination; a combination of penicillin G at 100 U/ml, amikacin sulfate at 30 μg/ml, and streptomycin sulfate at 100 μg/ml is suggested.

One hour prior to inoculation of the cell cultures with sample material, the medium in the tubes or plates should be substituted with an inoculation medium consisting of FBS-free Hanks balanced salt solution or MEM containing 5 ml of sodium bicarbonate solution per liter and antibiotics. Guanidine hydrochloride should be present in the inoculation medium at a final concentration of 100 μg/ml when cultures of HEp-2 cells are infected for detection of adenoviruses. The guanidine is used to selectively suppress the replication of enteroviruses while allowing unrestricted replication of adenoviruses (14). Guanidine should be present neither during routine passage of the cell lines nor during the production of cell monolayers in the tubes or multiple-well plates. Guanidine should not be used when one is attempting to isolate enteroviruses. Following addition of the inoculation medium, the cultures should be incubated for 1 h at 37°C and then inoculated by addition of 0.05 to 0.1 ml of environmental sample material to the inoculation medium contained in each tube or well. This inoculum may first be diluted as necessary in a diluent, such as inoculation medium, before being added to the cell cultures. After 4 h of postinoculation incubation at 37°C, the medium and inoculum contained in the inoculated tubes or wells should be removed and replaced with fresh inoculation medium that does not contain any sample material. The inoculated tubes or plates should then be incubated at 37°C, and the medium in the tubes or plates should be replaced on days 2 and 5 postinoculation (plus day 8 for adenovirus assays) with fresh inoculation medium containing FBS at a final concentration of 2% (vol/vol). Following inoculation with sample material, culture tubes can be incubated on a roller apparatus during the full postinoculation period in an effort to speed development of CPE. The cell cultures should be examined on days 2 to 7 postinoculation for the presence of characteristic enterovirus CPE or on days 4 to 10 postinoculation for the presence of characteristic adenovirus CPE. The viral titers from this type of assay are normally expressed in terms of tissue culture infectious dose as a 50% endpoint as determined by the technique of Reed and Muench (23) or as most probable number (a 100% endpoint which can be determined by using method 9510-G in reference 6) per gram (dry weight) of the soil sample that was examined. An estimation technique which can be used to convert 50% endpoint assay titers to 100% values has been published by Hurst et al. (14).

Method for Performance of Viral Plaque Assays

The BGM cell line can be used for detecting a large number of different *Enterovirus* species and also some other enteric viruses by plaque formation assay. Viral plaque formation assays can be performed by using a soft agar overlay technique, and the assay titers should be reported in terms of PFU (a 100% endpoint) per gram (dry weight) of the soil sample that was examined. The growth and maintenance media used for this cell line have been described above in the CPE assay section. The plaque formation assays can be performed with confluent monolayer cultures of cells grown in either clear glass prescription bottles, glass cell culture flasks, or polystyrene cell culture flasks. Growth of the cells and use of inoculation medium should be done using the type of procedure described for the CPE assay. The cell culture bottles or flasks should be tightly capped during all incubations. The inoculation medium should be removed from the bottles or flasks immediately prior to introduction of the inoculum onto the cell monolayers. The inoculum consists of an environmental sample that may have been diluted as necessary with a diluent such as inoculation medium. The volume of inoculum used per bottle or flask will vary depending on the surface area of the monolayer con-

tained in the individual bottles or flasks. Typically, one would add 0.1 to 0.2 ml of inoculum to a 25-cm² (cell surface area) culture bottle or flask. The inoculum should be evenly distributed over the surface of the cell monolayers by gently rocking (tilting) the bottles or flasks. After inoculation, the cultures should be incubated for 1 h at 37°C to allow viruses from the inoculum to adsorb (attach) to the cultured cells. The cultures may be kept on a gently tilting, rotating table during this adsorption period to aid even distribution of the inoculum and to prevent drying of the monolayer surface. Agar overlay medium should then be added to the cultures at a rate of 10 ml/25 cm² area of cell monolayer. A suggested overlay medium formulation is MEM lacking phenol red, containing 7.5 mg of neutral red dye per liter, and supplemented to final levels of 1% (vol/vol) FBS, 15 ml of 7.5% (wt/vol) stock sodium bicarbonate solution per liter, 7.5 g of agar per liter, and 50 mg of MgCl₂ per liter. It is advisable to add antibiotics to the overlay medium.

It is desirable to confirm the positive status of at least 10 plaques per sample, or all plaques if assaying the aliquoted sample yields less than that minimum number. Confirmation can be accomplished by individually collecting material from each of the 10 plaques and then inoculating the collected material from each plaque into separate, fresh cell cultures as described for the CPE assay. These fresh cultures should then be examined for the development of characteristic CPE as described for the CPE assay. Development of characteristic CPE in the inoculated cultures is considered a confirmation that the plaque was formed by viruses capable of replicating in that cell line. The percentage of confirmable plaques should be used to establish a confirmed virus level for each sample. It is those confirmed values which should be used for statistical analysis. The results of this type of assay are reported either as PFU or as confirmed PFU per gram (dry weight) of the soil sample that was examined.

Recent modifications that show promise for enhancing the sensitivity of detecting enteric viruses by plaque formation assays include pretreatment of the cell cultures with a medium that contains iododeoxyuridine and performance of sequential inoculation (2). Sequential inoculation is done by incubating the inoculum on a culture of an initial cell line for 2 h and then transferring that inoculum to a second culture of either the same or a different cell line. As an example, the best combination of cell lines for detecting species of *Enterovirus* might be BGM for the initial cell line and then either BGM or RD (a human rhabdomyosarcoma cell line) as the second cell line. The plaque assay titers from the two cultures are then summed to yield a total plaque assay titer (2). Concentrated environmental samples are sometimes toxic to the cell culture lines. This toxicity can be caused by the presence of organic compounds that are eluted from the soil and concentrated along with the viruses. Several approaches have been developed to reduce these cytotoxic effects (12). The simplest approach, and one which does not result in an overall loss of viral titer, is to dilute the filtered eluate or concentrated sample material before introducing it onto the cultured cells. A simple alternative is to wash the inoculum off the cell monolayers before adding the overlay medium. This is done after the sample is allowed to incubate on the monolayers for 1 h at either room temperature or 37°C, so that virus adsorption to the cells can occur. After this exposure period, the inoculum is poured out of the cell culture bottles or flasks and replaced with 10 ml (per 75-cm² cell culture area) of washing solution (0.85% [wt/vol] sodium chloride containing 2% [vol/vol] FBS). The culture bottles or flasks are then gently rocked twice from side to side, which allows the washing solution to resuspend any remaining inoculum. The washing solution is then poured out, and soft agar overlay medium is dispensed into the cell culture bottles or flasks. Some loss of virus titer can be expected to result from this washing procedure (12).

DETECTION OF BACTERIOPHAGES

Bacteriophages can be detected by direct assay of soil suspensions by using a plaque formation technique. If the presence of soil particles in the assay causes a problem, either because the resulting turbidity obscures assessment of the plaque assay results or because the number of contaminating soil bacteria and fungi carried along with the soil particles complicates plaque enumeration, then the bacteriophages can be eluted from the soil particles and the eluate can be assayed. The elution method described above for use with human enteric viruses can be used for elution of bacteriophages from soil, with assay of the filtered eluate. Some bacteriophages have a very limited resistance to pH 3.5 exposure. For this reason, it is preferable to avoid performing the low-pH concentration step if the processed eluate is to be analyzed for the presence of bacteriophages.

The major issue when performing a bacteriophage assay is understanding that candidate bacterial host strains differ with respect to the range of bacteriophages to which they are susceptible (9, 22). There are two general types of plaque assay methods which can be used to enumerate bacteriophages. The first is the traditional double-agar-layer technique (8). This approach relies on the use of a bottom layer of agar medium that has a high concentration of agar (10 to 15 g/liter) to help a top layer of agar medium, whose agar concentration is lower (5 to 8 g/liter), adhere to the inside bottom of the petri dish when the dishes are inverted during incubation. When this double-agar-layer technique is used, the layer of bottom agar medium is added to the petri plates and allowed to congeal. The bacteriophage plaque assay is then performed in the subsequently added layer of top agar medium. The host bacteria grow to form a turbid suspension, termed a lawn, within the top agar layer. The bacteriophages produce areas of lysis in the host lawn. These areas of lysis appear circular and are termed plaques. This traditional method can be used to assay approximately 1 ml of sample per 10-cm-diameter petri dish. The bottom agar layer is presumed to be inert, although dissolved compounds can diffuse from the bottom agar layer into the top agar layer. An example of intentional use of this diffusion action is the incorporation of crystal violet dye into the bottom agar layer (1), done to inhibit growth of contaminating gram-positive bacteria in the top agar layer. A typical medium formulation for this double-agar-layer assay would be, per plate, approximately 15 ml of a bottom agar medium composed of either tryptic soy broth or a tryptone-yeast extract medium such as modified LB broth (9) containing 10 g of agar per liter. The top agar layer for each 10-cm-diameter petri dish would consist of 3 ml of the same type of broth medium supplemented with salts of calcium, magnesium, or both to aid phage replication and containing 8 g of agar per liter, 1 ml of the sample or diluted sample that is being assayed, and 0.2 ml of a fresh overnight culture of host bacterial strain grown in the same broth medium (9). After the top agar layers congeal within the petri plates, the plates are inverted and subjected to a subsequent overnight incubation to allow formation of the lawn and development of visible plaques. The final agar concentration in the top agar layer with this formulation would be 5.7 g/liter. The composition of modified LB broth medium is, per liter, 10 of Bacto Tryptone (Difco Laboratories, Detroit, Mich.), 5 of Bacto Yeast Ex-

tract (Difco), 9 g of sodium chloride, 1 g of magnesium chloride heptahydrate, and 1 ml of 1 M calcium chloride. The pH of modified LB broth medium should be adjusted to 7.0 by addition of NaOH before the medium is sterilized by autoclaving prior to use.

An alternative bacteriophage assay approach is the single-agar-layer technique (reference 7 and method 9211-D in reference 6), which relies on use of a top agar layer that has a higher final agar concentration (approximately 8 g/ liter) to eliminate the need for a bottom agar layer. This allows the assay of as much as 5 ml of water sample or diluted sample per 10-cm-diameter petri dish. A typical medium formulation for this single-agar-layer technique would be 100 ml of a double-strength broth medium containing 16 g of agar per liter, combined with 100 ml of either sample or diluted sample plus 10 ml of a fresh overnight culture of host bacteria grown in the same type of broth medium. This mixture would be divided among 21 petri dishes at 10 ml per 10-cm-diameter dish. The final agar concentration in this single agar layer would be approximately 7.6 g/liter.

The final concentration of agar used in the single-agar-layer method or in the top agar layer of the traditional double-agar-layer method may affect the dispersion of phages through the plaque assay medium. For this reason, stock suspensions of some bacteriophage types seem to produce lower plaque assay titers with the single-agar-layer technique, which uses a higher agar concentration, than with the traditional double-agar-layer technique. An example of this, from personal observation, is the replication of phage 23356-B1 (American Type Culture Collection) when grown in its host, *Klebsiella pneumoniae* (ATCC 23356).

The addition of tetrazolium dyes to the agar medium can aid visualization of phage plaques. Dye concentrations of up to 150 μg/ml seem to be usable for this purpose without lowering the assay titers (9). The live bacterial cells in the lawn will appear colored because they reduce the dye to its colored formazan, while the plaque areas contain oxidized dye and thus will appear relatively colorless. This dye supplementation technique can be used with both the traditional double-agar-layer assay technique and the single-agar-layer technique. Dye supplementation can be accomplished by adding 0.75 ml of a 2% (wt/vol) filter-sterilized aqueous solution of either tetrazolium red dye (T-8877; Sigma Chemical Company, St. Louis, Mo.) or tetrazolium violet dye (T-0138; Sigma) per 100 ml of top agar medium used in the double-agar-layer technique. The amount of dye solution added per 100 ml of the double-strength medium used in the single-agar-layer technique would be 1.5 ml. A recent technique for reducing the level of contaminating bacteria in samples which are to be assayed for the presence of bacteriophage is to treat the samples with hydrogen peroxide and then add catalase to destroy any residual peroxide (1) prior to performing a viral assay on the samples.

Stock bacteriophage suspensions for use in comparative assays and to be used as positive controls in bacteriophage assays on material from environmental samples can be prepared by the following technique, which is similar to that described by Hershey et al. (8). First, a double-agar-layer assay is performed, and petri plates showing confluent lysis of the lawn by bacteriophages are each flooded with 5 ml of broth medium. Following 2 h of incubation at room temperature, the petri plates are swirled gently. The broth medium, now containing suspended phage, is aspirated off the top agar layer with a pipette and filtered through either a 0.25- or a 0.45-μm-pore-size (rated by absolute filtration) filter to remove contaminating bacteria. If the strain of bacteriophage being examined is stable when frozen, then the produced bacteriophage stock suspension (filtered aspirate) can be stored frozen either at or below − 70°C. If the strain of bacteriophage being examined is not stable when frozen, then the bacteriophage stock suspension should be stored under refrigeration at 1 to 4°C. The medium used for preparing stock bacteriophage suspensions should not contain any tetrazolium dye. The incubation temperature used for bacteriophage plaque formation assays varies depending on the host bacterial strain being used and the group of bacteriophages whose detection is being sought.

REFERENCES

1. **Asghari, A., S. R. Farrah, and G. Bitton.** 1992. Use of hydrogen peroxide treatment and crystal violet agar plates for selective recovery of bacteriophages from natural environments. *Appl. Environ. Microbiol* **58:**1159–1163.
2. **Benton, W. H., and C. J. Hurst.** 1990. Sequential inoculation as an adjunct in enteric virus plaque enumeration. *Water Res.* **24:**905–909.
3. **Bitton, G., M. J. Charles, and S. R. Farrah.** 1979. Virus detection in soils: a comparison of four recovery methods. *Can. J. Microbiol.* **25:**874–880.
4. **Chang, P. L., and T. F. Yen.** 1984. Interaction of *Escherichia coli* B and B/4 and bacteriophage T4D with Berea sandstone rock in relation to enhanced oil recovery. *Appl. Environ. Microbiol.* **47:**544–550.
5. **Duboise, S. M., B. E. Moore, C. A. Sorber, and B. P. Sagik.** 1979. Viruses in soil systems. *Crit. Rev. Microbiol.* **7:**245–285.
6. **Eaton, A. D., L. S. Clesceri, and A. E. Greenberg (ed.).** 1995. *Standard Methods for the Examination of Water and Wastewater,* 19th ed. American Public Health Association, Washington, D.C.
7. **Grabow, W. O. K., and P. Coubrough.** 1986. Practical direct plaque assay for coliphages in 100-ml samples of drinking water. *Appl. Environ. Microbiol.* **52:**430–433.
8. **Hershey, A. D., G. Kalmanson, and J. Bronfenbrenner.** 1943. Quantitative methods in the study of the phage-antiphage reaction. *J. Immunol.* **46:**267–279.
8a. **Hurst, C. J.** 1979. Viral detection and persistence during the land treatment of sludge and wastewater. Ph.D. dissertation. Baylor College of Medicine, Houston, Tex.
9. **Hurst, C. J., J. C. Blannon, R. L. Hardaway, and W. C. Jackson.** 1994. Differential effect of tetrazolium dyes upon bacteriophage plaque assay titers. *Appl. Environ. Microbiol.* **60:**3462–3465.
10. **Hurst, C. J., and C. P. Gerba.** 1979. Development of a quantitative method for the detection of enteroviruses in soil. *Appl. Environ. Microbiol.* **37:**626–632.
11. **Hurst, C. J., C. P. Gerba, and I. Cech.** 1980. Effects of environmental variables and soil characteristics on virus survival in soil. *Appl. Environ. Microbiol.* **40:**1067–1079.
12. **Hurst, C. J., and T. Goyke.** 1983. Reduction of interfering

cytotoxicity associated with wastewater sludge concentrates assayed for indigenous enteric viruses. *Appl. Environ. Microbiol.* **46:**133–139.
13. **Hurst, C. J., and T. Goyke.** 1986. Stability of viruses in waste water sludge eluates. *Can. J. Microbiol.* **32:**649–653.
14. **Hurst, C. J., K. A. McClellan, and W. H. Benton.** 1988. Comparison of cytopathogenicity, immunofluorescence and in situ DNA hybridization as methods for the detection of adenoviruses. *Water Res.* **22:**1547–1552.
15. **Hurst, C. J., and P. A. Murphy.** 1996. The transmission and prevention of infectious disease, p. 3–54. *In* C. J. Hurst (ed.), *Modeling Disease Transmission and Its Prevention by Disinfection.* Cambridge University Press, Cambridge.
16. **Hurst, C. J., S. A. Schaub, M. D. Sobsey, S. R. Farrah, C. P. Gerba, J. B. Rose, S. M. Goyal, E. P. Larkin, R. Sullivan, J. T. Tierney, R. T. O'Brien, R. S. Safferman, M. E. Morris, F. M. Wellings, A. L. Lewis, G. Berg, P. W. Britton, and J. A. Winter.** 1991. Multilaboratory evaluation of methods for detecting enteric viruses in soils. *Appl. Environ. Microbiol.* **57:**395–401.
17. **Lefler, E., and Y. Kott.** 1974. Enteric virus behavior in sand dunes. *Isr. J. Technol.* **12:**298–304.
18. **Moore, R. S., D. H. Taylor, M. M. Reddy, and L. S. Sturman.** 1982. Adsorption of reovirus by minerals and soils. *Appl. Environ. Microbiol.* **44:**852–859.
19. **Moore, R. S., D. H. Taylor, L. S. Sturman, M. M. Reddy, and G. W. Fuhs.** 1981. Poliovirus adsorption by 34 minerals and soils. *Appl. Environ. Microbiol.* **42:**963–975.
20. **Ostle, A. G., and J. G. Holt.** 1979. Elution and inactivation of bacteriophages on soil and cation-exchange resin. *Appl. Environ. Microbiol.* **38:**59–65.
21. **Preston, D. R., and S. R. Farrah.** 1988. Activation thermodynamics of virus adsorption to solids. *Appl. Environ. Microbiol.* **54:**2650–2654.
22. **Rajala-Mustonen, R. L., and H. Heinonen-Tanski.** 1994. Sensitivity of host strains and host range of coliphages isolated from Finnish and Nicaraguan wastewater. *Water Res.* **28:**1811–1815.
23. **Reed, L. J., and H. Muench.** 1938. A simple method of estimating fifty percent endpoints. *Am. J. Hyg.* **27:** 493–497.
24. **Straub, T. M., I. L. Pepper, M. Abbaszadegan, and C. P. Gerba.** 1994. A method to detect enteroviruses in sewage sludge-amended soil using the PCR. *Appl. Environ. Microbiol.* **60:**1014–1017.
25. **Taylor, D. H., A. R. Bellamy, and A. T. Wilson.** 1980. Interaction of bacteriophage R17 and reovirus type III with the clay mineral allophane. *Water Res.* **14:**339–346.
26. **Taylor, D. H., R. S. Moore, and L. S. Sturman.** 1981. Influence of pH and electrolyte composition on adsorption of poliovirus by soils and minerals. *Appl. Environ. Microbiol.* **42:**976–984.
27. **Williams, S. T., A. M. Mortimer, and L. Manchester.** 1987. Ecology of soil bacteriophages, p. 157–179. *In* S. M. Goyal, C. P. Gerba, and G. Bitton (ed.), *Phage Ecology.* John Wiley & Sons, New York.

Isolation, Culture, and Detection of Arbuscular Mycorrhizal Fungi

AMIEL G. JARSTFER AND DAVID M. SYLVIA

44

Mycorrhizae are mutualistic associations between beneficial soil fungi and plant roots and are common in natural soils. They have an important role in increasing plant uptake of phosphorus (P) and other poorly mobile nutrients (38). The arbuscular mycorrhizal (AM) fungi colonize more than 90% of all vascular plant families. The term vesicular-arbuscular mycorrhiza was originally applied to this symbiotic association, but since a major suborder lacks the ability to form vesicles in roots, AM is now the preferred acronym. Because of the obligate nature of these organisms, their manipulation is different than for most other soilborne fungi, and much is still unknown about their biology in natural and managed ecosystems. Disturbance can change AM fungus distribution, abundance, and species composition within the ecosystem (30). Although isolation of their spores from the environment is relatively simple, multiplication of spores generally requires several months of growth under high-light conditions. Advanced immunological, physiological, and molecular techniques to detect and identify soilborne organisms are now being applied to AM fungi and should lead to a better understanding of their distribution and function in the environment.

The purpose of this chapter is to summarize methods to (i) isolate and estimate soilborne propagules of AM fungi, (ii) propagate AM fungi by traditional and innovative methods, and (iii) biochemically detect and assess properties of these fungi by using more advanced methodology. A wide range of methods is currently used successfully by various research groups. Therefore, these methods may be modified for application in different situations. For further detail and explanation, the reader is referred to reviews of these methods (18, 24, 26, 37, 45) and to a more extensive, step-by-step treatment of many of the procedures presented here (51). Methods for manipulating ectomycorrhizal fungi have been detailed elsewhere (36).

ISOLATION AND ESTIMATION OF SOILBORNE PROPAGULES

AM fungi are among the most common fungi in soil, but they are often overlooked because they do not grow on standard dilution-plating media. Soilborne propagules of AM fungi may include chlamydospores or azygospores, colonized

Florida Agricultural Experiment Station Journal Series no. R- 4709.

roots, and hyphae. Isolating spores and quantifying root colonization are the most basic procedures for working with these fungi. Spores are required for establishing pure, single-species cultures, while detection of colonization in roots is necessary to verify a functional association by visualization of arbuscules. The spores of AM fungi are larger than those of most other fungi, ranging from 10 to 1,000 μm in diameter. Most spores are between 100 and 200 μm in diameter and can easily be observed with a dissecting microscope.

Hayman (18) and Schenck and Perez (45) have reviewed several methods for extracting spores of AM fungi from soil. The wet sieving and decanting, density-gradient-centrifugation method is the most widely used and problem-free method. The major variable in its application is in the use of single or multiple densities of sucrose. Multiple layers of different densities can provide cleaner spores and are useful for separating different species. Another method of separating spores from debris utilizes a series of sieves of various pore sizes. As with most of the techniques, it works best for sandy soils and less well for clay or organic soils. Soil samples with a significant clay content can be soaked in 6.3 mmol sodium hexametaphosphate to disperse the clay fraction (31).

Method for Spore Isolation

After collecting a soil sample, following suggestions for size and area coverage (see chapter 4 of this manual), place the sample (50 to 100 g is usually sufficient) into a 2-liter container and add 1.5 liters of water. Vigorously mix the suspension to free spores from soil and roots. For fungal species that form spores in roots (e.g., *Glomus intraradices* and *G. clarum*), blend the soil-root sample for 1 min in 300 ml of water to free spores from roots. Next, let the suspension settle for 15 to 45 s (times vary depending on soil texture) and decant the supernatant through standard sieves. Sieves should be selected so as to capture the spores of interest. The spore sizes of various species can be found in reference 45. Use a 425-μm sieve over a 45-μm sieve for unknown field samples. Examine the contents of the top sieve for sporocarps that may be up to 1 mm in diameter. For clay soils, it is advisable to repeat the decanting and sieving procedure on the settled soil. Roots can be collected from the larger-mesh sieve for evaluation of internal colonization. Transfer sievings to 50-ml centrifuge tubes with a fine stream of water from a wash bottle, and balance opposing tubes. Centrifuge the tubes at 1,200 to 1,300 × g in a swing-ing-bucket rotor for 3 min, allowing the centrifuge to stop without braking. Remove the supernatant carefully to avoid disturbing the pellet, and then with a finger remove the organic debris that adheres to the side of the tube. Suspend soil particles in chilled 1.17 M sucrose, mix the contents with a spatula, and centrifuge the samples immediately at 1,200 to 1,300 × g for 1.5 min, applying the brake to stop the centrifuge. Pour the supernatant through the small-mesh sieve, carefully rinse the spores held on the sieve with tap water, and wash the spores into a plastic petri dish scribed with parallel lines spaced 0.5 cm apart. Spores may be counted by scanning the dish under a dissecting microscope.

The identification of AM fungi can be difficult because the taxonomy of this group is based almost entirely on a limited number of morphological characteristics of the spores. Unless taxonomy is the major objective, it is recommended that the spores be keyed initially only to genus. Isolates of special interest should be given a unique isolate code and then classified to species at a later date. For authoritative detail on the identification of spores, the reader is directed elsewhere (34, 35). Maintenance of good germ plasm should be an essential part of any AM research program; therefore, careful notes need to be taken on the size, color, surface characteristics, and wall morphology of the spore types recovered. Initiate pot cultures with each spore type, keeping detailed records on the origin and subsequent pot culture history of the isolate.

The International Collection of Vesicular-Arbuscular Mycorrhizal Fungi (INVAM) maintains a collection of AM isolates. Samples of AM fungi may be submitted to INVAM for verification of classification and possible inclusion in their collection. For additional information, contact Dr. J. Morton, Division of Plant and Soil Sciences, West Virginia University, Morgantown, WV 26506-6057 (electronic mail address, INVAM@WVNVM.WVNET.EDU; World Wide Web site, http://invam.caf.wvu.edu/). Another culture collection, La Banque Européenne des Glomales, exists in Europe. For additional information, contact Dr. John Dodd, The International Institute of Biotechnology, University of Kent at Canterbury, P.O. Box 228, Canterbury, Kent CT2 7YW, United Kingdom (electronic mail address, J.C.DODD@UKC.AC.UK; World Wide Web site, http://kiwi.ukc.ac.uk/biolab/beg/index.html).

Root Colonization

The AM fungi do not cause obvious morphological changes of roots; however, they produce arbuscules and, in many cases, vesicles in roots. To observe AM structures within the root, it is necessary to clear cortical cells of cytoplasm and phenolic compounds and then to differentially stain the fungal tissue. Phillips and Hayman (40) published the oft-cited method to visualize AM fungi in roots by using 0.05% trypan blue in lactophenol, but the use of phenol is now discouraged (28). The clearing agent for nonpigmented roots is generally 10% KOH, but treatment with H_2O_2 (40) or NaOCl (15) may be necessary for pigmented roots. Decolorizing with hydrogen peroxide is slower than decolorizing with hypochlorite, but there is less danger of complete destruction of fungal and cortical tissue. However, hypochlorite is a very fast and effective bleaching agent, and the procedure requires no heating. Alternatives to trypan blue for staining are chlorazol black E (7) and acid fuchsin (27). For nonpigmented roots, it is also possible to observe colonization nondestructively by inducing autofluorescence (1).

Method for Clearing and Staining Roots

Place root samples (approximately 0.5 g) in perforated plastic holders (e.g., OmniSette tissue cassettes; Fisher Scientific, Pittsburgh, Pa.) and store them in cold water until they are processed. Place enough 1.8 M KOH into a beaker (without samples) to allow samples to be covered, and heat the samples to 80°C in a fume hood. Place samples in the heated KOH for the desired time—15 min for tender roots such as onion and 30 min for other roots. If samples are still pigmented after the initial treatment, rinse them with at least three changes of water and then place them in a beaker with either 30% (wt/wt) H_2O_2 at 50°C or 3% (wt/vol) NaOCl, acidified with several drops of 5 M HCl. Times can vary from several seconds to several minutes. Check roots frequently to avoid destruction of the cortex and fungal tissue. Rinse roots with copious amounts of water as soon as samples are bleached white or become transparent, and then rinse them with five changes of tap water. Cover the samples with tap water, add 5 ml of concentrated HCl for each 200 ml of water, stir, and drain. Repeat once. Dispense enough trypan blue stain into a beaker (without samples) to cover samples, and heat the samples to 80°C. To prepare the stain, add in order to a flask while stirring: 800 ml of glycerine, 800 ml of lactic acid, 800 ml of distilled water, and finally 1.2 g of trypan blue. Place samples in the stain for at least 30 min, cool them overnight, and drain the stain into the large flask for reuse. Rinse samples with two changes of tap water to destain. Additional destaining in water may be necessary for some roots.

Various methods have been used to estimate root colonization by AM fungi (18, 27). The gridline-intersect method has the advantage of providing an estimate of both the proportion of colonized root and total root length (14). This is important because some treatments affect root and fungal growth differently. For example, when P is applied, total root length may increase more rapidly than colonized root length, and thus the proportion of colonized root will decrease even though the actual length of colonized root is increasing.

McGonigle et al. (29) argued that the gridline-intersect method is somewhat subjective because arbuscules may be difficult to distinguish with a dissecting microscope. They proposed use of a magnified-intersect method whereby roots are observed at a magnification of ×200 and arbuscules are quantified separately from vesicles and hyphae. Another limitation of the gridline-intersect method is that the intensity of colonization at each location is not estimated. To obtain an estimate of intensity, one can use a morphometric technique (53) whereby a grid of dots is placed over an image of squashed roots and colonized cortical cells are counted.

Method for Root Colonization Assessment

Spread a cleared and stained root sample evenly in a scribed, 10-cm-diameter plastic petri dish. A grid of squares should be scribed on the underside of the dish as specified by Giovannetti and Mosse (14) so that the total number of root intersections will be equal to root length in centimeters. Using a dissecting microscope, scan only the grid lines and record the total number of root intersections with the grid as well as the number of intersects with colonized roots. Verify any questionable colonization with a compound microscope. To do this, cut out a small portion of the root with a scalpel, place it in water on a microscope slide, and look for AM structures at a magnification of ×100 to ×400. Remember that the stains are not specific for AM fungi—other fungi colonizing the root will also stain, and so it is important to verify the presence of arbuscules or vesicles in the root by using a compound microscope.

Propagule Assays

Spore counts often underestimate numbers of AM fungi, since colonized roots and hyphae can also serve as propagules. The most commonly used methods to obtain an estimate of total propagules are the most probable number (MPN) and infectivity assays. The MPN assay provides estimates of propagule numbers, but confidence limits are usually very large. The infectivity assay is less complex and time-consuming than the MPN assay, but the actual propagule numbers are not estimated. Rather, the infectivity assay provides a relative comparison of propagule density among various soils or treatments.

MPN Assay

The MPN assay was developed as a method for estimating the density of organisms in a liquid culture (8). It was first used to estimate the propagule density of AM fungi in soil by Porter (41). The general procedure for the MPN assay is to dilute natural soil with disinfested soil. Place equal portions of the dilution series into small containers (5 to 10 replications of each dilution), plant a susceptible host plant in each container, and grow the plants long enough to obtain good root colonization. Plants are then washed free of soil, and roots are assessed for the presence or absence of colonization. Values for an MPN assay can be obtained from published tables (12); however, these tables restrict experimental design, thereby reducing the accuracy that can be obtained. A better approach is to program the equations into a computer and directly solve for the MPN value on the basis of optimal experimental design—increased replication and decreased dilution factor improve accuracy and reduce confidence limits. Numerous factors affect the outcome of an MPN assay (2, 9, 33, 56); therefore, caution should be exercised when values from different experiments are compared. Nonetheless, this assay has been a useful tool for estimating propagule numbers in field soil, pot cultures, and various forms of inocula.

Important considerations for evaluating AM fungi with the MPN assay are as follows.

1. Dilution factor. Preliminary studies should be conducted so that the lowest possible dilutions are used to bracket actual numbers found in the soil.
2. Sample processing. Samples should be kept cool and processed as soon as possible after collection. The sample soil needs to be relatively dry, and root pieces >2 mm in diameter should be removed from the sample to allow thorough mixing with the diluent soil. These treatments will affect propagule numbers and viability, and so all samples must be treated similarly.
3. Diluent soil. The soil preferably should be the same as the original sample and should be pasteurized rather than sterilized. Controls with no sample added should be set up with the pasteurized soil to ensure that all AM propagules have been eliminated.
4. Host plant. The host must be highly susceptible to AM colonization, produce a rapidly growing, fibrous root system, and be readily cleared for observation of colonization. Zea mays L. is a good choice.
5. Length of assay. Plants need to be grown long enough so that roots fully exploit the soil in each container. It is better to err on the conservative side and grow plants until

they are pot bound. Roots with well-developed mycorrhizae are also more easily evaluated. A typical assay may run for 6 to 8 weeks.
6. Confirming negative colonization. The entire root system must be examined to confirm a negative reading.

Infectivity Assays

Plants are grown under standard conditions, and root colonization is estimated after 3 to 6 weeks (32). The amount of colonization is assumed to be proportional to the total number of AM propagules in the soil. The length of the assay is critical, and preliminary studies are needed to select the proper harvest time for a given plant-soil combination. If plants grow for too short a time, the full potential for colonization is not realized; however, plants grown for too long a time may become uniformly colonized despite differences in AM populations.

An infection-unit method may also be used to quantify mycorrhizal propagules (13). The principle is that a count of discrete points of infection is a more reliable measure of the number of viable propagules than are other methods. However, this method is applicable only in short-term experiments because infection units are discernible only during the initial stages (1 to 3 weeks) of colonization.

Quantification of Soilborne Hyphae

Even though the hyphae that grow into the soil matrix from the root are the functional organs for nutrient uptake and translocation, few researchers have obtained quantitative data on their growth and distribution. This is largely because of the technical difficulties in obtaining reliable data—there is no completely satisfactory method to quantify external hyphae of AM fungi in soil. Three major problems have yet to be overcome: (i) there is no reliable method to distinguish AM fungal hyphae from the myriad of other fungal hyphae in the soil, (ii) assessment of the viability and activity of hyphae is problematic, and (iii) meaningful quantification is very time-consuming. Nonetheless, clarification of the growth dynamics of external hyphae is essential to further understanding of their function in soil. Detailed methods for quantifying soilborne hyphae have been recently reviewed (50).

CULTURE METHODS
Traditional Culture Methods

The culture of AM fungi on plants in disinfested soil, using spores, roots, or infested soil as inocula, has been the most frequently used technique for increasing propagule numbers (45). Many host plants have been used under a variety of conditions (48, 52). Examples of plants that have been used successfully are alfalfa, maize, onion, Sudan grass, and wheat. Generally, the host should become well colonized (>50% of the root length), produce root mass quickly, and be able to tolerate the high-light conditions required for the fungus to reproduce rapidly. Hosts which can be propagated from seed are preferable to cuttings since they can be more easily disinfested. Most seeds may be disinfested with 10% household bleach (0.525% NaOCl) for 5 to 15 min followed by five washes of water.

Disinfesting the fungal propagules is a critical step because fungi, bacteria, actinomycetes, and nematodes may be propagated with or instead of the AM fungi. Hepper (20) reviewed procedures for disinfesting and germinating spores, and Williams (55) details a method for reducing contamination of colonized root pieces. The most effective methods use chlorine compounds, surfactants, and combinations of antibacterial agents. We routinely incubate pot culture-produced spores in 2% chloramine T, 200 ppm of streptomycin sulfate, and a trace of Tween 20 for 20 min followed by at least five changes of sterile water. Spores from the field usually have higher levels of contamination, and a thorough prewash with water containing the surfactant should be used prior to the disinfesting treatment.

All components of the culture system should be disinfested prior to initiation of a pot culture. The method of soil disinfestation is especially important—the objective is to kill existing AM fungi, pathogenic organisms, and weed seeds while preserving a portion of the nonpathogenic microbial community (24, 45). Several methods, including fumigation with biocides and exposure to ionizing radiation, have been used to eliminate AM fungi from soil. However, the safety and convenience of heat pasteurization make it the preferred method. Large batches (50 to 100 kg) may be treated by heating to 85°C for two 8-h periods with 48 h between treatments in a commercial soil pasteurizer. Alternatively, smaller batches (4 kg each) may be heated to the same temperature by three 2-min exposures to 700-W, 2,475-MHz microwave radiation. Prior to either treatment method, the soil must be passed through a 2-mm sieve and wetted to at least 10% (wt/wt) moisture.

Culturing AM fungi in soilless media avoids the detrimental organisms in nonsterile soil and allows control over many of the physical and chemical characteristics of the growth medium. Soilless media are more uniform in composition, weigh less, and provide aeration better than do soil media. Jarstfer and Sylvia (24) have reviewed the various conducive substrates. Most soilless media do not buffer P concentration, and so care must be taken to avoid high solution levels of P in the root zone. As discussed below for soil, several strategies may be used to regulate fertility and provide conducive conditions for the culture of AM fungi in soilless media.

Conducive environmental conditions for cultures of AM fungi are a balance of high light intensity, adequate moisture, and moderate soil temperature without detrimental additions of fertilizers or pesticides (24). Good light quality (λ = 400 to 700 nm) and high photosynthetic photon flux density are necessary for colonization and sporulation. Where natural light conditions are poor (photosynthetic photon flux density of <500 μmol m^{-2} s^{-1}), supplemental high-intensity lamps should be used. Soil moisture affects AM development directly and indirectly. Directly, excessive moisture can encourage the growth of hyperparasites on spores in the culture. Indirectly, any moisture condition that inhibits primary root growth will reduce the spread of colonization. The best strategy is to apply water regularly to a well-drained substrate. Likewise, soil temperature is also important directly for the fungus and indirectly as it affects the chosen hosts. Sporulation is positively correlated with soil temperature from 15 to near 30°C for many AM fungi; however, at higher temperatures, sporulation may decrease as the host is stressed. Generally, soil temperature and moisture conditions that are optimal for the host should also prove best for the fungus.

Chemical amendments can have both beneficial and detrimental effects on the development of colonized root systems and on sporulation. Responses to P and nitrogen (N) fertilization may be strain dependent (11) and are affected by the relative amounts of N and P supplied. Three approaches can be used to supply the plants with nutrients:

(i) apply balanced nutrients except for P, which is applied at a rate 10-fold more dilute than recommended, (ii) apply dilute but balanced nutrients frequently, or (iii) mix a time-release fertilizer in the substrate. Application of pesticides can also affect AM colonization and sporulation. Prior to testing of any pesticide with cultures of AM fungi, it is recommended that previous reviews (24, 45) and the manufacturer be consulted.

In the greenhouse, pot cultures should be isolated from nonsterile soil, splashing water, and crawling insects in order to prevent contamination. In addition, specific isolates of AM fungi should be kept well separated from each other. To initiate pot cultures, place a layer of inoculum 1 to 2 cm below the seed or cutting. Inoculum may consist of spores, colonized roots, or infested soil. Infested soil is often used to obtain initial isolates from the field; however, these mixed-species cultures should rapidly progress to single-species cultures initiated from 20 to 100 healthy, disinfested, and uniform spores. For critical taxonomic studies, single-spore cultures should be produced (45). For step-by-step methods of pot culturing and more extensive discussion of these methods, consult reference 24 or 51. Otherwise, cultures should be grown for 3 to 6 months under the conducive conditions stated above. At that time, soil cores of approximately 5% of the container volume should be assessed for colonization and spore production. Cultures with more than 10 spores per g may be stored in refrigeration at 4°C. Many isolates tolerate air drying and storage at room temperature for long periods (>5 years); however, as the shelf life of most of these fungi is uncertain, it is recommended that both methods be tested to ensure survival of isolated germ plasm. Spores may be conditioned to survive to −70°C by first being air dried with the host (10). New isolates should be deposited with either INVAM or La Banque Européenne des Glomales. The INVAM produces a newsletter that periodically updates cultural recommendations and state-of-the-art taxonomic directions for this group of fungi.

Aeroponic and Hydroponic Cultures

The primary benefit of aeroponic and hydroponic systems is that colonized roots and spores are produced free of any substrate, permitting more efficient production and distribution of inocula. At least seven species of AM fungi have been grown in various nutrient-solution systems on hosts representing at least 21 genera (24). Usually plants are inoculated with AM fungi and grown in sand or vermiculite before they are transferred into a culture system. The plants are grown for a period of 4 to 6 weeks under conditions conducive for colonization, after which they are washed and nondestructively checked for colonization (1); however, it is also possible to inoculate plants directly in the culture system (22). The P concentrations which have been reported to support AM growth in solution cultures range from <1 to 24 μmol.

A system which applies a fine nutrient mist to roots of intact plants (aeroponic culture) produces excellent AM inoculum and concentrations of spores greater than those produced in soil-based pot cultures of the same age. Because the colonized-root inoculum produced in this system is free of any substrate, it can be sheared with the very sharp blade of a food processor, resulting in very high propagule densities. A complete and detailed description of these methods and applications has been published recently (25).

Monoxenic Cultures

The growth of AM fungi in pure culture in the absence of a host has not been achieved. However, selected AM fungi colonize roots of intact plants or root-organ cultures to achieve monoxenic cultures that are useful for basic research on the symbiosis (21). More recently, Ri T-DNA-transformed roots have been used to obtain colonized root cultures (3). It should be noted, however, that colonization rates in these systems are slow and only limited amounts of colonized roots have been produced. Transformed-root culture offers the most efficient method to grow axenic colonized roots, as no plant growth regulators are required for sustained growth. Becard and Fortin (3) provide procedures for initiating transformed-root cultures.

BIOCHEMICAL DETECTION METHODS

Biochemical and molecular methods promise to expedite detection of AM fungi in the environment as well as to make AM taxonomy a more precise science. Serology, gene amplification, and fatty acid profiles all hold potential for better understanding of phylogenetic relationships among AM fungi. Protein analyses are allowing better understanding of the symbiosis and may lead to the discovery of unique molecules by which to tag AM fungi. Physiological stains and the chitin assay allow nonspecific assessment of vitality and biomass of AM fungi in roots and hyphae extracted from the soil, but proper controls are necessary for interpretation of results. None of these techniques are yet widely used or standardized, but they are developed to the point where they can be applied to environmental analyses with reasonable success. With all of them, purity of samples and controls are essential considerations.

Immunoassays

Serological techniques have the potential for specific detection of AM hyphae and spores in soil; however, few antibodies have been proven highly specific thus far. AM fungi, like many others, have poor antigenic properties (17). It appears that the most immunogenic components of both spores and hyphae are internal, cell membrane-associated components of the cell walls. In addition, associated bacteria and actinomycetes provide interfering and stronger antigenic sites on the external surfaces. These are especially difficult to remove from spores even with antibiotic agents, oxidants, or sonication because they are deeply embedded in the spore walls (6, 17). The development of hybridoma technology has opened the potential for production of large quantities of monoclonal antibodies specific for a single epitope; however, as with other fungi, low percentages (<20) of hybridomas produce antibiotics (17). Specific references to methods for immunizing animals, collecting sera, and producing and testing hybridoma products are provided elsewhere in this manual. When screening monoclonal products, the isolate and microbiological purity of the AM fungi should be the foremost concern. As suggested by Hahn et al. (17), immunogold labeling of thin sections will show where the antibody is attaching. Cultured material, using either soilless media or aeroponics, will provide fungal material that is more pure and abundant than that from field-collected fungi for both immunization and screening of antibody products. Hahn et al. (17) recommend intrasplenic immunization with 0.4 mg of intact spores. This should avoid exposure of cytoplasmic and inner wall antigens. If crushed or homogenized material is used, thorough washing and even mild enzymatic digestion may present more unique antigens to the immune system of the chosen animal. Purified cytoplasmic proteins make good antigens (57) but are of little value for labeling intact spores or hyphae from soil.

See reference 39 for a good introduction and specifics on the methods for use of monoclonal antibodies with mycorrhizae.

Gene Amplification

Although methods that detect and analyze DNA are not directly applicable to environmental samples, probes derived from this research should aid in development of specific methods for detection of AM fungi in field-collected soil samples. At present, methods allow the extraction, amplification, and cloning of AM-derived DNA. PCR allows amplification of DNA from single spores (46, 58) and from as little as 100 mg of fresh colonized roots (47). However, when cloning from larger quantities of spores, one should follow methods similar to those used by Zeze et al. (59) to free DNA from lipids, proteins, and RNAs. The methods that have been used are quite universal but can be improved by the identification of specific primers for PCR of 18S ribosomal DNA (47).

Fatty Acid and Protein Analyses

Fatty acid methyl ester (FAME) profiles of AM fungi have proven to be reliable measures of similarity below the family level (16). These profiles are stable across hosts and with storage and subculturing (4). Graham et al. (16) have found the fatty acid $16:1_{\omega 5}$ cis content of roots to provide a reproducible index of colonization. The FAME techniques for detecting and identifying AM fungi may be used with as little as 10 to 20 mg of spores (130 to 500 spores) or 15 to 30 mg of oven-dried root material (4, 16). However, roots colonized by fungi in the suborder Gigasporineae will be indistinguishable from noncolonized roots by this method (16). After collection and washing of spores or root material and air or oven drying, the methods detailed by Sasser (43) are used to extract and quantify the FAME content of AM fungi.

Qualities of proteins, in either spores or the colonized root, may also be used to detect and identify AM fungi. These assays can be based on total or enzyme-related protein molecules. Identification using isozymes depends on resolving differences between host enzyme and diagnostic fungal enzyme banding patterns after electrophoresis (21). The interactions between the particular hosts and fungi are extremely important, and some combinations yield protein bands that are indistinguishable. Careful preparation of root and hyphal materials according to protocols that prevent phenolic oxidation is necessary. With these methods, as with the other biochemical methods, isolate purity and positive controls are of utmost importance. It should also be noted that specific detection from the mycorrhizae requires an identical noncolonized root system as the control. Rosendahl and Sen (42) provide the specific methods for sample preparation and analysis as well as a review of previous applications.

Physiological Stains

Physiological stains provide a relative measure of the metabolic activity of spores and hyphae rather than detection of an individual species or isolate of AM fungus. At least three tetrazolium salts have been used as vital stains for AM spores (54). These compounds are 2,3,5-triphenyl tetrazolium chloride, 3-(4,5-dimethyl-2-thiazolyl)-2,5-diphenyl-2H-tetrazolium bromide (MTT), and 2-(p-iodophenyl)-3-(p-nitrophenyl)-5-phenyl-2H-tetrazolium chloride (INT). The assay depends on fungal reduction of these or similar colorless compounds to highly chromophoric ones. Problems associated with interpretation of the results involve uptake failure by thick-walled spores, minimal metabolic activity of quiescent spores, and inability to see a color change in very dark spores. To test spore viability, first isolate spores by the wet sieving and decanting, density-gradient-centrifugation method described above and then place the spores in a 2 mM solution for 72 h at 27°C (54). After this incubation period, a pink or red color indicates that at least 25 to 50% of these spores will infect a susceptible host under conducive conditions. This estimate is different from the number that will germinate, which we have experienced to be near 85% for spores reducing MTT. Roots containing vesicles and hyphae may also be treated with nitroblue tetrazolium chloride in a similar manner but requiring treatment with several buffers and a substrate, followed by counter staining with acid fuchsin for a determination of the live versus total colonization (44). Detailed methods for extracting soilborne hyphae and staining with INT-NADH to determine the length of active hyphae are given by Sylvia (49).

Fungal Biomass by Chitin Content

Chitin determination has been used to estimate fungal biomass in roots under controlled experimental conditions (5, 19). However, the utility of this method for natural soils is limited because chitin is ubiquitous in nature. It is found in the cell walls of many fungi and the exoskeletons of insects, and certain soils exhibit physical and chemical properties which interfere with the analysis (23). However, the content of chitin in roots may be used effectively to determine the fungal biomass when comparable noncolonized control plants are available. Chitin determination has also been used to estimate hyphal biomass in soil; however, the limitations discussed previously for this method under "Root Colonization" apply here. For this method, dried soil samples are mixed with concentrated KOH and autoclaved for 1 h to degrade chitin to chitosan. Subsamples of the soil-KOH suspension are transferred to centrifuge tubes and assayed for chitin as referenced above for roots. By subtracting the chitin content of control soil without AM fungi from soil containing AM fungi, the biomass of AM fungi can be estimated.

REFERENCES

1. **Ames, R. N., E. R. Ingham, and C. P. P. Reid.** 1982. Ultraviolet-induced autofluorescence of arbuscular mycorrhizal root infections: an alternative to clearing and staining methods for assessing infections. *Can. J. Microbiol.* **28:** 351–355.
2. **An, Z.-Q., J. W. Hendrix, D. E. Hershman, and G. T. Henson.** 1990. Evaluation of the "Most probable number" (MPN) and wet-sieving methods for determining soilborne populations of endogonaceous mycorrhizal fungi. *Mycologia* **82:**576–581.
3. **Becard, G., and J. A. Fortin.** 1988. Early events of vesicular-arbuscular mycorrhiza formation on Ri T-DNA transformed roots. *New Phytol.* **108:**211–218.
4. **Bentivenga, S. P., and J. B. Morton.** 1994. Stability and heritability of fatty acid methyl ester profiles of glomalean endomycorrhizal fungi. *Mycol. Res.* **98:**1419–1426.
5. **Bethlenfalvay, G. J., R. S. Pacovsky, and M. S. Brown.** 1981. Measurement of mycorrhizal infection in soybeans. *Soil Sci. Soc. Am. J.* **45:**871–874.
6. **Bonfante-Fasolo, P., and A. Schubert.** 1987. Spore wall architecture of *Glomus* spp. *Can. J. Bot.* **65:**539–546.
7. **Brundrett, M. C., Y. Piche, and R. L. Peterson.** 1984. A new method for observing the morphology of vesicular-arbuscular mycorrhizae. *Can. J. Bot.* **62:**2128–2134.
8. **Cochran, W. G.** 1950. Estimation of bacterial densities by means of the "most probable number." *Biometrics* **6:** 105–116.
9. **de Man, J. C.** 1975. The probability of most probable numbers. *Eur. J. Appl. Microbiol.* **1:**67–78.
10. **Douds, D. D., and N. C. Schenck.** 1990. Cryopreservation of spores of vesicular-arbuscular mycorrhizal fungi. *New Phytol.* **115:**667–674.
11. **Douds, D. D., and N. C. Schenck.** 1990. Increased sporulation of vesicular-arbuscular mycorrhizal fungi by manipulation of nutrient regimes. *Appl. Environ. Microbiol.* **56:** 413–418.
12. **Fisher, R. A., and F. Yates.** 1963. *Statistical Tables for Biological, Agricultural and Medical Research.* Oliver and Boyd, Edinburgh.
13. **Franson, R. L., and G. J. Bethlenfalvay.** 1989. Infection unit method of vesicular-arbuscular mycorrhizal propagule determination. *Soil Sci. Soc. Am. J.* **53:**754–756.
14. **Giovannetti, M., and B. Mosse.** 1980. An evaluation of techniques for measuring vesicular arbuscular mycorrhizal infection in roots. *New Phytol.* **84:**489–500.
15. **Graham, J. H., D. M. Eissenstat, and D. L. Drouillard.** 1991. On the relationship between a plant's mycorrhizal dependency and rate of vesicular-arbuscular mycorrhizal colonization. *Funct. Ecol.* **5:**773–779.
16. **Graham, J. H., N. C. Hodge, and J. B. Morton.** 1995. Fatty acid methyl ester profiles for characterization of Glomalean fungi and their endomycorrhizae. *Appl. Environ. Microbiol.* **61:**58–64.
17. **Hahn, A., K. Horn, and B. Hock.** 1995. Serological properties of mycorrhizas, p. 181–201. *In* A. Varma and B. Hock (ed.), *Mycorrhiza: Structure, Function, Molecular Biology and Biotechnology.* Springer-Verlag, Berlin.
18. **Hayman, D. S.** 1984. Methods for evaluating and manipulating vesicular-arbuscular mycorrhiza, p. 95–117. *In* J. M. Lynch and J. M. Grainger (ed.), *Microbiological Methods for Environmental Microbiology.* Academic Press, London.
19. **Hepper, C. M.** 1977. A colorimetric method for estimating vesicular-arbuscular mycorrhizal infection in roots. *Soil Biol. Biochem.* **9:**15–18.
20. **Hepper, C. M.** 1984. Isolation and culture of VA mycorrhizal (VAM) fungi, p. 95–112. *In* C. L. Powell and D. J. Bagyaraj (ed.), *VA Mycorrhiza.* CRC Press, Inc., Boca Raton, Fla.
21. **Hepper, C. M., R. Sen, and C. S. Maskall.** 1986. Identification of vesicular-arbuscular mycorrhizal fungi in roots of leek (*Allium porrum* L.) and maize (*Zea mays* L.) on the basis of enzyme mobility during polyacrylamide gel electrophoresis. *New Phytol.* **102:**529–539.
22. **Hung, L. L., D. M. O'Keefe, and D. M. Sylvia.** 1991. Use of hydrogel as a sticking agent and carrier for vesicular-arbuscular mycorrhizal fungi. *Mycol. Res.* **95:**427–429.
23. **Jarstfer, A. G., and R. M. Miller.** 1985. Progress in the development of a chitin assay technique for measuring extraradical soilborne mycelium of V-A mycorrhizal fungi, p. 410. *In* R. Molina (ed.), *Proceedings of the 6th North American Conference on Mycorrhizae.* Forest Research Laboratory, Corvallis, Oreg.
24. **Jarstfer, A. G., and D. M. Sylvia.** 1992. Inoculum production and inoculation strategies for vesicular-arbuscular mycorrhizal fungi, p. 349–377. *In* B. Metting (ed.), *Soil Microbial Ecology: Applications in Agriculture and Environmental Management.* Marcel Dekker, Inc., New York.
25. **Jarstfer, A. G., and D. M. Sylvia.** 1995. Aeroponic culture of VAM fungi, p. 427–441. *In* A. Varma and B. Hock (ed.), *Mycorrhiza: Structure, Function, Molecular Biology and Biotechnology.* Springer-Verlag, Berlin.
26. **Jeffries, P.** 1987. Use of mycorrhizae in agriculture. *Crit. Rev. Biotechnol.* **5:**319–357.

27. **Kormanik, P. P., and A.-C. McGraw.** 1982. Quantification of vesicular-arbuscular mycorrhizae in plant roots, p. 37–45. *In* N. C. Schenck (ed.), *Methods and Principles of Mycorrhizal Research.* American Phytopathology Society, St. Paul, Minn.

28. **Koske, R. E., and J. N. Gemma.** 1989. A modified procedure for staining roots to detect VA mycorrhizas. *Mycol. Res.* **92:**486–505.

29. **McGonigle, T. P., M. H. Miller, D. G. Evans, G. S. Fairchild, and J. A. Swan.** 1990. A new method which gives an objective measure of colonization of roots by vesicular-arbuscular mycorrhizal fungi. *New Phytol.* **115:**495–501.

30. **Miller, R. M.** 1987. The ecology of vesicular-arbuscular mycorrhizae in grass- and shrublands, p. 135–170. *In* G. R. Safir (ed.), *Ecophysiology of VA Mycorrhizal Plants.* CRC Press, Inc., Boca Raton, Fla.

31. **Miller, R. M., D. R. Reinhardt, and J. D. Jastrow.** 1995. External hyphal production of vesicular-arbuscular mycorrhizal fungi in pasture and tallgrass prairie communities. *Oecologia* **103:**17–23.

32. **Moorman, T., and F. B. Reeves.** 1979. The role of endomycorrhizae in revegetation practices in the semi-arid West. II. A bioassay to determine the effect of land disturbance on endomycorrhizal populations. *Am. J. Bot.* **66:**14–18.

33. **Morton, J. B.** 1985. Underestimation of most probable numbers of vesicular-arbuscular endophytes because of non-staining mycorrhizae. *Soil Biol. Biochem.* **17:**383–384.

34. **Morton, J. B.** 1988. Taxonomy of VA mycorrhizal fungi: classification, nomenclature, and identification. *Mycotaxon* **32:**267–324.

35. **Morton, J. B., and G. L. Benny.** 1990. Revised classification of arbuscular mycorrhizal fungi (Zygomycetes): a new order, Glomales, two new suborders, Glomineae and Gigasporineae, and two new families, Acaulosporaceae and Gigasporaceae, with an emendation of Glomaceae. *Mycotaxon* **37:**471–491.

36. **Norris, J. R., D. J. Read, and A. K. Varma (ed.).** 1991. *Methods in Microbiology,* vol. 23. *Techniques for the Study of Mycorrhiza.* Academic Press, London.

37. **Norris, J. R., D. J. Read, and A. K. Varma (ed.).** 1992. *Methods in Microbiology,* vol. 24. *Techniques for the Study of Mycorrhiza.* Academic Press, London.

38. **O'Keefe, D. M., and D. M. Sylvia.** 1991. Mechanisms of the vesicular-arbuscular mycorrhizal plant-growth response, p. 35–53. *In* D. K. Arora, B. Rai, K. G. Mukerji, and G. R. Knudsen (ed.), *Handbook of Applied Mycology.* Marcel Dekker, Inc., New York.

39. **Perotto, S., F. Malavasi, and G. W. Butcher.** 1992. Use of monoclonal antibodies to study mycorrhiza: present applications and perspectives, p. 221–248. *In* J. R. Norris, D. J. Read, and A. K. Varma (ed.), *Methods in Microbiology,* vol. 24. Academic Press, London.

40. **Phillips, J. M., and D. S. Hayman.** 1970. Improved procedures for clearing roots and staining parasitic and vesicular arbuscular mycorrhizal fungi for rapid assessment of infection. *Trans. Br. Mycol. Soc.* **55:**158–161.

41. **Porter, W. M.** 1979. The 'most probable number' method for enumerating infective propagules of vesicular-arbuscular mycorrhizal fungi in soil. *Aust. J. Soil Res.* **17:**515–519.

42. **Rosendahl, S., and R. Sen.** 1992. Isozyme analysis of mycorrhizal fungi and their mycorrhiza. *Methods Microbiol.* **24:**169–194.

43. **Sasser, M. J.** 1991. Identification of bacteria through fatty acid analysis, p. 199–204. *In* F. Klement, K. Rudolf, and D. C. Sands (ed.), *Methods of Phytobacteriology.* Akademmiai Kiado, Budapest.

44. **Schaffer, G. F., and R. L. Peterson.** 1993. Modifications to clearing methods used in combination with vital staining of roots colonized with vesicular-arbuscular mycorrhizal fungi. *Mycorrhiza* **4:**29–35.

45. **Schenck, N. C., and Y. Perez.** 1990. Isolation and culture of VA mycorrhizal fungi, p. 237–258. *In* D. P. Labeda (ed.), *Isolation of Biotechnological Organisms from Nature.* McGraw-Hill Publishing Company, New York.

46. **Simon, L., M. Lalonde, and T. D. Bruns.** 1992. Specific amplification of 18S fungal ribosomal genes from vesicular-arbuscular endomycorrhizal fungi colonizing roots. *Appl. Environ. Microbiol.* **58:**291–295.

47. **Simon, L., R. C. Levesque, and M. Lalonde.** 1993. Identification of endomycorrhizal fungi colonizing roots by flourescent single-strand conformation polymorphism-polymerase chain reaction. *Appl. Environ. Microbiol.* **59:**4211–4215.

48. **Sreenivasa, M. N., and D. J. Bagyaraj.** 1988. Selection of a suitable host for mass multiplication of *Glomus fasciculatum. Plant Soil* **106:**289–290.

49. **Sylvia, D. M.** 1988. Activity of external hyphae of vesicular-arbuscular mycorrhizal fungi. *Soil Biol. Biochem.* **20:**39–43.

50. **Sylvia, D. M.** 1992. Quantification of external hyphae of vesicular-arbuscular mycorrhizal fungi. *Methods Microbiol.* **24:**53–65.

51. **Sylvia, D. M.** 1994. Vesicular-arbuscular mycorrhizal fungi, p. 351–378. *In* R. W. Weaver, S. Angle, P. Bottomley, D. Bezdicek, S. Smith, A. Tabatabai, and A. Wollum (ed.), *Methods of Soil Analysis,* part 2. *Microbiological and Biochemical Properties.* Soil Science Society of America, Madison, Wis.

52. **Thompson, J. P.** 1986. Soilless cultures of vesicular-arbuscular mycorrhizae of cereals: effects of nutrient concentration and nitrogen source. *Can. J. Bot.* **64:**2282–2294.

53. **Toth, R., and D. Toth.** 1982. Quantifying vesicular-arbuscular mycorrhizae using a morphometric technique. *Mycologia* **74:**182–187.

54. **Walley, F. L., and J. J. Germida.** 1995. Estimating the viability of vesicular-arbuscular mycorrhzae fungal spores using tetrazolium salts as vital stains. *Mycologia* **87:**273–279.

55. **Williams, P. G.** 1990. Disinfecting vesicular-arbuscular mycorrhizas. *Mycol. Res.* **94:**995–997.

56. **Wilson, J. M., and M. J. Trinick.** 1982. Factors affecting the estimation of numbers of infective propagules of vesicular arbuscular mycorrhizal fungi by the most probable number method. *Aust. J. Soil Res.* **21:**73–81.

57. **Wright, S. F., J. B. Morton, and J. E. Sworobuk.** 1987. Identification of a vesicular-arbuscular mycorrhizal fungus by using monoclonal antibodies in an enzyme-linked immunosorbent assay. *Appl. Environ. Microbiol.* **53:**2222–2225.

58. **Wyss, P., and P. Bonfante.** 1993. Amplification of genomic DNA of arbuscular-mycorrhizal (AM) fungi by PCR using short arbitrary primers. *Mycol. Res.* **97:**1351–1357.

59. **Zeze, A., H. Dulieu, and V. Gianinazzi-Pearson.** 1994. DNA cloning and screening of a partial genomic library from an arbuscular mycorrhizal fungus, *Scutellospora castanea. Mycorrhiza* **4:**251–254.

Isolation and Culture of Endophytic Bacteria and Fungi

CHARLES W. BACON AND DOROTHY M. HINTON

45

Endophytic bacteria and fungi are defined as organisms that live in association with plants for most if not all of their life cycles. These organisms are therefore symbiotic and are further distinguished in that the bacterium or fungus lives within the plant or portions of it as a nonpathogen, although slight to moderate degrees of pathogenicity may be expressed. This distinction serves to separate from the following discussion the many obligate pathogenic microorganisms that form similar but detrimental associations with plants. Also not included in this discussion are mycorrhizal fungi and rhizobial bacteria, which are subjects of companion chapters in this volume. Fungal and bacterial endophytes live within the intercellular spaces of plants (Fig. 1), where they live off apoplastic nutrients (4, 27). The designation of a bacterium or fungus as an endophyte is based on the recovery of an organism from surface-sterilized plant material and seed. Thus, endophytes are pragmatically defined, but there is one error inherent with this definition. Endophytes must be isolated from plant material that has been adequately surface sterilized, and the isolated endophyte must be demonstrated to reside within the association by microscopic techniques. The successful detection of fungi and bacteria in plants is dependent on generalized procedures. Adequate surface sterilizing is important in delimiting the endophytic habit, and particular attention must be directed to achieve this. The literature is replete with descriptions of microorganisms as endophytic, but close scrutiny reveals that the plant material was not properly surfaced sterilized and no attempts were made to show the endophytic habit microscopically, a procedure that is particularly relevant when the recovered organism is a noted saprophyte. Finally, the plant material from which endophytic organisms are isolated must be disease free.

Endophytic fungi and bacteria may confer benefits to the plant, and the benefits may be reciprocal, resulting in an enhanced symbiotic system for specific plant characteristics. Therefore, the use of endophytic bacteria and fungi opens up new areas of biotechnological exploitations, which drive the necessity to isolate and culture these organisms. Endophytes are used for biological control of various plant diseases, for enhanced agronomic plant characteristics such as increased drought tolerances and nitrogen efficiency, for bioherbicides, and for pharmacological agents. Because these microorganisms are usually obligate biotrophic parasites, their culture, especially for fungi, if ever achieved, may be difficult and slow. The techniques for examining, culturing, and exploiting fungal endophytes are relatively new, and techniques for utilizing endophytic bacteria are even more recent than those designed for endophytic fungi. The techniques vary from simple to complex and have been shown to be fundamental for in planta studies of host-parasite relationships as well as for physiological and ecological studies. It is the intent of this review to briefly review endophytic microorganisms and outline procedures useful in establishing their presence. General isolation and cultivation procedures will be emphasized, and when available, precautions for specific microorganisms during use of a particular procedure will be indicated.

ENDOPHYTIC ORGANISMS
Fungi

Endophytic fungi include species in the family Clavicipitaceae (Clavicipitales; Ascomycotina) and other fungal taxa (Deuteromycotina, Zygomycotina, and Basidiomycotina), most of which are found associated with a diverse assemblage of plant hosts, including trees, shrubs, grasses, sedges, and rushes (16, 17, 21, 24, 46, 66). These fungi show a distinct range of host association patterns and evolutionary tendencies. Several fungi (Table 1) have casual endophytic associations, found in restricted areas of specific plant organs. Sometimes more than one species may be present. The endophytic habit may not be obligatory, as several endophytic species may also be found as saprophytes in nonendophytic habits (18, 28, 34, 54, 55). However, there are endophytic species that are found only as systemic fungi in nature. The majority of these fungi are associated with annual and perennial grasses (14, 24, 53, 66, 67). These grass endophytes are clavicipitaceous and related fungi (Table 1). The most casual of these associations is exemplified by one species, *Myriogenospora atramentosa* (Berk. and Curt.) Diehl, which forms an epibiotic association (38). The majority of endophytic fungi of grasses [species of *Balansia* and *Epichloe typhina* (Fr.) Tul.] show the true endosymbiotic habit and, depending on the species, produce external fructification on various grass organs. There are also several species of fungal endophytes that infect grasses and remain symptomless. These include species of *Neotyphodium* Glenn, Bacon, and Hanlin (29) (= *Acremonium* Morgan-Jones and Gams). The symptomless endophytes offer the greatest biotechnological chal-

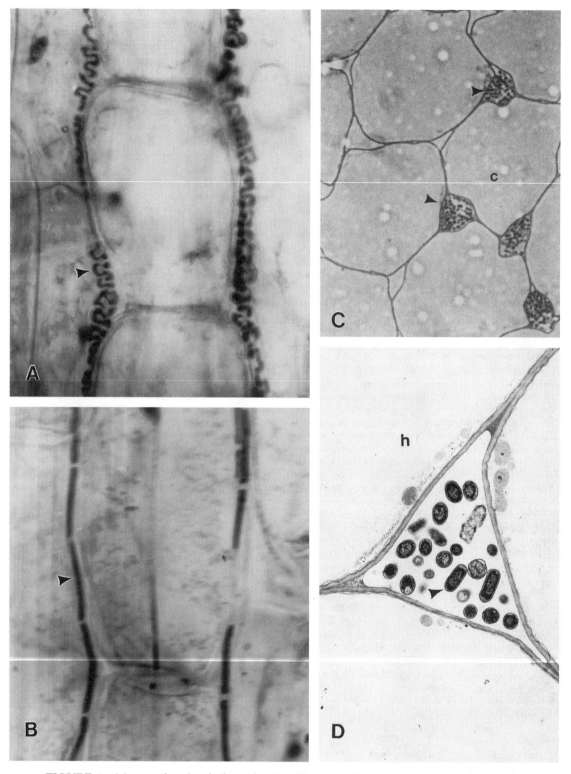

FIGURE 1 Micrographs of endophytic fungi and bacteria. (A) *Acremonium coenophialum*, the endophytic fungus of tall fescue. Coiled hypha can be seen between the intercellular spaces of plant cells (arrowhead) (magnification, ×100) (9). (B) Another view of this endophyte showing the straight nature of hyphae sometimes observed in tall fescue (magnification, ×100) (9). (C) *Enterobacter cloacae*, an endophytic bacterium of corn. Bacterial cells (arrowheads) are located between intercellular spaces of the cells (c) of the root cortex (32). (D) Transmission electron micrograph of *E. cloacae* in the intercellular spaces of host cells (h) of the root cortex.

TABLE 1 General methods for isolating endophytic fungi

Genus or group	Isolation method	Medium[a]	Incubation time (days), temp (°C)	References
Clavicipitaceous and related fungi				
Atkinsonella, Balansia, Epichloe, Myriogenospora	Ascospore discharge method (see text)	CMM, PDA	14–35, 24–28	3, 18
Acremonium, Atkinsonella, Balansia, Epichloe	Leaf sheath, inflorescence, and/or stem tissue	CMM, PDA, YEG	14–42, 24–28	2, 3, 9, 11
Neotyphodium (= *Acremonium*)	Seedling method: sterilize seed with full-strength bleach, place on agar tube and in darkness; isolate fungus that grows from etiolated seedling	CMM, PDA	14–50, 24–28	2, 3
	Liquid medium method: sterilize sheaths with full-strength bleach, aseptically macerate tissue, and place in flask of liquid medium	M102	14–90, 24–28	1, 8, 11
Nonclavicipitaceous and woody endophytes				
Numerous hyphomycetes, ascomycetes, and coelomycetes, e.g.:				
Fusarium moniliforme	Ascospore discharge and stem, leaf tissue method	PDA, CMM	7–14, 24–28	6, 32
Cladosporium sphaerospermum, Cryptosporiopsis sp., *Phomopsis* sp., *Septoria alni, Ophiovalsa suffusa, Phyllosticta, Mycosphaerella punctiformis, Melanconis alni, Rhabdocline parkeri, Cryptocline* sp., *Leptrostroma* sp.	Stem, leaf, twig	CMM	14–40, 20–24	17, 46, 55

[a] CMM, cornmeal-malt extract agar (see Table 3); PDA, potato dextrose agar; YEG, yeast extract-glucose agar (see Table 3); M102, liquid culture medium M102 (see Table 3).

lenge of all of the endophytes. The common feature of these fungi is that they are naturally and perennially associated with grasses in mutualistic associations called mutualisms (5, 16, 19). There are several extensive review papers on fungal endophytes, with detailed descriptions of species, grass hosts, toxicology, and biotechnological potentials (3, 7, 10, 19–21, 24, 28, 37, 57, 58, 66, 67). Nongrass or woody endophytes are found in association with conifers, deciduous trees, clover, alfalfa, and shrubby plants (15, 16, 24, 28, 37).

Bacteria

The bacterial endophytes include only a limited number of species. In the past, these associations have been described as involving endorhizosphere dwellers, but they also include infectious agents affecting other plant organs (Table 2). The number of bacterial endophytes is expected to increase as procedures for their detection improve. Most of the endophytic species belong to the families *Acetobacteriaceae, Enterobacteriaceae, Spirillaceae,* and *Bacillaceae* (22, 26, 36, 40). Thus, species of *Acetobacter, Achromobacter, Burkholderia, Campylobacter, Corynebacterium, Cytophaga, Bacillus, Brevibacterium, Flavobacterium, Leuconostoc, Klebsiella, Micrococcus, Enterobacter, Azospirillum,* and *Erwinia* have been associated with ovules, seeds, pods, roots, leaves, and stems of several plants of economic importance, and these bacteria have been shown to either have a positive role or cause no

apparent effects on the survival of several plant species (22, 25, 27, 33, 41, 43, 48, 52). Endophytic bacteria are anaerobic, aerobic, or microaerobic species. Many are diazotrophs. However, excluded as endophytes are those diazotrophic species that are not recovered from surface-sterilized plant parts, i.e., rhizobial bacteria. The distribution and endophytic nature of bacteria within plants have not been completely documented.

ISOLATION

Surface Sterilization and Sample Preparation

It is very important to always select vigorous and disease-free plant material for isolation work, as this prevents the isolation of localized pathogenic endophytic organisms. It is usually desirable to isolate several bacterial endophytes simultaneously. Therefore, the method used must allow for the recovery of mixed species of bacteria. The isolation of several taxa of fungi may present some problems, since fast-growing species prevent the growth of other species. However, selection of the appropriate incubation temperature and media can circumvent this problem. The general methods of surface sterilizing plant material for isolation utilize one or two sterilizers, such as commercial bleach or 1% chloramine-T in sequence, followed by ethanol or 50% sul-

TABLE 2 Identities and occurrence of selected bacterial endophytes

Bacterium	Plant(s) associated	Reference(s)
Azospirillum lipoferum, A. brasilense, A. amazonense	Corn, wheat, sorghum, sugarcane, pear millet, guinea grass, rice, and other tropical grasses[a]	26, 45, 47
Acetobacter diazotrophicus	Sugarcane	27
Campylobacter sp.	*Spartina alterniflora*	39
Klebsiella planticola (synonym, *K. pneumoniae*)	Sugarcane, rice	36, 48
Bacillus polymyxa	Sugarcane	48
B. pumilus	Corn, cotton	32, 41
Erwinia herbicola	Sugarcane	48
Burkholderia sp.	Cotton, corn	40
Enterobacter cloacae	Sugarcane, corn, grapevines	15, 32, 40, 48

[a] *A. lipoferum* is the most common isolate from surface-sterile plant parts of C-4 type grasses, while *A. brasilense* is the most common species isolated from C-3 grasses and sugarcane (22). Data are based on reported isolation of bacteria from surface-sterilized plant parts.

furic acid. Either one sterilizer or a combination of sterilizers accomplishes the important task of removing surface debris and microorganisms without interfering with endophytic organisms.

Prewash

A vigorous prewashing of plant material is necessary to reduce the number of surface microflora and remove debris. This procedure increases the effectiveness of the surface sterilizer. The prewash solution may be either sterile tap water, distilled water, or one of several dilute saline buffers. A pH of 5.6 for fungi and a pH of 7.2 for bacteria are recommended. Sterile distilled water may cause a decrease in some bacterial endophytes, although most endophytes should not be affected by water quality. The plant material should consist of the proximal portions of fresh roots, stems, and leaves. The material selected for isolation should be removed from the plant immediately before the prewash phase of this procedure. Excised roots, stems, or leaves are cut into 5- to 8-cm-long segments. It is not necessary to use detergents in any solution during the isolation of endophytes from green material.

The isolation of endophytes from seeds requires some special considerations, since seeds are dormant dehydrated structures, and most systemic seed-borne bacteria and fungi are similarly dormant. Improper rehydration of seeds affects the viability of bacteria (59) and fungi (60). Most dry seeds must first be wetted in sterile tap water briefly or allowed to imbibe for at least 2 to 4 h in sterile tap water or buffer on sterilized damp filter paper before any surface sterilizers are added. The temperature during the wetting or imbibing process may be room temperature for fungi or 4 to 10°C for bacteria. Alternatively, all seeds may be incubated at low temperatures, which prevents the multiplication of nonendophytic saprophytic flora on seeds. Dry seeds permitted to imbibe in sterilizers result in poor viability of systemic seed-borne microorganisms. Imbibed seeds can then be surface sterilized for endophyte isolation. If seeds have been treated with various pesticides, the prewash procedure should be preceded by water washes containing a detergent, followed by several washes in 75 to 90% ethanol and/or aqueous rinses with charcoal to remove pesticides (6). Consideration of a detergent must include the final staining procedure, since some identification is based on immunofluorescent stains, which would be affected by detergents that are also fluorescent.

Surface Sterilization

The container used to sterilize plants and seed should be the sterile disposable type with a screw cap. Specimen cups (4 to 6 oz [1 oz = ca. 29.6 ml]) are ideal for all washes and sterilizers. Surface sterilize most plant material by vigorous shaking on a rotary or reciprocal shaker in either a 1% chloramine-T solution for 30 min or full-strength bleach (5.25% sodium hypochlorite) solution containing 0.01% Tween 20 or 0.05% Triton X-100 for 3 to 5 min for fresh plant material or for as long as 15 min for seed and woody stems. Ovules, ovule-bearing pods, and tender plant parts may be surface sterilized in reduced concentrations of a sterilizer but for longer time periods. Thus, this type of material requires preliminary tests, and the procedure of Mundt and Hinkle (43) should be followed to determine the optimum conditions. The sterilizing solution is removed, and the plant material is rinsed three times (3 min each time) in sterile distilled water. Additional specific surface-sterilizing procedures for bacterial and fungal endophyte isolations are described below.

Fungal Endophyte Isolation

Before the plant material is plated on media for isolation, the 5- to 8-cm length of plant tissue should be cut again, with one-eighth to one-third of the length removed from each end. The final pieces are placed on a specific agar medium for isolation of fungi. The initial medium selected should be one that will be either semiselective for specific taxa or designed generally for fungal endophytes (Table 3). One of several antibiotics may be incorporated into media to prevent the growth of bacteria. Incubation should be between 24 and 28°C for 3 to 8 weeks.

Several species of the clavicipitaceous endophytes can be isolated from their ascospores that are produced on ascomata of infected plants (Table 1). The floral part containing the ascoma is taped to the lid of a petri dish, the lid is placed on the bottom dish of agar, and the dish is inverted, which allows discharge of ascospores onto the surface of the agar. Fungal colonies develop within 3 to 6 weeks, although it may take as long as 12 weeks for some species.

Isolation from Seed

Full-strength commercial bleach (5.25% NaOCl) is used to surface sterilize seed for endophyte isolation. Lesser concentrations of bleach and the 1% chloramine-T treatment described above will not sterilize most lots of seed. To isolate endophytes from seeds, deglume seeds by rubbing them vig-

TABLE 3 General isolation and storage media for endophytic fungi of grasses

Cornmeal-malt extract agar[a]
Cornmeal agar (Difco)	17.0 g
Malt extract	20.0 g
Yeast extract	2.0 g
Distilled water	1,000 ml

Yeast extract-glucose agar[b]
Yeast extract	2.0 g
KH_2PO_4	5.0 g
$MgSO_4$	0.5 g
Glucose	5.0 g
Agar	20.0 g
Streptomycin sulfate	0.15 g
Distilled water	1,000 ml

Liquid culture medium M102[c]
Sucrose	30.0 g
Malt extract	20.0 g
Bacto Peptone	2.0 g
Yeast extract	1.0 g
KCl	0.5 g
$MgSO_4·7H_2O$	0.5 g
Distilled water	1,000 ml

[a] Chloramphenicol (50 mg/liter) and other antibiotics may be added; the pH is adjusted to 5.6 with NaOH (9).

[b] Add filter-sterilized streptomycin after autoclaving and adjust the pH to 6.5.

[c] The pH 5.6 is adjusted; filter-sterilized chloramphenicol (50 mg/liter) may be added after autoclaving when the medium is used primarily for isolation.

orously between the hands for several minutes. Periodically collect the seeds that are freed of any adhering glumes. After 100 or more seeds have been collected, place them in a 4-oz sterile disposable cup and pour full-strength bleach solution into the cup. Cap the cup and agitate it continuously for 15 to 20 min. Pour off the bleach and replace it with 100 ml of sterile distilled water. Agitate the cup for 5 min, pour off the water, and replace it with another 100 ml of sterile distilled water. After agitation of seeds for 3 to 4 min more, pour off most of the water. Using sterile forceps, remove a seed and press it into potato dextrose agar, yeast extract-glucose agar, cornmeal-malt extract medium, or other suitable medium. A minimum of 20 plates with three seeds per plate is recommended. Seal plates with strips of Parafilm, masking tape, or other material and incubate them at room temperature. Colonies emerging from seeds before 1 week of incubation are likely to belong to seed surface-borne microorganisms. These colonies may be removed from plates as they develop. The characteristic white to pale buff colonies of *Neotyphodium* endophytes may become visible after 2 to 4 weeks. The time and appearance of colonies of other fungal endophytes (deuteromycetes, ascomycetes, zygomycetes, and basidiomycetes) will vary.

Isolation from Leaf or Stem

To isolate grass endophytes from leaf sheaths or culms, the procedure is similar. Young tissue should be obtained for isolation, as older tissues often contain many additional fungi that make isolation of slowly growing *Acremonium* endophytes difficult. Pieces of tissue approximately 5 mm or less in size should be obtained and placed in either full-strength or 50% Clorox solution as described above. The 1% chloramine solution works very well for most plant tissue and is preferable if this tissue is to be used for staining and

other histological preparations. Tissue should be agitated continuously for 15 min. However, after the first 5 min, two or three pieces of tissues are removed from the Clorox solution and every 2 min thereafter. The pieces are then rinsed vigorously in sterile distilled water and pressed into agar medium as described above. Four or five different surface sterilization times should be represented among resulting plates. Plates are sealed and incubated as described above.

Isolation from Ascospores

Several of the balansiae, i.e., *Balansia epichloe* (Weese) Diehl, and *E. typhina* produce fertile stromata that can be used as sources of isolates. The stromata must be mature and capable of discharging ascospores. Leaves or stems bearing mature stromata of the desired species are collected and used fresh; dried material usually will give poor results. The stromata are washed in sterile water, excess water is removed, and the stromata are taped to the lid of petri plates containing the desired medium. A satisfactory medium for this purpose is cornmeal-malt extract. To prevent bacterial contamination, an antibiotic such as streptomycin sulfate should be added to this medium. If the stroma is dry, a small amount of water is placed on its surface and the lid is replaced on the dish; ascospore discharge should proceed within a few minutes. The area under the leaf-bearing stroma may be marked on the bottom of the dish to identify the area to be viewed. Discharged ascospores may be observed with the ×200 objective of a light microscope. When enough ascospores are discharged, a new lid without stromata is placed on the plate and the petri plate is sealed with Parafilm. Plates are incubated at 24 to 26°C for 4 to 6 weeks. Colonies of endophytes, identified by their conidia and conidiophore morphology (42, 50), should be evident at least after 5 weeks following discharge. Colonies of *M. atramentosa* will not be visible until 6 to 7 weeks have passed, but this slow growth is host specific.

Woody and Nonclavicipitaceous Endophytes

Randomly selected whole healthy leaves or 2- to 3-year-old twigs from healthy trees or shrubs are collected and placed in 95% ethanol for exactly 1 min and then rinsed twice in sterile distilled water. The samples are then shaken in full-strength bleach for 5 min, rinsed in distilled water, and again shaken in 95% ethanol for exactly 30 s. The samples are rinsed in distilled water. Then, 2- to 4-mm-diameter disks from each leaf or 1.0- to 1.5-cm pieces from each twig are cut and plated on cornmeal-malt extract agar (Table 3). Conifer needles are similarly treated except that surface-sterilized leaves are cut into 1.0- to 2.5-cm sections. Plates are incubated for 2 to 8 weeks at 20 to 24°C in the dark.

Bacterial Endophyte Isolation

The incubation temperature for bacterial endophytes should be between 25 and 30°C. If several species of bacterial endophytes are desired, it may be necessary to incubate the tissue at several different temperatures. The incubation time varies and is medium dependent, but 24 to 72 h is common for bacterial endophytes.

Bacterial Endophytes from Seed and Woody Stem Tissue

Isolation of bacterial endophytes from seed onto agar media requires that the seed has not germinated. Germination of the seed is prevented by the addition of filter (0.23-nm pore size)-sterilized cycloheximide, 200 mg/liter (2 ml of a 100-

mg/ml stock solution in 75% ethanol), into cool medium. Some seed may require higher amounts of cycloheximide to retard germination, which should be determined experimentally. The cycloheximide also prevents most fungi from growing. Other fungicides that may be incorporated into media selective for bacterial endophyte isolation include benlate, chlorothalonil, pentachloronitrobenzene, and nystatin. Woody stems may be rinsed in 90 to 100% ethanol for 3 to 10 min, the stem ignited, and tissue aseptically removed from inner portions of the stem. Alternatively, full-strength bleach may be used as described for endophytic fungal isolation. Isolation techniques may be modified to include techniques for isolating xylem-inhabiting endophytic bacteria from woody stems (62, 68).

Bacterial Endophytes from Root and Leaf Tissue

Healthy roots are removed, immediately washed free of soil with sterile distilled water, and then soaked in sterile phosphate-buffered saline (pH 7.0) for 10 min to equilibrate osmotic pressure and prevent passive diffusion of sterilizing agents into roots (48). A 5- to 10-mm section of root is surface sterilized in a 1% aqueous solution of chloramine-T (45) for up to 30 min; a period of 1 h may be tried for woody and storage roots. The roots are then immersed for 30 min in sterile 0.05 M phosphate buffer (pH 7) and then rinsed several times in sterile distilled water. Roots are plated on one or more media for bacterial isolations. It may be necessary to divide the roots, omitting one set with the cut ends of each root from the chloramine-T solution; the other set is completely submerged. In the latter case, it will be necessary to cut sections of each root into smaller sections, discarding the upper one-quarter near the cut end. Leaves are detached and placed in the 1% chloramine-T solution for 30 min or in full-strength bleach for 5 min, washed for 3 min in sterile distilled water, and plated.

Isolation from Ground Seed and Plant Parts

A weighed amount, usually 10 g, of surface-sterilized, air-dried seeds or plant parts is blended in 90 ml of half-strength nutrient broth in a 125-ml flask. The flask is placed on a gyratory shaker for 2 to 4 h. Then, 1.0-, 0.75-, 0.50-, 0.25-, and 0.1-g samples of the suspension are introduced into screw-cap culture tubes containing 15 ml of full-strength nutrient broth and 0.20% cycloheximide. The tubes are incubated in the dark at 25 to 30°C for 3 to 4 days, after which 0.1 to 0.5 ml of the broth is spread over the surface of agar plates containing King's B medium. Alternatively, the broth may be spread over each of the three differential media as described by McInroy and Kloepper (40). The plates are observed daily for bacterial growth.

In Planta Visualization of Endophytes

Surface-sterilized plant material or seed may be examined with a light microscope for the occurrence of internal hyphae or bacterial cells. Fungal endophytes are stained with an aniline blue-lactic acid stain. The stain consists of 50 ml of 85% lactic acid, 0.1 g of aniline blue stain, and 100 ml of water. This stain is frequently used to examine scrapings of plant tissues from fresh and dried culms and leaf sheaths for the presence of endophytic mycelium (6, 7, 19). Tissues from sheaths or culms are removed, macerated if necessary, placed on a slide (exposed side up), covered with stain (three or more drops), and gently heated for a few seconds to aid penetration of the stain. Excess stain is then removed, water is added, and the tissues are examined at a magnification of ×400 or greater for the presence of non-

branching, intercellular, blue-stained hyphae. If dried culms are examined for endophytic mycelium, 0.1% aqueous aniline blue without lactic acid may be used and heating of the slides is omitted. Slides may be stored for 2 to 4 weeks by sealing the coverslip with clear nail polish before the stain decolorizes. Another stain used to detect endophytic fungi within plant tissue is rose bengal (52). Leaf blade tissue containing some species of endophytes, e.g., *Balansia* Spegazzini, must be cleared in order to examine endophytic hyphae (11), and the procedure of Hignight et al. (30) is recommended. It is not necessary to clear inner tissue removed from the inflorescence culm, leaf sheath, and similar poorly pigmented tissue.

Endophytic bacteria in plants may be visualized by the procedure of Patriquin and Dobereiner (45), using the tetrazolium stain. This stain, 2,3,5-triphenyltetrazolium chloride (PBMT), is prepared as a sterile solution and consists of 0.05 M potassium phosphate buffer (pH 7.0) containing 0.625 g of malic acid and 1.5 g of PBMT per liter. The buffer-malate mixture is autoclaved, and the PBMT is added after autoclaving. Chloramine-T surface-sterilized plant tissue (roots, shoots, and seed) is incubated in tubes overnight in the PBMT stain and is cut longitudinally or in cross section by hand or with a microtome. Sections of tissue can then be placed as wet mounts on slides and examined at a magnification of ×100 (Fig. 1). The bacterial cells are stained dark red to purple; with practice, endophytic bacteria can be easily observed at magnifications lower than ×100.

Both bacterial and fungal endophytes can be enhanced by counterstaining the PBMT-stained tissue with aniline blue for 1 min, removing the aniline blue, and then making a wet mount. This combined technique was used to establish the endophytic nature of the fungus *Fusarium moniliforme* Sheldon in corn plants (6, 32), which is not possible if either stain is used alone.

CULTURE

Most endophytic bacteria can be cultured in the laboratory on routine laboratory media. Endophytic fungi usually require complex, nondefined media. Usually their growth in vitro is very slow, suggesting that several requirements in the media are still absent. However, the amount of growth observed in culture is greater than that observed in planta, which might indicate that these fungi are naturally slowly growing.

Fungal Culture Media

Media used to isolate fungal endophytes are numerous and more complex than those used for culture after isolation (Table 3) (2, 3, 11, 35). All grass endophytes are biotrophic, and in combination with their hosts there is probably an exchange of certain classes of nutrients. Ultrastructural studies of endophytes indicate that there are no nutrient-absorbing structures, which suggests that all nutrients are derived from the apoplast (31, 49, 56). However, since both the qualitative and quantitative nature of nutrients in the apoplastic fluid and the nature by which the phloem is loaded are unknown but expected to vary for each plant species, isolation media, while intended to be generalized, may not be suitable for all species. Laboratory studies of grass endophytes have established that once isolated, the fungi can be cultured on a variety of media (4, 63). Following their isolation and growth, these fungi can be grown on media specific for several toxic and pharmacologically active compounds (Table 3) (3). The culture requirements of grass

TABLE 4 Isolation and culture media for endophytic bacteria[a]

Nutrient peptone agar (65)
Glucose	10.0 g
Malt extract	5.0 g
Yeast extract	1.0 g
Nutrient agar	15.0 g
Bacto Peptone	1.0 g
Distilled water	1,000 ml

MacConkey agar for *Klebsiella* sp.
MacConkey agar (Difco)	40.0 g
myo-Inositol	10.0 g
Carbenicillin	0.05 g
Distilled water	1,000 ml

Yeast mannitol agar (26)
Mannitol	0.5 g
Yeast extract	1.0 g
$MgSO_4·7H_2O$	0.2 g
NaCl	0.1 g
K_2HPO_4	0.5 g
Agar	15.0 g
Distilled water	1,000 ml

King's B medium
Proteose peptone no. 3	20.0 g
Glycerol	0.01 g
$MgSO_4$	1.5 g
K_2HPO_4	1.5 g
Distilled water	1,000 ml

Dobereiner and Day N-poor medium LGY (48)
DL-Malic acid	5.0 g
NaOH	3.0 g
$MgSO_4·H_2O$	0.2 g
NaCl	0.1 g
$CaCl_2$	0.02 g
Yeast extract	0.1 g
$FeCl_3$	10.0 mg
$NaMoO_4·2H_2O$	2.0 mg
$MnSO_2$	2.1 mg
H_3BO_3	2.8 mg
$Cu(NO_3)·3H_2O$	0.04 mg
$ZnSO_4·H_2O$	0.24 mg
NH_4Cl	1.0 g
Phosphate buffer[b]	100 ml
Distilled water	900 ml

SC medium for fastidious bacteria (22)
Papain digest of soy meal	8.0 g
Cornmeal agar	17.0 g
K_2HPO_4	1.0 g
KH_2PO_4	1.0 g
$MsSO_4·H_2O$	0.2 g
Bovine hemin·Cl[c]	15.0 ml
Bovine serum[c]	10.0 ml
Glucose[c]	1.0 ml
Cysteine[c]	1.0 g
Distilled water	1,000 ml

[a] Unless otherwise noted, the final pH is adjusted to between 6.8 and 7.0.
[b] Phosphate buffer consists of (per liter of water) 6.0 g of K_2HPO_2 and 4.0 g of KH_2PO_4; the pH of the medium is adjusted to 6.8.
[c] Bovine hemin chloride is prepared as a 0.1% solution in 0.05 N NaOH, bovine serum albumin fraction 5 is prepared as a 20% aqueous solution, and glucose is prepared as a 50% aqueous solution. All three solutions are filter sterilized and added to cool autoclaved medium. The final pH of the medium is adjusted to 6.6 with NaOH or HCl.

endophytes following isolation, especially species from wild annual ryegrass species and tall fescue grass maintained in culture for a year or more, are similar to those of any free-living fungus. There are two completely defined media for grass endophytes maintained under laboratory cultivation (1, 35). Fungal endophytes tolerate a wide variation in temperature but appear to have an optimum between 20 and 26°C (8, 23). The pH requirements of all fungal endophytes have not been determined, but the grass endophyte *Acremonium coenophialum* exhibited a wide tolerance to pH between 5.5 and 7.25 (23).

Bacterial Culture Media
The direct isolation of most bacterial endophytes from seed or plant tissue can be accomplished on nutrient agar, glucose-yeast extract agar, tryptic soy agar, the nitrogen-poor medium LGY, King's B medium, and MacConkey agar (Table 4) (12, 13, 26, 44, 45, 47, 64). If it is desired to isolate aerobic heterotrophic N_2-fixing bacteria, nutrient peptone agar (Table 4) or one of the media recommended by Watanabe et al. (65) can be used. Specific carbon sources and the nitrogen status of media encourage the growth of different groups of bacteria during isolation and initial culture. Such semiselective media are available for the isolation of *Azospirillum* and *Acetobacter* species and *Klebsiella planticola* (Table 4) (12, 13, 26, 36, 44, 61). McInroy and Kloepper (40) recommend the use of three media as a routine screening procedure during the isolation of endophytic bacteria (Table 4): medium R2A (Difco Laboratories, Detroit, Mich.) for oligotrophic bacteria, trytic soy agar (Difco) for culturable heterotrophic bacteria, and SC medium (22) for the growth of fastidious organisms.

REFERENCES
1. **Bacon, C. W.** 1985. A chemically defined medium for the growth and synthesis of ergot alkaloids by the species of *Balansia. Mycologia* **77:**418–423.
2. **Bacon, C. W.** 1988. Procedure for isolating the endophyte from tall fescue and screening isolates for ergot alkaloids. *Appl. Environ. Microbiol.* **54:**2615–2618.
3. **Bacon, C. W.** 1990. Isolation, culture, and maintenance of endophytic fungi of grasses, p. 259–282. *In* D. P. Labeda (ed.), *Isolation of Biotechnological Organisms from Nature.* McGraw-Hill, New York.
4. **Bacon, C. W.** 1992. Abiotic stress tolerances (moisture, nutrients) and photosynthesis in endophyte-infected tall fescue, p. 123–141. *In* S. Quisenberry and R. Joost (ed.), *Acremonium/Grass Interactions.* Elsevier Science Publishers, Amsterdam.
5. **Bacon, C. W.** 1995. Toxic endophyte-infected tall fescue and range grasses: historic perspectives. *J. Anim. Sci.* **73:**861–870.
6. **Bacon, C. W., D. M. Hinton, and M. D. Richardson.** 1994. A corn seedling test for resistance to *Fusarium moniliforme. Plant Dis.* **78:**302–305.
7. **Bacon, C. W., P. C. Lyons, J. K. Porter, and J. D. Robbins.** 1986. Ergot toxicity from endophyte-infected grasses: a review. *Agron. J.* **78:**106–116.
8. **Bacon, C. W., J. K. Porter, and J. D. Robbins.** 1979. Laboratory production of ergot alkaloids by species of *Balansia. J. Gen. Microbiol.* **113:**119–126.
9. **Bacon, C. W., J. K. Porter, J. D. Robbins, and E. S. Luttrell.** 1977. *Epichloe typhina* from toxic tall fescue grasses. *Appl. Environ. Microbiol.* **34:**576–581.
10. **Bacon, C. W., and M. R. Siegel.** 1988. The endophyte of tall fescue. *J. Prod. Agric.* **1:**45–55.
11. **Bacon, C. W., and J. F. White, Jr.** 1994. Stains, media,

and procedures for analyzing endophytes, p. 47–56. *In* C. W. Bacon and J. F. White, Jr. (ed.), *Biotechnology of Endophytic Fungi of Grasses*. CRC Press, Boca Raton, Fla.

12. **Bagley, S. T., and R. J. Seidler.** 1978. Primary *Klebsiella* identification with MacConkey-inositol carbenicillin agar. *Appl. Environ. Microbiol.* **36:**536–538.

13. **Balandreau, J.** 1983. Microbiology of the association. *Can. J. Bot.* **29:**851–859.

14. **Belesky, D. P., J. A. Stuedemann, and S. R. Wilkinson.** 1988. Ergopeptine alkaloids in grazed tall fescue. *Agron. J.* **80:**209–212.

15. **Bell, C. R., G. A. Dickie, W. L. G. Harvey, and Y. W. Y. F. Chan.** 1995. Endophytic bacteria in grapevine. *Can. J. Microbiol.* **41:**46–53.

16. **Carroll, G.** 1988. Fungal endophytes in stems and leaves: from latent pathogen to mutualistic symbiont. *Ecology* **69:**2–9.

17. **Carroll, G. C., and F. E. Carroll.** 1978. Studies on the incidence of coniferous needle endophytes in the Pacific Northwest. *Can. J. Bot.* **56:**3034–3043.

18. **Carroll, R. B., and E. S. Elliott.** 1964. Similarity of species of fungi isolated from roots of alfalfa and red clover. *Phytopathology* **54:**746.

19. **Clay, K.** 1988. Fungal endophytes of grasses: a defensive mutualism between plants and fungi. *Ecology* **69:**10–16.

20. **Clay, K.** 1990. Fungal endophytes of grasses. *Annu. Rev. Ecol. Syst.* **21:**275–297.

21. **Clement, S. L., W. J. Kaiser, and H. Eichenseer.** 1994. *Acremonium* endophytes in germplasms of major grasses and their utilization for insect resistance, p. 185–200. *In* C. W. Bacon and J. F. White, Jr. (ed.), *Biotechnology of Endophytic Fungi of Grasses*. CRC Press, Boca Raton, Fla.

22. **Davis, M. J., A. G. Gillaspie, Jr., R. W. Harris, and R. H. Lawson.** 1980. Ratoon stunting disease of sugar cane: isolation of the causal bacterium. *Science* **22:**1365–1367.

23. **Davis, N. D., E. M. Clark, K. A. Schrey, and U. L. Diener.** 1986. In vitro growth of *Acremonium coenophialum*, and endophyte of toxic tall fescue grass. *Appl. Environ. Microbiol.* **52:**888–891.

24. **Diehl, W. W.** 1950. *Balansia and Balansiae in America*, p. 1–82. USDA Agriculture Monograph 4. U.S. Government Printing Office, Washington, D.C.

25. **Dobereiner, J.** 1993. Recent changes in concepts of plant bacterial interactions: endophytic N_2 fixing bacteria. *Cienc. Cult.* **44:**310–313.

26. **Dobereiner, J., and F. O. Pedrosa.** 1987. *Nitrogen-Fixing Bacteria in Nonleguminous Crop Plants*. Springer-Verlag, New York.

27. **Dong, Z., M. J. Canny, M. E. McCully, M. R. Roboredo, C. F. Cabadilla, E. Ortega, and R. Rodés.** 1994. A nitrogen-fixing endophyte of sugarcane stems. A new role for the apoplast. *Plant Physiol.* **105:**1139–1147.

28. **Dorworth, C. E.** 1989. Mycoberbicides for forest weed biocontrol—the P.F.C. enhancement process, p. 116–119. *In* C. Bassett, L. J. Whitehouse, and J. A. Zabkiewicz (ed.), *Alternatives to the Chemical Control of Weeds*. FRI Bulletin. Forest Research Institute, Ministry of Forestry, Rotorua, New Zealand.

29. **Glenn, A. E., C. W. Bacon, R. Price, and R. T. Hanlin.** 1996. Molecular phylogeny of *Acremonium* and its taxonomic implications. *Mycologia* **88:**369–383.

30. **Hignight, K. W., G. A. Muilenburg, and A. J. P. Van Wijk.** 1993. A clearing technique for detecting the fungal endophyte *Acremonium* sp. in grasses. *Biotechnol. Histochem.* **68:**87–90.

31. **Hinton, D. M., and C. W. Bacon.** 1985. The distribution and ultrastructure of the endophyte of toxic tall fescue. *Can. J. Bot.* **63:**36–42.

32. **Hinton, D. M., and C. W. Bacon.** 1995. *Enterobacter cloacae* is an endophytic symbiont of corn. *Mycopathologia* **129:**117–125.

33. **Hollis, J. P.** 1951. Bacteria in healthy potato tissue. *Phytopathology* **41:**350–366.

34. **Johnson, J. A., and N. J. Whitney.** 1992. Isolation of fungal endophytes from black spruce (*Picea mariana*) dormant buds and needles from New Brunswick, Canada. *Can. J. Bot.* **70:**1754–1757.

35. **Kulkarni, R. K., and B. Nielsen.** 1986. Nutritional requirements for growth of fungus endophyte of tall fescue. *Mycologia* **78:**781–786.

36. **Ladha, J. K., W. L. Barraquio, and I. Watanabe.** 1983. Isolation and identification of nitrogen-fixing *Enterobacter cloacae* and *Klebsiella planticola* associated with rice plants. *Can. J. Microbiol.* **29:**1301–1308.

37. **Latch, G. C. M., M. J. Christensen, and R. E. Hickson.** 1988. Endophytes of annual and hybrid ryegrasses. *N. Z. J. Agric. Res.* **31:**57–63.

38. **Luttrell, E. S., and C. W. Bacon.** 1977. Classification of *Myriogenospora* in the Clavicipitaceae. *Can. J. Bot.* **55:**2090–2097.

39. **McClung, C. R., and D. G. Patriquin.** 1980. Isolation of a nitrogen-fixing *Campylobacter* species from the roots of *Spartina alterniflora* Loisel. *Can. J. Microbiol.* **26:**881–886.

40. **McInroy, J. A., and J. W. Kloepper.** 1995. Survey of indigenous bacterial endophytes from cotton and sweet corn. *Plant Soil* **173:**337–342.

41. **Misaghi, I. J., and C. R. Donndelinger.** 1990. Endophytic bacteria in symptom-free cotton plants. *Phytopathology* **80:**808–811.

42. **Morgan-Jones, G., and W. Gams.** 1982. Notes on Hyphomycetes. XLI. An endophyte of *Festuca arundinacea* and the anamorph of *Epichloe typhina*, new taxa in one of two new sections of *Acremonium*. *Mycotaxon* **15:**311–318.

43. **Mundt, J. O., and N. F. Hinkle.** 1976. Bacteria within ovules and seeds. *Appl. Environ. Microbiol.* **32:**694–698.

44. **Okon, Y., S. L. Albrecht, and R. H. Burris.** 1977. Methods for growing *Spirillium lipoferum* and for counting it in pure culture and in association with plants. *Appl. Environ. Microbiol.* **33:**85–88.

45. **Patriquin, D. G., and J. Dobereiner.** 1978. Light microscopy observations of tetrazolium-reducing bacteria in the endorhizosphere of maize and other grasses in Brazil. *Can. J. Microbiol.* **24:**734–742.

46. **Petrini, O., and G. C. Carroll.** 1981. Endophytic fungi in foliage of some Cuppressaceae in Oregon. *Can. J. Bot.* **59:**629–636.

47. **Rao, N. S. S.** 1983. Nitrogen-fixing bacteria associated with plantation and orchard plants. *Can. J. Microbiol.* **29:**863–866.

48. **Rennie, R. J., J. R. De Freitas, A. P. Ruschel, and P. B. Vose.** 1982. Isolation and identification of N_2-fixing bacteria associated with sugar cane (*Saccharum* sp.). *Can. J. Microbiol.* **28:**462–467.

49. **Rykard, D. M.** 1983. Comparative morphology of the conidial states and host-parasite relationship in members of Balansiae (Clavicipitaceae). Ph.D. thesis. University of Georgia, Athens.

50. **Rykard, D. M., E. S. Luttrell, and C. W. Bacon.** 1984. Conidiogenesis and conidiomata in the Clavicipitoideae. *Mycologia* **76:**1095–1103.

51. **Saha, D. C., M. A. Jackson, and R. L. Tate.** 1984. A rapid staining method for detection of endophytic fungus in turfgrasses. *Phytopathology* **74:**812.

52. **Samish, Z., R. Etinger-Tulezynska, and M. Bick.** 1961. Microflora within healthy tomatoes. *Appl. Microbiol.* **9:**20–25.

53. **Sampson, K.** 1937. Further observations on the systemic infection of *Lolium*. *Trans. Br. Mycol. Soc.* **21:**84–97.

54. **Sieber, T. N., F. Sieber-Canavesi, and C. E. Dorworth.** 1988. Endophytic fungi in four winter wheat cultivars (*Triticum aestivum*) differing in resistance against *Stagonospora nodorum* (Berk.) Cast. & Germ. = *Septoraia nodorum* (Berk.) Berk. *Phytopathology* **122:**289–306.

55. **Sieber, T. N., F. Sieber-Canavesi, and C. E. Dorworth.** 1991. Endophytic fungi of red alder (*Alnus rubra*) leaves and twigs in British Columbia. *Can. J. Bot.* **69:**407–411.

56. **Siegel, M. R., U. Jarlfors, G. C. M. Latch, and M. C. Johnson.** 1987. Ultrastructure of *Acremonium coenophialum*, *Acremonium lolii*, and *Epichloe typhina* endophytes in host and nonhost *Festuca* and *Lolium* species of grasses. *Can. J. Bot.* **65:**2357–2367.

57. **Siegel, M. R., G. C. M. Latch, L. P. Bush, N. F. Fammin, D. D. Rowen, B. A. Tapper, C. W. Bacon, and M. C. Johnson.** 1991. Alkaloids and insecticidal activity of grasses infected with fungal endophytes. *J. Chem. Ecol.* **16:**3301–3315.

58. **Siegel, M. R., G. C. M. Latch, and M. C. Johnson.** 1987. Fungal endophyte of grasses. *Annu. Rev. Phytopathol.* **25:**293–315.

59. **Sleesman, J. P., and C. Leben.** 1976. Bacterial desiccation: effects of temperature, relative humidity and culture age on survival. *Phytopathology* **66:**1334–1338.

60. **Sussman, A. S.** 1965. Longevity and survivability of fungi, p. 447–486. *In* G. C. Ainsworth and A. S. Sussman (ed.), *The Fungi*. Academic Press, New York.

61. **Talbot, H. W., and R. J. Seidler.** 1979. Cyclitol utilization associated with the presence of *Klebsielleae* in botanical environments. *Appl. Environ. Microbiol.* **37:**909–915.

62. **Timmer, L. W., R. H. Brlansky, R. F. Lee, and B. C. Raju.** 1983. A festidious, xylem-limited bacterium infecting ragweed. *Phytopathology* **73:**957–979.

63. **Ullasa, B. A.** 1969. *Balansia claviceps* in artificial culture. *Mycologia* **61:**572–579.

64. **Vincent, J. M.** 1970. *A Manual for the Practical Study of Root-Nodule Bacteria*. Blackwell Publications, Oxford.

65. **Watanabe, I., W. L. Barraquio, M. R. De Guzman, and D. A. Cabrera.** 1979. Nitrogen-fixing (acetylene reduction) activity and population of aerobic heterotrophic nitrogen-fixing bacteria associated with wetland rice. *Appl. Environ. Microbiol.* **37:**813–819.

66. **White, J. F., Jr.** 1987. The widespread distribution of endophytes in the Poaceae. *Plant Dis.* **71:**340–342.

67. **White, J. F., Jr., A. C. Morrow, and G. Morgan-Jones.** 1990. Endophyte-host associations in forage grasses. XII. A fungal endophyte of *Trichachne insularis* belonging to *Pseudocercosporella*. *Mycologia* **82:**218–226.

68. **Yance, C. E., and C. J. Chang.** 1987. Detection of xylelm-limited bacteria from sharpshooter leafhoppers and their feeding hosts in peach environ monitored by culture isolations and ELISA techniques. *Environ. Entomol.* **16:**68–71.

Methods of Soil Microbial Community Analysis

ANDREW OGRAM AND XIUHONG FENG

46

A knowledge of community structure is central to many ecological and environmental studies, including studies in such seemingly unrelated disciplines as chemical and environmental engineering, soil microbiology, biodegradation and bioremediation, and marine microbiology. A variety of techniques are currently available for analyzing various aspects of community structure, and many of these techniques may be applied with minor modifications to communities for which they were not originally developed. Even though analysis of microbial community structure is one of the most important aspects of environmental microbiology, it also is one of the most technically challenging.

The majority of microorganisms present in many environments may not be readily cultivated by current technologies and therefore may not be included in most analyses. Current estimates indicate that fewer than 1% of the microorganisms present in many environments are readily culturable, indicating that techniques based on laboratory cultivation may be significantly biased (41, 42). In fact, most species in many microbial communities may never have been described. Soil microbial communities probably are the most complex of natural communities, and one recent study has estimated that there may be as many as 4,000 species per gram of soil (41). Even if all species present in a community were cultivable, it is not feasible with current technologies to define and enumerate all species within any but rather simple communities because of time considerations. Many approaches either study small, well-defined groups of microorganisms, e.g., those involved in nitrification or the degradation of a particular xenobiotic, or use a broader-spectrum approach and define the relative numbers of individuals capable of utilizing an array of carbon sources (21) or possessing characteristic fatty acids (43).

To overcome difficulties and limitations associated with laboratory cultivation, some techniques that bypass cultivation and are based on the direct extraction of indicator molecules such as nucleic acids (36) or fatty acids (43) from environmental samples have been developed. The composition of extracted nucleic acids or fatty acids may be analyzed, and depending on the nature of the analysis, information regarding the structure, activity, and in some cases the nutritional state of the community may be obtained. No techniques are without limitations or biases, however, and the user should be familiar with the effects that a given bias will have on a particular application.

Strategies for community structure analysis may be divided into two general categories: (i) those relying on laboratory cultivation or incubation and (ii) those based on the direct extraction and analysis of indicator molecules such as nucleic and fatty acids. Since most environmental microbiological studies are primarily concerned with bacteria, only techniques concerned with characterization of bacteria will be covered here. The first section of this chapter covers approaches based on cultivation and the second addresses those based on analysis of indicator molecules that have been directly extracted from soil communities. Reduction and analysis of data by various means such as principal components analysis, hierarchical cluster analysis, and semivariance analysis (16) are useful for studying specific changes in communities (43) but are beyond the scope of this chapter.

SAMPLING CONSIDERATIONS

The first practical consideration encountered in analyzing the structures of soil microbial communities is the size and number of samples required. Sampling strategy has been discussed elsewhere in this book, but a basic understanding of the nature of soil microbial communities may aid in the design of an appropriate sampling strategy for a given application. An ecological community is loosely defined as "a unified assemblage of populations occurring and interacting at a given location" (2). Soil microbial communities may exist not as discrete units but rather as loosely bound units existing along a continuum of environmental variables that can extend laterally across a soil as well as downward through a soil profile. The minimum physical boundaries for a soil microbial community in terms of mass or volume have never been described, and communities in bulk soil are likely to consist of broad zones along environmental gradients that are specific to the individual study site. Soil community structures may, however, change dramatically on relatively small scales, such as may be observed in the top few centimeters of soil profiles, within soil aggregates, and in the rhizosphere. These microhabitats frequently are ignored in sampling soils for many studies, in which a kilogram of "bulk surface soil" may be collected from depths of between 0 and 15 cm and mixed together. This approach may result in combining relatively distinct communities present in these microzones with more homogeneous communities in bulk soil.

Microbial communities in bulk soil can be highly variable with regard to space and time (39), and the sampling strategy should be designed to account for this variability. An added level of variability is found when soils are sampled at various depths. Soil profiles are composed of a series of horizons that may vary with depth, and for most applications, sampling of subsoils should be conducted with attention to both depth and horizonation rather than to depth alone.

METHODS BASED ON LABORATORY CULTIVATION

Many commonly used methods of community analysis are based on the isolation of individual strains on solid medium, followed by characterization of the isolated colonies by one or more methods that will be discussed below. This general strategy is appropriate for studies requiring identification or characterization of individual isolates or for population genetics studies of a particular species. In addition, this may be the most sensitive approach available for enumeration and characterization of specific groups, if a suitable selective medium is available, such as selection of strains possessing particular growth characteristics (such as degradation of a particular xenobiotic) or resistances (such as to antibiotics or heavy metals). It may also be used in lieu of the more technically challenging and time-consuming approaches based on the characterization of nucleic acids or fatty acids directly isolated from the environmental sample. The greatest limitation of this general approach is that relatively few species present in an environmental sample are readily cultivated, and many strains of interest may not be included in the analysis.

Strain Isolation

Prior to characterization, strains must be isolated from the soil matrix and grown on laboratory medium. This is typically accomplished in three steps: (i) separation of cells from soil particulates, (ii) dilution of the cells to an appropriate level, and (iii) growth of the cells on appropriate growth medium.

Separation of bacterial cells from soil particulates may result in loss of a significant proportion of the community as a result of the close association of many cells with the organic matrix of soil particles. Many soil microorganisms produce extracellular polysaccharides that promote the irreversible adhesion of cells to soil particulates. Inefficient disruption of the bonds between the extracellular polysaccharide and the soil matrix decreases the efficiency of separation and may also contribute to a bias due to separation of those strains that are adhering loosely to soil particles. Separation is usually accomplished by vortex mixing 1 part of soil in 9 parts of an aqueous solution buffered to between pH 6.5 and 7.5 and containing small amounts of NaCl and a detergent such as sodium dodecyl sulfate. Suspension of the sample in this buffer serves two functions: it disperses the soil particles, thereby breaking up small aggregates that could decrease extraction efficiency, and at least partially disrupts the association between the cells and the soil particles.

Following suspension, the sample is serially diluted in sterile, osmotically balanced buffer such as phosphate-buffered saline such that between 20 and 200 colonies may be spread with a volume of 100 to 200 μl per plate of solid growth medium.

Because cultivation of environmental strains on laboratory media may be the single largest source of bias in community structure analysis, the method of cultivation becomes of primary importance. Standard plating media and incubation conditions are not appropriate for isolation and characterization of strict anaerobes or for those organisms that have unusual or unknown growth requirements. The researcher should be aware of the limitations of most plating media and may consider using a variety of different media and incubation conditions to optimize the diversity of isolates.

There is no single medium of choice for cultivation of heterotrophic aerobic bacteria isolated from soils, but many include various concentrations of yeast extract (37) or soil extracts (38) supplemented with inorganic nutrients. Soil bacteria may have a range of preferences for concentrations and types of nutrients, but many may be divided into two operationally defined classes: copiotrophs and oligotrophs. Copiotrophs have higher growth rates on rich complex media, while the growth of oligotrophs may be inhibited at these concentrations. Oligotrophs grow optimally on similar media that have been diluted to between 1 and 10% of the amounts used by copiotrophs. To ensure that the widest diversity of aerobic heterotrophs are isolated, it is best to spread cell suspensions on two sets of agar plates containing both nutrient concentrations.

Plating media can be used to intentionally select against the majority of species in order to isolate a specific group. Cycloheximide is commonly used as a selective agent to inhibit the growth of fungi. Fungi are very common in soils and can quickly overgrow agar plates, making the isolation of bacterial colonies very difficult.

The time allowed for incubation of soil isolates is of great importance because of the variability in growth rates among soil bacteria. Some colonies may appear on solid medium after only 1 or 2 days of incubation, while others may not appear before 1 or 2 weeks; consequently, plates should be examined daily for the appearance of new colonies. A limitation of this approach is that slowly growing bacteria may be overgrown by faster-growing bacteria, resulting in the inability to recover some strains.

Once bacterial strains have been cultivated, two general strategies may be used in selecting colonies for further characterization. The strategy of choice depends on the type of information desired. If the researcher is interested primarily in applications involving species richness (the number of different species present in a community), colonies that are different in morphology should be chosen. If, however, the researcher is interested primarily in species diversity (species richness and evenness), colonies should be chosen randomly. There are no rules for determining the number of colonies that should be selected for further analysis, and the decision usually is dictated by practical considerations based on plans for further analysis and the researcher's limitations and resources. After the appropriate strategy has been determined, strains should be purified by streaking each colony of interest at least two successive times for isolation.

Many environmental strains, particularly pseudomonads, secrete copious amounts of extracellular polysaccharides when grown on synthetic media, which may make separation of strains difficult. In addition, slowly growing strains may be overgrown by faster-growing strains and may be co-purified with faster-growing, and hence larger, colonies.

The limit of detection of individual species by this approach may be high, particularly if no system is used to select for the species of interest. The limit of detection may be calculated from the starting mass of soil, the final dilution, and the volume spread on the agar plates.

Colony Morphology

The simplest means of differentiating between species for the purposes of diversity measurements is by comparisons of colony morphology. Similar colony types may be grouped into morphotypes by careful observation of the shape, size, color, and level of mucoidy at various stages of colony age. This rather crude system does not allow identification of species, but it is suitable for accurate groupings of different species (18) and may be used for species diversity analyses or to reduce the number of isolates for further characterization by more sophisticated means.

Because of the great variety of soil microorganisms, it is not possible to give detailed descriptions of all specific colony morphologies that one might encounter. There are, however, qualitative guidelines that aid in identifying characteristics that differ among colony types. A note of caution should be interjected: colony morphologies are highly dependent on the growth substrate; colony characteristics observed for a strain on one medium may be quite different on another medium. This is particularly true when one is comparing colony morphologies on rich versus dilute media. Differences in morphology are often more easily spotted on rich media than on dilute media. Groupings of isolates into similar morphotypes should be conducted with colonies of similar age and from the same batch of media. There may also be some variation within the morphologies of colonies belonging to the same strain, and categorization of isolates based on specific colony characteristics should be somewhat broad and based on several aspects of morphology rather than on only one or two characteristics.

The most obvious colony characteristics are color and size. Many colonies typically appear somewhat off-white in color, but others exhibit various shades of colors, and close attention to these shades, in combination with other characteristics, may aid in discriminating between colony morphotypes. The relative sizes of colonies also are useful indicators of the relative growth rates of individual strains and can be used as a general comparison between strains. Precise sizing of colonies is not of great value because of possible small differences in growth rates between colonies, but colonies observed at a particular time may be classified as small, medium, or large (38).

The shape of colonies is likely to be the most important discriminant. Careful attention should be paid to the shape (round, irregular, or filamentous), the edges (smooth or small versus large lobes), the elevation (flat, embedded in the agar, concentric circles, convex, or with a raised area in the center), and the level of mucoidy (dry, powdery, wet, or very wet) (38).

METHODS FOR STRAIN IDENTIFICATION
Characterization Based on Growth Substrates

Environmental isolates are most commonly identified by growth on specific substrates and fermentative ability (21). This type of analysis typically is preceded by initial characterization of the isolated strains on the basis of Gram stain reaction, cell shape and size, and oxidase and catalase reactions. A number of systems are available from commercial sources, with those from Biolog (Hayward, Calif.) and API (Plainview, N.Y.) being among the most commonly used.

Lipid Analysis

Bacterial strains also may be identified by the presence of specific cellular lipids. There are several major classes and subclasses of lipids, including free acids, hydrocarbons, fatty alcohols, or other compounds of bound fatty acids (phospholipids, peptidolipids, glycolipids, etc.) (21). There are many different kinds of fatty acids in the lipids of microorganisms, and different organisms have different combinations of these fatty acids, thus forming distinct characteristic, stable patterns within taxa (9). These fatty acid patterns can be analyzed quantitatively to provide taxonomic information at the species level (46). Fatty acids can easily be extracted and esterified with methanol to form fatty acid methyl esters (FAMEs) (3). Phospholipids can be purified from mixtures of lipids by solid-phase extraction (48) and form phospholipid fatty acid methyl esters (PLFAs). These FAMEs and PLFAs can be analyzed quantitatively and qualitatively by high-resolution fused-silica capillary gas chromatography (27). This technique has been used widely and can be automated.

By comparison of the fatty acid pattern of an unknown isolate with those of reference strains in a computer database, one can identify the unknown isolate. This approach has been used widely in identifying medically important bacteria and plant pathogens (9). A few precautions must be considered when using FAMEs for identification of bacteria. First, purity of the isolate is very important. Second, the selection of reference strains is critical to the success of the identification. Third, standard environmental growth conditions for reference strains and isolates are needed for the determination of taxonomic relationships, because fatty acid patterns are influenced by many factors, including growth medium, incubation temperature and moisture, culture phase, and technique of analysis (9, 43). Finally, the database of environmental strains is limited, and a species identification may not be possible for many unknown environmental isolates.

Colony Hybridization

The presence or absence of particular genetic sequences within culturable isolates of a microbial community may be determined by hybridization of the isolated colonies with gene probes or cloned versions of the genes of interest (37). Probes may be directed toward specific functions, such as enzymes encoding the degradation of an organic contaminant or nitrogen fixation, or toward specific phylogenetic groups. An extensive list of gene probes that have been used is presented in an excellent review by Sayler and Layton (36). If the strains of interest are cultivable and if a strategy is available to select for these strains, colony hybridization may be quantitative and more sensitive than hybridization of DNA directly extracted from a soil sample (see below). In addition to all of the caveats and advantages concerning cultivation of soil bacteria, colony hybridization is subject to advantages and limitations associated with gene probing.

The greatest advantage of the application of gene probe technology to soil microbiology is that it can be a very specific means of detecting and enumerating individual genotypes. The greatest limitation is the current lack of knowledge concerning molecular genetics of soil microorganisms. In order to apply gene probe analysis to soil microorganisms, the molecular genetics of both the gene probe and the target sequences must be known. The gene of interest must have been cloned, and the nucleotide sequence of the target gene must be similar enough to that of the gene probe for hybridization to occur. If there is more than one type of gene responsible for a given function, colony hybridization may underestimate the total numbers of colonies capable of performing the function of interest. Conversely, if nontarget genes share enough sequence similarity with the gene probe for hybridization, overestimation of the numbers of target cells will occur. Underestimation due to the presence of alternative genes coding for similar functions is the most common source of error, and it is impossible to control for without a detailed knowledge of the genetics of soil microorganisms. Overestimation of target colonies by hybridization with similar genes may be controlled for to some extent by including closely related genes as negative controls during hybridization.

If one is interested in enumerating CFU having a particular genotype, cells should be quantitatively applied to the appropriate growth medium and incubated for up to 2 weeks, depending on the growth rates of the target populations. Appropriate positive and negative controls should be incubated under the same conditions as the colonies taken from soil. Positive control strains are typically those cells from which the gene probe was originally cloned, and negative controls are strains containing the cloning vectors and strains harboring genes that are related to, but different from, the gene of interest. Strains harboring the cloning vector are required negative controls because of the possibility of contamination of the probe with the cloning vector. This may occur if the probe was separated from the vector by digestion with restriction endonucleases and subsequently isolated from agarose gel. For a more complete description of probe preparation, the reader is advised to consult reference 30. The colonies are transferred to a DNA binding membrane and lysed with a detergent such as sodium dodecyl sulfate and sodium hydroxide. The released DNA is fixed to the membrane according to standard procedures (35). The membrane should be examined at this time to ensure that no cell debris that might participate in nonspecific binding of the probe remains on the membrane. If cell debris is visible, it may be easily washed off with a stream of sterile water. The membrane is then prehybridized to block any sites that may result in nonspecific binding of the probe and is hybridized with the labeled probe at a temperature that is approximately 10°C below the melting temperature of the hybrids. After 8 to 12 h, the excess probe is washed off in a solution containing an appropriate salt content and at a specific temperature. The lower the salt content and the higher the washing temperature, the greater the fidelity of the hybrids. Salt contents and temperatures for washing are typically determined empirically so that no hybridization is observed with the negative control; the reader is referred to reference 35 for appropriate conditions. The presence of the genes is usually detected by autoradiography (37).

rRNA Characterization

The most precise method of determining the phylogenetic placement of an isolated colony is by characterization of its rRNA. The characterization of rRNA genes (rDNA) is becoming a very powerful and commonly used means of accurately characterizing environmental isolates to phylogenetic levels from domain (*Bacteria*, *Archaea*, *Eucarya*) (47) to individual species.

The technique is based on the concept that rRNA molecules, particularly the 16S and 23S rRNA molecules of prokaryotes, are highly conserved throughout evolution and are therefore useful as specific indicators of the phylogenetic affiliation of environmental isolates (24). The nucleotide sequences of portions of the 16S and 23S rRNA molecules are subject to change at a faster rate over evolutionary time than are other portions of the same molecule. Those regions of the gene that change at a very slow rate over evolutionary time are characteristic of broad phylogenetic groupings, such

as domains, while those regions that change at a somewhat faster rate are characteristic of narrower phylogenetic groupings, such as the *Proteobacteria*, high G + C-content gram-positive bacteria, etc. The regions of these molecules that change most rapidly over evolutionary time (so-called hypervariable regions) are characteristic of individual species (24).

Two general strategies are usually used for rRNA analysis of isolated colonies: (i) colony hybridization with synthetic oligonucleotide probes directed toward specific phylogenetic groups and (ii) analysis of the nucleotide sequences of the 16S rRNA genes. The first strategy can be used to quickly categorize large numbers of isolates as to relatively broad phylogenetic groupings and can be quantitative if quantitative procedures are used in cultivation of the colonies. All of the advantages and disadvantages associated with cultivation of soil microorganisms apply to this procedure, and the general steps involved are similar to those described above for colony hybridization. An extensive list of probes used for identification of a range of phylogenetic groups is presented by Amann et al. (1).

The greatest limitation of this approach at the present time is the dearth of knowledge of the rRNA sequences of most environmental microorganisms. Synthetic oligonucleotide probes are designed on the basis of data available in databases such as the Ribosomal Database Project (RDP) maintained by the University of Illinois. Since the majority of soil microorganisms have never been isolated and their rRNA molecules have never been sequenced, many individual isolates may belong to as yet undescribed groups that do not hybridize with probes directed toward narrow phylogenetic groups.

Precise phylogenetic placement of isolates requires sequencing of individual 16S rRNA genes from DNA isolated from the colonies. While this procedure is much more informative than is colony hybridization with synthetic oligonucleotide probes, it also is much more time-consuming and technically demanding. In general, genomic DNA is isolated from the selected colonies, and primers specific to 16S rRNA genes are used to amplify these genes by PCR. The sizes of the amplification products depend on the primers selected but in general are between 1 and 1.5 kbp. The amplification products usually are separated by agarose gel electrophoresis, excised from the gel, and ligated into a cloning vector specifically designed for this purpose. A number of such systems, including the TA cloning vectors (Invitrogen, San Diego, Calif.) and the Cloneamp vectors (GIBCO BRL, Bethesda, Md.), are available.

Once the amplification products are cloned, their nucleotide sequences can be determined. Sequencing is now relatively convenient, with the availability of automated sequencers at many universities and commercial sites. Sequencing 16S rRNA genes may be complicated, however, by the presence of strong secondary structures that make denaturation of the template difficult under standard conditions. Sequencing of rRNA genes from some species may require alternative strategies for overcoming these problems, and the reader is referred to an excellent review on this subject by Lane (24).

A detailed description of phylogenetic analysis is presented elsewhere in this book, and only the general principles are presented here. A number of computer programs have been developed for this purpose. These programs, many of which are part of the Genetics Computer Group package and are available at most universities, are useful for comparing sequences with those in databases such as GenBank and for the construction of phylogenies. First approximations at phylogenies may be easily obtained with programs accessible from the RDP home page of the University of Illinois (http://rdp.life.uiuc.edu). The RDP home page also provides access to programs for rRNA approaches to strain identification, and the interested reader is encouraged to browse this site.

DNA Fingerprinting: RFLP, AP-PCR, and REP-PCR

Among the techniques that may be used to rapidly differentiate closely related environmental strains by DNA fingerprinting are restriction fragment length polymorphism (RFLP) analysis, arbitrarily primed PCR (AP-PCR), and repetitive extragenic palindromic PCR (REP-PCR). Regardless of the specific technique, the data obtained from DNA fingerprinting approaches are DNA fragments that, when separated on agarose or polyacrylamide gels, yield a banding pattern specific to the genome under investigation. These banding patterns are often likened to bar codes used in product merchandising. Fingerprinting techniques typically are not used to identify environmental isolates to the level of species but rather are used to demonstrate small genetic differences or similarities for population genetics studies or for studies of diversity within closely related species or strains. This type of approach is most appropriate for fine-structure analysis of specific components of microbial community structure rather than a broad-spectrum approach toward identifying specific taxonomic groups or describing the structures of entire communities.

The oldest of these approaches, RFLP analysis, uses a variety of restriction endonucleases to digest purified DNA from individual strains in order to identify polymorphisms within individual genes (44). Polymorphisms, or differences within a specific gene, may result in a different number of sites that are recognized by the restriction endonucleases used in the digestion. Such differences may be caused by single base pair changes or by larger changes such as insertions, deletions, and rearrangements. Fragments of different sizes within the digested DNA are separated by agarose gel electrophoresis, transferred to a DNA binding membrane by standard procedures, and hybridized with a labeled probe specific for the gene of interest. Differences or similarities between different isolates may then be identified. The similarity between numbers of restriction sites within individual genes also has been used to calculate evolutionary relatedness between genes from different organisms (29).

One of the primary requirements for RFLP analysis is that the target DNA be very pure; contaminants may partially inhibit restriction enzyme digestion, thereby yielding partial digestion products that confound the analysis. Chromosomal DNA should be subjected to multiple rounds of phenol-chloroform extraction and then at least two rounds of chloroform extraction. Depending on the strain, other steps may be required to remove contaminating substances (26). Perhaps the easiest approach to obtaining DNA suitable for RFLP analysis is by amplification of specific genes by PCR (chapter 48, this volume; 44): the DNA is amplified away from any contaminants, the amplification products are digested with the enzymes of choice, and the digestion products are visualized by gel electrophoresis. This approach has the added benefit that hybridization is not required; the digestion products result only from the gene of interest and are in sufficient concentrations to be visualized by staining with ethidium bromide.

The enzymes used in RFLP analyses typically are 6-bp recognition restriction endonucleases such as *Eco*RI, *Hin*dIII, and *Bam*HI. For relatively small (≤500-bp) segments of DNA or for highly conserved genes such as rRNA genes, restriction endonucleases that recognize 4 bp (and generate a greater number of fragments) such as *Taq*I, *Hae*III, and *Sau*3A are used.

A variety of different genes may be used as hybridization probes to increase the likelihood of identifying differences or similarities between environmental isolates. Examples of possible probes are genes encoding nitrogenase (*nifHDK*), rRNA, *recA*, or ribulose bisphosphate carboxylase (*rbcL*). A DNA binding membrane with digested DNA from several strains may be probed sequentially with different genes, stripping one probe off and then hybridizing with another. If membranes are to be reprobed, they should be checked prior to rehybridization for residual radioactivity by exposure to X-ray film.

Arbitrarily primed PCR (AP-PCR; a generic term incorporating randomly amplified polymorphic DNA and DNA amplification fingerprinting) is similar to RFLP analysis in that fingerprints specific to the target genome are generated, and the techniques may be used to differentiate between closely related strains (17). Unlike RFLP analysis, which is designed to identify differences within specific genes, these techniques are designed to screen for differences between entire genomes. The greater the number of primers used in individual amplification reactions, the greater the amount of the target genome scanned and the greater the likelihood of identifying differences between strains. For this reason, several AP-PCRs, each using a different primer, typically are performed. AP-PCR is much faster than RFLP analysis since it is PCR based, and no digestion with restriction endonucleases or hybridization is required. AP-PCR uses single PCR primers (rather than two different primers as is used in most PCR applications) that are typically of random base composition and usually between 8 and 10 bases in length. A number of companies synthesize such primers. The primers randomly amplify a variety of segments of the target DNA, typically yielding products between 200 and 2,000 bp in size that usually are separated on 2% agarose gels and visualized by staining with ethidium bromide (35).

The generation of amplification products is dependent on a number of variables, including primer sequence, primer concentration, template concentration, template quality, cycling parameters, and thermocycler. Because of the sensitivity of AP-PCR banding patterns to so many variables, a lack of reproducibility is frequently a problem. For this reason, three replicate amplifications should always be conducted. Reproducibility usually is not a problem between AP-PCR patterns generated by one operator who has taken care to be consistent between reactions, but comparisons of patterns generated by different laboratories or even by different operators in the same laboratory may not be possible (31).

REP-PCR resembles AP-PCR in that fingerprints characteristic of a particular strain are generated, but there may be fewer problems with variability than in AP-PCR. REP-PCR is based on the amplification of repeated sequences that are highly conserved and widely distributed among gram-negative genomes (8). The amplification is therefore not random and uses two primers (8). The approach has not been used as extensively as AP-PCR and RFLP for fingerprinting strains, but it is likely to become much more widely used in the future. An additional advantage of REP-PCR is that species-specific probes may be generated to identify individual species either by colony hybridizations or in extracts of community DNA (8).

ANALYSIS BASED ON DIRECT EXTRACTION OF INDICATOR MOLECULES

To bypass the current limitations and biases associated with laboratory cultivation, one can use alternative approaches to the analysis of community structure that are based on the analysis of indicator molecules extracted directly from environmental samples. The most commonly used indicator molecules are DNA, RNA, and phospholipids. These molecules yields different information regarding the structures and activities of microbial communities, and the user should choose the molecule and the method that best suit the specific application. Methods for the isolation of DNA, RNA, and phospholipids from environmental samples are presented elsewhere in this book.

Lipid Analysis

Total fatty acids extracted from environmental samples have been used to study microbial community structures and metabolic states and to compare similarities and differences among soil microbial communities (4, 12). Certain components of microbial populations are enriched in particular fatty acids. For example, the odd-number and branched-chain fatty acids are associated with gram-positive bacteria. Gram-negative bacteria contain higher proportions of even-number monounsaturated straight-chain and cyclopropane fatty acids, eubacteria generally do not contain polyunsaturated fatty acids, and plasmalogen phospholipids are enriched in anaerobic prokaryotic bacteria (43, 46).

The relative concentrations of different kinds of polar lipids in each membrane are characteristic for the type of membrane, the cell type, and the species; however, the fatty acid components of the individual membrane lipids are not fixed and may vary with nutritional state and environmental conditions in order to maintain membrane fluidity (22). By studying the changes of fatty acids profiles, especially the presence of certain phospholipid fatty acids as biomarkers for certain microbial types, an indication of the metabolic activity of the microbial community may be obtained (12, 15). The primary limitation of this method is that it is difficult to correlate the changes in phospholipid fatty acid patterns with the dynamics of specific groups of organisms, and more data are needed (12). Analysis of community structure by lipid analysis is discussed in greater detail in chapter 10.

Gene Probes

The type of information obtained by quantitative hybridization of gene probes to nucleic acids directly extracted from environmental samples is largely dependent on the target nucleic acid. Hybridization to DNA is useful for rapidly identifying specific aspects of community structure, such as the presence and relative concentrations of genes encoding a particular function. Hybridization to RNA is an indicator of the relative activities of the target groups. The activities of certain non-rRNA genes may be estimated by using mRNA as the target, although many mRNAs are too labile to be detected in this way. Analysis of rRNA will be discussed below, although many of the principles discussed here apply to rRNA and mRNA as well as DNA.

Many of the advantages and disadvantages of this approach are similar to those discussed for the use of gene probes in colony hybridization, although the quantitative aspects of hybridization of nucleic acids are not as straightforward as in colony hybridization. In colony hybridization, one merely counts the number of colonies that hybridize with the probe and relates this number to the number of CFU per gram of soil. In hybridization of extracted nucleic acids, probe specificity and probe length become important considerations. Precise determination of concentrations of specific sequences is difficult because little is known about the genetics of most soil microorganisms. It is conceivable that regions of the probe may hybridize with nontarget genes in the sample, resulting in an overestimation of the target gene. Alternatively, limited regions of the probe may hybridize with target genes in the sample because of evolutionary divergence of these genes, thereby underestimating the concentrations of the target sequences. For these reasons, probes should be relatively short (approximately 500-bp) segments of an internal region of a well-characterized gene, and the washing conditions should be kept relatively stringent (i.e., low salt and high temperature).

In general, a dilution series of known concentrations of the target gene (typically the probe DNA) is applied to a DNA binding membrane adjacent to the environmental samples. Triplicate samples are advised because of the variability of the procedure and variability associated with some membranes. Raskin et al. (32) conducted an extensive survey of variabilities in hybridization of rRNA bound to membranes from various manufacturers, and the reader is strongly encouraged to consult this article before purchasing nylon membranes for quantitative hybridizations. The inclusion of negative controls similar to those described for colony hybridization is very important.

DNAs are applied to the membrane as a grid through a vacuum manifold. Depending on the type of manifold, the DNA may be applied as circles (dot blots) or as narrow ellipses (slot blots). Slot blots are preferred for quantitative hybridization because the resulting autoradiograms are more easily analyzed by scanning densitometry.

Standard procedures for labeling the probe (either with radiolabels such as [^{32}P]dCTP or with nonradioactive labels), prehybridization, hybridization, and washing are followed (30, 35). The stringency of washing generally is kept high to avoid possible nonspecific hybridization with related but nontarget genes in the environmental sample (30).

The extent of hybridization usually is measured by exposure of the membrane to X-ray film (autoradiography). The X-ray film should be preflashed with a photographic strobe (35) in order to increase the sensitivity and the linear range of the autoradiogram. The membrane is wrapped in plastic wrap such as Saran Wrap, and the film and membrane are placed in an autoradiography cassette. The exposure of the film may be greatly enhanced by placing the film and membrane between two intensifying screens that may be purchased with the cassettes. The cassette should be kept at −80°C for between 3 h and 1 week, depending on the intensity of the signal emitted from the bound probe. The length of exposure must be determined empirically for each application. After an appropriate time, the film is developed and the densities of the exposed regions corresponding to the placement of the samples are measured by scanning densitometry. The relative amounts of hybridization in different samples may be determined from a standard curve relating density to mass of target generated from the dilution series of target genes. Alternatively, fingerprints of communities may be generated by reverse sample genome probing, in which a set of genes of interest are applied to the membrane and the sample DNA is labeled and used as a probe (45).

Comparisons of Communities by Percent G+C Profiles

An approach used to analyze shifts in microbial community structure with time or following some perturbation is comparison of the distributions of G + C contents in the community DNA (19). Bacterial chromosomes have G + C contents that are characteristic of their taxonomic groups, and the relative proportions of G + C contents in DNA extracted from a community should therefore be characteristic of the relative proportions of specific taxonomic groups within the community. Percent G + C profiles are generated by fractionating community DNA on CsCl-ethidium bromide gradients formed by ultracentrifugation. A + T-rich regions are less dense than G + C-rich regions, and a gradient of DNA of increasing G + C contents is formed downward through the ultracentrifuge tube. The tube is sampled by taking small fractions from the bottom of the gradient. The concentration of DNA (or A_{260}) is measured for each fraction and plotted as a function of fraction number.

Analysis of rRNA and rDNA

Analysis of the diversity of rRNAs or their genes in an environmental sample may be of great value in characterizing community structures and the relative activities of phylogenetic groups of interest. The strategies that can be used include sequencing and phylogenetic analysis (14, 25), hybridization with group-specific probes (1), and classification by RFLP patterns for the purpose of placement into different operational taxonomic units (28). Total community structure analyses are conducted on rDNA rather than rRNA molecules, since DNA includes all members of the community, regardless of activity. If, however, one is interested in analyzing the metabolically active component of the community, rRNA rather than rDNA should be analyzed.

The simplest approach to analysis of rRNA is hybridization with phylogenetic probes, using procedures similar to those described above in the section on gene probes. A variety of oligonucleotide probes that have been designed to hybridize exclusively to specific groups are listed in reference 1. This approach is useful for characterization of the active groups of a community, but it is less useful for identification of specific members of a community, since probes are directed toward broad phylogenetic groups.

For identification or characterization of individual members of a community, cloning and sequence analysis are required. Regardless of the specific approach to characterization of community rRNA or rDNA, the initial steps of these procedures require direct extraction of total RNA or DNA, reverse transcription of RNA to cDNA (if RNA is of interest), and PCR amplification of the rDNA or cDNA from the groups of interest by primers specific for these groups (24). Since the amplification products will have been derived from mixed communities, they are assumed to be mixtures of molecules of similar sizes originating from the various organisms present in the sample.

Prior to further analysis, population of molecules must be segregated by cloning (25), preferably in vectors specifically designed this purpose. The proportion of clones present in the PCR-generated library should not be assumed to be quantitatively representative of the proportions of the individual rRNA genes in the sample as a result of primer bias (40) and possible biases in the relative rates of PCR amplification of rRNA genes due to differences in copy number and proximity of the operons (10).

Once the clone library has been created, the number of clones to be analyzed further may be reduced by comparison of similar RFLP patterns. Since rRNA genes are only between 1.5 and 2 kb in size and are highly conserved, restriction endonucleases that recognize 4 bp, rather than 6, commonly are used (28). Reduction of the numbers of clones

for analysis by RFLP grouping may result in an overestimation of community diversity as a result of the presence of divergent rRNA (*rrn*) operons within some strains. Most bacterial species harbor between 1 and 10 *rrn* operons, and small differences between operons that result in different RFLP patterns have been observed in some strains (11).

Amplification of mixtures of molecules is subject to artifacts resulting from PCR, including the formation of amplification products composed of the 3′ end of one molecule and the 5′ end of another. These so-called chimeras are thought to present in fewer than 10% of the clones obtained by this method but can be very difficult to identify. Sequencing the amplification products and checking for chimeras with the Check Chimera program available through the RDP home page (http://www.rdp.life.uiuc.edu) is advised, but this program is not foolproof (34).

The most detailed information concerning the phylogenetic affiliations of the individual members of the community is obtained by sequencing each of the individual clones, but this may be an expensive and time-consuming procedure. Hence, it is usually conducted only on communities with low species diversities, such as those found in hot springs, bioreactors, and bacterioplankton. It should be noted that the widespread use of automated sequencers allows the characterization of many more clones than was once possible. A more suitable approach to analysis of complex communities, such as those in soils, is to determine the presence of phylogenetic groups of interest by hybridization of the individual clones with group-specific oligonucleotide probes (1, 25) as described in the section on colony hybridization.

REFERENCES

1. **Amann, R. I., W. Ludwig, and K.-H. Schleifer.** 1995. Phylogenetic identification and in situ detection of individual microbial cells without cultivation. *Microbiol. Rev.* **59:**143–169.
2. **Atlas, R. M., and R. Bartha.** 1993. *Microbial Ecology: Fundamentals and Applications,* 3rd ed. Benjamin/Cummings Co., New York.
3. **Bligh, E. G., and W. J. Dyer.** 1959. A rapid method of total lipid extraction and purification. *Can. J. Biochem. Physiol.* **37:**911–917.
4. **Bobbie, R. J., and D. C. White.** 1980. Characterization of benthic microbial community structure by high-resolution gas chromatography of fatty acid methyl esters. *Appl. Environ. Microbiol.* **39:**1212–1222.
5. **Bond, P. L., P. Hugenholtz, J. Keller, and L. Blackall.** 1995. Bacterial community structures of phosphate-removing and non-phosphate-removing activated sludges from sequencing batch reactors. *Appl. Environ. Microbiol.* **61:** 1910–1916.
6. **Bowman, J. P., J. H. Skerratt, P. D. Nichols, and L. I. Sly.** 1991. Phospholipid fatty acid and lipopolysaccharide fatty acid signature lipids in methane-utilizing bacteria. *FEMS Microbiol. Ecol.* **85:**15–22.
7. **Braun-Howland, E., P. Vescio, and S. Nierzwicki-Bauer.** 1993. Use of a simplified cell blot technique and 16S rRNA-directed probes for identification of common environmental isolates. *Appl. Environ. Microbiol.* **59:** 3219–3224.
8. **de Bruijn, F. J.** 1992. Use of repetitive (repetitive extragenic palindromic and enterobacterial repetitive intergenic consensus) sequences and the polymerase chain reaction to fingerprint the genomes of *Rhizobium meliloti* isolates and other soil bacteria. *Appl. Environ. Microbiol.* **58:**2180–2187.

9. **Dembitsky, V. M., E. E. Shubina, and A. G. Kashin.** 1992. Phospholipid and fatty acid composition of some basidiomycetes. *Phytochemistry* **31:**845–849.

10. **Farrelly, V., F. Rainey, and E. Stackebrandt.** 1995. Effect of genome size and *rrn* copy number on PCR amplification of 16S rRNA genes from a mixture of bacterial species. *Appl. Environ. Microbiol.* **61:**2798–2801.

11. **Feng, X. H., L. T. Ou, and A. Ogram.** Unpublished data.

12. **Frostegard, A., E. Baath, and A. Tunlid.** 1993. Shift in the structure of soil microbial communities in limed forests as revealed by phospholipid fatty acid analysis. *Soil Biol. Biochem.* **25:**723–730.

13. **Giesendorf, B., A. van Belkum, A. Koeken, H. Stegeman, M. Henkens, J. van der Plas, H. Goossens, H. Niesters, and W. Quint.** 1993. Development of species-specific DNA probes for *Campylobacter jejuni*, *Campylobacter coli*, and *Campylobacter lari* by polymerase chain reaction fingerprinting. *J. Clin. Microbiol.* **31:**1541–1546.

14. **Giovannoni, S. J., T. B. Britschgi, C. L. Moyer, and K. G. Field.** 1990. Genetic diversity in Sargasso Sea bacterioplankton. *Nature* (London) **345:**60–63.

15. **Guckert, J. B., M. A. Hood, and D. C. White.** 1986. Phopholipid ester-linked fatty acid profile changes during nutrient deprivation of *Vibrio cholerae*: increases in the *trans/cis* ratio and proportions of cyclopropyl fatty acids. *Appl. Environ. Microbiol.* **52:**794–801.

16. **Haack, S. K., H. Garchow, D. A. Odelson, L. J. Forney, and M. J. Klug.** 1994. Accuracy, reproducibility, and interpretation of fatty acid methyl ester profiles of model bacterial communities. *Appl. Environ. Microbiol.* **60:**2483–2493.

17. **Hadrys, H. M. Balick, and B. Schierwater.** 1992. Applications of random amplified polymorphic DNA (RAPD) in molecular ecology. *Mol. Ecol.* **1:**55–63.

18. **Haldeman, D. L., and P. S. Amy.** 1993. Diversity within a colony morphotype: implications for ecological research. *Appl. Environ. Microbiol.* **59:**933–935.

19. **Holben, W., and D. Harris.** 1991. Monitoring of changes in microbial community composition by fractionation of total community DNA based on G + C content, Q222. *In Abstracts of the 91st Annual Meeting of the American Society for Microbiology 1991.* American Society for Microbiology, Washington, D.C.

20. **Holben, W., J. Jansson, B. Chelm, and J. Tiedje.** 1988. DNA probe method for the detection of specific microorganisms in the soil bacterial community. *Appl. Environ. Microbiol.* **54:**703–711.

21. **Kennedy, A. C.** 1994. Carbon utilization and fatty acid profiles for characterization of bacteria, p. 543–553. *In* R. W. Weaver, S. Angle, P. Bottomley, D. Bezdicek, S. Smith, A. Tabatabai, and A. Wollum, (ed), *Methods of Soil Analysis*, part 2. *Microbiological and Biochemical Properties.* Soil Science Society of America, Inc., Madison, Wis.

22. **Kieft, T. L., D. B. Ringelberg, and D. C. White.** 1994. Changes in ester-linked phospholipid fatty acid profiles of subsurface bacteria during starvation and desiccation in a porous medium. *Appl. Environ. Microbiol.* **60:**3292–3299.

23. **Kopczynski, E. D., M. M. Bateson, and D. M. Ward.** 1994. Recognition of chimeric small-subunit ribosomal DNAs composed of genes from uncultivated microorganisms. *Appl. Environ. Microbiol.* **60:**746–748.

24. **Lane, D. J.** 1991. 16S/23S rRNA sequencing, p. 115–175. *In* E. Stackebrandt and M. Goodfellow (ed.), *Nucleic Acid Techniques in Bacterial Systematics.* John Wiley & Sons, New York.

25. **Liesack, W., and E. Stackebrandt.** 1992. Occurrence of novel groups of the domain *Bacteria* as revealed by analysis of genetic material isolated from an Australian terrestrial environment. *J. Bacteriol.* **174:**5072–5078.

26. **Malik, M., J. Kain, C. Pettigrew, and A. Ogram.** 1994. Purification and molecular analysis of microbial DNA from compost. *J. Microbiol. Methods* **20:**183–196.

27. **Moss, C. W.** 1981. Gas-liquid chromatography as an analytical tool in microbiology. *J. Chromatogr.* **203:**337–347.

28. **Moyer, C. L., F. C. Dobbs, and D. Karl.** 1994. Estimation of diversity and community structure through restriction fragment length polymorphism distribution analysis of bacterial 16S rRNA genes from a microbial mat at an active, hydrothermal vent system, Loihi Seamount, Hawaii. *Appl. Environ. Microbiol.* **60:**871–879.

29. **Nei, M., and W. H. Li.** 1979. Mathematical model for studying genetic variation in terms of restriction endonucleases. *Proc. Natl. Acad. Sci. USA* **76:**5269–5273.

30. **Ogram, A. V., and D. F. Bezdicek.** 1994. Nucleic acid probes, p. 665–687. *In* R. W. Weaver, S. Angle, P. Bottomley, D. Bezdicek, S. Smith, A. Tabatabai and A. Wollum (ed.), *Methods of Soil Analysis*, part 2. *Microbiological and Biochemical Properties.* Soil Science Society of America, Inc., Madison, Wis.

31. **Penner, G. A., A. Bush, R. Wise, W. Kim, L. Domier, K. Kasha, A. Laroche, G. Scoles, S. Molnar, and G. Fedak.** 1993. Reproducibility of random amplified polymorphic DNA (RAPD) analysis among laboratories. *PCR Methods Appl.* **2:**341–345.

32. **Raskin, L., W. C. Capman, M. D. Kane, B. E. Rittman, and D. A. Stahl.** 1996. Cultural evaluation of membrane supports for use in quantitative hybridizations. *Appl. Environ. Microbiol.* **62:**300–303.

33. **Ritz, K., and B. S. Griffiths.** 1994. Potential application of a community hybridization technique for assessing changes in the population structure of soil microbial communities. *Soil Biol. Biochem.* **26:**963–971.

34. **Robinson-Cox, J. F., M. M. Bateson, and D. M. Ward.** 1995. Evaluation of nearest-neighbor methods for detection of chimeric small-subunit rRNA sequences. *Appl. Environ. Microbiol.* **61:**1240–1245.

35. **Sambrook, J., E. F. Fritsch, and T. Maniatis.** 1989. *Molecular Cloning: a Laboratory Manual*, 2nd ed. Cold Spring Harbor Laboratory Press, Cold Spring Harbor, N.Y.

36. **Sayler, G. S., and A. C. Layton.** 1990. Environmental application of nucleic acid hybridization. *Annu. Rev. Microbiol.* **44:**625–648.

37. **Sayler, G. S., M. S. Shields, E. Tedford, A. Breen, S. Hooper, K. Sirotkin, and J. Davis.** 1985. Application of DNA-DNA colony hybridization to the detection of catabolic genotypes in environmental samples. *Appl. Environ. Microbiol.* **49:**1295–1303.

38. **Seeley, H. W., and P. VanDemark.** 1981. *Microbes in Action: a Laboratory Manual of Microbiology*, 3rd ed. W. H. Freeman & Co., New York.

39. **Smith, J. L., and J. Halvorson.** 1996. Personal communication.

40. **Suzuki, M. T., and S. J. Giovanonni.** 1996. Bias caused by template annealing in the amplification of mixtures of 16S rRNA genes by PCR. *Appl. Environ. Microbiol.* **62:** 625–630.

41. **Torsvik, V., J. Goksoyr, and F. Daae.** 1990. High diversity in DNA of soil bacteria. *Appl. Environ. Microbiol.* **56:** 782–787.

42. **Torsvik, V., K. Salte, R. Sorheim, and J. Goksoyr.** 1990. Comparison of phenotypic diversity and DNA heterogeneity in a population of soil bacteria. *Appl. Environ. Microbiol.* **56:**776–781.

43. **Vestal, J. R., and D. C. White.** 1989. Lipid analysis in microbial ecology. *Bioscience* **39:**535–541.

44. **Vilgalys, R., and M. Hester.** 1990. Rapid genetic identification and mapping of enzymatically amplified ribosomal DNA from several *Cryptococcus* species. *J. Bacteriol.* **172:** 4238–4246.

45. **Vourdouw, G., Y. Shen, C. Harrington, A. Telang, T. Jack, and D. Westlake.** 1993. Quantitative reverse sample genome probing of microbial communities and its application to oil field production waters. *Appl. Environ. Microbiol.* **59:**4101–4114.

46. **White, D. C., R. J. Bobbie, J. D. King, J. Nickels, and P. Amoe.** 1979. Lipid analysis of sediments for microbial biomass and community structure, p. 87–103. *In* C. D. Litchfield and P. L. Seyfried (ed.), *Methodology for Biomass Determinations and Microbial Activities in Sediments.* ASTM STP 673. American Society for Testing and Materials, Philadelphia.

47. **Woese, C. R., O. Kandler, and M. L. Wheelis.** 1990. Towards a natural system of organisms: proposal for the domains *Archaea, Bacteria,* and *Eucarya. Proc. Natl. Acad. Sci. USA* **87:**4576–4579.

48. **Zelles, L., and Q. Y. Bai.** 1993. Fractionation of fatty acids derived from soil lipids by solid phase extraction and their quantitative analysis by GC-MS. *Soil Biol. Biochem.* **25:**495–507.

Isolation and Purification of Bacterial Community DNA from Environmental Samples

WILLIAM E. HOLBEN

47

Note: *Although this chapter is located in the section on Soil, Rhizosphere, and Phyllosphere, the DNA purification protocols described here have been shown to be useful with samples from a wide variety of environments, including soils, sediments, sludges, bioreactors, and intestinal tracts. These protocols are thus presented for isolation and purification of bacterial community DNA from environmental samples* **in general.**

During the past decade, numerous DNA-based methodologies for the study of microbial populations and communities in various environments have been developed. Several reports describe the use of the techniques of molecular biology to distinguish, enumerate, and monitor individual bacterial populations within a complex microbial community. This is often accomplished by using appropriate molecular probes to detect DNA sequences specific to populations of interest. The specificity of detection possible with DNA-based approaches helps to overcome a major obstacle in environmental microbiology studies: the difficult task of specifically and directly monitoring an individual population in environmental samples and in the presence of the rest of the microbial community. DNA-based community-level analyses which describe microbial communities in terms of their structure, complexity, or similarity have more recently been reported. Studies at this broad level of resolution represent a means to monitor and compare entire, and often very complex, microbial communities in a single analysis. For an overview of DNA-based studies in environmental microbiology, the reader is referred to references 7, 8, and 13.

Among the advantages of DNA-based detection methods is that a particular DNA sequence can be detected directly; i.e., marker genes and selectable phenotypes are not required. Hence, DNA-based detection methods can be used to monitor either genetically modified or wild-type indigenous populations. Another advantage is that bacterial growth is not required for detection; thus, nonculturable and nonviable populations can be detected. This represents a major boon for studies in environmental microbiology, since typically only about 0.1 to 1% of the bacteria in environmental samples are readily cultured under laboratory conditions (2, 3). Obviating the need for culturing organisms also facilitates quantitative and comparative analyses because DNA obtained directly from environmental samples reflects the relative abundance of populations at the time of sampling. DNA-based (and RNA-based) detection

methods truly represent a new set of tools to expand the capabilities of investigators to detect, quantify, and monitor microorganisms and microbial communities in environmental samples.

Several protocols for the isolation and purification of bacterial community DNA from a variety of environmental samples have been reported in recent years (3–6, 10, 14–19). Most of these techniques derive from and can be grouped into two main methodological approaches: (i) direct lysis of bacteria in the presence of the environmental matrix followed by DNA recovery and purification and (ii) bacterial fractionation, whereby bacteria are first separated (fractionated) from the environmental matrix prior to cell lysis and DNA recovery. The earliest published protocols for bacterial community DNA recovery were relatively complex and labor-intensive, often requiring several days to obtain purified DNA from just a few samples. Subsequently published protocols, often representing substantive modifications of the earliest protocols, have made these procedures simpler and more generally useful than as originally published. Protocols for the recovery of bacterial community DNA by both direct lysis and bacterial fractionation are presented here. These methodologies are compared, and recommendations for the selection of either approach as appropriate for the characteristics of a given environmental sample are offered.

The protocols presented here currently are employed almost daily in my laboratory and have proven to be useful with a wide variety of environmental samples, including soils, sediments, bioreactors, animal intestinal tracts, granular activated sludges, and extracellular symbiotic bacterial assemblages (e.g., marine tubeworm symbionts). Both approaches feature postlysis purification of DNA by using cesium chloride-ethidium bromide equilibrium density gradients, which we have found to be the most satisfactory method for obtaining high-quality, high-molecular-weight DNA that is highly purified, regardless of sample origin. As such, these protocols represent good general approaches for the isolation of bacterial community DNA from most samples. The reader should note that other researchers have published protocols for bacterial DNA recovery from some of these, as well as from other, environmental samples. These other protocols often comprise modifications and optimization of previously published protocols, making them especially suitable for particular experimental systems or sample types. For example, Hilger and Myrold (4) describe a method effective for recovery of *Frankia* DNA from soil. Other protocols emphasize rapid processing of small-scale soil samples and are appropriate for analyses of large numbers of physically and biologically homogeneous samples (e.g., reference 18). Protocols optimized for recovery of bacterial community DNA from low-biomass soil and sediment samples have also been developed (9, 14, 16). Zhou et al. (19) recently presented a modified direct lysis protocol for soil bacterial communities which includes various postlysis DNA purification approaches that are selected depending on soil sample characteristics. Researchers are encouraged to consult these and other protocols where they apply to their own systems of study or for ideas to optimize DNA recovery from novel or problematic environmental samples.

GENERAL CONSIDERATIONS

Several parameters will affect the recovery of bacterial community DNA from environmental samples. Many of these either are common to both DNA recovery methods or control the selection of the approach used and thus can be discussed in general.

Bacterial Biomass

The bacterial biomass within a sample can determine whether usable amounts of bacterial community DNA will be recovered by a particular approach, given practical and logistical limitations of the protocols. In theory, one could take any usable protocol and simply scale it up to recover the desired amount of DNA from low-biomass samples. However, this is not always practical. Consider that the average bacterium contains 9×10^{-15} g (9×10^{-9} μg) of DNA. Thus, to obtain 1 μg of DNA, about 1.1×10^8 bacteria are needed. For samples such as soils, sludges, intestinal tracts, and some sediments, in which bacterial densities are generally in the range of 10^8 to 10^{10} bacteria per g, 1-g samples are sufficient to obtain 1 or more μg of DNA. On the other hand, samples such as marine water, groundwater, and subsurface soils and sediments, in which bacterial biomass is typically much lower, can require processing of large amounts of environmental sample to obtain 1 μg of DNA. For aqueous samples, concentration of biomass can often be accomplished by fairly simple filtration or centrifugation strategies. However, low-biomass soils and sediments can have such a low ratio of bacteria (and hence bacterial DNA) to solid matrix that processing sufficient sample to obtain just 1 μg of DNA may be difficult or impossible. Even assuming sufficient scale-up of protocols and completely efficient lysis, small amounts of DNA may be difficult to recover from large amounts of solid material in such samples. The direct lysis protocol, as described here, was developed for use with surface soils, sediments, sludges, and other high-biomass samples (or bacteria concentrated from aqueous samples) with total bacterial counts greater than 10^8 bacteria per g of solid material. The bacterial fractionation approach should be considered for other samples with lower biomass, since recovery and concentration of bacteria prior to lysis can overcome some of the problems associated with low-biomass samples.

Environmental Matrix

The nature of the environmental matrix can also have profound effects on the ability to recover bacterial community DNA and can be the driving force behind selection and optimization of techniques for DNA recovery. For example, aqueous samples with little or no particulate matter and low bacterial density can be processed by filtration or centrifugation of bacteria prior to lysis. Samples with large amounts of particulate material often are candidates for the bacterial fractionation approach, which has been used with soil and very successfully with chicken intestinal tract samples. Surface soils, sediments, and sludges, which typically have about 10^8 to 10^{10} bacteria per g, are generally amenable to the direct lysis approach. However, clay content in some soils and sediments interferes with bacterial DNA recovery, particularly with the direct lysis approach, probably as a result of adsorption of released DNA onto the clays (11, 19). Some measure of success in blocking DNA binding to clays in soil can be obtained by competition with excess phosphate or by altering the pH or ionic environment. However, no single method for effectively blocking DNA-binding sites in all clays, sediments, and soils has yet been reported.

Contaminants

Some compounds in environmental samples complicate the recovery and purification of bacterial DNA, often because they copurify with DNA. This is especially true for many of the polyphenolic compounds (e.g., humic and fulvic materials) that often are encountered in soils and sediments. They are of particular concern when the direct lysis approach is applied, since the DNA and significant amounts of these contaminants are in solution following bacterial lysis. Although such contaminants do not preclude the recovery of DNA, they can confound DNA quantitation because they typically have significant light absorbance at 260 nm, the wavelength generally used to quantitate DNA. Co-contaminating compounds also have been shown to interfere with subsequent manipulations and analyses of the DNA, particularly those involving enzymatic reactions (6, 10, 15, 19). It has been our experience that DNA isolated by either the bacterial fractionation or the direct lysis method described here is highly purified. In these protocols, a high degree of purity is achieved through removal of polyphenolic compounds from solution by using polyvinylpolypyrrolidone and through two rounds of equilibrium density centrifugation whereby the bacterial community DNA localizes in a very discrete band in the central region of the gradient while the contaminating proteins float, the RNA is pelleted, and the remaining contaminants are dispersed throughout the gradient. Other protocols which do not involve equilibrium density centrifugation can be found in the literature, and several have been cited in this chapter. Many of these employ preparative electrophoresis, exclusion chromatography, organic extractions, large-scale precipitation, or commercially available, prepacked chromatography media. While cesium chloride gradient purification of DNA appears to be generally useful for a wide variety of sample and contaminant types, final selection of the purification approach may be driven by the nature of the contaminants in the sample, logistical considerations, or equipment limitations.

Bacterial Adhesion

Another point to consider when selecting a method for bacterial DNA recovery is whether the target bacteria are tightly adhered to the particulate matrix of the sample. If they are, it may be difficult to efficiently recover bacteria from the sample prior to lysis, although some investigators have used mild detergents, ion-exchange resins, phosphate buffers, and other approaches to improve recoveries. The direct lysis approach is generally favored for such samples, since DNA from adhered bacteria can be liberated into solution.

Refractory Bacteria

The types of organisms encountered in samples can also affect DNA recovery from particular populations of bacteria.

For example, methanogens in general appear to be refractory to both enzymatic and mechanical lysis (unpublished observation). Selective lysis (or nonlysis) of certain populations is of particular concern when one is addressing community-level or diversity-based questions or when a population of interest is known to be difficult to lyse. The direct lysis protocol described here is mechanical in nature and appears to be quite effective with a wide variety of bacteria. Virtually no intact bacterial cells can be observed by microscopy after lysis. It is, of course, possible that DNA from specific populations that are refractory to lysis and present in low abundance is not being recovered. Investigators concerned about obtaining DNA from a particular population of interest are encouraged to design experiments to assess recovery, such as by monitoring pure cultures for lysis by microscopy, spiking environmental samples with labeled bacteria, assaying independently for the presence of the population (e.g., in an activity-based assay), or probing recovered DNA for the presence of sequences specific to the population of interest.

DIRECT LYSIS PROTOCOL FOR BACTERIAL COMMUNITY DNA RECOVERY

The direct lysis protocol employs high temperature, harsh detergent, and mechanical disruption for cell lysis. The liberated DNA is recovered and purified by cesium chloride-ethidium bromide equilibrium density centrifugation. Direct lysis is less labor-intensive than the bacterial fractionation approach and typically results in higher DNA yields. DNA recovered by this approach probably better represents the total bacterial community because recovery of bacteria from the matrix is not required and the mechanical nature of the lysis is probably less selective than with other enzymatic approaches. It appears that this direct lysis protocol approaches quantitative recovery of bacterial community DNA on the basis of predicted yields and microscopic analysis of environmental samples following lysis. This approach is generally favored over bacterial fractionation for experiments involving tightly adhered populations or when DNA that best represents the entire bacterial community is desired. DNA recovered by this approach typically is ≥25 kb in average length and thus is suitable for most subsequent molecular manipulations and analyses, including procedures involving size fractionation of restriction enzyme-digested DNA. The size range of newly isolated DNA can readily be assessed by agarose gel electrophoresis, including as standards DNA fragments of known size. This procedure recovers predominantly bacterial DNA from most soil and sediment samples (5). Samples obtained from other types of environments which have large populations of protozoa, other eukaryotic populations, or cell-free DNA may result in recovery of significant amounts of nonbacterial DNA which might or might not interfere with subsequent analyses and/or confound interpretation of data. It is advisable to assess the contribution of nonbacterial DNA for each new type of sample being analyzed, e.g., by hybridization analyses using eubacterial, archaebacterial, and eukaryotic cell-specific probes.

1. To each sample (typically 10 g for soils, sludges, and sediments; the amount can be adjusted depending on the nature of the sample) in an autoclaved 50-ml polycarbonate or polypropylene Oak Ridge tube, add 20 ml of NaPO$_4$ (100 mM, pH 7.0) and 2.5 ml of sterile 10% sodium dodecyl sulfate (SDS). In addition, 1.5 g of polyvinylpolypyrrolidone (acid washed and then neutralized) can be added for samples with high humic or fulvic content to help remove these contaminants from solution (6). Mix samples vigorously by vortexing, and then incubate them for 30 min at 70°C with mixing every 5 min.

2. Add 5 g of large glass beads (0.7- to 1.0-mm diameter; Sigma catalog no. G9393) and 5 g of small glass beads (0.2- to 0.3-mm diameter; Sigma catalog no. G9143). Place tubes horizontally on a reciprocal platform shaker, and shake them for 30 min at high speed (~100 oscillations/min) at room temperature.

3. Pellet particulates and cell debris by centrifugation at 7,796 × g for 10 min at 10°C (e.g., in a Sorvall SS34 rotor at 10,000 rpm).

4. Carefully transfer the supernatant to a clean, autoclaved Oak Ridge tube, and let the tube stand for 15 to 30 min on ice to precipitate the SDS. Clear the lysate by centrifugation at 7,796 × g for 10 min at 10°C (e.g., in a Sorvall SS34 rotor at 10,000 rpm). Carefully transfer the cleared lysate to a clean, autoclaved Oak Ridge tube.

5. Prepare cesium chloride-ethidium bromide equilibrium density gradients as follows (the amounts given below are for 18.5-ml gradients; all amounts and volumes of reagents should be adjusted proportionally for different sizes of gradient tubes). Adjust the volume of the lysate to 15.5 ml with sterile distilled H$_2$O, and then add 14.5 g (0.93 g/ml) of finely ground cesium chloride. Mix the contents by gentle inversion until the cesium chloride is dissolved, and let the tube stand at room temperature for 10 to 15 min to precipitate proteins. Clear the lysate by centrifugation at 2,000 × g for 10 min at 10°C (5,000 rpm in a Sorvall SS34 rotor). The precipitated proteins will form a floating layer that is foamy in appearance; this layer should be discarded.

6. Transfer the gradient mixture to an ultracentrifuge tube containing 0.65 ml of ethidium bromide (10 mg/ml); mix by inversion. Fill the remainder of the tube with cesium chloride balance solution ($R_f = 1.3870$; also containing 0.5 mg of ethidium bromide per ml). Balance and seal the tubes, and then band the bacterial community DNA by centrifugation at 255,800 × g (e.g., 52,000 rpm in a Sorvall TV865B rotor) for 9 to 16 h at 18°C.

7. Stop the ultracentrifuge (let the rotor coast to a stop at least from 3,000 rpm to zero), and gently remove the tubes from the rotor. Avoid unnecessary shaking of the tubes, as this can perturb the gradients.

8. Extract the DNA band in a 3- to 5-ml volume, using a 5-ml syringe and 18-gauge needle under UV illumination. (*Note:* Wear gloves and UV-protective goggles or glasses for protection from UV irradiation and ethidium bromide.) To extract DNA, first puncture the very top of the ultracentrifuge tube to allow air to enter. Next, insert the needle just below the visible DNA band with the beveled orifice pointing up, and slowly extract the DNA band by withdrawing the plunger of the syringe.

9. Transfer the DNA solution to a new ultracentrifuge tube, fill the remainder of the tube with cesium chloride balance solution ($R_f = 1.3870$; also containing 0.5 mg of ethidium bromide per ml), and repeat the ultracentrifugation step (step 6). This second ultracentrifugation step results in a substantial increase in DNA purity, since the contaminants disperse throughout the gradient while the DNA forms a discrete band in a small volume.

10. Repeat the DNA fractionation (steps 7 and 8), but this time withdraw the DNA in as small a volume as possible without leaving any behind (usually 1.5 to 3 ml).

11. Transfer the DNA solution to a small (~6-ml) capped disposable test tube. Extract the ethidium bromide from the DNA by using isopropanol saturated with 5 M

sodium chloride as follows: add an equal volume of isopropanol saturated with 5 M NaCl to the test tube, close the tube, and mix the contents by gentle inversion. Let the tube stand until the phases separate.

12. Remove the top (pink) layer (isopropanol-ethidium bromide) with a Pasteur pipette, and discard this solution appropriately.

13. Repeat this extraction process until all pink color is gone from the upper phase and then once more (typically a total of five extractions).

14. To precipitate the DNA from solution, transfer the solution to a labeled glass centrifuge tube (e.g., Corex tube), add 2 volumes of sterile distilled H_2O, and then add 1 volume of cold ($-20°C$) 100% isopropanol. Cover the tube securely with Parafilm, and mix the contents thoroughly by inversion or vortexing. A typical solution contains DNA (2.0 ml), 2 volumes of distilled H_2O (4.0 ml), and 1 volume (three times the original DNA volume) of isopropanol (6.0 ml). Incubate the solution for several hours to overnight at $-20°C$ (longer incubation times may facilitate recovery of DNA if low yields are anticipated). Note that this first precipitation step should not be longer than about 24 h or the cesium chloride may crystallize out of solution, complicating further purification.

15. Pellet the DNA by centrifugation at $4,400 \times g$ (e.g., at 7,500 rpm in a Sorvall SS34 rotor) at 4°C for 1 h. Position the tube with the label toward the outside of the rotor as a reference for the location of the DNA pellet, which may not be visible at this stage.

16. Discard (carefully pour off) the supernatant. Invert the tube on a paper towel, and drain it dry (~5 min). Complete drying under vacuum (complete air drying may be substituted for vacuum drying).

17. Dissolve the DNA pellet in 400 μl of sterile distilled H_2O. This can be accomplished by a combination of gentle vortexing and use of a pipettor (e.g., Pipetman P1000) to pipette the 400 μl up and down along the sides of the tube.

18. Transfer the DNA solution to a labeled 1.5-ml Eppendorf tube, and then collect the remaining liquid in the Corex tube by brief centrifugation and transfer it to the Eppendorf tube.

19. Add 40 μl of 3 M sodium acetate (pH 5.2) and 440 μl of cold ($-20°C$) 100% isopropanol to the tube, mix thoroughly by vortexing, and incubate the tube at $-20°C$ for at least 1 h (at this stage, the DNA may be stored for extended periods under these conditions [i.e., in 50% isopropanol and in the presence of monovalent cations]).

20. Collect the DNA by centrifugation in a microcentrifuge at full speed (~14,000 rpm) for 30 min at 4°C.

21. Remove and discard the supernatant with a Pasteur pipette. Wash the DNA pellet once with cold ($-20°C$) 70% ethanol by gentle inversion, centrifuge briefly (2 min) in the microcentrifuge, and then remove and discard the supernatant. Dry the DNA under vacuum (complete air drying may be substituted) and then resuspend it in a small volume (usually 100 μl) of sterile distilled water.

At this stage, purified DNA has been obtained from the environmental sample. This DNA sample can now be quantitated, manipulated, and analyzed by using the techniques of molecular biology. The reader is referred to references 1 and 12 for general procedures related to handling and analysis of purified DNA and to the current literature in microbial ecology and environmental microbiology for more specific descriptions of analyses of total bacterial community DNA.

Bacterial Fractionation Protocol for Bacterial Community DNA Recovery

The bacterial fractionation approach involves the separation of bacterial cells from the bulk of the environmental matrix prior to cell lysis and recovery of bacterial community DNA. Briefly, the entire environmental sample is brought into suspension by homogenization in buffer. Large particulates, fungal cells, and other debris are then removed by a low-speed centrifugation, leaving unattached bacterial cells in suspension. These bacterial cells are subsequently captured by high-speed centrifugation and lysed, and the bacterial DNA is recovered. Multiple rounds of homogenization and centrifugation can be performed on the same sample to enhance recovery of the bacterial fraction. The efficiency of recovery of the bacterial fraction with this approach is dependent on sample type. We have achieved recoveries ranging from ~34% for a mid-Michigan agricultural soil (6) to essentially quantitative recovery of bacteria from chicken intestinal tract contents (unpublished data). Bacterial lysis is accomplished by using a protocol which combines features of lysis protocols for various groups of bacteria, including the removal of humic contaminants from solution with polyvinylpolypyrrolidone, cell wall digestion by lysozyme and pronase, and incubation at high temperature (6). Bacterial DNA is recovered from the cell lysate by cesium chloride-ethidium bromide equilibrium density centrifugation and precipitation with isopropanol.

Bacterial community DNA recovered by this approach tends to be less contaminated with polyphenolic compounds and other contaminants because the bacteria are removed from the bulk of the environmental matrix prior to lysis. DNA recovered by bacterial fractionation has an average size of 50 kb, somewhat larger than typically is obtained by the direct lysis approach. This approach essentially recovers DNA only from bacteria (not fungal, protozoan, or cell-free DNA) because the bacteria are separated from the environmental matrix, larger, heavier cells, and any cell-free DNA (e.g., from the more labile protozoa) prior to lysis and DNA recovery. There is some concern, particularly with soil and sediment samples, that bacterial fractionation may preferentially recover rapidly growing bacterial cells (probably because they do not adhere as tightly to the matrix) and loosely adhered populations. This selectivity of recovery would be of particular consequence when DNA representing the entire bacterial community is desired. However, with other environments, such as the chicken intestinal tract, this approach has resulted in excellent and apparently quantitative recovery of all bacterial populations from the huge amount of plant and other material present in the raw sample. Another feature worth noting is that this approach makes it possible to recover live bacterial cells from large amounts of bulk material, which may be desirable for certain applications.

Note: The protocol described here is optimized for processing 50 g of soil, sludge, and sediment samples. Modifications in terms of starting sample size, buffer type and volume (e.g., to enhance recovery of tightly adhered cells, a mild detergent, surfactant, or phosphate buffer might be used), and centrifugation speeds (e.g., to recover microbacteria) may be desirable for other types of samples.

1. Combine each 50-g sample with 200 ml of homogenization solution and 15 g of polyvinylpolypyrrolidone (acid washed and then neutralized) in a blender jar. The homogenization buffer is based on Winogradsky's salt solution with added reductant and contains (per liter) 0.25 g of K_2HPO_4, 0.25 g of $MgSO_4 \cdot 7H_2O$, 0.125 g of NaCl, 2.5 mg of

$Fe_2(SO_4)3·H_2O$, 2.5 mg of $MnSO_4·4$ H_2O, and 0.2 M sodium ascorbate added as powder to achieve 0.2 M just prior to use (the reducing power of sodium ascorbate lessens with time). *Note:* If this protocol is being used to recover viable cells, the sodium ascorbate should be omitted because it appears to reduce the viability of cells during the fractionation process.

2. Homogenize the sample for three 1-min intervals, with 1 min of cooling in an ice-water bath between each interval.

3. Quickly transfer the homogenate (including all of the suspended solids) to a 250-ml centrifuge bottle, and pellet solids, fungi, and other debris by centrifugation at 640 $\times$ g (e.g., in a Sorvall GSA rotor at 2,500 rpm) for 15 min at 4°C.

4. Pour the supernatant into a clean 250-ml centrifuge bottle, being careful not to dislodge the large pellet. Collect the bacterial fraction by centrifugation at 15,000 $\times$ g (e.g., in a Sorvall GSA rotor at 12,000 rpm) for 20 min at 4°C.

5. Add 200 ml of homogenization buffer to the soil pellet from the last step, and repeat the homogenization and centrifugation steps two more times (i.e., repeat steps 2 to 4 twice more, combining the bacterial pellets in step 4).

6. Wash the cell pellet by carefully resuspending it in 200 ml of TE (33 mM Tris [pH 8.0], 1 mM EDTA), using a small, clean paintbrush (this approach is very effective at dispersing the tightly packed pellet). Collect the bacteria by centrifugation at 15,000 $\times$ g (e.g., in a Sorvall GSA rotor at 12,000 rpm) for 20 min at 4°C.

7. Resuspend the cell pellet in 20.0 ml of TE (again using the paintbrush), and then transfer the cell suspension to an autoclaved 50-ml Oak Ridge tube. Pretreat the cells for lysis by adding 5.0 ml of 5 M NaCl and 125 μl of 20% Sarkosyl, mixing, and then incubating the mixture at room temperature for 10 min. This pretreatment appears to increase the efficiency of lysis with lysozyme.

8. Collect the cells by centrifugation at 11,000 $\times$ g (e.g., in a Sorvall SS34 rotor at 12,000 rpm) for 20 min at 4°C.

9. Gently resuspend the cell pellet in 3.5 ml of 50 mM Tris (pH 8.0)–0.75 M sucrose–10 mM EDTA, again using the paintbrush.

10. Add 0.5 ml of lysozyme solution (grade 1 from chicken egg white [Sigma catalog no. L6876]; 40 mg/ml in TE), mix by vortexing, and then incubate the mixture at 37°C for 30 min.

11. Add 0.5 ml of pronase E (type XXV from *Streptomyces griseus* [Sigma catalog no. P6911]; 10 mg/ml in TE, predigested for 30 min at 37°C), mix by vortexing, and incubate the mixture at 37°C for 30 min.

12. Transfer the mixture to a 65°C water bath for 10 min, then add 250 μl of 20% Sarkosyl, mix by vortexing, and incubate the mixture at 65°C for 40 min.

13. Transfer the mixture to an ice bath, let it stand for at least 30 min, and then clear the lysate of cellular debris by centrifugation at 25,260 $\times$ g (e.g., in a Sorvall SS34 rotor at 18,000 rpm) for 1 h at 4°C.

14. Transfer the supernatant to a clean Oak Ridge tube, and then add 9.0 ml of sterile distilled H_2O, 12.7 g of finely ground cesium chloride, and 1.5 ml of ethidium bromide (10 mg/ml). Mix by gentle inversion until the cesium chloride is completely dissolved, and then adjust the refractive index to 1.3865 to 1.3885 (these values correspond to a density range of 1.55 to 1.58 g/ml) by adding cesium chloride or sterile distilled H_2O as needed.

15. Transfer the gradient mixture to an ultracentrifuge tube, and fill the remainder of the tube with cesium chloride balance solution (R_f = 1.3870; also containing 0.5 mg of ethidium bromide per ml). Balance and seal the tubes, and then band the bacterial community DNA by centrifugation at 255,800 $\times$ g (e.g., 52,000 rpm in a Sorvall TV865B rotor) for 9 to 16 h at 18°C.

16. Stop the ultracentrifuge (let the rotor coast to a stop at least from 3,000 rpm to zero), and gently remove the tubes from the rotor. Avoid unnecessary shaking of the tubes, as this can perturb the gradients.

17. Extract the DNA band in a 3- to 5-ml volume, using a 5-ml syringe and 18-gauge needle under UV illumination. (*Note:* Wear gloves or UV-protective goggles or glasses for protection from UV irradiation and ethidium bromide.) To extract DNA, first poke a hole at the very top of the ultracentrifuge tube to allow air to enter. Next, insert the needle just below the visible DNA band with the orifice pointing up, and slowly extract the DNA band by withdrawing the plunger of the syringe.

18. Transfer the DNA solution to a new ultracentrifuge tube, fill the remainder of the tube with cesium chloride balance solution (R_f = 1.3870, also containing 0.5 mg of ethidium bromide per ml), and repeat the ultracentrifugation step (step 6). This second ultracentrifugation step results in a substantial increase in DNA purity, since the contaminants disperse throughout the gradient while the DNA forms a discrete band in a small volume.

19. Repeat the DNA fractionation (steps 7 and 8), but this time withdraw the DNA in as small a volume as possible without leaving any behind (usually 1.5 to 3 ml).

20. Transfer the DNA solution to a small (~6-ml) capped disposable test tube, and then extract the ethidium bromide from the DNA by using isopropanol saturated with 5 M sodium chloride as follows: add an equal volume of isopropanol saturated with 5 M NaCl to the test tube, close the tube, and mix the contents by gentle inversion. Let the tube stand until the phases separate.

21. Remove the top (pink) layer (isopropanol-ethidium bromide) with a Pasteur pipette, and discard this solution appropriately.

22. Repeat this extraction process until all pink color is gone from the upper phase and then once more (typically a total of five extractions).

23. To precipitate the DNA from solution, transfer the solution to a labeled glass centrifuge tube (e.g., Corex tube), add 2 volumes of sterile distilled H_2O, and then add 1 volume of cold (−20°C) 100% isopropanol. Cover the tube securely with Parafilm, and mix the contents thoroughly by inversion or vortexing. A typical solution contains DNA (2.0 ml), 2 volumes of distilled H_2O (4.0 ml), and 1 volume (three times the original DNA volume) of isopropanol (6.0 ml). Incubate the solution overnight at −20°C. Note that this first precipitation step should not be longer than 24 h or the cesium chloride may crystallize out of solution, complicating further purification.

24. Pellet the DNA by centrifugation at 4,400 $\times$ g (e.g., at 7,500 rpm in a Sorvall SS34 rotor) at 4°C for 1 h. Position the tube with the label toward the outside of the rotor as a reference for the location of the pelleted DNA, which may not be visible at this stage.

25. Discard (carefully pour off) the supernatant. Invert the tube on a paper towel, and drain it dry (~5 min). Complete drying under vacuum (complete air drying may be substituted for vacuum drying).

26. Dissolve the DNA pellet in 400 μl of sterile distilled H_2O. This can be accomplished by a combination of vor-

texing and use of a pipettor (e.g., Pipetman P1000) to pipette the 400 μl up and down along the sides of the tube.

27. Transfer the DNA solution to a labeled 1.5-ml Eppendorf tube, and then collect the remaining liquid in the Corex tube by brief centrifugation and transfer it to the Eppendorf tube.

28. Add 40 μl of 3 M sodium acetate (pH 5.2) and 440 μl of cold ($-20°$C) 100% isopropanol to the tube, mix thoroughly by vortexing, and incubate the tube at $-20°$C for at least 1 h (at this stage, the DNA may be stored for extended periods under these conditions [i.e., in 50% isopropanol and in the presence of monovalent cations]).

29. Collect the DNA by centrifugation in a microcentrifuge at full speed (~14,000 rpm) for 30 min at 4°C.

30. Remove and discard the supernatant with a Pasteur pipette. Wash the DNA pellet once with cold ($-20°$C) 70% ethanol by gentle inversion, centrifuge briefly (2 min) in the microcentrifuge, and then remove and discard the supernatant. Dry the DNA under vacuum, and then resuspend it in a small volume (usually 100 μl) of sterile distilled water.

At this stage, purified DNA has been obtained from the environmental sample. This purified DNA can now be quantitated, manipulated, and analyzed by using the techniques of molecular biology. The reader is referred to reference 1 and 12 for general procedures related to handling and analysis of purified DNA and to the current literature for more specific descriptions of analyses of total bacterial community DNA.

REFERENCES

1. **Ausubel, F. M., R. Brent, R. E. Kingston, D. D. Moore, J. G. Seidman, J. A. Smith, and K. Struhl (ed.).** 1990. *Current Protocols in Molecular Biology*, vol. 2. Greene Publishing Associates and Wiley-Interscience, New York.

2. **Bakken, L. R.** 1985. Separation and purification of bacteria from soil. *Appl. Environ. Microbiol.* **49:**1482–1487.

3. **Fægri, A., V. L. Torsvik, J. Goksöyr.** 1977. Bacterial and fungal activities in soil: separation of bacteria and fungi by a rapid fractionated centrifugation technique. *Soil Biol. Biochem.* **9:**105–112.

4. **Hilger, A. B., and D. D. Myrold.** 1991. Method for extraction of *Frankia* DNA from soil, p. 107–113. *In* D. A. Crossley, Jr., D. C. Coleman, P. F. Hendrix, W. Cheng, D. H. Wright, M. H. Beare, and C. A. Edwards (ed.), *Modern Techniques in Soil Ecology*. Elsevier Science Publishing Co., New York.

5. **Holben, W. E.** 1994. Isolation and purification of bacterial DNA from soil, p. 727–751. *In* R. W. Weaver et al. (ed.), *Methods of Soil Analysis*, part 2. *Microbiological and Biochemical Properties.* SSSA Book Series no. 5. Soil Science Society of America, Madison, Wis.

6. **Holben, W. E., J. K. Jansson, B. K. Chelm, and J. M. Tiedje.** 1988. DNA probe method for the detection of specific microorganisms in the soil bacterial community. *Appl. Environ. Microbiol.* **54:**703–711.

7. **Holben, W. E., and J. M. Tiedje.** 1988. Applications of nucleic acid hybridization in microbial ecology. *Ecology* **69:** 561–568.

8. **Knight, I. T., W. E. Holben, J. M. Tiedje, and R. R. Colwell.** 1992. Nucleic acid hybridization techniques for detection, identification, and enumeration of microorganisms in the environment, p. 65–91. *In* M. A. Levin, R. J. Seidler, and M. Rogul (ed.), *Microbial Ecology: Principles, Methods, and Applications.* McGraw-Hill, Inc., New York.

9. **Ogram, A.** Personal communication.

10. **Ogram, A., G. S. Sayler, and T. Barkay.** 1987. The extraction and purification of microbial DNA from sediments. *J. Microbiol. Methods* **7:**57–66.

11. **Ogram, A., G. S. Sayler, D. Gustin, and R. J. Lewis.** 1988. DNA adsorption to soils and sediments. *Environ. Sci. Technol.* **22:**982–984.

12. **Sambrook, J., E. F. Fritsch, and T. Maniatis.** 1989. *Molecular Cloning: a Laboratory Manual*, 2nd ed. Cold Spring Harbor Laboratory Press, Cold Spring Harbor, N.Y.

13. **Sayler, G. S., and A. C. Layton.** 1990. Environmental application of nucleic acid hybridization. *Annu. Rev. Microbiol.* **44:**625–648.

14. **Smith, G. B., and J. M. Tiedje.** 1992. Isolation and characterization of a nitrite reductase gene and its use as a probe for denitrifying bacteria. *Appl. Environ. Microbiol.* **58:** 376–384.

15. **Steffan, R. J., J. Goksøyr, A. K. Bej, and R. M. Atlas.** 1988. Recovery of DNA from soils and sediments. *Appl. Environ. Microbiol.* **54:**2908–2915.

16. **Thiem, S.** Personal communication.

17. **Torvik, V., J. Goksøyr, and F. L. Daae.** 1990. High diversity in DNA of soil bacteria. *Appl. Environ. Microbiol.* **56:** 782–787.

18. **Tsai, Y.-L., and B. H. Olson.** 1991. Rapid method for direct extraction of DNA from soil and sediments. *Appl. Environ. Microbiol.* **57:**1070–1074.

19. **Zhou, J., M. A. Bruns, and J. M. Tiedje.** 1996. DNA recovery from soils of diverse composition. *Appl. Environ. Microbiol.* **62:**316–322.

PCR: Applications for Plant and Soil Microbes

IAN L. PEPPER

48

USES OF PCR

PCR amplifications involve a new technique that has revolutionized molecular biology methodologies. PCR is an enzymatic reaction that allows amplification of DNA through a repetitive process in vitro. During each cycle of PCR, any DNA that is present in the reaction is copied. Thus, during each cycle the amount of DNA theoretically doubles. In practice, 25 cycles of PCR result in approximately a 1 million-fold increase in the amount of DNA present. This amplification results in a large amount of a specific DNA sequence being produced, which can be purified via gel electrophoresis and visualized under UV light after ethidium bromide (EtBr) staining.

Amplification results in increased sensitivity of detection of a DNA sequence present in trace amounts in mixed populations. Although this technique was originally used for genetic and clinical purposes, there are increasing opportunities for its application in environmental microbiology. New applications include detection of microbial pathogens in environmental samples and estimates of microbial diversity and enzymatic activity.

Sample Sources

Although PCR is relatively easy to conduct in pure culture, it is often much more difficult to carry out on nucleic acid sequences derived from environmental samples. A two-step process is required: first, the environmental sample must be processed to remove substances that would inhibit PCR, and, second, the processed sample must be subjected to PCR amplification by use of appropriately designed primers. Normally primers are two 20- to 22-bp sequences that can be used to amplify at the species level. PCR can be inhibited either physically or chemically. Physical inhibition can result from the presence of soil colloids associated with DNA extracted from a soil or sediment sample. Such physical inhibition is thought to result from the colloids interfering with the annealing of primers to target DNA sequences. Chemical inhibition can result from the presence of inorganic substances, for example, an iron compound, or organic macromolecules such as humic acid and its related compounds (1, 27, 28). Compounds with the potential for inhibition can be present in any environmental sample. Examples of environmental samples include sewage sludges, soils, sediments, marine waters, groundwater, ultrapure water, and food. Because of the diverse nature of these samples, different PCR protocols have been developed for specific, sensitive detection of a variety of nucleic acid sequences.

Normal PCR

In normal PCR, the use of two unique primers results in a single amplification product. Often the amplification product is diagnostic for particular bacterial pathogens or indicator organisms. For example, *Escherichia coli* can be detected by the use of primers designed from the *lamB* gene, which codes for an outer membrane protein in *E. coli* (9).

Multiplex PCR

There is currently great interest in the detection of pathogens by using PCR, since the technique does not involve culturing the organisms prior to analysis. Important in this regard is the fact that many pathogens in environmental samples may be viable but nonculturable. Although such pathogens will not be detectable by conventional culture methodologies, they are capable of infecting humans and causing disease. Multiplex PCR involves the use of multiple sets of primers, which results in multiple products of amplification. Such a multiplex system has been used to distinguish *Salmonella* species from *E. coli* (30). The multiplex system is particularly useful when all of the products of amplification are diagnostic for a particular species, since there is then greater assurance that the pathogen of interest will be detected. In the case of *Salmonella* species, three sets of primers specific to the *Salmonella* genome were used. Multiplex PCR has also been used with multiple sets of primers specific to different genomes. In this case, more than one organism can be detected in one PCR.

Reverse Transcriptase-PCR

Microbial pathogens can be introduced into soil or marine samples through municipal sewage sludges. In the United States, it is common practice to apply treated sewage to agricultural land. Some cities (e.g., Los Angeles, Boston, and Honolulu) have direct sewage outfalls into the ocean. It is important that the fate of the pathogens in this material be determined, and PCR has enabled specific sensitive detection of introduced pathogens. In some cases, PCR can be used to detect specific viruses for which no other method of analysis is available. For example, there are no tissue culture methods available to detect Norwalk virus, which may become the leading cause of viral gastroenteritis in the United States.

Since many viruses exist as single-stranded complexes of RNA surrounded by a protein capsid, a different PCR protocol is necessary to detect these nucleic acid sequences. This protocol, known as reverse transcriptase-PCR (RT-PCR), involves the use of the enzyme reverse transcriptase. This enzyme is used to make a cDNA copy of the single-stranded RNA sequence. The cDNA is a direct copy of the bases in the RNA except that thymine replaces uracil. During the first cycle of PCR, a complementary strand to the cDNA is formed, and subsequently the reaction proceeds as in normal PCR.

Most frequently, RT-PCR is used to detect RNA viruses in environmental samples. Straub et al. (23, 24) used RT-PCR to detect enteroviruses in sewage sludge and sludge-amended soils. In this study, sensitivity of detection was 0.2 virus per ml of sludge. However, the authors noted that this may represent as many as 100 to 10,000 template copies, since in environmental samples such as sludge or soil, there are normally a number of noninfectious viruses accompanying each infectious virus.

Estimates of Enzymatic Potential

Detection of pathogens requires the use of highly specific primers that allow detection of nucleic acid sequences at the species level. However, there are other ways of using PCR on environmental samples with primers of lower specificity that amplify the DNA from a wider array of organisms. Primers derived from conserved sequences or plasmids that code for enzymatic activity can be used to estimate the potential for that enzymatic activity in an environmental sample, even when the activity is due to a diverse number of organisms. For example, primers derived from plasmid pJP4 have been used to estimate the potential degradation of 2,4-dichlorophenoxyacetic acid (15). This plasmid has been associated with a variety of different bacteria. PCR detection of plasmids has also been used to confirm potential transconjugants in gene transfer studies in soil (16). mRNA can also be RT-PCR amplified. Because mRNA has a half-life of only a few minutes and is produced only immediately prior to protein synthesis, detection of mRNA gives an estimate of actual expressed metabolic activity. Tsai and Olson (26) used guanidine isothiocyanate to extract mRNA from soil organisms and subsequently estimated the activity of toluene degradation. Such mRNA studies can be used in ecological studies to estimate specific rates of microbial transformations or rates of biodegradation.

Estimates of Microbial Diversity

Microbial diversity within an environmental sample can be determined by PCR analysis of 16S ribosomal DNA (rDNA) sequences. Approaches include the use of nested PCR with multiple primers or amplification of conserved 16S rDNA sequences followed by restriction fragment length polymorphism analysis and sequencing of specific bands. These analyses are critical to many recent phylogenic studies. PCR amplification of naphthalene-catabolic and 16S rRNA gene sequences has been used to characterize indigenous sediment bacteria (7). The 16S rDNA sequences of soil bacteria have also been PCR amplified and analyzed (5). A recent approach is to extract soil community DNA without culturing the participating microorganisms. PCR-generated 16S rDNA fragments can be produced by using universal primers from conserved regions of DNA. The 16S rDNA PCR products can be subcloned and further characterized by a restriction fragment length analysis termed amplified rDNA restriction analysis (ARDRA). Sequence analysis of fragments allows the diversity of uncultured organisms to be determined (31).

DNA Fingerprinting Using PCR

In recent years there has been increased interest in DNA fingerprinting methodologies to identify bacteria at the isolate level and to subtype pathogenic bacteria such as salmonellae or soil organisms such as rhizobia (6). PCR can be used in several different ways to fingerprint bacteria. One method uses a single arbitrary primer and is known as AP-PCR (32). A random or arbitrary primer (≈ 10 bp) can insert into a bacterial genome at multiple sites and thus give rise to multiple amplification products of variable size. When separated by electrophoresis, the products result in a banding pattern, or fingerprint, unique to the organism under investigation. However, when AP-PCR is used, it is critical that all PCR components and conditions be controlled rigorously, since variable conditions result in variable banding patterns (10). More recently, repetitive DNA sequences within bacteria have been used to generate DNA fingerprints. Primers designed from these repetitive sequences anneal to the repeat sequences and again result in multiple amplification products. ERIC (enterobacterial, repetitive, intergenic consensus sequence [6])-PCR and REP (repetitive intergenic palindromic sequence)-PCR are excellent examples of such protocols (29). ERIC-PCR and REP-PCR differ from AP-PCR in that two 20-bp primers are used in the assays. Overall, DNA fingerprinting methodologies have great potential to identify isolates of bacteria with similar DNA sequences. It is important to note, however, that the degree of discrimination with some of these methodologies is so great that isolates can often be placed into different groupings based on the fingerprints, even when no functional differences are observed between the organisms.

Quantification of PCR

In many cases, PCR technologies do not require quantification of the amount of amplified DNA. In certain kinds of investigations, however, a presence/absence result for a particular gene sequence is not sufficient and quantification is necessary. For example, an evaluation of the incidence of a particular pathogen within a soil sample would certainly benefit from quantification of amplified product. Therefore, it is important that the amount of amplified product can be related to the initial amount of bacterial DNA template that was present in the environmental sample to allow estimation of the incidence of the organism in question within the soil.

Standard curves utilizing known amounts of template can be PCR amplified, and the amplified product can be separated by gel electrophoresis. This results in a semiquantitative method of analysis in which unknown amounts of template within environmental samples can be amplified and compared visually with the standard curve. Densitometers can be used to improve the accuracy of the comparison.

A more recent method of quantification involves high-performance liquid chromatography (HPLC) detection of amplified DNA (11). Quantitation of PCR products is done by generating a standard curve correlating product DNA or HPLC peak response to template DNA in a linear response range. This standard curve can be used to determine the amount of template that was present in an original environmental sample.

PCR METHODOLOGIES

Design of Primers

It is probably already apparent that the choice of the primer sequences is critical for successful amplification of a specific DNA sequence. As in the case of gene probes, primer sequences can be deduced from known DNA sequences, and they can also be deduced by the use of computer programs. The overall choice of the primers must be guided by the aim of the researcher. If detection of a target DNA that is specific to a given species or genus of bacterium is required, then sequences unique to that bacterium must be used for the design of the primers. For example, the *lamB* gene codes for the production of an outer membrane protein in *E. coli*, and primers designed from this gene sequence will detect *E. coli* (9). However, in some cases it may be more prudent to design primers from conserved sequences. Thus, primers designed from "common *nod*" gene sequences could be used to detect all species of *Rhizobium*. Strategy is therefore an important aspect governing the choice of the sequence from which the primers will be designed. In addition, there are other criteria that determine the ultimate choice of primer sequences. In general, most primers are 17 to 30 bp, separated by a few as a 100 or as many as several thousand base pairs. Of course, the distance between the primers determines the size of the amplification product. It is also critical that the primers contain sequences that are different from each other, so that they anneal at different sites on the chromosome. If primers contain complementary sequences, they can hybridize to each other, producing a "primer dimer." Most primers also contain a high $G + C$ content ($>50\%$) so that the melting temperature of the primers is increased as a result of the three hydrogen bonds between these bases, compared with the two bonds between A and T. All known DNA sequences are now accessible via computer analysis using software such as Gene Bank. These programs allow unique sequences to be identified and evaluated as potential primers.

Specificity of Amplification

The degree of specificity of primers used in PCR can intentionally be varied. If unique primers are required to amplify specific DNA, then unique DNA sequences must be chosen as the target for amplification, e.g., a primer pair that allows amplification of *Salmonella*-specific DNA but not *E. coli*-specific DNA. Since these two groups of organisms are closely related, the design of these primers is critical to distinguish between the species of the two genera. However, if the detection of all or most bacterial DNAs in an environmental sample is required (e.g., in a bioassay), then primers termed universal primers are designed so as to allow amplification of a conserved DNA sequence present in all bacteria. Such conserved sequences exist within 16S rDNA.

The degree of specificity can be varied by primer design and also by changing the annealing temperature. In general, as the annealing temperature is increased, the number of base pair mismatches allowable for hybridization decreases, increasing the specificity of amplification. Increasing the annealing temperature from 50 to 55°C will often decrease nonspecific amplification. However, along with increased specificity, there is an associated decrease in sensitivity. The maximum allowable annealing temperature is generally 10°C less than the melting temperature, which depends on percent $G + C$.

Primers can sometimes result in several different amplification products as well as an amplification product specific to the target DNA. The amplification product from the correct target sequence can be identified by appropriate size markers of standard DNA but must be confirmed (see Confirmation of Product cDNA Sequences, below).

Sensitivity of Detection

Sensitivity of amplification is important when a given DNA sequence is in an environmental sample but in low concentration or copy number. In this instance, the amplification protocol must be sensitive enough to detect the target DNA sequence and not result in a false negative. The level of sensitivity must be defined so that a negative result can be quantified. For example, current Environmental Protection Agency drinking water standards require that <1 *Salmonella* organism per 100 ml be detectable; therefore, the use of PCR to detect salmonellae must somehow be capable of meeting these standards. Sensitivity can be increased by (i) optimization of amplification, (ii) increasing the number of cycles, or (iii) concentration of the amount of target DNA in an environmental sample, for example, by filtration of a large volume of water followed by elution of bacterial cells into a smaller final volume.

Sensitivity can be evaluated in terms of whole-cell lysates or pure genomic DNA preparations. In terms of whole cells, target bacteria are grown to late log phase in broth medium, serially diluted, and plated on agar to determine the number of CFUs. Concurrently, 500 μl of each dilution is added to a PCR tube and centrifuged for 10 min at 14,000 rpm. All but 10 μl of the supernatant is discarded, and the resulting pellet is resuspended in 89.5 μl of reaction mix lacking enzyme. Cells are boiled for 10 min, and 0.5 μl of *Taq* polymerase is added after tubes are cooled to room temperature. PCR amplification is then conducted. Sensitivity can also be determined in terms of pure genomic DNA by using 10-fold dilutions of stock DNA preparations as a template for PCR. DNA concentrations in the stock preparations are quantified by using A_{260} values.

Sensitivity in terms of cell counts based on CFUs can be misleading since the total number of copies of a target sequence is always greater than the number of CFUs. Broth cultures inevitably contain dead or lysed cells with target sequences, and each viable cell is likely to contain multiple copies of each genome (12). Therefore, equating sensitivity with CFUs tends to overestimate the actual sensitivity of the method. Sensitivity also depends on the method used to detect the amplified product. With EtBr staining of DNA, PCR can often detect 10^3 to 10^4 cells, whereas the use of ^{32}P-labeled gene probes usually increases the sensitivity by 2 orders of magnitude. For assays using PCR in soil, a sensitivity of one cell per gram of soil has been reported (18, 22).

In terms of pure DNA, 100 ag (10^{-18} g) of a 179-bp fragment has been amplified (9). Assuming a total genome of 4×10^6 bp, which is equivalent to approximately 9 fg of DNA, the target amplification product of 179 bp is equivalent to $(179/4 \times 10^6) \times 9 \times 10^{-15}$ g, or 0.4 ag. Therefore, one copy of the target DNA represents 0.4 ag, and the sensitivity of detection in this study was approximately 250 copies (100 ag).

Overall, the issue of sensitivity is complex and must be evaluated for each individual set of primers. Such an evaluation is critical if PCR amplifications are to be used for diagnostic purposes. In addition, it is important to note that sensitivity in pure culture is always greater than that in environmental samples as a result of the presence of PCR-inhibitory substances.

TABLE 1 Standard PCR mixture

Component	Amt (μl)
H$_2$O	61.5–66.5
10× reaction buffer[a]	10
Deoxynucleoside triphosphates (1.25 mM)	16
Primer 1 (0.1 μg/μl)	1.0
Primer 2 (0.1 μg/μl)	1.0
Template	5–10
Enzyme	0.5
Total vol	100

[a] Consists of 100 mM Tris-HCl (pH 8.3) (at 25°C), 500 mM KCl, 15 mM MgCl$_2$, and 0.01% (wt/vol) gelatin, autoclaved. This buffer provides preferred pH and ionic strength for PCR amplification reactions.

Basic PCR Protocol

A "standard" 100-μl PCR mixture that can routinely be used is shown in Table 1. The various components of the reaction mixture are added to a 0.5-ml polypropylene tube. The quality of the tube used in the reaction is important, since uniform and efficient heat transfer through the tube is critical for the cycling conditions. The components are added to the tube sequentially, using either a positive-displacement pipette or PCR-dedicated pipettes. To avoid nonspecific amplification products due to mispriming, the reaction is set up in ice baths. When a large number of reaction tubes are being set up, it is advisable to use so-called master mixes; e.g., if 10 amplifications are being set up, 10-fold amounts of all reagents (except the template) are mixed together in a single tube in bulk and then individually aliquoted into separate PCR tubes. The template is finally added to the tubes. The use of master mixes reduces errors when one is dispensing extremely small amounts of the reagents, e.g., *Taq* polymerase, to the individual tubes. To cut down on expenses, 50- or even 25-μl reactions can be run.

A typical standard PCR amplification consists of repeated cycling of temperatures to achieve (i) denaturation or separation of template double-stranded DNA, (ii) annealing of primers to the single-stranded template DNA, and (iii) extension from the 3' ends of primers to achieve the synthesis of a copy of each of the single strands. In a typical protocol, a starting point for the temperature cycling during denaturation is 94°C for 1.5 min, primer annealing at 37°C for 1 min, and primer extension step at 72°C for 1 min. This procedure is subsequently repeated for 25 cycles with the annealing temperature at 55 or 60°C. At the end of 25 cycles, all unextended target sequences are extended for 7 to 10 min at 72°C. At the end of the amplification cycles, the enzyme reaction is stopped by incubation at 4°C. Sensitivity of detection can be increased by using 10 μl of the amplified reaction mix as the template for a fresh PCR mixture and reamplifying for a further 25 cycles (double PCR).

Direct RT-PCR Protocol for Virus Detection

The method described is taken from reference 1. First, reverse transcriptase is used to synthesize a complementary DNA strand to the RNA template i.e., cDNA. The mix shown in Table 2 is heated for 5 min at 99°C to release viral RNA. After addition of 1 μl each of reverse transcriptase, random hexamers, and RNase inhibitor, samples are subjected to 1 cycle of 24°C for 10 min, 44°C for 50 min, 99°C for 5 min, and 5°C for 5 min to transcribe the RNA to

TABLE 2 Mixture for direct RT-PCR

Component	Amt (μl)
10× buffer	3
25 mM MgCl$_2$	7
10 mM deoxynucleoside triphosphate	8
Template	10

cDNA. All of this reaction mixture is now used as the template for cDNA amplification.

For cDNA amplification, the mix shown in Table 3 is made. The final volume of the PCR mixture is 100 μl. Typical PCR conditions are 94°C for 1 min, 55°C for 45 s, and 72°C for 45 s. After 25 cycles, the final extension cycle is 72°C for 7 min to complete all strands.

More recently, Reynolds et al. (20) used an integrated cell culture-PCR procedure to detect infectious enteroviruses. Here, PCR is done on cell culture lysates incubated for 24 to 48 h. By this method, infectious viruses are detected within 24 h. A further advantage is that the effects of PCR-inhibitory substances that are normally associated with environmental samples are reduced as a result of dilution in the cell culture.

Confirmation of Product DNA Sequences

The primers used in PCR delineate the amplified product size based on the positions of primer annealing within the microbial genome. The theoretical product size can be estimated from DNA size markers or ladders and gives an indication that the correct nucleic acid sequence has been amplified. However, it is useful to definitely prove that the actual product sequence is the desired target. Gene probe techniques such as Southern hybridizations have traditionally been used to identify target sequences. The correct amplification product can also be identified by the use of end-labeled gene probes specific to an internal region of the amplified product (2). An alternative approach that does not use probes is seminested PCR. In seminested PCR, the upstream external primer used in the original amplification and an internal downstream primer are utilized. The internal primer is designed from sequences contained within the theoretical original product sequence. Therefore, in the second seminested PCR protocol, amplification occurs only if the correct internal target sequence was initially present. Seminested PCR uses an additional 25 to 30 cycles, with the template being an aliquot of the original PCR mix. In essence, then, this is a double PCR and can increase sensitivity of detection by 1 to 2 orders of magnitude.

Limitations of PCR

The advantages of PCR are well known and include speed, sensitivity, and relatively low costs; however, there are some

TABLE 3 Mixture for cDNA amplification

Component	Amt (μl)
H$_2$O	57.5
10× buffer	7
25 mM MgCl$_2$	3
Taq enzyme (5 U/μl)	0.5
Primer 1 (0.5 μg/μl)	0.5
Primer 2 (0.5 μg/μl)	0.5

disadvantages. The major issue is that PCR will detect non-viable organisms. Josephson et al. (8) demonstrated that PCR would detect nonviable bacterial pathogens, and therefore the relevance of PCR-positive detection was brought into question. However, this study also showed that nucleic acids within nonviable bacteria degrade relatively quickly in the environment and therefore that detection of a PCR positive implied that the organisms either were viable or were recently viable. Analogous data for viruses were obtained by Straub et al. (24). Here, cell culture analyses of environmental samples for enteroviruses were negative, whereas PCR analyses gave positive identification. This study showed that noninfectious viruses could remain in the environment for long periods of time with intact nucleic acids protected by the viral capsids. Therefore, for viruses, the public health hazard of PCR positives needs to be evaluate. Note, however, that the integrated cell culture-PCR methodology (20) is much more likely to detect only infectious virus.

There are three other major limitations to PCR detection of gene sequences in environmental samples. One is small sample size; in other words, the PCR template is limited to only a few microliters, which can limit the amount of the environmental sample that can be processed. This problem can be overcome by innovative sample processing. The second limitation is the presence of PCR-inhibitory substances, and the third is laboratory contamination that results in false positives. These three issues are addressed below.

Preventing Contamination

False positives arise from contamination but can be reduced by taking the following steps.

1. Use pre- and post-PCR stations. To reduce carryover of amplified DNA, samples should be prepared at a pre-PCR station in a separate room or far away from where amplified products are analyzed. A different, post-PCR station is used to analyze products via electrophoresis or gene probes. The single most important step in avoiding contamination is completely separating PCR preparation from anything to do with the amplified PCR product.

2. If possible, dedicate sets of unique supplies and pipetting devices for each set of primers.

3. Autoclave buffer solutions and use HPLC-grade water. Note that primers, deoxynucleoside triphosphates, and *Taq* DNA polymerase cannot be autoclaved.

4. Aliquot reagents to minimize the number of repeated samplings from given reagents. Label aliquots so that if contamination occurs, it can be traced.

5. Use disposable gloves and change them frequently, particularly between pre- and post-PCR stations.

6. Spin down tubes before opening them to remove sample from tube caps and walls.

7. Use positive-displacement pipettes to avoid contamination via aerosols.

8. Premix reagents before dividing them into aliquots; i.e., use master mixes. Pipette reagents for a no-DNA negative control last.

9. Add all reagents before sample DNA; i.e., add DNA last, and cap each tube before proceeding to the next sample.

10. A method called hot start can also significantly reduce preamplification mispriming. This method involves setting up the reaction at approximately 80°C without one critical component such as the *Taq* polymerase. The reaction mixture is heated to 94 to 95°C to denature the template and then cooled to around 80°C. This procedure reduces the chance for the primers to anneal to nonspecific regions on the template, and further, there will be no extension due to the missing *Taq* polymerase.

11. Use the minimum number of PCR cycles possible for a given sample.

Contamination can be checked by the use of negative controls that are required with every assay. The negative control contains no added DNA as template.

PROCESSING SOIL SAMPLES FOR PCR AMPLIFICATIONS

Extraction of Cells from Soil

Two approaches that allow PCR detection of DNA sequences from soil have been developed. The first involves the extraction of intact cells from soil followed by cell lysis and subsequent PCR analysis. This can be done by adding surfactants to aid in the removal of cells that are sorbed to colloidal or humic substances and then extracting the soil with some kind of solvent. Many extracting solutions, including soluble sodium solutions, phosphate buffers, and calcium solutions, have been tried. In practice, the percent recovery of cells from soil depends on the nature of the soil as well as the nature of the organisms. Typically, higher recoveries are obtained from coarse-textured than fine-textured soils or soils high in organics. Autochthonous cells that have been present in soil for long periods of time are normally sorbed more strongly than young zymogenous cells growing rapidly as a result of a readily metabolizable substrate. Some cells are intrinsically more sticky than others because they secrete polysaccharides. All of these factors can result in the selective extraction of some bacterial populations over others. In addition, it is difficult to separate the cells from the soil particles themselves. Differential centrifugation has been used to overcome this obstacle (18, 22). However, despite these problems, cell extraction is still likely to result in a more representative estimate of the total bacterial community than are simple dilution and plating. Once the soil has been extracted, the total cell biomass can be lysed and then subjected to PCR analysis. Lysis is normally done directly in the thermocycler at a temperature of 98°C. Prior to lysis, the cells must be purified since PCR can be inhibited by humic materials (see Purification of DNA Extracted from Soil, below).

Direct Lysis of Bacteria in Soil

Several protocols for the extraction of community DNA have recently been published (14), but most rely on the use of lysozyme or sodium dodecyl sulfate (SDS) as the lysing agent. Once extracted, the DNA can be purified by cesium chloride (CsCl) density centrifugation or the use of commercial kits. An alternative and more frequently utilized approach is to use phenol extraction followed by ethanol precipitation as a means of purification. Once the DNA has been purified, it can be subjected to PCR analysis. Although cell lysis in situ is a very attractive approach, it is not without limitations. Soils high in colloids can adsorb DNA, with the result that the efficiencies of extraction are not identical in different soils (17). Also the efficiency of lysis is not always perfect. Finally, note that DNA extractions may include eukaryotic DNA as well as bacterial DNA.

Overall, the community (cells or DNA)-based extraction methods represent a higher proportion of the bacterial population than dilution and plating alone, and levels of recovery are often 20 to 30%. They also are likely to represent a segment of the population of bacteria different from that obtained with culture techniques. These methods also have the advantage of being well suited to subsequent PCR analysis.

Direct Lysis of Soil Samples

The direct lysis protocol described here is a modification of the procedure of Ogram et al. (17). Ten grams of wet soil is incubated in a 1.0 mM sodium phosphate-buffered solution (pH 7.0) with 0.25 g of SDS at 70°C for 30 min. The samples are vortexed every 5 min to aid in lysis. After incubation, 5 g of 200- to 300-μm-diameter and 5 g of 1,000-μm-diameter glass beads are added, and the suspension is shaken continuously for 30 min on a reciprocal platform shaker. Soil and debris are sedimented via centrifugation at 10,000 rpm for 10 min at 10°C, using a Sorvall SS-34 rotor. The supernatant is transferred to an Oak Ridge tube and placed on ice for 15 to 30 min to precipitate the SDS. The SDS precipitate is removed by centrifugation at 10,000 rpm for 10 min at 10°C, and the supernatant is saved. Other researchers have used physical methods to lyse bacteria; these can include thermal shock from freeze-heat cycles and sonication.

Purification of DNA Extracted from Soil

DNA that results from lysed soil bacteria will likely be contaminated with PCR-inhibitory substances, which include humic substances as well as metal ions (1, 27). Several approaches for the removal of these substances are outlined below.

Purification via CsCl Density Centrifugation

Extracted DNA in the supernatant is transferred to an Oak Ridge tube containing 14.5 g of CsCl. If necessary, sterile water is added to bring the supernatant volume to 15.5 ml before addition of the CsCl. The CsCl is dissolved by gently inverting the tubes, and the tubes are incubated at room temperature for 10 to 15 min. The precipitated proteins are removed by centrifugation at 7,500 rpm for 10 min at 10°C. The cleared supernatant is transferred to a Sorvall 18.5-ml Ultracrimp centrifuge tube containing 0.65 ml of 10-mg/ml EtBr. When necessary, any remaining volume is filled with CsCl-EtBr balance solution, and the tubes are balanced prior to sealing. A Sorvall TV865B rotor is used to band the DNA at 45,000 rpm at 18°C for at least 17 h in a Beckman L80-70 ultracentrifuge (Beckman Instruments).

After ultracentrifugation, a band is visible when viewed under long-wave UV illumination. Bands are fractionated by using 18-gauge needles with 3- or 6-ml syringes for extraction of the band and 23-gauge needles for making an air hole in the top of the tube. The extracted solution can be transferred to a 6-ml Beckman heat seal tube. The volume is adjusted by using CsCl-EtBr balance solution, and tubes are balanced and sealed. A second ultracentrifugation for further purification is performed, using the previous conditions, in a Beckman VTi65.2 rotor. The fractionation procedure is repeated, and the solution is transferred to a 14-ml Falcon tube for removal of the EtBr.

Isopropanol saturated with 5.0 M NaCl is added in volumes equal to the amount of fractionated DNA. The solutions are gently mixed and allowed to separate into distinct phases. The isopropanol-EtBr layer is discarded, and this procedure is repeated five times until all of the pink color due to the ethidium is removed. The remaining solution is measured and transferred to a clean Corex tube. Two volumes of sterile deionized H_2O (dH_2O) is added, followed by 3 volumes of ice-cold ($-20°C$) ethanol. The tubes are covered with Parafilm and mixed before overnight incubation at $-20°C$. Within 24 h, the precipitate is sedimented by centrifugation at 7,500 rpm at 4°C for 1 h. The supernatant is decanted, and the tube is inverted on Kim-Wipes and allowed to drain for 5 min. The pellets are air dried and resuspended in 400 μl of sterile dH_2O. After the DNA is resuspended, the solution is transferred to a sterile 1.5-ml Eppendorf tube, and 40 μl of 3.0 M sodium acetate and 880 μl of ice-cold ethanol are added. This solution is thoroughly mixed and incubated for at least 1 h at $-20°C$. An Eppendorf model 5415 Microfuge placed in a 4°C cold room is used to pellet the DNA at 12,000 rpm for 15 min. The supernatant is discarded, and the pellet is washed once with ice-cold 70% ethanol and then spun for 5 min at 12,000 rpm in the Microfuge. The tubes are inverted onto Kim-Wipes and allowed to air dry, after which the DNA is resuspended in 400 μl of sterile dH_2O.

The purity of the processed DNA is determined by measuring the A_{260}/A_{280} on a Beckman DU-6 spectrophotometer (Beckman Instruments). A ratio value greater than 1.8 is acceptable. Concentration of the DNA solution is measured on a TKO 100 Mini-fluorometer (Hoefer Scientific Instruments) as recommended by the manufacturer.

Purification via Elutip d Columns

Elutip d purification columns are prepared as instructed by the manufacturer (Schleicher & Schuell, Keene, N.H.). Supernatant containing DNA is passed through the Elutip d prefilter and column, after which the sample is recovered from the column with two 400-μl washes of high-salt buffer. The DNA is precipitated in 2.5 volumes of ethanol at $-20°C$ overnight ad pelleted by centrifugation in the cold for 20 min at 14,000. The pellet is washed in 70% ethanol, air dried, and resuspended in Tris-EDTA buffer with 1 mg of RNase per ml. The DNA is stored at 4°C.

Other Approaches to Purification

Several researchers have used other purification techniques, including Sephadex-Chelex columns (1), density gradient centrifugation (18), and hydroxyapatite columns (25).

PROCESSING PLANT SAMPLES FOR PCR

Basic Protocol

Microbes found within plant tissue include symbiotic microbes involved in nitrogen fixation. These organisms include *Rhizobium* and *Frankia* spp., which are found within nodules associated with plant roots. PCR amplification of these plant-associated microbes necessitates that the nodular material itself be treated prior to analysis. An outline for processing nodular material is presented below.

Ten micrograms of nodule material is weighed out, using a surface-sterilized pair of forceps. The nodules are rinsed in ethanol and later in sterile distilled water and then aseptically crushed (preferably under a hood) in the presence of dry ice on a spot plate, using a glass rod. The crushed material is placed in a 0.5-ml microcentrifuge tube containing the PCR mixture without the enzyme (Table 4), using a wide-mouth pipette or pipette tip. The sample (in the microcentrifuge tube) is placed on a heat block or a thermocycler and heated at 98°C for 20 min to lyse the "bacteroides." The sample is allowed to cool to room temperature, the *Taq* polymerase (0.5 μl) is added to the lysate, and the temperature cycling (25 cycles) is initiated for amplification.

TABLE 4 Mixture for PCR amplification of plant samples

Component	Amt (μl)
Sterile H_2O	60–65
$10\times$ PCR buffer	10
Deoxynucleoside triphosphates (1.25 mM)	16
Primer 1 (0.1 μg/μl)	0.5
Primer 2 (0.1 μg/μl)	0.5
Template	5–10

At the end of 25 amplification cycles (94°C-60°C-72°C), an additional 2 μl (0.2 μg) of each primer is added to the mixture, and the temperature cycling is continued for another 25 cycles of amplification. This "booster-PCR" protocol (21) is unique in that the ratio of primer to template is altered by adding a reduced amount of primer in the initial 25 cycles and the normal amount (0.2 μg) prior to the second round of amplification. This is done because excess primer relative to template DNA can result in decreased efficiency of PCR.

Primer Sequences for PCR Amplification of Rhizobia

To date there is limited published information on PCR primers to specifically detect rhizobia. However, there are numerous sequences that could serve as potential primer sequences (13). Primers can be designed for the detection of all rhizobia or could be specific enough to allow the detection of specific genera and species. For example, common *nod* and *fix* gene sequences can be used to detect all rhizobia. Bjourson et al. (4) designed primers and used a combined subtraction hybridization procedure coupled to PCR to detect specific strains of *Rhizobium leguminosarum* bv. trifolii. Earlier work by Bjourson and Cooper (3) identified strain-specific sequences of *Rhizobium loti*. More recently, REP and ERIC sequences have been used to design primers that allow identification of *Rhizobium meliloti* strains (6). Unique sequences such as those of the transposable Tn5 have been inserted into rhizobial bacterial genomes, and primers designed from Tn5 sequences have been used to detect specific strains of *R. leguminosarum* bv. phaseoli from within root nodules and soils (19).

CONCLUSIONS

PCR is a powerful new tool that can be used extensively to detect gene sequences associated with plant and soil microbes. Applications include detection of specific microbes and estimates of microbial diversity and enzymatic activity. The key to the successful use of PCR is in the design of the primers, which should be prompted by the research objectives of the scientist. However, although PCR on pure culture isolates is relatively straightforward, PCR detection of gene sequences in situ in environmental samples is not so simple. Such samples often contain PCR-inhibitory substances which must be removed prior to PCR amplification. Different protocols are available for microbes in either plant or soil environments. Overall, PCR adds a useful new technology to aid in the study of plant and soil microbes.

REFERENCES

1. **Abbaszadegan, M., M. S. Huber, C. P. Gerba, and I. L. Pepper.** 1993. Detection of enteroviruses in groundwater using polymerase chain reaction. *Appl. Environ. Microbiol.* **59:**1318–1324.
2. **Ausubel, F. M., R. Brent, R. E. Kingston, D. D. Moore, J. A. Smith, J. G. Seidman, and K. Struhl (ed.).** 1987. *Current Protocols in Molecular Biology.* John Wiley & Sons, New York.
3. **Bjourson, A. J., and J. E. Cooper.** 1988. Isolation of *Rhizobium loti* strain-specific DNA sequences by subtraction hybridization. *Appl. Environ. Microbiol.* **54:**2852–2855.
4. **Bjourson, A. J., C. E. Stone, and J. E. Cooper.** 1992. Combined subtraction hybridization and polymerase chain reaction amplification procedure for isolation of strain-specific *Rhizobium* DNA sequences. *Appl. Environ. Microbiol.* **58:**2296–2301.
5. **Bruce, K. D., W. D. Hiorns, J. L. Hobman, A. M. Osborn, D. Strike, and D. A. Ritchie.** 1992. Amplification of DNA from native populations of soil bacteria by using polymerase chain reaction. *Appl. Environ. Microbiol.* **58:**3413–3416.
6. **de Bruijn, F. J.** 1992. Use of repetitive (repetitive extragenic palindromic and enterobacterial repetitive intergenic consensus) sequences and the polymerase chain reaction to fingerprint the genomes of *Rhizobium meliloti* isolates and other soil bacteria. *Appl. Environ. Microbiol.* **58:**2180–2187.
7. **Herrick, J. B., E. L. Madsen, C. A. Batt, and W. G. Ghiorse.** 1993. Polymerase chain amplification of naphthalene catabolic and 16S rRNA gene sequences from indigenous sediment bacteria. *Appl. Environ. Microbiol.* **59:**687–694.
8. **Josephson, K. L., C. P. Gerba, and I. L. Pepper.** 1993. Polymerase chain reaction detection of nonviable bacterial pathogens. *Appl. Environ. Microbiol.* **59:**3513–3515.
9. **Josephson, K. L., S. D. Pillai, J. Way, C. P. Gerba, and I. L. Pepper.** 1991. Detection of fecal coliforms in soil by polymerase chain reaction and DNA : DNA hybridizations. *Soil Sci. Soc. Am. J.* **55:**1326–1332.
10. **Jutras, E. M., R. M. Miller, and I. L. Pepper.** 1995. Optimization of arbitrarily primed PCR for the identification of bacterial isolates. *J. Microbiol. Methods* **24:**55–63.
11. **Katz, E. D., W. Bloch, and J. Wages.** 1992. HPLC quantification and identification of DNA amplified by polymerase chain reaction. *Amplifications* **8:**10–13.
12. **Krawiec, S., and M. Riley.** 1990. Organization of the bacterial chromosome. *Microbiol. Rev.* **54:**502–539.
13. **Martinez, E., D. Romero, and R. Palacios.** 1990. The *Rhizobium* genome. *Crit. Rev. Plant Sci.* **9:**59–93.
14. **Moré, M. I., J. B. Herrick, M. C. Silva, W. G. Ghiorse, and E. L. Madsen.** 1994. Quantitative cell lysis of indigenous microorganisms and rapid extraction of microbial DNA from sediment. *Appl. Environ. Microbiol.* **60:**1572–1580.
15. **Neilson, J. W., K. L. Josephson, S. D. Pillai, and I. L. Pepper.** 1992. Polymerase chain reaction and gene probe detection of the 2,4-dichlorophenoxyacetic acid degradation plasmid, pJP4. *Appl. Environ. Microbiol.* **58:**1271–1275.
16. **Neilson, J. W., K. L. Josephson, I. L. Pepper, R. B. Arnold, G. D. DiGiovanni, and N. A. Sinclair.** 1994. Frequency of horizontal gene transfer of a large catabolic plasmid (pJP4) in soil. *Appl. Environ. Microbiol.* **60:**4053–4058.
17. **Ogram, A., G. S. Sayler, and T. Barkay.** 1987. The extraction and purification of microbial DNA from sediments. *J. Microbiol. Methods* **7:**57–66.
18. **Pillai, S. D., K. L. Josephson, R. L. Bailey, C. P. Gerba, and I. L. Pepper.** 1991. Rapid method for processing soil samples for polymerase chain reaction amplification of spe-

cific gene sequences. *Appl. Environ. Microbiol.* **57:** 2283–2286.

19. **Pillai, S. D., K. L. Josephson, R. L. Bailey, and I. L. Pepper.** 1992. Specific detection of rhizobia in root nodules and soil using the polymerase chain reaction. *Soil Biol. Biochem.* **24:**885–891.

20. **Reynolds, K. A., C. P. Gerba, and I. L. Pepper.** 1996. Detection of infectious enteroviruses by an integrated cell culture-PCR procedure. *Appl. Environ. Microbiol.* **62:** 1424–1427.

21. **Ruano, G., and K. K. Kidd.** 1990. Booster PCR: a biphasic paradigm for amplification of a few molecules of target. *Amplification* **3:**12–13.

22. **Steffan, R. J., and R. M. Atlas.** 1988. DNA amplification to enhance detection of generally engineered bacteria in environmental samples. *Appl. Environ. Microbiol.* **54:** 2185–2191.

23. **Straub, T. M., I. L. Pepper, M. Abbaszadegan, and C. P. Gerba.** 1994. A method to detect enteroviruses in sewage sludge-amended soil using the PCR. *Appl. Environ. Microbiol.* **60:**1014–1017.

24. **Straub, T. M., I. L. Pepper, and C. P. Gerba.** 1994. Detection of naturally occurring enteroviruses and hepatitis A virus in undigested and anaerobically digested sludge using the polymerase chain reaction. *Can. J. Microbiol.* **40:** 884–888.

25. **Torsvik, V. L.** 1980. Isolation of bacterial DNA from soil. *Soil Biol. Biochem.* **12:**15–21.

26. **Tsai, Y. L., and B. H. Olson.** 1991. Rapid method for direct extraction of DNA from soil and sediments. *Appl. Environ. Microbiol.* **57:**1070–1074.

27. **Tsai, Y. L., and B. H. Olson.** 1992. Rapid method for separation of bacterial DNA from humic substances in sediments for polymerase chain reaction. *Appl. Environ. Microbiol.* **58:**2292–2295.

28. **Tsai, Y. L., C. J. Palmer, and L. R. Sangermano.** 1993. Detection of *Escherichia coli* in sewage and sludge by polymerase chain reaction. *Appl. Environ. Microbiol.* **59:** 353–357.

29. **Versalovic, J., M. Schneider, F. J. de Bruijn, and J. R. Lupski.** 1994. Genomic fingerprinting of bacteria using repetitive sequence-based polymerase chain reaction. *Methods Mol. Cell. Biol.* **5:**25–40.

30. **Way, J. S., K. L. Josephson, S. D. Pillai, M. Abbaszadegan, C. P. Gerba, and I. L. Pepper.** 1993. Specific detection of *Salmonella* spp. by multiplex polymerase chain reaction. *Appl. Environ. Microbiol.* **59:**1473–1479.

31. **Weidner, S., W. Arnold, and A. Pühler.** 1996. Diversity of uncultured microorganisms associated with the seagrass *Halophila stipulacea* estimated by restriction fragment length polymorphism analysis of PCR-amplified 16S rRNA genes. *Appl. Environ. Microbiol.* **62:**766–771.

32. **Welsh, J., and M. McClelland.** 1990. Fingerprinting genomes using PCR with arbitrary primers. *Nucleic Acids Res.* **19:**861–866.

Quantification of Nitrogen Transformations

DAVID D. MYROLD

49

Nitrogen cycling in soils has been studied intensively for many years because of its importance to plant productivity. The complexity of the N cycle, with its many pools and processes (Fig. 1), has also caught the interest of soil scientists and microbiologists. In addition to the continued importance of N to agricultural production and the vigor of natural ecosystems, more recent concerns about soil acidification caused partly by atmospheric inputs of N, NO_3^- movement in surface and groundwaters, and contributions of N gases, such as N_2O and NO, to ozone destruction and the greenhouse effect have required N transformations to be quantified.

This chapter will describe some of the most useful methods for measuring the most important soil N transformations (Fig. 1): N_2 fixation, N mineralization, nitrification, and denitrification. It will be selective and not comprehensive, as entire books have been written about the techniques used to measure just one or two of the N cycle processes. Several of the chapters found in references 17 and 33 are especially recommended as additional sources of information about current methods for measuring N transformation rates in soil.

DINITROGEN FIXATION

The process of N_2 fixation involves the biological reduction of N_2 to the level of NH_3, which is subsequently incorporated into amino acids. It is a reaction that requires a large amount of energy to break the triple bond of N_2. The nitrogenase enzyme that catalyzes this reaction is oxygen sensitive and has been found only in bacteria.

Some bacteria fix N_2 in a free-living state (nonsymbiotic N_2 fixation), others are closely associated with plant roots (associative N_2 fixation), and others form a mutualistic symbiosis with plants (symbiotic N_2 fixation). Generally, the greatest rates of N_2 fixation occur in symbioses because the plants can directly supply the high energy demands of this process.

The first estimates of N_2 fixation came from measuring the increase in total N of the system under study. This method is not particularly sensitive but can work reasonably well for studying pure-culture syntheses of N_2-fixing symbioses. The method is more difficult to use in soil systems but has the advantage of being integrative over time and can be used to obtain field estimates (32, 34). Its sensitivity is probably limited to rates greater than about 20 kg of N ha^{-1},

and doing it well requires a significant amount of sampling and labor. Thus, this method will not be described in this chapter.

Greater sensitivity in measuring N_2 fixation can be obtained by using ^{15}N, a stable isotope of N that is present naturally as 0.3663% of N in the atmosphere. The definitive test for the presence of N_2-fixing activity is probably the incorporation of ^{15}N-labeled N_2. An alternative method is to label the soil with ^{15}N and estimate the amount of N_2 fixed by using the principle of isotope dilution. Each of these applications of ^{15}N techniques will be described.

The most recently developed method for estimating N_2 fixation is the use of acetylene as a surrogate substrate for the nitrogenase enzyme. This method is very sensitive and easy to do, which accounts for its widespread use. There are numerous pitfalls to its use, however; thus, the acetylene reduction assay must be used with care.

$^{15}N_2$ Gas as a Tracer

Nitrogen in the atmosphere, in soils, and in plants is largely made up of the ^{14}N isotope, with just a small proportion (about 0.37%) of the heavier ^{15}N isotope. Thus, when soil or a plant-soil system is exposed to an atmosphere that has been enriched with $^{15}N_2$, any N_2 fixed by the system will be enriched with ^{15}N, with the amount of $^{15}N_2$ fixed being proportional to the rate of N_2 fixation. This method is the definitive test for N_2 fixation and is particularly useful for calibrating the acetylene reduction method. In practice, $^{15}N_2$ fixation can be done only in the laboratory in closed systems. An example of a protocol for measuring N_2 fixation with $^{15}N_2$ is given below; additional details can be found in references 31 and 34.

Materials

A closed system of some sort is needed. For a soil system, this can be simply a sealed bottle, but more sophisticated designs are required for studying symbiotic N_2 fixation. Small plants can be enclosed entirely, whereas larger plants can have just their root systems sealed off from the atmosphere (Fig. 2). For all systems, it is important to strike a balance between minimizing the gas-phase volume to reduce the amount of $^{15}N_2$ needed while still having sufficient volume for adequate aeration and dilution of any gaseous products that might affect the measurement. A ratio of 0.3 liter

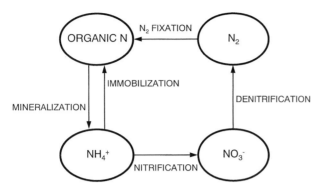

FIGURE 1 Diagram of the N cycle, showing major pools of N and dominant N transformations that occur in soil.

of gas-phase volume per g of plant tissue has been proposed as a rule of thumb (34).

$^{15}N_2$ is available from commercial sources, usually as 99 atom% $^{15}N_2$. This is the most convenient form; however, $^{15}N_2$ can also be made by oxidizing $^{15}NH_4{}^+$ with hypobromite (31).

Analysis of ^{15}N is by mass spectrometry (21).

Procedure

The system to be measured is sealed, and $^{15}N_2$ is added. A nonlabeled control should also be used. Often 10 atom% $^{15}N_2$ is a good working concentration; however, one can make more exact estimates for any given system on the basis of assumptions about N_2 fixation rates, system volumes, incubation length, and mass spectrometer sensitivities (31, 34). After a brief period to allow complete mixing of the added $^{15}N_2$, a sample of the gas phase should be taken to determine the actual ^{15}N enrichment. The length of the incubation period is dependent on the atom% ^{15}N enrichment and the N_2-fixing activity. The greater sensitivity attained with a long incubation period must be balanced against the need to maintain concentrations of the gases in the system close to atmospheric, e.g., 21% oxygen. Another gas sample should be taken at the end of the incubation period, and its atom% ^{15}N enrichment should be averaged with that of the initial sample. The soil or plant sample is then analyzed for atom% ^{15}N abundance. Grinding the entire plant for analysis is simplest if just an overall rate of N_2 fixation is desired; however, plant parts can be sampled separately, and a weighted average can be used. The atom% ^{15}N excess of the N_2-fixing system is the difference in atom% ^{15}N between the labeled and control systems. Divid-

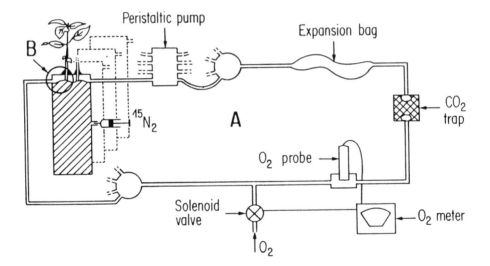

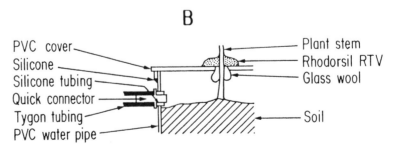

FIGURE 2 Example of a continuous-flow system for measuring N_2 fixation in plant-soil systems. This system could be used for either $^{15}N_2$ or acetylene reduction assays. (A) Closed-circuit system including an oxygen-regulating device; (B) detail of a plant container indicating sealing material for plant stems and aeration ports. PVC, polyvinyl chloride; RTV, a silicone rubber. Taken from reference 30 and used with permission of the publisher.

ing this atom% ^{15}N excess by the atom% ^{15}N excess of the atmosphere gives the fraction of N that came from N_2 fixation. Multiplying this fraction by the total N of the plant (or soil) and dividing by the incubation time will yield a rate of N_2 fixation.

^{15}N Isotope Dilution

^{15}N can be used to label the soil or nutrient solution rather than the atmosphere. In practice, this is easier to do and has the advantage that it can be done in the field or the laboratory. It also offers an integrative measure of N_2 fixation, something which is very difficult to do with either $^{15}N_2$ tracer or acetylene reduction methods. Especially with the increasing availability and reduced cost of ^{15}N isotope analysis, ^{15}N isotope dilution has become the method of choice for most N_2 fixation studies. It is not without its drawbacks and difficulties, however, and must be used carefully (30, 31, 34). The general principles for field studies of N_2 fixation by ^{15}N isotope dilution will be given.

Materials

^{15}N-labeled fertilizer, most commonly urea, ammonium salts, or nitrate salts, is used. ^{15}N enrichment of 5 to 10 atom% is usually adequate when N is added at a rate of 5 to 10 kg ha^{-1}.

Procedure

Well-replicated plots large enough to encompass the variability of the growing plants should be used. Reasonable plot sizes range from 1 to 2 m^2 for pasture legumes to 2 to 5 m^2 for grain legumes. Substantially larger plots are needed for tree-sized plants, although paired-tree plots are often used to reduce the cost of ^{15}N-labeled fertilizer material.

A critical aspect of all ^{15}N isotope dilution measurements of N_2 fixation is the selection of a non-N_2-fixing reference plant. For some legume species, this may be a nonnodulating variety of the same species; unfortunately, nonnodulating varieties are unavailable for most plants. A suitable reference plant should not fix N_2 but should have a similar rooting pattern to access the same pools of soil N, take up soil N in the same temporal pattern, and respond similarly to any treatments or environmental stresses. Sudan grass has been proposed as a reasonable choice for grain legumes; perennial ryegrass has been proposed for pasture legumes (31). There is no standard for shrubs or trees.

Soil is usually labeled by dissolving the ^{15}N-labeled fertilizer in water and sprinkling the solution uniformly over the treated plots. It is sometimes advantageous to split the fertilizer addition into two or more applications, particularly with perennial plants, in order to have the soil more uniformly labeled over the growing season. With annual N_2-fixing plants, it may be possible to label the soil organic N pool one year and then use this relatively uniformly labeled plot to measure N_2 fixation in subsequent years. Unamended control plots should also be used as a means of assessing the background atom% ^{15}N of the N_2-fixing and reference plant species.

Plants may be harvested periodically or just at the end of a growing season. Analysis is simplified if the entire plant can be ground; however, it is possible to sample different plant parts to obtain a weighted average. The fraction of N coming from N_2 fixation is equal to 1 minus the quotient of the atom% ^{15}N excess of the fixing plant over the atom% ^{15}N excess of the nonfixing plant. These ^{15}N excess volumes are relative to values for plants growing in control (nonlabeled) soils.

Acetylene Reduction

Measuring the ethylene produced from the reduction of acetylene is easy, inexpensive, and more sensitive than ^{15}N techniques (2, 16, 34). Although it has been used to estimate field rates over time in the field, it is not well suited for such work and is much better as a point-in-time measurement. Additional pitfalls are that the ratio of acetylene reduced to N_2 fixed is not always the 3:1 or 4:1 ratio given by the ideal stoichiometry of the reaction but is quite variable. This problem can be overcome by using $^{15}N_2$ fixation to determine the ratio for the system of interest, however. Other problems include the disturbance related to enclosing root nodules for the assay and the sensitivity of at least some root nodules to the so-called acetylene-induced decline (18). Excised root nodules should not be used. A key use of the acetylene reduction method is to screen for N_2-fixing activity in soil and root nodule samples.

Materials

A source of acetylene is needed. This is most easily generated by adding water to calcium carbide. Commercial acetylene can also be used but should be purified by passing the gas through concentrated sulfuric acid and water traps to remove contaminants. Some type of gastight system is needed to hold the sample. This can be as simple as a flask for soil samples or a complicated cuvette system for evaluating the activity of root nodules (Fig. 2). Ideally, a flowthrough system rather than a static incubation vessel is used with nodulated plants (26, 34). To assay the acetylene and ethylene, a gas chromatograph equipped with a flame ionization detector is required.

Procedure

The sample to be assayed is placed in a vessel large enough that there is an adequate supply of oxygen but small enough to enhance sensitivity. Acetylene is added to a final concentration of 10% (vol/vol) and well mixed throughout the gas phase. A control without acetylene should be used to measure background ethylene production. During the assay, gas samples are taken periodically for acetylene and ethylene analysis. Gas samples may be stored for a short period in syringes with the needles stuck into a rubber stopper; evacuated vials can be used for longer storage. Gas samples, typically 1 ml, are analyzed by gas chromatography, and concentrations of ethylene are calculated from a standard curve. Acetylene reduction rates are obtained by regression analysis of the ethylene concentration versus time data. The amount of N_2 fixed can be estimated by dividing the acetylene reduction rate by the measured (using $^{15}N_2$) or theoretical ratio of moles of acetylene reduced per mole of N fixed.

NITROGEN MINERALIZATION AND IMMOBILIZATION

The term N mineralization is most often used to describe the net accumulation of NH_4^+ plus NO_3^- in soil and is thought to represent the amount of N available for plant uptake. It is more accurate, however, to call this net N mineralization. Net N mineralization is the result of gross N mineralization (also called ammonification) plus gross nitrification minus any assimilation (often called immobilization) of NH_4^+ or NO_3^- into organic N. The concurrent processes of NH_4^+ production and consumption are done by all heterotrophic soil microorganisms.

Numerous laboratory and field methods have been devised to measure N mineralization (3, 6). A subset of these techniques for determining laboratory and field rates of net and gross N mineralization will be described below.

Net N Mineralization

The basic principle in determining net N mineralization rates is to measure the change in the sum of NH_4^+ plus NO_3^- concentrations during an incubation period. Soil conditions can be adjusted to optimize N mineralization or left unchanged to simulate natural conditions.

Laboratory Measurement

Laboratory measurement will determine net N mineralization under optimal conditions; however, it can be modified easily if other environmental conditions are desired.

Soil sampled from the field is sieved (<2 mm is the typical size used, although larger screens can be used for wet or clayey soils) and mixed to homogeneity, and soil water is adjusted to field capacity. The moist soil (about 25 g) is placed in a beaker or cup with a capacity of about 100 cm^3 and covered with polyethylene film that has a small hole to facilitate aeration. A minimum of three replicates should be used. A sample (equivalent to about 10 g [dry weight]) is taken at the beginning of the experiment for NH_4^+ and NO_3^- analysis. The soil is incubated at room temperature for 28 days. During the incubation, soil water content is kept within 5% of field capacity by addition of water as uniformly as possible. At the end of the incubation period, a sample of soil is taken for NH_4^+ and NO_3^- analysis, and the remainder is used to determine gravimetric water content.

Extraction of NH_4^+ and NO_3^- is done by shaking soil with 2 M KCl (10:1 extraction solution-to-soil ratio) for 1 h. The extract can be filtered (prerinse filters to remove any traces of inorganic N) or centrifuged. The clear solution is used for measuring the concentration of NH_4^+ and NO_3^- either colorimetrically or by distillation and titration (12).

Net N mineralization is calculated from the difference between NH_4^+ plus NO_3^-–N at the end and beginning of the incubation. (The difference in NO_3^-–N yields net nitrification.)

A common modification of this method is to sample at several intervals over the course of incubation. From the time course of NH_4^+ and NO_3^- production, it is possible to calculate the kinetics of net N mineralization (27).

Field Measurement

Hart et al. (6) present an excellent discussion of the advantages and disadvantages of several methods to measure net N mineralization in the field: buried bags, resin cores, open and closed cores, etc. All methods are disruptive to some degree. All sever roots, which may enhance N immobilization as the dead roots decompose and may alter soil water content by eliminating transpiration. Closed-top cores and buried bags eliminate leaching, whereas leaching can occur in open-top cores and resin cores but is measured only in the resin cores. Because it is one of the less complicated methods, the following closed-top, solid-cylinder method is perhaps as good as any.

Intact soil columns are collected in thin-walled steel or polyvinyl chloride cylinders which are sharpened at one end. Cylinders about 5 cm in diameter work well and can be cut to the desired length. A minimum of 20 cylinders per site should be used. Ten cylinders are used to measure initial NH_4^+ and NO_3^- concentrations; the other 10 are left in place and covered with a loosely fitting top. The cover prevents additions or removals of inorganic N by water. A 1-month incubation period is typically used.

Soil in the cylinders is removed and sieved (4 mm). The coarse and fine fractions are weighed and sampled for gravimetric water content. A sample of the fine fraction is extracted with 2 M KCl (10:1 extractant-to-soil ratio) and analyzed for NH_4^+ and NO_3^- (the amount of NH_4^+ and NO_3^- in the coarse fraction is assumed to be negligible). The difference in the amount of NH_4^+ and NO_3^-–N per core between the 1-month and initial sample gives the net N mineralization (the change in NO_3^-–N gives the net nitrification), which is usually expressed on an area basis.

The variability of field net N mineralization is always high. Some statistical efficiency can be gained by increasing the number of cylinders used and pooling samples from several cylinders to keep the number of laboratory analyses reasonable.

Gross N Mineralization

Measuring the gross rate of N mineralization (ammonification or NH_4^+ production) requires the use of $^{15}NH_4^+$ (6, 23). It also allows the gross rate of NH_4^+ consumption to be determined. The principle is the same for field or laboratory measurements; the major difference is the high variability in the field and greater difficulty in adding the ^{15}N label uniformly to field samples without disturbance. By analogy, gross nitrification and NO_3^- immobilization can be determined by using $^{15}NO_3^-$. Thus, only the method for gross N mineralization will be described.

Procedure

Add sufficient highly enriched (99 atom% ^{15}N) $^{15}NH_4^+$ to increase the native NH_4^+ pool size by about 1 mg of N per kg of soil, using a small volume of solution so as not to greatly disrupt the soil water status. This may be done conveniently by making several injections with a syringe and long needle. At least six samples should be treated in this way, half of which will be extracted immediately for initial measurement of NH_4^+ concentration and isotopic composition and sampled for gravimetric water content. The other half of the samples are incubated for 24 h prior to analysis.

A convenient method of preparing KCl extracts for ^{15}N analysis is to diffuse the NH_4^+ in the KCl extract onto an acidified paper filter disk (4), although distillation methods (7) also work. The atom% ^{15}N of the extracted NH_4^+ is measured by mass spectrometry (21).

Calculations

There are numerous ways to calculate gross N mineralization and NH_4^+ consumption rates from the concentration and atom% ^{15}N data, depending on the assumptions that one makes about reaction order and whether remineralization of immobilized ^{15}N is allowed (6, 23). Table 1 shows the calculation procedure assuming zero-order kinetics and no remineralization, which is usually reasonable for 24-h incubations.

Comments

Excellent analytical technique must be used in ^{15}N experiments because small amounts of contamination can greatly affect the final results. Hauck et al. (9) give sound advice on this topic.

TABLE 1 Calculation of gross N mineralization and NH_4^+ consumption rates by using the equation initially developed by Kirkham and Bartholomew (13)[a]

Time (days)	NH_4^+ conc (mg of N per kg of soil)	NH_4^+ isotopic composition (atom% ^{15}N excess)
0	10.0	50.0
1	5.0	12.5

[a] This method is valid for zero-order kinetics, no mineralization of immobilized ^{15}N, and cases when gross N mineralization and NH_4^+ consumption rates are not equal.

Definitions:

P_A = NH_4^+ production (gross N mineralization or ammonification) rate (mg of N per kg of soil per day)

c_A = NH_4^+ consumption rate (mg of N per kg of soil per day)

t = time (days)

$A_{XS,0}$ = atom% ^{15}N excess of the NH_4^+ pool at t = 0 (%)

$A_{XS,t}$ = atom% ^{15}N excess of the NH_4^+ pool at t = t (%)

atom% ^{15}N excess = atom% ^{15}N of a ^{15}N-enriched pool minus its atom% ^{15}N prior to ^{15}N addition

A_0 = NH_4^+ concentration at t = 0 (mg of N per kg of soil)

A_t = NH_4^+ concentration at t = t (mg of N per kg of soil)

Calculations:

P_A = $(A_0 - A_t/t) \cdot (\ln(A_{XS,0}/A_{XS,t})/\ln(A_0/A_t))$
= $((10 - 5)/1) \cdot (\ln(50/12.5)/\ln(10/5))$
= $5 \cdot (1.386/0.693)$
= 10 mg of N per kg of soil per day

c_A = $p_A - ((A_t - A_0)/t)$
= $10 - ((5 - 10)/1)$
= 15 mg of N per kg of soil per day

NITRIFICATION

Nitrification is most commonly used to describe the autotrophic oxidation of NH_4^+ to NO_2^- and further to NO_3^-. The first oxidation step from NH_4^+ to NO_2^- is carried out by the ammonia-oxidizing bacteria. It is usually the rate-limiting step. Ammonia oxidation can be inhibited by acetylene (10 Pa) and several other compounds, e.g., nitrapyrin (N-Serve). Nitrite oxidizers convert NO_2^- to NO_3^-. This process can be selectively inhibited by chlorate (ClO_3^-).

In most soils, autotrophic nitrification is the dominant NO_3^--producing transformation; however, NO_3^- production directly from organic N by heterotrophic microorganisms can be important in some soils. The following methods are for determining autotrophic nitrification, but modifications for heterotrophic nitrification will be noted.

Potential Nitrification Activity

Maximum, or potential, nitrification activity of a soil can be measured in a short-term laboratory assay by providing optimal conditions for nitrification to occur (25). Because the duration of this assay is too short for significant growth of autotrophic nitrifiers to occur, the potential nitrification activity is also an index of the size of the active ammonia-oxidizing population.

Procedure

An aerobic soil slurry (20 g of field moist soil in 90 ml of 0.5 mM potassium phosphate buffer [pH 7.2]) is amended with NH_4^+ (0.2 ml of a 0.25 M ammonium sulfate solution) and 1 ml of 1 M $KClO_3$. The soil slurry is shaken at room temperature, and four to five aliquots (5 ml each) are removed at 1- to 2-h intervals. The aliquots are measured for NO_2^-, which accumulates because ClO_3^- blocks its oxidation to NO_3^-. The rate of NO_2^- increase is calculated by linear regression and divided by the soil dry weight to give the potential nitrification activity.

Comments

An estimate of the size of the active ammonia-oxidizing population can be made by dividing the potential nitrification activity by the maximum per-cell activity measured in pure culture (60 to 320 pg of N cell^{-1} h^{-1} [25]). Sizes of ammonia-oxidizing populations estimated in this way are often 10-fold higher than most-probable-number population measurements.

Acetylene, at concentrations of >10 Pa, can selectively inhibit autotrophic nitrification without inhibiting heterotrophic nitrification. Thus, this soil slurry method can be used with acetylene in place of chlorate as an inhibitor. In this case, the production of NO_3^- with and without acetylene is measured, and the difference is the amount of heterotrophic nitrification.

Laboratory and Field Nitrification Rates

Net nitrification rates can be measured in laboratory incubations by using the methods used for measuring net N mineralization. The difference is that only the accumulation of NO_3^- is measured. Similarly, gross rates of nitrification can be determined by the isotope dilution method by adding ^{15}N as NO_3^- and measuring the dilution of the atom% ^{15}N of the NO_3^- pool.

DENITRIFICATION

Under anaerobic conditions, some bacteria can reduce NO_3^- to the gas NO, N_2O, or N_2 by the process known as denitrification. Under most conditions, N_2 is the primary product.

Denitrification typically has been the most poorly quantified N transformation. It often represented the N that was not accounted for in an N budget. The N budget approach probably is still the definitive standard; however, other methods have supplanted it as the method of choice in most studies. Losses of fertilizer N from denitrification are conveniently determined by using ^{15}N-labeled fertilizer. In this method, the amount of ^{15}N in all ecosystem components is tallied and subtracted from the ^{15}N that was added in the fertilizer. Dividing this missing ^{15}N by the atom% ^{15}N of the fertilizer yields the total amount of fertilizer N lost by denitrification. One advantage of this approach is that it gives an integrative measure of denitrification. A disadvantage is that leaching of ^{15}N is assumed to be negligible; however, this may be reasonable if little ^{15}N is found in deeper soil samples.

The two most commonly used methods to measure denitrification are not integrative (8). The simplest and least expensive is the so-called acetylene block method (15), which takes advantage of the fact that acetylene inhibits the last step in the denitrification process, the reduction of N_2O to N_2. Thus, the accumulation of N_2O reflects the denitrification rate. Excellent accounts of the advantages and disadvantages of using the acetylene block have been written (5, 11, 24). Two applications of this method will be presented: the assay for denitrifier enzyme activity and field measurements using soil cores.

Denitrification has also been measured by adding highly enriched $^{15}NO_3^-$ to soils and measuring the ^{15}N-labeled gaseous products (20, 22). The use of this technique in the field will be described.

TABLE 2 Use of the Bunsen absorption coefficient to calculate the total amount of N_2O in a system[a]

Temp (°C)	Bunsen coefficient (α)
5	1.06
10	0.88
15	0.74
20	0.63
25	0.54
30	0.47
35	0.41

[a] The total mass (M_t) of N_2O is equal to the gaseous concentration (C_g) of N_2O times the sum of the gas volume (V_g) plus the liquid volume (V_1) times the Bunsen coefficient (α): $M_t = C_g(V_g + V_1 \cdot \alpha)$.

Denitrifier Enzyme Activity

Measuring the denitrifier enzyme activity of a soil is a relatively quick and easy procedure. It is useful as an index of the denitrifying activity of a soil. Denitrifier enzyme activity has even been correlated to annual denitrification losses in some soils, although it is unlikely that there is a single, universal relationship. The procedure described is based on that given by Tiedje (28).

Procedure

Soil (10 to 25 g) is placed in a container that can be made gastight. Serum bottles or Erlenmeyer flasks work well. A solution containing readily available C (usually 1 mM glucose) and NO_3^- (1 mM KNO_3) is added in sufficient amounts to make a soil slurry. Chloramphenicol, which inhibits protein synthesis, can be added at a working concentration of 1 g liter^{-1}. The container is capped and flushed with an anaerobic gas. Acetylene is added to a final concentration of 10% (vol/vol). The soil slurry is shaken and incubated for 1 to 2 h. During this period, at least four gas samples are taken and analyzed for N_2O concentration. A gas chromatograph equipped with an electron capture detector works well for N_2O analysis. The slope of the increase in the total N_2O (gaseous and dissolved phases; Table 2) divided by the dry weight of the soil gives the value for denitrifier enzyme activity.

Field Measurements Using the Acetylene Block

Acetylene inhibition of N_2O reduction has been used to measure denitrification in static cores, cores with a recirculating gas phase, and soil covers (14, 19). Each of these methods has advantages and disadvantages (29); however, only the static core method will be detailed here because it is simpler and allows for more replication. Adequate replication is one of the most important considerations in measuring denitrification rates in soils because the coefficient of variation is often very large (>100%). The acetylene block method is also the most sensitive means of measuring denitrification, with rates of a few milligrams of N per hectare per day being detectable.

Procedure

Soil cores should be taken with a minimum of disturbance. This is often facilitated by using an impact coring device that houses a recessed sleeve, which serves to hold the soil core upon extraction. For soils with good structural stability, the sleeves can be removed when the cores are transferred to an incubation vessel. Otherwise, it is best to leave the soil in the sleeves. The sleeves may, however, be perforated to improve gas transport, which can be a limitation in the static core method. Soil cores 2.5 to 7.5 cm in diameter are quite typical. The larger-diameter cores tend to compact the soil less, and the greater soil volume helps to decrease sampling variability.

Soil cores are sealed gastight. It is important that the total volume of the gas phase be large enough that the aeration of the soil sample is not greatly altered. Purified acetylene (see the foregoing discussions of N_2 fixation methods) is added to a final concentration of 10% (vol/vol) and mixed by pumping with a large syringe. The sealed system is incubated for up to 24 h at in situ temperatures. Gas samples are taken after the gas phase is mixed with a large syringe. Often only two measurements are taken: the first shortly after the addition of acetylene, and the other at the end of the incubation. Additional samples can be taken at intermediate times to ensure that the production of N_2O is linear, which it should be unless substrate (either NO_3^- or C) becomes limiting during the incubation. Gas samples can be stored in evacuated tubes; however, one should always include controls for leakage and background contamination (19).

Nitrous oxide is analyzed by gas chromatography with an electron capture detector. The denitrification rate is calculated from the difference in the total amount of N_2O sampled at two time points divided by the difference in time and the weight (or surface area) of the soil. It is important to include the amount of dissolved N_2O, which is a fairly soluble gas. This can be done by using Henry's law or, as shown in Table 2, the Bunsen absorption coefficient.

Comments

The potential shortcomings of the acetylene inhibition and the static core techniques are related to gas diffusion and side effects from acetylene. For acetylene to inhibit N_2O reduction, it must reach the site of denitrification, and for N_2O to be measured, it must diffuse into the headspace. Gas transport is facilitated by exposing the sides and top of the soil core to the atmosphere and by mixing the headspace. These practices are probably adequate for soils low in clay and not water saturated. They may not be adequate for wet, highly structured clay soils. Unfortunately, there is no good way to circumvent the gas transport problems in such soils.

Acetylene can affect other processes in soil. Contaminants in the acetylene may inhibit microbial activity or serve as a substrate and enhance microbial activity, thus the need to use pure acetylene. Some soil microbes can use acetylene as a C source, and long exposure times can enrich these populations. This is minimized by short (<1-day) incubations. Perhaps the biggest problem is that acetylene inhibits ammonia oxidation, thereby inhibiting nitrification and the production of NO_3^-. For many soils, there is adequate NO_3^- so that denitrification does not become limiting over 1 day. Some soils, however, may become limited much sooner. This can be detected by measuring the time course of denitrification and looking for decreased activity over time. If a decline in denitrification activity is observed, a shorter incubation time can be used. Even so, in some soils nitrification and denitrification may be tightly coupled in space, and acetylene inhibition may underestimate denitrification rates.

^{15}N Gas Emission

Dinitrogen can be composed of two ^{14}N atoms, two ^{15}N atoms, or one ^{14}N atom and one ^{15}N atom. These molecular species are distributed according to a binomial distribution. Thus, the native atmosphere, which is about 0.37% ^{15}N, has very little (0.0013%) $^{30}N_2$. By adding NO_3^- with a

TABLE 3 Sample calculations for denitrification, using the principle of isotope distribution (1)[a]

Time (h)	Ion current ($^{29}N_2^+/^{28}N_2^+$; r'')	Ion current ($^{30}N_2^+/^{28}N_2^+$; r''')	Molecular fraction of $^{29}N_2$ (^{29}x)	Molecular fraction of $^{30}N_2$ (^{30}x)	Atom fraction of ^{15}N (a)
0	0.007353	0.00001352	0.007299	0.00001342 (x_a)	0.003663 (a_a)
24	0.0074205	0.00002927	0.007365	0.00002905 (x_m)	0.003712 (a_m)

[a] Definitions:

$^{29}x = r''/(1 + r'' + r''')$
$^{30}x = r'''/(1 + r'' + r''')$
$a = (^{29}x + 2\,^{30}x)/2$
d = fraction of N_2 from denitrification
a_p = atom fraction of $^{15}NO_3^-$ pool

Calculations:

$$d = (a_m - a_a)^2/(x_m + a_a^2 - 2a_a a_m)$$
$$= (0.003712 - 0.003663)^2/(0.00002905 + (0.003663)^2 - 2 \cdot 0.003663 \cdot 0.003712)$$
$$= 0.0001545$$
$$a_p = a_a + (x_m + a_a^2 - 2a_a a_m)/(a_m - a_a)$$
$$= 0.003663 + (0.00002905 + (0.003663)^2 - 2 \cdot 0.003663 \cdot 0.003712)/(0.003712 - 0.003663)$$
$$= 0.3182$$

high atom% ^{15}N enrichment to the soil, one can measure the increase in the amount of $^{30}N_2$ with good sensitivity. This was realized in the 1950s but has been applied with any regularity only since about 1980, largely because of the greater availability of sensitive mass spectrometers.

Although ^{15}N gas emission can in principle be used with soil cores and for laboratory incubations, its most useful application is for measuring rates of denitrification in relatively undisturbed field soils. The only major disturbance is the addition of highly labeled $^{15}NO_3^-$ (>20 atom% ^{15}N). This is probably not a major issue for soils that are normally fertilized with N and can be minimized in wildland soils. When the NO_3^- pool is labeled at 50 atom% ^{15}N or more, rates of 5 g of N ha^{-1} day^{-1} can be measured.

Procedure

Soil microplots are established at a size consistent with the gas collection chamber that will be used. A chamber with a diameter of 10 to 20 cm is fairly typical. Although the soil chamber can be pressed into the surface few centimeters of soil, it is also possible to use a permanent base that is implanted to a deeper depth. The soil chamber is then attached to this base when measurements are being made. The chamber itself is usually constructed of polyvinyl chloride with a rubber septum for sampling gases. Normally the headspace of the chamber is kept small to maximize sensitivity.

The microplot is labeled with NO_3^- enriched to 50 atom% ^{15}N or greater. Because of the high enrichment, only a small amount of label must be added, perhaps double or triple the native NO_3^- pool size. It is important, however, to add this label as evenly as possible. Thus the NO_3^- is usually dissolved in water and injected according to a grid pattern.

The soil chamber is set over the labeled soil, and gas samples are taken periodically during the incubation period, which might be a few hours. Samples taken from a cover over a nonlabeled plot also are taken to determine the background labeling of N_2 in the atmosphere. Gas samples are most conveniently stored in evacuated tubes for later mass spectroscopic analysis. A longer incubation time can be used to increase sensitivity; however, care must be taken that a long emplacement of the soil cover does not disturb the system by decreasing aeration, increasing temperatures, or affecting plant processes.

The ratios of $^{29}N_2$ and $^{30}N_2$ to $^{28}N_2$ are measured with

a mass spectrometer. These data are then used in rather complex calculations (Table 3) that give both the atom% ^{15}N of the NO_3^- pool that underwent denitrification and the fraction of the headspace gas that came from denitrification. By multiplying the fraction from denitrification by the total amount of N_2 gas in the chamber and dividing by the surface area and time, the denitrification flux is calculated. This flux is a function of both production (the denitrification rate) and transport; thus, further calculations are needed to calculate the denitrification rate (10).

Comments

The ^{15}N gas emission method suffers from gas diffusion problems similar to those inherent in the acetylene block method. Consequently, it is not surprising that the two methods have agreed quite well when comparisons have been made (22). In theory, it can be shown that denitrification may be under- or overestimated if the NO_3^- pool is not uniformly labeled; however, empirical evidence has not shown this to be a major problem.

SUMMARY

Measuring rates of N cycle transformations is not a simple task. Most measurements are disruptive in some way, and each has its advantages and disadvantages. As with all scientific analyses, the method should be selected in accordance with the experimental objectives and perhaps evaluated prior to use.

REFERENCES

1. **Arah, J. R. M.** 1992. New formulae for mass spectrometric analysis of nitrous oxide and dinitrogen emissions. *Soil Sci. Soc. Am. J.* **56**:795–800.
2. **Bergersen, F. J.** 1980. *Methods for Evaluating Biological Nitrogen Fixation.* John Wiley & Sons, New York.
3. **Binkley, D., and S. C. Hart.** 1989. The components of nitrogen availability assessments in forest soils. *Adv. Soil Sci.* **10**:57–112.
4. **Brooks, P. D., J. M. Stark, B. B. McInteer, and T. Preston.** 1989. Diffusion method to prepare soil extracts for automated nitrogen-15 analysis. *Soil Sci. Soc. Am. J.* **53**:1707–1711.

5. **Duxbury, J. M.** 1986. Advantages of the acetylene method of measuring denitrification, p. 73–92. *In* R. D. Hauck and R. W. Weaver (ed.), *Field Measurement of Dinitrogen Fixation and Denitrification.* Soil Science Society of America, Madison, Wis.

6. **Hart, S. C., J. M. Stark, E. A. Davidson, and M. K. Firestone.** 1994. Nitrogen mineralization, immobilization, and nitrification, p. 985–1018. *In* R. W. Weaver, J. S. Angle, and P. J. Bottomley (ed.), *Methods of Soil Analysis,* part 2. *Microbiological and Biochemical Properties.* Soil Science Society of America, Madison, Wis.

7. **Hauck, R. D.** 1982. Nitrogen–isotope-ratio analysis, p. 735–779. *In* A. L. Page, R. H. Miller, and D. R. Keeney (ed.), *Methods of Soil Analysis,* part 2. *Chemical and Microbiological Properties,* 2nd ed. American Society of Agronomy, Madison, Wis.

8. **Hauck, R. D.** 1986. Field measurement of denitrification—an overview, p. 59–72. *In* R. D. Hauck and R. W. Weaver (ed.), *Field Measurement of Dinitrogen Fixation and Denitrification.* Soil Science Society of America, Madison, Wis.

9. **Hauck, R. D., J. J. Meisinger, and R. L. Mulvaney.** 1994. Practical considerations in the use of nitrogen tracers in agricultural and environmental research, p. 907–950. *In* R. W. Weaver, J. S. Angle, and P. J. Bottomley (ed.), *Methods of Soil Analysis,* part 2. *Microbiological and Biochemical Properties.* Soil Science Society of America, Madison, Wis.

10. **Hutchinson, G. L., and A. R. Mosier.** 1981. Improved soil cover method for field measurement of nitrous oxide flux. *Soil Sci. Soc. Am. J.* **45**:311–316.

11. **Keeney, D. R.** 1986. Critique of the acetylene blockage technique for field measurement of denitrification, p. 103–115. *In* R. D. Hauck and R. W. Weaver (ed.), *Field Measurement of Dinitrogen Fixation and Denitrification.* Soil Science Society of America, Madison, Wis.

12. **Keeney, D. R., and D. W. Nelson.** 1982. Nitrogen—inorganic forms, p. 643–698. *In* A. L. Page, R. H. Miller, and D. R. Keeney (ed.), *Methods of Soil Analysis,* part 2. *Chemical and Microbiological Properties,* 2nd ed. American Society of Agronomy, Madison, Wis.

13. **Kirkham, D., and W. V. Bartholomew.** 1954. Equations for following nutrient transformations in soil utilizing tracer data. *Soil Sci. Soc. Am. Proc.* **18**:33–34.

14. **Klemedtsson, L., G. Hansson, and A. Mosier.** 1990. The use of acetylene for the quantification of N_2 and N_2O production from biological processes in soil, p. 167–180. *In* N. P. Revsbech and J. Sørensen (ed.), *Denitrification in Soil and Sediments.* Plenum Press, New York.

15. **Knowles, R.** 1990. Acetylene inhibition technique: development, advantages, and potential problems, p. 151–166. *In* N. P Revsbech and J. Sørensen (ed.), *Denitrification in Soil and Sediments.* Plenum Press, New York.

16. **Knowles, R., and W. L. Barraquio** 1994. Free-living dinitrogen-fixing bacteria, p. 179–197. *In* R. W. Weaver, J. S. Angle, and P. J. Bottomley (ed.), *Methods of Soil Analysis,* part 2. *Microbiological and Biochemical Properties.* Soil Science Society of America, Madison, Wis.

17. **Knowles, R., and T. H. Blackburn (ed.).** 1993. *Nitrogen Isotope Techniques.* Academic Press, Inc., San Diego, Calif.

18. **Minchin, F. R., and J. F. Witty.** 1989. Limitations and errors in gas exchange measurements with legume nodules, p. 79–96. *In* J. G. Torrey and L. J. Winship (ed.), *Applications of Continuous and Steady-State Methods to Root Biology.* Kluwer Academic Publishers, Dordrecht, The Netherlands.

19. **Mosier, A. R., and L. Klemedtsson.** 1994. Measuring denitrification in the field, p. 1047–1065. *In* R. W. Weaver, J. S. Angle, and P. J. Bottomley (ed.), *Methods of Soil Analysis,* part 2. *Microbiological and Biochemical Properties.* Soil Science Society of America, Madison, Wis.

20. **Mosier, A. R., and D. S. Schimel.** 1993. Nitrification and denitrification, p. 181–208. *In* R. Knowles and T. H. Blackburn (ed.), *Nitrogen Isotope Techniques.* Academic Press, Inc., San Diego, Calif.

21. **Mulvaney, R. L.** 1993. Mass spectrometry, p. 11–57. *In* R. Knowles and T. H. Blackburn (ed.), *Nitrogen Isotope Techniques.* Academic Press, Inc., San Diego, Calif.

22. **Myrold, D. D.** 1990. Measuring denitrification in soils using ^{15}N techniques, p. 181–198. *In* N. P Revsbech and J. Sørensen (ed.), *Denitrification in Soil and Sediments.* Plenum Press, New York.

23. **Powlson, D. S., and D. Barraclough.** 1993. Mineralization and assimilation in soil-plant systems, p. 209–242. *In* R. Knowles and T.H. Blackburn (ed.), *Nitrogen Isotope Techniques.* Academic Press, Inc., San Diego, Calif.

24. **Rolston, D. E.** 1986. Limitations of the acetylene blockage technique for field measurement of denitrification, p. 93–102. *In* R. D. Hauck and R. W. Weaver (ed.), *Field Measurement of Dinitrogen Fixation and Denitrification.* Soil Science Society of America, Madison, Wis.

25. **Schmidt, E. L., and L. W. Belser.** 1994. Autotrophic nitrifying bacteria, p. 159–177. *In* R. W. Weaver, J. S. Angle, and P. J. Bottomley (ed.), *Methods of Soil Analysis,* part 2. *Microbiological and Biochemical Properties.* Soil Science Society of America, Madison, Wis.

26. **Silvester, W. B., R. Parsons, F. R. Minchin, and J. F. Witty.** 1989. Simple apparatus for growth of nodulated plants and for continuous nitrogenase assay under defined gas phase, p. 55–66. *In* J. G. Torrey and L. J. Winship (ed.), *Applications of Continuous and Steady-State Methods to Root Biology.* Kluwer Academic Publishers, Dordrecht, The Netherlands.

27. **Stanford, G., and S. J. Smith.** 1972. Nitrogen mineralization potentials of soils. *Soil Sci. Soc. Am. Proc.* **38**:99–102.

28. **Tiedje, J. M.** 1994. Denitrifiers, p. 245–267. *In* R. W. Weaver, J. S. Angle, and P. J. Bottomley (ed.), *Methods of Soil Analysis,* part 2. *Microbiological and Biochemical Properties.* Soil Science Society of America, Madison, Wis.

29. **Tiedje, J. M., S. Simkins, and P. M. Groffman.** 1989. Perspectives on measurement of denitrification in the field including recommended protocols for acetylene based methods, p. 217–240. *In* M. Clarholm and L. Bergstrom (ed.), *Ecology of Arable Land.* Kluwer Academic Publishers, Dordrecht, The Netherlands.

30. **Vose, P. B., and R. L. Victoria.** 1986. Re-examination of the limitations of nitrogen-15 isotope dilution technique for the field measurement of dinitrogen fixation, p. 23–58. *In* R. D. Hauck and R. W. Weaver (ed.), *Field Measurement of Dinitrogen Fixation and Denitrification.* Soil Science Society of America, Madison, Wis.

31. **Warembourg, F. R.** 1993. Nitrogen fixation in soil and plant systems, p. 127–156. *In* R. Knowles and T. H. Blackburn (ed.), *Nitrogen Isotope Techniques.* Academic Press, Inc., San Diego, Calif.

32. **Weaver, R. W.** 1986. Measurement of biological dinitrogen fixation in the field, p. 1–10. *In* R. D. Hauck and R. W. Weaver (ed.), *Field Measurement of Dinitrogen Fixation and Denitrification.* Soil Science Society of America, Madison, Wis.

33. **Weaver, R. W., J. S. Angle, and P. J. Bottomley (ed.).** 1994. *Methods of Soil Analysis,* part 2. *Microbiological and Biochemical Properties.* Soil Science Society of America, Madison, Wis.

34. **Weaver, R. W., and S. K. A. Danso.** 1994. Dinitrogen fixation, p. 1019–1045. *In* R. W. Weaver, J. S. Angle, and P. J. Bottomley (ed.), *Methods of Soil Analysis,* part 2. *Microbiological and Biochemical Properties.* Soil Science Society of America, Madison, Wis.

Quantifying the Metabolic Activity of Microbes in Soil

G. STOTZKY

50

The overall metabolic activity of microbes in soil can be determined with respirometric techniques that monitor either CO_2 evolution or O_2 consumption. These methods, especially when CO_2 evolution is measured, probably provide the best and most easily measured index of the gross metabolic activity of the mixed microbial populations in soil (1, 13, 15). The "master jar" system (Fig. 1) (13, 19) enables not only the continuous collection of respired CO_2 but also the removal of subsamples of soil during an extended incubation for various analyses (e.g., transformation of substrates, species diversity, enzyme activities, and survival of introduced microorganisms) without disturbing the remaining soil. Sampling without disturbance eliminates artifactual peaks in CO_2 evolution that result from the physical disturbance of the soil. The soils are incubated at controlled temperatures and maintained at the -33-kPa water tension by continuous aeration with water-saturated, CO_2-free air. The amount of CO_2 trapped in NaOH collectors is determined, after precipitation of the CO_2 with $BaCl_2$, by automatic potentiometric titration with HCl. The amount of CO_2 evolved from the master jars is expressed on the basis of a constant amount of soil, usually 100 g (oven-dry equivalent), which normalizes the respiration rate regardless of the amount of soil remaining in the master jars.

The potential gross metabolic activity, both aerobic and anaerobic, of the heterotrophic soil microbiota can be measured by the addition of a nonspecific substrate (e.g., glucose), and the potential activity of specific populations can be evaluated by the addition of specific substrates (e.g., celluloses, starches, lipids, proteins, and xenobiotics) whose mineralization is dependent on the ability of these populations to synthesize the appropriate enzymes. In particular, aldehydes, which are highly selective substrates, can be used (3, 8, 9). Ratios of the gross metabolic activity (with glucose or other nonspecific substrates) to that of specific metabolic activities (e.g., with aldehydes or other selective substrates) can be used to indicate whether alterations in the physicochemical or biological characteristics of the soil (e.g., additions of pollutants, inocula, clay minerals, humic substances, etc., and fluctuations in aeration or humidity) exert an effect on the metabolism of all components of the indigenous microbial population or only on certain segments of the microbiota. These ratios also sharpen comparisons between control and treated soils (e.g., amended or inoculated).

When aldehydes are used as specific substrates, the data from studies with aldehydes should be correlated with those from nitrification studies. Both nitrification (an autotrophic process) and mineralization of aldehydes (a heterotrophic process) are restricted to specific, albeit different, microbial species, and the two processes show similar kinetics in soil, especially when soils are stressed (e.g., with heavy metals or acid precipitation) or altered (e.g., amended with different clay minerals) (16, 17).

When the effects of the inoculation of genetically modified organisms (GMO) are being studied, the soils should also be amended with the specific substrate (e.g., toluene, xylene, or 2,4-dichlorophenoxyacetate) on which the products of the novel gene(s) in a GMO function, to determine whether the substrate provides an ecological advantage to the GMO, whether intermediates are produced from the substrate, and how any advantages or intermediates affect both nonspecific and specific metabolic activities as well as other microbe-mediated ecological processes and microbial populations (5–7, 12). When the GMO contains genes that confer resistance to the toxicity of an antimicrobial agent, the soil should be amended with the agent to determine whether such a stress confers an ecological advantage on the GMO and whether this advantage, in turn, influences the activity and population dynamics of the indigenous microbiota.

SOIL PREPARATION

Sieve soil (top ~5 cm) collected in the field through a broad (e.g., 1-cm)-mesh screen to remove stones and plant debris and to disrupt large soil aggregates. Mix the sieved soil thoroughly to provide as uniform and representative a sample as possible. The sieved soil can be used immediately after collection or can be stored. Although soil used immediately probably reflects better the microbiological conditions that exist in situ, there are disadvantages: for example, if the same soil is to be used in subsequent experiments, collection from adjacent sites and in different seasons can result in both biotic and abiotic variability. Moreover, if the desired soil is located some distance from the laboratory, considerable time will be consumed in collection. The collection, sieving, and storage of quantities of soil from the same site

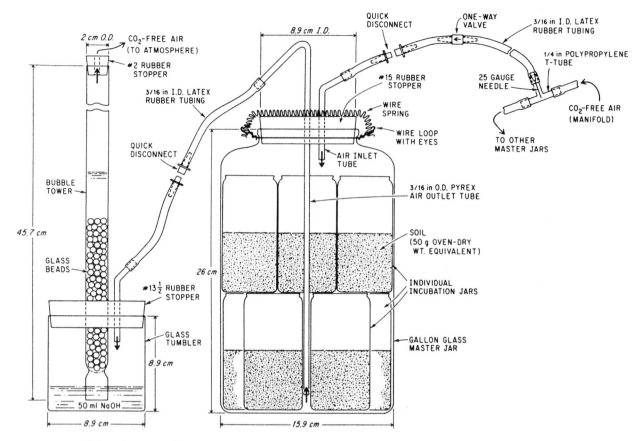

FIGURE 1 Incubation unit for measuring CO_2 evolution from soil. The master jar is used when subsamples of soil are to be removed during an incubation. When this is not required, soil can be placed directly into the master jar or into a smaller container. I.D., inside diameter; O.D., outside diameter.

sufficient for numerous experiments eliminate these disadvantages. Changes in the microbiota as the result of storage of the soil (20) can be rectified to some extent. For example, soils can be maintained in wooden flats (e.g., flats measuring 55 by 30 by 15 cm) in a greenhouse or a plant-growth chamber under a regimen of intermittent cropping and fluctuating temperatures and moisture (14). If a greenhouse or growth chamber is not available, soils can be stored at room temperature in large plastic or metal garbage cans lined with plastic garbage bags. Two weeks before the initiation of a study, pass the soil through a 2-mm-mesh sieve, and rejuvenate the soil by bringing it to its −33-kPa water tension and adding glucose (1% [wt/wt] in a mineral salts solution such as M9) and approximately 20 mg of fresh garden soil per g (oven-dry equivalent) of stored soil. Maintain the soil at room temperature, and mix it every few days (4, 19, 21).

When soil is to be amended with clay minerals, mix the soil after the initial sieving with the appropriate mined clay mineral. Use an electric-powered cement mixer for uniform and rapid mixing. The desired ratios of clay and soil can be achieved on a weight/weight or a volume/volume basis, although the latter (using buckets) is more conveniently conducted in the field, especially with large volumes of soil.

PROCEDURE

Before each experiment, mix the soil and desired additives in a plastic bag by kneading, and store the samples for 48

h at 4°C, with additional kneading at ca. 24 h, to enhance the uniform distribution of water, substrate, cells, and other additives. After the 48-h equilibration period, weigh 50 g (oven-dry equivalent) of soil into 100-cm^3 glass vials (baby food jars or commercially available vials). If cold-intolerant microorganisms are to be added, add the cells just before weighing the soils, and mix well. Keep the vials of soil cool (e.g., on ice or in a refrigerator) until all have been filled. Place the vials into a wide-mouth 1-gal (3.785-liter) jar (master jar) (Fig. 1) (pickle or mayonnaise jars are ideal and can be obtained inexpensively). Attach the master jar, via the air inlet tube, to the manifold of a respiration train that contains a scrubber system that removes oil, ambient CO_2, various nitrogen compounds, and other contaminants from the air source and then resaturates with water the air that continuously flushes the master jar (Fig. 2). Connect the air outlet tube of the master jar to the CO_2 collector (Fig. 1) (see below for details). At desired intervals, remove a soil vial from each master jar for microbiological, enzymatic, physical, chemical, and other analyses.

Immediately after placing the soil vials into the master jars, analyze the soil (most efficiently done with soil remaining after the vials have been filled) for the microbiological, enzymatic, physical, chemical, and other characteristics that will be analyzed during the experiments. These constitute the data for day 0.

Flush the soils in the master jars continuously with CO_2-free, water-saturated air to remove respired CO_2 and to

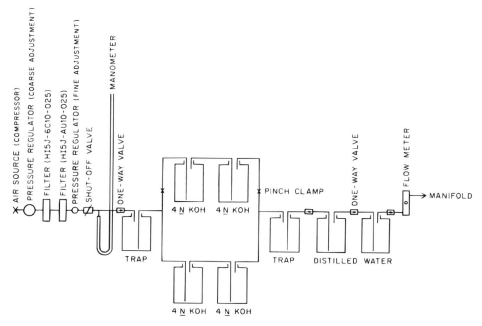

FIGURE 2 Schematic of the scrubber system used to remove CO_2 and contaminants from and to saturate with water the air that flushes the containers of soil (e.g., master jars).

maintain the soils at the −33-kPa water tension. Respiration is determined by trapping the evolved CO_2 in NaOH and periodically titrating the unneutralized NaOH with HCl, contained in a 10-ml self-filling burette, with an automatic titrator (e.g., Radiometer TTT80) connected to a pH meter (e.g., Radiometer PHM82 standard pH meter). (See below for details on reagents.)

Place into each master jar sufficient soil vials for the number of subsamples to be analyzed during the course of the study (it is advisable to place more soil vials than needed, in the event of any contingencies, such as dropping a vial or needing to continue the experiment longer than originally designed). Insert the rubber stopper (no. 15), with the air inlet and outlet tubes, and secure it with a wire spring, the ends of which are inserted into loops twisted in a wire circle that is fastened around the neck of the jar (Fig. 1). Attach the master jar, via the air inlet tube, to the manifold of the respiration train.

The air that flushes the master jars must be treated to remove inorganic and organic contaminants and ambient CO_2 and then saturated with water. The air scrubber system (Fig. 2) consists of (i) a pressure regulator with coarse adjustment; (ii) two filters (e.g., HI5J-6C10-025 and HI5J-AU10-025 [Finite Filter, Oxford, Mich.]) in series, to remove particulates, oil droplets, and other contaminants; (iii) a pressure regulator with fine adjustment (pressure range, 0 to 25 lb/in²); (iv) a shutoff valve; (v) a water-filled manometer constructed of Tygon tubing (e.g., with an inside diameter of 13 mm) that extends from the floor to the ceiling, is open to the atmosphere, and is vented to a sink or flask (this manometer serves as a pressure release valve in the event of a blockage in the airflow system; a water column of ~0.7 m is equivalent to 1 lb/in²); (vi) an empty 20-liter glass carboy that serves as a trap; (vii) two 20-liter glass carboys, each containing 5 liters of 4.0 N KOH (replace the KOH approximately every 2 weeks, as K_2CO_3 will eventually form in the bottom of the air inlet tube, disrupting the airflow, and the KOH will eventually be neutralized by the CO_2;

add a few drops of phenolphthalein to the KOH, and replace the KOH when the red color disappears; to prevent disruption of the airflow during replacement, two pairs of 20-liter carboys, each containing 5 liters of 4.0 N KOH, are placed in parallel, and the airflow is shunted between the pairs of carboys by the appropriate placement of pinch clamps); (viii) a second empty trap; (ix) two 20-liter glass carboys, each containing 6 liters of distilled water to rehumidify the air (replace the water periodically, as it may become contaminated with aerosolized KOH; add a few drops of phenolphthalein to the first carboy of water, and replace the water in both carboys when it turns red); and (x) a flowmeter.

The components of the scrubber system are connected with copper pipe to the manometer and then with glass or plastic, straight or T-shaped, tubing connectors or plastic one-way valves (outside diameter of 6.4 mm) and latex tubing (inside diameter of 6.4 mm; wall thickness of 1.6 mm). Each carboy has glass air inlet and outlet tubes attached through a rubber stopper in the mouth of the carboy. The air inlet tube extends 4 to 6 cm into the solution, and the outlet tube protrudes slightly below the rubber stopper. In the traps, the air inlet tube protrudes slightly below the stopper, and the outlet tube extends almost to the bottom of the carboy. This arrangement of inlet and outlet tubes in the traps enables the KOH and water to be pushed back into their carboys from the traps in the event that the airstream is disrupted and the one-way valves malfunction.

The air scrubber system is attached to a manifold that distributes the CO_2-free, water-saturated air to each master jar. The manifold consists of a series of glass or plastic T-shaped tubing connectors (outside diameter of 6.4 mm), the side arms of which are connected to adjacent T-shaped tubing connectors with latex tubing (inside diameter of 6.4 mm), and the perpendicular arm is attached to the air inlet tube of the master jar. A 25-gauge hypodermic needle, which serves to equalize the airflow to each master jar, is inserted into the perpendicular arm of the T-shaped tubing connector and is fixed in place by the latex tubing that covers the perpendicular arm at one end and extends to the female half of a plastic quick-disconnect connector at the other end. A one-way valve, to prevent backflow of air with a loss of CO_2 from the master jar in the event of a reduction in pressure in the airflow system, is inserted in this tubing. The male half of the quick-disconnect connector is inserted into the end of the tubing that is attached to the air inlet tube of the master jar.

The master jar is closed with a no. 15 rubber stopper that contains two glass tubes, each of which is attached to tubing containing the male half of plastic quick-disconnect connectors. One glass tube is the air inlet and terminates on the underside of the stopper; the other tube, which extends to the bottom of the master jar to prevent channeling of the airstream, is the air outlet. The manifold tubing (containing the female end of a quick-disconnect connector) is connected to the air inlet tubing of the master jar, and the outlet tubing of the master jar (containing the female end of a quick-disconnect connector) is connected to the air inlet tubing of the CO_2 collector.

The CO_2 collector is a 200-cm^3 glass tumbler (commercially available in restaurant supply stores) closed with a rubber stopper (no. $13\frac{1}{2}$) in which an adjustable glass column (chimney) (inside diameter of 17.5 mm and ca. 45-cm long), which extends into the NaOH in the tumbler, and a glass air inlet tube are inserted (Fig. 1). The base of the chimney is constricted (by simple glassblowing) to retain glass beads (6 mm), which disrupt air bubbles and increase the surface area of the NaOH, thereby maximizing the absorption of CO_2. The CO_2-free air is then vented to the atmosphere. Studies with ^{14}C-labeled substrates have shown that the absorption of CO_2 in the collectors is 100% efficient (13). The air inlet tube is attached to latex tubing (inside diameter of 6.4 mm) containing the male end of a plastic quick-disconnect connecter. The top of the chimney is closed with a no. 2 one-hole rubber stopper.

Before connecting the master jars to the manifold, set the air pressure gauge to about 3 lb/in^2 and open the air valve. Connect the master jars to the manifold, and tightly close with screw clamps any outlets on the manifold that are not connected to a master jar. Include several empty master jars (at least one for every eight experimental jars) interspersed among the jars containing soil to serve as blanks for nonrespired CO_2. Purge the master jars for approximately 30 min with CO_2-free air. Fill and empty a 50-ml stopcock-type automatic burette about four times with the NaOH solution (see below) to remove old NaOH and any precipitate. Fill the burette completely (50 ml) with NaOH, and dispense the NaOH into a glass tumbler. Insert a stopper with a chimney and air inlet tube into the tumbler, and connect the air inlet tubing to the outlet tubing of the master jar via the quick-disconnect connector. The continuous airstream will force the NaOH into the chimney. When all CO_2 collectors are connected, adjust the airflow through all master jars to comparable rates, based on the rate of bubbling, by raising or lowering the chimneys. Keep the outside of the chimneys in the vicinity of the stoppers lubricated with silicone grease, and wear a heavy glove when adjusting the chimneys (the glass chimneys become brittle after extensive use and can break during adjustment). Adjust the rate of airflow to 10 to 15 liters/h; monitor the airflow with the flowmeter.

The titration schedule will depend on when the substrate was added, the nature of the substrate, and the type of data desired. With glucose as the substrate, titrations are usually conducted daily after addition for the first 5 to 7 days, then

on alternate days for the subsequent week, and then on every third or fourth day for the remainder of the study. To determine the amount of CO_2 evolved, disconnect the tumbler from the master jar at the quick-disconnect connector, loosen the rubber stopper, and rinse the glass beads and the inside of the chimney, as well as the outside of the bottom of the chimney, with about 100 ml of distilled water into the tumbler. Add approximately 10 ml of the $BaCl_2$ solution (see below) to the tumbler to precipitate the adsorbed CO_2 as $BaCO_3$. (The amount of $BaCl_2$ added depends on the amount of CO_2 evolved. Add $BaCl_2$ until no more precipitate is formed.) Place the tumbler on a magnetic stir plate, add a magnetic stir bar, and insert the pH electrodes (glass and calomel or, preferably, a combination electrode) and the capillary HCl delivery tube attached to the magnetic valve of the automatic titrator. Fill the burette with HCl (see below), and adjust the capillary tube so that it dispenses the HCl onto the glass electrode. Turn on the magnetic stir plate, and start the titration. When the titration is complete, read on the burette the amount of HCl required to neutralize (pH 7.0) the NaOH in the tumbler, and record the reading.

Discard the neutralized NaOH, wash the tumbler with tap water, rinse it with distilled water, refill it with 50 ml of NaOH, insert a stopper with the chimney, and reattach the CO_2 collector to a master jar. Immediately after removing the CO_2 collector for titration and when not removing samples of soil, replace it with a fresh collector. Washing the beads and chimney and adding $BaCl_2$ to a collector to be titrated, as well as filling a new tumbler with fresh NaOH, are done while a CO_2 collector from another master jar is being titrated. Do not pour the neutralized NaOH down a sink drain, but pour it into a large bucket or carboy. Allow the $BaCO_3$ to accumulate on the bottom, decant the clear supernatant, and dispose of the concentrated $BaCO_3$ according to regulations for the disposal of hazardous wastes. When samples of soil are removed from a master jar after titration, purge the ambient air from the master jar for about 30 min before attaching a CO_2 collector.

Calculate the amount of carbon (C) respired by using the formula: mg of C = $(B - S)6N$, where B is the average amount (milliliters) of HCl required to neutralize the NaOH in the CO_2 collectors attached to empty master jars (blanks), S is the amount (milliliters) of HCl required to neutralize the NaOH in a CO_2 collector from an experimental master jar, 6 is the equivalent weight of C, and N is the normality of the HCl solution.

REAGENTS

NaOH (~1.5 N): Dissolve, with swirling, approximately 60 g of NaOH per liter of distilled water in a 20-liter borosilicate carboy. Fit into the mouth of the carboy a two-hole rubber stopper containing an air inlet tube and a solution outlet tube that is attached to a constant-volume 50-ml stopcock-type automatic burette. Connect the air inlet tube to a gas-drying tube containing Drierite and Ascarite to prevent ambient water vapor and CO_2 from entering the NaOH, and attach a rubber bulb to enable pressurization of the carboy. During the course of the study, adjust the normality of the NaOH solution to the amount of CO_2 produced.

NaOH (2.00 N standard): Commercially available.

HCl (~7.5 N): Dilute approximately 625 ml of concentrated HCl to 1 liter with distilled water (add the HCl to the water). As the normality of the NaOH in the CO_2 collector is adjusted to reflect decreasing or increasing (e.g., following pulsing with a carbon source) respiration rates, the normality of the HCl must also be changed to reflect the fivefold difference between the normality of the NaOH and that of the HCl. (The CO_2 collector contains 50 ml of NaOH, and the self-filling burette on the automatic titrator has a capacity of only 10 ml of HCl, to enhance precision.) Determine accurately the normality of the HCl by titrating it against the 2.00 N NaOH standard. Attach a Drierite-Ascarite column and rubber bulb to the air inlet tube as described for the NaOH carboy. Attach the outlet tube of the carboy containing HCl to the self-filling burette.

$BaCl_2$ (~3.5 M): Dissolve approximately 855 g of $BaCl_2 \cdot 2H_2O$ per liter of distilled water. Place the solution in a glass carboy equipped with a Drierite-Ascarite column and a rubber bulb on the air inlet tube, and connect the outlet tube to a 50- or 100-ml self-filling burette.

Distilled water: Fill a glass carboy with freshly distilled water. Attach a Drierite-Ascarite column and a rubber bulb to the air inlet tube and an eyedropper tube to the outlet tube. The flow of water to wash the chimney of the CO_2 collector is controlled with a pinch clamp on the outlet tubing just above the eyedropper tube.

KOH (~4.0 N): Dissolve approximately 224 g of KOH per liter of distilled water in a borosilicate flask. Transfer the KOH solution to the appropriate glass carboys in the scrubber system of the respiration train.

(*Note:* The dissolution of NaOH and KOH is an exothermic reaction and requires the use of a borosilicate container. The solutions must be allowed to cool before being transferred to other containers that are not heat tolerant.)

DISCUSSION

The method described above for quantifying the overall metabolic activity of soil, which has been used extensively in this laboratory to evaluate the effects on activity of various physicochemical and biological soil factors, including various environmental stresses (17, 18), is highly accurate and reproducible (e.g., variation between replicate master jars is usually less than 1%), relatively inexpensive (the apparatus is essentially handcrafted, except for the automatic titrator and the pH meter), simple (e.g., as it is an open system, fluctuations in pressure have a negligible effect), and rapid (the CO_2 evolved from 50 master jars can be measured in approximately 2.5 h).

Computer-controlled systems for measuring CO_2 evolution and O_2 consumption are commercially available. For example, the Micro-Oxymax (Columbus Instruments International Corporation, Columbus, Ohio) uses a single-beam, nondispersive infrared device to monitor CO_2 concentrations and an O_2 battery (fuel cell) to monitor O_2 concentrations. This apparatus can monitor up to 80 samples, using a microcomputer both as a controller and for the collection, storage, and presentation of data. It can also be modified to measure CH_4 and CO. However, it cannot apparently be used to distinguish $^{14}CO_2$ from $^{12}CO_2$, which may be necessary in some studies (e.g., on the degradation and priming action of an added substrate) (13). The approximate cost

of this system for monitoring 80 vessels for CO_2 and O_2 is $113,000. Another computerized system is the Respicond III (Nordgren Innovations, Umeå, Sweden), which can measure the respiration of 96 soil samples simultaneously for extended periods. This system is based on measuring decreases in the conductivity of a KOH solution as the CO_2 absorbed forms K_2CO_3 (10). The approximate cost of this system is $60,000.

One of the unrecognized problems associated with systems in which a hydroxide solution is present within the container of soil (e.g., as in the Biometer [2] or the Respicond III) is that the solution removes not only respired CO_2 but also water vapor, which eventually results in the water content of the soil being reduced below the -33-kPa tension that is optimum for microbial activity (13, 17). In the master jar system, this is not a problem, as the headspace above the soil is constantly flushed with a water-saturated airstream and the CO_2-trapping solution is external to the soil, thereby maintaining the soil at the -33-kPa water tension. Moreover, in some closed systems (e.g., the Biometer and Respicond III), the hydroxide solution and O_2 must be periodically replenished, and the systems do not enable the removal of subsamples of soil for other analyses without disturbance of the remaining soil.

Numerous other methods have been used to measure the amount of CO_2 evolved from soil. These methods are based on gravimetric, conductimetric, manometric, titrimetric, potentiometric, calorimetric, gas chromatographic, or optical (infrared) principles and have been reviewed (1, 13). Recently, a colorimetric method for measuring CO_2 evolved from soil has been suggested (11). Ultimately, the choice of methods is dictated by such practical factors as ease of use, cost, realistic sensitivity, duration of the study, sample size (e.g., some methods cannot deal adequately with relatively large samples of soil that may be necessary for statistical reasons), and the type of data required.

The overall metabolic activity of soil can also be estimated by measuring the consumption of O_2. However, this method will measure only the activity of aerobic and facultative organisms. Various methods for measuring the consumption of O_2 have been reviewed (1, 13). Computer-controlled systems (e.g., Micro-Oxymax) are also available.

REFERENCES

1. **Anderson, J. P. E.** 1982. Soil respiration, p. 831–871. *In* A. L. Page, R. H. Miller, and D. R. Keeney (ed.), *Methods of Soil Analysis*, part 2. American Society of Agronomy, Madison, Wis.
2. **Bartha, R., and D. Pramer.** 1965. Features of a flask for measuring the persistence and biological effects of pesticides. *Soil Sci.* **100:**68–70.
3. **Bewley, R. J. F., and G. Stotzky.** 1984. Degradation of vanillin in soil-clay mixtures treated with simulated acid rain. *Soil Sci.* **137:**415–417.
4. **Devanas, M. A., D. Rafaeli-Eshkol, and G. Stotzky.** 1986. Survival of plasmid-containing strains of *Escherichia coli* in soil: effect of plasmid size and nutrients on survival of hosts and maintenance of plasmids. *Curr. Microbiol.* **13:** 269–277.
5. **Doyle, J. D., K. A. Short, G. Stotzky, R. J. King, R. J. Seidler, and R. H. Olsen.** 1991. Ecologically significant effects of *Pseudomonas putida* PPO301(pRO103), genetically engineered to degrade 2,4-dichlorophenoxyacetate, on microbial populations and processes in soil. *Can. J. Microbiol.* **37:**682–691.
6. **Doyle, J. D., and G. Stotzky.** 1993. Methods for the detec-

tion of changes in the microbial ecology of soil caused by the introduction of microorganisms. *Microb. Releases* **2:** 63–72.

7. **Doyle, J. D., G. Stotzky, G. McClung, and C. W. Hendricks.** 1995. Effects of genetically engineered microorganisms on microbial populations and processes in natural habitats. *Adv. Appl. Microbiol.* **40:**237–287.

8. **Kunc, F., and G. Stotzky.** 1974. Effect of clay minerals on heterotrophic microbial activity in soil. *Soil Sci.* **118:** 186–195.

9. **Kunc, F., and G. Stotzky.** 1977. Acceleration of aldehyde decomposition in soil by montmorillonite. *Soil Sci.* **124:** 167–172.

10. **Nordgren, A.** 1988. Apparatus for the continuous, long-term monitoring of soil respiration rate in large numbers of samples. *Soil Biol. Biochem.* **20:**955–957.

11. **Rowell, M. J.** 1995. Colorimetric method for CO_2 measurement in soils. *Soil Biol. Biochem.* **27:**373–375.

12. **Short, K. A., J. D. Doyle, R. J. King, R. J. Seidler, G. Stotzky, and R. H. Olsen.** 1991. Effects of 2,4-dichlorophenol, a metabolite of a genetically engineered bacterium, and 2,4-dichlorophenoxyacetate on some microorganism-mediated ecological processes in soil. *Appl. Environ. Microbiol.* **57:**412–418.

13. **Stotzky, G.** 1965. Microbial respiration, p. 1550–1570. *In* C. A. Black et al. (ed.), *Methods of Soil Analysis.* American Society of Agronomy, Madison, Wis.

14. **Stotzky, G.** 1973. Techniques to study interactions between microorganisms and clay minerals *in vivo* and *in vitro*, p. 17–28. *In* T. Rosswall (ed.), *Modern Methods in the Study of Microbial Ecology. Bulletins from the Ecological Research Committee*, vol. 17.

15. **Stotzky, G.** 1974. Activity, ecology, and population dynamics of microorganisms in soil, p. 57–135. *In* A. I. Laskin and H. Lechevalier (ed.), *Microbial Ecology.* Chemical Rubber Co., Cleveland.

16. **Stotzky, G.** 1980. Surface interactions between clay minerals and microbes, viruses, and soluble organics, and the probable importance of these interactions to the ecology of microbes in soil, p. 231–249. *In* R. C. W. Berkeley, J. M. Lynch, J. Melling, P. R. Rutter, and B. Vincent (ed.), *Microbial Adhesion to Surfaces.* Ellis Horwood Limited, Chichester, England.

17. **Stotzky, G.** 1986. Influence of soil mineral colloids on metabolic processes, growth, adhesion, and ecology of microbes and viruses, p. 305–428. *In* P. M. Huang and M. Schnitzer (ed.), *Interactions of Soil Minerals with Natural Organics and Microbes.* Soil Science Society of America, Madison, Wis.

18. **Stotzky, G.** 1989. Gene transfer among bacteria in soil, p. 165–222. *In* S. B. Levy and R. V. Miller (ed.), *Gene Transfer in the Environment.* McGraw-Hill, New York.

19. **Stotzky, G., M. W. Broder, J. D. Doyle, and R. A. Jones.** 1993. Selected methods for the detection and assessment of ecological effects resulting from the release of genetically engineered microorganisms to the terrestrial environment. *Adv. Appl. Microbiol.* **38:**1–98.

20. **Stotzky, G., R. D. Goos, and M. I. Timonin.** 1962. Microbial changes occurring in soil as a result of storage. *Plant Soil* **16:**1–19.

21. **Tapp, H., L. Calamai, and G. Stotzky.** 1994. Adsorption and binding of the insecticidal proteins from *Bacillus thuringiensis* subsp. *kurstaki* and subsp. *tenebrionis* on clay minerals. *Soil Biol. Biochem.* **26:**663–679.

Assessment of Extracellular Enzymatic Activity in Soil

MATTHEW J. MORRA

51

INTRODUCTION TO SOIL ENZYMES

Extracellular enzymes are strictly defined as those enzymes that have passed through the cytoplasmic membrane of the cell originally responsible for synthesis (32). However, this definition becomes blurred within the context of a discussion on extracellular soil enzymes based on inability to precisely identify enzyme location within the soil matrix. A more realistic approach is to redefine the term in an operational manner as related to assay procedures used to monitor activity. Specifically, a soil enzyme is considered extracellular when it catalyzes substrate transformation in the absence of similar substrate metabolism by living cells. Implicit in this definition is the need to demonstrate that the catalyst is indeed a protein and not a mineral or nonproteinaceous organic compound.

It is precisely this complex, heterogeneous mixture of organic and inorganic components that makes the measurement of extracellular enzyme activities in soil difficult. However, these same soil components and heterogeneous microenvironments are also responsible for enzyme stabilization and preservation. An understanding of the chemical and physical characteristics of soil colloids is thus critical to development of a valid extracellular enzyme assay and interpretation of the derived data.

The soil solid phase consists of both inorganic and organic components which in many cases form heterogeneous mixtures defined as organo-mineral complexes. Soil inorganic components include both crystalline materials in the form of layer silicates and more poorly crystalline oxides, hydroxides, and oxyhydroxides (collectively termed hydrous oxides) (10). Those minerals in the clay-sized fraction (<2.0 μm in diameter) generally exert the greatest influence on enzymatic reactions because of their large surface areas and ability to act as cation and, in some cases, anion exchangers.

Natural organic materials in soils exist in two distinct forms: those that are recognizable as originating from a particular organism and those that have been chemically altered such that their origin is not easily determined (1, 38). Recognizable compounds include large polymeric materials (e.g., cellulose, protein, and lignin) as well as relatively simple molecules (e.g., sugars, organic acids, and amino acids). Large polymeric materials without a recognizable, highly organized structure are called humic materials. Humic materials are largely polar materials having a net negative charge and, like most layer silicates, a cation-exchange capacity. In soil, humic and nonhumic materials exist predominantly in association with mineral surfaces. These organo-mineral complexes bind both enzymes and substrates, thereby controlling the extent of catalysis.

The specific effect that soil colloids have on any one enzyme depends on the type of interaction. Mechanisms by which enzymes interact with soil constituents and the impact of this interaction on optimum reaction pH, susceptibility to proteolytic attack, heat stability, and kinetic parameters have been summarized (3, 6, 45) and continue to be the focus of research (37). A clear understanding of the impacts of these parameters is essential, since actual substrate transformation rates in soil are controlled by an enzyme's longevity and activity within its respective microenvironment.

Extracellular soil enzymes originate from bacteria, fungi, plants, and a variety of macroinvertebrates (18, 23). In many cases, the origin of the activity is unknown and may involve the participation of various forms of the same enzyme or isozymes (27, 28). A large number of extracellular enzyme activities have been identified in soils or soil extracts (9, 23, 43), and standardized methodologies for the measurement of many of these enzymes have been summarized (33, 40, 41).

Selection of an appropriate enzyme to assay depends on whether the objective centers on process-level investigations, soil characterization, or microbial activity indices. Process-level investigations include attempts to use enzyme activity measurements in (i) estimating soil fertility and fertilizer use efficiency, (ii) delineating temporal and spatial ecosystem differences, (iii) monitoring pollution, (iv) performing synthetic organic transformation studies, and (v) determining land use impacts (9, 34). The selection of appropriate enzyme assays in these cases is related to the degradation of specific substrates, many times without concern as to enzyme origin (e.g., fungi, earthworms, or plant root exudates). The source of the enzyme is likewise less important when one is characterizing soils to establish original location, as has been proposed for use in crime laboratories (47).

In contrast, if the objective centers on estimating and comparing microbial numbers and activities among different soils, enzyme origin and longevity in soil are of primary concern. The very nature of the desired data indicates a clear relationship of enzyme activity and living microbial cells. Dehydrogenase is inactive in extracellular form and thus frequently used as a measure of soil microbial activity (36). Extracellular enzyme activity as measured at any one time, however, reflects an enzyme pool which in many cases includes enzymes produced and released by the microorganism, stabilized by soil colloids, and accumulated in soil during an undefined time period. Temporal differences in microbial biomass or activity which alter enzyme synthesis could be buffered by the larger extracellular enzyme pool. Nonmicrobial sources of the same enzyme or isozymes included in this pool would cause additional difficulty in relating assay results to microbial numbers or activity indices.

The use of soil enzyme assays in microbial activity and biomass studies, therefore, must be approached with caution. In some cases assay results do correlate with microbial numbers or biomass estimates as well as microbial activities, but the results are mixed (7, 20, 29, 46). In a comprehensive evaluation, Frankenberger and Dick (14) determined the relationship of microbial respiration, biomass, and viable plate counts in 10 soils with the activities of 11 different soil enzymes. Alkaline phosphatase, amidase, α-glucosidase, and dehydrogenase correlated at the 1% level, and phosphodiesterase, arylsulfatase, invertase, α-galactosidase, and catalase correlated at the 5% level, with CO_2 evolution from glucose-amended soils. No correlation with CO_2 evolution from glucose-amended soils was observed with acid phosphatase and urease or with any of the enzymes in the absence of glucose. Phosphodiesterase and α-galactosidase levels were proportional to microbial numbers as determined by using selective media, and alkaline phosphatase, amidase, and catalase levels were proportional to microbial biomass as determined by using chloroform fumigation techniques.

SOIL ENZYME ASSAYS

Reaction rates as enhanced by soil enzymes ultimately control soil processes. Soil enzyme assays are thus valuable for evaluation of these processes, but the selection of an appropriate enzyme or battery of enzymes to measure is dependent on the investigator's specific objectives. Once this selection has been made, an appropriate assay method is devised, again with careful consideration of project objectives.

The investigator must first decide whether in situ substrate transformation rates are required or if the focus is more qualitative in nature. In situ results can be obtained only with intact soil, whereas soil extracts can yield fundamental information on the presence or absence of a specific extracellular enzyme, its catalytic properties, and fractionation among soil components. In any extraction procedure, cell damage and intracellular enzyme release must be avoided. Artifacts could conceivably result from the interaction of otherwise intracellular enzymes with soil constituents, potentially producing what appear to be catalytically active extracellular enzyme-soil complexes. Tabatabai and Fu (43) recently summarized extraction and purification procedures for soil enzymes including oxidoreductases, hydrolases, and a lyase. Although the procedures vary, they most commonly include enzyme extraction with a phosphate buffer or another solution which solubilizes humic materials. A compromise must be reached in selecting an extractant which effectively removes a large proportion of the soil activity without denaturing or inhibiting the enzyme. As a result, only <20% of the activity is extracted, possibly in a selective manner from various pools within the soil matrix (6). The extracts therefore include enzymatic activity, but the extraction is incomplete and the enzyme itself is by no means in a purified state.

Purification procedures have been used to increase specific activity of the enzyme within the extract (35, 43). Published procedures include one or more of the following: $(NH_4)_2SO_4$ precipitation, dialysis, ultrafiltration, ion-exchange chromatography, and size exclusion chromatography. Although increases in specific activity have been obtained, no soil enzyme has yet been purified to the extent possible for enzymes derived from other sources. This lack of purity must be considered when one is comparing the characteristics (e.g., molecular weight, electrophoretic mobility, and kinetic constants) of soil extract activities with purified enzymes.

Soil extracts cannot be used if the objective is to determine rates of substrate alteration approximating or correlating with in situ rates. Instead, entire soils are collected and used as the source of enzyme. Soil collection for enzymatic activity analysis presents the same problems as encountered when one is sampling for microorganisms (see chapter 41). A "representative" sample must be obtained from an environment that exhibits heterogeneity on a much smaller scale than can be physically sampled. Successfully obtaining this representative sample does not eliminate the fact that the data represent an average value and not true enzymatic activity as it occurs at the microsite level.

Following field collection, it is often desirable to store the soil sample for variable time periods prior to the assay. Typically, soil samples are air dried, passed through a 2-mm sieve, and stored at room temperature. The impact of drying the soil on extracellular enzyme activities varies with the specific enzyme. Those enzymes which exist primarily as stabilized complexes with soil colloids are not expected to be greatly affected by air drying, whereas enzyme activities associated with living cells will obviously be affected. Urease activity in soil is generally attributed to an extracellular enzyme pool and thus is not usually altered upon air drying of the soil (4). In contrast, dehydrogenase is inactivated in soil and as a result must be determined without soil drying (45). There is conflicting evidence concerning these generalizations, and soil storage for short time periods at field moisture concentrations and 4°C is recommended. If air drying is necessary, literature concerning the specific enzyme of interest must be consulted to determine potential effects on the activity to be measured.

Once soil samples have been collected and stored, critical decisions must be made with respect to assay parameters as they relate to in situ or maximal enzyme activities. Burns (6) has summarized the advantages and disadvantages of each approach, comparing enzyme activities found under conditions provided in conventional enzyme assays devised for purified proteins with enzyme activities occurring in soil under field conditions (Table 1). Adoption of standardized assay procedures has yet to be realized, although methodological problems have been addressed for nearly 30 years (25).

An attempt will be made here to describe the most commonly practiced methods for measuring soil enzyme activities. These methods most closely resemble the approach listed in the left column of Table 1 and are not necessarily representative of field conditions. Assays are most frequently conducted by amending soils with enough water to form a slurry, thereby facilitating homogeneous substrate distribution. The sample may or may not be shaken during incubation, but consistency is necessary because increased rates of substrate transformation often occur with shaking (42).

TABLE 1 Assay conditions and impacts on soil enzyme activities[a]

Conventional enzyme assay (maximum potential activity, highly reproducible)	Field conditions (submaximal activity, extremely variable)
Excess substrate	Usually limiting substrate
Homogeneous substrate	Heterogeneous substrate
Usually soluble substrate	Often insoluble substrate
Often artificial substrate	Natural substrate
Buffered at optimum pH	Soil pH
pH poised	pH may vary
Slurry conditions	Moisture level variable
Often shaken	Stationary
Constant temperature	Temperature variable
Flora and fauna absent	Flora and fauna present

[a] Reprinted from reference 6 with permission of the publisher.

Incubation in a controlled-temperature environment is required to ensure a constant rate of substrate transformation. Many published procedures specify 37°C, a practice questioned by Malcolm (25), who recommends an incubation temperature of either 30°C or field soil temperature. Incubation temperature, as with most assay parameters, can be selected to simulate field conditions or maximize substrate transformation in a highly reproducible laboratory assay.

Likewise, opinions differ with respect to the use of buffers in extracellular enzyme assays (5, 25, 45). Many standardized assays include the use of specific buffers (41), whereas substrate transformation in the absence of buffer is thought to better simulate potential activity of the enzyme under field conditions (5). Although this is principally a choice made on the basis of the objectives of the investigation, buffers should be used when substrate amendment or product formation alters slurry pH. In many cases, increased activity is achieved by determining the optimum reaction pH and buffering the soil slurry at this pH. This is the case for acid and alkaline phosphatases, in which cases buffer pHs of 6.5 and 11, respectively, were identified and recommended for use in standardized assays (Fig. 1). Phosphatase

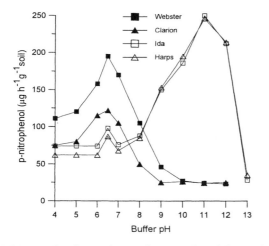

FIGURE 1 Catalytic release of p-nitrophenol by acid and alkaline phosphatases in soils as altered by buffer pH. Reprinted from reference 11 with kind permission from Elsevier Science Ltd., The Boulevard, Langford Lane, Kidlington 0X5 1GB, United Kingdom.

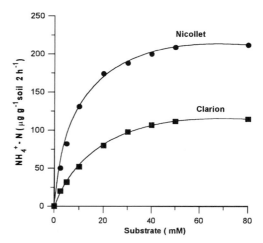

FIGURE 2 Ammonium release catalyzed by amidase in two soils incubated for 2 h with formamide as the substrate. Reprinted from reference 16 with permission of the publisher.

activity, as measured by product formation, decreases dramatically above and below these optimum pH values.

A soil enzyme assay usually includes an artificial, soluble substrate at a concentration sufficient to maintain zero-order kinetics, thus achieving a reaction rate proportional to enzyme concentration. One of the first steps in devising a valid soil enzyme assay is often identifying the amount of substrate necessary to obtain zero-order kinetics in soils having a range of enzyme concentrations. Figure 2 shows NH_4^+-N release catalyzed by amidase in two soils incubated for 2 h with formamide as the substrate (16). Burns (5) recommends a substrate concentration in excess of 5 times the K_m for the respective enzyme, which in this case was identified on average for eight soils as 12.3 mM formamide (17). An acceptable formamide concentration for measuring amidase activity in soil is thus approximately 63 mM, a concentration that is within the zero-order region as illustrated in Fig. 2. Verification that the assay accurately reflects enzyme activity and that catalysis rates are not affected by substrate concentration or inhibited by product formation is often performed by confirming a linear increase in enzyme activity with the amount of soil (8, 16, 24).

Assays include soil incubation for time periods ranging from 1 to 24 h or even longer before measurement of substrate disappearance or, more typically, product formation. Although assay sensitivity will dictate incubation time, shorter incubation periods (<5 h) are preferable for two reasons. Since zero-order kinetics with respect to the substrate are required, longer times increase the likelihood of developing a substrate limitation, thus causing a shift from zero- to first-order dependence on substrate concentration. The opposite effect, increased product formation, is also more likely at longer incubation times as a result of microbial proliferation and additional enzyme production. Both problems are detected by measuring the reaction rate at different incubation times and determining linearity of the relationship. Deviation toward the x axis indicates a possible substrate limitation or product inhibition of the reaction. Deviation toward the y axis suggests an increased reaction rate, possibly as a result of enzyme formation during the assay.

In most cases, microbial proliferation and enzyme synthesis during the incubation must be avoided, hence the emphasis on short incubation times. This is not always possible, and many assays require the use of inhibitors. An ideal inhibitor stops microbial proliferation and product or substrate assimilation by those microbes but does not lyse cells, destroy membrane integrity, or alter specific activity of the extracellular enzyme. This is a difficult if not impossible goal, but certain inhibitors offer acceptable results for specific enzymes. Toluene is the most often used inhibitor, producing reasonable microbial suppression when present in amounts 3 to 10% of the soil slurry liquid volume (2, 36, 41). A comparison of the effectiveness of toluene, ethanol, Triton X-100, and dimethyl sulfoxide in assays of soil amidase, acid phosphatase, alkaline phosphatase, catalase, α-glucosidase, α-galactosidase, invertase, dehydrogenase, arylsulfatase, and urease showed that no one compound provided satisfactory results for all enzymes (15). Toluene proved effective in most enzyme assays except those for catalase and dehydrogenase. However, increased activities of arylsulfatase and urease resulted, most likely as a result of membrane disruption and the release of intracellular enzymes. Ethanol functioned as a microbial inhibitor without significant inhibition of soil phosphatases, urease, catalase, and amidase. Triton X-100 demonstrated potential applicability in assays for urease, invertase, amidase, and α-galactosidase. Dimethyl sulfoxide amendment resulted in decreased activity for many of the enzymes without providing inhibition of microbial metabolism and proliferation.

All preceding considerations regarding the performance of a soil enzyme assay hinge on the ability to accurately and precisely measure the rate of the reaction as reflected in substrate disappearance or product formation. Most assays rely on the measurement of product because of analytical advantages with respect to sensitivity. Slight decreases in substrate concentration are difficult to detect, whereas lower background concentrations facilitate product measurement. However, in the case of soil assays, this advantage may be nullified by existing background concentrations of the product in soil or matrix interferences. For example, aminohydrolase activities are often quantified by measuring NH_4^+-N formation in soil amended with an organic substrate (41). Any preexisting NH_4^+-N in the respective soils will likewise be included in measurements performed at assay termination. Matrix interferences often result from a lack of specificity because colorimetric techniques include the absorbance of unknown organic compounds simultaneously extracted with the component of interest. A number of assays, including those for phosphatases, arylsulfatase, and β-glucosidase, rely on the colorimetric measurement of p-nitrophenol released as a result of substrate hydrolysis (41). Alkaline solutions of p-nitrophenol develop a yellow color with an A_{400} proportional to the phenol concentration. Unfortunately, soil extract A_{400} values also include a contribution from unknown soluble organics.

A control containing the same components as the treatment is thus essential to accurately quantify extracellular enzyme activity. In the preceding examples, background concentrations of NH_4^+-N or absorbances present in the control are subtracted from total values obtained from the respective treatments. Two approaches yield acceptable background estimates. During sample incubation of the first method, controls contain the same reagents as treatments except for substrate. Substrate is added to the control only after enzyme inhibition and reaction termination. Alternatively, some enzyme assay procedures, such as that for rhodanese, recommend pretreatment of control soils to achieve soil sterilization and enzyme denaturation (41). A sterilized control is essential in those instances in which inorganic or nonproteinaceous organics catalyze the same reaction as the enzyme of interest (19, 21, 49, 51). Possible sterilization methods include autoclaving, radiation treatment, and chemical amendment, but it is important to remember that all methods may potentially alter the soil and reduce legitimacy of the control (30, 35, 50).

DATA ANALYSIS AND INTERPRETATION

Investigators commonly interpret soil enzyme catalysis data by using techniques developed for purified enzyme systems (26, 41), although such an approach suffers from limitations in a highly heterogeneous soil environment. The basis for the approach relies on Michaelis-Menten theory, in which rate constants (k_1, k_2, k_3) describe single substrate (S) interaction with an enzyme (E), formation of an intermediate complex (ES), and eventual product formation (P).

$$S + E \underset{k_2}{\overset{k_1}{\rightleftharpoons}} ES \overset{k_3}{\longrightarrow} E + P$$

This theory is described by the relationship of reaction velocity with substrate concentration as shown in Fig. 2 by means of the Michaelis-Menten equation:

$$v = \frac{V_{max}[S]}{[S] + K_m}$$

where v is reaction velocity at a specific substrate concentration, V_{max} is the maximum velocity when the enzyme is substrate saturated, and K_m is the Michaelis constant.

To use the Michaelis-Menten equation, it is necessary to determine the values for V_{max} and K_m, a task that is difficult to do graphically using the model presented in Fig. 2. Instead, kineticists rearrange the classical Michaelis-Menten equation into linear forms as shown in Fig. 3 for inorganic pyrophosphatase in soil. Computer programs reduce the need for conversion to a linear form, but transformations are still of value for presentation purposes.

Investigators have reported K_m and V_{max} values for a number of enzymes and substrates in a variety of soils (8, 11–13, 17, 42). It is important to remember that the respective K_m and V_{max} values are only apparent kinetic constants. They are altered by the microenvironment of the enzyme as well as by the specific incubation parameters. Tabatabai and Bremner (42) showed that K_m values for arylsulfatase varied from 1.37×10^{-3} to 5.69×10^{-3} M and that those for phosphatase varied from 1.26×10^{-3} to 4.58×10^{-3} M in nine soils. Shaking the assay mixture during incubation decreased K_m and usually increased V_{max} for these two enzymes as well as urease (39).

Assay-specific changes in apparent K_m and V_{max} values inhibit the ability to compare data obtained by using differing procedures. When obtained under identical conditions, however, K_m is an index of optimal substrate concentration and V_{max} is the relative amount of enzyme. Lower K_m values imply a more efficient enzyme capable of functioning at lower substrate concentrations, whereas V_{max} is directly proportional to enzyme concentration.

Nannipieri et al. (27, 28) compared the kinetic parameters of urease, phosphatase, and protease activities in soil extracts. They were able to demonstrate, by using K_m values and Eadie-Scatchard plots, deviations from Michaelis-Menten kinetics and the possible presence of isozymes within the samples (Fig. 4). Eadie-Scatchard plots provide a graphical means to separate the activities of two or more enzymes (or multiple forms of the same enzyme) by using the number of resulting straight lines. The presence in soil of enzymes

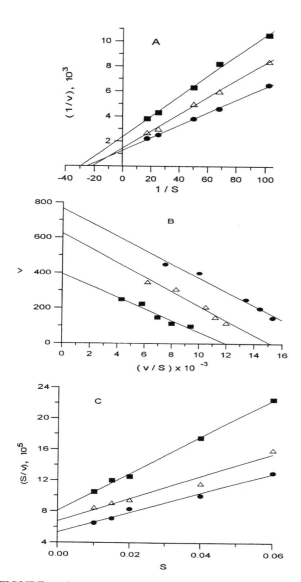

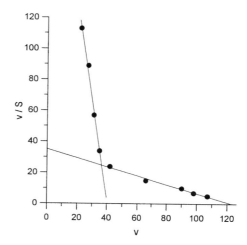

FIGURE 4 Eadie-Scatchard plot of phosphatases in soil extracts. The rate of the reaction (v) is expressed as nanomoles of p-nitrophenol released per hour per milliliter, and substrate concentration (S) is expressed as millimolar. Reprinted from reference 27 with kind permission from Elsevier Science Ltd., The Boulevard, Langford Lane, Kidlington 0X5 1GB, United Kingdom.

factor, R is the gas constant, and T is the temperature on the Kelvin scale. Plotting the natural logarithm of the initial reaction rate versus $1/T$ yields a straight line with a slope of $-E_a/R$. Investigators have also applied Eyring's transition-state theory to soil enzymes, thereby allowing them to calculate the activation parameters of free energy, entropy, and enthalpy (22, 31, 48). Although data collection and parameter calculation are easily performed by using rate and temperature relationships, data significance and interpretation are hindered by the complexity of the soil system.

CONCLUSIONS

A wide variety of detailed soil enzyme assay procedures are available, many of which are easily performed with basic pieces of laboratory equipment. These standardized assays are used to obtain fundamental information about soil enzymes and to predict in situ enzymatic catalysis rates. However, it remains difficult to extrapolate laboratory data to the prediction of actual biogeochemical cycling reactions, because subtle changes in the procedures alter absolute catalysis rates. Careful attention to methodology is necessary to ensure that assay procedures are appropriate for project objectives.

REFERENCES

1. **Aiken, G. R., D. M. McKnight, R. L. Wershaw, and P. MacCarthy (ed.).** 1982. *Humic Substances in Soil, Sediment, and Water.* John Wiley & Sons, New York.
2. **Boschker, H. T. S., S. A. Bertilsson, E. M. J. Dekkers, and T. E. Cappenberg.** 1995. An inhibitor-based method to measure initial decomposition of naturally occurring polysaccharides in sediments. *Appl. Environ. Microbiol.* **61:** 2186–2192.
3. **Boyd, S. A., and M. M. Mortland.** 1990. Enzyme interactions with clays and clay-organic matter complexes, p. 1–28. *In* J.-M. Bollag and G. Stotzky (ed.), *Soil Biochemistry,* vol. 6. Marcel Dekker, New York.
4. **Bremner, J. M., and R. L. Mulvaney.** 1978. Urease activity in soils, p. 149–196. *In* R. G. Burns (ed.), *Soil Enzymes.* Academic Press, London.

FIGURE 3 Linear transformations of the Michaelis-Menten equation for pyrophosphatase activity in soils. Velocity (v) is expressed as micrograms of P_i released per gram of soil per 5 h, and S is expressed as molar concentration. (A) Lineweaver-Burk plot; (B) Eadie-Hofstee plot; (C) Hanes-Woolf plot. ■, Okoboji soil; △, Clarion soil; ●, Webster soil. Reprinted from reference 8 with kind permission from Elsevier Science Ltd., The Boulevard, Langford Lane, Kidlington 0X5 1GB, United Kingdom.

with differing K_m values implies that substrate concentrations will control which enzyme is active and thus most important in biogeochemical cycling of the element in question.

Other investigators have gone beyond the calculation of the kinetic parameters K_m and V_{max} and have reported thermodynamic interpretations of their data in the form of activation energies (8, 17, 22, 44, 48). Thermodynamic relationships are developed by using the linearized version of the Arrhenius equation:

$$\ln k = \frac{-E_a}{RT} + \ln A$$

in which k is the rate constant for the reaction, E_a is the activation energy, A is the integration constant or frequency

5. **Burns, R. G.** 1978. Enzyme activity in soil: some theoretical and practical considerations, p. 295–340. *In* R. G. Burns (ed.), *Soil Enzymes.* Academic Press, London.

6. **Burns, R. G.** 1986. Interaction of enzymes with soil mineral and organic colloids, p. 429–451. *In* P. M. Huang and M. Schnitzer (ed.), *Interactions of Soil Minerals with Natural Organics and Microbes.* Soil Science Society of America, Madison, Wis.

7. **Casida, L. E., Jr.** 1977. Microbial metabolic activity in soil as measured by dehydrogenase determinations. *Appl. Environ. Microbiol.* **34:**630–636.

8. **Dick, W. A., and M. A. Tabatabai.** 1978. Inorganic pyrophosphatase activity in soils. *Soil Biol. Biochem.* **10:**59–65.

9. **Dick, W. A., and M. A. Tabatabai.** 1992. Significance and potential uses of soil enzymes, p. 95–127. *In* F. B. Metting, Jr. (ed.), *Soil Microbial Ecology: Applications in Agricultural and Environmental Management.* Marcel Dekker, New York.

10. **Dixon, J. B., and S. B. Weed (ed.).** 1989. *Minerals in Soil Environments,* 2nd ed. Soil Science Society of America, Madison, Wis.

11. **Eivazi, F., and M. A. Tabatabai.** 1977. Phosphatases in soils. *Soil Biol. Biochem.* **9:**167–172.

12. **Eivazi, F., and M. A. Tabatabai.** 1988. Glucosidases and galactosidases in soils. *Soil Biol. Biochem.* **20:**601–606.

13. **Frankenberger, W. T., Jr.** 1983. Kinetic properties of L-histidine ammonia-lyase activity in soils. *Soil Sci. Soc. Am. J.* **47:**71–74.

14. **Frankenberger, W. T., Jr., and W. A. Dick.** 1983. Relationships between enzyme activities and microbial growth and activity indices in soil. *Soil Sci. Soc. Am. J.* **47:** 945–951.

15. **Frankenberger, W. T., Jr., and J. B. Johanson.** 1986. Use of plasmolytic agents and antiseptics in soil enzyme assays. *Soil Biol. Biochem.* **18:**209–213.

16. **Frankenberger, W. T., Jr., and M. A. Tabatabai.** 1980. Amidase activity in soils. I. Method of assay. *Soil Sci. Soc. Am. J.* **44:**282–287.

17. **Frankenberger, W. T., Jr., and M. A. Tabatabai.** 1980. Amidase activity in soils. II. Kinetic parameters. *Soil Sci. Soc. Am. J.* **44:**532–536.

18. **Hartenstein, R.** 1982. Soil macroinvertebrates, aldehyde oxidase, catalase, cellulase and peroxidase. *Soil Biol. Biochem.* **14:**387–391.

19. **Hayes, M. H. B., and U. Mingelgrin.** 1991. Interactions between small organic chemicals and soil colloidal constituents, p. 323–407. *In* G. H. Bolt, M. F. DeBoodt, M. H. B. Hayes, and E. B. A. De Strooper (ed.), *Interactions at the Soil Colloid-Soil Solution Interface.* Kluwer Academic Publishers, Dordrecht, The Netherlands.

20. **Hersman, L. E., and K. L. Temple.** 1979. Comparison of ATP, phosphatase, pectinolyase, and respiration as indicators of microbial activity in reclaimed coal strip mine spoils. *Soil Sci.* **127:**70–73.

21. **Huang, P. M.** 1990. Role of soil minerals in transformations of natural organics and xenobiotics, p. 29–115. *In* J.-M. Bollag and G. Stotzky (ed.), *Soil Biochemistry,* vol. 6. Marcel Dekker, New York.

22. **Juma, N. G., and M. A. Tabatabai.** 1988. Comparison of kinetic and thermodynamic parameters of phosphomonoesterases of soils and of corn and soybean roots. *Soil Biol. Biochem.* **20:**533–539.

23. **Ladd, J. N.** 1978. Origin and range of enzymes in soil, p. 51–96. *In* R. G. Burns (ed.), *Soil Enzymes.* Academic Press, London.

24. **Ladd, J. N., and J. H. A. Butler.** 1972. Short-term assays of soil proteolytic enzyme activities using proteins and dipeptide derivatives as substrates. *Soil Biol. Biochem.* **4:** 19–30.

25. **Malcolm, R. E.** 1983. Assessment of phosphatase activity in soils. *Soil Biol. Biochem.* **15:**403–408.

26. **McLaren, A. D.** 1978. Kinetics and consecutive reactions of soil enzymes, p. 97–116. *In* R. G. Burns (ed.), *Soil Enzymes.* Academic Press, London.

27. **Nannipieri, P., B. Ceccanti, and D. Bianchi.** 1988. Characterization of humus-phosphatase complexes extracted from soil. *Soil Biol. Biochem.* **20:**683–691.

28. **Nannipieri, P., B. Ceccanti, S. Cervelli, and C. Conti.** 1982. Hydrolases extracted from soil: kinetic parameters of several enzymes catalysing the same reaction. *Soil Biol. Biochem.* **14:**429–432.

29. **Nannipieri, P., R. L. Johnson, and E. A. Paul.** 1978. Criteria for measurement of microbial growth and activity in soil. *Soil Biol. Biochem.* **10:**223–229.

30. **Negre, M., M. Gennari, C. Crecchio, and P. Ruggiero.** 1995. Effect of ethylene oxide sterilization on soil organic matter, spectroscopic analysis, and adsorption of acifluorfen. *Soil Sci.* **159:**199–206.

31. **Perucci, P., and L. Scarponi.** 1984. Arylsulphatase activity in soils amended with crop residues: kinetic and thermodynamic parameters. *Soil Biol. Biochem.* **16:**605–608.

32. **Priest, F. G.** 1984. *Extracellular Enzymes.* American Society for Microbiology, Washington, D.C.

33. **Roberge, M. R.** 1978. Methodology of soil enzyme measurement and extraction, p. 341–370. *In* R. G. Burns (ed.), *Soil Enzymes.* Academic Press, London.

34. **Schaffer, A.** 1993. Pesticide effects on enzyme activities in the soil ecosystem, p. 273–340. *In* J.-M. Bollag and G. Stotzky (ed.), *Soil Biochemistry,* vol. 8. Marcel Dekker, New York.

35. **Skujins, J.** 1976. Extracellular enzymes in soil. *Crit. Rev. Microbiol.* **4:**383–421.

36. **Skujins, J.** 1978. History of abiontic soil enzyme research, p. 1–49. *In* R. G. Burns (ed.), *Soil Enzymes.* Academic Press, London.

37. **Staunton, S., and H. Quiuampoix.** 1994. Adsorption and conformation of bovine serum albumin on montmorillonite: modification of the balance between hydrophobic and electrostatic interactions by protein methylation and pH variation. *J. Colloid Interface Sci.* **166:**89–94.

38. **Stevenson, F. J.** 1994. *Humus Chemistry,* 2nd ed. John Wiley & Sons, New York.

39. **Tabatabai, M. A.** 1973. Michaelis constants of urease in soils and soil fractions. *Soil Sci. Soc. Am. Proc.* **37:**707–710.

40. **Tabatabai, M. A.** 1986. Soil enzymes, p. 903–947. *In* A. L. Page, R. H. Miller, and D. R. Keeney (ed.), *Methods of Soil Analysis,* part 2. *Chemical and Microbiological Properties,* 2nd ed. American Society of Agronomy and Soil Science Society of America, Madison, Wis.

41. **Tabatabai, M. A.** 1994. Soil enzymes, p. 775–833. *In* R. W. Weaver, S. Angle, P. Bottomley, D. Bezdicek, S. Smith, A. Tabatabai, and A. Wollum (ed.), *Methods of Soil Analysis,* part 2. *Microbiological and Biochemical Properties.* Soil Science Society of America, Madison, Wis.

42. **Tabatabai, M. A., and J. M. Bremner.** 1971. Michaelis constants of soil enzymes. *Soil Biol. Biochem.* **3:**317–323.

43. **Tabatabai, M. A., and M. Fu.** 1992. Extraction of enzymes from soils, p. 197–227. *In* G. Stotzky and J.-M. Bollag (ed.), *Soil Biochemistry,* vol. 7. Marcel Dekker, New York.

44. **Tabatabai, M. A., and B. B. Singh.** 1976. Rhodanese activity of soils. *Soil Sci. Soc. Am. J.* **40:**381–385.

45. **Tate, R. L., III.** 1992. *Soil Organic Matter: Biological and Ecological Effects.* Krieger Publishing, Malabar, Fla.

46. **Tate, R. L., III, and R. E. Terry.** 1980. Variation in microbial activity in histosols and its relationship to soil moisture. *Appl. Environ. Microbiol.* **40:**313–317.

47. **Thornton, J. I., and A. D. McLaren.** 1975. Enzymic characterization of soil evidence. *J. Forensic Sci.* **20:**674–692.

48. **Tomar, J. S., and A. F. MacKenzie.** 1984. Effects of catechol and p-benzoquinone on the hydrolysis of urea and energy barriers of urease activity in soils. *Can. J. Soil Sci.* **64:**51–60.

49. **Voudrias, E. A., and M. Reinhard.** 1986. Abiotic organic reactions at mineral surfaces, p. 462–486. *In* J. A. Davis and K. F. Hayes (ed.), *Geochemical Processes at Mineral Surfaces.* American Chemical Society, Washington, D.C.

50. **Wolf, D. C., T. H. Dao, H. D. Scott, and T. L. Lavy.** 1989. Influence of sterilization methods on selected soil microbiological, physical, and chemical properties. *J. Environ. Qual.* **18:**39–44.

51. **Wolfe, N. L., U. Mingelgrin, and G. C. Miller.** 1990. Abiotic transformations in water, sediments, and soil, p. 103–168. *In* H. H. Cheng (ed.), *Pesticides in the Soil Environment: Processes, Impacts, and Modeling.* Soil Science Society of America, Madison, Wis.

Antibiotic, Chromogenic, and Luminescent Markers for Bacteria

JOE J. SHAW, S.-J. WU, N. P. SINGH, W. MAHAFFEE,
F. DANE, AND A. E. BROWN

52

Bacteria are ubiquitous in nature and unparalleled in their collective abilities to change environments, create or degrade exotic compounds, cause or prevent diseases, influence mineral and carbon cycles, colonize harsh environments, and perform myriad other tasks (48). The trend in recent years, since the first patenting of a genetically engineered microbe (GEM), has been to propose genetically engineered bacteria for ever more far-ranging and complex tasks. To achieve these ends, it is increasingly important to be able to release bacteria into microcosms or into the environment and be able to monitor their activities, or even recover them in order to assess efficacy and safety issues. A variety of methods have been employed, including the construction and use of bacteria that have distinctive and easily discerned phenotypes. These marked bacteria are generally similar to wild-type bacteria except for one or two characteristics that allow them to be easily identified or selectively cultured. Use of suitable markers permits the released strain to be differentiated from most or all other bacteria in a sample, especially when combined with other standard bacteriological methods. This chapter discusses antibiotic, chromogenic, and bioluminescent markers that have been or could be used to tag bacteria released into nonlaboratory settings. Comprehensive reviews of related topics have been prepared by others (4, 21, 38, 42, 43, 46, 59).

ANTIBIOTIC MARKERS

In 1929, Fleming (27) reported that staphylococci were lysed by contaminating *Penicillium* species, and the era of antibiotics was born. There are many thousands of antibiotics, and they are typically classified by either their chemical structures (50) or their modes of action (58). When considered by mode of action, antibiotics can be classified into five major families: inhibitors of cell wall synthesis, membrane-active agents, inhibitors of protein synthesis, inhibitors of nucleic acid synthesis, and inhibitors of metabolic function.

In spite of their powerful growth-inhibiting abilities, microorganisms are quite capable of resisting antibiotics. Resistance can be achieved through changes in permeability of cell wall or cell membrane to prevent access of antibiotics to their targets or through enhanced active efflux of the xenobiotic (56). Modification of the antibiotic target may occur, thus reducing its affinity for the drug (67). In addi-

tion, overproduction of the antibiotic target, such as hyperproduction of *p*-aminobenzoic acid in sulfonamide-resistant *Neisseria* species, may reduce antibiotic damage (36). The most direct way to combat antibiotics is to destroy them through enzymatic action. A recently proposed theory of resistance to antibiotics concerns the formation of biofilms (2, 17). Although cells embedded in biofilms may have difficulty obtaining food and other nourishment, they are protected from antibiotics by virtue of their position. Their slow metabolic and reproduction rates may also protect them from damage by some antibiotics (3).

Antibiotic resistance has been widely used as a genetic marker in recombinant DNA technology and microbial ecology. The attraction of antibiotic resistance markers is their ease of use. Antibiotic resistance determinants have been successfully applied to monitor bacteria in soil, leaves, and roots and even to assess aerial dispersal and epiphytic survival (37, 47, 63, 69, 73). Despite their convenience, several aspects should be considered when one is using antibiotic resistance markers. Because antibiotic resistance occurs at various levels in naturally occurring populations of microbes, background levels and specificity of selection for GEMs are concerns. Use of more than one kind of antibiotic resistance marker is commonly employed to address this problem. Combining resistance genes for antibiotics with genes for resistance to heavy metals may be employed. The study of microbes in a particular ecological niche is often done by introducing a GEM into that environment. The GEM is often genetically engineered to express a specific antibiotic resistance in order to be recovered on selective media. In this way, antibiotic resistance genes have been quite widely employed for a long time because of the ease with which selective media can be used to recover bacteria (14, 24, 73). Numerous plasmids and transposons have been used for such purposes (8, 21, 22, 29, 55).

Transposons are useful and are generally stable, but they typically carry the ability to move again, even if at low rates. Thus, there is always the possibility that the background population will become contaminated with the introduced DNA. Some disabled transposons that lack the ability to move without a trans-supplied protein, and thus are quite stable, have been constructed (5, 22, 55). Another means to introduce an antibiotic resistance gene with some stability is via homologous recombination, in which case it is preferable to identify an expendable region of the chromosome to receive the incoming DNA.

In contrast to the direct introduction of resistance genes, a spontaneous mutant that has a chromosomal mutation that confers antibiotic resistance may be selected. These mutants usually contain a still-functional gene that is nonetheless mutated in a way that confers resistance to a specific antibiotic. For instance, certain point mutations in 23S rRNA genes seem to cause no loss of function but do confer resistance to chloramphenicol (1, 49). Similarly, rifampin-resistant or nalidixic acid-resistant bacteria can also be isolated with RNA polymerase or DNA gyrase mutations. A simple procedure for selecting chromosomal antibiotic resistance has been described (68).

Once a spontaneous mutation has occurred, the mutated chromosomal gene can be cloned and used. It is often easier to recover and use a spontaneously occurring antibiotic-resistant mutant, especially when a strain proves refractory to transformation. However, the nature of the resistance mechanism is unknown at the time of isolation and seldom becomes fully known, and such mutations may be associated with pleiotropic effects. Thus, care should be taken when a spontaneously occurring antibiotic-resistant mutant is used, and efforts to characterize the mutant are required. For instance, is it an auxotroph, has its colony morphology changed, and does it seem to grow as well as the parental strain?

Some authors have suggested that the use of antibiotic resistance-tagged bacteria might contribute to increased antibiotic resistance in natural populations of bacteria (38), but it seems possible that profligate use of antibiotics, rather than use of resistance genes or strains, contributes to this problem. However, herbicide resistance and metal resistance genes have been proposed as alternative markers (21).

Another possible drawback with antibiotic resistance markers may be that the competitiveness of a GEM will be compromised (4). It has been reported that the survival times of rifampin-resistant pseudomonads are shorter than those of the rifampin-sensitive wild type in soil (16). Alterations in the symbiotic effectiveness of *Rhizobium trifolii* have been observed in a strain with engineered antibiotic resistance (31). Other researchers have reported a phenomenon called rifampin masking (51), wherein rifampin-resistant bacteria are apparently sensitive to rifampin immediately after recovery from soil. Some specific environmental conditions will make an antibiotic resistance marker inefficient. Cations can influence the activity of aminoglycoside antibiotics against *Pseudomonas aeruginosa* (72). These examples suggest that comparison of the marked strain with its parental strain should be made before the data from releases are interpreted.

CHROMOGENIC MARKERS

The product of some enzymatic reactions may be a pigment which imparts colors to colonies of bacteria. Genes encoding such enzymes have been cloned and used to detect GEMs, the premise being that the target organisms could be recovered from environmental samples and easily detected as colored colonies on agar media. Other chromogenic detection strategies do not employ visual inspection of agar plate cultures but instead utilize broth cultures and detect pigment formation spectrophotometrically.

The enzyme 2,4-dichlorophenoxyacetate causes the conversion of phenoxyacetate to phenol, which can be detected as red colonies after a specific stain is applied (41). Violacein

is a purple pigment, and the genes required for biosynthesis of this compound have been expressed in *Escherichia coli* (57). A DNA fragment apparently involved in actinorhodin synthesis causes the formation of brown pigment in *Streptomyces lividans* cells (34). The bright red pigment, prodigiosin, is produced by *Serratia marcescens*, and the trait can be introduced into other bacteria (20).

xylE Gene

The *xylE* gene product is catechol-2,3-dioxygenase, an enzyme that catalyzes the formation of hydroxymuconic acid, a compound that reacts with catechol to form the yellow-pigmented semialdehyde derivative. The resultant yellow colonies are easily detected in culture. The *xylE* gene has been suggested for use in detection of intact or viable cells because the gene product is easily rendered inactive by exposure to atmospheric oxygen. However, this feature could also lead to errors in estimates of GEM populations unless the stability of the enzyme in the target environment was well understood. This feature should not present a problem in plate culture. The *xylE* gene has been successfully expressed in a variety of gram-negative rods as well as *Streptomyces* species (35, 74).

β-Glucuronidase

An important enzyme, but one that is underutilized, is β-glucuronidase (39), which has good potential for chromogenic detection of GEMs. This enzyme catalyzes the hydrolysis of glucuronide compounds, and some of the resultant pigments are easily detectable. Most commonly, 5-bromo-4-chloro-3-indolyl-β-D-glucuronide is used as a substrate, and its hydrolysis yields a stable blue pigment. Fluorescent products result with the use of other glucuronide substrates (13). Since this enzyme is found in diverse enteric bacteria (28), background noise may create problems in certain applications.

LacZ Protein

The single most successful strategy using a chromogenic tag to monitor GEMs employed β-galactosidase (26). This enzyme is widely used in laboratory studies for other purposes, and it is well characterized and understood. Excellent spectrophotometric methods (12) and plate assays (54) exist for detection of enzyme activity. In the former case, a yellow pigment or a fluorescent product is produced; in the latter case, blue colonies are formed. Drahos et al. (26) reported excellent success in detecting the enzyme in genetically modified pseudomonads. Of 500 *Pseudomonas* strains tested, none were able to naturally express β-galactosidase, and thus expression of the enzyme in conjunction with other fluorescent *Pseudomonas* traits provided strong presumptive identification of GEMs. Also, concurrent expression of both β-galactosidase and the associated galactoside permease in *Pseudomonas fluorescens* permitted the bacteria to grow on lactose. A *Pseudomonas* isolate genetically engineered to express β-galactosidase was released into the environment in subsequent studies (44) and was easily and apparently accurately monitored by using visual plate assays in combination with various selection methods. However, as with β-glucuronidase, background activity could be a problem because of similar enzyme activities in other microbes, especially members of the family *Enterobacteriaceae*.

LUMINESCENT MARKERS

Bioluminescence is widespread in nature, occurring in diverse phyla including mollusks, insects, fungi, diatoms, and bacteria (32). Bioluminescence is thought to have evolved separately on several occasions. One of the most familiar and intriguing examples of bioluminescence concerns the luminous *Vibrio fischeri* that colonizes the light organs of marine fish (52, 53). These bacteria are provided with nutrients and a niche and, in return, provide the host fish with light that is used in communication. Insects too have interesting bioluminescent properties. Three related polypeptides were obtained from a single beetle species and shown to catalyze the production of three different colors of light in vitro: yellow, green, and orange (75).

Generally speaking, light production in biological systems requires two pivotal components: the luciferin and the luciferase. Luciferin is oxidized by the luciferase enzyme, with the concomitant release of light (rather than chemical) energy. Bioluminescence requires energy, reducing power, and oxygen, though the stage and form at which each is required in the process varies with the type of bioluminescence. Thus, the immediate light reaction in bacteria requires reduced flavin mononucleotide, whereas ATP is required for the insect reaction. These requirements might be expected to make genetically engineered bacteria less fit than their dark progenitors over the long term, but observable reductions in growth or pathogenic abilities have generally not been reported as a result of expression of the *lux* genes in diverse hosts such as *Anabaena* species (61), *Streptomyces callicolor* (60), and many gram-negative rods (65).

Bioluminescence

For monitoring GEMs, the most widely used luminescent marker is bacterial bioluminescence. The genes that encode bacterial bioluminescence are called *lux*, and five structural genes are required for the complete and unassisted production of light in bacteria, whereas only two cistrons need be expressed if luciferin is supplied to the cells (53). Thus, two different approaches can be taken. Bacteria may be constructed to emit light without requirement for exogenous addition of luciferin (expressing all five genes), or they may be constructed to synthesize only the luciferase enzyme and thus require exogenous luciferin. Both strategies have advantages.

The entire *lux* operon of *V. fischeri* (without the regulatory genes) is about 7 to 8 kb in size and contains the five structural genes, *luxCDABE*. The advantage of using this larger construct is that bacteria can bioluminesce unassisted, requiring no additional compounds if they correctly express the five genes. Three genes (*luxCDE*) encode enzymes involved in the synthesis of the luciferin, tetradecyl aldehyde. Two (*luxAB*) encode the subunits of the luciferase enzyme. Since no substrate is required, bacteria expressing all five structural genes can bioluminesce even in locations where it would generally be impossible or disruptive to add the luciferin, such as inside plant tissues or under high pressures (e.g., in deep-earth simulations).

In contrast, only the *luxAB* genes of either *V. fischeri* or *V. harveyi* may be introduced to mark bacteria. In these situations, an aldehyde substrate must be added. There are advantages to this situation. First, the smaller *luxAB* fragment is easier to use in construction of recombinant molecules. Second, simple withholding of substrate rather than dependency on a regulated promoter is enough to completely prevent light production (even though the luciferase may still be synthesized). Also, biosynthesis of the tetradecyl substrate is dependent on correct interaction of the *luxCDE* gene products with the fatty acid biosynthetic machinery of the host, and there is little information for most species

concerning this reaction. Thus, it could well be in some situations that biosynthesis of the substrate for the light reaction will be deficient and result in low levels of light production and subsequently low estimates of gene expression or bacterial numbers, etc. The major drawback in using only *luxAB* is that aldehyde must be added to observe bioluminescence, and this action can be very disruptive to the biology being studied. However, it is not a problem for many types of observations.

Other luciferase genes, including the firefly *luc* gene (23), have been cloned and can be used to tag bacteria. This single gene encodes a monomeric luciferase that utilizes a very different substrate than the bacterial luciferase. Nonetheless, this gene can be used essentially like the *luxAB* constructs, and substrate is added when needed. A *Rhizobium* isolate expressing the *luc* gene has been used in risk assessment studies (62). The luciferin is expensive, however, and is not readily taken up by cells.

Bioluminescence has been used to study gene expression, bacterial movement in plants, and bacterial growth (10, 40). A number of vectors to deliver the various bioluminescent systems have been described (9, 11, 45, 66). The particular vector or construct of choice is very case dependent, and a transposon that is easily delivered to one bacterial strain is less easily delivered to another. Major concerns are stable replication or insertion, strength of transcription, and how one will supply the luciferin (endogenous or exogenous).

Imaging methods vary, but generally X-ray film or electronic methods are used to produce an image. In the former case, a leaf or plant part is placed against film and held in place with paper clips. Alternately, the leaf and film may be held together by wrapping them both in aluminum foil. The plant and film are placed in secure darkness for one to many hours. Afterward, the film is recovered and developed. Drawbacks to this method are that the procedure is time-consuming and the handling of the leaf may cause tissue damage. Nonetheless, X-ray film methods were used successfully to study bacteria infecting the large sturdy leaves of *Anthurium* species (30). Advantages of this method are that it is inexpensive and requires no special equipment.

Electronic imaging provides the most satisfactory results, being quick and sensitive. However, low-light imaging systems are expensive. The results are generally superior to film methods because less light can be detected and software makes interpretation of the images relatively easy. Typically, a potted plant or a petri dish or other sample is placed in a chamber equipped with an overhead electronic camera. The camera is focused onto the desired area, and the bioluminescence is recorded. Thus, there is no need to actually handle the sample, and damage to leaves and other sensitive biological samples is reduced, allowing individual samples to be examined at multiple time points.

In our laboratory, we are using bioluminescent bacteria (expressing the entire *V. fischeri lux* operon) to study the gene expression of cabbage during infection by *Xanthomonas campestris* pv. campestris, which causes black rot disease. The bioluminescent bacteria are inoculated into plants, and as a result of their emission of light and the use of sensitive electronic imaging methods, the bacteria can be observed moving within the plants. The sensitivity is such that the bacteria can be observed in the very earliest stages of infection, when they have just moved into a new area of a leaf and have colonized only a single plant cell (estimated to be about 10^3 CFU). With this knowledge, the infected plant parts can be sampled and analyzed for gene expression days before the presence of the bacteria could be inferred by development of symptoms (25).

Electronic equipment varies, as does the technology for detecting low levels of light. We use a charge-coupled device camera; other equipment includes single-photon-counting cameras and image intensifiers. Price is a major consideration; when such equipment is purchased, it is desirable to arrange for on-site testing with actual samples.

Magnification is another concern. While a macro lens provides an enormous amount of information about the location of bacteria on leaf surfaces, it will yield no information about single bacterial cells, and microscopic objectives are needed for some studies. Also, the light emitted by bacteria embedded in the soil (such as those colonizing root surfaces) might be detectable only with fiber-optic cables, especially if there is a need to keep the sample intact.

Quite often it is not necessary to obtain an image of the bioluminescence, and a simple luminometer reading will provide valuable data. This is often the case when it is desirable to know only whether the bacteria are emitting detectable light or whether high levels (indicative of large numbers of bacteria) or low levels (indicative of few bacteria) of light are being produced. Thus, a small portion of a leaf or root or a small amount of soil can be placed into a luminometer for a reading. Essentially any clinical luminometer will suffice. A variation on this theme employs an enrichment procedure as follows. The sample (e.g., 0.1 g of soil) is placed into a test tube with several milliliters of growth medium and selective antibiotics and is aerated. At intervals, 100-μl aliquots of the enrichment culture are assayed for bioluminescence. By this means, a few bacteria in a sample can be induced to reproduce, increasing in numbers until bioluminescence is detectable. Alternately, the bacteria can be allowed to grow out of a sample onto agar, and the bioluminescence is detected visually (6, 7).

In limited field releases of genetically engineered xanthomonads, we have found use of the *lux* genes, coupled with the culture enrichment method, a powerful tool with which to study microbial movement and persistence in agroecosystems (64). The rate of movement of the bacteria from host plants onto weeds and into the soil, as well as the overall persistence of the GEMs on plants and inability to persist in the soil, has been measured (18, 19).

In another variation, bacteria are plated out onto selective or nonselective media and bioluminescent colonies are detected by eye or electronic imaging. Depending on the selective nature of the medium and the efficiency of recovery as well as the use of controls to monitor total recoverable bacteria, inferences can be made regarding the prominence of the bioluminescent bacteria in the soil.

Green Fluorescent Protein

The green fluorescent protein of the jellyfish, *Aequorea victoria*, is an unusual polypeptide that absorbs blue light and fluoresces green. The protein is not an enzyme and requires only an excitation light source, thus rendering moot the question of substrate addition. The cDNA has been subjected to mutational analysis (33), and variant proteins that emit colors other than green were obtained, suggesting possible multiple reporter systems activated by a single light source. In *E. coli* and *Caenorhabditis elegans*, the protein has been used to study gene expression and has caused no ill effects (15). As a marker gene, the green fluorescent protein gene offers promise and may be very useful for studying eukaryotes, for which either the expression of the five-gene *lux* operon or provision of an exogenous supply of substrate

for an enzymatic reaction would be difficult. It is quite possible that the excitation light source could penetrate several layers of plant cells, making the in planta study of genetically altered bacteria or viruses much easier. At present, however, no such studies have been reported.

MIXING AND MATCHING MARKERS

It is important to understand the limitations of markers as well as their advantages. For instance, the luciferase of *V. fischeri* is temperature sensitive, and bacteria growing at higher temperatures (37°C) emit far less light than bacteria producing similar amounts of enzyme but grown at lower temperatures (25°C). This may or may not present a problem, depending on the study conditions, the questions being addressed, and the nature and sensitivity of the detection equipment. In contrast, the luciferase enzyme of *V. harveyi* is much less sensitive to temperature fluctuations.

Some markers (such as luciferases) generally can be used to report the activity of only physiologically active cells, whereas others (e.g., the green fluorescent protein) would be expected to be useful for all cells, active or inactive. Still others (e.g., LacZ) might detect only culturable organisms. It is likely that a combination of markers will provide the most reliable approach.

All of the above-described markers have both strengths and weaknesses, and it is probable that no one system will be useful in all experiments. However, many of the advantages and disadvantages discussed above are often theoretical and have not been tested or are based on experimental data from artificial systems (i.e., not field conditions). In an attempt to examine the usefulness and limitations of these marker systems, we have taken an approach whereby several markers are used individually and in conjunction with each other. In our work, we are using bacterial bioluminescence, spontaneous antibiotic resistance, and a serological marker. Immunofluorescent colony staining (IFC) was developed by van Vuurde (70, 71). IFC is based on combining isolation of viable bacteria with serological differentiation of target colonies. The technique uses immunoglobulin G antibodies specific for a bacterial strain conjugated with fluorescein isothiocyanate to distinguish target colonies from nontarget colonies. Because bacteria identified by IFC remain viable, positive colonies may be subcultured for other uses such as identification by other methods, e.g., fatty acid methyl ester analysis or biochemical reactions. Since the use of IFC does not require any alteration in the phenotype or genotype of the wild type, this technique facilitates the comparison of an unaltered wild type with a genetically modified derivative of that wild type, such as the antibiotic-resistant and bioluminescent derivatives that we employ. Thus, direct assessment of effects on fitness of a GEM due to the addition or alteration of genetic material is possible. It is also feasible to determine the stability of the introduced genetic material in the environment because IFC allows for differentiation of the introduced bacterium from indigenous microflora without reliance on the introduced or altered genetic material. This could allow for rapid differentiation of genetically modified bacteria that have reverted to wild type (i.e., lost the introduced genetic material or spontaneous antibiotic resistance) from the indigenous microflora; however, IFC does not allow for differentiation of isolates that have reverted to wild type from those that have not reverted but are no longer culturable on antibiotic-amended media. Another drawback to the serological marker is that there is always the possibility of cross-reactions with bacteria that are not the target strain. Because of this possibility, controls (samples from the same environment that were not inoculated with the target strain) must be examined very carefully for cross-reacting strains, and a subset of positive isolates should be subcultured and identified by another means. These precautions should ensure the validity of the data collected.

A wild-type strain, a spontaneous rifampin-resistant mutant of the wild type, and a bioluminescent derivative of the wild type are individually inoculated onto cucumber seed and planted in the field. Root and soil samples are removed at various times during the growth of the plants and processed. This system allows for comparison of the population of the genetically modified derivatives with that of the unaltered wild type, thereby allowing us to assess the practicality of the genetic markers for monitoring our bacterial strain. Using IFC with samples inoculated with the genetic markers, we are also able to examine the effectiveness of those markers detecting populations of marked bacteria. Also, by subculturing IFC-positive colonies and performing fatty acid methyl ester analysis, we can check the specificity of our serological marker.

CONCLUDING REMARKS

In the future, new markers will become more widely available, as will the methodologies to apply them easily to microbiological questions. It seems likely that multiple colors of bioluminescence will be used to tag GEMs, perhaps one color to indicate location, another to report on metabolic activity, and perhaps a third to document transcription from a specific promoter. In addition, these remotely sensed markers will likely be combined with selectable markers.

Presumably, a greater understanding of marker effects on the host, regulated gene expression, and host chromosome structure will permit the design and routine use of GEMs tagged in very specific and desirable ways, with full understanding of effects on fitness. The green fluorescent protein shows great promise as a marker for a variety of reasons: (i) no substrate is required, (ii) background noise is not a problem in many environments, and (iii) multiple colors may be available in the future. It may be especially important for use in eukaryotic microbes (fungi and viruses) as these microbes become more widely utilized in release experiments.

Use of the multicistron *lux* operon will likely remain limited to prokaryotes because eukaryotes cannot generally express polycistrons and because they synthesize fatty acids via different means than bacteria. However, there are two exceptions to consider. First, the LuxA/B fusion protein (10) is expressed as a single gene and so can be expressed in eukaryotes, although luciferin will generally need to be supplied. Second, it is possible that the *lux* operon could be expressed in chloroplasts because of the prokaryotic natures of their gene expression and fatty acid biosynthesis.

REFERENCES

1. **Aagaard, C., H. Phan, S. Trevisanato, and R. A. Garrett.** 1994. A spontaneous point mutation in the single 23S rRNA gene of the thermophilic archaeon *Sulfolobus acidocaldarius* confers multiple drug resistance. *J. Bacteriol.* **176:** 7744–7747.
2. **Amábile-Cuevas, C. F.** 1993. *Origin, Evolution and Spread of Antibiotic Resistance Genes.* Landes, Austin, Tex.
3. **Anwar, H., and J. L. Strap.** 1992. Eradication of biofilm cells of *Staphylococcus aureus* with tobramycin and cephalexin. *Can. J. Microbiol.* **38:**618–625.
4. **Atlas, R. M., G. Sayler, R. S. Burlage, and A. K. Bej.** 1992. Molecular approaches for environmental monitoring of microorganisms. *BioTechniques* **5:**706–717.
5. **Barry, G.** 1986. Permanent insertion of foreign genes into the chromosomes of soil bacteria. *Bio/Technology* **4:** 446–449.
6. **Beauchamp, C. J., J. W. Kloepper, and H. Antoun.** 1993. Detection of genetically engineered bioluminescent pseudomonads in potato rhizosphere at different temperatures. *Microb. Releases* **1:**203–207.
7. **Beauchamp, C. J., J. W. Kloepper, and P. A. Lemke.** 1993. Luminometric analyses of plant root colonization by bioluminescent pseudomonads. *Can. J. Microbiol.* **39:** 434–441.
8. **Berg, C. M., D. E. Berg, and E. A. Groisman.** 1989. Transposable elements and the genetic engineering of bacteria, p. 879–925. *In* D. E. Berg and M. M. Howe (ed.), *Mobile DNA.* American Society for Microbiology, Washington, D.C.
9. **Boivin, R., F.-P. Chalifour, and P. Dion.** 1988. Construction of a Tn5 derivative encoding bioluminescence and its introduction in *Pseudomonas, Agrobacterium,* and *Rhizobium. Mol. Gen. Genet.* **213:**50–55.
10. **Boylan, M. O., J. Pelletier, S. Dhepagnon, S. Trudel, N. Sonenberg, and E. A. Meighen.** 1989. Construction of a fused LuxAB gene by site-directed mutagenesis. *J. Biolumin. Chemilumin.* **4:**310–316.
11. **Brennerova, M. V., and D. E. Crowley.** 1994. Direct detection of rhizosphere-colonizing *Pseudomonas* sp. using an *Escherichia coli* rRNA promoter in a Tn7-*lux* system. *FEMS Microbiol. Ecol.* **14:**319–330.
12. **Bronstein, I., J. Fortin, P. E. Stanley, G. Stewart, and L. J. Kricka.** 1994. Chemiluminescent and bioluminescent reporter gene assays. *Anal. Biochem.* **219:**169–181.
13. **Bronstein, I., J. J. Fortin, and J. C. Voyta.** 1994. Chemiluminescent reporter gene assays: sensitive detection of the *gus* and *seap* gene products. *BioTechniques* **17:**172–177.
14. **Campbell, J. N., D. D. Cass, and D. J. Peteya.** 1987. Colonization and penetration of intact canola seedling roots by an opportunistic fluorescent *Pseudomonas* sp. and the response of host tissue. *Phytopathology* **77:**1166–1173.
15. **Chalfie, M., Y. Tu, G. Euskirchen, W. W. Ward, and D. C. Prasher.** 1994. Green fluorescent protein as a marker for gene expression. *Science* **263:**802–805.
16. **Compeau, G. B., J. Al-Achi, E. Platsouka, and S. B. Levy.** 1988. Survival of rifampin-resistant mutants of *Pseudomonas fluorescens* and *Pseudomonas putida* in soil systems. *Appl. Environ. Microbiol.* **54:**2432–2438.
17. **Costerton, J. W., K.-J. Cheng, and G. G. Geesey.** 1987. Bacterial biofilms in nature and disease. *Annu. Rev. Microbiol.* **41:**435–464.
18. **Dane, F., and J. J. Shaw.** 1993. Growth of bioluminescent *Xanthomonas campestris* pv. *campestris* in and on susceptible and resistant host plants. *Mol. Plant-Microbe Interact.* **6:**786–789.
19. **Dane, F., and J. J. Shaw.** 1994. Endophytic and epiphytic growth of *Xanthomonas campestris* pv. *campestris* on susceptible and resistant host plants in the field environment: detection by bioluminescence. *Microb. Releases* **2:**223–229.
20. **Dauenhauer, S. A., R. A. Hull, and R. P. Williams.** 1984. Cloning and expression in *Escherichia coli* of *Serratia marcescens* genes encoding prodigiosin biosynthesis. *J. Bacteriol.* **158:**1128–1132.
21. **de Lorenzo, V.** 1994. Designing microbial systems for gene expression in the field. *Trends Biotechnol.* **12:**367–371.
22. **de Lorenzo, V., and K. N. Timmis.** 1994. Analysis and

construction of stable phenotypes in Gram-negative bacteria with Tn5- and Tn10-derived minitransposons. *Methods Enzymol.* **235:**386–405.

23. **de Wet, J. R., K. V. Wood, D. R. Helinski, and M. Deluca.** 1985. Cloning of firefly luciferase cDNA and the expression of active luciferase in *Escherichia coli. Proc. Natl. Acad. Sci. USA* **82:**7870–7873.

24. **Dijkstra, A. F., G. H. N. Scholten, and J. A. van Veen.** 1987. Colonization of wheat seedling (*Triticum aestivum*) roots by *Pseudomonas fluorescens* and *Bacillus subtilis. Biol. Fertil. Soils* **4:**41–46.

25. **Dodson, K. M.** 1995. Accumulation patterns of defense proteins of cabbage during compatible and incompatible interactions. M.S. thesis. Department of Plant Pathology, Auburn University, Auburn, Ala.

26. **Drahos, D. J., B. C. Hemming, and S. McPherson.** 1986. Tracking recombinant organisms in the environment: beta-galactosidase as a selectable non-antibiotic marker for fluorescent pseudomonads. *Bio/Technology* **4:**439–444.

27. **Fleming, A.** 1929. On the antibacterial action of cultures of a penicillium, with special reference to their use in the isolation of *B. influenzae. Br. J. Exp. Pathol.* **10:**226–236.

28. **Flemming, C. A., K. T. Leung, H. Lee, J. T. Trevors, and C. W. Greer.** 1994. Survival of *lux-lac*-marked biosurfactant-producing *Pseudomonas aeruginosa* UG2L in soil monitored by nonselective plating and PCR. *Appl. Environ. Microbiol.* **60:**1606–1613.

29. **Foster, T. J.** 1983. Plasmid-determined resistance to antimicrobial drugs and toxic metal ions in bacteria. *Microbiol. Rev.* **47:**361–409.

30. **Fukui, R., H. Fukui, R. McElhaney, F. C. Nelson, and A. M. Alvarez.** 1996. Relationship between symptom development and actual sites of infection in leaves of *Anthurium* inoculated with a bioluminescent strain of *Xanthomonas campestris* pv. diffenbachiae. *Appl. Environ. Microbiol.* **62:**1021–1028.

31. **Hagedorn, C.** 1979. Relationship of antibiotic resistance to effectiveness in *Rhizobium trifolii* populations. *Soil. Sci. Soc. Am. J.* **43:**921–925.

32. **Harvey, E. N.** 1952. *Bioluminescence.* Academic Press, New York.

33. **Heim, R., D. C. Prasher, and R. Y. Tsien.** 1994. Wavelength mutations and posttranslational autoxidation of green fluorescent protein. *Proc. Natl. Acad. Sci. USA* **91:**12501–12504.

34. **Horinouchi, S., and T. Beppu.** 1985. Construction and application of a promoter-probe plasmid that allows chromogenic identification of *Streptomyces lividans. J. Bacteriol.* **162:**406–412.

35. **Ingram, C., M. Brawner, P. Youngman, and J. Westpheling.** 1989. *xylE* functions as an efficient reporter gene in *Streptomyces* spp.: use for the study of *galP1*, a catabolite-controlled promoter. *J. Bacteriol.* **171:**6617–6624.

36. **Israili, Z. H.** 1987. Bacterial resistance to antimicrobial agent, p. 165–184. *In* S. S. Lamba and C. A. Walker (ed.), *Antibiotics and Microbial Transformations.* CRC Press, Inc., Boca Raton, Fla.

37. **Jacques, M.-A., L. K. Linda, and C. E. Morris.** 1995. Population sizes, immigration, and growth of epiphytic bacteria on leaves of different ages and positions of field-grown endive (*Cichorium endivia* var. *latifolia*). *Appl. Environ. Microbiol.* **61:**899–906.

38. **Jansson, J. K.** 1995. Tracking genetically engineered microorganisms in nature. *Curr. Opin. Biotechnol.* **6:**275–283.

39. **Jefferson, R. A.** 1989. The Gus reporter gene system. *Nature* (London) **342:**837–838.

40. **Keller, G. A., S. Gould, M. Deluca, and S. Subramani.** 1987. Firefly luciferase is targeted to peroxisomes in mammalian cells. *Proc. Natl. Acad. Sci. USA* **84:**3264–3268.

41. **King, R. J., K. A. Short, and R. J. Seidler.** 1991. Assay for detection and enumeration of genetically engineered microorganisms which is based on the activity of a deregulated 2,4-dichlorophenoxyacetate monooxygenase. *Appl. Environ. Microbiol.* **57:**1790–1792.

42. **Kloepper, J. W., and C. J. Beauchamp.** 1992. A review of issues related to measuring colonization of plant roots by bacteria. *Can. J. Microbiol.* **38:**1219–1232.

43. **Kluepfel, D. A.** 1993. The behavior and tracking of bacteria in the rhizosphere. *Annu. Rev. Phytopathol.* **31:**441–472.

44. **Kluepfel, D. A., E. L. Kline, H. D. Skipper, T. A. Hughes, D. T. Gordon, D. J. Drahos, G. F. Barry, B. C. Hemming, and E. J. Brandt.** 1991. The release and tracking of genetically engineered bacteria in the environment. *Phytopathology* **81:**348–352.

45. **Legocki, R. P., M. Legocki, T. O. Baldwin, and A. A. Szalay.** 1986. Bioluminescence in soybean root nodules: demonstration of a general approach to assay gene expression in vivo by using bacterial luciferase. *Proc. Natl. Acad. Sci. USA* **83:**9080–9084.

46. **Lindow, S. E.** 1995. The use of reporter genes in the study of microbial ecology. *Mol. Ecol.* **4:**555–566.

47. **Lindow, S. E., G. R. Knudsen, R. J. Seidler, M. V. Walter, V. W. Lambou, P. S. Amy, D. Schmedding, V. Prince, and S. Hern.** 1988. Aerial dispersal and epiphytic survival of *Pseudomonas syringae* during a pretest for the release of genetically engineered strains into the environment. *Appl. Environ. Microbiol.* **54:**1557–1563.

48. **Lindow, S. E., N. J. Panopoulos, and B. L. McFarland.** 1989. Genetic engineering of bacteria from managed and natural habitats. *Science* **244:**1300–1307.

49. **Mankin, A. S., and R. A. Garrett.** 1991. Chloramphenicol resistance mutations in the single 23S rRNA gene of the archaeon *Halobacterium halobium. J. Bacteriol.* **173:**3559–3563.

50. **Marshall, V. P., J. I. Cialdella, D. W. Elrod, and P. F. Wiley.** 1987. Microbial transformations of antibiotics, p. 7–45. *In* S. S. Lamba and C. A. Walker (ed.), *Antibiotics and Microbial Transformations.* CRC Press, Inc., Boca Raton, Fla.

51. **McInroy, J. A., G. Wei, G. Musson, and J. W. Kloepper.** 1992. Evidence for possible masking of rifampicin-resistance phenotype of marked bacteria in planta. *Phytopathology* **82:**1177. (Abstract.)

52. **Meighen, E. A.** 1988. Enzymes and genes from the lux operons of bioluminescent bacteria. *Annu. Rev. Microbiol.* **42:**151–176.

53. **Meighen, E. A.** 1994. Genetics of bacterial bioluminescence. *Annu. Rev. Genet.* **28:**117–139.

54. **Miller, J. H.** 1972. *Experiments in Molecular Genetics.* Cold Spring Harbor Laboratory, Cold Spring Harbor, N.Y.

55. **Musso, R. W., and T. Hodam.** 1989. Construction and characterization of versatile kanamycin-resistance cassettes derived from the Tn5 transposon. *Gene* **85:**205–207.

56. **Nikaido, H.** 1994. Prevention of drug access to bacterial targets: permeability barriers and active efflux. *Science* **264:**382–388.

57. **Pemberton, J. M., K. M. Vincent, and R. J. Penfold.** 1990. Cloning of heterologous expression of the violacein biosynthesis gene cluster from *Chromobacterium violaceum. Curr. Microbiol.* **22:**355–358.

58. **Quesnel, L. B., and A. D. Russell.** 1983. Introduction, p. 1–18. *In* L. B. Quesnel and A. D. Russell (ed.), *Antibiotics: Assessment of Antimicrobial Activity and Resistance.* Academic Press, Inc., New York.

59. **Saunders, J. R., and V. A. Saunders.** 1993. Genotypic and phenotypic methods for the detection of specific released microorganisms, p. 27–59. *In* C. Edwards (ed.), *Monitoring*

Genetically Manipulated Microorganisms in the Environment. John Wiley & Sons Ltd., Chichester, England.

60. **Schauer, A., M. Ranes, R. Santamaria, J. Guijarro, E. Lawlor, C. Mendez, K. Chater, and R. Losick.** 1988. Visualizing gene expression in time and space in the filamentous bacterium *Streptomyces coelicolor. Science* **240:**768–772.

61. **Schmetterer, G., C. P. Wolk, and J. Elhai.** 1986. Expression of luciferases from *Vibrio harveyi* and *Vibrio fischeri* in filamentous cyanobacteria. *J. Bacteriol.* **167:**411–414.

62. **Selbitschka, W., A. Puhler, and R. Simon.** 1992. The construction of *recA*-deficient *Rhizobium meliloti* and *R. leguminosarum* strains marked with *gusA* or *luc* cassettes for use in risk-assessment studies. *Mol. Ecol.* **1:**9–19.

63. **Seong, K.-Y., M. Höfte, J. Boelens, and W. Versraete.** 1991. Growth, survival, and root colonization of plant growth beneficial *Pseudomonas fluorescens* ANP15 and *Pseudomonas aeruginosa* 7NSK2 at different temperatures. *Soil Biol. Biochem.* **23:**423–428.

64. **Shaw, J. J., F. Dane, D. Geiger, and J. W. Kloepper.** 1992. Use of bioluminescence for detection of genetically engineered microorganisms released into the environment. *Appl. Environ. Microbiol.* **58:**267–273.

65. **Shaw, J. J., and C. I. Kado.** 1986. Development of a *Vibrio* bioluminescence gene-set to monitor phytopathogenic bacteria during the ongoing disease process in a nondisruptive manner. *Bio/Technology* **4:**560–564.

66. **Shaw, J. J., P. Rogowsky, T. J. Close, and C. I. Kado.** 1987. Working with bacterial bioluminescence. *Plant Mol. Biol. Rep.* **5:**225–236.

67. **Spratt, B. G.** 1994. Resistance to antibiotics mediated by target alterations. *Science* **264:**388–393.

68. **Stewart, G. J.** 1992. Natural transformation and its potential for gene transfer in the environment, p. 253–282. *In* M. A. Levin, R. J. Seidler, and M. Rogul (ed.), *Microbial Ecology: Principles, Methods, and Applications.* McGraw-Hill, Inc., New York.

69. **Vandenhove, H., R. Merckx, H. Wilmots, and K. Vlassak.** 1991. Survival of *Pseudomonas fluorescens* inocula of different physiological stages in soil. *Soil Biol. Biochem.* **23:**1133–1142.

70. **van Vuurde, J. W. L.** 1987. New approach in detecting phytopathogenic bacteria by combined immunoisolation and immunoidentification assays. *EPPO Bull.* **17:**139–148.

71. **van Vuurde, J. W. L.** 1990. Immunostaining of colonies for sensitive detection of viable bacteria in sample extracts and on plant roots, p. 907–912. *In* Z. Klement (ed.), *Plant Pathogenic Bacteria, Proceedings of the 7th International Conference on Plant Pathogenic Bacteria,* part B. Akadémiai Kiadò, Budapest.

72. **Washington, J. A., II, R. J. Snyder, P. C. Kohner, C. B. Wiltse, D. M. Ilstrup, and J. T. McCall.** 1978. Effect of cation content of agar on the activity of gentamicin, tobramycin, and amikacin against *Pseudomonas aeruginosa. J. Infect. Dis.* **137:**103–110.

73. **Williams, S. T., and F. L. Davies.** 1965. Use of antibiotics for selective isolation and enumeration of Actinomycetes in soil. *J. Gen. Microbiol.* **38:**251–261.

74. **Winstanley, C., J. A. W. Morgan, R. W. Pickup, and J. R. Saunders.** 1991. Use of a *xylE* marker gene to monitor survival of recombinant *Pseudomonas putida* populations in lake water by culture on nonselective media. *Appl. Environ. Microbiol.* **57:**1905–1913.

75. **Wood, K. V., Y. A. Lam, H. H. Seliger, and W. D. McElroy.** 1989. Complementary DNA coding click beetle luciferases can elicit bioluminescence of different colors. *Science* **244:**700–702.

Use of Fluorescent Antibodies for Studying the Ecology of Soil- and Plant-Associated Microbes

TIMOTHY R. McDERMOTT

53

The ability to detect and enumerate microbes in natural samples allows microbial ecologists to evaluate the suitability of various habitats for different microbes and assess microbial population responses to perturbations in the environment. Counting specific microorganisms is problematic in most natural samples because of the number and variety of nontarget organisms present. These problems are significant when one is interested in differentiating between strains or serotypes of the same species or between species of the same genus. In the latter example, it is difficult to devise a differential medium that will allow the cultivation and identification of wild-type bacteria at the strain level. Since each technique has its limitations, the microbial ecologist must choose techniques which are best suited for the organism(s), the environment being studied, and the specific questions being asked.

RATIONALE AND APPLICATION OF THE FA TECHNIQUE

While there are now several detection systems available, the fluorescent antibody (FA) technique is unique in that it allows studies of microbial ecological relationships at the strain level. It utilizes antibodies that have been tagged with a fluorescent dye to locate and identify bacteria or fungi in samples obtained from natural environments. The utility of this technique will vary somewhat from species to species, depending on relative serological complexity. However, even in cases of significant cross-reaction, measures that may still allow differentiation between strains can be taken. Such distinction is clearly useful when the objective is to probe population responses at a very specific level or when one is searching for specific released organisms (26). Conversely, one may take advantage of serological relatedness to monitor microbial population responses at the species or genus levels.

The FA technique was first used in biomedical applications (18). Its usefulness in soil microbiology was not realized, however, until it was exploited by E. L. Schmidt at the University of Minnesota. Initially, Schmidt and Bankole (59) combined contact slides with fluorochrome-labeled antibodies to demonstrate a specific detection system for *Aspergillus flavus*. Hill and Gray (34) later used FAs to study *Bacillus* species in soil. Schmidt and coworkers also used the FA technique for bacteria, extending its use to *Bradyrhizo-*

bium japonicum, *Beijerinckia* species, and *Nitrobacter winogradskyi* (8, 25, 58). Additional studies making use of FAs soon followed and included studies of other soil bacteria such as *Azotobacter* spp. (65) and *Azospirillum brasilense* (55).

Both beneficial and pathogenic plant-microbe interactions have been investigated with FAs. Examples include the rhizosphere ecology and interstrain competition of rhizobia for nodulation of their respective host legumes (24, 45, 47), the colonization of *Enterobacter* species on the root surfaces of grasses (4), and competition between strains of *Erwinia* during colonization of potato tubers (22). Others have extended the technique for studies designed to localize *Azospirillum* colonization in Kallar grass cortex tissue (51) and *Clavibacter michiganensis* (ring rot pathogen) in potato stems (64). An example of FA detection of specific bacteria from a rhizosphere sample is illustrated in Fig. 1, and examples of direct detection on the root are shown in Fig. 2. The utility of FAs for studying nonmycorrhizal (27, 50), ectomycorrhizal (61), and endomycorrhizal (29, 42, 70) fungal species also has been demonstrated.

The methods described in this chapter were developed by researchers working primarily with gram-negative bacteria to obtain polyclonal antibodies against whole cells. The reader can consult reference 44 for papers that describe work with other gram-negative bacteria as well as gram-positive organisms. Because of space limitations, no attempt was made to present a complete review of the topic. For an in-depth discussion of the applications of the FA technique, the reader is referred to references 57 and 9. Also, recent reviews of FA methods are offered by Dazzo and Wright (21), Hampton et al. (32), and Wright (72). The production of monoclonal antibodies is beyond the scope of this chapter, but Harlow and Lane (33) offer an extensive overview of the topic.

DEVELOPMENT OF ANTISERA
Antigen Preparation

Many components of a microorganism are antigenic. Proteins and polysaccharides contain numerous epitopes that will provoke an immune reaction, but peptides as small as a few amino acids can be antigenic as well. In most studies of plant- or soil-associated bacteria, researchers prepared cells in ways that favored the preservation of somatic antigens (O antigens associated with the outer membrane). However, flagellar antigens (H antigens) can also be very

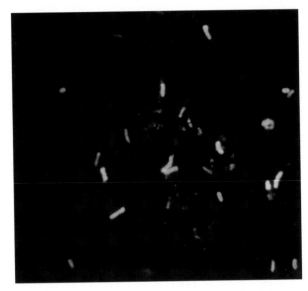

FIGURE 1 Detection of *B. japonicum* in samples from a soybean rhizosphere by using strain-specific FAs. FA-reactive and nonreactive cells appear as white and gray, respectively. As viewed with epifluorescence, reactive cells are brilliant green. Nonreactive cells are dull orange as a result of gelatin-RITC counterstain.

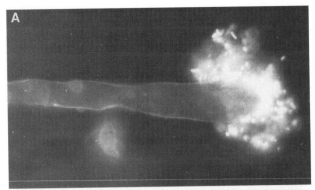

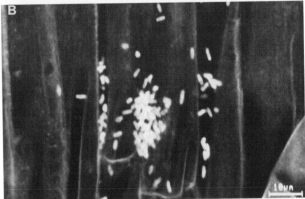

FIGURE 2 In situ detection of *Rhizobium leguminosarum* bv. trifolii on the root surface of its host, white clover (*Trifolium repens*), using indirect immunofluorescence microscopy with a monoclonal antibody specific for its lipopolysaccharide O antigen. Immunofluorescent cells in panel A (conventional epifluorescence) have aggregated at the tip of a root hair where the host lectin, trifoliin A, accumulates; cells in panel B (laser scanning confocal epifluorescence) have developed a microcolony on the rhizoplane. Reprinted from reference 21 with permission of Kluwer Academic Publishers.

useful. For fungi, workers have used both spores and hyphae. The preparation of viral antigens requires several special precautions and will not be covered here. The reader is referred to references 13 and 67 for detailed discussions of collecting and handling viruses.

To avoid extraneous fluorescence in the finished fluorochrome-labeled antibody preparation, care must be taken to eliminate any antigenic materials that may contaminate the antigen preparation and lead to nonspecific reactions. A typical protocol for bacteria is to grow the organism of interest in a defined medium (avoid complex media when possible, as background reactions may arise from yeast or protein extracts) and then wash the cells three times in 0.85% (wt/vol) saline solution that has been autoclaved and filtered through a 0.22-μm-pore-size filter. Resuspend the cells in autoclaved-filtered 0.85% saline to a concentration of 10^9 cells per ml (conveniently quantified with a Klett meter using McFarland standards [48]). For many bacteria, the highly immunoreactive O-antigen portion of the lipopolysaccharide is the basis for unique serological reactions, whereas flagellar antigens can substantially reduce the specificity of the resulting antiserum. The O antigens are stable to heat treatment, but flagellar antigens (along with other surface proteins) are inactivated by heat treatment. Place the cells in a sterile stoppered test tube, and vent the tube with a sterile syringe needle. Place the tube into a boiling water bath for 1 h. After cooling, add the preservative Merthiolate (thimerosal) to a final concentration of 1:10,000.

Some flagellar antigens offer a high degree of specificity (20) and may be used for detection of particular organisms. Flagellar antigens can be stabilized by mixing saline-suspended cells (nonheated) with an equal volume of 0.6% formalin in saline and incubating the cells at room temperature for 48 h. The cells are collected by centrifugation and resuspended to their original volume in 0.85% saline containing 0.3% formalin (23). The formalin treatment can significantly reduce the antigenicity of somatic antigens (20).

For fungi, spores and hyphae from liquid cultures are washed, homogenized with a Waring blender in sterile-filtered 0.85% saline, collected by centrifugation, and resuspended to 0.5 to 0.75 mg (wet weight) per ml of sterile-filtered saline (42, 43, 61, 70). Preparations are used undiluted and injected intravenously (i.v.). For vesicular-arbuscular mycorrhizae (VAM) fungi, which are difficult to manipulate in pure culture, spores can be harvested from soil material associated with aseptically grown plants (e.g., Leonard jar assembly [68]) that have been inoculated with a single isolate of VAM fungi. The spores are washed repeatedly in sterile water (29, 43) and pelleted by centrifugation. The spores are resuspended in a minimal volume (approximately 1,000 spores per ml), disrupted by sonication (on ice to avoid overheating), and then used as antigen after diluting 1:3 with phosphate-buffered saline (PBS) (29). Wilson et al. (70) described a method for harvesting VAM fungal hyphae for use as antigen. Surface-sterilized spores of *Gigaspora* sp. (WUM6) were germinated between sterile filter disks (0.45-μm pore size). The hyphae and vesicles were then separated from the spores and used as antigen.

Immunization

For the reader who is inexperienced in antibody production, Harlow and Lane (33) have prepared a comprehensive labo-

TABLE 1 Immunization schedules used for generation of polyclonal antibodies against bacteria (60) and fungi (61)

Bacteria		Fungi	
Day	Protocol	Day	Protocol
1	0.5-ml i.v. injection of cell (antigen) suspension	1	0.5-ml i.v. injection of homogenized suspension
2	1.0-ml i.v. injection	3	1.0-ml i.v. injection
3	1.5-ml i.v. injection	5	1.5-ml i.v. injection
4–6	Rest	6–9	Rest
7	1.5-ml i.v. injection	10	1.5-ml i.v. injection
8	2.0-ml i.v. injection	11	2.0-ml i.v. injection
9	2.0-ml i.v. injection	12–15	Rest
10–15	Rest	16	2.0-ml s.c. injection
16	Test bleed and titer determination	19	2.0-ml s.c. injection
18	Cardiac bleed (30–50 ml)	26	Test bleed; harvest blood if titer is sufficient

Additional bleeds can be made at 10-day intervals, but the titer should be checked between bleeds.

If titer is low at day 26, try increasing it with additional i.v. booster injections.

ratory manual for antibody production. A pure preparation of the antigen is presented to the animal's immune system such that antibodies specific to that antigen are produced. Prior to injection, preimmune blood must be sampled to determine if the animal's serum already contains antibodies reactive with the organism of interest. Titers can be determined by standard agglutinations tests (28). Animals with preimmune serum that contains antibodies against the target organism should not be used for serum production.

Rabbits typically are used for polyclonal antibody production. A variety of immunization schedules have been successful for bacteria (10, 16, 60) and fungi (29, 31, 36, 59) and usually involve injections administered subcutaneously (s.c.), intramuscularly (i.m.), or i.v. over a period of 4 to 5 weeks. Examples of i.v. immunization protocols used for bacteria and fungi are given in Table 1. Injections given s.c. and i.m. with adjuvants also can be used in conjunction with i.v. injections. They are often the initial doses and are followed by i.v. injections for titer-boosting purposes. Both Freund's complete and incomplete adjuvants have been used for s.c. and i.m. injections. The antigen suspension is thoroughly emulsified 1:1 with the adjuvant by drawing the mixture into a syringe (no needle) and then expelling it repeatedly until the mix is of a smooth consistency. Injections using adjuvants are always s.c. or i.m., never i.v. Booster injections of the antigen suspension alone are done i.v. to increase the antibody titer. They should be continued until the serum titer is at least 1:1,200.

Adjuvants are convenient in that their use reduces the number of injections required to elicit a satisfactory immune response. The complete adjuvant contains killed cells of *Mycobacterium tuberculosis* suspended in mineral oil. The killed *M. tuberculosis* aids in the stimulation of the immune system. The mineral oil functions as a slow-release mechanism, maintaining the antigen in the animal for extended periods to continue eliciting an immune response. Otherwise, the foreign matter will be quickly degraded and voided by the animal. The complete adjuvant can be used if the finished FA is intended for examinations of pure cultures or in vitro work but should be avoided or used only in the initial injections when one expects to use the FA to probe environmental samples. The presence of M. *tuberculosis* in the antigen solution used for booster injections can result in the finished serum containing significant titers against a second organism, which may introduce unwanted FA reactions.

Blood Harvest

Blood is collected when the titer is sufficient. The animal may be bled completely, which usually yields approximately 100 to 120 ml of blood (roughly 50 to 60 ml of crude serum). Alternatively, several 30-ml harvests can be taken every 10 days. If the latter method is used, one should verify that the titer remains at an adequate level for each bleed (booster shots may be required), and the animal should also receive sterile saline to replace body fluids lost to each bleed.

Two methods of harvest are used: cardiac puncture and ear bleed. Animal care facilities normally have trained personnel capable of either procedure, but the ear bleed procedure can be successfully carried out by previously untrained technicians without unneeded animal trauma. The ear bleed apparatus (Fig. 3) can either be constructed by a glassblower or purchased commercially (Bellco Biotechnology, Vineland, N.J.). For each bleeding, the animal is constrained, the ear is shaved and disinfected with 70% ethanol, and then the marginal ear vein is nicked with a sterile scalpel. The ear bleed apparatus is immediately placed over the ear, and a mild vacuum is gradually applied. As a seal is established at the base of the bleeding apparatus, blood will begin to flow, and 30 ml can be collected within a few minutes.

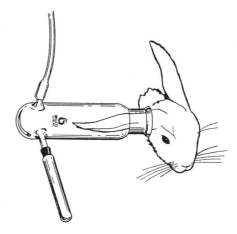

FIGURE 3 Ear bleed device as shown in the Bellco Glass catalog.

For successive bleeds, the ear is nicked at a new site slightly closer to the base, and the foregoing procedure is repeated.

A second method for bleeding rabbits uses oil of wintergreen (methyl salicylate) to dialyze the central ear artery. Apply the oil of wintergreen to the shaved ear and allow a few minutes for the artery to swell. Insert a sterile hypodermic needle into the artery (directed towards the base of the ear) and collect the blood that will begin to flow immediately. A 21-gauge needle can be used for blood samples of less than 5 ml, whereas a 19-gauge needle should be used to collect larger blood volumes. To stop blood flow, apply a piece of gauze and hold it in place with a tongue retractor. The blood should clot in a maximum of 30 min.

The harvested blood is clotted at room temperature for approximately 2 h and then placed in a refrigerator to allow the clot to shrink. The serum is decanted and cleared of unclotted blood cells by centrifugation at 5,000 × g for 10 min. The cleared serum can then be distributed in 20-ml volumes (scintillation vials work well) and frozen for future use.

FA PREPARATION
Serum Purification
Transfer 20 ml of cleared serum to a 50-ml beaker containing a small stir bar. The beaker is chilled by placing it within a larger (e.g., 600-ml) beaker containing a water-ice slush. Alternatively, the procedure can be carried out in a cold room facility. With slow stirring, slowly add an equal volume of cold saturated $(NH_4)_2SO_4$. Allow the serum proteins to precipitate over several hours (or overnight in a cold room), and then pellet them by centrifugation at 10,000 × g for 15 min. The supernatant is discarded, and the pellet is resuspended (gentle shaking) to the original volume with 0.85% saline. An equal volume of saturated $(NH_4)_2SO_4$ is again added, and the resulting precipitate is pelleted by centrifugation. This precipitation-centrifugation step can be repeated several times if the resuspended protein solution is still pink (contains free hemoglobin). The final precipitate should be resuspended in half the original volume, which will yield a convenient protein concentration. The partially purified globulins are then extensively dialyzed (molecular weight cutoff, 14,000- to 16,000) against pH 8.0 saline to remove the $(NH_4)_2SO_4$, which can interfere with the fluorochrome-labeling step (37).

Conjugating Antibodies to Fluorochrome
The following conjugation protocol is for use with the fluorescent dye fluorescein isothiocyanate (FITC). Slightly different protocols are offered by Somasegaran and Hoben (63) and Wright (72); Goldman (30) describes alternative procedures. Dialyze 10 ml of the partially purified globulin solution overnight against 1 liter of 80 mM Na_2CO_3–20 mM $NaHCO_3$ buffer (pH 9.5) containing Merthiolate as a preservative (final concentration of 1:10,000; do not use azide, as it will interfere with the conjugation reaction). Adjust the protein concentration to 10 mg/ml, and then for every 10 ml of globulin solution, add with slow stirring 3.0 mg of FITC previously dissolved in 3.0 ml of dimethyl sulfoxide. Incubate the mixture at room temperature without stirring for 3 h, and then adjust the pH of the conjugate mix to 7.4 with small increments of 1.0 N HCl.

Removal of Unreacted Fluorochrome
The product from the above-described labeling step will contain both conjugated and nonconjugated FITC. The nonconjugated FITC can be removed by either gel filtration or dialysis. Gel filtration is the preferred method, as it is considerably faster and yields a final product that should be totally free of nonreacted FITC. Sephadex G-25 normally is used for these separations, with bed dimensions of approximately 2.5-cm diameter by 30-cm length. Preequilibrate the Sephadex with PBS (0.85% [wt/vol] NaCl, 10 mM Na_2HPO_4 [pH adjusted to 7.2 with HCl], 0.01% [wt/vol] Merthiolate). Next, drain the column until the PBS is just at the bed surface, and then gently apply the FA mixture with a Pasteur pipette. Drain the FA mix into the column until the surface of the Sephadex is visible, and then add additional PBS, being careful not to disturb the layer of FITC-antibody conjugate. After adding at least 5 cm of head volume, begin filtration by allowing the column to slowly drain, adding additional PBS as required to maintain the head. Two yellow bands will be distinguishable; the upper band will be the nonreacted dye, whereas the lower, faster-running band will contain the dye-antibody conjugate. Collect the lower band, refrigerate it overnight, and then centrifuge it at 10,000 × g for 15 min. Save and filter the supernatant (FA conjugate) through a 0.45-μm-pore-size filter.

The specificity and the titer of the FA preparation should be checked. Mix the FA with an equal volume of glycerol and then, using 0.85% saline, dilute in twofold steps to 1/32 (gentle mixing, no vortexing). For each FA dilution, prepare a heat-fixed smear of the homologous organism. To evaluate the specificity of the FA, also prepare smears of as many related (strains) and unrelated organisms as is convenient. Apply 1 to 2 drops of the gelatin-rhodamine isothiocyanate (RITC) conjugate counterstain (described below) to the smear, and incubate it at room temperature for 20 min. Use a toothpick to spread 1 to 2 drops of the diluted FA on the smear, and then incubate it for 30 min at room temperature in the dark on a moistened paper towel and covered with an inverted petri dish (cover with aluminum foil). Gently rinse off excess FA with PBS, soak the preparation in PBS for 20 min and then in distilled water for 15 min, and air dry it. When viewing cells under epifluorescence, grade the fluorescing cells on a subjective scale: 0, no fluorescence; 1+ to 4+, increasingly good fluorescence (8). If the original serum had a titer of 1:1,200 or greater, one should be able to obtain 4+ reactions (brilliant yellow-green) with at least 1:8-diluted FA. If the nondiluted FA is graded at below 3+ (bright yellow-green), the conjugation step is suspect. One may be able to strengthen the fluorescence intensity of the FA by concentrating the sample. Concentration can be achieved by dialysis against polyethylene glycol. Transfer the conjugate to dialysis tubing (molecular weight cutoff, 12,000 to 14,000), and then place it on a bed of dry polyethylene glycol 20,000. Remove the dialysis bag when the volume has been reduced appropriately.

One may also be able to strengthen the fluorescence reaction by using anion-exchange chromatography to remove nonconjugated globulins (30). This separation is based on nonlabeled antibodies having less negative charge than those that are labeled; with increased label, elution requires higher ionic strength. However, since the antibodies will be variably labeled and thus variably charged, it is likely that the final volume of the yellow-colored eluted material will be considerably increased, and so it may have to be dialyzed against polyethylene glycol to reduce the volume.

CELL EXTRACTION AND RECOVERY
Counting efficiency with FAs, as with any enumeration technique, is dependent on quantitative extraction and recovery of the organism from the environmental sample. For soil and rhizosphere preparations, flocculation and centrifugation techniques have been developed for maximizing cell

recovery in a fashion that still allows for FA detection and counting. Unfortunately, given the heterogeneous nature of soils, optimization of either approach is an empirical exercise, and one may encounter some soils that are difficult (perhaps impossible) to work with in this regard. Soil characteristics that appear to be important are texture class and organic matter content. Soils with relatively high clay content tend to be good candidates for the flocculation technique, whereas sandy or silty soils may be clarified with centrifugation. While not always the case, soils that contain high levels of organic matter can present problems.

Counting efficiency for a particular soil can be determined by first sterilizing the soil, inoculating it with a known number of the organism of interest, and then conducting extraction experiments in which FA counts are compared with viable counts. Extraction techniques can also be optimized in nonsterilized soils in which FA counts are compared with viable counts obtained with an antibiotic-resistant mutant of the same strain used for antigen development. Three protocols developed to maximize extraction of cells from soil and plant roots are described below. All three can be modified as required for any soil type. Additionally, a method for direct detection on roots is provided.

Flocculation Method

The flocculation technique has been applied to both soil and rhizosphere samples (24, 40, 47). An extractant used with good success is a solution composed of hydrolyzed gelatin and $(NH_4)_2HPO_4$. A 1% (wt/vol in water) gelatin stock solution is adjusted to pH 10.5 and then partially hydrolyzed by autoclaving for 10 min. The gelatin stock solution is diluted 1:10 in 0.1 M $(NH_4)_2HPO_4$, with the final pH adjusted to 8.2, and used as the extractant.

Demezas and Bottomley (24) modified the extraction protocol of Kingsley and Bohlool (40). The cells are released from the soil particles by diluting 10 g of soil in 50 ml of the ammonium phosphate-gelatin extractant, adding 5 g of glass beads, and shaking the preparation for 15 min on a wrist action shaker. Another 45 ml of extractant is added, and the sample is shaken for an additional 15 min. A divalent cation is then added to flocculate the dispersed soil. Kingsley and Bohlool (40) achieved variable success by adding 0.2 g of Ca(OH)$_2$ and 0.5 g of MgCO$_3$ to the above-described soil-buffer mix. Demezas and Bottomley (24) compared various flocculants and found that a mixture of 0.8 g of CaCl$_2 \cdot$2H$_2$O and 0.5 g of 4MgCO$_3 \cdot$Mg(OH)$\cdot$4H$_2$O provided optimum flocculation, sample clarity, and cell recovery for the soil used in their work. After the flocculant is added and thoroughly mixed (5 min on the wrist action shaker), the suspension is allowed to settle for 1 h. Portions of the supernatant are then filtered (with suction) through Sudan black-treated polycarbonate filters (see Filter Preparation, below). A filtering manifold that accommodates at least six filters is preferred to optimize sample handling.

Centrifugation Technique

The density centrifugation technique (71) works well with a variety of soils, ranging from oxisol clays to sandy loam ultisols. This technique will likely prove superior for soils that respond poorly to flocculation treatment. Soils of this type would be expected to have textural classes dominated by the silt and sand fractions.

A 10-g soil sample is mixed with 95 ml of a 1% CaCl$_2$ solution, 1 drop of Tween 80 (Difco) and 3 drops of antifoam agent (Dow-B; Fisher Scientific) are added, and the solution is vigorously mixed for 5 min in a blender. A 10-ml aliquot is then diluted with 20 ml of a 1.33-g/cm^3 sucrose solution and layered on top of a 30-ml 1.33-g/cm^3 sucrose bed. The sample is centrifuged in a swinging-bucket rotor for 15 min. The centrifugal force used will vary depending on the soil. Centrifugation speeds generating forces of 252 × g to 715 × g yielded samples with sufficient clarity and without suffering extensive reduction in cell recovery. After centrifugation, 10 ml is removed from the sucrose layer and diluted with 90 ml of distilled H$_2$O. This suspension is then filtered through Sudan black-stained filters (see Filter Preparation, below).

Cell Recovery from Roots

A centrifugation technique to recover *B. japonicum* cells from the rhizoplane of soybean root tips also has been used (45). The soybean roots were excavated from the soil, the adhering soil was stripped away by pulling the root through the tips of sterile forceps, and the distal 2 cm of each of 10 to 15 roots was removed, weighed, and homogenized in PBS (0.15 M potassium phosphate, 0.85% [wt/vol] NaCl [pH 7.0]; 1 ml of buffer per root tip) in a Colworth-Stomacher (Sto-80; Tekmar Co., Cincinnati, Ohio). The homogenate was centrifuged for 5 min at 250 × g to sediment root debris and soil materials into a loose pellet. Portions (1.0 ml) of the supernatant were passed through filters, which were then treated with FAs.

Direct Examination of Roots

The FA technique can be useful for studying rhizosphere colonization and spatial relationships between microorganisms. Figure 2 demonstrates the use of FAs for locating sites of rhizobium colonization of root hairs and on the rhizoplane of clover roots. The following procedure for staining roots is modified from that of Bohlool (5). Shear forces resulting from vortexing or unnecessary turbulence should be avoided to minimize loss of attached microbes. Roots are first stained with an eriochrome black solution prepared by dissolving 9 mg of eriochrome black in 1.0 ml of dimethyl sulfoxide and then adding 5.0 ml of a solution containing 50 ml of dimethyl sulfoxide, 20 ml of distilled water, 10 ml of 0.1 M aluminum chloride, and 10 ml of 1.0 M acetic acid. The pH is adjusted to 5.2 with NaOH, and then distilled water is added to a final volume of 100 ml. The roots are soaked in the eriochrome solution until they appear red and then destained by soaking in 0.85% saline until the color no longer changes. Each root is placed on a glass slide and covered with gelatin-rhodamine counterstain (see below), which is allowed to dry at 65°C. FA is applied, and the root is incubated at room temperature (the incubation protocol for filters is described below) and then washed in three changes of 0.85% saline (for 10 min each time). Mounting fluid is applied along with a coverslip, and the sample is viewed.

FA STAINING OF FILTERS

Filter Preparation

The filter should be stained to obtain optimum background contrast for viewing the fluorescing cells. Polycarbonate filters (25-mm diameter) are preferred over cellulose filters, as their surface is more uniform and the cells tend not to become embedded and obscured from view. A pore size of 0.45 μm is commonly used, as 0.22-μm pores can clog quickly. This depends on the sample, and prefiltration can be used to remove organic materials and larger mineral particles, with small reductions in cell recovery (11). Prestained filters are available from vendors, but white filters can be easily stained in the laboratory (11). The staining solution is composed of 0.3% (wt/vol) Sudan black in 70% (vol/vol) ethanol. One hundred filters are placed in 500 ml of staining solution and soaked for 48 h. The solution is decanted, and the filters are washed with filtered distilled H_2O until free dye is no longer visible. The filters are transferred to aluminum foil, dried at room temperature, and conveniently stored with their spacers in their original container.

Gelatin-RITC Conjugate Counterstain

Bohlool and Schmidt (7) developed a gelatin RITC conjugate that is effective for controlling nonspecific staining (see Nonspecific Staining, below). The gelatin helps eliminate nonspecific binding of the FA and, when applied as a conjugate with RITC, provides an effective counterstain that gives good contrast to the green fluorescence of an FITC-labeled antibody. To prepare the gelatin-RITC conjugate, mix 2 g of gelatin with 100 ml of water, adjust the pH to 10 to 11 with 1 N NaOH, and then autoclave the sample for 10 min at 121°C. After cooling, readjust the sample to the original pH. In a separate container, dissolve 16 mg of RITC in a minimum volume of acetone, filter the sample through a 0.45-μm-pore-size filter to remove undissolved materials, and then add it to the gelatin solution. Conjugate the preparation overnight with slow stirring, and remove the nonconjugated RITC by dialysis against PBS (pH 7.2). Add Merthiolate (1 : 10,000), and divide the sample into 20-ml portions. Working stocks can be refrigerated, and excess volumes can be frozen (−20°C).

Sample Staining

After the sample has been extracted by one of the methods discussed above, an aliquot is collected on the Sudan black-stained filter, which is transferred to a glass slide. One to two drops of the gelatin-RITC conjugate is applied and gently spread with a sterile toothpick, and the sample is incubated at room temperature until nearly dry. The appropriately diluted FA is applied (2 to 3 drops per filter), and the filter and slide are placed on a moist paper towel, covered with a glass petri dish, and incubated at room temperature for 30 min in the dark. The stained filter is returned to the filtering manifold and washed (using vacuum) with at least 100 ml of 0.22-μm-pore-size filtered 0.85% saline.

The foregoing protocol is for the direct FA approach. The indirect FA technique is similar except that two antibodies are used. Primary antibodies against the target organism are made as described above but are not conjugated to a fluorochrome. The unlabeled primary antibody preparation is applied to the filter and incubated as for labeled antibodies. A secondary fluorochrome-labeled antibody that recognizes rabbit immunoglobulins (usually goat anti-rabbit antibodies) is then added. The secondary antibody is available commercially and is conjugated to the fluorescent dye as described for the direct approach. The advantage of the indirect FA approach is that one can use the same fluorochrome-labeled secondary antibody with any primary antibody preparation derived from serum of the same animal species. This avoids having to label each primary antibody used, which can be very convenient if large numbers of serologically different organisms are sought in the same environmental sample.

COUNTING

Different fluorescence microscope systems have been described (24, 60). In general, the microscope is equipped with a light source that provides the correct intensity and wavelength and with the filters necessary for the fluorochrome being used. The reader should contact sales representatives for information regarding the requirements for conversion of current microscopy equipment for use in epifluorescence work. It should also be pointed out that confocal microscopy offers distinct advantages over conventional microscopy for FA work (see below).

The stained and washed filter is transferred to a glass slide and marked with appropriate notations for identification purposes. Just prior to counting, 1 to 2 drops of FA mounting fluid (Difco) and a coverslip are applied to the filter. Allow a few minutes for the mounting fluid to spread evenly across the filter. At least 20 randomly chosen microscopic fields are examined for cells that brightly fluoresce. The number of bacteria in the sample is then estimated by the following equation (58): number/g of soil = $(N \times A \times D)/a \times V$, where N is the average number of cells per microscopic field, A is the effective filtering area of the filter (square centimeters), D is the dilution factor, a is the area of the microscopic field (square centimeters), and V is the volume (milliliters) of diluted soil solution passed through the filter. Both A and a can vary depending on the equipment used. See reference 11 for comments regarding detection limits and reproducibility of the FA technique with soil samples.

PROBLEMS AND LIMITATIONS OF THE FA TECHNIQUE

Potential problems with the FA technique include autofluorescence, nonspecific staining (antibody specificity), antigen stability, and determination of viability. Given the great heterogeneity of soils and rhizopheres, these problems may not be an issue in some samples, but they may present significant difficulties in others.

Autofluorescence

Autofluorescence results from plant tissue, soil, or organic debris fluorescing as a result of exposure to UV light. Often, the autofluorescing components of rhizosphere samples are readily distinguishable from the target organism by virtue of size, color, and/or intensity and do not interfere with organism counts or sample interpretations. However, autofluorescence can potentially render some samples unworkable. Normally, organic debris does not present problems unless the content is particularly high (1), but this problem can largely be suppressed by using counterstains (6). Plant tissue will autofluoresce, the intensity depending on species, tissue age, and tissue type, with each case perhaps requiring a unique treatment. Counterstains such as RITC (7, 55) and crystal violet (4) have been used to reduce plant autofluorescence to levels at which the target organism can be clearly distinguished. Most studies have used FITC as the fluorescent dye conjugated to the immunoglobulins, but plant tissues autofluoresce when exposed to light wavelengths typically used for FITC-labeled FAs. Brlansky et al. (15) avoided this problem by conjugating the antibodies to tetramethylrhodamine isothiocyanate, which has excitation and emission wavelengths different from those of FITC. Goldman (30) and Kawamura (38) provide lists of alternate fluorochromes.

The application of laser scanning confocal microscopy

to FA work (21, 56) promises to significantly enhance the power of the FA technique. The confocal imaging system allows for only signals from the focused plane to be detected (reviewed in references 62, 66, and 69), excluding background fluorescence arising from materials such as plant tissue, soil particles, or organic debris (see Fig. 2 for examples of confocal and conventional microscopy images). Confocal microscopy is noninvasive, provides a clear advantage over the conventional microscope with respect to definition and contrast, and greatly diminishes autofluorescence. Schloter et al. (56) demonstrated the usefulness of confocal microscopy for eliminating autofluorescence on wheat roots. In samples probed with tetramethylrhodamine isothiocyanate-labeled monoclonal antibodies raised against *Azospirillum brasilense*, they used a dual-laser system to excite the autofluorescence of the root to produce a green background upon which the distinctive red *Azospirillum* cells could easily be seen. They also used the optical sectioning ability of the confocal microscope to locate the *Azospirillum* cells within the root mucigel layer.

Nonspecific Staining

Nonspecific staining may result from binding of the antibody-dye conjugate to soil or plant components (60). Such apparent infidelity is not totally unexpected because regardless of whether polyclonal or monoclonal antibodies are used, the binding of proteins to soil clays and organic matter is well documented. The standard remedy for this problem is to prestain the sample with a gelatin-counterstain (typically RITC [7]) conjugate. This procedure effectively swamps potential nonspecific protein binding sites on soil or plant materials with a different proteinaceous material prior to application of the FA. Other possible sources of nonspecific staining may result from incomplete removal of nonconjugated fluorochrome, and when possible one should use the direct FA rather than indirect FA approach, as the latter requires a second layer of antibodies that may introduce additional nonspecific binding.

A potentially more serious type of nonspecific staining is due to antigenic cross-reactivity with nontarget organisms. Cross-reaction between totally unrelated organisms has been reported (8, 19). There also are instances of fungus spores behaving as "universal acceptors" for FAs, resulting in considerable cross-reaction or nonspecific staining (9). Cross-reaction also was observed between genera of fungi when hyphae were used as antigen (70). The investigator must determine the feasibility of FAs for each case. If the study involves introduction of an organism that has no history in the environment of interest, then FAs against the target organisms are applied to sample preparations prior to introduction to determine if there are cross-reactive organisms present. If nonspecific reactions are evident, then one may try isolating the cross-reactive organism and use it to absorb out the related antibodies (see reference 52 for absorption protocols).

If the immunoglobulin preparation is of suitable titer, it should be enriched with antibodies specific for particular epitopes of the target organism. Therefore, dilution of the FA to an appropriate working titer may reduce cross-reaction staining. Also, a minimum staining intensity must be established for which a positive score is recorded for the sample or cells in question (60). The use of low-titer FA and assignment of faintly staining cells as positive (19) can lead to confusion regarding the specificity of the FA.

Antigen Stability

It is important to determine that the antigenic determinants used to generate the antibodies from pure culture cells are sufficiently expressed by the organism in the natural setting. Significant changes in growth conditions may bring about changes in antigens or epitopes. Bohlool and Schmidt (9) cite several examples which suggest that antigen stability is likely to be not a problem. Also, some rhizobia are well known for the gross morphological changes that they undergo during the transition from free-living to symbiotic life styles, and there is evidence that the bacteroid outer membrane is structurally unstable (2, 3). Still, FAs are routinely used to identify bacteroids for the determination of nodule occupancy in interstrain competition studies (24, 46, 49). Monoclonal FAs also have been used in soil studies to differentiate between specific members of a serocluster; again, antigen stability was noted (41).

Even with the foregoing examples of antigen stability (and numerous others not cited), there are other reports in which antigen stability proved to be a problem. Sadowsky et al. (54) noted that antibodies generated from cells of different strains of *Rhizobium fredii* cultured on certain media would not react with cells cultured on some other media. Further, staining intensity was affected by growth phase, and cells apparently always required heat treatment prior to FA application. These peculiar features of FA staining suggest that some bacteria may have relatively few surface antigens that profoundly influence serological specificity, that heat treatment influences antigen structure such that reactive epitopes are exposed upon heat denaturation (but not in the native conformation), and that these dominant antigens are expressed transiently, depending on nutritional and physiological conditions. All of these points need to be taken into account when one is preparing antigen for injection and when one is preparing cells or samples for FA probing. If the organism of interest requires heat treatment for adequate FA binding, then root or rhizosphere studies may prove impractical, as significant plant protein denaturation can result, fouling the sample and making filtration difficult or impossible.

Viability

Until recently, a major drawback with FAs has been their inability to differentiate between live and dead cells. Some of the earliest reports pointed out problems associated with use of FAs in studies in which normal turnover of dead cells does not occur in a timely fashion (57). In sterilized soils, for example, dead cells may remain recognizable by FAs for several days. Kennedy and Wollum (39) demonstrated that the FA technique was substantially less sensitive than viable plate counts and the most-probable-number technique (based on plant infectivity) for estimating surviving cells after heat treatment of soil. Bottomley and Maggard (12) adapted an FA-direct viable count method developed for aquatic bacteria (14, 17, 35, 53) for assessing the viability of FA-detectable cells. After the sample has been extracted, yeast extract and nalidixic acid are added to the solution containing the recovered cells. The amended solution is incubated for 16 to 24 h and then processed for FA counting. This FA technique detects elongated cells that are responding to the addition of yeast extract as a growth-yielding substrate, and thus are deemed viable, but in the presence of nalidixic acid fail to complete division.

Brian Kinkle provided an excellent critical review of the manuscript, and Frank Dazzo contributed photographs for Fig. 2.

Research in my laboratory is supported by grants from the U.S. Department of Agriculture (94-37305-0574) and the National Science Foundation (IBN-941385 and IBN-9420798).

REFERENCES

1. **Apel, W. A., P. R. Dugan, J. A. Filppi, and M. S. Rheins.** 1976. Detection of *Thiobacillus ferrooxidans* in acid mine by indirect fluorescent antibody staining. *Appl. Environ. Microbiol.* **32:**159–165.

2. **Bal, A. K., S. Shantharam, and D. P. S. Verma.** 1980. Changes in the outer cell wall of *Rhizobium* during development of root nodule symbiosis in soybean. *Can. J. Microbiol.* **26:**1097–1103.

3. **Bal, A. K., and P. P. Wong.** 1982. Infection process and sloughing off of rhizobial outer membrane in effective nodules of lima bean. *Can. J. Microbiol.* **28:**890–896.

4. **Bilal, R., G. Rasul, M. Arshad, and K. A. Malik.** 1993. Attachment, colonization and proliferation of *Azospirillum brasilense* and *Enterobacter* spp. on root surfaces of grasses. *World J. Microbiol. Biotechnol.* **9:**63–69.

5. **Bohlool, B. B.** 1987. Fluorescence methods for study of *Rhizobium* in culture and in situ, p. 127–147. *In* G. H. Elkan (ed.), *Symbiotic Nitrogen Fixation Technology*. Marcel Dekker, Inc. New York.

6. **Bohlool, B. B., and T. D. Brock.** 1974. Immunofluorescence approach to the study of the ecology of *Thermoplasma acidophilum* in coal refuse material. *Appl. Microbiol.* **28:**11–16.

7. **Bohlool, B. B., and E. L. Schmidt.** 1968. Nonspecific staining: its control in immunofluorescence examination of soil. *Science* **162:**1012–1014.

8. **Bohlool, B. B., and E. L. Schmidt.** 1970. Immunofluorescent detection of *Rhizobium japonicum* in soil. *Soil Sci.* **110:**229–236.

9. **Bohlool, B. B., and E. L. Schmidt.** 1980. The immunofluorescence approach in microbial ecology. *Adv. Microb. Ecol.* **4:**203–241.

10. **Boothroyd, M., and D. L. Georgala.** 1964. Immunofluorescent identification of *Clostridium botulinum*. *Nature* (London) **202:**515–516.

11. **Bottomley, P. J.** 1994. Light microscopic methods for studying soil microorganisms, p. 81–105. *In* R. Weaver, J. S. Angle, and P. J. Bottomley (ed.), *Methods of Soil Analysis*. Soil Science Society of America, Madison, Wis.

12. **Bottomley, P. J., and S. P. Maggard.** 1990. Determination of viability within serotypes of a soil population of *Rhizobium leguminosarum* bv. trifolii. *Appl. Environ. Microbiol.* **56:**533–540.

13. **Brakke, M. K.** 1990. Preparation of antigens, viruses, p. 15–26. *In* R. Hampton, E. Ball, and S. De Boer (ed.), *Serological Methods for Detection and Identification of Viral and Bacterial Plant Pathogens. A Laboratory Manual*. The American Phytopathological Society, St. Paul, Minn.

14. **Brayton, P. R., M. L. Tamplin, A. Huq, and R. R. Colwell.** 1987. Enumeration of *Vibrio cholerae* O1 in Bangladesh waters by fluorescent antibody direct count. *Appl. Environ. Microbiol.* **53:**2862–2865.

15. **Brlansky, R. H., R. F. Lee, and L. W. Timmer.** 1981. Detection of plant rickettsialike bacteria *in situ* using immunofluorescence. *Phytopathology* **71:**863. (Abstract.)

16. **Chung, K. L., R. Z. Hawirko, and P. K. Isaac.** 1964. Cell wall replication. I. Cell wall growth of *B. cereus* and *B. megaterium*. *Can. J. Microbiol.* **10:**43–48.

17. **Colwell, R. R., P. R. Brayton, D. J. Grimes, D. B. Roszak, S. A. Huq, and L. M. Palmer.** 1985. Viable but nonculturable *Vibrio cholerae* and related pathogens in the environment. Implications for release of genetically-engineered microorganisms. *Bio Technology* **3:**817–820.

18. **Coons, A. H., H. J. Creech, and R. N. Jones.** 1941. Immunological properties of an antibody containing a fluorescent group. *Proc. Soc. Exp. Biol. Med.* **47:**200–202.

19. **Crowley, C. F., and S. H. De Boer.** 1982. Nonpathogenic bacteria associated with potato stems cross-react with *Corynebacterium sepedonicum* antisera in immunofluorescence. *Am. Potato J.* **59:**1–8.

20. **Davies, S. N.** 1951. The serology of *Bacillus polymyxa*. *J. Gen. Microbiol.* **5:**807–816.

21. **Dazzo, F. B., and S. F. Wright.** Production of anti-microbial antibodies and their use in immunofluorescence microscopy. *In* A. Akkermans, J. van Elsas, and F. de Bruijn (ed.), *Molecular Microbial Ecology Manual*, in press. Kluwer Academic Publishers, Dordrecht, The Netherlands.

22. **De Boer, S. H.** 1984. Enumeration of two competing *Erwinia carotovora* populations in potato tubers by a membrane filter-immunofluorescence procedure. *J. Appl. Bacteriol.* **57:**517–522.

23. **De Boer, S. H., and N. W. Schaad.** 1990. Preparation of antigens, bacteria, p. 27–31. *In* R. Hampton, E. Ball, and S. De Boer (ed.), *Serological Methods for Detection and Identification of Viral and Bacterial Plant Pathogens. A Laboratory Manual*. The American Phytopathological Society, St. Paul, Minn.

24. **Demezas, D. H., and P. J. Bottomley.** 1986. Autecology in rhizospheres and nodulating behavior of indigenous *Rhizobium trifolii*. *Appl. Environ. Microbiol.* **52:**1014–1019.

25. **Diem, H. G., G. Godbillon, and E. L. Schmidt.** 1977. Application of the fluorescent antibody technique to the study of an isolate of *Beijerinckia* in soil. *Can. J. Microbiol.* **23:**161–165.

26. **Ford, S., and B. H. Olson.** 1988. Methods for detecting genetically engineered microorganisms in the environment. *Adv. Microb. Ecol.* **10:**45–79.

27. **Frankland, J. C., A. D. Bailey, T. R. G. Gray, and A. A. Holland.** 1981. Development of an immunological technique for estimating mycelial biomass of *Mycena galopus* in leaf matter. *Soil Biol. Biochem.* **13:**87–92.

28. **Freter, R.** 1976. Agglutinin titration (Widal) for the diagnosis of enteric fever and other enterobacterial infections, p. 285–288. *In* N. R. Rose and H. Friedman (ed.), *Manual of Clinical Immunology*. American Society for Microbiology, Washington, D. C.

29. **Friese, C. F., and M. F. Allen.** 1991. Tracking the fates of exotic and local VA mycorrhizal fungi: methods and patterns. *Agric. Ecosyst. Environ.* **34:**87–96.

30. **Goldman, M.** 1968. *Fluorescent Antibody Methods*. Academic Press, New York.

31. **Gordon, M. A.** 1958. Differentiation of yeasts by means of fluorescent antibody. *Proc. Soc. Exp. Biol. Med.* **97:**694–698.

32. **Hampton, R., E. Ball, and S. De Boer (ed.).** 1990. *Serological Methods for Detection and Identification of Viral and Plant Pathogens*. American Phytopathological Society, St. Paul, Minn.

33. **Harlow, E., and D. Lane.** 1988. *Antibodies: a Laboratory Manual*. Cold Spring Harbor Laboratory, Cold Spring Harbor, N.Y.

34. **Hill, I. R., and T. R. G. Gray.** 1967. Application of the fluorescent antibody technique to an ecological study of bacteria in soil. *J. Bacteriol.* **93:**1888–1897.

35. **Hussong, D., R. R. Colwell, M. O'Brien, E. Weiss, A. D. Pearson, R. M. Weiner, and W. D. Burge.** 1987. Viable *Legionella pneumophila* not detectable by culture on agar media. *Bio/Technology* **5:**947–950.

36. **Kaufman, L., and B. Brandt.** 1964. Fluorescent-antibody studies of the mycelial form of *Histoplasma capsulatum* and morphologically similar fungi. *J. Bacteriol.* **87:**120–126.

37. **Kaufman, L., and W. B. Cherry.** 1961. Technical factors affecting the preparation of fluorescent antibody reagents. *J. Immunol.* **87:**72–79.

38. **Kawamura, A., Jr. (ed.).** 1969. *Fluorescent Antibody Techniques and Their Applications*. University of Tokyo Press, Tokyo, and University Park Press, Baltimore.

39. Kennedy, A. C., and A. G. Wollum II. 1988. Enumeration of *Bradyrhizobium japonicum* in soil subjected to high temperature: comparison of plate count, most probable number and fluorescent techniques. *Soil Biol. Biochem.* **20:** 933–937.

40. Kingsley, M. T., and B. B. Bohlool. 1981. Release of *Rhizobium* spp. from tropical soils and recovery for immunofluorescence enumeration. *Appl. Environ. Microbiol.* **42:** 241–248.

41. Kinkle, B. K., and E. L. Schmidt. 1992. Stability of a monoclonal antibody determinant in soil populations of *Bradyrhizobium japonicum*. *Soil Biol. Biochem.* **24:**819–820.

42. Kough, J. E., N. Malajczuk, and R. G. Linderman. 1983. Use of the indirect immunofluorescent technique to study the vesicular-arbuscular fungus *Gigaspora erthropa*, a new species forming arbuscular mycorrhizae. *Mycologia* **76:** 250–255.

43. Kough, J. E., N. Malajczuk, and R. G. Linderman. 1983. Use of the indirect immunofluorescent technique to study the vesicular-arbuscular fungus *Glomus epigaeum* and other *Glomus* species. *New Phytol.* **94:**57–62.

44. Macario, A. J. L., and E. Conway de Macario (ed.). 1985. *Monoclonal Antibodies against Bacteria*, vol. I and II. Academic Press, Orlando, Fla.

45. McDermott, T. R., and P. H. Graham. 1989. *Bradyrhizobium japonicum* inoculant mobility, nodule occupancy, and acetylene reduction in the soybean root system. *Appl. Environ. Microbiol.* **55:**2493–2498.

46. McDermott, T. R., P. H. Graham, and M. L. Ferrey. 1991. Competitiveness of indigenous populations of *Bradyrhizobium japonicum* serocluster 123 as determined using a root-tip marking procedure in growth pouches. *Plant Soil* **135:**245–250.

47. Moawad, H. A., W. R. Ellis, and E. L. Schmidt. 1984. Rhizosphere response as a factor in competition among three serogroups of indigenous *Rhizobium japonicum* for nodulation among of field-grown soybeans. *Appl. Environ. Microbiol.* **47:**607–612.

48. Paik, G., and M. T. Suggs. 1970. Reagents, stains, and miscellaneous test procedures, p. 930–950. *In* E. H. Lennette, E. H. Spaulding, and J. P. Truant (ed.), *Manual of Clinical Microbiology*. American Society for Microbiology, Washington, D.C.

49. Pepper, I. L., K. L. Josephson, C. S. Nautiyal, and D. P. Bourque. 1989. Strain identification of highly-competitive bean rhizobia isolated from root nodules: use of fluorescent antibodies, plasmid profiles and gene probes. *Soil Biol. Biochem.* **21:**749–753.

50. Preece, T. F., and D. J. Cooper. 1969. The preparation and use of a fluorescent antibody reaction for *Botrytis cinerea* grown on glass slides. *Trans. Br. Mycol. Soc.* **52:** 99–104.

51. Reinhold, B., T. Hurek, and I. Fendrik. 1987. Cross-reaction of predominant nitrogen-fixing bacteria with enveloped, round bodies in the root interior of Kallar grass. *Appl. Environ. Microbiol.* **53:**889–891.

52. Robert, F. M., and E. L. Schmidt. 1985. Somatic serogroups among 55 strains of *Rhizobium phaseoli*. *Can. J. Microbiol.* **31:**519–523.

53. Rollins, D. M., and R. R. Colwell. 1986. Viable but nonculturable stage of *Campylobacter jejuni* and its role in survival in the natural aquatic environment. *Appl. Environ. Microbiol.* **52:**531–538.

54. Sadowsky, M. J., B. B. Bohlool, and H. H. Keyser. 1987. Serological relatedness of *Rhizobium fredii* to other rhizobia and to the bradyrhizobia. *Appl. Environ. Microbiol.* **53:** 1785–1789.

55. Schank, S. C., R. L. Smith, G. C. Weiser, D. A. Zuberer, J. H. Bouton, K. H. Quesenberry, M. E. Tyler, J. R. Milam, and R. C. Littell. 1979. Fluorescent antibody technique to identify *Azospirillum brasilense* associated with roots of grasses. *Soil Biol. Biochem.* **11:**287–295.

56. Schloter, M., R. Borlinghaus, W. Bode, and A. Hartmann. 1993. Direct identification and localization of *Azospirillum* in the rhizosphere of wheat using fluorescence-labeled monoclonal antibodies and confocal scanning laser microscopy. *J. Microsc.* **171:**173–177.

57. Schmidt, E. L. 1973. Fluorescent antibody techniques for the study of microbial ecology. *Bull. Ecol. Comm.* (Stockholm) **17:**67–76.

58. Schmidt, E. L. 1974. Quantitative autecological study of microorganisms in soil by immunofluorescence. *Soil Sci.* **118:**141–149.

59. Schmidt, E. L., and R. O. Bankole. 1962. Detection of *Aspergillus flavus* in soil by immunofluorescent staining. *Science* **136:**776–777.

60. Schmidt, E. L., R. O. Bankole, and B. B. Bohlool. 1968. Fluorescent-antibody approach to study of rhizobia in soil. *J. Bacteriol.* **95:**1987–1992.

61. Schmidt, E. L., J. A. Biesbrock, B. B. Bohlool, and D. H. Marx. 1974. Study of mycorrhizae by means of fluorescent antibody. *Can. J. Microbiol.* **20:**137–139.

62. Shotton, D., and N. White. 1989. Confocal scanning microscopy: three-dimensional biological imaging. *Trends Biochem Sci.* **14:**435–439.

63. Somasegaran, P., and H. J. Hoben. 1994. *Handbook for Rhizobia*. Springer-Verlag, New York.

64. Stefani, E. 1989. In situ detection of *Cavibacter michiganensis* sp. *sepedonicus* in potato stems using the fluorescent antibody technique. *Phytopathol. Mediterr.* **28:**53–56.

65. Tchan, Y. T., and R. DeVille. 1970. Application de l'immunofluorescence a l'etude des *Azotobacter* du sol. *Ann. Inst. Pasteur* (Paris) **118:**665–673.

66. Tekola, P., Q. Zhu, and J. P. A. Baak. 1994. Confocal laser microscopy and image processing for three-dimensional microscopy. *Hum. Pathol.* **25:**12–21.

67. Van Regenmortel, M. H. V. 1982. *Serology and Immunochemistry of Plant Viruses*. Academic Press, New York.

68. Vincent, J. M. 1970. *A Manual for the Practical Study of Root-Nodule Bacteria*. IBP handbook no. 15. Blackwell Scientific Publications, Oxford.

69. White, J. G., W. B. Amos, and M. Fordham. 1987. An evaluation of confocal versus conventional imaging of biological structures by fluorescence light microscopy. *J. Cell Biol.* **105:**41–48.

70. Wilson, J. M., M. J. Trinick, and C. A. Parker. 1983. The identification of vesicular-arbuscular mycorrhizal fungi using immunofluorescence. *Soil Biol. Biochem.* **15:** 439–445.

71. Wollum, A. G., II, and R. H. Miller. 1980. Density centrifugation method for recovering *Rhizobium* spp. from soil for fluorescent-antibody studies. *Appl. Environ. Microbiol.* **39:**466–469.

72. Wright, S. F. 1994. Serology and conjugation of antibodies, p. 593–618. *In* R. Weaver, J. S. Angle, and P. J. Bottomley (ed.), *Methods of Soil Analysis*. Soil Science Society of America, Madison, Wis.

Reporter Gene Systems Useful in Evaluating In Situ Gene Expression by Soil- and Plant-Associated Bacteria

JOYCE E. LOPER AND STEVEN E. LINDOW

54

A reporter gene system is composed of a gene lacking its endogenous promoter, which will be transcribed only if placed downstream from an exogenous promoter. Consequently, the product of a reporter gene "reports" the transcriptional activity of a promoter to which it is fused. A gene is selected to serve as a reporter on the basis of several criteria, including the ease and sensitivity with which its product can be detected. Both reporter genes and marker genes encode a gene product that can be detected conveniently, but a marker gene is transcribed and translated from an endogenous promoter, whereas a reporter gene is transcribed from an exogenous promoter. The utility of marker genes in microbial ecology is discussed by Shaw et al. (chapter 52). All reporter genes lack their endogenous promoters, but some also lack a ribosome binding site and a translational start site. These reporter genes generate translational fusions, resulting in a chimeric protein. For studies of microbial ecology, reporter gene systems that generate transcriptional fusions have been used more commonly than those that generate translational fusions. This chapter focuses on those reporter gene systems that are useful in assessing the in situ transcriptional activity of promoters in bacterial cells in soil or plant tissues or on plant surfaces.

Most genes in microorganisms do not encode products that are conveniently detected or quantified in culture or especially in the habitats that microbes occupy in nature. The concept of using a gene that confers a readily assayable product to report the activity of another gene was a major breakthrough in molecular biology. Reporter genes have been indispensable tools used by molecular biologists to develop our present understanding of gene regulation in both prokaryotes and eukaryotes. More recently, reporter gene systems useful in assessing in situ gene expression by bacteria in natural habitats have been described (52).

CHARACTERISTICS OF REPORTER GENES USEFUL IN MICROBIAL ECOLOGY

To serve as useful tools for assessment of gene expression by bacteria inhabiting natural substrates, reporter genes must meet a number of criteria that are not required of those systems used exclusively to monitor gene expression by bacteria grown in culture. The most important criteria are listed below.

Detectability

The product of a reporter gene must be uncommon in the natural substrates to be evaluated. Background levels of the gene product contributed by organisms indigenous to the environment decrease the sensitivity of the reporter gene system and interfere with detection of even moderate levels of gene expression.

Biological or physical components of the environment can obscure the detection of certain reporter gene products. For example, soil particles can conceal products detected visually, and naturally occurring pigments can quench light reactions that otherwise could be detected and quantified easily.

Sensitivity

Bacteria in natural habitats are often found in microcolonies or even as individual cells rather than in the large cell aggregates typical of bacterial cultures. Therefore, sensitivity of detection is frequently an overriding consideration in selecting a reporter gene system for studies assessing in situ gene expression by bacteria inhabiting natural substrates. Sensitivity reflects the minimum concentration of a reporter gene product that can be detected in an environmental sample. At present, the most sensitive reporter gene systems can detect gene expression by a single bacterial cell if the target promoter is actively transcribed (82). Sensitive reporter genes provide an opportunity to evaluate temporal and spatial heterogeneity in gene expression.

Reliable Quantification

Bacteria inhabiting natural habitats may be subject to dramatic changes in nutrient availability and may periodically be exposed to stresses, such as those imposed by desiccation, anoxic environments, changes in osmolarity, or a variety of other factors. The optimal reporter gene system would be expressed by bacteria inhabiting variable and dynamic environments. In reality, all gene products are affected by at least some environmental or physiological factors. Therefore, the major factors influencing the stability and activity of reporter gene products must be defined and considered in selecting a reporter gene, designing the experiment, and interpreting experimental data.

Specific Activity

Quantitative estimates of reporter gene product concentrations are most useful if they can be normalized for the number of cells containing a reporter gene fusion. Differences in reporter gene product concentration can then be partitioned into effects contributed by changes in bacterial population size and by changes in gene expression. Reliable genetic markers, such as antibiotic resistance and others (44; chapter 52, this volume), facilitate quantification of culturable cells of the bacterial strain containing the reporter gene fusion. Certain bacteria, such as epiphytic strains of *Pseudomonas syringae* on leaf surfaces (93), exist in a culturable state in their natural habitats, but the nonculturability of organisms in many other habitats (77) complicates estimates of cell numbers that contribute the reporter gene product.

Responsiveness to Changes in Transcriptional Activity

Ideally, a reporter gene product should serve as a quantitative estimate of the immediate rate of transcription of the gene to which it is fused. A stable reporter gene product accumulates over time, however, and its concentration reflects cumulative transcriptional activity of the target promoter rather than the current transcriptional activity. For example, if transcription of a target promoter increases with time, then the reporter gene product encoded just prior to detection will constitute the majority present in the cell and will provide a good estimate of current transcriptional activity. If the target promoter is undergoing repression, however, a large proportion of a stable reporter gene product will have been produced earlier, and its concentration will not reflect the current transcriptional activity of the target promoter. We recognize that responsiveness comes at the cost of sensitivity: a labile reporter gene product may not accumulate to detectable concentrations. Therefore, a reporter gene should be selected to provide the best compromise between sensitivity and stability.

REPORTER GENE SYSTEMS USEFUL IN ENVIRONMENTAL MICROBIOLOGY

Unfortunately, the reporter gene systems used to assess in situ gene expression by bacteria in natural habitats have not been subjected to careful side-by-side comparisons according to the criteria listed above. Nevertheless, we highlight here the advantages and disadvantages of the major reporter genes that are useful for evaluating in situ gene expression by soil or plant-associated bacteria. We also refer the reader to a recent review of these and other reporter genes used in microbial ecology (52).

lacZ

The *lacZ* gene encodes the enzyme β-galactosidase, which cleaves the disaccharide lactose into glucose and galactose. β-Galactosidase can be quantified by a convenient colorimetric assay and can be detected by the development of a blue color in colonies grown on agar media supplemented with the substrate 5-bromo-4-chloro-3-indolyl-β-D-galactopyranoside (X-Gal). β-Galactosidase can also be quantified in extremely sensitive assays generating fluorescent and chemiluminescent products (7, 60, 96). Because of the ease and sensitivity of its detection and the large number of plasmid vectors and transposons available for making transcriptional and translational fusions, *lacZ* probably is the most common reporter gene used in studies of gene regulation by bacteria in culture (83). Background levels of β-galactosidase activity expressed by many indigenous bacteria (3, 11), plants, and animals limit the use of the *lacZ* reporter for assessing

in situ gene expression by bacteria inhabiting natural substrates. Fortunately, bacterial and eukaryotic β-galactosidases are differentially sensitive to high pH (7) and heat (95); therefore, eukaryotic enzymes can be selectively inactivated and the activities of microbial *lacZ* gene fusions can be assessed directly on plants (1). Pigmented and fluorescent compounds present in environmental samples (such as plant tissues or soil) may interfere with sensitive colorimetric or fluorometric assays of β-galactosidase activity (7). In spite of these difficulties, in situ gene transcription by many plant-associated bacteria has been assessed successfully by using *lacZ* fusions (1, 36, 65). The relatively large numbers of bacterial cells containing the *lacZ* fusion in plant tissues or on plant surfaces apparently allowed measurement of bacterial β-galactosidase despite background levels of enzymatic activity.

gusA

The *gusA* (also called *uidA*) gene of *Escherichia coli* encodes β-glucuronidase, which catalyzes the hydrolysis of a wide range of glucuronides (21, 39, 40). The most common substrate for this enzyme is 5-bromo-4-chloro-3-indolyl-β-D-glucuronide (X-Gluc), which produces a blue chromophore upon cleavage (39). Other compounds that produce a fluorescent pigment upon cleavage are also available (7, 8, 21, 60). *gusA* fusions are used extensively in studies of gene expression by viral plant pathogens, plants, yeasts, and filamentous fungi, all of which lack indigenous β-glucuronidase activity. *gusA* has also been useful in assessing gene expression by bacterial pathogens and symbionts of plants (55, 58, 92), especially in resolving spatial patterns of in situ transcriptional activity of bacterial cells (80). Its use in natural habitats containing mixed bacterial populations may be limited, however, by the presence of β-glucuronidase in certain indigenous bacteria (21).

xylE

The *xylE* gene encodes catechol-2,3-oxygenase, which converts catechol to 2-hydroxymuconic semialdehyde, a bright yellow pigment that can be visualized or measured spectrophotometrically (78, 97). The *xylE* gene, originally found on the TOL plasmid in *Pseudomonas putida*, is uncommon in other organisms. The stability of catechol-2,3-oxygenase varies with oxygen availability (66) and with the physiological status of the bacterial cell (74). In stationary-phase cells of *P. putida*, for example, specific activity of the enzyme decreased by 39% in 20 min, whereas the activity was relatively stable in exponentially growing cells. The enzyme also is inhibited dramatically by activated oxygen species (74). Therefore, quantification of *xylE* activity is subject to error unless the factors influencing stability of catechol-2,3-oxygenase are taken into account. Nevertheless, the *xylE* reporter was used successfully to study in situ gene expression by *P. putida* inhabiting the rhizosphere (9).

Genes Conferring Bioluminescence

Bacterial luciferases catalyze a light-emitting reaction that requires O_2, a long-chain aliphatic aldehyde substrate, and a source of reducing equivalents, usually reduced flavin mononucleotide ($FMNH_2$) (61). The active form of luciferase is a dimer encoded by the *luxAB* genes (61). In the marine organisms *Vibrio fischeri* and *Vibrio harveyi*, substrate biosynthesis is conferred by products of *luxCDE*. Both O_2 and $FMNH_2$ are available in most bacterial cells; therefore, virtually any bacterium can produce light in an aerobic environment if it contains the entire *luxCDABE* operon. In natural habitats, nutritional resources required for substrate production by bacteria may be limiting; therefore, many researchers use *luxAB* as a reporter and add the aldehyde substrate exogenously. *N*-Decanal, the most commonly used substrate, can be added to cells in the vapor phase (81, 84) or as an emulsion to cells suspended in aqueous media (13). The gene (*luc*) conferring luminescence in the North American firefly *Photinus pyralis* also has been used as a reporter. The *luc* reporter probably is of limited use in bacteria inhabiting natural substrates because the luciferin substrate must be added exogenously and does not readily penetrate into microbial cells (45). Several excellent reviews of the biochemistry and genetics of bacterial bioluminescence have been published (33, 61, 84).

Genes conferring bacterial bioluminescence are very useful reporters for studies of in situ gene expression for many reasons. (i) There is little background luciferase activity in most environmental samples. (ii) Luciferase can be quantified conveniently with scintillation counters, dedicated luminometers, or other photon-counting devices (11, 33). (iii) Luciferase can be detected very sensitively with charge-coupled device video cameras (6, 18, 33, 81). In fact, luminescence of an individual cell containing a highly expressed *lux* operon and growing under conditions in which metabolic activity is high and O_2 is abundant can be detected by this method (82). (iv) Because light emission by individual cells or, more commonly, cell aggregates in environmental substrates can be detected directly and nondestructively, spatial and temporal patterns of the in situ transcriptional activity of cells can be resolved (11, 70, 81).

Despite its virtues, the *lux* reporter system is not without limitations. Because bioluminescence requires abundant O_2 and $FMNH_2$ (61), it will not be expressed by cells growing in anaerobic environments or by metabolically inactive cells. More importantly, light emission will not be proportional to the rate of *lux* gene transcription unless cells produce $FMNH_2$ in nonlimiting amounts (i.e., unless cells are metabolically active). Light emission from bioluminescent cells suspended in translucent aqueous environments or present on the surfaces of different substrates can be measured, but the intensity of light emitted can be quenched by particulate material (i.e., soil particles) or soluble pigments. The extent of quenching must be determined if bioluminescence of bacteria in environmental samples is to be quantified accurately. Luciferase is a stable enzyme, and therefore the *lux* reporter will not respond rapidly to changes in transcriptional activity of the target promoter. Because luciferase is unstable at 37°C and can be denatured in vivo by mild temperature shock (84), its responsiveness can be maximized by a heat shock treatment.

GFP

The green fluorescent protein (GFP) from the jellyfish *Aequorea victoria* absorbs violet light with a maximum absorbance at 395 nm and emits green light with a maximum at 509 nm (12). The abundance of GFP in an individual cell can be quantified by fluorescence spectroscopy or confocal laser microscopy. Thus, this reporter gene system, like the *lux* system, has advantages over others (such as *lacZ*, *xylE*, and *inaZ*) in that the transcription of a gene in individual cells, rather than only the average transcription of a population of cells, can be measured. Cellular metabolism is not required for fluorescence of GFP, indicating that the GFP reporter will be useful in studies evaluating in situ gene expression by cells that are not actively growing, as may be the case in many natural habitats. Therefore, the GFP

reporter offers an attractive alternative to the use of *lux* fusions to visualize the transcription of genes in individual cells. GFP fluorescence is stable and persists after fixation of samples with glutaraldehyde or formaldehyde (12). GFP can be quantified over a large range of activity because the intensity of fluorescence can be controlled by the intensity of the incident irradiation. Red-shifted variant GFPs, which differ from the wild type in their optimal excitation wavelength (490 nm versus 395 nm) and their enhanced emission intensity, can be readily quantified by confocal laser microscopy with an argon laser light source ($\lambda = 488$ nm) (16, 28, 29). GFP and the red-shifted variant offer many exciting possibilities for the study of in situ gene expression by bacteria inhabiting natural substrates.

inaZ

Contrary to the common misconception that water freezes at the same temperature at which it melts (0°C), pure water can remain in the liquid state if cooled to −40°C (88). Supercooling to temperatures lower than −15°C rarely occurs in nature, however, because of the presence of heterologous ice nuclei. Almost all organic and inorganic substances can catalyze ice formation (i.e., serve as ice nuclei) at temperatures between −15 and −40°C, and certain compounds can serve as nuclei at temperatures as high as −6°C (88). In the absence of biological ice nuclei, however, water present in environmental samples can supercool to −6°C or lower. A number of bacterial and fungal species can serve as biological ice nuclei (35, 49, 73). The best characterized of these is *P. syringae*, a widely distributed bacterial epiphyte of plants (49). Ice nucleation activity (INA) of *P. syringae* is conferred by a single gene (*inaZ*) that encodes an outer membrane protein (InaZ) (90). Individual ice nucleation proteins do not serve as ice nuclei, but they form large, homogeneous aggregates that collectively orient water molecules into a conformation mimicking the crystalline structure of ice, thereby catalyzing ice formation (10, 19, 25, 41, 94). Oriented water molecules freeze at temperatures slightly below 0°C (i.e., −2 to −10°C) instead of supercooling.

The transcriptional activity of a promoter fused to a promoterless *inaZ* can be assessed by measuring INA. The characteristics of the *inaZ* reporter gene system are listed below.

Detection of Ice Nuclei in Environmental Samples

Ice nuclei can be detected by a tube freezing technique (34, 85) and a droplet freezing assay (87) (described below), both of which detect ice formation in a supercooled, aqueous solution in which an environmental sample is placed. Both methods detect ice nuclei directly from environmental samples and therefore can assess in situ gene expression by bacteria containing a transcriptional fusion to *inaZ*.

Background Levels of Ice Nuclei

A number of bacterial (35) and fungal (73) species can serve as ice nuclei, but these species generally are present in very low numbers in most environmental samples. At −5°C, there is very little background INA in the soil, rhizosphere, or water samples that we have tested. In contrast, epiphytic populations of ice nucleation-active (Ice⁺) bacteria are common on aerial surfaces of field-grown plants collected from favorable environments (i.e., those with cool temperatures and high relative humidity or free moisture) (35). Certainly, large populations of Ice⁺ bacteria on leaf surfaces of field-grown plants present background levels of ice nuclei that will interfere with detection of ice nuclei contributed by introduced bacteria containing the ice nucleation reporter gene. Although the ice nucleation reporter gene will not be useful in this situation, background levels of ice nuclei in most environmental samples rarely limit the usefulness of this reporter (32, 48).

Quantification of INA

INA, expressed as the number of ice nuclei per cell, can be quantified conveniently by a droplet freezing assay (87), which is described in detail below. Soil particles, plant material, and other constituents of environmental samples do not interfere with measurements of INA (34, 53). For example, INA expressed by *Pseudomonas fluorescens* was not altered when the bacterium was placed into soil suspensions (32). Therefore, INA of bacteria containing the *inaZ* reporter gene can be quantified accurately even from soil suspensions (46, 56).

INA is related quantitatively to the InaZ concentration in the cell. In *E. coli*, INA increases with the square of InaZ concentration until INA reaches 10^{-1} nuclei per cell (90, 91). Above that, INA increases with the third power (cube) of InaZ concentration until it reaches the maximum detectable level of one nucleus per cell. A similar second- to third-power relationship of INA to InaZ protein concentration also was established in *P. syringae* (47). This relationship is the basis of a hypothesis that a rate-limiting step of di- or trimerization of InaZ precedes formation of large protein aggregates on outer membranes of gram-negative bacteria that serve as ice nuclei (90). The second- to third-power relationship indicates that *inaZ* can serve as an amplifiable reporter of the transcriptional activities of promoters to which it is fused.

Sensitivity and Range of Detection

Two important and related advantages of the ice nucleation reporter system are the sensitivity with which INA can be detected and the tremendous range of INAs that can be quantified. Theoretically, a single ice nucleus can be detected simply by observing ice formation in a bacterial suspension cooled to $-5°C$. Because virtually every cell is ice nucleation active if the *inaZ* gene is transcribed from a very active promoter (47, 56, 90), it is conceivable that fewer than 10 Ice$^+$ cells could be detected in an environmental sample. This property coupled with the low levels of background INA present in most environmental samples imparts the exceptional sensitivity with which InaZ can be detected in environmental samples.

The lower limit of detection of INA is the inverse of the bacterial population size. If one ice nucleus is detected per ml of a bacterial culture containing 10^8 CFU/ml, then the INA is 10^{-8} ice nuclei per cell. It is therefore possible to estimate a range of INA varying by 10^8-fold, which corresponds to a 2×10^4-fold change in InaZ concentration. Few reporter systems can accurately estimate transcriptional activity over a 20,000-fold range of expression.

Lack of Dependence of INA on High Metabolic Activity

With one exception (72), researchers have reported that INA is expressed by bacterial cells in both exponential growth phase and stationary phase in culture (15, 27, 68), indicating that the *inaZ* reporter is useful for assessing gene expression of cells that are not actively growing, as may be common in natural habitats. INA of *P. syringae* containing *inaZ* transcribed from its native promoter commonly is 10- to 100-fold lower in exponential growth phase than in stationary phase (15, 27, 32, 68). A sequence similar to a "gear-box" promoter, which is typically more active in stationary-phase cells than in exponentially growing cells, is present 5' to *inaZ* in *P. syringae* (68). Therefore, the effect of physiological status on INA may be, at least partially, at the level of *inaZ* transcription and may vary among *inaZ* transcriptional fusions.

InaZ is relatively rare in an Ice$^+$ bacterial cell. A cell with an actively transcribed *inaZ* gene has approximately 500 to 1,000 InaZ proteins, which is far fewer than the 10^5 copies of major outer membrane proteins that are typically present in gram-negative bacteria (90). Therefore, expression of INA does not place a large metabolic demand on the bacterial cell. Differences in the relative growth and survival of near-isogenic Ice$^+$ and Ice$^-$ strains of *P. syringae* in culture or on leaf surfaces have not been detected (50), indicating that INA does not alter the bacterium's fitness in at least certain habitats.

Responsiveness to Changes in Transcriptional Activity

The responsiveness of INA to changes in transcription of the *inaZ* reporter appears to vary among bacterial genera. If an inoculum of *P. syringae* or *P. fluorescens* containing a transcriptional fusion to *inaZ* is grown under cultural conditions such that the target promoter is inactive, then gene induction can be detected within hours of inoculation of a natural substrate on which bacterial cells express *inaZ* (46, 56, 75, 76). In contrast, INA (at $-5°C$) expressed by *P. fluorescens* containing an *inaZ* fusion does not fully reflect the repressed level of gene expression until 12 to 24 h after transfer from conducive conditions to those that are not conducive to *inaZ* transcription (32, 56). Exponentially growing and stationary-phase cells of *P. fluorescens* are similar in the rate at which INA changes in response to a shift between conducive and nonconducive conditions (32). Therefore, INA of cells that are not actively growing, which may be a common physiological state of bacteria in many natural habitats, responds to changes in the transcription of *inaZ*. The relatively stable expression of INA by *P. fluorescens* is surprising when contrasted with reports demonstrating that ice nucleation proteins in their native hosts and in the heterologous host *E. coli* are subject to significant and rapid degradation (15, 91). More studies are needed to address this apparent discrepancy.

Usefulness of *inaZ* as a Reporter Gene

inaZ has many of the characteristics of a reporter gene system that is useful for assessing in situ gene expression by bacteria inhabiting natural habitats. Nevertheless, the *inaZ* reporter gene is not without limitations. Below, we provide background information on biological ice nucleation, elaborate on the characteristics of InaZ that are most relevant to its use as a reporter gene product, and give practical information on the methods and applications of the *inaZ* reporter gene system.

PROPERTIES OF ICE NUCLEATION RELEVANT TO ITS USE AS A REPORTER
inaZ, the InaZ Protein, and the INA Phenotype

Genes encoding InaZ have been cloned from a number of bacterial species (90). The largest portion of each gene is composed of a hierarchy of repeated motifs 24, 48, and 144 nucleotides in length. According to a structural model of InaZ aggregates (41), the repeated amino acid sequences of the InaZ protein produce a periodic array of hydrogen bond donor and acceptor groups on the protein surface that interact with neighboring InaZ proteins, membrane lipids, and water outside of the cell. Individual InaZ proteins have no fixed conformation; instead, they are elongated, flexible molecules that do not function as ice nuclei. A fixed conformation is acquired when the repeated sequences of neighboring InaZ proteins interdigitate to form large, nearly planar aggregates. The number of InaZ molecules in the planar aggregate determines the temperature at which an ice nucleus is active: an aggregate of only two or three monomers is required to form an ice nucleus that is active at $-10°C$, whereas an aggregate of at least 50 InaZ proteins is required to nucleate ice at $-2°C$. An ice nucleus that is active at $-5°C$ is formed when 20 to 40 monomers of the InaZ protein assemble cooperatively on the surface of the bacterial outer membrane (10, 25).

The relatively low abundance of InaZ in a bacterial cell (90) coupled with the requirement for aggregation of 20 to 40 InaZ proteins on the surface of the outer membrane may explain why each cell has a low probability of nucleating ice at temperatures of $-5°C$ or above. For example, only a fraction of cells (typically 10^{-1} or fewer) in a native population of Ice$^+$ P. syringae inhabiting leaf surfaces of plants in the field express ice nucleation activity at $-5°C$ at a given time (35, 49, 53).

Factors Influencing INA

Membrane Fluidity

Planar aggregates of InaZ proteins are stabilized through their interactions with lipids in the outer membrane of gram-negative bacteria (24). Factors that increase membrane fluidity (i.e., a shift to higher temperature [27] or treatment with phenethyl alcohol [51]) decrease INA, presumably by destabilizing InaZ protein aggregates.

Temperature

INA is affected by the temperature at which bacterial cells are grown as well as the temperature of the cells immediately before a freezing event. INA of gram-negative bacteria is maximal at growth temperatures of 18 to 24°C (27). INA of P. syringae at $-5°C$ is 1,000-fold lower in cultures grown at 30°C than in cultures grown at 20°C. Because the concentration of InaZ does not differ among cells grown within this temperature range, the influence of temperature appears to be at the posttranslational level, most likely on InaZ aggregation (27). Similarly, INA of cells that are not actively growing responds to temperature changes. For example, INA of P. syringae at $-5°C$ increased from 10^{-8} nuclei per cell to 1 nucleus per cell in response to a decrease in temperature from 30 to 16°C (68). These dramatic changes in INA, which occur very rapidly and do not require protein synthesis, are probably due to changes in membrane fluidity that occur when temperature is shifted (51).

Chemical Factors

INA expressed by P. syringae varies with the type of carbon source present in a culture medium (53, 68). Bacterial cultures grown in liquid media, even with vigorous aeration, generally do not express INA as efficiently as cultures grown on solid media (48, 72). These factors have less influence than temperature on INA, and their specific effects on inaZ transcription or translation or on InaZ stability or aggregation are unknown.

RECOMMENDATIONS FOR USE OF THE inaZ REPORT GENE SYSTEM

Plasmids and Transposons for Constructing Transcriptional Fusions to inaZ

To date, all reporter gene constructs are based on the inaZ gene cloned from P. syringae (26), which is 3,603 nucleotides in length and encodes a protein of 153 kDa. All constructs make use of a 3.8-kb EcoRI fragment (47), which contains DNA extending from the AhaIII site at nucleotide 775 through the EcoRI site at nucleotide 4453 of inaZ (GenBank database accession number X03035).

Use of the inaZ reporter for quantitative assessment of gene expression requires an understanding of the relationship between InaZ protein content and INA, which has been established in E. coli (90, 91) and P. syringae (47). Because this relationship may not apply to an InaZ fusion protein, we recommend inaZ only for generating transcriptional fusions, not for generating protein fusions. This discussion is limited to transcriptional fusions to the inaZ gene, which possesses a ribosome binding site and start site for translation of an InaZ protein of predictable size and activity. We refer the reader to a recent article by Panopoulos (71), in which methods for constructing inaZ transcriptional and protein fusions are described in some detail.

Transposons Useful for Making Fusions to Cloned Genes

The transposons Tn3-Spice (47) and Tn3-nice (46) are useful for mutagenesis of DNA cloned into a polA-independent plasmid vector. These transposons differ principally in the antibiotic resistances that they confer (i.e., resistance to streptomycin sulfate and spectinomycin is conferred by Tn3-Spice, and resistance to neomycin is conferred by Tn3-nice). In both transposons, the segment preceding the inaZ gene has translational stop codons in all three reading frames, thereby ensuring that only transcriptional fusions will be generated.

Plasmids Containing a Promoterless inaZ

The promoterless inaZ has been cloned in two plasmids to derive pLAFR6-inaZ (14) and pVSP61-inaZ (56).

The cosmid pLAFR6 (14) is a derivative of the broad-host-range plasmid pRK2. Transcriptional terminators flank inaZ, and so pLAFR6-inaZ confers very little basal INA on at least certain bacterial hosts (i.e., P. fluorescens and P. syringae). A multiple cloning site is located 5′ to inaZ to facilitate cloning of promoter sequences.

Plasmid pVSP61 is composed of two host-specific replicons for stable maintenance in Pseudomonas spp. and in E. coli and related enteric bacteria (86). A promoterless inaZ was cloned into the multiple cloning site of pVSP61 in opposite orientation to its lac promoter (56). Although inaZ is not flanked by transcriptional terminators, pVSP61-inaZ confers virtually no basal INA on P. fluorescens or P. syringae (56). Plasmid pVSP61-inaZ is particularly useful in studies evaluating gene expression by bacteria in natural habitats because it is stably maintained by strains of P. fluorescens, P. putida, P. syringae, and Erwinia carotovora inhabiting the rhizosphere, phyllosphere, or bulk soil (4, 46, 56). Unfortunately, the plasmid lacks useful cloning sites 5′ to inaZ. An improved version of the plasmid, which contains transcriptional and translational terminators and an array of useful cloning sites upstream of inaZ, is being derived (64).

Experimental Design

Evaluation of INA in Culture

The *inaZ* reporter is useful only if the test strain has no intrinsic INA and if it expresses INA from an introduced, transcribed *inaZ*. INA has been successfully expressed in all species of gram-negative bacteria into which genes conferring ice nucleation have been introduced, including many strains of *E. coli*, *P. syringae*, *P. fluorescens*, *Pseudomonas aureofaciens*, *Pseudomonas solanacearum*, *Xanthomonas campestris*, *Agrobacterium tumefaciens*, *Rhizobium meliloti*, *Erwinia amylovora*, *Erwina carotovora*, and a number of halophilic bacteria (2, 22, 46–48, 56). Nevertheless, the level of INA conferred by *inaZ* varies among bacteria, and the *inaZ* reporter will not be useful in bacteria that express INA inefficiently. Even strains within a given bacterial species vary in the efficiency in which they express INA. For example, most but not all strains of *Erwinia carotovora* express INA efficiently from an introduced plasmid containing *inaZ* transcribed from a *lac* promoter (4).

Evaluation of Sensitivity

A principal advantage of the *inaZ* reporter gene is its sensitivity and its range of detection. Nevertheless, certain promoters transcribe at levels that are too high for quantitative assessment by the *inaZ* reporter. For example, *P. syringae* containing a fusion of *inaZ* to an iron-regulated promoter expresses INA that is inversely related to iron availability of a culture medium (56). Virtually every cell expresses INA in the medium supplemented with 10^{-6} M $FeCl_3$, and higher levels of transcriptional activity expressed by cells growing in a medium with less iron cannot be accurately quantified with the *inaZ* reporter system. Therefore, it is important to select a reporter gene system with a range of detection that matches the transcriptional activity of the promoter to be evaluated.

Need for an InaZ⁺ Control

An *inaZ* gene transcribed from its native promoter or from any characterized promoter that is active in the test strain should be included as a control in experiments evaluating in situ gene expression by bacteria inhabiting natural substrates. This control strain allows one to separate effects attributable to transcription from those attributable to post-transcriptional effects on InaZ (i.e., translation, stability, or aggregation).

Quantification of INA by the Droplet Freezing Assay

In the droplet freezing assay (87), environmental samples (without prior preparation) are suspended in an aqueous solution; small volumes (i.e., droplets) of dilutions of the solution are cooled to $-5°C$, and the fraction of those that freeze is recorded. The number of ice nuclei in the sample is calculated from the fraction of droplets that freeze. INA of a bacterial population is calculated by dividing the number of ice nuclei by the number of culturable bacterial cells in the sample. A detailed description of the droplet freezing assay follows.

Preparation of Bacterial Suspensions from Environmental Samples

Place environmental samples, such as plant material or soil, in a dilute buffer such as 10 mM potassium phosphate buffer (pH 7.0). Bacterial INA is relatively insensitive to pH values of 3 to 9 but decreases outside this range. The aqueous solution should not contain high concentrations of salt or organic substances that could affect the freezing point of the solution. Bacterial cells can be dislodged from plant material by shaking or sonication in an ultrasonic cleaning bath as described previously (57, 69). In our experience, neither of these procedures affects bacterial INA significantly. It is essential that samples and dilution buffer tubes be maintained at a constant temperature (usually room temperature) throughout the entire assay procedure. For example, change the water with each use of a bath-type sonicator to avoid exposure of the samples to increased water temperatures.

Preparation of Foil

An ice nucleus-free surface on which samples of a bacterial suspension can be cooled is required for this assay. The smooth surface of aluminum foil coated with a thin film of paraffin is ideal for this purpose. Medium-weight foil (0.025 mm thick) should be cut into rectangular sheets sized to fit conveniently in the refrigerated bath that will be used in the assay. To prepare the foil surface, spray a nearly saturated solution of paraffin in xylene (approximately 5% [wt/vol]) on one side of the foil and place the foil sheets in an oven held at 65°C for 5 min. The heat will volatilize the solvent and melt the paraffin onto the sheet. After cooling to room temperature, the aluminum foil sheets appear slightly greasy; they can be stacked and stored indefinitely at room temperature.

Placement of Foil

The aluminum foil sheets are used as a solid surface to support droplets for the freezing assay. A refrigerated circulating bath is the heat sink for the assay. The temperature of the coolant in the bath should be high enough to reduce background levels of ice nuclei found in environmental samples and low enough to optimize detection of INA of the bacterial strain containing the reporter gene fusion. In our experience, $-5°C$ is the best compromise between these two factors. We prefer ethanol as a coolant because it has low viscosity (even at $-10°C$) and does not form a thick boundary layer underneath the aluminum foil. Polyethylene glycol also works well and does not pose a fire hazard, but it is messy and more viscous and has a greater tendency to foam than ethanol. Circulation of coolant in the bath must be maintained throughout the assay to ensure adequate and uniform heat transfer from the droplets to the heat sink.

To make a "boat" that can float on the surface of the coolant, fold up approximately 1 cm of each edge of the rectangular sheet of aluminum foil, with the paraffin-treated side up. Avoid trapping air bubbles underneath the foil sheets as you place them on the coolant surface. Droplets placed over air bubbles will not quickly cool to the bath temperatures, and differences in droplet temperature are important sources of error in the droplet freezing assay. Bubbles can be minimized by laying one end of the boat on the surface of the bath and slowly lowering the remainder of the foil sheet, in the same manner that one would use to place a coverslip on a microscope slide. To detect trapped air bubbles, direct a moist stream of air onto the floating foil; condensation will be visible on the foil surface but will not form directly over trapped air bubbles. Discard aluminum foil boats with trapped air bubbles.

Placement of Droplets on the Foil

Begin placing droplets of bacterial suspensions on the surface of the foil immediately after the foil is placed on the bath. Over time, a foamy, insulating layer of aerated coolant can collect under the foil. This insulating layer reduces heat transfer and causes false-negative results in the assay. Ideally, one should finish placing droplets on the foil within 5 min after the foil is placed on the coolant. Each droplet is a small volume (typically 10 μl), which enhances the rate of heat transfer between the droplet and the coolant. Plastic tipped micropipettes are convenient for delivering the droplets and are unlikely to scratch the surface of the aluminum foil sheet. The number of droplets that freeze is a probability function; therefore, estimates of ice nuclei in a sample become more accurate as the number of droplets evaluated increases. We typically evaluate 40 droplets per dilution, which represents a compromise between accuracy and expediency of the assay.

Observation of Frozen Droplets

Droplets from a bacterial suspension on a boat will freeze quickly at -5 or $-10°C$. If the freezing temperature of the sample is considerably higher than that of the coolant, then droplets will freeze within 10 to 20 s. If the nucleation temperature of the sample is similar to that of the coolant, then ice nucleation events may occur even 1 to 2 min after the droplets are placed on the boat. For this reason, observe the droplets for ice formation 1 to 2 min after you place droplets on the boat. If the room temperature is high or air movement over the bath is substantial, cover the bath with a piece of Plexiglas and observe ice formation through the Plexiglas cover. This will stabilize the temperature of the droplets and increase the reproducibility of the assay. Frozen droplets, which appear opaque and peaked, can easily be differentiated from clear, hemispherical unfrozen droplets. The distinction is less obvious if the unfrozen droplets are also opaque, as occurs if a soil or turbid bacterial suspension is being tested. In that case, touch droplets with a sterile toothpick or pipette tip to verify whether they are solid or liquid. Sometimes, a droplet that first appeared liquid will suddenly turn solid after it is touched. Such a drop must be considered unfrozen because it is likely that the probe triggered nucleation. Keep in mind that an insulating layer of aerated coolant can form underneath the foil, in which case frozen droplets will eventually begin to melt.

From a series of dilutions of bacterial suspensions, select those dilutions for which some but not all of the droplets freeze. The number of ice nuclei will be estimated most accurately from dilutions in which more than 5 but fewer than 30 of the 40 droplets freeze. As the proportion of frozen droplets increases, so does the probability that a given droplet contains more than one ice nucleus. This phenomenon is addressed in calculating the number of ice nuclei from the proportion of frozen droplets.

Calculation of INA

The number of ice nuclei in the sample is calculated as proposed by Vali (87), who recognized that freezing events are approximated by a Poisson distribution (because frozen droplets may have more than one ice catalyst present although only one catalyst is detected): $N = \ln[1/(1 - f)]/(V*D)$, where N is the concentration of active ice nuclei in the initial solution (i.e., nuclei per milliliter), $\ln[1/(1 - f)]$ is the calculated nucleation frequency based on a Poisson distribution of freezing events among the collection of droplets, f is the fraction of frozen droplets (number of droplets that froze/total number of droplets) (note: N should be calculated from dilutions in which the fraction of droplets that froze is greater than 0 and less than 1), V is the volume of each droplet (milliliter), and D is a factor describing the dilution of the solution from which droplets were tested from the initial solution.

The concentration of ice nuclei can be normalized for the density of culturable cells that contain the *inaZ* fusion, determined by the spread plate technique (23) on a selective agar medium. INA is calculated by dividing the concentration of ice nuclei (N) by the density of culturable bacteria (CFU per milliliter) present in the environmental sample. Calculations can be done with a variety of database or spreadsheet software packages or a computer program designed specifically to calculate numbers of ice nuclei and culturable bacterial cells (PlateCount; Taylor-made Software, Corvallis, Oreg.).

Interpretation of Results

The droplet freezing assay estimates the number of ice nuclei present in an aqueous solution. When this assay is combined with methods estimating numbers of culturable bacteria, one can calculate INA of a population of bacterial cells, expressed as the number of nuclei per bacterial cell. Theoretically, the maximum INA that can be detected is one nucleus per cell. (A bacterial cell cannot be divided between droplets, so even if the cell has more than one functional ice nucleus present in the outer membrane, the second nucleus will not be detected.) In certain situations, however, more than one nucleus per cell is observed. For example, cells can retain INA for a limited time after they lose viability (49), and some bacterial species (such as *Erwinia herbicola*) produce extracellular membrane vesicles that contain ice nuclei (19). Droplets containing nuclei contributed by nonviable cells and membrane vesicles will freeze even though these structures are not culturable cells.

INAs ranging from 1 to 10^{-8} nuclei per cell are expressed typically by cells containing transcriptional fusions to *inaZ*. Although one may incorrectly assume that INAs of less than 1 indicate heterogeneous gene expression among bacterial cells (i.e., *inaZ* is being transcribed by only a subset of cells in the population), this cannot be concluded from data obtained by the droplet freezing assay. *inaZ*, like most other reporter gene systems, reports the *average* level of transcriptional activity in a population of cells. In two extreme examples, a bacterial population expressing a mean INA of 10^{-2} ice nuclei per cell could be (i) a population of homogeneous cells, each of which expresses the same moderate activity, or (ii) a population in which 1% of the cells are fully expressed (1 nucleus per cell) and 99% express very low INA (i.e., 10^{-8} ice nuclei per cell). The droplet freezing assay does not differentiate between these extreme examples or intermediate cases that may exist within a population of cells.

APPLICATIONS OF REPORTER GENES IN ENVIRONMENTAL MICROBIOLOGY

Our current understanding of the activities of bacteria in natural habitats is limited by the lack of methods by which these activities can be assessed. Our concepts are grounded necessarily on extrapolations from the characteristics of bacteria grown in axenic culture, despite the general recognition that microbial habitats in nature differ profoundly from the microbial cultures that are amenable to scientific investigation. Reporter gene systems provide an opportunity for microbiologists to step beyond the confines of previous methodology in their attempts to understand the biology of microbes in their natural habitats. A few examples of the innovative ways that reporter gene systems are being used in environmental microbiology are described below.

Indirect Assessments of Phenotypic Expression

Many phenotypes of bacteria can be detected and quantified easily in culture but are difficult to detect or quantify in natural habitats. Phenotypes expressed by bacteria are commonly influenced by medium composition in culture and presumably by the chemical composition and physical environment of microhabitats that bacteria occupy in natural environments. For example, antibiotic production by bacteria grown in culture varies with carbon source, presence of micronutrients, and numerous other factors. Presumably, similar factors influence the production of antibiotics by bacteria inhabiting natural substrates, but this has rarely been established because of the difficulties encountered in extracting antibiotics from such substrates (see chapter 55). The product of a reporter gene, if fused to a promoter of an antibiotic biosynthesis operon, provides a reliable and convenient assessment of the expression of biosynthetic genes by bacteria in natural habitats. Reporter gene fusions also provide methods for assessing conditions that optimize or limit the expression of antibiotic biosynthesis genes in culture, thereby targeting certain edaphic factors that may limit the expression of such genes by bacteria in nature. Gene fusions to a *lacZ* or *inaZ* reporter have enabled researchers to confirm the in situ expression of genes determining the biosynthesis of oomycin A (36), pyoverdine (56), pyoluteorin (46), and phenazines (22). Because antibiotic biosynthesis genes can be transcribed weakly by bacterial populations occupying natural habitats relative to their expression by bacteria grown in culture (36, 46), sensitivity is a primary criterion in selecting a reporter gene for this purpose. Because of its sensitivity, the *inaZ* reporter is especially useful for this application; indeed, the low levels of transcriptional activity expressed by *P. fluorescens* inhabiting the rhizosphere (46, 56) would not have been detected with less sensitive reporters. Bacterial cells in natural habitats are not always growing exponentially but are thought to exist, at least periodically, in a state of starvation (67, 89). Therefore, a reporter gene system that functions only in actively growing cells is a poor choice for studies assessing in situ gene expression by bacteria in many natural habitats.

Estimation of Biologically Available Concentrations of Chemicals

Concentrations of chemicals can be determined accurately by analytical techniques, but these methods do not assess whether the chemical is present in a form that is available biologically. Numerous physicochemical factors influence the form of chemical compounds or elements in solution or soil (54). Because its form has a profound effect on the biological availability of a chemical, methods that evaluate only chemical concentration do not provide an accurate assessment of its biological relevance to an ecosystem. Biological sensors, constructed by fusing a reporter gene to a promoter that is transcribed at a rate proportional to the availability of an element or compound, can be used to assess biologically meaningful levels of specific chemicals in environmental samples. Sensors based on a bioluminescence (*lux*) reporter system have been developed for assessment of biologically available forms of Hg(II) (79) or naphthalene (11, 31, 43) in surface water amended with substrates required for bacterial growth. Naphthalene-responsive bioluminescence was observed when immobilized cells of *P. fluorescens* containing a transcriptional fusion of *luxCDABE* to a naphthalene-induced promoter were exposed to waste streams in which nutrients were supplied (30, 37). An array of sensors that estimate biologically available levels of a spectrum of environmental pollutants could provide tremendous insight into the biological risks imposed by such pollutants in the environment. Reporter gene products such as luciferase, InaZ, or GFP, which can be quantified accurately over a large range of activity, are good choices for this application. Because metabolic activity of the bacterial cells is maintained by nutrients supplied exogenously, a reporter gene system that functions optimally in actively growing cells is suitable for this purpose.

Characterization of Microbial Habitats

In many natural habitats, such as leaf and root surfaces, bacterial cells are not distributed uniformly but instead are found in microcolonies in sites that support their growth or survival (20). Despite this, our knowledge of microbial habitats is based on analysis of bulk environmental samples rather than analysis of the microsites in which bacteria exist in nature. Bacteria containing chemically responsive reporter gene constructs, similar to those described above, can be used to assess the chemical composition of bacterial microhabitats in nature. For example, a phosphate sensor, constructed by fusing a phosphate-responsive promoter to a promoterless *lacZ* gene, provides an assessment of phosphate availability to *P. putida* (17). Although the *lacZ* reporter gene was useful for assessing phosphate-mediated transcription by *P. putida* in the rhizosphere of plants grown in sand or a sterile soil, it was not of adequate sensitivity to assess transcription by the bacterium in the rhizosphere of plants grown in a nonsterile soil (17). Our laboratories constructed a sensor (termed *pvd-inaZ*) that is responsive to iron availability to *Pseudomonas* spp. in soil or on plant surfaces (56). The sensitivity of the *inaZ* reporter was needed to estimate concentrations of bioavailable iron in these habitats. In another application, bacterial bioluminescence could be detected within individual *Rhizobium*-infected plant cells (70), providing evidence that the availability of O_2 within bacteroids was sufficiently high for the bioluminescence reaction, in contrast to many expectations. In the future, a complement of biological sensors may be available to facilitate characterization of the chemical and physical constituents of the microhabitats that bacteria occupy in nature. Sensors based on reporter genes that confer visible products (i.e., the GFP gene, *gusA*, or *luxAB*) may enable researchers to evaluate the availability of chemicals to individual bacterial cells rather than to bacterial populations inhabiting natural substrates.

Evaluation of Biological Interactions

Pathogenesis, symbiosis, and colonization of plant surfaces are complex processes. In plant-associated bacteria, the capacity to interact with host plants is determined by multiple loci controlled through intricate regulatory cascades. The genetic events involved in these interactions cannot be elucidated in studies of cultured bacterial cells alone but instead must be evaluated in situ. Studies comparing bacteria grown in culture or on leaf surfaces, for example, indicate that patterns of gene expression differ markedly between cells grown in these habitats (13). Certain bacterial genes involved in plant-microbe interactions are transcribed actively in situ, such that mRNA levels are of adequate abundance to be detected directly. For the many genes that are critical to the interaction but are transcribed at much lower levels, however, reporter gene systems are essential tools enabling researchers to characterize the molecular events involved in processes of symbiosis and pathogenesis. For example, gene expression by plant-pathogenic or symbiotic bacteria in the tissues of plant hosts has been evaluated with *lux* (42, 70, 81), *gusA* (55, 58, 80, 92), *lacZ* (1, 5, 65), and *inaZ* (38, 59, 75, 76) reporters. *lux* and *gusA* reporters allow researchers to evaluate gene expression by bacterial pathogens and symbionts in specific host tissues microscopically, thereby providing a spatial and temporal analysis of the host-pathogen interaction. The *lacZ* and *inaZ* reporters provide information about the mean level of gene expression in a population of bacterial cells associated with the plant, and *inaZ* has been especially useful in situations in which additional sensitivity is required.

Estimation of Microbial Activity

Current methods for estimating in situ microbial activity, such as those that evaluate the rate of [^{3}H]thymidine incorporation or adenylate energy charge, measure the mean metabolic activity of entire taxa (for example, prokaryotic or eukaryotic microorganisms) but do not accurately assess the activity of a particular bacterial species or strain. A novel method for assessing the metabolic activity of an individual bacterial strain is based on the *lux* reporter system, which will emit light only if nonlimiting concentrations of O_2 and $FMNH_2$ are available. Because the intracellular concentration of $FMNH_2$ reflects the metabolic activity of the cell, light emission is an indirect indicator of metabolic activity. For example, when genetically modified cells of *P. fluorescens* that contained chromosomal *luxAB* fusions were inoculated into sterile soil, light production decreased with time. Amendment of the soil with nutrients resulted in increased bioluminescence, presumably because the metabolic activity of the cells, which was low in nonamended soil, was enhanced with addition of limiting nutrients (63). Therefore, metabolic activity of the cells could be deduced from differences in luminescence of *luxAB*-marked strains in nutrient-amended and nonamended soils (62, 63). Complementary constructs, in which metabolically responsive promoters are fused to metabolically insensitive reporter gene systems (e.g., the GFP or *inaZ* system), could also be useful in developing our understanding of the metabolic activities of bacteria in their natural habitats.

CONCLUDING REMARKS

Reporter gene systems provide an unprecedented opportunity for microbiologists to gain a new perspective on the activities of bacteria inhabiting natural substrates, including their expression of specific genes, their recognition of environmental signals, their metabolic activities, and the chemical composition of the habitats that they occupy. A number of innovative and complementary reporter gene systems that are useful in environmental microbiology are now available. Our discussion has focused on the ice nucleation reporter gene system, which has several unique attributes that make it extremely useful in studies evaluating in situ gene expression by bacteria inhabiting natural environments. We expect that the ice nucleation reporter gene system, coupled with complementary systems such as *lacZ*, *lux*, *gusA*, or GFP, will be an important tool in future inquiries into the ecology of microorganisms.

The comments and suggestions of Marcella Henkels and Virginia Stockwell during preparation of the manuscript are gratefully acknowledged.

The work in our laboratories on the ice nucleation reporter gene was supported in part by interagency agreement DW12935653-01-2 between the Biotechnology Program of the U.S. Environmental Protection Agency and the U.S. Department of Agriculture.

REFERENCES

1. **Arsène, F., S. Katupitiya, I. R. Kennedy, and C. Elmerich.** 1994. Use of *lacZ* fusions to study the expression of *nif* genes of *Azospirillum brasilense* in association with plants. *Mol. Plant-Microbe Interact.* **7:**748–757.
2. **Arvanitis, N., C. Vargas, G. Tegos, A. Perysinakis, J. J. Nieto, A. Ventosa, and C. Drainas.** 1995. Development of a gene reporter system in moderately halophilic bacteria by employing the ice nucleation gene of *Pseudomonas syringae*. *Appl. Environ. Microbiol.* **61:**3821–3825.
3. **Atlas, R. M., G. Sayler, R. S. Burlage, and A. K. Bej.** 1992. Molecular approaches for environmental monitoring of microorganisms. *BioTechniques* **12:**706–717.
4. **Boehm, M., and J. E. Loper.** Unpublished data.
5. **Bolton, G. W., E. W. Nester, and M. P. Gordon.** 1986. Plant phenolic compounds induce expression of the *Agrobacterium tumefaciens* loci needed for virulence. *Science* **232:**983–985.
6. **Brennerova, M. V., and D. E. Crowley.** 1994. Direct detection of rhizosphere-colonizing *Pseudomonas* sp. using an *Escherichia coli* rRNA promoter in a Tn*7-lux* system. *FEMS Microbiol. Ecol.* **14:**319–330.
7. **Bronstein, I., J. Fortin, P. E. Stanley, G. S. A. B. Stewart, and L. J. Kricka.** 1994. Chemiluminescent and bioluminescent reporter gene assays. *Anal. Biochem.* **219:**169–181.
8. **Bronstein, I., J. J. Fortin, J. C. Voyta, R.-R. Juo, B. Edwards, C. E. M. Olesen, N. Lijam, and L. J. Kricka.** 1994. Chemiluminescent reporter gene assays: sensitive detection of the GUS and SEAP gene products. *BioTechniques* **17:**172–177.
9. **Buell, C. R., and A. J. Anderson.** 1993. Expression of the *aggA* locus of *Pseudomonas putida* in vitro and in planta as detected by the reporter gene, *xylE*. *Mol. Plant-Microbe Interact.* **6:**331–340.
10. **Burke, M. J., and S. E. Lindow.** 1990. Surface properties and size of the ice nucleation site in ice nucleation active bacteria: theoretical considerations. *Cryobiology* **27:**80–84.
11. **Burlage, R. S., and C.-T. Kuo.** 1994. Living biosensors for the management and manipulation of microbial consortia. *Annu. Rev. Microbiol.* **48:**291–309.
12. **Chalfie, M., Y. Tu, G. Euskirchen, W. W. Ward, and D. C. Prasher.** 1994. Green fluorescent protein as a marker for gene expression. *Science* **263:**802–805.
13. **Cirvilleri, G., and S. E. Lindow.** 1994. Differential expression of genes of *Pseudomonas syringae* on leaves and in culture evaluated with random genomic *lux* fusions. *Mol. Ecol.* **3:**249–257.
14. **Dahlbeck, D., and B. J. Staskawicz.** Personal communication.
15. **Deininger, C. A., G. M. Mueller, and P. K. Wolber.** 1988. Immunological characterization of ice nucleation proteins from *Pseudomonas syringae*, *Pseudomonas fluorescens*, and *Erwinia herbicola*. *J. Bacteriol.* **170:**669–675.
16. **Delagrave, S., R. E. Hawtin, C. M. Silva, M. M. Yang, and D. C. Youvan.** 1995. Red-shifted excitation mutants of the green fluorescent protein. *Bio/Technology* **13:**151–154.
17. **de Weger, L. A., L. C. Dekkers, A. J. van der Bij, and B. J. J. Lugtenberg.** 1994. Use of phosphate-reporter bacteria to study phosphate limitation in the rhizosphere and in bulk soil. *Mol. Plant-Microbe Interact.* **7:**32–38.
18. **de Weger, L. A., P. Dunbar, W. F. Mahafee, B. J. J. Lugtenberg, and G. S. Sayler.** 1991. Use of bioluminescence markers to detect *Pseudomonas* spp. in the rhizosphere. *Appl. Environ. Microbiol.* **57:**3641–3644.
19. **Fall, R., and P. K. Wolber.** 1995. Biochemistry of bacterial ice nuclei, p. 63–83. *In* R. E. Lee, Jr., G. J. Warren, and L.

V. Gusta (ed.), *Biological Ice Nucleation and Its Applications*. APS Press, St. Paul, Minn.

20. **Foster, R. C., A. D. Rovira, and T. W. Cook.** 1983. *Ultrastructure of the Root-Soil Interface*. APS Press, St. Paul, Minn.

21. **Gallagher, S. R.** 1992. *GUS Protocols: Using the GUS Gene as a Reporter of Gene Expression*. Academic Press, New York.

22. **Georgakopoulos, D. G., M. Hendson, N. J. Panopoulos, and M. N. Schroth.** 1994. Analysis of expression of a phenazine biosynthesis locus of *Pseudomonas aureofaciens* PGS12 on seeds with a mutant carrying a phenazine biosynthesis locus-ice nucleation reporter gene fusion. *Appl. Environ. Microbiol.* **60:**4573–4579.

23. **Gerhardt, P., R. G. E. Murray, W. A. Wood, and N. R. Krieg** (ed.). 1994. *Methods for General and Molecular Bacteriology*, p. 255–257. American Society for Microbiology, Washington D.C.

24. **Govindarajan, A. G., and S. E. Lindow.** 1988. Phospholipid requirement for expression of ice nuclei in *Pseudomonas syringae* and *in vitro*. *J. Biol. Chem.* **263:**9333–9338.

25. **Govindarajan, A. G., and S. E. Lindow.** 1988. Size of bacterial ice-nucleation sites measured *in situ* by radiation inactivation analysis. *Proc. Natl. Acad. Sci. USA* **85:**1334–1338.

26. **Green, R. L., and G. J. Warren.** 1985. Physical and functional repetition in a bacterial ice nucleation gene. *Nature* (London) **317:**645–648.

27. **Gurian-Sherman, D., and S. E. Lindow.** 1995. Differential effects of growth temperature on ice nuclei active at different temperatures that are produced by cells of *Pseudomonas syringae*. *Cryobiology* **32:**129–138.

28. **Heim, R., A. B. Cubitt, and R. Y. Tsien.** 1995. Improved green fluorescence. *Nature* (London) **373:**663–664.

29. **Heim, R., D. C. Prasher, and R. Y. Tsien.** 1994. Wavelength mutations and posttranslational autooxidation of green fluorescent protein. *Proc. Natl. Acad. Sci. USA* **91:**12501–12504.

30. **Heitzer, A., K. Malachowsky, J. E. Thonnard, P. R. Bienkowski, D. C. White, and G. S. Sayler.** 1994. Optical biosensor for environmental on-line monitoring of naphthalene and salicylate bioavailability with an immobilized bioluminescent catabolic reporter bacterium. *Appl. Environ. Microbiol.* **60:**1487–1494.

31. **Heitzer, A., O. F. Webb, J. E. Thonnard, and G. S. Sayler.** 1992. Specific and quantitative assessment of naphthalene and salicylate bioavailability by using a bioluminescent catabolic reporter bacterium. *Appl. Environ. Microbiol.* **58:**1839–1846.

32. **Henkels, M. D., and J. E. Loper.** Unpublished data.

33. **Hill, P. J., C. E. D. Rees, M. K. Winson, and G. S. A. B. Stewart.** 1993. The application of *lux* genes. *Biotechnol. Appl. Biochem.* **17:**3–14.

34. **Hirano, S. S., L. S. Baker, and C. D. Upper.** 1985. Ice nucleation temperature of individual leaves in relation to population sizes of ice nucleation active bacteria and frost injury. *Plant Physiol.* **77:**259–265.

35. **Hirano, S. S., and C. D. Upper.** 1995. Ecology of ice nucleation-active bacteria, p. 41–61. *In* R. E. Lee, Jr., G. J. Warren, and L. V. Gusta (ed.), *Biological Ice Nucleation and Its Applications*. APS Press, St. Paul, Minn.

36. **Howie, W. J., and T. V. Suslow.** 1991. Role of antibiotic biosynthesis in the inhibition of *Pythium ultimum* in the cotton spermosphere and rhizosphere by *Pseudomonas fluorescens*. *Mol. Plant-Microbe Interact.* **4:**393–399.

37. **Huang, B., T. W. Wang, R. Burlage, and G. Sayler.** 1993. Development of an on-line sensor for bioreactor operation. *Appl. Biochem. Biotechnol.* **39:**371–382.

38. **Huynh, T. V., D. Dahlbeck, and B. J. Staskawicz.** 1989.

Bacterial blight of soybean: regulation of a pathogen gene determining host cultivar specificity. *Science* **245:**1374–1377.

39. **Jefferson, R. A.** 1989. The *gus* reporter gene system. *Nature* (London) **342:**837–838.

40. **Jefferson, R. A., S. M. Burgess, and D. Hirsh.** 1986. β-Glucuronidase from *Escherichia coli* as a gene-fusion marker. *Proc. Natl. Acad. Sci. USA* **83:**8447–8451.

41. **Kajava, A. V., and S. E. Lindow.** 1993. A molecular model of the three-dimensional structure of bacterial ice nucleation proteins. *J. Mol. Biol.* **232:**709–717.

42. **Kamoun, S., and C. I. Kado.** 1990. A plant-inducible gene of *Xanthomonas campestris* pv. *campestris* encodes an exocellular component required for growth in the host and hypersensitivity on non-hosts. *J. Bacteriol.* **172:**5165–5172.

43. **King, J. M. H., P. M. DiGrazia, B. Applegate, R. Buriage, J. Sanseverino, P. Dunbar, F. Larimer, and G. S. Sayler.** 1990. Rapid, sensitive bioluminescent reporter technology for naphthalene exposure and biodegradation. *Science* **249:**778–781.

44. **Kluepfel, D. A.** 1993. The behavior and tracking of bacteria in the rhizosphere. *Annu. Rev. Phytopathol.* **31:**441–472.

45. **Koncz, C., W. H. R. Langridge, O. Olsson, J. Schell, and A. A. Szalay.** 1990. Bacterial and firefly luciferase genes in transgenic plants: advantages and disadvantages of a reporter gene. *Dev. Genet.* **11:**224–232.

46. **Kraus, J., and J. E. Loper.** 1995. Characterization of a genomic region required for production of the antibiotic pyoluteorin by the biological control agent *Pseudomonas fluorescens* Pf-5. *Appl. Environ. Microbiol.* **61:**849–854.

47. **Lindgren, P. B., R. Frederick, A. G. Govindarajan, N. J. Panopoulos, B. J. Staskawicz, and S. E. Lindow.** 1989. An ice nucleation reporter system: identification of inducible pathogenicity genes in *Pseudomonas syringae* pv. *phaseolicola*. *EMBO J.* **8:**1291–1301.

48. **Lindow, S. E.** Unpublished data.

49. **Lindow, S. E.** 1983. The role of bacterial ice nucleation in frost injury to plants. *Annu. Rev. Phytopathol.* **21:**363–384.

50. **Lindow, S. E.** 1985. Ecology of *Pseudomonas syringae* relevant to the field use of Ice⁻ deletion constructed in vitro for plant frost control, p. 23–35. *In* H. O. Halvorson, D. Pramer, and M. Rogul (ed.), *Engineered Organisms in the Environment: Scientific Issues*. American Society for Microbiology, Washington, D.C.

51. **Lindow, S. E.** 1995. Membrane fluidity as a factor in production and stability of bacterial ice nuclei active at high subfreezing temperatures. *Cryobiology* **32:**247–258.

52. **Lindow, S. E.** 1995. The use of reporter genes in the study of microbial ecology. *Mol. Ecol.* **4:**555–566.

53. **Lindow, S. E., S. S. Hirano, W. R. Barchet, D. C. Arny, and C. D. Upper.** 1982. Relationship between ice nucleation frequency of bacteria and frost injury. *Plant Physiol.* **70:**1090–1093.

54. **Lindsay, W. L.** 1979. *Chemical Equilibria in Soil*. John Wiley & Sons, New York.

55. **Łojkowska, E., C. Dorel, P. Reignault, N. Hugouvieux-Cotte-Pattat, and J. Robert-Baudouy.** 1993. Use of GUS fusion to study the expression of *Erwinia chrysanthemi* pectinase genes during infection of potato tubers. *Mol. Plant-Microbe Interact.* **6:**488–494.

56. **Loper, J. E., and S. E. Lindow.** 1994. A biological sensor for iron available to bacteria in their habitats on plant surfaces. *Appl. Environ. Microbiol.* **60:**1934–1941.

57. **Loper, J. E., T. V. Suslow, and M. N. Schroth.** 1984. Lognormal distribution of bacterial populations in the rhizosphere. *Phytopathology* **74:**1454–1460.

58. **Lorang, J. M., and N. T. Keen.** 1995. Characterization of

avrE from *Pseudomonas syringae* pv. *tomato*: a *hrp*-linked avirulence locus consisting of at least two transcriptional units. *Mol. Plant-Microbe Interact.* 8:49–57.

59. **Ma, S.-W., V. L. Morris, and D. A. Cuppels.** 1991. Characterization of a DNA region required for production of the phytotoxin coronatine by *Pseudomonas syringae* pv. *tomato*. *Mol. Plant-Microbe Interact.* 4:69–74.

60. **Manafi, M., W. Kneifel, and S. Bascomb.** 1991. Fluorogenic and chromogenic substrates used in bacterial diagnostics. *Microbiol. Rev.* 55:335–348.

61. **Meighen, E. A.** 1993. Bacterial bioluminescence: organization, regulation, and application of the *lux* genes. *FASEB J.* 7:1016–1022.

62. **Meikle, A., L. A. Glover, K. Killham, and J. I. Prosser.** 1994. Potential luminescence as an indicator of activation of genetically-modified *Pseudomonas fluorescens* in liquid culture and in soil. *Soil Biol. Biochem.* 26:747–755.

63. **Meikle, A., K. Killham, J. I. Prosser, and L. A. Glover.** 1992. Luminometric measurement of population activity of genetically modified *Pseudomonas fluorescens* in the soil. *FEMS Microbiol. Lett.* 99:217–220.

64. **Miller, W., and S. E. Lindow.** Unpublished data.

65. **Mo, Y.-Y., and D. C. Gross.** 1991. Expression *in vitro* and during plant pathogenesis of the *syrB* gene required for syringomycin production by *Pseudomonas syringae* pv. *syringae*. *Mol. Plant-Microbe Interact.* 4:28–36.

66. **Morgan, J. A. W., C. Winstanley, R. W. Pickup, J. G. Jones, and J. R. Saunders.** 1989. Direct phenotypic and genotypic detection of a recombinant pseudomonad population released into lake water. *Appl. Environ. Microbiol.* 55:2537–2544.

67. **Morita, R. Y.** 1993. Bioavailability of energy and the starvation state, p. 1–23. *In* S. Kjelleberg (ed.), *Starvation in Bacteria.* Plenum Press, New York.

68. **Nemecek-Marshall, M., R. LaDuca, and R. Fall.** 1993. High-level expression of ice nuclei in a *Pseudomonas syringae* strain is induced by nutrient limitation and low temperature. *J. Bacteriol.* 175:4062–4070.

69. **O'Brien, R. D., and S. E. Lindow.** 1989. Effect of plant species and environmental conditions on epiphytic population sizes of *Pseudomonas syringae* and other bacteria. *Phytopathology* 79:619–627.

70. **O'Kane, D. J., W. L. Lingle, J. E. Wampler, M. Legocki, R. P. Legocki, and A. A. Szalay.** 1988. Visualization of bioluminescence as a marker of gene expression in *Rhizobium*-infected soybean root nodules. *Plant Mol. Biol.* 10:387–399.

71. **Panopoulos, N. J.** 1995. Ice nucleation genes as reporters, p. 271–281. *In* R. E. Lee, Jr., G. J. Warren, and L. V. Gusta (ed.), *Biological Ice Nucleation and Its Applications.* APS Press, St. Paul, Minn.

72. **Pooley, L., and T. A. Brown.** 1991. Effects of culture conditions on expression of the ice nucleation phenotype of *Pseudomonas syringae. FEMS Microbiol. Lett.* 77:229–232.

73. **Pouleur, S., C. Richard. J.-G. Martin, and H. Antoun.** 1992. Ice nucleation activity in *Fusarium acuminatum* and *Fusarium avenaceum. Appl. Environ. Microbiol.* 58:2960–2964.

74. **Pounder, J. I., C. R. Buell, and A. J. Anderson.** Personal communication.

75. **Rahme, L. G., M. N. Mindrinos, and N. J. Panopoulos.** 1992. Plant and environmental sensory signals control the expression of *hrp* genes in *Pseudomonas syringae* pv. *phaseolicola. J. Bacteriol.* 174:3499–3507.

76. **Ritter, C., and J. L. Dangl.** 1995. The *avrRpm1* gene of *Pseudomonas syringae* pv. *maculicola* is required for virulence on Arabidopsis. *Mol. Plant-Microbe Interact.* 8:444–453.

77. **Roszak, D. B., and R. R. Colwell.** 1987. Survival strategies of bacteria in the natural environment. *Microbiol. Rev.* **51:**365–379.

78. **Schweizer, H. P.** 1993. Two plasmids, X1918 and Z1918, for easy recovery of the *xylE* and *lacZ* reporter genes. *Gene* **134:**89–91.

79. **Selifonova, O., R. Burlage, and T. Barkay.** 1993. Bioluminescent sensors for detection of bioavailable Hg(II) in the environment. *Appl. Environ. Microbiol.* **59:**3083–3090.

80. **Sharma, S. B., and E. R. Signer.** 1990. Temporal and spatial regulation of the symbiotic genes of *Rhizobium meliloti* in planta revealed by transposon Tn5-*gusA. Genes Dev.* **4:**344–356.

81. **Shaw, J. J., and C. I. Kado.** 1986. Development of a *Vibrio* bioluminescence gene set to monitor phytopathogenic bacteria during the ongoing disease process in a non-disruptive manner. *Bio/Technology* **4:**560–564.

82. **Silcock, D. J., R. N. Waterhouse, L. A. Glover, J. I. Prosser, and K. Killham.** 1992. Detection of a single genetically modified bacterial cell in soil by using charge coupled device-enhanced microscopy. *Appl. Environ. Microbiol.* **58:**2444–2448.

83. **Slauch, J. M., and T. J. Silhavy.** 1991. Genetic fusions as experimental tools. *Methods Enzymol.* **204:**213–248.

84. **Stewart, G. S. A. B., and P. Williams.** 1992. *lux* genes and the applications of bacterial bioluminescence. *J. Gen. Microbiol.* **138:**1289–1300.

85. **Suslow, T. V., D. Matsubara, and M. Davies.** 1987. Application of tube nucleation assays to rapid population estimates of rhizobacteria expressing novel ice nucleation activity. *Curr. Plant Sci. Biotechnol. Agric.* **4:**1018–1024.

86. **Tucker, W.** Personal communication.

87. **Vali, G.** 1971. Quantitative evaluation of experimental results on the heterogeneous freezing nucleation of super-cooled liquids. *J. Atmos. Sci.* **28:**402–409.

88. **Vali, G.** 1995. Principles of ice nucleation, p. 1–39. *In* R. E. Lee, Jr., G. J. Warren, and L. V. Gusta (ed.), *Biological Ice Nucleation and Its Applications.* APS Press, St. Paul, Minn.

89. **van Elsas, J. D., and L. S. van Overbeek.** 1993. Bacterial responses to soil stimuli, p. 55–79. *In* S. Kjelleberg (ed.), *Starvation in Bacteria.* Plenum Press, New York.

90. **Warren, G. J.** 1995. Identification and analysis of *ina* genes and proteins, p. 85–99. *In* R. E. Lee, Jr., G. J. Warren, and L. V. Gusta (ed.), *Biological Ice Nucleation and Its Applications.* APS Press, St. Paul, Minn.

91. **Watanabe, N. M., M. W. Southworth, G. J. Warren, and P. K. Wolber.** 1990. Rates of assembly and degradation of bacterial ice nuclei. *Mol. Microbiol.* **4:**1871–1879.

92. **Wei, Z.-M., and S. V. Beer.** 1995. *hrpL* activates *Erwinia amylovora hrp* gene transcription and is a member of the ECF subfamily of σ factors. *J. Bacteriol.* **177:**6201–6210.

93. **Wilson, M., and S. E. Lindow.** 1992. Relationship of total and culturable cells in epiphytic populations of *Pseudomonas syringae. Appl. Environ. Microbiol.* **58:**3908–3913.

94. **Wolber, P. K.** 1993. Bacterial ice nucleation. *Adv. Microb. Physiol.* **34:**203–237.

95. **Young, D. C., S. D. Kingsley, K. A. Ryan, and F. J. Dutko.** 1993. Selective inactivation of eukaryotic β-galactosidase in assays for inhibitors of HIV-1 TAT using bacterial β-galactosidase as a reporter enzyme. *Anal. Biochem.* **215:**24–30.

96. **Zhang, Y.-Z., J. J. Naleway, K. D. Larison, Z. Huang, and R. P. Haugland.** 1991. Detecting *lacZ* gene expression in living cells with new lipophilic, fluorogenic β-galactosidase substrates. *FASEB J.* **5:**3108–3113.

97. **Zukowski, M. M., D. F. Gaffney, D. Speck, M. Kauffmann, A. Findeli, A. Wisecup, and J.-P. Lecocq.** 1983. Chromogenic identification of genetic regulatory signals in *Bacillus subtilis* based on expression of a cloned *Pseudomonas* gene. *Proc. Natl. Acad. Sci. USA* **80:**1101–1105.

Antibiotic Production by Soil and Rhizosphere Microbes In Situ

LINDA S. THOMASHOW, ROBERT F. BONSALL, AND DAVID M. WELLER

55

Antibiotics encompass a chemically heterogeneous group of small organic molecules of microbial origin that at low concentrations are deleterious to the growth or metabolic activities of other microorganisms (30). It has long been known that microorganisms capable of antibiotic production reside in soil, and in recent years it has been demonstrated unequivocally that antibiotics are synthesized in situ and can contribute to microbial antagonism and persistence. Genetic and molecular studies, coupled with sensitive and specific assay and detection systems, have confirmed the importance of antibiotics in plant defense, stimulating interest in the use of microorganisms as a practical means to improve the health and productivity of crop plants. The bioanalytical and genetic approaches are complementary, allowing not only the identification and quantification of antibiotics produced in situ but also evaluation of their activity and hence their biological significance. The techniques used are not new, but they have seldom been employed to evaluate microbial interactions in situ and may be applied in the future to optimize the production of desired metabolites under particular environmental conditions, to monitor the persistence or dissemination of bioactive metabolites of microbial origin in soil or plant tissues, and to evaluate the impact, on native microbial populations, of antibiotic-producing agents introduced for biological control, bioremediation, or biofertilization.

This chapter describes the use of well-characterized genes and mutant strains in indirect analyses of antibiotic production or activity in situ, discusses impediments to direct recovery and detection of antibiotics from natural sources, and reviews methods used to extract and quantify antibiotics produced by microorganisms in soil and plant materials.

INDIRECT AND DIRECT APPROACHES

Factors Affecting Antibiotic Production, Activity, and Detection

The quantity and quality of nutrients available and the ability of microbes to compete successfully for them are major determinants of microbial population size and metabolic activity and of antibiotic production in natural habitats. Nutrients are not uniformly dispersed throughout soil but rather are localized in the spermosphere and rhizosphere and in and around plant debris, wounds, lesions, and propa-

gules. When antibiotics have been detected in nature, it has been in material from these microhabitats, which are localized areas of intense microbial interaction (33, 34).

Whether an antibiotic will reach an ecologically significant or chemically detectable level in situ depends not only on the rate at which it is produced but also on the fraction of it available in a biologically active or extractable state. Soil particles and colloids, plant tissues, and microbial decomposition all act as sinks for biologically active molecules. The relative importance of physicochemical and biological processes in determining the fate of antibiotics in situ depends on the level of microbial activity and the rate and reversibility of sorption to soil surfaces, which in turn vary according to the chemistry of the compound and the soil and environmental conditions. In the upper soil horizons, however, microbial catabolism and perhaps plant uptake have greater impact on antibiotic availability than does soil quality (1, 2).

Activity Measurements In Situ: the Genetic Approach

Because antibiotics may reach threshold concentrations for activity within microsites while remaining at low or undetectable levels within a sample, it is important to evaluate not only the quantity of a substance that may be recovered but also its biological activity in situ. Such analyses are inherently indirect because the measured parameter is not the antibiotic itself but rather an effect attributed to its presence. The value of indirect approaches has been limited historically by uncertainty that the measured effect was related directly to the presence of the antibiotic in question. Many bacterial isolates produce multiple antibiotics with overlapping activity spectra, and thus it was not possible to correlate an effect in situ with a particular metabolite. In recent years this obstacle has been overcome by using mutant strains uniquely deficient in the production of specific metabolites. Studies in which such mutants are compared with their antibiotic-producing parental strains have a degree of specificity formerly lacking in indirect assays.

Mutants defective in antibiotic production may arise spontaneously or be induced chemically, by UV irradiation, or with molecular genetic approaches. The latter approach is preferred because the site of mutation can be identified and the mutated gene(s) can be recovered for further characterization. The strategy, which has been likened to Koch's postulates, consists of (i) mutagenesis to inactivate the trait of interest; (ii) screening of mutants and characterization of the mutant phenotype; (iii) complementation of mutants with DNA from a wild-type genomic library to genetically restore the trait; and (iv) comparison of the activities of the mutant, wild-type, and restored phenotypes. A fifth step, in which the complementing gene is mutated, used to replace the wild-type homolog, and shown to confer the mutant phenotype, is important to confirm the functional role of the mutated DNA but often is overlooked.

The value of the genetic approach is in direct proportion to the specificity of the mutation causing loss of antibiotic production. Assays used for phenotypic characterization must be able to distinguish among the various bioactive metabolites that may be present. The most useful mutants are those defective specifically in the synthesis of a particular metabolite, but until recently few such mutants were available. Antibiotic synthesis is a highly regulated process, and mutations in many genes other than those encoding biosynthetic enzymes can affect production (30). Such mutations also can influence the production of bioactive metabolites other than the one of interest or can adversely affect microbial fitness. Mutants used for in situ studies therefore must be indistinguishable from parental strains in traits other than the one of interest. When only limited characterization is practical, studies should include several independently derived mutants which are unlikely to have undetected or nonspecific mutations in common.

One advantage of the genetic approach is that mutant strains can be used to address a variety of experimental questions. Recent applications have focused on the effect of antibiotics on plant pathogens and on microbial survival and competitiveness (4, 30). Expression in situ of genes involved in antibiotic synthesis has been monitored with reporter genes (chapter 54, this volume). Less exploited is the use of genetically defined strains to develop models able to predict whether antibiotics will be produced or active in particular soils (30). Finally, the genetic approach remains a powerful option when an antibiotic is not sufficiently stable or abundant to be detected directly.

Extraction from Soil or Plant Material: Physical and Chemical Considerations

The isolation and detection of antibiotics from natural sources confirms that production has occurred and complements indirect evidence of antibiotic activity. Factors that will influence the efficiency of recovery include the stability of the antibiotic, its chemical and physical interactions with the sample matrix and the extraction solvent, and the handling of the sample before and during extraction. Properties such as solubility, thermolability, photosensitivity, and susceptibility to oxidation can be deduced from compounds produced in vitro. When the structure is known, insight can be gained into solubility, stability, and potential interactions with soil components. For example, methoxy groups are ring activators of phenolic compounds, increasing their susceptibility to oxidation, whereas carboxyl groups bonded directly to the ring reduce susceptibility to oxidation (2). Sorption of phenolic acids by soils is increased by the presence of methoxy groups or acrylic side chains on the aromatic ring (8).

Antibiotics produced in situ adsorb instantaneously upon contact with organic matter and charged groups on the surface of soil particles, and recovery declines continuously over time (2, 5, 32). The pH of soil and extractant solutions determines the charge of ionizable antibiotics, which in turn influences their solubility, affinity for soil colloids and or-

ganic matter, and uptake or diffusion into microorganisms and plants. In soil horizons where the humified organic content is high and microbial activity is maximal, the organic matter strongly influences the recovery of bioactive compounds such as antibiotics. In soils between pH 4.5 and 6.5, nonionic forms of organic acids and phenolic compounds are readily and irreversibly sorbed by soil organic matter (5, 12) or polymerized, even under abiotic conditions, into soil humic substances (7–9).

At pH values above the pK_a (approximately pH 4.5 for phenolic acids), important charge interactions occur with components of the inorganic soil fraction. Interactions may be direct, as when hydroxy acid anions bind to positively charged metal oxides, or indirect, as when such acids bind to negatively charged surfaces through divalent cationic "bridge" molecules that may themselves be either exchangeable or irreversibly embedded in mineral surfaces. Soils high in hydroxy-Al and -Fe compounds have a high adsorption rate and capacity for carboxyl and phenolic hydroxyl groups, and some Mn^{2+}-rich soils also have a high sorptive capacity for organic acids (8, 24). In comparison, clays are much less reactive (15), with adsorption related more to available surface area than to retention mechanism (8). However, adsorption through exchange ions can result in irreversible oxidation coupled with a reduction of Fe or Mn oxides (24) or can bring organic molecules into proximity with each other, favoring polymerization and irreversible retention (7).

Sample Collection and Preparation

The processing of soil and plant tissue samples, including collecting, storing, and sieving steps, can affect antibiotic recovery. For plant-associated samples, it is important to note how the soil and plant material are separated, how much plant tissue and soil are present, and whether specific portions of the plant or the root system are sampled. Quantitative variation can occur among replicated samples depending on whether macroscopic organic matter is distributed uniformly or removed by sieving.

Of particular importance are the measures taken to minimize microbial degradation of bioactive substances prior to and during extraction. If samples cannot be extracted immediately, freezing is preferable to air or oven drying because losses resulting from degradation, thermolability, and irreversible sorption can occur during the drying process.

The lower limit of sample size is determined by the efficiency of the extraction process and the sensitivity of detection, whereas the upper limit is set by how much material can conveniently be processed and how many samples and replications are required to obtain significant results. Soil sample sizes typically are 1 g (2, 9, 25, 28) or larger, and plant tissue samples of 80 seeds with glumes (20), root systems or root and crown segments from 50 to 200 seedlings (17, 19, 20, 31), or up to 17 g of potato tuber tissue (3) are representative. The efficiency of recovery is calculated from control samples to which the analyte has not or has been added in quantities spanning and exceeding the concentration range expected in test samples. Control samples should be of the same soil type and moisture content as test samples and should be incubated consistently (at least for a few minutes) prior to processing as for test samples. It also may be advantageous to add an internal standard to test samples to assess the efficiency of recovery (18). Such standards should have chemical properties similar to those of the substance being analyzed, should not occur naturally in the sample matrix, and must not interfere with subsequent analyses.

Solvent Considerations and Extraction

The choice of extractant will depend on the solubility and charge properties of the antibiotic and whether the objective is to determine the total amount present or the concentration available to interact with microorganisms and plants. The biologically available fraction is that which remains free in the soil solution or which can readily be released into the soil solution as environmental conditions permit. Compounds in the soil solution can be extracted with water, but water alone does not provide biologically meaningful estimates of available pool sizes because the reversibly bound fraction is not recovered and many antibiotics are sparingly soluble at ambient temperatures. Salt solutions (e.g., 0.5 M sodium acetate) and mild chelating agents are more effective

than water in recovering compounds retained through ionic interactions, presumably because the ions in solution compete with and displace the bound material (9). Even more effective are neutral to basic solutions (0.05 to 0.5 M, pH 7.0 to 8.0) of EDTA, which can displace organic anions bound through nonexchangeable cations to mineral surfaces (2, 7, 9). Some organic compounds are unstable in 0.5 M EDTA at pH 7.5 or 8.0, but neutral EDTA (0.25 M, pH 7.0) gave reliable estimates of free and reversibly bound phenolic acids in soils (2). Maximal recovery was achieved by 3 h with nonchelating aqueous solutions and by 6 to 12 h with chelating agents (9).

EDTA solutions do not extract biologically unavailable materials sorbed irreversibly to clay substrates or organic

TABLE 1 Isolation of antibiotics produced in situ

Compound	Source	Method			Recovery (%)	Detection limit[a]	Amt detected	Reference
		Extraction[b]	Chromatography	Detection				
Heterocyclic								
Phenazine-1-carboxylic acid	Rhizosphere	LLE, SPE	HPLC	UV	90–100		29–578 ng g of root^{-1}	31
Phenolic								
2,4-Diacetylphloroglucinol	Rhizosphere	LE	HPLC	UV	40–50		0.94–1.36 μ g of root^{-1}	17
	Soil	LE, SPE	HPLC	UV, amperometric	60–70		49–194 ng g of rhizosphere^{-1}	28
2-Hexyl-5-propylresorcinol	Seeds	LE, LLE	HPLC	UV	40–45		2.2–3.3 μg seed^{-1}	20
Phenylpyrrole								
Pyrrolnitrin	Seeds	LE, LLE	TLC	Bioautography	55–60		75–110 ng seed^{-1}	20
			GC	Amperometric	41–85		65–156 ng seed^{-1}	20
	Rhizosphere	LE, LLE, SPE	TLC	Bioautography	45–55		1 ng root^{-1}	20
			GC[c]	Amperometric	65–100		5.3–8.4 ng root^{-1}	20
	Potato	LE	TLC	Bioautography	4	0.87 ng wound^{-1}	≥0.87 ng wound^{-1}	3
			HPLC	UV		60 ng wound^{-1}	<60 ng wound^{-1}	3
Polyketide								
Pyoluteorin	Rhizosphere	LE	HPLC	UV	40–50		30–50 ng g of root^{-1}	26
Lipopeptide								
Herbicolin A	Crowns, roots	LE	TLC	Bioautography	70	25–50 ng seedling^{-1}	2.3–3.0 μg g of dry root^{-1}	19
Iturin A	Soil	LE	HPLC	UV	70–90		0.4–5.0 μg g of dry soil^{-1}	29a
Surfactin	Soil	LE	HPLC	UV	70–80	1.7–11.4 μg g of dry soil^{-1}		29a
Epidithiodiketopiperazine								
Chaetomin	Soil	LE	TLC	UV, bioautography				10
Gliotoxin	Soilless mix	LE	HPLC	UV	23		0.4–1.3 μg (cm^3)$^{-1}$	25
	Compost soil	LE	HPLC	UV	10–20		0.4–1.1 μg (cm^3)$^{-1}$	25
	Clay soil	LE	HPLC	UV	10–20		0.2–0.9 μg (cm^3)$^{-1}$	25
	Sandy soil	LE	HPLC	UV	10–20		0.02–0.14 μg (cm^3)$^{-1}$	25

[a] Detection limit for pure analyte added to sample matrix and subjected to fractionation as described.
[b] LLE, liquid-liquid phase extraction; SPE, solid-phase extraction; LE, liquid extraction.
[c] GC, gas chromatography.

matter and plant debris. Sodium hydroxide solutions, with or without autoclaving, can recover at least part of this material (9) and have been used to estimate the total quantity of phenolic acids present in situ (9, 18). A major disadvantage of strong alkaline extractions is the accompanying hydrolysis or dispersal of organic residues and humified material that can subsequently interfere with antibiotic detection (2, 9, 16).

To date, organic extractants have been used almost exclusively to recover antibiotics from natural sources and in our experience have proven more effective than salt, EDTA, or alkaline solutions. Samples, either hydrated or dry, are dispersed in the extractant; the solid residues are removed by settling or centrifugation; and the filtrate is collected and concentrated after passage through a solvent-compatible filter. A wide range of ratios of sample mass to solvent volume have been reported, but values of 1:1 to 1:5 are typical. Repeat extractions with smaller solvent volumes are more efficient than single extractions, but they are also more time-consuming and can prolong user exposure to toxic solvents.

Samples to be fractionated by high-performance liquid chromatography (HPLC) often require additional processing to remove soil residues that can foul chromatographic columns and interfere with UV detection. Humic acids usually are sedimented at acidic pH, although this results in salt accumulation and sample dilution and may cause significant loss of yield (18). Antibiotics with ionizable residues may be separated from contaminants by taking advantage of the pH-dependent, differential solubility of the neutral and charged forms in organic and aqueous solvents (27, 31). Most isolation procedures therefore include at least one liquid-liquid extraction step with immiscible solvents to partition antibiotics away from salt residues and interfering impurities and into organic solvents from which they can be concentrated readily.

Solid-phase extraction is being used increasingly to recover bioactive compounds from natural sources and offers many advantages over liquid-liquid solvent partitioning. Less solvent waste is generated, isolation is rapid and efficient, and detection limits may be increased because trace substances in large solvent volumes can be enriched. Waters Sep-Pak C_{18} cartridges are convenient for analytical-scale samples (28, 29, 31), and an octadecylsilica column was effective for preparative enrichment of the antibiotic 2,4-diacetylphloroglucinol (28). Aminoindoles and carboxylic indoles were recovered with high efficiency on Amberlite XAD-2 and C_{18} columns, respectively (23).

Chromatography and Detection

Thin-layer chromatography (TLC) is widely used to fractionate antibiotics recovered from natural materials (Tables 1 and 2). Advantages of TLC are that expensive instrumentation need not be available, compounds can be separated with good resolution, and methods are readily adaptable for preparative-scale work. Tissue extracts do not require extensive purification, and many samples can be run simultaneously. Both normal and reverse-phase adsorbants have been used with a variety of mobile-phase solvent systems (Table 2). Substances are visualized by UV absorption, chromogenic reaction with spray reagents, or bioautography, in which suspensions of sensitive or resistant indicator organisms in agar or broth are overlaid on chromatograms to detect bioactive spots (13). Antibiotic identity is confirmed

TABLE 2 Representative TLC systems for antibiotics produced in situ

Compound	Plate type	Solvent	R_f	Detection	Reference
Phenazine-1-carboxylic acid	Silica Gel G	Benzene-acetic acid, 95:5	0.17	UV, 365 nm	27
2-Hydroxyphenazine-1-carboxylic acid	Silica Gel G	Benzene-acetic acid, 95:5	0.19	UV, 365 nm	27
2-Hydroxyphenazine	Silica Gel G	Benzene-acetic acid, 95:5	0.04	UV, 365 nm	27
2,4-Diacetylphloroglucinol	Silica Gel 60 F_{254}	Chloroform-methanol, 19:1	0.20	UV, 254 nm; bioautography (*Bacillus subtilis*)	17
	Silica Gel GF	Dichloromethanol-hexane-methanol, 50:40:10	0.48	UV, 254 nm, 365 nm	29
	Reverse-phase KC18F	Acetonitrile-methanol-water, 1:1:1	0.85	UV, 254 nm; Pauly reagent	20a
Pyrrolnitrin	Silica Gel GHLF	Chloroform-acetone, 9:1	0.86	UV, 254 nm	21
	Silica Gel 60 F_{254}	Chloroform-acetone, 9:1	0.48	Erlich reagent, Pauly reagent	3
	Reverse-phase KC18F	Acetonitrile-methanol-water, 1:1:1	0.23	Erlich reagent, Pauly reagent	3
Pyoluteorin	Silica Gel GHLF	Chloroform-acetone, 9:1	0.36	UV, 254 nm	21
	Silica Gel 60 F_{254}	Chloroform-methanol, 19:1	0.50	UV, 254 nm; bioautography (*B. subtilis*)	17
	Reverse-phase, KC18F	Acetonitrile-methanol-water, 1:1:1	0.75	UV, 254 nm; Pauly reagent	20a
Herbicolin A	Silica Gel LK6DF	Chloroform-methanol-acetic acid-water, 65:25:4:3		Bioautography (*Candida albicans*)	20
Gliotoxin	Silica Gel LK6DF	Chloroform-acetone, 7:3	0.54	UV, 254 nm	25
Chaetomin	Silica Gel 60 F_{254}	Methylene chloride-methanol, 95:5		UV, 254 nm; bioautography (*Pythium ultimum*)	10

TABLE 3 Examples of HPLC systems for antibiotics produced in situ

Compound	Column type	Solvent	Profile	Detection	Sensitivity	Reference(s)
Phenazine-1-carboxylic acid	4-μm C$_{18}$ reverse-phase radial pack	Acetonitrile-water in 0.1% TFA[a]	Linear gradient	UV		31
2-Hydroxyphenazine-1-carboxylic acid	4-μm C$_{18}$ reverse-phase radial pack	Acetonitrile-water in 0.1% TFA	Linear gradient	UV		27
2-Hydroxyphenazine	4-μm C$_{18}$ reverse-phase radial pack	Acetonitrile-water in 0.1% TFA	Linear gradient	UV		27
2,4-Diacetylphloroglucinol	Nucleosil 120-5-C$_{18}$ reverse phase	Methanol-water in 0.43% phosphoric acid	Step gradient	UV, 270 nm	5 ng	17
	4-μm C$_{18}$ reverse-phase radial pack	Water-acetonitrile-methanol	Isocratic	UV, 278 nm		6
	5-μm ODS Hypersil	Water-methanol-tetrahydrofuran		UV, 254 nm	4 ng	28
		Water–methanol–tetrahydrofuran–0.05 M NaCl	Isocratic	Amperometric, 1.10 V	0.01 ng	28
Pyrrolnitrin	4-μm C$_{18}$ reverse-phase radial pack	Water-acetonitrile-methanol	Isocratic	UV, 225 nm		6
	5-μm Econosphere C$_{18}$	Acetonitrile-methanol-water	Isocratic	UV		3
Pyoluteorin	4-μm C$_{18}$ reverse-phase radial pack	Acetonitrile-methanol-water	Isocratic	UV, 310 nm	20 ng	6, 22
	Nucleosil 120-5-C$_{18}$ reverse phase	Methanol-water in 0.43% phosphoric acid	Step gradient	UV, 313 nm	50 ng	17, 26
Herbicolin A	5-μm Dynamax C$_{18}$ reverse phase	Acetonitrile-water in 0.1% TFA	Gradient	UV, 210 nm	1 μg	20
Iturin A	ODS-2 reverse phase	Acetonitrile-ammonium acetate (10 mM)	Isocratic	UV, 205 nm		14
Surfactin	ODS-2 reverse phase	Acetonitrile-TFA	Isocratic	UV, 205 nm		14
Gliotoxin	5-μm Ultrasphere ODS reverse phase	Water-acetonitrile-methanol-acetic acid	Isocratic	UV, 254 nm		25

[a] TFA, trifluoroacetic acid.

by appearance, distance traveled relative to the solvent front (R_f value), and cochromatography with pure standards in at least two different solvent systems. Quantities are estimated from spot size and intensity, or size of the inhibition zone for bioautography, at various dilutions relative to spots from known amounts of the standard run on the same plate (11).

Variations of up to 40% may be encountered between TLC data and published R_f values unless the same adsorbants (preferably from the same vendors) and preparative methods are used (see, for example, reference 3).

The versatility, resolving capability, and quantitative accuracy offered by HPLC make it the method of choice for

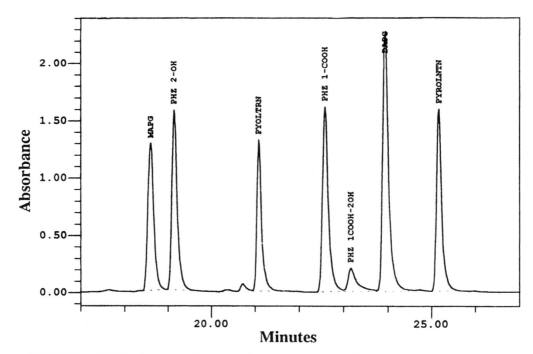

FIGURE 1 HPLC elution profile for antibiotics produced by fluorescent *Pseudomonas* spp. A mixture of reference antibiotics obtained from cultures grown in vitro was suspended in 35% acetonitrile with 0.1% trifluoroacetic acid and separated by reverse-phase chromatography on a Waters Nova-Pak C_{18} radial pack cartridge (4-μm particle diameter; 8 by 100 mm). A 2-min initial condition of 10% acetonitrile–0.1% trifluoroacetic acid at a flow rate of 1 ml/min was followed by a 20-min linear gradient to 100% acetonitrile–0.1% trifluoroacetic acid. Chromatography was on a Waters Millenium system. UV absorbance at 257 nm was detected with a Waters 996 photodiode array detector. Peaks shown (in order of elution): MAPG, monoacetylphloroglucinol; PHZ 2-OH, 2-hydroxyphenazine; PYOLTRN, pyoluteorin; PHZ 1-COOH, phenazine-1-carboxylic acid; PHZ 1COOH-2OH, 2-hydroxyphenazine-1-carboxylic acid; DAPG, 2,4-diacetylphloroglucinol; PYROLNTN, pyrrolnitrin.

most analyses of antibiotics produced in situ. Considerations in optimizing or developing a chromatographic system include selection of the column, the mobile phase, the elution profile, and the detector to be used. Reverse-phase columns have been used almost exclusively for antibiotics produced in situ, and a variety of liquid-phase systems and elution profiles have been described; representative examples are shown in Table 3 and Fig. 1. For applications that require processing of many samples, isocratic elution avoids time- and solvent-consuming column reequilibration between runs and is preferable to gradient elution.

Detection most frequently is by UV absorbance, and because photodiode array detectors concurrently monitor a range of wavelengths, they offer important advantages over fixed-wavelength units. Components within a mixture can be monitored simultaneously, each at its own unique absorption maximum, and subsequent spectral analyses can provide insight as to peak homogeneity and identity (31). Alternatively, significantly enhanced sensitivity and selectivity of detection can be obtained for phenolic compounds from some sources by amperometric detection (28), and fluorometric detection may offer similar advantages for compounds such as indoles (23) and some phenazine compounds.

REFERENCES

1. **Blum, U., S. B. Weed, and B. R. Dalton.** 1987. Influence of various soil factors on the effects of ferulic acid on leaf expansion of cucumber seedlings. *Plant Soil* **98:**111–130.

2. **Blum, U., A. D. Worsham, L. D. King, and T. M. Gerig.** 1994. Use of water and EDTA extractions to estimate available (free and reversibly bound) phenolic acids in Cecil soils. *J. Chem. Ecol.* **20:**341–359.

3. **Burkhead, K. D., D. A. Schisler, and P. J. Slininger.** 1994. Pyrrolnitrin production by biological control agent *Pseudomonas cepacia* B37w in culture and in colonized wounds of potatoes. *Appl. Environ. Microbiol.* **60:** 2031–2039.

4. **Carroll, H., Y. Moënne-Loccoz, D. N. Dowling, and F. O'Gara.** 1995. Mutational disruption of the biosynthesis genes coding for the antifungal metabolite 2,4-diacetylphloroglucinol does not influence the ecological fitness of *Pseudomonas fluorescens* F113 in the rhizosphere of sugarbeets. *Appl. Environ. Microbiol.* **61:**3002–3007.

5. **Chiou, C. T.** 1989. Theoretical considerations of the partition uptake of nonionic organic compounds by soil organic matter, p. 1–29. *In* B. L. Sawhney and K. Brown (ed.), *Reactions and Movement of Organic Chemicals in Soils.* SSSA Special Publication 22. Soil Science Society of America, Madison, Wis.

6. **Corbell, N., and J. E. Loper.** 1995. A global regulator of secondary metabolite production in *Pseudomonas fluorescens* Pf-5. *J. Bacteriol.* **177:**6230–6236.

7. **Dalton, B. R., U. Blum, and S. B. Weed.** 1983. Allelopathic substances in ecosystems: effectiveness of sterile soil components in altering recovery of ferulic acid. *J. Chem. Ecol.* **9:**1185–1201.

8. **Dalton, B. R., U. Blum, and S. B. Weed.** 1989. Differen-

tial sorption of exogenously applied ferulic, *p*-coumaric, *p*-hydroxybenzoic, and vanillic acids in soil. *Soil Sci. Soc. Am. J.* **53:**757–762.

9. **Dalton, B. R., S. B. Weed, and U. Blum.** 1987. Plant phenolic acids in soils: a comparison of extraction procedures. *Soil Sci. Soc. Am. J.* **51:**1515–1521.

10. **di Pietro, A., M. Gut-Rella, J. P. Pachlatko, and F. J. Schwinn.** 1992. Role of antibiotics produced by *Chaetomium globosum* in biocontrol of *Pythium ultimum*, a causal agent of damping-off. *Phytopathology* **82:**131–135.

11. **Fried, B., and J. Sherma.** 1982. *Thin-Layer Chromatography: Techniques and Applications.* Marcel Dekker, New York.

12. **Hasset, J. J., and W. L. Banwart.** 1989. The sorption of nonpolar organics by soils and sediments, p. 31–44. *In* B. L. Sawhney and K. Brown (ed.), *Reactions and Movement of Organic Chemicals in Soils.* SSSA Special Publication 22. Soil Science Society of America, Madison, Wis.

13. **Homans, A. L., and A. Fuchs.** 1970. Direct bioautography on thin-layer chromatograms as a method for detecting fungitoxic substances. *J. Chromatogr.* **51:**327–329.

14. **Huang, C.-C., T. Ano, and M. Shoda.** 1993. Nucleotide sequence and characteristics of the gene, *lpa-14*, responsible for biosynthesis of the lipoprotein antibiotics iturin A and surfactin from *Bacillus subtilis* RB14. *J. Ferment. Bioeng.* **76:**445–450.

15. **Huang, P. M., T. S. C. Want, M. K. Wang, M. H. Wu, and N. W. Hsu.** 1977. Retention of phenolic acids by noncrystalline hydroxy-aluminum and -iron compounds and clay minerals of soils. *Soil Sci.* **123:**213–219.

16. **Kaminsky, R., and W. H. Muller.** 1987. The extraction of soil phytotoxins using a neutral EDTA solution. *Soil Sci.* **124:**205–210.

17. **Keel, C., U. Schnider, M. Maurhofer, C. Voisard, J. Laville, U. Burger, P. Wirthner, D. Haas, and G. Défago.** 1992. Suppression of root diseases by *Pseudomonas fluorescens* CHA0: importance of the bacterial secondary metabolite 2,4-diacetylphloroglucinol. *Mol. Plant-Microbe Interact.* **5:**4–13.

18. **Kelley, W. T., D. L. Coffey, and T. C. Mueller.** 1994. Liquid chromatographic determination of phenolic acids in soil. *J. AOAC Int.* **4:**805–809.

19. **Kempf, H.-J., P. H. Bauer, and M. N. Schroth.** 1993. Herbicolin A associated with crown and roots of wheat after seed treatment with *Erwinia herbicola* B247. *Phytopathology* **83:**213–216.

20. **Kempf, H.-J., S. Sinterhauf, M. Müller, and P. Pachlatko.** 1994. Production of two antibiotics by a biocontrol bacterium in the spermosphere of barley and in the rhizosphere of cotton, p. 114–116. *In* M. H. Ryder, P. M. Stephens, and G. D. Bowen (ed.), *Improving Plant Productivity with Rhizobacteria.* CSIRO Division of Soils, Adelaide, South Australia, Australia.

20a. **Kraus, J.** Personal communication.

21. **Kraus, J., and J. E. Loper.** 1992. Lack of evidence for a role of antifungal metabolite production by *Pseudomonas fluorescens* Pf-5 in biological control of Pythium damping-off of cucumber. *Phytopathology* **82:**264–271.

22. **Kraus, J., and J. E. Loper.** 1995. Characterization of a genomic region required for production of the antibiotic pyoluteorin by the biological control agent *Pseudomonas fluorescens* Pf-5. *Appl. Environ. Microbiol.* **61:**849–854.

23. **Lebuhn, M., and A. Hartmann.** 1993. Method for the determination of indole-3-acetic acid and related compounds of L-tryptophan catabolism in soils. *J. Chromatogr.* **629:**255–266.

24. **Lehmann, R. G., H. H. Cheng, and J. B. Harsh.** 1987. Oxidation of phenolic acids by soil iron and manganese oxides. *Soil Sci. Soc. Am. J.* **51:**352–356.

25. **Lumsden, R. D., J. C. Locke, S. T. Adkins, J. F. Walter, and C. J. Rideout.** 1992. Isolation and localization of the antibiotic gliotoxin produced by *Gliocladium virens* from alginate prill in soil and soilless media. *Phytopathology* **82:**230–235.

26. **Maurhofer M., C. Keel, D. Haas, and G. Défago.** 1995. Influence of plant species on disease suppression by *Pseudomonas fluorescens* strain CHA0 with enhanced antibiotic production. *Plant Pathol.* **44:**40–50.

27. **Pierson, L. S., III, and L. S. Thomashow.** 1993. Cloning and heterologous expression of the phenazine biosynthetic locus from *Pseudomonas aureofaciens* 30–84. *Mol. Plant-Microbe Interact.* **5:**330–339.

28. **Shanahan, P., A. Borro, F. O'Gara, and J. D. Glennon.** 1992. Isolation, trace enrichment and liquid chromatographic analysis of diacetylphloroglucinol in culture and soil samples using UV and amperometric detection. *J. Chromatogr.* **606:**171–177.

29. **Shanahan, P., D. J. O'Sullivan, P. Simpson, J. D. Glennon, and F. O'Gara.** 1992. Isolation of 2,4-diacetylphloroglucinol from a fluorescent pseudomonad and investigation of physiological parameters influencing its production. *Appl. Environ. Microbiol.* **58:**353–358.

29a. **Shoda, M.** Personal communication.

30. **Thomashow, L. S., and D. M. Weller.** 1995. Current concepts in the use of introduced bacteria for biological disease control: mechanisms and antifungal metabolites, p. 187–235. *In* G. Stacey and N. Keen (ed.), *Plant-Microbe Interactions*, vol. 1. Chapman & Hall, New York.

31. **Thomashow, L. S., D. M. Weller, R. F. Bonsall, and L. S. Pierson III.** 1990. Production of the antibiotic phenazine-1-carboxylic acid by fluorescent *Pseudomonas* species in the rhizosphere of wheat. *Appl. Environ. Microbiol.* **56:**908–912.

32. **Weber, J. B., and C. T. Miller.** 1989. Organic chemical movement over and through soil, p. 305–334. *In* B. L. Sawhney and K. Brown (ed.), *Reactions and Movement of Organic Chemicals in Soils.* SSSA Special Publication 22. Soil Science Society of America, Madison, Wis.

33. **Weller, D. M., and L. S. Thomashow.** 1990. Antibiotics: evidence for their production and sites where they are produced, p. 703–711. *In* R. R. Baker and P. E. Dunn (ed.), *New Directions in Biological Control: Alternatives for Suppression of Agricultural Pests and Diseases.* Alan R. Liss, New York.

34. **Weller, D. M., and L. S. Thomashow.** 1993. Microbial metabolites with biological activity against plant pathogens, p. 172–180. *In* R. D. Lumsden and J. L. Vaughn (ed.), *Pest Management: Biologically Based Strategies.* American Chemical Society, Washington, D.C.

Quantification of Gene Transfer in Soil and the Rhizosphere

J. T. TREVORS AND J. D. VAN ELSAS

56

INTRODUCTION

This chapter contains a discussion of gene transfer, selected references, and protocols for studying and quantifying microbial gene transfer in soil. Standard protocols for studying gene transfer in soil are not available, but some have recently been suggested (46, 54).

The ability of microorganisms to transfer and/or acquire DNA and undergo genetic recombination (whereby DNA recombines with a replicon in recipient cells and then is stably inherited and expressed) allows soil microbial populations to adapt and evolve. Gene transfer, which requires recombination between homologous DNA sequences, generally occurs when donor and recipient cells are closely related (32, 45). However, recombination-independent gene transfer can occur between different bacterial species and even genera (32). The three mechanisms of gene transfer and gene flow that occur in soil are transformation, conjugation, and transduction. It is not known if protoplast fusion occurs in soil as a mechanism of gene transfer. Research on gene transfer in the environment (reviewed in references 13, 25, 50, 65, 80, and 81) is important from both environmental and human health perspectives related to the potential release of genetically engineered microorganisms into the environment for use in agriculture, forestry, biodegradation of toxic chemicals, and biomining.

The frequency of gene transfer in soil depends on a variety of interrelated, often changing, chemical, physical, and biological factors. Table 1 summarizes critical factors to consider when conducting gene transfer experiments in soil. Other factors may be added as required by the researchers. These factors should be carefully considered and incorporated into the experimental design during the planning stages of the research. Depending on the scientific hypothesis being tested, various factors may be more important than others.

In bacterial conjugation, survival of donor, recipient, and transconjugant cells, cell numbers, and the ratio of donor to recipient cells are important. In transduction, the presence of a critical density of transducing bacteriophages and susceptible bacterial host species is crucial. Environmental conditions such as temperature, pH, nutrient availability, and aerobic/anaerobic soil conditions that influence bacterial survival and activity also control gene transfer (17, 53).

It is also important to consider the carrier that will be used to introduce microbial cells into soils. The carrier may be water (distilled or deionized) or a nutrient solution, or the cells can be encapsulated in solid biodegradable carriers such as alginate or κ-carrageenan (6, 22; see review by Trevors et al. [71]). Other carriers include peat (67), mineral oil, and clay. Solid carriers such as alginate and κ-carrageenan may retard gene transfer in soil by physically separating donor cells from recipient cells. Alternatively, encapsulated cells may be protected from desiccation and predation by protozoans in soil. Therefore, survival of the encapsulated cells may be extended in soil.

It is also essential to use the most suitable plating medium for each gene transfer study. Most primary research papers contain formulations for media. The reader is also referred to *Handbook of Microbiological Media* by Atlas (1).

Importance of R-M Systems

Restriction-modification (R-M) systems, defined as endonucleases and methylases active at the same DNA sequences, also influence gene transfer (34, 60). Restriction enzymes recognize a specific DNA sequence and cleave the DNA unless the sequence is methylated by a modification enzyme (34). DNA in cells having an R-M system is methylated and acts as a substrate for restriction. However, the normal substrate for restriction may be foreign DNA entering cells as phages, conjugative plasmids, or naked DNA. R-M systems are widespread in prokaryotes, suggesting a central role in evolution. They are not essential for bacterial growth, DNA synthesis, recombination, or DNA repair. Price and Bickle (34) suggested that protection of a host cell by R-M is akin to protection by the immune system in higher eukaryotes. R-M assists in maintaining a particular genome while cleaving foreign DNA, making it accessible to genetic recombination. The ability of bacterial genomes to maintain constancy and still change gives rise to variants and diversity in populations (40, 61, 62). For additional information on microbial evolution, see references 5, 60 to 62, and 82; for additional information on gene transfer, see references 4, 10, 13, 14, 20, 23 to 25, 29, 36, 37, 43, 52, 54, 57 to 60, 68, 69, 74, 79, and 81.

Experimental Setup

Most studies on microbial gene transfer in soil have been conducted in soil microcosms under controlled laboratory conditions, which are not always representative of condi-

TABLE 1 Critical factors to consider when conducting gene transfer experiments in soil with microorganisms

1. Soil type, disturbed soils, non-disturbed soils, bulk density, pH, water content, aerobic or anaerobic conditions, presence of toxicant(s), nutrients, additional surfaces, presence of rhizophere

2. Microcosm, greenhouse, or field plot design and replication. In laboratory and greenhouse microcosms, the size of the microcosm and mass of soil used are important, especially when growing plants.

3. Donor-to-recipient cell numbers and ratios and their survival in conjugation experiments

4. Concentration and length of naked transforming DNA in transformation experiments

5. Host range of plasmid and duration of mating period, phage host range in transduction experiments, presence of competent cells in transformation experiments

6. If donor and recipient cells are added to soils, the distribution of the cells is critical for gene transfer.

7. Most suitable media for enumeration of donors, recipients, transconjugants, and transformants

8. Duration of the gene transfer experiment and careful control of environmental conditions such as temperature, relative humidity, soil water content, and light-dark cycles

9. Use of molecular techniques such as PCR, DNA-DNA hybridization, and reporter genes if possible

tions in field soils. Parameters such as soil temperature, pH, texture, organic matter content, presence of agrochemicals or contaminants (metals, organic contaminants), water-holding capacity, bulk density, and disturbed versus undisturbed soil exert a significant influence on bacterial survival, on persistence of DNA in soil, and consequently on gene transfer. The effects of such parameters may be studied by varying one parameter and determining the frequency of gene transfer (38, 77). It is also necessary to adequately determine if the gene transfer events occurred in soil and not on the plating media used to enumerate transconjugants in conjugation experiments (44, 78).

Soil masses from 10 to 100 g to several kilograms have been used in gene transfer experiments in soil microcosms. Therefore, microcosms vary significantly in size and design. Generally, containers or beakers made of glass, polyethylene, polypropylene, or polycarbonate are used. Glass and the latter two materials are autoclavable, which is often useful in preparing sterile control soils. Containers can be used to grow small plants, to produce a rhizosphere effect. Large containers must be used if plants are to be grown to maturity. In addition, plants remove soil water that must be replaced periodically if soil moisture is to be kept within a narrow range. The presence of plants in the soil microcosm may also require added fertilizer. Generally, the larger the microcosm and the more soil used, the more representative is the experiment of the natural environment. The use of larger microcosms also allows removal of subsamples without destroying the microcosm and influencing subsequent results. Alternatively, entire microcosms can be sacrificed if

the soil mass is low (e.g., 10 to 100 g). Soil samples as small as 1 g can be used to prepare serial dilutions for viable plating and direct DNA extraction, purification, and PCR amplification of a target DNA sequence (12, 23, 70). However, for viable plating, removal of 10 g of moist soil samples to prepare serial dilutions is generally advised. Also, a minimum of three 10-g moist soil samples should be dried at 105°C for 24 h or to constant mass, to determine the dry weight of soil necessary for subsequent calculation of the log CFU per gram (dry weight) of soil.

Often, experiments require the use of sterilized (autoclaved, tyndallized, or irradiated by using a ^{60}Co source at 4 Mrad [63, 73]) soil to allow the estimation of gene transfer and microbial survival of introduced cells in the absence of competing microorganisms or to prevent predation of introduced and indigenous bacterial cells by protozoans (11). It is difficult to sterilize even a small mass (10 to 100 g) of soil by repeated autoclaving with incubation periods between autoclaving to allow germination of spores. A comparative study on sterilization methods has been published by Tuominen et al. (72) and a review has been published by Trevors (63). In many laboratories, repeated autoclaving (usually done three times) is used, since it is the only means available. Chemical sterilization methods should not be used, as they may have a subsequent effect on the introduced microbial cells in a gene transfer study.

Transformation

Transformation is the uptake and integration of DNA by competent microbial cells. Competence is the ability of bacterial cells to bind and take up DNA in a form resistant to intracellular digestion by DNases. Successful transformation in microorganisms involves integration of foreign DNA into the recipient chromosome. Three types of transformation are known (32): (i) replacement transformation, in which donor DNA is substituted for homologous sequences in the genome of recipient cells, (ii) plasmid transformation, in which a new plasmid replicon is established in recipient cells without recombination events, and (iii) facilitated plasmid transformation, in which donor plasmid DNA contains homology to genomic DNA in recipient cells and recombination may establish a plasmid replicon. In an additional transformation mechanism known as ectopic transformation, foreign chromosomal genes are inserted into the chromosome at places other than their normal location, or foreign DNA is established in the genomic DNA of recipient cells (32).

Competence factors released by gram-positive *Bacillus subtilis* cells into the external medium induce competence in the population at high cell concentrations. However, gram-negative bacteria such as *Haemophilus influenzae* and *Pseudomonas stutzeri* do not release competence factors. Instead, competence is achieved by slowing the growth rate, switching from nutritionally rich to nutrient-depleted media, increasing cyclic AMP concentrations, or blocking DNA synthesis while permitting protein and RNA synthesis to proceed. These competence-inducing factors appear during the transition from the exponential to the stationary phase. Possibly, bacteria become competent during the stationary phase to obtain a greater catabolic range or resistance to waste metabolites and to obtain precursors for DNA synthesis (51). In soil, bacteria do not exhibit a growth curve like that observed in batch cultures. It is possible that in soil, a lack of nutrient(s) or exposure to increased ionic concentrations such as calcium may induce competence.

Conjugation

Conjugation is the transfer of plasmid or chromosomal DNA in a mating between donor and recipient cells, resulting in exconjugants after dissociation of cells involved in mating aggregates. The exconjugants are designated merozygotes after receiving DNA from donor cells, and merozygotes are called transconjugants after incoming DNA has been stabilized. Transconjugants can act as donors and the process can be repeated, producing secondary transconjugants and so on. Conjugation requires cell-to-cell contact by means of a conjugation bridge via the donor sex pilus, which is composed of a single protein, pilin. Contact initially occurs between the tip of the sex pilus and the cell envelope of recipient cells (32). Transfer-proficient plasmids are responsible for plasmid and chromosome transfer. Chromosomal transfer occurs when a conjugative plasmid integrates into and carries a segment of chromosomal DNA during conjugation. Nonconjugative plasmids can be mobilized directly or via integration with a conjugative plasmid. For instance, a conjugative plasmid may be transferred to a recipient containing a nontransmissible plasmid. Subsequently, both plasmids may be transferred to a recipient that contains neither plasmid. Triparental mobilization may occur when parent A contains a conjugative plasmid, parent B contains a nonconjugative plasmid, and parent C is the recipient of the nonconjugative plasmids. Although transfer frequencies are generally lower than those in biparental matings, triparental mating may be a mechanism for transfer of recombinant DNA in soil microorganisms. Conjugative plasmids may exhibit a narrow (limited to strains of one species) or wide (even trespassing genus boundaries) host range, and numerous examples of both types can be found in the literature.

Since the early 1970s, conjugation under different conditions using selected donor and recipient strains has been studied in soil microcosms. Although experiments under laboratory conditions with donor and recipient cells added to soils do not mimic the natural soil environment, these experiments allow gene transfer frequencies, as well as the conditions under which gene transfer occurs, to be determined. This information is of fundamental value and may be useful for assessing the frequency of gene transfer in soil from a regulatory perspective when genetically engineered microorganisms are to be released into soil. The gene transfer frequency can be expressed as the number of transconjugants per donor cell after a known period of time. Plasmid transfer frequencies should be calculated so as not to include division of transconjugants, since an increase in transconjugant numbers could be taken as a high frequency of plasmid transfer instead of division of cells after the initial plasmid transfer.

Low frequencies of transfer (e.g., 10^{-8} to 10^{-9}) are difficult to detect in soil. In many studies, selective markers used in conjugation experiments are antibiotic resistances carried on a conjugative plasmid. Upon acquiring the plasmid, transconjugant cells become resistant to the specific antibiotic(s). However, some antibiotic-resistant cells are present in soil, limiting the use of these resistance markers.

Transduction

Transduction is a gene transfer mechanism by which a bacteriophage transports genes from one bacterial cell to another. Bacteriophages, when replicating, can insert a certain amount of DNA, which may be of phage or bacterial origin, in their capsid. When the phage infects a bacterial host, it may insert the foreign DNA into the recipient genome, after

which the new host may express the foreign gene(s). Some phages can also reside in host bacterial cells as plasmids.

In generalized transduction, bacterial DNA is packaged erroneously into virions (32). Generalized transducing phages can contain any fragment of the bacterial genome. In specialized transduction, the DNA in transducing phages is a hybrid of bacterial and phage DNA (32). Specialized transducing phages often contain specific bacterial genome fragments.

Transduction may be a method by which bacteria such as *Escherichia coli* and *Salmonella typhimurium* acquire resistance to antibiotics, and it has been shown that R plasmids can be transferred by this method (31). Because transduction is phage-mediated genetic transfer, it is generally believed to occur between related microorganisms that the phage can infect. However, transduction can occur between different species. For instance, Ruhfel et al. (42) reported that bacteriophage CP-51 transduced *Bacillus anthracis*, *B. cereus*, and *B. thuringiensis*. Additionally, phage PBL1c transduced protoplasts of *B. larvae*, *B. subtilis*, and *B. popillae* (2).

METHODS FOR QUANTIFYING GENE TRANSFER IN SOIL

Bulk (Nonrhizosphere) and Rhizosphere Soil

Bulk (nonrhizosphere) soil can be operationally defined as the soil part not directly under the influence of roots, whereas the rhizosphere soil is directly influenced by roots. Bacteria and other microorganisms in bulk soil are often in a state of metabolic dormancy due to a lack of carbon and other nutrient(s). The surface of the root is referred to as the rhizoplane, the ectorhizosphere is defined as the area immediately surrounding the root, and the endorhizosphere is defined as the area within the root that can be colonized by some microorganisms (30). The endorhizosphere has also been defined as the outer soil zone of the root itself (83); it can extend a considerable distance from the plant roots and can include mycorrhizal fungal associations. In this situation, it is sometimes designated the mycorrhizosphere (30).

Bacteria can be dislodged from bulk and rhizosphere soil by shaking soil (200 to 300 rpm) in a dispersing agent such as 0.1% (wt/vol) sodium pyrophosphate (pH adjusted to 7.0) (66, 75). However, their extraction from the rhizosphere is more laborious. Shaking of washed roots with glass beads for 10 to 15 min or homogenization of roots by using a tissue grinder removes microbial cells from the rhizoplane for enumeration or analysis by PCR and DNA probing (70). Since bacterial cells can invade both epidermal and cortical plant cells, especially if the plant cells are damaged, their recovery may be difficult. Rhizoplane microbial cells can be detected within the roots by using staining and microscopic techniques.

Transformation

A pioneering study on transformation of *B. subtilis* in soil was reported by Graham and Istock (16). Frequencies of transformation of three markers (10^{-1} to 10^{-8} per recipient cell) depended on (i) whether mixed or single strain cultures were used and (ii) the time of sampling. The frequencies were significantly higher than the mutation rates. Bacteria were thought to be transformed by extracellular DNA produced by both *B. subtilis* strains in soil. Lorenz and Wackernagel (28) studied transformation of *P. stutzeri* in soil extract agar. Although the soil extract provided an environment in which transformation could occur, frequencies were 10- to 60-fold greater under limitations of nitrogen, phosphate, or pyruvate.

Extracellular DNA is stabilized when attached to sand, providing increased resistance to DNases (26, 41). The potential for sand-adsorbed DNA to transform both gram-positive and gram-negative bacteria has been examined (26, 27). *B. subtilis* was transformed at 25- to 50-fold-higher frequencies by sand-adsorbed DNA than by dissolved DNA (26). *P. stutzeri* was transformed at a less efficient rate by sand-adsorbed DNA. As *B. subtilis* cells attached (>10%) to DNA-coated sand better than *P. stutzeri* (0.8%), it was postulated that the former microorganism had a greater chance of contact with DNA to facilitate transformation. Transformation of *B. subtilis* was not inhibited by DNase I at concentrations of up to 1 μg/ml, whereas transformation of *P. stutzeri* cells was inhibited by 50 ng of DNase I per ml.

As far as we know, no reports of genetic transformation in the rhizosphere are available. The fate of naked DNA in rhizosphere soil may be assessed by addition of the DNA to an established rhizosphere or upon sowing or planting of seedlings. Alternatively, cells with a known reporter gene(s) could be added to soil, followed by establishment of a rhizosphere and the possibility of the cells providing transforming DNA to indigenous microbial species in the rhizosphere (for a review of molecular marker systems, see reference 35; for a review of survival and detection of bacteria, see reference 17). A naturally competent microbial strain that has been previously determined to act as a recipient in a standard non-soil transformation experiment can be added to the soils.

Procedure for Assessing and Quantifying Transformation in Soil

An excellent review on transformation in soil has been provided by Stotzky et al. (54), to which the reader is referred for additional details. In situ transformation can be set up by collecting soil samples from selected sites and passing them through a sieve (usually 2- or 4-mm mesh) to obtain a uniform soil. Soil may be used either directly or after short-term storage or sterilization for some treatments involving special procedures (54).

A realistic amount of naked DNA, e.g., 10 μg of chromosomal or 1 μg of plasmid, may be added to 2 g of soil, which is thoroughly mixed with a sterile stainless steel spatula or gently homogenized. If the DNA contains a unique reporter gene (see review by Prosser [35]) such as *lux* or Tn7::*lacZY* (18), the transformants containing the reporter gene can be probed. Also, the reporter gene can be extracted from soil samples as small as 1 g by direct soil DNA extraction, purified, and amplified by PCR for detection (70). The DNA concentration can be determined by using Hoechst 33258 (Calbiochem Diagnostics, La Jolla, Calif.), in which benzimidazole binds with DNA and enhances fluorescence. The fluorescence can be measured with a Hoefer 100 Mini-fluorometer (Hoefer Scientific Instruments, San Francisco, Calif.) calibrated with purified *E. coli* DNA (Sigma, St. Louis, Mo.) (70).

A second fluorometric method is based on a Tyler model SSF-600 solid-state fluorometer (Tyler Research, Edmonton, Alberta, Canada) with ethidium bromide as the fluorescent dye (70). Ethidium bromide is a nucleic acid-intercalating agent and frameshift mutagen and thus should be used with care. Consult the material safety data sheet (MSDS) for this chemical and use all appropriate safety measures. The SSF-600 fluorometer excites at 520 nm and measures the emission at 600 nm of the DNA-ethidium bromide complexes. A fluorometer blank not containing DNA is used to adjust the instrument to zero, while other cuvettes containing DNA standards (e.g., *E. coli* DNA; Sigma) are used to generate a standard curve to verify the linear response of the instrument.

After quantification and addition of DNA to soil, competent cells may be added, after which the soil-DNA-cell mix is incubated under the desired experimental conditions. During a time course study, soil samples may be serially diluted and plated on selective recipient- and marker-selective agar plates. The number of transformants and recipients and the transformation frequency per recipient are determined. Controls should be included to exclude background mutation (in soil without transforming DNA added) and to ensure that transformation did not occur on the agar plates (e.g., treat the agar plate with DNase).

Conjugation

The first study on bacterial conjugation in soil, using an *E. coli* mating system, was conducted by Weinberg and Stotzky (79). Frequencies of conjugation were lower in soil than in broth and were enhanced by amendment of soil with montmorillonite clay. However, one of the most common microorganisms considered for environmental release may be *Pseudomonas* species as a result of their ubiquitous nature and range of degradative abilities useful in bioremediation and in promotion of plant growth. Therefore, it is important to understand the nature of conjugative plasmids in *Pseudomonas* species to assess the risk of transfer of recombinant genes to indigenous bacteria. Lennon and DeCicco (21), examining 37 *Pseudomonas cepacia* strains from clinical, pharmaceutical-industrial, and environmental origins, found that large plasmids encoding antibiotic resistance(s) were a common feature in these strains. The plasmids were mobilizable by conjugation using the broad-host-range plasmid R751. In another study, 4 of 20 strains of *Pseudomonas syringae* pv. tomato were identified as conjugative by mobilization of the nonconjugative plasmid RSF1010 into *P. syringae* pv. syringae recipients (3). The ability to confer conjugation bridges might well be a widespread trait in bacterial communities in soil.

Top et al. (55) detected conjugal transfer of heavy metal resistance genes from *E. coli* to *Alcaligenes eutrophus* in sterile soil at frequencies of 10^{-5} per recipient cell. Transfer was the result of mobilization of a nonconjugative plasmid (pDN705) by a conjugative plasmid (RP4 or pULB113) present in another microorganism (triparental cross) or the recipient strain (retromobilization). Conjugation in nonsterile soils occurred at frequencies of 1.5×10^{-8} to 1.5×10^{-6} per recipient cell only when nutrients were added to soil.

In other experiments (39, 47, 48, 75), it was shown that the wheat rhizosphere exerts a significant gene transfer-enhancing effect on plasmid RP4 transfer between fluorescent pseudomonads. A marked version of RP4 also transferred to gram-negative members of the indigenous community, as evidenced by donor counterselection using a donor-specific phage (ΦR2f) followed by selective plating and probing (47). Roots also allowed for matings between cells which had been physically separated in soil, via the stimulation of movement toward the root, leading to cell-to-cell interactions (76). Knudsen et al. (19) developed a computer model to predict survival and conjugation of *P. cepacia* (containing recombinant plasmid R388:Tn*1721*) in rhizosphere and phyllosphere microcosms. The model was capable of predicting donor, recipient, and transconjugant populations at hourly intervals. Transconjugant numbers on leaf surfaces and in the rhizosphere of radish and bean plants were correctly predicted.

Procedure for Assessing and Quantifying Conjugation in Soil

Gene transfer by conjugation can be assessed in two different ways: (i) by introducing plasmid donor and recipient bacteria into soil and selecting for the recipient and the plasmid and (ii) by introducing the donor into soil and assessing transfer to indigenous bacteria by a donor counterselection method. Table 2, summarizes the different steps in the procedure; Table 3 summarizes a protocol for assessing gene transfer in a model rhizosphere. Alternatively, gene transfer can be studied in the rhizosphere of desired plants by growing plants in containers and removing roots at the desired time; any soil adhering to the roots is considered rhizosphere soil.

The procedure for studying plasmid transfer between donor and recipient strains is relatively simple. Transfer taking place on the selective agar instead of in soil should be checked for, preferably before an experiment is started as well as during the experiment. In a study using a *Pseudomonas fluorescens* strain, mating on the surface of the agar plate was inhibited by using nalidixic acid as one of the antibiotics for donor counterselection (44). Mutation controls of both donor and recipient should also be done to check for the occurrence of spontaneous resistant mutant CFU on transconjugant selective plates.

A problem in studying plasmid transfer to indigenous soil bacteria is counterselection of the used donor strain. Before transconjugants can be plated, the donor cell density should be reduced to avoid crowding of plates with donor colonies. Methods to eliminate the donor are based on (i) the use of markers that are not expressed in the donor, (ii) the use of an auxotrophic donor, (iii) the use of a donor with an inducible host-killing gene, (iv) the use of a donor which does not survive in the soil, and (v) the use of a bacteriophage to lyse the donor.

In experiments on plasmid transfer in soil, diverse bacteria are sometimes found on transconjugant-selective plates. These colonies are from bacteria naturally resistant to the selective antimicrobial agent(s), donor cells which escaped counterselection, or transconjugants. Colony filter hybridization with a specific DNA sequence present on the plasmid transferred can be used to distinguish the three. In addition, donor and transconjugant colonies can be distinguished by checking for a chromosomal marker in the donor.

Filter Mating Assay

Before a gene transfer experiment in soil is carried out, the frequency of gene transfer in a given system can be assessed in a filter mating. Transfer of the plasmid can be impaired because of endonuclease restriction activity in the recipient, or detection of transfer can fail because of lack of expression of the marker genes on the plasmid in the recipient. The following protocol is useful. Check plasmid transfer by adding 0.1 ml of washed cultures of donor and recipient on 0.22-μm-pore-size nitrocellulose filters on LB (or other appropriate) agar. Add strains separately on filters as controls for mutation. Incubate the cells overnight at a temperature adequate for growth of the strain with the lowest optimal temperature. Following incubation, place the filters into tubes containing 5 ml of sterile 0.85% NaCl solution and vortex them to dislodge bacterial cells from the filter. Plate

TABLE 2 Protocol for measuring plasmid transfer by conjugation in soil[a]

1. Select donor and recipient strains and plasmid. An example of a donor strain is *P. fluorescens* (76) containing the broad-host-range, self-transmissible plasmid RP4 encoding resistances to kanamycin, tetracycline, and ampicillin. A recipient strain can be produced by obtaining a rifampin- and nalidixic acid-resistant mutant of the donor strain without the plasmid. An excellent plating medium for fluorescent pseudomonads is King's B agar, composed of 2% peptone, 0.15% K_2HPO_4, 0.30% $MgSO_4 \cdot 7H_2O$, 1% glycerol, and 1.5% agar. The donors can be enumerated on King's B agar supplemented with kanamycin (80 μg/ml) and tetracycline (80 μg/ml). Recipients can be enumerated on 50 μg each of rifampin and nalidixic acid per ml. Transconjugants can be enumerated on King's B agar supplemented with all of the above-mentioned antibiotics. Culturable aerobic heterotrophic CFU can be obtained by plating on a variety of media (66).

2. Add donor and recipient cells to soil microcosm in buffer or distilled water suspension. Usually cell densities of 10^5 to 10^8/g of soil are added, the number depending on the purpose of the experiment. The moisture content of the soil should be known before and after addition of cells. The cells can be mixed throughout the soil microcosm with a sterile spatula or introduced onto the soil surface.

3. The number of viable donor and recipient cells should be determined in the inocula by viable plating on appropriate selective media for donor and recipient.

4. Controls can include an uninoculated soil or a soil inoculated with donor only or recipient only.

5. Periodically, remove soil samples from the microcosm and place in Erlenmeyer flasks containing sterile diluent such as 0.1% (wt/vol) sodium pyrophosphate (adjusted to pH 7) and 5 to 10 g of glass beads (5-mm diameter) or gravel (of the same size), stoppered with foam plugs. Samples are dispersed by vigorous shaking for 30 min to 1 h at 200 rpm. Serial dilutions can be prepared in the same diluent, and 100-μl aliquots can be plated onto selective agar medium for donor, recipient, and transconjugant CFU counts.

6. The frequency of transfer is determined as the number of transconjugants per recipient or donor cell and reported on dry-weight-of-soil basis. The dry weight of a known mass of soil can be determined by drying at 105°C for 24 h or until constant weight is achieved.

7. Further sampling at additional time intervals may also be conducted.

8. Plasmid(s) in CFU may be determined by agarose gel electrophoresis.

9. DNA probing may be carried out on transconjugant CFU, with the results compared with those for donor cells or donor plasmid.

[a] Based on RP4 as the donor plasmid and *P. fluorescens* as donor and recipient strains. Generally it is advisable to use a minimum of three replicates of each treatment.

TABLE 3 Procedure for studying gene transfer in the rhizosphere by using a model soil microcosm (Kuchenbuch soil chamber) (9)[a]

1. Germinate selected seeds, (e.g., *Triticum aestivum* [wheat], 2 to 3 days in the dark at room temperature) on sterile filter paper moistened with sterile distilled water.

2. Grow bacterial cultures to the selected phase of growth. Wash and prepare cells at the selected density in buffer or distilled or deionized water and use them to inoculate the soil in the model microcosm. One advantage of the model microcosm chamber is that the introduced bacterial cells (donors and/or recipients) and transconjugants can be placed at known distances from the root/membrane boundary. This permits the study of gene transfer at known distances form the model rhizosphere.

3. Set up soil microcosms as described by van Elsas et al. (76). Introduce donor and recipient bacteria cells into the soil. Layer the soil in the lower part of soil microcosms at the desired bulk density. Glue a nylon membrane and upper chamber on, and establish germinated seedlings on top of the nylon membrane.

4. Incubate microcosms in a controlled incubator (measure temperature and relative humidity and maintain them at the desired settings) under the selected light-dark cycle. Maintain the selected soil water content especially after plants start growing rapidly.

5. Destructively sample microcosms during plant development (e.g., after 7, 15, and 30 days), and analyze soil samples obtained by slicing layers from bottom to top of the chamber. This allows for assessment of the effect of root exudates on gene transfer frequencies and survival of cells. Report data as per gram (dry weight) of soil.

[a] See references 9, 75, and 76. Generally it is advisable to use a minimum of three replicates per treatment.

appropriate serial dilutions on media with correct antibiotics to enumerate donor, recipient, and transconjugant cells.

Conjugal Transfer Between Introduced Donor and Recipient Bacteria in Soil

The following protocol can be used to study plasmid transfer between donor and recipient *P. fluorescens* strains in soil (44). The medium and antibiotics to be used depend on the strains, plasmids, and markers. Soil samples are collected and dried to 10% below the water content at field capacity. Washed cell suspensions of the donor culture are added to the soil in such a volume that half of the lost water content is replaced, at final cell numbers of between 10^7 and 10^8/g of soil. Soil is mixed thoroughly and kept for 1 h before addition of the recipient strain in the same volume and numbers as the donor. Seedlings are planted or other additions are made to the soil samples, and the soil is incubated in the microcosms in a controlled environment chamber to prevent it from drying. For sampling, 10-g soil samples are removed and placed in Erlenmeyer flasks containing 95 ml of sterile 0.1% (wt/vol) sodium pyrophosphate (adjusted to pH 7.0) solution and 10 g of sterile gravel or sterile glass beads (diameter, 2 to 4 mm), and the flasks are shaken for 10 min at 200 rpm. Serial 10-fold dilutions of the soil slurry are made, and 100 μl is surface plated onto appropriate media with selective antibiotics to enumerate donor, recipient, and transconjugant cell numbers. Plates are usually in-

cubated for 2 to 5 days (check daily) at 20 to 37°C, depending on the strains. Numbers of CFU on transconjugant-selective plates are compared with those on the controls for mutation and plate mating. Confirmed transconjugants are enumerated and expressed per gram (dry weight) of soil. Conjugation frequencies per donor, per recipient, or per donor × recipient may be calculated.

Conjugal Transfer to Indigenous Bacteria in Soil by Using a Bacteriophage for Donor Counterselection

The procedure described below has been successfully used in studies on the transfer of different plasmids from *P. fluorescens* to indigenous bacteria in soil and the rhizosphere. Isolation of a specific phage and preparation of a crude phage lysate are described by Smit et al. (47). Soil samples are inoculated with the donor strain as described above. Ten-gram soil samples are added to Erlenmeyer flasks containing 95 ml of sterile 0.1% (wt/vol) sodium pyrophosphate solution (pH adjusted to 7.0) and 10 g of gravel. The flasks are shaken for 10 min at 200 rpm. Ten-fold dilutions are prepared for enumeration of donor and the total number of culturable bacteria. Subsamples (0.5 ml) from the Erlenmeyer flasks are pipetted into sterile microcentrifuge tubes, mixed by inverting the tubes with 0.5 ml of crude phage lysate (10^9 PFU/ml), and then incubated for 20 min at room temperature. Samples (0.1 to 0.2 ml) are plated onto transconjugant-selective plates. Selected transconjugants are checked for the presence of the plasmid by filter hybridization using a specific probe. Transconjugants which react positively with the probe are tested for not being phage-resistant donor CFU by reference to the antibiotic resistance of the donor or other characteristics. The species of selected transconjugants can be determined by using carbon utilization patterns (e.g., BIOLOG system; BIOLOG Inc., Hayward, Calif.) or some other method such as total cellular fatty acid analysis (MIDI, Newark, Del.) (8), and results are confirmed by plasmid isolations or filter matings to a suitable recipient. If microbial diversity is a component of the gene transfer experiments, the reader is referred to the text edited by Preist et al. (33), the excellent article by Torsvik et al. (56), reference 64, and chapter 46 of this volume.

Transduction

Transduction of *E. coli* in soil has been reported by Germida and Khachatourians (15). Zeph et al. (84) also reported transduction of *E. coli* by bacteriophage P1 in both sterile and nonsterile soils. In sterile soil, transduction was enhanced by amendments of montmorillonite or kaolinite clay. There were 3 to 4 log units more transductants in sterile than in nonsterile soil. However, there was no evidence of transduction to the indigenous soil microflora (84). In a subsequent study, these researchers used biotinylated DNA probes to detect transduction. No indigenous soil bacteria tested were transduced by phage P1 (85).

Transduction in the rhizosphere is so far unknown. Experiments could be conducted in a manner similar to that used for detection of transduction in bulk soils, by following procedures for establishing a rhizosphere and separating the rhizosphere soil from nonrhizosphere soil when sampling for transductants. Information forthcoming from these experiments will allow comparisons of the frequency of gene transfer between rhizosphere and nonrhizosphere soils and between the rhizospheres of different plants and with different phages and recipients.

Procedure for Assessing and Quantifying Transduction in Soil

Transduction can be assessed and quantified in soil microcosms by using known numbers of bacterial recipient cells as well as potentially transducing phage added to soil. To assess the effect of this variable, it is essential to set a variety of multiplicities of infection (MOIs) centered around 1. Smit et al. (49) showed that the MOI in soil can be artificially low because of sorption of phage to soil surfaces depending on soil type. For each soil, the optimal experimental conditions may need to be determined. Even at high MOIs transduction frequencies in microcosms with natural soil can be low because of the presence of competing or inhibiting microorganisms or lack of nutrient(s) and because only a fraction of a phage lysate is usually transducing. Most often transduction frequencies in soil decline below detection limits (often 10^{-8} at realistic host cell densities), and little can be done to increase this sensitivity.

Transduction in soil is quantified by determining the numbers of transductant CFU or recipient- and marker-selective agar plates and dividing these values by the number of potential recipients assayed on recipient-selective plates. Controls should consist of soil portions without additions, with phage only, and with recipient cells only. In addition, controls should account for the putative occurrence of plate transductions, e.g., by placing recipient strain and phage together on selective agar plates directly following their extraction from the separate soil portions.

Compartment Model for Gene Transfer

For a discussion of a compartment model approach to genetic change mechanisms in microorganisms and a specific example of a simulation model of gene transfer using the *mer* (mercury volatilization) operon, see reference 7. The compartment model simulates changes in cell numbers in different genetic configurations as a result of the mechanisms that confer genetic change. If models are to be representative of genetic diversity and the dynamic changes that occur, some essential conditions must be met. These have been summarized by Childress and Sharpe (7) as follows.

Genetic differences in bacteria represent different metabolic capabilities to tolerate or function under different conditions. The presence and distribution of genes necessary for metabolism of contaminants in a bacterial or microbial population are of considerable significance. The presence of the necessary genes in only a few cells in the bacterial population decreases the overall degradative capacity, as there may not be an increase in gene expression and degradation.

Inoculation of soil with microorganisms to enhance biodegradation requires information on gene transfer mechanisms and frequencies of gene transfer to indigenous microbial cells. Manipulation of a soil system by optimizing environmental conditions may require data on the initial distribution of degradative genes.

Degradative genes can be located on the chromosome, plasmid(s), and transposons. If the bacterial population is to be manipulated, the more data available on locations of genes, the easier it may be to optimize conditions for degradation. Factors such as the presence of pollutants, temperature, pH, and oxygen availability operate at molecular and cell levels. The effects that they produce are mediated by the genetic state of individual cells.

This work was sponsored by a grant from the EC-BIOTECH program (BIO2CT-92091), by a NATO collaborative grant to J.T.T. and J.D.V.E., and by a Natural Sciences and Engineering Research Council of Canada operating grant to J.T.T.

REFERENCES

1. **Atlas, R. M.** 1993. *Handbook of Microbiological Media.* CRC Press, Boca Raton, Fla.
2. **Bakhiet, N., and D. P. Stahly.** 1985. Studies on transfection and transformation of protoplasts of *Bacillus larvae*, *Bacillus subtilis* and *Bacillus popilliae*. *Appl. Environ. Microbiol.* **49:**577–581.
3. **Bender, C. L., and D. A. Cooksey.** 1986. Indigenous plasmids in *Pseudomonas syringae* pv. *tomato*: conjugative transfer and role in copper resistance. *J. Bacteriol.* **165:**534–541.
4. **Berg, G., and J. T. Trevors.** 1990. Bacterial conjugation between *Escherichia coli* and *Pseudomonas* spp. donor and recipient cells in soil. *J. Ind. Microbiol.* **5:**79–84.
5. **Cairns-Smith, A. G.** 1985. The first organisms. *Sci. Am.* **252:**90–100.
6. **Cassidy, M. B., K. Leung, H. Lee, and J. T. Trevors.** 1995. Survival of *lac-lux* marked *Pseudomonas aeruginosa* UG2Lr cells encapsulated in κ-carrageenan and alginate. *J. Microbiol. Methods* **23:**281–290.
7. **Childress, W. M., and P. J. H. Sharpe.** 1992. A compartment model approach to bacterial population genetics and biodegradation, p. 61–87. *In* C. J. Hurst (ed.), *Modeling the Metabolic and Physiologic Activities of Microorganisms.* John Wiley & Sons, Inc., New York.
8. **Colwell, R. R., and R. Grigorova (ed.).** 1987. *Methods in Microbiology*, vol. 19. *Current Methods for Classification and Identification of Microorganisms.* Academic Press, New York.
9. **Dijkstra, A. F., J. M. Govaert, G. H. N. Scholten, and J. D. van Elsas.** 1987. A soil chamber for studying the bacterial distribution in the vicinity of roots. *Soil Biol. Biochem.* **19:**351–352.
10. **Edwards, C. (ed.).** 1993. *Monitoring Genetically Manipulated Microorganisms in the Environment.* John Wiley & Sons, New York.
11. **England, L. S., H. Lee, and J. T. Trevors.** 1993. Bacterial survival in soil: effect of clays and protozoans. *Soil Biol. Biochem.* **25:**525–531.
12. **Flemming, C. A., K. Leung, H. Lee, J. T. Trevors, and C. Greer.** 1994. Survival of a *lux-lac* marked biosurfactant-producing *Pseudomonas aeruginosa* UG2 strain in soil: monitored by nonselective plating and PCR techniques. *Appl. Environ. Microbiol.* **60:**1606–1613.
13. **Fry, J. C., and M. J. Day (ed.).** 1990. *Bacterial Genetics in Natural Environments.* Chapman and Hall, London.
14. **Gauthier, M. J. (ed.).** 1992. *Gene Transfers and Environment.* Springer-Verlag, Heidelberg.
15. **Germida, J. J., and G. G. Khachatourians.** 1988. Transduction of *Escherichia coli* in soil. *Can. J. Microbiol.* **34:**190–193.
16. **Graham, J. B., and A. A. Istock.** 1978. Gene exchange in *Bacillus subtilis* in soil. *Mol. Gen. Genet.* **166:**287–290.
17. **Jackman, S. C., H. Lee, and J. T. Trevors.** 1992. Survival, detection and containment of bacteria. *Microb. Releases* **1:**125–154.
18. **Kluepfel, D. A., E. L. Kline, T. Hughes, H. Skipper, D. Gooden, D. J. Drahos, G. F. Barry, B. C. Hemming, and E. J. Brandt.** 1992. Field testing of a genetically engineered rhizosphere inhabiting pseudomonad: development of a model system, p. 189–200. *In* D. R. MacKenzie and S. C. Henry (ed.), *The Biosafety Results of Genetically Modified Plants and Microorganisms, Proceedings of the Kiawah Island Conference.* Agricultural Research Institute, Bethesda, Md.
19. **Knudsen, G. R., M. V. Walter, L. A. Porteous, V. J.**

Prince, J. L. Armstrong, and R. J. Seidler. 1988. Predictive model of conjugative plasmid transfer in the rhizosphere and phyllosphere. *Appl. Environ. Microbiol.* **54:** 343–347.

20. Lazcano, A., G. E. Fox, and J. F. Oro. 1992. Life before DNA: the origin and evolution of early Archean cells, p. 237–295. *In* R. P. Mortlock (ed.), *The Evolution of Metabolic Function.* CRC Press, Boca Raton, Fla.

21. Lennon, E., and B. T. DeCicco. 1991. Plasmids of *Pseudomonas cepacia* strains of diverse origins. *Appl. Environ. Microbiol.* **57:**2345–2350.

22. Leung, K., M. Cassidy, S. Holmes, H. Lee, and J. T. Trevors. 1995. Survival of carrageenan-encapsulated and unencapsulated *Pseudomonas aeruginosa* UG2Lr in a forest soil: monitored by polymerase chain reaction and viable plating. *FEMS Microbiol. Ecol.* **16:**71–82.

23. Leung, K., L. S. England, S. Weir, M. Cassidy, and J. T. Trevors. 1994. Microbial diversity in soil: effect of releasing genetically-engineered bacteria. *Mol. Ecol.* **3:**413–422.

24. Levin, M. A., R. J. Seidler, and M. Rogul (ed.). 1992. *Microbial Ecology: Principles, Methods and Applications.* McGraw-Hill, Inc., New York.

25. Levy, S. B., and R. V. Miller (ed.). 1989. *Gene Transfer in the Environment.* McGraw-Hill Publishing Company, New York.

26. Lorenz, M. G., B. W. Aardema, and W. Wakernagel. 1988. Highly efficient genetic transformation of *Bacillus subtilis* attached to sand grains. *J. Gen. Microbiol.* **134:** 107–112.

27. Lorenz, M. G., and W. Wackernagel. 1990. Natural genetic transformation of *Pseudomonas stutzeri* by sand-adsorbed DNA. *Arch. Microbiol.* **154:**380–385.

28. Lorenz, M. G., and W. Wackernagel. 1991. High frequency of natural genetic transformation of *Pseudomonas stutzeri* in soil extract supplemented with a carbon/energy and phosphorus source. *Appl. Environ. Microbiol.* **57:** 1246–1251.

29. Lorenz, M. G., and W. Wackernagel. 1994. Bacterial gene transfer by natural genetic transformation in the environment. *Microbiol. Rev.* **58:**563–602.

30. Lugtenberg, B. J. J., and L. A. de Weger. 1992. Plant root colonization by *Pseudomonas* spp., p. 13–19. *In* E. Galli, S. Silver, and B. Witholt (ed.), *Pseudomonas: Molecular Biology and Biotechnology.* American Society for Microbiology, Washington, D.C.

31. Mise, K., and R. Nakaya. 1977. Transduction of R plasmids by bacteriophages P1 and P22. *Mol. Gen. Genet.* **157:**131–138.

32. Porter, R. D. 1988. Modes of gene transfer in bacteria, p. 1–41. *In* R. Kucherlapati and G. R. Smith (ed.), *Genetic Recombination.* American Society for Microbiology, Washington, D.C.

33. Preist, F. G., A. Ramos-Cormezana, and B. J. Tindall (ed.). 1994. *Bacterial Diversity and Systematics.* Plenum Press, New York.

34. Price, C., and T. A. Bickle. 1986. A possible role for DNA restriction in bacterial evolution. *Microbiol. Sci.* **3:** 296–299.

35. Prosser, J. I. 1994. Molecular marker systems for detection of genetically engineered micro-organisms in the environment. *Microbiology* **140:**5–17.

36. Reanney, D. 1977. Gene transfer as a mechanism of microbial evolution. *BioScience* **27:**340–344.

37. Reanney, D. C., W. P. Roberts, and W. J. Kelly. 1982. Genetic interactions among microbial communities, p. 287–322. *In* A. T. Bull and J. H. Slater (ed.), *Microbial Interactions and Communities.* Academic Press, New York.

38. Richaume, A., J. S. Angle, and M. J. Sadowski. 1989. Influence of soil variables on in situ plasmid transfer from *Escherichia coli* to *Rhizobium fredii. Appl. Environ. Microbiol.* **55:**1730–1734.

39. Richaume, A., E. Smit, G. Faurie, and J. D. van Elsas. 1992. Influence of soil type on the transfer of RP4p from *Pseudomonas fluorescens* to indigenous bacteria. *FEMS Microbiol. Ecol.* **101:**281–292.

40. Riley, M. 1989. Constancy and change in bacterial genomes, p. 359–388. *In* J. S. Poindexter and E. R. Leadbetter (ed.), *Bacteria in Nature.* Plenum Press, New York.

41. Romanowski, G., M. G. Lorenz, and W. Wackernagel. 1991. Adsorption of plasmid DNA to mineral surfaces and protection against DNase I. *Appl. Environ. Microbiol.* **57:** 1057–1061.

42. Ruhfel, R. E., N. J. Robillard, and C. B. Thorne. 1984. Interspecies transduction of plasmids among *Bacillus anthracis, B. cereus,* and *B. thuringiensis. J. Bacteriol.* **157:** 708–711.

43. Seech, A., and J. T. Trevors. 1991. Environmental variables and evolution of xenobiotic catabolism in bacteria. *Trends Ecol. Evol.* **6:**79–83.

44. Smit, E., and J. D. van Elsas. 1990. Determination of plasmid transfer frequency in soil: consequences of bacterial mating on selective agar media. *Curr. Microbiol.* **21:** 151–157.

45. Smit, E., and J. D. van Elsas. 1992. Conjugal gene transfer in the soil environment: new approaches and developments, p. 79–94. *In* M. Gauthier (ed.), *Gene Transfers and Environment.* Springer-Verlag, New York.

46. Smit, E., and J. D. van Elsas. 1995. Detection of gene transfer in the environment: conjugation in soil. *In* A. D. L. Akkermans, J. D. van Elsas, and F. J. de Bruijn (ed.), *Molecular Microbial Ecology Manual.* Kluwer Academic Publishers, Dordrecht, The Netherlands.

47. Smit, E., J. D. van Elsas, J. A. van Veen, and W. M. de Vos. 1991. Detection of plasmid transfer from *Pseudomonas fluorescens* to indigenous bacteria in soil using bacteriophage *φ*R2f for donor counterselection. *Appl. Environ. Microbiol.* **57:**3482–3488.

48. Smit, E., D. Venne, and J. D. Van Elsas. 1993. Effects of cotransfer and retrotransfer on the mobilization of a genetically engineered IncQ plasmid between bacteria on filters and in soil. *Appl. Environ. Microbiol.* **59:**2257–2263.

49. Smit, E., A. C. Wolters, H. Lee, J. T. Trevors, and J. D. van Elsas. 1996. Interactions between a genetically engineered *Pseudomonas fluorescens* and bacteriophage R2f in soil; effects of nutrients, alginate encapsulation and rhizosphere. *Microb. Ecol.* **31:**125–140.

50. Stewart, G. J. 1992. Natural transformation and its potential for gene transfer in the environment, p. 253–282. *In* M. A. Levin, R. J. Seidler, and M. Rogul (ed.), *Microbial Ecology: Principles, Methods and Applications.* McGraw-Hill, Inc., New York.

51. Stewart, G. J., and C. A. Carlson. 1986. The biology of natural transformation. *Annu. Rev. Microbiol.* **40:** 211–235.

52. Stewart-Tull, D. E. S., and M. Sussman (ed.). 1992. *The Release of Genetically Modified Microorganisms-REGEM 2.* Plenum Press, New York.

53. Stotzky, G., and H. Babich. 1986. Survival of, and genetic transfer by, genetically engineered bacteria in natural environments. *Adv. Appl. Microbiol.* **31:**93–138.

54. Stotzky, G., E. Gallori, and M. Khanna. 1995. Transformation in soil. *In* A. D. L. Akkermans, J. D. van Elsas, and F. J. de Bruijn (ed.), *Molecular Microbial Ecology Manual.* Kluwer Academic Publishers, Dordrecht, The Netherlands.

55. Top, E., M. Mergeay, D. Springael, and W. Verstraete. 1990. Gene escape model: transfer of heavy metal resistance genes from *Escherichia coli* to *Alcaligenes eutrophus* on

agar plates and in soil samples. *Appl. Environ. Microbiol.* **56:**2471–2479.

56. **Torsvik, V., J. Goksoyr, F. L. Daae, R. Sorheim, J. Michalsen, and K. Salte.** 1994. Use of DNA analysis to determine the diversity of microbial communities, p. 39–48. *In* K. Ritz, J. Dighton, and K. E. Giller (ed.), *Beyond the Biomass.* British Society of Soil Science, Wiley-Sayce Publication, England.

57. **Trevors, J. T.** 1987. R-plasmid transfer in soil. *Bull. Environ. Contam. Toxicol.* **39:**74–77.

58. **Trevors, J. T.** 1987. Survival of *Escherichia coli* donor, recipient and tranconjugant cells in soil. *Water Air Soil Pollut.* **34:**409–414.

59. **Trevors, J. T.** 1988. Use of microcosms to study genetic interactions between microorganisms. *Microbiol. Sci.* **5:**132–136.

60. **Trevors, J. T.** 1995. Molecular evolution in bacteria. *Antonie van Leeuwenhoek* **67:**315–324.

61. **Trevors, J. T.** Genome size in bacteria. *Antonie van Leeuwenhoek,* in press.

62. **Trevors, J. T.** DNA in soil: adsorption, transformation and molecular evolution. *Antonie van Leeuweenhoek,* in press.

63. **Trevors, J. T.** Sterilization and inhibition of microbial activity in soil. *J. Microbiol. Methods,* in press.

64. **Trevors, J. T.** Bacterial biodiversity in chemically contaminated soils. Submitted for publication.

65. **Trevors, J. T., T. Barkay, and A. W. Bourquin.** 1987. Gene transfer among bacteria in soil and aquatic environments: a review. *Can. J. Microbiol.* **33:**191–198.

66. **Trevors, J. T., and S. Cook.** 1992. A comparison of plating media and diluents for enumeration of aerobic bacteria in a loam soil. *J. Microbiol. Methods* **14:**271–275.

67. **Trevors, J. T., and F. Grange.** 1992. Respiratory activity of a genetically-engineered *Pseudomonas fluorescens*—peat mixture introduced into agricultural soil. *J. Environ. Sci. Health* **A27:**879–887.

68. **Trevors, J. T., and K. M. Oddie.** 1986. R-plasmid transfer in soil and water. *Can. J. Microbiol.* **32:**610–613.

69. **Trevors, J. T., and M. E. Starodub.** 1987. R-plasmid transfer in non-sterile soil. *Syst. Appl. Microbiol.* **9:**312–315.

70. **Trevors, J. T., and J. D. van Elsas (ed.).** 1995. *Nucleic Acids in the Environment: Methods and Applications.* Springer-Verlag, Heidelberg.

71. **Trevors, J. T., J. D. van Elsas, H. Lee, and L. S. van Overbeek.** 1992. Use of alginate and other carriers for encapsulation of microbial cells for use in soil. *Microb. Releases* **1:**61–69.

72. **Tuominen, L., T. Kairesalo, and H. Hartikainen.** 1994.

Comparison of methods for inhibiting bacterial activity in sediment. *Appl. Environ. Microbiol.* **60:**3454–3457.

73. **van Elsas, J. D., J. M. Govaert, and J. A. van Veen.** 1987. Transfer of plasmid pFT30 between bacilli in soil as influenced by bacterial population dynamics and soil conditions. *Soil Biol. Biochem.* **19:**639–647.

74. **van Elsas, J. D., and J. T. Trevors.** 1991. Environmental risks and fate of genetically engineered microorganisms in soil. *J. Environ. Sci. Health* **A26:**981–1001.

75. **van Elsas, J. D., J. T. Trevors, and M. E. Starodub.** 1988. Bacterial conjugation between pseudomonads in the rhizosphere of wheat. *FEMS Microbiol. Ecol.* **53:**299–306.

76. **van Elsas, J. D., J. T. Trevors, M. E. Starodub, and L. S. van Overbeek.** 1990. Transfer of plasmid RP4 between pseudomonads after introduction into soil: influence of spatial and temporal aspects of inoculation. *FEMS Microbiol. Ecol.* **73:**1–12.

77. **van Veen, J. A., P. J. Kuikman, and J. D. van Elsas.** 1994. Modelling microbial interactions in soil; preliminary considerations and approaches, p. 29–46. *In* M. J. Bazin and J. M. Lynch (ed.), *Environmental Gene Release, Models, Experiments and Risk Assessment.* Chapman and Hall, London.

78. **Walter, M. V., L. A. Porteous, R. J. Seidler, and J. L. Armstrong.** 1991. Formation of transconjugants on plating media following in situ conjugation experiments. *Can. J. Microbiol.* **37:**703–707.

79. **Weinberg, S. R., and G. Stotzky.** 1972. Conjugation and genetic recombination of *Escherichia coli* in soil. *Soil Biol. Biochem.* **4:**171–180.

80. **Wellington, E. M. H., P. R. Herron, and N. Cresswell.** 1993. Gene transfer in terrestrial environments and the survival of bacterial inoculants in soil, p. 135–170. *In* C. Edwards (ed.), *Monitoring Genetically Manipulated Microorganisms in the Environment.* John Wiley & Sons, New York.

81. **Wellington, E. M. H., and J. D. Van Elsas.** 1992. *Gene Transfer between Microorganisms in the Natural Environment.* Pergamon Press, London.

82. **Woese, C. R.** 1987. Bacterial evolution. *Microbiol. Rev.* **51:**221–271.

83. **Wood, M.** 1989. *Soil Biology.* Chapman and Hall, New York.

84. **Zeph, L. R., M. A. Onaga, and G. Stotzky.** 1988. Transduction of *Escherichia coli* by bacteriophage P1 in soil. *Appl. Environ. Microbiol.* **54:**1731–1737.

85. **Zeph, L. R., and G. Stotzky.** 1989. Use of a biotinylated DNA probe to detect bacteria transduced by bacteriophage P1 in soil. *Appl. Environ. Microbiol.* **55:**661–665.

Microorganisms Interacting with Insects

JAMES R. FUXA AND YASUHISA KUNIMI

57

Most of the scientific literature and methodology for microorganisms interacting with invertebrates in soil deal with pathogens of insects, or entomopathogens. This particular group of microorganisms has been well studied because the class Insecta includes numerous pest species and because entomopathogens are widely researched as agents for insect pest management (153). The entomopathogens investigated for insect control include bacteria, viruses, fungi, protozoa, and nematodes.

Soil is an important reservoir for long-term survival of entomopathogens. Thus, methods dealing with soils often are incorporated into studies of these organisms. For example, new isolates of entomopathogens often are found directly or indirectly through soil sampling. Long-term methods of microbial control of insects (i.e., a single microbial release that results in entomopathogen replication and suppression of more than one pest generation) often include monitoring of the entomopathogen population in its soil reservoir. Current concerns with risk assessment of recombinant entomopathogens are necessitating a better understanding of environmental fate (including persistence and spread in soil), not only of the recombinants but also of the parental entomopathogens. Additionally, many insect pests live in soil; research on control of such pests requires introduction of the entomopathogen into soil and its subsequent sampling.

The purpose of this chapter is to review the basic methods used for studying entomopathogens in soils. These methods deal largely with detection, isolation, or quantification of the entomopathogen, either directly or indirectly through sampling the host insects. Methods also have been developed to inoculate entomopathogens into soil, release and sample marked entomopathogens, and study entomopathogens in soil microcosms.

SAMPLING INSECTS IN SOIL

Studies of interactions between insects and microorganisms often require sampling of the insects in soil. In many cases, studies of the interacting microorganism or entomopathogen similarly require initial sampling of the insects, with

Approved for publication by the Director of the Louisiana Agricultural Experiment Station as manuscript number 95-17-9166.

subsequent detection or isolation of the microorganism. Thus, methods of sampling insects and other invertebrates in soil are critical to studies of the microorganisms interacting with these animals. The basic methods of sampling insects in soil have been reviewed elsewhere (141) and will only be summarized here.

Insect Sampling

Absolute samples of insects in soil, particularly those including entomopathogens, virtually always involve collection of a soil sample and subsequent extraction of the animals. The extraction process can kill the insects or allow them to live; in the latter case, the sampled insects often are reared to permit pathogenesis. Live or dead insects can then be observed for signs or symptoms of the disease caused by an entomopathogen, which in turn can be identified or isolated.

Soil samples for absolute sampling of insects usually are collected with a corer (141), although tools as simple as shovels or trowels can be used to collect a specific volume of soil. Corers come in various sizes and shapes. Some are modified to facilitate sampling in extreme conditions, such as hard tropical or frozen tundra soils, or soft humus-rich materials such as manure heaps, in which collecting undisturbed samples can be difficult.

Insects are extracted from soil samples by one of two major types of methods, mechanical or behavioral-dynamic. Mechanical methods have the advantages of allowing storage of samples for long time periods (the insects do not have to be alive) and of finding all insect stages, including sedentary ones. A major disadvantage is that they are more labor-intensive. Behavioral-dynamic procedures, on the other hand, can be left unattended once set up; thus, in addition to requiring less labor, several samples can be handled simultaneously in batteries of extractors. However, these methods depend on the animal's behavior, which in turn can be influenced by the condition of the animal as well as environmental factors.

Mechanical extraction methods include simple visual search, dry sieving through screens of various meshes, soil washing-wet sieving whereby soil is washed through sieves, soil washing-flotation whereby insects float to the surface, centrifugation, sedimentation, and sectioning of soil cores impregnated with agar.

The basic apparatus for behavioral-dynamic methods is

the Berlese-Tullgren funnel; this device basically consists of a large funnel containing the soil sample, with a heat source at the top to drive the insects into a collecting jar or apparatus at the bottom. There have been many modifications and improvements on this basic design, including wet extractors (e.g., elutriators) in which the soil sample is flooded. Other extractors have used chemical fumes or electrical currents instead of heat as the stimulus for movement of insects or other invertebrates.

There also are relative sampling methods for insects in or on soil. The most common is the pitfall trap, basically consisting of a container sunk into the soil so that the open mouth is level with the soil surface. Like the Berlese-Tullgren funnel, this basic design has seen many modifications and improvements. Other methods include plowing a transect to expose soil insects, placing bottomless cages in certain areas to capture insects emerging from soil, and capturing insects with a bait placed in or on the soil.

Entomopathogens in Insects Sampled from Soil

Once the insects are removed from the soil sample, they can be examined for signs and symptoms of disease caused by entomopathogens. This often is done after the insects are reared in the laboratory for several days, to allow time for diseases to develop. The entomopathogens in diseased insects can be isolated or quantified.

Sampling of insects from soil and subsequent examination for entomopathogens can provide several levels of information. Insects from samples from a given area can be pooled, and then subsamples can be examined for entomopathogens (126); this technique is useful to determine simple presence or absence of a particular microorganism over a relatively large area. Perhaps the most common type of study is that in which the percentage of infected insects provides an estimate of prevalence of the entomopathogen in a given area or volume (30), though not in the form of pathogen units per unit area or volume. Sampling insects can provide an estimate of entomopathogen units per unit area or volume when numbers of pathogen units produced in insects of certain sizes or age categories can be determined and numbers of such infected hosts can be determined from absolute field samples (80).

Several of these methods of sampling insects in soil have been used in studies of entomopathogens. Examples include core samples with visual examination for insects infected with fungi (52, 91), nematodes (34, 82), bacteria (52, 82, 83), protozoa (52), or rickettsia (52, 82, 83); a rotary separator (similar in principle to sieving) to sample a fungus (116); core samples with separation by flotation for protozoa (74, 75), fungi (16), a yeast (74), or a bacterium (75); a modified Berlese-Tullgren funnel for studies of fungal transport (183); pitfall traps for insects with fungi (35) or nematodes (35, 114, 115); emergence cages to sample a fungus from insects (41, 122, 167); and a fungus in termites trapped with a bait (51).

Bias Associated with Sampling of Insects

Estimation of entomopathogen numbers in soil can become biased when these microorganisms are collected indirectly through sampling insects. This can happen as a result of altered behavior of infected insects or horizontal transmission (contamination) during or after sampling.

Sampling can be biased if the behavior of infected insects is affected to the degree that they become more or less likely than uninfected insects to be captured. For example, insects infected with a virus (79) or rickettsia (12) burrowed less deeply into soil than uninfected insects. When certain ants are infected with a fungus, they generally climb onto exposed positions on plants rather than maintain their typical ground-dwelling behavior; diseased ants also are known to stay away from their nests (28). If such aberrant behavior can be identified through preliminary research, it often can be counteracted by means of sampling design or mathematical transformation (37).

The sampling procedure or handling of the insects can cause errors in estimation of prevalence or pathogen units in soil. This subject has been reviewed previously (37) and will only be summarized here. The major concern is overestimation of prevalence of entomopathogens: uninfected insects can be exposed to entomopathogens by soil agitation or by contaminated equipment or hands, insects in one sample can spread disease to one another before they are separated, or stress due to sampling can increase susceptibility or induce latent infections that might not otherwise be activated. Underestimation also is possible if small, infected insects die and decompose before they are found. Countermeasures for such biases include rapid preservation of sampled insects by various methods for later examination, rapid separation of living insects, rearing of sampled insects for a shorter time than the entomopathogen's period of lethal infection, use of disinfectant techniques, and experimental evaluation of possible cross-infection.

DIRECT DETECTION OR QUANTIFICATION OF ENTOMOPATHOGENS IN SOIL

Entomopathogens can be sampled or isolated directly from soil samples, as opposed to being sampled with the insect host. Direct sampling methods include bioassay, physical methods, selective media, microscopy, molecular techniques, and serology.

Soil Bioassay

Bioassay is a valuable method for detection or quantification of insect pathogens. Other methods usually must be related to bioassay results at some point in a study because bioassay is the best indicator of biological activity. Bioassays in conjunction with standard curves can be used to estimate numbers of pathogen units per unit of soil.

Soil bioassays for entomopathogens have been limited to the use of live insects, but assays in cell cultures are also possible (174). Live insects have been used in two ways: as detection tools in the laboratory and as sentinel insects placed for a time in the host insect's habitat. With either method, however, the researcher must not assume that the bioassay insects are fully comparable to field insects with regard to susceptibility, behavior, and other attributes.

Laboratory insects have been used for detection and quantification of entomopathogens. Basically, the method involves the sampling of soil, the preparation of a soil suspension, and the exposure of bioassay insects to that suspension.

Soil bioassays for detecting viruses have had three variations: the aliquot of soil suspension is dispensed onto a disc of leaf (leaf disc method) (38–40, 69–71, 112), the aliquot of soil suspension is applied to the surface of solidified artificial diet (surface-treated method) (29, 127, 156, 157, 168, 178), or a soil sample is directly mixed with the artificial diet during its preparation (incorporating method) (65, 155). The leaf disc method also was used for an entomopathogenic fungus to determine stability of conidia (66) and for *Bacillus thuringiensis* to monitor insecticidal activity in soil (169, 172). The incorporating method was used to assess pathogenicity of *B. thuringiensis* in soil (121).

Bioassay insects can be exposed directly to soil samples. Ishibashi and Kondo (67, 68) investigated the persistence of nematodes in soil by placing insect larvae on the surface. Mracek (106) and Curran and Heng (22) used a similar assay for quantifying nematodes in soil. Krueger et al. (87) investigated the fate of a fungus in soil through direct exposure of bioassay insects. Entomopathogenic fungi generally invade the insect host through the integument, so test insects often are treated percutaneously with a soil sample containing entomopathogenic fungi.

In the case of sentinel (37) or bait (182) insects, appropriate host animals are placed for a time in soil and are observed there or in the laboratory for the development of disease. Bedding and Akhurst (10) developed the *Galleria* bait method, in which larvae of the greater wax moth, *Galleria mellonella*, are buried in the soil as a bait for entomopathogenic nematodes. Akhurst and Bedding (1) cited the various studies in which this method has been used. Other insects have been used in a similar manner to detect several genera and species of naturally occurring entomopathogenic fungi in soil (16, 86, 131, 182).

Entomopathogen Counts

Direct counts of entomopathogen units have the advantages of being more accurate than pathogen numbers estimated from sampled hosts and usually quicker and cheaper than bioassays. However, the presence of entomopathogen units does not necessarily signify infectivity, and such counts usually must be related to bioassay results. The four useful techniques that have been used for extraction and counting of entomopathogenic microorganisms present in soil are discussed below.

Physical Methods

Various physical approaches have been developed for sampling and quantifying microorganisms from soil.

Three methods for quantifying soilborne resting spores of fungi all use wet sieving and discontinuous density gradients. Macdonald and Spokes (97) developed a procedure for isolation of fungal resting spores by Percoll-based, discontinuous gradient centrifugation; recovery rates were estimated at 55 to 88%, and a total of 1.6×10^4 resting spores could be extracted from a 50-g sample of soil. Hajek and Wheeler (50) also applied this technique for quantification of soilborne, fungal resting spores. They reported that mean recoveries from soil range from 33 to 72% and that this method was less reliable at densities below 10^2 resting spores per g of soil. Since the Percoll-based method is too expensive for extensive use, Li et al. (92) developed an inexpensive, quantitative method for detecting resting spores in soil with sucrose and $CaCl_2$. This method yielded an average resting spore recovery of 79%.

Physical methods also have been used for entomopathogenic viruses. Thomas et al. (155) collected polyhedra of nuclear polyhedrosis virus (NPV) from soil by wet sieving and a continuous-flow system of centrifugation. Hukuhara (61) purified polyhedra of a cytoplasmic polyhedrosis virus from soil by multiple steps including desorption of the polyhedra with sodium pyrophosphate, aqueous two-phase separation with a dextran-polyethylene glycol system, and CsCl density gradient centrifugation. The recovery of polyhedra by this method was low (7%) because of soil aggregation during the centrifugation, so Hukuhara (62) improved the step by substituting metrizamide for CsCl. Evans et al. (29)

developed a simple method for extracting NPV from soil by ultrasonication in sodium dodecyl sulfate and differential centrifugation.

The Baermann funnel, centrifugal flotation, and flotation-sieving techniques have been used for enumeration of entomopathogenic nematodes in soil (22, 132). Physical methods have not been used to sample entomopathogenic bacteria or protozoa in soil.

Selective Media

A generally useful method for sampling, isolation, and counting of microorganisms (46) in soil is by means of selective media. These have been used extensively, mostly in studies of entomopathogen persistence. Generally, the method consists of suspending soil in sterile water and then using standard pour plate or streak plate isolation techniques.

Several selective media have been used to sample entomopathogenic fungi from insect cadavers and soil. The first medium (162) contained peptone, glucose, rose bengal, chloramphenicol, and cycloheximide. This medium was originally devised for the isolation of fungi from insect cadavers, but Doberski and Tribe (25) modified it for isolation from soil. A variety of fungicides and antibiotics have been used in selective media: streptomycin, erythrocine, oxytetracycline, chlortetracycline, penicillin, dodine (n-dodecylguanidine acetate), Morocide (binapacryl; 2-(2-butyl)-4,6-dinitrophenyl-3,3-dimethylactylate), and benomyl. Beilharz et al. (11) discovered that the incorporation of dodine into agar media made the media selective for the fungi *Beauveria bassiana* and *Metarhizium anisopliae*. Since then, dodine-based media have been used frequently for fungi in soil (94, 140, 143–147). Baath (8) reported that malt extract agar containing $CuSO_4$ selectively inhibited some soil fungi other than certain entomopathogenic fungi.

There have been many attempts to isolate *B. thuringiensis* from soil (26, 103). Competitive growth of *B. thuringiensis* relative to soil microorganisms was promoted by nutrient agar amended with polymyxin sulfate and penicillin G (129, 130). Another selective medium is peptone-yeast extract agar amended with nystatin as a fungicide and streptomycin as an antibiotic (169–171). A medium containing sodium acetate was used for isolating *B. thuringiensis* from soil with a background of 10^9 indigenous bacteria per g (160). Sodium acetate selectively inhibits the germination of *B. thuringiensis* spores; after spores of other bacteria germinate, these growing cells and other, non-spore-forming bacteria are eliminated by heat treatment. This method has been used for isolation of *B. thuringiensis* from soil samples from all over the world (19, 100). *B. thuringiensis* persists in soil predominantly as spores, so it is very difficult to enumerate vegetative cells. To overcome this problem, Akiba and Katoh (4) developed a selective medium based on the assimilation of carbon sources for the detection of a small number of *B. thuringiensis* vegetative cells in the soil; no growth of the bacterium was detected when spores were used as the inocula (2).

Milner (104) developed a semiquantitative procedure for detection of *Bacillus popilliae* subsp. *rhopaea* in soil. The procedure was based on the observation that spore germination of *B. popilliae* is much slower than that of most other sporeformers. The soil suspension in a germinating medium is subjected to seven cycles of incubation and heating to kill germinated spores and vegetative cells of most other sporeformers. This procedure reduced the number of contaminant spores by over 95% but was very complicated and time-consuming. An MYPGP medium supplemented with vancomycin is useful for the quantification of *B. popilliae* spores in soil (24, 142).

Other entomopathogenic bacteria also have been studied in soil with selective media. Hertlein et al. (56) developed a selective medium for *Bacillus sphaericus*, a mosquito pathogen. The medium was a reliable indicator for the presence of *B. sphaericus* in soil samples, but efficiency and selectivity of the medium were not reported. Yousten et al. (181) developed a defined medium, BATS, which allowed growth of 18 mosquito-pathogenic strains of *B. sphaericus* and inhibited the growth of 68% of the nonpathogenic *B. sphaericus* strains tested as well as other *Bacillus* spp. Guerineau et al. (48) developed adenosine and anthranilate media and compared these two media with the BATS medium for the isolation of *B. sphaericus* in soil samples. They found that BATS medium was the most effective for recovery of spores of mosquito-pathogenic strains of *B. sphaericus*. O'Callaghan and Jackson (110) reported several agar media for selective isolation and identification of *Serratia entomophila*.

Sampling, precision, and disadvantages of the use of selective media have been discussed previously (73, 161). The major disadvantages are that the method favors microorganisms with vigorous growth, that selection of methods and media heavily biases which microorganisms are sampled or counted, and that growth on media does not necessarily relate to pathogenicity. These disadvantages can be partially offset by the use of more than one medium or more than one technique (e.g, microscope count) for each sample (73).

Microscopy

Microscopy has been used to count microorganisms in soil. Usually the sample is suspended in water, though techniques such as embedding and sectioning or filtration and removal can be used, and then a count is made with electron, light, or fluorescence microscopy. Microscope counts have been used for viruses, a bacterium, and a protozoan in soil.

Hukuhara and Namura (64) microscopically demonstrated polyhedra of an NPV after staining of polyhedra with Buffalo black and preparation of thin sections of soil. They could readily distinguish polyhedra from soil particles. Taverner and Connor (154) used Giemsa stain instead of Buffalo black for another NPV. Hukuhara (60) detected polyhedra of a cytoplasmic polyhedrosis virus and capsules of a granulosis virus in soil with a scanning electron microscope.

There have been few microscopic examinations of soil for bacteria or protozoa. Immunofluorescence microscopy was used to detect *B. thuringiensis* (173). Germida (43) compared various stains, staining methods, and microscopy for detection of *Nosema locustae* spores in soil; fluorescein isothiocyanate was a better stain than aniline blue, acridine orange, fluorescein diacetate, crystal violet, safranin, Congo red, methyl red, erythrosin, phenolic rose bengal, malachite green, or eosin. Staining and microscopy were used for the assessment of survival and persistence of *N. locustae* in field soil (44).

Molecular Techniques

Some groups of microorganisms can be extracted from soil by selective media, whereas the others cannot. It can be difficult to detect the latter microorganisms in soil, especially when background counts are high. Application of molecular techniques can overcome this problem, and specific methods are now available for the detection of both culturable and nonculturable microorganisms. Among the entomopathogenic microorganisms, *B. thuringiensis* is the most promising for application of molecular techniques. Colony hybridization permits the sensitive detection of culturable bacteria with specific sequences in soil. Rupar et al. (128) used the cloned *cryIIIA* gene as a probe in colony hybridization and found two *cryIIIA*-hybridizing colonies in soil. Miteva et al. (105) applied DNA fingerprinting with M13 bacteriophage DNA as a probe for studying of inter- and intraserotypic variations of different *B. thuringiensis* strains.

Amounts of nucleic acids that can be extracted from soil may be small and below the level needed for detection by hybridization techniques. PCR is the most widely used method for the in vitro amplification of nucleic acids. Carozzi et al. (19) used PCR to identify *B. thuringiensis* strains containing *cryI*, *cryIII*, and *cryIV* genes in soil. Insecticidal activity predicted by the PCR screening corresponded with the insecticidal activity in insect bioassays. Bourque et al. (15) used PCR for distinguishing among strains of *B. thuringiensis* having the genes encoding CryIA(a), CryIA(b), and CryIA(c).

Molecular techniques are available but have not been used for research of other entomopathogen groups in soil. PCR has been successful in detecting DNA of NPVs (17); a single infected insect could be detected in a population of 20,000 by examination of the insect feces with this technique (109). There are many examples of restriction fragment length polymorphism analyses (RFLP) that have been used for baculoviruses (153).

With entomopathogenic fungi, molecular techniques are mainly used to identify polymorphisms. These include RFLP (55), rRNA-encoding DNA sequence (49), random amplification of polymorphic DNA by PCR (13, 32, 57, 150, 158), and construction of genomic DNA probes for hybridization (13, 55).

Entomopathogenic protozoa and nematodes are beginning to be subjected to molecular methods that could be useful in soil studies. For protozoa, comparative sequence analyses of RNAs and RFLP were used with microsporidia (9, 98, 99, 108, 149, 165). RFLP and rapid amplification of polymorphic DNA by PCR have been developed for nematodes (21, 78, 123–125).

Serology

Serological methods have been developed to the point where they can detect entomopathogens, but few studies have done so in the host's habitat.

The fluorescent antibody technique has been used to count entomopathogenic microorganisms in soil in conjunction with microscopy. West et al. (173) produced three kinds of antisera against purified crystals, autoclaved spores, and heat-killed vegetative cells of *B. thuringiensis* subsp. *aizawai*. They then monitored numbers of vegetative cells, parasporal crystals, and spores of *B. thuringiensis* in soil by immunofluorescence microscopy using these antisera. Hukuhara and Akami (63) observed polyhedra of an NPV in field soil by immunofluorescence microscopy. The method consisted of releasing polyhedra from soil in sodium pyrophosphate, separating them in an aqueous two-phase system, concentrating them by differential centrifugation, and indirect fluorescent antibody staining with guinea pig immunoglobulin G.

Fuxa et al. (40) used an enzyme-linked immunosorbent assay (ELISA) for detection of an NPV in soil. The ELISA detected as few as 360 polyhedra per g of soil, and absorbance values were inversely related to the amount of clay. Other techniques have the potential to be used in soil studies: single or double immunodiffusion for NPVs (135, 180); latex agglutination for viruses (138, 139); ELISA for cytoplasmic polyhedrosis virus (7, 117), iridescent virus (84), nonoc-cluded baculovirus (95), or a microsporidium (47, 81); plas-tic-bead ELISA for NPV (96); and a radioimmunoassay for NPV (113).

Comparisons of Quantification Methods

Some of the methods for quantification of entomopathogens have been directly or indirectly compared experimentally. Evans et al. (29) and Taverner and Connor (154) reported that polyhedra of NPVs in soil could be detected easily by light microscopy after staining with Buffalo black when numbers of polyhedra were relatively high, whereas bioassay was superior when numbers of polyhedra were low. Fuxa et al. (40) compared ELISA with bioassay for quantification of an NPV in soil and concluded that bioassay provided a better, more direct quantification than ELISA. Brownbridge et al. (16) compared the *Galleria* bait method with estimates made from sampled hosts infected with a fungal pathogen and found that *M. anisopliae* was the dominant entomopa-thogen detected by the former method whereas *Verticillium lecanii* was dominant in the latter. Curran and Heng (22) compared the Baermann funnel, flotation, and the *Galleria* bait methods for estimating the number of entomopatho-genic nematodes in soil; they suggested that a combination of flotation and *Galleria* bait techniques was best for this purpose. Saunders and All (132) compared three methods, flotation-sieving, centrifugation-flotation, and the Baer-mann funnel, for efficiency at extracting an entomopatho-genic nematode from soil samples; Baermann funnel extrac-tion was superior to flotation-sieving and centrifugation-flotation. On the other hand, Curran and Heng (22) re-ported that centrifugation-flotation recovered the greatest number of nematodes with a lower variance than the Baer-mann funnel technique. Such disparity in results seems to be related to differences in experimental methods, nematode strain, or soil type. Shimazu et al. (137) determined the infectious activity of fungal resting spores in soil during a winter season by using three kinds of bioassay techniques. West et al. (173) found that immunofluorescence micros-copy was more sensitive than a selective medium for detec-tion of *B. thuringiensis* in soil.

Whenever bioassay is compared with other techniques for sampling entomopathogens in soil or other components (37) of the insect's habitat, it usually is the most sensitive, the most biologically meaningful, and sometimes the most accurate method for detection or quantification. However, molecular techniques have been used so seldom that there has been virtually no direct comparison between them and bioassay for such purposes.

ECOLOGICAL TECHNIQUES FOR ENTOMOPATHOGENS IN SOIL

In addition to sampling methods for detection or quantifica-tion of entomopathogens, there are several techniques for studying the ecology of these microorganisms in soil. These techniques include the application of the microorganisms onto or into the soil in the field, elimination from the soil (decontamination) in the field, release and recovery, release of marked entomopathogens, and studies in soil microcosms.

Application onto or into Soil

There are numerous examples of application or inoculation of entomopathogens onto or into soil, because soil applica-tion is often attempted in research of microbial control of

insects. Several methods have been attempted, usually with at least moderate success.

The simplest means of soil inoculation is application to the soil surface without additional manipulation. Entomo-pathogens have been applied in aqueous suspension to the soil surface as a drench (76, 115), by sprayers or sprinklers (18, 72, 77, 133, 176, 177), and through drip irrigation (176). Dry formulations of entomopathogens have been ap-plied as a powder (85) or granule (41). These simple, surface applications have been used to inoculate viruses (72), nema-todes (76, 77, 115, 133, 176, 177), a fungus (18, 41), and a bacterium (85).

The soil surface sometimes is turned either immediately before or after surface application. The purpose generally is to provide the entomopathogen protection from sunlight or desiccation. In most cases, the soil is turned after, rather than before, application. For experimental purposes, soil has been turned with a hand trowel after fungal application as a spray or granule (41), with a rake after sprinkling of a fungal suspension (58), and with a tiller after spraying of virus (178). In one case, a fungus was sprayed onto soil after it was turned with a rake (167).

Application of an entomopathogen to the soil surface also can be followed by watering, presumably to wash the microorganism into the soil for protection from sunlight or desiccation. This technique has been used after sprinkler application of nematodes (34, 82) and after application of a bacterium formulated as a dust (82).

Entomopathogens occasionally are inoculated into soil with agricultural equipment. Suspensions of nematodes have been inoculated with a planter (120) and with a plow (114).

Certain techniques make use of the insect host of the entomopathogen. A nematode has been inoculated into soil by burying infected insects (115). A fungus has been applied to soil by spraying termites exposed by briefly removing a piece of their mound or by dusting termites similarly exposed by briefly turning fallen wood (51). Also, a fungus has been applied to the soil surface in a bait formulation (91), and nematodes have been applied on baits or to the soil beneath a bait (27, 102).

There has been very little direct comparison of the effica-cies of these various methods of soil inoculation. Applica-tion of a fungus as a granular formulation followed by tilling was significantly more efficacious than application in a water suspension following by tilling, which in turn was better than surface application of the granular formulation without tilling. Thus, efficacy of soil application was en-hanced by tilling the entomopathogen beneath the soil sur-face (41).

Elimination from Soil

Elimination of entomopathogens from soil in field situations has seldom been reported. Yet such techniques are poten-tially useful, not only in ecology but also in research for environmental risk assessment, in which it can be important to be able to eliminate a population of a released microor-ganism. In one case, a genetically marked virus was elimi-nated from soil in field plots with a formalin treatment (14).

Application to and Recovery from Soil

A common technique in studies of entomopathogens in soil is to apply pathogen units to soil in field situations and subsequently sample to recover these same pathogen units. The purpose of this technique generally is to test field persis-tence, or persistence and transport, in soil, although life cycle characteristics sometimes are monitored. The soil application and sampling methods for this type of research already have been discussed. This technique has been used with bacteria (3, 26, 119, 136), fungi (41, 53, 66, 107, 146, 148), viruses (69, 71, 112, 179), and nematodes (34, 42, 68, 166).

Marking Techniques

A problem encountered in release and recovery studies is the difficulty of distinguishing the released entomopathogen from indigenous microorganisms of the same species. This is particularly true if the entomopathogen can persist in soil for long time periods or if it has a wide geographical range. One way to counteract this difficulty is by marking the entomopathogen. In two cases, genetically altered insect viruses have been released in the field with subsequent soil sampling; virus detected by bioassay of the soil samples was then identified as progeny of the released, altered virus (14, 175). In another example, antibiotic-resistant marked strains of an entomopathogenic bacterium were used to study the persistence of the bacterium in soil (101).

Soil Microcosms

Microcosms often are used to answer specific questions about ecology of entomopathogens and, either directly or indirectly, their interactions with insects. Microcosms have been defined as "an attempt to reproduce, at least in part, the complex environmental interactions in controlled laboratory testing situations" (36) or, similarly, "a controlled reproducible laboratory system which attempts to simulate a portion of the real world" (45). Microcosms often are used in studies of entomopathogens and insects in soil because of the small sizes of the organisms and the relative ease of collecting and handling the soil medium, including its natural microflora and microfauna.

A wide variety of containers have been used to hold soil microcosms in research of entomopathogens. The most basic are simple containers, ranging in size from a few milliliters to 50 liters, perhaps placed in an incubator for temperature control. These have included simple cups (88), pots (111), flasks (93), petri dishes (118), culture tubes (76), glass or polyvinyl chloride tubes (33, 164), vials (43), plastic boxes (59), bottles (89), and bags (54). A degree of complexity can be added with multiple chambers (102) or with one container nested inside another, usually to maintain a certain humidity (31, 89, 90, 151). In certain designs, soil microcosms in various types and sizes of containers are set in greenhouses (23, 54, 145, 163). Microcosms often are constructed in the shape of columns, usually for studies of entomopathogen transport in soil (6, 43, 44, 134, 144).

Soil microcosms have been used for several purposes in studies of entomopathogens. Pathogenicity or virulence often is tested in these systems (54, 76, 88, 102, 163). Another common purpose for research with microcosms is to study the movement of nematodes or passive transport of other entomopathogens in soil columns (6, 20, 23, 44, 54, 134, 144, 145). Microcosms also have been useful to determine the effects on entomopathogens of abiotic (31, 59, 87, 89, 90, 111, 119, 151) and biotic (5, 43, 67, 93, 152, 159, 172) environmental factors. Persistence, measured by one or more of the soil procedures already discussed, is usually the entomopathogen characteristic monitored in these studies of environmental factors. Other subjects of study in soil microcosms have included saprobic growth (118) and various physiological processes (164).

Soil microcosms have been used in studies of all five

major groups of entomopathogens. The nematodes (6, 20, 59, 67, 76, 89, 90, 102, 111, 134, 159) and fungi (31, 54, 87, 88, 93, 118, 144, 145, 151, 152, 163) are studied most often in microcosms, although a bacterium (5, 119, 164, 172), protozoan (43, 44), and virus (23) also have been subjects of such research.

CONCLUSION

A body of methods has accumulated for studies of entomopathogens in soil. Such studies are frequent primarily because of the importance of soil as a reservoir for many of these microorganisms. The methods for these studies often are a combination of methods for studying invertebrates as well as the microorganisms in soil. Bioassay of soil samples remains a primary method for detecting or quantifying these microorganisms because it is both sensitive and biologically meaningful. However, molecular methods have great potential, particularly with respect to sensitivity for detection of entomopathogens. Methods for studying entomopathogens in soil are likely to become more dependent on molecular techniques in the future. Soil studies will become increasingly important for environmental risk assessment of recombinant entomopathogens.

REFERENCES

1. **Akhurst, R. J., and R. A. Bedding.** 1986. Natural occurrence of insect pathogenic nematodes (Steinernematidae and Heterorhabditidae) in soil in Australia. *J. Aust. Entomol. Soc.* **25:**241–244.
2. **Akiba, Y.** 1986. Microbial ecology of *Bacillus thuringiensis*. VI. Germination of *Bacillus thuringiensis* spores in the soil. *Appl. Entomol. Zool.* **21:**76–80.
3. **Akiba, Y.** 1991. Assessment of rainwater-mediated dispersion of field-sprayed *Bacillus thuringiensis* in the soil. *Appl. Entomol. Zool.* **26:**477–483.
4. **Akiba, Y., and K. Katoh.** 1986. Microbial ecology of *Bacillus thuringiensis*. V. Selective medium for *Bacillus thuringiensis* vegetative cells. *Appl. Entomol. Zool.* **21:**210–215.
5. **Akiba, Y., Y. Sekijima, K. Aizawa, and N. Fujiyoshi.** 1977. Microbial ecological studies on *Bacillus thuringiensis*. II. Dynamics of *Bacillus thuringiensis* in sterilized soil. *Jpn. J. Appl. Entomol. Zool.* **21:**41–46.
6. **Alatorre-Rosas, R., and H. K. Kaya.** 1990. Interspecific competition between entomopathogenic nematodes in the genera *Heterorhabditis* and *Steinernema* for an insect host in sand. *J. Invertebr. Pathol.* **55:**179–188.
7. **Arakawa, A.** 1991. Quantitative assay of cytoplasmic polyhedrosis virus in the feces of the silkworm, *Bombyx mori*, using enzyme linked immunosorbent assay. *J. Seric. Sci. Jpn.* **60:**105–111.
8. **Baath, E.** 1991. Tolerance of copper by entomogenous fungi and the use of copper-amended media for isolation of entomogenous fungi from soil. *Mycol. Res.* **95:**1140–1152.
9. **Baker, M. D., C. R. Vossbrinck, J. V. Maddox, and A. H. Undeen.** 1994. Phylogenetic relationships among *Vairimorpha* and *Nosema* species (Microspora) based on ribosomal RNA sequence data. *J. Invertebr. Pathol.* **64:**100–106.
10. **Bedding, R. A., and R. J. Akhurst.** 1975. A simple technique for the detection of insect parasitic rhabditid in soil. *Nematologica* **21:**109.
11. **Beilharz, V. C., D. J. Parbery, and H. J. Swart.** 1982. Dodine: a selective agent for certain soil fungi. *Trans. Br. Mycol. Soc.* **79:**507–511.
12. **Benz, G.** 1963. Physiopathology and histochemistry, p.

299–338. *In* E. A. Steinhaus (ed.), *Insect Pathology, an Advanced Treatise*, vol. 1. Academic Press, New York.

13. **Bidochka, M. J., M. A. McDonald, R. J. St. Leger, and D. W. Roberts.** 1994. Differentiation of species and strains of entomopathogenic fungi by random amplification of polymorphic DNA (RAPD). *Curr. Genet.* **25:** 107–113.

14. **Bishop, D. H. L., J. S. Cory, and R. D. Possee.** 1992. The use of genetically engineered virus insecticides to control insect pests, p. 137–146. *In* J. C. Fry and M. J. Day (ed.), *Release of Genetically Engineered and Other Micro-Organisms.* Cambridge University Press, Cambridge.

15. **Bourque, S. N., J. R. Valero, J. Mercier, M. C. Lavoie, and R. C. Levesque.** 1993. Multiplex polymerase chain reaction for detection and differentiation of the microbial insecticide *Bacillus thuringiensis. Appl. Environ. Microbiol.* **59:**523–527.

16. **Brownbridge, M., R. A. Humber, B. L. Parker, and M. Skinner.** 1993. Fungal entomopathogens recovered from Vermont forest soils. *Mycologia* **85:**358–361.

17. **Burand, J. P., H. M. Horton, S. Retnasami, and J. S. Elkinton.** 1992. The use of polymerase chain reaction and shortwave UV irradiation to detect baculovirus DNA on the surface of gypsy moth eggs. *J. Virol. Methods* **36:** 141–149.

18. **Cantwell, G. E., W. W. Cantelo, and R. F. W. Schroder.** 1986. Effect of *Beauveria bassiana* on underground stages of the Colorado potato beetle, *Leptinotarsa decemlineata* (Coleoptera: Chrysomelidae). *Great Lakes Entomol.* **19:** 81–84.

19. **Carozzi, N. B., V. C. Kramer, G. W. Warren, S. Evola, and M. G. Koziel.** 1991. Prediction of insecticidal activity of *Bacillus thuringiensis* strains by polymerase chain reaction product profiles. *Appl. Environ. Microbiol.* **57:** 3057–3061.

20. **Choo, H. Y., and H. K. Kaya.** 1991. Influence of soil texture and presence of roots on host finding by *Heterorhabditis bacteriophora. J. Invertebr. Pathol.* **58:**279–280.

21. **Curran, J., and F. Driver.** 1994. Molecular taxonomy of *Heterorhabditis*, p. 41–48. *In* A. M. Burnell, R. U. Ehlers, and J. P. Masson (ed.), *Genetics of Entomopathogenic Nematode-Bacterium Complexes.* European Commission, Brussels.

22. **Curran, J., and J. Heng.** 1992. Comparison of three methods for estimating the number of entomopathogenic nematodes present in soil samples. *J. Nematol.* **24:** 170–176.

23. **David, W. A. L., and B. O. C. Gardiner.** 1967. The persistence of a granulosis virus of *Pieris brassicae* in soil and in sand. *J. Invertebr. Pathol.* **9:**342–347.

24. **Dingman, D. W., and D. P. Stahly.** 1983. Medium promoting sporulation of *Bacillus larvae* and metabolism of medium components. *Appl. Environ. Microbiol.* **46:** 860–869.

25. **Doberski, J. W., and H. T. Tribe.** 1980. Isolation of entomogenous fungi from elm bark and soil with reference to ecology of *Beauveria bassiana* and *Metarhizium anisopliae. Trans. Br. Mycol. Soc.* **74:**95–100.

26. **Dulmage, H. T., and K. Aizawa.** 1982. Distribution of *Bacillus thuringiensis* in nature, p. 209–237. *In* E. Kurstak (ed.), *Microbial and Viral Pesticides.* Marcel Dekker, New York.

27. **Epsky, N. D., and J. L. Capinera.** 1988. Efficacy of the entomogenous nematode *Steinernema feltiae* against a subterranean termite, *Reticulitermes tibialis* (Isoptera: Rhinotermitidae). *J. Econ. Entomol.* **81:**1313–1317.

28. **Evans, H. C.** 1982. Entomogenous fungi in tropical forest ecosystems: an appraisal. *Ecol. Entomol.* **7:**47–60.

29. **Evans, H. F., J. M. Bishop, and E. A. Page.** 1980. Meth-
ods for the quantitative assessment of nuclear-polyhedrosis virus in soil. *J. Invertebr. Pathol.* **35:**1–8.

30. **Evans, H. F., and P. F. Entwistle.** 1982. Epizootiology of the nuclear polyhedrosis virus of European spruce sawfly with emphasis on persistence of virus outside the host, p. 449–461. *In* E. Kurstak (ed.), *Microbial and Viral Pesticides.* Marcel Dekker, New York.

31. **Fargues, J., O. Reisinger, P. H. Robert, and C. Aubart.** 1983. Biodegradation of entomopathogenic Hyphomycetes: influence of clay coating on *Beauveria bassiana* blastospore survival in soil. *J. Invertebr. Pathol.* **41:** 131–142.

32. **Fegan, M., J. M. Manners, D. J. Maclean, J. A. G. Irwin, K. D. Z. Samuels, D. G. Holdom, and D. P. Li.** 1993. Random amplified polymorphic DNA markers reveal a high degree of genetic diversity in the entomopathogenic fungus *Metarhizium anisopliae* var. *anisopliae. J. Gen. Microbiol.* **139:**2075–2081.

33. **Forschler, B. T., J. N. All, and W. A. Gardner.** 1990. *Steinernema feltiae* activity and infectivity in response to herbicide exposure in aqueous and soil environments. *J. Invertebr. Pathol.* **55:**375–379.

34. **Forschler, B. T., and W. A. Gardner.** 1991. Field efficacy and persistence of entomogenous nematodes in the management of white grubs (Coleoptera: Scarabeidae) in turf and pasture. *J. Econ. Entomol.* **84:**1454–1459.

35. **Fowler, H. G., and J. Justi, Jr.** 1987. Field collection techniques for genera *Neocurtilla* and *Scapteriscus* mole crickets and their effect upon the evaluation of their natural pathogens in Brazil. *Z. Angew. Entomol.* **104:**204–207.

36. **Frederick, R. J., and R. W. Pilsucki.** 1991. Nontarget species testing of microbial products intended for use in the environment, p. 32–50. *In* M. A. Levin and H. S. Strauss (ed.), *Risk Assessment in Genetic Engineering.* McGraw-Hill, New York.

37. **Fuxa, J. R.** 1987. Ecological methods, p. 23–41. *In* J. R. Fuxa and Y. Tanada (ed.), *Epizootiology of Insect Diseases.* John Wiley & Sons, New York.

38. **Fuxa, J. R., and A. R. Richter.** 1993. Lack of vertical transmission in *Anticarsia gemmatalis* (Lepidoptera: Noctuidae) nuclear polyhedrosis virus, a pathogen non-indigenous to Louisiana. *Environ. Entomol.* **22:**425–431.

39. **Fuxa, J. R., and A. R. Richter.** 1994. Distance and rate of spread of *Anticarsia gemmatalis* (Lepidoptera: Noctuidae) nuclear polyhedrosis virus released into soybean. *Environ. Entomol.* **23:**1308–1316.

40. **Fuxa, J. R., G. W. Warren, and C. Kawanishi.** 1985. Comparison of bioassay and enzyme-linked immunosorbent assay for quantification of *Spodoptera frugiperda* nuclear polyhedrosis virus in soil. *J. Invertebr. Pathol.* **46:** 133–138.

41. **Gaugler, R., S. D. Costa, and J. Lashomb.** 1989. Stability and efficacy of *Beauveria bassiana* soil inoculations. *Environ. Entomol.* **18:**412–417.

42. **Geden, C. J., and R. C. Axtell.** 1988. Effect of temperature on nematode (*Steinernema feltiae* [Nematoda: Steinernematidae]) treatment of soil for control of lesser mealworm (Coleoptera: Tenebrionidae) in turkey houses. *J. Econ. Entomol.* **81:**800–803.

43. **Germida, J. J.** 1984. Persistence of *Nosema locustae* spores in soil as determined by fluorescence microscopy. *Appl. Environ. Microbiol.* **47:**313–318.

44. **Germida, J. J., A. B. Ewen, and E. E. Onofriechuk.** 1987. *Nosema locustae* Canning (Microsporida) spore populations in treated field soils and resident grasshopper populations. *Can. Entomol.* **119:**355–360.

45. **Gillett, J. W., and J. M. Witt.** 1979. *Terrestrial Microcosms.* NSF-79-0024. National Science Foundation, Washington, D.C.

46. **Gray, T. R. G.** 1990. Methods for studying the microbial ecology of soil. *Methods Microbiol.* **22:**309–342.

47. **Greenstone, M. H.** 1983. An enzyme-linked immunosorbent assay for the *Amblyospora* sp. of *Culex salinarius* (Microspora: Amblyosporidae). *J. Invertebr. Pathol.* **41:**250–255.

48. **Guerineau, M., B. Alexander, and F. G. Priest.** 1991. Isolation and identification of *Bacillus sphaericus* strains pathogenic for mosquito larvae. *J. Invertebr. Pathol.* **57:**325–333.

49. **Hajek, A. E., R. A. Humber, J. S. Elkinton, B. May, S. R. A. Walsh, and J. C. Silver.** 1990. Allozyme and restriction fragment length polymorphism analyses confirm *Entomophaga maimaiga* responsible for 1989 epizootics in North American gypsy moth populations. *Proc. Natl. Acad. Sci. USA* **87:**6979–6982.

50. **Hajek, A. E., and M. M. Wheeler.** 1994. Application of techniques for quantification of soil-borne entomophthoralean resting spores. *J. Invertebr. Pathol.* **64:**71–73.

51. **Hänel, H., and J. A. L. Watson.** 1983. Preliminary field tests on the use of *Metarhizium anisopliae* for the control of *Nasutitermes exitiosus* (Hill) (Isoptera: Termitidae). *Bull. Entomol. Res.* **73:**305–313.

52. **Hanula, J. L., and T. G. Andreadis.** 1988. Parasitic microorganisms of Japanese beetle (Coleoptera: Scarabeidae) and associated scarab larvae in Connecticut soils. *Environ. Entomol.* **17:**709–714.

53. **Harrison, R. D., and W. A. Gardner.** 1992. Fungistasis of *Beauveria bassiana* by selected herbicides in soil. *J. Entomol. Sci.* **27:**233–238.

54. **Hartwig, J., and S. Oehmig.** 1992. BIO 1020—behaviour in the soil, and important factors affecting its action. *Pflanzenschutz-Nachr. Bayer* **45:**159–176.

55. **Hegedus, D. D., and G. G. Khachatourians.** 1993. Identification of molecular variants in mitochondrial DNAs of members of the genera *Beauveria, Verticillium, Paecilomyces, Tolypocladium,* and *Metarhizium. Appl. Environ. Microbiol.* **59:**4283–4288.

56. **Hertlein, B. C., R. Levy, and T. W. Miller, Jr.** 1979. Recycling potential and selective retrieval of *Bacillus sphaericus* from soil in a mosquito habitat. *J. Invertebr. Pathol.* **33:**217–221.

57. **Hodge, K. T., A. J. Sawyer, and R. A. Humber.** 1995. RAPD-PCR for identification of *Zoophthora radicans* isolates in biological control of the potato leafhopper. *J. Invertebr. Pathol.* **65:**1–9.

58. **Hokkanen, H. M. T.** 1993. Overwintering survival and spring emergence in *Meligethes aeneus:* effects of body weight, crowding, and soil treatment with *Beauveria bassiana. Entomol. Exp. Appl.* **67:**241–246.

59. **Hudson, W. G., and K. B. Nguyen.** 1989. Effects of soil moisture, exposure time, nematode age, and nematode density on laboratory infection of *Scapteriscus vicinus* and *S. acletus* (Orthoptera: Gryllotalpidae) by *Neoaplectana* sp. (Rhabditida: Steinernematidae). *Environ. Entomol.* **18:**719–722.

60. **Hukuhara, T.** 1972. Demonstration of polyhedra and capsules in soil with scanning electron microscope. *J. Invertebr. Pathol.* **20:**375–376.

61. **Hukuhara, T.** 1975. Purification of polyhedra of a cytoplasmic-polyhedrosis virus from soil. *J. Invertebr. Pathol.* **25:**337–342.

62. **Hukuhara, T.** 1977. Purification of polyhedra of a cytoplasmic polyhedrosis virus from soil using metrizamide. *J. Invertebr. Pathol.* **30:**270–272.

63. **Hukuhara, T., and K. Akami.** 1987. Demonstration of polyhedral inclusion bodies of a nuclear polyhedrosis virus in field soil by immunofluorescence microscopy. *J. Invertebr. Pathol.* **49:**130–132.

64. **Hukuhara, T., and H. Namura.** 1971. Microscopic demonstration of polyhedra in soil. *J. Invertebr. Pathol.* **18:**162–164.

65. **Hukuhara, T., and H. Namura.** 1972. Distribution of a nuclear-polyhedrosis virus of the fall webworm, *Hyphantria cunea,* in soil. *J. Invertebr. Pathol.* **19:**308–316.

66. **Ignoffo, C. M., C. Garcia, D. L. Hostetter, and R. E. Pinnell.** 1978. Stability of conidia of an entomopathogenic fungus, *Nomuraea rileyi,* in and on soil. *Environ. Entomol.* **7:**724–727.

67. **Ishibashi, N., and E. Kondo.** 1986. *Steinernema feltiae* (DD-136) and *S. glaseri:* persistence in soil and bark compost and their influence on native nematodes. *J. Nematol.* **18:**310–316.

68. **Ishibashi, N., and E. Kondo.** 1987. Dynamics of the entomogenous nematode *Steinernema feltiae* applied to soil with and without nematicide treatment. *J. Nematol.* **19:**404–412.

69. **Jaques, R. P.** 1964. The persistence of a nuclear-polyhedrosis virus in soil. *J. Invertebr. Pathol.* **6:**251–254.

70. **Jaques, R. P.** 1967. The persistence of a nuclear polyhedrosis virus in the habitat of the host insect, *Trichoplusia ni.* II. Polyhedra in soil. *Can. Entomol.* **99:**820–829.

71. **Jaques, R. P.** 1969. Leaching of the nuclear-polyhedrosis virus of *Trichoplusia ni* from soil. *J. Invertebr. Pathol.* **13:**256–263.

72. **Jaques, R. P.** 1970. Application of viruses to soil and foliage for control of the cabbage looper and imported cabbageworm. *J. Invertebr. Pathol.* **15:**328–340.

73. **Johnson, L. F., and E. A. Curl.** 1972. *Methods for Research on the Ecology of Soil-Borne Plant Pathogens.* Burgess Publishing, Minneapolis.

74. **Jouvenaz, D. P., G. E. Allen, W. A. Banks, and D. P. Wojcik.** 1977. A survey for pathogens of fire ants, *Solenopsis* spp., in the southeastern United States. *Fla. Entomol.* **60:**275–279.

75. **Jouvenaz, D. P., W. A. Banks, and J. D. Atwood.** 1980. Incidence of pathogens in fire ants, *Solenopsis* spp., in Brazil. *Fla. Entomol.* **63:**345–346.

76. **Jouvenaz, D. P., C. S. Lofgren, and R. W. Miller.** 1990. Steinernematid nematode drenches for control of fire ants, *Solenopsis invicta,* in Florida. *Fla. Entomol.* **73:**190–193.

77. **Jouvenaz, D. P., and W. R. Martin.** 1992. Evaluation of the nematode *Steinernema carpocapsae* to control fire ants in nursery stock. *Fla. Entomol.* **75:**148–151.

78. **Joyce, S. A., A. Reid, F. Driver, and J. Curran.** 1994. Application of polymerase chain reaction (PCR) methods to the identification of entomopathogenic nematodes, p. 178–187. *In* A. M. Burnell, R. U. Ehlers, and J. P. Masson (ed.), *Genetics of Entomopathogenic Nematode-Bacterium Complexes.* European Commission, Brussels.

79. **Kalmakoff, J., and A. M. Crawford.** 1982. Enzootic virus control of *Wiseana* spp. in the pasture environment, p. 435–448. *In* E. Kurstak (ed.), *Microbial and Viral Pesticides.* Marcel Dekker, New York.

80. **Kaupp, W. J.** 1983. Estimation of nuclear polyhedrosis virus produced in field populations of the European pine sawfly, *Neodiprion sertifer* (Geoff.) (Hymenoptera: Diprionidae). *Can. J. Zool.* **61:**1857–1861.

81. **Kawarabata, T., and S. Hayasaka.** 1987. An enzyme-linked immunosorbent assay to detect alkali-soluble spore surface antigens of strains of *Nosema bombycis* (Microspora: Nosematidae). *J. Invertebr. Pathol.* **50:**118–123.

82. **Kaya, H. K., M. G. Klein, and T. M. Burlando.** 1993. Impact of *Bacillus popilliae, Rickettsiella popilliae* and entomopathogenic nematodes on a population of the scarabeid, *Cyclocephala hirta. Biocontrol Sci. Technol.* **3:**443–453.

83. Kaya, H. K., M. G. Klein, T. M. Burlando, R. E. Harrison, and L. A. Lacey. 1992. Prevalence of two *Bacillus popilliae* Dutky morphotypes and blue disease in *Cyclocephala hirta* LeConte (Coleoptera: Scarabeidae) populations in California. *Pan-Pac. Entomol.* **68**:38–45.

84. Kelly, D. C., M. L. Edwards, H. F. Evans, and J. S. Robertson. 1978. The use of the enzyme linked immunosorbent assay to detect, and discriminate between, small iridescent viruses. *Ann. Appl. Biol.* **90**:369–374.

85. Klein, M. G. 1988. Pest management of soil-inhabiting insects with microorganisms. *Agric. Ecosystems Environ.* **24**:337–349.

86. Ko, W. H., J. K. Fujii, and K. M. Kanegawa. 1982. The nature of soil pernicious to *Coptotermes formosanus*. *J. Invertebr. Pathol.* **39**:38–40.

87. Krueger, S. R., J. R. Nechols, and W. A. Ramoska. 1991. Infection of chinch bug, *Blissus leucopterus leucopterus* (Hemiptera: Lygaeidae), adults from *Beauveria bassiana* (Deuteromycotina: Hyphomycetes) conidia in soil under controlled temperature and moisture conditions. *J. Invertebr. Pathol.* **58**:19–26.

88. Krueger, S. R., M. G. Villani, A. S. Martins, and D. W. Roberts. 1992. Efficacy of soil applications of *Metarhizium anisopliae* (Metsch.) Sorokin conidia, and standard and lyophilized mycelial particles against scarab grubs. *J. Invertebr. Pathol.* **59**:54–60.

89. Kung, S.-P., R. Gaugler, and H. K. Kaya. 1990. Influence of soil pH and oxygen on persistence of *Steinernema* spp. *J. Nematol.* **22**:440–445.

90. Kung, S.-P., R. Gaugler, and H. K. Kaya. 1990. Soil type and entomopathogenic nematode persistence. *J. Invertebr. Pathol.* **55**:401–406.

91. Latch, G. C. M., and W. M. Kain. 1983. Control of porina caterpillar (*Wiseana* spp.) in pasture by the fungus *Metarhizium anisopliae*. *N. Z. J. Exp. Agric.* **11**:351–354.

92. Li, Z., R. S. Soper, and A. E. Hajek. 1988. A method for recovering resting spores of entomophthorales. (Zygomycetes) from soil. *J. Invertebr. Pathol.* **52**:18–26.

93. Lingg, A. J., and M. D. Donaldson. 1981. Biotic and abiotic factors affecting stability of *Beauveria bassiana* conidia in soil. *J. Invertebr. Pathol.* **38**:191–200.

94. Liu, Z. Y., R. J. Milner, C. F. McRae, and G. G. Lutton. 1993. The use of dodine in selective media for isolation of *Metarhizium* spp. from soil. *J. Invertebr. Pathol.* **62**:248–251.

95. Longworth, J. F., and G. P. Carey. 1980. The use of an indirect enzyme-linked immunosorbent assay to detect baculovirus in larvae and adults of *Oryctes rhinoceros* from Tonga. *J. Gen. Virol.* **47**:431–438.

96. Ma, M., J. K. Burkholder, R. E. Webb, and H. T. Hsu. 1984. Plastic-bead ELISA: an inexpensive epidemiological tool for detecting gypsy moth (Lepidoptera: Lymantriidae) nuclear polyhedrosis virus. *J. Econ. Entomol.* **77**:537–540.

97. Macdonald, R. M., and J. R. Spokes. 1981. *Conidiobolus obscurus* in arable soil: a method for extracting and counting azygospores. *Soil Biol. Biochem.* **13**:551–553.

98. Malone, L. A., A. H. Broadwell, E. T. Lindridge, C. A. McIvor, and J. A. Ninham. 1994. Ribosomal RNA genes of two microsporidia, *Nosema apis* and *Vavraia oncoperae*, are very variable. *J. Invertebr. Pathol.* **64**:151–152.

99. Malone, L. A., and C. A. McIvor. 1993. Pulsed-field gel electrophoresis of DNA from four microsporidian isolates. *J. Invertebr. Pathol.* **61**:203–205.

100. Martin, P. A. W., and R. S. Travers. 1989. Worldwide abundance and distribution of *Bacillus thuringiensis* isolates. *Appl. Environ. Microbiol.* **55**:2437–2442.

101. Martin, W. F., and C. F. Reichelderfer. 1989. *Bacillus thuringiensis*: persistence and movement in field crops, p.

25. *In Program and Abstracts, Society for Invertebrate Pathology, XXII Annual Meeting*. Society for Invertebrate Pathology.

102. Mauldin, J. K., and R. H. Beal. 1989. Entomogenous nematodes for control of subterranean termites, *Reticulitermes* spp. (Isoptera: Rhinotermitidae). *J. Econ. Entomol.* **82**:1638–1642.

103. Meadows, M. P. 1993. *Bacillus thuringiensis* in the environment: ecology and risk assessment, p. 193–220. *In* P. F. Entwistle, J. S. Cory, M. J. Bailey, and S. Higgs (ed.), *Bacillus thuringiensis, an Environmental Biopesticide: Theory and Practice*. John Wiley & Sons, West Sussex, England.

104. Milner, R. J. 1977. A method for isolating milky disease, *Bacillus popilliae* var. *rhopaea*, spores from the soil. *J. Invertebr. Pathol.* **30**:283–287.

105. Miteva, V., A. Abadjieva, and R. Grigorova. 1991. Differentiation among strains and serotypes of *Bacillus thuringiensis* by M13 DNA fingerprinting. *J. Gen. Microbiol.* **137**:593–600.

106. Mracek, Z. 1982. Estimate of the number of infective larvae of *Neoaplectana carpocapsae* (Nematoda: Steinernematidae) in a soil sample. *Nematologica* **28**:303–306.

107. Müller-Kögler, E., and G. Zimmermann. 1986. Zur Lebensdauer von *Beauveria bassiana* in kontaminiertem Boden unter Freiland- und Laboratoriumsbedingungen. *Entomophaga* **31**:285–292.

108. Munderloh, U. G., T. J. Kurtti, and S. E. Ross. 1990. Electrophoretic characterization of chromosomal DNA from two microsporidia. *J. Invertebr. Pathol.* **56**:243–248.

109. Noguchi, Y., M. Kobayashi, and T. Shimada. 1994. An application of the polymerase chain reaction for practical diagnosis of the nuclear polyhedrosis in large-scale culture of *Bombyx mori*. *J. Seric. Sci. Jpn.* **63**:399–406.

110. O'Callaghan, M., and T. A. Jackson. 1993. Isolation and enumeration of *Serratia entomophila*, a bacterial pathogen of the New Zealand grass grub, *Costelytra zealandica*. *J. Appl. Bacteriol.* **75**:307–314.

111. Oetting, R. D., and J. G. Latimer. 1991. An entomogenous nematode *Steinernema carpocapsae* is compatible with potting media environments created by horticultural practices. *J. Entomol. Sci.* **26**:390–394.

112. Øgaard, L., C. F. Williams, C. C. Payne, and O. Zethner. 1988. Activity persistence of granulosis viruses [Baculoviridae] in soils in United Kingdom and Denmark. *Entomophaga* **33**:73–80.

113. Ohba, M., M. D. Summers, P. Hoops, and G. E. Smith. 1977. Immunoradiometric assay for baculovirus enveloped nucleocapsids and polyhedrin. *J. Invertebr. Pathol.* **30**:362–368.

114. Parkman, J. P., J. H. Frank, K. B. Nguyen, and G. C. Smart, Jr. 1993. Dispersal of *Steinernema scapterisci* (Rhabditida: Steinernematidae) after inoculative applications for mole cricket (Orthoptera: Gryllotalpidae) control in pastures. *Biol. Control* **3**:226–232.

115. Parkman, J. P., W. G. Hudson, J. H. Frank, K. B. Nguyen, and G. C. Smart, Jr. 1993. Establishment and persistence of *Steinernema scapterisci* (Rhabditida: Steinernematidae) in field populations of *Scapteriscus* spp. mole crickets (Orthoptera: Gryllotalpidae). *J. Entomol. Sci.* **28**:182–190.

116. Payah, W. S., and D. J. Boethel. 1986. Impact of *Beauveria bassiana* (Balsamo) Vuillemin on survival of overwintering bean leaf beetles, *Cerotoma trifurcata* (Forster), (Coleoptera, Chrysomelidae). *Z. Angew. Entomol.* **102**:295–303.

117. Payne, P., D. J. S. Arora, and S. Belloncik. 1982. An enzyme-linked immunosorbent assay for the detection of cytoplasmic polyhedrosis virus. *J. Invertebr. Pathol.* **40**:55–60.

118. Pereira, R. M., S. B. Alves, and J. L. Stimac. 1993. Growth of *Beauveria bassiana* in fire ant nest soil with amendments. *J. Invertebr. Pathol.* **62:**9–14.

119. Petras, S. F., and L. E. Casida, Jr. 1985. Survival of *Bacillus thuringiensis* spores in soil. *Appl. Environ. Microbiol.* **50:**1496–1501.

120. Poinar, G. O., Jr., J. S. Evans, and E. Schuster. 1983. Field test of the entomogenous nematode, *Neoaplectana carpocapsae*, for control of corn rootworm larvae (*Diabrotica* sp., Coleoptera). *Prot. Ecol.* **5:**337–342.

121. Pruett, C. J. H., H. D. Burges, and C. H. Wyborn. 1980. Effect of exposure to soil on potency and spore viability of *Bacillus thuringiensis*. *J. Invertebr. Pathol.* **35:**168–174.

122. Quinn, M. A., and A. A. Hower. 1985. Isolation of *Beauveria bassiana* (Deuteromycotina: Hyphomycetes) from alfalfa field soil and its effect on adult *Sitona hispidulus* (Coleoptera: Curculionidae). *Environ. Entomol.* **14:**620–623.

123. Reid, A. P. 1994. Molecular taxomony of *Steinernema*, p. 49–58. *In* A. M. Burnell, R. U. Ehlers, and J. P. Masson (ed.), *Genetics of Entomopathogenic Nematode-Bacterium Complexes*. European Commission, Brussels.

124. Reid, A. P., and W. M. Hominick. 1992. Restriction fragment length polymorphisms with the ribosomal DNA repeat unit of British entomopathogenic nematodes (Rhabditida: Steinernematidae). *Parasitology* **105:**317–323.

125. Reid, A. P., and W. M. Hominick. 1993. Cloning of the rDNA repeat unit from a British entomopathogenic nematode (Steinernematidae). *Parasitology* **107:**529–536.

126. Reinganum, C., S. J. Gagen, S. B. Sexton, and H. P. Vellacott. 1981. A survey for pathogens of the black field cricket, *Teleogryllus commodus*, in the western district of Victoria, Australia. *J. Invertebr. Pathol.* **38:**153–160.

127. Roome, R. E., and R. A. Daoust. 1976. Survival of the nuclear polyhedrosis virus of *Heliothis armigera* on crops and in soil in Botswana. *J. Invertebr. Pathol.* **27:**7–12.

128. Rupar, M. J., W. P. Donovan, R. G. Groat, A. C. Slaney, J. W. Mattison, T. B. Johnson, J. Charles, V. C. Dumanoir, and H. de Bajac. 1991. Two novel strains of *Bacillus thuringiensis* toxic to coleopterans. *Appl. Environ. Microbiol.* **57:**3337–3344.

129. Saleh, S. M., R. F. Harris, and O. N. Allen. 1969. Method for determining *Bacillus thuringiensis* var. *thuringiensis* Berliner in soil. *Can. J. Microbiol.* **15:**1101–1104.

130. Saleh, S. M., R. F. Harris, and O. N. Allen. 1970. Fate of *Bacillus thuringiensis* in soil: effect of soil pH and organic amendment. *Can. J. Microbiol.* **16:**677–680.

131. Sato, H., M. Shimazu, and N. Kamata. 1994. Detection of *Cordyceps militaris* Link (Clavicipitales; Clavicipitaceae) by burying pupae of *Quadricalcarifera punctatella* Motschlsky (Lepidoptera: Notodontidae). *Appl. Entomol. Zool.* **29:**130–132.

132. Saunders, M. C., and J. N. All. 1982. Laboratory extraction methods and field detection of entomophilic rhabditoid nematodes from soil. *Environ. Entomol.* **11:**1164–1165.

133. Schroeder, W. J. 1992. Entomopathogenic nematodes for control of root weevils of citus. *Fla. Entomol.* **75:**563–567.

134. Schroeder, W. J., and J. B. Beavers. 1987. Movement of the entomogenous nematodes of the families Heterorhabditidae and Steinernematidae in soil. *J. Nematol.* **19:**257–259.

135. Scott, H. A., W. C. Yearian, and S. Y. Young. 1976. Evaluation of the single radial diffusion technique for detection of nuclear polyhedrosis virus (NPV) infection in *Heliothis zea*. *J. Invertebr. Pathol.* **28:**229–232.

136. Sekijima, Y., Y. Akiba, K. Ono, K. Aizawa, and N. Fuji-

yoshi. 1977. Microbial ecological studies on *Bacillus thuringiensis*. II. Dynamics of *Bacillus thuringiensis* in soil of mulberry field. *Jpn. J. Appl. Entomol. Zool.* **21:**35–40.

137. Shimazu, M., C. Koizumi, T. Kushida, and J. Mitsuhashi. 1987. Infectivity of hibernated resting spores of *Entomophaga maimaiga* Humber, Shimazu et Soper (Entomophthorales: Entomophthoraceae). *Appl. Entomol. Zool.* **22:**216–221.

138. Shimizu, S., and A. Arakawa. 1986. Latex agglutination test for the detection of the cytoplasmic polyhedrosis virus and the densonucleosis virus of the silkworm, *Bombyx mori*. *J. Seric. Sci. Jpn.* **55:**153–157.

139. Shimizu, S., M. Ohba, K. Kanda, and K. Aizawa. 1983. Latex agglutination test for the detection of the flacherie virus of the silkworm, *Bombyx mori*. *J. Invertebr. Pathol.* **42:**151–155.

140. Sneh, B. 1991. Isolation of *Metarhizium anisopliae* from insects on an improved selective medium based on wheat germ. *J. Invertebr. Pathol.* **58:**269–273.

141. Southwood, T. R. E. 1978. *Ecological Methods, with Particular Reference to the Study of Insect Populations*, 2nd ed. Chapman and Hall, London.

142. Stahly, D. P., D. M. Takefman, C. A. Livasy, and D. W. Dingman. 1992. Selective medium for quantitation of *Bacillus popilliae* in soil and in commercial spore powders. *Appl. Environ. Microbiol.* **58:**740–743.

143. Storey, G. K., and W. A. Gardner. 1986. Sensitivity of the entomogenous fungus *Beauveria bassiana* to selected plant growth regulators and spray additives. *Appl. Environ. Microbiol.* **52:**1–3.

144. Storey, G. K., and W. A. Gardner. 1987. Vertical movement of commercially formulated *Beauveria bassiana* conidia through four Georgia soil types. *Environ. Entomol.* **16:**178–181.

145. Storey, G. K., and W. A. Gardner. 1988. Movement of an aqueous spray of *Beauveria bassiana* into the profile of four Georgia soils. *Environ. Entomol.* **17:**135–139.

146. Storey, G. K., W. A. Gardner, J. J. Hamm, and J. R. Young. 1989. Recovery of *Beauveria bassiana* propagules from soil following application of formulated conidia through an overhead irrigation system. *J. Entomol. Sci.* **22:**355–357.

147. Storey, G. K., W. A. Gardner, and E. W. Tollner. 1989. Penetration and persistence of commercially formulated *Beauveria bassiana* conidia in soil of two tillage systems. *Environ. Entomol.* **18:**835–839.

148. Storey, G. K., W. A. Gardner, and E. W. Tollner. 1989. Penetration and persistence of commercially formulated *Beauveria bassiana* conidia in soil of two tillage systems. *Environ. Entomol.* **18:**835–839.

149. Streett, D. A. 1994. Analysis of *Nosema locustae* (Microsporida: Nosematidae) chromosomal DNA with pulsed-field gel electrophoresis. *J. Invertebr. Pathol.* **63:**301–302.

150. Strongman, D. B., and R. M. MacKay. 1993. Discrimination between *Hirsutella longicolla* var. *longicolla* and *Hirsutella longicolla* var. *cornuta* using random amplified polymorphic DNA fingerprinting, *Mycologia* **85:**65–70.

151. Studdert, J. P., and H. K. Kaya. 1990. Water potential, temperature, and clay-coating of *Beauveria bassiana* conidia: effect on *Spodoptera exigua* pupal mortality in two soil types. *J. Invertebr. Pathol.* **56:**327–336.

152. Studdert, J. P., H. K. Kaya, and J. M. Duniway. 1990. Effect of water potential, temperature, and clay-coating on survival of *Beauveria bassiana* conidia in a loam and peat soil. *J. Invertebr. Pathol.* **55:**417–427.

153. Tanada, Y., and H. K. Kaya. 1993. *Insect Pathology*. Academic Press, San Diego, Calif.

154. Taverner, M. P., and E. F. Connor. 1992. Optical enu-

meration technique for detection of baculoviruses in the environment. *Environ. Entomol.* **21**:307–313.

155. **Thomas, E. D., C. F. Reichelderfer, and A. M. Heimpel.** 1972. Accumulation and persistence of a nuclear polyhedrosis virus of the cabbage looper in the field. *J. Invertebr. Pathol.* **20**:157–164.

156. **Thompson, C. G., and D. W. Scott.** 1979. Production and persistence of the nuclear polyhedrosis virus of the douglas-fir tussock moth, *Orgyia pseudotsugata* (Lepidoptera: Lymantriidae), in the forest ecosystem. *J. Invertebr. Pathol.* **33**:57–65.

157. **Thompson, C. G., D. W. Scott, and B. E. Wickman.** 1981. Long-term persistence of the nuclear polyhedrosis virus of the douglas-fir tussock moth, *Orgyia pseudotsugata* (Lepidoptera: Lymantriidae), in forest soil. *Environ. Entomol.* **10**:254–255.

158. **Tigano-Milani, M. S., A. C. M. M. Gomes, and B. W. S. Sobral.** 1995. Genetic variability among Brazilian isolates of the entomopathogenic fungus *Metarhizium anisopliae*. *J. Invertebr. Pathol.* **65**:206–210.

159. **Timper, P., H. K. Kaya, and B. A. Jaffee.** 1991. Survival of entomogenous nematodes in soil infested with the nematode-parasitic fungus *Hirsutella rhossiliensis* (Deuteromycotina: Hyphomycetes). *Biol. Control* **1**:42–50.

160. **Travers, R. S., P. A. W. Martin, and C. F. Reichelderfer.** 1987. Selective process for efficient isolation of soil *Bacillus* spp. *Appl. Environ. Microbiol.* **53**:1263–1266.

161. **Van Elsas, J. D., and C. Waalwijk.** 1991. Methods for the detection of specific bacteria and their genes in soil. *Agric. Ecosys. Environ.* **34**:97–105.

162. **Veen, K. H., and P. Ferron.** 1966. A selective medium for the isolation of *Beauveria tenella* and of *Metarhizium anisopliae*. *J. Invertebr. Pathol.* **8**:268–269.

163. **Villani, M. G., S. R. Krueger, P. C. Schroeder, F. Consolie, N. H. Consolie, L. M. Preston-Wilsey, and D. W. Roberts.** 1994. Soil application effects of *Metarhizium anisopliae* on Japanese beetle (Coleoptera: Scarabeidae) behavior and survival in turfgrass microcosms. *Environ. Entomol.* **23**:502–513.

164. **Visser, S., J. A. Addison, and S. B. Holmes.** 1994. Effects of Dipel® 176, a *Bacillus thuringiensis* subsp. *kurstaki* (*B.t.k.*) formulation, on the soil microflora and the fate of *B.t.k.* in an acid forest soil: a laboratory study. *Can. J. For. Res.* **24**:462–471.

165. **Vossbrinck, C. R., J. V. Maddox, S. Friedman, B. A. Debrunner-Vossbrinck, and C. R. Woese.** 1987. Ribosomal RNA sequence suggests microsporidia are extremely ancient eukaryotes. *Nature* (London) **326**: 411–414.

166. **Washino, R. K., and B. B. Westerdahl.** 1981. Influence of soil type and inoculation rate on population dynamics of *Romanomermis culicivorax*, p. 88–90. *In Mosquito Control Research Annual Report 1981, University of California.*

167. **Watt, B. A., and R. A. LeBrun.** 1984. Soil effects of *Beauveria bassiana* on pupal populations of the Colorado potato beetle (Coleoptera: Chrysomelidae). *Environ. Entomol.* **13**:15–18.

168. **Weseloh, R. M., and T. G. Andreadis.** 1986. Laboratory assessment of forest microhabitat substrates as sources of

the gypsy moth nuclear polyhedrosis virus. *J. Invertebr. Pathol.* **48**:27–33.

169. **West, A. W.** 1984. Fate of the insecticidal, proteinaceous parasporal crystal of *Bacillus thuringiensis* in soil. *Soil Biol. Biochem.* **16**:357–360.

170. **West, A. W., H. D. Burges, T. J. Dixon, and C. H. Wyborn.** 1985. Effect of incubation in non-sterilized and autoclaved arable soil on survival of *Bacillus thuringiensis* and *Bacillus cereus* spore inocula. *N. Z. J. Agric. Res.* **28**: 559–566.

171. **West, A. W., H. D. Burges, T. J. Dixon, and C. H. Wyborn.** 1985. Survival of *Bacillus thuringiensis* and *Bacillus cereus* spore inocula in soil: effects of pH, moisture, nutrient availability and indigenous microorganisms. *Soil Biol. Biochem.* **17**:657–665.

172. **West, A. W., H. D. Burges, R. J. White, and C. H. Wyborn.** 1984. Persistence of *Bacillus thuringiensis* parasporal crystal insecticidal activity in soil. *J. Invertebr. Pathol.* **44**: 128–133.

173. **West, A. W., N. E. Crook, and H. D. Burges.** 1984. Detection of *Bacillus thuringiensis* in soil by immunofluorescence. *J. Invertebr. Pathol.* **43**:150–155.

174. **Wigley, P. J., and P. D. Scott.** 1983. The seasonal incidence of cricket paralysis virus in a population of the New Zealand small field cricket, *Pteronemobius nigrovus* (Orthoptera: Gryllidae). *J. Invertebr. Pathol.* **41**:378–380.

175. **Wood, H. A., P. R. Hughes, and A. Shelton.** 1994. Field studies of the co-occlusion strategy with a genetically altered isolate of the *Autographa californica* nuclear polyhedrosis virus. *Environ. Entomol.* **23**:211–219.

176. **Woodring, J. L., and H. K. Kaya.** 1988. *Steinernematid and Heterorhabditid Nematodes: a Handbook of Techniques.* Southern Cooperative Series Bulletin 331. Arkansas Agricultural Experiment Station, Fayetteville.

177. **Wright, R. J., F. Agudelo-Silva, and R. Georgis.** 1987. Soil applications of steinernematid and heterorhabditid nematodes for control of Colorado potato beetles, *Leptinotarsa decemlineata* (Say). *J. Nematol.* **19**:201–206.

178. **Young, S. Y., and W. C. Yearian.** 1979. Soil application of *Pseudoplusia* NPV: persistence and incidence of infection in soybean looper caged on soybean. *Environ. Entomol.* **8**:860–864.

179. **Young, S. Y., and W. C. Yearian.** 1986. Movement of a nuclear polyhedrosis virus from soil to soybean and transmission in *Anticarsia gemmatalis* (Hübner) (Lepidoptera: Noctuidae) populations on soybean. *Environ. Entomol.* **15**:573–580.

180. **Young, S. Y., W. C. Yearian, and H. A. Scott.** 1975. Detection of nuclear polyhedrosis virus infection in *Heliothis* spp. by agar gel double diffusion. *J. Invertebr. Pathol.* **26**:309–312.

181. **Yousten, A. A., S. B. Fretz, and S. A. Jelley.** 1985. Selective medium for mosquito-pathogenic strains of *Bacillus sphaericus*. *Appl. Environ. Microbiol.* **49**:1532–1533.

182. **Zimmermann, G.** 1986. The 'Galleria bait method' for detection of entomopathogenic fungi in soil. *J. Appl. Entomol.* **102**:213–215.

183. **Zimmermann, G., and E. Bode.** 1983. Untersuchungen zur Verbreitung des insektenpathogen Pilzes *Metarhizium anisopliae* (Fungi imperfecti, Moniliales) durch Bodenarthropoden. *Pedobiologia* **25**:65–71.

SUBSURFACE AND LANDFILLS

VOLUME EDITOR
MICHAEL J. McINERNEY

SECTION EDITORS
RONALD W. HARVEY AND JOSEPH M. SUFLITA

Overview of Issues in Subsurface and Landfill Microbiology

RONALD W. HARVEY, JOSEPH M. SUFLITA, AND MICHAEL J. McINERNEY

58

Just over a decade ago, answering the question of whether microorganisms even existed in the terrestrial subsurface was a serious research endeavor. During the intervening years, this question has been rendered moot. The recognition of an active subsurface microflora led to a myriad of more sophisticated questions regarding the nature of the organisms inhabiting this environment. For instance, what are the limits of the biosphere? How diverse are subsurface microbial communities? What adaptations do the organisms possess that allow them to survive and proliferate in this environment? How were microorganisms introduced into the subsurface, how do they interact, and what factors govern their distribution? What is the role of subsurface microorganisms in the cycling of elements and the transfer of energy?

In considering such questions, it is entirely appropriate to question the "uniqueness" of the resident microflora. That is, are the organisms inhabiting the subsurface in any way out of the ordinary? It seems reasonable to hypothesize that once microorganisms are introduced into the subsurface, natural forces will select for those that are able to survive and disperse within the subsurface. Thus, those organisms that inhabit the subsurface are the ones that have the adaptive features necessary to maintain themselves and proliferate there. The uniqueness of the subsurface microbiota may be in the mechanisms that microorganisms have evolved to adapt to and withstand the selection forces inherent in the terrestrial subsurface.

To date, the majority of evidence indicates that most subsurface environments (exclusive of those controlled by a dominant abiotic factor such as temperature, pressure, or toxicants) possess climax ecological communities. Such ecosystems are characterized by a high degree of microbiological diversity, they possess trophic structure, and they exhibit materials cycling and energy transfer. Members of such ecosystems typically possess either structural, physiological, or reproductive adaptive features that allow them to become dispersed and survive in such ecosystems.

It may be that subsurface microbiota have great similarities with surface organisms, particularly when a metabolic basis is used for such an assessment. That is, similarities in metabolic pathways or in corresponding ecological niches may be found between specific surface and subsurface microorganisms. This should not be surprising and is also somewhat comforting. It is well known that metabolic principles represent a unifying feature of life. Diverse life forms exhibit a remarkable similarity at the biochemical or metabolic level of examination. By extension of this logic, it may be that subsurface microbial populations and community dynamics are also likely to be rather similar to those of surface environments. Consequently, it may also be that ecological principles can transcend the particular environment under consideration.

In spite of the proliferation of knowledge regarding the terrestrial subsurface and its tremendous economic importance, this environmental compartment remains one of the least understood microbial habitats. This is due, in part, to the high costs and technical problems associated with aseptically obtaining high-quality samples suitable for microbiological analysis. Advancement in our understanding of subsurface microbiology is also hampered by a lack of suitable methodologies for studying the activities of sparse, often ultraoligotrophic microbial populations that inhabit pristine subsurface environments. However, the last two decades have seen much progress in this important and developing field. The recent interest in subsurface microbiology is largely due to the need to control environmental problems caused by or affected by microorganisms in subsurface environments. These problems include souring of producing oil reservoirs, clogging of water supply aquifers by bacteria and/or products of bacterially mediated geochemical reactions, contamination of water supply wells by transport of pathogens, bacterial production of explosive gases in and adjacent to landfills, and the effects of microbial populations on the fate and transport of subsurface contaminants.

A fundamental impediment to the advancement of subsurface microbiology is the difficulty of obtaining representative samples of subsurface material that are free of contamination by nonindigenous organisms and thus suitable for microbiological study. Water supply wells have been in existence for several millennia. However, the act of constructing a well not only chemically alters the subsurface environment but also introduces nonresident microorganisms. Therefore, it is often unclear whether microbial communities observed in well water are truly representative of what is in the aquifer per se. Much of our current understanding of subsurface microbial populations derives from studies involving core samples. Core samples partially alleviated interpretational problems associated with the altered subsurface environment represented by wells. Also, most of the microbial populations in the subsurface are attached to solid surfaces and therefore may not be adequately sampled by wells. Chapter 59 discusses procedures for sampling groundwater and subsurface core material for subsequent microbiological studies. Included are descriptions of, advantages of, and problems associated with coring procedures used to sample shallow (<100 m) depths in unconsolidated formations (hollow-stem augers) and used to sample deeper and/or more consolidated formations (cable tools and rotary drills). In practice, it is often more accurate to assess the degree of microbial contamination of the recovered subsurface samples than to assume that the contamination did not occur. Tracer techniques for assessing this contamination and methods for minimizing further contamination during sample handling are also evaluated. It is evident that many advances have been made both in the method of collection and in the method of handling subsurface samples for microbiological study. However, it still difficult to say with absolute certainty that a given microorganism isolated from a sample of groundwater or subsurface core material originated from depth and is not merely an artifact of contamination.

In spite of the sampling and methodological problems inherent in subsurface microbiology, a number of important recent advances have been made in our understanding of microbial activities in groundwater environments. Much can be inferred about the activities of microorganisms in groundwater by assessing the changes in aquifer geochemistry along a given flow path. Flow paths in many aquifers are hundreds of kilometers in length, allowing synoptic observations of microbially mediated changes in geochemistry in low-nutrient environments that take hundreds or thousands of years to complete. Chapter 61 presents a number of studies in which observations of geochemical changes along flow paths are used to illustrate how microorganisms can alter both the chemical quality and the water-producing capacity of aquifers. Spatial patterns of sequential changes in the predominant terminal electron-accepting process for electron donor-limited (oligotrophic) and for electron acceptor-limited (e.g., organically contaminated) aquifers are discussed. Another important tool in assessing microbially mediated changes in aquifer geochemistry is stable isotope fractionation, as discussed in chapter 62. It is well known that microorganisms preferentially use lighter isotopes during many metabolic processes. Chapter 62 specifically focuses on the transformations of carbon to illustrate how microbial fractionation processes result in altered isotopic signatures in the resulting metabolic products. A recognition of the patterns observed in the end products relative to the starting substrates can lead to reasonable deductions regarding the types of microbial processes that may occur in the terrestrial subsurface.

Although much can be inferred from geochemical measurements about the activities of aquifer microorganisms involved in the predominant terminal electron-accepting processes, the nonconservative nature of many geochemical components can limit interpretation. A parallel approach involves incubation assays which are specific for a given electron-accepting process (chapter 63). The latter approach may be particularly well suited for analysis of organically contaminated aquifers in which the rates of microbial activity are fast enough to measure within a reasonable experimental time frame. These assays are performed in the laboratory with representative geochemical conditions and aquifer core material collected from the zone of interest. However, microbial or chemical contamination of the samples and deviations from in situ incubation conditions can result in interpretational difficulties, especially in evaluating

the rates at which the processes occur. Still another approach is the use of small-scale injection and recovery experiments that use unaltered in situ conditions and microbial populations in the aquifer itself. The in situ injection and recovery approach is best performed under natural-gradient (normal flow) conditions so that the flow rate is relatively constant and the rates of the terminal electron-accepting process can be calculated from a modified advection/dispersion model. A conservative tracer (usually bromide or chloride) is used to account for the flow velocity and the degrees of dispersion that take place during downgradient transport. A labeled terminal electron acceptor or a blocking compound that causes a buildup of a readily measured intermediate in the terminal electron-accepting process can be injected with the conservative tracer. The need to monitor the injectate in three-dimensional space over a relatively short time frame limits the method to shallow unconsolidated aquifer systems.

Small-scale injection and recovery experiments have also been used to delineate the subsurface transport behavior of microorganisms (chapter 64), which has important ramifications for microbially enhanced oil recovery processes, for the fate and transport of organic contaminants, and for the placement of water supply wells. Our ability to predict the subsurface transport of microorganisms on a large scale is hampered by the multitude of physiological, hydrological, and chemical factors that control their movement. Many of these factors are poorly understood and interrelated with other processes. Nevertheless, many advances have been made in our ability to influence and mathematically describe groundwater transport of microorganisms. An iterative approach which utilizes laboratory-scale, intermediate-scale, and field-scale experiments allows for an evaluation of specific processes under controlled laboratory conditions and tests of laboratory-derived transport hypotheses in situ. The use of well-characterized microbial-sized microspheres has facilitated the understanding of some of the abiotic factors that govern subsurface microbial transport behavior. However, there are a number of problems and areas of research that need to be addressed before we can reliably predict microbial transport over longer distances. In particular, there is a dearth of knowledge concerning the role of bactiverous subsurface eukaryotes, and methods of culturing subsurface microorganisms that do not alter the transport characteristics of the cell are needed. We also need to know how to mathematically describe the processes of microbial attachment and detachment in the presence of chemically complex surfaces and groundwaters and how to handle physical and chemical aquifer heterogeneity in transport models.

The aforementioned difficulties in accurately predicting microbial transport in the subsurface make it problematic to use microbial transport models to calculate how far a water supply well should be located from a contaminant source. Indeed, predictive transport models can over- or underpredict microbial movement in groundwater by several factors of 10. Most well protection models use the transport of viruses to calculate well setback distances and ignore pathogenic bacteria and protozoa whose transport behavior can be fundamentally different from that of viruses. The methods that are used to calculate setback distances for water supply wells are explained in chapter 66. Despite these limitations, for regional planning purposes, geostatistical models have been successfully applied to areas that are particularly susceptible to groundwater contamination. These models incorporate information about groundwater temperatures (which are used to calculate virus inactivation rates), hydraulic conductivities, and hydraulic gradients. However, viral inactivation can be affected by a number of factors other than temperature, such as interaction with solid surfaces and the amount of dissolved organic matter. It is clear that models used to predict subsurface microbial transport behavior must be improved to give more precise answers to these important public health questions. In particular, more information is needed on the transport behavior of microorganisms in the unsaturated zone, which is ignored in many of the current models.

The recognition that microorganisms inhabit both shallow and deep aquifers has led to the realization that petroleum reservoirs must also contain active microbial populations. However, the problems facing a petroleum microbiologist (chapter 65) are quite different from those of a microbiologist working with drinking water aquifers. Important concerns are the control of detrimental microbial activities such as hydrogen sulfide production (souring) and the use of microbes to enhance the oil production. The problems created by microorganisms during the souring process of an oil reservoir include corrosion of water-handling equipment, plugging of injection wells, and the safety issues and removal costs associated with the production of hydrogen sulfide. The key issue in the control of microbially induced souring is the early detection of the sulfide-producing bacteria that are responsible. The ability of microorganisms to improve oil production has been a controversial topic because of the lack of controlled laboratory and field studies. Chapter 65 describes the use of sandstone experimental systems to test the potential of microorganisms to release oil entrapped in rock. One way that bacteria can facilitate the release of oil is by producing biosurfactants that reduce the interfacial forces between the oil and aqueous phases. To be effective, the biosurfactant must reduce these forces by a factor of at least 10,000. Bacteria and the polymers that they produce can also enhance the recovery of oil by plugging the more conductive zones and larger pores in an oil-bearing formation which diverts the recovery fluid into previously bypassed regions of reservoir. Chapter 65 describes the procedures used to isolate and characterize polymer- or surfactant-producing bacteria from oil reservoirs. Although it is relatively easy to determine whether a microorganism can enhance oil recovery in controlled laboratory experiments, extrapolating this information to the recovery of oil in actual oil reservoirs, often at great depth, is nontrivial. A multitude of variables which are constantly changing throughout the exploitation process can affect the performance of oil recovery operations. For these reasons, it is difficult to determine whether a particular treatment has in fact caused an increase in oil production. Thus, in evaluating a microbial process in the field, it is critical to perform a control treatment that uses the same equipment and injection protocols as are used for the microbial treatment but lacks the key components of the microbial process itself, in order to verify that the microbial process is having the desired effect. Also, there may be a need to conduct tracer experiments to verify flow paths in the reservoir before a microbially enhanced oil recovery process is tested.

The microbiology of landfills (chapter 60) is also an important topic, as this waste disposal option is the most popular to date and is likely to remain so for the immediate future. It is arguable whether this topic should be included in a subsection dealing largely with subsurface microbiology rather than in the section on biodegradation. However, landfills are known to have serious environmental effects on the terrestrial subsurface, most notably the contamination of shallow groundwater reserves. In order to accurately assess this impact as well as other environmental threats, it is important to know what is actually buried in landfills and to what extent the microbial decomposition of interred refuse actually occurs. Chapter 60 provides a needed perspective for such evaluations.

Aside from the obvious connection of the impact of landfills on the terrestrial subsurface, there are other, more subtle similarities linking the two environments. First, the two habitats are strikingly similar with respect to the heterogeneous nature of microbial processes in them. In fact, one could hardly imagine a more heterogeneous environment than a municipal landfill. Such heterogeneity requires specialized sampling, handling, and analysis strategies with objectives clearly defined. Since sampling either habitat tends to be difficult, expensive, and time-consuming, there is an obvious need to obtain samples that are as representative as possible. For the results to be most meaningful, experimental systems and analytical techniques that are appropriate for the objectives of the study must be used. Guidance on these topics can also be found in chapter 60. Moreover, the microbial processes in each habitat are equally dependent on local hydrological conditions for a variety of important features, including the delivery of terminal electron donors and/or acceptors, the supply of nutrients for microbial growth, and the migration of accumulated metabolic intermediates or end products. However, preferential channeling of moisture through a landfill can severely limit refuse decomposition processes. An indication of the multiple factors that can potentially limit microbial refuse decomposition in landfills can be found in chapter 60.

A final similarity can also be noted. Current studies argue that microbial activities influence the geochemical processes in both landfills and subsurface environments and that the altered geochemistry, in turn, influences the selection of microorganisms proliferating in the habitat. However, while microbial activity may often be limited by the availability of electron donors in uncontaminated aquifer systems, this is rarely the case with landfills.

Lastly, it is hoped that most of the ensuing chapters will provide investigators with a greater understanding of the experimental approaches needed to study the microbiology of the terrestrial subsurface and an appreciation of interpretational limits imposed by the existing methodologies. As our knowledge of the types of metabolic interactions and the nutritional versatility of the indigenous microorganisms increases, we will be able to predict with greater confidence the effects that may be associated with perturbations of subsurface environments. This in turn will help provide a conceptual framework for biotechnological approaches for environmental restoration. The fact that subsurface habitats that have been physically separated from their surface counterparts for thousands to millions of years are now known to contain diverse populations of microorganisms has excited the interest of evolutionary biologists. The hope is that further study of subsurface microorganisms will provide insights into the process of microbial evolution and possibly into the origins of life itself.

Subsurface Drilling and Sampling

JAMES K. FREDRICKSON AND TOMMY J. PHELPS

59

Interest in the microbial ecology of terrestrial subsurface environments and the role of microorganisms in subsurface geochemical processes has grown rapidly since detailed studies demonstrated that a variety of shallow (5, 23) and deep (3, 4, 10, 36) environments contained substantial numbers of viable microorganisms. There has also been an increasing interest in utilizing the metabolic activities of microorganisms for in situ bioremediation of contaminated subsurface sediments and groundwater. Hence, scientific interest in naturally occurring biogeochemical processes and novel microorganisms (9, 19) in the subsurface along with practical needs for monitoring in situ bioremediation has increased the need for effective approaches for sampling the subsurface. Subsurface samples have been used to study the size and composition of the microbial community by using traditional and molecular biology-based methods, to assess the types and rates of microbial activities, and to measure the geochemical properties of the water and solids which comprise the microbial habitat.

The advancements in subsurface microbiology, especially of relatively deep environments (i.e., below 100 m), have been due at least in part to innovations in drilling and coring technologies. In spite of these innovations, however, obtaining representative samples of solids and water from the subsurface for microbiological and geochemical studies remains a major impediment to the study of subsurface microbiology.

The sampling of subsurface sediments, rocks, and water for microbiological and geochemical analyses presents unique challenges compared with the sampling of surface soils, waters, and sediments. Sampling of subsurface solids requires specialized technologies for drilling (advancing the borehole) and coring (collecting samples of subsurface materials). Different lithologies require different approaches for successfully collecting samples. For example, drilling and coring methods that are used to sample fine-grained sediments generally will not work for unconsolidated coarse-grained sediments that tend to flow or fall out of the core barrel upon retrieval. Also, the drilling methods used for sampling hard rock at depth generally require approaches such as rotary coring with circulating fluids that are distinct from hollow-stem augering, which is used for coring of shallow unconsolidated sediments.

The costs associated with drilling and coring often limit the number and type of samples that can be obtained. This can be problematic in that, not surprisingly, there can be considerable spatial heterogeneity in the distribution and function of microorganisms in subsurface environments (2, 40, 41). For soils, it is relatively simple to obtain sufficient numbers of samples to adequately characterize the microbiological properties of a given site. However, when one is sampling the subsurface, especially at depth, obtaining a sufficient number of samples to define horizontal and vertical variations in the microbiological properties can be technically difficult, labor-intensive, and costly.

It is generally accepted that subsurface solids must be sampled to adequately evaluate the microbiological characteristics of subsurface environments. The population densities and community composition are generally different in groundwater samples compared with subsurface sediments obtained from the same formation (28, 29), probably because of the selective attachment of some populations to solid surfaces and their low density in the pore waters. In addition, the process of drilling and developing a well leads to introduction of exogenous microorganisms and overall changes in the biogeochemical environment in and near the well (11, 23). Sampling groundwater for microbiological analysis is necessary when sampling solids is impractical or when the samples are being collected to check for water quality indicators and pathogens or to monitor changes in microbial populations and activities during bioremediation. Groundwater sampling is commonly used for monitoring geochemical parameters during in situ bioremediation and, in some cases, even for microbiological monitoring. Complications with sampling solids include the heterogeneity in the microbiological (and geochemical) properties of the subsurface (2, 40), perturbation of the surface site, and prohibitive costs. Sampling of groundwater from multiple wells can be an effective means for assessing microbiological and geochemical properties in aquifers at local (43) and regional (35) scales. In comparison with collecting core samples for geochemical and microbiological analyses, sampling groundwater from multipoint sampling wells provides considerable flexibility in terms of the location, number, and size of samples that can be collected.

In some cases, costs associated with the drilling and coring may be greater than the costs of analyses and investigations. At times, the drilling costs may be borne by other parties that are sampling the subsurface for other reasons, such as mineral or fossil fuel exploration or installation of water supply wells. Such "piggybacking" may enable access to otherwise unavailable samples but has disadvantages: the coring methods may not be appropriate for obtaining samples suitable for microbiological analysis, and investigators may not have control over the location and procedures used for coring. If drilling and coring are to be contracted, then selection of a drilling company must be carefully considered prior to sampling. Some companies have more experience and capabilities than others in scientific drilling and sampling. These are often the larger companies that have operations in many regions and even multiple countries. However, smaller locally owned companies are often more familiar with the local geology and have more experience at drilling and coring in the region. It is recommended that those unfamiliar with contracting for drilling and coring consult with an experienced geologist before proceeding. These are just a few examples of the challenges and issues that must be addressed when one is sampling the subsurface.

Approaches appropriate for drilling boreholes and collecting cores depend on the depths and lithologies to be sampled. Hollow-stem augering is the most common method used for shallow drilling in unconsolidated sediments, typically to depths less than 100 m, while rotary drilling is the most common method for drilling to depths greater than 100 m. Cable tool systems have also been used successfully for sampling subsurface sediments at depths of less than 300 m. These three commonly used approaches have been selected for inclusion in this chapter. The reader should consult references 11, 14, 15, 33, 38, 39, and 44 for additional details regarding these and other drilling and sampling methods.

Although many precautions must be taken to prevent contamination of samples during coring and processing, it is prudent to use one or more tracers so that an assessment of contamination can be made. Rigorous tracer regimens are particularly important for sampling deep subsurface environments, where circulating fluids are typically used to remove drill cuttings from the borehole. There are several sources of contamination that can occur during subsurface drilling and coring. One source is the circulating fluid that can enter the formation ahead of the drill bit or penetrate the core while it is being recovered from the borehole. Another source is smearing or mixing of part of the outer core, which has been in contact with drilling tools and fluids, into the uncontaminated inner core during sample paring and processing. After subsurface materials have been recovered, they are subjected to processing prior to analysis or use in experiments. Processing typically includes paring to remove outer core material and sectioning and disaggre-

gation of the inner portion of the core. Because many subsurface environments are anoxic, it is important to limit exposure of samples to air in order to prevent inhibition or inactivation of anaerobic and microaerophilic microorganisms by O_2 and to prevent oxidation of reduced geochemical species.

The goal of this chapter is to provide background information and general guidelines necessary for obtaining representative samples of subsurface solids and groundwater for microbiological and geochemical analyses. To this end, approaches for collecting representative solid and groundwater samples, processing of subsurface solids to reduce and assess microbiological and geochemical alterations during coring, and evaluating the quality of subsurface samples are described.

SAMPLING FORMATION MATERIALS

Hollow-Stem Augering: Shallow Sediment Sampling

Hollow-stem augering is typically used for sampling shallow, unconsolidated subsurface environments of less than 100 m in depth (Table 1). Typical uses of hollow-stem augering include drilling in shallow, unconsolidated lithologies for the installation of domestic water supply and monitoring wells in addition to the sampling of solids.

During hollow-stem augering, the borehole is advanced by rotating the auger flights, and the cuttings are brought to the surface by the auger action. When the desired sampling depth is reached, a split-spoon (Fig. 1) or other drive- or push-type core barrel is lowered through the auger flights to the bottom of the borehole. The core barrel is typically driven into the sediment by percussion, i.e., with a hammer. Once the core barrel is driven to the appropriate depth, it is brought to the surface. A hollow-stem auger drill rig is shown in Fig. 2, and a simplified diagram of coring through a hollow-stem auger is shown in Fig. 3.

It is common practice to thoroughly clean the surface of all auger flights and core barrels before each use. For additional information on decontamination of drilling tools, see the section in this chapter on sample processing and references 15, 38, 39, and 44. Once the core has been brought to the surface, the liner containing the intact core is removed from the core barrel and the ends are capped. The core can then be transferred to a location where it can be further processed.

The major advantages of coring through hollow-stem augers are that it is relatively simple, commonly available, and economical to use for obtaining nearly undisturbed unsatu-

TABLE 1 Comparison of various drilling methods

Drilling method	Applicable depth (m)	Lithology	Coring method	Advantages	Disadvantages
Hollow-stem augering	<100	Unconsolidated	Split spoon, Waterloo	Mobile, inexpensive	Shallow
Cable tool	<300	Unconsolidated or consolidated	Split spoon	Inexpensive, no drilling fluids	Slow
Rotary with mud or air	>1,000	Unconsolidated or consolidated	Split spoon, wireline	Deep access, fast	Costly, drilling fluids

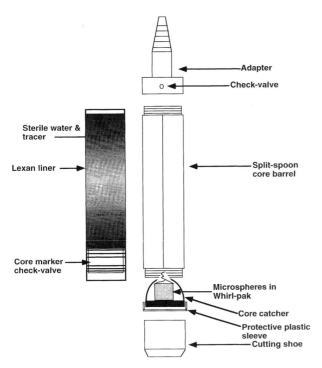

FIGURE 1 Diagram of split-spoon core barrel containing tracers and core liner for sampling saturated subsurface sediments by cable tool, hollow-stem auger, or rotary drilling methods.

rated or saturated subsurface samples. The drilling equipment is typically mounted on the back of a truck and is readily deployed. Depending on the total depth and frequency of coring, many boreholes can usually be cored over a few days. Another major advantage is that drilling fluids are not required, and hence a source of contamination is eliminated. However, for augering saturated sediments, nonnative water is often added to maintain fluid pressures on the formation to prevent the heaving of sediments into the augers. This water can be a potential source of microbial contamination, as it can move into the formation ahead of the auger flights. Before introducing nonnative water to a borehole from which microbiological samples are to be collected, it is important to disinfect the water by chlorination. The principal sources of contamination during hollow-stem augering, when nonnative fluids are not introduced intentionally to the borehole, are from overlying sediment and soil that may be carried down to the bottom of the borehole and from underlying fluidized sediments which may flow into the open borehole.

Sampling of unconsolidated sands by hollow-stem auger can be difficult because the samples lack cohesiveness and tend to fall out of the core barrel during the trip to the surface. To overcome this problem, a specialized sampling system referred to as the Waterloo sampler (47) was developed. The Waterloo sampler consists of an outer core barrel, an inner liner, and a plunger or piston. The piston is set at the bottom of the liner. As the core barrel is driven into the sediment, the piston slides upward inside the inner liner, creating a suction. The suction helps to retain the sample within the core liner during retrieval of the core barrel. Another advantage of the Waterloo system is that it can limit drainage of pore water. Drainage of pore water can

result in changing the solid-to-solution ratio of the sediments and replacement of water with air in some of the pores. Both of these processes can change the microbiological and geochemical properties of the sample.

A disadvantage of the Waterloo system is the requirement for an additional cable to hold the piston and the risk of the cables becoming entangled in the borehole. It is also difficult to core dense, consolidated, or stony materials with the hollow-stem auger. The depth to which hollow-stem augering can be used is limited as a result of increasing friction of the auger flights with depth, especially if rotation of the flights is stopped for any period of time. In spite of these disadvantages, hollow-stem augering is the method of choice because of its availability, simplicity, and low cost for sampling unconsolidated sediments at depths of less than 100 m.

Cable Tool Drilling: Shallow and Deep Sediment Sampling

Cable tool drilling has been used for centuries (17) for drilling water wells and is still commonly used in many regions of the world. Recently, it has been used to obtain high-quality samples suitable for microbiological analysis from unconsolidated sediments from depths as great as 200 m (20, 30).

Cable tool drilling is accomplished using a "drive" barrel that is attached to a set of "jars" or sliding hammers that advance the borehole by percussion action. A cable tool drill rig is shown in Fig. 4. The upper jar is raised by a cable

FIGURE 2 Photograph of a hollow-stem auger drill rig used for obtaining core samples from relatively shallow (<100 m) subsurface sediments.

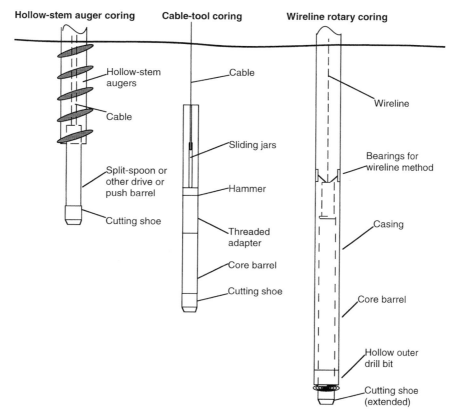

FIGURE 3 Simplified diagram of below-ground portion of coring process for collection of subsurface materials during hollow-stem auger, cable tool, and wireline rotary drilling.

attached to the drill rig and is then dropped onto the lower jar, which pushes the drive barrel deeper into the subsurface. The drive barrel is repeatedly lowered, driven, and retrieved to advance the hole. The materials obtained in this manner are generally not suitable for microbiological analysis because of the disturbance that occurs during the driving. In consolidated or dense formations, the hole is advanced via use of a hard tool or heavy bit that crushes the subsurface material, which is subsequently removed by bailing. It is usually necessary to advance metal casing in unconsolidated sediments to prevent the borehole from collapsing during and after drilling. The diameter of the casing, and hence of the borehole, can vary from <10 to >30 cm. It is often necessary to use smaller-diameter casing with increasing depth because of the greater friction associated with the larger-diameter casing. Telescoping of multiple casings eases the removal of casings after the borehole has been completed.

Core samples of formation materials suitable for microbiological analyses are obtained during cable tool drilling in a manner similar to that described for hollow-stem auger coring. A split-spoon or similar push-type core barrel replaces the drive barrel, and the split spoon is advanced by raising and dropping the upper jar and hammering the core barrel into the formation ahead of the borehole (Fig. 3). Split-spoon core barrels are the most common and flexible samplers to use in conjunction with cable tool. They are readily assembled and disassembled, can house an inner core liner, and are easily cleaned. To limit contact between the core liner and groundwater or slurry in the borehole or drill muds, it is necessary to use a core plug or marker (Fig. 1)

and to fill the liner with disinfected water. When the core barrel contacts and penetrates the formation, the core plug is pushed up through the core liner and the water that was in the liner exits through the check valve at the top of the assembly. The check valve in the core plug allows fluids and mud at the bottom of the borehole to be squeezed out of the liner as the core fills the liner.

A major advantage of cable tool drilling is that, as with hollow-stem auger drilling, no recirculating fluids are required, and hence the potential for contamination of samples is greatly reduced. Also, the equipment is widely available. An advantage over hollow-stem augering is that coring at depths of greater than 100 m can be readily achieved. In theory, cable tool drilling can be used to sample formations deeper than 100 m, but practical considerations limit the effective depth. A disadvantage is that cable tool drilling is slow; under ideal conditions, the borehole may be advanced 10 to 20 m/day. Depending on the geology, there can be considerable physical disruption of the sample as a result of the hammering, pulverization, and redistribution of subsurface materials. There can also be significant compaction of the cored materials. If there are gravels or cobbles, they will be fractured and ground up by the action of the jars. Fracturing of the solids can have a secondary impact on the microbiological and chemical characteristics of subsurface samples. For example, Bjornstad et al. (8) found that fracturing of cobbles was related to the generation and accumulation of hydrogen gas in the slurried drill cuttings in the hole. This production of H_2 led to an increase in H_2-utilizing bacteria in the borehole slurry, which provided a potential source of contamination of the subsurface materials.

FIGURE 4 Photograph of a cable tool drill rig.

Rotary Drilling: Deep Sediment and Rock Sampling

Rotary drilling and coring are commonly used technologies for subsurface mineral and fossil fuel exploration, scientific exploration, placement of deep wells for water supply, and monitoring groundwater contamination. Rotary drilling rigs range in size from small truck-mounted units used for shallow well drilling to large rigs that are used for deep (>1,500 m) gas and oil exploration and scientific drilling programs. Rotary drilling employs sections of hollow drill pipe that are threaded together as the borehole is deepened. The drill pipe is rotated at the surface by the drill rig. At the base of the drill pipe is a drill bit that provides the cutting action. For routine drilling operations, a tri-cone bit is used, which essentially homogenizes the materials into "cuttings" as the drill pipe is advanced. To prevent the borehole from becoming clogged with these cuttings, a circulating fluid is used to remove them.

Rotary drilling and coring require the use of drilling fluids, which typically consist of bentonite clay in combination with polymers. Polymers used in drilling fluids typically consist of polysaccharides such as guar gum (11). Another common commercial drilling fluid is Revert, a biodegradable grain-based mud that decomposes after drilling. The ingredients and their concentrations can be adjusted to alter the density, viscosity, or other characteristics of the drilling fluids. Alternatively, air or inert gases can serve as drilling fluids. In addition to removing cuttings from the borehole, drilling fluids also lubricate and cool the drill bit and help to stabilize the borehole. The drilling fluids also limit loss of drilling fluid to the formation and intrusion of formation water into the borehole. Loss of drilling fluid to the formation and collapse of the borehole are prevented by the formation of a filter cake on the sidewall of the borehole and by maintenance of proper hydrostatic pressures because of the drilling fluid's weight.

Drilling fluids are circulated through the borehole via a pump located at the surface. The drilling fluids are pumped downward through the hollow drill pipe and exit at the bottom of the borehole through ports in the drill bit. The fluids lubricate and cool the drill bit and remove the cuttings by returning them to the surface along the annulus between the drill pipe and the borehole. At the surface the cuttings are separated from the fluids, usually by settling or screening. Fluid weight and viscosity are routinely monitored and adjusted to correspond to changing conditions in the borehole. Drilling-fluid pressure is controlled by the pump and must be low enough to prevent the fluid pressure from greatly exceeding the formation pressures, which could destabilize the borehole and contaminate the formation ahead of the drill bit.

Shallow coring typically involves the removal of the entire drill pipe section by section from the borehole. The drill bit is then removed and replaced with a coring tool. The string of drill pipe is placed back into the hole, and coring proceeds. After coring, the drill pipe is removed from the borehole to retrieve the core. For shallow coring (<100 m), a split spoon, Shelby tube, or other type of driven or pushed core barrel can be used (38). For deeper coring (>100 m), sampling tools such as the Dennison or Pitcher barrels are often employed (38). At depths of >100 m when many cores are collected sequentially, the process of removing the entire drill string for each exchange of the drill bit and coring tool is time- and labor-intensive.

A significant improvement of coring during rotary drilling was the development of the wireline coring system. Rather than removing all of the drill pipe to switch from rotary drilling to coring, the upper section of drill pipe can be disconnected from the drill rig and a "messenger" can be lowered on a wireline through the inside of the drill pipe. An assembly used for wireline coring during rotary drilling is shown in Fig. 5. At the bottom of the borehole, the messenger engages and retrieves an inner drill bit through the inside of the hollow drill pipe. The wireline is used to bring the inner bit to the surface, where it is exchanged with a core barrel. At the bottom of the borehole, the core barrel is engaged to the drill pipe and an outer bit through a bearing which allows the drill pipe and outer bit to rotate while the sample tool and core liner remain stationary. The upper section of drill pipe is then reattached, and the pipe is rotated to advance the borehole. The outer drill bit at the bottom of the drill pipe is hollow and grinds away the rock outside the outer edges of the core barrel, allowing the core to enter the core barrel. After coring, a messenger is lowered on the wireline to retrieve the core barrel.

An extended shoe is often used with wireline coring to extend the cutting edge of the core barrel several centimeters in front of the drill bit/sediment interface in order to core into previously uncompromised subsurface materials. The assembly, with extended shoe, used for wireline coring during rotary drilling is shown in Fig. 5. The use of extended shoes is important for sampling permeable materials or when the drilling-fluid pressure exceeds the formation pressure, causing contamination of the sediments ahead of the drill bit. Wireline coring methods can save a substantial amount of time in coring deep rock and sediments compared with conventional rotary coring because the drill pipe is not removed during the operation. In addition, wireline coring

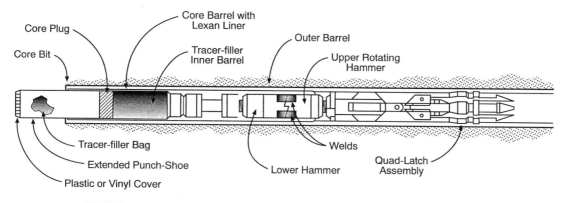

FIGURE 5 Diagram of assembly used for rotary drilling with wireline coring.

reduces the risk of borehole collapse, which can occur with frequent removal and replacement of the drill pipe.

Rotary wireline coring can be used to obtain intact core samples from a wide range of lithologies and at depths of kilometers. Because a rotating cutting action rather than percussion is used, rotary coring usually causes less physical disturbance of the core samples. The length of continuous core obtained can vary from less than 1 m to tens of meters. With this flexibility, relatively large quantities of sample can be obtained from a single core run. Rotary operations can also be used to core in angled or horizontal boreholes. In addition to rotary coring, a drive- or push-type sampler can be used in a rotary-drilled hole, although samples obtained in this manner will be subject to greater physical disturbance.

The principal disadvantage of rotary coring is the use of recirculating drilling fluids. Aqueous fluids can be problematic because of the large number of bacteria that they often contain. Beeman and Suflita (7) counted 10^9 total microorganisms per g (dry weight) of bentonite drilling muds during wireline rotary coring of southeast coastal plain subsurface sediments. Microbial contamination of subsurface samples during the coring process can occur via several routes (11). These routes include direct contamination of the formation where the mud pressure at the bottom of the bit exceeds the formation pressure, forcing mud and solutes into the formation ahead of the drill bit and core barrel. Seepage of drill mud that coats the outside of the core into pores or fractures of the core as it enters the core barrel is another potential source of contamination. Given the high density of bacteria in aqueous fluids compared with their population density in the subsurface, it is important to employ tracers to detect and quantify the extent of bacterial and drilling fluid contamination from drilling fluids whenever rotary operations are used.

Air or an inert gas such as argon or N_2 can be used as an alternative to aqueous drilling fluids to remove cuttings from the borehole. For example, pressurized argon was used to core deep basalt and associated interbedded sediments at a site on the Snake River Plain in Idaho (8). Because the core samples were being used for microbiological and geochemical analysis, argon was used rather than air to limit the exposure of the samples to O_2. Drilling with air or an inert gas is limited to the vadose zone and to moderate depths beneath the water table because of the excessive hydrostatic pressures that must be overcome when the borehole fills with groundwater. A major advantage of drilling with air or an inert gas is the reduced potential for microbiological contamination of core samples. One disadvantage of drilling with air is that significant drying of samples can occur during coring of permeable sediment or rock. Drying during air drilling can be reduced by using a Gel-Coring system (Baker-Hughes, Casper, Wyo.). For gel coring, core liners are prefilled with a viscous, inert gel. As the core advances, gel surrounds the core, forming platelets to isolate the core and minimize exposure of cored materials to O_2 and drilling fluids. This approach should also be useful for limiting intrusion of drilling fluids into cores.

Assessing Contamination

A principal concern in sampling subsurface environments is the chemical and biological contamination of samples either by extraneous sediments or by circulating drilling fluids during the coring, retrieval, and processing stages. These exogenous materials can contain microorganisms and solutes (or gases) that can affect the microbiological and geochemical properties of cores. Because the levels of microbial biomass and activity in the subsurface are often inherently low, even small amounts (i.e., milligrams per kilogram) of contamination can affect the results and their interpretation. Thus, it is often desirable to include tracers when samples from the subsurface are obtained.

Sources of microbial, dissolved, and gaseous contaminants include drilling fluids, groundwater-sediment slurry in the borehole, slough from overlying formations, and borehole sidewall materials. Contamination of sample materials can occur before, during, and after the coring operation. The intrusion of drilling fluids into core samples can introduce large numbers of bacteria and solutes that can alter sediment or groundwater chemistry. Contamination by drilling fluids can also affect the microbiological properties of the samples by altering pH, redox, and nutrient concentrations. During drilling with air or other gases, intrusion of gas can potentially result in displacement of pore water, resulting in drying and oxidation of the core. External sources of microorganisms, solutes, or gases from surface soils, the drilling operation, and sample processing should be traced when possible.

One illustration of how tracers can be used to estimate contamination by drilling fluids, such as can occur in permeable, unconsolidated sediments in advance of the drill bit during rotary coring, is shown in Fig. 6. In this laboratory column experiment (9a) with intact sandy aquifer material, it was demonstrated that solutes, in this case Br^-, can move in advance of bacteria and clay (Fig. 6A) under fluid pressures similar to those occurring during rotary coring. This example illustrates the differential transport of water and particles such as bacteria and clay and the buildup

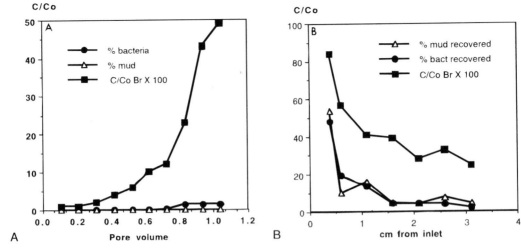

FIGURE 6 Transport of Br⁻, bacteria, and clay associated with drill muds through an intact core of aquifer sand (A) and distribution of these components over the length of the core after collection of approximately one pore volume (B). C/Co, ratio of the concentration at a given pore volume or distance to the starting concentration.

of a filter cake, as evidenced by the high concentration of clay near the column inlet (Fig. 6B). This filtration can greatly reduce contamination of sediments in advance of the drill bit with bacteria from the drilling mud.

Microbiological Tracers

A variety of tracers have been used to evaluate biological contamination of subsurface samples (Table 2). The opti-

mum microbiological tracer may be an organism with a rare but easily detected phenotype. For example, microorganisms that produce a distinguishing pigment or grow on a selective medium can be easily deployed and detected. *Saccharomyces cerevisiae* (baker's yeast) (46), *Serratia marcescens*, *Chromobacterium violaceum*, and *Bacillus subtilis* var. niger (ATCC 9372) (18) have previously been used to trace microbial transport in groundwater. However, the use of live organ-

TABLE 2 Techniques used to evaluate potential biological contamination of cored subsurface sediments

Technique (reference[s])	Method of introduction	Initial concn	Limit of detection
Latex fluorescent microspheres, 1.0 μm, carboxylated (20)	Dry mix with glass beads to hole bottom with bailer	5×10^9 g⁻¹ × >200 g	10^3 g⁻¹
(12, 39)	In bag in shoe/core catcher that breaks upon coring	3×10^{10} ml⁻¹ × 10 ml	10^3 g⁻¹
Serendipitous native microbes (7, 13)	Native, no introduction	10^3–10^7 ml⁻¹	10^0
Added microorganisms (18, 46)	Add to fluids	$<10^6$	10^0
Community-level physiological profiles[a] (33)	Comparative analysis	NA[b]	NA
Bacterial membrane lipid profiles[a] (33)	Comparative analysis	NA	NA
Bacterial CFU[a] (39)	Comparative analysis	NA	NA
Biochemical profiles of bacterial isolates[a] (33)	Comparative analysis	NA	NA

[a] Analysis based on pattern recognition.
[b] NA, not applicable.

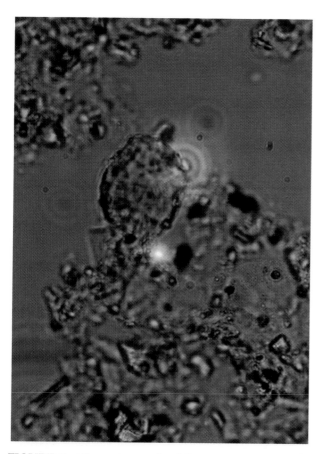

FIGURE 7 Photomicrograph of fluorescent microparticles among sediment particles, demonstrating easy detection of such particles when used as tracers. The photomicrograph was taken by using a combination of epifluorescent and phase-contrast microscopy.

isms as biological tracers may be regulated, and caution should be used in selecting and introducing such organisms. "Serendipitous" bacterial tracers that can be present in high concentrations in drilling fluids, such as *Gluconobacter* species (13) or coliforms (7), can also be used. Alternatively, fluorescent microspheres can be used as bacterial surrogates. Microspheres are available in a range of sizes, with different fluors, and with neutral, positive, or negative charges (Polysciences, Warrington, Pa.) and are readily detected by epifluorescence microscopy (Fig. 7). It has been our experience that some sediment and rock samples contain materials that fluoresce at the same excitation wavelength as the Polysciences NYO latex microspheres; therefore, the use of this type of particle should be avoided. Latex, carboxylated, 1-μm-diameter YG microspheres have been successfully used as bacterial tracers during coring with cable tool and wireline rotary operations (12, 20, 39). One disadvantage of using microspheres is that the cost can be prohibitive if large numbers are required.

Additional approaches to estimate the level and extent of biological contamination include comparative biological analyses of core samples and of potential sources of contamination, such as drilling fluids. These analyses can include membrane lipid profiles and community-level physiological profiles (33). Since these are community-level measures, they can be used only to determine if gross contamination

has occurred and will be relatively ineffective for detecting low-level microbial contamination. Analyses of individual microbial isolates can include biochemical profiles, lipid profiles, or genetic fingerprinting. Comparative approaches for assessing contamination must be used with caution because of the potential that the microbiota of the contaminant source may be indigenous to the sampled subsurface (i.e., enriched in the drilling fluids by mixing with the formation cuttings during drilling operations). This is of particular concern with recirculating drilling fluids. An additional precaution regarding comparison of isolates pertains to how representative the isolates are. For example, analysis of relatively few isolates among a population of 10^6 or more bacteria per cm^3 may not be adequate in all situations for addressing contamination.

Geochemical Tracers

It is desirable to include tracers of chemical contamination because contaminants could affect geochemical analyses and indirectly affect the microbiological properties of the sample by altering E_h, pH, or nutrient concentrations. Examples of the types of chemical tracers that can be used to trace contamination are provided in Table 3. Chemical tracers should be nonreactive and therefore conserved during the sampling process. Rhodamine and fluorescein (38) have been used to trace drilling fluids because they are highly colored even at low concentrations and therefore can be used as visual indicators of core contamination in the field. However, fluorescent dyes are typically sensitive to oxidizing agents and pH. Bromide is a useful solute tracer, as it is normally present at low concentrations in groundwater, is generally nonreactive, and is nontoxic. It can also be readily detected in the field by using an ion-specific electrode, although these electrodes must be calibrated often since they are subject to interferences from other ions and matrix effects. Potassium has also been used as an ionic tracer but has the potential disadvantages that it sorbs to clays in the drilling fluids via cation-exchange reactions and can be present at high background concentrations in some groundwaters. Sulfate and perchlorate can also be used as anionic tracers, but sulfate has the potential to be biologically reactive under anaerobic conditions (i.e., sulfate reduction). Inert perfluorocarbon compounds can be used as gaseous or aqueous tracers. Despite their limited solubility in aqueous solutions, they can be detected at extremely low concentrations by gas chromatography with an electron capture detector, making them very sensitive tracers. Also, different perfluorocarbons such as perfluorohexane and perfluoromethylcyclohexane can be used at different coring intervals in the same borehole as a check on the potential for contamination from the previous core run. Although perfluorocarbons are nonreactive, they are quite volatile. Therefore, precautions should be taken when samples are processed in an enclosed space such as a glove bag to prevent volatilization from and cross-contamination of samples.

Tracer Introduction

Methods for the introduction of tracers vary with the drilling technology and the sources of contamination to be traced. The potential for contamination of samples obtained during hollow-stem augering is less than for other coring methods. Because the action of the auger flights brings cuttings to the surface, there is little potential for downhole contamination with slough from the surface. A major source of microbial and solute contamination during hollow-stem auger coring beneath the water table is sediment and water in the borehole. In general, solute tracers such as bromide can be used

TABLE 3 Tracers used to evaluate chemical contamination in cores of subsurface sediments.

Tracer	Method of introduction	Initial concn (μg liter^{-1} or kg^{-1})	Limit of detection (μg liter^{-1} or kg^{-1})	Method(s) of detection	Notes
Rhodamine	Soluble in aqueous drilling fluids	20,000	0.5–2.0	Fluorimetry	Useful for on-site drilling-fluid analysis
Bromide	Soluble in aqueous drilling fluids	>500,000	10	Ion-specific electrode; ion chromatography	Liquid drilling-fluid analysis
	Concentrated solution directly to borehole	>500,000	10		Can be added to split-spoon core liners
Potassium	Soluble in aqueous drilling fluids	>500,000	10	Ion chromatography; inductively coupled plasma spectrometry	Liquid drilling-fluid analysis
Sulfate	Present in bentonite fluids or soluble	>150,000	10	Ion chromatography	Liquid drilling-fluid analysis
ClO$_4$ (as NaClO$_4$)	Sterile solution in Lexan liners	200,000	2,000	Ion chromatography	Can be added to split-spoon core liners
Tungsten carbide	Direct to borehole bottom	250,000,000	100,000,000	Ion microprobe	Add to bottom of borehole with a bailer
Glass beads	Direct to borehole bottom	>500,000,000	5,000,000	Microscopy	Tracer carrier/bulking agent
Noble gases	Direct to drilling fluid	>500,000	50	Gas chromatography	Gaseous drilling fluids
Perfluorocarbons	Add as methanol emulsion for solubility of 1 mg/liter	1,000	0.00001	Gas chromatography	Off-site aqueous fluid analysis
	Meter and vaporize solution	1,000	0.001		As gaseous fluid

to trace the intrusion of borehole slurry into cores. Bromide or other solute tracers as well as microbiological or inorganic particulate tracers can be prepared as a concentrated solution and added to the borehole by using a bailer. For example, at a Department of Energy field research site in New Mexico where wireline rotary coring was being used, a saturated LiBr salt was added directly to the mud tank during makeup of the drilling muds. This resulted in a concentration of 100 to 200 mg of Br$^-$ per liter in the circulating drill muds that was maintained throughout drilling and coring operations. Biological tracers can be introduced by similar means.

When the vadose zone is sampled, the need for tracers in borehole fluids is obviously eliminated. However, there is potential for contamination from particulates that fall into the borehole from overlying sediments and soil and from sidewall slough or drilling tools. These particles can penetrate subsurface materials while hammering advances the core barrel. A dry tracer mix consisting of glass beads as a bulking agent, fluorescent microspheres as a bacterial tracer, and tungsten carbide particles as a chemical tracer was developed for tracing particulate intrusion into vadose samples during cable tool coring (20). This mixture was placed at the bottom of the borehole by using a bailer with a check valve at the bottom that released the tracer mix upon contact with the bottom of the borehole (Fig. 8). In

core samples collected from the vadose zone, the concentration of fluorescent microspheres was found to increase as the number of hammer strikes required to advance the core barrel increased (18a).

Rotary drilling methods generally have greater potential for introducing contaminants into cores than do hollow-stem auger and cable tool coring methods because of the need for circulating drilling fluids and the greater depths at which these methods are generally used. Since the volumes of mud required for rotary drilling can exceed several hundred or even thousands of gallons, it is not economically feasible to mix fluorescent microspheres directly into drilling fluids. In contrast, microorganisms can easily and economically be cultured in large quantities and introduced directly into drilling fluids. Chemical tracers can also be introduced directly into the fluid. Because of the potential for loss of drill muds and dilution with groundwater, it is necessary to monitor the concentrations of tracers in the muds and adjust them accordingly. State and local regulatory agencies should be consulted prior to introduction of microbiological or chemical tracers.

One approach to using microspheres as biological tracers has been to deploy them at or near the interface between the coring tool and the formation. The microspheres can be placed in a Whirl-pak bag with the wire closures wrapped around the base of a core catcher in the shoe of a coring

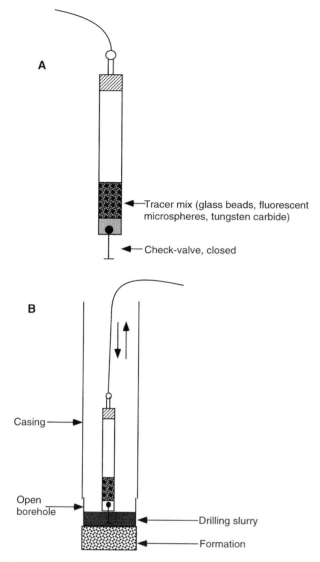

FIGURE 8 (A) Diagram of bailer with check valve for introducing tracers into a borehole for vadose- and saturated-zone coring. (B) The bailer is lowered to the bottom of the borehole on a cable, and the dry tracer is released, covering the exposed sediment.

tool (Fig. 1). As the core enters the shoe, the bag ruptures, releasing the microspheres. Another approach for introducing biological and dissolved tracers is to add them to an aqueous solution that is placed inside the core barrel liner (Fig. 1). This requires the use of a plug or marker that seals the bottom of the liner to prevent mud or borehole slurry from moving up into the core barrel and liner during the trip down the hole. As the core enters the barrel, the plug rises up the inside of the liner and the tracer is released through the check valve at the top of the core barrel. These approaches are effective for assessing the invasion of drilling fluid into cores but will not detect microbial contamination that occurs as the result of invasion of drilling fluid into the formation.

For drilling with air or other gases, perfluorocarbons can be used as tracers to evaluate the intrusion of gas into core samples. To introduce small quantities of perfluorocarbons

into the gas stream, perfluorocarbon can be metered with a high-pressure liquid chromatography pump and then vaporized by heating and injected directly into the circulated gas before it enters the borehole. Alternatively, for mud-rotary coring, the perfluorocarbon can be metered directly into the drilling-fluid line as a liquid.

Tracer Sampling

A critical aspect of using tracers to assess the quality of subsurface samples is collection and analysis of samples that allow for evaluation and interpretation of the results to determine if samples are representative of the subsurface. In addition to collecting pared core material for tracer analyses, it is important, where applicable, to collect samples of borehole slurry. This allows for quantification of the extent of contamination. Given that the tracer may not be distributed evenly along the outside of the core, it may also be necessary to sample the parings and pared core from different locations along the length of the core.

When live organisms are used as tracers or community-level biological analyses, it is necessary to prevent or limit microbial changes after sampling by performing the analyses immediately or by immediately freezing the samples, if this is acceptable. When inert particles such as fluorescent microspheres are used as tracers, biological changes after sampling are not a concern, although microspheres can decompose with time. Except for perfluorocarbons, no special precautions are required for samples to be analyzed for chemical tracers. Because of the volatile nature of perfluorocarbons, it is necessary to immediately add methanol to samples in a sealed container to preserve them for analysis.

The type of tracers and how they are deployed depend on the goals of the sampling, the types of lithologies being sampled, the type and source(s) of contaminants being traced, and the drilling and coring methods being used. These factors should be carefully considered prior to drilling and sampling to select the types of tracers and methods for introducing them that are best suited for the sampling requirements. Evaluations of tracer introduction, sampling, and analyses should generally be conducted prior to sample collection to ensure that the methods are working satisfactorily. For example, if microspheres are introduced properly during wireline coring, they should be detectable in the drilling fluids and on the outside and ends of the cores. Since it is generally neither practical nor economical to introduce tracers such as microspheres continuously during drilling, they are generally introduced only during coring for the purpose of collecting samples for microbiological analysis. Alternatively, if rotary drilling with circulating muds is being used to advance the borehole, Br^- maintained at concentrations of 100 to 200 mg liter^{-1} in the circulating fluids is effective for tracing drilling fluid penetration in advance of the borehole. In summary, tracers can provide valuable information on sample integrity when used properly.

Coring Procedures

Sample processing is typically conducted under an inert atmosphere to minimize chemical or biological changes in the cored materials. The main concern regarding chemical changes resulting from exposure to air is oxidation of reduced species such as Fe(II) and sulfides, inactivation of anaerobic and microaerophilic microorganisms, and drying of the sample. Stimulation of biological activities of the sample, which can occur as a result of coring disturbances (21, 24, 25, 29), is reduced by excluding potential electron donors and acceptors. The rate and extent of changes occurring in samples prior to analysis are minimized by processing subsurface materials under an inert atmosphere. An alternative to processing cores in the field that limits their exposure to air is to pare and segment the cores and then place them in containers and flush with an inert gas. Samples can then be transported to a laboratory for further processing.

Depending on the properties of the subsurface materials and the analyses and experiments to be conducted, it may be desirable to process cores in the field for several reasons. Timely paring and separation of representative core material from drilling fluids, surface contaminants, bottom-hole slough, and wall materials limit contamination. Furthermore, on-site processing also enables a rapid evaluation of sample quality and allows for modification of drilling procedures. For on-site processing of subsurface cores, a chamber similar to the modified glove bag designed and built by Coy (Ann Arbor, Mich.) (15) can be used.

At the field laboratory, liners containing cored materials are placed into the transfer tube, which is then flushed and evacuated for three or more cycles with an inert gas such as N_2 or Ar to displace air. After sufficient flushing, the interior airlock door is opened and materials are passed into the processing chamber. To achieve a completely anoxic atmosphere, it would be necessary to include H_2 in the gas mix, a desiccant, and a palladium catalyst. Desirability of a completely anoxic processing system must be weighed against the possible artificial stimulation of anaerobic bacteria by H_2. A slight positive gas pressure within the chamber helps to maintain anoxia.

Depending on the type of material cored, type of drilling fluid, and composition of the core liner, various techniques may be used for removing cores from liners. If the cores are short (<75 cm) and materials are consolidated, it may be possible to extrude the core from an intact liner or to section the core laterally into discrete sections prior to extrusion. Alternatively, if the cores are consolidated and the liner is Lexan or polycarbonate, it may be advantageous to score the liner on each side (along the core length) with utility knives. Core liners can be cut by using circular saws equipped with carbide-tipped blades. Although electrical outlets can be built into the chamber for running power tools, etc., caution is needed since moisture can build up over time, resulting in a potential shock hazard. Battery-operated handheld saws can be used, thus eliminating the need for electrical outlets for this purpose within the chamber.

After the cored materials are exposed, the core can be examined and the geological characteristics can be recorded. Evaluation of core quality, such as distinguishing slough or nonnative materials from the core and identifying segments for subsampling, can be made at this time. If the core material is physically disturbed, as evidenced by mixing or drilling-induced fractures, or if there are signs of gross chemical or biological contamination such as penetration of drilling muds or rhodamine if it is used as a tracer, then the core is generally not suitable for microbiological or chemical analyses. Removal of outer core material that has the greatest potential for being contaminated should be standard practice, especially for low-biomass subsurface materials. Drilling fluids are typically most concentrated at the ends and outermost edges of the core. These regions must be carefully pared with flame-sterilized or autoclaved tools. The amount of material to be pared depends on a variety of factors, including the drilling method and permeability of the material being sampled. If a visible tracer such as rhodamine was used, the distance of penetration can be used as a guide for

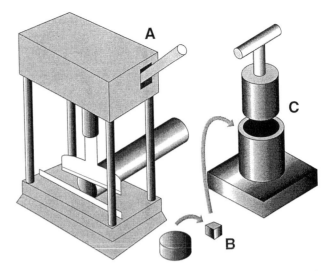

FIGURE 9 Diagram of hydraulic core splitter (A) and rock crusher (C) used for sampling consolidated subsurface materials. Careful paring of the discs produced by splitting of the core is required to produce rock fragments (B) from which outer, potentially contaminated materials have been removed. (Reprinted from reference 15 with kind permission of Elsevier Science-NL, Sara Burgerhartstraat 25, 1055 KV Amsterdam, The Netherlands.)

paring. Obviously, it is better to be conservative and to pare more than necessary than to risk including contaminated material in the sample. Subsamples of drilling fluids, outer portions of the core, and representative materials from inside the core should be collected and analyzed for tracers.

Most unconsolidated sediments are readily pared with sterile knives, spoons, spatulas, or chisels. Consolidated sediments or rocks may require use of hydraulic splitters to pare outer, potentially contaminated portions of a core from the representative materials in the center of the core. Hydraulic core splitters can be used to section a core into 2- to 5-cm-thick discs (Fig. 9A), which can be subsequently pared with a core splitter or a sterile chisel (Fig. 9B). Care must be taken to not contaminate inner material with outer core parings during this process. Between uses, the blades of the hydraulic splitters should be cleaned to remove particulates and fluids and disinfected. Pared materials can be further disaggregated by crushing in a sterilize Plattner mill (Fig. 9C). Careful consideration should also be given to subsampling and to whether sediments should be homogenized, especially if samples are being collected for multiple analyses. As there can be considerable vertical and horizontal heterogeneity in the sediment properties, samples collected only centimeters apart can have considerably different microbiological, geochemical, and physical properties.

Sample Transport and Disbursement

In the packaging chamber, samples are segregated, transferred to sterile Whirl-pak bags, and placed in sealable canning jars. To minimize exposure of sample to O_2 during transport, jars should be flushed with O_2-free N_2 or Ar via a cannula prior to sealing. This can be accomplished inside the anaerobic chamber. A novel approach for storage and transport of anaerobic sediment samples has been described by Cragg et al. (16). This approach uses heat-sealable "wine bags" consisting of a thin nylon-aluminum-polyethylene laminate with a thick inner polyethylene liner. Core segments or subsamples can be placed in these bags with a commercially available O_2-absorbing system used for culturing anaerobic bacteria and then sealed. Samples placed in bags in this manner can remain anaerobic for months. Sealed samples can then be transported to laboratories or shipped by overnight express carrier with cold packs to keep samples cool during transport. Although in situ temperatures may be above ambient air temperatures, the inclusion of cold packs is necessary to protect samples from elevated temperatures that can occur during transport. Alternatively, for some microbiological analyses that involve analysis of macromolecules such as lipids and nucleic acids, samples should be frozen immediately for processing at a later time. Samples for such analyses can be frozen on site in a standard $-20°C$ freezer, but if a freezer is not available, dry ice or liquid nitrogen can be used. Freezing samples in dry ice or liquid N_2 is advantageous in that nucleic acids and lipids are more stable at these lower temperatures.

Summary of Coring Procedures

Contaminants can be introduced before, during, and after drilling or during sample processing. After each use, all sampling equipment should be throughly cleaned with a steam cleaner or by other means to remove debris and drilling fluid. Equipment such as the core barrel, shoe, core catchers, and core liners should all be sterilized or, at a minimum, disinfected prior to use. Core liners are typically constructed of aluminum or Lexan and can be autoclaved prior to use. Precut Lexan liners can be wrapped with paper or foil and autoclaved prior to use in the field. In the field, core liners are removed from their wrapping and are placed directly in the core barrel. Retainer rings, core plugs, and core catchers are added if necessary, and the cutting shoe is screwed onto the end of the core barrel. Bags containing tracers such as microspheres are often placed inside the cutting shoe (Fig. 1 and 5), and the bottoms of cutting shoes may be covered with vinyl or plastic to limit fluid movement into the sampling barrel during descent into the borehole. After assembly, the sampler can be filled with sterile water or fluids containing tracers and transported to the drill rig for attachment to the coring assembly. During coring, the protective covering on the cutting shoe breaks and the core enters the liner and breaks the tracer bag, releasing tracers into the surrounding environment. After the core is retrieved, it should be capped to limit exposure of the core ends and transferred to an on-site processing facility or to the laboratory. Cores can then be put into a glove bag or similar chamber to be processed.

GROUNDWATER SAMPLING

Assuming proper drilling and installation, groundwater samples can be collected from wells that are useful for microbiological analysis. However, there are a number of factors that must be considered prior to sampling. A major question regarding collection of groundwater samples for microbiological analysis is whether the microorganisms associated with the water are representative of the in situ populations, many of which may be attached to solids. In addition, groundwater samples, depending on the well configuration and length of screened interval, can represent water collected over a relatively large volume of the subsurface, whereas solids are typically point samples. It is recognized that vertical integration of groundwater over a screened interval can result in considerable dilution of solutes and particulates. Consequently, it is generally better to collect sam-

ples from discrete zones. To better define spatial variations in microbial transport in a sand aquifer on Cape Cod, Harvey et al. (26) used a multilevel sampler (32) that consisted of a 3.2-cm-diameter polyvinyl chloride pipe that encased multiple screened 6.5-mm-diameter polyethylene tubes that exited the pipe to the surrounding aquifer. It is important to consider these and other factors prior to sampling groundwater from wells for microbiological or geochemical analyses. The purpose of this section is to identify and discuss some of the key issues associated with groundwater sampling and describe some of the more commonly used methods. The reader should consult references 6, 34, and 45 for additional information on groundwater sampling.

Methods for Groundwater Sampling

Because the local environment adjacent to a well can be substantially different from that in the aquifer, it is usually necessary to purge the standing water from a well before collecting water for microbiological or geochemical analyses. Stagnant water in the upper part of the casing is not in contact with the formation but rather is in contact with the casing and gases in the well bore. Hence, the geochemical and microbiological properties of this water rarely reflect properties of the formation waters. Biofilms of nonnative microorganisms can build up on well casings and well screens and can detach if the pumping rate is too high. Also, sediments in and around the well screen can be contaminated during well development. A general rule is to flush at least two to three well volumes before collecting water samples to ensure that the samples are representative of the formation. It is also desirable to monitor geochemical parameters, such as E_h and pH, and pump until stable values are obtained as indicators that formation water is being obtained. Although removal of stagnant water in the well bore is critical, purging itself can introduce artifacts if not conducted properly. For example, overpumping can increase turbidity of the water, cause dilution which can change the concentration of dissolved constituents, and dewater pores, which can alter both microbiological and chemical properties.

Once the well has been purged, there are multiple methods for obtaining groundwater. These include grab devices such as bailers (Fig. 8), suction or lift devices such as peristaltic or centrifugal pumps, positive-displacement systems such as gas-driven bladder pumps and electrical submersible pumps, and inertia pumps. Factors such as depth, type of analyses, and hydrological properties all influence the choice of collection device. For collecting microbiological samples, it is desirable to use inert, autoclavable materials for those components that come in contact with the water. Pumps that cannot be autoclaved can be sanitized by pumping a solution of disinfectant through them prior to use. The most common method of disinfection is chlorination because of its ease of use.

Actual collection of groundwater for microbiological analyses is relatively straightforward. The easiest approach is simply to flush and fill clean, sterile glass bottles with groundwater directly from the source tubing. For sampling anaerobic groundwater, it is important to fill the bottles completely to eliminate any air in the headspace that may oxidize reduced chemical species or inhibit strict anaerobes. When a suction-lift pump is used, a sterile vacuum flask can be used to aseptically collect groundwater samples while limiting exposure of the sample to air (34).

These methods are generally applicable for the collection of samples for the study of bacteria, protozoa, or viruses. For example, Harvey et al. have used stainless steel submersible

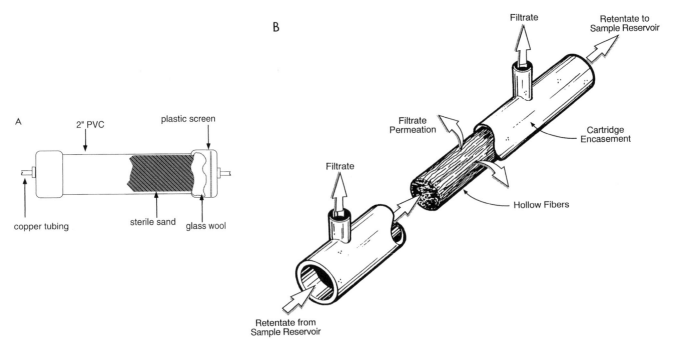

FIGURE 10 (A) Diagram of sand-trap sampling apparatus used to collect bacteria from the flow from an artesian well (reprinted from reference 42 with permission of the publisher). PVC, polyvinyl chloride. (B) Cutaway view of hollow-fiber tangential flow filter (reprinted, in part, from reference 31 with permission of the publisher).

pumps to collect groundwater samples suitable for studying the transport of bacteria (26) and protozoa (27) in a shallow, sandy aquifer on Cape Cod. Because the concentrations of viruses in groundwater are often quite low, their study often requires concentration by filtration. For example, Abbaszadegan et al. (1) used an electropositive cartridge filter to collect viruses from groundwater, which were subsequently eluted in a beef extract solution.

Stevens et al. (42) used a polyvinyl chloride cartridge packed with sand (Fig. 10A) that was connected in-line to the discharge line from an artesian well to select for bacteria that would attach to the sand surfaces during investigation of the microbiological properties of a deep, confined aquifer. The composition of the anaerobic microbiological community that developed on the sand particles was consistent with the geochemical properties of the well, indicating that this approach was effective for sampling bacteria that were a natural component of the aquifer. An alternative to the sand-cartridge method was used by Pedersen and Ekendahl (37) to study microbial communities in deep crystalline bedrock at a site in Sweden. Biofilm reactors containing slides, previously baked in a muffle furnace at 425°C for 4 h to remove organic particles, were used as the substratum for biofilm attachment.

Because the density of bacteria in groundwater is often low, between 10^3 and 10^5 cells ml^{-1} (28, 42), it may be desirable to concentrate the biomass for certain microbiological analyses. Kuwabara and Harvey (31) used a hollow-fiber, tangential-flow filtration system (Fig. 10B) to concentrate bacteria from groundwater. Fry et al. (22) also used hollow-fiber filtration to concentrate bacteria from groundwater so that nucleic acids could be extracted for analysis of the population by using nucleic acid probes of rRNA and sequence analysis of cloned 16S rRNA genes. Similarly, White et al. (43a) used tangential-flow filtration to concen-

trate groundwater biomass for lipid analyses during bioremediation.

While the recovery of bacteria from groundwater by hollow-fiber filtration can be efficient (>90%), especially when backpulsing is applied, there are a number of inherent problems and precautions. Hollow-fiber filters are not 100% efficient, and there will be some loss of cells during filtration. There may also be a tendency to lose more small-size cells than large-size cells. Another precaution is that filters need to be thoroughly cleaned and disinfected between uses to ensure that one is sampling groundwater bacteria and not bacteria attached to surfaces within the filter. A variety of chemical methods can be used to clean and disinfect the filter cartridges, including those recommended by the manufacturer. A variety of process control samples can also be collected to assess sampling procedures. Although most manufacturers do not guarantee the integrity of the filters after autoclaving, the materials withstand repeated autoclaving, and the integrity of the cartridges holds up well (31).

SUMMARY

This chapter provides background information and general guidelines necessary for obtaining samples from the subsurface environment for microbiological and geochemical analyses. It is recognized that not all microbiological investigations of the subsurface environment will be able to implement all of the procedures discussed in this chapter. However, by understanding the advantages and limitations of the various sampling technologies, by using tracers, and by employing rigorous procedures for sample processing, representative subsurface samples of defensible quality for scientific investigation and monitoring can be obtained.

REFERENCES

1. **Abbaszadegan, M., M. S. Huber, C. P. Gerba, and I. L. Pepper.** 1993. Detection of enteroviruses in groundwater with the polymerase chain reaction. *Appl. Environ. Microbiol.* **59:**318–1324.

2. **Adrian, N. R., J. A. Robinson, and J. M. Suflita.** 1994. Spatial variability in biodegradation rates as evidenced by methane production from an aquifer. *Appl. Environ. Microbiol.* **60:**3632–3639.

3. **Amy, P. S., D. L. Haldeman, D. Ringelberg, D. H. Hall, and C. Russell.** 1992. Comparison of identification systems for classification of bacteria isolated from water and endolithic habitats within the deep subsurface. *Appl. Environ. Microbiol.* **58:**3367–3373.

4. **Balkwill, D. L.** 1989. Numbers, diversity, and morphological characteristics of aerobic, chemoheterotrophic bacteria in deep subsurface sediments from a site in South Carolina. *Geomicrobiol. J.* **7:**33–51.

5. **Balkwill, D. L., and W. C. Ghiorse.** 1985. Characterization of subsurface bacteria associated with two shallow aquifers in Oklahoma. *Appl. Environ. Microbiol.* **50:**580–588.

6. **Barcelona, M. J., J. P. Gibb, J. A. Helfrich, and E. E. Garske.** 1986. *Practical Guide to Groundwater Sampling.* Contract report 374. Illinois State Geological Survey, Springfield.

7. **Beeman, R. E., and J. M. Suflita.** 1989. Evaluation of deep subsurface sampling procedures using serendipitous microbial contaminants as tracer organisms. *Geomicrobiol. J.* **7:**223–233.

8. **Bjornstad, B. N., J. P. McKinley, T. O. Stevens, S. A. Rawson, J. K. Fredrickson, and P. E. Long.** 1994. Generation of hydrogen gas as a result of drilling within the saturated zone. *Ground Water Monit. Remediation* **Fall:**140–147.

9. **Boone, D. R., Y. Liu, Z.-J. Zhao, D. L. Balkwill, G. R. Drake, T. O. Stevens, and H. C. Aldrich.** 1995. *Bacillus infernus* sp. nov., an Fe(III)- and Mn(IV)-reducing anaerobe from the deep terrestrial subsurface. *Int. J. Syst. Bacteriol.* **45:**441–448.

9a. **Brockman, F. J.** Unpublished data.

10. **Brockman, F. J., T. L. Kieft, J. K. Fredrickson, B. N. Bjornstad, S. W. Li, W. Spangenburg, and P. E. Long.** 1992. Microbiology of vadose zone paleosols in south-central Washington state. *Microb. Ecol.* **23:**279–301.

11. **Chapelle, F. H.** 1993. *Ground-Water Microbiology and Geochemistry.* John Wiley & Sons, Inc., New York.

12. **Chapelle, F. H., and P. B. McMahon.** 1991. Geochemistry of dissolved inorganic carbon in a coastal plain aquifer. 1. Sulfate from confining beds as an oxidant in microbial CO_2 production. *J. Hydrol.* **127:**85–108.

13. **Chapelle, F. H., J. T. Morris, P. B. McMahon, and J. L. Zelibor.** 1988. Bacterial metabolism and the del-13C composition of groundwater, Floridian aquifer, South Carolina. *Geology* **16:**117–121.

14. **Clark, R. R.** 1988. A new continuous sampling wireline system for acquisition of uncontaminated minimally disturbed soil samples. *Ground Water Monit. Rev.* **8**(4):66–72.

15. **Colwell, F. S., G. J. Stormberg, T. J. Phelps, S. A. Birnbaum, J. McKinley, S. A. Rawson, C. Veverka, S. Goodwin, P. E. Long, B. F. Russell, T. Garland, D. Thompson, P. Skinner, and S. Grover.** 1992. Innovative techniques for collection of saturated and unsaturated subsurface basalts and sediments for microbiological characterization. *J. Microbiol. Methods* **15:**279–292.

16. **Cragg, B. A., S. J. Bale, and R. J. Parkes.** 1992. A novel method for the transport and long-term storage of cultures and samples in an anaerobic atmosphere. *Lett. Appl. Microbiol.* **15:**125–128.

17. **Driscoll, F. G.** 1986. Well drilling methods, p. 268–339. In F. G. Driscoll (ed.), *Groundwater and Wells,* 2nd ed. Johnson Filtration Systems, St. Paul, Minn.

18. **Fournelle, H. J., E. K. Day, and W. B. Page.** 1957. Experimental ground water pollution at Anchorage, Alaska. *Public Health Rep.* **72:**203–209.

18a. **Fredrickson, J. K.** Unpublished data.

19. **Fredrickson, J. K., D. L. Balkwill, G. R. Drake, M. F. Romine, D. B. Ringelberg, and D. C. White.** 1995. Aromatic-degrading *Sphingomonas* isolates from the deep subsurface. *Appl. Environ. Microbiol.* **61:**1917–1922.

20. **Fredrickson, J. K., F. J. Brockman, B. N. Bjornstad, P. E. Long, S. W. Li, J. P. McKinley, J. V. Wright, J. L. Conca, T. L. Kieft, and D. L. Balkwill.** 1993. Microbiological characteristics of pristine and contaminated deep vadose sediments from an arid region. *Geomicrobiol. J.* **11:**95–107.

21. **Fredrickson, J. K., S. W. Li, F. J. Brockman, D. L. Haldeman, P. S. Amy, and D. L. Balkwill.** 1995. Time-dependent changes in viable numbers and activities of aerobic heterotrophic bacteria in subsurface samples. *J. Microbiol. Methods* **21:**253–265.

22. **Fry, N. K., J. K. Fredrickson, and D. A. Stahl.** 1994. Microbial ecology of deep anaerobic alkaline aquifers, abstr. N-150, p. 342. In *Abstracts of the 94th General Meeting of the American Society for Microbiology.* American Society for Microbiology, Washington, D.C.

23. **Ghiorse, W. C., and J. T. Wilson.** 1988. Microbial ecology of the terrestrial subsurface. *Adv. Appl. Microbiol.* **33:**107–173.

24. **Haldeman, D. L., P. S. Amy, D. C. White, and D. B. Ringelberg.** 1994. Changes in bacteria recoverable from subsurface volcanic rock samples during storage at 4°C. *Appl. Environ. Microbiol.* **60:**2697–2703.

25. **Haldeman, D. L., S. A. Penny, D. Ringelberg, D. C. White, R. E. Garen, and W. C. Ghiorse.** 1995. Microbial growth and resuscitation after community structure after perturbation. *FEMS Microbiol. Ecol.* **17:**27–37.

26. **Harvey, R. W., L. H. George, R. L. Smith, and D. R. LaBlanc.** 1989. Transport of microspheres and indigenous bacteria through a sandy aquifer: results of natural- and forced-gradient tracer experiments. *Environ. Sci. Technol.* **23:**51–56.

27. **Harvey, R. W., N. E. Kinner, A. Bunn, D. Macdonald, and D. Metge.** 1995. Transport behavior of groundwater protozoa and protozoan-sized microspheres in sandy aquifer sediments. *Appl. Environ. Microbiol.* **61:**209–217.

28. **Hazen, T. C., L. Jimenez, G. L. de Victoria, and C. B. Fliermans.** 1991. Comparison of bacteria from deep subsurface sediment and adjacent groundwater. *Microb. Ecol.* **22:**293–304.

29. **Hirsch, P., and E. Rades-Rohkolh.** 1988. Some special problems in the determination of viable counts of groundwater microorganisms. *Microb. Ecol.* **16:**99–113.

30. **Kieft, T. L., J. K. Fredrickson, J. P. McKinley, B. N. Bjornstad, S. A. Rawson, T. J. Phelps, F. J. Brockman, and S. M. Pfiffner.** 1995. Microbiological comparisons within and across contiguous lacustrine, paleosol, and fluvial subsurface sediments. *Appl. Environ. Microbiol.* **61:**749–757.

31. **Kuwabara, J. S., and R. W. Harvey.** 1990. Application of a hollow-fiber, tangential-flow device for sampling suspended bacteria and particles from natural waters. *J. Environ. Qual.* **19:**625–629.

32. **LeBlanc, D. R., S. P. Garabedian, K. M. Hess, L. W. Gelhar, R. D. Quadri, K. G. Stollenwerk, and W. W. Wood.** 1991. Large-scale natural gradient test in sand and gravel, Cape Cod, Massachusetts. *Water Resour. Res.* **27:**895–910.

33. **Lehman, R. M., F. S. Colwell, D. Ringelberg, and D. C. White.** 1995. Combined microbial community-level analyses for quality assurance of terrestrial subsurface cores. *J. Microbiol. Methods* **22:**263–281.

34. **McNabb, J. F., and G. E. Mallard.** 1984. Microbiological sampling in the assessment of groundwater pollution, p. 235–260. *In* G. Bitton and C. P. Gerba (ed.), *Groundwater Pollution Microbiology.* John Wiley & Sons, Inc., New York.

35. **Murphy, E. M., J. A. Schramke, J. K. Fredrickson, H. W. Bledsoe, A. J. Francis, D. S. Sklarew, and J. C. Linehan.** 1992. The influence of microbial activity and sedimentary organic carbon on the isotope geochemistry of the Middendorf Aquifer. *Water Resour. Res.* **28:**723–740.

36. **Pedersen, K., and S. Ekendahl.** 1990. Distribution and activity of bacteria in deep granitic groundwaters of southeastern Sweden. *Microb. Ecol.* **20:**37–52.

37. **Pedersen, K., and S. Ekendahl.** 1992. Assimilation of CO_2 and introduced organic compounds by bacterial communities in groundwater from Southeastern Sweden deep crystalline bedrock. *Microb. Ecol.* **23:**1–14.

38. **Phelps, T. J., C. B. Fliermans, T. R. Garland, S. M. Pfiffner, and D. C. White.** 1989. Recovery of deep subsurface sediments for microbiological studies. *J. Microbiol. Methods* **9:**267–280.

39. **Russell, B. F., T. J. Phelps, W. T. Griffin, and K. A. Sargent.** 1992. Procedures for sampling deep subsurface microbial communities in unconsolidated sediments. *Ground Water Monit. Rev.* **12:**96–104.

40. **Smith, R. L., R. W. Harvey, and D. R. Leblanc.** 1991. Importance of closely-spaced vertical sampling in delineating chemical and microbiological gradients in groundwater studies. *J. Contam. Hydrol.* **7:**285–300.

41. **Stevens, T. O., and B. S. Holbert.** 1995. Variability and density dependence of bacteria in terrestrial subsurface samples: implications for enumeration. *J. Microbiol. Methods* **21:**283–293.

42. **Stevens, T. O., J. P. McKinley, and J. K. Fredrickson.** 1993. Bacteria associated with deep, alkaline, anaerobic groundwaters in southeast Washington. *Microb. Ecol.* **25:**35–50.

43. **Vroblesky, D. A., and F. H. Chapelle.** 1994. Temporal and spatial changes of terminal electron-accepting processes in a petroleum hydrocarbon-contaminant aquifer and the significance for contaminated biodegradation. *Water Resour. Res.* **30:**1561–1570.

43a. **White, D. C., et al.** Personal communication.

44. **Wilson, J. T., J. F. McNabb, D. L. Balkwill, and W. C. Ghiorse.** 1983. Enumeration and characterization of bacteria indigenous to a shallow water-table aquifer. *Ground Water* **21:**134–142.

45. **Wilson, N.** 1995. *Soil Water and Ground Water Sampling.* CRC Press, Inc., Boca Raton, Fla.

46. **Wood, W. W., and G. G. Ehrlich.** 1979. Use of bakers yeast to trace microbial movement in ground water. *Ground Water* **16:**398–402.

47. **Zapico, M. M., S. Vales, and J. A. Cherry.** 1987. A wireline piston core barrel for sampling cohesionless sand and gravel below the water table. *Ground Water Monit. Rev.* **7:**74–82.

Microbial Studies of Landfills and Anaerobic Refuse Decomposition

MORTON A. BARLAZ

60

INTRODUCTION

A landfill is a disposal alternative for municipal solid waste (MSW) as well as certain industrial wastes, water and wastewater treatment sludges, and agricultural residues. Despite increases in recycling and composting, approximately 62% by weight of the MSW generated in the United States in 1993 was buried in sanitary landfills (71). There is a limit to the types of waste that can be recycled or composted, and combustion is typically more expensive than a landfill. Thus, landfills will be a significant waste repository for the foreseeable future. A complex series of biological and chemical reactions begins with the burial of refuse in a landfill, and landfills represent an active anaerobic ecosystem. In 1994, methane was recovered in commercial quantities from 119 landfills in the United States and Canada (70).

The objective of this chapter is to present techniques that can be used to study the biological reactions and microorganisms involved in anaerobic refuse decomposition as it occurs in a landfill. There are a number of other microbiological processes associated with solid waste management, including (i) aerobic decomposition and pathogen destruction during composting and (ii) the survival and release of refuse-associated viruses (64, 65). These processes are beyond the scope of this chapter, which is restricted to techniques designed to study MSW decomposition. However, many of the principles and techniques associated with sample collection and processing presented here are applicable to these other processes. A more thorough review of solid waste microbiology, which includes composting systems and viruses, was recently published (47).

This chapter emphasizes techniques as they have been applied to the study of MSW, which, as defined by the U.S. Environmental Protection Agency, includes residential, commercial, and nonhazardous industrial waste but excludes combustion ash, hazardous waste, sludges, and industrial process waste. The principles discussed in this chapter could be applied to these other waste types, which, with the exception of hazardous waste, are frequently buried in the same landfills that receive MSW.

This chapter begins with a brief description of the major components of a sanitary landfill followed by a discussion of MSW composition. Next, the manner in which cellulosic substrates are converted to methane and carbon dioxide is described. This section also discusses factors that limit de-composition in landfills. The next section presents information on systems that can be used to simulate refuse decomposition and techniques that can be used to measure refuse biodegradation. The remainder of the chapter is dedicated to presentation of techniques used to measure microbial activity, including viable cell counts, enzyme assays, and measurement of instantaneous rates of gas production.

Description of a Sanitary Landfill

While landfills have historically been the dominant alternative for MSW disposal, there has been substantial evolution in their design and operation. In the past, a landfill often represented little more than an open hole or marsh where refuse was dumped. The refuse was often not covered properly, sometimes it was burned for volume reduction, and there was little effort to control storm water runoff and downward migration of water that had come into contact with the refuse (leachate). With the implementation of subtitle D of the Resource Conservation and Recovery Act and regulations in many states, landfills have become highly engineered facilities designed to contain the refuse and separate it from the environment, capture leachate, and control gas migration.

The major components of a landfill are briefly summarized here; detailed design information is available elsewhere (21, 30, 54, 68). Prior to placement of waste in the ground, a site is typically excavated to increase the disposal volume-to-surface area ratio. The lowest component of a landfill is the liner system, which includes layers (i) to minimize the migration of leachate to the groundwater and (ii) to collect leachate for treatment. A common system used to minimize leachate migration begins with a layer of low-permeability soil, typically a 0.67- to 1-m-thick clay layer with a hydraulic conductivity of no more than 10^{-7} cm/s. A flexible membrane liner (FML) is often placed above the clay layer. The FML is typically 1.5-mm-thick polyethylene with an equivalent hydraulic conductivity (based on vapor diffusion) of about 10^{-12} cm/s (22). Together, the soil and FML are referred to as a composite liner. A drainage layer designed to promote the collection of leachate is placed above the composite or clay liner. This layer consists of sand with a conductivity of at least 10^{-2} cm/s. Slotted pipe is placed in the sand layer at intervals of 30 to 60 m to collect leachate and route it to a treatment system. A protective barrier is then installed above the leachate collection system to pro-

TABLE 1 Organic composition of residential refuse

Compound(s)	% of dry wt determined in reference:			
	6[a]	4	59	13
Cellulose	51.2	28.8	38.5	48.2
Hemicellulose	11.9	9.0	8.7	10.6
Lignin	15.2	23.1	28.0	14.5
Volatile solids	78.6	75.2	Not measured	71.4

[a] The following additional determinations were made for this sample: protein, 4.2%; soluble sugars, 0.35%; starch, 0.6%; and pectin, <3%.

tect it from the equipment used to place and compact the refuse. Refuse may then be placed above the protective layer. The waste is compacted and covered daily to minimize the attraction of rodents, wildlife, and disease-carrying insects; the blowing of the refuse away from the landfill; and contamination of storm water runoff. A 15-cm soil layer is typically used for daily cover. Synthetic materials that are rolled over the waste at the end of the working day and removed prior to burial of additional refuse are also available. The type of daily cover used is important because landfill samples may include substantial quantities of soil. Once refuse has reached the final design grade of the landfill, a final cover is applied. At a minimum, this cover includes a layer of low-permeability soil designed to minimize storm water infiltration, overlaid by a layer that will support vegetative growth. Vegetation serves to minimize erosion of the final cover and to promote evapotranspiration. Methane and carbon dioxide are the major terminal products of anaerobic decomposition in landfills. These and other gases are typically vented through wells placed in the refuse to minimize their migration off site. The gas may be vented, flared to reduce the release of methane, or recovered for use as energy.

Refuse Composition

Traditionally, MSW has been classified according to visual categories such as glass, paper, metals, etc., as determined from published data (71). However, the organic composition of MSW is more useful for biodegradation studies. Data on the organic composition of refuse are presented in Table 1. These data indicate that cellulose and hemicellulose are the principal biodegradable components of MSW. The other major organic component of MSW, lignin, is recalcitrant under anaerobic conditions (79). In addition, lignin interferes with the decomposition of cellulose and hemicellulose by physically impeding microbial access to these degradable carbohydrates. Other biodegradable organics present in smaller concentrations are protein and soluble sugars. Protein concentrations are typically obtained by multiplication of the organic nitrogen content by 6.25 (2). However, there are likely to be other forms of organic nitrogen in refuse that may be less degradable than protein, such as nitrogen-containing humic compounds. Thus, the reported protein concentrations are likely an upper limit.

BIOLOGICAL DECOMPOSITION IN A LANDFILL

The decomposition of MSW to methane in sanitary landfills is a microbially mediated process that requires the coordinated activity of several trophic groups of bacteria. In the first part of this section, the general pathway for anaerobic decomposition, as it has been documented to occur in other anaerobic ecosystems, is reviewed. Then, refuse decomposi-

tion is characterized in four phases defined by different gas and leachate characteristics and microbial population development.

The Microbiology of Anaerobic Decomposition

Several trophic groups of anaerobic bacteria are required for the production of methane from biological polymers such as cellulose, hemicellulose, and protein (Fig. 1) (15, 80). The first reaction (referred to as hydrolysis in Fig. 1) is the hydrolysis of polymers such as carbohydrates, fats, and proteins. The initial products of polymer hydrolysis are soluble sugars, amino acids, long-chain carboxylic acids, and glycerol. Fermentative microorganisms then ferment these hydrolysis products to short-chain carboxylic acids, ammonia, carbon dioxide, and hydrogen. Acetate, a direct precursor of methane, and alcohols are also formed. The next reaction is carried out by obligate proton-reducing (or H_2-producing) acetogens. They oxidize fermentation products, including propionate and butyrate, to acetate, carbon dioxide, and hydrogen. Oxidation of propionate and butyrate is thermodynamically favorable only at very low hydrogen concentrations (80). Thus, the obligate proton-reducing acetogenic bacteria function only in syntrophic association with a hydrogen scavenger such as a methanogen or a sulfate reducer. Typically, sulfate concentrations in landfills are minimal and methane is the major electron sink, although exceptions

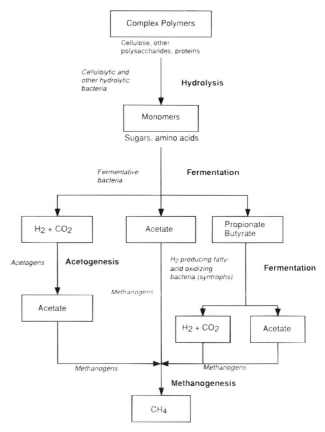

FIGURE 1 Overall process of anaerobic decomposition showing the manner in which various groups of fermentative anaerobes act together in the conversion of complex organic materials ultimately to methane and carbon dioxide. (Reprinted from reference 15 by permission of Prentice-Hall, Inc., Upper Saddle River, N.J.)

exist. The significance of another acetogenic reaction, the production of acetate from H_2 and CO_2, has not been established in the landfill ecosystem. The terminal step in the conversion of complex polymers to methane is carried out by the methanogenic bacteria. The most common methanogenic substrates are acetate and H_2 plus CO_2. Other substrates include formate, methanol, methylated amines, and methylated sulfides. Most methanogens have a pH optimum around 7 (81). Should the activity of the fermentative organisms exceed that of the acetogens and methanogens, there will be an imbalance in the ecosystem. Carboxylic acids and hydrogen will accumulate and the pH of the system will fall, thus inhibiting methanogenesis.

Microbiology of Refuse Decomposition

When refuse is placed in a landfill, biological decomposition resulting in methane formation as described above does not occur immediately. A period ranging from months to years is necessary for the proper growth conditions and the required microbiological system to become established. Thus, most research on refuse decomposition has been conducted by using laboratory simulations in which the rate of decomposition is accelerated. A characterization of refuse decomposition describing chemical and microbiological characteristics has been developed by using data from laboratory-scale reactors (9). Refuse decomposition is described as occurring in an aerobic phase, an anaerobic acid phase, an accelerated methane production phase, and a decelerated methane production phase. This description is summarized in Fig. 2.

In the aerobic phase (phase 1), both oxygen and nitrate are consumed, with soluble sugars serving as the carbon source for microbial activity. All of the trophic groups required for refuse methanogenesis (cellulolytics, acetogens, and methanogens) are present in fresh refuse, though there is little change in their populations. In the anaerobic acid phase (phase 2), carboxylic acids accumulate and the pH decreases as a result of an imbalance between fermentative activity and acetogenic and methanogenic activity. There is some cellulose and hemicellulose decomposition in phase 2. The methanogen population begins to increase, and methane is detected in the landfill gas. In phase 3, the accelerated methane production phase, there is a rapid increase in the rate of methane production to some maximum value. A methane concentration of 50 to 60% (vol/vol) is characteristic of this phase, as are a decrease in carboxylic acid concentrations, an increase in pH, little hydrolysis of solids, and increases in the populations of cellulolytic, acetogenic, and methanogenic bacteria. The accumulated carboxylic acids are the principal substrate supporting methane production in this phase. The fourth phase is termed the decelerated methane production phase. The methane concentration, pH, and cellulolytic and methanogenic populations remain at values similar to those in phase 3. Concurrently, the methane production rate decreases, the acetogenic population increases, and carboxylic acids are depleted. In addition, as carboxylic acid concentrations decrease, there is an increase in the rate of cellulose plus hemicellulose hydrolysis (Fig. 2). While acid utilization limits methane production in phases 2 and 3, solids hydrolysis limits methane production in phase 4.

There are limitations to the four-phase description of refuse decomposition presented here as it applies to full-scale landfills. First, the time required for the onset of each phase may be significantly longer than the times shown in Fig. 2. Second, gas and leachate samples from landfills may reflect a composite of refuse in several different states of decomposition, depending on the manner in which landfills are filled and leachate and gas are collected. Third, in the presence of significant sulfate concentrations, electrons would be diverted from methane production to sulfate reduction. Analysis of refuse excavated from the Fresh Kills landfill in New York indicated that construction and demolition debris accounted for up to 14% of the buried refuse volume, although the wallboard (primarily calcium sulfate) concentration was not measured (67). Later work (31) showed that inhibition of sulfate reduction resulted in increased rates of methane production in Fresh Kills landfill samples. Fourth, the presence of nitrate would stimulate denitrification and nitrogen gas production and would inhibit methanogenesis.

The presence of anaerobic protozoa in refuse excavated from landfills has been documented (28, 29). Many of the protozoa present in the excavated refuse contained symbiotic methanogenic bacteria that utilize hydrogen released by the host's hydrogenosomes. The dominant protozoan isolated from the samples was the ciliate *Metopus palaeformis*. The impact of *M. palaeformis* on methane production from refuse was evaluated in later work, and there was no evidence to suggest that protozoa were stimulating refuse decomposition through enhanced nutrient recycling (27). By using oligonucleotide probes, it was shown that the methanogenic symbiont was similar but not identical to *Methanobacterium formicicum* and that the symbiotic methanogen does not normally exist in refuse samples outside the host. Thus, protozoa do not appear to be a major factor in the transmission of methanogenic bacteria through aerobic environments into landfills.

While the focus here is on anaerobic decomposition, the fate of methane produced in landfills should be considered. Methane produced in landfills and not recovered for energy is not necessarily released to the environment. The potential for anaerobic methane oxidation in landfills and aerobic methane oxidation in landfill cover soils has been evaluated in recent work. Hocknull and Dalton (35) detected anaerobic methane oxidation activity in landfill leachate but not on the solid refuse. A number of electron acceptors, including nitrate, sulfate, fumarate, and iron(III), enhanced anaerobic methane oxidation, and additional experiments suggested that sulfate reducers played a key role in methane oxidation. However, the significance of anaerobic methane oxidation to overall methane production from landfills could not be assessed. Aerobic methane oxidation activity in landfill cover soil has been demonstrated in a number of studies (38, 41, 75) and may play a role in minimization of methane emissions from landfills that lack gas control.

Factors Limiting Decomposition in Landfills

Numerous researchers have tried to enhance refuse methanogenesis by manipulation of the landfill ecosystem. A number of factors, including moisture content and moisture flow, pH, particle size, inoculum addition, nutrient concentrations, and temperature, have been shown to influence the onset and rate of methane production. All of these variables will affect the environmental conditions present in a landfill, and these conditions in turn control microbial growth. This research has been summarized (7), and the two variables that appear to be most critical in controlling refuse methanogenesis are moisture content and pH. Adequate moisture and a pH about neutral are required for refuse methanogenesis. In addition, high moisture will promote the dissolution and mixing of soluble substrates and nutrients and will also provide a mechanism for microbial

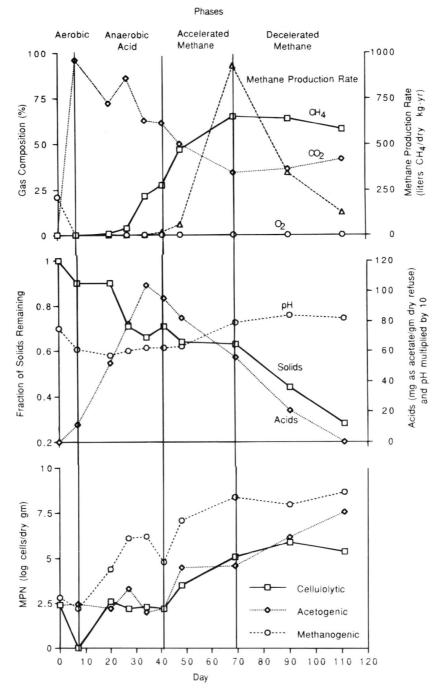

FIGURE 2 Summary of observed trends in refuse decomposition with leachate recycle. Gas volume data were corrected to dry gas at standard temperature and pressure. The acids are expressed as acetic acid equivalents. Solids remaining is the ratio of the weight of cellulose plus hemicellulose removed from a container divided by the weight of cellulose plus hemicellulose added to the container initially. Methanogen MPN data are the log of the average of the acetate- and H_2/CO_2-utilizing populations. (Reprinted from reference 9.)

transport within a landfill. While mixed refuse contains all of the microbes required for refuse decomposition, they are undoubtedly not well distributed among all of the degradable components of refuse.

Recent work has shown a positive correlation between increasing moisture content and the methane production rate of excavated refuse samples (31, 33). Methane production from excavated refuse samples was also limited to those with a pH around neutral (67). Leachate recycle and neutralization have been shown to enhance the onset and rate of methane production in laboratory-scale tests (5, 55), although more field experience is needed to fully document their value in full-scale landfills. Leachate composition data have recently been summarized (17).

MICROBIOLOGICAL STUDIES

A discussion of techniques applicable to landfill microbiology must begin with an acknowledgment that landfills are extremely heterogeneous as a result of the nature of the waste buried therein. It is probably not possible to obtain truly representative samples from landfills. Nonetheless, by collection of numerous samples from different parts of a landfill, one can obtain some indication of the status of the landfill. Given the heterogeneity issue, some researchers have studied landfill microbiology in laboratory-scale simulations, while others have focused on field sampling. Alternate systems for the study of landfill microbiology and refuse decomposition are discussed first, followed by presentation of techniques for sample collection and processing. Next, measurements that may be used to characterize the biological state of decomposing refuse are presented. The approaches discussed include traditional microbiological measurements such as enumeration and enzyme assays, use of emerging molecular tools, and chemical measurements such as the instantaneous rate of methane production and analysis of specific organic constituents in refuse.

Systems for the Study of Refuse Decomposition and Landfill Microbiology

A number of systems have been used to study refuse decomposition and landfill microbiology, with each system typically optimized for a specific objective. If the study objective is to obtain samples of refuse in various states of decomposition, then the researcher has some flexibility as to whether to work with samples from a full-scale landfill or from a laboratory simulation. When the objective is to evaluate the impact of a parameter on refuse decomposition, then a simulation of a landfill is required. The types of measurements to be made will also influence the system used to study refuse decomposition. When researchers are interested in collection of refuse samples for measurement of specific biological parameters (enumeration, hydrolytic enzyme activity, etc.), a mechanism for representative sampling is critical. When an overall measure of decomposition such as the methane production rate will suffice, a system in which solids sampling is difficult will still be acceptable. Sampling techniques for each type of system are discussed in the following section.

The ideal landfill simulation is a field-scale landfill containing several thousand kilograms of refuse. A limited number of studies on the impact of parameters such as moisture content and sludge addition on refuse decomposition have been conducted in field-scale test cells (16, 20, 32). While a system of field-scale test cells represents a near-perfect landfill simulation, the system has its drawbacks: it is expensive, and the nature of the simulation makes it difficult to obtain representative samples and to ensure 100% recovery of the gas produced. Furthermore, obtaining the desired data required several years of monitoring. (For a discussion of difficulties with respect to sampling, see Sample Collection and Processing, below.) Even when field-scale test cells are constructed, variability can be a problem. Croft and Fawcett (20) summarized the initial monitoring period for the Brogborough test cells in the United Kingdom and reported methane production rates of 4.3, 5.8, and 6.7 m³/metric ton-year for three replicate test cells, each containing 15,000 metric tons of refuse. A later report (16) concluded that the injec-

tion of either air or water into two of the test cells resulted in increased methane production. The air was hypothesized to increase the temperature of the refuse by providing a brief period of aerobic decomposition. These field-scale test cells are providing long-term data on the rate of methane production and serve to emphasize the need to monitor field-scale cells for periods in excess of 10 years. Given the time and expense associated with field-scale cells, the majority of work on parameter assessment has been conducted by using laboratory-scale simulations described later in this section.

As an alternative to the construction of field-scale test cells, refuse samples can be collected from full-scale landfills for certain study objectives. Such samples may be used for the isolation or enumeration of bacteria, enzyme assays, measurement of the methane production rate of an excavated sample, and measurement of chemical parameters such as pH, solids (cellulose, hemicellulose, lignin), and soluble organics. A number of researchers have collected samples of refuse from full-scale landfills to assess their methane production status. In this case, an excavated sample is collected and then placed in a laboratory-scale system for further study. Ham et al. (33) placed samples in 15-cm-diameter reactors (polyvinyl chloride pipe) capable of holding about 10 kg of refuse. The relatively large reactor allows collection of a more representative sample without elimination of sample components because they will not fit in the reactor. While such large reactors may be well suited to measurement of methane production activity at the time of sampling, manipulation of a large number of variables in this large system would prove difficult. Others worked with excavated refuse samples in reactors which hold only 0.2 to 0.3 kg (31). The smaller reactors allowed for manipulation of the ecosystem by molybdate and moisture addition with confidence that the amendments were in contact with the entire reactor. The same research group has also used wide-mouth bottles fitted with large rubber stoppers (44a). The stoppers were pierced with severed Balch tubes sealed with butyl rubber stoppers and aluminum crimp seals. This system was used successfully to monitor H₂ concentrations over time. Of course, the smaller sample size may require some component selection. Work to correlate sample moisture content and pH with the rate of methane production was also conducted by using excavated refuse samples placed in reactors (25, 61). The application of more traditional microbiological measurements to excavated refuse samples is discussed later in this section.

Where a laboratory-scale simulation of a landfill is desired, reactors ranging from 2 to 3,300 liters have been used. The largest reactors were about 4.2 m long and 1 m in diameter and held about 400 kg of MSW (56). These reactors are certainly large enough to obtain representative samples of MSW. However, their size makes them difficult to control and manipulate and expensive to replicate. In addition, it would not be possible to obtain representative subsamples over time. Others have worked with 110- to 210-liter steel drums that have advantages and disadvantages similar to those of the system described above (5, 23). Work has also been conducted in 2- to 4-liter reactors containing 1 to 2 kg of shredded refuse (6, 59). In the initial use of these reactors, multiple replicate reactors were set up and were destructively sampled over time to monitor microbial population development during decomposition (9). Recently, this concept has been expanded to add a ¹⁴C-radiolabeled material to the reactor to evaluate its conversion to ¹⁴CH₄ and ¹⁴CO₂ under simulated landfill conditions (13).

The use of shredded refuse makes it possible to work with

smaller quantities of refuse and still have a representative sample. Refuse is easily obtained from a municipal refuse collection service and may be shredded in a slow-speed, high-torque shredder such as a Shred Pax AZ-7H (Wood Dale, Ill.). However, shredding refuse provides a degree of mixing not typically found in landfills. This mixing, coupled with the decreased particle size, likely enhances biodegradation.

Whether one is working with 1 or 400 kg of refuse, the time to the onset of methane production varies. In the absence of some enhancement, 10 to 20 years may be required to complete the refuse decomposition cycle. This incubation period is not practical for many experiments, and strategies to enhance decomposition in laboratory-scale systems have been developed. Steps that can be taken to enhance decomposition include (i) incubation at about 40°C, (ii) the use of shredded refuse, and (iii) the use of leachate recycle and neutralization. It has been reported that 40°C is the optimum temperature for mesophilic refuse decomposition (34). Leachate recycle and neutralization involve addition of sufficient moisture to a refuse sample during the reactor filling operation to allow for the free drainage of liquid into a liquid collection system. This liquid can then be externally neutralized and recycled through the top of a reactor. If needed, or as part of the test protocol, nutrients could also be added prior to recycling. This process will move refuse from the anaerobic acid phase to the accelerated methane production phase in 1 to 2 months.

Another technique that has been used successfully to accelerate the onset of methane production in laboratory-scale systems is the addition of a seed. A seed of well-decomposed refuse has been shown to repeatedly initiate decomposition of fresh refuse or individual components thereof (3, 4, 8). It will eliminate the anaerobic acid phase of decomposition, and methane production will begin almost immediately. Because carboxylic acids will not accumulate during the anaerobic acid phase, they will not be present in large concentrations, and the increase in the methane production rate will not be as sharp as it would be in the absence of a seed. Seed should be excavated from a landfill known to be in an active state of methane production or developed in the laboratory by decomposing fresh refuse until its methane production rate is well into the decelerated methane production phase. Prior to use, the seed should be tested for viability. This can be done by placing a subsample in a 2-liter reactor and feeding it cellobiose to verify the activity of the methanogenic consortia. The seed may be stored in the laboratory for several years at room temperature prior to use to minimize the number of times that refuse must be excavated. Ideally, the methane potential of the seed will be relatively low because the refuse is well decomposed. However, control reactors may be required to measure background methane production associated with the seed. If excavated refuse is not sufficiently well decomposed, further decomposition can be enhanced in the laboratory with leachate recycle and neutralization to develop a more decomposed refuse seed.

Work with the use of anaerobic sewage sludge as a seed has been summarized previously (7). Anaerobic sewage sludge appears to stimulate the accumulation of carboxylic acids. Thus, its use requires substantial leachate neutralization, and it is not clear that the addition of anaerobic sewage sludge decreases the lag time relative to leachate recycle and neutralization in the absence of added sludge.

Enhancement of refuse decomposition is a suitable technique to evaluate the impact of a number of variables on decomposition and leachate quality. For example, this technique has been used to evaluate the impact of heavy metal sludges (56) and a lime-stabilized sludge cover soil on refuse decomposition (59). In the absence of such enhancement, tests to assess the impact of variables would take years to complete. Work has been conducted to show that there are no significant differences in the total anaerobic population or the subpopulations of cellulolytic, hemicellulolytic, acetogenic, or methanogenic (based on acetate or H_2/CO_2 utilization) bacteria in refuse incubated with or without leachate recycle (8). However, this work did not address whether the same species of organisms were active under the two incubation conditions.

Studies in which refuse decomposition is enhanced will provide data on the relative impacts of parameters on decomposition and whether a material has the potential to biodegrade under simulated landfill conditions. Work in small reactors filled with shredded refuse represents an idealized system, and degradation rates measured in this system cannot be extrapolated to the field.

The biochemical methane potential (BMP) technique has been modified for use in studies to assess the ultimate biodegradability of refuse samples. The BMP test was originally developed to measure the anaerobic biodegradability of a soluble organic chemical in a small batch reactor such as a 125-ml serum bottle (62). It has since been adapted in various ways for measurement of the biodegradability of solid samples (46, 72). In the BMP assay, small samples of refuse are added to a serum bottle with liquid medium and an inoculum. The samples are typically dried, shredded, and then ground in a Wiley mill to pass a 1-mm screen before use in the BMP assay. The methane potential of a sample is measured after incubation for 30 to 60 days. The BMP test provides a measure of ultimate biodegradability under idealized conditions that do not simulate a landfill. One application of the BMP assesses the potential of well-decomposed refuse excavated from a landfill to produce additional methane (72). Because only a few grams of sample may be added to a bottle, the sample must be homogeneous. Techniques for development of homogeneous samples from refuse are discussed in the following section. Given a homogeneous sample, the BMP test is quite repeatable.

Anaerobic biodegradation is mediated by the activity of bacterial consortia with a wide range of activities and syntrophic interactions. Because of the syntrophic interactions, the study of individual members of anaerobic consortia has always been difficult. A multistage continuous-culture technique has been applied to the study of landfill microbiology (18, 37, 51). In this system, reactor vessels of increasing size are connected in series, and the vessels are operated as a plug flow chemostat. As the reactor size increases, the imposed dilution rate decreases. This selects for faster-growing bacteria in the top vessel and slower-growing bacteria in the subsequent vessels. This technique is reported to separate habitat domains while still permitting the overlap of activity domains required for syntrophic activity. James and Watson-Craik (37) used three vessels to segregate hydrolytic or fermentative, acidogenic, and methanogenic activities. The technique was used to study the effects of environmental conditions, including temperature and carboxylic acid concentrations, on methane production.

The multistage continuous-culture technique is unique in its ability to separate different components of anaerobic consortia. Nonetheless, the system of reactors must be inoculated with a representative consortium. In the study by James and Watson-Craik (37), consortia were enriched on cellobiose, valerate, and butyrate. These consortia may or may not represent the composition of the microbial community in a landfill. In addition, environmental conditions in the continuous-culture system may select for a community that differs from that in a landfill. The researchers recognized the need to understand the role of surfaces in their system because growth surfaces are abundant in a landfill.

In summary, there are a number of potential systems available for the study of refuse biodegradation and microbiology. The system to be used must be selected in consideration of specific research objectives and sampling requirements. The ability to obtain samples will likely play a role in selection of the appropriate landfill simulation system. While field-scale tests are most realistic, they are expensive and difficult to reproduce. Also, it is difficult to obtain representative samples from them for microbiological analyses. Laboratory-scale systems allow for more rigorous experimental control and have more flexibility with respect to sampling, replication, and monitoring. Often, the time required to conduct a laboratory study can be shortened by the use of a seed. While direct comparisons have not been made, decomposed refuse is my own seed of choice because of its proven reliability. Serum bottle assays provide more limited information but are easy to conduct and are very repeatable. The multistage culture system is unique in its ability to separate the response of syntrophic consortia to environmental variables, but its operating requirements are relatively complex.

Sample Collection and Processing

Refuse samples are required for filling reactors to begin decomposition studies, for assessment of the decomposition status of refuse undergoing decomposition, and for microbiological assessment. This section begins with information on the collection and processing of fresh refuse samples for filling reactors and then discusses work required to obtain samples from full-scale landfills. Finally, the manner in which such samples should be processed for microbiological and chemical analyses is discussed.

Fresh refuse may be obtained from refuse collection vehicles en route or upon delivery at a landfill or combustion facility. The major precaution is to verify that the refuse originated in residential areas only and does not contain unusual materials. If the refuse is to be shredded, then approximately 1,000 kg of refuse should be obtained and shredded until it has a particle size small enough that representative samples can be obtained for use in laboratory-scale systems. The slow-speed, high-torque shredder referenced above will produce refuse less than 2 cm wide by 3 cm long. Prior to shredding, some presorting may be required to remove unusual materials and materials that cannot or should not be shredded. Once shredded, the refuse should be manually mixed and repetitively quartered until 5- to 10-kg piles remain. The refuse from one pile should then be selected at random for use in reactor filling. Refuse from multiple piles may be required, depending on the mass of refuse needed.

Refuse may be excavated from a landfill either for use as a seed or to obtain samples for microbiological and chemical analyses. If the refuse is required for use as a seed, it is not critical that it be representative of the landfill but only that it carry the requisite bacteria. Generally, landfill operating records will be available to identify a section of a landfill where the refuse is at least 15 years old. Ideally, it will be known whether the landfill is producing methane. Refuse for use as a seed may be excavated with construction equip-

ment. Excavated refuse will typically require shredding to obtain a homogeneous material for filling reactors. Exposure of the excavated refuse to oxygen during excavation and shredding will not destroy the methanogenic nature of the refuse.

Representative samples are more important when refuse is to be excavated from a landfill for microbiological and chemical analyses. Given the heterogeneous nature of landfills, the best approach is to obtain multiple samples over a preselected grid pattern. Samples may be obtained either with construction equipment or, preferably, with a bucket auger that has the capability to obtain deeper samples (67). The process of bringing refuse to the surface with an auger will likely reduce the particle size sufficiently that shredding is not required. Refuse excavated with construction equipment will probably require shredding. In either case, large samples (about 1,000 kg) should be excavated and then reduced by using quartering techniques to obtain as representative a subsample as possible. Westlake (73) attempted to study microbial metabolism in samples excavated from experimental landfill test cells in the United Kingdom and concluded that 25-g samples were inadequate because of sample heterogeneity. The 25-g samples were subsamples from 2 kg of shredded refuse.

When excavated samples are to be used for microbiological analyses, there is often concern regarding the effect of sample exposure to air. The microorganisms responsible for the terminal reactions in refuse decomposition—acetogens and methanogens—are obligate anaerobes. Those responsible for the hydrolytic and fermentative processes may be facultative or obligate anaerobes. In early work to measure the methane production status of refuse excavated from landfills, elaborate precautions were taken to minimize oxygen exposure. These included covering boreholes with tents and flooding them with an inert gas. It has since been shown that exposure of refuse samples in an active state of methane production to air for 2 to 4 h does not decrease the time required for these samples to resume methane production once replaced in an anaerobic system (3, 11, 12). While practical steps to minimize air exposure are warranted, extraordinary steps are not. In a recent study, samples were placed in a series of two oxygen-impermeable plastic bags (44).

Once refuse samples have been collected, they will require further processing specific to the required biological or chemical assay. Applicable techniques are described below.

Processing Samples for Microbiological Analyses

The first step in many microbiological measurements is formation of a liquid extract from the excavated solids. Until recently, only one study had evaluated alternative techniques for inoculum formation (10). The basic technique of this study was to blend refuse in anaerobic phosphate buffer (23.7 mM, pH 7.2) for about 60 s in a 4-liter blender. The buffer was boiled under nitrogen, but a reducing agent (cysteine hydrochloride) was not added because some oxidation occurs during the inoculum formation procedure and the oxidized form of cysteine exerts toxicity. Refuse near the blender blade became quite hot, but the bulk sample temperature did not increase in the 1-min blending period. The blended material was then squeezed by hand (covered with gloves), and the squeezings were used as the inoculum. Treatments evaluated to increase the efficiency of cell extraction included prechilling the refuse at 4°C, multiple blendings and hand squeezings, and use of blended refuse prior to hand squeezing. These treatments were based on a survey of the literature on rumen microbiology, in which there has been more extensive work on the enumeration of bacteria attached to cellulosic substrates (19, 42). The additional treatments did not increase the most probable number (MPN) of cellulolytic bacteria above the population measured by blending followed by hand squeezing. The inoculum formation technique was then validated by addition of a spike of rumen fluid to refuse followed by recovery of the spiked cellulolytic bacteria, using the inoculum formation technique and MPN enumerations. The results showed no evidence that the refuse was exerting a toxic effect on the rumen cellulolytic bacteria or that these bacteria were irreversibly attaching to refuse.

The use of a blender was appropriate in the study described above because all work was conducted with shredded refuse. By using a blender, it was possible to blend approximately 100 g (dry weight) of refuse. Palmisano et al. (48) worked with 10-g samples of refuse excavated from landfills. Samples were placed in a stomacher with prereduced anaerobic buffer and mixed for approximately 1 min. Westlake and Archer (74) added 1-g samples to Hungate tubes containing anaerobic diluent and agitated the tubes to form an inoculum. The use of small samples suggests that there was some self-selection of the components of refuse that were enumerated.

Recently, bacteria present on specific constituents of refuse were enumerated to identify which components of refuse carried refuse-decomposing microorganisms into landfills (57). Constituents tested included grass, leaves, branches, food waste, and mixed refuse. In this work, sample sizes were on the order of 10 kg. Thus, it was necessary to modify the previously developed inoculum formation technique. Solid samples were placed in a large (113-liter) plastic garbage can that had been wiped with ethanol and purged with sterile argon. A measured volume of filter (0.2-μm pore size)-sterilized anaerobic phosphate buffer (23.7 mM, pH 7.2) was then added to a sample to form a slurry. Large volumes of buffer were prepared by pumping the buffer through a 0.45-μm-pore-size filter and then boiling it under N_2. The sample was then stirred by hand (covered with arm-length gloves). Next, four samples were removed and placed in a 1-liter beaker. The contents of a beaker were poured through a second person's hands (covered with gloves) into a sterile 4-liter, argon-purged Erlenmeyer flask. Finally, solids caught in the person's hands were squeezed to release excess liquid into the flask. The liquid in this flask served as the inoculum for MPN enumerations. Inocula were serially diluted in phosphate buffer (23.7 mM, pH 7.2).

To evaluate the inoculum formation technique for the 10-kg sample, measured volumes of anaerobically digested sewage sludge were spiked into measured amounts of refuse (57). The cellulolytic and acetoclastic methanogenic bacteria in refuse, refuse plus sludge, and sludge were then enumerated by the MPN technique. Three spike treatments were evaluated: (i) refuse plus a low-level spike of sludge, (ii) refuse plus a high-level spike of sludge, and (iii) refuse plus a low-level spike without exposure to refuse particulate matter. For the cellulolytic bacteria, the ratio of the measured to expected MPN ranged from 0.72 to 2.22 in the various treatments. This range of measured to expected MPNs is not significantly different from 1.0 ($P = 0.05$). The ratio of measured to expected MPNs for the acetoclastic methanogenic bacteria was not significantly different from 1.0 for the high-level spike (0.23) and the low-level spike in the absence of refuse particulate matter (0.94). However, the measured population in the low-level spike was significantly lower ($P = 0.05$) than that of the low-level spike in the absence of refuse particulate matter. This finding suggests that sludge methanogenic bacteria attached to refuse particulate matter and that this attachment interfered with their enumeration. However, given the dependencies of methanogenic bacteria on soluble substrates and cellulolytic bacteria on insoluble substrates, this conclusion seems unlikely. If attachment was indeed a problem, then low spike recoveries would have been measured for the cellulolytic bacteria as well. The near-quantitative recovery of cellulolytic and methanogenic bacteria in this experiment suggested that the inoculum formation procedure was repeatable and did not cause a reduction in viable cells. The applicability of the inoculum formation procedure to the other trophic groups that participate in refuse decomposition was assumed.

Recently, the efficiency of a stomacher for extraction of microbial cells from refuse was evaluated (44). Duplicate 20-g samples of refuse were homogenized in 180 ml of a phosphate buffered diluent for 1 min. This extraction procedure was repeated six times, and its efficiency was judged on the basis of the number of aerobic bacteria that would grow on tryptone soya agar at 35°C. It was reported that 93.4% of the cells that were extractable were extracted in two cycles of the extraction procedure, and this treatment was recommended for future work.

Refuse extractions are also required to form a liquid inoculum for enzyme assays. This procedure has typically involved extraction of a 10-g sample in 30 ml of 0.2% Triton X-100 containing 0.75 M $MgSO_4$ (39, 49). After extraction, the samples were centrifuged at 10,000 × g, and the supernatant was used in the enzyme assays.

In summary, researchers have been consistent in their efforts to minimize exposure to oxygen during the inoculum formation process. Inocula have always been formed in an anaerobic buffer solution. However, there is variability in the sample size used and the specific method of microorganism extraction. As in other environmental samples that include solids, there is no measure of the extent to which attached microorganisms are separated during any inoculum formation technique. It is generally assumed that the fraction of microbes extracted is repeatable over a range of samples within the same study. Techniques for inoculum formation must be adapted to the objectives of a particular study.

Processing Samples for Chemical Analyses

Samples may also be excavated for analysis of the chemical composition of both the aqueous and solid phases. In all cases, a representative sample should be obtained by repetitive quartering of the total excavated sample. Quartering should occur after shredding or other size-reducing methods. Analysis of the aqueous phase is typically accomplished by adding sufficient deionized water to form a slurry. The pH should be measured in the field to prevent shifts due to atmospheric CO_2. The slurry may then be filtered and preserved by acidification prior to analysis of the soluble organic content by measurement of total organic carbon (TOC), chemical oxygen demand (COD), or other parameters such as carboxylic acids. When analysis of the solid phase is to be performed, the shredded subsample should be dried at 65 or 75°C and then ground in a Wiley mill to a 1-mm particle size. Particle size reduction is best accomplished by first performing a coarse grind (to 3 to 5 mm) and then reducing the sample mass by using a sample splitter (riffler). Next, the reduced mass should be ground to 1 mm. This finely ground refuse should then be redried at 65 or 75°C and stored in a tightly sealed mason jar. The finished product may be used for analysis of cellulose, hemicellulose, protein, lignin, volatile solids, and, perhaps, other organics. The significance of these compounds and appropriate analytical

TABLE 2 Media used for MPN tests

Addition (per liter of medium)	Total anaerobes	Cellulolytics	Xylanotics	Acetogens	Methanogens	
					Acetate	H$_2$/CO$_2$
PO$_4$ solution[a] (ml)	100	100	100	100	0.1[b]	0.2[b]
M3 solution[c] (ml)	100	100	100	100	100	100
Mineral solution[d] (ml)	10	10	10	10	10	10
Vitamin solution[e] (ml)	10	10	10	10	10	10
Resazurin (0.1%) (ml)	2	2	2	2	2	2
Volatile fatty acids[f] (ml)	10	10	10			
Yeast extract (g)	1	1	0.25		2	2
Trypticase peptones (g)	2	0.025	2		2	2
Hemin (0.01%)[g] (ml)	10	10	10			
NaHCO$_3$ (g)					0.5	0.5
Carbon source	See text	See text	See text	See text	See text	See text
Distilled water[h] (ml)	578	420	678	698	868	868
Boiled under:	N$_2$-CO$_2$ (80-20)	N$_2$-CO$_2$ (80-20)	N$_2$-CO$_2$ (80-20)	N$_2$-CO$_2$ (80-20)	N$_2$	N$_2$
NaHCO$_3$ (5% [wt/vol])[i,j] (ml)	70	70	70	70		
Cysteine hydrochloride (5% [wt/vol])[i] (ml)	10	10	10	10	10	10
Final pH	6.6	6.6	6.6	7.0	7.2	6.7

[a] The phosphate solution contained 16.1 g of KH$_2$PO$_4$ and 31.87 g of Na$_2$HPO$_4$·7H$_2$O per liter.

[b] A concentrated phosphate solution containing 15 g of KH$_2$PO$_4$ and 31.87 g of Na$_2$HPO$_4$·7H$_2$O per 100 ml was used to supply the phosphates. It was prepared under nitrogen and autoclaved separately. The indicated volume is aseptically added to individual tubes containing 9 ml of sterile medium.

[c] M3 solution contained, per liter, 10 g of NH$_4$Cl, 9 g of NaCl, 2 g of MgCl$_2$·6H$_2$O, and 1 g of CaCl$_2$·2H$_2$O.

[d] As described in reference 40, with the addition of 0.033 g of Na$_2$WO$_4$·2H$_2$O.

[e] As described in reference 78.

[f] As described by Leedle and Hespell (42), with the addition of phenylacetic acid (0.0068 g) and 3-phenylpropionic acid (0.0075 g) per liter of volatile fatty acid solution.

[g] Hemin was prepared by dissolving 0.01 g in 100 ml of deionized water that contained 0.1 g of NH$_4$Cl and 0.1 g of NaOH.

[h] After addition of distilled water, the pH was adjusted to 7.2 with NaOH, the medium was boiled, the remaining ingredients were added, and 9 ml of medium was dispensed into pressure tubes and autoclaved for 20 min at 121°C.

[i] Added after boiling.

[j] Boiled under an 80-20 N$_2$-CO$_2$ gas phase.

techniques are discussed below in the section on chemical measurements.

Microbiological Measurements

Techniques that may be used to characterize the microbiology of a refuse sample are discussed in this section. These techniques include enumeration methods, enzyme assays, measurement of the methane production rate of an excavated refuse sample, and emerging molecular techniques.

Enumeration of Microorganisms on Refuse

Techniques that have been used for the enumeration of landfill bacteria include acridine orange direct counts (AODC) (48), MPNs (9, 57), agar plate counts (44), and roll tubes (48, 74). Irrespective of the enumeration technique, the first step must be the formation of a liquid inoculum from a refuse sample as discussed above. Once a liquid inoculum is formed, microorganisms may be enumerated by any of the methods described below.

Adaptation of the AODC procedure to refuse has been presented by Palmisano et al. (48). Briefly, between 0.5 and 1 g of refuse was fixed in formaldehyde. Samples were then diluted to a known volume in sodium pyrophosphate and incubated for 5 to 10 min. Next, samples were cooled to 0°C and alternately sonicated and cooled. Samples were then stained, filtered, and counted.

Media used for the MPN enumeration of the total anaerobic population and the subpopulations of cellulolytic, hemi-cellulolytic, butyrate-catabolizing acetogenic, and acetoclastic and hydrogeneotrophic methanogenic bacteria are presented in Table 2. The medium for enumeration of the total anaerobic population contained 10 soluble carbon sources (cellobiose, glucose, maltose, xylose, galactose, arabinose, mannose, starch, glycerol, and galacturonic acid), each at a concentration of 2.5 mM. Carbon sources were representative of refuse hydrolysis products. Microbial growth on cellulose was detected by visible disappearance of ball-milled Whatman no. 1 filter paper. Filter paper strips would be more representative of the particle size of refuse. However, in preliminary work conducted prior to the adaptation of ball-milled cellulose, I never obtained repeatable results with filter paper. Another alternative substrate representative of the lignocellulosic components of MSW such as newspaper is ground wood. To my knowledge, no landfill researcher has used ground wood as an enumeration substrate. Xylan from oat spelts (Sigma, St. Louis, Mo.) was used for enumeration of the hemicellulolytic bacteria. Prior to use, the xylan was soaked in distilled water for 24 h to remove the soluble and nonsettleable material. After drying, it was ground in a small Wiley mill. Growth was determined by measurement of optical density (OD). Tubes were considered positive if their OD was significantly higher than the background OD present in uninoculated controls. These controls are required because of the presence of background OD associated with the ground xylan.

Acetogenic bacteria were enumerated on the basis of conversion of butyrate (40 mM) to acetate and hydrogen and subsequent conversion of the hydrogen to methane by a pure culture of *Methanobacterium formicicum* (43). The methane concentrations in tubes containing butyrate were compared with those in tubes without butyrate at each dilution. Tubes in which the methane concentration is significantly greater than that in the controls ($P = 0.01$) can be considered positive. Methanogen MPN tests were performed with either 80 mM acetate or 202.6 kPa of H_2/CO_2. Tubes were considered positive if they contained greater than 0.5% (vol/vol) methane. MPN tubes were incubated for 30 days except for the acetogen tubes, which were incubated for 60 days.

Palmisano et al. used a rich medium for enumeration of the total fermentative population (48). Their medium contained three carbon sources (glucose, cellobiose, and soluble starch) as well as several nutrient sources (yeast extract, Trypticase, hemin, and a carboxylic acid mixture). Anaerobic starch- and protein-degrading microbes were enumerated on peptone-yeast-glucose agar amended with soluble starch and gelatin, respectively. Starch-clearing zones were detected by the addition of potassium iodide, while proteolytic microbes were detected by addition of 0.05% (wt/vol) naphthol blue black in 7% (vol/vol) acetic acid. Colonies producing extracellular proteases were surrounded by clear zones. Maule et al. (44) used tryptone soya agar for enumeration of both total anaerobes and total aerobes by plate counts.

Roll tubes have also been used for the enumeration and isolation of cellulolytic bacteria from refuse (48, 74). In this procedure, the tubes are inspected visually or with the assistance of a microscope for a zone of clearing indicative of the consumption of insoluble cellulose. Researchers typically use ball-milled cellulose as the cellulose source, although other alternatives are available as discussed above.

There have been just a few reports of work identifying the presence of anaerobic fungi in refuse. In this work, bacteria were inhibited by the addition of streptomycin sulfate at 130 U/ml and penicillin K at 2,000 U/ml (10). Cycloheximide (0.05 mg/ml) was used to inhibit fungi in parallel tests. Theodorou et al. (69) used these same antibiotics and chloramphenicol when inhibition of methanogens was desired. Maule et al. (44) enumerated aerobic fungi in refuse samples by using malt extract agar in the absence of antibiotics. Researchers typically make extensive reference to the rumen microbiology literature for the appropriate techniques (77).

Only one research group has reported work with anaerobic protozoa from landfills (27–29). Liquid dilutions of the solid refuse sample were made either into liquid medium or onto nonnutrient agar plates streaked with *Escherichia coli*. A number of protozoa, including two *Mastigamoeba* species, *Heteromita* sp., *Chilomastix* sp., *Phreatamoeba* sp., and at least one other unidentified heterotrophic flagellate, were isolated from excavated refuse samples. The most frequently occurring isolate, *M. palaeformis* Kahl 1927, was found to carry methanogenic endosymbionts.

Habitat Simulation and Incubation Conditions

The use of culture techniques to enumerate or isolate microorganisms from environmental samples assumes that the desired populations will grow under the laboratory conditions provided. This is an imperfect assumption in any ecosystem. For example, the ratio of the total fermentative population measured in roll tubes to the AODC, for samples excavated from a landfill and incubated at 37°C, was between 2.2 ×

10^{-5} and 0.016 (48). This result is typical of other ecosystems and serves to emphasize the limitations of laboratory culture techniques.

In developing culture media, an attempt to simulate the habitat is typically made. For example, in studies of the rumen microflora, clarified rumen fluid is typically added to the medium. In landfill microbiology, adjustments can be made to the medium pH and incubation temperature. However, simulation of the organic matrix in a landfill is problematic. Leachate could be used for habitat simulation. However, leachate contains degradable organic carbon. Thus, if a medium is to support the growth of a specific trophic group, then the presence of uncharacterized carbon in the leachate will interfere with the enumeration assay. A second limitation of leachate is that a source of constant composition usually is not available. Leachate composition reflects the decomposition state of the refuse. Young leachate will likely be relatively high in carboxylic acid concentrations, while leachates from well-decomposed refuse will contain more humic-like compounds. Leachate composition from a landfill will also vary as it is affected by the overlying refuse and dilution with storm water. Even leachate from a laboratory-scale lysimeter will vary in composition over time unless the refuse is well decomposed prior to first use of the leachate. Preincubation of leachate from well-decomposed refuse to deplete any degradable carbon could be used to address these limitations. However, there is no report to date of a researcher having used leachate in a growth medium.

The incubation temperature for an enumeration assay should reflect the temperature of the ecosystem from which the inoculum is recovered. The optimum temperature for methane production in the mesophilic range has been reported as 41 to 42°C (34, 53). On the basis of these results, Barlaz et al. (9) conducted MPN assays at 40°C. The refuse used as the inoculum source for these tests was also incubated at 40°C. Palmisano et al. (48) incubated samples excavated from the Fresh Kills landfill at both 22 and 37°C. The average temperature of the excavated samples was 29°C, with a range of 10 to 63°C (67).

Direct Measures of Microbial Activity

Techniques that have been reported for the direct measurement of microbial activity and the presence of microbes in refuse include enzyme assays, turnover of added ^{14}C, measurement of the coenzyme F420, and scanning electron microscopy. These techniques are discussed below.

The activity of a number of hydrolytic enzymes in excavated refuse samples has been measured (49). Esterase activity was measured by the hydrolysis of fluorescein diacetate to fluorescein, the presence of which could be measured by fluorescence spectrophotometry. Protease activity was measured by using azocoll, a proteinaceous substrate that releases a blue dye on hydrolysis. Amylase activity was measured by using substrates labeled with a *p*-nitrophenol moiety. In this assay, *p*-nitrophenyl α-D-maltoheptaoside is converted to *p*-nitrophenylmaltotriose plus maltotetraose by α-amylase. The *p*-nitrophenylmaltotriose is ultimately hydrolyzed to glucose and to *p*-nitrophenol, which was measured spectrophotometrically. Others have measured lipase activity by a titrimetric measurement of fatty acids released from triglycerides (39). For a broader characterization, the API-ZYM system has been used to screen for the presence of 18 different enzymes (49).

Two methods have been published for measuring cellulase activity: one for decomposing refuse samples and a second for pure cultures. In work with decomposing refuse (49), cellulase activity was measured on the basis of the degradation of cellulose-azure, which releases a blue dye upon hydrolysis. In work with pure cultures of cellulolytic organisms isolated from decomposing refuse, a technique based on carboxymethylcellulose was presented (74). Cell extracts were obtained by first growing cells in liquid medium and then harvesting them in the late logarithmic phase. After centrifugation, bacterial pellets weighing 1 to 2 g were resuspended and then sonicated. The supernatant, which represented a cell-free whole-cell homogenate, was decanted and frozen until use. In addition, a cell-free supernatant fraction was produced by filter (0.2-μm pore size) sterilization of the original growth medium, after which ethanol was added to 70% by volume. The precipitated protein was then harvested by centrifugation, resuspended in buffer, and frozen until use. Carboxymethylcellulase activity was measured by the release of reducing sugars from a 0.5% (wt/vol) solution of carboxymethylcellulose. Cellobiohydrolase activity was measured by the release of p-nitrophenol from p-nitrophenylcellobioside. Total cellulase activity was measured as a decrease in turbidity of a 0.025% (wt/vol) cellulose suspension, using a spectrophotometer at 660 nm.

There are several limitations to the use of enzyme assays as described above. First, all of the analyses except the pure culture work were done with 10-g samples. If the results are to be used for general characterization of the presence or absence of activity, then this sample size may be sufficient. However, 10-g samples cannot be used to extrapolate activity to an entire landfill. Presumably, larger samples could be used for the initial extraction, although this might increase the opportunity for enzymes to attach to the solid matrix provided by the sample. As always, it is important to work with multiple replicate samples to explore spatial variability. Factors such as pH, refuse composition, temperature, and nutrient status will vary spatially and temporally. Thus, measurement of spatial variability will reflect variation in conditions throughout a landfill.

A second limitation to the use of enzyme assays is the effectiveness of the cell lysis and washing step. While 0.2% (wt/vol) Triton X-100 has been used, no work comparing its effectiveness with that of other detergents or lysis procedures has been published. In this respect, enzyme activity data are probably best evaluated as relative measures among samples within the same study.

A third limitation is specific to the cellulase assay. Jones and Grainger (39) evaluated their extraction technique by adding a commercial cellulase to both sterile and nonsterile refuse. Upon extraction, the cellulase was fully recovered from the sterile refuse, but only a fixed amount of enzyme activity was measured in the nonsterile refuse regardless of the amount of enzyme added. The authors suggested that the enzyme could have been deactivated by proteases.

A review of results of cellulase activity and cellulolytic population data suggests limitations with both measurements. The presence of cellulose as the dominant biodegradable polymer in refuse is not reflected in measurements of either cellulolytic populations or cellulase enzyme activity. Palmisano et al. studied hydrolytic enzyme activity (49) and population densities (48) in refuse samples excavated from three landfills. Cellulase activity was present in only 2 of 28 samples from the Fresh Kills landfill, 3 of 8 samples from a landfill in Arizona, and 1 of 17 samples from a landfill in Florida. In contrast, esterase, amylase, and protease activities were present in all 28 samples excavated from the Fresh Kills site. Similar problems arise on review of enumeration data. Palmisano et al. (48, 49) detected no cellulolytic organisms in several refuse samples. Barlaz et al. found cellulolytic populations to be about 3 orders of magnitude lower than methanogen populations in well-decomposed refuse (9) and found cellulolytic organisms in very low populations on individual components of refuse (57). It has been suggested that one or more of the enzymes responsible for cellulose hydrolysis was membrane bound and not extracted (49). This is consistent with a report that cellulose hydrolysis was greater in cell homogenates than in the extracellular fraction (74). However, it does not explain the seemingly low MPN results.

Researchers have used two other techniques, which are not formal enzyme assays, to assess the microbial activity of decomposing refuse: methane production and mineralization of [^{14}C]cellulose. A measure of methane-producing activity may be obtained by placing a sample of freshly excavated refuse in a reactor, sealing it, and sparging it with nitrogen to displace oxygen that entered during reactor loading (31, 67). The rate of methane production can then be measured, and samples can be amended to study the effects of various supplements. Because methane production from refuse requires the coordinated activities of several trophic groups, measurement of methane-producing activity provides a measure of the overall activity of the refuse sample as opposed to the activity of a specific organism or enzyme.

The metabolic ability of decomposing refuse to convert added [^{14}C]cellulose to $^{14}CO_2$ and $^{14}CH_4$ has also been measured (49). Ten-gram samples of refuse were placed in bottles to which [^{14}C]cellulose was added. Mineralization was measured by sparging the headspace of a bottle with nitrogen, thus forcing headspace gases through a system to trap $^{14}CO_2$ and $^{14}CH_4$. The system includes NaOH traps to dissolve the $^{14}CO_2$, followed by a combustion furnace to oxidize $^{14}CH_4$ to $^{14}CO_2$, and finally followed by another set of traps to dissolve the newly produced $^{14}CO_2$. Trapped radiolabeled gases are then quantified by scintillation counting. Here too, the measure reflects the combined activities of all trophic groups involved in cellulose conversion to methane.

The use of [^{14}C]cellulose has an important limitation. My experience with commercially purchased radiolabeled cellulose has shown that its specific activity is as much as 10 times greater than the level indicated by the vendor. Thus, if mass balances are involved, it is critical to measure the specific activity independently. The specific activity of commercially purchased [^{14}C]cellulose has been measured successfully by mixing radiolabeled cellulose with microcrystalline cellulose and then ball milling the combination for 5 days. Ball milling serves to dilute the specific activity of the radiolabeled material and to homogenize the small weight of labeled cellulose in a larger mass of unlabeled cellulose. After ball milling, triplicate 0.2-g samples were subjected to an acid hydrolysis designed to dissolve the cellulose (24, 52). This procedure includes hydrolysis in 72% (wt/vol) sulfuric acid followed by a secondary hydrolysis in 3% sulfuric acid. The hydrolysate will contain hydrolyzed cellulose as glucose. The activity of the hydrolysate may then be analyzed by scintillation counting, and the specific activity of the radiolabeled cellulose can be calculated.

F420 is a coenzyme found in methanogens and other archaebacteria and some eubacteria (74). Its fluorescence makes microscopic detection simple, and it may also be measured by high-pressure liquid chromatography (HPLC). However, its use as a quantitative indicator of methanogenic activity is limited because F420 extracted from environmental samples may include some F420 extracted from nonviable cells as well as extracellular F420. In addition, the cellular concentration of F420 varies with both species and environmental conditions, and F420 degrades upon exposure to oxygen.

Ether-linked lipids may be used for detection of total methanogen biomass, assuming that they are the sole archaebacteria present in a landfill. However, this technique does not provide information on the diversity and activity of the methanogen population (74).

Finally, one research group has used scanning electron microscopy and fluorescence light microscopy to study colonization on the surface of refuse samples (60). They reported 10^4 to 10^5 methanogens per cm^2 on excavated refuse samples and estimated that this was equivalent to 5×10^5 to 7×10^6 cells per g (wet weight).

In summary, microbial populations and activity in anaerobically decomposing refuse have been measured by using a number of viable cell count procedures, AODC, hydrolytic enzyme activities, methane production rates, the turnover of [^{14}C]cellulose, and both electron and fluorescence microscopy. Each measurement has its own set of strengths and weaknesses that must be considered in view of each study's objectives. Viable cell counts and enzyme activities are perhaps best used as a relative measure among different samples within the same study. The broadest measure, the methane production rate of an excavated refuse sample, is the most comparable among studies and often the most easily determined.

Molecular Techniques for the Study of Landfill Microbiology

The limitations of traditional laboratory culture techniques for the study of microorganisms in the environment are well known and are not unique to landfill microbiology. These limitations have stimulated the development of a number of molecular techniques as described elsewhere in this manual. The potential for the application of such techniques to landfill samples has been reviewed, although no actual data were presented (74). Antigenic fingerprinting has been used to characterize methanogens isolated from refuse samples. The relatedness of seven methanogenic isolates to a reference methanogen culture collection was reported (26). However, this technique can be applied only to previously isolated species and does not aid in the detection of uncultured species.

Maule et al. (44) were the first to apply DNA-based techniques to landfill samples. They prepared probes for 13 species of methanogens and used them to check for each species in landfill leachate. Five species, *Methanoculleus bourgense*, *Methanobacterium formicicum*, *Methanosarcina barkeri*, *Methanosphaera stadtmanae*, and *Methanobrevibacter ruminantium*, were shown to be present. The first two species were most common. The authors concluded that "[i]f suitable genetic markers for the detection and quantification of other microbiological groups such as the acetogens, cellulose degraders and sulfate reducers can be found, DNA-based systems may supersede culture-based methods for monitoring the microbiological changes in landfill." Raskin et al. (58) monitored the startup phase of an anaerobic digester by using oligonucleotide probes complementary to conserved tracts of the 16S rRNAs of phylogenetically defined groups of methanogens. The digesters were fed shredded MSW with certain components (glass, plastic, metal, yard waste) removed. Members of the order *Methanobacteriales* were the major hydrogen-utilizing methanogens present in both a

mesophilic and a thermophilic digester. *Methanosarcina* and *Methanosaeta* species were present in both digesters but at much lower concentrations. Silvey and Blackall (63) used denaturing gradient electrophoresis to monitor population shifts in leachate from a laboratory-scale landfill simulation. They reported that the population was largely stable over the 6-month monitoring period. They also presented a detailed description of their extraction and amplification procedures. In summary, while the use of molecular techniques holds much promise for advancing our understanding of landfill microbial ecology, results to date are only preliminary.

Chemical Measurements

Chemical measurements can be used to infer information about the microbiological state of refuse. The major biodegradable components of refuse are cellulose and hemicellulose. The other major organic present in refuse is lignin. As refuse decomposes, the cellulose and hemicellulose concentrations decrease, and since lignin is recalcitrant, its concentration increases. Techniques for the measurement of cellulose, hemicellulose, and lignin have been described (24, 52). The basis for the procedure is hydrolysis of a solid sample in 72% (wt/vol) sulfuric acid and then a secondary hydrolysis in 3% (wt/vol) sulfuric acid. These hydrolyses convert cellulose and hemicellulose to their respective monomers. HPLC is then used to quantify the concentrations of glucose, xylose, mannose, arabinose, and galactose in the acid hydrolysate. The glucose originates from cellulose; the other sugars originate from hemicellulose. The solids remaining after acid hydrolysis are used to measure the lignin concentration. Lignin is calculated from the weight loss of the dried solids after combustion at 625°C. Although recalcitrant, the chemical form of lignin may change during decomposition, resulting in apparent changes in the lignin concentration as measured by this digestion procedure (36).

Cellulose, hemicellulose, and lignin concentrations have been used to assess the degree of decomposition in a landfill (14, 72). In the 1980s, fresh refuse was reported to have a cellulose-to-lignin ratio of approximately 3.5 to 4, and this ratio was shown to decrease with age in excavated refuse samples (14). This ratio is particularly useful because it is not confounded by the presence of cover soil in samples. Such soil is typically present in excavated refuse samples and will dilute the concentrations of all solids. More recently, measurement of the hemicellulose concentration in refuse has become standard, and it may now be more appropriate to work with the ratio of cellulose plus hemicellulose to lignin. While this ratio cannot provide an absolute assessment of microbiological activity, it does provide information on the relative state of decomposition of refuse samples. This ratio has been used as an indicator of residual methane production potential in well-decomposed refuse with moderate success (72).

Protein is the next major degradable component of refuse. As explained earlier, protein measurements are subject to some error due to the presence of nonprotein organic nitrogen. Protein loss is difficult to quantify since protein that is present in refuse and is converted to cell mass during decomposition will still be measured as protein. Given the analytical problems and the relatively low concentration of protein in refuse, protein has not been routinely used as a measure of decomposition.

Volatile solids are measured on the basis of weight loss after combustion at 550°C (1). Volatile solids represent all organic matter in refuse including degradable forms (cellu-

TABLE 3 Summary of methods used to study refuse decomposition and landfill microbiology

Type of analysis or purpose	Method	Comments	Reference(s)
Decomposition studies	Actual landfill or field-scale test landfill	• Expensive • Hard to control • Must obtain large samples to assess overall methane production performance • Difficult to obtain representative samples for microbiological tests • Extended time required for complete decomposition cycle	16, 20, 32, 45, 73
	Reactors ranging from 2 to 3,300 liters	• Difficult to obtain representative samples from large reactors for microbiological assessment • Large systems hard to control • Smaller systems can be better controlled and sampled • Small systems require use of shredded refuse and some selection of what is added to reactor • Can accelerate decomposition	5, 6, 8, 9, 23, 31, 34, 53, 55, 56, 59
	Biochemical methane potential	• Measures ultimate anaerobic biodegradability of a small sample • Requires inoculum • Useful for measuring potential residual methane production if representative samples can be recovered • Inexpensive • Rapid (30 days) • Reproducible	11, 46, 62, 72
Decomposition studies, effect of environmental conditions on refuse microbes	Multistage chemostat	• Unique system for studying effects of environmental conditions on individual components of anaerobic consortia • Applicability to landfill dependent on similarity between active population in landfill and chemostat	18, 37, 51
Sample collection	Fresh refuse	• Should be representative of residential refuse • Requires shredding for homogenization	6, 9, 59
	Decomposed refuse excavated from landfill	• Must collect large samples and homogenize • Oxygen exposure not major problem with reasonable precaution during sampling	11, 12, 33, 44, 48, 49, 73
	Decomposed refuse from a laboratory reactor	• Sample history well known	4, 5, 9
Inoculum formation	Blend in phosphate buffer Stomacher Multiple extractions	• Shown to be repeatable, easy to implement • Limited to 10–50-g samples • Two extractions adequate in one study (44) • No difference on multiple blending in phosphate buffer (10)	9, 57, 44a 48 10, 19, 42, 44
	Triton X-100 and MgSO₄	• Used for enzyme assays • No comparative work done	39, 49
Enumeration	MPN	• Most common method • Researchers have been consistent in use of a rich medium for enumeration of the total anaerobic population, but the precise medium has varied • No comparative studies are available; media are available for enumeration of cellulolytics, hemicellulolytics, acetogens, and methanogens	9, 48, 57
	AODC	• Measures total viable plus nonviable cells • Gives count well above total anaerobic populations as measured by MPN	48
	Roll tubes	• Used primarily for cellulolytic bacteria • Ball-milled cellulose is most common cellulose form used in both roll tubes and liquid media • Used successfully for cellulolytic isolations	48, 74

(Table continued on next page)

TABLE 3 Summary of methods used to study refuse decomposition and landfill microbiology (*Continued*)

Type of analysis or purpose	Method	Comments	Reference(s)
Microbial activity	Hydrolytic enzymes	• Cellulase measurements apparently inconsistent with expected values based on cellulose decomposition in landfills • Effectiveness of cell lysis step unknown	39, 49
	[^{14}C]cellulose turnover	• Good measure of combined activity of microorganisms required for decomposition in landfill at time of sample excavation • Rapid results • Results only as representative as sample	49
	Methane production rate	• Instantaneous rate of methane production in sample • Not as sensitive as [^{14}C]cellulose • Rapid results • Results only as representative as sample	31, 67
Microbial identification and population dynamics	Antigenic fingerprinting	• Useful for comparison of methanogenic isolates with known isolates	26
	DNA probes	• Applied successfully in leachate • Results too preliminary to judge effectiveness	44, 63
Chemical measurements	Cellulose, hemicellulose, lignin	• Useful for assessment of the degree of decomposition • Requires representative sample	14, 33, 72

lose, hemicellulose) and recalcitrant forms (lignin, plastics, rubber). Given the wide range of compounds reported as volatile solids, they can provide only a rough estimate of the state of refuse decomposition.

Finally, cellulose, hemicellulose, and lignin concentrations have been used in laboratory-scale reactors to document the relationship between solids loss and methane production (6). This approach uses a mass balance based on methane potential. The methane potential of each component is calculated by using equation 1 (50):

$$C_nH_aO_bN_c +$$
$$[n - (a/4) - (b/2) + 3(c/4)]H_2O \rightarrow$$
$$[(n/2) - (a/8) + (b/4) + 3(c/8)]CO_2 \qquad (1)$$
$$+ [(n/2) + (a/8) - (b/4) - 3(c/8)]CH_4 + cNH_3$$

The mass balance is calculated from the methane potential of the solids removed from a reactor plus the measured methane, divided by the methane potential of the solids added to a reactor at the beginning of an experiment. On the basis of equation 1, 415 and 424 liters of methane at standard temperature and pressure would be expected for every kilogram of cellulose and hemicellulose degraded, respectively.

The soluble organic fraction of refuse may be used as a relative measure of organic accumulation. As explained earlier, carboxylic acids will accumulate during the anaerobic acid phase and decrease during the accelerated methane production phase until they reach steady-state pool levels during the decelerated methane production phase. Absolute concentrations of TOC, COD, and carboxylic acids are influenced by a number of system-specific parameters. Thus, concentrations representative of each phase of decomposition cannot be provided, and these data are best used when a time series is available to assess trends in decomposition. Either TOC or COD will serve as a measure of the total organic content. Even with the near depletion of carboxylic

acids in the final phase of decomposition, there will be other dissolved organics that are slowly degradable or recalcitrant. Thus, the biological oxygen demand/COD ratio is a good indicator of the presence of dissolved, degradable organic carbon. TOC is measured with a TOC analyzer, while COD is most easily measured by using kits containing preweighed reagents (Hach Co., Loveland, Colo.). Carboxylic acids can be measured by either a gas chromatograph equipped with a flame ionization detector or HPLC equipped with a conductivity detector. The biological oxygen demand measurement is described in reference 2.

Though an indirect measure of microbial activity, chemical parameters provide a measure of the results of biodegradative activity. While the issue of representative sampling is equally complex for chemical and microbiological measurements, chemical measurements circumvent the issue of cell or enzyme extraction efficiency.

SUMMARY

A landfill is a unique anaerobic ecosystem with an abundance of degradable organic carbon and a wide range of microbial activities. Its uniqueness is a result of (i) the long residence time of carbon in a landfill, (ii) the constant input of fresh organic carbon over the operational life, (iii) the large changes in environmental conditions (pH, soluble organic carbon) over the decomposition period, and (iv) changes in the presence and nature of insoluble polymeric substrates during decomposition. The microbiology of the landfill ecosystem has not been as thoroughly explored as that of other anaerobic systems such as anaerobic digesters and the rumen. Thus, it is likely that there are unique microorganisms that could be characterized given more emphasis on this ecosystem. As with other ecosystems, it is likely that the limitations of traditional culture techniques are limiting our ability to identify new microorganisms from landfills. Recently, researchers have begun to apply molecular techniques to the study of landfill microbes. As approaches that eliminate the need for culture techniques are developed, a wealth of new information on landfill microbiology can be expected. As suggested by Maule et al. (44), the development of genetic markers for all of the trophic groups required for refuse methanogenesis will greatly enhance our ability to characterize refuse decomposition and allow confirmation of the current models of microbial population dynamics in landfills.

A major challenge in studying landfill microbiology is the issue of system heterogeneity. To address this issue, researchers have developed a number of approaches, including small grab samples that are not representative, multiple replicate samples at full-scale landfills, and laboratory simulations with shredded refuse. The appropriate system for the study of landfill microbiology as well as the appropriate measures will be unique for each set of research objectives. Alternate systems and measures are summarized in Table 3. The types of measurements used to study landfill microbiology are analogous to those used in other ecosystems and include total and viable cell counts, enzyme activity, carbon turnover, and the production of terminal end products. The relationship between methane production and the biodegradation of the major organic constituents of refuse is well understood, and this knowledge provides an opportunity to use chemical measurements to evaluate the extent of refuse decomposition.

Landfills represent a large source of organic carbon, and the decomposition of this material is well documented. However, questions remain with respect to the rate and extent of decomposition. While older landfills have exhibited reduced rates of methane production, to my knowledge, there is not a documented case of a landfill ceasing to produce measurable methane. Thus, our knowledge is lacking with respect to ultimate landfill methane production, the amount of methane released after oxidation in the cover soil, and the amounts of organic carbon and other nutrients that are placed in long-term terrestrial storage. In addition to gaps in information on these macroscale indicators of refuse decomposition, the dominant microorganisms responsible for the decomposition process have been neither identified nor characterized.

REFERENCES

1. **Ackman, P.** 1981. *Cellulose and Lignin Testing Methods Research*. M.S. report. Department of Civil and Environmental Engineering, University of Wisconsin Madison.
2. **American Public Health Association.** 1985. *Standard Methods for the Examination of Water and Wastewater*, 16th ed. American Public Health Association, Washington, D.C.
3. **Barlaz, M. A.** Unpublished data.
4. **Barlaz, M. A., W. E. Eleazer, W. S. Odle III, X. Qian, and Y.-S. Wang.** *Biodegradative Analysis of Municipal Solid Waste in Laboratory-Scale Landfills*. EPA final report CR-818339. U.S. Environmental Protection Agency, Research Triangle Park, N.C. Submitted for publication.
5. **Barlaz, M. A., R. K. Ham, and M. W. Milke.** 1987. Gas production parameters in sanitary landfill simulators. *Waste Manage. Res.* **5:**27–39.
6. **Barlaz, M. A., R. K. Ham, and D. M. Schaefer.** 1989. Mass balance analysis of decomposed refuse in laboratory scale lysimeters. *J. Environ. Eng.* **115:**1088–1102.
7. **Barlaz, M. A., R. K. Ham, and D. M. Schaefer.** 1990. Methane production from municipal refuse: a review of enhancement techniques and microbial dynamics. *Crit. Rev. Environ. Control* **19:**557–584.
8. **Barlaz, M. A., R. K. Ham, and D. M. Schaefer.** 1992. Microbial, chemical and methane production characteristics of anaerobically decomposed refuse with and without leachate recycle. *Waste Manage. Res.* **10:**257–267.
9. **Barlaz, M. A., D. M. Schaefer, and R. K. Ham.** 1989. Bacterial population development and chemical characteristics of refuse decomposition in a simulated sanitary landfill. *Appl. Environ. Microbiol.* **55:**55–65.
10. **Barlaz, M. A., D. M. Schaefer, and R. K. Ham.** 1989. Effects of prechilling and sequential washing on the enumeration of microorganisms from refuse. *Appl. Environ. Microbiol.* **55:**50–54.
11. **Bogner, J. E.** 1990. Controlled study of landfill biodegradation rates using modified BMP assays. *Waste Manage. Res.* **8:**329–352.
12. **Bogner, J. E.** 1992. Anaerobic burial of refuse in landfills: increased atmospheric methane and implications for increased carbon storage. *Ecol. Bull.* **42:**98–108.
13. **Bonner, B. A., P. P. Calvert, M. A. Barlaz, and C. A. Pettigrew.** 1995. Polymer biodegradation under simulated landfill conditions. Presented at the International Congress of Pacific Basin Societies, December 17–22, Honolulu, Hawaii.
14. **Bookter, T. J., and R. K. Ham.** 1982. Stabilization of solid waste in landfills. *J. Environ. Eng.* **108:**1089–1100.
15. **Brock, T. D., M. T. Madigan, J. M. Martinko, and J. Parker.** 1994. *Biology of Microorganisms*, 7th ed. Prentice-Hall, Englewood Cliffs, N. J.
16. **Campbell, D., M. Caine, M. Meadows, and K. Knox.** 1995. Enhanced landfill gas production at large-scale test cells, p. 593–601. *In* T. H. Christensen, R. Cossu, and R. Stegmann (coordinators), *Proceedings of Sardinia 95, Fifth International Landfill Symposium*. CISA, Environmental Sanitary Engineering Center, Cagliari, Italy.
17. **Christensen, T. H., P. Kjeldsen, H. J. Albrechsten, G. Heron, P. H. Nielsen, P. L. Bjerg, and P. E. Holm.** 1994. Attenuation of landfill leachate pollutants in aquifers. *Crit. Rev. Environ. Sci. Technol.* **24:**119–202.
18. **Coutts, D. A. P., E. Senior, and M. T. M. Balba.** 1987. Multi-stage chemostat investigations of interspecies interactions in a hexanoate-catabolising microbial association isolated from anoxic landfill. *J. Appl. Bacteriol.* **62:**251–260.
19. **Craig, W. M., B. J. Hong, G. A. Broderick, and R. J. Bula.** 1984. In vitro inoculum enriched with particle-associated microorganisms for determining rates of fiber digestion and protein degradation. *J. Diary Sci.* **67:**2902–2909.
20. **Croft, B., and T. Fawcett.** 1993. *Landfill Gas Enhancement Studies. The Brogborough Test Cells.* ETSU B/B5/00080/REP. Energy Technology Support Unit, Department of Trade and Industry, Oxfordshire, England.
21. **Daniel, D. E. (ed.).** 1993. *Geotechnical Practise for Waste Disposal*. Chapman & Hall, New York.
22. **Daniel, D. E., and C. D. Shackelford.** 1989. Containment of landfill leachate with clay liners, p. 323–341. *In* T. H. Christensen, R. Cossu, and R. Stegmann (ed.), *Sanitary Landfilling: Process, Technology and Environmental Impact*. Academic Press, Inc., New York.
23. **Deipser, A., and R. Stegmann.** 1994. The origin and fate of volatile trace components in municipal solid waste landfills. *Waste Manage. Res.* **12:**128–139.
24. **Effland, M. J.** 1977. Modified procedure to determine acid soluble lignin in wood and pulp. *Tappi* **60:**143–144.
25. **Emberton, J. R.** 1986. The biological and chemical characterization of landfills. *In Proceedings of Energy from Landfill Gas*.
26. **Fielding, E. R., D. B. Archer, E. C. de Macario, and**

A. J. L. Macario. 1988. Isolation and characterization of methanogenic bacteria from landfills. *Appl. Environ. Microbiol.* **54:**835–836.

27. Finlay, B. J., K. J. Clarke, P. A. Cranwell, T. M. Embley, R. M. Hindle, and B. M. Simon. 1993. *Further Studies on the Role of Protozoa in Landfill.* ETSU B 1325. Energy Technology Support Unit, Department of Trade and Industry, Oxfordshire, England.

28. Finlay, B. J., K. J. Clarke, and B. M. Simon. 1990. *Protozoology of Landfill.* ETSU B 1256. Energy Technology Support Unit, Department of Trade and Industry, Oxfordshire, England.

29. Finlay, B. J., and T. Fenchel. 1991. An anaerobic protozoon, with symbiotic methanogens, living in municipal landfill material. *FEMS Microbiol. Ecol.* **85:**169–180.

30. Fluet, J. E., K. Badu-Tweneboah, and A. Khatami. 1992. A review of geosynthetic liner system technology. *Waste Manage. Res.* **10:**47–65.

31. Gurijala, K. R., and J. M. Suflita. 1993. Environmental factors influencing methanogenesis from refuse in landfills. *Environ. Sci. Technol.* **27:**1176–1181.

32. Halvadakis, C. P., A. N. Findikakis, C. Papelis, and J. O. Leckie. 1988. The mountain view controlled landfill project field experiment. *Waste Manage. Res.* **6:**103–114.

33. Ham, R. K., M. R. Norman, and P. R. Fritschel. 1993. Chemical characterization of fresh kills landfill refuse and extracts. *J. Environ. Eng.* **119:**1176–1195.

34. Hartz, K. E., R. E. Klink, and R. K. Ham. 1982. Temperature effects: methane generation from landfill samples. *J. Environ. Eng.* **108:**629–638.

35. Hocknull, M. D., and H. Dalton. 1993. *Anaerobic Methane Oxidation in Landfill.* ETSU B 1266. Energy Technology Support Unit, Department of Trade and Industry, Oxfordshire, England.

36. Iiyama, K., B. A. Stone, and B. J. Macauley. 1994. Compositional changes in compost during composting and growth of *Agaricus bisporus*. *Appl. Environ. Microbiol.* **60:**1538–1546.

37. James, A. G., and I. A. Watson-Craik. 1993. *Elucidation of Refuse Interspecies Interaction by Use of Laboratory Models.* ETSU B/B2/00148/REP. Energy Technology Support Unit, Department of Trade and Industry, Oxfordshire, England.

38. Jones, H. A., and D. B. Nedwell. 1993. Methane emission and methane oxidation in land-fill cover soil. *FEMS Microbiol. Ecol.* **102:**185–195.

39. Jones, K. L., and J. M. Grainger. 1983. The application of enzyme activity measurements to a study of factors affecting protein, starch and cellulose fermentation in a domestic landfill. *Eur. J. Appl. Microbiol. Biotechnol.* **18:**181–185.

40. Kenealy, W., and J. G. Zeikus. 1981. Influence of corrinoid antagonists on methanogen metabolism. *J. Bacteriol.* **146:**133–140.

41. Kightley, D., D. B. Nedwell, and M. Cooper. 1995. Capacity for methane oxidation in landfill cover soils measured in laboratory-scale soil microcosms. *Appl. Environ. Microbiol.* **61:**592–610.

42. Leedle, J. A. Z., and R. B. Hespell. 1980. Differential carbohydrate media and anaerobic replica plating techniques in delineating carbohydrate utilizing subgroups in rumen bacterial populations. *Appl. Environ. Microbiol.* **39:**709–719.

43. Mackie, R. I., and M. P. Bryant. 1981. Metabolic activity of fatty acid-oxidizing bacteria and the contribution of acetate, propionate, butyrate, and CO_2 to methanogenesis in cattle waste at 40 and 60°C. *Appl. Environ. Microbiol.* **41:**1363–1373.

44. Maule, A., P. Luton, and R. Sharp. 1994. *A Microbiological and Chemical Study of the Brogborough Test Cells.* ETSU

B/LF/00200/REP. Energy Technology Support Unit, Department of Trade and Industry, Oxfordshire, England.

44a. Mormile, M. R., K. R. Gurijala, J. A. Robinson, M. J. McInerney, and J. M. Suflita. 1996. The importance of hydrogen in landfill fermentations. *Appl. Environ. Microbiol.* **62:**1583–1588.

45. Nilsson, P., H. Karlsson, A. Lagerkvist, and J. E. Meijer. 1995. The coordinated test cell program in Sweden, p. 603–614. *In* T. H. Christensen, R. Cossu, and R. Stegmann (coordinators), *Proceedings of Sardinia 95, Fifth International Landfill Symposium.* CISA, Environmental Sanitary Engineering Center, Cagliari, Italy.

46. Owens, J. M., and D. P. Chynoweth. 1993. Biochemical methane potential of municipal solid waste (MSW) components. *Water Sci. Technol.* **27:**1–14.

47. Palmisano, A. C., and M. A. Barlaz (ed.). *Solid Waste Microbiology*, in press. CRC Press, Inc., Boca Raton, Fla.

48. Palmisano, A. C., D. A. Maruscik, and B. S. Schwab. 1993. Enumeration and hydrolytic microorganisms from three sanitary landfills. *J. Gen. Microbiol.* **139:**387–391.

49. Palmisano, A. C., B. S. Schwab, and D. A. Maruscik. 1993. Hydrolytic enzyme activity in landfilled refuse. *Appl. Microbiol. Biotechnol.* **38:**828–832.

50. Parkin, G. F., and W. F. Owen. 1986. Fundamentals of anaerobic digestion of wastewater sludges. *J. Environ. Eng.* **112:**867–920.

51. Parks, R. J., and E. Senior. 1988. Multistage chemostats and other models for studying anoxic ecosystems, p. 51–71. *In* J. W. T. Wimpenny (ed.), *CRC Handbook of Laboratory Model Systems for Microbial Ecosystems,* vol. 1. CRC Press, Inc., Boca Raton, Fla.

52. Pettersen, R. C., and V. H. Schwandt. 1991. Wood sugar analysis by anion chromatography. *J. Wood Chem. Technol.* **11:**495–501.

53. Pfeffer, J. T. 1974. Temperature effects on anaerobic fermentation of domestic refuse. *Biotechnol. Bioeng.* **16:**771–787.

54. Pfeffer, J. T. 1992. *Solid Waste Management Engineering.* Prentice-Hall, Englewood Cliffs, N.J.

55. Pohland, F. G. 1975. *Sanitary Landfill Stabilization with Leachate Recycle and Residual Treatment.* EPA grant no. R-801397. U.S. Environmental Protection Agency National Environmental Research Center, Cincinnati.

56. Pohland, F. G., and J. P. Gould. 1986. Co-disposal of municipal refuse and industrial waste sludge in landfills. *Water Sci. Technol.* **18:**177–192.

57. Qian, X., and M. A. Barlaz. 1996. Enumeration of anaerobic refuse decomposing microorganisms on refuse constituents. *Waste Manage. Res.* **14:**151–161.

58. Raskin, L., L. K. Poulsen, D. R. Noguera, B. E. Rittman, and D. A. Stahl. 1994. Quantification of methanogenic groups in anaerobic biological reactors by oligonucleotide probe hybridization. *Appl. Environ. Microbiol.* **60:**1241–1248.

59. Rhew, R., and M. A. Barlaz. 1995. The effect of lime stabilized sludge as a cover material on anaerobic refuse decomposition. *J. Environ. Eng.* **121:**499–506.

60. Robinson, J. P., and H. C. Sturz. 1993. *Nutrition and Inhibition of Methanogenic Bacteria in the Landfill Environment.* ETSU B 1271. Energy Technology Support Unit, Department of Trade and Industry, Oxfordshire, England.

61. Segal, J. P. 1987. Testing large landfill sites before construction of gas recovery facilities. *Waste Manage. Res.* **5:**123–131.

62. Shelton, D. R., and J. M. Tiedje. 1984. General method for determining anaerobic biodegradation potential. *Appl. Environ. Microbiol.* **47:**850–857.

63. Silvey, P., and L. L. Blackall. 1995. A study of the microbial ecology of MSW, p. 117–125. *In* T. H. Christensen,

R. Cossu, and R. Stegmann (coordinators), *Proceedings of Sardinia 95, Fifth International Landfill Symposium.* CISA, Environmental Sanitary Engineering Center, Cagliari, Italy.

64. **Strom, P. F.** 1985. Effect of temperature on bacterial species diversity in thermophilic solid-waste composting. *Appl. Environ. Microbiol.* **50:**899–905.

65. **Strom, P. F.** 1985. Identification of thermophilic bacteria in solid-waste composting. *Appl. Environ. Microbiol.* **50:** 906–913.

66. **Suflita, J. M.** 1996. Personal communication.

67. **Suflita, J. M., C. P. Gerba, R. K. Ham, A. C. Palmisano, W. L. Rathje, and J. A. Robinson.** 1992. The world's largest landfill: a multidisciplinary investigation. *Environ. Sci. Technol.* **26:**1486–1495.

68. **Tchobanoglous, G., H. Theisen, and S. Vigil.** 1993. *Integrated Solid Waste Management,* McGraw-Hill Inc., New York.

69. **Theodorou, M. K., C. King-Spooner, and D. E. Beever.** 1989. *Presence or Absence of Anaerobic Fungi in Landfill Refuse.* ETSU B 1246. Energy Technology Support Unit, Department of Energy, Harwell Laboratory, England.

70. **Thorneloe, S. A., and J. G. Pacey.** 1994. Landfill gas utilization—database of North American projects. Presented at the Solid Waste Association of North America 17th Landfill Gas Symposium, March 22–24, Long Beach, Calif.

71. **U. S. Environmental Protection Agency.** 1994. *Characterization of Municipal Solid Waste in the United States: 1994 Update.* EPA 530-R-94-042. U.S. Environmental Protection Agency, Washington, D.C.

72. **Wang, Y.-S., C. S. Byrd, and M. A. Barlaz.** 1994. Anaerobic biodegradability of cellulose and hemicellulose in excavated refuse samples. *J. Ind. Microbiol.* **13:**147–153.

73. **Westlake, K.** 1994. *Microbial Metabolism in the Brogborough Landfill Gas Enhancement Test Cells.* ETSU B/LF/00201/REP. Energy Technology Support Unit, Department of Trade and Industry, Oxfordshire, England.

74. **Westlake, K., and D. B. Archer.** 1990. Fundamental studies on cellulose degradation in *Landfills.* ETSU B 1228. Energy Technology Support Unit, Department of Energy, Harwell Laboratory, England.

75. **Whalen, S. C., W. S. Reeburgh, and K. A. Sandbeck.** 1990. Rapid methane oxidation in a landfill cover soil. *Appl. Environ. Microbiol.* **56:**3405–3411.

76. **Widdick, D. A., and T. M. Embley.** 1992. *Use of Nucleic Acid Technology in Landfill. A Feasibility Study.* ETSU B 1315. Energy Technology Support Unit, Department of Trade and Industry, Oxfordshire, England.

77. **Windam, W. R., and D. E. Akin.** 1984. Rumen fungi and forage fiber digestion. *Appl. Environ. Microbiol.* **48:** 473–476.

78. **Wolin, E. A., M. J. Wolin, and R. S. Wolfe.** 1963. Formation of methane by bacterial extracts. *Biol. Chem.* **238:** 2882–2886.

79. **Young, L. Y., and A. C. Frazer.** 1987. The fate of lignin and lignin derived compounds in anaerobic environments. *Geomicrobiol. J.* **5:**261–293.

80. **Zehnder, A. J. B.** 1978. Ecology of methane formation, p. 349–376. *In* R. Mitchell (ed.), *Water Pollution Microbiology,* vol. 2. John Wiley & Sons, Inc., New York.

81. **Zinder, S. H.** 1993. Physiological ecology of methanogenesis, p. 128–206. *In* J. G. Ferry (ed.), *Methanogenesis: Ecology, Physiology, Biochemistry and Genetics.* Chapman & Hall, New York.

Alteration of Aquifer Geochemistry by Microorganisms

FRANCIS H. CHAPELLE AND PAUL M. BRADLEY

61

It has long been known that microorganisms present in aquifer systems can alter groundwater geochemistry. As early as 1900 it was observed that groundwater associated with petroleum deposits often lacked dissolved sulfate, whereas groundwater not associated with petroleum contained sulfate. In 1917, it was first suggested that this effect was caused by the metabolism of sulfate-reducing bacteria (27). This hypothesis was confirmed in 1926 when sulfate-reducing bacteria were isolated from brines associated with petroleum deposits (2). Over the next 40 years, a number of investigators came to similar conclusions about the importance of microorganisms in altering groundwater geochemistry (3, 13). However, it was not until methods for aseptic sampling of subsurface sediments were developed (8, 12, 25, 29) and comprehensive evaluations of microorganisms present in these sediments were made (6, 7, 10, 11) that the effects of microbial metabolism on groundwater geochemistry became widely known.

It has recently been shown that microorganisms can affect the physical properties of aquifer systems as well as the chemistry of groundwater. Geologists have long known that secondary porosity (that is, porosity caused by the dissolution of solid aquifer matrix) can allow the accumulation of petroleum in otherwise nonporous rocks (24). In addition, it has been known that secondary porosity can enhance the water-bearing properties of some aquifer systems (22). However, the processes leading to secondary porosity development were not well understood. Painstaking isotopic and mass balance studies showed that decarboxylation of organic matter and other abiotic processes could not account for all of the observed secondary porosity in many systems (16). The realization that most aquifer systems contain active microorganisms raised the possibility that microbial metabolism could lead to secondary porosity development (6). Subsequent studies showed that microorganisms could serve to either destroy (19) or enhance (14, 21) porosity in aquifer sediments.

The central tenet of this chapter is that the metabolism and growth of microorganisms present in aquifers may alter the chemical composition of groundwater and the physical properties of aquifers. While the extents of these alterations vary widely between aquifers depending on rates of microbial metabolism, hydrologic setting, and aquifer mineralogy, the patterns of observed alterations exhibit certain similarities. The purpose of this chapter is to describe some common patterns of geochemical alterations observed in aquifer systems and to illustrate them with examples.

MICROBIAL ALTERATION OF GROUNDWATER GEOCHEMISTRY

The ways in which microbial processes alter the chemical composition of groundwater vary widely. However, two broad categories of water chemistry patterns are associated with conditions that limit microbial activity in a given aquifer system. The first and most common category of aquifers consists of those that are electron donor limited; i.e., microbial metabolism is inherently limited by the availability of organic carbon or other potential electron donors. Many aquifers systems in this category are composed of sediments or rocks tens to hundreds of million years old from which metabolizable organic carbon has been progressively removed over geologic time. The other broad category of aquifers consists of those that are electron acceptor limited. In these systems, there is an excess of electron donor available (usually but not necessarily organic carbon) such that microbial metabolism is constrained by the availability of electron acceptors. Examples of this category of aquifers include most natural petroleum reservoirs and, importantly, many aquifers that have been chemically contaminated by human activities. In this section, the alteration of groundwater geochemistry in each of these kinds of aquifers will be considered.

Electron Acceptor-Limited Aquifers

The groundwater geochemistry of the Black Creek aquifer in South Carolina was first described in 1950 (9) and has been extensively studied since that time (28). In addition, the effects of microbial processes on the groundwater chemistry of this aquifer have been described in detail (5, 17, 18). The Black Creek aquifer has attracted so much attention primarily because it exhibits a nearly complete sequence of classic water chemistry changes commonly observed in regional groundwater systems. In this hydrologic system, groundwater flows from recharge areas near the fall line to discharge areas near the Atlantic Ocean (Fig. 1). As groundwater moves along aquifer flow paths, there is a striking increase in concentrations of dissolved inorganic carbon (DIC) (Fig. 2A). Concentrations of DIC increase from less than 1 mmol/liter (mM) near recharge areas to more than 12 mM 150 km downgradient. This DIC is produced by

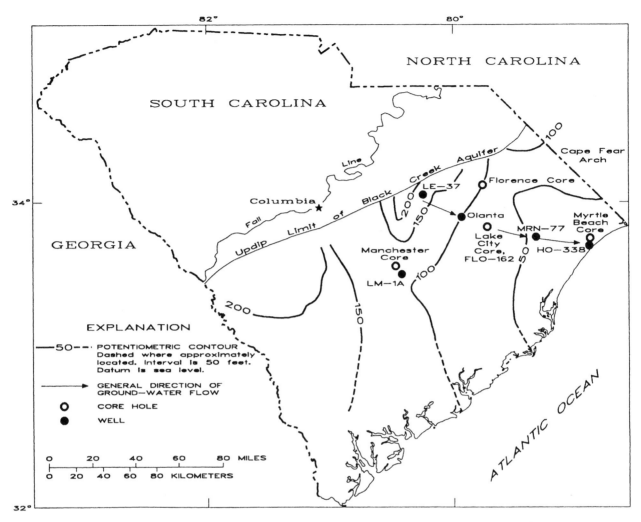

FIGURE 1 Map showing directions of groundwater flow, locations of wells used for water chemistry control, and locations of core holes penetrating the Black Creek aquifer in South Carolina.

microbial metabolism driving the dissolution of carbonate shell material in the aquifer according to the equation

$$CaCO_{3(carbonate)} + CO_{2(microbial)} \rightarrow Ca^{2+} + 2HCO_3^{-}$$

Because approximately half of the DIC produced comes directly from microbial metabolism (-6 mM) and because it takes approximately 150,000 years for groundwater to travel from the recharge area, it follows that microbial metabolism produces DIC at an overall rate of about 10^{-4} mmol/liter of water per year. Thus, even though increases of DIC concentration are large along the aquifer flow path, they reflect very low rates of microbial metabolism.

The principal reason rates of microbial metabolism are low in this system is that aquifers contain very little metabolizable organic carbon. It has been shown, for example, that Black Creek aquifer sediments contain between 0.1 and 1.0% organic carbon on a dry-weight basis (21). Because of these low rates of microbial metabolism, the amounts of available electron acceptors in the system [oxygen, Fe(III), sulfate, and CO_2] are large relative to the abundance of organic carbon. The electron donor-limited nature of this aquifer explains much of the behavior of other solutes in the system. Concentrations of dissolved oxygen drop below

detectable levels near the recharge area (Fig. 2B). Once groundwater becomes anoxic, concentrations of ferrous iron increase, indicating the initiation of active Fe(III) reduction (Fig. 2C). Downgradient of the high-iron zone, sulfate reduction becomes the principal electron acceptor (Fig. 2D), with sulfate in groundwater being continuously replenished from high-sulfate water trapped in clay beds that confine the aquifer (4, 26). It is not until the very end of the flow path that concentrations of sulfate decrease to the point that methanogenesis becomes an important process (Fig. 2E).

The tendency of microbial electron-accepting processes to proceed from O_2 reduction to Fe(III) reduction to sulfate reduction to methanogenesis in the direction of groundwater flow, as illustrated by the chemistry of the Black Creek aquifer, is the defining characteristic of electron donor-limited aquifers. This behavior follows directly from the abundance of potential electron acceptors relative to available electron donors.

Electron Acceptor-Limited Aquifers

While many ancient sediments have little available organic carbon (or other potential electron donors), geologic or hy-

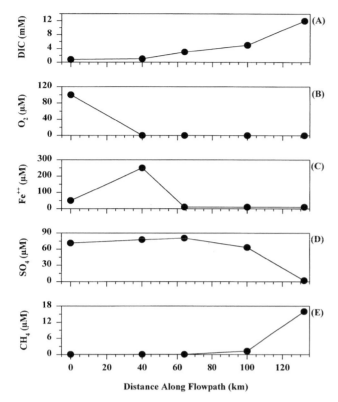

FIGURE 2 Concentrations of DIC (A), dissolved oxygen (B), ferrous iron (C), sulfate (D), and methane (E) in groundwater from the Black Creek aquifer.

drologic conditions can be such that there is an abundance of available carbon. When the supply of available carbon is very high, microbial metabolism can become limited by the lack of available electron acceptors. Examples of electron acceptor-limited groundwater systems include peat aquifers (which are common in northern latitudes), petroleum deposits, and aquifers that have been chemically contaminated by human contrivances such as petroleum spills, landfill leachates, or septic effluents. One of the best-documented examples of an electron acceptor-limited aquifer, however, is the result of natural hydrologic processes occurring near the town of Valdosta, Georgia.

The Floridan aquifer, which underlies much of the southeastern United States, is one of the most productive hydrologic systems in the world. The aquifer consists of Tertiary age (that is, about 50 million-year-old) limestones (23) characterized by the presence of nodular gypsum (a calcium sulfate mineral). Where Floridan limestones crop out at land surface, secondary dissolution features such as sinkholes and disappearing streams (karst topography) are common. These features contribute to an unusual groundwater quality problem near Valdosta.

The channel of the Withlacoochee River just north of Valdosta intersects a series of sinkholes (Fig. 3). These sinkholes capture a portion of the river's flow, and this water recharges the Floridan aquifer. Because the Withlacoochee River is a typical south Georgia blackwater stream, characterized by high concentrations of dissolved and particulate organic carbon (referred to as total organic carbon [TOC]), a large amount of TOC is delivered to the Floridan aquifer at this location (15). This TOC is a readily available electron

donor, and rates of microbial metabolism are very rapid in this system. Given the initial TOC concentrations of 6 mg/liter as C (-0.5 mM C), and given rapid rates of groundwater flow in this system (-1 km/year), overall rates of metabolism are on the order of 0.025 mmol of C per liter per year, which is 2 orders of magnitude faster than in the Black Creek aquifer. Because of such rapid rates of microbial metabolism, observed water chemistry patterns in the Floridan aquifer near Valdosta are much different from those in the Black Creek aquifer.

Dissolved oxygen is consumed rapidly (Fig. 4A), as was the case in the Black Creek aquifer. However, concentrations of methane increase rapidly (Fig. 4B) because of the initial lack of alternative electron acceptors such as sulfate (Fig. 4C). About 8 km downgradient, sulfate becomes available (Fig. 4D) and sulfide production becomes evident (Fig. 4E), indicating the initiation of active sulfate reduction. These patterns indicate that near the source of TOC recharge, available sulfate has been consumed by a combination of hydrologic (rapid gypsum dissolution) and microbial (rapid metabolism) factors. This has resulted in methanogenesis occurring *closest* to the recharge area, exactly opposite the pattern observed in the Black Creek aquifer, where methanogenesis occurred *farthest* from the recharge area. This pattern is characteristic of an electron acceptor-limited groundwater system.

Electron acceptor-limited groundwater systems are important because they are characteristic of groundwater contamination caused by human activities. One of the best-documented examples of this phenomenon has been given by Baedecker and coworkers at the site of a crude oil spill in Minnesota (1). At this site, it was shown that methanogenesis predominated near the crude oil, followed by Fe(III) reduction 10 to 20 m downgradient, followed by oxygen reduction 100 m downgradient. Sulfate was not available in sufficient quantity for sulfate reduction to be an important process in this system. This general pattern of water chemistry changes characteristic of electron acceptor limitations is common in chemically contaminated groundwater systems.

MICROBIAL ALTERATION OF AQUIFER PROPERTIES

Microbial processes not only systematically alter the geochemistry of groundwater but also can alter the hydrologic properties of the aquifers themselves. This is a topic of special interest to hydrologists, since groundwater can move freely only in sediments or rocks that have sufficient effective porosity (that is, interconnected pore space). This is also a topic of interest in contaminant hydrology because aquifer porosity can store oily-phase contaminants, making them difficult to remove. As before, it is useful to consider the microbial generation of aquifer porosity and permeability in the context of electron donor or electron acceptor limitations.

Electron Donor-Limited Aquifers

Groundwater flows in the Black Creek aquifer from recharge areas along the fall line to discharge areas along the Atlantic coast (Fig. 1). As discussed previously, the geochemistry of groundwater is significantly altered by microbial processes along this flow path (Fig. 2). In addition to altering water chemistry, microbial processes also alter the physical properties of this aquifer. Data from core holes oriented along flow paths have shown a marked lithologic

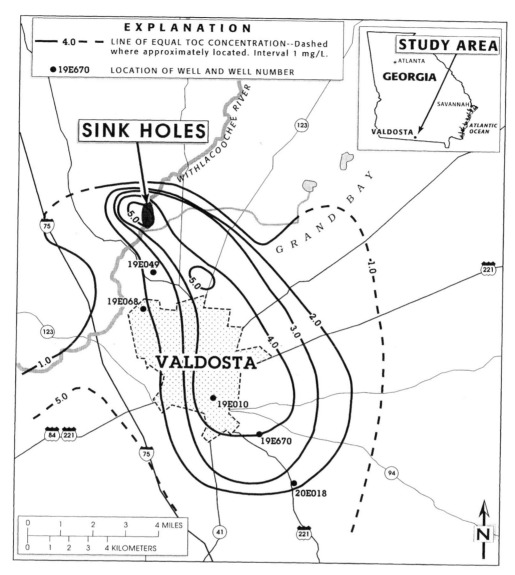

FIGURE 3 Map showing distribution of TOC in groundwater downgradient of sinkholes recharging the Floridan aquifer near Valdosta, Georgia.

change. In recharge areas near the fall line, secondary intergranular calcite cements are virtually nonexistent. At Lake City, which is intermediate between recharge areas and discharge areas, calcite cements are more common. In Myrtle Beach, which is near discharge areas, up to 50% of the total thickness of the aquifer has been cemented by intergranular calcite cement. This loss of porosity has a significant effect on the hydrology of this aquifer system. The transmissivity (that is, aquifer permeability multiplied by aquifer thickness) of the Black Creek aquifer is so low that groundwater cannot supply municipal water needs for the city of Myrtle Beach. This low transmissivity can be attributed, in part, to the abundance of intergranular cements that fill primary porosity in this aquifer system.

The microbial processes leading to the development of pore-filling cements in the Black Creek aquifer have been investigated in detail (19). This study showed that while sands of the Black Creek aquifer contained relatively little organic carbon, adjacent confining beds contained abundant organic carbon. The fermentation of this confining-bed organic carbon causes the accumulation of organic acids in confining-bed pore water. The diffusive flux of these organic acids to the Black Creek aquifer and the subsequent oxidation of these acids in the aquifer lead to a net mass transfer of carbon. As DIC accumulates in the Black Creek aquifer, the groundwater becomes oversaturated with respect to the mineral calcite, which readily precipitates according to the equation

$$Ca^{2+} + CO_3^- \rightarrow CaCO_3 \text{ (cement)}$$

The contribution of carbon derived from organic material to the pore-filling cements is recorded in the stable carbon isotope composition of the cements (Fig. 5). Primary carbonate shell material in these sediments has a $\delta^{13}C$ of about 0‰, whereas organic carbon in confining beds has a $\delta^{13}C$ of about 20‰. The pore-filling cements present at the Lake City site clearly show the signature of isotopically light carbon derived from organic matter. Because this organic mat-

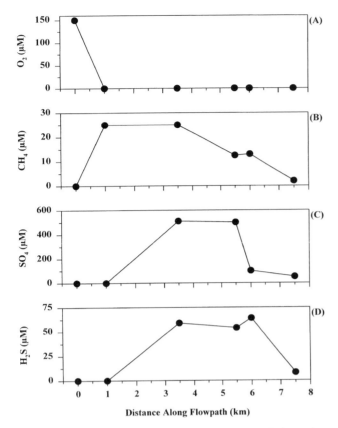

FIGURE 4 Concentrations of dissolved oxygen (A), methane (B), sulfate (C), and hydrogen sulfide (D) in groundwater from the Floridan aquifer near Valdosta, Georgia.

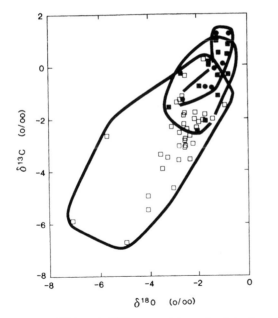

FIGURE 5 $\delta^{13}C$ and $\delta^{18}O$ compositions of calcite in sediments from the Black Creek aquifer. •, shell material in unconsolidated clays; ■, shell material in unconsolidated sands; □, intergranular calcite cements.

ter was oxidized by microbial activity, it follows that microbial activity is driving the production of intergranular cements. This is a particularly well-documented example of how microbial processes can alter the hydrologic properties of an aquifer system by destroying primary porosity. Many sedimentary rocks in the geologic record are cemented by secondary calcite cements, and it is likely that these cements also reflect microbial processes operating over long periods of geologic time.

Electron Acceptor-Limited Aquifers

In the previous example, we saw how microbial processes in an electron donor-limited aquifer can serve to decrease aquifer porosity by the production of intergranular cements. In electron acceptor-limited aquifers, this pattern is often reversed with the production of secondary porosity (that is, porosity caused by the dissolution of mineral grains).

This phenomenon was first studied systematically in a study of a shallow aquifer contaminated by petroleum hydrocarbons (14). These investigators were particularly interested in secondary porosity associated with silicate minerals such as quartz and feldspars. They prepared in situ microcosms by crushing feldspars, sieving them to obtain uniform size, cleaning them with low-power ultrasonication, placing them in porous polyethylene cylinders, and submerging them in wells tapping a petroleum-contaminated aquifer for 14 months. Split samples of material placed into each microcosm were left unreacted to serve as controls. On recovery, the mineral samples were examined with scanning electron microscopy.

It was found that bacterial cells exhibiting a variety of morphologies colonized mineral surfaces and that there was intense etching of the feldspars in particular. Some of this etching could be attributed to chemical dissolution due to the relatively high concentrations of organic acids in the groundwater. Other dissolution features were associated with individual cells or colonies of cells. It was observed that cells created a "reaction zone" in their immediate vicinity in which organic acids, and possibly exoenzymes produced within the cell, created dissolution "pits" and a net increase in secondary porosity.

In addition to experimental evidence that microorganisms can produce secondary porosity in aquifers (14), there is field evidence of this phenomenon as well. It has been shown that quartz grains within a petroleum hydrocarbon-contaminated aquifer (20) exhibit pitting and etching features (Fig. 6) that are not observed in quartz grains outside the contaminated zone. These investigators also showed that the greatest amounts of secondary porosity development in this aquifer were associated with high concentrations of organic acids. This finding confirms that microbially produced intermediate products such as organic acids are directly involved with the observed porosity development, as was suggested by experimental evidence (14).

The production of secondary porosity by microbial activity in electron acceptor-limited aquifers has several practical implications. In chemically contaminated aquifers, the production of porosity can lead to the entrapment of oily-phase contaminants, which can serve to complicate the process of aquifer remediation. On the other hand, the production of micropores and the subsequent diffusion of dissolved contaminants into them can serve to restrict the transport of dissolved contaminants (30). The most important economic effect of porosity production in electron acceptor-limited aquifers, however, occurs in petroleum reservoirs. Petrographic examination of rocks from petroleum reservoirs has

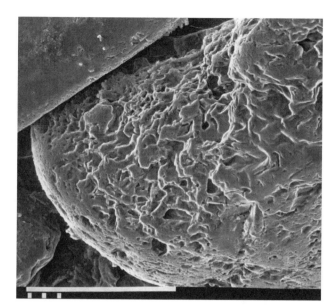

FIGURE 6 Scanning electron micrograph showing highly corroded quartz grain from a contaminated aquifer. Bar = 100 μm.

long suggested that the introduction of hydrocarbons can lead to porosity enhancement (24). This porosity enhancement, in turn, can lead to the accumulation of economically important quantities of petroleum or natural gas. It is probable that much of the porosity associated with important petroleum reservoirs is a direct result of microbial processes.

SUMMARY

The metabolism of microorganisms progressively alters the chemical composition of groundwater and the hydrologic properties of water-bearing strata in virtually all aquifer systems. The patterns of water chemistry changes observed depend on the relative abundance of electron donors and electron acceptors. When an aquifer is electron donor limited (that is, highly oligotrophic), the slow metabolism of available organic carbon results in the accumulation of DIC along aquifer flow paths, and available electron acceptors are consumed in the order dissolved oxygen > nitrate > Fe(III) > sulfate > CO_2 (methanogenesis). When an aquifer is electron acceptor limited (that is, there is an excess of available organic carbon), DIC also accumulates along aquifer flow paths but the order of electron acceptor consumption is reversed. The accumulation of intermediate and final products of microbial metabolism (such as organic acids, ferrous iron, hydrogen sulfide, and methane) in electron acceptor-limited aquifers often decreases the suitability of groundwater for human consumption. Concurrent with changes in groundwater geochemistry, the solid matrix of the aquifer can also be altered by microorganisms. The production of DIC in electron donor-limited aquifers can result in the precipitation of secondary carbonate cements that reduce primary porosity and permeability over geologic time. Conversely, the production of organic acids in electron acceptor-limited aquifers can lead to the production of secondary porosity. Thus, microorganisms can alter both the chemical quality of groundwater and the water-producing capacity of aquifer systems.

REFERENCES

1. **Baedecker, M. J., I. M. Cozzarelli, D. I. Siegel, P. C. Bennett, and R. P. Eganhouse.** 1993. Crude oil in a shallow sand and gravel aquifer. 3. Biogeochemical reactions and mass balance modeling in anoxic ground water. *Appl. Geochem.* **8:**569–586.
2. **Bastin, E. S.** 1926. The presence of sulphate-reducing bacteria in oil-field waters. *Science* **63:**21–24.
3. **Cedarstrom, D. J.** 1946. Genesis of groundwaters in the coastal plain of Virginia. *Econ. Geol.* **41(3):**218–245.
4. **Chapelle, F. H., and D. R. Lovley.** 1990. Rates of bacterial metabolism in deep coastal-plain aquifers. *Appl. Environ. Microbiol.* **56:**1865–1874.
5. **Chapelle, F. H., and P. B. McMahon.** 1991. Geochemistry of dissolved inorganic carbon in a coastal plain aquifer. 1. Sulfate from confining beds as an oxidant in microbial CO_2 production. *J. Hydrol.* **127:**85–108.
6. **Chapelle, F. H., J. T. Morris, P. B. McMahon, and J. L. Zelibor, Jr.** 1988. Bacterial metabolism and the del-13C composition of ground water, Floridan aquifer, South Carolina. *Geology* **16:**117–121.
7. **Chapelle, F. H., J. L. Zelibor, D. J. Grimes, and L. L. Knobel.** 1987. Bacteria in deep coastal plain sediments of Maryland: a possible source of CO_2 to ground water. *Water Resour. Res.* **23:**1625–1632.
8. **Dunlap, W. J., J. F. McNabb, M. R. Scalf, and R. L. Cosby.** 1977. *Sampling for Organic Chemicals and Microorganisms in the Subsurface.* Report EPA-600/2-77-176. U.S. Environmental Protection Agency.
9. **Foster, M. D.** 1950. The origin of high sodium bicarbonate waters in the Atlantic and Gulf Coastal Plains. *Geochim. Cosmochim. Acta* **1(1):**33–48.
10. **Fredrickson, J. F., D. L. Balkwill, J. M. Zachara, S. W. Li, F. J. Brockman, and M. A. Simmons.** 1991. Physiological diversity and distributions of heterotrophic bacteria in deep cretaceous sediments of the Atlantic Coastal Plain. *Appl. Environ. Microbiol.* **57:**402–411.
11. **Fredrickson, J. K., T. R. Garland, R. J. Hicks, J. M. Thomas, S. W. Li, and S. M. McFadden.** 1989. Lithotrophic and heterotrophic bacteria in deep subsurface sediments and their relation to sediment properties. *Geomicrobiol. J.* **7:**53–66.
12. **Ghiorse, W. C., and D. L. Balkwill.** 1983. Enumeration and characterization of bacteria indigenous to subsurface environments. *Dev. Ind. Microbiol.* **24:**213–224.
13. **Gurevich, M. S.** 1962. The role of microorganisms in producing the chemical composition of ground water. *Trans. Inst. Microbiol.* **9:**65–75.
14. **Hiebert, F. K., and P. Bennett.** 1992. Microbial control of silicate weathering in organic-rich ground water. *Science* **258:**278–281.
15. **Krause, R. E.** 1976. *Occurrence and Distribution of Color and Hydrogen Sulfide in Water from the Principal Artesian Aquifer in the Valdosta Area, Georgia.* U.S. Geological Survey Open File Report 76–378. U.S. Geological Survey, Columbia, S.C.
16. **Lundergard, P. D., and L. S. Land.** 1986. Carbon dioxide and organic acids: their role in porosity enhancement and cementation, Paleogene of the Texas Gulf Coast. *SEPM Spec. Publ.* **38:**129–146.
17. **McMahon, P. B., and F. H. Chapelle.** 1991. Microbial production of organic acids in aquitard sediments and its role in aquifer geochemistry. *Nature* (London) **349:**233–235.
18. **McMahon, P. B., and F. H. Chapelle.** 1991. Geochemistry of dissolved inorganic carbon in a Coastal-Plain aquifer. 2. Modeling carbon sources, sinks, and $\delta^{13}C$ evolution. *J. Hydrol.* **127:**109–135.
19. **McMahon, P. B., F. H. Chapelle, W. F. Falls, and P. M.**

Bradley. 1992. The role of microbial processes in linking sandstone diagenesis with organic-rich clays. *J. Sediment. Petrol.* **62**(1):1–10.

20. **McMahon, P. B., D. A. Vroblesky, P. M. Bradley, F. H. Chapelle, and C. D. Gullet.** 1995. Evidence for enhanced mineral dissolution in organic-acid-rich shallow ground water. *Ground Water* **33**:207–216.

21. **McMahon, P. B., D. F. Williams, and J. T. Morris.** 1990. Production and isotopic composition of bacterial CO_2 in deep coastal-plain sediments of South Carolina. *Ground Water* **28**:693–702.

22. **Meinzer, O. E.** 1923. *The Occurrence of Ground Water in the United States, with a Discussion of Principles.* U.S. Geological Survey Water-Supply Paper 489. U.S. Geological Survey, Columbia, S.C.

23. **Miller, J. A.** 1986. *Hydrogeologic Framework of the Floridan Aquifer System in Florida and Parts of Georgia, Alabama, and South Carolina.* U.S. Geological Survey Professional Paper 1403-B. U.S. Geological Survey, Columbia, S.C.

24. **Moncure, G. K., R. W. Lahann, and R. M. Siebert.** 1984. Origin of secondary porosity and cement distribution in a sandstone/shale sequence from the Frio Formation (Oligocene). *AAPG Mem.* **37**:151–161.

25. **Phelps, T. J., E. G. Raione, D. C. White, and C. B. Fliermans.** 1989. Microbial activities in deep subsurface environments. *Geomicrobiol. J.* **7**(1–2):79–92.

26. **Pucci, A. A., Jr., and J. P. Owens.** 1989. Geochemical variations in a core of hydrogeologic units near Freehold, New Jersey. *Ground Water* **27**:802–812.

27. **Rogers, G. S.** 1917. Chemical relations of the oil-field waters in San Joaquin Valley, California. *U.S. Geol. Surv. Bull.* **653**:93–99.

28. **Speiran, G. K.** 1987. Relation of aqueous geochemistry to sedimentary depositional environments. *Am. Water Resour. Assoc. Monogr. Ser.* **9**:79–96.

29. **Wilson, J. T., J. F. McNabb, D. L. Balkwill, and W. C. Ghiorse.** 1983. Enumeration and characterization of bacteria indigenous to a shallow water-table aquifer. *Ground Water* **21**:134–142.

30. **Wood, W. W., T. F. Kraemer, and P. P. Hearn.** 1990. Intragranular diffusion: an important mechanism influencing solute transport in clastic aquifers? *Science* **247**:1569–1572.

Stable Carbon Isotopes as Indicators of Microbial Activity in Aquifers

ETHAN L. GROSSMAN

62

Stable carbon isotopes (^{13}C/^{12}C) are sensitive indicators of microbial processes. Many important microbial reactions such as acetogenesis and CO_2 reduction are associated with large isotopic fractionations (i.e., enrichment of one isotope relative to another). These reactions can substantially affect the ^{13}C/^{12}C of dissolved bicarbonate, methane, and acetate in groundwater. Geochemists use chemical and isotopic analyses of these groundwater constituents as well as aquifer sediments to understand the processes active in aquifer systems. This approach is especially effective when applied to changes along groundwater flow paths. This chapter will discuss the principles and methodology behind the use of stable carbon isotopes as indicators of microbial activities in aquifers, along with specific applications. The treatment is far from exhaustive, and the reader is referred to the original studies for details.

ISOTOPE RATIO MASS SPECTROMETRY

Stable isotope ratios of the light elements (H, C, N, O, S) are measured as gases (H_2, CO_2, N_2, SO_2) with a dynamic source (gas) isotope ratio mass spectrometer (IRMS). This instrument consists of: (i) an inlet system to introduce the gas, (ii) an ion source to convert gas molecules into ions which are focused and accelerated into an ion beam, (iii) a mass analyzer (magnet) to separate the ion beam into beams of different mass/charge ratio, and (iv) ion detectors to determine the intensity of the different ion beams (35, 44). In conventional dual-inlet systems, sample and standard gases are alternately introduced into the mass spectrometer via capillary leaks. The gas chromatograph-combustion-IRMS (GC-C-IRMS), a recent advance, measures ^{13}C/^{12}C ratios by separating gas mixtures (e.g., N_2, CO_2, CH_4, C_2H_6) by gas chromatography, combusting reduced gases to CO_2, and then detecting the CO_2 in ion collectors (56, 76).

Carbon isotopic compositions are reported as the per mille (‰) difference in isotope ratio relative to a standard in δ notation:

$$\delta^{13}C(‰) = \left(\frac{^{13}C/^{12}C_{sample} - \,^{13}C/^{12}C_{standard}}{^{13}C/^{12}C_{standard}} \right) \times 1,000 \quad (1)$$

The standard for reporting carbon isotope data is PDB (Peedee belemnite), which is the carbonate skeleton of a belemnite from the Cretaceous Peedee Formation in South Carolina (31). Because PDB is no longer available, researchers analyze secondary standards such as NBS-20 (Solenhofen limestone) and NBS-19 (TS limestone) which have been calibrated to PDB.

SAMPLING AND SAMPLE PREPARATION

In sampling for dissolved carbon species in groundwater, care must be taken to collect samples representative of the aquifer water and to avoid exposure of sample to air. Conventional groundwater sampling methods are discussed in detail by Wood (96) and Herzog et al. (43), among others. Before sampling a well, stagnant water above the screened interval is removed by pumping or bailing. Researchers typically remove an amount equivalent to three times the volume of the stagnant water column (see reference 43 for a discussion of alternative approaches). Temperature, specific conductance, pH, and dissolved oxygen are monitored during pumping. Constant measurements are evidence that representative aquifer waters are being sampled. In recent years, scientists have employed innovative sampling techniques to study aquifer heterogeneity and dissolved gases in deep wells. One approach involves the use of a multilayer sampler in which aquifer water is allowed to equilibrate with deionized water in stacked dialysis cells spaced at 3-cm intervals, providing a fine-scale record of aquifer heterogeneity (74). Alternatively, packers may be used to isolate and sample discrete borehole intervals. Pressurized water samples can be collected in specially designed downhole sample vessels (79).

Sample preparation for carbon isotopic analyses involves conversion of the sample to CO_2, which is subsequently dried through a cold trap (e.g., dry ice-alcohol slurry) and introduced into the IRMS. CO_2 is transferred through the vacuum line using liquid nitrogen. Conventional IRMS can routinely analyze CO_2 sample sizes of 0.2 μmol. With high temperature combustion, somewhat larger samples may be required because of background CO_2 production from cupric oxide and copper reagents. Even smaller CO_2 samples (0.01 to 0.1 μmol) can be analyzed with the GC-C-IRMS (76).

DIC

To sample groundwater dissolved inorganic carbon (DIC) (aqueous CO_2 + HCO_3^- + CO_3^{2-}), common practice is to use Tygon tubing to fill glass bottles (>15 ml) from the bottom, overflowing by one sample volume. Samples with particulate organic matter can be poisoned by adding 0.5 ml of saturated $HgCl_2$ solution per 100 ml of water to eliminate postsampling microbial activity. Sample bottles are capped with Polyseal caps, stored on ice in the field, and refrigerated after returning to the laboratory.

DIC is extracted from groundwater by acidification of the sample in vacuo. The evolved CO_2 is then passed through a dry ice-alcohol slurry to trap water and then frozen with liquid nitrogen (38, 41). In an alternative method, ammoniacal strontium chloride solution is added to the water sample in the field, precipitating DIC as solid carbonate. The bottle is immediately capped and shaken (41). In the laboratory, the precipitate is filtered, dried, and then reacted with phosphoric acid on a vacuum line to produce CO_2. The precision of $\delta^{13}C$ determinations by either of these techniques is typically better than ± 0.1‰ (1σ).

Methane

Methane that occurs as head gas can be sampled by filling a basin with water, placing the outflow tube in an inverted 500-ml Mason jar in the basin, and letting the gas displace the water (22). Once an adequate gas sample has accumulated, the jar is capped under water. Samples are immediately stored on ice. A syringe is used to sample gas through a septum in the jar lid. For dissolved methane samples, liter bottles are filled from the bottom in a basin and overflowed by at least one sample volume. The bottle is sealed underwater with a Polyseal cap. Alternatively, soda or beer bottles and a bottle capper can be used.

To analyze methane on a standard IRMS, dissolved methane is stripped from the water using helium as a carrier gas. Prior to combustion, contaminant CO_2 is removed using a liquid nitrogen trap. Methane is then passed through an 850°C tube furnace filled with oxidized copper wire (CuO) (97). The resulting CO_2 is cryogenically dried and analyzed on the IRMS. Head-gas methane is similarly prepared, except that the sample can be directly injected through a septum into the carrier gas stream. Precision for methane ^{13}C analyses is typically ± 0.1‰ (1σ).

Methane can be analyzed quickly on a GC-C-IRMS. Head gas is directly injected into the instrument. Methane is chromatographically separated from air and CO_2, combusted, and then introduced into the IRMS as a pulse. The precision is on the order of ± 0.3‰. Dissolved methane can be analyzed by equilibration of the dissolved gas with a head gas (E. L. Grossman and D. J. Katz, unpublished data). The bottle is inverted in a bucket filled with methane-free water and opened. Twenty-five milliliters of He is added, displacing water, and then the bottle is capped with a shortened rubber stopper. The bottle is shaken and then gas is extracted with a syringe and injected into the GC-C-IRMS. For low-concentration samples, methane can be concentrated with a cryofocusing device, in which injected methane is first frozen on molecular sieve material using liquid nitrogen (67).

DOC

Methods for extracting dissolved organic carbon (DOC) from groundwater are still evolving. In earlier studies of groundwater DOC, samples were concentrated by adsorption chromatography using XAD-8 resin for the high-molecular-weight (HMW) fraction (mostly fulvic acid) and a Silicalite molecular sieve for the low-molecular-weight (LMW) fraction (58, 90). More recently, researchers are using re-

verse osmosis and ultrafiltration to concentrate and fractionate DOC (26). Application of these techniques to isotopic studies of groundwater DOC is in progress.

For isotopic analysis, the HMW DOC fraction is freeze-dried and isotopically analyzed by the sealed-tube combustion method (90; see reference 14 for details). Organic material, cupric oxide, and copper wire are added to Vycor or quartz tubing (9 mm outer diameter) which has been sealed at one end. The tube is evacuated and sealed with a torch. The sample and reagents are mixed and the tube is combusted in a furnace initially at 900°C for 2 h and then at 650°C for 2 h. Evolved CO_2 is released into the vacuum line using a tube cracker and then dried and transferred cryogenically. The LMW fraction is thermally extracted from Silicalite and combusted by passage through a cupric oxide furnace at 800 to 900°C (58, 90). Undoubtedly, future studies will involve the separation and isotopic analyses of LMW fractions by GC-C-IRMS.

FUNDAMENTALS OF ISOTOPE FRACTIONATION

Isotopes of the same element exhibit almost identical chemical behavior; however, the mass differences result in subtle differences in the velocity and thermodynamic properties of isotopically substituted molecules. There are two types of isotope fractionation, kinetic fractionation and equilibrium fractionation. Isotopic fractionation associated with equilibrium between coexistent phases is expressed by the fractionation factor α:

$$\alpha = R_A/R_B \qquad (2)$$

where R is the isotope ratio of interest (e.g., $^{13}C/^{12}C$, $^{34}S/^{32}S$) and subscripts A and B are the phases of interest. Fractionation factors approach unity at high temperature (e.g., 1,000°C). For convenience, isotopic fractionation is often reported as $1,000 \ln\alpha_{A-B}$, which approximates $\delta_A - \delta_B$. Researchers also use the enrichment factor ϵ to denote fractionation, where ϵ (‰) $= 1,000(\alpha - 1) \approx 1,000 \ln\alpha \approx \delta_A - \delta_B$. If α equals 1, both $1,000 \ln\alpha$ and ϵ equal 0‰.

For an ideal gas, the kinetic fractionation associated with velocity differences can be readily calculated. At a given temperature, the average kinetic energy per molecule, E_K, will be $1/2\ mv^2$. Since different molecules have the same E_K, molecules with less mass will have a greater average velocity. Thus, molecules with the light isotope will have greater velocities than molecules with the heavy isotope and will react and diffuse faster. Graham's Law describes the equation for the theoretical kinetic fractionation factor based on the velocity ratios:

$$\alpha = v_L/v_H = (m_H/m_L)^{1/2} \qquad (3)$$

where v is velocity, m is mass, and the subscripts L and H refer to the light- and heavy-isotope-bearing molecule. For CO_2, the theoretical kinetic fractionation factor for carbon isotopes is 1.011, equivalent to 11‰.

Kinetic fractionation will dominate in kinetic reactions (i.e., unidirectional reactions) and will result in products enriched in the light isotope. Conversely, the residual reactant will become progressively enriched in the heavy isotope. Kinetic fractionation is important in reactions such as C_3 photosynthesis (29, 62) and microbial methane oxidation (8, 24, 85). These reactions involve low-molecular-weight reactants (CO_2, CH_4) and relatively well-mixed systems. As seen from equation 3, kinetic fractionation decreases with increasing molecular weight. Thus, catabolism of large organic molecules results in little or no kinetic fractionation. For example, Trust et al. (86) saw no fractionation in the biodegradation of fluoranthene (molecular weight, 202.3) after 56% mineralization.

Equilibrium fractionation depends on the thermodynamic properties of the reactant and product. Substitution of heavy isotopes into molecules lowers the vibrational frequencies of the bonds, thus lowering zero point energy and producing a stronger bond (63). The heavy isotopes will favor the molecule for which substitution will result in the greatest decrease in the thermodynamic free energy of the system. Isotopic fractionation decreases as temperature increases, as the energy differences between isotopically substituted molecules become less important. In nature, reactions that occur abiotically at high temperature with small isotope fractionation are microbially mediated at low temperature with large isotope fractionation.

In general, isotope fractionation is proportional to the relative mass difference between the major and minor isotopes (3, 63). Relative mass difference equals the difference in mass divided by the average mass. For example, the relative mass difference between ^{13}C and ^{12}C is 1/12.5 or 0.08. In contrast, the relative mass difference between ^{1}H and ^{2}H (deuterium) is 0.67. Not surprisingly, $^{2}H/^{1}H$ ratios vary in nature by over 400‰, whereas $^{13}C/^{12}C$ ratios generally vary by less than 120‰. Another factor that influences the magnitude of isotopic fractionation observed in nature is the variety of valences and bond types (especially covalent bond types) associated with the element of interest. For example, $^{13}C/^{12}C$ and $^{34}S/^{32}S$ isotope ratios vary in nature to a greater extent than do $^{30}Si/^{28}Si$ and $^{44}Ca/^{40}Ca$ isotope ratios (44).

In cases where reaction products do not isotopically reequilibrate with the reactants, dramatic isotopic changes can occur in reactants and products as the reaction progresses. This process, called Rayleigh distillation, is described by the equation:

$$\frac{R}{R_0} = \frac{\delta + 1,000}{\delta_0 + 1,000} = f^{\alpha - 1} \quad (4)$$

where R_0 and R are the initial and final isotope ratios, respectively; δ_0 and δ are the initial and final δ values, respectively; f is the fraction of reactant remaining; and α is the fractionation factor of the reaction ($R_{products}/R_{reactants}$) at equilibrium. Examples where Rayleigh distillation may be important include ^{13}C enrichment of DIC with methanogenesis, ^{13}C enrichment of methane with methanotrophy, and ^{34}S enrichment in dissolved sulfate with sulfate reduction.

CARBON CYCLING AND THE $\delta^{13}C$ OF SEDIMENTARY CARBON

The average $\delta^{13}C$ of the Earth, estimated from deep-seated mantle carbon (e.g., diamonds), is -5 to -7‰. This primordial carbon has been acted upon by organisms and separated into two sedimentary reservoirs: (i) a ^{13}C-depleted organic carbon reservoir (about -23‰) and (ii) a ^{13}C-enriched carbonate reservoir (about 1‰) (Fig. 1) (3). Simple mass balance calculations incorporating the size and $\delta^{13}C$ of these reservoirs confirm an average Earth $\delta^{13}C$ of -6 ± 1‰.

Carbon isotopes are cycled between the two large sedimentary reservoirs through the small but geochemically important reservoirs of the biosphere, atmosphere, and ocean.

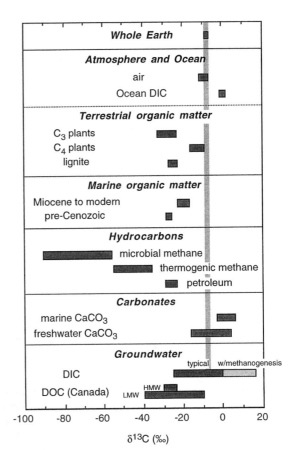

FIGURE 1 Typical $\delta^{13}C$ values for Earth material (from references 28 and 44, among others).

Photosynthesis of atmospheric CO_2 (-7 to -9‰) by terrestrial plants produces organic carbon with an average $\delta^{13}C$ of about -27‰ for C_3 plants and -13‰ for C_4 plants (Fig. 1 and 2) (28, 62). Crassulacean acid metabolism plants yield variable $\delta^{13}C$ values mostly in the range from -10 to -20‰. Marine algae use DIC in seawater to produce marine organic matter with a $\delta^{13}C$ of about -22‰ (in low to mid-latitudes).

A small fraction of terrestrial biomass (-12 to -28‰) is buried with sands, silts, and clays in fluvial and deltaic environments. Likewise, minor amounts of marine organic matter are deposited with marine sediments (-20 to -27‰, depending on age). This sedimentary organic matter (SOM) provides a substrate for microbial activity. Large portions of the continents have been intermittently flooded in the past, so the strata that form aquifers and aquitards often consist of interlayered marine and terrestrial sediments. In nearshore marine environments, SOM may be dominated by the influx of river-borne terrestrial organic matter. If the ancient shoreline lacked a source of clastic sediments, carbonate sediments (limestone, dolostone) may dominate. These sediments are usually low in organic carbon but form a large reservoir for ^{13}C-rich inorganic carbon (~1‰).

Carbon isotopes may be fractionated between different classes of organic compounds. For example, amino acids are ^{13}C-enriched and lipids are ^{13}C-depleted relative to bulk organic carbon (1, 27). Furthermore, the $\delta^{13}C$ values of cellulose and hemicellulose in vascular plants can be 4 to

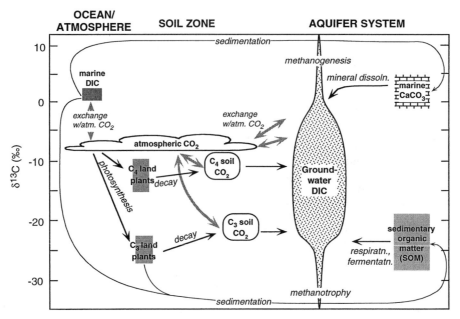

FIGURE 2 Schematic of atmosphere-ocean-land carbon cycling and its impact on groundwater DIC. Vertical ranges of polygons approximate $\delta^{13}C$ range of carbon reservoir.

7‰ higher (less negative) than that of lignin (11). Similar ^{13}C enrichments in cellulose and hemicellulose are observed in modern and ancient wood (83).

The isotopic composition of SOM can change with time in response to (i) preferential removal of compound groups that differ in $\delta^{13}C$ from the bulk organic matter, (ii) isotopic effects associated with bacterial degradation, and (iii) decarboxylation reactions which remove ^{13}C-enriched carboxyl groups (28, 53). Because cellulose and hemicellulose are more labile than lignin, degradation decreases the $\delta^{13}C$ of wood (and presumably SOM) by 1 to 2‰ (83).

CARBON ISOTOPES AND MICROBIAL PROCESSES

Carbon isotopes are useful indicators of those microbial processes that produce or consume isotopically distinct compounds. For example, respiration may add DIC with a $\delta^{13}C$ of −25‰ to groundwater with an initial $\delta^{13}C$ of −10‰. In such a case, mass balance equations can be used to determine the proportion of DIC added by respiration. Methanogenesis and methanotrophy are associated with large ^{13}C fractionation and impart a unique "fingerprint" to the $\delta^{13}C$ of DIC. In addition, the $\delta^{13}C$ of microbial biomass can be used to trace carbon sources and distinguish between autotrophs and heterotrophs.

Heterotrophy

It is common knowledge that the $\delta^{13}C$ of respired CO_2 approximates that of the substrate. Close examination, however, shows that microbial aerobic respiration can produce CO_2 depleted in ^{13}C by a few per mille relative to the substrate (n-alkanes [98], glucose [12]). Even larger ^{13}C depletions are suggested from experiments using lactose as the substrate (81). Carbon isotope fractionation between cell carbon in *Pseudomonas aeruginosa* and substrate averages −2.3 ± 0.6‰; similarly, fractionation between nucleic acids and substrate averages −2.4 ± 0.4‰ (21).

Methylotrophs preferentially utilize $^{12}CH_4$ over $^{13}CH_4$. CO_2 produced from methane oxidation is depleted in ^{13}C relative to methane by 4 to 30‰ (8, 24, 99). This fractionation causes progressive enrichment in residual CH_4, which subsequently results in progressive ^{13}C enrichment in CO_2 (although initially very depleted in ^{13}C; Fig. 3). On the other hand, if the CH_4 is completely oxidized, conservation of mass dictates that the net $\delta^{13}C$ of the respired CO_2 will equal that of the original CH_4. In experiments with *Methylomonas methanica*, Zyakun et al. (99) noted that carbon isotopes are fractionated between CH_4 and biomass as well as between CH_4 and CO_2. The net fractionation associated with methane consumption (−16‰) combines fractionations associated with biomass production (−27‰) and CO_2 production (−5‰). A further consideration is that ^{13}C

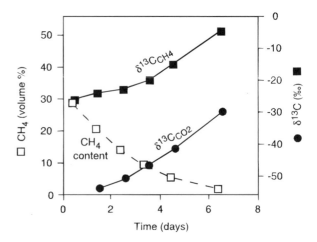

FIGURE 3 Laboratory experiment showing increases in the $\delta^{13}C$ values of CH_4 and CO_2 as CH_4 is consumed by methylotrophs. Reprinted with permission from *Nature* (8). © 1981 Macmillan Magazines Ltd.

fraction in biomass has been found to decrease as cells approach stationary-phase growth or when growth occurs in reduced methane concentrations (85). In natural sediments, estimates of fractionation associated with CH_4 consumption by methylotrophs range from -4 to $-26‰$ (47, 87, 93). Depending upon the $\delta^{13}C$ of the environmental methane and the amount of recycled biomass, CO_2 released into an aquifer by methylotrophs can initially have $\delta^{13}C$ values as low as $-100‰$.

Chemoautotrophy and Acetogenesis

Like photosynthesis, chemoautotrophy discriminates against heavy isotopes. In experiments with the sulfur-oxidizing bacteria *Thiomicrospira* sp. and *Thiobacillus neapolitanus*, Ruby et al. (77) observed a $-25‰$ fractionation in cell mass $\delta^{13}C$ relative to the DIC of the medium. Preuß et al. (68) observed that the ^{13}C fractionation between biomass and CO_2 depends on the CO_2 fixation pathway, varying from $-10‰$ for the reductive citric acid cycle to $-36‰$ for the reductive acetyl-coenzyme A pathway. Synthesized acetate is even more depleted in ^{13}C than biomass carbon. For the anaerobic synthesis of acetate from CO_2 and H_2 by *Acetobacterium woodii*, Preuß et al. (68) and Gelwicks et al. (36) obtained 1,000 $\ln\alpha_{acetate-CO_2}$ values of $-40‰$ and $-52‰$, respectively. In contrast, acetate produced aerobically by *Escherichia coli* was 12‰ *enriched* in ^{13}C relative to the glucose carbon source (12). Rinaldi et al. (73) found a smaller and opposite effect in the aerobic synthesis of acetate from ethanol by *Acetobacter suboxydans*. Acetate was depleted in ^{13}C relative to ethanol by about 5‰.

Methanogenesis

Isotope effects associated with methanogenesis are among the largest in nature. Not surprisingly, microbial methane represents the most ^{13}C-depleted substance on Earth (Fig. 1). Petroleum geochemists commonly measure the $\delta^{13}C$ of methane in the deep subsurface to determine whether the gas is biogenic (microbial) or thermocatalytic (abiotic) in origin (92) (Fig. 1). The magnitude of 1,000 $\ln\alpha_{CH_4-substrate}$ can depend upon the microorganism, pathway, temperature, and type of substrate and its concentration (references 34, 48, and 75, among others). Laboratory experiments with pure cultures of CO_2 reducers grown at 36 to 65°C, using CO_2 and H_2 as carbon and energy sources, yield 1,000 $\ln\alpha_{CH_4-CO_2}$ values of -25 to $-57‰$ (Table 1). Not surprisingly, the least fractionation ($-25‰$) occurred at the highest temperature (65°C). As $^{12}CO_2$ is preferentially removed from the system, residual CO_2 becomes progressively enriched in ^{13}C, approaching $\delta^{13}C$ values of 20‰.

Aceticlastic methanogenesis produces methane which is depleted in ^{13}C relative to methyl carbon and CO_2 which is depleted in ^{13}C relative to carboxyl carbon. For pure cultures of *Methanosarcina barkeri*, the fractionation between methane and methyl carbon is about $-21‰$ (37, 48). The fractionation between CO_2 and carboxyl carbon is also roughly $-21‰$ (37). Blair and Carter (13) estimated the fractionation associated with aceticlastic methanogenesis in natural systems based on ^{13}C measurements of acetate and methyl carbon in marine pore waters (Cape Lookout Bight). They obtained $-31 \pm 14‰$ for methane-methyl carbon fractionation and $-35 \pm 19‰$ for CO_2-carboxyl carbon fractionation. These results confirm that a large fractionation is associated with aceticlastic methanogenesis in nature; however, the magnitude is still uncertain. Also, these determinations were derived for systems with relatively large acetate pools (for example, up to 400 μM for Cape Lookout Bight). In natural systems where acetate is limiting, complete conversion of acetate to methane and CO_2 may result in considerably less fractionation.

CH_4-CO_2 fractionation determinations based on isotopic measurements of natural samples vary from -44 to $-75‰$ (Table 1) (33, 60, 61). The larger fractionation observed with natural samples compared with laboratory experiments may be due to the lower temperatures (2 to 17°C) of the former. Whiticar et al. (94) observed that $\delta^{13}C$ differences between CH_4 and CO_2 ($\Delta^{13}C_{CH_4-CO_2}$; not necessarily reflecting equilibrium) for marine environments are generally between $-50‰$ and $-90‰$, whereas $\Delta^{13}C_{CH_4-CO_2}$ values for freshwater environments are between $-40‰$ and $-60‰$. They proposed that this difference in ^{13}C fractionation reflects differences in the dominant methanogenic pathway, i.e., CO_2 reduction for marine environments and aceticlastic methanogenesis for freshwater environments.

TABLE 1 Carbon isotopic fractionation between CH_4 and CO_2 (1,000 $\ln\alpha_{CH_4-CO_2}$) in laboratory and field studies[a]

Environment	Taxon	Temp. (°C)	Substrate	1,000 $\ln\alpha_{CH_4-CO_2}$ (‰)	Reference
Lake mud		5–9		-44 to -75	61
Marine pore waters		10		-74	60
Sewage sludge		~35		-52	60
Laboratory (open)	*Methanosarcina barkeri*	40	CO_2/H_2	-43	34
Laboratory (open)	*Methanobacterium* sp. strain M.o.H.	40	CO_2/H_2	-57	34
Laboratory (closed)	*Methanobacterium thermoautotrophicum*	65	CO_2/H_2	-25	34
Landfill A		~20		-70	34
Landfill B		~20		-66	34
Sewage sludge tank		60		-45	34
Laboratory	*Methanosarcina barkeri*	40	CO_2/H_2	-43	8
Laboratory (open)	*Methanobacterium* sp.	37	CO_2/H_2	-35	10
Laboratory (open)	*Methanobacterium* sp.	46	CO_2/H_2	-34	10
Marine sediments		2		-72	33
Marine sediments		17		-63	33
Laboratory (closed)	*Methanobacterium formicicum*	34	CO_2/H_2	-48	6
Laboratory (open)	*Methanosarcina barkeri*	36	CO_2/H_2	-48	48

[a] Determinations for natural samples assume that the source of carbon for CH_4 is the ambient CO_2 at the time of collection.

This model appears valid for most systems, but may not accurately represent aquifer systems. Coleman et al. (23) and Grossman et al. (39) used hydrogen isotope data as evidence that CO_2 reduction is the primary methanogenic pathway in "freshwater" aquifer systems. Furthermore, Coleman et al. obtained $\Delta^{13}C_{CH_4\text{-}CO_2}$ values for groundwaters that are similar to those of marine systems (-60 to $-70‰$). Defining $\Delta^{13}C_{CH_4\text{-}CO_2}$ values using groundwater gases is tenuous, however, because groundwater samples average aquifer heterogeneity, which may include different anaerobic microenvironments.

Methanogenesis can cause $\delta^{13}C_{DIC}$ values to increase to as high as $+38‰$ (20). The impact of methanogenesis on groundwater DIC depends on $1,000 \ln\alpha$ (Table 1), substrate, and the effect of associated processes (i.e., fermentation). For CO_2 reduction, the $\delta^{13}C_{DIC}$-DIC trend depends on the ratio of fermentative CO_2 production to methanogenic CO_2 consumption. Assuming that fermentation is the only source of H_2, the overall stoichiometry of biodegradation with methanogenesis as the terminal process is (51):

$$C_nH_aO_b + (n - a/4 - b/2)H_2O \tag{5}$$
$$\rightarrow (n/2 - a/8 + b/4)CO_2 + (n/2 + a/8 - b/4)CH_4$$

For cellulose degradation, the CO_2/CH_4 ratio of the products is unity. The $\delta^{13}C_{DIC}$-DIC trends can be modeled assuming an open system with one input and one output (51, 95). Model inputs are the $1,000 \ln\alpha_{CH_4\text{-}CO_2}$, the composition and $\delta^{13}C$ of the initial substrate (equation 5), and the initial concentration and $\delta^{13}C$ of DIC.

Equation 5 shows that DIC concentration should not decrease with methanogenesis unless there is an independent H_2 source. Based on $\delta^{13}C_{DIC}$-DIC trends and other data, Stevens and McKinley (84) have proposed such a system in a basalt aquifer, where H_2 is produced by abiotic chemical reactions involving basalt. In this case, $\delta^{13}C_{DIC}$-DIC trends can be modeled as a closed system with no inputs and one output (i.e., Rayleigh distillation).

The isotopic fractionation associated with aceticlastic methanogenesis has not been adequately studied. On the basis of previous work, the ^{13}C composition of DIC contributed by aceticlastic methanogenesis can vary greatly depending on the concentration and isotopic composition of acetate. Other considerations include the taxa involved and temperature. Overall, aceticlastic methanogenesis should release somewhat ^{13}C-depleted DIC (-10 to $-40‰$?) derived from the carboxyl group of the acetate (13, 37).

^{13}C EVIDENCE FOR MICROBIAL ACTIVITY IN AQUIFER SYSTEMS

Microbial processes in aquifer systems are recorded in the $\delta^{13}C$ of DIC, dissolved methane, and DOC. Field studies typically focus on one or more of these groundwater constituents to track carbon cycling.

DIC

Early isotopic studies of aquifer systems were motivated by a need to correct DIC ^{14}C ages for the addition of DIC from carbonate mineral dissolution and "abiotic" coal and lignite degradation (45, 88). This work has led to a good understanding of the isotopic evolution of groundwater DIC.

Atmospheric CO_2 ($-7 \pm 1‰$) is the initial carbon source of groundwater, but the low atmospheric P_{CO_2} ($10^{-3.5}$ atm) provides only minor amounts of carbon (Fig. 2). A more important source is soil CO_2, which tends to be 3 to 7‰ enriched in ^{13}C relative to local vegetation (32, 72). Predominantly produced by root respiration, soil CO_2 is typically -19 to $-22‰$ in northern latitudes where C_3 plants dominate (70). Soil CO_2 $\delta^{13}C$ values are roughly 10‰ higher where C_4 plants dominate (L. Cifuentes, unpublished data). Reardon et al. (70) have shown that near the water table, groundwater $\delta^{13}C_{DIC}$ values are generally within 1‰ of those for soil CO_2. CO_2-charged groundwater will dissolve carbonate minerals if present, producing low-bicarbonate groundwaters (e.g., <7 mM) (30) with neutral pH and $\delta^{13}C_{DIC}$ values of -10 to $-15‰$. Subsequent microbial activity within the aquifer adds DIC, the $\delta^{13}C$ of which will depend upon the active microbial processes.

Respiration is the first microbial process to significantly affect groundwater DIC in the saturated zone. Early researchers were hesitant, however, to attribute these effects to microbial processes in aquifer systems. For example, Pearson and White (65) applied a simple ^{13}C mass balance model to $\delta^{13}C$ trends in the Carrizo aquifer in Texas and found evidence for the addition of ^{13}C-depleted carbon ($-25‰$ and $-30‰$), which they attributed to "CO_2 from lignite" and "petroleum-derived CO_2." Smith et al. (82) were among the first to integrate isotopic and microbiological approaches. They noted abundant nitrate-reducing bacteria and low counts of sulfate-reducing bacteria in groundwaters from the Chalk aquifer of the London Basin. Groundwater $\delta^{13}C_{DIC}$ data, however, showed progressively higher values along the flow path, a trend opposite that expected with anaerobic respiration. Thus, nitrate reduction and sulfate reduction were considered minor processes.

Like Smith et al. (82), Chapelle and Knobel (17) also observed an increase in $\delta^{13}C$ along a groundwater flow path (Aquia aquifer, Maryland); they suggested that this trend might reflect methanogenesis. However, dissolved methane concentration was low, requiring the authors to hypothesize that methane migrated out of the aquifer. Grossman et al. (39) found increasing $\delta^{13}C_{DIC}$ with increasing DIC content in high-bicarbonate waters in Texas (Fig. 4A). Assuming carbon addition with no mixing of waters, they estimated that carbon with a $\delta^{13}C$ as high as $+10‰$ was added to the groundwater, consistent with the combined effects of fermentation, CO_2 reduction, and calcium carbonate dissolution (Fig. 5). Furthermore, Grossman et al. found a correlation between high DIC content and high methane content, further evidence that the $\delta^{13}C$ trend and DIC addition are due to methanogenesis (39). The $\delta^{13}C_{DIC}$-DIC trend of Grossman et al. contrasts with that of Stevens and McKinley (84) (Fig. 4B). The former shows the expected trend when fermentation reactions are the H_2 source, whereas the latter can be explained best by an independent source of H_2 [in this case, reactions with Fe(II) silicates in basalts].

Few studies of aquifer systems have combined sediment microbiology and carbon isotopes. Using sediment cores and waters from the deep Coastal Plain aquifers in Maryland, Chapelle et al. (18) demonstrated that substantial counts of viable microorganisms exist in deep aquifers, including methanogens and sulfate reducers. These microorganisms could produce the observed down-gradient increase in DIC and $\delta^{13}C_{DIC}$; however, it was unclear whether the $\delta^{13}C_{DIC}$ trend reflected methanogenesis or some other poorly understood microbial process. To better understand the sources of DIC in groundwater, McMahon et al. (55) examined CO_2 production in sediments from a variety of Coastal Plain aquifers under aerobic and anaerobic conditions. After correction for carbonate dissolution, $\delta^{13}C$ values for evolved

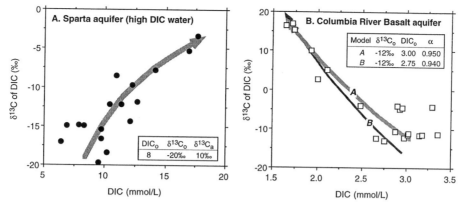

FIGURE 4 Evolution of the ^{13}C composition of DIC in two aquifer systems with microbial methane. DIC_0, initial DIC; $\delta^{13}C_0$, initial $\delta^{13}C$; $\delta^{13}C_a$, $\delta^{13}C$ of added carbon; α, fractionation factor between CH_4 and CO_2. (A) High-DIC waters from the Sparta aquifer in east-central Texas (39, 40, 97). Arrow shows trend for addition of DIC ($\delta^{13}C = 10‰$) through the combined action of fermentation, CO_2 reduction, and $CaCO_3$ dissolution (open model) (39). (B) Data for Columbia basalt aquifers in eastern Washington state (84). Arrows show trends for a closed system model (Rayleigh distillation) with consumption of CO_2 through CO_2 reduction. Trends were determined for two different sets of conditions (A and B).

CO_2 ranged from -18 to $-30‰$ and correlated with the $\delta^{13}C$ of the SOM. Furthermore, the data showed a down-dip trend toward higher $\delta^{13}C$ values for SOM and microbially produced CO_2, although not of the magnitude needed to explain the results of Chapelle et al. (18).

Comprehensive geochemical and isotopic models are

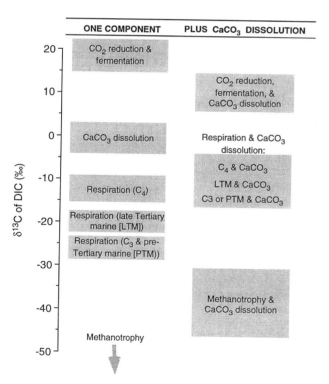

FIGURE 5 Carbon isotopic compositions of various sources of DIC. Ranges of compositions are rough and do not include fractionation associated with respiration. The effect of aceticlastic methanogenesis on the $\delta^{13}C$ of groundwater DIC is not well understood and is therefore not shown.

available to evaluate the simultaneous impact of microbial and abiotic processes on groundwater chemistry (e.g., Net-Path [66]). Lee and Strickland (52) modeled the geochemical and ^{13}C evolution in southeastern U.S. Coastal Plain aquifers and found evidence for organic matter oxidation by Fe(III) and sulfate reduction. Based on data from the Black Creek aquifer in South Carolina, McMahon and Chapelle (54) suggested that confining beds contribute DIC to the aquifer. They estimated that this DIC is derived from roughly 50% fermentation CO_2 and 50% $CaCO_3$ dissolution. Murphy et al. (59) modeled geochemical data for the Middendorf aquifer in South Carolina. The $\delta^{13}C$ values remained low ($\sim -21‰$) throughout tens of kilometers of flow path, suggesting microbial oxidation of SOM without significant $CaCO_3$ dissolution or methanogenesis.

Several studies have examined the use of carbon isotopes to monitor the biodegradation of petroleum hydrocarbons in contaminated aquifers. Aggarwal and Hinchee (2) observed unusually low $\delta^{13}C$ in soil CO_2 from contaminated sites, indicating mineralization of hydrocarbons. Jackson et al. (46) demonstrated in field and laboratory experiments that $\delta^{13}C_{CO_2}$ measurements can be sensitive indicators of rates of aerobic hydrocarbon mineralization in salt marshes, where the C_4 vegetation ($-16‰$) is enriched in ^{13}C relative to the hydrocarbons ($-29‰$). Several studies have reported increases in $\delta^{13}C_{DIC}$ associated with a contaminant plume. At the Bemidji, Minn., site, Baedecker et al. (5) observed a 17‰ increase in $\delta^{13}C_{DIC}$ over 3 years. Similar increases in $\delta^{13}C_{DIC}$ were observed by Herczeg et al. (42) and Landmeyer et al. (49). These $\delta^{13}C_{DIC}$ increases along with increases in CH_4 content clearly demonstrate contaminant degradation by methanogenesis. While these measurements aid geochemical modeling efforts and quantitative estimates of biodegradation, they are limited by an inadequate understanding of the isotopic fractionation associated with the biodegradation of organic substrates. Furthermore, mass balance approaches are compromised by the effects of CO_2 and CH_4 degassing, dissolution of carbonate minerals, and temporal and spatial heterogeneity. Nevertheless, this approach provides unique information on carbon cycling in contaminant plumes.

Methane

As previously mentioned, methanogenesis has a major impact on carbon isotope geochemistry because of the large fractionation between CO_2 and CH_4. Early ^{13}C studies of methane in the subsurface focused on economic occurrences of gas at depths where thermocatalytic methane is produced (>1,000 m). This methane typically yielded $\delta^{13}C$ values of greater than −55‰. Lower CH_4 $\delta^{13}C$ values ($\delta^{13}C_{CH_4}$) in deep samples, usually evidence for methanogenesis when found in shallow samples, were often attributed to abiotic processes such as migration (e.g., reference 25). However, subsequent isotopic studies of natural gas, including studies of deep production wells, have shown that microbial methane is ubiquitous. In fact, microbial methane has been found as deep as 3,350 m. Rice and Claypool (71) estimated that more than 20% of the discovered gas reserves are of biogenic origin.

Several studies have focused exclusively on groundwater methane. Coleman (22) analyzed methane from over 100 gas and water wells in Illinois and showed that gas wells in glacial sediments and water wells in glacial and shallow Paleozoic sediments produced methane with $\delta^{13}C_{CH_4}$ values of −66 to −91‰, indicative of a microbial origin. Coleman (22) confirmed the microbial origin of ^{13}C-depleted methane through ^{14}C analyses. Many of the gases had ^{14}C contents close to that of the glacial SOM from which the methane was presumably derived. Barker (7) and Barker and Fritz (9) analyzed over 260 groundwater samples from Ontario, Manitoba, and North Dakota. In addition to finding ubiquitous microbial methane ($\delta^{13}C$ values less than −60‰), they also found that sanitary landfill methane had $\delta^{13}C$ values intermediate between those of microbial and thermocatalytic methane (−38 to −56‰). Grossman et al. (39) observed an average $\delta^{13}C_{CH_4}$ difference of ~10‰ between Yegua (−65 ± 4‰) and Sparta (−55 ± 3‰) aquifers in east-central Texas. This difference corresponded to an 8‰ difference in the $\delta^{13}C$ of organic-rich sediments from each formation, evidence that Eocene SOM was the substrate for methanogenesis. Methanogenesis is not restricted to sedimentary aquifers. Sherwood Lollar et al. (78) found that bacterial methane was a common constituent in waters from the crystalline rocks of the Canadian and Fennoscandian Shields, and as mentioned earlier, Stevens and McKinley (84) showed isotopic evidence for methanogenesis in basalt aquifers.

Previous studies demonstrated that microbial methane is common in groundwater, but had not determined whether the process is ongoing. Simpkins and Parkin (80) used geochemical and isotopic evidence to demonstrate that methane in shallow Pleistocene and Holocene sediments in central Iowa was produced by CO_2 reduction. In a subsequent paper, Parkin and Simpkins (64) showed that the methane production rates in the sediments generally mirrored methane concentration profiles in groundwaters, indicating active CH_4 production.

Hydrogen isotope ratios ($^2H/^1H$ or D/H) have been used as an indicator of a methanogenic pathway (94). CO_2 reduction is believed to result in δD_{CH_4} values of −170 to −250‰, and aceticlastic methanogenesis presumably yields δD_{CH_4} values of −250 to −400‰. The validity of this model is still debated, with supporting (16) and contradicting (50) evidence. Groundwater methane δD values for Texas (−165 to −192‰ [39]) and Iowa and Illinois (−194 to −265‰ [23, 80]) are surprisingly consistent considering

the natural variability in methane δD values (< −170‰ to > −450‰ [78, 94]). The observed regional differences can be explained by latitudinal variations in the δD of meteoric water. Applying the Whiticar et al. model (94), these data argue for methanogenesis by CO_2 reduction in aquifer systems.

Methanotrophy imparts a distinctive isotopic signature on both the residual methane and the DIC. As mentioned earlier, methylotrophs preferentially utilize $^{12}CH_4$, leaving the residual methane enriched in ^{13}C. Figure 3 shows the variation in the isotopic composition of methane as methane is consumed by methylotrophs in a closed system. For fractionation factors of 0.996 and 0.974, the changes in isotopic composition after 50% of the methane has been consumed are 2.6‰ and 17.1‰, respectively (equation 4).

Barker and Fritz (8) used carbon isotopes as evidence for methane oxidation in shallow aquifers north of Lake Erie. The $\delta^{13}C_{CH_4}$ values, −29‰ to −37‰, were higher (more positive) than those of local biogenic methane (< −50‰) and thermocatalytic methane (−37‰ to −44‰). In a study of the Edwards, Trinity, Yegua, and Wilcox-Carrizo aquifers in Texas, Zhang (97) found trace methane (0.1 to 5 μM) with $\delta^{13}C$ values as high as −9‰ (Fig. 6). Many of the values are higher than that expected for an unaltered methane pool (probably < −25‰). Thus, the ^{13}C-enriched methane is likely the residual of a larger, partially oxidized methane reservoir. Figure 6 shows that the observed $\delta^{13}C_{CH_4}$ values and concentrations are consistent with CH_4 oxidation from CH_4-rich waters.

Microbial methane (−75‰) may undergo considerable oxidation and yet still retain $\delta^{13}C$ values that appear microbial (−60‰). With additional oxidation, the altered methane pool may appear thermocatalytic (e.g., −45‰) (Fig. 6). Fortunately, other evidence can be used to identify methane

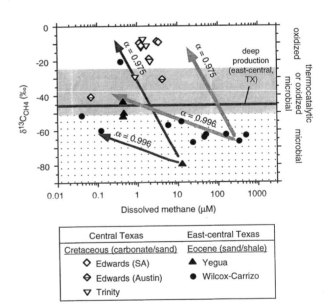

FIGURE 6 Carbon isotopic composition of methane versus dissolved methane concentration in Texas aquifers. Broad arrows show the methane oxidation trends assuming a closed system (equation 4), $\alpha_{CO_2-CH_4}$ values of 0.975 and 0.996, and initial compositions based on gas-rich Yegua and Wilcox-Carrizo waters. Interpretations at right are based on $\delta^{13}C$ ranges; boundaries between fields are gradational (modified from reference 97).

oxidation. Hydrogen isotopes are highly fractionated during methane oxidation (24). Covariant $\delta^{13}C$-δD trends in methane are strong evidence for oxidation. Unfortunately, $\delta^2 H$ measurements require much larger samples than do $\delta^{13}C$ measurements.

DIC $\delta^{13}C$ values may also provide evidence for methanotrophy. The $\delta^{13}C$ of CO_2 produced from methane oxidation can be as much as 100‰ lower than that of groundwater DIC. This effect has been observed in marine sediment cores. Pore water $\delta^{13}C_{DIC}$ values of < -30‰ directly above the zone of methanogenesis are compelling evidence for oxidation of upwardly diffusing methane (19, 60). Evidence for methane oxidation in the $\delta^{13}C$ of groundwater DIC is less apparent. For example, Zhang (97) found little or no $\delta^{13}C_{DIC}$ evidence for methane oxidation despite the fact that many methane samples showed unusual ^{13}C enrichment. Similarly, Canadian groundwaters with methane $\delta^{13}C$ values as high as -25‰ did not show unusually low $\delta^{13}C_{CO_2}$ values (8). This lack of significant impact on $\delta^{13}C_{DIC}$ probably reflects (i) the small fraction of CH_4 converted to CO_2 relative to biomass, (ii) the small size of the CH_4 pool compared to the DIC pool, and (iii) noise added by other sources of carbon (e.g., carbonate minerals, other respiration processes).

DOC

Groundwater DOC is a complex mixture of organic substances derived from biotic and abiotic reactions. It is both consumed and produced by microorganisms. Studies of groundwater DOC have been motivated by the role of DOC in pollutant transport and degradation and by the use of $^{14}C_{DOC}$ activities to date groundwater. ^{13}C compositions are analyzed to better understand sources of DOC. In their study of the Milk River aquifer in southern Alberta, Murphy et al. (57, 58) characterized the HMW and LMW DOC fractions and measured their $\delta^{13}C$ and ^{14}C activity. The HMW fraction was collected on XAD-8 resin and the LMW fraction was collected on a Silicalite molecular sieve. The HMW fraction was primarily composed of fulvic acids, whereas the LMW fraction consisted of short-chain aliphatic hydrocarbons and substituted alcohols. ^{14}C activities gave strong clues to the origin of the DOC. The ^{14}C activities of the HMW fraction followed the trend for DIC, suggesting a soil origin. In contrast, ^{14}C activities of the LMW fraction were low, suggesting derivation from kerogen (i.e., Cretaceous SOM). The $\delta^{13}C$ values of the HMW fraction exhibited a narrow range (-24 to -27‰) and approximated those of soil organic matter, thus supporting the conclusions based on ^{14}C. The $\delta^{13}C$ values of the LMW fraction were much more variable (-19 to -40‰), suggesting a complex origin. Samples near the recharge zone had the lowest $\delta^{13}C$ values and surprisingly low ^{14}C activities. Murphy et al. (58) hypothesized that the more labile LMW fraction is quickly recycled by microorganisms and does not survive transport from the soil zone, whereas the more recalcitrant HMW fraction is more likely to survive transport into and within the confined aquifer.

Wassenaar et al. (91) analyzed the ^{14}C and $\delta^{13}C$ of DOC and SOM in clayey tills from the same geographic region as the Murphy et al. (58) study. HMW and LMW fractions were also separated with XAD-8 resin and a Silicalite molecular sieve, respectively. Like Murphy et al., Wassenaar et al. found low variability in HMW fraction $\delta^{13}C$ values (-25 to -30‰) and high variability in LMW fraction $\delta^{13}C$ values (-10 to -46‰). These authors noted that the LMW samples with the lowest $\delta^{13}C$ values had the lowest ^{14}C activity and suggested that the LMW samples may be partly derived

from microbial activity on coal fragments in the tills. Murphy et al. (58) observed a similar pattern. Perhaps these low $\delta^{13}C$ values for the LMW fraction reflect acetogenesis (36, 68). The high $\delta^{13}C$ values in the LMW fraction (up to -10‰) may reflect fractionation associated with acetate consumption such as is observed with methanogenesis (37). In contrast to Murphy et al. (58), Aravena and Wassenaar (4) found $\delta^{13}C$ and ^{14}C evidence that HMW DOC (as well as CH_4) in the confined Alliston aquifer is derived in situ from Mid-Wisconsinan peats, rather than transported from the soil zone. The above studies show that isotopic measurements can be used to determine the source of DOC, which, depending upon the system, may be the soil zone, aquifer sediments, or both.

In a survey of humic substance $\delta^{13}C$ from a variety of Canadian groundwaters, Wassenaar et al. (89) observed geographic differences in $\delta^{13}C$ and, in some cases, $\delta^{13}C$ trends along flow paths. They suggested that these geographic differences and flow path trends reflect (i) eogenetic effects (e.g., decarboxylation), (ii) changes in carbon source due to climate change or human activity, (iii) multiple organic matter sources within a system (e.g., soil, peat, kerogen), and (iv) microbial processes. The ^{13}C data did not correlate with elemental composition but did correlate with DOC content. Unfortunately, DOC content also varied geographically. Thus, the process(es) dominating the ^{13}C signal for any one locality or sample could not be resolved. A subsequent study of a shallow, unconfined aquifer in Ontario evaluated the effect of DOC degradation on DIC content and $\delta^{13}C$ (90). Mass balance considerations suggest that carbonate mineral dissolution, rather than biodegradation of DOC, accounts for most of the DIC at this shallow (<20 m) site.

FUTURE DIRECTIONS

Future isotopic studies of microbial activity in aquifers will undoubtedly be linked to analyses of specific organic compounds in groundwater and sediments. With the development of the GC-C-IRMS, isotopic analyses of nanogram quantities of carbon are feasible. Future work will likely focus on microbial biomass and the isotopic composition of microbial biomarkers. Carbon isotope analyses of bacterial biomass or nucleic acids may be used to trace substrates of aquifer microorganisms (e.g., reference 21). We know from studies of hot and cold deep-sea vents that carbon isotopes can be used to trace carbon flow in ecosystems based upon sulfur-oxidizing chemoautotrophs (69) and methylotrophs (15). These same types of microorganisms are active in aquifer systems. Perhaps carbon isotope analyses of aquifer particulate organic carbon or DOC compounds can be used to determine the influence of chemoautotrophs and methylotrophs in aquifer systems.

A fruitful area of future research is the $\delta^{13}C$ of LMW DOC components, especially acetate. Studies of marine pore water acetate provide important information on microbial carbon cycling (13). Some technique development will be required to apply the same approach to pristine aquifers where DOC content is low, but it may be possible to study contaminated aquifers with current techniques. The role of stable carbon isotopes has been limited in groundwater DOC studies because soil zone SOM and aquifer SOM have similar ^{13}C compositions for the areas studied. Groundwater DOC studies need to be conducted at low latitudes to take advantage of the increased contribution of C_4 plants to soil zone organic matter. In such a case, groundwater DOC derived from the soil zone could be distinguished from that derived from the aquifer, especially in sediments that predate the C_4 pathway. ^{13}C analyses of microbial biomass could be used to determine the relative importance of aquifer organic carbon and soil zone organic carbon on microbial survival.

Stable carbon isotopes are a potentially powerful tool for determining the fate of organic contaminants. GC-C-IRMS analysis of specific compounds holds great promise for characterizing the proportion of contaminant degraded, but "tracer" compounds must be found that exhibit isotopic fractionation during biodegradation.

I thank P. McMahon, my students K. Clemence, D. Katz, and J. Routh, and three anonymous reviewers for reviewing the manuscript. Partial support was provided by the Subsurface Science Program of the Department of Energy (DE-FGO3-93ER61636).

REFERENCES

1. **Abelson, P. H., and T. C. Hoering.** 1960. The biogeochemistry of the stable isotopes of carbon. *Carnegie Inst. Washington Yearb.* **59:**158–165.

2. **Aggarwal, P. K., and R. E. Hinchee.** 1991. Monitoring in situ biodegradation of hydrocarbons by using stable carbon isotopes. *Environ. Sci. Technol.* **25:**1178–1180.

3. **Anderson, T. F., and M. A. Arthur.** 1983. Stable isotopes of oxygen and carbon and their application to sedimentologic and paleoenvironmental problems, p. 1-1–1-151. *In* M. A. Arthur, T. F. Anderson, I. R. Kaplan, J. Veizer, and L. S. Land (ed.), *Stable Isotopes in Sedimentary Geology.* Society for Economic Paleontology and Mineralogy, Tulsa, Okla.

4. **Aravena, R., and L. I. Wassenaar.** 1993. Dissolved organic carbon and methane in a regional confined aquifer, southern Ontario, Canada: carbon isotope evidence for associated subsurface sources. *Appl. Geochem.* **8:**483–493.

5. **Baedecker, M. J., I. M. Cozzarelli, R. P. Eganhouse, D. I. Siegel, and P. C. Bennett.** 1993. Crude oil in a shallow sand and gravel aquifer. III. Biogeochemical reactions and mass balance modeling in anoxic groundwater. *Appl. Geochem.* **8:**569–586.

6. **Balabane, M., E. Galimov, M. Hermann, and R. Létolle.** 1987. Hydrogen and carbon isotope fractionation during experimental production of bacterial methane. *Org. Geochem.* **11:**115–119.

7. **Barker, J. F.** 1979. Methane in groundwaters—a carbon isotope geochemical study. Ph.D. thesis. University of Waterloo, Waterloo, Canada.

8. **Barker, J. F., and P. Fritz.** 1981. Carbon isotope fractionation during microbial methane oxidation. *Nature* (London) **293:**289–291.

9. **Barker, J. F., and P. Fritz.** 1981. The occurrence and origin of methane in some groundwater flow systems. *Can. J. Earth Sci.* **18:**1802–1816.

10. **Belyaev, S. S., R. Wolkin, W. R. Kenealy, M. J. DeNiro, S. Epstein, and J. G. Zeikus.** 1983. Methanogenic bacteria from the Bondyuzhskoe oil field: general characterization and analysis of stable-carbon isotopic fractionation. *Appl. Environ. Microbiol.* **45:**691–697.

11. **Benner, R., M. L. Fogel, E. K. Sprague, and R. E. Hodson.** 1987. Depletion and ^{13}C in lignin and its implications for stable carbon isotope studies. *Nature* (London) **329:**708–710.

12. **Blair, N., A. Leu, E. Muñoz, J. Olsen, E. Kwong, and D. Des Marais.** 1985. Carbon isotopic fractionation in heterotrophic microbial metabolism. *Appl. Environ. Microbiol.* **50:**996–1001.

13. **Blair, N. E., and W. D. Carter, Jr.** 1992. The carbon isotope biogeochemistry of acetate from a methanogenic marine sediment. *Geochim. Cosmochim. Acta* **56:**1247–1258.

14. **Boutton, T.** 1991. Stable carbon isotope ratios of natural materials. 1. Sample preparation and mass spectrometric analysis, p. 155–171. *In* D. C. Coleman and B. Fry (ed.), *Carbon Isotope Techniques.* Academic Press, Inc., San Diego, Calif.

15. **Brooks, J. M., M. C. Kennicutt II, C. R. Fisher, S. A. Macko, K. Cole, J. J. Childress, R. R. Bidigare, and R. D. Vetter.** 1987. Deep-sea hydrocarbon seep communities: evidence for energy and nutritional carbon sources. *Science* **238:**1138–1142.

16. **Burke, R. A., Jr., C. S. Martens, and W. M. Sackett.** 1988. Seasonal variations of D/H and $^{13}C/^{12}C$ ratios of microbial methane in surface sediments. *Nature* (London) **332:**829–831.

17. **Chapelle, F. H., and L. L. Knobel.** 1985. Stable carbon isotopes of HCO_3^- in the Aquia aquifer, Maryland: evidence for an isotopically heavy source of CO_2. *Groundwater* **23:**592–599.

18. **Chapelle, F. H., J. L. Zelibor, Jr., D. J. Grimes, and L. L. Knobel.** 1987. Bacteria in deep Coastal Plain sediments of Maryland: a possible source of CO_2 to groundwater. *Water Resources Res.* **23:**1625–1632.

19. **Claypool, G. E., and C. N. Threlkeld.** 1983. Anoxic diagenesis and methane generation in sediments of the Blake Outer Ridge, Deep Sea Drilling Project Site 533, Leg 76, p. 391–402. *In Initial Reports of the Deep Sea Drilling Project*, vol. 76. U.S. Government Printing Office, Washington, D.C.

20. **Claypool, G. E., C. N. Threlkeld, P. N. Mankiewicz, M. A. Arthur, and T. F. Anderson.** 1985. Isotopic composition of interstitial fluids and origin of methane in slope sediment of the Middle America Trench, Deep Sea Drilling Project Leg 84, p. 683–691. *In Initial Reports of the Deep Sea Drilling Project*, vol. 84. U.S. Government Printing Office, Washington, D.C.

21. **Coffin, R. B., D. J. Velinsky, R. Devereux, W. A. Price, and L. A. Cifuentes.** 1990. Stable carbon isotope analysis of nucleic acids to trace source of dissolved substrates used by estuarine bacteria. *Appl. Environ. Microbiol.* **56:**2012–2020.

22. **Coleman, D. D.** 1976. Isotopic characterization of Illinois natural gas. Ph.D. thesis. University of Illinois at Urbana-Champaign, Urbana, Ill.

23. **Coleman, D. D., C.-L. Liu, and K. M. Riley.** 1988. Microbial methane in the shallow Paleozoic sediments and glacial deposits of Illinois, U.S.A. *Chem. Geol.* **71:**23–40.

24. **Coleman, D. D., J. B. Risatti, and M. Schoell.** 1981. Fractionation of carbon and hydrogen isotopes by methane-oxidizing bacteria. *Geochim. Cosmochim. Acta* **45:**1033–1037.

25. **Columbo, U., F. Gazzarrini, R. Gonfiantini, E. Tongiorgi, and L. Caflisch.** 1969. Carbon isotopic study of hydrocarbons in Italian natural gases, p. 499–516. *In* P. A. Schenk and I. Havenaar (ed.), *Advances in Organic Geochemistry.* Pergamon Press, Inc., Oxford.

26. **Crum, R. H., E. M. Murphy, and C. K. Keller.** 1996. A non-adsorptive method for the isolation and fractionation of natural dissolved organic carbon. *Water Res.* **30:**1304–1311.

27. **Degens, E. T., M. Behrendt, B. Gotthardt, and E. Reppmann.** 1968. Metabolic fractionation of carbon isotopes in marine plankton. II. Data on samples collected off the coasts of Peru and Ecuador. *Deep-Sea Res.* **15:**11–20.

28. **Deines, P.** 1980. The isotopic composition of reduced organic carbon, p. 329–406. *In* P. Fritz and J. C. Fontes (ed.), *Handbook of Environmental Isotope Geochemistry*, vol. 1. Elsevier, Amsterdam.

29. **Fogel, M. L., and L. A. Cifuentes.** 1993. Isotope fractionation during primary production, p. 73–98. *In* M. H. Engel and S. A. Macko (ed.), *Organic Geochemistry.* Plenum Press, New York.

30. **Foster, M. D.** 1950. The origin of high sodium bicarbonate waters in the Atlantic and Gulf Coastal Plains. *Geochim. Cosmochim. Acta* **1:**33–48.
31. **Friedman, I., and J. R. O'Neil.** 1977. Compilation of stable isotope fractionation factors of geochemical interest, p. KK1–KK12. *In Data of Geochemistry*, 6th ed. U.S. Geological Survey Professional Paper 440-KK. U.S. Government Printing Office, Washington, D.C.
32. **Fritz, P., E. J. Reardon, J. Barker, R. M. Brown, J. A. Cherry, R. W. D. Killey, and D. McNaughton.** 1978. The carbon isotope geochemistry of a small groundwater system in northeastern Ontario. *Water Resources Res.* **14:**1059–1067.
33. **Galimov, E. M., and K. A. Kvenvolden.** 1983. Concentrations and carbon isotopic compositions of CH_4 and CO_2 in gas from sediments of the Blake Outer Ridge, Deep Sea Drilling Project Leg 76, p. 403–407. *In Initial Reports of the Deep Sea Drilling Project*, vol. 76. U.S. Government Printing Office, Washington, D.C.
34. **Games, L. M., and J. M. Hayes.** 1976. On the mechanisms of CO_2 and CH_4 production in natural anaerobic environments, p. 51–73. *In J. O. Nriagu (ed.), Environmental Biogeochemistry*, vol. 1. *Carbon, Nitrogen, Phosphorus, Sulfur and Selenium Cycles.* Ann Arbor Science Publishers, Inc., Ann Arbor, Mich.
35. **Gat, J. R., and R. Gonfiantini.** 1981. *Stable Isotope Hydrology: Deuterium and Oxygen-18 in the Water Cycle.* Technical Report Series no. 210. International Atomic Energy Agency, Vienna.
36. **Gelwicks, J. T., J. B. Risatti, and J. M. Hayes.** 1989. Carbon isotope effects associated with autotrophic acetogenesis. *Appl. Environ. Microbiol.* **60:**467–472.
37. **Gelwicks, J. T., J. B. Risatti, and J. M. Hayes.** 1994. Carbon isotope effects associated with aceticlastic methanogenesis. *Org. Geochem.* **14:**441–446.
38. **Grossman, E. L.** 1982. Stable isotopes in live benthic foraminifera from the Southern California Borderland. Ph.D. thesis. University of Southern California, Los Angeles, Calif.
39. **Grossman, E. L., B. K. Coffman, S. J. Fritz, and H. Wada.** 1989. Bacterial production of methane and its influence on groundwater chemistry in east-central Texas aquifers. *Geology* **17:**495–499.
40. **Grossman, E. L., R. W. Hahn, and S. J. Fritz.** 1986. Origin of gaseous hydrocarbons in the Sparta Aquifer in Brazos and Burleson Counties, Texas. *Trans. Gulf Coast Assoc. Geol. Soc.* **36:**457–470.
41. **Hassan, A. A.** 1982. *Methodologies for Extraction of Dissolved Inorganic Carbon for Stable Carbon Isotope Studies: Evaluation and Alternatives.* U.S. Geological Survey Water Resources Investigations 82–6.
42. **Herczeg, A. L., S. B. Richardson, and P. J. Dillon.** 1991. Importance of methanogenesis for organic carbon mineralization in groundwater contaminated by liquid effluent, South Australia. *Appl. Geochem.* **6:**533–542.
43. **Herzog, B., J. Pennino, and G. Nielsen.** 1991. Groundwater sampling, p. 449–499. *In D. M. Nielsen (ed.), Practical Handbook of Ground-Water Monitoring.* Lewis Publishers, Chelsea, Mich.
44. **Hoefs, J.** 1987. *Stable Isotope Geochemistry*, 3rd ed. Springer-Verlag, Berlin.
45. **Ingerson, E., and F. J. Pearson, Jr.** 1964. Estimation of age and rate of motion of ground water by the ^{14}C-method, p. 263–283. *In Recent Researches in the Fields of the Hydrosphere, Atmosphere, and Nuclear Geochemistry.* Maruzen, Tokyo.
46. **Jackson, A. W., J. H. Pardue, and R. Araujo.** 1996. Monitoring crude oil mineralization in salt marshes: use of stable carbon isotope ratios. *Environ. Sci. Technol.* **30:**1139–1144.
47. **King, S. L., P. D. Quay, and J. M. Lansdown.** 1989. The $^{13}C/^{12}C$ kinetic isotope effect for soil oxidation of methane at ambient atmospheric concentrations. *J. Geophys. Res.* **94:**18,273–18,277.
48. **Krzycki, J. A., W. R. Kenealy, M. J. DeNiro, and J. G. Zeikus.** 1987. Stable carbon isotope fractionation by *Methanosarcina barkeri* during methanogenesis from acetate, methanol, or carbon dioxide-hydrogen. *Appl. Environ. Microbiol.* **53:**2597–2599.
49. **Landmeyer, J. E., D. A. Vroblesky, and F. H. Chapelle.** 1996. Stable carbon isotope evidence of biodegradation zonation in a shallow jet-fuel contaminated aquifer. *Environ. Sci. Technol.* **30:**1120–1128.
50. **Lansdown, J. M., P. D. Quay, and S. L. King.** 1992. CH_4 production via CO_2 reduction in a temperate bog: a source of ^{13}C-depleted CH_4. *Geochim. Cosmochim. Acta* **56:**3493–3503.
51. **LaZerte, B. D.** 1981. The relationship between total dissolved carbon dioxide and its stable carbon isotope ratio in aquatic sediments. *Geochim. Cosmochim. Acta* **45:**647–656.
52. **Lee, R. W., and D. J. Strickland.** 1988. Geochemistry of groundwater in Tertiary and Cretaceous sediments of the southeastern Coastal Plain in eastern Georgia, South Carolina, and southeastern North Carolina. *Water Resources Res.* **24:**291–303.
53. **Macko, S. A., M. H. Engel, and P. L. Parker.** 1993. Early diagenesis of organic matter in sediments: assessment of mechanisms and preservation by the use of isotopic molecular approaches, p. 211–224. *In M. H. Engel and S. A. Macko (ed.), Organic Geochemistry.* Plenum Press, New York.
54. **McMahon, P. B., and F. H. Chapelle.** 1991. Geochemistry of dissolved inorganic carbon in a Coastal Plain aquifer. 2. Modeling carbon sources, sinks, and $\delta^{13}C$ evolution. *J. Hydrol.* **127:**109–135.
55. **McMahon, P. B., D. F. Williams, and J. T. Morris.** 1990. Production and carbon isotopic composition of bacterial CO_2 in deep coastal plain sediments of South Carolina. *Ground Water* **28:**693–702.
56. **Merritt, D. A., and J. M. Hayes.** 1994. Factors controlling precision and accuracy in isotope-ratio-monitoring mass spectrometry. *Anal. Chem.* **66:**2336–2347.
57. **Murphy, E. M., S. N. Davis, A. Long, D. Donahue, and A. J. T. Jull.** 1989. ^{14}C in fractions of dissolved organic carbon in ground water. *Nature* (London) **337:**153–155.
58. **Murphy, E. M., S. N. Davis, A. Long, D. Donahue, and A. J. T. Jull.** 1989. Characterization and isotopic composition of organic and inorganic carbon in the Milk River Aquifer. *Water Resources Res.* **25:**1893–1905.
59. **Murphy, E. M., J. A. Schramke, J. K. Fredrickson, H. W. Bledsoe, A. J. Francis, D. S. Sklarew, and J. C. Linehan.** 1992. The influence of microbial activity and sedimentary organic carbon on the isotope geochemistry of the Middendorf aquifer. *Water Resources Res.* **28:**723–740.
60. **Nissenbaum, A., B. J. Presley, and I. R. Kaplan.** 1972. Early diagenesis in a reducing fjord, Saanich Inlet, British Columbia. I. Chemical and isotopic changes in major components of interstitial water. *Geochim. Cosmochim. Acta* **36:**1007–1027.
61. **Oana, S., and E. S. Deevey.** 1960. Carbon 13 in lake waters, and its possible bearing on paleolimnology. *Am. J. Sci.* **258a:**253–272.
62. **O'Leary, M. H.** 1988. Carbon isotopes in photosynthesis. *Bioscience* **38:**328–336.
63. **O'Neil, J. R.** 1986. Theoretical and experimental aspects of isotopic fractionation, p. 1–40. *In J. W. Valley, H. P. Taylor, Jr., and J. R. O'Neil (ed.), Reviews in Mineralogy,* vol. 16. Mineralogical Society of America, Chelsea, Mich.

64. **Parkin, T. B., and W. W. Simpkins.** 1995. Contemporary groundwater methane production from Pleistocene carbon. *J. Environ. Qual.* **24**:367–372.

65. **Pearson, F. J., and D. E. White.** 1967. Carbon 14 ages and flow rates of water in Carrizo Sand, Atascosa County, Texas. *Water Resources Res.* **3**:251–261.

66. **Plummer, L. N., E. C. Prestemon, and D. L. Parkhurst.** 1994. *An Interactive Code (NETPATH) for Modeling Net Geochemical Reactions along a Flow Path.* U.S. Geological Survey Water Resources Investigations Report 94-4169.

67. **Popp, B. N., F. J. Sansone, T. M. Rust, and D. A. Merritt.** 1995. Determination of concentration and carbon isotopic composition of dissolved methane in sediments and nearshore waters. *Anal. Chem.* **67**:405–411.

68. **Preuß, A., R. Schauder, G. Fuchs, and W. Stichler.** 1989. Carbon isotope fractionation by autotrophic bacteria with three different CO_2 fixation pathways. *Z. Naturforsch.* **44c**: 397–402.

69. **Rau, G. H., and J. I. Hedges.** 1979. Carbon-13 depletion in a hydrothermal vent mussel: suggestion of a chemosynthetic food source. *Science* **203**:648–649.

70. **Reardon, E. J., G. B. Allison, and P. Fritz.** 1979. Seasonal chemical and isotopic variations of soil CO_2 at Trout Creek, Ontario, p. 355–371. *In* W. Back and D. A. Stephenson (ed.), *Contemporary Hydrogeology—The George Burke Maxey Memorial Volume. J. Hydrol.* **43**.

71. **Rice, D. D., and G. E. Claypool.** 1981. Generation, accumulation, and resource potential of biogenic gas. *Am. Assoc. Petrol. Geol. Bull.* **65**:5–25.

72. **Rightmire, C. T., and B. B. Hanshaw.** 1973. Relationship between the carbon isotope composition of soil CO_2 and dissolved carbonate species in groundwater. *Water Resources Res.* **9**:958–967.

73. **Rinaldi, G., W. G. Meinschein, and J. M. Hayes.** 1974. Carbon isotopic fractionations associated with acetic acid production by *Acetobacter suboxydans. Biomed. Mass Spectrom.* **1**:412–414.

74. **Ronen, D., M. Magaritz, and I. Levy.** 1986. A multi-layer sampler for the study of detailed hydrochemical profiles in groundwater. *Water Res.* **20**:311–315.

75. **Rosenfield, W. D., and S. R. Silverman.** 1959. Carbon isotope fractionation in bacterial production of methane. *Science* **130**:1658–1659.

76. **Routh, J., and L. A. Cifuentes.** 1995. Application of compound-specific stable isotope measurements by GC-combustion-IRMS in geochemical research, p. 283–303. *In* M. Hyman and M. W. Rowe (ed.), *Advances in Analytical Geochemistry.* JAI Press Inc., Greenwich, Conn.

77. **Ruby, E. G., H. W. Jannasch, and W. G. Deuser.** 1987. Fractionation of stable carbon isotopes during chemoautotrophic growth of sulfur-oxidizing bacteria. *Appl. Environ. Microbiol.* **58**:1940–1943.

78. **Sherwood Lollar, B., S. K. Frape, P. Fritz, S. A. Macko, J. A. Welhan, R. Blomqvist, and P. W. Lahermo.** 1993. Evidence for bacterially generated hydrocarbon gas in Canadian Shield and Fennoscandian Shield rocks. *Geochim. Cosmochim. Acta* **57**:5073–5085.

79. **Sherwood Lollar, B., S. K. Frape, and S. M. Weise.** 1994. New sampling devices for environmental characterization of groundwater and dissolved gas chemistry (CH_4, N_2, He). *Environ. Sci. Technol.* **28**:2423–2427.

80. **Simpkins, W. W., and T. B. Parkin.** 1993. Hydrogeology and redox geochemistry of CH_4 in a Late Wisconsinan till and loess sequence in Central Iowa. *Water Resources Res.* **29**:3643–3657.

81. **Smejkal, V., F. D. Cook, and H. R. Krouse.** 1971. Studies of sulfur and carbon isotope fractionation with microorganisms isolated from springs of Western Canada. *Geochim. Cosmochim. Acta* **35**:787–800.

82. **Smith, D. B., R. A. Downing, R. A. Monkhouse, R. L. Otlet, and F. J. Pearson.** 1976. The age of groundwater in the Chalk of the London Basin. *Water Resources Res.* **12**: 392–404.

83. **Spiker, E. C., and P. G. Hatcher.** 1987. Carbon isotope fractionation of sapropelic organic matter during early diagenesis. *Org. Geochem.* **5**:283–290.

84. **Stevens, T. O., and J. P. McKinley.** 1995. Lithoautotrophic microbial ecosystems in deep basalt aquifers. *Science* **270**:450–454.

85. **Summons, R. E., L. L. Jahnke, and Z. Roksandic.** 1994. Carbon isotopic fractionation in lipids from methanotrophic bacteria: relevance for interpretation of the geochemical record of biomarkers. *Geochim. Cosmochim. Acta* **58**:2853–2863.

86. **Trust, B. A., J. G. Mueller, R. B. Coffin, and L. A. Cifuentes.** 1995. The biodegradation of fluoranthene as monitored using stable carbon isotopes, p. 233–239. *In* R. E. Hinchee, G. S. Douglas, and S. K. Ong (ed.), *Selected Papers from the Third International Symposium on In Situ and On-Site Bioreclamation.* Battelle Press, Columbus, Ohio.

87. **Tyler, S. C., P. M. Crill, and G. W. Brailsford.** 1994. $^{13}C/^{12}C$ fractionation of methane during oxidation in a temperate forested soil. *Geochim. Cosmochim. Acta* **58**: 1625–1633.

88. **Vogel, J. C.** 1967. Investigation of groundwater flow with radiocarbon, p. 355–369. *In Isotopes in Hydrology 1967.* International Atomic Energy Agency, Vienna.

89. **Wassenaar, L. I., R. Aravena, P. Fritz, and J. Barker.** 1990. Isotopic composition (^{13}C, ^{14}C, ^{2}H) and geochemistry of aquatic humic substances from groundwater. *Org. Geochem.* **15**:383–396.

90. **Wassenaar, L. I., R. Aravena, P. Fritz, and J. F. Barker.** 1991. Controls on the transport and carbon isotopic composition of dissolved organic carbon in a shallow groundwater system, Central Ontario, Canada. *Chem. Geol. (Isot. Geosci. Sect.)* **87**:39–57.

91. **Wassenaar, L. I., M. J. Hendry, R. Aravena, and P. Fritz.** 1990. Organic carbon isotope geochemistry of clayey deposits and their associated porewaters, southern Alberta. *J. Hydrol.* **120**:251–270.

92. **Whiticar, M. J.** 1994. Correlation of natural gases with their sources, p. 261–283. *In* L. B. Magoon and W. G. Dow (ed.), *The Petroleum System—from Source to Trap.* Am. Assoc. Petrol. Geol. Memoir 60.

93. **Whiticar, M. J., and E. Faber.** 1986. Methane oxidation in sediment and water column environments—isotope effects. *Org. Geochem.* **10**:759–768.

94. **Whiticar, M. J., E. Faber, and M. Schoell.** 1986. Biogenic methane formation in marine and freshwater environments: CO_2 reduction vs. acetate fermentation—isotope evidence. *Geochim. Cosmochim. Acta* **50**:693–709.

95. **Wigley, T. M. L., L. N. Plummer, and F. J. Pearson, Jr.** 1978. Mass transfer and carbon isotope evolution in natural water systems. *Geochim. Cosmochim. Acta* **42**:1117–1139.

96. **Wood, W. W.** 1976. Guidelines for collection and field analysis of ground-water samples for selected unstable constituents, p. 1–24. *In Techniques of Water-Resources Investigations of the U.S. Geological Survey.*

97. **Zhang, C.** 1994. Microbial geochemistry of groundwater in deep aquifers, central and east-central Texas. Ph.D. thesis. Texas A&M University, College Station, Tex.

98. **Zyakun, A. M., V. A. Bondar, V. V. Bezruchko, and A. N. Shkidchenko.** 1987. Separation of carbon isotopes by heterotrophic microorganisms growing on n-alkanes. *Mikrobiologiiya* **5**:953–957. (In Russian.)

99. **Zyakun, A. M., V. A. Bondar, and B. B. Namsaraev.** 1981. Fractionation of methane carbon isotopes by methane oxidizing bacteria. *Freib. Forsch.* **360**:19–27.

Determining the Terminal Electron-Accepting Reaction in the Saturated Subsurface

RICHARD L. SMITH

63

Microorganisms obtain their energy for metabolism by catalyzing a variety of oxidation-reduction reactions. In subsurface environments, where photosynthesis does not occur, the production and transfer of electrons is the driving force that governs all microbial processes. A wide range of microbially mediated oxidation-reduction couples are possible, many for which uniquely adapted groups of microorganisms have evolved. The electron source, or donor, for the oxidation-reduction couple can be either an organic or an inorganic compound. The same is also true for the electron-accepting compound. Within a given environment, the collective result of all microbial processes, most of which are oxidation-reduction reactions, is viewed as a microbial food chain. A compound that is a reduced end product for one microbe may be an oxidized substrate for another. By sequential coupling of the microbial processes, virtually all of the energy that is biologically available in a given substrate, or suite of substrates, will be extracted by the microbial population. In so doing, the microbial food chain serves as an electron conduit and channels electrons to the most oxidized, and therefore most energetically favorable, electron-accepting compound available within the localized environment.

This last step in the flow of electrons through the microbial food chain is called the terminal electron-accepting reaction; the last compound that is reduced is the terminal electron acceptor. The process itself is an important indicator of the nature of the microbial community within a habitat. Because it is the funnel through which electrons flow, the rate at which the terminal electron-accepting reaction is occurring is an integrated summary of the activity of the entire microbial food chain, and because there is a systematic progression of electron acceptors, the particular reaction that is occurring is indicative of the antecedent electron supply. Furthermore, the specialized nature of the microbes that facilitate each process, for example, methanogens or sulfate reducers, suggests the metabolic capacity or diversity that might be expected within the microbial community. Therefore, determining the terminal electron-accepting reaction is an important step in characterizing the microbial community in the subsurface.

SEQUENCE OF TERMINAL ELECTRON ACCEPTORS

The most commonly available terminal electron acceptors and their sequence of utilization are listed in Table 1. All are inorganic compounds that are relatively stable in an aqueous environment, and they represent various degrees of oxidation potential. There are key differences between the source, the solubility, and typical groundwater concentration ranges for the various electron acceptors, which are reflected in the nature of the various terminal electron-accepting reactions. Oxygen is the electron acceptor that provides the greatest energy yield; when it is present, aerobic metabolism will dominate. Indeed, the magnitude of this energetic advantage enables the final products of other electron-accepting reactions (e.g., Fe^{2+}, S^{2-}, and CH_4) to serve as electron donors when they migrate into oxygen-containing zones. However, oxygen availability in the subsurface is rather limited because an aquifer is, for the most part, a closed environment. Contact with the atmosphere occurs only at the water table. Thus, most commonly, as water enters an aquifer during recharge, oxygen concentrations are fixed via equilibrium with air or with the atmosphere in the unsaturated zone, and any additional oxygen entrainment along a flow path within the aquifer is usually insignificant. Travel distance and time along a flow path can be long (>100 km and $>1,000$ years), giving the subsurface microbial populations ample opportunity to exhaust the oxygen supply.

After oxygen, nitrate is the next electron acceptor in the progression sequence. In pristine situations, nitrate is not a prominent groundwater constituent and is found only in trace concentrations. This fact is directly related to the oligotrophic (nutrient-poor) nature of most groundwaters. However, nitrate is the most prevalent groundwater contaminant from anthropogenic sources and is found in increasingly wide geographic areas, with large ranges in concentration. Nitrate enters an aquifer from both localized (point) and areawide (nonpoint) sources (the latter stemming from agricultural practices). Therefore, the potential for encountering nitrate as a terminal electron acceptor in groundwater is not insignificant and is continually increasing, particularly for unconfined aquifers.

Iron, manganese, and sulfate sequentially follow nitrate as electron acceptors. All three are naturally abundant in many aquifers. Fe(III) most frequently occurs as insoluble oxides and hydroxides, which are often found as coatings on mineral grains and which have various degrees of crystallinity and structure. Thus, this electron acceptor is immobilized by its association with the aquifer solids. The large

TABLE 1 General characteristics of the electron acceptors most commonly found in groundwater, given in the order of their sequence of utilization

Electron acceptor	Reduced product	Primary source in groundwater	Concn range in groundwater[a]
O_2	H_2O	Atmosphere	0–0.4 mM
NO_3^-	N_2	Contamination	0–20 mM
Mn(IV)	Mn(II)	Mineral solids	Very low
Fe(III)	Fe(II)	Mineral solids	Very low
SO_4^{2-}	S^{2-}	Mineral solids, brines, marine sources, contamination	0–15 mM
CO_2	CH_4	Carbonate solids, organic matter degradation	0–4 mM

[a] Ranges given are concentrations generally found in groundwater and not meant to represent concentration extremes.

quantity of iron oxide coatings in an aquifer suggests that the oxidation capacity via Fe(III) reduction can be quite substantial. Mn(IV) is similar to Fe(III) in that it is quite insoluble in water yet may be present in large quantities in an aquifer. Very little is known about manganese reduction in the subsurface or even whether it is a respiratory terminal electron-accepting reaction. Fe(II) will abiotically reduce Mn(IV) at neutral pH; therefore, direct enzymatic reduction is not required (30). In contrast, sulfate is much more soluble and delivered to the microbial community via groundwater flow. In some cases, dissolution of sulfate-bearing minerals, such as gypsum, can serve as a long-term reservoir for this electron acceptor and should be considered in constructing sulfate mass balances along specific flow paths. In other settings, sulfate minerals are not abundant and sulfate is only a minor groundwater constituent. However, like nitrate, sulfate is present in many contaminant sources (e.g., landfills and septic systems), often in conjunction with increased electron acceptor demand. Therefore, sulfate reduction can be significant even where sulfate is naturally scarce.

When the other electron acceptors have been depleted, carbon dioxide becomes the terminal acceptor, being reduced to methane during methanogenesis. The supply of carbon dioxide for this process is essentially inexhaustible. A considerable electron supply, in the form of degradable organic carbon, must be available for the process to commence. As organic material is degraded by the microbial food chain, carbon dioxide is produced, which is then subsequently utilized by the methanogens. Hence, the electron acceptor in this case is internally generated. Additionally, carbon dioxide that was generated during reduction of all of the preceding electron acceptors and the carbonate solids that are present in many subsurface formations also contribute to the electron acceptor supply for methanogens. It should be noted that not all of the methane that is generated is actually derived from carbon dioxide reduction. A significant percentage comes from fermentation of acetate, which is also produced by the microbial food chain. For both carbon dioxide reduction and acetate fermentation, the electrons that are generated by microbial metabolism are transferred to carbon moieties, one organic and the other inorganic, and methane is subsequently produced. In keeping with the conceptual framework that the terminal electron-accepting reaction is a respiratory process, carbon dioxide is viewed as the terminal electron acceptor

when methanogenesis is occurring, but it is important to remember that not all of the methane is generated via this mechanism.

TERMINAL ELECTRON-ACCEPTING REACTIONS IN THE SUBSURFACE

The distribution of the terminal electron-accepting reactions in an aquifer is dictated by several factors, including the relative abundance of the various electron acceptors, the amount and availability (or degradability) of the electron supply, and the nature and rate of groundwater flow. Conceptually, there are two general distribution patterns. In the first case (Fig. 1), which is typical of most pristine groundwaters, the electron donor supply is the limiting factor. Naturally occurring organic matter is not readily abundant in the subsurface, and that which is present is usually refractory. This greatly limits the flow of carbon and energy through the microbial food chain. The sequence of terminal electron-accepting reactions occurs in the order shown in Table 1, but very long periods of time can be necessary before the supply of a given electron acceptor is depleted. The thickness of the reaction zones in the direction of groundwater flow depends on the supply of each individual electron acceptor, but the zones for the more abundant acceptors can be many kilometers long—so long that often the aquifer may discharge to a surface water system before oxygen or the other electron acceptors are depleted. The other distribution pattern results when the electron donor supply is plentiful and the microbial food chain is electron acceptor limited (Fig. 2). This is often (though not exclusively) the result of organic contamination and is usually a localized phenomenon. In this situation, all available electron acceptors are rapidly utilized, creating a central methanogenic zone. Then, as the organic compounds are consumed and diluted by transport downgradient, a point is reached at which sulfate, and then Fe(III), etc., is not limiting. As this happens, the other electron-accepting reactions become active. The end result is a double progression of processes in the direction of groundwater flow, first in the order shown in Table 1 and Fig. 1 and then in the reverse order. For any given case, pristine or contaminated, the exact pattern could differ from the two presented here. For example, some zones could be absent or insignificant if an electron acceptor is not present. But the key issue to consider for any groundwater situation is the relative ratio of electron donor and electron acceptor supplies (see chapter 61 for further discussion of this topic).

Another important factor is that an aquifer is not a homogeneous environment. There are many reasons for spatial variability, and most of them affect the subsurface microbial community. Geochemical gradients are particularly important variables with regard to electron-accepting reactions. Both horizontal and vertical geochemical gradients occur within an aquifer. The steepness of these gradients is dictated by the degree of dispersion (i.e., mixing) along each spatial axis. In general, dispersion is much greater in the direction of groundwater flow than in the transverse directions. Thus, when bulk flow is primarily horizontal, the sharper gradients would usually be found in the vertical direction. Vertical gradients can persist for many kilometers along a groundwater flow path. The existence of these gradients can restrict a terminal electron-accepting reaction to a relatively narrow vertical interval within an aquifer, depending on the electron acceptor supply. For example, in one nitrate-contaminated sand and gravel aquifer, denitrification occurs within a 5- to 7-m-thick zone (46). This means

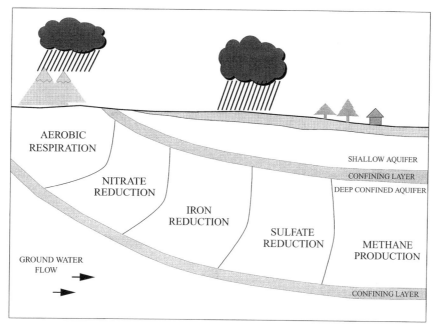

FIGURE 1   Diagram of the sequential succession of electron-accepting reactions in deep pristine groundwater. Modified from reference 31 with permission. © 1994 American Chemical Society.

that resolution in the vertical dimension must be carefully considered when one is conducting a study to determine the electron-accepting reactions in an aquifer. Short well screens or multilevel point samplers, spaced vertically to match the shape of the gradient, are best (45). The gradient can be determined to some extent by using screened augers during the installation process. Moreover, collection of core material must also reflect the existence of gradients. Cores collected from the same depth and vicinity as a well screen

can be combined with groundwater collected from the same depth. This ensures that the aquifer microbes are subjected to a representative geochemistry during incubation assays.

APPROACHES FOR DETERMINING THE ELECTRON-ACCEPTING REACTION

Microbial processes, especially the terminal electron-accepting reactions, alter the geochemistry of the aqueous en-

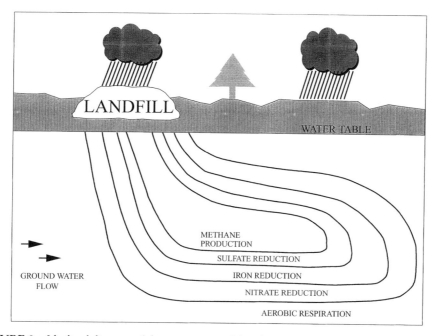

FIGURE 2   Idealized diagram of the orientation of the electron-accepting reactions in groundwater contaminated with landfill leachate.

vironment. Specific geochemical characteristics result when each of the various electron acceptors is reduced. These characteristics, or geochemical profiles, have been used to infer which process is active within any given region of an aquifer. This is a valid, acceptable approach. However, the conclusions may be limited to stating that process x either is occurring or has occurred, and there may be missing components of the geochemical profile that could confound the issue (e.g., sulfide removal in the presence of heavy metals). An alternative approach is to conduct incubation assays specific for a given electron-accepting reaction, because a demonstration that the process is active is the most direct proof that it is the terminal electron-accepting reaction. The latter approach assumes that the sample collection did not disrupt or alter the microbial community, that the incubation process can faithfully maintain the in situ geochemistry during the entire time course, and that the selected assay is sensitive enough to detect the low rates of activity that occur in the subsurface. The vast majority of microorganisms in an aquifer are attached to solid surfaces (18). This means that collection of core material is required for these activity assays to obtain a representative result, unless an alternative in situ approach is available (43).

In general, the best and most rigorous determination of the terminal electron-accepting reaction is obtained by combining the geochemical and microbiological approaches. Interpretation of the combined results involves fewer inherent assumptions and is less subject to error than interpretation of the results of each individual approach. An aquifer should actually be viewed from an ecosystem perspective. Therefore, an orientation that encompasses (i) the diversity and the nature of microbial processes that occur in the subsurface, (ii) the geochemical environment, and (iii) the hydrologic regime is the most appropriate when considering the function and the interaction of the terminal electron-accepting reactions.

AEROBIC RESPIRATION

Determining whether aerobic respiration is the dominant terminal process is relatively straightforward. When oxygen is present, it will be the electron acceptor, not only for the thermodynamic reasons mentioned previously but also because it is toxic to the obligately anaerobic processes (iron reduction, sulfate reduction, and methanogenesis) and in-

hibits the expression and the function of the denitrification enzymes (27). Only at very low oxygen concentrations, below ~30 μM (39, 50, 51), is there a potential for nitrate competing with oxygen for available electrons. Because an aquifer is water saturated, the oxygen concentration of the bulk groundwater is a good representation of oxygen available to the individual microorganism, even those that are attached. There is little likelihood that anaerobic microsites, which are common in many soils, are important in the groundwater environment.

In practice then, one can readily conclude that oxygen is the terminal electron acceptor when oxygen concentrations exceed 50 μM, unless there are no nitrogen oxides present, in which case the threshold may be 10 μM or lower. Oxygen is the one electron acceptor for which it is not necessary to conduct laboratory incubations as a confirmatory step. This is fortuitous, because oxygen consumption by pristine samples with low electron acceptor demand could be very difficult to detect. Thus, the pertinent methodology for determining aerobic respiration is that of sampling and quantifying dissolved oxygen in groundwater.

Collecting groundwater for quantification of dissolved oxygen must be done carefully; exposure to the atmosphere can result in overestimation, while stripping dissolved gases via suction lift pumping can result in underestimation. Given that the water table may be several meters below land surface, the challenge is to obtain a sample representative of the aquifer. Several techniques can be used (Table 2); each is best for specific conditions. First, the standing water in the well must be replaced with water from the formation. The standard procedure is to pump and discard at least 3 well casing volumes. Then the water sample is collected. A down-well pump with an inflatable packer can seal the screened interval from the atmosphere and can serve for both well purging and sample collection. On the other hand, a grab sampler can maintain in situ pressure for deep formations or gas-laden samples. For the small-bore multilevel samplers, suction lift is unavoidable. The best choice here is a peristaltic pump fitted with oxygen-impermeable tubing and a slow rate of pumping. Down-well oxygen determination is also possible (Table 2), though it can be quite time-consuming. Use of production wells for oxygen determinations can be considered only a qualitative approach and is best avoided because of turbulent mixing, long screened intervals, and the potential for entrainment of atmospheric oxygen.

TABLE 2 Sampling techniques to measure dissolved oxygen in groundwater

Technique	Advantage	Disadvantage	Consideration
Down-well, pressure lift pump	Single step for purge and sample	Specialized pump, not suitable for smaller pipes	Use O_2-impermeable tubing[a] and slow pump rate
Suction lift pump	Can be used with small-diameter wells	Water table must be above suction limit, potential outgassing	Use O_2-impermeable tubing and slow pump rate
Grab sampler[b]	Maintains pressure, no outgassing	Small volume collected	Avoid turbulent mixing when lowering into well
Down-well determination (lower O_2-measuring device[c] into well)	Prepump measurement, in situ temperature and pressure	Cumbersome and time-consuming, water table must be above suction limit	Preferable to continue pumping with suction lift pump during measurement

[a] See reference 23.
[b] See reference 25.
[c] Can be either an electrode or an ampoule filled with a colorimetric reagent (54).

The two most commonly used analytical procedures for oxygen are the Winkler iodometric technique (3) and use of oxygen-specific membrane electrodes. Both are suitable if calibrated properly but lack sensitivity for low oxygen concentrations. For the latter, a quick, field-friendly, colorimetric method uses Rhodazine-D in sealed glass ampoules (a proprietary product of CHEMetrics, Inc.) (54). A suggested approach is to collect groundwater samples containing >30 μM (1 mg liter^{-1}) in BOD (biological oxygen demand) bottles (3) for analysis in the laboratory using an oxygen-specific electrode that adapts to the neck of the BOD bottle. When oxygen is <30 μM, the ampoulated Rhodazine-D procedure is recommended.

DENITRIFICATION

When nitrate is an electron acceptor, it is reduced either to nitrogen gas by denitrification or to ammonium by dissimilatory nitrate reduction. Each process is respiratory in nature, each produces nitrite as an intermediate, yet each is catalyzed by a physiologically different group of microorganisms. Nitrate reduction to ammonium appears to be the preferred pathway when the electron supply greatly exceeds the amount of available nitrate (49). However, this is an uncommon situation for groundwater, so the process is probably also uncommon. In groundwater studies in which both processes were examined, denitrification was predominant (37, 46, 48), but in one study dissimilatory nitrate reduction was significant in microcosm experiments with aquifer core material (10). Aerobic (i.e., oxygen-indifferent) denitrification has also been reported (38), though it is not likely to be significant in groundwater.

Denitrification usually leaves a detectable impact on the nitrogen geochemistry of an aquifer. Nitrite and nitrous oxide concomitant with nitrate are often found when the process is occurring and therefore are good indicators. However, these compounds also result from nitrification. Concentrations of nitrogen gas exceeding atmospheric equilibrium are good additional evidence for denitrification, though care must be exercised in sampling and analysis to avoid loss of the excess nitrogen via bubble formation (46). Even stronger evidence would be the disappearance of nitrate and an increase of nitrogen gas along a known flow path in the aquifer.

Microbial processes, especially dissimilatory processes, commonly fractionate low-molecular-weight compounds because of the tendency of lighter isotopes to react faster than heavier isotopes. The result is an enrichment of the heavier isotopes in the substrate and the lighter isotopes in the products. This fractionation mechanism can be used to determine whether nitrate is being utilized as a terminal electron acceptor. As nitrate is consumed, the remaining nitrate pool is sequentially enriched in ^{15}N. The degree of enrichment is directly dependent on the amount of the original nitrate that has been reduced. Groundwater nitrate can be readily sampled, concentrated by freeze-drying, if necessary, and analyzed by mass spectrometry for the ^{15}N/^{14}N content. This natural abundance ratio (see reference 15 for an overview) can be used to infer the nitrate source in groundwater (4, 20, 29) and the extent of reduction that has occurred (11, 34, 46, 48). When coupled with ^{15}N determination of the products of reduction (N$_2$, NH$_4$$^+$) and a nitrogen mass balance, the stable isotope analysis will distinguish the relative extent of denitrification and/or dissimilatory nitrate reduction (46).

There are two incubation assays that have the sensitivity suitable to measure denitrification in groundwater samples. The first is the acetylene blockage technique. Acetylene inhibits the terminal enzyme of the denitrification pathway, nitrous oxide reductase, causing an accumulation of nitrous oxide when it is present (6, 56). A gas chromatograph equipped with an electron capture detector can readily detect low concentrations of nitrous oxide. Incubations are most easily conducted in flasks with an anoxic headspace (42). A good design is to slurry the core material with groundwater from the same site and incubate the samples at in situ temperatures. Production of nitrous oxide in the flasks is usually termed denitrification potential to clearly indicate that the incubations deviated from in situ conditions. The acetylene block technique has also been used for incubations in intact cores (44) and for in situ assays by injecting acetylene-laden water directly into an aquifer (44, 52). It should be noted that acetylene has been used almost exclusively for heterotrophic denitrification. Its utility for autotrophic denitrification, which can be significant in the subsurface (28), is unclear.

The other incubation assay uses ^{15}N as a tracer. Whereas the acetylene block technique is specific only for denitrification, incubation with ^{15}N-enriched nitrate can detect whether either or both denitrification and dissimilatory nitrate reduction to ammonium are occurring and the relative importance of each. In addition, this approach can be used to examine the turnover of nitrous oxide if sufficient quantities are present. The types of incubations are similar to those for the acetylene block; that is, they can be sediment slurries (42), within core (10), or in situ (41). Once the incubation is completed, the various nitrogen species are extracted and analyzed for ^{15}N content (10). Although this technique is much more involved and requires access to a mass spectrometer, a great deal more information can be obtained than with the acetylene block incubations.

IRON REDUCTION

In many aspects, iron reduction differs from the other terminal electron-accepting reactions. The substrate, Fe(III), is present in many forms, all as crystalline solids, and often in large quantities in the saturated subsurface. The very low aqueous solubility of Fe(III) requires that iron-reducing microorganisms be in direct contact with the mineral surface to use it as an electron acceptor. It also precludes quantifying the Fe(III) supply by collecting only groundwater samples. The product, Fe(II), is a common yet undesired constituent in groundwater drinking supplies. Its concentration in groundwater is controlled by a complicated set of precipitation/dissolution and ion-exchange reactions. The presence of Fe(II) in groundwater is a clear indication that iron reduction has occurred or is occurring at some point upgradient from the sampling point, but the concentration cannot be used to infer the extent of the reaction (2). In general, the reduction will likely be the result of microbial activity. Abiotic iron reduction appears to be much less common than was previously thought (30). The ferrozine colorimetric assay (19) has good sensitivity for Fe(II) and is most frequently the method of choice (2).

A demonstration that reducible forms of Fe(III) are present in the aquifer matrix will greatly strengthen the conclusion that iron reduction is occurring. However, not all forms of Fe(III) are biologically available (30). Those that are amorphous or less crystalline are more readily reducible. There appears to be some correlation between certain extraction techniques and the readily reducible fractions of

Fe(III) (22), but in actuality, the analytical procedures quantify an operationally defined fraction of the total iron present. Therefore, the prudent approach is to measure the amorphous Fe(III) oxides by using 0.5 M HCl extraction or Fe(III) oxidation capacity by using Ti(III)-EDTA extractions (22) and to couple that with sediment incubations (2). Production of Fe(II) during these incubations is clear evidence that both iron-reducing bacteria and the reducible forms of Fe(III) are present in the core material. Additional sediment incubations amended with synthetic Fe(III) oxides (33), with and without added acetate, can be used as positive controls to determine whether the samples have the potential capacity to reduce iron if it were present in an available form (2). Extraction and analysis of both solid and dissolved Fe(II) are necessary to determine the total amount of iron reduced during these incubations. Use of sediment slurries (see above) is the best technique for iron reduction incubations.

SULFATE REDUCTION

Despite the obvious presence of sulfides in a water sample, detecting the occurrence of sulfate reduction in an aquifer can be challenging. Low rates of activity can be coupled with relatively large quantities of metals that precipitate the sulfides as they are produced. In sulfate-rich formations, mineral dissolution and subsequent increase in groundwater sulfate content can greatly exceed and therefore mask sulfate removal via sulfate reduction (12). There are even situations in which sulfate entrainment into an aquifer via dissolution from confining layers can be matched by sulfate reduction. Hence the process could be occurring, though the substrate is apparently absent (12). Therefore, the geochemical evidence must be examined carefully to ascertain whether sulfate reduction is occurring. For this terminal electron-accepting reaction in particular, quantification of dissolved hydrogen concentrations could provide important ancillary information (see below).

Co-occurrence of sulfate and sulfide in groundwater would constitute a clear indication that sulfate reduction was occurring. This is the situation in marine sediments and likely extends to saltwater intrusions and brines in the saturated subsurface. However, more often, especially for freshwater aquifers, detection of acid volatile sulfides in core material will be necessary to arrive at the same conclusion. Pyrite and elemental sulfur, which are diagenetic end products of sulfide generation, are not particularly good indicators of recent sulfate reduction, though increases in extractable pyrite along a groundwater flow path can provide suggestive evidence for a sulfide source (13). Isotopic fractionation of sulfur isotopes has also been used to indicate sulfate reduction and is the geochemical tool used most frequently to determine sulfate reduction in groundwater. As with nitrate reduction, sulfate reduction favors reaction by the lighter isotope, resulting in an enrichment of ^{34}S in the unreacted sulfate. In some cases, the fractionation of ^{34}S in sulfate can be used to infer the occurrence of sulfate reduction along a flow path even when sulfate concentrations are increasing by dissolution (36). An accompanying depletion of ^{34}S in the sulfide pools would also be expected and has been used on occasion in groundwater studies as additional evidence for sulfate reduction (14). Sampling and assaying for sulfate are straightforward (3). For sulfide, avoid contact of the water sample with metal pipes and with oxygen, and avoid prolonged sample storage. Both colorimetric and titrimetric assays are appropriate for groundwater sulfide (3).

In surface water environments, sulfate reduction is most often assayed by using ^{35}S-labeled sulfate (26), either in microcosm incubations or by injecting the radiolabeled tracer into intact cores. Activity is determined by recovering and quantifying ^{35}S in the various reduced sulfur pools (16). However, only a very few groundwater studies have directly assessed sulfate reduction by using this technique (14), and more research in this area is clearly needed. Incubations could be conducted in the presence and absence of added electron donor (similar to the situation for iron reduction) to demonstrate sulfate-reducing potential. More common are enumeration and isolation of sulfate-reducing bacteria from groundwater and aquifer core material (8, 35, 53). This approach is based on the assumption that when sulfate is the terminal electron acceptor, larger numbers of sulfate-reducing bacteria will be present. While this is often the case, interpretations must be made with caution, using appropriate controls, and would be unarguably strengthened if supported with results from activity assays.

METHANOGENESIS

Methane is a common constituent of deep groundwater, especially in sedimentary basins, and in heavily contaminated shallow aquifers (17). It is a relatively stable compound (in the absence of oxygen) and can persist in the subsurface for long periods of time. Detection of methane does not necessarily denote active methanogenesis. Methane can originate from thermogenic as well as biogenic processes, entering groundwater from coal and petroleum deposits and from mining and drilling activities. In addition, biogenic methane is not necessarily contemporary. It may have been produced at a considerable distance upgradient from the sample site and/or much earlier in time.

Nonetheless, the first step for determining whether methane production is the terminal electron-accepting reaction is to quantify dissolved methane in groundwater. Sampling for methane involves the same considerations as sampling for any dissolved gas (see discussion above regarding oxygen); analysis is accomplished with headspace equilibration and gas chromatography (9, 40, 47). For shallow aquifers with relatively high rates of methane production, gas collection devices that cumulatively collect methane as a gas over a several-week period have also been described (1, 9). Thermogenic methane can be distinguished from biogenic methane by detecting the presence of higher hydrocarbons (e.g., ethane, butane, and propane) and by determining the carbon isotope composition ($^{13}C/^{12}C$) of the methane. Thermogenic methane tends to be lighter than biogenic methane, with -58 ‰ the approximate division between the two. However, methane oxidation can confound this interpretation (7). In addition, microbial methane production via carbon dioxide reduction fractionates the carbon dioxide (or, more appropriately, the dissolved inorganic carbon pool), enriching it in ^{13}C. Several studies have used these geochemical signatures to infer the source of methane in groundwater (5, 7, 21, 40). Complete isotopic characterization (C and H) may allow one to distinguish between methane produced from carbon dioxide reduction and that produced from acetate fermentation (55).

The relative simplicity and sensitivity of the methane assay greatly facilitate laboratory incubations to demonstrate methanogenesis. Aquifer core material can be incubated in sealed containers with an anoxic headspace for long periods of time, and the concentration of methane in the headspace can be assayed periodically. This method has been employed for many types of degradation experiments performed by using methane-producing conditions with subsurface material. It is also valid for simply demonstrating that methanogenesis is occurring. If background methane concentrations are high, it may be necessary to purge the headspace to improve the analytical sensitivity for quantifying small increases in methane concentration. Positive controls to which methane precursors have been added (e.g., acetate, formate, methanol, or trimethylamine) are useful to demonstrate whether the activity can be stimulated (8). ^{14}C-labeled substrates can be used to quantify the rates of the individual methane-producing reactions but are probably unnecessary to simply ascertain whether methanogenesis is the operative electron-accepting reaction. Enumeration of methanogenic bacteria would provide additional evidence for methane production. Methanogens would not be expected in significant numbers when other electron acceptors were functioning. However, like the case for sulfate reduction, suitable controls and samples from nonmethanogenic areas may be necessary to distinguish electron donor-poor situations in which methanogenic populations are small from those in which methanogens are present but inactive.

DISSOLVED HYDROGEN—AN ADDITIONAL TOOL

Recently, Lovley et al. (31) proposed that concentrations of dissolved hydrogen in groundwater can serve as an additional indicator of the predominant terminal electron-accepting reaction. Hydrogen is known to be an important intermediate for most anaerobic food chains (excepting perhaps denitrification) and serves as a direct link between fermentative and respiratory processes. This couple, termed interspecies hydrogen transfer (24), is characterized by a rapid production and consumption of dissolved hydrogen. It appears that the concentration of hydrogen is dictated by the hydrogen-consuming step or more explicitly by the electron-accepting reaction. A characteristic steady-state range of hydrogen concentrations can be attributed to each terminal electron acceptor and is independent of the electron donor supply (31, 32). In general, the greater the thermodynamic yield is for a given electron acceptor, the lower the predicted hydrogen concentration will be for environments in which it predominates (Table 3).

The approach has been applied by Chapelle et al. (13) at four groundwater field sites with different terminal electron acceptors. The approach has the greatest potential for distinguishing sulfate reduction and methane production because they are characterized by the highest hydrogen concentrations (Table 3). However, as noted throughout this chapter, there can be situations in which substrate consumption, product formation, or activity assays of the terminal

electron-accepting reaction are unequivocal. These are situations for which determination of dissolved hydrogen could be most valuable.

Environmental concentrations of hydrogen are extremely low. Therefore, sampling requires special precautions, and detection and quantification require specialized instrumentation. Metal surfaces catalyze the production of hydrogen in aqueous solutions and therefore must be avoided during sample collection. Hydrogen is also a very dynamic compound. No adequate sample storage or preservation procedure has been found, and analysis must be completed within 30 min of sample collection (13). Hydrogen gas is stripped from groundwater, separated from other gases on a packed chromatographic column, and quantified with a Reduction Gas Detector (proprietary product of Trace Analytical). Further details of sample collection and analysis are given by Chapelle et al. (13).

FINAL PERSPECTIVE
Determining the terminal electron-accepting reaction(s) at a groundwater study site can be a time-consuming and involved process, in large part because of the inaccessible nature of the subsurface and the slow rates at which the reactions occur. Yet whether from the basic microbiological, bioremediative, or drinking water perspective, this information is very important for understanding and interpreting the nature of the microbial food chain and the effects that it might have on groundwater geochemistry. There is also the global perspective: the saturated subsurface is the largest freshwater habitat on the planet. It is a unique habitat, one about which very little is known with respect to its microbial ecology. An integral part of future characterizations and insights regarding microbial communities in the subsurface must necessarily involve the context of the terminal electron-accepting reactions.

REFERENCES
1. **Adrian, N. R., J. A. Robinson, and J. M. Suflita.** 1994. Spatial variability in biodegradation rates as evidenced by methane production from an aquifer. *Appl. Environ. Microbiol.* **60:**3632–3639.
2. **Albrechtsen, H.-J., G. Heron, and T. H. Christensen.** 1995. Limiting factors for microbial Fe(III)-reduction in a landfill leachate polluted aquifer (Vejen, Denmark). *FEMS Microbiol. Ecol.* **16:**233–248.
3. **American Public Health Association.** 1981. *Standard Methods for the Examination of Water and Wastewater.* American Public Health Association, Washington, D.C.
4. **Aravena, R., M. L. Evans, and J. A. Cherry.** 1993. Stable isotopes of oxygen and nitrogen in source identification of nitrate from septic systems. *Ground Water* **31:**180–186.
5. **Baedecker, M. J., and W. Back.** 1979. Modern marine sediments as a natural analog to the chemically stressed environment of a landfill. *J. Hydrol.* **43:**393–414.
6. **Balderston, W. L., B. Sherr, and W. J. Payne.** 1976. Blockage by acetylene of nitrous oxide reduction in *Pseudomonas perfectomarinus. Appl. Environ. Microbiol.* **31:**504–508.
7. **Barker, J. F., and P. Fritz.** 1981. The occurrence and origin of methane in some groundwater flow systems. *Can. J. Earth Sci.* **18:**1802–1816.
8. **Beeman, R. E., and J. M. Suflita.** 1987. Microbial ecology of a shallow unconfined ground water aquifer polluted by municipal landfill leachate. *Microb. Ecol.* **14:**39–54.
9. **Beeman, R. E., and J. M. Suflita.** 1990. Environmental

TABLE 3 Anticipated hydrogen concentration for each terminal electron-accepting reaction[a]

Electron acceptor	Hydrogen concn (nM)
O_2	<0.1
NO_3^-	<0.1
Fe(III)	0.2–0.6
SO_4^{2-}	1–4
CO_2	>5

[a] From reference 13.

factors influencing methanogenesis in a shallow anoxic aquifer: a field and laboratory study. *J. Ind. Microbiol.* **5:** 45–58.

10. **Bengtsson, G., and H. Annadotter.** 1989. Nitrate reduction in a groundwater microcosm determined by ^{15}N gas chromatography-mass spectrometry. *Appl. Environ. Microbiol.* **55:**2861–2870.

11. **Böttcher, J., O. Strebel, S. Voerkelius, and H. L. Schmidt.** 1990. Using isotope fractionation of nitrate-nitrogen and nitrate-oxygen for evaluation of microbial denitrification in a sandy aquifer. *J. Hydrol.* **114:**413–424.

12. **Chapelle, F. H.** 1993. *Ground-Water Microbiology and Geochemistry.* John Wiley & Sons, Inc., New York.

13. **Chapelle, F. H., P. B. McMahon, N. M. Dubrovsky, R. F. Fujii, E. T. Oaksford, and D. A. Vroblesky.** 1995. Deducing the distribution of terminal electron-accepting processes in hydrologically diverse groundwater systems. *Water Resour. Res.* **31:**359–371.

14. **Dockins, W. S., G. J. Olson, G. A. McFeters, and S. C. Turbak.** 1980. Dissimilatory bacterial sulfate reduction in Montana groundwaters. *Geomicrobiol. J.* **2:**83–97.

15. **Ehleringer, J. R., and P. W. Rundel.** 1989. Stable isotopes: history, units, and instrumentation, p. 1–15. *In* P. W. Rundel, J. R. Ehleringer, and K. A. Nagy (ed.), *Stable Isotopes in Ecological Research.* Springer-Verlag, New York.

16. **Fossing, H., and B. B. Jørgensen.** 1990. Isotope exchange reactions with radiolabeled sulfur compounds in anoxic seawater. *Biogeochemistry* **9:**223–245.

17. **Freeze, R. A., and J. A. Cherry.** 1979. *Groundwater.* Prentice-Hall, Inc., Englewood Cliffs, N. J.

18. **Ghiorse, W. C., and J. T. Wilson.** 1988. Microbial ecology of the terrestrial subsurface. *Adv. Appl. Microbiol.* **33:** 107–172.

19. **Gibbs, M. M.** 1979. A simple method for the rapid determination of iron in natural waters. *Water Res.* **13:**295–297.

20. **Gormly, J. R., and R. F. Spalding.** 1979. Sources and concentrations of nitrate-nitrogen in ground water of the Central Platte region, Nebraska. *Ground Water* **17:**291–301.

21. **Grossman, E. L., B. K. Coffman, S. J. Fritz, and H. Wada.** 1989. Bacterial production of methane and its influence on ground-water chemistry in east-central Texas aquifers. *Geology* **17:**495–498.

22. **Heron, G., C. Crouzet, A. C. M. Bourg, and T. H. Christensen.** 1994. Speciation of Fe(II) and Fe(III) in contaminated aquifer sediments using chemical extraction techniques: *Environ. Sci. Technol.* **28:**1698–1705.

23. **Holm, T. R., G. K. George, and M. J. Barcelona.** 1988. Oxygen transfer through flexible tubing and its effects on ground water sampling results. *Ground Water Monit. Rev.* **8:**83–89.

24. **Ianotti, E. L., P. Kafkewitz, M. J. Wolin, and M. P. Bryant.** 1973. Glucose fermentation products of *Ruminococcus albus* grown in continuous culture with *Vibrio succinogenes*: changes caused by interspecies transer of H$_2$. *J. Bacteriol.* **114:**1231–1240.

25. **Johnson, R. L., J. F. Pankow, and J. A. Cherry.** 1987. Design of a ground-water sampler for collecting volatile organics and dissolved gases in small-diameter wells. *Ground Water* **25:**448–454.

26. **Jørgensen, B. B.** 1978. A comparison of methods for the quantification of bacterial sulfate reduction in coastal marine sediments. *Geomicrobiol. J.* **1:**29–64.

27. **Knowles, R.** 1982. Denitrification. *Microbiol. Rev.* **46:** 43–70.

28. **Korom, S. F.** 1992. Natural denitrification in the saturated zone: a review. *Water Resour. Res.* **28:**1657–1668.

29. **Kreitler, C. W., S. E. Ragone, and B. G. Katz.** 1978. N15/N14 ratios of groundwater nitrate, Long Island, New York. *Ground Water* **16:**404–409.

30. **Lovley, D. R.** 1991. Dissimilatory Fe(III) and Mn(IV) reduction. *Microbiol. Rev.* **55:**259–287.

31. **Lovley, D. R., F. H. Chapelle, and J. C. Woodward.** 1994. Use of dissolved H$_2$ concentrations to determine distribution of microbially catalyzed reactions in anoxic groundwater. *Environ. Sci. Technol.* **28:**1205–1210.

32. **Lovley, D. R., and S. Goodwin.** 1988. Hydrogen concentrations as an indicator of the predominant terminal electron-accepting reactions in aquatic sediments. *Geochim. Cosmochim. Acta* **52:**2993–3003.

33. **Lovley, D. R., and E. J. P. Phillips.** 1986. Availability of ferric iron for microbial reduction in bottom sediments of the freshwater tidal Potomac River. *Appl. Environ. Microbiol.* **52:**751–757.

34. **Mariotti, A., A. Landreau, and B. Simon.** 1988. ^{15}N isotope biogeochemistry and natural denitrification process in groundwater: application to the chalk aquifer of northern France. *Geochim. Cosmochim. Acta* **52:**1869–1878.

35. **Olson, G. J., W. S. Dockins, G. A. McFeters, and W. P. Iverson.** 1981. Sulfate-reducing and methanogenic bacteria from deep aquifers in Montana. *Geomicrobiol. J.* **2:** 327–340.

36. **Plummer, L. N., J. F. Busby, R. W. Lee, and B. B. Hanshaw.** 1990. Geochemical modeling of the Madison Aquifer in parts of Montana, Wyoming, and South Dakota. *Water Resour. Res.* **26:**1981–2014.

37. **Postma, D., C. Boesen, H. Kristiansen, and F. Larsen.** 1991. Nitrate reduction in an unconfined sandy aquifer: water chemistry, reduction processes, and geochemical modeling. *Water Resour. Res.* **27:**2027–2045.

38. **Robertson, L. A., and J. G. Kuenen.** 1984. Aerobic denitrification—old wine in new bottles? *Antonie van Leeuwenhoek* **50:**525–544.

39. **Sexstone, A. J., N. P. Revsbech, T. B. Parkin, and J. M. Tiedje.** 1985. Direct measurement of oxygen profiles and denitrification rates in soil aggregates. *Soil Sci. Soc. Am. J.* **49:**645–651.

40. **Simpkins, W. W., and T. B. Parkin.** 1993. Hydrogeology and redox geochemistry of CH$_4$ in a late Wisconsinan till and loess sequence in central Iowa. *Water Resour. Res.* **29:** 3643–3657.

41. **Smith, R. L., J. K. Böhlke, K. Revesz, S. P. Garabedian, and T. Yoshinari.** 1995. Assessing denitrification in nitrate-contaminated ground water using ^{15}N and natural gradient tracer tests. *Eos* **76:**F200.

42. **Smith, R. L., and J. H. Duff.** 1988. Denitrification in a sand and gravel aquifer. *Appl. Environ. Microbiol.* **54:** 1071–1078.

43. **Smith, R. L., and S. P. Garabedian.** Using transport model interpretations of tracer tests to study microbial processes in ground water. *In* A. L. Koch, J. A. Robinson, and G. A. Milliken (ed.), *Mathematical Models: Applications to Microbial Ecology*, in press. Chapman and Hall, New York.

44. **Smith, R. L., S. P. Garabedian, and M. H. Brooks.** Comparison of denitrification activity measurements in ground water using cores and natural gradient tracer tests. Submitted for publication.

45. **Smith, R. L., R. W. Harvey, and D. R. LeBlanc.** 1991. Importance of closely spaced vertical sampling in delineating chemical and microbiological gradients in groundwater studies. *J. Contam. Hydrol.* **7:**285–300.

46. **Smith, R. L., B. L. Howes, and J. H. Duff.** 1991. Denitrification in nitrate-contaminated groundwater: occurrence in steep vertical geochemical gradients. *Geochim. Cosmochim. Acta* **55:**1815–1825.

47. **Smith, R. L., B. L. Howes, and S. P. Garabedian.** 1991. In situ measurement of methane oxidation in groundwater by using natural-gradient tracer tests. *Appl. Environ. Microbiol.* **57:**1997–2004.

48. **Spalding, R. F., and J. D. Parrott.** 1994. Shallow groundwater denitrification. *Sci. Total Environ.* **141:**17–25.

49. **Tiedje, J. M., A. J. Sexstone, D. D. Myrold, and J. A. Robinson.** 1982. Denitrification: ecological niches, competition and survival. *Antonie van Leeuwenhoek* **48:** 569–583.

50. **Trevors, J. T.** 1985. The influence of oxygen concentrations on denitrification in soil. *Appl. Microbiol. Biotechnol.* **23:**152–155.

51. **Trevors, J. T., and M. E. Starodub.** 1987. Effect of oxygen concentration on denitrification in freshwater sediment. *J. Basic Microbiol.* **7:**387–391.

52. **Trudell, M. R., R. W. Gillham, and J. A. Cherry.** 1986. An in-situ study of the occurrence and rate of denitrification in a shallow unconfined sand aquifer. *J. Hydrol.* **83:** 251–268.

53. **van Beek, C. G. E. M., and D. van der Kooij.** 1982. Sulfate-reducing bacteria in ground water from clogging and nonclogging shallow wells in the Netherlands river region. *Ground Water* **20:**298–302.

54. **White, A. F., M. L. Peterson, and R. D. Solbau.** 1990. Measurement and interpretation of low levels of dissolved oxygen in ground water. *Ground Water* **28:**584–590.

55. **Whiticar, M. J., E. Faber, and M. Schoell.** 1986. Biogenic methane formation in marine and freshwater environments: CO_2 reduction vs. acetate fermentation—isotope evidence. *Geochim. Cosmochim. Acta* **50:**693–709.

56. **Yoshinari, T., R. Hynes, and R. Knowles.** 1977. Acetylene inhibition of nitrous oxide reduction and measurement of denitrification and nitrogen fixation in soil. *Soil Biol. Biochem.* **9:**177–183.

In Situ and Laboratory Methods To Study Subsurface Microbial Transport

RONALD W. HARVEY

64

Methodology for present-day studies of microbial transport through aquifers derives from earlier groundwater hydrology investigations in which microorganisms were used as markers to delineate groundwater flow. Bacteria which form red or yellow pigments were used as microbiological tracers in joint and pore aquifers in the latter part of the 19th century (1). Historically, nonindigenous microorganisms and spores were often preferred to nonreactive (conservative) solutes as groundwater tracers because of simplicity of detection (using light microscopy techniques or plating on selective media) and because the background level of nonindigenous microbes in the aquifer is generally negligible. Microorganisms and spores also have the advantage of being excluded on the basis of size from much of the fine, often dead-end porosity that does not contribute to flow. Therefore, biological tracers have been particularly useful in determining major flow paths in karstic (limestone) and highly fractured formations, in which the hydrology is dominated by the presence of preferred flow paths. Coliform bacteria (e.g., 11) and spores (e.g., 20) have often been used as tracers in karstic and fracture-flow aquifers. However, bacteriophage which can be traced for long distances and sorb very little under certain conditions (54) are becoming the particulate tracers of choice for delineation of flow in a number of aquifers dominated by preferred flow paths (57).

Although the use of microbial tracers has been instrumental in developing a better understanding of groundwater movement in certain types of aquifers, the importance of studying the transport behavior of the microorganisms themselves is now very apparent, largely because of an increasing dependence on limited and fragile groundwater resources. In particular, widespread contamination of shallow drinking water aquifers by microbial pathogens and chemical wastes has resulted in an increased interest in the factors that control subsurface microbial transport. The movement of nonindigenous bacteria, viruses, and protozoa through aquifers has been a public health concern in the United States (chapter 66), where contamination of water supply wells by microbial pathogens contributes significantly to the total number of waterborne disease outbreaks (15). Introduction of nonindigenous bacteria, viruses, and protozoa into groundwater often occurs as a result of various waste disposal practices. Deliberate injections of nonindigenous populations of bacteria into aquifers are now conducted to enhance in situ rates of bioremediation of organically contaminated groundwater and to increase oil recovery from less

transmissive zones by selective bacterial plugging of more-permeable strata (see chapter 65). It is well established that transport properties of introduced bacteria in groundwater can have major roles in transmission of some waterborne diseases (8), in the success of microbially enhanced oil recovery processes (40, 53), in the mobility of surface-active or hydrophobic groundwater contaminants (14, 41, 52), in pore clogging (78), in the potential dissemination of genetically engineered bacteria after release at aquifer restoration sites (76), and in the biorestoration of at least some organically contaminated aquifers (81). However, many of the factors controlling subsurface microbial transport are still poorly understood.

To better study the controls of microbial transport, it has been necessary to improve available methodology for investigating microbial transport behavior in the laboratory and in the field. The many physical, chemical, and biological factors that affect microbial transport through aquifer material (24) have necessitated refinement of techniques to enable observations to be made under more controlled experimental conditions than has been possible in the past. In this chapter, laboratory and field techniques for studying microbial transport behavior in aquifer materials are evaluated and selected studies resulting from this methodology are described.

FIELD VERSUS LABORATORY STUDIES

Although most studies delineating the controls of subsurface microbial transport are performed in the laboratory, advances in our understanding of microbial transport behavior would clearly benefit from an iterative approach that combines both laboratory and field investigation Fig. 1. Column studies provide a greater degree of control and therefore are useful in providing detailed, mechanistic information about specific processes affecting microbial transport. Also, columns can be designed to more nearly meet initial boundary conditions for microbial transport models. However, they cannot account for the many factors that control microbial mobility in situ. There are many factors in addition to the hydrological characteristics of the aquifer itself that control the movement of microorganisms in groundwater. For bacteria, these biotic and abiotic factors include growth, predation by protists, parasitism by bacteriophage and predatory bacteria (*Bdellovibrio* species), motility, lysis under unfavorable conditions, changes in cell size and propensity for attachment to solid surfaces in response to alterations in nutrient

FIELD WORK   **LAB WORK**

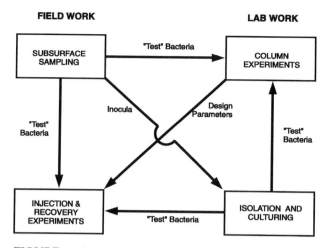

FIGURE 1 Integrated field and laboratory experimental approach to investigating transport behavior of microorganisms in groundwater environments.

conditions, spore formation in the case of some gram-positive species, reversible and irreversible attachment to solid surfaces, detachment from surfaces, and straining (24). Many of these processes are interrelated through other factors that are poorly understood or difficult to describe mathematically. Therefore, it is desirable to combine laboratory experiments that allow better delineation of individual processes with field studies that provide a framework in which the applicability of laboratory-derived results can be evaluated under natural conditions.

One problem with using flowthrough columns that are packed with aquifer material involves the difficulty in artificially recreating aquifer structure. Although flowthrough column experiments have provided useful information on many of the factors governing microbial transport through the subsurface (Table 1), a great deal of caution must be exercised when one is extrapolating laboratory results to the groundwater environment. Alterations in transport behavior that occur as a result of repacking aquifer sediments into flowthrough columns can be quite complex and difficult to sort out from natural transport behavior. It has been shown that transport of microbes through subsurface sediments that have been repacked into columns can differ from that observed in the field, even when flow velocity, porosity, and physicochemical conditions are similar (22). The effects on microbial transport behavior of macropore (secondary pore structure) destruction by repacking and sieving of unsaturated soil (72) and by mechanical disruption of whole saturated aquifer sediments (30) have been documented. The extent of spatial segregation of grain sizes resulting from wet manual (e.g., 44) or dry mechanical (68) column packing is not well known for subsurface sediments. However, it is clear that the repacking of whole aquifer sediments into large columns may have significant ramifications with regard to microbial transport behavior. In several recent microbial transport studies (16, 50, 69), problems associated with spatial fractionation of grain sizes have been ameliorated by the use of smaller-diameter, shorter-length columns packed with sieved (prefractionated) sand.

An important consideration in comparing transport results generated from natural-gradient and column experiments is the difference in flow velocities. Aquifers in which natural-gradient transport experiments have been run typically are characterized by slow groundwater velocities (<1 m/day) and steady temperatures. However, most column ex-

periments are run at flow velocities that are many times what is found in such aquifers, in part because the Mariot siphon and most pumps do not deliver accurately at low volumetric flow rates. The consequences of large differences in flow velocity between laboratory and field experiments lead to differences in the thickness of the hydrodynamic boundary layers (the quiescent layer of water immediately adjacent to grain surfaces), in the collector efficiency, and in attachment opportunities. Therefore, comparisons of column study results with those of the field can serve as an invaluable check.

IN SITU STUDIES
Tracers

Recent in situ investigations designed to examine microbial transport behavior in groundwater have involved the coinjection into the aquifer of the microorganism of interest with a conservative solute tracer (usually chloride, bromide, or iodide). Conservative tracers are typically used to monitor the velocity and direction or path of groundwater flow and the degree of dispersion (spreading of the injectate that is caused by diffusion and by mixing with adjacent parcels of water). Dimensionless concentration histories (breakthrough curves) of the conservative tracers are then obtained for sampling points that are downgradient and in the path of the injectate cloud. The characteristics of the conservative tracer breakthrough curve can then be used comparatively in determining some of the major transport parameters exhibited by the introduced microorganisms. When using halide tracers, it may be necessary to consider their effects upon microbial activity (20a) and upon the density of the injectate cloud (49a). For many small-scale injection and recovery experiments, halide concentrations in the 100 to 200 mg/liter range are sufficient. For a pulse (slug) input, conservative tracer transport data may be used to help calculate microbial transport parameters as described below.

Retardation (slowing down of transport relative to flow velocity) of the introduced microorganism can be determined by comparing velocities for the peak concentration of conservative tracer with that of the introduced microorganism at sampling points downgradient from point of injection: $R = V_{tracer}/V_{microbe}$, where V_{tracer} and $V_{microbe}$ are the observed average velocities of the peak concentrations of tracers and microorganisms, respectively, that appear at a downgradient well. If the breakthrough curves for the microorganism and conservative tracer are dissimilar in shape or the breakthrough curves are multipeaked, retardation may be approximated as the ratio of times required for the arrival of the centers of mass (estimated by numerical integration) for the introduced microorganism and the conservative tracer, respectively (29).

The degree of apparent dispersion of the introduced microorganism that occurs along the travel path through the aquifer can be obtained by comparing the degree of spreading in the breakthrough curve (at one-half the peak height) with that exhibited by the conservative tracer. In aquifer sediments, the apparent dispersion of the microorganism will usually be greater than that of the conservative tracer. The apparent spreading in the microorganism's concentration history is caused by reversible interactions with grain surfaces (which differentially retard portions of the population), diffusion (Brownian motion), and mixing with adjacent parcels of groundwater along the flow path. In contrast, only the latter two mechanisms contribute to the dispersion of the conservative solute because it does not react with

TABLE 1 Selected recent column studies delineating factors controlling subsurface microbial transport behavior[a]

Factor studied	Medium	Method	Conditions	Microorganism(s) or colloid	Reference
Bacteria					
Density-dependent attachment	Sandy soil	Downflow	Saturated, aerobic	Aquifer isolate	7
Flow velocity	1-mm glass beads	Upflow	Saturated, aerobic	*Pseudomonas fluorescens*	10
Structured heterogeneity	Clean quartz sand	Flowthrough	Saturated, aerobic	Aquifer isolates	16
Ionic strength	Aquifer sand	Upflow	Saturated, aerobic	*Pseudomonas* sp.	18
Various chemical treatments	Glass beads	Downflow	Saturated, aerobic	*Alcaligenes paradoxus*	21
Unsaturated flow conditions	Loamy soils	Downflow	Unsaturated, aerobic	*Burkholderia* and *Pseudomonas* spp.	34
Evaluation of models	Rounded quartz	Downflow	Saturated, aerobic	Aquifer isolates	37
Cell hydrophobicity and soil properties	Sandy and clay soil	Downflow	Unsaturated, aerobic	*Lactobacillus* spp.	39
Favorable transport properties	Berea sandstone	Horizontal flow	Saturated, aerobic	*Pseudomonas, Bacillus,* and *Clostridium* spp.	40
Facilitated transport of PAH	Aquifer sand	Flowthrough	Saturated, aerobic	*Bacillus subtilis*	41
Effect of sterilization techniques	Berea sandstone	Static	Saturated, aerobic	Landfill isolates	42
Permeability and motility	Berea sandstone	Static	Saturated, aerobic	*Bacillus* and *Enterobacter* spp.	43
Surface residence time	Subsurface sand	Downflow	Saturated, aerobic	Subsurface isolate	45
Survival	Aquifer sand	Static	Saturated, suboxic	*Pseudomonas* sp. strain B13 (GEM)	48
Facilitated transport of DDT	Sand	Flowthrough	Saturated, aerobic	*Pseudomonas* and *Bacillus* spp.	52
Rates of attachment and detachment	Sand	Upflow, pulse	Saturated, aerobic	Gram-negative aquifer isolates	55
Temperature-controlled motility	Aquifer sediment	Upflow	Saturated, aerobic	Subsurface isolate	55a
Selectivity and depth of plugging	Berea sandstone	Static	Saturated, aerobic		63
Motility with and without nutrients	Ottawa sand	Static	Saturated, anaerobic	*Escherichia coli*	66
Cell hydrophobicity and surface charge	Glass and Teflon	Static vs dynamic	Saturated, aerobic	Pseudomonads and coryneforms	67
Groundwater chemistry	Aquifer sand	Static	Saturated, aerobic	Uncultured aquifer population	69
Mineralogy and chemistry	Coated sand	Downflow	Saturated, aerobic	Aquifer isolates	70
Effect of macropores	Soil	Downflow	Unsaturated, aerobic	*E. coli*	72
Clogging efficiency	Quartz sand	Downflow	Saturated, aerobic	Aquifer and soil isolates	78
Viruses					
Matrix diffusion	Fractured tuff	Downflow	Saturated, aerobic	MS-2 (coliphage)	2
Attachment reversibility	(Un)bonded silica	Downflow	Saturated, aerobic	PRD-1 and MS-2 (phage)	3
Chemical perturbations	Silica beads	Downflow	Saturated, aerobic	MS-2 and poliovirus	4
Organic matter	Loamy fine sand	Downflow	Unsaturated, aerobic	MS-2 (coliphage)	62
Protozoa					
Field vs column transport behavior	Aquifer sand	Upflow	Saturated, aerobic	Unidentified flagellates	29
Microspheres					
Physical heterogeneity	Aquifer sand	Upflow	Saturated, aerobic	0.2–4.8-μm carboxylated microspheres	30
Optimal transport size	Aquifer sand	Upflow	Saturated, aerobic	0.7–6.2-μm carboxylated microspheres	29
Influence of oil	Berea sandstone	Horizontal flow	Saturated, aerobic	Unspecified, negative charge	40
Temperature effects	Aquifer sediment	Upflow	Saturated, aerobic	1.5-μm carboxylated microspheres	55a
Artificial fractures	Sand	Horizontal flow	Saturated, aerobic	1-μm latex microspheres	75

[a] PAH, polyaromatic hydrocarbons; DDT, dichlorodiphenyltrichlorethane; GEM, genetically engineered microorganism.

stationary surfaces. Apparent longitudinal (direction of flow) dispersion ($\mathcal{A}_L$) for the microorganisms can be estimated from the following equation (25):

$$\mathcal{A}_L = \frac{x_1(\Delta t/t_{\text{peak}})^2}{16 \ln 2}$$

where x_1 is the distance from point of injection, Δt is the duration of breakthrough when $C(t)$ is greater than one-half the peak concentration, and t_{peak} is the time to peak concentration. In many cases, such calculated values are only first approximations because the concentration histories of the microorganisms often do not conform to the assumed classical Gaussian-shaped breakthrough curves upon which the foregoing equation is based.

In granular aquifers, differences between the conservative tracer and microorganism breakthrough curves are due largely to the interactions between the microorganisms and grain surfaces. The simplest method of accounting for microbial attachment to surfaces is to assume that the microbial distribution between solution and solid surface is constant. This conditional approach assumes that attachment is always at equilibrium, and therefore the solid/solution partitioning of the microorganisms can be described by a single coefficient of distribution (K_d). However, judging from the protracted nature of the trailing ends of microorganism breakthrough curves relative to those of the conservative tracer in most field tests (Fig. 2A), an equilibrium assumption usually is not appropriate. A better approach involves consideration of the attachment and detachment kinetics of microorganisms being advected downgradient. It can be useful to assume that there are two types of attachment sites, one responsible for irreversible removal of the microorganisms and one responsible for reversible attachment. The former type results in an attenuation of the microorganism breakthrough curve (plotted as dimensionless concentration history versus time) relative to that of the conservative tracer. The latter type results in retardation of the center of mass of the unattached microbial population.

The fraction of the microorganisms injected into the aquifer that become irreversibly attached (within the experimental time frame) during transport to a downgradient sampler may be calculated from its relative breakthrough (RB). The RB is determined by comparing the numerical integration of the dimensionless concentration history of the microbe to that of the conservative tracer (28):

$$\text{RB} = \int_{t_0}^{t_f} \frac{C(t)}{C_0}\, dt \div \int_{t_0}^{t_f} \frac{\text{Tr}(t)}{\text{Tr}_0}\, dt$$

where C_0 and Tr_0 are the respective microorganism and conservative tracer concentrations in the injectate, $C(t)$ and $\text{Tr}(t)$ are concentrations downgradient at time t, t_0 is the time of injection, and t_f is the elapsed time from the beginning of injection until the termination of breakthrough.

The RB may, in turn, be used to estimate the collision efficiency factor. The collision efficiency factor, α, is a parameter in filtration theory that represents the physicochemical factors that determine irreversible microbial immobilization (25):

$$\alpha = \frac{d[[1 - 2(\mathcal{A}_L x_1)\ln(\text{RB})]^2 - 1]}{6(1 - \theta)\eta \mathcal{A}_L}$$

where d is the median grain size, x_1 is the travel distance from point of injection to the sampler at which the concentration history was monitored, θ is the porosity, and η is the single-collector efficiency. The latter parameter is the rate at which the microorganisms strike a single sand grain divided by the rate at which they move toward the grain and represents the physical factors determining collision. Its value can be

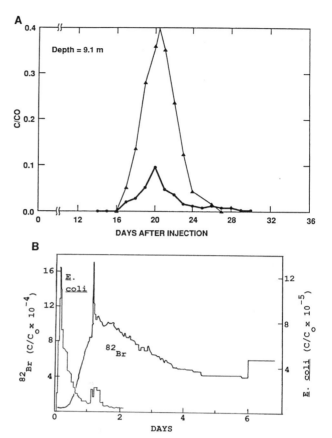

FIGURE 2 Dimensionless concentration histories for bacteria and bromide transported downgradient through aquifers from the point of coinjection. (A) Natural-gradient test in a well-sorted, sandy aquifer in Cape Cod, Massachusetts (reprinted from reference 8 with permission from John Wiley & Sons, Inc. ●; bacteria; ▲, bromide). (B) forced-gradient test in a fractured, granite aquifer at Chalk River Laboratory, Ontario (redrawn from reference 12 with kind permission from Elsevier Science Ltd., The Boulevard, Langford Lane, Kidlington 0X5 1GB, United Kingdom).

calculated by using the following expression (64): $\eta = 0.72 A_s N_{\text{Lo}}^{1/8} N_R^{15/8} + 2.4 \times 10^{-3} A_s N_G^{1.2} N_R^{-0.4} + 4 A_s^{1/3} N_{\text{Pe}}^{-2/3}$, where $A_s = 2(1 - p^5)/w$, $w = 2 - 3p + 3p^5 - 2p^6$, $p = (1 - \theta)^{1/3}$, $N_{\text{Lo}} = H/9\pi\mu a_p^2 U$, $N_R = a_p/a_s$, $N_G = 2a_p^2(\rho_p - \rho)g/9\mu U$, and $N_{\text{Pe}} = 48\pi\mu a_s a_p U/kT$. In this expression, a_p and a_s are the respective colloid and grain radii, μ is the fluid viscosity, U is the approach velocity of the groundwater, η is the porosity, ρ and ρ_p are the respective densities of the groundwater and the microorganism, g is the acceleration due to gravity, H is the Hamaker constant (~1 × 10^{-20} joules), k is the Boltzmann's constant (1.38 × 10^{-23} joules/mol °K), and T is temperature (°K).

Reversible types of attachment result in an increased travel time and apparent dispersion for the microorganisms relative to the conservative tracer. Although there is experimental evidence for the removal mechanisms (generally assumed irreversible) described by filtration theory (see above), there is also evidence that nonequilibrium reversible attachment is a common feature of microbial transport in aquifers (25). The nonequilibrium is caused by a rate of attachment that is different from the rate of detachment and can be accounted for by using different attachment and detachment rate constants in modified advection-dispersion models describing microbial transport in porous media. The development of specific models for describing microbial transport through aquifer material is beyond the scope of this chapter. The interested reader is referred to Lindqvist et al. (51) for a discussion of kinetic modeling of microbial transport through saturated subsurface sands, to Harvey and Garabedian (25) for a discussion of how filtration theory and mathematical descriptions of reversible attachment can be coupled with a hydrologic transport (advection-dispersion) model to describe small-scale field results, and to Corapcioglu and Haridas (13), Hornberger et al. (37), and Hurst (38) for discussions of subsurface microbial transport models.

In addition to providing a framework in which the transport behavior of microorganisms can be evaluated, conservative tracers can be useful for supplying information about movement of the injectate cloud within sampling arrays. This information facilitates a sampling protocol that will allow capture of most of the microbial breakthrough. However, the usefulness of conservative tracers in this regard depends on the type of aquifer in which the injection and recovery experiment is run. For homogeneous zones of sandy aquifers, dimensionless concentration histories of the conservative tracer can be similar to those of the bacteria or viruses coinjected with it (Fig. 2A) (5, 25). In general, physical heterogeneity of the system increases the dissimilarity between the breakthrough curves of the conservative tracer and the microorganisms (16, 30). For aquifers characterized by fracture flow, arrival of peak microbial abundance at a downgradient sampling point can greatly precede that of the conservative tracer (Fig. 2B). This is because the microorganisms are largely confined to the major flow paths (fractures) and therefore are subject to an average travel path that is much shorter than that of the conservative tracer. In extreme cases, microbial breakthrough at downgradient sampling points can terminate before arrival of measurable tracer concentrations. Because the pore volumes available for transport of microorganisms and solutes can be quite different in highly fractured aquifers, conservative tracers are less useful in providing real-time information about how and where to sample and in providing a comparative reference for determining microbial transport parameters.

The degrees of similarity between the microbial and sol-ute tracer breakthrough curves are also influenced by the size and surface chemistry of the microorganism being evaluated. For example, substantial retardation of aquifer flagellates (protozoa) relative to bromide was observed for transport in a well-sorted, sandy aquifer where retardation of indigenous bacteria is generally not observed (29). Earlier tests with different sizes (0.23 to 1.35 μm) and types of microspheres revealed that the reactivity of the surface was a much stronger determinant of retardation than microsphere size. This finding suggests that bacterial surface characteristics may be a more important determinant of retardation in some highly porous aquifer sediments than the size of the microbe (28). Retardation factors observed for various microorganisms and microbe-sized microspheres are listed in (Table 2) for small- and intermediate-scale injection-and-recovery studies involving different types of aquifers.

Microspheres have been quite useful as particulate tracers to study certain abiotic aspects of microbial transport behavior in situ. Because microspheres are physically and chemically well defined, they have been useful in determining the effects of cell size and surface characteristics on microbial transport behavior in sandy aquifer sediments. Microspheres also have been used to predict contamination by nonindigenous bacteria in core material obtained from deep subsurface sediments (see chapter 59). However, the transport properties of microspheres in porous media can differ from those of microorganisms (23). For example, it has been shown that the degrees of retardation, apparent dispersion, and immobilization exhibited by various types of commercially available microspheres differ from these exhibited by bacteria (28). On the other hand, transport properties of 2-μm-diameter carboxylated microspheres exhibit transport properties similar to those of groundwater flagellates (protozoa) in an injection-and-recovery experiment performed in a sand and gravel aquifer (29).

Natural- and Forced-Gradient Tracer Tests

Several different types of injection-and-recovery tests have been used to investigate microbial transport in situ (Fig. 3). Forced gradient tests are ones in which the flow field is controlled by high-volume pumping at the point of injection, point of withdrawal, or both. Forced-gradient tests have the advantage of substantially shortening the time frame required to obtain complete breakthrough of the in-

TABLE 2 Retardation in groundwater transport of microorganisms or microbe-sized microspheres relative to halide tracers

Aquifer	Colloid	Type of test	Distance (m)	Retardation factor	Reference
Fractured media					
Clay-rich till	MS-2 and PRD-1 (bacteriophage)	Natural gradient	4	0.01	57
Crystalline rock (granite)	*Escherichia coli*	Forced gradient	13	0.1	12
Layered basalt	Indigenous isolate (*Bacillus* sp.)	Forced gradient	27	0.6	33
Granular media					
Sand or gravel with clay and carbonate	*Saccharomyces cerevisiae* (yeast)	Forced gradient	1.5	0.7	83
Well-sorted sand and gravel	Indigenous bacterial community	Forced gradient	1.7	1.0	28
		Natural gradient	6.7	1.0	25
	PRD-1 (bacteriophage, 60 nm)	Natural gradient	13	1.0	5
	Uncharged 0.5-μm latex spheres	Natural gradient	6.7	1.0	28
	Polyacrolein 08-μm microspheres			1.3	28
	Carboxylated 0.5-μm microspheres			1.4	28
	Indigenous flagellates (protozoa)	Natural gradient	1.0	4.7	29

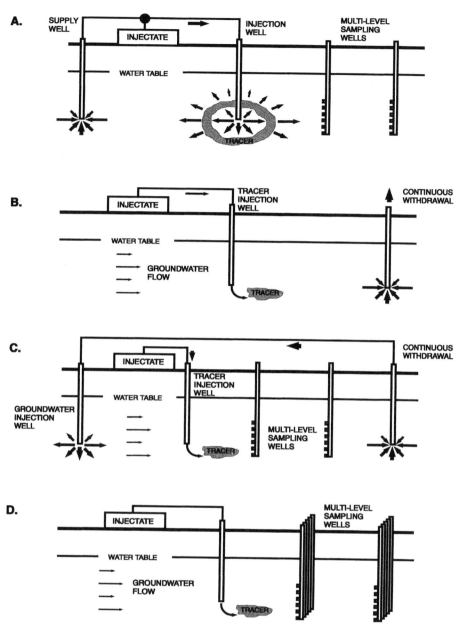

FIGURE 3 Schematic depiction of experimental designs for small- to intermediate-scale injection-and-recovery investigations examining microbial transport behavior in aquifers (modified from reference 8 with permission from John Wiley & Sons, Inc.). (A) Divergent, forced gradient; (B) convergent, forced gradient; (C) doublet cell, forced gradient; (D) natural gradient.

jected constituents. This type of test is particularly well suited for microbial transport studies in aquifers characterized by fracture flow, because the sampling well(s) do not need to be placed immediately down gradient. Hence, it is not necessary to completely understand the hydrology of the system before doing a transport experiment. A disadvantage of forced-gradient tests is that the flow field is nonuniform and therefore difficult to model. Divergent tracer tests involve a quick addition of a known quantity of microorganisms and conservative tracer into a continuous stream of groundwater being injected back into the aquifer. This setup produces radially divergent flow and forces the gradient in all directions (Fig. 3A). Because of the high degree of forced

dispersion, the distance over which the injectate can be monitored is limited. A convergent test can be accomplished by continuous withdrawal at the sampling well (Fig. 3B). The advantage of the convergent test is that all of the conservative tracer is recovered at the sampling well. This makes it possible to do a true mass balance on both the conservative tracer and the microorganisms. A variation of the simple convergent test is the double cell (Fig. 3C), in which an injection is made within the flow field created by a coupled continuous groundwater injection and withdrawal. The result is a flow field that can be controlled over moderate distances. The latter method is particularly useful for simulating conditions that may result during aquifer renovation.

In natural-gradient tests (Fig. 3D), the injectate is added slowly to the aquifer and the natural flow of groundwater advects the injectate cloud past rows of multilevel samplers (MLS) placed in rows perpendicular to the flow path. Natural-gradient tests are best suited for sandy aquifers where the flow direction can be predicted within reasonable limits and where the water table is shallow enough to permit the use of peristaltic pumps to sample water from each level of the MLS. Construction and use of MLS are discussed by Pickens et al. (60) and Smith et al. (73). For small-scale experiments, the individual MLS within the downgradient rows should be as close as possible. As a practical working limit, MLS installed by using hollow-stem auger drill rigs (see chapter 59) cannot be placed much closer than 1 m apart.

The quantity of microorganisms needed to conduct an injection and recovery experiment depends on the type of injection test to be run, travel distance, type of microorganism (and associated minimum level of detection), and aquifer characteristics. In general, the divergent (Fig. 3A) test will require the largest number of microorganisms. The radially divergent flow field in this type of test results in microbial decreases that are proportional to the square of the travel distance. This is, of course, in addition to losses from solution due to immobilization at solid surfaces. Natural-gradient experiments (Fig. 3D) often require the fewest number of microorganisms, because of the absence of forced dispersion caused by forced-gradient conditions, and/or, in the cases of continuous withdrawal at the sampling well (Fig. 3B and C), because of the absence of large dilution factors at the point of collection. If the physical parameters (flow direction, velocity, dispersivities, porosity, grain size distribution, injection and sampling point coordinates) are known and the aquifer is relatively homogeneous, a rough estimate of the number of microorganisms required for the test may be obtained by first calculating the degree of dilution that would occur along the experimental flow path as a result of dispersion and subsequently applying a filtration factor.

The degree of dilution due to dispersion or to sampling at some distance away from the travel path taken by the center of injectate mass may be roughly estimated from the following equation (17):

$$C_{(x,y,z,t)} = \left(\frac{C_0 V_0}{8\theta(\pi t)^{3/2}(D_x D_y D_z)^{1/2}} \right)$$
$$\exp\left(-\frac{(x - vt)^2}{4D_x t} - \frac{y^2}{4D_y t} - \frac{z^2}{4D_z t} \right)$$

where C is the concentration at time t, C_0 and V_0 are the respective initial concentration and volume at the point of injection, θ is the porosity, x is the distance in the direction of flow, y is the distance in the horizontal transverse direction, z is the distance in the vertical transverse direction, D_x, D_y, and D_z are the coefficients of dispersion in the respective x, y, and z directions, and v is the flow velocity. This expression is a modified solution of the well-known advection-dispersion equation (17):

$$\frac{\delta C}{\delta t} = -v\frac{\delta C}{\delta x} + D_x\frac{\delta^2 C}{\delta x^2} + D_y\frac{\delta^2 C}{\delta y^2} + D_z\frac{\delta^2 C}{\delta z^2}$$

Because local heterogeneities can substantively affect the values of the above-listed aquifer properties, it is almost always preferable to run an initial test with conservative tracers (as described in the preceding section). The resulting information on the flow direction and dispersivities test zone can be used to adjust the required volume of a subsequent coinjection involving microorganisms and tracers. Increasing the volume of the injectate results in both a decreased dispersion and a larger injectate cloud. The latter can often compensate for downgradient sampling points that do not lie immediately along the trajectory taken by the center of injectate mass.

The degree of immobilization experienced by microorganisms as they travel along the experimental flow path is difficult to estimate without a priori information derived from earlier in situ tests or from flowthrough column experiments. For bacteria, the loss of bacteria due to interaction with grain surfaces in sandy aquifer sediments can be at least one log unit per 10 m of travel (25). Consequently, for small-scale (≤10-m), natural-gradient injection-and-recovery tests, a minimum fluorescently stained bacterial population of 10^6 to 10^7/ml suspended in an injectate volume of 100 to 200 liters (25, 28) would be required to obtain an accurate concentration history at the downgradient samplers by using an epifluorescence direct-counting procedure (e.g., 36). The level of detection, which is an important determinant of the required number of labeled bacteria in the injectate, is controlled largely by the quantity of suspended particulate matter in the groundwater and, consequently, by the volume that may be filtered through the membrane filter used in the enumeration procedure.

Physiological Considerations

Most subsurface transport experiments referenced herein involved cultured microorganisms. Although data derived from these investigations provide useful information about the fate and transport of introduced microorganisms, the physical characteristics of cultured microorganisms in oligotrophic groundwater may change during the time frame of the experiment. In general, successive culturing in high-nutrient media results in enlarged cells that have a different propensity for attachment to solid surfaces than would be the case in their natural environment. It has been shown that groundwater protozoa that have been cultured in high-nutrient liquid media and reintroduced into the aquifer can undergo changes in cell size and in their propensity for solid surfaces with time and distance in the aquifer (29). This may result from their readaptation to low-nutrient, porous-medium conditions. It also appears that the physical properties of aquifer bacteria can undergo substantive changes during culturing. Most indigenous bacteria in the Cape Cod, Massachusetts, aquifer are transported as though they are nearly neutrally buoyant (26) and have apparent buoyant densities of ≤1.02 g/cm³. However, after culturing, these same bacteria exhibit buoyant densities of between 1.05 and 1.09 g/cm³ (32), similar to values reported for cultured bacteria isolated from other aquifers (79). It is clear that caution must be used when one is interpreting in situ transport experiments performed with cultured isolates.

Physiological changes during transport through the aquifer can be lessened by using whole communities of uncultured indigenous bacteria that have been concentrated directly from the zone in which the experiment is to be run. Using microorganisms that are already adapted to aquifer conditions can preclude interpretation problems resulting from physiology-induced changes in transport behavior. Bacteria may be concentrated from groundwater by dewatering techniques involving continuous-flow centrifugation or tangential-flow filtration and subsequently labeled for reinjection. An advantage of tangential-flow filtration is that the apparatus is field portable, allowing on-site concentration of aquifer microbial populations.

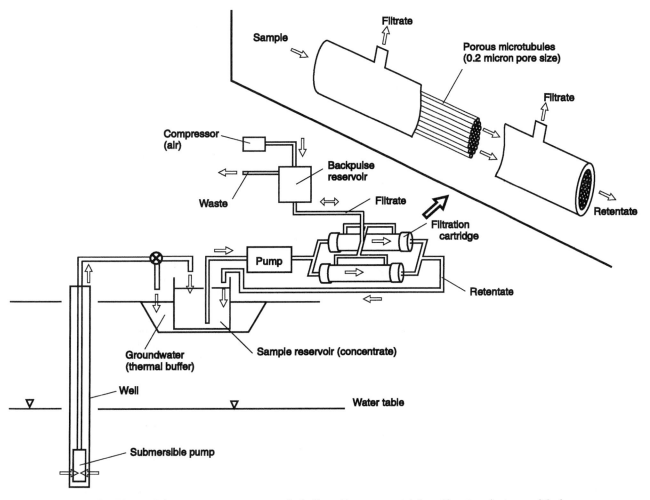

FIGURE 4 Schematic representation of a hollow-fiber tangential-flow filtration device modified for on-site concentration of indigenous microorganisms from groundwater (modified from reference 49 with permission from *Journal of Environmental Quality*).

Two methods of tangential-flow filtration have been used for this purpose. One method involves circulating a groundwater sample between parallel membrane filter sheets. The other method is one in which the groundwater sample is recirculated through hollow, porous fibers (microtubules). The principle behind tangential-flow filtration is that a recirculating water sample is forced at high velocity in a direction parallel to the filter surface. The scouring action provided by the high-velocity flow tends to prevent microorganisms and other colloidal particles from clogging the filter by keeping them in the recirculating stream. The transmembrane pressure causes a fraction of the liquid to exit the recirculating system through the membrane pores. A modified version of a commercially available microtubule-type tangential-flow device was found to exhibit a higher efficiency for concentrating bacteria from groundwater than either the parallel sheet-type tangential-flow device or the continuous-flow centrifuge (Fig. 4) (49). The advantage of a modified hollow-fiber device is that the design minimizes hydraulic dead space relative to the parallel sheet unit. It also allows periodic backpulse (reverse flow) of collected filtrate back through the porous microtubule walls. The backpulsing helps dislodge microorganisms attached to

the hollow-fiber walls that manage to escape the shearing action of the tangential-flow stream. This prevents clogging of the pores within the fiber walls and keeps the microbes in the recirculating groundwater stream.

When it is necessary to use cultured populations of a single groundwater isolate, it is desirable to maintain in situ characteristics. This may be facilitated by growth under conditions which mimic the environment from which the microbes were isolated. For example, it has been demonstrated that growing groundwater flagellates within porous media (sieved aquifer sediments) under low-nutrient and low-aquifer-pH conditions results in cells which appear to be similar in size to what is found in situ (46). Flagellates grown in this manner have been observed to exhibit transport behavior different from that of flagellates grown in high-nutrient liquid media (31).

It may be important to account for or control bacterial growth during subsurface transport in injection-and-recovery experiments. Uncontaminated aquifers are generally extremely oligotrophic. As a consequence, bacterial growth in groundwater is often too slow to be accurately measured by available methods (27). However, substantive bacterial growth can occur during transport downgradient in small-scale transport experiments as a result of the inadvertent introduction of dissolved organic matter (DOM) during the injection. The introduction of labile DOM is particularly problematic when cultured bacteria are used as an inoculum, because it is difficult to preclude residual broth nutrients and bacterial metabolites and lysates from contaminating the injectate. Inadvertent contamination with DOM during the introduction of cultured isolates may be lessened by a priori sequential resuspensions in nongrowth media or the groundwater into which the organisms will be added. Labeling the bacterial populations with the fluorochrome 4′,6-diamidino-2-phenylindole (DAPI), which is known to hamper bacterial metabolism (59), retards growth and provides a long-lived fluorescent tag that facilitates monitoring growth by the frequency-of-dividing-cells technique (27). It can also be helpful to monitor changes in abundance of labeled bacteria over several weeks in samples collected downgradient and maintained at groundwater temperature (25).

Heterogeneity

In most injection-and-recovery investigations involving microorganisms, it is important to be able to separate out heterogeneity-induced artifacts from true transport behavior. For bacteria, chemical heterogeneity and physical heterogeneity in the aquifer collectively affect most of the factors that influence the extent of travel in the aquifer (24). Even in relatively homogeneous aquifer sediments composed of well-sorted sand, physical and mineralogical heterogeneity in aquifer structure can cause apparent changes in microbial transport behavior. In a recent injection-and-recovery experiment in which a labeled indigenous bacterial population, a conservative tracer (bromide), and a particulate tracer (bacterium-sized microspheres) were coinjected into adjacent layers of a relatively homogeneous sandy aquifer, the relative order of appearance of peak abundance of the three constituents varied from layer to layer (30). Substantive changes in the rates of immobilization, retardation, and apparent dispersion over short distances in the aquifer can necessitate the need to verify that a given observed transport behavior is not merely an artifact of system heterogeneities.

In hydrologic investigations using conservative solute tracers, transport has been successfully predicted by using average aquifer properties and following large (kilograms) quantities of tracers for distances (over 100 m) sufficient for the average properties to be applicable (19). This approach is often impractical for field transport investigations involving microorganisms because of problems associated with high rates of immobilization, with maintaining viability over very long time periods, and with preventing clogging during point-source injections of large numbers of microbes into aquifer sediments. Another approach is to perform enough small-scale injection-and-recovery experiments at different points in the aquifer to ensure that observed trends in microbial transport behavior are reproducible.

LABORATORY STUDIES
Flowthrough Columns

Columns packed with subsurface or other porous material allow experiments to be conducted with much greater control than is possible in field studies. Such experiments have allowed a clearer delineation of many factors governing microbial transport through the saturated subsurface, including the effects of size-dependent exclusion from small pores (2),

attachment reversibility (3), chemotaxis (6, 9, 10), microbial abundance upon sorption kinetics (7), flow velocity (10), ionic strength (18), the influence of specific chemicals (21), sorption (35), rates of attachment and detachment (55), microbial surface residence times (45), cell hydrophobicity and surface charge (67), mineralogy (70), and clogging (78) (Table 1). Flowthrough saturated columns have also proven useful in studying transport behavior of genetically engineered bacteria before they are released into the subsurface (76).

As mentioned in a preceding section, the packing of large columns with subsurface sediments can be problematic because of a rearrangement of pore structure. Columns prepared with the aid of vibratory compaction can suffer from mechanically induced radial segregation (i.e., a greater proportion of coarser grains at the column periphery) (68), and large columns packed with a vibration-free packer can suffer from longitudinal grain-size segregation (30). Wet-packing procedures (e.g., 44) are most often used in flowthrough column experiments involving microbial transport. However, manual wet-packing procedures are not artifact free, judging from differences in apparent transport behavior of a polydispersed mixture of different-sized microspheres among replicate wet-packed columns (30). Sterilization of columns can also lead to alterations of pore structure. In particular, autoclaving of porous subsurface material can lead to a reduction in the surface area of clays available for adhesion and even alterations in pore surface charge. This can resulting in decreased microbial penetration rates within the column relative to those in which dry heat sterilization procedures were used (42).

In addition to the packing procedure, the importance of precluding the presence of air bubbles and maintaining saturated conditions cannot be overemphasized. This is because the presence of gas bubbles can change the hydraulic properties of the column and serve as highly efficient collectors for bacteria traveling through the column. The entrapment of bacteria within the gas-water interface is due to strong forces at the interface (surface tension) and is a well-documented phenomenon (80). Porous-medium columns may be voided of air pockets by an initial displacement of the intergranular gases with carbon dioxide, which can subsequently be scrubbed out by using a flowthrough solution of calcium sulfate followed by degassed groundwater (30). Although saturated columns of subsurface material have been successfully run in the downflow mode (2, 3, 82) (Fig. 5A), they are generally run in the upflow mode to prevent desaturation during the time course of the experiment. Upflow and downflow columns are often run in a vertical position. However, such an arrangement results in the flow direction and gravitational force acting along the same axis and may result in an overemphasis on sedimentation (56). A slight inclination (e.g., 10°) from the horizontal (Fig. 5B) can be used to allow closer simulation of typical groundwater flow in an unconfined aquifer (30). The Mariot siphon (Fig. 5A) can be used as an inexpensive means of providing constant head but is limited in terms of control. There are microprocessor-controlled piston pumps (e.g., ISCO model 5000) that deliver accurately and steadily at volumetric flow rates low enough (e.g., several milliliters per hour) to produce linear flow velocities within packed columns that are representative of field flow conditions (Fig. 5B).

Temperature can be an important factor in the attachment of microorganisms to surfaces. Therefore, it is important that column and fluid temperatures be reasonably close

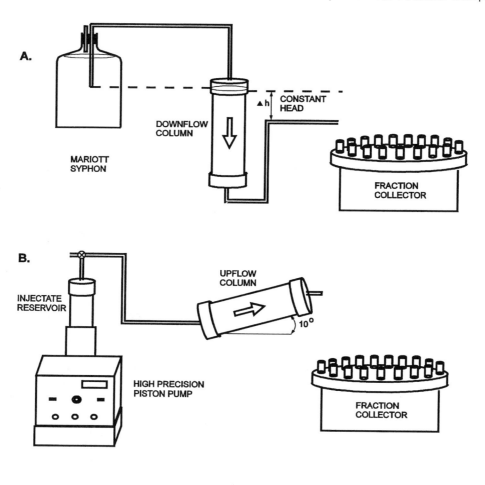

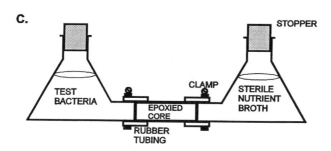

FIGURE 5 Experimental apparatuses for conducting laboratory-scale studies of microbial transport behavior in saturated subsurface media. (A) Downflow column with a Mariot siphon device and adjustable effluent tube elevation to provide constant head; (B) upflow column at a slight incline to the horizontal with a pressure-sensitive, high-precision piston pump to supply a constant head; (C) experimental setup for examining microbial transport behavior through consolidated materials under static conditions (modified from reference 43 with permission from the American Society for Microbiology).

to that of the aquifer in which the subsurface material was collected. Aquifers are usually thermally buffered and consequently are characterized by near-constant temperatures. The most effective method of conducting a column experiment at constant, in situ temperatures is to place the whole experimental assembly (column, reservoir, pump, collector, and tubing) in an environmental control room or large refrigerated incubator that is set to the temperature of the aquifer.

Flowthrough columns also are useful in investigations concerning microbial transport behavior in the vadose zone and in surface soils. However, maintaining constant degrees of saturation in subsurface sediments during unsaturated flow conditions in flowthrough columns can be problematic. Unsaturated flowthrough columns are run in a downflow mode and have been used to examine factors controlling microbial transport above the water table, i.e., unsaturated flow conditions (34), soil properties (39), and organic matter (62).

Because of the difficulty of maintaining sterile conditions in flowthrough columns and in the other associated apparatuses, it is generally desirable to either label a priori or use a method to differentially detect in the column eluant the microorganism(s) whose transport properties are being investigated. Several methods have been successfully used to label mixed and specific microbial populations prior to their addition to flowthrough, porous-medium columns. Subsurface bacterial isolates have been labeled with ^3H and ^{14}C prior to their addition to saturated sand columns, by amending separate cultures with [^3H]alanine, [^3H]glucose, [^3H]-adenosine, and [^{14}C]pyruvic acid (50). When a radiolabel is used to mark microorganisms, it is important to devise suitable controls that account for the microorganism's loss of label during the time course of the experiment. Prestaining methods have also given satisfactory results. DNA-specific fluorescent dyes (fluorochromes) such as DAPI allow labeling of whole bacterial and protozoan populations at very low concentrations (e.g., 5 μM). Because DAPI binds specifically to the DNA inside the cell, alterations of cell surface characteristics can be minimized relative to other stains that highlight structures within the cell envelope. However, DAPI has been shown to adversely affect bacterial metabolism (59) and therefore is best used in shorter experiments performed under nongrowth conditions.

Many flowthrough column studies with bacteria have used plate count procedures for detection of unlabeled bacteria appearing in the eluent (e.g., 10, 18, 47). The advantage of plate counting procedures is simplicity. The use of antibiotic-resistant bacteria in column experiments allows their detection on antibiotic-containing media, which offers specificity in addition to simplicity. The disadvantage of plate counting is that it often detects only a fraction of the total population added to a column. Also, culturability can change during the experimental time course as the bacteria adapt to conditions within the column. There are also a number of promising molecular techniques for labeling specific microorganisms for laboratory microcosm experiments that do not involve culturing. Some molecular techniques for marking microorganisms can be highly sensitive. By inserting *lux* genes from a marine vibrio, a single genetically engineered bacterium cell (*Pseudomonas syringae* pv. phaseolicola) was detected directly in soil by using charge-coupled device-enhanced microscopy, thereby avoiding problematic extractions of marker genes from a soil matrix (71). Methods involving DNA extraction and PCR have been used to detect low numbers of specific bacteria in sediment samples (77). Such a method was recently used to detect the presence of *Pseudomonas* sp. strain B13 in aquifer material (74). Monoclonal antibodies have also been used to detect specific bacteria in laboratory mesocosms (65). Because of the rapidity with which molecular methods and applications to subsurface microbiology are being developed, detailed discussions are not given here. A discussion of the development of molecular methods for detection of specific bacteria in environmental media is presented by Pickup (61).

Static Minicolumns

Static minicolumns have been useful in examining the role of taxis in bacterial transport through porous media (66), bacterial attachment behavior in the presence of model surfaces (67), the effect of groundwater chemistry on bacterial attachment to aquifer sediment grains (69), and survival of genetically engineered bacteria in aquifer sediments (48). Static minicolumns are particularly well suited for studies of microbial attachment behavior in the presence of saturated subsurface media. Their small size facilitates the use of replicate systems and helps avoid packing-related problems inherent with large columns. The absence of flow precludes the presence of a hydrodynamic boundary layer across which suspended microorganisms would otherwise have to cross to reach the surface. This simplifies the interpretation of microbial attachment kinetics, so that the effects of varying chemical, physical, and physiological conditions on attachment behavior in the presence of subsurface material can be more easily assessed.

Pore structure rearrangement during the sampling and repacking of subsurface material can be largely avoided in column experiments involving bacterial penetration through porous rock. This is because intact cores of consolidated material can be removed and handled intact for use in either flowthrough or static minicolumn experiments. A number of laboratory investigations involving bacterial penetration through low-permeability (<0.7-μm^2), consolidated subsurface material have been performed with Berea sandstone (12, 40, 43, 58, 63), which is relatively homogeneous and available in a variety of permeability classes (56). These experiments have demonstrated the ability of motile bacteria to penetrate porous rock at substantive rates (up to 3 cm/h) under conditions of very low flow. Assessments of bacterial growth or taxis through sandstone may be made by using a device consisting of a stagnant nutrient-saturated sandstone core connecting two flasks, one containing sterile nutrient broth and the other containing the bacterium of interest (40, 43) (Fig. 5C). The appearance of bacteria in the sterile flask establishes the time required for migration through the connecting sandstone, although differences in methods of detection (e.g., turbidity versus viable counts) can yield substantial differences in apparent travel times (12). A more thorough discussion of investigations involving growth, tactic, and chemotactic migration through sandstone is provide by McInerney (56).

CONCLUSIONS

Although microorganisms were first introduced into groundwater as tracers in order to gain information about the hydrology of complex groundwater environments, recent interest has clearly shifted to the transport behavior of the microorganisms themselves. The shift is due largely to the now recognized importance of subsurface microbial transport in aquifer restoration, migration of surface active and hydrophobic chemical contaminants, public health, and oil recovery. Much of the emphasis over the last two decades has focused on column experiments, because of interpretational and technical difficulties involved with conducting subsurface microbial transport investigations in the field. As a consequence, the development of methodology for conducting controlled field transport investigations has lagged behind that of laboratory experiments. Flowthrough and static column investigations offer the ability to achieve a degree of control not possible in the field. As described herein, methodology for studying specific controls of microbial transport in laboratory porous media is improving as techniques for accurately controlling flow at low (environmentally relevant) rates, isolating specific controlling factors, and quantifying the microorganisms of interest are refined. However, there is also a growing realization that processes controlling subsurface microbial transport behavior are interrelated and can be operative on spatial and temporal scales not conducive to laboratory study. This understanding is leading to more field-scale experiments and to research programs that involve a combination of field and laboratory studies. Most controlled field investigations of subsurface microbial transport are conducted on limited spatial scales relative to the scales of interest to those concerned with pathogen transport to water supply wells, with microbially enhanced oil recovery from petroleum reservoirs, and with the feasibility of using introduced bacteria for aquifer restoration. However, continued advances in our understanding of microbial transport behavior as a result of ongoing laboratory and small-scale field investigations will facilitate future larger-scale field experiments. This, in turn, will allow testing of more accurate models capable of describing subsurface microbial transport over considerable distances.

REFERENCES

1. **Abba, F., Orlandi, and A. Rondelli.** 1898. Über die Filtrationskraft des Bodens und die Fortschwemmung von Bakterien durch das Grundwasser. *Z. Hyg. Infektionskr. Krankh.* **31:**66–84.
2. **Bales, R. C., C. P. Gerba, G. H. Grondin, and S. L. Jensen.** 1989. Bacteriophage transport in sandy soil and fractured tuff. *Appl. Environ. Microbiol.* **55:**2061–2067.
3. **Bales, R. C., S. R. Hinkle, T. W. Kroeger, K. Stocking, and C. P. Gerba.** 1991. Bacteriophage adsorption during transport through porous media: chemical perturbations and reversibility. *Environ. Sci. Technol.* **25:**2088–2095.
4. **Bales, R. C., S. Li, K. M. Maquire, M. T. Yahya, and C. P. Gerba.** 1993. MS-2 and poliovirus transport in porous media: hydrophobic effects and chemical perturbations. *Water Resour. Res.* **29:**957–963.
5. **Bales, R. C., S. Li, K. M. Maquire, M. T. Yahya, C. P. Gerba, and R. W. Harvey.** 1995. Virus and bacteria transport in a sandy aquifer at Cape Cod, MA. *Ground Water* **33:**653–661.
6. **Barton, J. W., and R. M. Ford.** 1995. Determination of effective transport coefficients for bacterial migration in sand columns. *Appl. Environ. Microbiol.* **61:**3329–3335.
7. **Bengtsson, G., and R. Lindqvist.** 1995. Transport of soil bacteria controlled by density-dependent sorption kinetics. *Water Resour. Res.* **31:**1247–1256.
8. **Bitton, G., and R. W. Harvey.** 1992. Transport of pathogens through soils and aquifers, p. 103–124. *In* R. Mitchell (ed.), *Environmental Microbiology.* Wiley-Liss, New York.
9. **Bosma, T. N. P., J. J. L. Schnoor, G. Schraa, and A. J. B. Zehnder.** 1988. Simulation model for biotransformation of xenobiotics and chemotaxis in soil columns. *J. Contam. Hydrol.* **2:**225–236.
10. **Camper, A. K., J. T. Hayes, P. J. Surman, W. L. Jones, and A. B. Cunningham.** 1993. Effects of motility and adsorption rate coefficient on transport of bacteria through saturated porous media. *Appl. Environ. Microbiol.* **59:**3455–3462.
11. **Champ, D. R., and J. Schroeter.** 1988. Bacterial transport in fractured rock—a field-scale tracer test at the Chalk River Nuclear Laboratories. *Water Sci. Technol.* **20:**81–87.
12. **Chang, P. L., and T. F. Yen.** 1985. Interaction of *Pseudomonas putida* ATCC 12633 and bacteriophage gh-1 in Berea sandstone rock. *Appl. Environ. Microbiol.* **47:**544–550.
13. **Corapcioglu, M. Y., and A. Haridas.** 1985. Microbial transport in soils and groundwater: a numerical model. *Adv. Water Resour.* **8:**188–200.
14. **Corapcioglu, M. Y., and S. Kim.** 1985. Modeling facilitated contaminant transport by mobile bacteria. *Water Resour. Res.* **31:**2639–2647.
15. **Craun, G. F.** 1985. A summary of waterborne illness transmitted through contaminated groundwater. *J. Environ. Health* **48:**122.
16. **Fontes, D. E., A. L. Mills, G. M. Hornberger, and J. S. Herman.** 1991. Physical and chemical factors influencing transport of microorganisms through porous media. *Appl. Environ. Microbiol.* **57:**2473–2481.
17. **Freeze, R. A., and J. A. Cherry.** 1979. *Groundwater.* Prentice Hall, Englewood Cliffs, N.J.
18. **Gannon, J., Y. Tan, P. Baveye, and M. Alexander.** 1991. Effect of sodium chloride on transport of bacteria in a saturated aquifer material. *Appl. Environ. Microbiol.* **57:** 2497–2505.
19. **Garabedian, S. P., and D. R. LeBlanc.** 1991. Large-scale natural gradient tracer test in sand and gravel, Cape Cod, Massachusetts. 2. Analysis of spatial moments for a nonreactive tracer. *Water Resour. Res.* **27:**911–924.
20. **Gardner, G. D., and R. E. Gray.** 1976. Tracing subsurface flow in karst regions using artificially colored spores. *Bull. Assoc. Eng. Geol.* **13:**177–197.
20a. **Groffman, P. M., A. J. Gold, and G. Howard.** 1995. Hydrolic tracer effects on soil microbial activities. *Soil Sci. Soc. Am. J.* **59:**478–481.
21. **Gross, M. A., and B. E. Logan.** 1995. Influence of different chemical treatments on transport of *Alcaligenes paradoxus* in porous media. *Appl. Environ. Microbiol.* **61:**1750–1756.
22. **Harvey, R. W.** 1988. Transport of bacteria in a contaminated aquifer, p. 183–188. *In* G. E. Mallard and S. E. Ragone (ed.), *U.S. Geological Survey Water Resources Investigations Report* 88–4220.
23. **Harvey, R. W.** 1990. Evaluation of particulate and solute tracers for investigation of bacterial transport behavior in groundwater, p. 7-159–7-165. *In* C. B. Fliermans and T. C. Hazen (ed.), *Proceedings of the First International Symposium on Microbiology of the Deep Subsurface.* WSRC Information Services, Aiken, S.C.
24. **Harvey, R. W.** 1991. Parameters involved in modeling movement of bacteria in groundwater, p. 89–114. *In* C. J. Hurst (ed.), *Modeling the Environmental Fate of Microorganisms.* American Society for Microbiology, Washington, D.C.
25. **Harvey, R. W. and S. P. Garabedian.** 1991. Use of colloid filtration theory in modeling movement of bacteria through a contaminated sandy aquifer. *Environ. Sci. Technol.* **25:**178–185.
26. **Harvey, R. W., and S. P. Garabedian.** 1992. Reply to comment on "Use of colloid filtration theory in modeling movement of bacteria through a contaminated sandy aquifer." *Environ. Sci. Technol.* **26:**401–402.
27. **Harvey, R. W., and L. H. George.** 1987. Growth determinations for unattached bacteria in a contaminated aquifer. *Appl. Environ. Microbiol.* **53:**2992–2996.
28. **Harvey, R. W., L. H. George, R. L. Smith, and D. R. LeBlanc.** 1989. Transport of microspheres and indigenous bacteria through a sandy aquifer: results of natural and forced-gradient tracer experiments. *Environ. Sci. Technol.* **23:**51–56.
29. **Harvey, R. W., N. E. Kinner, A. Bunn, D. MacDonald, and D. W. Metge.** 1995. Transport behavior of groundwater protozoa and protozoan-sized microspheres in sandy aquifer sediments. *Appl. Environ. Microbiol.* **61:**209–217.
30. **Harvey, R. W., N. E. Kinner, D. MacDonald, D. W. Metge, and A. Bunn.** 1993. Role of physical heterogeneity in the interpretation of small-scale laboratory and field observations of microorganism, microsphere, and bromide transport through aquifer sediments. *Water Resour. Res.* **29:**2713–2721.
31. **Harvey, R. W., N. E. Kinner, N. Mayberry, K. Blakesly, D. W. Metge, and D. Kinner.** Effect of culturing procedure upon the subsurface transport of groundwater flagellates. Submitted for publication.
32. **Harvey, R. W., D. W. Metge, N. Kinner, and N. Mayberry.** Physiological consideration in applying laboratory-

determined buoyant densities to predictions of subsurface bacterial and protozoan transport. Submitted for publication.

33. **Harvey, R. W., C. Voss, and W. Souza.** Unpublished data.

34. **Hekman, W. E., C. E. Heijnen, S. L. G. E. Burgers, J. A. van Veen, and J. D. van Elsas.** 1995. Transport of bacterial inoculants through intact cores of two different soils as affected by water percolation and the presence of wheat plants. *FEMS Microb. Ecol.* **16:**143–158.

35. **Hendricks, D. W., F. J. Post, and D. R. Khairnar.** 1979. Sorption of bacteria on soils. *Water Air Soil Pollut.* **12:** 219–232.

36. **Hobbie, J. E., R. J. Daley, and S. Jasper.** 1977. Use of Nuclepore filters for counting bacteria by fluorescence microscopy. *Appl. Environ. Microbiol.* **33:**1225–1228.

37. **Hornberger, G. M., A. L. Mills, and J. S. Herman.** 1992. Bacterial transport in porous media: evaluation of a model using laboratory observations. *Water Resour. Res.* **28:** 915–938.

38. **Hurst, C. J. (ed.).** 1991. *Modeling the Environmental Fate of Microorganisms.* American Society for Microbiology, Washington, D.C.

39. **Huysman, F., and W. Verstraete.** 1993. Water-facilitated transport of bacteria in unsaturated soil columns: influence of cell surface hydrophobicity and soil properties. *Soil Biol. Biochem.* **25:**83–90.

40. **Jang, L. K., P. W. Chang, J. E. Findley, and T. F. Yen.** 1983. Selection of bacteria with favorable transport properties through porous rock for the application of microbially-enhanced oil recovery. *Appl. Environ. Microbiol.* **46:** 1066–1072.

41. **Jenkins, M. B., and L. W. Lion.** 1993. Mobile bacteria and transport of polynuclear aromatic hydrocarbons in porous media. *Appl. Environ. Microbiol.* **59:**3306–3313.

42. **Jenneman, G. L., M. J. McInerney, M. E. Crocker, and R. M. Knapp.** 1986. Effect of sterilization by dry heat or autoclaving on bacterial penetration through Berea sandstone. *Appl. Environ. Microbiol.* **51:**39–43.

43. **Jenneman, G. L., M. J. McInerney, and R. M. Knapp.** 1985. Microbial penetration through nutrient-saturated Berea sandstone. *Appl. Environ. Microbiol.* **50:**383–391.

44. **Johnson, M. J.** 1990. Relative permeabilities of gasoline, water, and air in sand. Master's thesis. Department of Civil Engineering, University of New Hampshire, Durham.

45. **Johnson, W. P., K. A. Blue, B. E. Logan, and R. G. Arnold.** 1995. Modeling bacterial detachment during transport through porous media as a residence-time-dependent process. *Water Resour. Res.* **31:**2649–2658.

46. **Kinner, N. E.** Personal communication.

47. **Kinoshita, T., R. C. Bales, M. T. Yahya, and C. P. Gerba.** 1993. Bacterial transport in a porous medium: retention of *Bacillus* and *Pseudomonas* on silica surfaces. *Water Res.* **27:** 1295–1301.

48. **Krumme, M. L., R. L., Smith, J. Egestorff, S. M. Thiem, J. M. Tiedje, K. N. Timmis, and D. F. Dwyer.** 1994. Behavior of pollutant-degrading microorganisms in aquifers: predictions for genetically engineered organisms. *Environ. Sci. Technol.* **28:**1134–1138.

49. **Kuwabara, J. S., and R. W. Harvey.** 1990. Application of a hollow-fiber, tangential-flow device for sampling suspended bacteria and particles from natural waters. *J. Environ. Qual.* **19:**625–629.

49a.**LeBlanc, D. R., S. P. Garabedian, K. M. Hess, L. W. Gelhar, R. D. Quadri, K. G. Stollenwerk, and W. W. Wood.** 1991. Large-scale natural gradient test in sand and gravel, Cape Cod, Massachusetts. I. Experimental design and observed tracer movement. *Water Resour. Res.* **27:** 895–910.

50. **Lindqvist, R., and G. Bengtsson.** 1991. Dispersal dynamics of groundwater bacteria. *Microb. Ecol.* **21:**49–72.

51. **Lindqvist, R., J. S. Cho, and C. G. Enfield.** 1994. A kinetic model for cell density dependent bacterial transport in porous media. *Water Resour. Res.* **30:**3291–3299.

52. **Lindqvist, R., and C. G. Enfield.** 1992. Biosorption of dichlorodiphenyltrichloroethane and hexachlorobenzene in groundwater and its implications for facilitated transport. *Appl. Environ. Microbiol.* **58:**2211–2218.

53. **Macleod, F. A., H. M. Lappin-Scott, and J. W. Costerton.** 1988. Plugging of a model rock system by using starved bacteria. *Appl. Environ. Microbiol.* **54:**1362–1372.

54. **Martin, R., and A. Thomas.** 1974. An example of the use of bacteriophage as a ground-water tracer. *J. Hydrol.* **23:** 73–78.

55. **McCaulou, D. R., R. C. Bales, and J. F. McCarthy.** 1994. Use of short-pulse experiments to study bacteria transport through porous media. *J. Contam. Hydrol.* **15:**1–14.

55a. **McCaulou, D. R., R. C. Bales, and R. G. Arnold.** 1995. Effect of temperature-controlled motility on transport of bacteria and microspheres through saturated sediment. *Water Resour. Res.* **31:**271–280.

56. **McInerney, M. J.** 1991. Use of models to predict bacterial penetration and movement within a subsurface matrix, p. 115–135. *In* C. J. Hurst (ed.), *Modeling the Environmental Fate of Microorganisms.* American Society for Microbiology, Washington, D.C.

57. **McKay, L. D., J. A. Cherry, R. C. Bales, M. T. Yahya, and C. P. Gerba.** 1993. A field example of bacteriophage as tracers of fracture flow. *Environ. Sci. Technol.* **27:** 1075–1079.

58. **Montgomery, A. D., M. J. McInerney, and K. L. Sublette.** 1990. Microbial control of the production of hydrogen sulfide by sulfate reducing bacteria. *Biotechnol. Bioeng.* **35:** 533–539.

59. **Parolin, C., A. Montecucco, G. Ciarrocchi, G. Pedrali-Noy, S. Valisena, M. Palumbo, and G. Palu.** 1990. The effect of the minor groove binding agent DAPI (2-amidino-diphenyl-indole) on DNA-directed enzymes: an attempt to explain inhibition of plasmid expression in *Escherichia coli.* *FEMS Microbiol. Lett.* **68:**341–346.

60. **Pickens, J. F., J. A. Cherry, G. E. Grisak, W. F. Merritt, and B. A. Risto.** 1978. A multilevel device for groundwater sampling and piezometric monitoring. *Ground Water* **16:** 322–327.

61. **Pickup, R. W.** 1991. Development of molecular methods for the detection of specific bacteria in the environment. *J. Gen. Microbiol.* **137:**1009–1019.

62. **Powelson, D. K., J. R. Simpson, and C. P. Gerba.** 1991. Effects of organic matter on virus transport in unsaturated flow. *Appl. Environ. Microbiol.* **57:**2192–2196.

63. **Raiders, R. A., M. J. McInerney, D. E. Revus, H. Torbati, R. M. Knapp, and G. E. Jenneman.** 1986. Selectivity and depth of microbial plugging in Berea sandstone cores. *J. Ind. Microbiol.* **1:**195–203.

64. **Rajagopalan, R., and C. Tien.** 1976. Trajectory analysis of deep-bed filtration with the sphere-in-cell porous media model. *J. Am. Inst. Chem. Eng.* **22:**523–533.

65. **Ramos-Gonzales, M. I., F. Ruizcabello, I. Brettar, F. Garrido, and J. L. Ramos.** 1992. Tracking genetically engineered bacteria: monoclonal antibodies against surface determinants of the soil bacterium *Pseudomonas putida* 2440. *J. Bacteriol.* **174:**2978–2985.

66. **Reynolds, P. J., P. Sharma, G. E. Jenneman, and J. J. McInerney.** 1989. Mechanisms of microbial movement in subsurface materials. *Appl. Environ. Microbiol.* **55:** 2280–2286.

67. **Rijnaarts, H. H. M., W. Norde, E. J. Bouwer, J. Lykema, and A. J. B. Zehnder.** 1993. Bacterial adhesion under static

and dynamic conditions. *Appl. Environ. Microbiol.* **59:** 3255–3265.

68. **Ripple, C. D., R. V. James, and J. Rubin.** 1974. Packing-induced radial particle-size segregation: influence on hydrodynamic dispersion and water transfer measurements. *Soil Sci. Soc. Am. Proc.* **38:**219–222.

69. **Scholl, M. A., and R. W. Harvey.** 1992. Laboratory investigations on the role of sediment surface and groundwater chemistry in transport of bacteria through a contaminated sandy aquifer. *Environ. Sci. Technol.* **26:**1410–1417.

70. **Scholl, M. A., A. L. Mills, J. S. Herman, and G. M. Hornberger.** 1990. The influence of mineralogy and solution chemistry on the attachment of bacteria to representative aquifer materials. *J. Contam. Hydrol.* **6:**321–336.

71. **Silcock, D. J., R. N. Waterhouse, L. A. Glover, J. L. Prosset, and K. Killham.** 1992. Detection of a single genetically modified bacterial cell in soil by using charge coupled device-enhanced microscopy. *Appl. Environ. Microbiol.* **58:** 2444–2448.

72. **Smith, M. S., G. W. Thomas, R. E. White, and D. Ritonga.** 1985. Transport of *Escherichia coli* through intact and disturbed soil columns. *J. Environ. Qual.* **14:**87–91.

73. **Smith, R. L., R. W. Harvey, and D. R. LeBlanc.** 1991. Importance of closely spaced vertical sampling in delineating chemical and microbiological gradients in groundwater studies. *J. Contam. Hydrol.* **7:**285–300.

74. **Thiem, S. M., M. L. Krumme, R. L. Smith, and J. M. Tiedje.** 1994. Use of molecular techniques to evaluate the survival of a microorganism injected into an aquifer. *Appl. Environ. Microbiol.* **60:**1059–1067.

75. **Toran, L., and A. V. Palumbo.** 1992. Colloid transport through fractured and unfractured laboratory sand columns. *J. Contam. Hydrol.* **9:**289–303.

76. **Trevors, J. T., and J. D. Vanelsas.** 1990. Transport of a genetically engineered *Pseudomonas fluorescens* strain through a soil microcosm. *Appl. Environ. Microbiol.* **56:** 401–408.

77. **Tsai, Y. L., and B. H. Olson.** 1992. Detection of low numbers of bacterial cells in soils and sediments by polymerase chain reaction. *Appl. Environ. Microbiol.* **58:** 754–757.

78. **Vandevivere, P., and P. Baveye.** 1992. Relationship between transport of bacteria and their clogging efficiency in sand columns. *Appl. Environ. Microbiol.* **58:**2523–2530.

79. **Wan, J., T. K. Tokunaga, and C. F. Tsang.** 1995. Bacterial sedimentation through porous medium. *Water Resour. Res.* **31:**1627–1636.

80. **Wan, J., J. L. Wilson, and T. L. Kieft.** 1994. Influence of the gas-water interface on transport of microorganisms through unsaturated porous media. *Appl. Environ. Microbiol.* **60:**509–516.

81. **Wilson, J. T., L. E. Leach, M. Henson, and J. N. Jones.** 1986. In situ biorestoration as a ground water remediation technique. *Ground Water Monit. Rev.* **6:**56–64.

82. **Wollum, A. G., II, and D. K. Cassel.** 1978. Transport of microorganisms in sand columns. *Soil Sci. Soc. Am. J.* **42:** 72–76.

83. **Wood, W. W., and G. G. Ehrlich.** 1978. Use of baker's yeast to trace microbial movement in ground water. *Ground Water* **16:**398–403.

Petroleum Microbiology: Biofouling, Souring, and Improved Oil Recovery

MICHAEL J. McINERNEY AND KERRY L. SUBLETTE

65

Beyond the general perception that sulfate-reducing bacteria are present and their activities can damage production equipment and sour oil reservoirs, the microbial ecology of petroleum reservoirs is poorly understood. Oil reservoirs were once considered too hostile for bacterial life because of extreme environmental factors such as temperature or salinity or considered inaccessible to life because the reservoir rock would prevent the penetration of microorganisms into the formation. However, it is now clear that many oil reservoirs have active and diverse populations of microorganisms (31), even in extremely thermophilic and hypersaline reservoirs (5, 6, 46). Also, bacteria can penetrate even rocks with very low permeabilities at appreciable rates (20, 50). Thus, oil reservoirs must not be viewed as sterile environments devoid of life. Microorganisms are present, and their activities will affect the day-to-day operations of the reservoir and, ultimately, the recovery of oil.

In this chapter, we discuss the methods used to study beneficial microbial activities such as enhanced oil recovery and detrimental activities such as souring.

OIL FIELD SAMPLING

To show unequivocally that a specific microorganism is present in the reservoir, aseptically obtained core material is needed (chapter 59). However, given the expense of drilling, such material is rarely available from oil reservoirs. Thus, the kinds of microorganisms present and their activities in the oil reservoir must be inferred from analysis of the fluids produced from the reservoir. Because of this, the selection of a reservoir with wells suitable for microbial sampling is an important concern. Preferably, a reservoir that has not been injected with fluid from another formation should be used to ensure that the fluid collected for analysis is native to the formation of interest. Also, the wells must produce fluid from only a single formation. Often, several oil-bearing formations may be encountered during drilling. In such cases, the same well is often used to produce fluids from each of these formations.

Samples should be collected at the wellhead if possible, and any stagnant liquid in the system should be removed prior to sample collection. Fittings and tubing connections should be sterilized prior to use. To reduce the exposure to oxygen, the collection bottle is filled and then flushed with fluid from the well prior to sample collection, or a bottle with an anoxic atmosphere is used (17). Even when these precautions are taken, interpretation of the data must always be tempered by the realization that any organisms detected in the fluid samples may be from biofilms growing on the surface of the pipes rather than directly from the reservovir.

CULTIVATION OF MICROBIAL POPULATIONS

Methods to enumerate and isolate microbial populations from oil reservoirs are similar to those used for other environments (chapter 6). The medium and cultural conditions must be adjusted to simulate the habitat as accurately as possible. Since most oil reservoirs have little or no oxygen (16), anaerobically prepared medium is used. A bicarbonate buffer system is used because many anaerobes require carbon dioxide for growth (9). The ionic strength of the medium is adjusted to match that of the produced water, or brine from the formation can be used. The pHs of most produced fluids are near neutral (16). However, on the basis of solubility calculations, Sheehy (45) suggests that the actual pH of thermophilic reservoirs may be significantly lower than that of the produced fluid. Atmospheric rather than the actual reservoir pressure is routinely used for isolation and enumeration studies (1, 5, 6, 46). Since pressures of 10 to 20 MPa do not adversely affect bacterial metabolism (27), it is assumed that microorganisms isolated at atmospheric pressure are similar to those that would have been isolated if the actual reservoir pressure was used. However, data to support or reject this hypothesis are not available. Thus, it is important to show that the isolates can grow and metabolize at actual reservoir conditions, including elevated pressures.

Microorganisms Useful for Oil Recovery

Microorganisms produce a variety of products that are potentially useful for oil recovery (31) (Table 1). The production of acids, solvents, gases, surfactants, and emulsifiers in or near the well bore prevents scale and paraffin deposition, alters the wettability of the rock, and changes the fluid saturations. All of these effects will improve oil drainage into the well. One of the major factors that limits oil production is the entrapment of oil in small pores by capillary forces. The reduction in the interfacial forces between the oil and brine by a biosurfactant will release the entrapped oil and may be the most important mechanism for microbially en-

TABLE 1 Types of microbial processes for oil recovery

Process	Production problem	Type of activity or product needed
Well bore cleanup	Paraffin and scale deposits	Emulsifiers, biosurfactants, solvents, acids, hydrocarbon degradation
Well stimulation	Formation damage, poor drainage	Gas, acids, solvents, biosurfactants
	High water production (coning)	Biomass and polymer production
Enhanced waterfloods	Poor displacement efficiency	Biosurfactants, solvents, polymers
	Poor sweep efficiency	Biomass and polymer production
	Souring	Nitrate reduction

hanced oil recovery (2, 10, 31). Finally, in situ growth and polymer production by microorganisms will reduce permeability variation and block water channels, which will improve the sweep efficiency of a recovery process and increase oil production (18, 24, 31, 38, 39). Most of these approaches rely on the stimulation of indigenous microorganisms in the reservoir by nutrient additions; therefore, the analysis of reservoir fluids for the presence of useful bacteria is essential. When a specific metabolic activity, such as biosurfactant production, is needed, the use an inoculum may be required.

Media with a readily utilizable organic carbon source are used to enumerate and isolate microorganisms that produce acids, solvents, and/or gas as end products (1, 6, 10). Polymer-producing bacteria are isolated by using a carbohydrate-based medium with a high carbon-to-nitrogen ratio, generally greater than 5:1 on a molar basis (36, 48). Polymer-producing bacteria will produce mucoid colonies on agar plates or viscous liquid cultures. The production of a cell-associated polymer rather than an excreted polymer may be more useful for microbial selective plugging processes since a cell-associated polymer may give more stable permeability reductions. Microorganisms that produce emulsifiers or biosurfactants are initially detected by the emulsifion of an oil layer in liquid medium (8, 21, 40). Isolates from these enrichments are then screened for emulsifying activity (8, 40) or for the ability to reduce the surface tension of water, indicating the production of a biosurfactant (21, 30). Screening for hemolytic zones around colonies on blood agar plates has also been used to detect biosurfactant-producing bacteria (32). Large reductions in capillary pressures are needed before oil is recovered. To be effective for enhanced oil recovery, the biosurfactant must reduce the interfacial tension between oil and brine by a factor of 10,000 or more. Interfacial tension is measured with a spinning drop tensiometer or by the pendant drop method (30, 49).

Detection of Detrimental Bacteria

Endpoint or most-probable-number enumerations for sulfate-reducing bacteria (SRB) are routinely done on reservoir fluids and equipment surfaces to assess the potential for souring and corrosion. Traditional cultivation methods are inexpensive and do not require much training of field personnel.

Another advantage of cultural methods is that SRB can be isolated for biocide testing and further physiological characterization. The organisms detected by amplification of SRB DNA sequences from a sulfate-reducing biofilm were similar to those routinely obtained by cultivation methods (22), suggesting that cultural methods detect most of the SRB present. The compositions of two media used to estimate the numbers of SRB from oil field samples are given in Table 2 (4, 47). The American Petroleum Institute recommended practice 38 uses a medium designated API RP-38 (4). Tanner (47) developed a modified version of this medium, designated (API)-RST, that gave higher SRB counts in shorter times from environmental samples than did API RP-38. To prepare the (API)-RST medium, the buffer is adjusted to pH 7.6 and all of the components except ferrous ammonium sulfate are added. The pH is readjusted if needed, and ferrous ammonium sulfate is added to form the precipitate. The medium is boiled under nitrogen and taken into an anaerobic chamber, where preweighed amounts of ascorbic acid and cysteine are added. The medium is dispensed into Hungate tubes (Belco Glass, Inc.) and then sterilized. This medium will detect lactate-using SRB, i.e., species within the genera *Desulfovibrio* and *Desulfotomaculum*, but can be modified to detect other SRB by using different electron donors.

In some oil field samples, *Shewanella putrefaciens* (43, 54) is the most numerous bacterium capable of reducing sulfur oxyanions to sulfide. However, this organism does not use sulfate as the electron acceptor and will not be detected in the media described above unless sulfite or thiosulfate is used as the electron acceptor. *S. putrefaciens* produces colonies with a salmon pink color on plate count agar (Difco), and this characteristic has been used to enumerate these organisms from oil field samples (43). The identity of isolates is confirmed by testing for the ability to reduce iron in medium with a nonfermentable electron donor and ferric iron as the electron acceptor (43). Selective enrichment with H_2 or formate as the electron donor and amorphous iron oxyhydroxide as the electron acceptor can also be used (34).

An enzyme-linked immunoabsorbent assay using polyclonal antibodies directed against adenosine 5'-phosphosul-

TABLE 2 Compositions of media used to detect SRB[a]

Component	Amt	
	API RP-38 (g/liter)	(API)-RST
Sodium lactate	2.4	4.2 g/liter
Yeast extract	1	0.7 g/liter
NaCl	10	10 g/liter
$(NH_4)SO_4$		0.3 g/liter
K_2HPO_4	0.01	0.6 g/liter
$MgSO_4 \cdot 7H_2O$	0.2	0.2 g/liter
$CaSO_4$		0.04 g/liter
TES[b]		1.5 g/liter
Vitamin solution[c]		10 ml
Trace metal solution[c]		5 ml
$Fe(NH_4)_2(SO_4)_2 \cdot 6H_2O$	0.2	0.3 g/liter
Iron nail	+	
Ascorbic acid	0.1	0.1 g/liter
Cysteine-HCl		0.2 g/liter

[a] From references 4 and 47.
[b] TES, N-tris(hydroxymethyl)methyl-2-aminoethanesulfonic acid.
[c] Composition is given in chapter 6.

fate reductase, one of the key enzymes involved in sulfate reduction, has been developed for rapid detection of SRB in oil field samples (15, 35). The assay is marketed under the trade name Rapidcheck (Conoco Specialty Products, Houston, Tex.). Since the enzyme is found in all known SRB, this test will detect almost all such organisms (35). The test is not affected by oil, sulfides, iron, or other constituents of produced brine and has a detection limit of 1,000 cells per ml. The main advantage of the test is that the results are available within 15 min. Problems with the detection of *Desulfobacter* species have been reported (7). Other immunological approaches using polyclonal antibodies directed against cell surface markers have been developed to detect specific strains of SRB (11, 41). Immunomagnetic beads coated with polyclonal antibodies against whole-cell antigens have been used to capture thermophilic SRB from North Sea oil fields (11).

With a few exceptions, the clustering of SRB based on phospholipid ester-linked fatty acids was the same as that predicted by 16S rRNA analysis (23). Thus, phospholipid profiles can be used not only to quantify SRB but also to study their diversity. Procedures for phospholipid analysis are discussed in chapter 10 of this manual.

A group-specific oligonucleotide probe (SRB385; 5'-CGGCGTCGCTGCGTCAGG-3') and genus-specific probes for gram-negative sulfate- and sulfur-reducing bacteria complementary to specific regions of rRNA have been developed (12). Fluorescent versions of these probes and epifluorescence microscopy have been used to visualize specific populations in biofilms (3). Group- and subgroup-specific primers have been synthesized to amplify 16S rRNA sequences of SRB (3, 22). A reverse sample genome probing method has been developed to study the microbial diversity of oil field production waters (52, 53). In this method, denatured chromosomal DNAs from bacteria isolated from the reservoir are spotted on a master filter. DNA extracted from an environmental sample is then labeled and hybridized with the filter to identify which of the bacterial genomes spotted on the filter are most prevalent in the sample.

Production fluids should also be routinely analyzed for the presence of general aerobic bacteria (GAB) to assess the level of contamination by slime-forming bacteria (4). These bacteria are important in developing biofilms that provide environments conducive for the growth of SRB. The following medium is recommended by the American Petroleum Institute (4) for use in waters containing less than 20,000 mg of total dissolved solids per liter: beef extract (3.0 g/liter), tryptone (5.0 g/liter), and dextrose (1.0 g/liter) at pH 7.0. Inoculated bottles are incubated at the temperature of the original water sample ±5°C.

DETECTION AND CONTROL OF MICROBIALLY INDUCED SOURING

The souring (increase in H_2S) of an oil reservoir can be directly linked to the growth of SRB and other sulfide-producing bacteria (e.g., *S. putrefaciens*) in the reservoir and associated water-handling equipment. It is common for reservoirs to become sour over a period of years. Souring is often a result of waterflooding or enhanced oil recovery processes in which sulfide-producing bacteria are either introduced into the reservoir or stimulated by nutrients introduced with the injected fluids. The consequences of souring are numerous and include (i) corrosion in production and injection wells and other water-handling equipment, (ii) plugging of injection wells by corrosion products and bio-

films, (iii) aquatic toxicity and odor associated with high H_2S levels in the produced water, (iv) safety hazards due to H_2S in retention ponds or other open vessels, and (v) increased costs to remove H_2S from natural gas before it enters the pipeline.

Recognition of Microbial Activity in an Oil Field Water System

The importance of early recognition of microbial activity in an oil field water system cannot be overemphasized. Costs in terms of replacement of equipment, well treatments, and shutdown time of wells can be very high. Certainly, monitoring of SRB in production fluids is a critical part of a program to control microbial activity. SRB are generally found associated with slime-forming bacteria in biofilms. The interiors of these films are anaerobic and highly conducive to the growth of SRB even when the surrounding environment is aerobic. The growth of SRB in the biofilm promotes corrosion by utilizing hydrogen and depolarizing cathodic sites on metal surfaces. In addition to cathodic depolarization, there is the direct attack of H_2S on metal surfaces. Both mechanisms give rise to black iron sulfides often associated with gelatinous mats of slime-forming bacteria. Corrosion also produces other chemical clues, discussed below, which can indicate an otherwise undetected problem (14).

The following are indicators of microbial activity in an oil field water-handling system: (i) plugging of filters, (ii) the presence of gelatinous, slimy, or hard deposits (black in color) in wash tanks, storage tanks, heater-treaters, free-water knockouts, or stagnant regions of flow lines, (iii) the appearance of H_2S in produced water, gas, or crude oil that was not previously sour, (iv) a pitting-type of corrosion accompanied by the presence of black iron sulfide, (v) an increase in the frequency of leaks, (vi) a decline in injectivity, and (vii) the production of black water and slime when injection wells are backflowed (33).

Examination of water samples, slime, or corroded materials can confirm the presence of SRB and other microorganisms. It is important that the sample be examined as soon as possible after removal from the oil field site since chemical clues such as the presence of H_2S may disappear and microbial populations may change.

Routine Monitoring of Microbial Activity

Many of the indicators of microbial activity given above become evident when SRB and slime-forming bacteria are well established in a water-handling system. It is obviously desirable to be able to detect and control microbial growth before a serious problem develops. Therefore, an oil field water system should be routinely monitored for microbial activity even when no problem is known to exist. Routine monitoring should include (i) enumeration of SRB and GAB, (ii) examination of metal surfaces and scrapings, and (iii) analysis of water quality.

Methods to enumerate SRB and GAB in water samples were discussed above. Water samples should be taken at various points in the system. The detection of SRB indicates a potential for corrosion and souring. An increase in the SRB count at any point in the system with time or an increase in the SRB count throughout the system demands immediate attention. GAB counts should be made on the same water samples. Counts that increase with time at any one location or throughout the system or that are greater than 10^4 cells per ml indicate significant contamination.

The analysis of water samples detects planktonic SRB

and GAB; therefore, these counts are only rough measures of the numbers of SRB and slime-forming bacteria in the system. The most reliable means of assessing the actual level of contamination of a water system is the direct examination of metal surfaces. Corrosion coupons installed at various sites in the system should be examined for evidence of microbial growth before being cleaned for corrosion analysis. Most corrosion coupons are mounted on the top of a pipe for easy removal. However, most microbial growth and corrosion occur in the bottom quadrant of the pipe. Therefore, coupons should be mounted flush with the bottom of the pipe. For monitoring microbial activity, coupons should be changed weekly, especially if high planktonic SRB or GAB counts have been obtained. In relatively clean systems, the sampling time may be extended to 1-month intervals if personnel constraints dictate. However, the statistical significance of the coupon analysis increases with increased sampling frequency. Several monitoring sites should be selected; for example, coupons might be placed in production wells, the free-water knockout, filters, the clarification system, junction points for commingled waters, and injection wells.

Immediately after removal, the coupons are placed in 10 ml of a sterile buffer in a capped plastic tube. The buffer is prepared with field water and contains a nontoxic surfactant and glass beads to help remove the biofilm from the coupons. The tubes are then shaken mechanically or by hand. After settling, the fluid is used to inoculate media for SRB or GAB counts. The results of the enumeration are recorded as the number of bacteria per cubic meter of coupon surface area. The detection of SRB indicates the potential for corrosion and souring. However, the key indicator of a serious problem is a progressive increase in the SRB count downstream from the initial source of corrosion (33).

In addition to the use of corrosion coupons, any scheduled or unscheduled downtime should be used to inspect the system for signs of microbial activity. Any change in the water chemistry such as an increase in the concentrations of sulfide, total iron, total suspended solids, or total organic carbon or a change in pH from the initial contamination site indicates microbial activity.

Control of Microbial Activity in Oil Field Waters

Physical or mechanical methods to control microbial activity are rather limited but can minimize souring and corrosion. Water systems must be kept cleaned. Water lines should be scraped and cleaned with slugs of surfactants and solvents. Whenever possible, the field operator should backflow water injection wells. A water system free of sludge and debris will reduce the number of sites on which SRB can become established. When alternate sources of injection water are used, each should be analyzed for SRB and GAB. If possible, an injection water that minimizes the contamination of the system and does not stimulate microbial growth and activity should be used. Whenever possible, avoid commingling of waters from different sources; inject them either separately or in alternating slugs. Commingling of different waters can stimulate growth of bacteria present in the waters. This is especially true when mixtures of fresh and saline waters are used. The water injection systems should be designed to eliminate stagnant points and minimize water-handling time on the surface. Low fluid flow velocities allow sludge to accumulate and produce environments suitable for the growth of SRB (33).

The most effective control of microbial activity in an oil field water system is currently by the use of biocides (15a). The use of the mechanical control methods discussed above will enhance the effectiveness of any biocide treatment. A number of factors must be considered in choosing a biocide for an individual treatment: the biocide must be active against the bacteria in the system under in situ conditions, it must remain active for a long enough period of time to treat sites far removed from the injection site of the chemical, it must be compatible with the reservoir fluids and any of the chemical agents that are used such as corrosion and scale inhibitors, oxygen scavengers, etc. and lastly, it must be economical.

Both inorganic and organic biocides are used. Inorganic biocides include strong oxidizing agents such as chlorine and chlorine dioxide. Chlorine and hypochlorous acid are generated on site by electrolysis of brines in a special electrolytic cell called a hypochlorite generator. Chlorine and hypochlorous acid are relatively low in cost and highly effective against planktonic bacteria but not as effective against biofilms. These compounds are inactivated by oxygen scavengers, and they oxidize corrosion and scale inhibitors. Highly contaminated systems require high doses to obtain the desired level of persistence. To reduce costs, chlorine treatment is often supplemented with the use of an organic biocide, e.g., quaternary ammonium compounds or formaldehyde. Chlorination of injection waters also has the disadvantage of producing chlorinated hydrocarbons, which can make the produced water unacceptable for discharge. The chlorinated hydrocarbons partition into the oil and may damage refinery catalysts.

Air mixtures of chlorine dioxide containing a partial pressure greater than 50 mm Hg can explode. Consequently, chlorine dioxide is never shipped but is produced on site by the reaction of sodium chlorate with sulfuric acid and sulfur dioxide. The yield of chlorine dioxide is dependent on the chemistry of the source water. The advantages of chlorine dioxide are that it is persistent and highly effective against biofilms and it oxidizes sulfur and iron deposits. However, the release of these materials may lead to plugging of injection wells and filters. Therefore, chlorine dioxide is best applied to a relatively clean system. It can increase corrosion rates, but it does not produce chlorinated hydrocarbons.

Organic biocides generally have a high degree of persistence in an oil field water system, but their effectiveness is dependent on the water chemistry and microbiology of the system. An organic biocide that may work well in one field may be totally ineffective in another. Therefore, it is important to screen a variety of biocides before choosing one or more for application. To be effective, the biocide must penetrate the biofilm in sufficient amounts to kill the bacteria. Therefore, the most meaningful biocide testing protocol will include biofilm testing. Screening of biocides against planktonic bacteria in the injection water can provide a wealth of information on chemical compatibility of the biocide with formation brine. Also, a biocide that is ineffective toward planktonic bacteria will generally also be ineffective against bacteria in biofilms. In practice, water samples from various sites in the water system are obtained, and each is exposed to several concentrations of the biocide. The manufacturer will recommend a minimum effective concentration and exposure time. After a predetermined time, SRB and GAB counts are obtained and compared with those of untreated samples. Only those biocides which exhibit total kill should be evaluated against biofilms.

The effectiveness of biocides against biofilms can be evaluated in situ, by using the whole or any part of the water treatment system, or in the laboratory, using a Robbins device (28, 42). A Robbins device is a tubular flow reactor with removable surfaces for the analysis of the biofilm. The device is designed so that surfaces can be removed aseptically without exposing the interior of the pipe to the outside environment and without altering the flow through the pipe. When water from a contaminated water system is circulated through the Robbins device, a biofilm containing SRB will develop on the removable surfaces called studs. The time required to develop a stable biofilm depends on the level of contamination in the system and the water chemistry. When a stable biofilm has developed as indicated by consistent SRB counts, the biocide treatment should begin. A control device and one or more devices treated with the biocide will be required for each biocide. Typically biocide testing will consist of the addition of daily slugs of the biocide followed by circulation of untreated water. The dosage and time of exposure should initially be based on the manufacturer's recommendation and can be increased if the killing efficiency is low. Studs should be examined visually, and the SRB counts should be conducted every 3 to 7 days.

It is difficult to kill all SRB in biofilms. Therefore, the question becomes, what is an acceptable number? The answer to this question is system dependent and firmly established only by relating operating experience with the levels of SRB and GAB contamination over time. Certainly, orders of magnitude decreases in SRB and GAB are desirable.

Another important question is whether to treat continuously or intermittently with slugs. Continuous treatment is usually more expensive than intermittent treatment; however, in some cases, SRB do not respond to slug treatment. In practice, slug treatments vary in frequency from two to three times per week to as few as once every 1 to 2 months. Significant cost savings can be obtained by treating contaminated segments separately with high dosages of the biocide while maintaining lower dosages elsewhere. However, the operator must make sure that the *entire* water handling system is treated.

It is not uncommon for the effectiveness of the biocide to decline with repeated use as a result of the development of resistance within the microbial populations in the system. This can be counteracted by using alternately two different biocides. When resistance to one biocide develops, the second biocide is used. If resistance to the second biocide develops, a return to the first biocide will generally be effective in controlling microbial activity.

The effectiveness of a biocide treatment program should be monitored continuously even if unwanted microbial growth has been controlled. The development of microbial resistance or seemingly minor changes in the water handling system can cause significant changes in the response of the system to the biocide. It is equally important that the treatment program be maintained once control of microbial activity has been achieved. It is tempting to save money by reducing biocide dosages once the problem is under control. However, the costs of reestablishing control after a biocide treatment program has been discontinued generally far outweigh any savings gained from a temporary cutback in the treatment program (26).

Microbial Control of Souring

Biocides are most effective in controlling the detrimental activities of SRB in surface facilities. However, controlling these activities in the reservoir with biocides is more difficult and often expensive. A different approach is to manipulate the ecology of the subsurface so that the terminal electron-accepting process is changed from the reduction of sulfur

oxyanions to nitrate reduction. Thus, even if sulfide producers are present in the reservoir, the production of sulfide is prevented. This can be done by adding nitrate and, if needed, a sulfide-tolerant, chemoautotrophic, nitrate-reducing bacterium (29).

MICROBIALLY ENHANCED OIL RECOVERY
Experimental Systems

The efficacy of microbial oil recovery processes has been demonstrated in laboratory experiments using sandstone cores connected to a flow system as shown in Fig. 1 (10, 18, 24, 25, 38, 49). Nutrients are injected into the core, and the system is incubated without fluid flow to allow time for in situ growth and metabolism to occur. The core is then flooded with an appropriate brine solution to determine if any additional oil is produced or if the permeability of the rock is changed. Since oil is entrapped in small pores in the rock, it is important to use consolidated rock rather than crushed rock or sand. Since it is often difficult and expensive to obtain core material from an oil reservoir, Berea sandstone is used as a model system and is commercially available in several permeability classes. The study of oil recovery from carbonate rock is difficult because this material has very small pore sizes and very low permeabilities (50), which makes it nearly impossible to inject fluids in the core. To avoid this problem, Adkins et al. (2) used a crushed limestone system to study microbially enhanced oil recovery from carbonates.

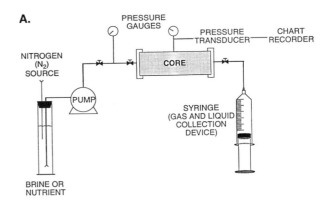

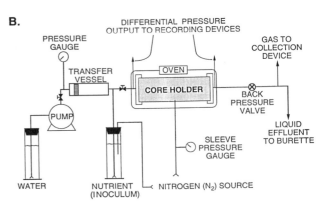

FIGURE 1 Consolidated and unconsolidated porous media systems to study microbial plugging and oil recovery. (A) Sandstone flow system; (B) high-pressure core system.

Cylindrical cores of Berea sandstone are cut by using diamond-tipped core barrels. Since the sandstone is obtained from a surface outcropping, the cores are steam cleaned for 2 weeks to remove organic compounds that may have percolated into the rock. Alternatively, the cores can be heated at 850°C for 4 to 24 h to destroy organic compounds in the core. The high-temperature treatment will also fuse clay particles and sterilize the core. The longitudinal surface of the core is coated with epoxy, and Plexiglas plates are glued to the ends of the core. The core is then wrapped in fiberglass cloth and coated with fiberglass resin before a final coat of epoxy is applied over the entire surface of the core. This procedure will force fluid though the core and provides a firm support for tubing connections. For high-pressure studies, cores are placed directly into rubber sleeves, which are then mounted inside of high-pressure core holders (Fig. 1) (17, 51).

The flow apparatus consists of reservoirs for nutrients, cell suspensions, and brine; a peristaltic pump; and a collection vessel. The effluent collection system should allow the measurement of gas, oil, and brine volumes. Flow lines are nylon or Teflon tubing with compression fittings, although stainless steel can also be used. The peristaltic pump must be able to maintain a nearly constant flow rate against a changing pressure gradient. For high-pressure work, a piston pump or a high-pressure liquid chromatography pump is used. All fluids are filtered through a 0.22-μm-pore-size filter prior to injection into the core to prevent the deposition of fines at the injection face. For the same reason, the cell concentration should be low, about 10^5/ml, although higher cell concentrations have been used (10). The flow rates must be maintained below the critical velocity for fines migration (about 0.5 cm·s^{-1}) to avoid permeability reductions (13). If heat-resistant materials are used, the flow system can be sterilized by steam or dry heat. Jenneman et al. (19) found that dry heat sterilization did not affect the clay morphology and the surface properties of the pores, whereas steam sterilization did, and they recommended that the former process be used to sterilize sandstone cores. Heat sterilization of the core after it has been wrapped in fiberglass may cause cracks and leaks. The core should be heat sterilized after the first coating of epoxy is applied or sterilized by the injection of disinfectants after it has been attached to the flow system.

The core is vacuum saturated with CO_2 to remove air before fluid saturation. Water-wet cores are prepared by flooding the core with a brine solution prior to the injection of oil. Oil-wet cores are prepared by injecting oil into the core prior to brine injection. The brine solution should match the ionic composition of the brine from the oil reservoir under study, or the actual brine from the reservoir can be used. If the cores were not fired to remove organic compounds and fuse the clays, a pretreatment with 10 to 100 pore volumes of a brine solution containing 0.1 M divalent cation such as calcium is required to prevent clay swelling. The calcium can be removed by flooding the core with a brine solution without calcium prior to nutrient addition. For subsequent injections, the ionic strength cannot be changed or clay swelling will occur.

For oil recovery studies, an oleic phase such as crude oil or a hydrocarbon of known composition such as hexadecane is injected into the core until brine is not detected in the effluent (10, 38, 39, 49). The core is then flooded with brine until no more oil is detected in the effluent. The amount of oil remaining inside the core after the brine flood is defined as the residual oil saturation. After the microbial treat-

ment, the core is again flooded with brine and the amount of oil in the effluent is quantified. It is important to note that changes in the flow rate or the viscosity of the fluids injected into the core will affect the residual oil saturation and make it difficult to determine whether oil recovery was the result of microbial action.

The effect of in situ microbial growth on the permeability of the rock is determined by measuring the change in the pressure gradient across the core before and after microbial treatment. If the flow rate and viscosity of the injected fluids are constant, then changes in the permeability of the core can be calculated from the pressure gradient across the core according to Darcy's law (18, 38):

$$K = (q\,\mu L)/(A\,\Delta P),$$

where K is the permeability in Darcy (Système International units of μm), q is the volumetric flow rate (cm^3 · s^{-1}), μ is the dynamic viscosity of the fluid [(mN · s) · (m^2)$^{-1}$], A is the cross-sectional area of the core (cm^2), L is the length of the core (cm), and ΔP is the differential pressure across the core (kPa). The pressure gradient is measured with transducers calibrated with a dead-weight tester. The sensitivity of the transducers can be adjusted by using diaphragms with the desired pressure range. Gauges or manometers can also be used to measure the pressure changes. Additional pressure taps can be placed along the core to determine the depth of microbial plugging. Flooding the core in the reverse direction is used to check the stability of the microbial plug and to remove any cells that accumulated at the injection face (38). Gas production can result in a free gas phase in the core which may account for up to 50% of the observed permeability reductions (17). Since most reservoirs have elevated pressures, which prevents the formation of a free gas phase, microbial studies should be conducted at reservoir pressures.

The selectively of a microbial plugging process is assessed by using a parallel core system in which two cores of different permeabilities are connected to a common injection line (39). A more realistic model of an actual oil reservoir consisted of two blocks of Berea sandstone which were separated from each other by a layer of filter paper (39). This system allowed crossflow of fluids between high and low permeability regions, as occurs in actual reservoirs.

The permeability and pore structure often vary between cores even if the cores are taken from the same block of sandstone. This makes it difficult to reproduce experimental findings from one core to the next. To avoid these complications, a model consolidated porous material in which glass beads are fused together by a gradual increase in temperature up to 700°C has been developed (44). The fused glass bead material has a reproducible pore size and surface area. The permeability of the fused glass beads is high (>6 Darcy), so gravity flow rather than a peristaltic pump is used for fluid injection. The change in permeability is calculated from change in the flow rate, since the hydraulic pressure gradient across the system is constant.

Field Applications

Because of the great variation in the physical and chemical properties among oil reservoirs and the numerous problems that may be encountered in oil production, it is difficult to recommend a specific process or procedure unless a detailed analysis of the well or the reservoir has been conducted. For these reasons, we will discuss the types of analyses needed to assess the effectiveness of a microbial oil recovery process rather than recommend a specific process or procedure. The choice of which process to use depends on the problem. Microbial processes have been developed to remove paraffins and scale in the well bore, rectify formation damage near the well bore, improve oil mobility, improve sweep efficiency of a recovery fluid, block water channels, etc. (Table 1). Several of these processes are now commercially available (37). The ultimate test of any oil recovery process is whether more oil was obtained and whether the process was cost-effective. However, these questions may be difficult to answer because of incomplete production records and the uncertainties encountered in demonstrating cause and effect in field studies. In addition to improving oil recovery, a microbial process can also be a success if it reduces the operating costs of a well or a field. If the amount of electricity needed to pump the oil or the number of chemical treatments needed to remove paraffin deposits can be reduced by microbial processes, the operating life of the well will be extended and it will be possible to recover oil that may not have been recovered if the well was prematurely closed for economic reasons.

Many factors, either associated with the well itself or within the reservoir, will affect oil production from a given well. These may include the cessation of production from a well, the replacement of worn or old equipment, the removal of scale and paraffin deposits, or variations in the stroke rate of the pump. The changes in production or injection rates in other wells will also affect oil production from the well under study. Thus, a thorough understanding of the production history and the current operational plan for the field is needed before a field study of any microbial process is begun.

To judge the effectiveness of a microbial treatment, the following analyses should be conducted. Samples of treated and untreated wells should be taken on a frequent basis before and after the implementation of the microbial process to determine whether the expected products of metabolism and the appropriate microorganisms increase in concentration. An increase indicates that the appropriate microbial activity was stimulated. A flowmeter installed on each production well is needed to measure total fluid production from that well. The amount of oil produced is then calculated from the water-to-oil ratio, which is determined on a frequent schedule. Preferably, data for each well in the field should be gathered over a period of several months to a year before and after the initiation of the microbial process. A control treatment using the same equipment and injection procedures but lacking key components of the microbial process should be conducted to determine how changes in equipment or the operation of the field affect oil recovery. For reservoir-wide processes, tracer experiments should be conducted to verify flow paths. The results of several successful microbial field trials are summarized in a book edited by Premuzic and Woodhead (37).

REFERENCES

1. **Adkins, J. P., L. A. Cornell, and R. S. Tanner.** 1992. Microbial composition of carbonate petroleum reservoir fluids. *Geomicrobiol. J.* **10:**87–97.
2. **Adkins, J. P., R. S. Tanner, E. O. Udegbunam, M. J. McInerney, and R. M. Knapp.** 1993. Microbially enhanced oil recovery from unconsolidated limestone cores. *Geomicrobiol. J.* **10:**77–86.
3. **Amann, R. I., J. Stromley, R. Devereux, R. Key, and D. A. Stahl.** 1992. Molecular and microscopic identification of sulfate-reducing bacteria in multispecies biofilms. *Appl. Environ. Microbiol.* **58:**614–623.
4. **American Petroleum Institute.** 1975. *Recommended Practice for the Biological Analysis of Subsurface Injection Waters.* American Petroleum Institute, Washington, D.C.
5. **Bernard, F. P., J. Connan, and M. Magot.** 1992. Indigenous microorganisms in connate water of many oil fields: a new tool in exploration and production techniques, p. 467–475. *In* SPE 24811. *Proceedings of the Society of Petroleum Engineers*, vol π. Society of Petroleum Engineers, Inc., Richardson, Tex.
6. **Bhupathiraju, V. K., M. J. McInerney, and R. M. Knapp.** 1993. Pretest studies for a microbially enhanced oil recovery field pilot in a hypersaline oil reservoir. *Geomicrobiol. J.* **11:**19–34.
7. **Brink, D. E., I. Vance, and D. C. White.** 1994. Detection of *Desulfobacter* in oil field environments by non-radioactive DNA probes. *Appl. Microbiol. Biotechnol.* **42:**469–475.
8. **Broderick, L. S., and J. J. Cooney.** 1992. Emulsification of hydrocarbons by bacteria from freshwater ecosystems. *Dev. Ind. Microbiol.* **23:**425–434.
9. **Bryant, M. P.** 1977. Microbiology of the rumen, p. 287–304. *In* M. J. Stevenson (ed.), *Duke's Physiology of Domestic Animals*, 9th ed. Cornell University Press, Ithaca, N.Y.
10. **Bryant, R. S., and J. Douglas.** 1988. Evaluation of microbial systems in porous media for EOR. *SPE Reservoir Eng.* **3:**489–495.
11. **Christensen, B. T. Torsvik, and T. Lien.** 1992. Immunomagnetically captured thermophilic sulfate-reducing bacteria from North Sea oil field waters. *Appl. Environ. Microbiol.* **58:**1244–1248.
12. **Devereux, R., M. D. Kane, J. Winfrey, and D. A. Stahl.** 1992. Genus- and group-specific hybridization probes for the determinative and environmental studies of sulfate-reducing bacteria. *Syst. Appl. Microbiol.* **15:**601–609.
13. **Grusbeck, C., and R. E. Collins.** 1982. Entrainment and deposition of fine particles in porous media. *Soc. Pet. Eng. J.* **22:**847–856.
14. **Hamilton, W. A.** 1985. Sulfate-reducing bacteria and anerobic corrosion. *Annu. Rev. Microbiol.* **39:**195–217.
15. **Horacek, G. L., and L. J. Gawel.** 1988. New kit for the rapid detection of SRB in the oil field, p. 189–192. *In* SPE 18199. *Proceedings of the Society of Petroleum Engineers*, vol π. Society of Petroleum Engineers, Inc., Richardson, Tex.
15a.**Jack, T. R., and D. W. S. Westlake.** 1995. Control in industrial settings, p. 265–292. *In* L. L. Barton (ed.), *Sulfate-Reducing Bacteria.* Plenum Press, New York.
16. **Jenneman, G. E.** 1989. The potential for in-situ microbial applications. *Dev. Pet. Sci.* **22:**37–74.
17. **Jenneman, G. E.** 1992. The effect of in-situ pressure on MEOR processes, p. 481–495. *In Proceedings of the Eighth Symposium on Enhanced Oil Recovery.* SPE/DOE 24203. Society of Petroleum Engineers, Richardson, Tex.
18. **Jenneman, G. E., R. M. Knapp, M. J. McInerney, D. E. Menzie, and D. E. Revus.** 1984. Experimental studies of in situ microbial enhanced oil recovery. *Soc. Pet. Eng. J.* **24:**33–37.
19. **Jenneman, G. E., M. J. McInerney, M. E. Crocker, and R. M. Knapp.** 1986. Effect of sterilization by dry heat or autoclaving on bacterial penetration through Berea sandstone. *Appl. Environ. Microbiol.* **51:**39–43.
20. **Jenneman, G. E., M. J. McInerney, and R. M. Knapp.** 1985. Microbial penetration through nutrient-saturated Berea sandstone. *Appl. Environ. Microbiol.* **50:**383–391.
21. **Jenneman, G. E., M. J. McInerney, R. M. Knapp, J. B. Clark, J. M. Ferro, D. E. Revus, and D. E. Menzie.** 1983. A halotolerant, biosurfactant-producing *Bacillus* species potentially useful for enhanced oil recovery. *Dev. Ind. Microbiol.* **25:**485–492.
22. **Kane, M. D., L. K. Poulsen, and D. A. Stahl.** 1993. Moni-

toring the enrichment and isolation of sulfate-reducing bacteria by using oligonucleotide hybridization probes designed from environmentally derived 16S rRNA sequences. *Appl. Environ. Microbiol.* **59**:682–686.

23. **Kohring, L. L., D. B. Ringelberg, R. Devereux, D. A. Stahl, M. W. Mittelman, and D. C. White.** 1994. Comparison of phylogenetic relationships based on phospholipid fatty acid profiles and ribosomal RNA sequence similarities among sulfate-reducing bacteria. *FEMS Microbiol. Lett.* **119**:303–308.

24. **Lappin-Scott, H. L., F. Cusack, and J. W. Costerton.** 1988. Nutrient resuscitation and growth of starved cells in sandstone cores: a novel approach to enhanced oil recovery. *Appl. Environ. Microbiol.* **54**:1373–1382.

25. **MacLeod, F. A., H. M. Lappin-Scott, and J. W. Costerton.** 1988. Plugging of a model rock system by using starved bacteria. *Appl. Environ. Microbiol.* **54**:1365–1372.

26. **Manning, F. S., and R. E. Thompson.** Unpublished data.

27. **Marquis, R. E., and P. Matsumura.** 1978. Microbial life under pressure, p. 105–158. *In* D. J. Kushner (ed.), *Microbial Life in Extreme Environments*. Academic Press, London.

28. **McCoy, W. F., J. D. Bryers, J. Robbins, and J. W. Costerton.** 1981. Observations of fouling biofilm formation. *Can. J. Microbiol.* **27**:910–917.

29. **McInerney, M. J., V. K. Bhupathiraju, and K. L. Sublette.** 1992. Evaluation of a microbial method to reduce hydrogen sulfide levels in a porous rock biofilm. *J. Ind. Microbiol.* **11**:53–58.

30. **McInerney, M. J., M. Javaheri, and D. Nagle, Jr.** 1990. Properties of the biosurfactant produced by *Bacillus licheniformis* strain JF-2. *J. Ind. Microbiol.* **5**:95–102.

31. **McInerney, M. J., and D. W. S. Westlake.** 1990. Microbially enhanced oil recovery, p. 409–445. *In* H. L. Ehrlich and C. L. Brierley (ed.), *Microbial Mineral Recovery*. McGraw-Hill Publishing Co., New York.

32. **Mulligan, C. N., D. G. Cooper, and R. J. Neufeld.** 1984. Selection of microbes producing biosurfactants in media without hydrocarbons. *J. Ferment. Technol.* **62**:311–314.

33. **National Association of Corrosion Engineers.** 1982. The role of bacteria in corrosion of oil field equipment. National Association of Corrosion Engineers, Houston.

34. **Nealson, K. H., and D. Saffarini.** 1994. Iron and manganese in anaerobic respiration: environmental significance, physiology, and regulation. *Annu. Rev. Microbiol.* **48**:311–343.

35. **Odum, J. M., K. Jessie, E. Knodel, and M. Emptage.** 1991. Immunological cross-reactivities of adenosine 5′-phosphosulfate reductases from sulfate-reducing and sulfide-oxidizing bacteria. *Appl. Environ. Microbiol.* **57**:727–733.

36. **Pfiffner, S. M., M. J. McInerney, G. E. Jenneman, and R. M. Knapp.** 1986. Isolation of halotolerant, thermotolerant, facultative polymer-producing bacteria and characterization of the exopolymer. *Appl. Environ. Microbiol.* **51**:1224–1229.

37. **Premuzic, E., and A. Woodhead (ed.).** 1993. *Microbial Enhancement of Oil Recovery—Recent Advances. Proceeding of the 1992 International Conference on Microbial Enhanced Oil Recovery*. Elsevier, Amsterdam.

38. **Raiders, R. A., M. J. McInerney, D. E. Revus, H. M. Torbati, R. M. Knapp, and G. E. Jenneman.** 1986. Selec-

tivity and depth of microbial plugging in Berea sandstone cores. *J. Ind. Microbiol.* **1**:195–203.

39. **Raiders, R. A., R. M. Knapp, and M. J. McInerney.** 1989. Microbial selective plugging and enhanced oil recovery. *J. Ind. Microbiol.* **4**:215–230.

40. **Rosenberg, E., A. Zuckerberg, C. Rubinowitz, and D. L. Gutnick.** 1979. Emulsifier of *Arthrobacter* RAG-1: isolation and emulsifying properties. *Appl. Environ. Microbiol.* **37**:402–408.

41. **Rosnes, J. T., T. Torsvik, and T. Lien.** 1991. Spore-forming thermophilic sulfate-reducing bacteria isolated from North Sea oil field waters. *Appl. Environ. Microbiol.* **57**:2302–2307.

42. **Ruseska, I., J. Robbins, J. W. Costerton, and E. S. Kashen.** 1982. Biocide testing against corrosion-causing oil field bacteria. *Oil Gas J.* **80**(10):253–264.

43. **Semple, K. M., and D. W. S. Westlake.** 1987. Characterization of iron-reducing *Alteromonas putrefaciens* strains from oil field fluids. *Can. J. Microbiol.* **33**:366–371.

44. **Shaw, J. C., B. Bramhill, N. C. Wardlaw, and J. W. Costerton.** 1985. Bacterial fouling in a model core system. *Appl. Environ. Microbiol.* **49**:693–701.

45. **Sheehy, A.** 1991. Microbial physiology and enhanced oil recovery. *Dev. Pet. Sci.* **31**:37–44.

46. **Stetter, K. O., R. Huber, E. Blochl, M. Kurr, R. D. Eden, M. Fielder, H. Cash, and I. Vance.** 1993. Hyperthermophilic archaea are thriving in deep North Sea and Alaskan oil reservoirs. *Nature* (London) **365**:743–745.

47. **Tanner, R. S.** 1989. Monitoring sulfate-reducing bacteria: comparison of enumeration media. *J. Microbiol. Methods* **10**:83–90.

48. **Tempest, D. W., and J. T. M. Wouters.** 1981. Properties and performance of microorganisms in chemostat culture. *Enzyme Microb. Technol.* **3**:283–290.

49. **Thomas, C. P., G. A. Bala, and M. L. Duvall.** 1991. Surfactant-based enhanced oil recovery mediated by naturally occurring microorganisms, p. 287–298. *In* SPE 22844. *Proceedings of the Society of Petroleum Engineers*, vol. π. Society of Petroleum Engineers, Inc., Richardson, Tex.

50. **Udegbunam, E. O., J. P. Adkins, R. M. Knapp, M. J. McInerney, and R. S. Tanner.** 1991. Assessing the effects of microbial metabolism and metabolites on reservoir pore structure, p. 309–316. *In* SPE 22846. *Proceedings of the Society of Petroleum Engineers*, vol. π. Society of Petroleum Engineers, Inc., Richardson, Tex.

51. **Vance, I., and D. E. Brink.** 1994. Propionate-driven sulfate reduction by oil field bacteria in a pressurized porous rock bioreactor. *Appl. Microbiol. Biotechnol.* **40**:920–925.

52. **Voordouw, G., Y. Shen, C. S. Harrington, A. J. Telang, T. R. Jack, and D. W. S. Westlake.** 1993. Quantitative reverse sample genome probing of microbial communities and its application to oil field production waters. *Appl. Environ. Microbiol.* **59**:4101–4114.

53. **Voordouw, G., J. K. Voordouw, R. R. Karkhof-Schweizer, P. M. Fedorak, and D. W. S. Westlake.** 1991. Reverse sample genome probing, a new technique for identification of bacteria in environmental samples by DNA hybridization, and its application to the identification of sulfate-reducing bacteria in oil field samples. *Appl. Environ. Microbiol.* **57**:3070–3078.

54. **Westlake, D. W. S.** 1991. Microbial ecology of corrosion and reservoir souring. *Dev. Pet. Sci.* **31**:257–263.

Placement and Protection of Drinking Water Wells

MARYLYNN V. YATES

66

INTRODUCTION

Groundwater supplies over 100 million Americans with their drinking water; in rural areas there is an even greater reliance on groundwater, as it comprises up to 95% of the water used (1). Traditionally, it has been assumed that groundwater is safe for consumption without treatment because the soil acts as a filter to remove pollutants. As a result, private wells generally do not receive treatment (7), nor do a large number of public water supply systems. The U.S. Environmental Protection Agency (EPA) has estimated that approximately 72% of the public water supply systems in the United States that use groundwater do not disinfect (22). However, the use of contaminated, untreated, or inadequately treated groundwater has been the major cause of waterborne disease outbreaks in this country since 1920 (3, 4). Between 1920 and 1994, 1,699 waterborne disease outbreaks, involving over 850,000 people and resulting in 1,214 deaths, were reported in the United States (5, 6, 9, 14). These data are summarized for 10-year periods in Fig. 1 (6, 9). The number of reported outbreaks and the number of associated cases of illness have risen dramatically since 1971 compared with the period from 1951 to 1970. During the period from 1971 through 1980, an average of 32.6 outbreaks per year was reported. From 1981 through 1990, the average was 27.6, compared with averages of 11.1 and 13.1 for the periods 1951 through 1960 and 1961 through 1970, respectively. The increase in reported numbers of outbreaks may be due to an improved reporting system implemented in 1971 (2); however, it is still believed that only a fraction of the total number of outbreaks is reported (13).

Causative agents of illness were identified in approximately one-half the disease outbreaks during the period from 1971 through 1994. The microorganisms most commonly identified as causative agents were *Giardia* and *Cryptosporidium* spp., hepatitis A virus, and *Shigella* sp. *Giardia lamblia* caused over 18% of the illness associated with waterborne disease outbreaks. Enteric viruses were identified as the causative agents of disease in 8.5% of the outbreaks during this period.

In the 1980s, use of untreated or inadequately treated groundwater was responsible for 43% of the outbreaks that occurred in the United States (Fig. 2) (6, 9). In outbreaks due to the consumption of contaminated, untreated groundwater that occurred from 1971 to 1985, sewage was most often identified as the contamination source. In groundwater systems, etiologic (disease-causing) agents were identified in only 38% of the outbreaks, with *Shigella* sp. and hepatitis A virus being the most commonly identified pathogens (5). In almost two-thirds (62%) of the outbreaks, no etiologic agent could be identified, and the illness was simply listed as gastroenteritis of unknown etiology, However, retrospective serological studies of outbreaks of acute nonbacterial gastroenteritis from 1976 through 1980 indicated that 42% of these outbreaks (i.e., the 62% for which no etiologic agent was identified) were caused by the Norwalk virus (10). Thus, it has been suggested that the Norwalk virus is responsible for approximately 23% of all reported waterborne disease outbreaks in the United States (12).

Drinking Water Regulations

The fact that microorganisms are responsible for numerous waterborne disease outbreaks every year led the EPA to reexamine the coliform standard, which has been used to indicate the microbiological quality of drinking water in the United States for more than 75 years. Increasing amounts of evidence collected during the past 15 to 20 years suggest that the coliform group may not be an adequate indicator of the presence of pathogenic viruses and possibly protozoan parasites in water. In 1985, the EPA proposed maximum contaminant level goals for viruses and *G. lamblia*, a protozoan parasite (20). These standards are in addition to the standard for the indicator microorganisms, total coliform bacteria. Rather than require public water systems to monitor the water for the presence of these pathogenic microorganisms, the EPA proposed treatment requirements for groundwater with the goal that the level of pathogenic viruses in the treated water would result in a risk of less than one infection per 10,000 persons per year (23). Water utilities may avoid chemical disinfection of the source water if they meet one of the EPA's "natural disinfection" criteria. The criteria include setback distance from a potential contamination source, depth to well screen or thickness of unsaturated zone, groundwater travel time from the contamination source to a supply well, and virus travel time from the contamination source to a supply well. The numerical values for the criteria are based on an acceptable virus concentration of two virus particles per 10^7 liters at the well-

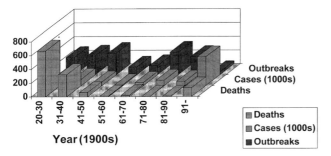

FIGURE 1 Waterborne disease outbreaks in the United States, 1920 to 1994.

head, which was calculated by using a risk of less than one infection per 10,000 people per year (17).

There are approximately 180,000 community and non-community public water supply systems with wells or well fields that are not groundwater under the direct influence of surface water wells. The draft proposed groundwater disinfection rule will allow such systems to avoid disinfecting provided they can demonstrate that adequate "natural disinfection" occurs between potential sources of fecal contamination (e.g., septic tanks and sewer lines) and wells.

Sources of Microorganisms

Microorganisms may be introduced into the subsurface environment in a variety of ways. In general, any practice that involves the application of domestic wastewater to the soil has the potential to cause microbiological contamination of groundwater. This is because the treatment processes to which the wastewater are subjected do not effect complete removal or inactivation of the disease-causing microorganisms present. Goyal et al. (8) isolated viruses from the groundwater beneath cropland being irrigated with sewage effluent. Viruses have been detected in the groundwater at several sites practicing land treatment of wastewater; these cases were reviewed by Keswick and Gerba (11). The burial of disposable diapers in sanitary landfills is a means by which pathogenic microorganisms in untreated human waste may be introduced into the subsurface. Vaughn et al. (27) detected viruses as far as 408 m downgradient of a landfill site in New York. Land application of treated sewage effluent for the purposes of groundwater recharge has also resulted in the introduction of viruses to the underlying groundwater (25, 26).

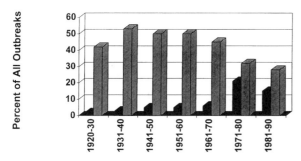

FIGURE 2 Waterborne outbreaks in untreated (▨) and disinfected-only (■) groundwater systems in the United States, 1920 to 1990.

METHODS TO DELINEATE WELLHEAD PROTECTION AREAS

There are several methods that can be used to establish placement of drinking water wells to minimize microbial contamination. The choice of the method to be used when establishing wellhead protection areas (WHPAs) depends on a variety of factors, including the vulnerability of the drinking water supply, the importance of the groundwater as a water supply, the number and types of contamination sources in the area, the availability of pertinent hydrogeologic information, and the availability of technical personnel and computer equipment. The six most common methods, listed in order of increasing technical complexity, are (i) arbitrary fixed radius, (ii) calculated fixed radius, (iii) simplified variable shapes, (iv) analytical methods, (v) hydrogeologic mapping, and (vi) numerical transport models. These methods, which are discussed in detail in reference 21, are described briefly below.

Arbitrary Fixed Radius

The arbitrary fixed radius method is the simplest method of delineating a WHPA. A circle of specified radius is drawn around a well, and no sources of potential contamination may be located within the circle. This method is widely used by states and local communities for establishing placement requirements for septic tank or other on-site wastewater treatment systems. The choice of radius varies widely, as shown in Table 1, from 7.62 to 152.4 m. The numerical value of the radius may be based on general hydrogeologic considerations as well as professional judgment. This method is very simple to establish and enforce. However, depending on the variability in soil conditions and source characteristics, it may not provide consistent protection throughout an area. In addition, the radius generally represents an average protective distance. Thus, areas upgradient of the well may be underprotected from contamination, while those downgradient may be overprotected.

Calculated Fixed Radius

The calculated fixed radius method is also a simple method in which a circle of a fixed radius is drawn around a well; however, the radius is calculated on the basis of certain criteria.

Source-Driven Radius

The radius of setback from a well can be based on the type of contaminant source. Many states have several setback requirements, depending on the contaminant source. For example, California requires a 15.2-m setback from a septic system, a 30.5-m setback from a drain field, and a 45.7-m setback from a cesspool (Table 1). In these cases, a number of concentric circles would be drawn around the well, and the specified contaminant sources would not be allowed in each of the rings.

Well-Driven Radius

Some setback distances are determined on the basis of characteristics such as the well discharge (pumping) rate or the type of well (Table 1). In the case of setting setback distances based on pumping rate, the time of travel of the contaminant from the source to the water supply well is considered. Larger radii require longer travel times; therefore, more inactivation and removal of pathogenic microorganisms can occur.

Different setback distances can also be specified on the basis of the presence of a confining layer between the soil

TABLE 1 State requirements for setback distances from public water supply wells

State	Setback radius (m)	Comments
Alabama	30.5	From well to septic system
Alaska	30.5–71	From community sewer lines
	45.7–71	From wastewater treatment and disposal systems
Arizona	15.2	From sewer
	30.5	From septic disposal field
California	15.2	From sewer or watertight septic tank
	30.5	From subsurface sewage leaching field
	45.7	From cesspool or seepage pit
Connecticut	22.9	Discharge of <10 gpm[a]
	45.7	Discharge of 10–50 gpm
	71	Discharge of >50 gpm
Delaware	30.5	All public water supply wells
Florida	30.5	From sanitary hazards
Georgia	15.2	No state statute in effect since 1986
Idaho	30.5	From septic drain field
Illinois	15.2	From septic tank
Indiana	71	From septic system
Iowa	30.5	From deep well
	71	From shallow well
Kansas	30.5	From all pollution sources
Kentucky	15.2	From septic system
Maryland	15.2	Confined aquifers
	30.5	Unconfined aquifers
Maine	91.4	All water supply wells
Massachusetts	15.2–121.9	Discharge <69 gpm
	76.2	Tubular wellfields, discharge of >69 gpm
	121.9	Gravel-packed wells, discharge of >69 gpm
Michigan	22.9–71	From septic tanks
	243.8–609.6	From known major sources of contamination
Minnesota	15.2	From septic system
Mississippi		Shall not be located so as to create a hazard
Missouri	30.5	Required; 91.4 m recommended
Montana		No specified distance
Nebraska	152.4	From septic system
Nevada	71	From septic system
New Hampshire	22.9	Discharge of 0.28 gpm
	121.9	Supplying >1,000 persons or having fire protection capability *or* supplying 25–1,000 persons and having no fire protection capability
	38.1	Discharge of 0.56 gpm
	45.7	Discharge of 0.84 gpm
	53.3	Discharge of 1.12 gpm
	71	Discharge of >1.39 gpm
North Carolina	45.7	Private wells
North Dakota	30.5	From septic system
Ohio	15.2	From known contamination source
Oregon	15.2	From septic tank
	30.5	From sewage disposal field or cesspool
Pennsylvania	30.5	Individual water supply wells
Rhode Island	71	Bedrock or driven wells
	121.9	Gravel-packed well
South Carolina	7.6	No written rule
South Dakota	15.2	From septic tank
	30.5	From drain field
Utah	3.0	From sewer lines for deep wells
	91.4	From specially constructed sewers for shallow wells
	457.2	Wellhead zone
Vermont	22.9–71	Transient, noncommunity wells
	30.5–71	Nontransient, noncommunity wells
	38.1–71	Community public water supply wells
Washington	30.5	From septic tank
	30.5	From sewage lagoon
West Virginia		Determined by director of State Department of Health
	15.2	Nonpublic water supply wells
Wisconsin	7.6	From septic tanks
	15.2	From drain fields
Wyoming		Documentation showing that discharge will not affect water quality required

[a] gpm, gallons per minute.

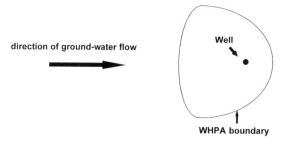

FIGURE 3 Simple variable shape method for delineating WHPAs.

surface and the aquifer from which the well draws water. For example, Maryland requires a 30.5-m setback for wells in unconfined aquifers but only 15.2 m for wells in confined aquifers. While requiring more information to develop than the arbitrary fixed radius method, the well-driven radius method has the potential to allow for some variability in local conditions. However, because a circle of fixed radius is drawn around the well, the upgradient area is likely to be underprotected and the downgradient area is likely to be overprotected.

Simplified Variable Shapes
The simplified variable shape method requires information about the pumping rate of the well in addition to information about the direction of groundwater flow at the site. The downgradient boundary of the WHPA is calculated by using the pumping rate of the well. Then, the upgradient boundary is calculated by using the time of travel from the contaminant. The orientation of the WHPA around the well is based on the direction of groundwater flow (Fig. 3).

An advantage of this method over the first two is that all areas of the WHPA should be equally protected. The additional information required to develop the variable shapes is generally available or can be obtained at a relatively modest cost. However, it is still a very simple method that does not take into account many of the physical heterogeneities that occur in subsurface geologic materials.

Analytical Methods
In analytical methods, WHPAs are delineated by using mathematical equations that describe the groundwater flow and contaminant transport in the area of the well. The uniform flow equation (19) can be used to determine the areal extent of the aquifer from which the pumping well draws water, the "zone of contribution" to the well (Fig. 4). Calcu-

lation of the zone of contribution requires knowledge of several hydrogeologic parameters, including the aquifer transmissivity, porosity, hydraulic gradient, hydraulic conductivity, and saturated thickness. By using this equation, the downgradient boundary and width of the WHPA can be delineated. The upgradient boundary can be calculated by using a time of contaminant travel criterion or some type of flow boundary criterion, such as the presence of a groundwater divide.

The analytical equations required by this method are generally simple to use for trained hydrologists and civil and environmental engineers. It has an advantage over the previously discussed methods in that it does take into account more site-specific hydrogeologic information. However, most analytic methods do not take into account hydrologic boundaries (e.g., streams), subsurface heterogeneities, and nonuniform rainfall or evapotranspiration.

Hydrogeologic Mapping
In some areas, the hydrogeologic characteristics can be mapped and used to delineate WHPAs. For example, topographic analysis of drainage basin divides can be combined with water table elevation data to delineate aquifer recharge areas. Potential contaminant sources in the recharge can then be sited by using time of travel calculations to prevent groundwater contamination. This method is most applicable to areas having flow boundaries located near the ground surface and areas with fractured bedrock or karst formations. Use of this method requires highly trained personnel with expertise in geologic and geomorphologic mapping.

Numerical Flow and Transport Models
The most technically complex method of delineating WHPAs involves the use of computer models that solve equations for groundwater flow and/or solute transport by numerical methods. A large number of numerical models have been developed to predict groundwater contamination by a variety of chemicals. Sixty-four of these models were reviewed by van der Heijde and Beljin (24).

The information required to use numerical models varies according to the specific model chosen but generally includes aquifer permeability, porosity, specific yield, saturated thickness, recharge rates, aquifer geometries, and the location of hydrogeologic boundaries. If the model has contaminant transport prediction capabilities in addition to groundwater flow prediction, several other parameters such as dispersivity, contaminant inactivation rate, and contaminant adsorption coefficient must also be known.

Although there are many models that can be used to delineate WHPAs, most of them were developed to protect the well from chemical contamination. To date, only one groundwater flow and contaminant transport model that has a user-friendly interface has been developed to delineate WHPAs for microorganisms. This model, VIRALT (16), was developed for the EPA to calculate virus transport to a pumping well from a contaminant source such as a septic tank.

EXAMPLES
Very few models that can be used to delineate WHPAs intended to protect from microbial contamination have been developed. Two models that have been developed will be discussed below.

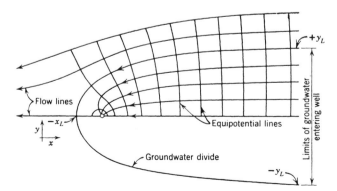

FIGURE 4 Analytical method for delineating WHPAs: plan view of flow to a pumping well. Reprinted from reference 21.

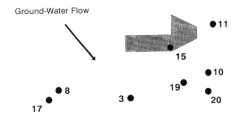

Ground-Water Flow

FIGURE 5 Virus tracer study site, Fort Devens, Massachusetts. ●, well; ■, tracer application area.

VIRALT

VIRALT (16) is a semianalytical and numerical model that simulates virus transport in the subsurface. It was developed for the EPA in support of the groundwater disinfection rule's natural disinfection criteria. VIRALT is a mathematical computer model that calculates the concentration of viruses at the water table and at a well after it has been transported through unsaturated and saturated subsurface media. The predictive capabilities of this model were tested recently (29), using data obtained at a field site where viruses were applied to the land and their transport to a series of wells was monitored.

Site Description

The location chosen for testing was a site in Fort Devens, Massachusetts, where an experiment to monitor the movement of coliphages during the rapid infiltration of wastewater was conducted by the U.S. Army (18) (Fig. 5). The rapid infiltration beds were composed of sand and gravel that had been replaced approximately 4 years prior to the tracer experiments. Several observation wells, 3 to 6 cm in diameter and cased to at least 1.2 m below the ground surface, were located on the site. High concentrations of f2 coliphage (approximately 10^8 PFU/liter) were added to the effluent as it was applied to the soil. The concentrations of the phage in several wells on and up to 53 m from the site were monitored for 21 days.

To test the ability of VIRALT to predict the measured values, several pieces of information characterizing the site had to be input into the model. These input data are shown in Table 2. In several instances, the required information was not reported by the investigators; in those cases, the default values provided by VIRALT were used. The sources of all input values (e.g., measured, default, assumed) are given in Table 2.

Results of Model Testing

VIRALT's prediction of f2 coliphage transport to well 15, located within the boundary of the wastewater infiltration beds, compared with the measured values is shown in Fig. 6. It can be seen that the model predicted that it would take longer for the virus to arrive at the well than it actually did. In addition, the maximum predicted concentration at well 15 was less than 10^3 PFU/liter, although measured concentrations reached 10^9 PFU/liter. The model predictions for wells 19, 10, and 20, located approximately 24, 28, and 53 m, respectively, from the infiltration site, also varied from the measured concentrations. In these cases, the model predicted that the maximum virus concentration at the well would be less than 1 PFU/liter. The investigators measured concentrations between 10^3 and 10^9 PFU/liter at these wells during the 21 days of the tracer study.

TABLE 2 Input parameters for Fort Devens, Massachusetts, site

Parameter	Value	Source of value
Saturated zone		
Transmissivity	104 m² day⁻¹	Measured
Aquifer thickness	12.2 m	Measured
Porosity	0.46	Measured
Hydraulic gradient	0.0148	Measured
Angle of flow	315°	Measured
Unsaturated zone		
No. of soil layers	1	Known
Soil type	Sand	Known
Thickness	19.5 m	Measured
Saturated hydraulic conductivity	8.5 m day⁻¹	Measured
Saturated moisture content	0.46	Measured
Residual moisture content	0.045	Default
Source		
Time period	7 days	Known
Leakage rate	0.46 m day⁻¹	Known
Virus concentration	2.5 × 10⁸ PFU liter⁻¹	Known
Source area	12,400 m²	Known
Contaminant transport		
Soil bulk density	1.43	Measured
Molecular diffusion	0	Default
Saturated dispersivity	1.2 m	Default
Unsaturated dispersivity	0.12 m	Assumed
Temperature	15°C	Default
Virus adsorption	0 cm³g⁻¹	Measured
Virus inactivation rate	0.29 day⁻¹	Default
Retardation coefficient	1	Default
Well pumping rate	0.012 m³ day⁻¹	Known

Geostatistical Model

Another model that has been developed is a regional screening model. This type of model is useful for regional planning purposes, because areas in a community with relatively higher vulnerability to groundwater contamination can be

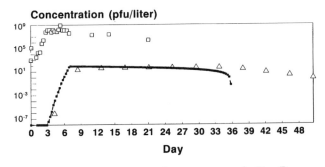

FIGURE 6 Virus transport during tracer study, Fort Devens, Massachusetts, well 15. □, measured; ■, predicted; △, predicted (ambient).

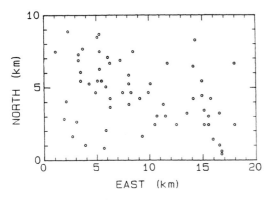

FIGURE 7 Sample collection sites.

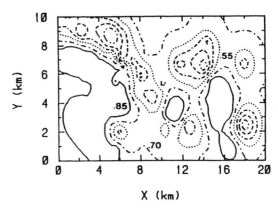

FIGURE 9 Contour diagram for the conditional probabilities that the setback distances are greater than 15 m.

distinguished from areas where contamination is less likely to be a problem.

A model of this type has been developed and used to predict setback distances between sources of viruses (in this case, septic tanks) and drinking water wells for a 200-km² area in the city of Tucson, Arizona (30, 31). It was assumed that virus inactivation rates were dependent only on the temperature of the groundwater and could be predicted from groundwater temperatures. Although septic tank setback distances are used here for illustrative purposes, this model could be used for determining separation distances between any potential source of contamination and a drinking water well.

Data Input Requirements

The data required to run this model include (i) virus inactivation rates in the groundwater (or groundwater temperatures) at 71 locations in the city Fig. 7 shows the relative locations of the samples used [28]), (ii) hydraulic conductivity of the aquifer at those 71 locations (at least), and (iii) hydraulic gradients at those 71 locations (at least).

Model Output

The output from this model is in the form of contour maps. Figure 8 shows the distances between contamination sources (e.g., septic tanks) and drinking water wells that would be required to achieve a 7-order-of-magnitude reduction in virus number (e.g., the removal of 10 million virus particles) in the time necessary for the water to move that distance.

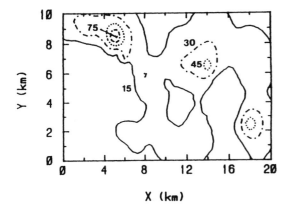

FIGURE 8 Septic tank setback distances (meters) estimated by disjunctive kriging.

For example, if a septic tank is placed on a contour marked 30, a well would have to be 30 m away for there to be a removal of 10 million virus particles. (The model can be run for any amount of virus reduction desired.) A wide range of septic tank setback distances (from less than 15 to more than 75 m) was calculated for a part of the city of Tucson.

The conditional probabilities associated with the estimated separation distances can also be calculated. In other words, the model can be used to answer the following questions.

1. Given a setback distance (e.g., specified by regulation), what is the probability that this would be adequate to protect the groundwater from viral contamination at different locations in the city?

2. Given a desired probability level, what setback distance would be necessary to be confident that the groundwater would be protected from contamination by viruses?

Case 1: Probabilities Associated with Specified Setback Distances

Probability maps were calculated for two setback distances for comparative purposes. Suppose that the local ordinance requires a minimum of 15-m separation distance between a septic tank and a drinking water well. Figure 9 shows the probability that there would be a 7-order-of-magnitude reduction in virus numbers in the time required for the water to travel 15 m. For the contour marked 0.85, one would be 85% sure that a 15-m separation distance will be adequate to meet the criterion of protection of the well water from virus contamination.

Figure 10 shows the probability contour map calculated on the basis of a 30-m separation distance between a septic tank and a well. In comparing this figure with Fig. 9, it can be seen that the contour which had a 70% probability in Fig. 9 now has an 85% probability of meeting the criterion. This is because 30 m has been set as the separation distance, which means that it will take longer for the viruses to travel to the well and there is more opportunity for natural virus inactivation. Thus, it follows that the probability that a 30-m separation distance would be adequate is higher (85%) than the probability estimated for a 15-m separation distance (70%).

Case 2: Setback Distances Associated with Specified Probabilities

In this case, rather than specifying a setback distance and calculating the associated probabilities, one specifies

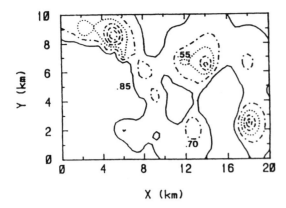

FIGURE 10 Contour diagram for the conditional probabilities that the setback distances are greater than 30 m.

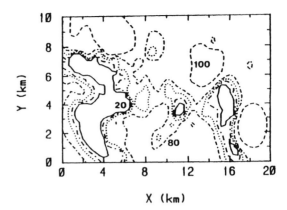

FIGURE 12 Contour diagrams for the setback distances (meters) given a conditional probability of 99%.

the desired probability level and calculates the associated setback distances. In the first example, a probability level of 0.9 was specified. The model calculates the setback distance necessary to be 90% certain that the actual setback distance required to achieve a 7-order-of-magnitude reduction in virus number is less than or equal to that distance. In Fig. 11, it can be seen that the required setback distances range from 20 to more than 100 m. If one wants to be 99% certain that the setback distances are adequate to prevent viral contamination, much larger separation distances are calculated (Fig. 12). For example, in some areas a 100-m setback distance would be required, whereas a 60-m setback distance was calculated when a 90% probability of achieving the removal criterion was specified.

To demonstrate the effect of adding pumping wells to the regional groundwater flow in the model calculations, a simple one-well case was used (30). The well chosen is pumped at a rate of 568 liters/min. In the former calculation, in which only regional groundwater flow was used in the setback distance calculation, this well was located on a 60-m contour (Fig. 8). When the 568-liter/min pumping rate is added to the travel time calculation, a setback distance of 156 m is required to achieve a 7-order-of-magnitude reduction in virus number (Fig. 13). If only 4 orders of magnitude of virus inactivation is required, the setback distance will be 93 m, which is still 1.5 times greater than that calculated without adding the effects of pumping. The actual

calculations would be more complicated than described here, as the effects of all of the wells' pumping would have to be included to get an accurate picture of the flow field in the basin. This simple example does show, however, that pumping has a large impact on the travel time, and thus setback distance, calculations and must be considered if the method is to be used for municipal planning purposes.

With the appropriate modifications to model the specific situation of interest, the methods could be used for community planning purposes. The first case described, namely, calculating the conditional probabilities given a specified setback distance, would be useful in a situation for which the minimum setback distance was specified by regulation. For example, a certain community has a regulation stating that 30 m is the minimum separation between a well and a septic tank. This model could be used to generate a conditional probability contour map. A decision to allow a septic tank to be placed in a certain location could then be based on the calculated probabilities. For example, it might be

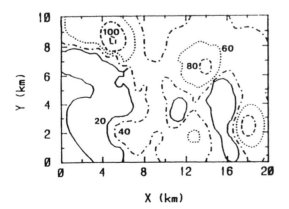

FIGURE 11 Contour diagrams for the setback distances (meters) given a conditional probability of 90%.

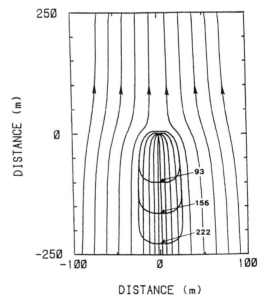

FIGURE 13 Setback distances (meters) calculated for a pumping well in a regional flow field.

decided that if the probability was 75% or greater, a septic tank would be permitted on any lot, provided that soil percolation test requirements were met. If the probability was between 50 and 75%, soil percolation test requirements could be made more stringent or the minimum lot size could be increased in order for a septic tank permit to be issued. If the probability were less than 50%, it might be decided that septic tanks would not be allowed at all.

The approach described in the second case could also be used for community planning purposes, in that a desired probability level could be specified (e.g., in a regulation), and the setback distances necessary to achieve that level would be calculated. One advantage of using this method is that the implicit assumption that the hydrogeologic characteristics of the area are constant would be avoided. The regulations would only have to specify a probability level to be met in order to allow a septic tank permit.

Limitations

There are several limitations in this model in its current form which must be recognized. (i) Only saturated zone transport is considered. There is no allowance for reduction in virus number as the water moves vertically through the unsaturated soil. This is a serious limitation because the greatest percentage of virus loss during subsurface transport is most likely to occur in the unsaturated zone. (ii) The influence of multiple pumping wells on the pattern and rate of groundwater flow has not been considered. Pumping wells act to increase the flow rate of water in some parts of the aquifer and slow it in others and may cause the direction of groundwater flow to be reversed in certain areas. This will have a profound effect on the calculated setback distances, as was illustrated above. The effects of multiple pumping wells will change the flow field even more and must be considered in actual practice. (iii) Inactivation of the viruses was the only removal mechanism included. It is well documented that adsorption to soil particles is an important removal mechanism, especially in the unsaturated zone.

CONCLUSIONS

There are many different methods that can be used to delineate zones around drinking water wells to protect the water supply from microbial contamination. However, in most of the United States, the simplest methods, i.e., the arbitrary fixed radius and calculated fixed radius methods, are used to delineate wellhead protection zones for microorganisms. There are several possible explanations for this. One is that the factors that control the fate and transport of enteric microorganisms in the subsurface are not understood well enough to allow the use of the numerical transport models. In addition, the amount of data required for these models is generally well beyond what is available in many communities. Another possible explanation for the reliance on the fixed radius methods is that it is still assumed that groundwater is safe from microbial contamination, and thus it is not necessary to expend efforts to accurately determine wellhead protection zones specifically for microorganisms. Finally, many of the wells that are used for drinking water are owned and operated by small communities and individual businesses. In most cases, these small systems or noncommunity systems do not have the resources available to obtain the information necessary to calculate wellhead protection zones by the more data-intensive methods.

REFERENCES

1. **Bitton, G., and C. P. Gerba.** 1984. Groundwater pollution microbiology: the emerging issue, p. 1–7. *In* G. Bitton and C. P. Gerba (ed.), *Groundwater Pollution Microbiology.* John Wiley & Sons, New York.
2. **Craun, G. F.** 1985. A summary of waterborne illness transmitted through contaminated groundwater. *J. Environ. Health* **48:**122–127.
3. **Craun, G. F.** 1986. Statistics of waterborne outbreaks in the U.S. (1920–1980), p. 73–159. *In* G. F. Craun (ed.), *Waterborne Diseases in the United States.* CRC Press, Boca Raton, Fla.
4. **Craun, G. F.** 1986. Recent statistics of waterborne disease outbreaks (1981–1983), p. 43–69. *In* G. F. Craun (ed.), *Waterborne Diseases in the United States.* CRC Press, Boca Raton, Fla.
5. **Craun, G. F.** 1990. *Methods for Investigation and Prevention of Waterborne Disease Outbreaks.* EPA-600/1-90/005a. U.S. Environmental Protection Agency Office of Research and Development, Cincinnati.
6. **Craun, G. F.** 1991. Causes of waterborne outbreaks in the United States. *Water Sci. Technol.* **24:**17–20.
7. **DiNovo, F., and M. Jaffe.** 1984. *Local Groundwater Protection, Midwest Region.* American Planning Association, Chicago.
8. **Goyal, S. M., B. H. Keswick, and C. P. Gerba.** 1984. Viruses in groundwater beneath sewage irrigated cropland. *Water Res.* **18:**299–302.
9. **Herwaldt, B. L., G. F. Craun, S. L. Stokes, and D. D. Juranek.** 1992. Outbreaks of waterborne disease in the United States: 1989–1990. *J. Am. Water Works Assoc.* **84:** 129–135.
10. **Kaplan, J. E., G. W. Gary, R. C. Baron, W. Singh, L. B. Schonberger, R. Feldman, and H. Greenberg.** 1982. Epidemiology of Norwalk gastroenteritis and the role of Norwalk virus in outbreaks of acute nonbacterial gastroenteritis. *Ann. Intern. Med.* **96:**756–761.
11. **Keswick, B. H., and C. P. Gerba.** 1980. Viruses in groundwater. *Environ. Sci. Technol.* **14:**1290–1297.
12. **Keswick, B. H., T. K. Satterwhite, P. C. Johnson, H. L. DuPont, S. L. Secor, J. A. Bitsura, G. W. Gary, and J. C. Hoff.** 1985. Inactivation of Norwalk virus in drinking water by chlorine. *Appl. Environ. Microbiol.* **50:**261–264.
13. **Lippy, E. C., and S. C. Waltrip.** 1984. Waterborne disease outbreaks—1946–1980: a thirty-five-year perspective. *J. Am. Water Works Assoc.* **76:**60–67.
14. **Moore, A. C., B. L. Herwaldt, G. F. Craun, R. L. Calderon, A. K. Highsmith, and D. D. Juranek.** 1994. Waterborne disease in the United States, 1991 and 1992. *J. Am. Water Works Assoc.* **86:**87–99.
15. **Olson, E., and D. Cameron.** 1995. *The Dirty Little Secret about Our Drinking Water.* Natural Resources Defense Council, Washington, D.C.
16. **Park, N.-S., T. N. Blanford, and P. S. Huyakorn.** 1991. *VIRALT—a Model for Simulating Viral Transport in Ground Water, Documentation and User's Guide, Version 2.0.* HydroGeoLogic, Inc., Herndon, Va.
17. **Regli, S., J. B. Rose, C. N. Haas, and C. P. Gerba.** 1991. Modeling the risk from *Giardia* and viruses in drinking water. *J. Am. Water Works Assoc.* **83:**76–84.
18. **Schaub, S. A., E. P. Meier, J. P. Kolmer, and C. A. Sorber.** 1975. *Land Application of Wastewater: the Fate of Viruses, Bacteria, and Heavy Metals at a Rapid Infiltration Site.* Report TR 7504. U.S. Army Medical Research and Development Command, Washington, D.C.
19. **Todd, D. K.** 1980. *Groundwater Hydrology.* John Wiley & Sons, New York.
20. **U.S. Environmental Protection Agency.** 1985. National primary drinking water regulations; synthetic organic

chemicals, inorganic chemicals, and microorganisms. *Fed. Regist.* **50:**46936–47022.

21. **U.S. Environmental Protection Agency.** 1987. *Guidelines for Delineation of Wellhead Protection Areas.* EPA440/6-87-010. U.S. Environmental Protection Agency Office of Ground-Water Protection, Washington, D.C.

22. **U.S. Environmental Protection Agency.** 1990. *Strawman Rule for Groundwater Disinfection.* Office of Drinking Water, Washington, D.C.

23. **U.S. Environmental Protection Agency.** 1992. *Draft Ground-Water Disinfection Rule.* Office of Ground Water and Drinking Water, Washington, D.C.

24. **van der Heijde, P., and M. S. Beljin.** 1988. *Model Assessment for Delineating Wellhead Protection Areas.* EPA440/6-88-002. U. S. Environmental Protection Agency Office of Ground-Water Protection, Washington, D.C.

25. **Vaughn, J. M., and E. F. Landry.** 1977. *Data Report: an Assessment of the Occurrence of Human Viruses in Long Island Aquatic Systems.* Brookhaven National Laboratory, Department of Energy and Environment, Upton, N.Y.

26. **Vaughn, J. M., and E. F. Landry.** 1978. The occurrence of human enteroviruses in a Long Island groundwater aquifer recharged with tertiary wastewater effluents, p. 233–245. *In* H. L. McKim (ed.), *State of Knowledge in Land Treatment of Wastewater,* vol. 2. U.S. Government Printing Office, Washington, D.C.

27. **Vaughn, J. M., E. F. Landry, L. J. Baranosky, C. A. Beckwith, M. C. Dahl, and N. C. Delihas.** 1978. Survey of human virus occurrence in wastewater-recharged groundwater on Long Island. *Appl. Environ. Microbiol.* **36:**47–51.

28. **Yates, M. V.** 1990. *The Use of Models for Granting Variances from Mandatory Disinfection of Ground Water Used as a Public Water Supply.* EPA/600/2-90/010. U.S. Environmental Protection Agency, Ada, Okla.

29. **Yates, M. V.** 1995. Field evaluation of the GWDR's natural disinfection criteria. *J. Am. Water Works Assoc.* **87:**76–85.

30. **Yates, M. V., and S. R. Yates.** 1989. Septic tank setback distances: a way to minimize virus contamination of drinking water. *Ground Water* **27:**202–208.

31. **Yates, M. V., S. R. Yates, A. W. Warrick, and C. P. Gerba.** 1986. Use of geostatistics to predict virus decay rates for determination of septic tank setback distances. *Appl. Environ. Microbiol.* **52:**479–483.

AEROBIOLOGY VII

VOLUME EDITOR
LINDA D. STETZENBACH

Introduction to Aerobiology

LINDA D. STETZENBACH

67

The study of airborne microorganisms has expanded from the traditional arena of transmission of disease via the respiratory route to include not only human pathogens but also plant pathogens, opportunistic and nonpathogenic organisms, and aerosolized microbial by-products. Airborne culturable and nonculturable bacteria, saprophytic fungi, free-living parasites, viruses, and algae that may result in adverse health effects or environmental impact are now studied within the field of aerobiology. Fragments of microbial agents (e.g., cell wall fragments, flagella, and genetic material) and microbial metabolites (e.g., volatile organic compounds, endotoxin, and mycotoxins) are also of interest.

This chapter introduces the field of aerobiology, discusses indoor and anthropogenic outdoor sources of airborne microorganisms that affect human health and the environment, and reviews the association of bioaerosols and indoor environmental quality issues. Subsequent chapters in this volume are focused on specific topics concerning airborne microbial agents.

BIOAEROSOLS

A collection of airborne biological particles is called a bioaerosol. Generally, bioaerosols are generated as polydispersed droplets or particles of different sizes ranging from 0.5 to 30 μm in diameter (76). Air serves as a mode of transport for the dispersal of bioaerosols from one location to another. The composition and concentration of the microorganisms comprising the bioaerosol vary with the source and the dispersal in the air until deposition (83, 103).

A variety of sources, including fresh and marine surface waters, soil, and plants, are found in the environment. Bioaerosols generated from water sources during splash and wave action are usually formed with a thin layer of moisture surrounding the microorganisms and often consist of aggregates of several organisms (120). Microorganisms released into the air from soil are often single units or associated with particles. These aerosols are characterized by the type of soil (44) and the presence of vegetation (80) and may contain particulate matter serving as "rafts" for microorganisms (78).

The transport and ultimate settling of a bioaerosol are affected by its physical properties and the environmental parameters that it encounters while airborne. The size, den-

sity, and shape of the droplets or particles comprise the most important physical characteristics, while the magnitude of air currents, relative humidity, and temperature are the significant environmental parameters (77, 83, 97). Temperature and relative humidity also contribute to the influx of airborne microorganisms. Increased concentrations of some fungal spores (e.g., *Nigrospora* and *Cladosporium* spp.) in outdoor air (22) and increased numbers of bacteria released from plant surfaces (80) have been associated with high temperatures and low relative humidity. These factors also influence the survival of airborne microorganisms and affect their ability to colonize on surfaces after deposition. While harsh environmental conditions tend to decrease the numbers of viable airborne organisms, there is variability in survival between groups of microorganisms and within genera. In general, fungal spores, enteric viruses, and amoebic cysts are somewhat resistant to environmental stresses encountered during transport through the air. Bacteria and algae are more susceptible, although bacterial endospores (e.g., *Bacillus* spp.) are quite resistant (67). Although it is generally believed that microbial numbers do not increase during transport, Dimmick et al. (31) reported one doubling of airborne bacterial cell numbers in a rotating-drum aerosol chamber with saturated-humidity conditions and tryptone added to the cell suspension prior to aerosolization. A detailed discussion of the factors affecting the survival and transport of bioaerosols is presented in chapter 69; information on the fate and transport of agricultural pathogens is presented in chapter 75.

Inhalation, ingestion, and dermal contact are routes of human exposure for airborne microorganisms, but inhalation is the predominant route resulting in adverse health effects. The average human inhales approximately 10 m^3 of air per day (83). Large airborne particles are lodged in the upper respiratory tract (nose and nasopharynx) (124). Generally, particles <6 μm in diameter are transported to the lung, but the greatest retention in the alveoli is of 1- to 2-μm particles (99, 103, 105). The airborne transmission of *Legionella pneumophila*, *Mycobacterium tuberculosis*, and newly recognized pathogens with an airborne transmission route (e.g., hantavirus) is known to cause severe infections. Asthma, hypersensitivity pneumonitis, and other respiratory illnesses are also associated with exposure to bioaerosols (Table 1) (41, 88, 112, 123).

Bioaerosols also may result in the spread of plant diseases,

TABLE 1 Adverse effects associated with exposure to airborne microorganisms

Microbial agent	Adverse effect(s)	
	Human health	Environmental
Algae	Allergic reactions	Odor problems
Bacteria	Hypersensitivity pneumonitis, infections, mucous membrane irritation	Deterioration of building materials, loss of agricultural productivity (crop and livestock diseases), odor problems
Endotoxin	Cough, headache, fever, malaise, muscle aches, nausea, respiratory distress	None reported
Fungi	Allergic reactions, asthma, dermal irritation, hypersensitivity pneumonitis, infections, mucous membrane irritation	Deterioration of building materials, loss of agricultural productivity (crop and livestock diseases), odor problems
Mycotoxin	Headache, muscle problems, neurologic disorders, respiratory distress, toxicosis	Loss of agricultural productivity (disease in livestock)
Protozoa	Encephalitis, hypersensitivity reactions, infections	May afford protection from biocide treatment to other microorganisms, loss of agricultural productivity (disease in livestock)
Virus	Infections	Loss of agricultural productivity (crop and livestock diseases)

loss of agricultural productivity, and deterioration of building materials (Table 1).

MAN-MADE SOURCES OF BIOAEROSOLS IN OUTDOOR ENVIRONMENTS

Numerous anthropogenic activities, especially agricultural practices and wastewater treatment processes, serve as the origin of bioaerosols in outdoor environments (Table 2). Increases in airborne concentrations of microorganisms during harvesting operations have been documented (17, 75), and several investigators have reported the presence of airborne bacteria and viruses resulting from wastewater treatment (34, 82, 99), sanitary landfill operations (98), and reuse-water irrigation practices (6, 113).

The release of biotechnology products (e.g., genetically engineered microorganisms and microbial pest control agents) developed to enhance agricultural productivity, mineral recovery, oil spill cleanup, and toxic waste disposal can also be a source of airborne microorganisms. The application of biotechnology products by aerosol mist increases the possibility for transport of the microbial product from the target crop to surrounding areas (81). Aerosolized, genetically modified cells have been monitored in a barn setting (85) and at biotechnology-based fermentors (62).

BIOAEROSOLS AND INDOOR ENVIRONMENTAL QUALITY

Deterioration of building materials, offensive odors, and adverse human health effects are associated with microbial contamination of indoor environments. Residences, offices, schools, health care facilities, enclosed agricultural structures (e.g., barns and crop storage areas), pharmaceutical and industrial facilities, and food processing plants are among the indoor environments where airborne microorga-

nisms have been studied (Table 2). Sources and reservoirs of microorganisms, including building materials and furnishings, pets, plants, and air conditioning systems, are present within these settings (78). Wallboard, ceiling tiles, carpeting and vinyl flooring, painted surfaces, upholstery and drapery, wallpaper, plastics, wood, cement, and brick are examples of building materials and components that can serve as sites for microbial colonization (Table 2).

Bacteria and algae generally grow in areas with standing water such as air-handling system components (e.g., water spray humidification systems and condensate pans) and sites where water intrusion or leaking (e.g., flooding and condensation) has occurred. However, with the exception of viruses, which require a living host cell for replication, microorganisms will colonize virtually any surface where there is sufficient moisture. Fungi, which have lower water activity (a_w) requirements than other microorganisms, tend to colonize a wide variety of building materials (48). *Penicillium* spp. and *Aspergillus versicolor* are primary colonizers that are often isolated on wallpaper and drier margins of wetted walls, while *Cladosporium* spp. proliferate as secondary colonizers. The a_ws for primary and secondary colonizers have been listed as <0.80 and 0.80 to 0.90, respectively (48). *Ulocladium* spp. and *Stachybotrys atra* are found when the a_w becomes >0.90 (48).

Air conditioning systems are often cited as amplification sites, and these systems have been associated with the dispersal of contaminants indoors (9, 38, 41), but naturally ventilated buildings are also affected by bioaerosols, as organisms can be transported via drafts through open windows and doors (78). When microbial amplification occurs, the indoor environment becomes a source of bioaerosol exposure to the occupants. Reynolds et al. (100) reported that indoor sources of bioaerosols may be significant when differences are noted between indoor and outdoor concentrations and/or populations. This exposure may be important because

TABLE 2 Anthropogenic sources and amplification sites of airborne microorganisms

Category	Site — Facility or activity	Microbial agent(s)	Concn/unit vol or area[a]	Reference(s)
Agricultural	Animal facilities	Bacteria	10^3–10^5 CFU/m^3	28
			10^4 CFU/m^3	114
		Endotoxin	100–500 pg/m^3	89
		Fungi	10^3 CFU/m^3	114
			10^6–10^7 CFU/m^3	32
	Composting	Actinomycetes—thermophilic	NR	74
		Bacteria—legionellae	10^3–10^6 CFU/g	56
		Fungi—thermophiles	10^2–10^6 CFU/g	111
	Crop processing	Fungi	150–69,000 CFU/m^3	60
	Farming, harvesting, baling, grain storage	Bacteria	46–6,500 CFU/m^3	79
			237–18,520 CFU/m^3	75
		Fungi	10^3–10^5 CFU/m^3	95
			10^3–10^7 CFU/m^3	17
			10^6–10^8 CFU/m^3	75
			10^8–10^9 CFU/m^3	32
			10^9 CFU/m^3	84
	Rural/suburban communities	Algae	102–228 CFU/m^3	15
			100–800 CFU/m^3	51
Air-handling systems	Cooling/heating systems	Bacteria		
		Bacilli	Air: 80 CFU/m^3	55
			Surfaces: 15–453 CFU/m^2	55
		Heterotrophic	Air: 140–1,649 CFU/m^2	55
			Surfaces: 10^2–10^6 CFU/m^2	55
		Pseudomonads	Air: 30–70 CFU/m^3	55
			Surfaces: 45–10^6 CFU/m^2	55
		Staphylococcus spp.	Air: 80–130 CFU/m^3	55
			Surfaces: 50 CFU/m^2	55
		Endotoxin	NR	41
		Fungi		
		Mesophilic	10^3 CFU/m^3	11
		Thermophilic	NR	18
		T. vulgaris	NR	7, 38
			10^2 CFU/m^3	33
		Parasites—amoebic cysts	NR	33, 118
	Cooling towers, water spray systems	Bacteria—*Legionella* spp.	NR	45
		Fungi		
		Aureobasidium pullulans	NR	123
		M. faeni	NR	38
	Humidifiers	Bacteria		
		L. pneumophila	10^2–10^4 CFU/ml	125
		Pseudomonas spp.	480–12,000 CFU/h	25
		Fungi—*T. vulgaris*	NR	116
Building materials and furnishings	Ceiling tile, insulation, painted surfaces, wallpaper	Fungi		
		Aspergillus sp.	200 CFU/m^3	69
		Cladosporium sp.	576 CFU/m^3	69
		Penicillium sp.	3,013 CFU/m^3	69
		S. atra	NR	27, 93
	Carpet	Bacteria	10^4 CFU/in^2	5
		Fungi		69
		Mesophilic	682 CFU/m^3	69
		Alternaria sp.	989 CFU/m^3	69
		S. atra	NR	
	Flooring (not carpet)	Bacteria	NR	37
	Garden soil	Bacteria—legionellae	NR	37
	Hot water heaters, hot water systems	Bacteria		
		Legionella sp.	2–12 CFU/m^3	13
		L. pneumophila	NR	53

(Table continued on next page)

TABLE 2 Anthropogenic sources and amplification sites of airborne microorganisms (*Continued*)

Site		Microbial agent(s)	Concn/unit vol or area[a]	Reference(s)
Category	Facility or activity			
	House dust	Algae	NR	8–10
		Fungi—not specified	10^3–10^6 CFU/g	30
			NR	49
		Alternaria sp.	1,133 CFU/g	121
			1.5×10^4 CFU/g	87
		Cladosporium sp.	1,700 CFU/g	121
			5.6×10^4 CFU/g	87
		Penicillium sp.	4,500 CFU/g	121
			8.1×10^4 CFU/g	87
		Rhodotorula sp.	1,166 CFU/g	121
		Yeasts—not specified	1,366 CFU/g	121
			5.9×10^5 CFU/g	87
Building occupants	House plants (greenhouses)	Fungi	6,881–62,590 CFU/m^3	19
	Humans (skin fragments)	Bacteria	400–2,790 CFU/m^3	82
	Mortuary (necropsies)	Bacteria		
		Coliforms	3–531 CFU/m^3	92
		Gram-positive cocci	20–163 CFU/m^3	92
Healthcare facilities	Operating room	Bacteria—not specified	10^2 CFU/m^3	115
			10^2–10^5 particles/m^3	108
	Patient rooms	Virus—varicella-zoster	NR	106
Industrial	Water system	Bacteria—*L. pneumophila*	10^4–10^6 CFU/ml	125
	Cellulose, wood chip factory, sawmill	Fungi—not specified	10^4–10^8 CFU/m^3	12
			10^8 CFU/m^3	32
		A. fumigatus	880–200,000 CFU/m^3	68
		Penicillium sp.	12–88,000 CFU/m^3	68
	Clean rooms	Viable particles	2.5–320 CFU/m^3	36
	Food processing, storage	Fungi—not specified	10^4 CFU/m^3	12
		Xerophiles	10^4 CFU/m^3	71
		Mycotoxin	NR	71, 73
	Manufacturing (cotton mill, tobacco processing, equipment cleaning)	Bacteria		
		Gram-negative rods	3,190–3,490 CFU/m^3	82
		Total	10–7,500 CFU/m^3	117
		Endotoxin	0.02–3.6 μg/m^3	117
		Fungi—not specified	10^5 CFU/m^3	71
		Total spores	600–10,600 spores/m^3	57
		A. fumigatus	20–410 CFU/m^3	57
	Packing boxes	Fungi—*A. pullulans*	NR	123
Wastewater treatment	Activated sludge	Bacteria—not specified	440 CFU/m^2	64
			10–10^5 CFU/m^3	72
			10^4 CFU/m^3	98
		Coliforms	0.27–5.17 CFU/m^3	34
		Enterics	10^4 CFU	99
		Endotoxin	1–350 ng/m^3	
		Fungi		
		Mesophilic	<DL to 7,220 CFU/m^3	61
		Thermophilic	<DL to 193 CFU/m^3	61
	Aeration tanks	Bacteria—not specified	120 CFU/m^2/s	64
			10^2–10^3 CFU/m^3	26
			86–7,143 CFU/m^3	14
		Fecal streptococci	10^2 CFU/m^3	26
		S. faecalis	<DL to 10^2 CFU/m^3	26
		Virus		
		Animal	ND	14
		Coliphage	0–9 PFU/m^3	14

(*Table continued on next page*)

TABLE 2 Anthropogenic sources and amplification sites of airborne microorganisms (*Continued*)

Site		Microbial agent(s)	Concn/unit vol or area[a]	Refer-ence(s)
Category	Facility or activity			
	Effluent irrigation	Bacteria—not specified	1,410 CFU/m^3	6
		E. coli	21–1,130 CFU/m^3	113
		Fecal coliforms	200 CFU/m^3	6
		Fecal streptococci	700 CFU/m^3	6
		Total coliforms	500 CFU/m^3	6
		Virus—enteric	NR	6, 35
	Not identified	Bacteria—gram-negative rods	9,900 CFU/m^3	82
	Sludge compost	Bacteria—coliforms	4–1,640 CFU/m^3	61
	Trickling filter	Bacteria		
		Coliforms	389–19,737 CFU/m^3	1
		Total	51–1,676 CFU/m^3	1

[a] NR, not reported; <DL, less than detection limit; ND, none detected. One inch = 2.54 cm.

people spend approximately 22 h/day in indoor environments (110).

Occupants of enclosed spaces are also a major source of bacterial bioaerosols, with human-to-human transmission often occurring in high-density indoor environments such as correctional facilities (54) and military training centers (16). Coughing and loud talking are reported to release approximately 10^4 droplets, while sneezing releases approximately 10^6 droplets (83). Microorganisms are also dispersed from surfaces as a result of activity by the occupants (21, 49, 83, 100).

Legionnaires' disease and tuberculosis are examples of defined illnesses resulting from exposure to a specific airborne microbial agent. The discovery of *L. pneumophila* as the cause of the outbreak of Legionnaires' disease in Philadelphia in 1976 increased the awareness of diseases caused by bioaerosols (45, 94, 122). A detailed discussion of *Legionella* spp. and Legionnaires' disease is presented in chapter 72. Similarly, the increased reporting of tuberculosis in both developing and industrialized countries has prompted renewed interest in the genus *Mycobacterium* and its airborne transmission. *M. tuberculosis* is spread via bioaerosols from an infected person and is recognized as a significant public health concern because of the low infectious dose (63). Nontuberculosis mycobacteria have also been associated with respiratory illness (24, 66). A discussion of airborne mycobacteria is presented in chapter 73. These illnesses are in contrast to sick building syndrome (SBS), which defines the often random, vague complaints reported by building occupants, such as nasal, eye, and mucous membrane irritation, lethargy, headache, rashes, dry skin, and shortness of breath (40, 70). The possible causes of SBS cited include comfort factors (temperature and relative humidity) (58), environmental tobacco smoke (45), chemical off-gassing from building materials (e.g., adhesives, paints, and particleboard) (70, 119), and microorganisms. Fungi and bacteria are most often cited as possible microbial contaminants leading to SBS complaints, particularly when moisture is present or water damage has occurred (29, 39, 40, 52).

MICROBIAL AGENTS

The environmental effects and human health complaints resulting from airborne microorganisms have renewed interest in a wide variety of microorganisms. Airborne fungi have been the focus of much concern because of their ability to cause serious respiratory infection and to elicit allergic reactions. Although few fungi actually cause infection in healthy individuals, immunosuppressed host defenses resulting from therapy (e.g., antibiotics, steroids, and drugs) or the presence of another disease-causing agent may increase the likelihood of fungal infection (90). More commonly, fungi initiate disease in the absence of infection by serving as aeroallergens resulting in asthma, allergic rhinitis, and respiratory distress. These maladies are often associated with microbe-contaminated environments (7, 45, 50, 71, 87). Numerous surveys have been conducted to determine exposure to airborne fungi resulting from a variety of outdoor and indoor sources (Table 2).

Adverse health effects may also result from exposure to mycotoxins which are produced by some fungi. Exposure of agricultural workers to mycotoxin-containing dust (71) and industrial exposure to mycotoxins which are produced by *Aspergillus fumigatus* (73) have been reported. Toxigenic *Penicillium* and mycotoxin-containing spores of *S. atra* have been found in contaminated buildings and have been associated with reports of illness (27). The trichothecene mycotoxin associated with *S. atra* spores (93, 96, 109) can produce toxicosis with low numbers of spores, causing severe health problems (109). β,1-3-D-Glucan, a polyglucose structure in fungal cell walls, has been associated with inflammatory response (101), and volatile products of fungal metabolism associated with characteristic moldy odors (e.g., geosmins) can result in occupant complaints (86, 104). An extensive review of fungal bioaerosols and mycotoxins is found in chapter 70; additional information on agricultural fungal pathogens is presented in chapter 75.

Airborne thermophilic actinomycetes (e.g., *Thermoactinomyces vulgaris* and *Micropolyspora faeni*) have been implicated in numerous cases of hypersensitivity pneumonitis and other allergic reactions (74). Exposure to thermophilic actinomycetes, whose optimal growth temperature is ≥40°C, may occur near compost or municipal landfills and during handling of decomposing organic matter in agricultural facilities (e.g., milling and sugar cane bagasse) (74, 98). Indoor exposure to thermophilic actinomycetes and resulting disease have been reported (7), implicating a heating system humidifier (38) and home humidifiers (116).

Surveys have reported the presence of airborne bacteria in indoor and outdoor environments (Table 2). However, the significance of these populations in offices, schools, residences, and outdoor environments has not been determined, in part because gram-positive cocci and *Bacillus* spp. are common airborne organisms and are isolated in the absence of adverse health effects. *Micrococcus* and *Staphylococcus* spp. commonly disseminate from nasal and oral surfaces, skin, clothing, and hair of building occupants (36). The finding of higher ratios of these bacteria isolated from indoor air than from outdoor air has been used as an indication of high occupancy rate, poor ventilation, or inadequate building maintenance (43), but additional research in this area is needed. While airborne transmission of bacteria and other microorganisms in health care facilities can cause nosocomial infections (107), hospitals adhere to strict infection control procedures and routinely monitor for airborne microorganisms. However, increasing numbers of elderly people are living in nursing homes and home health care facilities where maintenance and decontamination practices may not be as strictly followed. The risk of exposure to microorganisms that may act as opportunistic pathogens is, therefore, increased.

Exposure to airborne bacterial cells may also result in the inhalation of endotoxin, a lipopolysaccharide found in the cell walls of gram-negative bacteria. Exposure to endotoxin may result in fever, cough, headache, and respiratory impairment. Airborne endotoxin may be a major cause of illness in enclosed agricultural settings such as silage facilities, poultry processing houses, and cotton mills (23, 102). Chilled-water spray humidification systems used in textile manufacturing facilities have also been implicated as sources of aerosolized endotoxin associated with cases of humidifier fever and hypersensitivity pneumonitis (117). A review of airborne endotoxin is presented in chapter 71.

Although viruses do not replicate outside a susceptible host cell, they are readily transported through the air. Therefore, virus bioaerosols are a notable environmental concern. Numerous human viruses are transmitted via droplets and spread by the respiratory route from one person to another in the indoor environment (124). Enteric virus bioaerosols are produced at sewage treatment facilities (2), and aerosol transport of pathogenic viruses from infected plants has also been documented (47). An extensive review of airborne viruses is presented in chapter 74.

Blowing dust was reported as the source of 62 genera of airborne algae (15), and eutrophic lakes are cited as contributors to outdoor airborne algal concentrations. Dust (9) and aeration of aquariums (78) have been proposed as possible indoor sources. While exposure to algal extracts has been associated with adverse human health effects (8–10), the extent of allergic reactions due to algal bioaerosols has not been fully investigated. Additional research is needed to determine the environmental and human health effects of airborne algae.

Free-living amoebae (e.g., *Naegleria fowleri* and *Acanthamoeba* spp.) are indigenous to soil and water and can be aerosolized from natural and artificially heated waters such as power plant discharges, lakes, and hot springs. Cooling system waters have also been reported as potential sources of aerosolized amoebae (83). Humidifiers and ceiling dust were cited as an indoor source of airborne amoebic antigen (33). Although severe health effects can be elicited by exposure to these organisms (20), insufficient information is currently available on airborne protozoa.

Universal to the field of aerobiology is the need for mea-surement methods for detection and identification of the microorganism(s) of interest. Classical microbiological methods for measurement of airborne microorganisms rely on culture and/or microscopic assay using forced airflow sampling (59). Culture-based methods require appropriate conditions for growth of culturable organisms, but airborne microorganisms are stressed during transport and collection and may not respond to incubation conditions in the laboratory (65). Total count procedures, using direct microscopic enumeration, are tedious and often fail to distinguish genera and/or species. Methods to measure microbial contamination other than direct or culturable counts have recently been developed. These include assays for β,1-3-D-glucan (42, 46), ergosterol (42, 87, 91), and endotoxin (117). In addition, the use of PCR has recently been shown to be an effective means to detect and identify airborne bacteria (3, 4) and viruses (106). Chapter 68 details traditional sampling and analysis methods and discusses recently developed assays using biochemical and biotechnology-based techniques.

In summary, interest in the populations of airborne microorganisms in agricultural and industrial settings, health care facilities, residences, offices, and classroom environments has increased in recent years. The potential for adverse environmental and human health effects resulting from indoor and outdoor bioaerosol exposure has prompted renewed interest in aerobiology, and research activity in this rapidly expanding area of environmental microbiology has expanded. The following chapters in this section focus on specific topics related to bioaerosols, and the reader is directed to consult those chapters for additional information.

REFERENCES

1. **Adams, A. P., and J. C. Spendlove.** 1970. Coliform aerosols emitted by sewage treatment plants. *Science* **169:** 1218–1220.
2. **Adams, D. J., J. C. Spendlove, R. S. Spendlove, and B. B. Barnett.** 1982. Aerosol stability of infectious and potentially infectious reovirus particles. *Appl. Environ. Microbiol.* **44:**903–908.
3. **Alvarez, A. J., M. P. Buttner, and L. D. Stetzenbach.** 1995. PCR for bioaerosol monitoring: sensitivity and environmental interference. *Appl. Environ. Microbiol.* **61:** 3639–3644.
4. **Alvarez, A. J., M. P. Buttner, G. A. Toranzos, E. A. Dvorsky, A. Toro, T. B. Heikes, L. E. Mertikas, and L. D. Stetzenbach.** 1994. The use of solid-phase polymerase chain reaction for the enhanced detection of airborne microorganisms. *Appl. Environ. Microbiol.* **60:**374–376.
5. **Anderson, R. L.** 1969. Biological evaluation of carpeting. *Appl. Microbiol.* **18:**180–186.
6. **Applebaum, J., N. Guttman-Bass, M. Lugten, B. Teltsch, B. Fattal, and H. I. Shuval.** 1984. Dispersion of aerosolized enteric viruses and bacteria by sprinkler irrigation with wastewater. *Monogr. Virol.* **15:**193–201.
7. **Banaszak, E. F., J. Barboriak, J. Fink, G. Scanlon, D. P. Schlueter, A. Sosman, W. Thiede, and G. Unger.** 1974. Epidemiologic studies relating thermophilic fungi and hypersensitivity lung syndrome. *Am. Rev. Respir. Dis.* **110:**585–591.
8. **Bernstein, I. L., and R. S. Safferman.** 1966. Sensitivity of skin and bronchial mucosa to green algae. *J. Allergy* **38:**166–173.
9. **Bernstein, I. L., and R. S. Safferman.** 1973. Clinical sensitivity to green algae demonstrated by nasal challenge and in vitro tests of immediate hypersensitivity. *J. Allergy Clin. Immunol.* **51:**22–28.
10. **Bernstein, I. L., G. V. Villacorte, and R. S. Safferman.**

1969. Immunologic responses of experimental animals to green algae. *J. Allergy* **43**:191–199.

11. **Bernstein, R. S., W. G. Sorenson, D. Garabrant, C. Reaux, and R. D. Treitman.** 1983. Exposures to respirable, airborne *Penicillium* from a contaminated ventilation system: clinical, environmental and epidemiological aspects. *Am. Ind. Hyg. Assoc. J.* **44**:161–169.

12. **Blomquist, G., G. Strom, and L.-H. Stromquist.** 1984. Sampling of high concentrations of airborne fungi. *Scand. J. Work Environ. Health* **10**:109–113.

13. **Bollin, G. E., J. F. Plouffe, M. F. Para, and B. Hackman.** 1985. Aerosols containing *Legionella pneumophila* generated by shower heads and hot-water faucets. *Appl. Environ. Microbiol.* **50**:1128–1131.

14. **Brenner, K. P., P. V. Scarpino, and C. S. Clark.** 1988. Animal viruses, coliphage, and bacteria in aerosols and wastewater at a spray irrigation site. *Appl. Environ. Microbiol.* **54**:409–415.

15. **Brown, R. M., D. A. Larson, and H. C. Bold.** 1964. Airborne algae: their abundance and heterogeneity. *Science* **143**:583–585.

16. **Brundage, J. F., R. M. Scott, W. M. Lednar, D. W. Smith, and R. N. Miller.** 1988. Building-associated risk of febrile acute respiratory diseases in army trainees. *JAMA* **259**:2108–2112.

17. **Burge, H. A., M. L. Muilenberg, and J. A. Chapman.** 1991. Crop plants as a source of fungus spores of medical importance, p. 222–236. *In* J. H. Andrews and S. S. Hirano (ed.), *Microbial Ecology of Leaves.* Springer-Verlag, New York.

18. **Burge, H. A., W. R. Solomon, and J. R. Boise.** 1980. Microbial prevalence in domestic humidifiers. *Appl. Environ. Microbiol.* **39**:840–844.

19. **Burge, H. A., W. R. Solomon, and M. L. Muilenberg.** 1982. Evaluation of indoor plantings as allergen exposure sources. *Allergy Clin. Immunol.* **70**:101–108.

20. **Burrell, R.** 1991. Microbiological agents as health risks in indoor air. *Environ. Health Perspec.* **95**:29–34.

21. **Buttner, M. P., and L. D. Stetzenbach.** 1993. Monitoring of fungal spores in an experimental indoor environment to evaluate sampling methods and the effects of human activity on air sampling. *Appl. Environ. Microbiol.* **59**:219–226.

22. **Cammack, R. H.** 1955. Seasonal changes in three common constituents of the air spores of southern Nigeria. *Nature* (London) **176**:1270–1272.

23. **Castellan, R. M., S. A. Olenchock, J. L. Hankinson, P. D. Millner, J. B. Cooke, C. K. Bragg, H. H. Perkins, and R. R. Jacobs.** 1984. Acute bronchoconstriction induced by cotton dust: dose-related responses to endotoxin and other dust factors. *Ann. Intern. Med.* **101**:157–163.

24. **Contreras, M. A., O. T. Cheung, D. E. Sanders, and R. S. Goldstein.** 1988. Pulmonary infection with nontuberculous mycobacteria. *Am. Rev. Respir. Dis.* **137**:149–152.

25. **Couvelli, H. D., J. Kleeman, J. E. Martin, W. L. Landau, and R. L. Hughes.** 1973. Bacterial emission from both vapor and aerosol humidifiers. *Am. Rev. Respir. Dis.* **108**:698–701.

26. **Crawford, G. V., and P. H. Jones.** 1979. Sampling and differentiation techniques for airborne organisms emitted from wastewater. *Water Res.* **13**:393–399.

27. **Croft, W. S., B. B. Jarvis, and C. S. Yatawara.** 1986. Airborne outbreak of trichothecene toxicosis. *Atmos. Environ.* **20**:549–552.

28. **Curtis, S. E., R. K. Balsbaugh, and J. G. Drummond.** 1978. Comparison of Andersen eight-stage and two-stage viable air samplers. *Appl. Environ. Microbiol.* **35**:208–209.

29. **Dales, R. E., R. Burnett, and H. Awanenburg.** 1991.

Adverse health effects among adults exposed to home dampness and molds. *Am. Rev. Respir. Dis.* **143**:505–509.

30. **Davies, R. R.** 1960. Viable moulds in house dust. *Trans. Br. Mycol. Soc.* **43**:617–630.

31. **Dimmick, R. L., H. Wolochow, and M. A. Chatigny.** 1979. Evidence that bacteria can form new cells in airborne particles. *Appl. Environ. Microbiol.* **37**:924–927.

32. **Eduard, W., J. Lacey, K. Karlsson, U. Palmgren, G. Strom, and G. Blomquist.** 1990. Evaluation of methods for enumerating microorganisms in filter samples from highly contaminated occupational environments. *Am. Ind. Hyg. Assoc. J.* **51**:427–436.

33. **Edwards, J. H.** 1980. Microbial and immunological investigations and remedial action after an outbreak of humidifier fever. *Br. J. Ind. Med.* **37**:55–62.

34. **Fannin, K. F., S. C. Vana, and W. Jakubowski.** 1985. Effect of an activated sludge wastewater treatment plant on ambient air densities of aerosols containing bacteria and viruses. *Appl. Environ. Microbiol.* **49**:1191–1196.

35. **Fattal, B., E. Katzenelson, N. Guttman-Bass, and A. Sadovski.** 1984. Relative survival rates of enteric viruses and bacterial indicators in water, soil, and air. *Monogr. Virol.* **15**:184–192.

36. **Favero, M. S., J. R. Puleo, J. H. Marshall, and G. S. Oxborrow.** 1966. Comparative levels and types of microbial contamination detected in industrial clean rooms. *Appl. Microbiol.* **14**:539–551.

37. **Finch, J. E., J. Prince, and M. Hawksworth.** 1978. A bacteriological survey of the domestic environment. *J. Appl. Bacteriol.* **45**:357–364.

38. **Fink, J. N., E. F. Banazak, W. H. Thiede, and J. J. Barboriak.** 1971. Interstitial pneumonitis due to hypersensitivity to an organism contaminating a heating system. *Ann. Intern. Med.* **74**:80–83.

39. **Finnegan, M. J., and C. A. C. Pickering.** 1986. Building related illness. *Clin. Allergy* **16**:389–405.

40. **Finnegan, M. J., C. A. C. Pickering, and P. S. Burge.** 1984. The sick building syndrome: prevalence studies. *Br. Med. J.* **289**:1573–1575.

41. **Flaherty, D. K., F. H. Deck, J. Cooper, K. Bishop, P. A. Winzenburger, L. R. Smith, L. Bynum, and W. B. Witmer.** 1984. Bacterial endotoxin isolated from a water spray air humidification system as a putative agent of occupational-related lung disease. *Infect. Immun.* **43**:206–212.

42. **Fogelmark, B., H. Goto, K. Yuasa, B. Marchat, and R. Rylander.** 1992. Acute pulmonary toxicity of inhaled β-1,3-glucan and endotoxin. *Agents Actions* **35**:50–56.

43. **Gallup, J. M., J. Zanolli, and L. Olson.** 1993. Airborne bacterial exposure: preliminary results of volumetric studies performed in office buildings, schools, and homes in Southern California, p. 167–170. *In* P. Kalliokoski, M. Jantunen, and O. Seppanen (ed.), *Indoor Air '93, Proceedings of the 6th International Conference on Indoor Air Quality and Climate, Helsinki, Finland.* Indoor Air '93, Helsinki, Finland.

44. **Gillette, D. A., and I. H. Blifford, Jr.** 1972. Measurement of aerosol size distribution and vertical fluxes of aerosols on land subject to wind erosion. *J. Appl. Meteorol.* **11**:977–987.

45. **Gold, D. R.** 1992. Indoor air pollution. *Clin. Chest Med.* **13**:215–229.

46. **Goto, H., K. Yuasa, and R. Rylander.** 1994. (1-3)-β-D-Glucan in indoor air, its measurement and in vitro activity. *Am. J. Med.* **25**:81–83.

47. **Graham, D. C., C. E. Quinn, and L. F. Bradley.** 1977. Quantitative studies on the generation of aerosols of *Erwinia carotovora* var. *atroseptica* by simulated raindrop impaction on blackleg-infected potato stems. *J. Appl. Bacteriol.* **43**:413–424.

48. **Grant, C., C. A. Hunter, B. Flannigan, and A. F. Bravery.** 1989. The moisture requirements of moulds isolated from domestic dwellings. *Int. Biodeterior.* **25:**259–284.

49. **Gravesen, S.** 1978. Identification and prevalence of culturable mesophilic microfungi in house dust from 100 Danish homes. *Allergy* **33:**268–272.

50. **Gravesen, S.** 1979. Fungi as a cause of allergic disease. *Allergy* **34:**135–154.

51. **Gregory, P. H., E. D. Hamilton, and T. Sreeramulu.** 1955. Occurrence of the alga *Gloeocapsa* in air. *Nature* (London) **176:**1270.

52. **Harrison J., C. A. C. Pickering, E. B. Faragher, P. K. C. Austwick, S. A. Little, and L. Lawton.** 1992. An investigation of the relationship between microbial and particulate indoor air pollution and the sick building syndrome. *Respir. Med.* **86:**225–235.

53. **Helms, C. M., R. M. Massanari, R. Zeitler, S. Streed, M. J. R. Gilchrist, N. Hall, W. J. Hausler, Jr., J. Sywassink, W. Johnson, L. Wintermeyer, and W. J. Hierholzer.** 1983. Legionnaires' disease associated with a hospital water system: a cluster of 24 nosocomial cases. *Ann. Intern. Med.* **99:**172–178.

54. **Hoge, C. W., M. R. Reichler, E. A. Dominguez, J. C. Bremer, T. D. Mastro, K. A. Hendricks, D. M. Musher, J. A. Elliott, R. R. Facklam, and R. F. Breiman.** 1994. An epidemic of pneumococcal disease in an overcrowded, inadequately ventilated jail. *N. Engl. J. Med.* **331:** 643–648.

55. **Hugenholtz, P., and J. A. Fuerst.** 1992. Heterotrophic bacteria in an air-handling system. *Appl. Environ. Microbiol.* **58:**3914–3920.

56. **Hughes, M. S., and T. W. Steele.** 1994. Occurrence and distribution of *Legionella* species in composted plant materials. *Appl. Environ. Microbiol.* **60:**2003–2005.

57. **Huuskonen, M. S., K. Husman, J. Jarvisalo, O. Korhonen, M. Kotimaa, T. Kuusela, H. Nordman, A. Zitting, and R. Mantyjarvi.** 1984. Extrinsic allergic alveolitis in the tobacco industry. *Br. J. Ind. Med.* **41:**77–83.

58. **Jaakkola, J. J. K., and O. P. Heinonen.** 1989. Sick building syndrome, sensation of dryness and thermal comfort in relation to room temperature in an office building: need for individual control of temperature. *Environ. Int.* **15:** 163–168.

59. **Jensen, P. A., B. Lighthart, A. J. Mohr, and B. T. Shaffer.** 1994. Instrumentation used with microbial bioaerosols, p. 226–284. *In* B. Lighthart and A. J. Mohr (ed.), *Atmospheric Microbial Aerosols. Theory and Applications.* Chapman and Hall, New York.

60. **Jensen, P. A., W. F. Todd, M. E. Hart, R. L. Mickelsen, and D. M. O'Brien.** 1993. Evaluation and control of worker exposure to fungi in a beet sugar refinery. *Am. Ind. Hyg. Assoc. J.* **54:**742–748.

61. **Jones, B. L., and J. T. Cookson.** 1983. Natural atmospheric microbial conditions in a typical suburban area. *Appl. Environ. Microbiol.* **45:**919–934.

62. **Juozaitis, A., Y. Huang, K. Willeke, J. Donnelly, S. Kalatoor, A. Leeson, and R. Wyza.** 1994. Dispersion of respirable aerosols in a fermentor and their removal in an exhaust system. *Appl. Occup. Environ. Hyg.* **9:**552–559.

63. **Kaufmann, S. H. E., and J. D. A. van Embden.** 1993. Tuberculosis: a neglected disease strikes back. *Trends Microbiol.* **1:**2–5.

64. **Kenline, P. A., and P. V. Scarpino.** 1972. Bacterial air pollution from sewage treatment plants. *Am. Ind. Hyg. Assoc. J.* **May:**346–352.

65. **Kingston, D.** 1971. Selective media in air sampling: a review. *J. Appl. Bacteriol.* **34:**221–232.

66. **Kirschner, R. A., Jr., B. C. Parker, and J. O. Falkinham III.** 1992. Epidemiology of infection by nontuberculous mycobacteria. *Am. Rev. Respir. Dis.* **145:**271–275.

67. **Knudsen, G. R., and H. W. Spurr, Jr.** 1987. Field persistence and efficacy of five bacterial preparations for control of peanut leaf spot. *Plant Dis.* **71:**442–445.

68. **Kotimaa, M. H.** 1990. Occupational exposure to fungal and actinomycete spores during handling of wood chips. *Grana* **29:**153–156.

69. **Kozak, P. P., J. Gallup, L. H. Cummins, and S. A. Gillman.** 1980. Currently available methods for home mold surveys. II. Examples of problem homes surveyed. *Ann. Allergy* **45:**167–176.

70. **Kreiss, K., and M. J. Hodgson.** 1984. Building associated epidemics, p. 87–106. *In* P. J. Walsh (ed.), *Indoor Air Quality.* CRC Press, Boca Raton, Fla.

71. **Lacey, J., and B. Crook.** 1988. Fungal and actinomycete spores as pollutants of the workplace and occupational allergens. *Am. Occup. Hyg.* **32:**515–533.

72. **Laitinen, S., J. Kangas, M. Kotimaa, J. Leosovuori, P. J. Martikainen, A. Nevalainen, R. Sarantila, and K. Husman.** 1994. Workers' exposure to airborne bacteria and endotoxins at industrial wastewater treatment plants. *Am. Ind. Hyg. Assoc. J.* **55:**1055–1060.

73. **Land, C. J., K. Hult, R. Fuchs, S. Hagelberg, and H. Lundstrom.** 1987. Tremorgenic mycotoxins from *Aspergillus fumigatus* as a possible occupational health problem in sawmills. *Appl. Environ. Microbiol.* **53:**787–790.

74. **Land, G., M. R. McGinnis, J. Staneck, and A. Gatson.** 1991. Aerobic pathogenic *Actinomycetales*, p. 340–359. *In* A. Balows, W. J. Hausler, Jr., K. L. Herrmann, H. D. Isenberg, and H. J. Shadomy (ed.), *Manual of Clinical Microbiology,* 5th ed. American Society for Microbiology, Washington, D.C.

75. **Lighthart, B.** 1984. Microbial aerosols: estimated contribution of combine harvesting to an airshed. *Appl. Environ. Microbiol.* **47:**430–432.

76. **Lighthart, B.** 1994. Physics of microbial bioaerosols, p. 5–27. *In* B. Lighthart and A. J. Mohr (ed.), *Atmospheric Microbial Aerosols, Theory and Applications.* Chapman and Hall, New York.

77. **Lighthart, B., and A. J. Mohr.** 1987. Estimating downwind concentrations of viable airborne microorganisms in dynamic atmospheric conditions. *Appl. Environ. Microbiol.* **53:**1580–1583.

78. **Lighthart, B., and L. D. Stetzenbach.** 1994. Distribution of microbial bioaerosols, p. 68–98. *In* B. Lighthart and A. J. Mohr (ed.), *Atmospheric Microbial Aerosols, Theory and Applications.* Chapman and Hall, New York.

79. **Lindemann, J., H. A. Constantinidou, W. R. Barchet, and C. D. Upper.** 1982. Plants as sources of airborne bacteria, including ice nucleation-active bacteria. *Appl. Environ. Microbiol.* **44:**1059–1063.

80. **Lindemann, J., and C. D. Upper.** 1985. Aerial dispersal of epiphytic bacteria over bean plants. *Appl. Environ. Microbiol.* **50:**1229–1232.

81. **Lindow, S. E., G. R. Knudsen, R. J. Seidler, M. V. Walter, V. W. Lambou, P. S. Amy, V. Prince, and S. C. Hern.** 1988. Aerial dispersal and epiphytic survival of *Pseudomonas syringae* during a pretest for the release of genetically engineered strains into the environment. *Appl. Environ. Microbiol.* **54:**1557–1563.

82. **Lundholm, I. M.** 1982. Comparison of methods for quantitative determinations of airborne bacteria and evaluation of total viable counts. *Appl. Environ. Microbiol.* **44:** 179–183.

83. **Lynch, J. M., and N. J. Poole.** 1979. Aerial dispersal and the development of microbial communities, p. 140–170. *In* J. M. Lynch and N. J. Poole (ed.), *Microbial Ecology: a Conceptual Approach.* John Wiley & Sons, New York.

84. **Malmberg, P., A. Rask-Andersen, M. Lundholm, and U. Palmgren.** 1990. Can spores from molds and actinomy-

cetes cause organic dust toxic syndrome reaction? *Am. J. Ind. Med.* **17:**109–110.

85. **Marshall, B., P. Flynn, D. Kamely, and S. B. Levy.** 1988. Survival of *Escherichia coli* with and without ColE1::Tn5 after aerosol dispersal in a laboratory and a farm environment. *Appl. Environ. Microbiol.* **54:**1776–1783.

86. **Mattheis, J. P., and R. G. Roberts.** 1992. Identification of geosmin as a volatile metabolite of *Penicillium expansum. Appl. Environ. Microbiol.* **58:**3170–3172.

87. **Miller, J. D., A. M. Laflamme, Y. Sobol, P. Lafontaine, and R. Greenhalgh.** 1988. Fungi and fungal products in some Canadian houses. *Int. Biodeterior.* **24:**103–120.

88. **Miller, M. M., R. Patterson, J. N. Fink, and M. Roberts.** 1976. Chronic hypersensitivity lung disease with recurrent episodes of hypersensitivity pneumonitis due to a contaminated central humidifier. *Clin. Allergy* **6:**451–462.

89. **Milton, D. K., R. J. Gere, H. A. Feldman, and I. A. Greaves.** 1990. Endotoxin measurement: aerosol sampling and application of a new limulus method. *Am. Ind. Hyg. Assoc. J.* **51:**331–337.

90. **Mishra, S. K., L. Ajello, D. G. Ahearn, H. A. Burge, V. P. Kurup, D. L. Pierson, D. L. Price, R. S. Samson, R. S. Sandhu, B. Shelton, R. B. Simmons, and K. F. Switzer.** 1992. Environmental mycology and its importance to public health. *J. Med. Vet. Mycol.* **30:**287–305.

91. **Newell, S. Y., T. L. Arsuffi, and R. D. Fallon.** 1988. Fundamental procedures for determining ergosterol content of decaying plant material by liquid chromatography. *Appl. Environ. Microbiol.* **54:**1876–1879.

92. **Newson, S. W. B., C. Rowlands, J. Matthers, and C. J. Elliot.** 1983. Aerosols in the mortuary. *J. Clin. Pathol.* **36:**127–132.

93. **Nikulin, M., A. Pasanen, S. Berg, and E. Hintikka.** 1994. *Stachybotrys atra* growth and toxin production in some building materials and fodder under different relative humidities. *Appl. Environ. Microbiol.* **60:**3421–3424.

94. **Osterholm, M. T., T. D. Y. Chin, D. O. Osborne, H. B. Dull, A. G. Dean, D. W. Fraser, P. S. Hayes, and W. N. Hall.** 1983. A 1957 outbreak of legionnaires' disease associated with a meat packing plant. *Am. J. Epidemiol.* **117:**60–67.

95. **Pasanen, A., P. Kalliokoski, P. Pasanen, T. Salmi, and A. Tossavainen.** 1989. Fungi carried from farmers' work into farm homes. *Am. Ind. Hyg. Assoc. J.* **50:**631–633.

96. **Pasanen, A., M. Nikulin, M. Tuomainen, S. Berg, P. Parikka, and E. Hintikka.** 1993. Laboratory experiments on membrane filter sampling of airborne mycotoxins produced by *Stachybotrys atra* corda. *Atmos. Environ.* **27A:**9–13.

97. **Pedgley, D. E.** 1991. Aerobiology: the atmosphere as a source and sink for microbes, p. 43–59. *In* J. H. Andrews and S. S. Hirano (ed.), *Microbial Ecology of Leaves.* Springer-Verlag, New York.

98. **Rahkonen, P., M. Ettala, and I. Loikkanen.** 1987. Working conditions and hygiene at sanitary landfills in Finland. *Ann. Occup. Hyg.* **31:**505–513.

99. **Randall, C. W., and J. O. Ledbetter.** 1966. Bacterial air pollution from activated sludge units. *Am. Ind. Hyg. Assoc. J.* Nov.–Dec:506–519.

100. **Reynolds, S. J., A. J. Streifel, and C. E. McJilton.** 1990. Elevated airborne concentrations of fungi in residential and office environments. *Am. Ind. Hyg. Assoc. J.* **51:**601–604.

101. **Rylander, R., and B. Fogelmark.** 1994. Inflammatory responses by inhalation of endotoxin and (1-3)-β-D-glucan. *Am. J. Ind. Med.* **25:**101–102.

102. **Rylander, R., and J. Vesterlund.** 1982. Airborne endo-

toxins in various occupational environments. *Prog. Clin. Biol. Res.* **93:**399–409.

103. **Salem, H., and D. E. Gardner.** Health aspects of bioaerosols, p. 304–330. *In* B. Lighthart and A. J. Mohr (ed.), *Atmospheric Microbial Aerosols, Theory and Applications.* Chapman and Hall, New York.

104. **Samson, R. A.** 1985. Occurrence of moulds in modern living and working environments. *Eur. J. Epidemiol.* **1:**54–61.

105. **Sattar, S. A., and M. K. Ijaz.** 1987. Spread of viral infections by aerosols. *Crit. Rev. Environ. Control* **17:**89–131.

106. **Sawyer, M. H., C. J. Chamberlain, Y. N. Wu, N. Aintablian, and M. R. Wallace.** 1994. Detection of varicella-zoster virus DNA in air samples from hospital rooms. *J. Infect. Dis.* **169:**91–94.

107. **Schaal, K. P.** 1991. Medical and microbiological problems arising from airborne infection in hospitals. *J. Hosp. Infect.* **18:**451–459.

108. **Seal, D. V., and R. P. Clark.** 1990. Electronic particle counting for evaluating the quality of air in operating theatres: a potential basis for standards? *J. Appl. Bacteriol.* **68:**225–230.

109. **Sorenson, W. G., D. G. Frazer, B. B. Jarvis, J. Simpson, and V. A. Robinson.** 1987. Trichothecene mycotoxins in aerosolized conidia of *Stachybotrys atra. Appl. Environ. Microbiol.* **53:**1370–1375.

110. **Spangler, J. D., and K. Sexton.** 1983. Indoor air pollution: a public health perspective. *Science* **221:**9–17.

111. **Straatsma, G., R. A. Samson, T. W. Olijnsma, H. J. M. Opden Camp, J. P. G. Gerrits, and L. J. L. D. van Griensven.** 1994. Ecology of thermophilic fungi in mushroom compost with emphasis on *Scytalidium thermophilum* and growth stimulation of *Agaricus bisporus* mycelium. *Appl. Environ. Microbiol.* **60:**454–458.

112. **Strachan, D. P., B. Flannigan, E. M. McCabe, and F. McGarry.** 1990. Quantification of airborne moulds in the homes of children with and without wheeze. *Thorax* **45:**382–387.

113. **Teltsch, B., and E. Katzenelson.** 1978. Airborne enteric bacteria and viruses from spray irrigation with wastewater. *Appl. Environ. Microbiol.* **35:**290–296.

114. **Thorne, P. S., M. S. Kiekhaefer, P. Whitten, and K. J. Donham.** 1992. Comparison of bioaerosol sampling methods in barns housing swine. *Appl. Environ. Microbiol.* **58:**2543–2551.

115. **Tjade, O. H., and I. Gabor.** 1980. Evaluation of airborne operating room bacteria with a Biap slit sampler. *J. Hyg.* (Cambridge) **84:**37–40.

116. **Tourville, D. R., W. I. Weiss, P. T. Wertlake, and G. M. Leudemann.** 1972. Hypersensitivity pneumonitis due to contamination of home humidifier. *J. Allergy Clin. Immunol.* **49:**245–251.

117. **Walters, M., D. Milton, L. Larsson, and T. Ford.** 1994. Airborne environmental endotoxin: a cross validation of sampling and analysis techniques. *Appl. Environ. Microbiol.* **60:**996–1005.

118. **Warhurst, D.** 1977. Humidifier fever: amoebae and allergic lung disease. *Thorax* **32:**653–663.

119. **Weschler, C. J., A. T. Hodgson, and J. D. Woolsey.** 1992. Indoor chemistry: ozone, volatile organic compounds, and carpets. *Environ. Sci. Technol.* **26:**2371–2377.

120. **Wickman, H. H.** 1994. Deposition, adhesion, and release of bioaerosols, p. 99–165. *In* B. Lighthart and A. J. Mohr (ed.), *Atmospheric Microbial Aerosols, Theory and Applications.* Chapman and Hall, New York.

121. **Wickman, M., S. Gravesen, S. L. Nordvall, G. Pershagen, and J. Sundell.** 1992. Indoor viable dust-bound microfungi in relation to residential characteristics, living

habits, and symptoms in atopic and control children. *J. Allergy Clin. Immunol.* **89:**752–759.

122. **Winn, W. C., Jr.** 1988. Legionnaires disease: historical perspective. *Clin. Microbiol.* **1:**60–81.

123. **Woodard, E. D., B. Friedlander, R. J. Lesher W. Font, R. Kinsey, and F. T. Hearne.** 1988. Outbreak of hypersensitivity pneumonitis in an industrial setting. JAMA **259:**1965–1969.

124. **Zeterberg, J. M.** 1973. A review of respiratory virology and the spread of virulent and possibly antigenic viruses via air conditioning systems. *Ann. Allergy* **31:** 228–234.

125. **Zuravleff, J. J., V. L. Yu, J. W. Shonnard, J. D. Rihs, and M. Best.** 1983. *Legionella pneumophila* contamination of a hospital humidifier. *Am. Rev. Respir. Dis.* **128:** 657–661.

Sampling and Analysis of Airborne Microorganisms

MARK P. BUTTNER, KLAUS WILLEKE, AND SERGEY A. GRINSHPUN

Microbiologists have confronted the challenges of sampling and analysis of airborne microorganisms since the beginning of the 20th century. Today, the concentrations and compositions of airborne microorganisms are of interest in diverse areas such as agricultural and industrial settings, medicine, home and office environments, and military research. In all of these applications, the term "bioaerosol" is generally used to refer to airborne biological particles, such as bacterial cells, fungal spores, viruses, or pollen grains, and to their fragments and by-products. A wide variety of bioaerosol sampling and analysis methods have been used, and new methods are being developed. However, several problems remain to be solved. For instance, no single sampling method is suitable for the collection and analysis of all types of bioaerosols, and no standardized protocols are currently available. Therefore, data between studies are often difficult to compare because of differences in sampler design, collection time, airflow rate, and analysis method. In addition, human exposure limits have not been established for bioaerosols because of the lack of exposure, dose, and response data. This problem complicates the use of sampling results for risk assessment.

It is important for the investigator to carefully consider the objectives of sampling before any samples are taken. After determining what information is desired, an appropriate sampling and analysis method can be selected and incorporated into the monitoring design. The purpose of this chapter is to present various bioaerosol sampling and analysis methods in order to facilitate the selection of instrumentation and techniques. The principles of bioaerosol sampling are presented, followed by a review of sampling methods currently available, including the results of performance evaluations of the various sampler types. Equipment calibration and sampling considerations such as collection times and the number of samples are discussed. Analysis methods are addressed in the second section of the chapter, beginning with traditional culture and total count methods and concluding with some recently developed biochemical and molecular techniques.

BIOAEROSOL SAMPLING

The objective of bioaerosol sampling is the efficient removal and collection of biological particles from the air in a man-

ner which does not affect the ability to detect the organisms (e.g., alteration in culturability or biological integrity). This ability is dependent on the physical and biological characteristics of the organisms and on the physical features of the sampling instrument (21, 93). While a microorganism is airborne, its motion is governed by the same laws of physics as apply to biologically inert particles. The three principal collection methods used in quantitative bioaerosol sampling are impaction, impingement, and filtration.

Methods of Collection

Impaction

The impaction method separates particles from the airstream by utilizing the inertia of the particles to force their deposition onto a solid or semisolid collection surface. The collection surface is usually an agar medium for culture-based analysis or an adhesive-coated surface that can be analyzed microscopically. The impaction process depends on the inertial properties of the particle, such as size, density, and velocity, and on the physical parameters of the impactor, such as the inlet nozzle dimensions and the airflow pathway (69). The sampled air exits the impactor's inlet nozzle as an air jet directed at the collection surface, and particles with sufficient inertia impact. The lower-inertia particles remain airborne with the airflow (Fig. 1A). Centrifugal impaction also uses inertial forces to separate the particle from the airstream, but in a radial geometry (Fig. 1B).

Liquid Impingement

Liquid impingement is similar to impaction in that the inertial force of the particle is the principal force removing it from the air. However, the collection medium is a liquid, usually a dilute buffer solution, and the collected microorganisms move around freely in the bubbling liquid (Fig. 1C). As a result, aggregates of cells may be broken apart and particles remaining in the airstream may diffuse to the surface of a bubble and be transferred to the collection buffer in this manner. The collection of bioaerosol particles in a liquid medium allows division of the sample and the potential application of several analysis methods, as described below.

Filtration

Filtration achieves the separation of particles from the airstream by passage of the air through a porous medium, usu-

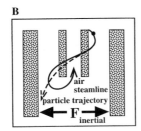
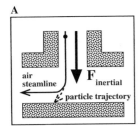
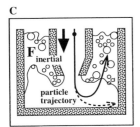
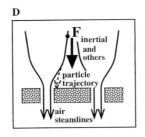

FIGURE 1 Mechanisms of collection utilized in bioaerosol sampling. A, solid plate impaction; B, centrifugal impaction; C, liquid impingement; D, filtration; $F_{inertial}$, inertial force. (Adapted from reference 68 with kind permission from Elsevier Science Ltd., The Boulevard, Langford Lane, Kidlington OX5 1GB, United Kingdom.)

ally a membrane filter. Collection of particles depends on their physical properties (size, shape, and density), the filter pore size, and the airflow rate (51). Inertial forces and other mechanisms such as interception, diffusion, and electrostatic attraction result in the collection of particles on the surface of the filter (Fig. 1D). Simultaneous action of all of these forces also removes particles smaller than the pore size of the filter (43).

Gravity

Gravity, or depositional sampling, is a nonquantitative collection method in which an agar medium is exposed to the environment and airborne organisms are collected primarily by gravity. This sampling method is often used because it is inexpensive and easily performed. However, collection of airborne microorganisms by this method is affected by the size and shape of the particles and by the motion of the surrounding air (69). As a result, large particles are more likely to be deposited onto the collection surface (13). This can lead to misrepresentation of the prevalence of airborne microorganisms and the exclusion of smaller particles from collection (80). In addition, the airborne concentration of the microorganisms cannot be determined by gravity sampling because the volume of air from which the particles originate is unknown. Gravity sampling has been compared with various methods that pass a known volume of air to the collection medium. The results show that the airborne concentrations derived from gravity sampling are not qualitatively or quantitatively accurate and do not compare favorably with those obtained by other sampling methods (16, 77, 78, 80).

Sampler Types

There are a wide variety of commercially available bioaerosol samplers (21, 43, 69). The selection of a sampler depends on a number of factors, such as sampler performance, expected bioaerosol concentration, and the analysis method.

Prior to initiating an investigation, the researcher should have an understanding of the specific objectives of air monitoring as well as the limitations of the various sampling methods. Although standardized sampling protocols are not available at the present time, guidelines have been published to assist in the selection of bioaerosol sampling protocols (21, 43). Some of the more widely used bioaerosol sampling methods are listed in Table 1 and discussed in the following sections.

Impactor Sampler

Impaction is the most commonly used method of collection for airborne microorganisms, and a variety of impactor samplers are commercially available (Table 1). They differ by the number of nozzles, the shape of the nozzles, and the number of stages. If air is drawn through a single nozzle, the shape of the nozzle is usually rectangular and the impactor is referred to as a slit sampler. If there are several nozzles, usually circular in shape, the plate with the impaction nozzles resembles a sieve and the impactor is sometimes referred to as a sieve sampler. If there are several stages with successively smaller nozzles, the sampler is referred to as a cascade impactor.

The Andersen six-stage impactor sampler (Graseby Andersen, Smyrna, Ga.) consists of six stages with decreasing nozzle diameter such that successive stages collect progressively smaller particles (3). Thus, the six-stage sampler measures the culturable bioaerosol concentrations in specific particle size ranges. One-stage and two-stage models of the sampler are also available. The Surface Air System (SAS; International PBI, Milan, Italy; distributed by Spiral Biotech, Inc., Bethesda, Md.) and Burkard (Burkard Manufacturing Co., Ltd., Rickmansworth, Hertfordshire, England) portable air samplers are battery-powered one-stage impactors which utilize agar-filled plates as the collection medium. The Reuter Centrifugal Sampler (RCS) and RCS Plus (Biotest Diagnostics Corp., Denville, N.J.) are portable, battery-powered samplers which centrifugally project the microorganisms onto agar strips. The portable samplers do not require external vacuum pumps and electrical outlets but are available only as single-stage devices.

Slit impactors deposit the bioaerosol onto an agar surface for the estimation of viable cells (e.g., Casella MK-II, distributed by BGI, Inc., Waltham, Mass.; Mattson/Garvin air sampler, from Barramundi Corp., Homosassa, Fla.; New Brunswick slit-to-agar sampler, from New Brunswick Scientific Co., Inc., Edison, N.J.) or onto an adhesive-coated surface for the microscopic enumeration of the collected particles, usually fungal spores or pollen grains (Burkard spore traps; Burkard Manufacturing Co., Ltd.). Many slit impactors have a moving collection surface to provide temporal discrimination of the bioaerosol concentration.

Liquid Impinger Sampler

Liquid impinger samplers are commonly used for the retrieval of bioaerosol particles over a wide range of airborne particle concentrations (Table 1). For analysis, the liquid sample can be concentrated by filtration or diluted by liquid addition, depending on the concentration of microorganisms. Several culture media can be inoculated with aliquots of the collection medium for the determination of groups of microorganisms with different culture requirements. The liquid samples may also be analyzed by biochemical, immunological, and molecular biological assays to detect the presence of specific microorganisms, culturable or nonculturable. The AGI-30 all-glass impinger sampler (Ace Glass, Inc.,

TABLE 1 General characteristics of several bioaerosol samplers

Sampler	Collection medium	Airflow rate (liters/min)	Sample analysis method(s)	Remarks
Impaction				
Andersen viable impactors, 1-, 2-, or 6-stage	Agar; 100-mm petri dishes	28.3	Culture	Particle size discrimination (2- and 6-stage models); vacuum pump required; counts corrected for multiple impaction
Biotest RCS and RCS Plus	Agar; plastic strips	40 (RCS) 50 (RCS Plus)	Culture	Portable, battery operated
Burkard spore traps, 24-h, 7-day, and personal samplers	Adhesive-coated surface, tape, or glass slide	10	Microscopic	Determination of total fungal spores and pollen
Burkard portable air sampler	Agar; 100-mm petri dishes	10–20	Culture	Portable, battery operated; counts corrected for multiple impaction
Casella MK-II	Agar petri dishes; 100 mm (small inlet), 150 mm (large inlet)	30 (small inlet), 700 (large inlet)	Culture	Time discrimination up to 6 min; limited availability
Mattson/Garvin 220 and P-320	Agar; 150-mm petri dishes	28.3	Culture	Time discrimination up to 1 h
New Brunswick slit-to-agar, STA-101, -203, and -303	Agar; 150-mm petri dishes	50	Culture	Time discrimination up to 1 h; vacuum pump required for some models
SAS, Super 90, Compact and High Flow	Agar; 55-mm contact plates, 84-mm contact plates	90 (Compact, Super 90), 180 (High Flow)	Culture	Portable, battery operated; counts corrected for multiple impaction
Impingement				
All-glass impingers (AGI-30 and AGI-4)	Liquid	12.5	Culture, microscopic, endotoxin assay, gene probes, PCR, immunoassay	High and low bioaerosol concentrations; vacuum pump required
Burkard multistage liquid impinger	Liquid	20	Culture, microscopic, endotoxin asay, gene probes, PCR, immunoassay	High and low bioaerosol concentrations; particle size discrimination; vacuum pump required
Filtration				
25-, 37-, or 47-mm filter cassettes	Filter, membrane	1–50	Microscopic, endotoxin assay, gene probes, PCR, immunoassay, culture (spores)	Loss of culturable vegetative cells; portable, useful for personal monitoring; vacuum pump required

Vineland, N.J.) is a widely used liquid impinger sampler that has a curved inlet tube designed to simulate the nasal passage, making this sampler useful for studying the respiratory infection potential of bioaerosols (23, 35, 43). For other applications, the inlet tube is washed with a known volume of collection fluid to recover nonrespirable airborne particles. The AGI-30 has an impaction distance of 30 mm from the jet to the bottom of the sampler. The AGI-4 model features a shorter distance of 4 mm to improve particle collection efficiency over the AGI-30. However, added sam-

pling stress may result from impaction against the glass bottom of the sampler, resulting in a loss of viability of cells. The Burkard multistage liquid impinger (Burkard Manufacturing Co., Ltd.) is a stainless steel sampler which collects particles in three size fractions: >10, 4 to 10, and <4 μm.

Filtration Sampling

The collection of airborne microorganisms onto a filter material is used in bioaerosol monitoring because of its simplicity, low cost, and versatility (Table 1). Air samples are usu-

TABLE 2 Calculated and reported d_{50} values for several bioaerosol samplers

Sampler	d_{50} (μm)		Performance evaluation studies
	Calculated (reference)	Reported	
Impaction			3, 7, 9, 12, 15, 16, 33, 44, 45, 48, 52, 82
Andersen 6-stage viable impactor			
Stage 1	6.24 (43), 6.61 (68)	7.0	
Stage 2	4.21	4.7	
Stage 3	2.86	3.3	
Stage 4	1.84	2.1	
Stage 5	0.94	1.1	
Stage 6	0.58	0.65	
Andersen 2-stage viable impactor			33, 44, 48, 87, 94
Stage 0	6.28	8.0	
Stage 1	0.83	0.95	
Andersen 1-stage viable impactor	0.58	0.65	44, 45, 87
Biotest RCS	7.5	3.8	44, 49, 60, 73
RCS Plus		0.82	
Burkard spore traps			
24-h, 7-day			12, 16, 81
Standard nozzle	3.70		
High-efficiency nozzle	2.17		
Personal sampler	2.52		
Burkard portable air sampler	4.18		
Casella MK-II, small inlet	0.67		9
Mattson/Garvin	0.53		44, 73
SAS			16, 44
Compact	1.97	2.0	
High flow	1.52 (43), 1.45 (68)	2.0	
Impingement			
All-glass impinger (AGI-30)	0.30		3, 7, 15, 44, 48, 52, 87, 92
Burkard multistage liquid impinger			26, 94
Stage 1		10	
Stage 2		4	
Stage 3			

ally collected onto 25-, 37-, or 47-mm-diameter filter membranes housed in disposable plastic cassettes, which are available from a variety of manufacturers (e.g., Gelman Sciences, Ann Arbor, Mich.; Millipore Corp., Bedford, Mass.; Nuclepore, Corning Costar Corp., Cambridge, Mass.; Poretics Corp., Livermore, Calif.). Polycarbonate, cellulose mixed ester, or polyvinyl chloride filter material may be used, depending on the nature of the bioaerosol and the method of sample analysis (43, 51). Filter membrane pore sizes range from 0.01 to 10 μm, with air sample flow rates from 1 to 50 liters/min. Filtration sampling is adaptable to a variety of assays (Table 1), but loss of viability of vegetative cells may occur, presumably as a result of desiccation stress during sampling (44, 58, 87).

Sampler Performance

Bioaerosol sampler performance has been reviewed (21, 35, 43, 48, 68, 86), and numerous laboratory and field studies comparing the utility of various sampling methods have been performed (Table 2). Data from these studies are often difficult to compare because of differences in samplers, the length of sampling time, the volume of air sampled, the sample analysis method, and the characteristics of the bioaerosol being measured. While no single sampler type currently available is ideally suited for the collection and analysis of all types of bioaerosols, consideration of the theoretical

aspects of air sampling and the experimental results of performance studies may facilitate the selection of the appropriate sampling method for a particular monitoring situation.

The performance of bioaerosol samplers can be divided into physical and biological components. Physical parameters include inlet sampling efficiency and collection efficiency, whereas biological sampling efficiency reflects primarily the effect of sampling on the culturable state of the microorganisms. Inlet sampling efficiency refers to the ability of the sampler inlet to extract particles from the ambient environment without bias due to the size, shape, or density of the particles. The inlet characteristics of several bioaerosol samplers have been calculated for different types of bioaerosol particles sampled under various conditions (35). Depending on the external wind direction and velocity relative to the inlet geometry and flow characteristics, increased or, more commonly, decreased particle concentration measurements may be obtained relative to the true concentration in the environment (10).

Collection efficiency is the ability of the sampler to remove particles from the airstream and transfer them to the collection medium. In an impactor sampler, the physical characteristics of the impaction nozzle(s) and the airflow rate are used to calculate the cut size, or d_{50}, of the impaction stage. The d_{50} refers literally to the particle diameter at which 50% of the particles are collected. However, because of the sharp cutoff characteristics of impactor samplers, the d_{50} is generally considered to be the particle diameter above which all particles are collected (62, 68). Theoretical d_{50} values have been calculated for several bioaerosol samplers (43, 68) and are shown in Table 2. For efficient collection, it is important to choose an impactor whose d_{50} is below the mean size of the microorganism being sampled. For membrane filters, the collection efficiency is approximately 100% for particles larger than the pore size (56). In an inertial device, such as the impactor, one usually refers to the "aerodynamic" diameter of the particle, which is the diameter of a unit-density sphere that has the same gravitational settling velocity as the particle in question (6). A nonspherical microorganism is thus described by a single dimension.

Biological sampling efficiencies differ among bioaerosol sampler types. Ideally, each sampler should collect all airborne microorganisms without altering the culturability or the biological integrity required for the detection and/or quantification of the microorganisms. Biological effects of sampling are difficult to assess because of the heterogeneity of bioaerosols with respect to particle size, biological composition, and environmental factors. Airborne microorganisms are subjected to a variety of environmental stressors, such as UV radiation, chemical pollutants, desiccation, and temperature extremes; consequently, many organisms may be in a nonculturable state (24). In addition, the stress of impaction may injure the collected microorganisms, depending on their physiological characteristics (84). The majority of analysis methods rely on culturing the collected microorganisms. Therefore, sampling stresses which can result in a reduction in the culturability of airborne microorganisms are important in assessing the overall bioaerosol sampler performance. Filtration sampling, for example, while highly efficient for the collection of airborne microorganisms, has the disadvantage of viability losses of vegetative cells, presumably as a result of desiccation (44, 58, 87). Therefore, filtration sampling in combination with culture analysis is generally used when bioaerosol concentrations are very high and sampling times are short. Filtration sampling is also used to monitor desiccation-resistant forms such as fungal spores or bacterial endospores, or in combination with a total count method of analysis (70).

The stress of sampling may also result in loss of culturability when agar or liquid collection media are used. However, it has been observed that the recovery of culturable cells can be improved by using certain culture media (14, 67, 89) or by the addition of certain compounds to the collection medium, such as osmoprotectants, which aid in the resuscitation of stressed or damaged cells (63). Impaction into different media also permits the differentiation of metabolic from structural injury (84). The length of collection time, discussed in the following section, also plays a major role in the efficacy of air sampling for the retrieval of culturable microorganisms.

Most of the bioaerosol sampler performance evaluations have not distinguished between physical and biological parameters in assessing efficiency, focusing instead on an overall measure of sampling efficacy. Although there is no single sampling method which is appropriate for all bioaerosols, the Andersen six-stage viable impactor and the AGI-30 sampler have been suggested as reference methods (21), and many sampler efficiency studies have included these samplers in side-by-side comparisons (Table 2). The principal advantage of the agar impactor samplers is collection of organisms directly onto the culture medium; however, one of the disadvantages of this sampling method is the noncult-

urable component of the bioaerosol, which is not measured and can be a significant percentage of the total composition (70). In addition, the overall performance of agar impactors can be adversely affected by overloading when bioaerosol concentrations are high, resulting in agar plates or strips which contain colonies that overlap and are too numerous to count (19). Other problems with agar impactor samplers which can produce erratic results and underestimation of the true bioaerosol concentration are (i) particle bounce (e.g., particles may rebound off the collection surface or another particle and reenter the airstream [69]), (ii) aggregation or clumping of microorganisms (e.g., multiple organisms impact at the same place on the agar surface and are enumerated as a single colony), and (iii) electrostatic forces (e.g., biological particles are attracted to the plastic rims or other surfaces of the collection device (3). In an impinger, microorganisms are collected by impaction onto a liquid-wetted surface and are also subjected to culturability losses through sampling stress (3, 88). In addition, the collected organisms may become reaerosolized in a manner similar to the aerosolization of microorganisms from bubbling liquids such as whirlpools and fermentors (5, 40, 47).

Collection Time

Sample collection time is an integral part of the bioaerosol sampling design, and guidelines for the selection of optimal sampling times for various bioaerosol samplers have been published (2, 69). Parameters which must be considered are the expected bioaerosol concentration, the quantitation range of the sampler, and the effect of sampling stress on the overall collection efficiency. For each sample, the sampling period must be sufficiently long to obtain a representative sample of the airborne microorganisms present without exceeding the upper quantitation limit of the sampler or causing losses in culturability of airborne organisms.

The selection of sample collection time is complicated by the fact that bioaerosol concentrations may vary greatly over time, often by several orders of magnitude within the same environment. Thus, air sample periods of short duration provide only a brief temporal and spatial glimpse of the environment, and several samples may be required to determine the average bioaerosol concentrations. Furthermore, the sampling environment may range from a relatively low bioaerosol concentration ($\leq 10^2$ CFU/m^3) in clean rooms and hospital operating rooms (29) to exceptionally high concentrations (10^5 to 10^{10} CFU/m^3) found in certain industrial and agricultural settings (28, 50).

While the ambient bioaerosol concentration is an unknown quantity, the quantitation range of a bioaerosol sampler can be determined if the airflow rate is known and the collection time is set (15). The lower quantitation limit (LQL) is obtained by assuming the detection of a single organism, whereas the upper quantitation limit (UQL) value is based on the maximum number of organisms which can be enumerated from a sample. When the collection medium is a liquid or a membrane filter from which organisms can be eluted, the sample can be serially diluted prior to analysis. Therefore, there is essentially no UQL for these sampling methods. For an impactor sampler, the UQL is determined by the area of the collection surface. In the case of an impactor with several nozzles in its stage(s), the UQL depends on the number of sampling nozzles. For example, the UQL of the Andersen single-stage sampler is reached when a CFU develops under each of the 400 sampling nozzles. When aerosol samples are collected near the UQL, errors in analysis, such as uncertainty associated with "positive-hole" corrections, may occur as a result of overcrowding on the impaction surface (see discussion of enumeration in Sample Analysis, below); other problems include difficulties in resolving distinct particles or colonies (19) and inhibition of the growth of microorganisms as a result of competition for space and nutrients. When a limited amount of overlap occurs, the true bioaerosol concentration can be statistically calculated from the number of colonies on the agar (19, 59). Conversely, air samples collected near the LQL of a sampler may contain an insufficient number of particles to accurately represent the true bioaerosol concentration.

One way to predict the optimal sampling time for a particular sampler is to determine the ideal surface density of microorganisms on the collection area and assume the order of magnitude of the bioaerosol. Nevalainen et al. (68) calculated the optimal sampling time for five bioaerosol samplers by using the formula $t = (\delta)(A)/(C_a)(Q)$, where t is the sampling time, δ is the desired surface density, A is the area of the sampling surface, C_a is the average expected bioaerosol concentration, and Q is the sampler flow rate.

Figure 2 illustrates the sampling times for an ideal surface density on a culture plate, δ_{macro}, of 1 colony per cm^2, or 10^4 particles per cm^2 for microscopic counts, δ_{micro}, with a sampler collection efficiency of 100%. The indicated sampling times for an airborne concentration of 10^3 particles per m^3 illustrate that the optimal sampling time depends on the sampling method and on the anticipated bioaerosol concentration. For instance, a high-volume sampler, such as the SAS sampler, may be best suited for sampling environments with low bioaerosol concentrations, while the Burkard personal sampler may be most effective at relatively high concentrations. Because the bioaerosol concentration is unknown and can only be estimated, often more than one collection time is used for sample collection to enhance the likelihood of obtaining useful data.

Another factor which influences the selection of the sample collection time is sampling stress, which can result in the loss of culturability of airborne microorganisms. It has been observed that increased sampling time has resulted in decreased viability for aerosolized vegetative bacterial

cells (15, 23, 36, 86). As air flows over the nutrient agar surface of an impactor, the agar may lose water content, resulting in a harder surface. Subsequently, sampled microorganisms may rebound from the hardened surface and not be collected (48). If they are collected, they may be less embedded in the agar, exposing the microorganisms to desiccation stresses. Thus, a doubling of sampling time may not result in a doubling of CFU, depending on the stress tolerance of the airborne microorganisms being sampled. Therefore, the investigator should consider taking several consecutive samples of short duration rather than a few samples for a long time interval.

Number of Samples

Given the heterogeneity and temporal variability exhibited by bioaerosols and the relatively short collection time for many air sampling methods, multiple samples are often taken to more accurately determine the concentration and composition of a bioaerosol. The number of air samples required depends on the statistical methods used to analyze the data, on the length of the sample collection time, and on the variability of the bioaerosol. Often pilot sampling studies are performed in a particular environment to arrive at the final sampling design. Simultaneous paired samples are often taken because of the high degree of variability which has been demonstrated between data from paired samplers of the same type (16, 52). Because of the changes in bioaerosol concentration over time and the small volume of air from which the sample is taken, a single air sample at a discrete point in time and space usually has limited value in the assessment of the airborne microorganisms in an environment of concern. Because the aerosolization of fungal spores and other microorganisms may be sporadic and the air environment may not be perfectly mixed, a negative result obtained from an air sample is not proof that a specific microorganism is not present.

Sampler Calibration

The concentration of airborne microorganisms is calculated by dividing the number of collected microorganisms by the sampled air volume (e.g., CFU per cubic meter). Therefore, the volumetric flow rate of each sampler should be calibrated and adjusted to a desired level, if necessary. This is particularly important for impactors, as an impactor's d_{50} is a function of the flow rate through the nozzle(s). A decreased flow rate (e.g., due to a weakened battery) increases the d_{50}; thus, collection is shifted toward larger particle sizes (62). If some of the nozzles in a multinozzle impactor are plugged, the flow velocity through the remaining nozzles is increased, resulting in a lower d_{50}, with increased collection and, possibly, increased stress on the microorganisms. In impactor sampling, the recommended distance between the exit plane of the impactor nozzle(s) and the top of the nutrient surface should be maintained. Otherwise, the impactor's collection characteristics, primarily the d_{50}, may be changed.

The flow rate of a battery-powered sampler is usually calibrated before and after sample collection. If the flow rate has decreased by less than 10%, the average of the two flow rate measurements is generally used for the calculations of the airborne concentration of microorganisms. If the difference exceeds 10%, the sample is considered invalid. Various calibration methods which measure the flow rate through a soap bubble meter or an electronic calibration instrument are available (22, 54, 55).

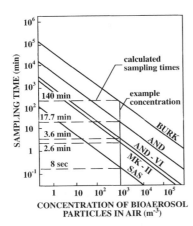

FIGURE 2 Collection times for selected bioaerosol samplers. BURK, Burkard spore trap (personal sampler); AND, Andersen six-stage viable impactor sampler; AND-VI, Andersen single-stage viable impactor sampler; MK-II, Casella MK-II sampler; SAS, Surface Air System high-flow sampler. (Adapted from reference 68 with kind permission from Elsevier Science Ltd., The Boulevard, Langford Lane, Kidlington 0X5 1GB, United Kingdom.)

SAMPLE ANALYSIS

The selection of an analysis method is a critical component of the sampling plan, and it should be designated before air sampling is conducted. Factors which influence the choice of an analytical method include the cost and length of time required for analysis, the sensitivity and specificity of the analysis method, the availability of sampling methods, and the expected characteristics of the bioaerosol of interest. Sample analyses are primarily cultural, microscopic, biochemical, immunological, or molecular biological assays. Traditionally, airborne microorganisms have been analyzed by culturable and total count (microscopic) determinations. However, limitations to both of these methods have led to the development of techniques that can increase the sensitivity and accuracy of bioaerosol monitoring.

Culture Methods

Culture Media and Incubation

Many of the currently available bioaerosol sampling methods rely on culture for the quantification and characterization of airborne bacteria and fungi. Microorganisms that are collected on a nutrient agar surface by impaction can be cultured directly, while organisms that are collected in a liquid or on a filter are transferred to a culture medium. Only those cells which survive and reproduce under the culture conditions to form visible colonies are enumerated. Because microorganisms exhibit a wide range of nutritional requirements, no formulation is capable of culturing every type of organism. Therefore, a common strategy in bioaerosol monitoring is to choose general media which promote the growth of the greatest diversity of species. This approach is especially useful in survey efforts directed toward the characterization of the airborne microbiota. Because of differences in temperature, incubation time, and nutritional requirements, no single culture medium is satisfactory for the simultaneous isolation of airborne bacteria and fungi. Generally it is necessary to perform replicate sampling using different culture media or to divide samples for inoculation onto multiple types of nutrient media.

Several conditions, such as pH, temperature, water activity, nutrients, antibiotics, light and aeration, can be manipulated to favor the growth of a select group of organisms (11). For example, incubation at a temperature of 55°C may be used to select for the growth of thermophilic actinomycetes, or antibiotic resistance markers may be used for the selective culture of a target species (1, 15, 90). However, the growth of microorganisms on selective media is hindered compared with that on general media. This may be especially true for culturing of airborne microorganisms, which are already stressed or damaged by aerosolization and sampling. For this reason, general media are often used for the initial culture of bioaerosol samples, followed by replication onto differential or selective media for identification (43).

Several broad-spectrum media have been evaluated for their utility in the retrieval of culturable airborne fungi (14, 67, 79, 89). The following media have been recommended, although the results vary between investigations. Malt extract agar (MEA), rose bengal-containing agars (e.g., rose bengal-streptomycin and dichloran-rose bengal-chloramphenicol), and dichloran glycerol-18 agar have been suggested for isolation of airborne fungi. MEA may be the most widely used fungal isolation medium. Of the formulations available, unamended 2% MEA was found to promote sporulation better than MEA amended with glucose and peptone (41, 85). Rose bengal- and dichloran-containing media have

the advantage of inhibiting the spread of rapidly growing fungal genera (e.g., *Rhizopus* and *Mucor* spp.), allowing the enumeration and identification of other fungi in the sample. However, rose bengal is photoactivated in direct sunlight, forming cytotoxic products that may inhibit fungal growth (4). Dichloran glycerol-18 is a low-water-activity medium ($a_w = 0.955$) developed for the isolation of xerophilic fungi (39), and it compares favorably with other media for culturing mesophilic airborne fungi (89). Other commonly used fungal media include potato dextrose agar and Sabouraud's agar (37, 80).

Incubation periods for fungi typically range from 3 to 7 days; however, fastidious xerophiles may require several weeks for mycelial development (8). Most airborne fungi are mesophilic and grow well at temperatures of 20 to 25°C, and medically important fungal pathogens grow well at 30°C (61). Incubation temperature can be also be used to select for certain species. *Aspergillus fumigatus* is a thermotolerant species which can be cultured at 37 to 45°C, above the temperature range of mesophilic fungi.

For the culture of bacteria, several broad-spectrum media, such as tryptic soy agar, nutrient agar, and casein soy peptone agar, may be used (11, 43). These media are often amended with antibiotics to restrict the growth of fungi. Incubation temperatures from 28 to 35°C for 1 to 7 days are usually used for environmental and human source bacteria, an important exception being thermophilic actinomycetes, which are cultured at 55°C.

Enumeration

Viable counts are determined after the appropriate incubation period by enumerating the CFU. The concentration of culturable airborne microorganisms (CFU per cubic meter) is determined by dividing the number of CFU per sample by the volume of air sampled. When microorganisms are collected with multiple-jet impactor samplers (e.g., Andersen and SAS impactor samplers; Table 1), positive-hole corrections are generally applied to the data (3, 16, 59). These corrections are estimations which account for the probability of multiple impactions of microorganisms through the same sampling jet, resulting in the enumeration of a single colony at that impaction site. The magnitude of the correction increases with the number of CFU, and the uncertainty of the estimate increases near the UQL of the sampler (3). The corrected CFU count for a sample is used to calculate the airborne concentration of microorganisms. When organisms are collected through a single nozzle onto a stationary or moving surface, colonies may overlap each other, but the resulting reduced number of CFU can be corrected (19).

Enumeration errors are associated with colony counts that are very low or very high. Colony counts that are too low can be nonrepresentative of the population and exhibit high variability. As colony counts increase, counting errors increase because of overlap of colonies and inhibitory effects of microorganisms on one another. For statistical accuracy, microbiologists traditionally enumerate only those plates with colony counts between 30 and 300 on a 100-mm-diameter plate. Recently, researchers have proposed that upper limits for efficient detection be determined by colony size and the resolution ability of the enumeration system (19). When filtration or impingement sampling is used, the wash solution or collection buffer can be serially diluted to obtain counts within an acceptable range. However, problems may arise when one collects air samples directly onto an agar surface without knowledge of the bioaerosol concentration being sampled and too few or too many colonies emerge on a plate. These difficulties can be lessened by setting more than one sample collection time, as discussed previously.

Identification

Fungal identification is based largely on the morphological characteristics of spores and spore-bearing structures by using direct microscopy. Misidentification of fungi is a significant source of error in culturable sample analysis; therefore, accurate fungal identification requires training and experience. Bacterial isolates may be identified by a variety of methods. Classical biochemical methods may be employed for identification to the genus and species level, or procedures such as Gram staining may be used to differentiate bacterial groups. A variety of identification systems which classify bacteria according to various biochemical characteristics, such as substrate utilization and cellular fatty acid profiles, are available (43). However, many of the databases used to identify isolates are composed mainly of clinical isolates; therefore, the identification of environmental microorganisms may be problematic.

Microscopy

Microscopic enumeration methods are used to obtain an estimate of the total number of microorganisms present in a sample. In contrast with culture techniques, microscopic analysis allows enumeration of both culturable and nonculturable microorganisms. However, identification of microorganisms to the species level is usually not possible without using a taxon-specific technique, such as immunospecific fluorescence staining (described below). A variety of stains may be used to differentiate biological particles from nonbiological material and respiring cells from nonrespiring cells.

Impaction onto a glass slide or tape and filtration sampling methods are generally used for the determination of total airborne fungal spores. Air samples collected with Burkard spore traps are analyzed via light microscopy by examining several longitudinal traverses of the collection slide or tape. The volume of air represented by a traverse can be calculated, and the average number of fungal spores per traverse is used to determine the total number of spores per cubic meter. When the sample is collected onto a moving slide, time discrimination of fungal spore concentration is possible by enumerating traverses between specified points along the collection surface. Although staining is not required for enumeration, several stains, including phenosaffranin, basic fuchsin, and lactophenol cotton blue, are commonly used to facilitate discrimination of spores from debris (11). Accurate identification to the genus level is possible for only a limited number of fungal spore types. Therefore, data are usually reported as total number of spores per cubic meter. Fungal spores collected by filtration may be washed from the filter membrane, stained with acridine orange, and enumerated by epifluorescence microscopy (70). However, some fungal spore types may resist staining or have dark pigmentation which masks fluorescence (11).

For the determination of total airborne bacteria by microscopic analysis, liquid impingement or filtration sampling is used. Aliquots of collection buffer or the filter membrane wash solution may be stained for epifluorescence microscopy by the acridine orange direct count method (38, 70, 91); however, >10^4 cells per sample are required to obtain useful results with this method. Cells and endospores may also be enumerated by bright-field or phase-contrast microscopy using a hemocytometer or counting chamber.

The advantage of microscopic analysis of air samples is the determination of total airborne microorganisms, both culturable and nonculturable. Because a significant fraction of airborne microorganisms may be nonculturable, microscopy is often used in conjunction with culture methods to analyze bioaerosol samples. The major disadvantage is the tedious and time-consuming nature of microscopic enumeration, although computerized image analysis systems may automate the process. Another drawback is the level of expertise required to identify fungal spores. Misidentification of fungal spores and the inability to distinguish microorganisms from nonbiological particles are common sources of error using this method (11).

Immunoassay

Immunoassay methods, used extensively in biomedical research, have been recognized for their potential application to bioaerosol analysis (13). Immunoassays rely on the binding of antibodies to a specific target antigen. Target antigens may be (i) cell surface-associated proteins or polysaccharides or (ii) human allergens. The development of a specific antibody is a critical component which affects the sensitivity and specificity of the assay; therefore, monoclonal antibodies are often used. Among the methods which may be applied to bioaerosol analysis are fluorescence immunoassay, enzyme immunoassay, and radioimmunoassay.

Fluorescence immunoassay consists of staining samples with a fluorescence-labeled antibody that binds specifically to the antigens on the surfaces of the target organisms and then enumeration by epifluorescence microscopy. Fluorescent labeling dyes include fluorescein, fluorescein isothiocyanate, and rhodamine isothiocyanate (43). Appropriate controls are necessary to determine levels of background and nonspecific fluorescence.

Radioimmunoassay consists of combining a specific antibody with a radioactive label, while enzyme immunoassay utilizes binding of the antibody or the antigen to an enzyme. The concentration of antigen is measured by radioactivity or enzyme activity. These methods have been applied for the measurement of airborne allergens such as dust mite allergen and animal dander (27, 57, 74). Impaction, impingement, and filtration sampling methods are compatible with the assay (20). The advantages of these methods for quantification of airborne allergen are their specificity and sensitivity. The major limitation of immunoassays is that specific antigens for microorganisms are difficult to define and standardize (61). As more microbial antigenic compounds are characterized, immunoassay analysis methods will provide data necessary for assessing environmental exposure to allergens.

Biochemical Assay

Biochemical methods of analysis are used to measure a biological compound of interest, such as endotoxins or mycotoxins. Because inhalation of these compounds may produce adverse health effects (17, 26), there have been efforts to measure these compounds directly in order to relate environmental exposure to human response. Endotoxins are lipopolysaccharides (LPS) in the cell walls of gram-negative bacterial cells. Airborne endotoxin is widespread because of the ubiquity of gram-negative bacteria in the environment, and elevated levels of endotoxin have been measured in bioaerosol samples from a variety of agricultural, industrial, and office environments (42). The most widely used method for measurement of endotoxin is the *Limulus* amebocyte lysate test. The lysate of amebocytes from the horseshoe crab, *Limulus polyphemus*, gels in the presence of LPS. This reaction forms the basis of endotoxin quantitation, and a variety of test systems are commercially available.

Airborne endotoxin data are expressed as nanograms per cubic meter or endotoxin units (EU) per cubic meter. EU is defined as the potency of 0.10 ng of a reference standard endotoxin (65). Because endotoxin potency varies among gram-negative bacteria, data are often expressed as EU to facilitate comparisons between studies. Filtration sampling is most often used to collect airborne endotoxin; however, various filter materials have been shown to inhibit the *Limulus* amebocyte lysate assay (66). Differences in collection and analysis methods between laboratories has made comparison of results difficult. In addition, interference with the *Limulus* amebocyte lysate reaction by inhibitors present in environmental samples must be accounted for in the measurement of endotoxin content (65).

Mycotoxins associated with dusts from fungus-contaminated grains have long been recognized as a source of illness (31). Toxicosis from exposure to *Stachybotrys atra* mycotoxin has also been found in a contaminated home (25). Filtration sampling using glass fiber filters and polycarbonate membranes has been used to measure aerosolized mycotoxins of *S. atra* in the laboratory (71, 83). Other biochemical methods which may be applied to the analysis of airborne fungi include ergosterol (64) and β-1,3-glucan (32, 34) assays.

PCR

PCR is a procedure used to rapidly amplify specific DNA sequences (75). This technique has been used successfully to enhance the detection of microorganisms in a variety of matrices, including air samples (1, 76). Application of the PCR technique to bioaerosol sampling eliminates the requirement for culture or microscopic examination. Therefore, the use of PCR seems particularly suited for detection of specific microorganisms that are difficult to culture, grow very slowly, or have never been cultured in vitro. The PCR method also provides an increase in sensitivity over traditional culture methods (46). Positive results with the PCR technique were demonstrated for the detection of bacteria aerosolized in a greenhouse when culture counts were negative (1).

Another advantage of PCR is the rapidity with which results are obtained compared with culture counts. It is possible to obtain results within hours of sample collection with PCR, compared with days or weeks for culture methods. It should be noted that PCR can be used for the direct detection of microorganisms provided that unique, non-cross-hybridizing sequences are available for the specific microorganisms to serve as primers for amplification. The authenticity of the amplified DNA can be confirmed by sequencing, restriction analysis, or gene probe hybridization. If there are multiple organisms of interest in a sample, it may be possible to use multiplex PCR on the same reaction mix to simultaneously amplify several DNA targets of interest (18). A concentration step is required for PCR to be effectively applied to air sample analysis. Two sample collection methods, liquid impingement and filtration, have been used successfully for air sample collection, with subsequent PCR detection of the target microorganism (1, 76).

A limitation of PCR is the inability to distinguish between culturable and nonculturable microorganisms. In addition, it is currently difficult to quantitate the concentration of the DNA segments which serve as templates for the amplification reaction. The major difficulty associated with PCR as a quantitative method is that no other technique yet matches its sensitivity and can be used to compare results. Thus, any data obtained by quantitative PCR need to be supported by carefully designed controls and compared with reliable standards (30). Quantitation of microbial populations in a given environment by PCR amplification has been demonstrated (53, 72).

With the wide variety of bioaerosol sampling and analysis methods in use, the investigator must have a clear understanding of the objectives of bioaerosol sampling before sampling is conducted. This information will assist in the selection of the appropriate sampling and analysis methodology for collecting meaningful bioaerosol data. As existing methods are improved and new techniques are developed, more accurate bioaerosol information may be obtained, which will further the understanding of human exposure to airborne microorganisms and aid in the development of standardized monitoring protocols.

REFERENCES

1. **Alvarez, A. J., M. P. Buttner, G. A. Toranzos, E. A. Dvorsky, A. Toro, T. B. Heikes, L. E. Mertikas, and L. D. Stetzenbach.** 1994. The use of solid-phase polymerase chain reaction for the enhanced detection of airborne microorganisms. *Appl. Environ. Microbiol.* **60:**374–376.
2. **American Conference of Governmental Industrial Hygienists.** 1989. *Guidelines for the Assessment of Bioaerosols in the Indoor Environment.* American Conference of Governmental Industrial Hygienists, Cincinnati.
3. **Andersen, A. A.** 1958. New sampler for the collection, sizing, and enumeration of viable airborne particles. *J. Bacteriol.* **76:**471–484.
4. **Banks, J. G., R. G. Board, J. Carger, and A. D. Dodge.** 1985. The cytotoxic and photodynamic inactivation of microorganisms by rose bengal. *J. Appl. Bacteriol.* **58:**392–400.
5. **Baron, P. A., and K. Willeke.** 1986. Respirable droplets from whirlpools: measurement of size distribution and estimation of disease potential. *Environ. Res.* **39:**8–18.
6. **Baron, P. A., and K. Willeke.** 1993. Gas and particle motion, p. 23–40. *In* K. Willeke and P. A. Baron (ed.), *Aerosol Measurement: Principles, Techniques and Applications.* Van Nostrand Reinhold, New York.
7. **Bausum, H. T., S. A. Schaub, K. F. Kenyon, and M. J. Small.** 1982. Comparison of coliphage and bacterial aerosols at a wastewater spray irrigation site. *Appl. Environ. Microbiol.* **43:**28–38.
8. **Beuchat, L. R.** 1992. Media for detecting and enumerating yeasts and moulds. *Int. J. Food Microbiol.* **17:**145–158.
9. **Blomquist, G., G. Strom, and L. H. Stromquist.** 1984. Sampling of high concentrations of airborne fungi. *Scand. J. Work Environ. Health* **10:**109–113.
10. **Brockmann, J. E.** 1993. Sampling and transport of aerosols, p. 77–111. *In* K. Willeke and P. A. Baron (ed.), *Aerosol Measurement: Principles, Techniques and Applications.* Van Nostrand Reinhold, New York.
11. **Burge, H. A.** 1995. Bioaerosol investigations, p. 1–23. *In* H. A. Burge (ed.), *Bioaerosols.* CRC Press, Boca Raton, Fla.
12. **Burge, H. A., J. R. Boise, J. A. Rutherford, and W. R. Solomon.** 1977. Comparative recoveries of airborne fungus spores by viable and non-viable modes of volumetric collection. *Mycopathologia.* **61:**27–33.
13. **Burge, H. A., and W. R. Solomon.** 1987. Sampling and analysis of biological aerosols. *Atmos. Environ.* **21:**451–456.
14. **Burge, H. P., W. R. Solomon, and J. R. Boise.** 1977. Comparative merits of eight popular media in aerometric studies of fungi. *J. Allergy Clin. Immunol.* **60:**199–203.
15. **Buttner, M. P., and L. D. Stetzenbach.** 1991. Evaluation

of four aerobiological sampling methods for the retrieval of aerosolized *Pseudomonas syringae*. *Appl. Environ. Microbiol.* **57:**1268–1270.

16. **Buttner, M. P., and L. D. Stetzenbach.** 1993. Monitoring airborne fungal spores in an experimental indoor environment to evaluate sampling methods and the effects of human activity on air sampling. *Appl. Environ. Microbiol.* **59:**219–226.

17. **Castellan, R. M., S. A. Olenchock, K. B. Kinsley, and J. L. Hankinson.** 1987. Inhaled endotoxin and decreased spirometric values: an exposure-response relation for cotton dust. *N. Engl. J. Med.* **317:**605–610.

18. **Chamberlain, J. S., R. A. Gibbs, J. E. Ranier, P. N. Nguyen, and C. T. Cashey.** 1988. Deletion screening of the Duchenne muscular dystrophy locus via multiplex DNA amplification. *Nucleic Acids Res.* **16:**11141–11156.

19. **Chang, C. W., Y. H. Hwang, S. A. Grinshpun, J. M. Macher, and K. Willeke.** 1994. Evaluation of counting error due to colony masking in bioaerosol sampling. *Appl. Environ. Microbiol.* **60:**3732–3738.

20. **Chapman, M. D.** 1995. Analytical methods: immunoassays, p. 235–248. *In* H. A. Burge (ed.), *Bioaerosols.* CRC Press, Boca Raton, Fla.

21. **Chatigny, M. A., J. M. Macher, H. A. Burge, and W. R. Solomon.** 1989. Sampling airborne microorganisms and aeroallergens, p. 199–220. *In* S. V. Hering (ed.), *Air Sampling Instruments for Evaluation of Atmospheric Contaminants,* 7th ed. American Conference of Governmental Industrial Hygienists, Cincinnati.

22. **Chen, B. T.** 1993. Instrument calibration, p. 493–520. *In* K. Willeke and P. A. Baron (ed.), *Aerosol Measurement: Principles, Techniques and Applications.* Van Nostrand Reinhold, New York.

23. **Cox, C. S.** 1987. *The Aerobiological Pathway of Microorganisms.* John Wiley and Sons Ltd., Chichester, England.

24. **Cox, C. S.** 1989. Airborne bacteria and viruses. *Sci. Prog.* (Oxford) **73:**469–500.

25. **Croft, W. S., B. B. Jarvis, and C. S. Yatawara.** 1986. Airborne outbreak of trichothecene toxicosis. *Atmos. Environ.* **20:**549–552.

26. **Crook, B., S. Higgins, and J. Lacey.** 1987. Methods for sampling microorganisms at solid waste disposal sites, p. 791–797. *In* D. R. Houghton, R. N. Smith, and H. O. W. Eggins (ed.), *Biodeterioration 7.* Elsevier, London.

27. **DeBlay, F., M. D. Chapman, and T. A. E. Platts-Mills.** 1991. Airborne cat allergen *Fel d* I: environmental control with the cat in situ. *Am. Rev. Respir. Dis.* **143:**1334.

28. **Eduard, W., J. Lacey, K. Karlsson, U. Palmgren, G. Strom, and G. Blomquist.** 1990. Evaluation of methods for enumerating microorganisms in filter samples from highly contaminated occupational environments. *Am. Ind. Hyg. Assoc. J.* **51:**427–436.

29. **Favero, M. S., J. R. Puleo, J. H. Marshall, and G. S. Oxborrow.** 1966. Comparative levels and types of microbial contamination detected in industrial clean rooms. *Appl. Microbiol.* **14:**539–551.

30. **Ferre, F.** 1992. Quantitative or semi-quantitative PCR: reality versus myth. *PCR Methods Appl.* **2:**1–9.

31. **Flannigan, B.** 1987. Mycotoxins in the air. *Int. Biodeterior.* **23:**73–78.

32. **Fogelmark, B., H. Goto, K. Yuasa, B. Marchat, and R. Rylander.** 1992. Acute pulmonary toxicity of inhaled β-1,3-glucan and endotoxin. *Agents Actions* **35:**50–56.

33. **Gillespie, V. L., C. S. Clark, H. S. Bjornson, S. J. Samuels, and J. W. Holland.** 1981. A comparison of two-stage and six-stage Andersen impactors for viable aerosols. *Am. Ind. Hyg. Assoc. J.* **42:**858–864.

34. **Goto, H., K. Yuasa, and R. Rylander.** 1994. (1-3)-β-D-Glucan in indoor air, its measurement and in vitro activity. *Am. J. Ind. Med.* **25:**81–83.

35. **Grinshpun, S. A., C. W. Chang, A. Nevalainen, and K. Willeke.** 1994. Inlet characteristics of bioaerosol sampler. *J. Aerosol Sci.* **25:**1503–1522.

36. **Hatch, M. T., and H. Wolochow.** 1969. Bacterial survival: consequences of the airborne state, p. 267–295. *In* R. L. Dimmick and A. B. Akers (ed.), *An Introduction to Experimental Aerobiology.* Wiley, New York.

37. **Hirsch, S. R., and J. A. Sosman.** 1976. A one-year survey of mold growth inside twelve homes. *Ann. Allergy* **36:**30–38.

38. **Hobbie, J. E., R. J. Daley, and S. Jasper.** 1977. Use of nuclepore filters for counting bacteria by fluorescence microscopy. *Appl. Environ. Microbiol.* **33:**1225–1228.

39. **Hocking, A. D., and J. I. Pitt.** 1980. Dichloran-glycerol medium for enumeration of xerophilic fungi from low-moisture foods. *Appl. Environ. Microbiol.* **39:**488–492.

40. **Huang, Y. L., K. Willeke, A. Juozaitis, J. Donnelly, A. Leeson, and R. Wyza.** 1994. Fermentation process monitoring through measurement of aerosol release. *Biotechnol. Prog.* **10:**32–38.

41. **Hunter, C. A., C. Grant, B. Flannigan, and A. F. Bravery.** 1988. Mould in buildings: the air spora of domestic buildings. *Int. Biodeterior.* **24:**81–101.

42. **Jacobs, R. R.** 1989. Airborne endotoxins: an association with occupational lung disease. *Appl. Ind. Hyg.* **4:**50–56.

43. **Jensen, P. A., B. Lighthart, A. J. Mohr, and B. T. Shaffer.** 1994. Instrumentation used with microbial bioaerosol, p. 226–284. *In* B. Lighthart and A. J. Mohr (ed.), *Atmospheric Microbial Aerosols, Theory and Applications.* Chapman and Hall, New York.

44. **Jensen, P. A., W. F. Todd, G. N. Davis, and P. V. Scarpino.** 1992. Evaluation of eight bioaerosol samplers challenged with aerosols of free bacteria. *Am. Ind. Hyg. Assoc. J.* **53:**660–667.

45. **Jones, W., K. Morring, P. Morey, and W. Sorenson.** 1985. Evaluation of the Andersen viable impactor for single stage sampling. *Am. Ind. Hyg. Assoc. J.* **46:**294–298.

46. **Josephson, K. L., C. P. Gerba, and I. L. Pepper.** 1993. Polymerase chain reaction detection of nonviable bacterial pathogens. *Appl. Environ. Microbiol.* **59:**3513–3515.

47. **Juozaitis, A., Y. L. Huang, K. Willeke, J. Donnelly, S. Kalatoor, A. Leeson, and R. Wyza.** 1994. Dispersion of respirable aerosols in a fermenter and their removal in an exhaust system. *Appl. Occup. Environ. Hyg.* **9:**552–559.

48. **Juozaitis, A., K. Willeke, S. A. Grinshpun, and J. Donnelly.** 1994. Impaction onto a glass slide or agar versus impingement into a liquid for the collection and recovery of airborne microorganisms. *Appl. Environ. Microbiol.* **60:**861–870.

49. **Kaye, S.** 1988. Efficiency of "Biotest RCS" as a sampler of airborne bacteria. *J. Parenteral Sci. Technol.* **42:**147–152.

50. **Kotimaa, M.** 1990. Occupational exposure to spores in the handling of wood chips. *Grana* **29:**153–156.

51. **Lee, K. W., and M. Ramamurthi.** 1993. Filter collection, p. 179–205. *In* K. Willeke and P. A. Baron (ed.), *Aerosol Measurement: Principles, Techniques and Applications.* Van Nostrand Reinhold, New York.

52. **Lembke, L. L., R. N. Kniseley, R. C. VanNostrand, and M. D. Hale.** 1981. Precision of the all-glass impinger and the Andersen microbial impactor for air sampling in solid-waste handling facilities. *Appl. Environ. Microbiol.* **42:**222–225.

53. **Leser, T. D., M. Boye, and N. B. Hendriksen.** 1995. Survival and activity of *Pseudomonas* sp. strain B13(FR1) in a marine microcosm determined by quantitative PCR and a rRNA-targeting probe and its effect on the indigenous bacterioplankton. *Appl. Environ. Microbiol.* **61:**1201–1207.

54. **Lippmann, M.** 1973. Instruments and techniques used in

calibrating sampling equipment, p. 101–122. *In The Industrial Environment—Its Evaluation and Control.* National Institute for Occupational Safety and Health, Washington, D.C.

55. **Lippmann, M.** 1989. Calibration of air sampling instruments, p. 73–110. *In* S. V. Hering (ed.), *Air Sampling Instruments for Evaluation of Atmospheric Contaminants,* 7th ed. American Conference of Governmental Industrial Hygienists, Cincinnati.

56. **Lippmann, M.** 1989. Sampling aerosols by filtration, p. 305–336. *In* S. V. Hering (ed.), *Air Sampling Instruments for Evaluation of Atmospheric Contaminants,* 7th ed. American Conference of Governmental Industrial Hygienists, Cincinnati.

57. **Luczynska, C. M., Y. Li, M. D. Chapman, and T. A. E. Platts-Mills.** 1990. Airborne concentration and particle size distribution of allergen derived from domestic cats (*Felis domesticus*): measurements using cascade impactor, liquid impinger and a two site monoclonal antibody assay for Fel d I. *Am. Rev. Respir. Dis.* **141:**361.

58. **Lundholm, I. M.** 1982. Comparison of methods for quantitative determinations of airborne bacteria and evaluation of total viable counts. *Appl. Environ. Microbiol.* **44:**179–183.

59. **Macher, J. M.** 1989. Positive-hole correction of multiple-jet impactors for collecting viable microorganisms. *Am. Ind. Hyg. Assoc. J.* **50:**561–568.

60. **Macher, J. M., and M. W. First.** 1983. Reuter centrifugal air sampler: measurement of effective airflow rate and collection efficiency. *Appl. Environ. Microbiol.* **45:**1960–1962.

61. **Madelin, T. M., and M. F. Madelin.** 1995. Biological analysis of fungi and associated molds, p. 361–386. *In* C. S. Cox and C. M. Wathes (ed.), *Bioaerosols Handbook.* CRC Press, Boca Raton, Fla.

62. **Marple, V. A., K. L. Rubow, and B. A. Olson.** 1993. Inertial, gravitational, centrifugal and thermal collection techniques, p. 206–232. *In* K. Willeke and P. A. Baron (ed.), *Aerosol Measurement: Principles, Techniques and Applications.* Van Nostrand Reinhold, New York.

63. **Marthi, B.** 1994. Resuscitation of microbial bioaerosols, p. 192–225. *In* B. Lighthart and A. J. Mohr (ed.), *Atmospheric Microbial Aerosols, Theory and Applications.* Chapman and Hall, New York.

64. **Miller, J. D., A. M. Laflamme, Y. Sobol, P. Lafontaine, and R. Greenhalgh.** 1988. Fungi and fungal products in some Canadian houses. *Int. Biodeterior.* **24:**103–120.

65. **Milton, D. K.** 1995. Endotoxin, p. 77–86. *In* H. A. Burge (ed.), *Bioaerosols.* CRC Press, Boca Raton, Fla.

66. **Milton, D. K., R. J. Gere, H. A. Feldman, and I. A. Greaves.** 1990. Endotoxin measurement: aerosol sampling and application of a new Limulus method. *Am. Ind. Hyg. Assoc. J.* **51:**331–337.

67. **Morring, K. L., W. G. Sorenson, and M. D. Attfield.** 1983. Sampling for airborne fungi: a statistical comparison of media. *Am. Ind. Hyg. Assoc. J.* **44:**662–664.

68. **Nevalainen, A., J. Pastuszka, F. Liebhaber, and K. Willeke.** 1992. Performance of bioaerosol samplers: collection characteristics and sampler design considerations. *Atmos. Environ.* **26A:**531–540.

69. **Nevalainen, A., K. Willeke, F. Liebhaber, J. Pastuszka, H. Burge, and E. Henningson.** 1993. Bioaerosol sampling, p. 471–492. *In* K. Willeke and P. A. Baron (ed.), *Aerosol Measurement: Principles, Techniques and Applications.* Van Nostrand Reinhold, New York.

70. **Palmgren, U., G. Strom, G. Blomquist, and P. Malmberg.** 1986. Collection of airborne micro-organisms on Nuclepore filters, estimation and analysis—CAMNEA method. *J. Appl. Bacteriol.* **61:**401–406.

71. **Pasanen, A., M. Nikulin, M. Tuomainen, S. Berg, P. Parikka, and E. Hintikka.** 1993. Laboratory experiments on membrane filter sampling of airborne mycotoxins produced by *Stachybotrys atra* corda. *Atmos. Environ.* **27A:**9–13.

72. **Picard, C., C. Ponsonnet, E. Paget, X. Nesme, and P. Simonet.** 1992. Detection and enumeration of bacteria in soil by direct DNA extraction and polymerase chain reaction. *Appl. Environ. Microbiol.* **58:**2717–2722.

73. **Placencia, A. M., J. T. Peeler, G. S. Oxborrow, and J. W. Danielson.** 1982. Comparison of bacterial recovery by Reuter centrifugal air sampler and slit-to-agar sampler. *Appl. Environ. Microbiol.* **44:**512–513.

74. **Platts-Mills, T. A. E., W. R. Thomas, R. C. Aalbersee, D. Vervloet, and M. D. Chapman.** 1992. Dust mite allergens and asthma: report of a second international workshop. *J. Allergy Clin. Immunol.* **89:**1046.

75. **Saiki, R. K., S. Scharf, F. Faloona, K. B. Mullis, G. T. Horn, H. Erlich, and N. Arnheim.** 1985. Enzymatic amplification of β-globin genomic sequence and restriction site analysis for diagnosis of sickle cell anemia. *Science* **230:**1350–1354.

76. **Sawyer, M. H., C. J. Chamberlin, Y. N. Wu, N. Aintablian, and M. R. Wallace.** 1994. Detection of varicella-zoster virus DNA in air samples from hospital rooms. *J. Infect. Dis.* **169:**91–94.

77. **Sayer, W. J., N. M. MacKnight, and H. W. Wilson.** 1972. Hospital airborne bacteria as estimated by the Andersen sampler versus gravity settling culture plate. *Am. J. Clin. Pathol.* **58:**558–562.

78. **Sayer, W. J., D. B. Shean, and J. Ghosseiri.** 1969. Estimation of airborne fungal flora by the Andersen sampler versus the gravity settling plate. *J. Allergy* **44:**214–227.

79. **Smid, T., E. Schokkin, J. S. M. Boleij, and D. Heederik.** 1989. Enumeration of viable fungi in occupational environments: a comparison of samplers and media. *Am. Ind. Hyg. Assoc. J.* **50:**235–239.

80. **Solomon, W. R.** 1975. Assessing fungus prevalence in domestic interiors. *J. Allergy Clin. Immunol.* **56:**235–242.

81. **Solomon, W. R., H. A. Burge, J. R. Boise, and M. Becker.** 1980. Comparative particle recoveries by the retracting rotorod, rotoslide and Burkard spore trap sampling in a compact array. *Int. J. Biometeor.* **24:**107–116.

82. **Sorber, C. A., H. T. Bausum, S. A. Schaub, and M. J. Small.** 1976. A study of bacterial aerosols at a wastewater irrigation site. *J. Water Pollut. Control Fed.* **48:**2367–2379.

83. **Sorenson, W. G., D. G. Frazer, B. B. Jarvis, J. Simpson, and V. A. Robinson.** 1987. Trichothecene mycotoxins in aerosolized conidia of *Stachybotrys atra. Appl. Environ. Microbiol.* **53:**1370–1375.

84. **Stewart, S. L., S. A. Grinshpun, K. Willeke, S. Terzieva, V. Ulevicius, and J. Donnelly.** 1995. Effect of impact stress on microbial recovery on an agar surface. *Appl. Environ. Microbiol.* **61:**1232–1239.

85. **Strachan, D. P., B. Flannigan, E. M. McCabe, and F. McGarry.** 1990. Quantification of airborne moulds in the homes of children with and without wheeze. *Thorax* **45:**382–387.

86. **Thompson, M. W., J. Donnelly, S. A. Grinshpun, A. Juozaitis, and K. Willeke.** 1994. Method and test system for evaluation of bioaerosol samplers. *J. Aerosol Sci.* **25:**1579–1593.

87. **Thorne, P. S., M. S. Kiekhaefer, P. Whitten, and K. J. Donham.** 1992. Comparison of bioaerosol sampling methods in barns housing swine. *Appl. Environ. Microbiol.* **58:**2543–2551.

88. **Tyler, M. E., and E. L. Shipe.** 1959. Bacterial aerosol samplers. I. Development and evaluation of the all-glass impinger. *Appl. Microbiol.* **7:**337–349.

89. **Verhoeff, A. P., J. H. van Wijnen, J. S. M. Boleij, B.**

Brunekreef, E. S. van Reenen-Hoekstra, and R. A. Samson. 1990. Enumeration and identification of airborne viable mould propagules in houses. *Allergy* **45:**275–284.

90. **Walter, M. V., B. Marthi, V. P. Fieland, and L. M. Ganio.** 1990. Effect of aerosolization on subsequent bacterial survival. *Appl. Environ. Microbiol.* **56:**3468–3472.

91. **Walters, M., D. Milton, L. Larsson, and T. Ford.** 1994. Airborne environmental endotoxin: a cross validation of sampling and analysis techniques. *Appl. Environ. Microbiol.* **60:**996–1005.

92. **White, L. A., D. J. Hadley, D. E. Davids, and R. Naylor.** 1975. Improved large-volume sampler for the collection of bacterial cells from aerosol. *Appl. Microbiol.* **29:** 335–339.

93. **Willeke, K., and P. A. Baron (ed.).** 1993. *Aerosol Measurement: Principles, Techniques and Applications.* Van Nostrand Reinhold, New York.

94. **Zimmerman, N. J., P. C. Reist, and A. G. Turner.** 1987. Comparison of two biological aerosol sampling methods. *Appl. Environ. Microbiol.* **53:**99–104.

Fate and Transport of Microorganisms in Air

ALAN JEFF MOHR

69

The fate and transport of microorganisms in the atmosphere are complicated issues involving many physical and biochemical factors. The transport of bioaerosols is primarily governed by hydrodynamic and kinetic factors, while their fate is dependent on their specific chemical makeup and the meteorological parameters to which they are exposed. When a particle approaches another surface, effects governed by the makeup of the cell wall influence deposition. Specific surface-surface interactions then dictate the release of the particle. The vast majority of airborne microorganisms are immediately inactivated is a result of environmental stresses (e.g., desiccation, temperature, and oxygen) that act to alter the makeup of the outer surface. The most significant environmental factors influencing viability are relative humidity (RH), temperature, and oxygen. Additional influences are exerted through air ions, solar irradiance, and open air factors (OAF). Some microorganisms have built-in mechanisms that act to repair damage inflicted during the aerosolization and transport phases. Although few generalities can be made concerning the aerosol stability or fate of microorganisms, bacteria tend to behave differently from viruses, which behave differently from molds and fungi.

Bioaerosol particles can be either solid or liquid and can come from a number of natural and anthropogenic sources.

Many of the physical and chemical processes that describe more general classifications of aerosols also apply to bioaerosols. Classic texts, which should be studied by all individuals interested in aerosols, include works by Fuchs (49), Hinds (61), and Hidy (60). Significant sources which address bioaerosols include texts by Cox (24), Anderson and Cox (6), and Lighthart and Mohr (76). An additional list of valuable references is compiled in the *Journal of Aerosol Science and Technology* (6a).

FATE OF BIOAEROSOLS
Physical and Chemical Factors Influencing Viability

Studies to determine the factors that influence viability in the airborne state have been performed for decades. Many of these studies have yielded differing results, primarily because different methodologies were used during their evaluations. Cox (24) outlines several factors which exert stresses on airborne microorganisms and lists water content or RH as being of primary interest (Table 1). He presents detailed information in chapters dealing with environmental parameters (RH and temperature, oxygen, OAF, etc.), which are broken down into sections dealing with phages, viruses, bacteria, and other microorganisms. The effects of temperature on aerosol stability are less well understood. Marthi (80) presents his survey of pertinent stresses as being either "primary" or "secondary" in effect (Table 1). He lists RH, temperature, radiation, and OAF as important primary stressors. For artificially generated aerosols, he lists secondary stressors that influence inactivation or protection provided to the primary stressors. Important secondary stressors include the method of aerosol generation, the makeup of generation fluid, the type of aerosol sampling method used, and the type of medium used to enumerate the collected microorganisms. It has been shown that the type of assay medium used can significantly influence the results for bioaerosol studies.

Water Content (RH)

The state of water and the water content associated with bioaerosols are fundamental factors influencing the fate or viability of these microorganisms. As the RH decreases, so does the water available to the exterior environment of the microorganism. Loss of water can cause dehydration, resulting in inactivation of many microorganisms. Of all of the measurable meteorological parameters, RH is the most important with respect to aerosol stability (24). Israeli et al. (65) studied freeze-dried microorganisms and showed the importance of water content to the viability of microorganisms. They concluded that the biomembrane, as phospholipid bilayers, undergoes conformation changes from crystalline to gel phases as a result of water loss. These transformations induced changes to cell proteins which in turn resulted in a loss of viability.

Not only is the water content of aerosolized microorganisms a principal factor contributing to aerosol viability (108–110), but the RH of a system will directly affect the density of the bioaerosol unit. The size, shape, and density of an aerosolized particle are directly related to the aerodynamic diameter, which determines settling velocity and location of deposition in the respiratory tract and influences the impaction characteristics of aerosol samplers.

The effects of RH can be influenced by the content of the suspension fluid used before aerosolization (7, 51), the content of the collection fluid (17), and prehumidification (107). For some microorganisms, shifts in RH after aerosol-

TABLE 1 Factors affecting viability of airborne microorganisms

Factors	Reference(s)
Primary	
RH	24, 80
Temperature	80
Radiation	80
OAF	80
Secondary	
Method of aerosol generation	80
Composition of generation fluid	80
Sampling method	80
Collection medium	80

ization have a more profound effect on aerosol stability than constant RH (55).

Temperature

The vapor pressure, and therefore the RH, of a system is dependent on the temperature. This relationship makes it very difficult to separate the effects of temperature and RH. Studies to determine the effect of temperature on aerosol stability have generally shown that increases in temperature tend to decrease the viability of airborne microorganisms (40).

Oxygen Concentration

Oxygen has an effect on the aerosol stability and infectivity of some bacteria (27). Free radicals of oxygen have been suggested as the cause of this inactivation, but negative correlations between oxygen concentration and viability have been observed by some investigators (26). Israeli et al. (65) notes that oxygen susceptibility increases with dehydration, increased oxygen concentration, and time of exposure.

Other Factors

Electromagnetic Radiation

Aerosol inactivation caused by electromagnetic radiation has been shown to be dependent on the wavelength and, hence, the intensity of the radiation. Shorter wavelengths contain more energy and are generally more deleterious to aerosolized microorganisms. RH (8), water activity (15, 68), oxygen concentration (15), age of the aerosol (24), and the presence of other gases all influence the effect that radiation exerts on airborne microorganisms. Short-wave ionizing radiation (X rays, gamma rays) can cause breaks in the DNA of microorganisms (15). Other studies showed that samplers monitoring bacteria from airborne emissions at sewage plants consistently yielded greater numbers at night (50).

OAF

Studies showing that inactivation rates for many biological aerosols were significantly increased upon challenge with HEPA-filtered air from outdoors compared with clean, inert laboratory-supplied air resulted in the term OAF (11, 43). While Cox (24) suggests that inactivation is caused primarily by reactions between ozone and olefins, OAF inactivation is probably caused by a combination of factors including pollutant concentration, RH, pressure fluctuations, and air ions (37).

Miscellaneous

Review articles by Cox (24), Strange and Cox (102), and Spendlove and Fannin (100) address the influences that pollutants, pressure fluctuations, air ions (90), and season may have on aerosol inactivation rates. Some atmospheric pollutants that which have been studied include NO_2, SO_2, and O_3. The deleterious effects of these pollutants are greatly influenced by RH. Some microorganisms display greater loss of viability at high RH, while others show the opposite effect at high RH. Some pollutants may react with water to form acids (46), and the pH of the environment may be involved in acid production and, hence, inactivation of aerosolized microorganisms. Other pollutants which have been studied include HCOH, CO, HCl, HF, C_2H_2, C_2H_4, and C_2H_8 (27, 36). These compounds are much less toxic than OAF.

A review (101) on the diurnal and annual variations of bioaerosol concentrations details some interesting phenomena. For a high desert climate, the upward flux of bacteria was shown to follow the solar irradiance/sensible heat cycle (77), with the maximum occurring at solar noon. It was also reported that even though the upward bacterial flux decreased after solar noon, the bacterial concentration increased as a result of a reduction in solar output (due to solar angle), which resulted in lower rates of bacterial inactivation. Annual variations in bioaerosol concentrations have been shown to exhibit minimums in the winter and to reach a primary peak in the late spring (101), with a secondary peak observed during midsummer. These maximums and minimums are related to favorable growth conditions (moderate temperatures, humidity, solar irradiance) and snow cover (low temperatures), respectively.

Lighthart and Shaffer (77) have recently designed a test apparatus to evaluate the flux of bacteria into the atmosphere. They found that in the summer, in a high desert location, the maximum upward flux of bacteria was equal to 1.7×10^4 CFU m^{-2} h^{-1}. It was also noted that the maximum occurred around solar noon. The upward flux was correlated with the daily solar irradiance/sensible heat cycle. Another study (78) performed over a grass seed field showed lower numbers of bacteria associated with the upward flux and that variations in airborne concentrations were linked to the local diurnal sea breezes.

Microbial Aerosol Stability

The experimental determination of aerosol stability is dependent on the ability to assay an entity for biological activity. Many sensitive immunological procedures which take advantage of the specificity of antigen-antibody reactions are available. The problem with these assays is that they reveal very little about the viability of the microorganism in question. Spendlove and Fannin (100) define virus survival by stating that stability or survival in the aerosol particle is really a measure of the ability of the virus to infect tissue culture cells of living hosts. For this reason, infectivity and stability should be treated as a single entity for collected bioaerosols.

This chapter presents data relating aerosol stability or infectivity studies with bacteria, animal viruses, bacterial viruses (phages), and other microorganisms. Cox (24) and Mohr (88) discuss in greater detail individual bacteria and viruses and provide rationales for the demonstrated stability. The results for many aerosol stability tests performed on the same microorganism often differ because testing techniques have not been standardized. Harper (54) noted that variations in results can be caused by (i) methods of aerosol generation, storage, and sampling; (ii) method of assay; (iii)

TABLE 2 Aerosol stability parameters for selected bacteria

Bacterium	Stability parameter(s)	Reference(s)
Bacillus subtilis	Death rate	108, 109
Enterobacter cloacae	RH, temperature, CO_2	70, 73, 81, 82
Erwinia herbicola	RH, temperature	81, 82, 105
Escherichia coli	RH, temperature, O_2, wet, dry	20–22, 25, 26, 111, 113
Klebsiella planticola	RH, temperature	81
Mycoplasma spp.	RH	118
M. pneumoniae	RH, temperature, sunlight	67, 69, 116–118
Pasteurella tularensis	RH, wet and dry generation, sunlight	8, 9, 17, 23, 29
Serratia marcescens	RH, O_2, freeze, time	27, 28, 31, 55, 59
Staphylococcus albus	Time, ambient temperature	108, 109
S. aureus	RH, temperature	114

differentiation of total, physical, and viability decay; (iv) presence or absence of light; (v) methods and extent to which RH and temperature are controlled; and (vi) method of data presentation. Additionally, variations can occur because of choice of suspending fluid, collection fluid, and content of atmosphere (presence or absence of oxygen and other gases) used for testing.

Bacteria

Some significant review papers addressing the aerosol characteristics of bacteria are available (3, 6, 24, 57, 76, 102). Table 2 shows some of the parameters that have been shown to influence the viability of selected bacteria. More variation in aerosol test results has been observed for bacteria than viruses, phages, and other microorganisms. This variation is due not only to differences in test procedures and data presentation but also to the greater structural and metabolic complexity of bacteria (cell walls, membranes, and metabolism). Much of the published data cannot be used for comparing aerosol stability because of the presence of several

stresses which could have acted synergistically in a detrimental manner or possibly enhanced survival in some unknown way. Some generalities concerning bacteria can be made. Aerosolized spores of many bacteria are extremely resistant to oxygen concentration, RH, and temperature (83).

Viruses

Several review papers addressing the aerosol characteristics of viruses are available (6, 24, 88, 100). Table 3 shows some of the parameters which have been shown to influence the viability of selected viruses. Viruses are normally very resistant to inactivation by oxygen. This fact and the relative simplicity of their structure explain why results of aerosol inactivation studies are more consistent for viruses than for bacteria. Generalities which can be made about aerosol inactivation rates for viruses include the following: (i) viruses with lipids in their outer coat or capsid are more stable at low RH than at high RH; (ii) viruses without lipids are more stable at high RH than at low RH; (iii) when viable viruses

TABLE 3 Aerosol stability parameters for selected viruses

Virus	Stability parameter(s)	Reference(s)
Adenovirus	RH	85
Bovine rhinotracheitis virus	RH, temperature, media	47, 97
Columbia SK virus	RH, temperature	4
Coxsackievirus	UV light	66
Encephalomyocarditis virus	RH, rehumidification, infectivity	32, 33
Enteric viruses	RH	106
Foot-and-mouth disease virus	Radiation, RH, temperature, weather factor	7, 41, 42, 95
Influenza virus	RH	79
Langat viruses A and B	RH, temperature	10
New Castle disease virus	RH, temperature	97
Picornavirus 37A	RH	5, 40
Pigeonpox virus	RH, inositol	112
Pono virus	Rh, temperature, environment, seasonal inactivation	10, 32, 35, 53, 58, 62
Reovirus	RH, temperature	2, 87
Rotavirus	RH, temperature	34, 62, 86, 94
Rous sarcoma virus	RH, inositol	112
Semliki Forest virus	RH	10, 34
Simian virus 40	RH, temperature	4
St. Louis encephalitis virus	RH temperature	91
Venezuelan equine encephalomyelitis virus	RH, temperature, sunlight	12, 46
Vesicular stomatitis virus	RH, temperature, O_3	48, 97
Yellow fever virus	RH, temperature	84
General	RH, drying	16

TABLE 4 Aerosol stability parameters for selected phages

Phage(s)	Stability parameter(s)	Reference(s)
T-series coliphages	RH	25
T1	Air ions	52, 104
T2, T3, T7	Mechanism of inactivation	10
T3, T7	Freeze	30
T3	RH, sample prehumidification	56
Bacteriophages	Sample prehumidification	107
MS 2	Prehumidification, RH, inactivation, temperature	45, 92, 104, 107
Pasteurella pestis phage	Prehumidification	56
S13	Prehumidification, RH, inactivation	45, 104, 107
ϕX174	RH, chemicals	44, 36

can no longer be detected after aerosol collection, the nucleic acid can be isolated and is still active (this evidence suggests that aerosol inactivation of viruses is not caused by nucleic acid inactivation but by denaturation of coat proteins); and (iv) prehumidification upon sampling increases recovery of viruses that lack lipids in their outer coat. Buckland and Tyrrell (16) first speculated that virus survival in aerosols might be related to the amounts of lipids in the capsid (envelope or outer shell). Many viruses exhibit higher survival rates at mid-range RH (54), regardless of temperature, while some display better survival at low and high RH and the lowest survival at mid-range RH (2, 87). However, survival may depend on the temperature and oxygen content of the test atmosphere at the time of testing. Some viruses are stable in the airborne state over broad temperature and RH ranges (4, 5).

Phages

Phages, like other viruses, are not inactivated by oxygen. Table 4 shows some of the parameters that are associated with the viability of selected phages. Most of the work has been performed on the T series of coliphages (phages which infect *Escherichia coli*). Generally, it has been determined that NaCl is toxic to phages, but that the effect can be reversed by addition of peptone to the suspending media. Aerosol inactivation of complex phages has been shown to be caused by the breakup of the head-tail complex, which could be brought about during the aerosol generation or sampling processes, and that prehumidification at sampling often increased recovery rates. Overall, the results for various workers are consistent because of the relative simplicity of the structure of phages.

Fungi

Fungi and their spores seem to be resistant to desiccation, but little work assessing their survival rates in aerosols has been completed. Interest has intensified recently with the phenomenon termed sick building syndrome. It is believed that fungi and bacteria are responsible for most cases of sick building syndrome even though most of the 100,000 known species of fungi do not cause disease in healthy people. More common effects of the presence of fungi in buildings are asthma or allergenic-type responses and sometimes significant respiratory distress. Many fungi produce toxins that, when inhaled in high concentrations (as can occur with farm workers or people working in granaries), can result in significant health consequences.

INACTIVATION MECHANISMS

Bioaerosols are subject to inactivation and transport the moment they become airborne. Desiccation of the droplet is the main factor responsible for inactivation. Particle sizes for these droplets are usually small (2 to 10 μm), and the particles tend to follow the streamlines of the local wind. Gravitational settling (for particles greater than 5.0 μm) and impaction are the leading causes of loss of these particles. Figure 1 illustrates that, depending on the aerodynamic diameter of a particle, it will either be small enough to follow the streamlines of the surrounding flows or large enough to cross streamline flow and impact on a surface.

Some relationships between aerosol stability and the biological composition of microorganisms have recently been identified (65, 115). Bacterial aerosol stability is considerably more complex than that observed for viruses. Cox (20–22) set up an elaborate series of stability experiments to examine the factors that influenced inactivation of *E. coli* B. When *E. coli* B was aerosolized from a suspension of distilled water into a highly purified nitrogen, argon, or helium atmosphere, it displayed similar patterns of survival as a function of RH. Under these conditions, survival was virtually complete at low RH but was critically dependent on RH above 80%. Under these conditions, Cox (24) speculated, the gases slightly modified the water structure through gas hydrate formation and water lattice modification, which affected the stability of biological structures. Cox (24) then

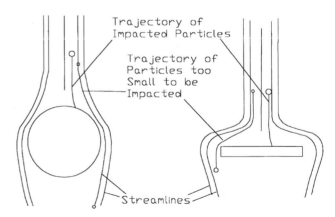

FIGURE 1 Particles less than 5 μm tend to follow flow streamlines around obstructions. Particles greater than 5 μm tend to cross streamlines and impact on obstructions.

explained the events that may take place as precursors to aerosol inactivation as follows. When *E. coli* is aerosolized into inert atmospheres at mid to high RH, the biological membrane constituents become destabilized through loss of water molecules. Additives such as polyhydroxyl compounds, which supersaturate, can stabilize these structures. The polyhydroxyl compounds, by binding to sites on proteins, cause conformational changes in them and thereby stabilize proteins, making them less susceptible to denaturation. This is convincing evidence that the state of proteins on the outer membrane of some microorganisms is critical to the resultant stability profile.

Cox (24) also explains that Maillard reactions between reducing sugars and amino acids are responsible for some denaturation and unfolding of the ovalbumin α-helix which occurs during drying. Additionally, he explained that dried proteins may be stabilized by similar reactions between other sugars and sodium glutamate. Soddu and Vieth (96) showed that sucrose can bind to collagen membranes with different affinities and produce conformational changes in the protein structure. Cox's (24) explanation of the sequence of events is as follows. There is little doubt that during the desiccation process, polyhydroxyl compounds and amino acids react together, causing conformational changes that strengthen the overall protein structure. The presence of sugar additives causes conformational changes in the coat proteins, and in the new configuration, coat proteins do not react (or react more slowly) with the polyhydroxy coat moieties. In the absence of these sugar additives and free molecules, the coat proteins may react irreversibly, through Maillard reactions, with polyhydroxy coat moieties and cause loss of viability. In addition, the sugar additives could compete with the polyhydroxy coat moieties for the reaction sites of the coat proteins or physically hinder that reaction's molecular collisions. The result in each case would be more aerostable microorganisms. When the proteins are in their more normal aqueous environment, the Maillard reactions leading to this inactivation may not occur because the reaction sites are separated either by bulk water molecules or by water molecules bound at the reactive sites. Removal of these water molecules by evaporation would then lead to the events proposed above. A possible mechanism for aerosol inactivation has been presented here by inference from mostly unrelated work. There have been many studies on solute concentration effects which may also be applied to help model the molecular events which take place during inactivation. In a practical sense, the process may be viewed as the increase of solute concentration which concomitantly occurs during drying of the aerosol droplet.

Freeze-drying, freezing, and aerosolization of microorganisms all act to remove water from the system. Elaborate studies have been performed with freeze-dried microorganisms as models to study the processes responsible for inactivation. Israeli et al. (65) studied freeze-dried microorganisms and showed the importance of water content to the viability of microorganisms. They concluded that some areas of the biomembrane, specifically the phospholipid bilayers, go through conformational changes, from crystalline to gel phases, as a result of water loss. These factors caused damage to cell proteins, which in turn resulted in a loss of viability.

BIOAEROSOL STABILIZERS

Several compounds have been shown to provide protection to aerosolized microorganisms. Generation fluid additives such as inositol and bovine serum albumin have been shown

to increase the recovery of some bacteria (51). Some polyhydric compounds such as raffinose, dextran, glycerol, glutamate, and inositol have been shown to increase recovery of bacteria, but the results are dependent on the RH of the system (20, 51, 112). Cox (24) hypothesizes that these compounds bind with available proteins and stabilize them against denaturization. Marthi and Lighthart (82) added betaine to the generating and collection fluid during field studies collecting airborne bacteria and observed increases from 21.6 to 61.3% compared with samples taken in the absence of betaine. The recovery was dependent on the concentration of betaine, but they found the optimum concentration to be between 2 and 5 mM. Betaine acts as an osmoprotectant. Additionally, trehalose has been shown to afford protection to both freeze-dried (65) and aerosolized bacteria. Trehalose is known to stabilize lipids, proteins, and phospholipids, important constituents of microorganisms.

REPAIR MECHANISMS AND RESUSCITATION

Microorganisms in the environment are constantly subjected to forces that cause inactivation. Some evidence indicates that cell growth occurs in airborne particles (39). The consequences of these forces are not always lethal. Marthi (80) and Cox (24) present interesting details involved with sublethal damage to microorganisms. Marthi (80) states that perhaps the most important effect of sublethal stress is the inability of stressed microorganisms to grow on different media, both selective and nonselective. The fact that microorganisms are present but cannot be detected by standard enumeration procedures will require additional planning by the investigator to confirm that the reported results are representative of the sample that has been collected. Cox (24) points out that although the microorganisms are stressed and not detectable by normal methods, they may be able to recover from the injury and initiate disease. Marthi (80) points out that the primary sites of initial stress include the outer membrane, cell wall, cytoplasmic membrane, RNA and ribosomes, and DNA. When injuries are inflicted on these bacterial cell constituents, the metabolic activities of microorganisms are concomitantly influenced. Cox (24) addresses repair mechanisms associated with surface structures, transport functions, and radiation damage in his review article.

VIABILITY MODELS
Exponential Decay Model

The loss of viability of aerosolized microorganisms is caused by complex physical, meteorological, and cellular interactions. Early attempts to explain aerosol viability relied on the exponential decay model, with mixed results. Even though exponential decay has been shown to be a simplification, application of intricate mathematical expressions often delivered conclusions which were no more accurate than simple expressions. If all of the known parameters were applied to a mathematical expression, there would be around 20 inputs, many of which would be dependent on other parameters during specific periods of flight. The expression for exponential decay is

$$V_t = V_0 e^{-K} \tag{1}$$

where V_t is viability at time t, V_0 is viability at time zero, and K is the decay rate constant.

Kinetic Model

Cox, in a series of articles (24), presents what he calls the kinetic model. He was dissatisfied with the lack of explanation of the exponential model, which did not account for the time-dependent decay observed for most microorganisms. Cox explains his kinetic model, which supposes that microorganisms contain a molecular species $B(n)H_2O$, the biological activity of which is essential for a microbial cell to replicate or be infectious. $B(n)H_2O$, when exposed to an environment of lowered water activity (or low RH), forms a series of hydrates similar to those formed by other biomolecules; some of these hydrates are unstable and spontaneously denature through a first-order process, i.e.,

$$-dx/dt = kx \qquad (2)$$

where x is the concentration of the species which denatures. The model form is then

$$B(n)H_2O \leftrightarrow B(n - x)H_2O + xH_2O$$

$$\overset{k_+}{\underset{k_-}{\leftrightarrow}} B(n - x - y)H_2O + yH_2O \qquad (3)$$

$$\leftrightarrow B + iH_2O$$

where $B(n - x)H_2O$ is the denatured form with a rate constant k_x and $B(n - x - y)H_2O$ is the denatured form with a rate constant k_y.

Cox (24) then applies probability theory to evaluate the likelihood of death, which is related to percent viability, and sets up appropriate boundary conditions and integrates the results. When denaturation follows a first-order pattern, the final form of the equation is written

$$\ln V = K_1[B(n - x)H_2O]_0(e^{-kt} - 1) + \ln 100 \qquad (4)$$

where k is a first-order denaturation constant, t is time, K_1 is the probability constant, and V is viability. Cox (24) has analyzed several hundred viability-time curves and has found very good agreement with experimental results for dehydration inactivation.

Catastrophe Model

Catastrophe theory was formulated to explain the nature of discontinuous events such as the breaking of waves on a beach, the crash of the stock market, or the sudden aggression displayed by an animal. Overall reactions involving large numbers of molecules appear to behave in a continuous fashion because discontinuities are smoothed, but on an individual level, the reactions are discontinuous. If the number of reactions is relatively low, as is the case when aerosol viability is being considered, discontinuity is appropriately represented and aerosol inactivation rates may be predicted. On an individual level, the loss of infectivity of an airborne microorganism is a sudden discontinuous event, and this change in state is termed the catastrophe. The basis of catastrophe theory is related to the potential energy and therefore equilibrium of the system. The potential energy is a function of what are termed control parameters, which govern the equilibrium of an event. Within a certain range, equilibrium will not be influenced by variations in the control parameters, but some small critical change can cause a shift in the potential energy and result in inactivation. Cox (24) combined catastrophe theory and kinetics to explain loss of viability caused by desiccation, temperature, oxygen, and OAF. The calculated viability curves agreed very well for inactivation caused by oxygen concentration and OAF and were more accurate than predictions based on probability theory for denaturation induced by desiccation.

Dispersion Models

Models have been developed to predict dispersion and deposition of microbial aerosols with respect to penetration of structures (99), to predict infectious microbial concentrations downwind from known sources (1, 69, 71, 98, 103, 119), and to evaluate the spread of plant pathogens outdoors (19). Many of the early models were based on the treatments of atmospheric diffusion which were created to predict the fate of air pollutants. This inert particle dispersion model by Pasquill (89) is empirical and based on observations of the deposition of inanimate particles. Results from its application yield average distributions of airborne microbes, have a somewhat limited downwind range, and should be applied to flat terrain with steady wind conditions. Assumptions which are made for this model include the following: (i) there is Gaussian distribution of the plume in the horizontal and vertical plane; (ii) there is total reflection of particles from the ground; (iii) particles are emitted from the source at a constant rate; (iv) wind velocity and direction are constant; (v) the terrain is flat; (vi) particles are smaller than 20 μm, making gravitational effects negligible; and (vii) diffusion downwind is negligible, as is the difference in wind velocity between the source and real wind. The classical form of the inert particle dispersion model is

$$X(x,y,z{:}H) = Q/\{2\pi(\sigma_y\sigma_z)u \times \exp[-0.5(Y/\sigma_y)^2]\}$$
$$\times \{\exp[-0.5((Z - H)/\sigma_y)^2] \qquad (5)$$
$$+ [-0.5((Z + H)/\sigma_z)^2]\}$$

where X is the number of particles per cubic meter at downwind location x, y, or z; Q is the number of particles emitted from the source per second; u is the mean air speed in meters per second; σ_y is the standard deviation of the horizontal concentration (at the downwind distance); σ_z is the standard deviation of the vertical concentration (at the downwind distance); and H is the height of the source plus the plume rise.

The equation can be simplified for ground level ($z = 0$) and becomes

$$X(x,y,0{:}H) = Q/\pi\sigma_y\sigma_zu \exp[-(y^2/2\sigma_y^2 + H^2/2\sigma_z^2)] \qquad (6)$$

For a ground-level source ($H = 0$) and analysis along the center line ($y = 0$), the equation becomes

$$X(x,y,z{:}H) = Q/\pi\sigma_y\sigma_zu \qquad (7)$$

As explained earlier, these equations are for inanimate particles and do not take into account microbial viability decay. Lighthart and Frisch (72) expanded on the inert particle diffusion model and added a biological death constant (α) and created a graphic method which would estimate ground-level concentrations when microbial death rate, mean wind speed, atmospheric stability class, source height, and downwind sampled distance were known. Microbial death rate was determined under laboratory conditions and applied to the model. The equation for biological death (BD), after modification, becomes

$$X(x,y,z{:}H)_{BD} = X(x,y,z{:}H) \exp(-\alpha t) \qquad (8)$$

where t was approximated by x/u.

Biological decay and the rate constants associated with inactivation must be determined under dynamic conditions in the laboratory. Teltsch et al. (103) presented methodology to estimate the death constant under field conditions to determine inactivation and fatal concentrations downwind from sprinklers using wastewater. Death constants must be evaluated over a wide range of environmental conditions for accurate results, and the determined decay constants cannot be applied for predictive uses because of the specific meteorological conditions under which the study took place.

Lighthart and Mohr (75) applied best-fit laboratory-generated decay constants to the Gaussian plume model, using microbial source strength and local hourly mean weather data to drive the model through a summer and winter day cycle. For near-source locations, higher wind speeds or short travel times exerted a major modulating effect during the day because time was inadequate for inactivation. Additionally, as travel time increased, because of low wind speed or long distances, modulation was due more to solar irradiance, RH, and temperature than to wind speed.

Spendlove (99) applied various models (box, test tube) to determine the penetration of structures by microbial aerosols. Models showed that for a single-story dwelling without air conditioning, the aerosol dose received by people was the same inside as outside and that air conditioning units would act to moderately decrease inside concentrations (depending on ventilation rates).

Recent developments may change the methods by which airborne microorganisms are modeled. These developments include changes in the application of computers to sort out numerous data inputs; new interests in the fate and deposition of genetically altered microorganisms to mediated destruction caused by climate, other microorganisms, and insects; and a better understanding of the parameters that influence aerosol inactivation and dispersion. Lighthart and Kim (74) have recently presented a simulation model which describes the dispersion of individual droplets of water containing viable microbes. By repeating the modeling process many times for individual droplets, an aerosol cloud can be simulated. The model accounts for the physical, chemical, biological, and measured meteorological parameters of each droplet for many time increments. The droplet model is separated into five submodels which include aerosol generation, evaporation, dispersion, deposition, and microbial death, all of which are calculated at chosen time intervals in the trajectory for each drop. The results show that evaporation is an important factor in determining deposition sites because of particle size dependence, chemical reactions, and protection offered by large droplets. Wind gust data are required because average wind velocity data tend to oversimplify what is occurring on a micrometeorological scale.

TRANSPORT OF BIOAEROSOLS

Naturally occurring bioaerosols are injected into the atmosphere either by chance (wind, rain, bursting bubbles, etc.) or by processes governed by natural selection. Wickman (115) presents an explanation based on the physical and biological (molecular processes) parameters and uses the terms deposition, adhesion, and release to explain the transport of bioaerosols. Cox (24) refers to the same mechanisms as "take-off processes" and "landing on surfaces," and his discussion is geared more to the consequences as they relate to the human respiratory system. Interest in the processes that influence bioaerosol transport has increased in recent years as a result of the application of genetically engineered microorganisms, and analyses of their release, dispersion, and possible effects, to surrounding areas. Chamberlain (18)

studied both the deposition and release of spores and pollens as they relate to biological surfaces.

Deposition and Adhesion

Wickman (115) presents a notable review of the forces which govern deposition and adhesion and presents several novel approaches that are now being used for these evaluations. Deposition is in part governed by the mass of the particle. Terminal velocity is governed by the mass of a particle, a cubic function of the particle diameter. Particles between 1.0 and 5.0 μm normally follow the streamlines of the surrounding air. Larger particles often have the momentum to deviate from the streamlines, impact surrounding surfaces, and be deposited. Particle bounce also increases with increasing mass. Additional forces include gravitational settling, convection (due to temperature variations), diffusion (thermal energy), and eddy diffusion. When particles enter the region near a surface, London-van der Waals forces (electrical dipole moments) and electrostatic forces (electrical charge forces of attraction and repulsion) can influence deposition.

Wickman (115) describes adhesive forces as being dominated by the molecular structure and organization near the contact surfaces. Two types of electrodynamic forces, London-van der Waals forces and electrostatic interactions, are primarily responsible for the at-a-distance interactions that occur between aerosol particles and surfaces. If liquid is present, interfacial reactions (meniscus formation) can develop. Adhesion is also governed by the atomic attraction between surfaces (Hamaker constants, electronic frequency) and geometrical factors (shape of surfaces, flat versus spherical).

Release

The energy required to overcome the forces of adhesion is deduced by analyzing and measuring particle release. These forces can be measured with either an atomic force microscope (13, 93) or a surface force apparatus (14, 38, 63, 64).

Adhesion can be overcome by mechanical or aerodynamic forces and initiate the release of biological particles on surfaces. The mechanical energy frequently originates from turbulent mixing of the wind. The force required to remove particles increases with decreasing particle size. The surfaces of bioaerosols are complex structures which include interactions between proteins, phospholipids, peptidoglycans, etc. Wickman (115) analyzes release characteristics by utilizing S-layers, a simpler exterior covering, present in some microorganisms. He concludes that in dry air, the S-layer surface is ridged and hydrophobic, but the surface groups are capable of reacting with adjacent S-layer surfaces. This tends to increase adhesion with increasing contact time. When water is more available, the surface becomes more hydrophilic, and adhesion forces steadily increase as a result of the reorientation and influences of exposed polar groups. As more water becomes available (higher RH), increased hydration ensures that the polar-hydrophilic groups are available at the surface so that equilibrium adhesion occurs. He concludes that RH and contact time are important factors that control the forces required to overcome adhesion which may result in release.

CONCLUSIONS

The study of bioaerosols has recently gained more attention because of the recognized importance of both indoor and outdoor events which have been shown to affect applicable populations. Significant additions to the mechanisms governing the fate and transport of bioaerosols have been identified. Hydrodynamic and physical factors have been studied in increasing detail so that the biochemical changes that are responsible for the inactivation of aerosolized microorganisms can be defined. Methods have been developed to study the mechanisms involved with the deposition, adhesion, and release of biological particles. It was pointed out that the water content and the state of water in the biological particle are the primary factors associated with inactivation and viability. Some microorganisms have been shown to multiply in aerosols, while others have displayed built-in repair mechanisms. The realization and application of standards for generating, storing, sampling, and presenting data will allow better comparison and reproducibility of results. Before a bioaerosol project is initiated, a substantial amount of study must be completed to determine methods to maximize the recovery of aerosolized microorganisms. These studies should include effects related to sampling media, counting media, generation media, and the makeup of test atmospheres.

A special thanks is extended to Frances Anderson for her help on this chapter. Her insights were essential to its success.

REFERENCES

1. **Adams, A. P., and J. C. Spendlove.** 1970. Coliform aerosols emitted by sewage treatment plants. *Science* **169:** 1218–1220.
2. **Adams, D. J., J. C. Spendlove, R. S. Spendlove, and B. B. Barnett.** 1982. Aerosol stability of infectious and potentially infectious reovirus particles. *Appl. Environ. Microbiol.* **44:**903–908.
3. **Akers, T. G.** 1973. Survival, damage and inactivation in aerosols, p. 73–86. *In* J. F. Hers and K. C. Winkler (ed.), *Airborne Transmission and Infection.* John Wiley & Sons, New York.
4. **Akers, T. G., S. Bond, and L. J. Goldberg.** 1966. Effect of temperature and relative humidity on the survival of airborne Colombia sk group viruses. *Appl. Microbiol.* **14:** 361–365.
5. **Akers, T. G., and M. T. Hatch.** 1968. Survival of a picornavirus and its infectious RNA after aerosolization. *Appl. Microbiol.* **16:**1811–1813.
6. **Anderson, J. D., and C. S. Cox.** 1976. Microbial survival, p. 203–226. *In* T. R. Gray and J. R. Postgate (ed.), *The Survival of Vegetative Microbes.* Cambridge University Press, Cambridge.
6a. **Anonymous.** 1991. *J. Aerosol. Sci. Technol.* **14(1):1–4.**
7. **Barlow, D. F.** 1972. The effects of various protecting agents on the inactivation of foot-and-mouth disease virus in aerosols and during freeze-drying. *J. Gen. Virol.* **17:** 281–288.
8. **Beebe, J. M.** 1959. Stability of disseminated aerosols of *Pasteurella tularensis* subjected to simulated solar radiations and various humidities. *J. Bacteriol.* **78:**18–24.
9. **Beebe, J. M., and G. W. Pirsch.** 1958. Response of airborne species of *Pasteurella* to artificial radiation simulating sunlight under different conditions of relative humidity. *Appl. Microbiol.* **6:**127–138.
10. **Benbough, J. E.** 1971. Some factors affecting the survival of airborne viruses. *J. Gen. Virol.* **10:**209–220.
11. **Benbough, J. E., and A. M. Hood.** 1971. Viricidal activity of open air. *J. Hyg.* (Cambridge) **69:**619–626.
12. **Berendt, R. F., and E. L. Dorsey.** 1971. Effect of simulated solar radiation and sodium fluorescein on the recovery of Venezuelan equine encephalomyelitis virus from aerosols. *Appl. Microbiol.* **21:**447–450.
13. **Binnig, G., and H. Rohrer.** 1986. Scanning tunneling microscopy. *IBM J. Res. Dev.* **30:**355–369.
14. **Bradley, R. S.** 1932. The cohesive force between solid surfaces and the surface energy of solids. *Philos. Mag.* **13:** 853–862.
15. **Bridges, B. A.** 1976. Survival of bacteria following exposure to ultraviolet and ionizing radiations. *Soc. Gen. Microbiol.* **26:**183–208.
16. **Buckland, F. E., and D. A. J. Tyrrell.** 1964. Loss of infectivity on drying various viruses. *Nature* (London) **195:**1063–1064.
17. **Cabelli, V. J.** 1962. The rehydration of aerosolized bacteria: compounds which enhance the survival of rehydrated *Pasteurella tularensis.* Technical Report no. 314. U.S. Army, Dugway Proving Ground, Dugway, Utah.
18. **Chamberlain, A. C.** 1967. Deposition of particles to natural surfaces, p. 138–164. *In* P. H. Gregory and J. L. Montieth (ed.), *Airborne Microbes.* Cambridge University Press, Cambridge.
19. **Cook, R. J., and K. F. Baker.** 1983. *The Nature and Practice of Biological Control of Plant Pathogens.* American Phytopathological Society Press, St. Paul, Minn.
20. **Cox, C. S.** 1966. The survival of *Escherichia coli* atomized into air and into nitrogen from distilled water and from solution of protecting agents as a function of relative humidity. *J. Gen. Microbiol.* **43:**383–399.
21. **Cox, C. S.** 1968. The aerosol survival of *Escherichia coli* B in nitrogen, argon, and helium atmospheres and the influence of relative humidity. *J. Gen. Microbiol.* **50:** 139–147.
22. **Cox, C. S.** 1970. Aerosol survival of *Escherichia coli* B disseminated from the dry state. *Appl. Microbiol.* **19:** 604–607.
23. **Cox, C. S.** 1971. Aerosol survival of *Pasteurella tularensis* disseminated from the wet and dry states. *Appl. Microbiol.* **21:**482–486.
24. **Cox, C. S.** 1987. *The Aerobiological Pathway of Microorganisms.* John Wiley & Sons, Chichester, England.
25. **Cox, C. S., and F. Baldwin.** 1966. The use of phage to study causes of loss of viability of *Escherichia coli* in aerosols. *J. Gen. Microbiol.* **44:**15–22.
26. **Cox, C. S., and F. Baldwin.** 1967. The toxic effect of oxygen upon the aerosol survival of *Escherichia coli* B. *J. Gen. Microbiol.* **49:**115–117.
27. **Cox, C. S., J. Baxter, and B. J. Maidment.** 1973. A mathematical expression for oxygen-induced death in dehydrated bacteria. *J. Gen. Microbiol.* **75:**179–185.
28. **Cox, C. S., S. J. Gagen, and J. Baxter.** 1974. Aerosol survival of *Serratia marcescens* as a function of oxygen concentration, relative humidity, and time. *Can. J. Microbiol.* **20:**1529–1534.
29. **Cox, C. S., and L. J. Goldberg.** 1972. Aerosol survival of *Pasteurella tularensis* and the influence of relative humidity. *Appl. Microbiol.* **23:**1–3.
30. **Cox, C. S., W. J. Harris, and J. Lee.** 1974. Viability and electron microscope studies of phages T3 and T7 subjected to freeze-drying, freeze-thawing and aerosolization. *J. Gen. Microbiol.* **81:**207–215.
31. **Cox, C. S., and R. J. Heckly.** 1973. Effects of oxygen upon freeze-dried and freeze-thawed bacteria: viability and free radical studies. *Can. J. Microbiol.* **19:**189–194.
32. **de Jong, J. C.** 1970. Decay mechanism of polio and EMC viruses in aerosols, p. 210–245. *In* I. H. Silver (ed.), *Proceedings of the Third International Symposium on Aerobiology.* Academic Press, New York.
33. **de Jong, J. C.** 1970. On the mechanism of the decay of poliomyelitis virus and encephalomyocarditis virus in aerosols, p. 210–211. *In* I. H. Silver (ed.), *Aerobiology,*

Third International Symposium on Aerobiology. Academic Press, Inc., London.

34. **de Jong, J. C., M. Harnsen, A. D. Plantinga, and T. Trouwborst.** 1976. Aerosol stability functions of Semliki Forest virus. *Appl. Environ. Microbiol.* **32:**315–319.

35. **de Jong, J. C., and K. C. Winkler.** 1968. The inactivation of poliovirus in aerosols. *J. Hyg.* (Cambridge) **66:** 557–565.

36. **de Mik, G., and L. de Groot.** 1973. Effect of gases on the aerosol stability of various microorganisms, p. 155–158. *In* J. F. P. Hers and K. C. Winkler (ed.), *Airborne Transmission and Infection.* John Wiley & Sons, New York.

37. **de Mik, G., and I. de Groot.** 1977. The germicidal effect of the open air in different parts of the Netherlands. *J. Hyg.* (Cambridge) **78:**175–180.

38. **Derjaguin, B. V., I. I. Abrikosov, and E. M. Lifshitz.** 1956. Direct measurement of molecular attraction between solids separated by a narrow gap. *Q. Rev. Chem. Soc.* **10:**295–329.

39. **Dimmick, R. L., H. Wolochow, and M. A. Chatigny.** 1979. Evidence that bacteria can form new cells in airborne particles. *Appl. Environ. Microbiol.* **37:**924–927.

40. **Dimmock, N. L.** 1967. Differences between the thermal inactivation of picornaviruses at "high" and "low" temperatures. *Virology* **31:**338–353.

41. **Donaldson, A. I.** 1972. Effect of radiation on selected virus. *Vet. Bull.* **48:**83–94.

42. **Donaldson, A. I., and N. P. Ferris.** 1975. The survival of foot-and-mouth disease virus in open air conditions. *J. Hyg.* (Cambridge) **74:**409–415.

43. **Druett, H. A.** 1973. The open air factor, p. 141–151. *In* J. F. P. Hers and K. C. Winkler (ed.), *Airborne Transmission and Infection.* John Wiley & Sons, New York.

44. **Dubovi, E. J.** 1971. Biological activity of the nucleic acid extracted from two bacterial viruses. *J. Appl. Microbiol.* **21:**624–630.

45. **Dubovi, E. J., and T. G. Akers.** 1970. Airborne stability of tailless bacterial viruses S-13 and MS-2. *J. Appl. Microbiol.* **19:**624–630.

46. **Ehrlich, R., and S. Miller.** 1971. Effect of relative humidity and temperature on airborne Venezuelan equine encephalitis virus. *J. Appl. Microbiol.* **22:**194–200.

47. **Elazhary, M. A. S. Y., and J. B. Derbyshire.** 1977. Effect of temperature, relative humidity and medium on the aerosol stability of infectious bovine rhinotracheitis virus. *Can. J. Comp. Med.* **43:**158–167.

48. **Fairchild, G. A.** 1974. Ozone effect on respiratory deposition of vesicular stomatitis virus aerosols. *Am. Rev. Respir. Dis.* **109:**446–451.

49. **Fuchs, N. A.** 1964. *The Mechanics of Aerosols.* Dover Publications Inc., New York.

50. **Goff, G. D., J. C. Spendlove, A. P. Adams, and P. S. Nicholes.** 1973. Emission of microbial aerosols from sewage treatment plants that use trickling filters. *Health Serv. Rep.* **88:**640–652.

51. **Goldberg, L. J., and I. Ford.** 1973. The function of chemical additives in enhancing microbial survival in aerosols, p. 86–89. *In* J. F. P. Hers and K. C. Winkler (ed.), *Airborne Transmission and Infection.* John Wiley & Sons, New York.

52. **Happ, J. W., J. B. Harstad, and L. M. Buchanan.** Effect of air ions on submicron T1 bacteriophage aerosols. *Appl. Microbiol.* **14:**888–891.

53. **Harper, G. J.** 1961. Airborne microorganisms: survival tests with four viruses. *J. Hyg.* (Cambridge) **59:**479–486.

54. **Harper, G. J.** 1963. The influence of environment on the survival of airborne virus particles in the laboratory. *Arch. Gesamte Virusforsch.* **13:**64–71.

55. **Hatch, M. T., and R. L. Dimmick.** 1966. Physiological responses of airborne bacteria to shifts in relative humidity. *Bacteriol. Rev.* **30:**597–603.

56. **Hatch, M. T., and J. C. Warren.** 1969. Enhanced recovery of airborne T$_3$ coliphage and *Pasteurella pestis* bacteriophage by means of a presampling humidification technique. *Appl. Microbiol.* **17:**685–689.

57. **Hatch, M. T., and H. Wolochow.** 1969. Bacterial aerosols, p. 267–295. *In* R. L. Dimmick and A. B. Akers (ed.), *An Introduction to Experimental Aerobiology.* Wiley Interscience, New York.

58. **Hemmes, J. H., K. C. Winkler, and S. M. Kool.** 1960. Virus survival as a seasonal factor in influenza and poliomyelitis. *Nature* (London) **188:**430–438.

59. **Hess, G. E.** 1965. Effects of oxygen on aerosolized *Serratia marcescens.* *Appl. Microbiol.* **13:**781–787.

60. **Hidy, G. M.** 1984. *Aerosols.* Academic Press Inc., New York.

61. **Hinds, W. C.** 1982. *Aerosol Technology.* John Wiley & Sons, New York.

62. **Ijaz, M. K., S. A. Sattar, C. M. Johnson-Lussenburg, and V. S. Springthorpe.** 1985. Comparison of the airborne survival of calf rotavirus and poliovirus type 1 (Sabin) aerosolized as a mixture. *Appl. Environ. Microbiol.* **49:** 289–293.

63. **Israelachvili, J. N., and G. E. Adams.** 1978. Measurement of forces between two mica surfaces in aqueous electrolyte solutions in the range 0–100 nm. *J. Chem. Soc. Faraday Trans. 1* **74:**975–1001.

64. **Israelachvili, J. N., and D. Tabor.** 1973. Van der Waals forces: theory and experiment. *Prog. Surf. Membr. Sci.* **7:** 1–55.

65. **Israeli, E., J. Gitelman, and B. Lighthart.** 1994. Death mechanisms in bioaerosols, p. 166–191. *In* B. Lighthart and A. J. Mohr (ed.), *Atmospheric Microbial Aerosols.* Chapman & Hall, New York.

66. **Jensen, M. M.** 1964. Inactivation of airborne viruses by ultraviolet irradiation. *Appl. Microbiol.* **12:**418–412.

67. **Kethley, T. W., W. B. Crown, and E. L. Fincher.** 1957. The nature and composition of experimental bacterial aerosols. *Appl. Microbiol.* **5:**1–17.

68. **Krinsky, N. I.** 1976. Cellular damage initiated by visible light. *Symp. Soc. Gen. Microbiol.* **26:**209–239.

69. **Lembke, L. L., and R. N. Kniseley.** 1980. Coliforms in aerosols generated by a municipal solid waste recovery system. *Appl. Environ. Microbiol.* **40:**888–891.

70. **Lighthart, B.** 1973. Survival of airborne bacteria in a high urban concentration of carbon monoxide. *Appl. Microbiol.* **25:**86–91.

71. **Lighthart, B.** 1984. Microbial aerosols: estimated contribution of combine harvesting to an airshed. *Appl. Environ. Microbiol.* **47:**430–432.

72. **Lighthart, B., and A. S. Frisch.** 1976. Estimation of viable airborne microbes downwind from a point source. *Appl. Environ. Microbiol.* **31:**700–704.

73. **Lighthart, B., V. E. Hiatt, and A. T. Rossano, Jr.** 1971. The survival of airborne *Serratia marcescens* in urban concentrations of sulfur dioxide. *J. Air. Pollut. Control Assoc.* **21:**539–642.

74. **Lighthart, B., and J. Kim.** 1989. Simulation of airborne microbial transport. *Appl. Environ. Microbiol.* **55:** 2349–2355.

75. **Lighthart, B., and A. J. Mohr.** 1987. Estimating downwind concentrations of viable airborne microorganisms in dynamic atmospheric conditions. *Appl. Environ. Microbiol.* **53:**1580–1583.

76. **Lighthart, B., and A. J. Mohr.** 1994. *Atmospheric Microbial Aerosols. Theory and Applications.* Chapman & Hall, New York.

77. **Lighthart, B., and B. T. Shaffer.** 1994. Bacterial flux

from chaparral into the atmosphere in mid-summer at a high desert location. *Atmos. Environ.* **28:**1267–1274.

78. **Lighthart, B., and B. T. Shaffer.** 1995. Airborne bacteria in the atmospheric surface layer: temporal distrubution above a grass seed field. *Appl. Environ. Microbiol.* **61:** 1492–1496.

79. **Loosli, C. G., H. M. Lemon, O. H. Robertson, and E. Appel.** 1943. Experimental air-borne influenza infection. I. Influence of humidity on survival of virus in air. *Proc. Soc. Exp. Biol. Med.* **53:**205–206.

80. **Marthi, B.** 1994. Resuscitation of microbial bioaerosols, p. 192–225. *In* B. Lighthart and A. J. Mohr (ed.), *Atmospheric Microbial Aerosols.* Chapman & Hall, New York.

81. **Marthi, B., V. P. Fieland, M. Walter, and R. J. Seidler.** 1990. Survival of bacteria during aerosolization. *Appl. Environ. Microbiol.* **56:**3436–3467.

82. **Marthi, B., and B. Lighthart.** 1990. Effects of betaine on enumeration of airborne bacteria. *Appl. Environ. Microbiol.* **56:**1286–1289.

83. **May, K. R., H. A. Druett, and L. P. Packman.** 1969. Toxicity of open air to a variety of microorganisms. *Nature* (London) **221:**1146–1147.

84. **Mayhew, C. J., W. D. Zimmerman, and N. Hahon.** 1968. Assessment of aerosol stability of yellow fever virus by fluorescent-cell counting. *Appl. Microbiol.* **16:**263–266.

85. **Miller, W. S., and M. S. Artenstein.** 1966. Aerosol stability of three acute respiratory disease viruses. *Proc. Soc. Exp. Biol. Med.* **123:**222–227.

86. **Moe, K., and G. J. Harper.** 1983. The effect of relative humidity and temperature on the survival of calf rotavirus in aerosol. *Arch. Virol.* **76:**211–216.

87. **Mohr, A. J.** 1984. Aerosol stability of reovirus. Dissertation. Utah State University, Logan.

88. **Mohr, A. J.** 1991. Development of models to explain the survival of viruses and bacteria in aerosols, p. 160–190. *In* C. J. Hurst (ed.), *Modeling the Environmental Fate of Microorganisms.* American Society for Microbiology, Washington, D.C.

89. **Pasquill, F.** 1961. The estimation of the dispersion of windborne material. *Meteorol. Mag.* **90:**33–49.

90. **Phillips, G., G. J. Harris, and M. W. Jones.** 1963. The effect of ions on microorganisms. *Int. J. Biometeorol.* **8:** 27–37.

91. **Rabey, F., R. J. Janssen, and L. M. Kelley.** 1969. Stability of St. Louis encephalitis virus in the airborne state. *Appl. Environ. Microbiol.* **18:**880–882.

92. **Resnick, I. G., A. J. Mohr, and B. G. Harper.** 1987. Production, purification and use of MS2 bacteriophage as a viral simulant, abstr. Q-23, p. 285. *In Abstracts of the 87th Annual Meeting of the American Society for Microbiology 1987.* American Society for Microbiology, Washington, D.C.

93. **Sarid, D.** 1991. *Scanning Force Microscopy.* Oxford University Press, New York.

94. **Sattar, S. A., M. K. Ijaz, C. M. Johnson-Lussenburg, and V. S. Springthorpe.** 1984. Effect of relative humidity on the airborne survival of rotavirus SA11. *Appl. Environ. Microbiol.* **47:**879–881.

95. **Smith, L. P., and M. E. Hugh-Jones.** 1969. The weather factor in foot-and-mouth disease epidemics. *Nature* (London) **223:**712–715.

96. **Soddu, A., and W. R. Vieth.** 1980. The effect of sugars on membranes. *J. Mol. Catal.* **7:**491–500.

97. **Songer, J. R.** 1966. Influence of relative humidity on the survival of some airborne viruses. *Appl. Microbiol.* **15:** 1–16.

98. **Sorber, C. A., H. T. Bausum, S. A. Schaub, and M. J. Small.** 1976. A study of bacterial aerosols at a wastewater

irrigation site. *J. Water Pollut. Control Fed.* **48:** 2367–2379.

99. **Spendlove, J. C.** 1975. Penetration of structures by microbial aerosols. *Dev. Ind. Microbiol.* **16:**427–436.

100. **Spendlove, J. C., and K. F. Fannin.** 1982. Methods of characterization of virus aerosols, p. 261–329. *In* C. P. Gerba and S. M. Goyal (ed.), *Methods in Environmental Virology.* Marcel Dekker, Inc., New York.

101. **Stetzenbach, L., and B. Lighthart.** 1994. Distribution of bioaerosols, p. 68–98. *In* B. Lighthart and A. J. Mohr (ed.), *Atmospheric Microbial Aerosols.* Chapman & Hall, New York.

102. **Strange, R. E., and C. S. Cox.** 1976. Survival of dried and airborne bacteria. *Symp. Soc. Gen. Microbiol.* **26:** 111–154.

103. **Teltsch, B., H. I. Shuval, and J. Tadmor.** 1980. Die-away kinetics of aerosolized bacteria from sprinkler application of wastewater. *Appl. Environ. Microbiol.* **39:** 1191–1197.

104. **Trouwborst, T., J. C. de Jong, and K. C. Winkler.** 1972. Mechanism of inactivation in aerosols of phage T_1. *J. Gen. Virol.* **15:**235.

105. **Walter, M. V., B. Marthi, V. P. Fieland, and L. M. Ganio.** 1990. Effect of aerosolization on subsequent bacterial survival. *Appl. Environ. Microbiol.* **56:**3468–3472.

106. **Ward, R. L., and C. S. Ashley.** 1977. Inactivation of enteric viruses in wastewater sludge through dewatering by evaporation. *Appl. Environ. Microbiol.* **34:**564.

107. **Warren J. C., T. G. Akers, and E. J. Dubovi.** 1969. Effect of prehumidification on sampling of selected airborne viruses. *Appl. Microbiol.* **18:**893–896.

108. **Webb, S. J.** 1959. Factors affecting the viability of airborne bacteria. I. Bacteria aerosolized from distilled water. *Can. J. Microbiol.* **5:**649–669.

109. **Webb, S. J.** 1960. Factors affecting the viability of airborne bacteria. III. The role of bonded water and protein structure in the death of air-borne cells. *Can. J. Microbiol.* **6:**89–105.

110. **Webb, S. J.** 1965. *The Role of Bound Water in the Maintenance of the Integrity of a Cell or Virus.* Charles C Thomas, Springfield, Ill.

111. **Webb, S. J.** 1969. The effects of oxygen on the possible repair of dehydration damage by *Escherichia coli. J. Gen. Microbiol.* **58:**317–326.

112. **Webb, S. J., R. Bather, and R. W. Hodges.** 1963. The effect of relative humidity and inositol on air-borne viruses. *Can. J. Microbiol.* **9:**87–94.

113. **Webb, S. J., and A. D. Booth.** 1969. The effect of radiation on *Escherichia coli. Nature* (London) **222:** 1199–1200.

114. **Wells, W. F.** 1934. Droplet and droplet nuclei. *Am. J. Hyg.* **20:**611–627.

115. **Wickman, H. H.** 1994. Deposition, adhesion, and release, p. 99–165. *In* B. Lighthart and A. J. Mohr (ed.), *Atmospheric Microbial Aerosols.* Chapman & Hall, New York.

116. **Wright, D. N., and G. D. Bailey.** 1969. Effect of relative humidity on the stability of *Mycoplasma pneumoniae* exposed to simulated solar ultraviolet and to visible radiation. *Can. J. Microbiol.* **15:**1449–1452.

117. **Wright, D. N., G. D. Bailey, and L. J. Goldberg.** 1969. Effect of temperature on survival of airborne *Mycoplasma pneumoniae. J. Bacteriol.* **99:**491–495.

118. **Wright, D. N., G. D. Bailey, and M. T. Hatch.** 1968. Survival of airborne mycoplasma as affected by relative humidity. *J. Bacteriol.* **95:**251–252.

119. **Zeterberg, J. M.** 1973. A review of respiratory virology and the spread of virulent and possibly antigenic viruses via air conditioning systems. *Ann. Allergy* **31:**228–234.

Airborne Fungi and Mycotoxins

CHIN S. YANG AND ECKARDT JOHANNING

70

Fungi are a heterogeneous group of organisms including the true slime molds (class Myxomycetes), the water molds (class Oomycetes), true fungi, and lichens. Some of these organisms, such as the true slime molds and the water molds, have been conveniently placed in the kingdom Fungi. They are, in many aspects, different from true fungi, although they are conventionally studied by mycologists (1). Fungi as a group inhabit a wide range of niches, such as exposed rock (some lichens), the sea (marine fungi), the North Pole, and the tropics (12). They have developed many different modes of obtaining nutrients. Some fungi, such as powdery mildews, are obligate parasites. Some fungi exist in symbiotic relationships with plant roots to form mycorrhizae or with algae to form lichens. A large number of fungi survive as saprophytes. Many of these saprophytes have been found to grow in the indoor environment of man-made buildings and have caused building-related complaints and illnesses (27, 59, 72).

Fungi produce a variety of secondary metabolites, including mycotoxins and fungal volatile organic compounds (VOCs). Mycotoxins are harmful to animals and humans. In addition to mycotoxins, some VOCs produced by actively growing fungi are known irritants or hazardous chemicals. They may pose a health risk to building occupants (3, 21, 59).

This chapter reviews the existing literature on airborne fungi, with emphasis on indoor fungal contamination and the health effects of mycotoxins and fungal organic volatiles. Chapter 75 discusses fungi of concern in agricultural settings. A wealth of literature on outdoor airborne fungi can also be found in reviews by Gregory (28), Flannigan et al. (22), Lacey (46, 47), and Levetin (51). It is important to keep in mind that outdoor airborne spores are often the source of indoor fungal contamination. Their impact on indoor airborne fungal populations could be immediate or delayed until they have settled and have colonized an indoor environment.

AIRBORNE FUNGAL POPULATIONS

It must be emphasized that fungal spores and occasionally hyphal fragments are dispersed and disseminated in air from one location to another. They cannot survive and complete their life cycles by staying afloat in air for an indefinite period of time. Therefore, with respect to fungal contamination, identifying and locating the source of fungal colonization is of higher importance than assessing airborne fungal data (7, 8).

The scientific literature contains numerous reports of studies assessing indoor fungal populations, the majority of which are based on air sampling data. These reports include some focus on hospitals and health care facilities (64, 81, 87), residential dwellings (10, 19, 86, 91), schools (20, 52), and office buildings (61). The focus of hospital sampling was often on *Aspergillus fumigatus*, an opportunistic pathogen. General fungal populations were identified in nonhospital sampling.

A comprehensive assessment of fungal contamination in the indoor environment should include consideration of environmental factors (such as outdoor air, ventilation mode, heating, occupant density, ventilation rate, and moisture), on-site inspection, air sampling, surface and source sampling, sample analysis, risk analysis, and finally remedial actions (7). Unfortunately, most investigations do not follow the approach of comprehensive assessment, often because of insufficient labor, time, and funds. An alternate approach to identifying indoor fungal contamination is to focus on inspection and source sampling.

Factors Affecting Airborne Fungal Populations

Two important factors that directly affect airborne fungal populations are the availability of food and free water for fungal growth and the method of spore dispersal. Other physical, chemical, and biological parameters affecting fungal growth, and hence airborne fungal populations, can be found in recent references (12, 29).

Substrates (Including Water Activity)

Fungi are achlorophyllous and require simple sugars, carbohydrates, and other organics, such as vitamins and amino acids, to survive. In the natural environment, fungi have developed a number of ways to obtain these nutrients (29), such as parasitic, symbiotic, and saprophytic relationships.

Humans share food and living space with fungi. Some fungi cause food spoilage or make food toxic to humans, yet we utilize fungi to produce edible mushrooms and useful by-products, such as enzymes and organic acids (27). *Botrytis cinerea* is a well-known disease-causing fungus on grapevines, strawberries, and other fruits and produce. Species of *Penicillium* and *Aspergillus* often cause spoilage in foodstuffs and make them inedible by animals and humans. Fungi are also known to cause wood stains, wood decay, and biodeterioration and biodegradation of polymers, carpet, plaster, wallpaper, paints and organic coatings, fuels and lubricants, leather, and paper products (27, 55, 62, 98). The references cited underscore the likelihood that fungi can and will grow in man-made environments. Consequently, controlling nutrient sources in order to limit fungal amplification is practically impossible.

One of the critical factors affecting fungal growth indoors is water. There are a number of ways to measure water availability in materials. Water or moisture content of a material is expressed as a percentage of the oven-dry weight (98). However, water content does not suggest the actual availability of free water in the material to fungi. A better measurement of water availability to fungi is water activity (a_w), which is expressed as

$$a_w = \frac{\text{vapor pressure of water in substrate}}{\text{vapor pressure of pure water}}$$

Detailed discussions of water activity in materials and food are presented by Troller and Scott (88) and Gravesen et al. (27).

Many common indoor fungi require an a_w near 1 for growth. Some xerophilic fungi, however, have an optimal a_w ranging from 0.65 to 0.90 (27). Both mesophilic and xerophilic fungi can be found in the indoor environment. Table 1 lists some mesophilic and xerophilic fungi and their minimum a_ws.

Fungal Spore Discharge Mechanisms

Fungal spores are released by two basic mechanisms: (i) active spore discharge and (ii) passive spore release. Various airborne fungal spores have been known to peak during certain hours of the day or night. This periodicity is related to spore discharge mechanisms and environmental factors in nature (47, 95). Details of these two mechanisms and environmental factors affecting spore release are presented by Lacey (47) and Levetin (51).

Fungi with active spore discharge include such common airborne fungi as *Sporobolomyces*, *Epicoccum*, and *Nigrospora* species and some smut-like yeasts. Many ascomycetes and

basidiomycetes also have active discharge mechanisms (51). Spores of *Sporobolomyces* species and some basidiospores are usually most abundant at night or in the predawn hours. Their spore release requires the absorption of moisture to build up release pressure. Dry-spore fungi, such as *Aspergillus*, *Penicillium*, and *Cladosporium* species, are often hydrophobic. They become airborne by passive force, such as air movement or rain drops. *Cladosporium* species usually dominates the airborne spore population during the day. Its spores stay airborne owing to the buoyancy of warmer air.

Some fungi, such as puffballs of basidiomycetes, release their spores in "spore clouds or puffs" when impacted by raindrops, humans, or small animals (51). The spore clouds may persist for a period of time until they are dispersed by air mixing. Results of air sampling can be greatly affected by whether the spore cloud has dispersed (46).

Many fungi frequently detected indoors and outdoors produce spores in a slimy mass. These fungi include such common indoor contaminants as *Acremonium* (although some species of *Acremonium* produce dry spores), *Stachybotrys*, *Fusarium*, and *Trichoderma* species. Slimy spores may be released into the air when they become dry or disturbed or when they are attached to other particles. Their dissemination may also be assisted by insects, animals, or water (96). Because slimy spores do not become airborne easily, their detection indoors should be considered significant. Any detection of *Stachybotrys* spores in air samples taken indoors should trigger further investigation and search for the fungus.

Airborne Fungal Populations

Fungi as a group produce a number of different spores, both sexual and asexual. Many types of spores are capable of becoming airborne. Some spores, such as chlamydospores (asexual) and zygospores (sexual), are not designed to be dispersed by air transmission, and there has been no report of recovering these spores in air. Hypogeous fungi (including such well-known fungi as truffles) produce subterranean sporocarps and disperse their spores through different mechanisms. However, both sexual and asexual spores of five major classes (Myxomycetes, Zygomycetes, Ascomycetes, Basidi-

TABLE 1 Selected fungi and their a_ws[a]

Fungus	a_w	Temp (°C)	Fungus	a_w	Temp (°C)
Absidia corymbifera	0.88	25	*Mucor circinelloides*	0.90	25
Alternaria citri	0.84	25	*Paecilomyces variotii*	0.84	25
Aspergillus candidus	0.75	25	*Penicillium brevicompactum*	0.81	23
A. flavus	0.78	33	*P. chrysogenum*	0.79	25
A. fumigatus	0.82	25	*P. citrinum*	0.80	25
A. niger	0.77	35	*P. expansum*	0.83	23
A. ochraceus	0.77	25	*P. frequentans*	0.81	23
A. restrictus	0.75	25	*P. griseofulvum*	0.81	23
A. sydowii	0.78	25	*P. spinulosum*	0.80	25
A. terreus	0.78	37	*Rhizopus microsporus*	0.90	25
A. versicolor	0.78	37	*R. stolonifer*	0.84	25
A. wentii	0.84	25	*R. oryzae*	0.88	25
Eurotium[b] *amstelodami*	0.70	25	*Stachybotrys chartarum*	0.94	NA[c]
E. chevalieri	0.71	33	*Syncephalastrum racemosum*	0.84	25
Emericella[b] *nidulans*	0.78	37	*Wallemia sebi*	0.70	25

[a] Data are from references 34, 77, and 88.
[b] Teleomorph of *Aspergillus* species.
[c] NA, not applicable.

omcyetes, and Deuteromycetes) have been isolated and reported from air.

Spores of ascomycetes and basidiomycetes have frequently been recovered from air samples by using spore trap samplers. Levetin (50) reported 18 genera of basidiospores from the atmosphere in Tulsa, Oklahoma. Many species of basidiospores have been demonstrated to be allergenic (4, 11, 18, 24, 49, 74). The importance of basidiomycetes in the indoor environment depends on the building construction. Wood-decaying basidiomycetes, such as polypores, are associated with wood and may be found in buildings constructed of wood. In fact, wood decay caused by polypores is a significant problem in the United Kingdom (76) and in the United States (98). Unfortunately, identification of airborne basidiomycetes collected by using culture techniques can be difficult, and these fungi are often missed or misidentified.

Ascospores have been identified from air samples collected using spore traps (78). Many ascomycetes, however, do not produce teleomorphs and ascospores in culture. This makes it difficult to determine the frequency of ascospores in air. Some ascomycetes, such as *Chaetomium* species and *Eurotium* species (teleomorph of some *Aspergillus* species), produce ascocarps in cultures, which suggests that ascospores may become airborne. Ascospores are suspected to be allergenic. Four species of *Chaetomium* were listed as licensed by the Food and Drug Administration for commercial production as allergens (78). Otherwise, little documentation is available on the allergenicity of ascospores.

The majority of airborne spores either trapped on samplers or grown from agar media are asexual spores of the subdivision Deuteromycotina and the class Zygomycetes. Members of the Deuteromycotina are anamorphs, or asexual states, of either Ascomycotina or Basidiomycotina. Asexual spores that belong to Deuteromycotina are called conidia; those of the class Zygomycetes are called sporangiospores. Many of these spores are known allergens. Some of them have been prepared into allergen extracts and approved by the Food and Drug Administration for medical uses (78). Common members of the Deuteromycotina found in air include *Alternaria, Aspergillus, Cladosporium, Epicoccum,* and *Penicillium* species. Species of *Mucor* and *Rhizopus,* both zygomycetes, are also frequently isolated from air.

Another group capable of producing and releasing spores into the air are the myxomycetes, or the true slime molds. They are considered to have similarities to both true fungi and animals (1) and have been placed in the kingdom Protista (6, 51). Spores of slime molds have been found in air samples (6, 28). Allergic reactions to extracts of slime molds have been reported. Giannini et al. (24) reported that 15.4% of patients tested showed positive skin reactions to extracts of *Fuligo septica, Lycogala epidendrum,* and *Stemonitis ferruginea.* Santilli et al. (74) and Benaim-Pinto (4) found that patients yielded positive skin responses to spore extracts of *F. septica.*

Outdoor airborne fungal populations may directly or indirectly affect the indoor population, since the pathways of infiltration are suspected to be from leaks and cracks or through doors, windows, and building air intake systems. Therefore, common outdoor fungal taxa are often the predominant fungal types detected indoors (52, 96). A review of the literature reveals some agreements and disagreements on the predominant fungi identified indoors. Yang et al. (96), in examining cultures of over 2,000 Andersen samples collected outdoors and in nonresidential buildings in the United States, found that *Cladosporium, Penicillium,* and *As-*

pergillus species, basidiomycetes, and *Alternaria* species were the top five fungal taxa found indoors as well as outdoors in frequency of occurrence. All five fungal taxa were detected in less than 40% of indoor samples. However, *Cladosporium* species was found in over 80% of outdoor samples, while *Penicillium* species was detected in 58%. These findings suggest that both *Cladosporium* and *Penicillium* species were common in outdoor air. The results were somewhat in agreement with those reported by Strachan et al. (86) from British homes. They found that *Penicillium* species, *Cladosporium* species, and basidiomycetes (including *Sistotrema brinkmanii*) were the common types of mold isolated and mold concentrations as well. Using a number of different types of samplers and sampling media, VerHoeff et al. (91) found that species of *Cladosporium, Penicillium,* and *Aspergillus* (including the teleomorph *Eurotium*) were common in homes in The Netherlands. In a survey using an Andersen sampler of 10 elementary schools in southern California, Dungy et al. (20) found that *Cladosporium, Alternaria,* and *Penicillium* species, sterile mycelia, and *Epicoccum* species were the top five fungal groups isolated indoors. The predominant fungi detected outdoors were slightly different in that *Aureobasidium* species was more frequently encountered than *Epicoccum* species. However, using a Roto Rod technique, they found that spores of *Alternaria* species, rust fungi, *Cladosporium* species, *Epicoccum* species, and smut fungi were predominant both indoors and outdoors. Through further comparison of airborne fungal populations in the same study, the authors found that the top seven fungal types were identical at schools and at homes. In a hospital sampling, Solomon et al. (81) found that *A. fumigatus, A. niger, Mucor* species, *Paecilomyces* species, and yeasts were the five most common fungi recovered in culture at 37°C. The agreements and disagreements in the findings may be attributed to sampling techniques, isolation media used, incubation temperatures, and geographical areas (56, 95).

FUNGI AS DISEASE AGENTS

Fungi are known to cause infections and allergies, which are well documented in the literature. They are briefly discussed below. Fungi and their by-products, such as $(1\text{-}3)\text{-}\beta\text{-D-glu-}$ can, mycotoxins, and VOCs, have also been implicated in other diseases and health effects.

Infections

Infections caused by fungi are called mycoses. They are categorized into endemic mycoses and opportunistic mycoses. Endemic mycoses are related to the geographical distribution of certain fungal pathogens. This type of infection is caused by the inhalation of airborne spores or conidia found in certain regions where there is a higher frequency of such fungi because of unique soil and plant/flora conditions (45, 69). Table 2 lists seven groups of fungi and the infections that occur through air transmission, the diseases they cause, and clinical manifestations.

Opportunistic infections are secondary complications that occur in patients with weakened immune systems. Patients usually have major systemic diseases or health-suppressed conditions such as complicated diabetes mellitus, some form of cancer, human immunodeficiency virus infection or AIDS, severe liver or kidney diseases, organ transplantation, burn injury, and immunosuppressive medication treatment. These infections are not contagious, and the fungi are not considered obligatory pathogens. Secondary fungal infections and medical complications related to air-

TABLE 2 Diseases transmitted by airborne fungi and affected tissues[a]

Fungal agent(s)	Disease	Affected tissues[b]
Histoplasma capsulatum	Histoplasmosis	Lung, eye, (skin and bone)
Cryptococcus neoformans	Cryptococcosis	Lung, central nervous system, meninges, skin, viscera
Coccidioides immitis	Coccidioidomycosis	Lung, multiorgan dissemination (skin, bone, meninges, joints)
Blastomyces dermatitis	Blastomycosis	Lung, skin and mucous membrane, bone, joints
Aspergillus spp.	Aspergillosis	Lung, bronchial airways and sinus cavities, ear canal, eye (cornea)
Mucorales, Zygomycetes	Mucormycosis	Nose, sinuses, eye, lung, (brain and other organs), gastrointestinal system

[a] Data are from references 45 and 69.
[b] Tissues in parentheses are less commonly affected.

borne fungal contamination in hospitals and transplant units have been reported sporadically. Immunocompromised patients and nursing home residents may be at an increased risk of opportunistic infections if elevated fungal levels and atypical fungi are present in indoor air surveys. Of most concern are *Aspergillus* spp., such as *A. fumigatus*, *A. flavus*, and *A. niger*. Soil, bird and bat droppings, water-damaged materials, or organic-rich substrates in buildings may be reservoirs for these fungi (5, 7, 45).

Prevention, diagnosis, and therapy of opportunistic infections may be difficult for many who are not well trained and experienced in this field. It is feasible that early recognition, preventive building engineering, and public health intervention can reduce the incidence of mycosis, especially the iatrogenic acquisition in health care facilities.

Allergies

Fungal material can cause an immunopathology with an exaggerated or inappropriate immune response, called hypersensitivity reaction or common allergy. The immune reactions of fungal allergies are listed in Table 3. The fungal spore is a known cause of allergic diseases (26) and was identified as one of the major indoor allergens (70). Several recent epidemiological studies have shown that long-duration indoor exposure to certain fungi can result in hypersensitivity reaction and chronic diseases. Mold levels less than or equal to outside background levels are usually well tolerated by most people. However, as atypical mold levels increase because of recurrent water leaks, home dampness, and high humidity, an increased frequency of respiratory problems (allergy and "sick building syndrome"-like symptoms) can be found (7, 8, 17, 22, 41, 71, 79, 84–86).

The reported percentages of the population with allergy to molds vary from 2% to 18%. Approximately 80% of asthmatics were reported to be allergic to molds (22). The inci-

dence and prevalence of allergic diseases is on the rise (70). In clinical allergy, patients can be specifically tested for mold allergy by a skin or serological test, and appropriate advice and treatment can then be prescribed. Patients with an atopy (a genetic trait of increased allergen sensitivity) are frequently allergic to multiple fungal species and manifest type I reactions (asthma, rhinitis, eczema, and hay fever).

All fungi may be allergenic depending on the exposure situation and dose (70), although the sensitivity of clinical tests may vary with the study population and individual immune system characteristics. Atopic individuals typically have a higher rate of positive skin reactions after provocation tests and serological allergy tests measuring antibody precipitins (immunoglobulin E [IgE]). Diseases such as allergic brochopulmonary aspergillosis and allergic fungal sinusitis possibly require additional host factors which are not well documented (70) and may be the result of the combined reaction of allergenic inflammation and the immunotoxic effect of fungal metabolites. The relevant route of exposure is inhalation, and the adverse effects are related to duration and intensity of fungal exposure. However, typical for allergic reactions is that once an individual develops an allergy to certain fungi, even small airborne concentrations can trigger an asthma attack or other allergic reactions. This response is different from fungal toxic-inflammatory health reactions, which depend on airborne concentrations and are similar for most people, whether they are sensitized or not. Allergy threshold levels to common molds of 100 CFU/m^3 for *Alternaria* species and of 3,000 CFU/m^3 for *Cladosporium* species have been reported (26).

Hypersensitivity Pneumonitis and Organic Dust Toxic Syndrome

Hypersensitivity pneumonitis (HP), also called extrinsic allergic alveolitis, is a well-recognized occupational disease

TABLE 3 Immunopathological responses caused by fungal hypersensitivity

Type	Immune response	Diseases
I Immediate hypersensitivity	IgE, mast cell	Asthma, rhinitis, eczema, hay fever
III Immune complex mediated	IgG, antigen-antibody complex deposition in the blood vessels and tissues	Hypersensitivity pneumonitis, Arthus reaction, extrinsic allergic alveolitis
IV Delayed-type hypersensitivity	Antigen-sensitized T lymphocyte	Allergic contact dermatitis, pneumonitis

TABLE 4 Fungal agents of HP and occupational dust exposure

Fungal agent	Source	Disease(s)
Aspergillus clavatus	Moldy malt	Malt worker's lung
Aureobasidium pullulans	Steam	Sauna-taker's lung
Alternaria spp.	Wood	Woodworker's lung
Botrytis cinerea	Moldy fruits	Winegrowers' lung
Cryptostroma corticale	Wood	Maple bark stripper's lung
Farnai rectivirgula	Straw	Potato riddler's lung
Serpula (Merilius) lacrymans	Moldy building	Dry rot lung
Penicillium spp.	Cork	Suberosis, woodman's disease
Mucor stolinifer	Moldy paprika	Paprika worker's lung
Trichosporon cutaneum	House dust	Japan summer pneumonitis

(Table 4). Organic dust toxic syndrome (ODTS), also called toxic pneumonitis, is a nonallergic, noninfectious form of an acute inflammatory lung reaction to fungal dust exposure. The differences between HP and ODTS can be difficult to distinguish. The clinical features, biochemistry, and pathophysiology of allergic reactions are difficult to separate from those of inflammatory toxic reactions. Table 5 lists comparative features of HP and ODTS. The significance of ODTS in occupational health is such that preventive measures for ODTS have been recommended for certain occupations (agriculture) by the National Institute for Occupational Safety and Health. The measures include the use of industrial hygiene controls, special protective equipment, ventilation, and respiratory protection (63).

Composting of biological waste is a new, growing technology in municipal waste management. Recent environmental monitoring suggests possible intense exposure risks to several fungi, bacteria, and viruses. High fungal levels of primarily *A. fumigatus*, *Penicillium* spp., and *Paecilomyces* spp. in the range of 2,000 to 20,000 CFU/m^3 were found in a sentinel health investigation of a young composting worker who developed alveolitis and lung fibrosis (42). Two of the 18 air samples contained *Stachybotrys chartarum* previously not identified in similar composting studies. A preliminary review and discussions with experts indicate that current hygiene practices and medical surveillance in many composting facilities appear to be inadequate. Furthermore, there is a paucity of good epidemiological studies related to health problems of workers in large-scale waste composting facilities, and there are no established guidelines for medical surveillance and monitoring of personnel working in such facilities (42).

Mycotoxins and Mycotoxicoses

Fungi produce a number of toxic chemicals such as the poisonous compounds found in some mushrooms and the toxic metabolites of some species of microfungi. Many of the mushroom poisons are polypeptides or amino acid-derived toxins (29). However, poisoning due to ingestion of poisonous mushrooms is excluded from the scope of this chapter. Lincoff and Mitchel (54) present a comprehensive discussion of mushroom toxins.

Some fungi have been known to produce secondary metabolites that are harmful to animals and humans when ingested (57), inhaled (16, 41, 59), or brought in contact with the skin (75). These toxic metabolites (Table 6), including alkaloids, cyclopeptides, and coumarins, are called mycotoxins (29). They have been called an "agent in search of a disease" (75) because of the multiple presentations and organs involved in the disease process. An overview (13) of clinically important health disorders based on various case reports and results of disease cluster investigations is presented for the most important mycotoxin producers (Table 6).

Some mycotoxins, such as lysergic acid, are derivatives of amino acids (such as tryptophan). Mycotoxins derived from other precursors are grouped into aromatic- and phenolic-related toxins and terpenoid toxins. Some well-known and potent mycotoxins in the aromatic- and phenolic-related toxin group are aflatoxin, zearalenone, and griseoful-

TABLE 5 Comparative features of HP and ODTS[a]

Condition	Immune responses	Affected tissue or organ	Exposure levels	Clinical features
HP (extrinsic allergic alveolitis)	Type IV delayed hypersensitivity, cell-mediated immune reaction	Lung alveoli, forming granulomas	10^6–10^{10} CFU of thermophilic actinomycetes or fungi/m^3	Dyspnea, cough, fatigue, poor appetite, weight loss, abnormal chest X ray, abnormal pulmonary functions, high antibody precipitins; may cause pulmonary fibrosis in long term
ODTS (toxic pneumonitis)	Nonallergenic, noninfectious, lack of IgG	Inflammatory lung reaction	High concentrations of fungi, >10^9 spores/m^3 of (1-3)-β-D-glucan or >1–2 μg/m^3 of endotoxins	Dyspnea, cough, headaches, fever, chills, malaise, acute inflammatory lung reaction, negative chest X ray, patient recovers after exposure cessation

[a] Data are from reference 92.

TABLE 6 Some toxigenic fungi and secondary chemical metabolites and associated health effects

Fungi	Chemical metabolite(s)	Health effect(s)
Penicillium (>150 species)	Patulin	Hemorrhage of lung, brain disease
	Citrinin	Renal damage, vasodilatation, bronchial constriction, increased muscular tone
	Ochratoxin A	Nephrotoxic, hepatotoxic
	Citroviridin	Neurotoxic
	Emodin	Reduced cellular oxygen uptake
	Gliotoxin	Lung disease
	Verraculogen	Neurotoxic: trembling in animals
	Secalonic acid D	Lung, teratogenic in rodents
Aspergillus species	Patulin	Hemorrhage of lung, brain disease
A. flavus and A. parasiticus	Aflatoxin B1	Liver cancer, respiratory system cancer, cytochrome P-450 monooxygenase disorder
A. versicolor	Sterigmatocystin	Carcinogen
A. ochraceus	Ochratoxin A	Nephrotoxic, hepatotoxic
Stachybotrys chartarum	Trichothecenes[a] (more than 170 derivatives known):	Immune suppression and dysfunction, cytotoxic, bleeding, dermal necrosis; high-dose ingestion lethal (human case reports); low-dose, chronic ingestion potentially lethal; teratogenic, abortogenic (in animals)
	T2	Alimentary toxic aleukia reported in Russia and Siberia
	Nivalenol	Staggering wheat in Siberia
	Deoxynivalenol	Red mold disease in Japan
	Diacetoxyscirpenol	Neurotoxic/nervous system and behavior abnormality
	Satratoxin H	
	Spirolactone	Anticomplement function
Fusarium spp.	Zearalenone	Phytoestrogen may alter immune function; stimulates growth of uterus and vulva, atrophy of ovary
Claviceps species	Ergot alkaloids	Prolactin inhibitor, vascular constriction, uterus contraction promoter

[a] Trichothecenes are also produced by species of *Myrothecium*, *Trichoderma*, *Trichothecium*, and *Gibberella* (teleomorph of some *Fusarium* species).

vin. The terpenoid toxins include trichothecenes and fusidanes (29). There are more than 200 mycotoxins produced by a variety of common fungi, according to the World Health Organization Environmental Health Criteria 105 on mycotoxins (94). Samson (73) suggested that there are currently more than 400 toxic metabolites and the number is increasing.

The International Agency for Research on Cancer (35) classifies aflatoxin, a toxin discovered in 1961 in *A. niger* and *A. parasiticus*, as yielding "sufficient evidence" for human and animal carcinogenicity. It may also be involved in occupational respiratory cancers among food and grain workers (82). Aflatoxin-induced disease has been well reviewed (30, 35, 43). Trichothecene toxins (T-2 toxin, *Fusarium* toxins) are listed with "limited evidence" for animals and "inadequate evidence" (no data available) for humans (35). Macrocyclic trichothecenes, such as satratoxin H, have not been classified. Epidemiological studies suggest a higher rate of upper respiratory tract and lung cancer in workers in the grain and food handling industry, which presents a high fungal product inhalation risk (82). The markedly high rate of lung cancer among uranium miners in Slesien, Germany (Schneeberg disease) may be related to combined effects of high radon and *Aspergillus* exposure in the underground mines (44).

The earliest known mycotoxin producers (toxigenic fungi), primarily *Claviceps purpurea*, produce the substance ergot. From a public health point of view, important toxigenic fungi (Table 6) are *Penicillium* species, *Aspergillus* spe-

cies, *Fusarium* species, *S. chartarum* (synonym, *S. atra*), *Paecilomyces* species, and *Trichoderma viridi*. These fungi can produce in humans and animals adverse health effects resulting in typical organ damage and disease which is neither allergic nor infectious in nature. In several disease outbreaks, human and animal deaths have been linked to exposure to toxigenic fungi, typically through ingestion. Current research, however, indicates that inhalation of certain mycotoxins has even stronger effects (15) and is more common in human disease outbreak. The occurrence of mycotoxins in the environment and related adverse health effects have been reviewed (13, 59, 75).

Mycotoxins may cause a variety of short-term as well as long-term adverse health effects (73, 75). Samson (73) divided the effects into four basic categories: acute, chronic, mutagenic, and teratogenic. These range from immediate toxic response and immunosuppression to potential long-term carcinogenic and teratogenic effects. Symptoms due to mycotoxins or toxin-containing spores (particularly those of *S. chartarum*) include dermatitis, recurring cold and flu-like symptoms, burning sore throat, headaches and excessive fatigue, diarrhea, and impaired or altered immune function (75). The ability of the body to resist infectious diseases may be weakened, resulting in opportunistic infections. Certain mycotoxins, such as zearalenone, have been found to cause infertility and stillbirths in pigs (57).

Historically, mycotoxins have been a problem to farmers and food industries and in Eastern European and Third

World countries. The higher-dose exposure to fungi and mycotoxins encountered by farmers and in food industries was generally considered unlikely to occur in nonfarming activities. However, many toxigenic fungi, such as *S. chartarum* and species of *Aspergillus*, *Penicillium*, and *Fusarium*, have been found to infest buildings with known indoor air- and building-related problems and with illnesses among the occupants (16, 22, 37, 41). It is suggested that inhalation exposure to mycotoxin-containing fungal spores is significant in the reported cases of building-related mycotoxicoses (37). Croft et al. (16) identified several cases of mycotoxicoses due to airborne exposure to the toxigenic fungus *S. atra* in a residential building. Additional cases of office building-associated *Stachybotrys* mycotoxicosis were reported by Johanning (39) and Johanning and colleagues (41). Satratoxin H was detected in the fungus isolated from the contaminated building. Symptoms and immunological tests for antibodies (IgE and IgG) specific to *S. atra* suggested exposure to the fungus.

Mycotoxins generally have low volatility; therefore, inhalation of volatile mycotoxins is not very likely (75). Rather, the toxins are an integral part of the fungus. Sorenson et al. (83) demonstrated in the laboratory that aerosolized conidia of *S. atra* contained trichothecene mycotoxins. The most common toxin was satratoxin H. Lesser amounts of satratoxin G and trichoverrols A and B were also detected but less frequently. They also found that most of the airborne particles were within the respirable range. Similar experiments conducted by Pasanen et al. (67) demonstrated that trichothecene mycotoxins were in airborne fungal propagules of *S. atra* and could be collected on membrane filters. Conidia of *A. flavus* and *A. parasiticus* were reported to contain aflatoxins (93). Miller (59) also reported detection of two mycotoxins, deoxynivalenol and T-2 toxin, in conidia of *Fusarium graminearum* and *F. sporotrichioides*, respectively. These references suggest that inhalation exposure to conidia may also increase the chance of exposure to mycotoxins.

Although relationships linking inhalation exposure to mycotoxin-containing fungal spores and symptoms of mycotoxicoses in fungus-infested indoor environments were established (16, 41, 67, 83), other possible exposure routes such as ingestion and skin contact are likely. Because fungal spores are ubiquitous in a contaminated environment, the chance of ingesting toxin-containing spores through eating, drinking, and smoking is likely to increase.

One important toxigenic fungus frequently detected in problem buildings is *S. chartarum*, which produces a series of potent cytotoxins as well as a variety of other compounds affecting the immune system (14, 38). In a recent case study of health effects and immunological laboratory changes related to indoor exposure to trichothecene (satratoxin H) and possibly other mycotoxins, disorders of the respiratory and central nervous systems were noted. Abnormal test results of the cellular and humoral immune systems were found (40). In earlier cases in Eastern Europe, typically in an agricultural setting, marked leukopenia or acute "radiation-mimetic" effects on the blood cell system with subsequent sepsis-like opportunistic infections after trichothecene ingestion were reported (36, 65).

Trichothecenes are considered to be the most potent small-molecule inhibitors of protein synthesis through inhibition of the peptidyltransferase activity (58, 90). These toxins can cause alveolar macrophage defects and may affect phagocytosis. They have been investigated for use in cancer treatment (25) but also in chemical-biological warfare. The presence of fungal chemical metabolites has been reported in several cases of animal and human ingestion-related mycotoxicosis, resulting sometimes in death (32, 37). Mycotoxins of the trichothecene group have been shown to cause depressed T- or B-lymphocyte activity, suppressed immunoglobulin and antibody production, reduced complement or interferon activity, and impaired macrophage-effector cell function of human neutrophils (97).

Laboratory changes of immunoglobulins (IgA, IgE, IgG, and IgM) in workers handling foodstuff contaminated with mycotoxins, primarily desoxinvalenol (vomitoxin), have been reported (89). An increase of IgA production and IgA nephropathy and a decrease of IgG and IgM after ingestion of vomitoxin was reported in a mice experiment (68). Renal failure and IgG deposition in the glomeruli after inhalation of ochratoxin produced by *Aspergillus ochraceus* was found in the case of a farmer (66).

A World Health Organization task group concluded in 1990 that an association between trichothecene exposure and human disease episodes is possible; however, only limited data are available (94). Immunotoxicological effects principally depend on the exposure conditions, dose, and timing. Some immunological effects may be only transient, of short duration, and difficult to detect in routine medical tests. Medical findings are often nonspecific, and other systemic diseases or causes need to be ruled out by an experienced clinician. Mycotoxicosis is still often not recognized by the treating physician, especially because exposure circumstances and presence of certain mycotoxins remain unknown. Better fungal exposure characterization and sampling techniques now available should improve the chances for better medical detection of mycotoxicosis, particularly by specialists in occupational and environmental medicine. Newer analytical methods involving immunoassays and cell line cytotoxicity analysis are able to provide relatively rapid and easy screening tests to detect the presence of mycotoxins in fungus-contaminated materials (23, 33, 53, 79, 80).

Fungal Volatile Organic Compounds

Fungi in active growth produce VOCs, which have been suggested as a possible health risk (59). A number of VOCs have been found in fungi common in indoor contamination. Most of these fungal VOCs are derivatives of alcohols, ketones, hydrocarbons, and aromatics. In a recent study, Larsen and Frisvad (48) examined the in vitro production of fungal volatiles from 47 *Penicillium* taxa and detected alcohols, ketones, esters, small alkenes, monterpenes, sesquiterpenes, and aromatics. However, aldehydes were not detected.

Some fungi produce VOCs with an unpleasant odor (27), while other fungi (such as mushrooms) produce VOCs with pleasant odors and flavors. 1-Octen-3-ol, one of the major fungal VOCs, has a characteristic mushroom odor. The musty, moldy, and earthy odors likely come from 2-octen-1-ol and geosmin (1,10-dimethyl-9 decalol) (22). Ezeonu et al. (21) identified ethanol, 2-ethyl hexanol, cyclohexane, and benzene in fiberglass air duct liners colonized by *Aspergillus versicolor*, *Acremonium obclavatum*, and *Cladosporium herbarum*. Acetone and 2-butanone were detected only on agar plate samples of *A. versicolor* and *Acremonium obclavatum*. The authors (21) indicated that both 2-ethyl hexanol and cyclohexane, detected in fiberglass air duct liners colonized by fungi, are eye and skin irritants. Benzene is a generally recognized hazardous chemical. Other fungal VOCs associated with two common indoor fungi, *Penicillium* and

Aspergillus spp., have been identified: 2-methyl-isoborneol, 2-methyl-1-propanol, 3-methyl-1-butanol, and 3-octanone (27). Additional fungal VOCs are compiled and listed by Batterman (3). Almost all of the published information regarding fungal VOCs concerns *Penicillium* and *Aspergillus* spp. Little is known about VOCs of other common indoor fungal contaminants.

Although musty, moldy odors are often associated with unhealthy environments, there is no cause-effect relationship directly linking fungal VOCs to human health effects (3). In fact, there is little documentation on the risk, irritancy, or short-term and long-term effects of human exposure to fungal VOCs. In a review, Flannigan et al. (22) suggested that fungal VOCs might be associated with symptoms from nausea to acute respiratory symptoms, but the effects of prolonged low-level exposure are not known. Increasing evidence exists to associate poor air quality with fungal odors and VOCs from contaminated building materials and ventilation systems (21, 27).

CONCLUSIONS

Fungi are ubiquitous in nature. They are considered to be allergenic agents and occasionally infectious and disease causing to susceptible people. Recent findings indicate that they have become a major problem in buildings where moisture control is poor or water intrusion is common. Some mycotoxins in fungal spores released into the surroundings are potentially irritating, toxic, teratogenic, carcinogenic, and immunosuppressive. In addition to being allergenic and infectious, mycotoxins in spores are now considered to be a major health threat in buildings where fungal infestation occurs. Miller (60) considered inhalation exposure to molds and their by-products to be well-characterized hazards.

Clinical diagnoses of mold allergies and fungal infections are generally much better established than emerging health concerns of such fungal metabolites as (1-3)-β-D-glucan, mycotoxins in airborne spores, and fungal VOCs. Furthermore, risk assessment of human exposure to these fungi and their by-products is complex and difficult (9) and in its infancy. Exposure to fungi and their by-products at various levels is ubiquitous in some indoor environments. Human sensitivity to them varies from individual to individual. Health implications from inhalation exposure of many fungal metabolites, particularly at low-level exposures, are poorly understood. Little is known concerning the consequences of short-term and long-term exposures to these fungal metabolites and whether the health effects are reversible.

Fungi and their growth in nonindustrial buildings should not be considered just a nuisance or minor problem. There is now significant evidence that fungi and their secondary metabolites are harmful to building occupants. It was recommended in a recent international workshop held in Baarn, The Netherlands, that "visible fungal growth in non-industrial indoor environments is not acceptable on medical and hygienic grounds" (2).

REFERENCES

1. **Alexopoulos, C. J., and C. W. Mims.** 1979. *Introductory Mycology*, 3rd ed. John Wiley & Sons, New York.
2. **Anonymous.** 1994. Recommendations, p. 531–538. *In* R. A. Samson, B. Flannigan, M. E. Flannigan, A. P. VerHoeff, O. C. G. Adan, and E. S. Hoekstra (ed.), *Health Implications of Fungi in Indoor Environments*. Elsevier, Amsterdam.
3. **Batterman, S. A.** 1995. Sampling and analysis of biological volatile organic compounds, p. 249–268. *In* H. A. Burge (ed.), *Bioaerosols*. Lewis Publishers, Boca Raton, Fla.
4. **Benaim-Pinto, C.** 1992. Sensitization to basidiomycetes and to *Fuligo septica* (Myxomycetae) in Venezuelan atopic patients suffering from respiratory allergy. *J. Allergy Clin. Immunol.* **89:**282–289.
5. **Benenson, A. S.** 1990. *Control of Communicable Disease in Man*, 15th ed. American Public Health Association, Washington, D.C.
6. **Blackwell, M.** 1990. Air dispersed spores of myxomycetes, p. 51–52. *In* E. G. Smith (ed.), *Sampling and Identifying Allergenic Pollens and Molds*. Blewstone Press, San Antonio, Tex.
7. **Burge, H. A.** 1990. The fungi, p. 136–162. *In* P. Morey, J. Feeley and J. Otten (ed.), *Biological Contaminants in Indoor Environments*. American Society for Testing and Materials, Philadelphia.
8. **Burge, H. A.** 1990. Bioaerosols: prevalence and health effects in the indoor environment. *J. Allergy Clin. Immunol.* **86:**687–701.
9. **Burge, H. A.** 1993. Characterization of bioaerosols in buildings in the United States, p. 131–137. *In IAQ '92: Environments for People*. ASHRAE, Atlanta.
10. **Burge, H. A.** 1995. Bioaerosols in the residential environment, p. 579–597. *In* C. S. Cox and C. M. Wathes (ed.), *Bioaerosols Handbook*. Lewis Publishers, Boca Raton, Fla.
11. **Butcher, B. T., C. E. O'Neil, M. A. Reed, L. C. Altman, M. Lopez, and S. B. Lehrer.** 1987. Basidiomycete allergy: measurement of spore-specific IgE antibodies. *J. Allergy Clin. Immunol.* **80:**803–809.
12. **Carroll, G. C., and D. T. Wicklow.** 1992. *The Fungal Community, Its Organization and Role in the Ecosystem*. Marcel Dekker, Inc., New York.
13. **Ciegler A., H. R. Burmeister, R. F. Vesonder, and C. W. Hesseltine.** 1981. Mycotoxins: occurrence in the environment. *In* R. C. Shank (ed.), *Mycotoxins and N-Nitroso Compound Environmental Risk*, vol. 1. CRC Press, Boca Raton, Fla.
14. **Corrier, D. E.** 1991. Mycotoxicosis: mechanism of immunosuppression. *Vet. Immunol. Immunopathol.* **30:**73–87.
15. **Creasia, D. A., J. D. Thurman, R. W. Wannemacher, Jr., and D. L. Bunnder.** 1990. Acute inhalation toxicity of T-2 mycotoxin in rat and guinea pig. *Fundam. Appl. Toxicol.* **14:**54–59.
16. **Croft, W. A., B. B. Jarvis, and C. S. Yatawara.** 1986. Airborne outbreak of trichothecene toxicosis. *Atmos. Environ.* **20:**549–552.
17. **Dales, R. E., R. Burnett, and H. Zwaneburg.** 1991. Adverse health effects among adults exposed to home dampness and molds. *Am. Rev. Respir. Dis.* **143:**505–509.
18. **Davis, W. E., W. E. Horner, J. E. Salvaggio, and S. B. Lehrer.** 1988. Basidiospore allergens: analysis of *Coprinus quadrifidus* spore, cap, and stalk extracts. *Clin. Allergy* **18:**261–267.
19. **DeKoster, J. A., and P. S. Thorne.** 1995. Bioaerosol concentrations in noncomplaint, complaint, and intervention homes in the midwest. *Am. Ind. Hyg. Assoc. J.* **56:**573–580.
20. **Dungy, C. I., P. P. Kozak, J. Gallup, and S. P. Galant.** 1986. Aeroallergen exposure in the elementary school setting. *Ann. Allergy* **56:**218–221.
21. **Ezeonu, I. M., D. L. Price, R. B. Simmons, S. A. Crow, and D. G. Ahearn.** 1994. Fungal production of volatiles during growth on fiberglass. *Appl. Environ. Microbiol.* **60:**4172–4173.
22. **Flannigan, B., E. M. McCabe, and F. McGarry.** 1991. Allergenic and toxigenic micro-organisms in houses. *J. Appl. Bacteriol.* **79:**61S–73S.
23. **Gareis, M.** 1995. Cytotoxicity testing of samples originat-

ing from problem building, p. 139–144. *In* E. Johanning and C. S. Yang (ed.), *Fungi and Bacteria in Indoor Air Environments: Health Effects, Detection and Remediation.* Eastern New York Occupational Health Program, Albany, N.Y.

24. **Giannini, E. H., W. T. Northey, and C. R. Leathers.** 1975. The allergenic significance of certain fungi rarely reported as allergens. *Ann. Allergy* **35:**372–376.

25. **Goodwin, W., C. D. Haas, C. Fabian, I. Heller-Bettinger, and B. Hoogstraten.** 1978. Phase I evaluation of anguidine (diacetoxyscirpenol, NSC-141537). *Cancer* **42:**23–26.

26. **Gravesen, S.** 1979. Fungi as a cause of allergic disease. *Allergy* **34:**135–154.

27. **Gravesen, S., J. C. Frisvad, and R. A. Samson.** 1994. *Microfungi.* Munksgaard, Copenhagen.

28. **Gregory, P. H.** 1973. *The Microbiology of Atmosphere,* 2nd edition. Halstead Press, New York.

29. **Griffin, D. H.** 1993. *Fungal Physiology.* Wiley-Liss, New York.

30. **Hendry, K. M., and E. C. Cole.** 1993. A review of mycotoxins in indoor air. *J. Toxicol. Environ. Health* **38:**183–198.

31. **Henson, P. M., and R. C. Murphy.** 1989. *Mediator of the Inflammatory Process.* Elsevier, New York.

32. **Hintikka, E.** 1987. Human stachybotryotoxicosis, p. 87–89. *In* T. D. Wyllie and L. G. Morehouse (ed.), *Mycotoxic Fungi, Mycotoxins, Mycotoxicosis,* vol. 3. Marcel Dekker, New York.

33. **Hintikka, E., and M. Nikulin.** 1995. Aerosol mycotoxins: a veterinary experience and perspective, p. 31–34. *In* E. Johanning and C. S. Yang (ed.), *Fungi and Bacteria in Indoor Air Environments: Health Effects, Detection and Remediation.* Eastern New York Occupational Health Program, Albany, N.Y.

34. **Hocking, A. D., and B. F. Miscamble.** 1995. Water relations of some Zygomycetes isolated from food. *Mycol. Res.* **99:**1113–1118.

35. **International Agency for Research on Cancer.** 1993. Monographs in the evaluation of carcinogenic risks to human. *Bull. W.H.O.* **56:**245–523.

36. **Jarmai, K.** 1929. Viskosusseptikamien bei alteren Fohlen und erwachsenen Pferden [Viscosussepsis in older colts and grown-up horses]. *Dtsch. Tierarzt. Wochenschr.* **33:**517–519.

37. **Jarvis, B. B.** 1990. Mycotoxins and indoor air quality, p. 201–214. *In* P. Morey, J. Feeley, and J. Otten (ed.), *Biological Contaminants in Indoor Environments.* American Society for Testing and Materials, Philadelphia.

38. **Jarvis, B. B., J. Salemme, and A. Morais.** 1995. Stachybotrys toxins. 1. *Nat. Toxins* **3:**10–16.

39. **Johanning, E.** 1995. Health problems related to fungal exposure—primarily of *Stachybotrys atra,* p. 169–182. *In* E. Johanning and C. S. Yang (ed.), *Fungi and Bacteria in Indoor Air Environments: Health Effects, Detection and Remediation.* Eastern New York Occupational Health Program, Albany, N.Y.

40. **Johanning, E., M. Gareis, M. Hintikka, M. Nikulin, and C. Yang.** 1995. Biological toxicity analyses of mycotoxin producing fungal samples from field studies, p. 32. *In Abstracts—American Industrial Hygiene Conference.* American Industrial Hygiene Association, Fairfax, Va.

41. **Johanning, E., P. R. Morey, and B. B. Jarvis.** 1993. Clinical-epidemiological investigation of health effects caused by *Stachybotrys atra* building contamination, p. 225–230. *In Indoor Air '93, Proceedings of the Sixth International Conference on Indoor Air Quality and Climate,* vol. 1. *Health Effects.* Helsinki, Finland.

42. **Johanning, E., E. Olmsted, and C. Yang.** 1995. Medical issues related to municipal waste composting, p. 24. *In Abstracts—American Industrial Hygiene Conference.* American Industrial Hygiene Association, Fairfax, Va.

43. **Krogh, P.** 1989. The role of mycotoxins in disease of animals and man. *J. Appl. Bacteriol. Symp. Suppl.* **79:**99S–204S.

44. **Kusak, V., S. Jelinek, and J. Ula.** 1990. Possible role of *Aspergillus flavus* in the pathogenesis of Schneeberg and Jachymov disease. *Neoplasm* **17:**441–449.

45. **Kwon-Chung, K. J., and J. E. Bennett.** 1992. *Medical Mycology.* Lea & Febiger, Philadelphia.

46. **Lacey, J.** 1981. Aerobiology of conidial fungi, p. 373–416. *In* G. C. Cole and B. Kendrick (ed.), *Biology of Conidial Fungi,* vol. 1. Academic Press, New York.

47. **Lacey, J.** 1991. Aerobiology and health, p. 157–185. *In* D. L. Hawksworth (ed.), *Frontiers in Mycology.* CAB International, Wallingford, England.

48. **Larsen, T. O., and J. C. Frisvad.** 1995. Characterization of volatile metabolites from 47 *Penicillium* taxa. *Mycol. Res.* **99:**1153–1166.

49. **Lehrer, S. B., M. Lopez, B. T. Butcher, J. Olson, M. Reed, and J. E. Salvaggio.** 1986. Basidiomycete mycelia and spore-allergen extracts: skin test reactivity in adults with symptoms or respiratory allergy. *J. Allergy Clin. Immunol.* **78:**478–485.

50. **Levetin, E.** 1991. Identifications and concentration of airborne basidiospores. *Grana* **30:**123–128.

51. **Levetin, E.** 1995. Fungi, p. 87–120. *In* H. A. Burge (ed.), *Bioaerosols.* Lewis Publishers, Boca Raton, Fla.

52. **Levetin, E., R. Shaughnessy, E. Fisher, B. Ligman, J. Harrison, and T. Brennan.** 1995. Indoor air quality in schools: exposure to fungal allergens. *Aerobiologia* **11:**27–34.

53. **Lewis, C. W., J. E. Smith, J. G. Anderson, and Y. M. Murad.** 1994. The presence of mycotoxin-associated fungal spores isolated from the indoor air of the damp domestic environment and cytotoxic to human cell lines. *Indoor Environ.* **3:**323–330.

54. **Lincoff, G., and D. H. Mitchel.** 1977. *Toxic and Hallucinogenic Mushroom Poisoning.* Van Nostrand Reinhold Co., New York.

55. **Llewellyn, G. C., and C. E. O'Rear.** 1990. *Biodeterioration Research 3: Mycotoxins, Biotoxins, Wood Decay, Air Quality, Cultural Properties, General Biodeterioration, and Degradation.* Plenum Press, New York.

56. **Macher, J. M., F.-Y. Huang, and M. Flores.** 1991. A two-year study of microbiological indoor air quality in a new apartment. *Arch. Environ. Health* **46:**25–29.

57. **Marasas, W. F. O., and P. E. Nelson.** 1987. Mycotoxicology, Pennsylvania State University Press, University Park, Pa.

58. **McLaughlin, C. S., M. H. Vaughan, I. M. Campbell, C. M. Wei, M. E. Stafford, and B. S. Hansen.** 1977. Inhibition of protein synthesis by trichothecenes, p. 263–273. *In* J. V. Rodrick, C. M. Hesseltine, and M. A. Mehlman (ed.), *Mycotoxins in Human and Animal Health.* Pathotox, Park Forest Sount, Ill.

59. **Miller, J. D.** 1992. Fungi as contaminants in indoor air. *Atmos. Environ.* **26A:**2163–2172.

60. **Miller, J. D.** 1993. Fungi and the building engineer, p. 147–158. *In IAQ '92: Environments for People.* ASHRAE, Atlanta.

61. **Morey, P. R., M. J. Hodgson, W. G. Sorenson, G. J. Kullman, W. W. Rhodes, and G. S. Visvesvara.** 1984. Environmental studies in moldy office buildings: biological agents, sources and preventive measures. *Ann. Am. Conf. Gov. Ind. Hyg.* **10:**21–35.

62. **Morgan-Jones, G., and B. J. Jacobsen.** 1988. Notes on hyphomycetes. LVIII. Some dematiaceous taxa, including two undescribed species of *Cladosporium,* associated with

biodeterioration of carpet, plaster and wallpaper. *Mycotaxon* **32:**223–236.
63. **National Institute for Occupational Safety and Health.** 1994. *Preventive Organic Duct Toxic Syndrome.* Publication 94-102. National Institute for Occupational Safety and Health, Cincinnati.
64. **Noble, W. C., and Y. M. Clayton.** 1963. Fungi in the air of hospital wards. *J. Gen. Microbiol.* **32:**397–402.
65. **Ozegovic, L.** 1971. Straw related disease in farm workers. *Veterinaria* **20:**263–267. (In Russian.)
66. **Paolo, N. D., A. Guarieri, F. Loi, G. Sacchi, A. M. Mangiorotti, and M. D. Paolo.** 1993. Acute renal failure from inhalation of mycotoxins. *Nephron* **64:**621–625.
67. **Pasanen, A.-L., M. Nikulin, M. Tuomainen, S. Berg, P. Parikka, and E.-L. Hintikka.** 1993. Laboratory experiments on membrane filter sampling of airborne mycotoxins produced by *Stachybotrys atra* Corda. *Atmos. Environ.* **27A:**9–13.
68. **Pestka, J. J., M. A. Moorman, and R. L. Warner.** 1989. Dysregulation of IgA production and IgA nephropathy induced trichothecene vomitoxin. *Food Chem. Toxicol.* **27:**361–368.
69. **Phaller, M. A.** 1992. Opportunistic fungal infections, p. 287–293. *In* J. M. and R. B. Wallace (ed.), *Maxcy-Rosenau Public Health and Preventive Medicine.* Appleton and Lange, East Norwalk, Conn.
70. **Pope, A. M., R. Patterson, and H. Burge.** 1993. *Indoor Allergens.* National Academy Press, Washington, D.C.
71. **Ruotsalainen, R., N. Jaakkola, and J. J. K. Jaakkola.** 1995. Dampness and molds in day-care centers as an occupational problem. *Int. Arch. Occup. Environ. Health* **66:**369–374.
72. **Rylander, R.** 1994. Office and domestic environments, p. 247–255. *In* R. Rylander and R. R. Jacobs (ed.), *Organic Dusts: Exposure, Effects, and Prevention.* Lewis Publishers, Boca Raton, Fla.
73. **Samson, R. A.** 1992. Mycotoxins: a mycologist's perspective. *J. Med. Vet. Mycol.* **30**(Suppl. 1):9–18.
74. **Santilli, J., W. J. Rockwell, and R. P. Collins.** 1985. The significance of the spores of the basidiomycetes (mushrooms and their allies) in bronchial asthma and allergic rhinitis. *Ann. Allergy* **55:**469–471.
75. **Schiefer, H. B.** 1990. Mycotoxins in indoor air: a critical toxicological viewpoint, p. 167–172. *In Indoor Air '90, Proceedings of the 5th International Conference on Indoor Air Quality and Climate,* vol. 1. Toronto.
76. **Singh, J.** 1994. Nature and extent of deterioration in buildings due to fungi, p. 34–53. *In* J. Singh (ed.), *Building Mycology.* Chapman and Hall, London.
77. **Smith, D., and A. H. S. Onions.** 1994. *The Preservation and Maintenance of Living Fungi.* CAB International, Wallingford, England.
78. **Smith, E. G.** 1990. *Sampling and Identifying Allergenic Pollens and Molds.* Blewstone Press, San Antonio, Tex.
79. **Smith, J. E., J. G. Anderson, C. W. Lewis, and Y. M. Murad.** 1992. Cytotoxic fungal spores in the indoor atmosphere of the damp domestic environment. *FEMS Microbiol. Lett.* **100:**337–344.
80. **Smoragiewicz, W., B. Cossette, A. Boutard, and K. Krzystyniak.** 1993. Trichothecene mycotoxins in the dust of ventilation systems in office buildings. *Int. Arch. Occup. Environ. Health* **65:**113–117.
81. **Solomon, W. R., H. P. Burge, and J. R. Boise.** 1978. Airborne *Aspergillus fumigatus* levels outside and with a large clinical center. *J. Allergy Clin. Immunol.* **62:**56–60.

82. **Sorenson, W. G.** 1990. Mycotoxins as potential occupational hazards. *Dev. Ind. Microbiol.* **31:**205–211.
83. **Sorenson, W. G., D. G. Frazier, B. B. Jarvis, J. Simpson, and V. A. Robinson.** 1987. Trichothecene mycotoxins in aerosolized conidia of *Stachybotrys atra. Appl. Environ. Microbiol.* **53:**1370–1375.
84. **Spengler, J., L. Neas, S. Nalai, D. Dockery, F. Speizer, J. Ware, and M. Raizenne.** 1993. Respiratory symptoms and housing characteristics, p. 165–170. *In Indoor Air '93. Proceedings of the Sixth International Conference on Indoor Air and Climate,* vol. 1. Helsinki, Finland.
85. **Strachan, D. P., and R. A. Elton.** 1986. Relationship between respiratory morbidity in children and the home environment. *Fam. Pract.* **3:**137–142.
86. **Strachan, D. P., B. Flannigan, E. M. McCabe, and F. McGarry.** 1990. Quantification of airborne moulds in the homes of children with and without wheeze. *Thorax* **45:**382–387.
87. **Streifel, A. J., and F. S. Rhame.** 1993. Hospital air filamentous fungal spore and particle counts in a specially designed hospital, p. 161–165. *In Indoor Air '93. Proceedings of the Sixth International Conference on Indoor Air and Climate,* vol. 4. Helsinki, Finland.
88. **Troller, J. A., and V. N. Scott.** 1992. Measurement of water activity (Aw) and acidity, p. 135–151. *In* C. Vanderzant and D. F. Splittstoesser (ed.), *Compendium of Methods for the Microbiological Examination of Foods.* American Public Health Association, Washington, D.C.
89. **Tutelyan, V. A., K. R. Dadiani, and N. E. Voitko.** 1992. Changes in aerological indicators of immune status of workers in contact with mycotoxin-contaminated foodstuffs. *Gig. Tr. Prof. Zabol.* [*Occup. Safety Health*] **8:**18–20. (In Russian.)
90. **Ueno, Y.** 1983. *Trichothecenes—Chemical, Biological and Toxicological Aspects,* p. 135–194. Elsevier, Amsterdam.
91. **VerHoeff, A. P., J. H. Van Wijnen, J. S. M. Boleij, B. Brunekreef, E. S. Van Reenen-Hoekstra, and R. A. Samson.** 1990. Enumeration and identification of airborne viable mould propagules in houses. *Allergy* **45:**275–284.
92. **Von Essen, S., R. A. Robbins, A. B. Thompson, and S. I. Rennard.** 1990. Organic dust toxic syndrome: an acute febrile reaction to organic dust exposure distinct from hypersensitivity pneumonitis. *Clin. Toxicol.* **28:**389–420.
93. **Wicklow, D. T., and O. Shotwell.** 1983. Intrafungal distribution of aflatoxins among conidia and sclerotia of *Aspergillus flavus* and *Aspergillus parasiticus. Can. J. Microbiol.* **29:**1–5.
94. **World Health Organization.** 1990. *Environmental Health Criteria 105, Selected Mycotoxins: Ochratoxins, Trichothecenes, Ergot.* World Health Organization, Geneva.
95. **Yang, C. S.** 1995. Understanding the biology of fungi found indoors, p. 131–137. *In* E. Johanning and C. S. Yang (ed.), *Fungi and Bacteria in Indoor Environments: Health Effects, Detection and Remediation.* Eastern New York Occupational Health Program, Albany, N.Y.
96. **Yang, C. S., L.-L. Hung, F. A. Lewis, and F. A. Zampiello.** 1993. Airborne fungal populations in non-residential buildings in the United States, p. 219–224. *In Indoor Air '93. Proceedings of the Sixth International Conference on Indoor Air and Climate,* vol. 4. Helsinki, Finland.
97. **Yarom, R., Y. Sherman, R. More, I. Ginsberg, R. Borinski, and B. Yagen.** 1984. T-2 toxin effect on bacterial infection and leukocyte functions. *Toxicol. Appl. Pharmacol.* **75:**60–68.
98. **Zabel, R. A., and J. J. Morrell.** 1992. *Wood Microbiology.* Academic Press, Inc., San Diego, Calif.

Airborne Endotoxin

STEPHEN A. OLENCHOCK

71

Airborne gram-negative bacteria are ubiquitous contaminants of soils, water, and other living organisms throughout the world. The outer membrane of gram-negative bacteria contains, as integral components, endotoxins which are heat-stable lipopolysaccharide complexes (62). Reviews of the chemical composition of endotoxins provide exhaustive descriptions of the molecular nature of these materials (19, 28, 36, 50). In the outermost membrane of the gram-negative bacterial cell, the lipopolysaccharide molecule consists of three regions: the O-specific polysaccharide (I), the core polysaccharide (II), and lipid A (III). The O-specific polysaccharide chain provides serologic specificity to the different serotypes of gram-negative bacteria. Substitutions, composed of sugars and their sequences on repeating oligosaccharide units, define the antigenic uniqueness for region I. This variability in composition is of importance to the differences in quality of toxicities found with different endotoxins (24, 25). Markedly less variability is found in the core polysaccharide, which can be identical for large groupings of bacteria. A unique ketose, 2-keto-3-deoxyoctonic acid, can be found in region II. Region III, the least variable chemical region, consists of lipid A. The predominant endotoxin activities ascribed to endotoxins can be found in lipid A (19, 36, 50).

Endotoxins are released into the environment after bacterial cell lysis or during active cell growth (1a). Release also occurs when intact bacterial cells are phagocytized by macrophages, in which case the liberated endotoxins contain increased toxicity (14). Ultrastructural analyses of gram-negative bacteria associated with environmental exposures demonstrate shedding of large amounts of endotoxin-containing membrane vesicles (17).

The data suggest that a significant portion of environmental dust-borne endotoxin occurs in the form of microvesicles (17). Endotoxins remain relatively stable in environmental dusts over time when the samples are stored in a manner that is not conducive to continuous growth of gram-negative bacteria. In a study of endotoxins in cotton dusts that were stored for a 6-year period, the concentrations in the dusts were not lower than the original levels (7).

ENVIRONMENTAL SOURCES OF ENDOTOXINS

Because of the ubiquitous nature of gram-negative bacteria and their endotoxins, exposures to these agents are com-

monplace. Reviews of environmental endotoxins and other bioaerosol exposures detail their presence in a variety of occupational environments (8, 15, 30). They are found in various materials associated with environments that contain large amounts of organic dusts or materials. In particular, activities related to agriculture and the processing of agricultural products provide sources of endotoxin exposures.

Endotoxins have been studied in particular depth in agricultural workplaces (Table 1), where they have been quantified in such diverse materials and areas as stored grains, silage, hays, straw, and animal bedding material (43); composted wood chips (45) and stored timber (16); tobacco; bulk cottons and cotton dust (40); *Agaricus bisporus* mushrooms, including processing materials such as manure, compost, and spawn (41); bulk and airborne dusts in swine confinement units (63) and poultry confinement and processing facilities (31); horse and dairy cow barns (44, 54); compost (34, 55); animal feed production (57); and biotechnology processes (37, 46).

However, exposures are not confined to traditionally dusty environments. Endotoxins can be found in many environments that are considered to be "nondusty" or "less dusty" as well, such as office buildings and libraries where humidification systems are operative (15). In more traditional industrial environments, endotoxins have been found in water spray humidification systems (18) and machining fluids (20).

HEALTH EFFECTS FROM ENDOTOXIN EXPOSURES

Endotoxins are biologically active materials that profoundly affect both humoral and cellular host mediation systems (1a, 3, 19, 36). Complement and coagulation systems are affected, and direct interactions of endotoxins with myriad human cell types, including basophils, mast cells, endothelial cells, macrophages, platelets, polymorphonuclear leukocytes, and T and B lymphocytes, have been reported.

Endotoxins affect many organ systems, each of which is subjected to a slightly different attack of mediators (26). The type and severity of organ damage are related to the type and amount of mediator generated, the time course, and the cellular receptors (26). Various human responses following intravenous and inhalation challenges with bacterial endotoxins (including fever), cardiovascular effects, and

TABLE 1 Selected sources of endotoxin contamination from agricultural materials

Source	Reference(s)
Animal-related sources	
Bedding (straw, hay)	43
Cow barns	54
Feed	57
Horse barns	44
Poultry confinement and processing units	31
Swine confinement units	63
Agricultural products	
Compost	34, 55
Cotton	40
Mushroom processing materials	41
Timber, stored	16
Wood chips, composted	45
Miscellaneous	
Biotechnology	37, 46

TABLE 2 Lung function effects in response to exposure to airborne endotoxins in agricultural dusts

Lung effects	Exposure	Endotoxin concn (ng/m^3)	Reference
Acute	Cotton dust	9	6
	Cotton dust	33	22
Chronic	Animal feed dust	0.2–470	56
	Cotton dust	1–20	29
	Swine confinement dust	11,332[a]	63

[a] Mean dust contaminant level in EU per meter cubed.

the immunopharmacological actions of such mediators as stress hormones, tumor necrosis factor, interleukins, and interferons are detailed elsewhere (4).

Research related to environmental exposures to endotoxins focuses on respiratory exposures to airborne dusts containing endotoxins. Endotoxin-induced neutrophil influx into the nasal tissues is suggested as a cause of mucous hypersecretion in a histochemical study in animals (23), and structural and functional changes in the lungs of intact animals and in isolated lung cells are reported (2). The primary cell responsible for initiating pulmonary reactions after inhalation of organic dusts laden with endotoxins is the pulmonary macrophage (51), and human alveolar macrophages are extremely sensitive to the effects of endotoxins in vitro (9). Additionally, lipopolysaccharide was shown to bind to pulmonary surfactant and alter its surface tension properties, which could result in impaired pulmonary function (10). Exogenous surfactant replacement in an experimental animal model restored the respiratory failure which was induced by intratracheal instillation of endotoxin (58).

Systemic signs and symptoms such as chest tightness, cough, shortness of breath, fever, and wheezing that are suggestive of inhalation of airborne endotoxins have been found in workers in sewage treatment facilities (32), swine confinement buildings (12), and poultry confinement units (59). In a study of asthmatic patients, it was suggested that inhalation of purified lipopolysaccharide induced local bronchial inflammation and a systemic inflammatory response (33). The International Commission on Occupational Health, through its Committee on Organic Dusts, suggests that endotoxin exposures may provoke different reactions, such as organic dust toxic syndrome, bronchoconstriction, or mucous membrane irritation, depending on the concentration and duration of exposure to the endotoxins (52).

Acute Lung Effects

A few large studies of endotoxin effects on lung function have been conducted. In those studies, the dust exposures were similar to exposures in workplace situations in that the exposures were not to endotoxins alone but to endotoxins in combination with actual dusts. Controlled exposures of human volunteers to cotton dusts laden with endotoxins led to the definition of an association between decreased

acute pulmonary function and the airborne level of endotoxins in the dusts (5, 6, 22). In these studies of acute pulmonary function effect as measured by the forced expiratory volume in 1 s, the threshold for zero pulmonary function change was defined for cotton workers who smoke as 33 ng/m^3 (22) (Table 2). The results from a larger study of a mixed population (with reference to prior cotton mill work and smoking history) indicated a calculated zero-change threshold of as little as 9 ng/m^3 (6) (Table 2). When these data from vertically elutriated dust samples are mathematically converted to contemporary endotoxin units (EU) and a conversion factor of 10 EU/ng is used (1), the thresholds are 330 and 90 EU/m^3, respectively. To put these calculated thresholds in perspective with other environmental exposures, levels of endotoxins from various sources were summarized as follows (39): total dust samples collected during bedding chopping at a dairy farm, 20,945 EU/m^3; silo unloading on a farm, 88,503 EU/m^3; Shanghai cotton mill opening area, 16,604 EU/m^3; and shackling line during poultry processing, 6,340 EU/m^3. Vertically elutriated dusts from the opening area in a Shanghai cotton mill contained 5,173 EU/m^3, and those from the hulling area in a rice commune in Shanghai contained 4,501 EU/m^3. Environmental exposures to airborne endotoxins in these agricultural settings are markedly greater than the calculated thresholds for zero pulmonary function change.

Chronic Lung Effects

Associations between airborne endotoxin concentrations and chronic lung disease have been reported also (Table 2). In a study of 443 cotton textile workers from two mills in Shanghai, a dose-response trend was observed between current endotoxin levels and the lung disease chronic bronchitis (29). This association was not observed with the concentration of vertically elutriated dust. Analysis of data from that study indicated that exposures of 1 to 20 ng/m^3 (approximately 10 to 200 EU/m^3) constituted an adverse respiratory health effect in the exposed workers.

In a separate epidemiologic study in The Netherlands of 315 animal feed workers, symptoms and lung function changes were related more to present and historic endotoxin exposures than to inspirable dust exposures (56). These investigators reported that lung function changes occurred at levels of endotoxin which ranged from 0.2 to 470 ng/m^3.

A study was conducted on the respiratory health status of 54 swine producers who worked an average of 10.7 years in the swine production industry in Canada (63). Chronic respiratory symptoms and baseline lung function, as measured by the forced vital capacity and forced expiratory volume in 1 s were associated most significantly with exposure to endotoxin. The mean total dust concentration in that study was 11,332 EU/m^3, and the effect of that exposure was much more important to chronic pulmonary function changes than was the gravimetric dust concentration. This occupational setting is one in which an exposure-effect relationship of airborne endotoxin exposure and (acute) work period pulmonary function decrements was shown as well (11).

In a separate longitudinal study of 168 swine confinement operators in the United States (53), it was found that longitudinal declines in spirometric measures of lung function were independently associated with cross-shift declines in lung function and higher concentrations of endotoxins in the bioaerosol. It was suggested by the study that minimizing exposure to agricultural aerosols may decrease the incidence of chronic lung disease among this population of workers.

SAMPLE COLLECTION AND TRANSPORT

Because of the ubiquity of gram-negative bacterial endotoxins, bulk materials, water, and airborne or settled dusts can be tested for their presence (8). Inhalable, respirable, and total dusts, vertically elutriated dusts, or multistage size-fractionated dusts can be analyzed also. Sampling times, flow rates, and sampling strategies will depend on the environment to be studied (8). For analyses of the endotoxin content in airborne dusts, polyvinyl chloride filters (29, 31, 40, 53) and glass fiber filters (27, 57, 63) are commonly used, although other filter media, such as polycarbonate membrane filters (60) or Teflon filters (48, 49), are also used. It has been suggested that the choice of appropriate filter medium may depend on the organic aerosol that is being sampled (21). Nonetheless, when standard safeguards in handling the filters before and after dust collection are followed, it has been traditionally accepted that the choice of filter material is of little consequence (47).

Because the container in which the collected sample is kept should not be an extraneous source of contamination, and dust may escape from the filter surface during transport, it is recommended (38) that after postweighing, the filters be placed in 50-ml sterile plastic conical centrifuge tubes with screw-on caps. With the filter sample placed with the dust side toward the lumen and the reverse side against the wall of the tube, the samples can be mailed or otherwise transported. The extraction of the filter sample will occur within the same transport tube; therefore, potential loss of dust from the filter surface will be of no consequence to the protocol. Depending on the distance to destination or the length of time required, the samples may require cool shipment to avoid any possibly significant growth of gram-negative bacteria. Other techniques such as placing the filter face down on laboratory Parafilm and folding a Parafilm "envelope" over it have proved adequate, although not ideal, for transporting samples through the mail (40).

ENDOTOXIN ANALYSIS

There is currently no internationally accepted and standardized method for extraction and analysis of gram-negative bacterial endotoxins from environmental samples. In general, bulk and airborne dusts with endotoxin contamination undergo aqueous extraction, and the supernatant fluids are analyzed by the *Limulus* amebocyte lysate assay. As a result of extraction validation studies (13, 42), current techniques use water or water with an added surfactant as the extraction method of choice. Further details on extraction are presented elsewhere (38). Because of its accuracy, sensitivity,

and reproducibility, the chromogenic modification of the *Limulus* amebocyte lysate assay has become the commonly accepted method of quantification of endotoxins (21, 38, 40). Continued evolution of that technique has resulted in the kinetic chromogenic modification (13, 27, 44), which includes the methodology to overcome sample-induced enhancement or inhibition of the test (27, 61). Other techniques for endotoxin quantification such as the kinetic-turbidimetric method (35) are used also, while comparisons of various methods continue (49).

SUMMARY

Gram-negative bacteria and their endotoxins are common contaminants of the environment, most commonly associated with organic or agricultural materials. Environmental and occupational exposures to these agents can have profound adverse effects on the host, especially on the structure and function of the respiratory tract. Bulk and airborne samples of the environment can be tested for the presence of endotoxins, and quantification of the exposures can be used to associate the levels of endotoxins with acute and chronic effects on pulmonary function. The commonly accepted method of analyzing endotoxins in environmental samples is the chromogenic modification of the *Limulus* amebocyte lysate test. The latest generation of the technique, the kinetic chromogenic modification, combines accuracy, reproducibility, and sensitivity of the chromogenic technique with the methodology to overcome sample-induced enhancement or inhibition of the test.

REFERENCES

1. **Bio Whittaker (Walkersville, Md.).** Personal communication.
1a. **Bradley, S. G.** 1979. Cellular and molecular mechanisms of action of bacterial endotoxins. *Annu. Rev. Microbiol.* **33:**67–94.
2. **Brigham, K. L., and B. Meyrick.** 1986. Endotoxin and lung injury. *Am. Rev. Respir. Dis.* **133:**913–927.
3. **Burrell, R.** 1990. Immunomodulation by bacterial endotoxin. *Crit. Rev. Microbiol.* **17:**189–208.
4. **Burrell, R.** 1994. Human responses to bacterial endotoxin. *Circ. Shock* **43:**137–153.
5. **Castellan, R. M., S. A. Olenchock, J. L. Hankinson, P. D. Millner, J. B. Cocke, C. K. Bragg, H. H. Perkins, Jr., and R. R. Jacobs.** 1984. Acute bronchoconstriction induced by cotton dust: dose-related responses to endotoxin and other dust factors. *Ann. Intern. Med.* **101:**157–163.
6. **Castellan, R. M., S. A. Olenchock, K. B. Kinsley, and J. L. Hankinson.** 1987. Inhaled endotoxin and decreased spirometric values: an exposure-response relation for cotton dust. *N. Engl. J. Med.* **317:**605–610.
7. **Castellan, R. M., S. A. Olenchock, A. Q. Wearden, K. B. Kinsley, S. S. Bajpayee, and H. H. Perkins, Jr.** 1992. Effect of storage time on endotoxin concentration in vertical elutriator cotton dust samples, p. 318–320. *In* L. N. Domelsmith, R. R. Jacobs, and P. J. Wakelyn (ed.), *Proceedings of the Sixteenth Cotton Dust Research Conference.* National Cotton Council of America, Memphis, Tenn.
8. **Crook, B., and S. A. Olenchock.** 1995. Industrial workplaces, p. 531–545. *In* C. S. Cox and C. M. Wathes (ed.), *Bioaerosols Handbook.* CRC Press, Inc., Lewis Publishers, Boca Raton, Fla.
9. **Davis, W. B., I. S. Barsoum, P. W. Ramwell, and H. Yeager, Jr.** 1980. Human alveolar macrophages: effects of endotoxin in vitro. *Infect Immun.* **30:**753–758.

10. **DeLucca, A. J., II, K. A. Brogden, and R. Engen.** 1988. Enterobacter agglomerans lipopolysaccharide-induced changes in pulmonary surfactant as a factor in the pathogenesis of byssinosis. *J. Clin. Microbiol.* **26:**778–780.

11. **Donham, K. J.** 1995. Health hazards of pork producers in livestock confinement buildings: from recognition to control, p. 43–48. *In* H. H. McDuffie, J. A. Dosman, K. M. Semchuk, S. A. Olenchock, and A. Senthilselvan (ed.), *Agricultural Health and Safety: Workplace, Environment, Sustainability.* CRC Press, Inc., Lewis Publishers, Boca Raton, Fla.

12. **Donham, K. J., D. C. Zavala, and J. A. Merchant.** 1984. Respiratory symptoms and lung function among workers in swine confinement buildings: a cross-sectional epidemiological study. *Arch. Environ. Health* **39:**96–101.

13. **Douwes, J., P. Versloot, A. Hollander, D. Heederik, and G. Doekes.** 1995. Influence of various dust sampling and extraction methods on the measurement of airborne endotoxin. *Appl. Environ. Microbiol.* **61:**1763–1769.

14. **Duncan, R. L., Jr., J. Hoffman, V. L. Tesh, and D. C. Morrison.** 1986. Immunologic activity of lipopolysaccharides released from macrophages after the uptake of intact E. coli in vitro. *J. Immunol.* **136:**2924–2929.

15. **Dutkiewicz, J., L. Jablonski, and S. A. Olenchock.** 1988. Occupational biohazards: a review. *Am. J. Ind. Med.* **14:**605–623.

16. **Dutkiewicz, J., W. G. Sorenson, D. M. Lewis, and S. A. Olenchock.** 1992. Levels of bacteria, fungi and endotoxin in stored timber. *Int. Biodeterior. Biodegrad.* **30:**29–46.

17. **Dutkiewicz, J., J. Tucker, R. Burrell, S. A. Olenchock, D. Schwegler-Berry, G. E. Keller III, B. Ochalska, F. Kaczmarski, and C. Skorska.** 1992. Ultrastructure of the endotoxin produced by gram-negative bacteria associated with organic dusts. *Syst. Appl. Microbiol.* **15:**272–285.

18. **Flaherty, D. K., F. H. Deck, J. Cooper, K. Bishop, P. A. Winzenburger, L. R. Smith, L. Bynum, and W. B. Witmer.** 1984. Bacterial endotoxin isolated from a water spray air humidification system as a putative agent of occupation-related lung disease. *Infect. Immun.* **43:**206–212.

19. **Galanos, C., M. A. Freudenberg, O. Luderitz, E. T. Tietschel, and O. Westphal.** 1979. Chemical, physicochemical and biological properties of bacterial lipopolysaccharides, p. 321–332. *In* E. Cohen (ed.), *Biomedical Applications of the Horseshoe Crab (Limulidae).* Alan R. Liss, Inc., New York.

20. **Gordon, T.** 1992. Acute respiratory effects of endotoxin-contaminated machining fluid aerosols in guinea pigs. *Fund. Am. Appl. Toxicol.* **19:**117–123.

21. **Gordon, T., K. Galdanes, and L. Brosseau.** 1992. Comparison of sampling media for endotoxin-contaminated aerosols. *Appl. Occup. Environ. Hyg.* **7:**472–477.

22. **Haglind, P., and R. Rylander.** 1984. Exposure to cotton dust in an experimental cardroom. *Br. J. Ind. Med.* **41:**340–345.

23. **Harkema, J. R., and J. A. Hotchkiss.** 1991. In vivo effects of endotoxin on nasal epithelial mucosubstances: quantitative histochemistry. *Exp. Lung Res.* **17:**743–761.

24. **Helander, I., M. Salkinoja-Salonen, and R. Rylander.** 1980. Chemical structure and inhalation toxicity of lipopolysaccharides from bacteria on cotton. *Infect. Immun.* **29:**859–866.

25. **Helander, I., M. Salkinoja-Salonen, and R. Rylander.** 1982. Pulmonary toxicity of endotoxins: comparison of lipopolysaccharides from various bacterial species. *Infect. Immun.* **35:**528–532.

26. **Hewett, J. A., and R. A. Roth.** 1993. Hepatic and extrahepatic pathobiology of bacterial lipopolysaccharides. *Pharmacol. Rev.* **45:**381–411.

27. **Hollander, A., D. Heederik, P. Versloot, and J. Douwes.**
1993. Inhibition and enhancement in the analysis of airborne endotoxin levels in various occupational environments. *Am. Ind. Hyg. Assoc. J.* **11:**647–653.

28. **Joklik, W. K., H. P. Willett, D. B. Amos, and C. M. Wilfert (ed.).** 1988. *Zinsser Microbiology,* 19th ed., p. 70–72. Appleton and Lange, Norwalk, Conn.

29. **Kennedy, S. M., D. C. Christiani, E. A. Eisen, D. H. Wegman, I. A. Greaves, S. A. Olenchock, T. T. Ye, and P.-L. Lu.** 1987. Cotton dust and endotoxin exposure-response relationships in cotton textile workers. *Am. Rev. Respir. Dis.* **135:**194–200.

30. **Lacy, J., and J. Dutkiewicz.** 1994. Bioaerosols and occupational lung disease. *J. Aerosol Sci.* **25:**1371–1404.

31. **Lenhart, S. W., P. D. Morris, R. E. Akin, S. A. Otenchock, W. S. Service, and W. P. Boone.** 1990. Organic dust, endotoxin, and ammonia exposures in the North Carolina poultry processing industry. *Appl. Occup. Environ. Hyg.* **5:**611–618.

32. **Lundholm, M., and R. Rylander.** 1980. Occupational symptoms among compost workers. *J. Occup. Med.* **22:**256–257.

33. **Michel, O., R. Ginanni, B. Le Bon, J. Content, J. Duchateau, and R. Sergysels.** 1992. Inflammatory response to acute inhalation of endotoxin in asthmatic patients. *Am. Rev. Respir. Dis.* **146:**352–357.

34. **Millner, P. D., S. A. Olenchock, E. Epstein, R. Rylander, J. Haines, J. Walker, B. L. Ooi, E. Horne, and M. Maritato.** 1994. Bioaerosols associated with composting facilities. *Compost Sci. Util.* **2:**6–57.

35. **Milton, D. K., R. J. Gere, H. A. Feldman, and I. A. Greaves.** 1990. Endotoxin measurement: aerosol sampling and application of a new limulus method. *Am. Ind. Hyg. Assoc. J.* **51:**331–337.

36. **Morrison, D. C., and R. J. Ulevitch.** 1978. The effects of bacterial endotoxins on host mediation systems. *Am. J. Pathol.* **93:**527–617.

37. **Olenchock, S. A.** 1988. Quantitation of airborne endotoxin levels in various occupational environments. *Scand. J. Work Environ. Health* **14:**72–73.

38. **Olenchock, S. A.** 1990. Endotoxins, p. 190–200. *In* P. R. Morey, J. C. Feeley, Sr., and J. A. Otten (ed.), *Biological Contaminants in Indoor Environments, ASTM STP 1071.* American Society for Testing and Materials, Philadelphia.

39. **Olenchock, S. A.** 1994. Health effects of biological agents: the role of endotoxins. *Appl. Occup. Environ. Hyg.* **9:**62–64.

40. **Olenchock, S. A., D. C. Christiani, J. C. Mull, T.-T. Ye, and P.-L. Lu.** 1990. Airborne endotoxin concentrations in various work areas within two cotton textile mills in the People's Republic of China. *Biomed. Environ. Sci.* **3:**443–451.

41. **Olenchock, S. A., D. M. Lewis, J. J. Marx, Jr., A. G. O'Campo, and G. J. Kullman.** 1989. Endotoxin contamination and immunological analyses of bulk samples from a mushroom farm, p. 139–150. *In* C. E. O'Rear and G. C. Llewellyn (ed.), *Biodeterioration Research 2.* Plenum Press, New York.

42. **Olenchock, S. A., D. M. Lewis, and J. C. Mull.** 1989. Effects of different extraction protocols on endotoxin analyses of airborne grain dusts. *Scand. J. Work Environ. Health* **15:**430–435.

43. **Olenchock, S. A., J. J. May, D. S. Pratt, L. A. Piacitelli, and J. E. Parker.** 1990. Presence of endotoxins in different agricultural environments. *Am. J. Ind. Med.* **18:**279–284.

44. **Olenchock, S. A., S. A. Murphy, J. C. Mull, and D. M. Lewis.** 1992. Endotoxin and complement activation in an analysis of environmental dusts from a horse barn. *Scand. J. Work Environ. Health* **18:**58–59.

45. **Olenchock, S. A., W. G. Sorenson, G. J. Kullman, and**

W. G. Jones. 1991. Biohazards in composted wood chips, p. 481–483. In H. R. Rossmoore (ed.), *Biodeterioration and Biodegradation 8*. Elsevier Applied Science, London.

46. **Palchak, R. B., R. Cohen, M. Ainslie, and C. L. Hoerner.** 1988. Airborne endotoxin associated with industrial-scale production of protein products in gram-negative bacteria. *Am. Ind. Hyg. Assoc. J.* **49:**420–421.

47. **Popendorf, W.** 1986. Report on agents. *Am. J. Ind. Med.* **10:**251–259.

48. **Preller, I., D. Heederik, H. Kromhout, J. S. M. Boleij, and M. J. M. Tielen.** 1995. Determinants of dust and endotoxin exposure of pig farmers: development of a control strategy using empirical modelling. *Am. Occup. Hyg.* **39:** 545–557.

49. **Reynolds, S. J., and D. K. Milton.** 1993. Comparison of methods for analysis of airborne endotoxin. *Appl. Occup. Environ. Hyg.* **8:**761–767.

50. **Rietschel, E. T., H. Brade, L. Brade, W. Kaca, K. Kawahara, B. Lindner, T. Luderitz, T. Tomita, U. Schade, U. Seydel, and U. Zaringer.** 1985. Newer aspects of the chemical structure and biological activity of bacterial endotoxins, p. 31–50. *In* J. W. ten Cate, H. R. Buller, A. Sturk, and J. Levin (ed.), *Bacterial Endotoxins: Structure, Biomedical Significance and Detection with the Limulus Amebocyte Lysate Test*. Alan R. Liss, Inc., New York.

51. **Rylander, R.** 1989. Role of endotoxin and glucan for the development of granulomatous disease in the lung. *Sarcoidosis* **6:**26–27.

52. **Rylander, R., D. C. Christiani, and Y. Peterson (ed.).** 1989. *Report of the Committee on Organic Dusts of the International Commission on Occupational Health*, p. 1–12. Institutionen for Hygien Goteborgs Universitet, Goteborg, Sweden.

53. **Schwartz D. A., K. J. Donham, S. A. Olenchock, W. J. Popendorf, D. S. Van Fossen, L. F. Burmeister, and J. A. Merchant.** 1995. Determinants of longitudinal changes in spirometric function among swine confinement operators and farmers. *Am. J. Respir. Crit. Care Med.* **151:** 47–53.

54. **Siegel, P. D., S. A. Olenchock, W. G. Sorenson, D. M.** Lewis, T. A. Bledsoe, J. J. May, and D. S. Pratt. 1991. Histamine and endotoxin contamination of hay and respirable hay dust. *Scand. J. Work Environ. Health* **17:**276–280.

55. **Sigsgaard, T., P. Malmros, L. Nersting, and C. Petersen.** 1994. Respiratory disorders and atopy in Danish refuse workers. *Am. J. Respir. Crit. Care Med.* **149:**1407–1412.

56. **Smid, T., D. Heederik, R. Houba, and P. H. Quanjer.** 1992. Dust- and endotoxin-related respiratory effects in the animal feed industry. *Am. Rev. Respir. Dis.* **146:** 1474–1479.

57. **Smid, T., D. Heederik, G. Mensink, R. Houba, and J. S. M. Boleij.** 1992. Exposure to dust, endotoxins, and fungi in the animal feed industry. *Am. Ind. Hyg. Assoc. J.* **53:** 362–368.

58. **Tashiro, K., W. Z. Li, T. Gakiya, T. Taniguchi, K. Nitta, and T. Kobayashi.** 1994. Effects of surfactant replacement on respiratory failure induced by intratracheal endotoxin injection. *Prog. Respir. Res.* **27:**212–215.

59. **Thelin, A., O. Tegler, and R. Rylander.** 1984. Lung reactions during poultry handling related to dust and bacterial endotoxin levels. *Eur. J. Respir. Dis.* **65:**266–271.

60. **Walters, M., D. Milton, L. Larsson, and T. Ford.** 1994. Airborne environmental endotoxin: a cross-validation of sampling and analysis techniques. *Appl. Environ. Microbiol.* **60:**996–1005.

61. **Whitmer, M. P., S. A. Olenchock, D. M. Lewis, and J. C. Mull.** 1992. Improved methodology to validate endotoxin levels in inhibitory agricultural samples, p. 576. *In* M. L. Myers, R. F. Herrick, S. A. Olenchock, J. R. Myers, J. E. Parker, D. L. Hard, and K. Wilson (ed.), *Papers and Proceedings of the Surgeon General's Conference on Agricultural Safety and Health*. U.S. Government Printing Office, Washington, D.C.

62. **Windholz, M., S. Budvari, L. Y. Stroumtsos, and M. N. Festig (ed.).** 1976. *The Merck Index*, 9th ed. Merck and Company, Rahway, N.J.

63. **Zejda, J. E., E. Barber, J. A. Dosman, S. A. Olenchock, H. H. McDuffie, C. Rhodes, and T. Hurst.** 1994. Respiratory health status in swine producers relates to endotoxin exposure in the presence of low dust levels. *J. Occup. Med.* **36:**49–56.

Legionellae and Legionnaires' Disease

BARRY S. FIELDS

72

INTRODUCTION

Bacteria of the family *Legionellaceae* are representatives of a group of organisms that survive as intracellular parasites or endosymbionts of free-living protozoa. The *Legionellaceae* maintained their anonymity until 1977, primarily because of their unique ecology. Initially the bacterium *Legionella pneumophila* was labeled the Legionnaires' disease bacterium. This nomenclature may have been misleading since one can easily assume that causing human disease is paramount to the organisms' existence. We now know that humans have coexisted with these bacteria for a very long time, and only recently has our industrial technology provided these organisms with a means for causing infection in humans.

Legionnaires' Disease

Legionnaires' disease is a consequence of altering the environment for human benefit. This illness continues to comprise a significant proportion of the many cases of pneumonia that occur in the developed world each year. Although the disease can be effectively treated with appropriate antimicrobial agents, current measures are much less efficient at preventing transmission of the bacteria from the environment to susceptible hosts.

Legionnaires' disease occurs when sufficient numbers of legionellae are aerosolized and subsequently inhaled by a susceptible host (9). Legionnaires' disease and legionellae have been included in the aerobiology section of this volume because the bacteria are transmitted to the host via aerosols. However, because these bacteria are primarily found in freshwater environments, many of the methods to be discussed are associated with microbiologic examination of water. Legionellae are relatively difficult to detect in the environment and even more difficult to detect clinically. The fastidious nature of these bacteria is primarily due to the fact that they derive their nutrients from and multiply in an intracellular environment. The bacteria multiply within single-cell protozoans in the environment and within alveolar macrophages in humans. The methods used to detect these organisms have evolved from research in both clinical and environmental microbiology.

Legionellosis

Legionellae were first isolated and identified as part of the investigation of respiratory illness in persons attending a Legionnaires' meeting in Philadelphia in 1976 (30, 46). This highly publicized investigation documented 239 cases and 34 deaths due to a previously unrecognized form of pneumonia. Legionellae are now associated with two forms of respiratory illness, collectively referred to as legionellosis (41, 33). Legionnaires' disease is the pneumonic and more severe form of legionellosis (Table 1). The other form of respiratory illness is named Pontiac fever after the first documented outbreak, which occurred at a health department in Pontiac, Michigan (41). Possible explanations for the manifestation of two disease syndromes caused by the same bacteria include the inability of some legionellae to multiply in human tissue (for a variety of reasons, including virulence, host range, or viability of the bacteria) and differences in host susceptibility (27, 48, 54). A recent community-based pneumonia incidence study has estimated that there are between 10,000 and 15,000 cases of legionellosis annually in the United States (44), approximately 25-fold higher than the number of cases annually reported to the Centers for Disease Control and Prevention. This study determined that the majority of cases of legionellosis are sporadic, with only about 4% outbreak related (44). The sources of community-acquired cases are difficult to identify, partly because of the ubiquitous nature of the bacterium. Although the organisms are relatively common in the environment, they infrequently cause disease (35).

A conceptual scheme of the chain of causation was proposed by Fraser in 1983 to describe the events that lead to an outbreak of legionellosis (29): (i) legionellae must be present in an environmental reservoir, (ii) amplifying factors must allow legionellae to multiply from low to high concentrations, (iii) there must be some means of disseminating legionellae from the reservoir so as to expose people, (iv) the legionellae must be virulent for humans, (v) the organisms must be inoculated at an appropriate site on the human host, and (vi) the host must be susceptible to infection by legionellae. This scheme serves as a model for understanding the epidemiology of legionellosis as well as development of prevention strategies.

THE *LEGIONELLACEAE* AND THEIR ECOLOGY

Bacteria of the genus *Legionella* are characterized as gram-negative, aerobic, rod-shaped bacteria. Cells are 0.3 to 0.9

TABLE 1 Characteristic of Legionnaires' disease and Pontiac fever

Legionnaires' disease	Pontiac fever
Progressive pneumonia, sometimes fatal	Influenza-like illness (nonpneumonic)
2–10-day incubation period	36-h (mean) incubation period
<5% attack rate	>95% attack rate
Risk factors: cigarette smoking, diabetes mellitus, cancer, end-stage renal disease, AIDS	No known risk factors

by 1 to 20 μm and motile, with one or more polar or lateral flagella (12). Legionellae use amino acids as their carbon and energy sources and do not oxidize or ferment carbohydrates. Currently, there are 39 species and 61 serologically distinct groups in the genus *Legionella* (19). Species identification and differentiation are performed serologically, although antisera for many species and serogroups are not available commercially (12). A single species of *Legionella*, *L. pneumophila*, causes approximately 90% of all documented cases of legionellosis (43). Although there are now 15 serogroups of *L. pneumophila*, 82% of all legionellosis cases are caused by *L. pneumophila* serogroup 1. Approximately one-half (i.e., 18) of the 39 species of *Legionella* have been associated with human disease. It is likely that most of the legionellae can cause human disease under the appropriate conditions; however, these infections are infrequently reported because they are rare and because of the lack of diagnostic reagents.

There appear to be a number of unidentified legionellae which cannot be grown on routine *Legionella* media. These organisms have been given the acronym LLAPs (*Legionella*-like amoebal pathogens) because they have been detected through their ability to grow intracellularly in protozoan cells (56). One of the organisms, *Sarcobium lyticum*, was isolated from protozoa in soil by Drozanski in 1954 (20). In 1992, comparative analysis of 16S rRNA genes of several species of legionellae and *Sarcobium* indicated that *S. lyticum* is very closely related to the legionellae (61). *S. lyticum* reacts with antisera specific for a number of legionellae and also reacts with a commercially available PCR kit designed to be specific for the genus *Legionella* (25a). It is likely that *Sarcobium* and *Legionella* belong in the same family or represent a single genus, although additional studies will be necessary to establish this relationship. One LLAP strain was isolated from the sputum of a pneumonia patient by enrichment in amoebae (56). A recent serologic study showed that approximately 20% of pneumonia patients demonstrated a fourfold rise in titer to 1 of 10 LLAPs (5). Collectively, these studies indicate the family *Legionellaceae* may contain a number of bacterial pathogens that cannot be detected by conventional techniques used for legionellae.

The Ecology of Legionellae
Water is the major reservoir for legionellae, and the bacteria are found in freshwater environments worldwide (28). Legionellae have been detected in as many as 40% of freshwater environments by culture and in up to 80% of freshwater sites tested by PCR (59, 64a). Several outbreaks of legionellosis have been associated with construction, and it was originally believed that the bacteria could survive and be transmitted to humans via soil. However, legionellae do not survive in dry environments, and these outbreaks are more

likely the result of massive descalement of plumbing systems due to changes in water pressure during construction (40, 47).

Initially, it was difficult to explain the pervasiveness of legionellae in aquatic environments. The bacteria are fastidious and require an unusual combination of nutrients in bacteriologic medium. These levels of nutrients would rarely be found in aquatic environments and, if present, would serve only to amplify faster-growing bacteria that would compete with the legionellae. However, the nutrients required by legionellae represent the need for an intracellular environment, not soluble nutrients commonly found in freshwater. Legionellae survive in aquatic and, possibly, in some soil environments as intracellular parasites of free-living protozoa (26, 53). Rowbotham first described the ability of *L. pneumophila* to infect protozoa (53) and later described these bacteria as "protozoonotic, i.e., naturally infecting protozoa" (55). Figure 1 shows *L. pneumophila* multiplying within the ciliated protozoan *Tetrahymena pyriformis*. Legionellae have been reported to multiply in 13 species of amoebae and 2 species of ciliated protozoa, while growth of legionellae in the absence of protozoa has been documented only on laboratory media (26). A number of studies have described the relationship between legionellae and protozoa in aquatic environments identified as potential or actual reservoirs of disease-causing strains (11, 34). Protozoa naturally present in these environments can support intracellular growth of legionellae in vitro (3). Legionellae can infect and multiply intracellularly in both protozoa and human phagocytic cells (26, 36). It appears that protozoa are the natural hosts of legionellae, whereas human phagocytic cells occasionally become ill-fated surrogates. Understanding that protozoa play a crucial role in the ecology of legionellae is critical to the development of successful prevention strategies. To understand the ecology of legionellae, these bacteria must be considered in the context of their microbial community, not as independent inhabitants of freshwater environments.

Aerosol Transmission from Aquatic Environments
Inhalation of legionellae in aerosolized droplets is the primary means of transmission for legionellosis (9). These aerosolized droplets must be of a respirable size (1 to 5 μm). No person-to-person transmission of Legionnaires' disease has been documented. A number of devices have been implicated as sources of aerosol transmission of legionellae (Table 2). These sources are of two general types: those producing aerosols of contaminated potable water, such as showers and faucets, and those from nonpotable water, such as cooling towers and whirlpool spas. Meaningful identification of sources of transmission requires a multidisciplinary approach including epidemiology, molecular epidemiology, and microbiologic techniques including water and, occasionally, air sampling (9).

DETECTION OF LEGIONELLAE IN THE ENVIRONMENT
There is considerable controversy concerning the public health benefits of monitoring certain environments for legionellae. Detection of legionellae in an environmental source is not necessarily evidence of the potential for disease. As previously stated, legionellae are ubiquitous and can be isolated from 20 to 40% of freshwater environments (1, 59). However, it is difficult to determine a course of action based on the presence of legionellae since it appears

that their presence rarely results in disease. There is general agreement that monitoring is warranted in order to identify the source of an outbreak of legionellosis or to evaluate the efficacy of biocides or prevention measures. Monitoring may be warranted in special settings where people are highly susceptible to illness due to *Legionella* infection, such as an organ transplant ward within a hospital. Most of the dissension revolves around the issue of regularly scheduled microbiologic monitoring for legionellae in public buildings or areas. The Centers for Disease Control and Prevention recommends aggressive maintenance and disinfection protocols for devices known to transmit legionellae but does not recommend regularly scheduled microbiologic assays for the bacteria (35). In the absence of associated disease, there is no clear evidence that basing interventions on microbiologic assays will lead to a reduction in Legionnaires' disease cases or outbreaks. Regular microbiologic assays may divert resources from proper maintenance protocols or engender a false sense of security. Others, however, have argued that monitoring of potable and recreational waters for other waterborne pathogens (primarily coliforms) is a major achievement of public health programs (38). In 1993, Joly wrote an excellent review of this debate (38), which should continue for the foreseeable future. The following sections describe procedures used to detect legionellae in the environment but do not address the relevance of these methods.

Collection of Water Samples

The number and types of sites that should be tested to detect legionellae must be determined on an individual basis. This is because of the diversity of plumbing and heating, ventilation, and air-conditioning (HVAC) systems in the variety of institutions that may be sampled. These institutions can include industrial facilities, hotels, hospitals, retirement homes, public facilities, and domestic environments. An environmental sampling protocol addressing selection of the appropriate sites to sample within a hospital was published in 1987 (4). This protocol can serve as a prototype for identifying sites that should be sampled in a variety of institutions. Generally, any water source that may be aerosolized should be considered a potential source for the transmission of legionellae. The bacteria are rarely found in municipal water supplies and tend to colonize plumbing systems and point-of-use devices. To colonize a system, the bacteria must multiply, and this requires temperatures above 25°C (12). Therefore, legionellae are most commonly found in hot water systems. The bacteria do not survive drying (40), and so condensate from air-conditioning equipment, which frequently evaporates, is not a likely source (Table 2).

Two primary sample types should be collected when one is sampling for legionellae: water samples and swabs of point-

of-use devices or system surfaces (16). Collection of at least 1 liter of water allows concentration of the sample if necessary. If the water source has recently been treated with chlorine or bromine, 0.5 ml of 0.1 N sodium thiosulfate may be added to each 1-liter sample to neutralize the disinfectant.

Swabs allow sampling of biofilms, which frequently contain legionellae. These can be taken from various points within plumbing systems or from surfaces of basins of cooling towers or spas. Swabs of faucet aerators and showerheads should be taken in conjunction with water samples from these sites and should be taken with the aerator or showerhead removed. The swabs can be streaked directly onto an agar plate or submerged in a small volume of water taken at the same time to prevent drying during transportation to the laboratory.

All samples should be transported to the laboratory in insulated coolers as protection against extreme heat or cold. Samples that will not be processed within 24 h from the time of collection should be refrigerated.

Culture of Water Samples

A schematic representation of methods for processing water samples for culture is shown in Fig. 2. The procedure chosen depends on the expected degree of total bacterial contamination in a particular sample. Potable waters generally have low bacterial concentrations and are either cultured directly or concentrated to detect legionellae. Nonpotable waters, such as those from cooling towers, generally do not require concentration because of their high bacterial concentrations.

Sample Concentration

Samples may be concentrated 10-fold or more by using either filtration or centrifugation. Filtration is used more frequently, although either procedure can be used successfully (16, 66). Water should be filter concentrated in a biological safety cabinet, using 0.2-μm-pore-size polycarbonate filters. Polycarbonate membranes allow suspended particles to collect on the filter surface without being trapped as with matrix-type filters. The filter membrane is then resuspended into a volume of the original sample water and vortexed for 30 s. Samples may be concentrated by centrifugation at 1,000 × g for 10 min, removing all but 10 ml of the supernatant, and vortexing (66).

Acid or Heat Pretreatment

A selective procedure is required to reduce the number of non-*Legionella* bacteria before culturing some water samples with high total bacterial concentrations. Non-*Legionella* bacteria can be selectively killed by either acid pretreatment or brief exposure to higher temperatures (6, 18). Legionellae are more resistant to lower pH and brief exposures to higher temperatures than many other freshwater bacteria. For acid pretreatment, the sample is mixed and incubated with an acid buffer (pH 2.2) for 3 to 30 min (6). The sample is usually neutralized with a 1.0 N KOH buffer before inoculation onto media. Heat pretreatment is accomplished by incubating 10 ml of sample in a 50°C water bath for 30 min (18).

Heat Enrichment

Heat enrichment or incubation of specimens at 35°C can improve recovery of legionellae by up to 30% (57). However, this procedure requires a considerable length of time before results can be obtained and may not be practical in many situations. Heat enrichment relies on autochthonous

---

TABLE 2 Sources known to transmit legionellae via aerosols[a]

Type of water	Transmitting devices
Potable	Showers, tap water faucets, respiratory therapy equipment
Nonpotable	Cooling towers and evaporative condensers, whirlpool spas, decorative fountains, ultrasonic mist machines, humidifiers

[a] Modified from reference 9, with permission.

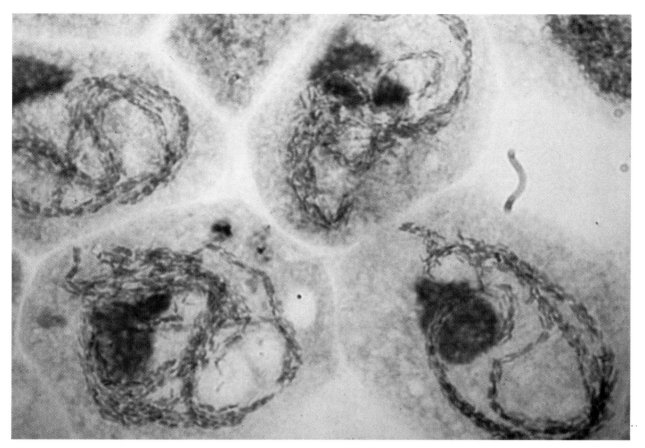

FIGURE 1   Gimenez stain of the ciliated protozoan *T. pyriformis* infected with *L. pneumophila*. Chains of multiplying *Legionella* cells are contained within vesicles as observed in human phagocytic cells infected with *L. pneumophila*. Magnification, ×1,650.

protozoa to amplify undetectable levels of legionellae. Aliquots of samples are incubated at 35°C. Incubated samples can be cultured after 2 to 6 weeks.

Culture Media

Legionellae were isolated by procedures used for the recovery of rickettsiae as early as 1943 (the TATLOCK strain, now *Legionella micdadei*), although these bacteria were not identified as legionellae until 1979 (8, 13, 32). The bacterium was first isolated on bacteriologic agar, using Mueller-Hinton agar supplemented with hemoglobin and IsoVitaleX (25). The medium used for the culture of legionellae has been improved several times, eventually resulting in the medium currently used, buffered charcoal yeast extract (BCYE) (21, 24, 51). The essential component in hemoglobin was found to be a soluble form of iron, and L-cysteine is the essential amino acid provided by the IsoVitaleX. These refinements led to the development of Feeley-Gorman agar, which provides better recovery of the organism from tissue (25). Later, starch was replaced with charcoal to detoxify the medium and the amino acid source was changed to yeast extract, resulting in charcoal yeast extract agar (24). Charcoal yeast extract agar is the base form for most media used to grow legionellae. The most widely used form, BYCE, contains a buffer and is supplemented with α-ketoglutarate (21, 51). Table 3 lists the primary components of BCYE agar and the supplements added for various purposes (24, 49, 65, 67).

Culture of environmental samples requires the use of selective and nonselective media in conjunction with previously described selection procedures. Most laboratories use multiple plates for each sample, including a BCYE agar plate, a BCYE agar plate containing three antimicrobial agents, and a BCYE agar plate containing the three antimicrobial agents plus glycine (Table 3). These media can be prepared with or without the indicator dyes, which impart a color specific for certain species of *Legionella* (65). Although the majority of *Legionella* spp. grow readily on BCYE agar, some require supplementation with bovine serum albumin to enhance growth. *L. micdadei* and several strains of *Legionella bozemanii* show a preference for BCYE with 1.0% albumin (49). All agar plates are inoculated with 0.1 ml of sample by the spread plate technique and incubated at 35°C in a humidified 2.5% CO_2 atmosphere or candle extinction jar.

Identification of Legionellae

Colonies of legionellae require approximately 72 h to appear on BCYE agar and may require 7 days or longer. Ideally, plates should be examined after 4 days of incubation and a second time before being discarded after between 7 and 10 days of incubation. Plates should be examined with a dissecting microscope and a light source to detect bacterial colonies resembling legionellae. After approximately 4 days of incubation, these colonies are 2 to 4 mm in diameter, convex, and round with entire edges (Fig. 3). The center

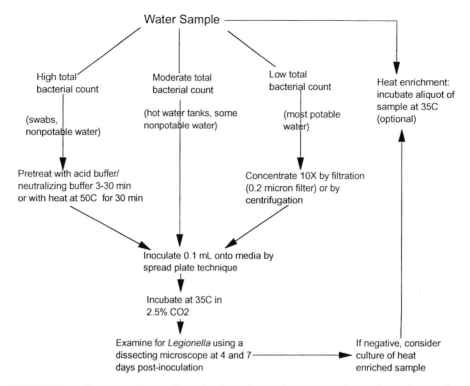

FIGURE 2 Overview of procedures for the culture of water samples to detect legionellae.

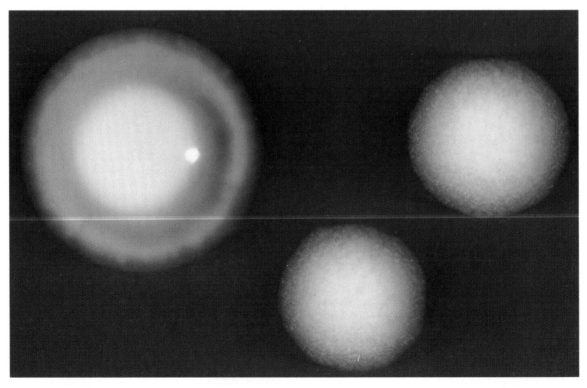

FIGURE 3 Two *Legionella* colonies and a non-*Legionella* bacterial colony as seen through a dissecting microscope upon primary isolation (4 days of incubation). Note the white "cut-glass" appearance of the center of the colony and the purple iridescence which borders it. The iridescence can be one of several colors; the significance of the color is unknown.

of the colony is usually a bright white with a textured appearance that has been described as "cut-glass like" or speckled. The white center of the colony is often bordered with blue, purple, green, or red iridescence. Some species of legionellae produce colonies that exhibit blue-white or red autofluorescence (18). The primary isolation plates can be examined with long-wave UV light to detect these autofluorescent colonies.

Colonies resembling legionellae can be presumptively identified on the basis of their requirement for L-cysteine by subculture on blood agar or BCYE agar without L-cysteine. Subcultured colonies that grow on BCYE agar, but not on blood agar or BCYE without L-cysteine, are presumed to be legionellae. Legionellae are relatively inert in many biochemical test media, so these tests are of limited value in identification of these bacteria. Definitive identification is usually accomplished by using a direct fluorescent antibody (DFA) or slide agglutination test with specific antisera (16). Identification can also be accomplished by fatty acid analysis and DNA hybridization (12).

Air Sampling

Examination of water samples is the most efficient microbiologic method for identifying sources of legionellae. Air sampling is an insensitive means of detecting these bacteria and therefore is of limited value in environmental sampling for legionellae. In certain instances, it may be beneficial to demonstrate the presence of legionellae in aerosol droplets associated with suspected reservoirs of the bacterium. Air sampling has been used to better define the roles of certain devices such as showers, faucets, and evaporative condensers in disease transmission (10). It is usually used to establish the presence of the legionellae in aerosol droplets and occasionally to quantitate or determine the sizes of particles containing legionellae. Information regarding particle size and numbers of viable bacteria can be calculated by using these procedures, but this approach requires much more stringent controls and calibration (16). Samplers should be placed in locations representative of human exposure, and investigators should wear an Occupational Safety and Health Administration-approved respirator if sampling involves exposure to potentially infectious aerosols.

The basic methods for sampling airborne bacteria include impingement in liquids, impaction of solid surfaces, filtration, sedimentation, centrifugation, and precipitation (70).

See chapter 68 for detailed discussion of sampling techniques for airborne microorganisms. Methods that have been used to sample air for legionellae include impingement in liquid by using an all-glass impinger (AGI), impaction of solid medium by using Andersen samplers, and the use of settle plates (10). Except for settle plates, these methods require a vacuum source and a means of controlling airflow. Several configurations of air sampling equipment can be used; they usually incorporate a device for controlling airflow (flowmeter-manometer) connected in the vacuum line between the sampler and vacuum source.

Chemical Corps-type AGIs with the stem 30 mm from the bottom of the flask have been used successfully to sample for legionellae (10). These samplers use the principle of impingement and washing of air, in which organisms are entrapped in a liquid medium. Because of the velocity at which samples are collected, clumps tend to be fragmented, leading to a more accurate count of bacteria present in the air. The disadvantages of this method are that this velocity tends to destroy some vegetative cells, it does not differentiate particle sizes, and AGIs are easily broken in the field. Yeast extract broth (0.25%) is the recommended liquid medium for AGI sampling of legionellae (16). Once the sample has been collected, the yeast extract broth may be processed by methods used for the culture of water samples.

Andersen samplers are viable particle samplers in which particles pass through jet orifices of decreasing size in cascade fashion until they hit an agar surface (70). The agar plates are then removed and incubated in order to culture any legionellae present. The stage distribution of the legionellae should indicate the extent to which the bacteria would have penetrated the respiratory system. The advantages of this sampling method are that the equipment is more durable, the sampler can determine the number and size of droplets containing legionellae, and agar plates can be placed directly in an incubator with no further manipulations. Both selective and nonselective BCYE agar can be used in an Andersen sampler. If the samples must be shipped to a laboratory, they should be packed and shipped without refrigeration as soon as possible.

Nonculture Methods for Detection of Legionellae

Several nonculture methods have been developed to detect legionellae in environmental samples. These methods offer the potential of greatly increased sensitivity. Culture re-

TABLE 3 Components and supplements of BCYE agar for culturing legionellae from the environment

Component	Concn	Purpose
Charcoal	2.0 g/liter	Base component
Yeast extract	10.0 g/liter	Base component
ACES[a] buffer	10.0 g/liter	Base component
Ferric pyrophosphate	0.25 g/liter	Base component
L-Cysteine	0.4 g/liter	Base component
Potassium α-ketoglutarate	1.0 g/liter	Base component
Agar	17.0 g/liter	Base component
Glycine	3.0 g/liter	Selective agent
Polymyxin B	50–100 U/ml	Selective agent (gram negative)
Vancomycin or cefamandole	1–5 g or 4 mg/liter	Selective agent (gram positive)
Anisomycin or cycloheximide	80 μg/ml (for either)	Selective agent (fungal)
Bromocresol blue	10 mg/liter	Indicator dye
Bromocresol purple	10 mg/liter	Indicator dye
Bovine serum albumin	10 g/liter	Supplement for some fastidious legionellae

[a] N-(2-Acetamido)-2-aminoethanesulfonic acid.

mains the method of choice for detecting legionellae, primarily because nonculture methods cannot provide information regarding the viability of the bacteria. These nonculture methods include detection of the organisms with specific antisera by DFA staining and procedures to detect nucleic acids of legionellae by PCR.

The use of DFA to detect legionellae is limited by the number of specific antisera that can be used. Since there are no antisera which specifically react with all *Legionella* species, a different antiserum must be used for each species or serogroup. Reports on the sensitivity and specificity of DFA testing of environmental specimens vary greatly, with most studies indicating that the test is relatively insensitive and nonspecific (38).

The use of PCR for detecting nucleic acids of legionellae in the environment has proved to be a valuable technique for investigations of legionellosis (48). A number of *Legionella* genes, including 5S rRNA, 16S rRNA, and *mip* genes, have been used as targets for PCR (42, 62). A commercially produced PCR kit (EnviroAmp kit; Perkin-Elmer) is available and contains two probes that detect amplified DNA from the genus *Legionella* and the species *L. pneumophila*. The EnviroAmp kit reports a sensitivity of 10 to 100 legionellae per ml. Results from the use of culture and the EnviroAmp kit for detection of one species, *L. pneumophila*, are similar. However, up to 80% of freshwaters are positive for the genus *Legionella* with the PCR kit (64a), while only approximately 20 to 40% of waters are positive by culture (1, 59). This discrepancy could be due to the presence of nonviable or injured organisms, a nonspecific reaction with unrelated organisms (although available data argue against this possibility), or the presence of related organisms, such as *S. lyticum*, that cannot be detected by conventional techniques used for legionellae.

Most investigations of epidemic legionellosis have used culture to detect legionellae in the environment. As a result, most of our epidemiologically relevant information concerning legionellosis is based on direct culture data. Until we possess a better understanding of the diversity and distribution of the legionellae, results from non-culture-based methods should be interpreted cautiously.

Subtyping Techniques

Associating an environmental isolate of *Legionella* with a clinical isolate from a patient with legionellosis usually requires a molecular subtyping procedure. *L. pneumophila* serogroup 1 (Lp1) accounts for most of the cases of legionellosis (9). However, Lp1 can be divided into a number of subtypes by using various techniques, indicating that this is a fairly heterogeneous serogroup (31). Identification of the bacterium, even to the serogroup level, is not sufficient to implicate an environmental isolate as the source of disease.

Initially legionellae were identified to the serogroup level during investigations of legionellosis. This form of serologic subtyping uses polyvalent or monoclonal antisera and may be adequate for identifying reservoirs of some of the more uncommon legionellae causing disease. The variety of strains and distribution of Lp1 necessitate more elaborate subtyping procedures to discriminate within these bacteria. Several groups of monoclonal antibodies have been developed for this purpose (2). An international panel of seven monoclonal antibodies was proposed in 1986 (2, 39). Although much information has been gained through the use of this panel, several of the cell lines have been lost and most of these reagents are no longer available. Use of these monoclonal antibodies has identified 10 type strains within

Lp1. The ability to differentiate Lp1 into 10 subtypes has greatly improved our ability to correctly identify sources of disease. For example, several outbreak investigations have found Lp1 in both the potable water system and cooling towers of a particular institution (10, 11). Differentiation with monoclonal antibodies indicated that the epidemic strain had colonized only one of these systems (either cooling tower or potable systems), indicating that the presence of Lp1 in the other system was not causing disease at that site. Without this level of subtyping, it would be much more difficult to confidently confirm the source of transmission.

Recent investigations have shown that the use of monoclonal antibodies may be inadequate for discriminating between disease-causing strains and other environmental isolates of Lp1 (68). Newer molecular techniques such as pulsed-field gel electrophoresis and arbitrarily primed PCR are able to discriminate within monoclonal subtypes of Lp1 and identify sources of disease-causing strains (31). These techniques appear to complement the use of monoclonal antibodies and are the most recent of several techniques that separate strains on the basis of DNA polymorphism (58). Other techniques used to discriminate between isolates of legionellae include restriction fragment length polymorphism analysis, plasmid analyses, electrophoretic alloenzyme typing, and RNA-DNA probing of DNA digests (64). Pulsed-field gel electrophoresis and arbitrarily primed PCR offer better discrimination and are less labor-intensive than these other techniques.

ENVIRONMENTAL APPROACHES TO CONTROLLING LEGIONELLOSIS

Practical information concerning treatment processes that effectively control legionellae is limited, especially in the United States. Various biocides and alternative disinfection methods, such as heat eradication, UV irradiation, and ozonation, have been tested to determine their abilities to kill legionellae (22, 50, 60, 63). Results obtained in these types of laboratory studies often fail to translate into effective prevention protocols (23). Several countries have produced guidelines or codes of practice relating to the control of legionellae; however, research to substantiate these practices is scarce, and the prevailing rationale for these recommendations is almost entirely empirical (14). At the time of this publication, guidelines for the minimization of legionellosis in building water systems in the United States are being drafted under the auspices of the American Society of Heating, Refrigerating and Air-Conditioning Engineers, Inc.

If one utilizes the chain of causation leading to an outbreak of legionellosis (as described earlier in this chapter), a control strategy would need only to interrupt this chain of events to be successful. The following discussion of control strategies is arranged according to this scheme.

Environmental Reservoirs

Attempting to control legionellae in water supplies would be an inefficient means of preventing legionellosis. It is highly unlikely that the legionellae could be eradicated, even on a limited scale, since they are integral members of the aquatic microbial community. The bacteria are ubiquitous in freshwater environments, and they are present in relatively low concentrations in most of these water supplies. They are rarely, if ever, isolated from water treatment facilities. The difficulty in detecting legionellae in water treatment plants and municipal water supplies is probably due to the lower temperatures of these waters. Legionellae are more frequently detected and are present in higher concentrations in warm or thermally altered environments (52).

Amplifying Reservoirs

As previously mentioned, temperature is a critical factor in the ability of legionellae to colonize reservoirs in which they are amplified. Other microorganisms and factors critical to the growth of legionellae are almost universally present in freshwater environments, and it is temperature that governs the numbers of these bacteria. An Australian study of cooling towers found that legionellae colonized or multiplied in towers with basin temperatures above 16°C, and multiplication became explosive at temperatures above 23°C (14). Conversely, legionellae are killed at temperatures above 55°C, and it has been suggested that potable hot water systems be maintained at temperatures between 55 and 60°C to prevent growth of the bacteria (63). This may be impractical for many institutions, especially hospitals, where the potential for scalding of patients exists.

Practices to control legionellae in amplifying reservoirs can be divided into two categories, routine maintenance and emergency decontamination procedures. Maintenance of respiratory therapy equipment requires that potable water not be used to rinse the equipment before patient use (45). Several documents address routine maintenance of cooling towers, evaporative condensers, and whirlpool spas (15, 17, 69). Generally, these documents suggest following manufacturers' recommendations regarding cleaning and biocide treatment of these devices. It is generally believed that well-maintained equipment does not permit excessive growth of legionellae (37). Emergency decontamination protocols describing hyperchlorination and cleaning procedures for cooling towers have been developed for towers implicated in the transmission of legionellosis (17, 69). There is no equivalent compilation of recommendations for potable water systems, although specific intervention strategies have been published (7, 45). The principal approaches to disinfection of potable systems are heat flushing at temperatures above 70°C, hyperchlorination, and physical cleaning of hot water tanks (45). Potable systems are easily recolonized and may require continuous intervention such as raising of hot water temperatures or continuous chlorination (45).

Aerosol Transmission

Some control strategies are intended to prevent exposure of susceptible individuals to aerosols which may contain legionellae. Manufacturers have improved the performance of drift eliminators for cooling towers in recent years (14). These newer eliminator modules can reduce the water loss from cooling towers by orders of magnitude. Early investigations of Legionnaires' disease associated with cooling towers resulted in a recommendation for relocation of cooling towers or air intake vents so that cooling tower exhaust would not be carried directly into the HVAC systems of buildings. Some documents have suggested eliminating air intake vents within 100 m of cooling towers, although data to support this recommended distance are limited.

Health care professionals, engineers, and industry continue to increase their awareness and understanding of legionellae and legionellosis. It is difficult to determine if this increased understanding has led to measures that have reduced the incidence of legionellosis. Fear of litigation has caused the administration of many institutions to implement prevention strategies that are unproven and lack sufficient scientific documentation. Effective prevention strate-

gies will require more effective decontamination techniques and approaches to prevent amplification of these bacteria in reservoirs.

REFERENCES

1. **Arnow, P. M., D. Weil, and M. F. Para.** 1985. Prevalence and significance of *Legionella pneumophila* contamination of residential hot-tap water systems. *J. Infect. Dis.* **152:** 145–151.

2. **Barbaree, J. M.** 1993. Selecting a subtyping technique for use in investigations of legionellosis epidemics, p. 169–172. *In* J. M. Barbaree, R. F. Breiman, and A. P. Dufour (ed.), *Legionella: Current Status and Emerging Perspectives.* American Society for Microbiology, Washington, D.C.

3. **Barbaree, J. M., B. S. Fields, J. C. Feeley, G. W. Gorman, and W. T. Martin.** 1986. Isolation of protozoa from water associated with a legionellosis outbreak and demonstration of intracellular multiplication of *Legionella pneumophila.* *Appl. Environ. Microbiol.* **51:**422–424.

4. **Barbaree, J. M., G. W. Gorman, W. T. Martin, B. S. Fields, and W. E. Morrill.** 1987. Protocol for sampling environmental sites for legionellae. *Appl. Environ. Microbiol.* **53:**1454–1458.

5. **Benson, R. F., T. J. Rowbotham, I. Bialkowska, D. Losos, J. C. Butler, H. B. Lipman, J. F. Plouffe, and B. S. Fields.** 1995. Serologic evidence of infection with nine *Legionella*-like amoebal pathogens in pneumonia patients, abstr. C-200, p. 35. *In Abstracts of the 95th General Meeting of the American Society for Microbiology 1995.* American Society for Microbiology, Washington, D.C.

6. **Bopp, C. A., J. W. Summer, G. K. Morris, and J. G. Wells.** 1981. Isolation of *Legionella* spp. from environmental water samples by low-pH treatment and use of a selective medium. *J. Clin. Microbiol.* **13:**714–719.

7. **Bornstein, N., C. Vieilly, M. Nowiki, J. C. Paucod, and J. Fleurette.** 1986. Epidemiological evidence of legionellosis transmission through domestic hot water supply systems and possibilities of control. *Isr. J. Med. Sci.* **13:**39–40.

8. **Bozeman, F. M., J. W. Humphries, and J. M. Campbell.** 1968. A new group of rickettsia-like agents recovered from guinea pigs. *Acta Virol.* **12:**87–93.

9. **Breiman, R. F.** 1993. State of the art lecture. Modes of transmission in epidemic and nonepidemic *Legionella* infection: directions for further study, p. 30–35. *In* J. M. Barbaree, R. F. Breiman, and A. P. Dufour (ed.), *Legionella: Current Status and Emerging Perspectives.* American Society for Microbiology, Washington, D.C.

10. **Breiman, R. F., W. Cozen, B. S. Fields, T. D. Mastro, S. J. Carr, J. S. Spika, and L. Mascola.** 1990. Role of air sampling in investigation of an outbreak of Legionnaires' disease associated with exposure to aerosols from an evaporative condenser. *J. Infect. Dis.* **161:**1257–1261.

11. **Breiman, R. F., B. S. Fields, G. N. Sanden, L. Volmer, A. Meier, and J. S. Spika.** 1990. Association of shower use with Legionnaires' disease: possible role of amoebae. *JAMA* **263:**2924–2926.

12. **Brenner, D. J., J. C. Feeley, and R. E. Weaver.** 1984. Family VII. *Legionellaceae,* p. 279. *In* N. R. Krieg and J. G. Holt (ed.), *Bergey's Manual of Systemic Bacteriology,* vol. 1. Williams & Wilkins, Baltimore.

13. **Brenner, D. J., A. G. Steigerwalt, and J. E. McDade.** 1979. Classification of the legionnaires' disease bacterium: *Legionella pneumophila,* genus novum, species nova, of the Family *Legionellaceae,* familia nova. *Ann. Intern. Med.* **90:** 656–658.

14. **Broadbent, C. R.** 1993. *Legionella* in cooling towers: practical research, design, treatment, and control guidelines, p. 217–222. *In* J. M. Barbaree, R. F. Breiman, and A. P. Du-

four (ed.), *Legionella: Current Status and Emerging Perspectives.* American Society for Microbiology, Washington, D.C.

15. **Centers for Disease Control.** 1985. Suggested health and safety guidelines for public spas and hot tubs. Publication no. 99-960. Centers for Disease Control, Atlanta.

16. **Centers for Disease Control and Prevention.** 1992. *Procedures for the Recovery of Legionella from the Environment.* Centers for Disease Control and Prevention, Atlanta.

17. **Cooling Tower Institute.** 1980. *Suggested Protocol for Emergency Cleaning of Cooling Towers and Related Equipment Suspected of Infection by Legionnaires' Disease Bacteria (Legionella pneumophila).* Cooling Tower Institute, Houston, Tex.

18. **Dennis, P. J. L.** 1988. Isolation of legionellae from environmental specimens, p. 31–44. *In* T. G. Harrison and A. G. Taylor (ed.), *A Laboratory Manual for Legionella.* John Wiley & Sons Ltd., New York.

19. **Dennis, P. J., D. J. Brenner, W. L. Thacker, R. Wait, G. Vesey, A. G. Steigerwalt, and R. F. Benson.** 1993. Five new *Legionella* species isolated from water. *Int. J. Syst. Bacteriol.* **43:**329–337.

20. **Drozanski, W. J.** 1991. *Sarcobium lyticum* gen. nov., sp. nov., an obligate intracellular bacterial parasite of small free-living amoebae. *Int. J. Syst. Bacteriol.* **41:**82–87.

21. **Edelstein, P. H.** 1981. Improved semi-selective medium for isolation of *Legionella pneumophila* from contaminated clinical and environmental specimens. *J. Clin. Microbiol.* **14:**298–303.

22. **Edelstein, P. H., R. E. Whittaker, R. L. Kreiling, and C. L. Howell.** 1982. Efficacy of ozone in eradication of *Legionella pneumophila* from hospital plumbing fixtures. *Appl. Environ. Microbiol.* **44:**1330–1334.

23. **England, A. C., III, D. W. Fraser, G. F. Mallison, D. C. Mackel, P. Skaliy, and G. W. Gorman.** 1982. Failure of *Legionella pneumophila* sensitivies to predict culture results from disinfectant-treated air-conditioning cooling towers. *Appl. Environ. Microbiol.* **43:**240–244.

24. **Feeley, J. C., R. J. Gibson, G. W. Gorman, N. C. Langford, J. K. Rasheed, D. C. Mackel, and W. B. Baine.** 1979. Charcoal yeast extract agar: primary isolation medium for *Legionella pneumophila. J. Clin. Microbiol.* **10:**437–441.

25. **Feeley, J. C., G. W. Gorman, R. E. Weaver, D. C. Mackel, and H. W. Smith.** 1978. Primary isolation medium for legionnaires' disease bacterium. *J. Clin. Microbiol.* **8:**320–328.

25a. **Fields, B.** Unpublished data.

26. **Fields, B. S.** 1993. *Legionella* and protozoa: interaction of a pathogen and its natural host, p. 129–136. *In* J. M. Barbaree, R. F. Breiman, and A. P. Dufour (ed.), *Legionella: Current Status and Emerging Perspectives.* American Society for Microbiology, Washington, D.C.

27. **Fields, B. S., J. M. Barbaree, G. N. Sanden, and W. E. Morrill.** 1990. Virulence of a *Legionella anisa* strain associated with Pontiac fever: an evaluation using protozoan, cell culture, and guinea pig models. *Infect. Immun.* **58:**3139–3142.

28. **Fliermans, C. B., W. B. Cherry, L. H. Orrison, S. J. Smith, D. L. Tison, and D. H. Pope.** 1981. Ecological distribution of *Legionella pneumophila. Appl. Environ. Microbiol.* **41:**9–16.

29. **Fraser, D. W.** 1984. Sources of legionellosis, p. 277–280. *In* C. Thornsberry, A. Balows, J. C. Feeley, and W. Jakubowski (ed.), *Proceedings of the Second International Symposium on Legionella.* American Society for Microbiology, Washington, D.C.

30. **Fraser, D. W., T. F. Tsai, W. Orenstein, W. E. Parkin, H. J. Beecham, R. G. Sharrar, H. Harris, G. F. Mallison, S. M. Martin, J. E. McDade, C. C. Shepard, P. S. Brachman, and the Field Investigation Team.** 1977. Legion-

naires' disease: description of an epidemic of pneumonia. *N. Engl. J. Med.* **297:**1189–1197.

31. **Gomez-Lus, P., B. S. Fields, R. F. Benson, W. T. Martin, S. P. O'Connor, and C. M. Black.** 1993. Comparison of arbitrarily primed polymerase chain reaction, ribotyping, and monoclonal antibody analysis for subtyping *Legionella pneumophila* serogroup 1. *J. Clin. Microbiol.* **31:**1940–1942.

32. **Hebert, G. A., C. W. Moss, L. K. McDougal, F. M. Bozeman, R. M. McKinney, and D. J. Brenner.** 1980. The rickettsia-like organism Tatlock (1943) and HEBA (1959): bacteria phenotypically similar but genetically distinct from *Legionella pneumophila* and the WIGA bacterium. *Ann. Intern. Med.* **92:**45–52.

33. **Helms, C. M., J. P. Viner, R. H. Sturm, E. S. Renner, and W. Johnson.** 1979. Comparative features of pneumococcal, mycoplasmal, and legionnaires' disease pneumonias. *Ann. Intern. Med.* **90:**543–547.

34. **Henke, M., and K. M. Seidel.** 1986. Association between *Legionella pneumophila* and amoebae in water. *Isr. J. Med. Sci.* **22:**690–695.

35. **Hoage, C. W., and R. F. Breiman.** 1991. Advances in the epidemiology and control of *Legionella* infections. *Epidemiol. Rev.* **13:**329–339.

36. **Horwitz, M. A., and S. C. Silverstein.** 1980. The legionnaires' disease bacterium (*Legionella pneumophila*) multiplies intracellularly in human monocytes. *J. Clin. Invest.* **66:**441–450.

37. **Jakubowski, W., C. V. Broome, E. E. Geldreich, and A. P. DuFour.** 1984. Round table discussion. Transmission and control, p. 351–355. *In* C. Thornsberry, A. Balows, J. C. Feeley, and W. Jakubowski (ed.), *Proceedings of the Second International Symposium on Legionella.* American Society for Microbiology, Washington, D.C.

38. **Joly, J. R.** 1993. Monitoring for the presence of *Legionella:* where, when, and how?, p. 211–216. *In* J. M. Barbaree, R. F. Breiman, and A. P. Dufour (ed.), *Legionella: Current Status and Emerging Perspectives.* American Society for Microbiology, Washington, D.C.

39. **Joly, J. R., R. M. McKinney, J. O. Tobin, W. F. Bibb, I. D. Watkins, and D. Ramsey.** 1985. Development of a standardized subgrouping scheme for *Legionella pneumophila* serogroup 1 using monoclonal antibodies. *J. Clin. Microbiol.* **23:**768–771.

40. **Katz, S. M., and J. M. Hammel.** 1987. The effect of drying, heat, and pH on the survival of *Legionella pneumophila. Ann. Clin. Lab. Sci.* **17:**150–156.

41. **Kaufmann, A. F., J. E. McDade, C. M. Patton, J. V. Bennett, P. Skaliy, J. C. Feeley, D. C. Anderson, M. E. Potter, V. F. Newhouse, M. B. Gregg, and P. S. Brachman.** 1981. Pontiac fever: isolation of the etiologic agent (*Legionella pneumophila*) and demonstration of its mode of transmission. *Am. J. Epidemiol.* **111:**337–339.

42. **Mahbubani, M. H., A. K. Bej, R. Miller, L. Haff, J. DiCesare, and R. M. Atlas.** 1990. Detection of *Legionella pneumophila* with polymerase chain reaction and gene probe methods. *Mol. Cell. Probes* **4:**175–187.

43. **Marston, B. J., H. B. Lipman, and R. F. Breiman.** 1994. Surveillance for Legionnaires' disease. *Arch. Intern. Med.* **154:**2417–2422.

44. **Marston, B. J., J. F. Plouffe, R. F. Breiman, T. M. File, R. F. Benson, M. Moyenuddin, W. L. Thacker, K. H. Wong, S. Skelton, B. Hackman, S. J. Salstrom, J. M. Barbaree, and The Community-Based Pneumonia Incidence Study Group.** 1993. Preliminary findings of a community-based pneumonia incidence study, p. 36–37. *In* J. M. Barbaree, R. F. Breiman, and A. P. Dufour (ed.), *Legionella: Current Status and Emerging Perspectives.* American Society for Microbiology, Washington, D.C.

45. **Mastro, T. D., B. S. Fields, R. F. Breiman, J. Campbell,**

B. D. Plikaytis, and J. S. Spika. 1991. Nosocomial Legionnaires' disease and use of medication nebulizers. *J. Infect. Dis.* **163:**667–671.

46. **McDade, J. E., C. C. Shepperd, D. W. Fraser, T. R. Tsai, M. A. Redus, W. R. Dowdle, and the Laboratory Investigation Team.** 1977. Legionnaires' disease. Isolation of the bacterium and demonstration of its role in other respiratory disease. *N. Engl. J. Med.* **297:**1197–1203.
47. **Mermel, L. A., S. L. Joesephson, C. H. Giorgio, J. Dempsey, and S. Parenteau.** 1995. Association of Legionnaires' disease with construction: contamination of potable water? *Infect. Control Hosp. Epidemiol.* **16:**76–80.
48. **Miller, L. A., J. L. Beebe, J. C. Butler, W. T. Martin, R. Benson, R. E. Hoffman, and B. S. Fields.** 1993. Use of polymerase chain reaction in an epidemic investigation of Pontiac fever. *J. Infect. Dis.* **168:**769–772.
49. **Morrill, W. E., J. M. B. S. Fields, G. N. Sanden, and W. T. Martin.** 1990. Increased recovery of *Legionella micdadei* and *Legionella bozemanii* on buffered charcoal yeast extract agar supplemented with albumin. *J. Clin. Microbiol.* **28:**616–618.
50. **Muraca, P., J. E. Stout, and V. L. Yu.** 1987. Comparative assessment of chlorine, heat, ozone, and UV light for killing *Legionella pneumophila* within a model plumbing system. *Appl. Environ. Microbiol.* **53:**447–553.
51. **Pasculle, A. W., J. C. Feeley, R. J. Gibson, L. G. Cordes, R. L. Myerowitz, C. M. Patton, G. W. Gorman, C. L. Carmack, J. W. Ezzell, and J. N. Dowling.** 1980. Pittsburgh pneumonia agent: direct isolation from human lung tissue. *J. Infect. Dis.* **141:**727–732.
52. **Plouffe, J. F., L. R. Webster, and B. Hackman.** 1983. Relationship between colonization of hospital buildings with *Legionella pneumophila* and hot water temperatures. *Appl. Environ. Microbiol.* **46:**769–770.
53. **Rowbotham, T. J.** 1980. Preliminary report on the pathogenicity of *Legionella pneumophila* for freshwater and soil amoebae. *J. Clin. Pathol.* **33:**1179–1183.
54. **Rowbotham, T. J.** 1980. Pontiac fever explained? *Lancet* **ii:**69.
55. **Rowbotham, T. J.** 1986. Current views on the relationships between amoebae, legionellae and man. *Isr. J. Med. Sci.* **22:**678–689.
56. **Rowbotham, T. J.** 1993. *Legionella*-like amoebal pathogens, p. 137–140. *In* J. M. Barbaree, R. F. Breiman, and A. P. Dufour (ed.), *Legionella: Current Status and Emerging Perspectives.* American Society for Microbiology, Washington, D.C.
57. **Sanden, G. N., W. E. Morrill, B. S. Fields, R. F. Breiman, and J. M. Barbaree.** 1992. Incubation of water samples containing amoebae improves detection of legionellae by the culture method. *Appl. Environ. Microbiol.* **58:**2001–2004.
58. **Schoonmaker, D., T. Heimberger, and G. Birkhead.** 1992. Comparison of ribotyping and restriction enzyme analysis using pulsed-field gel electrophoresis for distinguishing *Legionella pneumophila* isolates obtained during a nosocomial outbreak. *J. Clin. Microbiol.* **30:**1491–1498.
59. **Shelton, B. G., G. K. Morris, and G. W. Gorman.** 1993. Reducing risks associated with *Legionella* bacteria in building water systems, p. 279–281. *In* J. M. Barbaree, R. F. Breiman, and A. P. Dufour (ed.), *Legionella: Current Status and Emerging Perspectives.* American Society for Microbiology, Washington, D.C.
60. **Skaliy, P., T. A. Thompson, G. W. Gorman, G. K. Morris, H. V. McEachern, and D. C. Mackel.** 1980. Laboratory studies of disinfectants against *Legionella pneumophila*. *Appl. Environ. Microbiol.* **40:**697–700.
61. **Springer, N., W. Ludwig, W. Drozanski, R. Amann, and K. H. Schleifer.** 1992. The phylogenetic status of *Sarcobium lyticum*, an obligate intracellular bacterial parasite of small amoebae. *FEMS Microbiol. Lett.* **96:**199–202.
62. **Starnbach, M. N., S. Falkow, and L. S. Tompkins.** 1989. Species-specific detection of *Legionella pneumophila* in water by DNA amplification and hybridization. *J. Clin. Microbiol.* **27:**1257–1261.
63. **Stout, J. E., M. E. Best, and V. L. Yu.** 1986. Susceptibility of members of the family *Legionellacae* to thermal stress: implications for heat eradication methods in water distribution systems. *Appl. Environ. Microbiol.* **52:**396–399.
64. **Tompkins, L. S., and J. S. Loutit.** 1993. Detection of *Legionella* by molecular methods, p. 163–168. *In* J. M. Barbaree, R. F. Breiman, and A. P. Dufour (ed.), *Legionella: Current Status and Emerging Perspectives.* American Society for Microbiology, Washington, D.C.
64a. **Tyndall, D. (Oak Ridge, Tenn.).** Personal communication.
65. **Vickers, R. M., A. Brown, and G. M. Garrity.** 1981. Dye-containing buffered charcoal yeast extract medium for differentiation of members of the family *Legionellaceae*. *J. Clin. Microbiol.* **13:**380–382.
66. **Vickers, R. M., J. E. Stout, V. L. Yu, and J. D. Rihs.** 1987. Manual of culture methodology for *Legionella*. *Semin. Respir. Infect.* **2:**274–279.
67. **Wadowsky, R. M., and R. B. Yee.** 1981. Glycine-containing selective medium for isolation of legionellaceae from environmental specimens. *Appl. Environ. Microbiol.* **42:**768–772.
68. **Whitney, C. G., J. Hofmann, J. Pruckler, B. Matyas, R. Benson, B. Fields, L. Mermel, C. Giorgio, and R. Breiman.** 1994. A novel subtyping method to identify the source of an outbreak of Legionnaires' disease, abstr. J192, p. 193. *In Program and Abstracts of the 34th Interscience Conference on Antimicrobial Agents and Chemotherapy.* American Society for Microbiology, Washington, D.C.
69. **Wisconsin Division of Health.** 1987. *Control of Legionella in Cooling Towers: Summary Guidelines.* Wisconsin Division of Health, Madison, Wis.
70. **Wolf, H. W., P. Skaliy, L. B. Hall, M. M. Harris, H. M. Decker, L. M. Buchanan, and C. M. Dahlgren.** 1964. Sampling microbiological aerosols. Public Health Service publication no. 686. Government Printing Office, Washington, D.C.

Airborne *Mycobacterium* spp.

PAUL A. JENSEN

73

The only genus in the family Mycobacteriaceae is Mycobacte-rium. The mycobacteria are slightly curved or straight bacilli, 0.2 to 0.7 μm in diameter by 1.0 to 10 μm in length, and are sometimes branching. Mycobacteria have cell walls with a high lipid content that includes waxes having characteristic mycolic acids with long, branched chains. The minimal standards for including a species in the genus Mycobacterium are (i) acid-alcohol fastness, (ii) presence of mycolic acids containing 60 to 90 carbon atoms which are cleaved to C_{22} to C_{26} fatty acid methyl esters by pyrolysis, and (iii) a guanine-plus-cytosine content of the DNA of 61 to 71 mol% (51).

The genus Mycobacterium includes obligate parasites, saprophytes, and opportunistic pathogens. Most species are free living in soil and water. However, the more publicized mycobacteria, Mycobacterium tuberculosis complex and M. leprae, amplify in the tissues of humans and other warm-blooded animals (26).

Mycobacteria can be divided into two main groups based on growth rate. The slowly growing species require more than 7 days to form visible colonies on solid medium, while rapidly growing species require less than 7 days. In general, the slowly growing species are often pathogenic for humans or animals, while rapidly growing species are usually considered nonpathogenic for humans, although important exceptions exist (51). Mycobacterial species, complexes, and groups are listed in Table 1.

In general, environmental specimens are collected from the air, soil, and water. Clinical specimens include human sputum, urine, blood, feces, cerebrospinal fluid, tissue biopsy specimens (e.g., liver, bone marrow, and lymph nodes), or any suspected source of infection. In addition, specimens may be isolated from contaminated equipment, dressings, or other materials. Clinical specimens may be aerosolized, thus becoming airborne environmental specimens. Whether samples are taken from an environmental or clinical matrix, a mixed bacterial, and possibly fungal, flora should be expected.

BACTERIOLOGY, EPIDEMIOLOGY, AND PATHOGENESIS OF MYCOBACTERIAL INFECTIONS

M. tuberculosis Complex

The M. tuberculosis complex includes the species M. bovis, M. microti, M. africanum, and M. tuberculosis. M. bovis causes tuberculosis in cattle, humans, and other primates. M. bovis bacille Calmette-Guérin, usually referred to as M. bovis BCG or BCG, is an attenuated strain of M. bovis. This strain was attenuated by serial passage (22). This vaccine strain was found to be avirulent to guinea pigs, rabbits, and horses. It was first given to a human in 1921. BCG vaccines are the most extensively used human vaccines worldwide, with more than 3 billion doses administered (22). Because of the history of conflicting results from clinical trials, the efficacy of BCG remains unknown (13). M. microti causes naturally acquired generalized tuberculosis in warm-blooded animals. M. africanum is a cause of human tuberculosis in tropical Africa.

Tuberculosis was present in paleolithic times but remained an unimportant disease for humans until the necessary environmental and social changes occurred in feudal Europe, thus releasing the "great white plague." Koch first described the tubercle bacillus, M. tuberculosis, and demonstrated it to be the cause of tuberculosis.

Tuberculosis is spread almost exclusively by airborne residues of tiny droplets produced by infectious individuals while coughing, speaking, laughing, and sneezing (45). Using high-speed photography, Jennison estimated more than 20,000 droplets released during a sneeze and only a few dozen to a few hundred expelled while coughing (29). More recently, Loudon and Roberts collected aerosolized particles from subjects coughing and talking. Subjects expelled an average of 470 particles per cough (50% were 5 to 10 μm or smaller) and 1,800 particles while counting from 0 to 100 (50% were 30 μm or smaller) (33).

Nottrebart (39) described a situation in which a number of hospital employees were awarded compensation after developing active disease. Exposure to autopsy material and machines used to suction secretions from tuberculosis patients, the adjusting of tuberculosis patients' beds, and the application of an ointment to a tuberculosis ulcer were implicated. An additional incident of nosocomial transmission of tuberculosis occurred after exposure to a hospitalized patient who underwent surgical incision, drainage, and syringe irrigation of an abscess with debridement of the surrounding necrotic tissue. The patient spent 13 days in a positive-pressure patient room and 4 days in the intensive care unit. Four of five (80%) surgical suite employees, 28 of 33 (85%) general medical floor employees, and 6 of 20 (35%) intensive care unit employees had skin test conversions (27).

TABLE 1 Key biochemical tests for identification of mycobacteria[a]

Descriptive term	Species	Complex or group
Tuberculosis	M. tuberculosis	M. tuberculosis
	M. bovis	complex
	M. africanum	
	M. microti	
Hansen's disease	M. leprae	
Nonchromogens	M. avium	M. avium complex
	M. intracellulare	
	M. xenopi	
	M. haemophilum	
	M. malmoense	
	M. simoidei	
	M. genavense	
	M. celatum	
	M. ulcerans	
	M. terrae	M. terrae complex
	M. triviale	
	M. nonchromogenicum	
	M. gastri	
Photochromogens	M. kansasii	
	M. marinum	
	M. simiae	
	M. asiaticum	
Scotochromogens	M. gordonae	
	M. scrofulaceum	
	M. szulgai	
	M. flavescens	
Rapid growers	M. fortuitum	M. fortuitum group
	M. chelonae	M. chelonae group
	M. smegmatis	
	M. phlei	
	M. vaccae	

[a] Reprinted with permission from the *Manual of Clinical Microbiology*, 6th ed. (38).

More recently, transmission of M. *tuberculosis* was transmitted from an infectious crew member to other crew members on an aircraft. The risk of infection of frequent flyers also increased with increasing hours of exposure to the index case (19).

Despite the efforts of numerous health organizations worldwide, the eradication of tuberculosis has never been imminent. Fox (23) estimates that nearly half of the world's population is infected with M. *tuberculosis*, with approximately 8 million new cases and 3 million deaths attributable to tuberculosis yearly. After decades of decline, tuberculosis has risen in recent years, even in the United States, where up to 10 million individuals are believed to be infected (16).

Mycobacteria Other Than *M. tuberculosis*

Modern leprosy (Hansen's disease) is caused by M. *leprae*, first identified by Gerhard Armauer Hansen in Norway in 1874. There are two distinct types of Hansen's disease, lepromatous (skin) and tuberculoid (nerve) (11). Estimates by the World Health Organization put the global prevalence of Hansen's disease at 10 to 12 million (11). An individual with untreated lepromatous Hansen's disease may discharge up to 8×10^8 acid-fast bacilli in a single nose blow (57). Naturally occurring M. *leprae* infections of the nine-banded armadillo have been documented in Texas and Louisiana

(57). The mode of transmission has never been validated; some researchers suspect that it is transmitted through air, while others say that only direct contact will result in infection.

M. *avium* complex and M. *kansasii* are the most common pulmonary pathogens. M. *avium* complex organisms are ubiquitous and have been isolated from water, soil, and air (62). One of the most common types of M. *avium* complex isolated from humans and animals is serotype 8 (62). In general, these bacteria are of low pathogenicity and frequently infect immune-competent humans without causing disease. This lack of disease complicates the interpretation of air sampling results. In an aquatic study, airborne culturable mycobacteria in natural aerosols from the James River were collected by using six-stage Andersen samplers. M. *avium*, M. *intracellulare*, and M. *scrofulaceum* were detected. In general, the collected mycobacterial aerosols were associated with particles larger than 5 μm (21). In a study of acid, brown-water swamps of the southeastern United States, M. *avium* complex organisms were found in higher concentrations than in inland rivers (31). Kirschner et al. (31) concluded that swamp soils, waters, and aerosols may have played a role in the epidemiology of infections in the swamp area.

M. *kansasii* and M. *xenopi* have been found in hot and cold water systems (35). Several other species of mycobacteria are known to be pathogenic; however, their routes of transmission are either unknown or not suspected to be via aerosols (62).

SAMPLING CONSIDERATIONS

Bioaerosol sampling may be performed to verify airborne transmission of mycobacteria during epidemiological investigations and research studies to evaluate engineering controls. If air sampling is deemed appropriate, the user must keep in mind that false-negative results are quite possible and should interpret all negative findings with caution. False-positive results due to poor sampling and/or analytical technique are also possible.

Investigators should use appropriate personal protective equipment and practice good personal hygiene when conducting indoor environmental quality, infectious disease outbreak, and agricultural health investigations that have resulted in medically diagnosed symptoms. Such equipment may include respiratory protection gear to prevent inhalation of contaminants and microorganism-resistant clothing to prevent the transmission to investigators.

SAMPLING AND ANALYTICAL METHODS
Sampling for Airborne Mycobacteria

General guidelines for aerobiological sampling are discussed in chapter 68. Collection methods for airborne mycobacteria include culturable bioaerosol sampling and nonculturable and nonviable bioaerosol sampling. Because free bacteria (single cells) can be mycobacterial aerosols of interest in some environmental investigations, the sampling method must collect these droplet nuclei (5 μm) (4, 60). Often, however, the bioaerosols will be clumps of microorganisms or microorganisms attached to another particle such as a skin scale or sputum droplet. When one is sampling a clean room or other environment with extremely low levels of culturable bioaerosols, the lower limit of 30 CFU for statistical comparisons may not be achievable. In such a situation, a qualitative representation must be used without being able to accommodate statistical validity. When sampling for nonviable microorganisms or when culturability is not of concern, collection efficiency is the overriding concern.

Analytical Methods

Equally important, the analytical method chosen must be compatible with the sampling method and capable of detecting the organism of interest. The physiology of mycobacteria has been reviewed by Ratledge (42) and by Barksdale and Kim (6).

Microscopic Assay

Simple and differential staining may be performed; however, members of the genus *Mycobacterium* do not stain readily. The Ziehl-Neelsen (acid-alcohol) staining procedure was first proposed in 1883 and, with minor modifications, is used today to classify organisms as acid fast (25, 53). Fluorescence microscopy for the direct count of microorganisms has been described in a number of studies (40). In the fluorescence technique, the organisms are stained with auramine or some other fluorescent dye. The preparation is then examined within 2 h if not protected from light. In comparisons of clinical techniques, fluorescence microscopy techniques for *M. tuberculosis* gave results similar to or slightly poorer than those obtained by using PCR (9, 41).

Culture

The ideal culture medium for mycobacteria should promote rapid growth of small numbers and be simple to prepare from readily available ingredients. Many media have been recommended for the cultivation of *M. tuberculosis*; however, there is no general agreement as to which of them is best. Table 2 lists some suggested media for the cultivation of mycobacteria, and several references are available (18, 24, 37, 38). *M. leprae* has not been shown to grow in solid or liquid media. However, *M. leprae* grows naturally in armadillos and humans. Large quantities of *M. leprae* (10^6 organisms) may be cultivated in the footpads of BALB/c (nude) mice (57, 59).

Inoculated agar plates are incubated at the appropriate temperature for times ranging from days for rapidly growing mycobacteria to weeks for multidrug-resistant *M. tuberculosis* (4). Sometimes colonies of rapidly growing mycobacteria may take weeks to appear on initial isolation and reveal

TABLE 2 Suggested media for the cultivation of mycobacteria[a]

Solid
  Agar based
    1. Middlebrook 7H10
    2. Middlebrook 7H11
    3. Mitchison's selective 7H11
    4. Dubos oleic acid albumin
  Egg based
    1. Wallenstein
    2. Löwenstein-Jensen with RNA
    3. Löwenstein-Jensen with pyruvic acid
Liquid
  1. BACTEC 12B medium
  2. Middlebrook 7H9 broth
  3. Dubos albumin

[a] Reprinted from reference 8 with permission.

their shorter generation times only on subculture (28). This delay may occur as a result of bacterial injury (structural or metabolic) or the adaptation process to permit the organism to multiply on artificial medium. Petri dishes containing agar media may be conveniently incubated in gas-permeable bags that maintain moisture during lengthy incubation. An atmosphere enriched with 5 to 10% (vol/vol) CO_2 enhances the growth of many mycobacteria. However, CO_2 has not been shown conclusively to enhance the growth of mycobacteria on egg media (28). Most mycobacteria multiply at 35 to 37°C. A small group of mycobacteria (e.g., *M. marinum* and *M. ulcerans*) optimally grow at 30°C. A distinctive property of culturable mycobacteria includes their growth rates at 24, 31, 37, and 45°C. The incubator should also be humidified to minimize desiccation of the medium. If a liquid medium is used, the tubes should be well aerated but desiccation of the medium must be prevented. Tubes less than half filled have sufficient oxygen to support adequate growth. Laboratory medium blanks and field medium blanks to detect accidental contamination must be handled in the same manner as samples.

When sampling for mycobacteria with a modified six-stage sampler (stages 1 and 6 only) and Middlebrook 7H10, Macher et al. (34) lost 33 of 51 samples as a result of fungal overgrowth within the first 3 weeks after collection. Rapidly growing mycobacteria were recovered on three occasions at one location. If impinger samplers are used, the liquid collection medium may be filtered, directly inoculated on solid medium, or incubated directly if it will support the growth of the target mycobacterium. By using direct filtration techniques (3, 30) and the direct inoculation technique (3), the airborne concentrations may be estimated.

Distinctive properties of culturable mycobacteria include accumulation of niacin, susceptibility to thiophene-2-carboxylate hydrazide, nitrate reduction, production of catalase (37 and 68°C), Tween 80 hydrolysis, tellurite reduction, tolerance to 5% NaCl, iron uptake, arylsulfatase breakdown of phenolphthalein disulfate, growth on MacConkey agar, and ability to process pyrazinamidase; however, these individual assays are time-consuming (8). The BACTEC system has reduced the isolation time for mycobacteria to approximately 10 days. This system is based on the principle that mycobacteria multiply in Middlebrook 7H12 broth and metabolize ^{14}C-containing palmitic acid (52). $^{14}CO_2$ is released into the headspace of the vial. The instrument samples the headspace three times weekly for 6 weeks and measures the radioactivity. *M. tuberculosis*, *M. bovis*, and mycobacteria other than *M. tuberculosis* can be differentiated by using *p*-nitro-α-acetylamino-β-hydroxypropiophenone or thiophene-2-carboxylate hydrazide (8).

Mycolic acids (cellular fatty acids) of mycobacteria are structural in nature, occurring in the cell membrane or cell wall of all bacteria. When the bacteria are grown under standardized growth conditions, the mycolic acid profiles are reproducible within a genus, down to the subspecies or strain level in some microorganisms. The Microbial Identification System, developed by Microbial Identification, Inc. (Newark, Del.), provides a chromatographic technique and software libraries capable of identifying various microorganisms on the basis of mycolic acid composition (46, 47, 55, 58, 65). This chromatographic technique is also known as gas chromatography fatty acid methyl ester analysis, and a database containing the analysis libraries for culturable mycobacteria is available (55, 65).

Immunoassay

Many immunoassays are now readily available from commercial sources, permitting laboratories to rapidly develop in-house immunochemical analytical capability without lengthy antibody preparation. Some of the more widely used formats are radioimmunoassays, fluorescence immunoassays, and enzyme immunoassays. Witebsky and Conville (61) summarized recent developments in diagnostic mycobacteriology, including radioimmunoassays, with emphasis on laboratory capabilities relevant to the more rapid and accurate detection and identification of mycobacterial pathogens. Enzyme-linked immunosorbent assays (ELISAs) are now highly automated, and efforts are under way to commercially develop well-standardized kits containing appropriate controls and materials. ELISAs for M. *tuberculosis* (15, 32) and M. *leprae* (54) have been used in laboratories for identification.

Molecular Techniques

Diagnostic mycobacteriology is rapidly adapting molecular biology techniques in addition to classical identification methods to identify organisms. Genus-specific and species-specific DNA probes have been developed for mycobacteria (56). This approach has been successfully used to detect various organisms, including M. *tuberculosis* (10, 20, 36, 63, 64). Mycobacteria other than M. *tuberculosis* have been successfully identified (14), and Richter et al. (43) developed a three-primer PCR to detect M. *leprae*.

Membrane filters may be used to remove microorganisms from liquids and analyzed by PCR (7). If air samples are collected with a liquid impinger, then the collection fluid may be filtered and analyzed by PCR for a specific microorganism (1). Air samples were collected by Sawyer et al. (48) on filters and analyzed by PCR for varicella-zoster virus. Finally, Schafer (49) proposed sampling for airborne M. *tuberculosis* by using a method similar to that used by Sawyer et al. Restriction fragment length polymorphism analysis is widely used to distinguish genetic changes within a species. Such analysis has been used in epidemiological studies (12, 50) to track the transmission of M. *tuberculosis* in New York City. M. *leprae* isolates from different sources were found to have identical sequences of the spacer region between two RNA genes (17).

CONCLUSION

Dose-response data are not available for exposure to *Mycobacterium* spp. No occupational exposure limit for bioaerosols has been promulgated by the Occupational Safety and Health Administration. The American Conference of Governmental Industrial Hygienists has stated that "there are no numerical guidelines or threshold limit values (TLVs) that allow ready interpretation of bioaerosol data" (2). Riley (44) estimated the indoor M. *tuberculosis* droplet nuclei concentration in the air to be as low as 1 infectious unit per 310 m^3 (1 infectious unit per 12,000 ft^3) (5) and stated that the infectious dose is less than 10. Because indoor airborne mycobacterial concentrations are thought to be very low, bioaerosol monitoring of *Mycobacterium* spp. is problematic. Available methods include the measurement of viable (culturable and nonculturable) and nonviable microorganisms in both indoor (e.g., industrial, office, or residential) and outdoor (e.g., agricultural and general air quality) environments. Investigators must ensure that the purpose of sampling is clear, the organism of interest is identified, and the appropriate analytical technique is chosen prior to air sampling. Air sampling may be appropriate to include in epidemiological investigations, in research studies, or in situations indicated by a physician.

REFERENCES

1. **Alvarez, A. J., M. P. Buttner, G. A. Toranzos, E. A. Dvorsky, A. Toro, T. B. Heikes, L. E. Mertikas-Pifer, and L. D. Stetzenbach.** 1994. Use of solid-phase PCR for enhanced detection of airborne microorganisms. *Appl. Environ. Microbiol.* **60:**374–376.
2. **American Conference of Governmental Industrial Hygienists, Inc.** 1995. *1995–96 Threshold Limit Values (TLVs™) for Chemical Substances and Physical Agents and Biological Exposure Indices (BEIs),* p. 9–11. ACGIH, Cincinnati.
3. **American Public Health Association.** 1989. *Standard Methods for the Examination of Water and Waste Water,* 17th ed. American Public Health Association, Washington, D.C.
4. **American Thoracic Society.** 1990. Diagnostic standards and classification of tuberculosis. *Am. Rev. Respir. Dis.* **142:**725–735.
5. **American Thoracic Society.** 1992. Control of tuberculosis in the United States. *Am. Rev. Respir. Dis.* **146:**1623–1633.
6. **Barksdale, L., and K. S. Kim.** 1977. Mycobacterium. *Bacteriol. Rev.* **41:**217–372.
7. **Bej, A. K., M. H. Mahbubani, J. L. Dicesare, and R. L. Atlas.** 1991. Polymerase chain reaction-gene probe detection of microorganisms using filter-concentrated samples. *Appl. Environ. Microbiol.* **57:**3529–3534.
8. **Berlin, O. G. W.** 1990. Mycobacteria, p. 597–640. *In* E. J. Baron and S. M. Finegold (ed.), *Bailey & Scott's Diagnostic Microbiology.* The C. V. Mosby Co., St. Louis, Mo.
9. **Bodmer, T., A. Gurtner, K. Schopfer, and L. Matter.** 1994. Screening of respiratory tract specimens for the presence of *Mycobacterium tuberculosis* by using the Gen-Probe Amplified Mycobacterium Tuberculosis Direct Test. *J. Clin. Microbiol.* **32:**1483–1487.
10. **Brisson-Noel, A., D. Lecossier, X. Nassif, B. Gicquel, V. Levy-Frebault, and A. J. Hance.** 1989. Rapid diagnosis of tuberculosis by amplification of mycobacterial DNA in clinical samples. *Lancet* **ii:**1069–1071.
11. **Bryceson, A., and R. E. Pfaltzgraft.** 1990. *Leprosy.* Churchill Livingstone, London.
12. **Cleveland, J. L., J. Kent, B. F. Gooch, S. E. Valway, D. W. Marianos, W. R. Butler, and I. M. Onorato.** 1995. Multidrug-resistant Mycobacterium tuberculosis in an HIV dental clinic. *Infect. Control Hosp. Epidemiol.* **16:**7–11.
13. **Colditz, G. A., T. F. Brewer, C. S. Berkey, M. E. Wilson, E. Burdick, H. V. Fineberg, and F. Mosteller.** 1994. Efficacy of BCG vaccine in the prevention of tuberculosis. Meta-analysis of the published literature. *JAMA* **271:**698–702. (Comments in *ACP J. Club* **121**[Suppl. 1]:22 and *JAMA* **272:**765–766.)
14. **Cook, S. M., R. E. Bartos, C. L. Pierson, and T. S. Frank.** 1994. Detection and characterization of atypical mycobacteria by the polymerase chain reaction. *Diagn. Mol. Pathol.* **3:**53–58.
15. **Daftary, V. G., D. D. Banker, and G. V. Daftary.** 1994. ELISA test for tuberculosis. *Indian J. Med. Sci.* **48:**39–42.
16. **Daniel, T. M., J. H. Bates, and K. A. Downes.** 1994. History of tuberculosis, p. 13–24. *In* B. R. Bloom (ed.), *Tuberculosis: Pathogenesis, Protection, and Control.* ASM Press, Washington, D.C.
17. **de Wit, M. Y., and P. R. Klutzier.** 1994. *Mycobacterium leprae* isolates from different sources have identical sequences of the spacer region between the 16S and 23S ribosomal RNA genes. *Microbiology* **140:**1983–1987.
18. **Difco Laboratories.** 1984. *Difco Manual: Dehydrated Culture Media and Reagents for Microbiology.* Difco Laboratories, Inc., Detroit.
19. **Driver, C. R., S. E. Valway, W. M. Morgan, I. M. Onor-**

ato, and K. G. Castro. 1994. Transmission of *Mycobacterium tuberculosis* associated with air travel. *JAMA* **272:** 1031–1035.

20. **Eisenach, K. D., M. D. Cave, J. H. Bates, and J. T. Crawford.** 1991. Polymerase chain reaction amplification of a repetitive DNA sequence specific for *Mycobacterium tuberculosis. J. Infect. Dis.* **161:**977–981.

21. **Falkinham, J. O., K. L. George, M. A. Ford, and B. C. Parker.** 1990. Collection and characteristics of mycobacteria in aerosols, p. 71–83. *In* P. R. Morey, J. C. Freeley, and J. A. Otten (ed.), *Biological Contaminants in Indoor Environments. ASTM STP 1071.* American Society for Testing and Materials, Philadelphia.

22. **Fine, P. E., and L. C. Rodrigues.** 1990. Modern vaccines. Mycobacterial diseases. *Lancet* **335:**1016–1020.

23. **Fox, J. L.** 1990. TB: a grim disease of numbers. *ASM News* **56:**363–365.

24. **Gherna, R., P. Pienta, and R. Cote (ed.).** 1992. *Catalogue of Bacteria and Phages,* 18th ed. American Type Culture Collection, Rockville, Md.

25. **Harada, K., S. Gidoh, and S. Tsutsumi.** 1976. Staining mycobacteria with carbolfuchsin: properties of solution with different sample of basic fuchsin. *Microsc. Acta* **78:** 21–27.

26. **Holt, J. G., N. R. Krieg, P. H. A. Sneath, J. T. Staley, and S. T. Williams.** 1994. *Bergey's Manual of Determinative Bacteriology.* Williams & Wilkins, Baltimore.

27. **Hutton, M. D., W. W. Stead, G. M. Cauthen, A. B. Bloch, and W. M. Ewing.** 1990. Nosocomial transmission of tuberculosis associated with a draining abscess. *J. Infect. Dis.* **161:**286–295.

28. **Jenkins, P. A., S. R. Pattyn, and F. Portaels.** 1982. Diagnostic bacteriology, p. 441–470. *In* C. Ratledge and J. Stanford (ed.), *The Biology of the Mycobacteria.* Academic Press, New York.

29. **Jennison, M. W.** 1942. Atomizing of mouth and nose secretions into the air as revealed by high-speed photography, p. 106–128. *In* F. R. Moulton (ed.), *Aerobiology.* AAAS publication no. 17. American Association for the Advancement of Science, Washington, D.C.

30. **Jensen, P. A., W. F. Todd, G. N. Davis, and P. V. Scarpino.** 1992. Evaluation of eight bioaerosol samplers challenged with aerosols of free bacteria. *Am. Ind. Hyg. Assoc. J.* **53:**660–667.

31. **Kirschner, R. A., B. C. Parker, and J. O. Falkinham.** 1992. Epidemiology of infection by nontuberculous mycobacteria: *Mycobacterium avium, Mycobacterium intracellulare,* and *Mycobacterium scrofulaceum* in acid, brown-water swamps of the southeastern United States and their association with environmental variables. *Am. Rev. Respir. Dis.* **145:**271–275.

32. **Ling, M. L.** 1994. Update of the rapid diagnosis of infectious diseases. I. Bacteria, fungi and parasites. *Singapore Med. J.* **35:**316–318. (Review.)

33. **Loudon, R. G., and R. M. Roberts.** 1967. Droplet expulsion from the respiratory tract. *Am. Rev. Respir. Dis.* **95:** 435–442.

34. **Macher, J. M., L. E. Alevantis, Y.-L. Chang, and K.-S. Liu.** 1992. Effect of ultraviolet germicidal lamps on airborne microorganisms in an outpatient waiting room. *Appl. Occup. Environ. Hyg.* **7:**505–513.

35. **McSwiggan, D. A., and C. H. Collins.** 1974. The isolation of M. *kansasii* and M. *xenopi* from water systems. *Tubercle* **55:**291–297.

36. **Metchock, B., and L. Diem.** 1995. Algorithm for use of nucleic acid probes for identifying *Mycobacterium tuberculosis* from BACTEC 12B bottles. *J. Clin. Microbiol.* **33:** 1934–1937.

37. **Murray, P. R., E. J. Baron, M. A. Pfaller, F. C. Tenover,** and R. H. Yolken (ed.). 1995. *Manual of Clinical Microbiology,* 6th ed. ASM Press, Washington, D.C.

38. **Nolte, F. S., and B. Metchock.** 1995. Mycobacterium, p. 400–437. *In* P. R. Murray, E. J. Baron, M. A. Pfaller, F. C. Tenover, and R. H. Yolken (ed.), *Manual of Clinical Microbiology,* 6th ed., ASM Press, Washington, D.C.

39. **Nottrebart, H. C.** 1980. Nosocomial infections acquired by hospital employees. *Infect. Control* **1:**257–259.

40. **Palmgren, U., G. Ström, G. Blomquist, and P. Malmberg.** 1986. Collection of airborne micro-organisms on Nuclepore filters, estimation and analysis—CAMNEA method. *J. Appl. Bacteriol.* **61:**401–406.

41. **Pfyffer, G. E., P. Kissling, R. Wirth, and R. Weber.** 1994. Direct detection of *Mycobacterium tuberculosis* complex in respiratory specimens by a target-amplified test system. *J. Clin. Microbiol.* **32:**918–923.

42. **Ratledge, C.** 1976. The physiology of the mycobacteria. *Adv. Microb. Physiol.* **13:**115–244.

43. **Richter, E., M. Duchrow, C. Schluter, M. Hahn, H. D. Flad, and J. Gerdes.** 1994. Detection of Mycobacterium leprae by three-primer PCR. *Immunobiology* **191:**351–353.

44. **Riley, R. L.** 1961. Airborne pulmonary tuberculosis. *Bacteriol. Rev.* **25:**243–248.

45. **Rubin, J.** 1991. Mycobacterial disinfection and control, p. 377–384. *In* S. S. Block (ed.), *Disinfection, Sterilization, and Preservation,* 4th ed. Lea & Febiger, Philadelphia.

46. **Sasser, M.** 1990. Identification of bacteria through fatty acid analysis. *In* Z. Klement, K. Rudolph, and D. C. Sands (ed.), *Methods in Phytobacteriology.* Akademia Kiado, Budapest.

47. **Sasser, M.** 1990. *Identification of Bacteria by Gas Chromatography of Cellular Fatty Acids.* Technical note no. 101. Microbial Identification, Inc., Newark, Del.

48. **Sawyer, M. H., C. J. Chamberlin, Y. N. Wu, N. Aintablian, and M. R. Wallace.** 1994. Detection of varicella-zoster virus DNA in air samples from hospital rooms. *J. Infect. Dis.* **169:**91–94.

49. **Schafer, M. P.** 1994. *Sampling and Analytical Method Development for Airborne Mycobacterium tuberculosis. Study Protocol.* Division of Physical Sciences and Engineering, National Institute for Occupational Safety and Health, Cincinnati.

50. **Shafer, R. W., P. M. Small, C. Larkin, S. P. Singh, P. Kelly, M. F. Sierra, G. Schoolnik, and K. D. Chirgwin.** 1995. Temporal trends and transmission patterns during the emergence of multidrug-resistant tuberculosis in New York City: a molecular epidemiologic assessment. *J. Infect. Dis.* **171:**170–176.

51. **Shinnick, T. M., and R. C. Good.** 1994. Mycobacterial taxonomy. *Eur. J. Clin. Microbiol. Infect. Dis.* **13:**884–901. (Review.)

52. **Siddiqi, S. H., C. C. Hwangbo, V. Silcox, R. C. Good, D. E. Snider, and G. Middlebrook.** 1984. Rapid radiometric methods to detect and differentiate *Mycobacterium tuberculosis*/M. *bovis* from other mycobacterial species. *Am. Rev. Respir. Dis.* **130:**634–640.

53. **Silverton, R. E., and M. J. Anderson.** 1961. *Handbook of Medical Laboratory Formulae,* p. 175. Butterworths, London.

54. **Singh, N. B., S. Bhatnagar, A. Choudhary, H. P. Gupta, and S. M. Kaul.** 1994. Evaluation of diversified antigens for detection of *Mycobacterium leprae* antibodies from leprosy patients and contacts. *Indian J. Exp. Biol.* **32:**478–481.

55. **Smid, L., and M. Salfinger.** 1994. Mycobacterial identification by computer-aided gas-liquid chromatography. *Diagn. Microbiol. Infect. Dis.* **19:**81–88.

56. **Sritharan, V., J. V. Iralu, and R. H. Barker, Jr.** 1994. Comparison of genus- and species-specific probes for PCR

detection of mycobacterial infections. *Mol. Cell Probes* **8:** 409–416.

57. **Stewart-Tull, D. E. S.** 1982. *Mycobacterium leprae*—the bacteriologist's enigma, p. 273–307. *In* C. Ratledge and J. Stanford (ed.), *The Biology of the Mycobacteria.* Academic Press, New York.

58. **Welch, D. F.** 1991. Applications of cellular fatty acid analysis. *Clin. Microbiol. Rev.* **4:**422–438.

59. **Welch, T. M., R. H. Gerber, L. P. Murray, H. Ng, S. M. O'Neill, and L. Levy.** 1980. Viability of *Mycobacterium leprae* after multiplication in mice. *Infect. Immun.* **30:** 325–328.

60. **Wells, W. F.** 1955. Aerodynamics of droplet nuclei, p. 13–19. *In Airborne Contagion and Air Hygiene.* Harvard University Press, Cambridge, Mass.

61. **Witebsky, F. G., and P. S. Conville.** 1993. The laboratory diagnosis of mycobacterial diseases. *Infect. Dis. Clin. N. Am.* **7:**359–376. (Review.)

62. **Wolinsky, E.** 1979. Nontuberculous mycobacteria and associated diseases. *Am. Rev. Respir. Dis.* **119:**107–159.

63. **Wren, B., C. Clayton, and S. Tabaqchali.** 1990. Rapid identification of toxigenic *Clostridium difficile* by polymerase chain reaction. *Lancet* **335:**423. (Letter.)

64. **Yajko, D. M., C. Wagner, V. J. Tevere, T. Kocagöz, W. K. Hadley, and H. F. Chambers.** 1995. Quantitative culture of *Mycobacterium tuberculosis* from clinical sputum specimens and dilution endpoint of its detection by the Amplicor PCR assay. *J. Clin. Microbiol.* **33:**1944–1947.

65. **Yassin, A. F., H. Brzezinka, and K. P. Schaal.** 1993. Cellular fatty acid methyl ester profiles as a tool in the differentiation of members of the genus *Mycobacterium. Zentralbl. Bakteriol.* **279:**316–329.

Airborne Viruses

SYED A. SATTAR AND M. KHALID IJAZ

74

Viruses can become airborne through the release of contaminated liquids or dried material. Larger particles settle out rapidly on environmental surfaces in the immediate vicinity, but droplet nuclei (usually <5.0 μm in diameter) not only can remain suspended but can be transported over long distances by air currents. Wind-blown carriage of animal pathogenic viruses has been shown to cause outbreaks of disease considerable distances down wind from the source (16, 74). For example, an epizootic of pseudorabies in swine herds arose from airborne spread of virus across an area of nearly 150 km² (39), and retrospective studies of similar outbreaks indicated airborne spread of the virus up to 17 km (6). Intercontinental transport of human viruses through atmospheric dispersion of airborne particles has also been postulated, and Hammond et al. (43) suggest that such long-distance transport of airborne viruses may explain the pandemics of influenza.

Many common activities can contaminate air with viruses. Inhalation of air with such particles can lead to their retention in the respiratory tract, and airborne spread has been clearly documented for many viral infections of humans (26, 97) and animals (97). Figure 1 shows how susceptible hosts can be exposed to airborne viruses. However, infection through the inhalation and retention of droplet nuclei is generally regarded as true airborne spread. Infectious viruses (97) or their nucleic acids (100) have been detected in field samples of air. Airborne outbreaks of viral gastroenteritis (99) may be due to the translocation and ingestion of particles retained in the upper respiratory tract (103). In general, airborne spread of viruses is rapid as well as difficult to prevent and control.

METHODS FOR THE STUDY OF AIRBORNE VIRUSES

The methodology for generating, storing, and collecting viral aerosols has already been reviewed (18, 54, 97, 105). Chapter 68 also discusses the sampling techniques for airborne microorganisms. This chapter, therefore, is limited to a critical review of the information on the role of air in the spread of vertebrate viruses.

Proper study of airborne viruses requires specialized and custom-built equipment (111). Furthermore, virus suspensions with high titers (>10⁷ infective units/ml) are essential for such studies because virus infectivity can be lost during

generation and collection of aerosols and then there is dilution in air of the nebulized material. Work with airborne viruses also requires stringent safety precautions (30). Basically, one needs the following equipment and procedures for the study of airborne viruses and other types of microorganisms.

Aerosol Generation

To study the aerobiology of viruses, it is necessary to generate particles which are small enough (<5 μm in diameter) to remain suspended in air, thus permitting their retention in the respiratory tract upon inhalation. Several nebulizers are available for this purpose (18, 19, 97, 105). Nebulizers for the delivery of therapeutic drugs also produce particles in the respirable range (65) and can be adapted for generating microbial aerosols. The nature and composition of the virus suspending medium determine the size distribution of the particles aerosolized and protect the viability of the virus during nebulization and aging (97).

A physical tracer in the virus suspension to be sprayed is needed to differentiate between physical loss of the infectious virus due to the settling out of aerosols and its actual biological decay. Uranine, a fluorescent dye, added to the virus suspension has been found to be as effective as a radiolabeled virus (54); at the levels needed, the dye is harmless to viruses and cell cultures but also cheaper and safer to use than radioisotopes. However, dyes may affect the photosensitivity of viruses (18). Bacterial spores are also suitable tracers (18, 105). Nebulization fluids with mixtures of viruses have been used to directly compare their airborne stabilities (56). Antifoam agents in the virus suspension may also be needed to reduce frothing during nebulization (54).

Retention and Aging of Aerosols

Whereas the nebulized virus can be held as a static aerosol in any type of closed container (105), one generally uses a rotating stainless steel drum (37, 54, 105) to store virus aerosols in a dynamic state for studying the influence of various environmental factors on them. The drum is housed in a larger chamber which is vented through HEPA filters to prevent exposure to any accidental virus leakage. The drum is continuously rotated mechanically along its axis at a predetermined rate (e.g., 4 rpm) to keep the aerosolized material from settling out. This allows the virus to stay suspended in air for weeks if needed. The air inside the drum

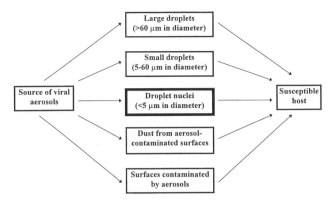

FIGURE 1 Direct or indirect exposure of susceptible hosts to aerosolized viruses.

can be preconditioned to the desired relative humidity (RH) level and temperature. Thorough air flushing of the drum is required between experiments. As far as we know, Goldberg drums or similar devices are not available commercially.

Ultrafine (<1 μm in diameter) natural (spider web) and artificial (tungsten) threads can be used as supports for virus-containing particles. Such anchored particles are still subject to the influence of various environmental factors around them (18, 105). Figure 2 is a micrograph of a virus-containing particle captured on a spider thread (68). In spite of their limitations (105), ultrafine threads are the best means to study the influence of atmospheric chemicals (19) and of light and irradiation (68, 76) on airborne viruses. Handley and Roe (45) have recently used an artificial fiber (2 to 3 μm in diameter) of ethylene vinyl acetate as an alternative to spider threads; tests with bacterial aerosols have found them to perform as well as spider threads. Their suitability for viruses remains to be evaluated.

Collection and Sizing of Aerosols

Most of the available aerosol collection devices (18, 105, 107) are unsuitable for working with viruses. All-glass impingers (AGI) are commonly used for collecting viral aerosols (97, 105) in relatively small volumes (about 10 liters/min) of air. The air to be sampled is sucked into the impinger, where it passes through a tube with a limiting orifice and impacts on the surface of the collecting fluid. An antifoaming agent in the collection fluid is usually needed to reduce frothing during the sampling process (54). The volume of air sampled depends on the sampling rate of the impinger and the time for which it is run. If the air is being sampled from an aerosol-holding device, its volume and the total number of samples to be collected for a given experiment will determine the volume of individual air samples. Prolonged operation of AGI for air sampling can also lead to the evaporation of the sampling fluid, thereby affecting the concentration of solutes in it. Use of preimpingers with AGI improves the collection of airborne viruses (97). The volume of air sampled must be replaced with fresh air to avoid creating a vacuum in the aerosol-holding device.

In the field, commercial large-volume air samplers (LVAS), which can easily process 10 m^3 of air per min, can be used to recover airborne viruses (18, 97, 105). The collection efficiencies of various LVAS vary considerably depending on the type of virus, the type and volume of air sampled, the nature of the collection fluid, and the rate of sampling. We have used the LVAS developed by White

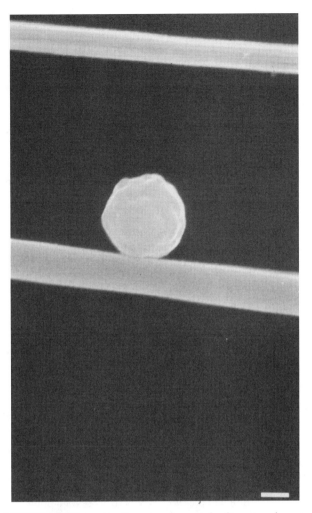

FIGURE 2 Scanning electron micrograph of an aerosol particle attached to a strand of spider web. Bar = 0.5 μm. (Reproduced with permission from B. Kournikakis, Defence Research Establishment, Suffield, Alberta, Canada.)

et al. (113) to collect respiratory and enteric viruses from artificially generated aerosols (89). Wallis et al. (109) have reported the use of acidic buffer-moistened cartridge filters to recover polioviruses aerosolized during the flushing of toilets; the air could be sampled with rates as high as 100 liters/min, but the virus eluted from the filters required further concentration.

Determination of the size distribution of airborne particles containing infectious viruses has required the adaptation (54) of the Andersen sieve sampler (2). Instead of a bacteriological agar medium, tryptose phosphate broth containing 3% gelatin is put into the sampler's petri plates. After sampling, the collection medium can be liquefied by holding the plates at 37°C for 1 h and poured out for virus titration and measurements of the physical tracer levels. In the three-stage glass impinger (75), virus retention corresponds to that in the human upper respiratory tract, bronchioles, and alveoli. A comparison of the three-stage Andersen sampler and the three-stage May AGI gave similar results in the sampling and sizing of artificially generated biological aerosols (119).

Exposing Experimental Animals or Human Subjects to Airborne Viruses

Table 1 summarizes information from reports published in the past 25 years. Studies on the susceptibility of animals to airborne viruses or determination of the minimal infective dose of viruses by the respiratory route requires exposing animals to standardized clouds of the test agent under controlled conditions. Traditionally, the Henderson apparatus (47) has been used for this purpose when one is working with small animals such as mice and rats. In this device, individual animals are placed in cylindrical holders with only the tip of the animal's face (including the nostrils) protruding through one end of the holder. The holders are then attached to ports on a larger tube connected to the aerosol source, and the only means of exposure of the animals to the aerosols is by inhalation of the air in the tube. Devices for working with cattle (79) or pigs (87) have also been described. Here it is important to note that the aerosolized virus must not be allowed to deposit on areas where the test animals could be exposed to it in ways other than through inhalation. Chung et al. (17) have reported the development of a low-cost wind tunnel for exposing human subjects to aerosols of >2 μm in diameter. Particles generated by a Collison-type nebulizer and introduced into the tunnel by using an aerosol injector are uniformly distributed to expose human subjects; the device is yet to be tested with airborne viruses.

SOURCES OF VIRAL AEROSOLS

Activities such as sneezing, coughing, flushing toilets, and changing diapers as well as shaking, homogenization, and sonication of virus-containing materials can generate infectious aerosols. Preventing the generation of and avoiding exposure to such aerosols are particularly important in laboratories and other settings where infectious material is handled. According to Pike (90), 27% of the cases of laboratory-acquired infections were due to airborne viruses; cases in research settings accounted for more than 67% of such infections. Whereas recent improvements in the design and construction of biohazard containment equipment and better enforcement of biosafety procedures have considerably reduced the risk of aerosol exposure, many laboratory workers do not appear to be fully aware of the dangers of infectious aerosols. Better training of laboratory workers, continued vigilance, and further improvements in equipment and procedures are needed to make these workplaces safer.

AIRBORNE SURVIVAL OF HUMAN AND ANIMAL PATHOGENIC VIRUSES

How long a given virus can remain infectious in air depends on the nature of the virus and the medium in which it was suspended before becoming airborne, ambient temperature, RH, atmospheric gases, lighting, and irradiation (97). The absence of standardization of the experimental protocols and wide variations in the system of reporting the results make direct comparisons of the findings from different studies extremely difficult.

The literature on this topic has already been reviewed (97), and the following is an update of the information from experimental and field studies on selected human and animal pathogenic viruses. Some of the viruses selected are known to spread by the airborne route, whereas for the others there appears to be a real or perceived risk for airborne

transmission. It should be noted that many of the viruses spread through aerosols are not normally associated with infections of the respiratory tract.

VZV

Investigations of chickenpox outbreaks clearly show airborne spread of varicella-zoster virus (VZV). Recently, Sawyer et al. (100) found VZV DNA in air samples from rooms housing patients with zoster or chickenpox, and the air remained positive up to 24 h after patient discharge. Whereas none of the VZV DNA-positive samples had infectious virus, PCR technology may prove very useful in studying the airborne transport of viruses in both indoor and outdoor settings.

Hepburn and Brooks (48) have described an outbreak of chickenpox in a military field hospital. Several patients and staff members in an isolation ward for cases of gastroenteritis were affected, and virus spread was most likely through air. This incident highlights the weaknesses in field hospital design and illustrates how easily biological warfare agents could spread in such settings (48).

Viruses Causing Acute Gastroenteritis

Caliciviruses

That Norwalk virus can spread by air is suggested from outbreak investigations (13, 35). The first such report was based on an outbreak in the emergency room of a hospital (99), and many who simply walked through the incriminated area also became infected. Contamination of air may occur by the aerosolization of the virus during vomiting (14, 15). Proper studies on the airborne spread of these viruses are not possible because the viruses cannot be cultured in the laboratory.

Rotaviruses

Rotaviruses are among the major causes of acute gastroenteritis in humans and animals (95). Apart from their spread by the fecal-oral route, epidemiological and experimental studies suggest airborne transmission of rotaviral infections (4). More recent studies on murine and other strains of bovine rotaviruses show their behavior in the airborne state to parallel that of previously studied human and animal rotaviruses (55).

Viruses Causing Infections of the Respiratory Tract

The behavior of aerosolized influenza and parainfluenza viruses has been studied in some detail (97). Recent studies in this regard have focused on challenging vaccinated animals with viral aerosols to study the protective effect of immunization (Table 1).

The relative importance of air in the spread of rhinovirus colds continues to be debated (24, 41). In our own studies (61), airborne human rhinovirus type 14 behaved in ways typical of other picornaviruses; its half-life was nearly 14 h when it was aerosolized from tryptose phosphate broth and the aerosols were held at 20°C with 80% RH. This finding suggests that rhinoviruses can remain infectious in air long enough to permit aerial spread.

Papillomaviruses

Reports of warts in the respiratory tracts of laser therapists (42) suggest exposure to papillomaviruses in the smoke plumes from vaporized verrucae. Garden et al. (34) first reported the detection of papillomavirus DNA in such

TABLE 1 Experimental challenge of animals or humans to viral aerosols: chronological list of selected studies published since 1970

Virus(es)	Host(s)	Remarks	Reference
Murine leukemia virus	Mice	First report on aerosol stability of the virus and its ability to spread by air.	69
	Mice	Nearly 40% of exposed mice developed leukemia within 25 mo.	77
Murine leukemia and Rous sarcoma viruses	Monkeys	Showed that tumor viruses could be readily transmitted via aerosols.	78
Adenovirus type 12	Hamsters	Airborne virus shown to be pathogenic to newborn animals.	22
Parainfluenzavirus type 1	Mice	Transmissibility rates did not increase after serial airborne challenge with the virus.	108
Adenovirus type 4, coxsackievirus A21	Human volunteers	Determined infectivity of these viruses by aerosols.	44
Marek's disease virus	Chickens	Exposure to effluent air from "donor cages" housing infected animals resulted in a high incidence of the infection in test chicks; passage of contaminated air through certain filters partially or completely prevented such infection.	12
Newcastle disease virus	Chickens	Unvaccinated birds shed much higher levels of virus than those previously vaccinated.	52
Vesicular stomatitis virus	Mice	Exposure of mice for 1 h to ozone resulted in a 70% increase in respiratory deposition of the virus.	28
Influenza A virus	Mice	Under conditions of aerosol inhalation, mice were found to be a suitable model for studies on pathogenesis.	32
	Mice	Extrapulmonary virus was in direct quantitative relationship to the extent of lung involvement.	33
Influenza virus	Mice	Mouse resistance to viral pneumonia was affected in the presence of manganese dioxide.	73
Moloney murine sarcoma and leukemia virus complex	Mice	When tumor extracts were aerosolized, both viruses survived for at least 2 h, but mice exposed to the aerosols did not develop an infection.	49
Bovine rhinotracheitis virus	Cattle	The study compared clinical and immunological responses after aerosol exposure or intramuscular inoculation; in both cases, the virus generally elicited comparable levels of serum antibody but not measurable nasal antibody; aerosol-exposed cattle shed virus from the nose, while the others did not.	31
Feline caliciviruses	Cats	Concluded that aerosol transmission probably plays little part in the spread of these viruses.	110
African swine fever virus (KWH/12)	Pigs	Concluded that the primary route of infection in pig was through the lower respiratory tract.	114
	Pigs	Animals became infected after challenge with aerosolized virus.	115
Newcastle disease virus (vaccine)	Chickens	Antibody response to aerosol challenge much better than that from administration in drinking water; no clinical disease; virus recovered from lungs for 10 days.	64
Rinderpest virus	Cattle	Low or high RH was shown to increase the probability of disease transmission by the respiratory route, but any aerial spread across distances greater than a few meters was believed to occur principally at night.	53
Feline caliciviruses	Cats	Animals could be infected by aerosol exposure or direct intranasal instillation.	86
Bovine respiratory syncytial virus	Holstein calves	Animals exposed to aerosols of the virus manifested moderate to severe signs of respiratory disease.	27
Japanese B encephalitis virus	Mice, rats, hamsters, guinea pigs, squirrel monkeys	Mice and hamsters were highly susceptible to aerosol challenge; guinea pigs and rats seroconverted but survived the infection; squirrel monkeys died after a high dose of infectious virus.	70
Rauscher murine leukemia virus	Mice	Infection of BALB/c mice through aerosols of leukemogenic virus was possible.	80
Influenza virus	Mice	The dynamics of B lymphocytes in lungs after aerosol challenge was studied; IgA-[a] and IgM-containing cells appeared first, followed by IgG-containing cells.	88

(Table continued on next page)

TABLE 1 Experimental challenge of animals or humans to viral aerosols: chronological list of selected studies published since 1970 *(Continued)*

Virus(es)	Host(s)	Remarks	Reference
Parainfluenzavirus type 3	Calves	First report of extensive purulent pneumonia in calves after exposure to aerosols of the virus and *Pasteurella haemolytica*.	57
Foot-and-mouth disease virus	Cattle	Cattle were infected with aerosolized virus.	72
Pseudorabies virus	Pigs	Virus was transmitted to seronegative pigs exposed to air from boxes containing infected pigs.	25
Infectious bronchitis virus (Australian T stain)	Chickens	Birds exposed to aerosols had earlier and slightly more severe respiratory symptoms.	93
Lassa fever virus	Guinea pigs and monkeys	Both species were susceptible when exposed to small-particle aerosols.	106
Mouse rotavirus	Mice	Neonatal mice developed acute gastroenteritis within 48 h of exposure to viral aerosols.	91
Rift Valley fever virus	Rats	When exposed to viral aerosols, 97% of the unvaccinated and 32% of the vaccinated animals died; levels of serum neutralizing antibody were predictive of the protective effect.	2
Fowlpox virus	Chickens	Day-old chicks were vaccinated by the aerosol route or by injection in the wing web; both methods induced immunity, but the aerosol method was more suitable for mass vaccinations.	23
Influenza virus	Mice	Intranasal inoculation was found to be superior to aerosol exposure in the immunization of mice.	59
Junin virus	Rhesus monkeys	All animals exposed to aerosolized virus became sick within 3 weeks and died; the clinical picture was similar to that seen in humans.	63
Influenza virus	Horses	Challenge of animals to aerosolized influenza virus was found to be a reliable means of assessing immunization against the disease.	84
Foot-and-mouth disease virus	Calves	In situ hybridization method showed presence of virus in target organs as early as 6 h after aerosol challenge.	11
Equid herpesvirus 1	Horses	Exposure of pregnant mares to viral aerosols resulted in abortions, but virus could not be demonstrated in the aborted fetuses.	104
Influenza virus	Humans	The vaccine was administered by either nose drops or large-particle aerosols; the aerosol route was more effective and well tolerated.	40
Respiratory coronavirus	Pigs	Aerosol challenge of young pigs induced a strong response in bronchial lymph nodes but not in the mesenteric lymph nodes.	10
Bovine herpesvirus strains 1.1 and 1.3	Calves	Strain 1.3 produced severe encephalitis with minimal respiratory lesions, whereas strain 1.1 gave respiratory lesions but no neurological disease; colostrum-fed animals became subclinically infected.	8
Influenza virus	Horses	Vaccinated and unvaccinated ponies were exposed to aerosols of a field isolate of influenza virus; vaccinated animals showed complete protection even after 15 mo of immunization.	83

[a] IgA, IgM, and IgG, immunoglobulins A, M, and G.

plumes from CO_2 laser vaporization of bovine fibropapillomas and human verrucae. Subsequently, PCR was used to confirm that the DNA in the laser plumes corresponded to the type in the lesion (62). In fact, infectious papillomavirus particles have also been detected in the vapors produced during CO_2 laser treatment as well as electrocoagulation of bovine warts (98). The same study also found that the amounts of DNA generated during the laser treatment of human and bovine warts were higher than those released during electrocoagulation; surgical masks were shown to be effective in filtering out the viruses in the vapors from both

types of treatment. Inoculation of cattle with debris from CO_2 laser plumes generated during the vaporization of bovine warts failed to infect the cattle (117). A recent PCR-based study (9) has reported widespread contamination of the facial areas of the therapists and the operating room with papillomavirus DNA released during CO_2 laser treatment and electrocoagulation of warts and neoplasia.

Retroviruses

Little is known about airborne spread of retroviruses (97). The advent of AIDS, however, has led to some recent studies on human immunodeficiency virus type 1 (HIV-1). Although airborne spread of AIDS is not known, survival of HIV-1 in air has been studied to assess the risk of exposure of health care personnel during orthopedic surgery. Infectious HIV-1 was detected in aerosols from certain types of surgical power tools (60), and the blood-containing aerosols produced were in the respirable range (58). Baggish et al. (5) claim to have detected HIV-1 in CO_2 laser smoke when pellets of experimentally infected cells were vaporized; we have found their findings difficult to interpret.

In our view, airborne spread of HIV-1 and other retroviruses could occur where high-titered suspensions of such viruses are handled, and strict adherence to safety precautions is necessary to eliminate the risk. It would be useful to know if cell-associated HIV-1 (in both infectious and proviral forms) behaves differently in air than does cell-free virus.

Viruses Causing Hemorrhagic Fevers

The ability of viruses causing hemorrhagic fevers to spread through air remains unclear. Their handling requires the highest level of biohazard containment (30); however, laboratories with such facilities are limited, and those that exist may not be equipped for aerobiological studies. In many cases, airborne spread of these viruses has been discovered through laboratory accidents or patterns of disease transmission in hospitals (21, 71) and animal-holding facilities.

The survival of Lassa fever virus in artificially generated aerosols was favored at low (30%) RH levels, and even at 32°C the virus survived long enough to permit its dispersal by air (106). Experimental aerosol exposure of monkeys and guinea pigs could infect and kill them, with the median 50% infectious dose for guinea pigs being as low as 15 virus PFU.

Laboratory-acquired infections in those handling hantaviruses or animals carrying these viruses are well documented (1, 72, 85, 118); the most likely means of exposure was by inhalation of infectious aerosols. Inhalation of artificially generated aerosols of hantaviruses can infect rats (94).

Rhesus monkeys became acutely ill and died when exposed to aerosols of Junin virus (63), the agent of Argentine hemorrhagic fever. The symptoms and pathology of the disease were very similar to those after parenteral exposure and also mimicked the clinical syndrome seen in humans.

The recent laboratory-acquired case of Sabia virus infection was probably due to aerosol exposure (7).

Viruses of Domestic Animals

FMDV

Foot-and-mouth disease virus (FMDV) has been studied extensively for its capacity to spread through air (97), and several airborne outbreaks of the disease have been documented (74). Models to forecast and analyze outbreaks of FMDV are now available (82). Sheep breathing air from cabinets with FMDV-infected pigs contracted the disease with amounts of virus as low as 10 50% tissue culture infective doses (36).

NDV

There is convincing evidence for the airborne spread of Newcastle disease virus (NDV), and air filters in poultry houses are highly effective in preventing outbreaks of the disease (51).

Kournikakis et al. (67, 68) used a vaccine (LaSota) strain of NDV to study the airborne survival and behavior of enve-

loped viruses as well as to field test protective equipment and methods for rapid collection and identification of viruses in air. The vaccine, available in the lyophilized form, is relatively easy to work with and safe. Each vaccine vial can yield nearly 10^8 PFU/ml. The influence of air temperature and RH on the airborne survival of this virus was very similar to that for other enveloped viruses (97). A combination of low RH (20 to 30%) and low temperature (about 10°C) is optimal for its survival in air in the dark (67); under such conditions, 56% of the aerosolized virus remained infectious even after 6 h, while the corresponding figure for 20°C was 39%. However, there was a nearly 99% drop in virus infectivity when NDV-containing aerosols were captured on spider threads and held under daylight with or without a cloud cover (Fig. 3).

Negative air ions may reduce the spread of NDV in chicken farms (81). Infected animals were placed upwind from the susceptibles ones and under controlled conditions of temperature (26.7°C), RH (50%), and ventilation rates (0.34 to 1.36 m^3/min); there was a 28% reduction in virus transmission when negative ion generators were used.

Pseudorabies Virus

Even though the herpesvirus that causes pseudorabies is relatively fragile, its capacity to spread through air parallels (39) that of FMDV, a picornavirus. The virus in air survived the best when RH was 55% and the air temperature was 4°C (101); the half-life of the virus under these optimal conditions was less than 1 h.

AEROSOL CHALLENGE STUDIES

Several studies have challenged human or animal hosts with artificially generated viral aerosols (97). Table 1 summarizes the information from selected studies published on this topic since 1970. The following conclusions are based on a critical analysis of these studies. (i) In most cases, susceptible hosts became infected upon exposure to the test virus, but often the amount of virus inhaled was unknown or may have been unrealistically high. (ii) In certain cases, the experimental design and setup did not exclude virus exposure of test subjects by means other than through inhalation. (iii) In spite of the importance of reporting RH and air temperature in such experiments, many of the investigators failed to mention these parameters. (iv) Of necessity, nearly all of these studies used laboratory-adapted strains of the test viruses; the extent to which these data apply to viruses in the field remains undetermined and perhaps undeterminable. (v) Any meaningful comparison of the data from these studies is difficult because there are no standard procedures for aerosol challenge studies. Therefore, much developmental work is needed before accurate and reliable information on the behavior of airborne viruses can be generated and such data can be applied to designing mathematical models and strategies to control and prevent disease spread through air.

COLLECTION OF NATURALLY OCCURRING VIRAL AEROSOLS

Collecting infectious viruses from naturally occurring aerosols continues to be difficult because the devices currently available for the purpose are generally noisy, bulky, expensive, and somewhat inefficient. As is true for many other types of environmental sampling, it is often too late to look for the suspected virus in air in an outbreak investigation.

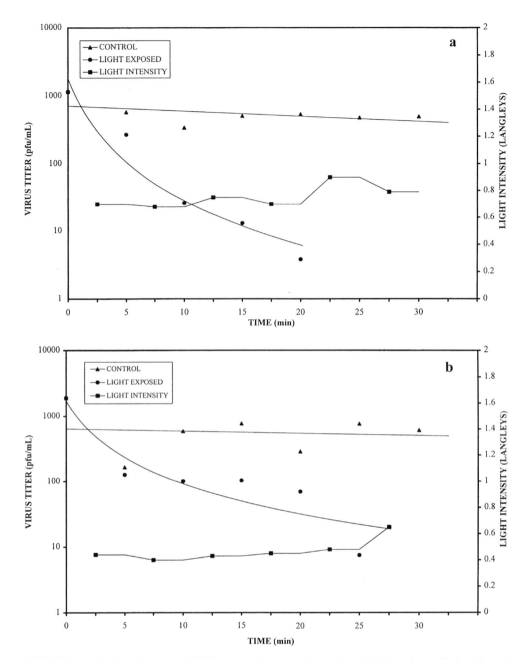

FIGURE 3 Biological decay of NDV captured on a spider web and held under sunlight (a) and under cloudy conditions (b). Control samples were held in the dark. (Reproduced with permission from B. Kournikakis, Defence Research Establishment, Suffield, Alberta, Canada.)

Even if it were feasible, regular monitoring of air for viruses is not recommended because of the limited significance of the findings. The data from molecular biological techniques (e.g., PCR) for viruses in air sample concentrates will be much more meaningful if the nucleic acid detected can reliably signal the presence or absence of virus infectivity. However, there are certain situations in which sampling of air for naturally occurring viral aerosols can be extremely valuable. This is exemplified by the studies of bat caves for airborne rabies virus. In this regard, not only was infectious rabies virus recovered from the air in the caves, but experimental animals exposed (in insectproof cages) to such air died of rabies (116).

Airborne spread of naturally occurring viral infections of humans and animals is known. In some of these cases (e.g., measles virus and influenza virus), air may be the chief vehicle of virus transfer. Others may be isolated instances of airborne spread of a virus which is normally transmitted by direct contact or by other vehicles. But it must be noted that any virus which can survive aerosolization has the potential for airborne transmission.

CONCLUDING REMARKS

Increasing use of recycled air will further enhance the risk of exposure of susceptible individuals to airborne viruses.

Such spread of viruses is also believed to exacerbate asthma and other ailments of the respiratory tract (66, 92). Therefore, there is an urgent need for improvements in the design and operation of air-handling/recycling systems, particularly in hospitals, clinics, offices, residential complexes, sports facilities, and aircraft. The continuing increase in the numbers of immunosuppressed and transplant patients also underscores this need. A recent U.S. survey on the association of outbreaks of infections with competitive sports has shown the infectious agents involved to be predominantly viruses, with several instances of airborne transmission (38). Innovative approaches are required to combine energy and resource conservation with much reduced risk of infection spread through conditioned and recycled air. Additional studies are also needed to establish better methods for disinfecting airborne viruses in recycled air.

Airborne dissemination of nonviable viruses or their components can lead to problems in laboratories using PCR or other such techniques, and presently available containment facilities may not be quite adequate to address this issue (96). Infectious aerosol generation by surgical tools (9, 20, 46, 98) and laboratory equipment (29) suggests that any such technology should be assessed at the design stage and corrective measures should be taken. This will require improved communication between design engineers and end users of the devices, including infection control practitioners.

The increased awareness of airborne infection, particularly in health care settings, has escalated the use of face masks, but the protective ability of such masks has been questioned in studies using respirable but nonmicrobial aerosols (112). There are no standardized methods to test the ability of face masks to withstand challenge with airborne infectious agents. This issue needs addressing.

Rapid inactivation (68) and higher levels of dilution of vertebrate viruses in open air suggest that they may be less likely to spread under outdoor conditions, but there is now enough evidence that certain types of animal viruses can be carried by air over several kilometers, leading to outbreaks at distant locations (6, 16, 39). While some suggest that this is true for human pathogens as well (43), much work is needed to substantiate this possibility. Perhaps long-distance transport of viruses can explain the peculiar transcontinental movement of outbreaks of rotaviral infections (50).

Airborne viruses continue to be a threat to human and animal health in spite of the sophisticated design and efficient protective functioning of the respiratory system (102). Much of the research thus far has documented outbreaks of airborne viral infections and identified to some degree the factors that influence virus survival in air. Now, greater emphasis is needed on reducing generation of infectious aerosols in indoor as well as outdoor settings and on enhancing the removal and/or inactivation of viruses and other infectious agents in air.

Sue Springthorpe's comments on the manuscript were very useful. We thank those who responded to our requests for relevant papers and reports. The help of K. Loewen (Office of Biosafety, Health Canada) in the literature search is much appreciated. We are grateful to B. Kournikakis and Canada's Defence Research Establishment for permission to reproduce the scanning electron micrograph and the graphs. We thank Elizabeth Kipp for the preparation of the manuscript.

REFERENCES

1. **Alexeyev, O. A., and B. A. Baranov.** 1993. Puumala virus infection without signs of renal involvement. *Scand. J. Infect. Dis.* **25:**525–527.

2. **Andersen, A. A.** 1958. New sampler for collection, sizing, and enumeration of viable airborne particles. *J. Bacteriol.* **76:**471–484.

3. **Anderson, G. W., Jr., J. O. Lee, A. O. Anderson, N. Powell, J. A. Mangiafico, and G. Meadors.** 1991. Efficacy of a Rift Valley fever virus vaccine against an aerosol infection in rats. *Vaccine* **9:**710–714.

4. **Ansari, S. A., V. S. Springthorpe, and S. A. Sattar.** 1991. Survival and vehicular spread of human rotaviruses: possible relation to seasonality of outbreaks. *Rev. Infect. Dis.* **13:**448–461.

5. **Baggish, M. S., B. J. Poiesz, D. Joret, P. Williamson, and A. Refai.** 1991. Presence of human immunodeficiency virus DNA in laser smoke. *Lasers Surg. Med.* **11:** 197–203.

6. **Banks, M.** 1993. DNA restriction fragment length polymorphism among British isolates of Aujeszky's disease virus: use of the polymerase chain reaction to discriminate among strains. *Br. Vet. J.* **149:**155–163.

7. **Barry, M., F. Bia, M. Cullen, L. Dembry, S. Fischer, D. Geller, W. Hierholzer, P. McPhedran, P. Rainey, M. Russi, E. Snyder, E. Wrone, J. P. Gonzalez, R. Rico-Hesse, R. Tesh, R. Ryder, R. Shope, W. P. Quinn, P. D. Galbraith, M. L. Cartter, J. L. Hadler, and A. DeMaria, Jr.** 1994. Arenavirus infection—Connecticut, 1994. *Morbid. Mortal. Weekly Rep.* **43:**635–636.

8. **Belknap, E. B., J. K. Collins, V. K. Ayers, and P. C. Schultheiss.** 1994. Experimental infection of neonatal calves with neurovirulent bovine herpesvirus type 1.3. *Vet. Pathol.* **31:**358–365.

9. **Bergbrant, I.-M., L. Samuelsson, S. Olofsson, F. Jonassen, and A. Ricksten.** 1994. Polymerase chain reaction for monitoring human papillomavirus contamination of medical personnel during treatment of genital warts with CO_2 laser and electrocoagulation. *Acta Derm. Venereol.* (Stockholm) **74:**393–395.

10. **Brim, T. A., J. L. VanCott, J. K. Lunney, and L. J. Saif.** 1994. Lymphocyte proliferation responses of pigs inoculated with transmissible gastroenteritis virus or porcine respiratory coronavirus. *Am. J. Vet. Res.* **55:**494–501.

11. **Brown, C. C., R. F. Meyer, H. J. Olander, C. House, and C. A. Mebus.** 1992. A pathogenesis study of foot-and-mouth disease in cattle, using in situ hybridization. *Can. J. Vet. Res.* **56:**189–193.

12. **Calnek, B. W., and S. B. Hitchner.** 1973. Survival and disinfection of Marek's disease virus and the effectiveness of filters in preventing airborne dissemination. *Poultry Sci.* **52:**35–43.

13. **Caul, E. O.** 1994. Small round structured viruses: airborne transmission and hospital control. *Lancet* **343:** 1240–1242.

14. **Chadwick, P. R., and R. McCann.** 1994. Transmission of a small round structured virus by vomiting during a hospital outbreak of gastroenteritis. *J. Hosp. Infect.* **26:** 251–259.

15. **Chadwick, P. R., M. Walker, and A. E. Rees.** 1994. Airborne transmission of a small round structured virus. *Lancet* **343:**171.

16. **Christensen, L. S., S. Mortensen, A. Botner, B. S. Strandbygaard, L. Ronsholt, C. A. Henriksen, and J. B. Andersen.** 1993. Further evidence of long distance airborne transmission of Aujeszky's disease (pseudorabies) virus. *Vet. Rec.* **132:**317–321.

17. **Chung, I. P., D. Dunn-Rankin, R. F. Phalen, and M. J. Oldham.** 1992. Low-cost wind tunnel for aerosol inhalation studies. *Am. Ind. Hyg. Assoc. J.* **53:**232–236.

18. **Cox, C. S.** 1987. *The Aerobiological Pathway of Microorganisms*, p. 1–293. John Wiley & Sons, Chichester, England.

19. **Cox, C. S.** 1989. Airborne bacteria and viruses. *Sci. Prog.* **73:**469–500.

20. **Cukier, J., M. F. Price, and L. O. Gentry.** 1989. Suction lipoplasty: biohazardous aerosols and exhaust mist—the clouded issue. *Plast. Reconstr. Surg.* **83:**494–497.

21. **Dalgard, D. W., R. J. Hardy, S. L. Pearson, G. J. Pucak, R. V. Quander, P. M. Zack, C. J. Peters, and P. B. Jahrling.** 1992. Combined simian hemorrhagic fever and Ebola virus infection in cynomolgus monkeys. *Lab. Anim. Sci.* **42:**152–157.

22. **Davis, W. G., A. R. Griesemer, A. J. Shadduck, and L. R. Farrelle.** 1971. Effect of relative humidity on dynamic aerosols of adenovirus 12. *Appl. Microbiol.* **21:**676–679.

23. **Deuter, A., D. J. Southee, and A. P. Mockett.** 1991. Fowlpox virus: pathogenicity and vaccination of day-old chickens via the aerosol route. *Res. Vet. Sci.* **50:**362–364.

24. **Dick, E. C., L. C. Jennings, K. A. Mink, C. D. Wartgow, and S. L. Inhorn.** 1987. Aerosol transmission of rhinovirus colds. *J. Infect. Dis.* **156:**442–448.

25. **Donaldson, A. I., R. C. Wardley, S. Martin, and N. P. Ferris.** 1983. Experimental Aujesky's disease in pigs: excretion, survival and transmission of the virus. *Vet. Rec.* **113:**490–494.

26. **Eickhoff, T. C.** 1994. Airborne nosocomial infection: a contemporary perspective. *Infect. Control Hosp. Epidemiol.* **15:**663–672.

27. **Elazhary, M. A. S. Y., M. Galina, R. S. Roy, M. Fontaine, and P. Lamothe.** 1980. Experimental infection of calves with bovine respiratory syncytial virus (Quebec strain). *Can. J. Comp. Med.* **44:**390–395.

28. **Fairchild, G. A.** 1974. Ozone effect on respiratory deposition of vesicular stomatitis virus aerosols. *Am. Rev. Respir. Dis.* **109:**446–451.

29. **Ferbas, J., K. R. Chadwick, A. Logar, A. E. Patterson, R. W. Gilpin, and J. B. Margolick.** 1995. Assessment of aerosol containment on the ELITE flow cytometer. *Cytometry* **22:**45–47.

30. **Fleming, D. O., J. H. Richardson, J. J. Tulis, and D. Vesley (ed.).** 1995. *Laboratory Safety: Principles and Practices*, 2nd ed. ASM Press, Washington, D.C.

31. **Frank, G. H., R. G. Marshall, and P. C. Smith.** 1977. Clinical and immunologic responses of cattle to infectious bovine rhinotracheitis virus after infection by viral aerosol or intramuscular inoculation. *Am. J. Vet. Res.* **38:**1497–1502.

32. **Frankova, V.** 1975. Inhalatory infection of mice with influenza A0/PR8 virus. I. The site of primary virus replication of its spread in the respiratory tract. *Acta Virol.* **19:**35–40.

33. **Frankova, V., and V. Rychterova.** 1975. Inhalatory infections of mice with influenza A0/PR8 virus. II. Detection of virus in the blood and extrapulmonary organs. *Acta Virol.* **19:**41–46.

34. **Garden, J. M., M. K. O'Banion, L. S. Shelnitz, K. S. Pinski, A. D. Bakus, M. E. Reichmann, and J. P. Sundberg.** 1988. Papillomavirus in the vapor of carbon dioxide laser-treated verrucae. *JAMA* **259:**1199–1202.

35. **Gellert, G. A., S. H. Waterman, D. Ewert, L. Oshiro, M. P. Giles, S. S. Monroe, L. Gorelkin, R. I. Glass.** 1990. An outbreak of acute gastroenteritis caused by a small round structured virus in a geriatric convalescent facility. *Infect. Control Hosp. Epidemiol.* **11:**459–464.

36. **Gibson, C. F., and A. I. Donaldson.** 1986. Exposure of sheep to natural aerosols of foot-and-mouth disease. *Res. Vet. Sci.* **41:**45–49.

37. **Goldberg, L. J., H. M. S. Watkins, E. E. Boerke, and M. A. Chatigny.** 1958. The use of a rotating drum for the study of aerosols over extended periods of time. *Am. J. Hyg.* **68:**85–93.

38. **Goodman, R. A., S. B. Thacker, S. L. Solomon, M. T. Osterholm, and J. M. Hughes.** 1994. Infectious diseases in competitive sports. *JAMA* **271:**862–867.

39. **Grant, R. H., A. B. Scheidt, and L. R. Rueff.** 1994. Aerosol transmission of a viable virus affecting swine: explanation of an epizootic of pseudorabies. *Int. J. Biometeorol.* **38:**33–39.

40. **Gruber, W. C., H. P. Hinson, K. L. Holland, J. M. Thompson, G. W. Reed, and P. F. Wright.** 1993. Comparative trial of large-particle aerosol and nose drop administration of live attenuated influenza vaccines. *J. Infect. Dis.* **168:**1282–1285.

41. **Gwaltney, J. M., Jr.** 1989. Rhinoviruses, p. 593–615. *In* A. S. Evans (ed.), *Viral Infections of Humans, Epidemiology and Control.* Plenum Medical Book Company, New York.

42. **Hallmo, P., and O. Naess.** 1991. Laryngeal papillomatosis with human papillomavirus DNA contracted by a laser surgeon. *Eur. Arch. Oto-Rhino-Laryngol.* **248:**425–427.

43. **Hammond, G. W., R. L. Raddatz, and D. E. Gelskey.** 1989. Impact of atmospheric dispersion and transport of viral aerosols on the epidemiology of influenza. *Rev. Infect. Dis.* **11:**494–497.

44. **Hamory, B. H., R. B. Couch, R. G. J. Douglas, S. H. Black, and V. Knight.** 1972. Characterization of the infective unit for man of two respiratory viruses. *Proc. Soc. Exp. Biol. Med.* **139:**890–893.

45. **Handley, B. A., and J. M. Roe.** 1994. An alternative microthread for the study of airborne survival of bacteria outdoors. *J. Appl. Bacteriol.* **77:**504–508.

46. **Heinsohn, P., D. L. Jewett, L. Balzer, C. H. Bennett, P. Seipel, and A. Rosen.** 1991. Aerosols created by some surgical power tools: particle size distribution and qualitative hemoglobin content. *Appl. Occup. Environ. Hyg.* **6:**773–776.

47. **Henderson, D. W.** 1952. An apparatus for the study of airborne infections. *J. Hyg.* **50:**53–88.

48. **Hepburn, N. C., and T. J. G. Brooks.** 1991. An outbreak of chickenpox in a military field hospital—the implications for biological warfare. *J. R. Soc. Med.* **84:**721–722.

49. **Hinshaw, V. S., F. L. Schaffer, and M. A. Chatigny.** 1976. Evaluation of Moloney murine sarcoma and leukemia virus complex as a model for airborne oncogenic virus biohazards: survival of airborne virus and exposure of mice. *J. Natl. Cancer Inst.* **57:**775–778.

50. **Ho, M.-S., R. I. Glass, P. F. Pinsky, and L. J. Anderson.** 1988. Rotavirus as a cause of diarrheal morbidity and mortality in the United States. *J. Infect. Dis.* **158:**1112–1116.

51. **Hopkins, S. R., and L. N. Drury.** 1971. Efficacy of air filters in preventing transmission of Newcastle disease. *Avian Dis.* **15:**596–603.

52. **Hugh-Jones, M. E., W. H. Allen, F. A. Dark, and J. G. Harper.** 1973. The evidence for the airborne spread of Newcastle disease. *J. Hyg.* **71:**325–339.

53. **Hyslop, N. S. G.** 1979. Observation on the survival and infectivity of airborne rinderpest virus. *Int. J. Biometeorol.* **23:**1–7.

54. **Ijaz, M. K., Y. G. Karim, S. A. Sattar, and C. M. Johnson-Lussenburg.** 1987. Development of methods to study the survival of airborne viruses. *J. Virol. Methods* **18:**87–106.

55. **Ijaz, M. K., S. A. Sattar, T. Alkarmi, F. K. Dar, A. R. Bhatti, and K. M. Elhag.** 1994. Studies on the survival of aerosolized bovine rotavirus (UK) and a murine rotavirus. *Comp. Immunol. Microbiol. Infect. Dis.* **17:**91–98.

56. **Ijaz, M. K., S. A. Sattar, C. M. Johnson-Lussenburg, and V. S. Springthorpe.** 1985. Comparison of the airborne survival of calf rotavirus and poliovirus type 1 (Sabin)

aerosolized as a mixture. *Appl. Environ. Microbiol.* **49:** 289–293.

57. **Jericho, K. W. F., C. le Q.-Darcel, and E. V. Langford.** 1982. Respiratory disease in calves produced with aerosols of parainfluenza-3 virus and *Pasteurella haemolytica*. *Can. J. Comp. Med.* **46:**293–301.

58. **Jewett, D. L., P. Heinsohn, C. Bennett, A. Rosen, and C. Neuilly.** 1992. Blood-containing aerosols generated by surgical techniques: a possible infectious hazard. *Am. Ind. Hyg. Assoc. J.* **53:**228–231.

59. **Johansson, B. E., and E. D. Kilbourne.** 1991. Comparison of intranasal and aerosol infection of mice in assessment of immunity to influenza virus infection. *J. Virol. Methods* **35:**109–114.

60. **Johnson, G. K., and W. S. Robinson.** 1991. Human immunodeficiency virus-1 (HIV-1) in the vapors of surgical power instruments. *J. Med. Virol.* **33:**47–50.

61. **Karim, Y. G., M. K. Ijaz, S. A. Sattar, and C. M. Johnson-Lussenburg.** 1985. Effect of relative humidity on the airborne survival of rhinovirus-14. *Can. J. Microbiol.* **31:** 1058–1061.

62. **Kashima, H. K., T. Kessis, P. Mounts, and K. Shah.** 1991. Polymerase chain reaction identification of human papillomavirus DNA in CO_2 laser plume from recurrent respiratory papillomatosis. *Otolaryngol. Head Neck Surg.* **104:**191–195.

63. **Kenyon, R. H., K. T. McKee, Jr., P. M. Zack, M. K. Rippy, A. P. Vogel, C. York, J. Meegan, C. Crabbs, and C. J. Peters.** 1992. Aerosol infection of rhesus macaques with Junin virus. *Intervirology* **33:**23–31.

64. **Kim, S. J., and P. B. Spradbrow.** 1978. Administration of a vaccine prepared from the Australian V4 strain of Newcastle disease virus by aerosol and drinking water. *Aust. Vet. J.* **54:**486–489.

65. **Knoch, M., E. Wunderlich, and S. Geldner.** 1994. A nebulizer system for highly reproducible aerosol delivery. *J. Aerosol Med.* **7:**229–237.

66. **Korppi, M.** 1988. Viruses and airborne allergens as precipitants of obstructive respiratory difficulties in children. *Ann. Clin. Res.* **20:**417–422.

67. **Kournikakis, B., D. Netolitsky, and J. Fildes.** 1987. Effects of temperature and relative humidity on the survival of Newcastle disease virus aerosols in the rotating drum (U), p. 1–17. Memorandum 1261. Defence Research Establishment, Suffield, Ralston, Alberta, Canada.

68. **Kournikakis, B., M. Simpson, and D. Netolitzky.** 1988. Photoinactivation of Newcastle disease virus in aerosol and in solution, p. 1–12. Memorandum 1263. Defence Research Establishment, Suffield, Ralston, Alberta, Canada.

69. **Larson, E. W.** 1970. Airborne transmission potential of murine leukemia infection, p. 213. *In* I. H. Silver (ed.), *Proceedings of the Third International Symposium on Aerobiology*. Academic Press, New York.

70. **Larson, E. W., J. W. Dominik, and T. W. Slove.** 1980. Aerosol stability and respiratory infectivity of Japanese B encephalitis virus. *Infect. Immun.* **30:**397–401.

71. **LeDuc, J. W.** 1989. Epidemiology of hemorrhagic fever viruses. *Rev. Infect. Dis.* **11:**S730–S735.

72. **Lee, H. W., and K. M. Johnson.** 1982. Laboratory-acquired infections with Hantaan virus, the etiological agent of Korean hemorrhagic fever. *J. Infect. Dis.* **146:** 645–651.

73. **Maigetter, R. Z., R. Ehrlich, J. D. Fenters, and D. E. Gardner.** 1976. Potentiating effects of manganese dioxide on experimental respiratory infections. *Environ. Res.* **11:** 386–391.

74. **Maragon, S., E. Facchin, F. Moutou, I. Massirio, G. Vincenzi, and G. Davies.** 1994. The 1993 Italian foot-and-mouth disease epidemic: epidemiological features of the four outbreaks identified in Verona province (Veneto region). *Vet. Rec.* **135:**53–57.

75. **May, K. R.** 1966. Multistage liquid impinger. *Bacteriol. Rev.* **30:**559–570.

76. **May, K. R., and H. A. Druett.** 1968. A microthread technique for studying the viability of microbes in a simulated airborne state. *J. Gen. Virol.* **51:**353–366.

77. **McKissick, G. E., R. A. Griesemer, and R. L. Farrell.** 1970. Aerosol transmission of Rauscher murine leukemia virus. *J. Natl. Cancer Inst.* **45:**625–636.

78. **McKissick, G. E., L. G. Wolf, R. L. Farrell, R. A. Griesemer, and A. Hellman.** 1970. Aerosol transmission of oncogenic viruses, p. 233–240. *In* I. H. Silver (ed.), *Proceedings of the Third International Symposium on Aerobiology*. Academic Press, New York.

79. **McVicar, J. W., and R. J. Eisner.** 1983. Aerosol exposure of cattle to foot-and-mouth disease virus. *J. Hyg.* **91:** 319–328.

80. **Merekalova, Z. I., N. M. Orekhova, and E. I. Zharona.** 1981. Possible ways of excreting murine leukemia virus into the environment and the so-called input gate for infection. *Eksp. Onkol.* **3:**59–62.

81. **Mitchell, B. W., and D. J. King.** 1994. Effect of negative air ionization on airborne transmission of Newcastle disease virus. *Avian Dis.* **38:**725–732.

82. **Moutou, F., and B. Durand.** 1994. Modelling the spread of foot-and-mouth disease virus. *Vet. Res.* **25:**279–285.

83. **Mumford, J. A., D. M. Jessett, E. A. Rollinson, D. Hannant, and M. E. Draper.** 1994. Duration of protective efficacy of equine influenza immunostimulating complex/tetanus vaccines. *Vet. Rec.* **134:**158–162.

84. **Mumford, J. A., and J. Wood.** 1992. Establishing an acceptability threshold for equine influenza vaccines. *Dev. Biol. Stand.* **79:**137–146.

85. **Niklasson, B. S.** 1992. Haemorrhagic fever with renal syndrome, virological and epidemiological aspects. *Pediatr. Nephrol.* **6:**201–204.

86. **Omerod, E., I. A. McCandlish, and O. Jarrett.** 1979. Disease produced by feline caliciviruses. *Vet. Rec.* **104:** 65–69.

87. **Osborne, A. D., J. R. Saunders, T. K. Sebunya, P. Willson, and G. H. Green.** 1985. A simple chamber for experimental reproduction of respiratory disease in pigs and other species. *Can. J. Comp. Med.* **49:**434–435.

88. **Owens, S. L., J. W. Osebold, and Y. C. Zee.** 1981. Dynamics of B-lymphocytes in the lungs of mice exposed to aerosolized influenza virus. *Infect. Immun.* **33:**231–238.

89. **Park, G. W.** 1980. The use of a large volume air sampler for the detection of airborne respiratory viruses: a feasibility study. M.Sc. thesis. University of Ottawa, Ottawa, Ontario, Canada.

90. **Pike, R. M.** 1979. Laboratory-associated infections: incidence, fatalities, causes and prevention. *Annu. Rev. Microbiol.* **33:**41–66.

91. **Prince, D. S., C. Astry, S. Vonderfecht, G. Jakab, F.-M. Shen, and R. H. Yolken.** 1986. Aerosol transmission of experimental rotavirus infection. *Pediatr. Infect. Dis.* **5:**218–222.

92. **Quintiliani, R.** 1994. The epidemiology of mild, moderate, and severe asthma. *Adv. Ther.* **11:**70–77.

93. **Ratanasethakul, C., and R. B. Cumming.** 1983. The effect of route of infection and strain of virus on the pathology of Australian infectious bronchitis. *Aust. Vet. J.* **60:** 209–213.

94. **Rossi, C. A., E. H. Stephenson, and J. W. LeDuc.** 1988. Aerosol transmission of Hantaan and related viruses to laboratory rats. *Am. J. Trop. Med. Hyg.* **38:**636–640.

95. **Saif, L. J., and K. W. Theil.** 1990. *Viral Diarrheas of Man and Animals.* CRC Press, Inc., Boca Raton, Fla.

96. **Saksena, N. K., D. Dwyer, and F. Barre-Sinoussi.** 1991. A "sentinel" technique for monitoring viral aerosol contamination. *J. Infect. Dis.* **164:**1021–1022.

97. **Sattar, S. A., and M. K. Ijaz.** 1987. Spread of viral infections by aerosols. *Crit. Rev. Environ. Control* **17:**89–131.

98. **Sawchuk, W. S., P. J. Weber, D. R. Lowy, and L. M. Dzubow.** 1989. Infectious papillomavirus in the vapor of warts treated with carbon dioxide laser or electrocoagulation: detection and protection. *J. Am. Acad. Dermatol.* **21:**41–49.

99. **Sawyer, L. A., J. J. Murphy, J. E. Kaplan, P. F. Pinsky, D. Chacon, S. Walmsley, L. B. Schonberger, A. Philips, K. Forward, C. Goldman, J. Brunton, R. A. Fralick, A. O. Carter, W. G. J. Gary, R. I. Glass, and D. E. Low.** 1988. 25- to 30-nm virus particle associated with a hospital outbreak of acute gastroenteritis with evidence for airborne transmission. *Am. J. Epidemiol.* **127:**1261–1271.

100. **Sawyer, M. H., C. J. Chamberlin, Y. N. Wu, N. Aintablian, and M. R. Wallace.** 1994. Detection of varicella-zoster virus DNA in air samples from hospital rooms. *J. Infect. Dis.* **169:**91–94.

101. **Schoenbaum, M. A., J. J. Zimmerman, G. W. Beran, and D. P. Murphy.** 1990. Survival of pseudorabies virus in aerosol. *Am. J. Vet. Res.* **51:**331–333.

102. **Skerrett, S. J.** 1994. Host defenses against respiratory infection. *Med. Clin. N. Am.* **78:**941–966.

103. **Slote, L.** 1976. Vital aerosols: a potential occupationally related health threat in aerated wastewater treatment systems. *J. Environ. Health* **38:**310–314.

104. **Smith, K. C., K. E. Whitwell, M. M. Binns, C. A. Dolby, D. Hannant, and J. A. Mumford.** 1992. Abortion of virologically negative foetuses following experimental challenge of pregnant pony mares with equid herpesvirus 1. *Equine Vet. J.* **24:**256–259.

105. **Spendlove, J. C., and K. F. Fannin.** 1982. Methods of characterization of virus aerosols, p. 261–329. *In* C. P. Gerba and S. M. Goyal (ed.), *Methods in Environmental Virology.* Marcel Dekker, Inc., New York.

106. **Stephenson, E. H., E. W. Larson, and J. W. Dominik.** 1984. Effect of environmental factors on aerosol-induced Lassa virus infection. *J. Med. Virol.* **14:**295–303.

107. **Stetzenbach, L. D.** 1992. Airborne microorganisms, p. 53–65. *In* J. Lederberg (ed.), *Encyclopedia of Microbiology,* vol. 1. Academic Press, Inc., New York.

108. **van der Veen, J., Y. Poort, and D. J. Birchfield.** 1972. Effect of relative humidity on experimental transmission of Sendai virus in mice. *Proc. Soc. Exp. Biol. Med.* **140:**1437–1440.

109. **Wallis, C., J. L. Melnick, V. C. Rao, and T. E. Sox.** 1985. Method for detecting viruses in aerosols. *Appl. Environ. Microbiol.* **50:**1181–1186.

110. **Wardley, R. C., and R. C. Poney.** 1977. Aerosol transmission of feline caliciviruses: an assessment of its epidemiological importance. *Br. Vet. J.* **133:**504–508.

111. **Wathes, C. M., and H. Johnson.** 1991. Physical protection against airborne pathogens and pollutants by a novel animal isolator in a level 3 containment laboratory. *Epidemiol. Infect.* **107:**157–170.

112. **Weber, A., K. Willeke, R. Marchioni, T. Myojo, R. McKay, J. Donnelly, and F. Liebhaber.** 1993. Aerosol penetration and leakage characteristics of masks used in the health care industry. *Am. J. Infect. Control* **21:**167–173.

113. **White, L. A., D. J. Hadley, D. E. Davids, and R. Naylor.** 1975. Improved large volume sampler for the collection of bacterial cells from aerosol. *Appl. Microbiol.* **29:**335–339.

114. **Wilkinson, P. J., and A. I. Donaldson.** 1977. Transmission studies with African swine fever virus. The early distribution of virus in pigs infected by airborne virus. *J. Comp. Pathol.* **87:**497–501.

115. **Wilkinson, P. J., A. I. Donaldson, A. Greig, and W. Bruce.** 1977. Transmission studies with African swine fever virus. Infection of pigs by airborne virus. *J. Comp. Pathol.* **87:**487–495.

116. **Winkler, W. G.** 1968. Airborne rabies virus isolation. *Bull. Wildl. Dis. Assoc.* **4:**37–40.

117. **Wisniewski, P. M., M. J. Warhol, R. F. Rando, T. V. Sedlacek, J. E. Kemp, and J. C. Fisher.** 1990. Studies on the transmission of viral disease via the CO_2 laser plume and ejecta. *J. Reprod. Med.* **35:**1117–1123.

118. **Wong, T. W., Y. C. Chan, E. H. Yap, Y. G. Joo, H. W. Lee, P. W. Lee, R. Yanagihara, C. J. Gibbs, Jr., and D. C. Gajdusek.** 1988. Serological evidence of hantavirus infection in laboratory rats and personnel. *Int. J. Epidemiol.* **17:**887–890.

119. **Zimmerman, N. J., P. C. Reist, and A. G. Turner.** 1987. Comparison of two biological aerosol sampling methods. *Appl. Environ. Microbiol.* **53:**99–104.

Aerobiology of Agricultural Pathogens

ESTELLE LEVETIN

75

All crop plants are subject to diseases caused by viruses, bacteria, and fungi. Although plant diseases had been recognized since ancient times, they were not connected with pathogenic microorganisms until the 19th century. In 1861, during the investigations of the Irish potato blight, the German botanist Anton de Bary proved experimentally that the fungus *Phytophthora infestans,* previously found on the infected potatoes, was actually the cause of the disease. This discovery by de Bary gave birth to the science of plant pathology. Soon, other plant pathogenic fungi were described; however, pathogenic bacteria and viruses were not identified until later in the 19th century. Even with the many advances in agriculture during the 20th century, plant diseases still pose a major threat to the world's crops, making the world's food supply highly vulnerable.

Studying the spread of plant disease within a crop and from field to field within a region is often referred to as botanical epidemiology. Dispersal of the pathogen is the necessary step for repeated cycles of infection and multiplication and, therefore, the spread of the epidemic. A thorough understanding of dispersal is necessary for predicting the onset and severity of the disease. Implicit within the scope of dispersal are the processes of release, transport, and deposition (3). Although propagules of plant pathogens are dispersed by wind, rain, soil water, insects, and even humans, this chapter is limited to discussion of the airborne spread of those pathogens, which may be the least-understood vehicle of dispersal.

RANGE OF AIRBORNE TRANSPORT

Plant pathogens dispersed by airborne transport generally produce enormous numbers of propagules that are passively carried through the atmosphere. The spread of plant disease through this aerobiological pathway can proceed over short distances through focal spread as well as over long distances (70).

An infection focus is an area of a crop with a contagious disease. Foci are often circular, but if they are strongly affected by wind, they may become comet-shaped or V-shaped. Foci generally have a constant radial expansion, with the rate varying with the scale of the infection from a few centimeters per day for a localized infection to hundreds of kilometers per year for a pandemic (70). The spread of the disease can be described as microscale, mesoscale, and synoptic scale, using meteorological terms that describe the horizontal movement of atmospheric particles.

Disease Spread

Microscale spread is usually limited to less than a few hundred meters within one field and occurs within one growing season. This roughly corresponds to a zero-order epidemic (27, 70). The focus begins with a single successful propagule causing an infection and creating a lesion. After several generations of localized pathogen spread (for a polycyclic disease with a repeating cycle), the focus may reach a detectable size. In an annual crop, the foci are often about 1 m in diameter around the initial source of infection but may be larger.

If growing conditions are unsuitable for the pathogen, the focus may stop expanding or even disappear with new growth in the canopy. However, when conditions are favorable for the pathogen, the disease will spread. As the primary focus continues to expand, secondary foci and later tertiary foci appear. This continued focal spread over a larger area is considered mesoscale spread and corresponds to a first-order epidemic. This may be restricted to one field but may spread over many fields or up to several hundred kilometers in size or even part of a continent during a single growing season (70).

Synoptic or macroscale spread occurs when the epidemic progresses for several years and spreads over an area several thousand kilometers in size. This is referred to as a second-order epidemic. This pandemic may even cover a whole continent after a certain number of years (70).

Long-Distance Transport

Long-range transport often explains the introduction of a pathogen to a new area or the reintroduction to areas where overwintering cannot occur (14). Since fungi are the most abundant plant pathogens, most of the well-studied examples of long-distance transport involve fungal spores. The transport of *Puccinia graminis* uredospores from the southern United States and Mexico to the wheat belt in the northern United States and Canada has been well documented, as have other instances of long-distance dispersal of rust spores (51).

Trajectory analysis is frequently used to trace the previous movement of a spore-laden parcel of air and locate the inoculum source. To calculate the trajectory, the average wind speed and direction in the transport layer are determined. Once the source is identified, forward trajectories are used to indicate further potential areas of fallout. In addition, various computer models have been used to trace the movement of spores from dissemination at a source to deposition at a sink (14). The atmospheric transport and dispersion and the branching atmospheric trajectory models are used to calculate trajectories based on upper air winds, temperature, and other parameters. The atmospheric transport and dispersion model will even calculate the concentration of the spores and their deposition amounts (14, 51). Recently the hybrid single-particle Legrangian integrated trajectory (HY-SPLIT) model has been used to successfully track the movement of *Peronospora tabacina* spores and predict the occurrence of blue mold (44).

Movement in the Atmosphere

Fungal spores are a normal component of the turbulent layer of the atmosphere, and airborne transport normally occurs here, with the dispersive power related to the intensity of turbulence (12, 24, 46). Airborne spores are present in large numbers any time that the ground is not covered with ice and snow. Spores are discharged from fungi growing as parasites or saprobes, and atmospheric concentrations may exceed 200,000 spores per m³ of air (35). Wind is the major factor, with gusts and lulls affecting takeoff, transport, and deposition (3, 51).

In a crop, spores have to pass from the laminar layer close to the leaf surface into the turbulent layer within the crop. Gusts and turbulence enhance spore removal from leaves by sweeping away the layer of slowly moving air next to the leaf surface (3). Before spores reach the free air above the crop, they must also pass through the boundary layer surrounding the crop. Possibly as many as 90% of the spores are deposited within the crop itself (20). The percentage that escapes from the canopy depends on the balance between deposition and turbulence, with greater escape during more turbulent winds. The position of the spores within the canopy also affects escape. Spores produced lower in the canopy will have lower rates of escape since they are exposed to slower winds and less turbulence. In the early stages of a disease, when the infection may be confined to the bottom or middle of the canopy, the spread of the disease may be limited. This may change as disease progresses. After the infection reaches the top of the canopy, the increased rate of spore escape may permit the disease to spread rapidly across the field (3).

Carried by wind, spores are transported both vertically and horizontally. They may be carried upward by convective activity or thermals and have been recovered from altitudes of over 5,000 m. In spring and summer, when the ground is heated to a greater degree, vertical movements of air masses are more common, as is greater turbulence. The atmosphere during this period is characterized by the maximum concentrations of airborne spores, including many pathogenic spores such as those of *Erysiphe*, *Drechslera*, and *Venturia* species. Horizontally, spores can be carried for thousands of kilometers, dispersing pathogens into new areas (23, 51).

Dispersal Mechanisms

Dispersal of spores into the atmosphere is dependent on the method of discharge by individual taxa as well as environmental factors such as temperature, humidity, and wind speed. Spores are released into the air either passively or by active discharge mechanisms.

Passive Mechanisms

Many pathogenic fungi such as conidia of the powdery mildews, *Alternaria* spp., and *Drechslera* spp. as well as rust uredospores and smut teliospores are passively dispersed in dry weather. For the majority of these dry-dispersed spores, the airborne concentration is dependent on the ease with which spores are detached from the parent mycelium (or fruiting structure) and by atmospheric conditions such as wind speed and turbulence. Wind gusts facilitate their removal, as does the movement of the leaf itself in the wind. The role of wind gusts in the dispersal of spores has recently been reviewed by Aylor (3). Since this method of dispersal (for both pathogenic and saprobic fungi) is promoted by warm dry weather, the entrained propagules are referred to as the dry air spora. They include the most abundant fungi in the environment and usually peak during the afternoon hours when humidity is low and wind speeds are increased (24). Although there are major similarities in the dry air spora worldwide, at any one time the air spora may be dominated by nearby sources of spores (35). During crop harvesting or mowing, incredible levels of spores (up to $10^9/m^3$) may be dispersed into the atmosphere, with *Alternaria*, *Cladosporium*, *Epicoccum*, or *Didymella* the most abundant taxa (35, 56). Dry-dispersed spores may also be removed by raindrops as they strike the leaf surface, causing a puffing action that may remove the conidia (17).

Rain splash can also propel spores into the atmosphere. While rain is the most important vehicle for splash dispersal, overhead irrigation can also disperse plant pathogens. Splash-borne spores or bacteria are usually produced in mucilage (a sticky or slimy polysaccharide layer), which inhibits their direct removal by wind. However, the mucilage also protects them from desiccation during dry weather. The first raindrops dissolve the mucilage and leave a spore suspension available for splash dispersal by additional raindrops. Recently, the dispersal of *Colletotrichum gloeosporioides* conidia by rain splash was studied, and the number of conidia was shown to vary from drop to drop (57). Splash droplets from the first water drop to strike the leaf released only a few conidia, and some mucilage with the highest number of conidia was found in the fifth water drop.

Rain splash confines dispersal to periods when the wet conditions on the new host are also favorable for germination and even reproduction (18, 29). Severity of bacterial brown spot of snap bean was found to increase after rain due to the rapid multiplication of the pathogen (28).

Many splash-dispersed spores have some adhesive properties that enable them to stick to the new host surface. These spores frequently have smooth, thin, hyaline walls and an elongate shape, such as conidia of *Fusarium* and *Pseudocercosporella* species. Although they lack the protective feature of thick-pigmented walls commonly found on members of the dry air spora, they are protected from desiccation by wet conditions during dispersal and deposition (18).

The effect of wind on dispersal of splash-borne spores depends on the size of the raindrops. While very large droplets carrying spores are unaffected by wind, small droplets may evaporate and allow the spore to be dispersed through the air as a true aerosol (17, 18, 36). Most splash droplets are too large to form aerosols. As a result, rain splash appears to be most important for local dispersal only. Net movement is often downward from younger to older leaves (28).

Raindrops are very efficient spore collectors, and the majority of spores in the 20- to 30-μm range are washed out by raindrops. Even a very light rain can effectively wash spores from the atmosphere; however, the significance of this rain scrubbing is difficult to quantify (17).

Active Mechanisms

Active discharge mechanisms propel spores into the turbulent layer, independent of wind. These mechanisms are widespread among the fungi, with the ballistics of some species quite spectacular. Many ascospores and basidiospores are actively discharged by mechanisms that require moisture or high humidity. Atmospheric concentrations of these spores often peak at night or during early morning hours when the humidity is high (3). Aerobiological studies have shown that ascospores are often abundant in the atmosphere during and after rainfall (1, 8). The explosive discharge of ascospores from the ascus is a widespread characteristic of the ascomycetes which is linked to available moisture. High osmotic pressure develops within the ascus either by the direct absorption of rainwater or by the swelling of mucilage within the ascus (49). The resulting pressure causes the ascus tip to rupture, forcing the spores out explosively into the turbulent layer. Discharge of basidiospores from mushrooms and bracket fungi also requires atmospheric moisture; however, the mechanism is not completely understood. High concentrations of atmospheric basidiospores have been reported from various locations, with the release occurring during times of high humidity (24, 26, 37, 48, 61). Although basidiospore dispersal is not directly tied to rainfall, fruiting bodies of basidiomycetes, such as mushrooms and puffballs, frequently develop following rain. Release of basidiospores from these fruiting bodies, therefore, results in high atmospheric concentrations during seasons when rainfall is frequent (37).

Survival in the Atmosphere

Fungal spores that are adapted to airborne dispersal are often much more resistant to environmental stress than the parent hyphae. However, they are still vulnerable to certain types of environmental damage while airborne. The exposure to harmful radiation and extremes of temperature and humidity can decrease the viability or infectivity of pathogenic species.

Change in relative humidity, often caused by changing wind speeds, may affect survival, especially for thin-walled spores, which may easily plasmolyze (12). The risks of desiccation are usually greatest in daytime and close to the ground. At night and at high altitudes, conditions are less stressful; spores have even been reported to germinate in the clouds (24). At the same time, spores may also serve as condensation nuclei for rain (51).

Radiation in the upper atmosphere, especially UV radiation, may also affect survival of spores carried into the upper air. The effects of radiation may outweigh the effects of temperature and humidity. Again, thin-walled colorless spores may be more vulnerable since they lack the protection provided by the melanin present in the cell walls of pigmented spores. Low temperatures in the upper atmosphere may be preservative, however, and protect spores from UV damage (24). Despite the environmental hazards, many spores are able to survive long-range transport, but the percentage of viable spores that actually reach a target and cause infection is low (51). Spores can be deposited on the crop surface by sedimentation, impaction, boundary layer exchange, turbulence, electrostatic deposition, and raindrops (51).

PATHOGENS SPREAD BY THE AEROBIOLOGICAL PATHWAY

A wide variety of plant pathogens, including viruses, bacteria, and fungi, are dispersed through the atmosphere. When they are deposited on a susceptible host, infection can occur, and when environmental conditions are favorable, the resulting disease spread may lead to widespread crop loss. A thorough understanding of the role of the aerobiological pathway in pathogen dispersal is necessary for the management and control of these diseases.

Viruses

Most plant viruses are spread by means of insect vectors, especially aphids. Barley yellow dwarf is often considered the most widespread and economically important virus disease of cereal crops; it affects more than 100 species, including wheat, rice, corn, barley, and oats. Many wild grasses are also infected and thereby serve as reservoirs of infection. Barley yellow dwarf is caused by a large complex of luteoviruses which share common aphid vectors. This topic has recently been reviewed by Irwin and Thresh (30) and others (9, 54). The aerobiological focus of barley yellow dwarf research concerns the atmospheric movement of the aphid vector and the importance of winds in long-distance aphid migration (30).

Plant viruses can become airborne when infected plants are damaged by high winds. This was described for tobacco mosaic virus (5), but in general, evidence on the aerosol dispersal of viruses is scant (36). The transport of virus particles by airborne pollen has been established for more than 15 plant viruses (see recent reviews by Cooper et al. [11] and Mink [47]).

Bacteria

Many species of bacteria are serious phytopathogens, causing devastating crop losses each year. Bacterial plant pathogens are generally gram-negative bacilli, with species of *Erwinia*, *Pseudomonas*, *Xanthomonas*, *Agrobacterium*, and *Corynebacterium* well-studied phytopathogens; however, nonpathogenic species in these genera also occur (6, 28, 36, 59, 62, 65). Recently Conn et al. (10) identified a coccoid bacterium, *Leuconostoc mesenteroides*, as the cause of a postharvest decay of tomato. *Clavibacter* species is the only recognized gram-positive phytopathogen (64).

Dispersal of pathogenic bacteria from plants is generally passive by water, wind-blown water, and animals. Agricultural practices and agricultural workers and their machinery and tools also play a major role in the spread of pathogenic bacteria, with humans typically responsible for most long-distance dispersal. In warm, humid climates, where dew and rain are common, dispersal by rain splash is the major means of disease spread (18, 22).

Phytopathogenic bacteria have no special mechanism for producing airborne propagules; however, they become airborne during rain or irrigation as well as during crop harvesting (36). Bacteria can be passively carried off the surfaces of plants or soil by air currents and may even become airborne as single cells, but they are usually carried on rafts of plant material or in droplet nuclei following splash dispersal (36, 40). In a recent review of air-sampling techniques, Lacey and Venette (36) cited many studies that document the airborne dispersal of bacterial pathogens.

None of the phytopathogenic bacteria are capable of forming endospores; thus, they remain susceptible to solar radiation and desiccation and therefore limited in their ability to survive long distance aerial dispersal. Survival requires rapid transport to an environment with adequate moisture, nutrition, and temperature (22). Loss of water, causing dehydration or desiccation, represents the major stress to airborne bacteria. Dehydration is generally caused by the evaporation of water from droplets carrying bacteria (45). This finding supports the suggestion by Hirano and Upper (28) that splash dispersal is significant only for local disease spread. However, it is also known that some bacteria can serve as cloud condensation nuclei. Franc (19) showed that viable *Erwinia carotovora* cells were recovered from precipitation in Colorado. The storm system that deposited the precipitation had originated in the Pacific Ocean off the coast of California. The ability of *E. carotovora* cells to function as cloud condensation nuclei and be protected from desiccation and radiation may explain their ability to survive this long-distance transport.

The airborne dispersal of bacteria which are not phytopathogens has been the focus of other agricultural research. Microorganisms are currently being used in various agricultural applications, including frost protection, control of plant diseases, and control of insect pests (66, 69). In some instances, the organisms are naturally occurring microbial pathogens of insects that are being used for pest control, or they are epiphytic competitors of foliar pathogens. In other cases, the organisms are genetically engineered microorganisms (GEMs) designed to perform specific actions.

A well-known example of GEMs is the ice⁻ strain of *Pseudomonas syringae*, which can prevent frost injury. Frost damage of leaves occurs when ice forms around specific ice-nucleating sites such as wall proteins of *P. syringae*, a common leaf surface bacterium (64). By using genetic engineering, the gene that controlled for the wall protein was deleted, thereby creating cells of *P. syringae* that would not serve as ice nucleation sites. Plants sprayed with cultures of this genetically engineered (ice⁻) bacterium could survive lower temperatures without frost damage (33, 64). While large-scale applications of ice⁻ strains of *P. syringae* are still being developed, a naturally occurring ice⁻ strain of *P. fluorescens*, A506, is now registered commercially to control frost injury of pear (69).

The movement and fate of GEMs have been the focus of recent study, and the literature is extensive. The topic has also been addressed in recent reviews (50, 66, 69). The safe utilization of GEMs necessitates that they be applied and retained in specific target areas. Field application of GEMs is generally through aerosol sprays. Dispersal occurs by the modes described for phytopathogenic bacteria, including splash, which is important for short-distance movement, and wind, which has greater potential for long-distance transport (45, 50). The possibility of long-distance dispersal of GEMs out of the target area always exists (45). As a result, it is necessary to carefully monitor the dispersal and account for the released population. Air sampling is one of the methods being used for the detection of GEMs as well as nonrecombinant organisms (36, 66). Monitoring the field release of ice⁻ bacteria showed that the results of field trials were closely predicted by laboratory and greenhouse experiments as well as previous field studies with nonrecombinant bacteria (69). Wilson and Lindow (69) concluded that the release of GEMs in carefully conducted studies should have few detrimental effects.

With the current progress in recombinant DNA technology, it is anticipated that agricultural applications of GEMs will increase in the coming decades (45). Aerobiological studies will continue to play a major role in assessing the

TABLE 1 Major fungal pathogens with airborne dispersal

Disease	Organism	Reference(s)
Late blight of potato	*Phytophthora infestans*	21, 42, 52, 55
Blue mold	*Peronospora tabacina*	3, 32, 43, 44, 68
Apple scab	*Venturia inaequalis*	4, 17, 18, 31, 55
Early blight, leaf spots	*Alternaria* spp.	2, 38, 60
Coffee rust	*Hemileia vastatrix*	51
Wheat stem rust	*Puccinia graminis*	15, 16, 51, 53, 58
Wheat leaf rust	*Puccinia recondita*	15, 16, 51
Smut fungi	*Ustilago* spp.	2, 13, 25, 26, 38, 48, 61
	Tilletia spp.	26

risks associated with the introduction of these recombinant microorganisms.

Fungi

Over 70% of all major crop diseases are caused by fungi, with thousands of fungal species recognized as plant pathogens. It is estimated that fungal diseases cost more than $3.5 billion to American farmers alone (34). In general, spores of most fungal pathogens are adapted for airborne transport; however, much of the aerobiological research of agricultural pathogens has focused on a limited number of fungi that cause economically important diseases (36). Air sampling can be a valuable tool, since the management of plant disease requires an understanding of the airborne dispersal of inoculum. The following discussion examines some of the significant fungal pathogens using the aerobiological pathway (Table 1).

Late Blight of Potato

The potato ranks as the world's fourth major food crop, but the continued increase of this crop in developing nations may be limited by the cost of disease control measures. Potatoes are attacked by many pathogens and pests, and the total amount of agricultural chemicals applied to the potato crop is greater than for any other food crop (52). Late blight of potato has been the most important disease of potato since the 1840s, when it caused the destruction of the potato crop in Ireland and the resulting widespread famine.

Phytophthora infestans is the pathogen responsible for late blight of potato (and also tomato in some areas). This oomycete occurs wherever potatoes are grown, and all potato cultivars are susceptible. Populations of the fungus are relatively short-lived, and a field with infected potatoes during a particular year may or may not show diseased plants the following year (21). The fungus invades host tissue, resulting in the rapid death of infected parts. Productivity of the potato plants is greatly reduced, and tuber destruction frequently occurs as well. Sporangia are produced on aerial hyphae which grow out of the stomata and are dispersed by wind, possibly carried for tens of kilometers (21). At low temperatures (10 to 15°C) and high humidity, sporangia germinate by producing numerous zoospores, while at higher temperatures each sporangium gives rise to a single germ tube that develops into hyphae. As a result, during cool, wet periods, the production of zoospores leads to an astonishingly rapid spread of the disease. Without fungicidal protection, a blighted field can be destroyed within a couple of days. Similar disease progression occurs in infected tomato fields (55).

The economic impact of fungicide application to control late blight (or other pathogens) must not be overlooked. Often the cost of repeated applications may actually outweigh the value of the crop at harvest time (64). Rather than relying on fungicides alone, one can use other techniques such as sanitation, crop rotation, and genetic resistance to reduce the need for fungicides (64). This multifaceted approach, referred to as integrated pest management, has been widely recommended for controlling plant disease. To reduce fungicide use, detailed information is needed about the pathogen life cycle, especially the dispersal phase. For pathogens with airborne dispersal, aerobiological studies and knowledge of the environmental conditions that promote dispersal can supply information on the optimum timing of fungicide applications to protect the crop.

Potato late blight forecasting has a 40-year history. In many potato-growing regions, when meteorological conditions indicate that *Phytophthora* spread is likely to occur, warnings are issued to apply fungicides (42). Forecasting models, such as BLITECAST, have been successful in reducing the number of fungicide applications. Sometimes as many as six applications can be omitted in a single season (42). Other models have successfully incorporated air-sampling data for blight forecasting. The aerobiological models could possibly lead to a 50% reduction in fungicide applications (55).

P. infestans overwinters as mycelia in infected tubers, but in areas where both mating types occur, it can also overwinter as oospores in the soil. When the fungus was introduced to Europe and the United States in the 1840s, only one mating type (A$_1$) was introduced. In 1950, the second mating type (A$_2$) was identified in central Mexico, the native home of the fungus. The A$_2$ mating type was confined to Mexico until 1980, when it spread through Europe. It is speculated that this mating type was introduced in a large shipment of potatoes from Mexico to Europe in the late 1970s. Outbreaks of the A$_2$ type began occurring in Europe in 1980; this type has subsequently spread throughout the world (52). The introduction of the A$_2$ mating type has increased concerns about the possibility of sexual reproduction and genetic recombination occurring in the field, with new strains developing. This possibility has broad implications for potato breeders searching for blight resistance. These concerns have been strengthened by the recent outbreaks of late blight in the Pacific Northwest states. The strain of *P. infestans* involved was resistant to Metalaxyl, the most widely used fungicide to control late blight (21).

Two types of long-distance dispersal of *P. infestans* apparently have taken place. Intercontinental migration has been associated with the transport of infected plants or tubers by humans. This occurred in the 1840s and again before the 1980 outbreak of the A$_2$ mating type. Long-distance dispersal over tens of kilometers is attributed to wind-blown sporangia. Sporangia can survive for hours at the reduced humidity encountered during transport (21). Maps showing the rapid progress of blight epidemics in the 1840s suggest that a second-order epidemic of late blight could possibly occur during a single growing season.

Blue Mold

Tobacco blue mold caused by the oomycete *Peronospora tabacina* (synonym, *P. hyoscyami* f. sp. *tabacina*) is an unpredictable disease of both wild and cultivated tobacco, causing devastation some years and not appearing at all during other years. Blue mold was first described in Australia during the 19th century, and the fungus was identified by Baily in 1890 (32). In North America, the disease was confined to seedbeds until 1979, when the first serious epidemic occurred. The infection rate was especially severe in both 1979 and 1980, with a second-order epidemic advancing at rates of 10 to 32 km/day northward in the eastern United States to southern Canada. Crop losses in the United States and Canada during these 2 years were estimated at approximately $350 million. Both host plants and pathogen exist year-round in tropical areas such as the Mediterranean and Caribbean basins. In temperate regions, tobacco is grown as an annual; in addition, *P. tabacina* is not able to overwinter in temperate zones. As a result, the long-distance transport of inocula from tropical regions, in the form of asexual sporangia, must be reintroduced each year.

Infection can occur within 4 h after a sporangium lands on the leaf surface. A symptom-free incubation period, typically 5 to 7 days, ends with the appearance of yellow lesions and the development of new sporangia. Unlike *Phyophthora infestans*, *P. tabacina* does not produce zoospores, and the sporangia themselves (often referred to as sporangiospores or conidia) are the only asexual propagules. Sporangia are released from the sporangiophore by the twisting movement of the sporangiophore that occurs during the drying of the leaf in the morning as humidity decreases and temperature increases (43). During the spring in Connecticut, the greatest number of sporangia were found in the air in the early morning, while in summer the highest concentrations occurred later in the morning, when turbulence may have been greater (3). Cool, wet, overcast weather favors the rapid advance of the fungus, while clear, hot, dry weather stops disease spread (43).

Each spring in the eastern United States, weather conditions are favorable for the northward transport of *Peronospora* sporangia from southern sources. Case studies of epidemics occurring from 1979 to 1986 suggest at least two likely pathways of disease spread (43). In some years, the pathogen spread northward from Florida and Georgia. In other years, the disease occurred in Kentucky and North Carolina without first occurring in the southeast. The source of the inoculum in this second pathway was believed to be south-central Texas on *Nicotiana repanda* or from cultivated tobacco in Mexico. These cases have shown that long-distant transport of sporangia can lead to severe epidemics of blue mold, and forecasting systems can potentially provide time for tobacco farmers to apply fungicides.

Wiglesworth et al. (68) reported on a forecasting system that uses molecular probes. They described the amplification of an oligonucleotide fragment of DNA from *P. tabacina* that could be used to detect the fungus in infected parts of tobacco plants. Spore traps along with the amplification of this DNA fragment were used in the prediction of a local blue mold outbreak. Other predictive models make use of extant disease outbreaks, weather fronts, and weather forecasts. The HY-SPLIT trajectory model (44) was successfully used during the spring and summer of 1995 to predict outbreaks of blue mold in the eastern United States. The trajectory model employs documented reports of blue mold along with meteorological data from the National Oceanic and Atmospheric Administration's nested grid model (44), which is used to predict short-term weather so as to plot trajectories of inoculum-laden parcels of air. Forecasters used the most recently reported and continuing sources of blue mold to develop daily trajectories. The daily forecast produced by HY-SPLIT describes future weather conditions at the source and along the anticipated pathway, with emphasis given to those atmospheric conditions that favor sporulation at the source, survival during transport, and deposition. An overall outlook describes the likelihood of blue mold spread over the subsequent 48 h. The forecasters anticipate that the use of this model will continue to provide valuable information in the efforts to control blue mold epidemics (44).

Apple Scab

Scab is the most important disease of apples, with all commercial varieties susceptible to attack by the ascomycete *Venturia inaequalis* (34). The disease occurs in every country where apple trees are grown, and similar scab diseases also affect pear and hawthorn. Scab lesions occur on both leaves and fruit; the disease can also cause premature defoliation (31). Control of apple scab requires repeated application of fungicides (8 to 20 times per season) to protect the crop. Without such treatment, 70 to 100% of the crop would be unsalable (34).

The fungus overwinters as immature fruiting bodies on dead leaves on the ground in the orchard. The ascocarps mature in the spring, and ascospores are actively discharged from ascocarps during rainfall. Although discharged during rain, the spores are dry airborne spores. The rain causes the asci to swell and release the spores (18). The airborne concentration of *V. inaequalis* ascospores decreased rapidly with height above ground. The concentrations at 3 m were only 6% of the values at 15 cm. A spore dispersal model suggests that this decrease was due to the rapid increase of wind speed and eddies above ground level (4). The spores become a major component of the air spora in orchards and are carried by wind to leaves, where they cause a primary infection (17, 55). During the spring there is usually a constant source of the ascospores. Ponti and Cavanni (55) showed that the ascospores could typically be found in the air of Italian apple orchards during periods of rainfall from mid-March to mid-June. Once the primary infection is established, conidia of the anamorphic stage (*Spilocaea pomi*) begin developing. The more primary inoculum present, the more rapidly the disease will build up and the more serious the epidemic will ultimately be.

Fungicide applications should be timed to coincide with periods of rainfall in the spring, when ascospores are released. Air sampling has been used to provide information about the duration of the ascospore season in a particular area and therefore help control the use of fungicides to the periods when it is absolutely necessary (55).

Alternaria Infection

Alternaria is a genus of asexual or imperfect fungi assigned to the form class Hyphomycetes. Fungi in this genus are anamorphs of ascomycetes, including members of the genus *Pleospora*. *Alternaria* species are characterized by very distinctive large multicellular dictyospores that have a beak and are produced in chains. Species of *Alternaria* occur as parasites on a number of crop plants, causing early blight or leaf spot diseases, or as saprobes on a large variety of organic substrates. Crop losses caused by *Alternaria* pathogens are less than losses due to more serious pathogens such as rusts or downy mildews; however, this genus is prominent in aerobiological literature since it is recognized as an important aeroallergen as well as a plant pathogen. Aerobiological surveys such as those conducted by the Aeroallergen Network of the American Academy of Allergy and Immunology (2) routinely report the presence of *Alternaria* conidia in the atmosphere. This cosmopolitan genus can frequently be found in atmospheric concentrations of several hundred to 1,000 or more spores per m³. In some areas, these spores may be in the atmosphere throughout the year. Peak concentrations are usually seen in the late summer or fall (38).

Many pathogenic species of *Alternaria*, including *A. solani* on potato and tomato, *A. brassicae* on members of the family *Brassicaceae*, *A. porri* on onions, and *A. alternata* on a variety of host species, have a worldwide distribution (60). In general, *Alternaria* pathogens often attack plants under stress, especially those stressed by drought, insect infestation, or senescence. Young seedlings are also susceptible to *Alternaria* infection but established to middle-aged plants are less vulnerable.

Alternaria spores are passively dispersed from infected leaves by moderate to strong gusty winds, with velocities of 2 to 3 m/s required for spore release. As a component of the dry air spora, dispersal typically occurs during midday, when conditions are warm and dry with high wind speeds. Greatest dispersal occurs during dry weather that immediately follows periods of rain or heavy dew. Prolonged dry windy periods will deplete spore reserves on the leaves and inhibit sporulation, which requires moist conditions. Spores can also disperse when washed or splashed from leaf surfaces by raindrops and by irrigation water; however, for *Alternaria* species, this method of dispersal is less important. By contrast, high humidity inhibits the release of spores from wet leaves, and airborne spores are washed from the atmosphere by rain and irrigation (60).

At present, there is no nationwide network of plant pathologists conducting air sampling for the detection of airborne pathogens. However, the Aeroallergen Network sponsored by the American Academy of Allergy and Immunology consists of a group of 74 certified air-sampling and reporting stations in the United States. The 1994 Aeroallergen Network report documented airborne fungal spore concentrations from 40 sampling stations. *Alternaria* conidia were among the top 10 spore types (in terms of atmospheric concentrations) reported from 37 stations, and at 15 stations *Alternaria* spores were among the top 5 (2). Although some of the airborne spores may be from saprobic species of *Alternaria*, these spores still represent a significant component of the air spora in the United States and may represent a potential threat to crops.

Rusts

There are about 6,000 species of rust fungi which attack a wide range of hosts among seed plants and cause some of the most destructive plant diseases. Millennia before scientists understood that pathogenic fungi could cause plant diseases, rust epidemics were studied, and the reddish lesions were noted on plants. These epidemics were recognized in ancient Greece and described in the writings of Aristotle and Theophrastus, who even noted that different plants varied in susceptibility to these diseases. Today, rust fungi still remain among the most serious agricultural pathogens. Some of the most important diseases of cereal crops are caused by these basidiomycetes, which have produced serious epidemics throughout history. In addition, coffee, apple, and pine trees as well as the economies dependent on these crops have also been devastated by rust fungi. The long-distance transport of rust fungi has been studied more extensively than that of any other fungal pathogen.

Coffee rust, caused by *Hemileia vastatrix*, destroyed the coffee plantations in Ceylon in the 1870s and 1880s and today threatens production wherever coffee is grown. In

1966, an outbreak of coffee rust in Angola produced spores that were apparently carried by favorable wind currents across the Atlantic Ocean. The spores were washed out by rainfall over coffee plantations in Brazil approximately 5 to 7 days later (51).

In North America, the most important rust pathogen is *Puccinia graminis* f. sp. *tritici*, the fungus responsible for stem rust of wheat (41). The organism has a complex life cycle that involves five spore stages on two separate host plants (wheat and barberry). Basidiospores which develop from overwintering teliospores are capable of infecting a young barberry leaf, eventually giving rise to spermatia and receptive hyphae and then aeciospores, also on the barberry leaf. The aeciospores transfer the infection to wheat, where the spores germinate, and hyphae enter the plant through stomata. Once the mycelium is established in wheat, the uredial stage develops within 2 weeks, appearing on the stem as long narrow lesions which produce dark red powdery masses of uredospores (also called urediniospores). Uredospores become airborne and reinfect new wheat plants to produce repeated generations of uredia. Near the end of the growing season, the uredia turn black as two-celled overwintering teliospores replace the uredospores (34, 39, 41).

The major vehicles for disease spread are the uredospores, which are easily carried by wind from one plant to another, giving rise to epidemics. In fact, they can be carried by prevailing winds for hundreds or thousands of kilometers (51). In the mild climate of northern Mexico and southern Texas, this stage can continue all winter and give rise to spring infections in northern states caused by spores carried by prevailing southerly winds. Uredospores can also be carried back to the south in late summer and fall. Initially demonstrated in 1923 by Stakman, the movement of rust uredospores along the "*Puccinia* pathway" in North America is one of the best-known examples of long-distance dispersal (53). In addition to *P. graminis*, *P. recondita*, *P. coronata*, and *Bipolaris maydis* as well as insect vectors of viral diseases may all be connected with dispersal along this pathway (51).

Wheat plants infected by *P. graminis* are severely weakened, but not destroyed, and the grain yield is significantly reduced. Wheat plants throughout the world are threatened, and it is estimated that worldwide over 1 million metric tons of wheat are lost annually as a result of stem rust. Evidence of the long-distance transport of *P. graminis* uredospores in other parts of the world was included in a review by Nagarajan and Singh (51). In Europe, two pathways have been studied: an eastern European path originating in Turkey and Romania and a western path from Morocco and Spain (51). Both pathways converge in the Scandinavian countries. In India, the uredospores of both *P. graminis* and *P. recondita* survive throughout the year in Nilgiri Hills in southern India. Dispersal to central and northern India occurs during November under the influence of tropical cyclones (51).

The long-distance dispersal of *P. graminis* uredospores between the eastern and western wheat-growing regions in Australia has been documented, and the overseas transport of stem rust has also been investigated (51). Two strains discovered in Australia were shown to be identical to strains found in South Africa in terms of pathogenicity and isozyme patterns. In addition, upper-air wind currents studied through the use of a weather balloon supported the possibility of this transport (51).

Successful long-distance dispersal of wheat rust uredospores is also dependent on source strength and viability. A mature uredium is capable of producing about 10,000 uredospores per day over a period of several weeks (58).

With a 5% disease severity of 50 uredia, a single plant would produce 500,000 uredospores per day. It has been estimated that a field of wheat with a moderate infection of stem rust would produce 4×10^{12} uredospores per day per ha (51). Nagarajan and Singh (51) reported that following long-distance transport, the spores had to be deposited within 120 h after takeoff to be infectious. Eversmeyer and Kramer (15) studied survival of *P. graminis* and *P. recondita* uredospores under a variety of temperature conditions in the field. At subfreezing temperatures during winter, no spores were viable after 96 h, but during spring, 10 to 20% of the inoculum was viable after 120 h and a fraction of 1% survived for 456 h. In environmental chamber experiments, spores remained viable for up to 864 h at constant temperatures of between 10 and 30°C. However, viability of spores exposed to freezing and subfreezing temperatures declined rapidly within a few hours of exposure (16). Although conditions in the atmosphere may be similarly harsh, the enormous numbers of spores produced will ensure that at least a small number of spores reach a suitable host.

Smuts

There are approximately 1,200 species of smut fungi classified in 50 genera of the Ustilaginales; however, the majority of species are classified into two large genera, *Tilletia* and *Ustilago* (67). Smuts are serious plant pathogens on cereal crops, causing millions of dollars in damages each year. Many native plants totaling approximately 4,000 host species are also affected (67).

The major dispersal phase of smut fungi consists of the asexual teliospores (64). Incredible numbers of teliospores develop from the mycelium, usually within galls, and are dispersed by wind. There is no repeating stage; each infected plant produces only one generation of teliospores. Many smuts overwinter as teliospores; however, other smut species overwinter as mycelia in infected grain (39).

Corn smut is a widespread disease occurring wherever corn is grown and is more prevalent on sweet corn than other varieties. The fungus *Ustilago maydis* can form galls on any aboveground plant part but is most conspicuous when the galls develop on the ear. The size and location of the galls reflect the degree of crop loss. Galls on the ear result in total loss, while those in other places reduce yield or cause stunted growth (64).

The fungus overwinters as teliospores on plant debris and in the soil. In spring, the teliospores germinate, each producing a basidium that forms basidiospores. These are carried by wind or splashed by rain to developing tissues of corn plants, either young seedlings or the growing tissue of older plants. Basidiospores germinate and invade the host tissue. The resulting mycelium grows prolifically and stimulates host cells to divide and form galls which may reach a size of 15 cm in diameter. As the galls mature, the interior darkens as the mycelium is converted into teliospores. The galls are initially covered by a membrane which later ruptures, releasing masses of dry spores (39).

Teliospores of various species are produced during different seasons. Hamilton (26) reported that airborne *Tilletia* spores peaked in late August and early September in England, while *Ustilago* spores were at their highest concentrations between mid-June and the end of July. The airborne concentrations of these two genera depend on a number of meteorological factors. Levels decrease during rainy or humid conditions and increase during periods of brilliant sun, strong winds, and high barometric pressure, as would be expected for members of the dry air spora (26).

Other studies have also documented the presence of smut spores in the atmosphere. Halwagy (25) reported that *Ustilago* spores were the second-most-common spore type identified in the Kuwait atmosphere. Misra (48) showed smut spores peaked in India during dry conditions during December and January, while Rubulis (61) reported that smut spores peaked during the late spring and fall in Sweden. In these two studies, the smuts were basically considered as one spore category, with no attempt to identify the species involved.

Crotzer and Levetin (13) examined the airborne concentration of smut teliospores in the Tulsa, Oklahoma, atmosphere during 1991 and 1992 and attempted to identify the most abundant smut spores. This study showed that teliospores occurred in the atmosphere on 100% of the days from May through October in 1991 and 1992. The average daily concentration generally ranged from 100 to 1,000 teliospores per m^3 during both seasons, with the mean for the 1991 season being 291 spores per m^3 and the peak being 1,874 spore per m^3, which occurred on 5 July. The mean for the 1992 season was 356 spores per m^3, with the peak of 5,906 spores per m^3 occurring on 12 May. Other data show that atmospheric smut spores are not limited to this 6-month period but are also prevalent earlier in spring and later in the fall (38). Although these spores were consistently present in the atmosphere during the period studied, their concentrations experienced many fluctuations due to daily variations in wind speed, precipitation, and relative humidity as well as phenologies of the host species and pathogens (13). Different species of smuts were observed at different times of the year. During May and June, prevalent teliospores included those species which infect Bermuda grass (*Ustilago cynodontis*), johnsongrass (*Sphacelotheca occidentalis*), oat (*U. kolleri*), and wheat (*U. tritici*). Smuts identified during September and October included *U. maydis*, which infects corn, and *U. brumivora* and *U. bullata*, which are pathogenic to several native Oklahoma grasses (13). Although some of the smuts identified in Tulsa were pathogens of native grasses, many others were important crop pathogens.

In addition to the significance of smut fungi as plant pathogens, it is recognized that smut teliospores can serve as potential aeroallergens since the atmosphere is often saturated with these spores for extended periods of time. In the 1994 report from the Aeroallergen Network (2), smut spores were among the top 10 spore types, in terms of atmospheric concentration, reported from 39 of 40 sampling stations that provide counts for fungal spores. At half of these stations, smut spores were in the top five (2). A few smut extracts are routinely used for diagnosis and desensitization, and many individuals have skin tested positive for these allergens and have elevated levels of immunoglobulin E antibodies (7, 63). However, the full extent of the allergenicity and clinical significance of various smut species is not known (13). At the 39 monitoring stations scattered across the United States, smut spores represent a significant component of the air spora and may represent a significant source of pathogen inoculum.

SUMMARY

Airborne fungal spores and other plant disease propagules are always present in the natural environment and pose a constant threat to food crops. With the rapidly expanding world population, diseases that threaten the food supply could result in widespread starvation. In addition, world food production must continue to increase in the coming decades in order to feed the projected population increases. For food production to increase, control of plant disease is essential. Aerobiological studies must be part of any effort to understand the distribution and epidemiology of agricultural pathogens that rely on air currents for dispersal.

Many aerobiological studies have been conducted on the local or long-distance spread of individual plant pathogens. Some of these studies are performed to develop mathematical or computer models for forecasting disease epidemics and predicting their onset and severity; for example, models of the spread of the tobacco blue mold pathogen have been successful in predicting the spread of this disease. For other pathogens, the lack of accurate spore dispersal data has often limited the development of epidemic models. These models require an understanding of all stages of transport, from spore takeoff to deposition within the crop, as well as knowledge of what percentage of the spores escape local deposition and what percentage are capable of surviving long-distance transport. Sensitivity of molecular techniques may help to answer some of the questions on long-distance transport.

Air sampling in the plant pathology community has generally been limited to individual studies for individual pathogens. There is no general monitoring for diverse pathogens. Although the importance of the *Puccinia* pathway for the dispersal of *P. graminis* uredospores is well known, there is no network of monitoring stations that could provide information of the advance of the pathogen or what other organisms utilize the same aerobiological pathway. Such networks exist in the aeroallergen research community. An important objective for the future might be a greater coordination among scientists interested in atmospheric transport.

REFERENCES

1. **Allitt, U.** 1986. Identity of airborne hyaline, one-septate ascospores and their relation to inhalant allergy. *Trans. Br. Mycol. Soc.* **87:**147–154.
2. **American Academy of Allergy and Immunology.** 1995. *1994 Pollen and Spore Report.* American Academy of Allergy and Immunology, Milwaukee, Wis.
3. **Aylor, D. E.** 1990. The role of intermittent wind in the dispersal of fungal pathogens. *Annu. Rev. Phytopathol.* **28:** 73–92.
4. **Aylor, D. E.** 1995. Vertical variation of aerial concentration of *Venturia inaequalis* ascospores in an apple orchard. *Phytopathology* **85:**175–181.
5. **Banttari, E. E., and J. R. Venette.** 1980. Aerosol spread of plant viruses: potential role in disease outbreaks. *Ann. N. Y. Acad. Sci.* **353:**167–173.
6. **Barras, F., F. van Gijsegem, and A. K. Chatterjee.** 1994. Extracellular enzymes and pathogenesis of soft-rot *Erwinia*. *Annu. Rev. Phytopathol.* **32:**201–234.
7. **Burge, H. A.** 1985. Fungus allergens. *Clin. Rev. Allergy* **3:** 319–329.
8. **Burge, H. A.** 1986. Some comments on the aerobiology of fungus spores. *Grana* **25:**143–146.
9. **Burnet, P. A. (ed.).** 1989. *Barley Yellow Dwarf Virus, the Yellow Plague of Cereals.* Centro Internacional de Mejoramiento de Maiz y Trigo, Mexico City, Mexico.
10. **Conn, K. E., J. M. Ogawa, B. T. Manji, and J. E. Adaskaveg.** 1995. *Leuconostoc mesenteroides*, the first report of a coccoid bacterium causing a plant disease. *Phytopathology* **85:**593–599.
11. **Cooper, J. I., S. E. Kelley, and P. R. Massalski.** 1988. Virus-pollen interactions. *Adv. Dis. Vector Res.* **5:** 221–249.

12. **Cox, C. S.** 1987. *The Aerobiological Pathway of Microorganisms.* John Wiley & Sons, New York.

13. **Crotzer, V., and E. Levetin.** The aerobiological significance of smut spores in Tulsa, Oklahoma. *Aerobiologia,* in press.

14. **Davis, J. M.** 1987. Modeling the long-range transport of plant pathogens in the atmosphere. *Annu. Rev. Phytopathol.* **25:**169–188.

15. **Eversmeyer, M. G., and C. L. Kramer.** 1994. Survival of *Puccinia recondita* and *P. graminis* urediniospores as affected by exposure to weather conditions at one meter. *Phytopathology* **84:**332–335.

16. **Eversmeyer, M. G., and C. L. Kramer.** 1995. Survival of *Puccinia recondita* and *P. graminis* urediniospores exposed to temperatures from subfreezing to 35°C. *Phytopathology* **85:**161–164.

17. **Fitt, B. D. L., and H. A. McCartney.** 1986. Spore dispersal in relation to epidemic models, p. 311–345. *In* K. J. Leonard and W. E. Fry (ed.), *Plant Disease Epidemiology,* vol. I. Macmillan Publishing Co., New York.

18. **Fitt, B. D. L., H. A. McCartney, and P. J. Walklate.** 1989. The role of rain in dispersal of pathogen inoculum. *Annu. Rev. Phytopathol.* **27:**241–270.

19. **Franc, G.** 1994. Atmospheric transport of *Erwinia carotovora,* preprint vol., p. 435–437. *In Proceedings of the 11th Conference of Biometeorology and Aerobiology.* American Meteorological Society, Boston.

20. **Frinking, H.** 1993. A historical perspective on aerobiology as a discipline, p. 23–28. *In* S. A. Isard (ed.), *Alliance for Aerobiology Research Workshop Report.* Illinois Natural History Survey, Champaign.

21. **Fry, W. E., S. B. Goodwin, J. M. Matuszak, L. J. Spielman, M. G. Milgroom, and A. Drenth.** 1992. Population genetics and intercontinental migrations of *Phytophthora infestans. Annu. Rev. Phytopathol.* **30:**107–129.

22. **Goodman, R. N., Z. Kiraly, and K. R. Wood.** 1986. *The Biochemistry and Physiology of Plant Disease.* University of Missouri Press, Columbia.

23. **Govi, G.** 1992. Aerial diffusion of phytopathogenic fungi. *Aerobiologia* **8:**84–93.

24. **Gregory, P. H.** 1973. *The Microbiology of the Atmosphere,* 2nd ed. Halstead Press, New York.

25. **Halwagy, M.** 1989. Seasonal airspora at three sites in Kuwait 1977–1982. *Mycol. Res.* **93:**208–213.

26. **Hamilton, E. D.** 1959. Studies on the air spora. *Acta Allergol.* **13:**143–175.

27. **Heesterbeek, J. A. P., and J. C. Zadoks.** 1987. Modelling pandemics of quarantine pests and diseases; problems and perspectives. *Crop Prot.* **6:**211–221.

28. **Hirano, S. S., and C. D. Upper.** 1990. Population biology and epidemiology of *Pseudomonas syringae. Annu. Rev. Phytopathol.* **28:**155–177.

29. **Huber, L., and T. J. Gillespie.** 1992. Modeling leaf wetness in relation to plant disease epidemiology. *Annu. Rev. Phytopathol.* **30:**553–577.

30. **Irwin, M. E., and J. M. Thresh.** 1990. Epidemiology of barley yellow dwarf: a study in ecological complexity. *Annu. Rev. Phytopathol.* **28:**393–424.

31. **Jeger, M. J.** 1984. Damage and loss in fruit orchards caused by airborne fungal pathogens, p. 225–236. *In* R. K. S. Wood and G. J. Jellis (ed.), *Plant Diseases: Infection, Damage and Loss.* Blackwell Scientific Publications, Oxford.

32. **Johnson, G. I.** 1989. *Peronospora hyoscyami* de Bary: taxonomic history, strains, and host range, p. 1–18. *In* W. E. McKeen (ed.), *Blue Mold of Tobacco.* APS Press, St. Paul, Minn.

33. **Kelman, A.** 1995. Contributions of plant pathology to the biological sciences and industry. *Annu. Rev. Phytopathol.* **33:**1–21.

34. **Kendrick, B.** 1992. *The Fifth Kingdom,* 2nd ed. Mycologue Publications, A Focus Text, Newburyport, Mass.

35. **Lacey, J.** 1991. Aerobiology and health: the role of airborne fungal spores in respiratory disease, p. 157–186. *In* D. L. Hawksworth (ed.), *Frontiers in Mycology.* C. A. B. International, Kew, United Kingdom.

36. **Lacey, J., and J. Venette.** 1995. Outdoor air sampling techniques, p. 407–471. *In* C. S. Cox and C. M. Wathes (ed.), *Bioaerosols Handbook.* Lewis Publishers, Boca Raton, Fla.

37. **Levetin, E.** 1991. Identification and concentration of airborne basidiospores. *Grana* **30:**123–128.

38. **Levetin, E.** 1995. Unpublished data.

39. **Levetin, E., and K. McMahon.** 1996. *Plants and Society.* Wm. C. Brown Publishers, Inc., Dubuque, Iowa.

40. **Lighthart, B., J. C. Spendlove, and T. G. Akers.** 1979. Factors in the production, release, and viability of biological particles: bacteria and viruses, p. 11–22. *In* R. L. Edmonds (ed.), *Aerobiology: the Ecological Systems Approach.* Dowden, Hutchinson & Ross, Stroudsburg, Pa.

41. **Littlefield, L. J.** 1981. *Biology of the Plant Rusts: an Introduction.* Iowa State University Press, Ames.

42. **MacKenzie, D. R., V. J. Elliott, B. A. Kidney, E. D. King, M. H. Royer, and R. L. Theberge.** 1983. Application of modern approaches to the study of the epidemiology of diseases caused by *Phytophthora,* p. 303–313. *In* D. C. Erwin, S. Bartnicki-Garcia, and P. H. Tsao (ed.), *Phytophthora: Its Biology, Taxonomy, Ecology, and Pathology.* APS Press, St. Paul, Minn.

43. **Main, C. E., and J. M. Davis.** 1989. Epidemiology and biometeorology of tobacco blue mold, p. 201–216. *In* W. E. McKeen (ed.), *Blue Mold of Tobacco.* APS Press, St. Paul, Minn.

44. **Main, C. E., T. Keever, and J. M. Davis.** 1995. Personal communication.

45. **Marthi, B.** 1994. Resuscitation of microbial bioaerosols, p. 192–225. *In* B. Lighthart and A. J. Mohr (ed.), *Atmospheric Microbial Aerosols: Theory and Applications.* Chapman and Hall, New York.

46. **Mason, C. J.** 1979. Atmospheric transport, p. 85–95. *In* R. L. Edmonds (ed.), *Aerobiology: the Ecological Systems Approach.* Dowden, Hutchinson & Ross, Stroudsburg, Pa.

47. **Mink, G. I.** 1993. Pollen- and seed-transmitted viruses and viroids. *Annu. Rev. Phytopathol.* **31:**375–402.

48. **Misra, R. P.** 1987. Studies on seasonal and diurnal variation in the occurrence of airborne spores of basidiomycetes. *Perspect. Mycol. Res.* **1:**243–252.

49. **Moore-Landecker, E.** 1990. *Fundamentals of the Fungi,* 3rd ed. Prentice-Hall, Englewood Cliffs, N.J.

50. **Mundt, C. C.** 1995. Models from plant pathology on the movement and fate of new genotypes of microorganisms in the environment. *Annu. Rev. Phytopathol.* **33:**467–488.

51. **Nagarajan, S., and D. V. Singh.** 1990. Long-distance dispersion of rust pathogens. *Annu. Rev. Phytopathol.* **28:**139–153.

52. **Niederhauser, J. S.** 1993. International cooperation in potato research and development. *Annu. Rev. Phytopathol.* **31:**1–21.

53. **Pedgley, D. E.** 1986. Long distance transport of spores, p. 346–365. *In* K. J. Leonard and W. E. Fry (ed.), *Plant Disease Epidemiology,* vol. I. Macmillan Publishing Co., New York.

54. **Plumb, R. T.** 1983. Barley yellow dwarf virus—a global problem, p. 185–198. *In* R. T. Plumb and J. M. Thresh (ed.), *Plant Virus Epidemiology.* Blackwell Science, London.

55. **Ponti, I., and P. Cavanni.** 1992. Aerobiology in plant protection. *Aerobiologia* **8:**94–101.

56. **Portnoy, J., J. Chapman, H. Burge, M. Muilenberg, and W. Solomon.** 1987. *Epicoccum* allergy: skin reaction patterns and spore/mycelium disparities recognized by IgG and IgE ELISA inhibition. *Ann. Allergy* **59:**39–43.

57. **Rajasab, A. H., and H. T. Chawda.** 1994. Dispersal of the conidia of *Colletotrichum gloeosporioides* by rain and the development of anthracnose on onion. *Grana* **33:**162–165.

58. **Roelfs, A. P.** 1985. Wheat and rye stem rust, p. 3–37. *In* A. P. Roelfs and W. R. Bushnell (ed.), *The Cereal Rusts*, vol. II. *Diseases, Distribution, Epidemiology, and Control*. Academic Press, Inc., Orlando, Fla.

59. **Romantschuk, M.** 1992. Attachment of plant pathogenic bacteria to plant surfaces. *Annu. Rev. Phytopathol.* **30:** 225–243.

60. **Rotem, Joseph.** 1994. *The Genus Alternaria*. APS Press, St. Paul, Minn.

61. **Rubulis, J.** 1984. Airborne fungal spores in Stockholm and Eskilstuna, central Sweden, *Nord. Aerobiol.* **1984:**85–93.

62. **Salmond, G. P. C.** 1994. Secretion of extracellular virulence factors by plant pathogenic bacteria. *Annu. Rev. Phytopathol.* **32:**181–200.

63. **Santilli, J., W. J. Rockwell, and R. P. Collins.** 1985. The significance of the spores of the basidiomycetes (mushrooms and their allies) in bronchial asthma and allergic rhinitis. *Ann. Allergy* **55:**469–471.

64. **Schumann, G.** 1991. *Plant Diseases: Their Biology and Social Impact*. APS Press, St. Paul, Minn.

65. **Starr, M. P.** 1984. Landmarks in the development of phytobacteriology. *Annu. Rev. Phytopathol.* **22:**169–188.

66. **Stetzenbach, L. D., S. C. Herm, and R. J. Seidler.** 1992. Field sampling design and experimental methods for the detection of airborne microorganisms, p. 543–555. *In* M. A. Levin, R. J. Seidler, and M. Rogul (ed.), *Microbial Ecology: Principles, Methods, and Applications*. McGraw-Hill, New York.

67. **Vanky, K.** 1987. *Illustrated Genera of Smut Fungi*. Gustav Fischer Verlag, Stuttgart, Germany.

68. **Wiglesworth, M. D., W. C. Nesmith, C. L. Schardl, S. Li, and M. R. Siegel.** 1994. Use of specific repetitive sequences in *Peronospora tabacina* for the early detection of the tobacco blue mold pathogen. *Phytopathology* **84:** 425–430.

69. **Wilson, M., and S. E. Lindow.** 1993. Release of recombinant microorganisms. *Annu. Rev. Microbiol.* **47:**913–944.

70. **Zadoks, J. C., and F. van den Bosch.** 1994. On the spread of plant disease: a theory of foci. *Annu. Rev. Phytopathol.* **32:**503–521.

BIOTRANSFORMATION AND BIODEGRADATION

VOLUME EDITOR
MICHAEL V. WALTER

SECTION EDITOR
RONALD L. CRAWFORD

Overview: Biotransformation and Biodegradation

MICHAEL V. WALTER AND RONALD L. CRAWFORD

This section on biotransformation and biodegradation is a compilation of some areas that the editors felt would be of interest, because of their relevance and importance, to those working in the field of bioremediation. The section consists of several chapters that describe general techniques or environments that are pertinent to the bioremediation of target pollutants. Other chapters discuss methods for measuring various aspects of bioremediation, such as determining biodegradability of a target pollutant or bioavailability of the contaminating compound(s), and how these factors affect the outcome of bioremediation. Three of the chapters describe some techniques that can be used to stimulate the process of biodegradation. A third of the chapters describe methods used in the biodegradation of specific classes of compounds and removal of metals from soil. While biodegradation is usually viewed in a positive light, one chapter describes the mechanisms of corrosion due to microbial activity and how to prevent or control it.

The hazard of pollutants can be reduced by conventional technologies which involve removal, alteration, or isolation of the pollutant. Such techniques typically consist of excavation followed by incineration or containment. However, these technologies are expensive, and in many cases they do not destroy the contaminating compound(s) but rather transfer them from one environment or form to another (Fig. 1). Bioremediation addresses the limitations of more conventional techniques by bringing about the actual destruction of many organic contaminants at reduced cost. As a result, over the last 20 years, bioremediation has grown from a virtually unknown technology to a technology that is considered for the cleanup of a wide range of contaminating compounds.

Bioremediation is the result of the biological breakdown, or biodegradation, of contaminating compounds. While a precise definition of biodegradation is nonexistent, the process generally involves the breakdown of organic compounds, usually by microorganisms, into more cell biomass and less complex compounds, and ultimately to water, and either carbon dioxide or methane (4). The extent of biodegradation and the rate at which it occurs depend on interactions between the environment, the number and type of microorganisms present, and the chemical structure of the contaminant(s) being degraded. Therefore, for a bioremediation effort to be successful, many factors must be taken into account (Fig. 2).

Biodegradation of most organic pollutants occurs at a faster rate under aerobic conditions, i.e., when oxygen is present for use as a final electron acceptor. A common misconception is that oxygen is readily available in soils. Because of low soil porosity or locations that are well below the surface, oxygen availability is very often the limiting factor. Developing methods of supplying oxygen to remote subsoil locations has become a very active research area within the field of bioremediation.

Biodegradation of many organic compounds will occur under anaerobic conditions, i.e., in the absence of oxygen, although the rate may not be as rapid as observed under aerobic conditions. Certain microorganisms are able to use compounds such as nitrate (NO_3), sulfate (SO_4^{2-}), or iron (Fe^{3+}) as final electron acceptors instead of oxygen. Research into this area has also identified microorganisms that use unique compounds such as manganese as final electron acceptors. However, biodegradation of some compounds, such as halogenated hydrocarbons, can be faster, at least initially, under anaerobic conditions. Techniques have been developed to measure the biodegradability of environmental pollutants under anaerobic conditions.

The biodegradation of compounds without excavation or removal is referred to as in situ bioremediation. In situ bioremediation, while often very difficult to sustain, is the preferred method of bioremediation because of lower costs. If the appropriate environmental conditions, nutrients (usually nitrogen, phosphate, and sulfur), and microorganisms are present, intrinsic bioremediation will occur. Intrinsic bioremediation is the biodegradation of a target pollutant without intervention. Unfortunately, because of time limits or suboptimal conditions, the rate of intrinsic bioremediation is usually not rapid enough from a regulatory standpoint. When this occurs, enhanced (engineered) bioremediation is required.

Enhanced bioremediation involves increasing the rate of biodegradation, which can be accomplished in two ways. One method is to supply required nutrients to the indigenous population (biostimulation). The other method involves inoculating microorganisms capable of degrading the target pollutant(s), either with or without nutrients, into the contaminated environment, thus augmenting the indigenous microbial population (bioaugmentation). While in situ bioremediation is often the preferred method, it is not always possible to achieve the desired results. In these situa-

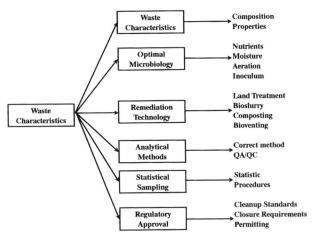

FIGURE 1 Parameters involved in the successful use of bioremediation and how they can influence successful implementation of the technology. QA/QC, quality assurance/quality control. Reprinted with permission from reference 5. Copyright Lewis Publishers, an imprint of CRC Press, Boca Raton, Fla., 1994.

tions, the contaminated soil is excavated and biodegraded by using solid-phase treatment or slurry reactors. Solid-phase treatment includes use of land farming, composting, soil windrows, static soil pile systems, or vessel reactors. Slurry-phase systems are carried out in lagoons or designed reactors (3).

Biodegradation is often a metabolic process which involves the complete breakdown of an organic compound into its inorganic components, a process commonly referred to as mineralization. Compounds having a molecular structure to which microorganisms have not been exposed (xenobiotic compounds) are resistant to bioremediation (recalcitrant) or are biodegraded incompletely.

Many xenobiotic compounds, such as haloaromatic com-

pounds, undergo biotransformation as opposed to biodegradation. Biotransformation is the chemical alteration of the molecular structure of organic or inorganic compounds, resulting in a different complexity or the loss of some characteristic property with no loss of molecular complexity (4). Biotransformation produces change in the structure of an organic compound, which may affect the toxicity and mobility of the original compound. Microbial populations growing on one compound may fortuitously transform a contaminating chemical, a process known as cometabolism. The microbial population does not increase in biomass or number as a result of the contaminating chemical and will not grow if the contaminant is the only substrate present (7).

Initially, biodegradation was most commonly associated with bacteria. However, fungi which are capable of degrading a wide range of environmental pollutants, most notably the polyaromatic hydrocarbons, DDT, trinitrotoluene, and pentachlorophenol, are currently being isolated (2). Many of the enzymes involved in contaminant biodegradation are produced by the fungi to degrade lignin, a naturally occurring plant polymer associated with cellulose and hemicellulose (1).

Research in biodegradation has demonstrated the existence of microorganisms capable of degrading chemicals such as some polychlorinated biphenyls and nitroaromatic compounds, once thought to be recalcitrant or subject only to biotransformation. The metabolic pathways and environmental conditions required to achieve biodegradation of these compounds are often unique to the particular compound. We have presented methods used to achieve biodegradation of these compounds in separate chapters. Unfortunately, practical considerations preclude discussing all of them. We feel that most of the prevalent compounds or classes of compounds are represented.

REFERENCES

1. **Atlas, R. M., and R. Bartha.** 1993. Biogeochemical cycling: carbon, hydrogen, and oxygen, p. 289–313. *In Microbial Ecology. Fundamentals and Applications,* 3rd ed. The Benjamin/Cummings Publishing Co., Redwood City, Calif.
2. **Bumpus, J. A.** 1993. White rot fungi and their potential use in soil bioremediation processes, p. 65–100. *In* J.-M. Bollag and G. Stotzky (ed.), *Soil Biochemistry,* vol. 8. Marcel Dekker, Inc., New York.
3. **Cookson, J. T.** 1995. Solid and slurry-phase bioremediation, p. 305–358. *In Bioremediation Engineering. Design and Application.* McGraw-Hill Inc., New York.
4. **Hoeppel, R. E., and R. E. Hinchee.** 1994. Enhanced biodegradation for on-site remediation of contaminated soils and ground water. *Environ. Sci. Pollut. Control Ser.* **6:**311–431.
5. **Huesemann, M.** 1994. Guidelines for land-treating petroleum hydrocarbon-contaminated soil. *J. Soil Contam.* **3:** 299–318.
6. **Leahy, M., and R. Brown.** 1994. Bioremediation: optimizing results. *Chem. Eng.* **101**(May):108–114.
7. **Reineke, W.** 1984. Microbial degradation of halogenated aromatic compounds, p. 319–353. *In* D. T. Gibson (ed.), *Microbial Degradation of Organic Compounds.* Marcel Dekker, Inc., New York.

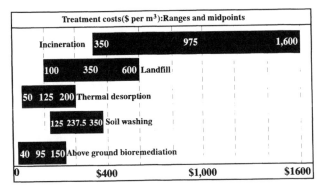

FIGURE 2 Relative costs for various remediation technologies. Reprinted from *Chemical Engineering* (6), with permission of the publisher.

Methods for Determining Biodegradability

EUGENE L. MADSEN

77

OVERVIEW

This chapter conveys basic principles pertinent to the design, implementation, and interpretation of protocols for determining biodegradability. Concepts presented in this chapter are applicable to all environmental contamination problems whose solutions can be addressed microbiologically. Although microbiological processes alter inorganic substances (i.e., heavy metals, acid mine drainage, cyanides, uranium, and metal alloy surfaces [chapters 37, 89, 93, and 94, this volume, and references 5, 7, 11, 28, 31, 32, 35, 43, 48, 62, 64, 66, 68]), the focus here is primarily on assessing the biodegradability of organic compounds. This emphasis reflects both my experience and a need to limit the scope of the chapter.

This chapter is complemented by others in this volume, including those that emphasize environmental sampling (chapters 4, 41–45, 59, and 68), growth of particular microorganisms (chapters 6–9, 16, 17, 43–45, 57, and 72–74), metabolic reactions in aquatic and subsurface habitats (chapters 26, 27, 33–36, 49–51, and 61–63), biodegradation of particular substances (chapters 79, 80, 84–86, 90, and 91), and the use of specific methodologies (chapters 5, 10–12, 31, 33, 46, 48, and 52–54). Chapters in *Methods for General and Molecular Bacteriology* (21a) that complement this chapter include those describing microbial growth and physiology (chapters 7, 8, and 11).

Evolutionary Role of Microorganisms in the Biosphere

The small size, ubiquitous distribution, high specific surface area, potentially high rate of metabolic activity, genetic malleability, potentially rapid growth rate, and unrivaled enzymatic and nutritional versatility of microorganisms cast them in the role of recycling agents for the biosphere. The vast diversity and quantities of inorganic (e.g., S^0, NH_3, H_2, and CH_4) and organic (e.g., carbohydrates, fats, proteins, lipids, nucleic acids, and hydrocarbons) materials present on Earth's surface are disseminated between a matching diversity of habitats whose physical and chemical characteristics span wide ranges of pH, temperature, salinity, oxygen tension, redox potential, water potential, etc. (13, 16a, 56, 70). It is appealing to speculate that this distribution of resources between a variety of environments was the source of selective pressure during the evolutionary diversification of microorganisms. The end product of this evolution is a microbial world capable of exploiting virtually all of the naturally occurring (and many of the synthetic) metabolic resources on Earth. Physiological exploitation of resources by microorganisms allows them to survive and grow. This simple growth and survival of microorganisms drives biogeochemical cycling of the elements and simultaneously maintains the conditions required for life by other inhabitants of the biosphere. As a result of microbial decay processes, essential nutrients (e.g., carbon, nitrogen, and phosphorus) present in the biomass of one generation of biota are made available to the next. Understanding the detailed microbiological mechanisms of the maintenance of ecosystems provides both practical and intellectual challenges for inquiries into environmental microbiology in general and biodegradation processes in particular.

Definitions of Biodegradability and Biodegradation

"Biodegradability" is a noun that embodies qualities representing the susceptibility of substances to alteration by microbial processes. The substances may be organic or inorganic. The alteration may be brought about by (i) intra- or extracellular enzymatic attack that is essential for growth of the microorganism(s) (e.g., the attacked substances are used as a source of carbon, energy, nitrogen, or other nutrients or as a final electron acceptor), (ii) enzymatic attack that is beneficial because it serves some protective purpose (e.g., mobilization of toxic mercury away from the vicinity of the cells), (iii) enzymatic attack that is of no detectable benefit to the microorganism (e.g., cometabolic reactions in which a physiologically useful primary substrate induces production of enzymes that fortuitously alter the molecular structure of another compound), and (iv) nonenzymatic reactions stemming from by-products of microbial physiology that cause geochemical change (e.g., consumption of O_2, production of fermentation by-products, or an alteration in pH).

Biodegradation of organic compounds is the partial simplification or complete destruction of their molecular structure by physiological reactions catalyzed by microorganisms (1, 2, 4, 23, 36, 69a). Biodegradation is routinely measured by applying chemical and physiological assays to laboratory incubations of flasks containing pure cultures of microorganisms, mixed cultures, or environmental samples (e.g., soil, water, sediment, or industrial sludges). When attempting to measure biodegradation or judge the biodegradability of

Environmental Context

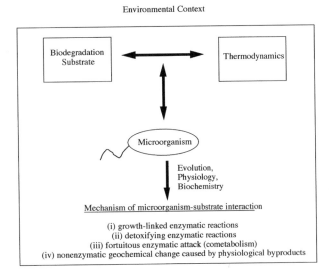

FIGURE 1 Conceptual overview of major determinants of biodegradability. The mechanisms by which microorganisms catalyze biodegradation reactions are defined by interactions between the context-dependent thermodynamic stability of the substrate and the evolution of physiological capabilities of microorganisms.

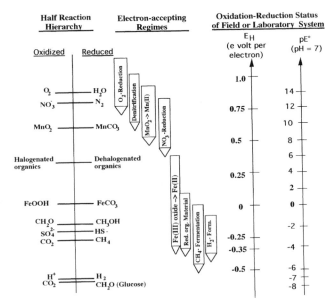

FIGURE 2 The hierarchy of final electron acceptors provides a simple means to integrate the thermodynamics, microbiology, and physiology of biogeochemical oxidation-reduction reactions. (See text for details. Modified from reference 70, Copyright © 1988. Reprinted by permission of John Wiley & Sons, Inc.)

substances, the investigator must define the environmental context so that potential reactants, reactions, and products can be identified. Microorganisms can catalyze only reactions that are thermodynamically possible. Furthermore, reaction mechanisms are largely constrained by precedents set during the evolution of physiological and biochemical functions. Because microbial evolution and biochemical research are ongoing, our understanding of mechanisms by which microorganisms degrade substrates continues to expand. Figure 1 provides a conceptual overview of major determinants of biodegradability.

Thermodynamics of Microbial Metabolic Processes

Zehnder and Stumm (70) presented an incisive and unifying means of understanding the diversity of microbial metabolic processes. Figure 2 graphically depicts the relationship between reduced and oxidized substrates as a vertically arranged hierarchy of oxidation-reduction half reactions. The vertical axes in Fig. 2 are E_h and, equivalently, pE. Compounds on the left of the half-reaction hierarchy are in an oxidized state, while those on the right are in the reduced form. Furthermore, the transition from oxidized to reduced forms is governed by the redox status of the system of interest and by catalytic mechanisms of microbially produced enzyme systems. Highly oxidizing conditions appear in the upper portion of the hierarchy in Fig. 2, while highly reducing conditions are listed in the lower portion. Figure 2 can be used to predict which combinations of half-reaction pairs are thermodynamically possible because under standard conditions the lower reaction proceeds leftward (electron producing) and the upper reaction proceeds rightward (electron accepting). Graphically, pairs of thermodynamically favorable half reactions can be linked simply by drawing arrows diagonally from the lower right to the upper left portion of the hierarchy. Fundamental reactions of the carbon cycle tie the oxidation of photosynthetically produced organic carbon (e.g., CH_2O; lower right of the hierarchy in

Fig. 2) to the variety of final electron acceptors that may be present in natural habitats (O_2, NO_3^-, Mn^{4+}, Fe^{3+}, SO_4^{2-}, CO_2). Each of these coupled half reactions is mediated by microorganisms. Moreover, when diagonal arrows directing carbohydrate oxidation to the reduction of these electron acceptors are drawn, the length of each arrow is proportional to the free energy gained by the microorganisms. Thus, microorganisms metabolizing carbohydrates with O_2 as a final electron acceptor are able to generate more ATP than those carrying out nitrate respiration. These microorganisms, in turn, gain more energy than those using Mn^{4+} and Fe^{3+} as final electron acceptors. This pattern continues down the hierarchy of electron-accepting regimes until methanogenesis (CO_2 as the final electron acceptor) is reached. There is a three-way convergence between the thermodynamics of half reactions, the physiology of microorganisms, and the presence of geochemical constituents actually found in field sites (or added to flasks used in biodegradation studies). It is notable that synthetic halogenated compounds (such as tetrachloroethylene and polychlorinated biphenyls) are also present in the hierarchy depicted in Fig. 2. These compounds are utilized as final electron acceptors by microorganisms. Knowledge of the reduction of halogenated compounds (and their locations in the hierarchy of final electron acceptors) plays an important role in strategies for testing biodegradability and eliminating environmental pollutants from field sites. Thus, final electron acceptors that dominate the physiological reactions of microorganisms provide useful criteria for categorizing both the biogeochemical regimes in field sites and physiological conditions examined in laboratory biodegradation experiments.

Biodegradation Contrasted with Traditional Microbial Physiology

Measurements of biodegradability routinely include cell growth, substrate loss, consumption of final electron acceptors, and production of both intermediary metabolites and final metabolic end products. These types of measures have been developed and traditionally applied by microbial physiologists to pure cultures of microorganisms in laboratory-prepared liquid media containing high concentrations (1 to 10 g/liter) of simple sugars and other growth substances (see chapters 7, 8, 10, and 11 in reference 21a). The term "biodegradability testing" (rather than "physiological assays") is warranted simply because (i) the measures are often applied to field sites or field site-derived samples (water, soil, sediment, industrial effluent, etc.); (ii) the substrates of interest are environmental pollutants that show little structural resemblance to substrates traditionally studied by physiologists; (iii) metabolism of the pollutants is often studied at low, environmentally relevant concentrations (≤ 1 ppm); (iv) when naturally occurring microbial communities are the objects of study, the populations of organisms responsible for the metabolic reactions are almost always unknown; and (v) when studies of pollutant metabolism by pure cultures are conducted, the organisms have usually been isolated and selected from field habitats on the basis of biodegradation capabilities of the cultures.

PRINCIPLES FOR MEASURING BIODEGRADABILITY

Biodegradation methodologies are designed to confirm, demonstrate, and explore both the net chemical changes and the associated intracellular details pertinent to how microorganisms influence the fate of organic contaminants. The procedures span a broad range of disciplines and sophisticated protocols. Figure 3 provides an overview of the variety of objectives, disciplines, and protocols that play key roles in biodegradation research. The two phases which serve as main divisions in Figure 3 result from the degree to which scientific detail is pursued. Phase 1 procedures treat samples of soil, sediments, water, or industrial effluents simply as "black boxes" which do or do not make contaminant compounds disappear, as judged by analytical chemical criteria. Phase 2 begins with the isolation of pure cultures of contaminant-degrading microorganisms. Once these have been obtained, then refined physiological and enzymatic assays may be performed. Furthermore, as DNA sequences of genes that code for metabolic pathways become increasingly available, molecular procedures will continue to gain prominence in biodegradation protocols. Among the final goals

Phase 1: Laboratory enrichment and demonstration of net metabolic activity

1. Soil, sediment, water, or industrial sludge in field site
↓
2. Aseptically remove, contain, transport to laboratory
↓
3. Divide into replicate live and abiotic treatments
↓
4. If appropriate, add radiolabeled or unlabeled organic compound of interest
↓
5. Use tools of analytical-chemistry or physiology to periodically measure consumption of parent compound and coreactants or production of metabolites or physiological endproducts
↓
6. Compare time-course data obtained from live and abiotic treatments
↓
7. Interpret and consider more reductionistic procedures shown in Phase 2

↓

Phase 2: Isolation of pure cultures and examination of physiological, biochemical, and molecular basis of pollutant metabolism

8. Isolate pure cultures capable of expressing metabolic activity determined in Phase 1
↓
9. Characterize growth, cell yield, sequential induction, and other physiological characteristics of the microorganisms during pollutant metabolism
↓
10. Extract and identify metabolites, enzymes, and cofactors associated with pollutant metabolism
↓
11. Cell-free examination of metabolites, enzymes, and cofactors
↓
12. Determine portion of genomic or plasmid DNA that codes for pollutant metabolism by screening a cloned DNA library, by transposon mutagenesis, or other procedures
↓
13. Conduct hybridization, restriction mapping, and sequencing DNA analyses seeking open reading frames, phylogenetic relationships to similar genes, and other key insights
↓
14. Elucidate details of gene expression and regulation via a variety of genetic and molecular techniques that include transposon mutagenesis, construction of expression clones, insertional inactivation, and inducer/reporter experiments

FIGURE 3 Two phases of procedures for understanding biodegradation processes. Phase 1 begins with environmental samples. Phase 2 proceeds through biochemical and molecular aspects of pollutant metabolism by single microorganisms.

of the procedures shown in phase 2 is understanding the molecular basis for gene expression and regulation.

Design and Implementation of Biodegradation Assays using Environmental Samples and Pure Cultures

The traditional black-box approach to biodegradation assays asks the question, Are microorganisms within this complex microbial community (e.g., derived from soil, water, sediment, or industrial sludge) able to metabolize the compound of interest? To answer this question, one aseptically gathers samples from a given field location, dispenses known weights or volumes of the samples to replicated vessels, handles the samples in a variety of ways that include a treatment that has been either sterilized or poisoned, incubates the test samples under laboratory conditions, and employs both chemical and physiological assays that monitor the fate of the test compound within experimental vessels over time (Fig. 3, phase 1). The objective of this general experimental design for biodegradation procedures using environmental samples or pure cultures is remarkably simple, yet there is a substantial series of obstacles that must be overcome before one obtains clear data that truly test a given set of specific hypotheses. Every design parameter selected for inclusion in a biodegradation assay can influence the resultant data. Therefore, decisions made in implementing biodegradation assays should be well reasoned. Table 1 summarizes many of the practical and theoretical decisions that must be made in developing biodegradation protocols. Step 1, a background issue considering information use, is fundamental to all related experimental decisions. The degree to which experimental minutiae of a given testing protocol must be initially considered is commensurate with the scrutiny that the final data will undergo. Artifacts and biases in data are virtually unavoidable in biodegradation assays (see below); thus, it may be wise to simply accept methodological limitations rather than worry about initial potential technical design flaws that may later have no practical impact.

Once the reason for conducting the biodegradation assay has been put in perspective (step 1), then another background issue, that of physiological conditions, should be confronted. Step 2 appears in Table 1 to acknowledge the fact that biodegradation is only a small portion of the perhaps thousands of physiological reactions simultaneously occurring when both pure cultures and mixed microbial populations in environmental samples are incubated in the laboratory. These physiological processes feed one another, interact in complex ways, and can be governed by many of the sometimes inadvertent physical and chemical manipulations made while preparing, incubating, and sampling assay vessels. Uncertainties become particularly striking when one is attempting to troubleshoot failed attempts to demonstrate biodegradation activity. The interplay between fundamental knowledge of physiology and experimental design parameters demands that a variety of issues be confronted: (i) the mechanism by which the compound is metabolized (e.g., as a carbon source, as a nitrogen source, or as a cometabolic substrate whose transformation will occur only when another compound is supplied); (ii) inclusion versus exclusion of potential growth-limiting vitamins and minerals; (iii) inclusion versus exclusion of air in the headspace of the reaction vessel; (iv) the solid-to-liquid ratios used in test vessels containing soil, sediments, or sludges; (v) the multiple roles of compounds in physiological reactions (for instance, nitrate can serve as both a nitrogen source and a final electron acceptor); and (vi) the fact that the compound whose biodegradation is being tested may be toxic at high concentrations or fall below some minimum threshold value for uptake and cell growth at low concentrations. Background considerations raised under steps 1 and 2 guide most of the practical steps needed for completing the implementation of the biodegradation assays (Table 1, steps 3 to 7).

Proper sampling and preparation of the source of microorganisms whose physiological activity is of interest (step 3 in Table 1) are critical for achieving valid biodegradation data. When mixed populations of naturally occurring microorganisms are being examined, the physiological information generated in biodegradation protocols is only as sound as was the investigators' skill in gathering microorganisms that truly represent the sampling site. Great care must be exercised during microbiological sampling. Aseptic techniques, such as the use of flame-sterilized implements and enclosure of samples within previously sterilized vessels, are often used (though these procedures may be unnecessary when one is sampling habitats [e.g., soil or sludges] likely to already contain the microflora present on surfaces of nonsterile implements and vessels). Because microbiological characteristics of environmental samples are prone to postsampling changes, procedures for sample fixation and storage must be scrutinized until deemed fully compatible with subsequent analyses that are planned. Immediate sample fixation (e.g., freezing in liquid nitrogen or addition of formalin, acid, or mercuric chloride to prevent metabolic and/or population shifts) is appropriate for a variety of microbiological characteristics (e.g., extraction of nucleic acids or phospholipid fatty acids) that do not require that cell viability be maintained. But biodegradation assays are, by definition, measures of dynamic chemical changes effected over time by live microorganisms. Therefore, cell fixation is impossible. Cooling of the samples on ice until laboratory processing is the most widely recommended sample-holding procedure. Cool temperatures retard the rate at which microbial populations change in response to microhabitat disturbances. However, by slowing metabolic activity, sample cooling will also cause dynamic gradients of diffusion-controlled environmental factors (such as O_2 concentration) to change. Thus, there is no perfect procedure for handling environmental samples. The microbial populations present in soil, sediment, water, and industrial effluent samples may begin to shift and change the moment an environmental sample is removed from a field site (37). These changes continue through cold storage, distribution of the samples to biodegradation vessels, and continued laboratory incubation during the biodegradation assays. It is this inevitable, intractable set of microbiological changes (as well as our inability to match laboratory to in situ field conditions) that makes it virtually impossible to extrapolate the results of laboratory biodegradation assays directly back to field sites (37).

When more reductionistic biodegradation information (Fig. 3) is being sought from populations of pure-cultured microorganisms (or from cell extracts or genes derived therefrom), then extreme care is still needed in scrutinizing the details required for the design and implementation of experiments (latter portion of step 3 in Table 1). Although a wide variety of possible hypotheses may be tested by the investigator, the outcome of the tests may be influenced by experimental manipulations that include (i) the selective pressures used to isolate the test microorganism(s), (ii) growth phase of the culture, (iii) enzyme systems induced, (iv) nutritional status of the cells, (v) how cells are broken

TABLE 1 Steps and decisions essential for implementing biodegradation assays

Step	Decisions
1. *Background*: Determine how the resultant data will be interpreted and used.	Objectives range from information about crude "biodegradation potential" to tests of specific hypotheses about physiological or biochemical factors governing biodegradation processes.
2. *Background*: Select the physiological conditions under which pollutant metabolism is to be measured.	The pivotal physiological concern is defining the mechanisms by which the compound(s) is metabolized. Of primary importance is discriminating among such possibilities as cometabolic reactions, use as an electron acceptor, and use as a carbon and energy source. Other concerns address conditions in experimental flasks such as nutrient sufficiency, which final electron acceptor regimen should dominate, what pollutant concentration ranges should be examined, and if conditions should be constantly changing (batch culture) or constant (continuous culture) during the assay.
3. *Practice*: Select and aseptically prepare or sample the microorganisms whose physiological activity is of interest.	For assays using environmental or industrial samples, aseptic sampling techniques involve use of tools (such as flame-sterilized scoops, spatulas, and knives) and sample placement within sterilized glass or plastic containers. For assays using pure cultures of microorganisms, the microorganisms must be aseptically grown under conditions that carefully define the cell physiological status (e.g., stage of growth, cell numbers, induced enzyme systems, nutritional state) desired by the investigator.
4. *Practice*: Select the physical apparatus and hence the physiological setting for biodegradation reactions to occur.	Glass (or plastic) vessels must be assembled. These contain the test compound(s), the microorganisms being studied, and any accompanying components of soils, sediments, sludges, and water in various ratios. The experimental hardware may be fitted with a variety of gas and water exchange assemblies for maintaining physiological conditions and assaying reaction progress.
5. *Practice*: Select a metabolic activity assay which is sensitive, effective, convenient, inexpensive, and compatible with experimental objectives.	The general assay categories are physiological assays (e.g., respirometry or growth) and chemical assays (which include gas chromatography, gas chromatography-mass spectrometry, HPLC, and radiotracer techniques).
6. *Practice*: Aseptically prepare stock solutions of ^{14}C-labeled organic compounds. Check radiopurity.	The validity of the results from biodegradation assays using ^{14}C-labeled substrates is dependent on substrate radiopurity and aseptic preparation of stock solutions.
7. *Practice*: Complete the experimental design parameters for the assay vessels and the assays themselves.	a. Concentration of the test substrate(s) b. Number of replicated flasks per treatment c. Whether flasks can be sampled repeatedly or if they require sacrifice at each sampling period d. Frequency of sampling e. Method of preparing abiotic controls f. Method for separating radioactive parent and product compounds from one another

for cell-free procedures, (vi) materials and conditions used in separating enzymes from cell extracts, (vii) addition of reagents to increase plasmid copy number, (viii) the G + C content of genomic DNA, etc. (14, 20, 22, 38, 52, 72; see also chapters 7, 8, 10, 11, 22, and 23 of reference 21a).

Depending on one's objectives, selecting the physical apparatus and hence the physical-chemical setting of the biodegradation assay (Table 1, step 4) can be a trivial matter or present severe logistical problems. Despite the fact that all biodegradation equipment has the same general goal—to contain both microorganisms and test organic compounds in ways that allow physiological modification of the organic compound to be measured—this goal has been manifest in hundreds of different ways in the published literature. The physical equipment used by a given investigator reflects a variety of factors, including the chemical properties of the test component; physiological conditions desired; the sensi-

tivity and accuracy needed to document the particular biodegradation reaction(s); the biochemical mechanism of biodegradation; the fastidiousness of the microorganism(s) carrying out the reaction; the amount of control sought by investigators over chemical, nutritional, or microscale gradients within experimental vessels; the degree that the investigator desires conditions during the biodegradation assay to match those in the field site of interest; and availability of personnel or hardware or other resources.

Selection of the physical apparatus is often intricately linked to how the metabolism of the organic compound will be documented. For instance, if the long-established biological oxygen demand (BOD) assay (traditionally used for sewage treatment purposes) is the assay of choice, then the widely used BOD bottle is likely to be the accompanying apparatus of choice: biodegradation will be gauged by using an indirect metabolic activity assay, oxygen consumption.

This and other indirect physiological biodegradation tests indicate chemical loss through a related vital microbiological parameter (cell growth, respiration, calorimetry, etc.) (Table 1, step 5). In contrast, methods which focus directly on demonstrating diminished mass of the contaminant and/or production of contaminant-specific metabolites are considered more rigorous. These direct procedures utilize specific chemical purification and/or detection instruments (including gas chromatography, gas chromatography-mass spectrometry, high-performance liquid chromatography [HPLC], spectrophotometry, and radioisotopic tracers) to monitor the abundance of test chemicals in experimental vessels. If a ^{14}C-labeled version of the test compound is added to the test apparatus and, later, steadily increasing amounts of ^{14}C are recovered from the assay vessel as $^{14}CO_2$, then microbial mineralization has been documented. Such radiotracer assays are often both elegantly rigorous and procedurally easy to complete.

Step 6 in Table 1, selection, preparation, and quality control for sterile ^{14}C-labeled stock solutions, represents critical procedures in implementing biodegradation studies involving radiotracers. The location of the ^{14}C-labeled atom(s) within the test molecule has a major bearing on experimental objectives and interpretation of data. If nuances of biochemical pathways are of interest, many refined hypotheses can be explained by varying the ^{14}C-labeled moieties. However, if complete mineralization is of interest, then central structural elements of the molecule (such as a benzene ring) should be uniformly labeled with ^{14}C. Sterility is required both for maintaining the expensive stock solutions and for allowing clear, confident interpretation of data from experiments in which the stock solutions are used. If microorganisms able to transform the ^{14}C-labeled substrate are present in the stock solutions, not only may the compound be altered prior to its intended use, but also the source of biodegradation activity found in experimental treatments may be uncertain. Both of these possibilities are disastrous for biodegradation testing. Thus, extreme care in aseptic handling of the vessels, solutions, and dilutions of ^{14}C-labeled stock solutions is essential. Furthermore, during subsequent biodegradation experiments, repeated tests should be carried out to ensure lack of microbial contamination of stock solutions and to assess the radiopurity of the substrate. Under most circumstances, an uninoculated, reagent-only control treatment serves to test for the presence of biodegradation activity in the stock solutions. Radiopurity of ^{14}C-labeled substrates is readily assayed by using HPLC separation techniques and a fraction collector (though thin-layer chromatography is suitable if the proper solvent mix is used in the mobile phase). The HPLC analysis system must be tested on unlabeled standards so that the precise elution time for given flow rates of the mobile phase is known. The total ^{14}C activity contained within a given volume of stock solution (determined via scintillation counting) is compared with the proportion of ^{14}C eluting at the expected time. In this way, radiopurity of the substrate standards can be discerned. In subsequent biodegradation tests, the proportion of transformed substrate (measured either via loss of the parent compound or conversion to $^{14}CO_2$ or daughter products) must far exceed (i.e., at least by a factor of 2 to 4) the level of ^{14}C impurities present in the stock solution. The manufacturer's specifications on product radiopurity (e.g., >98%) can generally be trusted for newly purchased materials.

When received from the manufacturer, the ^{14}C-labeled compound may be carrier free (i.e., as vanishingly small quantities of pure liquid, solid, or gas) or in a carrier solvent such as ethanol or acetone. In general, it is best to request the carrier-free form of substrate from the manufacturer. This avoids issues of solvent effects and/or cosubstrate influences on the results of microbial activity assays. The precise details of preparing and maintaining small volumes of sterile concentrated primary stock solutions are beyond the scope of this chapter. However, it should be noted that stock preparation approaches include addition of sterile water to the manufacturer's vessel (for solubilization and repeated collection of rinsates), dilution and washing of volatile ^{14}C-labeled solvents (originally supplied in sealed glass ampoules) with their unlabeled forms, preparation of saturator vessels that allow sparingly soluble substrates simply to float in sterile deionized water so that water drawn off contains fully dissolved aqueous-phase substrate at its saturated solubility, and filter sterilization using syringe-fitted units whose housings and filtration membranes fail to bind the radiolabeled substrate of interest. Other working stock solutions, derived from dilutions of the primary stock, can be readily prepared by aseptic transfer of known volumes of the primary stock to known volumes of suitable sterile solvents (often water). All stocks should be stored in a manner that stabilizes the material—usually in a glass vessel, sealed with a Teflon-lined cap, kept cold or frozen.

The seventh and final step in Table 1 is necessary for completing the logistical considerations implicit in all previous steps. Major issues are discussed below.

Concentration of the test substrate(s). Pollutant compounds should be added to test vessels at environmentally relevant concentrations. Ideally, the substrates should be free of carrier solvents. It is essential to understand a variety of properties of test compounds (e.g., aqueous solubility, volatility, and toxicity [to microorganisms and to the microbiologist]) before initiating biodegradation assays and the elective enrichment, isolation, and enumeration protocols that may follow (Fig. 3; 38) Table 2 presents chemical characteristics and related physiological considerations of selected environmental pollutants whose biodegradability is frequently examined.

Number of replicated flasks per treatment. Three replicates is generally considered the minimum, but the number may reach above 10 if experimental aims involve fine discrimination between treatment types (only you and your statistician know for sure).

Whether the replicate flasks can be sampled repeatedly or if each assay requires that the replicates be sacrificed at each sampling period. This issue is largely determined by the physical and chemical status of the test compound within the experimental apparatus. If the reaction mixture of cells, water, particles, gases, etc., can be subsampled in a reproducible manner that maintains the composition of the mixture as well as the integrity of the biodegradation process, then the repeated sampling strategy should be used. This obviates the need for the preparation of large numbers of replicate assay vessels at the beginning of the testing procedure. Furthermore, repeated sampling of the same vessels also ensures that the same microbiological populations (or their progeny) are sampled throughout an experiment, thus diminishing inoculum variability as a source of noise in the data.

It becomes necessary to sacrifice replicate vessels when subsampling cannot accurately quantify the compound of interest within test vessels. For example, in many studies of the metabolism of hydrophobic pesticides, not only may the compound's molecular structure be resistant to complete

TABLE 2 Characteristics of selected environmental contaminants and considerations for carrying out biodegradation assays and strain isolation procedures

Compound	Aqueous solubility (ppm)	Recommended starting concn in aqueous enrichment culture (ppm)	Mode of compound delivery to agar plates	Major considerations	References
Benzene	1,780	5	Vapor phase or embedded in wax	Toxicity, volatility; used aerobically and anaerobically as carbon and energy source	20, 24, 33
Toluene	535	5	Vapor phase	Toxicity, volatility; used aerobically and anaerobically as carbon and energy source	15, 17, 34, 61
Naphthalene	30	30	Vapor phase	Solubility, volatility; used aerobically as carbon and energy source	16, 55
Phenanthrene	1.3	1.3	Spray plate	Solubility; used aerobically as carbon and energy source	44, 67, 28a
Phenol	82,000	50	Dissolved in medium	Toxicity, volatility; used aerobically and anaerobically as carbon and energy source	6, 30
Tetrachloroethylene	150–200	0.5	Dissolved in medium and vapor phase	Toxicity, volatility; used anaerobically as final electron acceptor, not attacked aerobically; electron donors include H_2, methanol, and butyrate	21, 27a, 29, 41a, 63
		30,000	Dissolved in hexadecane		

mineralization, but the pesticide may also partition itself in complex patterns between a variety of solid components in environmental samples such as clays and humus. Uniformly sampling each of these components within test vessels is usually impossible. Thus, the analytical approach used at each sampling time may be to sacrifice three (or more) replicate vessels by subjecting the entire contents of each to an array of dissection, extraction, chromatographic, and detection procedures that quantify the disposition of ^{14}C-labeled parent and daughter compounds.

Frequency of sampling. Determining how often replicate biodegradation vessels should be sampled is no trivial issue. Plots of data showing loss of contaminants versus time generally convey the most information if several early datum points establish a stable initial concentration of the substrate, and then gradual loss begins and plateaus at some low level. Thus, three phases in a biodegradation curve are typically seen: lag, active biodegradation, and cessation. During lag and cessation periods, the sampling interval is often immaterial. The key issue about the active biodegradation phase is that it should not be missed. During the active biodegradation phase, the sampling intervals should be adjusted to one-third or one-fourth of that of the active period. For instance, if the substrate concentration drops from 100% to 20% in a 24-h period, the sampling interval should be 6 to 8 h. That sampling regimen produces clean, unambiguous substrate disappearance curves. However, depending on the statistical needs of the investigator, sampling intervals may be $1/10$ or $1/20$ of the active period (57).

With no prior knowledge of how long the lag time or active period will be, all that an investigator can do is be pragmatic: sample at convenient intervals that gradually increase with the duration of the test (e.g., first daily and then every 2, 4, and 7 days; for some anaerobic incubations, the sampling interval may be monthly [chapters 87 and 88, this volume, and references 33, 40]). Also, the investigator

should expect to repeat the experiments several times, using data from each successive assay to refine experimental design (especially sampling-interval) parameters. It is the relationship between the mass of the substrate and the mass and activity of the biodegrading populations that governs rates of substrate loss. If the substrate of interest is a carbon and energy source (e.g., phenol) added at a concentration of 1 ppm (1 μg/ml), and the initial phenol-degrading population is 10^4 cells per ml (or 10^{-8} g of cells per ml [this presumes that cells weigh 1 pg each]), then a substantial time lag is inevitable because an initial doubling of biomass will consume only 1% of the phenol (10^{-8} g of cells versus 10^{-6} g of phenol). But after several generations, as the mass of the active cells approaches 1 ppm (10^6 g of cells per ml), then in a single generation period nearly all of the phenol can theoretically be metabolized. This is a simple but insightful illustration of how both lag time and active biodegradation phases are governed by relative masses of test substrate and active cells present in test vessels.

Method of preparing abiotic controls. Methods for preparing abiotic controls include (i) sterilization, (ii) use of biological poisons, and (iii) use of uninoculated controls. These treatments are necessary so that the difference in behavior of the test compound in the live versus the abiotic treatments can be attributed to microbial activity. The ideal abiotic control treatment would simply eliminate all biological function while keeping the physical and chemical properties of environmental samples (composition and structure of everything from the test compound to microbial cells to the surfaces of clay minerals and humus polymers) completely unaltered. This is an impossibility. To avoid artifacts that may result from any single means of preparing abiotic controls, multiple procedures (described below) should be considered whenever possible.

Unquestionably the best sterilization procedure is filtration (membrane filter pore size of 0.22 μm). This procedure usually removes all microorganisms and retains dissolved chemical constituents in environmental samples but is appropriate only for relatively pristine aquatic samples. Since high-solids samples such as soil cannot be filter sterilized, alternative treatments must be used. Autoclaving soil for 1-h periods on 3 successive days effectively eliminates biological activity, as do γ irradiation (2 to 3 Mrads) and gas-phase sterilization (i.e., by ethylene oxide treatment). Soil sterilization techniques have recently been reviewed by Wolf and Skipper (69).

Chemical agents (poisons) can be easily introduced and highly effective in aqueous media and in aquatic samples for which the test medium is uniform and homogeneous and good cell-poison contact can be ensured. Because soil and sediments are a complex mixture of reactive components, the activity of chemical agents as metabolic poisons is often greatly diminished. Thus, high concentrations and multiple agents are often recommended when one is attempting to curtail microbial processes in soil. The chemical inhibitors of microbial activity that have been tested include $HgCl_2$, sodium azide, a variety of acids, strong base (NaOH), sodium cyanide, formaldehyde, and various antibiotics (12, 69). The effectiveness of an inhibitor varies with dosage and the specific properties of each environmental sample. Therefore, preliminary tests verifying process inhibition should be conducted while the design of biodegradation protocols is refined.

Uninoculated controls are treatments that, whenever possible, should be included in experimental designs. These are especially appropriate for tests using pure cultures or extracts derived therefrom (Fig. 3). Uninoculated controls verify that the reactants of interest are stable in the absence of microorganisms and that procedures taken to exclude microorganisms were effective. Uninoculated controls can be prepared only when a substantial portion of the biodegradation reaction medium is synthetic (e.g., a mixture of water, mineral salts, and the test compound) and has been sterilized. When the reaction medium consists primarily of the environmental sample itself (e.g., soil, sludge, or lake water), then clearly the inoculum is already present. In such cases, the uninoculated control cannot be prepared; instead, sterilization and/or poisoned treatments must be pursued.

Method for separating radioactive parent and product compounds from one another. In biodegradation procedures utilizing radiolabeled test compounds, it is generally unnecessary to use chromatographic separation measures for distinguishing the compounds of interest from others present in environmental samples. Instead, documentation of microbial attack on the parent labeled compound usually relies on changes in properties (volatility, molecular weight, solubility, hydrophobicity, etc.) of the radiotracer. These changes generally allow investigators to use simple partitioning procedures which quantify the radioactivity in different pools of the initial parent and intermediary or final daughter products. The simplest example of this is the conversion of a water-soluble, nonvolatile parent compound (such as nitrophenol) to CO_2. The parent remains in aqueous solution always, while CO_2 will leave aqueous solution at low pH and be partitioned into a separate headspace-located reservoir of base. Not all combinations of radiolabeled parent and daughter products differ unambiguously in their partitioning behaviors. For instance, ^{14}C-labeled benzene is both volatile and reasonably soluble in aqueous media (Table 2). Thus, it may be erroneous to presume at the end of a mineralization assay that 100% of the ^{14}C detected in a separate reservoir of base represents $^{14}CO_2$. A

variety of checks have been devised for ensuring that ^{14}C-labeled parent and transformation products are unambiguously partitioned. These take the form of additional procedures or control flasks that test alternative interpretations of the data (3, 8, 20, 38, 41, 71). Whenever one is uncertain about the partitioning behavior of ^{14}C-labeled pools of parent and daughter compounds in a given experimental apparatus, the trapping efficiencies of authentic standards must be tested after the standards are added to test vessels, alone and in mixtures, in the absence of microorganisms.

MANAGEMENT OF ORGANIC CONTAMINANTS IN FIELD SITES
Relationships between Biodegradation and Bioremediation

Bioremediation is the intentional use of biodegradation processes to eliminate environmental pollutants from sites where they have been intentionally or inadvertently released. Bioremediation technologies use the physiological potential of microorganisms, as documented most readily in laboratory assays, to eliminate environmental pollutants at field sites. In pursuing a remedy to environmental pollution, the concentration of pollutants must be reduced to levels that are acceptable to site owners and regulatory agencies that may be involved (46, 60).

The fundamental divisions in approaches to implementing bioremediation technology are based on two questions: Where will the contaminants be metabolized? How aggressively will site remediation be approached? Regarding location (Fig. 4), microbial processes may destroy environmental contaminants in situ, where they are found in the landscape, or ex situ, which requires that contaminants be moved from the landscape into some type of containment vessel (a bioreactor) for treatment. Regarding aggressiveness, intrinsic bioremediation is passive—it relies on the innate capacity of microorganisms present in field sites to respond to and metabolize the contaminants. Because intrinsic bioremediation occurs in the landscape where both indigenous microor-

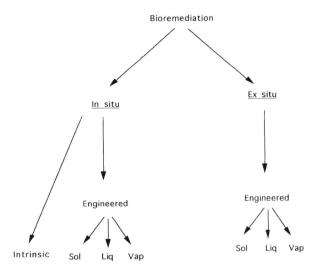

FIGURE 4 Overview of bioremediation approaches. Categories are based on where remediation will occur (in situ or ex situ), on how aggressively remediation is pursued (engineered or intrinsic), and on the physical status of the treatment system (Solid [Sol]-, liquid [Liq]-, or vapor [Vap]-phase treatment).

ganisms and contaminants reside, this type of bioremediation necessarily occurs in situ (Fig. 4). Alternatively, engineered bioremediation takes an active role in modifying a site to encourage and enhance the biodegradative capabilities of microorganisms. Each of the two major engineered bioremediation approaches may exploit solid-, slurry-, or vapor-phase systems for encouraging microorganisms to proliferate and metabolize the contaminant chemicals (Fig. 4).

Decisions for selecting the most effective bioremediation strategy are based on characteristics of the contaminants (e.g., toxicity, molecular structure, solubility, volatility, and susceptibility to microbial attack), the contaminated site (e.g., geology, hydrology, soil type, and climate and the legal, economic, and political pressures felt by the site owner), and the microbial process that will be exploited (e.g., pure culture, mixed cultures, and their respective growth conditions, and supplements) (Fig. 1; 36, 65). Engineered bioremediation relies on a variety of engineering procedures (control of water flow, aeration, chemical amendments, physical mixing, etc.) that influence both microbial populations and targeted contaminants (18, 42, 53). Furthermore, the efficacy of the remediation processes must be documented by chemical analysis of samples of water, air, and soil taken from the contaminated site. A full discussion of bioremediation technologies is beyond the scope of this chapter. A brief overview is presented below. However, readers are referred to references 2, 4, 19, 25–27, 46, 47, 49, 51, and 69a and chapters 76, 92, and 93 in this volume for additional details.

Intrinsic bioremediation is the management of contaminant biodegradation without taking any engineering steps to enhance the process. Intrinsic bioremediation utilizes the innate capabilities of naturally occurring microbial communities to metabolize environmental pollutants. The capacity of native microorganisms to carry out intrinsic bioremediation must be documented in laboratory biodegradation tests performed on site-specific samples. Furthermore, the effectiveness of intrinsic bioremediation must be proven with a site-monitoring regime that includes chemical analysis of contaminants, final electron acceptors, and/or other reactants or products indicative of biodegradation processes. This bioremediation strategy differs from no-action alternatives in that it requires adequate assessment of the existing biodegradation rates and adequate monitoring of the process. Intrinsic bioremediation may be used alone or in conjunction with other remediation techniques. In order for intrinsic bioremediation to be effective, the rate of contaminant destruction must be faster than the rate of contaminant migration. These relative rates depend on the type of contaminant, the microbial community, and the site hydrogeochemical conditions. Intrinsic bioremediation has been documented for a variety of contaminants and habitats, including low-molecular-weight polycyclic aromatic compounds in groundwater (39, 45), gasoline-related compounds in groundwater (54), crude oil in marine waters (10), and low-molecular-weight chlorinated solvents in groundwater (9, 40, 58).

Engineered bioremediation either accelerates intrinsic bioremediation or completely replaces it through the use of site modification procedures (such as excavation, hydrologic manipulations, and installation of bioreactors) that allow concentrations of nutrients, electron acceptors, or other materials to be managed in a manner that hastens biodegradation reactions. Engineered bioremediation is especially well suited for treating nonvolatile, sparingly soluble contaminants whose properties impede successful treatment by other technologies. There are a number of reasons why engineered bioremediation may be chosen over intrinsic bioremediation: time, cost, and liability. Because engineered bioremediation accelerates biodegradation reactions, this technology is appropriate for situations in which time frames for contaminant elimination are short or transport processes are causing the contaminant plume to advance rapidly. The need for rapid pollutant removal may be driven by an impending property transfer or by the impact of the contamination on the local community. A shortened cleanup time means a correspondingly lower cost of maintaining the site.

Engineered ex situ bioremediation has been used in municipal sewage treatment systems for over a century (42). In sewage treatment systems, wastewaters from municipalities are directed through an array of controlled environments which encourage microbial growth in filters, tanks, and digesters. Physical, chemical, and microbiological manipulations remove carbonaceous, nitrogenous, and other materials from water before it is discharged into rivers, lakes, or oceans. Engineered in situ bioremediation was implemented by Raymond and colleagues to clean up petroleum-contaminated groundwater as many as two decades ago (50). In the pioneering version of this technology, a groundwater circulation system was established which encouraged mixing of contaminants, microbial cells, and nutrients designed to encourage aerobic catabolic reactions. More recent engineered bioremediation case studies are described elsewhere (19, 47, 51).

Verification of Bioremediation in Field Sites

The reasons to establish sound scientific criteria for microbiological involvement in contaminant loss are as follows. (i) Biodegradation processes are often unique in their capacity to break intramolecular bonds of contaminant compounds; thus, contaminants can be destroyed and not simply transferred from one location to another, as is the case for many other pollution control technologies. (ii) When the mechanism of pollutant destruction is certain, key site management decisions about process enhancement can be made. (iii) In order for bioremediation to meet society's pollution control needs, the industry must adopt some standards for uniformity and quality control so that credibility and reliability can be attained (46).

But the question remains: What is adequate *proof* of bioremediation? The U.S. legal system provides a variety of categories of certainty in interpreting evidence. The categories depend on the type of case and the significance of the issues. Among the different burdens of proof are (i) proof beyond a reasonable doubt, (ii) proof in a clear and convincing manner, and (iii) proof beyond a preponderance of doubt. This chapter neither intends nor is able to dictate to regulatory or legal agencies which level of proof should be deemed adequate for bioremediation technology practitioners. Nonetheless, this section introduces approaches that can be used to distinguish biotic from abiotic reactions affecting contaminants at field sites where bioremediation technology is being applied (see references 36, 38, and 59 for additional discussions).

To answer the question What proves bioremediation?, it must be recognized that only under relatively rare circumstances is a proof of bioremediation unequivocal when a single piece of evidence is relied on. In the majority of cases, the complexities of contaminant mixtures, their hydrogeochemical settings, and accounting for competing abiotic mechanisms of contaminant loss make it a challenge to document biodegradation processes. Unlike controlled labora-

TABLE 3 Overview of steps and corresponding protocols required for demonstrating in situ biodegradation and bioremediation[a]

Step	Protocols
1. Develop historical records documenting loss of contaminants from sites.	1. Determine contaminant behavior in the field via a series of chemical analytical procedures performed in the field and in the laboratory on environmental samples.
2. Perform laboratory assays showing that microorganisms in site samples have the potential to transform the contaminants under expected site conditions.	2. Measure biodegradation activity in field samples incubated in the laboratory. Support this with information from pure cultures, cell-free preparations, and use of molecular procedures.
3. Obtain evidence that biodegradation potential is actually expressed in the field.	3. Demonstrate expression of biodegradation in situ (i) through the use of conservative tracers, (ii) by field detection of intermediary metabolites, (iii) through the use of replicate field plots to distinguish biotic from abiotic contaminant loss, (iv) by documenting adaptation of indigenous microorganisms to contaminants, using field gradients of specific contaminant-degrading populations and metabolic coreactants over time and/or space, or (v) via rigorous computer modeling strategies.

[a] From references 38 and 46.

tory experimentation wherein measurements can usually be easily interpreted, cause-and-effect relationships are often very difficult to establish at field sites. Furthermore, certain bioremediation data that may be convincing for some authorities may not be convincing for others. Thus, in seeking proof of bioremediation, several approaches described below should be independently pursued. The strategy should be to build a consistent, logical case relying on convergent lines of independent evidence.

The consensus of a recent National Research Council (46) committee in recommending criteria proving in situ bioremediation is as follows:

1. Develop historical records documenting loss of contaminants from field sites.

2. Perform laboratory assays unequivocally showing that microorganisms in site-derived samples have the potential to metabolize the contaminants under expected site conditions.

3. Demonstrate that the metabolic potential measured under point 2 is actually expressed in the field. To achieve this, microbiological mechanisms of contaminant attenuation must be distinguished from abiotic ones. Evidence deemed suitable for these purposes will vary according to the contaminants and conditions found at each site.

Table 3 provides an overview of the three steps for verifying bioremediation as well as the supporting protocols (36, 38).

During preparation of the manuscript, my research laboratory was supported by a grant from the Cornell Biotechnology Program, which is sponsored by the New York State Science and Technology Foundation (grant NYS CAT 92054), a consortium of industries, and the National Science Foundation. Additional support was provided by the Air Force Office of Scientific Research (grants AOFSR-91-0436, F49620-93-1-0414, and F49620-95-1-0346), Merck and Co., Inc. (grant PY-641308), the New York State Center for Hazardous Waste Management, the National Institute of Environmental Health Sciences (grants 08-P2ES05950A and ES05950-03), and USDA Hatch Formula Funds.

Expert manuscript preparation by P. Lisk is greatly appreciated.

REFERENCES

1. **Alexander, M.** 1981. Biodegradation of chemicals of environmental concern. *Science* **211**:132–138.
2. **Alexander, M.** 1994. *Biodegradation and Bioremediation.* Academic Press, New York.
3. **Anderson, J. P. E.** 1982. Soil respiration, p. 831–871. *In* A. L. Page, R. H. Miller, and D. R. Keeney (ed.), *Methods of Soil Analysis*, part 2. Soil Science Society of America, Inc., Madison, Wis.
4. **Atlas, R. M.** 1992. Oil spills: regulation and biotechnology. *Curr. Opin. Biotechnol.* **3**:220–223.
5. **Babu, G. R. V., J. H. Wolfram, and K. D. Chapatwala.** 1992. Conversion of sodium cyanide to carbon dioxide and ammonia by immobilized cells of *Pseudomonas putida. J. Ind. Microbiol.* **9**:235–238.
6. **Bak, F., and F. Widdel.** 1986. Anaerobic degradation of phenol and phenol derivatives by *Desulfobacterium phenolicum* new species. *Arch. Microbiol.* **146**:177–180.
7. **Barkay, T., C. Liebert, and M. Gillman.** 1989. Environmental significance of the potential for *mer*(Tn21)-mediated reduction of Hg^{2+} to Hg^0 in natural waters. *Appl. Environ. Microbiol.* **55**:1196–1202.
8. **Bartha, R., and D. Pramer.** 1965. Features of a flask and method for measuring the persistence and biological effects of pesticides in soil. *Soil Sci.* **100**:68–70.
9. **Beeman, R. E., J. E. Howell, S. H. Shoemaker, E. A. Salazar, and J. R. Buttram.** 1994. A field evaluation of in situ microbial reductive dehalogenation by the biotransformation of chlorinated ethenes, p. 14–27. *In* R. E. Hinchee, A. Leesen, L. Semprini, and S. K. Ong (ed.), *Bioremediation of Chlorinated and Polycyclic Aromatic Hydrocarbon Compounds.* Lewis Publishers, Inc., Boca Raton, Fla.
10. **Bragg, J. R., R. C. Prince, E. J. Harner, and R. M. Atlas.** 1994. Effectiveness of bioremediation for the Exxon Valdez oil spill. *Nature* (London) **368**:413–418.
11. **Brierley, C. L.** 1990. Bioremediation of metal-contaminated surface and groundwaters. *Geomicrobiol. J.* **8**:201–224.
12. **Brock, T. D.** 1978. The poisoned control in biogeochemical investigations, p. 717–726. *In* W. V. Krumbein (ed.), *Environmental Biogeochemistry and Geomicrobiology*, vol. 3. Ann Arbor Science Publishers Inc., Ann Arbor, Mich.
13. **Brock, T. D.** 1980. Environmental biogeochemistry, p.

93–103. In P. A. Trudinger, M. R. Walter, and B. J. Ralph (ed.), *Biogeochemistry of Ancient and Modern Environments.* Proceedings of the Fourth International Symposium on Environmental Biogeochemistry. Australian Academy of Sciences, Canberra, Australia.

14. **Dagley, S., and P. J. Chapman.** 1971. Evaluation of methods used to determine metabolic pathways. *Methods Microbiol.* **6A:**217–268.

15. **Dolfing, J., J. Zeyer, P. Binder-Eicher, and R. P. Schwarzenbach.** 1990. Isolation and characterization of a bacterium that mineralizes toluene in the absence of molecular oxygen. *Arch. Microbiol.* **154:**336–341.

16. **Eaton, R. W., and P. J. Chapman.** 1992. Bacterial metabolism of naphthalene: construction and use of recombinant bacteria to study ring cleavage of 1,2-dihydroxynaphthalene and subsequent reactions. *J. Bacteriol.* **174:** 7542–7554.

16a.**Ehrlich, H. L.** 1995. *Geomicrobiology,* 3rd ed. Marcel Dekker, Inc., New York.

17. **Evans, P. J., D. T. Mang, K. S. Kim, and L. Y. Young.** 1991. Anaerobic degradation of toluene by a denitrifying bacterium. *Appl. Environ. Microbiol.* **57:**1139–1145.

18. **Fiorenza, S., K. L. Duston, and C. H. Ward.** 1991. Decision making—is bioremediation a viable option. *J. Hazard. Mater.* **28:**171–183.

19. **Flathman, P. E., D. E. Jerger, and J. H. Exner.** 1994. *Bioremediation: Field Experience.* Lewis Publishers, Inc., Boca Raton, Fla.

20. **Focht, D. D.** 1994. Microbiological procedures for biodegradation research, p. 407–425. In R. W. Weaver, S. Angle, P. Bottomley, D. Bezdicek, S. Smith, A. Tabatabai, and A. Wollum (ed.), *Methods of Soil Analysis,* part 2. Soil Science Society of America, Inc., Madison, Wis.

21. **Freedman, D. L., and J. M. Gossett.** 1989. Biological reductive dechlorination of tetrachloroethylene and trichloroethylene to ethylene under methanogenic conditions. *Appl. Environ. Microbiol.* **55:**2144–2151.

21a.**Gerhardt, P., R. G. E. Murray, N. A. Wood, and N. R. Krieg (ed.).** 1994. *Methods for General and Molecular Bacteriology.* American Society for Microbiology, Washington, D.C.

22. **Gibson, D. T.** 1971. Assay of enzymes of aromatic metabolism. *Methods Microbiol.* **6A:**463–478.

23. **Gibson, D. T. (ed.).** 1984. *Microbial Degradation of Organic Compounds.* Marcel Dekker, Inc., New York.

24. **Gibson, D. T., G. E. Cardini, F. C. Maseles, and R. E. Kallio.** 1970. Incorporation of oxygen-18 into benzene by *Pseudomonas putida. Biochemistry* **9:**1631–1635.

25. **Hinchee, R. E. (ed.).** 1994. *Air Sparging.* Lewis Publishers, Inc., Boca Raton, Fla.

26. **Hinchee, R. E., and R. F. Olfenbuttel.** 1994. *On-Site Bioreclamation: Processes for Xenobiotic and Hydrocarbon Treatment.* Battelle Press, Columbus, Ohio.

27. **Hinchee, R. E., J. T. Wilson, and D. C. Downey (ed.).** 1995. *Intrinsic Bioremediation.* Battelle Press, Columbus, Ohio.

27a.**Holliger, C., G. Schraa, A. J. M. Stams, and A. J. B. Zehnder.** 1993. A highly purified enrichment culture couples the reductive dechlorination of tetrachloroethene to growth. *Appl. Environ. Microbiol.* **59:**2991–2997.

28. **Kalin, M., J. Cairns, and R. M. McCready.** 1991. Ecological engineering methods for acid mine drainage treatment of coal wastes. *Resour. Conserv. Recycl.* **5:**265–276.

28a.**Kiyohara, H., K. Nagao, and K. Yana.** 1982. Rapid screen for bacteria degrading water-insoluble, solid hydrocarbons on agar plates. *Appl. Environ. Microbiol.* **43:**454–457.

29. **Krumholz, L. R.** 1995. A new anaerobe that grows with tetrachloroethylene as an electron acceptor, abstr. Q-34, p. 403. *Abstracts of the 95th General Meeting of the American Society for Microbiology 1995.* American Society for Microbiology, Washington, D.C.

30. **Kukor, J. J., and R. H. Olsen.** 1992. Complete nucleotide sequence of TBUD, the gene encoding phenol-cresol hydroxylase from *Pseudomonas picketti* PKO1 and functional analysis of the encoded enzyme. *J. Bacteriol.* **174:** 6518–6526.

31. **Lovley, D. R.** 1991. Dissimilatory iron-III and manganese-IV reduction. *Microbiol. Rev.* **55:**259–287.

32. **Lovley, D. R.** 1993. Dissimilatory metal reduction. *Annu. Rev. Microbiol.* **47:**263–290.

33. **Lovley, D. R., J. D. Coates, J. C. Woodward, and E. J. P. Phillips.** 1995. Benzene oxidation coupled to sulfate reduction. *Appl. Environ. Microbiol.* **61:**953–958.

34. **Lovley, D. R., and D. J. Lonergan.** 1990. Anaerobic oxidation of toluene, phenol, and p-cresol by the dissimilatory iron-reducing organism, GS-15. *Appl. Environ. Microbiol.* **56:**1858–1864.

35. **Lovley, D. R., E. J. P. Phillips, Y. A. Gorby, and E. R. Landa.** 1991. Microbial reduction of uranium. *Nature* (London) **350:**413–416.

36. **Madsen, E. L.** 1991. Determining in situ biodegradation: facts and challenges. *Environ. Sci. Technol.* **25:**1662–1673.

37. **Madsen, E. L.** 1996. A critical analysis of methods for determining the composition and biogeochemical activities of soil microbial communities in situ, p. 287–370. In G. Stotzky and J.-M. Bollag (ed.), *Soil Biochemistry,* vol. 9. Marcel Dekker, Inc., New York.

38. **Madsen, E. L.** Theoretical and applied aspects of bioremediation: the influence of microbiological processes on organic compounds in field sites. In R. Burlage (ed.), *Techniques in Microbial Ecology,* in press, Oxford University Press, New York.

39. **Madsen, E. L., J. L. Sinclair, and W. C. Ghiorse.** 1991. In situ biodegradation: microbiological patterns in a contaminated aquifer. *Science* **252:**830–833.

40. **Major, D. W., E. W. Hodgins, and B. J. Butler.** 1991. Field and laboratory evidence of in situ biotransformation of tetrachloroethene to ethene and ethane at a chemical transfer facility in North Toronto, p. 147–172. In R. E. Hinchee and R. F. Olfenbuttel (ed), *On-Site Bioreclamation.* Butterworth-Heinemann, Stoneham, Mass.

41. **Marinucci, A. C., and R. Bartha.** 1979. Apparatus for monitoring the mineralization of volatile ^{14}C-labeled compounds. *Appl. Environ. Microbiol.* **38:**1020–1022.

41a.**Maymó-Gatell, X., V. Tandoi, J. M. Gossett, and S. H. Zinder.** 1995. Characterization of an H_2-utilizing enrichment culture that reductively dechlorinates tetrachloroethene to vinyl chloride and ethene in the absence of methanogenesis and acetogenesis. *Appl. Environ. Microbiol.* **61:** 3928–3933.

42. **McCarty, P. L.** 1991. Engineering concepts for in situ bioremediation. *J. Hazard. Mater.* **28:**1–11.

43. **McHale, A. P., and S. McHale.** 1994. Microbial biosorption of metals: potential in the treatment of metal pollution. *Biotechnol. Adv.* **12:**647–652.

44. **Mueller, J. G., P. J. Chapman, B. O. Blattmann, and P. H. Pritchard.** 1990. Isolation and characterization of a fluoranthene-utilizing strain of *Pseudomonas paucimobilis. Appl. Environ. Microbiol.* **56:**1079–1086.

45. **Murarka, I., E. Neuhauser, M. Sherman, B. B. Taylor, D. M. Mauro, J. Ripp, and T. Taylor.** 1992. Organic substances in the subsurface: delineation, migration, and remediation. *J. Hazard. Mater.* **32:**245–261.

46. **National Research Council.** 1993. *In Situ Bioremediation: When Does It Work?* National Academy Press, Washington, D.C.

47. **Norris, R. D., R. E. Hinchee, R. Brown, P. L. McCarty, L. Semprini, J. T. Wilson, D. G. Kampbell, M. Reinhard,**

D. J. Bouwer, R. C. Borden, T. M. Vogel, J. M. Thomas, and C. H. Ward. 1994. *Handbook of Bioremediation.* Lewis Publishers, Inc., Boca Raton, Fla.

48. **Oremland, R. S., C. W. Culbertson, and M. R. Winfrey.** 1991. Methylmercury decomposition in sediments and bacterial cultures: involvement of methanogens and sulfate reducers in oxidative demethylation. *Appl. Environ. Microbiol.* **57:**130–137.

49. **Pritchard, P. H.** 1992. Use of inoculation in bioremediation. *Curr. Opin. Biotechnol.* **3:**232–243.

50. **Raymond, R. L., J. O. Hudson, and V. W. Jamison.** 1976. Oil degradation in soil. *Appl. Environ. Microbiol.* **31:** 522–535.

51. **Rittmann, B. E., E. Seagren, B. A. Wrenn, A. J. Valocchi, C. Ray, and L. Raskin.** 1994. *In Situ Bioremediation,* 2nd ed. Noyes Publications, Park Ridge, N.J.

52. **Rogers, J. E., and D. T. Gibson.** 1977. Purification and properties of *cis*-toluene dihydrodiol dehydrogenase from *Pseudomonas putida. J. Bacteriol.* **130:**1117–1124.

53. **Ryan, J. R., R. C. Loehr, and E. Rucker.** 1991. Bioremediation of organic contaminated soils. *J. Hazard. Mater.* **28:** 159–169.

54. **Salanitro, J. P.** 1993. The role of bioattenuation in the management of aromatic hydrocarbon plumes in aquifers. *Ground Water Monit. Remediation* **13:**150–161.

55. **Sanseverino, J., B. M. Applegate, J. M. H. King, and G. S. Sayler.** 1993. Plasmid-mediated mineralization of naphthalene phenanthrene and anthracene. *Appl. Environ. Microbiol.* **59:**1931–1937.

56. **Schlegel, H. G., and H. W. Jannasch.** 1992. Prokaryotes and their habitats, p. 75–125. *In* A. Balows, H. G. Trüper, M. Dworkin, W. Harder, and K.-H. Schleifer (ed.), *The Prokaryotes,* 2nd ed. Springer-Verlag, New York.

57. **Scow, K. M., and M. A. Alexander.** 1992. Effect of diffusion on the kinetics of biodegradation experimental results with synthetic aggregates. *Soil Sci. Soc. Am. J.* **56:**128–134.

58. **Semprini, L., P. K. Kitanidis, D. H. Kampbell, and J. T. Wilson.** 1995. Anaerobic transformation of chlorinated aliphatic hydrocarbons in a sand aquifer based on spatial chemical distributions. *Water Resour. Res.* **31:**1051–1062.

59. **Shannon, M. J. R., and R. Unterman.** 1993. Evaluating bioremediation: distinguishing fact from fiction. *Annu. Rev. Microbiol.* **47:**715–738.

60. **Shauver, J. M.** 1993. A regulator's perspective on in situ bioremediation, p. 99–103. *In* National Research Council (ed.), *In Situ Bioremediation: When Does It Work?* National Academy Press, Washington, D.C.

61. **Shields, M. S., M. J. Reagin, R. R. Gerger, R. Campbell, and C. Somerville.** 1995. TOM, a new aromatic degradative plasmid from *Burkholderia (Pseudomonas) cepacia* G4. *Appl. Environ. Microbiol.* **61:**1352–1356.

62. **Summers, A. O.** 1992. The hard stuff: metals in bioremediation. *Curr. Opin. Biotechnol.* **3:**271–276.

63. **Tandoi, V., T. D. DiStefano, P. A. Bowser, J. M. Gossett, and S. H. Zinder.** 1994. Reductive dehalogenation of chlorinated ethenes and halogenated ethanes by a high-rate anaerobic enrichment culture. *Environ. Sci. Technol.* **28:**973–979.

64. **Thompson-Eagle, E. C., and W. T. Frankenberger, Jr.** 1992. Bioremediation of soils contaminated with selenium. *Adv. Soil Sci.* **17:**261–310.

65. **Tiedje, J. M.** 1993. Bioremediation from an ecological perspective, p. 110–120. *In* National Research Council (ed.), *In Situ Bioremediation: When Does It Work?* National Academy Press, Washington, D.C.

66. **Videla, H. A., and W. G. Characklis.** 1992. Biofouling and microbially influenced corrosion. *Int. Biodeterior. Biodegrad.* **29:**195–212.

67. **Weissenfels, W. D., M. Beyer, and J. Klein.** 1990. Degradation of phenanthrene, fluorene and fluoranthene by pure bacterial cultures. *Appl. Microbiol. Biotechnol.* **32:**479–484.

68. **Whitlock, J. L.** 1990. Biological detoxification of precious metal processing wastewaters. *Geomicrobiol. J.* **8:**241–249.

69. **Wolf, D. C., and H. D. Skipper.** 1994. Soil sterilization, p. 41–52. *In* R. W. Weaver, S. Angle, P. Bottomley, D. Bezdicek, S. Smith, A. Tabatabai, and A. Wollum (ed.), *Methods of Soil Analysis,* part 2. Soil Science Society of America, Inc., Madison, Wis.

69a.**Young, L. Y., and C. E. Cerniglia (ed.).** 1995. *Microbial Transformation and Degradation of Toxic Organic Chemicals.* Wiley-Liss, New York.

70. **Zehnder, A. J. B., and W. Stumm.** 1988. Geochemistry and biogeochemistry of anaerobic habitats, p. 1–38. *In* A. J. B. Zehnder (ed.), *Biology of Anaerobic Microorganisms.* John Wiley & Sons, Inc., New York.

71. **Zibilske, L. M.** 1994. Carbon mineralization, p. 835–863. *In* R. W. Weaver, S. Angle, P. Bottomley, D. Bezdicek, S. Smith, A. Tabatabai, and A. Wollum (ed.), *Methods of Soil Analysis,* part 2. Soil Science Society of America, Inc., Madison, Wis.

72. **Zylstra, G. J., and D. T. Gibson.** 1991. Aromatic hydrocarbon degradation: a molecular approach. *Genet. Eng.* **13:** 183–203.

Measuring and Modeling Physicochemical Limitations to Bioavailability and Biodegradation

ANURADHA RAMASWAMI AND RICHARD G. LUTHY

78

Bioprocesses associated with microbes have often been used to remediate soils and sludges contaminated with organic and inorganic pollutants (19). Biotreatment has the potential of reducing contaminant concentrations, as well as reducing toxicity and contaminant mobility, providing a low-cost alternative to physical-chemical treatment, incineration, or landfilling of contaminated media. The potential benefits of biotreatment can be offset by field observations that indicate limited efficiency of pollutant degradation and much slower rates of biotransformation compared with laboratory tests. Further, an initial more rapid rate of biotransformation typically is followed by a much slower biotransformation rate, often resulting in residues that are recalcitrant to further microbial attack despite favorable environmental conditions (8, 9, 18, 32, 34). These observations suggest that pollutants present in environmental systems may be unavailable to the microbes for degradation. Thus, the overall rate of biotransformation in environmental systems depends on (i) physicochemical phenomena that control the bioavailability of pollutants and (ii) biokinetic factors pertaining to electron acceptors, nutrient supply, toxicity, inhibition, and competitive substrate utilization.

In this chapter, we focus on physicochemical phenomena that control bioavailability and biotransformation rates; adequate electron acceptors and nutrients and optimal conditions for biotreatment are assumed to be present in the system. Systems comprising organic-phase liquids and sorbed contaminants are considered.

PHYSICOCHEMICAL PHENOMENA AND BIOAVAILABILITY

Several physicochemical processes can affect the bioavailability of pollutants to microbes. An understanding of bioavailability requires knowledge of the physical state of the pollutant substrate (e.g., solid, liquid, or sorbed) and its location relative to the microbes present in the environmental system. Consistent with the approach used by environmental engineers, two different environmental subsystems may be considered: (i) a source region containing soil, water, and a separate, non-aqueous-phase liquid (NAPL; e.g., diesel fuel or coal tar) that slowly releases organic pollutants over long periods of time and (ii) a soil-water region containing pollutants transported downstream from a contaminated zone. While several studies have focused on the soil-

water region, relatively little is known about physical, chemical, and microbial interactions in the NAPL region. Because of different dominant processes operating in the two regions, the NAPL system and the soil-water system are often discussed separately in the environmental engineering literature.

Soil-water systems are dominated by solid surfaces to which microbes are attached. Microbes, with a typical size of about 1 μm, are excluded from entering the smaller micropores of soil and porous media. Thus, pollutants present in the micropores of soil aggregates or solids may be unavailable to the bacteria and must diffuse through pore water to the external grain surface in order to be degraded. The intrapore diffusion may be retarded by sorption of pollutant to the solid surfaces of the micropores (36). In many cases of aerobic degradation, it appears that the pollutant mass sorbed on solid surfaces is largely unavailable to the microbes (16, 20, 25, 27). Thus, a working hypothesis for aerobic soil-water systems considers primarily the mobile, aqueous-phase substrate to be bioavailable for microbial degradation.

Consistent with the foregoing hypothesis, when a NAPL is present in the system and functions as a source of contaminants, only the fraction of pollutant mass dissolved in the aqueous phase may be assumed to be bioavailable (7, 11, 23, 33). Some exceptions include higher-molecular-weight aliphatic hydrocarbons and chlorinated polychlorinated biphenyl congeners that are essentially insoluble yet slowly undergo microbially induced transformations. In general, for environmental systems contaminated with NAPLs, it may be assumed that it is predominantly the dissolved aqueous-phase substrate that is readily available for microbial degradation.

Another factor that affects the bioavailability of organic pollutants in porous media is aging, i.e., the length of time the soil has been exposed to contamination. As contaminated soil ages, pollutants may diffuse into smaller, more tortuous micropores and chemically bind with soil and sediment material, becoming increasingly unavailable to microbes with the progress of time (4, 32). Aging may result also in weathered films that inhibit dissolution from complex NAPLs such as tars and crude petroleum (15). Thus, the bioavailability of organic pollutants in environmental systems may be controlled by physicochemical processes pertaining to (i) dissolution from a NAPL source to water; (ii) sorption and micropore diffusion in the soil-water region;

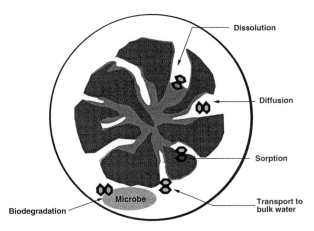

FIGURE 1 Schematic illustrating mass transfer and biodegradation of PAH solutes from liquid coal tar imbibed in microporous media.

and (iii) aging, i.e., the length of time a contaminant has been in contact with soil or sediment.

The schematic in Fig. 1 illustrates the effect of some of these processes on bioavailability. The figure depicts a slurry system with porous solids, the micropores of which are small and thus size exclude bacteria. The microporous solid shown in Fig. 1 is partially saturated with a NAPL, in this case coal tar, which functions as a source of polynuclear aromatic hydrocarbons (PAHs) such as naphthalene. Thus, the naphthalene present in coal tar undergoes dissolution at the intrapore NAPL-water interface, sorption-retarded diffusion in pore water, and external bulk-phase mass transfer before it is available for microbial degradation.

In studying the soil-water and NAPL-water interactions described above, it is important to distinguish between equilibrium partitioning processes and kinetic processes that control bioavailability. Equilibrium partitioning imposes a limit on the maximum aqueous-phase (bioavailable) contaminant concentration permissible in the presence of a sorbent solid or a separate organic-phase liquid (NAPL), while kinetic processes control the rate at which contaminants transfer to the aqueous phase, thereby becoming bioavailable.

In a soil-water system with no NAPL, maximum aqueous-phase pollutant concentrations are controlled by sorption-desorption equilibrium between the soil and water. At low pollutant concentrations, sorption equilibria may be represented by a linear sorption isotherm,

$$S = K_p C \qquad (1)$$

where S (mass/mass) is the mass of the pollutant sorbed per unit mass of sorbate at equilibrium and C is the corresponding equilibrium bulk aqueous-phase pollutant concentration (pollutant mass/volume water). K_p is a sorption coefficient that has been correlated with the fraction of organic carbon in the soil and the octanol-water partition coefficient of the pollutant (14). In a soil-water system with a solids loading of ρ (mass solids/volume water), the maximum bioavailable portion of the total pollutant mass may be represented by a sorption factor, sf, given by

$$sf = \frac{C}{[C + \rho S]} = \frac{1}{[1 + \rho K_p]} \qquad (2)$$

$$= \frac{\text{maximum pollutant mass in water}}{\text{total pollutant mass in system}}$$

Equation 2 indicates that maximum bioavailable pollutant concentrations, and hence biotransformation rates, may be expected to decrease with an increase in the solids loading, ρ, and with an increase in the sorption parameter, K_p. In experimental evaluations of naphthalene degradation, Mihelcic and Luthy (16) observed that an increase in indigenous microbial activity arising from an increase in soil loading in a bioreactor did not occur because of a decrease in bioavailable aqueous-phase naphthalene concentrations due to increased sorption. Weissenfels et al. (35) have demonstrated the impact of soil properties on bioavailability. The rate and extent of mineralization of anthracene oil were found to be inversely proportional to the sorption capacity of the soil; soils comprising a larger fraction of organic carbon showed a greater capacity for binding organic pollutants, rendering them unavailable for degradation. Thus, sorption equilibria can play a significant role in controlling the bioavailability of organic pollutants in soil-water systems.

Similarly, in the presence of a NAPL, equilibrium between water and the organic liquid controls the maximum permissible aqueous-phase pollutant concentration. Equilibrium partitioning between water and a multicomponent organic liquid may be described by Raoult's law (26). Consider ideal equilibrium dissolution of a PAH compound, e.g., naphthalene, from NAPL coal tar to water. The equilibrium aqueous-phase naphthalene concentration at any time, t, is given by

$$C_{eq,t} = x_t^{PAH} * C_{\text{pure liquid}}^{PAH} \qquad (3)$$

where x_t^{PAH} is the mole fraction of naphthalene in coal tar at time t and $C_{\text{pure liquid}}^{PAH}$ is the solubility of pure subcooled liquid naphthalene in water. Most coal tar constituents are much less soluble and much less degradable than naphthalene. Thus, as biotreatment proceeds and naphthalene is preferentially depleted from the organic phase, the mole fraction of naphthalene in coal tar will decrease, resulting in a proportional decrease in equilibrium aqueous-phase naphthalene concentrations. Thus, at any time t, equilibrium partitioning determines the maximum bioavailable fraction of the total pollutant mass initially present in the system. In NAPL systems, this fraction has been represented by a solubility factor, SF (24):

$$SF_{(t)} = \frac{V C_{eq,t}}{M_{total}^{PAH}} = \frac{V x_t^{PAH} C_{\text{pure liquid}}^{PAH}}{M_{total}^{PAH}} \qquad (4)$$

where V represents the bulk aqueous volume and M_{total}^{PAH} represents the total mass of naphthalene initially present in a system. Equation 4 indicates that the bioavailability of PAH compounds in NAPL systems is proportional to their solubility in water. Higher-molecular-weight PAH compounds that are more hydrophobic and less soluble in water may be expected to be less bioavailable for microbial degradation. A combination of low bioavailability and low intrinsic biodegradation rates may result in little or no removal of higher-molecular-weight PAH compounds from environmental matrices, as has been observed in field tests (9, 31). In addition, with the progress of time, the equilibrium concentrations of the more soluble PAHs such as naphthalene would decrease according to equation 4, reducing the solubility factor and thereby reducing the fraction of the initial total naphthalene that becomes bioavailable in the aqueous phase. Since only a portion of the total pollutant mass is available to the microbes, the overall bioavailability may be said to be limited by dissolution equilibria.

While equilibrium phenomena determine the maximum bioavailable (aqueous-phase) pollutant concentrations, observed aqueous-phase pollutant concentrations often exhibit departures from equilibrium due to slow rates of mass transfer. Thus, the overall rate of biotransformation also may be limited by slow physicochemical phenomena, e.g., slow desorption rates from solids, slow dissolution rates from liquids, or slow diffusion rates within micropores. The kinetics of mass transfer will affect overall biotransformation rates only if the rate of mass transfer is slower than the potential biodegradation rate. Thus, when mass transfer occurs more slowly than biodegradation, the system is said to be mass transfer limited, indicating physical constraints arising from slow pore diffusion, desorption, or dissolution. Efficient design and operation of biotreatment systems require knowledge of bioavailability and rate-limiting phenomena. These parameters may be evaluated through quantitative mathematical modeling techniques and experimental protocols described in the following sections.

MODELING TECHNIQUES

Mathematical models have been developed to couple equilibrium and mass transfer rate processes with biokinetic phenomena. Different models have been developed to capture the representative features of various environmental systems. A general, simplified representation of mass transport and biodegradation with one-dimensional groundwater flow in porous media containing macroporous NAPL blobs may be written as

$$\frac{\partial C}{\partial t} = D_x \frac{\partial^2 C}{\partial x^2} - v \frac{\partial C}{\partial x} + k_l a^{NAPL} [C_{eq,t} - C] - k_{bio} C \quad (5)$$

where C is the pollutant concentration in the macroporous (mobile) aqueous phase variable with location and time, i.e., (x, t); v is the pore water velocity; D_x is the longitudinal dispersion coefficient; k_{bio} is a first-order biokinetic coefficient; and $k_l a^{NAPL}$ is an external mass transfer coefficient for pollutant dissolution from the surface of NAPL blobs to bulk water. $k_l a^{NAPL}$ incorporates the specific NAPL-water interfacial area, a, i.e., the interfacial area per unit of bulk water in the system (17). The terms on the right-hand side of equation 5 represent the processes of macropore dispersion, advection, dissolution, and biodegradation, respectively. Simplified first-order biokinetic rate coefficients have often been used in modeling exercises to represent biodegradation; the simplification is appropriate for stable microbial populations and low substrate concentrations in the system (e.g., reference 28 and 29).

Sorption-retarded micropore diffusion in the immobile water trapped within soil aggregates has been found to result in nonequilibrium conditions in the field (3). Intra-aggregate sorption-diffusion is commonly modeled by using a radial-diffusion model (36):

$$\frac{\partial c_{(r,t)}}{\partial t} = D_{eff} \left[\frac{\partial^2 c_{(r,t)}}{\partial r^2} + \frac{2}{r} \frac{\partial c_{(r,t)}}{\partial r} \right] \text{ for } r < R \quad (6)$$

where $c_{(r,t)}$ is the micropore aqueous concentration of the pollutant at any radial location, r, within a soil aggregate of radius R. D_{eff} is an effective intra-aggregate diffusion coefficient that accounts for sorption-retarded diffusion within tortuous micropores, the tortuosity being inversely propor-

tional to the porosity of the soil grains. D_{eff} is given by $D_{eff} = D_m n^2/[n + K_p \rho_s(1 - n)]$. Here, D_m is the molecular diffusion coefficient of the pollutant in water, n is the porosity of the soil aggregate, K_p is the sorption coefficient, and ρ_s is the density of the dry solid material. Mass transfer between the immobile water held within microporous soil aggregates and the mobile water may be incorporated into the right-hand side of equation 5 by the additional term

$$\left[\frac{\partial C}{\partial t} \right]_{soil\ aggregate\ to\ bulk\ water} = k_l a^{soil} [c_{(r = R,t)} - C]$$

(7)

where $k_l a^{soil}$ is a mass transfer coefficient that describes external mass transfer from the surface of a microporous soil aggregate of radius R into the bulk fluid at concentration C. The rate of mass transfer is modeled to be proportional to a concentration driving force term given by the difference between the micropore aqueous concentration at the surface of the soil aggregate, $c_{(r=R,t)}$, and the macropore or bulk concentration, C.

Variants of the generalized model presented above have been used by several researchers. Chen et al. (5) have coupled advection-dispersion with biodegradation to describe the biotransformation of benzene and toluene in sandy aquifer material. Bosma (4) has, in addition, coupled radial sorption-retarded diffusion within microporous soil aggregates with biodegradation occurring in a one-dimensional column. Seagren et al. (28, 29) have combined advection, dispersion and biodegradation with dissolution from a NAPL source present in a flowthrough system packed with nonporous media. Equation 5 may be modified to represent batch mixed systems by deleting the first and second terms on the right-hand side, corresponding to dispersion and advection, respectively. Note that in mixed systems C will vary with time, t, but not with location. Batch soil-water systems focusing on micropore diffusion have been studied by Rijnaarts et al. (25), Mihelcic and Luthy (16), Scow and Alexander (27), and Chung et al. (6), while the additional impact of a NAPL source trapped within the micropores of porous aggregates has been studied by Ramaswami et al. (23).

The modeling framework presented in equation 5 yields several useful dimensionless parameters for assessing the impact of mass transfer kinetics on biotransformation rates. These parameters may be used to infer rate-limiting processes in environmental systems comprising soil, water, and/or a NAPL. The dimensionless parameters represent the rate of biodegradation relative to the rates of mass transfer due to advection, dissolution, and micropore diffusion and are summarized in Table 1. The significance of these parameters is described below.

Advection and biodegradation processes are compared by using a parameter, $D1$, that belongs to a family of dimensionless groups commonly referred to as the Damkohler number. A value of $D1 \gg 1$ indicates that advection occurs much more slowly than biodegradation, limiting overall biotransformation rates. Increasing the flow rate through porous media would increase the overall biotransformation rate only so long as $D1$ remains larger than 1. Once $D1$ becomes smaller than unity, water flows too rapidly through the system, resulting in insufficient time for biodegradation. In such cases, slower flow rates would enhance biotransformation.

Dissolution and biodegradation processes are compared by using a second type of Damkohler number, $D2$. A value of

TABLE 1 Dimensionless parameters comparing biodegradation and mass transfer rates

Definition and implication[a]	Reference(s)
$D1 = \dfrac{\text{biodegradation rate}}{\text{advection rate}} = \dfrac{k_{\text{bio}}L}{v}$ $D1 > 1$: slow advection limits biotransformation	28
$D2 = \dfrac{\text{biodegradation rate}}{\text{external mass transfer rate}} = \dfrac{k_{\text{bio}}}{k_l a^{\text{NAPL}}} \text{ or } \dfrac{k_{\text{bio}}}{k_l a^{\text{soil}}}$ $D2 > 1$: slow mass transfer from NAPL blobs or soil grains	4, 22, 24, 28
$D3 = \dfrac{\text{biodegradation rate}}{\text{micropore diffusion rate}} = \dfrac{k_{\text{bio}}R^2}{D_{\text{eff}}}$ $D3 > 1$: slow intra-aggregate micropore diffusion	6, 22, 24

[a] L is the characteristic length of the system in the flow direction; R is the grain size, representing the maximum diffusion path length within microporous soil aggregates. When the parameters $D1$, $D2$, and $D3$ are much larger than unity, mass transfer constraints are indicated, as shown. Conversely, values much smaller than unity indicate biokinetic limitations.

$D2 \gg 1$ indicates that external mass transfer, i.e., dissolution from macroporous NAPL blobs or desorption from the surface of soil aggregates, occurs much more slowly than biodegradation. Overall biotransformation rates in such systems may be enhanced by increasing the external mass transfer rate. Dissolution from NAPLs is dependent on the interfacial area between water and organic liquid and on the degree of mixing. In such cases, increasing the dispersion of the NAPL volume through generation of smaller NAPL globules with increased specific interfacial area will be beneficial, though this may be practically achieved only in ex situ processes employing some type of mixing (e.g., reference 23). Likewise, mass transfer in soil-water systems also depends on the surface area of contact between soil aggregates and water and on the tilling or mixing intensity. Thus, in soil systems and in NAPL systems, desorption and dissolution rates may be enhanced by increased mixing. Enhanced biotreatment rates upon mixing have been reported by Zehnder (37), as shown in Fig. 2, and by Wang and Bartha (34) and Griest et al. (12).

Micropore diffusion and biodegradation are compared by using a dimensionless parameter, $D3$, typically referred to as the Thiele modulus. A value of $D3 \gg 1$ indicates that slow pore diffusion rates limit overall biotransformation rates.

Diffusion rates may be increased by decreasing the diffusion path length within the grains, e.g., by particle size reduction. Particle size reduction may be achieved by pulverization, mixing, and grinding in ex situ slurry systems and by tilling in engineered land treatment systems. Pulverization has been found to be effective in enhancing desorption (32) and mineralization of organic contaminants (25).

The foregoing discussion illustrates the use of dimensionless parameters in assessing the impact of individual mass transfer processes on biotransformation rates. In environmental systems, however, several of these mass transfer phenomena occur simultaneously and sequentially (e.g., Fig. 1). Thus, determining the overall rate-limiting phenomena involves first detecting the slowest mass transfer process and then comparing the rate of that process with the biodegradation rate, using the parameters $D1$, $D2$, and $D3$ described above. The slowest mass transfer step may be ascertained through the use of dimensionless groups summarized in Table 2. To illustrate the use of the dimensionless parameters, we consider the slurry batch system shown on Fig. 1, for which the parameters $B1$, $D2$, and $D3$ are appropriate. The Biot number, $B1$, determines the slower of the two mass transfer processes: intrapore sorption-diffusion and external bulk-phase mass transfer. When $B1$ is larger than unity, the slower pore diffusion rate is compared with the biodegradation rate through the Thiele number, $D3$. Conversely, when $B1$ is smaller than unity, the slower boundary layer mass transfer rate is compared with the biodegradation rate through use of the Damkohler number, $D2$. Simultaneous analysis of the three parameters may be used to determine the overall rate-limiting phenomenon as illustrated in the flowchart in Fig. 3.

Thus, the dimensionless parameters derived from mathematical models provide quantitative criteria for identifying rate-limiting phenomena in environmental systems. A priori estimates of the dimensionless parameters may be made with knowledge of pollutant and medium properties (e.g., sorption capacity, diffusion coefficient, grain size, and micropore tortuosity), mass transfer characteristics, and biodegradation rates. Estimates of mass transfer coefficients for various system configurations may be obtained from chemical engineering mass transfer correlations. The validity of these correlations must be confirmed by experimental observations. Mass transfer studies of several soil and sediment systems demonstrate the successful adaptation and use of such correlations (e.g., references 10, 17, 21, and 23). Estimates of the biodegradation rate coefficients may be ob-

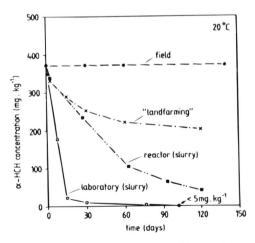

FIGURE 2 Biotransformation of hexachlorocyclohexane in systems ranging from the laboratory scale to field scale. The rate of biotransformation is greater in laboratory studies and is enhanced by mixing. (From reference 37.)

TABLE 2 Dimensionless parameters comparing various mass transfer rates

Definition and implication[a]	References
$B1 = \dfrac{\text{mass flux across external grain surface}}{\text{intra-aggregate micropore diffusive flux}} = \dfrac{k_l R}{D_{\text{eff}}\gamma}$	6, 24
$B1 > 1$: micropore diffusion is slower than external mass transfer	
$B2 = \dfrac{\text{external mass transfer rate}}{\text{advection rate}} = \dfrac{k_l a^{\text{NAPL}} L}{v}$ or $\dfrac{k_l a^{\text{soil}} L}{v}$	4, 28
$B2 > 1$: advection is slower than external mass transfer from NAPL blobs or soil grains	

[a] L is the characteristic length of the system in the flow direction; R is the grain size, representing the intraparticle diffusion path length; γ is a sorption factor given by $\gamma = n + K_p \rho_s (1 - n)$.

tained from the environmental microbiology literature; for example, some presumed first-order rate constants are summarized in references 13 and 30.

It should be noted that the uncertainties associated with estimating mass transfer and biokinetic parameters will result in uncertainties in the estimates of the dimensionless parameters summarized in Tables 1 and 2. Thus, dimensionless parameters significantly smaller or larger than unity can offer useful insights about likely rate-limiting phenomena. However, dimensionless parameters with estimated values closer to unity (e.g., 0.2 to 5) would not be conclusive. Further, with the progress of biotreatment, the mass transfer and biokinetic parameters can undergo significant changes, such as those due to aging effects that reduce mass transfer rates (e.g., reference 15), or be subjected to toxic, inhibitory, or competitive effects that can retard biodegradation rates (e.g., reference 1). For these reasons, it is necessary to complement the theoretical analysis of dimensionless parameters with test protocols that experimentally evaluate physicochemical limitations to biotransformation.

DETECTING AND MEASURING PHYSICOCHEMICAL CONSTRAINTS

Protocols for determining physicochemical constraints on biotransformation rates range from qualitative techniques to quantitative measurements. Qualitative tests suggest that mass transfer limitations may exist, while quantitative tests can establish the fact. Qualitative indicators include the following: (i) the pollutant persists in the environmental matrix despite the presence of viable microbes (8, 9, 32); (ii) freshly added compound degrades readily, while native compound in an aged field sample does not (8, 18, 32); (iii) results from biodegradation tests show incomplete conversion and a plateau effect (9, 37); (iv) leaching experiments

show low aqueous-phase substrate concentrations (8, 32); (v) pulverization or increased mixing enhances biotransformation rates (12, 25, 37); and (vi) addition of oxygen or nutrients does not enhance biotreatment rates.

More quantitative assessment of mass transfer limitations involves measuring equilibrium partitioning along with mass transfer and biotransformation rates in independent tests. The protocol for measuring physicochemical constraints on bioavailability and biodegradation rates is outlined below.

1. Abiotic batch tests are conducted to evaluate dissolution or desorption equilibria.

2. The batch equilibration tests are repeated with different contaminant loadings to generate a sorption isotherm to determine K_p. Alternatively, in NAPL systems, flowthrough mass transfer tests may be conducted at large residence times in order to evaluate dynamic changes in dissolution equilibria.

3. The solubility or sorption factor is computed on the basis of measured partitioning phenomena, quantifying maximum bioavailability in the system.

4. Additional abiotic flow through mass transfer tests are conducted at shorter residence times in order to assess the kinetics of mass transfer. The short residence time results in low bulk aqueous-phase pollutant concentrations and thus a large concentration driving force for mass transfer, enabling more accurate measurement of mass transfer rate coefficients.

5. The Damkohler number $D2$ is computed from measurements of the mass transfer coefficient.

6. Independent biodegradation or biomineralization tests are conducted to verify the utility of the Damkohler number in identifying rate-limiting phenomena.

The method described above is illustrated by two case studies, representing NAPL slurry and soil slurry systems, respectively.

Case Study 1: NAPL System

The first case study monitors initial rates of naphthalene biomineralization in a slurry system in which coal tar NAPL functions as the source of PAH contaminants. Details of this study may be found in references 23 and 24. Since a separate organic-phase liquid is present in this system, dissolution equilibria and kinetics are of interest. Two different NAPL coal tar slurry systems were tested by using the protocol summarized above: (i) a solid slurry system in which coal tar NAPL was imbibed in small (250-μm) microporous silica particles and dispersed in water (schematic in Fig. 1) and (ii) a liquid-liquid system in which a single large globule of coal tar NAPL was placed in a batch reactor. The solid slurry system creates a dense dispersion of small particles,

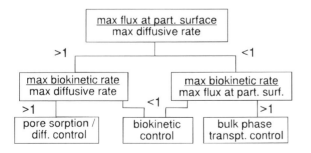

FIGURE 3 Framework for identifying rate-limiting phenomena in bioslurry systems by using dimensionless parameters. (Adapted from reference 24.)

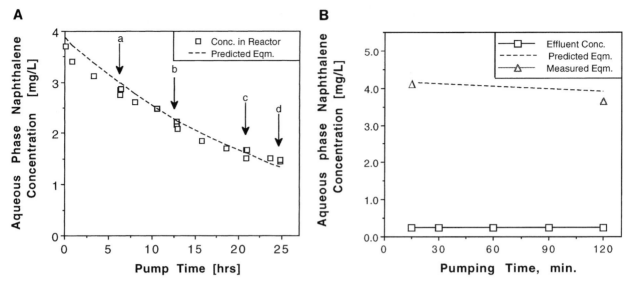

FIGURE 4 Evaluating mass transfer phenomena in coal tar NAPL-water systems. (A) Solid slurry system (0.25-mm diameter). Concentration profiles depicting equilibrium (Eqm.) conditions in a flowthrough system operated at a large residence time of 30 min. (B) Tar globule system (11-mm diameter). Concentration profiles showing a large departure from equilibrium in a flowthrough system operated at a smaller residence time of 6 min ($k_l a$ = 1.8/day).

thereby increasing the interfacial area between water and NAPL, resulting in rapid dissolution rates. In contrast, the globule system was characterized by low specific interfacial areas and slow dissolution rates.

Step 1

Batch tests were first conducted to evaluate the efficacy of Raoult's law in describing equilibrium partitioning in the two slurry systems. The two coal tar systems were maintained in closed vials for an extended period of time; the aqueous-phase naphthalene concentration was sampled periodically to ensure the approach of equilibrium. The aqueous-phase naphthalene concentrations measured at equilibrium compared well with Raoult's law predictions; thus, initial equilibrium partitioning between NAPL and water could be described by using the thermodynamic model of Raoult's law.

Step 2

In the second step, mass transfer tests were conducted in a flowthrough mode at a residence time of 30 min. Water was flushed through the two coal tar systems over an extended period of time. Continuous flushing depleted the naphthalene from the NAPL, resulting in a decreased mole fraction of naphthalene in coal tar and hence reduced equilibrium aqueous-phase concentrations with the progress of time. Equilibrium concentrations were measured by periodically discontinuing flushing (e.g., at points a to d in Fig. 4A) and allowing the contents of the vessel of reequilibrate. The measured variation in equilibrium concentrations due to flushing was found to be predicted fairly well from Raoult's law (Fig. 4). Thus, the batch and flowthrough equilibrium tests enable quantification of naphthalene partitioning in static and dynamic conditions, i.e., at the start of and during treatment.

Step 3

The equilibrium measurements were used to compute the solubility factor (equation 4) and hence the maximum frac-

tion of the naphthalene mass initially present in coal tar that partitions into the aqueous phase and becomes bioavailable to the microbes. For the two systems studied with a tar/water loading of 1 ml : 50 ml, the solubility factor at the start of biotreatment was found to be 0.012; this indicates that a maximum of 1.2% of the naphthalene initially present in coal tar can be bioavailable in the aqueous phase at the start of biodegradation. It is the bioavailable fraction of the pollutant mass that is assumed to undergo ready biotransformation. Hence, in a biokinetically limited system, the initial apparent fractional mineralization rate would be given by $k_{apparent}$ = SF · k_{bio}. Conversely, when mass transfer rates are slow, biomineralization proceeds at the same rate as mass transfer and the initial apparent fractional biomineralization rate is related to the mass transfer coefficient, $k_l a$, as $k_{apparent}$ = SF · $k_l a$. Thus, equilibrium phenomena represented by SF, as well as kinetic parameters determined by $k_l a$ and k_{bio}, play an important role in controlling observed biotransformation rates.

Step 4

The kinetics of dissolution of naphthalene from coal tar was assessed in flowthrough tests conducted at short residence times (e.g., 6 min), yielding nonequilibrium conditions. A sample concentration profile from one such test with the tar globule system is shown in Fig. 4B. The effluent naphthalene concentrations show a large departure from equilibrium, indicating slow mass transfer rates from the tar globule. The mass transfer coefficient, $k_l a$, was calculated from the observed concentration profiles, using the mass balance equation shown in equation 5 (modified for flowthrough, mixed systems). The tar globule system was characterized by slow mass transfer rates with experimentally determined $k_l a$ in the range of 1/day to 2/day. In a similar test, the solid slurry system exhibited rapid mass transfer with an average $k_l a$ of the order of 4,000/day.

Step 5

Damkohler numbers were then estimated by using experimentally measured values of the mass transfer coefficients and the range of first-order biokinetic rate coefficients for naphthalene obtained from the environmental microbiology literature (1/day to 25/day). Damkohler numbers in the range of 0.0003 to 0.006 were estimated for the solid slurry system, while Damkohler numbers in the range of 1 to 25 were estimated for the globule system.

Step 6

Analysis of the estimated Damkohler numbers indicates that the solid slurry may not be limited by mass transfer rate processes, while the tar globule system may be limited by slow rates of dissolution. Independent biomineralization tests conducted with the two systems confirmed this result (Fig. 5). Estimates of the initial biodegradation rate coefficient, k_{bio}, ranging from 12/day to 35/day were obtained from the biomineralization data. Experimentally derived estimates of k_{bio} and $k_l a$ were used to compute a more accurate range of Damkohler numbers, appropriate for the systems being studied. In the solid slurry system (Fig. 5A), for which the Damkohler number ranged from 0.003 to 0.008, miner-

alization proceeded rapidly at an apparent rate corresponding to $k_{apparent} = SF \cdot k_{bio}$, indicating biokinetic limitations. In the case of the globule system, for which Damkohler numbers ranged from 6 to 19, mineralization rates were retarded (Fig. 5B). The apparent initial mineralization rate corresponded to $k_{apparent} = SF \cdot k_l a$, indicating that slow mass transfer rates (with $k_l a \simeq 1$/day to 2/day) controlled the biotransformation rates. Aqueous-phase naphthalene concentrations in the bioreactors were also measured; near-equilibrium concentrations were observed in the solid slurry system, indicating rapid mass transfer, while departures from equilibrium were observed in the globule system, confirming the fact that dissolution rates were limiting biotransformation.

The study demonstrates that biotransformations in contaminated media are controlled by both (i) equilibrium partitioning phenomena that control the maximum bioavailable pollutant concentration in the aqueous phase and (ii) the slower of the mass transfer and biokinetic rates, which determines the overall rate of transformation of the bioavailable contaminant.

Similar quantitative measurements have been conducted in soil-water systems and demonstrate the combined effect of sorption equilibria and kinetics on biotransformation rates (e.g., references 16, 25, and 27). The study of Rijnaarts et al. (25) is described in the next case study.

Case Study 2: Soil-Water Desorption System

The study examined the desorption and degradation of α-hexachlorocyclohexane (HCH) from two soil samples recovered from a contaminated site (25). Two different soil slurry systems were considered: (i) an end-over-end mixed system with gentle mixing (system E) and (ii) a well-stirred system employing a magnetic stir bar (system S). Vigorous mixing in system S is believed to result in particle size reduction; the combination of greater mixing intensity and particle size reduction resulted in larger desorption rates for system S than for system E. The impact of desorption and mass transfer on biotransformation rates was assessed in the following manner.

Step 1

Desorption equilibria were investigated at different contaminant loadings. Nonsterile soil samples were placed in contact with water in the two systems described above. Natural microbial activity reduced HCH contaminant loadings in the two systems over time. Soil- and aqueous-phase HCH concentrations were measured at different time intervals, generating desorption isotherms from which the partition coefficient, K_p, was estimated as 171 kg/liter, and a sorption factor, SF $\simeq 1/18$, was obtained for a typical soil loading, ρ, of 100 g/liter.

Step 2

Desorption kinetics were evaluated by measuring changes in aqueous HCH concentrations over time. Lumped first-order mass transfer coefficients obtained from the desorption profiles were of the order of 0.42/day for system S and 0.05/day for system E. Thus, mass transfer rates were found to be an order of magnitude slower in system E compared with system S. The slow mass transfer rates were associated with slow intraparticle diffusion of HCH from the soil micropores.

Step 3

Maximum potential biokinetic rate coefficients for HCH degradation in slurry systems with similar soil loadings were

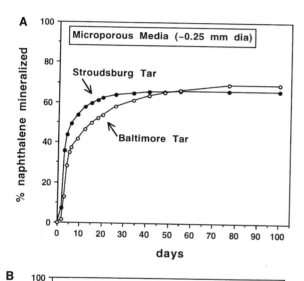

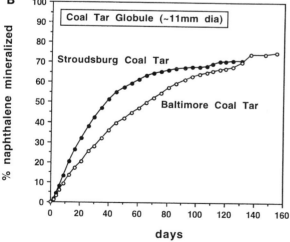

FIGURE 5 Biomineralization of naphthalene from coal tar. (A) Solid slurry system, with a Damkohler number much smaller than 1, is biokinetically limited. (B) Tar globule system, with a Damkohler number greater than unity, is mass transfer limited.

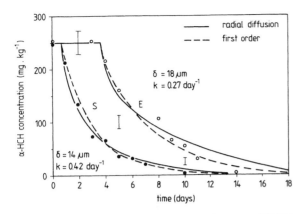

FIGURE 6 Observed biotransformation kinetics of HCH in systems E (gently mixed) and S (vigorously mixed). The lines depict HCH desorption kinetics. Biotransformation rates are observed to follow desorption kinetics, indicating mass transfer limitations. Reprinted with permission from reference 25. Copyright 1990 American Chemical Society.

obtained from the literature (2) and found to be of the order of 0.3/day. Thus, a priori estimates of the Damkohler number indicate a value in the range of 1 for system S and a value of 6 for system E. Analysis of the Damkohler number indicates that system E is likely to be mass transfer limited compared with system S.

Step 4

Biodegradation experiments were conducted to verify the presence of mass transfer constraints on biotransformation. The biotransformation of HCH in the two bioreactor systems was assessed by monitoring the change in sorbed-phase HCH concentration. It was observed that both stirred and end-over-end mixed systems were mass transfer limited. Thus, observed biotransformation rates could be described by desorption rates, modeled by empirically fitting either a radial-diffusion model or a first-order desorption model (Fig. 6). Stirring the slurry resulted in particle size reduction, more rapid mass transfer, and hence enhanced biotransformation rates. Thus, the impact of slow intraparticle mass transfer on biodegradation rates was demonstrated in this study.

The data and experimental protocols described above provide guidelines for measuring physicochemical limitations to biotransformations occurring in contaminated media. The experimental protocols, used in conjunction with the modeling framework, enable identification of rate-limiting phenomena and may be used to separate physicochemical constraints from biokinetic factors, resulting in more efficient and cost-effective management of biotreatment projects. The protocols described in this chapter focus on grain-scale phenomena and have been tested mainly in bench-scale systems. In addition to grain-scale processes, subsurface cleanup operations encounter large-scale spatial heterogeneities in soil and aquifer properties. More field tests are required to assess the impact of large-scale variations in hydraulic conductivity, contaminant entrapment, microbial density, and inhibition on bioavailability and biotransformation rates at contaminated sites.

REFERENCES

1. **Alvarez, P. J. J., and T. M. Vogel.** 1991. Substrate interactions of benzene, toluene and para-xylene during microbial degradation by pure cultures and mixed culture aquifer slurries. *Appl. Environ. Microbiol.* **57:**2981–2985.
2. **Bachmann, A., W. de Bruin, J. C. Jumelet, H. H. N. Rijnaarts, and J. B. Zehnder.** 1988. Aerobic biomineralization of alpha-hexachlorocyclohexane in contaminated soil. *Appl. Environ. Microbiol.* **54:**548–554.
3. **Ball, W. P., and P. V. Roberts.** 1991. Longterm sorption of halogenated organic chemicals by aquifer material. 1. Equilibrium. 2. Intraparticle diffusion. *Environ. Sci. Technol.* **25:**1223–1237.
4. **Bosma, T. N. P.** 1991. Ph.D. dissertation. Wageningen Agricultural University, Wageningen, The Netherlands.
5. **Chen, Y., L. M. Abriola, P. J. J. Alvarez, P. J. Anid, and T. M. Vogel,** 1992. Modeling transport and biodegradation of benzene and toluene in sandy aquifer material: comparison with experimental measurements. *Water Resour. Res.* **28:**1833–1847.
6. **Chung, G., B. J. McCoy, and K. M. Scow.** 1993. Criteria to assess when biodegradation is kinetically limited by intraparticle sorption diffusion. *Biotechnol. Bioeng.* **41:** 625–632.
7. **Efroymson, R. A., and M. Alexander.** 1994. Role of partitioning in biodegradation of phenanthrene dissolved in nonaqueous-phase liquids. *Environ. Sci. Technol.* **28:** 1172–1179.
8. **Erickson, D. C., R. C. Loehr, and E. F. Neuhauser.** 1993. PAH loss during bioremediation of manufactured gas plant site soils. *Water Res.* **27:**911–919.
9. **Gas Research Institute.** 1995. *Environmentally Acceptable Endpoints in Soil: Risk-Based Approach to Contaminated Site Management Based on Availability of Chemicals in Soil.* Draft report. Gas Research Institute, Chicago.
10. **Geller, J. T., and J. R. Hunt.** 1993. Mass transfer from nonaqueous phase organic liquids in water saturated porous media. *Water Resour. Res.* **29:**233–245.
11. **Ghoshal, S., A. Ramaswami, and R. G. Luthy.** 1996. Biodegradation of naphthalene from coal tar and heptamethylnonane in mixed batch systems. *Environ. Sci. Technol.* **30:** 1282–1291.
12. **Griest, W. H., A. J. Stewart, R. L. Tyndall, J. E. Caton, C. H. Ho, K. S. Ironside, W. M. Caldwell, and E. Tan.** 1993. Chemical and toxicological testing of composted explosives-contaminated soil. *Environ. Toxicol. Chem.* **12:** 1105–1116.
13. **Howard, P.** 1991. *Handbook of Environmental Degradation Rates.* Lewis Publishers, Chelsea, Mich.
14. **Karickhoff, S. W., D. S. Brown, and T. A. Scott.** 1979. Sorption of hydrophobic organic compounds on natural sediments. *Water Res.* **13:**241–248.
15. **Luthy, R. G., A. Ramaswami, S. Ghoshal, W. M. Merkel.** 1993. Interfacial films in coal tar NAPL-water systems. *Environ. Sci. Technol.* **27:**2914–2918.
16. **Mihelcic, J. R., and R. G. Luthy.** 1991. Sorption and microbial degradation of naphthalene in soil-water systems under denitrification conditions. *Environ. Sci. Technol.* **25:** 169–177.
17. **Miller, C. T., M. M. Poirier-McNeill, and A. S. Mayer.** 1990. Dissolution of trapped nonaqueous phase liquids. *Water Resour. Res.* **26:**2783–2796.
18. **Morgan, D. J., A. Battaglia, B. J. Hall, L. A. Vernieri, and M. A. Cushey.** 1992. *The GRI Accelerated Biotreatability Protocol for Assessing Conventional Biological Treatment of Soils: Development and Evaluation Using Soils from Manufactured Gas Plant Sites.* GRI-92/0499. Gas Research Institute, Chicago.
19. **National Research Council.** 1993. *In Situ Bioremediation: When Does It Work?* National Academy Press, Washington, D.C.
20. **Ogram, A. V., R. E. Jessup, L. T. Ou, and P. S. C. Rao.**

1985. Effects of sorption on biological degradation rates of acetic acid in soils. *Appl. Environ. Microbiol.* **49:**582–587.

21. **Powers, S. E., C. O. Loureiro, L. M. Abriola, and W. J. Weber.** 1991. Theoretical study of the significance of nonequilibrium dissolution of nonaqueous phase liquids in subsurface systems. *Water Resour. Res.* **27:**463–477.

22. **Ramaswami, A.** 1994. Ph.D. thesis. Carnegie Mellon University, Pittsburgh, Pa.

23. **Ramaswami, A., S. Ghoshal, and R. G. Luthy,** Mass transfer and bioavailability of PAH compounds in coal tar NAPL-slurry systems. 2. Experimental evaluations. Submitted for publication.

24. **Ramaswami, A., and R. G. Luthy.** Mass transfer and bioavailability of PAH compounds in coal tar NAPL-slurry systems. 1. Model development. Submitted for publication.

25. **Rijnaarts, H. H. M., A. Bachmann, J. C. Jumelet, and A. J. B. Zehnder.** 1990. Effect of desorption and intraparticle mass transfer on the aerobic mineralization of α-hexachlorocyclohexane in a contaminated calcareous soil. *Environ. Sci. Technol.* **24:**1349–1354.

26. **Schwarzenbach, R. P., P. M. Gschwend, and D. M. Imboden.** 1993. *Environmental Organic Chemistry.* John Wiley & Sons, Inc., New York.

27. **Scow, K. M., and M. Alexander.** 1992. Effect of diffusion on the kinetics of biodegradation: experimental results with synthetic aggregates. *Soil Sci. Soc. Am. J.* **51:**128–134.

28. **Seagren, E. A., B. E. Rittmann, and A. J. Valocchi,** 1993. Quantitative evaluation of flushing and biodegradation for enhancing in situ dissolution of nonaqueous phase liquids. *J. Contam. Hydrol.* **12**(1/2)**:**103–132.

29. **Seagren, E. A., B. E. Rittmann, and A. J. Valocchi.** 1994. Quantitative evaluation of the enhancement of NAPL-

30. **Sims, R. C., and M. R. Overcash.** 1988. *Fate of Polynuclear Aromatic Compounds in Soil-Plant Systems. Residue Reviews,* vol. 88. Springer-Verlag, New York.

31. **Smith, J. R., R. M. Tomicek, P. V. Swallow, R. L. Weightman, D. V. Nakles, and M. Helbling.** 1994. Definition of biodegradation endpoints for PAH contaminated soils using a risk-based approach. Presented at the Ninth Annual Conference on Contaminated Soils, University of Massachusetts at Amherst, Oct. 18–20.

32. **Steinberg, S. M., J. J. Pignatello, and B. L. Sawhney.** 1987. Persistence of 1,2-dibromoethane in soils: entrapment in intraparticle micropores. *Environ. Sci. Technol.* **21:** 1201–1213.

33. **Volkering, F., A. M. Breure, A. Sterkenberg, and J. G. Van Andel.** 1992. Microbial degradation of polycyclic aromatic hydrocarbons—effects of substrate bioavailability on bacterial growth kinetics. *Appl. Microbiol. Biotechnol.* **36:** 548–552.

34. **Wang, X., and R. Bartha.** 1990. Effects of bioremediation on residues, activity and toxicity in soil contaminated by fuel spills. *Soil Biol. Biochem.* **22:**501–505.

35. **Weissenfels, W. D., H. Klewer, and J. Langhoff.** 1992. Adsorption of polycyclic aromatic hydrocarbons (PAHs) by soil particles: influence on biodegradability and biotoxicity. *Appl. Microbiol. Biotechnol.* **36:**689–696.

36. **Wu, S., and P. M. Gschwend.** 1988. Numerical modeling of sorption kinetics of organic compounds to soil and sediment particles. *Water Resour. Res.* **24:**1373–1383.

37. **Zehnder, A. B. J.** 1991. *Mikrobiologische Reinigung eines kontaminierten Bodens,* p. 68–72. Beitrag des 9. Dechema-Fachgespräches Umweltschutz, Frankfurt/Main.

Lignocellulose Biodegradation

LEE A. DEOBALD AND DON L. CRAWFORD

79

LIGNOCELLULOSE STRUCTURE AND BIOSYNTHESIS

Lignocellulose is the term used to describe the composite of the predominant polymers of vascular plant biomass. The three major polymeric components are the polysaccharides cellulose and hemicellulose and the phenolic polymer of lignin. Cellulose is a linear polymer of glucose linked through β-1,4 linkages and is usually arranged into microcrystalline fibers. It makes up about 50% of the mass of vascular plants. Cellulose biosynthesis has been studied primarily in bacteria, whose synthetic pathway is thought to be representative of plant cellulose biosynthesis. Cellulose synthesis in *Acetobacterium xylinum* occurs through a condensation reaction between UDP-glucose and the 4' hydroxyl of the growing cellulose chain with liberation of UMP (20). The UDP-glucose is synthesized through a condensation reaction between glucose-1-phosphate and UTP. This is catalyzed by UTP:α-glucose-1-phosphate uridylyltransferase (18) and releases PP_i. There is, however, evidence that GDP-glucose is the cellulose precursor in plants (20). For more information on the biochemistry of polysaccharide biosynthesis, the reader should consult some of the reviews on the subject (18, 20).

Hemicellulose is a matrix polysaccharide of plant cell walls. It includes all of the plant cell polysaccharides excluding pectin and cellulose (49). Hemicelluloses are usually heteroglycans containing two to four, and occasionally more, different monosaccharides. Monosaccharides commonly found as a part of hemicelluloses include D-glucose, D-galactose, D-mannose, D-arabinose, D-xylose, α-D-glucuronic acid, and 4-O-methyl-α-D-glucuronic acid, with the monosaccharide composition depending on the plant source. There are, however, sources of homoglycan hemicelluloses. These include xylan, which is the polysaccharide containing β-D-xylopyranosyl linked through β-1,4 linkages, and mannan, which is D-mannopyranose residues linked through β-1,4 glycosidic linkages (18). Hemicelluloses are frequently branched, and sometimes hydroxyl groups of the main chain are acetylated. Hemicellulose chemistry and biosynthesis have been thoroughly reviewed elsewhere (20, 49).

Lignin is an amorphous aromatic polymer found in the cell walls and middle lamellae of vascular plants and surrounds cellulose microfibrils. It confers rigidity to cell walls and makes plants resistant to pathogen attack and mechanical stresses. Lignin makes up about 15 to 30% of the mass of vascular plants. It is synthesized from various substituted *p*-hydroxycinnamyl alcohol derivatives through a free radical reaction process initiated by peroxidases. Because of the free radical nature of the polymerization reaction, the radical intermediates undergo rearrangement reactions which produce mesomeric forms of the monomers with the unpaired electron located at various locations on the molecule. When these radical intermediates couple, a variety of nonchiral intermonomeric C-C and C-O bonds are formed. One of the most prevalent linkages is the β-O-4' bond, which is an ether between C-2 of an arylpropane side chain and the *para* position of another aromatic monomer (24). Also common are phenyl coumaran and biphenyl linkages. Numerous other intermonomeric linkages are present. All of the monomers are synthesized through the shikimic acid pathway through L-phenylalanine or L-tyrosine. The amino acids are converted to cinnamic acid and coumaric acid by L-phenylalanine ammonia lyase and L-tyrosine ammonia lyase, respectively. The intermediates are further modified by several more hydroxylations, methylations, and reductions to produce coniferyl alcohol, sinapyl alcohol, and coumaryl alcohol, which are polymerized to lignin (24). The monomers are stored in the form of glucosides which have a glycosidic linkage between the *para* hydroxyl group of the lignin monomer and the anomeric carbon of glucose. The glucosides are hydrolyzed, making them reactive to oxidation and free radical polymerization. Hardwood lignins contain mainly syringyl and guaiacyl residues, softwoods have mainly guaiacyl residues, and grass lignins have *p*-hydroxyphenyl residues in addition to guaiacyl and syringyl residues (24). In addition to C-C and ether bonds, grass lignins have coumaric acid and 4-hydroxy-3-methoxycinnamic (ferulic) acid esterified to lignin hydroxyl groups. These esters represent about 10% of the lignin (24).

INTRODUCTION TO BIODEGRADATION METHODS

Because of the abundance of lignocellulose in the biosphere and the importance of different components of lignocellulose in industrial processes and products, it is important to be knowledgeable about its metabolism in various environments. If one is to develop additional biotechnological uses of lignocellulose, it is appropriate to understand the nature and rates of metabolic processes acting upon lignocellulose. Some of the uses proposed for ligninolytic organisms are for upgrading of low-quality forage for use by ruminants, pretreatment of wood chips prior to pulping, and production of chemicals from lignin. Cellulolytic and hemicellulolytic organisms and enzymes have been proposed for use in hydrolytic pretreatment of lignocellulosic polysaccharides prior to their fermentation to ethanol or other solvents. The methods used to study biodegradation of lignocellulose will depend on the goals of the investigation. Various levels of complexity with regard to substrates, biocatalyst, and effects can be chosen for study. A researcher could choose to study a population, a small group of organisms, a single organism, or enzymes produced by a single organism. One might choose to study the effect of the biocatalyst on intact wood, ground and extracted lignocellulose, an isolated component of lignocellulose, such as hemicellulose or lignin, or an oligomer or dimer such as a dimeric lignin model compound. Some organisms that degrade lignin may do so to gain access to the polysaccharides encrusted by it. Therefore, because of the intimate association of the components, removal of one from the others alters their physical and chemical characteristics, making the results of studies using purified components not completely representative of naturally occurring processes. The effect that an investigator is interested in may be simple to measure, such as mineralization, or more complex, such as depolymerization or specific chemical transformations. Lignocellulose biodegradation has been reviewed by several authors in recent years (12, 35).

LIGNOCELLULOSE BIODEGRADATION METHODS

The effects of one or more microbial cultures on the different components of lignocellulose can be measured by exposing a substrate to the organism(s) of interest under ideal culture conditions. The conditions that might be controlled include temperature, aeration, culture moisture content, medium nutrient content, and pH. Although these conditions may be somewhat unrealistic with respect to natural degradative conditions, it may be hypothesized that a similar effect occurs under nonideal environmental conditions. After incubation of the substrate for a suitable period of time, the lignocellulosic substrate is freed from the microbial biomass as much as possible and is analyzed along with similarly treated control substrate, with differences being attributed to the effect of the microbial culture (26). Substrates that could be used include wood blocks or ground plant material from herbaceous or woody plants, such as sawdust. To minimize the effects of low-molecular-weight nonpolymeric materials, the ground lignocellulose is usually extracted with hot water and a mixture of benzene and ethanol (38).

Lignocellulolytic organisms can be grown on lignocellulose in either a liquid culture or a solid-state fermentation. In the liquid culture system, the bulk of the medium is liquid containing up to a few percent ground lignocellulose. The cultures may be grown in Erlenmeyer shake flasks and should not contain so much solid that it prevents mixing during shaking. The partially degraded lignocellulose can be recovered for analysis by filtration. Another liquid culture system makes use of air agitation of a culture in a test tube. This is especially suited to small cultures in which evolved carbon dioxide is trapped in respirometry experiments (see below) (9). Stationary liquid cultures can be used where

agitation is detrimental to enzyme induction, such as was found with *Phanerochaete chrysosporium* (29). The broth and solid substrate could be contained in standing Erlenmeyer or Roux flasks in a thin layer (0.5 to 1.0 cm deep) to allow sufficient aeration. Investigations of lignocellulose-degrading enzymes usually make use of liquid cultures containing lignocellulose or some other inducer. Cell mass and undegraded solid material can be removed by filtration or centrifugation prior to enzyme characterization. Solid-state culturing is used to produce modified lignocellulose for characterization. It contains a higher ratio of lignocellulose to water such that the slurry is maintained in a thin layer to ensure adequate aeration and heat transfer. Solid-state cultures are not usually agitated, although special mixing methods are used in large, experimental cultures.

There are many methods that can be used to analyze lignocellulose following its degradation by microorganisms. Quantitation of the individual components is usually done. Residual lignin can be quantitated by one of the methods used for plant material analysis, such as the Klason method (10). Klason lignin is the insoluble material remaining after sample digestion with sulfuric acid and heat. Although the Klason method is not recommended for sensitive quantitation of lignin in biodegraded wood because it slightly underestimates the lignin content (10, 25), it is useful for indicating lignin loss. In addition to the gravimetric Klason lignin analysis method, a spectrophotometric method can be used; in this approach, the lignin is dissolved in acetylbromide and acetic acid, and absorption of the sample at 280 nm is determined (47). The carbohydrate loss can be measured by analysis of the reducing sugar concentration in the hydrolysate resulting from the Klason analysis (5). Although this would not yield the identities of the reducing sugars or indicate whether they originated from cellulose or hemicellulose, the hydrolysate could also be analyzed chromatographically to give an estimate of the sugars' origins (41). These methods would produce quantitative estimates of lignin and carbohydrate fractions. Qualitative chemical changes in these fractions caused by microbial activity would require more sophisticated methods (described below).

RESPIROMETRY

Mineralization of lignocellulose or one of its components is one of the earliest and most common methods of assessing biodegradation. Mineralization is the complete degradation of an organic substrate to carbon dioxide and water under aerobic culture conditions or to carbon dioxide and methane for anaerobic cultures. Numerous methods have been developed and used to monitor mineralization processes in both the field and the laboratory (4). Most frequently, the carbon dioxide produced from a specific quantity of organic substrate is measured. The CO_2 is usually trapped in an aqueous solution of base and is then quantitated by titration with standard acid. Alternatively, the CO_2 can be adsorbed onto solid soda lime and measured gravimetrically. Another respirometric method is to measure the volume of gas consumed in a closed system with a carbon dioxide trap, using a manometer. One drawback of such a system is that the partial pressure of oxygen continually declines until the culture is reaerated. This problem can be overcome by the use of an electrolytic respirometer, in which case the oxygen is replaced as it is depleted. In this system, a decrease of the pressure inside the culture container causes a platinum electrode to come into contact with an aqueous salt solution which is hydrolyzed to yield oxygen. Hydrolysis continues in the electrolytic cell until the pressure in the flask returns to the starting pressure. The amount of oxygen consumed can be quantitated indirectly by trapping and measuring the hydrogen concomitantly produced during electrolysis or by measuring the current consumed during degradation. In addition to these techniques for measuring mineralization of unlabeled substrates, some techniques for quantitating CO_2 released from ^{14}C-labeled substrates have been developed. Carbon dioxide is trapped in an aqueous base and is transferred to an organic CO_2-trapping agent, such as monoethanolamine (8) or Carbo-Sorb (Packard Instruments, Meriden, Conn.). Scintillation cocktail is added, and the $^{14}CO_2$ is quantified by liquid scintillation counting. Respiration measurement techniques have been thoroughly covered elsewhere (4, 9). Measurement of respirometry in anaerobic cultures adds other challenges. In the absence of terminal electron acceptors for anaerobic respiration, methane can be expected as a product from organic compound degradation. [^{14}C]methane is quantified with a liquid proportional counter (39) or by direct liquid scintillation counting in sealed vials (51), or the methane must be separated from CO_2 and then combusted to CO_2, which is trapped and quantitated by liquid scintillation counting as described above (6).

CELLULOSE AND HEMICELLULOSE BIODEGRADATION

The enzymes involved in the degradation of cellulose have been thoroughly studied in a wide range of microorganisms. Two different enzymes hydrolyze the cellulose chains. One randomly cleaves glycosidic linkages within the polymer and is defined as a β-1,4-endoglucanase. Endoglucanases produce cellulose molecules of increasingly smaller molecular weights. The other is a β-1,4-exoglucanase which hydrolyzes the second glycosidic linkage from the nonreducing end of the chain to produce cellulose whose length is progressively decreased by two glucose residues and releases cellobiose. The third enzyme in this hydrolytic process is β-glucosidase, which is responsible for converting the disaccharide cellobiose to glucose. While the endo- and exoglucanases are generally extracellular enzymes, the β-glucosidases may be either intra- or extracellular, depending on the microorganism. Because of the much greater complexity of sugars and linkages in hemicelluloses, many more enzymes are involved in complete hydrolysis of the backbone and the branches. Because xylan is a common hemicellulose backbone constituent, much of the research on hemicellulases has centered around the xylanases. Endoxylanases that produce oligosaccharides from xylan are known. Also known are exoxylanases, which release xylose residues from the end of xylan (49). The ultimate products from these reactions are simple sugars that can be taken up and utilized by the organisms that hydrolyze the polysaccharides.

Methods frequently used to study cellulose and hemicellulose degradation have made use of the reducing capacity of the carbohydrates. For every glycosidic bond of the polysaccharides that is hydrolyzed, a new free reducing end, which can react with agents used to quantitate reducing sugars, is created. There are two commonly used assays to quantitate reducing sugars. One is based on the production of amino-nitrosalicylate from 3,5-dinitrosalicylate (37), while the other measures copper(I) with an arsenomolybdate reagent following reduction of copper(II) by the sugar (44). Both methods are spectrophotometric assays. Details of the application of these methods and others have been

thoroughly reviewed (50). In addition to these methods, measurement of soluble dye released from insoluble dyed polysaccharide has been used as a measure of hydrolytic activity. Cellulose powder, filter paper, or cotton can be easily dyed with Remazol brilliant blue R or some similar reactive dye (14, 36). Enzymatic activity is proportional to the amount of dye made soluble during hydrolysis of the insoluble product. This method can also be used with soluble polysaccharide substrates as long as the substrate can be separated, for example, by precipitation, prior to spectrophotometric measurement of the soluble product. Remazol brilliant blue R-dyed cellulose or cellulose azure is available from commercial sources (Sigma Chemical Co., St. Louis, Mo.).

Disaccharide-hydrolyzing enzymes are usually assayed by using chromogenic substrates that are similar to the natural substrates. p-Nitrophenol is commonly linked through the phenolic hydroxyl to a sugar in the appropriate glycosidic linkage. The substrate is normally colorless, but upon hydrolysis the nitrophenol that is released has a yellow color and absorbs maximally at around 420 nm. Many other natural and synthetic substrates which are used in a similar way are available. Sugars conjugated to 4-methylumbelliferone are another group of substrates commonly used. Hydrolysis of the bond releases 4-methylumbelliferone, which is much more intensely fluorescent than its precursor. These substrates are available from a variety of commercial sources (Sigma).

BIODEGRADATION OF THE LIGNIN COMPONENT

Lignin or lignocellulosic substrates can be subjected to biodegradation by organisms in culture media, and then the material remaining can be recovered for analysis following an appropriate incubation period. Because lignin, cellulose, and hemicellulose are intimately associated in nature, it is thought that the presence of a readily metabolizable carbohydrate is required to achieve significant lignin degradation in an experimental situation. This carbohydrate may be supplied in the form of glucose or some other simple sugar, but frequently the carbohydrate provided is in the form of cellulose and hemicellulose in intact lignocellulose or wood. This also provides results more representative of what occurs in nature. The lignocellulose or wood is degraded by the organism(s) for a period of time, at which point the lignocellulose is collected and dried. It is then powdered, and the lignin is extracted with cold, neutral organic solvents in a similar way to the preparation of milled wood lignin (25). Because much of the carbohydrate has been degraded and many of the bonds linking the hemicellulose to lignin have been hydrolyzed, yields of extractive lignins from degraded lignocellulose are higher than those from sound lignocellulose (25, 43). These extractive lignins are believed to be exclusively from the outward-facing surface of the lignocellulose particles that have had access to degradative enzymes and so are mostly biologically modified. The degraded lignins are analyzed and compared with undegraded lignin.

Substrates

In addition to methods described above whereby lignin, in association with the cellulose and hemicellulose, is biodegraded and then characterized, there are many methods to measure degradation of lignin or lignin-like substrates by themselves. Mineralization of lignin to carbon dioxide is one of the best measures of the ability of one or a group of microorganisms to degrade lignin. Many chemical and physical methods have been developed to characterize lignin and have been applied to partially degraded lignin to elucidate biotransformations mediated by lignin-degrading organisms.

A variety of lignin substrates wherein the lignin has been largely freed of the carbohydrate components of lignocellulose are available. Ideally, these should be relatively unmodified during isolation and should be representative of the bulk lignin. The preparation that best fits these criteria is the milled wood lignin of Björkman (7). It is extracted with neutral organic solvents from ball milled wood. Braun's native lignin, which is extracted with neutral ethanol, is similar to milled wood lignin (43). The yield of lignin from these extractive processes can be increased by prior treatment of the wood with either cellulase or brown-rot fungi (43). During growth of the brown-rot fungi on the polysaccharide components of wood, the lignin is only slightly modified. However, the lignin isolated by using these organisms is not completely representative of the intact lignin. Removal of the carbohydrate from lignin with concentrated mineral acid is very effective and yields lignin in high amounts. However, these harsh conditions yield lignin substrates that are considerably altered from the original lignin. Examples of materials from such processes include Klason lignin and acidolysis lignin. Despite the modifications, such materials do find use as substrates in lignin biodegradation research (40). The most widely used isolated lignins are those resulting as by-products from the pulping industry, such as kraft lignins. Because they are readily available in large quantities, their biodegradation is of considerable interest from a water quality maintenance viewpoint and because they would be the most likely starting materials for a biological process that might develop products from lignin. These materials are produced at relatively high temperatures with caustic chemicals and so may be considerably modified from the original lignin. However, methods of fractionating the pulp-derived lignins have been developed to yield material that has been less severely altered (33).

Lignin Model Compounds

Analysis of the mechanism of cleavage of intermonomeric linkages in lignin usually requires a substrate with a single type of linkage representative of the one under study to facilitate the analysis of products. The substrates primarily used for this purpose are lignin model or substructure compounds. These compounds, which are usually dimers, were first synthesized and used for analysis of lignin chemical processes, such as pulping chemistry (1). Monomeric lignin model compounds, such as syringic or ferulic acids, have occasionally been used in lignin biodegradation research, but because these compounds have many nonlignin sources in nature, their degradation is not very representative of lignin biodegradation.

Model compounds for study of lignin biodegradation are usually used after lignin degradation has been confirmed by a reliable measure of lignin mineralization, such as radiorespirometry with ^{14}C-labeled synthetic (DHP; described below) or natural lignin (8, 19). The model is subjected to biotransformation by microbial cultures grown under ligninolytic conditions, crude enzyme preparations, or isolated enzymes until most of the substrate has been metabolized. The resulting metabolites are extracted from the medium, separated chromatographically (by thin-layer, high-pressure liquid, or gas chromatography), and identified by using chemical identification techniques (mass spectrometry, infrared

[IR] spectroscopy, nuclear magnetic resonance [NMR] spectrometry, etc.). One can then hypothesize about the biochemical reactions that occurred that would convert the substrate to the observed products. It is assumed that the same enzymes are acting upon the lignin model substrate as are acting upon lignin in nature. The results that one obtains from model degradation analysis complement the information gained from analysis of partially degraded lignin isolated from degraded lignocellulose. Models can occasionally be taken up and transformed by intracellular enzymes that may or may not have a natural function of degrading lignin fragments. Therefore, it is not reliable to draw conclusions on the ability of an enzyme to degrade lignin in nature based solely on its ability to degrade a lignin model compound. To prove that an enzyme that degrades a model compound also degrades lignin in nature, it is necessary to show that it degrades isolated lignin (i.e., milled wood lignin).

When one is choosing to use lignin model compounds in research, there are many considerations that must be taken into account. Of most importance is how representative the compound and its intermonomeric linkage are of lignin. The most frequently used models have a β-O-4 ether linkage. Model compounds usually have the syringyl (3,5-dimethoxy-4-hydroxy) or guaiacyl (3-methoxy-4-hydroxy) orientation of aromatic ring substituents which are thought to be representative of lignin. The substituent groups that are found on the aromatic rings in the dimer will usually be different for the two rings, and occasionally ethoxy groups will be used as aromatic ring substituents in place of methoxyls. This allows a researcher to analyze the degradation products from the dimer and assign the origins of the metabolites to the various parts of the substrate with greater reliability. One can then formulate hypotheses about the biochemical reactions responsible for model compound biodegradation. Despite this approach, some intermediates observed cannot be assigned to one arylpropane group or the other. Frequently, it is necessary to use one of the observed degradation intermediates as a substrate in another experiment and determine what other products are formed to formulate a degradation pathway. Another important consideration is whether the compound should be phenolic or nonphenolic. Although there are free phenolic groups in lignin, they are not found at very high concentrations (24). The nonphenolic substrates are often used to avoid interference by laccase or tyrosinase. These enzymes have been long known to be produced by many wood-rotting fungi and other organisms and are known to oxidize the common phenolic lignin model compounds to a variety of products (3). Because these enzymes are thought to not play a major role in the early phases of lignin metabolism, the nonphenolic models are usually used to measure effects of other enzymes.

Lignin models are frequently used in attempts to measure enzymatic activities that may be responsible for lignin depolymerization. To facilitate detection of products, chromogenic substrates have been synthesized (48). This allows development of spectrophotometric, fluorometric, or agar plate assays for lignin metabolic activity.

Another important criterion in the choice of which lignin models to use is availability. Dimeric lignin model compounds are not available commercially and so must be custom synthesized. Many of these synthesis protocols, especially for the radiolabeled compounds, are very involved, entailing many steps, and require a well-equipped synthesis laboratory and considerable expertise. Veratrylglycerol-β-guaiacyl ether is one of the commonly used models (28) and can be synthesized in four steps from commercially available starting materials (30). Dehydrodivanillin can be synthesized in one step from vanillin (11).

Lacking the availability of these compounds, researchers may choose to use other, less lignin-like substrates that are available. One such group of compounds that have been used are the lignans. These have C-C bonds between two C_6-C_3 residues through the β-carbons, and they commonly have tetrahydrofuran rings (18). α-Conidendrin has been used as a substrate to isolate bacteria (45). Although these compounds are structurally different from lignin, they can be used to suggest possible degradation pathways.

Simple lignin-like model compounds can be produced through single-step condensation reactions starting with commercially available starting materials. The reactions are usually initiated in similar ways to lignin polymerization, using peroxidase or related chemical or biological oxidation catalysts. Substrates that can be produced in this way include dehydrodivanillin (11) and dehydrodiisoeugenol (30). Unfortunately, for some of the peroxidase-catalyzed reactions, yields of the desired condensation products are not very high and purification from by-products is usually required. Because two identical monomers condense to form a dimer, microbial degradation of the dimer may yield products whose origin cannot be unequivocally attributed to one half of the dimer or the other.

One of the most commonly used lignin model compounds is the dehydrogenative polymerizate of coniferyl alcohol, or DHP. It is prepared by reacting coniferyl alcohol with peroxide and hydrogen peroxide to produce a lignin-like polymer (16). The reaction conditions affect the molecular weight of the product and the degree to which the intermonomeric linkages resemble lignin (24). DHPs are frequently produced in ^{14}C-labeled form and are used in radiorespirometry experiments to demonstrate mineralization of lignin (27). This requires the synthesis of the coniferyl alcohol precursor with the label in the desired location.

Radioisotope-Labeled Lignin Substrates
One of the most useful lignin substrates is that in which the lignin in lignocellulose is labeled with ^{14}C. This can be used to demonstrate the mineralization of lignin in the presence of cellulose and hemicellulose and is representative of the lignin as it would be found in nature. It is prepared by feeding plants or fresh-cut twigs radiolabeled lignin precursors and then allowing the radioactivity to be incorporated into the lignin (8, 19). The precursors chosen for feeding should give specific labeling of the lignin without labeling other nonlignin materials. Ideally the precursors should be as close to coniferyl alcohol in the biosynthetic pathway as possible, and no compound earlier than L-phenylalanine in the pathway should be considered. It may be desired to have the lignin labeled at a specific carbon in the side chain, on the aromatic ring, or in the methoxyl group. This enables a researcher to determine if a particular portion of the lignin is degraded preferentially by a given organism. However, this also requires the feeding of coumaric acid, ferulic acid, or sinapic acid that has been synthesized with the label in the appropriate location in the precursor, starting with a limited selection of commercially available ^{14}C-labeled synthetic precursors. One means of avoiding the complicated process of synthesizing lignin precursors is to use commercially available ^{14}C-labeled L-phenylalanine as a precursor. This is usually available as the uniformly labeled substrate but can occasionally be found with the label in a specific location. However, because the

L-phenylalanine can also be incorporated into protein, it is essential to treat the resulting lignocellulose with protease to remove labeled protein or to at least hydrolyze a sample with acid and show that the label has not become incorporated into nonlignin material (8). A simple method has been developed to convert much of the L-phenylalanine to trans-cinnamic acid in vitro with L-phenylalanine ammonia lyase prior to feeding it to plants (42). This will increase the specificity of the lignin labeling. Another enzymatic method has been used to label the methoxyl group of caffeic acid to produce ferulic acid, using catechol-O-methyltransferase with the methyl group originating from commercially available S-[methyl-^{14}C]adenosyl-L-methionine (15).

^{14}C-labeled synthetic lignin or DHP is also used as a substrate to demonstrate lignin mineralization. Although it is a lignin model compound, it is regarded as having enough structural similarity to lignin and to have a high enough molecular weight that its metabolism is representative of natural lignin biodegradation (27). However, it lacks chemical linkages to carbohydrate, which may be important in the natural degradation of lignin.

LIGNIN ANALYSIS METHODS
Chemical changes occurring in the lignin component during microbial degradation can be determined by a variety of methods. The lignin fraction can be extracted from the lignocellulose and then be subjected to chemical degradative reactions to yield low-molecular-weight products that can be analyzed and identified by using chromatographic techniques (26). These methods are essentially those that have been used by chemists to characterize the structure of lignin. Oxidation of lignin with nitrobenzene at 180°C yields a variety of benzaldehyde and benzoic acid derivatives (46). The yields of these compounds give the researcher an indication of what chemical modifications occurred to the arylpropane side chain and whether methoxyl groups were removed from or additional hydroxyl groups were introduced onto aromatic rings. Other oxidation methods, including permanganate oxidation (26) and cupric oxide oxidation (22), have been used to characterize lignin. These methods yield aromatic acids and ketones whose profiles indicate chemical modifications. Prior to permanganate oxidation, it is necessary to alkylate free phenolic hydroxyl groups to prevent aromatic ring cleavage by the oxidant. If these groups are ethylated, it is also possible to estimate the number of methoxyl groups removed by the microbial culture. One of the unfortunate drawbacks of these methods is that a relatively small percentage of the original lignin is recovered in the form of identifiable products. This requires one to extrapolate the results to the rest of the lignin, which may be an inappropriate assumption.

Chemical Functional Group Analysis
Fungal degradation of lignin has been shown in some cases to be oxidative (26) and to remove methoxyl groups. These changes in lignin are demonstrated by quantitating the total hydroxyl, carboxyl, and methoxyl content. Hydroxyl content is measured by completely acetylating the lignin with acetic anhydride and an organic base such as pyridine. The acetylated lignin is separated, acetyl groups are saponified, and acetate is measured by titration with a base. Aliphatic hydroxyl concentration is taken as the difference between the total hydroxyl content from acetylation and phenolic hydroxyl content from UV spectrophotometry. Carboxyl groups can be quantitated by potentiometric titration (25).

The Zeisel method for alkoxyl determination has been widely used for determination of methoxyl groups in lignin (38). The methoxyl ether linkage is hydrolyzed with hydriodic acid to form methyl iodide, which is then distilled. The methyl iodide is reacted with bromine, and the resulting iodine is titrated with standard thiosulfate. This method, or variations of it, is still used for determining the methoxyl group content of lignin. The functional group values are often reported as the number of groups per aryl propane unit.

Spectrometric Methods

Spectrometric methods have also been used to characterize and quantify the chemical substituent groups in lignin. Those that have been widely used include UV-visible (UV-Vis) and IR spectroscopy and NMR spectrometry.

UV-Vis Spectroscopy

UV-Vis spectra do not yield much structural detail, but they are used because the spectrophotometers are readily available, and they do permit detection of gross changes in lignin. UV spectra are used primarily to quantitate substituent groups such as phenolic hydroxyls and α-carbonyl residues. Phenolic hydroxyls are measured by recording a difference spectra between samples in basic solvent and neutral solvent (17). Peaks at about 250 and 300 nm in the spectra can be used to quantitate phenolic hydroxyls. α-Carbonyl groups in lignin absorb at around 270 nm. The decrease in the absorbance of this peak upon sodium borohydride reduction is used to estimate the quantities of these groups (2).

IR Spectroscopy

Infrared spectroscopy has also been widely used to characterize the structure of both sound and microbially degraded lignin (13). As with other spectrometric methods, the assignment of peaks is based on numerous spectra of lignin model compounds. Unlike with many of the other spectrometric methods, it is not necessary to record spectra on isolated lignin. Because so many different functional groups contribute to an IR spectrum of lignin, many absorption bands overlap, making it difficult to attribute the bands to specific functional groups (23). It is possible to derivatize the lignin or chemically modify it and to compare spectra before and after modification to assign bands with more reliability. However, some ambiguity still exists. The utility of IR spectra in determining microbially mediated lignin transformations lies in quantitating functional groups that give intense bands in spectral regions that have few or no other interfering bands (23). Isolated carbonyl groups or conjugated carbonyl groups can be measured in this way (23). Loss of aromaticity can also be determined from changes in aromatic skeletal vibrations (13).

NMR Spectrometry

Characterization of degraded lignin by NMR spectrometry has also been done. Many workers have used NMR spectrometry to measure the spectra of various monomeric and dimeric model compounds with structural features representative of lignin. The chemical shift data obtained from these experiments can then be used to assign the peaks in spectra to structural features in lignin. While both ^{1}H and ^{13}C NMR spectra have been used to characterize the structure of lignin (31, 32), ^{13}C NMR spectrometry has advantages over ^{1}H NMR spectrometry. Proton decoupling and signals over a much wider range give ^{13}C NMR spectra that are easier to interpret than ^{1}H NMR spectra (31). Lignin isolated by

neutral solvent extraction from sound and microbially degraded lignocellulose can be analyzed by NMR spectrometry. Samples are usually dissolved in deuteroacetone or deuterodioxane-D_2O mixtures, and lignins for ^{1}H NMR spectrometry are frequently acetylated (34). Modern instruments are capable of recording ^{13}C spectra from solid samples, which overcomes the need to achieve high concentrations of lignin in solution. However, chemical shifts are altered slightly (21).

HYPOTHETICAL EXAMPLE PROBLEM

An example of how a researcher might study the microbial degradation of the components of lignocellulose is given here to demonstrate the order in which methods might be applied. The first step is to demonstrate that lignin and the polysaccharides are degraded by the organism of interest. This is done by radiorespirometry using ^{14}C-labeled substrates or by showing the disappearance of these components from lignocellulose incubated with the organism. The quantitative loss of substrate, Klason lignin, and reducing sugar can also be used to show degradation of the major components. To study carbohydrate degradative activity, concentrated and unconcentrated enzyme-containing culture broth from a test organism is incubated with lignocellulose or other cellulosic or hemicellulosic substrates, and the reducing sugars released are quantified by using the Somogyi-Nelson or a chromatographic method. Colored products released from chromogenic sugar conjugates can also be used to demonstrate disaccharide hydrolysis activity, such as β-glucosidase. Lignin can be isolated by neutral solvent extraction from sound and microbially degraded lignocellulose. These two lignin preparations would then be analyzed by UV-Vis and IR spectrophotometry and ^{13}C NMR spectrometry, and spectral differences would be noted. The isolated lignins should also be characterized for differences in functional groups such as methoxyl and carboxyl content. Chemical degradation of the isolated lignins and chromatographic analysis of the products further indicate what biochemical reactions ocurred during degradation of the lignin. To further study the lignin-degrading enzymes, a lignin model compound such as veratrylglycerol-β-guaiacyl ether could be incubated with the organism under ligninolytic conditions, and the metabolites would be analyzed to determine the biochemistry of degradation.

A multitude of methods have been applied to the chemical and physical characterization of degraded and undegraded lignin, cellulose, and hemicellulose. Many methods have also been used to study the microbial and biochemical degradation of the components of lignocellulose. This chapter was meant to be a sampling of a few of the most useful and frequently applied methods.

REFERENCES

1. **Adler, E., B. O. Lindgren, and U. Saeden.** 1952. The beta-guaiacyl ether of alpha-veratrylglycerol as a lignin model. *Sven. Papperstidn* **55:**245–253.
2. **Adler, E., and J. Marton.** 1959. Zur Kenntnis der Carbonylgruppen im Lignin. I. *Acta Chem. Scand.* **13:**75–96.
3. **Ander, P., and K.-E. Eriksson.** 1978. Lignin degradation and utilization by microorganisms, p. 1–58. *In* M. J. Bull (ed.), *Progress in Industrial Microbiology.* Elsevier Scientific Publishing Co., Amsterdam.
4. **Anderson, J. P. E.** 1982. Soil respiration, p. 831–871. *In* A. L. Page, R. H. Miller, and D. R. Keeney (ed.), *Methods of Soil Analysis,* Part 2. *Chemical and Microbiological Properties,*

2nd ed. Agronomy Society of America–Soil Science Society of America, Madison, Wis.

5. **Antai, S. P., and D. L. Crawford.** 1981. Degradation of softwood, hardwood, and grass lignocelluloses by two *Streptomyces* strains. *Appl. Environ. Microbiol.* **42:**378–380.

6. **Benner, R., A. E. MacCubbin, and R. E. Hodson.** 1984. Anaerobic biodegradation of the lignin and polysaccharide components of lignocellulose and synthetic lignin by sediment microflora. *Appl. Environ. Microbiol.* **47:**998–1004.

7. **Björkman, A.** 1956. Studies on finely divided wood. I. Extraction of lignin with neutral solvents. *Sven. Papperstidn.* **59:**477–485.

8. **Crawford, D. L., R. L. Crawford, and A. L. Pometto III.** 1977. Preparation of specifically labeled ^{14}C-(lignin)- and ^{14}C-(cellulose)-lignocelluloses and their decomposition by the microflora of soil. *Appl. Environ. Microbiol.* **33:**1247–1251.

9. **Crawford, R. L., and D. L. Crawford.** 1978. Radioisotopic methods for the study of lignin biodegradation. *Dev. Ind. Microbiol.* **19:**35–49.

10. **Effland, M. J.** 1977. Modified procedure to determine acid-insoluble lignin in wood and pulp. *Tappi* **60**(10):143–144.

11. **Elbs, K., and H. Lerch.** 1916. Über dehydrodivanillin. *J. Prakt. Chem.* **93:**1–9.

12. **Eriksson, K.-E., R. A. Blanchette, and P. Ander.** 1990. *Microbial and Enzymatic Degradation of Wood and Wood Components.* Springer-Verlag, Berlin.

13. **Faix, O., J. Bremer, O. Schmidt, and T. Stevanovic J.** 1991. Monitoring of chemical changes in white-rot degraded beech wood by pyrolysis-gas chromatography and Fourier-transform infrared spectroscopy. *J. Anal. Appl. Pyrol.* **21:**147–162.

14. **Fernly, H. N.** 1963. The use of reactive dyestuffs in enzymology: new substrates for cellulolytic enzymes. *Biochem. J.* **87:**90–95.

15. **Frazer, A. C., I. Bossert, and L. Y. Young.** 1986. Enzymatic aryl-O-methyl-^{14}C labeling of model lignin monomers. *Appl. Environ. Microbiol.* **51:**80–83.

16. **Freudenberg, K.** 1965. Lignin: its constitution and formation from p-hydroxycinnamyl alcohols. *Science* **148:**595–600.

17. **Goldschmid, O.** 1954. Determination of phenolic hydroxyl content of lignin preparations by ultraviolet spectrophotometry. *Anal. Chem.* **26:**1421–1423.

18. **Goodwin, T. W., and E. I. Mercer.** 1983. *Introduction to Plant Biochemistry,* 2nd ed. Pergamon Press, Oxford.

19. **Haider, K., and J. Trojanowski.** 1975. Decomposition of specifically ^{14}C-labelled phenols and dehydropolymers of coniferyl alcohol as models for lignin degradation by soft and white rot fungi. *Arch. Microbiol.* **105:**33–41.

20. **Hassid, W. Z.** 1970. Biosynthesis of sugars and polysaccharides, p. 301–373. *In* W. Pigman and D. Horton (ed.), *The Carbohydrates—Chemistry and Biochemistry,* 2nd ed., vol. IIA. Academic Press, New York.

21. **Hawkes, G. E., C. Z. Smith, J. H. P. Utley, R. R. Vargas, and H. Viertler.** 1993. A comparison of solution and solid state ^{13}C NMR spectra of lignins and lignin model compounds. *Holzforschung* **47:**302–312.

22. **Hedges, J. I., and J. R. Ertel.** 1982. Characterization of lignin by gas chromatography of cupric oxide oxidation products. *Anal. Chem.* **54:**174–178.

23. **Hergert, H. L.** 1971. Infrared spectra, p. 267–297. *In* K. V. Sarkanen and C. H. Ludwig (ed.), *Lignins: Occurrence, Formation, Structure, and Reactions.* Wiley-Interscience, New York.

24. **Higuchi, T.** 1985. Biosynthesis of lignin, p. 141–160. *In* T. Higuchi (ed.), *Biosynthesis and Biodegradation of Wood Components.* Academic Press, Orlando, Fla.

25. **Kirk, T. K., and H.-M. Chang.** 1974. Decomposition of lignin by white-rot fungi. I. Isolation of heavily degraded lignins from decayed spruce. *Holzforschung* **28:**217–222.

26. **Kirk, T. K., and H.-M. Chang.** 1975. Decomposition of lignin by white-rot fungi. II. Characterization of heavily degraded lignins from decayed spruce. *Holzforschung* **29:**56–64.

27. **Kirk, T. K., W. J. Connors, D. Bleam, W. F. Hackett, and J. G. Zeikus.** 1975. Preparation and microbial decomposition of synthetic [^{14}C] lignins. *Proc. Natl. Acad. Sci. USA* **72:**2515–2519.

28. **Kirk, T. K., J. M. Harkin, and E. B. Cowling.** 1968. Oxidation of guaiacyl- and veratryl-glycerol-β-guaiacyl ether by *Polyporus versicolor* and *Stereum frustulatum. Biochim. Biophys. Acta* **165:**134–144.

29. **Kirk, T. K., E. Schultz, W. J. Connors, L. F. Lorenz, and J. G. Zeikus.** 1978. Influence of culture parameters on lignin metabolism by *Phanerochaete chrysosporium. Arch. Microbiol.* **117:**277–285.

30. **Leopold, B.** 1950. Aromatic keto- and hydroxy-polyethers as lignin models. III. *Acta Chem. Scand.* **4:**1523–1537.

31. **Lüdemann, H.-D., and H. Nimz.** 1973. Carbon-13 nuclear magnetic resonance spectrea of lignins. *Biochem. Biophys. Res. Commun.* **52:**1162–1169.

32. **Ludwig, C. H., B. J. Nist, and J. L. McCarthy.** 1964. Lignin. XIII. The high resolution nuclear magnetic resonance spectroscopy of protons in acetylated lignins. *J. Am. Chem. Soc.* **86:**1196–1202.

33. **Lundquist, K., and T. K. Kirk.** 1980. Fractionation-purification of an industrial kraft lignin. *Tappi* **63**(1):80–82.

34. **Lundquist, K., and T. Olsson.** 1977. NMR studies of lignins. 1. Signals due to protons in formyl groups. *Acta Chem. Scand. Ser. B* **31:**788–792.

35. **McCarthy, A. J.** 1987. Lignocellulose-degrading actinomycetes. *FEMS Microbiol. Rev.* **46:**145–163.

36. **McCleary, B. V.** 1988. Soluble, dye-labeled polysaccharides for the assay of endohydrolases. *Methods Enzymol.* **160:**74–86.

37. **Miller, G. L.** 1959. Use of dinitrosalicylic acid reagent for determination of reducing sugar. *Anal. Chem.* **31:**426–428.

38. **Moore, W. E., and D. B. Johnson.** 1967. Determination of alkoxyl groups. *In Procedures for the Chemical Analysis of Wood and Wood Products.* U.S. Department of Agriculture, Madison, Wis.

39. **Nelson, D. R., and J. G. Zeikus.** 1974. Rapid method for the radioisotopic analysis of gaseous end products of anaerobic metabolism. *Appl. Microbiol.* **27:**258–261.

40. **Odier, E., and B. Monties.** 1978. Biodegradation of wheat lignin by *Xanthomonas* 23. *Ann. Microbiol. (Inst. Pasteur)* **129A:**361–377.

41. **Pettersen, R. C., V. H. Schwandt, and M. J. Effland.** 1984. An analysis of the wood sugar assay using HPLC: a comparison with paper chromatography. *J. Chromatogr. Sci.* **22:**478–484.

42. **Pometto, A. L., III, and D. L. Crawford.** 1981. Enzymatic production of the lignin precursor trans-[U-^{14}C]cinnamic acid from ʟ-[U-^{14}C]phenylalanine using ʟ-phenylalanine amonia-lyase. *Enzyme Microb. Technol.* **3:**73–75.

43. **Schubert, W. J., and F. F. Nord.** 1950. Investigation on lignin and lignification. II. The characterization of enzymatically liberated lignin. *J. Am. Chem. Soc.* **72:**3835–3838.

44. **Somogyi, M.** 1952. Notes on sugar determination. *J. Biol. Chem.* **195:**19–23.

45. **Sundman, V.** 1961. Experiments with bacterial degradation of lignin. *Paperi Puu* **43:**673–676.

46. **Tanahashi, M., and T. Higuchi.** 1988. Chemical degradation methods for characterization of lignins. *Methods Enzymol.* **161:**101–109.

47. **van Zyl, J. D.** 1978. Notes on the spectrophotometric determination of lignin in wood samples. *Wood Sci. Technol.* **12:**251–259.
48. **Weinstein, D. A., and M. H. Gold.** 1979. Synthesis of guaiacylglycol and glycerol-β-O-(β-methylumbelliferyl) ethers: lignin model substrates for the possible fluorometric assay of β-etherases. *Holzforschung* **33:**134–135.
49. **Whistler, R. L., and E. L. Richards.** 1970. Hemicelluloses,

p. 447–469. *In* W. Pigman and D. Horton (ed.), *The Carbohydrates—Chemistry and Biochemistry*, 2nd ed., vol. IIA. Academic Press, New York.
50. **Wood, T. M., and K. M. Bhat.** 1988. Methods for measuring cellulase activities. *Methods Enzymol.* **160:**87–112.
51. **Zehnder, A. J. B., B. A. Huser, and T. D. Brock.** 1979 Measuring methane with the liquid scintillation counter. *Appl. Environ. Microbiol.* **37:**897–899.

Methods for Measuring Hydrocarbon Biodegradation in Soils

INGEBORG D. BOSSERT AND DAVID S. KOSSON

80

Bioremediation is a cost-effective cleanup technology which has, within the past decade, gained wide acceptance for rehabilitating hydrocarbon (HC)-contaminated environments, including soils. In order to reliably implement and maximize the potential of this technology, a clear understanding of biodegradation processes in the environment and an assessment of their impact are essential. With the increasing use of bioremediation technologies, especially in situ applications, a growing repertoire of sampling and analytical techniques is providing more and new information for site characterization, process design, effective implementation, and evaluation. Concurrent with technological developments, greater oversight by regulatory agencies, especially in the United States and the European Community, has provided guidance and impetus for advancing the current state of the art. With these advances, however, may come an unrealistic expectation of cleanup goals. For example, many individual contaminants can now be measured in discrete soil samples at the parts-per-billion level or lower. A comprehensive cleanup strategy must consider not only actual contaminant concentration but also its representation in the matrix (i.e., heterogeneity), its flux (i.e., bioavailability), and the production and impact of metabolic intermediates in the environment. The focus of this chapter is on methodology for the measurement of HC biodegradation activity in soils. It is important to note, however, that such measurement provides merely one of the many tools for a complete environmental assessment and remediation.

HCs include a large and diverse group of organic compounds. By definition, HCs comprise solely those compounds containing carbon and hydrogen. In common parlance, however, the term HCs has grown to include related heteroatom organic compounds derived from petroleum and its products. The microbial degradation of HCs is a long-recognized process that has been measured by a variety of strategies. Early efforts at petroleum prospecting were based on the detection and enumeration of HC-degrading bacteria which were associated with soils overlying petroleum-bearing formations; large numbers of gaseous HC-degrading bacteria isolated from soils served as indicators for natural gas seepage (17, 26). The HC-degrading capability of microorganisms was also exploited during the 1970s and 1980s in the landfarming of oily wastes (7, 70), a process in which HC biodegradation was the predominant treatment and removal mechanism for the applied wastes. For high process effi-

ciency, methods for monitoring HC biodegradation were necessary, and they usually consisted of the direct measurement of HC losses in the well-mixed soil of a landfarm. Today, as remedial applications for microbial biodegradation activity have broadened to include in situ bioremediation technology in many types of soil environments, additional methods for monitoring and assessment are required. Optimization of bioremediation technologies requires methods which can clearly delineate the rate and extent of biodegradation in soils in order to fully assess and improve performance. In addition, the limitations of prior field applications must be addressed with the development, for example, of realistic cleanup criteria, real-time and field monitoring, improved sampling design for field heterogeneity, and the application of emerging technologies such as genetic probes and fiber optics.

INDICATORS OF HC BIODEGRADATION ACTIVITY

In a discussion of methods to measure HC biodegradation in soils, it is first necessary to consider what processes comprise the target activity. Biodegradation in soils is a process which relies on the combined metabolic activities of a diversity of microorganisms and their interactions in a physically, chemically, and biologically complex environment to break down organic contaminants. Because of this complexity, biodegradation activity is difficult to measure directly. The most direct means for assessing overall biodegradation activity relies on the analysis of reactants, e.g., HC contaminants, or their end products, e.g., CO_2. This information can be used to develop a mass balance for quantitation of the process. Other methods, such as microbial enumeration or the measurement of a specific activity, provide indirect evidence for biodegradation and are best used as indicators for the biodegradation potential of a particular soil. The specific methods chosen for measuring contaminant losses and biodegradation activity will vary, depending on the characteristics of the contaminated matrix and its assessment criteria. Numerous reports on HC biodegradation in soils are available in the current literature and include descriptions of many different approaches for measuring biodegradation activity. A recent review of HC biodegradation in soils provides a general overview of existing measurement techniques and their applications (13). More fundamental information can be found in a recommended microbial ecology text (4), and a compilation of detailed instruction on microbial methodology is available in a reference text (38). The focus of this chapter is to provide a condensation of the existing methodology for measuring HC biodegradation in soils.

Direct Measurement of HC Degradation

The direct measurement of HC residues in a soil determines overall losses and thus provides equivocal evidence for biodegradation. A decrease in concentration or a change in HC composition in a soil may be due either to abiotic mechanisms, such as volatilization, leaching, and complexation to soil organic matter, or to biotic processes, such as biodegradation. To ascertain whether biodegradation is a major mechanism for loss, proper controls for abiotic removal must be included in the analysis. In addition, other types of methodology which specifically measure microbial numbers and activity can augment HC analyses to determine what role biodegradation plays in degrading a contaminant in a particular environment.

A variety of quantitative and qualitative techniques are available to measure HC losses in soil. The choice of a particular method usually depends on the type of contamination and soil and on the information required. In most cases, HCs are present as a mixture of compounds, which may be analyzed in bulk, e.g., total petroleum HC or oil and grease (37), or by methods that delineate the loss of individual compounds or qualitative changes in the contaminant matrix (21, 24, 77). The latter strategy has been used successfully to assess biodegradation under poorly defined or controlled conditions in the field, as in the *Exxon Valdez* oil spill (71, 75). During environmental cleanup, changes in the ratio of easily biodegradable, straight-chain HCs, e.g., n-alkanes, to highly branched, recalcitrant species, e.g., pristane, phytane, and hopane, were used to assess HC losses in the field over time; no change in ratio indicated losses primarily through aging and abiotic processes, whereas a decreased ratio indicated loss through biodegradation (21).

HC analysis of a contaminated soil generally requires extraction from the matrix, separation of the HC components, and a means of detection (13). Extraction may be to the vapor phase, e.g., "purge and trap" or thermal desorption, or to organic solvents of different polarities, depending on the chemical nature of the contaminant(s) to be analyzed. Separation into individual components or classes of compounds is generally performed by either gas or liquid chromatographic techniques (31, 32, 40, 72, 85). Quantitative or analytical detection can be gravimetric or by a variety of spectrophotometric or potentiometric methods.

Several emerging technologies and applications show potential for directly measuring HC contaminants in their matrices, eliminating the need for HC extraction and separation. Such technologies range from remote sensing with fiber optics to immunoassay (3, 6, 20). Although promising for future use, their current applications are limited, primarily because of difficulties in quantitation or because of matrix and contaminant interferences.

Measurement of Microbial Indicators

Microorganisms are often the primary mediators of HC degradation and removal in soils. Measurement of their metabolic activities and numbers serves as a good indicator of the biodegradation potential of a soil. Under favorable environmental conditions, e.g., moderate temperature and pH, adequate moisture, and availability of inorganic nutrients and electron acceptors, organic contaminants serve as substrates for growth and energy to the soil microbiota and generally enhance metabolic activity and growth. Measurable microbial parameters provide an indication of how well a soil can support microbial activity and how well its microbiota has responded to contamination. It is important to note that the measurement of microbial activity and numbers affords only an indirect measure of overall degradation activity, but because of their broad applicability, relative simplicity, and low cost, these methods often are common measures for the assessment and prediction of bioremediation. An overview of existing methodologies for measuring soil microbial activity and numbers as a means for evaluating biodegradation potential in contaminated soils is presented later in this chapter.

Microbial Activity

Many different types of microorganisms, in particular bacteria and fungi, effect the biodegradation of organic contaminants in soils (13, 59). Along with their diversity occur a wide variety of metabolic processes which can be measured to monitor biodegradation activity, in either a broad or a selective manner. For example, some processes, such as respiration and energy production, are widespread and occur in many types of microbial populations, whereas other metabolic processes, e.g., reactions of specific enzymes such as ligninases or cellulases, may be limited to only a few types of microorganisms in a particular soil.

HC Metabolism and Respiration

The general scheme for the biodegradation of HC contaminants is an oxidative one, whereby the contaminant (electron donor) is oxidized in the presence of oxygen or an alternate electron acceptor; the electron acceptor, in turn, is reduced in the coupled reaction. If conditions do not limit biodegradation activity, the contaminant substrate ultimately is converted to CO_2, a process termed mineralization. The measurement of CO_2 production, therefore, assesses the end product of biodegradation, as mediated by a wide array of microorganisms, under a broad range of redox conditions (22, 33, 49, 53, 59). For these reasons, and because it can be quantitated by using nondestructive techniques, CO_2 production often serves as the preferred measure of microbial activity. However, it can be a poor surrogate in some soils, because CO_2 equilibria are easily affected by local conditions, especially pH, pressure, moisture, and inorganic carbonates; these impacts may skew the measurement of biological activity.

Many variations in methodology exist, but the principles of CO_2 measurement in soils are essentially the same for all approaches. Soil is incubated in a sealed vessel with an alkali sorbent in the headspace to trap the CO_2 produced in the soil; the trapped CO_2 is then quantified. The design of biometer flasks (8) exemplifies these principles, and such flasks are commonly used for soil biodegradation studies (14, 28, 71). Given the heterogeneous nature of soil, variability among replicate incubations has been reported to be remarkably small, less than within a 5% standard deviation, for well-mixed soils (81). A number of studies have demonstrated good correlation between CO_2 production and biomass or microbial numbers. For example, Anderson and Domsch (2) reported that approximately 40 mg of biomass produced 1 ml of CO_2 per h at 22°C. In their study, measured amounts varied between soils, but results were consistent within each soil type tested. In a recent report examining the use of biometers for treatability studies, mass balances of substrate loading to CO_2 production confirmed that within a noninhibitory concentration range, mineralization to CO_2 was proportional to the concentration of test compound, irrespective of its composition (81).

Other variations for measuring CO_2 production in soils include a modified gas train apparatus (64), which is especially useful for soil incubations containing volatile and semivolatile HCs (and other contaminants). In this approach, volatilized HCs in the headspace gases are selectively trapped in an organic solvent before CO_2 sorption into an alkali solution. By monitoring all contaminant species, including volatilized and mineralized components, complete mass balances are achievable. Radiotracer studies, in which ^{14}C-labeled (or ^{13}C-labeled) substrates are purposefully added to the soil, often employ this or similar methodologies to achieve good recovery and mass balances (63, 82). The use of radiolabeled substrates confers greater sensitivity and enables the fate of specific substrates, or even specific carbon configurations, to be determined in contaminant mixtures. Also, radiotracer methods are useful for measuring contaminant-derived CO_2 in soils with a high natural organic content or a large reserve of inorganic carbonates (96). However, as in the case of adding any exogenous substrate to the soil matrix, partitioning and bioavailability of the added material may differ markedly from weathered field contamination (23), so that the biodegradation activity observed with the added substrate may not necessarily represent that occurring under real field conditions.

Role of Electron Acceptors

Next to measurement of CO_2 production, one of the most common parameters used for measuring HC biodegradation activity in soils is O_2 consumption, or O_2 uptake rate. Because of the dynamic equilibria often existing between CO_2 and soils, especially those which are acidic in nature or which have large reserves of inorganic carbonate, the measurement of other biodegradation indicators, rather than mineralization to CO_2, is sometimes a preferred measure of activity. O_2 consumption is widely used to assess biodegradation activity in aerobic environments, both in the laboratory and in field soils (25, 49). Although direct detection methods such as thermal conductivity can be used, O_2 uptake is often measured manometrically. Current methods rely on electrolytic respirometry for monitoring soil respiration as a function of biodegradation activity. Because most electrolytic systems rely on pressure changes due to O_2 uptake, they are restricted to aerobic activity measurements and require high mass transfer rates, which can be attained only in well-mixed systems such as slurry reactors. Although the current technology generally does not reflect field conditions, it does provide useful information on the biodegradation potential of a contaminant in soil, especially over short time scales. Kinetic studies (25, 41) have shown good correlation between rates of oxygen uptake and chemical oxygen demand removal for a variety of organic compounds.

Until recently, because of the reduced nature of HCs, it was assumed that HC biodegradation occurred predominantly in aerobic environments, where O_2 was the primary reactant and electron acceptor. It is now recognized that many HCs are biodegraded under anoxic conditions, where other external electron acceptors drive the reaction (22, 33, 43). In aerobic environments, exogenous O_2 serves as the electron acceptor and is reduced to water. In anoxic environments, other chemical species, such as NO_3, Fe(III), SO_4, or HCO_3, serve as the electron sink during HC metabolism (22, 62, 76). Analogous to O_2 consumption as a measure of aerobic biodegradation activity, the depletion of external electron acceptors, or the production of their reduced

species, can serve as a measure of biodegradation activity in anoxic soils (42, 52, 87). Whereas CO_2 production provides an overall assessment of HC biodegradation in soil, mediated by a wide range of physiologically diverse microorganisms, the measurement of a specific electron acceptor or its reduced species provides a secondary measure which represents the activity of only a portion of the microbiota that may ultimately contribute to the overall CO_2 pool. The use of electron acceptors as a gauge for HC biodegradation in soils, therefore, is a selective technique. It can be very useful for measurements in the field, especially in soils which contain measurable concentrations of a appropriate species. The acceptor chosen for monitoring must be present at concentrations well above background soil levels (either naturally occurring or by amendment). For example, Albrechtson and coworkers (1) were able to demonstrate in situ microbial degradation of a landfill leachate coupled to the reduction of naturally occurring iron levels in the soil. In a field demonstration of controlled nitrate addition to enhance toluene biodegradation, Mester (68) demonstrated good correlation between nitrate utilization and toluene removal in both soil and gas phases. Methane also serves as an indicator for anaerobic activity in soils. However, observed methane levels must be interpreted with caution because of high background levels in soils rich in organic material or in areas with natural gas seepage.

Biochemical Assays

Closely coupled to metabolic and respiratory processes is energy generation via ATP. The measurement of ATP, or the overall adenylate charge, has also served to assess overall activity and biomass in a soil (34, 54, 74, 89, 92). A direct comparison between the ATP concentration in a soil and the corresponding biomass is often difficult because ATP levels can fluctuate widely (9, 48), depending on the metabolic state of the microbial population. The total adenylate energy charge is a more comparable method of analysis which normalizes ATP levels to the total pool of adenylates (65). By measuring their relative abundances, an estimation of biomass (e.g., numbers) or activity (e.g., respiration) of the degrading soil populations is possible (51, 60, 97).

Whereas respirometry and ATP measurements provide a broad assessment of microbial activity in soils, other biochemical assays may be used to target a smaller subpopulation of the degrading microbial community, by measuring a specific metabolic activity (44, 55, 78, 98). Unless activity is dependent on the contaminant, these measurements are generally not specific for HC-degrading activity. Many enzyme assays are colorimetric and can therefore serve a dual role as vital dye for visualizing living microorganisms. For example, fluorescein diacetate and sulfofluorescein diacetate, dyes used for microbial enumeration or biomass determinations, are hydrolyzed by esterases to the fluorescent product fluorescein (35, 91).

Substrate uptake assays using radiolabeled substrates provide another means for assessing the biodegradation potential, or microbial vitality, of a soil. The specificity of this technique relies primarily on the type of radiolabeled substrate chosen. A contaminant analog can provide specific information on the overall fate of a compound in a soil (15, 58, 88), while widely assimilated substrates such as ^{14}C-labeled glucose, acetate, or thymidine will give information on the generic metabolic activity, or heterotrophic potential, of a soil (27, 29, 98). In either case, the distribution of the labeled substrate, i.e., mineralization to CO_2, assimilation into biomass, sorption, and incorporation into the soil organic matter, provides details on the fate of a particular compound in the environment, including kinetic information and the determination of a mass balance. It is important to note that whereas substrates added to soils may be more easily quantified than in situ contamination, they may partition differently than the native, weathered species and may therefore not necessarily represent field contaminants.

Microbial Enumeration

Microbial enumeration is not a direct measure of activity in soils; however, it does provide a measure of microbial vitality, or biodegradation potential, in a contaminated soil by measuring how well the soil supports microbial growth (13). In addition, the distribution of type and the number of microorganisms at a site may help to characterize the site with respect to concentration and age of the contaminant. Fresh spills and/or high levels of contaminants often kill or inhibit a large sector of the soil microbiota, whereas soils with lower levels of aged contamination show a greater number and diversity of microorganisms (12, 27, 59, 95). In general, most enumeration techniques for evaluating the biodegradation potential of contaminated soils focus on specific bacterial populations in the soil. Recent attention has also been directed to other members of the soil microbiota that may play a significant role in biodegradation processes. For example, several researchers have reported on the enumeration of fungi and protozoa as a means of assessing biodegradation activity (63, 84). Increased numbers of protozoa in contaminated soils were attributed to predation on bacteria, whose numbers also increased in response to HC contamination; fungal populations demonstrated a smaller fluctuation.

As with activity measurements, enumeration techniques vary in specificity. Direct methods are generally the most inclusive (9, 10, 39, 66, 69), while culture methods, as well as biochemical and genetic techniques, are more selective and may be used to target specific populations (1, 5, 98, 99). Recent modifications in culturing techniques, developed specifically for the enumeration of HC-degrading populations, include an overlayer plating technique (11, 56) and a modified most-probable-number microtiter screening technique (18, 75). The overlayer plating technique enables the screening and isolation of microbial populations that degrade water-insoluble HCs; the method has been further developed to screen for cometabolic activity (83).

Genetic methods include the use of engineered reporter microorganisms that serve as in situ biosensors for target contaminants (46) and gene probe and hybridization procedures that target specific gene sequences or enzymes in a microbial community (67, 94). Improvements on the earlier methodology include better extraction and amplification techniques to remove inhibitory soil components and to increase sensitivity of the assay (47, 50, 90). However, further development and evaluation are still needed before these genetic techniques provide a reliable and quantitative means for microbial enumeration of soils.

A more detailed discussion of microbial enumeration can be found in chapter 41. It is important to note that although counting methods can serve as an indicator for *potential* biodegradation activity in a soil, each methodology introduces its own inherent bias and at best only approximates the actual numbers of microorganisms present in the environment.

ASSESSMENT OF HC BIODEGRADATION RATES

From the previous section, it can be seen that many strategies and methods are available to observe and quantify the rate and extent of HC biodegradation activity in soil. In addition to knowledge about contaminant profile and the local hydrogeological and climatic conditions, information gained from the measurement of HC biodegradation activity, especially rate parameters, provides critical input for the assessment of a site and for the design of strategies to remediate a contaminated soil. To accurately assess HC biodegradation activity and rates, especially as they occur in the field, it is important to use an appropriate test system for study, including proper procedures for sample collection and for preparation, analysis, and careful interpretation of the results. In this section, an overview of model systems and the development of kinetic parameters for the evaluation of HC biodegradation in soils is presented.

Test Systems for the Evaluation of HC Biodegradation Rates in Soils

Laboratory conditions generally allow for greater control of environmental parameters, including temperature, moisture, pH, and substrate concentration, by improved mixing and mass transfer. They thereby permit more precise measurement of the targeted activity in the matrix under study. The increased precision of laboratory measurements comes at some cost; incubations often do not adequately mimic real field conditions and therefore provide only an incomplete assessment of activity in the field. Moreover, transfer to the laboratory can disturb sample integrity and thereby affect activity. Results obtained from laboratory studies, therefore, must be interpreted with caution before extrapolation to the field.

Whereas field studies provide a more accurate representation of in situ conditions, the information gained is often limited as a result of inherent heterogeneities in the matrix, which affect site characterization and evaluation of biodegradation activity (13). In addition, poor mixing and mass transfer limitations restrict the ability to control or manipulate environmental parameters such as pH, temperature, and moisture, as well as electron acceptor and contaminant distribution, which often adversely affect biodegradation activity and rates.

Because they are easier and less costly to conduct, laboratory studies are routinely used to assess biodegradation activity in field samples. Despite their recognized limitations, with proper interpretation, laboratory studies still provide valuable information for field assessments and applications. Microcosms comprise well-mixed study systems that are often operated in a contained, static mode for good quantitation of reactants and products, in order to achieve a rigorous mass balance of contaminant and biodegradation activity. As already mentioned, the improved quantitation achievable in microcosms is offset by incubation conditions which are generally far removed from actual field conditions.

Soil columns provide a more dynamic representation of soil activity, including temporal and spatial compartmentalization of biodegradation activity, in a flowthrough gradient system. Columns containing premixed and repacked soils minimize variability of results due to matrix heterogeneity but, like microcosms, provide a less realistic representation of field activity, especially with regard to porosity and flow characteristics (36, 86). On the other hand, soil columns constructed from undisturbed core material provide a good representation of actual field conditions, but the results are more difficult to quantitate and evaluate and generally exhibit much greater variability between samples because of field heterogeneities.

Biodegradation Rates

Biodegradation rate information provides critical input for the assessment, design, and implementation of bioremediation for site cleanup, especially to project the time period and extent of cleanup. Rate information can be developed from both laboratory and field studies, but the use of biodegradation rates measured in the laboratory for the prediction of field bioremediation rates must be approached with caution. Laboratory measurements should be viewed as the limit case in which maximum achievable rates under the defined conditions are observed. These determinations may also include measurements in matrices other than soils, if appropriate corrections are made for mass transfer effects and chemical interactions as they occur in the soil-contaminant matrix. Because of its inherent complexity, prediction of the rates and extent of contaminant biodegradation in field soils is a challenging task. Mathematical models may help to overcome some of the current limitations, especially with the incorporation of independently measured transport and biodegradation parameters to simulate contaminant flux in soils.

In the simplest case, i.e., aqueous cell suspensions, the aqueous phase represents only a fraction of the total system volume. Translation and incorporation of biodegradation rates to a multiphasic soil system must also include consideration of the multiphase partitioning of contaminants in the soil matrix. If the specific biodegradation rate function observed in the laboratory aqueous system is defined as m, the rate observed in the whole soil system, assuming that interphase mass transfer is not limiting, will be equal to m/R, where R is a multiphase retardation factor dependent on numerous physical parameters.

This effect is illustrated by the work of Knaebel and coworkers (57) in their study on the biodegradation of surfactants, whereby they examined the effect of surfactant interactions with soil and soil components on aerobic biodegradation kinetics and extent. The desorption coefficients of the soil-bound surfactants negatively correlated with both the initial rates of degradation and the ultimate extent of mineralization achieved. A 3/2-order kinetic model originally proposed by Brunner and Focht (19) was found to best describe the overall coupled desorption and mineralization kinetics. Scow and coworkers have also demonstrated the dependency of biodegradation rates on contaminant diffusion and sorption in soils (79, 80).

Biodegradation rates measured in the laboratory and the field have been based on substrate disappearance, electron acceptor uptake, and/or CO_2 production. The biodegradation rates of several fuels were measured by using manually tilled laboratory soil columns (85). In this study, the observed ranges of half-lives were 1.7 to 6.0 weeks for jet fuel, 5.5 to >18 weeks for heating oil, 7 weeks for diesel oil, and >48 weeks for bunker C fuel. Gasoline removal during these studies was attributed primarily to volatilization. Loss rates of oil and grease have been measured in field plots during land treatment of a sludge from an oily waste lagoon (61). The half-life of the total oil and grease ranged from ca. 260 to 400 days. The half-lives for naphthalenes, alkanes, and specific aromatics were less than 30 days. Volatilization may have been a significant loss mechanism for the more volatile constituents; approximately 20 to 25% of the total oil and grease was considered refractory.

CO_2 evolution was used to quantify biodegradation rates in laboratory soil columns, using soils obtained from a coral reef atoll (73). The soils were contaminated with petroleum HCs from fuels associated with past military operations. Average total petroleum HC biodegradation rates were estimated to be between 25 and 50 mg/kg of soil per day. Field in situ respirometry has been used to evaluate the treatment effects of various management options during bioventing (30). Oxygen uptake was found to be more sensitive to process changes than to CO_2 evolution. Observed mean oxygen uptake rates varied between 0.016 and 0.030 day^{-1}.

For field studies, quantitative results can be obtained from soil, soil vapor, or groundwater sampling (68, 93). Hutchins and coworkers (53) measured the biodegradation rates of benzene, toluene, xylenes, ethylbenzene, and 1,2,4-trimethylbenzene under denitrifying conditions, using samples of contaminated and uncontaminated soil cores. First-order biodegradation rate constants ranged from 0.016 to 0.38 day^{-1} and 0.022 to 0.067 day^{-1} for contaminated and uncontaminated core material, respectively. Ranges in activity may be attributed to observed lag periods of up to 30 days in the uncontaminated core material. In addition, benzene and m-xylene contaminants were found to inhibit the basal rate of denitrification.

SUMMARY AND FUTURE DIRECTIONS

Reflecting the growing reliance on biodegradation processes to clean up HC-contaminated soils, numerous strategies and methods are currently available for measuring HC biodegradation in soils. Choice of a method will depend on the information required as determined by assessment criteria, applicability, and cost. Over the past decade, increased analytical capabilities have resulted in greater extraction efficiencies and diminishing limits of detection. With these expanded capabilities, it is of paramount importance to process and interpret this growing information in its proper environmental context. Emphasis must be placed on careful extrapolation to real field conditions, with an emphasis on contaminant flux and bioavailability. These two criteria are of particular importance because they ultimately will dictate the contaminant risk-benefit dynamics and play an important role in establishing realistic endpoint criteria for cleanup.

Great progress has been made in environmental, i.e., soil, analysis and assessment. However, several outstanding issues still remain: (i) validation and extrapolation of laboratory results to the field, (ii) accurate site characterization of an inherently heterogeneous matrix, and (iii) questions of mass transfer and bioavailability with regard to both electron donor, e.g., contaminant, and electron acceptor, e.g., O_2 or NO_3. Along with a continued refinement of existing and emerging technologies, a significant response to these issues may come from greater reliance on computer-assisted analyses, based on an evolving generation of models constructed with independently measured parameters, that realistically reflect and predict contaminant fate in the environment.

REFERENCES

1. **Albrechtson, H.-J., G. Heron, and T. H. Christensen.** 1995. Limiting factors of microbial Fe(III)-reduction in a landfill leachate polluted aquifer (Vejen, Denmark). *FEMS Microbiol. Ecol.* **16:**233–248.
2. **Anderson, J. P. E., and K. H. Domsch.** 1978. A physiological method for the quantitative measurement of microbial biomass in soils. *Soil Biol. Biochem.* **10:**215–221.
3. **Arnold, M. A.** 1990. Fiber-optic biosensors. *J. Biotechnol.* **15:**219–228.
4. **Atlas, R. M., and R. Bartha.** 1993. *Microbial Ecology: Fun-*

damentals and Applications, 3rd ed. Benjamin-Cummings, New York.

5. **Atlas, R. M., and G. S. Sayler.** 1989. Tracking microorganisms and genes in the environment, p. 31–50. *In* G. S. Omenn (ed.), *Environmental Biotechnology: Reducing Risks from Environmental Chemicals through Biotechnology*. Plenum Press, New York.

6. **Barnard, S. M., and D. R. Walt.** 1991. Fiber-optic organic vapor sensor. *Environ. Sci. Technol.* **25:**1301–1304.

7. **Bartha, R., and I. Bossert.** 1984. The treatment and disposal of petroleum wastes, p. 553–578. *In* R. Atlas (ed.), *Petroleum Microbiology*. Macmillan Publishing Co., New York.

8. **Bartha, R., and D. Pramer.** 1965. Features of a flask and method for measuring the persistence and biological effects of pesticides in soil. *Soil Sci.* **100:**68–70.

9. **Beloin, R. M., J. L. Sinclair, and W. C. Ghiorse.** 1988. Distribution and activity of microorganisms in subsurface sediments of a pristine study site in Oklahoma. *Microb. Ecol.* **16:**85–97.

10. **Bloem, J., M. Veninga, and J. Shepherd.** 1995. Fully automatic determination of soil bacterium numbers, cell volumes, and frequencies of dividing cells by confocal laser scanning microscopy and image analysis. *Appl. Environ. Microbiol.* **61:**926–936.

11. **Bogardt, A. H., and B. B. Hemmingsen.** 1992. Enumeration of phenanthrene-degrading bacteria by an overlayer technique and its use in evaluation of petroleum-contaminated sites. *Appl. Environ. Microbiol.* **58:**2579–2582.

12. **Bossert, I., and R. Bartha.** 1984. The fate of petroleum in soil ecosystems, p. 435–474. *In* R. Atlas (ed.), *Petroleum Microbiology*. Macmillan Publishing Co., New York.

13. **Bossert, I. D., and G. C. Compeau.** 1995. Cleanup of petroleum hydrocarbon contamination in soil, p. 77–126. *In* L. Y. Young and C. E. Cerniglia (ed.), *Microbial Transformation and Degradation of Toxic Organic Chemicals*. Wiley-Liss, Inc., New York.

14. **Bossert, I. D., W. M. Kachel, and R. Bartha.** 1984. Fate of hydrocarbons during oily sludge disposal in soil. *Appl. Environ. Microbiol.* **47:**763–767.

15. **Bouwer, E., N. Durant, L. Wilson, W. Zhang, and A. Cunningham.** 1994. Degradation of xenobiotic compounds *in-situ*: capabilities and limits. *FEMS Microbiol. Rev.* **15:**307–317.

16. **Bowlen, G., and D. S. Kosson.** 1995. In-situ processes for bioremediation of BTEX and petroleum fuel products, p. 515–544. *In* L. Y. Young and C. E. Cerniglia (ed.), *Microbial Transformation and Degradation of Toxic Organic Chemicals*. Wiley-Liss, Inc., New York.

17. **Brisbane, P. G., and J. N. Ladd.** 1965. The role of microorganisms in petroleum exploration. *Annu. Rev. Microbiol.* **19:**351–364.

18. **Brown, E. J., and J. F. Braddock.** 1990. Sheen screen, a miniaturized most-probable-number method for enumeration of oil-degrading microorganisms. *Appl. Environ. Microbiol.* **56:**3895–3896.

19. **Brunner, W., and D. D. Focht.** 1984. Deterministic three half-order kinetic model for degradation of added carbon substrates in soil. *Appl. Environ. Microbiol.* **47:**167–172.

20. **Carter, K. A.** 1992. Using immunoassays for soil analysis. *Environ. Prot.* **3:**43–44.

21. **Christensen, L. B., and T. H. Larsen.** 1993. Method for determining the age of diesel oil spills in the soil. *Ground Water Monit. Rev.* **13(4):**142–149.

22. **Colberg, P. J. S., and L. Y. Young.** 1995. Anaerobic degradation of nonhalogenated homocyclic aromatic compounds coupled with nitrate, iron, or sulfate reduction, p. 307–330. *In* L. Y. Young and C. E. Cerniglia (ed.), *Micro-*

bial Transformation and Degradation of Toxic Organic Chemicals. Wiley-Liss, Inc., New York.

23. **Connaughton, D. F., J. R. Stedinger, L. W. Lion, and M. L. Schuler.** 1993. Description of time-varying desorption kinetics: release of naphthalene from contaminated soils. *Environ. Sci. Technol.* **27:**2397–2403.

24. **Cozzarelli, I. M., R. P. Eganhouse, and M. J. Baedecker.** 1990. Transformation of monoaromatic HCs to organic acids in anoxic groundwater environment. *Environ. Geol. Water Sci.* **16:**135–141.

25. **Dang, J. S., D. M. Harvey, A. Jobbagy, and C. P. L. Grady, Jr.** 1989. Evaluation of biodegradation kinetics with respirometric data. *Res. J. Water Pollut. Control Fed.* **61:**1711–1721.

26. **Davis, J. B.** 1967. *Petroleum Microbiology*. Elsevier Publishing Co., New York.

27. **Dean-Ross, D.** 1989. Bacterial abundance and activity in hazardous waste-contaminated soil. *Bull. Environ. Contam. Toxicol.* **43:**511–517.

28. **Dibble, J. T., and R. Bartha.** 1979. Effect of environmental parameters on the biodegradation of oil sludge. *Appl. Environ. Microbiol.* **37:**729–739.

29. **Dobbins, D. C., and F. K. Pfaender.** 1988. Methodology for assessing respiration and cellular incorporation of radiolabeled substrates by soil microbial communities. *Microb. Ecol.* **15:**257–273.

30. **Dupont, R. R.** 1993. Fundamentals of bioventing applied to fuel contaminated sites. *Environ. Prog.* **12:**45–53.

31. **Eaganhouse, R. P., M. J. Baedecker, I. M. Cozzarelli, G. R. Alken, K. A. Thorn, and T. F. Dorsey.** 1993. Crude oil in a shallow sand and gravel aquifer. II. Organic geochemistry. *Appl. Geochem.* **8:**551–567.

32. **Eaganhouse, R. P., T. F. Dorsey, C. S. Phinney, and A. M. Westcott.** 1993. Determination of C6–C10 aromatic hydrocarbons in water by purge-and-trap capillary gas chromatography. *J. Chromatogr.* **628:**81–92.

33. **Edwards, E. A., and D. Grbic-Galic.** 1992. Complete mineralization of benzene by aquifer microorganisms under strictly anaerobic conditions. *Appl. Environ. Microbiol.* **58:**2663–2666.

34. **Elland, F.** 1983. A simple method for quantitative determination of ATP in soil. *Soil Biol. Biochem.* **15:**665–670.

35. **Federle, T. W., D. C. Dobbins, J. R. Thornton-Manning, and D. D. Jones.** 1986. Microbial biomass, activity, and community structure in subsurface soils. *Ground Water* **24:**365–374.

36. **Fontes, D. E., A. L. Mills, G. M. Hornberger, and J. S. Herman.** 1991. Physical and chemical factors influencing transport of microorganisms through porous media. *Appl. Environ. Microbiol.* **57:**2473–2481.

37. **Franson, M. A. (ed.).** 1992. *Standard Methods for the Examination of Water and Wastewater*. American Public Health Association, Washington, D.C.

38. **Gerhardt, P., R. G. E. Murray, W. A. Wood, and N. R. Krieg (ed.).** 1994. *Methods for General and Molecular Bacteriology*. ASM Press, Washington, D.C.

39. **Ghiorse, W. C., and D. L. Balkwill.** 1983. Enumeration and morphological characterization of bacteria indigenous to subsurface environments. *Dev. Ind. Microbiol.* **24:**213–224.

40. **Giger, W., and M. Blumer.** 1974. Polycyclic aromatic hydrocarbons in the environment: isolation and characterization by chromatography, visible, ultraviolet, and mass spectrometry. *Anal. Chem.* **46:**1663–1671.

41. **Grady, C. P. L., Jr., J. S. Dang, D. M. Harvey, A. Jobbagy, and X.-L. Wang.** 1989. Determination of biodegradation kinetics through use of electrolytic respirometry. *Water Sci. Technol.* **21:**957–968.

42. **Grbic-Galic, D.** 1990. Anaerobic microbial transformation

of nonoxygenated aromatic and acyclic compounds in soil, subsurface, and freshwater sediments. *Soil Biochem.* **6:** 117–189.

43. **Haggblom, M. M., M. D. Rivera, I. D. Bossert, J. E. Rogers, and L. Y. Young.** 1990. Anaerobic biodegradation of p-cresol under three reducing conditions. *Microb. Ecol.* **20:**141–150.

44. **Hammel, K. E.** 1995. Organopollutant degradation by ligninolytic fungi, p. 331–346. *In* L. Y. Young and C. E. Cerniglia (ed.), *Microbial Transformation and Degradation of Toxic Organic Chemicals.* Wiley-Liss, Inc., New York.

45. **Heitkamp, M. A., J. P. Freeman, and C. E. Cerniglia.** 1987. Naphthalene biodegradation in environmental microcosms: estimates of degradation rates and characterization of metabolites. *Appl. Environ. Microbiol.* **53:**129–136.

46. **Heltzer, A., K. Malachowsky, J. E. Thonnard, P. R. Bienkowski, D. C. White, and G. S. Sayler.** 1994. Optical biosensor for environmental on-line monitoring of naphthalene and salicylate bioavailability with an immobilized bioluminescent catabolic reporter bacterium. *Appl. Environ. Microbiol.* **60:**1487–1494.

47. **Herrick, J. B., E. L. Madsen, C. A. Batt, and W. C. Ghiorse.** 1993. Polymerase chain reaction amplification of naphthalene-catabolic and 16S rRNA gene sequences from indigenous sediment bacteria. *Appl. Environ. Microbiol.* **59:** 687–694.

48. **Hickman, G. T., and J. T. Novak.** 1989. Relationship between subsurface biodegradation rates and microbial density. *Environ. Sci. Technol.* **23:**525–532.

49. **Hinchee, R. E., and S. K. Ong.** 1992. A rapid *in-situ* respiration test for measuring aerobic biodegradation rates of hydrocarbons in soil. *J. Air Waste Manage. Assoc.* **42:** 1305–1312.

50. **Holben, W. E., J. K. Jansson, B. K. Chelm, and J. M. Tiedje.** 1988. DNA probe method for the detection of specific microorganisms in the soil bacterial community. *Appl. Environ. Microbiol.* **54:**703–711.

51. **Holm, P. E., P. H. Nielsen, H.-J. Albrechtsen, and T. H. Christensen.** 1992. Importance of unattached bacteria and bacteria attached to sediment in determining potentials for degradation of xenobiotic organic contaminants in an aerobic aquifer. *Appl. Environ. Microbiol.* **58:** 3020–3026.

52. **Hutchins, S. R.** 1991. Biodegradation of monoaromatic hydrocarbons by aquifer microorganisms using oxygen, nitrate, or nitrous oxide as the terminal electron acceptor. *Appl. Environ. Microbiol.* **57:**2403–2407.

53. **Hutchins, S. R., G. W. Sewell, D. A. Kovacs, and G. A. Smith.** 1991. Biodegradation of aromatic hydrocarbons by aquifer microorganisms under denitrifying conditions. *Environ. Sci. Technol.* **25:**68–76.

54. **Kieft, T., and L. Rosacker.** 1991. Application of respiration- and adenylate-based soil microbiological assays to deep subsurface terrestrial sediments. *Soil Biol. Biochem.* **23:**563–568.

55. **Kirchner, M. J., A. G. Wollum II, and L. D. King.** 1993. Soil microbial populations and activities in reduced chemical input agroecosystems. *Soil Sci. Soc. Am. J.* **57:** 1289–1295.

56. **Kiyohara, H., K. Nagao, and K. Yana.** 1982. Rapid screen for bacteria degrading water-insoluble, solid hydrocarbons on agar plates. *Appl. Environ. Microbiol.* **43:**458–461.

57. **Knaebel, D. B., T. W. Federle, D. C. McAvoy, and J. R. Vestal.** 1994. Effect of mineral and organic soil constituents on microbial mineralization of organic compounds in a natural soil. *Appl. Environ. Microbiol.* **60:**4500–4508.

58. **Konopka, A., and R. Turco.** 1991. Biodegradation of organic compounds in vadose zone and aquifer sediments. *Appl. Environ. Microbiol.* **57:**2260–2268.

59. **Leahy, J. G., and R. R. Colwell.** 1990. Microbial degradation of hydrocarbons in the environment. *Microbiol. Rev.* **54:**305–315.

60. **Lind, A. M., and F. Eiland.** 1989. Microbiological characterization and nitrate reduction in subsurface soils. *Biol. Fertil. Soils* **8:**197–203.

61. **Loehr, R. C., J. H. Martin, Jr., E. F. Neuhauser.** 1992. Land treatment of an aged oily sludge—organic loss and change in soil characteristics. *Water Res.* **26:**805–815.

62. **Lovley, D. R., M. J. Baedecker, D. J. Lonergan, I. M. Cozzarelli, E. J. Phillips, and D. I. Siegel.** 1989. Oxidation of aromatic contaminants coupled to microbial iron reduction. *Nature* (London) **329:**297–299.

63. **Madsen, E. L., J. L. Sinclair, and W. C. Ghiorse.** 1991. In-situ biodegradation: microbiological patterns in a contaminated aquifer. *Science* **252:**830–833.

64. **Marinucci, A. C., and R. Bartha.** 1979. Apparatus for monitoring the mineralization of volatile ^{14}C-labeled compounds. *Appl. Environ. Microbiol.* **38:**1020–1022.

65. **Martens, R.** 1985. Estimation of the adenylate energy charge in unamended and amended agricultural soils. *Soil Biol. Biochem.* **17:**765–772.

66. **Martin, K., L. L. Parsons, R. E. Murray, and M. S. Smith.** 1988. Dynamics of soil denitrifier populations: relationships between enzyme activity, most-probable number counts, and actual N gas loss. *Appl. Environ. Microbiol.* **54:** 2711–2716.

67. **McDonald, I. R., E. M. Kenna, and J. C. Murrell.** 1995. Detection of methanotrophic bacteria in environmental samples with the polymerase chain reaction. *Appl. Environ. Microbiol.* **61:**116–121.

68. **Mester, K.** 1995. Field evaluation of in-situ denitrification for bioremediation of contaminated soils and groundwater. Ph.D. dissertation. Rutgers University, New Brunswick, N. J.

69. **Mills, A. L., C. Breuil, and R. R. Colwell.** 1978. Enumeration of petroleum-degrading marine and estuarine microorganisms by the most probable number method. *Can. J. Microbiol.* **24:**552–557.

70. **Morgan, P., and R. J. Watkinson.** 1989. HC degradation in soils and methods for soil biotreatment. *Crit. Rev. Biotechnol.* **8:**305–333.

71. **Mueller, J. G., S. M. Resnick, M. E. Shelton, and P. H. Pritchard.** 1992. Effect of inoculation on the biodegradation of weathered Prudhoe Bay crude oil. *J. Ind. Microbiol.* **10:**95–102.

72. **Novotny, M., J. W. Strand, S. L. Smith, D. Wiesler, and F. J. Schwende.** 1990. Compositional studies of coal tar by capillary gas chromatography/mass spectrometry. *Fuel* **60:**213–220.

73. **Phelps, T. J., R. L. Siegrist, N. E. Korte, D. A. Pickering, J. M. Strong-Gunderson, A. V. Palumbo, J. F. Walker, C. M. Morrissey, and R. Mackowski.** 1994. Bioremediation of petroleum hydrocarbons in soil column lysimeters from Kwajalein Island. *Appl. Biochem. Biotechnol.* **45/46:** 835–845.

74. **Post, R. D., and A. N. Beeby.** 1993. Microbial biomass in suburban roadside soils: estimates based on extracted microbial C and ATP. *Soil Biol. Biochem.* **25:**199–204.

75. **Pritchard, P. H., J. G. Mueller, J. C. Rogers, F. V. Kremer, and J. A. Glaser.** 1992. Oil spill bioremediation: experiences, lessons, and results from the *Exxon Valdez* oil spill in Alaska. *Biodegradation* **3:**315–335.

76. **Ramanand, K., and J. Suflita.** 1991. Anaerobic degradation of m-cresol in anoxic aquifer slurries: carboxylation reactions in a sulfate-reducing bacterial enrichment. *Appl. Environ. Microbiol.* **57:**1689–1695.

77. **Russell, E., and C. Ohland.** 1993. In-situ bioremediation of petroleum solvent contaminated subsurface soils—Ona-

laska landfill, Wisconsin, p. 473–486. *In Hydrocarbon Contaminated Soils*. Lewis Publishers, Boca Raton, Fla.

78. **Schnuerer, J., and T. Rosswall.** 1982. Fluorescein diacetate hydrolysis as a measure of total microbial activity in soil and litter. *Appl. Environ. Microbiol.* **43:**1256–1261.

79. **Scow, K., and M. Alexander.** 1989. Kinetics of biodegradation in soil, p. 243–269. *In Reactions and Movement of Organic Chemicals in Soils*. Joint publication no. 22. Soil Society of America and American Society of Agronomy, Madison, Wis.

80. **Scow, K., and J. Hutson.** 1992. Effect of diffusion and sorption on the kinetics of biodegradation: theoretical considerations. *Soil Sci. Soc. Am.* **56:**119–127.

81. **Sharabi, N. E.-D., and R. Bartha.** 1993. Testing of some assumptions about biodegradability in soil as measured by carbon dioxide evolution. *Appl. Environ. Microbiol.* **59:** 1201–1205.

82. **Sheppard, M. I., L. L. Ewing, and J. L. Hawkins.** 1994. Soil degassing of carbon-14 dioxide: rates and factors. *J. Environ. Qual.* **23:**461–468.

83. **Shiaris, M. P., and J. J. Cooney.** 1983. Replica plating method for estimating phenanthrene-utilizing and phenanthrene-cometabolizing microorganisms. *Appl. Environ. Microbiol.* **45:**706–710.

84. **Sinclair, J. L., D. H. Kampbell, M. L. Cook, and J. T. Wilson.** 1993. Protozoa in subsurface sediments from sites contaminated with aviation gasoline or jet fuel. *Appl. Environ. Microbiol.* **59:**467–472.

85. **Song, H.-G., X. Wang, and R. Bartha.** 1990. Bioremediation potential of terrestrial fuel spills. *Appl. Environ. Microbiol.* **56:**652–656.

86. **Stuart, B.** 1995. Transport of microorganisms in soils and groundwater. Ph.D. dissertation. Rutgers University, New Brunswick, N.J.

87. **Suflita, J. M., L. Liang, and A. Saxena.** 1989. The anaerobic biodegradation of o-, m-, and p-cresol by sulfate-reducing enrichment cultures obtained from a shallow anoxic aquifer. *J. Ind. Microbiol.* **4:**255–266.

88. **Swindoll, C. M., C. M. Aellon, D. C. Dobbins, O. Jiang, S. C. Long, and F. K. Pfaender.** 1988. Aerobic biodegradation of natural and xenobiotic organic compounds by subsurface microbial communities. *Environ. Toxicol. Chem.* **7:** 291–299.

89. **Thierry, A., and R. Chicheportiche.** 1988. Use of ATP bioluminescence measurements for the estimation of biomass during biological humification. *Appl. Microbiol. Biotechnol.* **28:**199–202.

90. **Tsai, Y. L., and B. H. Olson.** 1992. Rapid method for separation of bacterial DNA from humic substances in sediments for polymerase chain reaction. *Appl. Environ. Microbiol.* **58:**2292–2295.

91. **Tsuji, T., Y. Kawasaki, S. Takeshima, T. Sekiya, and S. Tanaka.** 1995. A new fluorescence staining assay for visualizing living microorganisms in soil. *Appl. Environ. Microbiol.* **61:**3415–3421.

92. **Vaden, V. R., J. Webster, G. J. Hampton, M. S. Hall, and F. R. Leach.** 1987. Comparison of methods for extraction of ATP from soil. *J. Microbiol. Methods* **7:**211–217.

93. **Venkatraman, S., I. Bossert, D. Kosson, T. Boland, and J. Schuring.** 1995. *Integrated Pneumatic Fracturing for the In-Situ Treatment of Contaminated Soil.* EPA Cooperative Agreement no. CR818207-01-0. U.S. Environmental Protection Agency.

94. **Voordouw, G., J. K. Voordouw, T. R. Jack, J. Foght, P. M. Fedorak, and D. W. S. Westlake.** 1992. Identification of distinct communities of sulfate-reducing bacteria in oil fields by reverse sample genome probing. *Appl. Environ. Microbiol.* **58:**3542–3552.

95. **Walker, J. D., and R. R. Colwell.** 1976. Enumeration of petroleum-degrading microorganisms. *Appl. Environ. Microbiol.* **31:**198–207.

96. **Watwood, M. E., C. S. White, and C. N. Dahm.** 1991. Methodological modifications for accurate and efficient determination of contaminant biodegradation in unsaturated calcareous soils. *Appl. Environ. Microbiol.* **57:**717–720.

97. **Webster, J. J., B. S. Hall, and F. R. Leach.** 1992. ATP and adenylate energy charge determinations on core samples from an Av-fuel spill site at the Traverse City, Michigan Airport. *Bull. Environ. Contam. Toxicol.* **49:**232–237.

98. **White, D. C., W. M. Davis, J. S. Nickels, J. D. King, and R. J. Robbie.** 1979. Determination of the sedimentary microbial biomass by extractable lipid phosphate. *Oecologia* **40:**51–62.

99. **Wright, S. F.** 1992. Immunological techniques for detection, identification, and enumeration of microorganisms in the environment, p. 45–64. *In* M. Levin, R. Seidler, and M. Rogul (ed.), *Microbial Ecology: Principles, Methods, and Applications.* McGraw-Hill, Inc., New York.

Bioventilation and Modeling of Airflow in Soil

JOHN H. KRAMER AND STEPHEN J. CULLEN

81

This chapter includes an overview of modeling and simulating airflow through soils. The term "modeling" refers to the choice of equations and parameters used to represent reality; the term "simulations" refers to the results of running models. Summaries of air-water-soil interactions, the approaches taken to model them, and pertinent literature on available models are included. This chapter is intended to introduce readers with limited knowledge of the vadose zone (the subsurface where partially saturated soils occur) to the processes and problems confronted by modelers and to provide a starting point for understanding how the problems are being addressed. The subject is quickly evolving, with new approaches and codes being developed continuously, so the specific models mentioned are not the only codes available.

APPLICATION OF MODELS

The objective of applying models is to provide convincing and useful simulations of reality. This requires four crucial steps: (i) selecting the model most appropriate to the task, (ii) determining the most important input parameters for field measurement, (iii) calibrating the model to conform with initial observed site-specific conditions, and (iv) matching simulation results against subsequent field observations to gauge how accurately the model represents a known change in reality. The first two steps are technical in nature and require a good basic understanding of the physics driving airflow in soils and of the mathematics driving the models. The heart of model application is in the third and fourth steps: calibration and matching simulation results (validation).

Calibration involves assigning values to boundary conditions and input parameters that will drive the model to recreate a particular starting condition (usually a pressure field). The notion that the chosen model and boundary conditions apply is tenable only if the governing equations can be run to a stable solution that approximates an initial observed condition. The difficulty of this task increases with the complexity of the model and the number of observation points.

Matching simulation results to a known second condition is the only way to show that the potential use of simulation results for portraying changes in the modeled system is valid. The faithful response of a model to a stressed condition reinforces confidence in simulations at other times and conditions. Matching simulation results is frequently referred to as "validation," a term criticized in the recent literature because it connotes veracity and promotes unfounded confidence in the mathematical simulation which, at best, is only an approximation (39). "History matching" has been proposed as a more descriptive and less misleading term (9). Others feel that "validation" has a very specific meaning within the context of numerical mathematical modeling and should be retained (30). Whatever the name, any model application is unreliable unless predictive reliability is demonstrated in this way. Unfortunately, many problems to which vadose zone modeling must be applied do not lend themselves to a history-matching step. We may be attempting to predict the future with little history to match. Also, we are frequently working with spatial and temporal variability in the geologic environment that is too great. A cogent argument that vadose zone modeling in complex geology can never be "validated" is that it is impossible to sample enough vadose zone flow pathways.

It should not be assumed that application of a computer model is straightforward. To get up and running will require some involved hardware and software decisions and a lot of personnel time. Most computer source codes for powerful fate and transport groundwater flow models are written in FORTRAN, which will not run on popular desktop machines unless compiled into DOS- or Macintosh-compatible formats. Some models are available in compiled versions, but compatibility problems are not uncommon. Although model codes which have preprocessors that build control and input files are available, some of the most powerful codes do not. Most groundwater models are not directly adapted for airflow output, and results require transformation.

OVERVIEW OF MODELING AIRFLOW IN SOIL
Pertinent Literature

Airflow modeling in soils is analogous to groundwater flow modeling when reasonable assumptions are possible, and a well-developed groundwater modeling literature can be applied to the airflow problem. Application of groundwater flow models to airflow problems is in the nascent stage. An excellent summary of airflow modeling is provided by Joss and Baehr (22). Two important sources of up-to-date information on computer modeling codes, primarily aqueous flow and transport in the vadose and saturated zones but some airflow, are The International Ground Water Modeling Center in Golden, Colo., and The Center for Subsurface Modeling Support at the Robert S. Kerr Environmental Research Laboratory, Ada, Okla.

Public domain software, such as AIR2D and AIR3D (4, 21, 22), developed through public agencies tend to be well tested and have very usable documentation. AIR3D is based on the most widely used and well tested groundwater model, MODFLOW (31), and is being enhanced with graphical user interfaces to facilitate PC-based application (15). In addition to public domain codes, there are a number of computer models developed at research institutions or in the private sector which can be licensed or used through contracting services.

Usefulness of Airflow Models

Airflow modeling is typically used as a design or predictive tool. Simulations are used to locate wells and galleries for air injection or vacuum extraction and to predict either remediation times based on oxygen transport to in situ bioremediation zones or extraction of volatile organic contaminants.

Figure 1 illustrates two types of subsurface airflow: sparged airflow beneath the water table and pressure-driven airflow in the vadose zone. Sparged airflow is induced beneath the water table by injecting pressurized air into saturated soils (20). Sparged airflow is vertical and localized above the area of injection, often in continuous tubular channels which are held open by the pressure of the injected air (46). The exact location of the channels is extremely sensitive to minor variations in soil type and is typically neither uniform nor radially symmetric (1). Sparged air will contact groundwater (deliver oxygen) at the edges of bubbles and tubular channels and can be an effective means of delivering oxygen to organisms in saturated soils. Sparging

will affect the liquid flow system by entraining and mounding groundwater or by pushing water out of pore space normally available for liquid flow. Modeling of airflow in sparged systems is unknown us but could be used to simulate delivery of oxygen and removal of volatile organic contaminants from aquifers.

Available airflow computer models treat advective (pressure-driven) airflow in the vadose zone. For the remainder of this chapter, the term "airflow" will refer to advective airflow. Airflow is induced by either pushing air into the subsurface (bioventing) or pulling air out (vacuum extraction) as shown in Fig. 1. Recently there has been much interest in modeling airflow of soil remediation systems by using computer codes developed to simulate groundwater flow, primarily in relation to injection and extraction from wells (3, 4, 11, 17–20, 26, 33, 34, 37).

Modeling vadose zone processes is difficult because the temporal and spatial variability inherent in geologic systems makes any manageable mathematical representation an oversimplification of reality. Airflow is computationally cumbersome to model mathematically without assuming gas ideality (29) and ignoring slip effects (22) and because of nonlinear interdependencies in soil-water systems that control air-filled porosity (35). Another level of complication is introduced when one is modeling transport of specific gases in soil air (e.g., oxygen). The behavior of gases in soil involves numerous interdependent geochemical and biogeochemical reactions, including biological oxygen demand, chemical interferences, and phase changes (partitioning between liquid phase and gas phase). Therefore, one should not expect airflow modeling, especially that which deals with transport of particular gases, to be a foolproof or even a consistently reliable predictor of airflow and gas transport. Despite these recognized difficulties, simulating airflow has proven useful because there exists no better way of providing satisfactory estimates for projecting into the future or the subsurface, predicting the effects of environmental changes, evaluating the effects of various potential remedial alternatives, or analyzing the importance of field data through computer model sensitivity analysis.

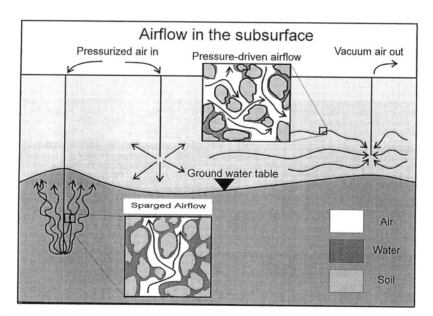

FIGURE 1 Aiflow in the subsurface and its effect on the groundwater table. Insets show the relationship of air, liquid, and solid for sparged airflow and pressure-driven airflow.

Airflow simulations are commonly used for aspects of environmental investigations which cannot be measured or analyzed in other ways. In fact, there exists no other tool for predicting airflow behavior in the subsurface over relevant time spans and areas.

Controls and Driving Force

It is necessary for airflow modelers to be familiar with the effects of geology and water content which can result in counterintuitive transport pathways. Geology, variable in space, defines the fixed geometry of potential airflow pathways. The most permeable soil materials are dry sands and gravels with relatively large pore spaces (>1 mm). The least permeable soil materials are moist silts and clays. These "bedded sedimentary deposits" generally occur in horizontal layers or elongated channels but can be tilted as well. Subsurface openings (fractures, cavities, cracks, pipeline backfill, etc.) can occur in any orientation and are very important components of the geologic framework (24).

Moisture content, a variable in time and space, modulates airflow by changing the cross-sectional area of openings (43). Fine-grained materials, despite having greater total porosity than coarse-grained materials, typically have less air-filled porosity because small pores hold more attached water. Airflow is zero in a saturated medium where all pores are occupied by liquid, and it increases only after moisture content decreases to the point where continuous openings occur (air entry value) (8). Noncontinuous air pockets are called entrapped air, typical of rapidly saturated or freshly drained soils. Entrapped air may fill up to 10% of the pore space, depending on soil type, and will provide a reservoir of gases which will dissolve into pore water or contribute oxygen to microbes in partially saturated soils. Entrapped air is difficult to measure and generally not considered in modeling airflow.

The force affecting airflow is pressure. Airflow will move in the direction of decreasing air pressure. The pressure gradients can be created by density contrasts from the volatilization of organic compounds (12), temperature-induced density contrasts, or the mixing of various gases. Pressure gradients can occur as a result of the subsurface production of gases such as methane (e.g., at landfills and swamps), release of pressurized natural gas deposits, human activities, or barometric pressure changes.

Transport Processes

Transport of gas components in airflow is affected by a host of biochemical processes, including biotransformation, mixing, sorption, vaporization, dissolution, and reaction. In general, airflow modeling does not attempt to simulate changes in air chemistry with the exception of simple first-order time decay (e.g., radon), equilibrium calculations of linear partitioning of volatile species from a liquid phase, or oxygen demand estimated as a linear function of pressure. Some types of models have been developed to predict the efficacy of vapor extraction by using assumptions about phase transformations. The program Hyperventilate (25) applies flow equations to soil air pressure data and calculates the flow volumes and extraction rates of specific gases on the basis of Henry's law. Different modes of gas-phase transport are treated only by the most powerful codes. A well-documented comprehensive one-dimensional (1-D) treatment for modeling of CO_2 production and transport in the vadose zone has been developed by Simunek and Suarez (41). This model, which treats both liquid and gaseous transport, is linked to production of CO_2 by microbial and root respiration and to uptake of CO_2 by root water.

Input Parameters

Input parameters are the Achilles' heel of vadose zone flow and transport model applications. They are critically important because they are the foundation upon which mathematical models rest, and they are vulnerable because problems associated with scale, the transient nature of airflow systems, and parameter heterogeneities make representative measurement elusive. Typical input parameters required for use in airflow models include the geometry and air-filled porosity of the domain to be modeled and boundary conditions. Airflow boundaries, paved surfaces in particular, are more safely modeled as leaky than impermeable (7).

Modelers must make practical adjustments to mathematical rigor. Mathematical rigor is compromised when no attempt is made to compute the effects of known processes (e.g., processes are assumed to be insignificant), when simplifying assumptions are made (e.g., geology is assumed to be homogeneous), or when empirical factors are introduced to simulate the results of complex phenomena. An example is the "interconnectedness index" used in the model SESOIL (16) to account for variable hydraulic properties of different soil types. The index is not measurable and can be assigned only on the basis of model performance; it is adjusted so that the simulation fits the known, anticipated, or desired field situation. The compromises work when the sensitivity of the result is insignificant.

The thoroughness of physical characterization is compromised when the mathematics requires parameters that cannot be practically measured (diffusion coefficients, soil-specific distribution coefficients, hydraulic functions, etc.) or when the inherent variability of parameter values is not sufficiently characterized. It is generally impractical to collect enough site-specific data to characterize the mean and variance of the many input measurements required for powerful codes.

Simplified, less powerful computer codes treat generalized volumes or simplified processes. Although they require fewer input parameters, making them easier to implement, they are not well suited for applications involving detailed modeling of transient airflow, complex geometries, irregular boundary conditions, or multiphase flow and transport. Powerful computer codes handle many conditions and processes but require hard-to-measure input parameters (species-specific diffusion coefficients, coupled relative hydraulic conductivity-phase saturation functions, soil-specific adsorption coefficients, etc.). The problem of impossible input requirements is usually handled by estimating average or "effective" parameters for unavailable measurements or by using default values from previous experience.

The sensitivity of a model to input parameters values can be evaluated by comparing the results of several model simulations of the same scenario run at different parameter values (36, 38). Sensitivity analysis is intended to provide an understanding of the range of reasonable results due to parameter heterogeneities for a specific modeled scenario. The most influential parameters can be identified for intensive and careful measurement in the field. A specific sensitivity analysis is therefore appropriate for each modeled scenario. Nofziger et al. (36) point out that "there is abundant evidence that the sensitivity and uncertainty are highly dependent on the scenario being modeled and the parameters used."

The most credible modeling applications incorporate

measured, site-specific hydraulic parameters. These parameters should be measured in situ where possible in order to minimize errors attributable to measurement scale or sampling and analytic techniques. It is most desirable to measure a suite of the most important parameters, such as moisture content, particle size distribution, and dry bulk density throughout the modeled volume. A well-characterized suite is one to which a statistical model can be applied and for which a measure of the mean, variance, and possible spatial correlation can be calculated. These characteristics aid in evaluating the uncertainty associated with model output through sensitivity analysis or stochastic (probability-based) treatment. Site-specific parameter measurements tie the model to the actual field setting, providing a link to reality.

A probabilistic argument can be advanced for generic modeling using databases of typical or estimated effective input parameters. The argument states that uncertainty in the simulation will be no greater if generic input is used than if measured site-specific input were used because uncertainty in site-specific input parameters is high. This argument is used to justify modeling in situations where it is impossible to collect sufficient field data and is often extended to situations where it is inconvenient to do so. The danger in this approach is that the simulation has no relation to site-specific conditions, and the inexperienced practitioner can be seduced to create a fabrication based on preconceived bias.

MODELING SOLUTION METHODS: ADAPTATION OF GROUNDWATER FLOW MODELS TO AIRFLOW

Modeling airflow is analogous to modeling groundwater flow. The flow equations are derived from conservation of mass and Darcy's law. Linearizing assumptions are used to generate an airflow equation of the form

$$\frac{\partial}{\partial x}\left(k_{xx}\frac{\partial\phi}{\partial x}\right) + \frac{\partial}{\partial y}\left(k_{xy}\frac{\partial\phi}{\partial y}\right) + \frac{\partial}{\partial z}\left(k_{xz}\frac{\partial\phi}{\partial z}\right) = S_a\frac{\partial\phi}{\partial t}$$

where ϕ is air pressure (M/LT^2) or pressure squared, depending on assumptions (29); k is air permeability (L^2); x, y, and z are Cartesian coordinates aligned along the major axes of the conductivity tensor with diagonal components k_{xx}, k_{yy}, and k_{zz}; S_a is specific storage of medium (L^{-1}); and t is time (T).

This is identical to the groundwater flow equation commonly employed in groundwater flow models, and the solutions from groundwater flow codes can be applied to airflow problems by converting output from any groundwater model in hydraulic head to air pressure by using one of two assumptions. Assumptions are reviewed and discussed in recent works (2, 22, 27, 29). The equation is used in a number of different types of models to calculate the pressure distribution of air throughout the modeled domain and/or to calculate the stream function distribution, as was done by Falta (11). Models employ different solution methods, with advantages and disadvantages as summarized below.

Analytical Method

Analytical models use exact mathematical solutions of partial differential equations to represent airflow. Soil properties are assumed to be homogeneous in the domain of interest. Approximate anisotropies can be treated through modification of axes scales (11). Air-filled porosity at any point in space is determined through precharacterized soil moisture relations for the soils of interest. Simulations are exact solutions to the continuously variable pressure field through time and space.

Advantages of analytic models include ease of use, time efficiencies, exact 3-D solutions, and others as discussed by McKee and Bumb (32). Analytical solutions can be used to quickly evaluate conditions at a site or to generate graphics for reports. Disadvantages are the unrealistic treatment of soil systems as homogeneous and the need for regular geometric boundaries amenable to exact mathematical representation. New developments in analytical solutions to airflow problems employ the analytic element method (27, 44, 45) which facilitates application to irregular domains by linking analytic solutions across element boundaries.

Finite-Difference Method

The numerical finite-difference technique is used in well-known, well-tested programs (e.g., AIR3D [21] and TOUGH2 [40]). The solution technique is well described as applied to saturated problems (5, 8, 14). Its advantage is that it is conceptually simple, employing mass balance (accounting for changes in air mass) in discrete spatial blocks progressing through time. A domain of specified dimension (1-D, 2-D, or 3-D) is divided (discretized) into n rectangular boxes, the centers of which are nodes of a grid. Each box can have uniquely variable hydraulic properties. Initial and boundary conditions, including air-filled porosity, air pressures, and constant pressure or constant flux boundaries, are set. The mass balance of air at each node is represented by writing the partial differential flow equations across each of the appropriate box boundaries (two for 1-D columns, four for 2-D grids, and six for 3-D grids). Since each node has as many neighboring nodes as boundaries (except at boundary nodes where conditions must be set), a system of n equations and n unknowns is generated. The pressure field at each node is found by solving the n "finite differences" for pressure simultaneously at preselected time steps. A steady-state starting pressure field can be found by using initial estimates of appropriate pressure heads at each node and employing iterative "relaxation" techniques to minimize residuals. The system can then be stressed by changing boundary conditions. Sources can be placed in the initial pressure field to simulate injection from point or line sources, and sinks can be defined to simulate extraction. Numerical solutions at different time steps represent snapshots in time of transient flow conditions. Discretization of the domain and selection of time steps are important for examining detail in the domain of interest, particularly in materials with distinct air entry pressures subjected to rapid wetting.

Advantages of the technique are that it is relatively efficient and versatile. Disadvantages are that node spacings are blocklike. Diagonal boundaries are represented in stair-step fashion, introducing error at irregular boundaries. The approach is limited when one is attempting to model anisotropic flow regimes in which the favored flow direction does not coincide with the grid axes. This situation would occur when one is attempting to model flow through deformed rocks, through folded or slumped soil horizons, or across tilted unconformities. In these cases, the simulated conductivities across the grid-cell boundaries may be less than the actual conductivity oblique to the boundaries. Geologic volumes with variable, nonorthogonal favored conductivity directions may require multiple adjoining modeled domains.

Finite-Element Method

Finite-element codes are conceptually and mathematically more difficult to understand than finite-difference models but are more versatile for modeling complex geometries and heterogeneous hydraulic conductivity fields (6). This is the technique most often applied in advanced vadose zone modeling of multiple processes and complex domains (10, 42, 47). The finite-element method breaks down the problem of spatial domain into polygonal elements with nodes at the corners. The polygonal elements are treated as regions for which approximate integrated values of the parameter of concern (usually pore pressure) are solved locally, using interpolating functions between the nodes. Flow equations are set up for each element, assuming parameter values at the nodes, and matrices of equations (both local and global) are solved simultaneously such that the changes in first derivatives (i.e, pressure gradients between nodes and between elements) are minimized. Finite-element solutions are standard practice in modern engineering design and are used to model problems analogous to subsurface flow such as stress distribution or heat transfer.

Advantages of finite-element models are that finite-element techniques can treat irregular boundaries and boundary conditions naturally. Finite-element techniques can also treat changes in principal flow directions within anisotropic formations. Disadvantages are the mathematical complexity and processing time, which can be longer than for finite-difference models.

Stochastic Methods

Any of the previously described deterministic mathematical solution methods can be used in a stochastic treatment. Stochastic treatments consider input parameters to be random variables and any particular solution to be only a single "realization" from a distribution of potential solutions. Any one realization is associated with a probability that it represents the average realization. The most commonly used stochastic treatment is called the Monte Carlo method. Numerous runs of a model are made, each with a different set of input parameters chosen randomly from a statistically characterized population of possible input parameter values (e.g., log-normal hydraulic conductivities). These numerous runs, in turn, build a population of solutions which can be characterized by a mean and standard deviation. The probability of any given outcome can be estimated from this sample of possible solutions (28). An alternative to the Monte Carlo procedure is the theoretical approach, in which the variance of the inputs can be carried through calculations and used to estimate the distributions of the possible solutions. This approach yields interesting insights. For example, Yeh et al. (48) predicted state-dependent anisotropy by correlating hydraulic conductivity anisotropies (K_x/K_z) to saturation percentage, variance in particle size distribution parameters, and layer thickness.

The advantage of a stochastic approach is that uncertainty in the input parameters is reflected in the solution, providing a measure of the uncertainty in the findings. A disadvantage is that the technique requires numerous simulations involving a well-organized effort and extensive computer time.

OTHER METHODS

Other approaches including non-theory-based empirical models (e.g., transfer function models) have been used with some success on a site-specific basis (23). An empirical model is based on observations through time or space which are fitted, usually involving a regression to a simplified mathematical expression. These can be scaled to extend past the domain of the original empirical observations. Empirical models have no theoretical basis and if transferred from one place to another may become invalid. These models cannot be used to interpret the importance of physically measurable properties. Projecting such models into the future is uncertain, particularly in transient flow regimes where input and through-flow may not bear any relation to the conditions for which the model was derived.

CONCLUSIONS

The state of the art in computer modeling of vadose zone flow and transport is that reliable prediction of contaminant flow in space and time has not been realized (13). Consequently, computer-modeled simulations alone are not a reliable basis upon which to design air delivery systems without data from monitoring networks. Nonetheless, computer modeling can be a useful design tool when combined with expert opinion. Monitoring and modeling in an iterative combination can be beneficial to understanding site-specific processes that affect airflow in the vadose zone. Furthermore, modeling represents the only available means of simulating potential contaminant transport into the future, between monitoring stations, or beyond monitoring networks.

Few groundwater computer models have been adapted to airflow; however, this is rapidly changing. Well-tested models such as AIR3D (21, 22) and Hyperventilate (25) exist in the public domain. Models and user interfaces suitable for a variety of vadose zone applications have also been and are being developed in private and research settings (15, 40).

Portions of the research for this chapter were supported by EPA Las Vegas cooperative agreement CR-816969-01-0 and the UCSB Institute for Crustal Studies (contribution no. 0202-40HW).

REFERENCES

1. **Ahlfeld, D., A. Dahmani, and W. Ji.** 1994. A conceptual model of field behavior of air sparging and its implications for application. *Ground Water Monit. Remediation* **14(4):** 132–139.
2. **American Society of Testing and Materials.** 1995. Standard guide to simulation of subsurface air flow using ground-water flow modeling codes. ASTM designation D18.21.93.17. American Society for Testing and Materials, Philadelphia.
3. **Baehr, A. L., and M. F. Hult.** 1991. Evaluation of unsaturated zone air permeability through pneumatic tests. *Water Resour. Res.* **27:**2605–2617.
4. **Baehr, A. L., and C. J. Joss.** 1995. An updated model of induced airflow in the unsaturated zone. *Water Resour. Res.* **31(2):**1–5.
5. **Bear, J.** 1972. *Dynamics of Fluids in Porous Media.* American Elsevier, New York.
6. **Bear, J.** 1979. *Hydraulics of Groundwater.* McGraw-Hill, New York.
7. **Beckett, G. D., and D. Huntley.** 1994. Characterization of flow parameters controlling soil vapor extraction. *Ground Water* **32:**239–247.
8. **Bouwer, H.** 1978. *Groundwater Hydrology,* p. 25–29. McGraw-Hill, New York.
9. **Bredehoeft, J. D., and L. F. Konikow.** 1993. Ground water models: validate or invalidate. *Ground Water* **31:**178. (Editorial.)

10. **Celia, M. A., and P. Binning.** 1992. A mass conservation numerical solution for two-phase flow in porous media with application to unsaturated flow. *Water Resour. Res.* **28:** 2819–2828.

11. **Falta, R. W.** 1995. Analytical solutions for gas flow due to gas injection and extraction from horizontal wells. *Ground Water* **33:**235–246.

12. **Falta, R. W., I. Javandel, K. Pruess, and P. Witherspoon.** 1989. Density-driven flow of gas in the unsaturated zone due to the evaporation of volatile organic compounds. *Water Resour. Res.* **25:**2159–2169.

13. **Fogg, G. E., D. R. Nielsen, and D. R. Shibberu.** 1994. Modeling contaminant transport in the vadose zone: perspective on state of the art, p. 249–267. *In* L. G. Wilson, S. J. Cullen, and L. G. Everett (ed.), *Handbook of Vadose Zone Characterization and Monitoring.* Lewis Publishers, Chelsea, Mich.

14. **Freeze, R. A., and J. A. Cherry.** 1979. *Groundwater.* Prentice-Hall, Englewood Cliffs, N.J.

15. **Geraghty and Miller Software.** 1994. American Petroleum Institute releases AIR3D, a vadose zone air flow model. *Geraghty Miller Software Newsl.* **7:**2–3.

16. **Hetrick, D. M., S. J. Scott, and M. J. Barden.** 1993. The new SESOIL user's guide. PUBL-SW-200-93. Wisconsin Department of Natural Resources, Madison.

17. **Johnson, P. C., and R. A. Ettinger.** 1994. Considerations for the design of in situ vapor extraction systems: radius of influence vs. zone of remediation. *Ground Water Monit. Remediation* **4**(3):123–128.

18. **Johnson, P. C., C. C. Stanley, M. W. Kemblowski, D. L. Byers, and J. D. Cohlhardt.** 1990. A practical approach to the design, operation, and monitoring of soil venting systems. *Ground Water Monit. Rev.* **9:**159–178.

19. **Johnson, P. C., C. C. Stanley, M. W. Kemblowski, and J. D. Cohlhardt.** 1990. Quantitative analysis for the cleanup of hydrocarbon-contaminated soils by in-situ venting. *Ground Water* **28:**413–429.

20. **Johnson, R. L., P. C. Johnson, D. B. McWhorter, R. E. Hinchee, and I. Goodman.** 1993. An overview of air sparging. *Ground Water Monit. Remediation* **13**(4): 127–135.

21. **Joss, C. J., and A. L. Baehr.** 1994. AIR3D—an adaptation of the ground-water-flow code MODFLOW to simulate three-dimensional air flow in the unsaturated zone. U.S. Geological Survey Open-file Report 94-533. U.S. Geological Survey, Reston, Va.

22. **Joss, C. J., and A. L. Baehr.** 1995. AIR2D—a computer program to simulate two dimensional axisymmetric air flow in the unsaturated zone. U.S. Geological Survey Open-file Report 95. U.S. Geological Survey, Reston, Va.

23. **Jury, W. A.** 1982. Simulation of solute transport using a transfer function model. *Water Resour. Res.* **18:**363–368.

24. **Kramer, J. H., and B. R. Keller.** 1994. Understanding the geologic framework of the vadose zone and its effect on storage and transmission of fluids, p. 137–158. *In* L. G. Wilson, S. J. Cullen, and L. G. Everett (ed.), *Handbook of Vadose Zone Characterization and Monitoring.* Lewis Publishers, Chelsea, Mich.

25. **Kruger, C. A., and J. G. Morse.** 1993. Decision-support software for soil vapor extraction technology application: Hyperventilate. EPA/600/R-93/028. Risk Reduction Engineering Laboratory, U.S. Environmental Protection Agency, Cincinnati.

26. **MacNeal, R. W.** 1994. Air flow modeling for bioventing applications. *Geraghty Miller Tech. Exchange Newsl.* **August:**1–4.

27. **MacNeal, R. W., S. W. Robertson, and K. Barr.** 1994. Gas flow modeling using the analytic element method,

p. 113–118. *In Proceedings of International Conference of Analytic Element Modeling of Groundwater Flow.* Available through Geraghty and Miller, Inc., Minneapolis.

28. **Massmann, J., R. A. Freeze, L. Smith, T. Sperling, and B. James.** 1991. Hydrogeological decision analysis. 2. Applications to ground water contamination. *Ground Water* **29:**536–548.

29. **Massmann, J. W.** 1989. Applying groundwater flow models in vapor extraction system design. *J. Environ. Eng.* **115:** 129–149.

30. **McCombie, C., and I. McKinley.** 1993. Validation—another perspective. *Ground Water* **31:**530. (Guest editorial.)

31. **McDonald, M. G., and A. L. Harbaugh.** 1988. A modular three-dimensional finite-difference ground-water flow model. *U.S. Geological Survey Techniques of Water-Resources Investigations,* book 6. U.S. Geological Survey, Reston Va.

32. **Mckee, C. R., and A. C. Bumb.** 1988. A three-dimensional analytical model for aid in selecting monitoring locations in the vadose zone. *Ground Water Monit. Rev.* **8:** 124–136.

33. **McWhorter, D. B.** 1990. Unsteady radial flow of gas in the vadose zone. *J. Contam. Hydrol.* **5:**297–314.

34. **Mohr, D. H., and P. H. Merz.** 1995. Application of a 2D air flow model to soil vapor extraction and bioventing case studies. *Ground Water* **33:**433–444.

35. **Nielsen, D. R., M. T. van Genuchten, and J. W. Biggar.** 1986. Water flow and solute transport processes in the unsaturated zone. *Water Resour. Res.* **22:**89S–108S.

36. **Nofziger, D. L., I.-S. Chen, and C. T. Haan.** 1993. Evaluation of unsaturated/vadose zone models for Superfund sites. EPA-600/R-93/184. R. S. Kerr Environmental Research Laboratory, U.S. Environmental Protection Agency, Ada, Okla.

37. **Nyer, E. K., R. Macneal, and D. C. Schafer.** 1994. VES design: using simple or sophisticated design methods. *Ground Water Monit. Remediation* **14**(3):101–104.

38. **Odencrantz, J. E., J. M. Farr, and C. Robinson.** 1992. Transport model parameter sensitivity for soil cleanup level determinations using SESOIL and AT123D in the context of the California Leaking Underground Fuel Tank Manual. *J. Soil Contam.* **1:**159–182.

39. **Oreskes, N., K. Shrader-Frechette, and K. Belitz.** 1994. Verification, validation, and confirmation of numerical models in the earth sciences. *Science* **263:**641–646.

40. **Pruess, F.** 1991. TOUGH2—a general-purpose numerical simulator for multiphase fluid and heat Flow. LBL-29400. Earth Sciences Division, Lawrence Berkeley Laboratory, Berkeley, Calif.

41. **Simunek, J., and D. L. Suarez.** 1992. The soil CO_2 code for simulating one-dimensional carbon dioxide production and transport in variably saturated porous media, version 1.1. Research Report no. 127. U.S. Salinity Laboratory, Agricultural Research Service, U.S. Department of Agriculture, Riverside, Calif.

42. **Simunek, J., T. Vogel, M. T. van Genuchten.** 1992. The SWMS 2D code for simulating water flow and solute transport in two-dimensional variably saturated media, version 1.1. Research Report no. 126. U.S. Salinity Laboratory, Agricultural Research Service, U.S. Department of Agriculture, Riverside, Calif.

43. **Springer, D. S., S. J. Cullen, and L. G. Everett.** 1994. Laboratory studies on air permeability, p. 217–248. *In* L. G. Wilson, S. J. Cullen, and L. G. Everett (ed.), *Handbook of Vadose Zone Characterization and Monitoring.* Lewis Publishers, Chelsea, Mich.

44. **Strack, O. D. L.** 1989. *Groundwater Mechanics.* Prentice-Hall, Englewood Cliffs, N.J.

45. **Strack, O. D. L.** 1994. *The Analytic Element Method: an Overview*. Department of Civil and Mining Engineering, University of Minnesota, Minneapolis.

46. **Wei, J., A. Dahmani, D. Ahlfeld, J. Lin, and E. Hill III.** 1993. Laboratory study of air sparging: flow visualization. *Ground Water Monit. Remediation* **13**(4):115–126.

47. **Yeh, J. T.-C., R. Srivastava, A. Guzman, and T. Harter.** 1993. A numerical model for water flow and chemical transport in variably saturated porous media. *Ground Water* **31**(4):634–644.

48. **Yeh, J. T.-C., L. W. Gelhar, and A. L. Gutjahr.** 1985. Stochastic analysis of unsaturated flow in heterogeneous soils. 2. Statistically anisotropic media with variable "a." *Water Resour. Res.* **21**:457–464.

Bioaugmentation

MICHAEL V. WALTER

82

The ability of microorganisms to degrade organic compounds and the rates at which degradation occurs depend on the interaction of numerous factors. These factors include chemical structure and concentration, the availability of the contaminant to the microorganisms, the size and nature of the microbial population, and the physical environment (4, 34, 36). Environmental factors such as moisture, pH, physical and chemical soil properties, and aeration will influence the activity of a microbial population in the biodegradation of contaminant compounds.

There are two methods of increasing the rate of biodegradation: stimulation of the indigenous population (biostimulation) and addition of microorganisms (bioaugmentation). Biostimulation involves supplying the required factors, i.e., nutrients such as nitrogen or oxygen, to the indigenous microorganisms. Bioaugmentation is the process of introducing microorganisms selected to perform a desired task, such as degrading the contaminant(s). The introduced microorganisms augment the indigenous population, hence the term bioaugmentation.

The practice of bioaugmentation is not new. Bioaugmentation is commonly used by the wastewater treatment industry at system startups, following an upset, or to improve treatment. The agricultural industry has added specific root nodule nitrogen-fixing bacteria to legume seeds (15). However, addition of microorganisms to facilitate biodegradation in the environment is for the most part an experimental and controversial method (15, 36).

Bioaugmentation is a useful tool that is appropriate for some but not all situations. Generally speaking bioaugmentation is most applicable for the treatment of compounds which are degraded very slowly, if at all, even under optimal conditions. Bioaugmentation has been successfully used to degrade pentachlorophenol (PCP) (5, 8, 29) and a mixture of aromatic hydrocarbons (19) in soil. Bioaugmentation may be useful when the concentration of the contaminant is toxic to indigenous microflora (4, 6, 15). Bioaugmentation may be appropriate after a chemical spill has occurred or when the environment is inhibitory to indigenous bioremediation (4, 15). Bioaugmentation is used most successfully when the contaminant is treated in a biological reactor where conditions can be controlled to maximize survival of the augmenting microbial population. Bioaugmentation generally involves the introduction of large numbers of nonindigenous microorganisms into the environment. For ex-

ample, a population of 10^{12} microorganisms per g (dry weight) was added off the Texas coast in 1990 (15). Six hundred gallons (1 gal = 3.785 liters) of trichloroethylene-degrading *Methylosinus trichosporium* at 10^9/ml was added to an aquifer at a rate of 1 gal/min (17, 33).

Successful application of bioaugmentation techniques requires the identification and isolation of useful microorganisms and the survival of those organisms after release into the environment. This chapter focuses on techniques for microbial strain selection, methods of evaluating microorganisms for biodegradative activity, the effect of inoculum size, and methods for releasing microorganisms into the environment.

STRAIN SELECTION

There are basically three methods by which a microorganism or group of microorganisms (consortium) capable of degrading a specific compound can be obtained: selective enrichment, the use of commercial products, or genetic engineering. This chapter focuses on the first two.

Selective Enrichment

Selective enrichment is designed to increase the population of a specific microorganism(s) relative to the initial inoculum. Inocula can be obtained from sludge, groundwater, or soil that has demonstrated degradative activity (7). The strategy involves providing conditions suitable for the growth of microorganisms capable of metabolizing the desired compound(s) (7, 9). For example, the desired microorganisms can be obtained by providing a particular substrate or suite of substrates that are the targets for biodegradation as the sole source(s) of carbon. Growth of the desired microorganisms may result in the loss of selective pressure in the growth media, allowing other organisms to grow as well. By inoculating the enrichment culture into fresh growth medium, the selective pressure can be reestablished or maintained (23, 30).

Techniques used for selective enrichment vary from simple batch-type methods, such as placing a sample of contaminated material that contains the target compound(s) and a population of indigenous microorganisms into a minimal medium and incubating it under appropriate conditions, to the use of highly sophisticated open systems (chemostats). The technique selected is based on time, available equipment, and ease of contaminant degradation. For example, Vecchioli et al. isolated a consortium of hydrocarbon-degrading microorganisms by adding 1 g of hydrocarbon-contaminated soil to 100 ml of minimal medium (36). Mueller et al. used a 2% inoculum of beach material contaminated with weathered Prudhoe Bay crude oil in Bushnell-Haas medium to establish an enrichment culture. The culture was maintained by routinely transferring a 2% inoculum into a fresh medium at weekly intervals (21).

Pfarl et al. used soil columns to enrich for strains capable of biodegrading 2-4 dichlorophenoxyacetic acid (2-4-D). Soil columns consisted of a glass tube with an inner diameter of 40 mm and length of 350 mm. Columns were packed with a mixture of 50 g of air-dried soil plus slightly moistened, prewashed quartz sand between two layers of glass wool. The columns were aerated by using compressed air bubbled through distilled water at 1 to 2 liters/h. Prior to the start of the enrichment process, the columns were rinsed with a minimal medium for 5 days. The 2-4-D was then added to the minimal medium (24).

To select microorganisms capable of degrading toxic or poorly soluble compounds, Ascon-Cabrera and Lebeault developed a biphasic method consisting of an organic layer and minimal medium. The biphasic components were 20 ml of silicon oil containing the target compounds and 80 ml of minimal medium into which an inoculum, activated sludge in this case, was added. The mixture was incubated at room temperature and stirred at 120 rpm. When turbidity was observed in the aqueous phase, 5 ml of culture was transferred to a new flask containing identical medium. Microorganisms were isolated by spreading culture samples on minimal medium agar plates with 100 to 200 mg of the appropriate compounds, various polychlorinated hydrocarbons, as sole carbon sources (2).

To select for microorganisms capable of biodegrading highly recalcitrant compounds, a continuous culture or chemostat may be used. Continuous culture uses a reaction vessel in an open system to which small amounts of sterile nutrient solution are added and from which equivalent amounts of spent nutrient solution containing microorganisms are removed. Continuous culture can require maintaining desired environmental conditions over a prolonged period of time while continuously adding small amounts of the target compound as a sole carbon soure. For example, Rozgaj and Glancer-Soljan used this technique over a 3-month period to selectively enrich for a microbial consortium capable of degrading 6-aminonaphthalene-2-sulfonic acid. The continuous culture was followed by another 3 months of batch culturing in 1,000-ml Erlenmeyer flasks containing 0.125 to 1.00 g of 6-aminonaphthalene-2-sulfonic acid per liter (28).

Commercial Strains

If one lacks the facilities or the time to undertake selective enrichment, it is possible to purchase highly adapted microbial strains, especially for hydrocarbon degradation. Bacteria are isolated from sites highly contaminated with organic material. By means of selection or induced mutation, an increase in the degradative activity is achieved. Pure strains are preserved by air- or freeze-drying techniques and then combined into mixed cultures (11). It is advisable to screen more than one commercial culture, using the target compound(s) as the sole carbon source, in shake flasks and in the environmental medium into which the microorganisms are to be introduced.

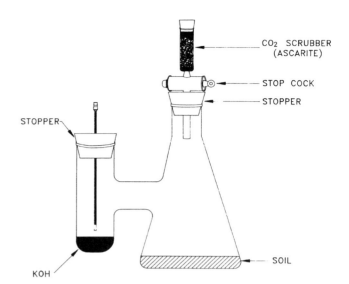

FIGURE 1 Biometer flask used for determining the extent of biodegradation by measuring CO_2, a product of biodegradation. The CO_2 is trapped in the KOH solution, which can be withdrawn and titrated with 1 N HCl to determine the amount of CO_2 produced.

SCREENING OF MICROBIAL STRAINS

Biodegradative activity can be evaluated directly by monitoring the loss of target compounds or indirectly by measuring the by-products of biodegradation or electron acceptors. A simple method requiring minimal equipment for screening microbial inocula in soil involves the use of biometer flasks. Mueller et al. modified this method to screen microorganisms in seawater (21). Biometer flasks (available from Bellco Biotechnology, Vineland, N.J., or equivalent) consist of a 250-ml Erlenmeyer flask with a side arm (Fig. 1). Potassium hydroxide is introduced into the side arm to trap CO_2. A thin layer of treated soil is added to the flask and aerated with CO_2-free air. Periodically, the KOH is withdrawn and titrated with a standard acid solution to determine the amount of CO_2 that has been produced (3). This is a useful screening technique because it is uncomplicated and inexpensive. This technique can be labor-intensive, however.

Dott et al. compared nine commercial inocula from three sources with activated sludge for degradation of summer fuel oil. Their screening protocol consisted of adding appropriate amounts of summer fuel oil to 100 ml of minimal medium. Biodegradation was monitored by solvent extraction followed by gas chromatography (11).

Fayad et al. compared the effectiveness of commercial bacterial products on degradation of oily mousse from the Gulf War oil spill by using 72-liter aquariums. Eighteen aquariums were separated into three groups of six each. Each group of aquariums received either 5, 10, or 20 g of the oily mousse per liter. In each group, two aquariums served as untreated controls, two received nutrients only, and two received nutrients and bacteria. Degradation of the oily mousse in the seawater was monitored by gas chromatography of methylene chloride extracts (13).

The U.S. Environmental Protection Agency commissioned the National Environmental Technology Applications Corporation to evaluate commercial products for use in Prince William Sound. The screening procedure was conducted in an electrolytic respirometer. An electrolytic respirometer is designed to measure the rate of respiration (oxygen uptake) by microorganisms in solution, soil, or sludge. Respirometers consist of a reactor module connected to an electrolytic oxygen generator. Depletion of oxygen within the reactor chamber, presumably by microbial activity, creates a vacuum which is sensed by a pressure indicator. This vacuum triggers the electrolytic oxygen generator, which creates oxygen by the process of electrolysis to counter the negative pressure. The electricity used to generate the oxygen is proportional to the amount (milligrams) of oxygen per liter (37). The CO_2 produced by microbial activity can be trapped in soda lime or potassium hydroxide.

In the Environmental Protection Agency-sponsored study, the decision of which products to field-test was based on rapid onset and high rate of oxygen uptake, substantial growth of oil degraders, and significant degradation of the aliphatic and aromatic hydrocarbon fractions (37).

EFFECT OF INOCULUM SIZE

The effect of initial population size on the rate of biodegradation may be dependent on the compound being degraded and environmental conditions present. The effects of different inoculum densities on various contaminants and some of the techniques used to enumerate the microbial populations are described below.

The effect of applying an *Alcaligenes* sp. capable of phenanthrene degradation was measured at three population densities ranging from 10^6 through 10^8 CFU/g of soil, using nonsterile soil. Results indicated enhanced biodegradation of phenanthrene to the same degree irrespective of inoculum level (19). The phenanthrene-degrading population was monitored by drop plating samples onto basal minimal salts agar, allowing the plates to dry overnight, and then spraying them with a 6% phenanthrene solution. The plates were incubated for 6 days at 25°C in the dark; colonies producing clear zones in the phenanthrene layer were counted.

In laboratory experiments, the degradation of PCP was found to be directly proportional to the initial cell density inoculated into the soil (8). However, after the first 2 days, soil that was inoculated with a minimum of 10^3 CFU/g had very similar PCP mineralization rates. PCP degraders were enumerated by using a radioisotopic most-probable-number technique. Samples were serially diluted in minimal medium, using a fivefold, five-replicate dilution series. Dilutions were inoculated with radioactive PCP, and replicates that evolved radioactive CO_2 above background levels were scored positive (8).

Comeau et al. measured the effect of inoculum density on 2-4-D biodegradation in soil. They found that the time for complete 2-4-D biodegradation was reduced by 0.5 day for each log increase in inoculum population when inoculum levels were above 10^5/ml (6).

The effect of inoculum size ranging from 10^4 to 10^8 CFU/liter on biodegradation of phenols was investigated by Vaishnav and Korthals Serial dilutions were plated on minimal salts agar containing 100 to 250 mg of phenol per liter. They found that the lag period decreased as the population increased. However, the overall rate of phenol degradation was independent of the microbial population (35).

Portier et al. compared the rates of degradation of polycyclic aromatic hydrocarbons (PAHs) in soil, using both mesocosms and field experiments. They compared PAH degradation in soil treated with commercial microorganisms with degradation in contaminated soil from which microorganisms were isolated. The rate of PAH biodegradation was correlated with microbial biomass and microbial density. Total microbial density was estimated by using media selective for bacteria, actinomycetes, yeasts, and filamentous fungi. Microbial biomass was measured by comparing levels of ATP. During the first 2 weeks of the study, mesocosms inoculated with commercial strains had higher rates of biodegradation, corresponding to higher populations of microorganisms. However, after 28 days, biodegradation rates from the two inocula were the same and corresponded to similar microbial populations. Field studies showed the same relationship between rate of bioremediation and population (26).

APPLICATION METHODS

One of the factors affecting the successful use of bioaugmentation is survival of the introduced strains. Research suggests that survival of microorganisms freely released into soil is influenced by numerous factors. These include nutrient availability, moisture, physical and chemical characteristics of the environment, predation, parasitism, and competition from the indigenous microbial population. Any one or a combination of these environmental factors can adversely affect the ability of an augmenting population to carry out biodegradation at useful rates over a prolonged period of time (12, 14, 26, 27). An important objective in bioaugmentation is to maximize the survival of microorganisms released into the environment. Different microbial cell carriers such as peat, mineral oil, agarose, clay, and wood chips are used to apply microorganisms into the environment to increase their survival. Another promising method for increasing microbial survival appears to be encapsulation of augmenting microorganisms.

Immobilization

Briglia et al. compared survival of PCP-degrading strains of *Rhodococcus* and *Flavobacterium* after both had been immobilized on polyurethane foam (PUR foam). The PUR foam (24 g, wet weight) was washed with distilled water and then sterilized in 250 ml of growth medium at 110°C for 20 min. The PUR foam growth medium was inoculated and incubated for 2 days at 28°C in the dark. The immobilized biomass was harvested by centrifugation. A total population of 7×10^{11} *Rhodococcus* and 9×10^{11} *Flavobacterium* cells was observed on 24 g of PUR foam. The inoculated PUR foam was mixed into soil to give an initial population of 9×10^9 cells per g of soil. Results indicated that the *Flavobacterium* population declined within 60 days after soil burial with no enhanced PCP degradation, while the *Rhodococcus* population degraded PCP in soil at a rate of 3.7 mg of PCP per day per kg of soil for 200 days (5).

Wiesel et al. demonstrated that a mixed bacterial culture capable of degrading PAHs could be immobilized onto either granular clay or lava slag. The inoculum was slowly percolated over the carrier material in a loop reactor and reached levels of 1 to 5×10^7 CFU/ml. They found that increasing the amount of inoculated carrier material increased the amount of biodegradation in shake flask experiments. They also observed biodegradation of PAHs in a model soil contained in a column using bacteria immobilized on granular clay. The biodegradation rate was not as high as in shake flask experiments. They concluded that the reduced rate may have been caused by heterogeneous distribution of PAH in the model soil (39).

Ogbonna et al. reported the use of a loofah sponge (*Luffa cylindrica*) as a support for microbial cell immobilization.

Loofah sponges are produced in large quantities in most African and Asian countries. They are light, cylindrical in shape, and made up of an interconnecting void within an open network of matrix support material. In comparison with other carriers, the density of a loofah sponge is very low and the porosity and specific pore volume are very high. The microorganisms were immobilized on the sponges by installing a "basket-like" device containing sponges in a 2.5-liter aerated fermentor. One impeller was located above the basket, while another impeller and an air sparger were placed below the basket. The fermentor was inoculated with sediment from a 24-h culture of a flocculating yeast. Results showed that immobilized cells grew well in the sponge, reaching cell densities of >3 g/g of sponge. They found that for fermentation purposes, an optimum cell/sponge weight ratio of 0.3 g : 1 g gave the best results (22). While not yet tested for bioremediation, supports like the loofah sponge offer intriguing possibilities.

Weber and Corseuil reported the use of biologically activated carbon (BAC) for enrichment of microorganisms and subsequent introduction into sand aquifers. Their system uses small BAC reactors in low-volume pumping schemes for the continuous growth, acclimation, and reintroduction of selectively enriched microorganisms into the subsurface. It was assumed that older cells would be sloughed from the BAC and were therefore less likely to reattach. The BAC reactors consisted of glass columns containing 1 g of 30-by-40 (U.S. standard sieve size) granular activated carbon. A seed culture of microorganisms selectively enriched for degradation of target compounds was isolated, and 1 ml of this culture, with a population of about 10^7/ml, was inoculated into the BAC reactors. The cells were allowed to attach for 12 h before the reactor was started. Synthetic groundwater and stock solutions of the target compounds were mixed and run in up-flow mode. The authors concluded that simultaneous contaminant destruction and enrichment for specific microorganisms occurred (38).

Otte et al. reported initial success with a technique called soil activation. Soil activation is the cultivation of biomass from contaminated soil for use in bioaugmentation. Active biomass was obtained from contaminated soil by using a batch bioreactor fed increasing concentrations of contaminants. The efficacy of soil activation was tested in static microcosms that received inoculum from the bioreactor. Results showed that within 36 h, 50% of the PCP had been mineralized in the inoculated soil, while no PCP mineralization was detected in the uninoculated control (23).

Encapsulation

Another method of increasing the survival of introduced microorganisms is the encapsulation of augmenting microbial strains. Encapsulation ensures the consistent presence of catabolically active biomass, provides for the slow release of nutrients, and provides a protective niche for the introduced microorganisms (18, 20).

Stormo and Crawford described a method for the encapsulation of microorganisms in microbeads composed of agar, alginate, or polyurethane. The beads ranged from 2 to 50 μm in diameter. Their process used a low-pressure ultrasonic nozzle which produced an aerosol of 2 to 50 μm regardless of the type of suspension that was being passed through it. The aqueous phase for the collection of the microspheres was 50 mM $CaCl_2$ for cells encapsulated in alginate or cold buffer for cells in agarose or prepolymer suspensions. A PCP-degrading *Flavobacterium* isolate that was encapsulated with this device retained high rates of PCP biodegradation (31).

Sumino et al. immobilized activated sludge by using acrylamide. They compared block polymerization with a tube polymerization method. In block polymerization, activated sludge was suspended in an acrylamide monomer, cross-linker (N,N'-methylenebisacrylamide), and promoter (β-dimethylamino-propionitrile), and then mixed with an initiator (potassium persulfate). Polymerization started within 1 min and was complete at 10 min. The gel block was passed through stainless steel mesh to yield small particles. The tube polymerization method was developed because of difficulty in obtaining uniform pellets by using block polymerization. The immobilization began as described; however, after a polymerization initiator was added, the mixture was immediately passed through a polyvinyl tube and incubated for 10 min at 20°C. The resulting elastic gel containing activated sludge was extruded from the polyvinyl tube and cut into pellets. In both procedures, monomeric acrylamide was found to be toxic to gram-negative bacteria. Improved survival was achieved by adding a macromolecular coagulant (Praestol 444K) to form particles prior to mixing with acrylamide monomer (32).

Wu and Wisecarver used a mixture of polyvinyl alcohol (PVA) and sodium alginate to encapsulate a phenol-degrading microorganism. They mixed bacteria suspended in growth medium with PVA. The PVA was added dropwise into a solution of saturated boric acid and 2% $CaCl_2$. The beads were gently stirred for 24 h and rinsed to remove excess boric acid. The addition of alginate to the PVA was necessary because initial attempts at cell encapsulation using only PVA produced agglomeration of the PVA beads into a mass of polymer unsuited for use in fluidized-bed reactors. Agglomeration of the beads was prevented by using PVA/alginate ratios of 5:1, 8:1, 10:1, and 12.5:1 (wt/wt) (40).

FUNGI

The ability of fungi, especially lignin-degrading fungi, to transform a wide variety of hazardous chemicals has generated interest in using these organisms in bioremediation (1, 16). Lamar et al. compared the abilities of three lignin-degrading fungi, *Phanerochaete chrysporium*, *Phanerochaete sordida*, and *Trametes hirsuta*, to degrade PCP and creosote in soil. They tested a total of seven control and fungal treatments in a pair of complementary field studies. The inocula were grown on a proprietary formulation consisting of a mixture of grain and sawdust inside autoclave bags each containing a microporous filter. The fungal inocula were applied 1 day after sterile aspen wood chips had been plowed into the soil. Results indicated that inoculation of soil with a 10% (wt/wt) loading of *P. sordida* resulted in the greatest decrease in PCP and creosote (16).

Davis et al. investigated the effect of *P. sordida* on the extent of biodegradation of 14 priority PAH pollutants under field conditions. After 56 days, compounds of five rings or more persisted at original concentrations. However, compounds containing four rings were depleted by 24 to 72%, while compounds containing three rings were depleted by 85 to 95%. Field application of the fungi was carried out as described above (10).

REFERENCES

1. **Alexander, M.** 1994. *Biodegradation and Bioremediation.* Academic Press, San Diego, Calif.
2. **Ascon-Cabrera, M., and J. M. Lebeault.** 1993. Selection

of xenobiotic-degrading microorganisms in a biphasic aqueous organic system. *Appl. Environ. Microbiol.* **59:** 1717–1724.

3. **Bartha, R., and C. Pramer.** 1965. Features of a flask and method for measuring the persistence and biological effects of pesticides in soil. *Soil Sci.* **100:**68–70.

4. **Bewley, R. J.** 1992. Bioremediation and waste management, p. 33–45. *In* D. E. S. Stewart-Tull and M. Sussman (ed.), *The Release of Genetically Modified Microorganisms.* Plenum Press, New York.

5. **Briglia, M., E. L. Nurmiaho-Lassila, G. Vallini, and M. Salkinoja-Salonen.** 1990. The survival of the pentachlorophenol-degrading *Rhodococcus cholorophenolicus* and *Flavobacterium* sp. in natural soil. *Biodegradation* **1:**273–281.

6. **Comeau, Y., C. W. Greer, and R. Samson.** 1993. Role of inoculum preparation and density of bioremediation of 2-4-D-contaminated soil by bioaugmentation. *Appl. Microbiol. Biotechnol.* **38:**681–687.

7. **Cookson, J. T.** 1994. *Bioremediation Engineering. Design and Application,* p. 433–460. McGraw-Hill, Inc., New York.

8. **Crawford, R. L., and W. W. Mohn.** 1985. Microbial removal of pentachlorophenol from soil using a *Flavobacterium. Enzyme Microb. Technol.* **7:**617–620.

9. **Crueger, W., and A. Creuger.** 1990. *Biotechnology, a Textbook of Industrial Biotechnology,* p. 4–8. Technical Publishers, Madison, Wis.

10. **Davis, M. W., J. A. Glaser, J. W. Evans, and R. T. Lamar.** 1993. Field evaluation of the lignin degrading fungus *Phanerocheate sordida* to treat creosote-contaminated soil. *Environ. Sci. Technol.* **27:**2572–2576.

11. **Dott, W., D. Feidierker, P. Kampfer, H. Schleibinger, and S. Strechel.** 1989. Comparison of autochthonous bacteria and commercially available cultures with respect to their effectiveness in fuel oil degradation. *J. Ind. Microbiol.* **4:**365–374.

12. **England, L. S., L. Hung, and J. Trevors.** 1993. Bacterial survival in soil: effect of clays and protozoa. *Soil Biol. Biochem.* **25:**525–531.

13. **Fayad, N. M., R. L. Endora, A. H. El-Mubarak, and A. B. Polancos, Jr.** 1992. Effectiveness of a bioremediation product in degrading oil spilled in the 1991 Arabian Gulf war. *Bull. Environ. Toxicol.* **49:**787–796.

14. **Goldstein, R. M., L. M. Mallory, and M. Alexander.** 1985. Reasons for the possible failure of inoculation to enhance biodegradation. *Appl. Environ. Microbiol.* **50:**977–983.

15. **Hoeppel, R. E., and R. E. Hinchee.** 1994. Enhanced bioremediation for on-site remediation of contaminated soils and ground water. *Environ. Sci. Pollut. Control Ser.* **6:** 311–431.

16. **Lamar, R. T., J. W. Evans, and J. Glaser.** 1993. Solid phase treatment of pentachlorophenol-contaminated soil using lignin-degrading fungi. *Environ. Sci. Technol.* **27:** 2566–2571.

17. **Lobenz, G.** 1995. Bioaugmentation clearing groundwater of TCE. *Bioremediation Rep.* **4:**1–2.

18. **Mattiasson, B.** 1987. *Immobilized Cells and Organelles,* vol. 1. CRC Press, Tampa, Fla.

19. **Moller, J., and H. Ingvorsen.** 1993. Biodegradation of phenanthrene in soil microcosms stimulated by an introduced *Alcaligenes* sp. *FEMS Microbiol. Ecol.* **102:**271–278.

20. **Mueller, J. G., J. Lin, S. E. Lantz, and P. H. Pritchard.** 1993. Recent developments in clean-up technologies. *Remediation* **3**(3):369–381.

21. **Mueller, J. G., S. M. Resnick, M. E. Shelton, and P. H. Pritchard.** 1992. Effect of inoculation on the biodegradation of weathered Prudhoe Bay crude oil. *J. Ind. Microbiol.* **10:**95–102.

22. **Ogbonna, J. C., Y. C. Liu, Y. K. Liu, and H. Tanaka.** 1994. Loofa (*Luffa cylindrica*) sponge as a carrier for microbial cell immobilization. *J. Ferment. Bioeng.* **78:**437–432.

23. **Otte, M. P., J. Gagnon, Y. Comeau, N. Matte, C. Greer, and R. Sampson.** 1994. Activation of an indigenous microbial consortium for bioaugmentation of pentachlorophenol/creosote contaminated soils. *Appl. Microbiol. Biotechnol.* **40:**926–932.

24. **Pfarl, C., G. Ditzelmuller, M. Loidl, and F. Streichsbier.** 1990. Microbial degradation of xenobiotic compounds. *FEMS Microbiol. Ecol.* **73:**255–262.

25. **Piotrowski, M. R.** 1991. Bioremediation of hydrocarbon contaminated surface water, groundwater, and soils: the microbial ecology approach. *Hydrocarb. Contam. Soils Groundwater* **1:**203–238.

26. **Portier, R., M. Bianchini, K. Fujisaki, C. Henery, and D. McMillin.** 1988. Comparison of effective toxicant biotransformation by autochthonous microorganisms and commercially available cultures in the in situ reclamation of abandoned industrial sites. *Schriftenr. Ver. Wasser Boden Lufthyg. Berlin-Dahlem* **80:**273–292.

27. **Recorbet, G., C. Steinberg, and G. Faurie.** 1992. Survival in soil of genetically engineered *Escherichia coli* as related to inoculum density, predation and competition. *FEMS Microbiol. Ecol.* **101:**251–260.

28. **Rozgaj, R., and M. Glancer-Soljan.** 1992. Total degradation of 6-aminonaphthalene-2-sulphonic acid by a mixed culture consisting of different bacterial genera. *FEMS Microbiol. Ecol.* **86:**229–235.

29. **Seech, A.** 1991. Biodegradation of pentachlorophenol in soil: response to physical, chemical, and biological treatments. *Can. J. Microbiol.* **37:**440–444.

30. **Stanbury, P. F., and A. Whitaker.** 1984. The isolation, preservation and improvement of industrial microorganisms, p. 26–73. *In Principles of Fermentation Technology.* Pergamon Press, Oxford.

31. **Stormo, K. E., and R. L. Crawford.** 1992. Preparation of encapsulated microbial cells for environmental applications. *Appl. Environ. Microbiol.* **58:**727–730.

32. **Sumino, T., H. Nakamura, and N. Mori.** 1991. Immobilization of activated sludge by the acrylamide method. *J. Ferment. Bioeng.* **72:**141–143.

33. **Texas General Land Office.** 1990. *Megabore Oil Spill off the Texas Coast. An Open Water Bioremediation Test.* Texas General Land Office, Austin.

34. **Torstensoon, L.** 1988. Microbial decomposition of herbicides in soil. *Outlook Agric.* **17:**120–124.

35. **Vaishnav, D. D., and E. Korthals.** 1989. Effect of microbial concentration on biodegradation rates of phenols. *J. Ind. Microbiol.* **4:**307–314.

36. **Vecchioli, G. I., M. T. Delpanno, and M. T. Painceira.** 1990. Use of selected autochthonous soil bacteria to enhance degradation of hydrocarbons in soil. *Environ Pollut.* **67:**249–258.

37. **Venosa, A., J. R. Haines, W. Nisamaneepong, R. Gouind, S. Pradhan, and B. Siddique.** 1992. Efficacy of commercial products in enhancing oil biodegradation in closed laboratory reactors. *J. Ind. Microbiol.* **10:**13–23.

38. **Weber, W. J., Jr., and H. X. Corseuil.** 1994. Inoculation of contaminated subsurface soils with enriched indigenous microbes to enhance bioremediation rates. *Water Res.* **28:** 1407–1414.

39. **Wiesel, I., S. M. Wubker, and H. J. Rehm.** 1993. Degradation of polycyclic aromatic hydrocarbons by an immobilized mixed bacterial culture. *Appl. Microbiol. Biotechnol.* **39:**110–116.

40. **Wu, K. A., and K. Wisecarver.** 1992. Cell immobilization using PVA cross linked with boric acid. *Biotechnol. Bioeng.* **39:**447–449.

Use of Fungi in Biodegradation

J. W. BENNETT AND B. D. FAISON

83

In nature, fungi do much of the dirty work. They are particularly efficient at degrading the major plant polymers, cellulose and lignin, but they also decompose a huge array of other organic molecules, including waxes, rubber, feathers, insect cuticles, and animal flesh. Although industrial microbiologists regularly harness fungal metabolism for brewing, baking, cheese preparation, and the production of antibiotics, commercial enzymes, and a number of commodity chemicals, fungi are best known for their dirty work. They spoil our foods, blight our crops, rot our buildings, contaminate our petri dishes, and cause some rather loathsome diseases. Paradoxically, despite this notoriety, the use of fungi in bioremediation has been limited compared with the use of bacteria. We here present a brief introduction to fungal taxonomy and mycological techniques, introduce methods for isolating fungi and for growing them in the laboratory, define some important terms, review examples of the successful applications of fungal organisms and enzymes for biodegradation, and point out the advantages and disadvantages of fungi as agents of bioremediation.

A LITTLE TAXONOMY

Like many other higher-order taxonomic units, the term "fungus" is difficult to define. It embraces a large group of nonphotosynthetic, lower eukaryotes once considered part of the plant kingdom but now usually afforded status in their own kingdom, the "Fifth Kingdom," on the basis of their characteristic absorptive mode of nutrition (71, 117). Traditionally, the Myxomycota, or slime molds, and the Eumycota, or "true fungi," comprise the two major subdivisions within this fungal kingdom. A plasmodium or pseudoplasmodium characterizes the Myxomycota, a group which includes well-known genera such as *Dictyostelium* and *Physarum*. Modern biologists believe that the slime molds are retained as fungi more out of tradition than on the basis of sound taxonomy (81), and they will not be considered further here.

Among the true fungi, the assimilative phase is usually filamentous or yeast-like. This group is divided into the lower fungi, which lack septation in the cell walls (the Mastigomycotina and Zygotina), and the higher fungi, which have septate cell walls (the Ascomycotina, Basidiomycotina, and Deuteromycotina). Most of the lower and all of the higher fungi have chitinized cell walls. A number of the lower fungi such as the hypochytrids and the oomycetes are now classified in the kingdom Protista (81).

In filamentous forms, the individual thread-like cells are called hyphae. A fungal colony, or portion of a colony, composed of many hyphae together is called a mycelium. The filamentous/mycelial growth form poses problems in determining the size of a single organism and in measuring the growth of fungi. In the older literature, the term "thallus" is often used to describe macroscopic mycelial formations.

Fungal taxonomy is based on reproductive morphology, which consists of both meiotic and mitotic spore-bearing structures. Both sexual and asexual spores are typically made in vast numbers. Many fungi have more than one morphologically distinct spore type at different phases in their life cycles. Further complicating this fungal pleomorphism is the fact that some fungi exist as either yeast or filamentous forms, depending on the environmental milieu, a phenomenon called dimorphism, best known from medically important species.

Like other eukaryotes, fungi have nuclei, mitochondria, 80S ribosomes, and chromosomes. Fungal cells may be haploid or diploid; the nuclei within a mycelium may all be genetically identical (monokaryotic) or a mixture of different genetic types (heterokaryotic). Basidiomycetes often have a special form of heterokaryon called a dikaryon. Although many fungi are microscopic, the best-known species, such as mushrooms and truffles, form macroscopic fruiting structures. Fungi are ubiquitous in terrestrial environments, and many fungi are capable of growing in environments hostile to most other forms of life (67). For example, fungi are the only eukaryotes known with thermophilic (60 to 62°C) optimal growth temperatures (109).

In summary, the fungi comprise an extremely heterogeneous group of eukaryotic, heterotrophic organisms that absorb their food. Mycology texts such as those by Alexopoulos and Mims (9), Ross (97), Moore-Landecker (85), and Carlile and Watkinson (24) are helpful introductions to this Fifth Kingdom. Another useful tool is *Ainsworth & Bisby's Dictionary of Fungi* (61). General taxonomic principles, as well as a guide to the sometimes arcane principles of fungal nomenclature, which are governed by the International Code of Botanical Nomenclature, are presented by Hawksworth (59). The last volumes (in two parts) of the classic series *The Fungi: an Advanced Treatise* provide comprehensive taxonomic coverage (1, 2), while the first volumes (3–5) give an almost encyclopedic review of the classical mycological literature. Fungal classification and taxonomy will also be covered in volume VII of a new comprehensive series, *The Mycota*, which is in preparation (38).

THE ACQUISITION, CARE, AND FEEDING OF FUNGI

Filamentous fungi have more described species than any other group of microorganisms, with about 70,000 named and approximately 1,500 new species published each year. As with bacteria, it is believed that only a small proportion of extant species are known to science. The total number of fungal species, both known and unknown, has been estimated at 250,000 (60).

About 170,000 pure strains of fungi, representing an estimated 7,000 different species, are maintained in culture collections internationally (60). Information about these resources can be accessed through the *World Directory of Collections of Cultures of Microorganisms* (107). One of the oldest and largest fungal collections is the Centraalbureau voor Schemmelcultures in The Netherlands. The major collection in the United States is the American Type Culture Collection (68). In addition, the collection at the Northern Regional Research Laboratory of the U.S. Department of Agriculture in Peoria, Ill., houses a large number of economically important fungi, and the Forest Products Laboratory in Madison, Wis., holds a major collection of wood rotters. As with bacterial and viral strains, a fee is usually charged for obtaining cultures.

In the laboratory, fungi and bacteria are treated similarly. They can be grown on agar media in petri plates or in liquid broths. Depending on the nutritional requirements of the individual species, either complex or defined media are used. Many fungi prefer media with acid pH. On petri plates, molds are readily distinguished from bacteria because they usually form dry colonies which may be brightly colored as a result of the pigmentation of their spores. With the naked eye, yeast colonies are not so easily distinguished from bacteria, but under the light microscope, their large size, compared with that of any run-of-the-mill prokaryote, is easy to discern. Recipes for common and specialty media for cultivating fungi are available in *Mycology Guidebook* (106) and *Handbook of Microbiological Media* (13). In isolating fungi from nature, antibiotics such as streptomycin or other bacterial growth inhibitors such as rose bengal are frequently added to media, thereby selectively enhancing growth of fungi (82).

When one is isolating culturable microorganisms from natural substrates, it is relatively easy to obtain a colony count based on the unicells of typical bacteria. Sampling for fungi is much harder because of their filamentous growth habit. There may be a large mass of fungal mycelium present, but the methods adapted from bacteriology may not detect it. Moreover, dormant fungal spores may produce numerous colonies, whereas thriving nonsporulating colonies may not be recovered at all. In liquid shake culture, many filamentous fungi form pellets, thus making direct turbidity assays impractical and making dry weights the most commonly used measure. In batch culture, the synthesis of many fungal products and enzymes is not correlated with growth but is triggered by the limitation of an essential nutrient. The terms "trophophase" and "idiophase," roughly comparable to bacterial log phase and stationary phase, have been used to describe filamentous growth. Both secondary metabolite production and the production of lignolytic enzymes are correlated with idiophase.

Enrichment cultures are a classical microbiological technique, commonly used for finding a specific microbe to degrade a certain toxic waste. Enrichment cultures favor the growth of a particular species based on its nutritional requirements. In the most common application of this method, aliquots of water, soil, leaf litter, or other mixed inocula are placed into a medium containing the targeted xenobiotic as the sole carbon source. Only organisms with degradative ability will grow. In liquid culture, competition for the substrate will lead to enrichment of the microbial strain that is able to grow fastest. On petri plates, colonies representing many species are usually isolated; these are then subcultured and tested further. With few exceptions. this approach leads to the isolation of bacteria. In general, fungi grow more slowly and produce fewer propagules than do bacteria. In addition, fungi are less likely than bacteria to have the capacity to use xenobiotics as sole carbon sources. Many fungi need a supplemental carbon source to sustain growth; i.e., their degradative potential is cometabolic. The much lauded capacity of white rot fungi to decompose lignin is a good example. Even though white rots are the major degraders of lignin in nature, they do not use the degraded lignin as a carbon source for growth (66, 75).

Thus, the key to successful isolation of fungi for xenobiotic degradation is twofold: (i) the recognition that fungi are easily outgrown by bacteria and (ii) the recognition that they produce many potent biodegradative enzymes capable of degrading toxic pollutants yet do not use these breakdown products to sustain growth. In order successfully to isolate fungi with potential for bioremediation, it is necessary to impose imaginative enrichment conditions, including the careful selection of supplementary carbon sources and bacteriostatic agents.

Serving as a manual to both methods and references, volume 4 of *Methods in Microbiology* (20) is specifically devoted to distinguishing mycological and bacteriological perspectives. Specialized techniques for collecting, isolating, cultivating, manipulating, and preserving fungi are given. Individual chapters are devoted to soil fungi, air sampling, and aquatic organisms. Although several decades old, this book is still a wonderful resource. See also volume 22 of *Methods in Microbiology* (*Techniques in Microbial Ecology* [56]), *Mycology Guidebook* (106), and chapters 6, 44, and 45 of this volume for guidance on the special handling of fungi. Other valuable publications concentrating on applied mycology are those by Arora et al. (12), Smith and Berry (102, 103), and Smith et al. (104).

DEFINITIONS AND DISTINCTIONS

Biodegradation, mineralization, bioremediation, biodeterioration, biotransformation, cometabolism, and bioaccumulation are terms not always used with appropriate sensitivity to subtle differences in meaning. The special role that fungi play in nature and in the human exploitation of their often unique metabolism can be clarified by thinking about the distinctions implied in this ecological lexicon.

Biodegradation is the biologically mediated breakdown of chemical compounds. It is an umbrella term, encompassing most of the other jargon addressed in this section, and generally implies a series of biochemical reactions. When biodegradation is complete, the process is called *mineralization*, i.e., the total breakdown of organic molecules into water, CO_2, and/or other inorganic end products. Not all authors are careful to distinguish between degrees of biodegradation; some use the term to describe almost any biologically mediated change in a substrate; others use it to describe mineralization (8).

Biotransformation, a word often used synonymously with *bioconversion,* usually refers to a single step in a biochemical pathway, in which a molecule (the precursor) is catalytically converted into a different one (the product). Many biotransformations in sequence constitute a metabolic pathway. Industrial processes frequently incorporate biotransformation reactions. A famous example is in the manufacture of steroids for birth control pills (95). When ecologists describe biotransformation, the environmental status of the transformed product is a primary consideration. Is it more water soluble and thus more easily excreted by the cell? Is it less toxic? Is it *more* hazardous than its precursor? When relatively innocuous precursors are converted into more toxic products, the process is called *activation.* The metabolic activation of mercury is a well-known example of a biotransformation with malign environmental impact.

Biodeterioration and *bioremediation* are the two aspects of biodegradation with an anthropomorphic emphasis. Biodeterioration is the breakdown of economically useful substances. Often the term is used narrowly to refer to the deterioration of substances that are normally resistant to biological attack, such as metals, plastics, drugs, cosmetics, paintings, sculpture, wood products, electrical equipment, fuels and oils, and other economically valuable objects (96).

In bioremediation, biological systems are used to transform and/or degrade toxic compounds or otherwise render them harmless. It can involve indigenous microbial populations with or without nutrient supplementation or exogenous organisms inoculated into the site. The goal is to disarm noxious chemicals without the formation of new toxins.

Bioaccumulation, sometimes used loosely synonymously with *biosorption,* is the concentration of substances without any metabolic transformation. Both living and dead cells may be involved. Bioaccumulation techniques can be used to concentrate substances such as copper, lead, silver, uranium, and certain radionuclides from aqueous environments, and the resultant loaded biomass can be recycled or contained.

Cometabolism is used in two ways. Usually, it describes the situation in which an organism is able to biotransform a substrate but is unable to grow on it. "Co-metabolism refers to any oxidation of substances without utilization of the energy derived from the oxidation to support microbial growth" (64). The phenomenon has also been called "cooxidation" and "gratuitous" or "fortuitous" metabolism. Many biochemists dislike the term (52, 65). Nevertheless, it has become well entrenched in the literature.

A second meaning of cometabolism is to describe the degradation of a given compound by the combined efforts of several organisms pooling their biochemical resources for mutual efforts (33).

THE FUNGAL WAY OF DEGRADATION

The placement of fungi into their own kingdom is based on their unique nutritional status; i.e., their cells secrete extracellular enzymes which break down potential food sources which are then absorbed back into the fungal colony. This way of life means that any discussion of fungal biodegradation must cover an extraordinary amount of catalytic capability. The decomposition of lignocellulose is prob-

ably the single most important degradative event in the Earth's carbon cycle. The utilization and transformation of the dead remains of other organisms is essential to the Earth's economy. An enormous ecological literature exists on the role of fungi as primary and secondary decomposers in these classic cycles of nature (see, for example, references 7, 25, 29, and 48).

From the human perspective, the power of fungal enzymes is Janus-faced. Molds destroy more food than any other group of microorganisms. They damage standing timber, finished wood products, and fibers as well as a wide range of noncellulosic products such as plastics, fuels, paints, glues, drugs, and other human artifacts (36, 96). On the other hand, many of the oldest biotechnological practices are also based on fungal catalytic power: baking, brewing, wine fermentation, the making of certain cheeses, and the koji process are ancient examples of the way humans have employed fungi for their own benefit. In the 20th century, numerous hydrolytic enzymes involved in the degradation of relatively simple biopolymers such as starch and protein have been purified, characterized, and utilized within industrial settings. These include fungal amylases, glucoamylases, lipases, pectinases, and proteases (see references 17 and 19 for reviews). Fungal cellulases provide a good example of the contrasting faces of a single enzymatic capability. During the Second World War, research by the U.S. Army on the microbial destruction of military clothing and tents led to the characterization of the cellulolytic mold *Trichoderma reesei.* Continuing research on *T. reesei* identified a complete set of cellulase enzymes required for the breakdown of cellulose to glucose (89, 93). These same enzymes now promise the potential of converting waste cellulosics into foods for our burgeoning population and have been subject of intense molecular biology research (39, 84). Although cellulase-produced glucose is not yet economically competitive, another traditional fungal process is: the conversion of lignocellulose into mushrooms (28, 105). These and other examples of economically advantageous uses of fungal biodegradation are displayed in Table 1.

Fungi are also good at bioaccumulation of heavy metals. Many species can absorb cadmium, copper, lead, mercury, and zinc onto their mycelia and spores. Sometimes the walls of dead fungi bind better than those of living ones. Systems using *Rhizopus arrhizus* have been developed for treating uranium and thorium (112). Spent fungal biomass from industrial fermentations is an available resource for the concentration of heavy-metal contamination (50, 51, 98).

What about fungal degradation of pollutants and toxic wastes? In some cases, traditional methods are being adopted for contemporary needs. For example, composting has been used to treat both pesticides (47) and munitions wastes (118). There is also rather a lot of descriptive biochemistry concerning the abilities of various fungi and their enzymes to biotransform pesticides (see reference 92 for enzymes and reference 101 for lists of specific compounds and organisms), but to date, the most sophisticated fungal approaches to environmental cleanup grow out of prior research on degradation of petroleum hydrocarbons (15, 26, 53) and on the adaptation of research on fungal treatment of lignocellulolytic wastes in the pulp and paper industry (73, 74, 76).

The ability to grow on petroleum hydrocarbons is widespread among the fungi (14, 27, 53). Jet crashes caused by blocked fuel lines due to the growth of *Cladosporium resinae,* first reported during World War II, are one of the more dramatic negative consequences of fungal ability to thrive in extreme habitats (79). Considerable information is available

TABLE 1 Examples of economically beneficial fungal degradation

Process	Substrate	Species	Representative reference(s)
Composting	Straw, manure, agricultural waste, bark	Consortia of bacteria and fungi, usually uncharacterized	18, 43
Mushroom cultivation	Lignocellulose, animal manure	*Agaricus bisoporus*	28, 62
	Straw, sawdust	*Pleurotus ostreatus* (oyster mushroom)	105
	Wood logs	*Lentinus odoides* (shiitake)	105
Single-cell protein	Alkanes	Yeasts, e.g., *Candida tropicalis*	34, 70
	Brewery wastes, molasses	*Saccharomyces cerevisiae, S. carlsbergensis*	80
	Sulfite waste liquid	*Candida utilis, Paecilomyces varioti*	80
Solid waste treatment	Sludge and sewage	Consortia of bacteria and fungi, usually uncharacterized	30, 32
	Pulp and paper mill	*Coriolus versicolor, Phanerochaete chrysosporium*	73, 74
Wastewater treatment	Distillery waste	Yeasts, especially *Candida utilis*	49
	Kraft bleaching effluent	*Phanerochaete chrysosporium*	37
	Tannery effluent	*Aspergillus* and *Penicillium* spp.	88

about the mycological flora associated with marine petroleum spills (27). On the industrial side, the years of research on single-cell protein, instituted with the goal of turning petroleum hydrocarbons into feed, have paid off in the study of the enzymatic mechanisms used by yeasts and other microorganisms in the biodegradation of petroleum wastes for environmental remediation (70, 94). Cytochrome P-450s are mixed-function oxidases (monooxygenases) derived from a superfamily of genes which are involved in many steps of petroleum degradation and in the biotransformation of a variety of environmental pollutants (116). Both detoxification and activation are associated with P-450 action. Fungal monooxygenases are more similar to mammalian than to bacterial cytochromes; Sariaslani (100) presents a particularly thorough review of these enzymes. Extensive biochemical and genetic data are available for several yeasts; there is also a large literature surrounding the aseptate filamentous species *Cunninghamella elegans* (27, 53).

Similarly, and to an even greater extent, research on pulp waste treatment, such as the decolorization of effluent from kraft pulp mills, and the subsequent mushrooming of research on white rot fungi have shown the power of wood-decaying species against a surprisingly large battery of environmental contaminants (23, 46).

THE WHITE ROT FUNGUS

Phanerochaete chrysosporium is a basidiomycete commonly known as the white rot fungus. It is the best studied of the ligninolytic fungi, a group whose natural habitat is forest litter and rotting wood. White rot fungi are unique among eukaryotes in having evolved nonspecific mechanisms for degrading lignin; curiously, they do not use lignin as a carbon source for their own growth (75). The biosynthesis of metabolites that do not directly contribute to the growth of the producing organism, after active growth has ceased, triggered by nutrient starvation, is reminiscent of the production of many small molecules, a process frequently called secondary metabolism. For this reason, lignin degradation is often described as a secondary metabolic process. Perhaps not coincidentally, lignin-degrading activity is induced by veratryl alcohol, a traditional low-molecular-weight secondary metabolite produced by ligninolytic cultures (41).

Lignin is unlike many natural polymers in that it consists of irregular phenyl propanoid units linked by nonhydrolyz-

able carbon-carbon and ether bonds. Lignin contains chiral carbons in both the L and D configurations, and this stereo irregularity further renders it resistant to attack by most microorganisms. Two families of heme peroxidases, lignin peroxidases (ligninases) and manganese-dependent peroxidases, as well as a peroxidase-generating system, are produced under conditions of nitrogen, sulfur, and carbon deprivation (41, 42, 54, 111). Using hydrogen peroxide, lignin peroxidases and manganese peroxidases promote the one-electron oxidation of lignin to free radicals, which then undergo spontaneous rearrangements. The enzyme systems involved in lignin biodegradation have been the subject of intense research, including the cloning and sequencing of a number of the relevant genes (for reviews, see references 6 and 63).

During the 1980s it became apparent that in addition to degrading lignin, *P. chrysosporium* is capable of mineralizing a wide variety of xenobiotics, including polyaromatic hydrocarbons, chlorinated phenols, nitroaromatics, and many other environmental toxins (Table 2), sometimes in complex mixtures (22). The ability to degrade such a broad spectrum of highly toxic and generally recalcitrant substrates puts white rot fungi in a class of their own. It is believed that this metabolic virtuosity is a function of their ecological niche as lignin degraders and that the enzymes of lignin degradation are largely responsible for the breakdown of these xenobiotics. Indeed, in the laboratory, most of the successful mineralization experiments have been conducted under ligninolytic conditions. Moreover, purified preparations of lignin peroxidases are capable of oxidizing a variety of the xenobiotics known to be mineralized by whole cell cultures of *P. chrysosporium* (e.g., polyaromatics, pentachlorophenols, and dichlorodibenzo-p-dioxin) although they are not involved in others {e.g., DDT [1,1,1-trichloro-2,2-bis(4-chlorophenyl)ethane] (77)}. For some fungal species, laccases, widely distributed copper-containing enzymes, may also be important (e.g., phenols, including chlorophenols [46, 100, 110]). In addition, a variety of intracellular enzymes, (e.g., reductases, methyltransferases, and cytochrome P-450 oxygenases), "plasma membrane potentials," and bioabsorption onto mycelia contribute to xenobiotic degradation in poorly understood ways (11, 69, 114, 115). Postulated mechanisms used by the white rot fungi to degrade pollutants have been summarized by Barr and Aust (16).

TABLE 2 Examples of xenobiotics degraded or transformed by *P. chrysosporium*

Type of compound	Example	Reference(s)
Aromatic hydrocarbons	Benzo[*a*]pyrene	57
	Phanthracene	21, 86, 108
	Pyrene	58
Chlorinated organics	Alkyl halide insecticides	72
	Atrazine	87
	Chloroanilines	10
	DDT	22, 77
	Pentachlorophenols	78, 83
	Trichlorophenol	69
	Polychlorinated biphenyls, Arochlor	35
	Polychlorinated dibenzo-*p*-dioxins	55, 58, 116
	Trichlorophenoxyacetic acid	99
Nitrogen aromatics	2,4-Dinitrotoluene (DNT)	113
	2,4,6-Trinitrotoluene (TNT)	45
	Hexahydro-1,3,5-trinitro-1,3,5-triazine (RDX)	44
Miscellaneous	Sulfanated azo dyes	31, 90, 91

MYCOREMEDIATION

Many workers divide bioremediation strategies into three general categories: (i) the target compound is used as a carbon source, (ii) the target compound is enzymatically attacked but is not used as a carbon source (cometabolism), and (iii) the target compound is not metabolized at all but is taken up and concentrated within the organism (bioaccumulation). Although fungi participate in all three strategies, they are often more proficient at cometabolism and bioaccumulation than at using xenobiotics as sole carbon sources.

The attributes that distinguish filamentous fungi from other life forms determine why they are good biodegraders. First, the mycelial growth habit gives a competitive advantage over single cells such as bacteria and yeasts, especially with respect to the colonization of insoluble substrates. In addition, fungi can rapidly ramify through substrates, literally digesting their way along by secreting a battery of extracellular degradative enzymes. Hyphal penetration provides a mechanical adjunct to the chemical breakdown effected by the secreted enzymes. The high surface-to-cell ratio characteristic of filaments maximizes both mechanical and enzymatic contact with the environment. Second, the extracellular nature of the degradative enzymes enables fungi to tolerate higher concentrations of toxic chemicals than would be possible if these compounds had to be brought into the cell. In addition, insoluble compounds that cannot cross a cell membrane are also susceptible to attack. Albeit at a low rate, fungi even have been shown to solubilize low-rank coal, a particularly persistent, irregular, and complex polymeric substrate that makes lignin look simple (27, 40). Finally, since the relevant enzymes are induced by nutritional signals independent of the target compound during secondary metabolism, they can act independently of the concentration of the substrate, and their frequently nonspecific nature means that they can act on chemically diverse substrates.

Among filamentous fungi, *P. chysosporium* is emerging as the model system for studying xenobiotic degradation. Since so little is understood about the fundamentals of how

this white rot mineralizes pollutants, it is not surprising that a great deal remains to be learned about the degradative mechanisms used by fungi in general. Oxidative enzymes play a major role, but in some cases organic acids and chelators excreted by the fungus also contribute to the process. Many of the toxic chemicals mineralized by fungi are already highly oxidized. Proton excretion, a mechanism that fungi use to lower the pH of their environment, appears to be involved in the reduction of some of these compounds (16).

What about the future? Aquatic fungi, anaerobic fungi, and mycorrhizal fungi in conjunction with pollutant-tolerant plants all provide opportunities for new research. Genetic engineering is another frontier. Fungal genes for degradative enzymes can be added to bacteria; alternatively, competent fungi can be modified to grow in an extended range of environments.

As we get better at recognizing what can and cannot be done with bioremediation, we will create a menu of choices using a broad range of organisms. In some situations, bioconcentration of a toxic waste is the best we can do. In others, the nonspecificity of the white rot fungi is ideally suited to treating low concentrations of mixed wastes in a nutrient-deficient habitat. In yet others, anaerobic bacteria are clearly the best candidates. Microbiologists know that pure cultures are rare in nature. Common sense tells us that in the real world, complete pathways of degradation are more likely to occur through the combined effects of many organisms. Judious combination of chemical and physical processes with biological schemes also offers promise. Filamentous fungi, yeasts, and nonphotosynthetic bacteria are the workhorses of biological degradation. Therefore, it is ironic that the popular press has chosen the word "green" to describe environmentally friendly technologies such as bioremediation. Decidedly not green in color (except for a few spore types) and most certainly underappreciated (even by most microbiologists), the fungi possess the most varied and most efficient battery of depolymerizing enzymes of all decomposers. They are guaranteed to be major players in the "greening" of toxic waste sites and other polluted habitats.

Research in J. W. B.'s laboratory is supported by grants from the U.S. Department of Defense (DNA 2-89, 116, 88-150) and the U.S. Department of Energy (DE-FG01-93EW532023); B.D.F. is supported by DOE contract OR 121120 from the U.S. Department of Energy's Office of Environmental Restoration and Waste Management with the In situ Remediation Integrated Program. Lockheed Martin manages Oak Ridge National Laboratory under contract no. DE-AC05-84OR21400 from the U.S. Department of Energy.

We thank Christine Murphey for manuscript preparation and Anna Loomis for help with the literature review.

REFERENCES

1. **Ainsworth, G. C., F. K. Sparrow, and A. S. Sussman.** 1973. *The Fungi: an Advanced Treatise,* vol. 4a. *A Taxonomic Review with Keys: Ascomycetes and Fungi Imperfecti.* Academic Press, New York.
2. **Ainsworth, G. C., F. K. Sparrow, and A. S. Sussman.** 1973. *The Fungi: an Advanced Treatise,* vol. 4b. *A Taxonomic Review with Keys: Basidiomycetes and Lower Fungi.* Academic Press, New York.
3. **Ainsworth, G. C., and A. S. Sussman.** 1965. *The Fungi: an Advanced Treatise,* vol. I. *The Fungal Cell.* Academic Press, New York.
4. **Ainsworth, G. C., and A. S. Sussman.** 1965. *The Fungi: an Advanced Treatise,* vol. II. *The Fungal Organism.* Academic Press, New York.
5. **Ainsworth, G. C., and A. S. Sussman.** 1965. *The Fungi:*

an Advanced Treatise, vol. III. *The Fungal Population.* Academic Press, New York.

6. **Alec, M., and M. H. Gold.** 1991. Genetics and molecular biology of the lignin-degrading basidiomycete *Phanerochaete chrysosporium*, p. 320–341. *In* J. W. Bennett and L. L. Lasure (ed.), *More Gene Manipulations in Fungi.* Academic Press, New York.

7. **Alexander, M.** 1977. *Introduction to Soil Microbiology*, 2nd ed. John Wiley & Sons, New York.

8. **Alexander, M.** 1981. Biodegradation of chemicals of environmental concern. *Science* **211:**132–138.

9. **Alexopoulos, C. J., and C. W. Mims.** 1979. *Introductory Mycology*, 3rd ed. John Wiley & Sons, New York.

10. **Arjmand, M., and H. Sandermann, Jr.** 1985. Mineralization of chloroaniline/lignin conjugates and of free chloroanilines by the white rot fungus *Phanerochaete chrysosporium. J. Agric. Food Chem.* **33:**1055–1060.

11. **Armenante, P. M., P. Nirupam, and G. Lewandowski.** 1994. Role of mycelium and extracellular protein in the biodegradation of 2,4,6-trichlorophenol by *Phanerochaete chrysosporium. Appl. Environ. Microbiol.* **60:**1711–1718.

12. **Arora, D. K., K. G. Mukerji, and E. H. Marth.** 1991. *Handbook of Applied Mycology*, vol. 3. *Foods and Feeds.* Marcel Dekker, Inc., New York.

13. **Atlas, R. M.** 1993. *Handbook of Microbiological Media.* CRC Press, Boca Raton, Fla.

14. **Atlas, R. M. (ed.).** 1984. *Petroleum Microbiology.* Macmillan, New York.

15. **Atlas, R. M., and R. Bartha.** 1992. Hydrocarbon biodegradation and oil spill bioremediation. *Adv. Microb. Ecol.* **12:**287–338.

16. **Barr, D. P., and S. D. Aust.** 1994. Mechanisms white rot fungi use to degrade pollutants. *Environ. Sci. Technol.* **28:**79A–87A.

17. **Berka, R. M., N. Dunn-Coleman, and M. Ward.** 1992. Industrial enzymes from Aspergillus species, p. 155–214. *In* J. W. Bennett and M. A. Klich (ed.), *Aspergillus. Biology and Industrial Applications.* Butterworth-Heinemann, Boston.

18. **Biddlestone, A. J., and K. R. Gray.** 1985. Composting, p. 1059–1070. *In* C. W. Robinson and J. A. Howell (ed.), *Comprehensive Biotechnology*, vol. 4. *The Practice of Biotechnology; Speciality Products and Service Activities.* Pergamon Press, Oxford.

19. **Bigelis, R.** 1992. Food enzymes, p. 361–415. *In* D. B. Finkelstein and C. Ball (ed.), *Biotechnology of Filamentous Fungi.* Butterworth-Heinemann, Boston.

20. **Booth, C. (ed.).** 1971. *Methods in Microbiology*, vol. 4. Academic Press, London.

21. **Bumpus, J. A.** 1989. Biodegradation of polycyclic aromatic hydrocarbons by *Phanerochaete chrysosporium. Appl. Environ. Microbiol.* **55:**154–158.

22. **Bumpus, J. A., and S. D. Aust.** 1987. Biodegradation of DDT [1,1,1-trichloro-2,2-bis(4-chlorophenyl)ethane] by the white rot fungus *Phanerochaete chrysosporium. Appl. Environ. Microbiol.* **53:**2001–2008.

23. **Bumpus, J. A., M. Tien, D. Wright, and S. D. Aust.** 1985. Oxidation of persistent environmental pollutants by white rot fungus. *Science* **228:**1434–1436.

24. **Carlile, M. J., and S. C. Watkinson.** 1994. *The Fungi.* Academic Press, London.

25. **Carroll, G. C., and D. T. Wicklow.** 1992. *The Fungal Community. Its Organization and Role in the Ecosystem*, 2nd ed. Marcel Dekker, Inc., New York.

26. **Cerniglia, C. E.** 1984. Microbial transformation of aromatic hydrocarbons, p. 99–128. *In* R. M. Atlas (ed.), *Petroleum Microbiology.* Macmillan, New York.

27. **Cerniglia, C. E., J. B. Sutherland, and S. A. Crow.** 1992. Fungal metabolism of aromatic hydrocarbons, p.

193–217. *In* G. Winkelmann (ed.), *Microbial Degradation of Natural Products.* VCH Press, Weinheim, Germany.

28. **Chang, S., J. A. Buswell, and P. G. Miles (ed.).** 1993. *Genetics and Breeding of Edible Mushrooms.* Gordon and Breach Science Publishers, Philadelphia.

29. **Cooke, R. C., and A. D. M. Rayner.** 1984. *Ecology of Saprotrophic Fungi.* Longman, London.

30. **Cooke, W. B.** 1976. Fungi in sewage, p. 389–434. *In* E. B. G. Jones (ed.), *Recent Advances in Aquatic Mycology.* John Wiley & Sons, New York.

31. **Cripps, C., J. A. Bumpus, and S. D. Aust.** 1990. Biodegradation of azo and heterocyclic dyes by *Phanerochaete chrysosporium. Appl. Environ. Microbiol.* **56:**1114–1118.

32. **Crueger, W., and A. Crueger.** 1982. *Biotechnology: a Textbook of Industrial Microbiology.* Sinauer Associates, Sunderland, Mass.

33. **Dagley, S.** 1987. Lessons from biodegradation. *Am. Rev. Microbiol.* **41:**1–23.

34. **Davis, P. (ed.).** 1974. *Single Cell Protein.* Academic Press, New York.

35. **Eaton, D. C.** 1985. Mineralization of polychlorinated biphenyls by *Phanerochaete chrysosporium*: a ligninolytic fungus. *Enzyme Microb. Technol.* **7:**194–195.

36. **Eggins, H. O. W., and W. D. Allsopp.** 1975. Biodeterioration and biodegradation by fungi, p. 301–319. *In* J. E. Smith and D. R. Berry (ed.), *The Filamentous Fungi*, vol. 1. *Industrial Fungi.* Edward Arnold, London.

37. **Eriksson, K.-E., and T. K. Kirk.** 1985. Biopulping, biobleaching and treatment of kraft bleaching effluents with white rot fungi, p. 271–294. *In* C. W. Robinson and J. A. Howell (ed.), *Comprehensive Biotechnology*, vol 4. *The Practice of Biotechnology, Speciality Products and Service Activities.* Pergamon Press, Oxford.

38. **Esser, K., and P. A. Lemke (ed.).** *The Mycota. A Comprehensive Treatise on Fungi as Experimental Systems for Basic and Applied Research*, vol. VII, in press. Springer-Verlag, New York.

39. **Eveleigh, D. E.** 1985. *Trichoderma*, pp. 487–510. *In* A. L. Demain and N. A. Solomon (ed.), *Biology of Industrial Microorganisms.* Benjamin/Cummings Publishing, Menlo Park, Calif.

40. **Faison, B. D.** 1991. Biological coal conversions. *Crit. Rev. Biotechnol.* **11:**347–366.

41. **Faison, B. D., and T. K. Kirk.** 1985. Factors involved in the regulation of a ligninase activity in *Phanerochaete chrysosporium. Appl. Environ. Microbiol.* **49:**299–304.

42. **Fenn, P., and T. K. Kirk.** 1981. Relationship of nitrogen to the onset and suppression of ligninolytic activity and secondary metabolism in *Phanerochaete chrysosporium. Arch. Microbiol.* **130:**59–65.

43. **Fermor, T. R.** 1993. Applied aspects of composting and bioconversion of lignocellulosic materials: an overview. *Int. Biodeterior. Biodegrad.* **3:**87–106.

44. **Fernando, T., and S. D. Aust.** 1991. Biodegradation of munition waste, TNT (2,4,6-trinitrotoluene), and RDX (hexahydro-1,3,5-trinitro-1,3,5-triazine) by *Phanerochaete chrysosporium*, p. 214–232. *In* D. W. Tedder and F. G. Pohland (ed.), *Emerging Technologies in Hazardous Waste Management II, ACS Symposium.* American Chemical Society, Washington, D.C.

45. **Fernando, T., J. A. Bumpus, and S. D. Aust.** 1990. Biodegradation of TNT (2,4,6-trinitrotoluene) by *Phanerochaete chrysosporium. Appl. Environ. Microbiol.* **56:**1666–1671.

46. **Field, J. A., E. de Jong, G. Feijoo-Costa, and J. A. M. de Bont.** 1993. Screening for ligninolytic fungi applicable to the biodegradation of xenobiotics. *Trends Biotechnol.* **11:**44–49.

47. **Fogarty, A. M., and O. H. Tuovinen.** 1991. Microbiolog-

ical degradation of pesticides in yard waste composting. *Microbiol. Rev.* **55:**225–233.

48. **Frankland, J. C., N. H. Hedger, and J. J. Swift (ed.).** 1982. *Decomposer Basidiomycetes: Their Biology and Ecology.* Cambridge University Press, Cambridge.

49. **Friedrich, J., A. Cimerman, and A. Perdih.** 1992. Use of fungi for bioconversion of distillery waste, p. 963–992. *In* D. K. Arora, R. P. Elander, and K. G. Mukerji (ed.), *Handbook of Applied Mycology,* vol. 4. *Fungal Biotechnology.* Marcel Dekker, Inc., New York.

50. **Gadd, G. M.** 1986. Fungal responses towards heavy metals, p. 83–110. *In* R. A. Herbert and G. A. Gadd (ed.), *Microbes in Extreme Environments.* Academic Press, London.

51. **Gadd, G. M.** 1992. Microbial control of heavy metal pollution, p. 59–88. *In* J. C. Fry. G. M. Gadd, R. A. Herbert, C. W. Jones, and I. A. Watson-Craik (ed.), *Microbial Control of Pollution.* Cambridge University Press, Cambridge.

52. **Gibson, D. T.** 1991. Biodegradation, biotransformation and the Belmont. *J. Ind. Microbiol.* **12:**1–12.

53. **Gibson, D. T. (ed.).** 1984. *Microbial Degradation of Organic Compounds.* Marcel Dekker, Inc., New York.

54. **Glenn, J. K., M. A. Morgan, M. B. Mayfield, M. Kumahara, and M. H. Gold.** 1983. An H_2O_2-requiring enzyme preparation involved in lignin biodegradation by the white rot basidiomycete *Phanerochaete chrysosporium. Biochem. Biophys. Res. Commun.* **114:**1077–1083.

55. **Gold, M. H., K. Valli, D. K. Joshi, and H. Wariishi.** 1992. Degradation of polychlorinated phenols and dichlorodibenzo-p-dioxin by *Phanerochaete chrysosporium.* p. 39–44. *In* 5th *International Conference on Biotechnology in the Pulp and Paper Industry.* Kyoto, Japan.

56. **Grigorova, R., and J. R. Norris.** 1990. *Methods in Microbiology,* vol. 22. *Techniques in Microbial Ecology.* Academic Press, London.

57. **Haemmerli, S. D., M. S. A. Leisola, D. Sanglard, and A. Feichter.** 1986. Oxidation of benzo(a)pyrene by extracellular ligninases of *Phanerochaete chrysosporium. J. Biol. Chem.* **261:**6900–6903.

58. **Hammel, K. E., B. Kalyanaraman, and T. K. Kirk.** 1986. Oxidation of polycyclic aromatic hydrocarbons and dibenzo[p]-dioxins by *Phanerochaete chrysosporium* ligninanse. *J. Biol. Chem.* **261:**16948–16952.

59. **Hawksworth, D. L.** 1974. *Mycologist's Handbook.* Commonwealth Mycological Institute, Kew, Surrey, England.

60. **Hawksworth, D. L., and B. E. Kirsop (ed.).** 1988. *Living Resource for Biotechnology. Filamentous Fungi.* Cambridge University Press, Cambridge.

61. **Hawksworth, D. L., B. C. Sutton, and G. C. Ainsworth.** 1983. *Ainsworth & Bisby's Dictionary of the Fungi,* 7th ed. Commonwealth Mycological Institute, Kew, Surrey, England.

62. **Hayes, W. A., and N. G. Nair.** 1975. The cultivation of *Agaricus bisporus* and other edible mushrooms, p. 212–248. *In* J. E. Smith and D. R. Berry (ed.), *The Filamentous Fungi,* vol. 1. *Industrial Mycology.* Edward Arnold, London.

63. **Holzbaur, E., L. F., A. Andrawis, and M. Tien.** 1991. Molecular biology of lignin peroxidases from *Phanerochaete chrysosporium,* p. 197–223. *In* S. A. Leong and R. M. Berka (ed.), *Molecular Industrial Mycology.* Marcel Dekker, Inc., New York.

64. **Horvath, R. S.** 1972. Microbial co-metabolism and the degradation of organic compounds in nature. *Bacteriol. Rev.* **36:**146–155.

65. **Hulbert, M. H., and S. Krawiec.** 1977. Cometabolism: a critique. *J. Theor. Biol.* **69:**287–291.

66. **Jeffries, T. W., S. Choi, and T. K. Kirk.** 1981. Nutri- tional regulation of lignin degradation by *Phanerochaete chrysosporium. Appl. Environ. Microbiol.* **42:**290–296.

67. **Jennings, D. H.** 1993. *Stress Tolerance of Fungi.* Marcel Dekker, Inc., New York.

68. **Jong, S. C., and M. J. Gantt (ed.).** 1987. *American Type Culture Collection Catalogue of Fungi/Yeast,* 17th ed. American Type Culture Collection, Rockville, Md.

69. **Joshi, D. K., and M. H. Gold.** 1993. Degradation of 2,4,5-trichlorophenol by the lignin-degrading basidiomycete *Phanerochaete chrysosporium. Appl. Environ. Microbiol.* **59:** 1779–1785.

70. **Kahlon, S. S.** 1991. Single-cell protein from molds and higher fungi, p. 499–540. *In* D. K. Arora, K. G. Mukerji, and E. H. Marth (ed.), *Handbook of Applied Mycology,* vol. 3. *Food and Feeds.* Marcel Dekker, Inc., New York.

71. **Kendrick, B.** 1985. *The Fifth Kingdom.* Mycologue Publications, Waterloo, Ontario, Canada.

72. **Kennedy, D. W., S. D. Aust, and J. A. Bumpus.** 1990. Comparative biodegradation of alkyl halide insecticides by the white rot fungus, *Phanerochaete chrysosporium* (BKM-F-1767). *Appl. Environ. Microbiol.* **56:**2347–2352.

73. **Kirk, T. K.** 1983. Degradation and conversion of lignocelluloses, p. 266–295. *In* J. Smith, D. R. Berry, and B. Kristiansen (ed.), *The Filamentous Fungi,* vol. IV. *Fungal Technology.* Edward Arnold, London.

74. **Kirk, T. K.** 1984. Degradation of lignin, p. 399–437. *In* D. T. Gibson (ed.), *Microbial Degradation of Organic Compounds.* Marcel Dekker, Inc., New York.

75. **Kirk, T. K., W. J. Connors, and J. G. Zeikus.** 1976. Requirement for a growth substrate during lignin decomposition by two wood-rotting fungi. *Appl. Environ. Microbiol.* **32:**192–194.

76. **Kirk, T. K., and P. Fenn.** 1982. Formation and action of the ligninolytic system in basidiomycetes, p. 67–90. *In* J. C. Frankland, N. H. Hedger, and J. J. Swift (ed.), *Decomposer Basidiomycetes: Their Biology and Ecology.* Cambridge University Press, Cambridge.

77. **Kohler, A., A. Jager, H. Willershausen, and H. Graf.** 1988. Extracellular ligninase of *Phanaerochaete chrysosporium* Burdsall has no role in the degradation of DDT. *Appl. Microbiol. Biotechnol.* **29:**618–620.

78. **Lamar, R. T., M. J. Larsen, and T. K. Kirk.** 1990. Sensitivity to and degradation of pentachlorophenol by *Phanerochaete chrysosporium. Appl. Environ. Microbiol.* **56:** 3519–3526.

79. **Lindley, N. D.** 1992. Hydrocarbon-degrading yeasts and filamentous fungi of biotechnological importance, p. 905–929. *In* D. K. Arora, R. P. Elander, and K. G. Mukerji (ed.), *Handbook of Applied Mycology,* vol. 4. *Fungal Biotechnology.* Marcel Dekker, Inc., New York.

80. **Litchfield, J. H.** 1979. Production of single-cell protein for use in food or feed, p. 93–155. *In* R. S. Porubcan and R. L. Sellars (ed.), *Microbial Technology,* 2nd ed., vol. 1. Academic Press, London.

81. **Margulis, L., and K. V. Schwartz.** 1982. *Five Kingdoms.* W. H. Freeman and Company, San Francisco.

82. **Martin, J. P.** 1950. Use of acid, rose bengal, and streptomycin in the plate method for estimating soil fungi. *Soil Sci.* **69:**215–232.

83. **Mileski, G. J., J. A. Bumpus, M. A. Jurek, and S. D. Aust.** 1988. Biodegradation of pentachlorophenol by the white rot fungus *Phanerochaete chrysosporium. Appl. Environ. Microbiol.* **54:**2885–2889.

84. **Montecourt, B. S., and D. E. Eveleigh.** 1985. Fungal carbohydrases: amylases and cellulases, p. 491–512. *In* J. W. Bennett and L. L. Lasure (ed.), *Gene Manipulations in Fungi.* Academic Press, New York.

85. **Moore-Landecker, E.** 1982. *Fundamentals of the Fungi.* Prentice-Hall, Inc., Englewood Cliffs, N.J.

86. **Morgan, P., S. T. Lewis, and R. J. Watkinson.** 1991. Comparison of abilities of white-rot fungi to mineralize selected xenobiotic compounds. *Appl. Microbiol. Biotechnol.* **34:**693–696.

87. **Mougin, C., C. Laugero, M. Asther, J. Dubroca, P. Frasse, and M. Asther.** 1994. Biotransformation of the herbicide atrazine by the white rot fungus *Phanerochaete chrysosporium*. *Appl. Environ. Microbiol.* **60:**705–708.

88. **Nandan, R., and S. Raisuddin.** 1992. Fungal degradation of industrial wastes and wastewater, p. 931–961. *In* D. K. Arora, R. P. Elander, and K. G. Mukerji (ed.), *Handbook of Applied Mycology*, vol. 4. *Fungal Biotechnology*. Marcel Dekker, Inc., New York.

89. **Nevalainen, K. M., M. E. Pentilla, A. Harkki, and T. T. Teeri.** 1991. The molecular biology of *Trichoderma* and its application to the expression of both homologous and heterologous genes. *In* S. A. Leong and R. M. Berka (ed.), *Molecular Industrial Mycology*. Marcel Dekker, Inc., New York.

90. **Pasti-Grigsby, M. B., A. Paszczynski, S. Goszczynski, D. L. Crawford, and R. L. Crawford.** 1992. Influence of aromatic substitution patterns on azo dye degradability by *Streptomyces* spp. and *Phanerochaete chrysosporium*. *Appl. Environ. Microbiol.* **58:**3605–3613.

91. **Paszczynski, A., M. B. Pasti-Grigsby, S. Goszczynski, R. L. Crawford, and D. L. Crawford.** 1992. Mineralization of sulfonated azo dyes and sulfanilic acid by *Phanerochaete chrysosporium* and *Streptomyces chromofuscus*. *Appl. Environ. Microbiol.* **58:**3598–3604.

92. **Raj, H. G., M. Saxena, A. Allameh, and K. G. Mukerji.** 1992. Metabolism of foreign compounds by fungi, p. 881–904. *In* K. K. Arora, R. P. Elander, and K. G. Mukerji (ed.), *Handbook of Applied Mycology*, vol. 4. *Fungal Biotechnology*. Marcel Dekker, Inc., New York.

93. **Reese, E. T., R. G. H. Sui, and H. S. Levinson.** 1950. The biological degradation of soluble cellulose derivatives and its relationship to the mechanism of cellulose hydrolysis. *J. Bacteriol.* **59:**485–497.

94. **Riser-Roberts, E.** 1992. *Bioremediation of Petroleum Contaminated Sites*. CRC Press, Boca Raton, Fla.

95. **Rosazza, J. P. (ed.).** 1982. *Microbial Transformations of Bioactive Compounds*, vol. 1 and 2. CRC Press, Boca Raton, Fla.

96. **Rose, A. H.** 1981. *Microbial Biodeterioration*, vol. 6. *Economic Microbiology*. Academic Press, London.

97. **Ross, I. K.** 1979. *Biology of the Fungi*. McGraw-Hill Book Co., New York.

98. **Ross, I. S.** 1975. Some effects of heavy metals on fungal cells. *Trans. Br. Mycol. Soc.* **64:**175–193.

99. **Ryan, T. P., and J. A. Bumpus.** 1989. Biodegradation of 2,4,5-trichlorophenoxyacetic acid in liquid culture and in soil by the white rot fungus *Phanerochaete chrysosporium*. *Appl. Microbiol. Biotechnol.* **31:**302–307.

100. **Sariaslani, F. S.** 1989. Microbial enzymes for oxidation of organic molecules. *Crit. Rev. Biotechnol.* **9:**172–257.

101. **Singh, U. D., N. Sethunathan, and K. Raghu.** 1991. Fungal degradation of pesticides, p. 541–588. *In* D. K. Arora, B. Rai, K. G. Mukerji, and G. R. Knudsen (ed.), *Handbook of Applied Mycology*, vol. 4. *Soil and Plants*. Marcel Dekker, Inc., New York.

102. **Smith, J. E., and D. R. Berry.** 1976. *The Filamentous Fungi*, vol. 2. *Biosynthesis and Metabolism*. Edward Arnold, London.

103. **Smith, J. E., and D. R. Berry.** 1978. *The Filamentous Fungi*, vol. 3. *Developmental Mycology*. Edward Arnold, London.

104. **Smith, J. E., D. R. Berry, and J. Kristiansen (ed.).** 1983. *The Filamentous Fungi*, vol. IV. *Fungal Technology*. Edward Arnold, London.

105. **Stamets, P.** 1993. *Growing Gourmet and Medicinal Mushrooms*. Ten Speed Press, Berkeley, Calif.

106. **Stevens, R. B. (ed.).** 1974. *Mycology Guidebook*. University of Washington Press, Seattle.

107. **Straines, J. E., V. F. McGowan, and V. B. D. Skerman.** 1986. *World Directory of Collections of Cultures of Microorganisms*. World Data Center, University of Brisbane, Queensland, Australia.

108. **Sutherland, J. D., A. L. Selby, J. P. Freeman, F. E. Evans, and C. E. Cerniglia.** 1991. Metabolism of phenanthrene by *Phanerochaete chrysosporium*. *Appl. Environ. Microbiol.* **57:**3310–3316.

109. **Tansey, M. R., and T. D. Brock.** 1972. The upper temperature limit for eukaryotic organisms. *Proc. Natl. Acad. Sci. USA* **69:**2426–2428.

110. **Thurston, C. F.** 1994. The structure and function of fungal laccases. *Microbiology* **140:**19–26.

111. **Tien, M., and T. K. Kirk.** 1983. Lignin-degrading enzyme from hymenomycete *Phanerochaete chrysosporium*. *Science* **221:**661–663.

112. **Treen-Sears, M. E., S. M. Martin, and B. Volesky.** 1984. Propagation of *Rhizopus javanicus* biosorbent. *Appl. Environ. Microbiol.* **48:**137–141.

113. **Valli, K., B. J. Brock, D. K. Joshi, and M. H. Gold.** 1992. Degradation of 2,4-dinitrotoluene by the lignin-degrading fungus *Phanerochaete chrysosporium*. *Appl. Environ. Microbiol.* **58:**221–228.

114. **Valli, K., and M. H. Gold.** 1991. Degradation of 2,4-dichlorophenol by the lignin-degrading fungus *Phanerochaete chrysosporium*. *J. Bacteriol.* **173:**345–352.

115. **Valli, K., H. Wariishi, and M. H. Gold.** 1992. Degradation of 2,7-dichlorodibenzo-p-dioxin by the lignin-degrading basidiomycete *Phaerochaete chrysosporium*. *J. Bacteriol.* **174:**2131–2137.

116. **Waterman, M. R., and E. J. Johnson (ed.).** 1991. *Methods in Enzymology*, vol. 206. *Cytochrome P-450*. Academic Press, Orlando, Fla.

117. **Whittaker, R. H.** 1969. New concepts of kingdoms of organisms. *Science* **163:**150–169.

118. **Williams, R. T., P. S. Ziegenfuss, and W. E. Sisk.** 1992. Composting of explosives and propellant contaminated soils under thermophilic and mesophilic conditions. *J. Ind. Microbiol.* **9:**137–144.

Practical Methods for the Isolation of Polycyclic Aromatic Hydrocarbon (PAH)-Degrading Microorganisms and the Determination of PAH Mineralization and Biodegradation Intermediates

KAY L. SHUTTLEWORTH AND CARL E. CERNIGLIA

84

Polycyclic aromatic hydrocarbons (PAHs) are widely distributed environmental contaminants that have detrimental biological effects, including acute and chronic toxicity, mutagenicity, and carcinogenicity. Because of their genotoxic potential, many PAHs have been included on the U.S. Environmental Protection Agency list of priority pollutants (9, 16). Because of the ubiquitous occurrence, recalcitrance, bioaccumulation potential, and suspected carcinogenic activity of PAHs, many researchers have developed procedures for the bioremediation of PAH-contaminated sites.

Previous reviews of the biodegradation of PAHs have emphasized the biochemistry of PAH transformation (24), ecological aspects (16), chemical pathways (9), genetic regulation (55), and bioremediation techniques (20, 41). In this review, the methods used for isolating PAH-degrading microorganisms and the techniques used to monitor PAH biodegradation are discussed. Consideration is given to the application of analytical methods used in the determination of PAHs and their metabolites in culture media and environmental samples.

TECHNIQUES FOR THE ISOLATION OF PAH DEGRADERS

Enrichment Cultures, Isolation of Microorganisms, and Maintenance of Stock Cultures

The best way to obtain PAH-utilizing bacteria is through the use of enrichment cultures. Some of the PAHs, particularly naphthalene, are volatile, and care must be taken to ensure that ample substrate is available for the duration of the enrichment period. This may be accomplished by using screw-cap flasks or by bubbling the medium with naphthalene-saturated air. Sufficient carbon, e.g., 100 ppm, must be available to allow for good microbial growth, but the limited aqueous solubility of PAHs (22) restricts the ways in which they can be added to the medium. PAHs may be solubilized in one of several water-miscible organic solvents, such as dimethyl sulfoxide, dimethylformamide, ethyl alcohol, or acetone. The PAHs should be filter sterilized with an organic-solvent-compatible filter and then added to the medium as a small amount of a concentrated stock solution. Many solvents, however, are toxic, and some are sources of carbon and energy for nontarget bacteria. For example, dimethyl sulfoxide is known to alter membrane permeability. Many bacteria can use ethanol as a carbon source, and

even dimethylformamide can be a carbon source for some bacteria (23, 50). The concentration of solvent should therefore be much less than 1% (vol/vol).

Alternatively, PAHs in volatile organic solvents, such as ether or methylene chloride, may be placed in sterile flasks. The organic solvent is allowed to evaporate, and then sterile mineral salts medium is added aseptically. This is more cumbersome than adding a small amount of solvent directly to the medium in a flask. Also, in comparison with the solvent evaporation technique, PAH addition via a nonvolatile solvent leads to better dispersion of many PAHs. In some cases, crystalline PAHs are added to the medium, but then the PAHs may float and cling to the glass above the air-liquid interface. Also, sterilization of crystalline PAHs is problematic, and it is difficult to accurately measure the small quantity needed for each flask. Most microorganisms grow very slowly when utilizing PAHs; therefore, enrichment cultures need to be given sufficient time to develop, e.g., 7 to 30 days, between transfers. Once isolates have been obtained from enrichment cultures by the plating techniques given below, it is best to maintain the bacteria in a PAH-containing medium. Growth of the bacteria in broth tubes can be substantially enhanced by agitating the tubes on a cell culture roller drum or similar device. Agitation ensures that the particulate PAHs are mixed throughout the medium and that the medium remains oxygenated, i.e., that the cells have ready access to both oxygen and carbon.

Spray Plates and Agar Overlay Plates

PAHs cannot be incorporated into agar by dissolving them in water; it is therefore more difficult than usual to isolate colonies which grow on the enrichment substrate. For naphthalene, the air can be saturated with naphthalene fumes by adding crystals to the plate lids. In other cases, agar plates can be sprayed with a 1 to 10% solution of a PAH in ether or acetone (33). The sprayer, such as those used for thin-layer chromatography (TLC), must produce a fine mist to obtain even coverage. The plates may be sprayed either before or after the bacteria are streaked or spread on the plates. The solvent(s) should evaporate quickly, and a thin, translucent film of PAH should remain on the plate. These plates are differential because colonies which can utilize the PAH will produce either zones of clearing or colored zones around them. If the plates are sprayed before streak or spread plating, the inoculating loop or spreader rod can disrupt the PAH film, making it difficult to distinguish zones of clearing. The adherence of PAHs to the agar varies; e.g., pyrene binds to agar better than phenanthrene. Alternatively, spread plates can be made with serial dilutions of the enrichment culture, microcolonies can be allowed to develop, and the plates can then be sprayed lightly with the PAH solution. For postinoculation spraying, great care must be taken not to overspray the plates because too much solvent can kill the bacteria or cause excessive colony spreading. In all cases, great care must be taken when spraying because the aerosol of a genotoxic PAH can easily contaminate a large area. Any spraying should be completed inside a cardboard box in a fume hood, and appropriate personal protection, e.g., gloves, laboratory coats, and respirators, must be worn. It is difficult to maintain strict sterile technique with spray plates; they should therefore be used only for initial isolation and not for final purification.

Agarose overlay plates (4) are an alternative to spray plates. Details of this technique are described below in the section on enumeration. When used as a pour plate technique, colony retrieval is complicated by the location of colonies within the agarose. This problem can be overcome by using a higher concentration (1.5%) of agarose and then performing standard streak or dilution plating on the hardened overlayer. Pyrene, fluorene, fluoranthene, or phenanthrene can be incorporated into agarose if 100% ethanol is used as the solvent. Because the PAHs are incorporated throughout the agarose, the zones of clearing are less pronounced than those formed on spray plates. The advantages to the agarose overlayers are the elimination of toxic aerosols and the ability to maintain strict sterile technique. Any bacterium sensitive to approximately 5% ethanol would not be isolated with this technique, but this quantity of solvent is necessary for the PAHs to be well dispersed so that they produce a uniform haze within the agarose.

Enumeration of PAH-Degrading Bacteria

The two basic methods which can be used to enumerate PAH degraders are the pour plate technique (4) and the most-probable-number (MPN) method (34); the general merits of the various microbial enumeration techniques are discussed elsewhere (34). There are several variations of the standard MPN method that can be used for PAHs (3). For any of the nonradiolabeled modifications, the PAH(s) must be the only carbon source; solvent-solubilized PAH(s) is added first, the solvent is evaporated aseptically, and mineral salts are then added aseptically. Positive results are recorded for visible growth (turbidity), but care must be taken to control for any turbidity related to the PAH itself. To save space and reagents, the assay can be conducted in 96-well microtiter plates (45). The well plates would not be a good alternative for volatile PAHs because of the potential for substrate loss and for cross-contamination. Studies with volatile PAHs, such as naphthalene, and with very low concentrations of semivolatile PAHs, such as phenanthrene, are best conducted in screw-cap tubes. Degradation of PAHs in the MPN tests can be evaluated by color formation, since colored ring fission products are frequently the result of PAH degradation. PAHs can, however, be used without appreciable buildup of these colored products, and therefore color should not be the sole method for assessing degradation. The turbidity-based MPN methods for PAH degraders are versatile because they can be used not only to enumerate the degraders of one specific PAH but also to estimate the total population of PAH degraders by concurrently incorporating several PAHs into the medium. In addition, radiolabeled MPN methods (37) can be used either to

enumerate the microorganisms that use the labeled PAH as a sole carbon source or to determine the number of microbes that cometabolize a PAH. There are several possible experimental designs for conducting ^{14}C MPN tests. One alternative is the double-vial method (37); the growth medium, radiolabeled PAH, and diluted sample are place in a minivial or small test tube, which is then placed in a scintillation vial containing NaOH or KOH as the CO_2 trap. The scintillation vial is capped, and after the desired incubation time, the minivial is removed. Radiolabeled MPN tests can also be completed by using standard 16-mm test tubes containing the desired growth medium, radiolabeled PAH, and diluted sample. Sodium or potassium hydroxide is added to center wells suspended from specifically designed rubber stoppers. The center wells are removed after the incubation is completed. For either the double-vial or the suspended-center-well method, scintillation fluid is added to the $^{14}CO_2$-containing hydroxide trap, and vials containing at least two to four times the radioactivity of the abiotic controls are scored as positive. Standard MPN tables are used for quantitation. Radiolabeled MPN methods have two main advantages over turbidity-based MPN methods. First, the procedures can be used to enumerate organisms that cometabolize PAHs because $^{14}CO_2$ will evolve from the radiolabeled PAH but not from the additional unlabeled substrates. Second, radiolabeling methods are extremely sensitive; PAH degradation can therefore be monitored at low PAH concentrations which are environmentally relevant (37). The disadvantages to ^{14}C MPN tests are the expense of the radiolabeled compounds and the lack of commercially available uniformly labeled PAHs. The production of $^{14}CO_2$ can be affected by the location of the radiolabeled carbons within the aromatic rings, since microbial enzymes catalyzing PAH degradation can be highly regiospecific.

An alternative to the MPN technique is the agarose overlay plate method (4). This method is not recommended for assessing the total number of PAH degraders because zones of clearing would develop only around colonies that could use the entire mixture of PAHs. It is, however, a very effective way to enumerate the microorganisms that degrade a single PAH. Also, the plates are differential; therefore, cometabolic degradation of PAHs can be detected by including small quantities of other nutrients, such as yeast extract or glucose, in the agarose. Colonies that do not utilize the PAH can be discounted because there will be no zone of clearing around them. Zones of clearing in the upper layer are easier to visualize if the mineral salts agar underlayer is kept as clear as possible by using high-purity agar or agarose. For the upper layer, 1% low-gel-temperature agarose is prepared with the chosen mineral salts; after cooling to 30 to 35°C, 0.2 ml of 8.5 mg of PAH per ml of ethanol is added to 3.5 ml of agarose. They are mixed immediately, and then 0.1 to 0.2 ml of the bacterial dilution is added and mixed. The agarose is then poured over the prepared underlayer agar. Plates are allowed to harden and are incubated for 1 to 3 weeks to allow zones of clearing to form around the PAH utilizers. This method is effective if the ethanolic PAH solution is well mixed into the agarose before addition of the bacterial dilution. If this is not done, high localized concentrations of ethanol will kill significant numbers of bacteria.

MONITORING METHODS FOR THE DEGRADATION OF PAHs
General

The degradation of PAHs can be monitored by measuring either parent compound removal or product formation. The method of choice will be determined by the information needed and by the nature of the sample. More options are available for analyzing pure cultures and laboratory-contaminated samples than exist for field samples. Most importantly, degradation in field samples cannot accurately be monitored with radiotracers because aging of a polluted soil sample greatly affects the potential biodegradation of the pollutants in that sample. Since aged PAHs in field samples will not be in equilibrium with freshly added radiolabeled PAHs (6), any freshly added PAHs will usually be degraded more rapidly than the aged PAHs. For example, Hatzinger and Alexander (25) found that the maximum rate of phenanthrene degradation in an unaged muck soil was 19.1%/day; the rate was only 2.7%/day, however, in samples in which the phenanthrene was abiotically laboratory aged with the soil for 315 days. The extent of phenanthrene mineralization was also affected by the aging time; in muck samples aged for 0 and 315 days, 59 and 42%, respectively, of the phenanthrene was mineralized. The degree to which aging may be expected to alter the rates and extent of biodegradation will depend on several factors, including aging time, soil structure, concentration of soil organic matter, concentration of the PAH, and the structure of the PAH.

Appropriate abiotic controls must always be included in the experimental design. This is more difficult for soil or sediment systems than for aqueous systems. If PAHs are added to previously uncontaminated soils, sterilization can be accomplished by using autoclaving techniques appropriate for soils (e.g., 1 h on two consecutive days) or by irradiation before addition of filter-sterilized PAHs. For soils obtained from contaminated sites, the production of suitable abiotic controls is more problematic. Microbial activity must be inhibited to the greatest degree possible while minimizing changes in the intrinsic association of the PAHs with the soil. The potential exists for the chemical structure or extractability of PAHs to be altered by autoclaving or irradiating the soil or by adding organic or inorganic inhibitors. Mercuric chloride is an extremely effective microbial inhibitor, but it will precipitate on the soil and may interfere with PAH extraction at the end of the experiment. Furthermore, proper disposal for mercuric chloride-contaminated samples can pose a problem. Since PAHs are not generally degraded anaerobically (39), an oxygen-free system could be an alternative control for a flowthrough system. This method has not been tested, however, and some PAHs can be effectively degraded under denitrifying conditions (39). Indeed, there has not been a rigorous comparison of the relative potentials for the various biological inhibitors to alter PAH extractability, and there are, unfortunately, drawbacks to every type of control. Choices must be made from among imperfect alternatives.

As with the enrichment cultures, the procedure for PAH addition can affect the experimental results. It is desirable to have any particulates distributed as evenly as possible, but the carrier solvent must be minimized to limit possible toxicity and the potential for an additional carbon source. If the system includes soil, it is particularly important for the PAHs to be uniformly dispersed. A dilute solution of PAH can be made in a solvent which evaporates quickly. The solution is evenly applied to the soil, the solvent is evaporated, and then the soil is mixed well. Alternatively, a dilute, water-miscible (acetone, dimethylformamide, dimethyl sulfoxide) solution of PAH(s) can be added dropwise to a soil- water slurry while it is being vigorously mixed; the slurry is allowed to equilibrate for a day or more, and then the aqueous phase is decanted (35). The volume of solvent should be high enough to provide for a good dispersion of the PAHs upon addition to the water but not so high as to extract native organic matter from the soil. The soil can be rinsed with water to remove excess unbound PAH and remaining solvent. A similar method involves adding the solubilized PAHs to the glass vessel, evaporating the solvent, adding soil and basal medium, equilibrating by shaking for several days, decanting the aqueous phase, and adding fresh medium (5). An "aged" sediment sample can be generated by prolonging the equilibration period or by holding the rinsed sediment for the desired aging time. In a field-aged soil, microbial activity can alter PAHs from the time of contamination to the time of sampling; this type of aging can be mimicked in the laboratory by controlling the incubation temperature and moisture. Alternatively, the soil can be aged at a low temperature to minimize microbial metabolism.

Pure Cultures and Laboratory-Contaminated Samples

Uptake

A rapid assessment of biodegradation potential can be made by using standard respirometry techniques (49). Oxygen uptake and/or CO_2 production is monitored with a differential respirometer or oxygen electrode, and PAH (substrate) utilization is proportional to the O_2 uptake. The exact procedures are provided by the manufacturers of the equipment. Endogenous respiration (controls) must be subtracted from respiration with the PAH. However, oxygen uptake alone does not provide detailed information on the percentage of the PAH mineralized to CO_2 or on the characteristics of metabolites.

All PAHs have characteristic UV absorption spectra, and removal of a parent compound from solution can be monitored by following changes in either the UV spectrum or a specific absorption maximum (40, 46). Although this is a rapid and simple method, great care must be taken to ensure that cells and PAH metabolites do not cause interference at the monitored wavelength(s). Dead-cell controls must also be used to ensure that loss of the PAH is due to active uptake and not to adsorption onto the cells. If the initial PAH concentration is below its aqueous solubility, UV data can be collected directly from the aqueous sample; for samples containing PAHs above the solubility limit, PAHs must be extracted into one of the less volatile organic solvents, such as ethyl acetate. For accurate quantitation, evaporative losses of the organic solvent must be avoided during the procedure and extraction efficiencies must be determined for the conditions used. The UV analysis can then be completed on the organic layer. Changes in PAH concentration can be determined by comparing the results with a standard curve of the PAH in the extracting solvent. The rapidity of this procedure can be offset by the need to verify that metabolites are not producing artifacts in the data. If corrections are not made for metabolites that absorb at the wavelength of interest, the actual uptake rate will be underestimated. For the greatest accuracy, the potential for artifact production needs to be assessed for each microorganism in combination with each PAH examined.

Mineralization

For bioremediation, it is important that PAHs be completely degraded, i.e., mineralized. There are several methods available for monitoring CO_2 production from radiolabeled PAHs. The basic principles are the same for all of the methods, but the configurations of the apparatus vary. In general, the PAH is added to a solution or soil, the microorganisms are given sufficient time to degrade the PAH, and the liberated CO_2 is trapped in an appropriate reagent. Reagents used for trapping CO_2 include NaOH, KOH, Ba(OH)$_2$, ethanolamine-ethylene glycol mixtures, and hyamine hydroxide. There is no single best trapping reagent; however, the following considerations will aid in making a decision. First, the trapping reagent must efficiently absorb all of the CO_2. The efficiency will depend not only on the trapping reagent but also on the configuration of the apparatus. For example, for flowthrough systems, the rate of CO_2 transfer will depend on the airflow rate and the size of the bubbles passing through the trapping solution. The smaller the bubbles, the better the gas-to-liquid transfer rate. When added to some scintillation cocktails, NaOH and KOH have a greater tendency than the organic traps to form two separate phases, which inhibit accurate counting. The volatility of organic traps can, however, substantially alter the pH of cultures in enclosed systems. With 12 ml of 30:70 (vol/vol) ethanolamine-ethylene glycol in the side arm of a biometer flask and 50 ml of a pH 7.5 medium in the flask, the pH of sterile medium can rise 0.5 to 3 units after 4 days of equilibration. The extent of pH change is a function of the type and concentration of buffer in the medium. Also, organic-based CO_2 traps readily absorb volatilized PAHs (1). This problem is avoided in flowthrough microcosm systems in which a special organic trap, e.g., polyurethane foam and Tenax resins designed to remove volatile organics but not CO_2, is placed between the reaction vessel and the CO_2 trap. The radioactivity of an aliquot of the trapping solution is measured in a scintillation counter. For all radioactive methods, the radiolabeled PAH must be mixed with the unlabeled PAH before addition to the aqueous medium. This is necessary because PAHs precipitate upon contact with the aqueous phase. If the labeled and unlabeled PAHs are not premixed, the radioactive and nonradioactive PAHs will be not be uniformly dispersed within the particles, and this can lead to preferential degradation of either labeled or unlabeled PAH, whichever is more available.

With all mineralization experiments, the systems must be well sealed to prevent loss of CO_2 through leaks. Systems can be checked for leaks by using bicarbonate (NaHCO$_3$) controls and checking for 100% recovery or, for small flowthrough systems, by submerging the vessels in water and checking for bubble formation. Leaks in flowthrough systems can also be checked by flushing the system with a noncombustible gas, such as helium, and monitoring all seals and joints with a gas leak detector like those used for gas chromatography (GC) systems.

For short-term degradation studies (i.e., a few hours), center-well incubation flasks are the most appropriate apparatus. These are small Erlenmeyer-type flasks with side arms and are suitable for the small volumes (<10 ml) needed for short-term work. Aluminum foil-covered rubber stoppers seal both the top and side arm, and the center well is suspended from the top stopper. For nonvolatile PAHs, the center wells can be made of either polypropylene or glass, but plastic center wells cannot be used with volatile PAHs such as phenanthrene because the plastic serves as a major sink for volatilized PAH. A small amount of trapping reagent, e.g., 5 M NaOH, is added to the center well, and the cells and PAH are added to the bottom of the flask. The stoppered side arm accommodates syringe needles, and the experiment can be started by injecting cells through the side arm. After the cells are incubated for the desired time, the reaction can be stopped by injecting sufficient sulfuric

acid through the side arm to lower the pH to 1 or 2. The addition of the sulfuric acid not only stops microbial metabolism but also drives CO_2 from the aqueous phase into the headspace. After 0.5 to 1 h, the CO_2 is equilibrated into the trap and the trapping reagent can be quantitatively removed from the well. The capacity of the CO_2 trap must exceed the expected production of CO_2 from the PAH. To ensure sufficient capacity, the system should be tested with a mixture of unlabeled and radiolabeled bicarbonate.

The biometer flasks are used in much the same way as the center-well flasks, but they accommodate larger volumes (20 to 75 ml) and are more suitable for studies lasting for several days. The basic design of a biometer flask is an Erlenmeyer flask with an attached side arm. The main flask contains the cells, PAH, and other growth components, whereas the side arm houses the CO_2 trap. The same experimental design principles as described for the center-well flasks hold for the biometer flasks. Unlike with the center-well flasks, CO_2 can be monitored over time in a single flask by sampling through a needle extending into the side arm. Fresh CO_2-trapping reagent is then added to replace the lost volume. When the CO_2 production rates are calculated, corrections must be made for dilution effects and for the amount of radiolabel removed during previous samplings. Because CO_2 is not actively driven out of the growth medium by either acid or airflow, the rate of accumulation in the CO_2 trap will depend on the rate of equilibrium between the compartments. The pH of the medium will have some influence on the rate of observed CO_2 release. Also, since this is a closed system, it can become oxygen limited over time.

Flowthrough systems have two main advantages over static systems. First, more precise rate data can be obtained, and second, volatile parent compounds and volatile metabolites can easily be separated from CO_2. There are many possible configurations of flowthrough systems, from the simple (28) to the complex (17, 29). Basically, each system consists of a growth vessel, a trap for volatile organics, and a CO_2 trap connected in sequence by tubes which carry clean, sterile, and preferably CO_2-free air. The organic trap may consist of an organic fluid, such as ethylene glycol monomethyl ether, or solids with a high organic adsorption capacity, such as Tenax or polyurethane foam. An organic fluid trap will evaporate with constant airflow, but it is easier to subsample an organic fluid than it is to subsample the solids. The microcosm system of Huckins et al. (29) can be monitored by inserting probes into ports; the disadvantage is that the more complex systems generally cost more than the simpler systems. Any flowthrough system must be leak-free, the trapping solutions and solids for both the CO_2 and the volatiles must not become saturated, and the gas flow rate to each vessel must be carefully regulated and metered so that CO_2 production rates can be compared between treatments. Volatilized PAHs and metabolites must not sorb anywhere but in the trap for volatiles; otherwise, a mass balance cannot be achieved.

Intermediate Metabolite Isolation

Liquid-Liquid Extraction
Metabolites and residual PAH(s) are extracted with an organic solvent, such as ethyl acetate or methylene chloride. Bacterial cultures can be extracted directly, but for fungi, mycelia must first be separated from the broth by filtering the culture through glass wool in a funnel. A volume of solvent equal to that of the medium is used to elute sorbed PAHs and metabolites from the retained mycelia. After all

of the solvent has drained by gravity, interstitial solvent should be forced out by pressing the cells with a spatula or glass rod. Any residual PAH which may have adsorbed to the glass of the experimental flasks can be recovered by rinsing the flasks with the chosen extracting solvent.

The bacterial culture or fungal filtrate and all of the solvent used as described above are transferred to a separatory funnel and shaken vigorously. After the contents have separated into two distinct layers, the funnel stopper is removed and the bottom, i.e., aqueous, layer is drained through the stopcock and retained. The solvent is then drained into a clean flask containing anhydrous sodium sulfate, which removes remaining water from the solvent. This must be swirled to ensure that all water is absorbed. Any cloudiness in the organic extract should disappear at this point. Carefully decant or pipette the solvent into a round-bottom flask with a ground-glass fitting. Since the sodium sulfate can interfere with later steps, a glass wool-packed funnel can be used to filter the solvent as it is being transferred. The extraction process should be performed twice more with equal volumes of solvent, and then all solvent fractions should be combined. This combined fraction contains the neutral organic-solvent-soluble metabolites and residual PAH.

Acidic metabolites should be extracted from the medium by lowering the pH of the aqueous phase to 2.0 to 3.0 with 1 M HCl. The acidified medium is extracted three more times as described above. The neutral and acid extracts should be kept separate to allow for individual analyses by TLC and high-performance liquid chromatography (HPLC).

Both extracts are evaporated to dryness under reduced pressure in a rotary evaporator with a water bath temperature of 36 to 38°C. For the smaller, more volatile PAHs, a lower temperature should be used to prevent excessive losses. A minimal volume of solvent, e.g., methanol or ethyl acetate, is then used to dissolve the dried sample. The resuspended sample should be pipetted into a small vial, and a second aliquot of solvent is added to the round-bottom flask. This wash is combined with the first, and the procedure is performed a third time. Better quantitative recovery is achieved by using this repetitive process instead of a single, larger volume of solvent. A gentle stream of argon or nitrogen can be used to evaporate the sample, which is held in an approximately 35°C water bath. The sample is finally dissolved in a known amount of methanol or acetonitrile for HPLC analysis or acetone for TLC analysis. The solvent volume must be large enough to ensure that all of the sample is actually resuspended; i.e., some PAHs have a relatively limited solubility in methanol. This consideration is particularly important if the initial PAH concentration was high.

Chromatographic Methods

In the analysis of PAH biodegradation intermediates, chromatography methods such as HPLC, TLC, and GC are now extensively used.

TLC has been widely used as a screening technique to monitor PAH degradation (Table 1). The major advantages of TLC over HPLC and GC are rapid and efficient detection, low cost, simplicity, and the ability to analyze multiple samples simultaneously. Generally, PAHs and their metabolites are separated on commercially prepared TLC sheets coated with silica gel containing a fluorescent indicator. A one-dimensional solvent system containing benzene-ethanol (9:1) or chloroform-acetone (8:2) is frequently used to obtain separation of residual PAH and aromatic ring-hydroxylated compounds. The compounds on the TLC plates can be visualized under UV light (254 nm) or by spraying a 2% solution of 2,6-dichloroquinone-4-chloroimide in methanol (Gibbs reagent) followed by exposure to ammonia vapor. A dark spot is produced under UV light, while the Gibbs reagent forms a purple color when it reacts with phenols.

For subsequent chemical analyses, the compounds can be extracted from the plate by either scraping the silica carefully with a spatula or using a vacuum-assisted spot collector. Compounds are then eluted from the silica with HPLC-grade methanol. In one TLC solvent system, benzene-hexane (1:1), the residual PAH migrates toward the solvent front and the polar metabolites remain at the origin of the TLC plate. This is a simple method to separate the PAH substrates from the polar compounds generated by microbial metabolism.

HPLC is an important tool for the determination of neutral and acidic compounds produced from the microbial transformation of PAHs (Table 2). Reverse-phase columns containing 5-μm-particle-size C_{18}-bonded silica are most commonly used to achieve separation. A mobile-phase gradient with either methanol or acetonitrile and water is typically used to separate phenols, quinones, dihydrodiols, and epoxides from the parent PAH. To separate acidic metabolites, generally either 1% acetic acid or 1% phosphoric acid is added to the mobile phase. Most analysts use a UV detector set at 254 nm to monitor PAH metabolites, although photodiode array detectors can be set at different wavelengths and the full spectrum of each peak can be acquired for comparison with authentic PAH spectra. Fluorescence and radiochemical detectors have also been used for detecting PAH metabolites, since the detectors are sensitive and can be used to obtain both structural and quantitative information.

For the determination of polar metabolites, GC and GC-mass spectroscopy (MS) have been used less often than other chromatographic methods, such as HPLC or TLC, because derivatization is needed for GC analysis. However, ring fission products derivatized with diazomethane have been determined by using a capillary GC. Acetylation is also commonly used to determine conjugated metabolites and dihydrodiols. Profiles of residual PAHs in environmental samples are routinely resolved by using capillary GC (Table 3).

Extraction and Analysis of PAHs from Soil Samples

For field samples, the only way to monitor PAH degradation is to monitor parent compound removal. To date, there has been no consensus on the best method for extracting PAHs from soils, and numerous methods have been used. A few examples are given in Table 3, which shows a variety of procedures available but is not exhaustive. The extraction methods for PAHs in soils include Soxhlet, supercritical fluid, and sonication. Until recently, the most common methods have been either Soxhlet extraction with methylene chloride or extended sonication or shaking in acetonitrile, methylene chloride, dimethyl sulfoxide, or other solvent. Recently, supercritical fluid extraction (SFE) has become increasingly popular because its extraction efficiency for PAHs is similar to that for Soxhlet extractions. SFE takes significantly less time and uses less hazardous solvents than Soxhlet (36, 53), but the SFE apparatus is relatively expensive. Moisture must be removed from samples for efficient extraction with water-immiscible solvents, e.g., methylene chloride. Soils can be dried by air, freeze-drying,

TABLE 1 Application of TLC for the analysis of PAH metabolites[a]

PAH studied	Metabolite(s)	Stationary phase	Solvent system	Detection method	Reference
AN	AN trans-1,2-dihydrodiol, 1-anthryl sulfate	Silica Gel UV$_{254}$	Benzene-ethanol (9:1)	2% Gibbs reagent, UV light (254 nm)	8
BA	2-Hydroxy-3-phenanthroic acid, 3-hydroxy-2-phenanthroic acid	Silica Gel 60 F$_{254}$	Benzene-acetic acid-H$_2$O (125:74:1)	2% Gibbs reagent, UV light (254 nm)	38
	1-Hydroxy-2-anthroic acid	Silica Gel 60 F$_{254}$	Chloroform-acetone (80:20)	2% Gibbs reagent, UV light (254 nm)	38
	BA metabolites separated from BA	Silica Gel F$_{254}$	Benzene-hexane (1:1)	UV light (254 nm)	14
	BA trans-3,4-, 8,9-, and 10,11-dihydrodiols separated from hydroxy BAs, epoxides, tetrols, and acetonides	Silica Gel F$_{254}$	Benzene-ethanol (9:1)	UV light (254 nm)	14
B(a)P	B(a)P trans-7,8- and 9,10-dihydrodiols from B(a)P 1,6- and 3,6-quinones and 3- and 9-hydroxy-B(a)P	Silica Gel F$_{254}$	Benzene-ethanol (9:1)	1% Gibbs reagent, UV light (254 and 365 nm)	12
FA	FA metabolites separated from FA	Silica Gel GF	Benzene-hexane (1:1)	UV light (254 nm)	31
	1-Fluorenone-1-carboxylic acid	Silica Gel IB2-F	Benzene-acetone-acetic acid (85:15:5)	UV light (254 nm)	31
	Further separation of polar and nonpolar compounds	Silica Gel GF	Benzene-ethanol (9:1)	UV light (254 nm)	31
NAPH	1-Naphthyl glucuronide, 1-naphthyl sulfate	Silica Gel GF$_{254}$	Ethyl acetate-hexane-acetic acid (17:5:1)	UV light (254 nm)	11
	NAPH trans-1,2-dihydrodiol from NAPH cis-1,2-dihydrodiol	Silica Gel 60F-254	Chloroform-acetone (8:2)	2% Gibbs reagent, UV light	15
	4-Hydroxy-1-tetralone, 1,2- and 1,4-naphthoquinones, 1- and 2-naphthols	Silica Gel 60F-254	Chloroform-acetone (8:2)	2% Gibbs reagent, UV light	15
PHEN	3-, 4-, and 9-phenanthrols, PHEN trans-3,4- and 9,10-dihydrodiols	Silica Gel LK6F Linear-K	Benzene-ethanol (9:1)	UV light	47
Pyrene	Pyrene metabolites from pyrene	Silica Gel GF	Benzene-hexane (1:1)	UV light	27
	1,6- and 1,8-pyrene quinones from 1-hydroxypyrene and 1,6- and 1,8-dihydroxypyrenes	Silica Gel 60 F$_{254}$	Benzene-ethanol (19:1)	2% Gibbs reagent, UV light (254 nm)	18
	Pyrene ring oxidation products separated	Silica Gel GF	Hexane-acetone (8:2)	UV light (254 nm)	27
	Highly polar acidic ring fission products separated	Silica Gel GF	Benzene-acetone-acetic acid (85:15:5)	UV light (254 nm)	27

[a] Abbreviations: AN, anthracene; BA, benz(a)anthracene; B(a)P, benz(a)pyrene; FA, fluoranthrene; NAPH, naphthalene; PHEN, phenanthrene.

or mixing with anhydrous sodium sulfate. Air drying or freeze-drying can reduce the concentration of volatile PAHs.

During the extraction of PAHs, other soil components, which can interfere with later analyses, are often coextracted. Various sample cleanup methods, including passage through silica gel, Florisil, or C$_{18}$ cartridges, have therefore been implemented (Table 3). Cleanup steps can,

however, cause appreciable losses of the analytes of interest (53). The PAH profiles can be obtained by GC, GC-MS, or HPLC with either UV or fluorescence detection. GC-MS is preferable to GC alone because extraneous coeluting compounds will have different molecular ions than the PAHs; i.e., GC-MS is more definitive than GC. Similarly, for HPLC analysis it is preferable to use a diode array detector rather than a single-wavelength UV

TABLE 2 Application of HPLC for the analysis of PAH metabolites[a]

PAH	Metabolite(s)	Column[b]	Mobile phase	Reference(s)
AN	1-Anthryl sulfate, AN trans-1,2-dihydrodiol	5-μm C$_{18}$ Ultrasphere (4.6 by 250 mm)	50–95% methanol-H$_2$O, 60 min, 1 ml/min	8, 19
	1-O-(2-Hydroxy-trans-1,2-dihydroanthryl)-β-D-xylopyranoside, 2-O-(1-hydroxy-trans-1,2-dihydroanthryl)-β-D-xylopyranoside, AN trans-1,2-dihydrodiol, 1-O-anthryl-β-D-xylopyranoside	Ultrasphere C$_{18}$ ODS (4.6 by 250 mm)	50–95% methanol-H$_2$O, 40 min, 1 ml/min	48
	AN cis-1,2-dihydrodiol, 2-hydroxy-3-naphthoic acid, salicylic acid, catechol	Ultrasphere C$_{18}$ ODS (4.6 by 250 mm)	50–95% methanol-H$_2$O (1% acetic acid), 40 min, 1 ml/min	Unpublished
BA	Phenanthroic acids separated from anthroic acid	5-μm Zorbax ODS (4.6 by 250 mm)	35–95% acetonitrile-H$_2$O (1% acetic acid), 30 min, 1 ml/min	38
	2-Hydroxy-3-phenanthroic acid, 3-hydroxy-2-phenanthroic acid	5-μm Spherisorb ODS (4.6 by 250 mm)	40:60 acetonitrile-H$_2$O (1% acetic acid), 1.5 ml/min	38
	BA trans-1,2-, 5,6-, and 8,9-dihydrodiols, 8- and 9-BA phenols	μBondapak C$_{18}$ (2 in series) (3.9 by 300 mm)	35–95% acetonitrile-H$_2$O, 60 min, 1 ml/min	14
	BA trans-1,2-, 5,6-, and 8,9-dihydrodiols	Zorbax ODS C$_{18}$ (6.2 by 250 mm)	65:35 methanol-H$_2$O, 1 ml/min	14
	BA trans-3,4-, 5,6-, and 8,9-dihydrodiols	Zorbax ODS (2 in series)	35–65% acetonitrile-H$_2$O, 60 min, 1 ml/min	14
	BA-tetraols and BA dihydrodiol-epoxides	Zorbax ODS C$_{18}$	65:35 methanol-H$_2$O, 0.8 ml/min	14
	Acetonides of BA-tetraols	Zorbax ODS C$_{18}$	65–100% methanol-H$_2$O, 30 min, 0.8 ml/min	14
B(a)P	B(a)P trans-9,10-, 4,5- and 7,8-dihydrodiols, 1,6- and 3,6-B(a)P quinones, 9- and 3-hydroxy-B(a)P	Two coupled μBondapak columns (3.9 by 300 mm)	60–85% methanol-H$_2$O, 0.8 ml/min	12
	B(a)P cis-9,10- and 7,8-dihydrodiols	μBondapak	30–70% acetonitrile-H$_2$O, 30 min, 1 ml/min	12
	B(a)P trans-7,8-dihydrodiol-9,10-epoxides, B(a)P-tetraols	Two coupled μBondapak C$_{18}$ columns	40–95% methanol-H$_2$O, 60 min, 0.8 ml/min	13
FA	9-Fluorenone-1-carboxylic acid, 8-hydroxy-7-methoxy-FA, 9-hydroxyfluorene, 9-fluorenone	5-μm C$_{18}$ Ultrasphere ODS (4.6 by 250 mm)	35–95% methanol-H$_2$O (1% acetic acid), 40 min, 1 ml/min	31, 32
	FA-2,3-dihydrodiol, 1- and 8-hydroxy-FA, 3-FA-β-glucopyranoside, 3-(8-OH-FA)-β-glucopyranoside	Zorbax ODS (4.6 by 250 mm)	30–95% methanol-H$_2$O, 40 min, 1 ml/min	44
Fluorene	9-Fluorenone, 9-fluorenol, 2-hydroxy-9-fluorenone	Ultrasphere C$_{18}$ (4.6 by 250 mm)	30–95% methanol-H$_2$O, 40 min, 1 ml/min	43
NAPH	NAPH trans-1,2-dihydrodiol, 4-hydroxy-1-tetralone, 1,2-naphthoquinone, 1,4-naphthoquinone, 1-naphthol, 2-naphthol	μBondapak C$_{18}$ (3.9 by 300 mm)	50–95% methanol-H$_2$O, 30 min, 0.4 ml/min	15
	1- and 2-naphthols, NAPH cis- and trans-1,2-dihydrodiols, 4-hydroxy-1-tetralone	Altex Ultrasphere ODS (4.6 by 250 mm)	50–95% methanol-H$_2$O, 30 min, 1 ml/min	11
	1-Naphthyl glucuronide, 1-naphthyl sulfate	C$_{18}$ μBondapak (3.9 by 300 mm)	0.01 M tetrabutyl ammonium bromide in methanol-H$_2$O (1:1), 1 ml/min	11
	NAPH cis-1,2-dihydrodiol, salicylic acid, catechol	5-μm Ultrasphere ODS (4.6 by 250 mm)	50–95% methanol-H$_2$O (1% acetic acid), 40 min, 1 ml/min	Unpublished
PHEN	PHEN trans-1,2- and 3,4-dihydrodiols	5-μm Ultrasphere cyano (4.6 by 250 mm)	Methanol-H$_2$O (1:1), 1 ml/min	19
	Purified PHEN dihydrodiols	10-μm μBondapak cyano (3.9 by 300 mm)	Hexane-ethanol (97:3), 1 ml/min	42
	3-, 4-, and 9-phenanthrols	Ultrasphere silica (4.6 by 250 mm)	Hexane-ethyl acetate (19:1), 2 ml/min	47
	PHEN trans-9,10-dihydrodiol, 9,10-phenanthrenequinone, 9-phenanthrol, 9-methoxyphenanthrene	10-μm Zorbax ODS (6.2 by 250 mm)	50–95% methanol-H$_2$O, 60 min, 1 ml/min	42

(Table continued on next page)

TABLE 2 Application of HPLC for the analysis of PAH metabolites[a] (*Continued*)

PAH	Metabolite(s)	Column	Mobile phase	Reference(s)
	Separate MTPA esters of PHEN 9,10-dihydrodiol	5-μm Spherisorb CN (4.6 by 250 mm)	Hexane-dioxane (98.9:1.1), 1 ml/min	42
	Phenanthrene 1-O-β-glucose	Altex Ultrasphere 5-μm ODS (4.6 by 250 mm)	50–95% methanol-H$_2$O, 30 min, 1 ml/min	10
	Phthalic acid, protocatechuic acid, 1-hydroxy-2-naphthoic acid, 3-phenanthrol, PHEN *cis*-1,2-, 3,4-, and 9,10-dihydrodiols	5-μm Spherisorb ODS (4.6 by 250 mm)	50–95% methanol-H$_2$O (1% acetic acid), 40 min, 1 ml/min	Unpublished
Pyrene	1,6- and 1,8-pyrene quinones, 1-hydroxypyrene, glucoside conjugates of 1-hydroxypyrene and 1,6- and 1,8-dihydroxypyrenes, 4-hydroxyperinaphthenone, 4-phenanthroic acid	5-μm ODS (4.6 by 250 mm)	50–95% methanol-H$_2$O, 30 min, 1 ml/min	18, 27

[a] See the footnote to Table 1. MTPA, methoxytrifluoromethyl-phenylacetic acid.
[b] ODS, octyldecyl silane.

TABLE 3 Techniques used for extracting PAHs from soils[a]

Contaminate	Extraction method	Solvent	Time	Cleanup step(s)	Analysis method	Detection method(s)	Recovery	Reference
Sewage sludge	Soxhlet	Methylene chloride	3 h	Florisil cartridge	HPLC	Fluorescence and photodiode array	80–100% with spiked soils	54
Diesel oil, laboratory spiked	Soxhlet	Methylene chloride	6 h	Silica gel eluted with hexane and benzene	GC	MS	Deuterated internal standard	52
Petrochemical wastes	Extraction	Methylene chloride	Not given	Silica gel eluted with hexane and 30% methylene chloride-hexane	GC	FID and MS	Not given	7
Pure PAHs, laboratory spiked	(1) Sonication[b] (2) Heated reflux	(1) Acetone (3 times)[b] (2) 2 N KOH-methanol (1:4, vol/vol), then extracted with hexane	(1) 30 min (2) 5 h	Coupled CN and C$_{18}$ cartridges rinsed with hexane and eluted with toluene	HPLC	UV (254 nm)	78–100% of spike	21
Certified reference material	SFE, 650 atm, 200°C	CO$_2$	40 min	None	GC	MS	97–207% of EPA-certified material based on liquid solvent extraction	36
Various wastes	Simple extraction	Acetone, then extracted with hexane	4 h	Silica gel eluted with hexane	(1) GC (2) HPLC	(1) FID (2) Fluorescence	60–80% of spiked samples	51
Petroleum waste sludge	SFE, 400 atm, 100°C	Freon-22	85 min	None	GC	MS	118–153% of 18-h methylene chloride sonication	26
Creosote	Simple extraction	Acetone	3 h	None	GC	FID	Not given	20
PAH contaminants, unspecified	Simple extraction	Dimethyl sulfoxide, then extracted with hexane-methylene chloride (10:1)	24 h	Silica gel eluted with hexane-methylene chloride (10:1)	GC	MS	Not given	30
Coal tar	SFE, 300 bar, 100°C	CO$_2$	40 min	None	GC	FID	79–110% of overnight methylene chloride Soxhlet	2

[a] FID, flame ionization detection; EPA, Environmental Protection Agency.
[b] Extraction procedures completed in series, not as two separate methods.

detector; the full spectrum of each peak can then be compared with authentic PAH spectra. Fluorescence detection is more sensitive than UV detection.

We thank Allison Luneau for invaluable help in preparation of the manuscript, Pat Fleischer for clerical assistance, and the NCTR Microbiology Division Research Staff for their comments.

REFERENCES

1. **Abbott, C. K., D. L. Sorensen, and R. C. Sims.** 1992. Use and efficiency of ethylene glycol monomethyl ether and monoethanolamine to trap volatized [7-^{14}C]naphthalene and $^{14}CO_2$. *Environ. Toxicol. Chem.* **11:**181–185.
2. **Anonymous.** 1992. *Supercritical Fluid Extraction for the Analysis of Contaminated Sites.* Electric Power Research Institute final report (TR-100754). Electric Power Research Institute, Palo Alto, Calif.
3. **Bagy, M. M. K., A. A. M. Shoreit, and S. K. Hemida.** 1992. Naphthalene-anthracene-utilizing microorganisms. *J. Basic Microbiol.* **32:**299–308.
4. **Bogardt, A. H., and B. B. Hemmingsen.** 1992. Enumeration of phenanthrene-degrading bacteria by an overlayer technique and its use in evaluation of petroleum-contaminated sites. *Appl. Environ. Microbiol.* **58:**2579–2582.
5. **Brannon, J. M., C. B. Rice, F. J. Reilly, Jr., J. C. Pennington, and V. A. McFarland.** 1993. Effects of sediment organic carbon on distribution of radiolabeled fluoranthene and PCBs among sediment, interstitial water, and biota. *Bull. Environ. Contam. Toxicol.* **51:**873–880.
6. **Burford, M. D., S. B. Hawthorne, and D. J. Miller.** 1993. Extraction rates of spiked versus native PAHs from heterogeneous environmental samples using supercritical fluid extraction and sonication in methylene chloride. *Anal. Chem.* **65:**1497–1505.
7. **Catallo, W. J., and R. J. Portier.** 1992. Use of indigenous and adapted microbial assemblages in the removal of organic chemicals from soils and sediments. *Water Sci. Technol.* **25:**229–237.
8. **Cerniglia, C. E.** 1982. Initial reactions in the oxidation of anthracene by *Cunninghamella elegans. J. Gen. Microbiol.* **128:**2055–2061.
9. **Cerniglia, C. E.** 1993. Biodegradation of polycyclic aromatic hydrocarbons. *Curr. Opin. Biotechnol.* **4:**331–338.
10. **Cerniglia, C. E., W. L. Campbell, J. P. Freeman, and F. E. Evans.** 1989. Identification of a novel metabolite in phenanthrene metabolism by the fungus *Cunninghamella elegans. Appl. Environ. Microbiol.* **55:**2275–2279.
11. **Cerniglia, C. E., J. P. Freeman, and R. K. Mitchum.** 1982. Glucuronide and sulfate conjugation in the fungal metabolism of aromatic hydrocarbons. *Appl. Environ. Microbiol.* **43:**1070–1075.
12. **Cerniglia, C. E., and D. T. Gibson.** 1979. Oxidation of benzo[a]pyrene by the filamentous fungus *Cunninghamella elegans. J. Biol. Chem.* **254:**12174–12180.
13. **Cerniglia, C. E., and D. T. Gibson.** 1980. Fungal oxidation of benzo[a]pyrene and (±)-trans-7,8-dihydroxy-7,8-dihydrobenzo[a]pyrene. *J. Biol. Chem.* **255:**5159–5163.
14. **Cerniglia, C. E., D. T. Gibson, and R. H. Dodge.** 1994. Metabolism of benz[a]anthracene by the filamentous fungus *Cunninghamella elegans. Appl. Environ. Microbiol.* **60:**3931–3938.
15. **Cerniglia, C. E., R. L. Hebert, P. J. Szaniszlo, and D. T. Gibson.** 1978. Fungal transformation of naphthalene. *Arch. Microbiol.* **117:**135–143.
16. **Cerniglia, C. E., and M. A. Heitkamp.** 1989. Microbial degradation of polycyclic aromatic hydrocarbons in the aquatic environments, p. 41–68. *In* U. Varanasi (ed.), *Metabolism of Polycyclic Aromatic Hydrocarbons in the Aquatic Environment.* CRC Press, Boca Raton, Fla.
17. **Cerniglia, C. E., and M. A. Heitkamp.** 1990. Polycyclic aromatic hydrocarbon degradation by *Mycobacterium. Methods Enzymol.* **188:**148–153.
18. **Cerniglia, C. E., D. W. Kelly, J. P. Freeman, and D. D. Miller.** 1986. Microbial metabolism of pyrene. *Chem. Biol. Interact.* **57:**203–216.
19. **Cerniglia, C. E., and S. K. Yang.** 1984. Stereoselective metabolism of anthracene and phenanthrene by the fungus *Cunninghamella elegans. Appl. Environ. Microbiol.* **47:**119–124.
20. **Ellis, B., P. Harold, and H. Kronberg.** 1991. Bioremediation of a creosote contaminated site. *Environ. Technol.* **12:**447–459.
21. **Eschenbach, A., M. Kästner, R. Bierl, G. Schaefer, and B. Mahro.** 1994. Evaluation of a new, effective method to extract polycyclic aromatic hydrocarbons from soil samples. *Chemosphere* **28:**683–692.
22. **Futoma, D. J., S. R. Smith, T. E. Smith, and J. Tanaka.** 1981. Solubility studies of PAH in water, p. 13–24. *In* D. J. Futoma, S. R. Smith, T. E. Smith, and J. Tanaka (ed.), *Polycyclic Aromatic Hydrocarbons in Water Systems.* CRC Press, Boca Raton, Fla.
23. **Ghisalba, O., P. Cevey, M. Küenzi, and H.-P. Schär.** 1985. Biodegradation of chemical waste by specialized methylotrophs, and alternatives to physical methods of waste disposal. *Conserv. Recycl.* **8:**47–71.
24. **Gibson, D. T., and V. Subramanian.** 1984. Microbial degradation of aromatic hydrocarbons, p. 181–252. *In* D. T. Gibson (ed.), *Microbial Degradation of Organic Compounds.* Marcel Dekker, New York.
25. **Hatzinger, P. B., and M. Alexander.** 1995. Effect of aging of chemicals in soil on their biodegradability and extractability. *Environ. Sci. Technol.* **29:**537–545.
26. **Hawthorne, S. B., J. J. Langenfeld, D. J. Miller, and M. D. Burford.** 1992. Comparison of supercritical $CHClF_2$, N_2O, and CO_2 for the extraction of polychlorinated biphenyls and polycyclic aromatic hydrocarbons. *Anal. Chem.* **64:**1614–1622.
27. **Heitkamp, M. A., J. P. Freeman, D. W. Miller, and C. E. Cerniglia.** 1988. Pyrene degradation by a *Mycobacterium* sp.: identification of ring oxidation and ring fission products. *Appl. Environ. Microbiol.* **54:**2556–2565.
28. **Hsu, T.-S., and R. Bartha.** 1979. Accelerated mineralization of two organophosphate insecticides in the rhizosphere. *Appl. Environ. Microbiol.* **37:**36–41.
29. **Huckins, J. N., J. D. Petty, and M. A. Heitkamp.** 1984. Modular containers for microcosm and process model studies on the fate and effects of aquatic contaminants. *Chemosphere* **13:**1329–1341.
30. **Hund, K., and W. Traunspurger.** 1994. Ecotox—evaluation strategy for soil bioremediation exemplified for a PAH-contaminated site. *Chemosphere* **29:**371–390.
31. **Kelley, I., J. P. Freeman, F. E. Evans, and C. E. Cerniglia.** 1991. Identification of a carboxylic acid metabolite from the catabolism of fluoranthene by a *Mycobacterium* sp. *Appl. Environ. Microbiol.* **57:**636–641.
32. **Kelley, I., J. P. Freeman, F. E. Evans, and C. E. Cerniglia.** 1993. Identification of metabolites from the degradation of fluoranthene by *Mycobacterium* sp. strain Pyr-1. *Appl. Environ. Microbiol.* **59:**800–806.
33. **Kiyohara, H., K. Nagao, and K. Yana.** 1982. Rapid screen for bacteria degrading water-insoluble, solid hydrocarbons on agar plates. *Appl. Environ. Microbiol.* **43:**454–457.
34. **Koch, A. L.** 1981. Growth measurement, p. 179–207. *In* P. Gerhardt (ed.), *Manual of Methods for General Bacteriology.* American Society for Microbiology, Washington, D.C.
35. **Landrum, P. F., W. S. Dupuis, and J. Kukkonen.** 1994. Toxicokinetics and toxicity of sediment-associated pyrene and phenanthrene in *Diporeia.* spp.: examination of equi-

librium-partitioning theory and residue-based effects for assessing hazard. *Environ. Toxicol. Chem.* **13:**1769–1780.

36. **Langenfeld, J. J., S. B. Hawthorne, D. J. Miller, and J. Pawliszyn.** 1993. Effects of temperature and pressure on supercritical fluid extraction efficiencies of polycyclic aromatic hydrocarbons and polychlorinated biphenyls. *Anal. Chem.* **65:**338–344.

37. **Lehmicke, L. G., R. T. Williams, and R. L. Crawford.** 1979. ^{14}C-most-probable-number method for enumeration of active heterotrophic microorganisms in natural waters. *Appl. Environ. Microbiol.* **38:**644–649.

38. **Mahaffey, W. R., D. T. Gibson, and C. E. Cerniglia.** 1988. Bacterial oxidation of chemical carcinogens: formation of polycyclic aromatic acids from benz[a]anthracene. *Appl. Environ. Microbiol.* **54:**2415–2423.

39. **Mihelcic, J. R., and R. G. Luthy.** 1988. Degradation of polycyclic aromatic hydrocarbon compounds under various redox conditions in soil-water systems. *Appl. Environ. Microbiol.* **54:**1182–1187.

40. **Moller, J., and H. Ingvorsen.** 1993. Biodegradation of phenanthrene in soil microcosms stimulated by an introduced *Alcaligenes* sp. *FEMS Microbiol. Ecol.* **102:**271–278.

41. **Mueller, J. G., P. J. Chapman, and P. H. Pritchard.** 1989. Creosote-contaminated sites: their potential for bioremediation. *Environ. Sci. Technol.* **23:**1197–1201.

42. **Narro, M. L., C. E. Cerniglia, C. Van Baalen, and D. T. Gibson.** 1992. Metabolism of phenanthrene by the marine cyanobacterium *Agmenellum quadruplicatum* PR-6. *Appl. Environ. Microbiol.* **58:**1351–1359.

43. **Pothuluri, J. V., J. P. Freeman, F. E. Evans, and C. E. Cerniglia.** 1993. Biotransformation of fluorene by the fungus *Cunninghamella elegans. Appl. Environ. Microbiol.* **59:**1977–1980.

44. **Pothuluri, J. V., R. H. Heflich, P. P. Fu, and C. E. Cerniglia.** 1992. Fungal metabolism and detoxification of fluoranthene. *Appl. Environ. Microbiol.* **58:**937–941.

45. **Stieber, M., F. Haeseler, P. Werner, and F. H. Frimmel.** 1994. A rapid screening method for micro-organisms degrading polycyclic aromatic hydrocarbons in microplates. *Appl. Microbiol. Biotechnol.* **40:**753–755.

46. **Stringfellow, W. T., and M. D. Aitken.** 1994. Comparative physiology of phenanthrene degradation by two dissimilar pseudomonads isolated from a creosote-contaminated soil. *Can. J. Microbiol.* **40:**432–438.

47. **Sutherland, J. B., A. L. Selby, J. P. Freeman, F. E. Evans, and C. E. Cerniglia.** 1991. Metabolism of phenanthrene by *Phanerochaete chrysosporium. Appl. Environ. Microbiol.* **57:**3310–3316.

48. **Sutherland, J. B., A. L. Selby, J. P. Freeman, P. P. Fu, D. D. Miller, and C. E. Cerniglia.** 1992. Identification of xyloside conjugates formed from anthracene by *Rhizoctonia solanii. Mycol. Res.* **96:**509–517.

49. **Umbreit, W. W., R. H. Burris, and J. F. Stauffer.** 1972. *Manometric and Biochemical Techniques, a Manual Describing Methods Applicable to the Study of Tissue Metabolism,* 5th ed. Burgess Publishing Co., Minneapolis.

50. **Urakami, T., H. Kobayashi, and H. Araki.** 1990. Isolation and identification of *N,N*-dimethylformamide-biodegrading bacteria. *J. Ferment. Bioeng.* **70:**45–47.

51. **van der Oost, R., F.-J. van Schooten, F. Ariese, and H. Heida.** 1994. Bioaccumulation, biotransformation and DNA binding of PAHs in feral eel (*Anguilla anguilla*) exposed to polluted sediments: a field survey. *Environ. Toxicol. Chem.* **13:**859–870.

52. **Wang, X., X. Yu, and R. Bartha.** 1990. Effect of bioremediation on polycyclic aromatic hydrocarbon residues in soil. *Environ. Sci. Technol.* **24:**1086–1089.

53. **Wenclawiak, B., C. Rathmann, and A. Teuber.** 1992. Supercritical-fluid extraction of soil samples and determination of polycyclic aromatic hydrocarbons (PAHs) by HPLC. *Fresenius J. Anal. Chem.* **344:**497–500.

54. **Wild, S. R., K. S. Waterhouse, S. P. McGrath, and K. C. Jones.** 1990. Organic contaminants in an agricultural soil with a known history of sewage sludge amendments: polynuclear aromatic hydrocarbons. *Environ. Sci. Technol.* **24:**1706–1711.

55. **Zylstra, G. J., and D. T. Gibson.** 1991. Aromatic hydrocarbon degradation: a molecular approach, p. 183–203. *In* J. K. Setlow (ed.), *Genetic Engineering: Principles and Methods,* vol. 13. Plenum Press, New York.

Biodegradation and Transformation of Nitroaromatic Compounds

SHIRLEY F. NISHINO AND JIM C. SPAIN

85

The addition of a nitro group to aromatic compounds dramatically alters the chemical properties of the molecule, and consequently the biodegradation pathways evolved by bacteria for the assimilation of the unsubstituted aromatic compounds are not capable of accommodating the corresponding nitro compounds. Few natural nitroaromatic compounds are known, and release of synthetic compounds into the biosphere has been a relatively recent event. Even so, microbes have evolved a variety of strategies for metabolism of aromatic nitro compounds.

METABOLISM AND PATHWAYS

Recently, a variety of novel oxidative and reductive mechanisms that lead to the complete mineralization of nitroaromatic compounds have been discovered. Extensive reviews of the recent research in these areas have been published (27, 46, 68, 69) and will be discussed only briefly here. What has emerged from this work is a dichotomy of how microbes respond to the presence of nitroaromatic compounds and how specific organisms can be isolated or exploited for their abilities to transform nitroaromatic compounds. In one scenario, the nitro compound is degraded (mineralized) and provides the organism with a source of carbon, nitrogen, or energy; therefore, enrichment and selection for microorganisms can be based on the ability of the nitro compound to support growth. In another scenario, there are other nitroaromatic compounds that will not support growth of microorganisms and thus provide no selective advantage that can be exploited for the isolation of organisms able to degrade the compounds. Nevertheless, microorganisms can often transform the aromatic nitro compounds through fortuitous reactions of nonspecific enzymes. These types of reactions are particularly common among the fungi and anaerobic bacteria. Although these transformations do not result in the complete removal of toxic compounds, novel strategies of bioprocess engineering offer the potential for use of biotransformation systems for the complete degradation or detoxification of nitroaromatic compounds.

Fungi

Various fungi, but in particular the white and brown rot fungi, under low-nutrient conditions produce extracellular ligninolytic enzymes that attack both the phenolic and nonphenolic components of lignin (21). The nonspecific nature of these systems allows the same enzymes to transform a wide variety of other substances, including nitro- and aminoaromatic compounds (5). The white rot fungus, *Phanerochaete chrysosporium*, has in recent years been found to degrade 2,4-dinitrotoluene (78) and 2,4,6-trinitrotoluene (TNT) (22, 49) under ligninolytic conditions. However, the ligninolytic system does not appear to be involved in the initial reduction reactions, which require the presence of live mycelia, but the ligninolytic enzymes are necessary for the subsequent reactions that lead to mineralization of the amino intermediates.

Anaerobic Bacteria

There have been several reports of degradation or transformation of nitro and aminoaromatic compounds under anaerobic conditions. Gorontzy et al. (28) found only nonspecific reduction of the nitro group to the corresponding amine by a variety of methanogenic bacteria, sulfate-reducing bacteria, and clostridia. None of the 17 strains examined could utilize the nitrophenols or nitrobenzoates as growth substrates, and the authors concluded that reduction of the nitro group was a nonspecific detoxification mechanism for those organisms. In contrast, sulfate-reducing bacteria use 2,4-dinitrophenol, 2,4- and 2,6-dinitrotoluene (8), and TNT (7, 59) as sources of nitrogen. The nitroaromatic compounds are reduced to the corresponding amines, whose subsequent fate is not clear, although reductive deamination has been proposed (8) as a possible mechanism for the elimination of the amino groups. Reductive deamination of 4-aminobenzoate metabolites before ring cleavage (64) and degradation of 3-aminobenzoate (65) have been demonstrated under anaerobic conditions. Nitroreductase activities in clostridia (47, 59, 60) produce nitroso-, hydroxylamino-, and aminoaromatic products from the parent nitroaromatic compounds.

Aerobic Bacteria

Aerobic bacteria can degrade nitroaromatic compounds through a variety of oxidative or reductive pathways. The known pathways for degradation of nitroaromatic compounds involve either the removal or the reduction of the nitro group and conversion of the resulting molecule into a substrate for oxidative ring fission. The cleavage products can be readily converted into the dicarboxylic acids of intermediary metabolic pathways.

Reductive pathways generally involve the conversion of

the nitroaromatic compound to hydroxylamino- or amino-aromatic intermediates (48). The reduced intermediates may be further transformed before ring cleavage and liberation of the nitrogen as ammonia. The reduction products are often the same as those produced by anaerobic bacteria (12), but anaerobic conditions are not necessary. The known transformations of the hydroxylamines include the action of a hydroxylaminolyase which converts hydroxyl-aminobenzoate to protocatechuate (31, 33, 62) and of hydroxylaminobenzene mutase, which converts hydroxylami-nobenzene to 2-aminophenol (54). A novel mechanism that results in the release of nitrite from the rearomatization of a reduced hydride-Meisenheimer complex has been demonstrated for picric acid (44) and TNT (81).

Oxidative catabolism of nitroaromatic compounds can be initiated by the action of either monooxygenase or dioxy-genase enzymes. Monooxygenases catalyze the replacement of the nitro group by a hydroxyl group in the degradation of 4-nitrophenol (73), 2-nitrophenol (86), and 4-methyl-5-nitrocatechol (32), resulting in the liberation of nitrite. Dioxygenases catalyze the initial attack on 2,4-dinitrotolu-ene (74), 1,3-dinitrobenzene (53), nitrobenzene (55), 3-ni-trobenzoate (51), and 2-nitrotoluene (3, 34). All of these oxidative reactions result in the release of the nitro group as nitrite and the formation of dihydroxy aromatic compounds able to serve as substrates for ring fission and subsequent metabolism.

ISOLATION (SELECTIVE ENRICHMENT) OF SPECIFIC BACTERIA ABLE TO GROW ON NITROAROMATIC COMPOUNDS

The current understanding of specific mechanisms for the degradation of nitroaromatic chemicals has been facilitated by the isolation of bacteria able to use nitroaromatic compounds as growth substrates rather than as nonspecific electron acceptors. Selective enrichment (38) to increase the relative abundance of bacteria with desired degradative abilities is the first step in this process. Factors to be considered prior to the initiation of an enrichment culture include the source of the inoculum, the concentration of the nitroaromatic compound, selection strategies for isolates, and methods to detect growth of the culture or metabolism of the nitroaromatic compound.

Inoculum

Source of Inoculum

Traditional strategies involving selection from activated sludges or a variety of ecosystems have not been useful for isolation of bacteria that degrade the more recalcitrant nitroaromatic compounds. Bacterial strains that use specific nitroaromatic compounds have been selected primarily from inocula previously exposed to the compound. Bacteria from a variety of sources can degrade compounds such as 4-ni-trophenol that have been widely distributed in the environment (1, 37, 72, 82). In contrast, bacteria which mineralize the more recalcitrant 2,4-dinitrotoluene have been isolated only from 2,4-dinitrotoluene-contaminated sites (56, 74). Nitrobenzene-degrading strains can be readily isolated from nitrobenzene-contaminated sites but not from uncontaminated sites (55). A similar result was observed for strains able to degrade s-triazines (14). These observations suggest that the presence of xenobiotic compounds provides the selective pressure for the development of the ability to degrade specific contaminants.

Adaptation to Degrade Nitroaromatic Compounds

Microbial communities can often adapt (acclimate) to degrade novel organic compounds. That is, the rate of degradation of a compound is increased by exposure of the microbial community to the particular compound (71). The mechanisms subsumed under the term "adaptation" are numerous but can be classified into three general categories: (i) growth of a small population that is capable of utilizing the compound, (ii) delayed induction of enzymes involved in the catabolic pathway, and (iii) genetic change(s) that enables the microorganisms to grow at the expense of the chemical. The mechanism involved in the adaptation of any particular microbial community to a novel substrate can be suggested by the length of the lag time from the initial exposure of the population to the chemical to the start of rapid disappearance of the chemical (67). Short lag periods of hours to a few days suggest that induction of a degradative pathway is the only requirement for the population to be able to degrade the compound. Lag periods of intermediate length, showing a gradual increase in rate of degradation, suggest that the lag period is the time required for the growth of a specific population of organisms capable of degrading a specific compound. Finally, extended lag periods followed by abrupt increases in degradation rate suggest that a genetic change allowed the microbial community to use the organic compound as a growth substrate. Support for the third mechanism is provided by the observation that nitro- and chloro-aromatic compounds previously thought to be resistant to biodegradation are now readily isolated from sites contaminated with such compounds but not from pristine sites.

Substrate Concentration

Careful attention must be given to the concentration of the nitroaromatic compounds available to the microbial population. Choice of an appropriate substrate concentration for selective enrichment involves consideration of toxicity, the concentration necessary to support growth, and the solubility of the nitroaromatic compound in the growth medium.

Toxicity

Toxicity both of the parent compound and of possible metabolic products must be considered. Nitroaromatic compounds typically serve as growth substrates for microorganisms at concentrations ranging from 10 to 100 mg/liter; however, some are toxic in that range as well. Concentrations in the lower part of the range may provide little selective pressure for isolation of bacteria able to degrade nitroaromatic compounds and may be difficult to measure; even so, initial concentrations of 10 to 20 mg/liter have been used successfully. When the initial amount is metabolized, the concentration of substrate supplied in subsequent additions or transfers can be increased until toxicity is exhibited. A lengthened lag period or slower rate of utilization can be taken as a sign of toxicity. If information about the concentrations of contaminants in the ecosystem chosen as a source of inoculum is available, it is often a good strategy to use a concentration of substrate similar to that found at the edges of the contaminated zone. If the concentration of substrate supplied is not enough to support extensive growth, the substrate can be added repeatedly.

Availability

With nonpolar nitroaromatic compounds, the solubility of the compound in aqueous solutions may be too low to support growth. A recent strategy successful with several synthetic compounds has been the use of dual-phase systems, in which the substrate of interest is dissolved at a high concentration in an inert organic solvent. The solvent serves as a reservoir of substrate, which is released to the aqueous phase as the substrate is utilized by the bacteria. A variety of solvents have been used with different levels of success (13, 19, 52, 61, 76). To date, this approach has not been used extensively with nitroaromatic compounds.

Procedures for Isolation

Initial Selection Conditions

The initial selection of cultures able to degrade nitroaromatic compounds generally is done in batch cultures. One gram of soil or 1 to 10 ml of water inoculated into 100 ml of a minimal medium supplemented with the appropriate nitroaromatic substrate as the sole source of nitrogen or carbon is a good starting point. The culture should be monitored for degradation of the nitroaromatic compound, and transfers should be made at appropriate intervals before the substrate is completely removed. As the nitroaromatic compound is degraded, a portion of the culture should be diluted 10-fold with fresh medium, and the process is repeated. If the degradative capability persists through multiple transfers, the inoculum size can be gradually decreased to increase the selection pressure and dilute out additional nutrients and substrates present in the initial inoculum. A variation on the batch culture technique is continuous perfusion and recirculation in small soil columns (4).

Isolation of Degradative Strains

When a mixed culture with the desired metabolic capability is available, individual strains can be isolated directly from the enrichment culture, or the enrichment culture may be placed in a chemostat to increase the selection pressure for the desired degradative capability. Individual strains can be isolated by spreading samples of the culture onto agar plates supplemented with the nitroaromatic compound or onto a complex medium such as nutrient agar. If the original inoculum came from an oligotrophic environment, use of diluted media such as $\frac{1}{10}$-strength nutrient agar will often be more successful than use of a full-strength rich medium. Isolation on complex media is useful when bacterial growth is very slow on minimal media or when the transformation of the nitroaromatic compound is only partial. Yeast extract (5 to 20 mg/liter) can be added to minimal media at this stage if the isolate seems to require growth factors. Individual colonies that grow on any of the plates are then tested for degradative ability. Testing of metabolic capabilities can be accomplished by auxanography (58) or by use of gradient plates (84).

Enhancement of Degradative Abilities

Enrichment cultures or isolated strains can be subjected to additional selection to enhance degradative abilities. Growth in chemostats at high dilution rates can be used to provide a continuous strong selection pressure for the desired degradative ability (36, 41). Various means of chemical or transposon mutagenesis (20) might also result in improved degradative capability. Molecular approaches to strain construction can be used to improve or expand the catabolic abilities of degradative strains (77). Strain construction in combination with chemostat selection has recently proved useful in isolating a *Pseudomonas* strain capable of mineralizing TNT (18).

Substrate Delivery

The volatility and water solubility of the nitroaromatic compound determine the means of delivery. Substrates that are readily soluble in aqueous solutions at concentrations useful for culturing bacteria can be dissolved directly in the culture medium. Such substrates include the isomeric nitrophenols, nitrobenzene, 2,4-dinitrotoluene, 1,3-dinitrobenzene, and TNT. The latter three nitroaromatic compounds, although soluble at the final working concentration, may be slow to dissolve and can be dissolved in a carrier solvent such as methanol, ethanol, acetone or dimethylformamide before addition to the culture medium. The nitroaromatic compound can also be dissolved in a volatile carrier such as acetone, and the resulting solution is used to coat the bottom of the culture vessel. The solvent is allowed to evaporate before addition of the culture. Volatile substrates may also be added to a culture in the vapor phase. For example, nitrobenzene can be provided to individual petri dishes by placing a cotton-plugged Durham tube containing nitrobenzene into the lid of the inverted agar plate, which is then sealed (26). Volatile substrates can also be provided by placing the plates into a sealed enclosure such as a desiccator with a small open container of the nitroaromatic substrate. Volatile substrates can be supplied to liquid cultures in the gas stream used to provide aeration. A single crystal of slightly water soluble compounds such as the nitrotoluenes can be placed in the center of a previously inoculated agar plate to produce a substrate gradient. Dual-phase cultures should be considered if high concentrations of a relatively insoluble nitroaromatic compound are required.

Nitroaromatic Compounds as Nutrient Sources

Nitroaromatic compounds may be used as either carbon or nitrogen sources. The decision as to whether to use the substrate as the sole source of carbon, nitrogen, or both may depend on the toxicity or solubility of the nitroaromatic compound. If the compound is used as the sole source of nitrogen, a source of carbon must be supplied in nitrogen-free minimal medium (10). A variety of easily metabolizable carbon sources such as glucose (9, 80), acetate (16, 42), succinate (10), lactate (2), glycerol, and pyruvate have been used. It has been suggested that an array of carbon sources be present in nitrogen-limited cultures so as not to limit the range of organisms able to grow in the culture (14); alternatively, multiple cultures, each with a different carbon source, can be inoculated.

In some instances the nitroaromatic compound can serve as the sole carbon and nitrogen source (51, 54, 55, 74). If there are multiple nitro groups on the aromatic ring, the elimination of nitrite can lead to toxicity. For example, during degradation of 2,4-dinitrotoluene by *Burkholderia* (formerly *Pseudomonas*) sp. strain DNT, growth is inhibited when nitrite concentrations approach 1 mM (56).

If bacteria have growth requirements other than the nitroaromatic compound, the addition of vitamins, amino acids, or more complex mixtures such as yeast or soil extracts might be necessary. Such factors must be supplied at levels that do not cause catabolite repression (23), which is expressed as increased lag times before growth on the nitroaromatic compound. Glucose, acetate, and succinate are frequent catabolite repressors because of the ease with which they are metabolized. Alternative substrates such as arginine, glycerol, or specific vitamins are less likely to act as catabolite repressors.

Detection of Degradation of Nitroaromatic Compounds

Substrate Disappearance

Substrate disappearance is a primary indication that the nitroaromatic compound is being utilized. High-performance liquid chromatography (HPLC) with UV detection is most commonly used to monitor substrate disappearance in liquid cultures. Capillary electrophoresis shows considerable promise for identification and quantitation of nitroaromatic compounds (57), but it is less sensitive than HPLC. Many nitroaromatic compounds are colored, so a visible decrease in the color of the culture can indicate substrate utilization. However, the color of some nitroaromatic compounds, such as nitrophenol, is pH dependent, and a slight shift in pH can give the appearance of substrate utilization. If the substrate is volatile, gas chromatography may be a convenient analytical method. However, disappearance of the substrate is only presumptive evidence that the compound is used for growth, and both uninoculated and killed controls must be included to ensure that sorption, volatilization, photolysis, or instability of the compound in aqueous solution is not confused with biodegradation. Disappearance of ^{14}C-labeled substrates indicates biodegradation only if a significant portion of the radiolabel is released as $^{14}CO_2$ and only if the fraction of the radiolabel in $^{14}CO_2$ is much larger than the level of radiochemical impurities.

Growth

If the nitroaromatic compound is used as a source of carbon or nitrogen, growth will accompany the disappearance of the substrate. Growth is easily recognized by increases in optical density in liquid cultures or by increase in colony size on agar plates. Growth can also be detected by increased protein concentration. An increase in cell counts is not reliable, because one of the possible starvation responses is an increase in cell numbers (50), that is, cell division without an increase in biomass. However, if cell numbers increase several orders of magnitude in the presence of the nitroaromatic compound but not in its absence (control cultures), then such increases can be taken as evidence of growth provided that cell yields are consistent and proportional to substrate disappearance.

Metabolite Accumulation

Metabolite accumulation can indicate that the nitroaromatic compound is a growth substrate and also provide insight into the degradative pathway. Steady accumulation of a metabolite indicates that it is a dead-end product and not part of a productive catabolic pathway. Transient accumulation of a metabolite during the early stages of culture growth suggests that the enzymes of the catabolic pathway are not induced simultaneously and that the metabolite is an intermediate of the pathway. Metabolite accumulation may be accompanied by either disappearance of a colored substrate or appearance of a colored intermediate, or both. For example, 2,4-dinitrotoluene is colorless in aqueous solutions, but the first metabolic intermediate produced by *Burkholderia* sp. strain DNT, 4-methyl-5-nitrocatechol, is bright yellow (74). Sequential appearance and disappearance of the metabolite during induction of *Burkholderia* sp. strain DNT on 2,4-dinitrotoluene turns the culture fluid from colorless to yellow to colorless again. Color changes or the lack of them can also be misleading. Two pathways for the degradation of 4-nitrophenol are known; one involves monooxygenation to hydroquinone (70), and the other involves dioxygenation to 4-nitrocatechol (39). Both 4-nitrophenol and 4-nitrocatechol are yellow; hydroquinone is colorless. Persistence of the yellow color in a culture containing 4-nitrophenol can indicate either that 4-nitrophenol is not degraded or that it has been converted to 4-nitrocatechol.

Ammonia and Nitrite Release

The known aerobic pathways for degradation of nitroaromatic compounds all result in the release of the nitro group as either ammonia or nitrite. The appearance of either of these in a culture would indicate the degradation of the nitroaromatic compound. To date, all pathways that result in the release of nitrite from a nitroaromatic compound, with the exception of the picric acid catabolic pathway (44), involve the oxidative release of all the nitro groups prior to ring cleavage. Catabolic pathways that result in the reduction of the nitro group can release the ammonia before or after ring cleavage. Release of these metabolites is concomitant with catabolism of the nitroaromatic substrate. Bacteria able to assimilate or transform nitrite or ammonia can mask the release of these metabolites from the nitroaromatic substrate.

Analytical Methods

Many of the analytical methods mentioned above are discussed elsewhere (25). HPLC is usually performed on C_{18} or C_8 reverse-phase columns with mobile phases consisting of mixtures of water and a less polar solvent such as methanol, acetonitrile, or tetrahydrofuran. Simple linear gradients from a more polar to a less polar mobile phase will separate many nitroaromatic compounds and their metabolites. Ion pair chromatography is often used to separate amino compounds with isocratic mobile phases. Common ion pair reagents such as tetrabutylammonium hydrogen sulfate and hexane sulfonic acid are available in commercial formulations. Photodiode array detectors offer a great improvement in convenience over variable-wavelength UV-visible light detectors. There are many simple but sensitive colorimetric assays for analysis of ammonia and nitrite (15). A commercially available assay for ammonia based on reductive amination (Sigma) is useful when metabolic intermediates interfere with standard colorimetric assays.

DESIGN AND OPERATION OF BIOREACTORS

Widespread environmental contamination by nitroaromatic compounds and explosives has created much of the impetus for recent research on their biodegradation. The best currently available technology for treatment of such materials involves incineration of soils or sorption to activated carbon, which is in turn incinerated. Incineration is a very costly treatment technology and draws much public criticism. Accordingly, much effort has been directed toward developing bioremediation systems, many of which include the use of bioreactors as a key component of the treatment system. All of the treatments described in this section involve metabolism of the nitroaromatic compound by microorganisms unable to use them as growth substrates. Therefore, they require additional sources of carbon and energy for growth and maintenance of the microbial cultures.

Composting

In its simplest form, the process of composting often does not require an actual container and is therefore easily expandable to accommodate large volumes of contaminated materials. It does, however, require large amounts of materials handling, space to accommodate bulk matter, and containment of leachates. Generally, not more than 10% of the composting pile can be contaminated soil (66); the rest is made up of compostable organic matter. Composting has been demonstrated for decontamination of soils contaminated with TNT and other explosives (83). Chemical and toxicological tests of composted explosive-contaminated soils indicate large reductions in the concentrations of explosives and their metabolites and the toxicity of leachates after composting (29, 30), but it is still not clear that acceptable levels of toxicity and mutagenicity are reached in composted soils. The ultimate fate of the nitroaromatic compounds in the residue is not known, and the volume of the hazardous material is increased considerably.

Anaerobic Treatment Systems

Anaerobic treatment systems have been proposed as a means of avoiding the accumulation of partially reduced intermediates during degradation of TNT. Under strictly anaerobic conditions ($E_h \leq 200$ mv), TNT can be completely reduced to triaminotoluene (59). Anaerobic treatment of explosive (primarily TNT)-contaminated (24) and herbicide (dinoseb)-contaminated (40) soils has been demonstrated in open bulk containers of soil, phosphate buffer, and potato starch. The potato starch served as a readily degradable carbon source which allowed the rapid establishment of anaerobiosis. In the case of the herbicides, no aromatic compounds remained in the bioreactors; however, with TNT, cresols and small organic acids remained as end products (24). To date, this is the only commercially available system specifically developed for bioremediation of nitroaromatic compounds.

Anaerobic/Aerobic Systems

To eliminate the hydroxytoluenes or aminotoluenes remaining in the anaerobic bioreactors following the disappearance of TNT, second-stage aerobic reactors have been proposed to hasten the removal of those intermediates which are more rapidly degraded under aerobic conditions (24, 63). Upon aeration, triaminotoluene bound to soil undergoes an oxidative polymerization, which immobilizes the chemical (63). However, further research is required to demonstrate whether the immobilization is permanent and whether any residual toxicity remains.

A two-stage anaerobic-aerobic process for degradation of nitrobenzene has also been described (17). Although nitrobenzene can be mineralized aerobically, there is potential for losses of nitrobenzene through air stripping. This problem can be avoided through a process which is initiated by a nonspecific anaerobic reduction of nitrobenzene to aniline, followed by aerobic degradation of the aniline. Glucose can serve as the carbon source and hydrogen donor for the anaerobic phase.

Fungus-Based Remediation Systems

The nonspecific, extracellular enzymes of the white rot fungi make them attractive candidates for bioremediation systems. Ligninolytic cultures of *P. chrysosporium* have been grown in a bench-scale fixed-film silicone membrane bioreactor (79) to study the potential for degradation of TNT (11). TNT in wastewater from munitions plants has also been successfully degraded by *P. chrysosporium* immobilized on a rotating biological contactor (75). Other proposals include the combination of fungi with bacteria in systems in which the fungi would first detoxify or modify the xenobiotic compound so that the bacterial population could mineralize the resulting metabolites (5).

Slurry-Phase Systems

Slurry-phase bioreactors in which soil is kept in suspension by mechanical mixers are being evaluated as a means of bioremediating contaminated soils and waters. Demonstrations of the biodegradation of polynuclear aromatic hydrocarbons by using a 30% (wt/vol) contaminated soil slurry have been reported (45). Other demonstrations have involved the treatment of explosive-contaminated soils. An aerobic reactor which used molasses as the cosubstrate reduced TNT from 1,300 to 10 mg/liter in 15 days (35); an anaerobic system achieved the same result more slowly. Other work indicated that the addition of a surfactant greatly enhanced the degradation rate of explosives by both acting as a cosubstrate and enhancing the bioavailability of the contaminants (85).

IN SITU APPLICATIONS

It has become clear that many nitroaromatic compounds can serve as growth substrates for bacteria. It is also evident that bacteria able to degrade nitroaromatic compounds are distributed in the environment at sites where contamination by nitroaromatic compounds has been chronic. Thus, it seems likely that treatment strategies based on in situ biodegradation of nitroaromatic compounds can be developed. Such treatment systems could avoid the necessity of providing alternate carbon and energy sources required by the treatment systems discussed in the preceding section. Many of the issues of materials handling and disposal of treated soils and effluents would also be minimized. Many in situ treatment systems for gasoline- and fuel-contaminated soils have been developed (6, 43) on the basis of the ready biodegradability of those compounds. With the discovery that bacteria can utilize many nitroaromatic compounds as growth substrates, it has become feasible to adapt many of the same treatment strategies to sites contaminated with nitroaromatic compounds. Compounds such as nitrobenzene, 3-nitrobenzoate, 2-nitrotoluene, 4-nitrotoluene, and 2,4-dinitrotoluene would be excellent candidates for in situ treatment.

REFERENCES

1. **Aelion, C. M., C. M. Swindoll, and F. K. Pfaender.** 1987. Adaptation to and biodegradation of xenobiotic compounds by microbial communities from a pristine aquifer. *Appl. Environ. Microbiol.* **53:**2212–2217.
2. **Allen, L. A.** 1949. The effect of nitro-compounds and some other substances on production of hydrogen sulphide by sulphate-reducing bacteria in sewage. *J. Appl. Bacteriol.* **12:** 26–38.
3. **An, D., D. T. Gibson, and J. C. Spain.** 1994. Oxidative release of nitrite from 2-nitrotoluene by a three-component enzyme system from *Pseudomonas* sp. strain JS42. *J. Bacteriol.* **176:**7462–7467.
4. **Audus, L. J.** 1952. The decomposition of 2,4-dichlorophenoxyacetic acid and 2-methyl-4-chlorophenoxyacetic acid in the soil. *J. Sci. Food Agric.* **3:**268–274.
5. **Barr, D. P., and S. D. Aust.** 1994. Mechanisms white rot fungi use to degrade pollutants. *Environ. Sci. Technol.* **28:** 78–87.
6. **Blackburn, J. W., and W. R. Hafker.** 1993. The impact of biochemistry, bioavailability and bioactivity on the se-

lection of bioremediation techniques. *Trends Biotechnol.* **11:**328–333.

7. **Boopathy, R., and C. F. Kulpa.** 1992. Trinitrotoluene (TNT) as a sole nitrogen source for a sulfate-reducing bacterium *Desulfovibrio* sp. (B strain) isolated from an anaerobic digester. *Curr. Microbiol.* **25:**235–241.

8. **Boopathy, R., and C. F. Kulpa.** 1993. Nitroaromatic compounds serve as nitrogen source for *Desulfovibrio* sp. (B strain). *Can. J. Microbiol.* **39:**430–433.

9. **Boopathy, R., J. Manning, C. Montemagno, and C. Kulpa.** 1994. Metabolism of 2,4,6-trinitrotoluene by a *Pseudomonas* consortium under aerobic conditions. *Curr. Microbiol.* **28:**131–137.

10. **Bruhn, C., H. Lenke, and H.-J. Knackmuss.** 1987. Nitrosubstituted aromatic compounds as nitrogen source for bacteria. *Appl. Environ. Microbiol.* **53:**208–210.

11. **Bumpus, J. A., and M. Tatarko.** 1994. Biodegradation of 2,4,6-trinitrotoluene by *Phanerochaete chrysosporium*: identification of initial degradation products and the discovery of a TNT metabolite that inhibits lignin peroxidases. *Curr. Microbiol.* **28:**185–190.

12. **Cerniglia, C. E., and C. C. Somerville.** 1995. Reductive metabolism of nitroaromatic and nitropolycyclic aromatic hydrocarbons, p. 99–115. *In* J. C. Spain (ed.), *Biodegradation of Nitroaromatic Compounds.* Plenum Publishing Corp., New York.

13. **Collins, A. M., J. M. Woodley, and J. M. Liddell.** 1995. Determination of reactor operation for the microbial hydroxylation of toluene in a two-liquid phase process. *J. Ind. Microbiol.* **14:**382–388.

14. **Cook, A. M., and R. Hütter.** 1981. Degradation of s-triazines: a critical view of biodegradation, p. 237–249. *In* T. Leisinger, A. M. Cook, R. Hütter, and J. Nüesch (ed.), *Microbial Degradation of Xenobiotics and Recalcitrant Compounds.* Academic Press, London.

15. **Daniels, L., R. S. Hanson, and J. A. Phillips.** 1994. Chemical analysis, p. 512–554. *In* P. Gerhardt, R. G. E. Murray, W. A. Wood, and N. R. Krieg (ed.), *Methods for General and Molecular Bacteriology.* American Society for Microbiology, Washington, D. C.

16. **de Bont, J. A. M., M. J. A. W. Vorage, S. Hartmans, and W. J. J. van den Tweel.** 1986. Microbial degradation of 1,3-dichlorobenzene. *Appl. Environ. Microbiol.* **52:**677–680.

17. **Dickel, O., W. Hang, and H.-J. Knackmuss.** 1993. Biodegradation of nitrobenzene by a sequential anaerobic-aerobic process. *Biodegradation* **4:**187–194.

18. **Duque, E., A. Haidour, F. Godoy, and J. L. Ramos.** 1993. Construction of a *Pseudomonas* hybrid strain that mineralizes 2,4,6-trinitrotoluene. *J. Bacteriol.* **175:**2278–2283.

19. **Efroymson, R. A., and M. Alexander.** 1991. Biodegradation by an *Arthrobacter* species of hydrocarbons partitioned into an organic solvent. *Appl. Environ. Microbiol.* **57:**1441–1447.

20. **Eisenstadt, E., B. C. Carlton, and B. J. Brown.** 1994. Gene mutation, p. 297–316. *In* P. Gerhardt, R. G. E. Murray, W. A. Wood, and N. R. Krieg (ed.), *Methods for General and Molecular Bacteriology.* American Society for Microbiology, Washington, D.C.

21. **Evans, C. S.** 1991. Enzymes of lignin degradation, p. 175–184. *In* W. B. Betts (ed.), *Biodegradation: Natural and Synthetic Materials.* Springer-Verlag, London.

22. **Fernando, T., J. A. Bumpus, and S. D. Aust.** 1990. Biodegradation of TNT (2,4,6-trinitrotoluene) by *Phanerochaete chrysosporium. Appl. Environ. Microbiol.* **56:**1666–1671.

23. **Fisher, S. H., and A. L. Sonenshein.** 1991. Control of carbon and nitrogen metabolism in *Bacillus subtilis. Annu. Rev. Microbiol.* **45:**107–135.

24. **Funk, S. B., D. J. Roberts, D. L. Crawford, and R. L. Crawford.** 1993. Initial-phase optimization for bioremediation of munition compound-contaminated soils. *Appl. Environ. Microbiol.* **59:**2171–2177.

25. **Gerhardt, P., R. G. E. Murray, W. A. Wood, and N. R. Krieg (ed.).** 1994. *Methods for General and Molecular Bacteriology.* American Society for Microbiology, Washington, D.C.

26. **Gibson, D. T.** 1976. Initial reactions in the bacterial degradation of aromatic hydrocarbons. *Zentralbl. Bakteriol. Hyg. Abt. 1 Orig. Reihe B* **162:**157–168.

27. **Gorontzy, T., O. Drzyzga, M. W. Kahl, D. Bruns-Nagel, J. Breitung, E. von Loew, and K.-H. Blotevogel.** 1994. Microbial degradation of explosives and related compounds. *Crit. Rev. Microbiol.* **20:**265–284.

28. **Gorontzy, T., J. Küver, and K.-H. Blotevogel.** 1993. Microbial transformation of nitroaromatic compounds under anaerobic conditions. *J. Gen. Microbiol.* **139:**1331–1336.

29. **Griest, W. H., A. J. Stewart, R. L. Tyndall, J. E. Caton, C.-H. Ho, K. S. Ironside, W. M. Caldwell, and E. Tan.** 1993. Chemical and toxicological testing of composted explosives-contaminated soil. *Environ. Toxicol. Chem.* **12:**1105–1116.

30. **Griest, W. H., R. L. Tyndall, A. J. Stewart, J. E. Caton, A. A. Vass, C.-H. Ho, and W. M. Caldwell.** 1994. Chemical characterization and toxicological testing of windrow composts from explosives-contaminated sediments. *Environ. Toxicol. Chem.* **14:**51–59.

31. **Groenewegen, P. E. J., and J. A. M. de Bont.** 1992. Degradation of 4-nitrobenzoate via 4-hydroxylaminobenzoate and 3,4-dihydroxybenzoate in *Comamonas acidovorans* NBA-10. *Arch. Microbiol.* **158:**381–386.

32. **Haigler, B. E., S. F. Nishino, and J. C. Spain.** 1994. Biodegradation of 4-methyl-5-nitrocatechol by *Pseudomonas* sp. strain DNT. *J. Bacteriol.* **176:**3433–3437.

33. **Haigler, B. E., and J. C. Spain.** 1993. Biodegradation of 4-nitrotoluene by *Pseudomonas* sp. strain 4NT. *Appl. Environ. Microbiol.* **59:**2239–2243.

34. **Haigler, B. E., W. H. Wallace, and J. C. Spain.** 1994. Biodegradation of 2-nitrotoluene by *Pseudomonas* sp. strain JS42. *Appl. Environ. Microbiol.* **60:**3466–3469.

35. **Hampton, M. L., and W. E. Sisk.** 1995. Field demonstration of soil slurry bioreactor technology for the remediation of explosives-contaminated soils. *In Platform Abstracts of the Third International Symposium In Situ and On-Site Bioreclamation.* Battelle Press, Columbus, Ohio.

36. **Harder, W.** 1981. Enrichment and characterization of degrading organisms, p. 77–96. *In* T. Leisinger, A. M. Cook, R. Hütter, and J. Nüesch (ed.), *Microbial Degradation of Xenobiotics and Recalcitrant Compounds.* Academic Press, London.

37. **Heitkamp, M. A., V. Camel, T. J. Reuter, and W. J. Adams.** 1990. Biodegradation of p-nitrophenol in an aqueous waste stream by immobilized bacteria. *Appl. Environ. Microbiol.* **56:**2967–2973.

38. **Holt, J. G., and N. R. Krieg.** 1994. Enrichment and isolation, p. 179–215. *In* P. Gerhardt, R. G. E. Murray, W. A. Wood, and N. R. Krieg (ed.), *Methods for General and Molecular Bacteriology.* American Society for Microbiology, Washington, D.C.

39. **Jain, R. K., J. H. Dreisbach, and J. C. Spain.** 1994. Biodegradation of p-nitrophenol via 1,2,4-benzenetriol by an *Arthrobacter* sp. *Appl. Environ. Microbiol.* **60:**3030–3032.

40. **Kaake, R. H., D. J. Roberts, T. O. Stevens, R. L. Crawford, and D. L. Crawford.** 1992. Bioremediation of soils contaminated with the herbicide 2-sec-butyl-4,6-dinitrophenol (dinoseb). *Appl. Environ. Microbiol.* **58:**1683–1689.

41. **Kellogg, S. T., D. K. Chatterjee, and A. M. Chakrabarty.**

1981. Plasmid-assisted molecular breeding: new technique for enhanced biodegradation of persistent toxic chemicals. *Science* **214:**1133–1135.

42. **Kitts, C. L., J. P. Lapointe, V. T. Lam, and R. A. Ludwig.** 1992. Elucidation of the complete *Azorhizobium* nicotinate catabolism pathway. *J. Bacteriol.* **174:**7791–7797.

43. **Lee, M. D., J. T. Wilson, and C. H. Ward.** 1987. In situ restoration techniques for aquifers contaminated with hazardous wastes. *J. Hazard. Mater.* **14:**71–82.

44. **Lenke, H., and H.-J. Knackmuss.** 1992. Initial hydrogenation during catabolism of picric acid by *Rhodococcus erythropolis* HL 24-2. *Appl. Environ. Microbiol.* **58:**2933–2937.

45. **Lewis, R. F.** 1993. SITE demonstration of slurry-phase bidegradation of PAH contaminated soil. *Air Waste* **43:**503–508.

46. **Marvin-Sikkema, F. D., and J. A. M. de Bont.** 1994. Degradation of nitroaromatic compounds by microorganisms. *Appl. Microbiol. Biotechnol.* **42:**499–507.

47. **McCormick, N. G., F. F. Feeherry, and H. S. Levinson.** 1976. Microbial transformation of 2,4,6-trinitrotoluene and other nitroaromatic compounds. *Appl. Environ. Microbiol.* **31:**949–958.

48. **Meulenberg, R., and J. A. M. de Bont.** 1995. Microbial production of catechols from nitroaromatic compounds, p. 37–52. *In* J. C. Spain (ed.), *Biodegradation of Nitroaromatic Compounds.* Plenum Publishing Corp., New York.

49. **Michels, J., and G. Gottschalk.** 1994. Inhibition of lignin peroxidase of *Phanerochaete chrysosporium* by hydroxylamino-dinitrotoluene, an early intermediate in the degradation of 2,4,6-trinitrotoluene. *Appl. Environ. Microbiol.* **60:**187–194.

50. **Morita, R. Y.** 1988. Bioavailability of energy and its relationship to growth and starvation survival in nature. *Can. J. Microbiol.* **34:**436–441.

51. **Nadeau, L. J., and J. C. Spain.** 1995. Bacterial degradation of *m*-nitrobenzoic acid. *Appl. Environ. Microbiol.* **61:**840–843.

52. **Nikolova, P., and O. P. Ward.** 1993. Whole cell biocatalysis in nonconventional media. *J. Ind. Microbiol.* **12:**76–86.

53. **Nishino, S. F., and J. C. Spain.** 1992. Initial steps in the bacterial degradation of 1,3-dinitrobenzene, abstr. Q-135, p. 358. *In Abstracts of the 92nd General Meeting of the American Society for Microbiology 1992.* American Society for Microbiology, Washington, D.C.

54. **Nishino, S. F., and J. C. Spain.** 1993. Degradation of nitrobenzene by a *Pseudomonas pseudoalcaligenes. Appl. Environ. Microbiol.* **59:**2520–2525.

55. **Nishino, S. F., and J. C. Spain.** 1995. Oxidative pathway for the biodegradation of nitrobenzene by *Comamonas* sp. strain JS765. *Appl. Environ. Microbiol.* **61:**2308–2313.

56. **Nishino, S. F., and J. C. Spain.** Unpublished results.

57. **Northrop, D. M., D. E. Martire, and W. A. MacCrehan.** 1991. Separation and identification of organic gunshot and explosive constituents by micellar electrokinetic capillary electrophoresis. *Anal. Chem.* **63:**1038–1042.

58. **Parke, D., and L. N. Ornston.** 1984. Nutritional diversity of *Rhizobiaceae* revealed by auxanography. *J. Gen. Microbiol.* **130:**1743–1750.

59. **Preuss, A., J. Fimpel, and G. Diekert.** 1993. Anaerobic transformation of 2,4,6-trinitrotoluene (TNT). *Arch. Microbiol.* **159:**345–353.

60. **Rafii, F., W. Franklin, R. H. Heflich, and C. E. Cerniglia.** 1991. Reduction of nitroaromatic compounds by anaerobic bacteria isolated from the human gastrointestinal tract. *Appl. Environ. Microbiol.* **57:**962–968.

61. **Rezessy-Szabó, J. M., G. N. M. Huijberts, and J. A. M. de Bont.** 1987. Potential of organic solvents in cultivating micro-organisms on toxic water-insoluble compounds, p.

295–301. *In* C. Laane, J. Tramper, and M. D. Lilly (ed.), *Biocatalysis in Organic Media.* Elsevier Science Publishers, Amsterdam.

62. **Rhys-Williams, W., S. C. Taylor, and P. A. Williams.** 1993. A novel pathway for the catabolism of 4-nitrotoluene by *Pseudomonas. J. Gen. Microbiol.* **139:**1967–1972.

63. **Rieger, P.-G., and H.-J. Knackmuss.** 1995. Basic knowledge and perspectives on biodegradation of 2,4,6-trinitrotoluene and related nitroaromatic compounds in contaminated soil, p. 1–18. *In* J. C. Spain (ed.), *Biodegradation of Nitroaromatic Compounds.* Plenum Publishing Corp., New York.

64. **Schnell, S., and B. Schink.** 1991. Anaerobic aniline degradation via reductive deamination of 4-aminobenzoyl-CoA in *Desulfobacterium anilini. Arch. Microbiol.* **155:**183–190.

65. **Schnell, S., and B. Schink.** 1992. Anaerobic degradation of 3-aminobenzoate by a newly isolated sulfate reducer and a methanogenic enrichment culture. *Arch. Microbiol.* **158:**328–334.

66. **Sims, R. C.** 1990. Soil remediation techniques at uncontrolled hazardous waste sites: a critical review. *J. Air Waste Manage. Assoc.* **40:**704–732.

67. **Spain, J. C.** 1990. Microbial adaptation in aquatic ecosystems, p. 181–190. *In* K. D. Racke and J. R. Coats (ed.), *Enhanced Biodegradation of Pesticides in the Environment.* American Chemical Society, Washington, D.C.

68. **Spain, J. C.** 1995. Biodegradation of nitroaromatic compounds. *Annu. Rev. Microbiol.* **49:**523–555.

69. **Spain, J. C. (ed.).** 1995. *Biodegradation of Nitroaromatic Compounds.* Plenum Publishing Corp., New York.

70. **Spain, J. C., and D. T. Gibson.** 1991. Pathway for biodegradation of *p*-nitrophenol in a *Moraxella* sp. *Appl. Environ. Microbiol.* **57:**812–819.

71. **Spain, J. C., and P. A. Van Veld.** 1983. Adaptation of natural microbial communities to degradation of xenobiotic compounds: effects of concentration, exposure time and chemical structure. *Appl. Environ. Microbiol.* **45:**428–435.

72. **Spain, J. C., P. A. Van Veld, C. A. Monti, P. H. Pritchard, and C. R. Cripe.** 1984. Comparison of *p*-nitrophenol biodegradation in field and laboratory test systems. *Appl. Environ. Microbiol.* **48:**944–950.

73. **Spain, J. C., O. Wyss, and D. T. Gibson.** 1979. Enzymatic oxidation of *p*-nitrophenol. *Biochem. Biophys. Res. Commun.* **88:**634–641.

74. **Spanggord, R. J., J. C. Spain, S. F. Nishino, and K. E. Mortelmans.** 1991. Biodegradation of 2,4-dinitrotoluene by a *Pseudomonas* sp. *Appl. Environ. Microbiol.* **57:**3200–3205.

75. **Sublette, K. L., E. V. Ganapathy, and S. Schwartz.** 1992. Degradation of munitions waste by *Phanerochaete chrysosporium. Appl. Biochem. Biotechnol.* **34/35:**709–723.

76. **Tiehm, A.** 1994. Degradation of polycyclic aromatic hydrocarbons in the presence of synthetic surfactants. *Appl. Environ. Microbiol.* **60:**258–263.

77. **Timmis, K. N., R. J. Steffan, and R. Unterman.** 1994. Designing microorganisms for the treatment of toxic wastes. *Annu. Rev. Microbiol.* **48:**525–557.

78. **Valli, K., B. J. Brock, D. K. Joshi, and M. H. Gold.** 1992. Degradation of 2,4-dinitrotoluene by the lignin-degrading fungus *Phanerochaete chrysosporium. Appl. Environ. Microbiol.* **58:**221–228.

79. **Venkatadri, R., and R. L. Irvine.** 1993. Cultivation of *Phanerochaete chrysosporium* and production of lignin peroxidase in novel biofilm reactor systems: hollow fiber reactor and silicone membrane reactor. *Water Res.* **27:**591–596.

80. **Villanueva, J. R.** 1964. Nitro-reductase from a *Nocardia* sp. *Antonie van Leeuwenhoek* **30:**17–32.

81. **Vorbeck, C., H. Lenke, P. Fischer, and H.-J. Knackmuss.** 1994. Identification of a hydride-Meisenheimer complex as a metabolite of 2,4,6-trinitrotoluene by a *Mycobacterium* strain. *J. Bacteriol.* **176:**932–934.

82. **Wiggins, B. A., and M. Alexander.** 1988. Role of chemical concentration and second carbon sources in acclimation of microbial communities for biodegradation. *Appl. Environ. Microbiol.* **54:**2803–2807.

83. **Williams, R. T., P. S. Ziegenfuss, and W. E. Sisk.** 1992. Composting of explosives and propellant contaminated soils under thermophilic and mesophilic conditions. *J. Ind. Microbiol.* **9:**137–144.

84. **Wolfaardt, G. M., J. R. Lawrence, M. J. Hendry, R. D. Robarts, and D. E. Caldwell.** 1993. Development of steady-state diffusion gradients for the cultivation of degradative microbial consortia. *Appl. Environ. Microbiol.* **59:** 2388–2396.

85. **Zappi, M. E., D. Gunnison, and H. L. Fredrickson.** 1995. Aerobic treatment of TNT contaminated soils using two engineering approaches, p. 281–287. *In* R. E. Hinchee, R. E. Hoeppel, and B. C. Alleman (ed.), *Bioremediation of Recalcitrant Organics.* Battelle Press, Columbus, Ohio.

86. **Zeyer, J., and P. C. Kearney.** 1984. Degradation of *o*-nitrophenol and *m*-nitrophenol by a *Pseudomonas putida*. *J. Agric. Food Chem.* **32:**238–242.

Biodegradation of Halogenated Solvents

LAWRENCE P. WACKETT

86

SIGNIFICANCE AND BIODEGRADABILITY OF CHLORINATED SOLVENTS

Chlorinated solvents have been used industrially in vast quantities as degreasing agents, heat transfer agents, and chemical intermediates in synthesis. Their usage has decreased because of public health concerns, but they are still widely used. Most heavily in use are C_1 compounds plus C_2 alkanes and alkenes containing chlorine and, in some cases, fluorine substituents in place of hydrogen atoms (Fig. 1). Most notable with respect to negative environmental impact are chlorofluorocarbons, implicated in ozone layer destruction (39), and CCl_4 and vinyl chloride, which have been demonstrated to be carcinogenic in mammals (35).

The biodegradability of the compounds shown in Fig. 1 varies widely, but a general rule is that biodegradability decreases with increasing halogen content. This reflects the relatively low chemical reactivity of polyhalogenated compounds and the necessity for multiple dehalogenation mechanisms for complete metabolism of the compound. In general, chlorinated alkenes are less biodegradable than the corresponding chloroalkanes.

It is sometimes erroneously stated that low-molecular-weight chlorinated compounds are only of human origin when, in fact, chlorinated natural products are found ubiquitously. For example, chloromethane is made in significant quantities by soil fungi (66). However, CCl_4 and chlorofluorocarbons are principally, if not exclusively, of anthropogenic origin. In this context, these latter compounds have been in the environment for only a short time on an evolutionary time scale, and this may reflect the rarity of microorganisms that can metabolize them.

MEASURING CHLORINATED SOLVENT BIODEGRADATION

An effective study on the microbiology of chlorinated solvent biodegradation requires proper handling and measurement of the compounds. In general, chlorinated solvents are highly volatile, necessitating the use of septum-sealed glass vials, or similar gastight containers, for growing cells or assessing the extent of metabolism. It is recommended that Teflon-lined rubber septa be used, as these compounds can dissolve in rubber, leading to experimental artifacts. The experimenter must also be cognizant of the somewhat low water solubility of the compounds and the time required for chlorinated solvents to reach equilibrium in dissolving. For example, erratic measurements of trichloroethene (TCE) concentrations in water led us to prepare 8 mM stock solutions of TCE in water and shake them overnight prior to the experiment. The appropriate volume of TCE was added by syringe to a Teflon septum-sealed glass vial filled with water and no headspace. The exclusion of a gas phase was important because of the relatively high Henry's law coefficient of chlorinated solvents, a quantitative measurement of their volatility. Chlorinated solvents readily partition between gas and liquid phases; thus use of an all-liquid-phase stock solution removes ambiguities in the amount of solvent transferred.

Gas chromatography is the method of choice for quantitative determination of chlorinated solvents. There are many suitable packed or capillary gas chromatography columns, and the choice can be made by consulting manufacturers' catalogs or literature examples. Electron capture detectors are very sensitive in their response to halogenated compounds. Electron capture detection is superior for polychlorinated compounds, but flame ionization detection is excellent for compounds such as chloromethane and vinyl chloride.

It is very important to complement measurements of chlorinated substrate disappearance with quantitative determination of products formed. The organic products may or may not be readily demonstrable in in vivo experiments. For example, if the substrate is serving as a carbon and energy source, then the carbon is assimilated. However, the halide ions will be released during metabolism and can be recovered stoichiometrically in a liquid medium. Fluoride may be determined with a fluoride-specific electrode; chloride and bromide may be determined by a colorimetric assay (6). It is necessary to ensure that the growth or incubation medium has a low background level of halide ions. For example, chloride is found in many salts and reagents. The colorimetric halide assay has also been performed on agar plates to screen clones expressing dehalogenase activity (29). In many cases, an organism or an enzyme does not assimilate or completely metabolize a chlorinated substrate, and a detectable organic product(s) is formed. Since the nature of these transformations is varied, it is impossible to give overall guidelines for organic product detection. To give one example, the oxidation of [^{14}C]TCE by *Pseudomonas putida*

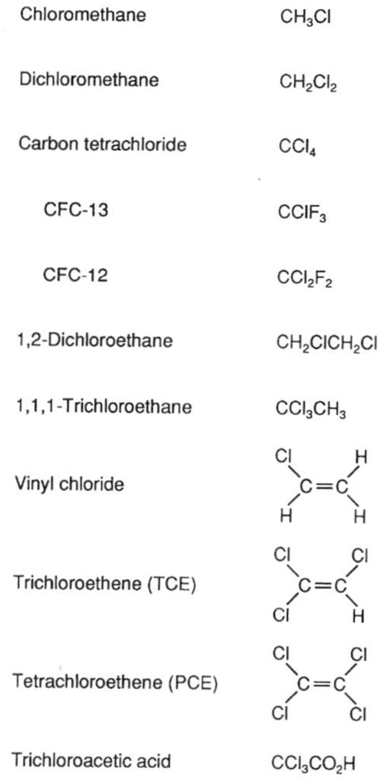

Chloromethane	CH₃Cl
Dichloromethane	CH₂Cl₂
Carbon tetrachloride	CCl₄
CFC-13	CClF₃
CFC-12	CCl₂F₂
1,2-Dichloroethane	CH₂ClCH₂Cl
1,1,1-Trichloroethane	CCl₃CH₃

Chloromethane — CH_3Cl

Dichloromethane — CH_2Cl_2

Carbon tetrachloride — CCl_4

CFC-13 — $CClF_3$

CFC-12 — CCl_2F_2

1,2-Dichloroethane — CH_2ClCH_2Cl

1,1,1-Trichloroethane — CCl_3CH_3

Vinyl chloride —

$$\underset{H}{\overset{Cl}{\diagdown}}C=C\underset{H}{\overset{H}{\diagup}}$$

Trichloroethene (TCE) —

$$\underset{Cl}{\overset{Cl}{\diagdown}}C=C\underset{H}{\overset{Cl}{\diagup}}$$

Tetrachloroethene (PCE) —

$$\underset{Cl}{\overset{Cl}{\diagdown}}C=C\underset{Cl}{\overset{Cl}{\diagup}}$$

Trichloroacetic acid — CCl_3CO_2H

FIGURE 1 Common names and chemical structures of representative halogenated solvents that are most widespread environmentally and most heavily studied in microbiological experiments. CFC, chlorofluorocarbon.

F1 or the responsible enzyme system, toluene dioxygenase, yields formic and glyoxylic acid that can be quantitatively determined by a high-pressure liquid chromatograph fitted with an organic acid column and radiochemical detection of eluting peaks (33).

MICROORGANISMS THAT METABOLIZE CHLORINATED SOLVENTS

Microorganisms that metabolize chlorinated solvents have been obtained (i) by enrichment culturing, (ii) from bioreactors established to degrade wastes, or (iii) from the ranks of known bacteria isolated for other purposes. An example of the latter is the biodegradation of hard-to-grade pollutants such as TCE or hydrochlorofluorocarbons that can be fortuitously oxidized by bacteria containing nonspecific oxygenases. To obtain new organisms, the enrichment culturing methods are established in consonance with the type of compound or biodegradation reaction desired. In general, perhalogenated compounds would be most likely metabolized by an initial reductive dehalogenation reaction by an anaerobic bacterium or consortium. A less halogenated compound such as vinyl chloride or chloromethane would more likely be metabolized via oxygenative mechanisms by aerobic bacteria. However, chloromethane and dichloromethane are known to be metabolized by both aerobic and anaerobic bacteria. This can be understood because both compounds may be dechlorinated by nucleophilic substitution mechanisms that can be operative in either aerobic or anaerobic bacteria (61).

Bacteria metabolizing chlorinated solvents can be grouped into three classes that require different conditions for supporting optimum dehalogenation activity: (i) those using chlorinated solvents as their sole carbon and energy sources, (ii) those using chlorinated solvents as their final electron acceptors, and (iii) those catalyzing fortuitous oxidative or reductive dehalogenation reactions. The discussion below is not an exhaustive treatment of all known bacteria, but illustrative examples are given for each of these three classes.

In general, the less chlorinated compounds are metabolized for carbon and energy because of the greater available energy derived from oxidation reactions (Table 1). This metabolism is not restricted to aerobic bacteria. Two different anaerobic acetogenic bacteria are known to grow on chloromethane (53) and dichloromethane (50). They carry out a net oxidation of the C_1 compounds that gives rise to the carboxyl group of acetate. The dichloromethane-fermenting organism forms a stable syntrophic association with other anaerobes.

Different aerobic bacteria that grow on C_1 and C_2 compounds containing two or fewer chlorine substituents have been identified. It is theoretically possible for a bacterium to grow on chloroform, TCE, and trichloroacetic acid, but well-documented examples do not exist at this time. A report of bacterial growth on TCE has not been reproduced, and the amount of TCE transformed in that study would not theoretically support significant growth (54).

Aerobic bacteria that metabolize chloromethane (24) and dichloromethane (47, 51) as their sole carbon and energy sources were obtained by enrichment culture. With both substrates, the bacteria obtained were methylotrophs and capable of growth on nonchlorinated C_1 compounds. However, many methylotrophs do not grow on chloromethane or dichloromethane. All methylotrophs yet characterized for growth on dichloromethane contain one of two distinct but evolutionarily related dichloromethane dehalogenases (4, 32, 47). The product of the dehalogenation reaction is formaldehyde, which can be oxidized to yield ATP and assimilated into cell carbon (Fig. 2).

A number of bacteria have been isolated for their ability to grow on larger chlorinated alkanes (5, 28, 30, 46, 48, 55). The largest-volume industrial chlorinated organic compound is 1,2-dichloroethane, and bacteria utilizing this compound have been extensively studied (44). A metabolic pathway in which 1,2-dichloroethane goes to 2-chloroethanol and 2-chloroacetate has been delineated (Fig. 2). The latter compound undergoes a second hydrolytic dechlorination reaction. The enzyme catalyzing the initial dechlorination of 1,2-dichloroethane has been crystallized, its structure has been solved to atomic resolution, and mechanistic details have been elucidated (57).

Bacteria that metabolize 2-haloacetates are long known

TABLE 1 Chlorinated solvents and organisms using them as the sole source of carbon and energy

Compounds	Organisms	Aerobic or anaerobic	Reference(s)
CH_3Cl	Methylotrophs	Aerobic	24
	Acetogens	Anaerobic	53
CH_2Cl_2	Methylotrophs	Aerobic	47, 50, 51
	Acetogens	Anaerobic	
CH_2ClCH_2Cl	*Xanthomonas* spp.	Aerobic	44
$CH_2{=}CHCl$	*Mycobacterium* spp.	Aerobic	23
$ClCH_2CO_2H$	Various	Aerobic	49
$Cl_2CH_2CO_2H$	Various	Aerobic	63

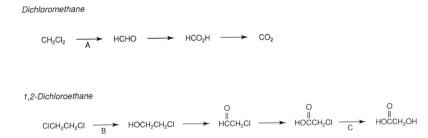

FIGURE 2 Bacterial metabolic pathways delineated for the aerobic metabolism of dichloromethane and 1,2-dichloroethane. Enzymatic steps in which C-Cl bonds are cleaved are highlighted by the letters A (dichloromethane dehalogenase), B (haloalkane dehalogenase), and C (chloroacetate halidohydrolase).

(19, 31), and numerous strains have been identified in the last 30 years (45, 49, 56). Some of those bacteria metabolize fluoroacetate (18), a natural product made by certain plants to be highly lethal to predatory animals (21). Other bacteria metabolize only haloacetates that are more chemically reactive than fluoroacetate; 2-chloroacetate is the substrate typically used. A subset of 2-chloroacetate-metabolizing organisms are able to metabolize 2,2-dichloroacetate (63). The metabolism of trichloroacetate is rare but has been reported (64).

Of the chloroethenes, only vinyl chloride, a major commodity chemical, is currently known to be a carbon and energy source for isolated bacterial strains. The bacteria are all *Mycobacterium* sp. isolates and are thought to use an oxygenase for initial metabolism of vinyl chloride (23).

A relatively recent, and very exciting, finding has been the identification of anaerobic bacteria that reduce chloroethenes in reactions apparently linked to ATP formation. Thus, tetrachloroethene and TCE serve as final electron acceptors and become reduced to less chlorinated products (25, 26, 40, 52). The use of chlorinated compounds as the final electron acceptor was previously established with *Desulfomonile tiedjei*, which reduces 3-chlorobenzoate to benzoate (38). Microbial reductive dechlorination occurs with a range of other compounds; for example, it has been well established that polychlorinated biphenyls undergo biologically catalyzed reduction (1, 8). The extent to which this reaction and other reactions observed with anaerobic consortia are energy linked is not rigorously established at present.

Enrichment culture techniques have failed to isolate bacteria that carry out well-documented energy-linked metabolism with some of the most environmentally widespread chlorinated solvents such as carbon tetrachloride and chloroform. As described above, TCE and tetrachloroethene are metabolized with concomitant ATP formation, but only by strictly anaerobic bacteria. To fill this void, aerobic bacteria have been identified that can oxidize chloroform and TCE when supplied with other organic compounds that support growth and, in some cases, induce the relevant enzyme system. The enzymes identified to cometabolize TCE are invariably nonspecific oxygenases. They are biosynthesized by different bacteria for the oxidation of methane (16), propane (58), propene (12), isoprene (13), phenol (22), toluene (15, 59, 65), isopropylbenzene (10), and ammonia (3). With soluble methane monooxygenase (16) and toluene dioxygenase (33), the oxidation of TCE has been investigated with purified enzyme components. These studies, and complementary in vivo experiments (2, 60), established that

oxidative cometabolism of TCE can generate reactive metabolites that are toxic to cells. This remains a major impediment to practical aerobic treatment of TCE when toxic effects limit the metabolic activity and survivability of the bacteria that are used in the process. A further knowledge of the reactions catalyzed by a wide spectrum of oxygenases might lead to the development of oxidative systems capable of counteracting toxic effects and, perhaps, assimilating TCE carbon into cell molecules.

Genetic engineering is beginning to be used for developing improved strains for chlorinated solvent biodegradation in three general ways: (i) to overcome problems of expressing the solvent-metabolizing enzyme, (ii) by protein engineering to obtain enzymes improved for chlorinated substrates, and (iii) by metabolic pathway engineering to combine genes that nature may not have spliced together via natural evolutionary processes. There are examples of combining TCE-oxidizing genes under control of a starvation-induced promoter (37) and generating mutants derepressed to make soluble methane monooxygenase, which rapidly oxidizes TCE, under a wider range of growth conditions (14). Variants of haloalkane dehalogenase, originally obtained from bacteria metabolizing 1,2-dichloroethane, are being generated with new substrate specificities. One approach used spontaneous mutation and a selective growth medium to obtain mutant enzymes more active with larger substrates (43). Site-directed mutagenesis experiments are also being conducted (44). Another series of protein engineering experiments use cytochrome $P450_{CAM}$ monooxygenase, an enzyme that apparently evolved for reactivity with nonchlorinated, natural product cyclic terpene substrates such as camphor (7). Molecular dynamics simulations have been performed with the wild-type cytochrome $P450_{CAM}$, and the data have been used to propose the generation of site-directed mutant enzymes that might be improved for binding and reacting with various chlorinated compounds (41, 42). Currently, a prediction has been made, using computational methods, that a double mutant of cytochrome $P450_{CAM}$ may show improved binding and catalysis with 1,1,1-trichloroethane, an important environmental pollutant (36).

Lastly, dechlorinating enzymes may be combined, even in unexpected ways, to metabolize polychlorinated compounds. Chloroform and carbon tetrachloride are targets for such an approach. In a recent study, pentachloroethane was a model compound shown to be completely dechlorinated via an engineered metabolic pathway (62). The recombinant bacterium, *P. putida* G786(pHG-2), metabolized pentachloroethane by sequential reductive and oxygenative dechlorination reactions. Pentachloroethane was reduced to TCE by cytochrome P450$_{CAM}$, and TCE was oxidized by the toluene dioxygenase enzyme system, from cloned *todABC$_1$C$_2$* genes, to glyoxylate and formate. *P. putida* G786(pHG-2) oxidized formate to CO$_2$. The engineered strain metabolizes a range of halogenated aliphatic compounds, including chlorofluorocarbons (27).

The most novel feature of the engineered pathway is that cytochrome P450$_{CAM}$ (20) and toluene dioxygenase (17) are known for catalyzing reactions other than dehalogenation. In the last few years, their ability to catalyze fortuitous dehalogenation reactions has been studied (9, 33, 34). This led to recruitment of their respective genes to engineer a dehalogenation pathway. It should be possible to engineer bacteria to metabolize other chlorinated compounds of environmental concern. This is being facilitated by the development of the University of Minnesota Biocatalysis/Biodegradation Database (UM-BBD), a computer resource currently accessible on the World Wide Web (URL = http://dragon.labmed.umn.edu/lynda/ [11]). The UM-BBD has information on microbial catabolic pathways, chemical compounds in those pathways, the organisms that transform the compounds, enzymes, and genes. This information is vital for optimal selection of enzymes and genes to design new pathways for metabolizing the most environmentally persistent chlorinated solvents.

Nature has designed an impressive array of approaches for metabolizing chlorinated solvents, but some compounds still largely resist microbial metabolism. Genetic techniques are becoming increasingly facile, information on catabolic genes and enzymes is increasing exponentially, and the means of cataloging that information are improving markedly with the advent of the World Wide Web. In this context, the rational design of metabolic pathways to handle chlorinated solvents will be used increasingly to augment the collection of bacteria that nature gives us by natural selection.

I acknowledge the many coworkers who have contributed to our knowledge of halogenated solvent biodegradation. I also thank Bonnie Allen for assistance in preparing the manuscript.

The research in my laboratory was supported by National Institutes of Health grant GM 41235 and Air Force Office of Scientific Research and Environmental Protection Agency grant CR820771-01-0.

REFERENCES

1. **Abramowicz, D. A.** 1990. Aerobic and anaerobic biodegradation of PCBs: a review. *Crit. Rev. Biotechnol.* **10:**241–251.
2. **Alvarez-Cohen, L., and P. L. McCarty.** 1991. Product toxicity and cometabolic competitive inhibition modeling of chloroform and trichloroethylene transformation by methanotrophic resting cells. *Appl. Environ. Microbiol.* **57:**1031–1037.
3. **Arciero, D., T. Vannelli, M. Logan, and A. B. Hooper.** 1989. Degradation of trichloroethylene by the ammonia-oxidizing bacterium Nitrosomonas europeae. *Biochem. Biophys. Res. Commun.* **159:**640–643.
4. **Bader, R., and T. Leisinger.** 1994. Isolation and characterization of the *Methylophilus* sp. strain DM11 gene encoding dichloromethane dehalogenase/glutathione S-transferase. *J. Bacteriol.* **176:**3466–3473.
5. **Belkin, S.** 1992. Biodegradation of haloalkanes. *Biodegradation* **3:**299–313.
6. **Bergmann, J. G., and J. Sanik.** 1957. Determination of trace amounts of chlorine in naphtha. *Anal. Chem.* **29:**241–243.

7. **Bradshaw, W. H., H. E. Conrad, E. J. Corey, I. C. Gunsalus, and D. Lednicer.** 1959. Microbial degradation of (+)-camphor. *J. Am. Chem. Soc.* **81:**5507.
8. **Brown, J. F., D. L. Bedard, M. J. Brennan, J. C. Carnahan, H. Feng, and R. E. Wagner.** 1987. Polychlorinated biphenyl dechlorination in aquatic sediments. *Science* **236:**709–712.
9. **Castro, C. E., R. S. Wade, and N. O. Belser.** 1985. Biodehalogenation: reactions of cytochrome P450 with polyhalomethanes. *Biochemistry* **24:**204–210.
10. **Dabrock, B., M. Kesseler, B. Averhoff, and G. Gottschalk.** 1994. Identification and characterization of a transmissable linear plasmid from *Rhodococcus erythropolis* BD2 that encodes isopropylbenzene and trichloroethene catabolism. *Appl. Environ. Microbiol.* **60:**853–860.
11. **Ellis, L. B. M., and L. P. Wackett.** 1995. A microbial biocatalysis database. *Soc. Ind. Microbiol. News* **45:**167–173.
12. **Ensign, S. A., M. R. Hyman, and D. J. Arp.** 1992. Cometabolic degradation of chlorinated alkenes by alkene monooxygenase in a propylene-grown *Xanthobacter* strain. *Appl. Environ. Microbiol.* **58:**3038–3046.
13. **Ewers, J., D. Freirer-Schröder, and H.-J. Knackmuss.** 1990. Selection of trichloroethane (TCE) degrading bacteria that resist inactivation by TCE. *Arch. Microbiol.* **154:**410–413.
14. **Fitch, M. W., D. W. Graham, R. G. Arnold, S. K. Agarwal, P. Phelps, G. E. Speitel, and G. Georgiou.** 1993. Phenotypic characterization of copper resistant mutants of *Methylosinus trichosporium* OB3b. *Appl. Environ. Microbiol.* **59:**2771–2776.
15. **Folsom, B. R., P. J. Chapman, and P. H. Pritchard.** 1990. Phenol and trichloroethylene degradation by *Pseudomonas cepacia* G4: kinetics and interactions between substrates. *Appl. Environ. Microbiol.* **56:**1279–1285.
16. **Fox, B. G., J. G. Bornemann, L. P. Wackett, and J. D. Lipscomb.** 1990. Haloalkane oxidation by the soluble methane monooxygenase from *Methylsinus trichosporium* OB3b: mechanistic and environmental implications. *Biochemistry* **29:**6419–6427.
17. **Gibson, D. T., G. J. Zylstra, and S. Chauhan.** 1990. Biotransformations catalyzed by toluene dioxygenase from *Pseudomonas putida* F1, p. 121–132. *In* S. Silver, A. M. Chakrabarty, B. Iglewski, and S. Kaplan (ed.), *Pseudomonas: Biotransformation, Pathogenesis, and Emerging Biotechnology.* ASM Press, Washington, D.C.
18. **Goldman, P.** 1965. The enzymatic cleavage of the carbon-fluorine bond in fluoroacetate. *J. Biol. Chem.* **240:**3434–3438.
19. **Goldman, P., G. W. A. Milne, and K. A. Keister.** 1968. Carbon-halogen bond cleavage: studies on bacterial halidohydrolases. *J. Biol. Chem.* **243:**428–434.
20. **Gunsalus, I. C., and G. C. Wagner.** 1978. Bacterial P450$_{CAM}$ methylene monooxygenase components: cytochrome m, putidaredoxin and putidaredoxin reductase. *Methods Enzymol.* **52:**166–188.
21. **Harborne, J. B.** 1988. *Introduction to Ecological Biochemistry.* Academic Press, London.
22. **Harker, A. R., and Y. Kim.** 1990. Trichloroethylene degradation by two independent aromatic-degrading pathways in *Alcaligenes eutrophus* JMP134. *Appl. Environ. Microbiol.* **56:**1179–1181.
23. **Hartmans, S., and J. A. M. de Bont.** 1992. Aerobic vinyl chloride metabolism in *Mycobacterium aurum* L1. *Appl. Environ. Microbiol.* **58:**1220–1226.
24. **Hartmans, S., A. Schmuckle, A. Cook, and T. Leisinger.** 1986. Methyl chloride: naturally occurring toxicant and C1-growth substrate. *J. Gen. Microbiol.* **132:**1139–1142.
25. **Holliger, C., and G. Schraa.** 1994. Physiological meaning

and potential for application of reductive dechlorination by anaerobic bacteria. *FEMS Microbiol. Rev.* **15:**297–305.

26. **Holliger, C., G. Schraa, A. J. M. Stams, and A. J. B. Zehnder.** 1993. A highly purified enrichment culture couples the reductive dechlorination of tetrachloroethene to growth. *Appl. Environ. Microbiol.* **59:**2991–2997.

27. **Hur, H.-G., M. J. Sadowsky, and L. P. Wackett.** 1994. Metabolism of chlorofluorocarbons and polybrominated compounds by *Pseudomonas putida* G786 (pHG-2) via an engineered metabolic pathway. *Appl. Environ. Microbiol.* **60:**4148–4158.

28. **Janssen, D. B., D. Jager, and B. Witholt.** 1987. Degradation of n-haloalkenes and α,ω-dihaloalkanes by wild-type and mutants of *Acinetobacter* sp. strain GJ70. *Appl. Environ. Microbiol.* **53:**561–566.

29. **Janssen, D. B., F. Pries, J. van der Ploeg, B. Kazemier, P. Terpstra, and B. Witholt.** 1989. Cloning of 1,2-dichloroethane degradation genes of *Xanthobacter autotrophicus* GJ110, and expression and sequencing of the *dhlA* gene. *J. Bacteriol.* **171:**6791–6799.

30. **Janssen, D. B., A. Scheper, L. Dijkhuizen, and B. Witholt.** 1985. Degradation of halogenated aliphatic compounds by *Xanthobacter autotrophicus* GJ10. *Appl. Environ. Microbiol.* **49:**673–677.

31. **Jensen, H. L.** 1957. Decomposition of chlorosubstituted aliphatic acids by soil bacteria. *Can. J. Microbiol.* **3:**151–164.

32. **La Roche, S. D., and T. Leisinger.** 1990. Sequence analysis and expression of the bacterial dichloromethane dehalogenase structural gene, a member of the glutathione S-transferase supergene family. *J. Bacteriol.* **172:**164–171.

33. **Li, S., and L. P. Wackett.** 1992. Trichloroethylene oxidation by toluene dioxygenase. *Biochem. Biophys. Res. Commun.* **185:**443–451.

34. **Li, S., and L. Wackett.** 1993. Reductive dehalogenation by cytochrome P450$_{CAM}$: substrate binding and catalysis. *Biochemistry* **32:**9355–9361.

35. **Maltoni, C., and G. Lefemine.** 1974. Carcinogenicity bioassays of vinylchloride: research plan and early results. *Environ. Res.* **7:**387–396.

36. **Manchester, J. I., and R. L. Ornstein.** 1995. Molecular dynamics simulations indicate that F87W, T185F-cytochrome P450$_{CAM}$ may reductively dehalogenate 1,1,1-trichloroethane. *J. Biomolec. Struct. Dyn.* **13:**413–422.

37. **Matin, A.** 1994. Starvation promoters of Escherichia coli. Their function, regulation, and use in bioprocessing and bioremediation. *Ann. N. Y. Acad. Sci.* **721:**277–291.

38. **Mohn, W. W., and J. M. Tiedje.** 1990. Strain DCB-1 conserves energy for growth from reductive dechlorination coupled to formate oxidation. *Arch. Microbiol.* **153:**267–271.

39. **Molina, M. J., and F. S. Rowland.** 1974. Stratospheric sink for chlorofluoromethanes: chloride atom-catalyzed destruction of ozone. *Nature* (London) **249:**810–812.

40. **Neumann, A., H. Scholz-Muramatsu, and G. Diekert.** 1994. Tetrachloroethene metabolism of *Dehalospirillium multivorans*. *Arch. Microbiol.* **162:**295–301.

41. **Paulsen, M. D., and R. L. Ornstein.** 1992. Predicting the product specificity and coupling of cytochrome P450$_{CAM}$. *J. Comput. Aided Mol. Des.* **6:**2077–2082.

42. **Paulsen, M. D., and R. L. Ornstein.** 1994. Active site mobility inhibits reductive dehalogenation of 1,1,1-trichloroethane by cytochrome P450$_{CAM}$. *J. Comput. Aided Mol. Des.* **8:**389–404.

43. **Pries, F., A. J. vanden Wijngaard, R. Bos, M. Pentenga, and D. B. Janssen.** 1994. The role of spontaneous cap domain mutations in haloalkane dehalogenase specificity and evolution. *J. Biol. Chem.* **269:**17490–17494.

44. **Pries, F., J. R. van der Ploeg, J. Dolfing, and D. B. Jans-**

sen. 1994. Degradation of halogenated aliphatic compounds: the role of adaption. *FEMS Microbiol. Rev.* **15:**279–295.

45. **Schneider, B., R. Müller, R. Frank, and F. Lingens.** 1991. Complete nucleotide sequences and comparison of the structural genes of two 2-haloalkanoic acid dehalogenases from *Pseudomonas* sp. strain CBS3. *J. Bacteriol.* **173:**1530–1535.

46. **Scholtz, R., A. Schmuckle, A. Cook, and T. Leisinger.** 1987. Degradation of eighteen 1-monohaloalkanes by *Arthrobacter* sp. strain HA1. *J. Gen. Microbiol.* **133:**267–274.

47. **Scholtz, R., L. P. Wackett, C. Egli, A. M. Cook, and T. Leisinger.** 1988. Dichloromethane dehalogenase with improved catalytic activity isolated from a fast-growing dichloromethane-utilizing bacterium. *J. Bacteriol.* **170:**5698–5704.

48. **Shochat, E., I. Hermoni, Z. Cohen, A. Abeliovich, and S. Belkin.** 1993. Bromoalkane-degrading *Pseudomonas* strains. *Appl. Environ. Microbiol.* **59:**1403–1409.

49. **Slater, J. H., D. Lovatt, A. J. Weightman, E. Senior, and A. T. Bull.** 1979. The growth of *Pseudomonas putida* on chlorinated aliphatic acids and its dehalogenase activity. *J. Gen. Microbiol.* **114:**125–136.

50. **Stromeyer, S. A., R. Hermann, A. M. Cook, and T. Leisinger.** 1993. Dichloromethane as the sole carbon source for an acetogenic mixed culture and isolation of a fermentative, dichloromethane-degrading bacterium. *Appl. Environ. Microbiol.* **59:**3790–3797.

51. **Stucki, G. R., R. Galli, H. Ebersold, and T. Leisinger.** 1981. Dehalogenation of dichloromethane by cell extracts of *Hyphomicrobium* DM2. *Arch. Microbiol.* **130:**366–370.

52. **Terzenbach, D. P., and M. Blaut.** 1994. Transformation of tetrachloroethylene to trichloroethylene by homoacetogenic bacteria. *FEMS Microbiol. Lett.* **123:**213–218.

53. **Traunecker, J., A. Preub, and G. Diekert.** 1991. Isolation and characterization of a methyl chloride utilizing strictly anaerobic bacterium. *Arch. Microbiol.* **156:**416–421.

54. **Vandenbergh, P. A., and B. S. Kunka.** 1988. Metabolism of volatile chlorinated aliphatic hydrocarbons by *Pseudomonas fluorescens*. *Appl. Environ. Microbiol.* **54:**2578–2579.

55. **Van den Wijngaard, A. J., K. van der Kamp, J. van der Ploeg, B. Kazemier, F. Pries, and D. B. Janssen.** 1992. Degradation of 1,2-dichloroethane by facultative methylotrophic bacteria. *Appl. Environ. Microbiol.* **58:**976–983.

56. **Van der Ploeg, J., G. van Hall, and D. B. Janssen.** 1991. *Xanthobacter autotrophicus* GJ10 and sequencing of the *dhlB* gene. *J. Bacteriol.* **173:**7925–7933.

57. **Verschueren, K. H., F. Seljee, H. J. Rozeboom, K. H. Kalk, and B. W. Dijkstra.** 1993. Crystallographic analysis of the catalytic mechanism of haloalkane dehalogenase. *Nature* (London) **363:**693–698.

58. **Wackett, L. P., G. A. Brusseau, S. R. Householder, and R. S. Hanson.** 1989. Survey of microbial oxygenases: trichloroethylene degradation by propane-oxidizing bacteria. *Appl. Environ. Microbiol.* **55:**2960–2964.

59. **Wackett, L. P., and D. T. Gibson.** 1988. Degradation of trichloroethylene by toluene dioxygenase in whole cell studies with *Pseudomonas putida* F1. *Appl. Environ. Microbiol.* **54:**1703–1708.

60. **Wackett, L. P., and S. R. Householder.** 1989. Toxicity of trichloroethylene to *Pseudomonas putida* F1 is mediated by toluene dioxygenase. *Appl. Environ. Microbiol.* **55:**2723–2725.

61. **Wackett, L. P., M. S. P. Logan, F. A. Blocki, and C. Baoli.** 1992. A mechanistic perspective on bacterial metabolism of chlorinated methanes. *Biodegradation* **3:**19–36.

62. **Wackett, L. P., M. J. Sadowski, L. M. Newman, H.-G. Hur, and S. Li.** 1994. Metabolism of polyhalogenated

compounds by a recombinant bacterium. *Nature* (London) **368:**627–629.

63. **Weightman, A. J., A. L. Weightman, and J. H. Slater.** 1985. Toxic effects of chlorinated and brominated alkanoic acids on *Pseudomonas putida* PP3: selection at high frequencies of mutations in genes encoding dehalogenases. *Appl. Environ. Microbiol.* **49:**1494–1501.

64. **Weightman, A. L., A. J. Weightman, and H. J. Slater.** 1992. Microbial dehalogenation of trichloroacetic acid. *World J. Microbiol. Biotechnol.* **8:**512–518.

65. **Winter, R. B., K.-M. Yen, and B. D. Ensley.** 1989. Efficient degradation of trichloroethylene by a recombinant *Escherichia coli. Bio/Technology* **7:**282–285.

66. **Wuosmaa, A. M., and L. P. Hager.** 1990. Methyl chloride transferase: a carbocation route for the synthesis of halometabolites. *Science* **249:**160–162.

Determination of Anaerobic Biodegradation Activity

JOSEPH M. SUFLITA, KATHLEEN L. LONDRY, AND GLENN A. ULRICH

87

The regulated interlinkage of oxidation and reduction reactions to release energy for biosynthetic purposes is a unifying feature of life. This undoubtedly reflects the fundamental biochemical unity of diverse life forms that share a common ancestry. It follows, then, that the tools to dissect and understand these metabolic interlinkages are applicable to the study of many types of organisms. It is also not surprising then, that the same rigorous criteria used to establish aerobic biodegradation mechanisms can also be applied to the study of anaerobic bioconversions. This chapter will explore these criteria and their interpretational limits. The principles herein are independent of the type of anaerobe or consortium catalyzing the transformation, the habitat under consideration, the particular type of substrate, or whether the metabolism is catalyzed in the field or in the laboratory. Precisely because metabolism is such a unifying feature of life, many of the criteria used to support the case for biodegradation can often be extrapolated from laboratory to field studies.

As is the case in the study of aerobic metabolism, there are many criteria that collectively argue for or against the prospects for anaerobic biodegradation. However, it is important to note that there is no single piece of evidence that by itself can definitively establish the role of anaerobes in any given transformation. Multiple lines of evidence are needed before a convincing argument to this effect can be made. Of course, simply garnering this type of evidence without reference to appropriate controls makes the resulting information useless. The appropriate controls include sterile, substrate-unamended, uninoculated, and ideally positive controls. While all such controls can and should be employed in laboratory investigations, suitable alterations for field studies are often required.

INITIAL ASSAYS OF ANAEROBIC BIODEGRADATION

To investigate the susceptibility of compounds to anaerobic decay, a screening protocol is typically an initial step. Such procedures are usually designed to quantify the end product of the reaction or the consumption of terminal electron acceptors. The results from such bioassays are compared with those from background controls and interpreted relative to the expected amount of end product formed or electron acceptor consumed, should complete mineralization of the parent substrate occur. These screening methods are reliable, inexpensive, and subject to standardization and provide a useful indication of the susceptibility of chemicals to anaerobic decay.

An example of a screening protocol is given in chapter 88. In chapter 88, a methanogenic bioassay is employed to measure the amount of end product (methane) formed in substrate-amended incubations relative to the amount formed in substrate-unamended controls. The results are then compared with the values predicted on the basis of the redox equation originally given by Symons and Buswell for compounds containing C, H, O, N, and S (80):

$$C_nH_aO_b N_cS_d + (n - a/4 - b/2 + 7c/4 + d/2)$$
$$H_2O \rightarrow (n/2 - a/8 + b/4 - 5c/8 + d/4)\ CO_2 \qquad (1)$$
$$+ (n/2 + a/8 - b/4 - 3c/8 - d/4)\ CH_4$$
$$+ cNH_4HCO_3 + dH_2S$$

The net amount of methane produced in incubations can reasonably be ascribed to the degree of substrate bioconversion. It is also advisable to incorporate a positive control in such assays, that is, to include a substrate that is known to be mineralized to the expected amount of methane in comparable incubations. For this purpose, substrates such as benzoic acid, acetate, or glucose are typically employed. When coupled with information on substrate decay (see below), such first-tier assays can provide powerful indications of the susceptibilities of a wide variety of substrates to anaerobic destruction (see for example references 1, 53, 54).

Methanogenesis is usually monitored by measurement of methane production by gas chromatography (GC) (see for example reference 7) and by measurement of increases in pressure, as indicated below. The proportion of the headspace of an incubation vessel consisting of methane is multiplied by the volume of gas (in milliliters) and then divided by the constant 22.4 ml/mmol (at STP) to give the millimoles of methane in the incubation vessel. Buswell's equation (see above) can be used to calculate the number of moles of methane and carbon dioxide that can theoretically be produced from a known amount of substrate. A yield of methane less than the amount stoichiometrically expected may indicate transformation of the substrate to metabolites, which accumulate.

Alternately, progress of methanogenic biodegradation assays can be monitored by examining the pressure that develops in substrate-amended incubations relative to the

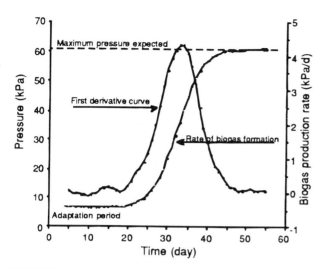

FIGURE 1 Accumulation of pressure with time in microcosms as a result of biogas formation from the methanogenic fermentation of benzoic acid and the first derivative of the resulting pressure curve. Reprinted from reference 78 with kind permission of Elsevier Science-NL, Sara Burgerhartstraat 25, 1055 KV Amsterdam, The Netherlands.

background controls, since the end products of methanogenic fermentation (biogas—CH_4 and, to a lesser extent, CO_2) partition to the headspace of sealed incubation vessels and create pressure. This pressure can be quantified in initial methanogenic screening protocols and interpreted successfully (Fig. 1). Generalized methods based on measurement of pressure accumulation for assessing the biodegradation of organic chemicals in a variety of anaerobic environments have been developed (5, 6, 22, 75, 78). Colleran et al. (22) reviewed the ways in which most of the procedures differ in quantification of methane, pressure, or both.

The sensitivity of a methanogenic biodegradation protocol is a function of a number of factors that require consideration before a reliable assay can be performed. Implicit in the use of the Buswell equation is knowledge of the mean oxidation state of the substrate carbon. More CH_4 than CO_2 would be expected from substrates that are relatively more reduced. That is, for the same amount of total carbon, it is expected that the methanogenic degradation of lipoidal materials would yield more methane than the mineralization of carbohydrates. There are also some compounds for which methane formation during anaerobic metabolism cannot reasonably be expected (e.g., urea and cyanuric acid). The use of the Buswell equation prior to the implementation of any screening protocol is strongly recommended.

Another consideration is the background rate of methane production relative to the amount of substrate added. If background methanogenesis is very high in the system under investigation, then detection of the net amount of methane due to the mineralization of the substrate may prove difficult. Options for this eventuality include diluting the inoculum to obtain a more reasonable background methanogenesis rate or adding a higher concentration of substrate. For the former option, diluted sewage sludge (10%) is often employed as a source of inoculum, while sedimentary environments are used for this purpose with little or no dilution. Use of higher substrate concentrations in the assay carries a greater risk of predisposing the technique to failure, because of the toxicity of the compound to the requisite microflora (see below). Chapter 88 of this manual indicates

how the toxicity of a substrate to methanogenesis can be evaluated prior to performing a biodegradation assay. For practical purposes, the substrate concentration employed should be one that is low enough that it does not inhibit the endogenous rate of methane production yet high enough to allow the clear detection of a methane signal above background levels should ultimate mineralization of the compound be possible. For many different types of chemicals, an initial substrate concentration of 50 to 250 ppm of C has been routinely employed. The highest possible substrate concentration that can be practically used is recommended, because it is sometimes possible to measure a metabolic intermediate (see below) that might otherwise escape detection when lower substrate amounts are employed.

The headspace-to-volume ratio of the incubation flask and temperature fluctuations also have the potential to reduce sensitivity and therefore reliability in initial anaerobic biodegradation screens. This is particularly true when pressure measurements are used instead of methane production. Temperature fluctuations can be reduced or eliminated by performing the assay in an incubator. However, room temperature incubations are more convenient. Variations in room temperature can limit the ability to detect biogas production to amounts that create a pressure of at least 3% in typical 160-ml serum bottle incubation systems (78). With a 60-ml headspace and the conservative assumption that only methane contributes to pressure, temperature fluctuations can limit the detection of substrate biodegradation to 12 to 20 ppm of C, depending on the mean oxidation state of the substrate carbon (Fig. 2). Figure 2 predicts that the sensitivity of methanogenic biodegradation assays is independent of the type of flask employed but is a function of the headspace-to-liquid-volume ratio. This indicates that an accurate assay of substrate biodegradation at lower substrate concentrations can be accomplished by decreasing the headspace volume in the containers. It is always advisable

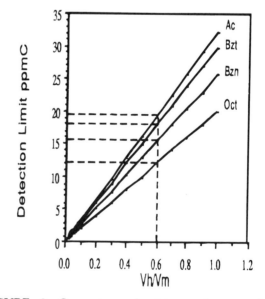

FIGURE 2 Comparison of minimum detection limits (dashed lines) for octane (Oct), benzene (Bzn), benzoate (Bzt), and acetate (Ac) using a pressure-based methanogenic biodegradation assay. Vh/Vm is the ratio of headspace to liquid volume. Reprinted from reference 78 with kind permission of Elsevier Science-NL, Sara Burgerhartstraat 25, 1055 KV Amsterdam, The Netherlands.

to confirm the results of screening assays based on pressure production by quantifying the amount of methane in the headspace of the incubation vessels by GC.

Similar screening protocols can also be developed for assessing anaerobic biodegradation of chemicals under a variety of other electron-accepting conditions (78). The procedures are analogous to the methanogenesis assay in that experimental incubations are constructed as described above and compared with sterile and substrate-unamended controls. However, instead of monitoring the end products of microbial metabolism, the consumption of nitrate or sulfate relative to that in control incubations is determined. In fact, this same logic can be employed to develop an assay for any terminal-electron-accepting process. An initial substrate concentration which does not inhibit the background rate of electron acceptor consumption but still allows the detection of electron acceptor removal in excess of background levels when biodegradation occurs is chosen. The amount of electron acceptor added to the incubations depends on the mean oxidation state of the substrate carbon, but amounts sufficient for the complete mineralization of the parent compound are typically employed. It is often advisable to focus on the consumption of electron acceptors rather than on the end products of the respective bioconversions, since the former is easier to analyze.

Each of these screening methods is able to provide information such as (i) the length of the lag or adaptation period, (ii) the rate of transformation once it does occur, (iii) the prospects for the accumulation of intermediates, and (iv) an indication of the completeness of degradation (Fig. 1). At least duplicate and usually triplicate incubations are typically employed for each compound under investigation, and, in our experience, the coefficients of variation in methane production or electron acceptor consumption rates are usually less than 5%.

Any screening procedure must be concerned with the possibility of both false-positive and false-negative indications of biodegradation. The former can be particularly problematic when fermentable organic substrates in the environment under study are unevenly distributed, resulting in a high degree of spatial heterogeneity. If this is the case, replicate incubations can exhibit significantly different background rates of metabolism. It is then recommended that the samples be composited to account for such differences and to allow a more reliable interpretation of the resulting screening assay. If biodegradation results are ambiguous, more definitive information can usually be obtained by reamending the incubations with the substrate of interest and additional electron acceptor if required.

False-negative indications of biodegradation are also part of any screening procedure. Some substrates can be biodegraded under anaerobic conditions yet not necessarily lead to the consumption of significant amounts of electron acceptor. For instance, methanol and methylamine can be completely mineralized to methane in many anaerobic environments, regardless of the sulfate concentration in the incubation mixture. Very little sulfate relative to the amount in substrate-unamended controls will be consumed in such experiments. If only sulfate depletion were monitored, an erroneous conclusion of recalcitrance would be made. It must be remembered that the tests outlined above are only screening procedures. It is always advisable to collect supporting evidence for or against conclusions regarding the susceptibility of various substrates to biodegradation.

CONSUMPTION OF ELECTRON-DONATING SUBSTRATES

The production of gaseous end products and the consumption of terminal electron acceptors in amounts close to those that are theoretically expected are powerful indications that the mineralization of the parent substrate occurred. However, this contention must be supported with other indications of biodegradation. For instance, a false indication of mineralization would be obtained if an abiotic transformation of the parent substrate preceded metabolism. In such a case, what is actually measured may be the mineralization of the transformed product rather than that of the substrate of interest. Furthermore, a negative result in the screening assays described above does not mean that the parent substrate is inherently recalcitrant under anaerobic conditions. The partial transformation of a parent substrate may lead to the formation of products that are relatively more resistant to microbial attack and are only slowly mineralized, if at all. It is therefore important to monitor the consumption of the electron donor in conjunction with performing the screening procedures outlined above.

Such assays monitor the reduction in electron donor concentration (usually the parent substrate of interest) as a function of time. This information is generally obtained from parallel incubations which contain the same inoculum and which are constructed at the same time as the other screening procedures. Obviously, this primary biodegradation information is easier to obtain in the laboratory than in the field. However, this most straightforward piece of evidence is in itself useless unless it is evaluated relative to abiotic fate processes occurring in the corresponding controls. In the laboratory, such determinations are made relative to uninoculated and sterile controls. In the field, no comparable set of controls exists. However, the destruction of a contaminant in the field can be evaluated among control and treatment areas. That is, substrate decay relative to that in control areas that have the contaminant of interest but do not receive the treatment under evaluation can be assessed. Often, such evaluations are made relative to the concentration of conservative organic or inorganic tracers. Implicit in the latter determination is a good knowledge of the nature and extent of the contaminant, the physical and chemical processes that control its distribution, and adequate sampling regimens.

Of course, the mass loss of a parent substrate in anaerobic incubations, even relative to that in appropriate controls, is not necessarily synonymous with microbial metabolism. Biodegradation assays are fraught with several difficulties, and it is foolish to rely solely on this type of information. Many contaminants may be initially transformed under anaerobic conditions and not subsequently metabolized. Such products are associated with their own environmental and health impacts, and substrate loss measurements alone would indicate complete biodegradation. Moreover, the assay of parent substrate decay requires a suite of analytical techniques that may change depending on the contaminant, its concentration, and the environmental matrix being assayed (see below). Often, it may not be possible to accurately monitor substrate decay without destructive sampling procedures. This eventuality makes it necessary to obtain relatively large quantities of inoculum at the start of the experiment. Factors such as these make standardization of a biodegradation protocol far more difficult. Generally, biodegradation assays based on substrate decay do not lend themselves to high-quality rate determinations, since the number of data points where biodegradation actually occurs can be limited. Lastly, there may be abiotic fate processes that occur in highly reduced nonsterile incubations that are less important in sterile or uninoculated controls. The judicious use of reducing agents to maintain highly reduced conditions in these incubations may help address such concerns (12). With these cautions in mind, the information on the consumption of electron donors can provide powerful evidence in support of anaerobic biodegradation.

Selection of Suitable Substrates

In the environment, organic pollutants are rarely found as sole contaminants. However, reductionist scientific reasoning dictates that only a single compound be studied at a time. An investigator must choose whether to study individual substrates or a mixture of chemical compounds, depending on the objective of the study, the interpretational limits, and other factors, such as the ability to analytically detect and measure the chemicals of interest. In some cases, substrates which are a composite of a wide variety of materials may be used, and the exact identity of each substrate may not be known. For instance, wastewaters, sludges, or leachates are frequently used as complex substrates for generalized assays of biodegradation, as indicated by the losses in volatile solids, biopolymer residues, etc., relative to those in sterile controls. Such assays provide very limited information, and attempts to determine which components in the mixtures are affected during the anaerobic biodegradation assay should be made.

In laboratory experiments, organic substrates are typically studied individually. The chosen compound is usually the dominant contaminant in an environment or is characteristic of a group of compounds of interest and therefore referred to as a model compound. The model compound approach facilitates comparison of results between related studies. While this approach may demonstrate that a type of compound is biodegradable, the results cannot be easily extrapolated to structurally related compounds. For example, the position of substituents on an aromatic nucleus affects the susceptibility of the compound to anaerobic decay, the biodegradation kinetics, and the dominant metabolic pathways involved in substrate destruction. Furthermore, co-metabolism will not be detected in the presence of a single substrate. Nevertheless, assays with single compounds have many advantages, such as simplifying detection of metabolites, and are among the most common studies done. Therefore, use of individual model compounds, while sometimes interpretationally limited, offers an excellent approach for initial screening assays as well as subsequent metabolic studies.

Structural features of potential substrates affect their susceptibility to anaerobic biodegradation. Under reducing conditions, more-oxidized substrates are generally more susceptible to microbial attack than their reduced counterparts. For example, under methanogenic conditions, long-chain alkanoic acids and alkanols are subject to microbial attack, whereas the degradation of the corresponding alkanes is much more difficult (personal observation). Under sulfate-reducing conditions, benzoate and, to a lesser extent, phenol are readily degraded (for a review, see reference 58), whereas extensive degradation of benzene is observed far less frequently (51). The complexity of an organic substrate also affects its susceptibility to microbial attack under anaerobic conditions. For instance, it has been known for decades that detergents and aliphatic hydrocarbons are more easily metabolized if the constituent alkyl chain is not highly branched (as summarized in reference 2). The same trend has been observed for the anaerobic biodegradation of gasoline additives, such as methyl-t-butyl ether. Methyl-t-butyl ether is far less susceptible to degradation than the straight-chain methyl butyl ether analog (64, 79).

Thermodynamic considerations are also useful when evaluating substrates for anaerobic biodegradation studies. This is especially true when a number of electron acceptors are potentially available to couple with substrate oxidation. However, the amount of energy available to the microorganisms will depend not only on the redox potential of the substrate and electron acceptor but also on the identity and concentration of all reaction products. For example, benzoic acid is readily degraded under most anaerobic conditions. A greater energetic benefit would be associated with benzoate decay coupled to nitrate reduction compared with other anaerobic conditions. As equation 2 illustrates, the fermentation of benzoate by syntrophic bacteria is thermodynamically unfavorable unless it is coupled with hydrogen consumption (equation 3) catalyzed by a methanogen (31, 66).

$$4 \, C_7H_5O_2^- + 28 \, H_2O \rightarrow 12 \, CH_3COO^- + 4 \, HCO_3^- + 12H_2 + 12H^+ \quad \Delta G^{o\prime} = 358.6 \text{ kJ/reaction} \tag{2}$$

$$4 \, C_7H_5O_2^- + 19 \, H_2O \rightarrow 12 \, CH_3COO^- + HCO_3^- + 3 \, CH_4 + 9H^+ \quad \Delta G^{o\prime} = -48.2 \text{ kJ/reaction} \tag{3}$$

Generally, the more energy there is that is potentially available from the oxidation of a substrate coupled to reduction of an electron acceptor, the more likely it is that the substrate will be degraded under those conditions. See the classic review of Thauer et al. (82) for an excellent primer relating to such thermodynamic considerations. Furthermore, thermodynamic constraints also influence the extent of substrate degradation. Studies of benzoate metabolism by a syntrophic bacterium suggest that end product buildup can create a thermodynamic constraint that precludes substrate metabolism below a critical Gibb's free energy value (83).

Special Concerns Regarding Electron Donor Consumption

While there are a myriad of issues that must be considered when designing metabolic experiments or evaluating environmental biodegradation results, three factors are paramount: bioavailability, toxicity, and the adaptation period.

Bioavailability refers to the accessibility of a substrate to the requisite microorganisms. In general, water-soluble substrates are more bioavailable and the likelihood of their being biodegraded is higher. Adsorption of substrates to solid matrices can also affect the susceptibility of the substrates to biodegradation. While covalent binding of substrates to solids, such as humic materials, may render the chemical more recalcitrant, adsorption of compounds to inert surfaces, such as granular activated carbon, may actually increase biodegradation. In natural systems, microorganisms often attach and grow at interfaces. Attached microorganisms may benefit from the adsorption of hydrophobic molecules to surfaces, since many potential substrates are concentrated there. The provision of a solid support for microbial and substrate attachment may increase the probability or rate of biodegradation, although adsorption must be corrected for when quantifying substrates. Poorly water-soluble compounds can be rendered more accessible by use of a carrier or cosolvent, such as ethanol, methanol, acetone, dimethylformamide, dimethyl sulfoxide, and mineral oil. Caution should be exercised with ethanol, since it is relatively rapidly converted to acetate, which could dramatically shift culture pH conditions.

As noted above, toxicity is another major concern in biodegradation studies, and both water-soluble and hydrophobic compounds can be toxic to anaerobes. The mechanisms of toxicity of hydrophobic compounds to microbes have recently been reviewed (76) and often involve interactions with cell membranes. The octanol/water partition coefficient ($\log P$) can be used to predict which substrates will exhibit solvent toxicity (44). Since one rarely knows a priori how toxic any substrate will be to a particular group of microorganisms, it is best to perform a toxicity survey prior to the actual biodegradation assay. By evaluating the effects of different concentrations of an organic compound on microbial activities, such as the degradation of a common substrate or the reduction of an electron acceptor, one can select an appropriate substrate level that will allow biodegradation to be reliably measured without predisposing the assay to failure.

Adaptation periods will frequently be encountered before the biodegradation of a particular substrate is observed. Under anaerobic conditions these periods may range from hours to over a year, and they may be due to a variety of reasons (55). Recognizing this fact allows investigators to choose suitable sampling and analysis intervals. Most often, a geometrically progressing time series, in which the frequency of sampling and analysis decreases with incubation duration, is chosen. Such a strategy helps minimize the effort and expense associated with monitoring the long-term incubations frequently associated with initial anaerobic bioconversions. A more frequent sampling effort can usually be reliably mounted when the parent substrate is reamended to the culture of interest.

Analytical Techniques

The analytical method used for monitoring substrates will depend on the physical and chemical characteristics of the compound. High-performance liquid chromatography (HPLC) is typically used for more-polar water-soluble substrates and electron acceptors, whereas GC is used for nonpolar volatile substrates and products. Other techniques, such as colorimetric assays, thin-layer chromatography, capillary electrophoresis, and nuclear magnetic resonance analysis, are also used but are more application specific. The selection of columns and analytical conditions for chromatography is compound specific; the literature should be consulted for details. Preparation of samples for analysis by any method must be appropriate and preferably quantitative.

A wide variety of detectors are available for both HPLC and GC, and a suitable one must be chosen for the compound of interest. For HPLC, a refractive index detector can serve most general purposes, but it suffers from relative insensitivity. Other detectors include fixed- or variable-wavelength UV absorbance, fluorescence, conductivity, or diode-array instruments. The last can facilitate decisions on analyte purity and can often suggest compound identity. For GC, two common nonspecific detectors are thermoconductivity and flame ionization detectors, while others are highly selective and sensitive for particular compound classes. These include electron capture, nitrogen-phosphorus, flame photometric, and chemiluminescence detectors. Mass spectrometers can be used as detectors with either liquid or gas chromatography. While relatively expensive, mass spectrometry (MS) offers outstanding sensitivity and can be used to identify analytes of interest.

Radiolabeled substrates are very useful for increasing the sensitivity of measurements, conducting rigorous mass balances, and confirming mineralization should it occur. For example, some substrates may be toxic at concentrations needed for reliable HPLC or GC analyses, but with the increased sensitivity of radiolabeled material, such substrates (as well as resulting metabolites and end products) can be monitored with an HPLC system equipped with a radioisotope detector. Of course, it is necessary to measure the radiochemical purity of a substrate prior to any metabolism study. In addition, caution must be exercised in interpreting the fate of a radiolabeled atom, as they may be exchanged or recycled through complex metabolic networks. As described later, $^{14}CO_2$ and $^{14}CH_4$ can be analyzed to determine whether mineralization of the parent substrate occurred.

Alternative Approaches for Non-Water-Soluble Compounds

The bulk of the literature on biodegradation under anaerobic conditions concerns substrates that are water soluble and relatively easily monitored by the techniques outlined above. However, many compounds of environmental concern are gaseous, hydrophobic, or insoluble. In addition to alternative analytical procedures, it is sometimes useful to consider alternate incubation conditions in order to assess the biodegradation of such substrates.

Anaerobic microbiology lends itself well to studies of gaseous substrates, as sealed incubation vessels are almost always used. Therefore, monitoring of gaseous or highly volatile chemicals usually involves an analysis of the headspace of incubation vessels by GC. Coupled with information on the solubility and partitioning characteristics of the analytes, headspace analysis is a reliable method of identifying and quantifying a variety of substrates (43).

Hydrophobic compounds can sometimes be added to cultures with a carrier. Inert hydrophobic carriers, such as 2,2,4,4,6,8,8-heptamethylnonane, can be used to create a two-phase incubation system (see for example reference 69). The hydrophobic substrate primarily partitions to the organic layer, while the organisms reside predominantly in the aqueous phase. Substrate transformation occurs primarily at the interface. If the substrate of interest is soluble in these carriers, direct analysis of the organic layer can be used to quantify substrate loss without the need for destructive analysis. The use of inert carriers may also reduce the toxicity often associated with hydrophobic substrates.

Many substrates cannot be dissolved in either aqueous solutions or common carriers but instead must be added as solids. In this case, it is advisable to maximize the surface area to improve the probability of microbial access and attack. For example, compounds can be dissolved in a volatile solvent and added to hot medium. As the medium cools, the pure substrate crystallizes (39a). Cultures can then be sacrificed with time for analysis of substrate decay. Typically, the culture fluid pH is adjusted if necessary, and then the culture fluid is extracted with an organic solvent. The latter is removed, concentrated, and analyzed by GC. In these cases it is advisable to add an internal standard to the culture fluid prior to extraction to account for extraction efficiency and analytical variation.

With large-molecular-weight materials (often solids), the use of radiolabeled substrates to detect labeled end products is a reliable method to assess anaerobic biodegradation. For example, anaerobic lignin degradation can be evaluated by using ^{14}C-labeled lignaceous materials (23, 87). Alternatively, for field studies in which radiolabeled substrates cannot be used, chemical treatment and/or pyrolysis followed by GC or GC-MS has been used for the evaluation of solid substrates, including lignin (57). Similarly, poly(3-hydroxyalkanoates) can be analyzed by GC after methanolysis (14).

Alternatively, microbial attack on polymeric material can sometimes be deduced by evaluating the change in average molecular weight of the material by size exclusion chromatography (14, 21, 89).

METABOLITES

Fundamental studies of the anaerobic metabolism of organic substances often lead to the discovery of intermediates that are characteristic of the particular pathway involved. Metabolites can be transient intermediates or end products resulting from the partial metabolism of substrates or analogs. Detection of these metabolites, either in the laboratory or in situ, provides strong evidence for microbial attack on the substrate of interest.

For example, aromatic hydrocarbons, such as benzene, toluene, and the xylene isomers, are dominant water-soluble components of refined petroleum products such as gasoline. Because of spills and leaks, these compounds are released into the environment. As aerobic microorganisms degrade these compounds, they can rapidly consume the available oxygen, and anaerobiosis develops. Thus, there is much concern over the fate of these compounds under anaerobic conditions (51).

Toluene and various xylene isomers are known to be attacked under anaerobic conditions by a variety of initial reactions, including ring and methyl group oxidation reactions (51). These hydrocarbons can also be transformed by addition reactions that can result in the transient formation of a variety of dioic acids that have so far been detected only under anaerobic conditions. This metabolism was first reported by Evans et al. (28) for a denitrifying isolate which transformed toluene to benzylsuccinic and benzylfumaric acid. These products were proposed to be dead-end metabolites formed by addition of succinyl-coenzyme A to the methyl moiety of toluene (28). By analogy, the same authors proposed a more productive route for toluene metabolism which involved addition of acetyl-coenzyme A to form hydrocinnamoyl-coenzyme A as an intermediate. Further evidence for this mechanism of initial transformation was provided by the identification of benzylsuccinic acid and E-phenylitaconic acid as products of toluene degradation by a denitrifying strain of Azoarcus tolulyticus (17). Benzylsuccinic acid and benzylfumaric acid were also reported to be dead-end metabolites of toluene degradation under sulfate-reducing (9, 10) and denitrifying (17, 74) conditions. Similarly, the anaerobic transformation of o-xylene has been reported to lead to (2-methylbenzyl)-succinic and -fumaric acids (10, 29), m-xylene metabolism has been linked to (3-methylbenzyl)-succinic and -fumaric acids (35), and p-xylene has been shown to be transformed to several products, including 4-methybenzylsuccinic acid (10).

These compounds are not normally present in hydrocarbon mixtures. To date, the only known mechanism for the production of these compounds is anaerobic biotransformation. Therefore, their detection in situ can be used as an indicator of anaerobic alkyl-benzene metabolism (8).

Organic compounds that are potential substrates for anaerobic biodegradation not only have distinctive chemical characteristics but also have a distinct carbon isotopic signature. This fact can be used to deduce whether putative metabolites are related to the original substrate. A powerful new technique of GC-isotope ratio mass spectrometry (GC-IRMS) (see below) can be used to determine the carbon isotopic signature of individual compounds present in complex mixtures that can be separated by GC. If the isotopic signature of a suspected metabolite matches that of the par-

ent substrate, then it is likely that the former originated from the latter. Furthermore, by using ^{13}C-enriched substrates, it may be possible to provide a clear isotopic signal over the background and to reliably assess a variety of biotransformations.

THE NEED TO MAINTAIN ANAEROBIC CONDITIONS

Microorganisms exhibit a wide range of tolerance to and differential requirements for oxygen. Even organisms considered strict anaerobes can metabolize oxygen when it is available, though they do not grow well in the presence of this gas (25, 50). It is beyond the scope of this chapter to consider the basic techniques associated with the cultivation and enumeration of anaerobic microorganisms. Familiarity with the importance of using O_2-free gases, redox indicators, and reducing agents, as well as the specialized anaerobic chambers, gassing stations, and equipment, is assumed. A general reference to these procedures can be found in chapter 6 of this manual and in Breznak and Costilow (12).

From a practical standpoint, the environmental microbiologist must often consider whether the transient exposure of a sample from an anaerobic environment to oxygen will adversely and irreversibly influence the resident microflora over and above the disturbance associated with any sampling. This consideration will often dictate whether extraordinary measures are needed to protect a sample from the effects of oxygen. It is generally assumed that the tolerance and growth range for microorganisms exposed to any abiotic factor are distinct and not necessarily the same. That is, the exposure of microorganisms to adverse ecological conditions may preclude microbial growth, but the cells do not necessarily die. Our experience indicates that particular care must be exercised when sampling environments that are only rarely exposed to oxygen, whereas much less diligence is necessary for environments that are regularly exposed. For instance, less concern would attend the sampling of a shallow anoxic aquifer that is regularly inundated with oxygenated rain than the sampling of the bovine rumen, where appreciable concentrations of oxygen are almost never evident.

Regardless of the oxygen status of the environment, it is generally wise to minimize sample exposure to this gas. In the field, we have found that metal ammunition boxes can be used for convenient sample transport to the laboratory and for sample storage under anaerobic conditions. Such boxes can be obtained at military surplus stores for a nominal fee. A hole can be made in the lid and stoppered to facilitate the exchange of the box atmosphere with O_2-free gases. The boxes are sturdy, will hold pressures for long periods of time (weeks), can be periodically flushed with the desired gas mixture, and can conveniently be taken into most anaerobic chambers. Once in the laboratory, strict anaerobic procedures can then be employed.

ANAEROBIC BIODEGRADATION LINKED TO VARIOUS ELECTRON ACCEPTORS

When oxygen is unavailable, the biodegradation of organic material can be coupled with the reduction of a variety of terminal electron acceptors. Consideration of all possible electron-accepting processes is beyond the scope of this chapter. In fact, to do so would prove redundant with other chapters in this manual. A good perspective on which electron acceptors are predominantly involved in microbial metabolism can be found in chapter 63. We therefore illustrate several major points of one of the more complex anaerobic processes: the reduction of carbon dioxide. Other electron-accepting conditions are discussed in a more cursory fashion, since they are more extensively covered in other sections of this book.

CARBON DIOXIDE REDUCTION

In the absence of more energetically favorable electron acceptors, carbon dioxide can be used as a terminal electron acceptor and reduced to form acetate or methane. Methanogenesis, the process by which carbon dioxide is reduced to methane, is carried out by a diverse group of strictly anaerobic *Archaea* known as methanogens. The ecology, taxonomy, physiology, and biochemistry of this diverse group have been reviewed elsewhere (47, 85). Methanogens generate methane and obtain energy for growth by transferring reducing equivalents from hydrogen to CO_2, by aceticlastic catabolism of acetate, or by disproportionating methyl-containing compounds such as trimethylamine or methanol. In biodegradation assays, it is seldom necessary to distinguish between methanogenesis by reduction of CO_2 versus cleavage of acetate derived from an organic substrate. In cases where the distinction must be made, appropriate procedures are available (46, 49, 77).

Although metabolically limited, methanogens abound in many anaerobic habitats because of their ability to participate in mutualistic relationships with other microorganisms. As terminal members of food webs, methanogens consume the products (e.g., H_2 and acetate) of syntrophic and fermentative bacteria that degrade a myriad of complex organic compounds. A wide variety of compounds have been shown to be susceptible to complete degradation by methanogenic consortia, and readers are referred to the literature for specific examples.

For biodegradation assays under methanogenic conditions, the recommended initial approach is to screen organic compounds by using the procedures outlined above and in chapter 88 of this manual. Strict anaerobic techniques and culture procedures are required for such assays. Ideal inocula for biodegradation assessment can be obtained from methanogenic environments that have a history of exposure to the compound(s) of interest. Fresh inoculum is generally recommended for such assays, but several studies have reported that storage for weeks or even months did not adversely affect metabolic activity (75, 78). It may be advantageous to dilute inocula to minimize the background rate of methane production or to minimize sulfide concentrations. Ideal incubation temperatures range from room temperature to 35°C. Background rates of methanogenesis must be accounted for by monitoring controls not amended with the substrates of interest. Sometimes, negative controls can be established in which methanogenesis is inhibited with bromoethanesulfonic acid (88). Enumeration and isolation of methanogens or isolation of compound degraders may follow the initial biodegradation screen. These procedures are outlined by Whitman et al. (85).

Radiolabeled Substrates

Additional evidence for substrate degradation and greater sensitivity for methane production can be obtained with the use of ^{14}C-labeled substrates. Substrate loss is monitored as the decrease of radioactivity in the aqueous phase. However,

the incorporation of radiolabel into biomass must be accounted for (36), and the production of the gaseous end products $^{14}CH_4$ and $^{14}CO_2$ can be measured by either of two methods. In one method, a gas proportional counter is used to quantify radioactive components that are separated by GC. Counting efficiencies of the gas proportional counter and total recoveries of radiolabel are determined by traditional means (70). Alternatively, the headspace of incubation vessels is purged with nitrogen into a series of alkaline CO_2 traps (68). Methane, which is not retained by the initial set of the traps, is mixed with oxygen and combusted over a catalyst to CO_2. The resulting CO_2 is subsequently trapped in a second set of alkali traps. The radioactivity associated with the two sets of traps is measured by liquid scintillation counting, and recoveries are calculated. When used with suitable controls, both methods provide reliable quantitative results.

Isotope Fractionation

Methanogenesis results in dramatic fractionations of carbon relative to that in abiotic methane production and microbial processes in general (see chapter 62, this volume). In both laboratory and field applications, isotope fractionations associated with methanogenesis can sometimes be used as evidence of this microbial process. The degree of fractionation depends on the substrate (52) as well as the methanogen (37). Methanogenesis also results in distinctive fractionation of hydrogen, which is once again dependent on the pathway by which methane is formed (84). The combination of carbon and hydrogen isotope signatures leads to predictions of contributions of methanogenic pathways in environments (63, 84). However, two factors must be considered in any interpretation of isotopic data: (i) isotope fractionation is very temperature dependent (3), with greater fractionation occurring at lower temperatures, and (ii) if any oxygen is available, methanotrophy can significantly affect methane isotopic values.

Measurement of isotope fractionation factors for methanogens and linking methane production to substrate biodegradation require careful experimentation. Previous studies have established procedures for controlled carbon dioxide incubations (38, 39). The isotopic compositions of carbon dioxide and methane in the incubation vessel can be determined by GC-IRMS or conventional oxidation of methane (11). If substrates other than carbon dioxide (such as acetate) are metabolized as a source of methane, analysis of these substrates by GC-IRMS or other means would be required to calculate fractionation factors (38). Expression of isotope fractionation factors should follow guidelines outlined by Grossman (chapter 62 of this manual) and Hayes (40). Isotopic fractionation offers a tool for detecting methanogenesis and for confirming that it contributes to the biogeochemical profile of an environment. However, this approach must be combined with more traditional methods to conclusively demonstrate methanogenic biodegradation.

Acetogenesis

An alternative form of carbon dioxide reduction involves the production of acetate rather than methane. Acetogenic bacteria (acetogens) form acetate from one-carbon substrates by the acetyl-coenzyme A pathway in a process referred to as acetogenesis. All acetogens are strictly anaerobic *Eubacteria*. While acetogens are noteworthy because of their ability to autotrophically fix carbon dioxide, they also have remarkable metabolic versatility. In addition to CO_2 plus H_2, most acetogens can use formate and many can use methanol, carbon monoxide, the methyl groups of methoxylated aromatic compounds, fermentable sugars, and even aromatic compounds (26). Transformations ascribed to homoacetogens include oxidation of primary alcohols, partial oxidation of organic acids, ether cleavage, and dehalogenation reactions. However, bipolymers that require extracellular hydrolysis are not attacked by these bacteria; homoacetogens depend on extracellular hydrolases produced by other fermenting anaerobes for utilization of these substrates (73). While at a disadvantage compared with other anaerobes such as sulfate-reducing bacteria and methanogens with respect to competition for individual substrates, acetogens are successful in natural anoxic environments because of their metabolic versatility. Acetogens use a wide variety of substrates, even simultaneously, and can form mutualistic associations, such as those involving interspecies hydrogen transfer (56, 73). The ecology, taxonomy, physiology, and biochemistry of acetogens have been reviewed recently by Drake (26, 26a). Details on the isolation and characterization of acetogens can also be found in other sources (24).

Biodegradation studies under acetogenic conditions per se are fairly rare. However, the quantity, ubiquity, and diversity of acetogens in anaerobic environments suggest their involvement in the anaerobic biodegradation of a myriad of natural materials. In particular, acetogens are capable of cleaving methyl groups from phenylmethylethers and other products of lignin cleavage. Transformations of aromatic compounds by acetogens are numerous; for further details, see the recent review by Frazer (34). Typical environments fruitful for studying such transformations include the hindguts of termites, bovine and ovine rumens, and anoxic sediments. Syringic acid and other phenylmethylether substrates can be analyzed by HPLC (56). Radiolabeled monomeric as well as polymeric lignaceous substrates are also available (13, 20, 23). Acetate production can be monitored by routine HPLC methods (79). However, the use of a ^{14}C label greatly facilitates detection of ^{14}C-acetate, which clearly attests to the transformation of the parent substrate but which may only be produced in trace quantities. Turnover of acetate in enrichment cultures by methanogens can be inhibited by the addition of bromoethanesulfonic acid, thereby allowing acetate to accumulate to detectable levels. The potential for anaerobic biodegradation of pollutants is seldom evaluated with respect to acetogens, yet they almost certainly play a role in this regard and deserve further study.

SULFATE REDUCTION

The sulfate-reducing bacteria (SRB) are capable of coupling organic matter oxidation with the reduction of sulfate as a terminal electron acceptor, and they are almost ubiquitously distributed in the environment. Not surprisingly, the majority of isolates have been obtained from marine habitats where sulfate is rarely limiting. However, SRB are known to tolerate diverse environmental conditions, including temperatures ranging from 0 to 110°C and pH values from 3 to about 9, as well as extremes of salinity, redox status, and degree of oxygenation. Depending on the particular organism, SRB may also employ thiosulfate, sulfite, tetrathionate, dithionate, elemental sulfur, nitrate, halo- or nitroaromatic compounds, malate, or fumarate as alternate electron acceptors (86). In addition, several isolates are known to couple the reduction of environmentally important metals, such as iron, chromium, and uranium, with the oxidation of electron donors (see chapter 89). Sulfate reduction is the dominant respiratory process in anoxic environments which contain sulfate but lack other thermodynamically more favorable electron acceptors. A review of the ecology of these organisms has recently been published (30).

Given the widespread distribution and metabolic diversity of SRB, it is not suprising that the activity of these organisms plays a pivotal role in the biogeochemical cycling of sulfur and carbon. They are known to catalyze the destruction of many pollutants (27). The sulfide product of their metabolism supports a variety of chemolithotrophic organisms capable of oxidizing reduced forms of inorganic sulfur to more oxidized sulfur species. The cycling of sulfur and descriptions of the organisms involved are detailed in chapter 34 of this manual.

The presence of rapid sulfate reduction in anoxic environments is evidenced by the blackening of sediments as a result of the precipitation of iron sulfide and by the characteristic odor of hydrogen sulfide gas. A chemical reaction between iron sulfide and elemental sulfur results in the formation of pyrite, an abundant iron sulfide mineral (for a review see reference 65). Although the consumption of sulfate and the production of sulfide are indicative of sulfate reduction, the absence of changes in these chemical species cannot be taken as evidence excluding this process. The oxidation of sulfide and the rapid precipitation of dissolved hydrogen sulfide with metals can modulate pool sizes of these species such that changes may be below detection limits. Estimates of the rate and extent of biodegradation which are based on sulfate depletion or the accumulation of sulfides can be compromised if the cycling of sulfur occurs.

The selection of an appropriate method for monitoring sulfate reduction depends on the anticipated rate and extent of sulfate reduction. For anaerobic biodegradation experiments in which relatively high concentrations of substrate are used, the depletion of sulfate from incubation mixtures can be monitored with an HPLC system equipped with a conductivity detector (see for example reference 78). In fact, HPLC methods are available for the quantification of sulfoxy anion intermediates involved in the cycling of sulfur, including thiosulfate, sulfite, and polythionates, even at low concentrations (1 μM) (4). Alternatively, the accumulation of dissolved sulfide can be monitored in incubations by the well-known methylene blue assay (18) or modifications thereof. However, sulfate reduction will likely be underestimated using this assay, since it detects only acid-volatile sulfides, including hydrogen sulfide and iron sulfide, but not other forms of reduced sulfur such as pyrite and elemental sulfur. Frequent monitoring of dissolved sulfide is advised, since high levels can inhibit biodegradation processes. Precipitation of soluble sulfide by reaction with $FeSO_4$ can sometimes be used to stimulate anaerobic biodegradation activity if sulfide toxicity is evident.

If small quantities of sulfate are to be reduced or if short incubation times will be employed (i.e., for in situ rate estimates), a radiotracer assay for sulfate reduction is recommended. This technique is orders of magnitude more sensitive than other sulfate reduction assays. Carrier-free $^{35}SO_4$ is relatively inexpensive and available with high specific activity and can easily be added to a wide variety of anaerobic incubation systems. The amount of sulfide formed as a result of sulfate reduction can be assessed by periodically taking subsamples (being careful to maintain anaerobic conditions) and fixing them in a cooled 10% zinc acetate solution. When slurries are sampled, it is essential to obtain both soluble and solid phases, since reduced sulfides will likely exist in both. The fixed samples are then stoppered and stored frozen prior to extraction and analysis of sulfides. Since exchange of the ^{35}S-isotope occurs readily, the pool sizes of labeled sulfides cannot be accurately determined (33). Thus, a simpler and more accurate single-step chromium extraction procedure, in which all reduced sulfur compounds are quantified, is preferred. The rate of sulfate reduction is based on the total amount of sulfide produced. Details of the extraction procedure, the factors affecting extraction efficiency, the protocols for making the reagents, and calculations for determining sulfate reduction rates are reviewed in chapter 34 of this manual and by Fossing and Jorgensen (32). This well-known technique has been used for measuring sulfate reduction rates in diverse environments, including coastal and deep sea sediments, marine microbial mats, sediments from lakes and rivers, and aquifers. Regardless of the method used to monitor sulfate-reducing activity, efforts should be made to confirm that the metabolism of the electron donor is truly coupled with electron acceptor decay. To garner such evidence, the sensitivity of the bioconversion to inhibition by molybdate is frequently evaluated. However, the same cautions attending the use of any inhibitor in metabolism studies are pertinent. While molybdate is known to interfere with sulfate reduction, it also can exhibit nonspecific inhibitory effects. More convincing evidence includes the cessation of the metabolism when the sulfate is either exhausted or removed from the incubation mixture and its resumption upon subsequent reamendment of the electron acceptor.

A potentially powerful approach for monitoring sulfate-reducing biodegradation activity in a variety of environmental matrices is the silver foil assay (19a). In this assay, a commercially available silver foil is cleaned with solvents and prepared by dipping in concentrated nitric acid for 10 to 20 s in order to cover the surface of the foil with a thin layer of silver oxide. The foil is then washed, air dried, autoclaved, and stored in an anaerobic chamber prior to use. A solution of Na^{35}SO$_4$ is then added to the environmental matrix under investigation. The prepared silver foil is then placed in contact with the environmental sample and incubated under anaerobic conditions. Sulfide produced as a consequence of sulfate reduction reacts rapidly with the silver oxide coating on the foil to form a radioactive precipitate. The foil is then recovered and washed overnight in distilled water to remove any unreacted and adherent ^{35}SO$_4$. [^{35}S]sulfide is then located on the foil and quantified by autoradiographic imaging. By maintaining the relative orientation of the foils, the exact locations in the environmental matrix that are associated with sulfate-reducing activity can be easily identified. In essence, the oxidized silver foil serves as a sulfide trap. The technique allows for the measurement of sulfate reduction with minimal disturbance to the normal distribution of the indigenous microflora.

IRON REDUCTION

The relatively high redox potential of Fe(III) suggests that it should serve as a favored electron acceptor for anaerobic respiration in a variety of environments. Indeed, this has been found for fresh, brackish, and marine habitats. Fe(III) reduction dominates over sulfate reduction and methanogenesis by keeping key intermediates of anaerobic decay, such as hydrogen and acetate, at concentrations too low to support the competing processes (61). However, exceptions to the above generalizations are notable (19). Further, the abiotic reduction of iron by reaction with reduced forms of sulfur such as hydrogen sulfide, iron sulfide, and pyrite is likely in sulfidogenic environments (16, 81).

A wide variety of microorganisms, including fungi, facultatively anaerobic bacteria, and strict anaerobes, are capable of reducing iron (59). Some of these organisms are capable of completely oxidizing organic compounds to CO$_2$, while others are incomplete oxidizers. The utilization of common fermentative intermediates such as hydrogen, acetate, and formate by dissimilatory iron-reducing bacteria likely accounts for the majority of Fe(III) reduction in the environment. Iron-reducing bacteria are also known to catalyze the oxidation of aromatic compounds, including important intermediates of anaerobic decay as well as several contaminant chemicals. Some iron reducers are capable of using a variety of electron acceptors, including Mn(IV), U(VI), NO$_3^-$, NO$_2^-$, S^0, SO$_4^{-2}$, and even oxygen (67).

The potential for iron reduction in an environment is usually determined by establishing the presence of oxidized iron and demonstrating that at least a fraction of it is capable of serving as an electron acceptor for the resident microflora. Iron(III) exists in a variety of insoluble chemical forms which vary in their degrees of crystalinity and their susceptibilities to reduction. Fe(III) sources with little or no crystalline structure, such as amorphous iron(III) oxyhydroxides, are relatively easily reduced. However, there is no doubt that crystalline forms of iron(III) can also serve as electron acceptors. For example, soluble Fe(II) was produced when magnetite was used as an electron acceptor by an iron-reducing pure culture (48). Nevertheless, a mild acid extraction procedure with hydroxylamine seems to correlate reasonably well with the amount of bioavailable Fe(III) (58a). Crystalline Fe(III) forms can be quantified by other procedures (41, 42).

Conclusive evidence for iron reduction is obtained by following the accumulation of reduced iron in incubations over time, as previously described (chapter 89 and reference 60). Both soluble and precipitated forms should be quantified in order to determine the total amount of Fe(II) produced. Unfortunately, in situ rate estimates are difficult to obtain, since radioactive ^{55}Fe is not useful as a tracer because of the rapid isotope exchange between reduced and oxidized iron forms (72). The addition of acetate or an easily reducible form of oxidized iron to selected incubations can serve as a control for determining whether carbon or Fe(III) is limiting iron reduction. Poorly crystalline Fe(III) oxide (60), Fe(III) nitrilotriacetic acid (71), Fe(III) pyrophosphate (15), and Fe(III) citrate (62) are all suitable sources of Fe(III) for this purpose. For biodegradation experiments, Fe(III) should be added in excess of the theoretical amount capable of being reduced via the oxidation of the electron donor. High concentrations of Fe(III) are desirable, since iron is reduced by a single electron. Substrate and iron(II) concentrations should be determined concomitantly so that iron reduction can be directly correlated with substrate disappearance. The amount of iron reduced in controls to which the compound of interest was not added must be accounted for so that a mass balance between Fe(II) production and substrate depletion can be obtained. Iron reduction catalyzed by sulfides or other reduced compounds which may be present in the samples of interest can be evaluated in heat- or chemically treated controls. The possibility of indirect Fe(III) reduction, catalyzed by sulfide production, can be addressed in controls in which sulfate reduction is inhibited by molybdate.

NITRATE REDUCTION

Anaerobic biodegradation can also be linked to the consumption of nitrate as a terminal electron acceptor. Nitrate can exhibit two major metabolic fate processes under anaerobic conditions which are distinguished on the basis of the resulting end products. Most commonly, nitrate (and nitrite) can be denitrified by a wide variety of microorganisms to result in the formation of nitrogen gas (N_2) and nitrous oxide (N_2O) as the principal products. Either organic or inorganic compounds can serve as electron donors supporting this metabolism. The other fate process for nitrate is dissimilatory nitrate reduction to NH_4^+. This process can occur in the same anaerobic habitats as those in which denitrification occurs. Most organisms with this capacity can ferment and can survive in the absence of nitrate, or they can couple their metabolism with the consumption of still other terminal electron acceptors.

Nitrate reduction can easily be evaluated by monitoring the consumption of the anion relative to that in substrate-unamended controls or by measuring the end products of the metabolism. In the former case, chromatographic analysis of nitrate or nitrite depletion from incubations by HPLC is a common procedure (see for example reference 78). However, interpretation of denitrification-linked biodegradation results based strictly on the disappearance of these anions can be confounded by their possible conversion to NH_4^+. Alternatively, denitrification-linked biodegradation can be evaluated by monitoring the appearance of N_2 and N_2O. In most assays, the reduction of N_2O is blocked by the addition of acetylene to the incubation mixture. The resulting N_2O is then easily quantified using a GC system equipped with an electron capture detector. A description of the intricacies associated with these and other nitrogen transformations can be found in chapters 35 and 49 of this manual.

In most anaerobic biodegradation assays, the redox dye resazurin is routinely included in the medium as an indicator of anaerobiosis. The same is true for assays that incorporate nitrate as a potential terminal electron acceptor. In its reduced state, the dye is colorless, but it becomes pink when it is oxidized. However, the intermediates of denitrification (NO and N_2O) can also oxidize resazurin and turn cultures pink (45). Therefore, one must resist the temptation to discard such cultures for fear of their being compromised by oxygen. Rather, if denitrification-linked biodegradation is an important fate process, the nitrogen intermediates will be consumed and the dye will eventually return to the colorless state. In fact, the transient appearance of the oxidized product of resazurin in such cultures (and not in appropriate controls) is often qualitative evidence that denitrification-linked biodegradation is occurring. Moreover, Jenneman et al. (45) noted that many denitrifiers growing on solid media containing resazurin could form pink colonies that could easily be isolated. This certainly represents an underutilized approach to the enumeration and eventual isolation of the denitrifying organisms that may be involved in biodegradation activities. No color alterations due to oxidation of resazurin would be evident in cultures in which dissimilatory nitrate reduction to NH_4^+ predominated.

CONCLUDING REMARK

Research over the past several decades has clearly illustrated that anaerobic microorganisms are much more metabolically diverse than was historically believed and that they are capable of catalyzing a wide variety of ecologically important biotransformations. This relatively recent appreciation of anaerobic microorganisms paralleled advances in microbiological theory, equipment, and methodology. This chapter has illustrated only a few of the common procedures for evaluating the biodegradation potential of anaerobic micro-

organisms. Many more techniques involving the enrichment and isolation of the requisite organisms, the generation of nutritionally exacting mutants, and the appropriate use of growing, resting, and cell-free preparations, as well as anaerobic biochemical and genetic methods for pathway analysis, are simply beyond the scope of this effort. It is hoped that investigators who employ the tools outlined herein are cognizant of the interpretational constraints and are equally aware that the discipline is still in its infancy. Much more remains to be learned.

REFERENCES

1. **Adrian, N. R., and J. M. Suflita.** 1994. Anaerobic biodegradation of halogenated and nonhalogenated N-, S-, and O-heterocyclic compounds in aquifer slurries. *Environ. Toxicol. Chem.* **13:**1551–1557.
2. **Alexander, M.** 1994. *Biodegradation and Bioremediation*, p. 159–176. Academic Press, San Diego, Calif.
3. **Alperin, M. J., N. E. Blair, D. B. Albert, T. M. Hoehler, and C. S. Martens.** 1992. Factors that control the stable carbon isotopic composition of methane produced in an anoxic marine sediment. *Global Biogeochem. Cycles* **6:**271–275.
4. **Bak, F., A. Schuhmann, and K. Jansen.** 1993. Determination of tetrathionate and thiosulfate in natural samples and microbial cultures by a new, fast and sensitive ion chromatographic technique. *FEMS Microbiol. Ecol.* **12:**257–264.
5. **Battersby, N. S., and V. Wilson.** 1988. Evaluation of a serum bottle technique for assessing the anaerobic biodegradability of organic chemicals under methanogenic conditions. *Chemosphere* **17:**2441–2460.
6. **Battersby, N. S., and V. Wilson.** 1989. Survey of the anaerobic biodegradation potential of organic chemicals in digesting sludge. *Appl. Environ. Microbiol.* **55:**433–439.
7. **Beeman, R. E., and J. M. Suflita.** 1987. Microbial ecology of a shallow unconfined ground water aquifer polluted by municipal landfill leachate. *Microb. Ecol.* **14:**39–54.
8. **Beller, H. R., W.-H. Ding, and M. Reinhard.** 1995. By-products of anaerobic alkylbenzene metabolism useful as indicators of *in situ* bioremediation. *Environ. Sci. Technol.* **29:**2864–2870.
9. **Beller, H. R., M. Reinhard, and D. Grbić-Galić.** 1992. Metabolic by-products of anaerobic toluene degradation by sulfate-reducing enrichment cultures. *Appl. Environ. Microbiol.* **58:**3192–3195.
10. **Beller, H. R., A. M. Spormann, P. K. Sharma, J. R. Cole, and M. Reinhard.** 1996. Isolation and characterization of a novel toluene-degrading, sulfate-reducing bacterium. *Appl. Environ. Microbiol.* **62:**1188–1196.
11. **Boutton, T. W.** 1991. Stable carbon isotope ratios of natural materials. I. Sample preparation and mass spectrometric analysis, p. 155–171. *In* D. C. Coleman and B. Fry (ed.), *Carbon Isotope Techniques.* Academic Press, Inc., San Diego.
12. **Breznak, J. A., and R. N. Costilow.** 1994. Physicochemical factors in growth, p. 137–154. *In* P. Gerhardt, R. G. E. Murray, W. A. Wood, and N. R. Krieg (ed.), *Methods for General and Molecular Bacteriology.* American Society for Microbiology, Washington, D.C.
13. **Brune, A., E. Miambi, and J. A. Breznak.** 1995. Roles of oxygen and the intestinal microflora in the metabolism of lignin-derived phenylpropanoids and other monoaromatic compounds by termites. *Appl. Environ. Microbiol.* **61:**2688–2695.
14. **Budwill, K., P. M. Fedorak, and W. J. Page.** 1992. Methanogenic degradation of poly(3-hydroxyalkanoates). *Appl. Environ. Microbiol.* **58:**1398–1401.

15. **Caccavo, F., D. J. Lonergan, D. R. Lovley, M. Davis, J. F. Stolz, and M. J. McInerney.** 1994. *Geobacter sulfurreducens* sp. nov., a hydrogen- and acetate-oxidizing dissimilatory metal-reducing microorganism. *Appl. Environ. Microbiol.* **60:**3752–3759.

16. **Canfield, D. E., R. Raiswell, and S. Bottrell.** 1992. The reactivity of sedimentary iron minerals towards sulfide. *Am. J. Sci.* **292:**659–683.

17. **Chee-Sanford, J. C., J. W. Frost, M. R. Fries, J. Zhou, and J. M. Tiedje.** 1996. Evidence for acetyl coenzyme A and cinnamoyl coenzyme A in the anaerobic toluene mineralization pathway in *Azoarcus tolulyticus* Tol-4. *Appl. Environ. Microbiol.* **62:**964–973.

18. **Cline, J. D.** 1969. Spectrophotometric determination of hydrogen sulfide in natural waters. *Limnol. Oceanogr.* **14:**454–458.

19. **Coates, J. D., R. T. Anderson, J. C. Woodward, E. J. P. Phillips, and D. R. Lovley.** Anaerobic hydrocarbon degradation in petroleum-contaminated harbor sediments under sulfate-reducing and artificially imposed iron-reducing conditions. *Environ. Sci. Technol.,* in press.

19a.**Cohen, Y.** Personal communication.

20. **Colberg, P. J.** 1988. Anaerobic microbial degradation of cellulose, lignin, oligolignols, and monoaromatic lignin derivatives, p. 333–372. *In* A. J. B. Zehnder (ed.), *Biology of Anaerobic Microorganisms.* Wiley, New York.

21. **Colberg, P. J., and L. Y. Young.** 1985. Anaerobic degradation of soluble fractions of [^{14}C-lignin] lignocellulose. *Appl. Environ. Microbiol.* **49:**345–349.

22. **Colleran, E., F. Concannon, T. Golden, F. Geoghegan, B. Crumlish, E. Killilea, M. Henry, and J. Coates.** 1992. Use of methanogenic activity tests to characterize anaerobic sludges, screen for anaerobic biodegradability and determine toxicity thresholds against individual trophic groups and species. *Water Sci. Technol.* **25:**31–40.

23. **Crawford, R. L.** 1981. *Lignin Biodegradation and Transformation,* p. 20–37. John Wiley & Sons, New York.

24. **Diekert, G.** 1992. The acetogenic bacteria, p. 517–533. *In* A. Balows, H. G. Truper, M. Dworkin, W. Harder, and K.-L. Schleifer (ed.), *The Prokaryotes,* 2nd ed. Springer Verlag, New York.

25. **Dilling, W., and H. Cypionka.** 1990. Aerobic respiration in sulfate-reducing bacteria. *FEMS Microbiol. Lett.* **71:**123–128.

26. **Drake, H. L.** 1994. Acetogenesis, acetogenic bacteria, and the acetyl-CoA "Wood/Ljungdahl" pathway: past and current perspectives, p. 3–60. *In* H. L. Drake (ed.), *Acetogenesis.* Chapman & Hall, New York.

26a.**Drake, H. L. (ed.).** 1994. *Acetogenesis.* Chapman & Hall, New York.

27. **Ensley, B. D., and J. M. Suflita.** 1995. Metabolism of environmental contaminants by mixed and pure cultures of sulfate-reducing bacteria, p. 293–332. *In* L. L Barton (ed.), *Sulfate-Reducing Bacteria.* Plenum Press, New York.

28. **Evans, P. J., W. Ling, B. Goldschmidt, E. R. Ritter, and L. Y. Young.** 1992. Metabolites formed during anaerobic transformation of toluene and *o*-xylene and their proposed relationship to the initial steps of toluene mineralization. *Appl. Environ. Microbiol.* **58:**496–501.

29. **Evans, P. J., D. T. Mang, K. S. Kim, and L. Y. Young.** 1991. Anaerobic degradation of toluene by a denitrifying bacterium. *Appl. Environ. Microbiol.* **57:**1139–1145.

30. **Fauque, G. D.** 1995. Ecology of sulfate-reducing bacteria, p. 217–241. *In* L. L. Barton (ed.), *Sulfate-Reducing Bacteria.* Plenum Press, New York.

31. **Ferry, J. G., and R. S. Wolfe.** 1976. Anaerobic degradation of benzoate to methane by a microbial consortium. *Arch. Microbiol.* **107:**33–40.

32. **Fossing, H., and B. B. Jorgensen.** 1989. Measurement of bacterial sulfate reduction in sediments: evaluation of a single-step chromium reduction method. *Biogeochemistry* **8:**205–222.

33. **Fossing, H., and B. B. Jorgensen.** 1990. Isotope exchange reactions with radiolabeled sulfur compounds in anoxic seawater. *Biogeochemistry* **9:**223–245.

34. **Frazer, A. C.** 1994. O-Demethylation and other transformations of aromatic compounds by acetogenic bacteria, p. 445–483. *In* H. L. Drake (ed.), *Acetogenesis.* Chapman & Hall, New York.

35. **Frazer, A. C., W. Ling, and L. Y. Young.** 1993. Substrate induction and metabolite accumulation during anaerobic toluene utilization by the denitrifying strain T1. *Appl. Environ. Microbiol.* **59:**3157–3160.

36. **Frazer, A. C., and L. Y. Young.** 1986. Anaerobic C$_1$ metabolism of the O-methyl-^{14}C-labeled substituent of vanillate. *Appl. Environ. Microbiol.* **51:**84–87.

37. **Games, L. M., J. M. Hayes, and R. P. Gunsalus.** 1978. Methane-producing bacteria: natural fractionations of the stable carbon isotopes. *Geochim. Cosmochim. Acta* **42:**1295–1297.

38. **Gelwicks, J. T., J. B. Risatti, and J. M. Hayes.** 1989. Carbon isotope effects associated with autotrophic acetogenesis. *Org. Geochem.* **14:**441–446.

39. **Gelwicks, J. T., J. B. Risatti, and J. M. Hayes.** 1994. Carbon isotope effects associated with aceticlastic methanogenesis. *Appl. Environ. Microbiol.* **60:**467–472.

39a.**Gieg, L.** Personal communication.

40. **Hayes, J. M.** 1993. Factors controlling ^{13}C contents of sedimentary organic compounds: principles and evidence. *Mar. Geol.* **113:**111–125.

41. **Heron, G., T. H. Christensen, and J. C. Tjell.** 1994. Oxidation capacity of aquifer sediments. *Environ. Sci. Technol.* **28:**153–158.

42. **Heron, G., C. Crouzet, A. C. M. Bourg, and T. H. Christensen.** 1994. Speciation of Fe(II) and Fe(III) in contaminated aquifer sediments using chemical extraction techniques. *Environ. Sci. Technol.* **28:**1698–1705.

43. **Hewitt, A. D., P. H. Mlyares, C. Leggett, and T. F. Jenkins.** 1992. Comparison of analytical methods for determination of volatile organic compounds in soils. *Environ. Sci. Technol.* **26:**1932–1938.

44. **Inoue, A., and K. Horikoshi.** 1989. A *Pseudomonas* thrives in high concentrations of toluene. *Nature* (London) **338:**264–266.

45. **Jenneman, G. E., A. D. Montgomery, and M. J. McInerney.** 1986. Method for detection of microorganisms that produce gaseous nitrogen oxides. *Appl. Environ. Microbiol.* **51:**776–780.

46. **Jeris, J. S., and P. L. McCarty.** 1965. The biochemistry of methane formation using ^{14}C-tracers. *J. Water Pollut. Control Fed.* **37:**178–192.

47. **Jones, W. J., D. P. Nagle, Jr., and W. B. Whitman.** 1987. Methanogens and the diversity of archaebacteria. *Microbiol. Rev.* **51:**135–177.

48. **Kosta, J. E., and K. H. Nealson.** 1995. Dissolution and reduction of magnetite by bacteria. *Environ. Sci. Technol.* **29:**2535–2540.

49. **Koyama, T.** 1963. Gaseous metabolism in lake sediments and paddy soils and the production of atmospheric methane and hydrogen. *J. Geophys. Res.* **68:**3971–3973.

50. **Krekeler, D., and H. Cypionka.** 1995. The preferred electron acceptor of *Desulfovibrio desulfuricans* CSN. *FEMS Microbiol. Ecol.* **17:**271–278.

51. **Krumholz, L., M. E. Caldwell, and J. M. Suflita.** 1996. Biodegradation of "BTEX" hydrocarbons under anaerobic conditions, p. 61–99. *In* R. Crawford and D. Crawford, (ed.), *Bioremediation: Principles and Applications.* Cambridge University Press, Cambridge.

52. **Krzycki, J. A., W. R. Kenealy, M. J. DeNiro, and J. G. Zeikus.** 1987. Stable carbon isotope fractionation by *Methanosarcina barkeri* during methanogenesis from acetate,

methanol, or carbon dioxide-hydrogen. *Appl. Environ. Microbiol.* **53:**2597–2599.

53. **Kuhn, E. P., and J. M. Suflita.** 1989. Anaerobic biodegradation of nitrogen-substituted and sulfonated benzene aquifer contaminants. *Hazard. Waste Hazard. Mater.* **6:** 121–133.

54. **Kuhn, E. P., and J. M. Suflita.** 1989. Microbial degradation of nitrogen, oxygen and sulfur heterocyclic compounds under anaerobic conditions: studies with aquifer samples. *Environ. Toxicol. Chem.* **8:**1149–1158.

55. **Linkfield, T. G., J. M. Suflita, and J. M. Tiedje.** 1989. Characterization of the acclimation period before anaerobic dehalogenation of halobenzoates. *Appl. Environ. Microbiol.* **55:**2773–2778.

56. **Liu, S., and J. M. Suflita.** 1993. H_2-CO_2-dependent anaerobic O-demethylation activity in subsurface sediments and by an isolated bacterium. *Appl. Environ. Microbiol.* **59:** 1325–1331.

57. **Logan, G. A., J. J. Boon, and G. Eglinton.** 1993. Structural biopolymer preservation in Miocene leaf fossils from the Clarkia site, northern Idaho. *Proc. Natl. Acad. Sci. USA* **90:**2246–2250.

58. **Londry, K. L., and P. M. Fedorak.** 1992. Benzoic acid intermediates in the anaerobic biodegradation of phenols. *Can. J. Microbiol.* **38:**1–11.

58a.**Lovley, D. R.** 1987. Rapid assay for microbially reducible ferric iron in aquatic sediments. *Appl. Environ. Microbiol.* **53:**1536–1540.

59. **Lovley, D. R.** 1990. Dissimilatory Fe(III) and Mn(IV) reduction. *Microbiol. Rev.* **55:**259–287.

60. **Lovley, D. R., and E. J. P. Phillips.** 1986. Organic matter mineralization with reduction of ferric iron in anaerobic sediments. *Appl. Environ. Microbiol.* **51:**683–689.

61. **Lovley, D. R., and E. J. P. Phillips.** 1987. Competitive mechanisms for inhibition of sulfate reduction and methane production in the zone of ferric iron reduction in sediments. *Appl. Environ. Microbiol.* **53:**2636–2641.

62. **Lovley, D. R., and E. J. P. Phillips.** 1988. Novel mode of microbial energy metabolism: organic carbon oxidation coupled to dissimilatory reduction of iron or manganese. *Appl. Environ. Microbiol.* **54:**1472–1480.

63. **Martens, C. S., N. E. Blair, C. D. Green, and D. J. Des Marais.** 1986. Seasonal variations in the stable carbon isotopic signature of biogenic methane in a coastal sediment. *Science* **233:**1300–1303.

64. **Mormile, M. R., S. Liu, and J. M. Suflita.** 1994. Anaerobic biodegradation of gasoline oxygenates: extrapolation of information to multiple sites and redox conditions. *Environ. Sci. Technol.* **28:**1727–1732.

65. **Morse, J. W., F. J. Millero, J. C. Cornwell, and D. Rickard.** 1987. The chemistry of the hydrogen sulfide and iron sulfide systems in natural waters. *Earth-Sci. Rev.* **24:**1–42.

66. **Mountfort, D. O., and M. P. Bryant.** 1982. Isolation and characterization of an anaerobic syntrophic benzoate-degrading bacterium from sewage sludge. *Arch. Microbiol.* **133:**249–256.

67. **Nealson, K. H., and D. Saffarini.** 1994. Iron and manganese in anaerobic respiration: environmental significance, physiology, and regulation. *Annu. Rev. Microbiol.* **48:**311–343.

68. **Nuck, B. A., and T. W. Federle.** A batch test for assessing the rate and extent of mineralization of ^{14}C-radiolabeled compounds under realistic anaerobic conditions. *Environ. Sci. Technol.*, in press.

69. **Rabus, R., R. Nordhaus, W. Ludwig, and F. Widdel.** 1993. Complete oxidation of toluene under strictly anoxic conditions by a new sulfate-reducing bacterium. *Appl. Environ. Microbiol.* **59:**1444–1451.

70. **Roberts, D. J., P. M. Fedorak, and S. E. Hrudey.** 1986.

Comparison of the fates of the methyl carbons of *m*-cresol and *p*-cresol in methanogenic consortia. *Can. J. Microbiol.* **33:**335–338.

71. **Roden, E. E., and D. R. Lovley.** 1993. Dissimilatory Fe(III) reduction by the marine microorganism *Desulfuromonas acetoxidans. Appl. Environ. Microbiol.* **59:**734–742.

72. **Roden, E. E., and D. R. Lovley.** 1993. Evaluation of ^{55}Fe as a tracer of Fe(III) reduction in aquatic sediments. *Geomicrobiol. J.* **11:**49–56.

73. **Schink, B.** 1994. Diversity, ecology, and isolation of acetogenic bacteria, p. 197–235. *In* H. L. Drake (ed.), *Acetogenesis.* Chapman & Hall, New York.

74. **Seyfried, B., G. Glod, R. Schocher, A. Tschech, and J. Zeyer.** 1994. Initial reactions in the anaerobic oxidation of toluene and *m*-xylene by denitrifying bacteria. *Appl. Environ. Microbiol.* **60:**4047–4052.

75. **Shelton, D. R., and J. M. Tiedje.** 1984. General method for determining anaerobic biodegradation potential. *Appl. Environ. Microbiol.* **47:**850–857.

76. **Sikkema, J., J. A. M. de Bont, and B. Poolman.** 1995. Mechanisms of membrane toxicity of hydrocarbons. *Microbiol. Rev.* **59:**201–222.

77. **Smith, P. H., and R. A. Mah.** 1966. Kinetics of acetate metabolism during sludge digestion. *Appl. Microbiol.* **14:** 368–371.

78. **Suflita, J. M., and F. Concannon.** 1995. Screening tests for assessing the anaerobic biodegradation of pollutant chemicals in subsurface environments. *J. Microbiol. Methods* **21:**267–281.

79. **Suflita, J. M., and M. R. Mormile.** 1993. Anaerobic biodegradation of known and potential gasoline oxygenates in the terrestrial subsurface. *Environ. Sci. Technol.* **27:** 976–978.

80. **Symons, G. E., and A. M. Buswell.** 1933. The methane fermentation of carbohydrates. *J. Am. Chem. Soc.* **55:** 2028–2036.

81. **Thamdrup, B., H. Fossing, and B. B. Jorgensen.** 1994. Manganese, iron, and sulfur cycling in a coastal marine sediment, Aarhus Bay, Denmark. *Geochim. Cosmochim. Acta* **58:**5115–5129.

82. **Thauer, R.K., K. Jungermann, and K. Dekker.** 1977. Energy conservation in chemotrophic anaerobic bacteria. *Bacteriol. Rev.* **41:**100–180.

83. **Warikoo, V., M. J. McInerney, J. A. Robinson, and J. M. Suflita.** 1996. Interspecies acetate transfer influences the extent of anaerobic benzoate degradation by syntrophic consortia. *Appl. Environ. Microbiol.* **62:**26–32.

84. **Whiticar, M. J., E. Faber, and M. Schoell.** 1986. Biogenic methane formation in marine and freshwater environments: CO_2 reduction vs. acetate fermentation—isotope evidence. *Geochim. Cosmochim. Acta* **50:**693–709.

85. **Whitman, W. B., T. L. Bowen, and D. R. Boone.** 1992. The methanogenic bacteria, p. 719–767. *In* A. Balows, H. G. Truper, M. Dworkin, W. Harder, and K.-L. Schleifer (ed.), *The Prokaryotes,* 2nd ed. Springer Verlag, New York.

86. **Widdel, F.** 1988. Microbiology and ecology of sulfate- and sulfur-reducing bacteria, p. 469–585. *In* A. J. B. Zehnder (ed.), *Biology of Anaerobic Microorganisms.* Wiley, New York.

87. **Young, L. Y., and A. C. Frazer.** 1987. The fate of lignin and lignin-derived compounds in anaerobic environments. *Geomicrobiol. J.* **5:**261–293.

88. **Zinder, S. H., T. Anguish, and S. C. Cardwell.** 1984. Selective inhibition by 2-bromoethanesulfonate of methanogenesis from acetate in a thermophilic anaerobic digestor. *Appl. Environ. Microbiol.* **47:**1343–1345.

89. **Ziomek, E., and R. E. Williams.** 1989. Modification of lignins by growing cells of the sulfate-reducing anaerobe *Desulfovibrio desulfuricans. Appl. Environ. Microbiol.* **55:** 2262–2266.

Anaerobic Biodegradability Assay

L. Y. YOUNG

88

BACKGROUND RATIONALE FOR PROCEDURE

A bioassay has a number of advantages over a chemical assay. For example, it is not necessary to know the composition of the material to be assayed. In addition, a bioassay permits the evaluation of the biologically available portion of the material. A chemical procedure or analysis would not be able to distinguish between the biodegradable and nonbiodegradable portions of organic matter. We first reported on an anaerobic biodegradation assay in 1979 (4) and termed it the biochemical methane potential (BMP) assay. In addition, along with the BMP, it is easy to use an adjunct test, the anaerobic toxicity assay (ATA), which is used if there is a likelihood that the test material is toxic. Our original assay has since been incorporated into the American Society for Testing and Materials manual *ASTM Standards on Materials and Environmental Microbiology* (1). These assays are simple and inexpensive and provide a means to determine relative anaerobic biodegradability of an effluent or a potential discharge to the environment.

The biodegradation assay, in many ways, is an anaerobic parallel to the biological oxygen demand assay. Standardized and optimal conditions are provided to facilitate relative comparisons between the materials being tested and a standard organic test mixture. To simplify interpretation, no nitrate, sulfate, or Fe(III) salts are added, as these could provide alternative anaerobic electron acceptors to the carbonate used in this methanogenic assay. On the other hand, if denitrifying or sulfate- or iron-reducing activity is specifically sought, the assay would need to be modified for that purpose (discussed later). Incubation temperatures are optimal for methanogenic activity, and incubation times are longer than in the standard biological oxygen demand tests. Activity is monitored as total gas and/or methane formation.

The disadvantages to this anaerobic bioassay also parallel those for the biological oxygen demand assay. For example, it is an empirical test which is dependent on the organisms' activity. This activity can vary depending on the source of the inoculum. Ideally, one wants an active heterogeneous inoculum and sufficient inoculum from the same batch for the entire array of bioassay bottles being evaluated. Furthermore, because the BMP is a bioassay, toxic compounds in the material to be tested will interfere, hence the need for the ATA. The assay is also enormously simplified by the

use of anaerobic serum bottles, first envisioned by Miller and Wolin (2).

WHEN TO USE

The assay can be used for screening the anaerobic biodegradation of specific organic compounds or organic wastes with unknown composition. It has been used to determine the relative biodegradability of feed sources to anaerobic treatment (4), landfill leacheate (3), and a series of organic chemicals (5). As noted in some of these earlier publications, assessment of the environmental fate of organic waste should include an anaerobic evaluation, given the facts that once in the environment, anoxic habitats are widespread (e.g., in anaerobic waste treatment systems, waterlogged soils, and freshwater and coastal marine sediments) and in many cases accessibility to them is limited (e.g., groundwater).

Whether the assay can be applied to in situ field monitoring depends on the questions to be addressed. If there is reason to believe that nutrients, organic or inorganic, may be limiting and that the contaminant is susceptible to anaerobic metabolism, then the contaminant can be subject to this assay. The results, however, do not reflect in situ activity but rather reflect the potential for biodegradation under the most optimal conditions. Methods for assessing in situ activity are addressed in section VI of this volume.

HOW TO USE
Materials

Most of the chemical constituents are readily available in a laboratory. Items which may need to be obtained ahead of time include serum bottles (usually 65 or 160 ml), thick (~0.5-inch [~1.27-cm]) black butyl rubber serum stoppers, aluminum crimps, glass syringe for gas measurements, gastight syringe for gas chromatographic injections, and gas chromatograph for carbon dioxide and/or methane measurements. Samples to be measured should be collected fresh, kept cool, filled to the top of the vessel, sealed or provided with an anaerobic headspace (nitrogen), and transported quickly to the laboratory. Samples should continue to be kept as anaerobic as possible during use, e.g., kept in sealed bottles with little or no headspace, under nitrogen or argon.

TABLE 1 Stock solutions for the anaerobic biodegradation test[a]

Stock solution	Compound	Concn (g/liter)	Amt added/4 liters	Concn in medium (mmol)
S-1	Resazurin	0.5	8 ml	
S-2	KH_2PO_4	69.0	8 ml	1.0
	K_2HPO_4	88.0		1.0
	$(NH_4)_2HPO_4$	10.0		0.15
	NH_4Cl	100.0		3.7
S-3[b]	$MgCl_2 \cdot 6H_2O$	60.0	40 ml	3.0
	$FeCl_2 \cdot 4H_2O$	20.0		1.0
	KCl	10.0		1.3
	$CaCl_2$	10.0		0.90
	KI	1.0		0.060
	$MnCl_2 \cdot 4H_2O$	0.40		0.020
	$CoCl_2 \cdot 6H_2O$	0.40		0.017
	$NiCl_2 \cdot 6H_2O$	0.050		0.0021
	$CuCl_2$	0.050		0.0037
	$ZnCl_2$	0.050		0.0037
	H_3BO_3	0.050		0.0081
	$Na_2MoO_4 \cdot 2H_2O$	0.050		0.0018
	$NaVO_3 \cdot nH_2O$	0.050		~0.0041
	Na_2SeO_3	0.010		0.00054
S-4	$Na_2S \cdot 9H_2O$	50.0	8 ml	0.40
Bicarbonate	$NaHCO_3$		16.8 g	50.0

[a] From reference 4 with kind permission from Elsevier Science Ltd., The Boulevard, Langford Lane, Kidlington OX5 1GB, United Kingdom.

[b] A small amount of precipitate may form in S-3 on standing. Shake well before using.

Media

Table 1 lists the constituents for preparation of the medium, which is a bicarbonate-buffered mineral salts medium with resazurin as a redox indicator and with NaS as a reductant. The medium is prepared and dispensed anaerobically; this is relatively easily done by providing a constant stream of anaerobic gas (either N_2 or Ar) over the medium. All constituents except the last two should be combined in water, boiled, and then cooled under an anaerobic gas stream before the last two constituents (S-4 and bicarbonate) are added. This step further reduces the medium. The medium should be dispensed into deoxygenated or pregassed serum bottles. It is easiest to preinoculate the medium with about 100 ml of digester sludge (or another appropriate inoculum with a heterogeneous anaerobic bacterial community) before dispensing it into replicate serum bottles. For these anaerobic assays, it is best to obtain anaerobic digester sludge as inocula. If that is not possible, another source, e.g., secondary sludge, though less desirable, can be used. The chemical or waste to be tested can be added before the stopper is placed or injected through the stopper by using a syringe. The concentration of the test material is chosen in order to minimize its toxicity yet provide sufficient organic carbon for measurable but not excessive gas production. Concentrations which have been reported for undefined waste material are in the range of 1 to 2 g of chemical oxygen demand per liter; for compounds of known composition, 50 mg of organic carbon per liter is suggested (1). The optimal incubation temperature is 35 to 37°C.

Measurements

The simplest measurement is that of cumulative total gas production over time, which is compared against background controls (without the test chemical or waste). If it is possible or desirable, gas composition (CO_2 and CH_4) in

each sample taken can be measured by gas chromatography. For readily degradable material, daily measurements should be taken for the first week, and then weekly measurements should be taken for a total of about 4 weeks. For material which degrades slowly, weekly measurements should be sufficient. Gas production is measured with a glass syringe which has been lubricated with water. The size of the syringe (5 to 50 ml) depends on the volume of gas being produced. The syringe is preflushed with anaerobic gas, inserted through the stopper, held in a horizontal position, and allowed to come to an equilibration point. The number of punctures of the thick, black rubber stoppers (~10 to 12) over the time course of the assay has not been a problem with respect to leakage. A problem can arise if the contaminant material being tested is volatile and/or sorbs into the butyl rubber. Alternatives then are to use Teflon-coated stoppers or stoppers composed of a more resistant formulation (e.g., viton). Since monitoring is based on gas measurements, some obvious precautions (e.g., make all measurements at the same temperature) should be taken. Atmospheric pressure changes have not generally been confounding. If specific carbon dioxide concentrations are needed, the amount in solution will have to be characterized either by calculation or by sacrificing samples by acidification.

Controls

Two types of controls should be considered and run with the samples. Background controls, consisting of medium and inoculum, provide background levels of gas production which is a result of endogenous microbial activity in the inoculum and the bioconversion of organics which may have been present in the inoculum source. Positive controls consist of medium inoculum and a known biodegradable substrate, such as an acetate-propionate mixture (3:1 by weight; concentration depends on that of the other materials being tested). This provides a positive measure of the activity of the inoculum and gives confidence in the data obtained from the test material. If the positive control is active and there is lower-than-predicted gas production from the test material, then there is confidence that the gas production is due to incomplete biodegradability and not an inactive inoculum.

INTERPRETATION, EVALUATION, AND REPORTING

There are several levels of interpretation which can be applied to the gas production data. The simplest is to compare total gas volume produced with that from the active controls. For this comparison, it is important that the concentrations of organic carbon used in the positive controls and the test material be the same. Hence, concentrations should be calculated on the basis of the number of carbon molecules so that the same amount of carbon is provided to the positive controls and the test bottles. Then theoretically, if the test material is completely degradable, the total gas (CO_2 and CH_4) should be the same as that from the positive acetate-propionate controls. If the material is incompletely degraded, the total gas formed will reflect that fact and be less than that from the positive controls. Data can be summarized graphically or in tabular form. For example, Table 2 summarizes the percent theoretical methane produced from different substrates, using a sludge inoculum from a waste treatment plant. Since inocula from different sources can give different results, broad comparisons of results obtained with different methods should be made cautiously.

If the test material is an undefined carbon source, the results can be measured as cumulative methane produced over time. Then, on the basis of the theoretical formation of 0.35 ml of methane per mg of chemical oxygen demand (4) for complete conversion of organic matter, the conversion efficiency of the unknown test material can be calculated. The methane yield is based on the idealized stoichiometry of complete conversion of the organic matter to carbon dioxide and methane. This approach works well with readily utilizable organic matter. As noted in the next section, there are several ways to evaluate the results.

THE ATA

If the total gas or methane production is less than expected, two interpretations are possible. One is that the matter is nondegradable or partially nondegradable, for a variety of

TABLE 2 Percentage of the theoretical methane produced from eight compounds with one sludge inoculum[a]

Substrate	% of theoretical methane
Ethanol	94
Polyethylene glycol 20,000	82
p-Cresol	94
m-Cresol	91
Phthalic acid	96
Di-n-butylphthalate	0
2-Octanol	87
m-Chlorobenzoate	0

[a] Excerpted from reference 5 with permission. Inoculum is from primary digester; incubation is over an 8-week period.

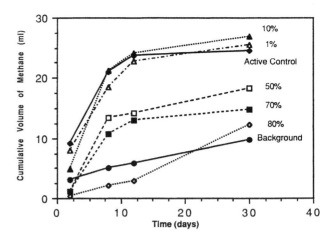

FIGURE 1 Cumulative volume of methane produced in various concentrations of leachate amended cultures in the ATA over time. (Reprinted from reference 3 with permission.)

reasons as noted in the next section. The second is that the matter is toxic. The difference between the two can be determined by the ATA, a simple means of determining if anaerobic degradation is being suppressed. The protocol is slightly different from that for the BMP. The ATA uses the same inoculum, concentrations of test material, background controls, positive controls, etc., as the BMP, plus an additional series of replicate bottles to which the acetate-propionate substrate is added. Different concentrations of the suspected toxic test material (e.g., 0.5, 0.1, and 0.01 of the original concentration suspected of being toxic) are added to the acetate-propionate replicate bottles. Toxicity is then observable as a reduction or suppression of gas (or methane) production compared with the uninhibited positive control. An example of an ATA taken from reference 3 is shown in Fig. 1.

VARIATIONS AND LIMITATIONS ON THE PROCEDURE

Several variations of the procedure have been described in the literature and can be implemented according to specific needs. For example, if a large number of samples are being routinely monitored on a regular basis, a pressure transducer system can be developed to automatically collect the gas measurements (5). If measurement of a denitrifying or sulfate- or iron-reducing activity is the goal, the method must be adapted. Because these anaerobic activities have not been the subject of an assay, appropriate tests would need to be developed and explored on a case-by-case basis. Since the BMP and ATA are both based on gas formation, the endpoints for evaluating these other anaerobic activities would likely be different (e.g., specifically assaying for substrate loss).

Finally, as noted at the beginning of this chapter, because the anaerobic biodegradability assay is a bioassay, the matter to be tested must be carbon based. If the material is not anaerobically biodegradable, the assay will not provide any insight as to why this is so. Nonbiodegradability may be due to a number of factors, such as the refractory nature of the compound, complexation or polymerization, toxicity to the microbial inoculum, incorporation into humic material, or lack of bioavailability. Determination of the specific factor responsible would require additional experimentation.

REFERENCES

1. **American Society for Testing and Materials.** 1993. *ASTM Standards on Materials and Environmental Microbiology*, 2nd ed., p. 166–170. American Society for Testing and Materials, Philadelphia.
2. **Miller, T. L., and M. J. Wolin.** 1974. A serum bottle modification of the Hungate technique for cultivating obligate anaerobes. *Appl. Environ. Microbiol.* **27:**985–987.
3. **O'Connor, O., R. Dewan, P. Galluzi, and L. Y. Young.** 1990. Landfill leacheate: a study of its anaerobic biodegradability and toxicity to methanogenesis. *Arch. Environ. Contam. Toxicol.* **19:**143–147.
4. **Owen, W. F., D. C. Stuckey, J. B. Healy, L. Y. Young, and P. L. McCarty.** 1979. Bioassay for monitoring biochemical methane potential and anaerobic toxicity. *Water Res.* **13:**485–492.
5. **Shelton, D. R., and J. M. Tiedje.** 1984. General method for determining anaerobic biodegradation potential. *Appl. Environ. Microbiol.* **47:**850–857.

Bioremediation and the Dissimilatory Reduction of Metals

GWEN LEE HARDING

89

While oxygen is the most energetically favorable electron acceptor for oxidizing organic contaminants, its availability is limited in numerous environments. Fortunately, many microorganisms have the capacity to use electron acceptors other than oxygen. Our understanding of microbial metabolism in anoxic environments has been greatly enhanced by recent investigations elucidating the role of metals as electron acceptors. We now know that the biological reduction of metals plays a significant role in the geochemistry of certain ecosystems (17, 18, 25–27, 31, 45, 49, 50). In addition, a role for dissimilatory reduction of metals has begun to unfold in the bioremediation arena either as a mechanism for reducing the toxicity of the metal itself or as an alternative electron acceptor for the concomitant oxidation of organic contaminants.

The dissimilatory reduction of a metal couples the oxidation of a substrate, the electron donor, to the reduction of a metal, the electron acceptor, thereby using the metal as an electron sink. When more than one electron acceptor is available, microorganisms will choose the one which will allow them to gain the most free energy and for which they are metabolically competent. The preferential order of the best-studied electron acceptors is oxygen, nitrate, sulfate, and carbon dioxide. The preferential reduction of the oxides and hydroxides of manganese and iron is between nitrate and sulfate (14, 39, 59). The primary focus of this chapter is on the dissimilatory reduction of iron. While manganese is also likely to be an important electron acceptor in some environments, our current understanding of its role in bioremediation is somewhat obscured by nonbiological reactions (50).

POTENTIAL USES OF DISSIMILATORY REDUCTION

Examples of metals other than iron or manganese which have been shown to undergo microbially mediated dissimilatory reduction include arsenic (1), chromate (51), cobalt (20), selenium oxyanions (47, 52, 55), and uranium (19, 41–43). The potential for reductive precipitation of uranium, chromium, selenium, lead, and technetium as a means of biorestoration has been reviewed recently (29). Promoting the microbial reduction of these metals may prove to be an effective technology for ameliorating their associated toxicity. However, while in some cases there are ongoing

efforts to move this technology to the field, much of the work to date has been confined to laboratory studies and not actual field demonstrations. An environment where metal reduction may play a significant role is wetlands. Constructed wetlands are becoming an acceptable technology for tertiary treatment of certain industrial and municipal wastewaters for which metals are a concern. One obstacle which needs to be addressed before precipitation becomes an acceptable in situ remediation technique is the potential for the metal to be remobilized, especially upon alterations in the redox potential of the environment.

Recent reviews detail our current understanding of the dissimilatory reduction of Fe(III) and Mn(IV) (27, 28, 50) as well as other metals (28). The potential importance of iron in particular in remediating organic contaminants either naturally or through engineered processes began to come into focus in the late 1980s and is continuing to gain momentum. The relative abundance of oxidized iron species in aquifers and sediments (2, 25, 36, 37) and the demonstration of microbial oxidation of pollutants coupled with iron reduction (7, 30, 35) have drawn particular attention to the possible role of iron as a significant source of reducing power in situ. Field studies of surface and subsurface contamination (2, 7), leachates from landfills (2, 3, 9, 14), and petroleum hydrocarbon plumes (6, 7, 10, 16, 57) support this concept.

DISSIMILATORY REDUCTION AS EVIDENCE FOR NATURAL ATTENUATION

In the above-mentioned field studies, it was observed that contaminated saturated environments exhibit a succession of redox zones, both horizontally and vertically, in which the electron acceptors, oxygen, nitrate, manganese oxides, iron oxides, hydroxides, and sulfate, are sequentially depleted, leaving ultimately a methanogenic environment. Vroblesky and Chapelle showed that the dominant terminal electron-accepting process (TEAP) can vary both temporally and spatially (57), emphasizing the dynamic nature of in situ processes. The availability and concentration of the various electron acceptors dictate the redox processes which dominate a particular environment, and the sum of all of the available electron acceptors dictates the capacity that an environment has for oxidizing organic contaminants. Understanding these concepts can provide a basis for predicting whether natural attenuation will contribute to the overall remediation at a site.

In an effort to capture these concepts in a quantitative manner, Barcelona and Holm (8) developed equations for calculating the oxidation capacity (OXC) of aquifers and their associated sediments. OXC is defined as the availability of oxidized species (moles per cubic meter) in an environmental compartment which can support the oxidation of contaminants. The equation which is currently used is the following (22, 23):

$$OXC = 4[O_2] + 5[NO_3^-] + [Fe(III)_{available}]$$
$$+ 2[Mn(IV)_{available}] + 8[SO_4^{2-}] \quad (1)$$
$$+ 4[CH_2O_{available}]$$

Examples of the application of this equation with use of data from several aquifers are documented in references 22 and 23.

While the above discussion presents the general theory for determining OXC, to actually measure OXC in sediments, a titanous EDTA (Ti^{3+}-EDTA) extraction method (22, 23) was developed. However, it should be noted that this method measures oxidized iron and manganese species and not other components which may be oxidizable. After anaerobic extraction of sediments with a solution of Ti^{3+}-EDTA, the unreacted Ti^{3+} is titrated and the OXC is calculated. Alternatively, the iron and manganese in the extract can be measured individually by atomic adsorption spectrophotometry. Measuring the OXC of a site provides at least an indication of the capacity that the specific environment has for oxidizing an organic contaminant.

The U.S. Environmental Protection Agency and the U.S. Air Force in collaboration with Engineering-Science, Inc., have drafted a protocol for identifying when "natural attenuation" or "intrinsic bioremediation" of a fuel contaminant plume in groundwater is occurring (58). Iron is one of the parameters included in the protocol and will likewise be included as an input parameter in soon to be released Bioplume III, the third generation of a model developed to evaluate the fate and transport of dissolved contaminants (58). The protocol suggests that observation of elevated concentrations of reduced iron within areas of dissolved BTEX (benzene, toluene, ethylbenze, and xylene) contamination is indicative of BTEX degradation by microbially mediated iron-reducing activity. Furthermore, the protocol suggests that the mass of BTEX lost to biodegradation can be estimated by knowing the volume of the contaminated water, the background Fe(II) concentration, and the concentration of Fe(II) in the contaminated area as indicated in the following equation (58):

$$BTEX_{Bio,Fe} = 0.05(Fe_M - Fe_B) \quad (2)$$

where $BTEX_{Bio,Fe}$ is the reduction in BTEX concentration via iron reduction, 0.05 is the concentration (milligrams per liter, average) of BTEX degraded per Fe(II) concentration (milligrams per liter) produced, Fe_M is the background Fe(II) concentration (milligrams per liter), and Fe_B is the measured Fe(II) concentration (milligrams per liter).

Several field studies in which ferrous iron concentrations contributed to the overall evidence supporting intrinsic bioremediation have been reported (6, 9, 10, 14, 16, 30, 57). It should be noted that equation 2 generates an extremely conservative estimation of the amount of BTEX degraded, since typically less than 5% of the Fe(II) produced during Fe(III) reduction is detectable as dissolved Fe(II) (27).

ADDITIONAL ROLES FOR IRON REDUCTION

Of the various species of oxidized iron, the more poorly crystalline structures have been observed to be more susceptible to microbial reduction (22, 36, 37, 48). However, in laboratory studies, chelating agents such nitrilotriacetate have been shown to enhance the rate of iron reduction and increase the availability of more bioresistant forms of Fe(III) (5, 46). These results suggest that it may be beneficial to investigate the use of chelating agents in field applications to increase the proportion of iron available for use as an electron sink.

Recent laboratory studies have also coupled the reduction of iron with degradation of nitroaromatic compounds. Heijman et al. (21) proposed a surface-bound iron redox cycle with bacteria reducing the iron, which in turn abiotically reduces nitroaromatic compounds. They also suggest a similar mechanism for reductive dehalogenations.

Strategies for remediating halogenated compounds often embrace a two-step process. The first step reductively dehalogenates the target compounds and is accomplished anaerobically. The second step is an aerobic step which mineralizes the carbon backbone. However, our understanding of the process of reductive dehalogenation as it occurs in natural environments is far from complete. Recent studies with two strains of *Shewanella putrefaciens* suggest that iron-reducing bacteria may directly dehalogenate compounds (53, 54) instead of contributing indirectly by reducing iron, which in turn abiotically dehalogenates the compound as suggested by Heijman et al. (21).

DEMONSTRATING THE PRESENCE OF IRON-REDUCING MICROBIAL POPULATIONS

Several iron-reducing populations contribute to the anaerobic degradation of organic material (24, 25, 27, 28, 40, 50) (Fig. 1). Some species can completely oxidize an organic substrate to CO_2 by using iron as the sole electron acceptor (31, 40). For example, *Geobacter metallireducens* (GS-15), an obligate anaerobe, is capable of growing on toluene, phenol, and p-cresol (30, 35). Some iron-reducing microorganisms, such as *S. putrefaciens*, contribute by oxidizing metabolic intermediates or products of fermentation (i.e., formate, lactate, pyruvate, and hydrogen) (11, 13, 33, 40, 44, 54). An interesting aspect of many of these organisms is the facultative nature of the iron reduction. Several iron reducers can reduce a variety of other electron acceptors, including nitrate, manganese oxides, sulfate, and even oxygen (4, 15, 18, 24, 40, 50).

To establish the presence of populations which have the potential to catalyze the reduction of iron, field samples are collected and incubated anaerobically and monitored for the production of Fe(II). Acetate and/or amorphous Fe(III) may be added to ensure that carbon and/or iron are not limiting (4). Synthetic amorphous iron can be generated by neutralizing a solution of $FeCl_3$ with NaOH as described by Lovley and Phillips (36). Strict anaerobic techniques should be employed throughout the processing and analyses of samples. If further enrichment for iron-reducing cultures is desired, recipes for media suitable for freshwater, estuarine, and sediment enrichments can be found in reference 36.

IRON DETERMINATION

Dissolved Fe(II) can be analyzed by atomic adsorption (2) or colorimetrically by the ferrozine method (36, 56). Field

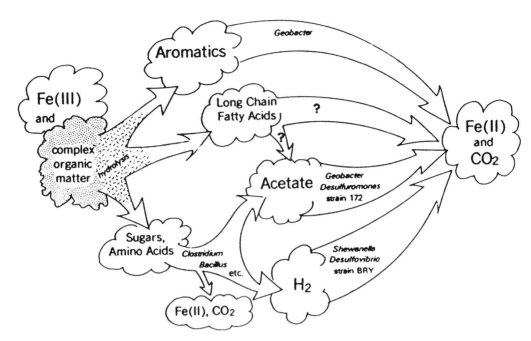

FIGURE 1 Model for oxidation of organic matter in sediments coupled to dissimilatory Fe(III) reduction showing examples of the microorganisms in pure culture known to catalyze the various reactions. Fermentation of sugars and amino acids has been simplified to designate the production of only the two major fermentation products, acetate and H_2. However, other short-chain fatty acids are produced in lesser amounts. These include propionate and formate, which may be directly oxidized to carbon dioxide by organisms such as G. *metallireducens* (propionate) or S. *putrefaciens* (formate), as well as lactate, which is oxidized to acetate and carbon dioxide by organisms such as S. *putrefaciens*. Reproduced with permission from the *Annual Review of Microbiology* (28), volume 47, ©1993 by Annual Reviews Inc.

kits based on the ferrozine method are now available from companies such as HACH. Total Fe(II), including that portion associated with the solid matrix, can be determined by first extracting the sample in 0.5 M HCl for 15 min (aqueous samples) or 60 min (sediments) before reacting it with the ferrozine reagent (36). While some procedures use 1,10-phenanthroline as the indicator reagent, this reagent is not as sensitive as ferrozine. Since Fe(II) is readily oxidized in air, analyses should be performed in the field or as quickly as possible after samples have been collected.

HYDROGEN CONCENTRATION AS AN INDICATOR OF IRON REDUCTION

Measurement of oxidized and reduced species as a means of delineating biological activities has its detractors. As pointed out by Chapelle et al. (13), simply knowing the patterns of electron acceptor consumption and final product accumulation does not necessarily delineate the true redox processes occurring at a specific point in time and place. Dilution along the groundwater flow may inadvertently lead to the misidentification of the boundary of a redox process.

An alternative method has been proposed for delineating TEAPs in general and can be used to identify when and where Fe(III) reduction is occurring (13, 32, 34). This method not only evaluates the presence or absence of various electron acceptors and their reduced counterparts but also includes measurement of hydrogen concentrations. The hypothesis is that hydrogen concentrations

more accurately reflect the predominate TEAP occurring at a specific location and time (13, 32, 34, 57). Conceptually, the hydrogen concentration associated with a particular TEAP is lower than for TEAPs which are less energetically efficient. For example, microorganisms which reduce iron inhibit sulfate reduction and methanogenesis by maintaining a localized hydrogen concentration which is too low to support the latter activities (13, 38, 57).

To determine dissolved hydrogen concentrations, groundwater samples are collected by using a "bubble strip" technique (12, 13, 34, 57) or a discrete depth water sampler (34). In both cases, dissolved hydrogen is allowed to equilibrate with a headspace for 15 min and then is analyzed by gas chromatography (34). Samples must be analyzed within 30 min of being collected. Iron-reducing environments are associated with H_2 concentrations of 0.1 to 0.8 nM (34).

SUMMARY

The concept of capitalizing on the microbially mediated dissimilatory reduction of metals for the purpose of bioremediation is still being developed. While field observations support a role for dissimilatory reduction of iron in the natural attenuation of organic contaminants, many of the proposed roles for dissimilatory reduction of metals described in this chapter are based on laboratory investigations and have not yet been proven to be fieldworthy. However, this chapter does summarize the tools presently available for evaluating whether biological reduction of iron is occurring at a site and the implications of the results with regard to the natural attenuation of organic contaminants.

REFERENCES

1. **Ahmann, D., A. L. Roberts, L. R. Krumholz, and F. M. M. Morel.** 1994. Microbe grows by reducing arsenic. *Nature* (London) **371:**750.
2. **Albrechtsen, H.-J.** 1994. Bacterial degradation under iron reducing conditions, p. 418–423. *In* R. E. Hinchee, B. C. Alleman, R. E. Hoeppel, and R. N. Miller (ed.), *Hydrocarbon Bioremediation.* Lewis Publishers, Boca Raton, Fla.
3. **Albrechtsen, H.-J., and T. H. Christensen.** 1994. Evidence for microbial iron-reduction in a landfill leachate polluted aquifer (Vejen, Denmark). *Appl. Environ. Microbiol.* **60:**3920–3925.
4. **Albrechtsen, H.-J., G. Heron, and T. H. Christensen.** 1995. Limiting factors for microbial Fe(III)—reduction in a landfill leachate polluted aquifer (Vejen, Denmark). *FEMS Microbiol. Ecol.* **16:**233–248.
5. **Arnold, R. G., T. M. Olson, and M. R. Hoffmann.** 1986. Kinetic and mechanism of dissimilative Fe(III) reduction by *Pseudomonas* sp. 200. *Biotechnol. Bioeng.* **28:**1657–1671.
6. **Baedecker, M. J., I. M. Cozzarelli, R. P. Eganhouse, D. I. Siegel, and P. C. Bennett.** 1993. Crude oil in a shallow sand and gravel aquifer. III. Biogeochemical reactions and mass balance modeling in anoxic groundwater. *Appl. Geochem.* **8:**569–586.
7. **Baedecker, M. J., D. E. Siegel, P. Bennett, and I. M. Cozzarelli.** 1989. The fate and effects of crude oil in a shallow aquifer. I. The distribution of chemical species and geochemical facies, p. 13–20. *In* G. E. Mallard and S. E. Ragone (ed.), *U.S. Geological Survey Water Resources Division Report 88-4220.* U.S. Geological Survey, Reston, Va.
8. **Barcelona, M. J., and T. R. Holm.** 1991. Oxidation-reduction capacities of aquifer solids. *Environ. Sci. Technol.* **25:**1565–1572.
9. **Bjerg, P. L., K. Rügge, J. K. Pedersen, and T. H. Christensen.** 1995. Distribution of redox-sensitive groundwater quality parameters downgradient of a landfill (Grindsted, Denmark). *Environ. Sci. Technol.* **29:**1387–1394.
10. **Borden, R. C., C. A. Gomez, and M. T. Becker.** 1995. Geochemical indicators of intrinsic bioremediation. *Ground Water* **33:**180–189.
11. **Caccavo, Jr., F., R. P. Blakemore, and D. R. Lovely.** 1992. A hydrogen-oxidizing, Fe(III)-reducing microorganism from the Great Bay Estuary, New Hampshire. *Appl. Environ. Microbiol.* **58:**3211–3216.
12. **Chapelle, F. H., and P. B. McMahon.** 1991. Geochemistry of dissolved inorganic carbon in a Coastal-Plain aquifer. 1. Sulfate from confining bead as an oxidant in microbial CO_2 production. *J. Hydrol.* **127:**85–108.
13. **Chapelle, F. H., P. B. McMahon, N. M. Dubrovsky, R. F. Fujii, E. T. Oaksford, and D. A. Vroblesky.** 1995. Deducing the distribution of terminal electron-accepting processes in hydrologically diverse groundwater systems. *Water Resour. Res.* **31:**359–371.
14. **Christensen, H. T., P. Kjeldsen, H.-I. Albrechtsen, G. Heron, P. H. Nielsen, P. L. Bjerg, and P. E. Holm.** 1994. Attenuation of landfill leachate pollutants in aquifers. *Crit. Rev. Environ. Sci. Technol.* **24:**119–202.
15. **Coleman, M. L., D. B. Hedrick, D. R. Lovley, D. C. White, and K. Pye.** 1993. Reduction of Fe(III) in sediments by sulphate-reducing bacteria. *Nature* (London) **361:**436–438.
16. **Ehrlich, G. G., E. M. Godsy, D. F. Goerlitz, and M. F. Jult.** 1983. Microbial ecology of a creosote-contaminated aquifer at St. Louis Park, Minnesota. *Dev. Ind. Microbiol.* **24:**235–245.
17. **Ehrlich, H. L.** 1990. *Geomicrobiology,* 2nd ed. Marcel Dekker, New York.
18. **Ghiorse, W. C.** 1988. Microbial reduction of manganese and iron, p. 305–331. *In* A. J. B. Zehnder (ed.), *Biology of Anaerobic Microorganisms.* John Wiley & Sons, Inc., New York.
19. **Gorby, Y., and D. R. Lovely.** 1992. Enzymatic uranium precipitation. *Environ. Sci. Technol.* **26:**205–207.
20. **Guenther, H., C. Frank, and H. Simon.** 1990. Use of cobalt complexes as redox mediators in microbial and electromicrobial reductions of different classes of compounds. *DECHEMA Biotechnol. Conf.* **4:**107–110.
21. **Heijman, C. G., E. Grieder, C. Holliger, and R. P. Schwarzenbach.** 1995. Reduction of nitroaromatic compounds coupled to microbial iron reduction in laboratory aquifer columns. *Environ. Sci. Technol.* **29:**775–783.
22. **Heron, G., T. H. Christensen, and J. C. Tjell.** 1994. Oxidation capacity of aquifer sediments. *Environ. Sci. Technol.* **28:**153–158.
23. **Heron, G., J. C. Tjell, and T. H. Christensen.** 1994. Oxidation capacity of aquifer sediment, p. 418–423. *In* R. E. Hinchee, B. C. Alleman, R. E. Hoeppel, and R. N. Miller (ed.), *Hydrocarbon Bioremediation.* Lewis Publishers, Boca Raton, Fla.
24. **Jones, J. G., W. Davison, and S. Gardener.** 1984. Iron reduction by bacteria: range of organisms involved and metals reduced. *FEMS Microbiol. Lett.* **21:**133–136.
25. **Lovley, D. R.** 1987. Organic matter mineralization with the reduction of ferric iron: a review. *Geomicrobiol. J.* **5:**375–399.
26. **Lovley, D. R.** 1990. Magnetite formation during microbial dissimilatory iron reduction, p. 151–166. *In* R. B. Frankel and R. P. Blakemore (ed.), *Iron Biominerals.* Plenum Publishing Corp., New York.
27. **Lovley, D. R.** 1991. Dissimilatory Fe(II) and Mn(IV) reduction. *Microbiol. Rev.* **55:**259–287.
28. **Lovley, D. R.** 1993. Dissimilatory metal reduction. *Annu. Rev. Microbiol.* **44:**263–290.
29. **Lovley, D. R.** 1995. Bioremediation of organic and metal contaminants with dissimilatory metal reduction. *J. Ind. Microbiol.* **14:**85–93.
30. **Lovley, D. R., M. J. Baedecker, D. J. Lonergan, I. M. Cozzarelli, E. J. P. Phillips, and D. I. Siegel.** 1989. Oxidation of aromatic contaminants coupled to microbial iron reduction. *Nature* (London) **339:**297–299.
31. **Lovley, D. R., F. H. Chapelle, and E. J. P. Phillips.** 1990. Fe(III)-reducing bacteria in deeply buried sediments of the Atlantic Coastal Plain. *Geology* **18:**954–957.
32. **Lovley, D. R., F. H. Chapelle, and J. C. Woodward.** 1994. Use of dissolved H_2 concentrations to determine distribution of microbially catalyzed redox reactions in anoxic groundwater. *Environ. Sci. Technol.* **28:**1205–1210.
33. **Lovley, D. R., S. J. Giovannoni, D. C. White, J. E. Champine, E. J. P. Phillips, Y. A. Gorby, and S. Goodwin.** 1993. *Geobacter metallireducens* gen. sp. nov., a microorganism capable of coupling the complete oxidation of organic compounds to the reduction of iron and other metals. *Arch. Microbiol.* **159:**336–344.
34. **Lovley, D. R., and S. Goodwin.** 1988. Hydrogen concentrations as an indicator of the predominant terminal electron-accepting reaction in aquatic sediments. *Geochim. Cosmochim. Acta* **52:**2993–3003.
35. **Lovley, D. R., and D. J. Lonergan.** 1990. Anaerobic oxidation of toluene, phenol, and *p*-cresol by the dissimilatory iron-reducing organism, GS-15. *Appl. Environ. Microbiol.* **56:**1858–1864.
36. **Lovley, D. R., and E. J. P. Phillips.** 1986. Organic matter mineralization with reduction of ferric iron in anaerobic sediments. *Appl. Environ. Microbiol.* **51:**683–689.

37. **Lovley, D. R., and E. J. P. Phillips.** 1986. Availability of ferric iron for microbial reduction in bottom sediments of the freshwater tidal Potomac River. *Appl. Environ. Microbiol.* **52:**751–757.

38. **Lovley, D. R., and E. J. P. Phillips.** 1987. Competitive mechanisms for inhibition of sulfate reduction and methane production in zones of ferric iron reduction in sediments. *Appl. Environ. Microbiol.* **53:**2636–2641.

39. **Lovley, D. R., and E. J. P. Phillips.** 1988. Manganese inhibition of microbial iron reduction in anaerobic sediments. *Geomicrobiol. J.* **6:**145–155.

40. **Lovley, D. R., and E. J. P. Phillips.** 1988. Novel mode of microbial energy metabolism: organic carbon oxidation coupled to dissimilatory reduction of iron or manganese. *Appl. Environ. Microbiol.* **54:**1427–1480.

41. **Lovley, D. R., and E. J. P. Phillips.** 1992. Reduction of uranium by *Desulfovibrio desulfuricans*. *Appl. Environ. Microbiol.* **58:**850–856.

42. **Lovley, D. R., and E. J. P. Phillips.** 1992. Bioremediation of uranium contamination with enzymatic uranium reduction. *Environ. Sci. Technol.* **26:**2228–2234.

43. **Lovley, D. R., E. J. P. Phillips, Y. A. Gorby, and E. R. Landa.** 1991. Microbial reduction of uranium. *Nature* (London) **350:**413–416.

44. **Lovley, D. R., E. J. P. Phillips, and D. J. Lonergan.** 1989. Hydrogen and formate oxidation coupled to dissimilatory reduction of iron or manganese by *Alteromonas putrefaciens*. *Appl. Environ. Microbiol.* **55:**700–706.

45. **Lovley, D. R., E. J. P. Phillips, and D. J. Lonergan.** 1991. Enzymatic versus nonenzymatic mechanisms for Fe(III) reduction in aquatic sediments. *Environ. Sci. Technol.* **25:**1062–1067.

46. **Lovley, D. R., J. C. Woodward, and F. H. Chapelle.** 1994. Stimulated anoxic biodegradation of aromatic hydrocarbons using Fe(III) ligands. *Nature* (London) **370:**128–131.

47. **Macy, J. M., T. A. Michel, and D. G. Kirsch.** 1989. Selenate reduction by a *Pseudomonas* species: a new mode of anaerobic respiration. *FEMS Microbiol. Lett.* **61:**195–198.

48. **Munch, J. C., and J. C. G. Ottow.** 1980. Preferential reduction of amorphous to crystalline iron oxides by bacterial activity. *Soil Sci.* **129:**15–21.

49. **Nealson, K. H., and C. R. Myers.** 1990. Iron reduction by bacteria: a potential role in the genesis of banded iron formations. *Am. J. Sci.* **290A:**35–45.

50. **Nealson, K. H., and C. R. Myers.** 1992. Microbial reduction of manganese and iron: new approaches to carbon cycling. *Appl. Environ. Microbiol.* **58:**439–443.

51. **Ohtake, H., and S. Silver.** 1994. Bacterial detoxification of toxic chromate, p. 403–415. *In* G. R. Chaudhry (ed.), *Biological Degradation and Bioremediation of Toxic Chemicals.* Dioscordes Press, Portland, Oreg.

52. **Oremland, R. S., J. T. Hollibaugh, A. S. Maest, T. S. Presser, L. G. Miller, and C. W. Culbertson.** 1989. Selenate reduction to elemental selenium by anaerobic bacteria in sediments and culture: biogeochemical significance of a novel, sulfate-independent respiration. *Appl. Environ. Microbiol.* **55:**2333–2343.

53. **Petrovskis, E. A., T. M. Vogel, and P. Adriaens.** 1994. Effects of electron acceptors and donors on transformation of tetrachloromethane by *Shewanella putrefaciens* MR-1. *FEMS Microbiol. Lett.* **121:**357–364.

54. **Picardal, F., R. G. Arnold, and B. B. Huey.** 1995. Effects of electron donor and acceptor conditions on reductive dehalogenation of tetrachloromethane by *Shewanella putrefaciens* 200. *Appl. Environ. Microbiol.* **61:**8–12.

55. **Steinberg, N. A., J. S. Blum, L. Hochstein, and R. S. Oremland.** 1992. Nitrate is a preferred electron acceptor for growth of freshwater selenate-respiring bacteria. *Appl. Environ. Microbiol.* **58:**426–428.

56. **Stookey, L. L.** 1970. Ferrozine—a new spectrophotometric reagent for iron. *Anal. Chem.* **42:**779–781.

57. **Vroblesky, D. A., and F. H. Chapelle.** 1994. Temporal and spatial changes of terminal electron-accepting processes in a petroleum hydrocarbon-contaminated aquifer and the significance for contaminant biodegradation. *Water Resour. Res.* **30:**1561–1570.

58. **Wiedemeier, T. H., J. T. Wilson, D. H. Kampbell, R. N. Miller, and J. E. Hansen.** 1995. Technical protocol for implementing intrinsic remediation with long-term monitoring for natural attenuation of fuel contamination dissolved in groundwater. Unpublished data.

59. **Zehnder, A. J. B.** 1988. Geochemistry and biogeochemistry of anaerobic habitats, p. 1–38. *In* A. J. Zehnder (ed.), *Biology of Anaerobic Microorganisms.* John Wiley & Sons, Inc., New York.

Aerobic Biotransformations of Polychlorinated Biphenyls

D. D. FOCHT

90

ENVIRONMENTAL TOXICOLOGY AND HISTORY

Polychlorinated biphenyls (PCBs) were introduced into the environment in the early 1920s from industrial processes requiring lubricants and heat capacitors with high thermal stability. Because of their persistence in the environment, biomagnification in the food chain, and suspected carcinogenic properties, they were banned from production and distribution in the United States in 1972. The chemical features that make these compounds so stable also make them resistant to decomposition in the environment. Thus, PCBs remain as a legacy from an era when soil and rivers were used as repositories of industrial wastes.

Commercial mixtures of PCBs generally contain 40 to 60 of the 209 possible congeners ranging from mono- to decachlorobiphenyl. They go under the trade names of Kanechlor (Mitsubishi-Monsanto; Japan), Phenoclor and Pyralene (Prodelec; France), Clophen (Beyer; Germany), Fenclor (Caffaro; Itaty), and Aroclor (Monsanto; United States, Canada, and United Kingdom). The trade name Aroclor is affixed with a number that represents the number of carbon atoms and the average percent molecular weight of chlorine. For example, Aroclor 1260 has 12 carbon atoms (biphenyl [BP]) and an average molecular weight of 60% chlorine. Not all congeners are toxic; 3,3',4,4'-tetrachlorobiphenyl and similar congeners having a planar structure, which resembles that of 2,3,7,8-tetrachlorodibenzodioxin, are among the most toxic (34).

BACTERIAL METABOLISM OF PCBs

The aerobic degradation of PCBs is synonymous with the BP pathway Fig. 1 BP-utilizing bacteria were first described by Lunt and Evans (28) and later shown to cometabolize chlorobiphenyls (3). Thus, the substrate of choice for isolating PCB degraders from the environment and growing them in culture has been BP. Claims have been made in the literature for isolation of bacteria able to utilize PCBs as growth substrates. However, these reports must be viewed as equivocal for one or more of the following reasons: the cultures are not available, the compositions of the growth media were inadequately defined, evidence for mineralization and dehalogenation was lacking, and contamination from BP was not considered.

Aerobic transformation of PCBs is generally less effective with increasing chlorine substitution, although transformation of some hexachlorobiphenyls has been reported (6, 22, 27). The extent of congener transformation is variable; in some cases, chlorobenzoates are produced, while in others, phenolic degradation products accumulate. Because dioxygenation occurs primarily at the 2,3 positions, it has generally been assumed that *ortho*-chlorines prevent dioxygenation, and thus the 2,2',5,5'-tetrachlorobiphenyl congener would not be attacked. However, Furukawa et al. (19) noted many years ago that this congener was oxidized to an unidentified phenolic product. More recently, Haddock et al. (20) have shown that this congener is attacked in two different ways by the same enzyme. One route results in dioxygenation at the 2,3 site to give the corresponding catechol by spontaneous elimination of HCl from the unstable chlorohydrodiol. The other route involved dioxygenation at the 3,4 positions to produce a diol, which was not further metabolized. The significance of 3,4-dioxygenation remains obscure, as 4-phenylcatechol is not a substrate for the phenylcatechol dioxygenase (9).

ISOLATION AND MAINTENANCE OF PCB-METABOLIZING BACTERIA

Enrichment culture of BP degraders is necessary when their numbers are below $10^3 \cdot g^{-1}$, which is not generally common in soil (14, 22). Use of enrichment culture, in which numbers are below this level (e.g., groundwater), will select for fast-growing prototrophic bacteria, such as pseudomonads, with defined media. Thus, direct isolation by spread plating yields a greater diversity of bacteria, including the slower-growing coryneforms. Small additions (50 mg · liter^{-1}) of yeast extract (YE) to defined mineral salts (MS) medium (12) are useful in accelerating colony formation of all bacteria, including pseudomonads, but high concentrations are to be avoided. For example, 0.1% YE, when used with 0.01% chlorobenzene and chlorotoluenes (37), does not select for growth on the chlorinated substrate but rather selects for growth of YE utilizers that are tolerant to it and may not metabolize it.

A small crystal of BP is placed on the bottom lid to allow the vapors to ascend to the top lid containing the MS agar. The two lids are sealed together with Parafilm or placed in a closed chamber (e.g., desiccator) and incubated at 28°C until colonies are visible. Incubations should never be car-

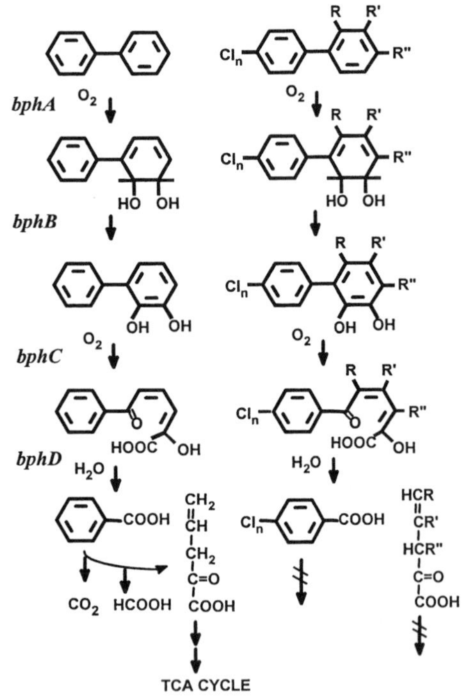

FIGURE 1 BP/PCB degradation pathway with the genes (bphABCD) encoding the four sequential steps in the pathway listed on the left. TCA, tricarboxylic acid. Redrawn from reference 7. Reprinted by permission of Kluwer Academic Publishers.

ried out at 37°C, as this temperature is lethal to many soil bacteria. Colonies can then be transferred to agar slants. Slants are prepared by adding a hexane solution of BP to each tube of medium to give a final concentration of 300 mg · liter^{-1} prior to autoclaving. Hexane will evaporate and BP will be deposited at the butt of the slant as it cools. The autoclave should be installed in a room having ventilation that meets public health standards or should be self-contained with its own condensing unit. If there is any doubt, the health and safety office of the institution should be consulted.

BP utilizers generally grow poorly in sterile liquid media. If BP has been added before the sample is autoclaved, the medium will form a large lump upon cooling and provide poor surface area for growth because of poor diffusion and low solubility (7 mg · liter^{-1}). Some cultures may grow well under these conditions, but most grow poorly in terms of rate and maximal density. Better growth is achieved by powdering BP with a mortar and pestle and adding it directly to the sterile MS medium. Airborne or substrate contamination is not a problem if the inoculant concentration is high (10^7 cells · ml^{-1}), as any contaminant could not outgrow the inoculant before the cells were harvested from the late-exponential-growth phase. Nevertheless, it cannot be stressed too strongly that this procedure, though useful for growing cells for metabolic studies, should never be used for maintenance or transfer of cultures. Subcultures and stock cultures should always be maintained aseptically on solid media.

BACTERIAL METABOLISM OF CHLOROBENZOATES

The production of chlorobenzoates by the BP/PCB pathway was first demonstrated 20 years ago (3) and later shown to occur in the environment (5, 11). Thus, chlorobenzoate degraders are relevant to complete biodegradation of PCBs. Construction of a PCB degradation pathway by combining the genes of BP and chlorobenzoate degraders was first suggested by Furukawa and Chakrabarty (18) and later demonstrated by others (1, 21, 24, 30). Axenic consortia have also been used to demonstrate mineralization of PCB congeners (2, 10, 33). Chlorobenzoates are metabolized by aerobic bacteria through four primary routes, which have been reviewed in detail elsewhere (13) but are summarized here as follows: (i) the *ortho*-chlorocatechol pathway, in which chlorine is removed after ring fission, (ii) dioxygenation of the aromatic ring with spontaneous removal of *ortho*-chlorines, (iii) dioxygenolytic removal of *meta*-chlorines, and (iv) nucleophilic hydroxyl displacement of *para*-substituted chlorines.

Unlike BP degraders, chlorobenzoate degraders are not as easily isolated from the environment. For example, a 3-chlorobenzoate utilizer described by Dorn et al. (8) required several months of adaptation to the substrate after first being grown on benzoate. A 4-chlorobenzoate degrader isolated by Marks et al. (29) required an acclimation of 4 months, and a dichlorobenzoate degrader (23) required 13 months for isolation. Moreover, 3-chlorobenzoate degradation does not readily occur in soil (15, 32) or unpolluted water (16) except when inoculated with degraders isolated from polluted environments. Transfer of the chlorobenzoate genes to indigenous bacteria has been well established in aquatic microcosms (17) and in soil (25). Schlömann (35) suggests that chlorobenzoate degradation may involve "recruitment" (i.e., genetic exchange) from other pathways. Consequently, direct isolation of chlorobenzoate degraders on selective media would not be a common occurrence.

AGAR PLATE METHODS

Auxotrophic bacteria that cometabolize PCBs cannot be cultured or enumerated on MS agar plates containing BP. Two methods can be used to assess metabolism of BP or PCBs. One involves a postincubation, after colony formation is observed. The other involves mixing of the substrate while the plates are solidifying. The first method, devised by Sylvestre (36) for metabolism of 4-chlorobiphenyl, involves spraying with a 5% (wt/vol) 4-chlorobiphenyl solution of diethyl ether to deposit a film. Plates are incubated for an additional 48 to 72 h. No incidence of toxicity from the solvent to 100 colonies picked at random was observed. The other method involves the addition of an acetone solution. When the plates have cooled to about 55°C, the solution is added and the plates are swirled manually until they start to set. At this time, the substrate is precipitated in small, grainy deposits throughout the agar. The translucent appearance of the plates enables BP or PCB metabolism to be observed by zones of clearing around colonies.

Dehalogenation of chlorobenzoates or other degradation products not utilized for growth can be detected by their addition (2 mM) and that of bromothymol blue (0.015%) to the MS agar medium. Positive colonies will be indicated by the formation of yellow zones on a green background, as a result of HCl production. For this method to work, the buffering capacity must be reduced by reducing the amount of phosphate salts in the MS medium (12) by half. This method is particularly useful for determining if hybrid strains or natural isolates can metabolize both BP and chlorobenzoate, which is uncommon. However, appropriate controls, lacking the chlorinated substrate, should be tested to ensure that acid production is not due to metabolism of the growth substrate.

EXTRACTION OF PCBs

Metabolism of PCBs and their metabolites can be studied with growing cells or resting cells. The latter are preferable, as incubations can be performed overnight for analysis the following day even though resting cell suspensions lose activity and may show lower specific activity than growing cells (27). Moreover, some strains of BP utilizers will not grow when incubated with PCBs because of substrate or product toxicity. Methods for preparations of resting cell suspensions are described in reference 12 and in other chapters of this book.

Assays are performed in 8-ml vials equipped with screw caps. PCBs are added from a hexane solution, and the solvent is evaporated in a fume hood to give the desired concentration (10 to 100 mg · liter^{-1}). Four milliliters of the washed cell suspension is added to the vial and incubated on a platform shaker. The cap should be loosely secured to permit adequate aeration. Because PCBs also adsorb to glassware and to cells, it is necessary to run a heat-killed control for each culture. Comparison between each culture and its control ensures that biological degradation has occurred.

At the end of the incubation, an internal standard is added to the flask in a hexane solution to give a concentration about 1/50 of the initial concentration of PCBs, which would be equivalent to a single congener. This congener (e.g., 2,2′,3,3′,5,5′-hexachlorobiphenyl) should be one that is not present in Aroclor 1242 or 1254 and is clearly separable from the other congeners in the gas chromatographic analysis. Retention times and characteristics of the congeners of commercial Aroclors can be found elsewhere (26, 31). The incubation mixture is acidified with 0.1 ml of 15 M H_3PO_4 to stop the reaction and to reduce emulsions upon extraction. Addition of Triton X-100 (1% final mixture) improves recovery, gives good separation of the hexane (top) and aqueous (bottom) phases, and is insensitive to an electron capture detector (ECD). An equivalent volume of hexane (4 ml) is added, and the vial is sealed and shaken for 1 h. Na_2SO_4 is added in excess of its solubility to reduce emulsions. The contents are placed into a 50-ml separatory funnel, and the aqueous (bottom) layer is drawn back to the vial for another extraction. The hexane layer is dried over anhydrous Na_2SO_4 and drawn off into a 10-ml volumetric flask. Extraction of the aqueous phase is performed two more times, and the volumetric flask is finally topped off with hexane.

PCBs have usually been analyzed by ECDs because they are 10- to 20-fold more sensitive than flame ionization detectors (FIDs). ECDs are relatively insensitive to hexane-extractable cellular components and to Triton X-100 (composed of long-chain alcohols). The disadvantage of ECDs is that they are not linear over a 25-fold change in concentration, require ultrahigh-purity gases, and require frequent cleaning and maintenance. Thus, standard curves need to be prepared from at least six samples over the concentration range of the samples. Moreover, licensing and regulatory requirements involved with the ^{63}Ni detector make them problematic in shipping to developing countries for use.

FIDs, on the other hand, give a linear response over 5 orders of magnitude and present fewer problems with maintenance. Sensitivity is not a problem, as 2 ng of individual congeners of PCBs is detectable. As FIDs have sensitivity more comparable to that of mass selective detectors on mass spectrometers, mass spectroscopists generally prefer an FID trace for evaluation prior to gas chromatography-mass spectrometry analysis. When an FID or mass selective detector is used, sodium dodecyl sulfate should be used instead of Triton X-100 because it does not partition into the hexane phase to interfere with the gas chromatographic analysis. The response mechanisms of ECDs and FIDs are completely opposite: ECDs are more sensitive to electronegative functional groups (e.g., halogens), which causes proportionally greater sensitivity to more highly chlorinated congeners.

For soil extraction, a 1:1 (vol/vol) hexane-acetone mixture should be used to disrupt water films that would otherwise prevent desorption of PCBs from soil. Extraction efficiency from spiked soil should first be determined, as recovery may vary as a result of organic matter content. The extract is dried over anhydrous Na_2SO_4 and passed through a silica gel no. 8 classic Sep-Pak (Millipore Corp., Bedford, Mass.) to remove soluble organic matter. Extracts that are yellow instead of clear may be indicative of a defective Sep-Pak. Because silica gel is very reactive to H_2O and CO_2, it is good practice to store unused Sep-Paks in a desiccator after the package has been exposed to the atmosphere. Exposed but unused Sep-Paks can be regenerated by heating at 110°C for 1 h.

EXTRACTION OF PCB METABOLITES

PCB degradation products (e.g., chlorobenzoates and chloroaliphatic acids) will not be extracted from soils or aqueous systems with hexane because they are too polar and will remain in the aqueous phase. Quantitative analysis of degradation products can be achieved by direct injection of aqueous extracts of cultures or an H_2O-methanol (1:1) extraction of soil into a high-pressure liquid chromatograph. The utility of high-pressure liquid chromatography is limited by the type of detector that is used; a UV detector is insensitive to saturated aliphatic acids (e.g., chloroacetic acid) lacking a chromophore group.

Analysis of degradation products by gas chromatography-mass spectrometry requires extraction with a more polar solvent than hexane. Ethyl acetate or diethyl ether (10 ml) is added to the aqueous phase (described in the previous section) in a 25-ml separatory funnel and rocked gently. The solvent layer is added to a vial or beaker and dried over 1 g of anhydrous Na_2SO_4 but not evaporated to dryness. Methylation of -OH groups is desirable, as degradation products may give very broad peaks or may be chemically and thermally unstable in the acid form. This is accomplished by concentrating the solvent to less than 1 ml with a rotary evaporator or with an N_2 stream. The sample is derivatized by adding an ethereal solution of diazomethane until a yellow color persists and is then concentrated under a gentle stream of N_2 to about 100 μl for analysis by gas chromatography-mass spectrometry (4).

REFERENCES

1. **Adams R. H., C. M. Huang, F. K. Higson, V. Brenner, and D. D. Focht.** 1992. Construction of a 3-chlorobiphenyl-utilizing recombinant from an intergeneric mating. *Appl. Environ. Microbiol.* **58:**647–654.
2. **Adriaens, P., H. P. Kohler, D. Kohler-Staub, and D. D. Focht.** 1989. Bacterial dehalogenation of chlorobenzoates and coculture biodegradation of 4,4′-dichlorobiphenyl. *Appl. Environ. Microbiol.* **55:**887–892.

3. **Ahmed, M., and D. D. Focht.** 1973. Degradation of polychlorinated biphenyls by two species of *Achromobacter*. *Can. J. Microbiol.* **19:**47–52.

4. **Arensdorf, J. J., and D. D. Focht.** 1995. A *meta* cleavage pathway for 4-chlorobenzoate, an intermediate in the metabolism of 4-chlorobiphenyl by *Pseudomonas cepacia* P166. *Appl. Environ. Microbiol.* **61:**443–447.

5. **Bailey, R. E., S. J. Gonsior, and W. L. Rhinehart.** 1983. Biodegradation of the monochlorobiphenyls and biphenyl in river water. *Environ. Sci. Technol.* **17:**617–621.

6. **Bedard, D. L., R. Unterman, L. H. Bopp, M. J. Brennan, M. L. Haberl, and C. Johnson.** 1986. Rapid assay for screening and characterizing microorganisms for the ability to degrade polychlorinated biphenyls. *Appl. Environ. Microbiol.* **51:**761–768.

7. **Brenner, V., J. J. Arensdorf, and D. D. Focht.** 1994. Genetic construction of PCB degraders. *Biodegradation* **5:**359–377.

8. **Dorn, E., M. Hellwig, W. Reineke, and H.-J. Knackmuss.** 1974. Isolation and characterization of a 3-chlorobenzoate degrading pseudomonad. *Arch. Microbiol.* **99:**61–70.

9. **Eltis, L. D., B. Hofmann, H.-J. Hecht, H. Luensdorf, and K. N. Timmis.** 1993. Purification and crystalization of 2,3-dihydroxybiphenyl 1,2-dioxygenase. *J. Biol. Chem.* **268:**2727–2732.

10. **Fava, F., and L. Marchetti.** 1991. Degradation and mineralization of 3-chlorobiphenyl by a mixed aerobic bacterial culture. *Appl. Microbiol. Biotechnol.* **36:**240–245.

11. **Flanagan, W. P., and R. J. May.** 1993. Metabolite detection as evidence for naturally occurring aerobic PCB biodegradation in Hudson River sediments. *Environ. Sci. Technol.* **27:**2207–2212.

12. **Focht, D. D.** 1994. Microbiological procedures for biodegradation research. p. 407–426. *In* R. W. Weaver, J. S. Angle, and P. S. Bottomley (ed.), *Methods of Soil Analysis, part 2. Microbiogical and Biochemical Properties.* Soil Science Society of America, Madison, Wis.

13. **Focht, D. D.** 1996. Biodegradation of chlorobenzoates, p. 71–80. *In* T. Nakazawa, K. Furukawa, D. Haas, and S. Silver (ed.), *Molecular Biology of Pseudomonads.* American Society for Microbiology, Washington, D.C.

14. **Focht, D. D., and W. Brunner.** 1985. Kinetics of biphenyl and polychlorinated biphenyl metabolism in soil. *Appl. Environ. Microbiol.* **50:**1058–1063.

15. **Focht, D. D., and D. Shelton.** 1987. Growth kinetics of *Pseudomonas alcaligenes* C-O relative to inoculation and 3-chlorobenzoate metabolism in soil. *Appl. Environ. Microbiol.* **53:**1846–1849.

16. **Fulthorpe, R. R., and R. C. Wyndham.** 1989. Survival and activity of a 3-chlorobenzoate-catabolic genotype in a natural system. *Appl. Environ. Microbiol.* **55:**1584–1590.

17. **Fulthorpe, R. R., and R. C. Wyndham.** 1991. Transfer and expression of the catabolic plasmid pBRC60 in wild bacterial recipients in a freshwater ecosystem. *Appl. Environ. Microbiol.* **57:**1546–1553.

18. **Furukawa, K., and A. M. Chakrabarty.** 1982. Involvement of plasmids in total degradation of chlorinated biphenyls. *Appl. Environ. Microbiol.* **44:**619–626.

19. **Furukawa, K., K. Tonomura, and A. Kamibayashi.** 1978. Effect of chlorine substitution on the biodegradability of polychlorinated biphenyls. *Appl. Environ. Microbiol.* **35:**223–227.

20. **Haddock, J. D., J. R. Horton, and D. T. Gibson.** 1995. Dihydroxylation and dechlorination of chlorinated biphenyls by purified biphenyl 2,3-dioxygenases from *Pseudomonas* sp. strain LB400. *J. Bacteriol.* **177:**20–26.

21. **Havel, J., and W. Reineke.** 1991. Total degradation of various chlorobiphenyls by cocultures and in vivo constructed hybrid pseudomonads. *FEMS Microbiol. Lett.* **78:**163–170.

22. **Hernandez, B. S., J. J. Arensdorf, and D. D. Focht.** 1995. Catabolic characteristics of biphenyl-utilizing isolates which cometabolize PCBs. *Biodegradation* **6:**75–82.

23. **Hernandez, B. S., F. K. Higson, R. Kondrat, and D. D. Focht.** 1991. Metabolism of and inhibition by chlorobenzoates in *Pseudomonas putida* P111. *Appl. Environ. Microbiol.* **57:**3361–3366.

24. **Hickey, W. J., V. Brenner, and D. D. Focht.** 1992. Mineralization of 2-chloro- and 2,5-dichlorobiphenyl by *Pseudomonas* sp. strain UCR2. *FEMS Microbiol. Lett.* **98:**175–180.

25. **Hickey, W. J., D. B. Searles, and D. D. Focht.** 1993. Enhanced mineralization of polychlorinated biphenyls in soil inoculated with chlorobenzoate-degrading bacteria. *Appl. Environ. Microbiol.* **59:**1194–1200.

26. **Hutzinger, O., S. Safe, and V. Zitko.** 1974. *The Chemistry of PCBs.* CRC Press, Cleveland.

27. **Kohler, H.-P. E., D. Kohler-Staub, and D. D. Focht.** 1988. Cometabolism of PCBs: enhanced transformation of Aroclor 1254 by growing bacterial cells. *Appl. Environ. Microbiol.* **54:**1940–1945.

28. **Lunt, D., and W. C. Evans.** 1970. The microbial metabolism of biphenyl. *Biochem. J.* **118:**54.

29. **Marks, T. S., A. R. W. Smith, and A. V. Quirk.** 1984. Degradation of 4-chlorobenzoic acid by *Arthrobacter* sp. *Appl. Environ. Microbiol.* **48:**1020–1025.

30. **Mokross, H., E. Schmidt, and W. Reineke.** 1990. Degradation of 3-chlorobiphenyl by in vivo constructed hybrid pseudomonads. *FEMS Microbiol. Lett.* **71:**179–185.

31. **Mullin, M. D., C. D. Pochini, S. McCrindle, M. Romkes, S. H. Safe, and L. H. Safe.** 1984. High-resolution PCB analysis: synthesis and chromatographic properties of all 209 PCB congeners. *Environ. Sci. Technol.* **18:**468–476.

32. **Pertsova, R. N., F. Kunc, and L. A. Golovleva.** 1984. Degradation of 3-chlorobenzoate in soil by pseudomonads carrying biodegradative plasmids. *Folia Microbiol.* **29:**242–247.

33. **Pettigrew, C. A., A. Breen, C. Corcoran, and G. S. Sayler.** 1990. Chlorinated biphenyl mineralization by individual populations and consortia of freshwater bacteria. *Appl. Environ. Microbiol.* **56:**2036–2045.

34. **Safe, S., S. Bandiera, T. Sawyer, L. Robertson, L. Safe, A. Parkinson, P. E. Thomas, D. E. Ryan, L. M. Reik, W. Levin, M. Denomme, and T. Fujita.** 1985. PCBs: structure-function relationships and mechanism of action. *Environ. Health Perspect.* **60:**47–56.

35. **Schlömann, M.** 1994. Evolution of chlorocatechol catabolic pathways. *Biodegradation* **5:**301–321.

36. **Sylvestre, M.** 1980. Isolation method for bacterial isolates capable of growth on *p*-chlorobiphenyl. *Appl. Environ. Microbiol.* **39:**1223–1224.

37. **Vandenbergh, P. A., R. H. Olsen, and J. F. Colaruotolo.** 1981. Isolation and genetic characterization of bacteria that degrade chloroaromatic compounds. *Appl. Environ. Microbiol.* **42:**737–739.

Biodegradation of Agricultural Chemicals

STEPHEN N. BRADLEY, TERRY B. HAMMILL, AND RONALD L. CRAWFORD

91

Pesticides and herbicides used in agriculture are the classic example of anthropogenic chemicals that enter the Earth's environment in large amounts via nonpoint sources. These compounds can adversely affect nontarget organisms and may be detrimental to human health if people are exposed to residues of the molecules in food, water, or agricultural products. The longer such chemicals persist in the environment, the greater the chance of their causing harm. Thus, scientists have long recognized the need to study degradative processes, particularly biodegradation, that transform these molecules in the environment. Mineralization—conversion of a chemical to carbon dioxide and other fully oxidized products—is the ultimate, and usually most desirable, level of biotransformation. Most anthropogenic compounds are mineralized by microbial processes in soil and water, but there are exceptions. Therefore, methods of measuring the biotransformation of agricultural chemicals by microorganisms, both pure cultures and mixed populations in soil or water, are essential in the repertoire of tools used by environmental microbiologists in studying the fate and effects of pesticides and herbicides in nature. Pure culture studies tell us about mechanisms of biotransformation, while mixed culture experiments tell us about the potential for microbial cotransformation processes that may exceed the capabilities of individual species.

In this chapter we discuss several specific methods that we have used to examine the biodegradability of herbicides by pure cultures and mixed populations of soil bacteria. We have chosen to use 2,4-dichlorophenoxy acetic acid (2,4-D) and 2,4-dinitro-6-sec-butylphenol (dinoseb) as model compounds. 2,4-D is a readily degradable herbicide of moderate toxicity to nontarget organisms, while dinoseb is a very persistent herbicide, now banned, that is highly toxic to virtually all living systems. We will report on both aerobic and anaerobic procedures for measuring the biodegradability of these molecules.

BIODEGRADATION OF 2,4-D AND DINOSEB BY PURE BACTERIAL CULTURES

The most convenient and unequivocal method for showing that a chemical is completely biodegraded to carbon dioxide uses radiorespirometric techniques. The compound of inter-

est, containing carbon in the form of radioactive ^{14}C, must first be purchased or synthesized. If, like our two examples, the compound is aromatic, the ^{14}C should be uniformly distributed in the rings of the benzene moiety, since this is usually the most recalcitrant portion of the molecule. Alternatively, the fate of side chain carbons can be studied by acquiring the chemical with ^{14}C in these moieties. U-[ring]-^{14}C-2,4-D is available commercially (e.g., Sigma Chemical Company, St. Louis, Mo.; 1 to 25 mCi/mmol as a solid). Phenyl-U-^{14}C-dinoseb can be acquired only by custom synthesis [3].

Aerobic Mineralization of 2,4-D by a Pure Culture

To demonstrate aerobic 2,4-D mineralization, agitated cultures in 250-ml flasks containing individual CO_2 traps are used. A trap consists of a small glass cup connected to a short glass rod passing through the bottom of the rubber stopper that seals the flask. Rubber can sometimes absorb volatile ^{14}C preliminary intermediates, so Teflon stoppers, although expensive, are better for use when volatile intermediates are produced during metabolism. Each cup contains 0.5 ml of sterile 1 N NaOH. For a bacterium such as *Alcaligenes eutrophus* JMP134(pJP4), isolated for its ability to grow on 2,4-D as a sole source of carbon and energy [1], we use a defined mineral salts medium (50 ml per flask) of the following final composition (grams per liter): $K_2HPO_4\cdot3H_2O$ (4.2), $NaH_2PO_4\cdot H_2O$ (1.0), NH_4Cl (2.0), nitrilotriacetic acid (0.1), $MgSO_4\cdot7H_2O$ (0.2), $FeSO_4\cdot7H_2O$ (0.012), $MnSO_4\cdot H_2O$ (0.003), $ZnSO_4\cdot7H_2O$ (0.003), and $CoSO_4$ (0.001). Substrate (2,4-D from a 5% neutralized stock solution in 1% dimethyl amine, sterilized by filtration) is added to the previously autoclaved basal medium to give a final diluted concentration of 100 mg/liter. The pH of the finished medium should be 6.5 to 7.0. The medium is dispensed aseptically (50 ml per flask) within a sterile transfer hood into previously sterilized flasks. A small volume (10 to 20 μl) of filter-sterilized radioactive 2,4-D (to provide at least 100,000 and preferably about 500,000 dpm per flask) is dispensed into a flask, which then is plugged with a rubber stopper previously sterilized by autoclaving and equipped with a CO_2 trap. A sufficient number of flasks should be prepared to examine at least four uninoculated controls, four inoculated but killed controls (0.02 to 0.03% [wt/vol] sodium azide),

Publication no. 95509 of the Idaho Agricultural Experiment Station.

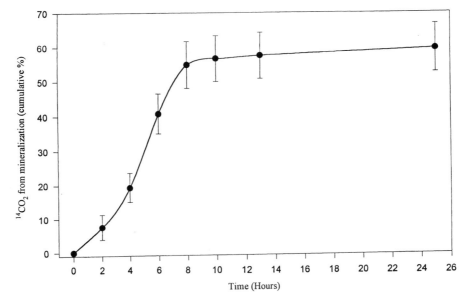

FIGURE 1 Mineralization of ^{14}C-2,4-D by A. *eutrophus*. 2,4-D was added at 100 mg/liter. Uninoculated controls and killed controls yielded no significant ^{14}CO$_2$. Datum points are averages of four replicate flasks ± standard deviations.

and four inoculated flasks. As an option, inoculated but nonradioactive controls can also be run.

At the start of the experiment, appropriate flasks are inoculated with a 5 to 10% inoculum of a log-phase culture of A. *eutrophus* grown on the defined 2,4-D medium without radioactive 2,4-D. The flasks are incubated at 30°C with shaking at 250 rpm. The NaOH in the trap is collected for counting the trapped ^{14}CO$_2$. The trap is washed twice with small volumes of sterile water, and the washes are mixed with the NaOH prior to counting. Scintillation fluid with high miscibility and low chemiluminescence must be used. Fresh sterile NaOH is added to the trap after each sampling. Separate sterile disposable pipettes are used for each flask and time point, and trapped NaOH is collected approximately every 2 h for microorganisms like A. *eutrophus*, which mineralize the 2,4-D quickly. To avoid contamination from too frequent opening of flasks, some investigators use an access port to the CO$_2$ trap in the form of a sterile needle to which a syringe can be attached to add and remove NaOH solutions or a more elaborate setup such as that described by Hsu and Bartha, which has a replaceable scintillation vial as a trap attached to a side arm (4). Biometers also can be constructed from serum bottles containing a 1-ml high-performance liquid chromatography (HPLC) autosampler vial suspended with a wire from a butyl rubber stopper. The vials are filled with 1 ml of 1 M NaOH or KOH trapping solution. Periodically the traps are removed and replaced with new, sterilized vials containing fresh alkali solution. In all cases, trapped ^{14}CO$_2$ is counted by placing the trapping solution and washes in scintillation vials containing 15 ml of scintillation cocktail (e.g., Bio-Safe II; Research Products International Corp.), and results are plotted cumulatively as percentage of added disintegrations per minute (carbon) recovered as ^{14}CO$_2$ with time, showing the standard deviations of replicate flasks (Fig. 1). Because of the substantial quenching activity of NaOH and KOH, external standards are used for conversion of

disintegrations per minute. An aerobic bacterium mineralizing a labeled substrate while using it as a growth substrate will typically convert about 50 to 60% of the substrate carbon into CO$_2$, with the remainder of the label being incorporated into cell material (Fig. 2). This can be confirmed, though generally this is not necessary, by collecting the cells by centrifugation or filtration, lysing them in Protosol (DuPont), and counting the lysate in a liquid scintillation counter.

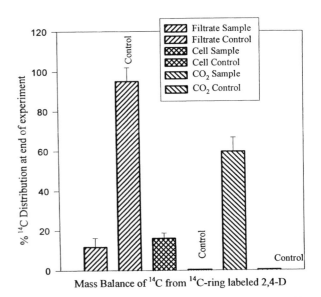

FIGURE 2 Recoveries of ^{14}C during mineralization of ^{14}C-2,4-D by A. *eutrophus*. Recoveries of ^{14}C were essentially 100% in both killed controls and inoculated flasks; values represent averages of four flasks ± standard deviations.

Biotransformation of 2,4-D by *A. eutrophus* as Measured by Removal of the Parent Molecule

It is often desirable to monitor biodegradation of a xenobiotic compound by other than isotopic procedures. HPLC is highly recommended, although it is the disappearance of the parent molecule, and not its mineralization, that is being measured. HPLC usually allows the investigator to discover something about intermediates and/or dead-end metabolites, since these compounds often show up in the chromatograms. Described below is a typical procedure using HPLC to monitor biodegradation of a readily degradable herbicide such as 2,4-D by a pure bacterial culture.

Medium Preparation

To show that a pure culture is able to use 2,4-D as a sole carbon source, a defined minimal medium is used. Such media usually contain some type of buffering system (such as a phosphate buffer) and trace elements. The composition (grams per liter) of the medium used for this experiment is as follows: $K_2HPO_4 \cdot 3H_2O$ (4.2), $NaH_2PO_4 \cdot H_2O$ (1.0), NH_4Cl (2.0), nitrilotriacetic acid (0.1), $MgSO_4 \cdot 7H_2O$ (0.2), $FeSO_4 \cdot 7H_2O$ (0.012), $MnSO_4 \cdot H_2O$ (0.003), $ZnSO_4 \cdot 7H_2O$ (0.003), and $CoSO_4$ (0.001). The pH of the medium is adjusted to 7.0 with either phosphoric acid or NaOH, whichever is appropriate, and the medium is dispensed into 125-ml Erlenmeyer flasks (50 ml per flask), which are plugged with cotton stoppers and autoclaved for 20 min at 126°C. 2,4-D from a 5% (wt/vol) stock solution in 1% (wt/vol) dimethylamine, filter sterilized, is added aseptically to each flask to give a final concentration of 100 mg/liter.

Inoculation and Sampling

The experiment is run in triplicate along with an uninoculated control flask. A 5 to 10% inoculum is added to each flask from an overnight culture of *A. eutrophus* JMP134(pJP4) grown on the defined medium. Cultures are incubated at 30°C on a rotary shaker at 250 rpm. The cultures are sampled by aseptically removing 1-ml portions of the medium, placing them in separate sterile 1.5-ml Eppendorf tubes, and immediately freezing them at −20°C. The first sample should be taken immediately after inoculation, with succeeding samples taken every 2 to 3 h for the duration of the study (usually 10 to 12 h). Samples are kept frozen until HPLC analysis.

HPLC Sample Preparation and Methodology

The frozen samples are thawed and centrifuged to remove cells and other particulates, using an Eppendorf 5415 C centrifuge (Brinkmann Instruments, Inc.) run for 5 min at 14,000 rpm. The supernatant is carefully removed to avoid disturbing the cell pellet and placed into HPLC vials sealed with screw caps and Teflon septa (SunBrokers). Depending on the mobile phase and sample medium components, we have found that it is sometimes necessary to mix the sample 1:1 with the mobile phase before injecting it onto the column to ensure that the sample is in solution, that it will be miscible with the mobile phase, and that it will not precipitate in the column during analysis.

HPLC analysis is conducted on a Hewlett-Packard model 1090 series II HPLC or equivalent machine. In the present example, an Alltech C_{18} 10-mm cartridge guard column is connected to a Microsorb C_{18} 3-μm-bead-diameter, 10-nm-pore-size, 4.6- by 50.0-mm HPLC column at room temperature. The diode array detector is set to monitor 284 nm,

which, although not the highest absorption value for 2,4-D, allows for detection and increased separation of other compounds generally associated with 2,4-D, such as dicamba (3,6-dichloro-2-methoxybenzoic acid) and MCPP [2-(4-chloro-*o*-tolyl)oxypropionic acid].

An isocratic mobile phase composed of 60% methanol (HPLC grade) and 40% of 4.4% formic acid is used. The formic acid solution is made by combining 50 ml of 88% formic acid with 950 ml of purified H_2O (Labconco Water Pro PS filter system or equivalent). The solution is mixed, filtered (0.22-μm pore size), and placed in an HPLC solvent reservoir. The methanol is also dispensed into a solvent reservoir. These solutions are sparged with helium before and during use.

The flow rate is set at 1.25 ml/min, and the injection volume is set at 20 μl. The typical run time of this method is 4 min, giving sufficient resolution of the 2,4-D peaks along with other compounds that may be found in a typical pesticide formulation that contains 2,4-D. If the run time is increased past the retention time of 2,4-D, then intermediate metabolites, if present, can also be detected. A calibration curve, in association with specified retention times, allows the concentration of 2,4-D to be automatically calculated by the instrument. Figure 3 shows a typical chromatogram produced by analyzing a sample containing 2,4-D by this method. Figure 4 shows degradation of 2,4-D by a pure culture of *A. eutrophus* JMP134(pJP4) as monitored by our HPLC procedure.

Pure Culture Biodegradation of Dinoseb

The following type of experiment has been used in our laboratory to study the biodegradation of dinoseb. It can easily be modified to examine how different nutrient conditions, temperature, and other conditions affect biodegradation kinetics. The pure cultures used in this example are strains of *Clostridium bifermentans*, including strain KMR-1 originally isolated by Regan and Crawford (7). Since dinoseb is a teratogenic compound, protective clothing, glasses, and gloves are required at all times.

Preparation of Medium

In a simple biodegradation experiment, brain heart infusion medium (BHI; Difco) is mixed at a concentration of 37 g/liter in 50 mM phosphate buffer (pH 7.0). The medium is prepared in an Erlenmeyer flask, and after it has been thoroughly mixed, dinoseb is added to a final concentration 50 mg/liter (50 ppm). The medium is then brought to a boil on a hot plate with magnetic stirrer or by heating the flask over a Bunsen burner with periodic agitation. When the flask is rapidly boiling, it is removed from the heat source, placed on ice, and immediately sparged with nitrogen gas until the medium is cool to replace the oxygen in the medium with nitrogen. When the medium is cool, 100-ml portions are dispensed into 125-ml serum bottles (SunBrokers) that have been flushed with nitrogen gas. The serum bottles are then capped with butyl rubber stoppers and aluminum crimp top caps (SunBrokers). The bottled medium is then autoclaved and is ready for use when cool. To test the effect of a rich medium on dinoseb degradation, control serum bottles are prepared by using a 50-mM phosphate buffer (pH 7.0) solution containing 50 mg of dinoseb per liter but no BHI. This medium is boiled and dispensed in the same manner as the BHI.

Inoculation and Sampling

Experiments are run in triplicate, and each bottle is inoculated with 0.5 ml of a heat-shocked spore suspension (60°C

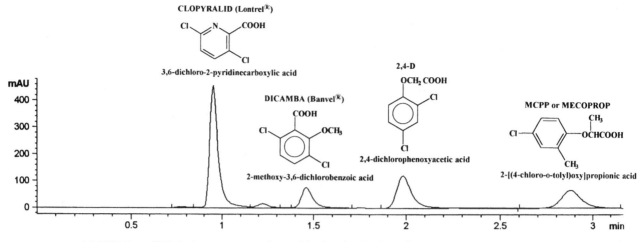

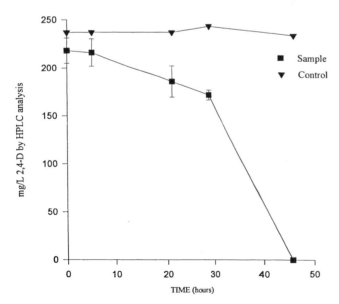

FIGURE 3 HPLC chromatogram of mixed herbicides. Peak 1 of retention time 1.96 min is 2,4-D. mAU, milli-absorbance units.

for 30 min) or 0.5 ml of an overnight vegetative culture grown in BHI. Before the bottles are inoculated, each butyl rubber top is flame sterilized with 95% ethanol. Inoculations are made with an appropriate sterile syringe and needle. Control serum bottles are inoculated in the same manner. Alternatively, control serum bottles can be inoculated with heat-killed cells. All bottles are incubated statically at 25°C in the dark.

Samples are taken before the initial injection of the inoculum and once daily thereafter. One-milliliter samples are removed with a 1-ml syringe and 1-in. (2.54-cm)-long, 22-gauge needle and placed in 1.5-ml polypropylene microcentrifuge tubes (Fisher). As the experiment progresses, gas (primarily CO_2) accumulates in the bottles. Caution should be used while removing samples, since the increased pressure tends to push the plunger in the syringe out of its barrel. The pressure changes when the needle is removed from the

FIGURE 4 Removal of 2,4-D from a defined medium by A. eutrophus as determined by HPLC. Points are averages of three replicates ± standard deviations.

bottle may also cause some of the sample to be evacuated from the syringe.

Samples may be frozen until analyzed by HPLC with little or no change in the dinoseb concentration. At the least, samples should be kept at 4°C to ensure that no additional biodegradation occurs after the sample has been drawn. Sample preparation for HPLC is discussed below.

BIODEGRADATION OF 2,4-D AND DINOSEB BY MIXED CULTURES OF BACTERIA

Absent or Slow Aerobic Mineralization of Dinoseb by Mixed Soil Microbial Populations

Dinoseb is much more recalcitrant than 2,4-D to mineralization under aerobic conditions, and it can persist in some soils apparently because few or no degraders of the molecule are present (9). There is little evidence to suggest that dinoseb can be mineralized aerobically by any pure microbial culture (8, 10), though many potential degrader strains have not been examined (e.g., white rot fungi). Dinoseb is therefore a good model for a recalcitrant herbicide. Its biodegradation under aerobic conditions is examined by essentially the same experiment as described above for 2,4-D, except that the replicates are inoculated with 1 ml of a soil suspension prepared by suspending 10 g of uniformly sieved (1-mm mesh) soil in 100 ml of sterile water. The suspension should be vigorously mixed before each subsample is taken for inoculating individual flasks. Also, since dinoseb is considerably more toxic than 2,4-D, it should be added at a lower concentration (e.g., 5 to 10 ppm), though concentration can be an experimental variable in experiments of this type. Other variables can include the addition of cosubstrates to stimulate indigenous consortia that might cometabolize the xenobiotic compound. However, using soil as an inoculum, we have not detected any aerobic dinoseb mineralization.

Biotransformation of Dinoseb under Anaerobic Conditions as Measured by Removal of the Parent Molecule

Although dinoseb is apparently not extensively degraded under aerobic conditions, it can be degraded anaerobically by microbial consortia (5). HPLC is an excellent technique for monitoring the process, since the formation and disappearance of biotransformation products can be observed along with the removal of the parent molecule. Using HPLC, we have found that pure cultures of *Clostridium* spp. (7) are able to degrade nitrated xenobiotic compounds under anaerobic conditions. Typical procedures are described below for using HPLC to monitor dinoseb biodegradation with mixed anaerobic consortia developed in soils supplemented with cosubstrates.

The biodegradation of dinoseb by consortia of anaerobic bacteria can also be studied by using HPLC. The experimental design can be modified to study the effects on biodegradation of carbon and nitrogen supplementation as well as temperature and other environmental factors. This type of experimentation can also be useful for predicting results in the natural environment and for determining the role of aerobic bacteria in the biodegradation process. Because of the teratogenic properties of dinoseb, gloves, glasses, and protective clothing are required at all times for those working with the compound.

Medium Preparation and Inoculation

Experiments with mixed anaerobic consortia are performed in triplicate in 500-ml Erlenmeyer flasks to which 100 g of soil per flask has been added. The soil used in our example was originally contaminated with dinoseb and had been bioremediated by using the technique developed by Kaake et al. (5). Such soil has been shown to contain a stable dinoseb-degrading microbial consortium. To each flask, 400 ml of a 50 mM phosphate buffer (pH 7.0) is added, and the resulting slurry is mixed well to ensure that the soil is saturated with buffer. Four grams of a carbon source such as glucose, fructose, starch, etc., is then added to each flask, along with dinoseb at 50 mg/liter (50 ppm). Control flasks are set up in the same manner, but no carbon source is added. The contents of flasks are mixed thoroughly to ensure a homogeneous distribution of both the carbon/nitrogen sources and the dinoseb. The flask is covered with aluminum foil, and the flasks are incubated statically at 25°C in the dark. During the course of the experiment, the flasks should be periodically swirled to ensure that no "dead spaces" develop in which dinoseb or carbon/nitrogen might become limiting.

Sampling the Aqueous Phase

During the initial stages of biodegradation, aerobic microorganisms consume the carbon/nitrogen source and act to lower the redox potential of the soil slurry. Therefore, both pH and redox potential (millivolts) are periodically measured, and aqueous samples are also taken. A general-purpose pH combination probe (Corning) and redox combination probe (Corning) attached to an Altex Φ60 pH meter (Beckman) work well for this task. Initial redox readings tend to be between +140 and +160 mV. As the culture becomes reduced and biodegradation occurs, redox potentials as low as −300 to −350 mV are seen. A reduction in pH will also be observed as the anaerobic members of the consortia begin to produce organic acids.

Aqueous samples from the flasks should be taken initially when the flasks are set up and once a day thereafter. The aqueous sample should be drawn from the highly reduced zone, as close as possible to the soil/aqueous-phase interface without drawing up excessive soil into the sample. Generally, 1 ml of sample from each flask is sufficient. Samples should be stored in 1.5-ml polypropylene microcentrifuge tubes (Fisher) and may be frozen until HPLC samples are

prepared, with little or no loss of dinoseb, for several weeks at least.

HPLC Sample Preparation and Methodology

If the samples have been frozen, they must be thawed and then centrifuged to remove any soil and/or cellular debris. Samples are centrifuged on an Eppendorf centrifuge 5415 C (Brinkmann Instruments) for 7 to 8 min at 14,000 rpm. A 1:1 dilution of the sample in acetonitrile is then prepared in a new 1.5-ml microcentrifuge tube for use in the HPLC analysis. The diluted sample is filtered into a 1.5-ml screw-top HPLC vial (SunBrokers), using a filter with a 1-ml syringe (Becton Dickinson) equipped with a 0.45-μm-pore-size, 4-mm-diameter nylon acrodisc (Gellman Sciences). Any pellet that may have formed during centrifugation must not be disturbed, since debris in the 1:1 dilution sample will make filtering very difficult. After filtering, the vial is capped with a screw cap and Teflon septum (SunBrokers).

HPLC Analysis

The HPLC method incorporates a gradient with acetonitrile and a lithium phosphate buffer made by adding 100 μl of concentrated phosphoric acid to 1 liter of Millipore-quality water generated from a Labconco Water Pro PS filter system. This solution is then brought to pH 4.0 by adding 1 M lithium hydroxide (usually between 300 and 600 μl). The buffer is then filtered through a 0.45-μm-pore-size filter membrane (Gellman Sciences) and placed in a clean HPLC solvent reservoir. Both the HPLC-grade acetonitrile and lithium phosphate buffer should be sparged with helium before use.

HPLC analysis is conducted on a Hewlett-Packard model 1090 series II HPLC or equivalent machine. A constant

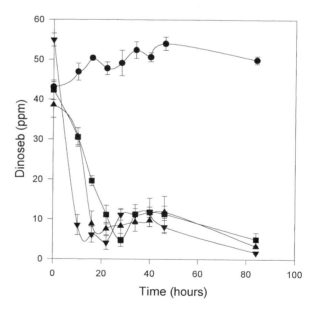

FIGURE 5 Biodegradation of dinoseb by C. *bifermentans* KMR-1 (■) TBH-1 (▼), and TBH-2 (▲) and by a control (●) that contained 100 ml of 50 mM phosphate buffer instead of BHI (no supplemental carbon). Each culture was grown in triplicate in 125-ml serum bottles containing 100 ml of anaerobically prepared BHI with 50 mg of dinoseb per liter. Cultures were injected with 0.5 ml of heat-shocked C. *bifermentans* spores at time zero. Error bars equal 1 standard deviation.

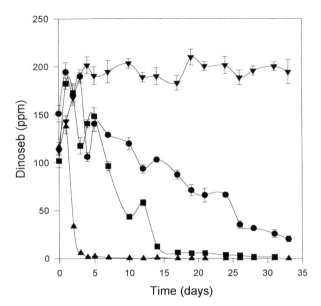

FIGURE 6 Degradation of dinoseb by mixed microbial populations in soil, using different supplemental carbon sources. Each experiment was conducted in triplicate in 500-ml Erlenmeyer flasks containing 400 ml of 50 mM phosphate buffer (pH 7.0), 200 mg of dinoseb per liter, and 100 g of a preacclimated soil. To each flask, 4 g of the desired carbon source was added. The carbon sources were as follows: potato starch (●), wheat starch 1 (■), and wheat starch 2 (▲). The control (▼) contained 100 g of soil, 400 ml of 50 mM phosphate buffer (pH 7.0), and 200 mg of dinoseb per liter but no supplemental carbon. Error bars equal 1 standard deviation.

phase 35% acetonitrile–65% lithium phosphate is maintained for the first minute of the sample run. Over the next 15 min the gradient is changed to 100% acetonitrile, and it is maintained at this concentration for 3 min (18 min into the run). The gradient is then returned to 35% acetonitrile–65% lithium phosphate over the next 2 min and maintained at this concentration for 2 more min. A 3-min postrun at the 35%–65% concentration is then performed,

giving a total run time per sample of 25 min. The flow rate during the sample run is kept at 0.40 ml/min, and the column temperature is kept at 42°C. A Phenomenex Spherex 5 C_{18} (250 by 2.0 mm) reverse-phase column is used for analyte separations. The diode array detector is set at 254 nm, with continuous scanning of the absorption spectrum of each peak from 190 to 600 nm. Concentration (parts of dinoseb per million) is calculated for each sample from a standard curve whose data are stored and used by the HPLC computer and its associated software. Because of the dilution made during the sample preparation, the peak area is multiplied by 2 in order to accurately estimate the concentration of dinoseb per sample.

This procedure has proven to be extremely reliable for tracking the biodegradation of dinoseb in samples taken from experiments with both pure cultures and consortia. Examples of typical experimental results are shown in Fig. 5 (three pure cultures) and Fig. 6 (soil consortium). A sample chromatograph can be seen Fig. 7.

OTHER AGRICULTURAL CHEMICALS

The two herbicides discussed above were offered as very specific examples of common agricultural chemicals. However, our methods for studying the biodegradation of these example compounds can be applied directly to many other compounds. For example, the techniques used for 2,4-D can be applied directly to the study of biodegradation of other phenoxy herbicides and related compounds like dicamba. The methods described for dinoseb can be applied directly to the study of virtually all nitroaromatic compounds. Of course, physiochemical characteristics may place some limitations on the use of these procedures for some chemicals. For example, adjustments would be needed for compounds that are highly insoluble, very volatile, or extremely toxic. Some general guidelines for making some of the most common adjustments follow.

A water-insoluble compound can be added to experimental flasks in a small volume (e.g., 0.1% solvent/medium final volumes) from a stock solution prepared in a water-miscible organic solvent such as dimethylformamide or dimethyl sulfoxide. Typically, this results in formation of a fine suspension or an emulsion of the compound in the

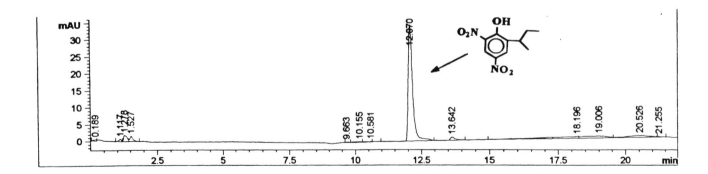

Retention Time

FIGURE 7 Sample chromatograph from a dinoseb biodegradation experiment. Dinoseb is identified via retention time (peak at 12.070 min) and alternatively by its spectral scan from 190 to 600 nm. mAU, milli-absorbance units.

microbial medium. The substrate is dissolved slowly as it is degraded. Extreme insolubility may result in very slow degradation rates, but solubility is one factor influencing the intrinsic biodegradability of all compounds. If the carrier solvent is dimethylformamide or dimethyl sulfoxide, it generally will not show toxicity toward, nor be degraded by, the microbial cultures under study.

Volatile compounds present special problems. To examine the conversion of a ^{14}C-labeled volatile compound to $^{14}CO_2$, one can use individual sealed flasks (three or more replicates) that are sacrificed at specific time intervals. We use Teflon Minimert valves or Teflon-faced gray butyl rubber stoppers (West Co., Phoenixville, Pa.) for sealing incubation vessels. As an alternative to using an in-flask CO_2 trap, the gas phase of a experimental vessel is flushed with a stream of CO_2-free air (scrubbed through 1 N NaOH) first through an organic vapor trap (e.g., a polyurethane filter or an organic solvent bubble trap) and then an NaOH CO_2 bubble trap. Radioactivity in the NaOH trap is counted by liquid scintillation procedures, as usual. For monitoring the removal of a highly volatile parent molecule, a technique such as headspace gas chromatography will usually be better than HPLC. Samples of incubation medium are removed periodically from sealed experimental vessels (sterile techniques are required) and placed in headspace sample vials that are immediately sealed with inert and impermeable stoppers and aluminum crimp seals. After equilibrium at a standard temperature, headspace samples are analyzed by gas chromatography. Total concentrations of volatile substrates (gas phase plus liquid phase) are calculated by using a dimensionless Henry's law constant (2). We have used headspace gas chromatography very successfully to monitor biodegradation and biotransformation of carbon tetrachloride (6). This technique requires fairly specialized equipment, and chromatography conditions will be specific to the compound(s) under study.

Highly toxic compounds may be inherently nonbiodegradable at the usual concentration levels employed in biodegradation experiments. In such cases, the use of radiolabeled substrates of very high specific radioactivity is recommended. The biodegradability of a chemical may then be examined at increasingly lower concentrations, down to the level of sensitivity (a few hundred disintegrations per minute of radioactivity) of the technique. Concentrations as low as a few micrograms per liter (parts per billion) may often be examined by using radioisotopic methods. Alternatively, gas chromatography techniques such as many of the standard U.S. Environmental Protection Agency protocols can be used to detect removal of parent compounds at parts-per-billion or even parts-per-trillion levels (e.g., electron capture detection of chlorinated compounds).

It is clear that there are no universal protocols for monitoring the biodegradation of agricultural chemicals. However, the techniques described here, or minor variations thereof, will be useful in most instances. As always, some common sense in choosing experimental methods and a thorough prior search of the relevant literature are required to produce results that are both trustworthy and reproducible.

REFERENCES

1. **Don, R. H., and J. M. Pemberton.** 1981. Properties of six pesticide degradation plasmids isolated from *Alcaligenes paradoxus* and *Alcaligenes eutrophus. J. Bacteriol.* **145:** 681–686.
2. **Gossett, J. M.** 1987. Measurement of Henry's Law constants for C_1 and C_2 chlorinated hydrocarbons. *Environ. Sci. Technol.* **21:**202–208.
3. **Goszczynski, S., and R. L. Crawford.** 1990. Isotopically labelled compounds for hazardous waste site cleanup investigations. Part I. Synthesis of [phenyl-U-^{14}C] labelled 2,4-dinitro-6-sec-butylphenol (dinoseb) and [phenyl-U-^{14}C] labelled 4-n-propylphenol. *J. Labelled Cmpd. Radiopharm.* **29:**35–42.
4. **Hsu, T. S., and R. Bartha.** 1979. Accelerated mineralization of two organophosphate insecticides in the rhizosphere. *Appl. Environ. Microbiol.* **37:**36–41.
5. **Kaake, R. H., D. L. Crawford, and R. L. Crawford.** Optimization of an anaerobic bioremediation process for soil contaminated with the nitroaromatic herbicide dinoseb (2-sec-butyl-4,6-dinitrophenol), p. 337–341. *In* R. E. Hinchee, D. B. Richardson, F. B. Metting, Jr., and G. Sayles (ed.), *Applied Biotechnology for Site Remediation.* Lewis Publishers, Boca Raton, Fla.
6. **Lewis, T. A., and R. L. Crawford.** 1993. Physiological factors affecting carbon tetrachloride dehalogenation by the denitrifying bacterium *Pseudomonas* sp. strain KC. *Appl. Environ. Microbiol.* **59:**1635–1641.
7. **Regan, K. M., and R. L. Crawford.** 1994. Characterization of *Clostridium bifermentans* and its biotransformation of 2,4,6-trinitrotoluene (TNT) and 1,3,5-triaza-1,3,5-trinitrocyclohexane (RDX). *Biotechnol. Lett.* **16:**1081–1086.
8. **Stevens, T. O.** 1989. Biodegradation of dinoseb (2-sec-butyl-4,6-dinitrophenol) and bioremediation of dinoseb-contaminated soils. Ph.D. dissertation. University of Idaho, Moscow.
9. **Stevens, T. O., R. L. Crawford, and D. L. Crawford.** 1990. Biodegradation of dinoseb (2-sec-butyl-4,6-dinitrophenol) in several Idaho soils with various dinoseb exposure histories. *Appl. Environ. Microbiol.* **56:**133–139.
10. **Stevens, T. O., R. L. Crawford, and D. L. Crawford.** 1991. Selection and isolation of bacteria capable of degrading dinoseb (2-sec-butyl-4,6-dinitrophenol). *Biodegradation* **2:**1–13.

Use of Soil Bioreactors and Microcosms in Bioremediation Research

STEVEN K. SCHMIDT AND KATE M. SCOW

92

Studying biodegradation in soil presents unique research challenges because of the heterogeneity of soil with respect to biological niches and the spatial distribution and availability of pollutants. A variety of complex biodegradation patterns result from physical interactions between pollutants and the soil matrix and from biological interactions among different organisms. The purpose of this chapter is to give an overview of methods to study the biodegradation of chemicals in soil microcosms and bioreactors. There have been several recent reviews of methods used to study the activity of soil microorganisms in microcosms (2, 14, 65). For example, Burns (14) reviewed the literature on microcosm studies that "explicitly attempted to understand how soil systems function in the field." In contrast, this chapter will focus only on the common methods used to study biodegradation processes in soil. Attempts have been made to standardize the methods used by researchers in this area (16, 30, 43), and we will refer to these standard methods whenever possible. Most of the methods described are used to determine the biodegradation potential of compounds in soil or are used to elucidate the effects of isolated variables, such as temperature, soil moisture, or sorption, on the kinetics of biodegradation. We will also discuss methods used to scale up soil microcosms to treat larger volumes of contaminated soil.

COLLECTION AND HANDLING OF SOILS

Sampling in the Field

Proper field sampling procedures should be designed so that representative samples of the study area can be obtained. For very heterogeneous and large sites, the spatial variability of soil characteristics should be considered in determining the location and number of representative soil samples (43, 51). Soils may be collected from a uniform depth, using a corer or shovel, and at various (e.g., 20 to 30) locations at the site and then pooled. Often the top 10 or 15 cm of soil is used, but vertical profiles of the soil may need to be sampled, especially at contaminated sites. If there is concern about contamination of samples by nonindigenous microorganisms, aseptic techniques such as flaming of the sampling device or push-tube sampling such as with the Shelby tube are used (15). Collected soils are stored in plastic bags on ice in a cooler until being placed in a cold room at 4°C for longer-term storage. Soil samples from contaminated sites should be stored in glass or other nonreactive containers. Long-term storage should be at field moisture levels, and air drying should be avoided. Zelles et al. (89) and others (e.g., Ross [50]) have reviewed information on changes in stored soil as a function of moisture and temperature.

Soil Preparation

Preparation of soil samples for biodegradation experiments can alter the physical, biological, and chemical properties of the soil as they exist in the field. There is a trade-off between having the samples be representative of field conditions and having relatively homogeneous samples so that differences can be attributed to the experimental treatments, not to soil heterogeneity. In preparation for experiments, it is common practice to pass soils through a 2- or 4-mm sieve. In some cases, samples are preincubated for up to a week at the experimental incubation temperature to overcome disturbance effects such as stimulation of respiration by release of soil carbon following sieving (3). Preincubation may not be advisable if the objective of the study is to measure biodegradation potential of soils at a contaminated site. Soil moisture levels should be determined in order to calculate the dry weight of the soil sample and so that additional experiments can be standardized for the same moisture conditions. Water content is usually determined by gravimetric methods (85). Depending on the goal of the experiment, other soil properties such as organic matter content, nutrient levels, pH, and texture can be determined by standard methods (85). References 51 and 68 provide information on methods for soil analysis.

Methods for Chemical Application

The test chemical may be added to soil in solution, as a gas, or, less commonly, as a solid. In calculating the amount of chemical to add, a decision must be made as to whether the amount or concentration added is to be calculated on the basis of the soil solution (per milliliter), solid phase (per gram [dry weight]), or gas phase (per milliliter or cubic centimeter). It may be necessary to know or estimate the chemical's Henry's coefficient and sorption partition coefficient to account for the partitioning of the chemical among the different phases of soil, in which case reference 38 provides useful information.

If the chemical is added as a solution, the water solubility may be so low that an organic solvent is required as a carrier.

In such cases, the organic solvent is usually highly volatile (e.g., methylene chloride, chloroform, or methanol) and, after being added, is allowed to vent to the atmosphere before the experiment is initiated. As little solvent as possible should be used because of the possibility of either stimulation or repression of the biodegradation rate by residual solvent. For example, Skidmore et al. (66) found that the rate of permethrin mineralization was increased if the chemical was added in a greater volume of methanol. The authors attributed the stimulation to an increase in the distribution of the chemical with more solvent (66); however, another explanation is that the solvent served as a cosubstrate that stimulated specific populations of microorganisms, in this case, methylotrophic bacteria. Another potential problem with the use of solvents is that many are toxic to a wide range of soil microorganisms. An alternative to using solvents is to add the chemical on a solid carrier, such as talc, sand, cellite, or a small subsample of the same soil, which is then mixed into the larger soil sample (83, 84).

If a chemical is added in a water solution, the soil moisture content may need to be reduced (e.g., through partial drying) so that the final moisture content after amendment is at the desired level. Soil with high clay content may form aggregates when the chemical is incorporated if the moisture content is too high. The chemical may be applied to the soil surface via small drops with a pipette or by spraying with a mister. Chemicals added to soil may be mixed in by stirring, blending, or placement on a roller for a period of time. At the start of the experiment, the initial concentration of the chemical should be measured if possible.

Incubation Conditions

Obviously, the incubation conditions will depend on the particular questions being investigated. Typically, soil is incubated at a constant temperature and in the dark or in opaque containers to prevent transformation of the chemical by photodecomposition. With respect to soil moisture, maximum microbial activity is at approximately 60% of the total pore volume (37) as a result of (i) reduced diffusion and access of microorganisms to their substrates at low moisture and (ii) reduction in diffusion of oxygen at high moisture conditions. Aerobic biodegradation studies are frequently conducted at 70 to 100% of 0.33 bar (33,000 Pa; the soil matric potential at "field capacity" [85]). Moisture loss is prevented by sealing the flasks or monitoring, e.g., by weight, and replacing lost moisture over time.

In aerobic studies, sufficient oxygen content is maintained by periodically opening sealed flasks to the atmosphere, or microcosms may be connected to a flowthrough system in which air is being continuously or intermittently supplied (see Biometers with Airflow, below). With very volatile chemicals, this may not be possible and flasks may have to remain sealed to the atmosphere. If aerobic conditions need to be confirmed, oxygen levels can be monitored by analyzing headspace samples.

For anaerobic studies, the headspace of the microcosm is composed of nitrogen, argon, or some other inert gas, and a reducing agent may be added to maintain anaerobic conditions (22). Soil may be flooded to reduce transfer of oxygen.

Controls

Sterile Controls

To show that a process is being carried out by biological activity, experimental treatments are frequently compared against sterile controls. Sterile controls also allow for the estimation of abiotic loss mechanisms, such as volatilization and sorption, that can occur simultaneously with biodegradation. Common methods of soil sterilization are gamma irradiation, chemical treatment (e.g., with sodium azide, mercuric chloride, or methyl bromide), and autoclaving. If inoculation of sterile material is required, then only gamma irradiation or autoclaving is an appropriate method. The most widely available method of sterilizing soil is double autoclaving, i.e., (i) autoclaving the soil for 90 min, (ii) allowing the soil to cool overnight, and (iii) reautoclaving the soil for 90 min. The first autoclaving induces the germination of heat-resistant endospores, which are then killed as vegetative cells during the second autoclaving.

Any method of sterilization will alter soil physical properties to some extent. Discussions of the merits and disadvantages of various soil sterilization techniques can be found in the literature (46, 52, 54, 67, 78, 87). Wolf et al. (87) compared the effects of treatments with dry heat, cobalt-60 gamma irradiation, propylene oxide, mercuric chloride, sodium azide, microwaves, chloroform, and antibiotics and found that mercuric chloride had the least effect on soil properties such as pH, surface area, and release of Mn.

Other Controls

As a general rule, there should be a control for every manipulation that could affect the process being measured. For example, in an experiment in which total CO_2 production from a chemical added in an organic solvent is being measured, a control receiving the same amount of solvent but no chemical should be measured to account for the conversion of solvent to CO_2. Many solvents are also toxic to soil microorganisms, and therefore a solvent-only control should be used in studies of the effects of chemicals on soil organisms or processes. In experiments in which an inoculum is added to soil, a killed inoculum treatment should be included as a control for possible fertilizer effects arising from death of the inoculum or from salts in the solution in which the inoculum is added.

Mass Balances

Performance of mass balances of added chemical at the end of experiments ensures that disappearance of a chemical is due to biodegradation and not to loss from adsorption, volatilization, or other abiotic mechanisms. If a radiolabeled compound is used, counts remaining in the soil may be in metabolites or incorporated into microbial biomass, and therefore it is recommended that analyses for the parent compound also be performed whenever possible. For volatile chemicals, measurement of volatilization is also recommended as described below.

SOIL SYSTEMS

Biometers

Measuring carbon dioxide evolution, preferably from a radiolabeled chemical, is one of the most common methods used in soil biodegradation studies because it demonstrates mineralization of the chemical and because it does not require destructive sampling. One of the simplest methods used, which is also recommended by the Organization for Economic Co-operation and Development (44) and the U. S. Environmental Protection Agency (65), is the biometer flask. Biometers are used in batch experiments that involve adding a single input of chemical and monitoring its disappearance over time. The basic biometer flask, first

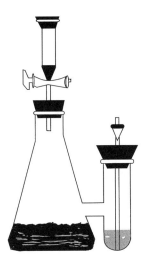

FIGURE 1 Organization for Economic Co-operation and Development-recommended biometer flask used to study the biodegradation of mineralizable pollutants (6, 44). Reprinted from reference 25 with permission.

described by Bartha and Pramer (6), is basically an Erlenmeyer flask with a side arm that contains fluid for trapping CO_2 (Bellco Glass Inc., Vineland, N.J.) (Fig. 1). The side arm can be used to trap ^{14}C-labeled CO_2 or total CO_2 as described in detail elsewhere (6, 57, 61, 62). Several authors have recommended changes to the biometer flask design, but the basic idea has remained the same. Haigh (25) described a modified biometer flask fitting that allows scintillation vials to be attached directly to the side arm of the flask. This modified design has the advantages of increased safety and of overall recovery of $^{14}CO_2$ but has the disadvantage of decreased rate of diffusion of the $^{14}CO_2$ from the soil to the trap (25). Also, many less expensive variants using standard glassware (such as unmodified Erlenmeyer flasks or Mason jars) have been developed; in such variants, a small vial containing a base trap is suspended from the top (28, 32, 59) or is placed on the soil surface. A mass balance of total ^{14}C should be calculated, whenever possible, at the end of the experiment.

The basic biometer approach has also been modified for a number of different specialized problems in the fields of biodegradation and bioremediation. For example, Yabannavar and Bartha (88) describe standardized methods for testing the degradability of plastics by using a combination of techniques including biometer studies, and Hsu and Bartha (29) used another modification of the biometer approach to look at effects of plant roots on biodegradation kinetics (see Plant-Soil Microcosms, below).

Biometers with Airflow

For semivolatile chemicals, Marinucci and Bartha (39) describe an adaptation of the biometer flask approach for carbon dioxide evolution. The system consists of a micro-Fernback flask connected to an exhaust manifold that allows the trapping of $^{14}CO_2$ in phenethylamine and trapping of the volatile organic compounds in either xylene or toluene. Sample are taken by using negative pressure to draw the flask atmosphere through a series of scintillation vials containing the various trapping solutions. Problems are sometimes encountered with backflow in such systems, but they can usually be avoided by properly disconnecting the system after each sampling (39).

A similar approach was used by Rüdel et al. (53) to study the degradation of various pesticides in soil. Their system had three types of traps: (i) ethylene glycol for volatile organic compounds, (ii) 1 M sulfuric acid for volatile basic compounds, and (iii) 0.5 M sodium hydroxide for $^{14}CO_2$. As with other such systems, the airflow can be passed through water to maintain a high relative humidity in the biometers.

Systems for Volatile Chemicals

By using highly volatile chemicals, a large fraction of the parent compound may be absorbed by traps for radiolabeled carbon dioxide; if a flowthrough system is used, there may be substantial removal of the initial concentration of volatile chemicals by the gas stream. Counts associated with volatile parent compounds or metabolites and counts associated with carbon dioxide can be partially discriminated by precipitation of the carbonate with barium chloride, washing, and resuspension of the precipitate (19).

A common approach used for very volatile chemicals is headspace analysis of airtight microcosms (21, 23, 31, 45). A small sample may be removed directly from the flask by using a gastight syringe, often through a Mini-Nert cap (Alltech Associates, Deerfield Ill.), and then injected into a gas chromatograph. Sometimes a gas sample is placed in a small vial containing an organic solvent or passed through a small cartridge containing an organic solid phase which functions as a vapor trap. In the latter two cases, the solvent is injected into the gas chromatograph, or the test chemical may be purged from the vapor trap by using a purge-and-trap system. For volatile chemicals, it is important that physical properties such as Henry's coefficients and sorption partition coefficients (38) be estimated or measured to account for the partitioning of the chemical among the different phases of soil and in order to determine the initial concentration. Without this information, it can be difficult to compare results from different studies.

Depending on the desired initial concentrations, the chemical may be added as a gas or liquid. With the latter, mixing may be a bit more problematic. Sample concentrations are compared against standards. If possible, a conservative, internal standard can be added to the experimental flask to determine sampling efficiency. Sometimes, a standard is added to the extraction solvent after the sample is collected (24).

Problems with headspace analysis are the sensitivity of the gas-phase concentration to temperature and pressure and leaking of the sample from the flask or syringe.

Slurry Bioreactors

The basic slurry approach is to suspend soil in an aqueous solution in a contained vessel that allows for mechanical mixing. The main advantage of slurry reactors is that they allow for optimization of conditions conducive to biodegradation. Mixing facilitates aeration and enhances the rate of chemical exchange between soil particles and the solution. Conditions can be further optimized for biodegradation by controlling pH and temperature and by providing nutrients and other specialized amendments, such as surfactants (77). Slurries can be maintained under aerobic or anaerobic conditions, and an inoculum can be added if the soil being treated has low populations of organisms capable of degrading the chemicals of interest (4, 9). For anaerobic biotreatability studies, the U.S. Environmental Protection Agency recommends the use of serum bottles as the reaction vessel and a 20% (wt/vol) mixture of soil and water (65, 80). As with all systems, it is recommended that a mass balance approach be used with slurry reactors whenever possible.

An example of the use of slurry reactors is the study of Mueller et al. (42). They used 1.5-liter slurry reactors with pH, temperature, and dissolved oxygen controls to treat soils contaminated with creosote and pentachlorophenol. They concluded that slurry-phase treatment was more effective for the cleanup of these sites than were solid-phase bioremediation techniques such as landfarming (42).

Soil Columns

The study of biodegradation in the presence of physical mass transfer limitations is usually conducted in soil columns. Movement of the chemical through the column may be primarily by advection and diffusion or by diffusion only, and movement may be through the solution (soluble chemicals) or gas phase (volatile chemicals). Columns may be maintained under saturated (all pores filled with water) or unsaturated conditions (18, 86). Soil column experiments require an in-depth understanding of the soil's physical parameters, e.g., bulk density, soil texture, and moisture. In addition, interpretation of biodegradation results from columns requires the use of mathematical models to separate the contributions of physical and biological phenomena to the measured data. Descriptions of protocols for conducting soil column experiments are found in the soil physics literature (e.g., reference 83).

Column experiments are conducted with sieved soil repacked into a column to a specific bulk density (12, 76) or in relatively undisturbed soil cores which are excavated and placed in columns (20, 36, 53, 69, 76). If a microbial inoculant is used, it is often physically mixed into soil before soil is added to columns. Introduction of the inoculum via the solution phase may result in nonuniform distribution. It is important that materials making up the column walls do not absorb the pollutant (75), and thus columns are often made of Teflon or stainless steel and tubing and fittings are made of Teflon. Clogging of soil columns can be a problem, especially in inlet tubing carrying metabolizable substrates. Filters are sometimes used to prevent movement of microorganisms into tubing, but the filters themselves may become clogged.

Pollutant movement through the column may be under steady-state (continuous flow), sequential batch (replacement of solution intermittently), or transient (single pulse of chemical is added) conditions. Saturated columns for studying the biodegradation of volatile organic chemicals are described in the literature (27, 35, 64). With volatile chemicals, a sequential batch reactor approach in which fresh solution containing oxygen, the chemical, and nutrients is used to replace the column solution at constant time intervals is commonly used. In unsaturated columns, it is important to maintain the same moisture content so that soil-water content within the column can be monitored by weight, by gamma ray attenuation, or by time domain reflectometry (74).

Pollutant concentrations, in either the gas or the solution phase, are monitored over the course of the experiment at the inlet and outlet of the column and sometimes from sample ports within the soil column. Small subcores of soil may be removed from the surface of the column for destructive sampling, in which case wooden or glass rods are used to refill the holes left by soil removal (76). To experimentally separate the influences of physical and biological processes on pollutant concentration, a conservative tracer such as chloride, bromide, or tritiated water is added to determine

the flow characteristics of the system. Sterile columns are usually run to determine the effect of adsorption to soil particles on the effluent concentration. Results of experiments are compared with those of simulations from models describing the system (e.g., reference 18) in order to determine the biodegradation kinetics uncoupled from physical and chemical influences.

Plant-Soil Microcosms

Recently there has been a revived interest in soil microcosms that include growing plants (e.g., reference 33). These types of studies are usually carried out to determine the effects of the rhizosphere microbial community on the biodegradation of chemicals or to determine the fate of chemicals in microcosms that simulate real systems. In studies of the former type, systems usually involve a two-compartment approach that allows the soil, which is emitting CO_2, to be isolated from the plant shoot, which absorbs CO_2 (29, 33). More complex microcosms for studying plant-soil interactions are reviewed elsewhere (8, 14).

EXPERIMENTAL DESIGN AND DATA ANALYSIS

The experimental design or data analysis scheme of a study will vary depending on the specific goals of the study. Whatever the goal of the study, experiments should be designed so that rigorous hypothesis-testing procedures can be applied to the data obtained from the experiment. All experimental treatments should be replicated at least in triplicate so that overall precision of the measurements can be determined and proper statistical analyses can be performed. Care should also be taken in designing systems that are to be subsampled over time to ensure that repeated sampling will not drastically reduce the volume of the soil or slurry in the reactor. Basic concepts of experimental design and statistics cannot be reviewed in this chapter, but many good texts are available (47, 70).

Of particular interest in biotreatability studies and studies of the factors controlling the rate of biodegradation in soil are recent advances in modeling of biodegradation kinetics in soil. The kinetics of biodegradation can be described by any number of mathematical models (13, 56, 58, 60, 61, 72). These models stress different aspects of biodegradation kinetics, including mass transfer constraints in the soil matrix, for example, the effects of sorption and diffusion (56), as well as the roles of other limiting factors and microbial population dynamics (60). These models should be viewed as aids for understanding the coupled processes controlling the fate of chemicals in soil and as analytical tools for testing specific hypotheses. For example, simple models and CO_2 evolution curves can be used to estimate population levels of specific biodegradative organisms in soil (55). Most commercially available statistics packages contain algorithms for performing nonlinear regression, and several reviews of this approach are available (7, 41, 49, 56).

Special caution should be exercised when one is analyzing carbon dioxide evolution data from soil incubations. Hess and Schmidt (26) and others (73) have stressed that analyzing accumulated CO_2 data leads to a number of statistical problems, including nonrandom, autocorrelated residuals and underestimation of experimental error. It is therefore preferable to use nonaccumulated raw data and the differential form of kinetic models to analyze soil respiration data. Examples of this approach to analyzing soil respiration data can be found in the literature (26, 71).

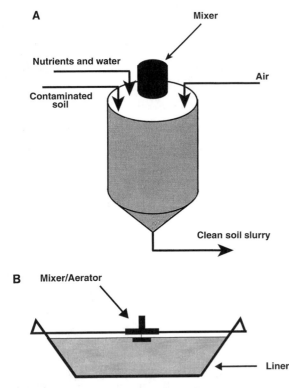

FIGURE 2 Two types of full-scale slurry bioreactors. (A) Slurry reactor; (B) pond or lagoon slurry reactor. Reprinted with permission from reference 10.

SCALE-UP

The methods described above can be used in studies of the factors controlling biodegradation rates in soil, as well in determining appropriate biotreatment strategies for a contaminated soil. Information from small-scale biodegradation studies can be used to design larger-scale biotreatment systems, and some small-scale systems can even be scaled up directly. The U.S. Environmental Protection Agency suggests three tiers of testing before full-scale application of a technology in the field (40, 65, 82). (i) Laboratory screening is performed to establish that chemical loss is actually due to biodegradation, to provide basic information on rates of biodegradation, and to control for variables for the second tier of the screening process. (ii) Bench-scale testing is conducted to establish performance and cost estimates, as well as to define design and operating parameters for scale-up of the technology. (iii) Pilot-scale testing should be carried out on the most promising technologies to obtain detailed cost and design data for the full-scale implementation of the biotreatment technology.

Of the small-scale systems described above, slurry reactors are probably the system most amenable to direct scale-up. The knowledge gained from laboratory and bench-scale slurry reactors can be used in the design of both large-scale bioreactors and pond or lagoon systems (Fig. 2) (10). The main advantage of using slurry reactors is that conditions in the system can be easily manipulated to maximize rates of biodegradation. Nutrients, microbial inoculants, electron acceptors, cosubstrates, and other amendments can be easily added to enhance the rate of biodegradation in these contained systems. Slurries can be maintained under aerobic or anaerobic conditions or even alternated between these

conditions, depending on the processes involved in biodegradation. Vigorous mixing can be done to increase the contact between cells and the pollutant. Slurry reactors can be especially effective in the remediation of soils containing highly sorbed chemicals (9, 17) because mixing breaks up soil aggregates and enhances the rate of exchange of the pollutant between soil particles and the solution.

Full-scale slurry reactors have been used successfully to clean up contaminated soil from a number of sites (81). As one example, ECOVA Corp. used 26,000-gallon (98,410-liter) slurry bioreactors capable of treating 60 yd^3 (45.9 m^3) of soil to bioremediate soils contaminated with 2,4-dichlorophenoxyacetic acid and 4-chloro-2-methyl-phenoxyacetic acid. More than 95% of both chemicals was removed from the slurries in 13 days (81).

Data from biotreatability and other laboratory studies can also be used in the design of other full-scale bioremediation technologies. In addition to slurry reactors, other bioreactors that are presently being used or considered for use in the field are contained landfarming operations and various forms of contained soil and compost piles (5). An advantage of landfarming is that large quantities of contaminated soil can be treated by using equipment normally employed in agricultural practices. Contained landfarming units differ from traditional landfarms in that they are constructed to contain the contaminated soil and leachates from the soil during treatment (5). Containment offers the advantages of allowing control of environmental variables such as moisture and temperature and ensuring that contamination of subsurface layers and groundwater will not take place during the treatment process. A major disadvantage of large contained landfarming systems is that they require expensive transport of contaminated soils to the treatment facility. Successful applications of landfarming in the cleanup of contaminated sites are documented in several recent publications (1, 5, 40, 81).

More information on scale-up of bioremediation technologies is available in numerous review articles (10, 17, 34, 63, 77, 81) and recent books (1, 5, 11). Also of interest for scale-up considerations are numerous studies that have compared the results from two systems or the results from a laboratory study with field trials (18, 27, 53, 76).

REFERENCES

1. **Alexander, M.** 1994. *Biodegradation and Bioremediation.* Academic Press, San Diego, Calif.
2. **Anderson, J. P. E.** 1990. Principles of and assay systems for biodegradation. *Adv. Appl. Technol. Ser.* **4:**129–145.
3. **Anderson, J. P. E., and K. H. Domsch.** 1978. A physiological method for the quantitative measurement of microbial biomass in soils. *Soil. Biol. Biochem.* **10:**215–221.
4. **Bachmann, A., P. Walet, P. Wijnen, W. de Bruin, J. L. M. Huntjens, W. Roelofsen, and A. J. B. Zehnder.** 1988. Biodegradation of alpha- and beta-hexachlorocyclohexane in soil slurry under different redox conditions. *Appl. Environ. Microbiol.* **54:**143–149.
5. **Baker, K. H., and D. S. Herson (ed.).** 1994. *Bioremediation.* McGraw-Hill Book Co., New York.
6. **Bartha, R., and D. Pramer.** 1965. Features of a flask and method for measuring the persistence and biological effects of pesticides in soil. *Soil Sci.* **100:**68–70.
7. **Bates, D. M., and D. G. Watts.** 1988. *Nonlinear Regression Analysis and Its Applications.* John Wiley & Sons, New York.
8. **Beyers, R. J., and H. T. Odum.** 1993. *Ecological Microcosms.* Springer-Verlag, New York.
9. **Black, W. V., O. A. O'Connor, D. S. Kosson, R. C.**

Ahlert, and J. E. Brugger. 1994. Slurry-based treatment of polyaryl contaminants sorbed onto soil. *J. Environ. Sci. Health* **A29:**833–843.

10. **Blackburn, J. W., and W. R. Hafker.** 1993. The impact of biochemistry, bioavailability and bioactivity on the selection of bioremediation techniques. *Trends Biotechnol.* **11:**328–333.

11. **Boulding, J. R.** 1995. *Practical Handbook of Soil, Vadose Zone and Groundwater Contamination: Assessment, Prevention and Remediation.* Lewis Publishers, Boca Raton, Fla.

12. **Bowman, B. T.** 1988. Mobility and persistence of metolachlor and aldicarb in field lysimeters. *J. Environ. Qual.* **17:**689–694.

13. **Brunner, W., and D. D. Focht.** 1984. Deterministic three-half-order kinetic model for microbial degradation of added carbon substrates in soil. *Appl. Environ. Microbiol.* **47:** 167–172.

14. **Burns, R. G.** 1988. Experimental models in the study of soil microbiology, p. 51–98. *In* J. W. T. Wimpenny (ed.), *Handbook of Laboratory Model Systems for Microbial Ecosystems.* CRC Press, Boca Raton, Fla.

15. **Chapelle, F.** 1993. *Groundwater Microbiology and Geochemistry.* John Wiley & Sons, New York.

16. **Code of Federal Regulations.** 1992. Inherent biodegradability in soil, p. 224–229. *Code of Federal Regulations,* vol. 40, part 796.3400. U.S. Government Printing Office, Washington D.C.

17. **Cutright, T. J.** 1995. A feasible approach to the bioremediation of contaminated soil: from lab-scale to field test. *Fresenius Environ. Bull.* **4:**67–73.

18. **Estrella, M. R., M. L. Brusseau, R. S. Maier, I. L. Pepper, P. J. Wierenga, and R. M. Miller.** 1993. Biodegradation, sorption, and transport of 2,4-dichlorophenoxyacetic acid in saturated and unsaturated soils. *Appl. Environ. Microbiol.* **59:**4266–4273.

19. **Fan, S., and K. M. Scow.** 1993. Biodegradation of TCE and toluene by indigenous microbial populations in soil. *Appl. Environ. Microbiol.* **59:**1911–1918.

20. **Fermanick, K. J., and T. C. Daniel.** 1991. Pesticide mobility and persistence in microlysimeter soil columns from a tilled and no-tilled plot. *J. Environ. Qual.* **20:**195–202.

21. **Freedman, D. L., and J. M. Gossett.** 1991. Biodegradation of dichloromethane and its utilization as a growth substrate under methanogenic conditions. *Appl. Environ. Microbiol.* **57:**2847–2857.

22. **Godsy, E. M., D. F. Goerlitz, and D. Grbic-Galic.** 1992. Methanogenic degradation kinetics of phenolic compounds in aquifer-derived microcosms. *Biodegradation.* **2:** 211–221.

23. **Godsy, E. M., D. F. Goerlitz, and D. Grbic-Galic.** 1992. Methanogenic biodegradation of creosote contaminants in natural and simulated ground water ecosystems. *Ground Water* **30:**232–242.

24. **Haag, F., M. Reinhard, and P. L. McCarty.** 1991. Degradation of toluene and p-xylene in anaerobic microcosms: evidence for sulfate as a terminal electron acceptor. *Environ. Toxicol. Chem.* **10:**1379–1389.

25. **Haigh, S. D.** 1993. A modified biometer flask for the measurement of mineralization of ^{14}C-labeled compounds in soil. *J. Soil Sci.* **44:**479–483.

26. **Hess, T. F., and S. K. Schmidt.** 1995. Improved procedure for obtaining statistically valid parameter estimates from soil respiration data. *Soil Biol. Biochem.* **27:**1–7.

27. **Hopkins, G. D., L. Semprini, and P. L. McCarty.** 1993. Microcosm and in situ field studies of enhanced biotransformation of trichloroethylene by phenol-utilizing microorganisms. *Appl. Environ. Microbiol.* **59:**2277–2285.

28. **Hoyle, B., K. M. Scow, G. Fogg, and J. Darby.** 1995. Effect of carbon:nitrogen ratio on kinetics of phenol bio-

degradation by *Acinetobacter* in saturated sand. *Biodegradation* **6:**283–293.

29. **Hsu, T. S., and R. Bartha.** 1979. Accelerated mineralization of two organophosphate insecticides in the rhizosphere. *Appl. Environ. Microbiol.* **37:**36–41.

30. **International Standards Organization.** 1993. *Soil Quality-Sampling,* part 6. *Guidance on the Collection, Handling and Storage of Soil for the Assessment of Aerobic Microbial Processes in the Laboratory.* ISO 10381/6-1993(E). International Standards Organization.

31. **Ioffe, B. V., and A. G. Vittenberg.** 1984. *Headspace Analysis and Related Methods in Gas Chromatography.* John Wiley & Sons, New York.

32. **Knaebel, D. B., and J. R. Vestal.** 1988. Comparison of double-vial to serum bottle radiorespirometry to measure microbial mineralization in soils. *J. Microbiol. Methods* **7:** 309–317.

33. **Knaebel, D. B., and J. R. Vestal.** 1992. Effects of intact rhizosphere microbial communities on the mineralization of surfactants in surface soils. *Can. J. Microbiol.* **38:** 643–653.

34. **LaBelle, B. E., and P. W. Hadley.** 1994. Bio beware!! Constraints and considerations when demonstrating bioremediation technologies in the field. *J. Soil Contam.* **3:** 119–126.

35. **Lanzarone, N., and P. L. McCarty.** 1990. Column studies on methanotrophic degradation of trichloroethene and 1,2-dichloroethane. *Ground Water* **28:**910–919.

36. **Leake, C. R.** 1991. Lysimeter studies. *Pestic. Sci.* **31:** 363–373.

37. **Linn, D. M., and J. W. Doran.** 1984. Effect of water-filled pore space on carbon dioxide and nitrous oxide production in tilled and non-tilled soils. *Soil Sci. Soc. Am. J.* **48:** 1267–1272.

38. **Lyman, W. J., W. F. Reehl, and D. H. Rosenblatt (ed.).** 1982. *Handbook of Chemical Property Estimation Methods.* McGraw-Hill Book Co., New York.

39. **Marinucci, A. C., and R. Bartha.** 1979. Apparatus for monitoring the mineralization of volatile ^{14}C-labeled compounds. *Appl. Environ. Microbiol.* **38:**1020–1022.

40. **McFarland, M. J., R. C. Sims, and J. W. Blackburn.** 1991. Use of treatability studies in developing remediation strategies for contaminated soils, p. 163–174. *In* G. S. Sayler, R. Fox, and J. W. Blackburn (ed.), *Environmental Biotechnology for Waste Treatment.* Plenum Press, New York.

41. **Motulsky, H. J., and L. A. Ransnas.** 1987. Fitting curves to data using nonlinear regression: a practical and non-mathematical review. *FASEB J.* **1:**365–374.

42. **Mueller, J. G., S. E. Lantz, B. O. Blattmann, and P. J. Chapman.** 1991. Bench-scale evaluation of alternative biological treatment processes for the remediation of pentachlorophenol- and creosote-contaminated materials: slurry-phase bioremediation. *Environ. Sci. Technol.* **25:** 1055–1061.

43. **Oliver, M. A., and R. Webster.** 1991. How geostatistics can help you. *Soil Use Manag.* **7:**206–217.

44. **Organization for Economic Co-operation and Development.** 1981. Inherent biodegradability in soil, p. 1–11. *In OECD Guidelines for Testing of Chemicals,* part 304A. Organization for Economic Co-operation and Development, Paris.

45. **Peterson, M. S., L. W. Lion, and C. A. Shoemaker.** 1988. Influence of vapor phase sorption and diffusion on the fate of trichloroethylene in an unsaturated aquifer system. *Environ. Sci. Technol.* **22:**571–578.

46. **Powlson, D. S., and D. S. Jenkinson.** 1976. The effects of biocidal treatments on metabolism in soil. II. Gamma irradiation, autoclaving, air-drying and fumigation. *Soil Biol. Biochem.* **8:**179–188.

47. **Richter, O., and D. Söndgerath.** 1990. *Parameter Estima-*

tion in Ecology: the Link between Data and Models. VCH Publishers, New York.

48. **Rijnaarts, H. H. M., A. Bachmann, J. C. Jumelet, and A. J. B. Zehnder.** 1991. Effect of desorption and intraparticle mass transfer on the aerobic biomineralization of alpha-hexachlorocyclohexane in a contaminated calcareous soil. *Environ. Sci. Technol.* **24:**1349–1354.

49. **Robinson, J. A.** 1985. Determining microbial kinetic parameters using nonlinear regression analysis: advantages and limitations in microbial ecology. *Adv. Microb. Ecol.* **8:**61–114.

50. **Ross, D. J.** 1991. Microbial biomass in a stored soil: a comparison of different estimation procedures. *Soil Biol. Biochem.* **10:**1005–1007.

51. **Rowell, D. L.** 1994. *Soil Science: Methods and Applications.* Longman Science & Technology and John Wiley & Sons, New York.

52. **Rozycki, M., and R. Bartha.** 1981. Problems associated with the use of azide as an inhibitor of microbial activity in soil. *Appl. Environ. Microbiol.* **41:**833–836.

53. **Rüdel, H., S. Schmidt, W. Kördel, and W. Klein.** 1993. Degradation of pesticides in soil: comparison of laboratory experiments in a biometer system and outdoor lysimeter experiments. *Sci. Total Environ.* **132:**181–200.

54. **Salonius, P. O., J. B. Robinson, and F. E. Chase.** 1967. A comparison of autoclaved and gamma-irradiated soils as media for microbial colonization experiments. *Plant Soil* **27:**239–248.

55. **Schmidt, S. K.** 1992. A substrate-induced growth-response (SIGR) method for estimating the biomass of microbial functional groups in soil and aquatic systems. *FEMS Microbiol. Ecol.* **101:**197–206.

56. **Schmidt, S. K.** 1992. Models for studying the population ecology of microorganisms in natural systems, p. 31–59. *In* C. J. Hurst (ed.), *Modeling the Metabolic and Physiologic Activities of Microorganisms.* John Wiley & Sons, New York.

57. **Schmidt, S. K., and M. J. Gier.** 1989. Dynamics of microbial populations in soil: indigenous microorganisms degrading 2,4-dinitrophenol. *Microb. Ecol.* **18:**285–296.

58. **Schmidt, S. K., S. Simkins, and M. Alexander.** 1985. Models for the kinetics of biodegradation of organic compounds not supporting growth. *Appl. Environ. Microbiol.* **50:**323–331.

59. **Scow, K. M., and M. Alexander.** 1992. Effect of diffusion and sorption on the kinetics of biodegradation: experimental results with synthetic aggregates. *Soil Sci. Soc. Am. J.* **56:**128–134.

60. **Scow, K. M., and J. Hutson.** 1992. Effect of diffusion and sorption on the kinetics of biodegradation: theoretical considerations. *Soil Sci. Soc. Am. J.* **56:**119–127.

61. **Scow, K. M., S. K. Schmidt, and M. Alexander.** 1989. Kinetics of biodegradation of mixtures of substrates in soil. *Soil Biol. Biochem.* **21:**703–708.

62. **Scow, K. M., S. Simkins, and M. Alexander.** 1986. Kinetics of mineralization of organic compounds at low concentrations in soil. *Appl. Environ. Microbiol.* **51:**1028–1035.

63. **Shannon, M. J. R., and R. Unterman.** 1993. Evaluating bioremediation: distinguishing fact from fiction. *Annu. Rev. Microbiol.* **47:**715–738.

64. **Siegrist, H., and P. L. McCarty.** 1987. Column methodologies for determining sorption and biotransformation potential for chlorinated aliphatic compounds in aquifers. *J. Contam. Hydrol.* **2:**31–50.

65. **Skladany, G. J., and K. H. Baker.** 1994. Laboratory biotreatability studies, p. 97–172. *In* K. H. Baker and D. S. Herson (ed.), *Bioremediation.* McGraw-Hill Book Co., New York.

66. **Skidmore, M. W., D. Kirkpatrick, and D. Shaw.** 1994. Influence of application methods on the degradation of permethrin in laboratory, soil aerobic metabolism studies. *Pestic. Sci.* **42:**101–107.

67. **Skipper, H. D., and D. T. Westermann.** 1973. Comparative effects of propylene oxide, sodium azide, and autoclaving on selected soil properties. *Soil Biol. Biochem.* **5:**409–414.

68. **Smith, K. A., and C. E. Mullins (ed.).** 1991. *Soil Analysis. Physical Methods.* Marcel Dekker, New York.

69. **Smith, W., S. Prasher, S. U. Khan, and N. Barthakur.** 1992. Leaching of ^{14}C-radiolabelled atrazine in long intact soil cores. *Trans. Am. Soc. Agric. Eng.* **35:**1213–1220.

70. **Sokal, R. R., and F. J. Rohif.** 1981. *Biometry,* 2nd ed. W. H. Freeman and Co., New York.

71. **Soulas, G.** 1993. Evidence for the existence of different physiological groups in the microbial community responsible for 2,4-D mineralization in soil. *Soil Biol. Biochem.* **25:**443–449.

72. **Stenström, J.** 1989. Kinetics of decomposition of 2,4-dichlorophenoxyacetic acid by *Alcaligenes eutrophus* JMP134 and in soil. *Toxic. Assess.* **4:**405–424.

73. **Taylor, B. R., and D. Parkinson.** 1988. Respiration and mass loss of aspen and pine leaf litter decomposing in laboratory microcosms. *Can. J. Bot.* **66:**1948–1959.

74. **Topp, E., and J. L. Davis.** 1985. Time-domain reflectometry (TDR) and its application to irrigation scheduling. *Adv. Irrig.* **3:**107–127.

75. **Topp, E., and W. Smith.** 1992. Sorption of the herbicides atrazine and metolachlor to selected plastics and silicone rubber. *J. Environ. Qual.* **21:**316–317.

76. **Topp, E., W. N. Smith, W. D. Reynolds, and S. U. Khan.** 1994. Atrazine and metolachlor dissipation in soils incubated in undisturbed cores, repacked cores, and flasks. *J. Environ. Qual.* **23:**693–700.

77. **Troy, M. A.** 1994. Bioengineering of soils and ground waters, p. 173–201. *In* K. H. Baker and D. S. Herson (ed.), *Bioremediation.* McGraw-Hill, Book Co., New York.

78. **Tuominem, L., T. Kairesalo, and H. Hartikainen.** 1994. Comparison of methods for inhibiting bacterial activity in sediment. *Appl. Environ. Microbiol.* **60:**3454–3457.

79. **Turco, R., and A. Konopka.** 1990. Biodegradation of carbofuran in enhanced and non-enhanced soils. *Soil Biol. Biochem.* **22:**195–201.

80. **U.S. Environmental Protection Agency.** 1988. Anaerobic microbiological transformation rate data for chemicals in the subsurface environment. *Fed. Regist.* **53:**22320–22323.

81. **U.S. Environmental Protection Agency.** 1990. *Engineering Bulletin, Slurry Biodegradation.* Publication no. EPA/540/2-90/016. U.S. Environmental Protection Agency, Washington, D.C.

82. **U.S. Environmental Protection Agency.** 1991. *Guide for Conducting Treatability Studies under CERCLA: Aerobic Biodegradation Remedy Screening–Interim Guidance.* Publication no. EPA/540/2-91/013A. U.S. Environmental Protection Agency, Washington, D.C.

83. **van Genuchten, M. T., and P. J. Wierenga.** 1986. Solute dispersion coefficients and retardation factors, p. 1025–1054. *In* A. Klute (ed.), *Methods of Soil Analysis,* part 1, 2nd ed. American Society of Agronomy and Soil Science Society of America, Madison, Wis.

84. **Watanabe, I.** 1978. Pentachlorophenol (PCP) decomposing activity of field soils treated annually with PCP. *Soil Biol. Biochem.* **10:**71–75.

85. **Weaver, R. W., S. Angle, P. Bottomley, D. Bezdicek, S. Smith, A. Tabatabai, and A. Wollum (ed.).** 1994. *Methods of Soil Analysis: Microbial and Biochemical Properties.* Soil Science Society of America, Madison, Wis.

86. **Wilson, J. T., and B. H. Wilson.** 1985. Biotransformation

of trichloroethylene in soil. *Appl. Environ. Microbiol.* **49:**
242–243.

87. **Wolf, D. C., T. H. Dao, H. D. Scott, and T. L. Lavy.**
1989. Influence of sterilization method on selected soil microbiological, physical and chemical properties. *J. Environ.
Qual.* **18:**39–44.

88. **Yabannavar, A. V., and R. Bartha.** 1994. Methods for

assessment of biodegradability of plastic films in soil. *Appl.
Environ. Microbiol.* **60:**3608–3614.

89. **Zelles, L., P. Adrian, Q. Y. Bai, K. Stepper, M. V. Adrian,
K. Fischer, A. Maier, and A. Ziegler.** 1991. Microbial
activity measured in soils stored under different temperature and humidity conditions. *Soil Biol. Biochem.* **10:**
955–962.

Microbiology for the Metal Mining Industry

CORALE L. BRIERLEY AND JAMES A. BRIERLEY

93

Microorganisms play both beneficial and detrimental roles in the mining and mineral processing of metals. On the one hand, microorganisms degrade certain toxic constituents used in mineral processing and concentrate and immobilize soluble heavy metals released as a result of mining and mineral processing activities. On the other hand, certain bacteria are responsible for one of the most persistent and destructive environmental problems—acid rock drainage (ARD). Yet the same bacteria responsible for ARD are also commercially exploited for cost-effective, efficient, and environmentally sound extraction of base metals such as copper, zinc, and nickel and for the pretreatment of ores and mineral concentrates in which precious metals (gold and silver) are embedded in sulfide minerals.

This chapter aims to provide an understanding of the microbiology and of the procedures used to assess ARD production, metals bioremediation, and commercial metals extraction. An overview of mining and mineral processing practices is provided as a foundation for appreciating the mining environment. The microorganisms and their reactions contributing to both beneficial and destructive activities in the mining environment are described. Sampling and assessment methods to predict and quantify microbial interactions with mineralized rock and soluble metal ions are outlined. Techniques to inhibit the bacterial production of ARD from rocks containing minerals composed of reduced sulfur and iron species are discussed. Practices using microorganisms to detoxify waste streams from mining and mineral processing operations and to responsibly process minerals for metals extraction are examined.

OVERVIEW OF METAL MINING AND MINERAL PROCESSING PRACTICES

Mining and processing of minerals involve many complex operations, some producing solid and aqueous wastes that must be properly managed to prevent environmental degradation (49). The mining industry's image has suffered significantly from its past actions, but stricter regulation, public reaction, improved technologies (including biotechnology), and the benefits associated with mitigation of environmental impacts have resulted in better stewardship of the land as well as more cost-effective mining practices (47).

Minerals are mined by a variety of surface and underground techniques. In surface mining, soil and rock (called overburden) overlying the ore body are removed in a process called stripping. Overburden is usually chemically inert. With the exception of turbidity, the runoff water from overburden usually does not present any risk to water quality. Turbidity is controlled by runoff containment and sedimentation. The ore body, exposed after stripping in surface mines and reached via shafts and tunneling in underground mines, is then drilled for placement of explosives and blasted to fragment the rock. Waste rock, which includes nonmineralized or low-grade mineralized rock, is removed from above or adjacent to the ore. This fragmented rock is usually placed in piles close to the mine. The nature and amount of mineralization in the waste rock, climatic conditions, and the ability of soil beneath the waste rock pile to attenuate contaminants determine the potential of runoff from the material to affect water quality. When the waste rock has acid-producing potential due to the presence of sulfide minerals and little or no acid-consuming carbonate minerals, special waste management practices are instituted. These practices are (i) using impermeable liners, (ii) comingling acid-generating with non-acid-generating materials, (iii) vegetating surfaces to minimize water and oxygen inflow, and (iv) collecting leachate.

In some cases, mines intercept groundwater. Dewatering is then required to maintain relatively dry conditions in the surface or underground mine. The dewatering product is pumped to the surface and used in processing facilities, discharged to receiving waters, used for agricultural purposes, or injected to groundwater. If the dewatering product contains constituents in excess of discharge standards for disposal, treatment may be required.

As mined, most ores contain valuable metals disseminated in a matrix of less valuable rock, called gangue. To separate metal-bearing minerals from the gangue, yielding a product higher in metal content, the ore must be crushed and/or ground small enough that each particle contains mostly the mineral to be recovered or mostly gangue. The separation of the particles on the basis of some difference between the ore mineral and the gangue yields a concentrate high in metal value and a finely ground waste rock (called tailings) containing very little metal. Tailings are disposed of in surface impoundments near the mineral processing site. The potential for tailings to affect water quality depends on the chemistry of the tailings material and the conditions at the disposal site. Excess solution from the tailings is collected and returned to the processing facility for reuse. If tailings pose a risk, liners may be required.

Properties used as the basis for separating valuable minerals from gangue include specific gravity, conductivity, magnetic susceptibility, affinity for certain chemicals, solubility, and the tendency to form chemical complexes. Processes for effecting the separation may be generally considered as gravity concentration, magnetic separation, electrostatic separation, flotation, and leaching. Solvent extraction, ion exchange, and carbon adsorption are widely applied techniques for concentrating metals from leaching solutions and for separating them from dissolved contaminants. Modern mineral processing also includes a method of piling ore, which may be fragmented material directly from the mine or may be crushed, onto impermeable pads and leaching the ore by allowing solutions to percolate through the material. This process, called heap leaching, is used to extract base and precious metals. Gold, silver, copper, zinc, and other metals are solubilized in the leach solutions. Bacteria are used in some commercial heap leach operations to solubilize base metals and to pretreat precious metal ores in which the gold and silver are occluded in iron sulfide minerals. Pretreatment substantially enhances gold and silver recoveries by degrading the mineral matrices and exposing the precious metals for dissolution with chemical reagents. Bacterial pretreatment will be discussed later in this chapter. Once metals of value are extracted from the heap-leached ore, the solid residue is rinsed with water and, if necessary, treated to meet constituent standards. The potential for water quality impacts from heap-leached, solid residue depends on the properties of the residue, reagents used, extent of water rinsing and treatment, and conditions specific to the site. Heap leach residues are disposed of in lined or unlined waste management units as required. Final closure may involve recontouring the piles, placing topsoil, revegetating, and providing erosion control.

Liquid wastes generated from flotation, leaching and solvent extraction practices, operation of mechanical equipment, storm water runoff from ore and waste rock storage areas, and wash-down water from metal recovery plants and maintenance buildings can contain elevated concentrations of regulated constituents. Depending on the characteristics and concentrations of the constituents, these wastes are contained in lined or unlined surface impoundments. Water is reused in processing facilities or may be treated before disposal.

In situ mining (also called solution mining) is a technique sometimes used to recover copper and uranium. In situ mining involves leaching the desired metal from mineralized ground in place. The ore body is penetrated and permeated with the leaching solution. The leach solution is contained within the mineralized area by impermeable geologic layers beneath the ore body and by controlled pumping. The leach solution containing the dissolved metal is recovered for processing at the surface. After in situ mining is completed, leached areas are restored to required standards by flushing the leached area with water until regulated constituent levels meet standards. Leach solutions and wash waters are contained in impoundments and evaporated or treated to meet disposal standards.

The microbial ecology of mining environments is far from being fully elucidated. However, even the harshest of these environments hosts a wide diversity of heterotrophic and autotrophic microorganisms whose identities and functions are largely unknown (45). Most microbial research and development efforts related to metal mining environments

have been directed toward the bacteria involved in acid rock formation and the commercial applications of bioleaching and minerals biooxidation. The reason for this is twofold. ARD is the most widespread, persistent, destructive, and least controllable environmental problem associated with base and precious metal mining. However, the same bacteria that cause ARD can be effectively applied in controlled situations to process base and precious metal deposits that are not technically and/or economically amenable to treatment by conventional mineral processing methods. Microorganisms that oxidize cyanide compounds and accumulate, precipitate, or otherwise transform heavy metals have been extensively studied in the laboratory but are underexploited commercially.

The remainder of this chapter considers the beneficial and destructive roles of microorganisms in mining, mineral processing, and mining waste management. Methods used to evaluate microbial processes in the mining environment are discussed. Commercial applications of microbiology in base and precious metal mining are accentuated in this chapter. Important microbial, metallurgical, and chemical interactions occurring in the mining environment are especially noted. Emphasis is placed on microbial activities related to ARD and minerals biooxidation because of their relative importance to the mining industry.

MICROBIOLOGY AND CHEMISTRY OF ARD
ARD Formation
ARD is the leachate resulting from the oxidation of sulfide minerals exposed to water, air, and bacteria and the resultant products from the interaction of acid, metal-bearing solutions reacting with alkaline rocks and water. ARD is the result of both bacterial and chemical activity. This section describes the key reactions causing ARD.

ARD initiates when sulfide minerals, such as pyrite (FeS_2), are exposed to air and water in rock outcrops, mining operations, or even springs:

$$4FeS_2 + 15O_2 + 2H_2O \rightarrow 4Fe^{3+} + 8SO_4^{2-} + 4H^+ \tag{1}$$

The pyrite chemically oxidizes, creating a slightly acidic environment conducive for the development of *Thiobacillus ferrooxidans* (53). This naturally occurring and ubiquitous chemolithotrophic microorganism colonizes the exposed mineral surfaces. *T. ferrooxidans* derives energy from the oxidation of inorganic sulfur- and iron-containing compounds, including mineral sulfides, ferrous iron, sulfur, thiosulfate, and tetrathionate, and assimilates CO_2 via the Calvin-Benson cycle (76). *T. ferrooxidans* accelerates the chemical rate of pyrite oxidation some 500,000 to 1 million times (8). Because pyrite (FeS_2) is common to sulfide mineral deposits, the bacterially catalyzed oxidation of pyrite is the single most important reaction contributing to ARD.

The bacterial oxidation of pyrite (reaction 1) produces ferric iron (Fe^{3+}), a strong oxidant that chemically oxidizes mineral sulfides including pyrite:

$$FeS_2 + 14Fe^{3+} + 8H_2O \rightarrow 15Fe^{2+} + 2SO_4^{2-} + 16H^+ \tag{2}$$

The ferrous iron (Fe^{2+}) resulting from this reaction is regenerated to ferric iron by *T. ferrooxidans*:

$$4Fe^{2+} + O_2 + 4H^+ \rightarrow 4Fe^{3+} + 2H_2O \tag{3}$$

The ferric iron is then available to oxidize more pyrite, and the cycle continues. Depending on the pH and ionic composition of the bulk solution, jarosite (basic ferric sulfate compounds) may form. Jarosite is the result of the chemical reaction between ferric iron, sulfate, and soluble potassium, although other monovalent ions, such as NH_4^+ and Ag^+, can induce jarosite formation:

$$K^+ + 3Fe^{3+} + 2SO_4^{2-}$$
$$+ 6H_2O \rightarrow KFe_3(SO_4)_2(OH)_6 + 6H^+ \quad \textbf{(4)}$$

$$K^+ + 3Fe3^{3+} + 2HSO_4^-$$
$$+ 6H_2O \rightarrow KFe_3(SO_4)_2(OH)_6 + 8H^+ \quad \textbf{(5)}$$

Jarosite formation is an acid-producing reaction (40, 51, 76). Under low-pH conditions, iron oxides such as goethite [αFeO(OH)] and lepidocrosite [γFeO(OH)] can also form (58).

As acidic, sulfurous, iron-bearing solution seeps from the sulfide-rich environment and encounters rocks, soils, and/or water of higher pH (pH > 2.5), the soluble ferric iron produced in reactions 1 and 3 will hydrolyze, generating additional acid:

$$Fe^{3+} + H_2O \rightarrow FeOH^{2+} + H^+ \quad \textbf{(6)}$$

$$Fe^{3+} + 2H_2O \rightarrow Fe(OH)_2^+ + 2H^+ \quad \textbf{(7)}$$

$$Fe^{3+} + 3H_2O \rightarrow Fe(OH)_3 + 3H^+ \quad \textbf{(8)}$$

The extent of ferric hydrolysis is dependent on the pH (40). This acid generation from iron hydrolysis must be quantitatively considered when ARD is treated with lime or limestone.

In the mining environment, a number of acid-consuming reactions occur (49). These include neutralization of acid by carbonate minerals such as calcite,

$$2H^+ + CaCO_3 \rightarrow Ca^{2+} + CO_2 + H_2O \quad \textbf{(9)}$$

and dolomite,

$$4H^+ + CaMg(CO_3)_2 \rightarrow Ca^{2+} + Mg^{2+}$$
$$+ 2CO_2 + 2H_2O \quad \textbf{(10)}$$

the reaction of acid with aluminosilicates (potassium and calcium feldspars),

$$2KAlSi_3O_8 + 2H^+ + H_2O \rightarrow Al_2Si_2O_5(OH)_4$$
$$+ 4SiO_2 + 2K^+ \quad \textbf{(11)}$$

$$CaAl_2Si_2O_8 + 2H^+ + H_2O \rightarrow Al_2Si_2O_5(OH)_4 + Ca^{2+}$$
$$\textbf{(12)}$$

the formation of gypsum,

$$CaCO_3 + SO_4^{2-} + 2H^+ + H_2O \rightarrow$$
$$CaSO_4 \cdot 2H_2O + CO_2 \quad \textbf{(13)}$$

and the bacterial oxidation of some divalent, base metal sulfides,

$$MS + 2.5O_2 + 2H^+ \rightarrow M^{2+} + SO_4^{2-} + H_2O$$
$$\textbf{(14)}$$

where MS is metal sulfide and M^{2+} is a divalent metal such as Cu^{2+}, Zn^{2+}, or Ni^{2+}.

Also present in the same environment as *T. ferrooxidans* are the chemoautotrophic, acidophilic *Thiobacillus thiooxidans*, which oxidizes reduced sulfur,

$$2S^0 + 2H_2O + 3O_2 \rightarrow 4H^+ + 2SO_4^{2-} \quad \textbf{(15)}$$

and *Leptospirillum ferrooxidans*, which oxidizes reduced iron compounds (reactions 2 and 3).

The balance between acid-generating and acid-consuming reactions dictates whether the system will be a net acid producer or net acid consumer. How this is actually determined will be discussed later.

The oxidation of pyrite (reaction 1) is an exothermic reaction with $\Delta H_r^0 = -1,440$ kJ mol^{-1}. In waste rock piles with active pyrite oxidation, sufficient heat is produced and retained within the pile for heat to build up. It is not unusual for temperatures to exceed 60°C (17, 53). Because *Thiobacillus* and *Leptospirillum* species are mesophilic, oxidizing inorganic substrates in the temperature range of approximately 10 to 40°C, the high temperatures that occur in some sulfidic waste piles eventually limit these organisms. However, it is now known that in sulfidic materials, a natural succession of chemoautotrophic bacteria will be established on the basis of temperature (53, 62). At lower temperatures (about 10 to 40°C), *Thiobacillus* and *Leptospirillum* species will predominate. At about 40°C, mesophilic chemoautotrophs will begin to die (63) and moderately thermophilic, acidophilic, chemoautotrophic bacteria will appear. At 30 to 40°C, mesophilic and moderately thermophilic bacteria will coexist. Oxidizing iron and sulfur compounds at a temperature range of about 40 to 60°C is a diverse group of organisms that are not yet well characterized. At a temperature of approximately 55°C, the moderate thermophiles are succeeded by the extremely thermophilic, acidophilic *Sulfolobus*, *Acidianus*, *Metallosphaera*, *Sulfobacillus*, and *Sulfurococcus* archaea. These bacteria oxidize iron and sulfur compounds under acid conditions at temperatures ranging from about 55 to near 85°C (53, 76). When the temperature exceeds the upper limits of the bacteria colonizing the moist and acidic areas of the sulfidic rock pile, all bacterial activity ceases and the temperature of the pile will decrease (62). Numerous papers describing the characteristics of the acidophilic thermophiles and methods of study have been published (1, 15, 64, 65).

Because of bacterial catalysis in ARD formation, efforts have been made to inhibit the growth of *Thiobacillus* and *Leptospirillum* species and the thermophilic bacteria by adding surfactants (72) and slow-release biocides (73). However, these chemical agents become diluted after a time and/or adsorbed to rock surfaces and eventually become ineffective. Frequent application is required. Applying reagents at depth in piles of sulfide-bearing rocks colonized by bacteria is nearly impossible. Once ARD is initiated, it is virtually unstoppable. To prevent the initiation of ARD, sulfide minerals must be isolated from air and water. This stops chemical oxidation and also inhibits the growth and activity of the bacteria that catalyze the reactions accelerating formation of ARD. Prevention is best implemented when mining activities are first started. Sulfide-bearing waste rocks fragmented by blasting, crushed sulfide ore residues from heap leach operations, and finely ground sulfidic tailings are especially vulnerable to chemical and biological oxidation because of the extensive exposure of sulfide surfaces to conditions conducive for ARD formation. Sulfide materials that have been identified as acid producing are subject to special disposal practices. Waste management options now used by the metal mining industry include

- Mixing the acid-producing minerals with acid-consuming rocks to neutralize acid that may be generated
- Encapsulating acid-producing rocks with low-permeability clay, minimizing air and water contact

- Recontouring waste piles, capping with clay, earth, and soil, vegetating surfaces to promote evapotranspiration, and implementing water control systems to prevent erosion

- Placing acid-producing rocks on impermeable pads and collecting, containing, and treating acid runoff

- Storing tailings and finely ground materials in lined impoundments and maintaining a water layer over the material, minimizing contact with air.

Mining wastes generated from the early days of mining activity until the middle of this century, before present-day environmental controls were established, have produced an environmental legacy (44) that not only is deleterious to the industry's image but has been difficult to technically rectify. Millions of tons of sulfide wastes exposed to air and water have produced ARD, contaminating thousands of miles of rivers and streams. Often in remote and virtually inaccessible locations, these exposed wastes cannot easily be reclaimed. These wastes will continue to produce ARD for hundreds of years and adversely affect lakes and streams.

Assessing Bacterial ARD Formation

Predicting ARD is both environmentally and economically important (77). If ARD production is predicted from testing of mine waste materials and ores, these materials can be specially handled to prevent ARD from starting. Prevention of the problem is far more cost-effective than maintaining collection and treatment facilities in perpetuity (20). Prediction testing can be used throughout the mine's life to assess the acid-generating potential of overburden, waste rock, tailings, and stockpiled ores. Predictive methods have recently been used to evaluate the rock materials making up the walls of open pit operations. When operations cease at some open pit mines, the pits will flood as a result of groundwater inflow. Permanent mine pit lakes possess a water quality that is dependent on complex chemical and biological interactions between the minerals making up the walls of the pit and the chemistry of inflowing groundwater (60).

Acid production and acid consumption of a mineralized material and the catalytic contribution of the chemolithotrophic bacteria to ARD formation can be quantified by the carbon-sulfur method, the biological acid-producing-potential (BAPP) test, the humidity cell test, and the large column leach test. None of the methods is approved by the Environmental Protection Agency. However, the carbon-sulfur method and the humidity cell test are under consideration by the American Society for Testing and Materials (ASTM) for standardization.

Carbon-Sulfur Method

Characterization of waste rock and ore is now required to environmentally permit a new mining operation. The carbon-sulfur analysis, a type of acid-base accounting, is being considered for standardization by ASTM. The carbon-sulfur method attempts to balance potentially acid-generating materials, which are generally sulfide minerals, with acid-consuming materials, which are primarily carbonate minerals, although hydroxides, silicates, and clays can also provide neutralization capacity (20). The method involves the following:

- Measuring total percent carbon and sulfur in the ore, waste rock, or tailings by infrared analysis

- Pyrolyzing the material, converting sulfide to SO_2. Sulfate-sulfur is assumed to be the sulfur remaining after

pyrolysis. The acid generation potential (AGP) in tons of calcium carbonate required per 1,000 tons (1 ton = 0.907 metric tons) of rock is calculated from the total sulfur and residual sulfur after pyrolysis (difference = sulfide-sulfur):

$$AGP = (\% \text{ sulfide-sulfur}) \times 10 \qquad (16)$$
$$\times 100.09/32.06$$

where AGP is expressed in tons of $CaCO_3$ per 1,000 tons, 10 represents (tons/1,000 tons) × 100, 100.09 is the molecular weight of $CaCO_3$ in grams per mole, and 32.06 is the molecular weight of sulfur in grams per mole.

- Treating the sample with 1 N HCl to solubilize carbonate minerals. The acid neutralization potential (ANP) is calculated as

$$ANP = (\% \text{ carbonate-carbon}) \times 10 \qquad (17)$$
$$\times 100.09/12.01$$

where ANP is expressed in tons of $CaCO_3$ per 1,000 tons, 10 represents (tons/1,000 tons) × 100, 100.09 is the molecular weight of $CaCO_3$ in grams per mole, and 12.01 is the molecular weight of carbon in grams per mole.

- Calculating the net acid generation (NAG):

$$NAG = AGP - ANP \qquad (18)$$

where NAG, AGP, and ANP are expressed in tons of $CaCO_3$ per 1,000 tons.

The initial hypothesis of this test was that a sample would generate acid only if the AGP exceeded the ANP. This is now known not to be true; in fact, some mineral samples with excessive ANP will in some cases be acid generating. Therefore, it is essential that other tests listed below be used in conjunction with the carbon-sulfur method to adequately predict the potential for ARD formation.

BAPP Test

The BAPP test (21) is based on oxidation of sulfide minerals by *T. ferrooxidans*. The objective of the test is to determine if the bacteria can generate enough sulfuric acid from oxidation of sulfides present in the sample to maintain acidic conditions suitable for their activity. If they can, microbial action will continue on a self-sustaining basis. The carbonate content of the rock is of paramount importance. Little or no carbonate mineralization results in acid generation and metal leaching. Excess carbonate prevents acid generation by maintaining conditions unfavorable for growth of the acidophilic, iron- and sulfide-oxidizing bacteria or by neutralization of any acid produced from sulfide oxidation.

To perform the BAPP test, a mineral sample is milled, yielding a particle size of less than 38 μm. A 30-g portion (a smaller amount of ore if the sulfide content exceeds 2%) of dry ore sample is placed in a flask with 70 ml of a basal salt solution [for example, 0.4 g of $(NH_4)_2SO_4$, 0.4 g of $MgSO_4 \cdot 7H_2O$, and 0.04 g of KH_2PO_4 per liter] suitable for *T. ferrooxidans*. The pH of the solution and mineral slurry is adjusted to 2.5 with sulfuric acid (10 to 36 N). The flask contents are equilibrated for 24 h on a gyratory shaker. If necessary, additional acid is added to maintain a pH between 2.5 and 2.8. The pH-adjusted slurry is inoculated with 5 ml of an active culture of *T. ferrooxidans*. The inoculum can be grown on a basal salts solution containing 33.0 g of $FeSO_4 \cdot 7H_2O$ per liter at pH 2.0. The weight of the flask and contents is recorded so that water can be added later to compensate for evaporative loss during the incubation period. The inoculated mineral suspension is incubated at 35°C on a gyratory shaker.

The pH is checked to ensure that it remains below 2.8. The concentration of a dissolved metal, usually iron but copper or other heavy metals if present in the material, is used as an indicator of bacterial activity. The flask contents are monitored every second day until microbiological activity ceases, i.e., until the pH and dissolved metal concentration remain constant. When microbiological activity ceases, 15 g of the finely ground rock is added and the culture is incubated for an additional 24 h. If the pH exceeds 3.5, the test is terminated. If the suspension is 3.5 or less, half of the weight of the sample (15 g) of rock material is added and incubation is continued for 24 h. After 24 h, if the pH is less than 3.5 or greater than 4, the test is terminated. If the pH is between 3.5 and 4.0, incubation is continued for an additional 48 h and the final pH value is recorded.

The acid demand of the sample is initially satisfied by addition of sulfuric acid to promote the growth of the bacteria. Once microbiological action has ceased, half of the original weight of the sample is added to the culture. If there has not been sufficient acid production, the pH will approach the natural pH of the sample (pH 3.5 or above). If the pH is 3.5 or less, another half-weight of sample is added and incubation is continued for up to 72 h. If the final pH is 3.5 or less, the sample is potentially acid producing. If the final pH is above 3.5, the results of the BAPP test are considered negative. A negative result indicates that the bacteria cannot sustain an oxidation reaction in the material. However, it is possible that the material is capable of generating acid at rates governed by chemical oxidation kinetics.

Humidity Cell Tests

Humidity cell tests are designed to model natural weathering of a waste material. A 1,000-g sample of waste rock, crushed to less than 6.3 mm, is placed in a special apparatus that provides control over air, temperature, and moisture. The device allows for collection and monitoring of oxidation products. Details of the humidity cell test are beyond the scope of this chapter; however, the test is being considered for ASTM standardization. Accelerated-weathering methods described by Caruccio (23) and Ferguson and Morin (31) are modifications of the proposed ASTM humidity cell test.

The test involves subjecting waste rock material in the humidity cell to alternating cycles of dry air (3 days) and water-saturated air (3 days) followed by a leach on day 7 with water. The leachate is collected and analyzed for pH, redox potential, acidity, alkalinity, sulfate, conductivity, and dissolved metals. The minimum period for this test is 20 weeks.

Modifications of the humidity cell include inoculation of the cell contents with *T. ferrooxidans*. If the waste rock material placed in the humidity cell has the potential for acid generation, as determined by the carbon-sulfur method and by the BAPP test, bacterial activity will accelerate acid production in the humidity cell.

The humidity cell test is intended to help identify the long-term potential of a solid waste sample to produce effluent that either meets or fails to meet established water quality standards under the conditions of the test. The aim of the method is to promote more rapid oxidation of solid waste constituents than would occur in nature and to maximize the loading of oxidation products in the weekly effluent. The combination of small volume of waste material,

temperature, excessive water, aeration, and inoculation with *T. ferrooxidans* ensures rapid oxidation rates.

Column Tests

Columns tests, which use cylinders ranging in size from 15 to 60 cm in diameter and 60 cm to 6 m in height, can be used to predict ARD from mine waste materials. Column tests are usually performed to complement the tests described above and to obtain data from larger-scale evaluations. For example, a larger particle size of waste material, amendments (e.g., lime or limestone) to neutralize potential acid production, or the effects of biocides may be evaluated. There are no defined protocols for columns tests. Column tests are custom-designed to achieve a specific objective, to simulate climatic conditions at a specific site, or both (20).

Columns are filled with rock, and deionized or distilled pH 6.0 to 6.5 water (pH adjusted with reagent-grade HNO_3) is applied to the waste rock. The water is dripped onto the ore at the top of the column and allowed to percolate through the waste rock. Alternatively, the entire column of waste rock is inundated with water. The columns are aerated from the base by using aquarium pumps or filtered, compressed air. If climatic conditions are being simulated, the water is dripped onto the rock in amounts that coincide with atmospheric precipitation at the mine site. For example, high rates of water application may be applied during a certain number of months to simulate the melting of snow. If inundation is used, the water is allowed to contact the ore for a period of 1 to 2 days each week and is then drained from the column and collected; for the remaining 5 or 6 days of each week, the column material is aerated. The waste rock in column tests is often inoculated with *T. ferrooxidans* to increase oxidation of sulfide minerals. Column tests are usually run for at least a year and often longer to obtain the necessary information.

The leachate from the waste rock column is collected and analyzed for pH, redox potential, acidity and alkalinity, conductivity, and sulfate, total iron, ferrous iron, and metal concentrations. The specific metals analyzed are based on regulated constituents. If the waste material is acid generating, several weeks to several months, depending on the particle size of the material, will elapse before oxidation begins. Once sulfides begin to oxidize and conditions within the column turn acid, the redox potential will rise, often to greater than 700 mV (standard hydrogen electrode), total iron will increase, and the Fe^{2+}/Fe^{3+} ratio will decrease as *T. ferrooxidans* oxidizes the Fe^{2+} to Fe^{3+}. Metal concentrations will increase.

Column tests provide extensive information on the rate of sulfide oxidation and the rate of release of metals from the rock of a certain particle size. Some models exist for relating the data derived from column tests to actual waste rock piles containing millions of tons of material. However, modeling is extremely complex because of the many biological, geochemical, geotechnical, and climatic variables that must be considered. The information gained from long and very costly column tests is used by the mining industry to make decisions on the long-term handling of waste materials. If the column tests predict that acid generation and concomitant metal release will occur, the waste material will be accorded special handling and disposition to minimize its contact with water and air (20).

Field Assessment of ARD

Often it is necessary to perform field assessments on stored waste materials or stockpiled ore to evaluate the potential for ARD or assess whether control measures that were implemented to prevent ARD are actually effective. Instrument monitoring and sampling are approaches applied to evaluating conditions within waste materials.

Instrumenting waste material piles to monitor temperature is an effective tool to detect sulfide oxidation. Because bacterial oxidation of pyrite and other sulfide minerals is an exothermic reaction ($\Delta H_r^0 = -1,440$ kJ mol^{-1}) and the heat is not well dissipated in waste rock piles, the temperature will rise. An increase in temperature is indicative of bacterial activity and hence ARD formation (42, 49).

Waste piles can be instrumented to detect gases. The chemolithotrophic bacteria consume O_2 in the oxidation of sulfides. If O_2 depletion is detected within waste piles, bacterial activity leading to ARD is likely (43). Detection of excessive CO_2 in the gaseous phase of the waste pile is also indicative of ARD because of carbonate mineral neutralization by acid:

$$2H^+ + CaCO_3 \rightarrow Ca^{2+} + CO_2 + H_2O \quad (19)$$

Lysimeters can be placed in waste rock piles to collect water for chemical and bacterial analyses. The pore water is analyzed for pH, redox potential, acidity and alkalinity, conductivity, and sulfate, total iron, ferrous iron, and metal concentrations. A decreasing pH and increasing acidity, redox potential, and cncentrations of total iron and soluble metals over time are indicative of ARD formation.

Enumerating *T. ferrooxidans* and the moderately and extremely thermophilic, iron-oxidizing bacteria (15) in water is accomplished by direct counts using a Petroff-Hauser countering chamber (45), agar plate colony counts (52), epifluorescence microscopy with acridine orange stain (2, 84), fluorescent antibody staining (38), dot-immunobinding (1, 4, 36), indirect measurements using nitrogen, protein (68), and ATP, DNA analyses (30, 69, 83), or the most probable number (MPN) (41) method. Direct counting, epifluorescence microscopy, and various staining and immunobinding assays do not distinguish between living and dead bacteria. Molecular probes, that is, the application of DNA technology, have considerable potential for assessing bacterial populations in mine wastes. However, the technology, which is still in its infancy relative to mining applications (69), is not widely applied. Agar plate colony counts lack accuracy because of the toxicity of gel components to the chemolithotrophic bacteria. Gels are difficult to use with thermophiles because of gel liquification. Mineral particles and high concentrations of metals interfere with indirect methods such as protein, nitrogen, and ATP analyses. The MPN method is frequently used because the procedure is specific for the iron-oxidizing bacteria responsible for ARD and it enumerates only viable bacteria. The MPN method involves serial dilution of the liquid sample containing the bacteria and culturing of the bacteria at each dilution in a liquid growth medium [e.g., 0.4 g of $(NH_4)_2SO_4$, 0.4 g of $MgSO_4\cdot7H_2O$, 0.04 g of KH_2PO_4, and 33.0 g of $FeSO_4\cdot7H_2O$ per liter, adjusted to pH 1.6 to 1.8 with H_2SO_4] that supports the growth of iron-oxidizing bacteria. The MPN method is a statistical method whose accuracy is increased by increasing the number of cultures made at each dilution of the original sample.

To assess ARD in the field, waste rock piles and tailings may be core drilled to obtain solid samples for detection of bacteria. Enumerating *T. ferrooxidans* in solids is highly inaccurate by any of the methods listed (15). Because the bacteria attach to the rock particles and cannot be easily dislodged from the solid particles, there is no simple way to adequately count the bacteria. The complexity of surface texture and particle clumping conceals some of the microorganisms associated with the particles. These characteristics preclude visual methods for counting bacteria. Bacterial counts can be obtained. However, the cell concentration data can be used for only relative comparisons, at best, as there is no adequate methodology for dislodging the entire microbial population from the mineral surface. The particle size effect, as it relates to surface area, must also be considered in attempts to determine population numbers on mineral samples. To decrease the number of variables in making a suspension of bacteria from minerals, a select particle size (e.g., 2 mm) can be produced by screening the mineral material. Nevertheless, the heterogeneity of particle size in waste rock and ore precludes determination of an absolute bacterial population.

Respirometry is increasingly being used as an indirect measurement of bacterial activity in solid mine wastes. Both O_2 and CO_2 uptake can be measured by this technique. Correlation of gas uptake with enumeration techniques provides an assessment of bacterial numbers and activity.

Bioleaching and Biooxidation

As destructive and as unstoppable as ARD is, the catalytic activity of the mesophilic and thermophilic chemoautotrophs has been harnessed for cost-effective, efficient, and environmentally acceptable commercial processing technologies called bioleaching and minerals biooxidation. Bioleaching is the bacterial oxidation of sulfide minerals, whereby metals of value (for example, copper, uranium, and zinc) are released into solution. Minerals biooxidation is a biological process in which iron sulfide minerals, such as pyrite and arsenopyrite (FeAsS), are degraded by bacteria and precious metals (gold and silver) are liberated for recovery by conventional metallurgical techniques. The best studied organisms in commercial bioleaching and minerals biooxidation are *T. ferrooxidans* and *L. ferrooxidans*. Increasingly, however, the moderately thermophilic, iron- and sulfide-oxidizing bacteria are being used or considered for use (13).

Since the 1950s, bacteria have been used to bioleach sulfidic, mineral-bearing waste rock from surface copper mines. The waste rock is piled near mining operations, a dilute sulfuric acid solution is applied to the top surfaces of the waste piles by using drip irrigation, and the solution is percolated through the pile. The moist, acid environment along with air entering from the tops and sides of the waste rock pile provides a conducive environment for naturally occurring *Thiobacillus* and *Leptospirillum* species to develop. Bacterial numbers reach 10^6 to 10^7 per ml of leach solution. The bacteria oxidize copper sulfide minerals (chalcocite [Cu_2S], covellite [CuS], and chalcopyrite [$CuFeS_2$]) and pyrite, releasing soluble copper and ferric iron. Direct bacterial action on the copper sulfide minerals and oxidation of these minerals by ferric iron solubilize the copper, which is carried from the waste pile by the percolating acid solution. The copper is recovered by solvent extraction, and high-grade cathode copper is produced by electrowinning. Some 20% of the world's copper is estimated to be produced by bioleaching (13). With environmental restraints placed on smelter operations and a global abundance of sulfidic copper ore deposits, copper production from bioleaching is anticipated to increase. Evidence of this is plentiful in Chile, where even higher-grade copper sulfide ores are bioleached in a process called bacterial thin-layer leaching (22). The copper sulfide ore is crushed to less than 6.3 mm and placed on an impermeable pad to a height of 3 to 6 m. A dilute sulfuric acid solution is applied along with a mixed culture of mesophilic iron- and sulfide-oxidizing bacteria, and bacterial catalysis commences. After 7 to 9 months, bioleaching is complete, with about 80% of the copper extracted from the ore.

A heap-leaching method has been developed to process precious metal ores in which elemental gold and silver are encased in sulfide minerals. Processing of these ores with cyanide, but without any treatment to expose the precious metals, results in very poor (often less than 30%) gold or silver recovery. Precious metal ores, amenable to minerals biooxidation, are called refractory-sulfidic precious metal ores. As with bacterial leaching of copper sulfide ores, the refractory-sulfidic precious metal ores are crushed, but usually to a larger size (less than 19 mm), and stacked on impermeable pads to heights of approximately 6 to 12 m. Bacteria can be added to the crushed ore as it is stacked onto lined pads (16). With drip irrigation of the heap, using dilute sulfuric acid containing low concentrations of ammonium and phosphate ions, the bacteria in close proximity to the sulfide minerals rapidly oxidize the pyrite and arsenopyrite in which the gold and silver are embedded. Within several months, depending on the ore characteristics, the bacteria and ferric iron have oxidized the sulfide minerals, exposing the elemental gold and silver. The ore is then washed with water to remove acid, soluble heavy metals, and iron. The ore is neutralized and treated with a dilute sodium cyanide solution or other reagents (e.g., thiourea or thiosulfate) that solubilize the precious metals. Biooxidation of refractory-sulfidic precious metal ores is called pretreatment because the ore is subjected to an additional treatment process before undergoing conventional metallurgical extraction with gold-solubilizing reagents. Minerals biooxidation pretreatment of refractory-sulfidic precious metal ores is revolutionizing the gold industry. This technology allows ores that were previously uneconomic to process by conventional methods to be profitably mined and processed (18).

Some refractory-sulfidic precious metal ores are amenable to concentration (that is, separation of the sulfides occluding the gold and silver from other gangue constituents such as calcite or silica minerals). This upgrading yields a concentrate that has a much higher gold content and therefore can be economically processed with technology that has a higher intrinsic capital and operating cost. Mineral biooxidation, using mesophilic and moderately thermophilic iron- and sulfide-oxidizing bacteria, is now used in commercial plants around the world to pretreat sulfidic-refractory precious metal concentrates. This process entails grinding the gold- or silver-bearing pyrite and/or arsenopyrite to a size of less than 100 μm. This material is fed continuously to a large stainless steel or rubber-lined steel tank, which is aerated and agitated. Ammonium and phosphate salts are fed to the tank containing a dilute solution of sulfuric acid. *Thiobacillus* and *Leptospirillum* species, adapted to the mineral and high concentrations of heavy metals that are solubilized during the minerals biooxidation process, are grown in the tank on the mineral concentrate. Because the conditions (nutrients, mineral sulfide energy source, O_2 and CO_2, temperature) in the tank are highly conducive for the bacteria, the pyrite and arsenopyrite are rapidly oxidized by the bacteria and the ferric iron oxidant. In commercial minerals biooxidation plants, 60 to 80% of the sulfide mineral at a solid's density of about 20% (wt/vol) is solubilized in the first tank in 48 to 60 h. The solution and partially oxidized mineral slurry are usually directed to two smaller, aerated and agitated tanks in series for further sulfide min-

eral biodegradation. After sufficient sulfide mineral is biooxidized, the biologically pretreated material is washed with water and subjected to a cyanide leach to extract the gold and silver. With biooxidation pretreatment, gold recoveries of 98% and greater are achieved. Because the mineral concentrates contain high sulfide content ($>20\%$ S^{2-}, wt/wt) and oxidation is rapid, heat from this exothermic reaction is excessive and is not adequately dissipated by radiation, stirring, and aeration. Therefore, reactors are maintained at about 40°C with internal cooling coils or external water curtains. In 1994, the first commercial plant using mixed cultures of moderately thermophilic bacteria, which have not been identified or extensively characterized, was commissioned. This plant operates at between 48 and 53°C. With less cooling required, treatment costs are decreased (13). The largest commercial minerals biooxidation plant is located in Ghana, on the African continent. With recent expansion, the plant processes 1,000 metric tons of mineral concentrate every 24 h.

Bioleaching and minerals biooxidation are alternative processes to smelting and roasting, which discharge large amounts of SO_2 and As_2O_3. These gases are difficult to contain, and the sulfuric acid product is in global oversupply. Recovered arsenic is often contaminated with heavy metals, necessitating disposal in a hazardous waste landfill. In contrast, biological processing degrades only the sulfide minerals occluding the gold and silver, so less sulfuric acid is produced. Lime or limestone is added to the dilute sulfuric acid solution containing the ferric iron and arsenate, increasing the pH to between 3 and 4 and precipitating ferric arsenate. Because this precipitate is not leached by the Environmental Protection Agency's toxic characteristics leach procedure, it is not considered a hazardous waste and can be disposed of in tailing impounds.

The same tests used to assess the microbial activity in the formation of ARD are used to assess amenability of ores and mineral concentrates to bioleaching and minerals biooxidation. The BAPP test is used to evaluate the potential of the sample to be biooxidized. Column tests are mandatory for evaluating bioleaching and biooxidation of ores. Ore in the columns is inoculated with *T. ferrooxidans* by intimate mixing of the bacteria with the ore particles (16) or by applying a bacterial inoculum to the top surface of the ore column. Column tests are used to optimize ore particle size for maximum sulfide oxidation and metal recovery and to analyze other conditions, such as the best heap height and the amount of nutrients required by the bacteria. The engineering design, costs, and anticipated profit of commercial bioleaching and biooxidation operations involving millions of tons of ore are obtained from a series of tests using columns often no larger than 15 cm in diameter by 2.0 m in height.

BIOREMEDIATION OF INORGANIC CONTAMINANTS

Bioremediation of Cyanide

Dilute cyanide solutions are used frequently in mineral processing, and discharges from these operations must eventually be treated. Chemical methods are most frequently used by the mining industry for cyanide destruction. However, biological processes have become increasingly more attractive to the industry because they are cost effective and can be used in diverse environments.

Cyanide oxidation by microorganisms was detailed in

1965 by Howe (48), reviewed in 1976 by Knowles (54), and further considered in the 1990s as public interest in "green" technologies increased (6, 59). However, for effective bioremediation of mining wastes, microorganisms must degrade thiocyanate (66) and metal-cyanide complexes (24, 32, 71).

In the mid-1980s, Homestake Mining Company developed a full-scale biological process for degradation of CN^-, SCN^-, NH_4^+, and weak-acid-dissociable metal-cyanide complexes [$Zn(CN)_2$, $Ni(CN)_2$, and $Cu(CN)_2$] (81):

$$M_x(CN)_y + 4H_2O + O_2 \rightarrow M^{2+} + 2HCO_3 + 2NH_3^- \quad (20)$$

where M represents = Zn^{2+}, Ni^{2+}, Ni^{2+}, or Cu^{2-},

$$SCN^- + 2.5O_2 + 2H_2O \rightarrow SO_4^{2-} + HCO_3^- + NH_3 \quad (21)$$

$$NH_4^+ + 1.5O_2 \rightarrow NO_2^- + 2H^+ + H_2O \quad (22)$$

$$NO_2^- + 1/2O_2 \rightarrow NO_3^- \quad (23)$$

After bacteria capable of performing these oxidation reactions were obtained, the kinetics were enhanced by optimizing conditions and the full-scale system was engineered. A 21,000-m^3 24 h^{-1} cyanide treatment plant at Homestake Mining Company's Lead, South Dakota, precious metal mine commenced operation in 1984. The plant consists of 48 rotating biological contactors (RBCs) followed by clarification and pressure sand filters with dual-medium beds. The 48 RBCs, 3.65 m in diameter and 7.6 m in length, are arranged in a mirror image plant design with 24 discs per side. The wastewater flow is perpendicular to the discs, with five discs per train. The first two RBCs per train are principally involved with cyanide and thiocyanate oxidation (reactions 20 and 21) using a mixed culture of Pseudomonas species. Metals, released from the cyanide complex, are adsorbed to the bacteria. Homestake's biooxidation plant removes about 92% of the total cyanide, 99% of the cyanide associated with the weak-acid-dissociable metal-cyanide complexes, and 95% of the Cu^{2+} and other heavy metals. The remaining three RBCs accomplish nitrification (reactions 22 and 23) (80).

There is increasing interest in biologically degrading cyanide retained in ore heaps following cyanide leaching to solubilize gold and silver. The present remediation treatment, after cyanide leaching is completed, entails repeated washing of the spent heaps with water. This requires large quantities of water and considerable time because of hydrologic constraints. Microorganisms are expected to hasten cyanide degradation, enhance heavy metal mobilization, and decrease overall water consumption. Although this technology is relatively new, encouraging results have been reported (3, 37, 74).

Obtaining microorganisms for cyanide degradation in water or in spent cyanide, heap leach operations begins by enriching for or isolating the organisms from tailings impoundments where cyanide effluents are stored. The samples are returned to the laboratory and incubated with an organic energy source in a medium containing free cyanide and metal-cyanide complexes in concentrations similar to those noted at the sample site. By subculturing on increasing concentrations of free and complexed cyanide, an isolate or mixed culture of organisms can be selected to perform in the target environment. This was the method used by Lien and Altringer (55), who isolated a cyanide-degrading Pseudomonas pseudoalcaligenes, and by Whitlock and Mudder

(81), who developed Homestake Mining Company's full-scale cyanide degradation plant.

Testing the effectiveness of the P. pseudoalcaligenes culture (55) for degradation of cyanide in tailings pond water was conducted in 2.5-cm-diameter glass columns packed with either quartz chips (0.635 cm) or activated carbon (0.2 to 2 mm), that provided surface area for colonization of the bacteria. Bacterial film development was promoted by recirculating PGY broth (5 g of peptone, 2.5 g of glycerol, and 0.5 g of yeast extract per liter) and an inoculum of P. pseudoalcaligenes through the column for 2 to 3 days. Following establishment of the bacterial culture in the column, tailings pond water containing 280 mg of CN per liter at pH 10.5 was passed through the column system. Over 90% of the cyanide was removed from the solution. Note that the pH of the cyanide solution was sufficiently high to prevent volatilization of HCN. The removal of the cyanide was attributed to bacterial metabolism.

Metals Biosorption

Bacteria (9, 10), fungi (7, 35), and algae (26) bind heavy metal ions. The initial mechanism for metal binding by microorganisms is electrostatic attraction between charged metal ions in solution and charged functional groups on microbial cell walls (70). The cell walls are composed of macromolecules with functional groups (principally carboxylate, amine, imidazole, phosphate, sulfhydryl, and sulfate) that contribute a net negative charge to the microorganism's surface. These functional groups remain active even when the microorganism is nonliving. Electrostatic binding is a rapid reaction and, depending on the binding constant between the functional group and the metal ion, can be quite stable. Metal binding to cell wall functional groups is highly dependent on water chemistry (9, 10) and the chemical characteristics of the metals (5). If conditions are excessively acidic, H^+ preferentially binds to the negatively charged sites, excluding the binding of heavy metals (25, 67). Most heavy metals form negatively charged complexes at certain pH values. Because the overall net charge on the cell wall is negative, these anionic complexes are not bound by the biomass.

Once bound to the cell wall, metals are actively transported across the cell membrane by living microorganisms. This process, which is much slower than electrostatic binding, requires expenditure of energy by the microorganisms. Although not definitively understood, the mechanism for transporting toxic heavy metals into the cell is believed to be the same mechanism that organisms use for accumulating essential metals such as sodium, calcium, and magnesium. It is likely that the transport mechanism is unable to differentiate between essential metal ions and toxic metals of similar atomic radius and charge.

Some microorganisms precipitate metals directly on the cell surfaces. Select species of Citrobacter produce a phosphate compound that precipitates metals directly on the cell surface. This process has been studied extensively for the immobilization of heavy metals and radionuclides (75).

The binding of heavy metals onto cell surfaces for concentration from waste streams containing dilute metal concentrations and for possible recovery and reuse of the metals has been commercially developed by using nonliving bacteria, algae, fungi, and other biomass, such as duckweed and sphagnum moss. The Bioclaim process used a bacterial biomass, produced as a by-product from the manufacturing of industrial enzymes. The nonliving bacteria were immobilized in polyethyleneimine and glutaraldehyde, extruded to form permeable beads, and applied in fixed- and fluidized-bed reactors (14). Various types of algae, immobilized in silica gel to form a product called Alga SORB, were used in fixed-bed reactors for removing soluble metals from waste streams and metal-contaminated groundwater (28). Jeffers and Corwin (50) immobilized a variety of biological materials, including microorganisms, in polysulfone, producing a metal adsorbent. Metals bound to any immobilized biomass products can be eluted by using acid, base for amphoteric metals, or complexing agents such as EDTA. Metals concentrated in the eluate can potentially be reclaimed by metallurgical techniques. Biosorbents are regenerated for reuse by washing with $NaCO_3$ or $NaOH$ solutions, affixing Na^+ to the charged sites of the cell wall functional groups.

Although researched extensively, metals biosorption using living microorganisms has not achieved extensive commercial following because of the problems attendant with keeping microorganisms alive in waste streams containing variable concentrations of heavy metals and other substances deleterious to living organisms. One application that is used principally to treat groundwater or other streams with very low concentrations of heavy metals is aerobic wetlands (10). Although largely relying on abundant growth of aquatic vegetation for uptake of soluble heavy metals, these specially constructed wetlands (19) foster growth of large algal populations. These algae actively accumulate metals (12).

Microorganisms isolated from the environment or obtained as by-products of fermentation or microorganisms immobilized in organic or inorganic matrices (14) are quantitatively evaluated for their capacity to accumulate metals from experimental biosorption equilibrium isotherms similar to those used to evaluate activated carbon (78). A dilute concentration of microorganisms or biosorbent material is contacted with a series of solutions containing increasing concentrations of soluble metal. An equilibrium is established whereby a certain amount of metal bound to the biomass is in equilibrium with the metal remaining free in solution. After agitation for a given period of time (12 to 24 h) and at a specific temperature, the solution is analyzed for the residual metal concentration and the filtered biomass is analyzed for metal binding to allow a mass balance. Equilibrium isotherms are graphically presented by plotting the metal uptake by the biosorbent (in weight weight^{-1} or mole weight^{-1}) against the residual metal concentration in solution (weight volume^{-1}). The graphs produced are hyperbolic as the metal uptake by the biomass levels off as the metal binding sites on the biomass become saturated at high metal concentrations. The maximum uptake of metal by the biomass, as well as the shape of the equilibrium isotherm, is important. The steeper the curve is at low metal concentrations, the more desirable the biomass is as an adsorbent, because the functional groups have a high affinity for the metal species (78).

Because the speciation of metals in solutions is crucial to the binding of the metals to functional groups on the biomass, the effluent to be bioremediated is characterized. Parameters examined are major ions, thermodynamic properties, minor ions, redox status, acid-base components, and complexing agents. Modeling of the waste stream, using the properties and equilibria among the different components, will aid in understanding the speciation of the metals in solutions (27, 61).

Immobilized biosorbents possess a particle size, mechanical strength, and rigidity sufficient for column testing. The biosorbents are loaded into small (2 by 15 cm) or large columns, and a metal-bearing solution is passed down-flow or pumped up-flow through the column. The solution is not recycled. Parameters evaluated in column tests are residence time of the solution in the column to achieve the desirable effluent concentration and maximum metal loading on the biosorbent. Other factors, such as optimum particle size and different solution chemistries, can also be tested.

Bacterially Mediated Metals Precipitation

Sulfate-reducing bacteria (SRBs) are used in highly controlled reactor systems and in so-called engineered or constructed anaerobic wetlands for removal of sulfate and heavy metals from ARD and other aqueous, metal-contaminated streams. The SRBs, species of *Desulfovibrio* and *Desulfotomaculum*, oxidize organic matter or H_2 by using sulfate as an electron acceptor to produce hydrogen sulfide and bicarbonate as products:

$$2CH_2O + SO_4^{2-} \rightarrow H_2S + 2HCO_3^- \qquad (24)$$

$$5H_2 + SO_4^{2-} \rightarrow H_2S + 4H_2O \qquad (25)$$

The sulfide immediately reacts with soluble heavy metal ions to form highly insoluble metal sulfides:

$$M^{2+} + S^{2-} \rightarrow MS\downarrow \qquad (26)$$

where M^{2+} represents a divalent metal such as Zn^{2+}, Cu^{2+}, or Ni^{2+}.

The most sophisticated plant using SRBs is located at a zinc refinery in The Netherlands. Each day, this plant treats 5,000 m^3 of groundwater containing heavy metals and sulfate (29). The H_2S can be produced (reactions 24 and 25) in a separate reactor or in the same reactor with the waste stream to be treated. Excess H_2S is directed to an aerobic bioreactor, where sulfide is oxidized to elemental sulfur by *Thiobacillus* species functioning at near-neutral pH values:

$$2H_2S + O_2 \rightarrow 2S^0\downarrow + 2H_2O \qquad (27)$$

The metal sulfides and elemental sulfur are recovered and recycled.

Engineered or constructed anaerobic wetlands, also called passive treatment systems, are increasingly being used by the mining industry to treat metal-contaminated and sulfate-bearing waste streams. Passive systems are relatively efficient and highly cost-effective. These systems usually consist of several unit operations. If the water is too acidic for growth of SRBs, the first step is passage of the waste stream through an anoxic drain or tank filled with limestone. The pH is increased, and some iron precipitates. The waste stream is then directed through cells, which are usually tanks or high-density polyethylene (HDPE)-lined pits filled with mushroom compost or manure, which serve as an inoculum and energy source. Straw, soil, wood chips, or rocks are added to increase hydraulic conductivity. The lined pits are often covered with topsoil and planted for aesthetics, as the vegetation serves no role in metal removal. The sulfate in the waste stream is reduced to H_2S, precipitating metals within the passive system. Bicarbonate (reaction 24) neutralizes acid. Design criteria for passive treatment systems are typically based on a volumetric loading factor (moles of metal per cubic meter of organic matter oxidized per day) (46, 82).

Sulfate reduction has been considered as a method for permanent stabilization of sulfidic mine tailings. When sulfide tailings are submerged to prevent sulfide biooxidation by excluding air, SRBs and some organic matter are present naturally. The SRBs present in this anoxic environment produce H_2S (reaction 24), causing metal precipitation and stabilization (reaction 26) (33).

A staged system to evaluate anaerobic and aerobic treatment of mine effluents containing excessive sulfate and heavy metals is described by Maree et al. (57). The continuous system consists of primary anaerobic, aerobic, and secondary anaerobic stages. The primary anaerobic stage is a columnar reactor filled with dolomite pebbles. At the top of the column is a stripping tower to remove H_2S with a gas mixture of N_2 and CO_2. The purpose of the primary anaerobic stage is to produce H_2S and precipitate metal sulfides from a waste stream amended with an organic energy source. The H_2S is produced by the SRB biofilm on the dolomite pebbles. Sludge is regularly removed from the tapered base of the anaerobic column. The aerobic stage is a completely mixed reactor and settling tank. The purpose of this stage is to oxidize residual H_2S and biodegrade residual organics from the primary anaerobic column. Although most effluents do not require a final treatment, the secondary anaerobic treatment is to remove residual chemical oxygen demand.

Microbial Metal Reduction, Methylation, and Demethylation

Microorganisms transform metals by reduction, methylation, and demethylation. Some methylation reactions, such as the methylation of mercury, produce a metal species that is readily bioaccumulated and is toxic. Some methylation reactions result in volatilization of the metals. Volatilization of selenium (34) has been studied in environments similar to those found in the mining industry. Demethylation is the cleavage of methyl groups from methylated metals and metalloids, resulting in ionic metal species.

Microbial reduction of metals and metalloids has been researched extensively by Lovley (56), who has collected samples from mining-type environments. Among the metal bioreductions that could potentially be important to remediation in mining are the microbial reduction of Fe^{3+}, Mn^{4+}, U^{6+}, SeO_4^{2-}, SeO_3^{2-} (39), and Cr^{6+} (79).

Little is known about the extent and remedial importance of bioreduction and volatilization of metals and metalloids in the mining environment. When aerobic wetlands and passive treatment systems are used to attenuate metals, conditions in the sediments may support the activity of organisms capable of these metal transformations (11). However, the extent of this activity in the mobilization or immobilization of metals in the mining environment is not known.

SUMMARY

In the last decade, biotechnology has become very important to the mining industry. Microorganisms catalyze the oxidation of sulfide minerals, elemental sulfur, and ferrous iron, producing ARD, which contaminates surface and groundwaters with high concentrations of acid, sulfate, and toxic heavy metals. Methods to predict ARD and to assess preventive measures have been developed and are applied throughout the world by the mining industry.

Biotechnical processes involving SRBs, metals biosorption, and cyanide degradation are now used to cost-effectively achieve discharge standards for regulated constituents. Bioleaching to extract valuable base metals and minerals biooxidation to enhance recovery of gold and silver are now used globally in commercial processes that are less

damaging to the environment and more cost-effective than many conventional processes.

REFERENCES

1. **Amaro, A. M., K. B. Hallberg, E. B. Lindstrom, and C. A. Jerez.** 1994. An immunological assay for detection and enumeration of thermophilic biomining microorganisms. *Appl. Environ. Microbiol.* **60:**3470–3473.
2. **American Society for Testing and Materials.** 1990. *Standard Test Method for Enumeration of Aquatic Bacteria by Epifluorescence Microscopy Counting Procedure.* ASTM D 4455-85. American Society for Testing and Materials, Philadelphia.
3. **Aronstein, B. N., A. Maka, and V. J. Srivastava.** 1994. Chemical and biological removal of cyanides from aqueous and soil-containing systems. *Appl. Microbiol. Biotechnol.* **41:**700–707.
4. **Arredondo, R., and C. A. Jerez.** 1989. Specific dot-immunobinding assay for detection and enumeration of *Thiobacillus ferrooxidans. Appl. Environ. Microbiol.* **55:**2025–2029.
5. **Avery, S. V., and J. M. Tobin.** 1993. Mechanism of adsorption of hard and soft metal ions to *Saccharomyces cerevisiae* and influence of hard and soft anions. *Appl. Environ. Microbiol.* **59:**2851–2856.
6. **Babu, G. R. V., J. H. Wolfram, and K. D. Chapatwala.** 1992. Conversion of sodium cyanide to carbon dioxide and ammonia by immobilized cells of *Pseudomonas putida. J. Ind. Microbiol.* **9:**235–238.
7. **Brady, D., S. Stoll, and J. R. Duncan.** 1994. Biosorption of heavy metal cations by non-viable yeast biomass. *Environ. Technol.* **15:**429–438.
8. **Brierley, C. L.** 1978. Bacterial leaching. *Crit. Rev. Microbiol.* **6:**207–262.
9. **Brierley, C. L.** 1990. Metal immobilization using bacteria, p. 303–323. *In* H. L. Ehrlich and C. L. Brierley (ed.), *Microbial Mineral Recovery.* McGraw-Hill Publishing Co., New York.
10. **Brierley, C. L.** 1991. Bioremediation of metal-contaminated surface and groundwaters. *Geomicrobiol. J.* **8:**201–223.
11. **Brierley, C. L.** 1992. Selenium in mine and mill environments, p. 175–179. *In Randol at MinEXPO 292.* Randol International, Ltd., Golden, Colo.
12. **Brierley, C. L.** 1993. Environmental biotechnology applications in mining. *In BIOMINE '93.* Australian Mineral Foundation, Inc., Adelaide, South Australia.
13. **Brierley, C. L.** 1995. Bacterial oxidation. *Eng. Mining J.* **196:**42–44.
14. **Brierley, C. L., and J. A. Brierley.** 1993. Immobilization of biomass for industrial application of biosorption, p. 35–44. *In* A. E. Torma, M. L. Apel, and C. L. Brierley (ed.), *Biohydrometallurgical Technologies.* The Minerals, Metals & Materials Society, Warrendale, Pa.
15. **Brierley, C. L., J. A. Brierley, P. R. Norris, and D. P. Kelly.** 1980. Metal-tolerant, microorganisms of hot, acid environments, p. 39–51. *In* G. W. Gould and J. E. L. Corry (ed.), *Microbial Growth and Survival in Extremes of Environment.* Academic Press, London.
16. **Brierley, J. A.** 1994. Biooxidation-heap technology for pretreatment of refractory sulfidic gold ore. *In BIOMINE '94.* Australian Mineral Foundation, Adelaide, South Australia.
17. **Brierley, J. A., and S. J. Lockwood.** 1977. The occurrence of thermophilic iron-oxidizing bacteria in a copper leaching system. *FEMS Microbiol. Lett.* **2:**163–165.
18. **Brierley, J. A., R. Y. Wan, D. L. Hill, and T. C. Logan.** 1995. Biooxidation-heap pretreatment technology for processing lower grade refractory gold ore, p. 253–262. *In* T.

Vargas, C. A. Jerez, J. V. Wiertz, and H. Toledo (ed.), *Biohydrometallurgical Processing*, vol. I. University of Chile, Santiago.

19. **Brodie, G. A.** 1991. Staged aerobic wetlands systems to treat acid drainage. Presented at Reclamation 2000: Technologies for Success. 8th National Meeting of the American Society of Surface Mining and Reclamation, Durango, Colo., 14–17 May 1991.

20. **Broughton, L. M., R. W. Chambers, and A. M. Robertson.** 1992. *Mine Rock Guidelines: Design and Control of Drainage Water Quality*. Report 93301. Saskatchewan Environment and Public Safety, Mines Pollution Control Branch, Prince Albert, Saskatchewan, Canada

21. **Bruynesteyn, A., and D. W. Duncan.** 1979. *Determination of Acid Production Potential of Waste Materials*. Paper A-79-29. The Metallurgical Society of the American Institute of Mining Engineers, New York.

22. **Bustos, S., S. Castro, and R. Montealegre.** 1993. The Sociedad Minera Pudahuel bacterial thin-layer leaching process at Lo Aguirre. *FEMS Microbiol. Rev.* **11:**231–236.

23. **Caruccio, F. T.** 1968. An evaluation of factors affecting acid mine drainage production and the ground water interactions in selected areas of western Pennsylvania, p. 107–151. *In Proceedings of the Second Symposium on Coal Mine Drainage Research*. Bituminous Coal Research, Monroeville, Pa.

24. **Chapatwala, K. D., G. R. V. Babu, and J. H. Wolfram.** 1993. Screening of encapsulated microbial cells for the degradation of inorganic cyanides. *J. Ind. Microbiol.* **11:**69–72.

25. **Crist, D. R., R. H. Crist, J. R. Martin, and J. R. Watson.** 1994. Ion exchange systems in proton-metal reactions with algal cell walls. *FEMS Microbiol. Rev.* **14:**309–314.

26. **Crist, R. H., K. Oberholser, D. Schwartz, J. Marzoff, D. Ryder, and D. R. Crist.** 1988. Interactions of metals and protons with algae. *Environ. Sci. Technol.* **22:**755–760.

27. **Darimont, A., and J. Frenay.** 1990. Metals in aqueous solutions, pp. 65–80. *In B. Volesky (ed.), Biosorption of Heavy Metals*. CRC Press, Boca Raton, Fla.

28. **Darnall, D. W., R. M. McPherson, and J. Gardea-Torresdey.** 1989. Metal recovery from geothermal waters and groundwaters using immobilized algae, p. 341–348. *In J. Salley, R. G. L. McCready, and P. L. Wichlacz (ed.), Biohydrometallurgy 1989*. CANMET SP89-10. Canada Centre for Mineral and Energy Technology, Ottawa.

29. **de Vegt, A. L.** (Paques Environmental Technology, Exton, Pa.) 1995. Personal communication.

30. **Espejo, R. T., J. Pizarro, E. Jedliki, O. Orellana, and J. Romero.** 1995. Bacterial population in the bioleaching of copper as revealed by analysis of DNA obtained from leached ores and leaching solutions, p. 1–8. *In C. A. Jerez, T. Vargas, H. Toledo, and J. V. Wiertz (ed.), Biohydrometallurgical Processing*, vol. II. University of Chile, Santiago.

31. **Ferguson, K. D., and K. A. Morin.** 1991. The prediction of acid rock drainage: lessons from the database, p. 83–86. *In Proceedings of the Second International Conference on the Abatement of Acidic Drainage*. Mine Environment Neutral Drainage Program, Canada Centre for Mineral and Energy Technology, Ottawa. Canada.

32. **Figueira, M. M., V. S. T. Cifninelli, and V. R. Linardi.** 1995. Bacterial degradation of metal cyanide complexes, p. 333–339. *In C. A. Jerez, T. Vargas, H. Toledo, and J. V. Wiertz (ed.), Biohydrometallurgical Processing*, vol. II. University of Chile, Santiago.

33. **Fortin, D., B. Davis, G. Southam, and T. J. Beveridge.** 1995. Biogeochemical phenomena induced by bacteria within sulfidic mine tailings. *J. Ind. Microbiol.* **14:**178–185.

34. **Frankenberger, W. T., Jr., and U. Karlson.** 1995. Volatilization of selenium from a dewatered seleniferous sediment: a field study. *J. Ind. Microbiol.* **14:**226–232.

35. **Gadd, G. M.** 1993. Interaction of fungi with toxic metals. *New Phytol.* **124:**25–60.

36. **Garcia, A., and C. A. Jerez.** 1995. Changes of the solid-adhered populations of *Thiobacillus ferrooxidans*, *Leptospirillum ferroxidans* and *Thiobacillus thiooxidans* in leaching ores as determined by immunological analysis, p. 19–30. *In C. A. Jerez, T. Vargas, H. Toledo, and J. V. Wiertz (ed.), Biohydrometallurgical Processing*, vol. II. University of Chile, Santiago.

37. **Garcia, H. J., M. C. Fuerstenau, and J. L. Hendrix.** 1993. Biodegradation of cyanide under anaerobic conditions, p. 377–389. *In J. Hager, B. Hansen, W. Imrie, J. Pusatori, and V. Ramachandran (ed.), Extraction and Processing for the Treatment and Minimization of Wastes*. The Minerals, Metals & Materials Society, Warrendale, Pa.

38. **Gates, J. E., and K. D. Pham.** 1979. An indirect fluorescent antibody staining technique for determining population levels of *Thiobacillus ferrooxidans* in acid mine drainage waters. *Microb. Ecol.* **5:**121–127.

39. **Gharieb, M. M., S. C. Wilinson, and G. M. Gadd.** 1995. Reduction of selenium oxyanions by unicellular, polymorphic and filamentous fungi: cellular location of reduced selenium and implications for tolerance. *J. Ind. Microbiol.* **14:**300–311.

40. **Grishin, S. I., J. M. Bigham, and O. H. Tuovinen.** 1988. Characterization of jarosite formed upon bacterial oxidation of ferrous sulfate in a packed-bed reactor. *Appl. Environ. Microbiol.* **54:**3101–3106.

41. **Guay, R.** 1993. *Development of a Modified MPN Procedure to Enumerate Iron Oxidizing Bacteria*. MEND Report 1.14.2. Mine Environment Neutral Drainage Program, Canada Centre for Mineral and Energy Technology, Ottawa.

42. **Harries, J. R., and A. I. M. Ritchie.** 1980. The use of temperature profiles to estimate the pyritic oxidation rate in a waste rock dump from an opencut mine. *Water Air Soil Pollut.* **15:**405–423.

43. **Harries, J. R., and A. I. M. Ritchie.** 1985. Pore gas composition in waste rock dumps undergoing pyritic oxidation. *Soil Sci.* **140:**143–152.

44. **Harris, L.** 1992. Mining and the well-informed citizen. *Mining Eng.* **44:**999–1002.

45. **Harrison, A. P., Jr.** 1984. The acidophilic thiobacilli and other acidophilic bacteria that share their habitat. *Ann. Rev. Microbiol.* **38:**265–292.

46. **Hedin, R. S., R. W. Narin, and R. L. P. Kleinmann.** 1994. *Passive Treatment of Coal Mine Drainage*. Bureau of Mines Information Circular 9389. U.S. Department of the Interior, Washington, D.C.

47. **Hodges, C. A.** 1995. Mineral resources, environmental issues, and land use. *Science* **268:**1305–1312.

48. **Howe, R. H. L.** 1965. Bio-destruction of cyanide wastes—advantages and disadvantages. *Int. J. Air Water Pollut.* **9:**463–478.

49. **Hutchison, I. P. G., and R. D. Ellison (ed.).** 1992. *Mine Waste Management*. Lewis Publishers, Boca Raton, Fla.

50. **Jeffers, T. H., and R. R. Corwin.** 1993. Waste water remediation using immobilized biological extractants, p. 1–13. *In A. E. Torma, M. L. Apel, and C. L. Brierley (ed.), Biohydrometallurgical Technologies*. The Minerals, Metals & Materials Society, Warrendale, Pa.

51. **Jennings, S. R., and D. J. Dollbopf.** 1995. *Geochemical Characterization of Sulfide Mineral Weathering for Remediation of Acid Producing Mine Wastes*. Reclamation Research Publication 9502. Reclamation Research Unit, Montana State University, Bozeman.

52. **Johnson, D. B., and S. McGinness.** 1991. A highly efficient and universal solid medium for growing mesophilic and moderately thermophilic, iron-oxidizing, acidophilic bacteria. *J. Microbiol. Methods* **13:**113–122.

53. **Kelley, B. C., and O. H. Tuovinen.** 1988. Microbiological oxidations of minerals in mine tailings, p. 33–53. *In* W. Salomons and U. Forsnter (ed.), *Chemistry and Biology of Solid Waste: Dredged Material and Mine Tailings*. Springer-Verlag, Berlin.

54. **Knowles, C. J.** 1976. Microorganisms and cyanide. *Bacteriol. Rev.* **40:**652–680.

55. **Lien, R. H., and P. B. Altringer.** 1993. Case study: bacterial cyanide detoxification during closure of the Green Springs gold heap leach operation, p. 219–227. *In* A. E. Torma, M. L. Apel, and C. L. Brierley (ed.), *Biohydrometallurgy Technologies*. The Minerals, Metals & Materials Society, Warrendale, Pa.

56. **Lovley, D. R.** 1995. Bioremediation of organic and metal contaminants with dissimilatory metal reduction. *J. Ind. Microbiol.* **14:**85–93.

57. **Maree, J. P., A. Gerber, and E. Hill.** 1987. An integrated process for biological treatment of sulfate-containing industrial effluents. *J. Water Pollut. Control Fed.* **59:**1069–1074.

58. **Melluish, J. M., L. A. Groat, D. B. Dreisinger, R. Branion, B. J. Y Leong, and D. R. Crombie.** 1993. Mineralogical characterization of residues from a microbiologically leached copper sulphide ore, p. 127–136. *In* A. E. Torma, J. E Wey, and V. I. Lakshmanan (ed.), *Biohydrometallurgical Technologies*. The Minerals, Metals & Materials Society, Warrendale, Pa.

59. **Meyers, P. R., P. Gokool, D. E. Rawlings, and D. R. Woods.** 1991. An efficient cyanide-degrading *Bacillus pumilus* strain. *J. Gen. Microbiol.* **137:**1397–1400.

60. **Miller, G., B. Lyons, and A. Davis.** Pit lake water quality in open-pit metal mines. *Environ. Sci. Technol.*, in press.

61. **Modak, J. M., and K. A. Natarajan.** 1995. Biosorption of metals using non-living biomass—a review. *Miner. Metall. Process.* **12:**189–196.

62. **Murr, L. E., and J. A. Brierley.** 1978. The use of large-scale test facilities in studies of the role of microorganisms in commercial leaching operations, p. 491–520. *In* L. E. Murr, A. E. Torma, and J. A. Brierley (ed.), *Metallurgical Applications of Bacterial Leaching and Related Microbiological Phenomena*. Academic Press, New York.

63. **Niemela, S. I., C. Sivela, T. Luoma, and O. H. Tuovinen.** 1994. Maximum temperature limits for acidophilic, mesophilic bacteria in biological leaching systems. *Appl. Environ. Microbiol.* **60:**3444–3446.

64. **Norris, P. R.** 1993. Thermoacidophilic archaebacteria: potential applications. *Biochem. Soc. Symp.* **58:**171–180.

65. **Norris, P. R., and W. J. Ingledew.** 1992. Acidophilic bacteria: adaptations and applications, p. 115–142. *In* R. A. Herbert and R. J. Sharp (ed.), *Molecular Biology and Biotechnology of Extremophiles*. Chapman and Hall, New York.

66. **Paruchuri, Y. L., N. Shivaraman, and P. Kumaran.** 1990. Microbial transformation of thiocyanate. *Environ. Pollut.* **68:**15–28.

67. **Pirszel, J., B. Pawlik, and T. Skowronski.** 1995. Cation-exchange capacity of algae and cyanobacteria: a parameter of their metal sorption abilities. *J. Ind. Microbiol.* **14:**319–322.

68. **Ramsay, B., J. Ramsay, M. De Tremblay, and C. Chavarie.** 1988. A method for the quantification of bacterial protein in the presence of jarosite. *Geomicrobiol. J.* **6:**171–177.

69. **Rawlings, D. E.** 1995. Restriction enzyme analysis of 16S rRNA genes for the rapid identification of *Thiobacillus ferrooxidans*, *Thiobacillus thiooxidans* and *Leptospirillum ferrooxidans* strains in leaching environments, p. 9–17. *In* C. A. Jerez, T. Vargas, H. Toledo, and J. V. Wiertz (ed.), *Biohydrometallurgical Processing*, vol. II. University of Chile, Santiago.

70. **Remacle, J.** 1990. The cell wall and metal binding, p. 83–92. *In* B. Volesky (ed.), *Biosorption of Heavy Metals*. CRC Press, Boca Raton, Fla.

71. **Silva-Avalos, J., M. G. Richmond, O. Nagppan, and D. A. Kunz.** 1990. Degradation of the metal-cyano complex tetracyanonickelate(II) by cyanide-utilizing bacterial isolates. *Appl. Environ. Microbiol.* **56:**3664–3670.

72. **Siwik, R., S. Payant, and K. Wheeland.** 1989. Control of acid generation from reactive waste rock with the use of chemicals, p. 181–193. *In* M. E. Chalkley, B. R. Conard, V. I. Lakshmanan, and K. G. Wheeland (ed.), *Tailings and Effluent Management*. Pergamon Press, New York.

73. **Sobek, A. A., E. Reutern, and J. B. Pausch.** 1989. Method of making and a slow release composition for abating acid water formation. U.S. Patent 4,869,905.

74. **Thompson, L. C., and E. Jones.** 1993. Bio-detoxification of spent ore and heap leach solutions, p. 343–346. *In Randol Gold Conference at Beaver Creek 93*. Randol International, Ltd., Golden, Colo.

75. **Tolley, M. R., L. R. Strachan, and L. E. Macaskie.** 1995. Lanthanum accumulation from acidic solutions using a *Citrobacter* sp. immobilized in a flow-through bioreactor. *J. Ind. Microbiol.* **14:**271–280.

76. **Tuovinen, O. H., B. C. Kelley, and S. N. Groudev.** 1991. Mixed cultures in biological leaching processes and mineral biotechnology, p. 373–427. *In* J. G. Zeikus and E. A. Johnson (ed.), *Mixed Cultures in Biotechnology*. McGraw-Hill, New York.

77. **Turney, W. R., and B. M. Thomson.** 1993. Minerals and mine drainage. *Water Environ. Res.* **65:**410–413.

78. **Volesky, B.** 1990. Removal and recovery of heavy metals by biosorption, p. 8–43. *In* B. Volesky (ed.), *Biosorption of Heavy Metals*. CRC Press, Boca Raton, Fla.

79. **Wang, Y. T., and H. Shen.** 1995. Bacterial reduction of hexavalent chromium. *J. Ind. Microbiol.* **14:**159–163.

80. **Whitlock, J. L.** 1990. Biological detoxification of precious metal processing wastewaters. *Geomicrobiol. J.* **8:**241–249.

81. **Whitlock, J. L., and T. I. Mudder.** 1986. The Homestake wastewater treatment process: biological removal of toxic parameters from cyanidation wastewaters and bioassay effluent monitoring, p. 327–339. *In* R. W. Lawrence, R. M. R. Branion, and H. G. Ebner (ed.), *Fundamental and Applied Biohydrometallurgy*. Elsevier, Amsterdam.

82. **Wildeman, T., G. Brodie, and J. Gusek.** 1993. *Wetland Design for Mining Operations*. BiTech Publishers, Ltd., Richmond, B.C., Canada.

83. **Yates, J. R., J. H. Lobos, and D. S. Holmes.** 1986. The use of genetic probes to detect microorganisms in biomining operations. *J. Ind. Microbiol.* **1:**129–135.

84. **Yeh, T. Y., R. M. Kelly, J. C. Cox, and G. J. Olson.** 1988. Significance of cell fluorescence color of acridine orange-stained *Thiobacillus ferrooxidans* under epifluorescence microscopy, p. 145–150. *In* P. R. Norris and D. P. Kelly (ed.), *BioHydroMetallurgy*. Science and Technology Letters, Kew, Surrey, England.

Microbiologically Influenced Corrosion

NICHOLAS J. E. DOWLING AND JEAN GUEZENNEC

94

Over the last decade, the deterioration of engineering materials by microorganisms has received unprecedented attention (4, 16, 22, 44a, 50), the impetus for which has been provided by significant industrial economic losses. More steel (the material of reference) is subject to corrosion in natural, untreated waters than in any other, more exotic environment. In consequence, considerable metal surface areas are brought into contact with microorganisms, which results in rapid colonization and possible deterioration. Any significant corrosion event thus occurs in the presence of, and often through, a biofilm. In practical terms, the concentration of chloride, the presence of inorganic deposits, and the concentration of oxygen are equally important; however, these parameters are frequently modified by the microbiota in a way that is deleterious to the underlying material.

The economic repercussions of corrosion in general, and microbiologically influenced corrosion (MIC) in particular, are alarming. The losses in heavy industry are probably more evident since the postfailure investigations are more rigorous. Examples are numerous. The Canadian deuterium uranium reactors rely on the heat-exchanger alloy N08800. In the years preceding 1990, the systematic penetration of these tubes by lake water sulfate-reducing bacteria under a calcareous deposit led to some $13 million in replacement costs, with $300,000/day in lost energy costs (11). Estimates for unscheduled outages (down time) of U.S. nuclear utilities per 1.3 megawatt unit easily absorb $1 million/day in 1990 dollars. Other high-cost heavy industries such as gas transmission estimate 70% of their corrosion cases to be heavily influenced by bacterial effects (2). If biofouling of engineering surfaces is also taken into account, then the cost of American petroleum refining losses alone are estimated to run at $1.4 billion/year (48). Establishing rational estimates is difficult under the best of circumstances; however, the figures that are available in terms of biocide costs, replacement materials (upgrades), and preventative maintenance in off-line plants indicate that MIC has become a problem of truly international proportions. The industries with clearly perceptible microbiological problems include the nuclear and fossil fuel electric power generating companies (57), armaments (60), pipelines (46), pulp and paper (95), oil fields (28), and the offshore industry (43).

Despite the identification of these not inconsiderable areas of economic interest, few original advances of a theoretical nature have been made since early in this century (103). For lack of theoretical proposals, this obscure form of corrosion has thus become phenomenological in the sense that observations have been made in the laboratory and in the field concerning the microbiological, electrochemical, and metallurgical aspects with little attempt at real integration. The clear retarding factor is the lack of technical overlap between these disciplines.

DETERMINATION OF MIC IN SITU

For the purposes of this chapter, it is convenient to limit the definition of MIC to the loss in engineering properties of a metallic alloy as a result of microbial activities. Notably, this does not exclude the problems generated as a result of attempts to remove biofilms from surfaces. Thus manganese-oxidizing bacteria which deposit encrusting films on stainless steel (59) cannot be removed by hypochlorite treatment since the Mn^{2+}-Mn^{4+} mixture will oxidize to permanganate (Mn^{7+}), an agent which is chemically corrosive. The form of corrosion differs considerably among alloys, as is the case with chemical corrosion. Criteria for identifying MIC in situ are controversial and vary considerably with the material under test. It seems reasonable, however, to specify the following conditions: (i) the presence of chloride or other penetrative anion, even in low concentration (approximately 7 ppm); (ii) the presence of an organic carbon source, although autotrophic organisms such as *Gallionella* sp. appear capable of propagating corrosion; (iii) localized corrosion morphology; (iv) localized binding of the corrosion products around the pit with copious extracellular polysaccharide; (v) a variation in the redox across the corrosion product layer (manifested in terms of a color change); (vi) pyrite-type corrosion products; (vii) helical morphology of bacterial cells with encrusted iron oxide as determined by optical microscopy; and (viii) excess of 10^7 cell per g (wet weight) of corrosion product as measured by acridine orange direct counts (37). These very loose criteria rarely apply simultaneously, nor are they specific for a particular environment; they are simply indications that the propagation of the corrosion observed in the field is in part controlled by microorganisms.

Electrochemical and Physical Models

Models of microbial corrosion usually integrate the perception that biofilms maintain inhomogeneities on metal sur-

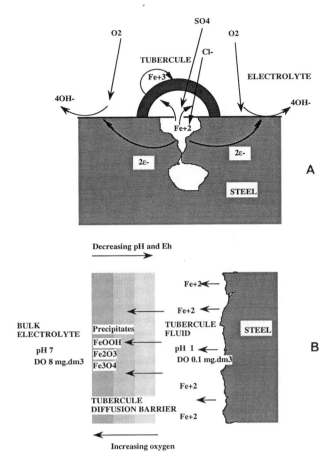

FIGURE 1 General schema of microbial corrosion (propagation) with spatial separation of anodic and cathodic sites.

faces (21). Corrosion initiates in a localized zone delimited by the extent of a surface defect or some inhomogeneity of the environment immediately adjacent to the metal surface. As the metal oxidizes, anions such as chloride and sulfate (sulfide) migrate to the corroding zone to counterbalance the charge deficit. The local environment subsequently becomes strongly reduced, and the pH can easily fall to 1.0. Dissolution of steel into the immediate acidic surroundings is followed by radial diffusion of the ferrous ions a small distance away from the corrosion site to an environment where the pH and E_h are relatively high and precipitation of a ferric oxide-hydroxide is thermodynamically favorable (83). This rapid accumulation of corrosion product forms what is eventually known as a tubercule, situated above and around the developing pit. Figure 1A shows an idealized form of a tubercule, of a kind that appears to be frequent on both carbon and stainless steels. The exterior of the tubercule is characterized by the presence of oxygen and a high pH (slightly higher than that of the bulk fluid). Under these conditions, bacteria are present in quantity in the wall of the tubercule, which is situated above the deepening hole in the metal known as a propagating pit. By whatever means that the corrosion initiates, the tubercule maintains a barrier sealing in an acidic zone which has a high concentration of reduced ions and an absence of oxygen. In electrochemical terms, the iron atoms (Fe^0) lose two electrons and pass into solution. The electrons liberated in the anodic reaction are consumed in the cathodic reaction outside the tubercule,

where O_2 molecules are reduced to hydroxyl anions, thus increasing the pH locally. In mineral terms, lepidocrocite (α-FeOOH), hematite (Fe_2O_3), and magnetite (Fe_3O_4) are bound in the tubercule wall by the bacteria and their extracellular polysaccharide. Obligately anaerobic organisms such as sulfate-reducing bacteria profit from the redox gradient across the tubercule wall by removing sulfate from the bulk solution as it diffuses to the inner reduced part of the tubercule. The HS- produced stimulates the anodic (dissolutive) half reaction.

It is clear that a similar but not identical form of (chemical) corrosion can occur without biological intervention (26). In both chemical and microbiological cases, when the film covering the occluded zone is mechanically breached, the acid contents are released, oxygen enters the hitherto anodic (reduced) zone, and the steel repassivates. Under conditions in which the tubercule was sustained by microorganisms, however, sufficient bacteria remain at the corroded site to propagate corrosion in the same manner as previously. This is not the case under a purely chemical corrosion regime.

The tubercular form of corrosion has frequently been observed on the welded structures of stainless steels (8, 10), where initiation occurs above the heat-affected zone. This area, common to all gas-tungsten arc welds, exhibits distortions in the size and shape of the metal grains as a result of the uneven postweld cooling process (9). Tubercules of a more generalized nature occur in carbon steel freshwater conduits, with the result that the internal diameter of these pipes becomes occluded with the quantity of corrosion products (72).

Field Identification Techniques

Until recently, the absence of good microbiological techniques directly applicable in the field severely hampered field diagnosis of MIC. Corrosion products provide fractal surfaces from which it is difficult to remove bacterial cells for enumeration and the classical procedures necessary for identification. Field samples are frequently composed of iron oxide suspensions or soil in oil-water emulsions. Nevertheless, a series of morphological and mineralogical observations have been made which are consistent with MIC. In the case of weld deterioration of medium-grade stainless steels, the duplex structure is revealed in the case of the American Iron and Steel Institute (AISI) 300 series. The most frequently reported failures (8) concern the 304L and 316L welded structures, the general elemental compositions of which are provided in Table 1.

The corrosion resistance of these austenitic (face-centered cubic) alloys is usually considered quite adequate in freshwater (80). When welded, however, at least two crystal structures are present in the metal matrix, composed of austenite and ferrite (body-centered cubic). Energy-dispersive spectroscopy coupled with scanning electron microscopy enables the pinpoint distinction between the two phases. The quantity of ferrite is controlled by several parameters, including the type of weld rod used and arc energy input/distance traveled, and is essentially due to the migration of certain elements during the cooling process. The ferrite has a higher concentration of chromium and molybdenum than the austenite, while the latter has an elevated level of nickel. Since chromium and molybdenum contribute to the corrosion resistance of both phases, the ferrite thus has slightly more noble properties than the austenite. MIC of these welds accumulates chloride anions from the bulk water at the weld site and promotes corrosion, usually by the selec-

TABLE 1 Elemental compositions of two AISI stainless steels subject to MIC

Alloy	Cr	Ni	Mo	Si	Mn	C	S	P
304L	18.2	8.2	0.02	0.3	0.8	0.02	0.003	0.01
316L	17.1	9.3	2.1	0.5	1.4	0.02	0.008	0.02

tive removal of the austenite. Figure 2 presents an example of the tubercule composed of the corrosion products, bacterial cell mass, and extracellular polysaccharide found above a welded pipe. The classic metallurgical signature of the ferrite skeleton that is found underneath the tubercule is shown in Fig. 3.

Mineralogical Observations

The corrosion products associated with MIC differ in some circumstances from those produced in chemical corrosion. Powder X-ray diffraction has proved quite useful in distinguishing specific corrosion products characteristic of MIC. The mineral mackinawite (FeS_{1-x}) has been described as a signature for the intervention of sulfate-reducing bacteria in carbon steel corrosion (68, 98). Other sulfides of potential interest are greigite (Fe_3S_4) and marcasite (FeS_2), although they are not exclusive to microbiological formation. The presence of these minerals in nongeologically produced freshwaters has been taken as an indication that sulfate-reducing bacteria support the corrosion reactions. Further evidence of the uniqueness of MIC corrosion products has been provided by Mössbauer spectroscopy. Under conditions in which a steel structure is cathodically protected and the pH is relatively high, a rare mineral named green rust II [$4Fe(OH)_2, 2FeOOH, FeSO_4, 4H_2O$] (78) is formed in the presence of the sulfate-reducing bacteria. Moulin et al. (74) have described the appearance of this mineral on excessively corroded harbor wall sheet steel pilings at sites in several European ports. The predicted service lifetime of the steel in these cases was sharply reduced.

Other materials are known to have unusual minerals appearing in microbiological corrosion products. Sulfate-reducing bacteria in contact with the nickel-based alloy Monel 400 (66% Ni, 30.5% Cu, 1.3% Fe) develop bright green $CuCl_2·3Cu(OH)_2$ tubercules overlaid with black

$(Fe·Ni)_xS_y$ in the reduced zones. In fact, copper-nickel alloys appear to be particularly subject to corrosion initiated and propagated by free sulfide species in solution. Other copper-bearing alloys such as the common heat-exchanger tube alloy 90/10 Cu/Ni, and indeed copper metal, form the corrosion products chalcocite (Cu_2S) and djurleite ($Cu_{1.93}S$-$Cu_{1.97}S$) (67), the formation of which, by purely chemical means at room temperature, is doubtful.

Free Corrosion Potential Measurements

Considerable effort has been devoted to electrochemical analyses of microbiological corrosion. The open circuit potential (or free corrosion potential [Ecorr]) is a mixed potential resulting from the combined oxidation and reduction reactions which occur on the metal surface. Usually the predominating reduction reaction of stainless steel in seawater is the reduction of oxygen:

$$O_2 + 2H_2O + 4e \Rightarrow 4OH^- \tag{1}$$

which according to the Nernst equation has the reversible potential $E°$:

$$E = E° \left(\frac{[O_2]}{[OH^-]} \right) + \frac{RT}{nF} \cdot \ln \frac{[O_2]}{[OH^-]^4} \tag{2}$$

In seawater, marine biofilms push the open circuit potential of stainless steels to near the pitting or breakdown potential predicted for these materials in chloride-containing water (94). Figure 4 shows the effect of accumulating biofilm on a 23% Cr–3% Mo stainless steel as a function of time, with respect to a sterile autoclaved control. After 70 days, the Ecorr had reached approximately + 350 mV/SCE as a result of surface-biofilm interaction. The addition of the electron transport chain inhibitor sodium azide rapidly halted all respiration on the steel surface, and a sharp fall was observed in the open circuit potential to sterile values at about + 120 mV/SCE. Recent results (5, 17) have indicated that this increase in potential was due to an acceleration of the cathodic reaction when coupled to an iron anode. Active corrosion in seawater under these conditions would therefore be driven by microbiological phenomena, the exact nature of which is the subject of some dispute. The foremost argument against aerobic bacterial acceleration of the cathodic half-reaction is provided by the oxygen reduction by the microorganisms themselves. If microorganisms remove oxygen, then a reduced quantity is available for cathodic reduction, and the corrosion rate should be correspondingly lower (equation 1). Laboratory studies of corrosion in aerobic neutral-pH systems show that uniform removal of oxygen by aerobic bacteria reduces the availability of the cathodic reactant and lowers the corrosion rate in consequence (81). Laboratory studies, however, leave a lot to be desired in recreating localized corrosion under true environmental conditions.

The identification of microorganisms on in-service corroded structures has been considered a first step in the identification of MIC. Analytical methods for the determination of specific genera are severely hampered by sampling problems, however. For example, carbon steel and copper-based

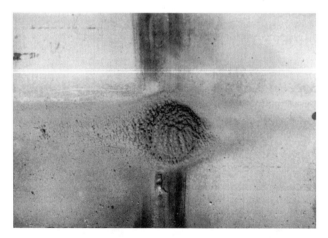

FIGURE 2 Bacterial tubercule above a stainless steel weld in a lower pipe section. Note that well water flowed from left to right, perpendicular to the weld. (Reprinted with kind permission of S. Borenstein, S. I., Palo Alto, Calif.)

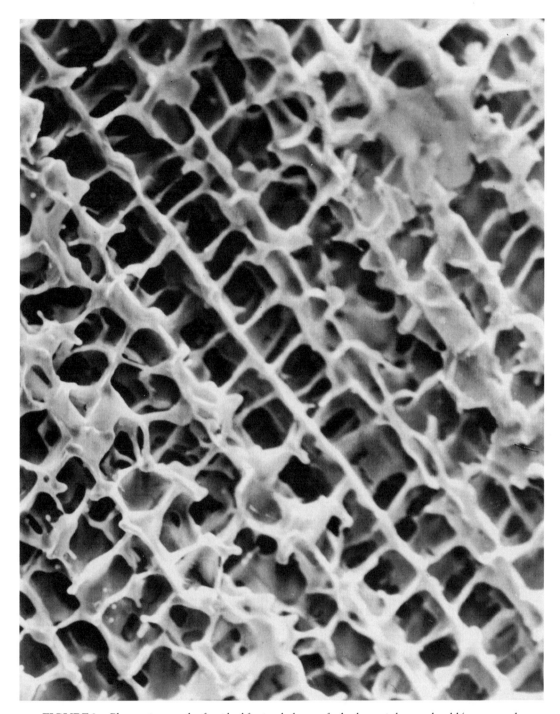

FIGURE 3 Photomicrograph of residual ferrite skeleton of a duplex stainless steel weld (corresponding to Fig. 2) after microbiological attack whereby the austenite phase was selectively dissolved.

alloys are sensitive to HS- corrosion in sour (H_2S-rich) oil and gas environments (23). Samples from these sites contain oil, secondary produced water, and suspended solids, which render enumeration and isolation procedures inefficient (107). Several field tests specific for sulfate-reducing bacteria, and in particular members of the frequently cited genus *Desulfovibrio*, have been developed. Variously they test for the presence of the hydrogenase and APS reductase enzymes involved in sulfate reduction or offer culture methods for specific classes of bacteria. An analysis of the various merits

of several tests has been published (93). Another approach has been to use DNA probes which are extremely specific and are less affected by multiphase samples (104). Remarkably, none of these commercial methods attempts to assay for in situ bacterial activity in any class. Methods using the stable isotopic ratio $^{32}S/^{34}S$ (61) or radioactive labels such as ^{35}S (66) on corroding metal surfaces appear to offer excellent methods of determining rates of sulfate reduction in an area where the actual species performing the process is probably immaterial. The nonselective radiolabels (82) ^{14}C

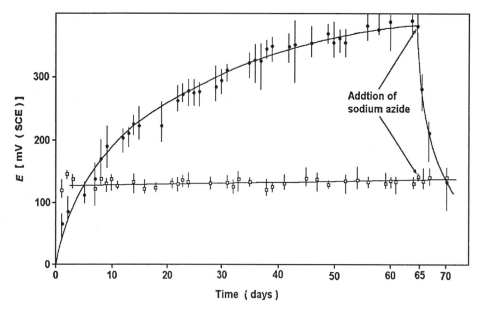

FIGURE 4 Evolution of the Ecorr of a 23% Cr–3% Mo stainless steel in seawater in the absence (open squares) and presence (closed circles) of a natural biofilm, showing the effect of biocide injection. (Reprinted with kind permission of V. Scotto, ICMM, Italy.)

and ^{3}H have been used to implicate hydrogenophilic anaerobic bacteria in the corrosion of gas transmission lines. These latter isotopes have the advantage of being less troubled by the tendency to precipitate like the sulfur tracers, which are removed as iron pyrite in the presence of ferrous ions.

MECHANISTIC STUDIES

While the engineering questions appear to be somewhat tractable, the microbiological aspects of MIC are complicated by the impact of corrosion processes on metabolic rates. In "simple" systems in which there is a relatively homogeneous biofilm, the nature of the surface affects the bacterial metabolism (100). Initial electrostatic attachment of the cell to the surface rapidly evolves to the production of extracellular polymers which fix the cell irreversibly to the substrate. This event is accompanied by a change in the cell metabolic activity which differs significantly from that of the planktonic state. The mature biofilm, generally considered in terms of a three-dimensional structure several cells thick, exhibits discontinuities in the diffusion characteristics of different molecules (50, 54, 85) which affect the corrosion processes. While biofilms appearing on metal surfaces immersed in natural waters are composed of a wide diversity of organisms, the contribution to the corrosion rate by sulfate-reducing bacteria appears greater than that of most other metabolic types of organisms. Under ordinary chemical corrosion conditions, the dissolution reactions (oxidation of the metal) are in electrochemical equilibrium with the reduction of some atomic or molecular species available in solution. Indeed, the electrons generated in the former reaction are consumed in the latter. In the previous section, which was concerned with oxic systems, the cathodic reactant was assumed to be oxygen. The much-cited hypothesis of Von Volzogen Kuhr and van der Vlught (103) states that in anoxic systems, protons constitute the principal cathodic reactant and that the primary corrosive effect of sulfate-reducing bacteria involves the consumption of hydrogen generated in the cathodic half-reaction. Thus, the anodic (dissolutive) half-reaction,

$$M^0 \Rightarrow M^{+2} + 2e \qquad (3)$$

where M represents any metallic element losing two electrons to pass into the oxidized state, is balanced by the cathodic (reductive) half-reaction in which the two electrons are simultaneously consumed:

$$2H^+ + 2e \Rightarrow H_2 \qquad (4)$$

The hydrogen so formed is then available for bacterial oxidation. The hypothesis introduced the idea that the removal of hydrogen to low levels would displace the equilibrium of the forward reaction of equation 4 such that the anodic dissolution (equation 3) would accelerate in consequence. Certain hydrogenophilic sulfate-reducing bacteria (notably *Desulfovibrio* spp.) have a very low K_m for hydrogen (86), and these same species can indeed grow on hydrogen generated in the corrosion cathodic half-reaction (12). This information, however, does not constitute proof that hydrogenophilic bacteria accelerate the anodic process in equation 3. The separation of a corrosive *Desulfovibrio desulfuricans* strain from the metal surface by virtue of a membrane showed that unchecked accumulation of molecular hydrogen at the metal surface resulted in a corrosion rate lower than that observed with the cells present on the surface (15). Excluding any corrosion effect that HS- has on corroding metal surfaces, the cathodic depolarization hypothesis thus appears feasible for the methanogenic bacterium *Methanococcus thermolithotrophicus* (14) despite a substantially lower affinity for hydrogen by methanogenic bacteria.

The importance of consortia involving sulfate-reducing bacteria with other metabolic types and in aerobic systems has been neglected despite being a common environment. A study with *Bacillus* sp., *Hafnia alvei*, and *Desulfovibrio gigas* demonstrated a clear increase in corrosion rate with *Desulfovibrio* sp. in nominally aerobic coculture (45). Electrochemical measurements (presented in admittance units, i.e., square centimeters per ohm) showed that the biculture *Hafnia* sp. with the *Desulfovibrio* sp. (2.5×10^{-3} mhos·cm²) and the triculture of these organisms with the *Bacillus* sp. provoked relatively high corrosion rates (1.4×10^{-3} mhos·cm⁻²) with respect to the sterile control (0.5×10^{-3} mhos·cm²) after 16 days of continuous culture. It is deduced that (i) active sulfate-reducing bacteria in "aerobic" media were necessary for the increased corrosion rates and (ii) bacterially influenced corrosion was accompanied by a reduced metal surface (Ecorr of -0.6 V/SCE). Sulfate-reducing bacteria are obligate anaerobes but oxygen tolerant (13) and withstand small quantities of oxygen in the presence of oxygen-respiring aerobes (*Bacillus* sp.) and the oxygen-consuming cathodic half-reaction present under neutral-pH corrosion. Studies at low (53) and high (52) oxygen concentrations show that optimal corrosion conditions occur in the presence of both oxygen and sulfide, possibly as a result of the presence of an Fe-FeS galvanic cell on the metal surface. Aerobic bacteria in both cases reduce the oxygen concentration in microniches on the metal surface, permitting sulfate reduction. Meanwhile, sufficient oxygen is available at a slightly remote site (some millimeters distant from the microniche) to sustain the cathodic half-reaction necessary for corrosion.

In the absence of oxygen, gas transmission pipelines are subject to considerable microbial corrosion (18) whereby the cathodic half-reaction may be provided by the reduction of protons. Undersea pipelines which transport mixed gases and liquid hydrocarbons are subject to deposit corrosion (at low pH) due to bacterial fermentation coupled with sulfate reduction. Notably, the volatile fatty acids formate, acetate, propionate and butyrate have been detected at levels of between 25 and 2,000 ppm, and under these conditions, corrosion deposits around the corroded sites yielded the obligately anaerobic fermenting bacteria *Clostridium* spp. and *Butyribacterium* spp. This basic system has been modeled with a triculture of *Desulfovibrio* sp., *Desulfobacter* sp., and the fermenter *Eubacterium limosum* (20). These organisms represent different metabolic groups which are capable of lithotrophic growth on hydrogen, respiration at the expense of sulfate, and fermentation of butanol or glucose. The principal reactions of interest with respect to corrosion were

$$4H_2 + 2CO_2 \Rightarrow CH_3COO^- + H^+ + 2H_2O$$
$$G^{0\prime} = -22.7 \text{ kJ/mol(H}_2) \quad (5)$$

$$CH_3COO^- + 3H^+ + SO_4^{2-} \Rightarrow 2CO_2 + H_2S + 2H_2O$$
$$G^{0\prime} = -63 \text{ kJ/mol(SO}_4^{2-}) \quad (6)$$

$$4H_2 + SO_4^{2-} + H^+ \Rightarrow HS^- + 4H_2O$$
$$G^{0\prime} = -152 \text{ kJ/mol(SO}_4^{2-}) \quad (7)$$

Figure 5 shows the reduction of sulfate with time as a function of bacterial population complexity with a *Eubacterium* sp., *Desulfobacter* sp., and *Desulfovibrio* sp. consortium in a system with multiple sources of carbon and electron donors. Sulfate reduction was principally conducted by the nonhydrogenophilic *Desulfobacter* sp., while the other two organisms favored fermentative reactions over dissimilatory sulfate reduction. Briefly, the consumption of H_2 by acetate fermentation is inefficient with respect to sulfate reduction (equations 5 and 7). *Desulfobacter* sp. appeared to perform the majority of the sulfate reduction purely because this species is nutritionally limited and incapable of fermentation. The *Desulfovibrio* sp., however, obtained most of its

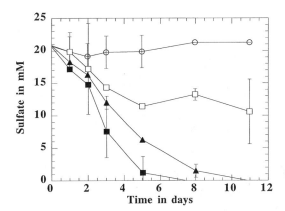

FIGURE 5 Sulfate removal during corrosion of C1020 carbon steel in the presence of *E. limosum* (○), *E. limosum* and *Desulfovibrio* sp., (□), *E. limosum* and *Desulfobacter* sp. (▲), and the triculture (■).

energy by fermentation of butanol via fumarate to succinate (-86 kJ/mol of H_2) rather than by sulfate respiration at the expense of hydrogen cathodically produced or otherwise. In corrosion terms, the situation is complex. Sulfides precipitate ferrous ions which may form an intact film on the metal surface, thus stopping all corrosion. Concurrently, the galvanic couple Fe-FeS tends to accelerate the primary corrosion reaction (Fe^0 to Fe^{2+}), aided by the reduced pH provided in volatile fatty acid fermentation. Scanning electron microscopy of the carbon steel surface (Fig. 6) shows the heterogeneous aspect of the surface with corrosion products and bacterial cells. Recent studies of the physiology of several sulfate-reducing bacteria show that the oxidation of hydrogen at the expense of Fe^{3+} reduction is also feasible (63), which introduces the possibility that the cathodic reactions exist inside the tubercule depicted in Fig. 1. Other mineral transformations performed by bacteria may also affect the corrosion resistance of a material. Certain sulfate-reducing bacteria, among other organisms (25), are capable of reducing Cr^{6+} to Cr^{3+} in the presence of hydrogen (34, 64). Steels containing above 9% Cr, including stainless steels, tend to form corrosion-resistant surfaces which are enriched in chromium hydroxides (77). The effect of chromate-reducing bacteria on these materials has yet to be established.

The interaction of bacteria and metal structures under mechanical stress is a source of considerable concern. Analyses of anaerobic bacterial activity by *Clostridium acetobutylicum* and a marine sulfate-reducing consortium have shown that stress corrosion cracking (SCC) can be accelerated by the ingress of hydrogen of bacterial origin into palladium (29) and carbon steels (87). SCC by hydrogen infiltration into the metal matrix is a well-known phenomenon whereby alloys crack at stresses which are nominally well within their elastic limit. Gangloff and Kelly (35) have shown that *Desulfovibrio vulgaris* greatly increased the rate of ambient-temperature fatigue cracking. While the accumulation of hydrogen at the crack tip was correlated with bacterial activity, the actual presence of the organisms at the tip was not necessary. Very little is known about these phenomena other than the high prevalence of SCC in sulfide-rich, reduced environments.

Corrosion of concrete and the embedded rebar skeleton also occurs by microbiological oxidation and reduction reactions. *Thiobacillus* spp. promote concrete corrosion by

FIGURE 6 Consortium of fermenting bacteria and *Desulfovibrio* sp. on the corroding surface of C1020 carbon steel.

leaching out the aggregate binder, which is pH sensitive. Milde et al. (71) isolated *Thiobacillus neapolitanus*, *T. thiooxidans*, *T. intermedius*, and *T. novellus* from sewers subject to crown corrosion in Hamburg, Germany. Sulfate-reducing bacteria present in sewage water release HS- which passes into the headspace as H_2S. The thiobacilli oxidize the sulfide to sulfuric acid on the upper zone (the crown) in contact with the headspace air. Laboratory simulations with Portland cement (initial pH of 9) showed very rapid deterioration uniquely in the zone where an active thiobacillus-containing consortium oxidized H_2S (91).

Microbial corrosion studies have also been hampered by the problems in developing suitable electrolytes for laboratory corrosion experiments which closely resemble environmental conditions. Certain nutritional supplements present in yeast extract, Trypticase soy (amines and nitrates) (65), minerals such as phosphates (106), and morpholinepropanesulfonic acid (MOPS) (20) commonly introduced into microbiological media are chemical corrosion inhibitors which skew results. Thus, media conceived for isolation procedures are rarely useful for corrosion experiments, except when bacteria metabolically remove this "inhibitor" and allow corrosion to proceed by default. Furthermore, it is probable that the isolation procedure itself produces only opportunistic organisms (108) which play little part in the original corrosion reaction from which the inoculum was taken.

CONTROL METHODS

Existing structures can be efficiently protected from microbial corrosion with biocides of numerous categories (49); however, the strategy of choice must conform to a battery of environmental protection laws. Quaternary ammonium salts, chlorine, carbamates, and glutaraldehyde all have persistence or toxicity problems which vary in acceptability (84). Filtration and UV radiation with ozone are efficient for controlling the microorganisms' access to the surfaces at risk but not decontamination (1, 30). It thus becomes apparent that chemical treatments for industrial biofouling involve a plethora of considerations not always related to the primary task of eliminating bacteria.

Oxidizing Biocides

Chlorination of both freshwater (42) and seawater (101) systems is both effective and less persistent than other biocide systems. Chlorine has the advantage of ease of application as a gas or as salts of hypochlorous acid such as sodium and calcium hypochlorite and removes 99% of planktonic bacterial cells at levels of 0.08 mg of HOCl/liter (51). The concentrations required for bacteriostasis or complete disinfection of biofilms, however, depend on, among other parameters, the pH and the availability of oxidizable materials (42). Chlorination of seawater systems displaces the naturally occurring bromide to form hypobromous acid, which some consider to be the active molecular species. The acid then dissociates, depending on the pH,

$$HOCl + Br^- \Rightarrow HOBr + Cl^- \qquad (8)$$

$$HOBr \Rightarrow OBr^- + H^+ \qquad (9)$$

and is considered more reactive against biofilms (99) than $HOCl^-$. Franklin et al. (32), however, showed that biofilms occurring on corroding carbon steel surfaces were not sensitive to either HOCl or HOBr at a concentration of 2 ppm. The absence of biostasis due to chlorine on corroded surfaces

has been confirmed by Freiter (33) and may be due to consumption of the oxidized molecular species by the corrosion products. At 16 ppm of HOCl, the number of sessile bacteria was reduced by 3 orders of magnitude (31). At this concentration, however, HOCl has an adverse affect on the corrosion rate.

Chlorination is usually used in continuous treatment at 0.1 to 0.2 mg/liter, but with intermittent doses at between 0.5 and 1 mg/liter. Chlorine ($HClO^-$) has a high redox potential, which increases the risk of pit initiation on ordinarily corrosion-resistant materials such as stainless steels. The addition of 0.1 to 0.2 ppm of residual chlorine, however (73), reduced the corrosion rate of stainless steels in seawater, primarily by affecting oxygen reduction in the biofilm. It is thus imperative to establish the threshold concentration of $HClO^-$ necessary to achieve bacteriostasis without excess chlorine affecting the chemical corrosion rate. While chlorine has no significant toxic effects, its discharge without treatment into the environment may oxidize natural compounds to carcinogenic haloamines, trihalomethanes, and other halogenated organic derivatives (27). This effect may be offset by the application of very low levels of chlorine with low concentrations of cupric ions (47), which has been reported to give synergistic results without increasing the corrosion rate. Chlorine dioxide (ClO_2) is another oxidizing biocide which does not form hypochlorous acid in seawater and can be used in the presence of phenols or ammonia. Chlorine dioxide oxidizes without "chlorination" and kills microorganisms by reaction with the cell structure, accelerating the metabolism to the detriment of cell growth or by inhibiting protein synthesis (7). Organic compounds that deliver oxidizing halogens (chloroisocyanates, 1-bromo-3-chloro-5,5-dimethylhydantoin) have sometimes been used as substitutes for chlorine gas for safety reasons (58).

Another oxidizing biocide, but one with a more limited use, is ozone (O_3). Ozone has a very short half-life and apparently produces no undesirable compounds. A distinct advantage of ozone is the rapidity of bacterial inactivation (76).

Nonoxidizing Biocides

Other chemical species can be just as effective as chlorine, but with persistence and toxicity problems which are difficult to ignore. Monochloramine is more penetrative into biofilms (109) than $HClO^-$, apparently has different physiological effects on the bacteria (96), but has no corrosion side effects. The compounds 2,2-dibromo-3-nitrilopropionamide, organosulfur (MBT), and organonitrogen-sulfur (DMTT) are frequently used biocides in closed-circuit systems (6, 69, 89). Quaternary ammonium salts are generally effective against algae and bacteria, but their use as well as that of the organotin products is strictly controlled in many countries (70). Aldehydes, and glutaraldehyde in particular, are used to reduce bacterial numbers in down-hole petroleum deposits (24) which would otherwise contribute to high levels of corrosion.

MATERIALS SELECTION

It is almost axiomatic that the materials found to be subject to MIC were selected for specific properties such as mechanical strength, thermal conductivity, weldability, chemical corrosion resistance, and cost. Invariably, MIC has never been considered to exist under special conditions which merited a category of corrosion distinct from the general "corrosivity" of natural water. "Abnormal" corrosion rates in otherwise benign environments can thus be identified as microbial corrosion whereby target sites usually involve some structural heterogeneity. The effects of manufacturing specific components, forged, wrought plate steel, welding, etc., introduce inhomogeneities in the final metallurgy in terms of crystal structure, welding slag, heated tinted surfaces, and intermetallic precipitates, which increase the susceptibility to chemical corrosion (80). The reduction in resistance to MIC is less well defined. Heat-tinted surfaces introduced in welding appear to reduce corrosion resistance of 304L welds in freshwater (56). Laboratory studies have been able to simulate this effect with a mixed microbial consortium including *Acinetobacter* spp., *Vibrio* sp., *Pseudomonas* sp., *Shigella* sp., three members of the *Bacillus* genus, and two sulfate-reducing bacteria (55). These organisms represent members of several metabolic groups commonly found in weld metal corrosion products: obligately aerobic bacteria, fermenting facultative anaerobes, and finally sulfidogenic bacteria. Nonwelded structures are also subject to selective attack. The stainless steel AISI 316L (Table 1) can undergo selective depletion of the elements Cr, Fe, and Ni at metal grain boundaries (Fig. 7) as a result of attachment by *Citrobacter freundii* (36). The attack of specific metallurgical defects is not restricted to stainless steels, however: microorganisms preferentially attach to high-strength low-alloy steels at partially melted zones where local increases in sulfur content at the grain boundaries of the steel have been observed (105). Unfortunately, no further microbiological analyses have been conducted to elucidate why the

bacteria were so attached and the link between the organisms and the consequent corrosion.

Corrosion in Fe-Ni-based alloys is mitigated in a general sense by the increase in alloying elements such as chromium and molybdenum. As a result, it is common to describe the corrosion resistance in terms of a pitting resistance equivalence (PRE = % Cr + 3.3% Mo + 16% N). This coefficient reflects a chemical ranking of each stainless steel with respect to the general corrosion resistance in acid electrolyte. The high (6%)-molybdenum stainless steels are considered to be among the most corrosion-resistant materials available; nevertheless, *D. desulfuricans* has reportedly initiated corrosion in several of these alloys under laboratory conditions (92). While no industrial failures of these materials have been reported, the laboratory study raises the possibility that the corrosion resistances of different alloys under a microbiological regime may not be ranked in the same order as under a chemical corrosion regime. The selection of corrosion-resistant materials has an important economic component which is subject to the raw mineral fluctuations at the London metal exchange (79). The 6% Mo alloys with a PRE of about 43 cost perhaps 2.5 times more than the 2% Mo alloy AISI 316 with a PRE of 23. If the addition of higher-alloying elements cannot be justified due to the effects of commonly occurring microorganisms by a significant gain in corrosion resistance, then other strategies must be used. For example, materials such as titanium (62) may be considered, but they are accompanied by prohibitive cost.

Not all alloys are specifically designed for corrosion resis-

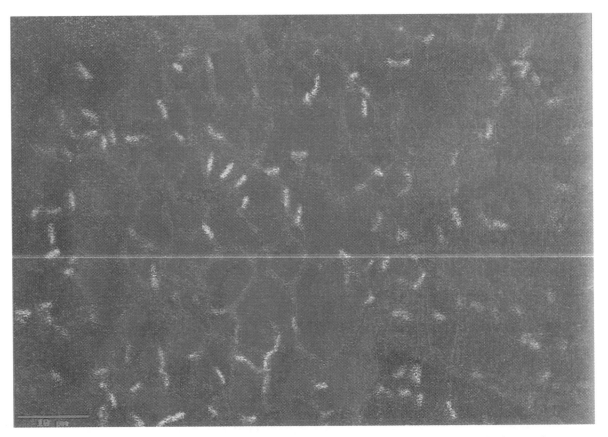

FIGURE 7 Attachment of *C. freundii* to AISI 316L stainless steel and subsequent depletion of the metal oxide grain boundaries. (Reprinted with kind permission of G. Geesey, Montana State University.)

tance. Carbon steels used in cutting and drilling machinery are subject to corrosion by *Pseudomonas* spp. and on occasion sulfate-reducing bacteria (39). In terms of world tonnage, carbon steels are by far the most widely used; however, they are rather easily corroded. Usual compensative measures involve a combination of painting and cathodic protection, which are also subject to microbiological deterioration (44). Polymers covering the surface of the steel provide a barrier to the ingress of water and salts, while sacrificial anodes of materials higher in the electroactivity series (Mg, Al, Zn, etc.) polarize the iron into the cathodic domain. In consequence, the local pH of the surface increases as a result of oxygen reduction to hydroxyl ions (Fig. 1A). The local alkaline pH precipitates calcium and magnesium cations from seawater or "hard" freshwater in a hard film onto the iron surface. The calcareous film so formed reduces the current flowing between the sacrificial anode and the protected iron. Under normal circumstances, this reduction in current is a desirable property since the anodes need to be replaced less frequently. Bacteria interfere with this process by reducing the local pH on the surface (75) and destabilizing the magnesium and calcium deposits (40). X-ray diffraction has identified changes in bruceite [hexagonal $Mg(OH)_2$], aragonite (orthorhombic $CaCO_3$), and calcite (rhombohedral $CaCO_3$) which lead to a current increase and fast dissolution of the anodes. A more severe complication of cathodic protection is due to sulfate-reducing bacteria present in reduced sediments. The hydrogenophilic members of this group consume the hydrogen produced at the surface, thus depolarizing the steel and increasing the likelihood of corrosion (41).

Materials other than copper- and iron-based compositions are also subject to microbial deterioration. Aluminum is extremely sensitive to corrosion by the fungus *Hormoconis* (formerly *Cladosporium*) *resinae* (3, 90, 102). Growth of this organism at the interface between water condensates and the fuel oils and kerosene in jet aircraft fuel tanks and radiators promotes the corrosion of lightweight aluminum alloys used in these applications. Figure 8 shows the corrosion imprint of fungal mycelia on the aluminum surface after clean-

ing. Aluminum is a passivatable material in the sense that surface oxidation to Al^{3+} forms a relatively intact film of a highly ordered porous nature (Al_2O_3) which retards further oxidation. A corrosive cell is created by the fungus, which partially oxidizes the alkanes to dodecanoic acid. This fatty acid combined with the local concentration effect of chloride, both of which partition to the aqueous phase from the kerosene, promotes corrosion of the aluminum directly in contact with the water which condenses out of the alkane fraction. Prior to the advent of nonpolar biocides, considerable MIC was detected in aircraft wings as a result of the degradation of the jet fuel kerosene. The local temperature due to air friction increases substantially and promoted both mycelial growth and subsequent corrosion.

Concerns for microbial corrosion have also extended to the behavior of modern alloys in very long term contact with natural untreated waters. Several western countries, notably France, Finland, Sweden, the United Kingdom, and the United States, possess civil nuclear energy programs which have produced significant quantities of high-level nuclear waste over a number of decades. Disposal of this waste in a rational manner requires very long term storage (minimum 1,000 years) in materials which are, of necessity, corrosion resistant. Assessments of potentially corrosive genera present in deep wells have identified the thiobacilli, the nitrifying and denitrifying bacteria, and sulfate-reducing and iron-oxidizing bacteria (97) as potential problems. Candidate container materials under test by the U.S. Nuclear Regulatory Commission include nickel-based alloys, stainless steels, and copper-nickel alloys. Extensive modeling has shown that initial radiation flux to the package materials will be in the order of 5×10^5 rad/h and that a maximum temperature of 200°C will develop at the wall. While this will initially tend to sterilize and dry out the immediate surroundings, the seep-back of water in the last 250-year period of the containment may transport microorganisms to the package surface. Under these latter conditions, microbial corrosion may well be viable (38). Chemical analysis of water originating from well J-13 at the proposed Yucca Mountain, Arizona, burial site has demonstrated sufficient

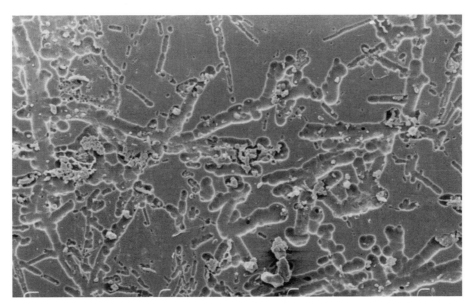

FIGURE 8 Selective attack of aluminum alloy after removal of *H. resinae* mycelia. (Reprinted with kind permission of H. Vidéla, INIFTA, Argentina.)

mineral nutrients with bicarbonate to support autotrophic growth. Microbial corrosion by autochthonous iron-oxidizing bacteria of 304 and 316 stainless steel welds has indeed been demonstrated at another field test site in Arizona (19). It is envisaged that construction of a TUFF repository would unavoidably involve the introduction of large quantities of hydraulic and fuel oils among other organic materials to the site, which would also support heterotrophic microorganisms of a more corrosive nature.

CONCLUSIONS

Corrosion as a direct result of contact between microorganisms and metals is not implicit. No study has conclusively shown that corrosion initiation (an event which occurs on the scale of a few micrometers) is the result of direct microbiological intervention. The arrangement previously described, with spatial separation between anodic and cathodic sites, represents corrosion propagation, a quasi-stable state exacerbated by microorganisms. The question of spatial resolution of the anodic and cathodic sites has been examined by Franklin et al. (32), using the scanning vibrating electrode technique; they established that viable and active bacteria (*Pseudomonas* sp.) are necessary for the biological propagation of corrosion in common carbon steels. Under sterile conditions, anodic zones are initially evenly distributed across the steel surface but passivate with time as a thin film of oxide and phosphate covers the surface. In the presence of bacteria, most of the anodic zones repassivate; however, one or two sites develop intense anodic currents from which corrosion propagates, i.e., becomes self-sustaining. Since the anodic sites clearly exist under sterile conditions, exclusive initiation by microorganisms seems unrealistic. The only real case for bacterial initiation appears to involve *Thiobacillus ferrooxidans*, which secretes an organic film around iron pyrite (FeS_2), thus increasing its dissolution rate (88). Steel of many grades contains iron and manganese sulfide inclusions which dissolve preferentially and give rise to metal attack. The relationship of this phenomenon with bacterial activity has yet to be established.

Despite the profusion of articles on microbiological corrosion, the relationship between free-developing complex biofilms and their effect on underlying metal substrates is poorly understood. The fault lies in part with the difficulty in recreating a corrosive environment under laboratory conditions in terms of the microbiota and local electrochemical surface discontinuities. The pursuit of these goals is further impeded by the remarkable absence of multidisciplinary laboratories. Researchers are usually schooled within a single discipline, which encourages fidelity and punishes expeditions to other areas. Despite occasional comprehensive research programs organized by government agencies and heavy industry, plant engineers in many countries appear to be ill informed on the countermeasures available. This has led to either neglect through ignorance or overprotection, which can be expensive and deleterious to plant and environment. Clearly a measured response is in order. Proposals such as the development of new alloys which are specifically resistant to microorganisms or the synthesis of new nonpersistent chemical or biological treatments may well provide some relief.

A precise definition for microbial corrosion has yet to be established, since chemical corrosion cannot be separated from the MIC phenomenon. In seawater, a considerable corrosive force is constantly provided by the 30 g of salt per liter. Corrosion acceleration by sulfate-reducing bacteria in seawater biofilms is therefore due to a combination of the salt and the metabolic activities of the organisms. The convenient term "microbiological corrosion" indicates that the corrosion rate-limiting step is controlled by sulfidogenic bacteria, for example, and not the chloride ion concentration.

After consideration of the broad and often contradictory mass of information currently available, two distinct rational approaches appear viable: a practical agenda would require that justification for measures against microbiological corrosion reside in the identification of corrosion products, rates, and microflora which are consistent with those reported for that alloy. These activities would, by definition, be short term in nature. In contrast, a scientific agenda would require, under strictly defined conditions, the elucidation of the energetic relationship between Faraday's law (essentially the loss of metal with the passage of charge) and the metabolic activities of a well-defined consortium. This nontrivial task will probably require finite element analysis in order to facilitate the integration of both biological and electrochemical processes.

REFERENCES

1. **Abshire, R. L.** 1986. The use of UV radiation as a method of sterilization in the pharmaceutical industry, p. D1.1–D1.19 *In Proceedings of the International Ozone Association European Committee.* Paris.
2. **Allred, R., J. Sudbury, and D. C. Olson.** 1959. Corrosion is controlled by bactericide treatment. *World Oil* **149:** 111–112.
3. **Allyon, E. S., and B. M. Rosales.** 1994. Electrochemical test for predicting microbiologically influenced corrosion of aluminium and AA 7005 alloy. *Corrosion* **50:**571–575.
4. **Angell, P., S. Borenstein, R. Buchanan, S. Dexter, N. Dowling, B. Little, C. Lundin, M. McNeil, D. Pope, R. Tatnall, D. White, and G. Ziegenfuss (ed.).** 1995. *International Conference on Microbially Influenced Corrosion.* National Association of Corrosion Engineers, Houston.
5. **Audouard, J. P., C. Compère, N. J. E. Dowling, D. Feron, D. Festy, A. Mollica, T. Rogne, V. Scotto, U. Steinsmo, K. Taxen, and D. Thierry.** 1995. Effect of marine biofilms on high performance stainless steels exposed in European coastal waters, p. 198–210. *In Microbial Corrosion. Proceedings of the 3rd International European Federation of Corrosion (EFC) Workshop.* EFC Publication no. 15. European Federation of Corrosion.
6. **Bendiksen, B., and J. Parsons.** 1990. Impact of biocides on polymer stabilisation of iron and manganese in recirculating cooling water systems, article 83. *In Proceedings of Conference, Corrosion '90.* National Association of Corrosion Engineers, Houston.
7. **Bernarde, M., W. Snow, V. Olivieri, and B. Davidson.** 1967. Kinetics and mechanism of bacterial disinfection by chlorine dioxide. *Appl. Microbiol.* **15:**257–265.
8. **Borenstein, S.** 1988. Microbiologically influenced corrosion failure analyses. *Mater. Perform.* **27**(8):62–66.
9. **Borenstein, S.** 1994. *Microbiologically Influenced Corrosion Handbook.* Woodhead Publishing Ltd., Cambridge.
10. **Borenstein, S., and P. B. Lindsay.** 1988. Microbiologically influenced corrosion failure analyses. *Mater. Perform.* **27**(3):51–54.
11. **Brennenstuhl, A. M., and P. E. Doherty.** 1990. The economic impact of microbiologically influenced corrosion at Ontario hydro's nuclear power plants, p. 7/5–7/10. *In N. J. E. Dowling, M. W. Mittelman, and J. C. Danko (ed.), Proceedings of Conference: Microbially Influ-*

enced Corrosion and Deterioration. Center for Materials Processing, University of Tennessee, Knoxville.

12. **Cord-Ruwisch, R., and F. Widdel.** 1986. Corroding iron as a hydrogen source for sulfate reduction in growing cultures of sulfate-reducing bacteria. *Appl. Microbiol. Biotechnol.* **25:**169–174.

13. **Cypionka, H., F. Widdel, and N. Pfennig.** 1985. Survival of sulfate-reducing bacteria after oxygen-stress, and growth in sulfate-free oxygen-sulfide gradients. *FEMS Microbiol. Ecol.* **31:**39–45.

14. **Daniels, L., N. Belay, B. Rajagopal, and P. Weimer.** 1987. Bacterial methanogenesis and growth from CO_2 with elemental iron as the sole source of electrons. *Science* **237:**509–511.

15. **Daumas, S., Y. Massiani, and J. Crousier.** 1988. Microbiological battery induced by sulphate-reducing bacteria. *Corros. Sci.* **28:**1041–1050.

16. **Dexter, S. (ed.).** 1986. *Biologically Induced Corrosion NACE-8 Proceedings of the International Conference on Biologically Induced Corrosion.* National Association of Corrosion Engineers, Houston.

17. **Dexter, S., and G. Gao.** 1988. Effect of seawater biofilms on corrosion potential and oxygen reduction of stainless steel. *Corrosion* **44(10):**717–723.

18. **Dias, O. C., and M. C. Bromel.** 1990. Microbially induced organic acid underdeposit attack in a gas pipeline. *Mater. Perform.* **April:**53–56.

19. **Dowling, N., C. Lundin, D. Sachs, J. Bullen, and D. White.** 1992. Principles of the selection of effective and economic corrosion-resistant alloys in contact with biologically active environments, p. 361–367. *In* C. Interrante and R. Pablan (ed.), *Proceedings of Conference: Scientific Basis for Nuclear Waste Management XVI.* Materials Research Society, Boston.

20. **Dowling, N. J. E., S. T. Brooks, T. Phelps, and D. White.** 1992. Effects of selection and fate of substrates supplied to anaerobic bacteria involved in the corrosion of pipe-line steel. *J. Ind. Microbiol.* **10:**207–215.

21. **Dowling, N. J. E., M. Mittelman, and D. C. White.** 1991. The role of consortia in microbially influenced corrosion, p. 341–372. *In* G. Zeikus (ed.), *Mixed Cultures in Biotechnology.*

22. **Dowling, N. J. E., M. W. Mittelman, and J. C. Danko (ed.).** 1990. *Proceedings of Congress, Microbially Influenced Corrosion and Deterioration.* University of Tennessee, Knoxville.

23. **Dziewulski, D., S. Franleigh, and D. Pope.** 1990. Microbial production of hydrogen sulfide in gas storage and production fields: field studies, preliminary modeling and control, article no. 35. *In Proceedings of Conference, Corrosion '90.* National Association of Corrosion Engineers, Houston.

24. **Eager, R., J. Leder, J. Stabnley, and A. Theis.** 1988. The use of glutaraldehyde for microbiological control in waterflood systems, article 84. *In Proceedings of Conference, Corrosion '88.* National Association of Corrosion Engineers, Houston.

25. **Ehrlich, H.** 1990. *Geomicrobiology,* 2nd ed. Marcel Dekker, New York.

26. **Eklund, G.** 1974. Initiation of pitting at sulfide inclusions in stainless steel. *J. Electrochem. Soc.* **121(4):**469–473.

27. **Electric Power Research Institute.** 1981. Biofouling control practice and assessment. EPRI-CS1796. Electric Power Research Institute, Palo Alto, Calif.

28. **Farquhar, G.** 1974. The sulfate-reducing bacteria and oilfield bacterial corrosion. A review of the current state of the art, article 1. *In Proceedings of Conference, Corrosion '74.* National Association of Corrosion Engineers, Houston.

29. **Ford, T., P. Searson, T. Harris, and R. Mitchell.** 1990. Investigation of microbiologically produced hydrogen permeation through palladium. *J. Electrochem. Soc.* **137(4):**1175–1179.

30. **Francis, P. D.** 1987. The use of ultraviolet light and ozone to remove organic contaminants in ultrapure water. *In Proceedings of Infrared Circuit Manufacturers Consortium Conference.*

31. **Franklin, M., D. Nivens, A. Vass, M. Mittelman, R. Jack, N. Dowling, and D. White.** 1991. Effect of chlorine and chlorine/bromine biocide treatments on the number and activity of biofilm bacteria and on carbon steel corrosion. *Corrosion* **47(2):**128–134.

32. **Franklin, M. J., D. C. White, and H. S. Isaacs.** 1991. Spatial and temporal relationships between localized microbial metabolic activity and localized electrochemical activity on steel, *In Proceedings of Conference, Corrosion '91.* National Association of Corrosion Engineers, Houston.

33. **Freiter, E.** 1992. Effects of a corrosion inhibitor on bacterial and microbiologically influenced corrosion. *Corrosion* **48(4):**266–276.

34. **Fude, L., B. Harris, M. Urritia, and T. Beveridge.** 1994. Reduction of Cr(VI) by a consortium of sulfate-reducing bacteria (SRBIII). *Appl. Environ. Microbiol.* **60:**1525–1531.

35. **Gangloff, R. P., and R. G. Kelly.** 1994. Microbe-enhanced environmental fatigue crack propagation in HY130 steel. *Corrosion* **50(5):**345–354.

36. **Geesey, G. G., R. Avci, D. Daly, M. Hamilton, P. Shope, and G. Harkin.** 1995. *Corros. Sci.* **38(1):**73–95.

37. **Geesey, G. G., and W. Costerton.** 1979. Microbiology of a northern river, bacterial distribution and relationship to suspended sediment and organic carbon. *Can. J. Microbiol.* **25:**1058–1062.

38. **Geesey, G. G., and G. Cragnolino.** 1995. A review of the potential for microbially influenced corrosion of high level waste containers in an unsaturated repository site, p. 76/1-76/20. *In* P. Angell, S. Borenstein, R. Buchanan, S. Dexter, N. Dowling, B. Little, C. Lundin, M. McNeil, D. Pope, R. Tatnall, D. White, and G. Ziegenfuss (ed.), *Proceedings of International Conference on Microbially Influenced Corrosion.* National Association of Corrosion Engineers, Houston.

39. **Gomez de Saravia, S. G., P. S. Guiamet, M. F. L. de Mele, and H. A. Videla.** 1991. Biofilm effects and MIC of carbon steel in electrolyte media and contaminated with microbial strains isolated from cutting-oil emulsions. *Corrosion* **47(9):**687–692.

40. **Guezennec, J., N. Dowling, J. Bullen, and D. C. White.** 1994. Relationship between bacterial colonization and cathodic current density associated with mild steel surfaces. *Biofouling* **8:**133–146.

41. **Guezennec, J. G., and M. Therené.** 1988. A study of the influence of cathodic protection on the growth of SRB in marine sediments by electrochemical techniques, p. 256–265. *In* C. A. C. Sequeria and A. K. Tiller (ed.), *Microbial Corrosion.* Elsevier Science, Amsterdam.

42. **Hassan, R., and L. Oh.** 1989. Effect of sodium hypochlorite (chlorox) and its mode of application of biofilm development. *Biofouling* **1:**353–361.

43. **Hill, E. C.** 1987. *Microbial Problems in the Off-Shore Oil Industry.* Institute of Petroleum, London.

44. **Holthe, R., E. Bardal, and P. O. Gartland.** 1989. Time dependence of cathodic properties of materials in seawater. *Mater. Perform.* **June:**16–23.

44a. **Howsam, P. (ed.).** 1990. *Microbiology in Civil Engineering.* FEMS symposium 59. Chapman & Hall, London.

45. **Jack, R., D. Ringelberg, and D. White.** 1992. Differential

corrosion rates of carbon steel by combinations of *Bacillus* sp., *Hafnia alvei* and *Desulfovibrio gigas* established by phospholipid analysis of electrode biofilm. *Corros. Sci.* **33**(12): 1843–1853.

46. **Kasahara, K., and F. Kajiyama.** 1986. Role of sulfate-reducing bacteria in the localized corrosion of buried pipes, p. 172–183. *In* S. Dexter (ed.), *Biologically Induced Corrosion. NACE-8. Proceedings of the International Conference on Biologically Induced Corrosion.* National Association of Corrosion Engineers. Houston.

47. **Knox-Holmes, B.** 1993. Biofouling control with low levels of copper and chlorine. *Biofouling* **7**:157–166.

48. **Knudsen, J. G.** 1981. Fouling of heat transfer surfaces, p. 57–82. *In Power Condenser Heat Transfer Technology.* Hemisphere Publishing Co., New York.

49. **Lamot, J.** 1988. Role of biocides in controlling microbial corrosion, p. 224–234. *In* A. Sequeira and A. Tiller (ed.), *Proceedings of Conference, European Federation of Corrosion Workshop on Microbial Corrosion.* Portugal.

50. **Lawrence, J. R., G. M. Wolfaardt, and D. R. Korber.** Determination of diffusion coefficients in biofilms by confocal laser microscopy. *Appl. Environ. Microbiol.* **60:** 1166–1173.

51. **Lechevalier, M., C. Cawthorn, and R. Lee.** 1988. Inactivation of biofilm bacteria. *Appl. Environ. Microbiol.* **54:** 2492–2499.

52. **Lee, W., Z. Lewandowski, M. Morrison, W. Charaklis, R. Avci, and P. Nielsen.** 1993. Corrosion of mild steel underneath aerobic biofilms containing sulfate-reducing bacteria. Part II. At high dissolved oxygen concentration. *Biofouling* **7:**197–216.

53. **Lee, W., Z. Lewandowski, S. Okabe, W. Charaklis, and R. Avci.** 1993. Corrosion of mild steel underneath aerobic biofilms containing sulfate-reducing bacteria. Part I. At low dissolved oxygen concentration. *Biofouling* **7:**197–216.

54. **Lewandowski, Z., W. Lee, W. Charaklis, and B. Little.** 1989. Dissolved oxygen and pH microelectrode measurements at water-immersed metal surface. *Corrosion* **45**(2): 92–98.

55. **Li, P., R. Buchanan, C. Lundin, P. Angell, A. Tuthill, and R. Avery.** 1995. Laboratory studies of microbially influenced corrosion of stainless steel weldments in simulated Tennessee river water, p. 40/1–40/14. *In* P. Angell S. Borenstein, R. Buchanan, S. Dexter, N. Dowling, B. Little, C. Lundin, M. McNeil, D. Pope, R. Tatnall, D. White, and G. Ziegenfuss (ed.), *Proceedings of International Conference on Microbially Influenced Corrosion.* National Association of Corrosion Engineers, Houston.

56. **Li, P., R. Buchanan, C. Lundin, D. Sachs, A. Tuthill, R. Avery, and J. Bullen.** 1995. Corrosion studies of stainless steel weldments in a microorganism-containing Arizona ground water system, p. 41/1–41/13. *In* P. Angell, S. Borenstein, R. Buchanan, S. Dexter, N. Dowling, B. Little, C. Lundin, M. McNeil, D. Pope, R. Tatnall, D. White, and G. Ziegenfuss (ed.), *Proceedings of International Conference on Microbially Influenced Corrosion.* National Association of Corrosion Engineers, Houston.

57. **Licina, G.** 1989. *Microbial Corrosion: 1988 Workshop Proceedings.* ER-6345. Electric Power Research Institute, Palo Alto, Calif.

58. **Liden, L., D. Burton, L. Bongers, and A. Holand.** 1980. Effects of chlorobrominated and chlorinated cooling water on estuarine organisms. *J. Water Pollut. Control Fed.* **53:**173–182.

59. **Linhardt, P.** 1994. Manganoxidierende bakterien und lochkorrosion an turbinenteilen aus CrNi-stahl in einem laufkraftwerk. *Werkst. Korros.* **45:**79–83.

60. **Little, B., M. Walch, P. Wagner, S. Gerchakov, and R.**

Mitchell. 1984. The impact of extreme obligate bacteria on corrosion processes, p. 511–520. *In Proceedings of 6th International Congress on Marine Corrosion and Fouling.*

61. **Little, B. J., P. Wagner, and J. Jones-Meehan.** 1994. Sulfur isotope fractionation in sulfide corrosion products as an indicator for microbiologically influence corrosion, p. 180–187. *In* J. Kearns and B. J. Little (ed.), *Microbiologically Influenced Corrosion Testing.* ASTM STP 1232. American Society for Testing and Materials, Philadelphia.

62. **Little, B. J., P. Wagner, and R. Ray.** 1993. An evaluation of titanium exposed to thermophilic and marine biofilms, paper 308. *In Proceedings of Conference.* National Association of Corrosion Engineers, Houston.

63. **Lovely, D., and E. Phillips.** 1994. Novel processes for anaerobic sulfate production from elemental sulfur by sulfate-reducing bacteria. *Appl. Environ. Microbiol.* **60:** 2394–2399.

64. **Lovely, D., and E. Phillips.** 1994. Reduction of chromate by *Desulfovibrio vulgaris* and its c_3 cytochrome. *Appl. Environ. Microbiol.* **60:**726–728.

65. **Mara, D., and D. Williams.** 1971. Corrosion of mild steel by nitrate-reducing bacteria. *Chem. Ind.* **May:**566–567.

66. **Maxwell, S., and W. A. Hamilton.** 1986. Modified radio-respirometric method for determining the sulfate reduction activity of biofilms on metal surfaces. *J. Microbiol. Methods* **5:**83–91.

67. **McNeil, M. B., J. M. Jones, and B. J. Little.** 1991. Production of sulfide minerals by sulfate-reducing bacteria during microbiologically influenced corrosion of copper. *Corrosion* **47**(9):674–677.

68. **McNeil, M. B., and B. J. Little.** 1990. Mackinawite formation during microbial corrosion. *Corrosion* **46**(7): 599–600.

69. **Meitz, A.** 1984. Efficacy and decomposition of DBNPA in two cooling water systems, p. 243–298. *In Proceedings of the International Water Conference.* IWC/84-50.

70. **Merrill, R.** 1986. Regulatory toxicology, p. 917–932. *In* C. Klassen, M. Amdur, and J. Doul (ed.), *Toxicology*, 3rd ed.

71. **Milde, K., W. Sand, W. Wolff, and E. Bock.** 1983. Thiobacilli of the corroded concrete walls of the Hamburg sewer system. *J. Gen. Microbiol.* **129:**1327–1333.

72. **Miller, J. D. A.** 1981. Metals, p. 149–202. *In* A. H. Rose (ed.), *Microbial Biodeterioration.* Academic Press, London.

73. **Mollica, A., G. Ventura, and E. Traverso.** 1990. Electrochemical monitoring of the biofilm growth on active-passive alloy tubes of heat exchanger using seawater as cooling medium, p. 341–349. *In Proceedings of 11th Corrosion Congress,* vol. 4. AIM, Florence, Italy.

74. **Moulin, J. M., K. Johnson, R. Karius, and B. Resiak.** 1995. Special corrosion of steel sheet piles in temperate seawater, p. 53/1–53/9. *In* P. Angell et al. (ed.), *Proceedings of International Conference on Microbially Influenced Corrosion.* National Association of Corrosion Engineers, Houston.

75. **National Association of Corrosion Engineers.** 1972. *Control of External Corrosion on Underground or Submerged Metallic Piping Systems.* NACE standard RP-01-69. National Association of Corrosion Engineers, Houston.

76. **Nebel, C., and T. Nebel.** 1984. Ozone: the process water sterilant. *Pharmacol. Manuf.* **1**(2):16–22.

77. **Okamoto, G.** 1974. Passive film of 18-8 stainless steel structure and its function. *Corros. Sci.* **13:**471–489.

78. **Olowe, A. A., P. Bauer, J. M. R. Génin, and J. Guezennec.** 1989. Mössbauer effect evidence of the existence of green rust 2 transient compound from bacterial corrosion in marine sediments. *Corrosion* **45**(3):229–234.

79. **Osozawa, K.** 1991. Materials development influenced by price of raw materials. *Corros. Eng.* (Japan) **40:**455–456.

80. **Peckner, D., and I. M. Burnstein.** 1977. *Handbook of Stainless Steels.* McGraw-Hill, New York.

81. **Pedersen, A., and M. Hermansson.** 1989. The effects on metal corrosion by *Serratia marcescens* and a *Pseudomonas* sp. *Biofouling* **1:**313–322.

82. **Phelps, T. J., R. M. Schram, D. Ringleberg, N. Dowling, and D. C. White.** 1991. Anaerobic microbial activities including hydrogen-mediated acetogenesis within natural gas transmission lines. *Biofouling* **3:**265–276.

83. **Pourbaix, M.** 1974. *Atlas of Electrochemical Equilibria.* National Association of Corrosion Engineers, Houston.

84. **Puckorius, P., and T. Harris.** 1981. Zero discharge via water reuse in cooling towers, p. 799–812. *In Proceedings of Water Reuse Symposium II.* Washington, D.C.

85. **Revsbech, N. P.** 1989. Diffusion characteristics of microbial communities determined by use of oxygen microsensors. *J. Microbiol. Methods* **9:**111–122.

86. **Robinson, J., and J. Tiedje.** 1984. Competition between sulfate-reducing and methanogenic bacteria for H_2 under resting and growing conditions. *Arch. Microbiol.* **137:**26–32.

87. **Robinson, M., and P. Kilgallon.** 1994. Hydrogen embrittlement of cathodically protected high-strength, low alloy steels exposed to sulfate-reducing bacteria. *Corrosion* **50**(8):626–635.

88. **Rodriguez-Leiva, M., and H. Tributsch.** 1988. Morphology of bacterial leaching patterns by Thiobacillus ferroxidans on synthetic pyrite. *Arch. Microbiol.* **149:**401–405.

89. **Rueska, I., J. Robbins, W. Costerton, and E. Lashen.** 1982. Biocide testing against corrosion-causing oil-field bacteria helps control plugging. *Oil Gas J.* **3:**8–13.

90. **Salvarezza, R., M. F. L. de Mele, and H. Videla.** 1983. Mechanisms of the microbial corrosion of aluminum alloys. *Corrosion* **39**(1):26–32.

91. **Sand, W., E. Bock, and D. White.** 1984. Role of sulfur oxidizing bacteria in the degradation of concrete, article 96. *In Proceedings of Conference, NACE '84.* National Association of Corrosion Engineers, Houston.

92. **Scott, P. J., J. Goldie, and M. Davies.** 1991. Ranking alloys for susceptibility to MIC—a preliminary report on high-Mo alloys. *Mater. Perform.* **January:**55–57.

93. **Scott, P. J. B., and M. Davies.** 1992. Survey of field kits for sulfate-reducing bacteria. *Mater. Perform.* **May:**64–68.

94. **Scotto, V., R. Di Centio, and G. Marcenaro.** 1985. The influence of marine aerobic microbial film on stainless steel corrosion behaviour. *Corros. Sci.* **25**(3):185–194.

95. **Soimajärvi, J., M. Pursiainen, and J. Korhonen.** 1978. Sulphate-reducing bacteria in paper machine waters and in suction roll perforations. *Eur. J. Appl. Microbiol. Biotechnol.* **5:**87–93.

96. **Stewart, P., T. Griebe, R. Srivnivasan, C. Chen, F. Yu, D. deBeer, and G. McFeters.** 1994. Comparison of respiratory activity and culturability during monochloramine disinfection of binary population biofilms. *Appl. Environ. Microbiol.* **60:**1690–1692.

97. **Stroes-Gascoyne, S.** 1989. *The Potential for Microbial Life in a Canadian High-Level Nuclear Fuel Waste Disposal Vault: a Nutrient and Energy Source Analysis.* Atomic Energy of Canada Ltd. report AECL-9574. Whiteshell Nuclear Research Establishment, Pinawa, Canada.

98. **Tiller, A. K.** 1990. Biocorrosion in civil engineering, p. 24–38. *In* P. Howsam (ed.), *Microbiology in Civil Engineering.* FEMS symposium 59. Chapman & Hall, London.

99. **Trulear, M., and C. Wiatr.** 1988. Recent advances in halogen based biocontrol, article 19. *In Proceedings of Conference, Corrosion '88.* National Association of Corrosion Engineers, Houston.

100. **Van Loosdrecht, M., J. Lyklema, W. Norde, and A. Zehnder.** 1990. Influence of interfaces on microbial activity. *Microbiol. Rev.* **54:**75–87.

101. **Ventura, G., E. Traverso, and A. Mollica.** 1989. Effect of NaCLO biocide additions in natural seawater on stainless steel corrosion resistance. *Corrosion* **45**(4):319–325.

102. **Videla, H., and W. G. Charaklis.** 1992. Biofouling and microbially influenced corrosion. *Int. Biodeterior. Biodegrad.* **29:**195–212.

103. **Von Volzogen Kuhr, C. A. H., and I. S. van der Vlught.** 1934. Graphatization of cast iron as an electrochemical process in anaerobic soils. *Water* **18:**147–165.

104. **Voordouw, G., Y. Shen, C. Harrington, A. Telang, T. Jack, and D. Westlake.** 1993. Quantitative reverse sample genome probing of microbial communities and its application to oil field production waters. *Appl. Environ. Microbiol.* **59:**4101–4114.

105. **Walsh, D., E. R. Willis, and T. VanDiepen.** 1995. Susceptibility of low alloy weldments to microbiologically influenced corrosion: attachment and pitting initiation, p. 61/1–61/18. *In* P. Angell, S. Borenstein, R. Buchanan, S. Dexter, N. Dowling, B. Little, C. Lundin, M. McNeil, D. Pope, R. Tatnall, D. White, and G. Ziegenfuss (ed.), *Proceedings of International Conference on Microbially Influenced Corrosion.* National Association of Corrosion Engineers, Houston.

106. **Weimer, P., M. Van Kavelaar, C. Michel, and T. Ng.** 1988. Effect of phosphate on the corrosion of carbon steel and on the composition of corrosion products in two-stage continuous cultures of *Desulfovibrio desulfuricans. Appl. Environ. Microbiol.* **54:**386–396.

107. **Widdel, F.** 1988. Microbiology and ecology of sulfate and sulfur-reducing bacteria, p. 469–585. *In* A. Zehnder (ed.), *Biology of Anaerobic Microorganisms.* Wiley Interscience, New York.

108. **Winogradsky, S.** 1949. *Microbiologie du Sol.* Oeuvres Complètes. Mason, Paris.

109. **Yu, F., and G. McFeters.** 1994. Physiological responses of bacteria in biofilms to disinfection. *Appl. Environ. Microbiol.* **60:**2462–2466.

Scale-Up of Processes for Bioremediation

ROGER A. KORUS

95

Scale-up is a procedure for designing and building a large-scale system based on experimental results from small-scale model systems. Few bioremediation processes can be scaled up by simply building a large version of the model system. Typically key kinetic and transport effects are scale dependent, and different effects can dominate design and operation at different scales. Often kinetic effects dominate on a small scale, and transport effects dominate in a large-scale system.

Engineering scale-up methods are more difficult to apply to in situ than to ex situ bioremediation. Field applications of in situ bioremediation are governed by complex factors that are not well defined. Pollutant toxicity, physical properties of the soil matrix, microbial composition, and nutrient availability affect the rates of contaminant degradation, and the heterogeneity of a field site can be extremely difficult to characterize. Usually the rate-controlling steps in a large-scale bioremediation process are mass transport steps. Often oxygen supply and desorption of contaminants from the soil matrix into the aqueous phase control the rate of in situ bioremediation. Oxygen supply is determined by injector spacing, rates of injection, and soil permeability. Contaminant desorption rates are determined by partition coefficients and the nature of the soil matrix. Engineering scale-up methods other than geometric similarity are difficult to apply to in situ bioremediation because of the complex nature of these processes and the limited ability to engineer the processes.

In contrast to in situ bioremediation, ex situ bioremediation rates can be dominated by mixing processes, and ex situ processes can be engineered to optimize mixing. Scale-up methods have been developed and tested for a wide range of industrial processes, and these methods are well suited to the scale-up of ex situ bioremediation processes. Applications of scale-up methods to ex situ bioremediation is the focus of this chapter, with some discussion of in situ processes.

For ex situ, and to a lesser extent for in situ, bioremediation, kinetic and sorption effects that influence bioremediation rates may be determined by small-scale treatability tests. The presence of active indigenous organisms and tight binding of contaminants to the organic or clay fraction of the soil matrix may limit contaminant degradability at any scale. Such kinetic and sorption limitations are relatively insensitive to scale. The initial step in scale-up is identifying poten-tial rate-limiting steps for biodegradation as scale increases. While kinetic rates and equilibrium partitioning are readily determined on a laboratory scale, transport effects are best analyzed by theoretical models. Since transport phenomena become rate controlling as scale increases, theory must be combined with small-scale experimental data to successfully design and predict the performance of large-scale systems.

Most scale-up problems result from the dependence of transport processes on scale. Fluid flow and mass transfer are the principal mechanisms for scale-up considerations. Scale-up problems are more pronounced for aerobic than for anaerobic processes, since oxygen transport is very sensitive to scale. Scale-up problems are more pronounced for continuous than for batch processes because of the fluid flow complexity of continuous systems.

SCALE-UP METHODS

Laboratory data cannot be used to directly simulate a full-scale bioremediation process or to design unit operations when key transport effects cannot be accurately scaled down for laboratory experiments. Therefore, the practice of using a scale factor and geometric similarity to design a full-scale process from laboratory data must be closely examined. Some operations such as metal precipitation can be accurately simulated in the laboratory. Other operations involving interfacial transport or mixing phenomena cannot be accurately simulated in a laboratory.

In the laboratory, there is typically a rapid and often a nearly complete degradation of contaminants under optimum conditions. This high rate and complete contaminant degradation can be difficult to achieve in the field (Fig. 1). The reasons for this difference are the dominance of transport effects at the large scale, especially the slow rate of contaminant desorption from soils which tightly bind the contaminants. Laboratory testing is typically done with high agitation, rates of oxygen and nutrient supply sufficiently high to maximize the rate of contaminant degradation, and uniform mixing conditions. This difference between laboratory and field results illustrates the difficulty in applying scale-up methods. If laboratory experiments cannot duplicate the transport effects of the field, these transport effects must be correctly modeled as a guide to scale-up.

If the rate-controlling mechanism is expected to be different in the field than in the laboratory, scale-up methods

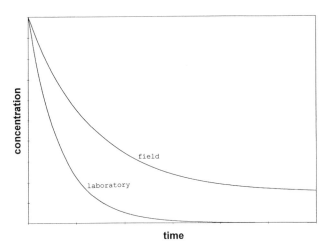

FIGURE 1 Contaminant degradation commonly observed in the laboratory and in the field.

must be carefully considered. If in situ field conditions or ex situ reactor conditions are known, it may be possible to scale down these conditions for laboratory tests (see "Scale-Down," below). If field conditions cannot be duplicated in the laboratory, mathematical models and simulation algorithms can use laboratory testing data to predict field or reactor performance (see "Scale-Up Models," below). In either case, well-designed laboratory testing will give a better understanding of system performance and improved predictions for large-scale system performance (see "Process Improvements," below).

Aerobic bioremediation processes are especially difficult to scale up. Mixing, hydrodynamics, and bubble coalescence effects cannot be simulated on a small scale. Laboratory tests give rates as a function of oxygen concentrations, but full-scale design must be based on experience, theory, and methods for scale-up. Typically the alternatives are to perform pilot plant tests at a scale large enough to simulate field conditions or perform a theoretical simulation of the bioremediation operations. Simulation models have been developed for many of the unit operations of bioremediation, including air stripping and activated carbon adsorption (21). These computer models rely on an accurate mathematical model of the operation and the availability of good model parameter estimates. Laboratory experiments can be useful for determining parameters in the model, but these experiments should be designed for parameter estimation rather than small-scale simulation of the bioremediation process.

A good example of the difficulty in scale-up is given by Oldshue for scale-up from an 80-liter to a 10,000-liter agitated bioreactor (22). Four different properties affecting $k_l a$ (the oxygen mass transfer coefficient times the gas-liquid interfacial area) were selected as bases for scale-up: constant power/volume (P/V), constant liquid circulation time, constant shear, and constant Reynolds number for the impeller. Scale-up based on constant P/V could be used to maintain constant oxygen supply, but shear rate will be increased by 70% with scale-up and circulation time will be increased about threefold. Scale-up based on circulation time requires 25 times more power at the large scale than for P/V scale-up, while constant shear scale-up requires 5 times less power.

Scale-up methods give predictions of process performance at a larger scale. Verification of these scale-up predictions can be made at pilot plant scales, intermediate be-

tween laboratory and field scales. However, pilot plant studies can be very expensive, often approaching the costs of a small commercial-scale operation. Therefore, there are large incentives for moving from bench scale to commercial scale with at most one intermediate step and at a high scale-up ratio. Solids-handling operations are difficult to confidently scale up with scale-up ratios greater than 50 to 100, while fluid processes can be scaled up with ratios 1,000 and more (6).

A critical question is whether the benefits derived from the additional pilot-scale studies required with low scale-up ratios result in a significant improvement in commercial performance. Limited experience at the pilot scale often results in inadequate commercial performance as first seen in an extension of the time required to start up a large-scale process. Experience in the chemical industry suggests that a pilot program whose total cost is less than the costs incurred in 3 months of start-up expenditures for a commercial unit can be a good investment (6). However, when full-scale equipment is available and process experience is extensive, full-scale tests can be carried out with existing full-scale equipment. Costs for such full-scale tests can be prohibitive, and only a limited range of variables can be studied.

Most scale-up methods can be divided into three approaches: rules of thumb, dimensional or time-scale analysis, and fundamental or semifundamental methods. A method is selected on the basis of the knowledge available and the related ability to model the behavior of the bioremediation process. Rules of thumb are used when little modeling information is available. Fundamental and semifundamental methods are used when complex mathematical models can be confidently applied. Dimensional analysis is applicable when known and tested correlations are available and is applicable to operations such as air stripping, biopiles, and slurry reactors for which degradation rates are controlled by well-defined physical mechanisms.

Rules of Thumb

Rules of thumb are most commonly applied when a rate-controlling transport step is identified but modeling is not well developed or realistic parameter estimates are not available for modeling and computer simulation. Rules of thumb are based on performance indices that govern the dominant transfer rate. For example, if oxygen transfer from gas to liquid is rate controlling, appropriate performance indices could be specific power input into an agitated reactor (P/V) or $k_l a$. These are the two most widely used scale-up criteria for industrial fermentors, with P/V used when suspension of solids, blending of liquids, dispersion of nonmiscible liquids, heat transfer of the liquid phase, or gas adsorption is the scale-up criterion (3).

The principle of similarity is the most widely used rule of thumb. Two systems are geometrically similar when the ratios of corresponding dimensions in one system are equal to the corresponding ratios of the other system. Hence, geometrical similarity exists when equipment at different scales has the same shape. Geometric similarity as a rule of thumb method should be used with great caution. Few operations are independent of scale, since transport rates will greatly vary with scale. Also, there often occurs a shift in the rate-controlling step above a certain scale. Initially, this is a shift from rate control to transport control at the pilot scale or a smaller scale. A safer approach for maintaining a constant value for a transport step is dimensional analysis.

TABLE 1 Dimensionless numbers useful in bioremediation analysis

Phenomenon	Dimensionless no.[a]	Physical meaning
Fluid flow	Reynolds no., $\text{Re} = \dfrac{\rho v D}{\eta}$	$\dfrac{\text{Inertial forces}}{\text{Viscous forces}}$
	Power no., $P_o = \dfrac{P}{\rho N^3 D^5}$	$\dfrac{\text{Total dissipated power}}{\text{Power due to inertia}}$
Mass transfer	Sherwood no., $\text{Sh} = \dfrac{kD}{D_e}$	$\dfrac{\text{Total mass transfer}}{\text{Mass transfer by diffusion}}$
	Schmidt no., $\text{Sc} = \dfrac{v}{D_e}$	$\left(\dfrac{\text{Hydrodynamic boundary layer}}{\text{Mass transfer boundary layer}}\right)^3$
	Peclet no., $\text{Pe} = \dfrac{vL}{D_e}$	$\dfrac{\text{Mass transfer by convection}}{\text{Mass transfer by diffusion}}$
Reaction	Damkohler I, $\text{Da}_\text{I} = \dfrac{rL}{vC}$	$\dfrac{\text{Reaction rate}}{\text{Mass transport by convection}}$
	Damkohler II, $\text{Da}_\text{II} = \dfrac{rL^2}{D_e C}$	$\dfrac{\text{Reaction rate}}{\text{Mass transport by diffusion}}$
	Thiele modulus, $\phi = R\sqrt{\dfrac{r}{D_e C}}$	$\left(\dfrac{\text{Reaction rate in particle}}{\text{Diffusion rate in particle}}\right)^{1/2}$

[a] Definitions: C, concentration; D, diameter; D_e, effective diffusivity; k, transfer coefficient; L, length; N, rotational speed; P, power input; r, reaction rate; R, radius; v, velocity; ρ, density; η, viscosity.

Dimensional and Regime Analysis

Dimensional analysis is an intermediate method of analysis between detailed mathematical modeling and empirical analysis. This method is based on maintaining a dimensionless group constant during scale-up. Even if bioremediation systems or operations cannot be completely described by mathematical models, the key variables can often be identified. Dimensionless groups arise in the modeling of fluid flow, heat transfer, mass transfer, and chemical and biological reactions (Table 1). To maintain transport and reaction rates during scale-up, it is sufficient that the key dimensionless numbers be held constant.

To apply dimensional analysis, the rate-limiting factors must be identified. These factors are typically mixing, mass transfer, and the rate of the biological reaction. Also, it is important to identify the limiting factor which is under the control of the process designer or operator. For example, the overall rate of biodegradation may be limited by oxygen supply. This could be determined by small-scale tests which indicate that degradation rates are proportional to oxygen supply or based on scale-up experience indicating that oxygen cannot easily be supplied at the rate necessary to support the microbial activity measured in the field or predicted for ex situ reactors. Oxygen supply increases with increased air supply, power input, and impeller speed for a stirred reactor. If power input and impeller speed can be controlled during scale-up, the power number might be selected for scale-up. Often such a scale-up criterion gives unreasonably large values for energy input or impeller speed. With increasing scale, energy input per unit volume and impeller speeds must be reduced. In such cases, dimensionless numbers cannot be maintained constant, and scale-up consists of operation at maximum values of oxygen supply, power input, and impeller speed in order to minimize limiting rates.

A dimensionless analysis cannot be made unless enough is known about the bioremediation process to decide which variables are important and what types of equations will model the process. It is often sufficient to identify the rate-controlling step to select the appropriate dimensionless numbers (Table 1). To identify the most important dimensionless groups, a time-scale or regime analysis may be valuable (20). Regime refers to the mechanism which controls system performance. Bioremediation rates can be controlled by kinetics, transport, or equilibrium phenomena. This analysis identifies which regime is rate determining.

The relative time scales for mixing and reaction are important in determining scale-up criteria. Typically there is a shift from kinetic control at a small scale to transport limitation at a large scale. Characteristic time constants for kinetic and transport processes (Table 2) can be used to predict the importance of these effects. For a first-order process, the time constant is defined as the reciprocal of the rate constant. For processes which are not first order, the time constant is calculated as the ratio between capacity and flow. Time constants that are small indicate that the corresponding effect is at equilibrium and does not influence bioremediation rates.

Fundamental and Semifundamental Methods

Fundamental methods are based on the application of turbulence models to describe the influence of process design and operating conditions on mixing and fluid flow (20). These models include a complete set of momentum and mass balances and are extremely complex, and many simplifications must be made to obtain solutions. In the semifundamental methods, simpler flow models are combined with a kinetic model to give a reasonably precise process description. The parameters in the flow model are scale dependent, and thus the influence of scale can be examined by model simulations. Because of the modeling and solution difficulties of multiphase, heterogeneous bioremediation processes, the fundamental and semifundamental approaches are not yet applicable to bioremediation scale-up.

SCALE-UP MODELS

The objective of scale-up is to design the operation of a bioremediation process from theoretical principles and using laboratory and pilot-scale data. It is crucial to determine on a small scale the factors which will control the bioremediation rate under application conditions. Bioremediation process rates are nearly always limited by mass transfer rather than

TABLE 2 Time constants for bioremediation processes[a]

Transport or kinetic process	Time constant
Gas	$(1 - \epsilon)V/v_g$, gas volume/gas flow rate
Liquid flow	V/Q, volume/liquid flow rate
Diffusion	L^2/D_e, length²/diffusivity
Oxygen supply	$1/k_1a$, 1/oxygen mass transfer coefficient
Microbial growth	$1/\mu$, 1/specific growth rate
Nutrient consumption	
Zero-order reaction ($C \gg K_s$)	C/r_{max}, concentration/consumption rate
1st-order reaction ($C \ll K_s$)	K_s/r_{max}, saturation constant/consumption rate
Heat transfer	$\rho V C_p/UA$, heat capacity/heat removal rate
Mixing with mechanical agitation	$4V/(1.5ND^3)$, volume/mixing speed · diameter³

[a] Definitions not provided in the footnote to Table 1 or above: C_p, heat capacity; U, heat transfer coefficient; ϵ, void fraction.

the intrinsic biological degradation kinetics. Consequently, the transfer rates of oxygen, nutrients, and contaminants are usually limiting. These transfer rates are strongly affected by hydraulic transport for in situ bioremediation and by mixing and aeration for ex situ bioremediation.

Modeling is essential to good scale-up since models must guide the application of laboratory data in scale-up predictions. A large fraction of the materials describing scale-up are dedicated not to scale-up as such but to a review of the state of the art of our understanding of the biological, physical, and chemical phenomena taking place in any given process. A better understanding of these phenomena is crucial to successful scale-up. Also, we must accept that larger-scale systems are less well defined, and we cannot carry the more accurate description of laboratory systems to a larger scale. With increasing system size, our model predictions will be less accurate.

Process Modeling

An accurate and quantitative process model is essential for scale-up with a high scale-up ratio. For scale-up it is necessary to identify the rate-controlling step for bioremediation and to model that step. The mathematical model can be used to guide scale-up by maximizing the rate of this rate-controlling step. Table 3 presents rate-limiting factors for biodegradation of soil contaminants.

A mathematical model has been developed for the bioremediation of hydrocarbons in a soil matrix to predict the rate-controlling step and the remediation rate during the bioremediation of a contaminated soil (14). This model is

TABLE 3 Potential rate-limiting factors for biodegradation of soil contaminants

Class of factor	Measurable effect	Key parameter(s)
Kinetic	Activity of indigenous organisms	k, X, K_m Inhibition constants
	Presence of inhibitory or toxic compounds	Activity decay constant
Equilibrium	Partitioning of contaminants	Partition coefficients
Transport	Desorption of contaminants	k_sA_s k_1a
	Oxygen supply	
	Nutrient supply	

based on mass transfer of oxygen and oil into the aqueous solution in the soil matrix and biodegradation of the hydrocarbons in the aqueous solution. Monod's equation was used to describe the biodegradation rate in aqueous solution, reducing to first-order kinetics at low contaminant concentrations, and degradation rates are proportional to the concentration of active microorganisms. Mass transfer equations were used to describe oxygen supply to the microorganisms and oil desorption from the soil matrix to the aqueous solution.

Biodegradation of contaminants can be viewed as a serial process in which the rates of contaminant desorption, oxygen supply, and biodegradation kinetics are equal when expressed in equivalent units, for example milligrams of biological oxygen demand (BOD) per second (14). These three rates can be expressed as

$$r_c = k_sA_s (C_{cs} - C_c) \qquad (1)$$

where r_c is the contaminant desorption rate (milligrams of BOD per second), k_s is the contaminant mass transfer coefficient (centimeters per second), A_s is the solid-liquid interfacial area for contaminant mass transfer (square centimeters), C_{cs} is the solubility of the contaminant in water (milligrams of BOD per cubic centimeter), and C_c is the aqueous concentration of contaminant (milligrams of BOD per cubic centimeter), and as

$$r_o = k_1a (C_g^* - C_1) \qquad (2)$$

where r_o is the rate of oxygen supply (milligrams of BOD per second), k_1 is the oxygen mass transfer coefficient (centimeter per second), a is the gas-liquid interfacial area for oxygen mass transfer (square centimeters), C_g^* is the solubility of oxygen in the aqueous phase in equilibrium with the gas-phase oxygen concentration (milligrams of BOD per cubic centimeter), and C_1 is the aqueous-phase oxygen concentration. For mass transfer of contaminants and oxygen the individual parameters k_s, A_s, k_1, and a are difficult to evaluate separately and therefore are grouped as single parameters, k_sA_s and k_1a.

The rate at which microorganisms degrade contaminants, r_b, can be expressed as

$$r_b = V \frac{kXC_c}{K_m + C_c} \cdot \frac{C_1}{K'_m + C_1} \qquad (3)$$

where V is the volume (cubic centimeters), k is the contaminant biodegradation rate constant (1/mg · s), X is the concentration of active microorganisms (milligrams per cubic centimeter), and K_m and K'_m (milligrams of BOD per centimeter) are the concentrations of contaminant and oxygen, respectively, where the biodegradation rate is half of the maximum rate. K'_m is typically between 0.1 and 1 mg of O_2 per liter.

Process modeling can be based on rates as shown above or on equilibrium models. Equilibrium modeling of contaminant biodegradation was recently reviewed (19). From analyses of data reported for both natural and engineered systems, it is clear that results are highly dependent on the contaminant, the identities and concentrations of organisms, and the nature of the sorbent. Complex physical, chemical, and biological parameters affecting bioavailability and biodegradation are extremely difficult to define. The effects of surfactants on the biodegradation of sorbed, hydrophobic pollutants is an excellent example of the difficulty in modeling. The presence of surfactants has been shown to have no effect, a stimulatory effect, or an inhibitory effect on the degradation of target substrates, depending on the nature of the surfactant and organisms used (19). However, no guiding principles allow the prediction of biodegradation rates based on surfactant types or concentrations. Surfactant toxicity, surfactant degradation, efficiency of substrate utilization by microorganisms, and surfactant contaminant effects on cell growth influence biodegradation rates but are generally unknown.

Because of system complexities, transport or rate modeling must be simplified by identifying a rate-limiting step. For example, with forced aeration of soil piles, this rate-limiting step is either mass transfer of oxygen into the aqueous phase, desorption of contaminant, or the biodegradation kinetics. The most successful modeling has been for situations in which oxygen supply is rate limiting (5). Oxygen supply models are physical models based on the minimum required oxygen demand. The geometry of the system and the soil permeability are the main parameters needed to describe the system. For desorption rate control models, nonequilibrium sorption parameters such as the effective pore diffusion coefficient must be estimated, and hydrodynamic parameters must be determined. For biodegradation rate control models, the distribution and composition of the microorganisms must be determined, and an accurate kinetic model which may contain toxic or inhibitory parameters must be developed.

Contaminant Desorption

In soil bioremediation, the rate-limiting step is often the desorption of contaminants. Sorption to soil particles and organic matter in soils often determines the bioavailability of organic pollutants. Also, this bioavailability is an important toxicity characteristic, and the U.S. Environmental Protection Agency has established the toxicity characteristic leaching procedure (12). Rates of in situ soil bioremediation are governed by mass transfer of contaminants (desorption and diffusion), the convective-dispersive flux of oxygen and nutrients, and the microbiological content of the soil. On-site testing can determine the rate and extent of biodegradation and ex situ process feasibility (4). The extent of biodegradation is often limited by contaminant desorption.

Transport modeling is done either with an equilibrium model usually based on batch equilibrium sorption measurements or with mass transfer models describing contaminant transport. Although equilibrium models do not describe the complex hydrodynamics, diffusional processes, and adsorption and desorption rates, equilibrium models can describe and predict the bioavailability of sorbed contaminants. The basic assumption of an equilibrium model is that sorbed-phase solute is not degraded and that sorption is reversible and rapid compared with the rate of microbial degradation. Therefore, sorption affects the biodegradation kinetics only by decreasing the aqueous contaminant concentration. This approach has been the method of choice for complex systems with hydrophobic pollutants that are slowly degraded. However, equilibrium will not be established if desorption from the solid phase is slow so that the rates of desorption and degradation are similar; in that case, aqueous concentrations are determined by desorption rates rather than equilibrium conditions.

Abiotic soil desorption and equilibrium partition coefficient measurements are probably the most important tests to complement microbial degradation measurements. Desorption tests measure the site-specific soil-water partition coefficients for the contaminants of interest. Several experimental protocols are available for measuring partition coefficients (24, 28). At two extremes, contaminants can be detected near their solubility limits or can be undetectable in the aqueous phase. These can represent the two extremes of bioavailability and biodegradability. Measurements of aqueous-phase and soil-phase concentrations in equilibrium may be sufficient to indicate potential problems with soil sorption. Many organic contaminants are hydrophobic, have a low water solubility, and are tightly held to the soil phase. Desorption of such contaminants is likely to be rate limiting especially for ex situ bioremediation, in which case supplementation with nutrients and microorganisms and careful control of environmental parameters can minimize these potential rate-limiting effects. Failure in the bioremediation of polyaromatic hydrocarbon compounds has been attributed to strong sorption to soil at former manufactured gas plant sites where gaseous fuels were produced from soft coal (24).

The use of surfactants and cosolvents has been investigated in attempts to increase bioavailability of contaminants that are strongly bound to soil. However, high surfactant concentrations can be required to achieve small increases in solubility. Typically 2% surfactant solutions are needed to remove a high percentage of compounds such as higher-ringed polyaromatic hydrocarbons, polychlorinated biphenyls, and higher-molecular-weight hydrocarbons from soils. Similar enhanced removal can be obtained with biosurfactants (25).

Values of $k_s A_s$ are more difficult to obtain. This desorption rate is strongly influenced by agitation, surfactant concentration, soil composition and structure, soil particle or aggregate size, contaminant partition coefficient, and age of contamination. Desorption experiments can determine $k_s A_s$, but care must be taken to properly control the variables that affect desorption. Adsorption isotherms and desorption rates can be determined by sequential batch washing (15) or column desorption tests (16). Desorption rates can be determined as a function of particle size (8).

An accurate evaluation of $k_s A_s$ is complicated by the heterogeneous nature and poor definition of contaminant-soil systems. Some success has been achieved in modeling mass transfer from a separate contaminant phase. During degradation, these non-aqueous-phase liquids often dissolve under conditions such that phase equilibrium is not achieved and dissolution is proportional to $k_s A_s$. Experimental determinations and correlations for $k_s A_s$ depend on interfacial area of the non-aqueous-phase liquid and liquid velocity at the interface (10). For adsorbed contaminants, $k_s A_s$ varies with soil composition and structure, with concentration and age of contamination, and therefore with time.

For example, slurry reactor tests indicate that the rate of naphthalene mass transfer decreases with time, with size of the medium, and with aging of the tar prior to testing (16).

Although the majority of published studies suggest that sorption reduces bioavailability and biodegradation occurs predominantly in the aqueous phase, some studies suggest that sorbed compounds are available to microorganisms without prior desorption (19). Bacteria have different abilities to attach to surfaces, but this relationship is complex. With surface attachment and for in situ bioremediation, physical and mathematical models can be very complex and difficult to verify. For in situ contaminant transport modeling, the mathematical description of groundwater flow, contaminant transport, and diffusion through porous media can be found in several texts (29). Model parameters describing sorption, solute transport, groundwater flow, microbial activity, and soil properties must be determined by laboratory testing. Care must be taken in selecting the type of testing and in obtaining representative samples.

Gas-Liquid Mass Transfer Basis

Most aerobic processes are controlled by the rate of oxygen transfer from the gas phase to the liquid phase (see equation 2). Oxygen supply can be estimated from oxygen transfer correlations and the physical properties of the slurry (26), but k_1a is difficult to predict for land farming tillage, and no published correlations exist.

In aerobic soil slurry reactors, it is difficult to maintain high oxygen concentrations because of the tendency for gas bubbles to coalesce (2). Also, since the reactors are usually low in profile, there is a very short liquid-gas contact time and a small surface area-to-volume ratio for the bubbles. Mechanical agitation is required to disperse gas bubbles and give smaller gas bubbles, but as the solids concentration increases, this agitation effect decreases (2). Operational problems such as diffuser clogging, solids settling, and materials corrosion must be avoided. The design of agitators and aeration diffusers is critical to the performance of soil slurry reactors, especially at high solids loading. Solid loading is usually in the 20 to 40% range.

The parameter k_1a is frequently used as a basis for scale-up of aerobic processes. However, k_1a is dependent on input airflow rate per reactor volume, power input per volume for mixing, and height of reactor. Usually scale-up is guided by a mathematical model relating k_1a to system variables rather than by dimensionless numbers. Operational or design variables are adjusted to maintain a constant k_1a and therefore a constant volumetric supply of oxygen. This approach has been successfully used to scale-up large wastewater treatment aerated reactors. An example of a model used for scale-up is (11)

$$k_1a = 2.37(1.02)^{(T-20)}(Q/S)^{1.07}(h)^{-0.45} \quad (4)$$

where T is temperature (degrees Centigrade), Q is the airflow rate to the reactor at 0°C and 1 atm (1 atm = 101.29 kPa) (cubic meters per hour), S is the cross-sectional area of the tank (square meters), and h is the liquid depth in the tank (meters). Additionally, oxygen supply can be increased by the mixing action occurring in a wide tank, which can produce supersaturation. Upwelling of the higher-oxygen-concentration liquid from the lower depths can occur without sufficient time for complete desorption at the surface. The percent above saturation, %SS, can be related to tank height by %SS = 1.50 $(h)^{1.35}$ (11).

In situ forced aeration and ex situ forced aeration of soil piles are modeled by physical models in which the geometry of the system and the soil permeability are the main parameters. Equations for such complex systems cannot be solved analytically, and numerical solutions have been developed for well-defined geometry and homogeneous physical properties (5).

Fluid Flow Basis

Scale-up of ex situ bioremediation processes is typically limited by oxygen supply or contaminant desorption. The key parameter that must be optimized in such instances is the degree of mixing. Oxygen transfer rate, contaminant desorption rate, shear rate, and heat transfer rate all increase with degree of mixing. A common scale-up criterion is constant power input per unit volume if a rate-limiting step is controlled by mixing. For ex situ processes, there should be a process minimum for P/V, and this minimum will be more apparent at larger scales. For aerobic processes, this P/V requirement could be associated with oxygen supply, contaminant desorption and aqueous-phase transport, nutrient transport, or cell suspension.

Degree of mixing is often judged by the extent of solids suspension. Three classes of solids suspension are complete uniform suspension, off-bottom suspension, and on-bottom motion (18). Off-bottom suspension is appropriate when contaminant desorption is rate limiting, since good circulation is required in the aqueous phase. The key to desorption is to provide an adequate flow velocity around the particles as a driving force for desorption.

The power required for mixing depends on solids concentration, particle size and size distribution, and particle density. These effects are represented in part by the settling velocity of the solid particles. At very low solids concentration, solid particles will settle at a velocity given by Stokes' law (17). For example, 200- to 300-μm soil particles will settle at 1 to 5 cm/s. However, settling velocity will be reduced in solid slurries because of hindered settling and increased slurry viscosity. Hindered settling velocity, u_s, can be estimated from the free settling velocity, u_t, by

$$u_s = u_t(\epsilon_1)^n \quad (5)$$

where ϵ_1 is the aqueous-phase volume fraction and n is approximately 4 (17). Settling velocity is inversely proportional to the slurry viscosity, and slurry viscosity increases by approximately a factor of 5 relative to water for ϵ_1 = 0.7.

Slurry Reactors

Slurry reactors were initially developed in the 1950s for use in the chemical process industries. Major developments in engineering modeling and design of three-phase slurry reactors were achieved in the 1960s and 1970s. Correlations were developed to determine the mechanical and gas-induced agitation necessary to suspend solid particles and to minimize intraparticle diffusion. Overall reaction rates were derived on the basis of the intrinsic reaction kinetics and mass transfer effects for reactant desorption and diffusional resistances within solid particles and between phases. Complex reaction pathways, including product inhibition, consecutive reactions, and complex kinetics, were modeled.

The majority of this development and modeling work was done for chemical processes in which mass transfer effects were the primary limiting steps. In biological degradation of recalcitrant compounds such as nitroaromatics, other factors may be rate limiting. Therefore, high-energy mixing and/or complete suspension of the soil particles may not be necessary.

Typically, catalytic slurry reactors in the chemical processing industries operate with 5 to 10% solids and 100- to 200-μm particles. Soil slurry reactors differ in having larger particles with strongly adsorbed pollutants operated at a higher solids loading. Some reactor designs require fine pulverization of the soil by multiple grinding steps prior to introduction into the reactor. This allows complete mixing, but there is a significant capital and operational cost associated with such a completely fluidized system. Soil slurry reactors are operated much like the slurry reactors used for coal cleaning and the leaching of metal ores.

Considerable literature describing anaerobic slurry reactors for municipal and farm sewage sludge digestion is available. The mass of organics degraded per unit volume of reactor per time can often be much higher in anaerobic treatment because the limitations of oxygen transfer do not apply. Some anaerobic digester designs are unstirred, and the predominant mixing mechanism is by gas generated in the reactor. Propeller-type mixers are often added for more thorough mixing and to maintain the solids in suspension. The current process design for most soil slurry reactors is to finely pulverize the material and try to keep it in suspension with significant power input to shaft stirrers, aerators, recirculation pumps, or a combination thereof. The alternative approach is to not mix at all or just mix occasionally. A twofold reduction in the time required for the removal of 2-sec-butyl-4,6-nitrophenol (dinoseb) and trinitrololuene intermediates was observed with a continuous stirred tank reactor in comparison with an unstirred batch reactor for a 50% soil slurry (23). Daniel et al. tested the effect of agitation on the degradation of trinitrotoluene and reduced intermediates and found that agitation increased the rate of 2,4-diamino-6-nitrotoluene degradation (9). With the extended residence time in soil slurry reactors, there is probably no need for a high shear or complete suspension agitation, especially in anaerobic design.

The operational and design variables having a major influence on soil slurry reactor performance are the desorption rate from the solid phase, diffusion in the liquid phase, and oxygen supply for aerobic systems. In both aerobic and anaerobic soil slurry reactors, the rate-limiting steps should be identified and their degradation rates minimized in the design, with minimum capital and operational costs. The following factors are major considerations in bioreactor design and operation:

1. Desorption rate depends strongly on soil and pollutant chemical structure. Adsorption isotherms and desorption rates can be determined by sequential batch washing.

2. Diffusion of pollutants to the aqueous phase depends on particle size and porosity. Pollutant distribution should be determined as a function of particle size. Particles greater than 400 μm cannot be easily suspended. The adsorption and diffusion of intermediates, additional reactants, and consortium bacteria are of primary importance.

3. Oxygen supply is dependent on aeration rate, solids loading, and agitation. Rates of oxygen should be determined by measuring oxygen transfer uptake rates for complex reactor designs, since oxygen transfer correlations are difficult to apply. Operational problems such as diffuser clogging, solids settling, and materials corrosion must be avoided. The design of agitators and aeration diffusers is critical to the performance of soil slurry reactors, especially at high solids loading.

SCALE-DOWN

Often small-scale experiments which provide the basis for scale-up can be designed by scaling down the anticipated large-scale system, and in this way scale-up problems can be minimized. However, at smaller scales it may be impossible to reproduce large-scale effects. Still, the small-scale system should duplicate as closely as possible the heterogeneities, flow nonidealities, and oxygen transport environment that will exist at the large scale. Often scale-up will require the use of existing facilities or a committed process design, so it is important to mimic this design at a smaller scale for initial tests.

Typically scale-down considerations precede scale-up since initial laboratory experiments must be planned to achieve data which will provide the basis for scale-up. Also, a realistic system can be used to evaluate proposed process changes for an existing operating process. Also, many of the unit operations of bioremediation can or will soon be modeled for computer simulation, and parameter values can be determined in a properly scaled-down experiment.

Shuler and Kargi give three examples of small-scale biochemical engineering experiments and models that mimic a large known piece of equipment (27). The first example is a small-scale apparatus that approximates the variations in substrate and dissolved oxygen concentrations that might be expected from a mixing-time or time constant analysis. A real continuous stirred tank reactor is modeled by a two-compartment system to approximate the residence time distributions expected in a large reactor. One vessel is aerated and one is purged with nitrogen to simulate a dead zone in the large-scale system. Such a scale-down apparatus could be used to estimate the system's response (for example, microbial growth rate or contaminant degradation) to changes in medium composition, use of surfactants, aeration rate, or use of different inoculum preparations.

As fermentors are scaled up, the mixing time usually increases. Compartmental models can be used to adjust laboratory data to describe a large-scale system in which mixing time is greater (27). Since scale-down cannot adequately represent the inhomogeneities of the large-scale system, a model of the mixing effects with scale is necessary to guide scale-up.

PROCESS IMPROVEMENTS
Reactor Intensification and Economy of Scale

In scale-up, there should be a learning curve reflecting the increasing process understanding with time. With scale-up, this increased understanding should provide more efficient operation as design, operating conditions, and control are improved. In the development of large-scale systems, the learning curve begins with the discovery of a reaction system (e.g., degradation of specific contaminants by a microbial consortium) and the subsequent path for process development has three broadly identifiable areas of activity (3): (i) economies of scale with an increase in throughput per system volume as the scale increases; (ii) process intensification with a reduction in the size of individual process units while maintaining performance; and (iii) discovery of an alternative reaction system, after which the scale-up cycle is repeated.

The predictive ability of mathematical models is crucial to the evaluation of process concepts for the design and operation of bioremediation systems. During scale-up from the laboratory scale to the pilot scale and full scale, improved models and more accurate parameter estimates guide process optimization. Process intensification and profitability are closely linked, as organizations must show continuous improvements in process design and operation to remain competitive.

Kinetic Characteristics

The measurement of biodegradation rates by indigenous microorganisms is the first step in microbiological characterization. These measurements can be complicated by low microbial populations or by the absence of species capable of degrading contaminants. Also, optimum conditions of temperature, oxygen nutrient supply, and contaminant availability due to low solubility and sorption can limit degradation rates, especially in early tests for which these limiting factors are not well defined.

Published biodegradation rate constants do not provide a good comparative basis for process design since experimental conditions vary greatly. Many factors can influence biodegradation rates and may not be reported; these factors include transport effects, acclimation of microorganisms to toxic chemicals, inhibition, and cometabolism. Also, zero-order kinetics, rather than the first-order kinetics with respect to substrate as given in Table 2, have been frequently reported for biodegradation (1). Zero-order kinetics may indicate rate limitation by another factor such as oxygen supply, another nutrient, or limiting solubility of the target compound. Also, documentation for less than 1% of the more than 2,000 new compounds submitted to the U.S. Environmental Protection Agency each year for regulatory review contains data on biodegradability (7). Therefore, degradation rate constants are used mainly for order-of-magnitude estimates, and other approaches to predicting biodegradability that are based on physical or chemical properties of compounds have been developed.

The main objective of microbial degradation tests is to determine if the indigenous microorganisms are capable of bioremediation when conditions are optimized or if inoculation by nonindigenous microorganisms is required. For example, in the bioremediation of dinoseb-contaminated sites, one site was remediated with indigenous microorganisms while a second site required inoculation and was remediated with inoculum from the first site (13). Whereas pure cultures or consortia of bacteria do not in general persist when introduced into a natural environment, soil and groundwater reactors can be operated to maintain a population of introduced microorganisms.

Bioremediation must compete economically and functionally with alternate remediation technologies. Competing technologies are often incineration and chemical treatments. Bioremediation usually competes well on a cost basis, especially for petroleum products and many solvents. However, bioremediation suffers from the large amount of preliminary information necessary to support process design. Lack of information on waste characteristics and microbial physiology and the complex options for process design and operation can make bioremediation more difficult to apply than alternate technologies. The variable end results of bioremediation are due in large part to these complexities of process design.

REFERENCES

1. **Alexander, M.** 1994. *Biodegradation and Bioremediation.* Academic Press, San Diego, Calif.
2. **Andrews, G.** 1990. Large-scale bioprocessing of solids. *Biotechnol. Prog.* **6:**225–230.
3. **Atkinson, B.** 1991. *Biochemical Engineering and Biotechnology Handbook,* 2nd ed. Stockton Press, New York.
4. **Autry, A. R., and G. M. Ellis.** 1992. Bioremediation: an effective remedial alternative for petroleum hydrocarbon-contaminated soil. *Environ. Prog.* **11:**318–323.
5. **Battaglia, A., and D. J. Morgan.** 1994. Ex-situ forced aeration of soil piles: a physical model. *Environ. Prog.* **13:** 178–187.
6. **Bisio, A., and R. L. Kabel.** 1985. *Scaleup of Chemical Processes.* John Wiley & Sons, New York.
7. **Boethling, R. S., and A. Sabljic.** 1989. Screening-level model for aerobic biodegradability based on a survey of expert knowledge. *Environ. Sci. Technol.* **23:**672–679.
8. **Compeau, G. C., W. D. Mahaffey, and L. Patras.** 1991. Full-scale bioremediation of contaminated soil and water, p. 91–109. *In* G. S. Sayler, R. Fox, and J. W. Blackburn (ed.), *Environmental Biotechnology for Waste Treatment.* Plenum Press, New York.
9. **Daniel, B. P., R. A. Korus, and D. L. Crawford.** 1995. Anaerobic bioremediation of munitions-contaminated soil, p. 161–175. *In* B. S. Schepart (ed.), *Bioremediation of Pollutants in Soil and Water.* ASTM Press, Philadelphia.
10. **Geller, J. T., and J. R. Hunt.** 1993. Mass transfer from nonaqueous phase organic liquids in water-saturated porous media. *Water Resour. Res.* **29:**833–845.
11. **Jackson, M. L., and C.-C. Shen.** 1978. Aeration and mixing in deep tank fermentation systems. *AIChE J.* **24:** 63–71.
12. **Johnson, L. D., and R. H. James.** 1989. Sampling and analysis of hazardous wastes, p. 13.3–13.44. *In* H. M. Freeman (ed.), *Standard Handbook of Hazardous Waste Treatment and Disposal.* McGraw-Hill Book Co., New York.
13. **Kaake, R. H., D. J. Roberts, T. O. Stevens, R. L. Crawford, and D. L. Crawford.** 1992. Bioremediation of soils contaminated with the herbicide 2-*sec*-butyl-4,6-dinitrophenol (dinoseb). *Appl. Environ. Microbiol.* **58:** 1683–1689.
14. **Li, K. Y., S. N. Annamalai, and J. R. Hopper.** 1993. Rate controlling model for bioremediation of oil contaminated soil. *Environ. Prog.* **12:**257–261.
15. **Linz, D. G., E. F. Neuhauser, and A. C. Middleton.** 1991. Perspectives on bioremediation in the gas industry, p. 25–36. *In* G. S. Sayler, R. Fox, and J. W. Blackburn (ed.), *Environmental Biotechnology for Waste Treatment.* Plenum Press, New York.
16. **Luthy, R. G., D. A. Dzombak, C. A. Peters, S. B. Roy, A. Ramaswami, D. V. Nakles, and B. R. Nott.** 1994. Remediating tar-contaminated soils at manufactured gas plant sites. *Environ. Sci. Technol.* **28:**266A–276A.
17. **McCabe, W. L., J. C. Smith, and P. Harriott.** 1993. *Unit Operations of Chemical Engineering,* 5th ed. McGraw-Hill, Inc., New York.
18. **McDonough, R. J.** 1992. *Mixing for the Process Industries.* Van Nostrand Reinhold, New York.
19. **Mihelcic, J. R., D. R. Lueking, R. J. Mitzell, and J. M. Stapleton.** 1993. Bioavailability of sorbed- and separate-phase chemicals. *Biodegradation* **4:**141–153.
20. **Nielsen, J. H.** 1994. *Bioreaction Engineering Principles.* Plenum Press, New York.
21. **Nyer, E. K.** 1993. *Practical Techniques for Groundwater and Soil Remediation.* CRC Press, Inc., Boca Raton, Fla.
22. **Oldshue, J. Y.** 1966. Fermentation mixing scale-up techniques. *Biotechnol. Bioeng.* **8:**3–24.
23. **Roberts, D. J., R. H. Kaake, S. B. Funk, D. L. Crawford, and R. L. Crawford.** 1993. Field scale anaerobic bioremediation of dinoseb-contaminated soils, p. 219–244. *In* M. A. Levin and M. Gealt (ed.), *Biotreatment of Industrial and Hazardous Waste.* McGraw-Hill, Inc., New York.
24. **Rogers, J. A., D. J. Tedaldi, and M. C. Kavanaugh.** 1993. A screening protocol for bioremediation of contaminated soil. *Environ. Prog.* **12:**146–156.
25. **Scheibenbogen, K., R. G. Zytner, H. Lee, and J. T. Trevors.** 1994. Enhanced removal of selected hydrocarbons from soil by *Pseudomonas aeruginosa* UG2 biosurfactants

and some chemical surfactants. *J. Chem. Tech. Biotechnol.* **59:**53–59.

26. **Shah, Y. T.** 1979. *Gas-Liquid-Solid Reactor Design.* McGraw-Hill, Inc., New York.

27. **Shuler, M. L., and F. Kargi.** 1992. *Bioprocess Engineering.* Prentice-Hall, Inc., Englewood Cliffs, N.J.

28. **Wu, S., and P. M. Gschwend.** 1986. Sorption kinetics of hydrophobic compounds to natural sediments and soils. *Environ. Sci. Technol.* **20:**717–725.

29. **Yong, R. N., A. M. O. Mohamed, and B. P. Warkentin.** 1992. *Principles of Contaminant Transport in Soils.* Elsevier, New York.

Author Index

Subject Index